NOTATION

	$\lceil x \rceil$	the smallest integer greater than or equal to the real number x: the *ceiling* of x
	$a \equiv b \pmod{n}$	a is congruent to b modulo n
RELATIONS	$A \times B$	the Cartesian, or cross, product of sets A, B: $\{(a, b) \mid a \in A, b \in B\}$
	$\mathcal{R} \subseteq A \times B$	$\mathcal{R}$ is a relation from A to B
	$a \mathrel{\mathcal{R}} b; (a, b) \in \mathcal{R}$	a is related to b
	$a \mathrel{\not{\mathcal{R}}} b; (a, b) \notin \mathcal{R}$	a is not related to b
	$\mathcal{R}^c$	the converse of relation $\mathcal{R}$: $(a, b) \in \mathcal{R}$ iff $(b, a) \in \mathcal{R}^c$
	$\mathcal{R} \circ \mathcal{S}$	the composite relation for $\mathcal{R} \subseteq A \times B$, $\mathcal{S} \subseteq B \times C$: $(a, c) \in \mathcal{R} \circ \mathcal{S}$ if $(a, b) \in \mathcal{R}$, $(b, c) \in \mathcal{S}$ for some $b \in B$
	$\mathrm{lub}\{a, b\}$	the least upper bound of a and b
	$\mathrm{glb}\{a, b\}$	the greatest lower bound of a and b
	$[a]$	the equivalence class of element a (relative to an equivalence relation $\mathcal{R}$ on a set A): $\{x \in A \mid x \mathrel{\mathcal{R}} a\}$
FUNCTIONS	$f: A \to B$	f is a function from A to B
	$f(A_1)$	for $f: A \to B$ and $A_1 \subseteq A$, $f(A_1)$ is the image of A_1 under f — that is, $\{f(a) \mid a \in A_1\}$
	$f(A)$	for $f: A \to B$, $f(A)$ is the range of f
	$f: A \times A \to B$	f is a binary operation on A
	$f: A \times A \to B \ (\subseteq A)$	f is a closed binary operation on A
	$1_A: A \to A$	the identity function on A: $1_A(a) = a$ for each $a \in A$
	$f\vert_{A_1}$	the restriction of $f: A \to B$ to $A_1 \subseteq A$
	$g \circ f$	the composite function for $f: A \to B$, $g: B \to C$: $(g \circ f)a = g(f(a))$, for $a \in A$
	f^{-1}	the inverse of function f
	$f^{-1}(B_1)$	the preimage of $B_1 \subseteq B$ for $f: A \to B$
	$f \in O(g)$	f is "big Oh" of g; f is of order g
THE ALGEBRA OF STRINGS	Σ	a finite set of symbols called an alphabet
	λ	the empty string
	$\|x\|$	the length of string x
	Σ^n	$\{x_1 x_2 \cdots x_n \mid x_i \in \Sigma\}, n \in \mathbf{Z}^+$
	Σ^0	$\{\lambda\}$
	Σ^+	$\bigcup_{n \in \mathbf{Z}^+} \Sigma^n$: the set of all strings of positive length
	Σ^*	$\bigcup_{n \geq 0} \Sigma^n$: the set of all finite strings
	$A \subseteq \Sigma^*$	A is a language
	AB	the concatenation of languages $A, B \subseteq \Sigma^*$: $\{ab \mid a \in A, b \in B\}$
	A^n	$\{a_1 a_2 \cdots a_n \mid a_i \in A \subseteq \Sigma^*\}, n \in \mathbf{Z}^+$
	A^0	$\{\lambda\}$
	A^+	$\bigcup_{n \in \mathbf{Z}^+} A^n$
	A^*	$\bigcup_{n \geq 0} A^n$: the Kleene closure of language A
	$M = (S, \mathcal{S}, \mathbb{O}, \nu, \omega)$	a finite state machine M with internal states S, input alphabet $\mathcal{S}$, output alphabet $\mathbb{O}$, next state function $\nu: S \times \mathcal{S} \to S$ and output function $\omega: S \times \mathcal{S} \to \mathbb{O}$

DISCRETE AND COMBINATORIAL MATHEMATICS

An Applied Introduction

FIFTH EDITION

RALPH P. GRIMALDI
Rose-Hulman Institute of Technology

PEARSON

Addison
Wesley

Boston San Francisco New York
London Toronto Sydney Tokyo Singapore Madrid
Mexico City Munich Paris Cape Town Hong Kong Montreal

Publisher:	Greg Tobin
Senior Acquisitions Editor:	William Hoffman
Assistant Editor:	RoseAnne Johnson
Executive Marketing Manager:	Yolanda Cossio
Senior Marketing Manager:	Pamela Laskey
Marketing Assistant:	Heather Peck
Managing Editor:	Karen Guardino
Senior Production Supervisor:	Peggy McMahon
Senior Manufacturing Buyer:	Hugh Crawford
Composition and Technical Art Rendering:	Techsetters, Inc.
Production Services:	Barbara Pendergast
Design Supervisor:	Barbara T. Atkinson
Cover Designer:	Dennis Schaefer
Photo Research and Design Specifications:	Beth Anderson
Cover Illustration:	George V. Kelvin

Photographs of Blaise Pascal, Aristotle, Lord Bertrand Arthur William Russell, Euclid, Augusta Ada Byron (Countess of Lovelace), Gottfried Wilhelm Leibniz, Carl Friedrich Gauss, Leonhard Euler, Arthur Cayley, Pierre de Fermat, Niels Henrik Abel, and Evariste Galois are reproduced courtesy of the Bettman Archive (Corbis). Photographs of George Boole, Peter Gustav Lejeune Dirichlet, David Hilbert, Giuseppe Peano, James Joseph Sylvester, Sophie Germain, and Emmy Noether are reproduced courtesy of Historical Pictures/Stock Montage. The photograph of Claude Elwood Shannon is reproduced courtesy of the MIT Museum. The photograph of Edsger W. Dijkstra is reproduced courtesy of the University of Texas at Austin. The photographs of Andrew John Wiles and Rear Admiral Grace Murray Hopper are reproduced courtesy of AP/Wide World. The photographs of Georg Cantor, Alan Mathison Turing, William Rowan Hamilton, and Leonardo Fibonacci are reproduced courtesy of The Granger Collection. The photograph of Paul Erdös is reproduced courtesy of Christopher Barker. The photographs of Andrei Nikolayevich Kolmogorov, Thomas Bayes, and Al-Khowârizmî are reproduced courtesy of the St. Andrews University MacTutor Archive. The photograph of David A. Huffman is reproduced courtesy of Manuel Enrique Bermúdez of the Department of Computer and Information Science and Engineering at the University of Florida. The photograph of Joseph P. Kruskal is reproduced courtesy of Leiden University.

Library of Congress Cataloging-in-Publication Data

Grimaldi, Ralph P.
   A review of discrete and combinatorial mathematics / by Ralph P. Grimaldi. – 5th ed.
     p. cm.
   Includes index.
   Rev. ed of: Discrete and combinatorial mathematics, c1999.
   ISBN 0-201-72634-3
    1. Mathematics. 2. Computer science–Mathematics. 3. Combinatorial analysis. I. Grimaldi, Ralph P. Discrete and combinatorial mathematics. II. Title.

QA39.2.G748 2003
510–dc21

2002038383

ISBN 0-201-72634-3

12 13 14 15 V092 16 15

*Alla memoria
di mia madre
e mio padre
con affetto e stima*

Preface

It has been more than twenty years since September 2, 1982, when I signed the contract to develop what turned into the first edition of this present textbook. At that time the idea of further editions never crossed my mind. Consequently, I continue to find myself simultaneously very humbled and very pleased with the way this textbook has been received by so many instructors and especially students. The first four editions of this textbook have found their way into many colleges and universities here in the United States. They have also been used in other nations such as Australia, Canada, England, Ireland, Japan, Mexico, the Netherlands, Scotland, Singapore, South Africa, and Sweden. I can only hope that this fifth edition will continue to enlighten and challenge all those who wish to learn about some of the many facets of the fascinating area of mathematics called discrete mathematics.

The technological advances of the last four decades have resulted in many changes in the undergraduate curriculum. These changes have fostered the development of many single-semester and multiple-semester courses where some of the following are introduced:

1. Discrete methods that stress the finite nature inherent in many problems and structures;

2. Combinatorics — the algebra of enumeration, or counting, with its fascinating inter-relations with so many finite structures;

3. Graph theory with its applications and interrelations with areas such as data structures and methods of optimization; and

4. Finite algebraic structures that arise in conjunction with disciplines such as coding theory, methods of enumeration, gating networks, and combinatorial designs.

A primary reason for studying the material in any or all of these four major topics is the abundance of applications one finds in the study of computer science — especially in the areas of data structures, the theory of computer languages, and the analysis of algorithms. In addition, there are also applications in engineering and the physical and life sciences, as well as in statistics and the social sciences. Consequently, the subject matter of discrete and combinatorial mathematics provides valuable material for students in many majors — not just for those majoring in mathematics or computer science.

The major purpose of this new edition is to continue to provide an introductory survey in both discrete and combinatorial mathematics. The coverage is intended for the beginning student, so there are a great number of examples with detailed explanations. (The examples are numbered separately and a thick line is used to denote the end of each example.) In addition, wherever proofs are given, they too are presented with sufficient detail (with the novice in mind).

The text strives to accomplish the following objectives:

1. To introduce the student at the sophomore-junior level, if not earlier, to the topics and techniques of discrete methods and combinatorial reasoning. Problems in counting, or enumeration, require a careful analysis of structure (for example, whether or not order and repetition are relevant) and logical possibilities. There may even be a question of existence for some situations. Following such a careful analysis, we often find that the solution of a problem requires simple techniques for counting the possible outcomes that evolve from the breakdown of the given problem into smaller subproblems.

2. To introduce a wide variety of applications. In this regard, whenever data structures (from computer science) or structures from abstract algebra are required, only the basic theory needed for the application is developed. Furthermore, the solutions of some applications lend themselves to iterative procedures that lead to specific algorithms. The algorithmic approach to the solution of problems is fundamental in discrete mathematics, and this approach reinforces the close ties between this discipline and the area of computer science.

3. To develop the mathematical maturity of the student through the study of an area that is so different from the traditional coverage in calculus and differential equations. Here, for example, there is the opportunity to establish results by counting a certain collection of objects in more than one way. This provides what are called combinatorial identities; it also introduces a novel proof technique. In this edition the nature of proof, along with what constitutes a valid argument, is developed in Chapter 2, in conjunction with the laws of logic and rules of inference. The coverage is extensive, keeping the student (with minimal background) in mind. [For the reader with a logic course (or something comparable) in his or her background, this material can be skipped over with little or no difficulty.] Proofs by mathematical induction (along with recursive definitions) are introduced in Chapter 4 and then used throughout the subsequent chapters.

With regard to theorems and their proofs, in many instances an attempt has been made to motivate theorems from observations on specific examples. In addition, whenever a finite situation provides a result that is not true for the infinite case, this situation is singled out for attention. Proofs that are extremely long and/or rather special in nature are omitted. However, for the very small number of proofs that are omitted, references are supplied for the reader interested in seeing the validation of these results. (The amount of emphasis placed on proofs will depend on the goals of the individual instructor and on those of his or her student audience.)

4. To present an adequate survey of topics for the computer science student who will be taking more advanced courses in areas such as data structures, the theory of computer languages, and the analysis of algorithms. The coverage here on groups, rings, fields, and Boolean algebras will also provide an applied introduction for mathematics majors who wish to continue their study of abstract algebra.

The prerequisites for using this book are primarily a sound background in high school mathematics and an interest in attacking and solving a variety of problems. No particular programming ability is assumed. Program segments and procedures are given in pseudocode, and these are designed and explained in order to reinforce particular examples. With regard to calculus, we shall mention later in this preface its extent in Chapters 9 and 10.

My primary motivation for writing the first four editions of this book has been the encouragement I had received over the years from my students and colleagues, as well as from the students and instructors who used the first four editions of the textbook at many different colleges and universities. Those four editions reflected both my interests and concerns and those of my students, as well as the recommendations of the Committee on the Undergraduate Program in Mathematics and of the Association of Computing Machinery. This fifth edition continues along the same lines, reflecting the suggestions and recommendations made by the instructors and especially the students who have used or are using the fourth edition.

Features

Following are brief descriptions of some of the major features of this newest edition. These are designed to assist the reader (student or otherwise) in learning the fundamentals of discrete and combinatorial mathematics.

Emphasis on algorithms and applications. Algorithms and applications in many areas are presented throughout the text. For example:

1. Chapter 1 includes several instances where the introductory topics on enumeration are needed — one example, in particular, addresses the issue of over-counting.

2. Section 7 of Chapter 5 provides an introduction to computational complexity. This material is then used in Section 8 of this chapter in order to analyze the running times of some elementary pseudocode procedures.

3. The material in Chapter 6 covers languages and finite state machines. This introduces the reader to an important area in computer science — the theory of computer languages.

4. Chapters 7 and 12 include discussions on the applications and algorithms dealing with topological sorting and the searching techniques known as the depth-first search and the breadth-first search.

5. In Chapter 10 we find the topic of recurrence relations. The coverage here includes applications on (a) the bubble sort, (b) binary search, (c) the Fibonacci numbers, (d) the Koch snowflake, (e) Hasse diagrams, (f) the data structure called the stack, (g) binary trees, and (h) tilings.

6. Chapter 16 introduces the fundamental properties of the algebraic structure called the group. The coverage here shows how this structure is used in the study of algebraic coding theory and in counting problems that require Polya's method of enumeration.

Detailed explanations. Whether it is an example or the proof of a theorem, explanations are designed to be careful and thorough. The presentation is primarily focused on improving understanding on the part of the reader who is seeing this type of material for the first time.

Exercises. The role of the exercises in any mathematics text is a crucial one. The amount of time spent on the exercises greatly influences the pace of the course. Depending on the interest and mathematical background of the student audience, an instructor should find that the class time spent on discussing exercises will vary.

There are over 1900 exercises in the 17 chapters. Those that appear at the end of each section generally follow the order in which the section material is developed. These exercises are designed to (a) review the basic concepts in the section; (b) tie together ideas presented in earlier sections of the chapter; and (c) introduce additional concepts that are related to the material in the section. Some exercises call for the development of an algorithm, or the writing of a computer program, often to solve a certain instance of a general problem. These usually require only a minimal amount of programming experience.

Each chapter concludes with a set of supplementary exercises. These provide further review of the ideas presented in the chapter, and also use material developed in earlier chapters.

Solutions are provided at the back of the text for almost all parts of all the odd-numbered exercises.

Chapter summaries. The last numbered section in each chapter provides a summary and historical review of the major ideas covered in that chapter. This is intended to give the reader an overview of the contents of the chapter and provide information for further study and applications. Such further study can be readily assisted by the list of references that is supplied.

In particular, the summaries at the ends of Chapters 1, 5, and 9 include tables on the enumeration formulas developed within each of these chapters. Sometimes these tables include results from earlier chapters in order to make comparisons and to show how the new results extend the prior ones.

Organization

The areas of discrete and combinatorial mathematics are somewhat new to the undergraduate curriculum, so there are several options as to which topics should be covered in these courses. Each instructor and each student may have different interests. Consequently, the coverage here is fairly broad, as a survey course mandates. Yet there will always be further topics that some readers may feel should be included. Furthermore, there will also be some differences of opinion with regard to the order in which some topics are presented in this text.

The nature and importance of the algorithmic approach to problem solving is stressed throughout the text. Ideas and approaches on problem solving are further strengthened by the interrelations between enumeration and structure, two other major topics that provide unifying threads for the material developed in the book.

The material is subdivided into four major areas. The first seven chapters form the underlying core of the book and present the fundamentals of discrete mathematics. The coverage here provides enough material for a one-quarter or one-semester course in discrete mathematics. The material in Chapter 2 can be reviewed by those with a background in logic. For those interested in developing and writing proofs, this material should be examined very carefully. A second course — one that emphasizes combinatorics — should include Chapters 8, 9, and 10 (and, time permitting, sections 1, 2, 3, 10, 11, and 12 of Chapter 16). In Chapter 9 some results from calculus are used; namely, fundamentals on differentiation and partial fraction decompositions. However, for those who wish to skip this chapter, sections 1, 2, 3, 6, and 7 of Chapter 10 can still be covered. A course that emphasizes the theory and applications of finite graphs can be developed from Chapters 11, 12, and 13. These chapters form the third major subdivision of the text. For a course in applied algebra, Chapters 14, 15, 16, and 17 (the fourth, and final, subdivision) deal with the algebraic structures — group, ring, Boolean algebra, and field — and include applications on cryptology, switching functions, algebraic coding theory, and combinatorial designs. Finally, a course on the role of discrete structures in computer science can be developed from the material in Chapters 11, 12, 13, 15, and sections 1–9 of Chapter 16. For here we find applications on switching functions, the RSA cryptosystem, and algebraic coding theory, as well as an introduction to graph theory and trees, and their role in optimization.

Other possible courses can be developed by considering the following chapter dependencies.

Chapter	Dependence on Prior Chapters
1	No dependence
2	No dependence (Hence an instructor can start a course in discrete mathematics with either the study of logic or an introduction to enumeration.)
3	1, 2
4	1, 2, 3
5	1, 2, 3, 4
6	1, 2, 3, 5 (Minor dependence in Section 6.1 on Sections 4.1, 4.2)
7	1, 2, 3, 5, 6 (Minor dependence in Section 7.2 on Sections 4.1, 4.2)
8	1, 3 (Minor dependence in Example 8.6 on Section 5.3)
9	1, 3
10	1, 3, 4, 5, 9 (Minor dependence in Example 10.33 on Section 7.3)
11	1, 2, 3, 4, 5 (Although some graph-theoretic ideas are mentioned in Chapters 5, 6, 7, 8, and 10, the material in this chapter is developed with no dependence on the graph-theoretic material given in these earlier results.)
12	1, 2, 3, 4, 5, 11
13	3, 5, 11, 12
14	2, 3, 4, 5, 7 (The Euler phi function (ϕ) is used in Section 14.3. This function is derived in Example 8.8 of Section 8.1 but the result can be used here in Chapter 14 without covering Chapter 8.)
15	2, 3, 5, 7
16	1, 2, 3, 4, 5, 7
17	2, 3, 4, 5, 7, 14

In addition, the index has been very carefully developed in order to make the text even more flexible. Terms are presented with primary listings and several secondary listings. Also there is a great deal of cross referencing. This is designed to help the instructor who may want to change the order of presentation and deviate from the straight and narrow.

Changes in the Fifth Edition

The changes here in the fifth edition of *Discrete and Combinatorial Mathematics* reflect the observations and recommendations of students and instructors who have used earlier editions of the text. As with the first four editions, the tone and purpose of the text remain intact. The author's goal is still the same: to provide within these pages a sound, readable, and understandable introduction to the foundations of discrete and combinatorial mathematics — for the beginning student or reader. Among the changes one will find in this fifth edition we mention the following:

• The examples in Section 4 of Chapter 1 now include material on *runs*, a concept that arises in the study of statistics — in particular, in the area of quality control.

• Exercise 13 for Section 3 of Chapter 2 develops the rule of inference known as *resolution*, a rule that serves as the basis for many computer programs designed to automate a reasoning system.

• The earlier editions of this text included a section that introduced the notion of probability. This section has now been expanded and three additional optional sections have been added for those who wish to further examine some of the introductory ideas associated with discrete probability — in particular, the axioms of probability, conditional probability, independence, Bayes' Theorem, and discrete random variables.

- The coverage on partial orders and total orders in Section 3 of Chapter 7 now includes an optional example where the Catalan numbers arise in this context.

- The introductory material in Section 1 of Chapter 8 has been rewritten to provide a more readable transition between the coverage on counting and Venn diagrams in Section 3 of Chapter 3 and the more general technique known as the Principle of Inclusion and Exclusion.

- One of the fascinating features of discrete and combinatorial mathematics is the variety of ways a given problem can be solved. In the fourth edition (in Chapters 1 and 3) the reader learned, in two different contexts, that a positive integer n had 2^{n-1} compositions — that is, there are 2^{n-1} ways to write n as an ordered sum of positive-integer summands. This result is now established in three other ways: (i) by the Principle of Mathematical Induction in Chapter 4; (ii) using generating functions in Chapter 9; and (iii) by solving a recurrence relation in Chapter 10.

- For those who want even more on discrete probability, Section 2 of Chapter 9 includes an example that deals with the geometric random variable.

- Section 2 of Chapter 10 now includes a discussion of the work by Gabriel Lamé in estimating the number of divisions used in the Euclidean algorithm to find the greatest common divisor of two positive integers.

- The Master theorem (of importance in the analysis of algorithms) is introduced and developed in an exercise for Section 6 of Chapter 10.

- The material on transport networks (in Section 3 of Chapter 13) has been updated and now incorporates the Edmonds-Karp algorithm in the procedure originally developed by Lester Ford and Delbert Fulkerson.

- The coverage on modular arithmetic in Section 3 of Chapter 14 now includes applications dealing with the linear congruential pseudorandom number generator, private-key cryptosystems, and modular exponentiation. Further, in Section 4 of Chapter 14, the material dealing with the Chinese Remainder Theorem, which was only stated in previous editions, now includes a proof of this result as well as an example dealing with how it is applied.

- Section 4 of Chapter 16 is new and optional. The material here provides an introduction to the RSA public-key cryptosystem and shows how one can apply some of the theoretical results developed in prior sections of the text.

- As with the second, third, and fourth editions, a great deal of effort has been applied in updating the summary and historical review at the end of each chapter. Consequently, new references and/or new editions are provided where appropriate.

- For this fifth edition, the following pictures and photographs have been added to the summary and historical review of certain chapters: a picture of Thomas Bayes and a photograph of Andrei Nikolayevich Kolmogorov in Chapter 3; a picture of Al-Khowârizmî in Chapter 4; a photograph of David A. Huffman in Chapter 12; and a photograph of Joseph B. Kruskal in Chapter 13.

Ancillaries

- There is an *Instructor's Solutions Manual* that is available, from the publisher, for those instructors who adopt the textbook for their classes. It contains the solutions and/or answers for all of the exercises within the 17 chapters and the three appendices of this textbook.

- There is also a *Student's Solutions Manual* that is available separately. It contains the solutions and/or answers for all of the odd-numbered exercises in the textbook. In some cases more than one solution is presented.

- The following Web site provides additional resources for learning more about discrete and combinatorial mathematics. In addition it also provides a way for readers to contact the author with comments, suggestions, or possible errors they have found.

www.aw.com/grimaldi

Acknowledgments

If space permitted, I should like to mention each of the students who provided help and encouragement when I was writing the five editions of this book. Their suggestions helped to remove many mistakes and ambiguities, thus improving the exposition. Most helpful in this category were Paul Griffith, Meredith Vannauker, Paul Barloon, Byron Bishop, Lee Beckham, Brett Hunsaker, Tom Vanderlaan, Michael Bryan, John Breitenbach, Dan Johnson, Brian Wilson, Allen Schneider, John Dowell, Charles Wilson, Richard Nichols, Charles Brads, Jonathan Atkins, Kenneth Schmidt, Donald Stanton, Mark Stremler, Stephen Smalley, Anthony Hinrichs, Kevin O'Bryant, and Nathan Terpstra.

I thank Larry Alldredge, Claude Anderson, David Rader, Matt Hopkins, John Rickert, and Martin Rivers for their comments on the computer science material, and Barry Farbrother, Paul Hogan, Dennis Lewis, Charles Kyker, Keith Hoover, Matthew Saltzman, and Jerome Wagner for their enlightening remarks on some of the applications.

I gratefully acknowledge the persistent enthusiasm and encouragement of the staff at Addison-Wesley (both past and present), especially Wayne Yuhasz, Thomas Taylor, Michael Payne, Charles Glaser, Mary Crittendon, Herb Merritt, Maria Szmauz, Adeline Ruggles, Stephanie Botvin, Jack Casteel, Jennifer Wall, Joanne Sousa Foster, Karen Guardino, Peggy McMahon, Deborah Schneider, Laurie Rosatone, Carolyn Lee-Davis, and Jennifer Albanese. William Hoffman, and especially RoseAnne Johnson and Barbara Pendergast, deserve the most recognition for their outstanding contributions to this fifth edition. The efforts put forth by Steven Finch in proofreading the text and that of Paul Lorczak who checked the accuracy of the answers to the exercises are also greatly appreciated.

I am also indebted to my colleagues John Kinney, Robert Lopez, Allen Broughton, Gary Sherman, George Berzsenyi, and especially Alfred Schmidt, for their interest and encouragement throughout the writing of this and/or earlier editions.

Thanks and appreciation are due the following reviewers of the first, second, third, fourth, and/or fifth editions.

Norma E. Abel	*Digital Equipment Corporation*
Larry Alldredge	*Qualcomm, Inc.*
Charles Anderson	*University of Colorado, Denver*
Claude W. Anderson III	*Rose-Hulman Institute of Technology*
David Arnold	*Baylor University*
V. K. Balakrishnan	*University of Maine at Orono*
Robert Barnhill	*University of Utah*
Dale Bedgood	*East Texas State University*
Jerry Beehler	*Tri-State University*
Katalin Bencsath	*Manhattan College*
Allan Bishop	*Western Illinois University*
Monte Boisen	*Virginia Polytechnic Institute*

Samuel Councilman	*California State University at Long Beach*
Robert Crawford	*Western Kentucky University*
Ellen Cunningham, SP	*Saint Mary-of-the-Woods College*
Carl DeVito	*Naval Postgraduate School*
Vladimir Drobot	*San Jose State University*
John Dye	*California State University at Northridge*
Carl Eckberg	*San Diego State University*
Michael Falk	*Northern Arizona University*
Marvin Freedman	*Boston University*
Robert Geitz	*Oberlin College*
James A. Glasenapp	*Rochester Institute of Technology*
Gary Gordon	*Lafayette College*
Harvey Greenberg	*University of Colorado, Denver*
Laxmi Gupta	*Rochester Institute of Technology*
Eleanor O. Hare	*Clemson University*
James Harper	*Central Washington University*
David S. Hart	*Rochester Institute of Technology*
Maryann Hastings	*Marymount College*
W. Mack Hill	*Worcester State College*
Stephen Hirtle	*University of Pittsburgh*
Arthur Hobbs	*Texas A&M University*
Dean Hoffman	*Auburn University*
Richard Iltis	*Willamette University*
David P. Jacobs	*Clemson University*
Robert Jajcay	*Indiana State University*
Akihiro Kanamori	*Boston University*
John Konvalina	*University of Nebraska at Omaha*
Rochelle Leibowitz	*Wheaton College*
James T. Lewis	*University of Rhode Island*
Y-Hsin Liu	*University of Nebraska at Omaha*
Joseph Malkevitch	*York College (CUNY)*
Brian Martensen	*The University of Texas at Austin*
Hugh Montgomery	*University of Michigan*
Thomas Morley	*Georgia Institute of Technology*
Richard Orr	*Rochester Institute of Technology*
Edwin P. Oxford	*Baylor University*
John Rausen	*New Jersey Institute of Technology*
Martin Rivers	*Lexmark International, Inc.*
Gabriel Robins	*University of Virginia*
Chris Rodger	*Auburn University*
James H. Schmerl	*University of Connecticut*
Paul S. Schnare	*Eastern Kentucky University*
Leo Schneider	*John Carroll University*
Debra Diny Scott	*University of Wisconsin at Green Bay*
Gary E. Stevens	*Hartwick College*
Dalton Tarwater	*Texas Tech University*
Jeff Tecosky-Feldman	*Harvard University*
W. L. Terwilliger	*Bowling Green State University*
Donald Thompson	*Pepperdine University*

Thomas Upson	*Rochester Institute of Technology*
W. D. Wallis	*Southern Illinois University*
Larry West	*Virginia Commonwealth University*
Yixin Zhang	*University of Nebraska at Omaha*

Special thanks are due to Douglas Shier of Clemson University for the outstanding work he did in reviewing the manuscripts of all five editions. Thanks are also due to Joan Shier for letting Doug review the fourth and fifth editions.

The translation for the dedication is due to Dr. Yvonne Panaro of Northern Virginia Community College. Thank you, Yvonne, and thank you, Patter (Patricia Wickes Thurston), for your role in obtaining the translation.

A text of this length requires the use of many references. The members of the library staff of Rose-Hulman Institute of Technology were always available when books and articles were needed, so it is only fitting to express one's appreciation for the efforts of John Robson, Sondra Nelson, Dong Chao, Jan Jerrell, and especially Amy Harshbarger and Margaret Ying. In addition, Keith Hoover and Raymond Bland are thanked for rescuing the author from the perils of many hardware problems.

The last, and surely the most important, note of thanks belongs once again to the ever-patient and encouraging now-retired secretary of the Rose-Hulman mathematics department — Mrs. Mary Lou McCullough. Thank you for the fifth time, Mary Lou, for all of your work!

Alas, the remaining errors, ambiguities, and misleading comments are once again the sole responsibility of the author.

R.P.G.
Terre Haute, Indiana

Contents

FUNDAMENTALS
OF DISCRETE
MATHEMATICS

1

Fundamental Principles of Counting

Enumeration, or counting, may strike one as an obvious process that a student learns when first studying arithmetic. But then, it seems, very little attention is paid to further development in counting as the student turns to "more difficult" areas in mathematics, such as algebra, geometry, trigonometry, and calculus. Consequently, this first chapter should provide some warning about the seriousness and difficulty of "mere" counting.

Enumeration does not end with arithmetic. It also has applications in such areas as coding theory, probability and statistics, and in the analysis of algorithms. Later chapters will offer some specific examples of these applications.

As we enter this fascinating field of mathematics, we shall come upon many problems that are very simple to state but somewhat "sticky" to solve. Thus, be sure to learn and understand the basic formulas — but do *not* rely on them too heavily. For without an analysis of each problem, a mere knowledge of formulas is next to useless. Instead, welcome the challenge to solve unusual problems or those that are different from problems you have encountered in the past. Seek solutions based on your own scrutiny, regardless of whether it reproduces what the author provides. There are often several ways to solve a given problem.

1.1
The Rules of Sum and Product

Our study of discrete and combinatorial mathematics begins with two basic principles of counting: the rules of sum and product. The statements and initial applications of these rules appear quite simple. In analyzing more complicated problems, one is often able to break down such problems into parts that can be solved using these basic principles. We want to develop the ability to "decompose" such problems and piece together our partial solutions in order to arrive at the final answer. A good way to do this is to analyze and solve many diverse enumeration problems, taking note of the principles being used. This is the approach we shall follow here.

Our first principle of counting can be stated as follows:

> **The Rule of Sum:** If a first task can be performed in m ways, while a second task can be performed in n ways, and the two tasks cannot be performed simultaneously, then performing either task can be accomplished in any one of $m + n$ ways.

Note that when we say that a particular occurrence, such as a first task, can come about in m ways, these m ways are assumed to be distinct, unless a statement is made to the contrary. This will be true throughout the entire text.

EXAMPLE 1.1	A college library has 40 textbooks on sociology and 50 textbooks dealing with anthropology. By the rule of sum, a student at this college can select among $40 + 50 = 90$ textbooks in order to learn more about one or the other of these two subjects.

EXAMPLE 1.2	The rule can be extended beyond two tasks as long as no pair of tasks can occur simultaneously. For instance, a computer science instructor who has, say, seven different introductory books each on C++, Java, and Perl can recommend any one of these 21 books to a student who is interested in learning a first programming language.

EXAMPLE 1.3	The computer science instructor of Example 1.2 has two colleagues. One of these colleagues has three textbooks on the analysis of algorithms, and the other has five such textbooks. If n denotes the maximum number of different books on this topic that this instructor can borrow from them, then $5 \leq n \leq 8$, for here both colleagues *may* own copies of the same textbook(s).

The following example introduces our second principle of counting.

EXAMPLE 1.4	In trying to reach a decision on plant expansion, an administrator assigns 12 of her employees to two committees. Committee A consists of five members and is to investigate possible favorable results from such an expansion. The other seven employees, committee B, will scrutinize possible unfavorable repercussions. Should the administrator decide to speak to just one committee member before making her decision, then by the rule of sum there are 12 employees she can call upon for input. However, to be a bit more unbiased, she decides to speak with a member of committee A on Monday, and then with a member of committee B on Tuesday, before reaching a decision. Using the following principle, we find that she can select two such employees to speak with in $5 \times 7 = 35$ ways.

> **The Rule of Product:** If a procedure can be broken down into first and second stages, and if there are m possible outcomes for the first stage and if, for each of these outcomes, there are n possible outcomes for the second stage, then the total procedure can be carried out, in the designated order, in mn ways.

EXAMPLE 1.5	The drama club of Central University is holding tryouts for a spring play. With six men and eight women auditioning for the leading male and female roles, by the rule of product the director can cast his leading couple in $6 \times 8 = 48$ ways.

EXAMPLE 1.6	Here various extensions of the rule are illustrated by considering the manufacture of license plates consisting of two letters followed by four digits.

a) If no letter or digit can be repeated, there are $26 \times 25 \times 10 \times 9 \times 8 \times 7 = 3{,}276{,}000$ different possible plates.

b) With repetitions of letters and digits allowed, $26 \times 26 \times 10 \times 10 \times 10 \times 10 = 6{,}760{,}000$ different license plates are possible.

c) If repetitions are allowed, as in part (b), how many of the plates have only vowels (A, E, I, O, U) and even digits? (0 is an even integer.)

EXAMPLE 1.7

In order to store data, a computer's main memory contains a large collection of circuits, each of which is capable of storing a *bit* — that is, one of the *bi*nary dig*its* 0 or 1. These storage circuits are arranged in units called (memory) cells. To identify the cells in a computer's main memory, each is assigned a unique name called its *address*. For some computers, such as embedded microcontrollers (as found in the ignition system for an automobile), an address is represented by an ordered list of eight bits, collectively referred to as a *byte*. Using the rule of product, there are $2 \times 2 \times 2 \times 2 \times 2 \times 2 \times 2 \times 2 = 2^8 = 256$ such bytes. So we have 256 addresses that may be used for cells where certain information may be stored.

A kitchen appliance, such as a microwave oven, incorporates an embedded microcontroller. These "small computers" (such as the PICmicro microcontroller) contain thousands of memory cells and use two-byte addresses to identify these cells in their main memory. Such addresses are made up of two consecutive bytes, or 16 consecutive bits. Thus there are $256 \times 256 = 2^8 \times 2^8 = 2^{16} = 65{,}536$ available addresses that could be used to identify cells in the main memory. Other computers use addressing systems of four bytes. This 32-bit architecture is presently used in the Pentium[†] processor, where there are as many as $2^8 \times 2^8 \times 2^8 \times 2^8 = 2^{32} = 4{,}294{,}967{,}296$ addresses for use in identifying the cells in main memory. When a programmer deals with the UltraSPARC[‡] or Itanium[§] processors, he or she considers memory cells with eight-byte addresses. Each of these addresses comprises $8 \times 8 = 64$ bits, and there are $2^{64} = 18{,}446{,}744{,}073{,}709{,}551{,}616$ possible addresses for this architecture. (Of course, not all of these possibilities are actually used.)

EXAMPLE 1.8

At times it is necessary to combine several different counting principles in the solution of one problem. Here we find that the rules of both sum and product are needed to attain the answer.

At the AWL corporation Mrs. Foster operates the Quick Snack Coffee Shop. The menu at her shop is limited: six kinds of muffins, eight kinds of sandwiches, and five beverages (hot coffee, hot tea, iced tea, cola, and orange juice). Ms. Dodd, an editor at AWL, sends her assistant Carl to the shop to get her lunch — either a muffin and a hot beverage or a sandwich and a cold beverage.

By the rule of product, there are $6 \times 2 = 12$ ways in which Carl can purchase a muffin and hot beverage. A second application of this rule shows that there are $8 \times 3 = 24$ possibilities for a sandwich and cold beverage. So by the rule of sum, there are $12 + 24 = 36$ ways in which Carl can purchase Ms. Dodd's lunch.

[†]Pentium (R) is a registered trademark of the Intel Corporation.

[‡]The UltraSPARC processor is manufactured by Sun (R) Microsystems, Inc.

[§]Itanium (TM) is a trademark of the Intel Corporation.

1.2
Permutations

Continuing to examine applications of the rule of product, we turn now to counting linear arrangements of objects. These arrangements are often called *permutations* when the objects are distinct. We shall develop some systematic methods for dealing with linear arrangements, starting with a typical example.

EXAMPLE 1.9

In a class of 10 students, five are to be chosen and seated in a row for a picture. How many such linear arrangements are possible?

The key word here is *arrangement*, which designates the importance of *order*. If A, B, C, . . . , I, J denote the 10 students, then BCEFI, CEFIB, and ABCFG are three such different arrangements, even though the first two involve the same five students.

To answer this question, we consider the positions and possible numbers of students we can choose from in order to fill each position. The filling of a position is a stage of our procedure.

$$10 \quad \times \quad 9 \quad \times \quad 8 \quad \times \quad 7 \quad \times \quad 6$$

1st position	2nd position	3rd position	4th position	5th position

Each of the 10 students can occupy the 1st position in the row. Because repetitions are not possible here, we can select only one of the nine remaining students to fill the 2nd position. Continuing in this way, we find only six students to select from in order to fill the 5th and final position. This yields a total of 30,240 possible arrangements of five students selected from the class of 10.

Exactly the same answer is obtained if the positions are filled from right to left — namely, $6 \times 7 \times 8 \times 9 \times 10$. If the 3rd position is filled first, the 1st position second, the 4th position third, the 5th position fourth, and the 2nd position fifth, then the answer is $9 \times 6 \times 10 \times 8 \times 7$, still the same value, 30,240.

As in Example 1.9, the product of certain consecutive positive integers often comes into play in enumeration problems. Consequently, the following notation proves to be quite useful when we are dealing with such counting problems. It will frequently allow us to express our answers in a more convenient form.

Definition 1.1

For an integer $n \geq 0$, *n factorial* (denoted $n!$) is defined by

$$0! = 1,$$
$$n! = (n)(n-1)(n-2) \cdots (3)(2)(1), \quad \text{for} \quad n \geq 1.$$

One finds that $1! = 1, 2! = 2, 3! = 6, 4! = 24$, and $5! = 120$. In addition, for each $n \geq 0$, $(n+1)! = (n+1)(n!)$.

Before we proceed any further, let us try to get a somewhat better appreciation for how fast $n!$ grows. We can calculate that $10! = 3,628,800$, and it just so happens that this is exactly the number of *seconds* in six *weeks*. Consequently, 11! exceeds the number of seconds in one *year*, 12! exceeds the number in 12 years, and 13! surpasses the number of seconds in a *century*.

If we make use of the factorial notation, the answer in Example 1.9 can be expressed in the following more compact form:

$$10 \times 9 \times 8 \times 7 \times 6 = 10 \times 9 \times 8 \times 7 \times 6 \times \frac{5 \times 4 \times 3 \times 2 \times 1}{5 \times 4 \times 3 \times 2 \times 1} = \frac{10!}{5!}.$$

Definition 1.2

Given a collection of n distinct objects, any (linear) arrangement of these objects is called a *permutation* of the collection.

Starting with the letters a, b, c, there are six ways to arrange, or permute, all of the letters: abc, acb, bac, bca, cab, cba. If we are interested in arranging only two of the letters at a time, there are six such size-2 permutations: ab, ba, ac, ca, bc, cb.

If there are n distinct objects and r is an integer, with $1 \leq r \leq n$, then by the rule of product, the number of permutations of size r for the n objects is

$$P(n, r) = \underset{\substack{\text{1st} \\ \text{position}}}{n} \times \underset{\substack{\text{2nd} \\ \text{position}}}{(n-1)} \times \underset{\substack{\text{3rd} \\ \text{position}}}{(n-2)} \times \cdots \times \underset{\substack{\text{rth} \\ \text{position}}}{(n-r+1)}$$

$$= (n)(n-1)(n-2) \cdots (n-r+1) \times \frac{(n-r)(n-r-1) \cdots (3)(2)(1)}{(n-r)(n-r-1) \cdots (3)(2)(1)}$$

$$= \frac{n!}{(n-r)!}.$$

For $r = 0$, $P(n, 0) = 1 = n!/(n-0)!$, so $P(n, r) = n!/(n-r)!$ holds for all $0 \leq r \leq n$. A special case of this result is Example 1.9, where $n = 10$, $r = 5$, and $P(10, 5) = 30{,}240$. When permuting all of the n objects in the collection, we have $r = n$ and find that $P(n, n) = n!/0! = n!$.

Note, for example, that if $n \geq 2$, then $P(n, 2) = n!/(n-2)! = n(n-1)$. When $n > 3$ one finds that $P(n, n-3) = n!/[n-(n-3)]! = n!/3! = (n)(n-1)(n-2) \cdots (5)(4)$.

The number of permutations of size r, where $0 \leq r \leq n$, from a collection of n objects, is $P(n, r) = n!/(n-r)!$. (Remember that $P(n, r)$ counts (linear) arrangements in which the objects can*not* be repeated.) However, if repetitions are allowed, then by the rule of product there are n^r possible arrangements, with $r \geq 0$.

EXAMPLE 1.10

The number of permutations of the letters in the word COMPUTER is 8!. If only five of the letters are used, the number of permutations (of size 5) is $P(8, 5) = 8!/(8-5)! = 8!/3! = 6720$. If repetitions of letters are allowed, the number of possible 12-letter sequences is $8^{12} \doteq 6.872 \times 10^{10}$.[†]

EXAMPLE 1.11

Unlike Example 1.10, the number of (linear) arrangements of the four letters in BALL is 12, not 4! (= 24). The reason is that we do not have four distinct letters to arrange. To get the 12 arrangements, we can list them as in Table 1.1(a).

[†]The symbol "$\doteq$" is read "is approximately equal to."

Table 1.1

A	B	L	L	A	B	L_1	L_2	A	B	L_2	L_1
A	L	B	L	A	L_1	B	L_2	A	L_2	B	L_1
A	L	L	B	A	L_1	L_2	B	A	L_2	L_1	B
B	A	L	L	B	A	L_1	L_2	B	A	L_2	L_1
B	L	A	L	B	L_1	A	L_2	B	L_2	A	L_1
B	L	L	A	B	L_1	L_2	A	B	L_2	L_1	A
L	A	B	L	L_1	A	B	L_2	L_2	A	B	L_1
L	A	L	B	L_1	A	L_2	B	L_2	A	L_1	B
L	B	A	L	L_1	B	A	L_2	L_2	B	A	L_1
L	B	L	A	L_1	B	L_2	A	L_2	B	L_1	A
L	L	A	B	L_1	L_2	A	B	L_2	L_1	A	B
L	L	B	A	L_1	L_2	B	A	L_2	L_1	B	A

(a)   (b)

If the two L's are distinguished as L_1, L_2, then we can use our previous ideas on permutations of distinct objects; with the four distinct symbols B, A, L_1, L_2, we have 4! = 24 permutations. These are listed in Table 1.1(b). Table 1.1 reveals that for each arrangement in which the L's are indistinguishable there corresponds a *pair* of permutations with distinct L's. Consequently,

2 × (Number of arrangements of the letters B, A, L, L)

= (Number of permutations of the symbols B, A, L_1, L_2),

and the answer to the original problem of finding all the arrangements of the four letters in BALL is 4!/2 = 12.

EXAMPLE 1.12

Using the idea developed in Example 1.11, we now consider the arrangements of all nine letters in DATABASES.

There are 3! = 6 arrangements with the A's distinguished for each arrangement in which the A's are not distinguished. For example, $DA_1TA_2BA_3SES$, $DA_1TA_3BA_2SES$, $DA_2TA_1BA_3SES$, $DA_2TA_3BA_1SES$, $DA_3TA_1BA_2SES$, and $DA_3TA_2BA_1SES$ all correspond to DATABASES, when we remove the subscripts on the A's. In addition, to the arrangement $DA_1TA_2BA_3SES$ there corresponds the pair of permutations $DA_1TA_2BA_3S_1ES_2$ and $DA_1TA_2BA_3S_2ES_1$, when the S's are distinguished. Consequently,

(2!)(3!)(Number of arrangements of the letters in DATABASES)

= (Number of permutations of the symbols D, A_1, T, A_2, B, A_3, S_1, E, S_2),

so the number of arrangements of the nine letters in DATABASES is 9!/(2! 3!) = 30,240.

Before stating a general principle for arrangements with repeated symbols, note that in our prior two examples we solved a new type of problem by relating it to previous enumeration principles. This practice is common in mathematics in general, and often occurs in the derivations of discrete and combinatorial formulas.

If there are n objects with n_1 indistinguishable objects of a first type, n_2 indistinguishable objects of a second type, ..., and n_r indistinguishable objects of an rth type, where $n_1 + n_2 + \cdots + n_r = n$, then there are $\dfrac{n!}{n_1!\, n_2! \cdots n_r!}$ (linear) arrangements of the given n objects.

EXAMPLE 1.13

The MASSASAUGA is a brown and white venomous snake indigenous to North America. Arranging all of the letters in MASSASAUGA, we find that there are

$$\frac{10!}{4!\, 3!\, 1!\, 1!\, 1!} = 25{,}200$$

possible arrangements. Among these are

$$\frac{7!}{3!\, 1!\, 1!\, 1!\, 1!} = 840$$

in which all four A's are together. To get this last result, we considered all arrangements of the seven symbols AAAA (one symbol), S, S, S, M, U, G.

EXAMPLE 1.14

Determine the number of (staircase) paths in the xy-plane from $(2, 1)$ to $(7, 4)$, where each such path is made up of individual steps going one unit to the right (R) or one unit upward (U). The blue lines in Fig. 1.1 show two of these paths.

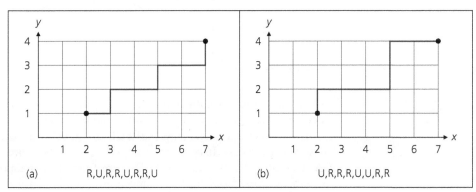

(a)   R,U,R,R,U,R,R,U   (b)   U,R,R,R,U,U,R,R

Figure 1.1

Beneath each path in Fig. 1.1 we have listed the individual steps. For example, in part (a) the list R, U, R, R, U, R, R, U indicates that starting at the point $(2, 1)$, we first move one unit to the right [to $(3, 1)$], then one unit upward [to $(3, 2)$], followed by two units to the right [to $(5, 2)$], and so on, until we reach the point $(7, 4)$. The path consists of five R's for moves to the right and three U's for moves upward.

The path in part (b) of the figure is also made up of five R's and three U's. In general, the overall trip from $(2, 1)$ to $(7, 4)$ requires $7 - 2 = 5$ horizontal moves to the right and $4 - 1 = 3$ vertical moves upward. Consequently, each path corresponds to a list of five R's and three U's, and the solution for the number of paths emerges as the number of arrangements of the five R's and three U's, which is $8!/(5!\, 3!) = 56$.

EXAMPLE 1.15

We now do something a bit more abstract and prove that if n and k are positive integers with $n = 2k$, then $n!/2^k$ is an integer. Because our argument relies on counting, it is an example of a *combinatorial proof*.

Consider the n symbols $x_1, x_1, x_2, x_2, \ldots, x_k, x_k$. The number of ways in which we can arrange all of these $n = 2k$ symbols is an integer that equals

$$\underbrace{\frac{n!}{2!\, 2! \cdots 2!}}_{k \text{ factors of } 2!} = \frac{n!}{2^k}.$$

Finally, we will apply what has been developed so far to a situation in which the arrangements are no longer linear.

EXAMPLE 1.16

If six people, designated as A, B, . . . , F, are seated about a round table, how many different circular arrangements are possible, if arrangements are considered the same when one can be obtained from the other by rotation? [In Fig. 1.2, arrangements (a) and (b) are considered identical, whereas (b), (c), and (d) are three distinct arrangements.]

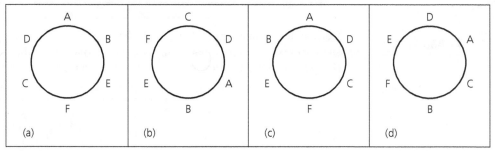

Figure 1.2

We shall try to relate this problem to previous ones we have already encountered. Consider Figs. 1.2(a) and (b). Starting at the top of the circle and moving clockwise, we list the distinct linear arrangements ABEFCD and CDABEF, which correspond to the same circular arrangement. In addition to these two, four other linear arrangements — BEFCDA, DABEFC, EFCDAB, and FCDABE — are found to correspond to the same circular arrangement as in (a) or (b). So inasmuch as each circular arrangement corresponds to six linear arrangements, we have $6 \times$ (Number of circular arrangements of A, B, . . . , F) = (Number of linear arrangements of A, B, . . . , F) = 6!.

Consequently, there are $6!/6 = 5! = 120$ arrangements of A, B, . . . , F around the circular table.

EXAMPLE 1.17

Suppose now that the six people of Example 1.16 are three married couples and that A, B, and C are the females. We want to arrange the six people around the table so that the sexes alternate. (Once again, arrangements are considered identical if one can be obtained from the other by rotation.)

Before we solve this problem, let us solve Example 1.16 by an alternative method, which will assist us in solving our present problem. If we place A at the table as shown in Fig. 1.3(a), five locations (clockwise from A) remain to be filled. Using B, C, . . . , F to fill

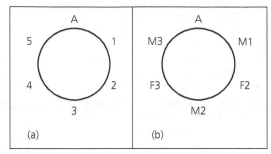

Figure 1.3

these five positions is the problem of permuting B, C, . . . , F in a linear manner, and this can be done in 5! = 120 ways.

To solve the new problem of alternating the sexes, consider the method shown in Fig. 1.3(b). A (a female) is placed as before. The next position, clockwise from A, is marked M1 (Male 1) and can be filled in three ways. Continuing clockwise from A, position F2 (Female 2) can be filled in two ways. Proceeding in this manner, by the rule of product, there are $3 \times 2 \times 2 \times 1 \times 1 = 12$ ways in which these six people can be arranged with no two men or women seated next to each other.

EXERCISES 1.1 AND 1.2

1. During a local campaign, eight Republican and five Democratic candidates are nominated for president of the school board.

a) If the president is to be one of these candidates, how many possibilities are there for the eventual winner?

b) How many possibilities exist for a pair of candidates (one from each party) to oppose each other for the eventual election?

c) Which counting principle is used in part (a)? in part (b)?

2. Answer part (c) of Example 1.6.

3. Buick automobiles come in four models, 12 colors, three engine sizes, and two transmission types. (a) How many distinct Buicks can be manufactured? (b) If one of the available colors is blue, how many different blue Buicks can be manufactured?

4. The board of directors of a pharmaceutical corporation has 10 members. An upcoming stockholders' meeting is scheduled to approve a new slate of company officers (chosen from the 10 board members).

a) How many different slates consisting of a president, vice president, secretary, and treasurer can the board present to the stockholders for their approval?

b) Three members of the board of directors are physicians. How many slates from part (a) have (i) a physician nominated for the presidency? (ii) exactly one physician appearing on the slate? (iii) at least one physician appearing on the slate?

5. While on a Saturday shopping spree Jennifer and Tiffany witnessed two men driving away from the front of a jewelry shop, just before a burglar alarm started to sound. Although everything happened rather quickly, when the two young ladies were questioned they were able to give the police the following information about the license plate (which consisted of two letters followed by four digits) on the get-away car. Tiffany was sure that the second letter on the plate was either an O or a Q and the last digit was either a 3 or an 8. Jennifer told the investigator that the first letter on the plate was either a C or a G and that the first digit was definitely a 7. How many different license plates will the police have to check out?

6. To raise money for a new municipal pool, the chamber of commerce in a certain city sponsors a race. Each participant pays a $5 entrance fee and has a chance to win one of the different-sized trophies that are to be awarded to the first eight runners who finish.

a) If 30 people enter the race, in how many ways will it be possible to award the trophies?

b) If Roberta and Candice are two participants in the race, in how many ways can the trophies be awarded with these two runners among the top three?

7. A certain "Burger Joint" advertises that a customer can have his or her hamburger with or without any or all of the following: catsup, mustard, mayonnaise, lettuce, tomato, onion, pickle, cheese, or mushrooms. How many different kinds of hamburger orders are possible?

8. Matthew works as a computer operator at a small university. One evening he finds that 12 computer programs have been submitted earlier that day for batch processing. In how many ways can Matthew order the processing of these programs if (a) there are no restrictions? (b) he considers four of the programs higher in priority than the other eight and wants to process those four first? (c) he first separates the programs into four of top priority, five of lesser priority, and three of least priority, and he wishes to process the 12 programs in such a way that the top-priority programs are processed first and the three programs of least priority are processed last?

9. Patter's Pastry Parlor offers eight different kinds of pastry and six different kinds of muffins. In addition to bakery items one can purchase small, medium, or large containers of the following beverages: coffee (black, with cream, with sugar, or with cream and sugar), tea (plain, with cream, with sugar, with cream and sugar, with lemon, or with lemon and sugar), hot cocoa, and orange juice. When Carol comes to Patter's, in how many ways can she order

a) one bakery item and one medium-sized beverage for herself?

b) one bakery item and one container of coffee for herself and one muffin and one container of tea for her boss, Ms. Didio?

c) one piece of pastry and one container of tea for herself, one muffin and a container of orange juice for Ms. Didio, and one bakery item and one container of coffee for each of her two assistants, Mr. Talbot and Mrs. Gillis?

10. Pamela has 15 different books. In how many ways can she place her books on two shelves so that there is at least one book on each shelf? (Consider the books in each arrangement to be stacked one next to the other, with the first book on each shelf at the left of the shelf.)

11. Three small towns, designated by A, B, and C, are interconnected by a system of two-way roads, as shown in Fig. 1.4.

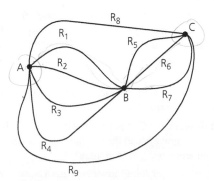

Figure 1.4

a) In how many ways can Linda travel from town A to town C?

b) How many different round trips can Linda travel from town A to town C and back to town A?

c) How many of the round trips in part (b) are such that the return trip (from town C to town A) is at least partially different from the route Linda takes from town A to town C? (For example, if Linda travels from town A to town C along roads R_1 and R_6, then on her return she might take roads R_6 and R_3, or roads R_7 and R_2, or road R_9, among other possibilities, but she does *not* travel on roads R_6 *and* R_1.)

12. List all the permutations for the letters a, c, t.

13. a) How many permutations are there for the eight letters a, c, f, g, i, t, w, x?

b) Consider the permutations in part (a). How many start with the letter t? How many start with the letter t and end with the letter c?

14. Evaluate each of the following.

a) $P(7, 2)$ **b)** $P(8, 4)$ **c)** $P(10, 7)$ **d)** $P(12, 3)$

15. In how many ways can the symbols a, b, c, d, e, e, e, e, e be arranged so that no e is adjacent to another e?

16. An alphabet of 40 symbols is used for transmitting messages in a communication system. How many distinct messages (lists of symbols) of 25 symbols can the transmitter generate if symbols can be repeated in the message? How many if 10 of the 40 symbols can appear only as the first and/or last symbols of the message, the other 30 symbols can appear anywhere, and repetitions of all symbols are allowed?

17. In the Internet each network interface of a computer is assigned one, or more, Internet addresses. The nature of these Internet addresses is dependent on network size. For the Internet Standard regarding reserved network numbers (STD 2), each address is a 32-bit string which falls into one of the following three classes: (1) A class A address, used for the largest networks, begins with a 0 which is then followed by a seven-bit *network number*, and then a 24-bit *local address*. However, one is restricted from using the network numbers of all 0's or all 1's and the local addresses of all 0's or all 1's. (2) The class B address is meant for an intermediate-sized network. This address starts with the two-bit string 10, which is followed by a 14-bit network number and then a 16-bit local address. But the local addresses of all 0's or all 1's are not permitted. (3) Class C addresses are used for the smallest networks. These addresses consist of the three-bit string 110, followed by a 21-bit network number, and then an eight-bit local address. Once again the local addresses of all 0's or all 1's are excluded. How many different addresses of each class are available on the Internet, for this Internet Standard?

18. Morgan is considering the purchase of a low-end computer system. After some careful investigating, she finds that there are seven basic systems (each consisting of a monitor, CPU, keyboard, and mouse) that meet her requirements. Furthermore, she also plans to buy one of four modems, one of three CD ROM drives, and one of six printers. (Here each peripheral device of a given type, such as the modem, is compatible with all seven basic systems.) In how many ways can Morgan configure her low-end computer system?

19. A computer science professor has seven different programming books on a bookshelf. Three of the books deal with C++, the other four with Java. In how many ways can the professor arrange these books on the shelf (a) if there are no restrictions? (b) if the languages should alternate? (c) if all the C++ books must be next to each other? (d) if all the C++ books must be next to each other and all the Java books must be next to each other?

20. Over the Internet, data are transmitted in structured blocks of bits called *datagrams*.

   a) In how many ways can the letters in DATAGRAM be arranged?

   b) For the arrangements of part (a), how many have all three A's together?

21. a) How many arrangements are there of all the letters in SOCIOLOGICAL?

   b) In how many of the arrangements in part (a) are A and G adjacent?

   c) In how many of the arrangements in part (a) are all the vowels adjacent?

22. How many positive integers n can we form using the digits 3, 4, 4, 5, 5, 6, 7 if we want n to exceed 5,000,000?

23. Twelve clay targets (identical in shape) are arranged in four hanging columns, as shown in Fig. 1.5. There are four red targets in the first column, three white ones in the second column, two green targets in the third column, and three blue ones in the fourth column. To join her college drill team, Deborah must break all 12 of these targets (using her pistol and only 12 bullets) and in so doing must always break the existing target at the bottom of a column. Under these conditions, in how many different orders can Deborah shoot down (and break) the 12 targets?

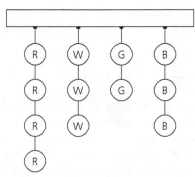

Figure 1.5

24. Show that for all integers $n, r \geq 0$, if $n + 1 > r$, then
$$P(n + 1, r) = \left(\frac{n+1}{n+1-r} \right) P(n, r).$$

25. Find the value(s) of n in each of the following:
(a) $P(n, 2) = 90$, (b) $P(n, 3) = 3P(n, 2)$, and
(c) $2P(n, 2) + 50 = P(2n, 2)$.

26. How many different paths in the xy-plane are there from $(0, 0)$ to $(7, 7)$ if a path proceeds one step at a time by going either one space to the right (R) or one space upward (U)? How many such paths are there from $(2, 7)$ to $(9, 14)$? Can any general statement be made that incorporates these two results?

27. a) How many distinct paths are there from $(-1, 2, 0)$ to $(1, 3, 7)$ in Euclidean three-space if each move is one of the following types?

   (H): $(x, y, z) \rightarrow (x + 1, y, z)$;

   (V): $(x, y, z) \rightarrow (x, y + 1, z)$;

   (A): $(x, y, z) \rightarrow (x, y, z + 1)$

   b) How many such paths are there from $(1, 0, 5)$ to $(8, 1, 7)$?

   c) Generalize the results in parts (a) and (b).

28. a) Determine the value of the integer variable *counter* after execution of the following program segment. (Here i, j, and k are integer variables.)

```
counter := 0
for i := 1 to 12 do
    counter := counter + 1
for j := 5 to 10 do
    counter := counter + 2
for k := 15 downto 8 do
    counter := counter + 3
```

   b) Which counting principle is at play in part (a)?

29. Consider the following program segment where i, j, and k are integer variables.

```
for i := 1 to 12 do
    for j := 5 to 10 do
        for k := 15 downto 8 do
            print (i - j)*k
```

   a) How many times is the **print** statement executed?

   b) Which counting principle is used in part (a)?

30. A sequence of letters of the form abcba, where the expression is unchanged upon reversing order, is an example of a *palindrome* (of five letters). (a) If a letter may appear more than twice, how many palindromes of five letters are there? of six letters? (b) Repeat part (a) under the condition that no letter appears more than twice.

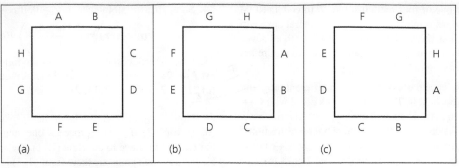

Figure 1.6

31. Determine the number of six-digit integers (no leading zeros) in which (a) no digit may be repeated; (b) digits may be repeated. Answer parts (a) and (b) with the extra condition that the six-digit integer is (i) even; (ii) divisible by 5; (iii) divisible by 4.

32. a) Provide a combinatorial argument to show that if n and k are positive integers with $n = 3k$, then $n!/(3!)^k$ is an integer.

b) Generalize the result of part (a).

33. a) In how many possible ways could a student answer a 10-question true-false test?

b) In how many ways can the student answer the test in part (a) if it is possible to leave a question unanswered in order to avoid an extra penalty for a wrong answer?

34. How many distinct four-digit integers can one make from the digits 1, 3, 3, 7, 7, and 8?

35. a) In how many ways can seven people be arranged about a circular table?

b) If two of the people insist on sitting next to each other, how many arrangements are possible?

36. a) In how many ways can eight people, denoted A, B, . . . , H be seated about the square table shown in Fig. 1.6, where Figs. 1.6(a) and 1.6(b) are considered the same but are distinct from Fig. 1.6(c)?

b) If two of the eight people, say A and B, do not get along well, how many different seatings are possible with A and B not sitting next to each other?

37. Sixteen people are to be seated at two circular tables, one of which seats 10 while the other seats six. How many different seating arrangements are possible?

38. A committee of 15 — nine women and six men — is to be seated at a circular table (with 15 seats). In how many ways can the seats be assigned so that no two men are seated next to each other?

39. Write a computer program (or develop an algorithm) to determine whether there is a three-digit integer $abc \ (= 100a + 10b + c)$ where $abc = a! + b! + c!$.

1.3
Combinations: The Binomial Theorem

The standard deck of playing cards consists of 52 cards comprising four suits: clubs, diamonds, hearts, and spades. Each suit has 13 cards: ace, 2, 3, . . . , 9, 10, jack, queen, king. If we are asked to draw three cards from a standard deck, in succession and without replacement, then by the rule of product there are

$$52 \times 51 \times 50 = \frac{52!}{49!} = P(52, 3)$$

possibilities, one of which is AH (ace of hearts), 9C (nine of clubs), KD (king of diamonds). If instead we simply select three cards at one time from the deck so that the order of selection of the cards is no longer important, then the six permutations AH–9C–KD, AH–KD–9C, 9C–AH–KD, 9C–KD–AH, KD–9C–AH, and KD–AH–9C all correspond to just one (unordered) selection. Consequently, each selection, or combination, of three cards, *with no reference to order*, corresponds to 3! permutations of three cards. In equation form this translates into

(3!) × (Number of selections of size 3 from a deck of 52)

$$= \text{Number of permutations of size 3 for the 52 cards}$$

$$= P(52, 3) = \frac{52!}{49!}.$$

Consequently, three cards can be drawn, without replacement, from a standard deck in $52!/(3! \, 49!) = 22,100$ ways.

> If we start with n distinct objects, each *selection*, or *combination*, of r of these objects, with no reference to order, corresponds to $r!$ permutations of size r from the n objects. Thus the number of combinations of size r from a collection of size n is
>
> $$C(n, r) = \frac{P(n, r)}{r!} = \frac{n!}{r!(n - r)!}, \qquad 0 \le r \le n.$$

In addition to $C(n, r)$ the symbol $\binom{n}{r}$ is also frequently used. Both $C(n, r)$ and $\binom{n}{r}$ are sometimes read "n choose r." Note that for all $n \ge 0$, $C(n, 0) = C(n, n) = 1$. Further, for all $n \ge 1$, $C(n, 1) = C(n, n - 1) = n$. When $0 \le n < r$, then $C(n, r) = \binom{n}{r} = 0$.

A word to the wise! When dealing with any counting problem, we should ask ourselves about the importance of order in the problem. When order is relevant, we think in terms of permutations and arrangements and the rule of product. When order is not relevant, combinations could play a key role in solving the problem.

EXAMPLE 1.18

A hostess is having a dinner party for some members of her charity committee. Because of the size of her home, she can invite only 11 of the 20 committee members. Order is not important, so she can invite "the lucky 11" in $C(20, 11) = \binom{20}{11} = 20!/(11! \, 9!) = 167,960$ ways. However, once the 11 arrive, how she arranges them around her rectangular dining table is an arrangement problem. Unfortunately, no part of the theory of combinations and permutations can help our hostess deal with "the offended nine" who were not invited.

EXAMPLE 1.19

Lynn and Patti decide to buy a PowerBall ticket. To win the grand prize for PowerBall one must match five numbers selected from 1 to 49 inclusive and then must also match the powerball, an integer from 1 to 42 inclusive. Lynn selects the five numbers (between 1 and 49 inclusive). This she can do in $\binom{49}{5}$ ways (since matching does *not* involve order). Meanwhile Patti selects the powerball—here there are $\binom{42}{1}$ possibilities. Consequently, by the rule of product, Lynn and Patti can select the six numbers for their PowerBall ticket in $\binom{49}{5}\binom{42}{1} = 80,089,128$ ways.

EXAMPLE 1.20

a) A student taking a history examination is directed to answer any seven of 10 essay questions. There is no concern about order here, so the student can answer the examination in

$$\binom{10}{7} = \frac{10!}{7! \, 3!} = \frac{10 \times 9 \times 8}{3 \times 2 \times 1} = 120 \text{ ways.}$$

b) If the student must answer three questions from the first five and four questions from the last five, three questions can be selected from the first five in $\binom{5}{3} = 10$ ways, and the other four questions can be selected in $\binom{5}{4} = 5$ ways. Hence, by the rule of product, the student can complete the examination in $\binom{5}{3}\binom{5}{4} = 10 \times 5 = 50$ ways.

c) Finally, should the directions on this examination indicate that the student must answer seven of the 10 questions where at least three are selected from the first five, then there are three cases to consider:

 i) The student answers three of the first five questions and four of the last five: By the rule of product this can happen in $\binom{5}{3}\binom{5}{4} = 10 \times 5 = 50$ ways, as in part (b).

 ii) Four of the first five questions and three of the last five questions are selected by the student: This can come about in $\binom{5}{4}\binom{5}{3} = 5 \times 10 = 50$ ways — again by the rule of product.

 iii) The student decides to answer all five of the first five questions and two of the last five: The rule of product tells us that this last case can occur in $\binom{5}{5}\binom{5}{2} = 1 \times 10 = 10$ ways.

Combining the results for cases (i), (ii), and (iii), by the rule of sum we find that the student can make $\binom{5}{3}\binom{5}{4} + \binom{5}{4}\binom{5}{3} + \binom{5}{5}\binom{5}{2} = 50 + 50 + 10 = 110$ selections of seven (out of 10) questions where each selection includes at least three of the first five questions.

EXAMPLE 1.21

a) At Rydell High School, the gym teacher must select nine girls from the junior and senior classes for a volleyball team. If there are 28 juniors and 25 seniors, she can make the selection in $\binom{53}{9} = 4,431,613,550$ ways.

b) If two juniors and one senior are the best spikers and must be on the team, then the rest of the team can be chosen in $\binom{50}{6} = 15,890,700$ ways.

c) For a certain tournament the team must comprise four juniors and five seniors. The teacher can select the four juniors in $\binom{28}{4}$ ways. For each of these selections she has $\binom{25}{5}$ ways to choose the five seniors. Consequently, by the rule of product, she can select her team in $\binom{28}{4}\binom{25}{5} = 1,087,836,750$ ways for this particular tournament.

Some problems can be treated from the viewpoint of either arrangements or combinations, depending on how one analyzes the situation. The following example demonstrates this.

EXAMPLE 1.22

The gym teacher of Example 1.21 must make up four volleyball teams of nine girls each from the 36 freshman girls in her P.E. class. In how many ways can she select these four teams? Call the teams A, B, C, and D.

a) To form team A, she can select any nine girls from the 36 enrolled in $\binom{36}{9}$ ways. For team B the selection process yields $\binom{27}{9}$ possibilities. This leaves $\binom{18}{9}$ and $\binom{9}{9}$ possible ways to select teams C and D, respectively. So by the rule of product, the four teams can be chosen in

$$\binom{36}{9}\binom{27}{9}\binom{18}{9}\binom{9}{9} = \left(\frac{36!}{9!\,27!}\right)\left(\frac{27!}{9!\,18!}\right)\left(\frac{18!}{9!\,9!}\right)\left(\frac{9!}{9!\,0!}\right)$$

$$= \frac{36!}{9!\,9!\,9!\,9!} \doteq 2.145 \times 10^{19} \text{ ways.}$$

b) For an alternative solution, consider the 36 students lined up as follows:

1st	2nd	3rd		35th	36th
student	student	student	$\cdots$	student	student

To select the four teams, we must distribute nine A's, nine B's, nine C's, and nine D's in the 36 spaces. The number of ways in which this can be done is the number of arrangements of 36 letters comprising nine each of A, B, C, and D. This is now the familiar problem of arrangements of nondistinct objects, and the answer is

$$\frac{36!}{9!\,9!\,9!\,9!}, \quad \text{as in part (a).}$$

Our next example points out how some problems require the concepts of both arrangements and combinations for their solutions.

EXAMPLE 1.23

The number of arrangements of the letters in TALLAHASSEE is

$$\frac{11!}{3!\,2!\,2!\,2!\,1!\,1!} = 831,600.$$

How many of these arrangements have no adjacent A's?

When we disregard the A's, there are

$$\frac{8!}{2!\,2!\,2!\,1!\,1!} = 5040$$

ways to arrange the remaining letters. One of these 5040 ways is shown in the following figure, where the arrows indicate nine possible locations for the three A's.

Three of these locations can be selected in $\binom{9}{3} = 84$ ways, and because this is also possible for all the other 5039 arrangements of E, E, S, T, L, L, S, H, by the rule of product there are $5040 \times 84 = 423,360$ arrangements of the letters in TALLAHASSEE with no consecutive A's.

Before proceeding we need to introduce a concise way of writing the sum of a list of $n + 1$ terms like $a_m, a_{m+1}, a_{m+2}, \ldots, a_{m+n}$, where m and n are integers and $n \geq 0$. This notation is called the *Sigma notation* because it involves the capital Greek letter Σ; we use it to represent a summation by writing

$$a_m + a_{m+1} + a_{m+2} + \cdots + a_{m+n} = \sum_{i=m}^{m+n} a_i.$$

Here, the letter i is called the *index* of the summation, and this index accounts for all integers starting with the *lower limit* m and continuing on up to (and including) the *upper limit* $m + n$.

We may use this notation as follows.

1) $\displaystyle\sum_{i=3}^{7} a_i = a_3 + a_4 + a_5 + a_6 + a_7 = \sum_{j=3}^{7} a_j$, for there is nothing special about the letter i.

2) $\displaystyle\sum_{i=1}^{4} i^2 = 1^2 + 2^2 + 3^2 + 4^2 = 30 = \sum_{k=0}^{4} k^2$, because $0^2 = 0$.

3) $\displaystyle\sum_{i=11}^{100} i^3 = 11^3 + 12^3 + 13^3 + \cdots + 100^3 = \sum_{j=12}^{101} (j-1)^3 = \sum_{k=10}^{99} (k+1)^3$.

4) $\displaystyle\sum_{i=7}^{10} 2i = 2(7) + 2(8) + 2(9) + 2(10) = 68 = 2(34) = 2(7 + 8 + 9 + 10) = 2\sum_{i=7}^{10} i$.

5) $\displaystyle\sum_{i=3}^{3} a_i = a_3 = \sum_{i=4}^{4} a_{i-1} = \sum_{i=2}^{2} a_{i+1}$.

6) $\displaystyle\sum_{i=1}^{5} a = a + a + a + a + a = 5a$.

Furthermore, using this summation notation, we see that one can express the answer to part (c) of Example 1.20 as

$$\binom{5}{3}\binom{5}{4} + \binom{5}{4}\binom{5}{3} + \binom{5}{5}\binom{5}{2} = \sum_{i=3}^{5} \binom{5}{i}\binom{5}{7-i} = \sum_{j=2}^{4} \binom{5}{7-j}\binom{5}{j}.$$

We shall find use for this new notation in the following example and in many other places throughout the remainder of this book.

EXAMPLE 1.24

In the studies of algebraic coding theory and the theory of computer languages, we consider certain arrangements, called *strings*, made up from a prescribed *alphabet* of symbols. If the prescribed alphabet consists of the symbols 0, 1, and 2, for example, then 01, 11, 21, 12, and 20 are five of the nine strings of *length* 2. Among the 27 strings of length 3 are 000, 012, 202, and 110.

In general, if n is any positive integer, then by the rule of product there are 3^n strings of length n for the alphabet 0, 1, and 2. If $x = x_1 x_2 x_3 \cdots x_n$ is one of these strings, we define the *weight* of x, denoted wt(x), by wt$(x) = x_1 + x_2 + x_3 + \cdots + x_n$. For example, wt$(12) = 3$ and wt$(22) = 4$ for the case where $n = 2$; wt$(101) = 2$, wt$(210) = 3$, and wt$(222) = 6$ for $n = 3$.

Among the 3^{10} strings of length 10, we wish to determine how many have even weight. Such a string has even weight precisely when the number of 1's in the string is even.

There are six different cases to consider. If the string x contains no 1's, then each of the 10 locations in x can be filled with either 0 or 2, and by the rule of product there are 2^{10} such strings. When the string contains two 1's, the locations for these two 1's can be selected in $\binom{10}{2}$ ways. Once these two locations have been specified, there are 2^8 ways to place either 0 or 2 in the other eight positions. Hence there are $\binom{10}{2}2^8$ strings of even weight that contain two 1's. The numbers of strings for the other four cases are given in Table 1.2.

Table 1.2

Number of 1's	Number of Strings	Number of 1's	Number of Strings
4	$\binom{10}{4}2^6$	8	$\binom{10}{8}2^2$
6	$\binom{10}{6}2^4$	10	$\binom{10}{10}$

Consequently, by the rule of sum, the number of strings of length 10 that have even weight is $2^{10} + \binom{10}{2}2^8 + \binom{10}{4}2^6 + \binom{10}{6}2^4 + \binom{10}{8}2^2 + \binom{10}{10} = \sum_{n=0}^{5} \binom{10}{2n}2^{10-2n}$.

Often we must be careful of *overcounting* — a situation that seems to arise in what may appear to be rather easy enumeration problems. The next example demonstrates how overcounting may come about.

EXAMPLE 1.25

a) Suppose that Ellen draws five cards from a standard deck of 52 cards. In how many ways can her selection result in a hand with no clubs? Here we are interested in counting all five-card selections such as

  i) ace of hearts, three of spades, four of spades, six of diamonds, and the jack of diamonds.
  ii) five of spades, seven of spades, ten of spades, seven of diamonds, and the king of diamonds.
  iii) two of diamonds, three of diamonds, six of diamonds, ten of diamonds, and the jack of diamonds.

If we examine this more closely we see that Ellen is restricted to selecting her five cards from the 39 cards in the deck that are not clubs. Consequently, she can make her selection in $\binom{39}{5}$ ways.

b) Now suppose we want to count the number of Ellen's five-card selections that contain at least one club. These are precisely the selections that were *not* counted in part (a). And since there are $\binom{52}{5}$ possible five-card hands in total, we find that

$$\binom{52}{5} - \binom{39}{5} = 2{,}598{,}960 - 575{,}757 = 2{,}023{,}203$$

of all five-card hands contain at least one club.

c) Can we obtain the result in part (b) in another way? For example, since Ellen wants to have at least one club in the five-card hand, let her first select a club. This she can do in $\binom{13}{1}$ ways. And now she doesn't care what comes up for the other four cards. So after she eliminates the one club chosen from her standard deck, she can then select the other four cards in $\binom{51}{4}$ ways. Therefore, by the rule of product, we count the number of selections here as

$$\binom{13}{1}\binom{51}{4} = 13 \times 249{,}900 = 3{,}248{,}700.$$

Something here is definitely *wrong*! This answer is larger than that in part (b) by more than one million hands. Did we make a mistake in part (b)? Or is something wrong with our present reasoning?

For example, suppose that Ellen first selects

the three of clubs

and then selects

the five of clubs,

king of clubs,

seven of hearts, and

jack of spades.

If, however, she first selects

the five of clubs

and then selects

the three of clubs,

king of clubs,

seven of hearts, and

jack of spades,

is her selection here really different from the prior selection we mentioned? Unfortunately, no! And the case where she first selects

the king of clubs

and then follows this by selecting

the three of clubs,

five of clubs,

seven of hearts, and

jack of spades

is not different from the other two selections mentioned earlier.

Consequently, this approach is *wrong* because we are overcounting — by considering like selections as if they were distinct.

d) But is there any other way to arrive at the answer in part (b)? Yes! Since the five-card hands must each contain at least one club, there are five cases to consider. These are given in Table 1.3. From the results in Table 1.3 we see, for example, that there are $\binom{13}{2}\binom{39}{3}$ five-card hands that contain exactly two clubs. If we are interested in having exactly three clubs in the hand, then the results in the table indicate that there are $\binom{13}{3}\binom{39}{2}$ such hands.

Table 1.3

Number of Clubs	Number of Ways to Select This Number of Clubs	Number of Cards That Are Not Clubs	Number of Ways to Select This Number of Nonclubs
1	$\binom{13}{1}$	4	$\binom{39}{4}$
2	$\binom{13}{2}$	3	$\binom{39}{3}$
3	$\binom{13}{3}$	2	$\binom{39}{2}$
4	$\binom{13}{4}$	1	$\binom{39}{1}$
5	$\binom{13}{5}$	0	$\binom{39}{0}$

Since no two of the cases in Table 1.3 have any five-card hand in common, the number of hands that Ellen can select with at least one club is

$$\binom{13}{1}\binom{39}{4} + \binom{13}{2}\binom{39}{3} + \binom{13}{3}\binom{39}{2} + \binom{13}{4}\binom{39}{1} + \binom{13}{5}\binom{39}{0}$$

$$= \sum_{i=1}^{5} \binom{13}{i}\binom{39}{5-i}$$

$$= (13)(82{,}251) + (78)(9139) + (286)(741) + (715)(39) + (1287)(1)$$

$$= 2{,}023{,}203.$$

We shall close this section with three results related to the concept of combinations.

First we note that for integers n, r, with $n \geq r \geq 0$, $\binom{n}{r} = \binom{n}{n-r}$. This can be established algebraically from the formula for $\binom{n}{r}$, but we prefer to observe that when dealing with a selection of size r from a collection of n distinct objects, the selection process leaves behind $n - r$ objects. Consequently, $\binom{n}{r} = \binom{n}{n-r}$ affirms the existence of a correspondence between the selections of size r (objects chosen) and the selections of size $n - r$ (objects left behind). An example of this correspondence is shown in Table 1.4, where $n = 5, r = 2$, and the distinct objects are 1, 2, 3, 4, and 5. This type of correspondence will be more formally defined in Chapter 5 and used in other counting situations.

Table 1.4

Selections of Size $r = 2$ (Objects Chosen)		Selections of Size $n - r = 3$ (Objects Left Behind)	
1. 1, 2	6. 2, 4	1. 3, 4, 5	6. 1, 3, 5
2. 1, 3	7. 2, 5	2. 2, 4, 5	7. 1, 3, 4
3. 1, 4	8. 3, 4	3. 2, 3, 5	8. 1, 2, 5
4. 1, 5	9. 3, 5	4. 2, 3, 4	9. 1, 2, 4
5. 2, 3	10. 4, 5	5. 1, 4, 5	10. 1, 2, 3

Our second result is a theorem from our past experience in algebra.

THEOREM 1.1

The Binomial Theorem. If x and y are variables and n is a positive integer, then

$$(x + y)^n = \binom{n}{0}x^0 y^n + \binom{n}{1}x^1 y^{n-1} + \binom{n}{2}x^2 y^{n-2} + \cdots$$

$$+ \binom{n}{n-1}x^{n-1}y^1 + \binom{n}{n}x^n y^0 = \sum_{k=0}^{n} \binom{n}{k}x^k y^{n-k}.$$

Before considering the general proof, we examine a special case. If $n = 4$, the coefficient of $x^2 y^2$ in the expansion of the product

$$(x + y)\ (x + y)\ (x + y)\ (x + y)$$

1st   2nd   3rd   4th
factor factor factor factor

is the number of ways in which we can select two x's from the four x's, one of which is available in each factor. (Although the x's are the same in appearance, we distinguish them as the x in the first factor, the x in the second factor, $\ldots$, and the x in the fourth factor. Also, we note that when we select two x's, we use two factors, leaving us with two other factors from which we can select the two y's that are needed.) For example, among the possibilities, we can select (1) x from the first two factors and y from the last two or (2) x from the first and third factors and y from the second and fourth. Table 1.5 summarizes the six possible selections.

Table 1.5

Factors Selected for x		Factors Selected for y	
(1)	1, 2	(1)	3, 4
(2)	1, 3	(2)	2, 4
(3)	1, 4	(3)	2, 3
(4)	2, 3	(4)	1, 4
(5)	2, 4	(5)	1, 3
(6)	3, 4	(6)	1, 2

Consequently, the coefficient of x^2y^2 in the expansion of $(x + y)^4$ is $\binom{4}{2} = 6$, the number of ways to select two distinct objects from a collection of four distinct objects.

Now we turn to the proof of the general case.

Proof: In the expansion of the product

$$(x + y)\,(x + y)\,(x + y) \cdots (x + y)$$

<table>
<tr><td>1st
factor</td><td>2nd
factor</td><td>3rd
factor</td><td>nth
factor</td></tr>
</table>

the coefficient of $x^k y^{n-k}$, where $0 \le k \le n$, is the number of different ways in which we can select k x's [and consequently $(n - k)$ y's] from the n available factors. (One way, for example, is to choose x from the first k factors and y from the last $n - k$ factors.) The total number of such selections of size k from a collection of size n is $C(n, k) = \binom{n}{k}$, and from this the binomial theorem follows.

In view of this theorem, $\binom{n}{k}$ is often referred to as a *binomial coefficient*. Notice that it is also possible to express the result of Theorem 1.1 as

$$(x + y)^n = \sum_{k=0}^{n} \binom{n}{n - k} x^k y^{n-k}.$$

EXAMPLE 1.26

a) From the binomial theorem it follows that the coefficient of $x^5 y^2$ in the expansion of $(x + y)^7$ is $\binom{7}{5} = \binom{7}{2} = 21$.

b) To obtain the coefficient of $a^5 b^2$ in the expansion of $(2a - 3b)^7$, replace $2a$ by x and $-3b$ by y. From the binomial theorem the coefficient of $x^5 y^2$ in $(x + y)^7$ is $\binom{7}{5}$, and $\binom{7}{5} x^5 y^2 = \binom{7}{5}(2a)^5(-3b)^2 = \binom{7}{5}(2)^5(-3)^2 a^5 b^2 = 6048 a^5 b^2$.

COROLLARY 1.1

For each integer $n > 0$,

    a) $\binom{n}{0} + \binom{n}{1} + \binom{n}{2} + \cdots + \binom{n}{n} = 2^n$, and

    b) $\binom{n}{0} - \binom{n}{1} + \binom{n}{2} - \cdots + (-1)^n \binom{n}{n} = 0$.

Proof: Part (a) follows from the binomial theorem when we set $x = y = 1$. When $x = -1$ and $y = 1$, part (b) results.

Our third and final result generalizes the binomial theorem and is called the *multinomial theorem*.

THEOREM 1.2

For positive integers n, t, the coefficient of $x_1^{n_1} x_2^{n_2} x_3^{n_3} \cdots x_t^{n_t}$ in the expansion of $(x_1 + x_2 + x_3 + \cdots + x_t)^n$ is

$$\frac{n!}{n_1! \, n_2! \, n_3! \cdots n_t!},$$

where each n_i is an integer with $0 \le n_i \le n$, for all $1 \le i \le t$, and $n_1 + n_2 + n_3 + \cdots + n_t = n$.

Proof: As in the proof of the binomial theorem, the coefficient of $x_1^{n_1} x_2^{n_2} x_3^{n_3} \cdots x_t^{n_t}$ is the number of ways we can select x_1 from n_1 of the n factors, x_2 from n_2 of the $n - n_1$ remaining factors, x_3 from n_3 of the $n - n_1 - n_2$ now remaining factors, $\ldots$, and x_t from n_t of the last $n - n_1 - n_2 - n_3 - \cdots - n_{t-1} = n_t$ remaining factors. This can be carried out, as in part (a) of Example 1.22, in

$$\binom{n}{n_1} \binom{n - n_1}{n_2} \binom{n - n_1 - n_2}{n_3} \cdots \binom{n - n_1 - n_2 - n_3 - \cdots - n_{t-1}}{n_t}$$

ways. We leave to the reader the details of showing that this product is equal to

$$\frac{n!}{n_1! \, n_2! \, n_3! \cdots n_t!},$$

which is also written as

$$\binom{n}{n_1, n_2, n_3, \ldots, n_t}$$

and is called a *multinomial coefficient*. (When $t = 2$ this reduces to a binomial coefficient.)

EXAMPLE 1.27

    a) In the expansion of $(x + y + z)^7$ it follows from the multinomial theorem that the coefficient of $x^2 y^2 z^3$ is $\binom{7}{2,2,3} = \frac{7!}{2! \, 2! \, 3!} = 210$, while the coefficient of xyz^5 is $\binom{7}{1,1,5} = 42$ and that of $x^3 z^4$ is $\binom{7}{3,0,4} = \frac{7!}{3! \, 0! \, 4!} = 35$.

    b) Suppose we need to know the coefficient of $a^2 b^3 c^2 d^5$ in the expansion of $(a + 2b - 3c + 2d + 5)^{16}$. If we replace a by v, $2b$ by w, $-3c$ by x, $2d$ by y, and 5 by z, then we can apply the multinomial theorem to $(v + w + x + y + z)^{16}$ and determine the coefficient of $v^2 w^3 x^2 y^5 z^4$ as $\binom{16}{2,3,2,5,4} = 302{,}702{,}400$. But $\binom{16}{2,3,2,5,4} (a)^2 (2b)^3 (-3c)^2 (2d)^5 (5)^4 = \binom{16}{2,3,2,5,4} (1)^2 (2)^3 (-3)^2 (2)^5 (5)^4 (a^2 b^3 c^2 d^5) = 435{,}891{,}456{,}000{,}000 \, a^2 b^3 c^2 d^5$.

EXERCISES 1.3

1. Calculate $\binom{6}{2}$ and check your answer by listing all the selections of size 2 that can be made from the letters a, b, c, d, e, and f.

2. Facing a four-hour bus trip back to college, Diane decides to take along five magazines from the 12 that her sister Ann Marie has recently acquired. In how many ways can Diane make her selection?

3. Evaluate each of the following.

    a) $C(10, 4)$ **b)** $\binom{12}{7}$ **c)** $C(14, 12)$ **d)** $\binom{15}{10}$

4. In the Braille system a symbol, such as a lowercase letter, punctuation mark, suffix, and so on, is given by raising at least one of the dots in the six-dot arrangement shown in part (a) of Fig. 1.7. (The six Braille positions are labeled in this part of the figure.) For example, in part (b) of the figure the dots in positions 1 and 4 are raised and this six-dot arrangement represents the letter c. In parts (c) and (d) of the figure we have the representations for the letters m and t, respectively. The definite article "the" is shown in part (e) of the figure, while part (f) contains the form for the suffix "ow." Finally, the semicolon, ; , is given by the six-dot arrangement in part (g), where the dots at positions 2 and 3 are raised.

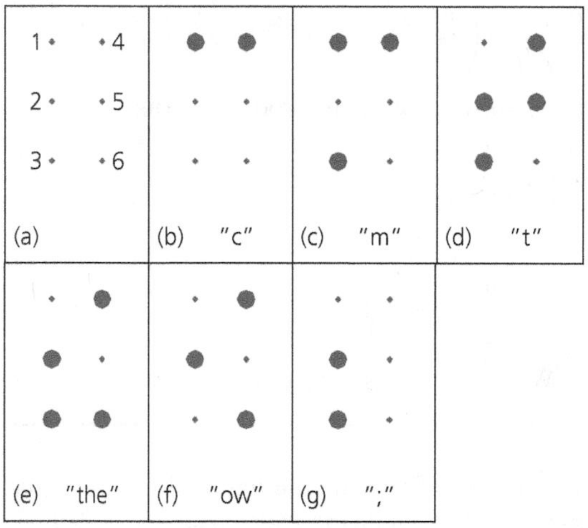

Figure 1.7

  a) How many different symbols can we represent in the Braille system?

  b) How many symbols have exactly three raised dots?

  c) How many symbols have an even number of raised dots?

5. a) How many *permutations* of size 3 can one produce with the letters m, r, a, f, and t?

  b) List all the *combinations* of size 3 that result for the letters m, r, a, f, and t.

6. If n is a positive integer and $n > 1$, prove that $\binom{n}{2} + \binom{n-1}{2}$ is a perfect square.

7. A committee of 12 is to be selected from 10 men and 10 women. In how many ways can the selection be carried out if (a) there are no restrictions? (b) there must be six men and six women? (c) there must be an even number of women? (d) there must be more women than men? (e) there must be at least eight men?

8. In how many ways can a gambler draw five cards from a standard deck and get (a) a flush (five cards of the same suit)? (b) four aces? (c) four of a kind? (d) three aces and two jacks? (e) three aces and a pair? (f) a full house (three of a kind and a pair)? (g) three of a kind? (h) two pairs?

9. How many bytes contain (a) exactly two 1's; (b) exactly four 1's; (c) exactly six 1's; (d) at least six 1's?

10. How many ways are there to pick a five-person basketball team from 12 possible players? How many selections include the weakest and the strongest players?

11. A student is to answer seven out of 10 questions on an examination. In how many ways can he make his selection if (a) there are no restrictions? (b) he must answer the first two questions? (c) he must answer at least four of the first six questions?

12. In how many ways can 12 different books be distributed among four children so that (a) each child gets three books? (b) the two oldest children get four books each and the two youngest get two books each?

13. How many arrangements of the letters in MISSISSIPPI have no consecutive S's?

14. A gym coach must select 11 seniors to play on a football team. If he can make his selection in 12,376 ways, how many seniors are eligible to play?

15. a) Fifteen points, no three of which are collinear, are given on a plane. How many lines do they determine?

  b) Twenty-five points, no four of which are coplanar, are given in space. How many triangles do they determine? How many planes? How many tetrahedra (pyramidlike solids with four triangular faces)?

16. Determine the value of each of the following summations.

  a) $\displaystyle\sum_{i=1}^{6}(i^2 + 1)$ **b)** $\displaystyle\sum_{j=-2}^{2}(j^3 - 1)$ **c)** $\displaystyle\sum_{i=0}^{10}[1 + (-1)^i]$

  d) $\displaystyle\sum_{k=n}^{2n}(-1)^k$, where n is an odd positive integer

  e) $\displaystyle\sum_{i=1}^{6} i(-1)^i$

17. Express each of the following using the summation (or Sigma) notation. In parts (a), (d), and (e), n denotes a positive integer.

  a) $\dfrac{1}{2!} + \dfrac{1}{3!} + \dfrac{1}{4!} + \cdots + \dfrac{1}{n!}$, $n \geq 2$

b) $1 + 4 + 9 + 16 + 25 + 36 + 49$

c) $1^3 - 2^3 + 3^3 - 4^3 + 5^3 - 6^3 + 7^3$

d) $\dfrac{1}{n} + \dfrac{2}{n+1} + \dfrac{3}{n+2} + \cdots + \dfrac{n+1}{2n}$

e) $n - \left(\dfrac{n+1}{2!}\right) + \left(\dfrac{n+2}{4!}\right) - \left(\dfrac{n+3}{6!}\right) + \cdots$
$+ (-1)^n \left(\dfrac{2n}{(2n)!}\right)$

18. For the strings of length 10 in Example 1.24, how many have (a) four 0's, three 1's, and three 2's; (b) at least eight 1's; (c) weight 4?

19. Consider the collection of all strings of length 10 made up from the alphabet 0, 1, 2, and 3. How many of these strings have weight 3? How many have weight 4? How many have even weight?

20. In the three parts of Fig. 1.8, eight points are equally spaced and marked on the circumference of a given circle.

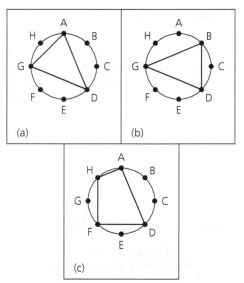

Figure 1.8

a) For parts (a) and (b) of Fig. 1.8 we have two different (though congruent) triangles. These two triangles (distinguished by their vertices) result from two selections of size 3 from the vertices A, B, C, D, E, F, G, H. How many different (whether congruent or not) triangles can we inscribe in the circle in this way?

b) How many different quadrilaterals can we inscribe in the circle, using the marked vertices? [One such quadrilateral appears in part (c) of Fig. 1.8.]

c) How many different polygons of three or more sides can we inscribe in the given circle by using three or more of the marked vertices?

21. How many triangles are determined by the vertices of a regular polygon of n sides? How many if no side of the polygon is to be a side of any triangle?

22. a) In the complete expansion of $(a + b + c + d) \cdot (e + f + g + h)(u + v + w + x + y + z)$ one obtains the sum of terms such as agw, cfx, and dgv. How many such terms appear in this complete expansion?

b) Which of the following terms do *not* appear in the complete expansion from part (a)?

   i) afx **ii)** bvx **iii)** chz
   iv) cgw **v)** egu **vi)** dfz

23. Determine the coefficient of $x^9 y^3$ in the expansions of (a) $(x + y)^{12}$, (b) $(x + 2y)^{12}$, and (c) $(2x - 3y)^{12}$.

24. Complete the details in the proof of the multinomial theorem.

25. Determine the coefficient of

a) xyz^2 in $(x + y + z)^4$

b) xyz^2 in $(w + x + y + z)^4$

c) xyz^2 in $(2x - y - z)^4$

d) xyz^{-2} in $(x - 2y + 3z^{-1})^4$

e) $w^3 x^2 y z^2$ in $(2w - x + 3y - 2z)^8$

26. Find the coefficient of $w^2 x^2 y^2 z^2$ in the expansion of (a) $(w + x + y + z + 1)^{10}$, (b) $(2w - x + 3y + z - 2)^{12}$, and (c) $(v + w - 2x + y + 5z + 3)^{12}$.

27. Determine the sum of all the coefficients in the expansions of

a) $(x + y)^3$ **b)** $(x + y)^{10}$ **c)** $(x + y + z)^{10}$

d) $(w + x + y + z)^5$

e) $(2s - 3t + 5u + 6v - 11w + 3x + 2y)^{10}$

28. For any positive integer n determine

a) $\displaystyle\sum_{i=0}^{n} \dfrac{1}{i!(n-i)!}$ **b)** $\displaystyle\sum_{i=0}^{n} \dfrac{(-1)^i}{i!(n-i)!}$

29. Show that for all positive integers m and n,
$$n \binom{m+n}{m} = (m+1) \binom{m+n}{m+1}.$$

30. With n a positive integer, evaluate the sum
$$\binom{n}{0} + 2\binom{n}{1} + 2^2\binom{n}{2} + \cdots + 2^k\binom{n}{k} + \cdots + 2^n\binom{n}{n}.$$

31. For x a real number and n a positive integer, show that

a) $1 = (1+x)^n - \binom{n}{1} x^1 (1+x)^{n-1}$
$+ \binom{n}{2} x^2 (1+x)^{n-2} - \cdots + (-1)^n \binom{n}{n} x^n$

b) $1 = (2+x)^n - \binom{n}{1}(x+1)(2+x)^{n-1}$
$+ \binom{n}{2}(x+1)^2 (2+x)^{n-2} - \cdots + (-1)^n \binom{n}{n}(x+1)^n$

c) $2^n = (2+x)^n - \binom{n}{1}x^1(2+x)^{n-1}$
$+ \binom{n}{2}x^2(2+x)^{n-2} - \cdots + (-1)^n \binom{n}{n}x^n$

32. Determine x if $\sum_{i=0}^{50} \binom{50}{i}8^i = x^{100}$.

33. a) If a_0, a_1, a_2, a_3 is a list of four real numbers, what is $\sum_{i=1}^{3}(a_i - a_{i-1})$?

b) Given a list—$a_0, a_1, a_2, \ldots, a_n$—of $n+1$ real numbers, where n is a positive integer, determine $\sum_{i=1}^{n}(a_i - a_{i-1})$.

c) Determine the value of $\sum_{i=1}^{100}\left(\frac{1}{i+2} - \frac{1}{i+1}\right)$.

34. a) Write a computer program (or develop an algorithm) that lists all selections of size 2 from the objects 1, 2, 3, 4, 5, 6.

b) Repeat part (a) for selections of size 3.

1.4
Combinations with Repetition

When repetitions are allowed, we have seen that for n distinct objects an arrangement of size r of these objects can be obtained in n^r ways, for an integer $r \geq 0$. We now turn to the comparable problem for combinations and once again obtain a related problem whose solution follows from our previous enumeration principles.

EXAMPLE 1.28

On their way home from track practice, seven high school freshmen stop at a restaurant, where each of them has one of the following: a cheeseburger, a hot dog, a taco, or a fish sandwich. How many different purchases are possible (from the viewpoint of the restaurant)?

Let c, h, t, and f represent cheeseburger, hot dog, taco, and fish sandwich, respectively. Here we are concerned with how many of each item are purchased, not with the order in which they are purchased, so the problem is one of selections, or combinations, with repetition.

In Table 1.6 we list some possible purchases in column (a) and another means of representing each purchase in column (b).

Table 1.6

(a)	(b)
1. c, c, h, h, t, t, f	1. x x | x x | x x | x
2. c, c, c, c, h, t, f	2. x x x x | x | x | x
3. c, c, c, c, c, c, f	3. x x x x x x | | | x
4. h, t, t, f, f, f, f	4. | x | x x | x x x x
5. t, t, t, t, t, f, f	5. | | x x x x x | x x
6. t, t, t, t, t, t, t	6. | | x x x x x x x |
7. f, f, f, f, f, f, f	7. | | | x x x x x x x

For a purchase in column (b) of Table 1.6 we realize that each x to the left of the first bar (|) represents a c, each x between the first and second bars represents an h, the x's between the second and third bars stand for t's, and each x to the right of the third bar stands for an f. The third purchase, for example, has three consecutive bars because no one bought a hot dog or taco; the bar at the start of the fourth purchase indicates that there were no cheeseburgers in that purchase.

Once again a correspondence has been established between two collections of objects, where we know how to count the number in one collection. For the representations in column (b) of Table 1.6, we are enumerating all arrangements of 10 symbols consisting of seven x's and three |'s, so by our correspondence the number of different purchases for column (a) is

$$\frac{10!}{7!\,3!} = \binom{10}{7}.$$

In this example we note that the seven x's (one for each freshman) correspond to the size of the selection and that the three bars are needed to separate the $3 + 1 = 4$ possible food items that can be chosen.

When we wish to select, *with repetition*, r of n distinct objects, we find (as in Table 1.6) that we are considering all arrangements of r x's and $n - 1$ |'s and that their number is

$$\frac{(n + r - 1)!}{r!(n - 1)!} = \binom{n + r - 1}{r}.$$

Consequently, the number of combinations of n objects taken r at a time, *with repetition*, is $C(n + r - 1, r)$.

(In Example 1.28, $n = 4$, $r = 7$, so it is possible for r to exceed n when repetitions are allowed.)

EXAMPLE 1.29

A donut shop offers 20 kinds of donuts. Assuming that there are at least a dozen of each kind when we enter the shop, we can select a dozen donuts in $C(20 + 12 - 1, 12) = C(31, 12) = 141{,}120{,}525$ ways. (Here $n = 20$, $r = 12$.)

EXAMPLE 1.30

President Helen has four vice presidents: (1) Betty, (2) Goldie, (3) Mary Lou, and (4) Mona. She wishes to distribute among them $1000 in Christmas bonus checks, where each check will be written for a multiple of $100.

   a) Allowing the situation in which one or more of the vice presidents get nothing, President Helen is making a selection of size 10 (one for each unit of $100) from a collection of size 4 (four vice presidents), with repetition. This can be done in $C(4 + 10 - 1, 10) = C(13, 10) = 286$ ways.

   b) If there are to be no hard feelings, each vice president should receive at least $100. With this restriction, President Helen is now faced with making a selection of size 6 (the remaining six units of $100) from the same collection of size 4, and the choices now number $C(4 + 6 - 1, 6) = C(9, 6) = 84$. [For example, here the selection 2, 3, 3, 4, 4, 4 is interpreted as follows: Betty does not get anything extra — for there is no 1 in the selection. The one 2 in the selection indicates that Goldie gets an additional $100. Mary Lou receives an additional $200 ($100 for each of the two 3's in the selection). Due to the three 4's, Mona's bonus check will total $100 + 3($100) = $400.]

c) If each vice president must get at least $100 and Mona, as executive vice president, gets at least $500, then the number of ways President Helen can distribute the bonus checks is

$$\underbrace{C(3+2-1, 2)}_{\substack{\textbf{Mona gets}\\\textbf{exactly \$500}}} + \underbrace{C(3+1-1, 1)}_{\substack{\textbf{Mona gets}\\\textbf{exactly \$600}}} + \underbrace{C(3+0-1, 0)}_{\substack{\textbf{Mona gets}\\\textbf{exactly \$700}}} = 10 = \underbrace{C(4+2-1, 2)}_{\substack{\textbf{Using the}\\\textbf{technique in part (b)}}}$$

Having worked examples utilizing combinations with repetition, we now consider two examples involving other counting principles as well.

EXAMPLE 1.31

In how many ways can we distribute seven bananas and six oranges among four children so that each child receives at least one banana?

After giving each child one banana, consider the number of ways the remaining three bananas can be distributed among these four children. Table 1.7 shows four of the distributions we are considering here. For example, the second distribution in part (a) of Table 1.7 — namely, 1, 3, 3 — indicates that we have given the first child (designated by 1) one additional banana and the third child (designated by 3) two additional bananas. The corresponding arrangement in part (b) of Table 1.7 represents this distribution in terms of three b's and three bars. These six symbols — three of one type (the b's) and three others of a second type (the bars) — can be arranged in $6!/(3!\,3!) = C(6, 3) = C(4+3-1, 3) = 20$ ways. [Here $n = 4$, $r = 3$.] Consequently, there are 20 ways in which we can distribute the three additional bananas among these four children. Table 1.8 provides the comparable situation for distributing the six oranges. In this case we are arranging nine symbols — six of one type (the o's) and three of a second type (the bars). So now we learn that the number of ways we can distribute the six oranges among these four children is $9!/(6!\,3!) = C(9, 6) = C(4+6-1, 6) = 84$ ways. [Here $n = 4$, $r = 6$.] Therefore, by the rule of product, there are $20 \times 84 = 1680$ ways to distribute the fruit under the stated conditions.

Table 1.7

1)	1, 2, 3	1)	$b\,\|\,b\,\|\,b\,\|$
2)	1, 3, 3	2)	$b\,\|\,\|\,b\,b\,\|$
3)	3, 4, 4	3)	$\|\,\|\,b\,\|\,b\,b$
4)	4, 4, 4	4)	$\|\,\|\,\|\,b\,b\,b$
(a)		(b)	

Table 1.8

1)	1, 2, 2, 3, 3, 4	1)	$o\,\|\,o\,o\,\|\,o\,o\,\|\,o$
2)	1, 2, 2, 4, 4, 4	2)	$o\,\|\,o\,o\,\|\,\|\,o\,o\,o$
3)	2, 2, 2, 3, 3, 3	3)	$\|\,o\,o\,o\,\|\,o\,o\,o\,\|$
4)	4, 4, 4, 4, 4, 4	4)	$\|\,\|\,\|\,o\,o\,o\,o\,o\,o$
(a)		(b)	

EXAMPLE 1.32

A message is made up of 12 different symbols and is to be transmitted through a communication channel. In addition to the 12 symbols, the transmitter will also send a total of 45 (blank) spaces between the symbols, with at least three spaces between each pair of consecutive symbols. In how many ways can the transmitter send such a message?

There are 12! ways to arrange the 12 different symbols, and for each of these arrangements there are 11 positions between the 12 symbols. Because there must be at least three spaces between successive symbols, we use up 33 of the 45 spaces and must now locate the remaining 12 spaces. This is now a selection, with repetition, of size 12 (the spaces) from a collection of size 11 (the locations), and this can be accomplished in $C(11+12-1, 12) = 646,646$ ways.

Consequently, by the rule of product the transmitter can send such messages with the required spacing in $(12!)\binom{22}{12} \doteq 3.097 \times 10^{14}$ ways.

In the next example an idea is introduced that appears to have more to do with number theory than with combinations or arrangements. Nonetheless, the solution of this example will turn out to be equivalent to counting combinations with repetitions.

EXAMPLE 1.33

Determine all integer solutions to the equation

$$x_1 + x_2 + x_3 + x_4 = 7, \qquad \text{where } x_i \geq 0 \quad \text{for all } 1 \leq i \leq 4.$$

One solution of the equation is $x_1 = 3$, $x_2 = 3$, $x_3 = 0$, $x_4 = 1$. (This is different from a solution such as $x_1 = 1, x_2 = 0, x_3 = 3, x_4 = 3$, even though the same four integers are being used.) A possible interpretation for the solution $x_1 = 3, x_2 = 3, x_3 = 0, x_4 = 1$ is that we are distributing seven pennies (identical objects) among four children (distinct containers), and here we have given three pennies to each of the first two children, nothing to the third child, and the last penny to the fourth child. Continuing with this interpretation, we see that each nonnegative integer solution of the equation corresponds to a selection, with repetition, of size 7 (the *identical* pennies) from a collection of size 4 (the *distinct* children), so there are $C(4 + 7 - 1, 7) = 120$ solutions.

At this point it is crucial that we recognize the equivalence of the following:

a) The number of integer solutions of the equation

$$x_1 + x_2 + \cdots + x_n = r, \qquad x_i \geq 0, \qquad 1 \leq i \leq n.$$

b) The number of selections, with repetition, of size r from a collection of size n.

c) The number of ways r identical objects can be distributed among n distinct containers.

In terms of distributions, part (c) is valid only when the r objects being distributed are identical and the n containers are distinct. When both the r objects and the n containers are distinct, we can select any of the n containers for each one of the objects and get n^r distributions by the rule of product.

When the objects are distinct but the containers are identical, we shall solve the problem using the Stirling numbers of the second kind (Chapter 5). For the final case, in which both objects and containers are identical, the theory of partitions of integers (Chapter 9) will provide some necessary results.

EXAMPLE 1.34

In how many ways can one distribute 10 (identical) white marbles among six distinct containers?

Solving this problem is equivalent to finding the number of nonnegative integer solutions to the equation $x_1 + x_2 + \cdots + x_6 = 10$. That number is the number of selections of size 10, with repetition, from a collection of size 6. Hence the answer is $C(6 + 10 - 1, 10) = 3003$.

We now examine two other examples related to the theme of this section.

EXAMPLE 1.35

From Example 1.34 we know that there are 3003 nonnegative integer solutions to the equation $x_1 + x_2 + \cdots + x_6 = 10$. How many such solutions are there to the inequality $x_1 + x_2 + \cdots + x_6 < 10$?

One approach that may seem feasible in dealing with this inequality is to determine the number of such solutions to $x_1 + x_2 + \cdots + x_6 = k$, where k is an integer and $0 \leq k \leq 9$. Although feasible now, the technique becomes unrealistic if 10 is replaced by a somewhat larger number, say 100. In Example 3.12 of Chapter 3, however, we shall establish a combinatorial identity that will help us obtain an alternative solution to the problem by using this approach.

For the present we transform the problem by noting the correspondence between the nonnegative integer solutions of

$$x_1 + x_2 + \cdots + x_6 < 10 \tag{1}$$

and the integer solutions of

$$x_1 + x_2 + \cdots + x_6 + x_7 = 10, \qquad 0 \leq x_i, \qquad 1 \leq i \leq 6, \qquad 0 < x_7. \tag{2}$$

The number of solutions of Eq. (2) is the same as the number of nonnegative integer solutions of $y_1 + y_2 + \cdots + y_6 + y_7 = 9$, where $y_i = x_i$ for $1 \leq i \leq 6$, and $y_7 = x_7 - 1$. This is $C(7 + 9 - 1, 9) = 5005$.

Our next result takes us back to the binomial and multinomial expansions.

EXAMPLE 1.36

In the binomial expansion for $(x + y)^n$, each term is of the form $\binom{n}{k} x^k y^{n-k}$, so the total number of terms in the expansion is the number of nonnegative integer solutions of $n_1 + n_2 = n$ (n_1 is the exponent for x, n_2 the exponent for y). This number is $C(2 + n - 1, n) = n + 1$.

Perhaps it seems that we have used a rather long-winded argument to get this result. Many of us would probably be willing to believe the result on the basis of our experiences in expanding $(x + y)^n$ for various small values of n.

Although experience is worthwhile in pattern recognition, it is not always enough to find a general principle. Here it would prove of little value if we wanted to know how many terms there are in the expansion of $(w + x + y + z)^{10}$.

Each distinct term here is of the form $\binom{10}{n_1, n_2, n_3, n_4} w^{n_1} x^{n_2} y^{n_3} z^{n_4}$, where $0 \leq n_i$ for $1 \leq i \leq 4$, and $n_1 + n_2 + n_3 + n_4 = 10$. This last equation can be solved in $C(4 + 10 - 1, 10) = 286$ ways, so there are 286 terms in the expansion of $(w + x + y + z)^{10}$.

And now once again the binomial expansion will come into play, as we find ourselves using part (a) of Corollary 1.1

EXAMPLE 1.37

a) Let us determine all the different ways in which we can write the number 4 as a sum of positive integers, where the order of the summands is considered relevant. These representations are called the *compositions* of 4 and may be listed as follows:

1) 4   5) $2 + 1 + 1$
2) $3 + 1$ 6) $1 + 2 + 1$
3) $1 + 3$ 7) $1 + 1 + 2$
4) $2 + 2$ 8) $1 + 1 + 1 + 1$

Here we include the sum consisting of only one summand — namely, 4. We find that for the number 4 there are eight compositions in total. (If we do *not* care about the order of the summands, then the representations in (2) and (3) are no longer considered to be different — nor are the representations in (5), (6), and (7). Under these circumstances we find that there are five *partitions* for the number 4 — namely, 4; $3 + 1$; $2 + 2$; $2 + 1 + 1$; and $1 + 1 + 1 + 1$. We shall learn more about partitions of positive integers in Section 9.3.)

b) Now suppose that we wish to *count* the number of compositions for the number 7. Here we do *not* want to list all of the possibilities — which include 7; $6 + 1$; $1 + 6$; $5 + 2$; $1 + 2 + 4$; $2 + 4 + 1$; and $3 + 1 + 2 + 1$. To count all of these compositions, let us consider the number of possible summands.

i) For one summand there is only one composition — namely, 7.

ii) If there are two (positive) summands, we want to count the number of integer solutions for

$$w_1 + w_2 = 7, \qquad \text{where } w_1, w_2 > 0.$$

This is equal to the number of integer solutions for

$$x_1 + x_2 = 5, \qquad \text{where } x_1, x_2 \geq 0.$$

The number of such solutions is $\binom{2 + 5 - 1}{5} = \binom{6}{5}$.

iii) Continuing with our next case, we examine the compositions with three (positive) summands. So now we want to count the number of *positive* integer solutions for

$$y_1 + y_2 + y_3 = 7.$$

This is equal to the number of *nonnegative* integer solutions for

$$z_1 + z_2 + z_3 = 4,$$

and that number is $\binom{3 + 4 - 1}{4} = \binom{6}{4}$.

We summarize cases (i), (ii), and (iii), and the other four cases in Table 1.9, where we recall for case (i) that $1 = \binom{6}{6}$.

Table 1.9

n = The Number of Summands in a Composition of 7		The Number of Compositions of 7 with n Summands	
(i)	$n = 1$	(i)	$\binom{6}{6}$
(ii)	$n = 2$	(ii)	$\binom{6}{5}$
(iii)	$n = 3$	(iii)	$\binom{6}{4}$
(iv)	$n = 4$	(iv)	$\binom{6}{3}$
(v)	$n = 5$	(v)	$\binom{6}{2}$
(vi)	$n = 6$	(vi)	$\binom{6}{1}$
(vii)	$n = 7$	(vii)	$\binom{6}{0}$

Consequently, the results from the right-hand side of our table tell us that the (total) number of compositions of 7 is

$$\binom{6}{6} + \binom{6}{5} + \binom{6}{4} + \binom{6}{3} + \binom{6}{2} + \binom{6}{1} + \binom{6}{0} = \sum_{k=0}^{6} \binom{6}{k}.$$

From part (a) of Corollary 1.1 this reduces to 2^6.

In general, one finds that for each positive integer m, there are $\sum_{k=0}^{m-1} \binom{m-1}{k} = 2^{m-1}$ compositions.

EXAMPLE 1.38

From Example 1.37 we know that there are $2^{12-1} = 2^{11} = 2048$ compositions of 12. If our interest is in those compositions where each summand is even, then we consider, for instance, compositions such as

$$2 + 4 + 6 = 2(1 + 2 + 3) \qquad 2 + 8 + 2 = 2(1 + 4 + 1)$$
$$8 + 2 + 2 = 2(4 + 1 + 1) \qquad 6 + 6 = 2(3 + 3).$$

In each of these four examples, the parenthesized expression is a composition of 6. This observation indicates that the number of compositions of 12, where each summand is even, equals the number of (all) compositions of 6, which is $2^{6-1} = 2^5 = 32$.

Our next two examples provide applications from the area of computer science. Furthermore, the second example will lead to an important summation formula that we shall use in many later chapters.

EXAMPLE 1.39

Consider the following program segment, where i, j, and k are integer variables.

```
for i := 1 to 20 do
    for j := 1 to i do
        for k := 1 to j do
            print (i * j + k)
```

How many times is the **print** statement executed in this program segment?

Among the possible choices for i, j, and k (in the order i–first, j–second, k–third) that will lead to execution of the **print** statement, we list (1) 1, 1, 1; (2) 2, 1, 1; (3) 15, 10, 1; and (4) 15, 10, 7. We note that $i = 10$, $j = 12$, $k = 5$ is not one of the selections to be considered, because $j = 12 > 10 = i$; this violates the condition set forth in the second **for** loop. Each of the above four selections where the **print** statement is executed satisfies the condition $1 \leq k \leq j \leq i \leq 20$. In fact, any selection a, b, c ($a \leq b \leq c$) of size 3, with repetitions allowed, from the list 1, 2, 3, ..., 20 results in one of the correct selections: here, $k = a$, $j = b$, $i = c$. Consequently the **print** statement is executed

$$\binom{20 + 3 - 1}{3} = \binom{22}{3} = 1540 \text{ times.}$$

If there had been r (≥ 1) **for** loops instead of three, the **print** statement would have been executed $\binom{20 + r - 1}{r}$ times.

EXAMPLE 1.40

Here we use a program segment to derive a summation formula. In this program segment, the variables i, j, n, and *counter* are integer variables. Furthermore, we assume that the value of n has been set prior to this segment.

```
counter := 0
for i := 1 to n do
    for j := 1 to i do
        counter := counter + 1
```

From the results in Example 1.39, after this segment is executed the value of (the variable) *counter* will be $\binom{n+2-1}{2} = \binom{n+1}{2}$. (This is also the number of times that the statement

(*)   counter := counter + 1

is executed.)

This result can also be obtained as follows: When $i := 1$, then j varies from 1 to 1 and (*) is executed once; when i is assigned the value 2, then j varies from 1 to 2 and (*) is executed twice; j varies from 1 to 3 when i is assigned the value 3, and (*) is executed three times; in general, for $1 \leq k \leq n$, when $i := k$, then j varies from 1 to k and (*) is executed k times. In total, the variable *counter* is incremented [and the statement (*) is executed] $1 + 2 + 3 + \cdots + n$ times.

Consequently,

$$\sum_{i=1}^{n} i = 1 + 2 + 3 + \cdots + n = \binom{n+1}{2} = \frac{n(n+1)}{2}.$$

The derivation of this summation formula, obtained by counting the same result in two different ways, constitutes a combinatorial proof.

Our last example for this section introduces the idea of a run, a notion that arises in statistics — in particular, in the detecting of trends in a statistical process.

EXAMPLE 1.41

The counter at Patti and Terri's Bar has 15 bar stools. Upon entering the bar Darrell finds the stools occupied as follows:

O O E O O O O E E E O O O E O,

where O indicates an occupied stool and E an empty one. (Here we are not concerned with the occupants of the stools, just whether or not a stool is occupied.) In this case we say that the occupancy of the 15 stools determines seven runs, as shown:

$$\underset{\text{Run}}{\underline{OO}} \quad \underset{\text{Run}}{\underline{E}} \quad \underset{\text{Run}}{\underline{OOOO}} \quad \underset{\text{Run}}{\underline{EEE}} \quad \underset{\text{Run}}{\underline{OOO}} \quad \underset{\text{Run}}{\underline{E}} \quad \underset{\text{Run}}{\underline{O}}.$$

In general, a *run* is a consecutive list of identical entries that are preceded and followed by different entries or no entries at all.

A second way in which five E's and 10 O's can be arranged to provide seven runs is

E O O O E E O O E O O O O O E.

We want to find the total number of ways five E's and 10 O's can determine seven runs. If the first run starts with an E, then there must be four runs of E's and three runs of O's. Consequently, the last run must end with an E.

Let x_1 count the number of E's in the first run, x_2 the number of O's in the second run, x_3 the number of E's in the third run, . . . , and x_7 the number of E's in the seventh run. We want to find the number of integer solutions for

$$x_1 + x_3 + x_5 + x_7 = 5, \qquad x_1, x_3, x_5, x_7 > 0 \tag{3}$$

and

$$x_2 + x_4 + x_6 = 10, \qquad x_2, x_4, x_6 > 0. \qquad (4)$$

The number of integer solutions for Eq. (3) equals the number of integer solutions for

$$y_1 + y_3 + y_5 + y_7 = 1, \qquad y_1, y_3, y_5, y_7 \geq 0.$$

This number is $\binom{4+1-1}{1} = \binom{4}{1} = 4$. Similarly, for Eq. (4), the number of solutions is $\binom{3+7-1}{7} = \binom{9}{7} = 36$. Consequently, by the rule of product there are $4 \cdot 36 = 144$ arrangements of five E's and 10 O's that determine seven runs, the first run starting with E.

The seven runs may also have the first run starting with an O and the last run ending with an O. So now let w_1 count the number of O's in the first run, w_2 the number of E's in the second run, w_3 the number of O's in the third run, . . . , and w_7 the number of O's in the seventh run. Here we want the number of integer solutions for

$$w_1 + w_3 + w_5 + w_7 = 10, \qquad w_1, w_3, w_5, w_7 > 0$$

and

$$w_2 + w_4 + w_6 = 5, \qquad w_2, w_4, w_6 > 0.$$

Arguing as above, we find that the number of ways to arrange five E's and 10 O's, resulting in seven runs where the first run starts with an O, is $\binom{4+6-1}{6}\binom{3+2-1}{2} = \binom{9}{6}\binom{4}{2} = 504$.

Consequently, by the rule of sum, the five E's and 10 O's can be arranged in $144 + 504 = 648$ ways to produce seven runs.

EXERCISES 1.4

1. In how many ways can 10 (identical) dimes be distributed among five children if (a) there are no restrictions? (b) each child gets at least one dime? (c) the oldest child gets at least two dimes?

2. In how many ways can 15 (identical) candy bars be distributed among five children so that the youngest gets only one or two of them?

3. Determine how many ways 20 coins can be selected from four large containers filled with pennies, nickels, dimes, and quarters. (Each container is filled with only one type of coin.)

4. A certain ice cream store has 31 flavors of ice cream available. In how many ways can we order a dozen ice cream cones if (a) we do not want the same flavor more than once? (b) a flavor may be ordered as many as 12 times? (c) a flavor may be ordered no more than 11 times?

5. a) In how many ways can we select five coins from a collection of 10 consisting of one penny, one nickel, one dime, one quarter, one half-dollar, and five (identical) Susan B. Anthony dollars?

b) In how many ways can we select n objects from a collection of size $2n$ that consists of n distinct and n identical objects?

6. Answer Example 1.32, where the 12 symbols being transmitted are four A's, four B's, and four C's.

7. Determine the number of integer solutions of

$$x_1 + x_2 + x_3 + x_4 = 32,$$

where

a) $x_i \geq 0, \quad 1 \leq i \leq 4$ **b)** $x_i > 0, \quad 1 \leq i \leq 4$

c) $x_1, x_2 \geq 5, \quad x_3, x_4 \geq 7$

d) $x_i \geq 8, \quad 1 \leq i \leq 4$ **e)** $x_i \geq -2, \quad 1 \leq i \leq 4$

f) $x_1, x_2, x_3 > 0, \quad 0 < x_4 \leq 25$

8. In how many ways can a teacher distribute eight chocolate donuts and seven jelly donuts among three student helpers if each helper wants at least one donut of each kind?

9. Columba has two dozen each of n different colored beads. If she can select 20 beads (with repetitions of colors allowed) in 230,230 ways, what is the value of n?

10. In how many ways can Lisa toss 100 (identical) dice so that at least three of each type of face will be showing?

11. Two n-digit integers (leading zeros allowed) are considered equivalent if one is a rearrangement of the other. (For example, 12033, 20331, and 01332 are considered equivalent five-digit integers.) (a) How many five-digit integers are not equivalent? (b) If the digits 1, 3, and 7 can appear at most once, how many nonequivalent five-digit integers are there?

12. Determine the number of integer solutions for

$$x_1 + x_2 + x_3 + x_4 + x_5 < 40,$$

where

a) $x_i \geq 0, \quad 1 \leq i \leq 5$

b) $x_i \geq -3, \quad 1 \leq i \leq 5$

13. In how many ways can we distribute eight identical white balls into four distinct containers so that (a) no container is left empty? (b) the fourth container has an odd number of balls in it?

14. a) Find the coefficient of $v^2 w^4 xz$ in the expansion of $(3v + 2w + x + y + z)^8$.

b) How many distinct terms arise in the expansion in part (a)?

15. In how many ways can Beth place 24 different books on four shelves so that there is at least one book on each shelf? (For any of these arrangements consider the books on each shelf to be placed one next to the other, with the first book at the left of the shelf.)

16. For which positive integer n will the equations

(1) $x_1 + x_2 + x_3 + \cdots + x_{19} = n$, and

(2) $y_1 + y_2 + y_3 + \cdots + y_{64} = n$

have the same number of positive integer solutions?

17. How many ways are there to place 12 marbles of the same size in five distinct jars if (a) the marbles are all black? (b) each marble is a different color?

18. a) How many nonnegative integer solutions are there to the pair of equations $x_1 + x_2 + x_3 + \cdots + x_7 = 37$, $x_1 + x_2 + x_3 = 6$?

b) How many solutions in part (a) have $x_1, x_2, x_3 > 0$?

19. How many times is the **print** statement executed for the following program segment? (Here, $i, j, k,$ and m are integer variables.)

```
for i := 1 to 20 do
  for j := 1 to i do
    for k := 1 to j do
      for m := 1 to k do
        print (i * j) + (k * m)
```

20. In the following program segment, $i, j, k,$ and *counter* are integer variables. Determine the value that the variable *counter* will have after the segment is executed.

```
counter := 10
for i := 1 to 15 do
  for j := i to 15 do
    for k := j to 15 do
      counter := counter + 1
```

21. Find the value of *sum* after the given program segment is executed. (Here $i, j, k,$ *increment*, and *sum* are integer variables.)

```
increment := 0
sum := 0
for i := 1 to 10 do
  for j := 1 to i do
    for k := 1 to j do
      begin
        increment := increment + 1
        sum := sum + increment
      end
```

22. Consider the following program segment, where $i, j, k, n,$ and *counter* are integer variables and the value of n (a positive integer) is set prior to this segment.

```
counter := 0
for i := 1 to n do
  for j := 1 to i do
    for k := 1 to j do
      counter := counter + 1
```

We shall determine, in two different ways, the number of times the statement

```
counter := counter + 1
```

is executed. (This is also the value of *counter* after execution of the program segment.) From the result in Example 1.39, we know that the statement is executed $\binom{n+3-1}{3} = \binom{n+2}{3}$ times. For a fixed value of i, the **for** loops involving j and k result in $\binom{i+1}{2}$ executions of the counter increment statement. Consequently, $\binom{n+2}{3} = \sum_{i=1}^{n} \binom{i+1}{2}$. Use this result to obtain a summation formula for

$$1^2 + 2^2 + 3^2 + \cdots + n^2 = \sum_{i=1}^{n} i^2.$$

23. a) Given positive integers m, n with $m \geq n$, show that the number of ways to distribute m identical objects into n distinct containers with no container left empty is

$$C(m-1, m-n) = C(m-1, n-1).$$

b) Show that the number of distributions in part (a) where each container holds at least r objects ($m \geq nr$) is

$$C(m-1+(1-r)n, n-1).$$

24. Write a computer program (or develop an algorithm) to list the integer solutions for

a) $x_1 + x_2 + x_3 = 10, \quad 0 \leq x_i, \quad 1 \leq i \leq 3$

b) $x_1 + x_2 + x_3 + x_4 = 4, \quad -2 \leq x_i, \quad 1 \leq i \leq 4$

25. Consider the 2^{19} compositions of 20. (a) How many have each summand even? (b) How many have each summand a multiple of 4?

26. Let n, m, k be positive integers with $n = mk$. How many compositions of n have each summand a multiple of k?

27. Frannie tosses a coin 12 times and gets five heads and seven tails. In how many ways can these tosses result in (a) two runs of heads and one run of tails; (b) three runs; (c) four runs;

(d) five runs; (e) six runs; and (f) equal numbers of runs of heads and runs of tails?

28. a) For $n \geq 4$, consider the strings made up of n bits — that is, a total of n 0's and 1's. In particular, consider those strings where there are (exactly) two occurrences of 01. For example, if $n = 6$ we want to include strings such as 010010 and 100101, but not 101111 or 010101. How many such strings are there?

b) For $n \geq 6$, how many strings of n 0's and 1's contain (exactly) three occurrences of 01?

c) Provide a combinatorial proof for the following: For $n \geq 1$,

$$2^n = \binom{n+1}{1} + \binom{n+1}{3} + \cdots + \begin{cases} \binom{n+1}{n}, & n \text{ odd} \\ \binom{n+1}{n+1}, & n \text{ even.} \end{cases}$$

1.5
The Catalan Numbers (Optional)

In this section a very prominent sequence of numbers is introduced. This sequence arises in a wide variety of combinatorial situations. We'll begin by examining one specific instance where it is found.

EXAMPLE 1.42

Let us start at the point $(0, 0)$ in the xy-plane and consider two kinds of moves:

$$\text{R}: (x, y) \to (x + 1, y) \qquad \text{U}: (x, y) \to (x, y + 1).$$

We want to know how we can move from $(0, 0)$ to $(5, 5)$ using such moves — one unit to the right or one unit up. So we'll need five R's and five U's. At this point we have a situation like that in Example 1.14, so we know there are $10!/(5!\,5!) = \binom{10}{5}$ such paths. But now we'll add a twist! In going from $(0, 0)$ to $(5, 5)$ one may touch but *never* rise above the line $y = x$. Consequently, we want to include paths such as those shown in parts (a) and (b) of Fig. 1.9 but not the path shown in part (c).

The first thing that is evident is that each such arrangement of five R's and five U's must start with an R (and end with a U). Then as we move across this type of arrangement — going from left to right — the number of R's at any point must equal or exceed the number of U's. Note how this happens in parts (a) and (b) of Fig. 1.9 but not in part (c). Now we can solve the problem at hand if we can count the paths [like the one in part (c)] that go from $(0, 0)$ to $(5, 5)$ but rise above the line $y = x$. Look again at the path in part (c) of Fig. 1.9. Where does the situation there break down for the first time? After all, we start with the requisite R — then follow it by a U. So far, so good! But then there is a second U and, at this (first) time, the number of U's exceeds the number of R's.

Now let us consider the following transformation:

$$\text{R, U, U,} \mid \text{U, R, R, R, U, U, R} \leftrightarrow \text{R, U, U,} \mid \text{R, U, U, U, R, R, U.}$$

What have we done here? For the path on the left-hand side of the transformation, we located the first move (the second U) where the path rose above the line $y = x$. The moves up to and including this move (the second U) remain as is, but the moves that follow are interchanged — each U is replaced by an R and each R by a U. The result is the path on the right-hand side of the transformation — an arrangement of four R's and six U's, as seen in part (d) of Fig. 1.9. Part (e) of that figure provides another path to be avoided; part (f) shows what happens when this path is transformed by the method described above. Now suppose we start with an arrangement of six U's and four R's, say

$$\text{R, U, R, R, U, U, U,} \mid \text{U, U, R.}$$

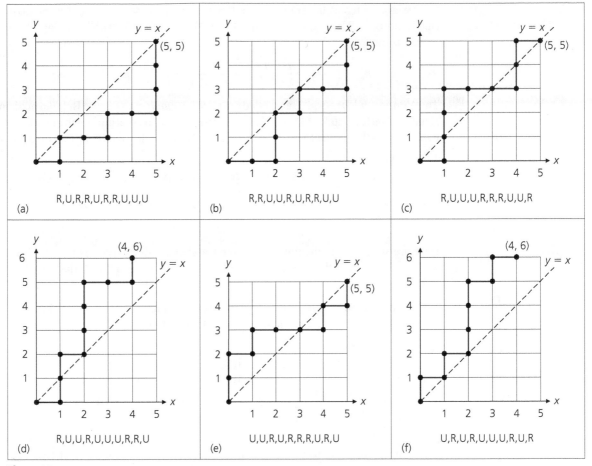

Figure 1.9

Focus on the first place where the number of U's exceeds the number of R's. Here it is in the seventh position, the location of the fourth U. This arrangement is now transformed as follows: The moves up to and including the fourth U remain as they are; the last three moves are interchanged—each U is replaced by an R, each R by a U. This results in the arrangement

$$R, U, R, R, U, U, U, \mid R, R, U.$$

—one of the *bad* arrangements (of five R's and five U's) we wish to avoid as we go from (0, 0) to (5, 5). The correspondence established by these transformations gives us a way to count the number of bad arrangements. We alternatively count the number of ways to arrange four R's and six U's—this is $10!/(4!\,6!) = \binom{10}{4}$. Consequently, the number of ways to go from (0, 0) to (5, 5) without rising above the line $y = x$ is

$$\binom{10}{5} - \binom{10}{4} = \frac{10!}{5!\,5!} - \frac{10!}{4!\,6!} = \frac{6(10)! - 5(10)!}{6!\,5!}$$

$$= \left(\frac{1}{6}\right)\left(\frac{10!}{5!\,5!}\right) = \frac{1}{(5+1)}\binom{10}{5} = \frac{1}{(5+1)}\binom{2 \cdot 5}{5} = 42.$$

The above result generalizes as follows. For any integer $n \geq 0$, the number of paths (made up of n R's and n U's) going from $(0, 0)$ to (n, n), without rising above the line $y = x$, is

$$b_n = \binom{2n}{n} - \binom{2n}{n-1} = \frac{1}{n+1}\binom{2n}{n}, \qquad n \geq 1, \qquad b_0 = 1.$$

The numbers $b_0, b_1, b_2, \ldots$ are called the *Catalan numbers*, after the Belgian mathematician Eugène Charles Catalan (1814–1894), who used them in determining the number of ways to parenthesize the product $x_1 x_2 x_3 x_4 \cdots x_n$. For instance, the five $(= b_3)$ ways to parenthesize $x_1 x_2 x_3 x_4$ are:

$$(((x_1 x_2) x_3) x_4) \quad ((x_1 (x_2 x_3)) x_4) \quad ((x_1 x_2)(x_3 x_4)) \quad (x_1 ((x_2 x_3) x_4)) \quad (x_1 (x_2 (x_3 x_4))).$$

The first seven Catalan numbers are $b_0 = 1$, $b_1 = 1$, $b_2 = 2$, $b_3 = 5$, $b_4 = 14$, $b_5 = 42$, and $b_6 = 132$.

EXAMPLE 1.43

Here are some other situations where the Catalan numbers arise. Some of these examples are very much like the result in Example 1.42. A change in vocabulary is often the only difference.

a) In how many ways can one arrange three 1's and three -1's so that all six partial sums (starting with the first summand) are nonnegative? There are five $(= b_3)$ such arrangements:

$$1, 1, 1, -1, -1, -1 \qquad 1, 1, -1, -1, 1, -1 \qquad 1, -1, 1, 1, -1, -1$$
$$1, 1, -1, 1, -1, -1 \qquad 1, -1, 1, -1, 1, -1$$

In general, for $n \geq 0$, one can arrange n 1's and n -1's, with all $2n$ partial sums nonnegative, in b_n ways.

b) Given four 1's and four 0's, there are 14 $(= b_4)$ ways to list these eight symbols so that in each list the number of 0's never exceeds the number of 1's (as a list is read from left to right). The following shows these 14 lists:

10101010	11001010	11100010
10101100	11001100	11100100
10110010	11010010	11101000
10110100	11010100	
10111000	11011000	11110000

For $n \geq 0$, there are b_n such lists of n 1's and n 0's.

c)

Table 1.10

$(((ab)c)d)$	$(((abc$	111000
$((a(bc))d)$	$((a(bc$	110100
$((ab)(cd))$	$((ab(c$	110010
$(a((bc)d))$	$(a((bc$	101100
$(a(b(cd)))$	$(a(b(c$	101010

Consider the first column in Table 1.10. Here we find five ways to parenthesize the product $abcd$. The first of these is $(((ab)c)d)$. Reading left to right, we list the three occurrences of the left parenthesis "(" and the letters a, b, c — maintaining the order in which these six symbols occur. This results in $(((abc$, the first expression in column 2 of Table 1.10. Likewise, $((a(bc))d)$ in column 1 corresponds to $((a(bc$ in column 2 — and so on, for the other three entries in each of columns 1 and 2. Now one can also go backward, from column 2 to column 1. Take an expression in column 2 and append "d)" to the right end. For instance, $((ab(c$ becomes $((ab(cd)$. Reading this new expression from left to right, we now insert a right parenthesis ")" whenever a product of two results arises. So, for example, $((ab(cd)$ becomes

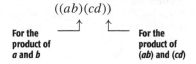

The correspondence between the entries in columns 2 and 3 is more immediate. For an entry in column 2 replace each "(" by a "1" and each letter by a "0". Reversing this process, we replace each "1" by a "(", the first 0 by a, the second by b, and the third by c. This takes us from the entries in column 3 to those in column 2.

Now consider the correspondence between columns 1 and 3. (This correspondence arises from the correspondence between columns 1 and 2 and the one between columns 2 and 3.) It shows us that the number of ways to parenthesize the product $abcd$ equals the number of ways to list three 1's and three 0's so that, as such a list is read from left to right, the number of 1's always equals or exceeds the number of 0's. The number of ways here is 5 ($= b_3$).

In general, one can parenthesize the product $x_1 x_2 x_3 \cdots x_n$ in b_{n-1} ways.

d) Let us arrange the integers 1, 2, 3, 4, 5, 6 in two rows of three so that (1) the integers increase in value as each row is read, from left to right, and (2) in any column the smaller integer is on top. For example, one way to do this is

$$\begin{array}{ccc} 1 & 2 & 4 \\ 3 & 5 & 6 \end{array}$$

Now consider three 1's and three 0's. Arrange these six symbols in a list so that the 1's are in positions 1, 2, 4 (the top row) and the 0's are in positions 3, 5, 6 (the bottom row). The result is 110100. Reversing the process, start with another list, say 101100 (where the number of 0's never exceeds the number of 1's, as the list is read from left to right). The 1's are in positions 1, 3, 4 and the 0's are in positions 2, 5, 6. This corresponds to the arrangement

$$\begin{array}{ccc} 1 & 3 & 4 \\ 2 & 5 & 6 \end{array}$$

which satisfies conditions (1) and (2), as stated above. From this correspondence we learn that the number of ways to arrange 1, 2, 3, 4, 5, 6, so that conditions (1) and (2) are satisfied, is the number of ways to arrange three 1's and three 0's in a list so that as the six symbols are read, from left to right, the number of 0's never exceeds the number of 1's. Consequently, one can arrange 1, 2, 3, 4, 5, 6 and satisfy conditions (1) and (2) in b_3 ($= 5$) ways.

In closing let us mention that the Catalan numbers will come up in other sections — in particular, Section 5 of Chapter 10. Further examples can be found in reference [3] by M. Gardner. For even more results about these numbers one should consult the references for Chapter 10.

EXERCISES 1.5

1. Verify that for each integer $n \geq 1$,
$$\binom{2n}{n} - \binom{2n}{n-1} = \frac{1}{n+1}\binom{2n}{n}.$$

2. Determine the value of b_7, b_8, b_9, and b_{10}.

3. a) In how many ways can one travel in the xy-plane from $(0, 0)$ to $(3, 3)$ using the moves R: $(x, y) \to (x + 1, y)$ and U: $(x, y) \to (x, y + 1)$, if the path taken may touch but *never* fall below the line $y = x$? In how many ways from $(0, 0)$ to $(4, 4)$?

b) Generalize the results in part (a).

c) What can one say about the first and last moves of the paths in parts (a) and (b)?

4. Consider the moves

R: $(x, y) \to (x + 1, y)$ and   U: $(x, y) \to (x, y + 1)$,

as in Example 1.42. In how many ways can one go

a) from $(0, 0)$ to $(6, 6)$ and not rise above the line $y = x$?

b) from $(2, 1)$ to $(7, 6)$ and not rise above the line $y = x - 1$?

c) from $(3, 8)$ to $(10, 15)$ and not rise above the line $y = x + 5$?

5. Find the other three ways to arrange 1, 2, 3, 4, 5, 6 in two rows of three so that the conditions in part (d) of Example 1.43 are satisfied.

6. There are b_4 $(= 14)$ ways to arrange 1, 2, 3, . . . , 8 in two rows of four so that (1) the integers increase in value as each row is read, from left to right, and (2) in any column the smaller integer is on top. Find, as in part (d) of Example 1.43,

a) the arrangements that correspond to each of the following.

    i) 10110010   **ii)** 11001010   **iii)** 11101000

b) the lists of four 1's and four 0's that correspond to each of these arrangements of 1, 2, 3, . . . , 8.

    i) 1 3 4 5   **ii)** 1 2 3 7   **iii)** 1 2 4 5
       2 6 7 8   4 5 6 8   3 6 7 8

7. In how many ways can one parenthesize the product $abcdef$?

8. There are 132 ways in which one can parenthesize the product $abcdefg$.

a) Determine, as in part (c) of Example 1.43, the list of five 1's and five 0's that corresponds to each of the following.

    i) $(((ab)c)(d(ef)))$
    ii) $(a(b(c(d(ef)))))$
    iii) $((((ab)(cd))e)f)$

b) Find, as in Example 1.43, the way to parenthesize $abcdef$ that corresponds to each given list of five 1's and five 0's.

    i) 1110010100
    ii) 1100110010
    iii) 1011100100

9. Consider drawing n semicircles on and above a horizontal line, with no two semicircles intersecting. In parts (a) and (b) of Fig. 1.10 we find the two ways this can be done for $n = 2$; the results for $n = 3$ are shown in parts (c)–(g).

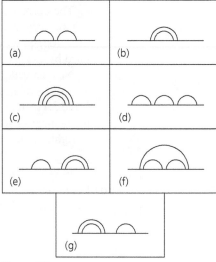

Figure 1.10

i) How many different drawings are there for four semicircles?

ii) How many for any $n \geq 0$? Explain why.

10. a) In how many ways can one go from $(0, 0)$ to $(7, 3)$ if the only moves permitted are R: $(x, y) \to (x + 1, y)$ and U: $(x, y) \to (x, y + 1)$, and the number of U's may never exceed the number of R's along the path taken?

b) Let m, n be positive integers with $m > n$. Answer the question posed in part (a), upon replacing 7 by m and 3 by n.

11. Twelve patrons, six each with a $5 bill and the other six each with a $10 bill, are the first to arrive at a movie theater, where the price of admission is five dollars. In how many ways can these 12 individuals (all loners) line up so that the number with a $5 bill is never exceeded by the number with a $10 bill (and, as a result, the ticket seller is always able to make any necessary change from the bills taken in from the first 11 of these 12 patrons)?

1.6
Summary and Historical Review

In this first chapter we introduced the fundamentals for counting combinations, permutations, and arrangements in a large variety of problems. The breakdown of problems into components requiring the same or different formulas for their solutions provided a key insight into the areas of discrete and combinatorial mathematics. This is somewhat similar to the *top-down approach* for developing algorithms in a structured programming language. Here one develops the algorithm for the solution of a difficult problem by first considering major subproblems that need to be solved. These subproblems are then further *refined* — subdivided into more easily workable programming tasks. Each level of refinement improves on the clarity, precision, and thoroughness of the algorithm until it is readily translatable into the code of the programming language.

Table 1.11 summarizes the major counting formulas we have developed so far. Here we are dealing with a collection of n distinct objects. The formulas count the number of ways to select, or order, with or without repetitions, r of these n objects. The summaries of Chapters 5 and 9 include other such charts that evolve as we extend our investigations into other counting methods.

Table 1.11

Order Is Relevant	Repetitions Are Allowed	Type of Result	Formula	Location in Text
Yes	No	Permutation	$P(n, r) = n!/(n-r)!,$ $\quad 0 \leq r \leq n$	Page 7
Yes	Yes	Arrangement	$n^r, \quad n, r \geq 0$	Page 7
No	No	Combination	$C(n, r) = n!/[r!(n-r)!] = \binom{n}{r},$ $\quad 0 \leq r \leq n$	Page 15
No	Yes	Combination with repetition	$\binom{n+r-1}{r}, \quad n, r \geq 0$	Page 27

As we continue to investigate further principles of enumeration, as well as discrete mathematical structures for applications in coding theory, enumeration, optimization, and sorting schemes in computer science, we shall rely on the fundamental ideas introduced in this chapter.

The notion of permutation can be found in the Hebrew work *Sefer Yetzirah* (*The Book of Creation*), a manuscript written by a mystic sometime between 200 and 600. However, even earlier, it is of interest to note that a result of Xenocrates of Chalcedon (396–314 B.C.) may possibly contain "the first attempt on record to solve a difficult problem in permutations and combinations." For further details consult page 319 of the text by T. L. Heath [4], as well as page 113 of the article by N. L. Biggs [1], a valuable source on the history of enumeration. The first textbook dealing with some of the material we discussed in this chapter was *Ars Conjectandi* by the Swiss mathematician Jakob Bernoulli (1654–1705). The text was published posthumously in 1713 and contained a reprint of the first formal treatise on probability. This treatise had been written in 1657 by Christiaan Huygens (1629–1695), the Dutch physicist, mathematician, and astronomer who discovered the rings of Saturn.

The binomial theorem for $n = 2$ appears in the work of Euclid (300 B.C.), but it was not until the sixteenth century that the term "binomial coefficient" was actually introduced by Michel Stifel (1486–1567). In his *Arithmetica Integra* (1544) he gives the binomial coefficients up to the order of $n = 17$. Blaise Pascal (1623–1662), in his research on probability, published in the 1650s a treatise dealing with the relationships among binomial coefficients, combinations, and polynomials. These results were used by Jakob Bernoulli in proving the general form of the binomial theorem in a manner analogous to that presented in this chapter. Actual use of the symbol $\binom{n}{r}$ did not begin until the nineteenth century, when it was used by Andreas von Ettinghausen (1796–1878).

Blaise Pascal (1623–1662)

It was not until the twentieth century, however, that the advent of the computer made possible the systematic analysis of processes and algorithms used to generate permutations and combinations. We shall examine one such algorithm in Section 10.1.

The first comprehensive textbook dealing with topics in combinations and permutations was written by W. A. Whitworth [10]. Also dealing with the material of this chapter are Chapter 2 of D. I. Cohen [2], Chapter 1 of C. L. Liu [5], Chapter 2 of F. S. Roberts [6], Chapter 4 of K. H. Rosen [7], Chapter 1 of H. J. Ryser [8], and Chapter 5 of A. Tucker [9].

REFERENCES

1. Biggs, Norman L. "The Roots of Combinatorics." *Historia Mathematica* 6 (1979): pp. 109–136.
2. Cohen, Daniel I. A. *Basic Techniques of Combinatorial Theory*. New York: Wiley, 1978.
3. Gardner, Martin. "Mathematical Games, Catalan Numbers: An Integer Sequence that Materializes in Unexpected Places." *Scientific American* 234, no. 6 (June 1976): pp. 120–125.
4. Heath, Thomas Little. *A History of Greek Mathematics*, vol. 1. Reprint of the 1921 edition. New York: Dover Publications, 1981.
5. Liu, C. L. *Introduction to Combinatorial Mathematics*. New York: McGraw-Hill, 1968.
6. Roberts, Fred S. *Applied Combinatorics*. Englewood Cliffs, N.J.: Prentice-Hall, 1984.
7. Rosen, Kenneth H. *Discrete Mathematics and Its Applications*, 5th ed. New York: McGraw-Hill, 2003.
8. Ryser, H. J. *Combinatorial Mathematics*. Published by the Mathematical Association of America. New York: Wiley, 1963.

9. Tucker, Alan. *Applied Combinatorics*, 4th ed. New York: Wiley, 2002.

10. Whitworth, W. A. *Choice and Chance*. Reprint of the 1901 edition. New York: Hafner, 1965.

SUPPLEMENTARY EXERCISES

1. In the manufacture of a certain type of automobile, four kinds of major defects and seven kinds of minor defects can occur. For those situations in which defects do occur, in how many ways can there be twice as many minor defects as there are major ones?

2. A machine has nine different dials, each with five settings labeled 0, 1, 2, 3, and 4.

a) In how many ways can all the dials on the machine be set?

b) If the nine dials are arranged in a line at the top of the machine, how many of the machine settings have no two adjacent dials with the same setting?

3. Twelve points are placed on the circumference of a circle and all the chords connecting these points are drawn. What is the largest number of points of intersection for these chords?

4. A choir director must select six hymns for a Sunday church service. She has three hymn books, each containing 25 hymns (there are 75 different hymns in all). In how many ways can she select the hymns if she wishes to select (a) two hymns from each book? (b) at least one hymn from each book?

5. How many ways are there to place 25 different flags on 10 numbered flagpoles if the order of the flags on a flagpole is (a) not relevant? (b) relevant? (c) relevant and every flagpole flies at least one flag?

6. A penny is tossed 60 times yielding 45 heads and 15 tails. In how many ways could this have happened so that there were no consecutive tails?

7. There are 12 men at a dance. (a) In how many ways can eight of them be selected to form a cleanup crew? (b) How many ways are there to pair off eight women at the dance with eight of these 12 men?

8. In how many ways can the letters in WONDERING be arranged with exactly two consecutive vowels?

9. Dustin has a set of 180 distinct blocks. Each of these blocks is made of either wood or plastic and comes in one of three sizes (small, medium, large), five colors (red, white, blue, yellow, green), and six shapes (triangular, square, rectangular, hexagonal, octagonal, circular). How many of the blocks in this set differ from

a) the *small red wooden square* block in exactly one way? (For example, the *small red plastic square* block is one such block.)

b) the *large blue plastic hexagonal* block in exactly two ways? (For example, the *small red plastic hexagonal* block is one such block.)

10. Mr. and Mrs. Richardson want to name their new daughter so that her initials (first, middle, and last) will be in alphabetical order with no repeated initial. How many such triples of initials can occur under these circumstances?

11. In how many ways can the 11 identical horses on a carousel be painted so that three are brown, three are white, and five are black?

12. In how many ways can a teacher distribute 12 different science books among 16 students if (a) no student gets more than one book? (b) the oldest student gets two books but no other student gets more than one book?

13. Four numbers are selected from the following list of numbers: $-5, -4, -3, -2, -1, 1, 2, 3, 4$. (a) In how many ways can the selections be made so that the product of the four numbers is positive and (i) the numbers are distinct? (ii) each number may be selected as many as four times? (iii) each number may be selected at most three times? (b) Answer part (a) with the product of the four numbers negative.

14. Waterbury Hall, a university residence hall for men, is operated under the supervision of Mr. Kelly. The residence has three floors, each of which is divided into four sections. This coming fall Mr. Kelly will have 12 resident assistants (one for each of the 12 sections). Among these 12 assistants are the four senior assistants — Mr. DiRocco, Mr. Fairbanks, Mr. Hyland, and Mr. Thornhill. (The other eight assistants will be new this fall and are designated as junior assistants.) In how many ways can Mr. Kelly assign his 12 assistants if

a) there are no restrictions?

b) Mr. DiRocco and Mr. Fairbanks must both be assigned to the first floor?

c) Mr. Hyland and Mr. Thornhill must be assigned to different floors?

15. a) How many of the 9000 four-digit integers 1000, 1001, 1002, . . . , 9998, 9999 have four distinct digits that are either increasing (as in 1347 and 6789) or decreasing (as in 6421 and 8653)?

b) How many of the 9000 four-digit integers 1000, 1001, 1002, . . . , 9998, 9999 have four digits that are either nondecreasing (as in 1347, 1226, and 7778) or nonincreasing (as in 6421, 6622, and 9888)?

16. a) Find the coefficient of x^2yz^2 in the expansion of $[(x/2) + y - 3z]^5$.

b) How many distinct terms are there in the complete expansion of

$$\left(\frac{x}{2} + y - 3z\right)^5 ?$$

c) What is the sum of all coefficients in the complete expansion?

17. a) In how many ways can 10 people, denoted A, B, . . . , I, J, be seated about the rectangular table shown in Fig. 1.11, where Figs. 1.11(a) and 1.11(b) are considered the same but are considered different from Fig. 1.11(c)?

b) In how many of the arrangements of part (a) are A and B seated on longer sides of the table across from each other?

18. a) Determine the number of nonnegative integer solutions to the pair of equations

$$x_1 + x_2 + x_3 = 6, \qquad x_1 + x_2 + \cdots + x_5 = 15,$$
$$x_i \geq 0, \quad 1 \leq i \leq 5.$$

b) Answer part (a) with the pair of equations replaced by the pair of inequalities

$$x_1 + x_2 + x_3 \leq 6, \qquad x_1 + x_2 + \cdots + x_5 \leq 15,$$
$$x_i \geq 0, \quad 1 \leq i \leq 5.$$

19. For any given set in a tennis tournament, opponent A can beat opponent B in seven different ways. (At 6–6 they play a tie breaker.) The first opponent to win three sets wins the tournament. (a) In how many ways can scores be recorded with A winning in five sets? (b) In how many ways can scores be recorded with the tournament requiring at least four sets?

20. Given n distinct objects, determine in how many ways r of these objects can be arranged in a circle, where arrangements are considered the same if one can be obtained from the other by rotation.

21. For every positive integer n, show that

$$\binom{n}{0} + \binom{n}{2} + \binom{n}{4} + \cdots = \binom{n}{1} + \binom{n}{3} + \binom{n}{5} + \cdots$$

22. a) In how many ways can the letters in UNUSUAL be arranged?

b) For the arrangements in part (a), how many have all three U's together?

c) How many of the arrangements in part (a) have no consecutive U's?

23. Francesca has 20 different books but the shelf in her dormitory residence will hold only 12 of them.

a) In how many ways can Francesca line up 12 of these books on her bookshelf?

b) How many of the arrangements in part (a) include Francesca's three books on tennis?

24. Determine the value of the integer variable *counter* after execution of the following program segment. (Here i, j, k, l, m, and n are integer variables. The variables r, s, and t are also integer variables; their values — where $r \geq 1$, $s \geq 5$, and $t \geq 7$ — have been set prior to this segment.)

```
counter := 10
for i := 1 to 12 do
   for j := 1 to r do
      counter := counter + 2
for k := 5 to s do
   for l := 3 to k do
      counter := counter + 4
for m := 3 to 12 do
   counter := counter + 6
for n := t downto 7 do
   counter := counter + 8
```

25. a) Find the number of ways to write 17 as a sum of 1's and 2's if order is relevant.

b) Answer part (a) for 18 in place of 17.

c) Generalize the results in parts (a) and (b) for n odd and for n even.

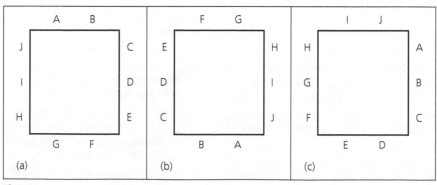

Figure 1.11

26. a) In how many ways can 17 be written as a sum of 2's and 3's if the order of the summands is (i) not relevant? (ii) relevant?

b) Answer part (a) for 18 in place of 17.

27. a) If n and r are positive integers with $n \geq r$, how many solutions are there to

$$x_1 + x_2 + \cdots + x_r = n,$$

where each x_i is a positive integer, for $1 \leq i \leq r$?

b) In how many ways can a positive integer n be written as a sum of r positive integer summands ($1 \leq r \leq n$) if the order of the summands is relevant?

28. a) In how many ways can one travel in the xy-plane from $(1, 2)$ to $(5, 9)$ if each move is one of the following types:

(R): $(x, y) \rightarrow (x + 1, y)$;   (U): $(x, y) \rightarrow (x, y + 1)$?

b) Answer part (a) if a third (diagonal) move

(D): $(x, y) \rightarrow (x + 1, y + 1)$

is also possible.

29. a) In how many ways can a particle move in the xy-plane from the origin to the point $(7, 4)$ if the moves that are allowed are of the form:

(R): $(x, y) \rightarrow (x + 1, y)$;   (U): $(x, y) \rightarrow (x, y + 1)$?

b) How many of the paths in part (a) do not use the path from $(2, 2)$ to $(3, 2)$ to $(4, 2)$ to $(4, 3)$ shown in Fig. 1.12?

c) Answer parts (a) and (b) if a third type of move

(D): $(x, y) \rightarrow (x + 1, y + 1)$

is also allowed.

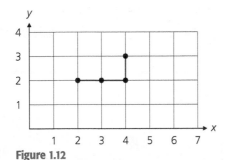

Figure 1.12

30. Due to their outstanding academic records, Donna and Katalin are the finalists for the outstanding physics student (in their college graduating class). A committee of 14 faculty mem-

bers will each select one of the candidates to be the winner and place his or her choice (checked off on a ballot) into the ballot box. Suppose that Katalin receives nine votes and Donna receives five. In how many ways can the ballots be selected, one at a time, from the ballot box so that there are always more votes in favor of Katalin? [This is a special case of a general problem called, appropriately, *the ballot problem*. This problem was solved by Joseph Louis François Bertrand (1822–1900).]

31. Consider the 8×5 grid shown in Fig. 1.13. How many different rectangles (with integer-coordinate corners) does this grid contain? [For example, there is a rectangle (square) with corners $(1, 1)$, $(2, 1)$, $(2, 2)$, $(1, 2)$, a second rectangle with corners $(3, 2)$, $(4, 2)$, $(4, 4)$, $(3, 4)$, and a third with corners $(5, 0)$, $(7, 0)$, $(7, 3)$ $(5, 3)$.]

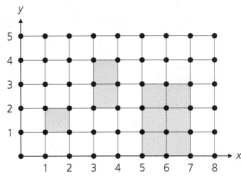

Figure 1.13

32. As head of quality control, Silvia examined 15 motors, one at a time, and found six defective (D) motors and nine in good (G) working condition. If she listed each finding (of D or G) after examining each individual motor, in how many ways could Silvia's list start with a run of three G's and have six runs in total?

33. In order to graduate on schedule, Hunter must take (and pass) four mathematics electives during his final six quarters. If he may select these electives from a list of 12 (that are offered every quarter) and he does not want to take more than one of these electives in any given quarter, in how many ways can he select and schedule these four electives?

34. In how many ways can a family of four (mother, father, and two children) be seated at a round table, with eight other people, so that the parents are seated next to each other and there is one child on a side of each parent? (Two seatings are considered the same if one can be rotated to look like the other.)

2
Fundamentals of Logic

In the first chapter we derived a summation formula in Example 1.40 (Section 1.4). We obtained this formula by counting the same collection of objects (the statements that were executed in a certain program segment) in two different ways and then equating the results. Consequently, we say that the formula was established by a combinatorial proof. This is one of many different techniques for arriving at a proof.

In this chapter we take a close look at what constitutes a valid argument and a more conventional proof. When a mathematician wishes to provide a proof for a given situation, he or she must use a system of logic. This is also true when a computer scientist develops the algorithms needed for a program or system of programs. The logic of mathematics is applied to decide whether one statement follows from, or is a logical consequence of, one or more other statements.

Some of the rules that govern this process are described in this chapter. We shall use these rules in proofs (provided in the text and required in the exercises) throughout subsequent chapters. However, at no time can we hope to arrive at a point at which we can apply the rules in an automatic fashion. As in applying the counting ideas discussed in Chapter 1, we should always analyze and seek to understand the situation given. This often calls for attributes we cannot learn in a book, such as insight and creativity. Merely trying to apply formulas or invoke rules will not get us very far either in proving results (such as theorems) or in doing enumeration problems.

2.1
Basic Connectives and Truth Tables

In the development of any mathematical theory, assertions are made in the form of sentences. Such verbal or written assertions, called *statements* (or *propositions*), are declarative sentences that are either true or false — but *not* both. For example, the following are statements, and we use the lowercase letters of the alphabet (such as p, q, and r) to represent these statements.

p: Combinatorics is a required course for sophomores.

q: Margaret Mitchell wrote *Gone with the Wind*.

r: $2 + 3 = 5$.

On the other hand, we do not regard sentences such as the exclamation

"What a beautiful evening!"

or the command

"Get up and do your exercises."

as statements since they do not have *truth values* (true or false).

The preceding statements represented by the letters p, q, and r are considered to be *primitive* statements, for there is really no way to break them down into anything simpler. New statements can be obtained from existing ones in two ways.

1) Transform a given statement p into the statement $\neg p$, which denotes its *negation* and is read "Not p."

For the statement p above, $\neg p$ is the statement "Combinatorics is not a required course for sophomores." (We do not consider the negation of a primitive statement to be a primitive statement.)

2) Combine two or more statements into a *compound* statement, using the following *logical connectives*.

a) Conjunction: The *conjunction* of the statements p, q is denoted by $p \wedge q$, which is read "p and q." In our example the compound statement $p \wedge q$ is read "Combinatorics is a required course for sophomores, **and** Margaret Mitchell wrote *Gone with the Wind*."

b) Disjunction: The expression $p \vee q$ denotes the *disjunction* of the statements p, q and is read "p or q." Hence "Combinatorics is a required course for sophomores, **or** Margaret Mitchell wrote *Gone with the Wind*" is the verbal translation for $p \vee q$, when p, q are as above. We use the word "or" in the *inclusive* sense here. Consequently, $p \vee q$ is true if one or the other of p, q is true or if *both* of the statements p, q are true. In English we sometimes write "and/or" to point this out. The *exclusive* "or" is denoted by $p \veebar q$. The compound statement $p \veebar q$ is true if one or the other of p, q is true but *not both* of the statements p, q are true. One way to express $p \veebar q$ for the example here is "Combinatorics is a required course for sophomores, or Margaret Mitchell wrote *Gone with the Wind*, but not both."

c) Implication: We say that "p implies q" and write $p \rightarrow q$ to designate the statement, which is the *implication* of q by p. Alternatively, we can also say

(i) "If p, then q."	**(ii)** "p is *sufficient* for q."
(iii) "p is a *sufficient condition* for q."	**(iv)** "q is *necessary* for p."
(v) "q is a *necessary condition* for p."	**(vi)** "p only if q."

A verbal translation of $p \rightarrow q$ for our example is "If combinatorics is a required course for sophomores, then Margaret Mitchell wrote *Gone with the Wind*." The statement p is called the *hypothesis* of the implication; q is called the *conclusion*. When statements are combined in this manner, there need not be any causal relationship between the statements for the implication to be true.

d) Biconditional: Last, the *biconditional* of two statements p, q, is denoted by $p \leftrightarrow q$, which is read "p if and only if q," or "p is necessary and sufficient for q." For our p, q, "Combinatorics is a required course for sophomores if and only if Margaret Mitchell wrote *Gone with the Wind*" conveys the meaning of $p \leftrightarrow q$. We sometimes abbreviate "p if and only if q" as "p iff q."

Throughout our discussion on logic we must realize that a sentence such as

"The number x is an integer."

is *not* a statement because its truth value (true or false) cannot be determined until a numerical value is assigned for x. If x were assigned the value 7, the result would be a true statement. Assigning x a value such as $\frac{1}{2}$, $\sqrt{2}$, or π, however, would make the resulting statement false. (We shall encounter this type of situation again in Sections 2.4 and 2.5 of this chapter.)

In the foregoing discussion, we mentioned the circumstances under which the *compound* statements $p \vee q$, $p \veebar q$ are considered true, on the basis of the truth of their components p, q. This idea of the truth or falsity of a compound statement being dependent only on the truth values of its components is worth further investigation. Tables 2.1 and 2.2 summarize the truth and falsity of the negation and the different kinds of compound statements on the basis of the truth values of their components. In constructing such *truth tables*, we write "0" for false and "1" for true.

Table 2.1

p	$\neg p$
0	1
1	0

Table 2.2

p	q	$p \wedge q$	$p \vee q$	$p \veebar q$	$p \rightarrow q$	$p \leftrightarrow q$
0	0	0	0	0	1	1
0	1	0	1	1	1	0
1	0	0	1	1	0	0
1	1	1	1	0	1	1

The four possible truth assignments for p, q can be listed in any order. For later work, the particular order presented here will prove useful.

We see that the columns of truth values for p and $\neg p$ are the opposite of each other. The statement $p \wedge q$ is true only when both p, q are true, whereas $p \vee q$ is false only when both the component statements p, q are false. As we noted before, $p \veebar q$ is true when exactly one of p, q is true.

For the implication $p \rightarrow q$, the result is true in all cases except where p is true and q is false. We do not want a true statement to lead us into believing something that is false. However, we regard as true a statement such as "If $2 + 3 = 6$, then $2 + 4 = 7$," even though the statements "$2 + 3 = 6$" and "$2 + 4 = 7$" are both false.

Finally, the biconditional $p \leftrightarrow q$ is true when the statements p, q have the same truth value and is false otherwise.

Now that we have been introduced to certain concepts, let us investigate a little further some of these initial ideas about connectives. Our first two examples should prove useful for such an investigation.

EXAMPLE 2.1

Let s, t, and u denote the following primitive statements:

> s: Phyllis goes out for a walk.
>
> t: The moon is out.
>
> u: It is snowing.

The following English sentences provide possible translations for the given (symbolic) compound statements.

a) $(t \wedge \neg u) \rightarrow s$: If the moon is out and it is not snowing, then Phyllis goes out for a walk.

b) $t \to (\neg u \to s)$: If the moon is out, then if it is not snowing Phyllis goes out for a walk. [So $\neg u \to s$ is understood to mean $(\neg u) \to s$ as opposed to $\neg(u \to s)$.]

c) $\neg(s \leftrightarrow (u \vee t))$: It is not the case that Phyllis goes out for a walk if and only if it is snowing or the moon is out.

Now we will work in reverse order and examine the logical (or symbolic) notation for three given English sentences:

d) "Phyllis will go out walking if and only if the moon is out." Here the words "if and only if" indicate that we are dealing with a biconditional. In symbolic form this becomes $s \leftrightarrow t$.

e) "If it is snowing and the moon is not out, then Phyllis will not go out for a walk." This compound statement is an implication where the hypothesis is also a compound statement. One may express this statement in symbolic form as $(u \wedge \neg t) \to \neg s$.

f) "It is snowing but Phyllis will still go out for a walk." Now we come across a new connective — namely, *but*. In our study of logic we shall follow the convention that the connectives *but* and *and* convey the same meaning. Consequently, this sentence may be represented as $u \wedge s$.

Now let us return to the results in Table 2.2, particularly the sixth column. For if this is one's first encounter with the truth table for the implication $p \to q$, then it may be somewhat difficult to accept the stated entries — especially the results in the first two rows (where p has the truth value 0). The following example should help make these truth value assignments easier to grasp.

| **EXAMPLE 2.2** | Consider the following scenario. It is almost the week before Christmas and Penny will be attending several parties that week. Ever conscious of her weight, she plans not to weigh herself until the day after Christmas. Considering what those parties may do to her waistline by then, she makes the following resolution for the December 26 outcome: "If I weigh more than 120 pounds, then I shall enroll in an exercise class."

Here we let p and q denote the (primitive) statements

p: I weigh more than 120 pounds.

q: I shall enroll in an exercise class.

Then Penny's statement (implication) is given by $p \to q$.

We shall consider the truth values of this particular example of $p \to q$ for the rows of Table 2.2. Consider first the easier cases in rows 4 and 3.

- Row 4: p and q both have the truth value 1. On December 26 Penny finds that she weighs more than 120 pounds and promptly enrolls in an exercise class, just as she said she would. Here we consider $p \to q$ to be true and assign it the truth value 1.

- Row 3: p has the truth value 1, q has the truth value 0. Now that December 26 has arrived, Penny finds her weight to be over 120 pounds, but she makes no attempt to enroll in an exercise class. In this case we feel that Penny has broken her resolution — in other words, the implication $p \to q$ is false (and has the truth value 0).

The cases in rows 1 and 2 may not immediately agree with our intuition, but the example should make these results a little easier to accept.

- Row 1: p and q both have the truth value 0. Here Penny finds that on December 26 her weight is 120 pounds or less and she does not enroll in an exercise class. She has not violated her resolution; we take her statement $p \rightarrow q$ to be true and assign it the truth value 1.

- Row 2: p has the truth value 0, q has the truth value 1. This last case finds Penny weighing 120 pounds or less on December 26 but still enrolling in an exercise class. Perhaps her weight is 119 or 120 pounds and she feels this is still too high. Or maybe she wants to join an exercise class because she thinks it will be good for her health. No matter what the reason, she has not gone against her resolution $p \rightarrow q$. Once again, we accept this compound statement as true, assigning it the truth value 1.

Our next example discusses a related notion: the *decision* (or *selection*) structure in computer programming.

EXAMPLE 2.3

In computer science the **if-then** and **if-then-else** decision structures arise (in various formats) in high-level programming languages such as Java and C++. The hypothesis p is often a relational expression such as $x > 2$. This expression then becomes a (logical) statement that has the truth value 0 or 1, depending on the value of the variable x at that point in the program. The conclusion q is usually an "executable statement." (So q is not one of the logical statements that we have been discussing.) When dealing with "**if** p **then** q," in this context, the computer executes q only on the condition that p is true. For p false, the computer goes to the next instruction in the program sequence. For the decision structure "**if** p **then** q **else** r," q is executed when p is true and r is executed when p is false.

Before continuing, a word of caution: Be careful when using the symbols $\rightarrow$ and $\leftrightarrow$. The implication and the biconditional are not the same, as evidenced by the last two columns of Table 2.2.

In our everyday language, however, we often find situations where an implication is used when the intention actually calls for a biconditional. For example, consider the following implications that a certain parent might direct to his or her child.

$s \rightarrow t$:   If you do your homework, then you will get to watch the baseball game.

$t \rightarrow s$:   You will get to watch the baseball game only if you do your homework.

- Case 1: The implication $s \rightarrow t$. When the parent says to the child, "If you do your homework, then you will get to watch the baseball game," he or she is trying a positive approach by emphasizing the enjoyment in watching the baseball game.

- Case 2: The implication $t \rightarrow s$. Here we find the negative approach and the parent who warns the child in saying, "You will get to watch the baseball game only if you do your homework." This parent places the emphasis on the punishment (lack of enjoyment) to be incurred.

In either case, the parent probably wants his or her implication — be it $s \rightarrow t$ or $t \rightarrow s$ — to be understood as the biconditional $s \leftrightarrow t$. For in case 1 the parent wants to hint at the punishment while promising the enjoyment; in case 2, where the punishment has been used (perhaps, to threaten), if the child does in fact do the homework, then that child will definitely be given the opportunity to enjoy watching the baseball game.

In scientific writing one must make every effort to be unambiguous — when an implication is given, it ordinarily cannot, and should not, be interpreted as a biconditional. Definitions are a notable exception, which we shall discuss in Section 2.5.

Before we continue let us take a step back. When we summarized the material that gave us Tables 2.1 and 2.2, we may not have stressed enough that the results were for any statements p, q — *not* just primitive statements p, q. Examples 2.4 through 2.6 should help to reinforce this.

EXAMPLE 2.4

Let us examine the truth table for the compound statement "Margaret Mitchell wrote *Gone with the Wind*, and if $2 + 3 \neq 5$, then combinatorics is a required course for sophomores." In symbolic notation this statement is written as $q \wedge (\neg r \to p)$, where p, q, and r represent the primitive statements introduced at the start of this section. The last column of Table 2.3 contains the truth values for this result. We obtained these truth values by using the fact that the conjunction of any two statements is true if and only if both statements are true. This is what we said earlier in Table 2.2, and now one of our statements — namely, the implication $\neg r \to p$ — is definitely a compound statement, not a primitive one. Columns 4, 5, and 6 in this table show how we build the truth table up by considering smaller parts of the compound statement and by using the results from Tables 2.1 and 2.2.

Table 2.3

p	q	r	$\neg r$	$\neg r \to p$	$q \wedge (\neg r \to p)$
0	0	0	1	0	0
0	0	1	0	1	0
0	1	0	1	0	0
0	1	1	0	1	1
1	0	0	1	1	0
1	0	1	0	1	0
1	1	0	1	1	1
1	1	1	0	1	1

EXAMPLE 2.5

In Table 2.4 we develop the truth tables for the compound statements $p \vee (q \wedge r)$ (column 5) and $(p \vee q) \wedge r$ (column 7).

Table 2.4

p	q	r	$q \wedge r$	$p \vee (q \wedge r)$	$p \vee q$	$(p \vee q) \wedge r$
0	0	0	0	0	0	0
0	0	1	0	0	0	0
0	1	0	0	0	1	0
0	1	1	1	1	1	1
1	0	0	0	1	1	0
1	0	1	0	1	1	1
1	1	0	0	1	1	0
1	1	1	1	1	1	1

Because the truth values in columns 5 and 7 differ (in rows 5 and 7), we must avoid writing a compound statement such as $p \vee q \wedge r$. Without parentheses to indicate which of the connectives $\vee$ and $\wedge$ should be applied first, we have no idea whether we are dealing with $p \vee (q \wedge r)$ or $(p \vee q) \wedge r$.

Our last example for this section illustrates two special types of statements.

EXAMPLE 2.6

The results in columns 4 and 7 of Table 2.5 reveal that the statement $p \rightarrow (p \vee q)$ is true and that the statement $p \wedge (\neg p \wedge q)$ is false for all truth value assignments for the component statements p, q.

Table 2.5

p	q	$p \vee q$	$p \rightarrow (p \vee q)$	$\neg p$	$\neg p \wedge q$	$p \wedge (\neg p \wedge q)$
0	0	0	1	1	0	0
0	1	1	1	1	1	0
1	0	1	1	0	0	0
1	1	1	1	0	0	0

Definition 2.1

A compound statement is called a *tautology* if it is true for all truth value assignments for its component statements. If a compound statement is false for all such assignments, then it is called a *contradiction*.

Throughout this chapter we shall use the symbol T_0 to denote any tautology and the symbol F_0 to denote any contradiction.

We can use the ideas of tautology and implication to describe what we mean by a valid argument. This will be of primary interest to us in Section 2.3, and it will help us develop needed skills for proving mathematical theorems. In general, an argument starts with a list of *given* statements called *premises* and a statement called the *conclusion* of the argument. We examine these premises, say $p_1, p_2, p_3, \ldots, p_n$, and try to show that the conclusion q follows logically from these given statements — that is, we try to show that if each of $p_1, p_2, p_3, \ldots, p_n$ is a true statement, then the statement q is also true. To do so one way is to examine the implication

$$(p_1 \wedge p_2 \wedge p_3 \wedge \cdots \wedge p_n)^{\dagger} \rightarrow q,$$

where the hypothesis is the conjunction of the n premises. If any one of $p_1, p_2, p_3, \ldots, p_n$ is false, then no matter what truth value q has, the implication $(p_1 \wedge p_2 \wedge p_3 \wedge \cdots \wedge p_n) \rightarrow q$ is true. Consequently, if we start with the premises $p_1, p_2, p_3, \ldots, p_n$ — each with truth value 1 — and find that under these circumstances q also has the value 1, then the implication

$$(p_1 \wedge p_2 \wedge p_3 \wedge \cdots \wedge p_n) \rightarrow q$$

is a *tautology* and we have a *valid argument*.

[†]At this point we have dealt only with the conjunction of two statements, so we must point out that the conjunction $p_1 \wedge p_2 \wedge p_3 \wedge \cdots \wedge p_n$ of n statements is true if and only if each p_i, $1 \leq i \leq n$, is true. We shall deal with this generalized conjunction in detail in Example 4.16 of Section 4.2.

1. Determine whether each of the following sentences is a statement.

a) In 2003 George W. Bush was the president of the United States.

b) $x + 3$ is a positive integer.

c) Fifteen is an even number.

d) If Jennifer is late for the party, then her cousin Zachary will be quite angry.

e) What time is it?

f) As of June 30, 2003, Christine Marie Evert had won the French Open a record seven times.

2. Identify the primitive statements in Exercise 1.

3. Let p, q be primitive statements for which the implication $p \to q$ is false. Determine the truth values for each of the following.

a) $p \wedge q$ **b)** $\neg p \vee q$ **c)** $q \to p$ **d)** $\neg q \to \neg p$

4. Let p, q, r, s denote the following statements:

p: I finish writing my computer program before lunch.
q: I shall play tennis in the afternoon.
r: The sun is shining.
s: The humidity is low.

Write the following in symbolic form.

a) If the sun is shining, I shall play tennis this afternoon.

b) Finishing the writing of my computer program before lunch is necessary for my playing tennis this afternoon.

c) Low humidity and sunshine are sufficient for me to play tennis this afternoon.

5. Let p, q, r denote the following statements about a particular triangle ABC.

p: Triangle ABC is isosceles.
q: Triangle ABC is equilateral.
r: Triangle ABC is equiangular.

Translate each of the following into an English sentence.

a) $q \to p$ **b)** $\neg p \to \neg q$
c) $q \leftrightarrow r$ **d)** $p \wedge \neg q$
e) $r \to p$

6. Determine the truth value of each of the following implications.

a) If $3 + 4 = 12$, then $3 + 2 = 6$.

b) If $3 + 3 = 6$, then $3 + 4 = 9$.

c) If Thomas Jefferson was the third president of the United States, then $2 + 3 = 5$.

7. Rewrite each of the following statements as an implication in the **if-then** form.

a) Practicing her serve daily is a sufficient condition for Darci to have a good chance of winning the tennis tournament.

b) Fix my air conditioner or I won't pay the rent.

c) Mary will be allowed on Larry's motorcycle only if she wears her helmet.

8. Construct a truth table for each of the following compound statements, where p, q, r denote primitive statements.

a) $\neg(p \vee \neg q) \to \neg p$ **b)** $p \to (q \to r)$
c) $(p \to q) \to r$ **d)** $(p \to q) \to (q \to p)$
e) $[p \wedge (p \to q)] \to q$ **f)** $(p \wedge q) \to p$
g) $q \leftrightarrow (\neg p \vee \neg q)$
h) $[(p \to q) \wedge (q \to r)] \to (p \to r)$

9. Which of the compound statements in Exercise 8 are tautologies?

10. Verify that $[p \to (q \to r)] \to [(p \to q) \to (p \to r)]$ is a tautology.

11. a) How many rows are needed for the truth table of the compound statement $(p \vee \neg q) \leftrightarrow [(\neg r \wedge s) \to t]$, where p, q, r, s, and t are primitive statements?

b) Let $p_1, p_2, \ldots, p_n$ denote n primitive statements. Let p be a compound statement that contains at least one occurrence each of p_i, for $1 \le i \le n$ — and p contains no other primitive statement. How many rows are needed to construct the truth table for p?

12. Determine all truth value assignments, if any, for the primitive statements p, q, r, s, t that make each of the following compound statements false.

a) $[(p \wedge q) \wedge r] \to (s \vee t)$

b) $[p \wedge (q \wedge r)] \to (s \veebar t)$

13. If statement q has the truth value 1, determine all truth value assignments for the primitive statements, p, r, and s for which the truth value of the statement

$$(q \to [(\neg p \vee r) \wedge \neg s]) \wedge [\neg s \to (\neg r \wedge q)]$$

is 1.

14. At the start of a program (written in pseudocode) the integer variable n is assigned the value 7. Determine the value of n after each of the following *successive* statements is encountered during the execution of this program. [Here the value of n following the execution of the statement in part (a) becomes the value of n for the statement in part (b), and so on, through the statement in part (d). For positive integers $a, b, \lfloor a/b \rfloor$ returns the integer part of the quotient — for example, $\lfloor 6/2 \rfloor = 3$, $\lfloor 7/2 \rfloor = 3$, $\lfloor 2/5 \rfloor = 0$, and $\lfloor 8/3 \rfloor = 2$.]

a) `if` $n > 5$ `then` $n := n + 2$

b) **if** $((n + 2 = 8)$ **or** $(n - 3 = 6))$ **then**
    $n := 2 * n + 1$

c) **if** $((n - 3 = 16)$ **and** $(\lfloor n/6 \rfloor = 1))$ **then**
    $n := n + 3$

d) **if** $((n \neq 21)$ **and** $(n - 7 = 15))$ **then**
    $n := n - 4$

15. The integer variables m and n are assigned the values 3 and 8, respectively, during the execution of a program (written in pseudocode). Each of the following *successive* statements is then encountered during program execution. [Here the values of m, n following the execution of the statement in part (a) become the values of m, n for the statement in part (b), and so on, through the statement in part (e).] What are the values of m, n after each of these statements is encountered?

a) **if** $n - m = 5$ **then** $n := n - 2$

b) **if** $((2 * m = n)$ **and** $(\lfloor n/4 \rfloor = 1))$ **then**
    $n := 4 * m - 3$

c) **if** $((n < 8)$ **or** $(\lfloor m/2 \rfloor = 2))$ **then** $n := 2 * m$
    else $m := 2 * n$

d) **if** $((m < 20)$ **and** $(\lfloor n/6 \rfloor = 1))$ **then**
    $m := m - n - 5$

e) **if** $((n = 2 * m)$ **or** $(\lfloor n/2 \rfloor = 5))$ **then**
    $m := m + 2$

16. In the following program segment i, j, m, and n are integer variables. The values of m and n are supplied by the user earlier in the execution of the total program.

for $i := 1$ **to** m **do**
    for $j := 1$ **to** n **do**
        if $i \neq j$ **then**
            print $i + j$

How many times is the **print** statement in the segment executed when (a) $m = 10$, $n = 10$; (b) $m = 20$, $n = 20$; (c) $m = 10$, $n = 20$; (d) $m = 20$, $n = 10$?

17. After baking a pie for the two nieces and two nephews who are visiting her, Aunt Nellie leaves the pie on her kitchen table to cool. Then she drives to the mall to close her boutique for the day. Upon her return she finds that someone has eaten one-quarter of the pie. Since no one was in her house that day — except for the four visitors — Aunt Nellie questions each niece and nephew about who ate the piece of pie. The four "suspects" tell her the following:

Charles:	Kelly ate the piece of pie.
Dawn:	I did not eat the piece of pie.
Kelly:	Tyler ate the pie.
Tyler:	Kelly lied when she said I ate the pie.

If only one of these four statements is true and only one of the four committed this heinous crime, who is the vile culprit that Aunt Nellie will have to punish severely?

2.2
Logical Equivalence: The Laws of Logic

In all areas of mathematics we need to know when the entities we are studying are equal or essentially the same. For example, in arithmetic and algebra we know that two nonzero real numbers are equal when they have the same magnitude and algebraic sign. Hence, for two nonzero real numbers x, y, we have $x = y$ if $|x| = |y|$ and $xy > 0$, and conversely (that is, if $x = y$, then $|x| = |y|$ and $xy > 0$). When we deal with triangles in geometry, the notion of congruence arises. Here triangle ABC and triangle DEF are congruent if, for instance, they have equal corresponding sides — that is, the length of side AB = the length of side DE, the length of side BC = the length of side EF, and the length of side CA = the length of side FD.

Our study of logic is often referred to as the *algebra of propositions* (as opposed to the algebra of real numbers). In this algebra we shall use the truth tables of the statements, or propositions, to develop an idea of when two such entities are essentially the same. We begin with an example.

EXAMPLE 2.7

For primitive statements p and q, Table 2.6 provides the truth tables for the compound statements $\neg p \lor q$ and $p \to q$. Here we see that the corresponding truth tables for the two statements $\neg p \lor q$ and $p \to q$ are exactly the same.

Table 2.6

p	q	$\neg p$	$\neg p \vee q$	$p \rightarrow q$
0	0	1	1	1
0	1	1	1	1
1	0	0	0	0
1	1	0	1	1

This situation leads us to the following idea.

Definition 2.2   Two statements s_1, s_2 are said to be *logically equivalent*, and we write $s_1 \Longleftrightarrow s_2$, when the statement s_1 is true (respectively, false) if and only if the statement s_2 is true (respectively, false).

Note that when $s_1 \Longleftrightarrow s_2$ the statements s_1 and s_2 provide the same truth tables because s_1, s_2 have the same truth values for *all* choices of truth values for their primitive components.

As a result of this concept we see that we can express the connective for the implication (of primitive statements) in terms of negation and disjunction — that is, $(p \rightarrow q) \Longleftrightarrow \neg p \vee q$. In the same manner, from the result in Table 2.7 we have $(p \leftrightarrow q) \Longleftrightarrow (p \rightarrow q) \wedge (q \rightarrow p)$, and this helps validate the use of the term *biconditional*. Using the logical equivalence from Table 2.6, we find that we can also write $(p \leftrightarrow q) \Longleftrightarrow (\neg p \vee q) \wedge (\neg q \vee p)$. Consequently, if we so choose, we can eliminate the connectives $\rightarrow$ and $\leftrightarrow$ from compound statements.

Table 2.7

p	q	$p \rightarrow q$	$q \rightarrow p$	$(p \rightarrow q) \wedge (q \rightarrow p)$	$p \leftrightarrow q$
0	0	1	1	1	1
0	1	1	0	0	0
1	0	0	1	0	0
1	1	1	1	1	1

Examining Table 2.8, we find that negation, along with the connectives $\wedge$ and $\vee$, are all we need to replace the *exclusive or* connective, $\underline{\vee}$. In fact, we may even eliminate either $\wedge$ or $\vee$. However, for the related applications we want to study later in the text, we shall need both $\wedge$ and $\vee$ as well as negation.

Table 2.8

p	q	$p \underline{\vee} q$	$p \vee q$	$p \wedge q$	$\neg(p \wedge q)$	$(p \vee q) \wedge \neg(p \wedge q)$
0	0	0	0	0	1	0
0	1	1	1	0	1	1
1	0	1	1	0	1	1
1	1	0	1	1	0	0

We now use the idea of logical equivalence to examine some of the important properties that hold for the algebra of propositions.

For all real numbers a, b, we know that $-(a + b) = (-a) + (-b)$. Is there a comparable result for primitive statements p, q?

EXAMPLE 2.8

In Table 2.9 we have constructed the truth tables for the statements $\neg(p \wedge q)$, $\neg p \vee \neg q$, $\neg(p \vee q)$, and $\neg p \wedge \neg q$, where p, q are primitive statements. Columns 4 and 7 reveal that $\neg(p \wedge q) \Longleftrightarrow \neg p \vee \neg q$; columns 9 and 10 reveal that $\neg(p \vee q) \Longleftrightarrow \neg p \wedge \neg q$. These results are known as *DeMorgan's Laws*. They are similar to the familiar law for real numbers,

$$-(a + b) = (-a) + (-b),$$

already noted, which shows the negative of a sum to be equal to the sum of the negatives. Here, however, a crucial difference emerges: The negation of the *conjunction* of two primitive statements p, q results in the *disjunction* of their negations $\neg p$, $\neg q$, whereas the negation of the *disjunction* of these same statements p, q is logically equivalent to the *conjunction* of their negations $\neg p$, $\neg q$.

Table 2.9

p	q	$p \wedge q$	$\neg(p \wedge q)$	$\neg p$	$\neg q$	$\neg p \vee \neg q$	$p \vee q$	$\neg(p \vee q)$	$\neg p \wedge \neg q$
0	0	0	1	1	1	1	0	1	1
0	1	0	1	1	0	1	1	0	0
1	0	0	1	0	1	1	1	0	0
1	1	1	0	0	0	0	1	0	0

Although p, q were primitive statements in the preceding example we shall soon learn that DeMorgan's Laws hold for any two arbitrary statements.

In the arithmetic of real numbers, the operations of addition and multiplication are both involved in the principle called the Distributive Law of Multiplication over Addition: For all real numbers a, b, c,

$$a \times (b + c) = (a \times b) + (a \times c).$$

The next example shows that there is a similar law for primitive statements. There is also a second related law (for primitive statements) that has no counterpart in the arithmetic of real numbers.

EXAMPLE 2.9

Table 2.10 contains the truth tables for the statements $p \wedge (q \vee r)$, $(p \wedge q) \vee (p \wedge r)$, $p \vee (q \wedge r)$, and $(p \vee q) \wedge (p \vee r)$. From the table it follows that for all primitive statements p, q, and r,

$$p \wedge (q \vee r) \Longleftrightarrow (p \wedge q) \vee (p \wedge r) \qquad \text{The Distributive Law of } \wedge \text{ over } \vee$$
$$p \vee (q \wedge r) \Longleftrightarrow (p \vee q) \wedge (p \vee r) \qquad \text{The Distributive Law of } \vee \text{ over } \wedge$$

The second distributive law has no counterpart in the arithmetic of real numbers. That is, it is not true for all real numbers a, b, and c that the following holds: $a + (b \times c) = (a + b) \times (a + c)$. For $a = 2$, $b = 3$, and $c = 5$, for instance, $a + (b \times c) = 17$ but $(a + b) \times (a + c) = 35$.

Table 2.10

p	q	r	$p \wedge (q \vee r)$	$(p \wedge q) \vee (p \wedge r)$	$p \vee (q \wedge r)$	$(p \vee q) \wedge (p \vee r)$
0	0	0	0	0	0	0
0	0	1	0	0	0	0
0	1	0	0	0	0	0
0	1	1	0	0	1	1
1	0	0	0	0	1	1
1	0	1	1	1	1	1
1	1	0	1	1	1	1
1	1	1	1	1	1	1

Before going any further, we note that, in general, if s_1, s_2 are statements and $s_1 \leftrightarrow s_2$ is a tautology, then s_1, s_2 must have the same corresponding truth values (that is, for each assignment of truth values to the primitive statements in s_1 and s_2, s_1 is true if and only if s_2 is true and s_1 is false if and only if s_2 is false) and $s_1 \Longleftrightarrow s_2$. When s_1 and s_2 are logically equivalent statements (that is, $s_1 \Longleftrightarrow s_2$), then the compound statement $s_1 \leftrightarrow s_2$ is a tautology. Under these circumstances it is also true that $\neg s_1 \Longleftrightarrow \neg s_2$, and $\neg s_1 \leftrightarrow \neg s_2$ is a tautology.

If s_1, s_2, and s_3 are statements where $s_1 \Longleftrightarrow s_2$ and $s_2 \Longleftrightarrow s_3$ then $s_1 \Longleftrightarrow s_3$. When two statements s_1 and s_2 are not logically equivalent, we may write $s_1 \not\Longleftrightarrow s_2$ to designate this situation.

Using the concepts of logical equivalence, tautology, and contradiction, we state the following list of laws for the algebra of propositions.

The Laws of Logic

For any primitive statements p, q, r, any tautology T_0, and any contradiction F_0,

1) $\neg\neg p \Longleftrightarrow p$ Law of *Double Negation*

2) $\neg(p \vee q) \Longleftrightarrow \neg p \wedge \neg q$ *DeMorgan's* Laws
$\neg(p \wedge q) \Longleftrightarrow \neg p \vee \neg q$

3) $p \vee q \Longleftrightarrow q \vee p$ *Commutative* Laws
$p \wedge q \Longleftrightarrow q \wedge p$

4) $p \vee (q \vee r) \Longleftrightarrow (p \vee q) \vee r$[†]   *Associative* Laws
$p \wedge (q \wedge r) \Longleftrightarrow (p \wedge q) \wedge r$

5) $p \vee (q \wedge r) \Longleftrightarrow (p \vee q) \wedge (p \vee r)$ *Distributive* Laws
$p \wedge (q \vee r) \Longleftrightarrow (p \wedge q) \vee (p \wedge r)$

6) $p \vee p \Longleftrightarrow p$ *Idempotent* Laws
$p \wedge p \Longleftrightarrow p$

7) $p \vee F_0 \Longleftrightarrow p$ *Identity* Laws
$p \wedge T_0 \Longleftrightarrow p$

[†]We note that because of the Associative Laws, there is no ambiguity in statements of the form $p \vee q \vee r$ or $p \wedge q \wedge r$.

8) $p \vee \neg p \Longleftrightarrow T_0$ *Inverse* Laws
     $p \wedge \neg p \Longleftrightarrow F_0$

9) $p \vee T_0 \Longleftrightarrow T_0$ *Domination* Laws
     $p \wedge F_0 \Longleftrightarrow F_0$

10) $p \vee (p \wedge q) \Longleftrightarrow p$ *Absorption* Laws
      $p \wedge (p \vee q) \Longleftrightarrow p$

We now turn our attention to proving all of these properties. In so doing we realize that we could simply construct the truth tables and compare the results for the corresponding truth values in each case — as we did in Examples 2.8 and 2.9. However, before we start writing, let us take one more look at this list of 19 laws, which, aside from the Law of Double Negation, fall naturally into pairs. This pairing idea will help us after we examine the following concept.

Definition 2.3   Let s be a statement. If s contains no logical connectives other than $\wedge$ and $\vee$, then the *dual* of s, denoted s^d, is the statement obtained from s by replacing each occurrence of $\wedge$ and $\vee$ by $\vee$ and $\wedge$, respectively, and each occurrence of T_0 and F_0 by F_0 and T_0, respectively.

If p is any primitive statement, then p^d is the same as p — that is, the dual of a primitive statement is simply the same primitive statement. And $(\neg p)^d$ is the same as $\neg p$. The statements $p \vee \neg p$ and $p \wedge \neg p$ are duals of each other whenever p is primitive — and so are the statements $p \vee T_0$ and $p \wedge F_0$.

Given the primitive statements p, q, r and the compound statement

$$s: \quad (p \wedge \neg q) \vee (r \wedge T_0),$$

we find that the dual of s is

$$s^d: \quad (p \vee \neg q) \wedge (r \vee F_0).$$

(Note that $\neg q$ is unchanged as we go from s to s^d.)

We now state and use a theorem without proving it. However, in Chapter 15 we shall justify the result that appears here.

THEOREM 2.1   *The Principle of Duality.* Let s and t be statements that contain no logical connectives other than $\wedge$ and $\vee$. If $s \Longleftrightarrow t$, then $s^d \Longleftrightarrow t^d$.

As a result, laws 2 through 10 in our list can be established by proving one of the laws in each pair and then invoking this principle.

We also find that it is possible to derive many other logical equivalences. For example, if q, r, s are primitive statements, the results in columns 5 and 7 of Table 2.11 show us that

$$(r \wedge s) \rightarrow q \Longleftrightarrow \neg(r \wedge s) \vee q$$

or that $[(r \wedge s) \rightarrow q] \leftrightarrow [\neg(r \wedge s) \vee q]$ is a tautology. However, instead of always constructing more (and, unfortunately, larger) truth tables it might be a good idea to recall from Example 2.7 that for primitive statements p, q, the compound statement

$$(p \rightarrow q) \leftrightarrow (\neg p \vee q)$$

Table 2.11

q	r	s	$r \wedge s$	$(r \wedge s) \rightarrow q$	$\neg(r \wedge s)$	$\neg(r \wedge s) \vee q$
0	0	0	0	1	1	1
0	0	1	0	1	1	1
0	1	0	0	1	1	1
0	1	1	1	0	0	0
1	0	0	0	1	1	1
1	0	1	0	1	1	1
1	1	0	0	1	1	1
1	1	1	1	1	0	1

is a tautology. If we were to *replace* each occurrence of this primitive statement p by the compound statement $r \wedge s$, then we would obtain the earlier tautology

$$[(r \wedge s) \rightarrow q] \leftrightarrow [\neg(r \wedge s) \vee q].$$

What has happened here illustrates the first of the following two *substitution rules*:

1) Suppose that the compound statement P is a tautology. If p is a *primitive* statement that appears in P and we replace *each* occurrence of p by the *same* statement q, then the resulting compound statement P_1 is also a tautology.

2) Let P be a compound statement where p is an arbitrary statement that appears in P, and let q be a statement such that $q \Longleftrightarrow p$. Suppose that in P we replace one or more occurrences of p by q. Then this replacement yields the compound statement P_1. Under these circumstances $P_1 \Longleftrightarrow P$.

These rules are further illustrated in the following two examples.

EXAMPLE 2.10

a) From the first of DeMorgan's Laws we know that for all primitive statements p, q, the compound statement

$$P: \quad \neg(p \vee q) \leftrightarrow (\neg p \wedge \neg q)$$

is a tautology. When we replace each occurrence of p by $r \wedge s$, it follows from the first substitution rule that

$$P_1: \quad \neg[(r \wedge s) \vee q] \leftrightarrow [\neg(r \wedge s) \wedge \neg q]$$

is also a tautology. Extending this result one step further, we may replace each occurrence of q by $t \rightarrow u$. The same substitution rule now yields the tautology

$$P_2: \quad \neg[(r \wedge s) \vee (t \rightarrow u)] \leftrightarrow [\neg(r \wedge s) \wedge \neg(t \rightarrow u)],$$

and hence, by the remarks following shortly after Example 2.9, the logical equivalence

$$\neg[(r \wedge s) \vee (t \rightarrow u)] \Longleftrightarrow [\neg(r \wedge s) \wedge \neg(t \rightarrow u)].$$

b) For primitive statements p, q, we learn from the last column of Table 2.12 that the compound statement $[p \wedge (p \rightarrow q)] \rightarrow q$ is a tautology. Consequently, if r, s, t, u are any statements, then by the first substitution rule we obtain the new tautology

$$[(r \rightarrow s) \wedge [(r \rightarrow s) \rightarrow (\neg t \vee u)]] \rightarrow (\neg t \vee u)$$

when we replace each occurrence of p by $r \rightarrow s$ and each occurrence of q by $\neg t \vee u$.

Table 2.12

p	q	$p \to q$	$p \wedge (p \to q)$	$[p \wedge (p \to q)] \to q$
0	0	1	0	1
0	1	1	0	1
1	0	0	0	1
1	1	1	1	1

EXAMPLE 2.11

a) For an application of the second substitution rule, let P denote the compound statement $(p \to q) \to r$. Because $(p \to q) \Longleftrightarrow \neg p \vee q$ (as shown in Example 2.7 and Table 2.6), if P_1 denotes the compound statement $(\neg p \vee q) \to r$, then $P_1 \Longleftrightarrow P$. (We also find that $[(p \to q) \to r] \leftrightarrow [(\neg p \vee q) \to r]$ is a tautology.)

b) Now let P represent the compound statement (actually a tautology) $p \to (p \vee q)$. Since $\neg \neg p \Longleftrightarrow p$, the compound statement P_1: $p \to (\neg \neg p \vee q)$ is derived from P by replacing *only the second occurrence* (but *not* the first occurrence) of p by $\neg \neg p$. The second substitution rule still implies that $P_1 \Longleftrightarrow P$. [Note that P_2: $\neg \neg p \to (\neg \neg p \vee q)$, derived by replacing *both* occurrences of p by $\neg \neg p$, is also logically equivalent to P.]

Our next example demonstrates how we can use the idea of logical equivalence together with the laws of logic and the substitution rules.

EXAMPLE 2.12

Negate and simplify the compound statement $(p \vee q) \to r$.
We organize our explanation as follows:

1) $(p \vee q) \to r \Longleftrightarrow \neg(p \vee q) \vee r$ [by the first substitution rule because $(s \to t) \leftrightarrow (\neg s \vee t)$ is a tautology for primitive statements s, t].

2) Negating the statements in step (1), we have $\neg[(p \vee q) \to r] \Longleftrightarrow \neg[\neg(p \vee q) \vee r]$.

3) From the first of DeMorgan's Laws and the first substitution rule, $\neg[\neg(p \vee q) \vee r] \Longleftrightarrow \neg \neg(p \vee q) \wedge \neg r$.

4) The Law of Double Negation and the second substitution rule now gives us $\neg \neg(p \vee q) \wedge \neg r \Longleftrightarrow (p \vee q) \wedge \neg r$.

From steps (1) through (4) we have $\neg[(p \vee q) \to r] \Longleftrightarrow (p \vee q) \wedge \neg r$.

When we wanted to write the negation of an implication, as in Example 2.12, we found that the concept of logical equivalence played a key role — in conjunction with the laws of logic and the substitution rules. This idea is important enough to warrant a second look.

EXAMPLE 2.13

Let p, q denote the primitive statements

p: Joan goes to Lake George. q: Mary pays for Joan's shopping spree.

and consider the implication

$p \to q$:   If Joan goes to Lake George, then Mary will pay for Joan's shopping spree.

Here we want to write the negation of $p \to q$ in a way other than simply $\neg(p \to q)$. We want to avoid writing the negation as "It is not the case that if Joan goes to Lake George, then Mary will pay for Joan's shopping spree."

To accomplish this we consider the following. Since $p \to q \Leftrightarrow \neg p \vee q$, it follows that $\neg(p \to q) \Leftrightarrow \neg(\neg p \vee q)$. Then by DeMorgan's Law we have $\neg(\neg p \vee q) \Leftrightarrow \neg\neg p \wedge \neg q$, and from the Law of Double Negation and the second substitution rule it follows that $\neg\neg p \wedge \neg q \Leftrightarrow p \wedge \neg q$. Consequently,

$$\neg(p \to q) \Leftrightarrow \neg(\neg p \vee q) \Leftrightarrow \neg\neg p \wedge \neg q \Leftrightarrow p \wedge \neg q,$$

and we may write the negation of $p \to q$ in this case as

$\neg(p \to q)$:   Joan goes to Lake George, but Mary does not
pay for Joan's shopping spree.

(*Note*: The negation of an if-then statement does *not* begin with the word *if*. It is *not* another *implication*.)

EXAMPLE 2.14

In Definition 2.3 the dual s^d of a statement s was defined only for statements involving negation and the basic connectives $\wedge$ and $\vee$. How does one determine the dual of a statement such as s: $p \to q$, where p, q are primitive?

Because $(p \to q) \Leftrightarrow \neg p \vee q$, s^d is logically equivalent to the statement $(\neg p \vee q)^d$, which is $\neg p \wedge q$.

The implication $p \to q$ and certain statements related to it are now examined in the following example.

EXAMPLE 2.15

Table 2.13 gives the truth tables for the statements $p \to q$, $\neg q \to \neg p$, $q \to p$, and $\neg p \to \neg q$. The third and fourth columns of the table reveal that

$$(p \to q) \Leftrightarrow (\neg q \to \neg p).$$

Table 2.13

p	q	$p \to q$	$\neg q \to \neg p$	$q \to p$	$\neg p \to \neg q$
0	0	1	1	1	1
0	1	1	1	0	0
1	0	0	0	1	1
1	1	1	1	1	1

The statement $\neg q \to \neg p$ is called the *contrapositive* of the implication $p \to q$. Columns 5 and 6 of the table show that

$$(q \to p) \Leftrightarrow (\neg p \to \neg q).$$

The statement $q \to p$ is called the *converse* of $p \to q$; $\neg p \to \neg q$ is called the *inverse* of $p \to q$. We also see from Table 2.13 that

$$(p \to q) \not\Leftrightarrow (q \to p) \quad \text{and} \quad (\neg p \to \neg q) \not\Leftrightarrow (\neg q \to \neg p).$$

Consequently, we must keep the implication and its converse straight. The fact that a certain implication $p \to q$ is true (in particular, as in row 2 of the table) does *not* require that the converse $q \rightarrow p$ also be true. However, it does necessitate the truth of the contrapositive $\neg q \rightarrow \neg p$.

Let us consider a specific example where p, q represent the statements

p: Jeff is concerned about his cholesterol (HDL and LDL) levels.

q: Jeff walks at least two miles three times a week.

Then we obtain

- (The implication: $p \rightarrow q$). If Jeff is concerned about his cholesterol levels, then he will walk at least two miles three times a week.
- (The contrapositive: $\neg q \rightarrow \neg p$). If Jeff does not walk at least two miles three times a week, then he is not concerned about his cholesterol levels.
- (The converse: $q \rightarrow p$). If Jeff walks at least two miles three times a week, then he is concerned about his cholesterol levels.
- (The inverse: $\neg p \rightarrow \neg q$). If Jeff is not concerned about his cholesterol levels, then he will not walk at least two miles three times a week.

If p is true and q is false, then the implication $p \rightarrow q$ and the contrapositive $\neg q \rightarrow \neg p$ are false, while the converse $q \rightarrow p$ and the inverse $\neg p \rightarrow \neg q$ are true. For the case where p is false and q is true, the implication $p \rightarrow q$ and the contrapositive $\neg q \rightarrow \neg p$ are now true, while the converse $q \rightarrow p$ and the inverse $\neg p \rightarrow \neg q$ are false. When p, q are both true or both false, then the implication is true, as are the contrapositive, converse, and inverse.

We turn now to two examples involving the simplification of compound statements. For simplicity, we shall list the major laws of logic being used, but we shall not mention any applications of our two substitution rules.

EXAMPLE 2.16

For primitive statements p, q, is there any simpler way to express the compound statement $(p \vee q) \wedge \neg(\neg p \wedge q)$ — that is, can we find a simpler statement that is logically equivalent to the one given?

Here one finds that

	Reasons
$(p \vee q) \wedge \neg(\neg p \wedge q)$	
$\Longleftrightarrow (p \vee q) \wedge (\neg\neg p \vee \neg q)$	DeMorgan's Law
$\Longleftrightarrow (p \vee q) \wedge (p \vee \neg q)$	Law of Double Negation
$\Longleftrightarrow (p \vee (q \wedge \neg q)$	Distributive Law of $\vee$ over $\wedge$
$\Longleftrightarrow p \vee F_0$	Inverse Law
$\Longleftrightarrow p$	Identity Law

Consequently, we see that

$$(p \vee q) \wedge \neg(\neg p \wedge q) \Longleftrightarrow p,$$

so we can express the given compound statement by the simpler logically equivalent statement p.

EXAMPLE 2.17

Consider the compound statement

$$\neg[\neg[(p \vee q) \wedge r] \vee \neg q],$$

where p, q, r are primitive statements. This statement contains four occurrences of primitive statements, three negation symbols, and three connectives.

From the laws of logic it follows that

	Reasons
$\neg[\neg[(p \lor q) \land r] \lor \neg q]$	
$\Longleftrightarrow \neg\neg[(p \lor q) \land r] \land \neg\neg q$	DeMorgan's Law
$\Longleftrightarrow [(p \lor q) \land r] \land q$	Law of Double Negation
$\Longleftrightarrow (p \lor q) \land (r \land q)$	Associative Law of $\land$
$\Longleftrightarrow (p \lor q) \land (q \land r)$	Commutative Law of $\land$
$\Longleftrightarrow [(p \lor q) \land q] \land r$	Associative Law of $\land$
$\Longleftrightarrow q \land r$	Absorption Law (as well as the Commutative Laws for $\land$ and $\lor$)

Consequently, the original statement

$$\neg[\neg[(p \lor q) \land r] \lor \neg q]$$

is logically equivalent to the much simpler statement

$$q \land r,$$

where we find only two primitive statements, no negation symbols, and only one connective.

Note further that from Example 2.7 we have

$$\neg[[(p \lor q) \land r] \to \neg q] \Longleftrightarrow \neg[\neg[(p \lor q) \land r] \lor \neg q],$$

so it follows that

$$\neg[[(p \lor q) \land r] \to \neg q] \Longleftrightarrow q \land r.$$

We close this section with an application on how the ideas in Examples 2.16 and 2.17 can be used in simplifying switching networks.

EXAMPLE 2.18

A switching network is made up of wires and switches connecting two terminals T_1 and T_2. In such a network, each switch is either open (0), so that no current flows through it, or closed (1), so that current does flow through it.

In Fig. 2.1(a) we have a network with one switch. Each of parts (b) and (c) contains two (independent) switches.

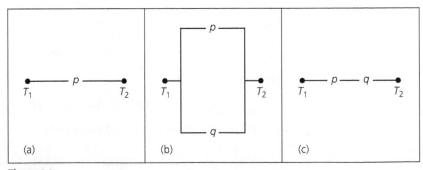

Figure 2.1

For the network in part (b), current flows from T_1 to T_2 if either of the switches p, q is closed. We call this a *parallel* network and represent it by $p \lor q$. The network in part (c) requires that each of the switches p, q be closed in order for current to flow from T_1 to T_2. Here the switches are in *series*; this network is represented by $p \wedge q$.

The switches in a network need not act independently of each other. Consider the network shown in Fig. 2.2(a). Here the switches labeled t and $\neg t$ are not independent. We have coupled these two switches so that t is open (closed) if and only if $\neg t$ is simultaneously closed (open). The same is true for the switches at q, $\neg q$. (Also, for example, the three switches labeled p are not independent.)

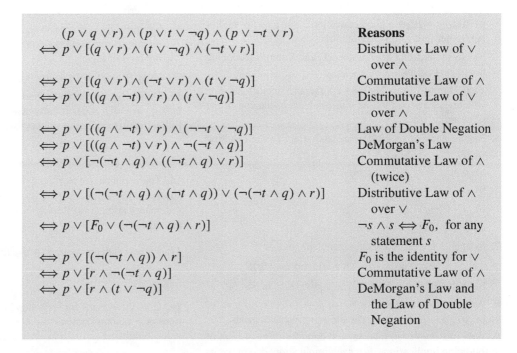

(a)   (b)

Figure 2.2

This network is represented by the statement $(p \vee q \vee r) \wedge (p \vee t \vee \neg q) \wedge (p \vee \neg t \vee r)$. Using the laws of logic, we may simplify this statement as follows.

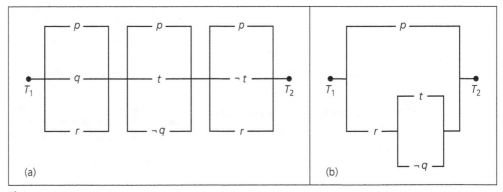

	Reasons
$(p \vee q \vee r) \wedge (p \vee t \vee \neg q) \wedge (p \vee \neg t \vee r)$	
$\Longleftrightarrow p \vee [(q \vee r) \wedge (t \vee \neg q) \wedge (\neg t \vee r)]$	Distributive Law of $\vee$ over $\wedge$
$\Longleftrightarrow p \vee [(q \vee r) \wedge (\neg t \vee r) \wedge (t \vee \neg q)]$	Commutative Law of $\wedge$
$\Longleftrightarrow p \vee [((q \wedge \neg t) \vee r) \wedge (t \vee \neg q)]$	Distributive Law of $\vee$ over $\wedge$
$\Longleftrightarrow p \vee [((q \wedge \neg t) \vee r) \wedge (\neg\neg t \vee \neg q)]$	Law of Double Negation
$\Longleftrightarrow p \vee [((q \wedge \neg t) \vee r) \wedge \neg(\neg t \wedge q)]$	DeMorgan's Law
$\Longleftrightarrow p \vee [\neg(\neg t \wedge q) \wedge ((\neg t \wedge q) \vee r)]$	Commutative Law of $\wedge$ (twice)
$\Longleftrightarrow p \vee [(\neg(\neg t \wedge q) \wedge (\neg t \wedge q)) \vee (\neg(\neg t \wedge q) \wedge r)]$	Distributive Law of $\wedge$ over $\vee$
$\Longleftrightarrow p \vee [F_0 \vee (\neg(\neg t \wedge q) \wedge r)]$	$\neg s \wedge s \Longleftrightarrow F_0$, for any statement s
$\Longleftrightarrow p \vee [(\neg(\neg t \wedge q)) \wedge r]$	F_0 is the identity for $\vee$
$\Longleftrightarrow p \vee [r \wedge \neg(\neg t \wedge q)]$	Commutative Law of $\wedge$
$\Longleftrightarrow p \vee [r \wedge (t \vee \neg q)]$	DeMorgan's Law and the Law of Double Negation

Hence $(p \vee q \vee r) \wedge (p \vee t \vee \neg q) \wedge (p \vee \neg t \vee r) \Longleftrightarrow p \vee [r \wedge (t \vee \neg q)]$, and the network shown in Fig. 2.2(b) is equivalent to the original network in the sense that current flows from T_1 to T_2 in network (a) exactly when it does so in network (b). But network (b) has only four switches, five fewer than network (a).

EXERCISES 2.2

1. Let p, q, r denote primitive statements.

   a) Use truth tables to verify the following logical equivalences.

   i) $p \rightarrow (q \wedge r) \Longleftrightarrow (p \rightarrow q) \wedge (p \rightarrow r)$
   ii) $[(p \vee q) \rightarrow r] \Longleftrightarrow [(p \rightarrow r) \wedge (q \rightarrow r)]$
   iii) $[p \rightarrow (q \vee r)] \Longleftrightarrow [\neg r \rightarrow (p \rightarrow q)]$

   b) Use the substitution rules to show that
   $$[p \rightarrow (q \vee r)] \Longleftrightarrow [(p \wedge \neg q) \rightarrow r].$$

2. Verify the first Absorption Law by means of a truth table.

3. Use the substitution rules to verify that each of the following is a tautology. (Here p, q, and r are primitive statements.)

   a) $[p \vee (q \wedge r)] \vee \neg[p \vee (q \wedge r)]$

   b) $[(p \vee q) \rightarrow r] \leftrightarrow [\neg r \rightarrow \neg(p \vee q)]$

4. For primitive statements p, q, r, and s, simplify the compound statement
   $$[[[(p \wedge q) \wedge r] \vee [(p \wedge q) \wedge \neg r]] \vee \neg q] \rightarrow s.$$

5. Negate and express each of the following statements in smooth English.

   a) Kelsey will get a good education if she puts her studies before her interest in cheerleading.

   b) Norma is doing her homework, and Karen is practicing her piano lessons.

   c) If Harold passes his C++ course and finishes his data structures project, then he will graduate at the end of the semester.

6. Negate each of the following and simplify the resulting statement.

   a) $p \wedge (q \vee r) \wedge (\neg p \vee \neg q \vee r)$

   b) $(p \wedge q) \rightarrow r$

   c) $p \rightarrow (\neg q \wedge r)$

   d) $p \vee q \vee (\neg p \wedge \neg q \wedge r)$

7. **a)** If p, q are primitive statements, prove that
   $$(\neg p \vee q) \wedge (p \wedge (p \wedge q)) \Longleftrightarrow (p \wedge q).$$

   b) Write the dual of the logical equivalence in part (a).

8. Write the dual for (a) $q \rightarrow p$, (b) $p \rightarrow (q \wedge r)$, (c) $p \leftrightarrow q$, and (d) $p \veebar q$, where p, q, and r are primitive statements.

9. Write the converse, inverse, and contrapositive of each of the following implications. For each implication, determine its truth value as well as the truth values of its corresponding converse, inverse, and contrapositive.

   a) If $0 + 0 = 0$, then $1 + 1 = 1$.

   b) If $-1 < 3$ and $3 + 7 = 10$, then $\sin(\frac{3\pi}{2}) = -1$.

10. Determine whether each of the following is true or false. Here p, q are arbitrary statements.

    a) An equivalent way to express the converse of "p is sufficient for q" is "p is necessary for q."

    b) An equivalent way to express the inverse of "p is necessary for q" is "$\neg q$ is sufficient for $\neg p$."

    c) An equivalent way to express the contrapositive of "p is necessary for q" is "$\neg q$ is necessary for $\neg p$."

11. Let p, q, and r denote primitive statements. Find a form of the contrapositive of $p \rightarrow (q \rightarrow r)$ with (a) only one occurrence of the connective $\rightarrow$; (b) no occurrences of the connective $\rightarrow$.

12. Show that for primitive statements p, q,
    $$p \veebar q \Longleftrightarrow [(p \wedge \neg q) \vee (\neg p \wedge q)] \Longleftrightarrow \neg(p \leftrightarrow q).$$

13. Verify that $[(p \leftrightarrow q) \wedge (q \leftrightarrow r) \wedge (r \leftrightarrow p)] \Longleftrightarrow [(p \rightarrow q) \wedge (q \rightarrow r) \wedge (r \rightarrow p)]$, for primitive statements p, q, and r.

14. For primitive statements p, q,

    a) verify that $p \rightarrow [q \rightarrow (p \wedge q)]$ is a tautology.

    b) verify that $(p \vee q) \rightarrow [q \rightarrow q]$ is a tautology by using the result from part (a) along with the substitution rules and the laws of logic.

    c) is $(p \vee q) \rightarrow [q \rightarrow (p \wedge q)]$ a tautology?

15. Define the connective "Nand" or "Not ... and ..." by $(p \uparrow q) \Longleftrightarrow \neg(p \wedge q)$, for any statements p, q. Represent the following using only this connective.

    a) $\neg p$ **b)** $p \vee q$ **c)** $p \wedge q$
    d) $p \rightarrow q$ **e)** $p \leftrightarrow q$

16. The connective "Nor" or "Not ... or ..." is defined for any statements p, q by $(p \downarrow q) \Longleftrightarrow \neg(p \vee q)$. Represent the statements in parts (a) through (e) of Exercise 15, using only this connective.

17. For any statements p, q, prove that

    a) $\neg(p \downarrow q) \Longleftrightarrow (\neg p \uparrow \neg q)$

    b) $\neg(p \uparrow q) \Longleftrightarrow (\neg p \downarrow \neg q)$

18. Give the reasons for each step in the following simplifications of compound statements.

    a) $[(p \vee q) \wedge (p \vee \neg q)] \vee q$ **Reasons**
    $\Longleftrightarrow [p \vee (q \wedge \neg q)] \vee q$
    $\Longleftrightarrow (p \vee F_0) \vee q$
    $\Longleftrightarrow p \vee q$

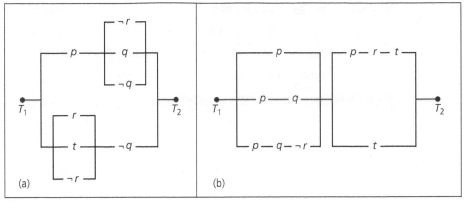

Figure 2.3

b) $(p \rightarrow q) \wedge [\neg q \wedge (r \vee \neg q)]$ **Reasons**
$\Leftrightarrow (p \rightarrow q) \wedge \neg q$
$\Leftrightarrow (\neg p \vee q) \wedge \neg q$
$\Leftrightarrow \neg q \wedge (\neg p \vee q)$
$\Leftrightarrow (\neg q \wedge \neg p) \vee (\neg q \wedge q)$
$\Leftrightarrow (\neg q \wedge \neg p) \vee F_0$
$\Leftrightarrow \neg q \wedge \neg p$
$\Leftrightarrow \neg (q \vee p)$

19. Provide the steps and reasons, as in Exercise 18, to establish the following logical equivalences.

   a) $p \vee [p \wedge (p \vee q)] \Leftrightarrow p$

   b) $p \vee q \vee (\neg p \wedge \neg q \wedge r) \Leftrightarrow p \vee q \vee r$

   c) $[(\neg p \vee \neg q) \rightarrow (p \wedge q \wedge r)] \Leftrightarrow p \wedge q$

20. Simplify each of the networks shown in Fig. 2.3.

2.3
Logical Implication: Rules of Inference

At the end of Section 2.1 we mentioned the notion of a valid argument. Now we will begin a formal study of what we shall mean by an argument and when such an argument is valid. This in turn will help us when we investigate how to prove theorems throughout the text.

We start by considering the general form of an argument, one we wish to show is valid. So let us consider the implication

$$(p_1 \wedge p_2 \wedge p_3 \wedge \cdots \wedge p_n) \rightarrow q.$$

Here n is a positive integer, the statements $p_1, p_2, p_3, \ldots, p_n$ are called the *premises* of the argument, and the statement q is the *conclusion* for the argument.

The preceding argument is called *valid* if whenever each of the premises $p_1, p_2, p_3, \ldots, p_n$ is true, then the conclusion q is likewise true. [Note that if any one of $p_1, p_2, p_3, \ldots, p_n$ is false, then the hypothesis $p_1 \wedge p_2 \wedge p_3 \wedge \cdots \wedge p_n$ is false and the implication $(p_1 \wedge p_2 \wedge p_3 \wedge \cdots \wedge p_n) \rightarrow q$ is automatically true, regardless of the truth value of q.] Consequently, one way to establish the validity of a given argument is to show that the statement $(p_1 \wedge p_2 \wedge p_3 \wedge \cdots \wedge p_n) \rightarrow q$ is a tautology.

The following examples illustrate this particular approach.

EXAMPLE 2.19

Let p, q, r denote the primitive statements given as

   p: Roger studies.

   q: Roger plays racketball.

   r: Roger passes discrete mathematics.

Now let p_1, p_2, p_3 denote the premises

p_1:   If Roger studies, then he will pass discrete mathematics.

p_2:   If Roger doesn't play racketball, then he'll study.

p_3:   Roger failed discrete mathematics.

We want to determine whether the argument

$$(p_1 \wedge p_2 \wedge p_3) \rightarrow q$$

is valid. To do so, we rewrite p_1, p_2, p_3 as

$$p_1:\quad p \rightarrow r \qquad p_2:\quad \neg q \rightarrow p \qquad p_3:\quad \neg r$$

and examine the truth table for the implication

$$[(p \rightarrow r) \wedge (\neg q \rightarrow p) \wedge \neg r] \rightarrow q$$

given in Table 2.14. Because the final column in Table 2.14 contains all 1's, the implication is a tautology. Hence we can say that $(p_1 \wedge p_2 \wedge p_3) \rightarrow q$ is a valid argument.

Table 2.14

			p_1	p_2	p_3	$(p_1 \wedge p_2 \wedge p_3) \rightarrow q$
p	q	r	$p \rightarrow r$	$\neg q \rightarrow p$	$\neg r$	$[(p \rightarrow r) \wedge (\neg q \rightarrow p) \wedge \neg r] \rightarrow q$
0	0	0	1	0	1	1
0	0	1	1	0	0	1
0	1	0	1	1	1	1
0	1	1	1	1	0	1
1	0	0	0	1	1	1
1	0	1	1	1	0	1
1	1	0	0	1	1	1
1	1	1	1	1	0	1

EXAMPLE 2.20

Let us now consider the truth table in Table 2.15. The results in the last column of this table show that for any primitive statements p, r, and s, the implication

$$[p \wedge ((p \wedge r) \rightarrow s)] \rightarrow (r \rightarrow s)$$

Table 2.15

p_1				p_2	q	$(p_1 \wedge p_2) \rightarrow q$
p	r	s	$p \wedge r$	$(p \wedge r) \rightarrow s$	$r \rightarrow s$	$[(p \wedge ((p \wedge r) \rightarrow s)] \rightarrow (r \rightarrow s)$
0	0	0	0	1	1	1
0	0	1	0	1	1	1
0	1	0	0	1	0	1
0	1	1	0	1	1	1
1	0	0	0	1	1	1
1	0	1	0	1	1	1
1	1	0	1	0	0	1
1	1	1	1	1	1	1

is a tautology. Consequently, for premises

$$p_1: \quad p \qquad p_2: \quad (p \wedge r) \to s$$

and conclusion $q: (r \to s)$, we know that $(p_1 \wedge p_2) \to q$ is a valid argument, and we may say that the truth of the conclusion q is *deduced* or *inferred* from the truth of the premises p_1, p_2.

The idea presented in the preceding two examples leads to the following.

Definition 2.4 If p, q are arbitrary statements such that $p \to q$ is a tautology, then we say that p *logically implies* q and we write $p \Rightarrow q$ to denote this situation.

When p, q are statements and $p \Rightarrow q$, the implication $p \to q$ is a tautology and we refer to $p \to q$ as a *logical implication*. Note that we can avoid dealing with the idea of a tautology here by saying that $p \Rightarrow q$ (that is, p logically implies q) if q is true whenever p is true.

In Example 2.6 we found that for primitive statements p, q, the implication $p \to (p \vee q)$ is a tautology. In this case, therefore, we can say that p logically implies $p \vee q$ and write $p \Rightarrow (p \vee q)$. Furthermore, because of the first substitution rule, we also find that $p \Rightarrow (p \vee q)$ for any statements p, q — that is, $p \to (p \vee q)$ is a tautology for any statements p, q, whether or not they are primitive statements.

Let p, q be arbitrary statements.

1) If $p \Longleftrightarrow q$, then the statement $p \leftrightarrow q$ is a tautology, so the statements p, q have the same (corresponding) truth values. Under these conditions the statements $p \to q$, $q \to p$ are tautologies, and we have $p \Rightarrow q$ and $q \Rightarrow p$.

2) Conversely, suppose that $p \Rightarrow q$ and $q \Rightarrow p$. The logical implication $p \to q$ tells us that we never have statement p with the truth value 1 and statement q with the truth value 0. But could we have q with the truth value 1 and p with the truth value 0? If this occurred, we could not have the logical implication $q \to p$. Therefore, when $p \Rightarrow q$ and $q \Rightarrow p$, the statements p, q have the same (corresponding) truth values and $p \Longleftrightarrow q$.

Finally, the notation $p \nRightarrow q$ is used to indicate that $p \to q$ is *not* a tautology — so the given implication (namely, $p \to q$) is *not* a logical implication.

EXAMPLE 2.21 From the results in Example 2.8 (Table 2.9) and the first substitution rule, we know that for statements p, q,

$$\neg(p \wedge q) \Longleftrightarrow \neg p \vee \neg q.$$

Consequently,

$$\neg(p \wedge q) \Rightarrow (\neg p \vee \neg q) \quad \text{and} \quad (\neg p \vee \neg q) \Rightarrow \neg(p \wedge q)$$

for all statements p, q. Alternatively, because each of the implications

$$\neg(p \wedge q) \to (\neg p \vee \neg q) \quad \text{and} \quad (\neg p \vee \neg q) \to \neg(p \wedge q)$$

is a tautology, we may also write

$$[\neg(p \wedge q) \to (\neg p \vee \neg q)] \Longleftrightarrow T_0 \quad \text{and} \quad [(\neg p \vee \neg q) \to \neg(p \wedge q)] \Longleftrightarrow T_0.$$

Returning now to our study of techniques for establishing the validity of an argument, we must take a careful look at the size of Tables 2.14 and 2.15. Each table has eight rows. For Table 2.14 we were able to express the three premises p_1, p_2, and p_3, and the conclusion q, in terms of the three primitive statements p, q, and r. A similar situation arose for the argument we analyzed in Table 2.15, where we had only two premises. But if we were confronted, for example, with establishing whether

$$[(p \rightarrow r) \wedge (r \rightarrow s) \wedge (t \vee \neg s) \wedge (\neg t \vee u) \wedge \neg u] \rightarrow \neg p$$

is a logical implication (or presents a valid argument), the needed table would require $2^5 = 32$ rows. As the number of premises gets larger and our truth tables grow to 64, 128, 256, or more rows, this first technique for establishing the validity of an argument rapidly loses its appeal.

Furthermore, looking at Table 2.14 once again, we realize that in order to establish whether

$$[(p \rightarrow r) \wedge (\neg q \rightarrow p) \wedge \neg r] \rightarrow q$$

is a valid argument, we need to consider only those rows of the table where each of the three premises $p \rightarrow r$, $\neg q \rightarrow p$, and $\neg r$ has the truth value 1. (Remember that if the hypothesis — consisting of the conjunction of all of the premises — is false, then the implication is true regardless of the truth value of the conclusion.) This happens only in the third row, so a good deal of Table 2.14 is not really necessary. (It is not always the case that only one row has all of the premises true. Note that in Table 2.15 we would be concerned with the results in rows 5, 6, and 8.)

Consequently, what these observations are telling us is that we can possibly eliminate a great deal of the effort put into constructing the truth tables in Table 2.14 and Table 2.15. And since we want to avoid even larger tables, we are persuaded to develop a list of techniques called *rules of inference* that will help us as follows:

1) Using these techniques will enable us to consider only the cases wherein all the premises are true. Hence we consider the conclusion only for those rows of a truth table wherein each premise has the truth value 1 — and we do *not* construct the truth table.

2) The rules of inference are fundamental in the development of a step-by-step validation of how the conclusion q logically follows from the premises p_1, p_2, p_3, ..., p_n in an implication of the form

$$(p_1 \wedge p_2 \wedge p_3 \wedge \cdots \wedge p_n) \rightarrow q.$$

Such a development will establish the validity of the given argument, for it will show how the truth of the conclusion can be deduced from the truth of the premises.

Each rule of inference arises from a logical implication. In some cases, the logical implication is stated without proof. (However, several of these proofs will be dealt with in the Section Exercises.)

Many rules of inference arise in the study of logic. We concentrate on those that we need to help us validate the arguments that arise in our study of logic. These rules will also help us later when we turn to methods for proving theorems throughout the remainder of the text. Table 2.19 (on p. 78) summarizes the rules we shall now start to investigate.

EXAMPLE 2.22

For a first example we consider the rule of inference called *Modus Ponens,* or the *Rule of Detachment.* (*Modus Ponens* comes from Latin and may be translated as "the method of affirming.") In symbolic form this rule is expressed by the logical implication

$$[p \wedge (p \to q)] \to q,$$

which is verified in Table 2.16, where we find that the fourth row is the only one where both of the premises p and $p \to q$ (and the conclusion q) are true.

Table 2.16

p	q	$p \to q$	$p \wedge (p \to q)$	$[p \wedge (p \to q)] \to q$
0	0	1	0	1
0	1	1	0	1
1	0	0	0	1
1	1	1	1	1

The actual rule will be written in the tabular form

$$\frac{\begin{array}{c} p \\ p \to q \end{array}}{\therefore q}$$

where the three dots ($\therefore$) stand for the word "therefore," indicating that q is the conclusion for the premises p and $p \to q$, which appear above the horizontal line.

This rule arises when we argue that *if* (1) p is true, *and* (2) $p \to q$ is true (or $p \Rightarrow q$), *then* the conclusion q must also be true. (After all, if q were false and p were true, then we could not have $p \to q$ true.)

The following valid arguments show us how to apply the Rule of Detachment.

a) 1) Lydia wins a ten-million-dollar lottery. $\qquad$ p
   2) If Lydia wins a ten-million-dollar lottery, then Kay will quit her job. $\quad$ $\underline{p \to q}$
   3) Therefore Kay will quit her job. $\qquad$ $\therefore q$

b) 1) If Allison vacations in Paris, then she will have to win a scholarship. $\quad$ $p \to q$
   2) Allison is vacationing in Paris. $\qquad$ $\underline{p}$
   3) Therefore Allison won a scholarship. $\qquad$ $\therefore q$

Before closing the discussion on our first rule of inference let us make one final observation. The two examples in (a) and (b) might suggest that the valid argument $[p \wedge (p \to q)] \to q$ is appropriate only for primitive statements p, q. However, since $[p \wedge (p \to q)] \to q$ is a tautology for primitive statements p, q, it follows from the first substitution rule that (all occurrences of) p or q may be replaced by compound statements — and the resulting implication will also be a tautology. Consequently, if r, s, t, and u are primitive statements, then

$$\frac{\begin{array}{c} r \vee s \\ (r \vee s) \to (\neg t \wedge u) \end{array}}{\therefore \neg t \wedge u}$$

is a valid argument, by the Rule of Detachment — just as $[(r \vee s) \wedge [(r \vee s) \to (\neg t \wedge u)]] \to (\neg t \wedge u)$ is a tautology.

A similar situation — in which we can apply the first substitution rule — occurs for each of the rules of inference we shall study. However, we shall not mention this so explicitly with these other rules of inference.

EXAMPLE 2.23

A second rule of inference is given by the logical implication

$$[(p \rightarrow q) \wedge (q \rightarrow r)] \rightarrow (p \rightarrow r),$$

where p, q, and r are any statements. In tabular form it is written

$$\begin{array}{c} p \rightarrow q \\ q \rightarrow r \\ \hline \therefore p \rightarrow r \end{array}$$

This rule, which is referred to as the *Law of the Syllogism,* arises in many arguments. For example, we may use it as follows:

1) If the integer 35244 is divisible by 396, then the integer 35244 is divisible by 66. $\qquad p \rightarrow q$

2) If the integer 35244 is divisible by 66, then the integer 35244 is divisible by 3. $\qquad q \rightarrow r$

3) Therefore, if the integer 35244 is divisible by 396, then the integer 35244 is divisible by 3. $\qquad \therefore p \rightarrow r$

The next example involves a slightly longer argument that uses the rules of inference developed in Examples 2.22 and 2.23. In fact, we find here that there may be more than one way to establish the validity of an argument.

EXAMPLE 2.24

Consider the following argument.

1) Rita is baking a cake.

2) If Rita is baking a cake, then she is not practicing her flute.

3) If Rita is not practicing her flute, then her father will not buy her a car.

4) Therefore Rita's father will not buy her a car.

Concentrating on the forms of the statements in the preceding argument, we may write the argument as

$$\begin{array}{c} p \\ p \rightarrow \neg q \\ \neg q \rightarrow \neg r \\ \hline \therefore \neg r \end{array} \qquad (*)$$

Now we need no longer worry about what the statements actually stand for. Our objective is to use the two rules of inference that we have studied so far in order to deduce the truth of the statement $\neg r$ from the truth of the three premises p, $p \rightarrow \neg q$, and $\neg q \rightarrow \neg r$.

We establish the validity of the argument as follows:

Steps	**Reasons**
1) $p \to \neg q$	Premise
2) $\neg q \to \neg r$	Premise
3) $p \to \neg r$	This follows from steps (1) and (2) and the Law of the Syllogism
4) p	Premise
5) $\therefore \neg r$	This follows from steps (4) and (3) and the Rule of Detachment

Before continuing with a third rule of inference we shall show that the argument presented at (*) can be validated in a second way. Here our "reasons" will be shortened to the form we shall use for the rest of the section. However, we shall always list whatever is needed to demonstrate how each step in an argument comes about, or follows, from prior steps.

A second way to validate the argument follows.

Steps	**Reasons**
1) p	Premise
2) $p \to \neg q$	Premise
3) $\neg q$	Steps (1) and (2) and the Rule of Detachment
4) $\neg q \to \neg r$	Premise
5) $\therefore \neg r$	Steps (3) and (4) and the Rule of Detachment

---

EXAMPLE 2.25

The rule of inference called *Modus Tollens* is given by

$$p \to q$$
$$\frac{\neg q}{\therefore \neg p}$$

This follows from the logical implication $[(p \to q) \land \neg q] \to \neg p$. *Modus Tollens* comes from Latin and can be translated as "method of denying." This is appropriate because we deny the conclusion, q, so as to prove $\neg p$. (Note that we can also obtain this rule from the one for Modus Ponens by using the fact that $p \to q \iff \neg q \to \neg p$.)

The following exemplifies the use of Modus Tollens is making a valid inference:

1) If Connie is elected president of Phi Delta sorority, then Helen will pledge that sorority.　　　　　　　　　　　　　　　　　　　　　$p \to q$

2) Helen did not pledge Phi Delta sorority.　　　　　　　　　　　　$\neg q$

3) Therefore Connie was not elected president of Phi Delta sorority.　　$\therefore \neg p$

And now we shall use Modus Tollens to show that the following argument is valid (for primitive statements p, r, s, t, and u).

$$p \to r$$
$$r \to s$$
$$t \lor \neg s$$
$$\neg t \lor u$$
$$\frac{\neg u}{\therefore \neg p}$$

Both Modus Tollens and the Law of the Syllogism come into play, along with the logical equivalence we developed in Example 2.7.

Steps	Reasons
1) $p \rightarrow r, r \rightarrow s$	Premises
2) $p \rightarrow s$	Step (1) and the Law of the Syllogism
3) $t \vee \neg s$	Premise
4) $\neg s \vee t$	Step (3) and the Commutative Law of $\vee$
5) $s \rightarrow t$	Step (4) and the fact that $\neg s \vee t \Longleftrightarrow s \rightarrow t$
6) $p \rightarrow t$	Steps (2) and (5) and the Law of the Syllogism
7) $\neg t \vee u$	Premise
8) $t \rightarrow u$	Step (7) and the fact that $\neg t \vee u \Longleftrightarrow t \rightarrow u$
9) $p \rightarrow u$	Steps (6) and (8) and the Law of the Syllogism
10) $\neg u$	Premise
11) $\therefore \neg p$	Steps (9) and (10) and Modus Tollens

Before continuing with another rule of inference let us summarize what we have just accomplished (and *not* accomplished). The preceding argument shows that

$$[(p \rightarrow r) \wedge (r \rightarrow s) \wedge (t \vee \neg s) \wedge (\neg t \vee u) \wedge \neg u] \Rightarrow \neg p.$$

We have *not* used the laws of logic, as in Section 2.2, to express the statement

$$(p \rightarrow r) \wedge (r \rightarrow s) \wedge (t \vee \neg s) \wedge (\neg t \vee u) \wedge \neg u$$

as a simpler logically equivalent statement. Note that

$$[(p \rightarrow r) \wedge (r \rightarrow s) \wedge (t \vee \neg s) \wedge (\neg t \vee u) \wedge \neg u] \not\Longleftrightarrow \neg p.$$

For when p has the truth value 0 and u has the truth value 1, the truth value of $\neg p$ is 1 while that of $\neg u$ and $(p \rightarrow r) \wedge (r \rightarrow s) \wedge (t \vee \neg s) \wedge (\neg t \vee u) \wedge \neg u$ is 0.

Let us once more examine a tabular form for each of the two related rules of inference, Modus Ponens and Modus Tollens.

$$\text{Modus Ponens:} \quad \begin{array}{c} p \rightarrow q \\ \underline{ p } \\ \therefore q \end{array} \qquad \text{Modus Tollens:} \quad \begin{array}{c} p \rightarrow q \\ \underline{ \neg q } \\ \therefore \neg p \end{array}$$

The reason we wish to do this is that there are other tabular forms that may arise — and these are similar in appearance but present *invalid* arguments — where each of the premises is true but the conclusion is false.

a) Consider the following argument:

1) If Margaret Thatcher is the president of the United States, then she is at least 35 years old. $p \rightarrow q$
2) Margaret Thatcher is at least 35 years old. $\underline{ q }$
3) Therefore Margaret Thatcher is the president of the United States. $\therefore p$

Here we find that $[(p \rightarrow q) \wedge q] \rightarrow p$ is *not* a tautology. For if we consider the truth value assignments p: 0 and q: 1, then each of the premises $p \rightarrow q$ and q is true while the conclusion p is false. This *invalid* argument results from the *fallacy* (error in reasoning) where we try to argue by the converse — that is, while $[(p \rightarrow q) \wedge p] \Rightarrow q$, it is *not the case* that $[(p \rightarrow q) \wedge q] \Rightarrow p$.

b) A second argument where the conclusion doesn't necessarily follow from the premises may be given by:

1) If $2 + 3 = 6$, then $2 + 4 = 6$. $p \to q$

2) $2 + 3 \neq 6$. $\neg p$

3) Therefore $2 + 4 \neq 6$. $\therefore \neg q$

In this case we find that $[(p \to q) \wedge \neg p] \to \neg q$ is *not* a tautology. Once again the truth value assignments p: 0 and q: 1 show us that the premises $p \to q$ and $\neg p$ can both be true while the conclusion $\neg q$ is false. The fallacy behind this invalid argument arises from our attempt to argue by the inverse — for although $[(p \to q) \wedge \neg q] \Rightarrow \neg p$, it does *not* follow that $[(p \to q) \wedge \neg p] \Rightarrow \neg q$.

Before proceeding further we now mention a rather simple but important rule of inference.

EXAMPLE 2.26

The following rule of inference arises from the observation that if p, q are true statements, then $p \wedge q$ is a true statement.

Now suppose that statements p, q occur in the development of an argument. These statements may be (given) premises or results that are derived from premises and/or from results developed earlier in the argument. Then under these circumstances the two statements p, q can be combined into their conjunction $p \wedge q$, and this new statement can be used in later steps as the argument continues.

We call this rule the *Rule of Conjunction* and write it in tabular form as

$$\begin{array}{c} p \\ q \\ \hline \therefore p \wedge q \end{array}$$

As we proceed further with our study of rules of inference, we find another fairly simple but important rule.

EXAMPLE 2.27

The following rule of inference — one we may feel just illustrates good old common sense — is called the *Rule of Disjunctive Syllogism*. This rule comes about from the logical implication

$$[(p \vee q) \wedge \neg p] \to q,$$

which we can derive from Modus Ponens by observing that $p \vee q \Longleftrightarrow \neg p \to q$.

In tabular form we write

$$\begin{array}{c} p \vee q \\ \neg p \\ \hline \therefore q \end{array}$$

This rule of inference arises when there are exactly two possibilities to consider and we are able to eliminate one of them as being true. Then the other possibility has to be true. The following illustrates one such application of this rule.

1) Bart's wallet is in his back pocket or it is on his desk. $p \vee q$

2) Bart's wallet is not in his back pocket. $\neg p$

3) Therefore Bart's wallet is on his desk. $\therefore q$

At this point we have examined five rules of inference. But before we try to validate any more arguments like the one (with 11 steps) in Example 2.25, we shall look at one more of these rules. This one underlies a method of proof that is sometimes confused with the contrapositive method (or proof) given in Modus Tollens. The confusion arises because both methods involve the negation of a statement. However, we will soon realize that these are two distinct methods. (Toward the end of Section 2.5 we shall compare and contrast these two methods once again.)

EXAMPLE 2.28

Let p denote an arbitrary statement, and F_0 a contradiction. The results in column 5 of Table 2.17 show that the implication $(\neg p \to F_0) \to p$ is a tautology, and this provides us with the rule of inference called the *Rule of Contradiction*. In tabular form this rule is written as

$$\frac{\neg p \to F_0}{\therefore p}$$

Table 2.17

p	$\neg p$	F_0	$\neg p \to F_0$	$(\neg p \to F_0) \to p$
1	0	0	1	1
0	1	0	0	1

This rule tells us that if p is a statement and $\neg p \to F_0$ is true, then $\neg p$ must be false because F_0 is false. So then we have p true.

The Rule of Contradiction is the basis of a method for establishing the validity of an argument — namely, the method of *Proof by Contradiction*, or *Reductio ad Absurdum*. The idea behind the method of Proof by Contradiction is to establish a statement (namely, the conclusion of an argument) by showing that, if this statement were false, then we would be able to deduce an impossible consequence. The use of this method arises in certain arguments which we shall now describe.

In general, when we want to establish the validity of the argument

$$(p_1 \land p_2 \land \cdots \land p_n) \to q,$$

we can establish the validity of the logically equivalent argument

$$(p_1 \land p_2 \land \cdots \land p_n \land \neg q) \to F_0.$$

[This follows from the tautology in column 7 of Table 2.18 and the first substitution rule — where we replace the primitive statement p by the statement $(p_1 \land p_2 \land \cdots \land p_n)^{\dagger}$.]

Table 2.18

p	q	F_0	$p \land \neg q$	$(p \land \neg q) \to F_0$	$p \to q$	$(p \to q) \leftrightarrow [(p \land \neg q) \to F_0]$
0	0	0	0	1	1	1
0	1	0	0	1	1	1
1	0	0	1	0	0	1
1	1	0	0	1	1	1

---

†In Section 4.2 we shall provide the reason why we know that for any statements $p_1, p_2, \ldots, p_n$, and q, it follows that $(p_1 \land p_2 \land \cdots \land p_n) \land \neg q \iff p_1 \land p_2 \land \cdots \land p_n \land \neg q$.

When we apply the method of Proof by Contradiction, we first assume that what we are trying to validate (or prove) is actually false. Then we use this assumption as an additional premise in order to produce a contradiction (or impossible situation) of the form $s \wedge \neg s$, for some statement s. Once we have derived this contradiction we may then conclude that the statement we were given was in fact true — and this validates the argument (or completes the proof).

We shall turn to the method of Proof by Contradiction when it is (or appears to be) easier to use $\neg q$ *in conjunction with* the premises $p_1, p_2, \ldots, p_n$ in order to deduce a contradiction than it is to deduce the conclusion q directly from the premises $p_1, p_2, \ldots, p_n$. The method of Proof by Contradiction will be used in some of the later examples for this section — namely, Examples 2.32 and 2.35. We shall also find it frequently reappearing in other chapters in the text.

Now that we have examined six rules of inference, we summarize these rules and introduce several others in Table 2.19 (on the following page).

The next five examples will present valid arguments. In so doing, these examples will show us how to apply the rules listed in Table 2.19 in conjunction with other results, such as the laws of logic.

EXAMPLE 2.29

Our first example demonstrates the validity of the argument

$$p \to r$$
$$\neg p \to q$$
$$\underline{q \to s}$$
$$\therefore \neg r \to s$$

Steps	**Reasons**
1) $p \to r$	Premise
2) $\neg r \to \neg p$	Step (1) and $p \to r \Longleftrightarrow \neg r \to \neg p$
3) $\neg p \to q$	Premise
4) $\neg r \to q$	Steps (2) and (3) and the Law of the Syllogism
5) $q \to s$	Premise
6) $\therefore \neg r \to s$	Steps (4) and (5) and the Law of the Syllogism

A second way to validate the given argument proceeds as follows.

Steps	**Reasons**
1) $p \to r$	Premise
2) $q \to s$	Premise
3) $\neg p \to q$	Premise
4) $p \vee q$	Step (3) and $(\neg p \to q) \Longleftrightarrow (\neg\neg p \vee q) \Longleftrightarrow (p \vee q)$, where the second logical equivalence follows by the Law of Double Negation
5) $r \vee s$	Steps (1), (2), and (4) and the Rule of the Constructive Dilemma
6) $\therefore \neg r \to s$	Step (5) and $(r \vee s) \Longleftrightarrow (\neg\neg r \vee s) \Longleftrightarrow (\neg r \to s)$, where the Law of Double Negation is used in the first logical equivalence

---

The next example is somewhat more involved.

Table 2.19

Rule of Inference	Related Logical Implication	Name of Rule

1) p
   $p \to q$
   $\therefore q$
	$[p \wedge (p \to q)] \to q$	Rule of Detachment
		(Modus Ponens)

2) $p \to q$
   $q \to r$
   $\therefore p \to r$
	$[(p \to q) \wedge (q \to r)] \to (p \to r)$	Law of the Syllogism

3) $p \to q$
   $\neg q$
   $\therefore \neg p$
	$[(p \to q) \wedge \neg q] \to \neg p$	Modus Tollens

4) p
   q
   $\therefore p \wedge q$
		Rule of Conjunction

5) $p \vee q$
   $\neg p$
   $\therefore q$
	$[(p \vee q) \wedge \neg p] \to q$	Rule of Disjunctive
		Syllogism

6) $\neg p \to F_0$ $\therefore p$	$(\neg p \to F_0) \to p$	Rule of Contradiction
7) $p \wedge q$ $\therefore p$	$(p \wedge q) \to p$	Rule of Conjunctive Simplification
8) p $\therefore p \vee q$	$p \to p \vee q$	Rule of Disjunctive Amplification

9) $p \wedge q$
   $p \to (q \to r)$
   $\therefore r$
	$[(p \wedge q) \wedge [p \to (q \to r)]] \to r$	Rule of Conditional
		Proof

10) $p \to r$
   $q \to r$
   $\therefore (p \vee q) \to r$
	$[(p \to r) \wedge (q \to r)] \to [(p \vee q) \to r]$	Rule for Proof
		by Cases

11) $p \to q$
   $r \to s$
   $p \vee r$
   $\therefore q \vee s$
	$[(p \to q) \wedge (r \to s) \wedge (p \vee r)] \to (q \vee s)$	Rule of the
		Constructive
		Dilemma

12) $p \to q$
   $r \to s$
   $\neg q \vee \neg s$
   $\therefore \neg p \vee \neg r$
	$[(p \to q) \wedge (r \to s) \wedge (\neg q \vee \neg s)] \to (\neg p \vee \neg r)$	Rule of the
		Destructive
		Dilemma

EXAMPLE 2.30   Establish the validity of the argument

$$p \to q$$
$$q \to (r \wedge s)$$
$$\neg r \vee (\neg t \vee u)$$
$$\underline{p \wedge t}$$
$$\therefore u$$

Steps	Reasons
1) $p \rightarrow q$	Premise
2) $q \rightarrow (r \wedge s)$	Premise
3) $p \rightarrow (r \wedge s)$	Steps (1) and (2) and the Law of the Syllogism
4) $p \wedge t$	Premise
5) p	Step (4) and the Rule of Conjunctive Simplification
6) $r \wedge s$	Steps (5) and (3) and the Rule of Detachment
7) r	Step (6) and the Rule of Conjunctive Simplification
8) $\neg r \vee (\neg t \vee u)$	Premise
9) $\neg(r \wedge t) \vee u$	Step (8), the Associative Law of $\vee$, and DeMorgan's Laws
10) t	Step (4) and the Rule of Conjunctive Simplification
11) $r \wedge t$	Steps (7) and (10) and the Rule of Conjunction
12) $\therefore u$	Steps (9) and (11), the Law of Double Negation, and the Rule of Disjunctive Syllogism

EXAMPLE 2.31

This example will provide a way to show that the following argument is valid.

If the band could not play rock music or the refreshments were not delivered on time, then the New Year's party would have been canceled and Alicia would have been angry. If the party were canceled, then refunds would have had to be made. No refunds were made.

Therefore the band could play rock music.

First we convert the given argument into symbolic form by using the following statement assignments:

p: The band could play rock music.
q: The refreshments were delivered on time.
r: The New Year's party was canceled.
s: Alicia was angry.
t: Refunds had to be made.

The argument above now becomes

$$(\neg p \vee \neg q) \rightarrow (r \wedge s)$$
$$r \rightarrow t$$
$$\underline{\neg t}$$
$$\therefore p$$

We can establish the validity of this argument as follows.

Steps	Reasons
1) $r \rightarrow t$	Premise
2) $\neg t$	Premise
3) $\neg r$	Steps (1) and (2) and Modus Tollens
4) $\neg r \vee \neg s$	Step (3) and the Rule of Disjunctive Amplification
5) $\neg(r \wedge s)$	Step (4) and DeMorgan's Laws
6) $(\neg p \vee \neg q) \rightarrow (r \wedge s)$	Premise
7) $\neg(\neg p \vee \neg q)$	Steps (6) and (5) and Modus Tollens
8) $p \wedge q$	Step (7), DeMorgan's Laws, and the Law of Double Negation
9) $\therefore p$	Step (8) and the Rule of Conjunctive Simplification

EXAMPLE 2.32   In this instance we shall use the method of Proof by Contradiction. Consider the argument

$$\neg p \leftrightarrow q$$
$$q \rightarrow r$$
$$\underline{\neg r}$$
$$\therefore p$$

To establish the validity for this argument, we assume the negation $\neg p$ of the conclusion p as another premise. The objective now is to use these four premises to derive a contradiction F_0. Our derivation follows.

Steps	**Reasons**
1) $\neg p \leftrightarrow q$	Premise
2) $(\neg p \rightarrow q) \wedge (q \rightarrow \neg p)$	Step (1) and $(\neg p \leftrightarrow q) \Longleftrightarrow [(\neg p \rightarrow q) \wedge (q \rightarrow \neg p)]$
3) $\neg p \rightarrow q$	Step (2) and the Rule of Conjunctive Simplification
4) $q \rightarrow r$	Premise
5) $\neg p \rightarrow r$	Steps (3) and (4) and the Law of the Syllogism
6) $\neg p$	Premise (the one assumed)
7) r	Steps (5) and (6) and the Rule of Detachment
8) $\neg r$	Premise
9) $r \wedge \neg r (\Longleftrightarrow F_0)$	Steps (7) and (8) and the Rule of Conjunction
10) $\therefore p$	Steps (6) and (9) and the method of Proof by Contradiction

If we examine further what has happened here, we find that

$$[(\neg p \leftrightarrow q) \wedge (q \rightarrow r) \wedge \neg r \wedge \neg p] \Rightarrow F_0.$$

This requires the truth value of $[(\neg p \leftrightarrow q) \wedge (q \rightarrow r) \wedge \neg r \wedge \neg p]$ to be 0. Because $\neg p \leftrightarrow q$, $q \rightarrow r$, and $\neg r$ are the given premises, each of these statements has the truth value 1. Consequently, for $[(\neg p \leftrightarrow q) \wedge (q \rightarrow r) \wedge \neg r \wedge \neg p]$ to have the truth value 0, the statement $\neg p$ must have the truth value 0. Therefore p has the truth value 1, and the conclusion p of the argument is true.

Before we consider our next example, we need to examine columns 5 and 7 of Table 2.20. These identical columns tell us that for primitive statements p, q, and r,

$$[p \rightarrow (q \rightarrow r)] \Longleftrightarrow [(p \wedge q) \rightarrow r].$$

Using the first substitution rule, let us replace each occurrence of p by the compound statement $(p_1 \wedge p_2 \wedge \cdots \wedge p_n)$. Then we obtain the new result

$$[(p_1 \wedge p_2 \wedge \cdots \wedge p_n) \rightarrow (q \rightarrow r)] \Longleftrightarrow [(p_1 \wedge p_2 \wedge \cdots \wedge p_n \wedge q)^\dagger \rightarrow r].$$

---

† In Section 4.2 we shall present a formal proof of why

$$(p_1 \wedge p_2 \wedge \cdots \wedge p_n) \wedge q \Longleftrightarrow p_1 \wedge p_2 \wedge \cdots \wedge p_n \wedge q.$$

Table 2.20

p	q	r	$p \wedge q$	$(p \wedge q) \rightarrow r$	$q \rightarrow r$	$p \rightarrow (q \rightarrow r)$
0	0	0	0	1	1	1
0	0	1	0	1	1	1
0	1	0	0	1	0	1
0	1	1	0	1	1	1
1	0	0	0	1	1	1
1	0	1	0	1	1	1
1	1	0	1	0	0	0
1	1	1	1	1	1	1

This result tells us that if we wish to establish the validity of the argument (*) we may be able to do so by establishing the validity of the corresponding argument (**).

$$
\begin{array}{ll}
(*) \qquad p_1 & \qquad\qquad (**) \qquad p_1 \\
 p_2 & \qquad\qquad p_2 \\
 \vdots & \qquad\qquad \vdots \\
 \underline{p_n} & \qquad\qquad p_n \\
 \therefore q \rightarrow r & \qquad\qquad \underline{q} \\
 & \qquad\qquad \therefore r
\end{array}
$$

After all, suppose we want to show that $q \rightarrow r$ has the truth value 1, when each of $p_1, p_2, \ldots, p_n$ does. If the truth value for q is 0, then there is nothing left to do, since the truth value for $q \rightarrow r$ is 1. Hence the real problem is to show that $q \rightarrow r$ has truth value 1, when each of $p_1, p_2, \ldots, p_n$, *and* q does — that is, we need to show that when $p_1, p_2, \ldots, p_n, q$ each have truth value 1, then the truth value of r is 1.

We demonstrate this principle in the next example.

EXAMPLE 2.33

In order to establish the validity of the argument

$$
(*) \qquad\qquad
\begin{array}{l}
u \rightarrow r \\
(r \wedge s) \rightarrow (p \vee t) \\
q \rightarrow (u \wedge s) \\
\underline{\neg t} \\
\therefore q \rightarrow p
\end{array}
$$

we consider the corresponding argument

$$
(**) \qquad\qquad
\begin{array}{l}
u \rightarrow r \\
(r \wedge s) \rightarrow (p \vee t) \\
q \rightarrow (u \wedge s) \\
\neg t \\
\underline{q} \\
\therefore p
\end{array}
$$

[Note that q is the hypothesis of the conclusion $q \rightarrow p$ for argument (*) and that it becomes another premise for argument (**) where the conclusion is p.]

To validate the argument (**) we proceed as follows.

Steps	**Reasons**
1) q	Premise
2) $q \to (u \wedge s)$	Premise
3) $u \wedge s$	Steps (1) and (2) and the Rule of Detachment
4) u	Step (3) and the Rule of Conjunctive Simplification
5) $u \to r$	Premise
6) r	Steps (4) and (5) and the Rule of Detachment
7) s	Step (3) and the Rule of Conjunctive Simplification
8) $r \wedge s$	Steps (6) and (7) and the Rule of Conjunction
9) $(r \wedge s) \to (p \vee t)$	Premise
10) $p \vee t$	Steps (8) and (9) and the Rule of Detachment
11) $\neg t$	Premise
12) $\therefore p$	Steps (10) and (11) and the Rule of Disjunctive Syllogism

We now know that for argument (**)

$$[(u \to r) \wedge [(r \wedge s) \to (p \vee t)] \wedge [q \to (u \wedge s)] \wedge \neg t \wedge q] \Rightarrow p,$$

and for argument (*) it follows that

$$[(u \to r) \wedge [(r \wedge s) \to (p \vee t)] \wedge [q \to (u \wedge s)] \wedge \neg t] \Rightarrow (q \to p).$$

Examples 2.29 through 2.33 have given us some idea of how to establish the validity of an argument. Following Example 2.25 we discussed two situations indicating when an argument is invalid — namely, when we try to argue by the converse or the inverse. So now it is time for us to learn a little more about how to determine when an argument is invalid.

Given an argument

$$
\begin{array}{l}
p_1 \\
p_2 \\
p_3 \\
\quad \vdots \\
\underline{p_n} \\
\therefore q
\end{array}
$$

we say that the argument is invalid if it is possible for each of the premises $p_1, p_2, p_3, \ldots, p_n$ to be true (with truth value 1), while the conclusion q is false (with truth value 0).

The next example illustrates an indirect method whereby we may be able to show that an argument we *feel* is invalid (perhaps because we cannot find a way to show that it is valid) actually *is* invalid.

EXAMPLE 2.34   Consider the primitive statements $p, q, r, s,$ and t and the argument

$$
\begin{array}{l}
p \\
p \vee q \\
q \to (r \to s) \\
\underline{t \to r} \\
\therefore \neg s \to \neg t
\end{array}
$$

To show that this is an invalid argument, we need *one* assignment of truth values for each of the statements $p, q, r, s,$ and t such that the conclusion $\neg s \to \neg t$ is false (has the truth value 0) while the four premises are all true (have the truth value 1). The only time the conclusion $\neg s \rightarrow \neg t$ is false is when $\neg s$ is true and $\neg t$ is false. This implies that the truth value for s is 0 and that the truth value for t is 1.

Because p is one of the premises, its truth value must be 1. For the premise $p \vee q$ to have the truth value 1, q may be either true (1) or false (0). So let us consider the premise $t \rightarrow r$ where we know that t is true. If $t \rightarrow r$ is to be true, then r must be true (have the truth value 1). Now with r true (1) and s false (0), it follows that $r \rightarrow s$ is false (0), and that the truth value of the premise $q \rightarrow (r \rightarrow s)$ will be 1 only when q is false (0).

Consequently, under the truth value assignments

$$p: \quad 1 \qquad q: \quad 0 \qquad r: \quad 1 \qquad s: \quad 0 \qquad t: \quad 1,$$

the four premises

$$p \qquad p \vee q \qquad q \rightarrow (r \rightarrow s) \qquad t \rightarrow r$$

all have the truth value 1, while the conclusion

$$\neg s \rightarrow \neg t$$

has the truth value 0. In this case we have shown the given argument to be invalid.

The truth value assignments $p: 1$, $q: 0$, $r: 1$, $s: 0$, and $t: 1$ of Example 2.34 provide one case that *disproves* what we thought might have been a valid argument. We should now start to realize that in trying to show that an implication of the form

$$(p_1 \wedge p_2 \wedge p_3 \wedge \cdots \wedge p_n) \rightarrow q$$

presents a valid argument, we need to consider *all* cases where the premises $p_1, p_2, p_3, \ldots,$ p_n are true. [Each such case is an assignment of truth values for the primitive statements (that make up the premises) where $p_1, p_2, p_3, \ldots, p_n$ are true.] In order to do so — namely, to cover the cases without writing out the truth table — we have been using the rules of inference together with the laws of logic and other logical equivalences. To cover all the necessary cases, we cannot use one specific example (or case) as a means of establishing the validity of the argument (for all possible cases). However, whenever we wish to show that an implication (of the preceding form) is not a tautology, all we need to find is one case for which the implication is false — that is, one case in which all the premises are true but the conclusion is false. This *one* case provides a *counterexample* for the argument and shows it to be invalid.

Let us consider a second example wherein we try the indirect approach of Example 2.34.

EXAMPLE 2.35

What can we say about the validity or invalidity of the following argument? Here $p, q, r,$ and s denote primitive statements.)

$$\begin{array}{c} p \rightarrow q \\ q \rightarrow s \\ r \rightarrow \neg s \\ \underline{\neg p \vee r} \\ \therefore \neg p \end{array}$$

Can the conclusion $\neg p$ be false while the four premises are all true? The conclusion $\neg p$ is false when p has the truth value 1. So for the premise $p \rightarrow q$ to be true, the truth value of q must be 1. From the truth of the premise $q \rightarrow s$, the truth of q forces the truth of s. Consequently, at this point we have statements p, q, and s all with the truth value 1.

Continuing with the premise $r \rightarrow \neg s$, we find that because s has the truth value 1, the truth value of r must be 0. Hence r is false. But with $\neg p$ false and the premise $\neg p \veebar r$ true, we also have r true. Therefore we find that $p \Rightarrow (\neg r \wedge r)$.

We have failed in our attempt to find a counterexample to the validity of the given argument. However, this failure has shown us that the given argument is valid — and the validity follows by using the method of Proof by Contradiction.

This introduction to the rules of inference has been far from exhaustive. Several of the books cited among the references listed near the end of this chapter offer additional material for the reader who wishes to pursue this topic further. In Section 2.5 we shall apply the ideas developed in this section to statements of a more mathematical nature. For we shall want to learn how to develop a proof for a theorem. And then in Chapter 4 another very important proof technique called *mathematical induction* will be added to our arsenal of weapons for proving mathematical theorems. First, however, the reader should carefully complete the exercises for this section.

EXERCISES 2.3

1. The following are three valid arguments. Establish the validity of each by means of a truth table. In each case, determine which rows of the table are crucial for assessing the validity of the argument and which rows can be ignored.

a) $[p \wedge (p \rightarrow q) \wedge r] \rightarrow [(p \vee q) \rightarrow r]$

b) $[[(p \wedge q) \rightarrow r] \wedge \neg q \wedge (p \rightarrow \neg r)] \rightarrow (\neg p \vee \neg q)$

c) $[[p \vee (q \vee r)] \wedge \neg q] \rightarrow (p \vee r)$

2. Use truth tables to verify that each of the following is a logical implication.

a) $[(p \rightarrow q) \wedge (q \rightarrow r)] \rightarrow (p \rightarrow r)$

b) $[(p \rightarrow q) \wedge \neg q] \rightarrow \neg p$

c) $[(p \vee q) \wedge \neg p] \rightarrow q$

d) $[(p \rightarrow r) \wedge (q \rightarrow r)] \rightarrow [(p \vee q) \rightarrow r]$

3. Verify that each of the following is a logical implication by showing that it is impossible for the conclusion to have the truth value 0 while the hypothesis has the truth value 1.

a) $(p \wedge q) \rightarrow p$

b) $p \rightarrow (p \vee q)$

c) $[(p \vee q) \wedge \neg p] \rightarrow q$

d) $[(p \rightarrow q) \wedge (r \rightarrow s) \wedge (p \vee r)] \rightarrow (q \vee s)$

e) $[(p \rightarrow q) \wedge (r \rightarrow s) \wedge (\neg q \vee \neg s)] \rightarrow (\neg p \vee \neg r)$

4. For each of the following pairs of statements, use Modus Ponens or Modus Tollens to fill in the blank line so that a valid argument is presented.

a) If Janice has trouble starting her car, then her daughter Angela will check Janice's spark plugs.
Janice had trouble starting her car.
∴ _____

b) If Brady solved the first problem correctly, then the answer he obtained is 137.
Brady's answer to the first problem is not 137.
∴ _____

c) If this is a **repeat-until** loop, then the body of this loop is executed at least once.

_____
∴ The body of the loop is executed at least once.

d) If Tim plays basketball in the afternoon, then he will not watch television in the evening.

_____
∴ Tim didn't play basketball in the afternoon.

5. Consider each of the following arguments. If the argument is valid, identify the rule of inference that establishes its validity. If not, indicate whether the error is due to an attempt to argue by the converse or by the inverse.

a) Andrea can program in C++, and she can program in Java.
Therefore Andrea can program in C++.

b) A sufficient condition for Bubbles to win the golf tournament is that her opponent Meg not sink a birdie on the last hole.
Bubbles won the golf tournament.
Therefore Bubbles' opponent Meg did not sink a birdie on the last hole.

c) If Ron's computer program is correct, then he'll be able to complete his computer science assignment in at most two hours.
It takes Ron over two hours to complete his computer science assignment.
Therefore Ron's computer program is not correct.

d) Eileen's car keys are in her purse, or they are on the kitchen table.

Eileen's car keys are not on the kitchen table.
Therefore Eileen's car keys are in her purse.

  e) If interest rates fall, then the stock market will rise.
Interest rates are not falling.
Therefore the stock market will not rise.

6. For primitive statements p, q, and r, let P denote the statement

$$[p \wedge (q \wedge r)] \vee \neg[p \vee (q \wedge r)],$$

while P_1 denotes the statement

$$[p \wedge (q \vee r)] \vee \neg[p \vee (q \vee r)].$$

  a) Use the rules of inference to show that

$$q \wedge r \Rightarrow q \vee r.$$

  b) Is it true that $P \Rightarrow P_1$?

7. Give the reason(s) for each step needed to show that the following argument is valid.

$$[p \wedge (p \rightarrow q) \wedge (s \vee r) \wedge (r \rightarrow \neg q)] \rightarrow (s \vee t)$$

Steps	Reasons
1) p	
2) $p \rightarrow q$	
3) q	
4) $r \rightarrow \neg q$	
5) $q \rightarrow \neg r$	
6) $\neg r$	
7) $s \vee r$	
8) s	
9) $\therefore s \vee t$	

8. Give the reasons for the steps verifying the following argument.

$$(\neg p \vee q) \rightarrow r$$
$$r \rightarrow (s \vee t)$$
$$\neg s \wedge \neg u$$
$$\underline{\neg u \rightarrow \neg t}$$
$$\therefore p$$

Steps	Reasons
1) $\neg s \wedge \neg u$	
2) $\neg u$	
3) $\neg u \rightarrow \neg t$	
4) $\neg t$	
5) $\neg s$	
6) $\neg s \wedge \neg t$	
7) $r \rightarrow (s \vee t)$	
8) $\neg(s \vee t) \rightarrow \neg r$	
9) $(\neg s \wedge \neg t) \rightarrow \neg r$	
10) $\neg r$	
11) $(\neg p \vee q) \rightarrow r$	
12) $\neg r \rightarrow \neg(\neg p \vee q)$	
13) $\neg r \rightarrow (p \wedge \neg q)$	
14) $p \wedge \neg q$	
15) $\therefore p$	

9. a) Give the reasons for the steps given to validate the argument

$$[(p \rightarrow q) \wedge (\neg r \vee s) \wedge (p \vee r)] \rightarrow (\neg q \rightarrow s).$$

Steps	Reasons
1) $\neg(\neg q \rightarrow s)$	
2) $\neg q \wedge \neg s$	
3) $\neg s$	
4) $\neg r \vee s$	
5) $\neg r$	
6) $p \rightarrow q$	
7) $\neg q$	
8) $\neg p$	
9) $p \vee r$	
10) r	
11) $\neg r \wedge r$	
12) $\therefore \neg q \rightarrow s$	

  b) Give a direct proof for the result in part (a).

  c) Give a direct proof for the result in Example 2.32.

10. Establish the validity of the following arguments.

  a) $[(p \wedge \neg q) \wedge r] \rightarrow [(p \wedge r) \vee q]$

  b) $[p \wedge (p \rightarrow q) \wedge (\neg q \vee r)] \rightarrow r$

  c) $p \rightarrow q$ **d)** $p \rightarrow q$
       $\neg q$ $r \rightarrow \neg q$
      $\underline{\quad \neg r \quad}$ $\underline{\quad r \quad}$
    $\therefore \neg(p \vee r)$ $\therefore \neg p$

  e) $p \rightarrow (q \rightarrow r)$ **f)** $p \wedge q$
      $\neg q \rightarrow \neg p$ $p \rightarrow (r \wedge q)$
       $\underline{\quad p \quad}$ $r \rightarrow (s \vee t)$
     $\therefore r$ $\underline{\quad \neg s \quad}$
                            $\therefore t$

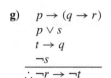

  g) $p \rightarrow (q \rightarrow r)$ **h)** $p \vee q$
      $p \vee s$ $\neg p \vee r$
      $t \rightarrow q$ $\underline{\quad \neg r \quad}$
      $\underline{\quad \neg s \quad}$ $\therefore q$
   $\therefore \neg r \rightarrow \neg t$

11. Show that each of the following arguments is invalid by providing a counterexample — that is, an assignment of truth values for the given primitive statements p, q, r, and s such that all premises are true (have the truth value 1) while the conclusion is false (has the truth value 0).

  a) $[(p \wedge \neg q) \wedge [p \rightarrow (q \rightarrow r)]] \rightarrow \neg r$

  b) $[[(p \wedge q) \rightarrow r] \wedge (\neg q \vee r)] \rightarrow p$

  c) $p \leftrightarrow q$ **d)** p
      $q \rightarrow r$ $p \rightarrow r$
      $r \vee \neg s$ $p \rightarrow (q \vee \neg r)$
     $\underline{\neg s \rightarrow q}$ $\underline{\neg q \vee \neg s}$
       $\therefore s$ $\therefore s$

12. Write each of the following arguments in symbolic form. Then establish the validity of the argument or give a counterexample to show that it is invalid.

a) If Rochelle gets the supervisor's position and works hard, then she'll get a raise. If she gets the raise, then she'll buy a new car. She has not purchased a new car. Therefore either Rochelle did not get the supervisor's position or she did not work hard.

b) If Dominic goes to the racetrack, then Helen will be mad. If Ralph plays cards all night, then Carmela will be mad. If either Helen or Carmela gets mad, then Veronica (their attorney) will be notified. Veronica has not heard from either of these two clients. Consequently, Dominic didn't make it to the racetrack and Ralph didn't play cards all night.

c) If there is a chance of rain or her red headband is missing, then Lois will not mow her lawn. Whenever the temperature is over 80°F, there is no chance for rain. Today the temperature is 85°F and Lois is wearing her red headband. Therefore (sometime today) Lois will mow her lawn.

13. a) Given primitive statements p, q, r, show that the implication

$$[(p \vee q) \wedge (\neg p \vee r)] \rightarrow (q \vee r)$$

is a tautology.

b) The tautology in part (a) provides the rule of inference known as *resolution*, where the conclusion $(q \vee r)$ is called the *resolvent*. This rule was proposed in 1965 by J. A. Robinson and is the basis of many computer programs designed to automate a reasoning system.

In applying resolution each premise (in the hypothesis) and the conclusion are written as *clauses*. A clause is a primitive statement or its negation, or it is the disjunction of terms each of which is a primitive statement or the negation of such a statement. Hence the given rule has the clauses $(p \vee q)$ and $(\neg p \vee r)$ as premises and the clause $(q \vee r)$ as its conclusion (or, resolvent). Should we have the premise $\neg(p \wedge q)$, we replace this by the logically equivalent clause $\neg p \vee \neg q$, by the first of DeMorgan's Laws. The premise $\neg(p \vee q)$ can be replaced by the two clauses $\neg p$, $\neg q$. This is due to the second DeMorgan Law and the Rule of Conjunctive Simplification. For the premise $p \vee (q \wedge r)$, we apply the Distributive Law of $\vee$ over $\wedge$ and the Rule of Conjunctive Simplification to arrive at either of the two clauses $p \vee q$, $p \vee r$. Finally, the premise $p \rightarrow q$ becomes the clause $\neg p \vee q$.

Establish the validity of the following arguments, using resolution (along with the rules of inference and the laws of logic).

(i)
$$p \vee (q \wedge r)$$
$$\underline{p \rightarrow s}$$
$$\therefore r \vee s$$

(ii)
$$p$$
$$\underline{p \leftrightarrow q}$$
$$\therefore q$$

(iii)
$$p \vee q$$
$$p \rightarrow r$$
$$\underline{r \rightarrow s}$$
$$\therefore q \vee s$$

(iv)
$$\neg p \vee q \vee r$$
$$\neg q$$
$$\underline{\neg r}$$
$$\therefore \neg p$$

(v)
$$\neg p \vee s$$
$$\neg t \vee (s \wedge r)$$
$$\neg q \vee r$$
$$\underline{p \vee q \vee t}$$
$$\therefore r \vee s$$

c) Write the following argument in symbolic form, then use resolution (along with the rules of inference and the laws of logic) to establish its validity.

Jonathan does not have his driver's license or his new car is out of gas. Jonathan has his driver's license or he does not like to drive his new car. Jonathan's new car is not out of gas or he does not like to drive his new car. Therefore, Jonathan does not like to drive his new car.

2.4
The Use of Quantifiers

In Section 2.1, we mentioned how sentences that involve a variable, such as x, need not be statements. For example, the sentence "The number $x + 2$ is an even integer" is not necessarily true or false unless we know what value is substituted for x. If we restrict our choices to integers, then when x is replaced by -5, -1, or 3, for instance, the resulting statement is false. In fact, it is false whenever x is replaced by an odd integer. When an even integer is substituted for x, however, the resulting statement is true.

We refer to the sentence "The number $x + 2$ is an even integer" as an *open statement*, which we formally define as follows.

Definition 2.5   A declarative sentence is an *open statement* if

1) it contains one or more variables, and

2) it is not a statement, but

3) it becomes a statement when the variables in it are replaced by certain allowable choices.

When we examine the sentence "The number $x + 2$ is an even integer" in light of this definition, we find it is an open statement that contains the single variable x. With regard to the third element of the definition, in our earlier discussion we restricted the "certain allowable choices" to integers. These allowable choices constitute what is called the *universe* or *universe of discourse* for the open statement. The universe comprises the choices we wish to consider or allow for the variable(s) in the open statement. (The universe is an example of a *set*, a concept we shall examine in some detail in the next chapter.)

In dealing with open statements, we use the following notation:

The open statement "The number $x + 2$ is an even integer" is denoted by $p(x)$ [or $q(x)$, $r(x)$, etc.]. Then $\neg p(x)$ may be read "The number $x + 2$ is *not* an even integer."

We shall use $q(x, y)$ to represent an open statement that contains two variables. For example, consider

$q(x, y)$:   The numbers $y + 2$, $x - y$, and $x + 2y$ are even integers.

In the case of $q(x, y)$, there is more than one occurrence of each of the variables x, y. It is understood that when we replace one of the x's by a choice from our universe, we replace the other x by the same choice. Likewise, when a substitution (from the universe) is made for one occurrence of y, that same substitution is made for all other occurrences of the variable y.

With $p(x)$ and $q(x, y)$ as above, and the universe still stipulating the integers as our only allowable choices, we get the following results when we make some replacements for the variables x, y.

$p(5)$:   The number $7(= 5 + 2)$ is an even integer. (FALSE)

$\neg p(7)$:   The number 9 is not an even integer. (TRUE)

$q(4, 2)$:   The numbers 4, 2, and 8 are even integers. (TRUE)

We also note, for example, that $q(5, 2)$ and $q(4, 7)$ are both false statements, whereas $\neg q(5, 2)$ and $\neg q(4, 7)$ are true.

Consequently, we see that for both $p(x)$ and $q(x, y)$, as already given, some substitutions result in true statements and others in false statements. Therefore we can make the following true statements.

1)   For some x, $p(x)$.

2)   For some x, y, $q(x, y)$.

Note that in this situation, the statements "For some x, $\neg p(x)$" and "For some x, y, $\neg q(x, y)$" are also true. [Since the statements "For some x, $p(x)$" and "For some x, $\neg p(x)$" are both true, we realize that the second statement is *not* the negation of the first — even though the open statement $\neg p(x)$ is the negation of the open statement $p(x)$. And a similar result is true for the statements involving $q(x, y)$ and $\neg q(x, y)$.]

The phrases "For some x" and "For some x, y" are said to *quantify* the open statements $p(x)$ and $q(x, y)$, respectively. Many postulates, definitions, and theorems in mathematics involve statements that are quantified open statements. These result from the two types of *quantifiers*, which are called the *existential* and the *universal quantifiers*.

Statement (1) uses the *existential quantifier* "For some x," which can also be expressed as "For at least one x" or "There exists an x such that." This quantifier is written in symbolic form as $\exists x$. Hence the statement "For some x, $p(x)$" becomes $\exists x \; p(x)$, in symbolic form.

Statement (2) becomes $\exists x \; \exists y \; q(x, y)$ in symbolic form. The notation $\exists x, y$ can be used to abbreviate $\exists x \; \exists y \; q(x, y)$ to $\exists x, y \; q(x, y)$.

The *universal quantifier* is denoted by $\forall x$ and is read "For all x," "For any x," "For each x," or "For every x." "For all x, y," "For any x, y," "For every x, y," or "For all x and y" is denoted by $\forall x \; \forall y$, which can be abbreviated to $\forall x, y$.

Taking $p(x)$ as defined earlier and using the universal quantifier, we can change the open statement $p(x)$ into the (quantified) statement $\forall x \; p(x)$, a false statement.

If we consider the open statement $r(x)$: "$2x$ is an even integer" with the same universe (of all integers), then the (quantified) statement $\forall x \; r(x)$ is a true statement. When we say that $\forall x \; r(x)$ is true, we mean that no matter which integer (from our universe) is substituted for x in $r(x)$, the resulting statement is true. Also note that the statement $\exists x \; r(x)$ is a true statement, whereas $\forall x \; \neg r(x)$ and $\exists x \; \neg r(x)$ are both false.

The variable x in each of open statements $p(x)$ and $r(x)$ is called a *free variable* (of the open statement). As x varies over the universe for an open statement, the truth value of the statement (that results upon the replacement of each occurrence of x) may vary. For instance, in the case of $p(x)$, we found $p(5)$ to be false—while $p(6)$ turns out to be a true statement. The open statement $r(x)$, however, becomes a true statement for every replacement (for x) taken from the universe of all integers. In contrast to the open statement $p(x)$ the statement $\exists x \; p(x)$ has a fixed truth value—namely, true. And in the symbolic representation $\exists x \; p(x)$ the variable x is said to be a *bound* variable—it is bound by the existential quantifier $\exists$. This is also the case for the statements $\forall x \; r(x)$ and $\forall x \; \neg r(x)$, where in each case the variable x is bound by the universal quantifier $\forall$.

For the open statement $q(x, y)$ we have two free variables, each of which is bound by the quantifier $\exists$ in either of the statements $\exists x \; \exists y \; q(x, y)$ or $\exists x, y \; q(x, y)$.

The following example shows how these new ideas about quantifiers can be used in conjunction with the logical connectives.

EXAMPLE 2.36   Here the universe comprises all real numbers. The open statements $p(x)$, $q(x)$, $r(x)$, and $s(x)$ are given by

$$p(x): \quad x \geq 0 \qquad r(x): \quad x^2 - 3x - 4 = 0$$
$$q(x): \quad x^2 \geq 0 \qquad s(x): \quad x^2 - 3 > 0.$$

Then the following statements are true.

1)
$$\exists x \; [p(x) \wedge r(x)]$$

This follows because the real number 4, for example, is a member of the universe and is such that both of the statements $p(4)$ and $r(4)$ are true.

2)
$$\forall x \; [p(x) \rightarrow q(x)]$$

If we replace x in $p(x)$ by a negative real number a, then $p(a)$ is false, but $p(a) \rightarrow q(a)$ is true regardless of the truth value of $q(a)$. Replacing x in $p(x)$ by a nonnegative real number b, we find that $p(b)$ and $q(b)$ are both true, as is $p(b) \rightarrow q(b)$. Consequently, $p(x) \rightarrow q(x)$ is true for all replacements x taken from the universe of all real numbers, and the (quantified) statement $\forall x \; [p(x) \rightarrow q(x)]$ is true.

This statement may be translated into any of the following:

a) For every real number x, if $x \geq 0$, then $x^2 \geq 0$.

b) Every nonnegative real number has a nonnegative square.

c) The square of any nonnegative real number is a nonnegative real number.

d) All nonnegative real numbers have nonnegative squares.

Also, the statement $\exists x \, [p(x) \rightarrow q(x)]$ is true.

The next statements we examine are false.

1') $$\forall x \, [q(x) \rightarrow \overline{s}(x)]$$

We want to show that the statement is false, so we need exhibit only one *counterexample* — that is, *one value of* x for which $q(x) \rightarrow s(x)$ is false — rather than prove something for all x as we did for statement (2). Replacing x by 1, we find that $q(1)$ is true and $s(1)$ is false. Therefore $q(1) \rightarrow s(1)$ is false, and consequently the (quantified) statement $\forall x \, [q(x) \rightarrow s(x)]$ is false. [Note that $x = 1$ does not produce the only counterexample: Every real number a between $-\sqrt{3}$ and $\sqrt{3}$ will make $q(a)$ true and $s(a)$ false.]

2') $$\forall x \, [r(x) \vee s(x)]$$

Here there are many values for x, such as 1, $\frac{1}{2}$, $-\frac{3}{2}$, and 0, that produce counterexamples. Upon changing quantifiers, however, we find that the statement $\exists x \, [r(x) \vee s(x)]$ is true.

3') $$\forall x \, [r(x) \rightarrow p(x)]$$

The real number -1 is a solution of the equation $x^2 - 3x - 4 = 0$, so $r(-1)$ is true while $p(-1)$ is false. Therefore the choice of -1 provides the unique counterexample we need to show that this (quantified) statement is false.

Statement (3') may be translated into either of the following:

a) For every real number x, if $x^2 - 3x - 4 = 0$, then $x \geq 0$.

b) For every real number x, if x is a solution of the equation $x^2 - 3x - 4 = 0$, then $x \geq 0$.

Now we make the following observations. Let $p(x)$ denote any open statement (in the variable x) with a prescribed *nonempty* universe (that is, the universe contains at least one member). Then if $\forall x \, p(x)$ is true, so is $\exists x \, p(x)$, or

$$\forall x \, p(x) \Rightarrow \exists x \, p(x).$$

When we write $\forall x \, p(x) \Rightarrow \exists x \, p(x)$ we are saying that the implication $\forall x \, p(x) \rightarrow \exists x \, p(x)$ is a logical implication — that is, $\exists x \, p(x)$ is true whenever $\forall x \, p(x)$ is true. Also, we realize that the hypothesis of this implication is the quantified *statement* $\forall x \, p(x)$, and the conclusion is $\exists x \, p(x)$, another quantified *statement*. On the other hand, it does not follow that if $\exists x \, p(x)$ is true, then $\forall x \, p(x)$ must be true. Hence $\exists x \, p(x)$ does not logically imply $\forall x \, p(x)$, in general.

Our next example brings out the fact that the quantification of an open statement may not be as explicit as we might prefer.

EXAMPLE 2.37

a) Let us consider the universe of all real numbers and examine the sentences:

1) If a number is rational, then it is a real number.

2) If x is rational, then x is real.

We should agree that these sentences convey the same information. But we should also question whether the sentences are statements or open statements. In the case of sentence (2) we at least have the presence of the variable x. But neither sentence contains an expression such as "For all," or "For every," or "For each." Our one and only clue to indicate that we are dealing with universally quantified statements here is the presence of the indefinite article "a" in the first sentence. In situations like these the use of the universal quantifier is *implicit* as opposed to *explicit*.

If we let $p(x), q(x)$ be the open statements

$$p(x): \quad x \text{ is a rational number} \qquad q(x): \quad x \text{ is a real number,}$$

then we must recognize the fact that both of the given sentences are somewhat informal ways of expressing the quantified statement

$$\forall x \, [p(x) \to q(x)].$$

b) For the universe of all triangles in the plane, the sentence

> "An equilateral triangle has three angles of 60°, and conversely."

provides another instance of implicit quantification. Here the indefinite article "An" is the only indication that we might be able to express this sentence as a statement with a universal quantifier. If the open statements

$$e(t): \quad \text{Triangle } t \text{ is equilateral.}$$

$$a(t): \quad \text{Triangle } t \text{ has three angles of 60°.}$$

are defined for this universe, then the given sentence can be written in the explicit quantified form

$$\forall t \, [e(t) \leftrightarrow a(t)].$$

c) In the typical trigonometry textbook one often comes across the trigonometric identity

$$\sin^2 x + \cos^2 x = 1.$$

This identify contains no explicit quantification, and the reader must understand or be told that it is defined for all real numbers x. When the universe of all real numbers is specified (or at least understood), then the identity can be expressed by the (explicitly) quantified statement

$$\forall x \, [\sin^2 x + \cos^2 x = 1].$$

d) Finally, consider the universe of all positive integers and the sentence

> "The integer 41 is equal to the sum of two perfect squares."

Here we have one more example where the quantification is implicit — but this time the quantification is existential. We may express the result here in a more formal (and symbolic) manner as

$$\exists m \, \exists n \, [41 = m^2 + n^2].$$

The next example demonstrates that the truth value of a quantified statement may depend on the universe prescribed.

EXAMPLE 2.38

Consider the open statement $p(x)$: $x^2 \geq 1$.

1) If the universe consists of all positive integers, then the quantified statement $\forall x \; p(x)$ is true.

2) For the universe of all positive real numbers, however, the same quantified statement $\forall x \; p(x)$ is false. The positive real number $1/2$ provides one of many possible counterexamples.

Yet for either universe, the quantified statement $\exists x \; p(x)$ is true.

One use of quantifiers in a computer science setting is illustrated in the following example.

EXAMPLE 2.39

In the following program segment, n is an integer variable and the variable A is an array $A[1], A[2], \ldots, A[20]$ of 20 integer values.

```
for n := 1 to 20 do
    A[n] := n * n - n
```

The following statements about the array A can be represented in quantified form, where the universe consists of all integers from 1 to 20, inclusive.

1) Every entry in the array is nonnegative:

$$\forall n \; (A[n] \geq 0).$$

2) There exist two consecutive entries in A where the larger entry is twice the smaller:

$$\exists n \; (A[n + 1] = 2A[n]).$$

3) The entries in the array are sorted in (strictly) ascending order:

$$\forall n \; [(1 \leq n \leq 19) \rightarrow (A[n] < A[n + 1])].$$

Our last statement requires the use of two integer variables m, n.

4) The entries in the array are distinct:

$$\forall m \; \forall n \; [(m \neq n) \rightarrow (A[m] \neq A[n])], \quad \text{or}$$
$$\forall m, n \; [(m < n) \rightarrow (A[m] \neq A[n])].$$

Before continuing, we summarize and somewhat extend, in Table 2.21, what we have learned about quantifiers.

The results in Table 2.21 may appear to involve only one open statement. However, we should realize that the open statement $p(x)$ in the table may stand for a conjunction of open statements, such as $q(x) \wedge r(x)$, or an implication of open statements, such as $s(x) \rightarrow t(x)$. If, for example, we want to know when the statement $\exists x \; [s(x) \rightarrow t(x)]$ is true, then we look at the table for $\exists x \; p(x)$ and use the information provided there. The table tells us that $\exists x \; [s(x) \rightarrow t(x)]$ is true when $s(a) \rightarrow t(a)$ is true for some (at least one) a in the prescribed universe.

We will look further into quantified statements involving more than one open statement. Before doing so, however, we need to examine the following definition. This definition is comparable to Definitions 2.2 and 2.4 where we defined the ideas of logically equivalent statements and logical implication. It settles the same types of questions for open statements.

Table 2.21

Statement	When Is It True?	When Is It False?
$\exists x\ p(x)$	For some (at least one) a in the universe, $p(a)$ is true.	For every a in the universe, $p(a)$ is false.
$\forall x\ p(x)$	For every replacement a from the universe, $p(a)$ is true.	There is at least one replacement a from the universe for which $p(a)$ is false.
$\exists x\ \neg p(x)$	For at least one choice a in the universe, $p(a)$ is false, so its negation $\neg p(a)$ is true.	For every replacement a in the universe, $p(a)$ is true.
$\forall x\ \neg p(x)$	For every replacement a from the universe, $p(a)$ is false and its negation $\neg p(a)$ is true.	There is at least one replacement a from the universe for which $\neg p(a)$ is false and $p(a)$ is true.

Definition 2.6

Let $p(x), q(x)$ be open statements defined for a given universe.

The open statements $p(x)$ and $q(x)$ are called (*logically*) *equivalent*, and we write $\forall x\ [p(x) \Longleftrightarrow q(x)]$ when the biconditional $p(a) \leftrightarrow q(a)$ is true for each replacement a from the universe (that is, $p(a) \Longleftrightarrow q(a)$ for each a in the universe). If the implication $p(a) \rightarrow q(a)$ is true for each a in the universe (that is, $p(a) \Rightarrow q(a)$ for each a in the universe), then we write $\forall x\ [p(x) \Rightarrow q(x)]$ and say that $p(x)$ *logically implies* $q(x)$.

For the universe of all triangles in the plane, let $p(x), q(x)$ denote the open statements

$$p(x): \quad x \text{ is equiangular} \qquad q(x): \quad x \text{ is equilateral.}$$

Then for every particular triangle a (a replacement for x) we know that $p(a) \leftrightarrow q(a)$ is true (that is, $p(a) \Longleftrightarrow q(a)$, for every triangle in the plane). Consequently, $\forall x\ [p(x) \Longleftrightarrow q(x)]$.

Observe that here and, in general, $\forall x\ [p(x) \Longleftrightarrow q(x)]$ if and only if $\forall x\ [p(x) \Rightarrow q(x)]$ and $\forall x\ [q(x) \Rightarrow p(x)]$.

We also realize that a definition similar to Definition 2.6 can be given for two open statements that involve two or more variables.

Now we take another look at the logical equivalence of statements (not open statements) as we examine the converse, inverse, and contrapositive of a statement of the form $\forall x\ [p(x) \rightarrow q(x)]$.

Definition 2.7

For open statements $p(x), q(x)$ — defined for a prescribed universe — and the universally quantified statement $\forall x\ [p(x) \rightarrow q(x)]$, we define:

1) The *contrapositive* of $\forall x\ [p(x) \rightarrow q(x)]$ to be $\forall x\ [\neg q(x) \rightarrow \neg p(x)]$.
2) The *converse* of $\forall x\ [p(x) \rightarrow q(x)]$ to be $\forall x\ [q(x) \rightarrow p(x)]$.
3) The *inverse* of $\forall x\ [p(x) \rightarrow q(x)]$ to be $\forall x\ [\neg p(x) \rightarrow \neg q(x)]$.

The following two examples illustrate Definition 2.7.

EXAMPLE 2.40

For the universe of all quadrilaterals in the plane let $s(x)$ and $e(x)$ denote the open statements

$$s(x): \quad x \text{ is a square} \qquad e(x): \quad x \text{ is equilateral.}$$

a) The statement

$$\forall x \, [s(x) \rightarrow e(x)]$$

is a true statement and is logically equivalent to its contrapositive

$$\forall x \, [\neg e(x) \rightarrow \neg s(x)]$$

because $[s(a) \rightarrow e(a)] \Longleftrightarrow [\neg e(a) \rightarrow \neg s(a)]$ for each replacement a. Hence

$$\forall x \, [s(x) \rightarrow e(x)] \Longleftrightarrow \forall x \, [\neg e(x) \rightarrow \neg s(x)].$$

b) The statement

$$\forall x \, [e(x) \rightarrow s(x)]$$

is a false statement and is the converse of the true statement

$$\forall x \, [s(x) \rightarrow e(x)].$$

The false statement

$$\forall x \, [\neg s(x) \rightarrow \neg e(x)]$$

is the inverse of the given statement $\forall x \, [s(x) \rightarrow e(x)]$.

Since $[e(a) \rightarrow s(a)] \Longleftrightarrow [\neg s(a) \rightarrow \neg e(a)]$ for each specific quadrilateral a, we find that the converse and inverse are logically equivalent — that is,

$$\forall x \, [e(x) \rightarrow s(x)] \Longleftrightarrow \forall x \, [\neg s(x) \rightarrow \neg e(x)].$$

EXAMPLE 2.41

Here $p(x)$ and $q(x)$ are the open statements

$$p(x): \quad |x| > 3 \qquad q(x): \quad x > 3$$

and the universe consists of all real numbers.

a) The statement $\forall x \, [p(x) \rightarrow q(x)]$ is a false statement. For example, if $x = -5$, then $p(-5)$ is true while $q(-5)$ is false. Consequently, $p(-5) \rightarrow q(-5)$ is false, and so is $\forall x \, [p(x) \rightarrow q(x)]$.

b) We can express the converse of the given statement [in part (a)] as follows:

> Every real number greater than 3 has magnitude
> (or, absolute value) greater than 3.

In symbolic form this true statement is written $\forall x \, [q(x) \rightarrow p(x)]$.

c) The inverse of the given statement is also a true statement. In symbolic form we have $\forall x \, [\neg p(x) \rightarrow \neg q(x)]$, which can be expressed in words by

> If the magnitude of a real number is less than or equal to 3,
> then the number itself is less than or equal to 3.

And this is logically equivalent to the (converse) statement given in part (b).

d) Here the contrapositive of the statement in part (a) is given by $\forall x \, [\neg q(x) \rightarrow \neg p(x)]$. This false statement is logically equivalent to $\forall x \, [p(x) \rightarrow q(x)]$ and can be expressed as follows:

If a real number is less than or equal to 3, then so is its magnitude.

e) Together with $p(x)$ and $q(x)$ as above, consider the open statement

$$r(x): \quad x < -3,$$

which is also defined for the universe of all real numbers. The following four statements are all true:

Statement:	$\forall x \, [p(x) \to (r(x) \vee q(x))]$
Contrapositive:	$\forall x \, [\neg(r(x) \vee q(x)) \to \neg p(x)]$
Converse:	$\forall x \, [(r(x) \vee q(x)) \to p(x)]$
Inverse:	$\forall x \, [\neg p(x) \to \neg(r(x) \vee q(x))]$

In this case (because the statement and its converse are both true) we find that the statement $\forall x \, [p(x) \leftrightarrow (r(x) \vee q(x))]$ is true.

Now we use the results of Table 2.21 once again as we examine the next example.

EXAMPLE 2.42

Here the universe consists of all the integers, and the open statements $r(x)$, $s(x)$ are given by

$$r(x): \quad 2x + 1 = 5 \qquad s(x): \quad x^2 = 9.$$

We see that the statement $\exists x \, [r(x) \wedge s(x)]$ is false because there is no one integer a such that $2a + 1 = 5$ and $a^2 = 9$. However, there is an integer $b \, (= 2)$ such that $2b + 1 = 5$, and there is a second integer $c \, (= 3 \text{ or } -3)$ such that $c^2 = 9$. Therefore the statement $\exists x \, r(x) \wedge \exists x \, s(x)$ is true. Consequently, the existential quantifier $\exists x$ does not distribute over the logical connective $\wedge$. This one counterexample is enough to show that

$$\exists x \, [r(x) \wedge s(x)] \not\Leftrightarrow [\exists x \, r(x) \wedge \exists x \, s(x)],$$

where $\not\Leftrightarrow$ is read "is *not* logically equivalent to." It also demonstrates that

$$[\exists x \, r(x) \wedge \exists x \, s(x)] \not\Rightarrow \exists x \, [r(x) \wedge s(x)],$$

where $\not\Rightarrow$ is read "does *not* logically imply." So the statement

$$[\exists x \, r(x) \wedge \exists x \, s(x)] \to \exists x \, [r(x) \wedge s(x)]$$

is *not* a tautology.

What, however, can we say about the converse of a quantified statement of this form? At this point we present a general argument for *any* (arbitrary) open statements $p(x), q(x)$ and *any* (arbitrary) prescribed universe.

Examining the statement

$$\exists x \, [p(x) \wedge q(x)] \to [\exists x \, p(x) \wedge \exists x \, q(x)],$$

we find that when the hypothesis $\exists x \, [p(x) \wedge q(x)]$ is true, there is at least one element c in the universe for which the statement $p(c) \wedge q(c)$ is true. By the Rule of Conjunctive Simplification (see Section 2.3), $[p(c) \wedge q(c)] \Rightarrow p(c)$. From the truth of $p(c)$ we have the true statement $\exists x \, p(x)$. Similarly we obtain $\exists x \, q(x)$, another true statement. So $\exists x \, p(x) \wedge$

$\exists x \, q(x)$ is a true statement. Since $\exists x \, p(x) \wedge \exists x \, q(x)$ is true whenever $\exists x \, [p(x) \wedge q(x)]$ is true, it follows that

$$\exists x \, [p(x) \wedge q(x)] \Rightarrow [\exists x \, p(x) \wedge \exists x \, q(x)].$$

Arguments similar to the one for Example 2.42 provide the logical equivalences and logical implications listed in Table 2.22. In addition to those listed in Table 2.22 many other logical equivalences and logical implications can be derived.

Table 2.22 Logical Equivalences and Logical Implications for Quantified Statements in One Variable

For a prescribed universe and any open statements $p(x)$, $q(x)$ in the variable x:

$$\exists x \, [p(x) \wedge q(x)] \Rightarrow [\exists x \, p(x) \wedge \exists x \, q(x)]$$

$$\exists x \, [p(x) \vee q(x)] \Longleftrightarrow [\exists x \, p(x) \vee \exists x \, q(x)]$$

$$\forall x \, [p(x) \wedge q(x)] \Longleftrightarrow [\forall x \, p(x) \wedge \forall x \, q(x)]$$

$$[\forall x \, p(x) \vee \forall x \, q(x)] \Rightarrow \forall x \, [p(x) \vee q(x)]$$

Our next example lists several of these and demonstrates how two of them are verified.

EXAMPLE 2.43

Let $p(x)$, $q(x)$, and $r(x)$ denote open statements for a given universe. We find the following logical equivalences. (Many more are also possible.)

1) $\forall x \, [p(x) \wedge (q(x) \wedge r(x))] \Longleftrightarrow \forall x \, [(p(x) \wedge q(x)) \wedge r(x)]$
To show that this statement is a logical equivalence we proceed as follows:
For each a in the universe, consider the *statements* $p(a) \wedge (q(a) \wedge r(a))$ and $(p(a) \wedge q(a)) \wedge r(a)$. By the Associative Law for $\wedge$, we have

$$p(a) \wedge (q(a) \wedge r(a)) \Longleftrightarrow (p(a) \wedge q(a)) \wedge r(a).$$

Consequently, for the *open statements* $p(x) \wedge (q(x) \wedge r(x))$ and $(p(x) \wedge q(x)) \wedge r(x)$, it follows that

$$\forall x \, [p(x) \wedge (q(x) \wedge r(x))] \Longleftrightarrow \forall x \, [(p(x) \wedge q(x)) \wedge r(x)].$$

2) $\exists x \, [p(x) \rightarrow q(x)] \Longleftrightarrow \exists x \, [\neg p(x) \vee q(x)]$
For each c in the universe, it follows from Example 2.7 that

$$[p(c) \rightarrow q(c)] \Longleftrightarrow [\neg p(c) \vee q(c)].$$

Therefore the statement $\exists x \, [p(x) \rightarrow q(x)]$ is true (respectively, false) if and only if the statement $\exists x \, [\neg p(x) \vee q(x)]$ is true (respectively, false), so

$$\exists x \, [p(x) \rightarrow q(x)] \Longleftrightarrow \exists x \, [\neg p(x) \vee q(x)].$$

3) Other logical equivalences that we shall often find useful include the following.
   a) $\forall x \, \neg\neg p(x) \Longleftrightarrow \forall x \, p(x)$
   b) $\forall x \, \neg[p(x) \wedge q(x)] \Longleftrightarrow \forall x \, [\neg p(x) \vee \neg q(x)]$
   c) $\forall x \, \neg[p(x) \vee q(x)] \Longleftrightarrow \forall x \, [\neg p(x) \wedge \neg q(x)]$

4) The results for the logical equivalences in 3(a), (b), and (c) remain valid when all of the universal quantifiers are replaced by existential quantifiers.

The results of Tables 2.21 and 2.22 and Examples 2.42 and 2.43 will now help us with a very important concept. How do we negate quantified statements that involve a single variable?

Consider the statement $\forall x\ p(x)$. Its negation — namely, $\neg[\forall x\ p(x)]$ — can be stated as "It is not the case that for all x, $p(x)$ holds." This is not a very useful remark, so we consider $\neg[\forall x\ p(x)]$ further. When $\neg[\forall x\ p(x)]$ is true, then $\forall x\ p(x)$ is false, and so for some replacement a from the universe $\neg p(a)$ is true and $\exists x\ \neg p(x)$ is true. Conversely, whenever the statement $\exists x\ \neg p(x)$ is true we know that $\neg p(b)$ is true for some member b of the universe. Hence $\forall x\ p(x)$ is false and $\neg[\forall x\ p(x)]$ is true. So the statement $\neg[\forall x\ p(x)]$ is true if and only if the statement $\exists x\ \neg p(x)$ is true. (Similar considerations also tell us that $\neg[\forall x\ p(x)]$ is false if and only if $\exists x\ \neg p(x)$ is false.)

These observations lead to the following rule for negating the statement $\forall x\ p(x)$:

$$\neg[\forall x\ p(x)] \Longleftrightarrow \exists x\ \neg p(x).$$

In a similar way, Table 2.21 shows us that the statement $\exists x\ p(x)$ is true (false) precisely when the statement $\forall x\ \neg p(x)$ is false (true). This observation then motivates a rule for negating the statement $\exists x\ p(x)$:

$$\neg[\exists x\ p(x)] \Longleftrightarrow \forall x\ \neg p(x).$$

These two rules for negation, and two others that follow from them, are given in Table 2.23 for convenient reference.

Table 2.23 Rules for Negating Statements with One Quantifier

$$\neg[\forall x\ p(x)] \Longleftrightarrow \exists x\ \neg p(x)$$
$$\neg[\exists x\ p(x)] \Longleftrightarrow \forall x\ \neg p(x)$$
$$\neg[\forall x\ \neg p(x)] \Longleftrightarrow \exists x\ \neg\neg p(x) \Longleftrightarrow \exists x\ p(x)$$
$$\neg[\exists x\ \neg p(x)] \Longleftrightarrow \forall x\ \neg\neg p(x) \Longleftrightarrow \forall x\ p(x)$$

We use the rules for negating quantified statements in the following example.

EXAMPLE 2.44

Here we find the negation of two statements, where the universe comprises all of the integers.

1) Let $p(x)$ and $q(x)$ be given by

$$p(x):\quad x \text{ is odd} \qquad q(x):\quad x^2 - 1 \text{ is even.}$$

The statement "If x is odd, then $x^2 - 1$ is even" can be symbolized as $\forall x\ [p(x) \to q(x)]$. (This is a true statement.)

The negation of this statement is determined as follows:

$$\neg[\forall x\ (p(x) \to q(x))] \Longleftrightarrow \exists x\ [\neg(p(x) \to q(x))]$$
$$\Longleftrightarrow \exists x\ [\neg(\neg p(x) \vee q(x))] \Longleftrightarrow \exists x\ [\neg\neg p(x) \wedge \neg q(x)]$$
$$\Longleftrightarrow \exists x\ [p(x) \wedge \neg q(x)]$$

In words, the negation says, "There exists an integer x such that x is odd and $x^2 - 1$ is odd (that is, not even)." (This statement is false.)

2) As in Example 2.42, let $r(x)$ and $s(x)$ be the open statements

$$r(x): \quad 2x + 1 = 5 \qquad s(x): \quad x^2 = 9.$$

The quantified statement $\exists x \, [r(x) \wedge s(x)]$ is false because it asserts the existence of at least one integer a such that $2a + 1 = 5$ ($a = 2$) and $a^2 = 9$ ($a = 3$ or -3). Consequently, its negation

$$\neg[\exists x \, (r(x) \wedge s(x))] \Longleftrightarrow \forall x \, [\neg(r(x) \wedge s(x))] \Longleftrightarrow \forall x \, [\neg r(x) \vee \neg s(x)]$$

is true. This negation may be given in words as "For every integer x, $2x + 1 \neq 5$ or $x^2 \neq 9$."

Because a mathematical statement may involve more than one quantifier, we continue this section by offering some examples and making some observations on these types of statements.

EXAMPLE 2.45

Here we have two real variables x, y, so the universe consists of all real numbers. The commutative law for the addition of real numbers may be expressed by

$$\forall x \, \forall y \, (x + y = y + x).$$

This statement may also be given as

$$\forall y \, \forall x \, (x + y = y + x).$$

Likewise, in the case of the multiplication of real numbers, we may write

$$\forall x \, \forall y \, (xy = yx) \quad \text{or} \quad \forall y \, \forall x \, (xy = yx).$$

These two examples suggest the following general result. If $p(x, y)$ is an open statement in the two variables x, y (with either a prescribed universe for both x and y or one prescribed universe for x and a second for y), then the statements $\forall x \, \forall y \, p(x, y)$ and $\forall y \, \forall x \, p(x, y)$ are logically equivalent — that is, the statement $\forall x \, \forall y \, p(x, y)$ is true (respectively, false) if and only if the statement $\forall y \, \forall x \, p(x, y)$ is true (respectively, false). Hence

$$\forall x \, \forall y \, p(x, y) \Longleftrightarrow \forall y \, \forall x \, p(x, y).$$

EXAMPLE 2.46

When dealing with the associative law for the addition of real numbers, we find that for all real numbers x, y, and z,

$$x + (y + z) = (x + y) + z.$$

Using universal quantifiers (with the universe of all real numbers), we may express this by

$$\forall x \, \forall y \, \forall z \, [x + (y + z) = (x + y) + z] \quad \text{or} \quad \forall y \, \forall x \, \forall z \, [x + (y + z) = (x + y) + z].$$

In fact, there are $3! = 6$ ways to order these three universal quantifiers, and all six of these quantified statements are logically equivalent to one another.

This is actually true for all open statements $p(x, y, z)$, and to shorten the notation, one may write, for example,

$$\forall x, y, z \ p(x, y, z) \Longleftrightarrow \forall y, x, z \ p(x, y, z) \Longleftrightarrow \forall x, z, y \ p(x, y, z),$$

describing the logical equivalence for three of the six statements.

In Examples 2.45 and 2.46 we encountered quantified statements with two and three bound variables — each such variable bound by a universal quantifier. Our next example examines a situation in which there are two bound variables — and this time each of these variables is bound by an existential quantifier.

EXAMPLE 2.47

For the universe of all integers, consider the true statement "There exist integers x, y such that $x + y = 6$." We may represent this in symbolic form by

$$\exists x \, \exists y \, (x + y = 6).$$

If we let $p(x, y)$ denote the open statement "$x + y = 6$," then an equivalent statement can be given by $\exists y \, \exists x \, p(x, y)$.

In general, for any open statement $p(x, y)$ and universe(s) prescribed for the variables x, y,

$$\exists x \, \exists y \, p(x, y) \Longleftrightarrow \exists y \, \exists x \, p(x, y).$$

Similar results follow for statements involving three or more such quantifiers.

When a statement involves both existential and universal quantifiers, however, we must be careful about the order in which the quantifiers are written. Example 2.48 illustrates this case.

EXAMPLE 2.48

We restrict ourselves here to the universe of all integers and let $p(x, y)$ denote the open statement "$x + y = 17$."

1) The statement

$$\forall x \, \exists y \, p(x, y)$$

says that "For every integer x, there exists an integer y such that $x + y = 17$." (We read the quantifiers from left to right.)

This statement is true; once we select *any* x, the integer $y = 17 - x$ does *exist* and $x + y = x + (17 - x) = 17$. But we realize that each value of x gives rise to a different value of y.

2) Now consider the statement

$$\exists y \, \forall x \, p(x, y).$$

This statement is read "There exists an integer y so that for all integers x, $x + y = 17$." This statement is false. Once *an* integer y is selected, the *only* value that x can have (and still satisfy $x + y = 17$) is $17 - y$.

If the statement $\exists y \, \forall x \, p(x, y)$ were true, then every integer (x) would equal $17 - y$ (for some one fixed y). This says, in effect, that all integers are equal!

Consequently, the statements $\forall x \, \exists y \, p(x, y)$ and $\exists y \, \forall x \, p(x, y)$ are generally not logically equivalent.

Translating mathematical statements — be they postulates, definitions, or theorems — into symbolic form can be helpful for two important reasons.

1) Doing so forces us to be very careful and precise about the meanings of statements, the meanings of phrases such as "For all x" and "There exists an x," and the order in which such phrases appear.

2) After we translate a mathematical statement into symbolic form, the rules we have learned should then apply when we want to determine such related statements as the negation or, if appropriate, the contrapositive, converse, or inverse.

Our last two examples illustrate this, and in so doing, extend the results in Table 2.23.

EXAMPLE 2.49

Let $p(x, y)$, $q(x, y)$, and $r(x, y)$ represent three open statements, with replacements for the variables x, y chosen from some prescribed universe(s). What is the negation of the following statement?

$$\forall x \; \exists y \; [(p(x, y) \wedge q(x, y)) \to r(x, y)]$$

We find that

$$\neg[\forall x \; \exists y \; [(p(x, y) \wedge q(x, y)) \to r(x, y)]]$$
$$\Longleftrightarrow \exists x \; [\neg\exists y \; [(p(x, y) \wedge q(x, y)) \to r(x, y)]]$$
$$\Longleftrightarrow \exists x \; \forall y \; \neg[(p(x, y) \wedge q(x, y)) \to r(x, y)]$$
$$\Longleftrightarrow \exists x \; \forall y \; \neg[\neg[p(x, y) \wedge q(x, y)] \vee r(x, y)]$$
$$\Longleftrightarrow \exists x \; \forall y \; [\neg\neg[p(x, y) \wedge q(x, y)] \wedge \neg r(x, y)]$$
$$\Longleftrightarrow \exists x \; \forall y \; [(p(x, y) \wedge q(x, y)) \wedge \neg r(x, y)].$$

Now suppose that we are trying to establish the validity of an argument (or a mathematical theorem) for which

$$\forall x \; \exists y \; [(p(x, y) \wedge q(x, y)) \to r(x, y)]$$

is the conclusion. Should we want to try to prove the result by the method of Proof by Contradiction, we would assume as an additional premise the negation of this conclusion. Consequently, our additional premise would be the statement

$$\exists x \; \forall y \; [(p(x, y) \wedge q(x, y)) \wedge \neg r(x, y)].$$

Finally, we consider how to negate the definition of *limit*, a fundamental concept in calculus.

EXAMPLE 2.50

In calculus, one studies the properties of real-valued functions of a real variable. (Functions will be examined in Chapter 5 of this text.) Among these properties is the existence of limits, and one finds the following definition: Let I be an open interval[†] containing the real number a and suppose the function f is defined throughout I, except possibly at a. We say that f has the *limit* L as x approaches a, and write $\lim_{x \to a} f(x) = L$, if (and only if) for every $\epsilon > 0$ there exists a $\delta > 0$ so that, for all x in I, $(0 < |x - a| < \delta) \to (|f(x) - L| < \epsilon)$. This can be expressed in symbolic form as

$$\lim_{x \to a} f(x) = L \Longleftrightarrow \forall \epsilon > 0 \; \exists \delta > 0 \; \forall x \; [(0 < |x - a| < \delta) \to (|f(x) - L| < \epsilon)].$$

[†]The concept of an open interval is defined at the end of Section 3.1.

[Here the universe comprises the real numbers in the open interval I, except possibly a. Also, the quantifiers $\forall \epsilon > 0$ and $\exists \delta > 0$ now contain some restrictive information.] Then, to negate this definition, we do the following (in which certain steps have been combined):

$$\lim_{x \to a} f(x) \neq L$$

$$\Longleftrightarrow \neg[\forall \epsilon > 0 \ \exists \delta > 0 \ \forall x \ [(0 < |x - a| < \delta) \to (|f(x) - L| < \epsilon)]]$$

$$\Longleftrightarrow \exists \epsilon > 0 \ \forall \delta > 0 \ \exists x \ \neg[(0 < |x - a| < \delta) \to (|f(x) - L| < \epsilon)]$$

$$\Longleftrightarrow \exists \epsilon > 0 \ \forall \delta > 0 \ \exists x \ \neg[\neg(0 < |x - a| < \delta) \lor (|f(x) - L| < \epsilon)]$$

$$\Longleftrightarrow \exists \epsilon > 0 \ \forall \delta > 0 \ \exists x \ [\neg\neg(0 < |x - a| < \delta) \land \neg(|f(x) - L| < \epsilon)]$$

$$\Longleftrightarrow \exists \epsilon > 0 \ \forall \delta > 0 \ \exists x \ [(0 < |x - a| < \delta) \land (|f(x) - L| \geq \epsilon)]$$

Translating into words, we find that $\lim_{x \to a} f(x) \neq L$ if (and only if) there exists a positive (real) number ϵ such that for every positive (real) number δ, there is an x in I such that $0 < |x - a| < \delta$ (that is, $x \neq a$ and its distance from a is less than δ) but $|f(x) - L| \geq \epsilon$ [that is, the value of $f(x)$ differs from L by at least ϵ].

EXERCISES 2.4

1. Let $p(x), q(x)$ denote the following open statements.

$$p(x): \quad x \leq 3 \qquad q(x): \quad x + 1 \text{ is odd}$$

If the universe consists of all integers, what are the truth values of the following statements?

a) $q(1)$ **b)** $\neg p(3)$ **c)** $p(7) \lor q(7)$

d) $p(3) \land q(4)$ **e)** $\neg(p(-4) \lor q(-3))$

f) $\neg p(-4) \land \neg q(-3)$

2. Let $p(x), q(x)$ be defined as in Exercise 1. Let $r(x)$ be the open statement "$x > 0$." Once again the universe comprises all integers.

a) Determine the truth values of the following statements.

   i) $p(3) \lor [q(3) \lor \neg r(3)]$
  ii) $p(2) \to [q(2) \to r(2)]$
 iii) $[p(2) \land q(2)] \to r(2)$
  iv) $p(0) \to [\neg q(-1) \leftrightarrow r(1)]$

b) Determine all values of x for which $[p(x) \land q(x)] \land r(x)$ results in a true statement.

3. Let $p(x)$ be the open statement "$x^2 = 2x$," where the universe comprises all integers. Determine whether each of the following statements is true or false.

a) $p(0)$ **b)** $p(1)$ **c)** $p(2)$

d) $p(-2)$ **e)** $\exists x \ p(x)$ **f)** $\forall x \ p(x)$

4. Consider the universe of all polygons with three or four sides, and define the following open statements for this universe.

$a(x)$:   all interior angles of x are equal

$e(x)$: x is an equilateral triangle

$h(x)$:   all sides of x are equal

$i(x)$: x is an isosceles triangle

$p(x)$: x has an interior angle that exceeds $180°$

$q(x)$: x is a quadrilateral

$r(x)$: x is a rectangle

$s(x)$: x is a square

$t(x)$: x is a triangle

Translate each of the following statements into an English sentence, and determine whether the statement is true or false.

a) $\forall x \ [q(x) \veebar t(x)]$ **b)** $\forall x \ [i(x) \to e(x)]$

c) $\exists x \ [t(x) \land p(x)]$ **d)** $\forall x \ [(a(x) \land t(x)) \leftrightarrow e(x)]$

e) $\exists x \ [q(x) \land \neg r(x)]$ **f)** $\exists x \ [r(x) \land \neg s(x)]$

g) $\forall x \ [h(x) \to e(x)]$ **h)** $\forall x \ [t(x) \to \neg p(x)]$

i) $\forall x \ [s(x) \leftrightarrow (a(x) \land h(x))]$

j) $\forall x \ [t(x) \to (a(x) \leftrightarrow h(x))]$

5. Professor Carlson's class in mechanics is comprised of 29 students of which exactly

1) three physics majors are juniors;

2) two electrical engineering majors are juniors;

3) four mathematics majors are juniors;

4) twelve physics majors are seniors;

5) four electrical engineering majors are seniors;

6) two electrical engineering majors are graduate students; and

7) two mathematics majors are graduate students.

Consider the following open statements.

$c(x)$:   Student x is in the class (that is, Professor Carlson's mechanics class as already described).

$j(x)$: Student x is a junior.

$s(x)$: Student x is a senior.

$g(x)$: Student x is a graduate student.

$p(x)$: Student x is a physics major.

$e(x)$: Student x is an electrical engineering major.

$m(x)$: Student x is a mathematics major.

Write each of the following statements in terms of quantifiers and the open statements $c(x)$, $j(x)$, $s(x)$, $g(x)$, $p(x)$, $e(x)$, and $m(x)$, and determine whether the given statement is true or false. Here the universe comprises all of the 12,500 students enrolled at the university where Professor Carlson teaches. Furthermore, at this university each student has only one major.

a) There is a mathematics major in the class who is a junior.

b) There is a senior in the class who is not a mathematics major.

c) Every student in the class is majoring in mathematics or physics.

d) No graduate student in the class is a physics major.

e) Every senior in the class is majoring in either physics or electrical engineering.

6. Let $p(x, y), q(x, y)$ denote the following open statements.

$$p(x, y): \quad x^2 \geq y \qquad q(x, y): \quad x + 2 < y$$

If the universe for each of x, y consists of all real numbers, determine the truth value for each of the following statements.

a) $p(2, 4)$

b) $q(1, \pi)$

c) $p(-3, 8) \wedge q(1, 3)$

d) $p\left(\frac{1}{2}, \frac{1}{3}\right) \vee \neg q(-2, -3)$

e) $p(2, 2) \rightarrow q(1, 1)$

f) $p(1, 2) \leftrightarrow \neg q(1, 2)$

7. For the universe of all integers, let $p(x), q(x), r(x), s(x),$ and $t(x)$ be the following open statements.

$p(x)$: $x > 0$

$q(x)$: x is even

$r(x)$: x is a perfect square

$s(x)$: x is (exactly) divisible by 4

$t(x)$: x is (exactly) divisible by 5

a) Write the following statements in symbolic form.

  i) At least one integer is even.
  ii) There exists a positive integer that is even.
  iii) If x is even, then x is not divisible by 5.
  iv) No even integer is divisible by 5.
  v) There exists an even integer divisible by 5.
  vi) If x is even and x is a perfect square, then x is divisible by 4.

b) Determine whether each of the six statements in part (a) is true or false. For each false statement, provide a counterexample.

c) Express each of the following symbolic representations in words.

  i) $\forall x\,[r(x) \rightarrow p(x)]$
  ii) $\forall x\,[s(x) \rightarrow q(x)]$
  iii) $\forall x\,[s(x) \rightarrow \neg t(x)]$
  iv) $\exists x\,[s(x) \wedge \neg r(x)]$

d) Provide a counterexample for each false statement in part (c).

8. Let $p(x)$, $q(x)$, and $r(x)$ denote the following open statements.

$$p(x): \quad x^2 - 8x + 15 = 0$$
$$q(x): \quad x \text{ is odd}$$
$$r(x): \quad x > 0$$

For the universe of all integers, determine the truth or falsity of each of the following statements. If a statement is false, give a counterexample.

a) $\forall x\,[p(x) \rightarrow q(x)]$

b) $\forall x\,[q(x) \rightarrow p(x)]$

c) $\exists x\,[p(x) \rightarrow q(x)]$

d) $\exists x\,[q(x) \rightarrow p(x)]$

e) $\exists x\,[r(x) \rightarrow p(x)]$

f) $\forall x\,[\neg q(x) \rightarrow \neg p(x)]$

g) $\exists x\,[p(x) \rightarrow (q(x) \wedge r(x))]$

h) $\forall x\,[(p(x) \vee q(x)) \rightarrow r(x)]$

9. Let $p(x), q(x),$ and $r(x)$ be the following open statements.

$$p(x): \quad x^2 - 7x + 10 = 0$$
$$q(x): \quad x^2 - 2x - 3 = 0$$
$$r(x): \quad x < 0$$

a) Determine the truth or falsity of the following statements, where the universe is all integers. If a statement is false, provide a counterexample or explanation.

  i) $\forall x\,[p(x) \rightarrow \neg r(x)]$
  ii) $\forall x\,[q(x) \rightarrow r(x)]$
  iii) $\exists x\,[q(x) \rightarrow r(x)]$
  iv) $\exists x\,[p(x) \rightarrow r(x)]$

b) Find the answers to part (a) when the universe consists of all positive integers.

c) Find the answers to part (a) when the universe contains only the integers 2 and 5.

10. For the following program segment, m and n are integer variables. The variable A is a two-dimensional array $A[1, 1]$, $A[1, 2], \ldots, A[1, 20], \ldots, A[10, 1], \ldots, A[10, 20]$, with 10 rows (indexed from 1 to 10) and 20 columns (indexed from 1 to 20).

```
for m := 1 to 10 do
    for n := 1 to 20 do
        A[m, n] := m + 3 * n
```

Write the following statements in symbolic form. (The universe for the variable m contains only the integers from 1 to 10 inclusive; for n the universe consists of the integers from 1 to 20 inclusive.)

a) All entries of A are positive.

b) All entries of A are positive and less than or equal to 70.

c) Some of the entries of A exceed 60.

d) The entries in each row of A are sorted into (strictly) ascending order.

e) The entries in each column of A are sorted into (strictly) ascending order.

f) The entries in the first three rows of A are distinct.

11. Identify the bound variables and the free variables in each of the following expressions (or statements). In both cases the universe comprises all real numbers.

a) $\forall y \,\exists z \,[\cos(x + y) = \sin(z - x)]$

b) $\exists x \,\exists y \,[x^2 - y^2 = z]$

12. a) Let $p(x, y)$ denote the open statement "x divides y," where the universe for each of the variables x, y comprises all integers. (In this context "divides" means "exactly divides" or "divides evenly.") Determine the truth value of each of the following statements; if a quantified statement is false, provide an explanation or a counterexample.

    i) $p(3, 7)$ **ii)** $p(3, 27)$

    iii) $\forall y \, p(1, y)$ **iv)** $\forall x \, p(x, 0)$

    v) $\forall x \, p(x, x)$ **vi)** $\forall y \,\exists x \, p(x, y)$

    vii) $\exists y \,\forall x \, p(x, y)$

    viii) $\forall x \,\forall y \,[(p(x, y) \wedge p(y, x)) \rightarrow (x = y)]$

b) Determine which of the eight statements in part (a) will change in truth value if the universe for each of the variables x, y were restricted to just the positive integers.

c) Determine the truth value of each of the following statements. If the statement is false, provide an explanation or a counterexample. [The universe for each of x, y is as in part (b).]

    i) $\forall x \,\exists y \, p(x, y)$ **ii)** $\forall y \,\exists x \, p(x, y)$

    iii) $\exists x \,\forall y \, p(x, y)$ **iv)** $\exists y \,\forall x \, p(x, y)$

13. Suppose that $p(x, y)$ is an open statement where the universe for each of x, y consists of only three integers: 2, 3, 5. Then the quantified statement $\exists y \, p(2, y)$ is logically equivalent to $p(2, 2) \vee p(2, 3) \vee p(2, 5)$. The quantified statement $\exists x \,\forall y \, p(x, y)$ is logically equivalent to $[p(2, 2) \wedge p(2, 3) \wedge p(2, 5)] \vee [p(3, 2) \wedge p(3, 3) \wedge p(3, 5)] \vee [p(5, 2) \wedge p(5, 3) \wedge p(5, 5)]$. Use conjunctions and/or disjunctions to express the following statements without quantifiers.

    a) $\forall x \, p(x, 3)$ **b)** $\exists x \,\exists y \, p(x, y)$ **c)** $\forall y \,\exists x \, p(x, y)$

14. Let $p(n), q(n)$ represent the open statements

$$p(n): \quad n \text{ is odd} \qquad q(n): \quad n^2 \text{ is odd}$$

for the universe of all integers. Which of the following statements are logically equivalent to each other?

a) If the square of an integer is odd, then the integer is odd.

b) $\forall n \, [p(n)$ is necessary for $q(n)]$

c) The square of an odd integer is odd.

d) There are some integers whose squares are odd.

e) Given an integer whose square is odd, that integer is likewise odd.

f) $\forall n \, [\neg p(x) \rightarrow \neg q(n)]$

g) $\forall n \, [p(n)$ is sufficient for $q(n)]$

15. For each of the following pairs of statements determine whether the proposed negation is correct. If correct, determine which is true: the original statement or the proposed negation. If the proposed negation is wrong, write a correct version of the negation and then determine whether the original statement or your corrected version of the negation is true.

a) Statement: For all real numbers x, y, if $x^2 > y^2$, then $x > y$.
Proposed negation: There exist real numbers x, y such that $x^2 > y^2$ but $x \le y$.

b) Statement: There exist real numbers x, y such that x and y are rational but $x + y$ is irrational.
Proposed negation: For all real numbers x, y, if $x + y$ is rational, then each of x, y is rational.

c) Statement: For all real numbers x, if x is not 0, then x has a multiplicative inverse.
Proposed negation: There exists a nonzero real number that does not have a multiplicative inverse.

d) Statement: There exist odd integers whose product is odd.
Proposed negation: The product of any two odd integers is odd.

16. Write the negation of each of the following statements as an English sentence — without symbolic notation. (Here the universe consists of all the students at the university where Professor Lenhart teaches.)

a) Every student in Professor Lenhart's C++ class is majoring in computer science or mathematics.

b) At least one student in Professor Lenhart's C++ class is a history major.

17. Write the negation of each of the following true statements. For parts (a) and (b) the universe consists of all integers; for parts (c) and (d) the universe comprises all real numbers.

a) For all integers n, if n is not (exactly) divisible by 2, then n is odd.

b) If k, m, n are any integers where $k - m$ and $m - n$ are odd, then $k - n$ is even.

c) If x is a real number where $x^2 > 16$, then $x < -4$ or $x > 4$.

d) For all real numbers x, if $|x - 3| < 7$, then $-4 < x < 10$.

18. Negate and simplify each of the following.

    a) $\exists x \, [p(x) \vee q(x)]$ **b)** $\forall x \, [p(x) \wedge \neg q(x)]$

    c) $\forall x \, [p(x) \rightarrow q(x)]$

    d) $\exists x \, [(p(x) \vee q(x)) \rightarrow r(x)]$

19. For each of the following statements state the converse, inverse, and contrapositive. Also determine the truth value for each given statement, as well as the truth values for its converse, inverse, and contrapositive. (Here "divides" means "exactly divides.")

a) [The universe comprises all positive integers.]
If $m > n$, then $m^2 > n^2$.

b) [The universe comprises all integers.]
If $a > b$, then $a^2 > b^2$.

c) [The universe comprises all integers.]
If m divides n and n divides p, then m divides p.

d) [The universe consists of all real numbers.]
$\forall x \, [(x > 3) \rightarrow (x^2 > 9)]$

e) [The universe consists of all real numbers.]
For all real numbers x, if $x^2 + 4x - 21 > 0$, then $x > 3$ or $x < -7$.

20. Rewrite each of the following statements in the *if-then* form. Then write the converse, inverse, and contrapositive of your implication. For each result in parts (a) and (c) give the truth value for the implication and the truth values for its converse, inverse, and contrapositive. [In part (a) "divisibility" requires a remainder of 0.]

a) [The universe comprises all positive integers.]
Divisibility by 21 is a sufficient condition for divisibility by 7.

b) [The universe comprises all snakes presently slithering about the jungles of Asia.]
Being a cobra is a sufficient condition for a snake to be dangerous.

c) [The universe consists of all complex numbers.]
For every complex number z, z being real is necessary for z^2 to be real.

21. For the following statements the universe comprises all nonzero integers. Determine the truth value of each statement.

a) $\exists x \, \exists y \, [xy = 1]$ **b)** $\exists x \, \forall y \, [xy = 1]$

c) $\forall x \, \exists y \, [xy = 1]$

d) $\exists x \, \exists y \, [(2x + y = 5) \wedge (x - 3y = -8)]$

e) $\exists x \, \exists y \, [(3x - y = 7) \wedge (2x + 4y = 3)]$

22. Answer Exercise 21 for the universe of all nonzero real numbers.

23. In the arithmetic of real numbers, there is a real number, namely 0, called the identity of addition because $a + 0 = 0 + a = a$ for every real number a. This may be expressed in symbolic form by

$$\exists z \, \forall a \, [a + z = z + a = a].$$

(We agree that the universe comprises all real numbers.)

a) In conjunction with the existence of an additive identity is the existence of additive inverses. Write a quantified statement that expresses "Every real number has an additive inverse." (Do not use the minus sign anywhere in your statement.)

b) Write a quantified statement dealing with the existence of a multiplicative identity for the arithmetic of real numbers.

c) Write a quantified statement covering the existence of multiplicative inverses for the nonzero real numbers. (Do not use the exponent -1 anywhere in your statement.)

d) Do the results in parts (b) and (c) change in any way when the universe is restricted to the integers?

24. Consider the quantified statement $\forall x \, \exists y \, [x + y = 17]$. Determine whether this statement is true or false for each of the following universes: (a) the integers; (b) the positive integers; (c) the integers for x, the positive integers for y; (d) the positive integers for x, the integers for y.

25. Let the universe for the variables in the following statements consist of all real numbers. In each case negate and simplify the given statement.

a) $\forall x \, \forall y \, [(x > y) \rightarrow (x - y > 0)]$

b) $\forall x \, \forall y \, [(x < y) \rightarrow \exists z \, (x < z < y)]$

c) $\forall x \, \forall y \, [(|x| = |y|) \rightarrow (y = \pm x)]$

26. In calculus the definition of the limit L of a sequence of real numbers $r_1, r_2, r_3, \ldots$ can be given as

$$\lim_{n \to \infty} r_n = L$$

if (and only if) for every $\epsilon > 0$ there exists a positive integer k so that for all integers n, if $n > k$ then $|r_n - L| < \epsilon$.

In symbolic form this can be expressed as

$$\lim_{n \to \infty} r_n = L \iff \forall \epsilon > 0 \; \exists k > 0 \; \forall n \, [(n > k) \rightarrow |r_n - L| < \epsilon].$$

Express $\lim_{n \to \infty} r_n \neq L$ in symbolic form.

2.5
Quantifiers, Definitions, and the Proofs of Theorems

In this section we shall combine some of the ideas we have already studied in the prior two sections. Although Section 2.3 introduced rules and methods for establishing the validity of an argument, unfortunately the arguments presented there seemed to have little to do with anything mathematical. [The rare exceptions are in Example 2.23 and the erroneous argument in part (b) of the material preceding Example 2.26.] Most of the arguments dealt with certain individuals and predicaments they were either in or about to face.

But now that we have learned some of the properties of quantifiers and quantified statements, we are better equipped to handle arguments that will help us to prove mathematical theorems. Before dealing with theorems, however, we shall consider how mathematical definitions are traditionally presented in scientific writing.

Following Example 2.3 in Section 2.1, the discussion concerned how an implication might be used in place of a biconditional in everyday conversation. But in scientific writing, it was noted, we should avoid any and all situations where an ambiguous interpretation might come about — in particular, an implication should not be used when a biconditional is intended. However, there is one major exception to that rule and it concerns the way that mathematical definitions are traditionally presented in mathematics textbooks and other scientific literature. Example 2.51 demonstrates this exception.

EXAMPLE 2.51

a) Let us start with the universe of all quadrilaterals in the plane and try to identify those that are called rectangles.

One person might say that

"If a quadrilateral is a rectangle then it has four equal angles."

Another individual might identify these special quadrilaterals by observing that

"If a quadrilateral has four equal angles, then it is a rectangle."

(Here both people are making implicitly quantified statements, where the quantifier is universal.)

Given the open statements

$p(x)$: x is a rectangle $q(x)$: x has four equal angles,

we can express what the first person says as

$$\forall x \, [p(x) \rightarrow q(x)],$$

while for the second person we would write

$$\forall x \, [q(x) \rightarrow p(x)].$$

So which of the preceding (quantified) statements identifies or defines a rectangle? Perhaps we feel that they both do. But how can that be, since one statement is the converse of the other and, in general, the converse of an implication is *not* logically equivalent to the implication.

Here the reader must consider what is intended — not just what each of the two people has said, or the symbolic expressions we have written to represent these statements. In this situation each person is using an implication with the meaning of a biconditional. They are both intending (though not stating)

$$\forall x \, [p(x) \leftrightarrow q(x)],$$

— that is, each is really telling us that

"A quadrilateral is a rectangle *if and only if* it has four equal angles."

b) Within the universe of all integers we can distinguish the even integers by means of a certain property and so we may define them as follows:

For every integer n we call n even if it is divisible by 2.

(By the expression "divisible by 2" we mean "exactly divisible by 2" — that is, there is no remainder upon division of the dividend n by the divisor 2.)

If we consider the open statements

$$p(n): \quad n \text{ is an even integer} \qquad q(n): \quad n \text{ is divisible by 2},$$

then it *appears* that the preceding definition may be written symbolically as

$$\forall n \, [q(n) \rightarrow p(n)].$$

After all, the given quantified statement (in the preceding definition) is an implication. However, the situation here is similar to that given in part (a). What appears to be stated is not what is intended. The intention is for the reader to interpret the given definition as

$$\forall n \, [q(n) \leftrightarrow p(n)],$$

that is,

"For every integer n, we call n even *if and only if* n is divisible by 2."

(Note that the open statement "n is divisible by 2" can also be expressed by the *open* statement "$n = 2k$, for some integer k." Don't be misled here by the presence of the quantifier "for some integer k" — for the expression $\exists k \, [n = 2k]$ is still an open statement because n remains a free variable.)

So now we see how quantifiers may enter into the way we state mathematical definitions — and that the traditional way in which such a definition appears is as an implication. But beware and remember: It is *only in definitions* that an implication can be (mis)read and correctly interpreted as a biconditional.

Note how we defined the limit concept in Example 2.50. There we wrote "if (and only if)" since we wanted to let the reader know our intention. Now we are free to replace "if (and only if)" by simply "if."

Having settled our discussion on the nature of mathematical definitions, we continue now with an investigation of arguments involving quantified statements.

| EXAMPLE 2.52 |

Suppose that we start with the universe that comprises only the 13 integers 2, 4, 6, 8, . . . , 24, 26. Then we can establish the statement:

For all n (meaning $n = 2, 4, 6, \ldots, 26$),

we can write n as the sum of at most three perfect squares.

The results in Table 2.24 provide a case-by-case verification showing the given (quantified) statement to be true. (We might call this statement a theorem.)

Table 2.24

$2 = 1 + 1$	$10 = 9 + 1$	$20 = 16 + 4$
$4 = 4$	$12 = 4 + 4 + 4$	$22 = 9 + 9 + 4$
$6 = 4 + 1 + 1$	$14 = 9 + 4 + 1$	$24 = 16 + 4 + 4$
$8 = 4 + 4$	$16 = 16$	$26 = 25 + 1$
	$18 = 16 + 1 + 1$	

This exhaustive listing is an example of a proof using the technique we call, rather appropriately, the *method of exhaustion*. This method is reasonable when we are dealing with a fairly small universe. If we are confronted with a situation in which the universe is larger but within the range of a computer that is available to us, then we might write a program to check all of the individual cases.

(Note that for certain cases in Table 2.24 more than one answer may be possible. For example, we could have written $18 = 9 + 9$ and $26 = 16 + 9 + 1$. But this is all right. We were told that each positive even integer less than or equal to 26 could be written as the sum of one, two, or three perfect squares. We were *not* told that each such representation had to be unique, so more than one possibility could occur. What we had to check in each case was that there was at least one possibility.)

In the previous example we mentioned the word *theorem*. We also found this term used in Chapter 1 — for example, in results like the binomial theorem and the multinomial theorem where we were introduced to certain types of enumeration problems. Without getting overly technical, we shall consider *theorems* to be statements of mathematical interest, statements that are known to be true. Sometimes the term *theorem* is used only to describe major results that have many and varied consequences. Certain of these consequences that follow rather immediately from a theorem are termed *corollaries* (as in the case of Corollary 1.1 in Section 1.3). In this text, however, we shall not be so particular in our use of the word theorem.

Example 2.52 is a nice starting point to examine the proof of a quantified statement. Unfortunately, a great number of mathematical statements and theorems often deal with universes that do not lend themselves to the method of exhaustion. When faced with establishing or proving a result for all integers, for example, or for all real numbers, then we cannot use a case-by-case method like the one in Example 2.52. So what can we do?

We start by considering the following rule.

The Rule of Universal Specification: If an open statement becomes true for *all* replacements by the members in a given universe, then that open statement is true for *each specific* individual member in that universe. (A bit more symbolically — if $p(x)$ is an open statement for a given universe, and if $\forall x \ p(x)$ is true, then $p(a)$ is true for each a in the universe.)

This rule indicates that the truth of an open statement in one particular instance follows (as a special case) from the more general (for the entire universe) truth of that universally quantified open statement. The following examples will show us how to apply this idea.

EXAMPLE 2.53

a) For the universe of all people, consider the open statements

$m(x)$: x is a mathematics professor $c(x)$: x has studied calculus.

Now consider the following argument.

All mathematics professors have studied calculus.

Leona is a mathematics professor.

Therefore Leona has studied calculus.

If we let l represent this particular woman (in our universe) named Leona, then we can rewrite this argument in symbolic form as

$$\forall x \, [m(x) \rightarrow c(x)]$$
$$\underline{m(l)}$$
$$\therefore c(l)$$

Here the two statements above the line are the premises of the argument, and the statement $c(l)$ below the line is its conclusion. This is comparable to what we saw in Section 2.3, except now we have a premise that is a universally quantified statement. As was the case in Section 2.3, the premises are all assumed to be true and we must try to establish that the conclusion is also true under these circumstances. Now, to establish the validity of the given argument, we proceed as follows.

Steps	**Reasons**
1) $\forall x \, [m(x) \rightarrow c(x)]$	Premise
2) $m(l)$	Premise
3) $m(l) \rightarrow c(l)$	Step (1) and the Rule of Universal Specification
4) $\therefore c(l)$	Steps (2) and (3) and the Rule of Detachment

Note that the statements in steps (2) and (3) are *not* quantified statements. They are the types of statements we studied earlier in the chapter. In particular, we can apply the rules of inference we learned in Section 2.3 to these two statements to deduce the conclusion in step (4).

We see here that the Rule of Universal Specification enables us to take a universally quantified premise and deduce from it an ordinary statement (that is, one that is not quantified). This (ordinary) statement — namely, $m(l) \rightarrow c(l)$ — is one specific true instance of the universally quantified true premise $\forall x \, [m(x) \rightarrow c(x)]$.

b) For an example of a more mathematical nature let us consider the universe of all triangles in the plane in conjunction with the open statements

$$p(t): \quad t \text{ has two sides of equal length.}$$
$$q(t): \quad t \text{ is an isosceles triangle.}$$
$$r(t): \quad t \text{ has two angles of equal measure.}$$

Let us also focus our attention on one specific triangle with no two angles of equal measure. This triangle will be called triangle XYZ and will be designated by c. Then we find that the argument

In triangle XYZ there is no pair of angles of equal measure.	$\neg r(c)$
If a triangle has two sides of equal length, then it is isosceles.	$\forall t \, [p(t) \rightarrow q(t)]$
If a triangle is isosceles, then it has two angles of equal measure.	$\underline{\forall t \, [q(t) \rightarrow r(t)]}$
Therefore triangle XYZ has no two sides of equal length.	$\therefore \neg p(c)$

is a valid one — as evidenced by the following.

Steps	**Reasons**
1) $\forall t\, [p(t) \to q(t)]$	Premise
2) $p(c) \to q(c)$	Step (1) and the Rule of Universal Specification
3) $\forall t\, [q(t) \to r(t)]$	Premise
4) $q(c) \to r(c)$	Step (3) and the Rule of Universal Specification
5) $p(c) \to r(c)$	Steps (2) and (4) and the Law of the Syllogism
6) $\neg r(c)$	Premise
7) $\therefore \neg p(c)$	Steps (5) and (6) and Modus Tollens

Once again we see how the Rule of Universal Specification helps us. Here it has taken the universally quantified statements at steps (1) and (3) and has provided us with the (ordinary) statements at steps (2) and (4), respectively. Then at this point we were able to apply the rules of inference we learned in Section 2.3 (namely, the Law of the Syllogism and Modus Tollens) to derive the conclusion $\neg p(c)$ in step (7).

c) Now for one last argument to drive the point home! Here we'll consider the universe to be made up of the entire student body at a particular college. One specific student, Mary Gusberti, will be designated by m.

For this universe and the open statements

$$j(x): \quad x \text{ is a junior} \qquad s(x): \quad x \text{ is a senior}$$

$$p(x): \quad x \text{ is enrolled in a physical education class}$$

we consider the argument:

No junior or senior is enrolled in a physical education class.

Mary Gusberti is enrolled in a physical education class.

Therefore Mary Gusberti is not a senior.

In symbolic form this argument becomes

$$\forall x\, [(j(x) \vee s(x)) \to \neg p(x)]$$
$$\underline{p(m)}$$
$$\therefore \neg s(m)$$

Now the following steps (and reasons) establish the validity of this argument.

Steps	**Reasons**
1) $\forall x\, [(j(x) \vee s(x)) \to \neg p(x)]$	Premise
2) $p(m)$	Premise
3) $(j(m) \vee s(m)) \to \neg p(m)$	Step (1) and the Rule of Universal Specification
4) $p(m) \to \neg(j(m) \vee s(m))$	Step (3), $(q \to t) \Longleftrightarrow (\neg t \to \neg q)$, and the Law of Double Negation
5) $p(m) \to (\neg j(m) \wedge \neg s(m))$	Step (4) and DeMorgan's Law
6) $\neg j(m) \wedge \neg s(m)$	Steps (2) and (5) and the Rule of Detachment (or Modus Ponens)
7) $\therefore \neg s(m)$	Step (6) and the Rule of Conjunctive Simplification

In Example 2.53 we have had our first opportunity to apply the Rule of Universal Specification. Using the rule in conjunction with Modus Ponens (or the Rule of Detachment) and

Modus Tollens, we are able to state the following corresponding analogs, each of which involves a universally quantified premise. In either case we consider a fixed universe that includes a specific member c and make use of the open statements $p(x)$, $q(x)$ defined for this universe.

$$
\begin{array}{ll}
(1) \quad \forall x\,[p(x) \to q(x)] & (2) \quad \forall x\,[p(x) \to q(x)] \\
\qquad \underline{p(c)} & \qquad \underline{\neg q(c)} \\
\qquad \therefore q(c) & \qquad \therefore \neg p(c)
\end{array}
$$

These two valid arguments are presented here for the same reason we presented them for the rules of inference — Modus Ponens and Modus Tollens — in Section 2.3 (in the discussion between Examples 2.25 and 2.26). We want to examine some possible errors that may arise when the results in (1) and (2) are not used correctly.

Let us start with the universe of all polygons in the plane. Within this universe we shall let c denote one specific polygon — the quadrilateral $EFGH$, where the measure of angle E is 91°. For the open statements

$$p(x): \quad x \text{ is a square} \qquad q(x): \quad x \text{ has four sides,}$$

the following argument is *invalid*.

(**1′**)   All squares have four sides.

          Quadrilateral $EFGH$ has four sides.

          Therefore quadrilateral $EFGH$ is a square.

In symbolic form this argument translates into

$$
(\mathbf{1''}) \qquad\qquad
\begin{array}{l}
\forall x\,[p(x) \to q(x)] \\
\underline{q(c)} \\
\therefore\, p(c)
\end{array}
$$

Unfortunately, although the premises are true, the conclusion is false. (For a square has no angle of measure 91°.) We admit that there might be some confusion between this argument and the valid one in (1) above. For when we apply the Rule of Universal Specification to the quantified premise in (1″), in this instance we arrive at the *invalid* argument

$$
\begin{array}{l}
p(c) \to q(c) \\
\underline{q(c)} \\
\therefore\, p(c)
\end{array}
$$

And here, as in Section 2.3, the error in reasoning lies in our attempt to argue by the converse.

A second invalid argument — from the misuse of argument (2) above — can also be given, as shown in the following.

(**2′**)   All squares have four sides.

          Quadrilateral $EFGH$ is not a square.

          Therefore quadrilateral $EFGH$ does not have four sides.

Translating (2′) into symbolic form results in

$$
(\mathbf{2''}) \qquad\qquad
\begin{array}{l}
\forall x\,[p(x) \to q(x)] \\
\underline{\neg p(c)} \\
\therefore\, \neg q(c)
\end{array}
$$

This time the Rule of Universal Specification leads us to

$$p(c) \rightarrow q(c)$$
$$\underline{\neg p(c)}$$
$$\therefore \neg q(c)$$

where the fallacy arises because we are trying to argue by the inverse.

And now let us look back at the three parts of Example 2.53. Although the arguments presented there involved premises that were universally quantified statements, there was never any instance where a universally quantified statement appeared in the conclusion. We now want to remedy this situation, since many theorems in mathematics have the form of a universally quantified statement. To do so we need the following considerations.

Start with a given universe and the open statement $p(x)$. To establish the truth of the statement $\forall x \; p(x)$, we must establish the truth of $p(c)$ for each member c in the given universe. But if the universe has many members or, for example, contains all the positive integers, then this exhaustive, if not exhausting, task of validating the truth of each $p(c)$ becomes difficult, if not impossible. To get around this situation we shall prove that $p(c)$ is true — but now we do it for the case where c denotes a *specific but arbitrarily chosen* member from the prescribed universe.

Should the preceding open statement $p(x)$ have the form $q(x) \rightarrow r(x)$, for open statements $q(x)$ and $r(x)$, then we shall assume the truth of $q(c)$ as an additional premise and try to deduce the truth of $r(c)$ — by using definitions, axioms, previously proven theorems, and the logical principles we have studied. For when $q(c)$ is false, the implication $q(c) \rightarrow r(c)$ is true, regardless of the truth value of $r(c)$.

The reason that the element c must be arbitrary (or generic) is to make sure that what we do and prove about c is applicable for *all* the other elements in the universe. If we are dealing with the universe of all integers, for example, we cannot choose c in an arbitrary manner by selecting c as 4, or by selecting c as an even integer. In general, we cannot make any assumptions about our choice for c unless these assumptions are valid for *all* the other elements of the universe. The word *generic* is applied to the element c here because it indicates that our choice (for c) must share all of the common characteristics of the elements for the given universe.

The principle we have described in the preceding three paragraphs is named and summarized as follows.

> **The Rule of Universal Generalization:** If an open statement $p(x)$ is proved to be true when x is replaced by any *arbitrarily chosen* element c from our universe, then the universally quantified statement $\forall x \; p(x)$ is true. Furthermore, the rule extends beyond a single variable. So if, for example, we have an open statement $q(x, y)$ that is proved to be true when x and y are replaced by *arbitrarily chosen* elements from the same universe, or their own respective universes, then the universally quantified statement $\forall x \; \forall y \; q(x, y)$ [or, $\forall x, y \; q(x, y)$] is true. Similar results hold for the cases of three or more variables.

Before we demonstrate the use of this rule in any examples, we wish to look back at part (1) of Example 2.43 in Section 2.4. It turns out that the explanation given there to establish that

$$\forall x \; [p(x) \wedge (q(x) \wedge r(x))] \iff \forall x \; [(p(x) \wedge q(x)) \wedge r(x)]$$

anticipated what we have now described in detail as the Rules of Universal Specification and Universal Generalization.

Now we'll turn to an example which is strictly symbolic. This example provides an opportunity to apply the Rule of Universal Generalization.

EXAMPLE 2.54

Let $p(x)$, $q(x)$, and $r(x)$ be open statements that are defined for a given universe. We show that the argument

$$\forall x \,[p(x) \to q(x)]$$
$$\frac{\forall x \,[q(x) \to r(x)]}{\therefore \forall x \,[p(x) \to r(x)]}$$

is valid by considering the following.

Steps	Reasons
1) $\forall x \,[(p(x) \to q(x)]$	Premise
2) $p(c) \to q(c)$	Step (1) and the Rule of Universal Specification
3) $\forall x \,[q(x) \to r(x)]$	Premise
4) $q(c) \to r(c)$	Step (3) and the Rule of Universal Specification
5) $p(c) \to r(c)$	Steps (2) and (4) and the Law of the Syllogism
6) $\therefore \forall x \,[p(x) \to r(x)]$	Step (5) and the Rule of Universal Generalization

Here the element c introduced in steps (2) and (4) is the same *specific* but *arbitrarily chosen* element from the universe. Since this element c has no *special or distinguishing properties* but does share all of the common features of every other element in this universe, we can use the Rule of Universal Generalization to go from step (5) to step (6).

And so at last we have dealt with a valid argument where a universally quantified statement appears as the conclusion, as well as among the premises.

The question that now may be at the back of the reader's mind is one of practicality. Namely, when would we ever need to use the argument that we validated in Example 2.54? We may find that we have already used it (perhaps, unknowingly) in earlier algebra and geometry courses, as we demonstrate in the following example.

EXAMPLE 2.55

a) For the universe of all real numbers, consider the open statements

$$p(x): \quad 3x - 7 = 20 \qquad q(x): \quad 3x = 27 \qquad r(x): \quad x = 9.$$

The following solution of an algebraic equation parallels the valid argument from Example 2.54.

1) If $3x - 7 = 20$, then $3x = 27$. $\forall x \,[p(x) \to q(x)]$
2) If $3x = 27$, then $x = 9$. $\forall x \,[q(x) \to r(x)]$
3) Therefore, if $3x - 7 = 20$, then $x = 9$. $\therefore \forall x \,[p(x) \to r(x)]$

b) When we dealt with the universe of all quadrilaterals in plane geometry, we may have found ourselves relating something like this:

"Since every square is a rectangle, and every rectangle
is a parallelogram, it follows that every square is a parallelogram."

In this case we are using the argument in Example 2.54 for the open statements

$p(x): \quad x$ is a square $q(x): \quad x$ is a rectangle $r(x): \quad x$ is a parallelogram.

Now we continue with one more argument to validate.

EXAMPLE 2.56

The steps and reasons needed to establish the validity of the argument

$$\forall x \, [p(x) \vee q(x)]$$
$$\underline{\forall x \, [(\neg p(x) \wedge q(x)) \rightarrow r(x)]}$$
$$\therefore \forall x \, [\neg r(x) \rightarrow p(x)]$$

are given as follows. [Here the element c is in the universe assigned for the argument. Also, since the conclusion is a universally quantified implication, we can assume $\neg r(c)$ as an additional premise — as was mentioned earlier when the Rule of Universal Generalization was first introduced.]

Steps	**Reasons**
1) $\forall x \, [p(x) \vee q(x)]$	Premise
2) $p(c) \vee q(c)$	Step (1) and the Rule of Universal Specification
3) $\forall x \, [(\neg p(x) \wedge q(x)) \rightarrow r(x)]$	Premise
4) $[\neg p(c) \wedge q(c)] \rightarrow r(c)$	Step (3) and the Rule of Universal Specification
5) $\neg r(c) \rightarrow \neg[\neg p(c) \wedge q(c)]$	Step (4) and $s \rightarrow t \Longleftrightarrow \neg t \rightarrow \neg s$
6) $\neg r(c) \rightarrow [p(c) \vee \neg q(c)]$	Step (5), DeMorgan's Law, and the Law of Double Negation
7) $\neg r(c)$	Premise (assumed)
8) $p(c) \vee \neg q(c)$	Steps (7) and (6) and Modus Ponens
9) $[p(c) \vee q(c)] \wedge [p(c) \vee \neg q(c)]$	Steps (2) and (8) and the Rule of Conjunction
10) $p(c) \vee [q(c) \wedge \neg q(c)]$	Step (9) and the Distributive Law of $\vee$ over $\wedge$
11) $p(c)$	Step (10), $q(c) \wedge \neg q(c) \Longleftrightarrow F_0$, and $p(c) \vee F_0 \Longleftrightarrow p(c)$
12) $\therefore \forall x \, [\neg r(x) \rightarrow p(x)]$	Steps (7) and (11) and the Rule of Universal Generalization

Before going on we want to point out a convention that the reader may not like but will have to get used to. It concerns our coverage of the Rules of Universal Specification and Universal Generalization. In the first case we started with the statement $\forall x \, p(x)$ and then dealt with $p(c)$ for some specific element c in our universe. For the Rule of Universal Generalization we obtained the truth of $\forall x \, p(x)$ from that of $p(c)$, where c was arbitrarily selected from the universe. Unfortunately, we'll often find ourselves using the letter x instead of c to denote the element — but as long as we understand what is happening we shall soon find the convention easy enough to work with.

The results of Example 2.54 and especially Example 2.56 lead us to believe that we can use universally quantified statements and the rules of inference — including the Rules of Universal Specification and Universal Generalization — to formalize and prove a variety of arguments and, hopefully, theorems. When we do so it appears that the validation of some rather short arguments requires quite a number of steps, because we have been very meticulous and included all the steps and reasons — we left little, if anything, to the imagination. The reader should rest assured that when we start to prove mathematical theorems, we shall present the proofs in the more conventional paragraph style. We shall no longer mention each and every application of the laws of logic and the other tautologies or the rules of inference. On occasion we may single out a certain rule of inference, but our attention will be primarily directed to the use of definitions, mathematical axioms and principles (other than those we have found in our study of logic), and other (earlier) theorems we have been able to prove. Why then have we been learning all of this material on validating arguments? Because it will provide us with a framework to fall back on whenever we doubt whether a given attempt at a proof really does the job. If in doubt, we have our study of logic to supply us with a somewhat mechanical but strictly objective means to help us decide.

And now we present paragraph-style proofs for some results about the integers. (These results may be considered rather obvious to us — in fact, we may find we have already seen and used some of them. But they provide an excellent setting for writing some simple proofs.) The proofs we shall presently introduce use the following ideas, which we now formally define. [The first idea was mentioned earlier in part (b) of Example 2.51.]

Definition 2.8   Let n be an integer. We call n *even* if n is divisible by 2 — that is, if there exists an integer r so that $n = 2r$. If n is not even, then we call n *odd* and find for this case that there exists an integer s where $n = 2s + 1$.

THEOREM 2.2   For all integers k and l, if k, l are both odd, then $k + l$ is even.

Proof: In this proof we shall number the steps so that we may refer to them for some later remarks. After this we shall no longer number the steps.

1) Since k and l are odd, we may write $k = 2a + 1$ and $l = 2b + 1$, for some integers a, b. This is due to Definition 2.8.

2) Then

$$k + l = (2a + 1) + (2b + 1) = 2(a + b + 1),$$

by virtue of the Commutative and Associative Laws of Addition and the Distributive Law of Multiplication over Addition — all of which hold for integers.

3) Since a, b are integers, $a + b + 1 = c$ is an integer; with $k + l = 2c$, it follows from Definition 2.8 that $k + l$ is even.

Remarks

1) In step (1) of the preceding proof k and l were chosen in an arbitrary manner, so we know by the Rule of Universal Generalization that the result obtained is true for all odd integers.

2) Although we may not realize it, we are using the Rule of Universal Specification (twice) in step (1). The first argument implicit in this step reads as follows.

   i) If n is an odd integer, then $n = 2r + 1$ for some integer r.
   ii) The integer k is a specific (but arbitrarily chosen) odd integer.
   iii) Therefore we may write $k = 2a + 1$ for some (specific) integer a.

3) In step (1) we do not have $k = 2a + 1$ and $l = 2a + 1$. Since k, l are arbitrarily chosen, it *may* be the case that $k = l$ — and when this happens we have $2a + 1 = k = l = 2b + 1$, from which it follows that $a = b$. [Since k may not equal l, it follows

that $(k-1)/2 = a$ may not equal $b = (l-1)/2$. Thus we should use the different variables a and b.]

Before we proceed with another theorem — written in the more conventional manner — let us examine the following.

EXAMPLE 2.57	Consider the following statement for the universe of integers.

$$\text{If } n \text{ is an integer, then } n^2 = n \text{ — or, } \forall n \ [n^2 = n].$$

Now for $n = 0$ it is true that $n^2 = 0^2 = 0 = n$. And if $n = 1$, it is also true that $n^2 = 1^2 = 1 = n$. However, we *cannot* conclude $n^2 = n$ for every integer n. The Rule of Universal Generalization does *not* apply here, for we cannot consider the choice of 0 (or 1) as an arbitrarily chosen integer. If $n = 2$, we have $n^2 = 4 \neq 2 = n$, and this one counterexample is enough to tell us that the given statement is false. However, either replacement — namely, $n = 0$ or $n - 1$ — is enough to establish the truth of the statement:

$$\text{For some integer } n, \ n^2 = n \text{ — or, } \exists n \ [n^2 = n].$$

We close — at last — with three results to demonstrate how we shall write proofs throughout the remainder of the text.

THEOREM 2.3	For all integers k and l, if k and l are both odd, then their product kl is also odd.

Proof: Since k and l are both odd, we may write $k = 2a + 1$ and $l = 2b + 1$, for some integers a and b — because of Definition 2.8. Then the product $kl = (2a + 1)(2b + 1) = 4ab + 2a + 2b + 1 = 2(2ab + a + b) + 1$, where $2ab + a + b$ is an integer. Therefore, by Definition 2.8 once again, it follows that kl is odd.

The preceding proof is an example of a direct proof. In our next example we shall prove a result in three ways: first by a direct argument (or proof), then by the contrapositive method, and finally by the method of proof by contradiction. [For the (method of) proof by contradiction we put in some extra details, since this is our first opportunity to use this technique.] The reader should not assume, however, that every theorem can be so readily proved in a variety of ways.

THEOREM 2.4	If m is an even integer, then $m + 7$ is odd.

Proof:

1) Since m is even, we have $m = 2a$ for some integer a. Then $m + 7 = 2a + 7 = 2a + 6 + 1 = 2(a + 3) + 1$. Since $a + 3$ is an integer, we know that $m + 7$ is odd.

2) Suppose that $m + 7$ is not odd, hence even. Then $m + 7 = 2b$ for some integer b and $m = 2b - 7 = 2b - 8 + 1 = 2(b - 4) + 1$, where $b - 4$ is an integer. Hence m is odd. [The result follows because the statements $\forall m \ [p(m) \to q(m)]$ and $\forall m [\neg q(m) \to \neg p(m)]$ are logically equivalent.]

3) Now assume that m is even and that $m + 7$ is also even. (This assumption is the negation of what we want to prove.) Then $m + 7$ even implies that $m + 7 = 2c$ for some integer c. And, consequently, $m = 2c - 7 = 2c - 8 + 1 = 2(c - 4) + 1$ with $c - 4$ an integer, so m is odd. Now we have our contradiction. We started with m even and deduced m odd — an impossible situation, since no integer can be both even and odd. How did we arrive at this dilemma? Simple — we made a mistake! This mistake is the false assumption — namely, $m + 7$ is even — that we wanted to believe at the start of the proof. Since the assumption is false, its negation is true, and so we now have $m + 7$ odd.

The second and third proofs for Theorem 2.4 appear to be somewhat similar. This is because the contradiction we derived in the third proof arises from the hypothesis of the theorem and its negation. We shall see as we progress (as early as the next chapter) that a contradiction may also be obtained by deriving the negation of a known fact — a fact that is *not* the hypothesis of the theorem we are attempting to prove. For now, however, let us think about this similarity a little more. Suppose we start with the open statements $p(m)$ and $q(m)$ — for a prescribed universe — and consider a theorem of the form $\forall m \, [p(m) \rightarrow q(m)]$. If we try to prove this result by the contrapositive method, then we shall actually prove the logically equivalent statement $\forall m \, [\neg q(m) \rightarrow \neg p(m)]$. To do so we assume the truth of $\neg q(m)$ (for any specific but arbitrarily chosen m in the universe) and show that this leads to the truth of $\neg p(m)$. On the other hand, if we wish to prove the theorem $\forall m \, [p(m) \rightarrow q(m)]$ by the method of proof by contradiction, then we assume that the statement $\forall m \, [p(m) \rightarrow q(m)]$ is false. This amounts to the fact that $p(m) \rightarrow q(m)$ is false for at least one replacement for m from the universe — that is, there is some element m in the universe for which $p(m)$ is true and $q(m)$ is false [or $\neg q(m)$ is true]. We then use the truth of $p(m)$ and $\neg q(m)$ to derive a contradiction. [In the third proof of Theorem 2.4 we obtained $p(m) \wedge \neg p(m)$.] These two methods can be compared symbolically in the following — where m is specific but arbitrarily chosen for the method of contraposition.

	Assumption	**Result Derived**
Contraposition	$\neg q(m)$	$\neg p(m)$
Contradiction	$p(m)$ and $\neg q(m)$	F_0

In general, when we are able to establish a theorem by either a direct proof or an indirect proof, the direct approach is less cumbersome than an indirect approach. (This certainly appears to be the case for the three proofs presented for Theorem 2.4.) When we do not have any prescribed directions given for attempting the proof of a certain theorem, we might start with a direct approach. If we succeed, then all is well. If not, then we might consider trying to find a counterexample to what we thought was a theorem. Should our search for a counterexample fail, then we might consider an indirect approach. We might prove the contrapositive of the theorem, or obtain a contradiction, as we did in the third proof of Theorem 2.4, by assuming the truth of the hypothesis and the truth of the negation of the conclusion (for some element m in the universe) in the given theorem.

We close this section with one more indirect proof by the method of contraposition.

THEOREM 2.5   For all positive real numbers x and y, if the product xy exceeds 25, then $x > 5$ or $y > 5$.

Proof: Consider the negation of the conclusion — that is, suppose that $0 < x \leq 5$ and $0 < y \leq 5$. Under these circumstances we find that $0 = 0 \cdot 0 < x \cdot y \leq 5 \cdot 5 = 25$, so the product

xy does *not* exceed 25. (This indirect method of proof now establishes the given statement, since we know that an implication is logically equivalent to its contrapositive.)

EXERCISES 2.5

1. In Example 2.52 why did we stop at 26 and not at 28?

2. In Example 2.52 why didn't we include the odd integers between 2 and 26?

3. Use the method of exhaustion to show that every even integer between 30 and 58 (including 30 and 58) can be written as a sum of at most three perfect squares.

4. Let n be a positive integer greater than 1. We call n *prime* if the only positive integers that (exactly) divide n are 1 and n itself. For example, the first seven primes are 2, 3, 5, 7, 11, 13, and 17. (We shall learn more about primes in Chapter 4.) Use the method of exhaustion to show that every integer in the universe 4, 6, 8, . . . , 36, 38 can be written as the sum of two primes.

5. For each of the following (universes and) pairs of statements, use the Rule of Universal Specification, in conjunction with Modus Ponens and Modus Tollens, in order to fill in the blank line so that a valid argument results.

a) [The universe comprises all real numbers.]
All integers are rational numbers.
The real number π is not a rational number.
∴ _____

b) [The universe comprises the present population of the United States.]
All librarians know the Library of Congress Classification System.

∴ Margaret knows the Library of Congress Classification System.

c) [The same universe as in part (b).]

Sondra is an administrative director.
∴ Sondra knows how to delegate authority.

d) [The universe consists of all quadrilaterals in the plane.]
All rectangles are equiangular.

∴ Quadrilateral $MNPQ$ is not a rectangle.

6. Determine which of the following arguments are valid and which are invalid. Provide an explanation for each answer. (Let the universe consist of all people presently residing in the United States.)

a) All mail carriers carry a can of mace.
Mrs. Bacon is a mail carrier.
Therefore Mrs. Bacon carries a can of mace.

b) All law-abiding citizens pay their taxes.
Mr. Pelosi pays his taxes.
Therefore Mr. Pelosi is a law-abiding citizen.

c) All people who are concerned about the environment recycle their plastic containers.
Margarita is not concerned about the environment.
Therefore Margarita does not recycle her plastic containers.

7. For a prescribed universe and any open statements $p(x)$, $q(x)$ in the variable x, prove that

a) $\exists x\, [p(x) \vee q(x)] \Longleftrightarrow \exists x\, p(x) \vee \exists x\, q(x)$

b) $\forall x\, [p(x) \wedge q(x)] \Longleftrightarrow \forall x\, p(x) \wedge \forall x\, q(x)$

8. **a)** Let $p(x), q(x)$ be open statements in the variable x, with a given universe. Prove that

$$\forall x\, p(x) \vee \forall x\, q(x) \Rightarrow \forall x\, [p(x) \vee q(x)].$$

[That is, prove that when the statement $\forall x\, p(x) \vee \forall x\, q(x)$ is true, then the statement $\forall x\, [p(x) \vee q(x)]$ is true.]

b) Find a counterexample for the converse in part (a). That is, find open statements $p(x), q(x)$ and a universe such that $\forall x\, [p(x) \vee q(x)]$ is true, while $\forall x\, p(x) \vee \forall x\, q(x)$ is false.

9. Provide the reasons for the steps verifying the following argument. (Here a denotes a specific but arbitrarily chosen element from the given universe.)

$$\forall x\, [p(x) \rightarrow (q(x) \wedge r(x))]$$
$$\underline{\forall x\, [p(x) \wedge s(x)]}$$
$$\therefore \forall x\, [r(x) \wedge s(x)]$$

Steps	Reasons
1) $\forall x\, [p(x) \rightarrow (q(x) \wedge r(x))]$	
2) $\forall x\, [p(x) \wedge s(x)]$	
3) $p(a) \rightarrow (q(a) \wedge r(a))$	
4) $p(a) \wedge s(a)$	
5) $p(a)$	
6) $q(a) \wedge r(a)$	
7) $r(a)$	
8) $s(a)$	
9) $r(a) \wedge s(a)$	
10) $\therefore \forall x\, [r(x) \wedge s(x)]$	

10. Provide the missing reasons for the steps verifying the following argument:

$$\forall x\, [p(x) \vee q(x)]$$
$$\exists x\, \neg p(x)$$
$$\forall x\, [\neg q(x) \vee r(x)]$$
$$\underline{\forall x\, [s(x) \rightarrow \neg r(x)]}$$
$$\therefore \exists x\, \neg s(x)$$

Steps	Reasons
1) $\forall x\,[p(x) \vee q(x)]$	Premise
2) $\exists x\,\neg p(x)$	Premise
3) $\neg p(a)$	Step (2) and the definition of the truth for $\exists x\,\neg p(x)$. [Here a is an element (replacement) from the universe for which $\neg p(x)$ is true.] The reason for this step is also referred to as the *Rule of Existential Specification.*
4) $p(a) \vee q(a)$	
5) $q(a)$	
6) $\forall x\,[\neg q(x) \vee r(x)]$	
7) $\neg q(a) \vee r(a)$	
8) $q(a) \rightarrow r(a)$	
9) $r(a)$	
10) $\forall x\,[s(x) \rightarrow \neg r(x)]$	
11) $s(a) \rightarrow \neg r(a)$	
12) $r(a) \rightarrow \neg s(a)$	
13) $\neg s(a)$	
14) $\therefore\,\exists x\,\neg s(x)$	Step (13) and the definition of the truth for $\exists x\,\neg s(x)$. The reason for this step is also referred to as the *Rule of Existential Generalization.*

11. Write the following argument in symbolic form. Then either verify the validity of the argument or explain why it is invalid. [Assume here that the universe comprises all adults (18 or over) who are presently residing in the city of Las Cruces (in New Mexico). Two of these individuals are Roxe and Imogene.]

All credit union employees must know COBOL. All credit union employees who write loan applications must know Excel.[†] Roxe works for the credit union, but she doesn't know Excel. Imogene knows Excel but doesn't know COBOL. Therefore Roxe doesn't write loan applications and Imogene doesn't work for the credit union.

12. Give a direct proof (as in Theorem 2.3) for each of the following.

a) For all integers k and l, if k, l are both even, then $k + l$ is even.

b) For all integers k and l, if k, l are both even, then kl is even.

13. For each of the following statements provide an indirect proof [as in part (2) of Theorem 2.4] by stating and proving the contrapositive of the given statement.

a) For all integers k and l, if kl is odd, then k, l are both odd.

b) For all integers k and l, if $k + l$ is even, then k and l are both even or both odd.

14. Prove that for every integer n, if n is odd, then n^2 is odd.

15. Provide a proof by contradiction for the following: For every integer n, if n^2 is odd, then n is odd.

16. Prove that for every integer n, n^2 is even if and only if n is even.

17. Prove the following result in three ways (as in Theorem 2.4): If n is an odd integer, then $n + 11$ is even.

18. Let m, n be two positive integers. Prove that if m, n are perfect squares, then the product mn is also a perfect square.

19. Prove or disprove: If m, n are positive integers and m, n are perfect squares, then $m + n$ is a perfect square.

20. Prove or disprove: There exist positive integers m, n, where m, n, and $m + n$ are all perfect squares.

21. Prove that for all real numbers x and y, if $x + y \geq 100$, then $x \geq 50$ or $y \geq 50$.

22. Prove that for every integer n, $4n + 7$ is odd.

23. Let n be an integer. Prove that n is odd if and only if $7n + 8$ is odd.

24. Let n be an integer. Prove that n is even if and only if $31n + 12$ is even.

2.6
Summary and Historical Review

This second chapter has introduced some of the fundamentals of logic — in particular, some of the rules of inference and methods of proof necessary for establishing mathematical theorems.

The first systematic study of logical reasoning is found in the work of the Greek philosopher Aristotle (384–322 B.C.). In his treatises on logic Aristotle presented a collection of principles for deductive reasoning. These principles were designed to provide a foundation

[†] The Excel spreadsheet is a product of Microsoft, Inc.

for the study of all branches of knowledge. In a modified form, this type of logic was taught up to and throughout the Middle Ages.

Aristotle (384–322 B.C.)

The German mathematician Gottfried Wilhelm Leibniz (1646–1716) is often considered the first scholar who seriously pursued the development of symbolic logic as a universal scientific language. This he professed in his essay *De Arte Combinatoria*, published in 1666. His research in the area of symbolic logic, carried out from 1679 to 1690, gave considerable impetus to the creation of this mathematical discipline.

Following the work by Leibniz, little change took place until the nineteenth century, when the English mathematician George Boole (1815–1864) created a system of mathematical logic that he introduced in 1847 in the pamphlet *The Mathematical Analysis of Logic, Being an Essay Towards a Calculus of Deductive Reasoning*. In the same year, Boole's countryman Augustus DeMorgan (1806–1871) published *Formal Logic; or, the Calculus of Inference, Necessary and Probable*. In some ways this treatise extended Boole's work

George Boole (1815–1864)

considerably. Then, in 1854, Boole detailed his ideas and further research in the notable work *An Investigation in the Laws of Thought, on Which Are Founded the Mathematical Theories of Logic and Probability*. The American logician Charles Sanders Peirce (1839–1914), who was also an engineer and philosopher, introduced the formal concept of the *quantifier* into the study of symbolic logic.

The concepts formulated by Boole were thoroughly examined in the work of another German scholar, Ernst Schröder (1841–1902). These results are known collectively as *Vorlesungen über die Algebra der Logik*; they were published in the period from 1890 to 1895.

Further developments in the area saw an even more modern approach evolve in the work of the German logician Gottlieb Frege (1848–1925) between 1879 and 1903. This work significantly influenced the monumental *Principia Mathematica* (1910–1913) by England's Alfred North Whitehead (1861–1947) and Bertrand Russell (1872–1970). Here what was begun by Boole was finally brought to fruition. Thanks to this remarkable effort and the work of other twentieth-century mathematicians and logicians, in particular the comprehensive *Grundlagen der Mathematik* (1934–1939) of David Hilbert (1862–1943) and Paul Bernays (1888–1977), the more polished techniques of contemporary mathematical logic are now available.

Several sections of this chapter stressed the importance of proof. In mathematics a proof bestows authority on what might otherwise be dismissed as mere opinion. Proof embodies the power and majesty of pure reason. But even more than that, it suggests new mathematical ideas. Our concept of proof goes hand in hand with the notion of a *theorem* — a mathematical statement the truth of which has been confirmed by means of a logical argument, namely, a *proof*. For those who feel they can ignore the importance of logic and the rules of inference, we submit the following words of wisdom spoken by Achilles in Lewis Carroll's *What the Tortoise Said to Achilles*: "Then Logic would take you by the throat, and force you to do it!"

Comparable coverage of the material presented in this chapter can be found in Chapters 2 and 11 of the text by K. A. Ross and C. R. B. Wright [11]. The first two chapters of the text by S. S. Epp [3] provide many examples and some computer science applications for those who wish to see more on logic and proof at a very readable introductory level. The text by H. Delong [2] provides an historical survey of mathematical logic, together with an examination of the nature of its results and the philosophical consequences of these results. This is also the case with the texts by H. Eves and C. V. Newsom [4], R. R. Stoll [13], and R. L. Wilder [14], wherein the relationships among logic, proof, and set theory (the topic of our next chapter) are examined in their roles in the foundations of mathematics.

For more on resolution (introduced in Exercise 13 of Section 2.3) and automated reasoning, the reader should examine the texts by J. H. Gallier [6] and M. R. Genesereth and N. J. Nilsson [7].

The text by E. Mendelson [9] provides an interesting intermediate introduction for those readers who wish to pursue additional topics in mathematical logic. A somewhat more advanced treatment is given in the work of S. C. Kleene [8]. Accounts of other work in mathematical logic are presented in the compendium edited by J. Barwise [1].

The objective of the works by D. Fendel and D. Resek [5] and R. P. Morash [10] is to prepare the student with a calculus background for the more theoretical mathematics found in abstract algebra and real analysis. Each of these texts provides an excellent introduction to the basic methods of proof. The unique text by D. Solow [12] is devoted entirely to introducing the reader who has a background in high school mathematics to the primary techniques used in writing mathematical proofs.

REFERENCES

1. Barwise, Jon (editor). *Handbook of Mathematical Logic*. Amsterdam: North Holland, 1977.
2. Delong, Howard. *A Profile of Mathematical Logic*. Reading, Mass.: Addison-Wesley, 1970.
3. Epp, Susanna S. *Discrete Mathematics with Applications*, 2nd ed. Boston, Mass.: PWS Publishing Co., 1995.
4. Eves, Howard, and Newsom, Carroll V. *An Introduction to the Foundations and Fundamental Concepts of Mathematics*, rev. ed. New York: Holt, 1965.
5. Fendel, Daniel, and Resek, Diane. *Foundations of Higher Mathematics*. Reading, Mass.: Addison-Wesley, 1990.
6. Gallier, Jean H. *Logic for Computer Science*. New York: Harper & Row, 1986.
7. Genesereth, Michael R., and Nilsson, Nils J. *Logical Foundations of Artificial Intelligence*. Los Altos, Calif: Morgan Kaufmann, 1987.
8. Kleene, Stephen C. *Mathematical Logic*. New York: Wiley, 1967.
9. Mendelson, Elliott. *Introduction to Mathematical Logic*, 3rd ed. Monterey, Calif.: Wadsworth and Brooks/Cole, 1987.
10. Morash, Ronald P. *Bridge to Abstract Mathematics: Mathematical Proof and Structures*. New York: Random House/Birkhäuser, 1987.
11. Ross, Kenneth A., and Wright, Charles R. B. *Discrete Mathematics*, 4th ed. Upper Saddle River, N.J.: Prentice-Hall, 1999.
12. Solow, Daniel. *How to Read and Do Proofs*, 3rd ed. New York: Wiley, 2001.
13. Stoll, Robert R. *Set Theory and Logic*. San Francisco: Freeman, 1963.
14. Wilder, Raymond L. *Introduction to the Foundations of Mathematics*, 2nd ed. New York: Wiley, 1965.

SUPPLEMENTARY EXERCISES

1. Construct the truth table for

$$p \leftrightarrow [(q \wedge r) \rightarrow \neg(s \vee r)].$$

2. a) Construct the truth table for

$$(p \rightarrow q) \wedge (\neg p \rightarrow r).$$

   b) Translate the statement in part (a) into words such that the word "not" does not appear in the translation.

3. Let p, q, and r denote primitive statements. Prove or disprove (provide a counterexample for) each of the following.

   a) $[p \leftrightarrow (q \leftrightarrow r)] \Longleftrightarrow [(p \leftrightarrow q) \leftrightarrow r]$

   b) $[p \rightarrow (q \rightarrow r)] \Longleftrightarrow [(p \rightarrow q) \rightarrow r]$

4. Express the negation of the statement $p \leftrightarrow q$ in terms of the connectives $\wedge$ and $\vee$.

5. Write the following statement as an implication in two ways, each in the *if-then* form: Either Kaylyn practices her piano lessons or she will not go to the movies.

6. Let p, q, r denote primitive statements. Write the converse, inverse, and contrapositive of

   a) $p \rightarrow (q \wedge r)$ b) $(p \vee q) \rightarrow r$

7. a) For primitive statements p, q, find the dual of the statement $(\neg p \wedge \neg q) \vee (T_0 \wedge p) \vee p$.

   b) Use the laws of logic to show that your result from part (a) is logically equivalent to $p \wedge \neg q$.

8. Let p, q, r, and s be primitive statements. Write the dual of each of the following compound statements.

   a) $(p \vee \neg q) \wedge (\neg r \vee s)$

   b) $p \rightarrow (q \wedge \neg r \wedge s)$

   c) $[(p \vee T_0) \wedge (q \vee F_0)] \vee [r \wedge s \wedge T_0]$

9. For each of the following, fill in the blank with the word *converse*, *inverse*, or *contrapositive* so that the result is a true statement.

   a) The converse of the inverse of $p \rightarrow q$ is the _____ of $p \rightarrow q$.

   b) The converse of the inverse of $p \rightarrow q$ is the _____ of $q \rightarrow p$.

   c) The inverse of the converse of $p \rightarrow q$ is the _____ of $p \rightarrow q$.

   d) The inverse of the converse of $p \rightarrow q$ is the _____ of $q \rightarrow p$.

   e) The inverse of the contrapositive of $p \rightarrow q$ is the _____ of $p \rightarrow q$.

10. Establish the validity of the argument

$$[(p \rightarrow q) \wedge [(q \wedge r) \rightarrow s] \wedge r] \rightarrow (p \rightarrow s).$$

11. Prove or disprove each of the following, where p, q, and r are any statements.

a) $[(p \veebar q) \veebar r] \Longleftrightarrow [p \veebar (q \veebar r)]$

b) $[p \veebar (q \to r)] \Longleftrightarrow [(p \veebar q) \to (p \veebar r)]$

12. Write the following argument in symbolic form. Then either establish the validity of the argument or provide a counterexample to show that it is invalid.

> If it is cool this Friday, then Craig will wear his suede jacket if the pockets are mended. The forecast for Friday calls for cool weather, but the pockets have not been mended. Therefore Craig won't be wearing his suede jacket this Friday.

13. Consider the open statement

$$p(x, y): \quad y - x = y + x^2$$

where the universe for each of the variables x, y comprises all integers. Determine the truth value for each of the following statements.

a) $p(0, 0)$ **b)** $p(1, 1)$

c) $p(0, 1)$ **d)** $\forall y \; p(0, y)$

e) $\exists y \; p(1, y)$ **f)** $\forall x \; \exists y \; p(x, y)$

g) $\exists y \; \forall x \; p(x, y)$ **h)** $\forall y \; \exists x \; p(x, y)$

14. Determine whether each of the following statements is true or false. If false, provide a counterexample. The universe comprises all integers.

a) $\forall x \; \exists y \; \exists z \; (x = 7y + 5z)$

b) $\forall x \; \exists y \; \exists z \; (x = 4y + 6z)$

15. Suppose two opposite corner squares are removed from an 8×8 chessboard — as in part (a) of Fig. 2.4. Can the remaining 62 squares be covered by 31 dominos (rectangles consisting of two adjacent squares — one white and the other blue, as shown in the figure)? (When a domino is placed on the chessboard, a square of a given color need not be placed on a square of the same color.)

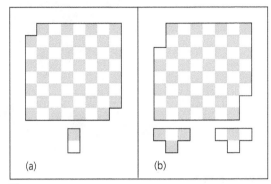

Figure 2.4

16. In part (b) of Fig. 2.4 we have an 8×8 chessboard where two squares (one blue and one white) have been removed from each of two opposite corners. Can the remaining 60 squares be covered by 15 T-shaped figures (of three white squares and one blue one, or three blue squares and one white one — as shown in the figure)? [The reader may wish to verify that a 4×4 chessboard (of all 16 squares) can be covered by four of the T-shaped figures. Then it follows that an 8×8 chessboard (of all 64 squares) can be covered by 16 of the T-shaped figures.]

3

Set Theory

Underlying the mathematics we study in algebra, geometry, combinatorics, probability, and almost every other area of contemporary mathematics is the notion of a set. Very often this concept provides an underlying structure for a concise formulation of the mathematical topic being investigated. Consequently, many books on mathematics have an introductory chapter on set theory or mention in an appendix those parts of the theory that are needed in the text. Here it may appear that, in opening the book with a chapter on fundamentals of counting, we have neglected set theory. Actually we have relied on intuition; each time the word *collection* appeared in Chapter 1, we were dealing with a set. Also, in Sections 2.4 and 2.5, the notion of a set (if not the term itself) was invoked when we dealt with the universe (of discourse) for an open statement.

Trying to define a set is rather difficult and often results in the circular use of such synonyms as "class," "collection," and "aggregate." When we first began the study of geometry, we used our intuition to grasp the ideas of point, line, and incidence. Then we started to define new terms and prove theorems, relying on these intuitive notions along with certain axioms and postulates. In our study of set theory, intuition is invoked once again, this time for the comparable ideas of element, set, and membership.

We shall find that the ideas we developed in Chapter 2 on logic are closely tied to set theory. Furthermore, many of the proofs we shall study in this chapter draw on the ideas developed in Chapter 2.

3.1
Sets and Subsets

We have a "gut feeling" that a set should be a well-defined collection of objects. These objects are called *elements* and are said to be *members* of the set.

The adjective *well-defined* implies that for any element we care to consider, we are able to determine whether it is in the set under scrutiny. Consequently, we avoid dealing with sets that depend on opinion, such as the set of outstanding major league pitchers for the 1990s.

We use capital letters, such as $A, B, C, \ldots$, to represent sets and lowercase letters to represent elements. For a set A we write $x \in A$ if x is an element of A; $y \notin A$ indicates that y is not a member of A.

EXAMPLE 3.1

A set can be designated by listing its elements within *set braces*. For example, if A is the set consisting of the first five positive integers, then we write $A = \{1, 2, 3, 4, 5\}$. Here $2 \in A$ but $6 \notin A$.

Another standard notation for this set provides us with $A = \{x | x$ is an integer and $1 \le x \le 5\}$. Here the vertical line | within the set braces is read "such that." The symbols $\{x | \ldots\}$ are read "the set of all x such that. . . ." The properties following | help us determine the elements of the set that is being described.

Beware! The notation $\{x | 1 \le x \le 5\}$ is not an adequate description of the set A unless we have agreed in advance that the elements we are considering are integers. When such an agreement is adopted, we say that we are specifying a *universe*, or *universe of discourse*, which is usually denoted by $\mathcal{U}$. We then select only elements from $\mathcal{U}$ to form our sets. In this particular problem, if $\mathcal{U}$ denotes the set of all integers or the set of all positive integers, then $\{x | 1 \le x \le 5\}$ adequately describes A. If $\mathcal{U}$ is the set of all real numbers, then $\{x | 1 \le x \le 5\}$ would contain all of the real numbers between 1 and 5 inclusive; if $\mathcal{U}$ consists of only even integers, then the only members of $\{x | 1 \le x \le 5\}$ would be 2 and 4.

EXAMPLE 3.2

For $\mathcal{U} = \{1, 2, 3, \ldots\}$, the set of positive integers, we consider the following sets. At the same time we introduce various notations one may use to describe such sets.

a) $A = \{1, 4, 9, \ldots, 64, 81\} = \{x^2 | x \in \mathcal{U}, x^2 < 100\} = \{x^2 | x \in \mathcal{U} \wedge x^2 < 100\}$

b) $B = \{1, 4, 9, 16\} = \{y^2 | y \in \mathcal{U}, y^2 < 20\} = \{y^2 | y \in \mathcal{U}, y^2 < 23\}$
$= \{y^2 | y \in \mathcal{U} \wedge y^2 \le 16\}$.

c) $C = \{2, 4, 6, 8, \ldots\} = \{2k | k \in \mathcal{U}\}$.

Sets A and B are examples of *finite* sets, whereas C is an *infinite* set. When dealing with sets like A or C, we can either describe the sets in terms of properties the elements must satisfy or list enough elements to indicate what is, we hope, an obvious pattern. For any finite set A, $|A|$ denotes the number of elements in A and is referred to as the *cardinality*, or *size*, of A. In this example we find that $|A| = 9$ and $|B| = 4$.

Here the sets B and A are such that every element of B is also an element of A. This important relationship occurs throughout set theory and its applications, and it leads to the following definition.

Definition 3.1

If C, D are sets from a universe $\mathcal{U}$, we say that C is a *subset* of D and write $C \subseteq D$, or $D \supseteq C$, if every element of C is an element of D. If, in addition, D contains an element that is not in C, then C is called a *proper subset* of D, and this is denoted by $C \subset D$ or $D \supset C$.

Note that for all sets C, D from a universe $\mathcal{U}$, if $C \subseteq D$, then

$$\forall x \, [x \in C \Rightarrow x \in D],$$

and if $\forall x \, [x \in C \Rightarrow x \in D]$, then $C \subseteq D$.

Here the universal quantifier $\forall x$ indicates that we should have to consider every element x in the prescribed universe $\mathcal{U}$. However, for each replacement c (from $\mathcal{U}$) where the statement $c \in C$ is false, we know that the implication $c \in C \rightarrow c \in D$ is true, regardless of the truth value of the statement $c \in D$. Consequently, we actually need to consider only those replacements c' (from $\mathcal{U}$) where the statement $c' \in C$ is true. If for each such c' we find that the statement $c' \in D$ is also true, then we know that $\forall x \, [x \in C \Rightarrow x \in D]$ or, equivalently, $C \subseteq D$.

Also, we find that for all subsets C, D of $\mathcal{U}$,

$$C \subset D \Rightarrow C \subseteq D,$$

and when C, D are finite,

$$C \subseteq D \Rightarrow |C| \leq |D|, \quad \text{and} \quad C \subset D \Rightarrow |C| < |D|.$$

However, for $\mathcal{U} = \{1, 2, 3, 4, 5\}$, $C = \{1, 2\}$, and $D = \{1, 2\}$, we see that C is a subset of D (that is, $C \subseteq D$), but it is not a proper subset of D (or, $C \not\subset D$). So, in general, we do *not* find that $C \subseteq D \Rightarrow C \subset D$.

EXAMPLE 3.3

In an early version of ANSI (American National Standards Institute) FORTRAN, no distinction was made between uppercase and lowercase letters, and a variable name consisted of a single letter followed by at most five characters (letters or digits). If $\mathcal{U}$ denotes the set of all such variable names, then by the rules of sum and product, $|\mathcal{U}| = 26 + 26(36) + 26(36)^2 + \cdots + 26(36)^5 = 26 \sum_{i=0}^{5} 36^i = 1,617,038,306$. Thus, $\mathcal{U}$ is large, but still finite. An integer variable in this programming language had to start with one of the letters I, J, K, L, M, N. So if A denotes the subset of all integer variables in this early version of ANSI FORTRAN, then $|A| = 6 + 6(36) + 6(36)^2 + \cdots + 6(36)^5 = 6 \sum_{i=0}^{5} 36^i = 373,162,686$.

The subset concept may now be used to develop the idea of set equality. First we consider the following example.

EXAMPLE 3.4

For the universe $\mathcal{U} = \{1, 2, 3, 4, 5\}$, consider the set $A = \{1, 2\}$. If $B = \{x \mid x^2 \in \mathcal{U}\}$, then the members of B are 1, 2. Here A and B contain the same elements — and no other element(s) — leading us to feel that the sets A and B are *equal*.

However, it is also true here that $A \subseteq B$ and $B \subseteq A$, and we prefer to formally define the idea of set equality by using these subset relations.

Definition 3.2

For a given universe $\mathcal{U}$, the sets C and D (taken from $\mathcal{U}$) are said to be *equal*, and we write $C = D$, when $C \subseteq D$ and $D \subseteq C$.

From these ideas on set equality, we find that neither order nor repetition is relevant for a general set. Consequently, we find, for example, that $\{1, 2, 3\} = \{3, 1, 2\} = \{2, 2, 1, 3\} = \{1, 2, 1, 3, 1\}$.

Now that we have defined the concepts of subset and set equality, we shall use the quantifiers of Section 2.4 to examine the negations of these ideas.

For a given universe $\mathcal{U}$, let A, B be sets taken from $\mathcal{U}$. Then we may write

$$A \subseteq B \Longleftrightarrow \forall x \, [x \in A \Rightarrow x \in B].$$

From the (quantified) definition of $A \subseteq B$, we find that

$$A \not\subseteq B \text{ (that is, } A \text{ is } not \text{ a subset of } B)$$
$$\Longleftrightarrow \neg\forall x \, [x \in A \Rightarrow x \in B]$$
$$\Longleftrightarrow \exists x \, \neg[x \in A \Rightarrow x \in B]$$
$$\Longleftrightarrow \exists x \, \neg[\neg(x \in A) \vee x \in B]$$
$$\Longleftrightarrow \exists x \, [x \in A \wedge \neg(x \in B)]$$
$$\Longleftrightarrow \exists x \, [x \in A \wedge x \notin B].$$

Hence $A \nsubseteq B$ if there is at least one element x in the universe where x is a member of A but x is not a member of B.

In a similar way, because $A = B \Longleftrightarrow A \subseteq B \wedge B \subseteq A$, then

$$A \neq B \Longleftrightarrow \neg(A \subseteq B \wedge B \subseteq A) \Longleftrightarrow \neg(A \subseteq B) \vee \neg(B \subseteq A) \Longleftrightarrow A \nsubseteq B \vee B \nsubseteq A.$$

Therefore two sets A and B are not equal if and only if (1) there exists at least one element x in $\mathcal{U}$ where $x \in A$ but $x \notin B$ or (2) there exists at least one element y in $\mathcal{U}$ where $y \in B$ and $y \notin A$ — or perhaps both (1) and (2) occur.

We also note that for any sets $C, D \subseteq \mathcal{U}$ (that is, $C \subseteq \mathcal{U}$ and $D \subseteq \mathcal{U}$),

$$C \subset D \Longleftrightarrow C \subseteq D \wedge C \neq D.$$

Now that we have introduced the four ideas of set membership, set equality, subset, and proper subset, we shall consider one more example to see what these concepts tell us, as well as what they do *not* tell us. Following this example, the proof of our first theorem for this chapter will be fairly straightforward — because it readily follows from some of these ideas.

EXAMPLE 3.5

Let $\mathcal{U} = \{1, 2, 3, 4, 5, 6, x, y, \{1, 2\}, \{1, 2, 3\}, \{1, 2, 3, 4\}\}$ (where x, y are the 24th, 25th lowercase letters of the alphabet and do not represent anything else, such as $3, 5$, or $\{1, 2\}$). Then $|\mathcal{U}| = 11$.

a) If $A = \{1, 2, 3, 4\}$, then $|A| = 4$ and here we have

  i) $A \subseteq \mathcal{U}$; ii) $A \subset \mathcal{U}$; iii) $A \in \mathcal{U}$;
  iv) $\{A\} \subseteq \mathcal{U}$; v) $\{A\} \subset \mathcal{U}$; but vi) $\{A\} \notin \mathcal{U}$.

b) Now let $B = \{5, 6, x, y, A\} = \{5, 6, x, y, \{1, 2, 3, 4\}\}$. Then $|B| = 5$, *not* 8. And now we find that

  i) $A \in B$; ii) $\{A\} \subseteq B$; and iii) $\{A\} \subset B$.

But

  iv) $\{A\} \notin B$;
  v) $A \nsubseteq B$ (that is, A is not a subset of B); and
  vi) $A \not\subset B$ (that is, A is not a proper subset of B).

THEOREM 3.1

Let $A, B, C \subseteq \mathcal{U}$.

a) If $A \subseteq B$ and $B \subseteq C$, then $A \subseteq C$. b) If $A \subset B$ and $B \subseteq C$, then $A \subset C$.
c) If $A \subseteq B$ and $B \subset C$, then $A \subset C$. d) If $A \subset B$ and $B \subset C$, then $A \subset C$.

Before we prove this theorem we want to recall a comment we made back in Section 2.5. It concerns our coverage of the Rules of Universal Specification and Universal Generalization and appears after Example 2.56. For now it is appropriate in this new area on set theory. When we want to prove, for example, that $x \in A \Rightarrow x \in C$, we shall start by considering any fixed but arbitrarily chosen element x in $\mathcal{U}$ — but we shall want this element x to be such that "$x \in A$" is a true statement (*not* an open statement). Then we must show that this same fixed but arbitrarily chosen element x is also in C. The proofs we present are consequently referred to as *element arguments*. Always remember that in these proofs x represents a fixed but arbitrarily chosen element of A — and though x is generic (since it is *not* a specifically named element in A), it does remain the same throughout each proof.

Proof: We shall prove parts (a) and (b) and leave the remaining parts for the exercises.

a) To prove that $A \subseteq C$, we need to verify that for all $x \in \mathcal{U}$, if $x \in A$ then $x \in C$. We start with an element x from A. Since $A \subseteq B$, $x \in A$ implies $x \in B$. Then with $B \subseteq C$, $x \in B$ implies $x \in C$. So $x \in A$ implies $x \in C$ (by the Law of the Syllogism — Rule 2 in Table 2.19 — since $x \in A$, $x \in B$, and $x \in C$ are statements), and $A \subseteq C$.

b) Since $A \subset B$, if $x \in A$ then $x \in B$. With $B \subseteq C$, it then follows that $x \in C$, so $A \subseteq C$. However, $A \subset B \Rightarrow$ there exists an element $b \in B$ such that $b \notin A$. Because $B \subseteq C$, $b \in B \Rightarrow b \in C$. Thus $A \subseteq C$ and there exists an element $b \in C$ with $b \notin A$, so $A \subset C$.

Our next example involves several subset relations.

EXAMPLE 3.6	Let $\mathcal{U} = \{1, 2, 3, 4, 5\}$ with $A = \{1, 2, 3\}$, $B = \{3, 4\}$, and $C = \{1, 2, 3, 4\}$. Then the following subset relations hold:

a) $A \subseteq C$ **b)** $A \subset C$

c) $B \subset C$ **d)** $A \subseteq A$

e) $B \nsubseteq A$

f) $A \not\subset A$ (that is, A is not a proper subset of A)

The sets A, B are just two of the subsets of C. We are interested in determining how many subsets C has in total. Before answering, however, we need to introduce the set with no members.

Definition 3.3	The *null set*, or *empty set*, is the (unique) set containing no elements. It is denoted by $\emptyset$ or $\{\ \}$.

We note that $|\emptyset| = 0$ but $\{0\} \neq \emptyset$. Also, $\emptyset \neq \{\emptyset\}$ because $\{\emptyset\}$ is a set with one element, namely, the null set.

The empty set satisfies the following property given in Theorem 3.2. To establish this property we use the method of proof by contradiction (or *reductio ad absurdum*). Following the proof of Theorem 2.4 (in Section 2.5), we said that in establishing a theorem by this method, we assumed the negation of the result and arrived at a contradiction. In our prior work (as found in Example 2.32 and the third proof of Theorem 2.4), we arrived at a contradiction of the form $r \wedge \neg r$ or $p(m) \wedge \neg p(m)$, respectively — where $\neg r$ was a premise in Example 2.32 and $p(m)$ a specific instance of the hypothesis in Theorem 2.4. In proving Theorem 3.2 things are now a little different. This time we shall find ourselves denying (or contradicting) an earlier result we have accepted as true, namely, the definition of the null set.

THEOREM 3.2	For any universe $\mathcal{U}$, let $A \subseteq \mathcal{U}$. Then $\emptyset \subseteq A$, and if $A \neq \emptyset$, then $\emptyset \subset A$.

Proof: If the first result is not true, then $\emptyset \nsubseteq A$, so there is an element x from the universe with $x \in \emptyset$ but $x \notin A$. But $x \in \emptyset$ is impossible. So we reject the assumption $\emptyset \nsubseteq A$ and find that $\emptyset \subseteq A$. In addition, if $A \neq \emptyset$, then there is an element $a \in A$ (and $a \notin \emptyset$), so $\emptyset \subset A$.

EXAMPLE 3.7

Returning now to Example 3.6 we determine the number of subsets of the set $C = \{1, 2, 3, 4\}$. In constructing a subset of C, we have, for each member x of C, two distinct choices: Either include it in the subset or exclude it. Consequently, there are $2 \times 2 \times 2 \times 2$ choices, resulting in $2^4 = 16$ subsets of C. These include the empty set $\emptyset$ and the set C itself. Should we need the number of subsets of two elements from C, the result is the number of ways two objects can be selected from a set of four objects, namely, $C(4, 2)$ or $\binom{4}{2}$. As a result, the total number of subsets of C, 2^4, is also the sum $\binom{4}{0} + \binom{4}{1} + \binom{4}{2} + \binom{4}{3} + \binom{4}{4}$, where the first summand is for the empty set, the second summand for the four *singleton* subsets, the third summand for the six subsets of size 2, and so on. So $2^4 = \sum_{k=0}^{4} \binom{4}{k}$.

Definition 3.4

If A is a set from universe $\mathcal{U}$, the *power* set of A, denoted $\mathscr{P}(A)$,[†] is the collection (or set) of all subsets of A.

EXAMPLE 3.8

For the set C of Example 3.7, $\mathscr{P}(C) = \{\emptyset, \{1\}, \{2\}, \{3\}, \{4\}, \{1, 2\}, \{1, 3\}, \{1, 4\}, \{2, 3\}, \{2, 4\}, \{3, 4\}, \{1, 2, 3\}, \{1, 2, 4\}, \{1, 3, 4\}, \{2, 3, 4\}, C\}$.

> For any finite set A with $|A| = n \geq 0$, we find that A has 2^n subsets and that $|\mathscr{P}(A)| = 2^n$. For any $0 \leq k \leq n$, there are $\binom{n}{k}$ subsets of size k. Counting the subsets of A according to the number, k, of elements in a subset, we have the combinatorial identity
>
> $$\binom{n}{0} + \binom{n}{1} + \binom{n}{2} + \cdots + \binom{n}{n} = \sum_{k=0}^{n} \binom{n}{k} = 2^n, \text{ for } n \geq 0.$$

This identity was established earlier in Corollary 1.1(a). The presentation here is another example of a combinatorial proof because the identity is established by counting the same collection of objects (subsets of A) in two different ways.

A systematic way to represent the subsets of a given nonempty set can be accomplished by using a coding scheme known as a *Gray code*. This is demonstrated in our next example.

EXAMPLE 3.9

Consider the binary strings (of 0's and 1's) in Fig. 3.1. In particular, examine the first column of the strings in part (b). How did this column come about? First we see 0, then 1 — as in part (a) of the figure. Then we see 1 followed by 0 — the reverse order (from bottom to top) of the two binary strings in part (a). Once we obtain the first column for the binary strings in part (b), we then list two 0's followed by two 1's.

Continuing with the strings in part (c) of the figure, now we concentrate on the first two columns. The first four entries (binary strings of length 2) are precisely the four strings in part (b). The last four entries (again, binary strings of length 2) are likewise the binary strings in part (b) — now in reverse order (from bottom to top). For these eight strings of length 2, we append 0 to the right of the first four and 1 to the right of the last four.

For each Gray code in parts (a), (b), (c) of the figure, as we go from one binary string (in a column) to the next binary string (in that column), there is exactly one bit that changes. For instance, in part (b), in going from 10 to 11, we find one change (from 0 to 1) in the second position. Furthermore, for the third and fourth strings in part (c), as we go from

---

[†]In some computer science textbooks the reader may find the notation 2^A used for $\mathscr{P}(A)$.

(a)			(c)				(d)	(e)	(f)
0		∅	0 0	0		∅	000	000	000
1		{x}	1 0	0		{x}	100	010	001
			1 1	0		{x, y}	110	011	101
			0 1	0		{y}	010	001	100
0	0	∅	0 1	1		{y, z}	011	101	110
1	0	{x}	1 1	1		{x, y, z}	111	111	010
1	1	{x, y}	1 0	1		{x, z}	101	110	011
0	1	{y}	0 0	1		{z}	001	100	111
(b)									

Figure 3.1

110 to 010, there is exactly one change—from 1 to 0 in the first position. The fourth and fifth strings have the one change from 0 to 1—this time in the third position. Also notice how the first and last strings for each code differ in the last position. Part (d) of the figure demonstrates this for the strings of length 3.

This technique, for constructing a Gray code for the strings of length 2 from those of length 1 and the strings of length 3 from those of length 2, is an example of a *recursive* construction. (This idea will be examined in more detail in Section 4.2.)

When we examine each Gray code in parts (a), (b), (c) of Fig. 3.1, we see a listing of subsets to the right of each of these codes. For example, in part (b), if we start with the set $A = \{x, y\}$ and keep the order of the elements fixed,[†] then we can list the subsets of A in terms of binary strings of length 2. We write 0 for an element when it is not in the subset and 1 when it is. Hence the subset $\{x\}$ is encoded as 10 because the "first" element x (of ordered set A) is in the subset, while the "second" element y (of ordered set A) is not present—as the 0, in 10, indicates. For part (c), the (ordered) set $B = \{x, y, z\}$ has its eight subsets listed next to the elements of the Gray code. As we go from one subset to the next (in a given column), we see that there is exactly one change in the makeup of the subset. For instance, in going from $\{x, y\}$ (110) to $\{y\}$ (010), exactly one element is deleted—as indicated by the change from 1 to 0 in the first positions of 110 and 010. Likewise, as we go from $\{z\}$ (001) to ∅ (000), exactly one element is deleted—the change from 1 to 0, in the third bits of 001 and 000, indicates this. Examining the change from $\{y, z\}$ (011) to $\{x, y, z\}$ (111), we see that one new element is added—here it is x. The change from 0 to 1 as we go from 011 to 111 takes this into account.

Note that the first four subsets in part (c) are the four subsets in part (b). Further, the last four subsets in part (c) come about from the same four subsets in part (b)—this time in reverse order and with the element z included in each subset.

The recursive construction given here shows how we can continue to develop Gray codes for binary strings of longer length. When this coding scheme was introduced—just prior to the start of this example—we spoke of it as *a* Gray code, not as *the* Gray code. Other Gray codes are possible. The code in part (e) of Fig. 3.1 provides a second Gray code for the eight binary strings of length 3. Furthermore, if we no longer require the first and last entries in a code to differ in only one position, then the code in part (f) of Fig. 3.1 would also serve as a Gray code for the eight binary strings of length 3.

[†] Originally we considered the elements of a set as unordered, so we are making an exception here. In textbooks dealing with data structures, such ordered sets are often referred to as *lists* and one finds, for instance, the ordered set $\{x, y, z\}$ denoted by $[x, y, z]$ or $\langle x, y, z \rangle$.

The ability to count certain, or all, subsets of a given set provides a second approach for the solution of two of our earlier examples.

EXAMPLE 3.10

In Example 1.14, we counted the number of (staircase) paths in the xy-plane from $(2, 1)$ to $(7, 4)$ where each such path is made up of individual steps going one unit to the right (R) or one unit upward (U). Figure 3.2 is the same as Fig. 1.1, where two of the possible paths are indicated.

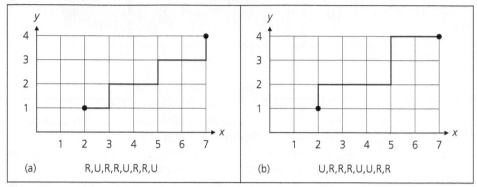

(a)   R,U,R,R,U,R,R,U   (b)   U,R,R,R,U,U,R,R

Figure 3.2

The path in Fig. 3.2(a) has its three upward (U) moves located in positions 2, 5, and 8 of the list at the bottom of the figure. Consequently, this path determines the three-element subset $\{2, 5, 8\}$ of the set $\{1, 2, 3, \ldots, 8\}$. In Fig. 3.2(b) the path determines the three-element subset $\{1, 5, 6\}$. Conversely, if we start, for example, with the subset $\{1, 3, 7\}$ of $\{1, 2, 3, \ldots, 8\}$, then the path that determines this subset is given by U, R, U, R, R, R, U, R.

Consequently, the number of paths sought here equals the number of subsets A of $\{1, 2, 3, \ldots, 8\}$, where $|A| = 3$. There are $\binom{8}{3} = \frac{8!}{3!\,5!} = 56$ such paths (and subsets), as we found in Example 1.14.

If we had considered the moves R to the right, instead of the upward moves U, we would have found the answer to be the number of subsets B of $\{1, 2, 3, \ldots, 8\}$, where $|B| = 5$. There are $\binom{8}{5} = \frac{8!}{5!\,3!} = 56$ such subsets. (The idea presented here was examined earlier for the result developed in Table 1.4.)

EXAMPLE 3.11

In part (b) of Example 1.37 of Section 1.4 we learned that there are 2^6 compositions for the integer 7 — that is, there are 2^6 ways to write 7 as a sum of one or more positive integers, where the order of the summands is relevant. The result we obtained there used the binomial theorem in conjunction with the answers for seven cases that were summarized in Table 1.9. Now we shall obtain this result in a somewhat different and easier way.

First consider the following composition of 7:

Here we have seven summands, each of which is 1, and six plus signs.

For the set $\{1, 2, 3, 4, 5, 6\}$ there are 2^6 subsets. But what does this have to do with the compositions of 7?

Consider a subset of $\{1, 2, 3, 4, 5, 6\}$, say $\{1, 4, 6\}$. Now form the following composition of 7:

$$(1 + 1) + 1 + (1 + 1) + (1 + 1)$$

1st plus sign   4th plus sign   6th plus sign

Here the subset $\{1, 4, 6\}$ indicates that we should place parentheses around the 1's on either side of the first, fourth, and sixth plus signs. This results in the composition

$$2 + 1 + 2 + 2.$$

If the same way we find that the subset $\{1, 2, 5, 6\}$ indicates the use of the first, second, fifth, and sixth plus signs, giving us

$$(1 + 1 + 1) + 1 + (1 + 1 + 1)$$

1st plus sign   2nd plus sign   5th plus sign   6th plus sign

or the composition $3 + 1 + 3$.

Going in reverse we see that the composition $1 + 1 + 5$ comes from

$$1 + 1 + (1 + 1 + 1 + 1 + 1)$$

and is determined by the subset $\{3, 4, 5, 6\}$ of $\{1, 2, 3, 4, 5, 6\}$. In Table 3.1 we have listed six compositions of 7 along with the corresponding subset of $\{1, 2, 3, 4, 5, 6\}$ that determines each of them.

Table 3.1

	Composition of 7		Determining Subset of $\{1, 2, 3, 4, 5, 6\}$
(i)	$1 + 1 + 1 + 1 + 1 + 1 + 1$	(i)	$\emptyset$
(ii)	$1 + 2 + 1 + 1 + 1 + 1$	(ii)	$\{2\}$
(iii)	$1 + 1 + 3 + 1 + 1$	(iii)	$\{3, 4\}$
(iv)	$2 + 3 + 2$	(iv)	$\{1, 3, 4, 6\}$
(v)	$4 + 3$	(v)	$\{1, 2, 3, 5, 6\}$
(vi)	7	(vi)	$\{1, 2, 3, 4, 5, 6\}$

The examples we have obtained here indicate a correspondence between the compositions of 7 and the subsets of $\{1, 2, 3, 4, 5, 6\}$. Consequently, once again we find that there are 2^6 compositions of 7. In fact, for each positive integer m, there are 2^{m-1} compositions of m.

Out next example yields another important combinatorial identity.

EXAMPLE 3.12

For integers n, r with $n \geq r \geq 1$,

$$\binom{n+1}{r} = \binom{n}{r} + \binom{n}{r-1}.$$

Although this result can be established algebraically from the definition of $\binom{n}{r}$ as $n!/(r!(n-r)!)$, we use a combinatorial approach. Let $A = \{x, a_1, a_2, \ldots, a_n\}$ and consider all subsets of A that contain r elements. There are $\binom{n+1}{r}$ such subsets. Each of these falls into exactly one of the following two cases: those subsets that contain the element x and those that do not. To obtain a subset C of A, where $x \in C$ and $|C| = r$, place x in C and then select $r - 1$ of the elements $a_1, a_2, \ldots, a_n$. This can be done in $\binom{n}{r-1}$ ways. For the other case we want a subset B of A with $|B| = r$ and $x \notin B$. So we select r elements from among $a_1, a_2, \ldots, a_n$, which we can do in $\binom{n}{r}$ ways. It then follows by the rule of sum that $\binom{n+1}{r} = \binom{n}{r} + \binom{n}{r-1}$.

Before we proceed any further let us reconsider the result of Example 3.12, but this time we shall do it in light of what we learned in Example 3.10.

Once again we let n, r be positive integers where $n \geq r \geq 1$. Then $\binom{n+1}{r}$ counts the number of (staircase) paths in the xy-plane from $(0, 0)$ to $(n + 1 - r, r)$, where, as in Example 3.10, each such path has

$$(n + 1) - r \quad \text{horizontal moves of the form } (x, y) \to (x + 1, y), \quad \text{and}$$

$$r \quad \text{vertical moves of the form } (x, y) \to (x, y + 1).$$

The last edge in each of these (staircase) paths terminates at the point $(n + 1 - r, r)$ and starts at either (i) the point $(n - r, r)$ or (ii) the point $(n + 1 - r, r - 1)$.

In case (i) we have the last edge horizontal, namely, $(n - r, r) \to (n + 1 - r, r)$; the number of (staircase) paths from $(0, 0)$ to $(n - r, r)$ is $\binom{(n-r)+r}{r} = \binom{n}{r}$. For case (ii) the last edge is vertical, namely, $(n + 1 - r, r - 1) \to (n + 1 - r, r)$; the number of (staircase) paths from $(0, 0)$ to $(n + 1 - r, r - 1)$ is $\binom{(n+1-r)+(r-1)}{r-1} = \binom{n}{r-1}$. Since these two cases exhaust all possibilities and have nothing in common, it follows that

$$\binom{n+1}{r} = \binom{n}{r} + \binom{n}{r-1}.$$

EXAMPLE 3.13

We now investigate how the identity of Example 3.12 can help us solve Example 1.35, where we sought the number of nonnegative integer solutions of the inequality $x_1 + x_2 + \cdots + x_6 < 10$.

For each integer k, $0 \leq k \leq 9$, the number of solutions to $x_1 + x_2 + \cdots + x_6 = k$ is $\binom{6+k-1}{k} = \binom{5+k}{k}$. So the number of nonnegative integer solutions to $x_1 + x_2 + \cdots + x_6 < 10$ is

$$\binom{5}{0} + \binom{6}{1} + \binom{7}{2} + \binom{8}{3} + \cdots + \binom{14}{9}$$

$$= \left[\binom{6}{0} + \binom{6}{1}\right] + \binom{7}{2} + \binom{8}{3} + \cdots + \binom{14}{9}, \quad \text{since } \binom{5}{0} = 1 = \binom{6}{0}$$

$$= \left[\binom{7}{1} + \binom{7}{2}\right] + \binom{8}{3} + \cdots + \binom{14}{9}, \quad \text{since } \binom{6}{0} + \binom{6}{1} = \binom{7}{1}$$

$$= \left[\binom{8}{2} + \binom{8}{3}\right] + \binom{9}{4} + \cdots + \binom{14}{9}, \quad \text{since } \binom{7}{1} + \binom{7}{2} = \binom{8}{2}$$

$$= \left[\binom{9}{3} + \binom{9}{4}\right] + \cdots + \binom{14}{9} = \cdots = \binom{14}{8} + \binom{14}{9} = \binom{15}{9} = 5005.$$

EXAMPLE 3.14

In Fig. 3.3 we find a part of the useful and interesting array of numbers called *Pascal's triangle*

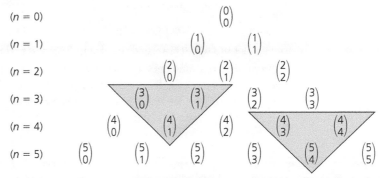

Figure 3.3

Note that in this partial listing the two triangles shown satisfy the condition that the binomial coefficient at the bottom of the inverted triangle is the sum of the other two terms in the triangle. This result follows from the identity in Example 3.12.

When we replace each of the binomial coefficients by its numerical value, the Pascal triangle appears as shown in Fig. 3.4.

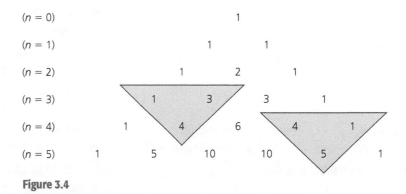

Figure 3.4

There are certain sets of numbers that appear frequently throughout the text. Consequently, we close this section by assigning them the following designations.

a) **Z** = the set of *integers* = {0, 1, −1, 2, −2, 3, −3, . . .}

b) **N** = the set of *nonnegative integers* or *natural numbers* = {0, 1, 2, 3, . . .}

c) **Z**⁺ = the set of *positive integers* = {1, 2, 3, . . .} = $\{x \in \mathbf{Z} \mid x > 0\}$

d) **Q** = the set of *rational numbers* = $\{a/b \mid a, b \in \mathbf{Z}, b \neq 0\}$

e) **Q**⁺ = the set of *positive rational numbers* = $\{r \in \mathbf{Q} \mid r > 0\}$

f) **Q*** = the set of *nonzero rational numbers*

g) **R** = the set of *real numbers*

h) $\mathbf{R}^+$ = the set of *positive real numbers*

i) $\mathbf{R}^*$ = the set of *nonzero real numbers*

j) $\mathbf{C}$ = the set of *complex numbers* = $\{x + yi \mid x, y \in \mathbf{R}, i^2 = -1\}$

k) $\mathbf{C}^*$ = the set of *nonzero complex numbers*

l) For each $n \in \mathbf{Z}^+$, $\mathbf{Z}_n = \{0, 1, 2, \ldots, n - 1\}$

m) For real numbers a, b with $a < b$, $[a, b] = \{x \in \mathbf{R} \mid a \le x \le b\}$, $(a, b) = \{x \in \mathbf{R} \mid a < x < b\}$, $[a, b) = \{x \in \mathbf{R} \mid a \le x < b\}$, $(a, b] = \{x \in \mathbf{R} \mid a < x \le b\}$. The first set is called a *closed interval*, the second set an *open interval*, and the other two sets *half-open intervals*.

EXERCISES 3.1

1. Which of the following sets are equal?

a) $\{1, 2, 3\}$ **b)** $\{3, 2, 1, 3\}$

c) $\{3, 1, 2, 3\}$ **d)** $\{1, 2, 2, 3\}$

2. Let $A = \{1, \{1\}, \{2\}\}$. Which of the following statements are true?

a) $1 \in A$ **b)** $\{1\} \in A$

c) $\{1\} \subseteq A$ **d)** $\{\{1\}\} \subseteq A$

e) $\{2\} \in A$ **f)** $\{2\} \subseteq A$

g) $\{\{2\}\} \subseteq A$ **h)** $\{\{2\}\} \subset A$

3. For $A = \{1, 2, \{2\}\}$, which of the eight statements in Exercise 2 are true?

4. Which of the following statements are true?

a) $\emptyset \in \emptyset$ **b)** $\emptyset \subset \emptyset$ **c)** $\emptyset \subseteq \emptyset$

d) $\emptyset \in \{\emptyset\}$ **e)** $\emptyset \subset \{\emptyset\}$ **f)** $\emptyset \subseteq \{\emptyset\}$

5. Determine all of the elements in each of the following sets.

a) $\{1 + (-1)^n \mid n \in \mathbf{N}\}$

b) $\{n + (1/n) \mid n \in \{1, 2, 3, 5, 7\}\}$

c) $\{n^3 + n^2 \mid n \in \{0, 1, 2, 3, 4\}\}$

6. Consider the following six subsets of $\mathbf{Z}$.

$A = \{2m + 1 \mid m \in \mathbf{Z}\}$ $B = \{2n + 3 \mid n \in \mathbf{Z}\}$
$C = \{2p - 3 \mid p \in \mathbf{Z}\}$ $D = \{3r + 1 \mid r \in \mathbf{Z}\}$
$E = \{3s + 2 \mid s \in \mathbf{Z}\}$ $F = \{3t - 2 \mid t \in \mathbf{Z}\}$

Which of the following statements are true and which are false?

a) $A = B$ **b)** $A = C$ **c)** $B = C$

d) $D = E$ **e)** $D = F$ **f)** $E = F$

7. Let A, B be sets from a universe $\mathcal{U}$. (a) Write a quantified statement to express the proper subset relation $A \subset B$. (b) Negate the result in part (a) to determine when $A \not\subset B$.

8. For $A = \{1, 2, 3, 4, 5, 6, 7\}$, determine the number of

a) subsets of A

b) nonempty subsets of A

c) proper subsets of A

d) nonempty proper subsets of A

e) subsets of A containing three elements

f) subsets of A containing 1, 2

g) subsets of A containing five elements, including 1, 2

h) subsets of A with an even number of elements

i) subsets of A with an odd number of elements

9. a) If a set A has 63 proper subsets, what is $|A|$?

b) If a set B has 64 subsets of odd cardinality, what is $|B|$?

c) Generalize the result of part (b)

10. Which of the following sets are nonempty?

a) $\{x \mid x \in \mathbf{N}, 2x + 7 = 3\}$

b) $\{x \in \mathbf{Z} \mid 3x + 5 = 9\}$

c) $\{x \mid x \in \mathbf{Q}, x^2 + 4 = 6\}$

d) $\{x \in \mathbf{R} \mid x^2 + 4 = 6\}$

e) $\{x \in \mathbf{R} \mid x^2 + 3x + 3 = 0\}$

f) $\{x \mid x \in \mathbf{C}, x^2 + 3x + 3 = 0\}$

11. When she is about to leave a restaurant counter, Mrs. Albanese sees that she has one penny, one nickel, one dime, one quarter, and one half-dollar. In how many ways can she leave some (at least one) of her coins for a tip if (a) there are no restrictions? (b) she wants to have some change left? (c) she wants to leave at least 10 cents?

12. Let $A = \{1, 2, 3, 4, 5, 7, 8, 10, 11, 14, 17, 18\}$.

a) How many subsets of A contain six elements?

b) How many six-element subsets of A contain four even integers and two odd integers?

c) How many subsets of A contain only odd integers?

13. Let $S = \{1, 2, 3, \ldots, 29, 30\}$. How many subsets A of S satisfy (a) $|A| = 5$? (b) $|A| = 5$ and the smallest element in A is 5? (c) $|A| = 5$ and the smallest element in A is less than 5?

14. a) How many subsets of $\{1, 2, 3, \ldots, 11\}$ contain at least one even integer?

b) How many subsets of $\{1, 2, 3, \ldots, 12\}$ contain at least one even integer?

c) Generalize the results of parts (a) and (b).

15. Give an example of three sets W, X, Y such that $W \in X$ and $X \in Y$ but $W \notin Y$.

16. Write the next three rows for the Pascal triangle shown in Fig. 3.4

17. Complete the proof of Theorem 3.1.

18. For sets $A, B, C \subseteq \mathcal{U}$, prove or disprove (with a counterexample), the following: If $A \subseteq B$, $B \nsubseteq C$, then $A \nsubseteq C$.

19. In part (i) of Fig. 3.5 we have the first six rows of Pascal's triangle, where a hexagon centered at 4 appears in the last three rows. If we consider the six numbers (around 4) at the vertices of this hexagon, we find that the two alternating triples — namely, 3, 1, 10 and 1, 5, 6 — satisfy $3 \cdot 1 \cdot 10 = 30 = 1 \cdot 5 \cdot 6$. Part (ii) of the figure contains rows 4 through 7 of Pascal's triangle. Here we find a hexagon centered at 10, and the alternating triples at the vertices — in this case, 4, 10, 15 and 6, 20, 5 — satisfy $4 \cdot 10 \cdot 15 = 600 = 6 \cdot 20 \cdot 5$.

a) Conjecture the general result suggested by these two examples.

b) Verify the conjecture in part (a).

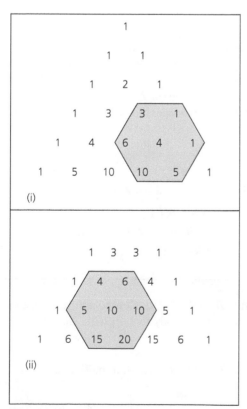

Figure 3.5

20. a) Among the strictly increasing sequences of integers that start with 1 and end with 7 are:

i) 1, 7   **ii)** 1, 3, 4, 7   **iii)** 1, 2, 4, 5, 6, 7

How many such strictly increasing sequences of integers start with 1 and end with 7?

b) How many strictly increasing sequences of integers start with 3 and end with 9?

c) How many strictly increasing sequences of integers start with 1 and end with 37? How many start with 62 and end with 98?

d) Generalize the results in parts (a) through (c).

21. One quarter of the five-element subsets of $\{1, 2, 3, \ldots, n\}$ contain the element 7. Determine n (≥ 5).

22. For a given universe $\mathcal{U}$, let $A \subseteq \mathcal{U}$ where A is finite with $|\mathcal{P}(A)| = n$. If $B \subseteq \mathcal{U}$, how many subsets does B have, if (a) $B = A \cup \{x\}$, where $x \in \mathcal{U} - A$? (b) $B = A \cup \{x, y\}$, where $x, y \in \mathcal{U} - A$? (c) $B = A \cup \{x_1, x_2, \ldots, x_k\}$, where $x_1, x_2, \ldots, x_k \in \mathcal{U} - A$?

23. Determine which row of Pascal's triangle contains three consecutive entries that are in the ratio $1 : 2 : 3$.

24. Use the recursive technique of Example 3.9 to develop a Gray code for the 16 binary strings of length 4. Then list each of the 16 subsets of the ordered set $\{w, x, y, z\}$ next to its corresponding binary string.

25. Suppose that A contains the elements v, w, x, y, z and no others. If a given Gray code for the 32 subsets of A encodes the ordered set $\{v, w\}$ as 01100 and the ordered set $\{x, y\}$ as 10001, write A as the corresponding ordered set.

26. For positive integers n, r show that

$$\binom{n+r+1}{r} = \binom{n+r}{r} + \binom{n+r-1}{r-1} + \cdots$$
$$+ \binom{n+2}{2} + \binom{n+1}{1} + \binom{n}{0}$$
$$= \binom{n+r}{n} + \binom{n+r-1}{n} + \cdots$$
$$+ \binom{n+2}{n} + \binom{n+1}{n} + \binom{n}{n}.$$

27. In the original abstract set theory formulated by Georg Cantor (1845–1918), a set was defined as "any collection into a whole of definite and separate objects of our intuition or our thought." Unfortunately, in 1901, this definition led Bertrand Russell (1872–1970) to the discovery of a contradiction — a result now known as *Russell's paradox* — and this struck at the very heart of the theory of sets. (But since then several ways have been found to define the basic ideas of set theory so that this contradiction no longer comes about.)

Russell's paradox arises when we concern ourselves with whether a set can be an element of itself. For example, the set of all positive integers is not a positive integer — or $\mathbf{Z}^+ \notin \mathbf{Z}^+$. But the set of all abstractions is an abstraction.

Now in order to develop the paradox let S be the set of all sets A that are not members of themselves — that is, $S = \{A | A$ is a set $\wedge A \notin A\}$.

    a) Show that if $S \in S$, then $S \notin S$.

    b) Show that if $S \notin S$, then $S \in S$.

The results in parts (a) and (b) show us that we must avoid trying to define sets like S. To do so we must restrict the types of elements that can be members of a set. (More about this is mentioned in the Summary and Historical Review in Section 3.8.)

28. Let $A = \{1, 2, 3, \ldots, 39, 40\}$.

    a) Write a computer program (or develop an algorithm) to generate a random six-element subset of A.

    b) For $B = \{2, 3, 5, 7, 11, 13, 17, 19, 23, 29, 31, 37\}$, write a computer program (or develop an algorithm) to generate a random six-element subset of A and then determine whether it is a subset of B.

29. Let $A = \{1, 2, 3, \ldots, 7\}$. Write a computer program (or develop an algorithm) that lists all the subsets B of A, where $|B| = 4$.

30. Write a computer program (or develop an algorithm) that lists all the subsets of $\{1, 2, 3, \ldots, n\}$, where $1 \leq n \leq 10$. (The value of n should be supplied during program execution.)

3.2
Set Operations and the Laws of Set Theory

After learning how to count, a student usually faces methods for combining counting numbers. First this is accomplished through addition. Usually the student's world of arithmetic revolves about the set $\mathbf{Z}^+$ (or a subset of $\mathbf{Z}^+$ that can be spoken and written about, as well as punched out on a hand-held calculator) wherein the addition of two elements from $\mathbf{Z}^+$ results in a third element of $\mathbf{Z}^+$, called the sum. Hence the student can concentrate on addition without having to enlarge his or her arithmetic world beyond $\mathbf{Z}^+$. This is also true for the operation of multiplication.

The addition and multiplication of positive integers are said to be *closed binary operations* on $\mathbf{Z}^+$. For example, when we compute $a + b$, for $a, b \in \mathbf{Z}^+$, there are two *operands*, namely, a and b. Hence the operation is called *binary*. And since $a + b \in \mathbf{Z}^+$ when $a, b \in \mathbf{Z}^+$, we say that the binary operation of addition (on $\mathbf{Z}^+$) is *closed*. The binary operation of (nonzero) division, however, is *not* closed for $\mathbf{Z}^+$ — we find, for example, that $1/2 \,(= 1 \div 2) \notin \mathbf{Z}^+$, even though $1, 2 \in \mathbf{Z}^+$. Yet this operation is closed when we consider the set $\mathbf{Q}^+$ instead of the set $\mathbf{Z}^+$.

We now introduce the following binary operations for sets.

Definition 3.5

For $A, B, \subseteq \mathcal{U}$ we define the following:

    a) $A \cup B$ (the *union* of A and B) $= \{x | x \in A \vee x \in B\}$.

    b) $A \cap B$ (the *intersection* of A and B) $= \{x | x \in A \wedge x \in B\}$.

    c) $A \bigtriangleup B$ (the *symmetric difference* of A and B) $= \{x | (x \in A \vee x \in B) \wedge x \notin A \cap B\} = \{x | x \in A \cup B \wedge x \notin A \cap B\}$.

Note that if $A, B \subseteq \mathcal{U}$, then $A \cup B, A \cap B, A \bigtriangleup B \subseteq \mathcal{U}$. Consequently, $\cup, \cap$, and $\bigtriangleup$ are closed binary operations on $\mathcal{P}(\mathcal{U})$, and we may also say that $\mathcal{P}(\mathcal{U})$ is *closed* under these (binary) operations.

EXAMPLE 3.15

With $\mathcal{U} = \{1, 2, 3, \ldots, 9, 10\}$, $A = \{1, 2, 3, 4, 5\}$, $B = \{3, 4, 5, 6, 7\}$, and $C = \{7, 8, 9\}$, we have:

    a) $A \cap B = \{3, 4, 5\}$ **b)** $A \cup B = \{1, 2, 3, 4, 5, 6, 7\}$

c) $B \cap C = \{7\}$ **d)** $A \cap C = \emptyset$

e) $A \triangle B = \{1, 2, 6, 7\}$ **f)** $A \cup C = \{1, 2, 3, 4, 5, 7, 8, 9\}$

g) $A \triangle C = \{1, 2, 3, 4, 5, 7, 8, 9\}$

In Example 3.15 we see that $A \cap B \subseteq A \subseteq A \cup B$. This result is not special for just this example but is true in general. The result follows because

$$x \in A \cap B \Rightarrow (x \in A \land x \in B) \Rightarrow x \in A$$

(by the Rule of Conjunctive Simplification — Rule 7 of Table 2.19), and

$$x \in A \Rightarrow (x \in A \lor x \in B) \Rightarrow x \in A \cup B$$

(where the first logical implication is a result of the Rule of Disjunctive Amplification — Rule 8 of Table 2.19).

Motivated by parts (d), (f), and (g) of Example 3.15, we introduce the following general ideas.

Definition 3.6

Let $S, T \subseteq \mathcal{U}$. The sets S and T are called *disjoint*, or *mutually disjoint*, when $S \cap T = \emptyset$.

THEOREM 3.3

If $S, T \subseteq \mathcal{U}$, then S and T are disjoint if and only if $S \cup T = S \triangle T$.

Proof: We start with S, T disjoint. (To prove that $S \cup T = S \triangle T$ we use Definition 3.2. In particular, we shall provide two element arguments, one for each inclusion.) Consider each x in $\mathcal{U}$. If $x \in S \cup T$, then $x \in S$ or $x \in T$ (or perhaps both). But with S and T disjoint, $x \notin S \cap T$ so $x \in S \triangle T$. Consequently, because $x \in S \cup T$ implies $x \in S \triangle T$, we have $S \cup T \subseteq S \triangle T$. For the opposite inclusion, if $y \in S \triangle T$, then $y \in S$ or $y \in T$. (But $y \notin S \cap T$; we don't actually use this here.) So $y \in S \cup T$. Therefore $S \triangle T \subseteq S \cup T$. And now that we have $S \cup T \subseteq S \triangle T$ and $S \triangle T \subseteq S \cup T$, it follows from Definition 3.2 that $S \triangle T = S \cup T$.

We prove the converse by the method of proof by contradiction. To do so we consider any $S, T \subseteq \mathcal{U}$ and keep the hypothesis (that is, that $S \cup T = S \triangle T$) as is, but we assume the negation of the conclusion (that is, we assume that S and T are *not* disjoint). So if $S \cap T \neq \emptyset$, let $x \in S \cap T$. Then $x \in S$ and $x \in T$, so $x \in S \cup T$ and

$$x \in S \triangle T (= S \cup T).$$

But when $x \in S \cup T$ and $x \in S \cap T$, then

$$x \notin S \triangle T.$$

From this contradiction — namely, $x \in S \triangle T \land x \notin S \triangle T$ — we realize that our original assumption was incorrect. Consequently, we have S and T disjoint.

In proving the first part of Theorem 3.3 we showed that if S, T are any sets, then $S \triangle T \subseteq S \cup T$. The disjointness of S and T was needed only for the opposite inclusion.

After mastering the skill of addition, one usually comes next to subtraction. Here the set **N** causes some difficulty. For example, **N** contains 2 and 5 but $2 - 5 = -3$, and $-3 \notin \mathbf{N}$. Therefore the binary operation of subtraction is not closed for **N**, although it is closed for the *superset* **Z** of **N**. So for **Z** we can introduce the *unary*, or *monary, operation* of negation where we take the "minus" or "negative" of a number such as 3, getting -3.

We now introduce a comparable unary operation for sets.

Definition 3.7   For a set $A \subseteq \mathcal{U}$, the *complement* of A, denoted $\mathcal{U} - A$, or $\overline{A}$, is given by $\{x \mid x \in \mathcal{U} \land x \notin A\}$.

EXAMPLE 3.16   For the sets of Example 3.15, $\overline{A} = \{6, 7, 8, 9, 10\}$, $\overline{B} = \{1, 2, 8, 9, 10\}$, and $\overline{C} = \{1, 2, 3, 4, 5, 6, 10\}$.

For every universe $\mathcal{U}$ and every set $A \subseteq \mathcal{U}$, we find that $\overline{A} \subseteq \mathcal{U}$. Therefore $\mathcal{P}(\mathcal{U})$ is *closed* under the unary operation defined by the complement.

The following concept is related to the concept of the complement.

Definition 3.8   For $A, B \subseteq \mathcal{U}$, the *(relative) complement* of A in B, denoted $B - A$, is given by $\{x \mid x \in B \land x \notin A\}$.

EXAMPLE 3.17   For the sets of Example 3.15 we have:

a) $B - A = \{6, 7\}$ b) $A - B = \{1, 2\}$ c) $A - C = A$

d) $C - A = C$ e) $A - A = \emptyset$ f) $\mathcal{U} - A = \overline{A}$

In order to motivate our next theorem, we first consider the following.

EXAMPLE 3.18   For $\mathcal{U} = \mathbf{R}$, let $A = [1, 2]$ and $B = [1, 3)$. Then we find that

a) $A = \{x \mid 1 \leq x \leq 2\} \subseteq \{x \mid 1 \leq x < 3\} = B$

b) $A \cup B = \{x \mid 1 \leq x < 3\} = B$

c) $A \cap B = \{x \mid 1 \leq x \leq 2\} = A$

d) $\overline{B} = (-\infty, 1) \cup [3, +\infty) \subseteq (-\infty, 1) \cup (2, +\infty) = \overline{A}$

This next theorem now shows us that the four results in Example 3.18 are related in general. In order to prove this theorem we again make use of Definition 3.2, as we discover the interplay between the notions of subset, union, intersection, and complement.

THEOREM 3.4   For any universe $\mathcal{U}$ and any sets $A, B \subseteq \mathcal{U}$, the following statements are equivalent:

a) $A \subseteq B$ b) $A \cup B = B$

c) $A \cap B = A$ d) $\overline{B} \subseteq \overline{A}$

Proof: In order to prove the theorem, we prove that (a) $\Rightarrow$ (b), (b) $\Rightarrow$ (c), (c) $\Rightarrow$ (d), and (d) $\Rightarrow$ (a). (The reason this suffices to prove this theorem is based on the idea presented in Exercise 13 at the end of Section 2.2.)

i) (a) $\Rightarrow$ (b)   If A, B are any sets, then $B \subseteq A \cup B$ (as mentioned after Example 3.15). For the opposite inclusion, if $x \in A \cup B$, then $x \in A$ or $x \in B$, but since $A \subseteq B$, in either case we have $x \in B$. So $A \cup B \subseteq B$ and, since we now have both inclusions, it follows (once again from Definition 3.2) that $A \cup B = B$.

ii) (b) $\Rightarrow$ (c)   Given sets A, B, we always have $A \supseteq A \cap B$ (as mentioned after Example 3.15). For the opposite inclusion, let $y \in A$. With $A \cup B = B$, $y \in A \Rightarrow y \in A \cup B \Rightarrow y \in B$ (since $A \cup B = B$) $\Rightarrow y \in A \cap B$, so $A \subseteq A \cap B$ and we conclude that $A = A \cap B$.

iii) (c) $\Rightarrow$ (d)   We know that $z \in \overline{B} \Rightarrow z \notin B$. Now if $z \in A \cap B$, then $z \in B$, since $A \cap B \subseteq B$. The contradiction — namely, $z \notin B \wedge z \in B$ — tells us that $z \notin A \cap B$. Therefore, $z \notin A$ because $A \cap B = A$. But $z \notin A \Rightarrow z \in \overline{A}$, so $\overline{B} \subseteq \overline{A}$.

iv) (d) $\Rightarrow$ (a)   Last, $w \in A \Rightarrow w \notin \overline{A}$. If $w \notin B$, then $w \in \overline{B}$. With $\overline{B} \subseteq \overline{A}$ it then follows that $w \in \overline{A}$. This time we get the contradiction $w \notin \overline{A} \wedge w \in \overline{A}$, and this tells us that $w \in B$. Hence $A \subseteq B$.

With a bit of theorem proving under our belts, we now introduce some of the major laws that govern set theory. These bear a marked resemblance to the laws of logic given in Section 2.2. In many instances these set theoretic laws are similar to the arithmetic properties of the real numbers, where "$\cup$" plays the role of "$+$" and "$\cap$" the role of "$\times$." However, there are several differences.

The Laws of Set Theory

For any sets A, B, and C taken from a universe $\mathcal{U}$

1) $\overline{\overline{A}} = A$	Law of *Double Complement*
2) $\overline{A \cup B} = \overline{A} \cap \overline{B}$	*DeMorgan's* Laws
$\overline{A \cap B} = \overline{A} \cup \overline{B}$	
3) $A \cup B = B \cup A$	*Commutative* Laws
$A \cap B = B \cap A$	
4) $A \cup (B \cup C) = (A \cup B) \cup C$	*Associative* Laws
$A \cap (B \cap C) = (A \cap B) \cap C$	
5) $A \cup (B \cap C) = (A \cup B) \cap (A \cup C)$	*Distributive* Laws
$A \cap (B \cup C) = (A \cap B) \cup (A \cap C)$	
6) $A \cup A = A$	*Idempotent* Laws
$A \cap A = A$	
7) $A \cup \emptyset = A$	*Identity* Laws
$A \cap \mathcal{U} = A$	
8) $A \cup \overline{A} = \mathcal{U}$	*Inverse* Laws
$A \cap \overline{A} = \emptyset$	
9) $A \cup \mathcal{U} = \mathcal{U}$	*Domination* Laws
$A \cap \emptyset = \emptyset$	
10) $A \cup (A \cap B) = A$	*Absorption* Laws
$A \cap (A \cup B) = A$	

All these laws can be established by element arguments, as in the first part of the proof of Theorem 3.3. We demonstrate this by establishing the first of DeMorgan's Laws and the second Distributive Law, that of intersection over union.

Proof: Let $x \in \mathcal{U}$. Then

$$x \in \overline{A \cup B} \Rightarrow x \notin A \cup B$$
$$\Rightarrow x \notin A \text{ and } x \notin B$$
$$\Rightarrow x \in \overline{A} \text{ and } x \in \overline{B}$$
$$\Rightarrow x \in \overline{A} \cap \overline{B},$$

so $\overline{A \cup B} \subseteq \overline{A} \cap \overline{B}$. To establish the opposite inclusion, we check to see that the converse of each logical implication is also a logical implication (that is, that each logical implication is, in fact, a logical equivalence). As a result we find that

$$x \in \overline{A} \cap \overline{B} \Rightarrow x \in \overline{A} \text{ and } x \in \overline{B}$$
$$\Rightarrow x \notin A \text{ and } x \notin B$$
$$\Rightarrow x \notin A \cup B$$
$$\Rightarrow x \in \overline{A \cup B}.$$

Therefore $\overline{A} \cap \overline{B} \subseteq \overline{A \cup B}$. Consequently, with $\overline{A \cup B} \subseteq \overline{A} \cap \overline{B}$ and $\overline{A} \cap \overline{B} \subseteq \overline{A \cup B}$, it follows from Definition 3.2 that $\overline{A \cup B} = \overline{A} \cap \overline{B}$.

In our second proof, we shall establish both subset relations simultaneously by using the logical equivalence ($\Longleftrightarrow$) as opposed to the logical implications ($\Rightarrow$ and $\Leftarrow$).

Proof: For each $x \in \mathcal{U}$,

$$x \in A \cap (B \cup C) \Longleftrightarrow (x \in A) \text{ and } (x \in B \cup C)$$
$$\Longleftrightarrow (x \in A) \text{ and } (x \in B \text{ or } x \in C)$$
$$\Longleftrightarrow (x \in A \text{ and } x \in B) \text{ or } (x \in A \text{ and } x \in C)$$
$$\Longleftrightarrow (x \in A \cap B) \text{ or } (x \in A \cap C)$$
$$\Longleftrightarrow x \in (A \cap B) \cup (A \cap C).$$

As we have equivalent statements throughout, we have established both subset relations simultaneously, so $A \cap (B \cup C) = (A \cap B) \cup (A \cap C)$. (The equivalence of the third and fourth statements follows from the comparable principle in the laws of logic — namely, the Distributive Law of conjunction over disjunction.)

The reader undoubtedly expects the pairing of the laws in items 2 through 10 to have some importance. As with the laws of logic, these pairs of statements are called *duals*. One statement can be obtained from the other by replacing all occurrences of $\cup$ by $\cap$ and vice versa, and all occurrences of $\mathcal{U}$ by $\emptyset$ and vice versa.

This leads us to the following formal idea.

Definition 3.9   Let s be a (general) statement dealing with the equality of two set expressions. Each such expression may involve one or more occurrences of sets (such as A, $\overline{A}$, B, $\overline{B}$, etc.), one or more occurrences of $\emptyset$ and $\mathcal{U}$, and only the set operation symbols $\cap$ and $\cup$. The *dual* of s, denoted s^d, is obtained from s by replacing (1) each occurrence of $\emptyset$ and $\mathcal{U}$ (in s) by $\mathcal{U}$ and $\emptyset$, respectively; and (2) each occurrence of $\cap$ and $\cup$ (in s) by $\cup$ and $\cap$, respectively.

As in Section 2.2, we shall state and use the following theorem. We shall prove a more general result in Chapter 15.

THEOREM 3.5   *The Principle of Duality.* Let s denote a theorem dealing with the equality of two set expressions (involving only the set operations $\cap$ and $\cup$ as described in Definition 3.9). Then s^d, the dual of s, is also a theorem.

Using this principle cuts our work down considerably. For each pair of laws in items 2 through 10, one need prove only one of the statements and then invoke this principle to obtain the other statement in the pair.

We must be careful about applying Theorem 3.5. This result cannot be applied to particular situations but only to results (theorems) about sets in general. For example, let us consider the *particular* situation where $\mathcal{U} = \{1, 2, 3, 4, 5\}$ and $A = \{1, 2, 3, 4\}$, $B = \{1, 2, 3, 5\}$, $C = \{1, 2\}$, and $D = \{1, 3\}$. Under these circumstances

$$A \cap B = \{1, 2, 3\} = C \cup D.$$

However, we cannot infer that $s: A \cap B = C \cup D \Rightarrow s^d: A \cup B = C \cap D$. For here $A \cup B = \{1, 2, 3, 4, 5\}$, whereas $C \cap D = \{1\}$. The reason why Theorem 3.5 is not applicable here is that although $A \cap B = C \cup D$ in this *particular* example, it is not true in general (that is, for any sets A, B, C, and D taken from a universe $\mathcal{U}$).

EXAMPLE 3.19   Inasmuch as Definition 3.9 and Theorem 3.5 do not mention anything about subsets, can we find a dual for the statement $A \subseteq B$ (where A, $B \subseteq \mathcal{U}$)?

Here we get an opportunity to use some of the results in Theorem 3.4. We can deal with the statement $A \subseteq B$ by using the equivalent statement $A \cup B = B$.

The dual of $A \cup B = B$ gives us $A \cap B = B$. But $A \cap B = B \Longleftrightarrow B \subseteq A$. Consequently, the dual of the statement $A \subseteq B$ is the statement $B \subseteq A$. (We could also have obtained this result by using $A \subseteq B \Longleftrightarrow A \cap B = A$.)

When we consider the relations that may exist among the sets that are involved in a set-equality or subset statement, we can investigate the situation graphically.

Named in honor of the English logician John Venn (1834–1923), a *Venn diagram* is constructed as follows: $\mathcal{U}$ is depicted as the interior of a rectangle, while subsets of $\mathcal{U}$ are represented by the interiors of circles and other closed curves. Figure 3.6 shows four Venn diagrams. The (blue) shaded region in Fig. 3.6(a) represents the set A, whereas $\overline{A}$ is represented by the unshaded area. The shaded region in Fig. 3.6(b) comprises $A \cup B$; the set $A \cap B$ is represented by the shaded region in Fig. 3.6(c). The Venn diagram for $A - B$ is given in part (d) of this figure.

In Fig. 3.7 Venn diagrams are used to establish the second of DeMorgan's Laws. Figure 3.7(a) has everything except $A \cap B$ shaded, so the shaded portion represents $\overline{A \cap B}$. We now develop a Venn diagram to depict $\overline{A} \cup \overline{B}$. In Fig. 3.7(b), $\overline{A}$ is the shaded region (outside the circle representing set A). Likewise, $\overline{B}$ is the shaded region shown in Fig. 3.7(c). When the results from Fig. 3.7(b) and Fig. 3.7(c) are put together, we get the Venn diagram for their union in Fig. 3.7(d). Since the shaded region in part (d) is the same as that in part (a), it follows that $\overline{A \cap B} = \overline{A} \cup \overline{B}$.

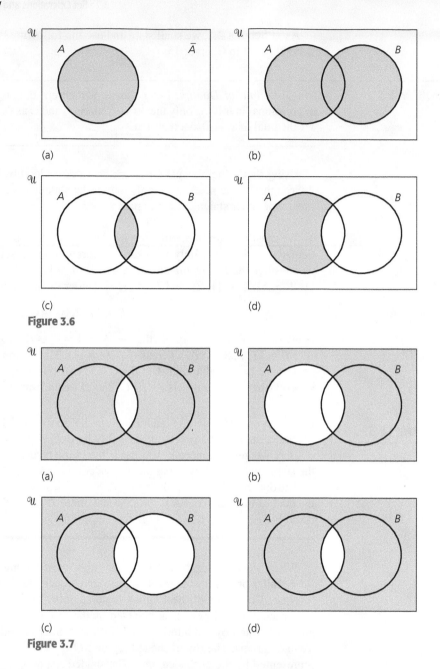

Figure 3.6

Figure 3.7

We further illustrate the use of these diagrams by showing that for any sets A, B, $C \subseteq \mathcal{U}$,

$$\overline{(A \cup B) \cap C} = (\overline{A} \cap \overline{B}) \cup \overline{C}.$$

Instead of shading regions, another approach that also uses Venn diagrams numbers the regions as shown in Fig. 3.8 where, for example, region 3 is $\overline{A} \cap B \cap \overline{C}$ and region 7 is $A \cap \overline{B} \cap C$. Each region is a set of the form $S_1 \cap S_2 \cap S_3$, where S_1 is replaced by A or $\overline{A}$, S_2 by B or $\overline{B}$, and S_3 by C or $\overline{C}$. Consequently, by the rule of product, there are eight possible regions.

Consulting Fig. 3.8, we see that $A \cup B$ comprises regions 2, 3, 5, 6, 7, 8 and that regions 4, 6, 7, 8 make up set C. Therefore $(A \cup B) \cap C$ comprises the regions common to $A \cup B$

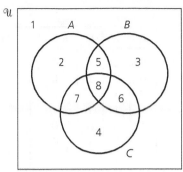

Figure 3.8

and C: namely, regions 6, 7, 8. Consequently, $\overline{(A \cup B)} \cap \overline{C}$ is made up of regions 1, 2, 3, 4, 5. The set $\overline{A}$ consists of regions 1, 3, 4, 6, while regions 1, 2, 4, 7 make up $\overline{B}$. Consequently, $\overline{A} \cap \overline{B}$ comprises regions 1 and 4. Since regions 4, 6, 7, 8 comprise C, the set $\overline{C}$ is made up of regions 1, 2, 3, 5. Taking the union of $\overline{A} \cap \overline{B}$ with $\overline{C}$, we then finish with regions 1, 2, 3, 4, 5, as we did for $\overline{(A \cup B)} \cap \overline{C}$.

One more technique for establishing set equalities is the *membership table*. (This method is akin to using the truth table introduced in Section 2.1.)

We observe that for sets A, $B \subseteq \mathcal{U}$, an element $x \in \mathcal{U}$ satisfies exactly one of the following four situations:

a) $x \notin A$, $x \notin B$ **b)** $x \notin A$, $x \in B$.

c) $x \in A$, $x \notin B$ **d)** $x \in A$, $x \in B$.

When x is an element of a given set, we write a 1 in the column representing that set in the membership table; when x is not in the set, we enter a 0. Table 3.2 gives the membership tables for $A \cap B$, $A \cup B$, $\overline{A}$ in this notation. Here, for example, the third row in part (a) of the table tells us that when an element $x \in \mathcal{U}$ is in set A but not in B, then it is not in $A \cap B$ but it is in $A \cup B$.

Table 3.2

A	B	$A \cap B$	$A \cup B$
0	0	0	0
0	1	0	1
1	0	0	1
1	1	1	1

(a)

A	$\overline{A}$
0	1
1	0

(b)

These binary operations on 0 and 1 are the same as in ordinary arithmetic (relative to $\cdot$ and $+$) except that $1 \cup 1 = 1$.

Using membership tables, we can establish the equality of two sets by comparing their respective columns in the table. Table 3.3 demonstrates this for the Distributive Law of union over intersection. We see here how each of the eight rows corresponds with exactly one of the eight regions in the Venn diagram of Fig. 3.8. For example, row 1 corresponds with region 1: $\overline{A} \cap \overline{B} \cap \overline{C}$; and row 6 corresponds with region 7: $A \cap \overline{B} \cap C$.

Table 3.3

A	B	C	$B \cap C$	$A \cup (B \cap C)$	$A \cup B$	$A \cup C$	$(A \cup B) \cap (A \cup C)$
0	0	0	0	0	0	0	0
0	0	1	0	0	0	1	0
0	1	0	0	0	1	0	0
0	1	1	1	1	1	1	1
1	0	0	0	1	1	1	1
1	0	1	0	1	1	1	1
1	1	0	0	1	1	1	1
1	1	1	1	1	1	1	1

↑   ↑

**Since these columns are identical, we conclude
that** $A \cup (B \cap C) = (A \cup B) \cap (A \cup C)$.

Before we continue let us make two points. (1) A Venn diagram is simply a graphical representation of a membership table. (2) The use of Venn diagrams and/or membership tables may be appealing, especially to the reader who presently does not appreciate writing proofs. However, neither one of these techniques specifies the logic and reasoning displayed in the element arguments we presented, for instance, to prove that for any A, B, $C \subseteq \mathcal{U}$,

$$\overline{A \cup B} = \overline{A} \cap \overline{B}, \quad \text{and} \quad A \cap (B \cup C) = (A \cap B) \cup (A \cap C).$$

We feel that Venn diagrams may help us to understand certain mathematical situations —but when the number of sets involved exceeds three, the diagram could be difficult to draw.

In summary, let us agree that the element argument (especially with its detailed explanations) is more rigorous than these two techniques and is the preferred method for proving results in set theory.

Now that we have the laws of set theory, what can we do with them? The following examples will demonstrate how the laws are used to simplify a complicated set expression or to derive new set equalities. (When more than one law is used in a given step, we list the principal law as the reason.)

EXAMPLE 3.20

Simplify the expression $\overline{\overline{(A \cup B) \cap C} \cup \overline{B}}$.

$\overline{\overline{(A \cup B) \cap C} \cup \overline{B}}$	**Reasons**
$= \overline{\overline{((A \cup B) \cap C)}} \cap \overline{\overline{B}}$	DeMorgan's Law
$= ((A \cup B) \cap C) \cap B$	Law of Double Complement
$= (A \cup B) \cap (C \cap B)$	Associative Law of Intersection
$= (A \cup B) \cap (B \cap C)$	Commutative Law of Intersection
$= [(A \cup B) \cap B] \cap C$	Associative Law of Intersection
$= B \cap C$	Absorption Law

The reader should note the similarity between the steps and reasons in this example and those for simplifying the statement

$$\neg[\neg[(p \vee q) \wedge r] \vee \neg q]$$

to the statement

$$q \wedge r$$

in Example 2.17.

EXAMPLE 3.21

Express $\overline{A - B}$ in terms of $\cup$ and $\overline{}$.

From the definition of relative complement, $A - B = \{x \mid x \in A \wedge x \notin B\} = A \cap \overline{B}$. Therefore,

	Reasons
$\overline{A - B} = \overline{A \cap \overline{B}}$	
$\quad = \overline{A} \cup \overline{\overline{B}}$	DeMorgan's Law
$\quad = \overline{A} \cup B$	Law of Double Complement

EXAMPLE 3.22

From the observation made in Example 3.21, we have $A \triangle B = \{x \mid x \in A \cup B \wedge x \notin A \cap B\} = (A \cup B) - (A \cap B) = (A \cup B) \cap \overline{(A \cap B)}$, so

	Reasons
$\overline{A \triangle B} = \overline{(A \cup B) \cap \overline{(A \cap B)}}$	
$\quad = \overline{(A \cup B)} \cup \overline{\overline{(A \cap B)}}$	DeMorgan's Law
$\quad = \overline{(A \cup B)} \cup (A \cap B)$	Law of Double Complement
$\quad = (A \cap B) \cup \overline{(A \cup B)}$	Commutative Law of $\cup$
$\quad = (A \cap B) \cup (\overline{A} \cap \overline{B})$	DeMorgan's Law
$\quad = [(A \cap B) \cup \overline{A}] \cap [(A \cap B) \cup \overline{B}]$	Distributive Law of $\cup$ over $\cap$
$\quad = [(A \cup \overline{A}) \cap (B \cup \overline{A})] \cap [(A \cup \overline{B}) \cap (B \cup \overline{B})]$	Distributive Law of $\cup$ over $\cap$
$\quad = [\mathcal{U} \cap (B \cup \overline{A})] \cap [(A \cup \overline{B}) \cap \mathcal{U}]$	Inverse Law
$\quad = (B \cup \overline{A}) \cap (A \cup \overline{B})$	Identity Law
$\quad = (\overline{A} \cup B) \cap (A \cup \overline{B})$	Commutative Law of $\cup$
$\quad = (\overline{A} \cup B) \cap \overline{(\overline{A} \cap B)}$	DeMorgan's Law
$\quad = \overline{A} \triangle B$	
$\quad = (A \cup \overline{B}) \cap (\overline{A} \cup B)$	Commutative Law of $\cap$
$\quad = (A \cup \overline{B}) \cap \overline{(A \cap \overline{B})}$	DeMorgan's Law
$\quad = A \triangle \overline{B}$	

In closing this section we extend the set operations of $\cup$ and $\cap$ beyond three sets.

Definition 3.10

Let I be a nonempty set and $\mathcal{U}$ a universe. For each $i \in I$ let $A_i \subseteq \mathcal{U}$. Then I is called an *index set* (or *set of indices*), and each $i \in I$ is called an *index*.

Under these conditions,

$$\bigcup_{i \in I} A_i = \{x \mid x \in A_i \quad \text{for at least one } i \in I\}, \qquad \text{and}$$

$$\bigcap_{i \in I} A_i = \{x \mid x \in A_i \quad \text{for every } i \in I\}.$$

We can rephrase Definition 3.10 by using quantifiers:

$$x \in \bigcup_{i \in I} A_i \iff \exists i \in I \, (x \in A_i) \qquad x \in \bigcap_{i \in I} A_i \iff \forall i \in I \, (x \in A_i)$$

Then $x \notin \bigcup_{i \in I} A_i \iff \neg [\exists i \in I \, (x \in A_i)] \iff \forall i \in I \, (x \notin A_i)$; that is, $x \notin \bigcup_{i \in I} A_i$ if and only if $x \notin A_i$ for *every* index $i \in I$. Similarly, $x \notin \bigcap_{i \in I} A_i \iff \neg [\forall i \in I \, (x \in A_i)] \iff \exists i \in I \, (x \notin A_i)$; that is, $x \notin \bigcap_{i \in I} A_i$ if and only if $x \notin A_i$ for *at least one* index $i \in I$.

If the index set I is the set $\mathbf{Z}^+$, we can write

$$\bigcup_{i \in \mathbf{Z}^+} A_i = A_1 \cup A_2 \cup \cdots = \bigcup_{i=1}^{\infty} A_i, \qquad \bigcap_{i \in \mathbf{Z}^+} A_i = A_1 \cap A_2 \cap \cdots = \bigcap_{i=1}^{\infty} A_i.$$

EXAMPLE 3.23

Let $I = \{3, 4, 5, 6, 7\}$, and for each $i \in I$ let $A_i = \{1, 2, 3, \ldots, i\} \subseteq \mathcal{U} = \mathbf{Z}^+$. Then $\bigcup_{i \in I} A_i = \bigcup_{i=3}^{7} A_i = \{1, 2, 3, \ldots, 7\} = A_7$, whereas $\bigcap_{i \in I} A_i = \{1, 2, 3\} = A_3$.

EXAMPLE 3.24

Let $\mathcal{U} = \mathbf{R}$ and $I = \mathbf{R}^+$. If for each $r \in \mathbf{R}^+$, $A_r = [-r, r]$, then $\bigcup_{r \in I} A_r = \mathbf{R}$ and $\bigcap_{r \in I} A_r = \{0\}$.

When dealing with generalized unions and intersections, membership tables and Venn diagrams are unfortunately next to useless, but the rigorous element approach, as demonstrated in the first part of the proof of Theorem 3.3, is still available.

THEOREM 3.6

Generalized DeMorgan's Laws. Let I be an index set where for each $i \in I$, $A_i \subseteq \mathcal{U}$. Then

a) $\overline{\bigcup_{i \in I} A_i} = \bigcap_{i \in I} \overline{A_i}$

b) $\overline{\bigcap_{i \in I} A_i} = \bigcup_{i \in I} \overline{A_i}$

Proof: We shall prove Theorem 3.6(a) and leave the proof of part (b) for the reader. For each $x \in \mathcal{U}$, $x \in \overline{\bigcup_{i \in I} A_i} \iff x \notin \bigcup_{i \in I} A_i \iff x \notin A_i$, for all $i \in I \iff x \in \overline{A_i}$, for all $i \in I \iff x \in \bigcap_{i \in I} \overline{A_i}$.

EXERCISES 3.2

1. For $\mathcal{U} = \{1, 2, 3, \ldots, 9, 10\}$ let $A = \{1, 2, 3, 4, 5\}$, $B = \{1, 2, 4, 8\}$, $C = \{1, 2, 3, 5, 7\}$, and $D = \{2, 4, 6, 8\}$. Determine each of the following:

a) $(A \cup B) \cap C$

b) $A \cup (B \cap C)$

c) $\overline{C} \cup \overline{D}$

d) $\overline{C \cap D}$

e) $(A \cup B) - C$

f) $A \cup (B - C)$

g) $(B - C) - D$

h) $B - (C - D)$

i) $(A \cup B) - (C \cap D)$

2. If $A = [0, 3]$, $B = [2, 7)$, with $\mathcal{U} = \mathbf{R}$, determine each of the following:

a) $A \cap B$

b) $A \cup B$

c) $\overline{A}$

d) $A \triangle B$

e) $A - B$

f) $B - A$

3. a) Determine the sets A, B where $A - B = \{1, 3, 7, 11\}$, $B - A = \{2, 6, 8\}$, and $A \cap B = \{4, 9\}$.

b) Determine the sets C, D where $C - D = \{1, 2, 4\}$, $D - C = \{7, 8\}$, and $C \cup D = \{1, 2, 4, 5, 7, 8, 9\}$.

4. Let $A, B, C, D, E \subseteq \mathbf{Z}$ be defined as follows:

$A = \{2n \,|\, n \in \mathbf{Z}\}$ — that is, A is the set of all (integer) multiples of 2;

$B = \{3n \,|\, n \in \mathbf{Z}\}$; $\qquad C = \{4n \,|\, n \in \mathbf{Z}\}$;

$D = \{6n \,|\, n \in \mathbf{Z}\}$; and $\quad E = \{8n \,|\, n \in \mathbf{Z}\}$.

a) Which of the following statements are true and which are false?

i) $E \subseteq C \subseteq A$

ii) $A \subseteq C \subseteq E$

iii) $B \subseteq D$

iv) $D \subseteq B$

v) $D \subseteq A$

vi) $\overline{D} \subseteq \overline{A}$

b) Determine each of the following sets.

i) $C \cap E$ ii) $B \cup D$ iii) $A \cap B$

iv) $B \cap D$ v) $\overline{A}$ vi) $A \cap E$

5. Determine which of the following statements are true and which are false.

a) $\mathbf{Z}^+ \subseteq \mathbf{Q}^+$ **b)** $\mathbf{Z}^+ \subseteq \mathbf{Q}$

c) $\mathbf{Q}^+ \subseteq \mathbf{R}$ **d)** $\mathbf{R}^+ \subseteq \mathbf{Q}$

e) $\mathbf{Q}^+ \cap \mathbf{R}^+ = \mathbf{Q}^+$ **f)** $\mathbf{Z}^+ \cup \mathbf{R}^+ = \mathbf{R}^+$

g) $\mathbf{R}^+ \cap \mathbf{C} = \mathbf{R}^+$ **h)** $\mathbf{C} \cup \mathbf{R} = \mathbf{R}$

i) $\mathbf{Q}^* \cap \mathbf{Z} = \mathbf{Z}$

6. Prove each of the following results without using Venn diagrams or membership tables. (Assume a universe $\mathcal{U}$.)

a) If $A \subseteq B$ and $C \subseteq D$, then $A \cap C \subseteq B \cap D$ and $A \cup C \subseteq B \cup D$.

b) $A \subseteq B$ if and only if $A \cap \overline{B} = \emptyset$.

c) $A \subseteq B$ if and only if $\overline{A} \cup B = \mathcal{U}$.

7. Prove or disprove each of the following:

a) For sets $A, B, C \subseteq \mathcal{U}$, $A \cap C = B \cap C \Rightarrow A = B$.

b) For sets $A, B, C \subseteq \mathcal{U}$, $A \cup C = B \cup C \Rightarrow A = B$.

c) For sets $A, B, C \subseteq \mathcal{U}$,
$[(A \cap C = B \cap C) \wedge (A \cup C = B \cup C)] \Rightarrow A = B$.

d) For sets, $A, B, C \subseteq \mathcal{U}$, $A \triangle C = B \triangle C \Rightarrow A = B$.

8. Using Venn diagrams, investigate the truth or falsity of each of the following, for sets $A, B, C \subseteq \mathcal{U}$.

a) $A \triangle (B \cap C) = (A \triangle B) \cap (A \triangle C)$

b) $A - (B \cup C) = (A - B) \cap (A - C)$

c) $A \triangle (B \triangle C) = (A \triangle B) \triangle C$

9. If $A = \{a, b, d\}$, $B = \{d, x, y\}$, and $C = \{x, z\}$, how many proper subsets are there for the set $(A \cap B) \cup C$? How many for the set $A \cap (B \cup C)$?

10. For a given universal set $\mathcal{U}$, each subset A of $\mathcal{U}$ satisfies the idempotent laws of union and intersection. (a) Are there any real numbers that satisfy an idempotent property for addition? (That is, can we find any real number(s) x such that $x + x = x$?) (b) Answer part (a) upon replacing addition by multiplication.

11. Write the dual statement for each of the following set-theoretic results.

a) $\mathcal{U} = (A \cap B) \cup (A \cap \overline{B}) \cup (\overline{A} \cap B) \cup (\overline{A} \cap \overline{B})$

b) $A = A \cap (A \cup B)$

c) $A \cup B = (A \cap B) \cup (A \cap \overline{B}) \cup (\overline{A} \cap B)$

d) $A = (A \cup B) \cap (A \cup \emptyset)$

12. Let $A, B \subseteq \mathcal{U}$. Use the equivalence $A \subseteq B \Leftrightarrow A \cap B = A$ to show that the dual statement of $A \subseteq B$ is the statement $B \subseteq A$.

13. Prove or disprove each of the following for sets $A, B \subseteq \mathcal{U}$.

a) $\mathcal{P}(A \cup B) = \mathcal{P}(A) \cup \mathcal{P}(B)$

b) $\mathcal{P}(A \cap B) = \mathcal{P}(A) \cap \mathcal{P}(B)$

14. Use membership tables to establish each of the following:

a) $\overline{A \cap B} = \overline{A} \cup \overline{B}$ **b)** $A \cup A = A$

c) $A \cup (A \cap B) = A$

d) $\overline{(A \cap B) \cup (\overline{A} \cap C)} = (A \cap \overline{B}) \cup (\overline{A} \cap \overline{C})$

15. **a)** How many rows are needed to construct the membership table for $A \cap (B \cup C) \cap (D \cup \overline{E} \cup \overline{F})$?

b) How many rows are needed to construct the membership table for a set made up from the sets $A_1, A_2, \ldots, A_n$, using $\cap$, $\cup$, and $\overline{}$?

c) Given the membership tables for two sets A, B, how can the relation $A \subseteq B$ be recognized?

d) Use membership tables to determine whether or not $(A \cap B) \cup \overline{(B \cap C)} \supseteq A \cup \overline{B}$.

16. Provide the justifications (selected from the laws of set theory) for the steps that are needed to simplify the set

$$(A \cap B) \cup [B \cap ((C \cap D) \cup (C \cap \overline{D}))],$$

where $A, B, C, D \subseteq \mathcal{U}$.

Steps	Reasons
$(A \cap B) \cup [B \cap ((C \cap D) \cup (C \cap \overline{D}))]$	
$= (A \cap B) \cup [B \cap (C \cap (D \cup \overline{D}))]$	
$= (A \cap B) \cup [B \cap (C \cap \mathcal{U})]$	
$= (A \cap B) \cup (B \cap C)$	
$= (B \cap A) \cup (B \cap C)$	
$= B \cap (A \cup C)$	

17. Using the laws of set theory, simplify each of the following:

a) $A \cap (B - A)$

b) $(A \cap B) \cup (A \cap B \cap \overline{C} \cap D) \cup (\overline{A} \cap B)$

c) $(A - B) \cup (A \cap B)$

d) $\overline{A} \cup \overline{B} \cup (A \cap B \cap \overline{C})$

18. For each $n \in \mathbf{Z}^+$ let $A_n = \{1, 2, 3, \ldots, n - 1, n\}$. (Here $\mathcal{U} = \mathbf{Z}^+$ and the index set $I = \mathbf{Z}^+$.) Determine

$$\bigcup_{n=1}^{7} A_n, \quad \bigcap_{n=1}^{11} A_n, \quad \bigcup_{n=1}^{m} A_n, \quad \text{and} \quad \bigcap_{n=1}^{m} A_n,$$

where m is a fixed positive integer.

19. Let $\mathcal{U} = \mathbf{R}$ and let $I = \mathbf{Z}^+$. For each $n \in \mathbf{Z}^+$ let $A_n = [-2n, 3n]$. Determine each of the following:

a) A_3 **b)** A_4

c) $A_3 - A_4$ **d)** $A_3 \triangle A_4$

e) $\bigcup_{n=1}^{7} A_n$ **f)** $\bigcap_{n=1}^{7} A_n$

g) $\bigcup_{n \in \mathbf{Z}^+} A_n$ **h)** $\bigcap_{n=1}^{\infty} A_n$

20. Provide the details for the proof of Theorem 3.6(b).

3.3
Counting and Venn Diagrams

With all of the theoretical work and theorem proving we did in the last section, now is a good time to examine some additional counting problems.

For sets A, B from a finite universe $\mathcal{U}$, the following Venn diagrams will help us obtain counting formulas for $|\overline{A}|$ and $|A \cup B|$ in terms of $|\mathcal{U}|$, $|A|$, $|B|$, and $|A \cap B|$.

As Fig. 3.9 demonstrates, $A \cup \overline{A} = \mathcal{U}$ and $A \cap \overline{A} = \emptyset$, so by the rule of sum, $|A| + |\overline{A}| = |\mathcal{U}|$ or $|\overline{A}| = |\mathcal{U}| - |A|$. The sets A, B, in Fig. 3.10, have empty intersection, so here the rule of sum leads us to $|A \cup B| = |A| + |B|$ and necessitates that A, B be finite but does not require any condition on the cardinality of $\mathcal{U}$.

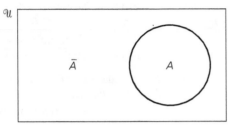

Figure 3.9

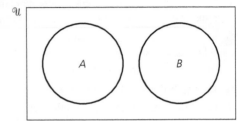

Figure 3.10

Turning to the case where A, B are not disjoint, we motivate the formula for $|A \cup B|$ with the following example.

EXAMPLE 3.25

In a class of 50 college freshmen, 30 are studying C++, 25 are studying Java, and 10 are studying both languages. How many freshmen are studying either computer language?

We let $\mathcal{U}$ be the class of 50 freshmen, A the subset of those students studying C++, and B the subset of those studying Java. To answer the question, we need $|A \cup B|$. In Fig. 3.11 the numbers in the regions are obtained from the given information: $|A| = 30$, $|B| = 25$, $|A \cap B| = 10$. Consequently, $|A \cup B| = 45 \neq 55 = 30 + 25 = |A| + |B|$, because $|A| + |B|$ counts the students in $A \cap B$ twice. To remedy this overcount, we subtract $|A \cap B|$ from $|A| + |B|$ to obtain the correct formula: $|A \cup B| = |A| + |B| - |A \cap B|$.

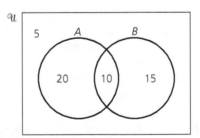

Figure 3.11

If A and B are finite sets, then $|A \cup B| = |A| + |B| - |A \cap B|$. Consequently, finite sets A and B are (mutually) disjoint if and only if $|A \cup B| = |A| + |B|$.

In addition, when $\mathcal{U}$ is finite, from DeMorgan's Law we have $|\overline{A} \cap \overline{B}| = |\overline{A \cup B}| = |\mathcal{U}| - |A \cup B| = |\mathcal{U}| - |A| - |B| + |A \cap B|$.

This situation extends to three sets, as the following example illustrates.

EXAMPLE 3.26

An AND gate in an ASIC (Application Specific Integrated Circuit) has two inputs: I_1, I_2, and one output: O. (See Fig. 3.12). Such an AND gate can have any or all of the following defects:

D_1:   The input I_1 is stuck at 0.

D_2:   The input I_2 is stuck at 0.

D_3:   The output O is stuck at 1.

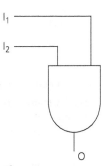

Figure 3.12

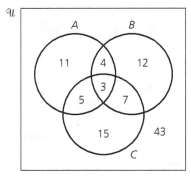

Figure 3.13

For a sample of 100 such gates we let A, B, and C be the subsets (of these 100 gates) having defects D_1, D_2, and D_3, respectively. With $|A| = 23$, $|B| = 26$, $|C| = 30$, $|A \cap B| = 7$, $|A \cap C| = 8$, $|B \cap C| = 10$, and $|A \cap B \cap C| = 3$, how many gates in the sample have at least one of the defects D_1, D_2, D_3?

Working backward from $|A \cap B \cap C| = 3$ to $|A| = 23$, we label the regions as shown in Fig. 3.13 and find that $|A \cup B \cup C| = |A| + |B| + |C| - |A \cap B| - |A \cap C| - |B \cap C| + |A \cap B \cap C| = 23 + 26 + 30 - 7 - 8 - 10 + 3 = 57$. Thus the sample contains 57 AND gates with at least one of the defects and $100 - 57 = 43$ AND gates with no defect.

> If A, B, C are finite sets, then $|A \cup B \cup C| = |A| + |B| + |C| - |A \cap B| - |A \cap C| - |B \cap C| + |A \cap B \cap C|$.
>
> From the formula for $|A \cup B \cup C|$ and DeMorgan's Law, we find that if the universe $\mathcal{U}$ is finite, then $|\overline{A} \cap \overline{B} \cap \overline{C}| = |\overline{A \cup B \cup C}| = |\mathcal{U}| - |A \cup B \cup C| = |\mathcal{U}| - |A| - |B| - |C| + |A \cap B| + |A \cap C| + |B \cap C| - |A \cap B \cap C|$.

We close this section with a problem that uses this last result.

EXAMPLE 3.27

A student visits an arcade each day after school and plays one game of either Laser Man, Millipede, or Space Conquerors. In how many ways can he play one game each day so that he plays each of the three types at least once during a given school week?

Here there is a slight twist. The set $\mathcal{U}$ consists of all arrangements of size 5 taken from the set of three games, with repetitions allowed. The set A represents the subset of all sequences of five games played during the week *without playing Laser Man*. The sets B and C are defined similarly, leaving out Millipede and Space Conquerors, respectively. The enumeration techniques of Chapter 1 give $|\mathcal{U}| = 3^5$, $|A| = |B| = |C| = 2^5$, $|A \cap B| =$

$|A \cap C| = |B \cap C| = 1^5 = 1$ and $|A \cap B \cap C| = 0$, so by the preceding formula there are $|\overline{A} \cap \overline{B} \cap \overline{C}| = 3^5 - 3 \cdot 2^5 + 3 \cdot 1^5 - 0 = 150$ ways the student can select his daily games during a school week and play each type of game at least once.

This example can be expressed in an equivalent distribution form, since we are seeking the number of ways to distribute five distinct objects (Monday, Tuesday, . . . , Friday) among three distinct containers (the computer games) with no container left empty. More will be said about this in Chapter 5.

EXERCISES 3.3

1. During freshman orientation at a small liberal arts college, two showings of the latest James Bond movie were presented. Among the 600 freshmen, 80 attended the first showing and 125 attended the second showing, while 450 didn't make it to either showing. How many of the 600 freshmen attended twice?

2. A manufacturer of 2000 automobile batteries is concerned about defective terminals and defective plates. If 1920 of her batteries have neither defect, 60 have defective plates, and 20 have both defects, how many batteries have defective terminals?

3. A binary string of length 12 is made up of 12 bits (that is, 12 symbols, each of which is a 0 or a 1). How many such strings either start with three 1's or end in four 0's?

4. Determine $|A \cup B \cup C|$ when $|A| = 50$, $|B| = 500$, and $|C| = 5000$, if (a) $A \subseteq B \subseteq C$; (b) $A \cap B = A \cap C = B \cap C = \emptyset$; and (c) $|A \cap B| = |A \cap C| = |B \cap C| = 3$ and $|A \cap B \cap C| = 1$.

5. How many permutations of the digits 0, 1, 2, . . . , 9 either start with a 3 or end with a 7?

6. A professor has two dozen introductory textbooks on computer science and is concerned about their coverage of the topics (A) compilers, (B) data structures, and (C) operating systems.

The following data are the numbers of books that contain material on these topics:

$|A| = 8$	$|B| = 13$	$|C| = 13$
$|A \cap B| = 5$	$|A \cap C| = 3$	$|B \cap C| = 6$
$|A \cap B \cap C| = 2$		

(a) How many of the textbooks include material on exactly one of these topics? (b) How many do not deal with any of the topics? (c) How many have no material on compilers?

7. How many permutations of the 26 different letters of the alphabet contain (a) either the pattern "OUT" or the pattern "DIG"? (b) neither the pattern "MAN" nor the pattern "ANT"?

8. A six-character variable name in a certain version of ANSI FORTRAN starts with a letter of the alphabet. Each of the other five characters can be either a letter or a digit. (Repetitions are allowed.) How many six-character variable names contain the pattern "FUN" or the pattern "TIP"?

9. How many arrangements of the letters in MISCELLANEOUS have no pair of consecutive identical letters?

10. How many arrangements of the letters in CHEMIST have H before E, or E before T, or T before M? (Here "before" means anywhere before, not just immediately before.)

3.4
A First Word on Probability

When one performs an *experiment* such as tossing a single fair coin, rolling a single fair die, or selecting two students at random from a class of 20 to work on a project, a set of all possible outcomes for each situation is called a *sample space*. Consequently, {H, T} serves as a sample space for the first experiment mentioned and {1, 2, 3, 4, 5, 6} is a sample space for the roll of a single fair die. Moreover, $\{\{a_i, a_j\} | 1 \le i \le 20, 1 \le j \le 20, i \ne j\}$ can be used for the last experiment, with a_i denoting the ith student, for each $1 \le i \le 20$.

In dealing with the sample space $\mathcal{S} = \{1, 2, 3, 4, 5, 6\}$ for the roll of a single fair die, we feel that each of the six possible outcomes has the *same*, or *equal, likelihood* of occurrence. Using this assumption of equal likelihood, we shall start our study of probability theory with a definition for probability that was first given by the French mathematician Pierre-Simon de Laplace (1749–1827) in his *Analytic Theory of Probability*.

Under the assumption of equal likelihood, let $\mathscr{S}$ be the sample space for an experiment $\mathscr{E}$. Each subset A of $\mathscr{S}$, including the empty subset, is called an *event*. Each element of $\mathscr{S}$ determines an *outcome*, so if $|\mathscr{S}| = n$ and $a \in \mathscr{S}$, $A \subseteq \mathscr{S}$, then

$$Pr(\{a\}) = \textit{The probability that } \{a\} \textit{ (or, a) occurs} = \frac{|\{a\}|}{\mathscr{S}} = \frac{1}{n}, \textit{ and}$$

$$Pr(A) = \textit{The probability that } A \textit{ occurs} = \frac{|A|}{|\mathscr{S}|} = \frac{|A|}{n}.$$

[*Note:* We often write $Pr(a)$ for $Pr(\{a\})$.]

We demonstrate these ideas in the following four examples.

EXAMPLE 3.28

When Daphne tosses a fair coin, what is the probability she gets a head? Here the sample space $\mathscr{S} = \{H, T\}$ with $A = \{H\}$ and we find that

$$Pr(A) = \frac{|A|}{|\mathscr{S}|} = \frac{1}{2}.$$

EXAMPLE 3.29

If Dillon rolls a fair die, what is the probability he gets (a) a 5 or a 6, (b) an even number? For either part the sample space $\mathscr{S} = \{1, 2, 3, 4, 5, 6\}$. In part (a) we have event $A = \{5, 6\}$ and $Pr(A) = \frac{|A|}{|\mathscr{S}|} = \frac{2}{6} = \frac{1}{3}$. For part (b) we consider event $B = \{2, 4, 6\}$ and find that $Pr(B) = \frac{|B|}{|\mathscr{S}|} = \frac{3}{6} = \frac{1}{2}$.

Furthermore we also notice here that

i) $Pr(\mathscr{S}) = \frac{|\mathscr{S}|}{|\mathscr{S}|} = \frac{6}{6} = 1$ — after all, the occurrence of the event $\mathscr{S}$ is a certainty; and

ii) $Pr(\overline{A}) = Pr(\{1, 2, 3, 4\}) = \frac{|\overline{A}|}{\mathscr{S}} = \frac{4}{6} = \frac{2}{3} = 1 - \frac{1}{3} = 1 - Pr(A)$.

EXAMPLE 3.30

There are 20 students enrolled in Mrs. Arnold's fourth-grade class. Hence, if she wants to select two of her students, at random, to take care of the class rabbit, she may make her selection in $\binom{20}{2} = 190$ ways, so $|\mathscr{S}| = 190$.

Now suppose that Kyle and Kody are two of the 20 students in the class and we let A be the event that Kyle is one of the students selected and B be the event that the selection includes Kody. Consequently, upon choosing the students, at random, the probability that Mrs. Arnold selects

a) both Kyle and Kody is $Pr(A \cap B) = \binom{2}{2}/\binom{20}{2} = 1/190$;

b) neither Kyle nor Kody is $Pr(\overline{A} \cap \overline{B}) = \binom{18}{2}/\binom{20}{2} = 153/190$;

c) Kyle but not Kody is $Pr(A \cap \overline{B}) = \binom{1}{1}\binom{18}{1}/\binom{20}{2} = 18/190 = 9/95$.

EXAMPLE 3.31

Consider drawing five cards from a standard deck of 52 cards. This can be done in $\binom{52}{5} = 2{,}598{,}960$ ways. Now suppose that Tanya draws five cards, at random, from a standard deck. What is the probability she gets (a) three aces and two jacks; (b) three aces and a pair; (c) a full house (that is, three of one kind and a pair)?

In all three cases we have $|\mathscr{S}| = 2{,}598{,}960$.

a) There are $\binom{4}{3} = 4$ ways in which one can select three aces and $\binom{4}{2} = 6$ ways in which two jacks may be selected. Consequently, if A is the event where Tanya draws three aces and two jacks, then $|A| = \binom{4}{3}\binom{4}{2} = 4 \cdot 6 = 24$ and $Pr(A) = 24/2{,}598{,}960 \doteq 0.000009234$.

b) Once again there are $\binom{4}{3} = 4$ ways to select the aces, and there are $\binom{4}{2} = 6$ ways to select a pair of deuces, or a pair of threes, ..., or a pair of tens, or a pair of jacks, ..., or a pair of kings. So the pair can be selected in $\binom{12}{1}\binom{4}{2} = 12 \cdot 6 = 72$ ways. If B is the event where three aces and a pair are drawn, then $Pr(B) = (4 \cdot 72)/2{,}598{,}960 \doteq 0.000110814$.

c) From part (b) we know there are $4 \cdot 72 = 288$ full houses with three aces. Likewise, there are 288 full houses with three deuces, 288 with three threes, ..., and 288 with three kings. So the probability that Tanya draws a full house is $\binom{13}{1}\binom{4}{3}\binom{12}{1}\binom{4}{2}/\binom{52}{5} = 3744/2{,}598{,}960 \doteq 0.001440576$.

If these three probabilities appear on the slim side, consider the chances of Tanya drawing a royal flush — that is, the ten, jack, queen, king, and ace of one given suit. For this five-card hand the probability is only $4/\binom{52}{5} = 4/2{,}598{,}960 \doteq 0.000001539$.

To study some additional sample spaces we need to introduce the idea of the ordered pair. This arises in the following structure.

Definition 3.11

For sets A, B, the *Cartesian product*, or *cross product*, of A and B is denoted by $A \times B$ and equals $\{(a, b) | a \in A, b \in B\}$.

We call the elements of $A \times B$ ordered pairs. For $(a, b), (c, d) \in A \times B$, we have $(a, b) = (c, d)$ if and only if $a = c$ and $b = d$.[†]

EXAMPLE 3.32

If $A = \{1, 2, 3\}$ and $B = \{x, y\}$, then $A \times B = \{(1, x), (1, y), (2, x), (2, y), (3, x), (3, y)\}$ while $B \times A = \{(x, 1), (x, 2), (x, 3), (y, 1), (y, 2), (y, 3)\}$. Here $(1, x) \in A \times B$ but $(1, x) \notin B \times A$, although $(x, 1) \in B \times A$. So $A \times B \neq B \times A$, but $|A \times B| = 6 = 2 \cdot 3 = |A||B| = |B||A| = |B \times A|$.

Now let us see how the Cartesian product can arise in a probability problem.

EXAMPLE 3.33

Suppose Concetta rolls two fair dice. This experiment can be decomposed as follows. Let $\mathscr{E}_1$ be the experiment where the first die is rolled — with sample space $\mathscr{S}_1 = \{1, 2, 3, 4, 5, 6\}$. Likewise we let $\mathscr{E}_2$ account for the second die rolled — also with sample space $\mathscr{S}_2 = \{1, 2, 3, 4, 5, 6\}$. (To keep the two dice distinct we can imagine the first die rolled with the left hand and the second with the right. Or we can have the first die colored red and the

---

[†]More about ordered pairs and the Cartesian product is given in Section 5.1.

second green — in order to distinguish them.) Consequently, when Concetta rolls these dice the sample space

$$\mathcal{S} = \mathcal{S}_1 \times \mathcal{S}_2 = \{(1, 1), (1, 2), (1, 3), (1, 4), (1, 5), (1, 6), (2, 1), (2, 2), (2, 3), (2, 4),$$
$$(2, 5), (2, 6), (3, 1), (3, 2), (3, 3), (3, 4), (3, 5), (3, 6), (4, 1), (4, 2),$$
$$(4, 3), (4, 4), (4, 5), (4, 6), (5, 1), (5, 2), (5, 3), (5, 4), (5, 5), (5, 6),$$
$$(6, 1), (6, 2), (6, 3), (6, 4), (6, 5), (6, 6)\}$$
$$= \{(x, y) | x, y = 1, 2, 3, 4, 5, 6\}.$$

Now consider the following events:

A: Concetta rolls a 6 (that is, the top faces of the dice sum to 6);

B: The sum of the dice is at least 7;

C: Concetta rolls an even sum; and

D: The sum of the dice is 6 or less.

a) Here

  i) $A = \{(1, 5), (2, 4), (3, 3), (4, 2), (5, 1)\}$ with $Pr(A) = |A|/|\mathcal{S}| = 5/36$;

  ii) $B = \{(1, 6), (2, 5), (3, 4), (4, 3), (5, 2), (6, 1), (2, 6), (3, 5), (4, 4),$ $(5, 3), (6, 2), (3, 6), (4, 5), (5, 4), (6, 3), (4, 6), (5, 5), (6, 4), (5, 6),$ $(6, 5), (6, 6)\} = \{(x, y) | x, y = 1, 2, 3, 4, 5, 6; x + y \geq 7\}$ with $Pr(B) = |B|/|\mathcal{S}| = 21/36 = 7/12$;

  iii) $C = \{(1, 1), (1, 3), (2, 2), (3, 1), (1, 5), (2, 4), (3, 3), (4, 2), (5, 1),$ $(2, 6), (3, 5), (4, 4), (5, 3), (6, 2), (4, 6), (5, 5), (6, 4), (6, 6)\}$ with $Pr(C) = |C|/|\mathcal{S}| = 18/36 = 1/2$; and

  iv) $D = \{(1, 1), (1, 2), (2, 1), (1, 3), (2, 2), (3, 1), (1, 4), (2, 3), (3, 2), (4, 1),$ $(1, 5), (2, 4), (3, 3), (4, 2), (5, 1)\}$ with $Pr(D) = |D|/|\mathcal{S}| = 15/36 = 5/12$.

b) We notice the following:

  i) $A \cup B = \{(x, y) | x, y = 1, 2, 3, 4, 5, 6; x + y \geq 6\}$, so $|A \cup B| = 26$ and $Pr(A \cup B) = |A \cup B|/|\mathcal{S}| = \frac{26}{36} = \frac{5}{36} + \frac{21}{36} = Pr(A) + Pr(B)$;

  ii) $C \cup D = \{(1, 1), (1, 2), (2, 1), (1, 3), (2, 2), (3, 1), (1, 4), (2, 3), (3, 2),$ $(4, 1), (1, 5), (2, 4), (3, 3), (4, 2), (5, 1), (2, 6), (3, 5), (4, 4), (5, 3),$ $(6, 2), (4, 6), (5, 5), (6, 4), (6, 6)\}$ so $|C \cup D| = 24$ and $Pr(C \cup D) = |C \cup D|/|\mathcal{S}| = 24/36 = 2/3$.

Here, however,

$$Pr(C \cup D) = 24/36 \neq 33/36 = 18/36 + 15/36 = Pr(C) + Pr(D), \text{ although}$$
$$Pr(C \cup D) = 24/36 = 18/36 + 15/36 - 9/36 = Pr(C) + Pr(D) - P(C \cap D).$$

The result here and that in part (i) [of (b)] mirror the ideas we saw earlier in the formulas following Example 3.25.

  iii) Finally, $Pr(\overline{B}) = Pr(D) = 15/36 = 1 - 21/36 = 1 - Pr(B)$.

Let us consider a second example where the Cartesian product is used. This time we'll also learn about another important structure.

EXAMPLE 3.34   An experiment $\mathcal{E}$ is conducted as follows: A single die is rolled and its outcome noted, and then a coin is flipped and its outcome noted. Determine a sample space $\mathcal{S}$ for $\mathcal{E}$.

Let $\mathcal{E}_1$ denote the first part of experiment $\mathcal{E}$, and let $\mathcal{S}_1 = \{1, 2, 3, 4, 5, 6\}$ be a sample space for $\mathcal{E}_1$. Likewise let $\mathcal{S}_2 = \{H, T\}$ be a sample space for $\mathcal{E}_2$, the second part of the experiment. Then $\mathcal{S} = \mathcal{S}_1 \times \mathcal{S}_2$ is a sample space for $\mathcal{E}$.

This sample space can be represented pictorially with a *tree diagram* that exhibits all the possible outcomes of experiment $\mathcal{E}$. In Fig. 3.14 we have such a tree diagram, which proceeds from left to right. From the left-most endpoint, six branches originate for the six outcomes of the first stage of the experiment $\mathcal{E}$. From each point, numbered $1, 2, \ldots, 6$, two branches indicate the subsequent outcomes for tossing the coin. The 12 ordered pairs at the right endpoints constitute the sample space $\mathcal{S}$.

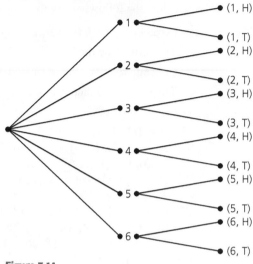

Figure 3.14

Now for this experiment $\mathcal{E}$ consider the events

A: A head appears when the coin is tossed.

B: A 3 appears when the die is rolled.

Then $A = \{(1, H), (2, H), (3, H), (4, H), (5, H), (6, H)\}$ and $B = \{(3, H), (3, T)\}$. So $Pr(A) = |A|/|\mathcal{S}| = 6/12 = 1/2$, $Pr(B) = |B|/|\mathcal{S}| = 2/12 = 1/6$, and

$$P(A \cup B) = \frac{7}{12} = \frac{6}{12} + \frac{2}{12} - \frac{1}{12} = Pr(A) + Pr(B) - Pr(A \cap B).$$

Before we continue let us look back at Examples 3.33 and 3.34. We may not realize it, but we have been making a certain assumption. In Example 3.33 we assumed that the outcome for the first die had no influence on the outcome for the second die. Likewise, in Example 3.34 we assumed that the outcome for the die had no bearing on the outcome for the coin. This concept of *independence* will be examined more closely in Section 3.6.

In our next example we extend the idea of the Cartesian (or, cross) product to more than two sets.

EXAMPLE 3.35

If Charles tosses a fair coin four times, what is the probability that he gets two heads and two tails?

Here the sample space for the first toss is $\mathcal{S}_1 = \{H, T\}$. Likewise, for the second, third, and fourth tosses, we have $\mathcal{S}_2 = \mathcal{S}_3 = \mathcal{S}_4 = \{H, T\}$. So, for this experiment of tossing a fair coin four times, we have the sample space $\mathscr{S} = \mathscr{S}_1 \times \mathscr{S}_2 \times \mathscr{S}_3 \times \mathscr{S}_4$, where a typical element of $\mathscr{S}$ is an ordered quadruple. For example, one such ordered quadruple is (H, T, T, T) (which may also be denoted HTTT). In this problem $|\mathscr{S}| = |\mathscr{S}_1||\mathscr{S}_2||\mathscr{S}_3||\mathscr{S}_4| = 2^4 = 16$. The event A we are concerned about contains all arrangements of H, H, T, T, so $|A| = 4!/(2!\,2!) = 6$. Consequently, $Pr(A) = |A|/|\mathscr{S}| = 6/16 = 3/8$.

(Comparable to Examples 3.33 and 3.34, here the result of each toss is *independent* of the outcome of any previous toss.)

The next example also requires some of the formulas developed in Chapter 1 for arrangements.

EXAMPLE 3.36

The acronym WYSIWYG (for, What you see is what you get!) is used to describe a user-interface. This user-interface presents material on a VDT (Video-display terminal) in precisely the same format the material appears on hard copy.

There are $7!/(2!\,2!) = 1260$ ways in which the letters in the acronym WYSIWYG can be arranged. Of these, $120(= 5!)$ arrangements have both consecutive W's and consecutive Y's. Consequently, if the letters for this acronym are arranged in a random manner, then we find the probability that the arrangement has both consecutive W's and consecutive Y's is $120/1260 \doteq 0.0952$.

The probability that a random arrangement of these seven letters starts and ends with the letter W is $[(5!/2!)]/[(7!/(2!\,2!))] = 60/1260 \doteq 0.0476$.

In our final example we shall use the concept of a Venn diagram.

EXAMPLE 3.37

In a survey of 120 passengers, an airline found that 48 enjoyed wine with their meals, 78 enjoyed mixed drinks, and 66 enjoyed iced tea. In addition, 36 enjoyed any given pair of these beverages and 24 passengers enjoyed them all. If two passengers are selected at random from the survey sample of 120, what is the probability that

a) (Event A) they both want only iced tea with their meals?

b) (Event B) they both enjoy exactly two of the three beverage offerings?

From the information provided, we construct the Venn diagram shown in Fig. 3.15. The sample space $\mathscr{S}$ consists of the pairs of passengers we can select from the sample of 120, so $|\mathscr{S}| = \binom{120}{2} = 7140$. The Venn diagram indicates that there are 18 passengers who drink only iced tea, so $|A| = \binom{18}{2}$ and $Pr(A) = 51/2380$. The reader should verify that $Pr(B) = 3/34$.

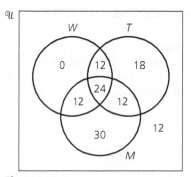

Figure 3.15

EXERCISES 3.4

1. The sample space for an experiment is $\mathcal{S} = \{a, b, c, d, e, f, g, h\}$, where each outcome is equally likely. If event $A = \{a, b, c\}$ and event $B = \{a, c, e, g\}$, determine (a) $Pr(A)$; (b) $Pr(B)$; (c) $Pr(A \cap B)$; (d) $Pr(A \cup B)$; (e) $Pr(\overline{A})$; (f) $Pr(\overline{A} \cup B)$; and (g) $Pr(A \cap \overline{B})$.

2. Joshua draws two ping-pong balls from a bowl of twenty ping-pong balls numbered 1 to 20. Provide a sample space for this experiment if

　a) the first ball drawn is replaced before the second ball is drawn.

　b) the first ball drawn is not replaced before the second ball is drawn.

3. A sample space $\mathcal{S}$ (for an experiment $\mathcal{E}$) contains 25 equally likely outcomes. If an event A (for this experiment $\mathcal{E}$) is such that $Pr(A) = 0.24$, how many outcomes are there in A?

4. A sample space $\mathcal{S}$ (for an experiment $\mathcal{E}$) contains n equally likely outcomes. If an event A (for this experiment $\mathcal{E}$) contains 7 of these outcomes and $Pr(A) = 0.14$, what is n?

5. The Tuesday night dance club is made up of six married couples and two of these twelve members must be chosen to find a dance hall for an upcoming fund raiser. (a) If the two members are selected at random, what is the probability they are both women? (b) If Joan and Douglas are one of the couples in the club, what is the probability at least one of them is among the two who are chosen?

6. If two integers are selected, at random and without replacement, from $\{1, 2, 3, \ldots, 99, 100\}$, what is the probability the integers are consecutive?

7. Two integers are selected, at random and without replacement, from $\{1, 2, 3, \ldots, 99, 100\}$. What is the probability their sum is even?

8. If three integers are selected, at random and without replacement, from $\{1, 2, 3, \ldots, 99, 100\}$, what is the probability their sum is even?

9. Jerry tosses a fair coin six times. What is the probability he gets (a) all heads; (b) one head; (c) two heads; (d) an even number of heads; and (e) at least four heads?

10. Twenty-five slips of paper, numbered 1, 2, 3, ..., 25, are placed in a box. If Amy draws six of these slips, without replacement, what is the probability that (a) the second smallest number drawn is 5? (b) the fourth largest number drawn is 15?

(c) the second smallest number drawn is 5 and the fourth largest number drawn is 15?

11. Darci rolls a fair die three times. What is the probability that (a) her second and third rolls are both larger than her first roll? (b) the result of her second roll is greater than that of her first roll and the result of her third roll is greater than the second?

12. In selecting a new server for its computing center, a college examines 15 different models, paying attention to the following considerations: (A) cartridge tape drive, (B) DVD Burner, and (C) SCSI RAID Array (a type of failure-tolerant disk-storage device). The numbers of servers with any or all of these features are as follows: $|A| = |B| = |C| = 6$, $|A \cap B| = |B \cap C| = 1$, $|A \cap C| = 2$, $|A \cap B \cap C| = 0$. (a) How many of the models have exactly one of the features being considered? (b) How many have none of the features? (c) If a model is selected at random, what is the probability that it has exactly two of these features?

13. At the Gamma Kappa Phi sorority the 15 sisters who are seniors line up in a random manner for a graduation picture. Two of these sisters are Columba and Piret. What is the probability that this graduation picture will find (a) Piret at the center position in the line? (b) Piret and Columba standing next to each other? (c) exactly five sisters standing between Columba and Piret?

14. The freshman class of a private engineering college has 300 students. It is known that 180 can program in Java, 120 in Visual BASIC[†], 30 in C++, 12 in Java and C++, 18 in Visual BASIC and C++, 12 in Java and Visual BASIC, and 6 in all three languages.

　a) A student is selected at random. What is the probability that she can program in exactly two languages?

　b) Two students are selected at random. What is the probability that they can (i) both program in Java? (ii) both program only in Java?

15. An integer is selected at random from 3 through 17 inclusive. If A is the event that a number divisible by 3 is chosen and B is the event that the number exceeds 10, determine $Pr(A)$, $Pr(B)$, $Pr(A \cap B)$, and $Pr(A \cup B)$. How is $Pr(A \cup B)$ related to $Pr(A)$, $Pr(B)$, and $Pr(A \cap B)$?

16. **a)** If the letters in the acronym WYSIWYG are arranged in a random manner, what is the probability the arrangement starts and ends with the same letter?

　b) What is the probability that a randomly generated arrangement of the letters in WYSIWYG has no pair of consecutive identical letters?

[†]Visual BASIC is a trademark of the Microsoft Corporation.

3.5
The Axioms of Probability (Optional)

In Section 3.4 our typical experiment had a sample space where each outcome had the same likelihood, or probability, of occurrence. If this does not happen, what do we do? Let us start by considering the following examples.

EXAMPLE 3.38

Suppose Trudy tosses a single coin but it is *not* fair — for instance, suppose this coin is loaded to come up heads twice as often as it comes up tails. Here the sample space $\mathscr{S} = \{H, T\}$, as in Example 3.28, but unlike that example where $Pr(H)^{\dagger} = Pr(T)$, in this situation we have $Pr(H) \neq Pr(T)$. With H, T as the only outcomes, we have $1 = Pr(\mathscr{S}) = Pr(\{H\} \cup \{T\}) = Pr(H) + Pr(T)$. Since $Pr(H) = 2Pr(T)$, it follows that $1 = Pr(H) + Pr(T) = 2Pr(T) + Pr(T)$, so $Pr(T) = 1/3$ and $Pr(H) = 2/3$.

EXAMPLE 3.39

A warehouse contains 10 motors, three of which are defective (D). The other seven are in good (G) working condition. A first inspector enters the warehouse and selects (and inspects) one of the motors. For this experiment $\mathscr{E}_1$, we have the sample space $\mathscr{S}_1 = \{D, G\}$ where $Pr(D) = 3/10$ and $Pr(G) = 7/10$. The next day a second inspector enters this same warehouse and selects (and inspects) a motor. For this second experiment — call it $\mathscr{E}_2$ — we likewise have $\mathscr{S}_2 = \{D, G\}$. But how do we define $Pr(D)$, $Pr(G)$ in this case? The answer depends on whether the first motor selected remained in the warehouse, or was removed.

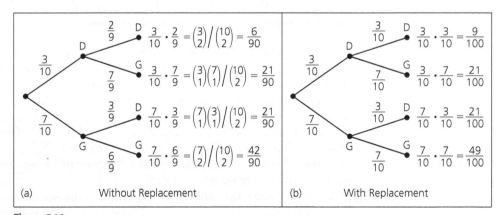

Figure 3.16

The tree diagrams in Fig. 3.16 deal with the two possibilities. For part (a) of the figure consider, for example, the case where the first motor selected is defective (D), with probability 3/10, and then the second motor selected is also defective (D). Since motors are not replaced here, when selecting the second motor the inspector is dealing with nine motors — two defective (D) and seven in good (G) working condition. Hence the probability of selecting a defective motor here is 2/9, *not* 3/10. So this situation, as shown by the top branching, has probability $\frac{3}{10} \cdot \frac{2}{9} = \binom{3}{2}/\binom{10}{2} = \frac{6}{90} = \frac{1}{15}$. The comparable case in part (b) of the figure has probability $\frac{3}{10} \cdot \frac{3}{10} = \frac{9}{100}$.

---

†Recall that when an event consists of a single outcome — say a, we may abbreviate $Pr(\{a\})$ as $Pr(a)$.

When selecting two motors, either with or without replacement, the sample space is $\mathscr{S} = \{DD, DG, GD, GG\}$ where, for instance, DG is used to abbreviate (D, G). Yet in neither situation do the outcomes have the same likelihood of occurrence. If the selections are done without replacement [as in Fig. 3.16(a)], then $Pr(DD) = \frac{6}{90}$, $Pr(DG) = \frac{21}{90}$, $Pr(GD) = \frac{21}{90}$, $Pr(GG) = \frac{42}{90}$, with $\frac{6}{90} + \frac{21}{90} + \frac{21}{90} + \frac{42}{90} = 1 = P(\mathscr{S})$. When the first motor is replaced [as in Fig. 3.16(b)], we have $Pr(DD) = \frac{9}{100}$, $Pr(DG) = \frac{21}{100}$, $Pr(GD) = \frac{21}{100}$, $Pr(GG) = \frac{49}{100}$, with $\frac{9}{100} + \frac{21}{100} + \frac{21}{100} + \frac{49}{100} = 1 = Pr(\mathscr{S})$.

From this point on we'll deal exclusively with the case where the two selections are made without replacement. Consider the following events:

A: One (that is exactly one) motor is defective: $\{DG, GD\}$;

B: At least one motor is defective: $\{DG, GD, DD\}$;

C: Both motors are defective: $\{DD\}$;

E: Both motors are in good working condition: $\{GG\}$.

Here

$$Pr(A) = \frac{21}{90} + \frac{21}{90} = \frac{7}{15}, \qquad Pr(B) = \frac{21}{90} + \frac{21}{90} + \frac{6}{90} = \frac{8}{15},$$

$$Pr(C) = \frac{6}{90} = \frac{1}{15}, \qquad\qquad Pr(E) = \frac{42}{90} = \frac{7}{15}.$$

Further, (i) $\overline{B} = E$ and $Pr(\overline{B}) = Pr(E) = \frac{7}{15} = 1 - \frac{8}{15} = 1 - Pr(B)$; and (ii) $A \cup C = B$ with $A \cap C = \emptyset$, so $Pr(A \cup C) = Pr(B) = \frac{8}{15} = \frac{7}{15} + \frac{1}{15} = Pr(A) + Pr(C)$.

What we did in the latter part of Example 3.39 now motivates our next observation. This observation extends our earlier results in Section 3.4 where each outcome of the sample space had the same likelihood, or probability, of occurring.

Let $\mathscr{S}$ be the sample space for an experiment $\mathscr{E}$. Each element $a \in \mathscr{S}$ is called an *outcome*, or *elementary event*, and we let $Pr(\{a\}) = Pr(a)$ denote the probability that this outcome occurs. Each nonempty subset A of $\mathscr{S}$ is still called an *event*. If event $A = \{a_1, a_2 \ldots, a_n\}$, where a_i is an outcome, for all $1 \le i \le n$, then $Pr(A) = \sum_{i=1}^{n} Pr(a_i)$. (*Note:* When $A = \emptyset$ we assign $Pr(A) = 0$, a result we shall actually establish later in this section.)

However, before we get to our axioms of probability, there is a point that needs to be clarified. We know that when a fair die is rolled, the sample space $\mathscr{S} = \{1, 2, 3, 4, 5, 6\}$, where each outcome has the same likelihood, or probability, of occurrence — namely, 1/6. However, if this die is rolled six times we should *not* expect to see one occurrence of each of the possible outcomes 1, 2, . . . , 6. Should this die be rolled 60 times we want each roll (after the first) to be unaffected by any previous roll — that is, each roll (after the first) is to be independent of any previous roll. Further, we cannot expect each of the six possible outcomes to occur ten times. In fact, if the 1 comes up 20 times and this die is then rolled 60 more times we cannot expect to see 1 come up 20 times again. So what can we expect? If, in rolling this fair die n times, the outcome of 1 occurs m times, then as n grows larger we expect the *relative frequency m/n* to approach 1/6.

So far this discussion has dealt with a sample space where each outcome has the same likelihood, or probability, of occurrence. However, the idea is still appropriate if we consider any sample space — for example, the sample space of Example 3.38. Equally important is how one can use the idea of relative frequency in modeling an experiment. For suppose we have a coin that we believe to be biased — perhaps because it is heavier than other similar coins that we have weighed. In tossing this coin the sample space is $\mathcal{S} = \{H, T\}$, but how can we determine $Pr(H)$, $Pr(T)$? We might toss the coin n times, assuming the outcome of each toss (after the first) is not affected by any previous outcome. If H comes up m times, then we can assign $Pr(H) = m/n$ and $Pr(T) = (n - m)/n = 1 - (m/n)$, where the accuracy of these assigned probabilities improves as n grows larger.

Having addressed the issue of probabilities as relative frequencies, now it is time to focus on the topic of this section — namely, the axioms of probability. One should find these axioms rather intuitive, especially when we look back at some of the results in Example 3.29 and part (b) of Example 3.33. The axioms were first introduced in 1933 by Andrei Kolmogorov and they apply to the case when the sample space $\mathcal{S}$ is finite.

The Axioms of Probability

Let $\mathcal{S}$ be the sample space for an experiment $\mathcal{E}$. If A, B are any events — that is, $\emptyset \subseteq A, B \subseteq \mathcal{S}$ (so we now allow the empty set to be an event), then

1) $Pr(A) \geq 0$

2) $Pr(\mathcal{S}) = 1$

3) if A, B are disjoint (or, mutually disjoint) then $Pr(A \cup B) = Pr(A) + Pr(B)$.[†]

Using these axioms we shall now establish a number of applicable results.

THEOREM 3.7

The Rule of Complement. Let $\mathcal{S}$ be the sample space for an experiment $\mathcal{E}$. If A is an event (that is, $A \subseteq \mathcal{S}$), then

$$Pr(\overline{A}) = 1 - Pr(A).$$

Proof: We know that $\mathcal{S} = A \cup \overline{A}$ with $A \cap \overline{A} = \emptyset$. So from axioms (2) and (3) it follows that $1 = Pr(\mathcal{S}) = Pr(A \cup \overline{A}) = Pr(A) + Pr(\overline{A})$, and $Pr(\overline{A}) = 1 - Pr(A)$.

Note that when $A = \emptyset$ in Theorem 3.7 we have $1 = Pr(\mathcal{S}) = Pr(\overline{A}) = 1 - Pr(A)$, so $Pr(\emptyset) = Pr(A) = 0$, in agreement with our earlier assignment.

The result of Theorem 3.7 can help cut down on our calculations in solving certain probability problems. This is demonstrated in the next two examples.

EXAMPLE 3.40

Suppose the letters in the word PROBABILITY are arranged in a random manner. Determine $Pr(A)$ for the event

A: The arrangement begins with one letter and ends in a different letter.

[†]Although our major concern in this chapter (if not the entire text) deals with $\mathcal{S}$ finite, when $\mathcal{S}$ is infinite Kolmogorov provided the fourth axiom:

4) if $A_1, A_2, A_3, \ldots$ are events (taken from $\mathcal{S}$) and $A_i \cap A_j = \emptyset$ for all $1 \leq i < j$, then

$$Pr\left(\bigcup_{n=1}^{\infty} A_n\right) = \sum_{n=1}^{\infty} Pr(A_n).$$

We consider four cases:

1) Start with the situation where neither B nor I appears at the start or finish. There are seven remaining (distinct) letters. Any one of them can be used at the start of the arrangement and there are six choices then for the last letter. For the nine letters in between there are $\frac{9!}{2!\,2!}$ arrangements. So for this case there are $(7)\left(\frac{9!}{2!\,2!}\right)(6) = 3{,}810{,}240$ possibilities.

2) Now suppose that B is used as the first or last letter (but not in both positions) and I only appears among the nine letters in the center. With one B so placed there are seven other (distinct) letters that can be used at the opposite end of the arrangement. The nine remaining letters in between can be arranged in $\frac{9!}{2!}$ ways, so this case accounts for $(2)(7)\frac{9!}{2!} = 2{,}540{,}160$ arrangements.

3) If we use one of the I's and none of the B's to start or end an arrangement, then there are again $2{,}540{,}160$ arrangements, as we had in case (2).

4) Finally, if one of B, I is used at the start and the other letter at the end, we can arrange the remaining nine letters in between in $9!$ ways. So here we have the final $2(9!) = 725{,}760$ arrangements.

Here $|\mathscr{S}| = \frac{11!}{2!\,2!} = 9{,}979{,}200$, so $Pr(A) = \frac{9{,}616{,}320}{9{,}979{,}200} = \frac{53}{55}$.

This result took quite a lot of calculations. So instead of the event A let us consider the event $\overline{A}$ — that is, the event where the arrangement begins and ends with the same letter. How many such arrangements are there? Say we use the letter B at the start and finish of the arrangement. Then the other nine letters in between can be arranged in $\frac{9!}{2!}$ ways. If I is used in place of B another $\frac{9!}{2!}$ arrangements result. So $|\overline{A}| = 9!$ and $Pr(\overline{A}) = \frac{9!}{(11!/(2!\,2!))} = \frac{2}{55}$.

With much less effort Theorem 3.7 shows us that $Pr(A) = 1 - Pr(\overline{A}) = \frac{53}{55}$.

EXAMPLE 3.41

Due to an intense preseason workout schedule, Coach Davis has honed her volleyball team into a major contender. Consequently, the probability her team will win any given tournament is 0.7, regardless of any previous win or loss. Suppose the team is slated to play eight tournaments.

a) The probability the women will win all eight tournaments is $(0.7)^8 \doteq 0.057648$. Could they possibly lose all eight tournaments? Yes, with probability $(0.3)^8 \doteq 0.000066$.

b) What is the probability the team wins exactly five of the eight tournaments? One way this can happen is if the team wins the first and second tournaments, loses the next three, and then wins the last three. We represent this by WWLLLWWW. The probability for this outcome is $(0.7)^2(0.3)^3(0.7)^3 = (0.7)^5(0.3)^3$. Another possibility that results in five tournament wins can be represented by WWLLWWLW. The probability here is $(0.7)^2(0.3)^2(0.7)^2(0.3)(0.7) = (0.7)^5(0.3)^3$. At this point we see that the probability Coach Davis's team wins five of the eight tournaments is

(The number of arrangements of five W's and three L's) $\times (0.7)^5(0.3)^3$.

From the material in Sections 1.2 and 1.3, especially Example 1.22, we know that there are $\frac{8!}{5!\,3!} = \binom{8}{5}$ ways to arrange five W's and three L's. Consequently, the probability the team wins five tournaments is

$$\binom{8}{5}(0.7)^5(0.3)^3 \doteq 0.254122.$$

c) Finally, what is the probability the team wins at least one tournament? Let us not do here what we did in Example 3.40. If we let A be the given event, then $Pr(A) = \sum_{i=1}^{8} \binom{8}{i}(0.7)^i(0.3)^{8-i}$. But $Pr(A)$ is more readily determined as $1 - Pr(\overline{A})$, where $Pr(\overline{A}) =$ the probability the team loses all eight tournaments $= (0.3)^8 \doteq 0.000066$ [as in part (a)]. Consequently, $Pr(A) = 1 - (0.3)^8 \doteq 0.999934$.

Before we go on we want to examine the structure of the answer at the end of part (b) of Example 3.41. Each tournament in the example results in either a win (success) or loss (failure). Further, after the first tournament, the outcome of each later tournament is *independent* of the outcome of any previous tournament. Such a two-outcome occurrence is called a *Bernoulli trial*. If there are n such trials and each trial has probability p of success and probability q $(= 1 - p)$ of failure, then the probability that there are (exactly) k successes among these n trials is

$$\binom{n}{k}p^k q^{n-k}, \quad 0 \leq k \leq n.$$

(We shall come upon this idea again in Section 16.5 when we study the application of Abelian groups in coding theory.)

Returning now to the axioms of probability, we know from axiom (3) that, for A, $B \subseteq \mathscr{S}$, if $A \cap B = \emptyset$ then $Pr(A \cup B) = Pr(A) + Pr(B)$. But what can we say if $A \cap B \neq \emptyset$?

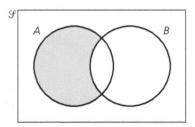

Figure 3.17

For the Venn diagram in Fig. 3.17 the interior of the rectangle represents the universe — here the sample space $\mathscr{S}$. The shaded region in the diagram denotes the event $A - B = A \cap \overline{B}$. Further,

i) the events $A \cap \overline{B}$ and B are disjoint, since $(A \cap \overline{B}) \cap B = A \cap (\overline{B} \cap B) = A \cap \emptyset = \emptyset$; and

ii) $(A \cap \overline{B}) \cup B = (A \cup B) \cap (\overline{B} \cup B) = (A \cup B) \cap \mathscr{S} = A \cup B$.

From these two observations and axiom (3) it follows that

(*) $Pr(A \cup B) = Pr((A \cap \overline{B}) \cup B) = Pr(A \cap \overline{B}) + Pr(B)$.

Next note that $A = A \cap \mathscr{S} = A \cap (B \cup \overline{B}) = (A \cap B) \cup (A \cap \overline{B})$ where $(A \cap B) \cap (A \cap \overline{B}) = (A \cap A) \cap (B \cap \overline{B}) = A \cap \emptyset = \emptyset$. So once again axiom (3) gives us

$$Pr(A) = Pr(A \cap B) + Pr(A \cap \overline{B}), \quad \text{or}$$
(**) $$Pr(A \cap \overline{B}) = Pr(A) - Pr(A \cap B).$$

The results in Eqs. (*) and (**) now establish the following.

THEOREM 3.8   *The Additive Rule.* If $\mathscr{S}$ is the sample space for an experiment $\mathscr{E}$, and $A, B \subseteq \mathscr{S}$, then

$$Pr(A \cup B) = Pr(A \cap \overline{B}) + P(B) = Pr(A) + Pr(B) - Pr(A \cap B).$$

At this point we use the result in Theorem 3.8 in the following two examples.

EXAMPLE 3.42   Yosi selects a card from a well-shuffled standard deck. What is the probability his card is a club or a card whose face value is between 3 and 7 inclusive?

Start by defining the events A, B as follows:

A: The card drawn is a club.

B: The face value of the card drawn is between 3 and 7 inclusive.

The answer to the problem is $Pr(A \cup B)$.

Here $Pr(A) = 13/52$ and $Pr(B) = 20/52$. Also $Pr(A \cap B) = 5/52$—for the 3 of clubs, 4 of clubs, ... , and 7 of clubs. Consequently, by Theorem 3.8, we have

$$Pr(A \cup B) = Pr(A) + Pr(B) - Pr(A \cap B) = \frac{13}{52} + \frac{20}{52} - \frac{5}{52} = \frac{28}{52} = \frac{7}{13}.$$

EXAMPLE 3.43   Diane inspects 120 cast aluminum rods and classifies the diameter and surface finish of each rod as adequate or superior. Her findings are summarized in Table 3.4.

Table 3.4

		Diameter	
		adequate	superior
Surface	adequate	10	18
Finish	superior	12	80

Define the events A, B as follows:

A: The diameter of the rod is classified as superior.

B: The surface finish of the rod is classified as superior.

Then

$$Pr(A) = (18 + 80)/120 = 98/120 = 49/60 \doteq 0.816667$$
$$Pr(B) = (12 + 80)/120 = 92/120 = 23/30 \doteq 0.766667$$
$$Pr(A \cap B) = 80/120 = 2/3 \doteq 0.666667.$$

By Theorem 3.8

$$Pr(A \cup B) = Pr(A) + Pr(B) - Pr(A \cap B)$$
$$= \frac{98}{120} + \frac{92}{120} - \frac{80}{120} = \frac{110}{120} = \frac{11}{12} \doteq 0.916667.$$

So 110 $[\doteq 110.40 = (0.92)(120)]$ of these 120 rods have either a superior diameter or a superior surface finish, or perhaps both.

In addition,

$Pr(\overline{A})$ = the probability the diameter of the rod is classified as adequate = $\frac{(10 + 12)}{120}$ = $\frac{22}{120} = 1 - \frac{98}{120} = 1 - Pr(A)$, and

$Pr(\overline{B})$ = the probability the surface finish of the rod is classified as adequate = $\frac{(10 + 18)}{120}$ = $\frac{28}{120} = 1 - \frac{92}{120} = 1 - Pr(B)$.

Using DeMorgan's Laws we also find that $Pr(\overline{A} \cup \overline{B}) = Pr(\overline{A \cap B}) = 1 - Pr(A \cap B) = 1 - \frac{2}{3} = \frac{1}{3}$, and $Pr(\overline{A} \cap \overline{B}) = Pr(\overline{A \cup B}) = 1 - Pr(A \cup B) = 1 - \frac{11}{12} = \frac{1}{12}$.

Now we want to extend the result of Theorem 3.8 to more than two events. The following theorem deals with three events and suggests the pattern for four or more.

THEOREM 3.9

Let $\mathcal{S}$ be the sample space for an experiment $\mathcal{E}$. For events $A, B, C \subseteq \mathcal{S}$,

$$Pr(A \cup B \cup C) =$$
$$Pr(A) + Pr(B) + Pr(C) - Pr(A \cap B) - Pr(A \cap C) - Pr(B \cap C) + Pr(A \cap B \cap C).$$

Proof: The Laws of Set Theory from Section 3.2 validate what follows:

$$\begin{aligned}
Pr(A \cup B \cup C) &= Pr((A \cup B) \cup C) = Pr(A \cup B) + Pr(C) - Pr((A \cup B) \cap C) \\
&= Pr(A) + Pr(B) - Pr(A \cap B) + Pr(C) - Pr((A \cap C) \cup (B \cap C)) \\
&= Pr(A) + Pr(B) + Pr(C) - Pr(A \cap B) \\
&\quad - [Pr(A \cap C) + Pr(B \cap C) - Pr((A \cap C) \cap (B \cap C))] \\
&= Pr(A) + Pr(B) + Pr(C) - Pr(A \cap B) \\
&\quad - Pr(A \cap C) - Pr(B \cap C) + Pr(A \cap B \cap C).
\end{aligned}$$

Note that the last equality follows because $(A \cap C) \cap (B \cap C) = A \cap B \cap C$ by the Associative, Commutative, and Idempotent Laws of Intersection. Also note the similarity between the formula for $Pr(A \cup B \cup C)$ and that for $|A \cup B \cup C|$ (given prior to Example 3.27).

Further, we see that the formula for $Pr(A \cup B \cup C)$ involves 7 ($= 2^3 - 1$) summands. For four events we would have 15 ($= 2^4 - 1$) summands: (i) $4 = \binom{4}{1}$ summands—one for each single event; (ii) $6 = \binom{4}{2}$ summands—one for each pair of events; (iii) $4 = \binom{4}{3}$ summands—one for each triple of events; and (iv) $1 = \binom{4}{4}$ summand for all four of the events. When dealing with n events, $A_1, A_2, \ldots, A_n$, where $n \geq 2$, the formula for $Pr(A_1 \cup A_2 \cup \cdots \cup A_n)$ has a total of $\sum_{r=1}^{n} \binom{n}{r} = \sum_{r=0}^{n} \binom{n}{r} - \binom{n}{0} = 2^n - 1$ summands, by Corollary 1.1. For $1 \leq r \leq n$, there are $\binom{n}{r}$ summands—one for each way we can select r of the n events. Each of these summands is preceded by a plus sign, for r odd, or a minus sign, for r even.

We'll see more formulas like the one in Theorem 3.9 in Section 8.1. For now let us apply the result of this theorem in the following example.

EXAMPLE 3.44

The game of Roulette is played by initially spinning a small white ball on a circular wheel that is divided into 38 sections of equal area. These sections are labeled 00, 0, 1, 2, 3, ..., 36.

As the wheel slows down, the number of the section where the ball comes to rest is the outcome for that one play of the game.

The numbers on the wheel are colored as follows.

Green:	00	0							
Red:	1	3	5	7	9	12	14	16	18
	19	21	23	25	27	30	32	34	36
Black:	2	4	6	8	10	11	13	15	17
	20	22	24	26	28	29	31	33	35

A player may place bets in various ways, such as (i) odd, even (here 00 and 0 are considered neither even nor odd); (ii) low (1–18), high (19–36); or (iii) red, black.

Gary enjoys Roulette and decides to place bets according to the events.

    A: The outcome is low. B: The outcome is red. C: The outcome is odd.

What is the probability Gary wins at least one of his bets — that is, what is $Pr(A \cup B \cup C)$?

Here $Pr(A) = Pr(B) = Pr(C) = 18/38$, $Pr(A \cap B) = Pr(A \cap C) = 9/38$, $Pr(B \cap C) = 10/38$, $Pr(A \cap B \cap C) = 5/38$, and by Theorem 3.9

$$Pr(A \cup B \cup C) = \frac{18}{38} + \frac{18}{38} + \frac{18}{38} - \frac{9}{38} - \frac{9}{38} - \frac{10}{38} + \frac{5}{38} = \frac{31}{38} \doteq 0.815789.$$

In closing this section we need to make one more point. The examples we've seen here and in the previous section have all dealt with finite sample spaces. Yet it is possible to have situations where a sample space is infinite. For instance, suppose a man takes a driver's test until he passes it. If he passes the test on his first try, we write P for this outcome. Should he need three attempts to pass the test, then we write FFP to denote the first and second failures followed by his passing of the test. Hence the sample space may be given here as $\mathcal{S} = \{P, FP, FFP, FFFP, \ldots\}$, an example of a *countably infinite*[†] set.

When dealing with sample spaces that are finite or countably infinite, we call the sample space *discrete*. The coverage here in Chapter 3 deals strictly with discrete sample spaces that are finite. However, in Section 9.2, we'll consider an example where the sample space is countably infinite.

Finally, suppose an experiment calls for a technician to record the temperature, in degrees Fahrenheit, of a heated iron rod. Theoretically, the sample space here could comprise an open interval of real numbers — for instance, $\mathcal{S} = \{t \,|\, 180°F < t < 190°F\}$. Here the sample space is again infinite, but this time it is *uncountably*[‡] *infinite*. In this case the sample space is called *continuous* and now one needs calculus to solve the related probability problems. We will not pursue this here but will direct the interested reader to the chapter references — especially, the text by J. J. Kinney [7].

EXERCISES 3.5

1. Let $\mathcal{S}$ be the sample space for an experiment $\mathcal{E}$ and let A, B be events from $\mathcal{S}$, where $Pr(A) = 0.4$, $Pr(B) = 0.3$, and $Pr(A \cap B) = 0.2$. Determine $Pr(\overline{A})$, $Pr(\overline{B})$, $Pr(A \cup B)$, $Pr(\overline{A \cup B})$, $Pr(A \cap \overline{B})$, $Pr(\overline{A} \cap B)$, $Pr(A \cup \overline{B})$, and $Pr(\overline{A} \cup B)$.

2. Ashley tosses a fair coin eight times. What is the probability she gets (a) six heads; (b) at least six heads; (c) two heads; and (d) at most two heads?

[†] The interested reader can find more on countable sets in Appendix 3.

[‡] More on uncountable sets can be found in Appendix 3.

3. Ten ping-pong balls labeled 1 to 10 are placed in a box. Two of these balls are then drawn, in succession and without replacement, from the box.

a) Find the sample space for this experiment.

b) Find the probability that the label on the second ball drawn is smaller than the label on the first.

c) Find the probability that the label on one ball is even while the label on the other is odd.

4. Russell draws one card from a standard deck. If A, B, C denote the events

A: The card is a spade.

B: The card is red.

C: The card is a picture card (that is, a jack, queen, or king).

Find $Pr(A \cup B \cup C)$.

5. Let $\mathscr{S}$ be the sample space for an experiment $\mathscr{E}$. If A, B are disjoint events from $\mathscr{S}$ with $Pr(A) = 0.3$ and $Pr(A \cup B) = 0.7$, what is $Pr(B)$?

6. If $\mathscr{S}$ is the sample space for an experiment $\mathscr{E}$ and $A, B \subseteq \mathscr{S}$, how is $Pr(A \triangle B)$ related to $Pr(A), Pr(B),$ and $Pr(A \cap B)$? [*Note:* $Pr(A \triangle B)$ is the probability that exactly one of the events A, B occurs.]

7. A die is loaded so that the probability a given number turns up is proportional to that number. So, for example, the outcome 4 is twice as likely as the outcome 2, and the outcome 3 is three times as likely as that of 1. If this die is rolled, what is the probability the outcome is (a) 5 or 6; (b) even; (c) odd?

8. Suppose we have two dice — each loaded as described in the previous exercise. If these dice are rolled, what is the probability the outcome is (a) 10; (b) at least 10; (c) a double?

9. Juan tosses a fair coin five times. What is the probability the number of heads always exceeds the number of tails as each outcome is observed?

10. Three types of foam are tested to see if they meet specifications. Table 3.5 summarizes the results for the 125 samples tested.

Table 3.5

		Specifications Are Met	
		No	Yes
Foam	1	5	60
Type	2	7	30
	3	8	15

Let A, B denote the events

A: The sample has foam type 1.

B: The sample meets specifications.

Determine $Pr(A), Pr(B), Pr(A \cap B), Pr(A \cup B), Pr(\overline{A}), Pr(\overline{B}), Pr(\overline{A} \cup \overline{B}), Pr(\overline{A} \cap \overline{B}), Pr(A \triangle B)$.

11. Consider the game of Roulette as described in Example 3.44.

a) If the game is played once, what is the probability the outcome is (i) high or odd; (ii) low or black?

b) If the game is played twice, what is the probability (i) both outcomes are black; (ii) one outcome is red and the other green?

12. Let $\mathscr{S}$ be the sample space for an experiment $\mathscr{E}$ and let $A, B \subseteq \mathscr{S}$. If $Pr(A) = Pr(B), Pr(A \cap B) = 1/5,$ and $Pr(\overline{A \cup B}) = 1/5,$ determine $Pr(A \cup B), Pr(A), Pr(A - B), Pr(A \triangle B)$.

13. The following data give the age and gender of 14 science professors at a small junior college.

25 M   39 F   27 F   53 M   36 F   37 F   30 M

29 F   32 M   31 M   38 F   26 M   24 F   40 F

One professor will be chosen at random to represent the faculty on the board of trustees. What is the probability that the professor chosen is a man or over 35?

14. The nine members of a coed intramural volleyball team are to be randomly selected from nine college men and ten college women. To be classified as coed the team must include at least one player of each gender. What is the probability the selected team includes more women than men?

15. While traveling through Pennsylvania, Ann decides to buy a lottery ticket for which she selects seven integers from 1 to 80 inclusive. The state lottery commission then selects 11 of these 80 integers. If Ann's selection matches seven of these 11 integers she is a winner. What is the probability Ann is a winner?

16. Let S be the sample space for an experiment $\mathscr{E}$ and let $A, B \subseteq \mathscr{S}$ with $A \subseteq B$. Prove that $Pr(A) \leq Pr(B)$.

17. Let $\mathscr{S}$ be the sample space for an experiment $\mathscr{E}$, and let $A, B \subseteq \mathscr{S}$. If $Pr(A) = 0.7$ and $Pr(B) = 0.5$, prove that $Pr(A \cap B) \geq 0.2$.

3.6
Conditional Probability: Independence
(Optional)

Throughout Sections 3.4 and 3.5 — especially prior to and at the end of Example 3.35, as well as in and after Example 3.41 — we mentioned the idea of the independence of outcomes. There we questioned whether the occurrence of a certain outcome might somehow affect the occurrence of another outcome. In this section we extend this idea from a single outcome to an event and make it more mathematically precise. To do so we proceed with the following.

EXAMPLE 3.45

Vincent rolls a pair of fair dice. The sample space $\mathscr{S}$ for this experiment is shown in Fig. 3.18, along with the events

A: The sum (on the faces) is at least 9.

B: A double is rolled.

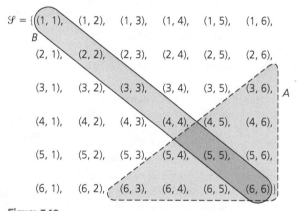

Figure 3.18

We see that $Pr(A) = \frac{10}{36} = \frac{5}{18}$, $Pr(B) = \frac{6}{36} = \frac{1}{6}$, and $Pr(A \cap B) = Pr(B \cap A) = \frac{2}{36} = \frac{1}{18}$.

But now, instead of just asking about the probability of the occurrence of event B, we go one step further. Here we want to determine the probability of the occurrence of event B given the *condition* that event A has occurred. This *conditional probability* is denoted by $Pr(B|A)$ and may be determined as follows.

The occurrence of event A reduces the sample space from the 36 equally-likely ordered pairs in $\mathscr{S}$ to the 10 equally-likely ordered pairs in A. Among the ordered pairs in A, two are also doubles — namely, $(5, 5)$ and $(6, 6)$. Consequently, the probability of B given $A = Pr(B|A) = \frac{2}{10}$, and we notice that $\frac{2}{10} = \frac{(2/36)}{(10/36)} = \frac{Pr(B \cap A)}{Pr(A)}$.

Before we suggest the result at the end of Example 3.45 as a general formula, let us consider a second example — one where the outcomes are not equally likely.

EXAMPLE 3.46

Lindsay has a coin that is biased with $Pr(\text{H}) = \frac{2}{3}$ and $Pr(\text{T}) = \frac{1}{3}$. She tosses this coin three times, where the result of each toss is independent of any preceding result. The eight possible outcomes in the sample space have the following probabilities:

$$Pr(\text{HHH}) = \left(\tfrac{2}{3}\right)^3 = \tfrac{8}{27}$$

$$Pr(\text{HHT}) = Pr(\text{HTH}) = Pr(\text{THH}) = \left(\tfrac{2}{3}\right)^2 \left(\tfrac{1}{3}\right) = \tfrac{4}{27}$$

$$Pr(\text{HTT}) = Pr(\text{THT}) = Pr(\text{TTH}) = \left(\tfrac{2}{3}\right) \left(\tfrac{1}{3}\right)^2 = \tfrac{2}{27}$$

$$Pr(\text{TTT}) = \left(\tfrac{1}{3}\right)^3 = \tfrac{1}{27}.$$

[Note that the sum of these probabilities is $\tfrac{8}{27} + 3\left(\tfrac{4}{27}\right) + 3\left(\tfrac{2}{27}\right) + \tfrac{1}{27} = \tfrac{8+12+6+1}{27} = 1$.]
Consider the events

A: The first toss results in a head [so $A = \{\text{HTT, HTH, HHT, HHH}\}$ and $Pr(A) = \tfrac{2}{27} + 2\left(\tfrac{4}{27}\right) + \tfrac{8}{27} = \tfrac{18}{27}$].

B: The number of heads is even [so $B = \{\text{TTT, HHT, HTH, THH}\}$ and $Pr(B) = \tfrac{1}{27} + 3\left(\tfrac{4}{27}\right) = \tfrac{13}{27}$].

Furthermore, $A \cap B = \{\text{HTH, HHT}\}$ and $Pr(B \cap A) = Pr(A \cap B) = \tfrac{4}{27} + \tfrac{4}{27} = \tfrac{8}{27}$.

To determine the conditional probability of B given A — that is, $Pr(B|A)$ — we'll make A our new sample space and redefine the probability of the four outcomes in A as follows:

$$Pr'(\text{HTT}) = \frac{Pr(\text{HTT})}{Pr(A)} = \frac{(2/27)}{(18/27)} = \frac{1}{9} \qquad Pr'(\text{HTH}) = \frac{Pr(\text{HTH})}{Pr(A)} = \frac{(4/27)}{(18/27)} = \frac{2}{9}$$

$$Pr'(\text{HHT}) = \frac{Pr(\text{HHT})}{Pr(A)} = \frac{(4/27)}{(18/27)} = \frac{2}{9} \qquad Pr'(\text{HHH}) = \frac{Pr(\text{HHH})}{Pr(A)} = \frac{(8/27)}{(18/27)} = \frac{4}{9}$$

[We see that $Pr'(\text{HTT}) + Pr'(\text{HTH}) + Pr'(\text{HHT}) + Pr'(\text{HHH}) = \tfrac{1}{9} + \tfrac{2}{9} + \tfrac{2}{9} + \tfrac{4}{9} = 1$.]
Among the four outcomes in A, two of them satisfy the condition given in event B — namely, HTH and HHT, the outcomes in $B \cap A$. Consequently, $Pr(B|A) = Pr'(\text{HTH}) + Pr'(\text{HHT}) = \tfrac{2}{9} + \tfrac{2}{9} = \tfrac{4}{9} = \tfrac{8}{18} = \tfrac{8/27}{18/27} = \frac{Pr(B \cap A)}{Pr(A)}$.

Motivated by the final result in each of the last two examples, we now summarize the underlying general procedure. We want a formula for $Pr(B|A)$, the conditional probability of the occurrence of event B given the occurrence of event A. Further, this formula should help us avoid unnecessary calculations such as those in Example 3.46, where we recalculated the probability of each outcome in A.

Now once we know that the event A has occurred, the sample space $\mathcal{S}$ shrinks to the outcomes in A. If we divide the probability of each outcome in A by $Pr(A)$, as in Example 3.46, the sum of these new probabilities sums to 1, so A can serve as the new sample space. Further, suppose e_1, e_2 are two outcomes in $\mathcal{S}$ with $Pr(e_2) = kPr(e_1)$, where k is a constant. If e_1, $e_2 \in A$, then within the new sample space A the probability of e_1 is still k times that of e_2.

To calculate $Pr(B|A)$ we now consider those outcomes in event A that are in event B. This gives us the outcomes in event $B \cap A$ and leads us to the following.

If $\mathcal{S}$ is the sample space for an experiment $\mathcal{E}$ and A, $B \subseteq \mathcal{S}$, then

the *conditional probability of B given A* $= Pr(B|A) = \dfrac{Pr(B \cap A)}{Pr(A)}$,

so long as $Pr(A) \neq 0$.

Further,

$$Pr(B \cap A) = Pr(A \cap B) = Pr(A)Pr(B|A),$$

and upon changing the roles of A and B we have

$$Pr(A \cap B) = Pr(B \cap A) = Pr(B)Pr(A|B).$$

The result

$$Pr(A)Pr(B|A) = Pr(A \cap B) = Pr(B)Pr(A|B)$$

is often called the *multiplicative rule*.

Without realizing it, we actually used the multiplicative rule in Example 3.39 — in the case where the motors were not replaced after inspection. The first part of our next example now reinforces how we use this rule.

EXAMPLE 3.47

A cooler contains seven cans of cola and three cans of root beer. Without looking at the contents, Gustavo reaches in and withdraws one can for his friend Jody. Then he reaches in again to get a can for himself.

Let A, B denote the events

A: The first selection is a can of cola.

B: The second selection is a can of cola.

a) Using the multiplicative rule, the probability that Gustavo chooses two cans of cola is

$$Pr(A \cap B) = Pr(A)Pr(B|A) = \left(\frac{7}{10}\right)\left(\frac{6}{9}\right) = \frac{7}{15}.$$

[Here $Pr(B|A) = 6/9$ because after the first can of cola is removed, the cooler then contains six cans of cola and three of root beer.]

b) The multiplicative rule and the additive rule (of Theorem 3.8) tell us that the probability Gustavo selects two cans of cola or two cans of root beer is

$$Pr(A \cap B) + Pr(\overline{A} \cap \overline{B}) = P(A)P(B|A) + Pr(\overline{A})Pr(\overline{B}|\overline{A})$$

$$= \left(\frac{7}{10}\right)\left(\frac{6}{9}\right) + \left(\frac{3}{10}\right)\left(\frac{2}{9}\right) = \frac{8}{15}.$$

c) Finally, let us determine $Pr(B)$. To do so we develop a new formula with the help of the Venn diagram (for a sample space $\mathcal{S}$ and events A, B) in Fig. 3.19. From the figure (and the laws of set theory) we see that $B = B \cap \mathcal{S} = B \cap (A \cup \overline{A}) = (B \cap A) \cup (B \cap \overline{A})$, where $(B \cap A) \cap (B \cap \overline{A}) = B \cap (A \cap \overline{A}) = B \cap \emptyset = \emptyset$.

$$Pr(B) = Pr(B \cap A) + Pr(B \cap \overline{A})$$

$$= Pr(A)Pr(B|A) + Pr(\overline{A})Pr(B|\overline{A})$$

$$= \left(\frac{7}{10}\right)\left(\frac{6}{9}\right) + \left(\frac{3}{10}\right)\left(\frac{7}{9}\right) = \frac{63}{90} = \frac{7}{10}.$$

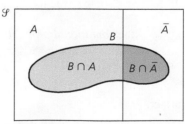

Figure 3.19

The result at the end of Example 3.47 — namely, for A, $B \subseteq \mathcal{S}$

$$Pr(B) = Pr(A)Pr(B|A) + Pr(\overline{A})Pr(B|\overline{A})$$

is referred to as the *Law of Total Probability*. Our next example shows how this result can be generalized.

EXAMPLE 3.48

Emilio is a system integrator for personal computers. As such he finds himself using keyboards from three companies. Company 1 supplies 60% of the keyboards, company 2 supplies 30% of the keyboards, and the remaining 10% comes from company 3. From past experience Emilio knows that 2% of company 1's keyboards are defective, while the percentages of defective keyboards for companies 2, 3 are 3% and 5%, respectively. If one of Emilio's computers is selected, at random, and then tested, what is the probability it has a defective keyboard?

Let A denote the event

A: The keyboard comes from company 1.

Events B, C are defined similarly for companies 2, 3, respectively. Event D, meanwhile, is

D: The keyboard is defective.

Here we are interested in $Pr(D)$. Guided by the Venn diagram in Fig. 3.20, we see that $D = D \cap \mathcal{S} = D \cap (A \cup B \cup C) = (D \cap A) \cup (D \cap B) \cup (D \cap C)$. But here $A \cap B = A \cap C = B \cap C = \emptyset$. So now, for example, the Laws of Set Theory show us that $(D \cap A) \cap (D \cap B) = D \cap (A \cap B) = D \cap \emptyset = \emptyset$. Likewise, $(D \cap A) \cap (D \cap C) = (D \cap B) \cap (D \cap C) = \emptyset$, and $(D \cap A) \cap (D \cap B) \cap (D \cap C) = \emptyset$. Consequently, by Theorem 3.9, we have

$$Pr(D) = Pr(D \cap A) + Pr(D \cap B) + Pr(D \cap C)$$
$$= Pr(A)Pr(D|A) + Pr(B)P(D|B) + Pr(C)Pr(D|C).$$

(Here we have the Law of Total Probability for three sets; that is, the sample space $\mathcal{S}$ is the union of three sets, any two of which are disjoint.)

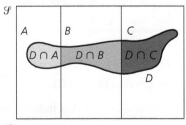

Figure 3.20

From the information given at the start of this example we know that

$Pr(A) = 0.6$	$Pr(B) = 0.3$	$Pr(C) = 0.1$
$Pr(D|A) = 0.02$	$Pr(D|B) = 0.03$	$Pr(D|C) = 0.05.$

So $Pr(D) = (0.6)(0.02) + (0.3)(0.03) + (0.1)(0.05) = 0.026$, and this tells us that 2.6% of the personal computers integrated by Emilio will have defective keyboards.

The next example takes us back to the situation in Example 3.48 and introduces us to *Bayes' Theorem*. As with the Law of Total Probability, the situation here likewise generalizes — that is, when appropriate, Bayes' Theorem may be applied to any sample space $\mathscr{S}$ that is decomposed into two or more events that are disjoint in pairs.

EXAMPLE 3.49

Referring back to the information in the preceding example, now we ask the question "If one of Emilio's personal computers is found to have a defective keyboard, what is the probability that keyboard came from company 3?"

Using the notation in Example 3.48 we see that here the given condition is D and that we want to find $Pr(C|D)$.

$$Pr(C|D) = \frac{Pr(C \cap D)}{Pr(D)} = \frac{Pr(C)Pr(D|C)}{Pr(A)Pr(D|A) + Pr(B)Pr(D|B) + Pr(C)Pr(D|C)}$$

$$= \frac{(0.1)(0.05)}{(0.6)(0.02) + (0.3)(0.03) + (0.1)(0.05)} = \frac{0.005}{0.026} = \frac{5}{26} \doteq 0.192308.$$

[Before leaving this example let us observe a small point. Since we have a choice on how to rewrite the numerator of $\frac{Pr(C \cap D)}{Pr(D)}$, do we know we've made the correct choice? Yes! The other choice, namely, $Pr(C \cap D) = Pr(D)Pr(C|D)$, would tell us that $Pr(C|D) = \frac{Pr(C \cap D)}{Pr(D)} = \frac{Pr(D)Pr(C|D)}{Pr(D)} = Pr(C|D)$, a correct but not very useful result.]

Having dealt with the Law of Total Probability and Bayes' Theorem, it is now time to settle the issue of independence. In our work on conditional probability we learned earlier that for events A, B, taken from a sample space $\mathscr{S}$, $Pr(A \cap B) = Pr(A)Pr(B|A)$. Should the occurrence of event A have no effect on that of B, we have $Pr(B|A) = Pr(B)$ — and so event B is independent of event A. These considerations now guide us to the following.

Definition 3.12

Given a sample space $\mathscr{S}$ with events A, $B \subseteq \mathscr{S}$, we call A, B *independent* when

$$Pr(A \cap B) = Pr(A)Pr(B).$$

For A, $B \subseteq \mathscr{S}$, the general situation has $Pr(B)Pr(A|B) = Pr(B \cap A) = Pr(A \cap B) = Pr(A)Pr(B|A)$. Using this and the result in Definition 3.12 we now have three ways to decide when A, B are independent:

1) $Pr(A \cap B) = Pr(A)Pr(B)$;

2) $Pr(A|B) = Pr(A)$; or

3) $Pr(B|A) = Pr(B)$.

We also realize that A is independent of B if and only if B is independent of A.

Our next example uses the preceding discussion to decide whether two events are independent.

EXAMPLE 3.50

Suppose Arantxa tosses a fair coin three times. Here the sample space $\mathscr{S}$ = {HHH, HHT, HTH, THH, HTT, THT, TTH, TTT}, where each outcome has probability $\frac{1}{8}$.

Consider the events

A: The first toss is H: A = {HHH, HHT, HTH, HTT} and $Pr(A) = \frac{1}{2}$;

B: The second toss is H: B = {HHH, HHT, THH, THT} and $Pr(B) = \frac{1}{2}$;

C: There are at least two H's: $\quad C = \{\text{HHT, HTH, THH, HHH}\}$ and $Pr(C) = \frac{1}{2}$.

a) $A \cap B = \{\text{HHH, HHT}\}$, so $\quad Pr(A \cap B) = \frac{1}{4} = \left(\frac{1}{2}\right)\left(\frac{1}{2}\right) = Pr(A)Pr(B)$. Consequently, the events A, B are independent.

b) $A \cap C = \{\text{HHH, HHT, HTH}\}$, so $P(A \cap C) = \frac{3}{8} \neq \left(\frac{1}{2}\right)\left(\frac{1}{2}\right) = Pr(A)Pr(C)$. Therefore, the events A, C are *not* independent.

c) Likewise, $\quad Pr(B \cap C) = \frac{3}{8} \neq \left(\frac{1}{2}\right)\left(\frac{1}{2}\right) = Pr(B)Pr(C)$ so $\quad B, C$ are also *not* independent.

d) The event $\overline{B} = \{\text{TTT, TTH, HTT, HTH}\}$ and $Pr(\overline{B}) = \frac{1}{2}$. Further $A \cap \overline{B} = \{\text{HTH, HTT}\}$ with $Pr(A \cap \overline{B}) = \frac{1}{4} = \left(\frac{1}{2}\right)\left(\frac{1}{2}\right) = Pr(A)Pr(\overline{B})$. So not only are the events A, B independent but the events A, $\overline{B}$ are also independent.

The first part of the following theorem shows us that what has happened here in parts (a) and (d) is not an isolated instance.

THEOREM 3.10

Let A, B be events taken from a sample space $\mathscr{S}$. If A, B are independent, then (a) A, $\overline{B}$ are independent; (b) $\overline{A}$, B are independent; and (c) $\overline{A}$, $\overline{B}$ are independent.

Proof: [We shall prove part (a) and leave the proofs of parts (b), (c) for the Section Exercises.] Since $A = A \cap \mathscr{S} = A \cap (B \cup \overline{B}) = (A \cap B) \cup (A \cap \overline{B})$ and $(A \cap B) \cap (A \cap \overline{B}) = A \cap (B \cap \overline{B}) = A \cap \emptyset = \emptyset$, we have $Pr(A) = Pr(A \cap B) + Pr(A \cap \overline{B})$. With A, B independent, it follows that $Pr(A \cap B) = Pr(A)Pr(B)$. The last two equations imply that $Pr(A \cap \overline{B}) = Pr(A) - Pr(A \cap B) = Pr(A) - Pr(A)Pr(B) = Pr(A)[1 - Pr(B)] = Pr(A)Pr(\overline{B})$. Consequently, from Definition 3.12 we know that A, $\overline{B}$ are independent.

Our next example will help motivate the idea of independence for three events.

EXAMPLE 3.51

Tino and Monica each roll a fair die. If we let x denote the result of Tino's roll and y that of Monica's, then once again $\mathscr{S} = \{(x, y) | 1 \leq x, y \leq 6\}$. Now consider the events A, B, C:

A: Tino rolls a 1, 2, or 6.

B: Monica rolls a 3, 4, 5, or 6.

C: The sum of Tino's and Monica's rolls is 7.

Here $Pr(A) = \frac{18}{36} = \frac{1}{2}$, $Pr(B) = \frac{24}{36} = \frac{2}{3}$, and $Pr(C) = \frac{6}{36} = \frac{1}{6}$. Further,

$A \cap B = \{(a, b) | a \in \{1, 2, 6\}, b \in \{3, 4, 5, 6\}\}$, so $|A \cap B| = 12$ and $Pr(A \cap B) = \frac{12}{36} = \frac{1}{3} = \left(\frac{1}{2}\right)\left(\frac{2}{3}\right) = Pr(A)Pr(B)$, so A, B are independent;

$A \cap C = \{(1, 6), (2, 5), (6, 1)\}$ and $Pr(A \cap C) = \frac{3}{36} = \frac{1}{12} = \left(\frac{1}{2}\right)\left(\frac{1}{6}\right) = Pr(A)Pr(C)$, making A, C independent;

$B \cap C = \{(4, 3), (3, 4), (2, 5), (1, 6)\}$ and $Pr(B \cap C) = \frac{4}{36} = \frac{1}{9} = \left(\frac{2}{3}\right)\left(\frac{1}{6}\right) = Pr(B)Pr(C)$, so B, C are also independent.

Finally,

$A \cap B \cap C = \{(1, 6), (2, 5)\}$ and $Pr(A \cap B \cap C) = \frac{2}{36} = \frac{1}{18} = \left(\frac{1}{2}\right)\left(\frac{2}{3}\right)\left(\frac{1}{6}\right) = Pr(A)Pr(B)Pr(C)$.

What has happened in Example 3.51 leads us to the following.

Definition 3.13   For a sample space $\mathscr{S}$ and events A, B, $C \subseteq \mathscr{S}$, we say that A, B, C are *independent* if

1) $Pr(A \cap B) = Pr(A)Pr(B)$;

2) $Pr(A \cap C) = Pr(A)Pr(C)$;

3) $Pr(B \cap C) = Pr(B)Pr(C)$; and

4) $Pr(A \cap B \cap C) = Pr(A)Pr(B)Pr(C)$.

Looking back now at Example 3.51 we see that there we verified the independence of the events A, B, C. But did we do too much? In particular, do we really need condition (4) in Definition 3.13? Perhaps we may feel that the first three conditions are enough to insure the fourth condition. But, perhaps, they are not enough. The next example will help us settle this issue.

EXAMPLE 3.52   Adira tosses a fair coin four times. So in this case the sample space $\mathscr{S} = \{x_1 x_2 x_3 x_4 | x_i \in \{H, T\}, 1 \le i \le 4\}$.

Let A, B, $C \subseteq \mathscr{S}$ be the events:

A: Adira's first toss is a tail (T);

B: Adira's last toss is a tail (T); and

C: The four tosses yield two heads and two tails.

For these events we find that $Pr(A) = \frac{8}{16} = \frac{1}{2}$, $Pr(B) = \frac{8}{16} = \frac{1}{2}$, and $Pr(C) = \frac{1}{16}\binom{4}{2} = \frac{6}{16} = \frac{3}{8}$.

In addition,

$$Pr(A \cap B) = \frac{4}{16} = \frac{1}{4} = \left(\frac{1}{2}\right)\left(\frac{1}{2}\right) = Pr(A)Pr(B),$$
$$Pr(A \cap C) = \frac{3}{16} = \left(\frac{1}{2}\right)\left(\frac{3}{8}\right) = Pr(A)Pr(C), \text{ and}$$
$$Pr(B \cap C) = \frac{3}{16} = \left(\frac{1}{2}\right)\left(\frac{3}{8}\right) = Pr(B)Pr(C).$$

However, $A \cap B \cap C = \{THHT\}$ and $Pr(A \cap B \cap C) = \frac{1}{16} = \frac{2}{32} \ne \frac{3}{32} = \left(\frac{1}{2}\right)\left(\frac{1}{2}\right)\left(\frac{3}{8}\right) = Pr(A)Pr(B)Pr(C)$. So while the three events in Example 3.51 are independent, the three events in this example are *(mutually) independent in pairs* — but *not* independent.

In closing this section we provide a summary of the probability rules and laws we have learned in this and the preceding section.

Summary of Probability Rules and Laws

1) *The Rule of Complement:* $Pr(\overline{A}) = 1 - Pr(A)$

2) *The Additive Rule:* $Pr(A \cup B) = Pr(A) + Pr(B) - Pr(A \cap B)$.
When A, B are disjoint, $Pr(A \cup B) = Pr(A) + Pr(B)$.

3) *Conditional Probability:* $Pr(A|B) = \dfrac{Pr(A \cap B)}{Pr(B)}$, $Pr(B) \ne 0$

4) *Multiplicative Rule:* $Pr(A)Pr(B|A) = Pr(A \cap B) = Pr(B)Pr(A|B)$.
When A, B are independent, $Pr(A \cap B) = Pr(A)Pr(B)$.

5) *The Law of Total Probability:* $Pr(B) = Pr(A)Pr(B|A) + Pr(\overline{A})Pr(B|\overline{A})$

6) *The Law of Total Probability (Extended Version):* If $A_1, A_2, \ldots, A_n \subseteq \mathcal{S}$, where $n \geq 3$, $A_i \cap A_j = \emptyset$ for all $1 \leq i < j \leq n$, and $\mathcal{S} = \cup_{i=1}^n A_i$, then for any event B,

$$Pr(B) = Pr(A_1)Pr(B|A_1) + \cdots + Pr(A_n)Pr(B|A_n) = \sum_{i=1}^{n} Pr(A_i)Pr(B|A_i).$$

7) *Bayes' Theorem:* $Pr(A|B) = \dfrac{Pr(A \cap B)}{Pr(B)} = \dfrac{Pr(A)Pr(B|A)}{Pr(A)Pr(B|A) + Pr(\overline{A})Pr(B|\overline{A})}$

8) *Bayes' Theorem (Extended Version):* If $A_1, A_2, \ldots, A_n \subseteq \mathcal{S}$, where $n \geq 3$, $A_i \cap A_j = \emptyset$ for all $1 \leq i < j \leq n$, and $\mathcal{S} = \cup_{i=1}^n A_i$, then for any event B, and each $1 \leq k \leq n$,

$$Pr(A_k|B) = \frac{Pr(A_k \cap B)}{Pr(B)} = \frac{Pr(A_k)Pr(B|A_k)}{Pr(A_1)Pr(B|A_1) + \cdots + Pr(A_n)Pr(B|A_n)}$$
$$= \frac{Pr(A_k)Pr(B|A_k)}{\sum_{i=1}^{n} Pr(A_i)Pr(B|A_i)}.$$

EXERCISES 3.6

1. Recall that in a standard deck of 52 cards there are 12 picture cards — four each of jacks, queens, and kings. Kevin draws one card from the deck. Find the probability his card is a king if we know that the card drawn is an ace or a picture card.

2. Let A, B be events taken from a sample space $\mathcal{S}$. If $Pr(A) = 0.6$, $Pr(B) = 0.4$, and $Pr(A \cup B) = 0.7$, find $Pr(A|B)$ and $Pr(A|\overline{B})$.

3. If Coach Mollet works his football team throughout August, then the probability the team will be the division champion is 0.75. The probability the coach will work his team throughout August is 0.80. What is the probability Coach Mollet works his team throughout August and the team finishes as the division champion?

4. The 420 freshmen at an engineering college take either calculus or discrete mathematics (but not both). Further, both courses are offered providing either an introduction to a CAS (computer algebra system) or using such a system extensively throughout the course. The results in Table 3.6 summarize how the 420 freshmen are distributed.

Table 3.6

	CAS (Introduction)	CAS (Extensive Coverage)
Calculus	170	120
Discrete Mathematics	80	50

a) If Sandrine is taking calculus, what is the probability her class is only being introduced to the use of a CAS?

b) Derek's class is making extensive use of the CAS. What is the probability Derek is taking discrete mathematics?

5. Let $\mathcal{S}$ be the sample space for an experiment $\mathcal{E}$ and let A, B be events from $\mathcal{S}$. If A, B are independent, prove that

$$Pr(A \cup B) = Pr(A) + Pr(\overline{A})Pr(B)$$
$$= Pr(B) + Pr(\overline{B})Pr(A).$$

6. Ceilia tosses a fair coin five times. What is the probability she gets three heads, if the first toss results in (a) a head; (b) a tail?

7. One bag contains 15 identical (in shape) coins — nine of silver and six of gold. A second bag contains 16 more of these coins — six silver and 10 gold. Bruno reaches in and selects one coin from the first bag and then places it in the second bag. Then Madeleine selects one coin from this second bag.

a) What is the probability Madeleine selected a gold coin?

b) If Madeleine's coin is gold, what is the probability Bruno had selected a gold coin?

8. A coin is loaded so that $Pr(H) = 2/3$ and $Pr(T) = 1/3$. Todd tosses this coin twice.

Let A, B be the events

A: The first toss is a tail. B: Both tosses are the same.

Are A, B independent?

9. Suppose that A, B are independent with $Pr(A \cup B) = 0.6$ and $Pr(A) = 0.3$. Find $Pr(B)$.

10. Alice tosses a fair coin seven times. Find the probability she gets four heads given that (a) her first toss is a head; (b) her first and last tosses are heads.

11. Paulo tosses a fair coin five times. If A, B denote the events

  A: Paulo gets an odd number of tails.

  B: Paulo's first toss is a tail.

are A, B independent?

12. The probability that a certain mechanical component fails when first used is 0.05. If the component does not fail immediately, the probability it will function correctly for at least one year is 0.98. What is the probability that a new component functions correctly for at least one year?

13. Paul has two coolers. The first contains eight cans of cola and three cans of lemonade. The second cooler contains five cans of cola and seven cans of lemonade. Paul randomly selects one can from the first cooler and puts it into the second cooler. Five minutes later Betty randomly selects two cans from the second cooler. If both of Betty's selections are cans of cola, what is the probability Paul initially selected a can of lemonade?

14. Let $\mathcal{S}$ be the sample space for an experiment $\mathcal{E}$ and let $A, B, C \subseteq \mathcal{S}$. If events A, B are independent, events A, C are disjoint, and events B, C are independent, find $Pr(B)$ if $Pr(A) = 0.2$, $Pr(C) = 0.4$, and $Pr(A \cup B \cup C) = 0.8$.

15. An electronic system is made up of two components connected in parallel. Consequently, the system fails only when both of the components fail. The probability the first component fails is 0.05 and, when this happens, the probability the second component fails is 0.02. What is the probability the electronic system fails?

16. Gayla has a bag of 19 marbles of the same size. Nine of these marbles are red, six blue, and four white. She randomly selects three of the marbles, without replacement, from the bag. What is the probability Gayla has withdrawn more red than white marbles?

17. Let A, B, C be independent events taken from a sample space $\mathcal{S}$. If $Pr(A) = 1/8$, $Pr(B) = 1/4$, and $Pr(A \cup B \cup C) = 1/2$, find $Pr(C)$.

18. A company involved in the integration of personal computers gets its graphics cards from three sources. The first source provides 20% of the cards, the second source 35%, and the third source 45%. Past experience has shown that 5% of the cards from the first source are found to be defective, while those from the second and third sources are found to be defective 3% and 2%, respectively, of the time.

  a) What percentage of the company's graphics cards are defective?

  b) If a graphics card is selected and found to be defective, what is the probability it was provided by the third source?

19. Gustavo tosses a fair coin twice. For this experiment consider the following events:

  A: The first toss is a head.

  B: The second toss is a tail.

  C: The tosses result in one head and one tail.

Are the events A, B, and C independent?

20. Three missiles are fired at an enemy arsenal. The probabilities the individual missiles will hit the arsenal are 0.75, 0.85, and 0.9. Find the probability that at least two of the missiles hit the arsenal.

21. Dustin and Jennifer each toss three fair coins. What is the probability (a) each of them gets the same number of heads? (b) Dustin gets more heads than Jennifer? (c) Jennifer gets more heads than Dustin?

22. Tiffany and four of her cousins play the game of "odd person out" to determine who will rake up the leaves at their grandmother Mary Lou's home. Each cousin tosses a fair coin. If the outcome for one cousin is different from that of the other four, then this cousin has to rake the leaves. What is the probability that a "lucky" cousin is determined after the coins are flipped only once?

23. Ninety percent of new airport-security personnel have had prior training in weapon detection. During their first month on the job, personnel without prior training fail to detect a weapon 3% of the time, while those with prior training fail only 0.5% of the time. What is the probability a new airport-security employee, who fails to detect a weapon during the first month on the job, has had prior training in weapon detection?

24. The binary string 101101, where the string is unchanged upon reversing order, is called a *palindrome* (of length 6). Suppose a binary string of length 6 is randomly generated, with 0, 1 equally likely for each of the six positions in the string. What is the probability the string is a palindrome if the first and sixth bits (a) are both 1; (b) are the same?

25. In defining the notion of independence for three events we found (in Definition 3.13) that we had to check four conditions. If there are four events, say E_1, E_2, E_3, E_4, then we have to check 11 conditions — six of the form $Pr(E_i \cap E_j) = Pr(E_i)Pr(E_j)$, $1 \le i < j \le 4$; four of the form $Pr(E_i \cap E_j \cap E_k) = Pr(E_i)Pr(E_j)Pr(E_k)$, $1 \le i < j < k \le 4$; and $Pr(E_1 \cap E_2 \cap E_3 \cap E_4) = Pr(E_1)Pr(E_2)Pr(E_3)Pr(E_4)$.
(a) How many conditions need to be checked for the independence of five events? (b) How many for n events, where $n \ge 2$?

26. Let A, B be events taken from a sample space $\mathcal{S}$. If $Pr(A \cap B) = 0.1$ and $Pr(\overline{A} \cap \overline{B}) = 0.3$, what is $Pr(A \bigtriangleup B | A \cup B)$?

27. Urn 1 contains 14 envelopes (of the same size) — six each contain $1 and the other eight each contain $5. Urn 2 contains eight envelopes (of the same size as those in urn 1) — three each contain $1 and the other five each contain $5. Three envelopes are randomly selected from urn 1 and transferred to urn 2. If Carmen now draws one envelope from urn 2, what is the probability her selection contains $1?

28. Let A, B be events taken from a sample space $\mathcal{S}$ (with $Pr(A) > 0$ and $Pr(B) > 0$). If $Pr(B|A) < Pr(B)$, prove that $Pr(A|B) < Pr(A)$.

29. Let A, B be events taken from a sample space $\mathcal{S}$. If $Pr(A) = 0.5$, $Pr(B) = 0.3$, and $Pr(A|B) + Pr(B|A) = 0.8$, what is $Pr(A \cap B)$?

30. Let $\mathcal{S}$ be the sample space for an experiment $\mathcal{E}$, with events A, $B \subseteq \mathcal{S}$. If $Pr(A|B) = Pr(A \,\triangle\, B) = 0.5$ and $Pr(A \cup B) = 0.7$, determine $Pr(A)$ and $Pr(B)$.

3.7
Discrete Random Variables (Optional)

In this section we introduce a fundamental idea in the study of probability and statistics — namely, the random variable. Since we are dealing exclusively with discrete sample spaces, we shall deal only with discrete random variables. Consequently, whenever the term random variable arises, it is understood that it is a discrete random variable — that is, a random variable defined for a discrete sample space. [Those interested in continuous random variables should consult the chapter references. Chapter 3 of the text by John J. Kinney [7] is an excellent starting point.]

We introduce the concept of a random variable in an informal way. The following example will help us do this.

EXAMPLE 3.53   If Keshia tosses a fair coin four times, the sample space for this random experiment may be given as

$\mathcal{S} = \{$HHHH,

HHHT, HHTH, HTHH, THHH,

HHTT, HTHT, HTTH, THHT, THTH, TTHH,

HTTT, THTT, TTHT, TTTH,

TTTT$\}$.

Now, for each of the 16 strings of H's and T's in $\mathcal{S}$, we define the *random variable* X as follows:

For $x_1 x_2 x_3 x_4 \in \mathcal{S}$, $X(x_1 x_2 x_3 x_4)$ counts the number of H's that appear among the four components x_1, x_2, x_3, x_4. Consequently,

$X(\text{HHHH}) = 4$,

$X(\text{HHHT}) = X(\text{HHTH}) = X(\text{HTHH}) = X(\text{THHH}) = 3$,

$X(\text{HHTT}) = X(\text{HTHT}) = X(\text{HTTH}) = X(\text{THHT}) = X(\text{THTH}) = X(\text{TTHH}) = 2$,

$X(\text{HTTT}) = X(\text{THTT}) = X(\text{TTHT}) = X(\text{TTTH}) = 1$, and

$X(\text{TTTT}) = 0$.

We see that X associates[†] each of the 16 strings of H's and T's in $\mathcal{S}$ with one of the nonnegative integers in $\{0, 1, 2, 3, 4\}$ (a subset of $\mathbf{R}$). This allows us to think of an outcome in $\mathcal{S}$ in terms of a real number. Further, suppose we are interested in the event

A: the four tosses result in two H's and two T's.

[†]This association by X between the strings in $\mathcal{S}$ and the nonnegative integers 0, 1, 2, 3, 4 is an example of a function — an idea to be covered in detail in Chapter 5. In general, a *random variable* is a function from the sample space $\mathcal{S}$ of an experiment $\mathcal{E}$ to $\mathbf{R}$, the set of real numbers. The *domain* of any random variable X is $\mathcal{S}$ and the *codomain* is always $\mathbf{R}$. The *range* in this case is $\{0, 1, 2, 3, 4\}$. (The concepts of domain, codomain, and range are formally defined in Section 5.2.)

In our earlier work we might have described this event by writing

$$A = \{HHTT, HTHT, HTTH, THHT, THTH, TTHH\}.$$

Now we can summarize the six outcomes in this event by writing $A = \{x_1 x_2 x_3 x_4 | X(x_1 x_2 x_3 x_4) = 2\}$, and this may be abbreviated to $A = \{x_1 x_2 x_3 x_4 | X = 2\}$. Also, we express $Pr(A)$, in terms of the random variable X, as $Pr(X = 2)$. So here we have $Pr(A) = Pr(X = 2) = 6/16 = 3/8$. Similarly, it follows that $Pr(X = 4) = 1/16$ since there is only one outcome for this case — namely, HHHH.

The following provides what we call the *probability distribution* for this particular random variable X.

x	$Pr(X = x)$
0	1/16
1	4/16 = 1/4
2	6/16 = 3/8
3	4/16 = 1/4
4	1/16

Observe how $\sum_{x=0}^{4} Pr(X = x) = 1$ in agreement with axiom (2) of Section 3.5. Also, it is understood that $Pr(X = x) = 0$ for $x \neq 0, 1, 2, 3, 4$.

Let us now reinforce what we have learned by considering a second example.

EXAMPLE 3.54

Suppose Giorgio rolls a pair of fair dice. This experiment was examined earlier — for instance, in Examples 3.33 and 3.45. The sample space here comprises 36 ordered pairs and may be expressed as $\mathscr{S} = \{(x, y) | 1 \leq x \leq 6, 1 \leq y \leq 6\}$.

We define the random variable X, for each ordered pair (x, y) in $\mathscr{S}$, by $X((x, y)) = x + y$, the sum of the numbers that appear on the (tops of) two fair dice. Then X takes on the following values:

$$X((1, 1)) = 2$$
$$X((1, 2)) = X((2, 1)) = 3$$
$$X((1, 3)) = X((2, 2)) = X((3, 1)) = 4$$
$$X((1, 4)) = X((2, 3)) = X((3, 2)) = X((4, 1)) = 5$$
$$X((1, 5)) = X((2, 4)) = X((3, 3)) = X((4, 2)) = X((5, 1)) = 6$$
$$X((1, 6)) = X((2, 5)) = X((3, 4)) = X((4, 3)) = X((5, 2)) = X((6, 1)) = 7$$
$$X((2, 6)) = X((3, 5)) = X((4, 4)) = X((5, 3)) = X((6, 2)) = 8$$
$$X((3, 6)) = X((4, 5)) = X((5, 4)) = X((6, 3)) = 9$$
$$X((4, 6)) = X((5, 5)) = X((6, 4)) = 10$$
$$X((5, 6)) = X((6, 5)) = 11$$
$$X((6, 6)) = 12$$

The probability distribution for X is as follows:

$Pr(X = 2) = 1/36$	$Pr(X = 6) = 5/36$	$Pr(X = 10) = 3/36$
$Pr(X = 3) = 2/36$	$Pr(X = 7) = 6/36$	$Pr(X = 11) = 2/36$
$Pr(X = 4) = 3/36$	$Pr(X = 8) = 5/36$	$Pr(X = 12) = 1/36$
$Pr(X = 5) = 4/36$	$Pr(X = 9) = 4/36$	

This can be abbreviated somewhat by

$$Pr(X = x) = \begin{cases} \dfrac{x-1}{36}, & x = 2, 3, 4, 5, 6, 7 \\[2mm] \dfrac{12-(x-1)}{36}, & x = 8, 9, 10, 11, 12. \end{cases}$$

Note that $\sum_{x=2}^{12} Pr(X = x) = 1$.

Having finished with describing X and its probability distribution, now let us consider the events:

B: Giorgio rolls an 8 — that is, the sum of the two dice is 8.

C: Giorgio rolls at least a 10.

The event $B = \{(2, 6), (3, 5), (4, 4), (5, 3), (6, 2)\}$ and $Pr(B) = Pr(X = 8) = 5/36$. Meanwhile $C = \{(4, 6), (5, 5), (6, 4), (5, 6), (6, 5), (6, 6)\}$ and $Pr(C) = 6/36 = 3/36 + 2/36 + 1/36 = Pr(X = 10) + Pr(X = 11) + Pr(X = 12) = Pr(10 \le X \le 12) = \sum_{x=10}^{12} Pr(X = x) = \sum_{x \ge 10} Pr(X = x)$.

The preceding two examples have shown us how a random variable may be described by its probability distribution. Now we shall see how a random variable can be characterized by means of two measures — its *expected value*, a measure of central tendency, and its *variance*, a measure of dispersion.

When a fair coin is tossed 10 times, our intuition may suggest that we expect to get five heads and five tails. Yet we know that we could actually see 10 heads, although the probability for this outcome is only $\binom{10}{10}\left(\frac{1}{2}\right)^{10} = \frac{1}{1024} \doteq 0.000977$, while the probability for five heads and five tails is substantially higher as $\binom{10}{5}\left(\frac{1}{2}\right)^5\left(\frac{1}{2}\right)^5 = \frac{252}{1024} \doteq 0.246094$. Similarly, we may want to know how many times we might expect to see a 6 when a fair die is rolled 50 times. To deal with such concerns we introduce the following idea.

Definition 3.14   Let X be a random variable defined for the outcomes in a sample space $\mathscr{S}$. The *mean*, or *expected value*, of X is

$$E(X) = \sum_{x} x \cdot Pr(X = x),$$

where the sum is taken over all the values x determined by the random variable $X^{\dagger}$.

The following example deals with $E(X)$ in several different situations.

[†]One finds the terms *mean (value)* and *expectation* also used to describe $E(X)$, as well as the alternate notation μ_X. Further, although our discussion deals solely with finite sample spaces, the above formula is valid for countably infinite sample spaces, so long as the infinite sum converges.

EXAMPLE 3.55

a) If a fair coin is tossed once and X counts the number of heads that appear, then

$$\mathscr{S} = \{H, T\}, \qquad X(H) = 1, \qquad X(T) = 0, \qquad Pr(X = 0) = Pr(X = 1) = \frac{1}{2},$$

and
$$E(X) = \sum_{x=0}^{1} x \cdot Pr(X = x) = 0 \cdot \frac{1}{2} + 1 \cdot \frac{1}{2} = \frac{1}{2}.$$

Note that $E(X)$ is neither 0 nor 1.

b) If one fair die is rolled, then $\mathscr{S} = \{1, 2, 3, 4, 5, 6\}$. Further, for each $1 \leq i \leq 6$, we have $X(i) = i$ and $Pr(X = i) = 1/6$. So here

$$E(X) = \sum_{x=1}^{6} x \cdot Pr(X = x) = 1 \cdot \frac{1}{6} + 2 \cdot \frac{1}{6} + 3 \cdot \frac{1}{6} + 4 \cdot \frac{1}{6} + 5 \cdot \frac{1}{6} + 6 \cdot \frac{1}{6}$$

$$= \left(\frac{1}{6}\right)(1 + 2 + \cdots + 6) = \frac{21}{6} = \frac{7}{2}.$$

Note, once again, that $E(X)$ is not among the values determined by the random variable X.

c) Suppose now we have a loaded die, where the probability of rolling the number i is proportional to i. As in part (b), $\mathscr{S} = \{1, 2, 3, 4, 5, 6\}$ and $X(i) = i$ for $1 \leq i \leq 6$. However, here, if p is the probability of rolling 1, then ip is the probability of rolling i, for each of the other five outcomes i, where $2 \leq i \leq 6$. From axiom (2), $1 = \sum_{i=1}^{6} ip = p(1 + 2 + \cdots + 6) = 21p$, so $p = 1/21$ and $Pr(X = i) = i/21$, $1 \leq i \leq 6$. Consequently,

$$E(X) = \sum_{x=1}^{6} x \cdot Pr(X = x) = 1 \cdot \frac{1}{21} + 2 \cdot \frac{2}{21} + \cdots + 6 \cdot \frac{6}{21}$$

$$= \frac{1 + 4 + 9 + 16 + 25 + 36}{21} = \frac{91}{21} = \frac{13}{3}.$$

d) Consider the random variable X in Example 3.53, where a fair coin was tossed four times. Then here

$$E(X) = \sum_{x=0}^{4} x \cdot Pr(X = x) = 0 \cdot \frac{1}{16} + 1 \cdot \frac{4}{16} + 2 \cdot \frac{6}{16} + 3 \cdot \frac{4}{16} + 4 \cdot \frac{1}{16}$$

$$= \frac{0 + 4 + 12 + 12 + 4}{16} = 2.$$

In this case $E(X)$ is found to be among the values determined by the random variable X.

e) Finally, for Example 3.54, where Giorgio rolled a pair of fair dice, we find that

$$E(X) = 2 \cdot \frac{1}{36} + 3 \cdot \frac{2}{36} + \cdots + 7 \cdot \frac{6}{36} + 8 \cdot \frac{5}{36} + \cdots + 11 \cdot \frac{2}{36} + 12 \cdot \frac{1}{36}$$

$$= \frac{252}{36} = 7.$$

Before continuing let us recall from Section 3.5 that a Bernoulli trial is an experiment with exactly two outcomes — success, with probability p, and failure, with probability $q = 1 - p$. When such an experiment is performed n times, and the outcome of any one trial is independent of the outcomes of any previous trials, then the probability that there are (exactly) k successes among the n trials is $\binom{n}{k}p^k q^{n-k}$, $0 \leq k \leq n$.

Now if we consider the sample space of all 2^n possibilities for the n outcomes of these n Bernoulli trials, then we can define the random variable X, where X counts the number of successes among the n trials. Under these circumstances X is called a *binomial random variable* and

$$Pr(X = x) = \binom{n}{x}p^x q^{n-x}, \qquad x = 0, 1, 2, \ldots, n.$$

This probability distribution is called the *binomial probability distribution* and it is completely determined by the values of n and p. Further, it is precisely the type of probability distribution that occurs in Example 3.53, where we regard an H as a success and find that

$$Pr(X = 0) = \frac{1}{16} = \binom{4}{0}\left(\frac{1}{2}\right)^0\left(\frac{1}{2}\right)^4 \qquad Pr(X = 3) = \frac{1}{4} = \binom{4}{3}\left(\frac{1}{2}\right)^3\left(\frac{1}{2}\right)^1$$

$$Pr(X = 1) = \frac{1}{4} = \binom{4}{1}\left(\frac{1}{2}\right)^1\left(\frac{1}{2}\right)^3 \qquad Pr(X = 4) = \frac{1}{16} = \binom{4}{4}\left(\frac{1}{2}\right)^4\left(\frac{1}{2}\right)^0$$

$$Pr(X = 2) = \frac{3}{8} = \binom{4}{2}\left(\frac{1}{2}\right)^2\left(\frac{1}{2}\right)^2$$

The five previous results can be summarized by

$$Pr(X = x) = \binom{4}{x}\left(\frac{1}{2}\right)^x\left(\frac{1}{2}\right)^{4-x}, \qquad x = 0, 1, \ldots, 4.$$

But why should we be bringing all of this up here in the discussion on the expected value of a random variable? At this point notice that in part (d) of Example 3.55 we found $E(X) = 2$, where X is the binomial random variable described above. For this binomial random variable X we have $n = 4$ and $p = 1/2$. Is it just a coincidence here that $E(X) = 2 = (4)(1/2) = np$?

Suppose we were to roll a fair die 12 times and ask for the number of times we *expect* to see a 5 come up. Here the binomial random variable X would count the number of times a 5 is rolled among the 12 rolls. Our intuition might *suggest* the answer is $2 = (12)(1/6) = np$. But is this once again $E(X)$ for this binomial random variable X? Instead of verifying this result directly — by using the formula in Definition 3.14, we shall obtain the result from the following theorem.

THEOREM 3.11   Let X be the binomial random variable that counts the number of successes, each with probability p, among n Bernoulli trials. Then $E(X) = np$.

Proof: From Definition 3.14 we have

$$E(X) = \sum_{x=0}^{n} x \cdot Pr(X = x) = \sum_{x=0}^{n} x\binom{n}{x}p^x q^{n-x},$$

where $q = 1 - p$. Since $x\binom{n}{x}p^x q^{n-x} = 0$ when $x = 0$, it follows that

$$E(X) = \sum_{x=1}^{n} x\binom{n}{x}p^x q^{n-x} = \sum_{x=1}^{n} x\frac{n!}{x!(n-x)!}p^x q^{n-x}$$

$$= \sum_{x=1}^{n} \frac{n!}{(x-1)!(n-x)!}p^x q^{n-x} = np\sum_{x=1}^{n} \frac{(n-1)!}{(x-1)!(n-x)!}p^{x-1}q^{n-x}$$

$$= np \sum_{y=0}^{n-1} \frac{(n-1)!}{y![n-(y+1)]!} p^y q^{n-(y+1)}, \qquad \text{upon substituting } y = x - 1,$$

upon substituting $y = x - 1$, and realizing that y varies from 0 to $n-1$ when x varies from 1 to n

$$= np \sum_{y=0}^{n-1} \binom{n-1}{y} p^y q^{(n-1)-y} = np(p+q)^{n-1}, \qquad \text{by the binomial theorem}$$

$$= np, \quad \text{since } p + q = 1.$$

As a result of Theorem 3.11 we now know that upon rolling a fair die 12 times the number of 5's we *expect* to see is $(12)(1/6) = 2$, as our intuition suggested earlier. Better still, should we roll this fair die 1200 times and let the random variable Y count the number of 5's that appear, then Y is a binomial random variable with $n = 1200$, $p = 1/6$, and $Pr(Y = y) = \binom{1200}{y}\left(\frac{1}{6}\right)^y \left(\frac{5}{6}\right)^{1200-y}$, $y = 0, 1, 2, \ldots, 1200$. Further, instead of trying to determine $E(Y)$ by actually calculating $\sum_{y=0}^{1200} y\binom{1200}{y}\left(\frac{1}{6}\right)^y \left(\frac{5}{6}\right)^{1200-y}$, we obtain $E(Y) = np = (1200)(1/6) = 200$, quite readily from Theorem 3.11.

Having dealt with the concept of the mean, or expected value, of a random variable X, we turn now to the variance of X — a measure of how widely the values determined by X are dispersed or spread out. If X is the random variable defined on the sample space $\mathcal{S}_X = \{a, b, c\}$, where $X(a) = -1$, $X(b) = 0$, $X(c) = 1$, and $Pr(X = x) = 1/3$, for $x = -1, 0, 1$, then $E(X) = 0$. But then if Y is the random variable defined on the sample space $\mathcal{S}_Y = \{r, s, t, u, v\}$, where $Y(r) = -4$, $Y(s) = -2$, $Y(t) = 0$, $Y(u) = 2$, $Y(v) = 4$, and $Pr(Y = y) = 1/5$, for $y = -4, -2, 0, 2, 4$, we get the same mean — that is, $E(Y) = 0$. However, although $E(X) = E(Y)$, we can see that the values determined by Y are more spread out about the mean of 0 than the values determined by X. To measure this notion of dispersion we introduce the following.

Definition 3.15

Let $\mathcal{S}$ be the sample space for an experiment $\mathcal{E}$ and let X be a random variable defined on the outcomes in $\mathcal{S}$. Suppose further that $E(X)$ is the mean, or expected value, of X. Then the *variance* of X, denoted σ_X^2, or Var(X), is defined by

$$\sigma_X^2 = \text{Var}(X) = E(X - E(X))^2 = \sum_x (x - E(X))^2 \cdot Pr(X = x),$$

where the sum is taken over all the values of x determined by the random variable X.

The *standard deviation* of X, denoted σ_X, is defined by

$$\sigma_X = \sqrt{\text{Var}(X)}.$$

Now let us apply Definition 3.15 in the following.

EXAMPLE 3.56

Let X be the random variable defined on the outcomes of the sample space $\mathcal{S} = \{a, b, c, d\}$, with $X(a) = 1$, $X(b) = 3$, $X(c) = 4$, and $X(d) = 6$.

Suppose the probability distribution for X is

x	$Pr(X = x)$
1	1/5
3	2/5
4	1/5
6	1/5.

Then

$$E(X) = 1 \cdot \frac{1}{5} + 3 \cdot \frac{2}{5} + 4 \cdot \frac{1}{5} + 6 \cdot \frac{1}{5} = \frac{17}{5}$$

and

$$\begin{aligned}
\text{Var}(X) &= E(X - E(X))^2 \\
&= \sum_x \left(x - \frac{17}{5}\right)^2 \cdot Pr(X = x) \\
&= \left(1 - \frac{17}{5}\right)^2 \left(\frac{1}{5}\right) + \left(3 - \frac{17}{5}\right)^2 \left(\frac{2}{5}\right) \\
&\quad + \left(4 - \frac{17}{5}\right)^2 \left(\frac{1}{5}\right) + \left(6 - \frac{17}{5}\right)^2 \left(\frac{1}{5}\right) \\
&= \left(\frac{144}{25}\right)\left(\frac{1}{5}\right) + \left(\frac{4}{25}\right)\left(\frac{2}{5}\right) + \left(\frac{9}{25}\right)\left(\frac{1}{5}\right) + \left(\frac{169}{25}\right)\left(\frac{1}{5}\right) \\
&= \left(\frac{330}{25}\right)\left(\frac{1}{5}\right) = \frac{66}{25},
\end{aligned}$$

so $\quad \sigma_X = \sqrt{\text{Var}(X)} = \sqrt{\frac{66}{25}} = \frac{1}{5}\sqrt{66} \doteq 1.624808.$

Our next result provides a second way by which we can compute $\text{Var}(X)$.

THEOREM 3.12

If X is a random variable defined on the outcomes of a sample space $\mathscr{S}$, then

$$\text{Var}(X) = E(X^2) - [E(X)]^2.$$

Proof: From Definition 3.15 we know that

$$\text{Var}(X) = E(X - E(X))^2 = \sum_x (x - E(X))^2 \cdot Pr(X = x).$$

Expanding within the summation we have

$$\begin{aligned}
\text{Var}(X) &= \sum_x (x^2 - 2x E(X) + [E(X)]^2) \cdot Pr(X = x) \\
&= \sum_x x^2 Pr(X = x) - 2E(X) \sum_x x \cdot Pr(X = x) \\
&\quad + [E(X)]^2 \sum_x Pr(X = x), \quad \text{because } E(X) \text{ is a constant} \\
&= E(X)^2 - 2E(X)E(X) + [E(X)]^2, \quad \text{because } \sum_x x \cdot Pr(X = x) = E(X) \\
&\qquad\qquad\qquad\qquad\qquad\qquad\qquad \text{and } \sum_x Pr(X = x) = 1 \\
&= E(X^2) - [E(X)]^2.
\end{aligned}$$

Let us check the result for $\text{Var}(X)$ in Example 3.56 by using Theorem 3.12.

EXAMPLE 3.57

The information in Example 3.56 provides the following:

x	x^2	$Pr(X = x)$
1	1	1/5
3	9	2/5
4	16	1/5
6	36	1/5

So $E(X^2) = \sum_x x^2 Pr(X = x) = (1)\left(\frac{1}{5}\right) + (9)\left(\frac{2}{5}\right) + (16)\left(\frac{1}{5}\right) + (36)\left(\frac{1}{5}\right) = \frac{71}{5}$.

Earlier in Example 3.56 we learned that $E(X) = 17/5$. Consequently, from Theorem 3.12 we have $\text{Var}(X) = E(X^2) - [E(X)]^2 = \frac{71}{5} - \left(\frac{17}{5}\right)^2 = \left(\frac{1}{25}\right)(355 - 289) = \frac{66}{25}$, as we found earlier.

We'll use the formula of Theorem 3.12 a second time in the following.

EXAMPLE 3.58

In Example 3.53 we studied the random variable X, which counted the number of heads that result when a fair coin is tossed four times. Soon thereafter we learned that X was a binomial random variable with $n = 4$, $p = 1/2$, $q = 1 - p = 1/2$, and $Pr(X = x) = \binom{4}{x}\left(\frac{1}{2}\right)^x\left(\frac{1}{2}\right)^{4-x}$, $x = 0, 1, 2, 3, 4$. Further, in part (d) of Example 3.55 we found that $E(X) = 2$ ($= np$, as we learned later in Theorem 3.11). To compute $\text{Var}(X)$ we use the formula in Theorem 3.12, but first we consider the following.

x	x^2	$Pr(X = x)$
0	0	1/16
1	1	4/16
2	4	6/16
3	9	4/16
4	16	1/16

Using these results we find that $E(X^2) = \sum_{x=0}^{4} x^2 Pr(X = x) = 0 \cdot \frac{1}{16} + 1 \cdot \frac{4}{16} + 4 \cdot \frac{6}{16} + 9 \cdot \frac{4}{16} + 16 \cdot \frac{1}{16} = \frac{80}{16} = 5$. So $\text{Var}(X) = E(X^2) - [E(X)]^2 = 5 - (2)^2 = 1 = 4\left(\frac{1}{2}\right)\left(\frac{1}{2}\right) = npq$, a result that is true in general. Further, for this random variable, the standard deviation $\sigma_X = 1$.

As we mentioned, the preceding example contains an instance of a more general result. We state that result now in our next theorem and outline a proof for this theorem in the Section Exercises.

THEOREM 3.13

Let X be the binomial random variable that counts the number of successes, each with probability p, among n independent Bernoulli trials. Then $\text{Var}(X) = npq$ and $\sigma_X = \sqrt{npq}$, where $q = 1 - p$.

As a result of Theorems 3.11 and 3.13 we now find that our next example requires little calculation.

EXAMPLE 3.59

Due to top-notch recruiting, Coach Jenkins' baseball team has probability 0.85 of winning each of the 12 baseball games it will play during the spring semester. (Here the outcome of each game is independent of the outcome of any previous game.)

Let X be the random variable that counts the number of games Coach Jenkins' team wins during the spring semester. Then $Pr(X = x) = \binom{12}{x}(0.85)^x(0.15)^{12-x}$, $x = 0, 1, 2, \ldots, 12$. Further, with $n = 12$ and $p = 0.85$, we readily see that $E(X) = \sum_{x=0}^{12} x\binom{12}{x}(0.85)^x(0.15)^{12-x} = np = 12(0.85) = 10.2$ and $\text{Var}(X) = \sum_{x=0}^{12}(x - 10.2)^2\binom{12}{x}(0.85)^x(0.15)^{12-x} = \sum_{x=0}^{12} x^2\binom{12}{x}(0.85)^x(0.15)^{12-x} - (10.2)^2 = npq = (12)(0.85)(0.15) = 1.53$.

A word of warning! The preceding example shows how easy it is to compute $E(X)$ and $\text{Var}(X)$ for a binomial random variable X, once we know the values of n and p. But remember, the formulas in Theorems 3.11 and 3.13 are valid only when the random variable X is *binomial*.

Before we introduce the last idea for this section we shall consider an example in order to motivate and illustrate the idea.

EXAMPLE 3.60

Referring back to Example 3.59, at this point we want to determine a *lower bound* for the probability that the random variable X is within k standard deviations σ_X of the mean $E(X)$, for $k = 2, 3$. When $k = 2$ we find that $Pr(E(X) - 2\sigma_X \leq X \leq E(X) + 2\sigma_X) = Pr(|X - E(X)| \leq 2\sigma_X)$. From the calculations in Example 3.59 we know that $E(X) = 10.2$ and $\text{Var}(X) = 1.53$, so $\sigma_X = \sqrt{1.53} \doteq 1.236932$. Consequently, $Pr(|X - E(X)| \leq 2\sigma_X) \doteq Pr(10.2 - 2(1.236932) \leq X \leq 10.2 + 2(1.236932)) \doteq Pr(7.726136 \leq X \leq 12.673864) = Pr(X = 8) + Pr(X = 9) + \cdots + Pr(X = 12) = \sum_{x=8}^{12}\binom{12}{x}(0.85)^x(0.15)^{12-x} \doteq 0.068284 + 0.171976 + 0.292358 + 0.301218 + 0.142242 = 0.976078$.

Likewise, for $k = 3$, $Pr(|X - E(X)| \leq 3\sigma_X) = Pr(6.489204 \leq X \leq 13.910796) = Pr(X = 7) + Pr(X = 8) + \cdots + Pr(X = 12) \doteq 0.019280 + 0.068284 + \cdots + 0.142242 = 0.995358$.

But where is this lower bound that we mentioned at the start of our discussion? Looking at the results for $k = 2, 3$ once more, we see that $Pr(|X - E(X)| \leq 2\sigma_X) \doteq 0.976078 \geq \frac{3}{4} = 1 - \frac{1}{2^2}$ and $Pr(|X - E(X)| \leq 3\sigma_X) \doteq 0.995358 \geq \frac{8}{9} = 1 - \frac{1}{3^2}$. So

$$Pr(|X - E(X)| \leq k\sigma_X) \geq 1 - \frac{1}{k^2}, \quad \text{for } k = 2, 3.$$

Further, although this lower bound is on the crude side, our next result will show that it is true for any positive real number k. In addition, the result is true for any random variable X, not just a binomial random variable like the one we have used here.

THEOREM 3.14

Chebyshev's Inequality. Let $\mathscr{S}$ be the sample space for an experiment $\mathscr{E}$ and let X be a random variable defined on the outcomes in $\mathscr{S}$. If $E(X)$ is the mean of X and σ_X its standard deviation, then for any $k > 0$,

$$Pr(E(X) - k\sigma_X \leq X \leq E(X) + k\sigma_X) = Pr(|X - E(X)| \leq k\sigma_X) \geq 1 - \frac{1}{k^2}.$$

[Here, as in Example 3.60, X accounts for those x values where $x = X(s)$ for some $s \in \mathscr{S}$ and $|x - E(X)| \leq k\sigma_X$.]

Proof: The proof presented here is for X discrete.[†] However, the result is also true for continuous random variables.

[†]The proof presented here is valid for the case where the sample space is countably infinite, so long as all the summations converge.

Let A, B be the following subsets of **R**.

$$A = \{x\,|\,|x - E(X)| > k\sigma_X\} \qquad B = \{x\,|\,|x - E(X)| \le k\sigma_X\}$$

(Note that A, B are not necessarily events for they need not be subsets of $\mathscr{S}$. They are subsets of the set of real numbers determined by the random variable X.)

We know that

$$\text{Var}(X) = \sigma_X^2 = \sum_x (x - E(X))^2 Pr(X = x)$$

$$= \sum_{x \in A}(x - E(X))^2 Pr(X = x) + \sum_{x \in B}(x - E(X))^2 Pr(X = x)$$

$$\ge \sum_{x \in A}(x - E(X))^2 Pr(X = x), \quad \text{as } \sum_{x \in B}(x - E(X))^2 Pr(X = x) \ge 0.$$

For $x \in A$, $|x - E(X)| > k\sigma_X$ and so it follows that here $|x - E(X)| \ge k\sigma_X$. Since $(x - E(X))^2 = |x - E(X)|^2$ we now have

$$\sigma_X^2 \ge \sum_{x \in A}|x - E(X)|^2 Pr(X = x) \ge k^2\sigma_X^2 \sum_{x \in A} Pr(X = x), \quad \text{and}$$

$$\sigma_X^2 \ge k^2\sigma_X^2 \sum_{x \in A} Pr(X = x) \Rightarrow \sigma_X^2 \ge k^2\sigma_X^2 Pr(|X - E(X)| > k\sigma_X)$$

$$\Rightarrow \frac{1}{k^2} \ge Pr(|X - E(X)| > k\sigma_X) \Rightarrow -\frac{1}{k^2} \le -Pr(|X - E(X)| > k\sigma_X)$$

$$\Rightarrow 1 - \frac{1}{k^2} \le 1 - Pr(|X - E(X)| > k\sigma_X)$$

$$\Rightarrow 1 - \frac{1}{k^2} \le Pr(|X - E(X)| \le k\sigma_X).$$

Our last example for this section shows how one might apply Chebyshev's Inequality.

| **EXAMPLE 3.61** | Angelica is selling boxes of candy for her choir's Christmas fund raiser. The pieces of candy are packed into each box so that the mean number of pieces is 125 with a standard deviation of 5 pieces. To find a lower bound on the probability that a box of Angelica's candy contains between 118 and 132 pieces we proceed as follows. |

Here the random variable X counts the number of pieces of candy in a box, with $E(X) = 125$ and $\sigma_X = 5$. Applying Chebyshev's Inequality we have

$$Pr(118 \le X \le 132) = Pr(118 - 125 \le X - 125 \le 132 - 125)$$

$$= Pr(-7 \le X - 125 \le 7) = Pr(|X - 125| \le 7)$$

$$= Pr\left(|X - E(X)| \le \left(\frac{7}{5}\right)\sigma_X\right) \ge 1 - \frac{1}{\left(\frac{7}{5}\right)^2} = 1 - \frac{25}{49} = \frac{24}{49}.$$

Consequently, the probability that a box of Angelica's candy contains between 118 and 132 pieces is at least $24/49 \doteq 0.489796$. (Note here that the value of k in Chebyshev's Inequality is $7/5$, which is *not* an integer.)

EXERCISES 3.7

1. Let X be a random variable with the following probability distribution.

x	0	1	2	3	4
$Pr(X = x)$	$\frac{1}{8}$	$\frac{1}{4}$	$\frac{1}{4}$	$\frac{1}{4}$	$\frac{1}{8}$

Determine (a) $Pr(X = 3)$; (b) $Pr(X \leq 4)$; (c) $Pr(X > 0)$; (d) $Pr(1 \leq X \leq 3)$; (e) $Pr(X = 2 | X \leq 3)$; and (f) $Pr(X \leq 1 \text{ or } X = 4)$.

2. The probability distribution for a random variable X is given by $Pr(X = x) = (3x + 1)/22$, $x = 0, 1, 2, 3$. Determine (a) $Pr(X = 3)$; (b) $Pr(X \leq 1)$; (c) $Pr(1 \leq X < 3)$; (d) $Pr(X > -2)$; and (e) $Pr(X = 1 | X \leq 2)$.

3. A shipment of 120 graphics cards contains 10 that are defective. Serena selects five of these cards, without replacement, and inspects them to see which, if any, are defective. If the random variable X counts the number of defective graphics cards in Serena's selection, determine (a) $Pr(X = x)$, $x = 0, 1, 2, \ldots, 5$; (b) $Pr(X = 4)$; (c) $Pr(X \geq 4)$; and (d) $Pr(X = 1 | X \leq 2)$.

4. Connie tosses a fair coin three times. If $X = X_1 - X_2$, where X_1 counts the number of heads that result and X_2 counts the number of tails that result, determine (a) the probability distributions for X_1, X_2, and X; and (b) the means $E(X_1)$, $E(X_2)$, and $E(X)$.

5. Let X be the random variable where $Pr(X = x) = 1/6$ for $x = 1, 2, 3, \ldots, 6$. (Here X is a *uniform discrete* random variable.) Determine (a) $Pr(X \geq 3)$; (b) $Pr(2 \leq X \leq 5)$; (c) $Pr(X = 4 | X \geq 3)$; (d) $E(X)$; and (e) $Var(X)$.

6. A computer dealer finds that the number of laptop computers her dealership sells each day is a random variable X where the probability distribution for X is given by

$$Pr(X = x) = \begin{cases} \dfrac{cx^2}{x!}, & x = 1, 2, 3, 4, 5 \\ 0, & \text{otherwise,} \end{cases}$$

where c is a constant. Determine (a) the value of c; (b) $Pr(X \geq 3)$; (c) $Pr(X = 4 | X \geq 3)$; (d) $E(X)$; and (e) $Var(X)$.

7. A random variable X has probability distribution given by

$$Pr(X = x) = \begin{cases} c(6 - x), & x = 1, 2, 3, 4, 5 \\ 0, & \text{otherwise,} \end{cases}$$

where c is a constant. Determine (a) the value of c; (b) $Pr(X \leq 2)$; (c) $E(X)$; and (d) $Var(X)$.

8. Wayne tosses an unfair coin — one that is biased so that a head is three times as likely to occur as a tail. How many heads should Wayne expect to see if he tosses the coin 100 times?

9. Suppose that X is a binomial random variable where $Pr(X = x) = \binom{n}{x} p^x (1 - p)^{n-x}$, $x = 0, 1, 2, \ldots, n$. If $E(X) = 70$ and $Var(X) = 45.5$, determine n, p.

10. A carnival game invites a player to select one card from a standard deck of 52 cards. If the card is a seven or a jack the player is given five dollars. For a king or an ace the player is given eight dollars. The other 36 cards result in the player losing. How much should one be willing to pay to play this game so that it is fair — that is, so that the expected value of the player's net winnings is 0?

11. The route that Jackie follows to school each day includes eight stoplights. When she reaches each stoplight, the probability that the stoplight is red is 0.25 and it is assumed that the stoplights are spaced far enough apart so as to operate independently. If the random variable X counts the number of red stoplights Jackie encounters one particular day on her ride to school, determine (a) $Pr(X = 0)$; (b) $Pr(X = 3)$; (c) $Pr(X \geq 6)$; (d) $Pr(X \geq 6 | X \geq 4)$; (e) $E(X)$; and (f) $Var(X)$.

12. Suppose that a random variable X has mean $E(X) = 17$ and variance $Var(X) = 9$, but its probability distribution is unknown. Use Chebyshev's Inequality to estimate a lower bound for (a) $Pr(11 \leq X \leq 23)$; (b) $Pr(10 \leq X \leq 24)$; and (c) $Pr(8 \leq X \leq 26)$.

13. Suppose that a random variable X has mean $E(X) = 15$ and variance $Var(X) = 4$, but its probability distribution is unknown. Use Chebyshev's Inequality to find the value of the constant c where $Pr(|X - 15| \leq c) \geq 0.96$.

14. Fred rolls a fair die 20 times. If X is the random variable that counts the number of 6's that come up during the 20 rolls, determine $E(X)$ and $Var(X)$.

15. A carton contains 20 computer chips, four of which are defective. Isaac tests these chips — one at a time and without replacement — until he either finds a defective chip or has tested three chips. If the random variable X counts the number of chips Isaac tests, find (a) the probability distribution for X; (b) $Pr(X \leq 2)$; (c) $Pr(X = 1 | X \leq 2)$; (d) $E(X)$; and (e) $Var(X)$.

16. Suppose that X is a random variable defined on a sample space $\mathcal{S}$ and that a, b are constants. Show that (a) $E(aX + b) = aE(x) + b$ and (b) $Var(aX + b) = a^2 Var(X)$.

17. Let X be a binomial random variable with $Pr(X = x) = \binom{n}{x} p^x q^{n-x}$, $x = 0, 1, 2, \ldots, n$, where n (≥ 2) is the number of Bernoulli trials, p is the probability of success for each trial, and $q = 1 - p$.

  a) Show that $E(X(X - 1)) = n^2 p^2 - np^2$.

  b) Using the fact that $E(X(X - 1)) = E(X^2 - X) = E(X^2) - E(X)$ and that $E(X) = np$, show that $Var(X) = npq$.

18. In alpha testing a new software package, a software engineer finds that the number of defects per 100 lines of code is a random variable X with probability distribution:

x	1	2	3	4
$Pr(X = x)$	0.4	0.3	0.2	0.1

Find (a) $Pr(X > 1)$; (b) $Pr(X = 3|X \geq 2)$; (c) $E(X)$; and (d) $\text{Var}(X)$.

19. In Mario Puzo's novel *The Godfather*, at the wedding reception for his daughter Constanzia, Don Vito Corleone discusses with his godson Johnny Fontane how he will deal with the movie mogul Jack Woltz. And in this context he speaks the famous line

"I'll make him an offer he can't refuse."

If we let the random variable X count the number of letters and apostrophes in a randomly selected word (from the above quotation) and we assume that each of the eight words has the same probability of being selected, determine (a) the probability distribution for X; (b) $E(X)$; and (c) $\text{Var}(X)$.

20. An assembly comprises three electrical components that operate independently. The probabilities that these components function according to specifications are 0.95, 0.9, and 0.88. If the random variable X counts the number of components that function according to specifications, determine (a) the probability distribution for X; (b) $Pr(X \geq 2|X \geq 1)$; (c) $E(X)$; and (d) $\text{Var}(X)$.

21. An urn contains five chips numbered 1, 2, 3, 4, and 5. When two chips are drawn (without replacement) from the urn, the random variable X records the higher value. Find $E(X)$ and σ_X.

3.8
Summary and Historical Review

In this chapter we introduced some of the fundamentals of set theory, together with certain relationships to enumeration problems and probability theory.

The algebra of set theory evolved during the nineteenth and early twentieth centuries. In England, George Peacock (1791–1858) was a pioneer in mathematical reforms and was among the first, in his *Treatise on Algebra*, to revolutionize the entire conception of algebra and arithmetic. His ideas were further developed by Duncan Gregory (1813–1844), William Rowan Hamilton (1805–1865), and Augustus DeMorgan (1806–1871), who attempted to remove ambiguity from elementary algebra and cast it in the strict postulational form. Not until 1854, however, when Boole published his *Investigation of the Laws of Thought*, was an algebra dealing with sets and logic formalized and the work of Peacock and his contemporaries extended.

The presentation here is primarily concerned with finite sets. However, the investigation of infinite sets and their cardinalities has occupied the minds of many mathematicians and philosophers. (More about this can be found in Appendix 3. However, the reader may want to learn more about functions — as presented in Chapter 5 — before looking into the material in this appendix.) The intuitive approach to set theory was taken until the time of the Russian-born mathematician Georg Cantor (1845–1918), who defined a set, in 1895, in a way comparable to the "gut feeling" we mentioned at the start of Section 3.1. His definition, however, was one of the obstacles he was never able to entirely remove from his theory of sets.

In the 1870s, when Cantor was researching trigonometric series and series of real numbers, he needed a device to compare the sizes of infinite sets of numbers. His treatment of the infinite as an actuality, on the same level as the finite, was quite revolutionary. Some of his work was rejected because it proved to be much more abstract than what many mathematicians of his time were accustomed to. However, his work won wide enough acceptance so that by 1890 the theory of sets, both finite and infinite, was considered a branch of mathematics in its own right.

By the turn of the century the theory was widely accepted, but in 1901 the paradox now known as Russell's paradox (which was discussed in Exercise 27 of Section 3.1) showed that set theory, as originally proposed, was internally inconsistent. The difficulty seemed to be in the unrestricted way in which sets could be defined; the idea of a set's being a

Georg Cantor (1845–1918)

Reproduced courtesy of The Granger Collection, New York

member of itself was considered particularly suspect. In their work *Principia Mathematica*, the British mathematicians Lord Bertrand Arthur William Russell (1872–1970) and Alfred North Whitehead (1861–1947) developed a hierarchy in the theory of sets known as the *theory of types*. This axiomatic set theory, among other twentieth-century formulations, avoided the Russell paradox. In addition to his work in mathematics, Lord Russell wrote books dealing with philosophy, physics, and his political views. His remarkable literary talent was recognized in 1950 when he was awarded the Nobel prize for literature.

Lord Bertrand Arthur William Russell (1872–1970)

The discovery of Russell's paradox — even though it could be remedied — had a profound impact on the mathematical community, for many began to wonder if other contradictions were still lurking. Then in 1931 the Austrian-born mathematician (and logician) Kurt Gödel (1906–1978) formulated that "under a specified consistency condition, any sufficiently strong formal axiomatic system must contain a proposition such that neither it nor its negation is provable and that any consistency proof for the system must use ideas and methods beyond those of the system itself." And unfortunately, from this we learn that we cannot establish — in a mathematically rigorous manner — that there are no contradictions in mathematics. Yet despite "Gödel's proof," mathematical research continues on — in fact, to the point where the amount of research since 1931 has surpassed that in any other period in history.

The use of the set membership symbol $\in$ (a stylized form of the Greek letter epsilon) was introduced in 1889 by the Italian mathematician Giuseppe Peano (1858–1932). The symbol "$\in$" is an abbreviation for the Greek word "$\epsilon\sigma\tau\iota$" meaning "is."

The Venn diagrams of Section 3.2 were introduced by the English logician John Venn (1834–1923) in 1881. In his book *Symbolic Logic*, Venn clarified ideas previously developed by his countryman George Boole (1815–1864). Furthermore, Venn contributed to the development of probability theory — as described in the widely read textbook he wrote on this subject. The Gray code, which we used in Section 3.1 to store the subsets of a finite set as binary strings, was developed in the 1940s by Frank Gray at the AT&T Bell Laboratories. Originally, such codes were used to minimize the effect of errors in the transmission of digital signals.

If we wish to summarize the importance of the role of set theory in the development of twentieth-century mathematics, the following quote attributed to the German mathematician David Hilbert (1862–1943) is worth pondering: "No one shall expel us from the paradise which Cantor has created for us."

In Section 3.1 we mentioned the array of numbers known as Pascal's triangle. We could have introduced this array in Chapter 1 with the binomial theorem, but we waited until we had some combinatorial identities that we needed to verify how the triangle is constructed. The array appears in the work of the Chinese algebraist Chu Shi-kie (1303), but its first appearance in Europe was not until the sixteenth century, on the title page of a book by Petrus Apianus (1495–1552). Niccolo Tartaglia (1499–1559) used the triangle in computing powers of $(x + y)$. Because of his work on the properties and applications of this triangle, the array has been named in honor of the French mathematician Blaise Pascal (1623–1662).

Although probability theory originated with games of chance and enumeration problems, we included it here because set theory has evolved as the exact medium needed to state and solve problems in this important contemporary area of applied mathematics. In the decade following 1660, probability entered European thought as a way of understanding stable frequencies in random processes. Ideas, which exemplify this consideration, were put forth by Blaise Pascal, and these led to the first systematic treatise on probability, written in 1657 by Christian Huygens (1629–1695). In 1812 Pierre-Simon de Laplace (1749–1827) collected all the ideas developed on probability theory at that time — starting with the definition in which each individual outcome is equally likely — and published them in his *Analytic Theory of Probability*. Among other ideas, this text includes the *Central Limit Theorem* — a fundamental result at the heart of hypothesis testing (in statistics). Along with Pierre-Simon de Laplace, Thomas Bayes (1702–1761) also showed how to determine probabilities by examining certain empirical data. Bayes' Theorem honors the name of this English Presbyterian minister and mathematician. Chebyshev's Inequality (of Section 3.7) is named for the Russian mathematician Pafnuty Lvovich Chebyshev (1821–1894), who may be better remembered for his work in number theory and interest in mechanics. Finally, the axiomatic approach to probability was first given in 1933 by the Russian mathematician Andrei Nikolayevich Kolmogorov (1903–1987) in his monograph *Grundbegriffe der Wahrscheinlichkeitsrechnung* (*Foundations of the Theory of Probability*).

More on the history and development of set theory can be found in Chapter 26 of C. B. Boyer [1]. Formal developments of set theory, including results on infinite sets, can be found in H. B. Enderton [3], P. R. Halmos [4], J. M. Henle [5], and P. C. Suppes [8]. An interesting history of the origins of probability and statistical ideas, up to the Newtonian era, can be found in F. N. David [2]. A more contemporary coverage is given in the text by V. J. Katz [6]. Chapters 1 and 2 of J. J. Kinney [7] are an excellent source for those interested in learning more about discrete probability.

Andrei Nikolayevich Kolmogorov (1903–1987)

Thomas Bayes (1702–1761)

REFERENCES

1. Boyer, Carl B. *History of Mathematics*. New York: Wiley, 1968.
2. David, Florence Nightingale. *Games, Gods, and Gambling*. New York: Hafner, 1962.
3. Enderton, Herbert B. *Elements of Set Theory*. New York: Academic Press, 1977.
4. Halmos, Paul R. *Naive Set Theory*. New York: Van Nostrand, 1960.
5. Henle, James M. *An Outline of Set Theory*. New York: Springer-Verlag, 1986.
6. Katz, Victor J. *A History of Mathematics (An Introduction)*. New York: Harper Collins, 1993.
7. Kinney, John J. *Probability: An Introduction with Statistical Applications*. New York: Wiley, 1997.
8. Suppes, Patrick C. *Axiomatic Set Theory*. New York: Van Nostrand, 1960.

SUPPLEMENTARY EXERCISES

1. Let A, B, $C \subseteq \mathcal{U}$. Prove that $(A - B) \subseteq C$ if and only if $(A - C) \subseteq B$.

2. Give a combinatorial argument to show that for integers n, r with $n \geq r \geq 2$,

$$\binom{n+2}{r} = \binom{n}{r} + 2\binom{n}{r-1} + \binom{n}{r-2}.$$

3. Let A, B, $C \subseteq \mathcal{U}$. Prove or disprove (with a counterexample) each of the following:

a) $A - C = B - C \Rightarrow A = B$

b) $[(A \cap C = B \cap C) \wedge (A - C = B - C)] \Rightarrow A = B$

c) $[(A \cup C = B \cup C) \wedge (A - C = B - C)] \Rightarrow A = B$

4. a) For positive integers m, n, r, with $r \leq \min\{m, n\}$, show that

$$\binom{m+n}{r} = \binom{m}{0}\binom{n}{r} + \binom{m}{1}\binom{n}{r-1} + \binom{m}{2}\binom{n}{r-2}$$
$$+ \cdots + \binom{m}{r}\binom{n}{0} = \sum_{k=0}^{r} \binom{m}{k}\binom{n}{r-k}.$$

b) For n a positive integer, show that

$$\binom{2n}{n} = \sum_{k=0}^{n} \binom{n}{k}^2.$$

5. a) In how many ways can a teacher divide a group of seven students into two teams each containing at least one student? two students?

b) Answer part (a) upon replacing seven with a positive integer $n \geq 4$.

6. Determine whether each of the following statements is true or false. For each false statement, give a counterexample.

a) If A and B are infinite sets, then $A \cap B$ is infinite.

b) If B is infinite and $A \subseteq B$, then A is infinite.

c) If $A \subseteq B$ with B finite, then A is finite.

d) If $A \subseteq B$ with A finite, then B is finite.

7. A set A has 128 subsets of even cardinality. (a) How many subsets of A have odd cardinality? (b) What is $|A|$?

8. Let $A = \{1, 2, 3, \ldots, 15\}$.

a) How many subsets of A contain all of the odd integers in A?

b) How many subsets of A contain exactly three odd integers?

c) How many eight-element subsets of A contain exactly three odd integers?

d) Write a computer program (or develop an algorithm) to generate a random eight-element subset of A and have it print out how many of the eight elements are odd.

9. Let $A, B, C \subseteq \mathcal{U}$. Prove that

$$(A \cap B) \cup C = A \cap (B \cup C) \text{ if and only if } C \subseteq A.$$

10. Let $\mathcal{U}$ be a given universe with $A, B \subseteq \mathcal{U}$, $|A \cap B| = 3$, $|A \cup B| = 8$, and $|\mathcal{U}| = 12$.

a) How many subsets $C \subseteq \mathcal{U}$ satisfy $A \cap B \subseteq C \subseteq A \cup B$? How many of these subsets C contain an even number of elements?

b) How many subsets $D \subseteq \mathcal{U}$ satisfy $\overline{A \cup B} \subseteq D \subseteq \overline{A \cup B}$? How many of these subsets D contain an even number of elements?

11. Let $\mathcal{U} = \mathbf{R}$ and let the index set $I = \mathbf{Q}^+$. For each $q \in \mathbf{Q}^+$, let $A_q = [0, 2q]$ and $B_q = (0, 3q]$. Determine

a) $A_{7/3}$ **b)** $A_3 \triangle B_4$

c) $\bigcup_{q \in I} A_q$ **d)** $\bigcap_{q \in I} B_q$

12. For a universe $\mathcal{U}$ and sets $A, B \subseteq \mathcal{U}$, prove that

a) $A \triangle B = B \triangle A$ **b)** $A \triangle \overline{A} = \mathcal{U}$

c) $A \triangle \mathcal{U} = \overline{A}$

d) $A \triangle \emptyset = A$, so $\emptyset$ is the identity for $\triangle$, as well as for $\cup$

13. Consider the membership table (Table 3.7). If we are given the condition that $A \subseteq B$, then we need consider only those

rows of the table for which this is true — rows 1, 2, and 4, as indicated by the arrows. For these rows, the columns for B and $A \cup B$ are exactly the same, so this membership table shows that $A \subseteq B \Rightarrow A \cup B = B$.

Table 3.7

	A	B	$A \cup B$
$\rightarrow$	0	0	0
$\rightarrow$	0	1	1
	1	0	1
$\rightarrow$	1	1	1

Use membership tables to verify each of the following:

a) $A \subseteq B \Rightarrow A \cap B = A$

b) $[(A \cap B = A) \wedge (B \cup C = C)] \Rightarrow A \cup B \cup C = C$

c) $C \subseteq B \subseteq A \Rightarrow (A \cap \overline{B}) \cup (B \cap \overline{C}) = A \cap \overline{C}$

d) $A \triangle B = C \Rightarrow A \triangle C = B$ and $B \triangle C = A$

14. State the dual of each theorem in Exercise 13. (Here you will want to use the result of Example 3.19 in conjunction with Theorem 3.5.)

15. a) Determine the number of linear arrangements of m 1's and r 0's with no adjacent 1's. (State any needed condition(s) for m, r.)

b) If $\mathcal{U} = \{1, 2, 3, \ldots, n\}$, how many sets $A \subseteq \mathcal{U}$ are such that $|A| = k$ with A containing no consecutive integers? [State any needed condition(s) for n, k.]

16. If the letters in the word BOOLEAN are arranged at random, what is the probability that the two O's remain together in the arrangement?

17. At a high school science fair, 34 students received awards for scientific projects. Fourteen awards were given for projects in biology, 13 in chemistry, and 21 in physics. If three students received awards in all three subject areas, how many received awards for exactly (a) one subject area? (b) two subject areas?

18. Fifty students, each with 75¢, visited the arcade of Example 3.27. Seventeen of the students played each of the three computer games, and 37 of them played at least two of them. No student played any other game at the arcade, nor did any student play a given game more than once. Each game costs 25¢ to play, and the total proceeds from the student visit were $24.25. How many of these students preferred to watch and played none of the games?

19. In how many ways can 15 laboratory assistants be assigned to work on one, two, or three different experiments so that each experiment has at least one person spending some time on it?

20. Professor Diane gave her chemistry class a test consisting of three questions. There are 21 students in her class, and every student answered at least one question. Five students did not answer the first question, seven failed to answer the second question, and six did not answer the third question. If nine students answered all three questions, how many answered exactly one question?

21. Let $\mathcal{U}$ be a given universe with $A, B \subseteq \mathcal{U}$, $A \cap B = \emptyset$, $|A| = 12$, and $|B| = 10$. If seven elements are selected from $A \cup B$, what is the probability the selection contains four elements from A and three from B?

22. For a finite set A of integers, let $\sigma(A)$ denote the sum of the elements of A. Then if $\mathcal{U}$ is a finite universe taken from $\mathbf{Z}^+$, $\Sigma_{A \in \mathcal{P}(\mathcal{U})} \sigma(A)$ denotes the sum of all elements of all subsets of $\mathcal{U}$. Determine $\Sigma_{A \in \mathcal{P}(\mathcal{U})} \sigma(A)$ for

   a) $\mathcal{U} = \{1, 2, 3\}$ **b)** $\mathcal{U} = \{1, 2, 3, 4\}$

   c) $\mathcal{U} = \{1, 2, 3, 4, 5\}$ **d)** $\mathcal{U} = \{1, 2, 3, \ldots, n\}$

   e) $\mathcal{U} = \{a_1, a_2, a_3, \ldots, a_n\}$, where
      $s = a_1 + a_2 + a_3 + \cdots + a_n$

23. a) In chess, the king can move one position in any direction. Assuming that the king is moved only in a forward manner (one position up, to the right, or diagonally northeast), along how many different paths can a king be moved from the lower-left corner position to the upper-right corner position on the standard 8×8 chessboard?

   b) For the paths in part (a), what is the probability that a path contains (i) exactly two diagonal moves? (ii) exactly two diagonal moves that are consecutive? (iii) an even number of diagonal moves?

24. Let $A, B \subseteq \mathbf{R}$, where $A = \{x \mid x^2 - 7x = -12\}$ and $B = \{x \mid x^2 - x = 6\}$. Determine $A \cup B$ and $A \cap B$.

25. Let $A, B \subseteq \mathbf{R}$, where $A = \{x \mid x^2 - 7x \leq -12\}$ and $B = \{x \mid x^2 - x \leq 6\}$. Determine $A \cup B$ and $A \cap B$.

26. Four torpedoes, whose probabilities of destroying an enemy ship are 0.75, 0.80, 0.85, and 0.90, are fired at such a vessel. Assuming the torpedoes operate independently, what is the probability the enemy ship is destroyed?

27. Travis tosses a fair coin twice. Then he tosses a biased coin, one where the probability of a head is 3/4, four times. What is the probability Travis's six tosses result in five heads and one tail?

28. Let $\mathcal{S}$ be the sample space for an experiment $\mathcal{E}$, with events $A, B \subseteq \mathcal{S}$. Prove that

$$Pr(A|B) \geq \frac{Pr(A) + Pr(B) - 1}{Pr(B)}.$$

29. Let A, B, C be independent events taken from a sample space $\mathcal{S}$. Prove that the events A and $B \cup C$ are independent.

30. What is the minimum number of times we must toss a fair coin so that the probability that we get at least two heads is at least 0.95?

31. A large jet aircraft has two wheels per landing gear for added safety. The tires are rated so that even with a "hard landing" the probability of any single tire blowing out is only 0.10. (a) What is the probability that a landing gear (with two tires) will survive even a hard landing with at least one good tire? (b) In order for the plane to land safely, all three landing gears (the nose and both wing landing gears) must have at least one good tire. What is the probability that the jet will be able to land safely even on a hard landing?

32. Let $\mathcal{S}$ be the sample space for an experiment $\mathcal{E}$ and let A, B be events — that is, $A, B \subseteq \mathcal{S}$. Prove that $Pr(A \cap B) \geq Pr(A) + Pr(B) - 1$. (This result is known as *Bonferroni's Inequality*.)

33. The exit door at the end of a hallway is open half of the time. On a table by the entrance to this hallway is a box containing 10 keys, but only one of these keys opens the exit door at the end of the hallway. Upon entering the hallway Marlo selects two of the keys from the box. What is the probability she will be able to leave the hallway via the exit door, without returning to the box for more keys?

34. Dustin tosses a fair coin eight times. Given that his first and last outcomes are the same, what is the probability he tossed five heads and three tails?

35. The probability Coach Sears' basketball team wins any given game is 0.8, regardless of any prior win or loss. If her team plays five games, what is the probability it wins more games than it loses?

36. Suppose that the number of boxes of cereal packaged each day at a certain packaging plant is a random variable — call it X — with $E(X) = 20,000$ boxes and $\mathrm{Var}(X) = 40,000$ boxes^2. Use Chebyshev's Inequality to find a lower bound on the probability that the plant will package between 19,000 and 21,000 boxes of cereal on a particular day.

37. Find the probability of getting one head (exactly) two times when three fair coins are tossed four times.

38. Devon has a bag containing 22 poker chips — eight red, eight white, and six blue. Aileen reaches in and withdraws three of the chips, without replacement. Find the probability that Aileen has selected (a) no blue chips; (b) one chip of each color; or (c) at least two red chips.

39. Let X be a random variable with probability distribution

$$Pr(X = x) = \begin{cases} c(x^2 + 4), & x = 0, 1, 2, 3, 4 \\ 0, & \text{otherwise,} \end{cases}$$

where c is a constant. Determine (a) the value of c; (b) $Pr(X > 1)$; (c) $Pr(X = 3 | X \geq 2)$; (d) $E(X)$; and (e) $\mathrm{Var}(X)$.

40. A dozen urns each contain four red marbles and seven green ones. (All 132 marbles are of the same size.) If a dozen students each select a different urn and then draw (with replacement) five marbles, what is the probability that at least one student draws at least one red marble?

41. Maureen draws five cards from a standard deck: the 6 of diamonds, 7 of diamonds, 8 of diamonds, jack of hearts, and king of spades. She discards the jack and king and then draws two cards from the remaining 47. What is the probability Maureen finishes with (a) a straight flush; (b) a flush (but not a straight flush); and (c) a straight (but not a straight flush)?

42. In the game of *pinochle* the deck consists of 48 cards — two each of the 9, 10, jack, queen, king, and ace for each of the four suits. There are four players and each is dealt 12 cards. What is the probability a given player is dealt four kings (one of each suit), four queens (one of each suit), and four other cards none of which is a king or queen? (Such a hand is referred to as a *bare roundhouse*.)

43. A grab bag contains one chip with the number 1, two chips each with the number 2, three chips each with the number 3, . . . , and n chips each with the number n, where $n \in \mathbf{Z}^+$. All chips are of the same size, those numbered 1 to m are red, and those numbered $m + 1$ to n are blue, where $m \in \mathbf{Z}^+$ and $m \leq n$. If Casey draws one chip, what is the probability it is the chip with 1 on it, given that the chip is red?

44. A fair die is rolled three times and the random variable X records the number of different outcomes that result. For example, if two 5's and one 4 are rolled, then X records two different outcomes. Determine (a) the probability distribution for X; (b) $E(X)$; and (c) Var(X).

45. When a coin is tossed three times, for the outcome HHT we say that two *runs* have occurred — namely, HH and T. Likewise, for the outcome THT we find three runs: T, H, and T. (The notion of a *run* was first introduced in Example 1.41.) Now suppose a biased coin, with $Pr(\mathrm{H}) = 3/4$, is tossed three times and the random variable X counts the number of runs that result. Determine (a) the probability distribution for X; (b) $E(X)$; and (c) σ_X.

4

Properties of the Integers: Mathematical Induction

Having known about the integers since our first encounters with arithmetic, in this chapter we examine a special property exhibited by the subset of positive integers. This property will enable us to establish certain mathematical formulas and theorems by using a technique called *mathematical induction*. This method of proof will play a key role in many of the results we shall obtain in the later chapters of this text. Furthermore, this chapter will provide us with an introduction to five sets of numbers that are very important in the study of discrete mathematics and combinatorics — namely, the triangular numbers, the harmonic numbers, the Fibonacci numbers, the Lucas numbers, and the Eulerian numbers.

When $x, y \in \mathbf{Z}$, we know that $x + y, xy, x - y \in \mathbf{Z}$. Thus we say that the set $\mathbf{Z}$ is *closed* under (the binary operations of) addition, multiplication, and subtraction. Turning to division, however, we find, for example, that $2, 3 \in \mathbf{Z}$ but that the rational number $\frac{2}{3}$ is *not* a member of $\mathbf{Z}$. So the set $\mathbf{Z}$ of all integers is *not closed* under the binary operation of *nonzero* division. To cope with this situation, we shall introduce a somewhat restricted form of division for $\mathbf{Z}$ and shall concentrate on special elements of $\mathbf{Z}^+$ called *primes*. These primes turn out to be the "building blocks" of the integers, and they provide our first example of a representation theorem — in this case the Fundamental Theorem of Arithmetic.

4.1
The Well-Ordering Principle: Mathematical Induction

Given any two distinct integers x, y, we know that we must have either $x < y$ or $y < x$. However, this is also true if, instead of being integers, x and y are rational numbers or real numbers. What makes $\mathbf{Z}$ special in this situation?

Suppose we try to express the subset $\mathbf{Z}^+$ of $\mathbf{Z}$, using the inequality symbols $>$ and $\geq$. We find that we can define the set of positive elements of $\mathbf{Z}$ as

$$\mathbf{Z}^+ = \{x \in \mathbf{Z} | x > 0\} = \{x \in \mathbf{Z} | x \geq 1\}.$$

When we try to do likewise for the rational and real numbers, however, we find that

$$\mathbf{Q}^+ = \{x \in \mathbf{Q} | x > 0\} \qquad \text{and} \qquad \mathbf{R}^+ = \{x \in \mathbf{R} | x > 0\},$$

but we cannot represent $\mathbf{Q}^+$ or $\mathbf{R}^+$ using $\geq$ as we did for $\mathbf{Z}^+$.

The set $\mathbf{Z}^+$ is different from the sets $\mathbf{Q}^+$ and $\mathbf{R}^+$ in that *every* nonempty subset X of $\mathbf{Z}^+$ contains an integer a such that $a \leq x$, for all $x \in X$ — that is, X contains a *least* (or *smallest*) element. This is not so for either $\mathbf{Q}^+$ or $\mathbf{R}^+$. The sets themselves do not contain least elements. There is no smallest positive rational number or smallest positive real number. If q is a positive rational number, then since $0 < q/2 < q$, we would have the smaller positive rational number $q/2$.

These observations lead us to the following property of the set $\mathbf{Z}^+ \subset \mathbf{Z}$.

The Well-Ordering Principle: Every *nonempty* subset of $\mathbf{Z}^+$ contains a smallest element. (We often express this by saying that $\mathbf{Z}^+$ is *well ordered*.)

This principle serves to distinguish $\mathbf{Z}^+$ from $\mathbf{Q}^+$ and $\mathbf{R}^+$. But does it lead anywhere that is mathematically interesting or useful? The answer is a resounding "Yes!" It is the basis of a proof technique known as mathematical induction. This technique will often help us to prove a general mathematical statement involving positive integers when certain instances of that statement suggest a general pattern.

We now establish the basis for this induction technique.

THEOREM 4.1

The Principle of Mathematical Induction. Let $S(n)$ denote an open mathematical statement (or set of such open statements) that involves one or more occurrences of the variable n, which represents a positive integer.

   a) If $S(1)$ is true; and

   b) If whenever $S(k)$ is true (for some particular, but arbitrarily chosen, $k \in \mathbf{Z}^+$), then $S(k+1)$ is true;

then $S(n)$ is true for all $n \in \mathbf{Z}^+$.

Proof: Let $S(n)$ be such an open statement satisfying conditions (a) and (b), and let $F = \{t \in \mathbf{Z}^+ | S(t) \text{ is false}\}$. We wish to prove that $F = \emptyset$, so to obtain a contradiction we assume that $F \neq \emptyset$. Then by the Well-Ordering Principle, F has a least element m. Since $S(1)$ is true, it follows that $m \neq 1$, so $m > 1$, and consequently $m - 1 \in \mathbf{Z}^+$. With $m - 1 \notin F$, we have $S(m - 1)$ true. So by condition (b) it follows that $S((m - 1) + 1) = S(m)$ is true, contradicting $m \in F$. This contradiction arose from the assumption that $F \neq \emptyset$. Consequently, $F = \emptyset$.

We have now seen how the Well-Ordering Principle is used in the proof of the Principle of Mathematical Induction. It is also true that the Principle of Mathematical Induction is useful if one wants to prove the Well-Ordering Principle. However, we shall not concern ourselves with that fact right now. In this section our major goal will center on understanding and using the Principle of Mathematical Induction. (But in the exercises for Section 4.2 we shall examine how the Principle of Mathematical Induction is used to prove the Well-Ordering Principle.)

In the statement of Theorem 4.1 the condition in part (a) is referred to as the *basis step*, while that in part (b) is called the *inductive step*.

The choice of 1 in the first condition of Theorem 4.1 is not mandatory. All that is needed is for the open statement $S(n)$ to be true for some *first* element $n_0 \in \mathbf{Z}$ so that the induction process has a starting place. We need the truth of $S(n_0)$ for our basis step. The integer n_0 could be 5 just as well as 1. It could even be zero or negative because the set $\mathbf{Z}^+$ in union with $\{0\}$ or any *finite* set of negative integers is well ordered. (When we do an induction proof and start with $n_0 < 0$, we are considering the set of all *consecutive* negative integers $\geq n_0$ in union with $\{0\}$ and $\mathbf{Z}^+$.)

Under these circumstances, we may express the Principle of Mathematical Induction, using quantifiers, as

$$[S(n_0) \wedge [\forall\, k \geq n_0 \; [S(k) \Rightarrow S(k+1)]]] \Rightarrow \forall\, n \geq n_0 \;\; S(n).$$

We may get a somewhat better understanding of why this method of proof is valid by using our intuition in conjunction with the situation presented in Fig. 4.1.

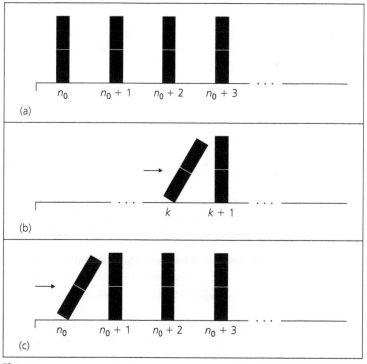

Figure 4.1

In part (a) of the figure we see the first four of an infinite (ordered) arrangement of dominos, each standing on end. The spacing between any two consecutive dominos is always the same, and it is such that if any one domino (say the kth) is pushed over to the right, then it will knock over the next $((k+1)$st) domino. This process is suggested in Fig. 4.1(b). Our intuition leads us to feel that this process will continue, the $(k+1)$st domino toppling and knocking over (to the right) the $(k+2)$nd domino, and so on. Part (c) of the figure indicates how the truth of $S(n_0)$ provides the push (to the right) to the first domino (at n_0). This provides the basis step and sets the process in motion. The truth of $S(k)$

forcing the truth of $S(k + 1)$ gives us the inductive step and continues the toppling process. We then infer the fact that $S(n)$ is true for all $n \geq n_0$ as we imagine *all* the successive dominos toppling (to the right.)

We shall now demonstrate several results that call for the use of Theorem 4.1.

EXAMPLE 4.1

For all $n \in \mathbf{Z}^+$, $\sum_{i=1}^{n} i = 1 + 2 + 3 + \cdots + n = \frac{n(n+1)}{2}$.

Proof: For $n = 1$ the open statement

$$S(n): \quad \sum_{i=1}^{n} i = 1 + 2 + 3 + \cdots + n = \frac{n(n + 1)}{2}$$

becomes $S(1)$: $\sum_{i=1}^{1} i = 1 = (1)(1 + 1)/2$. So $S(1)$ is true and we have our *basis step*—and a starting point from which to begin the induction. Assuming the result true for $n = k$ (for some $k \in \mathbf{Z}^+$), we want to establish our *inductive step* by showing how the truth of $S(k)$ "forces" us to accept the truth of $S(k + 1)$. [The assumption of the truth of $S(k)$ is our *induction hypothesis*.] To establish the truth of $S(k + 1)$, we need to show that

$$\sum_{i=1}^{k+1} i = \frac{(k + 1)(k + 2)}{2}.$$

We proceed as follows.

$$\sum_{i=1}^{k+1} i = 1 + 2 + \cdots + k + (k + 1) = \left(\sum_{i=1}^{k} i \right) + (k + 1) = \frac{k(k + 1)}{2} + (k + 1),$$

for we are assuming the truth of $S(k)$. But

$$\frac{k(k + 1)}{2} + (k + 1) = \frac{k(k + 1)}{2} + \frac{2(k + 1)}{2} = \frac{(k + 1)(k + 2)}{2},$$

establishing the inductive step [condition (b)] of the theorem.

Consequently, by the Principle of Mathematical Induction, $S(n)$ is true for all $n \in \mathbf{Z}^+$.

Now that we have obtained the summation formula for $\sum_{i=1}^{n} i$ in two ways (see Example 1.40), we shall digress from our main topic and consider two examples that use this summation formula.

EXAMPLE 4.2

A wheel of fortune has the numbers from 1 to 36 painted on it in a random manner. Show that regardless of how the numbers are situated, there are three consecutive (on the wheel) numbers whose total is 55 or more.

Let x_1 be any number on the wheel. Counting clockwise from x_1, label the other numbers $x_2, x_3, \ldots, x_{36}$. For the result to be false, we must have $x_1 + x_2 + x_3 < 55$, $x_2 + x_3 + x_4 < 55$, $\ldots$, $x_{34} + x_{35} + x_{36} < 55$, $x_{35} + x_{36} + x_1 < 55$, and $x_{36} + x_1 + x_2 < 55$. In these 36 inequalities, each of the terms $x_1, x_2, \ldots, x_{36}$ appears (exactly) three times, so each of the integers $1, 2, \ldots, 36$ appears (exactly) three times. Adding all 36 inequalities, we find that $3 \sum_{i=1}^{36} x_i = 3 \sum_{i=1}^{36} i < 36(55) = 1980$. But $\sum_{i=1}^{36} i = (36)(37)/2 = 666$, and this gives us the contradiction that $1998 = 3(666) < 1980$.

EXAMPLE 4.3

Among the 900 three-digit integers (from 100 to 999) those such as 131, 222, 303, 717, 848, and 969, where the integer is the same whether it is read from left to right or from right to left, are called *palindromes*. Without actually determining all of these three-digit palindromes, we would like to determine their sum.

The typical palindrome under study here has the form $aba = 100a + 10b + a = 101a + 10b$, where $1 \le a \le 9$ and $0 \le b \le 9$. With nine choices for a and ten for b, it follows from the rule of product that there are 90 such three-digit palindromes. Their sum is

$$\sum_{a=1}^{9} \left(\sum_{b=0}^{9} aba \right) = \sum_{a=1}^{9} \sum_{b=0}^{9} aba = \sum_{a=1}^{9} \sum_{b=0}^{9} (101a + 10b)$$

$$= \sum_{a=1}^{9} \left[10(101a) + 10 \sum_{b=0}^{9} b \right] = \sum_{a=1}^{9} \left[10(101a) + 10 \sum_{b=1}^{9} b \right]$$

$$= \sum_{a=1}^{9} \left[1010a + \frac{10(9 \cdot 10)}{2} \right] = \sum_{a=1}^{9} (1010a + 450)$$

$$= 1010 \sum_{a=1}^{9} a + 9(450)$$

$$= \frac{1010(9 \cdot 10)}{2} + 4050 = 49{,}500.$$

The next summation formula takes us from first powers to squares.

EXAMPLE 4.4

Prove that for each $n \in \mathbf{Z}^+$,

$$\sum_{i=1}^{n} i^2 = \frac{n(n+1)(2n+1)}{6}.$$

Proof: Here we are dealing with the open statement

$$S(n): \quad \sum_{i=1}^{n} i^2 = \frac{n(n+1)(2n+1)}{6}.$$

Basis Step: We start with the statement $S(1)$ and find that

$$\sum_{i=1}^{1} i^2 = 1^2 = \frac{1(1+1)(2(1)+1)}{6}.$$

so $S(1)$ is true.

Inductive Step: Now we assume the truth of $S(k)$, for some (particular) $k \in \mathbf{Z}^+$ — that is, we assume that

$$\sum_{i=1}^{k} i^2 = \frac{k(k+1)(2k+1)}{6}$$

is a true statement (when n is replaced by k). From this assumption we want to deduce the truth of

$$S(k+1): \quad \sum_{i=1}^{k+1} i^2 = \frac{(k+1)((k+1)+1)(2(k+1)+1)}{6}$$

$$= \frac{(k+1)(k+2)(2k+3)}{6}.$$

Using the induction hypothesis $S(k)$, we find that

$$\sum_{i=1}^{k+1} i^2 = 1^2 + 2^2 + \cdots + k^2 + (k+1)^2 = \sum_{i=1}^{k} i^2 + (k+1)^2$$

$$= \left[\frac{k(k+1)(2k+1)}{6}\right] + (k+1)^2$$

$$= (k+1)\left[\frac{k(2k+1)}{6} + (k+1)\right] = (k+1)\left[\frac{2k^2 + 7k + 6}{6}\right]$$

$$= \frac{(k+1)(k+2)(2k+3)}{6},$$

and the general result follows by the Principle of Mathematical Induction.

The formulas from Examples 4.1 and 4.4 prove handy in deriving our next result.

EXAMPLE 4.5

Figure 4.2 provides the first four entries of the sequence of *triangular* numbers. We see that $t_1 = 1, t_2 = 3, t_3 = 6, t_4 = 10$, and, in general, $t_i = 1 + 2 + \cdots + i = i(i+1)/2$, for each $i \in \mathbf{Z}^+$. For a fixed $n \in \mathbf{Z}^+$ we want a formula for the sum of the first n triangular numbers — that is, $t_1 + t_2 + \cdots + t_n = \sum_{i=1}^{n} t_i$. When $n = 2$ we have $t_1 + t_2 = 4$. For $n = 3$ the sum is 10. Considering n fixed (but arbitrary) we find that

$$\sum_{i=1}^{n} t_i = \sum_{i=1}^{n} \frac{i(i+1)}{2} = \frac{1}{2}\sum_{i=1}^{n}(i^2 + i) = \frac{1}{2}\sum_{i=1}^{n}i^2 + \frac{1}{2}\sum_{i=1}^{n}i$$

$$= \frac{1}{2}\left[\frac{n(n+1)(2n+1)}{6}\right] + \frac{1}{2}\left[\frac{n(n+1)}{2}\right] = n(n+1)\left[\frac{2n+1}{12} + \frac{1}{4}\right]$$

$$= \frac{n(n+1)(n+2)}{6}.$$

Consequently, if we wish to know the sum of the first 100 triangular numbers, we have

$$t_1 + t_2 + \cdots + t_{100} = \frac{100(101)(102)}{6} = 171{,}700.$$

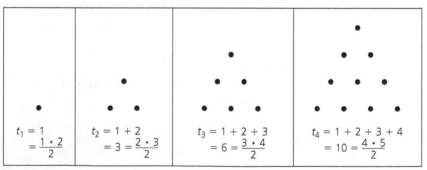

Figure 4.2

Before we present any more results, let us note how we started the proofs in Examples 4.1 and 4.4. In both cases we simply replaced the variable n by 1 and verified the truth of some rather easy equalities. Considering how the inductive step in each of these proofs was definitely more complicated to establish, we might question the need for bothering with these basis steps. So let us examine the following example.

EXAMPLE 4.6

If $n \in \mathbf{Z}^+$, establish the validity of the open statement

$$S(n): \quad \sum_{i=1}^{n} i = 1 + 2 + 3 + \cdots + n = \frac{n^2 + n + 2}{2}.$$

This time we shall go directly to the inductive step. Assuming the truth of the statement

$$S(k): \quad \sum_{i=1}^{k} i = 1 + 2 + 3 + \cdots + k = \frac{k^2 + k + 2}{2}$$

for some (particular) $k \in \mathbf{Z}^+$, we want to infer the truth of the statement

$$S(k+1): \quad \sum_{i=1}^{k+1} i = 1 + 2 + 3 + \cdots + k + (k+1) = \frac{(k+1)^2 + (k+1) + 2}{2}$$
$$= \frac{k^2 + 3k + 4}{2}.$$

As we did previously, we use the induction hypothesis and calculate as follows:

$$\sum_{i=1}^{k+1} i = 1 + 2 + 3 + \cdots + k + (k+1) = \left(\sum_{i=1}^{k} i \right) + (k+1)$$
$$= \frac{k^2 + k + 2}{2} + (k+1)$$
$$= \frac{k^2 + k + 2}{2} + \frac{2k + 2}{2} = \frac{k^2 + 3k + 4}{2}.$$

Hence, for each $k \in \mathbf{Z}^+$, it follows that $S(k) \Rightarrow S(k+1)$. But before we decide to accept the statement $\forall n \, S(n)$ as a true statement, let us reconsider Example 4.1. From that example we learned that $\sum_{i=1}^{n} i = n(n+1)/2$, for all $n \in \mathbf{Z}^+$. Therefore, we can use these two results (from Example 4.1 and the one already "established" here) to conclude that for all $n \in \mathbf{Z}^+$,

$$\frac{n(n+1)}{2} = \sum_{i=1}^{n} i = \frac{n^2 + n + 2}{2},$$

which implies that $n(n+1) = n^2 + n + 2$ and $0 = 2$. (Something is wrong somewhere!)

If $n = 1$, then $\sum_{i=1}^{1} 1 = 1$, but $(n^2 + n + 2)/2 = (1 + 1 + 2)/2 = 2$. So $S(1)$ is not true. But we may feel that this result just indicates that we have the wrong starting point. Perhaps $S(n)$ is true for all $n \geq 7$, or all $n \geq 137$. Using the preceding argument, however, we know that for any starting point $n_0 \in \mathbf{Z}^+$, if $S(n_0)$ were true, then

$$\frac{n_0^2 + n_0 + 2}{2} = \sum_{i=1}^{n_0} i = 1 + 2 + 3 + \cdots + n_0.$$

From the result in Example 4.1 we have $\sum_{i=1}^{n_0} i = n_0(n_0 + 1)/2$, so it follows once again that $0 = 2$, and we have no possible starting point.

This example should indicate to the reader the need to establish the basis step—no matter how easy it may be to verify it.

Now consider the following pseudocode procedures. The procedure in Fig. 4.3 uses a **for** loop to accumulate the sum of the squares. The second procedure (Fig. 4.4) demonstrates how the result of Example 4.4 can be used in place of such a loop. In both procedures the input is a positive integer n and the output is $\sum_{i=1}^{n} i^2$. However, whereas the pseudocode within the **for** loop of the procedure in Fig. 4.3 entails a total of n additions and n multiplications (not to mention the $n - 1$ additions for incrementing the counter variable i), the procedure in Fig. 4.4 requires only two additions, three multiplications, and one (integer) division. And this total number of additions, multiplications, and (integer) divisions is still 6 as the value of n increases. Consequently, the procedure in Fig. 4.4 is considered more efficient. (This idea of a *more efficient* procedure will be examined further in Sections 5.7 and 5.8.)

```
procedure SumOfSquares1 (n: positive integer)
begin
   sum := 0
   for i := 1 to n do
      sum := sum + i²
end
```

Figure 4.3

```
procedure SumOfSquares2 (n: positive integer)
begin
   sum := n * (n + 1) * (2 * n + 1)/6
end
```

Figure 4.4

Looking back at our first two applications of mathematical induction (in Examples 4.1 and 4.4), we might wonder whether this principle applies only to the verification of *known* summation formulas. The next seven examples show that mathematical induction is a vital tool in many other circumstances as well.

EXAMPLE 4.7

Let us consider the sums of consecutive odd positive integers.

1) 1 $= 1$ $(= 1^2)$
2) $1 + 3$ $= 4$ $(= 2^2)$
3) $1 + 3 + 5$ $= 9$ $(= 3^2)$
4) $1 + 3 + 5 + 7$ $= 16$ $(= 4^2)$

From these first four cases we *conjecture* the following result: The sum of the first n consecutive odd positive integers is n^2; that is, for all $n \in \mathbf{Z}^+$,

$$S(n): \quad \sum_{i=1}^{n} (2i - 1) = n^2.$$

Now that we have developed what we feel is a true summation formula, we use the Principle of Mathematical Induction to verify its truth for *all* $n \geq 1$.

From the preceding calculations, we see that $S(1)$ is true [as are $S(2)$, $S(3)$, and $S(4)$], and so we have our basis step. For the inductive step we assume the truth of $S(k)$ for some k (≥ 1) and have

$$\sum_{i=1}^{k}(2i - 1) = k^2.$$

We now deduce the truth of $S(k + 1)$: $\sum_{i=1}^{k+1}(2i - 1) = (k + 1)^2$. Since we have assumed the truth of $S(k)$, our induction hypothesis, we may now write

$$\sum_{i=1}^{k+1}(2i - 1) = \sum_{i=1}^{k}(2i - 1) + [2(k + 1) - 1] = k^2 + [2(k + 1) - 1]$$

$$= k^2 + 2k + 1 = (k + 1)^2.$$

Consequently, the result $S(n)$ is true for all $n \geq 1$, by the Principle of Mathematical Induction.

Now it is time to investigate some results that are not summation formulas.

EXAMPLE 4.8

In Table 4.1, we have listed in adjacent columns the values of $4n$ and $n^2 - 7$ for the positive integers n, where $1 \leq n \leq 8$. From the table, we see that $(n^2 - 7) < 4n$ for $n = 1, 2, 3, 4, 5$; but when $n = 6, 7, 8$, we have $4n < (n^2 - 7)$. These last three observations lead us to conjecture: For all $n \geq 6$, $4n < (n^2 - 7)$.

Table 4.1

n	$4n$	$n^2 - 7$	n	$4n$	$n^2 - 7$
1	4	−6	5	20	18
2	8	−3	6	24	29
3	12	2	7	28	42
4	16	9	8	32	57

Once again, the Principle of Mathematical Induction is the proof technique we need to verify our conjecture. Let $S(n)$ denote the open statement: $4n < (n^2 - 7)$. Then Table 4.1 confirms that $S(6)$ is true [as are $S(7)$ and $S(8)$], and we have our basis step. (At last we have an example wherein the starting point is an integer $n_0 \neq 1$.)

In this example, the induction hypothesis is $S(k)$: $4k < (k^2 - 7)$, where $k \in \mathbf{Z}^+$ and $k \geq 6$. In order to establish the inductive step, we need to obtain the truth of $S(k + 1)$ from that of $S(k)$. That is, from $4k < (k^2 - 7)$ we must conclude that $4(k + 1) < [(k + 1)^2 - 7]$. Here are the necessary steps:

$$4k < (k^2 - 7) \Rightarrow 4k + 4 < (k^2 - 7) + 4 < (k^2 - 7) + (2k + 1)$$

(because for $k \geq 6$, we find $2k + 1 \geq 13 > 4$), and

$$4k + 4 < (k^2 - 7) + (2k + 1) \Rightarrow 4(k + 1) < (k^2 + 2k + 1) - 7 = (k + 1)^2 - 7.$$

Therefore, by the Principle of Mathematical Induction, $S(n)$ is true for all $n \geq 6$.

EXAMPLE 4.9

Among the many interesting sequences of numbers encountered in discrete mathematics and combinatorics, one finds the *harmonic numbers* $H_1, H_2, H_3, \ldots$, where

$$H_1 = 1$$

$$H_2 = 1 + \frac{1}{2}$$

$$H_3 = 1 + \frac{1}{2} + \frac{1}{3},$$

$$\ldots,$$

and, in general, $H_n = 1 + \frac{1}{2} + \frac{1}{3} + \cdots + \frac{1}{n}$, for each $n \in \mathbf{Z}^+$.

The following property of the harmonic numbers provides one more opportunity for us to apply the Principle of Mathematical Induction.

$$\text{For all } n \in \mathbf{Z}^+, \sum_{j=1}^{n} H_j = (n+1)H_n - n.$$

Proof: As we have done in the earlier examples (that is, Examples 4.1, 4.4, and 4.7), we verify the basis step at $n = 1$ for the open statement $S(n)$: $\sum_{j=1}^{n} H_j = (n+1)H_n - n$. This result follows readily from

$$\sum_{j=1}^{1} H_j = H_1 = 1 = 2 \cdot 1 - 1 = (1+1)H_1 - 1.$$

To verify the inductive step, we assume the truth of $S(k)$, that is,

$$\sum_{j=1}^{k} H_j = (k+1)H_k - k.$$

This assumption then leads us to the following:

$$\begin{aligned}
\sum_{j=1}^{k+1} H_j = \sum_{j=1}^{k} H_j + H_{k+1} &= [(k+1)H_k - k] + H_{k+1} \\
&= (k+1)H_k - k + H_{k+1} \\
&= (k+1)[H_{k+1} - (1/(k+1))] - k + H_{k+1} \\
&= (k+2)H_{k+1} - 1 - k \\
&= (k+2)H_{k+1} - (k+1).
\end{aligned}$$

Consequently, we now know from the Principle of Mathematical Induction that $S(n)$ is true for all positive integers n.

EXAMPLE 4.10

For all $n \geq 0$ let $A_n \subset \mathbf{R}$, where $|A_n| = 2^n$ and the elements of A_n are listed in ascending order. If $r \in \mathbf{R}$, prove that in order to determine whether $r \in A_n$ (by the procedure developed below), we must compare r with no more than $n + 1$ elements in A_n.

When $n = 0$, $A_0 = \{a\}$ and only one comparison is needed. So the result is true for $n = 0$ (and we have our basis step). For $n = 1$, $A_1 = \{a_1, a_2\}$ with $a_1 < a_2$. In order to determine whether $r \in A_1$, at most two comparisons must be made. Hence the result follows when $n = 1$. Now if $n = 2$, we write $A_2 = \{b_1, b_2, c_1, c_2\} = B_1 \cup C_1$, where $b_1 < b_2 < c_1 < c_2$, $B_1 = \{b_1, b_2\}$, and $C_1 = \{c_1, c_2\}$. Comparing r with b_2, we determine which of the two possibilities — (i) $r \in B_1$; or (ii) $r \in C_1$ — can occur. Since $|B_1| = |C_1| = 2$, either one of the two possibilities requires at most two more comparisons (from the prior case where $n = 1$). Consequently, we can determine whether $r \in A_2$ by making no more than $2 + 1 = n + 1$ comparisons.

We now argue in general. Assume the result true for some $k \geq 0$ and consider the case for A_{k+1}, where $|A_{k+1}| = 2^{k+1}$. In order to establish our inductive step, let $A_{k+1} = B_k \cup C_k$, where $|B_k| = |C_k| = 2^k$, and the elements of B_k, C_k are in ascending order with the largest element x in B_k smaller than the least element in C_k. Let $r \in \mathbf{R}$. To determine whether $r \in A_{k+1}$, we consider whether $r \in B_k$ or $r \in C_k$.

a) First we compare r and x. (One comparison)

b) If $r \leq x$, then because $|B_k| = 2^k$, it follows by the induction hypothesis that we can determine whether $r \in B_k$ by making no more than $k + 1$ additional comparisons.

c) If $r > x$, we do likewise with the elements in C_k. We make at most $k + 1$ additional comparisons to see whether $r \in C_k$.

In any event, at most $(k + 1) + 1$ comparisons are made.

The general result now follows by the Principle of Mathematical Induction.

EXAMPLE 4.11

One of our first concerns when we evaluate the quality of a computer program is whether the program does what it is supposed to do. Just as we cannot prove a theorem by checking specific cases, so we cannot establish the correctness of a program simply by testing various sets of data. (Furthermore, doing this would be quite difficult if our program were to become a part of a larger software package wherein, perhaps, a data set is internally generated.) Since software development places a great deal of emphasis on structured programming, this has brought about the need for *program verification*. Here the programmer or the programming team must prove that the program being developed is correct *regardless* of the data set supplied. The effort invested at this stage considerably reduces the time that must be spent in debugging the program (or software package). One of the methods that can play a major role in such program verification is mathematical induction. Let us see how.

The pseudocode program segment shown in Fig. 4.5 is supposed to produce the answer $x(y^n)$ for real variables x, y with n a nonnegative integer. (The values for these three variables are assigned earlier in the program.) We shall verify the correctness of this program segment by mathematical induction for the open statement.

$S(n)$: For *all* x, $y \in \mathbf{R}$, if the program reaches the top of the **while** loop with $n \in \mathbf{N}$, after the loop is bypassed (for $n = 0$) or the two loop instructions are executed n (> 0) times, then the value of the real variable *answer* is $x(y^n)$.

```
while n ≠ 0 do
    begin
        x := x * y
        n := n - 1
    end
answer := x
```

Figure 4.5

The flowchart for this program segment is shown in Fig. 4.6. Referring to it will help us as we develop our proof.

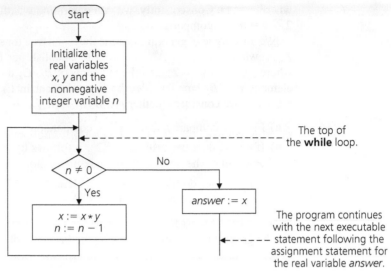

Figure 4.6

First consider $S(0)$, the statement for the case where $n = 0$. Here the program reaches the top of the **while** loop, but since $n = 0$, it follows the No branch in the flowchart and assigns the value $x = x(1) = x(y^0)$ to the real variable *answer*. Consequently, the statement $S(0)$ is true and the basis step of our induction argument is established.

Now we assume the truth of $S(k)$, for some nonnegative integer k. This provides us with the induction hypothesis.

$S(k)$: For *all* x, $y \in \mathbf{R}$, if the program reaches the top of the **while** loop with $k \in \mathbf{N}$, after the loop is bypassed (for $k = 0$) or the two loop instructions are executed k (> 0) times, then the value of the real variable *answer* is $x(y^k)$.

Continuing with the inductive step of the proof, when dealing with the statement $S(k + 1)$, we note that because $k + 1 \geq 1$, the program will not simply follow the No branch and bypass the instructions in the **while** loop. Those two instructions (in the **while** loop) will be executed at least once. When the program reaches the top of the **while** loop for the first time, $n = k + 1 > 0$, so the loop instructions are executed and the program returns to the top of the **while** loop where now we find that

- The value of y is unchanged.
- The value of x is $x_1 = x(y^1) = xy$.
- The value of n is $(k + 1) - 1 = k$.

But now, by our induction hypothesis (applied to the real numbers x_1, y), we know that after the **while** loop for x_1, y and $n = k$ is bypassed (for $k = 0$) or the two loop instructions are executed k (> 0) times, then the value assigned to the real variable *answer* is

$$x_1(y^k) = (xy)(y^k) = x(y^{k+1}).$$

So by the Principle of Mathematical Induction, $S(n)$ is true for all $n \geq 0$ and the correctness of the program segment is established.

EXAMPLE 4.12

Recall (from Examples 1.37 and 3.11) that for a given $n \in \mathbf{Z}^+$, a *composition* of n is an *ordered* sum of positive-integer summands summing to n. In Fig. 4.7 we find the compositions of 1, 2, 3, and 4. We see that

a) 1 has $1 = 2^0 = 2^{1-1}$ composition, 2 has $2 = 2^1 = 2^{2-1}$ compositions, 3 has $4 = 2^2 = 2^{3-1}$ compositions, and 4 has $8 = 2^3 = 2^{4-1}$ compositions; and

b) the eight compositions of 4 arise from the four compositions of 3 in two ways: (i) Compositions $(1')$–$(4')$ result by increasing the last summand (in each corresponding composition of 3) by 1; (ii) Each of compositions $(1'')$–$(4'')$ is obtained by appending "+1" to the corresponding composition of 3.

$(n = 1)$	1			$(n = 4)$	$(1')$	4
					$(2')$	$1 + 3$
$(n = 2)$	2				$(3')$	$2 + 2$
	$1 + 1$				$(4')$	$1 + 1 + 2$
$(n = 3)$	(1)	3			$(1'')$	$3 + 1$
	(2)	$1 + 2$			$(2'')$	$1 + 2 + 1$
	(3)	$2 + 1$			$(3'')$	$2 + 1 + 1$
	(4)	$1 + 1 + 1$			$(4'')$	$1 + 1 + 1 + 1$

Figure 4.7

The observations in part (a) suggest that for all $n \in \mathbf{Z}^+$, $S(n)$: n has 2^{n-1} compositions. The result [in part (a)] for $n = 1$ provides our basis step, $S(1)$. So now let us assume the result true for some (fixed) $k \in \mathbf{Z}^+$ — namely, $S(k)$: k has 2^{k-1} compositions. At this point consider $S(k + 1)$. One can develop the compositions of $k + 1$ from those of k as in part (b) above (where $k = 3$). For $k \geq 1$, we find that the compositions of $k + 1$ fall into two distinct cases:

1) The compositions of $k + 1$, where the last summand is an integer $t > 1$: Here this last summand t is replaced by $t - 1$, and this type of replacement provides a correspondence between all of the compositions of k and all those compositions of $k + 1$, where the last summand exceeds 1.

2) The compositions of $k + 1$, where the last summand is 1: In this case we delete "+1" from the right side of this type of composition of $k + 1$. Once again we get a correspondence between all the compositions of k and all those compositions of $k + 1$, where the last summand is 1.

Therefore, the number of compositions of $k + 1$ is twice the number for k. Consequently, it follows from the induction hypothesis that the number of compositions of $k + 1$ is $2(2^{k-1}) = 2^k$. The Principle of Mathematical Induction now tells us that for all $n \in \mathbf{Z}^+$, $S(n)$: n has 2^{n-1} compositions (as we learned earlier in Examples 1.37 and 3.11).

EXAMPLE 4.13

We learn from the equation $14 = 3 + 3 + 8$ that we can express 14 using only 3's and 8's as summands. But what may prove to be surprising is that for all $n \geq 14$,

$S(n)$: n can be written as a sum of 3's and/or 8's (with no regard to order).

As we start to verify $S(n)$ for all $n \geq 14$, we realize that the given introductory sentence shows us that the basis step $S(14)$ is true. For the inductive step we assume the truth of $S(k)$ for some $k \in \mathbf{Z}^+$, where $k \geq 14$, and then consider what can happen for $S(k + 1)$. If there is at least one 8 in the sum (of 3's and/or 8's) that equals k, then we can replace this 8 by three 3's and obtain $k + 1$ as a sum of 3's and/or 8's. But suppose that no 8 appears as a summand of k. Then the only summand used is a 3, and, since $k \geq 14$, we must have at least five 3's as summands. And now if we replace five of these 3's by two 8's, we obtain the sum $k + 1$, where the only summands are 3's and/or 8's. Consequently, we have shown how $S(k) \Rightarrow S(k + 1)$ and so the result follows for all $n \geq 14$ by the Principle of Mathematical Induction.

Now that we have seen several applications of the Principle of Mathematical Induction, we shall close this section by introducing another form of mathematical induction. This second form is sometimes referred to as the *Alternative Form of the Principle of Mathematical Induction* or the *Principle of Strong Mathematical Induction*.

Once again we shall consider a statement of the form $\forall \, n \geq n_0 \, S(n)$, where $n_0 \in \mathbf{Z}^+$, and we shall establish both a basis step and an inductive step. However, this time the basis step may require proving more than just the first case — where $n = n_0$. And in the inductive step we shall assume the truth of all the statements $S(n_0)$, $S(n_0 + 1)$, ..., $S(k - 1)$, and $S(k)$, in order to establish the truth of the statement $S(k + 1)$. We formally present this second Principle of Mathematical Induction in the following theorem.

THEOREM 4.2   *The Principle of Mathematical Induction — Alternative Form.* Let $S(n)$ denote an open mathematical statement (or set of such open statements) that involves one or more occurrences of the variable n, which represents a positive integer. Also let $n_0, n_1 \in \mathbf{Z}^+$ with $n_0 \leq n_1$.

   a) If $S(n_0)$, $S(n_0 + 1)$, $S(n_0 + 2)$, ..., $S(n_1 - 1)$, and $S(n_1)$ are true; and

   b) If whenever $S(n_0)$, $S(n_0 + 1)$, ..., $S(k - 1)$, and $S(k)$ are true for some (particular but arbitrarily chosen) $k \in \mathbf{Z}^+$, where $k \geq n_1$, then the statement $S(k + 1)$ is also true;

   then $S(n)$ is true for all $n \geq n_0$.

As in Theorem 4.1, condition (a) is called the *basis step* and condition (b) is called the *inductive step*.

The proof of Theorem 4.2 is similar to that of Theorem 4.1 and will be requested in the Section Exercises. We shall also learn in the exercises for Section 4.2 that the two forms of mathematical induction (given in Theorems 4.1 and 4.2) are equivalent, for each can be shown to be a valid proof technique when we assume the truth of the other.

Before we give any examples where Theorem 4.2 is applied, let us mention, as we did for Theorem 4.1, that n_0 need not actually be a positive integer — it may, in reality, be 0 or even possibly a negative integer. And now that we have taken care of that point once again, let us see how we might apply this new proof technique.

Our first example should be familiar. We shall simply apply Theorem 4.2 in order to obtain the result in Example 4.13 in a second way.

EXAMPLE 4.14

The following calculations indicate that it is possible to write (without regard to order) the integers 14, 15, 16 using only 3's and/or 8's as summands:

$$14 = 3 + 3 + 8 \qquad 15 = 3 + 3 + 3 + 3 + 3 \qquad 16 = 8 + 8$$

On the basis of these three results, we make the conjecture

For every $n \in \mathbf{Z}^+$ where $n \geq 14$,

$$S(n): \quad n \text{ can be written as a sum of 3's and/or 8's.}$$

Proof: It is apparent that the statements $S(14)$, $S(15)$, and $S(16)$ are true — and this establishes our basis step. (Here $n_0 = 14$ and $n_1 = 16$.)

For the inductive step we assume the truth of the statements

$$S(14), S(15), \ldots, S(k-2), S(k-1), \text{ and } S(k)$$

for some $k \in \mathbf{Z}^+$, where $k \geq 16$. [The assumption of the truth of these $(k - 14) + 1$ statements constitutes our induction hypothesis.] And now if $n = k + 1$, then $n \geq 17$ and $k + 1 = (k - 2) + 3$. But since $14 \leq k - 2 \leq k$, from the truth of $S(k - 2)$ we know that $(k - 2)$ can be written as a sum of 3's and/or 8's; so $(k + 1) = (k - 2) + 3$ can also be written in this form. Consequently, $S(n)$ is true for all $n \geq 14$ by the alternative form of the Principle of Mathematical Induction.

In Example 4.14 we saw how the truth of $S(k + 1)$ was deduced by using the truth of the one prior result $S(k - 2)$. Our last example presents a situation wherein the truth of more than one prior result is needed.

EXAMPLE 4.15

Let us consider the integer sequence $a_0, a_1, a_2, a_3, \ldots$, where

$$a_0 = 1, a_1 = 2, a_2 = 3, \qquad \text{and}$$
$$a_n = a_{n-1} + a_{n-2} + a_{n-3}, \quad \text{for all } n \in \mathbf{Z}^+ \text{ where } n \geq 3.$$

(Then, for instance, we find that $a_3 = a_2 + a_1 + a_0 = 3 + 2 + 1 = 6$; $a_4 = a_3 + a_2 + a_1 = 6 + 3 + 2 = 11$; and $a_5 = a_4 + a_3 + a_2 = 11 + 6 + 3 = 20$.)

We claim that the entries in this sequence are such that $a_n \leq 3^n$ for all $n \in \mathbf{N}$ — that is, $\forall n \in \mathbf{N} \ S'(n)$, where $S'(n)$ is the open statement: $a_n \leq 3^n$.

For the basis step, we observe that

i) $a_0 = 1 = 3^0 \leq 3^0$;

ii) $a_1 = 2 \leq 3 = 3^1$; and

iii) $a_2 = 3 \leq 9 = 3^2$.

Consequently, we know that $S'(0)$, $S'(1)$, and $S'(2)$ are true statements.

So now we turn our attention to the inductive step where we assume the truth of the statements $S'(0), S'(1), S'(2), \ldots, S'(k-1), S'(k)$, for some $k \in \mathbf{Z}^+$ where $k \geq 2$. For the case where $n = k + 1 \geq 3$ we see that

$$a_{k+1} = a_k + a_{k-1} + a_{k-2}$$
$$\leq 3^k + 3^{k-1} + 3^{k-2}$$
$$\leq 3^k + 3^k + 3^k = 3(3^k) = 3^{k+1},$$

so $[S'(k-2) \wedge S'(k-1) \wedge S'(k)] \Rightarrow S'(k+1)$.

Therefore it follows from the alternative form of the Principle of Mathematical Induction that $a_n \le 3^n$ for all $n \in \mathbf{N}$.

Before we close this section, let us take a second look at the preceding two results. In both Example 4.14 and Example 4.15 we established the basis step by verifying the truth of three statements: $S(14)$, $S(15)$, and $S(16)$ in Example 4.14; and, $S'(0)$, $S'(1)$, and $S'(2)$ in Example 4.15. However, to obtain the truth of $S(k + 1)$ in Example 4.14, we actually used only one of the $(k - 14) + 1$ statements in the induction hypothesis — namely, the statement $S(k - 2)$. For Example 4.15 we used three of the $k + 1$ statements in the induction hypothesis — in this case, the statements $S'(k - 2)$, $S'(k - 1)$, and $S'(k)$.

EXERCISES 4.1

1. Prove each of the following for all $n \ge 1$ by the Principle of Mathematical Induction.

a) $1^2 + 3^2 + 5^2 + \cdots + (2n - 1)^2 = \dfrac{n(2n - 1)(2n + 1)}{3}$

b) $1 \cdot 3 + 2 \cdot 4 + 3 \cdot 5 + \cdots + n(n + 2) =$
$\dfrac{n(n + 1)(2n + 7)}{6}$

c) $\displaystyle\sum_{i=1}^{n} \frac{1}{i(i + 1)} = \frac{n}{n + 1}$

d) $\displaystyle\sum_{i=1}^{n} i^3 = \frac{n^2(n + 1)^2}{4} = \left(\sum_{i=1}^{n} i\right)^2$

2. Establish each of the following for all $n \ge 1$ by the Principle of Mathematical Induction.

a) $\displaystyle\sum_{i=1}^{n} 2^{i-1} = \sum_{i=0}^{n-1} 2^i = 2^n - 1$

b) $\displaystyle\sum_{i=1}^{n} i(2^i) = 2 + (n - 1)2^{n+1}$

c) $\displaystyle\sum_{i=1}^{n} (i)(i!) = (n + 1)! - 1$

3. a) Note how $\sum_{i=1}^{n} i^3 + (n + 1)^3 = \sum_{i=0}^{n}(i + 1)^3 = \sum_{i=0}^{n}(i^3 + 3i^2 + 3i + 1)$. Use this result to obtain a formula for $\sum_{i=1}^{n} i^2$. (Compare with the formula given in Example 4.4.)

b) Use the idea presented in part (a) to find a formula for $\sum_{i=1}^{n} i^3$ and one for $\sum_{i=1}^{n} i^4$. [Compare the result for $\sum_{i=1}^{n} i^3$ with the formula in part (d) of Exercise 1 for this section.]

4. A wheel of fortune has the integers from 1 to 25 placed on it in a random manner. Show that regardless of how the numbers are positioned on the wheel, there are three adjacent numbers whose sum is at least 39.

5. Consider the following program segment (written in pseudocode):

```
for i := 1 to 123 do
    for j := 1 to i do
        print i * j
```

a) How many times is the print statement of the third line executed?

b) Replace i in the second line by i^2, and answer the question in part (a).

6. a) For the four-digit integers (from 1000 to 9999) how many are palindromes and what is their sum?

b) Write a computer program to check the answer for the sum in part (a).

7. A lumberjack has $4n + 110$ logs in a pile consisting of n layers. Each layer has two more logs than the layer directly above it. If the top layer has six logs, how many layers are there?

8. Determine the positive integer n for which

$$\sum_{i=1}^{2n} i = \sum_{i=1}^{n} i^2.$$

9. Evaluate each of the following:

a) $\sum_{i=11}^{33} i$ **b)** $\sum_{i=11}^{33} i^2$.

10. Determine $\sum_{i=51}^{100} t_i$, where t_i denotes the ith triangular number, for $51 \le i \le 100$.

11. a) Derive a formula for $\sum_{i=1}^{n} t_{2i}$, where t_{2i} denotes the $2i$th triangular number for $1 \le i \le n$.

b) Determine $\sum_{i=1}^{100} t_{2i}$.

c) Write a computer program to check the result in part (b).

12. a) Prove that $(\cos \theta + i \sin \theta)^2 = \cos 2\theta + i \sin 2\theta$, where $i \in \mathbf{C}$ and $i^2 = -1$.

b) Using induction, prove that for all $n \in \mathbf{Z}^+$,

$$(\cos \theta + i \sin \theta)^n = \cos n\theta + i \sin n\theta.$$

(This result is known as *DeMoivre's Theorem*.)

c) Verify that $1 + i = \sqrt{2}(\cos 45° + i \sin 45°)$, and compute $(1 + i)^{100}$.

13. a) Consider an 8×8 chessboard. It contains sixty-four 1×1 squares and one 8×8 square. How many 2×2

squares does it contain? How many 3×3 squares? How many squares in total?

b) Now consider an $n \times n$ chessboard for some fixed $n \in \mathbf{Z}^+$. For $1 \leq k \leq n$, how many $k \times k$ squares are contained in this chessboard? How many squares in total?

14. Prove that for all $n \in \mathbf{Z}^+$, $n > 3 \Rightarrow 2^n < n!$.

15. Prove that for all $n \in \mathbf{Z}^+$, $n > 4 \Rightarrow n^2 < 2^n$.

16. a) For $n = 3$ let $X_3 = \{1, 2, 3\}$. Now consider the sum

$$s_3 = \frac{1}{1} + \frac{1}{2} + \frac{1}{3} + \frac{1}{1 \cdot 2} + \frac{1}{1 \cdot 3} + \frac{1}{2 \cdot 3} + \frac{1}{1 \cdot 2 \cdot 3}$$

$$= \sum_{\emptyset \neq A \subseteq X_3} \frac{1}{p_A},$$

where p_A denotes the product of all elements in a nonempty subset A of X_3. Note that the sum is taken over all the nonempty subsets of X_3. Evaluate this sum.

b) Repeat the calculation in part (a) for s_2 (where $n = 2$ and $X_2 = \{1, 2\}$) and s_4 (where $n = 4$ and $X_4 = \{1, 2, 3, 4\}$).

c) Conjecture the general result suggested by the calculations from parts (a) and (b). Prove your conjecture using the Principle of Mathematical Induction.

17. For $n \in \mathbf{Z}^+$, let H_n denote the nth harmonic number (as defined in Example 4.9).

a) For all $n \in \mathbf{N}$ prove that $1 + \left(\frac{n}{2}\right) \leq H_{2^n}$.

b) Prove that for all $n \in \mathbf{Z}^+$,

$$\sum_{j=1}^{n} j H_j = \left[\frac{n(n+1)}{2}\right] H_{n+1} - \left[\frac{n(n+1)}{4}\right].$$

18. Consider the following four equations:

1) $$1 = 1$$
2) $$2 + 3 + 4 = 1 + 8$$
3) $$5 + 6 + 7 + 8 + 9 = 8 + 27$$
4) $$10 + 11 + 12 + 13 + 14 + 15 + 16 = 27 + 64$$

Conjecture the general formula suggested by these four equations, and prove your conjecture.

19. For $n \in \mathbf{Z}^+$, let $S(n)$ be the open statement

$$\sum_{i=1}^{n} i = \frac{(n + (1/2))^2}{2}.$$

Show that the truth of $S(k)$ implies the truth of $S(k+1)$ for all $k \in \mathbf{Z}^+$. Is $S(n)$ true for all $n \in \mathbf{Z}^+$?

20. Let S_1 and S_2 be two sets where $|S_1| = m$, $|S_2| = r$, for $m, r \in \mathbf{Z}^+$, and the elements in each of S_1, S_2 are in ascending order. It can be shown that the elements in S_1 and S_2 can be merged into ascending order by making no more than $m + r - 1$ comparisons. (See Lemma 12.1.) Use this result to establish the following.

For $n \geq 0$, let S be a set with $|S| = 2^n$. Prove that the number of comparisons needed to place the elements of S in ascending order is bounded above by $n \cdot 2^n$.

21. During the execution of a certain program segment (written in pseudocode), the user assigns to the integer variables x and n any (possibly different) positive integers. The segment shown in Fig. 4.8 immediately follows these assignments. If the program reaches the top of the **while** loop, state and prove (by mathematical induction) what the value assigned to *answer* will be after the two loop instructions are executed $n \ (> 0)$ times.

```
while n ≠ 0 do
  begin
    x := x * n
    n := n - 1
  end
answer := x
```

Figure 4.8

22. In the program segment shown in Fig. 4.9, x, y, and *answer* are real variables, and n is an integer variable. Prior to execution of this **while** loop, the user supplies real values for x and y and a nonnegative integer value for n. Prove (by mathematical induction) that for all $x, y \in \mathbf{R}$, if the program reaches the top of the **while** loop with $n \in \mathbf{N}$, after the loop is bypassed (for $n = 0$) or the two loop instructions are executed $n \ (> 0)$ times, then the value assigned to *answer* is $x + ny$.

```
while n ≠ 0 do
  begin
    x := x + y
    n := n - 1
  end
answer := x
```

Figure 4.9

23. a) Let $n \in \mathbf{Z}^+$, where $n \neq 1, 3$. Prove that n can be expressed as a sum of 2's and/or 5's.

b) For all $n \in \mathbf{Z}^+$ show that if $n \geq 24$, then n can be written as a sum of 5's and/or 7's.

24. A sequence of numbers $a_1, a_2, a_3, \ldots$ is defined by

$$a_1 = 1 \qquad a_2 = 2 \qquad a_n = a_{n-1} + a_{n-2}, n \geq 3.$$

a) Determine the values of a_3, a_4, a_5, a_6, and a_7.

b) Prove that for all $n \geq 1$, $a_n < (7/4)^n$.

25. For a fixed $n \in \mathbf{Z}^+$, let X be the random variable where $Pr(X = x) = \frac{1}{n}$, $x = 1, 2, 3, \ldots, n$. (Here X is called a *uniform discrete* random variable.) Determine $E(X)$ and $Var(X)$.

26. Let a_0 be a fixed constant and, for $n \geq 1$, let $a_n = \sum_{i=0}^{n-1} \binom{n-1}{i} a_i a_{(n-1)-i}$.

a) Show that $a_1 = a_0^2$ and that $a_2 = 2a_0^3$.

b) Determine a_3 and a_4 in terms of a_0.

c) Conjecture a formula for a_n in terms of a_0 when $n \geq 0$. Prove your conjecture using the Principle of Mathematical Induction.

27. Verify Theorem 4.2.

28. a) Of the $2^{5-1} = 2^4 = 16$ compositions of 5, determine how many start with (i) 1; (ii) 2; (iii) 3; (iv) 4; and (v) 5.

b) Provide a combinatorial proof for the result in part (a) of Exercise 2.

4.2
Recursive Definitions

Let us start this section by considering the integer sequence $b_0, b_1, b_2, b_3, \ldots,$ where $b_n = 2n$ for all $n \in \mathbf{N}$. Here we find that $b_0 = 2 \cdot 0 = 0$, $b_1 = 2 \cdot 1 = 2$, $b_2 = 2 \cdot 2 = 4$, and $b_3 = 2 \cdot 3 = 6$. If, for instance, we need to determine b_6, we simply calculate $b_6 = 2 \cdot 6 = 12$ — without the need to calculate the value of b_n for any other $n \in \mathbf{N}$. We can perform such calculations because we have an *explicit* formula — namely, $b_n = 2n$ — that tells us how b_n is determined from n (alone).

In Example 4.15 of the preceding section, however, we considered the integer sequence $a_0, a_1, a_2, a_3, \ldots,$ where

$$a_0 = 1, a_1 = 2, a_2 = 3, \quad \text{and}$$
$$a_n = a_{n-1} + a_{n-2} + a_{n-3}, \quad \text{for all } n \in \mathbf{Z}^+ \text{ where } n \geq 3.$$

Here we do *not* have an *explicit* formula that defines each a_n in terms of n for all $n \in \mathbf{N}$. If we want the value of a_6, for example, we need to know the values of a_5, a_4, and a_3. And these values (of a_5, a_4, and a_3) require that we also know the values of a_2, a_1, and a_0. Unlike the rather easy situation where we determined $b_6 = 2 \cdot 6 = 12$, in order to calculate a_6, here we might find ourselves writing

$$\begin{aligned}
a_6 &= a_5 + a_4 + a_3 \\
&= (a_4 + a_3 + a_2) + (a_3 + a_2 + a_1) + (a_2 + a_1 + a_0) \\
&= [(a_3 + a_2 + a_1) + (a_2 + a_1 + a_0) + a_2] \\
&\quad + [(a_2 + a_1 + a_0) + a_2 + a_1] + (a_2 + a_1 + a_0) \\
&= [[(a_2 + a_1 + a_0) + a_2 + a_1] + (a_2 + a_1 + a_0) + a_2] \\
&\quad + [(a_2 + a_1 + a_0) + a_2 + a_1] + (a_2 + a_1 + a_0) \\
&= [[(3 + 2 + 1) + 3 + 2] + (3 + 2 + 1) + 3] \\
&\quad + [(3 + 2 + 1) + 3 + 2] + (3 + 2 + 1) \\
&= 37.
\end{aligned}$$

Or, in a somewhat easier manner, we could have gone in the opposite direction with these considerations:

$$\begin{aligned}
a_3 &= a_2 + a_1 + a_0 = 3 + 2 + 1 = 6 \\
a_4 &= a_3 + a_2 + a_1 = 6 + 3 + 2 = 11 \\
a_5 &= a_4 + a_3 + a_2 = 11 + 6 + 3 = 20 \\
a_6 &= a_5 + a_4 + a_3 = 20 + 11 + 6 = 37.
\end{aligned}$$

No matter how we arrive at a_6, we realize that the two integer sequences — $b_0, b_1, b_2, b_3, \ldots,$ and $a_0, a_1, a_2, a_3, \ldots$ — are more than just numerically different. The integers $b_0, b_1, b_2, b_3, \ldots,$ can be very readily listed as $0, 2, 4, 6, \ldots,$ and for any $n \in \mathbf{N}$ we have the *explicit* formula $b_n = 2n$. On the other hand, we might find it rather difficult (if not impossible) to determine such an explicit formula for the integers $a_0, a_1, a_2, a_3, \ldots$.

What is happening here for a sequence of integers can also occur for other mathematical concepts — such as sets and binary operations [as well as functions (in Chapter 5), languages (in Chapter 6), and relations (in Chapter 7)]. Sometimes it is difficult to define a mathematical concept in an explicit manner. But, as for the sequence $a_0, a_1, a_2, a_3, \ldots$, we may be able to define what we need in terms of similar prior results. (We shall examine what we mean by this in several examples in this section.) When we do so we say that the concept is defined *recursively*, using the method, or process, of *recursion*. In this way we obtain the concept we are interested in studying — by means of a *recursive definition*. Hence, although we do not have an explicit formula here for the sequence $a_0, a_1, a_2, a_3, \ldots$, we do have a way of defining the integers a_n, for $n \in \mathbf{N}$, by recursion. The assignments

$$a_0 = 1, \qquad a_1 = 2, \qquad a_2 = 3$$

provide a base for the recursion.

The equation

$$a_n = a_{n-1} + a_{n-2} + a_{n-3}, \quad \text{for } n \in \mathbf{Z}^+ \text{ where } n \geq 3, \tag{*}$$

provides the recursive process; it indicates how to obtain new entries in the sequence from those prior results we already know (or can calculate). [*Note*: The integers computed from Eq. (*) may also be computed from the equation $a_{n+3} = a_{n+2} + a_{n+1} + a_n$, for $n \in \mathbf{N}$.]

We now use the concept of the *recursive definition* to settle something that was mentioned in three footnotes in Sections 2.1 and 2.3. After studying Section 2.2 we knew (from the laws of logic) that for any statements p_1, p_2, and p_3, we had

$$p_1 \wedge (p_2 \wedge p_3) \Longleftrightarrow (p_1 \wedge p_2) \wedge p_3,$$

and, consequently, we could write $p_1 \wedge p_2 \wedge p_3$ without any chance of ambiguity. This is because the truth value for the conjunction of three statements does not depend on the way parentheses might be introduced to direct the order of forming the conjunctions of pairs of (given or resultant) statements. But we were concerned about what meaning we should attach to an expression such as $p_1 \wedge p_2 \wedge p_3 \wedge p_4$. The following example now settles that issue.

EXAMPLE 4.16

The logical connective $\wedge$ was defined (in Section 2.1) for only two statements at a time. How, then, does one deal with an expression such as $p_1 \wedge p_2 \wedge p_3 \wedge p_4$, where p_1, p_2, p_3, and p_4 are statements? In order to answer this question we introduce the following recursive definition, wherein the concept at a certain $[(n+1)\text{st}]$ stage is developed from the comparable concept at an earlier $[n\text{th}]$ stage.

Given any statements $p_1, p_2, \ldots, p_n, p_{n+1}$, we define

1) the conjunction of p_1, p_2 by $p_1 \wedge p_2$ (as we did in Section 2.1), and

2) the conjunction of $p_1, p_2, \ldots, p_n, p_{n+1}$, for $n \geq 2$, by

$$p_1 \wedge p_2 \wedge \cdots \wedge p_n \wedge p_{n+1} \Longleftrightarrow (p_1 \wedge p_2 \wedge \cdots \wedge p_n) \wedge p_{n+1}.$$

[The result in (1) establishes the base for the recursion, while the logical equivalence in (2) is used to provide the recursive process. Note that the statement on the right-hand side of the logical equivalence in (2) is the conjunction of *two* statements: p_{n+1} and the previously determined statement $(p_1 \wedge p_2 \wedge \cdots \wedge p_n)$.]

Therefore, we define the conjunction of p_1, p_2, p_3, p_4 by

$$p_1 \wedge p_2 \wedge p_3 \wedge p_4 \Longleftrightarrow (p_1 \wedge p_2 \wedge p_3) \wedge p_4.$$

Then, by the associative law of $\wedge$, we find that

$$
\begin{aligned}
(p_1 \wedge p_2 \wedge p_3) \wedge p_4 &\Longleftrightarrow [(p_1 \wedge p_2) \wedge p_3] \wedge p_4 \\
&\Longleftrightarrow (p_1 \wedge p_2) \wedge (p_3 \wedge p_4) \\
&\Longleftrightarrow p_1 \wedge [p_2 \wedge (p_3 \wedge p_4)] \\
&\Longleftrightarrow p_1 \wedge [(p_2 \wedge p_3) \wedge p_4] \\
&\Longleftrightarrow p_1 \wedge (p_2 \wedge p_3 \wedge p_4).
\end{aligned}
$$

These logical equivalences show that the truth value for the conjunction of four statements is also independent of the way parentheses might be introduced to indicate how to associate the given statements.

Using the above definition, we now extend our results to the following "Generalized Associative Law for $\wedge$."

Let $n \in \mathbf{Z}^+$ where $n \geq 3$, and let $r \in \mathbf{Z}^+$ with $1 \leq r < n$. Then

$S(n)$:   For any statements $p_1, p_2, \ldots, p_r, p_{r+1}, \ldots, p_n$,

$$(p_1 \wedge p_2 \wedge \cdots \wedge p_r) \wedge (p_{r+1} \wedge \cdots \wedge p_n) \Longleftrightarrow p_1 \wedge p_2 \wedge \cdots \wedge p_r \wedge p_{r+1} \wedge \cdots \wedge p_n.$$

Proof: The truth of the statement $S(3)$ follows from the associative law for $\wedge$ — and this establishes the basis step for our inductive proof. For the inductive step we assume that $S(k)$ is true for some $k \geq 3$ and all $1 \leq r < k$. That is, we assume the truth of

$S(k)$: $(p_1 \wedge p_2 \wedge \cdots \wedge p_r) \wedge (p_{r+1} \wedge \cdots \wedge p_k)$

$$\Longleftrightarrow p_1 \wedge p_2 \wedge \cdots \wedge p_r \wedge p_{r+1} \wedge \cdots \wedge p_k.$$

Then we show that $S(k) \Rightarrow S(k+1)$. When we consider $k+1$ statements, then we must account for all $1 \leq r < k+1$.

1) If $r = k$, then

$$(p_1 \wedge p_2 \wedge \cdots \wedge p_k) \wedge p_{k+1} \Longleftrightarrow p_1 \wedge p_2 \wedge \cdots \wedge p_k \wedge p_{k+1},$$

from our recursive definition.

2) For $1 \leq r < k$, we have

$(p_1 \wedge p_2 \wedge \cdots \wedge p_r) \wedge (p_{r+1} \wedge \cdots \wedge p_k \wedge p_{k+1})$

$$
\begin{aligned}
&\Longleftrightarrow (p_1 \wedge p_2 \wedge \cdots \wedge p_r) \wedge [(p_{r+1} \wedge \cdots \wedge p_k) \wedge p_{k+1}] \\
&\Longleftrightarrow [(p_1 \wedge p_2 \wedge \cdots \wedge p_r) \wedge (p_{r+1} \wedge \cdots \wedge p_k)] \wedge p_{k+1} \\
&\Longleftrightarrow (p_1 \wedge p_2 \wedge \cdots \wedge p_r \wedge p_{r+1} \wedge \cdots \wedge p_k) \wedge p_{k+1} \\
&\Longleftrightarrow p_1 \wedge p_2 \wedge \cdots \wedge p_r \wedge p_{r+1} \wedge \cdots \wedge p_k \wedge p_{k+1}.
\end{aligned}
$$

So it follows by the Principle of Mathematical Induction (Theorem 4.1) that the open statement $S(n)$ is true for all $n \in \mathbf{Z}^+$ where $n \geq 3$.

Our next example provides us with a second opportunity to generalize an associative law — but this time we shall deal with sets instead of statements.

EXAMPLE 4.17

In Definition 3.10 we extended the binary operations of $\cup$ and $\cap$ to an arbitrary (finite or infinite) number of subsets from a given universe $\mathcal{U}$. However, these definitions do not rely on the binary nature of the operations involved, and they do not provide a *systematic* way of determining the union or intersection of any finite number of sets.

To overcome this difficulty, we consider the sets $A_1, A_2, \ldots, A_n, A_{n+1}$, where $A_i \subseteq \mathcal{U}$ for all $1 \leq i \leq n + 1$, and we define their union *recursively* as follows:

1) The union of A_1, A_2 is $A_1 \cup A_2$. (This is the base for our recursive definition.)

2) The union of $A_1, A_2, \ldots, A_n, A_{n+1}$, for $n \geq 2$, is given by

$$A_1 \cup A_2 \cup \cdots \cup A_n \cup A_{n+1} = (A_1 \cup A_2 \cup \cdots \cup A_n) \cup A_{n+1},$$

where the set on the right-hand side of the set equality is the union of *two* sets, namely, $A_1 \cup A_2 \cup \cdots \cup A_n$ and A_{n+1}. (Here we have the recursive process needed to complete our recursive definition.)

From this definition we obtain the following "Generalized Associative Law for $\cup$." If $n, r \in \mathbf{Z}^+$ with $n \geq 3$ and $1 \leq r < n$, then

$S(n)$: $(A_1 \cup A_2 \cup \cdots \cup A_r) \cup (A_{r+1} \cup \cdots \cup A_n)$

$$= A_1 \cup A_2 \cup \cdots \cup A_r \cup A_{r+1} \cup \cdots \cup A_n,$$

where $A_i \subseteq \mathcal{U}$ for all $1 \leq i \leq n$.

Proof: The truth of $S(n)$ for $n = 3$ follows from the associative law of $\cup$, thereby providing the basis step needed for this inductive proof. Assuming the truth of $S(k)$ for some $k \in \mathbf{Z}^+$, where $k \geq 3$ and $1 \leq r < k$, we shall now establish our inductive step by showing that $S(k) \Rightarrow S(k + 1)$. When dealing with $k + 1$ (≥ 4) sets we need to consider all $1 \leq r < k + 1$. We find that

1) For $r = k$ we have

$$(A_1 \cup A_2 \cup \cdots \cup A_k) \cup A_{k+1} = A_1 \cup A_2 \cup \cdots \cup A_k \cup A_{k+1}.$$

This follows from the given recursive definition.

2) If $1 \leq r < k$, then

$(A_1 \cup A_2 \cup \cdots \cup A_r) \cup (A_{r+1} \cup \cdots \cup A_k \cup A_{k+1})$

$$= (A_1 \cup A_2 \cup \cdots \cup A_r) \cup [(A_{r+1} \cup \cdots \cup A_k) \cup A_{k+1}]$$
$$= [(A_1 \cup A_2 \cup \cdots \cup A_r) \cup (A_{r+1} \cup \cdots \cup A_k)] \cup A_{k+1}$$
$$= (A_1 \cup A_2 \cup \cdots \cup A_r \cup A_{r+1} \cup \cdots \cup A_k) \cup A_{k+1}$$
$$= A_1 \cup A_2 \cup \cdots \cup A_r \cup A_{r+1} \cup \cdots \cup A_k \cup A_{k+1}.$$

So it follows by the Principle of Mathematical Induction that $S(n)$ is true for all integers $n \geq 3$.

Similar to the result in Example 4.17, the intersection of the $n + 1$ sets $A_1, A_2, \ldots, A_n$, A_{n+1} (each taken from the same universe $\mathcal{U}$) is defined recursively by:

1) The intersection of A_1, A_2 is $A_1 \cap A_2$.

2) For $n \geq 2$, the intersection of $A_1, A_2, \ldots, A_n, A_{n+1}$ is given by

$$A_1 \cap A_2 \cap \cdots \cap A_n \cap A_{n+1} = (A_1 \cap A_2 \cap \cdots \cap A_n) \cap A_{n+1},$$

the intersection of the *two* sets $A_1 \cap A_2 \cap \cdots \cap A_n$ and A_{n+1}.

We find that the recursive definitions for the union and intersection of any finite number of sets provide the means by which we can extend the DeMorgan Laws of Set Theory. We shall establish (by using mathematical induction) one of these extensions in the next example and request a proof of the other extension in the Section Exercises.

EXAMPLE 4.18

Let $n \in \mathbf{Z}^+$ where $n \geq 2$, and let $A_1, A_2, \ldots, A_n \subseteq \mathcal{U}$ for each $1 \leq i \leq n$. Then

$$\overline{A_1 \cap A_2 \cap \cdots \cap A_n} = \overline{A_1} \cup \overline{A_2} \cup \cdots \cup \overline{A_n}.$$

Proof: The basis step of this proof is given for $n = 2$. It follows from the fact that $\overline{A_1 \cap A_2} = \overline{A_1} \cup \overline{A_2}$ — by the second of DeMorgan's Laws (listed in the Laws of Set Theory in Section 3.2).

Assuming the truth of the result for some k, where $k \geq 2$, we have

$$\overline{A_1 \cap A_2 \cap \cdots \cap A_k} = \overline{A_1} \cup \overline{A_2} \cup \cdots \cup \overline{A_k}.$$

And when we consider $k + 1 \ (\geq 3)$ sets, the induction hypothesis is used to obtain the third set equality in the following:

$$\overline{A_1 \cap A_2 \cap \cdots \cap A_k \cap A_{k+1}} = \overline{(A_1 \cap A_2 \cap \cdots \cap A_k) \cap A_{k+1}}$$

$$= \overline{(A_1 \cap A_2 \cap \cdots \cap A_k)} \cup \overline{A_{k+1}} = (\overline{A_1} \cup \overline{A_2} \cup \cdots \cup \overline{A_k}) \cup \overline{A_{k+1}}$$

$$= \overline{A_1} \cup \overline{A_2} \cup \cdots \cup \overline{A_k} \cup \overline{A_{k+1}}.$$

This then establishes the inductive step in our proof and so we obtain this generalized DeMorgan Law for all $n \geq 2$ by the Principle of Mathematical Induction.

Now that we have seen the two recursive definitions (in Examples 4.16 and 4.17), as we continue to investigate situations where this type of definition arises, we shall generally refrain from labeling the base and recursive parts. Likewise, we may not always designate the basis and inductive steps in a proof by mathematical induction.

As we look back at Examples 4.16 and 4.17, the recursive definitions in these two examples should seem similar to us. For if we interchange the statement p_i with the set A_i, for all $1 \leq i \leq n + 1$, and if we interchange each occurrence of $\wedge$ with $\cup$ and replace $\Longleftrightarrow$ with $=$, then we can obtain the recursive definition in Example 4.17 from the one given in Example 4.16.

In a similar way one can recursively define the sum and product of n real numbers, where $n \in \mathbf{Z}^+$ and $n \geq 2$. Then we can obtain (by the Principle of Mathematical Induction) generalized associative laws for the addition and multiplication of real numbers. (In the

Section Exercises the reader will be requested to do this.) We want to be aware of such generalized associative laws because we have been using them and will continue to use them. The reader may be surprised to learn that we have already used the generalized associative law of addition. In each of Examples 4.1 and 4.4, for instance, the generalized associative law of addition was used to establish the inductive step (in the proof by mathematical induction). Furthermore, now that we are more aware of it, the generalized associative law of addition can be used (usually, in an implicit manner) in recursive definitions — for now there will be no chance for ambiguity if one wants to add four or more summands. For example, we could define the sequence of harmonic numbers H_1, H_2, H_3, ..., by

1) $H_1 = 1$; and
2) For $n \geq 1$, $H_{n+1} = H_n + \left(\frac{1}{n+1}\right)$.

Turning from addition to multiplication, we may use the generalized associative law of multiplication to provide a recursive definition of $n!$. In this case we write

1) $0! = 1$; and
2) For $n \geq 0$, $(n + 1)! = (n + 1)(n!)$.

(This was suggested in the paragraph following Definition 1.1 in Section 1.2.) Also, the integer sequence b_0, b_1, b_2, b_3, ..., given explicitly (at the start of this section) by the formula $b_n = 2n$, $n \in \mathbf{N}$, can now be defined recursively by

1) $b_0 = 0$; and
2) For $n \geq 0$, $b_{n+1} = b_n + 2$.

When we investigate the sequences in our next two examples, we shall once again find recursive definitions. In addition we shall establish results where the generalized associative law of addition will be used — although in an implicit manner.

EXAMPLE 4.19

In Section 4.1 we introduced the sequence of rational numbers called the harmonic numbers. Now we introduce an integer sequence that is prominent in combinatorics and graph theory (and that we shall study further in Chapters 10, 11, and 12). The *Fibonacci numbers* may be defined recursively by

1) $F_0 = 0$, $F_1 = 1$; and
2) $F_n = F_{n-1} + F_{n-2}$, for $n \in \mathbf{Z}^+$ with $n \geq 2$.

Hence, from the recursive part of this definition, it follows that

$$F_2 = F_1 + F_0 = 1 + 0 = 1 \qquad F_4 = F_3 + F_2 = 2 + 1 = 3$$
$$F_3 = F_2 + F_1 = 1 + 1 = 2 \qquad F_5 = F_4 + F_3 = 3 + 2 = 5.$$

We also find that $F_6 = 8$, $F_7 = 13$, $F_8 = 21$, $F_9 = 34$, $F_{10} = 55$, $F_{11} = 89$, and $F_{12} = 144$.

The recursive definition of the Fibonacci numbers can be used (in conjunction with the Principle of Mathematical Induction) to establish many of the interesting properties that these numbers exhibit. We investigate one of these properties now.

Let us consider the following five results that deal with sums of squares of the Fibonacci numbers.

1) $F_0^2 + F_1^2 = 0^2 + 1^2 = 1 = 1 \times 1$
2) $F_0^2 + F_1^2 + F_2^2 = 0^2 + 1^2 + 1^2 = 2 = 1 \times 2$
3) $F_0^2 + F_1^2 + F_2^2 + F_3^2 = 0^2 + 1^2 + 1^2 + 2^2 = 6 = 2 \times 3$

4) $F_0^2 + F_1^2 + F_2^2 + F_3^2 + F_4^2 = 0^2 + 1^2 + 1^2 + 2^2 + 3^2 = 15 = 3 \times 5$

5) $F_0^2 + F_1^2 + F_2^2 + F_3^2 + F_4^2 + F_5^2 = 0^2 + 1^2 + 1^2 + 2^2 + 3^2 + 5^2 = 40 = 5 \times 8$

From what is suggested in these calculations, we conjecture that

$$\forall n \in \mathbf{Z}^+ \sum_{i=0}^{n} F_i^2 = F_n \times F_{n+1}.$$

Proof: For $n = 1$, the result in Eq. (1) — namely, $F_0^2 + F_1^2 = 1 \times 1$ — shows us that the conjecture is true in this first case.

Assuming the truth of the conjecture for some $k \geq 1$, we obtain the induction hypothesis:

$$\sum_{i=0}^{k} F_i^2 = F_k \times F_{k+1}.$$

Turning now to the case where $n = k + 1 \; (\geq 2)$ we find that

$$\sum_{i=0}^{k+1} F_i^2 = \sum_{i=0}^{k} F_i^2 + F_{k+1}^2 = (F_k \times F_{k+1}) + F_{k+1}^2 = F_{k+1} \times (F_k + F_{k+1}) = F_{k+1} \times F_{k+2}.$$

Hence the truth of the case for $n = k + 1$ follows from the case for $n = k$. So the given conjecture is true for all $n \in \mathbf{Z}^+$ by the Principle of Mathematical Induction. (The reader may wish to note that the prior calculation uses the generalized associative law of addition. Furthermore we employ the recursive definition of the Fibonacci numbers; it allows us to replace $F_k + F_{k+1}$ by F_{k+2}.)

EXAMPLE 4.20

Closely related to the Fibonacci numbers is the sequence known as the *Lucas numbers*. This sequence is defined recursively by

1) $L_0 = 2, L_1 = 1$; and

2) $L_n = L_{n-1} + L_{n-2}$, for $n \in \mathbf{Z}^+$ with $n \geq 2$.

The first eight Lucas numbers are given in Table 4.2

Table 4.2

n	0	1	2	3	4	5	6	7
L_n	2	1	3	4	7	11	18	29

Although they are not as prominent as the Fibonacci numbers, the Lucas numbers also possess many interesting properties. One of the interrelations between the Fibonacci and Lucas numbers is illustrated in the fact that

$$\forall n \in \mathbf{Z}^+ \; L_n = F_{n-1} + F_{n+1}.$$

Proof: Here we need to consider what happens when $n = 1$ and $n = 2$. We find that

$$L_1 = 1 = 0 + 1 = F_0 + F_2 = F_{1-1} + F_{1+1}, \quad \text{and}$$

$$L_2 = 3 = 1 + 2 = F_1 + F_3 = F_{2-1} + F_{2+1},$$

so the result is true in these first two cases.

Next we assume that $L_n = F_{n-1} + F_{n+1}$ for the integers $n = 1, 2, 3, \ldots, k-1, k$, where $k \geq 2$, and then we consider the Lucas number L_{k+1}. It turns out that

$$L_{k+1} = L_k + L_{k-1} = (F_{k-1} + F_{k+1}) + (F_{k-2} + F_k) \qquad (*)$$
$$= (F_{k-1} + F_{k-2}) + (F_{k+1} + F_k) = F_k + F_{k+2} = F_{(k+1)-1} + F_{(k+1)+1}.$$

Therefore, it follows from the alternative form of the Principle of Mathematical Induction that $L_n = F_{n-1} + F_{n+1}$ for all $n \in \mathbf{Z}^+$. [The reader should observe how we used the recursive definitions for both the Fibonacci numbers and the Lucas numbers in the calculations at (*).]

EXAMPLE 4.21

In Section 1.3 we introduced the binomial coefficients $\binom{n}{r}$ for $n, r \in \mathbf{N}$, where $n \geq r \geq 0$. Corollary 1.1 in that section revealed that $\sum_{r=0}^{n} \binom{n}{r} = \sum_{r=0}^{n} C(n, r) = 2^n$, the total number of subsets for a set of size n. With the help of the result in Example 3.12 we can now define these binomial coefficients recursively by

$$\binom{n+1}{r} = \binom{n}{r} + \binom{n}{r-1}, \quad n \geq r \geq 0,$$

$$\binom{0}{0} = 1, \qquad \binom{n}{r} = 0, \quad r > n, \qquad \binom{n}{r} = 0, \quad r < 0.$$

At this time we present a second set of numbers, each of which is also dependent on two integers. For $m, k \in \mathbf{N}$, the *Eulerian numbers* $a_{m,k}$ are defined recursively by

$$a_{m,k} = (m-k)a_{m-1,k-1} + (k+1)a_{m-1,k}, \quad 0 \leq k \leq m-1, \qquad (*)$$
$$a_{0,0} = 1, \qquad a_{m,k} = 0, \quad k \geq m, \qquad a_{m,k} = 0, \quad k < 0.$$

(In Exercise 18 of the Section Exercises we shall examine a situation that shows how this recursive definition may arise.) The values for $a_{m,k}$, where $1 \leq m \leq 5$ and $0 \leq k \leq m-1$, are given as follows:

						Row Sum
($m = 1$)			1			$1 = 1!$
($m = 2$)		1		1		$2 = 2!$
($m = 3$)		1	4	1		$6 = 3!$
($m = 4$)	1	11	11	1		$24 = 4!$
($m = 5$)	1	26	66	26	1	$120 = 5!$

These results suggest that for a fixed $m \in \mathbf{Z}^+$, $\sum_{k=0}^{m-1} a_{m,k} = m!$, the number of permutations of m objects taken m at a time. We see that the result is true for $1 \leq m \leq 5$. Assuming the result true for some fixed $m \,(\geq 1)$, upon using the recursive definition at (*), we find that

$$\sum_{k=0}^{m} a_{m+1,k} = \sum_{k=0}^{m} [(m+1-k)a_{m,k-1} + (k+1)a_{m,k}]$$
$$= [(m+1)a_{m,-1} + a_{m,0}] + [ma_{m,0} + 2a_{m,1}] + [(m-1)a_{m,1} + 3a_{m,2}] + \cdots$$
$$+ [3a_{m,m-3} + (m-1)a_{m,m-2}] + [2a_{m,m-2} + ma_{m,m-1}]$$
$$+ [a_{m,m-1} + (m+1)a_{m,m}].$$

Since $a_{m,-1} = 0 = a_{m,m}$ we can write

$$\sum_{k=0}^{m} a_{m+1,k} = [a_{m,0} + m a_{m,0}] + [2a_{m,1} + (m-1)a_{m,1}] + \cdots$$

$$+ [(m-1)a_{m,m-2} + 2a_{m,m-2}] + [m a_{m,m-1} + a_{m,m-1}]$$

$$= (m+1) \sum_{k=0}^{m-1} a_{m,k} = (m+1)m! = (m+1)!$$

Consequently, the result is true for all $m \geq 1$ — by the Principle of Mathematical Induction. (We'll see the Eulerian numbers again in Section 9.2.)

In closing this section we shall introduce the idea of a *recursively defined set X*. Here we start with an initial collection of elements that are in X — and this provides the base of the recursion. Then we provide a rule or list of rules that tell us how to find new elements in X from other elements already known to be in X. This rule (or list of rules) constitutes the recursive process. But now (and this part is new) we are also given an *implicit* restriction — that is, a statement to the effect that no element can be found in the set X except for those that were given in the initial collection or those that were formed using the prescribed rule(s) provided in the recursive process.

We demonstrate the ideas given here in the following example.

EXAMPLE 4.22

Define the set X recursively by

1) $1 \in X$; and

2) For each $a \in X$, $a + 2 \in X$.

Then we claim that X consists (precisely) of all positive odd integers.

Proof: If we let Y denote the set of all positive odd integers — that is, $Y = \{2n + 1 | n \in \mathbf{N}\}$ — then we want to show that $Y = X$. This means, as we learned in Section 3.1, that we must verify both $Y \subseteq X$ and $X \subseteq Y$.

In order to establish that $Y \subseteq X$, we must prove that every positive odd integer is in X. This will be accomplished through the Principle of Mathematical Induction. We start by considering the open statement

$$S(n): \quad 2n + 1 \in X,$$

which is defined for the universe $\mathbf{N}$. The basis step — that is, $S(0)$ — is true here because $1 = 2(0) + 1 \in X$ by part (1) of the recursive definition of X. For the inductive step we assume the truth of $S(k)$ for some $k \geq 0$; this tells us $2k + 1$ is an element in X. With $2k + 1 \in X$ it then follows by part (2) of the recursive definition of X that $(2k + 1) + 2 = (2k + 2) + 1 = 2(k + 1) + 1 \in X$, so $S(k + 1)$ is also true. Consequently, $S(n)$ is true (by the Principle of Mathematical Induction) for all $n \in \mathbf{N}$ and we have $Y \subseteq X$.

For the proof of the opposite inclusion (namely, $X \subseteq Y$) we use the recursive definition of X. First we consider part (1) of the definition. Since $1 \ (= 2 \cdot 0 + 1)$ is a positive odd integer, we have $1 \in Y$. To complete the proof, we must verify that any integer in X that results from part (2) of the recursive definition is also in Y. This is done by showing that $a + 2 \in Y$ whenever the element a in X is also an element in Y. For if $a \in Y$, then $a = 2r + 1$, where $r \in \mathbf{N}$ — this by the definition of a positive odd integer. Thus

$a + 2 = (2r + 1) + 2 = (2r + 2) + 1 = 2(r + 1) + 1$, where $r + 1 \in \mathbf{N}$ (actually, $\mathbf{Z}^+$), and so $a + 2$ is a positive odd integer. This places $a + 2$ in Y and now shows that $X \subseteq Y$.

From the preceding two inclusions — that is, $Y \subseteq X$ and $X \subseteq Y$ — it follows that $X = Y$.

EXERCISES 4.2

1. The integer sequence $a_1, a_2, a_3, \ldots$, defined explicitly by the formula $a_n = 5n$ for $n \in \mathbf{Z}^+$, can also be defined recursively by

1) $a_1 = 5$; and

2) $a_{n+1} = a_n + 5$, for $n \geq 1$.

For the integer sequence $b_1, b_2, b_3, \ldots$, where $b_n = n(n + 2)$ for all $n \in \mathbf{Z}^+$, we can also provide the recursive definition:

1)' $b_1 = 3$; and

2)' $b_{n+1} = b_n + 2n + 3$, for $n \geq 1$.

Give a recursive definition for each of the following integer sequences $c_1, c_2, c_3, \ldots$, where for all $n \in \mathbf{Z}^+$ we have

a) $c_n = 7n$ **b)** $c_n = 7^n$

c) $c_n = 3n + 7$ **d)** $c_n = 7$

e) $c_n = n^2$ **f)** $c_n = 2 - (-1)^n$

2. a) Give a recursive definition for the disjunction of statements $p_1, p_2, \ldots, p_n, p_{n+1}, n \geq 1$.

b) Show that if $n, r \in \mathbf{Z}^+$, with $n \geq 3$ and $1 \leq r < n$, then

$$(p_1 \vee p_2 \vee \cdots \vee p_r) \vee (p_{r+1} \vee \cdots \vee p_n)$$
$$\Longleftrightarrow p_1 \vee p_2 \vee \cdots \vee p_r \vee p_{r+1} \vee \cdots \vee p_n.$$

3. Use the result of Example 4.16 to prove that if $p, q_1, q_2, \ldots, q_n$ are statements and $n \geq 2$, then

$$p \vee (q_1 \wedge q_2 \wedge \cdots \wedge q_n)$$
$$\Longleftrightarrow (p \vee q_1) \wedge (p \vee q_2) \wedge \cdots \wedge (p \vee q_n).$$

4. For $n \in \mathbf{Z}^+$, $n \geq 2$, prove that for any statements $p_1, p_2, \ldots, p_n$,

a) $\neg(p_1 \vee p_2 \vee \cdots \vee p_n) \Longleftrightarrow \neg p_1 \wedge \neg p_2 \wedge \cdots \wedge \neg p_n$.

b) $\neg(p_1 \wedge p_2 \wedge \cdots \wedge p_n) \Longleftrightarrow \neg p_1 \vee \neg p_2 \vee \cdots \vee \neg p_n$.

5. a) Give a recursive definition for the intersection of the sets $A_1, A_2, \ldots, A_n, A_{n+1} \subseteq \mathcal{U}, n \geq 1$.

b) Use the result in part (a) to show that for all $n, r \in \mathbf{Z}^+$ with $n \geq 3$ and $1 \leq r < n$,

$$(A_1 \cap A_2 \cap \cdots \cap A_r) \cap (A_{r+1} \cap \cdots \cap A_n)$$
$$= A_1 \cap A_2 \cap \cdots \cap A_r \cap A_{r+1} \cap \cdots \cap A_n.$$

6. For $n \geq 2$ and any sets $A_1, A_2, \ldots, A_n \subseteq \mathcal{U}$, prove that $\overline{A_1 \cup A_2 \cup \cdots \cup A_n} = \overline{A_1} \cap \overline{A_2} \cap \cdots \cap \overline{A_n}$.

7. Use the result of Example 4.17 to show that if sets $A, B_1, B_2, \ldots, B_n \subseteq \mathcal{U}$ and $n \geq 2$, then

$$A \cap (B_1 \cup B_2 \cup \cdots \cup B_n)$$
$$= (A \cap B_1) \cup (A \cap B_2) \cup \cdots \cup (A \cap B_n).$$

8. a) Develop a recursive definition for the addition of n real numbers $x_1, x_2, \ldots, x_n$, where $n \geq 2$.

b) For all real numbers x_1, x_2, and x_3, the associative law of addition states that $x_1 + (x_2 + x_3) = (x_1 + x_2) + x_3$. Prove that if $n, r \in \mathbf{Z}^+$, where $n \geq 3$ and $1 \leq r < n$, then

$$(x_1 + x_2 + \cdots + x_r) + (x_{r+1} + \cdots + x_n)$$
$$= x_1 + x_2 + \cdots + x_r + x_{r+1} + \cdots + x_n.$$

9. a) Develop a recursive definition for the multiplication of n real numbers $x_1, x_2, \ldots, x_n$, where $n \geq 2$.

b) For all real numbers x_1, x_2, and x_3, the associative law of multiplication states that $x_1(x_2 x_3) = (x_1 x_2)x_3$. Prove that if $n, r \in \mathbf{Z}^+$, where $n \geq 3$ and $1 \leq r < n$, then

$$(x_1 x_2 \cdots x_r)(x_{r+1} \cdots x_n) = x_1 x_2 \cdots x_r x_{r+1} \cdots x_n.$$

10. For all $x \in \mathbf{R}$,

$$|x| = \sqrt{x^2} = \begin{cases} x, & \text{if } x \geq 0 \\ -x, & \text{if } x < 0 \end{cases}, \quad \text{and}$$

$-|x| \leq x \leq |x|$. Consequently, $|x + y|^2 = (x + y)^2 = x^2 + 2xy + y^2 \leq x^2 + 2|x||y| + y^2 = |x|^2 + 2|x||y| + |y|^2 = (|x| + |y|)^2$, and $|x + y|^2 \leq (|x| + |y|)^2 \Rightarrow |x + y| \leq |x| + |y|$, for all $x, y \in \mathbf{R}$.

Prove that if $n \in \mathbf{Z}^+$, $n \geq 2$, and $x_1, x_2, \ldots, x_n \in \mathbf{R}$, then

$$|x_1 + x_2 + \cdots + x_n| \leq |x_1| + |x_2| + \cdots + |x_n|.$$

11. Define the integer sequence $a_0, a_1, a_2, a_3, \ldots$, recursively by

1) $a_0 = 1, a_1 = 1, a_2 = 1$; and

2) For $n \geq 3$, $a_n = a_{n-1} + a_{n-3}$.

Prove that $a_{n+2} \geq (\sqrt{2})^n$ for all $n \geq 0$.

12. For $n \geq 0$ let F_n denote the nth Fibonacci number. Prove that

$$F_0 + F_1 + F_2 + \cdots + F_n = \sum_{i=0}^{n} F_i = F_{n+2} - 1.$$

13. Prove that for any positive integer n,

$$\sum_{i=1}^{n} \frac{F_{i-1}}{2^i} = 1 - \frac{F_{n+2}}{2^n}.$$

14. As in Example 4.20 let $L_0, L_1, L_2, \ldots$ denote the Lucas numbers, where (1) $L_0 = 2$, $L_1 = 1$; and (2) $L_{n+2} = L_{n+1} + L_n$, for $n \geq 0$. When $n \geq 1$, prove that

$$L_1^2 + L_2^2 + L_3^2 + \cdots + L_n^2 = L_n L_{n+1} - 2.$$

15. If $n \in \mathbf{N}$, prove that $5F_{n+2} = L_{n+4} - L_n$.

16. Give a recursive definition for the set of all

a) positive even integers

b) nonnegative even integers

17. One of the most common uses for the recursive definition of sets is to define the *well-formed formulae* in various mathematical systems. For example, in the study of logic we can define the well-formed formulae as follows:

1) Each primitive statement p, the tautology T_0, and the contradiction F_0 are well-formed formulae; and

2) If p, q are well-formed formulae, then so are

  i) $(\neg p)$ **ii)** $(p \vee q)$ **iii)** $(p \wedge q)$

  iv) $(p \rightarrow q)$ **v)** $(p \leftrightarrow q)$

Using this recursive definition, we find that for the primitive statements p, q, r, the compound statement $((p \wedge (\neg q)) \rightarrow (r \vee T_0))$ is a well-formed formula. We can derive this well-formed formula as follows:

Steps	Reasons
1) p, q, r, T_0	Part (1) of the definition
2) $(\neg q)$	Step (1) and part (2i) of the definition
3) $(p \wedge (\neg q))$	Steps (1) and (2) and part (2iii) of the definition
4) $(r \vee T_0)$	Step (1) and part (2ii) of the definition
5) $((p \wedge (\neg q)) \rightarrow (r \vee T_0))$	Steps (3) and (4) and part (2iv) of the definition

For the primitive statements p, q, r, and s, provide derivations showing that each of the following is a well-formed formula.

  a) $((p \vee q) \rightarrow (T_0 \wedge (\neg r)))$

  b) $(((\neg p) \leftrightarrow q) \rightarrow (r \wedge (s \vee F_0)))$

18. Consider the permutations of 1, 2, 3, 4. The permutation 1432, for instance, is said to have one *ascent* — namely, 14 (since $1 < 4$). This same permutation also has two *descents* — namely, 43 (since $4 > 3$) and 32 (since $3 > 2$). The permutation 1423, on the other hand, has two ascents, at 14 and 23 — and the one descent 42.

  a) How many permutations of 1, 2, 3 have k ascents, for $k = 0, 1, 2$?

  b) How many permutations of 1, 2, 3, 4 have k ascents, for $k = 0, 1, 2, 3$?

  c) If a permutation of 1, 2, 3, 4, 5, 6, 7 has four ascents, how many descents does it have?

d) Suppose a permutation of 1, 2, 3, $\ldots$, m has k ascents, for $0 \leq k \leq m - 1$. How many descents does the permutation have?

e) Consider the permutation $p = 12436587$. This permutation of 1, 2, 3, $\ldots$, 8 has four ascents. In how many of the nine locations (at the start, end, or between two numbers) in p can we place 9 so that the result is a permutation of 1, 2, 3, $\ldots$, 8, 9 with (i) four ascents; (ii) five ascents?

f) Let $\pi_{m,k}$ denote the number of permutations of 1, 2, 3, $\ldots$, m with k ascents. Note how $\pi_{4,2} = 11 = 2(4) + 3(1) = (4 - 2)\pi_{3,1} + (2 + 1)\pi_{3,2}$. How is $\pi_{m,k}$ related to $\pi_{m-1,k-1}$ and $\pi_{m-1,k}$?

19. a) For $k \in \mathbf{Z}^+$ verify that $k^2 = \binom{k}{2} + \binom{k+1}{2}$.

b) Fix n in $\mathbf{Z}^+$. Since the result in part (a) is true for all $k = 1, 2, 3, \ldots, n$, summing the n equations

$$1^2 = \binom{1}{2} + \binom{2}{2}$$

$$2^2 = \binom{2}{2} + \binom{3}{2}$$

$$\vdots \qquad \vdots \qquad \vdots$$

$$n^2 = \binom{n}{2} + \binom{n+1}{2}$$

we have $\sum_{k=1}^{n} k^2 = \sum_{k=1}^{n} \binom{k}{2} + \sum_{k=1}^{n} \binom{k+1}{2} = \binom{n+1}{3} + \binom{n+2}{3}$. [The last equality follows from Exercise 26 for Section 3.1 because $\sum_{k=1}^{n} \binom{k}{2} = \binom{1}{2} + \binom{2}{2} + \binom{3}{2} + \cdots + \binom{n}{2} = 0 + \binom{2}{0} + \binom{3}{1} + \cdots + \binom{n}{n-2} = \binom{n+1}{n-2} = \binom{n+1}{3}$ and $\sum_{k=1}^{n} \binom{k+1}{2} = \binom{2}{2} + \binom{3}{2} + \binom{4}{2} + \cdots + \binom{n+1}{2} = \binom{2}{0} + \binom{3}{1} + \binom{4}{2} + \cdots + \binom{n+1}{n-1} = \binom{n+2}{n-1} = \binom{n+2}{3}$. Show that

$$\binom{n+1}{3} + \binom{n+2}{3} = \frac{n(n+1)(2n+1)}{6}.$$

c) For $k \in \mathbf{Z}^+$ verify that $k^3 = \binom{k}{3} + 4\binom{k+1}{3} + \binom{k+2}{3}$.

d) Use part (c) and the results from Exercise 26 for Section 3.1 to show that

$$\sum_{k=1}^{n} k^3 = \binom{n+1}{4} + 4\binom{n+2}{4} + \binom{n+3}{4} = \frac{n^2(n+1)^2}{4}$$

e) Find $a, b, c, d \in \mathbf{Z}^+$ so that for any $k \in \mathbf{Z}^+$, $k^4 = a\binom{k}{4} + b\binom{k+1}{4} + c\binom{k+2}{4} + d\binom{k+3}{4}$.

20. a) For $n \geq 2$, if $p_1, p_2, p_3, \ldots, p_n, p_{n+1}$ are statements, prove that

$$[(p_1 \rightarrow p_2) \wedge (p_2 \rightarrow p_3) \wedge \cdots \wedge (p_n \rightarrow p_{n+1})]$$
$$\Rightarrow [(p_1 \wedge p_2 \wedge p_3 \wedge \cdots \wedge p_n) \rightarrow p_{n+1}].$$

b) Prove that Theorem 4.2 implies Theorem 4.1.

c) Use Theorem 4.1 to establish the following: If $\emptyset \neq S \subseteq \mathbf{Z}^+$, so that $n \in S$ for some $n \in \mathbf{Z}^+$, then S contains a least element.

d) Show that Theorem 4.1 implies Theorem 4.2.

4.3
The Division Algorithm: Prime Numbers

Although the set $\mathbf{Z}$ is not closed under nonzero division, in many instances one integer (exactly) divides another. For example, 2 divides 6 and 7 divides 21. Here the division is exact and there is no remainder. Thus 2 dividing 6 implies the existence of a quotient — namely, 3 — such that $6 = 2 \cdot 3$. We formalize this idea as follows.

Definition 4.1

If $a, b \in \mathbf{Z}$ and $b \neq 0$, we say that b *divides* a, and we write $b|a$, if there is an integer n such that $a = bn$. When this occurs we say that b is a *divisor* of a, or a is a *multiple* of b.

With this definition we are able to speak of division inside of $\mathbf{Z}$ without going to $\mathbf{Q}$. Furthermore, when $ab = 0$ for $a, b \in \mathbf{Z}$, then either $a = 0$ or $b = 0$, and we say that $\mathbf{Z}$ has no *proper divisors of* 0. This property enables us to *cancel* as in $2x = 2y \Rightarrow x = y$, for $x, y \in \mathbf{Z}$, because $2x = 2y \Rightarrow 2(x - y) = 0 \Rightarrow 2 = 0$ or $x - y = 0 \Rightarrow x = y$. (Note that at no time did we mention multiplying both sides of the equation $2x = 2y$ by $\frac{1}{2}$. The number $\frac{1}{2}$ is outside the system $\mathbf{Z}$.)

We now summarize some properties of this division operation. Whenever we divide by an integer a, we assume that $a \neq 0$.

THEOREM 4.3

For all $a, b, c \in \mathbf{Z}$

   a) $1|a$ and $a|0$. **b)** $[(a|b) \wedge (b|a)] \Rightarrow a = \pm b$.
   c) $[(a|b) \wedge (b|c)] \Rightarrow a|c$. **d)** $a|b \Rightarrow a|bx$ for all $x \in \mathbf{Z}$.
   e) If $x = y + z$, for some $x, y, z \in \mathbf{Z}$, and a divides two of the three integers x, y, and z, then a divides the remaining integer.
   f) $[(a|b) \wedge (a|c)] \Rightarrow a|(bx + cy)$, for all $x, y \in \mathbf{Z}$. (The expression $bx + cy$ is called a *linear combination of b, c*.)
   g) For $1 \leq i \leq n$, let $c_i \in \mathbf{Z}$. If a divides each c_i, then $a|(c_1x_1 + c_2x_2 + \cdots + c_nx_n)$, where $x_i \in \mathbf{Z}$ for all $1 \leq i \leq n$.

Proof: We prove part (f) and leave the remaining parts for the reader.

If $a|b$ and $a|c$, then $b = am$ and $c = an$, for some $m, n \in \mathbf{Z}$. So $bx + cy = (am)x + (an)y = a(mx + ny)$ (by the Associative Law of Multiplication and the Distributive Law of Multiplication over Addition — since the elements in $\mathbf{Z}$ satisfy both of these laws). Since $bx + cy = a(mx + ny)$, with $mx + ny \in \mathbf{Z}$, it follows that $a|(bx + cy)$.

We find part (g) of the theorem useful when we consider the following question.

EXAMPLE 4.23

Do there exist integers x, y, z (positive, negative, or zero) so that $6x + 9y + 15z = 107$? Suppose that such integers did exist. Then since $3|6$, $3|9$, and $3|15$, it would follow from part (g) of Theorem 4.3 that 3 is a divisor of $6x + 9y + 15z$ and, consequently, 3 is a divisor of 107 — but this is not so. Hence there do not exist such integers x, y, z.

Several parts of Theorem 4.3 help us in the following

EXAMPLE 4.24	Let a, $b \in \mathbf{Z}$ so that $2a + 3b$ is a multiple of 17. (For example, we could have $a = 7$ and $b = 1$ — and $a = 4$, $b = 3$ also works.) Prove that 17 divides $9a + 5b$.

Proof: We observe that $17|(2a + 3b) \Rightarrow 17|(-4)(2a + 3b)$, by part (d) of Theorem 4.3. Also, since $17|17$, it follows from part (f) of the theorem that $17|(17a + 17b)$. Hence, $17|[(17a + 17b) + (-4)(2a + 3b)]$, by part (e) of the theorem. Consequently, as $[(17a + 17b) + (-4)(2a + 3b)] = [(17 - 8)a + (17 - 12)b] = 9a + 5b$, we have $17|(9a + 5b)$.

Using this binary operation of integer division we find ourselves in the area of mathematics called *number theory*, which examines the properties of integers and other sets of numbers. Once considered an area of strictly pure (abstract) mathematics, number theory is now an essential applicable tool — especially, in dealing with computer and Internet security. But for now, as we continue to examine the set $\mathbf{Z}^+$ further, we notice that for all $n \in \mathbf{Z}^+$ where $n > 1$, the integer n has at least two positive divisors, namely, 1 and n itself. Some integers, such as 2, 3, 5, 7, 11, 13, and 17 have exactly two positive divisors. These integers are called *primes*. All other positive integers (greater than 1 and not prime) are called *composite*. An immediate connection between prime and composite integers is expressed in the following lemma.

LEMMA 4.1	If $n \in \mathbf{Z}^+$ and n is composite, then there is a prime p such that $p|n$.

Proof: If not, let S be the set of all composite integers that have no prime divisor(s). If $S \neq \emptyset$, then by the Well-Ordering Principle, S has a least element m. But if m is composite, then $m = m_1 m_2$, where m_1, $m_2 \in \mathbf{Z}^+$ with $1 < m_1 < m$ and $1 < m_2 < m$. Since $m_1 \notin S$, m_1 is prime or divisible by a prime — so, there exists a prime p such that $p|m_1$. Since $m = m_1 m_2$, it now follows from part (d) of Theorem 4.3 that $p|m$, and so $S = \emptyset$.

Now why did we call the preceding result a *lemma* instead of a theorem? After all, it had to be proved like all other theorems in the book so far. The reason is that although a lemma is itself a theorem, its major role is to help prove other theorems.

In listing the primes we are inclined to believe that there are infinitely many such numbers. We now verify that this is true.

THEOREM 4.4	(Euclid) There are infinitely many primes.

Proof: If not, let p_1, p_2, ..., p_k be the finite list of all primes, and let $B = p_1 p_2 \cdots p_k + 1$. Since $B > p_i$ for all $1 \leq i \leq k$, B cannot be a prime. Hence B is composite. So by Lemma 4.1 there is a prime p_j, where $1 \leq j \leq k$ and $p_j|B$. Since $p_j|B$ and $p_j|p_1 p_2 \cdots p_k$, by Theorem 4.3(e) it follows that $p_j|1$. This contradiction arises from the assumption that there are only finitely many primes; the result follows.

Yes, this is the same Euclid from the fourth century B.C. whose *Elements*, written on 13 parchment scrolls, included the first organized coverage of the geometry we studied in high school. One finds, however, that these 13 books are also concerned with number theory. In particular, Books VII, VIII, and IX dwell on this topic. The preceding theorem (with proof) is found in Book IX.

We turn now to the major idea of this section. This result enables us to deal with nonzero division in $\mathbf{Z}$ when that division is not exact.

THEOREM 4.5

The Division Algorithm. If $a, b \in \mathbf{Z}$, with $b > 0$, then there exist unique $q, r \in \mathbf{Z}$ with $a = qb + r, 0 \leq r < b$.

Proof: If $b \mid a$ the result follows with $r = 0$, so consider the case where $b \nmid a$ (that is, b does not divide a).

Let $S = \{a - tb \mid t \in \mathbf{Z}, a - tb > 0\}$. If $a > 0$ and $t = 0$, then $a \in S$ and $S \neq \emptyset$. For $a \leq 0$, let $t = a - 1$. Then $a - tb = a - (a - 1)b = a(1 - b) + b$, with $(1 - b) \leq 0$, because $b \geq 1$. So $a - tb > 0$ and $S \neq \emptyset$. Hence, for any $a \in \mathbf{Z}$, S is a nonempty subset of $\mathbf{Z}^+$. By the Well-Ordering Principle, S has a least element r, where $0 < r = a - qb$, for some $q \in \mathbf{Z}$. If $r = b$, then $a = (q + 1)b$ and $b \mid a$, contradicting $b \nmid a$. If $r > b$, then $r = b + c$, for some $c \in \mathbf{Z}^+$, and $a - qb = r = b + c \Rightarrow c = a - (q + 1)b \in S$, contradicting r being the least element of S. Hence, $r < b$.

This now establishes a quotient q and remainder r, where $0 \leq r < b$, for the theorem. But are there other q's and r's that also work? If so, let $q_1, q_2, r_1, r_2 \in \mathbf{Z}$ with $a = q_1 b + r_1$, for $0 \leq r_1 < b$, and $a = q_2 b + r_2$, for $0 \leq r_2 < b$. Then $q_1 b + r_1 = q_2 b + r_2 \Rightarrow b \mid q_1 - q_2 \mid = |r_2 - r_1| < b$, because $0 \leq r_1, r_2 < b$. If $q_1 \neq q_2$, we have the contradiction $b \mid q_1 - q_2 \mid < b$. Hence $q_1 = q_2, r_1 = r_2$, and the quotient and remainder are unique.

As we mentioned in the preceding proof, when $a, b \in \mathbf{Z}$ with $b > 0$, then there exists a unique *quotient* q and a unique *remainder* r where $a = qb + r$, with $0 \leq r < b$. Furthermore, under these circumstances, the integer b is called the *divisor* while a is termed the *dividend*.

EXAMPLE 4.25

a) When $a = 170$ and $b = 11$ in the division algorithm, we find that $170 = 15 \cdot 11 + 5$, where $0 \leq 5 < 11$. So when 170 is divided by 11, the quotient is 15 and the remainder is 5.

b) If the dividend is 98 and the divisor is 7, then we find that $98 = 14 \cdot 7$. So in this case the quotient is 14 and the remainder is 0, and 7 (exactly) divides 98.

c) For the case of $a = -45$ and $b = 8$ we have $-45 = (-6)8 + 3$, where $0 \leq 3 < 8$. Consequently, the quotient is -6 and the remainder is 3 when the dividend is -45 and the divisor is 8.

d) Let $a, b \in \mathbf{Z}^+$.

1) If $a = qb$ for some $q \in \mathbf{Z}^+$, then $-a = (-q)b$. So, in this case, when $-a \ (< 0)$ is divided by $b \ (> 0)$ the quotient is $-q \ (< 0)$ and the remainder is 0.

2) If $a = qb + r$ for some $q \in \mathbf{N}$ and $0 < r < b$, then $-a = (-q)b - r = (-q)b - b + b - r = (-q - 1)b + (b - r)$. For this case, when $-a \ (< 0)$ is divided by $b \ (> 0)$ the quotient is $-q - 1 \ (< 0)$ and the remainder is $b - r$, where $0 < b - r < b$.

Despite the proof of Theorem 4.5 and the results in Example 4.25, we really do *not* have any systematic way to calculate the quotient q and remainder r when we divide an integer a (the dividend) by the positive integer b (the divisor). The proof of Theorem 4.5 guarantees the existence of such integers q and r, but the proof is *not* constructive. It does not appear to tell us how to actually calculate q and r, and it does not mention anything about the ability to use multiplication tables or perform long division. To remedy this situation we provide the procedure (written in pseudocode) in Fig. 4.10. Our next example illustrates the idea presented in part of this procedure.

```
procedure IntegerDivision (a, b: integers)
begin
  if a = 0 then
    begin
      quotient := 0
      remainder := 0
    end
  else
    begin
      r := abs(a) {the absolute value of a}
      q := 0
      while r ≥ b do
        begin
          r := r - b
          q := q + 1
        end
      if a > 0 then
        begin
          quotient := q
          remainder := r
        end
      else if r = 0 then
        begin
          quotient := -q
          remainder := 0
        end
      else
        begin
          quotient := -q - 1
          remainder := b - r
        end
    end
end
```

Figure 4.10

<table>
<tr><td>**EXAMPLE 4.26**</td></tr>
</table>

Just as the multiplication of positive integers may be viewed as repeated addition, so too can we view (integer) division as repeated subtraction. We see that subtraction does play a role in the definition of the set S in the proof of Theorem 4.5.

When calculating $4 \cdot 7$, for example, we can think in terms of repeated addition and write

$$2 \cdot 7 = 7 + 7 = 14$$

$$3 \cdot 7 = (2 + 1) \cdot 7 = 2 \cdot 7 + 1 \cdot 7 = (7 + 7) + 7 = 14 + 7 = 21$$

$$4 \cdot 7 = (3 + 1) \cdot 7 = 3 \cdot 7 + 1 \cdot 7 = ((7 + 7) + 7) + 7 = 21 + 7 = 28.$$

If, on the other hand, we wish to divide 37 by 8, then we should think of the quotient q as the number of 8's contained in 37. When each one of these 8's is removed (that is, subtracted) and no other 8 can be removed without giving us a negative result, then the integer that is left (remaining) is the remainder r. So we can calculate q and r by thinking in terms of repeated subtraction as follows:

$$37 - 8 = 29 \geq 8,$$
$$29 - 8 = (37 - 8) - 8 = 37 - 2 \cdot 8 = 21 \geq 8,$$
$$21 - 8 = ((37 - 8) - 8) - 8 = 37 - 3 \cdot 8 = 13 \geq 8,$$
$$13 - 8 = (((37 - 8) - 8) - 8) - 8 = 37 - 4 \cdot 8 = 5 < 8.$$

The last line shows that four 8's can be subtracted from 37 before we obtain a nonnegative result — namely, 5 — that is smaller than 8. Therefore, in this example we have $q = 4$ and $r = 5$.

Using the division algorithm, we consider some results on representing integers in bases other than 10.

EXAMPLE 4.27

Write 6137 in the octal system (base 8). Here we seek nonnegative integers $r_0, r_1, r_2, \ldots, r_k$, with $0 < r_k < 8$, such that $6137 = (r_k \cdots r_2 r_1 r_0)_8$.

With $6137 = r_0 + r_1 \cdot 8 + r_2 \cdot 8^2 + \cdots + r_k \cdot 8^k = r_0 + 8(r_1 + r_2 \cdot 8 + \cdots + r_k \cdot 8^{k-1})$, r_0 is the remainder obtained in the division algorithm when 6137 is divided by 8.

Consequently, since $6137 = 1 + 8 \cdot 767$, we have $r_0 = 1$ and $767 = r_1 + r_2 \cdot 8 + \cdots + r_k \cdot 8^{k-1} = r_1 + 8(r_2 + r_3 \cdot 8 + \cdots + r_k \cdot 8^{k-2})$. This yields $r_1 = 7$ (the remainder when 767 is divided by 8) and $95 = r_2 + r_3 \cdot 8 + \cdots + r_k \cdot 8^{k-2}$. Continuing in this manner, we find $r_2 = 7$, $r_3 = 3$, $r_4 = 1$, and $r_i = 0$ for all $i \geq 5$, so

$$6137 = 1 \cdot 8^4 + 3 \cdot 8^3 + 7 \cdot 8^2 + 7 \cdot 8 + 1 = (13771)_8.$$

We can arrange the successive divisions by 8 as follows:

		Remainders
8	6137	
8	767	$1 (r_0)$
8	95	$7 (r_1)$
8	11	$7 (r_2)$
8	1	$3 (r_3)$
	0	$1 (r_4)$

EXAMPLE 4.28

In the field of computer science, the binary number system (base 2) is very important. Here the only symbols that one may use are the bits 0 and 1. In Table 4.3 we have listed the binary representations of the (base-10) integers from 0 to 15. Here we have included leading zeros and find that we need four bits because of the leading 1 in the representations for the integers from 8 to 15. With five bits we can continue up to 31 ($= 32 - 1 = 2^5 - 1$); six bits are necessary to proceed to 63 ($= 64 - 1 = 2^6 - 1$). In general, if $x \in \mathbf{Z}$ and $0 \leq x < 2^n$, for $n \in \mathbf{Z}^+$, then we can write x in base 2 by using n bits. Leading zeros appear when $0 \leq x \leq 2^{n-1} - 1$, and for $2^{n-1} \leq x \leq 2^n - 1$ the first (most significant) bit is 1.

Information is generally stored in machines in units of eight bits called bytes, so for machines with memory cells of one byte we can store in a single cell any one of the binary

Table 4.3

Base 10	Base 2	Base 10	Base 2
0	0 0 0 0	8	1 0 0 0
1	0 0 0 1	9	1 0 0 1
2	0 0 1 0	10	1 0 1 0
3	0 0 1 1	11	1 0 1 1
4	0 1 0 0	12	1 1 0 0
5	0 1 0 1	13	1 1 0 1
6	0 1 1 0	14	1 1 1 0
7	0 1 1 1	15	1 1 1 1

equivalents of the integers from 0 to $2^8 - 1 = 255$. For a machine with two-byte cells, any one of the integers from 0 to $2^{16} - 1 = 65{,}535$ can be stored in binary form in each cell. A machine with four-byte cells would take us up to $2^{32} - 1 = 4{,}294{,}967{,}295$.

When a human deals with long sequences of 0's and 1's, the job soon becomes very tedious and the chance for error increases with the tedium. Consequently, it is common (especially in the study of machine and assembly languages) to represent such long sequences of bits in another notation. One such notation is the *hexadecimal (base*-16) *notation.* Here there are 16 symbols, and because we have only 10 symbols in the standard base-10 system, we introduce the following six additional symbols:

A	(Alfa)	C	(Charlie)	E	(Echo)
B	(Bravo)	D	(Delta)	F	(Foxtrot)

In Table 4.4 the integers from 0 to 15 are given in terms of both the binary and the hexadecimal number systems.

Table 4.4

Base 10	Base 2	Base 16	Base 10	Base 2	Base 16
0	0 0 0 0	0	8	1 0 0 0	8
1	0 0 0 1	1	9	1 0 0 1	9
2	0 0 1 0	2	10	1 0 1 0	A
3	0 0 1 1	3	11	1 0 1 1	B
4	0 1 0 0	4	12	1 1 0 0	C
5	0 1 0 1	5	13	1 1 0 1	D
6	0 1 1 0	6	14	1 1 1 0	E
7	0 1 1 1	7	15	1 1 1 1	F

To convert from base 10 to base 16, we follow a procedure like the one outlined in Example 4.27. Here we are interested in the remainders upon successive divisions by 16. Therefore, if we want to represent the (base-10) integer 13,874,945 in the hexadecimal system, we do the following calculations:

Remainders

$$
\begin{array}{r|l}
16 & 13{,}874{,}945 \\
16 & 867{,}184 \\
16 & 54{,}199 \\
16 & 3{,}387 \\
16 & 211 \\
16 & 13 \\
& 0
\end{array}
$$

$$
\begin{array}{ll}
1 & (r_0) \\
0 & (r_1) \\
7 & (r_2) \\
11\ (= \text{B}) & (r_3) \\
3 & (r_4) \\
13\ (= \text{D}) & (r_5)
\end{array}
$$

Consequently, $13{,}874{,}945 = (\text{D3B701})_{16}$.

There is, however, an easier approach for converting between base 2 and base 16. For example, if we want to convert the binary (one-byte) integer 01001101 to its base-16 counterpart, we break the number into blocks of four bits:

$$
\underbrace{0100}_{4} \quad \underbrace{1101}_{\text{D}}
$$

We then convert each block of four bits to its base-16 representation (as shown in Table 4.4), and we have $(01001101)_2 = (\text{4D})_{16}$. If we start with the (two-byte) number $(\text{A13F})_{16}$ and want to convert in the other direction, we replace each hexadecimal symbol by its (four-bit) binary equivalent (also as shown in Table 4.4):

$$
\underbrace{\text{A}}_{1010} \quad \underbrace{1}_{0001} \quad \underbrace{3}_{0011} \quad \underbrace{\text{F}}_{1111}
$$

This results in $(\text{A13F})_{16} = (1010000100111111)_2$.

EXAMPLE 4.29

We need negative integers in order to perform the binary operation of subtraction in terms of addition [that is, $(a - b) = a + (-b)$]. When we are dealing with the binary representation of integers, we can use a popular method that enables us to perform addition, subtraction, multiplication, and (integer) division: the *two's complement method*. The method's popularity rests on its implementation by only two electronic circuits — one to invert and the second to add.

In Table 4.5 the integers from -8 to 7 are represented by the four-bit patterns shown. The nonnegative integers are represented as they were in Tables 4.3 and 4.4. To obtain the results for $-8 \le n \le -1$, first consider the binary representation of $|n|$, the absolute value of n. Then do the following:

1) Replace each 0(1) in the binary representation of $|n|$ by 1(0); this result is called the *one's complement* of (the given representation of) $|n|$.

2) Add 1 ($= 0001$ in this case) to the result in step (1). This result is called the *two's complement* of n.

For example, to obtain the two's complement (representation) of -6, we proceed as follows.

6
↓
1) Start with the binary   0110
   representation of 6.   ↓
2) Interchange the 0's and 1's; this   1001
   result is the one's complement of 0110.   ↓
3) Add 1 to the prior result. $1001 + 0001 = 1010$

We can also obtain the four-bit patterns for the values $-8 \le n \le -1$ by using the four-bit patterns for the integers from 0 to 7 and complementing (interchanging 0's and 1's) these patterns as shown by four such pairs of patterns in Table 4.5. Note in Table 4.5 that the four-bit patterns for the nonnegative integers start with 0, whereas 1 is the first bit for the negative integers in the table.

Table 4.5

Two's Complement Notation				
Value Represented	**Four-Bit Pattern**			
7	0	1	1	1
6	0	1	1	0
5	0	1	0	1
4	0	1	0	0
3	0	0	1	1
2	0	0	1	0
1	0	0	0	1
0	0	0	0	0
-1	1	1	1	1
-2	1	1	1	0
-3	1	1	0	1
-4	1	1	0	0
-5	1	0	1	1
-6	1	0	1	0
-7	1	0	0	1
-8	1	0	0	0

EXAMPLE 4.30

How do we perform the subtraction $33 - 15$ in base 2, using the two's complement method with patterns of eight bits (= one byte)?

We want to determine $33 - 15 = 33 + (-15)$. We find that $33 = (00100001)_2$, and $15 = (00001111)_2$. Therefore we represent -15 by

$$11110000 + 00000001 = 11110001.$$

The addition of integers represented in two's complement notation is the same as ordinary binary addition, except that all results must have the same size bit patterns. This means that when two integers are added by the two's complement method, any extra bit that results on the left of the answer (by a final carry) must be discarded. We illustrate this in the following calculations.

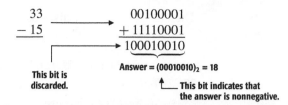

$$\begin{array}{r} 33 \\ -\ 15 \end{array} \qquad \begin{array}{r} 00100001 \\ +\ 11110001 \\ \hline 100010010 \end{array}$$

Answer = $(00010010)_2$ = 18

This bit is discarded.

This bit indicates that the answer is nonnegative.

To find $15 - 33$ we use $15 = (00001111)_2$ and $33 = (00100001)_2$. Then, to calculate $15 - 33$ as $15 + (-33)$, we represent -33 by $11011110 + 00000001 = 11011111$. This gives us the results

$$\begin{array}{r} 15 \\ -\ 33 \end{array} \qquad \begin{array}{r} 00001111 \\ +\ 11011111 \\ \hline 11101110 \end{array}$$

This bit indicates that the answer is negative.

In order to get the positive form of the answer, we proceed as follows:

11101110

1) Take the one's complement. $\downarrow$ 00010001

2) Add 1 to the prior result. $\downarrow$ 00010010

Since $(00010010)_2 = 18$, the answer is -18.

One problem we have avoided in the two preceding calculations involves the size of the integers that we can represent by eight-bit patterns. No matter what size patterns we use, the size of the integers that can be represented is limited. When we exceed this size, an *overflow error* results. For example, if we are working with eight-bit patterns and try to add 117 and 88, we obtain

$$\begin{array}{r} 117 \\ +\ 88 \end{array} \qquad \begin{array}{r} 01110101 \\ +\ 01011000 \\ \hline 11001101 \end{array}$$

This bit indicates that the answer is negative.

This result shows how we can detect an overflow error when adding two numbers. Here an overflow error is indicated: The sum of the eight-bit patterns for two positive integers has resulted in the eight-bit pattern for a negative integer. Similarly, when the addition of (the eight-bit patterns of) two negative integers results in the eight-bit pattern of a positive integer, an overflow error is detected.

To see why the procedure in Example 4.30 works in general, let $x, y \in \mathbf{Z}^+$ with $x > y$.

Let $2^{n-1} \le x < 2^n$. Then the binary representation for x is made up of n bits (with the leading bit 1). The binary representation for 2^n consists of $n + 1$ bits: a leading bit 1 followed by n 0's. The binary representation for $2^n - 1$ consists of n 1's.

When we subtract y from $2^n - 1$, we have

$$(2^n - 1) - y = \underbrace{11\ldots1}_{n\ 1\text{'s}} - y, \text{ the one's complement of } y.$$

Then $(2^n - 1) - y + 1$ gives us the two's complement of $-y$, and

$$x - y = x + [(2^n - 1) - y + 1] - 2^n,$$

where the final term, -2^n, results in the removal of the extra bit that arises on the left of the answer.

We close this section with one final result on composite integers.

EXAMPLE 4.31

If $n \in \mathbf{Z}^+$ and n is composite, then there exists a prime p such that $p|n$ and $p \le \sqrt{n}$.

Proof: Since n is composite, we can write $n = n_1 n_2$, where $1 < n_1 < n$ and $1 < n_2 < n$. We claim that one of the integers n_1, n_2 must be less than or equal to $\sqrt{n}$. If not, then $n_1 > \sqrt{n}$ and $n_2 > \sqrt{n}$ give us the contradiction $n = n_1 n_2 > (\sqrt{n})(\sqrt{n}) = n$. Without loss of generality, we shall assume that $n_1 \le \sqrt{n}$. If n_1 is prime, the result follows. If n_1 is not prime, then by Lemma 4.1 there exists a prime $p < n_1$ where $p|n_1$. So $p|n$ and $p \le \sqrt{n}$.

EXERCISES 4.3

1. Verify the remaining parts of Theorem 4.3.

2. Let $a, b, c, d \in \mathbf{Z}^+$. Prove that (a) $[(a|b) \wedge (c|d)] \Rightarrow ac|bd$; (b) $a|b \Rightarrow ac|bc$; and (c) $ac|bc \Rightarrow a|b$.

3. If p, q are primes, prove that $p|q$ if and only if $p = q$.

4. If $a, b, c \in \mathbf{Z}^+$ and $a|bc$, does it follow that $a|b$ or $a|c$?

5. For all integers a, b, and c, prove that if $a \nmid bc$, then $a \nmid b$ and $a \nmid c$.

6. Let $n \in \mathbf{Z}^+$ where $n \ge 2$. Prove that if $a_1, a_2, \ldots, a_n$, $b_1, b_2, \ldots, b_n \in \mathbf{Z}^+$ and $a_i|b_i$ for all $1 \le i \le n$, then $(a_1 a_2 \cdots a_n)|(b_1 b_2 \cdots b_n)$.

7. a) Find three positive integers a, b, c such that $31|(5a + 7b + 11c)$.

b) If $a, b, c \in \mathbf{Z}$ and $31|(5a + 7b + 11c)$, prove that $31|(21a + 17b + 9c)$.

8. A grocery store runs a weekly contest to promote sales. Each customer who purchases more than \$20 worth of groceries receives a game card with 12 numbers on it; if any of these numbers sum to exactly 500, then that customer receives a \$500 shopping spree (at the grocery store). After purchasing \$22.83 worth of groceries at this store, Eleanor receives her game card on which are printed the following 12 numbers: 144, 336, 30, 66, 138, 162, 318, 54, 84, 288, 126, and 456. Has Eleanor won a \$500 shopping spree?

9. Let $a, b \in \mathbf{Z}^+$. If $b|a$ and $b|(a + 2)$, prove that $b = 1$ or $b = 2$.

10. If $n \in \mathbf{Z}^+$, and n is odd, prove that $8|(n^2 - 1)$.

11. If $a, b \in \mathbf{Z}^+$, and both are odd, prove that $2|(a^2 + b^2)$ but $4 \nmid (a^2 + b^2)$.

12. Determine the quotient q and remainder r for each of the following, where a is the dividend and b is the divisor.

a) $a = 23$, $b = 7$ **b)** $a = -115$, $b = 12$

c) $a = 0$, $b = 42$ **d)** $a = 434$, $b = 31$

13. If $n \in \mathbf{N}$, prove that $3|(7^n - 4^n)$.

14. Write each of the following (base-10) integers in base 2, base 4, and base 8.

a) 137   **b)** 6243   **c)** 12,345

15. Write each of the following (base-10) integers in base 2 and base 16.

a) 22   **b)** 527   **c)** 1234   **d)** 6923

16. Convert each of the following hexadecimal numbers to base 2 and base 10.

a) A7   **b)** 4C2   **c)** 1C2B   **d)** A2DFE

17. Convert each of the following binary numbers to base 10 and base 16.

a) 11001110   **b)** 00110001

c) 11110000   **d)** 01010111

18. For what base do we find that $251 + 445 = 1026$?

19. Find all $n \in \mathbf{Z}^+$ where n divides $5n + 18$.

20. Write each of the following integers in two's complement representation. Here the results are eight-bit patterns.

a) 15   **b)** -15 **c)** 100

d) -65 **e)** 127   **f)** -128

21. If a machine stores integers by the two's complement method, what are the largest and smallest integers that it can store if it uses bit patterns of (a) 4 bits? (b) 8 bits? (c) 16 bits? (d) 32 bits? (e) 2^n bits, $n \in \mathbf{Z}^+$?

22. In each of the following problems, we are using four-bit patterns for the two's complement representations of the integers from -8 to 7. Solve each problem (if possible), and then convert the results to base 10 to check your answers. Watch for any overflow errors.

a)   0101
    $+\ 0001$

b)   1101
    $+\ 1110$

c)   0111
        + 1000

d)   1101
        + 1010

23. If $a, x, y \in \mathbf{Z}$, and $a \neq 0$, prove that $ax = ay \Rightarrow x = y$.

24. Write a computer program (or develop an algorithm) to convert a positive integer in base 10 to base b, where $2 \leq b \leq 9$.

25. The Division Algorithm can be generalized as follows: For $a, b \in \mathbf{Z}$, $b \neq 0$, there exist unique $q, r \in \mathbf{Z}$ with $a = qb + r$, $0 \leq r < |b|$. Using Theorem 4.5, verify this generalized form of the algorithm for $b < 0$.

26. Write a computer program (or develop an algorithm) to convert a positive integer in base 10 to base 16.

27. For $n \in \mathbf{Z}^+$, write a computer program (or develop an algorithm) that lists all positive divisors of n.

28. Define the set $X \subseteq \mathbf{Z}^+$ recursively as follows:

1) $3 \in X$; and

2) If $a, b \in X$, then $a + b \in X$.

Prove that $X = \{3k \mid k \in \mathbf{Z}^+\}$, the set of all positive integers divisible by 3.

29. Let $n \in \mathbf{Z}^+$ with $n = r_k \cdot 10^k + \cdots + r_2 \cdot 10^2 + r_1 \cdot 10 + r_0$ (the base-10 representation of n). Prove that

a) $2|n$ if and only if $2|r_0$

b) $4|n$ if and only if $4|(r_1 \cdot 10 + r_0)$

c) $8|n$ if and only if $8|(r_2 \cdot 10^2 + r_1 \cdot 10 + r_0)$

State a general theorem suggested by these results.

4.4
The Greatest Common Divisor:
The Euclidean Algorithm

Continuing with the division operation developed in Section 4.3, we turn our attention to the divisors of a pair of integers.

Definition 4.2   For $a, b \in \mathbf{Z}$, a positive integer c is said to be a *common divisor of a and b* if $c|a$ and $c|b$.

EXAMPLE 4.32   The common divisors of 42 and 70 are 1, 2, 7, and 14, and 14 is the *greatest* of the common divisors.

Definition 4.3   Let $a, b \in \mathbf{Z}$, where either $a \neq 0$ or $b \neq 0$. Then $c \in \mathbf{Z}^+$ is called a *greatest common divisor* of a, b if

a) $c|a$ and $c|b$ (that is, c is a common divisor of a, b), and

b) for any common divisor d of a and b, we have $d|c$.

The result in Example 4.32 satisfies these conditions. That is, 14 divides both 42 and 70, and any common divisor of 42 and 70 — namely, 1, 2, 7, and 14 — divides 14. However, this example deals with two small integers. What would we do with two integers each having 20 digits? We consider the following questions.

1) Given $a, b \in \mathbf{Z}$, where at least one of a, b is not 0, does a greatest common divisor of a and b always exist? If so, how does one find such an integer?

2) How many greatest common divisors can a pair of integers have?

In dealing with these questions, we concentrate on $a, b \in \mathbf{Z}^+$.

THEOREM 4.6   For all $a, b \in \mathbf{Z}^+$, there exists a unique $c \in \mathbf{Z}^+$ that is *the* greatest common divisor of a, b.

Proof: Given $a, b \in \mathbf{Z}^+$, let $S = \{as + bt \mid s, t \in \mathbf{Z}, as + bt > 0\}$. Since $S \neq \emptyset$, by the Well-Ordering Principle S has a least element c. We claim that c is a greatest common divisor of a, b.

Since $c \in S$, $c = ax + by$, for some $x, y \in \mathbf{Z}$. Consequently, if $d \in \mathbf{Z}$ and $d \mid a$ and $d \mid b$, then by Theorem 4.3(f) $d \mid (ax + by)$, so $d \mid c$.

If $c \nmid a$, we can use the division algorithm to write $a = qc + r$, with $q, r \in \mathbf{Z}^+$ and $0 < r < c$. Then $r = a - qc = a - q(ax + by) = (1 - qx)a + (-qy)b$, so $r \in S$, contradicting the choice of c as the least element of S. Consequently, $c \mid a$, and by a similar argument, $c \mid b$.

Hence all $a, b \in \mathbf{Z}^+$ have a greatest common divisor. If c_1, c_2 both satisfy the two conditions of Definition 4.3, then with c_1 as a greatest common divisor, and c_2 as a common divisor, it follows that $c_2 \mid c_1$. Reversing roles, we find that $c_1 \mid c_2$, and so we conclude from Theorem 4.3(b) that $c_1 = c_2$ because $c_1, c_2 \in \mathbf{Z}^+$.

We now know that for all $a, b \in \mathbf{Z}^+$, the greatest common divisor of a, b exists — and it is unique. This number will be denoted by $\gcd(a, b)$. Here $\gcd(a, b) = \gcd(b, a)$; and for each $a \in \mathbf{Z}$, if $a \neq 0$, then $\gcd(a, 0) = |a|$. Also when $a, b \in \mathbf{Z}^+$, we have $\gcd(-a, b) = \gcd(a, -b) = \gcd(-a, -b) = \gcd(a, b)$. Finally, $\gcd(0, 0)$ is not defined and is of no interest to us.

From Theorem 4.6 we see that not only does $\gcd(a, b)$ exist but that $\gcd(a, b)$ is also the *smallest positive integer* we can write as a *linear combination* of a and b. However, we must realize that if $a, b, c \in \mathbf{Z}^+$ and $c = ax + by$ for some $x, y \in \mathbf{Z}$, then we do *not* necessarily know that c is $\gcd(a, b)$ — unless we somehow also know that c is the smallest positive integer that can be written as such a linear combination of a and b.

Finally, integers a and b are called *relatively prime* when $\gcd(a, b) = 1$ — that is, when there exist $x, y \in \mathbf{Z}$ with $ax + by = 1$.

EXAMPLE 4.33

Since $\gcd(42, 70) = 14$, we can find $x, y \in \mathbf{Z}$ with $42x + 70y = 14$, or $3x + 5y = 1$. By inspection, $x = 2$, $y = -1$ is a solution; $3(2) + 5(-1) = 1$. But for $k \in \mathbf{Z}$, $1 = 3(2 - 5k) + 5(-1 + 3k)$, so $14 = 42(2 - 5k) + 70(-1 + 3k)$, and the solutions for x, y are not unique.

In general, if $\gcd(a, b) = d$, then $\gcd((a/d), (b/d)) = 1$. (Verify this!) If $(a/d)x_0 + (b/d)y_0 = 1$, then $1 = (a/d)(x_0 - (b/d)k) + (b/d)(y_0 + (a/d)k)$, for each $k \in \mathbf{Z}$. So $d = a(x_0 - (b/d)k) + b(y_0 + (a/d)k)$, yielding infinitely many solutions to $ax + by = d$.

The preceding example and the prior observations work well enough when a, b are fairly small. But how does one find $\gcd(a, b)$ for some arbitrary $a, b \in \mathbf{Z}^+$? If $a \mid b$, then $\gcd(a, b) = a$; and if $b \mid a$, then $\gcd(a, b) = b$ — otherwise, we turn to the following result, which we owe to Euclid.

THEOREM 4.7

Euclidean Algorithm. Let $a, b \in \mathbf{Z}^+$. Set $r_0 = a$ and $r_1 = b$ and apply the division algorithm n times as follows:

$$r_0 = q_1 r_1 + r_2, \qquad\qquad 0 < r_2 < r_1$$
$$r_1 = q_2 r_2 + r_3, \qquad\qquad 0 < r_3 < r_2$$
$$r_2 = q_3 r_3 + r_4, \qquad\qquad 0 < r_4 < r_3$$

$$\cdots \cdots \cdots \cdots \cdots$$

$$r_i = q_{i+1} r_{i+1} + r_{i+2}, \qquad 0 < r_{i+2} < r_{i+1}$$

$$\cdots \cdots \cdots \cdots \cdots$$

$$r_{n-2} = q_{n-1} r_{n-1} + r_n, \qquad 0 < r_n < r_{n-1}$$
$$r_{n-1} = q_n r_n.$$

Then r_n, the last nonzero remainder, equals $\gcd(a, b)$.

Proof: To verify that $r_n = \gcd(a, b)$, we establish the two conditions of Definition 4.3.

Start with the first division process listed (where $r_0 = a$ and $r_1 = b$). If $c|r_0$ and $c|r_1$, then as $r_0 = q_1 r_1 + r_2$, it follows that $c|r_2$. Next $[(c|r_1) \wedge (c|r_2)] \Rightarrow c|r_3$, because $r_1 = q_2 r_2 + r_3$. Continuing down through the division processes, we get to where $c|r_{n-2}$ and $c|r_{n-1}$. From the next-to-last equation, we conclude that $c|r_n$, and this verifies condition (b) of Definition 4.3.

To establish condition (a) we go in reverse order. From the last equation, $r_n|r_{n-1}$, and so $r_n|r_{n-2}$, because $r_{n-2} = q_{n-1} r_{n-1} + r_n$. Continuing up through the equations, we get to where $r_n|r_4$ and $r_n|r_3$, so $r_n|r_2$. Then $[(r_n|r_3) \wedge (r_n|r_2)] \Rightarrow r_n|r_1$ (that is, $r_n|b$), and finally $[(r_n|r_2) \wedge (r_n|r_1)] \Rightarrow r_n|r_0$, (that is, $r_n|a$). Hence $r_n = \gcd(a, b)$.

We have now used the word *algorithm* in describing the statements set forth in Theorems 4.5 and 4.7. This term will recur frequently throughout other chapters of this text, so it may be a good idea to consider just what it connotes.

First and foremost, an *algorithm* is a list of *precise* instructions designed to solve a particular *type* of problem—not just one special case. In general, we expect all of our algorithms to receive *input* and provide the needed result(s) as *output*. Also, an algorithm should provide the same result whenever we repeat the value(s) for the input. This happens when the list of instructions is such that each intermediate result that comes about from the execution of each instruction is unique, depending on only the (initial) input and on any results that may have been derived at any preceding instructions. In order to accomplish this any possible vagueness must be eliminated from the algorithm; the instructions must be described in a simple yet unambiguous manner, a manner that can be executed by a machine. Finally, our algorithms cannot go on indefinitely. They must terminate after the execution of a *finite* number of instructions.

In Theorem 4.7 we are confronted with the determination of the greatest common divisor of any two positive integers. Hence this algorithm receives the two positive integers a, b as its input and generates their greatest common divisor as the output.

The use of the word *algorithm* in Theorem 4.5 is based on tradition. As stated, it does not provide the precise instructions we need to determine the output we want. (We mentioned this fact prior to Example 4.26.) To eliminate this shortcoming of Theorem 4.5, however, we set forth the instructions in the pseudocode procedure of Fig. 4.10.

We now apply the Euclidean algorithm in the following five examples.

EXAMPLE 4.34

Find the greatest common divisor of 250 and 111, and express the result as a linear combination of these integers.

$$250 = 2(111) + 28, \qquad 0 < 28 < 111$$
$$111 = 3(28) + 27, \qquad 0 < 27 < 28$$
$$28 = 1(27) + 1, \qquad 0 < 1 < 27$$
$$27 = 27(1) + 0.$$

So 1 is the last nonzero remainder. Therefore $\gcd(250, 111) = 1$, and 250 and 111 are relatively prime. Working backward from the third equation, we have $1 = 28 - 1(27) = 28 - 1[111 - 3(28)] = (-1)(111) + 4(28) = (-1)(111) + 4[250 - 2(111)] = 4(250) - 9(111) = 250(4) + 111(-9)$, a linear combination of 250 and 111.

This expression of 1 as a linear combination of 250 and 111 is not unique, for $1 = 250[4 - 111k] + 111[-9 + 250k]$, for any $k \in \mathbf{Z}$.

We also have

$$\gcd(-250, 111) = \gcd(250, -111) = \gcd(-250, -111) = \gcd(250, 111) = 1.$$

Our next example is somewhat more general, as it concerns the greatest common divisor for an infinite number of pairs of integers.

EXAMPLE 4.35

For any $n \in \mathbf{Z}^+$, prove that the integers $8n + 3$ and $5n + 2$ are relatively prime.

When $n = 1$ we find that $\gcd(8n + 3, 5n + 2) = \gcd(11, 7) = 1$.

For $n \geq 2$ we have $8n + 3 > 5n + 2$, and as in the previous example, we may write

$$8n + 3 = 1(5n + 2) + (3n + 1), \qquad 0 < 3n + 1 < 5n + 2$$
$$5n + 2 = 1(3n + 1) + (2n + 1), \qquad 0 < 2n + 1 < 3n + 1$$
$$3n + 1 = 1(2n + 1) + n, \qquad 0 < n < 2n + 1$$
$$2n + 1 = 2(n) + 1, \qquad 0 < 1 < n$$
$$n = n(1) + 0.$$

Consequently, the last nonzero remainder is 1, so $\gcd(8n + 3, 5n + 2) = 1$ for all $n \geq 1$. But we could also have arrived at this conclusion if we had noticed that

$$(8n + 3)(-5) + (5n + 2)(8) = -15 + 16 = 1.$$

And since 1 is expressed as a linear combination of $8n + 3$ and $5n + 2$, and no *smaller* positive integer can have this property, it follows that the greatest common divisor of $8n + 3$ and $5n + 2$ is 1, for any positive integer n.

EXAMPLE 4.36

At this point we shall use the Euclidean algorithm to develop a procedure (in pseudocode) that will find $\gcd(a, b)$ for all $a, b \in \mathbf{Z}^+$. The procedure in Fig. 4.11 employs the binary operation **mod,** where for $x, y \in \mathbf{Z}^+$, x **mod** $y = $ the remainder after x is divided by y. For example, 7 **mod** 3 is 1, and 18 **mod** 5 is 3. (We shall deal with "the arithmetic of remainders" in more detail in Chapter 14.)

```
procedure gcd (a, b: positive integers)
begin
  r := a mod b
  d := b
  while r > 0 do
    begin
      c := d
      d := r
      r := c mod d
    end
end {gcd(a, b) is d, the last nonzero remainder}
```

Figure 4.11

Meanwhile, if we call this procedure for $a = 168$ and $b = 456$, the procedure first assigns r the value 168 **mod** 456 = 168 and d the value 456. Since $r > 0$ the code in the **while** loop is executed (for the first time) and we obtain the following: $c = 456$, $d = 168$,

$r = 456 \bmod 168 = 120$. We then find that the code in the **while** loop is executed three more times with the following results:

(2nd pass): $c = 168, d = 120, r = 168 \bmod 120 = 48$

(3rd pass): $c = 120, d = 48, r = 120 \bmod 48 = 24$

(4th pass): $c = 48, d = 24, r = 48 \bmod 24 = 0$.

Since r is now 0, the procedure tells us that $\gcd(a, b) = \gcd(168, 456) = 24$, the final value of d (the last nonzero remainder).

EXAMPLE 4.37

Griffin has two unmarked containers. One container holds 17 ounces and the other holds 55 ounces. Explain how Griffin can use his two containers to measure exactly one ounce.

From the Euclidean algorithm we find that

$$55 = 3(17) + 4, \qquad 0 < 4 < 17$$
$$17 = 4(4) + 1, \qquad 0 < 1 < 4.$$

Therefore $1 = 17 - 4(4) = 17 - 4[55 - 3(17)] = 13(17) - 4(55)$. Consequently, Griffin must fill his smaller (17-ounce) container 13 times and empty the contents (for the first 12 times) into the larger container. (Griffin empties the larger container whenever it is full.) Before he fills the smaller container for the thirteenth time, Griffin has $12(17) - 3(55) = 204 - 165 = 39$ ounces of water in the larger (55-ounce) container. After he fills the smaller container for the thirteenth time, he will empty 16 $(= 55 - 39)$ ounces from this container, filling the larger container. Exactly one ounce will be left in the smaller container.

EXAMPLE 4.38

Assisting students in programming classes, Brian finds that on the average he can help a student debug a Java program in six minutes, but it takes 10 minutes to debug a program written in C++. If he works continuously for 104 minutes and doesn't waste any time, how many programs can he debug in each language?

Here we seek integers $x, y \geq 0$, where $6x + 10y = 104$, or $3x + 5y = 52$. As $\gcd(3, 5) = 1$, we can write $1 = 3(2) + 5(-1)$, so $52 = 3(104) + 5(-52) = 3(104 - 5k) + 5(-52 + 3k)$, $k \in \mathbf{Z}$. In order to obtain $0 \leq x = 104 - 5k$ and $0 \leq y = -52 + 3k$, we must have $(52/3) \leq k \leq (104/5)$. So $k = 18, 19, 20$ and there are three possible solutions:

a) $(k = 18)$: $x = 14,$ $y = 2$ **b)** $(k = 19)$: $x = 9,$ $y = 5$

c) $(k = 20)$: $x = 4,$ $y = 8$

The equation in Example 4.38 is an example of a *Diophantine equation*: a linear equation requiring integer solutions. This type of equation was first investigated by the Greek algebraist Diophantus, who lived in the third century A.D.

Having solved one such equation, we seek to discover when a Diophantine equation has a solution. The proof is left to the reader.

THEOREM 4.8

If $a, b, c \in \mathbf{Z}^+$, the Diophantine equation $ax + by = c$ has an integer solution $x = x_0, y = y_0$ if and only if $\gcd(a, b)$ divides c.

We close this section with a concept that is related to the greatest common divisor.

Definition 4.4

For $a, b, c \in \mathbf{Z}^+$, c is called a *common multiple* of a, b if c is a multiple of both a and b. Furthermore, c is the *least common multiple* of a, b if it is the smallest of all positive integers that are common multiples of a, b. We denote c by $\text{lcm}(a, b)$.

If $a, b \in \mathbf{Z}^+$, then the product ab is a common multiple of both a and b. Consequently, the set of all (positive) common multiples of a, b is nonempty. So it follows from the Well-Ordering Principle that the $\text{lcm}(a, b)$ does exist.

EXAMPLE 4.39

a) Since $12 = 3 \cdot 4$ and no other smaller positive integer is a multiple of both 3 and 4, we have $\text{lcm}(3, 4) = 12 = \text{lcm}(4, 3)$. However, $\text{lcm}(6, 15) \neq 90$ — for although 90 is a multiple of both 6 and 15, there is a smaller multiple, namely, 30. And since no other common multiple of 6 and 15 is smaller than 30, it follows that $\text{lcm}(6, 15) = 30$.

b) For all $n \in \mathbf{Z}^+$, we find that $\text{lcm}(1, n) = \text{lcm}(n, 1) = n$.

c) When $a, n \in \mathbf{Z}^+$, we have $\text{lcm}(a, na) = na$. [This statement is a generalization of part (b). The earlier statement follows from this one when $a = 1$.]

d) If $a, m, n \in \mathbf{Z}^+$ with $m \leq n$, then $\text{lcm}(a^m, a^n) = a^n$. [And $\gcd(a^m, a^n) = a^m$.]

THEOREM 4.9

Let $a, b, c \in \mathbf{Z}^+$, with $c = \text{lcm}(a, b)$. If d is a common multiple of a and b, then $c \mid d$.

Proof: If not, then because of the division algorithm we can write $d = qc + r$, where $0 < r < c$. Since $c = \text{lcm}(a, b)$, it follows that $c = ma$ for some $m \in \mathbf{Z}^+$. Also, $d = na$ for some $n \in \mathbf{Z}^+$, because d is a multiple of a. Consequently, $na = qma + r \Rightarrow (n - qm)a = r > 0$, and r is a multiple of a. In a similar way r is seen to be a multiple of b, so r is a common multiple of a, b. But with $0 < r < c$, we contradict the claim that c is the least common multiple of a, b. Hence $c \mid d$.

Our last result for this section ties together the concepts of the greatest common divisor and the least common multiple. Furthermore, it provides us with a way to calculate $\text{lcm}(a, b)$ for all $a, b \in \mathbf{Z}^+$. The proof of this result is left to the reader.

THEOREM 4.10

For all $a, b \in \mathbf{Z}^+$, $ab = \text{lcm}(a, b) \cdot \gcd(a, b)$.

EXAMPLE 4.40

By virtue of Theorem 4.10 we have the following:

a) For all $a, b \in \mathbf{Z}^+$, if a, b are relatively prime, then $\text{lcm}(a, b) = ab$.

b) The computations in Example 4.36 establish the fact that $\gcd(168, 456) = 24$. As a result we find that

$$\text{lcm}(168, 456) = \frac{(168)(456)}{24} = 3{,}192.$$

EXERCISES 4.4

1. For each of the following pairs $a, b \in \mathbf{Z}^+$, determine $\gcd(a, b)$ and express it as a linear combination of a, b.

a) $231, 1820$ b) $1369, 2597$ c) $2689, 4001$

2. For $a, b \in \mathbf{Z}^+$ and $s, t \in \mathbf{Z}$, what can we say about $\gcd(a, b)$ if

a) $as + bt = 2$? b) $as + bt = 3$?

c) $as + bt = 4$? d) $as + bt = 6$?

3. For $a, b \in \mathbf{Z}^+$ and $d = \gcd(a, b)$, prove that

$$\gcd\left(\frac{a}{d}, \frac{b}{d}\right) = 1.$$

4. For $a, b, n \in \mathbf{Z}^+$, prove that $\gcd(na, nb) = n \gcd(a, b)$.

5. Let $a, b, c \in \mathbf{Z}^+$ with $c = \gcd(a, b)$. Prove that c^2 divides ab.

6. Let $n \in \mathbf{Z}^+$.

a) Prove that $\gcd(n, n + 2) = 1$ or 2.

b) What possible values can $\gcd(n, n + 3)$ have? What about $\gcd(n, n + 4)$?

c) If $k \in \mathbf{Z}^+$, what can we say about $\gcd(n, n + k)$?

7. For $a, b, c, d \in \mathbf{Z}^+$, prove that if $d = a + bc$, then

$$\gcd(b, d) = \gcd(a, b).$$

8. Let $a, b, c \in \mathbf{Z}^+$ with $\gcd(a, b) = 1$. If $a|c$ and $b|c$, prove that $ab|c$. Does the result hold if $\gcd(a, b) \neq 1$?

9. Let $a, b \in \mathbf{Z}$, where at least one of a, b is nonzero.

a) Using quantifiers, restate the definition for $c = \gcd(a, b)$, where $c \in \mathbf{Z}^+$.

b) Use the result in part (a) in order to decide when $c \neq \gcd(a, b)$ for some $c \in \mathbf{Z}^+$.

10. If a, b are relatively prime and $a > b$, prove that $\gcd(a - b, a + b) = 1$ or 2.

11. Let $a, b, c \in \mathbf{Z}^+$ with $\gcd(a, b) = 1$. If $a|bc$, prove that $a|c$.

12. Let $a, b \in \mathbf{Z}^+$ where $a \geq b$. Prove that $\gcd(a, b) = \gcd(a - b, b)$.

13. Prove that for any $n \in \mathbf{Z}^+$, $\gcd(5n + 3, 7n + 4) = 1$.

14. An executive buys \$2490 worth of presents for the children of her employees. For each girl she gets an art kit costing \$33; each boy receives a set of tools costing \$29. How many presents of each type did she buy?

15. After a weekend at the Mohegan Sun Casino, Gary finds that he has won \$1020 — in \$20 and \$50 chips. If he has more \$50 chips than \$20 chips, how many chips of each denomination could he possibly have?

16. Let $a, b \in \mathbf{Z}^+$. Prove that there exist $c, d \in \mathbf{Z}^+$ such that $cd = a$ and $\gcd(c, d) = b$ if and only if $b^2|a$.

17. Determine those values of $c \in \mathbf{Z}^+$, $10 < c < 20$, for which the Diophantine equation $84x + 990y = c$ has no solution. Determine the solutions for the remaining values of c.

18. Verify Theorems 4.8 and 4.10.

19. If $a, b \in \mathbf{Z}^+$ with $a = 630$, $\gcd(a, b) = 105$, and $\mathrm{lcm}(a, b) = 242$, 550, what is b?

20. For each pair a, b in Exercise 1, find $\mathrm{lcm}(a, b)$.

21. For each $n \in \mathbf{Z}^+$, what are $\gcd(n, n + 1)$ and $\mathrm{lcm}(n, n + 1)$?

22. Prove that $\mathrm{lcm}(na, nb) = n \, \mathrm{lcm}(a, b)$ for all $n, a, b \in \mathbf{Z}^+$.

4.5
The Fundamental Theorem of Arithmetic

In this section we extend Lemma 4.1 and show that for each $n \in \mathbf{Z}^+$, $n > 1$, either n is prime or n can be written as a product of primes, where the representation is unique up to order. This result, known as the *Fundamental Theorem of Arithmetic*, can be found in an equivalent form in Book IX of Euclid's *Elements*.

The following two lemmas will help us accomplish our goal.

LEMMA 4.2

If $a, b \in \mathbf{Z}^+$ and p is prime, then $p|ab \Rightarrow p|a$ or $p|b$.

Proof: If $p|a$, then we are finished. If not, then because p is prime, it follows that $\gcd(p, a) = 1$, and so there exist integers x, y with $px + ay = 1$. Then $b = p(bx) + (ab)y$, where $p|p$ and $p|ab$. So it follows from parts (d) and (e) of Theorem 4.3 that $p|b$.

LEMMA 4.3

Let $a_i \in \mathbf{Z}^+$ for all $1 \leq i \leq n$. If p is prime and $p|a_1 a_2 \cdots a_n$, then $p|a_i$ for some $1 \leq i \leq n$.

Proof: We leave the proof of this result to the reader.

Using Lemma 4.2 we now have another opportunity to establish a result by the method of proof by contradiction.

EXAMPLE 4.41

We want to show that $\sqrt{2}$ is irrational.

If not, we can write $\sqrt{2} = a/b$, where $a, b \in \mathbf{Z}^+$ and $\gcd(a, b) = 1$. Then $\sqrt{2} = a/b \Rightarrow 2 = a^2/b^2 \Rightarrow 2b^2 = a^2 \Rightarrow 2|a^2 \Rightarrow 2|a$. (Why?) Also, $2|a \Rightarrow a = 2c$ for some $c \in \mathbf{Z}^+$, so $2b^2 = a^2 = (2c)^2 = 4c^2$ and $b^2 = 2c^2$. But then $2|b^2 \Rightarrow 2|b$. Since 2 divides both a and b, it follows that $\gcd(a, b) \geq 2$ — but this contradicts the earlier claim that $\gcd(a, b) = 1$. [*Note:* The preceding proof for the irrationality of $\sqrt{2}$ was known to Aristotle (384–322 B.C.) and is similar to that given in Book X of Euclid's *Elements*.]

Before we turn to the main result for this section, let us point out that the integer 2 in the preceding example is not that special. The reader will be asked to show in the Section Exercises that in fact $\sqrt{p}$ is irrational for every prime p. Now that we have mentioned this fact, it is time to present the Fundamental Theorem of Arithmetic.

THEOREM 4.11

Every integer $n > 1$ can be written as a product of primes uniquely, up to the order of the primes. (Here a single prime is considered a product of one factor.)

Proof: The proof consists of two parts: The first part covers the existence of a prime factorization, and the second part deals with its uniqueness.

If the first part is not true, let $m > 1$ be the smallest integer not expressible as a product of primes. Since m is not a prime, we are able to write $m = m_1 m_2$, where $1 < m_1 < m$, $1 < m_2 < m$. But then m_1, m_2 can be written as products of primes, because they are less than m. Consequently, with $m = m_1 m_2$ we can obtain a prime factorization of m.

In order to establish the uniqueness of a prime factorization, we shall use the alternative form of the Principle of Mathematical Induction (Theorem 4.2). For the integer 2, we have a unique prime factorization, and assuming uniqueness of representation for $3, 4, 5, \ldots,$ $n - 1$, we suppose that $n = p_1^{s_1} p_2^{s_2} \cdots p_k^{s_k} = q_1^{t_1} q_2^{t_2} \cdots q_r^{t_r}$, where each p_i, $1 \leq i \leq k$, and each q_j, $1 \leq j \leq r$, is a prime. Also $p_1 < p_2 < \cdots < p_k$, and $q_1 < q_2 < \cdots < q_r$, and $s_i > 0$ for all $1 \leq i \leq k$, $t_j > 0$ for all $1 \leq j \leq r$.

Since $p_1|n$, we have $p_1|q_1^{t_1} q_2^{t_2} \cdots q_r^{t_r}$. By Lemma 4.3, $p_1|q_j$ for some $1 \leq j \leq r$. Because p_1 and q_j are primes, we have $p_1 = q_j$. In fact $j = 1$, for otherwise $q_1|n \Rightarrow q_1 = p_e$ for some $1 < e \leq k$ and $p_1 < p_e = q_1 < q_j = p_1$. With $p_1 = q_1$, we find that $n_1 = n/p_1 = p_1^{s_1-1} p_2^{s_2} \cdots p_k^{s_k} = q_1^{t_1-1} q_2^{t_2} \cdots q_r^{t_r}$. Since $n_1 < n$, by the induction hypothesis it follows that $k = r$, $p_i = q_i$ for $1 \leq i \leq k$, $s_1 - 1 = t_1 - 1$ (so $s_1 = t_1$), and $s_i = t_i$ for $2 \leq i \leq k$. Hence the prime factorization of n is unique.

This result is now used in the following five examples.

EXAMPLE 4.42

For the integer 980,220 we can determine the prime factorization as follows:

$$980{,}220 = 2^1(490{,}110) = 2^2(245{,}055) = 2^2 3^1(81{,}685) = 2^2 3^1 5^1(16{,}337)$$
$$= 2^2 3^1 5^1 17^1(961) = 2^2 \cdot 3 \cdot 5 \cdot 17 \cdot 31^2$$

EXAMPLE 4.43

Suppose that $n \in \mathbf{Z}^+$ and that

(*) $10 \cdot 9 \cdot 8 \cdot 7 \cdot 6 \cdot 5 \cdot 4 \cdot 3 \cdot 2 \cdot n = 21 \cdot 20 \cdot 19 \cdot 18 \cdot 17 \cdot 16 \cdot 15 \cdot 14.$

Since 17 is a prime factor of the integer on the right-hand side of Eq. (*) it must also be a factor for the left-hand side (by the uniqueness part of the Fundamental Theorem of

Arithmetic). But 17 does not divide any of the factors 10, 9, 8, . . . , 3 or 2, so it follows that $17|n$. (A similar argument shows us that $19|n$).

EXAMPLE 4.44

For $n \in \mathbf{Z}^+$, we want to count the number of positive divisors of n. For example, the number 2 has two positive divisors: 1 and itself. Likewise, 1 and 3 are the only positive divisors of 3. In the case of 4, we find the three positive divisors 1, 2, and 4.

To determine the result for each $n \in \mathbf{Z}^+$, $n > 1$, we use Theorem 4.11 and write $n = p_1^{e_1} p_2^{e_2} \cdots p_k^{e_k}$, where for each $1 \le i \le k$, p_i is a prime and $e_i > 0$. If $m|n$, then $m = p_1^{f_1} p_2^{f_2} \cdots p_k^{f_k}$, where $0 \le f_i \le e_i$ for all $1 \le i \le k$. So by the rule of product, the number of positive divisors of n is

$$(e_1 + 1)(e_2 + 1) \cdots (e_k + 1).$$

For example, since $29{,}338{,}848{,}000 = 2^8 3^5 5^3 7^3 11$, we find that $29{,}338{,}848{,}000$ has $(8 + 1)(5 + 1)(3 + 1)(3 + 1)(1 + 1) = (9)(6)(4)(4)(2) = 1728$ positive divisors.

Should we want to know how many of these 1728 divisors are multiples of $360 = 2^3 3^2 5$, then we must realize that we want to count the integers of the form $2^{t_1} 3^{t_2} 5^{t_3} 7^{t_4} 11^{t_5}$ where

$$3 \le t_1 \le 8, \qquad 2 \le t_2 \le 5, \qquad 1 \le t_3 \le 3, \qquad 0 \le t_4 \le 3, \qquad \text{and} \qquad 0 \le t_5 \le 1.$$

Consequently, the number of positive divisors of $29{,}338{,}848{,}000$ that are divisible by 360 is

$$[(8 - 3) + 1][(5 - 2) + 1][(3 - 1) + 1][(3 - 0) + 1][(1 - 0) + 1]$$
$$= (6)(4)(3)(4)(2) = 576.$$

To determine how many of the 1728 positive divisors of $29{,}338{,}848{,}000$ are perfect squares, we need to consider all divisors of the form $2^{s_1} 3^{s_2} 5^{s_3} 7^{s_4} 11^{s_5}$, where each of s_1, s_2, s_3, s_4, s_5 is an even nonnegative integer. Consequently, here we have

5 choices for s_1 — namely, 0, 2, 4, 6, 8;

3 choices for s_2 — namely, 0, 2, 4;

2 choices for each of s_3, s_4 — namely, 0, 2; and

1 choice for s_5 — namely, 0.

It then follows that the number of positive divisors of $29{,}338{,}848{,}000$ that are perfect squares is $(5)(3)(2)(2)(1) = 60$.

For our next example we shall need the multiplicative counterpart of the Sigma-notation (for addition) that we first observed in Section 1.3. Here we use the capital Greek letter Π for the Pi-notation.

We can use the Pi-notation to express the product $x_1 x_2 x_3 x_4 x_5 x_6$, for example, as $\prod_{i=1}^{6} x_i$. In general, one can express the product of the $n - m + 1$ terms $x_m, x_{m+1}, x_{m+2}, \ldots, x_n$, where $m, n \in \mathbf{Z}$ and $m \le n$, as $\prod_{i=m}^{n} x_i$. As with the Sigma-notation the letter i is called the *index* of the product, and here this index accounts for all $n - m + 1$ integers starting with the *lower limit* m and continuing on up to (and including) the *upper limit* n.

This notation is demonstrated in the following:

1) $\prod_{i=3}^{7} x_i = x_3 x_4 x_5 x_6 x_7 = \prod_{j=3}^{7} x_j$, since there is nothing special about the letter i;

2) $\prod_{i=3}^{6} i = 3 \cdot 4 \cdot 5 \cdot 6 = 6!/2!$;

3) $\prod_{i=m}^{n} i = m(m+1)(m+2)\cdots(n-1)(n) = n!/(m-1)!$, for all $m, n \in \mathbf{Z}^+$ with $m \le n$; and

4) $\prod_{i=7}^{11} x_i = x_7 x_8 x_9 x_{10} x_{11} = \prod_{j=0}^{4} x_{7+j} = \prod_{j=0}^{4} x_{11-j}$.

EXAMPLE 4.45

If $m, n \in \mathbf{Z}^+$, let $m = p_1^{e_1} p_2^{e_2} \cdots p_t^{e_t}$ and $n = p_1^{f_1} p_2^{f_2} \cdots p_t^{f_t}$, with each p_i prime and $0 \le e_i$ and $0 \le f_i$ for all $1 \le i \le t$. Then if $a_i = \min\{e_i, f_i\}$, the minimum (or smaller) of e_i and f_i, and $b_i = \max\{e_i, f_i\}$, the maximum (or larger) of e_i and f_i, for all $1 \le i \le t$, we have

$$\gcd(m, n) = p_1^{a_1} p_2^{a_2} \cdots p_t^{a_t} = \prod_{i=1}^{t} p_i^{a_i} \quad \text{and} \quad \text{lcm}(m, n) = p_1^{b_1} p_2^{b_2} \cdots p_t^{b_t} = \prod_{i=1}^{t} p_i^{b_i}.$$

For example, let $m = 491{,}891{,}400 = 2^3 3^3 5^2 7^2 11^1 13^2$ and let $n = 1{,}138{,}845{,}708 = 2^2 3^2 7^1 11^2 13^3 17^1$. Then with $p_1 = 2$, $p_2 = 3$, $p_3 = 5$, $p_4 = 7$, $p_5 = 11$, $p_6 = 13$, and $p_7 = 17$, we find $a_1 = 2$, $a_2 = 2$, $a_3 = 0$ (the exponent of 5 in the prime factorization of n must be 0, because 5 does not appear in the prime factorization), $a_4 = 1$, $a_5 = 1$, $a_6 = 2$, and $a_7 = 0$. So

$$\gcd(m, n) = 2^2 3^2 5^0 7^1 11^1 13^2 17^0 = 468{,}468.$$

We also have

$$\text{lcm}(m, n) = 2^3 3^3 5^2 7^2 11^2 13^3 17^1 = 1{,}195{,}787{,}993{,}400.$$

Our final result for this section ties together the Fundamental Theorem of Arithmetic with the fact that any two consecutive integers are relatively prime (as observed in Exercise 21 for Section 4.4).

EXAMPLE 4.46

Here we seek an answer to the following question. Can we find three consecutive positive integers whose product is a perfect square — that is, do there exist $m, n \in \mathbf{Z}^+$ with $m(m+1)(m+2) = n^2$?

Suppose that such positive integers m, n do exist. We recall that $\gcd(m, m+1) = 1 = \gcd(m+1, m+2)$, so for any prime p, if $p \mid (m+1)$, then $p \nmid m$ and $p \nmid (m+2)$. Furthermore, if $p \mid (m+1)$, it follows that $p \mid n^2$. And since n^2 is a perfect square, by the Fundamental Theorem of Arithmetic, we find that the exponents on p in the prime factorizations of both $m+1$ and n^2 must be the same *even* integer. This is true for each prime divisor of $m+1$, so $m+1$ is a perfect square. With n^2 and $m+1$ both being perfect squares, we conclude that the product $m(m+2)$ is also a perfect square. However, the product $m(m+2)$ is such that $m^2 < m^2 + 2m = m(m+2) < m^2 + 2m + 1 = (m+1)^2$. Consequently, we find that $m(m+2)$ is wedged between two *consecutive* perfect squares — and is not equal to either of them. So $m(m+2)$ cannot be a perfect square, and there are no three consecutive positive integers whose product is a perfect square.

EXERCISES 4.5

1. Write each of the following integers as a product of primes $p_1^{n_1} p_2^{n_2} \cdots p_k^{n_k}$, where $0 < n_i$ for all $1 \le i \le k$

$$\text{and } p_1 < p_2 < \cdots < p_k.$$

a) 148,500   b) 7,114,800   c) 7,882,875

2. Determine the greatest common divisor and the least common multiple for each pair of integers in the preceding exercise.

3. Let $t \in \mathbf{Z}^+$ and $p_1, p_2, p_3, \ldots, p_t$ be distinct primes. If $m \in \mathbf{Z}^+$ has prime factorization $p_1^{e_1} p_2^{e_2} p_3^{e_3} \cdots p_t^{e_t}$, what is the prime factorization of (a) m^2? (b) m^3?

4. Verify Lemma 4.3.

5. Prove that $\sqrt{p}$ is irrational for any prime p.

6. The change machine at Cheryll's laundromat contains n quarters, $2n$ nickels, and $4n$ dimes, where $n \in \mathbf{Z}^+$. Find all values of n so that these coins total k dollars, where $k \in \mathbf{Z}^+$.

7. Find the number of positive divisors for each integer in Exercise 1.

8. a) How many positive divisors are there for

$$n = 2^{14}3^95^87^{10}11^313^537^{10}?$$

b) For the divisors in part (a), how many are

   i) divisible by $2^33^45^711^237^2$?
   ii) divisible by 1,166,400,000?
   iii) perfect squares?
   iv) perfect squares that are divisible by $2^23^45^211^2$?
   v) perfect cubes?
   vi) perfect cubes that are multiples of $2^{10}3^95^27^511^213^237^2$?
   vii) perfect squares and perfect cubes?

9. Let $m, n \in \mathbf{Z}^+$ with $mn = 2^43^45^37^111^313^1$. If $\text{lcm}(m, n) = 2^23^35^27^111^213^1$, what is $\gcd(m, n)$?

10. Extend the results in Example 4.45 and find the greatest common divisor and least common multiple for the three integers in Exercise 1.

11. How many positive integers n divide $100137n + 248396544$?

12. Let $a \in \mathbf{Z}^+$. Find the smallest value of a for which $2a$ is a perfect square and $3a$ is a perfect cube.

13. a) Let $a \in \mathbf{Z}^+$. Prove or disprove: (i) If $10|a^2$, then $10|a$; and (ii) If $4|a^2$, then $4|a$.

b) Generalize the true result(s) in part (a).

14. Let $a, b, c \in \{0, 1, 2, \ldots, 9\}$ with at least one of a, b, c nonzero. Prove that the six-digit integer $abcabc$ is divisible by at least three distinct primes.

15. Determine the smallest perfect square that is divisible by 7!

16. For all $n \in \mathbf{Z}^+$, prove that n is a perfect square if and only if n has an odd number of positive divisors.

17. Find the smallest positive integer n for which the product $1260 \times n$ is a perfect cube.

18. Two hundred coins numbered 1 to 200 are put in a row across the top of a cafeteria table. Two hundred students are assigned numbers (from 1 to 200) and are asked to turn over certain coins. The student assigned number 1 is supposed to turn over all the coins. The student assigned number 2 is supposed to turn over every other coin, starting with the second coin. In general, the student assigned the number n, for each $1 \le n \le 200$, is supposed to turn over every nth coin, starting with the nth coin.

a) How many times will the 200th coin be turned over?

b) Will any other coin(s) be turned over as many times as the 200th coin?

c) Will any coin be turned over more times than the 200th coin?

19. How many different products can one obtain by multiplying any two (distinct) integers in the set

a) $\{4, 8, 16, 32\}$?   **b)** $\{4, 8, 16, 32, 64\}$?

c) $\{4, 8, 9, 16, 27, 32, 64, 81, 243\}$?

d) $\{4, 8, 9, 16, 25, 27, 32, 64, 81, 125, 243, 625, 729, 3125\}$?

e) $\{p^2, p^3, p^4, p^5, p^6, q^2, q^3, q^4, q^5, q^6, r^2, r^3, r^4, r^5\}$, where p, q, and r are distinct primes?

20. Write a computer program (or develop an algorithm) to find the prime factorization of an integer $n > 1$.

21. In triangle ABC the length of side BC is 293. If the length of side AB is a perfect square, the length of side AC is a power of 2, and the length of side AC is twice the length of side AB, determine the perimeter of the triangle.

22. Express each of the following in simplest form.

a) $\displaystyle\prod_{i=1}^{10}(-1)^i$

b) $\displaystyle\prod_{i=1}^{2n+1}(-1)^i$, where $n \in \mathbf{Z}^+$

c) $\displaystyle\prod_{i=4}^{8}\frac{(i+1)(i+2)}{(i-1)(i)}$

d) $\displaystyle\prod_{i=n}^{2n}\frac{i}{2n-i+1}$, where $n \in \mathbf{Z}^+$

23. a) Let $n = 88,200$. In how many ways can one factor n as ab where $1 < a < n$, $1 < b < n$, and $\gcd(a, b) = 1$? (*Note:* Here order is not relevant. So, for example, $a = 8$, $b = 11,025$, and $a = 11,025$, $b = 8$ result in the same unordered factorization.)

b) Answer part (a) for $n = 970,200$.

c) Generalize the results in parts (a) and (b).

24. Use the Pi-notation to write each of the following.

a) $(1^2 + 1)(2^2 + 2)(3^2 + 3)(4^2 + 4)(5^2 + 5)$

b) $(1 + x)(1 + x^2)(1 + x^3)(1 + x^4)(1 + x^5)$

c) $(1 + x)(1 + x^3)(1 + x^5)(1 + x^7)(1 + x^9)(1 + x^{11})$

25. Prove that if $n \in \mathbf{Z}^+$ and $n \ge 2$, then

$$\prod_{i=2}^{n}\left(1 - \frac{1}{i^2}\right) = \frac{n+1}{2n}.$$

26. When does a positive integer n have exactly

a) two positive divisors?   **b)** three positive divisors?

c) four positive divisors?   **d)** five positive divisors?

27. Let $n \in \mathbf{Z}^+$. We say that n is a *perfect* integer if $2n$ equals the sum of all the positive divisors of n. For example, since $2(6) = 12 = 1 + 2 + 3 + 6$, it follows that 6 is a perfect integer.

a) Verify that 28 and 496 are perfect integers.

b) If $m \in \mathbf{Z}^+$ and $2^m - 1$ is prime, prove that $2^{m-1}(2^m - 1)$ is a perfect integer. [You may find the result from part (a) of Exercise 2 for Section 4.1 useful here.]

4.6
Summary and Historical Review

According to the Prussian mathematician Leopold Kronecker (1823–1891), "God made the integers, all the rest is the work of man . . . . All results of the profoundest mathematical investigation must ultimately be expressible in the simple form of properties of the integers." In the spirit of this quotation, we find in this chapter how the handiwork of the Almighty has been further developed by men and women over the last 24 centuries.

Starting in the fourth century B.C. we find in Euclid's *Elements* not only the geometry of our high school experience but also the fundamental ideas of number theory. Propositions 1 and 2 of Euclid's Book VII include an example of an algorithm to determine the greatest common divisor of two positive integers by using an efficient technique to solve, in a *finite* number of steps, a specific type of problem.

The term *algorithm*, like its predecessor *algorism*, was unknown to Euclid. In fact, this term did not enter the vocabulary of most people until the late 1950s when the computer revolution began to make its impact on society. The word comes from the name of the famous Islamic mathematician, astronomer, and textbook writer Abu Ja'far Mohammed ibn Mûsâ al-Khowârizmî (c. 780–850). The last part of his name, al-Khowârizmî, which is translated as "a man from the town of Khowârizm," gave rise to the term *algorism*. The word *algebra* comes from al-jabr, which is contained in the title of al-Khowârizmî's textbook *Kitab al-jabr w'al muquabala*. Translated into Latin during the thirteenth century, this book had a profound impact on the mathematics developed during the European Renaissance.

Euclid (c. 400 B.C.)

Al-Khowârizmî (c. 780–850)

As mentioned in Section 4.4, our use of the word *algorithm* connotes a precise step-by-step method for solving a problem in a finite number of steps. The first person credited with developing the concept of a computer algorithm was Augusta Ada Byron (1815–1852), the Countess of Lovelace. The only child of the famous poet Lord Byron and Annabella Millbanke, Augusta Ada was raised by a mother who encouraged her intellectual talents. Trained in mathematics by the likes of Augustus DeMorgan (1806–1871), she continued her studies by assisting the gifted English mathematician Charles Babbage (1792–1871) in the development of his design for an early computing machine — the "Analytical Engine."

The most complete accounts of this machine are found in her writings, wherein one finds a great deal of literary talent along with the essence of the modern computer algorithm. Further details on the work of Charles Babbage and Augusta Ada Byron Lovelace can be found in Chapter 2 of the work by S. Augarten [1].

Augusta Ada Byron, Countess of Lovelace (1815–1852)

In the century following Euclid, we find some number theory in the work of Eratosthenes. However, it was not until five centuries later that the first major new accomplishments in the field were made by Diophantus of Alexandria. In his work *Arithmetica*, his integer solutions of linear (and higher-order) equations stood as a mathematical beacon in number theory until the French mathematician Pierre de Fermat (1601–1665) came on the scene.

The problem we stated in Theorem 4.8 was investigated by Diophantus and further analyzed during the seventh century by Hindu mathematicians, but it was not actually solved completely until the 1860s, by Henry John Stephen Smith (1826–1883).

For more on some of these mathematicians and others who have worked in the theory of numbers, consult L. Dickson [4]. Chapter 5 in I. Niven, H. S. Zuckerman, and H. L. Montgomery [10] deals with the solutions of Diophantine equations and their applications.

In the work *Formulario Matematico*, published in 1889, Giuseppe Peano (1858–1932) formulated the set of nonnegative integers on the basis of three undefined terms: zero, number, and successor. His formulation is as follows:

a) Zero is a number.

b) For each number n, its successor is a number.

c) No number has zero as its successor.

d) If two numbers m, n have the same successor, then $m = n$.

e) If T is a set of numbers where $0 \in T$, and where the successor of n is in T whenever n is in T, then T is the set of all numbers.

In these postulates the notion of order (successor) and the technique called mathematical induction are seen to be intimately related to the idea of number (that is, nonnegative integer). Peano attributed the formulation to Richard Dedekind (1831–1916), who was the first to develop these ideas; nonetheless, these postulates are generally known as "Peano's postulates."

The first European to apply the Principle of Mathematical Induction in proofs was the Venetian scientist Francisco Maurocylus (1491–1575). His book, *Arithmeticorum Libri Duo* (published in 1575), contains a proof, by mathematical induction, that the sum of the first n positive odd integers is n^2. In the next century, Pierre de Fermat made further improvements on the technique in his work involving "the method of infinite descent." Blaise Pascal (c. 1653), in proving such combinatorial results as $C(n, k)/C(n, k + 1) = (k + 1)/(n - k)$, $0 \le k \le n - 1$, used induction and referred to the technique as the work of Maurocylus. The actual term *mathematical induction* was not used, however, until the nineteenth century when it appeared in the work of Augustus DeMorgan (1806–1871). In 1838 he described the process with great care and gave it the name *mathematical induction*. (An interesting survey on this topic is found in the article by W. H. Bussey [2].)

The text by B. K. Youse [13] illustrates many varied applications of the Principle of Mathematical Induction in algebra, geometry, and trigonometry. For more on the relevance of this method of proof to the problems of programming and the development of algorithms, the text by M. Wand [12] (especially Chapter 2) provides ample background and examples.

More on the theory of numbers can be found in the texts by G. H. Hardy and E. M. Wright [5], W. J. LeVeque [7, 8], and I. Niven, H. S. Zuckerman, and H. L. Montgomery [10]. At a level comparable to that of this chapter, Chapter 3 of V. H. Larney [6] provides an enjoyable introduction to this material. The text by K. H. Rosen [11] integrates applications in cryptography and computer science in its development of the subject. The journal article by M. J. Collison [3] examines the history of the Fundamental Theorem of Arithmetic. The articles in [9] recount some interesting developments in number theory.

REFERENCES

1. Augarten, Stan. *BIT by BIT, An Illustrated History of Computers*. New York: Ticknor & Fields, 1984.

2. Bussey, W. H. "Origins of Mathematical Induction." *American Mathematical Monthly* 24 (1917): pp. 199–207.

3. Collison, Mary Joan. "The Unique Factorization Theorem: From Euclid to Gauss." *Mathematics Magazine* 53 (1980): pp. 96–100.

4. Dickson, L. *History of the Theory of Numbers*. Washington, D.C.: Carnegie Institution of Washington, 1919. Reprinted by Chelsea, in New York, in 1950.

5. Hardy, Godfrey Harold, and Wright, Edward Maitland. *An Introduction to the Theory of Numbers*, 5th ed. Oxford: Oxford University Press, 1979.

6. Larney, Violet Hachmeister. *Abstract Algebra: A First Course*. Boston: Prindle, Weber & Schmidt, 1975.

7. LeVeque, William J. *Elementary Theory of Numbers*. Reading, Mass.: Addison-Wesley, 1962.

8. LeVeque, William J. *Topics in Number Theory*, Vols. I and II. Reading, Mass.: Addison-Wesley, 1956.

9. LeVeque, William J., ed. *Studies in Number Theory*. MAA Studies in Mathematics, Vol. 6. Englewood Cliffs, N.J.: Prentice-Hall, 1969. Published by the Mathematical Association of America.

10. Niven, Ivan, Zuckerman, Herbert S., and Montgomery, Hugh L. *An Introduction to the Theory of Numbers*, 5th ed. New York: Wiley, 1991.

11. Rosen, Kenneth H. *Elementary Number Theory*, 4th ed. Reading, Mass.: Addison-Wesley, 2000.

12. Wand, Mitchell. *Induction, Recursion, and Programming*. New York: Elsevier North Holland, 1980.

13. Youse, Bevan K. *Mathematical Induction*. Englewood Cliffs, N.J.: Prentice-Hall, 1964.

SUPPLEMENTARY EXERCISES

1. Let a, d be fixed integers. Determine a summation formula for $a + (a + d) + (a + 2d) + \cdots + (a + (n - 1)d)$, for $n \in \mathbf{Z}^+$. Verify your result by mathematical induction.

2. In the following pseudocode program segment the variables n and sum are integer variables. Following the execution of this program segment, which value of n is printed?

```
n := 3
sum := 0
while sum < 10,000 do
  begin
    n := n + 7
    sum := sum + n
  end
print n
```

3. Consider the following five equations.

1) $\qquad\qquad 1 = 1$

2) $\qquad 1 - 4 = -(1 + 2)$

3) $\qquad 1 - 4 + 9 = 1 + 2 + 3$

4) $\qquad 1 - 4 + 9 - 16 = -(1 + 2 + 3 + 4)$

5) $1 - 4 + 9 - 16 + 25 = 1 + 2 + 3 + 4 + 5$

Conjecture the general formula suggested by these five equations, and prove your conjecture.

4. For $n \in \mathbf{Z}^+$, prove each of the following by mathematical induction:

a) $5 \mid (n^5 - n)$ **b)** $6 \mid (n^3 + 5n)$

5. For all $n \in \mathbf{Z}^+$, let $S(n)$ be the open statement: $n^2 + n + 41$ is prime.

a) Verify that $S(n)$ is true for all $1 \le n \le 9$.

b) Does the truth of $S(k)$ imply that of $S(k + 1)$ for all $k \in \mathbf{Z}^+$?

6. For $n \in \mathbf{Z}^+$ define the sum s_n by the formula

$$s_n = \frac{1}{2!} + \frac{2}{3!} + \frac{3}{4!} + \cdots + \frac{(n - 1)}{n!} + \frac{n}{(n + 1)!}.$$

a) Verify that $s_1 = \frac{1}{2}$, $s_2 = \frac{5}{6}$, and $s_3 = \frac{23}{24}$.

b) Compute s_4, s_5, and s_6.

c) On the basis of your results in parts (a) and (b), conjecture a formula for the sum of the terms in s_n.

d) Verify your conjecture in part (c) for all $n \in \mathbf{Z}^+$ by the Principle of Mathematical Induction.

7. For all $n \in \mathbf{Z}$, $n \ge 0$, prove that

a) $2^{2n+1} + 1$ is divisible by 3.

b) $n^3 + (n + 1)^3 + (n + 2)^3$ is divisible by 9.

8. Let $n \in \mathbf{Z}^+$ where n is odd and n is not divisible by 5. Prove that there is a power of n whose units digit is 1.

9. Find the digits x, y, z where $(xyz)_9 = (zyx)_6$.

10. If $n \in \mathbf{Z}^+$, how many possible values are there for $\gcd(n, n + 3000)$?

11. If $n \in \mathbf{Z}^+$ and $n \ge 2$, prove that $2^n < \binom{2n}{n} < 4^n$.

12. If $n \in \mathbf{Z}^+$, prove that 57 divides $7^{n+2} + 8^{2n+1}$.

13. For all $n \in \mathbf{Z}^+$, show that if $n \ge 64$, then n can be written as a sum of 5's and/or 17's.

14. Determine all $a, b \in \mathbf{Z}$ such that $\frac{a}{7} + \frac{b}{12} = \frac{1}{84}$.

15. Given $r \in \mathbf{Z}^+$, write $r = r_0 + r_1 \cdot 10 + r_2 \cdot 10^2 + \cdots + r_n \cdot 10^n$, where $0 \le r_i \le 9$ for $0 \le i \le n - 1$, and $0 < r_n \le 9$.

a) Prove that $9 \mid r$ if and only if $9 \mid (r_n + r_{n-1} + \cdots + r_2 + r_1 + r_0)$.

b) Prove that $3 \mid r$ if and only if $3 \mid (r_n + r_{n-1} + \cdots + r_2 + r_1 + r_0)$.

c) If $t = 137486\underline{x}225$, where x is a single digit, determine the value(s) of x such that $3 \mid t$. Which values of x make t divisible by 9?

16. Frances spends $6.20 on candy for prizes in a contest. If a 10-ounce box of this candy costs $.50 and a 3-ounce box costs $.20, how many boxes of each size did she purchase?

17. a) How many positive integers can we express as a product of nine primes (repetitions allowed and order not relevant) where the primes may be chosen from $\{2, 3, 5, 7, 11\}$?

b) How many of the positive integers in part (a) have at least one occurrence of each of the five primes?

18. Find the product of all (positive) divisors of (a) 1000; (b) 5000; (c) 7000; (d) 9000; (e) $p^m q^n$, where p, q are distinct primes and $m, n \in \mathbf{Z}^+$; and (f) $p^m q^n r^k$, where p, q, r are distinct primes and $m, n, k \in \mathbf{Z}^+$.

19. a) Ten students enter a locker room that contains 10 lockers. The first student opens all the lockers. The second student changes the status (from closed to open, or vice versa) of every other locker, starting with the second locker. The third student then changes the status of every third locker, starting at the third locker. In general, for $1 < k \le 10$, the kth student changes the status of every kth locker, starting with the kth locker. After the tenth student has gone through the lockers, which lockers are left open?

b) Answer part (a) if 10 is replaced by $n \in \mathbf{Z}^+$, $n \ge 2$.

20. Let $A = \{a_1, a_2, a_3, a_4, a_5\} \subseteq \mathbf{Z}^+$. Prove that A contains a nonempty subset S where the sum of the elements in S is a multiple of 5. (Here it is possible to have a sum consisting of only one summand.)

21. Consider the set $\{1, 2, 3\}$. Here we may write $\{1, 2, 3\} = \{1, 2\} \cup \{3\}$, where $1 + 2 = 3$. For the set $\{1, 2, 3, 4\}$ we find that $\{1, 2, 3, 4\} = \{1, 4\} \cup \{2, 3\}$, where $1 + 4 = 2 + 3$.

However, things change when we examine the set $\{1, 2, 3, 4, 5\}$. In this case, if $C \subseteq \{1, 2, 3, 4, 5\}$ and we let s_C denote the sum of the elements in C, then we find that there is no way to write $\{1, 2, 3, 4, 5\} = A \cup B$, with $A \cap B = \emptyset$ and $s_A = s_B$.

a) For which $n \in \mathbf{Z}^+$, $n \geq 3$, can we write $\{1, 2, 3, \ldots, n\} = A \cup B$, with $A \cap B = \emptyset$ and $s_A = s_B$? (As above, s_A and s_B denote the sums of the elements in A and B, respectively.)

b) Let $n \in \mathbf{Z}^+$ with $n \geq 3$. If we can write $\{1, 2, 3, \ldots, n\} = A \cup B$ with $A \cap B = \emptyset$ and $s_A = s_B$, describe how such sets A and B can be determined.

22. Determine those integers n for which $\frac{5n-4}{6}$ and $\frac{7n+1}{4}$ are also integers.

23. Let $a, b \in \mathbf{Z}^+$.

a) Prove that if $a^2 | b^2$ then $a | b$.

b) Is it true that if $a^2 | b^3$ then $a | b$?

24. Let n be a fixed positive integer that satisfies the property: For all $a, b \in \mathbf{Z}^+$, if $n | ab$ then $n | a$ or $n | b$. Prove that $n = 1$ or n is prime.

25. Suppose that $a, b, k \in \mathbf{Z}^+$ and that k is not a power of 2.

a) Prove that if $a^k + b^k \neq 2$, then $a^k + b^k$ is composite.

b) If $n \in \mathbf{Z}^+$ and n is not a power of 2, prove that if $2^n + 1$ is prime, then n is prime.

For the next three exercises, recall that H_n, F_n, and L_n denote the nth harmonic, Fibonacci, and Lucas numbers, respectively.

26. Prove that for all $n \in \mathbf{N}$, $H_{2^n} \leq 1 + n$.

27. Prove that $F_n \leq (5/3)^n$ for all $n \in \mathbf{N}$.

28. For $n \in \mathbf{N}$, prove that

$$L_0 + L_1 + L_2 + \cdots + L_n = \sum_{i=0}^{n} L_i = L_{n+2} - 1.$$

29. a) For the five-digit integers (from 10000 to 99999) how many are palindromes and what is their sum?

b) Write a computer program to check the answer for the sum in part (a).

30. Let a, b be odd with $a > b$. Prove that $\gcd(a, b) = \gcd\left(\frac{a-b}{2}, b\right)$.

31. Let $n \in \mathbf{Z}^+$ with u the units digit of n. Prove that $7 | n$ if and only if $7 | (\frac{n-u}{10} - 2u)$.

32. Let $m, n \in \mathbf{Z}^+$ with $19m + 90 + 8n = 1998$. Determine m, n so that (a) n is minimal; (b) m is minimal.

33. Catrina selects three integers from $\{0, 1, 2, 3, 4, 5, 6, 7, 8, 9\}$ and then forms the six possible three-digit integers (leading zero allowed) they determine. For instance, for the selection 1, 3, and 7, she would form the integers 137, 173, 317, 371, 713, and 731. Prove that no matter which three integers she initially selects, it is not possible for all six of the resulting three-digit integers to be prime.

34. Consider the three-row and four-column table shown in Fig. 4.12. Show that it is possible to place eight of the nine integers 2, 3, 4, 7, 10, 11, 12, 13, 15 in the remaining eight cells of the table so that the average of the integers in each row is the same integer and the average of the integers in each column is the same integer. Specify which of the nine integers given cannot be used and show how the other eight integers are placed in the table.

Figure 4.12

35. Allen writes the consecutive integers $1, 2, 3, \ldots, n$ on a blackboard. Then Barbara erases one of these integers. If the average of the remaining integers is $35\frac{7}{17}$, what is n and what integer was erased?

36. Leslie selects a random integer between 1 and 100 (inclusive). Find the probability her selection is divisible by (a) 2 or 3; (b) 2, 3, or 5.

37. Let $m = p_1^{e_1} p_2^{e_2} p_3^{e_3} p_4^{e_4}$ and $n = p_1^{f_1} p_2^{f_2} p_3^{f_3} p_5^{f_5}$, where p_1, p_2, p_3, p_4, p_5 are distinct primes, and $e_1, e_2, e_3, e_4, f_1, f_2, f_3, f_5 \in \mathbf{Z}^+$. How many common divisors are there for m, n?

5

Relations and
Functions

In this chapter we extend the set theory of Chapter 3 to include the concepts of relation and function. Algebra, trigonometry, and calculus all involve functions. Here, however, we shall study functions from a set-theoretic approach that includes finite functions, and we shall introduce some new counting ideas in the study. Furthermore, we shall examine the concept of function complexity and its role in the study of the analysis of algorithms.

We take a path along which we shall find the answers to the following (closely related) six problems:

1) The Defense Department has seven different contracts that deal with a high-security project. Four companies can manufacture the distinct parts called for in each contract, and in order to maximize the security of the overall project, it is best to have all four companies working on some part. In how many ways can the contracts be awarded so that every company is involved?

2) How many seven-symbol quaternary (0, 1, 2, 3) sequences have at least one occurrence of each of the symbols 0, 1, 2, and 3?

3) An $m \times n$ *zero-one matrix* is a matrix A with m rows and n columns, such that in row i, for all $1 \le i \le m$, and column j, for all $1 \le j \le n$, the entry a_{ij} that appears is either 0 or 1. How many 7×4 zero-one matrices have exactly one 1 in each row and at least one 1 in each column? (The zero-one matrix is a data structure that arises in computer science. We shall learn more about it in later chapters.)

4) Seven (unrelated) people enter the lobby of a building which has four additional floors, and they all get on an elevator. What is the probability that the elevator must stop at every floor in order to let passengers off?

5) For positive integers m, n with $m < n$, prove that

$$\sum_{k=0}^{n}(-1)^k \binom{n}{n-k}(n-k)^m = 0.$$

6) For every positive integer n, verify that

$$n! = \sum_{k=0}^{n}(-1)^k \binom{n}{n-k}(n-k)^n.$$

Do you recognize the connection among the first four problems? The first three are the same problem in different settings. However, it is not obvious that the last two problems are related or that there is a connection between them and the first four. These identities, however, will be established using the same counting technique that we develop to solve the first four problems.

5.1
Cartesian Products and Relations

We start with an idea that was introduced earlier in Definition 3.11. However, we repeat the definition now in order to make the presentation here independent of this prior encounter.

Definition 5.1   For sets A, B the *Cartesian product*, or *cross product*, of A and B is denoted by $A \times B$ and equals $\{(a, b) | a \in A, b \in B\}$.

We say that the elements of $A \times B$ are *ordered pairs*. For (a, b), $(c, d) \in A \times B$, we have $(a, b) = (c, d)$ if and only if $a = c$ and $b = d$.

If A, B are finite, it follows from the rule of product that $|A \times B| = |A| \cdot |B|$. Although we generally will not have $A \times B = B \times A$, we will have $|A \times B| = |B \times A|$.

Here $A \subseteq \mathcal{U}_1$ and $B \subseteq \mathcal{U}_2$, and we may find that the universes are different — that is, $\mathcal{U}_1 \neq \mathcal{U}_2$. Also, even if A, $B \subseteq \mathcal{U}$, it is not necessary that $A \times B \subseteq \mathcal{U}$, so unlike the cases for union and intersection, here $\mathcal{P}(\mathcal{U})$ is not necessarily closed under this binary operation.

We can extend the definition of the Cartesian product, or cross product, to more than two sets. Let $n \in \mathbf{Z}^+$, $n \geq 3$. For sets $A_1, A_2, \ldots, A_n$, the *(n-fold) product* of $A_1, A_2, \ldots, A_n$ is denoted by $A_1 \times A_2 \times \cdots \times A_n$ and equals $\{(a_1, a_2, \ldots, a_n) | a_i \in A_i, 1 \leq i \leq n\}$.[†] The elements of $A_1 \times A_2 \times \cdots \times A_n$ are called ordered *n-tuples*, although we generally use the term *triple* in place of 3-tuple. As with ordered pairs, if $(a_1, a_2, \ldots, a_n)$, $(b_1, b_2, \ldots, b_n) \in A_1 \times A_2 \times \cdots \times A_n$, then $(a_1, a_2, \ldots, a_n) = (b_1, b_2, \ldots, b_n)$ if and only if $a_i = b_i$ for all $1 \leq i \leq n$.

EXAMPLE 5.1   Let $A = \{2, 3, 4\}$, $B = \{4, 5\}$. Then

a) $A \times B = \{(2, 4), (2, 5), (3, 4), (3, 5), (4, 4), (4, 5)\}$.

b) $B \times A = \{(4, 2), (4, 3), (4, 4), (5, 2), (5, 3), (5, 4)\}$.

c) $B^2 = B \times B = \{(4, 4), (4, 5), (5, 4), (5, 5)\}$.

d) $B^3 = B \times B \times B = \{(a, b, c) | a, b, c \in B\}$; for instance, $(4, 5, 5) \in B^3$.

EXAMPLE 5.2   The set $\mathbf{R} \times \mathbf{R} = \{(x, y) | x, y \in \mathbf{R}\}$ is recognized as the real plane of coordinate geometry and two-dimensional calculus. The subset $\mathbf{R}^+ \times \mathbf{R}^+$ is the interior of the first quadrant of this plane. Likewise $\mathbf{R}^3$ represents Euclidean three-space, where the three-dimensional interior of any sphere (of positive radius), two-dimensional planes, and one-dimensional lines are subsets of importance.

EXAMPLE 5.3   Once again let $A = \{2, 3, 4\}$ and $B = \{4, 5\}$, as in Example 5.1, and let $C = \{x, y\}$. The construction of the Cartesian product $A \times B$ can be represented pictorially with the aid of a *tree diagram*, as in part (a) of Fig. 5.1. This diagram proceeds from left to right. From

---

[†]When dealing with the Cartesian product of three or more sets, we must be careful about the lack of associativity. In the case of three sets, for example, there is a difference between any two of the sets $A_1 \times A_2 \times A_3$, $(A_1 \times A_2) \times A_3$, and $A_1 \times (A_2 \times A_3)$ because their respective elements are ordered triples (a_1, a_2, a_3), and the distinct ordered pairs $((a_1, a_2), a_3)$ and $(a_1, (a_2, a_3))$. Although such differences are important in certain instances, we shall not concentrate on them here and shall always use the nonparenthesized form $A_1 \times A_2 \times A_3$. This will also be our convention when dealing with the Cartesian product of four or more sets.

the left-most endpoint, three branches originate — one for each of the elements of A. Then from each point, labeled 2, 3, 4, two branches emanate — one for each of the elements 4, 5 of B. The six ordered pairs at the right endpoints constitute the elements (ordered pairs) of $A \times B$. Part (b) of the figure provides a tree diagram to demonstrate the construction of $B \times A$. Finally, the tree diagram in Fig. 5.1 (c) shows us how to envision the construction of $A \times B \times C$, and demonstrates that $|A \times B \times C| = 12 = 3 \times 2 \times 2 = |A||B||C|$.

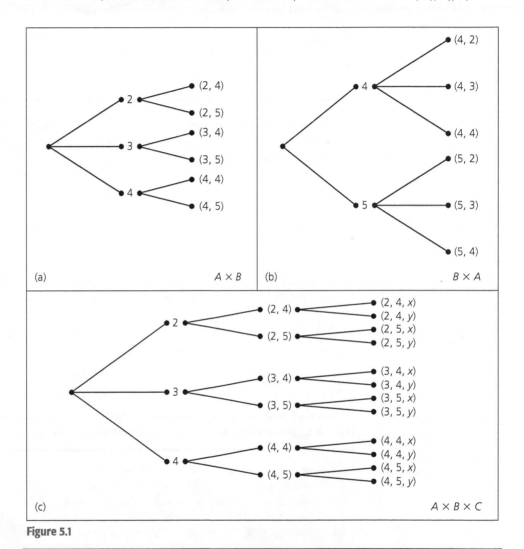

Figure 5.1

In addition to their tie-in with Cartesian products, tree diagrams also arise in other situations.

EXAMPLE 5.4   At the Wimbledon Tennis Championships, women play at most three sets in a match. The winner is the first to win two sets. If we let N and E denote the two players, the tree diagram in Fig. 5.2 indicates the six ways in which this match can be won. For example, the starred line segment (edge) indicates that player E won the first set. The double-starred edge indicates that player N has won the match by winning the first and third sets.

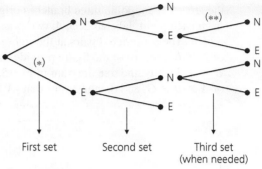

First set   Second set   Third set
(when needed)

Figure 5.2

Tree diagrams are examples of a general structure called a *tree*. Trees and graphs are important structures that arise in computer science and optimization theory. These will be investigated in later chapters.

For the cross product of two sets, we find the subsets of this structure of great interest.

Definition 5.2	For sets A, B, any subset of $A \times B$ is called a *(binary) relation* from A to B. Any subset of $A \times A$ is called a *(binary) relation* on A.

Since we will primarily deal with binary relations, for us the word "relation" will mean binary relation, unless something otherwise is specified.

EXAMPLE 5.5	With A, B as in Example 5.1, the following are some of the relations from A to B.

    a) $\emptyset$ **b)** $\{(2, 4)\}$

    c) $\{(2, 4), (2, 5)\}$ **d)** $\{(2, 4), (3, 4), (4, 4)\}$

    e) $\{(2, 4), (3, 4), (4, 5)\}$ **f)** $A \times B$

Since $|A \times B| = 6$, it follows from Definition 5.2 that there are 2^6 possible relations from A to B (for there are 2^6 possible subsets of $A \times B$).

For finite sets A, B with $|A| = m$ and $|B| = n$, there are 2^{mn} relations from A to B, including the empty relation as well as the relation $A \times B$ itself.

    There are also $2^{nm}(= 2^{mn})$ relations from B to A, one of which is also $\emptyset$ and another of which is $B \times A$. The reason we get the same number of relations from B to A as we have from A to B is that any relation $\mathcal{R}_1$ from B to A can be obtained from a unique relation $\mathcal{R}_2$ from A to B by simply reversing the components of each ordered pair in $\mathcal{R}_2$ (and vice versa).

EXAMPLE 5.6	For $B = \{1, 2\}$, let $A = \mathcal{P}(B) = \{\emptyset, \{1\}, \{2\}, \{1, 2\}\}$. The following is an example of a *relation* on A: $\mathcal{R} = \{(\emptyset, \emptyset), (\emptyset, \{1\}), (\emptyset, \{2\}), (\emptyset, \{1, 2\}), (\{1\}, \{1\}), (\{1\}, \{1, 2\}),$ $(\{2\}, \{2\}), (\{2\}, \{1, 2\}), (\{1, 2\}, \{1, 2\})\}$. We can say that the relation $\mathcal{R}$ is the *subset relation* where $(C, D) \in \mathcal{R}$ if and only if $C, D \subseteq B$ and $C \subseteq D$.

EXAMPLE 5.7

With $A = \mathbf{Z}^+$, we may define a relation $\mathcal{R}$ on set A as $\{(x, y) | x \le y\}$. This is the familiar "is less than or equal to" relation for the set of positive integers. It can be represented graphically as the set of points, with positive integer components, located on or above the line $y = x$ in the Euclidean plane, as partially shown in Fig. 5.3. Here we cannot list the entire relation as we did in Example 5.6, but we note, for example, that $(7, 7)$, $(7, 11) \in \mathcal{R}$, but $(8, 2) \notin \mathcal{R}$. The fact that $(7, 11) \in \mathcal{R}$ can also be denoted by $7\,\mathcal{R}\,11$; $(8, 2) \notin \mathcal{R}$ becomes $8\,\cancel{\mathcal{R}}\,2$. Here $7\,\mathcal{R}\,11$ and $8\,\cancel{\mathcal{R}}\,2$ are examples of the *infix* notation for a relation.

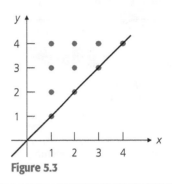

Figure 5.3

Our last example helps us to review the idea of a recursively defined set.

EXAMPLE 5.8

Let $\mathcal{R}$ be the subset of $\mathbf{N} \times \mathbf{N}$ where $\mathcal{R} = \{(m, n) | n = 7m\}$. Consequently, among the ordered pairs in $\mathcal{R}$ one finds $(0, 0)$, $(1, 7)$, $(11, 77)$, and $(15, 105)$. This relation $\mathcal{R}$ on $\mathbf{N}$ can also be given recursively by

1) $(0, 0) \in \mathcal{R}$; and

2) If $(s, t) \in \mathcal{R}$, then $(s + 1, t + 7) \in \mathcal{R}$.

We use the recursive definition to show that the ordered pair $(3, 21)$ (from $\mathbf{N} \times \mathbf{N}$) is in $\mathcal{R}$. Our derivation is as follows: From part (1) of the recursive definition we start with $(0, 0) \in \mathcal{R}$. Then part (2) of the definition gives us

i) $(0, 0) \in \mathcal{R} \Rightarrow (0 + 1, 0 + 7) = (1, 7) \in \mathcal{R}$;

ii) $(1, 7) \in \mathcal{R} \Rightarrow (1 + 1, 7 + 7) = (2, 14) \in \mathcal{R}$; and

iii) $(2, 14) \in \mathcal{R} \Rightarrow (2 + 1, 14 + 7) = (3, 21) \in \mathcal{R}$.

We close this section with these final observations.

1) For any set A, $A \times \emptyset = \emptyset$. (If $A \times \emptyset \ne \emptyset$, let $(a, b) \in A \times \emptyset$. Then $a \in A$ and $b \in \emptyset$. Impossible!) Likewise, $\emptyset \times A = \emptyset$.

2) The Cartesian product and the binary operations of union and intersection are interrelated in the following theorem.

THEOREM 5.1

For any sets A, B, $C \subseteq \mathcal{U}$:

a) $A \times (B \cap C) = (A \times B) \cap (A \times C)$

b) $A \times (B \cup C) = (A \times B) \cup (A \times C)$

c) $(A \cap B) \times C = (A \times C) \cap (B \times C)$

d) $(A \cup B) \times C = (A \times C) \cup (B \times C)$

Proof: We prove part (a) and leave the other parts for the reader. We use the same concept of set equality (as in Definition 3.2 of Section 3.1) even though the elements here are ordered pairs. For all $a, b \in \mathcal{U}$, $(a, b) \in A \times (B \cap C) \Longleftrightarrow a \in A$ and $b \in B \cap C \Longleftrightarrow a \in A$ and $b \in B, C \Longleftrightarrow a \in A, b \in B$ and $a \in A, b \in C \Longleftrightarrow (a, b) \in A \times B$ and $(a, b) \in A \times C \Longleftrightarrow (a, b) \in (A \times B) \cap (A \times C)$.

EXERCISES 5.1

1. If $A = \{1, 2, 3, 4\}$, $B = \{2, 5\}$, and $C = \{3, 4, 7\}$, determine $A \times B$; $B \times A$; $A \cup (B \times C)$; $(A \cup B) \times C$; $(A \times C) \cup (B \times C)$.

2. If $A = \{1, 2, 3\}$, and $B = \{2, 4, 5\}$, give examples of (a) three nonempty relations from A to B; (b) three nonempty relations on A.

3. For A, B as in Exercise 2, determine the following: (a) $|A \times B|$; (b) the number of relations from A to B; (c) the number of relations on A; (d) the number of relations from A to B that contain $(1, 2)$ and $(1, 5)$; (e) the number of relations from A to B that contain exactly five ordered pairs; and (f) the number of relations on A that contain at least seven elements.

4. For which sets A, B is it true that $A \times B = B \times A$?

5. Let A, B, C, D be nonempty sets.

a) Prove that $A \times B \subseteq C \times D$ if and only if $A \subseteq C$ and $B \subseteq D$.

b) What happens to the result in part (a) if any of the sets A, B, C, D is empty?

6. The men's final at Wimbledon is won by the first player to win three sets of the five-set match. Let C and M denote the players. Draw a tree diagram to show all the ways in which the match can be decided.

7. a) If $A = \{1, 2, 3, 4, 5\}$ and $B = \{w, x, y, z\}$, how many elements are there in $\mathcal{P}(A \times B)$?

b) Generalize the result in part (a).

8. Logic chips are taken from a container, tested individually, and labeled defective or good. The testing process is continued until either two defective chips are found or five chips are tested in total. Using a tree diagram, exhibit a sample space for this process.

9. Complete the proof of Theorem 5.1.

10. A rumor is spread as follows. The originator calls two people. Each of these people phones three friends, each of whom in turn calls five associates. If no one receives more than one call, and no one calls the originator, how many people now know the rumor? How many phone calls were made?

11. For $A, B, C \subseteq \mathcal{U}$, prove that

$$A \times (B - C) = (A \times B) - (A \times C).$$

12. Let A, B be sets with $|B| = 3$. If there are 4096 relations from A to B, what is $|A|$?

13. Let $\mathcal{R} \subseteq \mathbf{N} \times \mathbf{N}$ where $(m, n) \in \mathcal{R}$ if (and only if) $n = 5m + 2$. (a) Give a recursive definition for $\mathcal{R}$. (b) Use the recursive definition from part (a) to show that $(4, 22) \in \mathcal{R}$.

14. a) Give a recursive definition for the relation $\mathcal{R} \subseteq \mathbf{Z}^+ \times \mathbf{Z}^+$ where $(m, n) \in \mathcal{R}$ if (and only if) $m \geq n$.

b) From the definition in part (a) verify that $(5, 2)$ and $(4, 4)$ are in $\mathcal{R}$.

5.2
Functions: Plain and One-to-One

In this section we concentrate on a special kind of relation called a *function*. One finds functions in many different settings throughout mathematics and computer science. As for general relations, they will reappear in Chapter 7, where we shall examine them much more thoroughly.

Definition 5.3

For nonempty sets A, B, a *function*, or *mapping, f from A to B*, denoted $f: A \to B$, is a relation from A to B in which every element of A appears exactly once as the first component of an ordered pair in the relation.

We often write $f(a) = b$ when (a, b) is an ordered pair in the function f. For $(a, b) \in f$, b is called *the image* of a under f, whereas a is a *preimage* of b. In addition, the definition suggests that f is a method for *associating* with each $a \in A$ the *unique* element $f(a) = b \in B$. Consequently, $(a, b), (a, c) \in f$ implies $b = c$.

EXAMPLE 5.9

For $A = \{1, 2, 3\}$ and $B = \{w, x, y, z\}$, $f = \{(1, w), (2, x), (3, x)\}$ is a function, and consequently a relation, from A to B. $\mathscr{R}_1 = \{(1, w), (2, x)\}$ and $\mathscr{R}_2 = \{(1, w), (2, w), (2, x), (3, z)\}$ are relations, but not functions, from A to B. (Why?)

Definition 5.4

For the function $f: A \rightarrow B$, A is called the *domain* of f and B the *codomain* of f. The subset of B consisting of those elements that appear as second components in the ordered pairs of f is called the *range* of f and is also denoted by $f(A)$ because it is the set of images (of the elements of A) under f.

In Example 5.9, the domain of $f = \{1, 2, 3\}$, the codomain of $f = \{w, x, y, z\}$, and the range of $f = f(A) = \{w, x\}$.

A pictorial representation of these ideas appears in Fig. 5.4. This diagram suggests that a may be regarded as an *input* that is *transformed* by f into the corresponding *output*, $f(a)$. In this context, a C++ compiler can be thought of as a function that transforms a source program (the input) into its corresponding object program (the output).

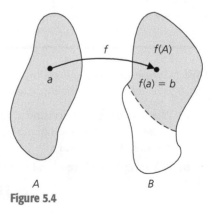

Figure 5.4

EXAMPLE 5.10

Many interesting functions arise in computer science.

a) A common function encountered is the *greatest integer function,* or *floor function.* This function $f: \mathbf{R} \rightarrow \mathbf{Z}$, is given by

$$f(x) = \lfloor x \rfloor = \text{the greatest integer less than or equal to } x.$$

Consequently, $f(x) = x$, if $x \in \mathbf{Z}$; and, when $x \in \mathbf{R} - \mathbf{Z}$, $f(x)$ is the integer to the immediate left of x on the real number line.
For this function we find that

1) $\lfloor 3.8 \rfloor = 3, \lfloor 3 \rfloor = 3, \lfloor -3.8 \rfloor = -4, \lfloor -3 \rfloor = -3;$

2) $\lfloor 7.1 + 8.2 \rfloor = \lfloor 15.3 \rfloor = 15 = 7 + 8 = \lfloor 7.1 \rfloor + \lfloor 8.2 \rfloor;$ and

3) $\lfloor 7.7 + 8.4 \rfloor = \lfloor 16.1 \rfloor = 16 \neq 15 = 7 + 8 = \lfloor 7.7 \rfloor + \lfloor 8.4 \rfloor.$

b) A second function — one related to the floor function in part (a) — is the *ceiling function*. This function $g: \mathbf{R} \to \mathbf{Z}$ is defined by

$$g(x) = \lceil x \rceil = \text{the least integer greater than or equal to } x.$$

So $g(x) = x$ when $x \in \mathbf{Z}$, but when $x \in \mathbf{R} - \mathbf{Z}$, then $g(x)$ is the integer to the immediate right of x on the real number line. In dealing with the ceiling function one finds that

1) $\lceil 3 \rceil = 3$, $\lceil 3.01 \rceil = \lceil 3.7 \rceil = 4 = \lceil 4 \rceil$, $\lceil -3 \rceil = -3$, $\lceil -3.01 \rceil = \lceil -3.7 \rceil = -3$;

2) $\lceil 3.6 + 4.5 \rceil = \lceil 8.1 \rceil = 9 = 4 + 5 = \lceil 3.6 \rceil + \lceil 4.5 \rceil$; and

3) $\lceil 3.3 + 4.2 \rceil = \lceil 7.5 \rceil = 8 \neq 9 = 4 + 5 = \lceil 3.3 \rceil + \lceil 4.2 \rceil$.

c) The function trunc (for truncation) is another integer-valued function defined on $\mathbf{R}$. This function deletes the fractional part of a real number. For example, trunc(3.78) $= 3$, trunc(5) $= 5$, trunc(-7.22) $= -7$. Note that trunc(3.78) $= \lfloor 3.78 \rfloor = 3$ while trunc(-3.78) $= \lceil -3.78 \rceil = -3$.

d) In storing a matrix in a one-dimensional array, many computer languages use the *row major* implementation. Here, if $A = (a_{ij})_{m \times n}$ is an $m \times n$ matrix, the first row of A is stored in locations 1, 2, 3, ..., n of the array if we start with a_{11} in location 1. The entry a_{21} is then found in position $n + 1$, while entry a_{34} occupies position $2n + 4$ in the array. In order to determine the location of an entry a_{ij} from A, where $1 \leq i \leq m$, $1 \leq j \leq n$, one defines the *access function* f from the entries of A to the positions 1, 2, 3, ..., mn of the array. A formula for the access function here is $f(a_{ij}) = (i - 1)n + j$.

a_{11}	a_{12}	$\cdots$	a_{1n}	a_{21}	a_{22}	$\cdots$	a_{2n}	a_{31}	$\cdots$		a_{ij}		$\cdots$		a_{mn}

1　2　$\cdots$　n　$n+1$　$n+2$　$\cdots$　$2n$　$2n+1$　$\cdots$　$(i-1)n+j$　$\cdots$　$(m-1)n+n$ $(= mn)$

EXAMPLE 5.11

We may use the floor and ceiling functions in parts (a) and (b), respectively, of Example 5.10 to restate some of the ideas we examined in Chapter 4.

a) When studying the division algorithm, we learned that for all $a, b \in \mathbf{Z}$, where $b > 0$, it was possible to find unique $q, r \in \mathbf{Z}$ with $a = qb + r$ and $0 \leq r < b$. Now we may add that $q = \lfloor \frac{a}{b} \rfloor$ and $r = a - \lfloor \frac{a}{b} \rfloor b$.

b) In Example 4.44 we found that the positive integer

$$29{,}338{,}848{,}000 = 2^8 3^5 5^3 7^3 11$$

has

$$60 = (5)(3)(2)(2)(1) = \left\lceil \frac{(8+1)}{2} \right\rceil \left\lceil \frac{(5+1)}{2} \right\rceil \left\lceil \frac{(3+1)}{2} \right\rceil \left\lceil \frac{(3+1)}{2} \right\rceil \left\lceil \frac{(1+1)}{2} \right\rceil$$

positive divisors that are perfect squares. In general, if $n \in \mathbf{Z}^+$ with $n > 1$, we know that we can write

$$n = p_1^{e_1} p_2^{e_2} \cdots p_k^{e_k}$$

where $k \in \mathbf{Z}^+$, p_i is prime for all $1 \leq i \leq k$, $p_i \neq p_j$ for all $1 \leq i < j \leq k$, and $e_i \in \mathbf{Z}^+$ for all $1 \leq i \leq k$. This is due to the Fundamental Theorem of Arithmetic. Then if $r \in \mathbf{Z}^+$, we find that the number of positive divisors of n that are perfect rth powers is $\prod_{i=1}^{k} \left\lceil \frac{e_i + 1}{r} \right\rceil$. When $r = 1$ we get $\prod_{i=1}^{k} \lceil e_i + 1 \rceil = \prod_{i=1}^{k} (e_i + 1)$, which is the number of positive divisors of n.

EXAMPLE 5.12

In Sections 4.1 and 4.2 we were introduced to the concept of a sequence in conjunction with our study of recursive definitions. We should now realize that a sequence of real numbers $r_1, r_2, r_3, \ldots$ can be thought of as a function $f: \mathbf{Z}^+ \to \mathbf{R}$ where $f(n) = r_n$, for all $n \in \mathbf{Z}^+$. Likewise, an integer sequence $a_0, a_1, a_2, \ldots$ can be defined by means of a function $g: \mathbf{N} \to \mathbf{Z}$ where $g(n) = a_n$, for all $n \in \mathbf{N}$.

In Example 5.9 there are $2^{12} = 4096$ relations from A to B. We have examined one function among these relations, and now we wish to count the total number of functions from A to B.

For the general case, let A, B be nonempty sets with $|A| = m$, $|B| = n$. Consequently, if $A = \{a_1, a_2, a_3, \ldots, a_m\}$ and $B = \{b_1, b_2, b_3, \ldots, b_n\}$, then a typical function $f: A \to B$ can be described by $\{(a_1, x_1), (a_2, x_2), (a_3, x_3), \ldots, (a_m, x_m)\}$. We can select any of the n elements of B for x_1 and then do the same for x_2. (We can select any element of B for x_2 so that the same element of B may be selected for both x_1 and x_2.) We continue this selection process until one of the n elements of B is finally selected for x_m. In this way, using the rule of product, there are $n^m = |B|^{|A|}$ functions from A to B.

Therefore, for A, B in Example 5.9, there are $4^3 = |B|^{|A|} = 64$ functions from A to B, and $3^4 = |A|^{|B|} = 81$ functions from B to A. In general, we do not expect $|A|^{|B|}$ to equal $|B|^{|A|}$. Unlike the situation for relations, we cannot always obtain a function from B to A by simply interchanging the components in the ordered pairs of a function from A to B (or vice versa).

Now that we have the concept of a function as a special type of relation, we turn our attention to a special type of function.

Definition 5.5

A function $f: A \to B$ is called *one-to-one*, or *injective*, if each element of B appears at most once as the image of an element of A.

If $f: A \to B$ is one-to-one, with A, B finite, we must have $|A| \le |B|$. For arbitrary sets A, B, $f: A \to B$ is one-to-one if and only if for all $a_1, a_2 \in A$, $f(a_1) = f(a_2) \Rightarrow a_1 = a_2$.

EXAMPLE 5.13

Consider the function $f: \mathbf{R} \to \mathbf{R}$ where $f(x) = 3x + 7$ for all $x \in \mathbf{R}$. Then for all $x_1, x_2 \in \mathbf{R}$, we find that

$$f(x_1) = f(x_2) \Rightarrow 3x_1 + 7 = 3x_2 + 7 \Rightarrow 3x_1 = 3x_2 \Rightarrow x_1 = x_2,$$

so the given function f is one-to-one.

On the other hand, suppose that $g: \mathbf{R} \to \mathbf{R}$ is the function defined by $g(x) = x^4 - x$ for each real number x. Then

$$g(0) = (0)^4 - 0 = 0 \quad \text{and} \quad g(1) = (1)^4 - (1) = 1 - 1 = 0.$$

Consequently, g is *not* one-to-one, since $g(0) = g(1)$ but $0 \ne 1$ — that is, g is *not* one-to-one because there exist real numbers x_1, x_2 where $g(x_1) = g(x_2) \not\Rightarrow x_1 = x_2$.

EXAMPLE 5.14

Let $A = \{1, 2, 3\}$ and $B = \{1, 2, 3, 4, 5\}$. The function

$$f = \{(1, 1), (2, 3), (3, 4)\}$$

is a one-to-one function from A to B;

$$g = \{(1, 1), (2, 3), (3, 3)\}$$

is a function from A to B, but it fails to be one-to-one because $g(2) = g(3)$ but $2 \neq 3$.

For A, B in Example 5.14 there are 2^{15} relations from A to B and 5^3 of these are functions from A to B. The next question we want to answer is how many functions $f: A \to B$ are one-to-one. Again we argue for general finite sets.

With $A = \{a_1, a_2, a_3, \ldots, a_m\}$, $B = \{b_1, b_2, b_3, \ldots, b_n\}$, and $m \leq n$, a one-to-one function $f: A \to B$ has the form $\{(a_1, x_1), (a_2, x_2), (a_3, x_3), \ldots, (a_m, x_m)\}$, where there are n choices for x_1 (that is, any element of B), $n - 1$ choices for x_2 (that is, any element of B except the one chosen for x_1), $n - 2$ choices for x_3, and so on, finishing with $n - (m - 1) = n - m + 1$ choices for x_m. By the rule of product, the number of one-to-one functions from A to B is

$$n(n - 1)(n - 2) \cdots (n - m + 1) = \frac{n!}{(n - m)!} = P(n, m) = P(|B|, |A|).$$

Consequently, for A, B in Example 5.14, there are $5 \cdot 4 \cdot 3 = 60$ one-to-one functions $f: A \to B$.

Definition 5.6

If $f: A \to B$ and $A_1 \subseteq A$, then

$$f(A_1) = \{b \in B | b = f(a), \text{ for some } a \in A_1\},$$

and $f(A_1)$ is called the *image of A_1 under f*.

EXAMPLE 5.15

For $A = \{1, 2, 3, 4, 5\}$ and $B = \{w, x, y, z\}$, let $f: A \to B$ be given by $f = \{(1, w), (2, x), (3, x), (4, y), (5, y)\}$. Then for $A_1 = \{1\}$, $A_2 = \{1, 2\}$, $A_3 = \{1, 2, 3\}$, $A_4 = \{2, 3\}$, and $A_5 = \{2, 3, 4, 5\}$, we find the following corresponding images under f:

$f(A_1) = \{f(a) | a \in A_1\} = \{f(a) | a \in \{1\}\} = \{f(a) | a = 1\} = \{f(1)\} = \{w\};$

$f(A_2) = \{f(a) | a \in A_2\} = \{f(a) | a \in \{1, 2\}\} = \{f(a) | a = 1 \text{ or } 2\}$
$\qquad = \{f(1), f(2)\} = \{w, x\};$

$f(A_3) = \{f(1), f(2), f(3)\} = \{w, x\}$, and $f(A_3) = f(A_2)$ because $f(2) = x = f(3);$

$f(A_4) = \{x\};$ and $f(A_5) = \{x, y\}.$

EXAMPLE 5.16

a) Let $g: \mathbf{R} \to \mathbf{R}$ be given by $g(x) = x^2$. Then $g(\mathbf{R}) =$ the range of $g = [0, +\infty)$. The image of $\mathbf{Z}$ under g is $g(\mathbf{Z}) = \{0, 1, 4, 9, 16, \ldots\}$, and for $A_1 = [-2, 1]$ we get $g(A_1) = [0, 4]$.

b) Let $h: \mathbf{Z} \times \mathbf{Z} \to \mathbf{Z}$ where $h(x, y) = 2x + 3y$. The domain of h is $\mathbf{Z} \times \mathbf{Z}$, not $\mathbf{Z}$, and the codomain is $\mathbf{Z}$. We find, for example, that $h(0, 0) = 2(0) + 3(0) = 0$ and $h(-3, 7) = 2(-3) + 3(7) = 15$. In addition, $h(2, -1) = 2(2) + 3(-1) = 1$, and for each $n \in \mathbf{Z}$, $h(2n, -n) = 2(2n) + 3(-n) = 4n - 3n = n$. Consequently, $h(\mathbf{Z} \times \mathbf{Z})$ = the range of $h = \mathbf{Z}$. For $A_1 = \{(0, n)|n \in \mathbf{Z}^+\} = \{0\} \times \mathbf{Z}^+ \subseteq \mathbf{Z} \times \mathbf{Z}$, the image of A_1 under h is $h(A_1) = \{3, 6, 9, \ldots\} = \{3n|n \in \mathbf{Z}^+\}$.

Our next result deals with the interplay between the images of subsets (of the domain) under a function f and the set operations of union and intersection.

THEOREM 5.2

Let $f: A \to B$, with $A_1, A_2 \subseteq A$. Then

a) $f(A_1 \cup A_2) = f(A_1) \cup f(A_2)$;   **b)** $f(A_1 \cap A_2) \subseteq f(A_1) \cap f(A_2)$;

c) $f(A_1 \cap A_2) = f(A_1) \cap f(A_2)$ when f is one-to-one.

Proof: We prove part (b) and leave the remaining parts for the reader.

For each $b \in B$, $b \in f(A_1 \cap A_2) \Rightarrow b = f(a)$, for some $a \in A_1 \cap A_2 \Rightarrow [b = f(a)$ for some $a \in A_1]$ and $[b = f(a)$ for some $a \in A_2] \Rightarrow b \in f(A_1)$ and $b \in f(A_2) \Rightarrow b \in f(A_1) \cap f(A_2)$, so $f(A_1 \cap A_2) \subseteq f(A_1) \cap f(A_2)$.

Definition 5.7

If $f: A \to B$ and $A_1 \subseteq A$, then $f|_{A_1}: A_1 \to B$ is called the *restriction of f to A_1* if $f|_{A_1}(a) = f(a)$ for all $a \in A_1$.

Definition 5.8

Let $A_1 \subseteq A$ and $f: A_1 \to B$. If $g: A \to B$ and $g(a) = f(a)$ for all $a \in A_1$, then we call g an *extension of f to A.*

EXAMPLE 5.17

For $A = \{1, 2, 3, 4, 5\}$, let $f: A \to \mathbf{R}$ be defined by $f = \{(1, 10), (2, 13), (3, 16), (4, 19), (5, 22)\}$. Let $g: \mathbf{Q} \to \mathbf{R}$ where $g(q) = 3q + 7$ for all $q \in \mathbf{Q}$. Finally, let $h: \mathbf{R} \to \mathbf{R}$ with $h(r) = 3r + 7$ for all $r \in \mathbf{R}$. Then

  i) g is *an* extension of f (from A) to $\mathbf{Q}$;

 ii) f is *the* restriction of g (from $\mathbf{Q}$) to A;

iii) h is *an* extension of f (from A) to $\mathbf{R}$;

 iv) f is *the* restriction of h (from $\mathbf{R}$) to A;

  v) h is *an* extension of g (from $\mathbf{Q}$) to $\mathbf{R}$; and

 vi) g is *the* restriction of h (from $\mathbf{R}$) to $\mathbf{Q}$.

EXAMPLE 5.18

Let $A = \{w, x, y, z\}$, $B = \{1, 2, 3, 4, 5\}$, and $A_1 = \{w, y, z\}$. Let $f: A \to B$, $g: A_1 \to B$ be represented by the diagrams in Fig. 5.5. Then $g = f|_{A_1}$ and f is an extension of g from A_1 to A. We note that for the given function $g: A_1 \to B$, there are five ways to extend g from A_1 to A.

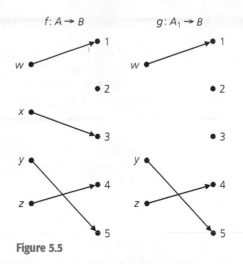

Figure 5.5

EXERCISES 5.2

1. Determine whether or not each of the following relations is a function. If a relation is a function, find its range.

a) $\{(x, y) | x, y \in \mathbf{Z}, y = x^2 + 7\}$, a relation from $\mathbf{Z}$ to $\mathbf{Z}$

b) $\{(x, y) | x, y \in \mathbf{R}, y^2 = x\}$, a relation from $\mathbf{R}$ to $\mathbf{R}$

c) $\{(x, y) | x, y \in \mathbf{R}, y = 3x + 1\}$, a relation from $\mathbf{R}$ to $\mathbf{R}$

d) $\{(x, y) | x, y \in \mathbf{Q}, x^2 + y^2 = 1\}$, a relation from $\mathbf{Q}$ to $\mathbf{Q}$

e) $\mathcal{R}$ is a relation from A to B where $|A| = 5$, $|B| = 6$, and $|\mathcal{R}| = 6$.

2. Does the formula $f(x) = 1/(x^2 - 2)$ define a function $f: \mathbf{R} \to \mathbf{R}$? A function $f: \mathbf{Z} \to \mathbf{R}$?

3. Let $A = \{1, 2, 3, 4\}$ and $B = \{x, y, z\}$. (a) List five functions from A to B. (b) How many functions $f: A \to B$ are there? (c) How many functions $f: A \to B$ are one-to-one? (d) How many functions $g: B \to A$ are there? (e) How many functions $g: B \to A$ are one-to-one? (f) How many functions $f: A \to B$ satisfy $f(1) = x$? (g) How many functions $f: A \to B$ satisfy $f(1) = f(2) = x$? (h) How many functions $f: A \to B$ satisfy $f(1) = x$ and $f(2) = y$?

4. If there are 2187 functions $f: A \to B$ and $|B| = 3$, what is $|A|$?

5. Let $A, B, C \subseteq \mathbf{R}^2$ where $A = \{(x, y) | y = 2x + 1\}$, $B = \{(x, y) | y = 3x\}$, and $C = \{(x, y) | x - y = 7\}$. Determine each of the following:

a) $A \cap B$ **b)** $B \cap C$
c) $\overline{A \cup \overline{C}}$ **d)** $\overline{B} \cup \overline{C}$

6. Let $A, B, C \subseteq \mathbf{Z}^2$ where $A = \{(x, y) | y = 2x + 1\}$, $B = \{(x, y) | y = 3x\}$, and $C = \{(x, y) | x - y = 7\}$.

a) Determine

  i) $A \cap B$ **ii)** $B \cap C$
  iii) $\overline{A \cup \overline{C}}$ **iv)** $\overline{B} \cup \overline{C}$

b) How are the answers for (i)–(iv) affected if $A, B, C \subseteq \mathbf{Z}^+ \times \mathbf{Z}^+$?

7. Determine each of the following:

a) $\lfloor 2.3 - 1.6 \rfloor$ **b)** $\lfloor 2.3 \rfloor - \lfloor 1.6 \rfloor$ **c)** $\lceil 3.4 \rceil \lfloor 6.2 \rfloor$
d) $\lfloor 3.4 \rfloor \lceil 6.2 \rceil$ **e)** $\lfloor 2\pi \rfloor$ **f)** $2\lceil \pi \rceil$

8. Determine whether each of the following statements is true or false. If the statement is false, provide a counterexample.

a) $\lfloor a \rfloor = \lceil a \rceil$ for all $a \in \mathbf{Z}$.

b) $\lfloor a \rfloor = \lceil a \rceil$ for all $a \in \mathbf{R}$.

c) $\lfloor a \rfloor = \lceil a \rceil - 1$ for all $a \in \mathbf{R} - \mathbf{Z}$.

d) $-\lceil a \rceil = \lceil -a \rceil$ for all $a \in \mathbf{R}$.

9. Find all real numbers x such that

a) $7\lfloor x \rfloor = \lfloor 7x \rfloor$ **b)** $\lfloor 7x \rfloor = 7$
c) $\lfloor x + 7 \rfloor = x + 7$ **d)** $\lfloor x + 7 \rfloor = \lfloor x \rfloor + 7$

10. Determine all $x \in \mathbf{R}$ such that $\lfloor x \rfloor + \lfloor x + \frac{1}{2} \rfloor = \lfloor 2x \rfloor$.

11. a) Find all real numbers x where $\lceil 3x \rceil = 3\lceil x \rceil$.

b) Let $n \in \mathbf{Z}^+$ where $n > 1$. Determine all $x \in \mathbf{R}$ such that $\lceil nx \rceil = n\lceil x \rceil$.

12. For $n, k \in \mathbf{Z}^+$, prove that $\lceil n/k \rceil = \lfloor (n-1)/k \rfloor + 1$.

13. a) Let $a \in \mathbf{R}^+$ where $a \geq 1$. Prove that (i) $\lfloor \lceil a \rceil /a \rfloor = 1$; and (ii) $\lceil \lfloor a \rfloor /a \rceil = 1$.

b) If $a \in \mathbf{R}^+$ and $0 < a < 1$, which result(s) in part (a) is (are) true?

14. Let $a_1, a_2, a_3, \ldots$ be the integer sequence defined recursively by

1) $a_1 = 1$; and

2) For all $n \in \mathbf{Z}^+$ where $n \geq 2$, $a_n = 2a_{\lfloor n/2 \rfloor}$.

    a) Determine a_n for all $2 \leq n \leq 8$.

    b) Prove that $a_n \leq n$ for all $n \in \mathbf{Z}^+$.

15. For each of the following functions, determine whether it is one-to-one and determine its range.

    a) $f: \mathbf{Z} \to \mathbf{Z}$, $f(x) = 2x + 1$

    b) $f: \mathbf{Q} \to \mathbf{Q}$, $f(x) = 2x + 1$

    c) $f: \mathbf{Z} \to \mathbf{Z}$, $f(x) = x^3 - x$

    d) $f: \mathbf{R} \to \mathbf{R}$, $f(x) = e^x$

    e) $f: [-\pi/2, \pi/2] \to \mathbf{R}$, $f(x) = \sin x$

    f) $f: [0, \pi] \to \mathbf{R}$, $f(x) = \sin x$

16. Let $f: \mathbf{R} \to \mathbf{R}$ where $f(x) = x^2$. Determine $f(A)$ for the following subsets A taken from the domain $\mathbf{R}$.

    a) $A = \{2, 3\}$
    b) $A = \{-3, -2, 2, 3\}$

    c) $A = (-3, 3)$
    d) $A = (-3, 2]$

    e) $A = [-7, 2]$
    f) $A = (-4, -3] \cup [5, 6]$

17. Let $A = \{1, 2, 3, 4, 5\}$, $B = \{w, x, y, z\}$, $A_1 = \{2, 3, 5\} \subseteq A$, and $g: A_1 \to B$. In how many ways can g be extended to a function $f: A \to B$?

18. Give an example of a function $f: A \to B$ and $A_1, A_2 \subseteq A$ for which $f(A_1 \cap A_2) \neq f(A_1) \cap f(A_2)$. [Thus the inclusion in Theorem 5.2(b) may be proper.]

19. Prove parts (a) and (c) of Theorem 5.2.

20. If $A = \{1, 2, 3, 4, 5\}$ and there are 6720 injective functions $f: A \to B$, what is $|B|$?

21. Let $f: A \to B$, where $A = X \cup Y$ with $X \cap Y = \emptyset$. If $f|_X$ and $f|_Y$ are one-to-one, does it follow that f is one-to-one?

22. For $n \in \mathbf{Z}^+$ define $X_n = \{1, 2, 3, \ldots, n\}$. Given $m, n \in \mathbf{Z}^+$, $f: X_m \to X_n$ is called *monotone increasing* if for all $i, j \in X_m$, $1 \leq i < j \leq m \Rightarrow f(i) \leq f(j)$. (a) How many monotone increasing functions are there with domain X_7 and codomain X_5? (b) Answer part (a) for the domain X_6 and codomain X_9. (c) Generalize the results in parts (a) and (b). (d) Determine the number of monotone increasing functions $f: X_{10} \to X_8$ where $f(4) = 4$. (e) How many monotone increasing functions $f: X_7 \to X_{12}$ satisfy $f(5) = 9$? (f) Generalize the results in parts (d) and (e).

23. Determine the access function $f(a_{ij})$, as described in Example 5.10(d), for a matrix $A = (a_{ij})_{m \times n}$, where (a) $m = 12$, $n = 12$; (b) $m = 7$, $n = 10$; (c) $m = 10$, $n = 7$.

24. For the access function developed in Example 5.10(d), the matrix $A = (a_{ij})_{m \times n}$ was stored in a one-dimensional array using the row major implementation. It is also possible to store this matrix using the column major implementation, where each entry a_{i1}, $1 \leq i \leq m$, in the first column

of A is stored in locations $1, 2, 3, \ldots, m$, respectively, of the array, when a_{11} is stored in location 1. Then the entries a_{i2}, $1 \leq i \leq m$, of the second column of A are stored in locations $m + 1, m + 2, m + 3, \ldots, 2m$, respectively, of the array, and so on. Find a formula for the access function $g(a_{ij})$ under these conditions.

25. a) Let A be an $m \times n$ matrix that is to be stored (in a contiguous manner) in a one-dimensional array of r entries. Find a formula for the access function if a_{11} is to be stored in location k (≥ 1) of the array [as opposed to location 1 as in Example 5.10(d)] and we use (i) the row major implementation; (ii) the column major implementation.

    b) State any conditions involving m, n, r, and k that must be satisfied in order for the results in part (a) to be valid.

26. The following exercise provides a combinatorial proof for a summation formula we have seen in four earlier results: (1) Exercise 22 in Section 1.4; (2) Example 4.4; (3) Exercise 3 in Section 4.1; and (4) Exercise 19 in Section 4.2.

Let $A = \{a, b, c\}$, $B = \{1, 2, 3, \ldots, n, n+1\}$, and $S = \{f: A \to B \mid f(a) < f(c) \text{ and } f(b) < f(c)\}$.

    a) If $S_1 = \{f: A \to B \mid f \in S \text{ and } f(c) = 2\}$, what is $|S_1|$?

    b) If $S_2 = \{f: A \to B \mid f \in S \text{ and } f(c) = 3\}$, what is $|S_2|$?

    c) For $1 \leq i \leq n$, let $S_i = \{f: A \to B \mid f \in S \text{ and } f(c) = i + 1\}$. What is $|S_i|$?

    d) Let $T_1 = \{f: A \to B \mid f \in S \text{ and } f(a) = f(b)\}$. Explain why $|T_1| = \binom{n+1}{2}$.

    e) Let $T_2 = \{f: A \to B \mid f \in S \text{ and } f(a) < f(b)\}$ and $T_3 = \{f: A \to B \mid f \in S \text{ and } f(a) > f(b)\}$. Explain why $|T_2| = |T_3| = \binom{n+1}{3}$.

    f) What can we conclude about the sets

$$S_1 \cup S_2 \cup S_3 \cup \cdots \cup S_n \text{ and } T_1 \cup T_2 \cup T_3?$$

    g) Use the results from parts (c), (d), (e), and (f) to verify that

$$\sum_{i=1}^{n} i^2 = \frac{n(n+1)(2n+1)}{6}.$$

27. One version of *Ackermann's function* $A(m,n)$ is defined recursively for $m, n \in \mathbf{N}$ by

$A(0, n) = n + 1$, $n \geq 0$;
$A(m, 0) = A(m - 1, 1)$, $m > 0$; and
$A(m, n) = A(m - 1, A(m, n - 1))$, $m, n > 0$.

[Such functions were defined in the 1920s by the German mathematician and logician Wilhelm Ackermann (1896–1962), who was a student of David Hilbert (1862–1943). These functions play an important role in computer science — in the theory of recursive functions and in the analysis of algorithms that involve the union of sets.]

    a) Calculate $A(1, 3)$ and $A(2, 3)$.

    b) Prove that $A(1, n) = n + 2$ for all $n \in \mathbf{N}$.

c) For all $n \in \mathbf{N}$ show that $A(2, n) = 3 + 2n$.

d) Verify that $A(3, n) = 2^{n+3} - 3$ for all $n \in \mathbf{N}$.

28. Given sets A, B, we define a *partial function f* with domain A and codomain B as a function from A' to B, where $\emptyset \neq A' \subset A$. [Here $f(x)$ is not defined for $x \in A - A'$.] For example, $f: \mathbf{R}^* \to \mathbf{R}$, where $f(x) = 1/x$, is a partial function on $\mathbf{R}$ since $f(0)$ is not defined. On the finite side, $\{(1, x), (2, x), (3, y)\}$ is a partial function for domain $A = \{1, 2, 3, 4, 5\}$ and codomain $B = \{w, x, y, z\}$. Furthermore, a computer program may be thought of as a partial function. The program's input is the input for the partial function and the program's output is the output of the function. Should the program fail to terminate, or terminate abnormally (perhaps, because of an attempt to divide by 0), then the partial function is considered to be undefined for that input. (a) For $A = \{1, 2, 3, 4, 5\}$, $B = \{w, x, y, z\}$, how many partial functions have domain A and codomain B? (b) Let A, B be sets where $|A| = m > 0$, $|B| = n > 0$. How many partial functions have domain A and codomain B?

5.3
Onto Functions: Stirling Numbers of the Second Kind

The results we develop in this section will provide the answers to the first five problems stated at the beginning of this chapter. We find that the *onto* function is the key to all of the answers.

Definition 5.9   A function $f: A \to B$ is called *onto*, or *surjective*, if $f(A) = B$ — that is, if for all $b \in B$ there is at least one $a \in A$ with $f(a) = b$.

EXAMPLE 5.19   The function $f: \mathbf{R} \to \mathbf{R}$ defined by $f(x) = x^3$ is an onto function. For here we find that if r is any real number in the codomain of f, then the real number $\sqrt[3]{r}$ is in the domain of f and $f(\sqrt[3]{r}) = (\sqrt[3]{r})^3 = r$. Hence the codomain of $f = \mathbf{R} =$ the range of f, and the function f is onto.

The function $g: \mathbf{R} \to \mathbf{R}$, where $g(x) = x^2$ for each real number x, is *not* an onto function. In this case no negative real number appears in the range of g. For example, for -9 to be in the range of g, we would have to be able to find a *real* number r with $g(r) = r^2 = -9$. Unfortunately, $r^2 = -9 \Rightarrow r = 3i$ or $r = -3i$, where $3i, -3i \in \mathbf{C}$, but $3i, -3i \notin \mathbf{R}$. So here the range of $g = g(\mathbf{R}) = [0, +\infty) \subset \mathbf{R}$, and the function g is *not* onto. Note, however, that the function $h: \mathbf{R} \to [0, +\infty)$ defined by $h(x) = x^2$ *is* an onto function.

EXAMPLE 5.20   Consider the function $f: \mathbf{Z} \to \mathbf{Z}$ where $f(x) = 3x + 1$ for each $x \in \mathbf{Z}$. Here the range of $f = \{\ldots, -8, -5, -2, 1, 4, 7, \ldots\} \subset \mathbf{Z}$, so f is *not* an onto function. If we examine the situation here a little more closely, we find that the integer 8, for example, is not in the range of f even though the equation

$$3x + 1 = 8$$

can be easily solved — giving us $x = 7/3$. But that is the problem, for the rational number $7/3$ is *not* an integer — so there is no x in the domain $\mathbf{Z}$ with $f(x) = 8$.

On the other hand, each of the functions

1) $g: \mathbf{Q} \to \mathbf{Q}$, where $g(x) = 3x + 1$ for $x \in \mathbf{Q}$; and

2) $h: \mathbf{R} \to \mathbf{R}$, where $h(x) = 3x + 1$ for $x \in \mathbf{R}$

is an onto function. Furthermore, $3x_1 + 1 = 3x_2 + 1 \Rightarrow 3x_1 = 3x_2 \Rightarrow x_1 = x_2$, regardless of whether x_1 and x_2 are integers, rational numbers, or real numbers. Consequently, all three of the functions f, g, and h are one-to-one.

EXAMPLE 5.21

If $A = \{1, 2, 3, 4\}$ and $B = \{x, y, z\}$, then

$$f_1 = \{(1, z), (2, y), (3, x), (4, y)\} \quad \text{and} \quad f_2 = \{(1, x), (2, x), (3, y), (4, z)\}$$

are both functions from A *onto* B. However, the function $g = \{(1, x), (2, x), (3, y), (4, y)\}$ is not onto, because $g(A) = \{x, y\} \subset B$.

If A, B are finite sets, then for an onto function $f: A \to B$ to possibly exist we must have $|A| \geq |B|$. Considering the development in the first two sections of this chapter, the reader undoubtedly feels it is time once again to use the rule of product and count the number of onto functions $f: A \to B$ where $|A| = m \geq n = |B|$. Unfortunately, the rule of product proves inadequate here. We shall obtain the needed result for some specific examples and then conjecture a general formula. In Chapter 8 we shall establish the conjecture using the Principle of Inclusion and Exclusion.

EXAMPLE 5.22

If $A = \{x, y, z\}$ and $B = \{1, 2\}$, then all functions $f: A \to B$ are onto except $f_1 = \{(x, 1), (y, 1), (z, 1)\}$, and $f_2 = \{(x, 2), (y, 2), (z, 2)\}$, the *constant* functions. So there are $|B|^{|A|} - 2 = 2^3 - 2 = 6$ onto functions from A to B.

In general, if $|A| = m \geq 2$ and $|B| = 2$, then there are $2^m - 2$ onto functions from A to B. (Does this formula tell us anything when $m = 1$?)

EXAMPLE 5.23

For $A = \{w, x, y, z\}$ and $B = \{1, 2, 3\}$, there are 3^4 functions from A to B. Considering subsets of B of size 2, there are 2^4 functions from A to $\{1, 2\}$, 2^4 functions from A to $\{2, 3\}$, and 2^4 functions from A to $\{1, 3\}$. So we have $3(2^4) = \binom{3}{2}2^4$ functions from A to B that are definitely not onto. However, before we acknowledge $3^4 - \binom{3}{2}2^4$ as the final answer, we must realize that not all of these $\binom{3}{2}2^4$ functions are distinct. For when we consider all the functions from A to $\{1, 2\}$, we are removing, among these, the function $\{(w, 2), (x, 2), (y, 2), (z, 2)\}$. Then, considering the functions from A to $\{2, 3\}$, we remove the same function: $\{(w, 2), (x, 2), (y, 2), (z, 2)\}$. Consequently, in the result $3^4 - \binom{3}{2}2^4$, we have twice removed each of the constant functions $f: A \to B$, where $f(A)$ is one of the sets $\{1\}$, $\{2\}$, or $\{3\}$. Adjusting our present result for this, we find that there are $3^4 - \binom{3}{2}2^4 + 3 = \binom{3}{3}3^4 - \binom{3}{2}2^4 + \binom{3}{1}1^4 = 36$ onto functions from A to B.

Keeping $B = \{1, 2, 3\}$, for any set A with $|A| = m \geq 3$, there are $\binom{3}{3}3^m - \binom{3}{2}2^m + \binom{3}{1}1^m$ functions from A onto B. (What result does this formula yield when $m = 1$? when $m = 2$?)

The last two examples suggest a pattern that we now state, without proof, as our general formula.

For finite sets A, B with $|A| = m$ and $|B| = n$, there are

$$\binom{n}{n}n^m - \binom{n}{n-1}(n-1)^m + \binom{n}{n-2}(n-2)^m - \cdots$$

$$+(-1)^{n-2}\binom{n}{2}2^m + (-1)^{n-1}\binom{n}{1}1^m = \sum_{k=0}^{n-1}(-1)^k\binom{n}{n-k}(n-k)^m$$

$$= \sum_{k=0}^{n}(-1)^k\binom{n}{n-k}(n-k)^m$$

onto functions from A to B.

EXAMPLE 5.24

Let $A = \{1, 2, 3, 4, 5, 6, 7\}$ and $B = \{w, x, y, z\}$. Applying the general formula with $m = 7$ and $n = 4$, we find that there are

$$\binom{4}{4}4^7 - \binom{4}{3}3^7 + \binom{4}{2}2^7 - \binom{4}{1}1^7 = \sum_{k=0}^{3}(-1)^k\binom{4}{4-k}(4-k)^7$$

$$= \sum_{k=0}^{4}(-1)^k\binom{4}{4-k}(4-k)^7 = 8400 \text{ functions from } A \text{ onto } B.$$

The result in Example 5.24 is also the answer to the first three questions proposed at the start of this chapter. Once we remove the unnecessary vocabulary, we recognize that in all three cases we want to distribute seven different objects into four distinct containers with no container left empty. We can do this in terms of onto functions.

For Problem 4 we have a sample space $\mathcal{S}$ consisting of the $4^7 = 16,384$ ways in which seven people can each select one of the four floors. (Note that 4^7 is also the total number of functions $f: A \to B$ where $|A| = 7$, $|B| = 4$.) The event that we are concerned with contains 8400 of those selections, so the probability that the elevator must stop at every floor is $8400/16384 \doteq 0.5127$, slightly more than half of the time.

Finally, for Problem 5, since $\sum_{k=0}^{n}(-1)^k\binom{n}{n-k}(n-k)^m$ is the number of onto functions $f: A \to B$ for $|A| = m$, $|B| = n$, for the case where $m < n$ there are no such functions and the summation is 0.

Problem 6 will be addressed in Section 5.6.

Before going on to anything new, however, we consider one more problem.

EXAMPLE 5.25

At the CH Company, Joan, the supervisor, has a secretary, Teresa, and three other administrative assistants. If seven accounts must be processed, in how many ways can Joan assign the accounts so that each assistant works on at least one account and Teresa's work includes the most expensive account?

First and foremost, the answer is not 8400 as in Example 5.24. Here we must consider two disjoint subcases and then apply the rule of sum.

  a) If Teresa, the secretary, works only on the most expensive account, then the other six accounts can be distributed among the three administrative assistants in $\sum_{k=0}^{3}(-1)^k\binom{3}{3-k}(3-k)^6 = 540$ ways. ($540 =$ the number of onto functions $f: A \to B$ with $|A| = 6$, $|B| = 3$.)

b) If Teresa does more than just the most expensive account, the assignments can be made in $\sum_{k=0}^{4}(-1)^k\binom{4}{4-k}(4-k)^6 = 1560$ ways. ($1560 =$ the number of onto functions $g: C \to D$ with $|C| = 6, |D| = 4$.)

Consequently, the assignments can be given under the prescribed conditions in $540 + 1560 = 2100$ ways. [We mentioned earlier that the answer would not be 8400, but it is $(1/4)(8400) = (1/|B|)(8400)$, where 8400 is the number of onto functions $f: A \to B$, with $|A| = 7$ and $|B| = 4$. This is no coincidence, as we shall learn when we discuss Theorem 5.3.]

We now continue our discussion with the distribution of distinct objects into containers with none left empty, but now the containers become identical.

EXAMPLE 5.26

If $A = \{a, b, c, d\}$ and $B = \{1, 2, 3\}$, then there are 36 onto functions from A to B or, equivalently, 36 ways to distribute four distinct objects into three distinguishable containers, with no container empty (and no regard for the location of objects in a given container). Among these 36 distributions we find the following collection of six (one of six such possible collections of six):

1) $\{a, b\}_1$ $\{c\}_2$ $\{d\}_3$ **2)** $\{a, b\}_1$ $\{d\}_2$ $\{c\}_3$

3) $\{c\}_1$ $\{a, b\}_2$ $\{d\}_3$ **4)** $\{c\}_1$ $\{d\}_2$ $\{a, b\}_3$

5) $\{d\}_1$ $\{a, b\}_2$ $\{c\}_3$ **6)** $\{d\}_1$ $\{c\}_2$ $\{a, b\}_3$,

where, for example, the notation $\{c\}_2$ means that c is in the second container. Now if we no longer distinguish the containers, these $6 = 3!$ distributions become identical, so there are $36/(3!) = 6$ ways to distribute the distinct objects a, b, c, d among three identical containers, leaving no container empty.

For $m \geq n$ there are $\sum_{k=0}^{n}(-1)^k\binom{n}{n-k}(n-k)^m$ ways to distribute m distinct objects into n numbered (but otherwise identical) containers with no container left empty. Removing the numbers on the containers, so that they are now identical in appearance, we find that one distribution into these n (nonempty) identical containers corresponds with $n!$ such distributions into the numbered containers. So the number of ways in which it is possible to distribute the m distinct objects into n identical containers, with no container left empty, is

$$\frac{1}{n!}\sum_{k=0}^{n}(-1)^k\binom{n}{n-k}(n-k)^m.$$

This will be denoted by $S(m, n)$ and is called a *Stirling number of the second kind*.

We note that for $|A| = m \geq n = |B|$, there are $n! \cdot S(m, n)$ onto functions from A to B.

Table 5.1 lists some Stirling numbers of the second kind.

EXAMPLE 5.27

For $m \geq n$, $\sum_{i=1}^{n} S(m, i)$ is the number of possible ways to distribute m distinct objects into n identical containers with empty containers allowed. From the fourth row of Table 5.1

Table 5.1

					$S(m, n)$			
m \ n	1	2	3	4	5	6	7	8
1	1							
2	1	1						
3	1	3	1					
4	1	7	6	1				
5	1	15	25	10	1			
6	1	31	90	65	15	1		
7	1	63	301	350	140	21	1	
8	1	127	966	1701	1050	266	28	1

we see that there are $1 + 7 + 6 = 14$ ways to distribute the objects a, b, c, d among three identical containers, with some container(s) possibly empty.

We continue now with the derivation of an identity involving Stirling numbers of the second kind. The proof is combinatorial in nature.

THEOREM 5.3   Let m, n be positive integers with $1 < n \leq m$. Then

$$S(m + 1, n) = S(m, n - 1) + nS(m, n).$$

Proof: Let $A = \{a_1, a_2, \ldots, a_m, a_{m+1}\}$. Then $S(m + 1, n)$ counts the number of ways in which the objects of A can be distributed among n identical containers, with no container left empty.

There are $S(m, n - 1)$ ways of distributing $a_1, a_2, \ldots, a_m$ among $n - 1$ identical containers, with none left empty. Then, placing a_{m+1} in the remaining empty container results in $S(m, n - 1)$ of the distributions counted in $S(m + 1, n)$ — namely, those distributions where a_{m+1} is in a container by itself. Alternatively, distributing $a_1, a_2, \ldots, a_m$ among the n identical containers with none left empty, we have $S(m, n)$ distributions. Now, however, for each of these $S(m, n)$ distributions the n containers become distinguished by their contents. Selecting one of the n distinct containers for a_{m+1}, we have $nS(m, n)$ distributions of the total $S(m + 1, n)$ — namely, those where a_{m+1} is in the same container as another object from A. The result then follows by the rule of sum.

To illustrate Theorem 5.3 consider the triangle shown in Table 5.1. Here the largest number corresponds with $S(m + 1, n)$, for $m = 7$ and $n = 3$, and we see that $S(7 + 1, 3) = 966 = 63 + 3(301) = S(7, 2) + 3S(7, 3)$. The identity in Theorem 5.3 can be used to extend Table 5.1 if necessary.

If we multiply the result in Theorem 5.3 by $(n - 1)!$ we have

$$\left(\frac{1}{n}\right)[n!S(m + 1, n)] = [(n - 1)!S(m, n - 1)] + [n!S(m, n)].$$

This new form of the equation tells us something about numbers of onto functions. If $A = \{a_1, a_2, \ldots, a_m, a_{m+1}\}$ and $B = \{b_1, b_2, \ldots, b_{n-1}, b_n\}$ with $m \geq n - 1$, then

$$\left(\frac{1}{n}\right) \text{ (The number of onto functions } h: A \to B)$$

$$= \text{(The number of onto functions } f: A - \{a_{m+1}\} \to B - \{b_n\})$$

$$+ \text{(The number of onto functions } g: A - \{a_{m+1}\} \to B).$$

Thus the relationship at the end of Example 5.25 is not just a coincidence.

We close this section with an application that deals with a counting problem in which the Stirling numbers of the second kind are used in conjunction with the Fundamental Theorem of Arithmetic.

EXAMPLE 5.28

Consider the positive integer $30{,}030 = 2 \times 3 \times 5 \times 7 \times 11 \times 13$. Among the unordered factorizations of this number one finds

   i) $30 \times 1001 = (2 \times 3 \times 5)(7 \times 11 \times 13)$

   ii) $110 \times 273 = (2 \times 5 \times 11)(3 \times 7 \times 13)$

   iii) $2310 \times 13 = (2 \times 3 \times 5 \times 7 \times 11)(13)$

   iv) $14 \times 33 \times 65 = (2 \times 7)(3 \times 11)(5 \times 13)$

   v) $22 \times 35 \times 39 = (2 \times 11)(5 \times 7)(3 \times 13)$

The results given in (i), (ii), and (iii) demonstrate three of the ways to distribute the six distinct objects $2, 3, 5, 7, 11, 13$ into two identical containers with no container left empty. So these first three examples are three of the $S(6, 2) = 31$ unordered two-factor factorizations of $30{,}030$ — that is, there are $S(6, 2)$ ways to factor $30{,}030$ as mn where $m, n \in \mathbf{Z}^+$ for $1 < m, n < 30{,}030$ and where order is not relevant. Likewise, the results in (iv) and (v) are two of the $S(6, 3) = 90$ unordered ways to factor $30{,}030$ into three integer factors, each greater than 1. If we want at least two factors (greater than 1) in each of these unordered factorizations, then we find that there are $\sum_{i=2}^{6} S(6, i) = 202$ such factorizations. If we want to include the *one-factor* factorization $30{,}030$ — where we distribute the six distinct objects $2, 3, 5, 7, 11, 13$ into one (identical) container — then we have 203 such factorizations in total.

---

1. Give an example of finite sets A and B with $|A|, |B| \geq 4$ and a function $f: A \to B$ such that (a) f is neither one-to-one nor onto; (b) f is one-to-one but not onto; (c) f is onto but not one-to-one; (d) f is onto and one-to-one.

2. For each of the following functions $f: \mathbf{Z} \to \mathbf{Z}$, determine whether the function is one-to-one and whether it is onto. If the function is not onto, determine the range $f(\mathbf{Z})$.

   a) $f(x) = x + 7$ **b)** $f(x) = 2x - 3$

   c) $f(x) = -x + 5$ **d)** $f(x) = x^2$

   e) $f(x) = x^2 + x$ **f)** $f(x) = x^3$

3. For each of the following functions $g: \mathbf{R} \to \mathbf{R}$, determine whether the function is one-to-one and whether it is onto. If the function is not onto, determine the range $g(\mathbf{R})$.

   a) $g(x) = x + 7$ **b)** $g(x) = 2x - 3$

   c) $g(x) = -x + 5$ **d)** $g(x) = x^2$

   e) $g(x) = x^2 + x$ **f)** $g(x) = x^3$

4. Let $A = \{1, 2, 3, 4\}$ and $B = \{1, 2, 3, 4, 5, 6\}$. (a) How many functions are there from A to B? How many of these are one-to-one? How many are onto? (b) How many functions are there from B to A? How many of these are onto? How many are one-to-one?

5. Verify that $\sum_{k=0}^{n} (-1)^k \binom{n}{n-k} (n - k)^m = 0$ for $n = 5$ and $m = 2, 3, 4$.

6. a) Verify that $5^7 = \sum_{i=1}^5 \binom{5}{i}(i!)S(7, i)$.

b) Provide a combinatorial argument to prove that for all $m, n \in \mathbf{Z}^+$,

$$m^n = \sum_{i=1}^m \binom{m}{i}(i!)S(n, i).$$

7. a) Let $A = \{1, 2, 3, 4, 5, 6, 7\}$ and $B = \{v, w, x, y, z\}$. Determine the number of functions $f: A \to B$ where (i) $f(A) = \{v, x\}$; (ii) $|f(A)| = 2$; (iii) $f(A) = \{w, x, y\}$; (iv) $|f(A)| = 3$; (v) $f(A) = \{v, x, y, z\}$; and (vi) $|f(A)| = 4$.

b) Let A, B be sets with $|A| = m \geq n = |B|$. If $k \in \mathbf{Z}^+$ with $1 \leq k \leq n$, how many functions $f: A \to B$ are such that $|f(A)| = k$?

8. A chemist who has five assistants is engaged in a research project that calls for nine compounds that must be synthesized. In how many ways can the chemist assign these syntheses to the five assistants so that each is working on at least one synthesis?

9. Use the fact that every polynomial equation having real-number coefficients and odd degree has a real root in order to show that the function $f: \mathbf{R} \to \mathbf{R}$, defined by $f(x) = x^5 - 2x^2 + x$, is an onto function. Is f one-to-one?

10. Suppose we have seven different colored balls and four containers numbered I, II, III, and IV. (a) In how many ways can we distribute the balls so that no container is left empty? (b) In this collection of seven colored balls, one of them is blue. In how many ways can we distribute the balls so that no container is empty and the blue ball is in container II? (c) If we remove the numbers from the containers so that we can no longer distinguish them, in how many ways can we distribute the seven colored balls among the four identical containers, with some container(s) possibly empty?

11. Determine the next two rows ($m = 9, 10$) of Table 5.1 for the Stirling numbers $S(m, n)$, where $1 \leq n \leq m$.

12. a) In how many ways can 31,100,905 be factored into three factors, each greater than 1, if the order of the factors is not relevant?

b) Answer part (a), assuming the order of the three factors is relevant.

c) In how many ways can one factor 31,100,905 into two or more factors where each factor is greater than 1 and no regard is paid to the order of the factors?

d) Answer part (c), assuming the order of the factors is to be taken into consideration.

13. a) How many two-factor unordered factorizations, where each factor is greater than 1, are there for 156,009?

b) In how many ways can 156,009 be factored into two or more factors, each greater than 1, with no regard to the order of the factors?

c) Let $p_1, p_2, p_3, \ldots, p_n$ be n distinct primes. In how many ways can one factor the product $\prod_{i=1}^n p_i$ into two or more factors, each greater than 1, where the order of the factors is not relevant?

14. Write a computer program (or develop an algorithm) to compute the Stirling numbers $S(m, n)$ when $1 \leq m \leq 12$ and $1 \leq n \leq m$.

15. A lock has n buttons labeled $1, 2, \ldots, n$. To open this lock we press each of the n buttons exactly once. If no two or more buttons may be pressed simultaneously, then there are $n!$ ways to do this. However, if one may press two or more buttons simultaneously, then there are more than $n!$ ways to press all of the buttons. For instance, if $n = 3$ there are six ways to press the buttons one at a time. But if one may also press two or more buttons simultaneously, then we find 13 cases — namely,

(1) 1, 2, 3	(2) 1, 3, 2	(3) 2, 1, 3
(4) 2, 3, 1	(5) 3, 1, 2	(6) 3, 2, 1
(7) {1, 2}, 3	(8) 3, {1, 2}	(9) {1, 3}, 2
(10) 2, {1, 3}	(11) {2, 3}, 1	(12) 1, {2, 3}
(13) {1, 2, 3}.		

[Here, for example, case (12) indicates that one presses button 1 first and then buttons 2, 3 (together) second.] (a) How many ways are there to press the buttons when $n = 4$? $n = 5$? How many for n in general? (b) Suppose a lock has 15 buttons. To open this lock one must press 12 different buttons (one at a time, or simultaneously in sets of two or more). In how many ways can this be done?

16. At St. Xavier High School ten candidates $C_1, C_2, \ldots, C_{10}$, run for senior class president.

a) How many outcomes are possible where (i) there are no ties (that is, no two, or more, candidates receive the same number of votes? (ii) ties are permitted? [Here we may have an outcome such as $\{C_2, C_3, C_7\}$, $\{C_1, C_4, C_9, C_{10}\}$, $\{C_5\}$, $\{C_6, C_8\}$, where C_2, C_3, C_7 tie for first place, C_1, C_4, C_9, C_{10} tie for fourth place, C_5 is in eighth place, and C_6, C_8 are tied for ninth place.] (iii) three candidates tie for first place (and other ties are permitted)?

b) How many of the outcomes in section (iii) of part (a) have C_3 as one of the first-place candidates?

c) How many outcomes have C_3 in first place (alone, or tied with others)?

17. For $m, n, r \in \mathbf{Z}^+$ with $m \geq rn$, let $S_r(m, n)$ denote the number of ways to distribute m distinct objects among n identical containers where each container receives at least r of the objects. Verify that

$$S_r(m + 1, n) = nS_r(m, n) + \binom{m}{r - 1}S_r(m + 1 - r, n - 1).$$

18. We use $s(m, n)$ to denote the number of ways to seat m people at n circular tables with at least one person at each table. The arrangements at any one table are not distinguished if one can be rotated into another (as in Example 1.16). The ordering of the tables is *not* taken into account. For instance, the arrangements in parts (a), (b), (c) of Fig. 5.6 are considered the same; those in parts (a), (d), (e) are distinct (in pairs).

The numbers $s(m, n)$ are referred to as the *Stirling numbers of the first kind*.

a) If $n > m$, what is $s(m, n)$?

b) For $m \geq 1$, what are $s(m, m)$ and $s(m, 1)$?

c) Determine $s(m, m - 1)$ for $m \geq 2$.

d) Show that for $m \geq 3$,

$$s(m, m - 2) = \left(\frac{1}{24}\right) m(m - 1)(m - 2)(3m - 1).$$

19. As in the previous exercise, $s(m, n)$ denotes a Stirling number of the first kind.

a) For $m \geq n > 1$ prove that

$$s(m, n) = (m - 1)s(m - 1, n) + s(m - 1, n - 1).$$

b) Verify that for $m \geq 2$,

$$s(m, 2) = (m - 1)! \sum_{i=1}^{m-1} \frac{1}{i}.$$

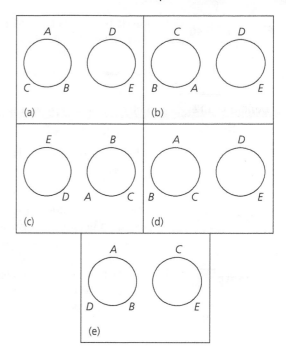

Figure 5.6

5.4
Special Functions

In Section 2 of Chapter 3 we mentioned that addition is a closed binary operation on the set $\mathbf{Z}^+$, whereas $\cap$ is a closed binary operation on $\mathcal{P}(\mathcal{U})$ for any given universe $\mathcal{U}$. We also noted in that section that "taking the minus" of an integer is a unary operation on $\mathbf{Z}$. Now it is time to make these notions of (closed) binary and unary operations more precise in terms of functions.

Definition 5.10 — For any nonempty sets A, B, any function $f: A \times A \to B$ is called a *binary operation* on A. If $B \subseteq A$, then the binary operation is said to be *closed* (*on A*). (When $B \subseteq A$ we may also say that *A is closed under f*.)

Definition 5.11 — A function $g: A \to A$ is called a *unary*, or *monary*, operation on A.

EXAMPLE 5.29

a) The function $f: \mathbf{Z} \times \mathbf{Z} \to \mathbf{Z}$, defined by $f(a, b) = a - b$, is a closed binary operation on $\mathbf{Z}$.

b) If $g: \mathbf{Z}^+ \times \mathbf{Z}^+ \to \mathbf{Z}$ is the function where $g(a, b) = a - b$, then g is a binary operation on $\mathbf{Z}^+$, but it is *not* closed. For example, we find that $3, 7 \in \mathbf{Z}^+$, but $g(3, 7) = 3 - 7 = -4 \notin \mathbf{Z}^+$.

c) The function $h: \mathbf{R}^+ \to \mathbf{R}^+$ defined by $h(a) = 1/a$ is a unary operation on $\mathbf{R}^+$.

EXAMPLE 5.30

Let $\mathcal{U}$ be a universe, and let $A, B \subseteq \mathcal{U}$. (a) If $f: \mathcal{P}(\mathcal{U}) \times \mathcal{P}(\mathcal{U}) \to \mathcal{P}(\mathcal{U})$ is defined by $f(A, B) = A \cup B$, then f is a closed binary operation on $\mathcal{P}(\mathcal{U})$. (b) The function $g: \mathcal{P}(\mathcal{U}) \to \mathcal{P}(\mathcal{U})$ defined by $g(A) = \overline{A}$ is a unary operation on $\mathcal{P}(\mathcal{U})$.

Definition 5.12

Let $f: A \times A \to B$; that is, f is a binary operation on A.

   a) f is said to be *commutative* if $f(a, b) = f(b, a)$ for all $(a, b) \in A \times A$.

   b) When $B \subseteq A$ (that is, when f is closed), f is said to be *associative* if for all $a, b, c \in A$, $f(f(a, b), c) = f(a, f(b, c))$.

EXAMPLE 5.31

The binary operation of Example 5.30 is commutative and associative, whereas the binary operation in part (a) of Example 5.29 is neither.

EXAMPLE 5.32

   a) Define the closed binary operation $f: \mathbf{Z} \times \mathbf{Z} \to \mathbf{Z}$ by $f(a, b) = a + b - 3ab$. Since both the addition and the multiplication of integers are commutative binary operations, it follows that

$$f(a, b) = a + b - 3ab = b + a - 3ba = f(b, a),$$

so f is commutative.

      To determine whether f is associative, consider $a, b, c \in \mathbf{Z}$. Then

$$f(a, b) = a + b - 3ab \quad \text{and} \quad f(f(a, b), c) = f(a, b) + c - 3f(a, b)c$$
$$= (a + b - 3ab) + c - 3(a + b - 3ab)c$$
$$= a + b + c - 3ab - 3ac - 3bc + 9abc,$$

whereas

$$f(b, c) = b + c - 3bc \quad \text{and} \quad f(a, f(b, c)) = a + f(b, c) - 3af(b, c)$$
$$= a + (b + c - 3bc) - 3a(b + c - 3bc)$$
$$= a + b + c - 3ab - 3ac - 3bc + 9abc.$$

Since $f(f(a, b), c) = f(a, f(b, c))$ for all $a, b, c \in \mathbf{Z}$, the closed binary operation f is associative as well as commutative.

   b) Consider the closed binary operation $h: \mathbf{Z} \times \mathbf{Z} \to \mathbf{Z}$, where $h(a, b) = a|b|$. Then $h(3, -2) = 3|-2| = 3(2) = 6$, but $h(-2, 3) = -2|3| = -6$. Consequently, h is *not* commutative. However, with regard to the associative property, if $a, b, c \in \mathbf{Z}$, we find that

$$h(h(a, b), c) = h(a, b)|c| = a|b||c| \quad \text{and}$$
$$h(a, h(b, c)) = a|h(b, c)| = a|b|c|| = a|b||c|,$$

so the closed binary operation h is associative.

EXAMPLE 5.33

If $A = \{a, b, c, d\}$, then $|A \times A| = 16$. Consequently, there are 4^{16} functions $f: A \times A \to A$; that is, there are 4^{16} closed binary operations on A.

    To determine the number of commutative closed binary operations g on A, we realize that there are four choices for each of the assignments $g(a, a)$, $g(b, b)$, $g(c, c)$, and $g(d, d)$.

We are then left with the $4^2 - 4 = 16 - 4 = 12$ other ordered pairs (in $A \times A$) of the form (x, y), $x \neq y$. These 12 ordered pairs must be considered in sets of two in order to insure commutativity. For example, we need $g(a, b) = g(b, a)$ and may select any one of the four elements of A for $g(a, b)$. But then this choice must also be assigned to $g(b, a)$. Therefore, since there are four choices for each of these $12/2 = 6$ sets of two ordered pairs, we find that the number of commutative closed binary operations g on A is $4^4 \cdot 4^6 = 4^{10}$.

Definition 5.13

Let $f: A \times A \rightarrow B$ be a binary operation on A. An element $x \in A$ is called an *identity* (or *identity element*) for f if $f(a, x) = f(x, a) = a$, for all $a \in A$.

EXAMPLE 5.34

a) Consider the (closed) binary operation $f: \mathbf{Z} \times \mathbf{Z} \rightarrow \mathbf{Z}$, where $f(a, b) = a + b$. Here the integer 0 is an identity since $f(a, 0) = a + 0 = 0 + a = f(0, a) = a$, for each integer a.

b) We find that there is no identity for the function in part (a) of Example 5.29. For if f had an identity x, then for *any* $a \in \mathbf{Z}$, $f(a, x) = a \Rightarrow a - x = a \Rightarrow x = 0$. But then $f(x, a) = f(0, a) = 0 - a \neq a$, unless $a = 0$.

c) Let $A = \{1, 2, 3, 4, 5, 6, 7\}$, and let $g: A \times A \rightarrow A$ be the (closed) binary operation defined by $g(a, b) = \min\{a, b\}$—that is, the minimum (or smaller) of a, b. This binary operation is commutative and associative, and for any $a \in A$ we have $g(a, 7) = \min\{a, 7\} = a = \min\{7, a\} = g(7, a)$. So 7 is an identity element for g.

In parts (a) and (c) of Example 5.34 we examined two (closed) binary operations, each of which has an identity. Part (b) of that example showed that such an operation need not have an identity element. Could a binary operation have more than one identity? We find that the answer is no when we consider the following theorem.

THEOREM 5.4

Let $f: A \times A \rightarrow B$ be a binary operation. If f has an identity, then that identity is unique.
Proof: If f has more than one identity, let $x_1, x_2 \in A$ with

$$f(a, x_1) = a = f(x_1, a), \quad \text{for all } a \in A, \qquad \text{and}$$
$$f(a, x_2) = a = f(x_2, a), \quad \text{for all } a \in A.$$

Consider x_1 as an element of A and x_2 as an identity. Then $f(x_1, x_2) = x_1$. Now reverse the roles of x_1 and x_2—that is, consider x_2 as an element of A and x_1 as an identity. We find that $f(x_1, x_2) = x_2$. Consequently, $x_1 = x_2$, and f has at most one identity.

Now that we have settled the issue of the uniqueness of the identity element, let us see how this type of element enters into one more enumeration problem.

EXAMPLE 5.35

If $A = \{x, a, b, c, d\}$, how many closed binary operations on A have x as the identity?
Let $f: A \times A \rightarrow A$ with $f(x, y) = y = f(y, x)$ for all $y \in A$. Then we may represent f by a table as in Table 5.2. Here the nine values, where x is the first component—as in (x, c), or the second component—as in (d, x), are determined by the fact that x is the identity element. Each of the 16 remaining (vacant) entries in Table 5.2 can be filled with any one of the five elements in A.

Table 5.2

f	x	a	b	c	d
x	x	a	b	c	d
a	a	$-$	$-$	$-$	$-$
b	b	$-$	$-$	$-$	$-$
c	c	$-$	$-$	$-$	$-$
d	d	$-$	$-$	$-$	$-$

Hence there are 5^{16} closed binary operations on A where x is the identity. Of these $5^{10} = 5^4 \cdot 5^{(4^2-4)/2}$ are commutative. We also realize that there are 5^{16} closed binary operations on A where b is the identity. So there are $5^{17} = \binom{5}{1}5^{16} = \binom{5}{1}5^{5^2-[2(5)-1]} = \binom{5}{1}5^{(5-1)^2}$ closed binary operations on A that have an identity, and of these $5^{11} = \binom{5}{1}5^{10} = \binom{5}{1}5^45^{(4^2-4)/2}$ are commutative.

Having seen several examples of functions (in Examples 5.16(b), 5.29, 5.30, 5.32, 5.33, 5.34, and 5.35) where the domain is a cross product of sets, we now investigate functions where the domain is a subset of a cross product.

Definition 5.14

For sets A and B, if $D \subseteq A \times B$, then $\pi_A: D \to A$, defined by $\pi_A(a, b) = a$, is called the *projection* on the first coordinate. The function $\pi_B: D \to B$, defined by $\pi_B(a, b) = b$, is called the *projection* on the second coordinate.

We note that if $D = A \times B$ then π_A and π_B are both onto.

EXAMPLE 5.36

If $A = \{w, x, y\}$ and $B = \{1, 2, 3, 4\}$, let $D = \{(x, 1), (x, 2), (x, 3), (y, 1), (y, 4)\}$. Then the projection $\pi_A: D \to A$ satisfies $\pi_A(x, 1) = \pi_A(x, 2) = \pi_A(x, 3) = x$, and $\pi_A(y, 1) = \pi_A(y, 4) = y$. Since $\pi_A(D) = \{x, y\} \subset A$, this function is *not* onto.

For $\pi_B: D \to B$ we find that $\pi_B(x, 1) = \pi_B(y, 1) = 1$, $\pi_B(x, 2) = 2$, $\pi_B(x, 3) = 3$, and $\pi_B(y, 4) = 4$, so $\pi_B(D) = B$ and this projection is an onto function.

EXAMPLE 5.37

Let $A = B = \mathbf{R}$ and consider the set $D \subseteq A \times B$ where $D = \{(x, y) | y = x^2\}$. Then D represents the subset of the Euclidean plane that contains the points on the parabola $y = x^2$.

Among the infinite number of points in D we find the point $(3, 9)$. Here $\pi_A(3, 9) = 3$, the x-coordinate of $(3, 9)$, whereas $\pi_B(3, 9) = 9$, the y-coordinate of the point.

For this example, $\pi_A(D) = \mathbf{R} = A$, so π_A is onto. (The projection π_A is also one-to-one.) However, $\pi_B(D) = [0, +\infty) \subset \mathbf{R}$, so π_B is *not* onto. [Nor is it one-to-one — for example, $\pi_B(2, 4) = 4 = \pi_B(-2, 4)$.]

We now extend the notion of projection as follows. Let $A_1, A_2, \ldots, A_n$ be sets, and $\{i_1, i_2, \ldots, i_m\} \subseteq \{1, 2, \ldots, n\}$ with $i_1 < i_2 < \cdots < i_m$ and $m \leq n$. If $D \subseteq A_1 \times A_2 \times \cdots \times A_n = \times_{i=1}^n A_i$, then the function $\pi: D \to A_{i_1} \times A_{i_2} \times \cdots \times A_{i_m}$ defined by $\pi(a_1, a_2, \ldots, a_n) = (a_{i_1}, a_{i_2}, \ldots, a_{i_m})$ is the projection of D on the i_1th, i_2th, $\ldots$, i_mth coordinates. The elements of D are called (ordered) *n-tuples*; an element in $\pi(D)$ is an (ordered) *m-tuple*.

These projections arise in a natural way in the study of *relational data bases*, a standard technique for organizing and describing large quantities of data by modern large-scale computing systems. In situations like credit card transactions, not only must existing data be organized but new data must be inserted, as when credit cards are processed for new cardholders. When bills on existing accounts are paid, or when new purchases are made on these accounts, data must be updated. Another example arises when records are searched for special considerations, as when a college admissions office searches educational records seeking, for its mailing lists, high school students who have demonstrated certain levels of mathematical achievement.

The following example demonstrates the use of projections in a method for organizing and describing data on a somewhat smaller scale.

EXAMPLE 5.38   At a certain university the following sets are related for purposes of registration:

A_1 = the set of course numbers for courses offered in mathematics.

A_2 = the set of course titles offered in mathematics.

A_3 = the set of mathematics faculty.

A_4 = the set of letters of the alphabet.

Consider the *table*, or relation,[†] $D \subseteq A_1 \times A_2 \times A_3 \times A_4$ given in Table 5.3.

Table 5.3

Course Number	Course Title	Professor	Section Letter
MA 111	Calculus I	P. Z. Chinn	A
MA 111	Calculus I	V. Larney	B
MA 112	Calculus II	J. Kinney	A
MA 112	Calculus II	A. Schmidt	B
MA 112	Calculus II	R. Mines	C
MA 113	Calculus III	J. Kinney	A

The sets A_1, A_2, A_3, A_4 are called the *domains of the relational data base*, and *table D* is said to have *degree* 4. Each element of D is often called a *list*.

The projection of D on $A_1 \times A_3 \times A_4$ is shown in Table 5.4. Table 5.5 shows the results for the projection of D on $A_1 \times A_2$.

Table 5.4

Course Number	Professor	Section Letter
MA 111	P. Z. Chinn	A
MA 111	V. Larney	B
MA 112	J. Kinney	A
MA 112	A. Schmidt	B
MA 112	R. Mines	C
MA 113	J. Kinney	A

Table 5.5

Course Number	Course Title
MA 111	Calculus I
MA 112	Calculus II
MA 113	Calculus III

[†]Here the relation D is *not* binary. In fact, D is a *quaternary* relation.

Tables 5.4 and 5.5 are another way of representing the same data that appear in Table 5.3. Given Tables 5.4 and 5.5, one can recapture Table 5.3.

The theory of relational data bases is concerned with representing data in different ways and with the operations, such as projections, needed for such representations. The computer implementation of such techniques is also considered. More on this topic is mentioned in the exercises and chapter references.

EXERCISES 5.4

1. For $A = \{a, b, c\}$, let $f: A \times A \to A$ be the closed binary operation given in Table 5.6. Give an example to show that f is *not* associative.

Table 5.6

f	a	b	c
a	b	a	c
b	a	c	b
c	c	b	a

2. Let $f: \mathbf{R} \times \mathbf{R} \to \mathbf{Z}$ be the closed binary operation defined by $f(a, b) = \lceil a + b \rceil$. (a) Is f commutative? (b) Is f associative? (c) Does f have an identity element?

3. Each of the following functions $f: \mathbf{Z} \times \mathbf{Z} \to \mathbf{Z}$ is a closed binary operation on $\mathbf{Z}$. Determine in each case whether f is commutative and/or associative.

  a) $f(x, y) = x + y - xy$

  b) $f(x, y) = \max\{x, y\}$, the maximum (or larger) of x, y

  c) $f(x, y) = x^y$

  d) $f(x, y) = x + y - 3$

4. Which of the closed binary operations in Exercise 3 have an identity?

5. Let $|A| = 5$. (a) What is $|A \times A|$? (b) How many functions $f: A \times A \to A$ are there? (c) How many closed binary operations are there on A? (d) How many of these closed binary operations are commutative?

6. Let $A = \{x, a, b, c, d\}$.

  a) How many closed binary operations f on A satisfy $f(a, b) = c$?

  b) How many of the functions f in part (a) have x as an identity?

  c) How many of the functions f in part (a) have an identity?

  d) How many of the functions f in part (c) are commutative?

7. Let $f: \mathbf{Z}^+ \times \mathbf{Z}^+ \to \mathbf{Z}^+$ be the closed binary operation defined by $f(a, b) = \gcd(a, b)$. (a) Is f commutative? (b) Is f associative? (c) Does f have an identity element?

8. Let $A = \{2, 4, 8, 16, 32\}$, and consider the closed binary operation $f: A \times A \to A$ where $f(a, b) = \gcd(a, b)$. Does f have an identity element?

9. For distinct primes p, q let $A = \{p^m q^n \mid 0 \le m \le 31, 0 \le n \le 37\}$. (a) What is $|A|$? (b) If $f: A \times A \to A$ is the closed binary operation defined by $f(a, b) = \gcd(a, b)$, does f have an identity element?

10. State a result that generalizes the ideas presented in the previous two exercises.

11. For $\emptyset \neq A \subseteq \mathbf{Z}^+$, let $f, g: A \times A \to A$ be the closed binary operations defined by $f(a, b) = \min\{a, b\}$ and $g(a, b) = \max\{a, b\}$. Does f have an identity element? Does g?

12. Let $A = B = \mathbf{R}$. Determine $\pi_A(D)$ and $\pi_B(D)$ for each of the following sets $D \subseteq A \times B$.

  a) $D = \{(x, y) \mid x = y^2\}$

  b) $D = \{(x, y) \mid y = \sin x\}$

  c) $D = \{(x, y) \mid x^2 + y^2 = 1\}$

13. Let A_i, $1 \le i \le 5$, be the domains for a table $D \subseteq A_1 \times A_2 \times A_3 \times A_4 \times A_5$, where $A_1 = \{U, V, W, X, Y, Z\}$ (used as code names for different cereals in a test), and $A_2 = A_3 = A_4 = A_5 = \mathbf{Z}^+$. The table D is given as Table 5.7.

  a) What is the degree of the table?

  b) Find the projection of D on $A_3 \times A_4 \times A_5$.

  c) A domain of a table is called a *primary key* for the table if its value uniquely identifies each list of D. Determine the primary key(s) for this table.

14. Let A_i, $1 \le i \le 5$, be the domains for a table $D \subseteq A_1 \times A_2 \times A_3 \times A_4 \times A_5$, where $A_1 = \{1, 2\}$ (used to identify the daily vitamin capsule produced by two pharmaceutical companies), $A_2 = \{A, D, E\}$, and $A_3 = A_4 = A_5 = \mathbf{Z}^+$. The table D is given as Table 5.8.

  a) What is the degree of the table?

  b) What is the projection of D on $A_1 \times A_2$? on $A_3 \times A_4 \times A_5$?

  c) This table has no primary key. (See Exercise 13.) We can, however, define a *composite primary key* as the cross product of a *minimal* number of domains of the table, whose components, taken collectively, uniquely identify each list of D. Determine some composite primary keys for this table.

Table 5.7

Code Name of Cereal	Grams of Sugar per 1-oz Serving	% of RDA[a] of Vitamin A per 1-oz Serving	% of RDA of Vitamin C per 1-oz Serving	% of RDA of Protein per 1-oz Serving
U	1	25	25	6
V	7	25	2	4
W	12	25	2	4
X	0	60	40	20
Y	3	25	40	10
Z	2	25	40	10

[a]RDA = recommended daily allowance

Table 5.8

Vitamin Capsule	Vitamin Present in Capsule	Amount of Vitamin in Capsule in IU[a]	Dosage: Capsules / Day	No. of Capsules per Bottle
1	A	10,000	1	100
1	D	400	1	100
1	E	30	1	100
2	A	4,000	1	250
2	D	400	1	250
2	E	15	1	250

[a]IU = international units

5.5
The Pigeonhole Principle

A change of pace is in order as we introduce an interesting distribution principle. This principle may seem to have nothing in common with what we have been doing so far, but it will prove to be helpful nonetheless.

In mathematics one sometimes finds that an almost obvious idea, when applied in a rather subtle manner, is the key needed to solve a troublesome problem. On the list of such obvious ideas many would undoubtedly place the following rule, known as the *pigeonhole principle*.

The Pigeonhole Principle: If m pigeons occupy n pigeonholes and $m > n$, then at least one pigeonhole has two or more pigeons roosting in it.

One situation for $6 (= m)$ pigeons and $4 (= n)$ pigeonholes (actually birdhouses) is shown in Fig. 5.7. The general result readily follows by the method of proof by contradiction. If the result is not true, then each pigeonhole has at most one pigeon roosting in it—for a total of at most $n (< m)$ pigeons. (Somewhere we have lost at least $m - n$ pigeons!)

But now what can pigeons roosting in pigeonholes have to do with mathematics—discrete, combinatorial, or otherwise? Actually, this principle can be applied in various problems in which we seek to establish whether a certain situation can actually occur. We

Figure 5.7

illustrate this principle in the following examples and shall find it useful in Section 5.6 and at other points in the text.

EXAMPLE 5.39

An office employs 13 file clerks, so at least two of them must have birthdays during the same month. Here we have 13 pigeons (the file clerks) and 12 pigeonholes (the months of the year).

Here is a second rather immediate application of our principle.

EXAMPLE 5.40

Larry returns from the laundromat with 12 pairs of socks (each pair a different color) in a laundry bag. Drawing the socks from the bag randomly, he'll have to draw at most 13 of them to get a matched pair.

From this point on, application of the pigeonhole principle may be more subtle.

EXAMPLE 5.41

Wilma operates a computer with a magnetic tape drive. One day she is given a tape that contains 500,000 "words" of four or fewer lowercase letters. (Consecutive words on the tape are separated by a blank character.) Can it be that the 500,000 words are all distinct?

From the rules of sum and product, the total number of different possible words, using four or fewer letters, is

$$26^4 + 26^3 + 26^2 + 26 = 475{,}254.$$

With these 475,254 words as the pigeonholes, and the 500,000 words on the tape as the pigeons, it follows that at least one word is repeated on the tape.

EXAMPLE 5.42

Let $S \subset \mathbf{Z}^+$, where $|S| = 37$. Then S contains two elements that have the same remainder upon division by 36.

Here the pigeons are the 37 positive integers in S. We know from the division algorithm (of Theorem 4.5) that when any positive integer n is divided by 36, there exists a unique quotient q and unique remainder r, where

$$n = 36q + r, \qquad 0 \le r < 36.$$

The 36 possible values of r constitute the pigeonholes, and the result is now established by the pigeonhole principle.

EXAMPLE 5.43

Prove that if 101 integers are selected from the set $S = \{1, 2, 3, \ldots, 200\}$, then there are two integers such that one divides the other.

For each $x \in S$, we may write $x = 2^k y$, with $k \geq 0$, and $\gcd(2, y) = 1$. (This result follows from the Fundamental Theorem of Arithmetic.) Then y must be odd, so $y \in T = \{1, 3, 5, \ldots, 199\}$, where $|T| = 100$. Since 101 integers are selected from S, by the pigeonhole principle there are two distinct integers of the form $a = 2^m y$, $b = 2^n y$ for some (the same) $y \in T$. If $m < n$, then $a|b$; otherwise, we have $m > n$ and then $b|a$.

EXAMPLE 5.44

Any subset of size 6 from the set $S = \{1, 2, 3, \ldots, 9\}$ must contain two elements whose sum is 10.

Here the pigeons constitute a six-element subset of $\{1, 2, 3, \ldots, 9\}$, and the pigeonholes are the subsets $\{1, 9\}, \{2, 8\}, \{3, 7\}, \{4, 6\}, \{5\}$. When the six pigeons go to their respective pigeonholes, they must fill at least one of the two-element subsets whose members sum to 10.

EXAMPLE 5.45

Triangle ACE is equilateral with $AC = 1$. If five points are selected from the interior of the triangle, there are at least two whose distance apart is less than $1/2$.

For the triangle in Fig. 5.8, the four smaller triangles are congruent equilateral triangles and $AB = 1/2$. We break up the interior of triangle ACE into the following four regions, which are mutually disjoint in pairs:

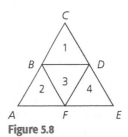

Figure 5.8

R_1: the interior of triangle BCD together with the points on the segment BD, excluding B and D.

R_2: the interior of triangle ABF.

R_3: the interior of triangle BDF together with the points on the segments BF and DF, excluding B, D, and F.

R_4: the interior of triangle FDE.

Now we apply the pigeonhole principle. Five points in the interior of triangle ACE must be such that at least two of them are in one of the four regions R_i, $1 \leq i \leq 4$, where any two points are separated by a distance less than $1/2$.

EXAMPLE 5.46

Let S be a set of six positive integers whose maximum is at most 14. Show that the sums of the elements in all the nonempty subsets of S cannot all be distinct.

For each nonempty subset A of S, the sum of the elements in A, denoted s_A, satisfies $1 \leq s_A \leq 9 + 10 + \cdots + 14 = 69$, and there are $2^6 - 1 = 63$ nonempty subsets of S. We should like to draw the conclusion from the pigeonhole principle by letting the possible sums, from 1 to 69, be the pigeonholes, with the 63 nonempty subsets of S as the pigeons, but then we have too few pigeons.

So instead of considering all nonempty subsets of S, we cut back to those nonempty subsets A of S where $|A| \leq 5$. Then for each such subset A it follows that $1 \leq s_A \leq 10 + 11 + \cdots + 14 = 60$. There are 62 nonempty subsets A of S with $|A| \leq 5$ — namely, all the subsets of S except for $\emptyset$ and the set S itself. With 62 pigeons (the nonempty subsets A of S where $|A| \leq 5$) and 60 pigeonholes (the possible sums s_A), it follows by the pigeonhole principle that the elements of at least two of these 62 subsets must yield the same sum.

EXAMPLE 5.47

Let $m \in \mathbf{Z}^+$ with m odd. Prove that there exists a positive integer n such that m divides $2^n - 1$.

Consider the $m + 1$ positive integers $2^1 - 1, 2^2 - 1, 2^3 - 1, \ldots, 2^m - 1, 2^{m+1} - 1$. By the pigeonhole principle and the division algorithm there exist $s, t \in \mathbf{Z}^+$ with $1 \leq s < t \leq m + 1$, where $2^s - 1$ and $2^t - 1$ have the same remainder upon division by m. Hence $2^s - 1 = q_1 m + r$ and $2^t - 1 = q_2 m + r$, for $q_1, q_2 \in \mathbf{N}$, and $(2^t - 1) - (2^s - 1) = (q_2 m + r) - (q_1 m + r)$, so $2^t - 2^s = (q_2 - q_1)m$. But $2^t - 2^s = 2^s(2^{t-s} - 1)$; and since m is odd, we have $\gcd(2^s, m) = 1$. Hence $m | (2^{t-s} - 1)$, and the result follows with $n = t - s$.

EXAMPLE 5.48

While on a four-week vacation, Herbert will play at least one set of tennis each day, but he won't play more than 40 sets total during this time. Prove that no matter how he distributes his sets during the four weeks, there is a span of consecutive days during which he will play exactly 15 sets.

For $1 \leq i \leq 28$, let x_i be the total number of sets Herbert will play from the start of the vacation to the end of the ith day. Then $1 \leq x_1 < x_2 < \cdots < x_{28} \leq 40$, and $x_1 + 15 < \cdots < x_{28} + 15 \leq 55$. We now have the 28 distinct numbers $x_1, x_2, \ldots, x_{28}$ and the 28 distinct numbers $x_1 + 15, x_2 + 15, \ldots, x_{28} + 15$. These 56 numbers can take on only 55 different values, so at least two of them must be equal, and we conclude that there exist $1 \leq j < i \leq 28$ with $x_i = x_j + 15$. Hence, from the start of day $j + 1$ to the end of day i, Herbert will play exactly 15 sets of tennis.

Our last example for this section deals with a classic result that was first discovered in 1935 by Paul Erdös and George Szekeres.

EXAMPLE 5.49

Let us start by considering two particular examples:

1) Note how the sequence 6, 5, 8, 3, 7 (of length 5) contains the decreasing subsequence 6, 5, 3 (of length 3).

2) Now note how the sequence 11, 8, 7, 1, 9, 6, 5, 10, 3, 12 (of length 10) contains the increasing subsequence 8, 9, 10, 12 (of length 4).

These two instances demonstrate the general result: For each $n \in \mathbf{Z}^+$, a sequence of $n^2 + 1$ distinct real numbers contains a decreasing or increasing subsequence of length $n + 1$.

To verify this claim let $a_1, a_2, \ldots, a_{n^2+1}$ be a sequence of $n^2 + 1$ distinct real numbers. For $1 \leq k \leq n^2 + 1$, let

$x_k =$ the maximum length of a decreasing subsequence that ends with a_k, and

$y_k =$ the maximum length of an increasing subsequence that ends with a_k.

For instance, our second particular example would provide

k	1	2	3	4	5	6	7	8	9	10
a_k	11	8	7	1	9	6	5	10	3	12
x_k	1	2	3	4	2	4	5	2	6	1
y_k	1	1	1	1	2	2	2	3	2	4

If, in general, there is no decreasing or increasing subsequence of length $n + 1$, then $1 \leq x_k \leq n$ and $1 \leq y_k \leq n$ for all $1 \leq k \leq n^2 + 1$. Consequently, there are at most n^2 *distinct* ordered pairs (x_k, y_k). But we have $n^2 + 1$ ordered pairs (x_k, y_k), since $1 \leq k \leq n^2 + 1$. So the pigeonhole principle implies that there are two identical ordered pairs (x_i, y_i), (x_j, y_j), where $i \neq j$ — say $i < j$. Now the real numbers $a_1, a_2, \ldots, a_{n^2+1}$ are distinct, so if $a_i < a_j$ then $y_i < y_j$, while if $a_j < a_i$ then $x_j > x_i$. In either case we no longer have $(x_i, y_i) = (x_j, y_j)$. This contradiction tells us that $x_k = n + 1$ or $y_k = n + 1$ for some $n + 1 \leq k \leq n^2 + 1$; the result then follows.

For an interesting application of this result, consider $n^2 + 1$ sumo wrestlers facing forward and standing shoulder to shoulder. (Here no two wrestlers have the same weight.) We can select $n + 1$ of these wrestlers to take one step forward so that, as they are scanned from left to right, their successive weights either decrease or increase.

EXERCISES 5.5

1. In Example 5.40, what plays the roles of the pigeons and of the pigeonholes?

2. Show that if eight people are in a room, at least two of them have birthdays that occur on the same day of the week.

3. An auditorium has a seating capacity of 800. How many seats must be occupied to guarantee that at least two people seated in the auditorium have the same first and last initials?

4. Let $S = \{3, 7, 11, 15, 19, \ldots, 95, 99, 103\}$. How many elements must we select from S to insure that there will be at least two whose sum is 110?

5. a) Prove that if 151 integers are selected from $\{1, 2, 3, \ldots, 300\}$, then the selection must include two integers x, y where $x|y$ or $y|x$.

b) Write a statement that generalizes the results of part (a) and Example 5.43.

6. Prove that if we select 101 integers from the set $S = \{1, 2, 3, \ldots, 200\}$, there exist m, n in the selection where $\gcd(m, n) = 1$.

7. a) Show that if any 14 integers are selected from the set $S = \{1, 2, 3, \ldots, 25\}$, there are at least two whose sum is 26.

b) Write a statement that generalizes the results of part (a) and Example 5.44.

8. a) If $S \subseteq \mathbf{Z}^+$ and $|S| \geq 3$, prove that there exist distinct $x, y \in S$ where $x + y$ is even.

b) Let $S \subseteq \mathbf{Z}^+ \times \mathbf{Z}^+$. Find the minimal value of $|S|$ that guarantees the existence of distinct ordered pairs $(x_1, x_2), (y_1, y_2) \in S$ such that $x_1 + y_1$ and $x_2 + y_2$ are both even.

c) Extending the ideas in parts (a) and (b), consider $S \subseteq \mathbf{Z}^+ \times \mathbf{Z}^+ \times \mathbf{Z}^+$. What size must $|S|$ be to guarantee the existence of distinct ordered triples $(x_1, x_2, x_3), (y_1, y_2, y_3) \in S$ where $x_1 + y_1$, $x_2 + y_2$, and $x_3 + y_3$ are all even?

d) Generalize the results of parts (a), (b), and (c).

e) A point $P(x, y)$ in the Cartesian plane is called a *lattice point* if $x, y \in \mathbf{Z}$. Given distinct lattice points $P_1(x_1, y_1), P_2(x_2, y_2), \ldots, P_n(x_n, y_n)$, determine the smallest value of n that guarantees the existence of $P_i(x_i, y_i), P_j(x_j, y_j)$, $1 \leq i < j \leq n$, such that the midpoint of the line segment connecting $P_i(x_i, y_i)$ and $P_j(x_j, y_j)$ is also a lattice point.

9. a) If 11 integers are selected from $\{1, 2, 3, \ldots, 100\}$, prove that there are at least two, say x and y, such that $0 < |\sqrt{x} - \sqrt{y}| < 1$.

b) Write a statement that generalizes the result of part (a).

10. Let triangle ABC be equilateral, with $AB = 1$. Show that if we select 10 points in the interior of this triangle, there must be at least two whose distance apart is less than $1/3$.

11. Let $ABCD$ be a square with $AB = 1$. Show that if we select five points in the interior of this square, there are at least two whose distance apart is less than $1/\sqrt{2}$.

12. Let $A \subseteq \{1, 2, 3, \ldots, 25\}$ where $|A| = 9$. For any subset B of A let s_B denote the sum of the elements in B. Prove that there are distinct subsets C, D of A such that $|C| = |D| = 5$ and $s_C = s_D$.

13. Let S be a set of five positive integers the maximum of which is at most 9. Prove that the sums of the elements in all the nonempty subsets of S cannot all be distinct.

14. During the first six weeks of his senior year in college, Brace sends out at least one resumé each day but no more than 60 resumés in total. Show that there is a period of consecutive days during which he sends out exactly 23 resumés.

15. Let $S \subset \mathbf{Z}^+$ with $|S| = 7$. For $\emptyset \neq A \subseteq S$, let s_A denote the sum of the elements in A. If m is the maximum element in S, find the possible values of m so that there will exist distinct subsets B, C of S with $s_B = s_C$.

16. Let $k \in \mathbf{Z}^+$. Prove that there exists a positive integer n such that $k \mid n$ and the only digits in n are 0's and 3's.

17. a) Find a sequence of four distinct real numbers with no decreasing or increasing subsequence of length 3.

b) Find a sequence of nine distinct real numbers with no decreasing or increasing subsequence of length 4.

c) Generalize the results in parts (a) and (b).

d) What do the preceding parts of this exercise tell us about Example 5.49?

18. The 50 members of Nardine's aerobics class line up to get their equipment. Assuming that no two of these people have the same height, show that eight of them (as the line is equipped from first to last) have successive heights that either decrease or increase.

19. For $k, n \in \mathbf{Z}^+$, prove that if $kn + 1$ pigeons occupy n pigeonholes, then at least one pigeonhole has $k + 1$ or more pigeons roosting in it.

20. How many times must we roll a single die in order to get the same score (a) at least twice? (b) at least three times? (c) at least n times, for $n \geq 4$?

21. a) Let $S \subset \mathbf{Z}^+$. What is the smallest value for $|S|$ that guarantees the existence of two elements x, $y \in S$ where x and y have the same remainder upon division by 1000?

b) What is the smallest value of n such that whenever $S \subseteq \mathbf{Z}^+$ and $|S| = n$, then there exist three elements x, y, $z \in S$ where all three have the same remainder upon division by 1000?

c) Write a statement that generalizes the results of parts (a) and (b) and Example 5.42.

22. For $m, n \in \mathbf{Z}^+$, prove that if m pigeons occupy n pigeonholes, then at least one pigeonhole has $\lfloor (m-1)/n \rfloor + 1$ or more pigeons roosting in it.

23. Let $p_1, p_2, \ldots, p_n \in \mathbf{Z}^+$. Prove that if $p_1 + p_2 + \cdots + p_n - n + 1$ pigeons occupy n pigeonholes, then either the first pigeonhole has p_1 or more pigeons roosting in it, or the second pigeonhole has p_2 or more pigeons roosting in it, $\ldots$, or the nth pigeonhole has p_n or more pigeons roosting in it.

24. Given 8 Perl books, 17 Visual BASIC[†] books, 6 Java books, 12 SQL books, and 20 C++ books, how many of these books must we select to insure that we have 10 books dealing with the same computer language?

5.6
Function Composition
and Inverse Functions

When computing with the elements of $\mathbf{Z}$, we find that the (closed binary) operation of addition provides a method for combining two integers, say a and b, into a third integer, namely $a + b$. Furthermore, for each integer c there is a second integer d where $c + d = d + c = 0$, and we call d the additive *inverse* of c. (It is also true that c is the additive *inverse* of d.)

Turning to the elements of $\mathbf{R}$ and the (closed binary) operation of multiplication, we have a method for combining any r, $s \in \mathbf{R}$ into their product rs. And here, for each $t \in \mathbf{R}$, if $t \neq 0$, then there is a real number u such that $ut = tu = 1$. The real number u is called the multiplicative *inverse* of t. (The real number t is also the multiplicative *inverse* of u.)

In this section we first study a method for combining two functions into a single function. Then we develop the concept of the inverse (of a function) for functions with certain properties. To accomplish these objectives, we need the following preliminary ideas.

[†]Visual BASIC is a trademark of the Microsoft Corporation.

Having examined functions that are one-to-one and those that are onto, we turn now to functions with both of these properties.

Definition 5.15

If $f: A \to B$, then f is said to be *bijective*, or to be a *one-to-one correspondence*, if f is both one-to-one and onto.

EXAMPLE 5.50

If $A = \{1, 2, 3, 4\}$ and $B = \{w, x, y, z\}$, then $f = \{(1, w), (2, x), (3, y), (4, z)\}$ is a one-to-one correspondence from A (on)to B, and $g = \{(w, 1), (x, 2), (y, 3), (z, 4)\}$ is a one-to-one correspondence from B (on)to A.

It should be pointed out that whenever the term *correspondence* was used in Chapter 1 and in Examples 3.11 and 4.12, the adjective *one-to-one* was implied though never stated.

For any nonempty set A there is always a very simple but important one-to-one correspondence, as seen in the following definition.

Definition 5.16

The function $1_A: A \to A$, defined by $1_A(a) = a$ for all $a \in A$, is called the *identity function* for A.

Definition 5.17

If $f, g: A \to B$, we say that f and g are *equal* and write $f = g$, if $f(a) = g(a)$ for all $a \in A$.

A common pitfall in dealing with the equality of functions occurs when f and g are functions with a common domain A and $f(a) = g(a)$ for all $a \in A$. It may *not* be the case that $f = g$. The pitfall results from not paying attention to the codomains of the functions.

EXAMPLE 5.51

Let $f: \mathbf{Z} \to \mathbf{Z}$, $g: \mathbf{Z} \to \mathbf{Q}$ where $f(x) = x = g(x)$, for all $x \in \mathbf{Z}$. Then f, g share the common domain $\mathbf{Z}$, have the same range $\mathbf{Z}$, and act the same on every element of $\mathbf{Z}$. Yet $f \neq g$! Here f is a one-to-one correspondence, whereas g is one-to-one but not onto; so the codomains do make a difference.

EXAMPLE 5.52

Consider the functions $f, g: \mathbf{R} \to \mathbf{Z}$ defined as follows:

$$f(x) = \begin{cases} x, & \text{if } x \in \mathbf{Z} \\ \lfloor x \rfloor + 1, & \text{if } x \in \mathbf{R} - \mathbf{Z} \end{cases} \qquad g(x) = \lceil x \rceil, \text{ for all } x \in \mathbf{R}$$

If $x \in \mathbf{Z}$, then $f(x) = x = \lceil x \rceil = g(x)$.

For $x \in \mathbf{R} - \mathbf{Z}$, write $x = n + r$ where $n \in \mathbf{Z}$ and $0 < r < 1$. (For example, if $x = 2.3$, we write $2.3 = 2 + 0.3$, with $n = 2$ and $r = 0.3$; for $x = -7.3$ we have $-7.3 = -8 + 0.7$, with $n = -8$ and $r = 0.7$.) Then

$$f(x) = \lfloor x \rfloor + 1 = n + 1 = \lceil x \rceil = g(x).$$

Consequently, even though the functions f, g are defined by *different* formulas, we realize that they are the *same* function — because they have the same domain and codomain and $f(x) = g(x)$ for all x in the domain $\mathbf{R}$.

Now that we have dispensed with the necessary preliminaries, it is time to examine an operation for combining two appropriate functions.

Definition 5.18

If $f: A \to B$ and $g: B \to C$, we define the *composite function*, which is denoted $g \circ f: A \to C$, by $(g \circ f)(a) = g(f(a))$, for each $a \in A$.

EXAMPLE 5.53

Let $A = \{1, 2, 3, 4\}$, $B = \{a, b, c\}$, and $C = \{w, x, y, z\}$ with $f: A \to B$ and $g: B \to C$ given by $f = \{(1, a), (2, a), (3, b), (4, c)\}$ and $g = \{(a, x), (b, y), (c, z)\}$. For each element of A we find:

$$(g \circ f)(1) = g(f(1)) = g(a) = x \qquad (g \circ f)(3) = g(f(3)) = g(b) = y$$
$$(g \circ f)(2) = g(f(2)) = g(a) = x \qquad (g \circ f)(4) = g(f(4)) = g(c) = z$$

So

$$g \circ f = \{(1, x), (2, x), (3, y), (4, z)\}.$$

Note: The composition $f \circ g$ is *not* defined.

EXAMPLE 5.54

Let $f: \mathbf{R} \to \mathbf{R}$, $g: \mathbf{R} \to \mathbf{R}$ be defined by $f(x) = x^2$, $g(x) = x + 5$. Then

$$(g \circ f)(x) = g(f(x)) = g(x^2) = x^2 + 5,$$

whereas

$$(f \circ g)(x) = f(g(x)) = f(x + 5) = (x + 5)^2 = x^2 + 10x + 25.$$

Here $g \circ f: \mathbf{R} \to \mathbf{R}$ and $f \circ g: \mathbf{R} \to \mathbf{R}$, but $(g \circ f)(1) = 6 \neq 36 = (f \circ g)(1)$, so even though both composites $f \circ g$ and $g \circ f$ can be formed, we do not have $f \circ g = g \circ f$. Consequently, the composition of functions is not, in general, a commutative operation.

The definition and examples for composite functions required that the codomain of f = domain of g. If range of $f \subseteq$ domain of g, this will actually be enough to yield the composite function $g \circ f: A \to C$. Also, for any $f: A \to B$, we observe that $f \circ 1_A = f = 1_B \circ f$.

An important recurring idea in mathematics is the investigation of whether combining two entities with a common property yields a result with this property. For example, if A and B are finite sets, then $A \cap B$ and $A \cup B$ are also finite. However, for infinite sets A and B, we have $A \cup B$ infinite but $A \cap B$ could be finite.

For the composition of functions we have the following result.

THEOREM 5.5

Let $f: A \to B$ and $g: B \to C$.

a) If f and g are one-to-one, then $g \circ f$ is one-to-one.

b) If f and g are onto, then $g \circ f$ is onto.

Proof:

a) To prove that $g \circ f: A \to C$ is one-to-one, let $a_1, a_2 \in A$ with $(g \circ f)(a_1) = (g \circ f)(a_2)$. Then $(g \circ f)(a_1) = (g \circ f)(a_2) \Rightarrow g(f(a_1)) = g(f(a_2)) \Rightarrow f(a_1) = f(a_2)$, because g is one-to-one. Also, $f(a_1) = f(a_2) \Rightarrow a_1 = a_2$, because f is one-to-one. Consequently, $g \circ f$ is one-to-one.

b) For $g \circ f \colon A \to C$, let $z \in C$. Since g is onto, there exists $y \in B$ with $g(y) = z$. With f onto and $y \in B$, there exists $x \in A$ with $f(x) = y$. Hence $z = g(y) = g(f(x)) = (g \circ f)(x)$, so the range of $g \circ f = C =$ the codomain of $g \circ f$, and $g \circ f$ is onto.

Although function composition is not commutative, if $f \colon A \to B$, $g \colon B \to C$, and $h \colon C \to D$, what can we say about the functions $(h \circ g) \circ f$ and $h \circ (g \circ f)$? Specifically, is $(h \circ g) \circ f = h \circ (g \circ f)$? That is, is function composition associative?

Before considering the general result, let us first investigate a particular example.

| **EXAMPLE 5.55** | Let $f, g, h \colon \mathbf{R} \to \mathbf{R}$, where $f(x) = x^2$, $g(x) = x + 5$, and $h(x) = \sqrt{x^2 + 2}$. |

Then $((h \circ g) \circ f)(x) = (h \circ g)(f(x)) = (h \circ g)(x^2) = h(g(x^2)) = h(x^2 + 5) = \sqrt{(x^2 + 5)^2 + 2} = \sqrt{x^4 + 10x^2 + 27}$.

On the other hand, we see that $(h \circ (g \circ f))(x) = h((g \circ f)(x)) = h(g(f(x))) = h(g(x^2)) = h(x^2 + 5) = \sqrt{(x^2 + 5)^2 + 2} = \sqrt{x^4 + 10x^2 + 27}$, as above.

So in this particular example, $(h \circ g) \circ f$ and $h \circ (g \circ f)$ are two functions with the same domain and codomain, and for all $x \in \mathbf{R}$, $((h \circ g) \circ f)(x) = \sqrt{x^4 + 10x^2 + 27} = (h \circ (g \circ f))(x)$. Consequently, $(h \circ g) \circ f = h \circ (g \circ f)$.

We now find that the result in Example 5.55 is true in general.

THEOREM 5.6   If $f \colon A \to B$, $g \colon B \to C$, and $h \colon C \to D$, then $(h \circ g) \circ f = h \circ (g \circ f)$.

Proof: Since the two functions have the same domain, A, and codomain, D, the result will follow by showing that for every $x \in A$, $((h \circ g) \circ f)(x) = (h \circ (g \circ f))(x)$. (See the diagram shown in Fig. 5.9.)

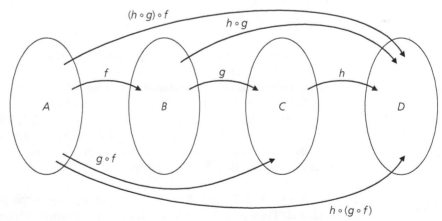

Figure 5.9

Using the definition of the composite function we know that for each $x \in A$ it takes two steps to determine $(g \circ f)(x)$. First we find $f(x)$, the image of x under f. This is an element of B. Then we apply the function g to the element $f(x)$ to determine $g(f(x))$, the image of $f(x)$ under g. This results in an element of C. At this point we apply the function h to the element $g(f(x))$ to determine $h(g(f(x))) = h((g \circ f)(x)) = (h \circ (g \circ f))(x)$. This result is an element of D. Similarly, starting once again with x in A, we have $f(x)$ in B, and now we apply the composite function $h \circ g$ to $f(x)$. This gives us $((h \circ g) \circ f)(x) = (h \circ g)(f(x)) = h(g(f(x)))$.

Since $((h \circ g) \circ f)(x) = h(g(f(x))) = (h \circ (g \circ f))(x)$, for each x in A, it now follows that

$$(h \circ g) \circ f = h \circ (g \circ f).$$

Consequently, the composition of functions is an associative operation.

By virtue of the associative property for function composition, we can write $h \circ g \circ f$, $(h \circ g) \circ f$ or $h \circ (g \circ f)$ without any problem of ambiguity. In addition, this property enables us to define powers of functions, where appropriate.

Definition 5.19 If $f: A \to A$ we define $f^1 = f$, and for $n \in \mathbf{Z}^+$, $f^{n+1} = f \circ (f^n)$.

This definition is another example wherein the result is defined *recursively*. With $f^{n+1} = f \circ (f^n)$, we see the dependence of f^{n+1} on a previous power, namely, f^n.

EXAMPLE 5.56 With $A = \{1, 2, 3, 4\}$ and $f: A \to A$ defined by $f = \{(1, 2), (2, 2), (3, 1), (4, 3)\}$, we have $f^2 = f \circ f = \{(1, 2), (2, 2), (3, 2), (4, 1)\}$ and $f^3 = f \circ f^2 = f \circ f \circ f = \{(1, 2), (2, 2), (3, 2), (4, 2)\}$. (What are f^4, f^5?)

We now come to the last new idea for this section: the existence of the invertible function and some of its properties.

Definition 5.20 For sets A, B, if $\mathcal{R}$ is a relation from A to B, then the *converse* of $\mathcal{R}$, denoted $\mathcal{R}^c$, is the relation from B to A defined by $\mathcal{R}^c = \{(b, a) | (a, b) \in \mathcal{R}\}$.

To get $\mathcal{R}^c$ from $\mathcal{R}$, we simply interchange the components of each ordered pair in $\mathcal{R}$. So if $A = \{1, 2, 3, 4\}$, $B = \{w, x, y\}$, and $\mathcal{R} = \{(1, w), (2, w), (3, x)\}$, then $\mathcal{R}^c = \{(w, 1), (w, 2), (x, 3)\}$, a relation from B to A.

Since a function is a relation we can also form the converse of a function. For the same preceding sets A, B, let $f: A \to B$ where $f = \{(1, w), (2, x), (3, y), (4, x)\}$. Then $f^c = \{(w, 1), (x, 2), (y, 3), (x, 4)\}$, a relation, but not a function, from B to A. We wish to investigate when the converse of a function yields a function, but before getting too abstract let us consider the following example.

EXAMPLE 5.57 For $A = \{1, 2, 3\}$ and $B = \{w, x, y\}$, let $f: A \to B$ be given by $f = \{(1, w), (2, x), (3, y)\}$. Then $f^c = \{(w, 1), (x, 2), (y, 3)\}$ is a function from B to A, and we observe that $f^c \circ f = 1_A$ and $f \circ f^c = 1_B$.

This finite example leads us to the following definition.

Definition 5.21 If $f: A \to B$, then f is said to be *invertible* if there is a function $g: B \to A$ such that $g \circ f = 1_A$ and $f \circ g = 1_B$.

Note that the function g in Definition 5.21 is also invertible.

EXAMPLE 5.58

Let $f, g: \mathbf{R} \to \mathbf{R}$ be defined by $f(x) = 2x + 5$, $g(x) = (1/2)(x - 5)$. Then $(g \circ f)(x) = g(f(x)) = g(2x + 5) = (1/2)[(2x + 5) - 5] = x$, and $(f \circ g)(x) = f(g(x)) = f((1/2)(x - 5)) = 2[(1/2)(x - 5)] + 5 = x$, so $f \circ g = 1_{\mathbf{R}}$ and $g \circ f = 1_{\mathbf{R}}$.
Consequently, f and g are both invertible functions.

Having seen some examples of invertible functions, we now wish to show that the function g of Definition 5.21 is unique. Then we shall find the means to identify an invertible function.

THEOREM 5.7

If a function $f: A \to B$ is invertible and a function $g: B \to A$ satisfies $g \circ f = 1_A$ and $f \circ g = 1_B$, then this function g is unique.

Proof: If g is not unique, then there is another function $h: B \to A$ with $h \circ f = 1_A$ and $f \circ h = 1_B$. Consequently, $h = h \circ 1_B = h \circ (f \circ g) = (h \circ f) \circ g = 1_A \circ g = g$.

As a result of this theorem we shall call the function g *the inverse* of f and shall adopt the notation $g = f^{-1}$. Theorem 5.7 also implies that $f^{-1} = f^c$.

We also see that whenever f is an invertible function, so is the function f^{-1}, and $(f^{-1})^{-1} = f$, again by the uniqueness in Theorem 5.7. But we still do not know what conditions on f insure that f is invertible.

Before stating our next theorem we note that the invertible functions of Examples 5.57 and 5.58 are all bijective. Consequently, these examples provide some motivation for the following result.

THEOREM 5.8

A function $f: A \to B$ is invertible if and only if it is one-to-one and onto.

Proof: Assuming that $f: A \to B$ is invertible, we have a unique function $g: B \to A$ with $g \circ f = 1_A$, $f \circ g = 1_B$. If $a_1, a_2 \in A$ with $f(a_1) = f(a_2)$, then $g(f(a_1)) = g(f(a_2))$, or $(g \circ f)(a_1) = (g \circ f)(a_2)$. With $g \circ f = 1_A$ it follows that $a_1 = a_2$, so f is one-to-one. For the onto property, let $b \in B$. Then $g(b) \in A$, so we can talk about $f(g(b))$. Since $f \circ g = 1_B$, we have $b = 1_B(b) = (f \circ g)(b) = f(g(b))$, so f is onto.

Conversely, suppose $f: A \to B$ is bijective. Since f is onto, for each $b \in B$ there is an $a \in A$ with $f(a) = b$. Consequently, we define the function $g: B \to A$ by $g(b) = a$, where $f(a) = b$. This definition yields a unique function. The only problem that could arise is if $g(b) = a_1 \neq a_2 = g(b)$ because $f(a_1) = b = f(a_2)$. However, this situation cannot arise because f is one-to-one. Our definition of g is such that $g \circ f = 1_A$ and $f \circ g = 1_B$, so we find that f is invertible, with $g = f^{-1}$.

EXAMPLE 5.59

From Theorem 5.8 it follows that the function $f_1: \mathbf{R} \to \mathbf{R}$ defined by $f_1(x) = x^2$ is not invertible (it is neither one-to-one nor onto), but $f_2: [0, +\infty) \to [0, +\infty)$ defined by $f_2(x) = x^2$ is invertible with $f_2^{-1}(x) = \sqrt{x}$.

The next result combines the ideas of function composition and inverse functions. The proof is left to the reader.

THEOREM 5.9	If $f: A \to B$, $g: B \to C$ are invertible functions, then $g \circ f: A \to C$ is invertible and $(g \circ f)^{-1} = f^{-1} \circ g^{-1}$.

Having seen some examples of functions and their inverses, one might wonder whether there is an algebraic method to determine the inverse of an invertible function. If the function is finite, we simply interchange the components of the given ordered pairs. But what if the function is defined by a formula, as in Example 5.59? Fortunately, the algebraic manipulations prove to be little more than a careful analysis of "interchanging the components of the ordered pairs." This is demonstrated in the following examples.

EXAMPLE 5.60	For $m, b \in \mathbf{R}$, $m \neq 0$, the function $f: \mathbf{R} \to \mathbf{R}$ defined by $f = \{(x, y) \mid y = mx + b\}$ is an invertible function, because it is one-to-one and onto.

To get f^{-1} we note that

$$\begin{aligned} f^{-1} &= \{(x, y) \mid y = mx + b\}^c = \{(y, x) \mid y = mx + b\} \\ &= \underbrace{\{(x, y) \mid x = my + b\}}_{} = \{(x, y) \mid y = (1/m)(x - b)\}. \end{aligned}$$

This is where we rename the variables (replacing x by y and y by x) in order to change the components of the ordered pairs of f.

So $f: \mathbf{R} \to \mathbf{R}$ is defined by $f(x) = mx + b$, and $f^{-1}: \mathbf{R} \to \mathbf{R}$ is defined by $f^{-1}(x) = (1/m)(x - b)$.

EXAMPLE 5.61	Let $f: \mathbf{R} \to \mathbf{R}^+$ be defined by $f(x) = e^x$, where $e \doteq 2.7183$, the base for the natural logarithm. From the graph in Fig. 5.10 we see that f is one-to-one and onto, so $f^{-1}: \mathbf{R}^+ \to \mathbf{R}$ does exist and $f^{-1} = \{(x, y) \mid y = e^x\}^c = \{(x, y) \mid x = e^y\} = \{(x, y) \mid y = \ln x\}$, so $f^{-1}(x) = \ln x$.

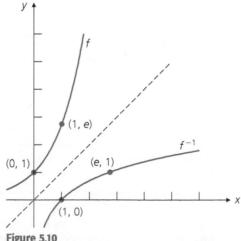

Figure 5.10

We should note that what happens in Fig. 5.10 happens in general. That is, the graphs of f and f^{-1} are symmetric about the line $y = x$. For example, the line segment connecting the points $(1, e)$ and $(e, 1)$ would be bisected by the line $y = x$. This is true for any corresponding pair of points $(x, f(x))$ and $(f(x), f^{-1}(f(x)))$.

This example also yields the following formulas:

$$x = 1_{\mathbf{R}}(x) = (f^{-1} \circ f)(x) = \ln(e^x), \quad \text{for all } x \in \mathbf{R}.$$
$$x = 1_{\mathbf{R}^+}(x) = (f \circ f^{-1})(x) = e^{\ln x}, \quad \text{for all } x > 0.$$

Even when a function $f: A \to B$ is not invertible, we find use for the symbol f^{-1} in the following sense.

Definition 5.22   If $f: A \to B$ and $B_1 \subseteq B$, then $f^{-1}(B_1) = \{x \in A \mid f(x) \in B_1\}$. The set $f^{-1}(B_1)$ is called the *preimage of B_1 under f.*

Be careful! We are now using the symbol f^{-1} in two different ways. Although we have the concept of a preimage for any function, not every function has an inverse function. Consequently, we cannot assume the existence of an inverse for a function f just because we find the symbol f^{-1} being used. A little caution is needed here.

EXAMPLE 5.62   Let $A = \{1, 2, 3, 4, 5, 6\}$ and $B = \{6, 7, 8, 9, 10\}$. If $f: A \to B$ with $f = \{(1, 7), (2, 7), (3, 8), (4, 6), (5, 9), (6, 9)\}$, then the following results are obtained.

a) For $B_1 = \{6, 8\} \subseteq B$, we have $f^{-1}(B_1) = \{3, 4\}$, since $f(3) = 8$ and $f(4) = 6$, and for any $a \in A$, $f(a) \notin B_1$ unless $a = 3$ or $a = 4$. Here we also note that $|f^{-1}(B_1)| = 2 = |B_1|$.

b) In the case of $B_2 = \{7, 8\} \subseteq B$, since $f(1) = f(2) = 7$ and $f(3) = 8$, we find that the preimage of B_2 under f is $\{1, 2, 3\}$. And here $|f^{-1}(B_2)| = 3 > 2 = |B_2|$.

c) Now consider the subset $B_3 = \{8, 9\}$ of B. For this case it follows that $f^{-1}(B_3) = \{3, 5, 6\}$ because $f(3) = 8$ and $f(5) = f(6) = 9$. Also we find that $|f^{-1}(B_3)| = 3 > 2 = |B_3|$.

d) If $B_4 = \{8, 9, 10\} \subseteq B$, then with $f(3) = 8$ and $f(5) = f(6) = 9$, we have $f^{-1}(B_4) = \{3, 5, 6\}$. So $f^{-1}(B_4) = f^{-1}(B_3)$ even though $B_4 \supset B_3$. This result follows because there is no element a in the domain A where $f(a) = 10$ — that is, $f^{-1}(\{10\}) = \emptyset$.

e) Finally, when $B_5 = \{8, 10\}$ we find that $f^{-1}(B_5) = \{3\}$ since $f(3) = 8$ and, as in part (d), $f^{-1}(\{10\}) = \emptyset$. In this case $|f^{-1}(B_5)| = 1 < 2 = |B_5|$.

Whenever $f: A \to B$, then for each $b \in B$ we shall write $f^{-1}(b)$ instead of $f^{-1}(\{b\})$. For the function in Example 5.62, we find that

$$f^{-1}(6) = \{4\} \quad f^{-1}(7) = \{1, 2\} \quad f^{-1}(8) = \{3\} \quad f^{-1}(9) = \{5, 6\} \quad f^{-1}(10) = \emptyset.$$

EXAMPLE 5.63   Let $f: \mathbf{R} \to \mathbf{R}$ be defined by

$$f(x) = \begin{cases} 3x - 5, & x > 0 \\ -3x + 1, & x \le 0. \end{cases}$$

a) Determine $f(0)$, $f(1)$, $f(-1)$, $f(5/3)$, and $f(-5/3)$.

b) Find $f^{-1}(0)$, $f^{-1}(1)$, $f^{-1}(-1)$, $f^{-1}(3)$, $f^{-1}(-3)$, and $f^{-1}(-6)$.

c) What are $f^{-1}([-5, 5])$ and $f^{-1}([-6, 5])$?

a) $f(0) = -3(0) + 1 = 1$ $f(5/3) = 3(5/3) - 5 = 0$
$f(1) = 3(1) - 5 = -2$ $f(-5/3) = -3(-5/3) + 1 = 6$
$f(-1) = -3(-1) + 1 = 4$

b) $f^{-1}(0) = \{x \in \mathbf{R} \mid f(x) \in \{0\}\} = \{x \in \mathbf{R} \mid f(x) = 0\}$
     $= \{x \in \mathbf{R} \mid x > 0 \text{ and } 3x - 5 = 0\} \cup \{x \in \mathbf{R} \mid x \leq 0 \text{ and } -3x + 1 = 0\}$
     $= \{x \in \mathbf{R} \mid x > 0 \text{ and } x = 5/3\} \cup \{x \in \mathbf{R} \mid x \leq 0 \text{ and } x = 1/3\}$
     $= \{5/3\} \cup \emptyset = \{5/3\}$

[Note how the horizontal line $y = 0$ — that is, the x-axis — intersects the graph in Fig. 5.11 only at the point $(5/3, 0)$.]

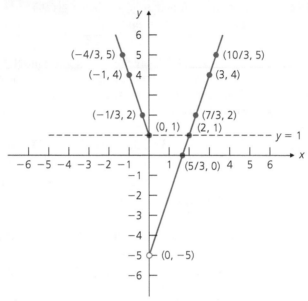

Figure 5.11

$f^{-1}(1) = \{x \in \mathbf{R} \mid f(x) \in \{1\}\} = \{x \in \mathbf{R} \mid f(x) = 1\}$
     $= \{x \in \mathbf{R} \mid x > 0 \text{ and } 3x - 5 = 1\} \cup \{x \in \mathbf{R} \mid x \leq 0 \text{ and } -3x + 1 = 1\}$
     $= \{x \in \mathbf{R} \mid x > 0 \text{ and } x = 2\} \cup \{x \in \mathbf{R} \mid x \leq 0 \text{ and } x = 0\}$
     $= \{2\} \cup \{0\} = \{0, 2\}$

[Here we note how the dashed line $y = 1$ intersects the graph in Fig. 5.11 at the points $(0, 1)$ and $(2, 1)$.]

$f^{-1}(-1) = \{x \in \mathbf{R} \mid x > 0 \text{ and } 3x - 5 = -1\} \cup \{x \in \mathbf{R} \mid x \leq 0 \text{ and } -3x + 1 = -1\}$
     $= \{x \in \mathbf{R} \mid x > 0 \text{ and } x = 4/3\} \cup \{x \in \mathbf{R} \mid x \leq 0 \text{ and } x = 2/3\}$
     $= \{4/3\} \cup \emptyset = \{4/3\}$

$$f^{-1}(3) = \{-2/3, 8/3\} \qquad f^{-1}(-3) = \{2/3\}$$

$f^{-1}(-6) = \{x \in \mathbf{R} \mid x > 0 \text{ and } 3x - 5 = -6\} \cup \{x \in \mathbf{R} \mid x \leq 0 \text{ and } -3x + 1 = -6\}$
     $= \{x \in \mathbf{R} \mid x > 0 \text{ and } x = -1/3\} \cup \{x \in \mathbf{R} \mid x \leq 0 \text{ and } x = 7/3\}$
     $= \emptyset \cup \emptyset = \emptyset$

c) $f^{-1}([-5, 5]) = \{x \mid f(x) \in [-5, 5]\} = \{x \mid -5 \leq f(x) \leq 5\}$.

     (Case 1) $x > 0$: $-5 \leq 3x - 5 \leq 5$

                          $0 \leq 3x \leq 10$

                          $0 \leq x \leq 10/3$ — so we use $0 < x \leq 10/3$.

(Case 2) $x \leq 0$: $-5 \leq -3x + 1 \leq 5$

$$-6 \leq -3x \leq 4$$

$2 \geq x \geq -4/3$ —here we use $-4/3 \leq x \leq 0$.

Hence $f^{-1}([-5, 5]) = \{x \mid -4/3 \leq x \leq 0 \text{ or } 0 < x \leq 10/3\} = [-4/3, 10/3]$. Since there are no points (x, y) on the graph (in Fig. 5.11) where $y \leq -5$, it follows from our prior calculations that $f^{-1}([-6, 5]) = f^{-1}([-5, 5]) = [-4/3, 10/3]$.

EXAMPLE 5.64

a) Let $f: \mathbf{Z} \to \mathbf{R}$ be defined by $f(x) = x^2 + 5$. Table 5.9 lists $f^{-1}(B)$ for various subsets B of the codomain $\mathbf{R}$.

b) If $g: \mathbf{R} \to \mathbf{R}$ is defined by $g(x) = x^2 + 5$, the results in Table 5.10 show how a change in domain (from $\mathbf{Z}$ to $\mathbf{R}$) affects the preimages (in Table 5.9).

Table 5.9

B	$f^{-1}(B)$
$\{6\}$	$\{-1, 1\}$
$[6, 7]$	$\{-1, 1\}$
$[6, 10]$	$\{-2, -1, 1, 2\}$
$[-4, 5)$	$\emptyset$
$[-4, 5]$	$\{0\}$
$[5, +\infty)$	$\mathbf{Z}$

Table 5.10

B	$g^{-1}(B)$
$\{6\}$	$\{-1, 1\}$
$[6, 7]$	$[-\sqrt{2}, -1] \cup [1, \sqrt{2}]$
$[6, 10]$	$[-\sqrt{5}, -1] \cup [1, \sqrt{5}]$
$[-4, 5)$	$\emptyset$
$[-4, 5]$	$\{0\}$
$[5, +\infty)$	$\mathbf{R}$

The concept of a preimage appears in conjunction with the set operations of intersection, union, and complementation in our next result. The reader should note the difference between part (a) of this theorem and part (b) of Theorem 5.2.

THEOREM 5.10

If $f: A \to B$ and $B_1, B_2 \subseteq B$, then (a) $f^{-1}(B_1 \cap B_2) = f^{-1}(B_1) \cap f^{-1}(B_2)$; (b) $f^{-1}(B_1 \cup B_2) = f^{-1}(B_1) \cup f^{-1}(B_2)$; and (c) $f^{-1}(\overline{B_1}) = \overline{f^{-1}(B_1)}$.

Proof: We prove part (b) and leave parts (a) and (c) for the reader.

For $a \in A$, $a \in f^{-1}(B_1 \cup B_2) \iff f(a) \in B_1 \cup B_2 \iff f(a) \in B_1$ or $f(a) \in B_2 \iff a \in f^{-1}(B_1)$ or $a \in f^{-1}(B_2) \iff a \in f^{-1}(B_1) \cup f^{-1}(B_2)$.

Using the notation of the preimage, we see that a function $f: A \to B$ is one-to-one if and only if $|f^{-1}(b)| \leq 1$ for each $b \in B$.

Discrete mathematics is primarily concerned with finite sets, and the last result of this section demonstrates how the property of finiteness can yield results that fail to be true in general. In addition, it provides an application of the pigeonhole principle.

THEOREM 5.11

Let $f: A \to B$ for finite sets A and B, where $|A| = |B|$. Then the following statements are equivalent: (a) f is one-to-one; (b) f is onto; and (c) f is invertible.

Proof: We have already shown in Theorem 5.8 that (c) $\Rightarrow$ (a) and (b), and that together (a), (b) $\Rightarrow$ (c). Consequently, this theorem will follow when we show that for these conditions on A, B, (a) $\Longleftrightarrow$ (b). Assuming (b), if f is not one-to-one, then there are elements a_1, $a_2 \in A$, with $a_1 \neq a_2$, but $f(a_1) = f(a_2)$. Then $|A| > |f(A)| = |B|$, contradicting $|A| = |B|$. Conversely, if f is not onto, then $|f(A)| < |B|$. With $|A| = |B|$ we have $|A| > |f(A)|$, and it follows from the pigeonhole principle that f is not one-to-one.

Using Theorem 5.11 we now verify the combinatorial identity introduced in Problem 6 at the start of this chapter. For if $n \in \mathbf{Z}^+$ and $|A| = |B| = n$, there are $n!$ one-to-one functions from A to B and $\sum_{k=0}^{n}(-1)^k \binom{n}{n-k}(n-k)^n$ onto functions from A to B. The equality $n! = \sum_{k=0}^{n}(-1)^k \binom{n}{n-k}(n-k)^n$ is then the numerical equivalent of parts (a) and (b) of Theorem 5.11. [This is also the reason why the diagonal elements $S(n, n)$, $1 \leq n \leq 8$, shown in Table 5.1 all equal 1.]

EXERCISES 5.6

1. a) For $A = \{1, 2, 3, 4, \ldots, 7\}$, how many bijective functions $f: A \to A$ satisfy $f(1) \neq 1$?

b) Answer part (a) where $A = \{x \mid x \in \mathbf{Z}^+, 1 \leq x \leq n\}$, for some fixed $n \in \mathbf{Z}^+$.

2. a) For $A = (-2, 7] \subseteq \mathbf{R}$ define the functions $f, g: A \to \mathbf{R}$ by

$$f(x) = 2x - 4 \quad \text{and} \quad g(x) = \frac{2x^2 - 8}{x + 2}.$$

Verify that $f = g$.

b) Is the result in part (a) affected if we change A to $[-7, 2)$?

3. Let $f, g: \mathbf{R} \to \mathbf{R}$, where $g(x) = 1 - x + x^2$ and $f(x) = ax + b$. If $(g \circ f)(x) = 9x^2 - 9x + 3$, determine a, b.

4. Let $g: \mathbf{N} \to \mathbf{N}$ be defined by $g(n) = 2n$. If $A = \{1, 2, 3, 4\}$ and $f: A \to \mathbf{N}$ is given by $f = \{(1, 2), (2, 3), (3, 5), (4, 7)\}$, find $g \circ f$.

5. If $\mathcal{U}$ is a given universe with (fixed) $S, T \subseteq \mathcal{U}$, define $g: \mathcal{P}(\mathcal{U}) \to \mathcal{P}(\mathcal{U})$ by $g(A) = T \cap (S \cup A)$ for $A \subseteq \mathcal{U}$. Prove that $g^2 = g$.

6. Let $f, g: \mathbf{R} \to \mathbf{R}$ where $f(x) = ax + b$ and $g(x) = cx + d$ for all $x \in \mathbf{R}$, with a, b, c, d real constants. What relationship(s) must be satisfied by a, b, c, d if $(f \circ g)(x) = (g \circ f)(x)$ for all $x \in \mathbf{R}$?

7. Let $f, g, h: \mathbf{Z} \to \mathbf{Z}$ be defined by $f(x) = x - 1$, $g(x) = 3x$,

$$h(x) = \begin{cases} 0, & x \text{ even} \\ 1, & x \text{ odd.} \end{cases}$$

Determine (a) $f \circ g$, $g \circ f$, $g \circ h$, $h \circ g$, $f \circ (g \circ h)$, $(f \circ g) \circ h$; (b) f^2, f^3, g^2, g^3, h^2, h^3, h^{500}.

8. Let $f: A \to B$, $g: B \to C$. Prove that (a) if $g \circ f: A \to C$ is onto, then g is onto; and (b) if $g \circ f: A \to C$ is one-to-one, then f is one-to-one.

9. a) Find the inverse of the function $f: \mathbf{R} \to \mathbf{R}^+$ defined by $f(x) = e^{2x+5}$.

b) Show that $f \circ f^{-1} = 1_{\mathbf{R}^+}$ and $f^{-1} \circ f = 1_{\mathbf{R}}$.

10. For each of the following functions $f: \mathbf{R} \to \mathbf{R}$, determine whether f is invertible, and, if so, determine f^{-1}.

a) $f = \{(x, y) \mid 2x + 3y = 7\}$

b) $f = \{(x, y) \mid ax + by = c, b \neq 0\}$

c) $f = \{(x, y) \mid y = x^3\}$

d) $f = \{(x, y) \mid y = x^4 + x\}$

11. Prove Theorem 5.9.

12. If $A = \{1, 2, 3, 4, 5, 6, 7\}$, $B = \{2, 4, 6, 8, 10, 12\}$, and $f: A \to B$ where $f = \{(1, 2), (2, 6), (3, 6), (4, 8), (5, 6), (6, 8), (7, 12)\}$, determine the preimage of B_1 under f in each of the following cases.

a) $B_1 = \{2\}$ **b)** $B_1 = \{6\}$

c) $B_1 = \{6, 8\}$ **d)** $B_1 = \{6, 8, 10\}$

e) $B_1 = \{6, 8, 10, 12\}$ **f)** $B_1 = \{10, 12\}$

13. Let $f: \mathbf{R} \to \mathbf{R}$ be defined by

$$f(x) = \begin{cases} x + 7, & x \leq 0 \\ -2x + 5, & 0 < x < 3 \\ x - 1, & 3 \leq x \end{cases}$$

a) Find $f^{-1}(-10)$, $f^{-1}(0)$, $f^{-1}(4)$, $f^{-1}(6)$, $f^{-1}(7)$, and $f^{-1}(8)$.

b) Determine the preimage under f for each of the intervals (i) $[-5, -1]$; (ii) $[-5, 0]$; (iii) $[-2, 4]$; (iv) $(5, 10)$; and (v) $[11, 17)$.

14. Let $f: \mathbf{R} \to \mathbf{R}$ be defined by $f(x) = x^2$. For each of the following subsets B of $\mathbf{R}$, find $f^{-1}(B)$.

a) $B = \{0, 1\}$ **b)** $B = \{-1, 0, 1\}$

c) $B = [0, 1]$ **d)** $B = [0, 1)$

e) $B = [0, 4]$ **f)** $B = (0, 1] \cup (4, 9)$

15. Let $A = \{1, 2, 3, 4, 5\}$ and $B = \{6, 7, 8, 9, 10, 11, 12\}$. How many functions $f: A \to B$ are such that $f^{-1}(\{6, 7, 8\}) = \{1, 2\}$?

16. Let $f: \mathbf{R} \to \mathbf{R}$ be defined by $f(x) = \lfloor x \rfloor$, the greatest integer in x. Find $f^{-1}(B)$ for each of the following subsets B of $\mathbf{R}$.

    a) $B = \{0, 1\}$ **b)** $B = \{-1, 0, 1\}$

    c) $B = [0, 1)$ **d)** $B = [0, 2)$

    e) $B = [-1, 2]$ **f)** $B = [-1, 0) \cup (1, 3]$

17. Let $f, g: \mathbf{Z}^+ \to \mathbf{Z}^+$ where for all $x \in \mathbf{Z}^+$, $f(x) = x + 1$ and $g(x) = \max\{1, x - 1\}$, the maximum of 1 and $x - 1$.

    a) What is the range of f?

    b) Is f an onto function?

    c) Is the function f one-to-one?

    d) What is the range of g?

    e) Is g an onto function?

    f) Is the function g one-to-one?

    g) Show that $g \circ f = 1_{\mathbf{Z}^+}$.

    h) Determine $(f \circ g)(x)$ for $x = 2, 3, 4, 7, 12$, and 25.

    i) Do the answers for parts (b), (g), and (h) contradict the result in Theorem 5.8?

18. Let f, g, h denote the following closed binary operations on $\mathscr{P}(\mathbf{Z}^+)$. For $A, B \subseteq \mathbf{Z}^+$, $f(A, B) = A \cap B$, $g(A, B) = A \cup B$, $h(A, B) = A \bigtriangleup B$.

    a) Are any of the functions one-to-one?

    b) Are any of f, g, and h onto functions?

    c) Is any one of the given functions invertible?

    d) Are any of the following sets infinite?

        (1) $f^{-1}(\emptyset)$ (2) $g^{-1}(\emptyset)$

        (3) $h^{-1}(\emptyset)$ (4) $f^{-1}(\{1\})$

        (5) $g^{-1}(\{2\})$ (6) $h^{-1}(\{3\})$

        (7) $f^{-1}(\{4, 7\})$ (8) $g^{-1}(\{8, 12\})$

        (9) $h^{-1}(\{5, 9\})$

    e) Determine the number of elements in each of the finite sets in part (d).

19. Prove parts (a) and (c) of Theorem 5.10.

20. **a)** Give an example of a function $f: \mathbf{Z} \to \mathbf{Z}$ where (i) f is one-to-one but not onto; and (ii) f is onto but not one-to-one.

    b) Do the examples in part (a) contradict Theorem 5.11?

21. Let $f: \mathbf{Z} \to \mathbf{N}$ be defined by

$$f(x) = \begin{cases} 2x - 1, & \text{if } x > 0 \\ -2x, & \text{for } x \le 0. \end{cases}$$

    a) Prove that f is one-to-one and onto.

    b) Determine f^{-1}.

22. If $|A| = |B| = 5$, how many functions $f: A \to B$ are invertible?

23. Let $f, g, h, k: \mathbf{N} \to \mathbf{N}$ where $f(n) = 3n$, $g(n) = \lfloor n/3 \rfloor$, $h(n) = \lfloor (n + 1)/3 \rfloor$, and $k(n) = \lfloor (n + 2)/3 \rfloor$, for each $n \in \mathbf{N}$. (a) For each $n \in \mathbf{N}$ what are $(g \circ f)(n)$, $(h \circ f)(n)$, and $(k \circ f)(n)$? (b) Do the results in part (a) contradict Theorem 5.7?

5.7
Computational Complexity†

In Section 4.4 we introduced the concept of an algorithm, following the examples set forth by the division algorithm (of Section 4.3) and the Euclidean algorithm (of Section 4.4). At that time we were concerned with certain properties of a general algorithm:

- The precision of the individual step-by-step instructions

- The input provided to the algorithm, and the output the algorithm then provides

- The ability of the algorithm to solve a certain type of problem, not just specific instances of the problem

- The uniqueness of the intermediate and final results, based on the input

---

†The material in Sections 5.7 and 5.8 may be skipped at this point. It will not be used very much until Chapter 10. The only place where this material appears before Chapter 10 is in Example 7.13, but that example can be omitted without any loss of continuity.

- The finite nature of the algorithm in that it terminates after the execution of a finite number of instructions

When an algorithm correctly solves a certain type of problem and satisfies these five conditions, then we may find ourselves examining it further in the following ways.

1) Can we somehow measure how long it takes the algorithm to solve a problem of a certain size? Whether we can may very well depend, for example, on the compiler being used, so we want to develop a measure that doesn't actually depend on such considerations as compilers, execution speeds, or other characteristics of a given computer.

   For example, if we want to compute a^n for $a \in \mathbf{R}$ and $n \in \mathbf{Z}^+$, is there some "function of n" that can describe how fast a given algorithm for such exponentiation accomplishes this?

2) Suppose we can answer questions such as the one set forth at the start of item 1. Then if we have two (or more) algorithms that solve a given problem, is there perhaps a way to determine whether one algorithm is "better" than another?

In particular, suppose we consider the problem of determining whether a certain real number x is present in the list of n real numbers $a_1, a_2, \ldots, a_n$. Here we have a problem of size n.

If there is an algorithm that solves this problem, how long does it take to do so? To measure this we seek a function $f(n)$, called the *time-complexity function*[†] of the algorithm. We expect (both here and in general) that the value of $f(n)$ will increase as n increases. Also, our major concern in dealing with any algorithm is how the algorithm performs for *large* values of n.

In order to study what has now been described in a somewhat informal manner, we need to introduce the following fundamental idea.

Definition 5.23   Let $f, g \colon \mathbf{Z}^+ \to \mathbf{R}$. We say that g *dominates* f (or f is *dominated* by g) if there exist constants $m \in \mathbf{R}^+$ and $k \in \mathbf{Z}^+$ such that $|f(n)| \le m|g(n)|$ for all $n \in \mathbf{Z}^+$, where $n \ge k$.

Note that as we consider the values of $f(1), g(1), f(2), g(2), \ldots$, there is a point (namely, k) after which the size of $f(n)$ is bounded above by a positive multiple (m) of the size of $g(n)$. Also, when g dominates f, then $|f(n)/g(n)| \le m$ [that is, the size of the quotient $f(n)/g(n)$ is bounded by m], for those $n \in \mathbf{Z}^+$ where $n \ge k$ and $g(n) \ne 0$.

When f is dominated by g we say that f is of *order* (*at most*) g and we use what is called "big-Oh" notation to designate this. We write $f \in O(g)$, where $O(g)$ is read "order g" or "big-Oh of g." As suggested by the notation "$f \in O(g)$," $O(g)$ represents the set of all functions with domain $\mathbf{Z}^+$ and codomain $\mathbf{R}$ that are dominated by g. These ideas are demonstrated in the following examples.

EXAMPLE 5.65   Let $f, g \colon \mathbf{Z}^+ \to \mathbf{R}$ be given by $f(n) = 5n$, $g(n) = n^2$, for $n \in \mathbf{Z}^+$. If we compute $f(n)$ and $g(n)$ for $1 \le n \le 4$, we find that $f(1) = 5$, $g(1) = 1$; $f(2) = 10$, $g(2) = 4$; $f(3) =$

[†]We could also study the *space-complexity function* of an algorithm, which we need when we attempt to measure the amount of memory required for the execution of an algorithm on a problem of size n. In this text, however, we limit our study to the time-complexity function.

15, $g(3) = 9$; and $f(4) = 20$, $g(4) = 16$. However, $n \geq 5 \Rightarrow n^2 \geq 5n$, and we have $|f(n)| = 5n \leq n^2 = |g(n)|$. So with $m = 1$ and $k = 5$, we find that for $n \geq k$, $|f(n)| \leq m|g(n)|$. Consequently, g dominates f and $f \in O(g)$. [Note that $|f(n)/g(n)|$ is bounded by 1 for all $n \geq 5$.]

We also realize that for all $n \in \mathbf{Z}^+$, $|f(n)| = 5n \leq 5n^2 = 5|g(n)|$. So the dominance of f by g is shown here with $k = 1$ and $m = 5$. This is enough to demonstrate that the constants k and m of Definition 5.23 need *not* be unique.

Furthermore, we can generalize this result if we now consider functions $f_1, g_1 : \mathbf{Z}^+ \to \mathbf{R}$ defined by $f_1(n) = an$, $g_1(n) = bn^2$, where a, b are nonzero real numbers. For if $m \in \mathbf{R}^+$ with $m|b| \geq |a|$, then for all $n \geq 1 (= k)$, $|f_1(n)| = |an| = |a|n \leq m|b|n \leq m|b|n^2 = m|bn^2| = m|g_1(n)|$, and so $f_1 \in O(g_1)$.

In Example 5.65 we observed that $f \in O(g)$. Taking a second look at the functions f and g, we now want to show that $g \notin O(f)$.

EXAMPLE 5.66

Once again let $f, g : \mathbf{Z}^+ \to \mathbf{R}$ be defined by $f(n) = 5n$, $g(n) = n^2$, for $n \in \mathbf{Z}^+$.

If $g \in O(f)$, then in terms of quantifiers, we would have

$$\exists m \in \mathbf{R}^+ \; \exists k \in \mathbf{Z}^+ \; \forall n \in \mathbf{Z}^+ \; [(n \geq k) \Rightarrow |g(n)| \leq m|f(n)|].$$

Consequently, to show that $g \notin O(f)$, we need to verify that

$$\forall m \in \mathbf{R}^+ \; \forall k \in \mathbf{Z}^+ \; \exists n \in \mathbf{Z}^+ \; [(n \geq k) \wedge (|g(n)| > m|f(n)|)].$$

To accomplish this, we first should realize that m and k are arbitrary, so we have no control over their values. The only number over which we have control is the positive integer n that we select. Now no matter what the values of m and k happen to be, we can select $n \in \mathbf{Z}^+$ such that $n > \max\{5m, k\}$. Then $n \geq k$ (actually $n > k$) and $n > 5m \Rightarrow n^2 > 5mn$, so $|g(n)| = n^2 > 5mn = m|5n| = m|f(n)|$ and $g \notin O(f)$.

For those who prefer the method of proof by contradiction, we present a second approach. If $g \in O(f)$, then we would have

$$n^2 = |g(n)| \leq m|f(n)| = mn$$

for all $n \geq k$, where k is some fixed positive integer and m is a (real) *constant*. But then from $n^2 \leq mn$ we deduce that $n \leq m$. This is impossible because $n(\in \mathbf{Z}^+)$ is a variable that can increase without bound while m is still a constant.

EXAMPLE 5.67

a) Let $f, g : \mathbf{Z}^+ \to \mathbf{R}$ with $f(n) = 5n^2 + 3n + 1$ and $g(n) = n^2$. Then $|f(n)| = |5n^2 + 3n + 1| = 5n^2 + 3n + 1 \leq 5n^2 + 3n^2 + n^2 = 9n^2 = 9|g(n)|$. Hence for all $n \geq 1$ ($= k$), $|f(n)| \leq m|g(n)|$ for any $m \geq 9$, and $f \in O(g)$. We can also write $f \in O(n^2)$ in this case.

In addition, $|g(n)| = n^2 \leq 5n^2 \leq 5n^2 + 3n + 1 = |f(n)|$ for all $n \geq 1$. So $|g(n)| \leq m|f(n)|$ for any $m \geq 1$ and all $n \geq k \geq 1$. Consequently $g \in O(f)$. [In fact, $O(g) = O(f)$; that is, any function from $\mathbf{Z}^+$ to $\mathbf{R}$ that is dominated by one of f, g is also dominated by the other. We shall examine this result for the general case in the Section Exercises.]

b) Now consider $f, g: \mathbf{Z}^+ \to \mathbf{R}$ with $f(n) = 3n^3 + 7n^2 - 4n + 2$ and $g(n) = n^3$. Here we have $|f(n)| = |3n^3 + 7n^2 - 4n + 2| \leq |3n^3| + |7n^2| + |-4n| + |2| \leq 3n^3 + 7n^3 + 4n^3 + 2n^3 = 16n^3 = 16|g(n)|$, for all $n \geq 1$. So with $m = 16$ and $k = 1$, we find that f is dominated by g, and $f \in O(g)$, or $f \in O(n^3)$.

Since $7n - 4 > 0$ for all $n \geq 1$, we can write $n^3 \leq 3n^3 \leq 3n^3 + (7n - 4)n + 2$ whenever $n \geq 1$. Then $|g(n)| \leq |f(n)|$ for all $n \geq 1$, and $g \in O(f)$. [As in part (a), we also have $O(f) = O(g) = O(n^3)$ in this case.]

We generalize the results of Example 5.67 as follows. Let $f: \mathbf{Z}^+ \to \mathbf{R}$ be the polynomial function where $f(n) = a_t n^t + a_{t-1} n^{t-1} + \cdots + a_2 n^2 + a_1 n + a_0$, for $a_t, a_{t-1}, \ldots, a_2, a_1, a_0 \in \mathbf{R}$, $a_t \neq 0$, $t \in \mathbf{N}$. Then

$$|f(n)| = |a_t n^t + a_{t-1} n^{t-1} + \cdots + a_2 n^2 + a_1 n + a_0|$$
$$\leq |a_t n^t| + |a_{t-1} n^{t-1}| + \cdots + |a_2 n^2| + |a_1 n| + |a_0|$$
$$= |a_t| n^t + |a_{t-1}| n^{t-1} + \cdots + |a_2| n^2 + |a_1| n + |a_0|$$
$$\leq |a_t| n^t + |a_{t-1}| n^t + \cdots + |a_2| n^t + |a_1| n^t + |a_0| n^t$$
$$= (|a_t| + |a_{t-1}| + \cdots + |a_2| + |a_1| + |a_0|) n^t.$$

In Definition 5.23, let $m = |a_t| + |a_{t-1}| + \cdots + |a_2| + |a_1| + |a_0|$ and $k = 1$, and let $g: \mathbf{Z}^+ \to \mathbf{R}$ be given by $g(n) = n^t$. Then $|f(n)| \leq m|g(n)|$ for all $n \geq k$, so f is dominated by g, or $f \in O(n^t)$.

It is also true that $g \in O(f)$ and that $O(f) = O(g) = O(n^t)$.

This generalization provides the following special results on summations.

EXAMPLE 5.68

a) Let $f: \mathbf{Z}^+ \to \mathbf{R}$ be given by $f(n) = 1 + 2 + 3 + \cdots + n$. Then (from Examples 1.40 and 4.1) $f(n) = \left(\frac{1}{2}\right)(n)(n+1) = \left(\frac{1}{2}\right) n^2 + \left(\frac{1}{2}\right) n$, so $f \in O(n^2)$.

b) If $g: \mathbf{Z}^+ \to \mathbf{R}$ with $g(n) = 1^2 + 2^2 + 3^2 + \cdots + n^2 = \left(\frac{1}{6}\right)(n)(n+1)(2n+1)$ (from Example 4.4), then $g(n) = \left(\frac{1}{3}\right) n^3 + \left(\frac{1}{2}\right) n^2 + \left(\frac{1}{6}\right) n$ and $g \in O(n^3)$.

c) If $t \in \mathbf{Z}^+$, and $h: \mathbf{Z}^+ \to \mathbf{R}$ is defined by $h(n) = \sum_{i=1}^{n} i^t$, then $h(n) = 1^t + 2^t + 3^t + \cdots + n^t \leq n^t + n^t + n^t + \cdots + n^t = n(n^t) = n^{t+1}$ so $h \in O(n^{t+1})$.

Now that we have examined several examples of function dominance, we shall close this section with two final observations. In the next section we shall apply the idea of function dominance in the analysis of algorithms.

1) When dealing with the concept of function dominance, we seek the best (or tightest) bound in the following sense. Suppose that $f, g, h: \mathbf{Z}^+ \to \mathbf{R}$, where $f \in O(g)$ and $g \in O(h)$. Then we also have $f \in O(h)$. (A proof for this is requested in the Section Exercises.) If $h \notin O(g)$, however, the statement $f \in O(g)$ provides a "better" bound on $|f(n)|$ than the statement $f \in O(h)$. For example, if $f(n) = 5$, $g(n) = 5n$, and $h(n) = n^2$, for all $n \in \mathbf{Z}^+$, then $f \in O(g)$, $g \in O(h)$, and $f \in O(h)$, but $h \notin O(g)$. Therefore, we are provided with more information by the statement $f \in O(g)$ than by the statement $f \in O(h)$.

2) Certain orders, such as $O(n)$ and $O(n^2)$, often occur when we deal with function dominance. Therefore they have come to be designated by special names. Some of the most important of these orders are listed in Table 5.11.

Table 5.11

Big-Oh Form	Name
$O(1)$	Constant
$O(\log_2 n)$	Logarithmic
$O(n)$	Linear
$O(n \log_2 n)$	$n \log_2 n$
$O(n^2)$	Quadratic
$O(n^3)$	Cubic
$O(n^m),\ m = 0, 1, 2, 3, \ldots$	Polynomial
$O(c^n),\ c > 1$	Exponential
$O(n!)$	Factorial

EXERCISES 5.7

1. Use the results of Table 5.11 to determine the best "big-Oh" form for each of the following functions $f: \mathbf{Z}^+ \to \mathbf{R}$.

a) $f(n) = 3n + 7$ **b)** $f(n) = 3 + \sin(1/n)$

c) $f(n) = n^3 - 5n^2 + 25n - 165$

d) $f(n) = 5n^2 + 3n \log_2 n$

e) $f(n) = n^2 + (n - 1)^3$

f) $f(n) = \dfrac{n(n + 1)(n + 2)}{(n + 3)}$

g) $f(n) = 2 + 4 + 6 + \cdots + 2n$

2. Let $f, g: \mathbf{Z}^+ \to \mathbf{R}$, where $f(n) = n$ and $g(n) = n + (1/n)$, for $n \in \mathbf{Z}^+$. Use Definition 5.23 to show that $f \in O(g)$ and $g \in O(f)$.

3. In each of the following, $f, g: \mathbf{Z}^+ \to \mathbf{R}$. Use Definition 5.23 to show that g dominates f.

a) $f(n) = 100 \log_2 n$, $g(n) = \left(\frac{1}{2}\right) n$

b) $f(n) = 2^n$, $g(n) = 2^{2n} - 1000$

c) $f(n) = 3n^2$, $g(n) = 2^n + 2n$

4. Let $f, g: \mathbf{Z}^+ \to \mathbf{R}$ be defined by $f(n) = n + 100$, $g(n) = n^2$. Use Definition 5.23 to show that $f \in O(g)$ but $g \notin O(f)$.

5. Let $f, g: \mathbf{Z}^+ \to \mathbf{R}$, where $f(n) = n^2 + n$ and $g(n) = \left(\frac{1}{2}\right) n^3$, for $n \in \mathbf{Z}^+$. Use Definition 5.23 to show that $f \in O(g)$ but $g \notin O(f)$.

6. Let $f, g: \mathbf{Z}^+ \to \mathbf{R}$ be defined as follows:

$$f(n) = \begin{cases} n, & \text{for } n \text{ odd} \\ 1, & \text{for } n \text{ even} \end{cases} \qquad g(n) = \begin{cases} 1, & \text{for } n \text{ odd} \\ n, & \text{for } n \text{ even} \end{cases}$$

Verify that $f \notin O(g)$ and $g \notin O(f)$.

7. Let $f, g: \mathbf{Z}^+ \to \mathbf{R}$ where $f(n) = n$ and $g(n) = \log_2 n$, for $n \in \mathbf{Z}^+$. Show that $g \in O(f)$ but $f \notin O(g)$.

(*Hint:*

$$\lim_{n \to \infty} \frac{n}{\log_2 n} = +\infty.$$

This requires the use of calculus.)

8. Let $f, g, h: \mathbf{Z}^+ \to \mathbf{R}$ where $f \in O(g)$ and $g \in O(h)$. Prove that $f \in O(h)$.

9. If $g: \mathbf{Z}^+ \to \mathbf{R}$ and $c \in \mathbf{R}$, we define the function $cg: \mathbf{Z}^+ \to \mathbf{R}$ by $(cg)(n) = c(g(n))$, for each $n \in \mathbf{Z}^+$. Prove that if $f, g: \mathbf{Z}^+ \to \mathbf{R}$ with $f \in O(g)$, then $f \in O(cg)$ for all $c \in \mathbf{R}$, $c \neq 0$.

10. a) Prove that $f \in O(f)$ for all $f: \mathbf{Z}^+ \to \mathbf{R}$.

b) Let $f, g: \mathbf{Z}^+ \to \mathbf{R}$. If $f \in O(g)$ and $g \in O(f)$, prove that $O(f) = O(g)$. That is, prove that for all $h: \mathbf{Z}^+ \to \mathbf{R}$, if h is dominated by f, then h is dominated by g, and conversely.

c) If $f, g: \mathbf{Z}^+ \to \mathbf{R}$, prove that if $O(f) = O(g)$, then $f \in O(g)$ and $g \in O(f)$.

11. The following is analogous to the "big-Oh" notation introduced in conjunction with Definition 5.23.

For $f, g: \mathbf{Z}^+ \to \mathbf{R}$ we say that f is of *order at least* g if there exist constants $M \in \mathbf{R}^+$ and $k \in \mathbf{Z}^+$ such that $|f(n)| \geq M|g(n)|$ for all $n \in \mathbf{Z}^+$, where $n \geq k$. In this case we write $f \in \Omega(g)$ and say that f is "big Omega of g." So $\Omega(g)$ represents the set of all functions with domain $\mathbf{Z}^+$ and codomain $\mathbf{R}$ that dominate g.

Suppose that $f, g, h: \mathbf{Z}^+ \to \mathbf{R}$, where $f(n) = 5n^2 + 3n$, $g(n) = n^2$, $h(n) = n$, for all $n \in \mathbf{Z}^+$. Prove that (a) $f \in \Omega(g)$; (b) $g \in \Omega(f)$; (c) $f \in \Omega(h)$; and (d) $h \notin \Omega(f)$ —that is, h is *not* "big Omega of f."

12. Let $f, g: \mathbf{Z}^+ \to \mathbf{R}$. Prove that $f \in \Omega(g)$ if and only if $g \in O(f)$.

13. a) Let $f: \mathbf{Z}^+ \to \mathbf{R}$ where $f(n) = \sum_{i=1}^{n} i$. When $n = 4$, for example, we have $f(n) = f(4) = 1 + 2 + 3 + 4 > 2 + 3 + 4 > 2 + 2 + 2 = 3 \cdot 2 = \lceil (4 + 1)/2 \rceil 2 = 6 >$

$(4/2)^2 = (n/2)^2$. For $n = 5$, we find $f(n) = f(5) = 1 + 2 + 3 + 4 + 5 > 3 + 4 + 5 > 3 + 3 + 3 = 3 \cdot 3 = \lceil(5 + 1)/2\rceil 3 = 9 > (5/2)^2 = (n/2)^2$. In general, $f(n) = 1 + 2 + \cdots + n > \lceil n/2 \rceil + \cdots + n > \lceil n/2 \rceil + \cdots + \lceil n/2 \rceil = \lceil (n + 1)/2 \rceil \lceil n/2 \rceil > n^2/4$. Consequently, $f \in \Omega(n^2)$.

Use

$$\sum_{i=1}^{n} i = \frac{n(n + 1)}{2}$$

to provide an alternative proof that $f \in \Omega(n^2)$.

b) Let $g: \mathbf{Z}^+ \to \mathbf{R}$ where $g(n) = \sum_{i=1}^{n} i^2$. Prove that $g \in \Omega(n^3)$.

c) For $t \in \mathbf{Z}^+$, let $h: \mathbf{Z}^+ \to \mathbf{R}$ where $h(n) = \sum_{i=1}^{n} i^t$. Prove that $h \in \Omega(n^{t+1})$.

14. For $f, g: \mathbf{Z}^+ \to \mathbf{R}$, we say that f is "big Theta of g," and write $f \in \Theta(g)$, when there exist constants $m_1, m_2 \in \mathbf{R}^+$ and $k \in \mathbf{Z}^+$ such that $m_1|g(n)| \leq |f(n)| \leq m_2|g(n)|$, for all $n \in \mathbf{Z}^+$, where $n \geq k$. Prove that $f \in \Theta(g)$ if and only if $f \in \Omega(g)$ and $f \in O(g)$.

15. Let $f, g: \mathbf{Z}^+ \to \mathbf{R}$. Prove that

$$f \in \Theta(g) \text{ if and only if } g \in \Theta(f).$$

16. a) Let $f: \mathbf{Z}^+ \to \mathbf{R}$ where $f(n) = \sum_{i=1}^{n} i$. Prove that $f \in \Theta(n^2)$.

b) Let $g: \mathbf{Z}^+ \to \mathbf{R}$ where $g(n) = \sum_{i=1}^{n} i^2$. Prove that $g \in \Theta(n^3)$.

c) For $t \in \mathbf{Z}^+$, let $h: \mathbf{Z}^+ \to \mathbf{R}$ where $h(n) = \sum_{i=1}^{n} i^t$. Prove that $h \in \Theta(n^{t+1})$.

5.8
Analysis of Algorithms

Now that the reader has been introduced to the concept of function dominance, it is time to see how this idea is used in the study of algorithms. In this section we present our algorithms as pseudocode procedures. (We shall also present algorithms as lists of instructions. The reader will find this to be the case in later chapters.)

We start with a procedure to determine the balance in a savings account.

EXAMPLE 5.69

In Fig. 5.12 we have a procedure (written in pseudocode) for computing the balance in a savings account n months (for $n \in \mathbf{Z}^+$) after it has been opened. (This balance is the procedure's output.) Here the user supplies the value of n, the input for the program. The variables *deposit, balance*, and *rate* are real variables, while i is an integer variable. (The annual interest rate is 0.06.)

```
procedure AccountBalance(n: integer)
begin
   deposit := 50.00      {The monthly deposit}
   i := 1                {Initializes the counter}
   rate := 0.005         {The monthly interest rate}
   balance := 100.00     {Initializes the balance}
   while i ≤ n do
      begin
         balance := deposit + balance + balance * rate
         i := i + 1
      end
end
```

Figure 5.12

Consider the following specific situation. Nathan puts $100.00 in a new account on January 1. Each month the bank adds the interest (*balance * rate*) to Nathan's account — on the first of the month. In addition, Nathan deposits an additional $50.00 on the first of each month (starting on February 1). This program tells Nathan the balance in his account after n months have gone by (assuming that the interest rate does not change). [*Note*: After one month, $n = 1$ and the balance is $50.00 (new deposit) + $100.00 (initial deposit) + ($100.00)(0.005) (the interest) = $150.50. When $n = 2$ the new balance is $50.00 (new deposit) + $150.50 (previous balance) + ($150.50) (0.005) (new interest) = $201.25.]

Our objective is to count (measure) the total number of operations (such as assignments, additions, multiplications, and comparisons) this program segment takes to compute the balance in Nathan's account n months after he opened it. We shall let $f(n)$ denote the total number of these operations. [Then $f: \mathbf{Z}^+ \to \mathbf{R}$. (Actually, $f(\mathbf{Z}^+) \subseteq \mathbf{Z}^+$.)]

The program segment begins with four assignment statements, where the integer variable i and the real variable *balance* are initialized, and the values of the real variables *deposit* and *rate* are declared. Then the **while** loop is executed n times. Each execution of the loop involves the following seven operations:

1) Comparing the present value of the counter i with n.

2) Increasing the present value of *balance* to *deposit* + *balance* + *balance* * *rate*; this involves one multiplication, two additions, and one assignment.

3) Incrementing the value of the counter by 1; this involves one addition and one assignment.

Finally, there is one more comparison. This is made when $i = n + 1$, so the **while** loop is terminated and the other six operations (in steps 2 and 3) are not performed.

Therefore, $f(n) = 4 + 7n + 1 = 7n + 5 \in O(n)$. Consequently, we say that $f \in O(n)$. For as n gets larger, the "order of magnitude" of $7n + 5$ depends primarily on the value n, the number of times the **while** loop is executed. Therefore, we could have obtained $f \in O(n)$ by simply counting the number of times the **while** loop was executed. Such shortcuts will be used in our calculations for the remaining examples.

Our next example introduces us to a situation where three types of complexity are determined. These measures are called the *best-case* complexity, the *worst-case* complexity, and the *average-case* complexity.

EXAMPLE 5.70   In this example we examine a typical *searching* process. Here an array of n (≥ 1) integers $a_1, a_2, a_3, \ldots, a_n$ is to be searched for the presence of an integer called *key*. If the integer is found, the value of *location* indicates its first location in the array; if it is not found the value of *location* is 0, indicating an unsuccessful search.

We cannot assume that the entries in the array are in any particular order. (If they were, the problem would be easier and a more efficient algorithm could be developed.) The input for this algorithm consists of the array (which is read in by the user or provided, perhaps, as a file from an external source), along with the number n of elements in the array, and the value of the integer *key*.

The algorithm is provided in the pseudocode procedure in Fig. 5.13.

We shall define the complexity function $f(n)$ for this algorithm to be the number of elements in the array that are examined until the value *key* is found (for the first time) or the array is exhausted (that is, the number of times the **while** loop is executed).

What is the best thing that can happen in our search for *key*? If *key* = a_1, we find that *key* is the first entry of the array, and we had to compare *key* with only one element of the array. In this case we have $f(n) = 1$, and we say that the *best-case complexity* for our algorithm

```
procedure LinearSearch(key, n: integer; a₁,a₂,a₃,...,aₙ: integers)
begin
  i := 1                            {initializes the counter}
  while (i ≤ n and key ≠ aᵢ) do
    i := i + 1
  if i ≤ n then location := i       {successful search}
  else location := 0                {unsuccessful search}
end {location is the subscript of the first array entry that equals key;
      location is 0 if key is not found}
```

Figure 5.13

is $O(1)$ (that is, it is constant and independent of the size of the array). Unfortunately, we cannot expect such a situation to occur very often.

From the best situation we turn now to the worst. We see that we have to examine all n entries of the array if (1) the first occurrence of *key* is a_n or (2) *key* is not found in the array. In either case we have $f(n) = n$, and the *worst-case complexity* here is $O(n)$. (The worst-case complexity will typically be considered throughout the text.)

Finally, we wish to obtain an estimate of the average number of array entries examined. We shall assume that the n entries of the array are distinct and are all equally likely (with probability p) to contain the value *key*, and that the probability that *key* is not in the array is equal to q. Consequently, we have $np + q = 1$ and $p = (1 - q)/n$.

For each $1 \le i \le n$, if *key* equals a_i, then i elements of the array have been examined. If *key* is not in the array, then all n array elements are examined. Therefore, the *average-case complexity* is determined by the average number of array elements examined, which is

$$f(n) = (1 \cdot p + 2 \cdot p + 3 \cdot p + \cdots + n \cdot p) + n \cdot q = p(1 + 2 + 3 + \cdots + n) + nq$$

$$= \frac{pn(n + 1)}{2} + nq.$$

If $q = 0$, then *key* is in the array, $p = 1/n$ and $f(n) = (n + 1)/2 \in O(n)$. For $q = 1/2$, we have an even chance that *key* is in the array and $f(n) = (1/(2n))[n(n + 1)/2] + (n/2) = (n + 1)/4 + (n/2) \in O(n)$. [In general, for all $0 \le q \le 1$, we have $f(n) \in O(n)$.]

EXAMPLE 5.71[†]

The result in Example 5.70 for the average number of array elements examined in the linear search algorithm may also be calculated using the idea of the random variable. When the algorithm is applied to the array $a_1, a_2, a_3, \ldots, a_n$ (of n distinct integers), we let the discrete random variable X count the number of array elements examined in the search for the integer *key*. Here the sample space can be considered as $\{1, 2, 3, \ldots, n, n^*\}$, where for $1 \le i \le n$, we have the case where *key* is found to be a_i — so that the i elements $a_1, a_2, a_3, \ldots, a_i$ have been examined. The entry n^* denotes the situation where all n elements are examined but *key* is not found among any of the array elements $a_1, a_2, a_3, \ldots, a_n$.

Once again we assume that each array entry has the same probability p of containing the value *key* and that q is the probability that *key* is not in the array. Then $np + q = 1$ and

---

[†]This example uses the concept of the discrete random variable which was introduced in the optional material in Section 3.7. It may be skipped without loss of continuity.

we have $Pr(X = i) = p$, for $1 \leq i \leq n$, and $Pr(X = n^*) = q$. Consequently, the average number of array elements examined during the execution of the linear search algorithm is

$$E(X) = \sum_{i=1}^{n} i Pr(X = i) + n Pr(X = n^*)$$

$$= \sum_{i=1}^{n} ip + np = p(1 + 2 + 3 + \cdots + n) + nq = \frac{pn(n+1)}{2} + nq.$$

Early in the discussion of the previous section, we mentioned how we might want to compare two algorithms that both correctly solve a given type of problem. Such a comparison can be accomplished by using the time-complexity functions for the algorithms. We demonstrate this in the next two examples.

EXAMPLE 5.72

The algorithm implemented in the pseudocode procedure of Fig. 5.14 outputs the value of a^n for the input a, n, where a is a real number and n is a positive integer. The real variable x is initialized as 1.0 and then used to store the values a, a^2, a^3, ..., a^n during execution of the **for** loop. Here we define the time-complexity function $f(n)$ for the algorithm as the number of multiplications that occur in the **for** loop. Consequently, we have $f(n) = n \in O(n)$.

```
procedure Power1(a: real; n: positive integer)
begin
  x := 1.0
  for i := 1 to n do
    x := x * a
end
```

Figure 5.14

EXAMPLE 5.73

In Fig. 5.15 we have a second pseudocode procedure for evaluating a^n for all $a \in \mathbf{R}$, $n \in \mathbf{Z}^+$. Recall that $\lfloor i/2 \rfloor$ is the greatest integer in (or the floor of) $i/2$.

```
procedure Power2(a: real; n: positive integer)
begin
  x := 1.0
  i := n
  while i > 0 do
    begin
      if i ≠ 2 * ⌊i/2⌋ then     {i is odd}
        x := x * a
      i := ⌊i/2⌋
      if i > 0 then
        a := a * a
    end
end
```

Figure 5.15

For this procedure the real variable x is initialized as 1.0 and then used to store the appropriate powers of a until it contains the value of a^n. The results shown in Fig. 5.16 demonstrate what is happening to x (and a) for the cases where $n = 7$ and 8. The numbers 1, 2, 3, and 4 indicate the first, second, third, and fourth times the statements in the **while** loop (in particular, the statement $i := \lfloor i/2 \rfloor$) are executed. If $n = 7$, then because $2^2 < 7 < 2^3$, we have $2 < \log_2 7 < 3$. Here the **while** loop is executed three times and

$$3 = \lfloor \log_2 7 \rfloor + 1 < \log_2 7 + 1,$$

where $\lfloor \log_2 7 \rfloor$ denotes the greatest integer in $\log_2 7$, which is 2. Also, when $n = 8$, the number of times the **while** loop is executed is

$$4 = \lfloor \log_2 8 \rfloor + 1 = \log_2 8 + 1,$$

since $\log_2 8 = 3$.

```
n = 7                          n = 8
x := 1.0                       x := 1.0
i := 7                         i := 8

    ⎧ x := x * a   {x = a}       ⎧ i := 4
1 ⎨ i := 3                   1 ⎨ a := a * a
    ⎩ a := a * a

                               ⎧ i := 2
                             2 ⎨ a := a * a

    ⎧ x := x * a   {x = a³}
2 ⎨ i := 1                       ⎧ i := 1
    ⎩ a := a * a              3 ⎨ a := a * a

    ⎧ x := x * a   {x = a⁷}       ⎧ x := x * a   {x = a⁸}
3 ⎨ i := 0                   4 ⎨ i := 0

    [x = a⁷ = a · a² · a⁴]       [x = (((a)²)²)²]
```

Figure 5.16

We shall define the time-complexity function $g(n)$ for (the implementation of) this exponentiation algorithm as the number of times the **while** loop is executed. This is also the number of times the statement $i := \lfloor i/2 \rfloor$ is executed. (Here we assume that the time interval for the computation of each $\lfloor i/2 \rfloor$ is independent of the magnitude of i.) On the basis of the foregoing two observations, we want to establish that for all $n \geq 1$, $g(n) \leq \log_2 n + 1 \in O(\log_2 n)$. We shall establish this by the Principle of Mathematical Induction (the alternative form — Theorem 4.2) on the value of n.

When $n = 1$, we see in Fig. 5.15 that i is odd, x is assigned the value of $a = a^1$, and a^1 is determined after only $1 = \log_2 1 + 1$ execution of the **while** loop. So $g(1) = 1 \leq \log_2 1 + 1$.

Now assume that for all $1 \leq n \leq k$, $g(n) \leq \log_2 n + 1$. Then for $n = k + 1$, during the first pass through the **while** loop the value of i is changed to $\left\lfloor \dfrac{k+1}{2} \right\rfloor$. Since $1 \leq \left\lfloor \dfrac{k+1}{2} \right\rfloor \leq k$, by the induction hypothesis we shall execute the **while** loop $g\left(\left\lfloor \dfrac{k+1}{2} \right\rfloor\right)$ more times, where $g\left(\left\lfloor \dfrac{k+1}{2} \right\rfloor\right) \leq \log_2 \left\lfloor \dfrac{k+1}{2} \right\rfloor + 1$.

Therefore

$$g(k + 1) \leq 1 + \left[\log_2 \left\lfloor \frac{k + 1}{2} \right\rfloor + 1 \right] \leq 1 + \left[\log_2 \left(\frac{k + 1}{2} \right) + 1 \right]$$

$$= 1 + [\log_2(k + 1) - \log_2 2 + 1] = \log_2(k + 1) + 1.$$

For the time-complexity function of Example 5.72, we found that $f(n) \in O(n)$. Here we have $g(n) \in O(\log_2 n)$. It can be verified that g is dominated by f but f is not dominated by g. Therefore, *for large n*, this second algorithm is considered more efficient than the first algorithm (of Example 5.72). (However, note how much easier the pseudocode in Fig. 5.14 is than that of the procedure in Fig. 5.15.)

In closing this section, we shall summarize what we have learned by making the following observations.

1) The results we established in Examples 5.69, 5.70, 5.72, and 5.73 are useful when we are dealing with moderate to large values of n. For small values of n, such considerations about time-complexity functions have little purpose.

2) Suppose that algorithms A_1 and A_2 have time-complexity functions $f(n)$ and $g(n)$, respectively, where $f(n) \in O(n)$ and $g(n) \in O(n^2)$. We must be cautious here. We might expect an algorithm with linear complexity to be "perhaps more efficient" than one with quadratic complexity. But we really need more information. If $f(n) = 1000n$ and $g(n) = n^2$, then algorithm A_2 is fine until the problem size n exceeds 1000. If the problem size is such that we never exceed 1000, then algorithm A_2 is the better choice. However, as we mentioned in observation 1, as n grows larger, the algorithm of linear complexity becomes the better alternative.

3) In Fig. 5.17 we have graphed a log-linear plot for the functions associated with some of the orders given in Table 5.11. [Here we have replaced the (discrete) integer variable n by the (continuous) real variable n.] This should help us to develop some feeling for their relative growth rates (especially for large values of n).

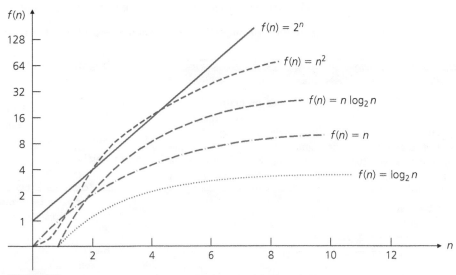

Figure 5.17

The data in Table 5.12 provide estimates of the running times of algorithms for certain orders of complexity. Here we have the problem sizes $n = 2$, 16, and 64, and we assume that the computer can perform one operation every 10^{-6} second $= 1$ microsecond (on the average). The entries in the table then estimate the running times in microseconds. For example, when the problem size is 16 and the order of complexity is $n \log_2 n$, then the running time is a very brief $16 \log_2 16 = 16 \cdot 4 = 64$ microseconds; for the order of complexity 2^n, the running time is 6.5×10^4 microseconds $= 0.065$ seconds. Since both of these time intervals are so short, it is difficult for a human to observe much of a difference in execution times. Results appear to be instantaneous in either case.

Table 5.12

Problem size n	Order of Complexity					
	$\log_2 n$	n	$n \log_2 n$	n^2	2^n	$n!$
2	1	2	2	4	4	2
16	4	16	64	256	6.5×10^4	2.1×10^{13}
64	6	64	384	4096	1.84×10^{19}	$> 10^{89}$

However, such estimates can grow rather rapidly. For instance, suppose we run a program for which the input is an array A of n different integers. The results from this program are generated in two parts:

1) First the program implements an algorithm that determines the subsets of A of size 1. There are n such subsets.

2) Then a second algorithm is implemented to determine all the subsets of A. There are 2^n such subsets.

Let us assume that we have a computer that can determine each subset of A in a microsecond. For the case where $|A| = 64$, the first part of the output is executed almost instantaneously — in approximately 64 microseconds. For the second part, however, Table 5.12 indicates that the amount of time needed to determine all the subsets of A will be about 1.84×10^{19} microseconds. We cannot be too content with this result, however, since

$$1.84 \times 10^{19} \text{ microseconds} \doteq 2.14 \times 10^8 \text{ days} \doteq 5845 \text{ centuries.}$$

EXERCISES 5.8

1. In each of the following pseudocode program segments, the integer variables i, j, n, and *sum* are declared earlier in the program. The value of n (a positive integer) is supplied by the user prior to execution of the segment. In each case we define the time-complexity function $f(n)$ to be the number of times the statement *sum* := *sum* + 1 is executed. Determine the best "big-Oh" form for f.

a)
```
begin
    sum := 0
    for i := 1 to n do
        for j := 1 to n do
            sum := sum + 1
end
```

b)
```
begin
    sum := 0
    for i := 1 to n do
        for j := 1 to n * n do
            sum := sum + 1
end
```

c)
```
begin
    sum := 0;
    for i := 1 to n do
        for j := i to n do
            sum := sum + 1
end
```

d)
```
begin
    sum := 0
    i := n
```

```
    while i > 0 do
      begin
        sum := sum + 1
        i := ⌊i/2⌋
      end
  end
e) begin
    sum := 0
    for i := 1 to n do
      begin
        j := n
        while j > 0 do
          begin
            sum := sum + 1
            j := ⌊j/2⌋
          end
      end
  end
```

2. The following pseudocode procedure implements an algorithm for determining the maximum value in an array $a_1, a_2, a_3, \ldots, a_n$ of integers. Here $n \geq 2$ and the entries in the array need not be distinct.

```
procedure Maximum (n: integer;
    a₁,a₂,a₃,...,aₙ: integers)
begin
  max := a₁
  for i := 2 to n do
    if aᵢ > max then
      max := aᵢ
end
```

a) If the worst-case complexity function $f(n)$ for this segment is determined by the number of times the comparison $a_i > max$ is executed, find the appropriate "big-Oh" form for f.

b) What can we say about the best-case and average-case complexities for this implementation?

3. a) Write a computer program (or develop an algorithm) to locate the first occurrence of the maximum value in an array $a_1, a_2, a_3, \ldots, a_n$ of integers. (Here $n \in \mathbf{Z}^+$ and the entries in the array need not be distinct.)

b) Determine the worst-case complexity function for the implementation developed in part (a).

4. a) Write a computer program (or develop an algorithm) to determine the minimum and maximum values in an array $a_1, a_2, a_3, \ldots, a_n$ of integers. (Here $n \in \mathbf{Z}^+$ with $n \geq 2$, and the entries in the array need not be distinct.)

b) Determine the worst-case complexity function for the implementation developed in part (a).

5. The following pseudocode procedure can be used to evaluate the polynomial

$$8 - 10x + 7x^2 - 2x^3 + 3x^4 + 12x^5,$$

when x is replaced by an arbitrary (but fixed) real number r.

For this particular instance, $n = 5$ and $a_0 = 8$, $a_1 = -10$, $a_2 = 7$, $a_3 = -2$, $a_4 = 3$, and $a_5 = 12$.

```
procedure PolynomialEvaluation1
    (n: nonnegative integer;
     r,a₀,a₁,a₂,...,aₙ: real)
begin
  product := 1.0
  value := a₀
  for i := 1 to n do
    begin
      product := product * r
      value := value + aᵢ * product
    end
end
```

a) How many additions take place in the evaluation of the given polynomial? (Do not include the $n - 1$ additions needed to increment the loop variable i.) How many multiplications?

b) Answer the questions in part (a) for the general polynomial

$$a_0 + a_1 x + a_2 x^2 + a_3 x^3 + \cdots + a_{n-1} x^{n-1} + a_n x^n,$$

where $a_0, a_1, a_2, a_3, \ldots, a_{n-1}, a_n$ are real numbers and n is a positive integer.

6. We first note how the polynomial in the previous exercise can be written in the *nested multiplication method*:

$$8 + x(-10 + x(7 + x(-2 + x(3 + 12x)))).$$

Using this representation, the following pseudocode procedure (implementing *Horner's method*) can be used to evaluate the given polynomial.

```
procedure PolynomialEvaluation2
    (n: nonnegative integer;
     r,a₀,a₁,a₂,...,aₙ: real)
begin
  value := aₙ
  for j := n - 1 down to 0 do
    value := aⱼ + r * value
end
```

Answer the questions in parts (a) and (b) of Exercise 5 for the new procedure given here.

7. Let $a_1, a_2, a_3, \ldots$ be the integer sequence defined recursively by

1) $a_1 = 0$; and

2) For $n > 1$, $a_n = 1 + a_{\lfloor n/2 \rfloor}$.

Prove that $a_n = \lfloor \log_2 n \rfloor$ for all $n \in \mathbf{Z}^+$.

8. Let $a_1, a_2, a_3, \ldots$ be the integer sequence defined recursively by

    1) $a_1 = 0$; and

    2) For $n > 1$, $a_n = 1 + a_{\lceil n/2 \rceil}$.

Find an explicit formula for a_n and prove that your formula is correct.

9. Suppose the probability that the integer *key* is in the array $a_1, a_2, a_3, \ldots, a_n$ (of n distinct integers) is 3/4 and that each array element has the same probability of containing this value. If the linear search algorithm of Example 5.70 is applied to this array and value of *key*, what is the average number of array elements that are examined?

10. When the linear search algorithm is applied to the array $a_1, a_2, a_3, \ldots, a_n$ (of n distinct integers) for the integer *key*,

suppose the probability that *key* has the value a_i is $i / [n(n + 1)]$, for $1 \le i \le n$. Under these circumstances, what is the average number of array elements examined?

11. a) Write a computer program (or develop an algorithm) to determine the location of the first entry in an array $a_1, a_2, a_3, \ldots, a_n$ of integers that repeats a previous entry in the array.

    b) Determine the worst-case complexity for the implementation developed in part (a).

12. a) Write a computer program (or develop an algorithm) to determine the location of the first entry a_i in an array $a_1, a_2, a_3, \ldots, a_n$ of integers, where $a_i < a_{i-1}$.

    b) Determine the worst-case complexity for the implementation developed in part (a).

5.9
Summary and Historical Review

In this chapter we developed the function concept, which is of great importance in all areas of mathematics. Although we were primarily concerned with finite functions, the definition applies equally well to infinite sets and includes the functions of trigonometry and calculus. However, we did emphasize the role of a finite function when we transformed a finite set into a finite set. In this setting, computer output (that terminates) can be thought of as a function of computer input, and a compiler can be regarded as a function that transforms a (source) program into a set of machine-language instructions (object program).

The actual word *function*, in its Latin form, was introduced in 1694 by Gottfried Wilhelm Leibniz (1646–1716) to denote a quantity associated with a curve (such as the slope of the curve or the coordinates of a point of the curve). By 1718, under the direction of Johann Bernoulli (1667–1748), a function was regarded as an algebraic expression made up of constants and a variable. Equations or formulas involving constants and variables

Gottfried Wilhelm Leibniz (1646–1716)

came later with Leonhard Euler (1707–1783). His is the definition of "function" generally found in high school mathematics. Also, in about 1734, we find in the work of Euler and Alexis Clairaut (1713–1765) the notation $f(x)$, which is still in use today.

Euler's idea remained intact until the time of Jean Baptiste Joseph Fourier (1768–1830), who found the need for a more general type of function in his investigation of trigonometric series. In 1837, Peter Gustav Lejeune Dirichlet (1805–1859) set down a more rigorous formulation of the concepts of variable, function, and the correspondence between the independent variable x and the dependent variable y, when $y = f(x)$. Dirichlet's work emphasized the relationship between two sets of numbers and did not call for the existence of a formula or expression connecting the two sets. With the developments in set theory during the nineteenth and twentieth centuries came the generalization of the function as a particular type of relation.

Peter Gustav Lejeune Dirichlet (1805–1859)

In addition to his fundamental work on the definition of a function, Dirichlet was also quite active in applied mathematics and in number theory, where he found need for, and was the first to formally state, the pigeonhole principle. Consequently, this principle is sometimes referred to as the Dirichlet drawer principle or the Dirichlet box principle.

The nineteenth and twentieth centuries saw the use of the special function, one-to-one correspondence, in the study of the infinite. In about 1888, Richard Dedekind (1831–1916) defined an infinite set as one that can be placed into a one-to-one correspondence with a proper subset of itself. [Galileo (1564–1642) had observed this for the set $\mathbf{Z}^+$.] Two infinite sets that could be placed in a one-to-one correspondence with each other were said to have the same *transfinite cardinal number*. In a series of articles, Georg Cantor (1845–1918) developed the idea of levels of infinity and showed that $|\mathbf{Z}| = |\mathbf{Q}|$ but $|\mathbf{Z}| < |\mathbf{R}|$. A set A with $|A| = |\mathbf{Z}|$ is called *countable*, or *denumerable*, and we write $|\mathbf{Z}| = \aleph_0$ as Cantor did, using the Hebrew letter aleph, with the subscripted 0, to denote the first level of infinity. To show that $|\mathbf{Z}| < |\mathbf{R}|$, or that the real numbers were *uncountable*, Cantor devised a technique now referred to as the Cantor diagonal method. (More about the theory of countable and uncountable sets can be found in Appendix 3.)

The Stirling numbers of the second kind (in Section 5.3) are named in honor of James Stirling (1692–1770), a pioneer in the development of generating functions, a topic we will investigate later in the text. These numbers appear in his work *Methodus Differentialis*, published in London in 1730. Stirling was an associate of Sir Isaac Newton (1642–1727) and was using the Maclaurin series in his work 25 years before Colin Maclaurin (1698–1746). However, although his name is not attached to this series, it appears in the approximation known as Stirling's formula: $n! \doteq (2\pi n)^{1/2} e^{-n} n^n$, which, as justice would have it, was actually developed by Abraham DeMoivre (1667–1754).

Using the counting principles developed in Section 5.3, the results in Table 5.13 extend the ideas that were summarized in Table 1.11. Here we count the number of ways it is possible to distribute m objects into n containers, under the conditions prescribed in the first three columns of the table. (The cases wherein neither the objects nor the containers are distinct will be covered in Chapter 9.)

Table 5.13

Objects Are Distinct	Containers Are Distinct	Some Container(s) May Be Empty	Number of Distributions
Yes	Yes	Yes	n^m
Yes	Yes	No	$n!\, S(m, n)$
Yes	No	Yes	$S(m, 1) + S(m, 2) + \cdots + S(m, n)$
Yes	No	No	$S(m, n)$
No	Yes	Yes	$\dbinom{n + m - 1}{m}$
No	Yes	No	$\dbinom{n + (m - n) - 1}{(m - n)} = \dbinom{m - 1}{m - n} = \dbinom{m - 1}{n - 1}$

Finally, the "big-Oh" notation of Section 5.7 was introduced by Paul Gustav Heinrich Bachmann (1837–1920) in his book *Analytische Zahlentheorie*, an important work on number theory, published in 1892. This notation has become prominent in approximation theory, in such areas as numerical analysis and the analysis of algorithms. In general, the notation $f \in O(g)$ denotes that we do not know the function f explicitly but do know an upper bound on its order of magnitude. The "big-Oh" symbol is sometimes referred to as the Landau symbol, in honor of Edmund Landau (1877–1938), who used this notation throughout his work.

Further properties of the Stirling numbers of the second kind are given in Chapter 4 of D. I. A. Cohen [3] and in Chapter 6 of the text by R. L. Graham, D. E. Knuth, and O. Patashnik [7]. The article by D. J. Velleman and G. S. Call [11] provides a very interesting introduction to the Stirling numbers of the second kind — as well as the Eulerian numbers introduced in Example 4.21. For more on infinite sets and the work of Georg Cantor, consult Chapter 8 of H. Eves and C. V. Newsom [6] or Chapter IV of R. L. Wilder [12]. The presentation in the book by J. W. Dauben [5] covers the controversy surrounding set theory at the turn of the century and shows how certain aspects of Cantor's personal life played such an integral part in his understanding and defense of set theory.

More examples that demonstrate how to apply the pigeonhole principle are given in the articles by K. R. Rebman [9] and A. Soifer and E. Lozansky [10]. Further results and extensions on problems arising from the principle are covered in the article by D. S. Clark and J. T. Lewis [2]. During the twentieth century a great deal of research has been devoted to generalizations of the pigeonhole principle, culminating in the subject of Ramsey theory, named for Frank Plumpton Ramsey (1903–1930). An interesting introduction to Ramsey theory can be found in Chapter 5 of D. I. A. Cohen [3]. The text by R. L. Graham, B. L. Rothschild, and J. H. Spencer [8] provides further worthwhile information.

Extensive coverage on the topic of relational data bases can be found in the work of C. J. Date [4]. Finally, the text by S. Baase and A. Van Gelder [1] is an excellent place to continue the study of the analysis of algorithms.

REFERENCES

1. Baase, Sara, and Van Gelder, Allen. *Computer Algorithms: Introduction to Design & Analysis*, 3rd ed. Reading, Mass.: Addison-Wesley, 2000.
2. Clark, Dean S., and Lewis, James T. "Herbert and the Hungarian Mathematician: Avoiding Certain Subsequence Sums." *The College Mathematics Journal* 21 (March 1990): pp. 100–104.
3. Cohen, Daniel I. A. *Basic Techniques of Combinatorial Theory*. New York: Wiley, 1978.
4. Date, C. J. *An Introduction to Database Systems*, 7th ed. Boston, Mass.: Addison-Wesley, 2002.
5. Dauben, Joseph Warren. *Georg Cantor: His Mathematics and Philosophy of the Infinite*. Lawrenceville, N. J.: Princeton University Press, 1990.
6. Eves, Howard, and Newsom, Carroll V. *An Introduction to the Foundations and Fundamental Concepts of Mathematics*, rev. ed. New York: Holt, 1965.
7. Graham, Ronald L., Knuth, Donald E., and Patashnik, Oren. *Concrete Mathematics*, 2nd ed. Reading, Mass.: Addison-Wesley, 1994.
8. Graham, Ronald L., Rothschild, Bruce L., and Spencer, Joel H. *Ramsey Theory*, 2nd ed. New York: Wiley, 1980.
9. Rebman, Kenneth R. "The Pigeonhole Principle (What it is, how it works, and how it applies to map coloring)." *The Two-Year College Mathematics Journal*, vol. 10, no. 1 (January 1979): pp. 3–13.
10. Soifer, Alexander, and Lozansky, Edward, "Pigeons in Every Pigeonhole." *Quantum* (January 1990): pp. 25–26, 32.
11. Velleman, Daniel J., and Call, Gregory S. "Permutations and Combination Locks." *Mathematics Magazine* 68 (October 1995): pp. 243–253.
12. Wilder, Raymond L. *Introduction to the Foundations of Mathematics*, 2nd ed. New York: Wiley, 1965.

SUPPLEMENTARY EXERCISES

1. Let $A, B \subseteq \mathcal{U}$. Prove that

  a) $(A \times B) \cap (B \times A) = (A \cap B) \times (A \cap B)$; and

  b) $(A \times B) \cup (B \times A) \subseteq (A \cup B) \times (A \cup B)$.

2. Determine whether each of the following statements is true or false. For each false statement give a counterexample.

  a) If $f: A \to B$ and $(a, b), (a, c) \in f$, then $b = c$.

  b) If $f: A \to B$ is a one-to-one correspondence and A, B are finite, then $A = B$.

  c) If $f: A \to B$ is one-to-one, then f is invertible.

  d) If $f: A \to B$ is invertible, then f is one-to-one.

  e) If $f: A \to B$ is one-to-one and $g, h: B \to C$ with $g \circ f = h \circ f$, then $g = h$.

  f) If $f: A \to B$ and $A_1, A_2 \subseteq A$, then $f(A_1 \cap A_2) = f(A_1) \cap f(A_2)$.

  g) If $f: A \to B$ and $B_1, B_2 \subseteq B$, then $f^{-1}(B_1 \cap B_2) = f^{-1}(B_1) \cap f^{-1}(B_2)$.

3. Let $f: \mathbf{R} \to \mathbf{R}$ where $f(ab) = af(b) + bf(a)$, for all $a, b \in \mathbf{R}$. (a) What is $f(1)$? (b) What is $f(0)$? (c) If $n \in \mathbf{Z}^+$, $a \in \mathbf{R}$, prove that $f(a^n) = na^{n-1}f(a)$.

4. Let $A, B \subseteq \mathbf{N}$ with $1 < |A| < |B|$. If there are 262,144 relations from A to B, determine all possibilities for $|A|$ and $|B|$.

5. If $\mathcal{U}_1, \mathcal{U}_2$ are universal sets with $A, B \subseteq \mathcal{U}_1$, and $C, D \subseteq \mathcal{U}_2$, prove that $(A \cap B) \times (C \cap D) = (A \times C) \cap (B \times D)$.

6. Let $A = \{1, 2, 3, 4, 5\}$ and $B = \{1, 2, 3, 4, 5, 6\}$. How many one-to-one functions $f: A \to B$ satisfy (a) $f(1) = 3$? (b) $f(1) = 3$, $f(2) = 6$?

7. Determine all real numbers x for which

$$x^2 - \lfloor x \rfloor = 1/2.$$

8. Let $\mathcal{R} \subseteq \mathbf{Z}^+ \times \mathbf{Z}^+$ be the relation given by the following recursive definition.

    1) $(1, 1) \in \mathcal{R}$; and

    2) For all $(a, b) \in \mathcal{R}$, the three ordered pairs $(a + 1, b)$, $(a + 1, b + 1)$, and $(a + 1, b + 2)$ are also in $\mathcal{R}$.

Prove that $2a \geq b$ for all $(a, b) \in \mathcal{R}$.

9. Let a, b denote fixed real numbers and suppose that $f: \mathbf{R} \to \mathbf{R}$ is defined by $f(x) = a(x + b) - b$, $x \in \mathbf{R}$. (a) Determine $f^2(x)$ and $f^3(x)$. (b) Conjecture a formula for $f^n(x)$, where $n \in \mathbf{Z}^+$. Now establish the validity of your conjecture.

10. Let A_1, A and B be sets with $\{1, 2, 3, 4, 5\} = A_1 \subset A$, $B = \{s, t, u, v, w, x\}$, and $f: A_1 \to B$. If f can be extended to A in 216 ways, what is $|A|$?

11. Let $A = \{1, 2, 3, 4, 5\}$ and $B = \{t, u, v, w, x, y, z\}$. (a) If a function $f: A \to B$ is randomly generated, what is the probability that it is one-to-one? (b) Write a computer program (or develop an algorithm) to generate random functions $f: A \to B$ and have the program print out how many functions it generates until it generates one that is one-to-one.

12. Let S be a set of seven positive integers the maximum of which is at most 24. Prove that the sums of the elements in all the nonempty subsets of S cannot be distinct.

13. In a ten-day period Ms. Rosatone typed 84 letters to different clients. She typed 12 of these letters on the first day, seven on the second day, and three on the ninth day, and she finished the last eight on the tenth day. Show that for a period of three consecutive days Ms. Rosatone typed at least 25 letters.

14. If $\{x_1, x_2, \ldots, x_7\} \subseteq \mathbf{Z}^+$, show that for some $i \neq j$, either $x_i + x_j$ or $x_i - x_j$ is divisible by 10.

15. Let $n \in \mathbf{Z}^+$, n odd. If $i_1, i_2, \ldots, i_n$ is a permutation of the integers $1, 2, \ldots, n$, prove that $(1 - i_1)(2 - i_2) \cdots (n - i_n)$ is an even integer. (Which counting principle is at work here?)

16. With both of their parents working, Thomas, Stuart, and Craig must handle ten weekly chores among themselves. (a) In how many ways can they divide up the work so that everyone is responsible for at least one chore? (b) In how many ways can

the chores be assigned if Thomas, as the eldest, must mow the lawn (one of the ten weekly chores) and no one is allowed to be idle?

17. Let $n \in \mathbf{N}$, $n \geq 2$. Show that $S(n, 2) = 2^{n-1} - 1$.

18. Mrs. Blasi has five sons (Michael, Rick, David, Kenneth, and Donald) who enjoy reading books about sports. With Christmas approaching, she visits a bookstore where she finds 12 different books on sports.

    a) In how many ways can she select nine of these books?

    b) Having made her purchase, in how many ways can she distribute the books among her sons so that each of them gets at least one book?

    c) Two of the nine books Mrs. Blasi purchased deal with basketball, Donald's favorite sport. In how many ways can she distribute the books among her sons so that Donald gets at least the two books on basketball?

19. Let $m, n \in \mathbf{Z}^+$ with $n \geq m$. (a) In how many ways can one distribute n distinct objects among m different containers with no container left empty? (b) In the expansion of $(x_1 + x_2 + \cdots + x_m)^n$, what is the sum of all the multinomial coefficients $\binom{n}{n_1, n_2, \ldots, n_m}$ where $n_1 + n_2 + \cdots + n_m = n$ and $n_i > 0$ for all $1 \leq i \leq m$?

20. If $n \in \mathbf{Z}^+$ with $n \geq 4$, verify that $S(n, n - 2) = \binom{n}{3} + 3\binom{n}{4}$.

21. If $f: A \to A$, prove that for all $m, n \in \mathbf{Z}^+$, $f^m \circ f^n = f^n \circ f^m$. (First let $m = 1$ and induct on n. Then induct on m. This technique is known as *double induction*.)

22. Let $f: X \to Y$, and for each $i \in I$, let $A_i \subseteq X$. Prove that

    a) $f\left(\bigcup_{i \in I} A_i\right) = \bigcup_{i \in I} f(A_i)$.

    b) $f\left(\bigcap_{i \in I} A_i\right) \subseteq \bigcap_{i \in I} f(A_i)$.

    c) $f\left(\bigcap_{i \in I} A_i\right) = \bigcap_{i \in I} f(A_i)$,   for f one-to-one.

23. Given a nonempty set A, let $f: A \to A$ and $g: A \to A$ where

$$f(a) = g(f(f(a))) \quad \text{and} \quad g(a) = f(g(f(a)))$$

for all a in A. Prove that $f = g$.

24. Let A be a set with $|A| = n$.

    a) How many closed binary operations are there on A?

    b) A closed ternary (3-ary) operation on A is a function $f: A \times A \times A \to A$. How many closed ternary operations are there on A?

    c) A closed k-ary operation on A is a function $f: A_1 \times A_2 \times \cdots \times A_k \to A$, where $A_i = A$, for all $1 \leq i \leq k$. How many closed k-ary operations are there on A?

    d) A closed k-ary operation for A is called *commutative* if

$$f(a_1, a_2, \ldots, a_k) = f(\pi(a_1), \pi(a_2), \ldots, \pi(a_k)),$$

where $a_1, a_2, \ldots, a_k \in A$ (repetitions allowed), and

$\pi(a_1), \pi(a_2), \ldots, \pi(a_k)$ is any rearrangement of $a_1, a_2, \ldots, a_k$. How many of the closed k-ary operations on A are commutative?

25. a) Let $S = \{2, 16, 128, 1024, 8192, 65536\}$. If four numbers are selected from S, prove that two of them must have the product 131072.

b) Generalize the result in part (a).

26. If $\mathcal{U}$ is a universe and $A \subseteq \mathcal{U}$, we define the *characteristic function* of A by $\chi_A: \mathcal{U} \to \{0, 1\}$, where

$$\chi_A(x) = \begin{cases} 1, & x \in A \\ 0, & x \notin A \end{cases}$$

For sets $A, B \subseteq \mathcal{U}$, prove each of the following:

a) $\chi_{A \cap B} = \chi_A \cdot \chi_B$, where $(\chi_A \cdot \chi_B)(x) = \chi_A(x) \cdot \chi_B(x)$

b) $\chi_{A \cup B} = \chi_A + \chi_B - \chi_{A \cap B}$

c) $\chi_{\overline{A}} = 1 - \chi_A$, where $(1 - \chi_A)(x) = 1(x) - \chi_A(x) = 1 - \chi_A(x)$

(For $\mathcal{U}$ finite, placing the elements of $\mathcal{U}$ in a fixed order results in a one-to-one correspondence between subsets A of $\mathcal{U}$ and the arrays of 0's and 1's obtained as the images of $\mathcal{U}$ under χ_A. These arrays can then be used for the computer storage and manipulation of certain subsets of $\mathcal{U}$.)

27. With $A = \{x, y, z\}$, let $f, g: A \to A$ be given by $f = \{(x, y), (y, z), (z, x)\}$, $g = \{(x, y), (y, x), (z, z)\}$. Determine each of the following: $f \circ g$, $g \circ f$, f^{-1}, g^{-1}, $(g \circ f)^{-1}$, $f^{-1} \circ g^{-1}$, and $g^{-1} \circ f^{-1}$.

28. a) If $f: \mathbf{R} \to \mathbf{R}$ is defined by $f(x) = 5x + 3$, find $f^{-1}(8)$.

b) If $g: \mathbf{R} \to \mathbf{R}$, where $g(x) = |x^2 + 3x + 1|$, find $g^{-1}(1)$.

c) For $h: \mathbf{R} \to \mathbf{R}$, given by

$$h(x) = \left| \frac{x}{x + 2} \right|,$$

find $h^{-1}(4)$.

29. If $A = \{1, 2, 3, \ldots, 10\}$, how many functions $f: A \to A$ (simultaneously) satisfy $f^{-1}(\{1, 2, 3\}) = \emptyset$, $f^{-1}(\{4, 5\}) = \{1, 3, 7\}$, and $f^{-1}(\{8, 10\}) = \{8, 10\}$?

30. Let $f: A \to A$ be an invertible function. For $n \in \mathbf{Z}^+$ prove that $(f^n)^{-1} = (f^{-1})^n$. [This result can be used to define f^{-n} as either $(f^n)^{-1}$ or $(f^{-1})^n$.]

31. In certain programming languages, the functions pred and succ (for predecessor and successor, respectively) are functions from $\mathbf{Z}$ to $\mathbf{Z}$ where $\text{pred}(x) = \pi(x) = x - 1$ and $\text{succ}(x) = \sigma(x) = x + 1$.

a) Determine $(\pi \circ \sigma)(x)$, $(\sigma \circ \pi)(x)$.

b) Determine π^2, π^3, $\pi^n (n \geq 2)$, σ^2, σ^3, $\sigma^n (n \geq 2)$.

c) Determine π^{-2}, π^{-3}, π^{-n} $(n \geq 2)$, σ^{-2}, σ^{-3}, $\sigma^{-n}(n \geq 2)$, where, for example, $\sigma^{-2} = \sigma^{-1} \circ \sigma^{-1} = (\sigma \circ \sigma)^{-1} = (\sigma^2)^{-1}$. (See Supplementary Exercise 30.)

32. For $n \in \mathbf{Z}^+$, define $\tau: \mathbf{Z}^+ \to \mathbf{Z}^+$ by $\tau(n) = $ the number of positive-integer divisors of n.

a) Let $n = p_1^{e_1} p_2^{e_2} p_3^{e_3} \ldots p_k^{e_k}$, where $p_1, p_2, p_3, \ldots, p_k$ are distinct primes and e_i is a positive integer for all $1 \leq i \leq k$. What is $\tau(n)$?

b) Determine the three smallest values of $n \in \mathbf{Z}^+$ for which $\tau(n) = k$, where $k = 2, 3, 4, 5, 6$.

c) For all $k \in \mathbf{Z}^+$, $k > 1$, prove that $\tau^{-1}(k)$ is infinite.

d) If $a, b \in \mathbf{Z}^+$ with $\gcd(a, b) = 1$, prove that $\tau(ab) = \tau(a)\tau(b)$.

33. a) How many subsets $A = \{a, b, c, d\} \subseteq \mathbf{Z}^+$, where $a, b, c, d > 1$, satisfy the property $a \cdot b \cdot c \cdot d = 2 \cdot 3 \cdot 5 \cdot 7 \cdot 11 \cdot 13 \cdot 17 \cdot 19$?

b) How many subsets $A = \{a_1, a_2, \ldots, a_m\} \subseteq \mathbf{Z}^+$, where $a_i > 1$, $1 \leq i \leq m$, satisfy the property $\prod_{i=1}^{m} a_i = \prod_{j=1}^{n} p_j$, where the p_j, $1 \leq j \leq n$, are distinct primes and $n \geq m$?

34. Give an example of a function $f: \mathbf{Z}^+ \to \mathbf{R}$ where $f \in O(1)$ and f is one-to-one. (Hence f is not constant.)

35. Let $f, g: \mathbf{Z}^+ \to \mathbf{R}$ where

$$f(n) = \begin{cases} 2, & \text{for } n \text{ even} \\ 1, & \text{for } n \text{ odd} \end{cases} \qquad g(n) = \begin{cases} 3, & \text{for } n \text{ even} \\ 4, & \text{for } n \text{ odd} \end{cases}$$

Prove or disprove each of the following: (a) $f \in O(g)$; and (b) $g \in O(f)$.

36. For $f, g: \mathbf{Z}^+ \to \mathbf{R}$ we define $f + g: \mathbf{Z}^+ \to \mathbf{R}$ by $(f + g)(n) = f(n) + g(n)$, for $n \in \mathbf{Z}^+$. [*Note*: The plus sign in $f + g$ is for the addition of the functions f and g, while the plus sign in $f(n) + g(n)$ is for the addition of the real numbers $f(n)$ and $g(n)$.]

a) Let $f_1, g_1: \mathbf{Z}^+ \to \mathbf{R}$ with $f \in O(f_1)$ and $g \in O(g_1)$. If $f_1(n) \geq 0$, $g_1(n) \geq 0$, for all $n \in \mathbf{Z}^+$, prove that $(f + g) \in O(f_1 + g_1)$.

b) If the conditions $f_1(n) \geq 0$, $g_1(n) \geq 0$, for all $n \in \mathbf{Z}^+$, are not satisfied, as in part (a), provide a counterexample to show that

$$f \in O(f_1), g \in O(g_1) \not\Rightarrow (f + g) \in O(f_1 + g_1).$$

37. Let $a, b \in \mathbf{R}^+$, with $a, b > 1$. Let $f, g: \mathbf{Z}^+ \to \mathbf{R}$ be defined by $f(n) = \log_a n$, $g(n) = \log_b n$. Prove that $f \in O(g)$ and $g \in O(f)$. [Hence $O(\log_a n) = O(\log_b n)$.]

6

Languages: Finite State Machines

In this era of computers and telecommunications, we find ourselves confronted every day with input-output situations. For example, in purchasing a package of chewing gum from a vending machine, we *input* some coins and then press a button to get our expected *output*, the package of chewing gum we desire. The first coin that we input sets the machine in motion. Although we usually don't care about what happens inside the machine (unless some kind of breakdown occurs and we suffer a loss), we should realize that somehow the machine is keeping track of the coins we insert, until the correct total has been inserted. Only then, and not before, does the vending machine output the desired package of chewing gum. Consequently, for the vendor to make the expected profit per package of chewing gum, the machine must *internally remember*, as each coin is inserted, what sum of money has been deposited.

A computer is another example of an input-output device. Here the input is generally some type of information, and the output is the result obtained after processing this information. How the input is processed depends on the internal workings of the computer; it must have the ability to remember past information as it works on the information it is currently processing.

Using the concepts we developed earlier on sets and functions, in this chapter we shall investigate an abstract model called a *finite state machine*, or *sequential circuit*. Such circuits are one of two basic types of control circuits found in digital computers. (The other type is a *combinational circuit* or *gating network*, which is examined in Chapter 15.) They are also found in other systems such as our vending machine, as well as in the controls for elevators and in traffic-light systems.

As the name indicates, a finite state machine has a finite number of internal states where the machine remembers certain information when it is in a particular state. However, before getting into this concept we need some set-theoretic material in order to talk about what constitutes valid input for such a machine.

6.1
Language: The Set Theory of Strings

Sequences of symbols, or characters, play a key role in the processing of information by a computer. Inasmuch as computer programs are representable in terms of finite sequences of characters, some algebraic way is needed for handling such finite sequences, or *strings*.

Throughout this section we use Σ to denote a nonempty *finite* set of symbols, collectively called an *alphabet*. For example, we may have $\Sigma = \{0, 1\}$ or $\Sigma = \{a, b, c, d, e\}$.

In any alphabet Σ, we do not list elements that can be formed from other elements of Σ by juxtaposition (that is, if $a, b \in \Sigma$, then the string ab is the juxtaposition of the symbols a and b). As a result of this convention, alphabets such as $\Sigma = \{0, 1, 2, 11, 12\}$ and $\Sigma = \{a, b, c, ba, aa\}$ are not considered. (In addition, this convention will help us later in Definition 6.5, when we talk about the length of a string.)

Using an alphabet Σ as the starting place, we can construct strings from the symbols of Σ in a systematic manner by using the following idea.

Definition 6.1

If Σ is an alphabet and $n \in \mathbf{Z}^+$, we define the *powers of* Σ recursively as follows:

1) $\Sigma^1 = \Sigma$; and
2) $\Sigma^{n+1} = \{xy | x \in \Sigma, y \in \Sigma^n\}$, where xy denotes the juxtaposition of x and y.

EXAMPLE 6.1

Let Σ be an alphabet.

If $n = 2$, then $\Sigma^2 = \{xy | x, y \in \Sigma\}$. For instance, with $\Sigma = \{0, 1\}$ we find $\Sigma^2 = \{00, 01, 10, 11\}$.

When $n = 3$, the elements of Σ^3 have the form uv, where $u \in \Sigma$ and $v \in \Sigma^2$. But since we know the form of the elements in Σ^2, we may also regard the strings in Σ^3 as sequences of the form uxy, where $u, x, y \in \Sigma$. As an example for this case, suppose that $\Sigma = \{a, b, c, d, e\}$. Then Σ^3 would contain $5^3 = 125$ three-symbol strings — among them aaa, acb, ace, cdd, and eda.

In general, for all $n \in \mathbf{Z}^+$ we find that $|\Sigma^n| = |\Sigma|^n$ because we are dealing with arrangements (of size n) where we are allowed to repeat any of the $|\Sigma|$ objects.

Now that we have examined Σ^n for $n \in \mathbf{Z}^+$, we shall look into one more power of Σ.

Definition 6.2

For an alphabet Σ we define $\Sigma^0 = \{\lambda\}$, where λ denotes the *empty string* — that is, the string consisting of *no* symbols taken from Σ.

The symbol λ is never an element in our alphabet Σ, and we should not mistake it for the blank (space) that is found in many alphabets.

However, although $\lambda \notin \Sigma$, we do have $\emptyset \subseteq \Sigma$, so we need to be cautious here. We observe that (1) $\{\lambda\} \not\subseteq \Sigma$ since $\lambda \notin \Sigma$; and (2) $\{\lambda\} \neq \emptyset$ because $|\{\lambda\}| = 1 \neq 0 = |\emptyset|$.

In order to speak collectively about the sets $\Sigma^0, \Sigma^1, \Sigma^2, \ldots$, we introduce the following notation for unions of such sets.

Definition 6.3

If Σ is an alphabet, then

a) $\Sigma^+ = \bigcup_{n=1}^{\infty} \Sigma^n = \bigcup_{n \in \mathbf{Z}^+} \Sigma^n$; and b) $\Sigma^* = \bigcup_{n=0}^{\infty} \Sigma^n$.

We see that the only difference between the sets Σ^+ and Σ^* is the presence of the element λ because $\lambda \in \Sigma^n$ only when $n = 0$. Also $\Sigma^* = \Sigma^+ \cup \Sigma^0$.

In addition to using the term *string*, we shall also refer to the elements of Σ^+ or Σ^* as *words* and sometimes as *sentences*. For $\Sigma = \{0, 1, 2\}$, we find such words as 0, 01, 102, and 1112 in both Σ^+ and Σ^*.

Finally, we note that even though the sets Σ^+ and Σ^* are *infinite*, the elements of these sets are *finite* strings of symbols.

EXAMPLE 6.2

For $\Sigma = \{0, 1\}$ the set Σ^* consists of all finite strings of 0's and 1's together with the empty string. For n reasonably small, we could actually list all strings in Σ^n.

If $\Sigma = \{\beta, 0, 1, 2, \ldots, 9, +, -, \times, /, (,)\}$, where β denotes the blank (or space), it is harder to describe Σ^* and, for $n > 2$, there are too many strings to list in Σ^n. Here in Σ^* we find familiar arithmetic expressions such as $(7 + 5)/(2 \times (3 - 10))$ as well as gibberish such as $+)((7/\times + 3/($.

We are now confronted with a familiar situation. As with statements (Chapter 2), sets (Chapter 3), and functions (Chapter 5), once again we need to be able to decide when two objects under study — in this case strings — are to be considered the same. We investigate this issue next.

Definition 6.4

If $w_1, w_2 \in \Sigma^+$, then we may write $w_1 = x_1 x_2 \cdots x_m$ and $w_2 = y_1 y_2 \cdots y_n$, for $m, n \in \mathbf{Z}^+$, and $x_1, x_2, \ldots, x_m, y_1, y_2, \ldots, y_n \in \Sigma$. We say that the strings w_1 and w_2 are *equal*, and we write $w_1 = w_2$, if $m = n$, and $x_i = y_i$ for all $1 \leq i \leq m$.

It follows from this definition that two strings in Σ^+ are *equal* only when each is formed from the same number of symbols from Σ and the corresponding symbols in the two strings match identically.

The number of symbols in a string is also needed to define another property.

Definition 6.5

Let $w = x_1 x_2 \cdots x_n \in \Sigma^+$, where $x_i \in \Sigma$ for each $1 \leq i \leq n$. We define the *length* of w, which is denoted by $\|w\|$, as the value n. For the case of λ, we have $\|\lambda\| = 0$.

As a result of Definition 6.5, we find that for any alphabet Σ, if $w \in \Sigma^*$ and $\|w\| \geq 1$, then $w \in \Sigma^+$, and conversely. Also, for all $y \in \Sigma^*$, $\|y\| = 1$ if and only if $y \in \Sigma$. Should Σ contain the symbol β (for the blank), it is still the case that $\|\beta\| = 1$.

If we use a particular alphabet, say $\Sigma = \{0, 1, 2\}$, and examine the elements $x = 01$, $y = 212$, and $z = 01212$ (in Σ^*), we find that

$$\|z\| = \|01212\| = 5 = 2 + 3 = \|01\| + \|212\| = \|x\| + \|y\|.$$

In order to continue our study of the properties of strings and alphabets, we need to extend the idea of juxtaposition a little further.

Definition 6.6

Let $x, y \in \Sigma^+$ with $x = x_1 x_2 \cdots x_m$ and $y = y_1 y_2 \cdots y_n$, so that each x_i, for $1 \leq i \leq m$, and each y_j, for $1 \leq j \leq n$, is in Σ. The *concatenation* of x and y, which we write as xy, is the string $x_1 x_2 \cdots x_m y_1 y_2 \cdots y_n$.

The *concatenation* of x and λ is $x\lambda = x_1 x_2 \cdots x_m \lambda = x_1 x_2 \cdots x_m = x$, and the *concatenation* of λ and x is $\lambda x = \lambda x_1 x_2 \cdots x_m = x_1 x_2 \cdots x_m = x$. Finally, the *concatenation* of λ and λ is $\lambda\lambda = \lambda$.

Here we have defined a closed binary operation on Σ^* (and Σ^+). This operation is associative but not commutative (unless $|\Sigma| = 1$), and since $x\lambda = \lambda x = x$ for all $x \in \Sigma^*$, the element $\lambda \in \Sigma^*$ is the identity for the operation of concatenation. The ideas embodied in the last two definitions (the length of a string and the operation of concatenation) are interrelated in the result

$$\|xy\| = \|x\| + \|y\|, \quad \text{for all } x, y \in \Sigma^*,$$

from which we obtain the special case

$$\|x\| = \|x\| + 0 = \|x\| + \|\lambda\| = \|x\lambda\| \ (= \|\lambda x\|).$$

Finally, for each $z \in \Sigma$, we have $\|z\| = \|z\lambda\| = \|\lambda z\| = 1$, whereas $\|zz\| = 2$.

The closed binary operation of concatenation now leads us to another recursive definition. Earlier we looked at powers of an alphabet Σ. Now we examine powers of strings.

Definition 6.7   For each $x \in \Sigma^*$, we define the *powers* of x by $x^0 = \lambda$, $x^1 = x$, $x^2 = xx$, $x^3 = xx^2$, ..., $x^{n+1} = xx^n$, ..., where $n \in \mathbf{N}$.

This definition is another illustration of how a mathematical entity is given in a recursive manner: The mathematical entity we presently seek is derived from previously derived entities. The definition provides a way for us to deal with the n-fold concatenation [an $(n + 1)$st power] of a string as the concatenation of the string with its $(n - 1)$-fold concatenation (an nth power). In so doing, the definition includes the special case where the string is just one symbol.

EXAMPLE 6.3   If $\Sigma = \{0, 1\}$ and $x = 01$, then $x^0 = \lambda$, $x^1 = 01$, $x^2 = 0101$, and $x^3 = 010101$. For all $n > 0$, x^n consists of a string of n 0's and n 1's where the first symbol is 0 and the symbols alternate. Here $\|x^2\| = 4 = 2\|x\|$, $\|x^3\| = 6 = 3\|x\|$, and, for all $n \in \mathbf{N}$, $\|x^n\| = n\|x\|$.

We are just about ready to tackle the major theme of this section, the concept of a language. Before we do so, however, we need to inquire about three other ideas. These ideas involve special subsections of strings.

Definition 6.8   If $x, y \in \Sigma^*$ and $w = xy$, then the string x is called a *prefix* of w, and if $y \neq \lambda$, then x is said to be a *proper prefix*. Similarly, the string y is called *a suffix* of w; it is a *proper suffix* when $x \neq \lambda$.

EXAMPLE 6.4   Let $\Sigma = \{a, b, c\}$, and consider the string $w = abbcc$. Then each of the strings λ, a, ab, abb, $abbc$, and $abbcc$ is a prefix of w, and except for $abbcc$ itself, each is a proper prefix. On the other hand, each of the strings λ, c, cc, bcc, $bbcc$, and $abbcc$ is a suffix of w, where the first five strings are proper suffixes.

In general, for an alphabet Σ, if $n \in \mathbf{Z}^+$ and $x_i \in \Sigma$, for all $1 \leq i \leq n$, then each of λ, x_1, $x_1 x_2$, $x_1 x_2 x_3$, ..., and $x_1 x_2 x_3 \cdots x_n$ is a prefix of the string $x = x_1 x_2 x_3 \cdots x_n$. And λ, x_n, $x_{n-1} x_n$, $x_{n-2} x_{n-1} x_n$, ..., and $x_1 x_2 x_3 \cdots x_n$ are all suffixes of x. So x has $n + 1$ prefixes, n of which are proper—and the situation is the same for suffixes.

EXAMPLE 6.5   If $\|x\| = 5$, $\|y\| = 4$, and $w = xy$, then w has x as a proper prefix and y as a proper suffix. In all, w has nine proper prefixes and nine proper suffixes because λ is both a proper prefix and a proper suffix for every string in Σ^+. Here xy is both a prefix and a suffix, but in neither case is it proper.

EXAMPLE 6.6	For a given alphabet Σ, let $w, a, b, c, d \in \Sigma^*$. If $w = ab = cd$, then

    1) a is a prefix of c, or c is a prefix of a; and

    2) b is a suffix of d, or d is a suffix of b.

Definition 6.9   If $x, y, z \in \Sigma^*$ and $w = xyz$, then y is called a *substring* of w. When at least one of x and z is different from λ (so that y is different from w), we call y a *proper substring*.

EXAMPLE 6.7	For $\Sigma = \{0, 1\}$, let $w = 00101110 \in \Sigma^*$. We find the following substrings in w:

    1) 1011: This arises in only one way — when $w = xyz$, with $x = 00$, $y = 1011$, and $z = 10$.

    2) 10: This example comes about in two ways:

        a) $w = xyz$ where $x = 00$, $y = 10$, and $z = 1110$; and

        b) $w = xyz$ for $x = 001011$, $y = 10$, and $z = \lambda$.

In case (b) the substring is also a (proper) suffix of w.

Now that we are familiar with the necessary definitions, it is time to think about the concept of language. When we consider the standard alphabet, including the blank, many strings such as *qxio, the wxxy red atzl*, and *aeytl* do not represent words or parts of sentences that appear in the English language, even though they are elements of Σ^*. Consequently, in order to consider only those words and expressions that make sense in the English language, we concentrate on a subset of Σ^*. This leads us to the following generalization.

Definition 6.10   For a given alphabet Σ, any subset of Σ^* is called a *language* over Σ. This includes the subset $\emptyset$, which we call the *empty language*.

EXAMPLE 6.8	With $\Sigma = \{0, 1\}$, the sets $A = \{0, 01, 001\}$ and $B = \{0, 01, 001, 0001, \ldots\}$ are examples of languages over Σ.

EXAMPLE 6.9	With Σ the alphabet of 26 letters, 10 digits, and the special symbols used in a given implementation of C++, the collection of executable programs for that implementation constitutes a language. In the same situation, each executable program could be considered a language, as could a particular set of such programs.

Since languages are sets, we can form the union, intersection, and symmetric difference of two languages. However, for the work here, an extension of the closed binary operation defined (in Definition 6.6) for strings is more useful.

Definition 6.11   For an alphabet Σ and languages $A, B \subseteq \Sigma^*$, the *concatenation* of A and B, denoted AB, is $\{ab \mid a \in A, b \in B\}$.

We might compare concatenation with the cross product. We shall see that just as $A \times B \neq B \times A$ in general, we also have $AB \neq BA$ in general. For A, B finite we did have $|A \times B| = |B \times A|$, but here $|AB| \neq |BA|$ is possible for finite languages.

EXAMPLE 6.10

Let $\Sigma = \{x, y, z\}$, and let A, B be the finite languages $A = \{x, xy, z\}$, $B = \{\lambda, y\}$. Then $AB = \{x, xy, z, xyy, zy\}$ and $BA = \{x, xy, z, yx, yxy, yz\}$, so

1) $|AB| = 5 \neq 6 = |BA|$; and
2) $|AB| = 5 \neq 6 = 3 \cdot 2 = |A\|B|$.

The differences arise because there are two ways to represent xy: (1) xy for $x \in A$, $y \in B$ and (2) $xy\lambda$ where $xy \in A$ and $\lambda \in B$. [The concept of uniqueness of representation is something we cannot take for granted. Although it does not hold here, it is a key to the success of many mathematical ideas. We saw this, for example, in the Fundamental Theorem of Arithmetic (Theorem 4.11).]

The preceding example suggests that for finite languages A and B, $|AB| \leq |A\|B|$. This can be shown to be true in general.

The following theorem deals with some of the properties satisfied by the concatenation of languages.

THEOREM 6.1

For an alphabet Σ, let A, B, $C \subseteq \Sigma^*$. Then

a) $A\{\lambda\} = \{\lambda\}A = A$ b) $(AB)C = A(BC)$
c) $A(B \cup C) = AB \cup AC$ d) $(B \cup C)A = BA \cup CA$
e) $A(B \cap C) \subseteq AB \cap AC$ f) $(B \cap C)A \subseteq BA \cap CA$

Proof: We prove parts (d) and (f) and leave the other parts for the reader.

(d) Since we are trying to prove that two sets are equal, once again we use the idea of set equality that we first found in Definition 3.2. Starting with x in Σ^* we find that $x \in (B \cup C)A \Rightarrow x = yz$ for $y \in B \cup C$ and $z \in A \Rightarrow (x = yz$ for $y \in B, z \in A)$ or $(x = yz$ for $y \in C, z \in A) \Rightarrow x \in BA$ or $x \in CA \Rightarrow x \in BA \cup CA$, so $(B \cup C)A \subseteq BA \cup CA$. Conversely, it follows that $x \in BA \cup CA \Rightarrow x \in BA$ or $x \in CA \Rightarrow (x = ba_1$ where $b \in B$ and $a_1 \in A)$ or $(x = ca_2$ where $c \in C$ and $a_2 \in A)$. Assume $x = ba_1$ for $b \in B$, $a_1 \in A$. Since $B \subseteq B \cup C$, we have $x = ba_1$, where $b \in B \cup C$ and $a_1 \in A$. Then $x \in (B \cup C)A$, so $BA \cup CA \subseteq (B \cup C)A$. (The argument is similar if $x = ca_2$.) With both inclusions established, it follows that $(B \cup C)A = BA \cup CA$.

(f) For $x \in \Sigma^*$, we see that $x \in (B \cap C)A \Rightarrow x = yz$ where $y \in B \cap C$ and $z \in A \Rightarrow (x = yz$ for $y \in B$ and $z \in A)$ and $(x = yz$ for $y \in C$ and $z \in A) \Rightarrow x \in BA$ and $x \in CA \Rightarrow x \in BA \cap CA$, so $(B \cap C)A \subseteq BA \cap CA$.

With $\Sigma = \{x, y, z\}$, let $B = \{x, xx, y\}$, $C = \{y, xy\}$, and $A = \{y, yy\}$. Then $xyy \in BA \cap CA$, but $xyy \notin (B \cap C)A$. Consequently, $(B \cap C)A \subset BA \cap CA$ for these particular languages.

Comparable to the concepts of Σ^n, Σ^*, Σ^+, the following definitions are given for an arbitrary language $A \subseteq \Sigma^*$.

Definition 6.12

For a given language $A \subseteq \Sigma^*$ we can construct other languages as follows:

a) $A^0 = \{\lambda\}$, $A^1 = A$, and for all $n \in \mathbf{Z}^+$, $A^{n+1} = \{ab | a \in A, b \in A^n\}$.

b) $A^+ = \bigcup_{n \in \mathbf{Z}^+} A^n$, the *positive closure* of A.

c) $A^* = A^+ \cup \{\lambda\}$. The language A^* is called the *Kleene closure* of A, in honor of the American logician Stephen Cole Kleene (1909–1994).

EXAMPLE 6.11

If $\Sigma = \{x, y, z\}$ and $A = \{x\}$, then (1) $A^0 = \{\lambda\}$; (2) $A^n = \{x^n\}$, for each $n \in \mathbf{N}$; (3) $A^+ = \{x^n | n \geq 1\}$; and (4) $A^* = \{x^n | n \geq 0\}$.

EXAMPLE 6.12

Let $\Sigma = \{x, y\}$.

a) If $A = \{xx, xy, yx, yy\} = \Sigma^2$, then A^* is the language of all strings w in Σ^* where the length of w is even.

b) With A as in part (a) and $B = \{x, y\}$, the language BA^* contains all the strings in Σ^* of odd length. In this case we also find that $BA^* = A^*B$ and that $\Sigma^* = A^* \cup BA^*$.

c) The language $\{x\}\{x, y\}^*$ (the concatenation of the languages $\{x\}$ and $\{x, y\}^*$) contains every string in Σ^* for which x is a prefix. The language $\{x\}\{x, y\}^+$ (the concatenation of the languages $\{x\}$ and $\{xy\}^+$) contains every string in Σ^+ for which x is a proper prefix.

The language containing all strings in Σ^* for which yy is a suffix can be defined by $\{x, y\}^*\{yy\}$.

Every string in the language $\{x, y\}^*\{xxy\}\{x, y\}^*$ has xxy as a substring.

d) Each string in the language $\{x\}^*\{y\}^*$ consists of a finite number (possibly zero) of x's followed by a finite number (also possibly zero) of y's. And although $\{x\}^*\{y\}^* \subseteq \{x, y\}^*$, the string $w = xyx$ is in $\{x, y\}^*$ but not in $\{x\}^*\{y\}^*$. Hence $\{x\}^*\{y\}^* \subset \{x, y\}^*$.

EXAMPLE 6.13

In the algebra of real numbers, if $a, b \in \mathbf{R}$ and $a, b > 0$, then $a^2 = b^2 \Rightarrow a = b$. However, in the case of languages, if $\Sigma = \{x, y\}$, $A = \{\lambda, x, x^3, x^4, \ldots\} = \{x^n | n \geq 0\} - \{x^2\}$ and $B = \{x^n | n \geq 0\}$, then $A^2 = B^2 (= B)$, but $A \neq B$. (*Note*: We never have $\lambda \in \Sigma$, but it is possible to have $\lambda \in A \subseteq \Sigma^*$.)

We continue this section with a lemma and a second theorem that deal with the properties of languages.

LEMMA 6.1

Let Σ be an alphabet, with languages $A, B \subseteq \Sigma^*$. If $A \subseteq B$, then for all $n \in \mathbf{Z}^+$, $A^n \subseteq B^n$.

Proof: Since $A^1 = A \subseteq B = B^1$, it follows that the result is true in the case for $n = 1$. Assuming the truth for $n = k$, we have $A \subseteq B \Rightarrow A^k \subseteq B^k$. Now consider a string x from A^{k+1}. From part (a) of Definition 6.12 we know that $x = x_1 x_k$, where $x_1 \in A$, $x_k \in A^k$. If $A \subseteq B$ then $A^k \subseteq B^k$ (by the induction hypothesis), and we have $x_1 \in B$, $x_k \in B^k$. Consequently, $x = x_1 x_k \in BB^k = B^{k+1}$ and $A^{k+1} \subseteq B^{k+1}$. By the Principle of Mathematical Induction, it now follows that if $A \subseteq B$, then for all $n \in \mathbf{Z}^+$, $A^n \subseteq B^n$.

Note: Lemma 6.1 does *not* establish that $A^+ \subseteq B^+$ or that $A^* \subseteq B^*$. These results are part of our next theorem.

THEOREM 6.2

For an alphabet Σ and languages $A, B \subseteq \Sigma^*$,

a) $A \subseteq AB^*$ **b)** $A \subseteq B^*A$

c) $A \subseteq B \Rightarrow A^+ \subseteq B^+$ **d)** $A \subseteq B \Rightarrow A^* \subseteq B^*$

e) $AA^* = A^*A = A^+$ **f)** $A^*A^* = A^* = (A^*)^* = (A^*)^+ = (A^+)^*$

g) $(A \cup B)^* = (A^* \cup B^*)^* = (A^*B^*)^*$

Proof: We provide the proofs for parts (c) and (g).

(c) Let $A \subseteq B$ and $x \in A^+$. Then $x \in A^+ \Rightarrow x \in A^n$, for some $n \in \mathbf{Z}^+$. From Lemma 6.1 it then follows that $x \in B^n \subseteq B^+$, and we have shown that $A^+ \subseteq B^+$.

(g) $[(A \cup B)^* = (A^* \cup B^*)^*]$. We know that $A \subseteq A^*, B \subseteq B^* \Rightarrow (A \cup B) \subseteq (A^* \cup B^*)$ $\Rightarrow (A \cup B)^* \subseteq (A^* \cup B^*)^*$ [by part (d)]. Conversely, we also see that $A, B \subseteq A \cup B \Rightarrow$ $A^*, B^* \subseteq (A \cup B)^*$ [by part (d)] $\Rightarrow (A^* \cup B^*) \subseteq (A \cup B)^* \Rightarrow (A^* \cup B^*)^* \subseteq (A \cup B)^*$ [by parts (d) and (f)]. From both inclusions it follows that $(A \cup B)^* = (A^* \cup B^*)^*$.

$[(A^* \cup B^*)^* = (A^*B^*)^*]$. First we find that $A^*, B^* \subseteq A^*B^*$ [by parts (a) and (b)] $\Rightarrow$ $(A^* \cup B^*) \subseteq A^*B^* \Rightarrow (A^* \cup B^*)^* \subseteq (A^*B^*)^*$ [by part (d)]. Conversely, if $xy \in A^*B^*$ where $x \in A^*$ and $y \in B^*$, then $x, y \in A^* \cup B^*$, so $xy \in (A^* \cup B^*)^*$, and $A^*B^* \subseteq$ $(A^* \cup B^*)^*$. Using parts (d) and (f) again, $(A^*B^*)^* \subseteq (A^* \cup B^*)^*$, and so the result follows.

As we close this first section we further examine the idea of a recursively defined set (given in Section 4.2), as demonstrated in the following three examples.

EXAMPLE 6.14

For the alphabet $\Sigma = \{0, 1\}$ consider the language $A \subseteq \Sigma^*$ where each word in A contains exactly one occurrence of the symbol 0. Then A is an infinite set, and among the words in A one finds 0, 01, 10, 01111, 11110111, and 11111111110. There are also infinitely many words in Σ^* that are not in A—such as, 1, 11, 00, 000, 010, and 011111111110. We can define this language A recursively as follows:

1) Our base step tells us that $0 \in A$; and

2) For the recursive process we want to include in A the words $1x$ and $x1$, for each word $x \in A$.

Using this definition, the following discussion shows us that the word 1011 is in A.

From part (1) of our definition, we know that $0 \in A$. Then by applying part (2) of our definition—three times—we find:

i) $01 \in A$, because $0 \in A$;

ii) $011 \in A$, because $01 \in A$; and

iii) Since $011 \in A$, we have $1011 \in A$.

EXAMPLE 6.15

For $\Sigma = \{(,)\}$—the alphabet containing the left and right parentheses—we want to consider the language $A \subseteq \Sigma^*$ consisting of those nonempty strings of parentheses that are grammatically correct for algebraic expressions. Hence we find, for example, the three strings (( )), ((( ) ( ))), and ( ) ( ) ( ) in this language, but we do not find strings such as (( ) ( ), ( )) (( ), or)( )(( ))). We see that if a string $x(\neq \lambda)$ is to be in A, then

i) we must have the same number of left parentheses in x as there are right parentheses; and

ii) the number of left parentheses must (always) be greater than or equal to the number of right parentheses, as we examine each of the parentheses in x — reading them consecutively from left to right.

The language A may be given recursively as follows:

1) ( ) is in A; and

2) For all x, $y \in A$ we have (i) $xy \in A$ and (ii) $(x) \in A$.

[As we mentioned prior to Example 4.22, we also have an implicit restriction here — that no string of parentheses is in A unless it can be derived through steps (1) and (2) above.]

Using this recursive definition, the following shows us how to establish that the string $(( )( ))$ in Σ^* is in the language A.

Steps	Reasons
1) ( ) is in A.	Part (1) of the recursive definition
2) ( )( ) is in A.	Step (1) and part (2i) of the definition
3) $(( )( ))$ is in A.	Step (2) and part (2ii) of the definition

---

EXAMPLE 6.16

Given an alphabet Σ, consider the string $x = x_1 x_2 x_3 \cdots x_{n-1} x_n$ in Σ^* where $x_i \in \Sigma$ for each $1 \le i \le n$ and $n \in \mathbf{Z}^+$. The *reversal* of x, denoted x^R, is the string obtained from x by reading the symbols (in x) from right to left — that is, $x^R = x_n x_{n-1} \cdots x_3 x_2 x_1$. For example, if $\Sigma = \{0, 1\}$ and $x = 01101$, then $x^R = 10110$ and for $w = 101101$ we find $w^R = 101101 = w$. In general, we can define the reversal of a string (from Σ^*) recursively as follows:

1) $\lambda^R = \lambda$; and

2) For each $n \in \mathbf{N}$, if $x \in \Sigma^{n+1}$, then we can write $x = zy$ where $z \in \Sigma$ and $y \in \Sigma^n$ — and here we define $x^R = (zy)^R = (y^R)z$.

Using this recursive definition we shall now prove that if Σ is an alphabet and x_1, $x_2 \in \Sigma^*$, then $(x_1 x_2)^R = x_2^R x_1^R$.

Proof: Here the proof is by mathematical induction — on the value of $\|x_1\|$. If $\|x_1\| = 0$, then $x_1 = \lambda$ and $(x_1 x_2)^R = (\lambda x_2)^R = x_2^R = x_2^R \lambda = x_2^R \lambda^R = x_2^R x_1^R$ because $\lambda^R = \lambda$ from part (1) of the recursive definition. Consequently, the result is true in this first case and this establishes the basis step. For the inductive step we shall assume the result is true for all y, $x_2 \in \Sigma^*$ where $\|y\| = k$ for some $k \in \mathbf{N}$. Now consider what happens for x_1, $x_2 \in \Sigma^*$ where $x_1 = zy_1$, with $\|z\| = 1$ and $\|y_1\| = k$. Here we find that $(x_1 x_2)^R = (zy_1 x_2)^R = (y_1 x_2)^R z$ [from part (2) of the recursive definition] $= x_2^R y_1^R z$ (from the induction hypothesis) $= x_2^R (zy_1)^R$ [again by part (2) of the recursive definition] $= x_2^R x_1^R$. Therefore the result is true for all x_1, $x_2 \in \Sigma^*$ by the Principle of Mathematical Induction.

EXERCISES 6.1

1. Let $\Sigma = \{a, b, c, d, e\}$. (a) What is $|\Sigma^2|$? $|\Sigma^3|$? (b) How many strings in Σ^* have length at most 5?

2. For $\Sigma = \{w, x, y, z\}$ determine the number of strings in Σ^* of length 5 (a) that start with w; (b) with precisely two w's; (c) with no w's; (d) with an even number of w's.

3. If $x \in \Sigma^*$ and $\|x^3\| = 36$, what is $\|x\|$?

4. Let $\Sigma = \{\beta, x, y, z\}$ where β denotes a blank, so $x\beta \neq x$, $\beta\beta \neq \beta$, and $x\beta y \neq xy$ but $x\lambda y = xy$. Compute each of the following:

a) $\|\lambda\|$ **b)** $\|\lambda\lambda\|$ **c)** $\|\beta\|$

d) $\|\beta\beta\|$ **e)** $\|\beta^3\|$ **f)** $\|x\beta\beta y\|$

g) $\|\beta\lambda\|$ **h)** $\|\lambda^{10}\|$

5. Let $\Sigma = \{v, w, x, y, z\}$ and $A = \bigcup_{n=1}^{6} \Sigma^n$. How many strings in A have xy as a proper prefix?

6. Let Σ be an alphabet. Let $x_i \in \Sigma$ for $1 \leq i \leq 100$ (where $x_i \neq x_j$ for all $1 \leq i < j \leq 100$). How many nonempty substrings are there for the string $s = x_1 x_2 \cdots x_{100}$?

7. For the alphabet $\Sigma = \{0, 1\}$, let $A, B, C \subseteq \Sigma^*$ be the following languages:

$$A = \{0, 1, 00, 11, 000, 111, 0000, 1111\},$$

$$B = \{w \in \Sigma^* \mid 2 \leq \|w\|\},$$

$$C = \{w \in \Sigma^* \mid 2 \geq \|w\|\}.$$

Determine the following subsets (languages) of Σ^*.

a) $A \cap B$ **b)** $A - B$ **c)** $A \triangle B$

d) $A \cap C$ **e)** $B \cup C$ **f)** $\overline{(A \cap C)}$

8. Let $A = \{10, 11\}$, $B = \{00, 1\}$ be languages for the alphabet $\Sigma = \{0, 1\}$. Determine each of the following: (a) AB; (b) BA; (c) A^3; (d) B^2.

9. If $A, B, C,$ and D are languages over Σ, prove that (a) $(A \subseteq B \wedge C \subseteq D) \Rightarrow AC \subseteq BD$; and (b) $A\emptyset = \emptyset A = \emptyset$.

10. For $\Sigma = \{x, y, z\}$, let $A, B \subseteq \Sigma^*$ be given by $A = \{xy\}$ and $B = \{\lambda, x\}$. Determine (a) AB; (b) BA; (c) B^3; (d) B^+; (e) A^*.

11. Given an alphabet Σ, is there a language $A \subseteq \Sigma^*$ where $A^* = A$?

12. For $\Sigma = \{0, 1\}$ determine whether the string 00010 is in each of the following languages (taken from Σ^*).

a) $\{0, 1\}^*$ **b)** $\{000, 101\}\{10, 11\}$

c) $\{00\}\{0\}^*\{10\}$ **d)** $\{000\}^*\{1\}^*\{0\}$

e) $\{00\}^*\{10\}^*$ **f)** $\{0\}^*\{1\}^*\{0\}^*$

13. For $\Sigma = \{0, 1\}$ describe the strings in A^* for each of the following languages $A \subseteq \Sigma^*$.

a) $\{01\}$ **b)** $\{000\}$

c) $\{0, 010\}$ **d)** $\{1, 10\}$

14. For $\Sigma = \{0, 1\}$ determine all possible languages $A, B \subseteq \Sigma^*$ where $AB = \{01, 000, 0101, 0111, 01000, 010111\}$.

15. Given a nonempty language $A \subseteq \Sigma^*$, prove that if $A^2 = A$, then $\lambda \in A$.

16. For a given alphabet Σ, let $a \in \Sigma$ — with a fixed. Define the functions $p_a, s_a, r: \Sigma^* \to \Sigma^*$ and the function $d: \Sigma^+ \to \Sigma^*$ as follows:

i) The prefix (by a) function: $p_a(x) = ax$, $x \in \Sigma^*$.

ii) The suffix (by a) function: $s_a(x) = xa$, $x \in \Sigma^*$.

iii) The reversal function: $r(\lambda) = \lambda$; for $x \in \Sigma^+$, if $x = x_1 x_2 \cdots x_{n-1} x_n$, where $x_i \in \Sigma$ for all $1 \leq i \leq n$, then $r(x) = x_n x_{n-1} \cdots x_2 x_1 = x^R$ (as defined in Example 6.16).

iv) The front deletion function: for $x \in \Sigma^+$, if $x = x_1 x_2 x_3 \cdots x_n$, then $d(x) = x_2 x_3 \cdots x_n$.

a) Which of these four functions is (or are) one-to-one?

b) Determine which of these four functions is (or are) onto. If a function is not onto, determine its range.

c) Are any of these four functions invertible? If so, determine their inverse functions.

d) Suppose that $\Sigma = \{a, e, i, o, u\}$. How many words x in Σ^4 satisfy $r(x) = x$? How many in Σ^5? How many in Σ^n, where $n \in \mathbf{N}$?

e) For $x \in \Sigma^*$, determine

$$(d \circ p_a)(x) \quad \text{and} \quad (r \circ d \circ r \circ s_a)(x).$$

f) If $\Sigma = \{a, e, i, o, u\}$ and $B = \{ae, ai, ao, oo, eio, eiouu\} \subseteq \Sigma^*$, find $r^{-1}(B)$, $p_a^{-1}(B)$, $s_a^{-1}(B)$, and $|d^{-1}(B)|$.

17. If $A(\neq \emptyset)$ is a language and $A^2 = A$, prove that $A = A^*$.

18. Provide the proofs for the remaining parts of Theorems 6.1 and 6.2.

19. Prove that for all finite languages $A, B \subseteq \Sigma^*$, $|AB| \leq |A\|B|$.

20. For $\Sigma = \{x, y\}$, use finite languages from Σ^* (as in Example 6.12), together with set operations, to describe the set of strings in Σ^* that (a) contain exactly one occurrence of x; (b) contain exactly two occurrences of x; (c) begin with x; (d) end in yxy; (e) begin with x or end in yxy or both; (f) begin with x or end in yxy but not both.

21. For $\Sigma = \{0, 1\}$, let $A \subseteq \Sigma^*$ be the language defined recursively as follows:

1) The symbols 0, 1 are both in A — this is the base for our definition; and

2) For each word x in A, the word $0x1$ is also in A — this constitutes the recursive process.

a) Find four different words — two of length 3 and two of length 5 — in A.

b) Use the given recursive definition to show that 0001111 is in A.

c) Explain why 00001111 is not in A.

22. Provide a recursive definition for each of the following languages $A \subseteq \Sigma^*$ where $\Sigma = \{0, 1\}$.

a) $x \in A$ if (and only if) the number of 0's in x is even.

b) $x \in A$ if (and only if) all of the 1's in x precede all of the 0's.

23. Use the recursive definition given in Example 6.15 to verify that each of the following strings is in the language A of that example.

a) (())() **b)** (())()() **c)** ()(()())

24. For an alphabet Σ a string x in Σ^* is called a *palindrome* if $x = x^R$ — that is, x is equal to its reversal. If $A \subseteq \Sigma^*$ where $A = \{x \in \Sigma^* | x = x^R\}$, how can we define the language A recursively?

25. For $\Sigma = \{0, 1\}$, let $A \subseteq \Sigma^*$, where $A = \{00, 1\}$. How many strings in A^* have length 3? length 4? length 5? length 6?

26. For $\Sigma = \{0, 1\}$, let $A \subseteq \Sigma^*$, where $A = \{00, 111\}$. How many strings in A^* have length 19?

27. For $\Sigma = \{0, 1\}$, let $A, B \subseteq \Sigma^*$, where A is the language of all strings in Σ^* of even length, while B is the language of all strings in Σ^* of odd length. Give a recursive definition for each of the languages A, B.

28. Let $\Sigma = \{a, b, c\}$. Determine the smallest number of words one must select from Σ^4 to guarantee that at least two of the words start and end with the same letter.

6.2
Finite State Machines: A First Encounter

We return now to the vending machine mentioned at the start of this chapter and analyze it in the following circumstance.

At a metropolitan office, a vending machine dispenses two flavors of chewing gum (each flavor in a package of five pieces): peppermint (P) and spearmint (S). The cost of a package of either flavor is 20¢. The machine accepts nickels, dimes, and quarters and returns the necessary change. One day Mary Jo decides she'd like a package of peppermint-flavored chewing gum. She goes to the vending machine, inserts two nickels and a dime, in that order, and presses the white button, denoted W. Out comes her package of peppermint-flavored chewing gum. (To get a package of spearmint-flavored chewing gum one presses the black button, denoted B.)

What Mary Jo has done, in making her purchase, can be represented as shown in Table 6.1, where t_0 is the initial time, when she inserts her first nickel, and t_1, t_2, t_3, t_4 are later moments in time, with $t_1 < t_2 < t_3 < t_4$.

Table 6.1

	t_0	t_1	t_2	t_3	t_4
State	(1) s_0	(4) s_1 (5¢)	(7) s_2 (10¢)	(10) s_3 (20¢)	(13) s_0
Input	(2) 5¢	(5) 5¢	(8) 10¢	(11) W	
Output	(3) Nothing	(6) Nothing	(9) Nothing	(12) P	

The numbers (1), (2), ..., (12), (13) in this table indicate the order of events in the purchase of Mary Jo's package of peppermint chewing gum. For each input at time t_i, $0 \le i \le 3$, there is *at that time* a corresponding output and then a change in state. The new state at time t_{i+1} depends on both the input and the (present) state at time t_i.

The machine is in a state of readiness at state s_0. It waits for a customer to start inserting coins that will total 20¢ or more and then press a button to get a package of chewing gum. If at any time the total of the coins inserted exceeds 20¢, the machine provides the needed change (before the customer presses the button to get the package of chewing gum).

At time t_0 Mary Jo provides the machine with her first input, 5¢. She receives nothing at this time, but at the later time t_1 the machine is in state s_1, where it *remembers* her total of 5¢ and waits for her second input (of 5¢ at time t_1). The machine again (at time t_1) provides no output, but at the next time, t_2, it is in state s_2, remembering a total of 10¢ = 5¢ (remembered at state s_1) + 5¢ (inserted at time t_1). Providing her dime (at time t_2) as the next input to the machine, Mary Jo does not receive a package of chewing gum at this time because the machine doesn't "know" which flavor Mary Jo prefers, but it does "know" now (t_3) that she has inserted the necessary total of $20\cancel{c} = 10\cancel{c}$ (remembered at state s_2) + $10\cancel{c}$ (inserted at time t_2). At last Mary Jo presses the white button, and at time t_3 the machine dispenses the output (her package of peppermint chewing gum) and then returns, at time t_4, to the starting state s_0, just in time for Mary Jo's friend Rizzo to deposit a quarter, receive her nickel change, press the black button, and obtain the package of spearmint chewing gum she desires. The purchase made by Rizzo is analyzed in Table 6.2.

Table 6.2

	t_0	t_1	t_2
State	(1) s_0	(4) s_3 (20¢)	(7) s_0
Input	(2) 25¢	(5) B	
Output	(3) 5¢ change	(6) S	

What has happened in the case of this vending machine can be abstracted to help in the analysis of certain aspects of digital computers and telephone communication systems.

The major features of such a machine are as follows:

1) The machine can be in only one of *finitely many states* at a given time. These states are called the *internal states* of the machine, and at a given time the total memory available to the machine is the knowledge of which internal state it is in at that moment.

2) The machine will accept as *input* only a finite number of symbols, which collectively are referred to as the *input alphabet* $\mathcal{I}$. In the vending machine example, the input alphabet is {nickel, dime, quarter, W, B}, each item of which is recognized by each internal state.

3) An *output* and a *next state* are determined by each combination of inputs and internal states. The finite set of all possible outputs constitutes the *output alphabet* $\mathcal{O}$ for the machine.

4) We assume that the sequential processings of the machine are *synchronized* by separate and distinct clock pulses and that the machine operates in a *deterministic* manner, where the output is completely determined by the total input provided and the starting state of the machine.

These observations lead us to the following definition.

Definition 6.13

A *finite state machine* is a five-tuple $M = (S, \mathcal{I}, \mathcal{O}, v, \omega)$, where S = the set of internal states for M; $\mathcal{I}$ = the input alphabet for M; $\mathcal{O}$ = the output alphabet for M; $v: S \times \mathcal{I} \rightarrow S$ is the *next state function*; and $\omega: S \times \mathcal{I} \rightarrow \mathcal{O}$ is the *output function*.

Using the notation of this definition, if the machine is in state s at time t_i and we input x at this time, then the output at time t_i is $\omega(s, x)$. This output is followed by a transition of the machine at time t_{i+1} to the next internal state given by $v(s, x)$.

We assume that when a finite state machine receives its first input, we are at time $t_0 = 0$ and the machine is in a designated starting state denoted by s_0. Our development will concentrate primarily on the output and state transitions that take place sequentially, with little or no reference to the sequence of clock pulses at times $t_0, t_1, t_2, \ldots$.

Since the sets S, $\mathscr{I}$, and $\mathbb{O}$ are finite, it is possible to represent v and ω, for a given finite state machine, by means of a table that lists $v(s, x)$ and $\omega(s, x)$ for all $s \in S$ and all $x \in \mathscr{I}$. Such a table is referred to as the *state table* or *transition table* for the given machine. A second representation of the machine is made by means of a *state diagram*.

We demonstrate the state table and state diagram in the following examples.

EXAMPLE 6.17

Consider the finite state machine $M = (S, \mathscr{I}, \mathbb{O}, v, \omega)$, where $S = \{s_0, s_1, s_2\}$, $\mathscr{I} = \mathbb{O} = \{0, 1\}$, and v, ω are given by the *state table* in Table 6.3. The first column of the table lists the (*present*) *states* for the machine. The entries in the second row are the elements of the input alphabet $\mathscr{I}$, listed once under v and then again under ω. The six numbers in the last two columns (and last three rows) are elements of the output alphabet $\mathbb{O}$.

Table 6.3

	v		ω	
	0	1	0	1
s_0	s_0	s_1	0	0
s_1	s_2	s_1	0	0
s_2	s_0	s_1	0	1

To calculate $v(s_1, 1)$, for example, we find s_1 in the column of present states and proceed horizontally over from s_1 until we are below the entry 1 in the section of the table for v. This entry gives $v(s_1, 1) = s_1$. In the same way we find $\omega(s_1, 1) = 0$.

With s_0 designated as the starting state, if the input provided to M is the string 1010, then the output is 0010, as demonstrated in Table 6.4. Here the machine is left in state s_2, so that if we had another input string, we would provide the first character of that string, here 0, at state s_2 unless the machine is *reset* to start once again at s_0.

Table 6.4

State	s_0	$v(s_0, 1) = s_1$	$v(s_1, 0) = s_2$	$v(s_2, 1) = s_1$	$v(s_1, 0) = s_2$
Input	1	0	1	0	0
Output	$\omega(s_0, 1) = 0$	$\omega(s_1, 0) = 0$	$\omega(s_2, 1) = 1$	$\omega(s_1, 0) = 0$	

Since we are primarily interested in the output, not in the sequence of transition states, the same machine can be represented by means of a *state diagram*. Here we can obtain the output string without actually listing the transition states. In such a diagram each internal state s is represented by a circle with s inside of it. For states s_i and s_j, if $v(s_i, x) = s_j$ for $x \in \mathscr{I}$, and $\omega(s_i, x) = y$ for $y \in \mathbb{O}$, we represent this in the state diagram by drawing a *directed edge* (or *arc*) from the circle for s_i to the circle for s_j and labeling the arc with the input x and output y as shown in Fig. 6.1.

With these conventions, the state diagram for the machine M of Table 6.3 is shown in Fig. 6.2. Although the table is more compact, the diagram enables us to follow an input string through each transition state it determines, picking up each of the corresponding

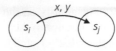

Figure 6.1

output symbols before each transition. Here if the input string is 00110101, then starting at state s_0, the first input of 0 yields an output of 0 and returns us to s_0. The next input of 0 yields the same result, but for the third input, 1, the output is 0 and we are now in state s_1. Continuing in this manner, we arrive at the output string 00000101 and finish in state s_1. (We note that the input string 00110101 is an element of $\mathcal{I}^*$, the Kleene closure of $\mathcal{I}$, and that the output string is in $\mathbb{O}^*$, the Kleene closure of $\mathbb{O}$.)

Starting at s_0, what is the output string for the input string 1100101101?

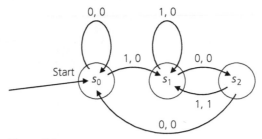

Figure 6.2

EXAMPLE 6.18

For the vending machine described earlier in this section, we have the state table, Table 6.5, with

1) $S = \{s_0, s_1, s_2, s_3, s_4\}$, where at state s_k, for each $0 \le k \le 4$, the machine remembers retaining $5k$ cents.

2) $\mathcal{I} = \{5¢, 10¢, 25¢, B, W\}$, where B denotes the black button one presses for a package of spearmint-flavored chewing gum and W the white button for a package of peppermint-flavored chewing gum.

3) $\mathbb{O} = \{n$ (nothing), P (peppermint chewing gum), S (spearmint chewing gum), 5¢, 10¢, 15¢, 20¢, 25¢\}$.

Table 6.5

	ν					ω				
	5¢	10¢	25¢	B	W	5¢	10¢	25¢	B	W
s_0	s_1	s_2	s_4	s_0	s_0	n	n	5¢	n	n
s_1	s_2	s_3	s_4	s_1	s_1	n	n	10¢	n	n
s_2	s_3	s_4	s_4	s_2	s_2	n	n	15¢	n	n
s_3	s_4	s_4	s_4	s_3	s_3	n	5¢	20¢	n	n
s_4	s_4	s_4	s_4	s_0	s_0	5¢	10¢	25¢	S	P

As we observed in the discussion just prior to Example 6.18, for a general finite state machine $M = (S, \mathcal{I}, \mathbb{O}, \nu, \omega)$, the input can be realized as an element of $\mathcal{I}^*$, with the output from $\mathbb{O}^*$. Consequently, it is to our advantage to extend the domains of v and ω from $S \times \mathcal{I}$ to $S \times \mathcal{I}^*$. For ω we enlarge the codomain to $\mathbb{O}^*$, recalling, should the need arise, that both $\mathcal{I}^*$ and $\mathbb{O}^*$ contain the empty string, λ. With these extensions, if $x_1 x_2 \cdots x_k \in \mathcal{I}^*$, for $k \in \mathbf{Z}^+$, then starting at any state $s_1 \in S$, we have

$$v(s_1, x_1) = s_2{}^{\dagger}$$

$$v(s_1, x_1 x_2) = v(v(s_1, x_1), x_2) = v(s_2, x_2) = s_3$$

$$v(s_1, x_1 x_2 x_3) = v(v(\underbrace{v(s_1, x_1)}_{s_2}, x_2), x_3) = v(s_3, x_3) = s_4$$

$$\underbrace{\qquad}_{v(s_2, x_2) = s_3}$$

$$\cdots\cdots\cdots$$

$$v(s_1, x_1 x_2 \cdots x_k) = v(s_k, x_k) = s_{k+1}, \quad \text{and}$$

$$\omega(s_1, x_1) = y_1$$

$$\omega(s_1, x_1 x_2) = \omega(s_1, x_1)\omega(v(s_1, x_1), x_2) = \omega(s_1, x_1)\omega(s_2, x_2) = y_1 y_2$$

$$\omega(s_1, x_1 x_2 x_3) = \omega(s_1, x_1)\omega(s_2, x_2)\omega(s_3, x_3) = y_1 y_2 y_3$$

$$\cdots\cdots\cdots$$

$$\omega(s_1, x_1 x_2 \cdots x_k) = \omega(s_1, x_1)\omega(s_2, x_2) \cdots \omega(s_k, x_k) = y_1 y_2 \cdots y_k \in \mathbb{O}^*$$

Also, $v(s_1, \lambda) = s_1$ for all $s_1 \in S$.

(We shall use these extensions again in Chapter 7.)

We close this section with an example that is relevant in computer science.

EXAMPLE 6.19

Let $x = x_5 x_4 x_3 x_2 x_1 = 00111$ and $y = y_5 y_4 y_3 y_2 y_1 = 01101$ be binary numbers where x_1 and y_1 are the least significant bits. The leading 0's in x and y are there to make the strings for x and y of equal length and to guarantee enough places to complete the sum. A *serial binary adder* is a finite state machine that we can use to obtain $x + y$. The diagram in Fig. 6.3 illustrates this, where $z = z_5 z_4 z_3 z_2 z_1$ has the least significant bit z_1.

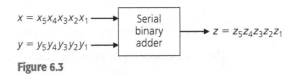

Figure 6.3

In the addition $z = x + y$, we have

$$
\begin{array}{rccccc}
x = & 0 & 0 & 1 & 1 & 1 \\
+ y = +\, & 0 & 1 & 1 & 0 & 1 \\
\hline
z = & 1 & 0 & 1 & 0 & 0 \\
\end{array}
$$

third
addition   first
addition

We note that for the first addition $x_1 = y_1 = 1$ and $z_1 = 0$, whereas for the third addition we have $x_3 = y_3 = 1$ and $z_3 = 1$ because of a *carry* from the addition of x_2 and y_2 (and the

---

†The state s_2 is determined by s_1 and x_1. It is not simply the second in a predetermined list of states.

carry from $x_1 + y_1$). Consequently, each output depends on the sum of two inputs and the ability to *remember* a carry of 0 or 1, which is crucial when the carry is 1.

The serial binary adder is modeled by a finite state machine $M = (S, \mathscr{I}, \mathbb{O}, \nu, \omega)$ as follows. The set $S = \{s_0, s_1\}$, where s_i indicates a carry of i; $\mathscr{I} = \{00, 01, 10, 11\}$, so there is a pair of inputs, depending on whether we are seeking $0 + 0$, $0 + 1$, $1 + 0$, or $1 + 1$, respectively; and $\mathbb{O} = \{0, 1\}$. The functions ν, ω are given in the state table (Table 6.6) and the state diagram (Fig. 6.4).

Table 6.6

	ν				ω			
	00	01	10	11	00	01	10	11
s_0	s_0	s_0	s_0	s_1	0	1	1	0
s_1	s_0	s_1	s_1	s_1	1	0	0	1

In Table 6.6 we find, for example, that $\nu(s_1, 01) = s_1$ and $\omega(s_1, 01) = 0$, because s_1 indicates a carry of 1 from the addition of the previous bits. The 01 input indicates that we are adding 0 and 1 (and carrying a 1). Hence the sum is 10 and $\omega(s_1, 01) = 0$ for the 0 in 10. The carry is again remembered in $s_1 = \nu(s_1, 01)$.

From the state diagram (Fig. 6.4) we see that the starting state must be s_0 because there is no carry prior to the addition of the least significant bits.

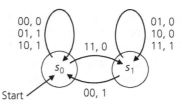

Figure 6.4

The state diagrams in Figs. 6.2 and 6.4 are examples of *labeled directed graphs*. We shall see more about graph theory throughout the text, for it has applications not only in computer science and electrical engineering but also in coding theory (prefix codes) and optimization (transport networks).

EXERCISES 6.2

1. Using the finite state machine of Example 6.17, find the output for each of the following input strings $x \in \mathscr{I}^*$, and determine the last internal state in the transition process. (Assume that we always start at s_0.)

   a) $x = 1010101$ **b)** $x = 1001001$ **c)** $x = 101001000$

2. For the finite state machine of Example 6.17, an input string x, starting at state s_0, produces the output string 00101. Determine x.

3. Let $M = (S, \mathscr{I}, \mathbb{O}, \nu, \omega)$ be a finite state machine where $S = \{s_0, s_1, s_2, s_3\}$, $\mathscr{I} = \{a, b, c\}$, $\mathbb{O} = \{0, 1\}$, and ν, ω are determined by Table 6.7.

Table 6.7

	ν			ω		
	a	b	c	a	b	c
s_0	s_0	s_3	s_2	0	1	1
s_1	s_1	s_1	s_3	0	0	1
s_2	s_1	s_1	s_3	1	1	0
s_3	s_2	s_3	s_0	1	0	1

a) Starting at s_0, what is the output for the input string *abbccc*?

b) Draw the state diagram for this finite state machine.

4. Give the state table and the state diagram for the vending machine of Example 6.18 if the cost of a package of chewing gum (peppermint or spearmint) is increased to 25¢.

5. A finite state machine $M = (S, \mathcal{I}, \mathcal{O}, \nu, \omega)$ has $\mathcal{I} = \mathcal{O} = \{0, 1\}$ and is determined by the state diagram shown in Fig. 6.5.

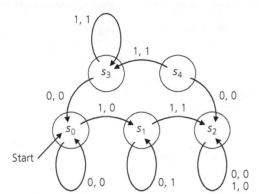

Figure 6.5

a) Determine the output string for the input string 110111, starting at s_0. What is the last transition state?

b) Answer part (a) for the same string but with s_1 as the starting state. What about s_2 and s_3 as starting states?

c) Find the state table for this machine.

d) In which state should we start so that the input string 10010 produces the output 10000?

e) Determine an input string $x \in \mathcal{I}^*$ of minimal length, such that $\nu(s_4, x) = s_1$. Is x unique?

6. Machine M has $\mathcal{I} = \{0, 1\} = \mathcal{O}$ and is determined by the state diagram shown in Fig. 6.6.

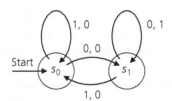

Figure 6.6

a) Describe in words what this finite state machine does.

b) What must state s_1 remember?

c) Find two languages $A, B \subseteq \mathcal{I}^*$ such that for every $x \in AB$, $\omega(s_0, x)$ has 1 as a suffix.

7. a) If S, $\mathcal{I}$, and $\mathcal{O}$ are finite sets, with $|S| = 3$, $|\mathcal{I}| = 5$, and $|\mathcal{O}| = 2$, determine (i) $|S \times \mathcal{I}|$; (ii) the number of functions $\nu: S \times \mathcal{I} \to S$; and (iii) the number of functions $\omega: S \times \mathcal{I} \to \mathcal{O}$.

b) For S, $\mathcal{I}$, and $\mathcal{O}$ in part (a), how many finite state machines do they determine?

8. Let $M = (S, \mathcal{I}, \mathcal{O}, \nu, \omega)$ be a finite state machine with $\mathcal{I} = \mathcal{O} = \{0, 1\}$ and S, ν, and ω determined by the state diagram shown in Fig. 6.7.

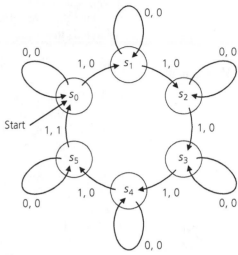

Figure 6.7

a) Find the output for the input string $x = 0110111011$.

b) Give the transition table for this finite state machine.

c) Starting in state s_0, if the output for an input string x is 0000001, determine all possibilities for x.

d) Describe in words what this finite state machine does.

9. a) Find the state table for the finite state machine in Fig. 6.8, where $\mathcal{I} = \mathcal{O} = \{0, 1\}$.

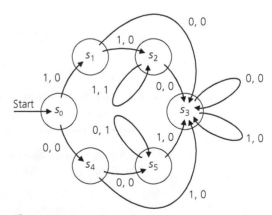

Figure 6.8

b) Let $x \in \mathcal{I}^*$ with $\|x\| = 4$. If 1 is a suffix of $\omega(s_0, x)$, what are the possibilities for the string x?

c) Let $A \subseteq \{0, 1\}^*$ be the language where $\omega(s_0, x)$ has 1 as a suffix for all x in A. Determine A.

d) Find the language $A \subseteq \{0, 1\}^*$ where $\omega(s_0, x)$ has 111 as a suffix for all x in A.

6.3
Finite State Machines: A Second Encounter

Having seen some examples of finite state machines, we turn to the study of some additional machines that are relevant to the design of computer hardware. One important type of machine is the *sequence recognizer*.

Here, $\mathcal{I} = \mathbb{O} = \{0, 1\}$, and we want to construct a machine that recognizes each occurrence of the sequence 111 as it is encountered in an input string $x \in \mathcal{I}^*$. For example, if $x = 1110101111$, then the corresponding output should be 0010000011, where a 1 in the ith position of the output indicates that a 1 can be found in positions i, $i - 1$, and $i - 2$ of x. Here overlapping of sequences of 111 can occur, so some characters in the input string can be thought of as characters in more than one triple of 1's.

Letting s_0 denote the starting state, we realize that we must have a state to remember 1 (the possible start of 111) and a state to remember 11. In addition, any time our input symbol is 0, we go back to s_0 and start the search for three successive 1's over again.

In Fig. 6.9, s_1 remembers a single 1, and s_2 remembers the string 11. If s_2 is reached, then a third "1" indicates the occurrence of the triple in the input string, and the output 1 *recognizes* this occurrence. But this third "1" also means that we have the first two 1's of another possible triple coming up in the string (as happens in 11101011 "1" 1). So after recognizing the occurrence of 111 with an output of 1, we return to state s_2 to remember the two inputs of 1 "1".

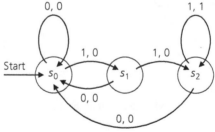

Figure 6.9

If we are concerned with recognizing all strings that end in 111, then for each $x \in \mathcal{I}^*$, the machine will recognize such a sequence with final output 1. This machine is then a recognizer of the language $A = \{0, 1\}^*\{111\}$.

Another finite state machine that recognizes the same triple 111 is shown in Fig. 6.10. The finite state machines represented by the state diagrams in Figs. 6.9 and 6.10 perform

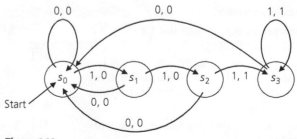

Figure 6.10

the same task and are said to be *equivalent*. The state diagram in Fig. 6.10 has one more state than that in Fig. 6.9, but at this stage we are not overly concerned with getting a finite state machine with a minimal number of states. In Chapter 7 we shall develop a technique to take a given finite state machine M and find one that is equivalent to it and has the smallest number of internal states needed.

The next example is a bit more selective.

EXAMPLE 6.21

Now we want to not only recognize the occurrence of 111 but we want to recognize only those occurrences that end in a position that is a multiple of three. Consequently, with $\mathcal{I} = \mathbb{O} = \{0, 1\}$, if $x \in \mathcal{I}^*$, where $x = 1110111$, then we want $\omega(s_0, x) = 0010000$, not 0010001. In addition, for $x \in \mathcal{I}^*$, where $x = 111100111$, the output $\omega(s_0, x)$ is to be 001000001, not 001100001, for here, because of length considerations, overlapping of sequences of 111 is not allowed.

Again we start at s_0 (Fig. 6.11), but now s_1 must remember a first 1 only if it occurs in x in position 1, 4, 7, ... If the input at s_0 is 0, we cannot simply return to s_0 as in Example 6.20. We must remember that this 0 is the *first* of three symbols of no interest. Hence from s_0 we go to s_3 and then to s_4, processing any triple of the form $0yz$ where 0 occurs in x in position $3k + 1$, $k \geq 0$. The same type of situation happens at s_1 if the input is 0. Finally, at s_2 the sequence 111 is recognized with an output of 1, if it occurs. The machine then returns to s_0 to input the next symbol of the input string.

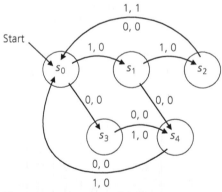

Figure 6.11

EXAMPLE 6.22

Figure 6.12 shows the state diagrams for finite state machines that will recognize the occurrence of the sequence 0101 in an input string $x \in \mathcal{I}^*$, where $\mathcal{I} = \mathbb{O} = \{0, 1\}$. The machine in Fig. 6.12(a) recognizes with an output of 1 each occurrence of 0101 in an input string, regardless of where it occurs. In Fig. 6.12(b) the machine recognizes with an output of 1 only those prefixes of x whose length is a multiple of four and that end in 0101. (Hence no overlapping is allowed here.) Consequently, for $x = 01010100101$, $\omega(s_0, x) = 00010100001$ for (a), whereas for (b), $\omega(s_0, x) = 00010000000$.

Now that we have examined some finite state machines that serve as sequence recognizers, it is only fair to consider a set of sequences that *cannot* be recognized by a finite state machine. This example gives us another opportunity to apply the pigeonhole principle.

EXAMPLE 6.23

Let $\mathcal{I} = \mathbb{O} = \{0, 1\}$. Can we construct a finite state machine that recognizes precisely those strings in the language $A = \{01, 0011, 000111, \ldots\} = \{0^i 1^i \mid i \in \mathbf{Z}^+\}$? If we can, then if s_0

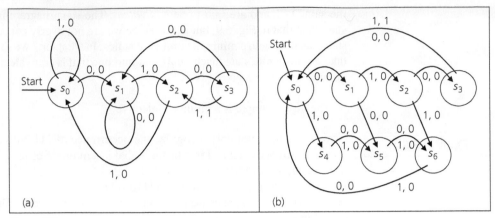

Figure 6.12

denotes the starting state, we shall expect $\omega(s_0, 01) = 01$, $\omega(s_0, 0011) = 0011$, and, in general, $\omega(s_0, 0^i 1^i) = 0^i 1^i$, for all $i \in \mathbf{Z}^+$. [*Note:* Here, for example, we want $\omega(s_0, 0011) = 0011$, where the first 1 in the output is for recognition of the substring 01 and the second 1 is for recognition of the string 0011.]

Suppose that there is a finite state machine $M = (S, \mathcal{I}, \mathcal{O}, \nu, \omega)$ that can recognize precisely those strings in A. Let $s_0 \in S$, where s_0 is the starting state, and let $|S| = n \geq 1$. Now consider the string $0^{n+1} 1^{n+1}$ in the language A. If our machine M is to operate correctly, then we want $\omega(s_0, 0^{n+1} 1^{n+1}) = 0^{n+1} 1^{n+1}$. Therefore, we see in Table 6.8 how this finite state machine will process the $n + 1$ 0's, starting at the state s_0, then continuing at the n states $s_1 = \nu(s_0, 0)$, $s_2 = \nu(s_1, 0)$, ..., and $s_n = \nu(s_{n-1}, 0)$. Since $|S| = n$, by applying the pigeonhole principle to the $n + 1$ states $s_0, s_1, s_2, \ldots, s_{n-1}, s_n$, we realize that there are two states s_i and s_j where $i < j$ but $s_i = s_j$.

Table 6.8

State	s_0	s_1	s_2	...	s_{n-1}	s_n	s_{n+1}	...	s_{2n}	s_{2n+1}
Input	0	0	0	...	0	0	1	...	1	1
Output	0	0	0	...	0	0	1	...	1	1

Now in Table 6.9 we see how the removal of the $j - i$ columns — for states $s_{i+1}, \ldots, s_j$ — results in Table 6.10. This table shows us that the finite state machine M recognizes the string $x = 0^{(n+1)-(j-i)} 1^{n+1}$, where $n + 1 - (j - i) < n + 1$. Unfortunately $x \notin A$, so M recognizes a string that it is *not* supposed to recognize. This demonstrates that we cannot construct a finite state machine that recognizes precisely those strings in the language $A = \{0^i 1^i | i \in \mathbf{Z}^+\}$.

Table 6.9

State	s_0	s_1	s_2	...	s_i	s_{i+1}	...	s_j	s_{j+1}	...	s_n	s_{n+1}	...	s_{2n}	s_{2n+1}
Input	0	0	0	...	0	0	...	0	0	...	0	1	...	1	1
Output	0	0	0	...	0	0	...	0	0	...	0	1	...	1	1

Table 6.10

State	s_0	s_1	s_2	$\cdots$	s_i	s_{j+1}	$\cdots$	s_n	s_{n+1}	$\cdots$	s_{2n}	s_{2n+1}
Input	0	0	0	$\cdots$	0	0	$\cdots$	0	1	$\cdots$	1	1
Output	0	0	0	$\cdots$	0	0	$\cdots$	0	1	$\cdots$	1	1

A class of finite state machines that is important in the design of digital devices consists of the *k-unit delay machines*, where $k \in \mathbf{Z}^+$. For $k = 1$, we want to construct a machine M such that if $x = x_1x_2 \cdots x_{m-1}x_m$, then for starting state s_0, $\omega(s_0, x) = 0x_1x_2 \cdots x_{m-1}$, so that the output is the input delayed one time unit (clock pulse). [The use of 0 as the first symbol in $\omega(s_0, x)$ is conventional.]

EXAMPLE 6.24

Let $\mathcal{I} = \mathbb{O} = \{0, 1\}$. With starting state s_0, $\omega(s_0, x) = 0$ for $x = 0$ or 1 because the first output is 0; the states s_1 and s_2 (in Fig. 6.13) remember a prior input of 0 and 1, respectively. In the figure, we label, for example, the arc from s_1 to s_2 with 1, 0 because with an input of 1 we need to go to s_2 where inputs of 1 at time t_i are remembered so that they can become outputs of 1 at time t_{i+1}. The 0 in the label 1, 0 is the output because starting in s_1 indicates that the prior input was 0, which becomes the present output. The labels on the other arcs are obtained by the same type of reasoning.

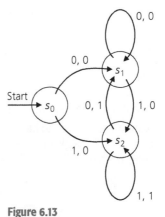

Figure 6.13

EXAMPLE 6.25

Observing the structure of a one-unit delay, we extend our ideas to the two-unit delay machine shown in Fig. 6.14. If $x \in \mathcal{I}^*$, let $x = x_1x_2 \cdots x_m$ where $m > 2$; if s_0 is the starting state, then $\omega(s_0, x) = 00x_1 \cdots x_{m-2}$. For states s_0, s_1, s_2 the output is 0 for all possible inputs. States s_3, s_4, s_5, and s_6 must remember the two prior inputs 00, 01, 10, and 11, respectively. To get the other arcs in the diagram, we shall consider one such arc and then use similar reasoning for the others. For the arc from s_5 to s_3 in Fig. 6.14(a), let the input be 0. Since the prior input to s_5 from s_2 is 0, we must go to the state that remembers the two prior inputs 00. This is state s_3. Going back two states from s_5 to s_2 to s_0, we see that the input is 1 (from s_0 to s_2). This then becomes the output (delayed two units) for the arc from s_5 to s_3. The complete machine is shown in part (b) of Fig. 6.14.

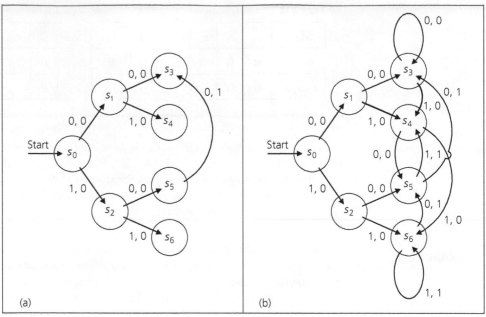

Figure 6.14

We turn now to some additional properties that arise in the study of finite state machines. The machine in Fig. 6.15 will be used for examples of the terms defined.

Definition 6.14   Let $M = (S, \mathcal{I}, \mathbb{O}, \nu, \omega)$ be a finite state machine.

a) For $s_i, s_j \in S$, s_j is said to be *reachable* from s_i if $s_i = s_j$ or if there is an input string $x \in \mathcal{I}^+$ such that $\nu(s_i, x) = s_j$. (In Fig. 6.15, state s_3 is reachable from s_0, s_1, s_2, and s_3 but not from s_4, s_5, s_6, or s_7. No state is reachable from s_3 except s_3 itself.)

b) A state $s \in S$ is said to be *transient* if $\nu(s, x) = s$ for $x \in \mathcal{I}^*$ implies $x = \lambda$; that is, there is no $x \in \mathcal{I}^+$ with $\nu(s, x) = s$. (For the machine in Fig. 6.15, s_2 is the only transient state.)

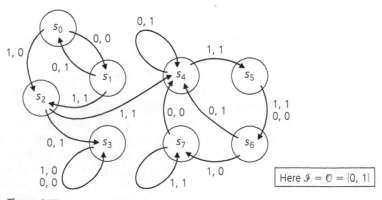

Figure 6.15

c) A state $s \in S$ is called a *sink*, or *sink state*, if $\nu(s, x) = s$, for all $x \in \mathscr{I}^*$. (s_3 is the only sink in Fig. 6.15.)

d) Let $S_1 \subseteq S$, $\mathscr{I}_1 \subseteq \mathscr{I}$. If $\nu_1 = \nu|_{S_1 \times \mathscr{I}_1} : S_1 \times \mathscr{I}_1 \to S$ (that is, the restriction of ν to $S_1 \times \mathscr{I}_1 \subseteq S \times \mathscr{I}$) has its range within S_1, then with $\omega_1 = \omega|_{S_1 \times \mathscr{I}_1}$, $M_1 = (S_1, \mathscr{I}_1, \mathbb{O}, \nu_1, \omega_1)$ is called a *submachine* of M. (With $S_1 = \{s_4, s_5, s_6, s_7\}$, and $\mathscr{I}_1 = \{0, 1\}$, we get a submachine M_1 of the machine M in Fig. 6.15.)

e) A machine is said to be *strongly connected* if for any states $s_i, s_j \in S$, s_j is reachable from s_i. (The machine in Fig. 6.15 is not strongly connected, but the submachine M_1 in part (d) has this property.)

We close this section with a concept that uses a tree diagram.

Definition 6.15

For a finite state machine M, let s_i, s_j be two distinct states in S. An input string $x \in \mathscr{I}^+$ is called a *transfer* (or *transition*) *sequence* from s_i to s_j if

a) $\nu(s_i, x) = s_j$, and

b) $y \in \mathscr{I}^+$ with $\nu(s_i, y) = s_j \Rightarrow \|y\| \geq \|x\|$.

There can be more than one such (shortest) sequence for two states s_i, s_j.

EXAMPLE 6.26

Find a transfer sequence from state s_0 to state s_2 for the finite state machine M given by the state table in Table 6.11, where $\mathscr{I} = \mathbb{O} = \{0, 1\}$.

Table 6.11

	ν		ω	
	0	1	0	1
s_0	s_6	s_1	0	1
s_1	s_5	s_0	0	1
s_2	s_1	s_2	0	1
s_3	s_4	s_0	0	1
s_4	s_2	s_1	0	1
s_5	s_3	s_5	1	1
s_6	s_3	s_6	1	1

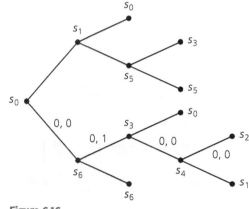

Figure 6.16

In constructing the tree diagram of Fig. 6.16, we start at state s_0 and find those states that can be reached from s_0 by using strings of length 1. Here we find s_1 and s_6. Then we do the same thing with s_1 and s_6, finding, as a result, those states reachable from s_0 with input strings of length 2. Continuing to expand the tree from left to right, we get to a vertex labeled with the desired state, s_2. Each time we reach a vertex labeled with a state used previously, we terminate that part of the expansion because we cannot reach any new states. After we arrive at the state we want, we backtrack to s_0 and use the state table to label the branches, as shown in Fig. 6.16. Hence, for $x = 0000$, $\nu(s_0, x) = s_2$ with $\omega(s_0, x) = 0100$. (Here x is unique.)

1. Let $\mathcal{I} = \mathcal{O} = \{0, 1\}$. (a) Construct a state diagram for a finite state machine that recognizes each occurrence of 0000 in a string $x \in \mathcal{I}^*$. (Here overlapping is allowed.) (b) Construct a state diagram for a finite state machine that recognizes each string $x \in \mathcal{I}^*$ that ends in 0000 and has length $4k$, $k \in \mathbf{Z}^+$. (Here overlapping is not permitted.)

2. Answer Exercise 1 for each of the sequences 0110 and 1010.

3. Construct a state diagram for a finite state machine with $\mathcal{I} = \mathcal{O} = \{0, 1\}$ that recognizes all strings in the language $\{0, 1\}^*\{00\} \cup \{0, 1\}^*\{11\}$.

4. For $\mathcal{I} = \mathcal{O} = \{0, 1\}$ a string $x \in \mathcal{I}^*$ is said to have *even parity* if it contains an even number of 1's. Construct a state diagram for a finite state machine that recognizes all nonempty strings of even parity.

5. Table 6.12 defines ν and ω for a finite state machine M where $\mathcal{I} = \mathcal{O} = \{0, 1\}$.

a) Draw the state diagram for M.

b) Determine the output for the following input sequences, starting at s_0 in each case: (i) $x = 111$; (ii) $x = 1010$; (iii) $x = 00011$.

Table 6.12

	ν		ω	
	0	1	0	1
s_0	s_0	s_1	0	0
s_1	s_0	s_1	1	1

c) Describe in words what machine M does.

d) How is this machine related to that shown in Fig. 6.13?

6. Show that it is not possible to construct a finite state machine that recognizes precisely those sequences in the language $A = \{0^i 1^j | i, j \in \mathbf{Z}^+, i > j\}$. (Here the alphabet for A is $\Sigma = \{0, 1\}$.)

7. For each of the machines in Table 6.13, determine the transient states, sink states, submachines (where $\mathcal{I}_1 = \{0, 1\}$), and strongly connected submachines (where $\mathcal{I}_1 = \{0, 1\}$).

8. Determine a transfer sequence from state s_2 to state s_5 in finite state machine (c) of Exercise 7. Is your sequence unique?

Table 6.13

	ν		ω	
	0	1	0	1
s_0	s_4	s_1	0	0
s_1	s_4	s_2	0	1
s_2	s_3	s_5	0	0
s_3	s_2	s_5	1	0
s_4	s_4	s_4	1	1
s_5	s_2	s_3	0	1

(a)

	ν		ω	
	0	1	0	1
s_0	s_0	s_1	1	0
s_1	s_0	s_1	0	1
s_2	s_1	s_3	0	0
s_3	s_0	s_4	0	0
s_4	s_4	s_4	1	1

(b)

	ν		ω	
	0	1	0	1
s_0	s_1	s_2	0	1
s_1	s_0	s_2	1	1
s_2	s_2	s_3	1	1
s_3	s_6	s_4	0	0
s_4	s_5	s_5	1	0
s_5	s_3	s_4	1	0
s_6	s_6	s_6	0	0

(c)

6.4
Summary and Historical Review

In this chapter we have been introduced to the theory of languages and to a discrete structure called a *finite state machine*. Using our prior development of elementary set theory and finite functions, we were able to combine some abstract notions and to model digital devices such as sequence recognizers and delays. Comparable coverage of this material appears in

Chapter 1 of L. L. Dornhoff and F. E. Hohn [3] and in Chapter 2 of D. F. Stanat and D. F. McAllister [15].

The finite state machine we developed is based on the model put forth in 1955 by G. H. Mealy in [11] and is consequently referred to as the "Mealy machine." The model is based on earlier concepts found in the work of D. A. Huffman [8] and E. F. Moore [13]. For further reading on the pioneering work dealing with various aspects and applications of the finite state machine, consult the material edited by E. F. Moore [14]. Additional information on the actual synthesis of such machines and on related hardware considerations, along with an extensive coverage of many related ideas, can be found in Chapters 9–15 of Z. Kohavi [9].

For more on languages and their relation to finite state machines, one should look into the UMAP module by W. J. Barnier [1], Chapter 8 of J. L. Gersting [4], and Chapters 7 and 8 of A. Gill [5]. A comprehensive coverage of these (and related) topics is given in the texts by J. G. Brookshear [2], J. E. Hopcroft and J. D. Ullman [7], H. R. Lewis and C. H. Papadimitriou [10], M. Minsky [12], and D. Wood [16].

David Hilbert
(1862–1943)

Alan Mathison Turing (1912–1954)
Reproduced courtesy of The Granger Collection, New York

One may be surprised to learn that the basic ideas of automata theory were developed to solve rather theoretical questions in the foundations of mathematics — as posed in 1900 by the German mathematician David Hilbert (1862–1943). In 1935 the English mathematician and logician Alan Mathison Turing (1912–1954) became interested in Hilbert's decision problem, which asked if there could be a general method one could apply to a given statement in order to determine if that statement were true. Turing's approach to the solution of this problem led him to develop what is now known as a *Turing machine*, the most general model for a computing machine. By using this model, he was able to establish very profound theoretical results about how computers should have to operate — before any such machines were actually built. During World War II Turing worked for the Foreign Office at Bletchley Park, where he did extensive work on the cryptanalysis of Nazi ciphers. His efforts contributed to the breaking of the mechanical cipher machine *Enigma*, a breakthrough that helped to bring about the defeat of the Third Reich. Following the war (and up to the time of his death), Turing's interest in the ability of machines to think led him to play a major role in the development of actual (not just theoretical) computers. For more on the life of this interesting scholar one should look into the biography by A. Hodges [6].

REFERENCES

1. Barnier, William J. "Finite-State Machines as Recognizers" (UMAP Module 671). *The UMAP Journal* 7, no. 3 (1986): pp. 209–232.

2. Brookshear, J. Glenn. *Theory of Computation: Formal Languages, Automata, and Complexity.* Reading, Mass.: Benjamin/Cummings, 1989.

3. Dornhoff, Larry L., and Hohn, Franz E. *Applied Modern Algebra.* New York: Macmillan, 1978.

4. Gersting, Judith L. *Mathematical Structures for Computer Science*, 5th ed. New York: Freeman, 2003.

5. Gill, Arthur. *Applied Algebra for the Computer Sciences*, Prentice-Hall Series in Automatic Computation. Englewood Cliffs, N.J.: Prentice-Hall, 1976.

6. Hodges, Andrew. *Alan Turing: The Enigma.* New York: Simon and Schuster, 1983.

7. Hopcroft, John E., and Ullman, Jeffrey D. *Introduction to Automata Theory, Languages, and Computation.* Reading, Mass.: Addison-Wesley, 1979.

8. Huffman, D. A. "The Synthesis of Sequential Switching Circuits." *Journal of the Franklin Institute* 257 (March 1954): pp. 161–190, (April 1954): pp. 275–303. Reprinted in Moore [14].

9. Kohavi, Zvi. *Switching and Finite Automata Theory*, 2nd ed. New York: McGraw-Hill, 1978.

10. Lewis, Harry R., and Papadimitriou, Christos H. *Elements of the Theory of Computation*, 2nd ed. Englewood Cliffs, N.J.: Prentice-Hall, 1997.

11. Mealy, G. H. "A Method for Synthesizing Sequential Circuits." *Bell System Technical Journal* 34 (September 1955): pp. 1045–1079.

12. Minsky, Marvin. *Computation: Finite and Infinite Machines.* Englewood Cliffs, N.J.: Prentice-Hall, 1967.

13. Moore, E. F. "Gedanken-experiments on Sequential Machines." *Automata Studies, Annals of Mathematical Studies*, no. 34: pp. 129–153. Princeton, N.J.: Princeton University Press, 1956.

14. Moore, E. F., ed., *Sequential Machines: Selected Papers.* Reading, Mass.: Addison-Wesley, 1964.

15. Stanat, Donald F., and McAllister, David F. *Discrete Mathematics in Computer Science.* Englewood Cliffs, N.J.: Prentice-Hall, 1977.

16. Wood, Derick. *Theory of Computation.* New York: Wiley, 1987.

SUPPLEMENTARY EXERCISES

1. Let $\Sigma_1 = \{w, x, y\}$ and $\Sigma_2 = \{x, y, z\}$ be alphabets. If $A_1 = \{x^i y^j \mid i, j \in \mathbf{Z}^+, j > i \geq 1\}$, $A_2 = \{w^i y^j \mid i, j \in \mathbf{Z}^+, i > j \geq 1\}$, $A_3 = \{w^i x^j y^i z^j \mid i, j \in \mathbf{Z}^+, j > i \geq 1\}$, and $A_4 = \{z^j (wz)^i w^j \mid i, j \in \mathbf{Z}^+, i \geq 1, j \geq 2\}$, determine whether each of the following statements is true or false.

a) A_1 is a language over Σ_1.

b) A_2 is a language over Σ_2.

c) A_3 is a language over $\Sigma_1 \cup \Sigma_2$.

d) A_1 is a language over $\Sigma_1 \cap \Sigma_2$.

e) A_4 is a language over $\Sigma_1 \triangle \Sigma_2$.

f) $A_1 \cup A_2$ is a language over Σ_1.

2. For languages $A, B \subseteq \Sigma^*$, does $A^* \subseteq B^* \Rightarrow A \subseteq B$?

3. Give an example of a language A over an alphabet Σ, where $(A^2)^* \neq (A^*)^2$.

4. For $\Sigma = \{0, 1\}$ consider the languages $A, B, C \subseteq \Sigma^*$ where $A = \{01, 11\}$, $B = \{01, 11, 111\}$, and $C = \{01, 11, 1111\}$. (a) How are A^* and B^* related? (b) How about A^* and C^*?

5. Let M be the finite state machine shown in Fig. 6.17. For states s_i, s_j, where $0 \leq i, j \leq 2$, let $\mathbb{O}_{ij}$ denote the set of all nonempty output strings that M can produce as it goes from state s_i to state s_j. If $i = 2$, $j = 0$, for example, $\mathbb{O}_{20} = \{0\}\{1, 00\}^*$.

Find $\mathbb{O}_{02}$, $\mathbb{O}_{22}$, $\mathbb{O}_{11}$, $\mathbb{O}_{00}$, and $\mathbb{O}_{10}$.

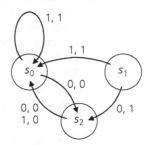

Figure 6.17

6. Let M be the finite state machine in Fig. 6.18.

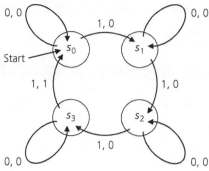

Figure 6.18

a) Find the state table for this machine.

b) Explain what this machine does.

c) How many distinct input strings x are there such that $\|x\| = 8$ and $\nu(s_0, x) = s_0$? How many are there with $\|x\| = 12$?

7. Let $M = (S, \mathcal{I}, \mathbb{O}, \nu, \omega)$ be a finite state machine with $|S| = n$, and let $0 \in \mathcal{I}$.

a) Show that for the input string $0000\ldots$, the output is eventually periodic.

b) What is the maximum number of 0's we can input before the periodic output starts?

c) What is the length of the maximum period that can occur?

8. For $\mathcal{I} = \mathbb{O} = \{0, 1\}$, let M be the finite state machine given in Table 6.14. If the starting state for M is *not* s_1, find an input string x (of smallest length) such that $\nu(s_i, x) = s_1$, for all $i = 2, 3, 4$. (Hence x gets the machine M to state s_1 regardless of the starting state.)

Table 6.14

	ν		ω	
	0	1	0	1
s_1	s_4	s_3	0	0
s_2	s_2	s_4	0	1
s_3	s_1	s_2	1	0
s_4	s_1	s_4	1	1

9. Let $\mathcal{I} = \mathbb{O} = \{0, 1\}$. Construct a state diagram for a finite state machine that reverses (from 0 to 1 or from 1 to 0) the symbols appearing in the 4th, in the 8th, in the 12th, ..., positions of an input string $x \in \mathcal{I}^+$. For example, if s_0 is the starting state, then $\omega(s_0, 0000) = 0001$, $\omega(s_0, 000111) = 000011$, and $\omega(s_0, 000000111) = 000100101$.

10. With $\mathcal{I} = \mathbb{O} = \{0, 1\}$, let M be the finite state machine given in Table 6.15. Here s_0 is the starting state. Let $A \subseteq \mathcal{I}^+$ where $x \in A$ if and only if the last symbol in $\omega(s_0, x)$ is 1. [There may be more than one 1 in the output string $\omega(s_0, x)$.] Construct a finite state machine wherein the last symbol of the output string is 1 for all $y \in \mathcal{I}^+ - A$.

Table 6.15

	ν		ω	
	0	1	0	1
s_0	s_1	s_2	1	0
s_1	s_2	s_1	0	1
s_2	s_2	s_3	0	1
s_3	s_1	s_0	1	0

11. Let $\mathcal{I} = \mathbb{O} = \{0, 1\}$ for the two finite state machines M_1 and M_2, given in Tables 6.16 and 6.17, respectively. The starting state for M_1 is s_0, whereas s_3 is the starting state for M_2.

Table 6.16

	ν_1		ω_1	
	0	1	0	1
s_0	s_0	s_1	1	0
s_1	s_1	s_2	0	0
s_2	s_2	s_0	0	1

Table 6.17

	ν_2		ω_2	
	0	1	0	1
s_3	s_3	s_4	1	1
s_4	s_4	s_3	1	0

We connect these machines as shown in Fig. 6.19. Here each output symbol from M_1 becomes an input symbol for M_2. For example, if we input 0 to M_1, then $\omega_1(s_0, 0) = 1$ and $\nu_1(s_0, 0) = s_0$. As a result, we then input 1 $(= \omega_1(s_0, 0))$ to M_2 to get $\omega_2(s_3, 1) = 1$ and $\nu_2(s_3, 1) = s_4$.

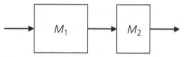

Figure 6.19

We construct a machine $M = (S, \mathcal{I}, \mathbb{O}, \nu, \omega)$ that represents this connection of M_1 and M_2 as follows:

$$\mathcal{I} = \mathbb{O} = \{0, 1\}.$$

$S = S_1 \times S_2, \quad$ where S_i is the set of internal states for M_i, for $i = 1, 2$.

$\nu: S \times \mathcal{I} \to S, \quad$ where

$\nu((s, t), x) = (\nu_1(s, x), \nu_2(t, \omega_1(s, x))), \quad$ for $s \in S_1, t \in S_2,$ and $x \in \mathcal{I}$.

$\omega: S \times \mathcal{I} \to \mathbb{O}$, where

$$\omega((s, t), x) = \omega_2(t, \omega_1(s, x)), \quad \text{for } s \in S_1, t \in S_2, \text{ and } x \in \mathcal{I}.$$

a) Find a state table for machine M.

b) Determine the output string for the input string 1101. After this string is processed, in which state do we find (i) machine M_1? (ii) machine M_2?

12. Although the state diagram seems more convenient than the state table when we are dealing with a finite state machine $M = (S, \mathcal{I}, \mathbb{O}, \nu, \omega)$, as the input strings get longer and the sizes of S, $\mathcal{I}$, and $\mathbb{O}$ increase, the state table proves useful when simulating the machine on a computer. The block form of the table suggests the use of a matrix or two-dimensional array for storing ν, ω. Use this observation to write a program (or develop an algorithm) that will simulate the machine in Table 6.18.

Table 6.18

	ν		ω	
	0	1	0	1
s_1	s_2	s_1	0	0
s_2	s_3	s_1	0	0
s_3	s_3	s_1	1	1

7

Relations: The Second Time Around

In Chapter 5 we introduced the concept of a (binary) relation. Returning to relations in this chapter, we shall emphasize the study of relations on a set A — that is, subsets of $A \times A$. Within the theory of languages and finite state machines from Chapter 6, we find many examples of relations on a set A, where A represents a set of strings from a given alphabet or a set of internal states from a finite state machine. Various properties of relations are developed, along with ways to represent finite relations for computer manipulation. Directed graphs reappear as a way to represent such relations. Finally, two types of relations on a set A are especially important: equivalence relations and partial orders. Equivalence relations, in particular, arise in many areas of mathematics. For the present we shall use an equivalence relation on the set of internal states in a finite state machine M in order to find a machine M_1, with as few internal states as possible, that performs whatever tasks M is capable of performing. The procedure is known as the minimization process.

7.1
Relations Revisited: Properties of Relations

We start by recalling some fundamental ideas considered earlier.

| **Definition 7.1** | For sets A, B, any subset of $A \times B$ is called a (*binary*) *relation* from A to B. Any subset of $A \times A$ is called a (*binary*) *relation* on A. |

As mentioned in the sentence following Definition 5.2, our primary concern is with binary relations. Consequently, for us the word "relation" will once again mean binary relation, unless something otherwise is specified.

| **EXAMPLE 7.1** | **a)** Define the relation $\mathcal{R}$ on the set $\mathbf{Z}$ by $a\,\mathcal{R}\,b$, or $(a, b) \in \mathcal{R}$, if $a \leq b$. This subset of $\mathbf{Z} \times \mathbf{Z}$ is the ordinary "less than or equal to" relation on the set $\mathbf{Z}$, and it can also be defined on $\mathbf{Q}$ or $\mathbf{R}$, but not on $\mathbf{C}$. |

b) Let $n \in \mathbf{Z}^+$. For $x, y \in \mathbf{Z}$, the *modulo n relation* $\mathcal{R}$ is defined by $x\,\mathcal{R}\,y$ if $x - y$ is a multiple of n. With $n = 7$, we find, for instance, that $9\,\mathcal{R}\,2$, $-3\,\mathcal{R}\,11$, $(14, 0) \in \mathcal{R}$, but $3\,\cancel{\mathcal{R}}\,7$ (that is, 3 is *not* related to 7).

c) For the universe $\mathcal{U} = \{1, 2, 3, 4, 5, 6, 7\}$ consider the (fixed) set $C \subseteq \mathcal{U}$ where $C = \{1, 2, 3, 6\}$. Define the relation $\mathcal{R}$ on $\mathcal{P}(\mathcal{U})$ by $A \mathcal{R} B$ when $A \cap C = B \cap C$. Then the sets $\{1, 2, 4, 5\}$ and $\{1, 2, 5, 7\}$ are related since $\{1, 2, 4, 5\} \cap C = \{1, 2\} = \{1, 2, 5, 7\} \cap C$. Likewise we find that $X = \{4, 5\}$ and $Y = \{7\}$ are so related because $X \cap C = \emptyset = Y \cap C$. However, the sets $S = \{1, 2, 3, 4, 5\}$ and $T = \{1, 2, 3, 6, 7\}$ are *not* related — that is, $S \not\mathcal{R} T$ — since $S \cap C = \{1, 2, 3\} \neq \{1, 2, 3, 6\} = T \cap C$.

| **EXAMPLE 7.2** | Let Σ be an alphabet, with language $A \subseteq \Sigma^*$. For $x, y \in A$, define $x \mathcal{R} y$ if x is a prefix of y. Other relations can be defined on A by replacing "prefix" with either "suffix" or "substring." |

| **EXAMPLE 7.3** | Consider a finite state machine $M = (S, \mathcal{I}, \mathcal{O}, \nu, \omega)$. |

    a) For $s_1, s_2 \in S$, define $s_1 \mathcal{R} s_2$ if $\nu(s_1, x) = s_2$, for some $x \in \mathcal{I}$. Relation $\mathcal{R}$ establishes the *first level of reachability*.

    b) The relation for the *second level of reachability* can also be given for S. Here $s_1 \mathcal{R} s_2$ if $\nu(s_1, x_1 x_2) = s_2$, for some $x_1 x_2 \in \mathcal{I}^2$. This can be extended to higher levels if the need arises. For the general *reachability* relation we have $\nu(s_1, y) = s_2$, for some $y \in \mathcal{I}^*$.

    c) Given $s_1, s_2 \in S$ the relation of *1-equivalence*, which is denoted by $s_1 E_1 s_2$ and is read "s_1 is 1-equivalent to s_2", is defined when $\omega(s_1, x) = \omega(s_2, x)$ for all $x \in \mathcal{I}$. Consequently, $s_1 E_1 s_2$ indicates that if machine M starts in either state s_1 or s_2, the output is the same for each element of $\mathcal{I}$. This idea can be extended to states being *k-equivalent*, where we write $s_1 E_k s_2$ if $\omega(s_1, y) = \omega(s_2, y)$, for all $y \in \mathcal{I}^k$. Here the same output string is obtained for each input string in $\mathcal{I}^k$ if we start at either s_1 or s_2.

        If two states are k-equivalent for all $k \in \mathbf{Z}^+$, then they are called *equivalent*. We shall look further into this idea later in the chapter.

We now start to examine some of the properties a relation can satisfy.

| **Definition 7.2** | A relation $\mathcal{R}$ on a set A is called *reflexive* if for all $x \in A$, $(x, x) \in \mathcal{R}$. |

    To say that a relation $\mathcal{R}$ is reflexive simply means that each element x of A is related to itself. All the relations in Examples 7.1 and 7.2 are reflexive. The general reachability relation in Example 7.3(b) and all of the relations mentioned in part (c) of that example are also reflexive. [What goes wrong with the relations for the first and second levels of reachability given in parts (a) and (b) of Example 7.3?]

| **EXAMPLE 7.4** | For $A = \{1, 2, 3, 4\}$, a relation $\mathcal{R} \subseteq A \times A$ will be reflexive if and only if $\mathcal{R} \supseteq \{(1, 1), (2, 2), (3, 3), (4, 4)\}$. Consequently, $\mathcal{R}_1 = \{(1, 1), (2, 2), (3, 3)\}$ is not a reflexive relation on A, whereas $\mathcal{R}_2 = \{(x, y) \mid x, y \in A, x \leq y\}$ is reflexive on A. |

| **EXAMPLE 7.5** | Given a finite set A with $|A| = n$, we have $|A \times A| = n^2$, so there are 2^{n^2} relations on A. How many of these are reflexive? |

    If $A = \{a_1, a_2, \ldots, a_n\}$, a relation $\mathcal{R}$ on A is reflexive if and only if $\{(a_i, a_i) \mid 1 \leq i \leq n\} \subseteq \mathcal{R}$. Considering the other $n^2 - n$ ordered pairs in $A \times A$ [those of the form (a_i, a_j), where $i \neq j$ for $1 \leq i, j \leq n$] as we construct a reflexive relation $\mathcal{R}$ on A, we either include or exclude each of these ordered pairs, so by the rule of product there are $2^{(n^2-n)}$ reflexive relations on A.

Definition 7.3 Relation $\mathcal{R}$ on set A is called *symmetric* if $(x, y) \in \mathcal{R} \Rightarrow (y, x) \in \mathcal{R}$, for all $x, y \in A$.

EXAMPLE 7.6 With $A = \{1, 2, 3\}$, we have:

a) $\mathcal{R}_1 = \{(1, 2), (2, 1), (1, 3), (3, 1)\}$, a symmetric, but not reflexive, relation on A;

b) $\mathcal{R}_2 = \{(1, 1), (2, 2), (3, 3), (2, 3)\}$, a reflexive, but not symmetric, relation on A;

c) $\mathcal{R}_3 = \{(1, 1), (2, 2), (3, 3)\}$ and $\mathcal{R}_4 = \{(1, 1), (2, 2), (3, 3), (2, 3), (3, 2)\}$, two relations on A that are both reflexive and symmetric; and

d) $\mathcal{R}_5 = \{(1, 1), (2, 3), (3, 3)\}$, a relation on A that is neither reflexive nor symmetric.

To count the symmetric relations on $A = \{a_1, a_2, \ldots, a_n\}$, we write $A \times A$ as $A_1 \cup A_2$, where $A_1 = \{(a_i, a_i) | 1 \leq i \leq n\}$ and $A_2 = \{(a_i, a_j) | 1 \leq i, j \leq n, i \neq j\}$, so that every ordered pair in $A \times A$ is in exactly one of A_1, A_2. For A_2, $|A_2| = |A \times A| - |A_1| = n^2 - n = n(n-1)$, an even integer. The set A_2 contains $(1/2)(n^2 - n)$ subsets S_{ij} of the form $\{(a_i, a_j), (a_j, a_i)\}$ where $1 \leq i < j \leq n$. In constructing a symmetric relation $\mathcal{R}$ on A, for each ordered pair in A_1 we have our usual choice of exclusion or inclusion. For each of the $(1/2)(n^2 - n)$ subsets $S_{ij} (1 \leq i < j \leq n)$ taken from A_2 we have the same two choices. So by the rule of product there are $2^n \cdot 2^{(1/2)(n^2-n)} = 2^{(1/2)(n^2+n)}$ symmetric relations on A.

In counting those relations on A that are both reflexive and symmetric, we have only one choice for each ordered pair in A_1. So we have $2^{(1/2)(n^2-n)}$ relations on A that are both reflexive and symmetric.

Definition 7.4 For a set A, a relation $\mathcal{R}$ on A is called *transitive* if, for all $x, y, z \in A$, $(x, y), (y, z) \in \mathcal{R} \Rightarrow (x, z) \in \mathcal{R}$. (So if x "is related to" y, and y "is related to" z, we want x "related to" z, with y playing the role of "intermediary.")

EXAMPLE 7.7 All the relations in Examples 7.1 and 7.2 are transitive, as are the relations in Example 7.3(c).

EXAMPLE 7.8 Define the relation $\mathcal{R}$ on the set $\mathbf{Z}^+$ by $a \mathcal{R} b$ if a (exactly) divides b — that is, $b = ca$ for some $c \in \mathbf{Z}^+$. Now if $x \mathcal{R} y$ and $y \mathcal{R} z$, do we have $x \mathcal{R} z$? We know that $x \mathcal{R} y \Rightarrow y = sx$ for some $s \in \mathbf{Z}^+$ and $y \mathcal{R} z \Rightarrow z = ty$ where $t \in \mathbf{Z}^+$. Consequently, $z = ty = t(sx) = (ts)x$ for $ts \in \mathbf{Z}^+$, so $x \mathcal{R} z$ and $\mathcal{R}$ is transitive. In addition, $\mathcal{R}$ is reflexive, but not symmetric, because, for example, $2 \mathcal{R} 6$ but $6 \not{\mathcal{R}} 2$.

EXAMPLE 7.9 Consider the relation $\mathcal{R}$ on the set $\mathbf{Z}$ where we define $a \mathcal{R} b$ when $ab \geq 0$. For all integers x we have $xx = x^2 \geq 0$, so $x \mathcal{R} x$ and $\mathcal{R}$ is reflexive. Also, if $x, y \in \mathbf{Z}$ and $x \mathcal{R} y$, then

$$x \mathcal{R} y \Rightarrow xy \geq 0 \Rightarrow yx \geq 0 \Rightarrow y \mathcal{R} x,$$

so the relation $\mathcal{R}$ is symmetric as well. However, here we find that $(3, 0), (0, -7) \in \mathcal{R}$ — since $(3)(0) \geq 0$ and $(0)(-7) \geq 0$ — but $(3, -7) \notin \mathcal{R}$ because $(3)(-7) < 0$. Consequently, this relation is *not* transitive.

EXAMPLE 7.10

If $A = \{1, 2, 3, 4\}$, then $\mathcal{R}_1 = \{(1, 1), (2, 3), (3, 4), (2, 4)\}$ is a transitive relation on A, whereas $\mathcal{R}_2 = \{(1, 3), (3, 2)\}$ is not transitive because $(1, 3), (3, 2) \in \mathcal{R}_2$ but $(1, 2) \notin \mathcal{R}_2$.

At this point the reader is probably ready to start counting the number of transitive relations on a finite set. But this is not possible here. For unlike the cases dealing with the reflexive and symmetric properties, there is no known general formula for the total number of transitive relations on a finite set. However, at a later point in this chapter we shall have the necessary ideas to count the relations $\mathcal{R}$ on a finite set, where $\mathcal{R}$ is (simultaneously) reflexive, symmetric, and transitive.

For now we consider one last property for relations.

Definition 7.5

Given a relation $\mathcal{R}$ on a set A, $\mathcal{R}$ is called *antisymmetric* if for all $a, b \in A$, $(a \mathcal{R} b$ and $b \mathcal{R} a) \Rightarrow a = b$. (Here the only way we can have both a "related to" b and b "related to" a is if a and b are one and the same element from A.)

EXAMPLE 7.11

For a given universe $\mathcal{U}$, define the relation $\mathcal{R}$ on $\mathcal{P}(\mathcal{U})$ by $(A, B) \in \mathcal{R}$ if $A \subseteq B$, for $A, B \subseteq \mathcal{U}$. So $\mathcal{R}$ is the subset relation of Chapter 3 and if $A \mathcal{R} B$ and $B \mathcal{R} A$, then we have $A \subseteq B$ and $B \subseteq A$, which gives us $A = B$. Consequently, this relation is antisymmetric, as well as reflexive and transitive, but it is not symmetric.

Before we are led astray into thinking that "not symmetric" is synonymous with "antisymmetric", let us consider the following.

EXAMPLE 7.12

For $A = \{1, 2, 3\}$, the relation $\mathcal{R}$ on A given by $\mathcal{R} = \{(1, 2), (2, 1), (2, 3)\}$ is not symmetric because $(3, 2) \notin \mathcal{R}$, and it is not antisymmetric because $(1, 2), (2, 1) \in \mathcal{R}$ but $1 \neq 2$. The relation $\mathcal{R}_1 = \{(1, 1), (2, 2)\}$ is both symmetric and antisymmetric.

How many relations on A are antisymmetric? Writing

$$A \times A = \{(1, 1), (2, 2), (3, 3)\} \cup \{(1, 2), (2, 1), (1, 3), (3, 1), (2, 3), (3, 2)\},$$

we make two observations as we try to construct an antisymmetric relation $\mathcal{R}$ on A.

1) Each element $(x, x) \in A \times A$ can be either included or excluded with no concern about whether or not $\mathcal{R}$ is antisymmetric.

2) For an element of the form (x, y), $x \neq y$, we must consider both (x, y) and (y, x) and we note that for $\mathcal{R}$ to remain antisymmetric we have three alternatives: (a) place (x, y) in $\mathcal{R}$; (b) place (y, x) in $\mathcal{R}$; or (c) place neither (x, y) nor (y, x) in $\mathcal{R}$. [What happens if we place both (x, y) and (y, x) in $\mathcal{R}$?]

So by the rule of product, the number of antisymmetric relations on A is $(2^3)(3^3) = (2^3)(3^{(3^2-3)/2})$. If $|A| = n > 0$, then there are $(2^n)(3^{(n^2-n)/2})$ antisymmetric relations on A.

For our next example we return to the concept of function dominance, which we first defined in Section 5.7.

EXAMPLE 7.13

Let $\mathscr{F}$ denote the set of all functions with domain $\mathbf{Z}^+$ and codomain $\mathbf{R}$; that is, $\mathscr{F} = \{f \mid f: \mathbf{Z}^+ \to \mathbf{R}\}$. For $f, g \in \mathscr{F}$, define the relation $\mathscr{R}$ on $\mathscr{F}$ by $f \mathscr{R} g$ if f is dominated by g (or $f \in O(g)$). Then $\mathscr{R}$ is reflexive and transitive.

If $f, g: \mathbf{Z}^+ \to \mathbf{R}$ are defined by $f(n) = n$ and $g(n) = n + 5$, then $f \mathscr{R} g$ and $g \mathscr{R} f$ but $f \neq g$, so $\mathscr{R}$ is *not antisymmetric*. In addition, if $h: \mathbf{Z}^+ \to \mathbf{R}$ is given by $h(n) = n^2$, then $(f, h), (g, h) \in \mathscr{R}$, but neither (h, f) nor (h, g) is in $\mathscr{R}$. Consequently, the relation $\mathscr{R}$ is also *not symmetric*.

At this point we have seen the four major properties that arise in the study of relations. Before closing this section we define two more notions, each of which involves three of these four properties.

Definition 7.6

A relation $\mathscr{R}$ on a set A is called a *partial order*, or a *partial ordering relation*, if $\mathscr{R}$ is reflexive, antisymmetric, and transitive.

EXAMPLE 7.14

The relation in Example 7.1(a) is a partial order, but the relation in part (b) of that example is not because it is not antisymmetric. All the relations of Example 7.2 are partial orders, as is the subset relation of Example 7.11.

Our next example provides us with the opportunity to relate this new idea of a partial order with results we studied in Chapters 1 and 4.

EXAMPLE 7.15

We start with the set $A = \{1, 2, 3, 4, 6, 12\}$ — the set of positive integer divisors of 12 — and define the relation $\mathscr{R}$ on A by $x \mathscr{R} y$ if x (exactly) divides y. As in Example 7.8 we find that $\mathscr{R}$ is reflexive and transitive. In addition, if $x, y \in A$ and we have both $x \mathscr{R} y$ and $y \mathscr{R} x$, then

$$x \mathscr{R} y \Rightarrow y = ax, \text{ for some } a \in \mathbf{Z}^+, \text{ and}$$

$$y \mathscr{R} x \Rightarrow x = by, \text{ for some } b \in \mathbf{Z}^+.$$

Consequently, it follows that $y = ax = a(by) = (ab)y$, and since $y \neq 0$, we have $ab = 1$. Because $a, b \in \mathbf{Z}^+$, $ab = 1 \Rightarrow a = b = 1$, so $y = x$ and $\mathscr{R}$ is antisymmetric — hence it defines a partial order for the set A.

Now suppose we wish to know how many ordered pairs occur in this relation $\mathscr{R}$. We may simply list the ordered pairs from $A \times A$ that comprise $\mathscr{R}$:

$$\mathscr{R} = \{(1, 1), (1, 2), (1, 3), (1, 4), (1, 6), (1, 12), (2, 2), (2, 4), (2, 6),$$
$$(2, 12), (3, 3), (3, 6), (3, 12), (4, 4), (4, 12), (6, 6), (6, 12), (12, 12)\}$$

In this way we learn that there are 18 ordered pairs in the relation. But if we then wanted to consider the same type of partial order for the set of positive integer divisors of 1800, we should definitely be discouraged by this method of simply *listing* all the ordered pairs. So let us examine the relation $\mathcal{R}$ a little closer. By the Fundamental Theorem of Arithmetic we may write $12 = 2^2 \cdot 3$ and then realize that if $(c, d) \in \mathcal{R}$, then

$$c = 2^m \cdot 3^n \quad \text{and} \quad d = 2^p \cdot 3^q,$$

where $m, n, p, q \in \mathbf{N}$ with $0 \leq m \leq p \leq 2$ and $0 \leq n \leq q \leq 1$.

When we consider the fact that $0 \leq m \leq p \leq 2$, we find that each possibility for m, p is simply a selection of size 2 from a set of size 3 — namely, the set $\{0, 1, 2\}$ — where repetitions are allowed. (In any such selection, if there is a smaller nonnegative integer, then it is assigned to m.) In Chapter 1 we learned that such a selection can be made in $\binom{3+2-1}{2} = \binom{4}{2} = 6$ ways. And, in like manner, n and q can be selected in $\binom{2+2-1}{2} = \binom{3}{2} = 3$ ways. So by the rule of product there should be $(6)(3) = 18$ ordered pairs in $\mathcal{R}$ — as we found earlier by actually listing all of them.

Now suppose we examine a similar situation, the set of positive integer divisors of $1800 = 2^3 \cdot 3^2 \cdot 5^2$. Here we are dealing with $(3 + 1)(2 + 1)(2 + 1) = (4)(3)(3) = 36$ divisors, and a typical ordered pair for this partial order (given by division) looks like $(2^r \cdot 3^s \cdot 5^t, 2^u \cdot 3^v \cdot 5^w)$, where $r, s, t, u, v, w \in \mathbf{N}$ with $0 \leq r \leq u \leq 3$, $0 \leq s \leq v \leq 2$, and $0 \leq t \leq w \leq 2$. So the number of ordered pairs in the relation is

$$\binom{4+2-1}{2}\binom{3+2-1}{2}\binom{3+2-1}{2} = \binom{5}{2}\binom{4}{2}\binom{4}{2} = (10)(6)(6) = 360,$$

and we definitely should *not* want to have to list all of the ordered pairs in the relation in order to obtain this result.

In general, for $n \in \mathbf{Z}^+$ with $n > 1$, use the Fundamental Theorem of Arithmetic to write $n = p_1^{e_1} p_2^{e_2} p_3^{e_3} \cdots p_k^{e_k}$, where $k \in \mathbf{Z}^+$, $p_1 < p_2 < p_3 < \cdots < p_k$, and p_i is prime and $e_i \in \mathbf{Z}^+$ for each $1 \leq i \leq k$. Then n has $\prod_{i=1}^{k}(e_i + 1)$ positive integer divisors. And when we consider the same type of partial order for this set (of positive integer divisors of n), we find that the number of ordered pairs in the relation is

$$\prod_{i=1}^{k}\binom{(e_i + 1) + 2 - 1}{2} = \prod_{i=1}^{k}\binom{e_i + 2}{2}.$$

In closing this section we introduce the equivalence relation — a concept that is very important in the study of mathematics.

Definition 7.7   An *equivalence relation* $\mathcal{R}$ on a set A is a relation that is reflexive, symmetric, and transitive.

EXAMPLE 7.16

a) The relation in Example 7.1(b) and all the relations in Example 7.3(c) are equivalence relations.

b) If $A = \{1, 2, 3\}$, then

$\mathcal{R}_1 = \{(1, 1), (2, 2), (3, 3)\}$,

$\mathcal{R}_2 = \{(1, 1), (2, 2), (2, 3), (3, 2), (3, 3)\}$,

$\mathcal{R}_3 = \{(1, 1), (1, 3), (2, 2), (3, 1), (3, 3)\}$, and

$\mathcal{R}_4 = \{(1, 1), (1, 2), (1, 3), (2, 1), (2, 2), (2, 3), (3, 1), (3, 2), (3, 3)\} = A \times A$

are all equivalence relations on A.

c) For a given finite set A, $A \times A$ is the largest equivalence relation on A, and if $A = \{a_1, a_2, \ldots, a_n\}$, then the equality relation $\mathcal{R} = \{(a_i, a_i) \mid 1 \leq i \leq n\}$ is the smallest equivalence relation on A.

d) Let $A = \{1, 2, 3, 4, 5, 6, 7\}$, $B = \{x, y, z\}$, and $f: A \to B$ be the onto function

$$f = \{(1, x), (2, z), (3, x), (4, y), (5, z), (6, y), (7, x)\}.$$

Define the relation $\mathcal{R}$ on A by $a\,\mathcal{R}\,b$ if $f(a) = f(b)$. Then, for instance, we find here that $1\,\mathcal{R}\,1$, $1\,\mathcal{R}\,3$, $2\,\mathcal{R}\,5$, $3\,\mathcal{R}\,1$, and $4\,\mathcal{R}\,6$.

For each $a \in A$, $f(a) = f(a)$ because f is a function — so $a\,\mathcal{R}\,a$, and $\mathcal{R}$ is reflexive. Now suppose that $a, b \in A$ and $a\,\mathcal{R}\,b$. Then $a\,\mathcal{R}\,b \Rightarrow f(a) = f(b) \Rightarrow f(b) = f(a) \Rightarrow b\,\mathcal{R}\,a$, so $\mathcal{R}$ is symmetric. Finally, if $a, b, c \in A$ with $a\,\mathcal{R}\,b$ and $b\,\mathcal{R}\,c$, then $f(a) = f(b)$ and $f(b) = f(c)$. Consequently, $f(a) = f(c)$, and we see that $(a\,\mathcal{R}\,b \wedge b\,\mathcal{R}\,c) \Rightarrow a\,\mathcal{R}\,c$. So $\mathcal{R}$ is transitive. Since $\mathcal{R}$ is reflexive, symmetric, and transitive, it is an equivalence relation.

Here $\mathcal{R} = \{(1, 1), (1, 3), (1, 7), (2, 2), (2, 5), (3, 1), (3, 3), (3, 7), (4, 4), (4, 6), (5, 2), (5, 5), (6, 4), (6, 6), (7, 1), (7, 3), (7, 7)\}$.

e) If $\mathcal{R}$ is a relation on a set A, then $\mathcal{R}$ is both an equivalence relation and a partial order on A if and only if $\mathcal{R}$ is the equality relation on A.

EXERCISES 7.1

1. If $A = \{1, 2, 3, 4\}$, give an example of a relation $\mathcal{R}$ on A that is

a) reflexive and symmetric, but not transitive

b) reflexive and transitive, but not symmetric

c) symmetric and transitive, but not reflexive

2. For relation (b) in Example 7.1, determine five values of x for which $(x, 5) \in \mathcal{R}$.

3. For the relation $\mathcal{R}$ in Example 7.13, let $f: \mathbf{Z}^+ \to \mathbf{R}$ where $f(n) = n$.

a) Find three elements f_1, f_2, $f_3 \in \mathcal{F}$ such that $f_i\,\mathcal{R}\,f$ and $f\,\mathcal{R}\,f_i$, for all $1 \le i \le 3$.

b) Find three elements g_1, g_2, $g_3 \in \mathcal{F}$ such that $g_i\,\mathcal{R}\,f$ but $f\,\mathcal{\not{R}}\,g_i$, for all $1 \le i \le 3$.

4. a) Rephrase the definitions for the reflexive, symmetric, transitive, and antisymmetric properties of a relation $\mathcal{R}$ (on a set A), using quantifiers.

b) Use the results of part (a) to specify when a relation $\mathcal{R}$ (on a set A) is (i) *not* reflexive; (ii) *not* symmetric; (iii) *not* transitive; and (iv) *not* antisymmetric.

5. For each of the following relations, determine whether the relation is reflexive, symmetric, antisymmetric, or transitive.

a) $\mathcal{R} \subseteq \mathbf{Z}^+ \times \mathbf{Z}^+$ where $a\,\mathcal{R}\,b$ if $a|b$ (read "a divides b," as defined in Section 4.3).

b) $\mathcal{R}$ is the relation on $\mathbf{Z}$ where $a\,\mathcal{R}\,b$ if $a|b$.

c) For a given universe $\mathcal{U}$ and a fixed subset C of $\mathcal{U}$, define $\mathcal{R}$ on $\mathcal{P}(\mathcal{U})$ as follows: For $A, B \subseteq \mathcal{U}$ we have $A\,\mathcal{R}\,B$ if $A \cap C = B \cap C$.

d) On the set A of all lines in $\mathbf{R}^2$, define the relation $\mathcal{R}$ for two lines ℓ_1, ℓ_2 by $\ell_1\,\mathcal{R}\,\ell_2$ if ℓ_1 is perpendicular to ℓ_2.

e) $\mathcal{R}$ is the relation on $\mathbf{Z}$ where $x\,\mathcal{R}\,y$ if $x + y$ is odd.

f) $\mathcal{R}$ is the relation on $\mathbf{Z}$ where $x\,\mathcal{R}\,y$ if $x - y$ is even.

g) Let T be the set of all triangles in $\mathbf{R}^2$. Define $\mathcal{R}$ on T by $t_1\,\mathcal{R}\,t_2$ if t_1 and t_2 have an angle of the same measure.

h) $\mathcal{R}$ is the relation on $\mathbf{Z} \times \mathbf{Z}$ where $(a, b)\mathcal{R}(c, d)$ if $a \le c$. [*Note*: $\mathcal{R} \subseteq (\mathbf{Z} \times \mathbf{Z}) \times (\mathbf{Z} \times \mathbf{Z})$.]

6. Which relations in Exercise 5 are partial orders? Which are equivalence relations?

7. Let $\mathcal{R}_1$, $\mathcal{R}_2$ be relations on a set A. (a) Prove or disprove that $\mathcal{R}_1$, $\mathcal{R}_2$ reflexive $\Rightarrow \mathcal{R}_1 \cap \mathcal{R}_2$ reflexive. (b) Answer part (a) when each occurrence of "reflexive" is replaced by (i) symmetric; (ii) antisymmetric; and (iii) transitive.

8. Answer Exercise 7, replacing each occurrence of $\cap$ by $\cup$.

9. For each of the following statements about relations on a set A, where $|A| = n$, determine whether the statement is true or false. If it is false, give a counterexample.

a) If $\mathcal{R}$ is a relation on A and $|\mathcal{R}| \ge n$, then $\mathcal{R}$ is reflexive.

b) If $\mathcal{R}_1$, $\mathcal{R}_2$ are relations on A and $\mathcal{R}_2 \supseteq \mathcal{R}_1$, then $\mathcal{R}_1$ reflexive (symmetric, antisymmetric, transitive) $\Rightarrow \mathcal{R}_2$ reflexive (symmetric, antisymmetric, transitive).

c) If $\mathcal{R}_1$, $\mathcal{R}_2$ are relations on A and $\mathcal{R}_2 \supseteq \mathcal{R}_1$, then $\mathcal{R}_2$ reflexive (symmetric, antisymmetric, transitive) $\Rightarrow \mathcal{R}_1$ reflexive (symmetric, antisymmetric, transitive).

d) If $\mathcal{R}$ is an equivalence relation on A, then $n \le |\mathcal{R}| \le n^2$.

10. If $A = \{w, x, y, z\}$, determine the number of relations on A that are (a) reflexive; (b) symmetric; (c) reflexive and symmetric; (d) reflexive and contain (x, y); (e) symmetric and contain (x, y); (f) antisymmetric; (g) antisymmetric and contain (x, y); (h) symmetric and antisymmetric; and (i) reflexive, symmetric, and antisymmetric.

11. Let $n \in \mathbf{Z}^+$ with $n > 1$, and let A be the set of positive integer divisors of n. Define the relation $\mathcal{R}$ on A by $x\,\mathcal{R}\,y$ if x

(exactly) divides y. Determine how many ordered pairs are in the relation $\mathcal{R}$ when n is (a) 10; (b) 20; (c) 40; (d) 200; (e) 210; and (f) 13860.

12. Suppose that p_1, p_2, p_3 are distinct primes and that $n, k \in \mathbf{Z}^+$ with $n = p_1^5 p_2^3 p_3^k$. Let A be the set of positive integer divisors of n and define the relation $\mathcal{R}$ on A by $x \, \mathcal{R} \, y$ if x (exactly) divides y. If there are 5880 ordered pairs in $\mathcal{R}$, determine k and $|A|$.

13. What is wrong with the following argument?

Let A be a set with $\mathcal{R}$ a relation on A. If $\mathcal{R}$ is symmetric and transitive, then $\mathcal{R}$ is reflexive.

Proof: Let $(x, y) \in \mathcal{R}$. By the symmetric property, $(y, x) \in \mathcal{R}$. Then with (x, y), $(y, x) \in \mathcal{R}$, it follows by the transitive property that $(x, x) \in \mathcal{R}$. Consequently, $\mathcal{R}$ is reflexive.

14. Let A be a set with $|A| = n$, and let $\mathcal{R}$ be a relation on A that is antisymmetric. What is the maximum value for $|\mathcal{R}|$? How many antisymmetric relations can have this size?

15. Let A be a set with $|A| = n$, and let $\mathcal{R}$ be an equivalence relation on A with $|\mathcal{R}| = r$. Why is $r - n$ always even?

16. A relation $\mathcal{R}$ on a set A is called *irreflexive* if for all $a \in A$, $(a, a) \notin \mathcal{R}$.

a) Give an example of a relation $\mathcal{R}$ on $\mathbf{Z}$ where $\mathcal{R}$ is irreflexive and transitive but not symmetric.

b) Let $\mathcal{R}$ be a nonempty relation on a set A. Prove that if $\mathcal{R}$ satisfies any two of the following properties — irreflexive, symmetric, and transitive — then it cannot satisfy the third.

c) If $|A| = n \geq 1$, how many different relations on A are irreflexive? How many are neither reflexive nor irreflexive?

17. Let $A = \{1, 2, 3, 4, 5, 6, 7\}$. How many symmetric relations on A contain exactly (a) four ordered pairs? (b) five ordered pairs? (c) seven ordered pairs? (d) eight ordered pairs?

18. a) Let $f: A \to B$, where $|A| = 25$, $B = \{x, y, z\}$, and $|f^{-1}(x)| = 10$, $|f^{-1}(y)| = 10$, $|f^{-1}(z)| = 5$. If we define the relation $\mathcal{R}$ on A by $a \, \mathcal{R} \, b$ if $a, b \in A$ and $f(a) = f(b)$, how many ordered pairs are there in the relation $\mathcal{R}$?

b) For $n, n_1, n_2, n_3, n_4 \in \mathbf{Z}^+$, let $f: A \to B$, where $|A| = n$, $B = \{w, x, y, z\}$, $|f^{-1}(w)| = n_1$, $|f^{-1}(x)| = n_2$, $|f^{-1}(y)| = n_3$, $|f^{-1}(z)| = n_4$, and $n_1 + n_2 + n_3 + n_4 = n$. If we define the relation $\mathcal{R}$ on A by $a \, \mathcal{R} \, b$ if $a, b \in A$ and $f(a) = f(b)$, how many ordered pairs are there in the relation $\mathcal{R}$?

7.2
Computer Recognition: Zero-One Matrices and Directed Graphs

Since our interest in relations is focused on those for finite sets, we are concerned with ways of representing such relations so that the properties of Section 7.1 can be easily verified. For this reason we now develop the necessary tools: relation composition, zero-one matrices, and directed graphs.

In a manner analogous to the composition of functions, relations can be combined in the following circumstances.

Definition 7.8   If A, B, and C are sets with $\mathcal{R}_1 \subseteq A \times B$ and $\mathcal{R}_2 \subseteq B \times C$, then the *composite relation* $\mathcal{R}_1 \circ \mathcal{R}_2$ is a relation from A to C defined by $\mathcal{R}_1 \circ \mathcal{R}_2 = \{(x, z) \mid x \in A, z \in C,$ and there exists $y \in B$ with $(x, y) \in \mathcal{R}_1, (y, z) \in \mathcal{R}_2\}$.

Beware! The composition of two relations is written in an order opposite to that for function composition. We shall see why in Example 7.21.

EXAMPLE 7.17   Let $A = \{1, 2, 3, 4\}$, $B = \{w, x, y, z\}$, and $C = \{5, 6, 7\}$. Consider $\mathcal{R}_1 = \{(1, x), (2, x),$ $(3, y), (3, z)\}$, a relation from A to B, and $\mathcal{R}_2 = \{(w, 5), (x, 6)\}$, a relation from B to C. Then $\mathcal{R}_1 \circ \mathcal{R}_2 = \{(1, 6), (2, 6)\}$ is a relation from A to C. If $\mathcal{R}_3 = \{(w, 5), (w, 6)\}$ is another relation from B to C, then $\mathcal{R}_1 \circ \mathcal{R}_3 = \emptyset$.

EXAMPLE 7.18

Let A be the set of employees at a computing center, while B denotes a set of high-level programming languages, and C is a set of projects $\{p_1, p_2, \ldots, p_8\}$ for which managers must make work assignments using the people in A. Consider $\mathcal{R}_1 \subseteq A \times B$, where an ordered pair of the form (L. Alldredge, Java) indicates that employee L. Alldredge is proficient in Java (and perhaps other programming languages). The relation $\mathcal{R}_2 \subseteq B \times C$ consists of ordered pairs such as (Java, p_2), indicating that Java is considered an essential language needed by anyone who works on project p_2. In the composite relation $\mathcal{R}_1 \circ \mathcal{R}_2$ we find (L. Alldredge, p_2). If no other ordered pair in $\mathcal{R}_2$ has p_2 as its second component, we know that if L. Alldredge was assigned to p_2 it was solely on the basis of his proficiency in Java. (Here $\mathcal{R}_1 \circ \mathcal{R}_2$ has been used to set up a matching process between employees and projects on the basis of employee knowledge of specific programming languages.)

Comparable to the associative law for function composition, the following result holds for relations.

THEOREM 7.1

Let A, B, C, and D be sets with $\mathcal{R}_1 \subseteq A \times B$, $\mathcal{R}_2 \subseteq B \times C$, and $\mathcal{R}_3 \subseteq C \times D$. Then $\mathcal{R}_1 \circ (\mathcal{R}_2 \circ \mathcal{R}_3) = (\mathcal{R}_1 \circ \mathcal{R}_2) \circ \mathcal{R}_3$.

Proof: Since both $\mathcal{R}_1 \circ (\mathcal{R}_2 \circ \mathcal{R}_3)$ and $(\mathcal{R}_1 \circ \mathcal{R}_2) \circ \mathcal{R}_3$ are relations from A to D, there is some reason to believe they are equal. If $(a, d) \in \mathcal{R}_1 \circ (\mathcal{R}_2 \circ \mathcal{R}_3)$, then there is an element $b \in B$ with $(a, b) \in \mathcal{R}_1$ and $(b, d) \in (\mathcal{R}_2 \circ \mathcal{R}_3)$. Also, $(b, d) \in (\mathcal{R}_2 \circ \mathcal{R}_3) \Rightarrow (b, c) \in \mathcal{R}_2$ and $(c, d) \in \mathcal{R}_3$ for some $c \in C$. Then $(a, b) \in \mathcal{R}_1$ and $(b, c) \in \mathcal{R}_2 \Rightarrow (a, c) \in \mathcal{R}_1 \circ \mathcal{R}_2$. Finally, $(a, c) \in \mathcal{R}_1 \circ \mathcal{R}_2$ and $(c, d) \in \mathcal{R}_3 \Rightarrow (a, d) \in (\mathcal{R}_1 \circ \mathcal{R}_2) \circ \mathcal{R}_3$, and $\mathcal{R}_1 \circ (\mathcal{R}_2 \circ \mathcal{R}_3) \subseteq (\mathcal{R}_1 \circ \mathcal{R}_2) \circ \mathcal{R}_3$. The opposite inclusion follows by similar reasoning.

As a result of this theorem no ambiguity arises when we write $\mathcal{R}_1 \circ \mathcal{R}_2 \circ \mathcal{R}_3$ for either of the relations in Theorem 7.1. In addition, we can now define the powers of a relation $\mathcal{R}$ on a set.

Definition 7.9

Given a set A and a relation $\mathcal{R}$ on A, we define the *powers of* $\mathcal{R}$ recursively by (a) $\mathcal{R}^1 = \mathcal{R}$; and (b) for $n \in \mathbf{Z}^+$, $\mathcal{R}^{n+1} = \mathcal{R} \circ \mathcal{R}^n$.

Note that for $n \in \mathbf{Z}^+$, $\mathcal{R}^n$ is a relation on A.

EXAMPLE 7.19

If $A = \{1, 2, 3, 4\}$ and $\mathcal{R} = \{(1, 2), (1, 3), (2, 4), (3, 2)\}$, then $\mathcal{R}^2 = \{(1, 4), (1, 2), (3, 4)\}$, $\mathcal{R}^3 = \{(1, 4)\}$, and for $n \geq 4$, $\mathcal{R}^n = \varnothing$.

As the set A and the relation $\mathcal{R}$ on A grow larger, calculations such as those in Example 7.19 become tedious. To avoid this tedium, the tool we need is the computer, once a way can be found to tell the machine about the set A and the relation $\mathcal{R}$ on A.

Definition 7.10

An $m \times n$ *zero-one matrix* $E = (e_{ij})_{m \times n}$ is a rectangular array of numbers arranged in m rows and n columns, where each e_{ij}, for $1 \leq i \leq m$ and $1 \leq j \leq n$, denotes the entry in the ith row and jth column of E, and each such entry is 0 or 1. [We can also write $(0, 1)$-matrix for this type of matrix.]

EXAMPLE 7.20

The matrix

$$E = \begin{bmatrix} 1 & 0 & 0 & 1 \\ 0 & 1 & 0 & 1 \\ 1 & 0 & 0 & 0 \end{bmatrix}$$

is a 3×4 $(0, 1)$-matrix where, for example, $e_{11} = 1$, $e_{23} = 0$, and $e_{31} = 1$.

In working with these matrices, we use the standard operations of matrix addition and multiplication *with the stipulation that* $1 + 1 = 1$. (Hence the addition is called Boolean.)

EXAMPLE 7.21

Consider the sets A, B, and C and the relations $\mathcal{R}_1$, $\mathcal{R}_2$ of Example 7.17. With the orders of the elements in A, B, and C fixed as in that example, we define the *relation matrices* for $\mathcal{R}_1$, $\mathcal{R}_2$ as follows:

$$M(\mathcal{R}_1) = \begin{array}{c} \\ (1) \\ (2) \\ (3) \\ (4) \end{array} \begin{array}{cccc} (w) & (x) & (y) & (z) \\ \begin{bmatrix} 0 & 1 & 0 & 0 \\ 0 & 1 & 0 & 0 \\ 0 & 0 & 1 & 1 \\ 0 & 0 & 0 & 0 \end{bmatrix} \end{array}, \qquad M(\mathcal{R}_2) = \begin{array}{c} \\ (w) \\ (x) \\ (y) \\ (z) \end{array} \begin{array}{ccc} (5) & (6) & (7) \\ \begin{bmatrix} 1 & 0 & 0 \\ 0 & 1 & 0 \\ 0 & 0 & 0 \\ 0 & 0 & 0 \end{bmatrix} \end{array}$$

In constructing $M(\mathcal{R}_1)$, we are dealing with a relation from A to B, so the elements of A are used to mark the rows of $M(\mathcal{R}_1)$ and the elements of B designate the columns. Then to denote, for example, that $(2, x) \in \mathcal{R}_1$, we place a 1 in the row marked (2) and the column marked (x). Each 0 in this matrix indicates an ordered pair in $A \times B$ that is missing from $\mathcal{R}_1$. For example, since $(3, w) \notin \mathcal{R}_1$, there is a 0 for the entry in row (3) and column (w) of the matrix $M(\mathcal{R}_1)$. The same process is used to obtain $M(\mathcal{R}_2)$.

Multiplying these matrices,[†] we find that

$$M(\mathcal{R}_1) \cdot M(\mathcal{R}_2) = \begin{bmatrix} 0 & 1 & 0 & 0 \\ 0 & 1 & 0 & 0 \\ 0 & 0 & 1 & 1 \\ 0 & 0 & 0 & 0 \end{bmatrix} \begin{bmatrix} 1 & 0 & 0 \\ 0 & 1 & 0 \\ 0 & 0 & 0 \\ 0 & 0 & 0 \end{bmatrix} = \begin{array}{c} (1) \\ (2) \\ (3) \\ (4) \end{array} \begin{array}{ccc} (5) & (6) & (7) \\ \begin{bmatrix} 0 & 1 & 0 \\ 0 & 1 & 0 \\ 0 & 0 & 0 \\ 0 & 0 & 0 \end{bmatrix} \end{array} = M(\mathcal{R}_1 \circ \mathcal{R}_2),$$

where the rows of the 4×3 matrix $M(\mathcal{R}_1 \circ \mathcal{R}_2)$ are marked by the elements of A while its columns are marked by the elements of C. In general we have: If $\mathcal{R}_1$ is a relation from A to B and $\mathcal{R}_2$ is a relation from B to C, then $M(\mathcal{R}_1) \cdot M(\mathcal{R}_2) = M(\mathcal{R}_1 \circ \mathcal{R}_2)$. That is, the product of the relation matrices for $\mathcal{R}_1$, $\mathcal{R}_2$, in that order, equals the relation matrix of the composite relation $\mathcal{R}_1 \circ \mathcal{R}_2$. (This is why the composition of two relations was written in the order specified in Definition 7.8.)

The reader will be asked to prove the general result of Example 7.21, along with some results from our next example, in Exercises 11 and 12 at the end of this section.

Further properties of relation matrices are exhibited in the following example.

[†]The reader who is not familiar with matrix multiplication or simply wishes a brief review should consult Appendix 2.

EXAMPLE 7.22

Let $A = \{1, 2, 3, 4\}$ and $\mathcal{R} = \{(1, 2), (1, 3), (2, 4), (3, 2)\}$, as in Example 7.19. Keeping the order of the elements in A fixed, we define the *relation matrix* for $\mathcal{R}$ as follows: $M(\mathcal{R})$ is the 4×4 (0, 1)-matrix whose entries m_{ij}, for $1 \leq i, j \leq 4$, are given by

$$m_{ij} = \begin{cases} 1, & \text{if } (i, j) \in \mathcal{R}, \\ 0, & \text{otherwise.} \end{cases}$$

In this case we find that

$$M(\mathcal{R}) = \begin{bmatrix} 0 & 1 & 1 & 0 \\ 0 & 0 & 0 & 1 \\ 0 & 1 & 0 & 0 \\ 0 & 0 & 0 & 0 \end{bmatrix}.$$

Now how can this be of any use? If we compute $(M(\mathcal{R}))^2$ using the convention that $1 + 1 = 1$, then we find that

$$(M(\mathcal{R}))^2 = \begin{bmatrix} 0 & 1 & 0 & 1 \\ 0 & 0 & 0 & 0 \\ 0 & 0 & 0 & 1 \\ 0 & 0 & 0 & 0 \end{bmatrix},$$

which happens to be the relation matrix for $\mathcal{R} \circ \mathcal{R} = \mathcal{R}^2$. (Check Example 7.19.) Furthermore,

$$(M(\mathcal{R}))^4 = \begin{bmatrix} 0 & 0 & 0 & 0 \\ 0 & 0 & 0 & 0 \\ 0 & 0 & 0 & 0 \\ 0 & 0 & 0 & 0 \end{bmatrix},$$

which is also the relation matrix for the relation $\mathcal{R}^4$ — that is, $(M(\mathcal{R}))^4 = M(\mathcal{R}^4)$. Also, recall that $\mathcal{R}^4 = \emptyset$, as we learned in Example 7.19.

What has happened here carries over to the general situation. We now state some results about relation matrices and their use in studying relations.

Let A be a set with $|A| = n$ and $\mathcal{R}$ a relation on A. If $M(\mathcal{R})$ is the relation matrix for $\mathcal{R}$, then

a) $M(\mathcal{R}) = \mathbf{0}$ (the matrix of all 0's) if and only if $\mathcal{R} = \emptyset$

b) $M(\mathcal{R}) = \mathbf{1}$ (the matrix of all 1's) if and only if $\mathcal{R} = A \times A$

c) $M(\mathcal{R}^m) = [M(\mathcal{R})]^m$, for $m \in \mathbf{Z}^+$

Using the (0, 1)-matrix for a relation, we now turn to the recognition of the reflexive, symmetric, antisymmetric, and transitive properties. To accomplish this we need the concepts introduced in the following three definitions.

Definition 7.11

Let $E = (e_{ij})_{m \times n}$, $F = (f_{ij})_{m \times n}$ be two $m \times n$ (0, 1)-matrices. We say that E *precedes*, or *is less than*, F, and we write $E \leq F$, if $e_{ij} \leq f_{ij}$, for all $1 \leq i \leq m$, $1 \leq j \leq n$.

EXAMPLE 7.23

With $E = \begin{bmatrix} 1 & 0 & 1 \\ 0 & 0 & 1 \end{bmatrix}$ and $F = \begin{bmatrix} 1 & 0 & 1 \\ 0 & 1 & 1 \end{bmatrix}$, we have $E \leq F$. In fact, there are eight $(0, 1)$-matrices G for which $E \leq G$.

Definition 7.12

For $n \in \mathbf{Z}^+$, $I_n = (\delta_{ij})_{n \times n}$ is the $n \times n$ $(0, 1)$-matrix where

$$\delta_{ij} = \begin{cases} 1, & \text{if } i = j, \\ 0, & \text{if } i \neq j. \end{cases}$$

Definition 7.13

Let $A = (a_{ij})_{m \times n}$ be a $(0, 1)$-matrix. The *transpose* of A, written A^{tr}, is the matrix $(a_{ji}^*)_{n \times m}$ where $a_{ji}^* = a_{ij}$, for all $1 \leq j \leq n$, $1 \leq i \leq m$.

EXAMPLE 7.24

For $A = \begin{bmatrix} 0 & 1 \\ 0 & 0 \\ 1 & 1 \end{bmatrix}$, we find that $A^{\text{tr}} = \begin{bmatrix} 0 & 0 & 1 \\ 1 & 0 & 1 \end{bmatrix}$.

As this example demonstrates, the ith row (column) of A equals the ith column (row) of A^{tr}. This indicates a method we can use in order to obtain the matrix A^{tr} from the matrix A.

THEOREM 7.2

Given a set A with $|A| = n$ and a relation $\mathcal{R}$ on A, let M denote the relation matrix for $\mathcal{R}$. Then

a) $\mathcal{R}$ is reflexive if and only if $I_n \leq M$.

b) $\mathcal{R}$ is symmetric if and only if $M = M^{\text{tr}}$.

c) $\mathcal{R}$ is transitive if and only if $M \cdot M = M^2 \leq M$.

d) $\mathcal{R}$ is antisymmetric if and only if $M \cap M^{\text{tr}} \leq I_n$. (The matrix $M \cap M^{\text{tr}}$ is formed by operating on corresponding entries in M and M^{tr} according to the rules $0 \cap 0 = 0 \cap 1 = 1 \cap 0 = 0$ and $1 \cap 1 = 1$ — that is, the usual multiplication for 0's and /or 1's.)

Proof: The results follow from the definitions of the relation properties and the $(0, 1)$-matrix. We demonstrate this for part (c), using the elements of A to designate the rows and columns in M, as in Examples 7.21 and 7.22.

Let $M^2 \leq M$. If $(x, y), (y, z) \in \mathcal{R}$, then there are 1's in row (x), column (y) and in row (y), column (z) of M. Consequently, in row (x), column (z) of M^2 there is a 1. This 1 must also occur in row (x), column (z) of M because $M^2 \leq M$. Hence $(x, z) \in \mathcal{R}$ and $\mathcal{R}$ is transitive.

Conversely, if $\mathcal{R}$ is transitive and M is the relation matrix for $\mathcal{R}$, let s_{xz} be the entry in row (x) and column (z) of M^2, with $s_{xz} = 1$. For s_{xz} to equal 1 in M^2, there must exist at least one $y \in A$ where $m_{xy} = m_{yz} = 1$ in M. This happens only if $x \mathcal{R} y$ and $y \mathcal{R} z$. With $\mathcal{R}$ transitive, it then follows that $x \mathcal{R} z$. So $m_{xz} = 1$ and $M^2 \leq M$.

The proofs of the remaining parts are left to the reader.

The relation matrix is a useful tool for the computer recognition of certain properties of relations. Storing information as described here, this matrix is an example of a *data structure*. Also of interest is how the relation matrix is used in the study of graph theory[†] and how graph theory is used in the recognition of certain properties of relations.

At this point we shall introduce some fundamental concepts in graph theory. Often these concepts will be given within examples and not in terms of formal definitions. In Chapter 11, however, the presentation will not assume what is given here and will be more rigorous and comprehensive.

Definition 7.14

Let V be a finite nonempty set. A *directed graph* (or *digraph*) G *on* V is made up of the elements of V, called the *vertices* or *nodes* of G, and a subset E, of $V \times V$, that contains the *(directed) edges*, or *arcs*, of G. The set V is called the *vertex set* of G, and the set E is called the *edge set*. We then write $G = (V, E)$ to denote the graph.

If $a, b \in V$ and $(a, b) \in E$[‡], then there is an edge *from a to b*. Vertex a is called the *origin* or *source* of the edge, with b the *terminus*, or *terminating vertex*, and we say that b is *adjacent from a* and that a is *adjacent to b*. In addition, if $a \neq b$, then $(a, b) \neq (b, a)$. An edge of the form (a, a) is called a *loop* (at a).

EXAMPLE 7.25

For $V = \{1, 2, 3, 4, 5\}$, the diagram in Fig. 7.1 is a directed graph G on V with edge set $\{(1, 1), (1, 2), (1, 4), (3, 2)\}$. Vertex 5 is a part of this graph even though it is not the origin or terminus of an edge. It is referred to as an *isolated* vertex. As we see here, edges need not be straight line segments, and there is no concern about the length of an edge.

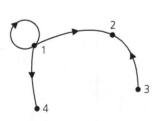

Figure 7.1

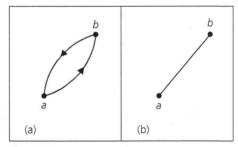

Figure 7.2

When we develop a *flowchart* to study a computer program or algorithm, we deal with a special type of directed graph where the shapes of the vertices may be important in the analysis of the algorithm. Road maps are directed graphs, where the cities and towns are represented by vertices and the highways linking any two localities are given by edges. In road maps, an edge is often directed in both directions. Consequently, if G is a directed graph and $a, b \in V$, with $a \neq b$, and both $(a, b), (b, a) \in E$, then the single undirected edge $\{a, b\} = \{b, a\}$ in Fig. 7.2(b) is used to represent the two directed edges shown in Fig. 7.2(a). In this case, a and b are called *adjacent* vertices. (Directions may also be disregarded for loops.)

[†]Since the terminology of graph theory is not standardized, the reader may find some differences between definitions given here and in other texts.

[‡]In this chapter we allow only one edge from a to b. Situations where multiple edges occur are called *multigraphs*. These are discussed in Chapter 11.

Directed graphs play an important role in many situations in computer science. The following example demonstrates one of these.

Computer programs can be processed more rapidly when certain statements in the program are executed concurrently. But in order to accomplish this we must be aware of the dependence of some statements on earlier statements in the program. For we cannot execute a statement that needs results from other statements — statements that have not yet been executed.

In Fig. 7.3(a) we have eight assignment statements that constitute the beginning of a computer program. We represent these statements by the eight corresponding vertices $s_1, s_2, s_3, \ldots, s_8$ in part (b) of the figure, where a directed edge such as (s_1, s_5) indicates that statement s_5 cannot be executed until statement s_1 has been executed. The resulting directed graph is called the *precedence graph* for the given lines of the computer program. Note how this graph indicates, for example, that statement s_7 cannot be executed until after each of the statements s_1, s_2, s_3, and s_4 has been executed. Also, we see how a statement such as s_1 must be executed before it is possible to execute any of the statements s_2, s_4, s_5, s_7, or s_8. In general, if a vertex (statement) s is adjacent *from* m other vertices (and no others), then the corresponding statements for these m vertices must be executed before statement s can be executed. Similarly, should a vertex (statement) s be adjacent *to* n other vertices, then each of the corresponding statements for these vertices requires the execution of statement s before it can be executed. Finally, from the precedence graph we see that the statements s_1, s_3, and s_6 can be processed concurrently. Following this, the statements s_2, s_4, and s_8 can be executed at the same time, and then the statements s_5 and s_7. (Or we could process statements s_2 and s_4 concurrently, and then the statements s_5, s_7, and s_8.)

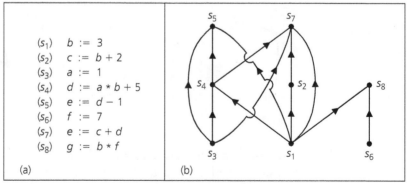

(s_1)	$b := 3$
(s_2)	$c := b + 2$
(s_3)	$a := 1$
(s_4)	$d := a * b + 5$
(s_5)	$e := d - 1$
(s_6)	$f := 7$
(s_7)	$e := c + d$
(s_8)	$g := b * f$

(a)   (b)

Figure 7.3

Now we want to consider how relations and directed graphs are interrelated. For a start, given a set A and a relation $\mathcal{R}$ on A, we can construct a directed graph G with vertex set A and edge set $E \subseteq A \times A$, where $(a, b) \in E$ if $a, b \in A$ and $a \mathcal{R} b$. This is demonstrated in the following example.

For $A = \{1, 2, 3, 4\}$, let $\mathcal{R} = \{(1, 1), (1, 2), (2, 3), (3, 2), (3, 3), (3, 4), (4, 2)\}$ be a relation on A. The directed graph associated with $\mathcal{R}$ is shown in Fig. 7.4(a), where the undirected edge $\{2, 3\}(= \{3, 2\})$ is used in place of the pair of distinct directed edges $(2, 3)$ and $(3, 2)$. If the directions in Fig. 7.4(a) are ignored, we get the *associated undirected graph* shown in part (b) of the figure. Here we see that the graph is *connected* in the sense that for any two vertices x, y, with $x \neq y$, there is a *path* starting at x and ending at y. Such a path consists of a *finite sequence of undirected edges*, so the edges $\{1, 2\}$, $\{2, 4\}$ provide a path from 1 to 4, and the edges $\{3, 4\}$, $\{4, 2\}$, and $\{2, 1\}$ provide a path from 3 to 1. The sequence of edges $\{3, 4\}$, $\{4, 2\}$, and $\{2, 3\}$ provides a path from 3 to 3. Such a *closed* path is called a *cycle*. This is an example of an undirected cycle of *length* 3, because it has three edges in it.

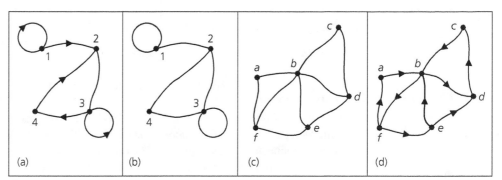

Figure 7.4

When we are dealing with paths (in both directed and undirected graphs), no vertex may be repeated. Therefore, the sequence of edges $\{a, b\}$, $\{b, e\}$, $\{e, f\}$, $\{f, b\}$, $\{b, d\}$ in Fig. 7.4(c) is *not* considered to be a path (from a to d) because we pass through the vertex b more than once. In the case of cycles, the path starts and terminates at the same vertex and has *at least three edges*. In Fig. 7.4(d) the sequence of edges (b, f), (f, e), (e, d), (d, c), (c, b) provides a *directed cycle* of length 5. The six edges (b, f), (f, e), (e, b), (b, d), (d, c), (c, b) do *not* yield a directed cycle in the figure because of the repetition of vertex b. If their directions are ignored, the corresponding six edges, in part (c) of the figure, likewise pass through vertex b more than once. Consequently, these edges are not considered to form a cycle for the undirected graph in Fig. 7.4(c).

Now since we require a cycle to have *length* at least 3, we shall not consider loops to be cycles. We also note that loops have no bearing on graph connectivity.

We choose to define the next idea formally because of its relevance to what we did earlier in Section 6.3.

Definition 7.15   A directed graph G on V is called *strongly connected* if for all x, $y \in V$, where $x \neq y$, there is a path (in G) of directed edges from x to y — that is, either the directed edge (x, y) is in G or, for some $n \in \mathbf{Z}^+$ and distinct vertices $v_1, v_2, \ldots, v_n \in V$, the directed edges (x, v_1), (v_1, v_2), $\ldots$, (v_n, y) are in G.

It is in this sense that we talked about strongly connected machines in Chapter 6. The graph in Fig. 7.4(a) is connected but not strongly connected. For example, there is no directed path from 3 to 1. In Fig. 7.5 the directed graph on $V = \{1, 2, 3, 4\}$ is strongly connected and *loop-free*. This is also true of the directed graph in Fig. 7.4(d).

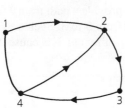

Figure 7.5

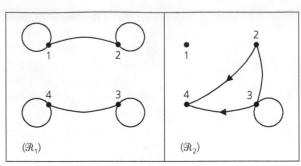

Figure 7.6

| EXAMPLE 7.28 |

For $A = \{1, 2, 3, 4\}$, consider the relations $\mathcal{R}_1 = \{(1, 1), (1, 2), (2, 1), (2, 2), (3, 3), (3, 4), (4, 3), (4, 4)\}$ and $\mathcal{R}_2 = \{(2, 4), (2, 3), (3, 2), (3, 3), (3, 4)\}$. As Fig. 7.6 illustrates, the graphs of these relations are *disconnected*. However, each graph is the union of two connected pieces called the *components* of the graph. For $\mathcal{R}_1$ the graph is made up of two strongly connected components. For $\mathcal{R}_2$, one component consists of an isolated vertex, and the other component is connected but not strongly connected.

| EXAMPLE 7.29 |

The graphs in Fig. 7.7 are examples of undirected graphs that are loop-free and have an edge for every pair of distinct vertices. These graphs illustrate the *complete graphs* on n vertices which are denoted by K_n. In Fig. 7.7 we have examples of the complete graphs on three, four, and five vertices, respectively. The complete graph K_2 consists of two vertices x, y and an edge connecting them, whereas the complete graph K_1 consists of one vertex and no edges because loops are not allowed.

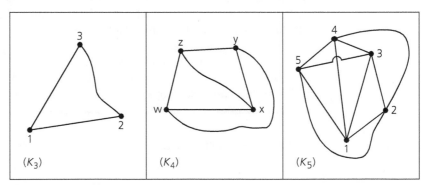

Figure 7.7

In this drawing of K_5 two edges cross, namely, $\{3, 5\}$ and $\{1, 4\}$. However, there is no point of intersection creating a new vertex. If we try to avoid the crossing of edges by drawing the graph differently, we run into the same problem all over again. This difficulty will be examined in Chapter 11 when we deal with the planarity of graphs.

A digraph G on a vertex set V gives rise to a relation $\mathcal{R}$ on V where $x \mathcal{R} y$ if (x, y) is an edge in G. Consequently, there is a $(0, 1)$-matrix for G, and since this relation matrix comes about from the adjacencies of pairs of vertices, it is referred to as the *adjacency matrix* for G as well as the relation matrix for $\mathcal{R}$.

At this point we tie together the properties of relations and the structure of directed graphs.

EXAMPLE 7.30

If $A = \{1, 2, 3\}$ and $\mathcal{R} = \{(1, 1), (1, 2), (2, 2), (3, 3), (3, 1)\}$, then $\mathcal{R}$ is a reflexive antisymmetric relation on A, but it is neither symmetric nor transitive. The directed graph associated with $\mathcal{R}$ consists of five edges. Three of these edges are loops that result from the reflexive property of $\mathcal{R}$. (See Fig. 7.8.) In general, if $\mathcal{R}$ is a relation on a finite set A, then $\mathcal{R}$ is reflexive if and only if its directed graph contains a loop at each vertex (element of A).

EXAMPLE 7.31

The relation $\mathcal{R} = \{(1, 1), (1, 2), (2, 1), (2, 3), (3, 2)\}$ is symmetric on $A = \{1, 2, 3\}$, but it is not reflexive, antisymmetric, or transitive. The directed graph for $\mathcal{R}$ is found in Fig. 7.9. In general, a relation $\mathcal{R}$ on a finite set A is symmetric if and only if its directed graph may be drawn so that it contains only loops and undirected edges.

EXAMPLE 7.32

For $A = \{1, 2, 3\}$, consider $\mathcal{R} = \{(1, 1), (1, 2), (2, 3), (1, 3)\}$. The directed graph for $\mathcal{R}$ is shown in Fig. 7.10. Here $\mathcal{R}$ is transitive and antisymmetric but not reflexive or symmetric. The directed graph indicates that a relation on a set A is transitive if and only if it satisfies the following: For all $x, y \in A$, if there is a (directed) path from x to y in the associated graph, then there is an edge (x, y) also. [Here $(1, 2), (2, 3)$ is a (directed) path from 1 to 3, and we also have the edge $(1, 3)$ for transitivity.] Notice that the directed graph in Fig. 7.3 of Example 7.26 also has this property.

The relation $\mathcal{R}$ is antisymmetric because there are no ordered pairs in $\mathcal{R}$ of the form (x, y) and (y, x) with $x \neq y$. To use the directed graph of Fig. 7.10 to characterize antisymmetry, we observe that for any two vertices x, y, with $x \neq y$, the graph contains at most one of the edges (x, y) or (y, x). Hence there are no undirected edges aside from loops.

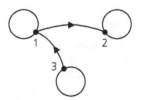

Figure 7.8

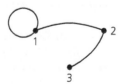

Figure 7.9

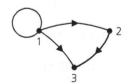

Figure 7.10

Our final example deals with equivalence relations.

EXAMPLE 7.33

For $A = \{1, 2, 3, 4, 5\}$, the following are equivalence relations on A:

$$\mathcal{R}_1 = \{(1, 1), (1, 2), (2, 1), (2, 2), (3, 3), (3, 4), (4, 3), (4, 4), (5, 5)\},$$
$$\mathcal{R}_2 = \{(1, 1), (1, 2), (1, 3), (2, 1), (2, 2), (2, 3), (3, 1), (3, 2), (3, 3),$$
$$(4, 4), (4, 5), (5, 4), (5, 5)\}.$$

Their associated graphs are shown in Fig. 7.11. If we ignore the loops in each graph, we find the graph decomposed into components such as K_1, K_2, and K_3. In general, a relation on a finite set A is an equivalence relation if and only if its associated graph is one complete graph augmented by loops at every vertex or consists of the disjoint union of complete graphs augmented by loops at every vertex.

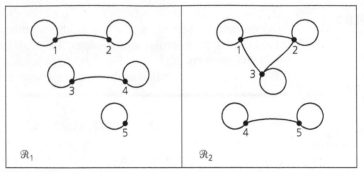

Figure 7.11

---

1. For $A = \{1, 2, 3, 4\}$, let $\mathcal{R}$ and $\mathcal{S}$ be the relations on A defined by $\mathcal{R} = \{(1, 2), (1, 3), (2, 4), (4, 4)\}$ and $\mathcal{S} = \{(1, 1), (1, 2), (1, 3), (2, 3), (2, 4)\}$. Find $\mathcal{R} \circ \mathcal{S}$, $\mathcal{S} \circ \mathcal{R}$, $\mathcal{R}^2$, $\mathcal{R}^3$, $\mathcal{S}^2$, and $\mathcal{S}^3$.

2. If $\mathcal{R}$ is a reflexive relation on a set A, prove that $\mathcal{R}^2$ is also reflexive on A.

3. Provide a proof for the opposite inclusion in Theorem 7.1.

4. Let $A = \{1, 2, 3\}$, $B = \{w, x, y, z\}$, and $C = \{4, 5, 6\}$. Define the relations $\mathcal{R}_1 \subseteq A \times B$, $\mathcal{R}_2 \subseteq B \times C$, and $\mathcal{R}_3 \subseteq B \times C$, where $\mathcal{R}_1 = \{(1, w), (3, w), (2, x), (1, y)\}$, $\mathcal{R}_2 = \{(w, 5), (x, 6), (y, 4), (y, 6)\}$, and $\mathcal{R}_3 = \{(w, 4), (w, 5), (y, 5)\}$. (a) Determine $\mathcal{R}_1 \circ (\mathcal{R}_2 \cup \mathcal{R}_3)$ and $(\mathcal{R}_1 \circ \mathcal{R}_2) \cup (\mathcal{R}_1 \circ \mathcal{R}_3)$. (b) Determine $\mathcal{R}_1 \circ (\mathcal{R}_2 \cap \mathcal{R}_3)$ and $(\mathcal{R}_1 \circ \mathcal{R}_2) \cap (\mathcal{R}_1 \circ \mathcal{R}_3)$.

5. Let $A = \{1, 2\}$, $B = \{m, n, p\}$, and $C = \{3, 4\}$. Define the relations $\mathcal{R}_1 \subseteq A \times B$, $\mathcal{R}_2 \subseteq B \times C$, and $\mathcal{R}_3 \subseteq B \times C$ by $\mathcal{R}_1 = \{(1, m), (1, n), (1, p)\}$, $\mathcal{R}_2 = \{(m, 3), (m, 4), (p, 4)\}$, and $\mathcal{R}_3 = \{(m, 3), (m, 4), (p, 3)\}$. Determine $\mathcal{R}_1 \circ (\mathcal{R}_2 \cap \mathcal{R}_3)$ and $(\mathcal{R}_1 \circ \mathcal{R}_2) \cap (\mathcal{R}_1 \circ \mathcal{R}_3)$.

6. For sets A, B, and C, consider relations $\mathcal{R}_1 \subseteq A \times B$, $\mathcal{R}_2 \subseteq B \times C$, and $\mathcal{R}_3 \subseteq B \times C$. Prove that (a) $\mathcal{R}_1 \circ (\mathcal{R}_2 \cup \mathcal{R}_3) = (\mathcal{R}_1 \circ \mathcal{R}_2) \cup (\mathcal{R}_1 \circ \mathcal{R}_3)$; and (b) $\mathcal{R}_1 \circ (\mathcal{R}_2 \cap \mathcal{R}_3) \subseteq (\mathcal{R}_1 \circ \mathcal{R}_2) \cap (\mathcal{R}_1 \circ \mathcal{R}_3)$.

7. For a relation $\mathcal{R}$ on a set A, define $\mathcal{R}^0 = \{(a, a) | a \in A\}$. If $|A| = n$, prove that there exist $s, t \in \mathbf{N}$ with $0 \le s < t \le 2^{n^2}$ such that $\mathcal{R}^s = \mathcal{R}^t$.

8. With $A = \{1, 2, 3, 4\}$, let $\mathcal{R} = \{(1, 1), (1, 2), (2, 3), (3, 3), (3, 4), (4, 4)\}$ be a relation on A. Find two relations $\mathcal{S}$, $\mathcal{T}$ on A where $\mathcal{S} \neq \mathcal{T}$ but $\mathcal{R} \circ \mathcal{S} = \mathcal{R} \circ \mathcal{T} = \{(1, 1), (1, 2), (1, 4)\}$.

9. How many 6×6 (0, 1)-matrices A are there with $A = A^{\mathrm{tr}}$?

10. If $E = \begin{bmatrix} 1 & 0 & 1 & 1 \\ 0 & 0 & 0 & 1 \\ 1 & 0 & 0 & 0 \end{bmatrix}$, how many (0, 1)-matrices F satisfy $E \le F$? How many (0, 1)-matrices G satisfy $G \le E$?

11. Consider the sets $A = \{a_1, a_2, \ldots, a_m\}$, $B = \{b_1, b_2, \ldots, b_n\}$, and $C = \{c_1, c_2, \ldots, c_p\}$, where the elements in each set remain fixed in the order given here. Let $\mathcal{R}_1$ be a relation from A to B, and let $\mathcal{R}_2$ be a relation from B to C. The relation matrix for $\mathcal{R}_i$ is $M(\mathcal{R}_i)$, where $i = 1, 2$. The rows and columns of these matrices are indexed by the elements from the appropriate sets A, B, and C according to the orders already prescribed. The matrix for $\mathcal{R}_1 \circ \mathcal{R}_2$ is the $m \times p$ matrix $M(\mathcal{R}_1 \circ \mathcal{R}_2)$, where the elements of A (in the order given) index the rows and the elements of C (also in the order given) index the columns.

Show that for all $1 \le i \le m$ and $1 \le j \le p$, the entries in the ith row and jth column of $M(\mathcal{R}_1) \cdot M(\mathcal{R}_2)$ and $M(\mathcal{R}_1 \circ \mathcal{R}_2)$ are equal. [Hence $M(\mathcal{R}_1) \cdot M(\mathcal{R}_2) = M(\mathcal{R}_1 \circ \mathcal{R}_2)$.]

12. Let A be a set with $|A| = n$, and consider the order for the listing of its elements as fixed. For $\mathcal{R} \subseteq A \times A$, let $M(\mathcal{R})$ denote the corresponding relation matrix.

a) Prove that $M(\mathcal{R}) = \mathbf{0}$ (the $n \times n$ matrix of all 0's) if and only if $\mathcal{R} = \emptyset$.

b) Prove that $M(\mathcal{R}) = \mathbf{1}$ (the $n \times n$ matrix of all 1's) if and only if $\mathcal{R} = A \times A$.

c) Use the result of Exercise 11, along with the Principle of Mathematical Induction, to prove that $M(\mathcal{R}^m) = [M(\mathcal{R})]^m$, for all $m \in \mathbf{Z}^+$.

13. Provide the proofs for Theorem 7.2(a), (b), and (d).

14. Use Theorem 7.2 to write a computer program (or to develop an algorithm) for the recognition of equivalence relations on a finite set.

15. a) Draw the digraph $G_1 = (V_1, E_1)$ where $V_1 = \{a, b, c, d, e, f\}$ and $E_1 = \{(a, b), (a, d), (b, c), (b, e), (d, b), (d, e), (e, c), (e, f), (f, d)\}$.

b) Draw the undirected graph $G_2 = (V_2, E_2)$ where $V_2 = \{s, t, u, v, w, x, y, z\}$ and $E_2 = \{\{s, t\}, \{s, u\}, \{s, x\}, \{t, u\}, \{t, w\}, \{u, w\}, \{u, x\}, \{v, w\}, \{v, x\}, \{v, y\}, \{w, z\}, \{x, y\}\}$.

16. For the directed graph $G = (V, E)$ in Fig. 7.12, classify each of the following statements as true or false.

a) Vertex c is the origin of two edges in G.

b) Vertex g is adjacent to vertex h.

c) There is a directed path in G from d to b.

d) There are two directed cycles in G.

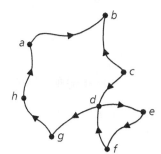

Figure 7.12

17. For $A = \{a, b, c, d, e, f\}$, each graph, or digraph, in Fig. 7.13 represents a relation $\mathcal{R}$ on A. Determine the rela-

tion $\mathcal{R} \subseteq A \times A$ in each case, as well as its associated relation matrix $M(\mathcal{R})$.

18. For $A = \{v, w, x, y, z\}$, each of the following is the $(0, 1)$-matrix for a relation $\mathcal{R}$ on A. Here the rows (from top to bottom) and the columns (from left to right) are indexed in the order v, w, x, y, z. Determine the relation $\mathcal{R} \subseteq A \times A$ in each case, and draw the directed graph G associated with $\mathcal{R}$.

a) $M(\mathcal{R}) = \begin{bmatrix} 0 & 1 & 1 & 0 & 0 \\ 1 & 0 & 1 & 1 & 1 \\ 0 & 0 & 0 & 0 & 1 \\ 0 & 0 & 0 & 0 & 1 \\ 0 & 0 & 0 & 0 & 0 \end{bmatrix}$

b) $M(\mathcal{R}) = \begin{bmatrix} 0 & 1 & 1 & 1 & 0 \\ 1 & 0 & 1 & 0 & 0 \\ 1 & 1 & 0 & 0 & 1 \\ 1 & 0 & 0 & 0 & 1 \\ 0 & 0 & 1 & 1 & 0 \end{bmatrix}$

19. For $A = \{1, 2, 3, 4\}$, let $\mathcal{R} = \{(1, 1), (1, 2), (2, 3), (3, 3), (3, 4)\}$ be a relation on A. Draw the directed graph G on A that is associated with $\mathcal{R}$. Do likewise for $\mathcal{R}^2$, $\mathcal{R}^3$, and $\mathcal{R}^4$.

20. a) Let $G = (V, E)$ be the directed graph where $V = \{1, 2, 3, 4, 5, 6, 7\}$ and $E = \{(i, j) | 1 \le i < j \le 7\}$.

    i) How many edges are there for this graph?

    ii) Four of the directed paths in G from 1 to 7 may be given as:

      1) $(1, 7)$;

      2) $(1, 3), (3, 5), (5, 6), (6, 7)$;

      3) $(1, 2), (2, 3), (3, 7)$; and

      4) $(1, 4), (4, 7)$.

      How many directed paths (in total) exist in G from 1 to 7?

b) Now let $n \in \mathbf{Z}^+$ where $n \ge 2$, and consider the directed graph $G = (V, E)$ with $V = \{1, 2, 3, \dots, n\}$ and $E = \{(i, j) | 1 \le i < j \le n\}$.

    i) Determine $|E|$.

    ii) How many directed paths exist in G from 1 to n?

    iii) If $a, b \in \mathbf{Z}^+$ with $1 \le a < b \le n$, how many directed paths exist in G from a to b?

    (The reader may wish to refer back to Exercise 20 in Section 3.1.)

21. Let $|A| = 5$. (a) How many directed graphs can one construct on A? (b) How many of the graphs in part (a) are actually undirected?

22. For $|A| = 5$, how many relations $\mathcal{R}$ on A are there? How many of these relations are symmetric?

23. a) Keeping the order of the elements fixed as 1, 2, 3, 4, 5, determine the $(0, 1)$ relation matrix for each of the equivalence relations in Example 7.33.

b) Do the results of part (a) lead to any generalization?

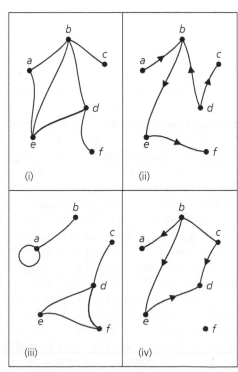

Figure 7.13

24. How many (undirected) edges are there in the complete graphs K_6, K_7, and K_n, where $n \in \mathbf{Z}^+$?

25. Draw a precedence graph for the following segment found at the start of a computer program:

$$
\begin{array}{ll}
(s_1) & a := 1 \\
(s_2) & b := 2 \\
(s_3) & a := a + 3 \\
(s_4) & c := b \\
(s_5) & a := 2 * a - 1 \\
(s_6) & b := a * c \\
(s_7) & c := 7 \\
(s_8) & d := c + 2
\end{array}
$$

26. a) Let $\mathcal{R}$ be the relation on $A = \{1, 2, 3, 4, 5, 6, 7\}$, where the directed graph associated with $\mathcal{R}$ consists of the two components, each a directed cycle, shown in Fig. 7.14. Find

Figure 7.14

the smallest integer $n > 1$, such that $\mathcal{R}^n = \mathcal{R}$. What is the smallest value of $n > 1$ for which the graph of $\mathcal{R}^n$ contains some loops? Does it ever happen that the graph of $\mathcal{R}^n$ consists of only loops?

b) Answer the same questions from part (a) for the relation $\mathcal{R}$ on $A = \{1, 2, 3, \ldots, 9, 10\}$, if the directed graph associated with $\mathcal{R}$ is as shown in Fig. 7.15.

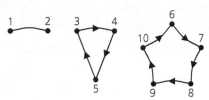

Figure 7.15

c) Do the results in parts (a) and (b) indicate anything in general?

27. If the complete graph K_n has 703 edges, how many vertices does it have?

7.3
Partial Orders: Hasse Diagrams

If you ask children to recite the numbers they know, you'll hear a uniform response of "1, 2, 3, ..." Without paying attention to it, they list these numbers in increasing order. In this section we take a closer look at this idea of order, something we may have taken for granted. We start with some observations about the sets $\mathbf{N}$, $\mathbf{Z}$, $\mathbf{Q}$, $\mathbf{R}$, and $\mathbf{C}$.

The set $\mathbf{N}$ is closed under the binary operations of (ordinary) addition and multiplication, but if we seek an answer to the equation $x + 5 = 2$, we find that no element of $\mathbf{N}$ provides a solution. So we enlarge $\mathbf{N}$ to $\mathbf{Z}$, where we can perform subtraction as well as addition and multiplication. However, we soon run into trouble trying to solve the equation $2x + 3 = 4$. Enlarging to $\mathbf{Q}$, we can perform nonzero division in addition to the other operations. Yet this soon proves to be inadequate; the equation $x^2 - 2 = 0$ necessitates the introduction of the real but irrational numbers $\pm \sqrt{2}$. Even after we expand from $\mathbf{Q}$ to $\mathbf{R}$, more trouble arises when we try to solve $x^2 + 1 = 0$. Finally we arrive at $\mathbf{C}$, the complex numbers, where any polynomial equation of the form $c_n x^n + c_{n-1} x^{n-1} + \cdots + c_2 x^2 + c_1 x + c_0 = 0$, where $c_i \in \mathbf{C}$ for $0 \leq i \leq n$, $n > 0$ and $c_n \neq 0$, can be solved. (This result is known as the Fundamental Theorem of Algebra. Its proof requires material on functions of a complex variable, so no proof is given here.) As we kept building up from $\mathbf{N}$ to $\mathbf{C}$, gaining more ability to solve polynomial equations, something was lost when we went from $\mathbf{R}$ to $\mathbf{C}$. In $\mathbf{R}$, given numbers r_1, r_2, with $r_1 \neq r_2$, we know that either $r_1 < r_2$ or $r_2 < r_1$. However, in $\mathbf{C}$ we have $(2 + i) \neq (1 + 2i)$, but what meaning can we attach to a statement such as "$(2 + i) < (1 + 2i)$"? We have lost the ability to "order" the elements in this number system!

As we start to take a closer look at the notion of order we proceed as in Section 7.1 and let A be a set with $\mathcal{R}$ a relation on A. The pair $(A, \mathcal{R})$ is called a *partially ordered set*, or *poset*, if relation $\mathcal{R}$ on A is a partial order, or a partial ordering relation (as given in Definition 7.6). If A is called a poset, we understand that there is a partial order $\mathcal{R}$ on A that makes A into this poset. Examples 7.1(a), 7.2, 7.11, and 7.15 are posets.

EXAMPLE 7.34

Let A be the set of courses offered at a college. Define the relation $\mathcal{R}$ on A by $x \mathcal{R} y$ if x, y are the same course or if x is a prerequisite for y. Then $\mathcal{R}$ makes A into a poset.

EXAMPLE 7.35

Define $\mathcal{R}$ on $A = \{1, 2, 3, 4\}$ by $x \mathcal{R} y$ if $x | y$ — that is, x (exactly) divides y. Then $\mathcal{R} = \{(1, 1), (2, 2), (3, 3), (4, 4), (1, 2), (1, 3), (1, 4), (2, 4)\}$ is a partial order, and $(A, \mathcal{R})$ is a poset. (This is similar to what we learned in Example 7.15.)

EXAMPLE 7.36

In the construction of a house certain jobs, such as digging the foundation, must be performed before other phases of the construction can be undertaken. If A is a set of tasks that must be performed in building a house, we can define a relation $\mathcal{R}$ on A by $x \mathcal{R} y$ if x, y denote the same task or if task x must be performed before the start of task y. In this way we place an order on the elements of A, making it into a poset that is sometimes referred to as a PERT (Program Evaluation and Review Technique) network. (Such networks came into play during the 1950s in order to handle the complexities that arose in organizing the many individual activities required for the completion of projects on a very large scale. This technique was actually developed and first used by the U.S. Navy in order to coordinate the many projects that were necessary for the building of the Polaris submarine.)

Consider the diagrams given in Fig. 7.16. If part (a) were part of the directed graph associated with a relation $\mathcal{R}$, then because $(1, 2), (2, 1) \in \mathcal{R}$ with $1 \neq 2$, $\mathcal{R}$ could not be antisymmetric. For part (b), if the diagram were part of the graph of a transitive relation $\mathcal{R}$, then $(1, 2), (2, 3) \in \mathcal{R} \Rightarrow (1, 3) \in \mathcal{R}$. Since $(3, 1) \in \mathcal{R}$ and $1 \neq 3$, $\mathcal{R}$ is not antisymmetric, so it cannot be a partial order.

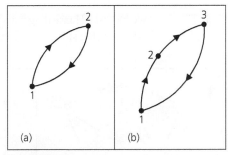

Figure 7.16

From these observations, if we are given a relation $\mathcal{R}$ on a set A, and we let G be the directed graph associated with $\mathcal{R}$, then we find that:

i) If G contains a pair of edges of the form (a, b), (b, a), for $a, b \in A$ with $a \neq b$, or

ii) If $\mathscr{R}$ is transitive and G contains a directed cycle (of length greater than or equal to three),

then the relation $\mathscr{R}$ cannot be antisymmetric, so $(A, \mathscr{R})$ fails to be a partial order.

EXAMPLE 7.37

Consider the directed graph for the partial order in Example 7.35. Figure 7.17(a) is the graphical representation of $\mathscr{R}$. In part (b) of the figure, we have a somewhat simpler diagram, which is called the *Hasse diagram* for $\mathscr{R}$.

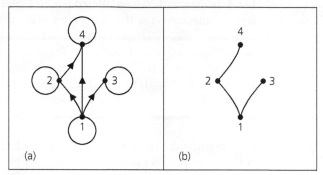

Figure 7.17

When we know that a relation $\mathscr{R}$ is a partial order on a set A, we can eliminate the loops at the vertices of its directed graph. Since $\mathscr{R}$ is also transitive, having the edges $(1, 2)$ and $(2, 4)$ is enough to insure the existence of edge $(1, 4)$, so we need not include that edge. In this way we obtain the diagram in Fig. 7.17(b), where we have not lost the directions on the edges — the directions are assumed to go from the bottom to the top.

In general, if $\mathscr{R}$ is a partial order on a finite set A, we construct a Hasse diagram for $\mathscr{R}$ on A by drawing a line segment from x *up to* y, if $x, y \in A$ with $x \mathscr{R} y$ and, most important, if there is no other element $z \in A$ such that $x \mathscr{R} z$ and $z \mathscr{R} y$. (So there is nothing "in between" x and y.) If we adopt the convention of reading the diagram from bottom to top, then it is not necessary to direct any edges.

EXAMPLE 7.38

In Fig. 7.18 we have the Hasse diagrams for the following four posets. (a) With $\mathscr{U} = \{1, 2, 3\}$ and $A = \mathscr{P}(\mathscr{U})$, $\mathscr{R}$ is the subset relation on A. (b) Here $\mathscr{R}$ is the "(exactly) divides" relation

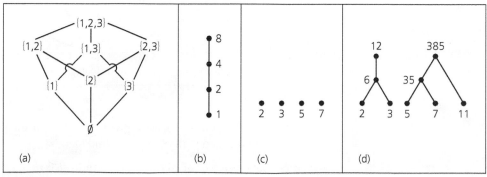

Figure 7.18

applied to $A = \{1, 2, 4, 8\}$. (c) and (d) Here the same relation as in part (b) is applied to $\{2, 3, 5, 7\}$ in part (c) and to $\{2, 3, 5, 6, 7, 11, 12, 35, 385\}$ in part (d). In part (c) we note that a Hasse diagram can have all isolated vertices; it can also have two (or more) connected pieces, as shown in part (d).

EXAMPLE 7.39

Let $A = \{1, 2, 3, 4, 5\}$. The relation $\mathcal{R}$ on A, defined by $x \mathcal{R} y$ if $x \leq y$, is a partial order. This makes A into a poset that we can denote by $(A, \leq)$. If $B = \{1, 2, 4\} \subset A$, then the set $(B \times B) \cap \mathcal{R} = \{(1, 1), (2, 2), (4, 4), (1, 2), (1, 4), (2, 4)\}$ is a partial order on B.

In general if $\mathcal{R}$ is a partial order on A, then for each subset B of A, $(B \times B) \cap \mathcal{R}$ makes B into a poset where the partial order on B is induced from $\mathcal{R}$.

We turn now to a special type of partial order.

Definition 7.16

If $(A, \mathcal{R})$ is a poset, we say that A is *totally ordered* (or, *linearly ordered*) if for all $x, y \in A$ either $x \mathcal{R} y$ or $y \mathcal{R} x$. In this case $\mathcal{R}$ is called a *total order* (or, a *linear order*).

EXAMPLE 7.40

a) On the set **N**, the relation $\mathcal{R}$ defined by $x \mathcal{R} y$ if $x \leq y$ is a total order.

b) The subset relation applied to $A = \mathcal{P}(\mathcal{U})$, where $\mathcal{U} = \{1, 2, 3\}$, is a partial, but not total, order: $\{1, 2\}, \{1, 3\} \in A$ but we have neither $\{1, 2\} \subseteq \{1, 3\}$ nor $\{1, 3\} \subseteq \{1, 2\}$.

c) The Hasse diagram in part (b) of Fig. 7.18 shows a total order. In Fig. 7.19(a) we have the directed graph for this total order — alongside its Hasse diagram in part (b).

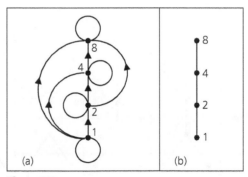

Figure 7.19

Could these notions of partial and total order ever arise in an industrial problem?

Say a toy manufacturer is about to market a new product and must include a set of instructions for its assembly. In order to assemble the new toy, there are seven tasks, denoted $A, B, C, \ldots, G$, that one must perform in the partial order given by the Hasse diagram of Fig. 7.20. Here we see, for example, that all of the tasks B, A, and E must be completed before we can work on task C. Since the set of instructions is to consist of a listing of these tasks, numbered $1, 2, 3, \ldots, 7$, how can the manufacturer write the listing and make sure that the partial order of the Hasse diagram is maintained?

What we are really asking for here is whether we can take the partial order $\mathcal{R}$, given by the Hasse diagram, and find a total order $\mathcal{T}$ on these tasks for which $\mathcal{R} \subseteq \mathcal{T}$. The answer is yes, and the technique that we need is known as *topological sorting*.

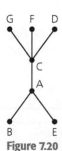

Figure 7.20

Topological Sorting Algorithm

(for a partial order $\mathcal{R}$ on a set A with $|A| = n$)

Step 1: Set $k = 1$. Let H_1 be the Hasse diagram of the partial order.

Step 2: Select a vertex v_k in H_k such that no (implicitly directed) edge in H_k starts at v_k.

Step 3: If $k = n$, the process is completed and we have a total order

$$\mathcal{T}: v_n < v_{n-1} < \cdots < v_2 < v_1$$

that contains $\mathcal{R}$.

If $k < n$, then remove from H_k the vertex v_k and all (implicitly directed) edges of H_k that terminate at v_k. Call the result H_{k+1}. Increase k by 1 and return to step (2).

Here we have presented our algorithm as a precise list of instructions, with no concern about the particulars of the pseudocode used in earlier chapters and with no reference to its implementation in a particular computer language.

Before we apply this algorithm[†] to the problem at hand, we should observe the deliberate use of "a" before the word "vertex" in step (2). This implies that the selection need not be unique and that we can get several different total orders $\mathcal{T}$ containing $\mathcal{R}$. Also, in step (3), for vertices v_{i-1} where $2 \leq i \leq n$, the notation $v_i < v_{i-1}$ is used because it is more suggestive of "v_i before v_{i-1}" than is the notation $v_i \, \mathcal{T} \, v_{i-1}$.

In Fig. 7.21, we show the Hasse diagrams that evolve as we apply the topological sorting algorithm to the partial order in Fig. 7.20. Below each diagram, the total order is listed as it evolves.

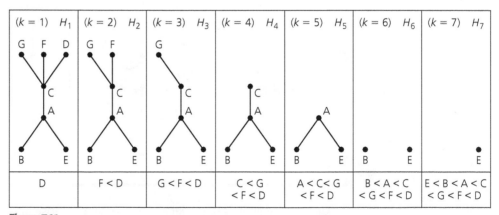

Figure 7.21

If the toy manufacturer writes the instructions in a list as 1-E, 2-B, 3-A, 4-C, 5-G, 6-F, 7-D, he or she will have a total order that preserves the partial order needed for correct assembly. This total order is one of 12 possible answers.

[†] Here we are only concerned with applying this algorithm. Hence we are assuming that it works and we shall not present a proof of that fact. Furthermore, we may operate similarly with other algorithms we encounter.

As is typical in discrete and combinatorial mathematics, this algorithm provides a procedure that reduces the size of the problem with each successive application.

The next example provides a situation where the number of distinct total orders for a particular partial order is determined.

EXAMPLE 7.41[†]

Let p, q be distinct primes. In part (a) of Fig. 7.22 we have the Hasse diagram for the partial order $\mathcal{R}$ of all positive-integer divisors of p^2q. Applying the topological sorting algorithm to this Hasse diagram, we find in Fig. 7.22(b) the five total orders $\mathcal{T}_i$, where $\mathcal{R} \subseteq \mathcal{T}_i$, for $1 \leq i \leq 5$.

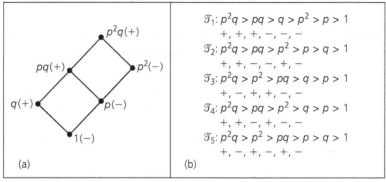

$\mathcal{T}_1: p^2q > pq > q > p^2 > p > 1$
  $+, +, +, -, -, -$
$\mathcal{T}_2: p^2q > pq > p^2 > p > q > 1$
  $+, +, -, -, +, -$
$\mathcal{T}_3: p^2q > p^2 > pq > q > p > 1$
  $+, -, +, +, -, -$
$\mathcal{T}_4: p^2q > pq > p^2 > q > p > 1$
  $+, +, -, +, -, -$
$\mathcal{T}_5: p^2q > p^2 > pq > p > q > 1$
  $+, -, +, -, +, -$

(a)   (b)

Figure 7.22

Now look at Fig. 7.22 again. This time focus on the three plus signs and three minus signs in part (a) of the figure and in the list below each total order in part (b). When we apply the topological sorting algorithm to the given partial order $\mathcal{R}$, step (2) of the algorithm implies that the first divisor selected is always p^2q. This accounts for the first plus sign in each $\mathcal{T}_i$, $1 \leq i \leq 5$. Continuing to apply the algorithm we get two more plus signs and the three minus signs.

Could there ever be more minus signs than plus signs in our corresponding list, as a total order is developed? For example, could we start with $+, -, -,$? If so, we have failed to correctly apply step (2) of the topological sorting algorithm — we should have recognized pq as the unique candidate to select after p^2q and p^2. In fact, for $0 \leq k \leq 2$, p^kq must be selected before p^k can be. Consequently, for each list of three plus signs and three minus signs, there is always at least as many plus signs as minus signs, as the list is read from left to right. Comparing now with the result in part (a) of Example 1.43, we see that the number of total orders for the given partial order is $5 = \frac{1}{3+1}\binom{2 \cdot 3}{3}$. Further, for $n \geq 1$, the topological sorting algorithm can be applied to the partial order of all positive divisors of $p^{n-1}q$ to yield $\frac{1}{n+1}\binom{2n}{n}$ total orders, another instance where the Catalan numbers arise.

In the topological sorting algorithm, we saw how the Hasse diagram was used in determining a total order containing a given poset $(A, \mathcal{R})$. This algorithm now prompts us to examine further properties of a partial order. At the start, particular emphasis will be given

---

[†]This example refers back to the optional material on Catalan numbers in Section 1.5. It may be skipped with no loss of continuity.

to a vertex like the vertex v_k in step (2) of the algorithm. The special property exhibited by such a vertex is now considered in the following.

Definition 7.17

If $(A, \mathcal{R})$ is a poset, then an element $x \in A$ is called a *maximal* element of A if for all $a \in A$, $a \neq x \Rightarrow x \mathcal{R} a$. An element $y \in A$ is called a *minimal* element of A if whenever $b \in A$ and $b \neq y$, then $b \mathcal{\not{R}} y$.

If we use the contrapositive of the first statement in Definition 7.17, then we can state that $x (\in A)$ is a maximal element if for each $a \in A$, $x \mathcal{R} a \Rightarrow x = a$. In a similar manner, $y \in A$ is a minimal element if for each $b \in A$, $b \mathcal{R} y \Rightarrow b = y$.

EXAMPLE 7.42

Let $\mathcal{U} = \{1, 2, 3\}$ and $A = \mathcal{P}(\mathcal{U})$.

    a) Let $\mathcal{R}$ be the subset relation on A. Then $\mathcal{U}$ is maximal and $\emptyset$ is minimal for the poset $(A, \subseteq)$.

    b) For B, the collection of proper subsets of $\{1, 2, 3\}$, let $\mathcal{R}$ be the subset relation on B. In the poset $(B, \subseteq)$, the sets $\{1, 2\}$, $\{1, 3\}$, and $\{2, 3\}$ are all maximal elements; $\emptyset$ is still the only minimal element.

EXAMPLE 7.43

With $\mathcal{R}$ the "less than or equal to" relation on the set $\mathbf{Z}$, we find that $(\mathbf{Z}, \leq)$ is a poset with neither a maximal nor a minimal element. The poset $(\mathbf{N}, \leq)$, however, has minimal element 0 but no maximal element.

EXAMPLE 7.44

When we look back at the partial orders in parts (b), (c), and (d) of Example 7.38, the following observations come to light.

    1) The partial order in part (b) has the unique maximal element 8 and the unique minimal element 1.

    2) Each of the four elements — 2, 3, 5, and 7 — is both a maximal element and a minimal element for the poset in part (c) of Example 7.38.

    3) In part (d) the elements 12 and 385 are both maximal. Each of the elements 2, 3, 5, 7, and 11 is a minimal element for this partial order.

Are there any conditions indicating when a poset must have a maximal or minimal element?

THEOREM 7.3

If $(A, \mathcal{R})$ is a poset and A is finite, then A has both a maximal and a minimal element.

Proof: Let $a_1 \in A$. If there is no element $a \in A$ where $a \neq a_1$ and $a_1 \mathcal{R} a$, then a_1 is maximal. Otherwise there is an element $a_2 \in A$ with $a_2 \neq a_1$ and $a_1 \mathcal{R} a_2$. If no element $a \in A$, $a \neq a_2$, satisfies $a_2 \mathcal{R} a$, then a_2 is maximal. Otherwise we can find $a_3 \in A$ so that $a_3 \neq a_2$, $a_3 \neq a_1$ (Why?) while $a_1 \mathcal{R} a_2$ and $a_2 \mathcal{R} a_3$. Continuing in this manner, since A is finite, we get to an element $a_n \in A$ with $a_n \mathcal{\not{R}} a$ for all $a \in A$ where $a \neq a_n$, so a_n is maximal.

The proof for a minimal element follows in a similar way.

Returning now to the topological sorting algorithm, we see that in each iteration of step (2) of the algorithm, we are selecting a maximal element from the original poset $(A, \mathcal{R})$, or a poset of the form $(B, \mathcal{R}')$ where $\emptyset \neq B \subset A$ and $\mathcal{R}' = (B \times B) \cap \mathcal{R}$. At least one such element exists (in each iteration) by virtue of Theorem 7.3. Then in the second part of step (3), if x is the maximal element selected [in step (2)], we remove from the present poset all elements of the form (a, x). This results in a smaller poset.

We turn now to the study of some additional concepts involving posets.

Definition 7.18

If $(A, \mathcal{R})$ is a poset, then an element $x \in A$ is called a *least* element if $x \mathcal{R} a$ for all $a \in A$. Element $y \in A$ is called a *greatest* element if $a \mathcal{R} y$ for all $a \in A$.

EXAMPLE 7.45

Let $\mathcal{U} = \{1, 2, 3\}$, and let $\mathcal{R}$ be the subset relation.

a) With $A = \mathcal{P}(\mathcal{U})$, the poset $(A, \subseteq)$ has $\emptyset$ as a least element and $\mathcal{U}$ as a greatest element.

b) For $B =$ the collection of nonempty subsets of $\mathcal{U}$, the poset $(B, \subseteq)$ has $\mathcal{U}$ as a greatest element. There is no least element here, but there are three minimal elements.

EXAMPLE 7.46

For the partial orders in Example 7.38, we find that

1) The partial order in part (b) has a greatest element 8 and a least element 1.

2) There is no greatest element or least element for the poset in part (c).

3) No greatest element or least element exists for the partial order in part (d).

We have seen that it is possible for a poset to have several maximal and minimal elements. What about least and greatest elements?

THEOREM 7.4

If the poset $(A, \mathcal{R})$ has a greatest (least) element, then that element is unique.

Proof: Suppose that $x, y \in A$ and that both are greatest elements. Since x is a greatest element, $y \mathcal{R} x$. Likewise, $x \mathcal{R} y$ because y is a greatest element. As $\mathcal{R}$ is antisymmetric, it follows that $x = y$.

The proof for the least element is similar.

Definition 7.19

Let $(A, \mathcal{R})$ be a poset with $B \subseteq A$. An element $x \in A$ is called a *lower bound* of B if $x \mathcal{R} b$ for all $b \in B$. Likewise, an element $y \in A$ is called an *upper bound* of B if $b \mathcal{R} y$ for all $b \in B$.

An element $x' \in A$ is called a *greatest lower bound* (glb) of B if it is a lower bound of B and if for all other lower bounds x'' of B we have $x'' \mathcal{R} x'$. Similarly $y' \in A$ is a *least upper bound* (lub) of B if it is an upper bound of B and if $y' \mathcal{R} y''$ for all other upper bounds y'' of B.

EXAMPLE 7.47

Let $\mathcal{U} = \{1, 2, 3, 4\}$, with $A = \mathcal{P}(\mathcal{U})$, and let $\mathcal{R}$ be the subset relation on A. If $B = \{\{1\}, \{2\}, \{1, 2\}\}$, then $\{1, 2\}$, $\{1, 2, 3\}$, $\{1, 2, 4\}$, and $\{1, 2, 3, 4\}$ are all upper bounds for

B (in $(A, \mathcal{R})$), whereas $\{1, 2\}$ is a least upper bound (and is in B). Meanwhile, a greatest lower bound for B is $\emptyset$, which is not in B.

EXAMPLE 7.48

Let $\mathcal{R}$ be the "less than or equal to" relation for the poset $(A, \mathcal{R})$.

a) If $A = \mathbf{R}$ and $B = [0, 1]$, then B has glb 0 and lub 1. Note that $0, 1 \in B$. For $C = (0, 1]$, C has glb 0 and lub 1, and $1 \in C$ but $0 \notin C$.

b) Keeping $A = \mathbf{R}$, let $B = \{q \in \mathbf{Q} | q^2 < 2\}$. Then B has $\sqrt{2}$ as a lub and $-\sqrt{2}$ as a glb, and neither of these real numbers is in B.

c) Now let $A = \mathbf{Q}$, with B as in part (b). Here B has no lub or glb.

These examples lead us to the following result.

THEOREM 7.5

If $(A, \mathcal{R})$ is a poset and $B \subseteq A$, then B has at most one lub (glb).

Proof: We leave the proof to the reader.

We close this section with one last ordered structure.

Definition 7.20

The poset $(A, \mathcal{R})$ is called a *lattice* if for all $x, y \in A$ the elements lub$\{x, y\}$ and glb$\{x, y\}$ both exist in A.

EXAMPLE 7.49

For $A = \mathbf{N}$ and $x, y \in \mathbf{N}$, define $x \, \mathcal{R} \, y$ by $x \le y$. Then lub$\{x, y\} = \max\{x, y\}$, glb$\{x, y\} = \min\{x, y\}$, and $(\mathbf{N}, \le)$ is a lattice.

EXAMPLE 7.50

For the poset in Example 7.45(a), if $S, T \subseteq \mathcal{U}$, with lub$\{S, T\} = S \cup T$ and glb$\{S, T\} = S \cap T$, then $(\mathcal{P}(\mathcal{U}), \subseteq)$ is a lattice.

EXAMPLE 7.51

Consider the poset in Example 7.38(d). Here we find, for example, that

$$\text{lub}\{2, 3\} = 6, \quad \text{lub}\{3, 6\} = 6, \quad \text{lub}\{5, 7\} = 35, \quad \text{lub}\{7, 11\} = 385, \quad \text{lub}\{11, 35\} = 385,$$

and

$$\text{glb}\{3, 6\} = 3, \quad \text{glb}\{2, 12\} = 2, \quad \text{glb}\{35, 385\} = 35.$$

However, even though lub$\{2, 3\}$ exists, there is no glb for the elements 2 and 3. In addition, we are also lacking (among other considerations) glb$\{5, 7\}$, glb$\{11, 35\}$, glb$\{3, 35\}$, and lub$\{3, 35\}$. Consequently, this partial order is not a lattice.

EXERCISES 7.3

1. Draw the Hasse diagram for the poset $(\mathcal{P}(\mathcal{U}), \subseteq)$, where $\mathcal{U} = \{1, 2, 3, 4\}$.

2. Let $A = \{1, 2, 3, 6, 9, 18\}$, and define $\mathcal{R}$ on A by $x \, \mathcal{R} \, y$ if $x | y$. Draw the Hasse diagram for the poset $(A, \mathcal{R})$.

3. Let $(A, \mathcal{R}_1)$, $(B, \mathcal{R}_2)$ be two posets. On $A \times B$, define relation $\mathcal{R}$ by $(a, b) \, \mathcal{R} \, (x, y)$ if $a \, \mathcal{R}_1 \, x$ and $b \, \mathcal{R}_2 \, y$. Prove that $\mathcal{R}$ is a partial order.

4. If $\mathcal{R}_1$, $\mathcal{R}_2$ in Exercise 3 are total orders, is $\mathcal{R}$ a total order?

5. Topologically sort the Hasse diagram in part (a) of Example 7.38.

6. For $A = \{a, b, c, d, e\}$, the Hasse diagram for the poset $(A, \mathcal{R})$ is shown in Fig. 7.23. (a) Determine the relation matrix for $\mathcal{R}$. (b) Construct the directed graph G (on A) that is associated with $\mathcal{R}$. (c) Topologically sort the poset $(A, \mathcal{R})$.

7. The directed graph G for a relation $\mathcal{R}$ on set $A = \{1, 2, 3, 4\}$ is shown in Fig. 7.24. (a) Verify that $(A, \mathcal{R})$ is a poset and find its Hasse diagram. (b) Topologically sort $(A, \mathcal{R})$. (c) How many more directed edges are needed in Fig. 7.24 to extend $(A, \mathcal{R})$ to a total order?

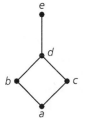

Figure 7.23

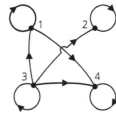

Figure 7.24

8. Prove that if a poset $(A, \mathcal{R})$ has a least element, it is unique.

9. Prove Theorem 7.5.

10. Give an example of a poset with four maximal elements but no greatest element.

11. If $(A, \mathcal{R})$ is a poset but not a total order, and $\emptyset \neq B \subset A$, does it follow that $(B \times B) \cap \mathcal{R}$ makes B into a poset but not a total order?

12. If $\mathcal{R}$ is a relation on A, and G is the associated directed graph, how can one recognize from G that $(A, \mathcal{R})$ is a total order?

13. If G is the directed graph for a relation $\mathcal{R}$ on A, with $|A| = n$, and $(A, \mathcal{R})$ is a total order, how many edges (including loops) are there in G?

14. Let $M(\mathcal{R})$ be the relation matrix for relation $\mathcal{R}$ on A, with $|A| = n$. If $(A, \mathcal{R})$ is a total order, how many 1's appear in $M(\mathcal{R})$?

15. a) Describe the structure of the Hasse diagram for a totally ordered poset $(A, \mathcal{R})$, where $|A| = n \geq 1$.

b) For a set A where $|A| = n \geq 1$, how many relations on A are total orders?

16. a) For $A = \{a_1, a_2, \ldots, a_n\}$, let $(A, \mathcal{R})$ be a poset. If $M(\mathcal{R})$ is the corresponding relation matrix, how can we recognize a maximal or minimal element of the poset from $M(\mathcal{R})$?

b) How can one recognize the existence of a greatest or least element in $(A, \mathcal{R})$ from the relation matrix $M(\mathcal{R})$?

17. Let $\mathcal{U} = \{1, 2, 3, 4\}$, with $A = \mathcal{P}(\mathcal{U})$, and let $\mathcal{R}$ be the subset relation on A. For each of the following subsets B (of A), determine the lub and glb of B.

a) $B = \{\{1\}, \{2\}\}$

b) $B = \{\{1\}, \{2\}, \{3\}, \{1, 2\}\}$

c) $B = \{\emptyset, \{1\}, \{2\}, \{1, 2\}\}$

d) $B = \{\{1\}, \{1, 2\}, \{1, 3\}, \{1, 2, 3\}\}$

e) $B = \{\{1\}, \{2\}, \{3\}, \{1, 2\}, \{1, 3\}, \{2, 3\}\}$

18. Let $\mathcal{U} = \{1, 2, 3, 4, 5, 6, 7\}$, with $A = \mathcal{P}(\mathcal{U})$, and let $\mathcal{R}$ be the subset relation on A. For $B = \{\{1\}, \{2\}, \{2, 3\}\} \subseteq A$, determine each of the following.

a) The number of upper bounds of B that contain (i) three elements of $\mathcal{U}$; (ii) four elements of $\mathcal{U}$; (iii) five elements of $\mathcal{U}$

b) The number of upper bounds that exist for B

c) The lub for B

d) The number of lower bounds that exist for B

e) The glb for B

19. Define the relation $\mathcal{R}$ on the set $\mathbf{Z}$ by $a \mathcal{R} b$ if $a - b$ is a nonnegative even integer. Verify that $\mathcal{R}$ defines a partial order for $\mathbf{Z}$. Is this partial order a total order?

20. For $X = \{0, 1\}$, let $A = X \times X$. Define the relation $\mathcal{R}$ on A by $(a, b) \mathcal{R} (c, d)$ if (i) $a < c$; or (ii) $a = c$ and $b \leq d$. (a) Prove that $\mathcal{R}$ is a partial order for A. (b) Determine all minimal and maximal elements for this partial order. (c) Is there a least element? Is there a greatest element? (d) Is this partial order a total order?

21. Let $X = \{0, 1, 2\}$ and $A = X \times X$. Define the relation $\mathcal{R}$ on A as in Exercise 20. Answer the same questions posed in Exercise 20 for this relation $\mathcal{R}$ and set A.

22. For $n \in \mathbf{Z}^+$, let $X = \{0, 1, 2, \ldots, n - 1, n\}$ and $A = X \times X$. Define the relation $\mathcal{R}$ on A as in Exercise 20. Remember that each element in this total order $\mathcal{R}$ is an ordered pair whose components are themselves ordered pairs. How many such elements are there in $\mathcal{R}$?

23. Let $(A, \mathcal{R})$ be a poset. Prove or disprove each of the following statements.

a) If $(A, \mathcal{R})$ is a lattice, then it is a total order.

b) If $(A, \mathcal{R})$ is a total order, then it is a lattice.

24. If $(A, \mathcal{R})$ is a lattice, with A finite, prove that $(A, \mathcal{R})$ has a greatest element and a least element.

25. For $A = \{a, b, c, d, e, v, w, x, y, z\}$, consider the poset $(A, \mathcal{R})$ whose Hasse diagram is shown in Fig. 7.25. Find

a) glb$\{b, c\}$ **b)** glb$\{b, w\}$

c) glb$\{e, x\}$ **d)** lub$\{c, b\}$

e) lub$\{d, x\}$ **f)** lub$\{c, e\}$

g) lub$\{a, v\}$

Is $(A, \mathcal{R})$ a lattice? Is there a maximal element? a minimal element? a greatest element? a least element?

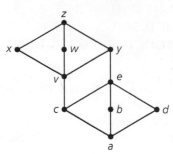

Figure 7.25

26. Given partial orders $(A, \mathcal{R})$ and $(B, \mathcal{S})$, a function f: $A \to B$ is called *order-preserving* if for all $x, y \in A$, $x \mathcal{R} y \Rightarrow$ $f(x) \mathcal{S} f(y)$. How many such order-preserving functions are there for each of the following, where $\mathcal{R}$, $\mathcal{S}$ both denote $\leq$ (the usual "less than or equal to" relation)?

a) $A = \{1, 2, 3, 4\}$, $B = \{1, 2\}$;

b) $A = \{1, \ldots, n\}$, $n \geq 1$, $B = \{1, 2\}$;

c) $A = \{a_1, a_2, \ldots, a_n\} \subset \mathbf{Z}^+$, $n \geq 1$, $a_1 < a_2 < \cdots < a_n$, $B = \{1, 2\}$;

d) $A = \{1, 2\}$, $B = \{1, 2, 3, 4\}$;

e) $A = \{1, 2\}$, $B = \{1, \ldots, n\}$, $n \geq 1$; and

f) $A = \{1, 2\}$, $B = \{b_1, b_2, \ldots, b_n\} \subset \mathbf{Z}^+$, $n \geq 1$, $b_1 < b_2 < \cdots < b_n$.

27. Let p, q, r, s be four distinct primes and $m, n, k, \ell \in \mathbf{Z}^+$. How many edges are there in the Hasse diagram of all positive divisors of (a) p^3; (b) p^m; (c) p^3q^2; (d) p^mq^n; (e) $p^3q^2r^4$; (f) $p^mq^nr^k$; (g) $p^3q^2r^4s^7$; and (h) $p^mq^nr^ks^\ell$?

28. Find the number of ways to totally order the partial order of all positive-integer divisors of (a) 24; (b) 75; and (c) 1701.

29. Let p, q be distinct primes and $k \in \mathbf{Z}^+$. If there are 429 ways to totally order the partial order of positive-integer divisors of p^kq, how many positive-integer divisors are there for this partial order?

30. For $m, n \in \mathbf{Z}^+$, let A be the set of all $m \times n$ $(0, 1)$-matrices. Prove that the "precedes" relation of Definition 7.11 makes A into a poset.

7.4
Equivalence Relations and Partitions

As we noted earlier in Definition 7.7, a relation $\mathcal{R}$ on a set A is an equivalence relation if it is reflexive, symmetric, and transitive. For any set $A \neq \emptyset$, the relation of equality is an equivalence relation on A, where two elements of A are related if they are identical; equality thus establishes the property of "sameness" among the elements of A.

If we consider the relation $\mathcal{R}$ on $\mathbf{Z}$ defined by $x \mathcal{R} y$ if $x - y$ is a multiple of 2, then $\mathcal{R}$ is an equivalence relation on $\mathbf{Z}$ where all even integers are related, as are all odd integers. Here, for example, we do not have $4 = 8$, but we do have $4 \mathcal{R} 8$, for we no longer care about the size of a number but are concerned with only two properties: "evenness" and "oddness." This relation splits $\mathbf{Z}$ into two subsets consisting of the odd and even integers: $\mathbf{Z} = \{\ldots, -3, -1, 1, 3, \ldots\} \cup \{\ldots, -4, -2, 0, 2, 4, \ldots\}$. This splitting up of $\mathbf{Z}$ is an example of a partition, a concept closely related to the equivalence relation. In this section we investigate this relationship and see how it helps us count the number of equivalence relations on a finite set.

Definition 7.21

Given a set A and index set I, let $\emptyset \neq A_i \subseteq A$ for each $i \in I$. Then $\{A_i\}_{i \in I}$ is a *partition* of A if

a) $A = \bigcup_{i \in I} A_i$ and   **b)** $A_i \cap A_j = \emptyset$,   for all $i, j \in I$ where $i \neq j$.

Each subset A_i is called a *cell* or *block* of the partition.

EXAMPLE 7.52

If $A = \{1, 2, 3, \ldots, 10\}$, then each of the following determines a partition of A:

a) $A_1 = \{1, 2, 3, 4, 5\}$, $A_2 = \{6, 7, 8, 9, 10\}$

b) $A_1 = \{1, 2, 3\}$, $A_2 = \{4, 6, 7, 9\}$, $A_3 = \{5, 8, 10\}$

c) $A_i = \{i, i + 5\}$, $1 \le i \le 5$

In these three examples we note how each element of A belongs to *exactly one* cell in each partition.

EXAMPLE 7.53

Let $A = \mathbf{R}$ and, for each $i \in \mathbf{Z}$, let $A_i = [i, i + 1)$. Then $\{A_i\}_{i \in \mathbf{Z}}$ is a partition of $\mathbf{R}$.

Now just how do partitions come into play with equivalence relations?

Definition 7.22

Let $\mathcal{R}$ be an equivalence relation on a set A. For each $x \in A$, the *equivalence class* of x, denoted $[x]$, is defined by $[x] = \{y \in A \,|\, y \,\mathcal{R}\, x\}$.

EXAMPLE 7.54

Define the relation $\mathcal{R}$ on $\mathbf{Z}$ by $x \,\mathcal{R}\, y$ if $4|(x - y)$. Since $\mathcal{R}$ is reflexive, symmetric, and transitive, it is an equivalence relation and we find that

$$[0] = \{\ldots, -8, -4, 0, 4, 8, 12, \ldots\} = \{4k | k \in \mathbf{Z}\}$$
$$[1] = \{\ldots, -7, -3, 1, 5, 9, 13, \ldots\} = \{4k + 1 | k \in \mathbf{Z}\}$$
$$[2] = \{\ldots, -6, -2, 2, 6, 10, 14, \ldots\} = \{4k + 2 | k \in \mathbf{Z}\}$$
$$[3] = \{\ldots, -5, -1, 3, 7, 11, 15, \ldots\} = \{4k + 3 | k \in \mathbf{Z}\}.$$

But what about $[n]$, where n is an integer other than 0, 1, 2, or 3? For example, what is $[6]$? We claim that $[6] = [2]$ and to prove this we use Definition 3.2 (for the equality of sets) as follows. If $x \in [6]$, then from Definition 7.22 we know that $x \,\mathcal{R}\, 6$. Here this means that 4 divides $(x - 6)$, so $x - 6 = 4k$ for some $k \in \mathbf{Z}$. But then $x - 6 = 4k \Rightarrow x - 2 = 4(k + 1) \Rightarrow 4$ divides $(x - 2) \Rightarrow x \,\mathcal{R}\, 2 \Rightarrow x \in [2]$, so $[6] \subseteq [2]$. For the opposite inclusion start with an element y in $[2]$. Then $y \in [2] \Rightarrow y \,\mathcal{R}\, 2 \Rightarrow 4$ divides $(y - 2) \Rightarrow y - 2 = 4l$ for some $l \in \mathbf{Z} \Rightarrow y - 6 = 4(l - 1)$, where $l - 1 \in \mathbf{Z} \Rightarrow 4$ divides $y - 6 \Rightarrow y \,\mathcal{R}\, 6 \Rightarrow y \in [6]$, so $[2] \subseteq [6]$. From the two inclusions it now follows that $[6] = [2]$, as claimed.

Further, we also find, for example, that $[2] = [-2] = [-6]$, $[51] = [3]$, and $[17] = [1]$. Most important, $\{[0], [1], [2], [3]\}$ provides a partition of $\mathbf{Z}$.

[*Note*: Here the index set for the partition is implicit. If, for instance, we let $A_0 = [0]$, $A_1 = [1]$, $A_2 = [2]$, and $A_3 = [3]$, then one possible index set I (as in Definition 7.21) is $\{0, 1, 2, 3\}$. When a collection of sets is called a partition (of a given set) but no index set is specified, the reader should realize that the situation is like the one given here — where the index set is implicit.]

EXAMPLE 7.55

Define the relation $\mathcal{R}$ on the set $\mathbf{Z}$ by $a \,\mathcal{R}\, b$ if $a^2 = b^2$ (or, $a = \pm b$). For all $a \in \mathbf{Z}$, we have $a^2 = a^2$ — so $a \,\mathcal{R}\, a$ and $\mathcal{R}$ is reflexive. Should $a, b \in \mathbf{Z}$ with $a \,\mathcal{R}\, b$, then $a^2 = b^2$ and it follows that $b^2 = a^2$, or $b \,\mathcal{R}\, a$. Consequently, relation $\mathcal{R}$ is symmetric. Finally, suppose that $a, b, c \in \mathbf{Z}$ with $a \,\mathcal{R}\, b$ and $b \,\mathcal{R}\, c$. Then $a^2 = b^2$ and $b^2 = c^2$, so $a^2 = c^2$ and $a \,\mathcal{R}\, c$. This makes the given relation transitive. Having established the three needed properties, we now know that $\mathcal{R}$ is an equivalence relation.

What can we say about the corresponding partition of $\mathbf{Z}$?

Here one finds that $[0] = \{0\}$, $[1] = [-1] = \{-1, 1\}$, $[2] = [-2] = \{-2, 2\}$, and, in general, for each $n \in \mathbf{Z}^+$, $[n] = [-n] = \{-n, n\}$. Furthermore, we have the *partition*

$$\mathbf{Z} = \bigcup_{n=0}^{\infty} [n] = \bigcup_{n \in \mathbf{N}} [n] = \{0\} \cup \left(\bigcup_{n=1}^{\infty} \{-n, n\} \right) = \{0\} \cup \left(\bigcup_{n \in \mathbf{Z}^+} \{-n, n\} \right).$$

These examples lead us to the following general situation.

THEOREM 7.6

If $\mathcal{R}$ is an equivalence relation on a set A, and x, $y \in A$, then (a) $x \in [x]$; (b) $x \mathcal{R} y$ if and only if $[x] = [y]$; and (c) $[x] = [y]$ or $[x] \cap [y] = \emptyset$.

Proof:

a) This result follows from the reflexive property of $\mathcal{R}$.

b) The proof here is somewhat reminiscent of what was done in Example 7.54.

   If $x \mathcal{R} y$, let $w \in [x]$. Then $w \mathcal{R} x$ and because $\mathcal{R}$ is transitive, $w \mathcal{R} y$. Hence $w \in [y]$ and $[x] \subseteq [y]$. With $\mathcal{R}$ symmetric, $x \mathcal{R} y \Rightarrow y \mathcal{R} x$. So if $t \in [y]$, then $t \mathcal{R} y$ and by the transitive property, $t \mathcal{R} x$. Hence $t \in [x]$ and $[y] \subseteq [x]$. Consequently, $[x] = [y]$. Conversely, let $[x] = [y]$. Since $x \in [x]$ by part (a), then $x \in [y]$ or $x \mathcal{R} y$.

c) This property tells us that two equivalence classes can be related in only one of two possible ways. Either they are identical or they are disjoint.

   We assume that $[x] \neq [y]$ and show how it then follows that $[x] \cap [y] = \emptyset$. If $[x] \cap [y] \neq \emptyset$, then let $v \in A$ with $v \in [x]$ and $v \in [y]$. Then $v \mathcal{R} x$, $v \mathcal{R} y$, and, since $\mathcal{R}$ is symmetric, $x \mathcal{R} v$. Now $(x \mathcal{R} v$ and $v \mathcal{R} y) \Rightarrow x \mathcal{R} y$, by the transitive property. Also $x \mathcal{R} y \Rightarrow [x] = [y]$ by part (b). This contradicts the assumption that $[x] \neq [y]$, so we reject the supposition that $[x] \cap [y] \neq \emptyset$, and the result follows.

Note that if $\mathcal{R}$ is an equivalence relation on A, then by parts (a) and (c) of Theorem 7.6 the distinct equivalence classes determined by $\mathcal{R}$ provide us with a partition of A.

EXAMPLE 7.56

a) If $A = \{1, 2, 3, 4, 5\}$ and $\mathcal{R} = \{(1, 1), (2, 2), (2, 3), (3, 2), (3, 3), (4, 4), (4, 5),$ $(5, 4), (5, 5)\}$, then $\mathcal{R}$ is an equivalence relation on A. Here $[1] = \{1\}$, $[2] = \{2, 3\} =$ $[3]$, $[4] = \{4, 5\} = [5]$, and $A = [1] \cup [2] \cup [4]$ with $[1] \cap [2] = \emptyset$, $[1] \cap [4] = \emptyset$, and $[2] \cap [4] = \emptyset$. So $\{[1], [2], [4]\}$ determines a partition of A.

b) Consider part (d) of Example 7.16 once again. We have $A = \{1, 2, 3, 4, 5, 6, 7\}$, $B = \{x, y, z\}$, and $f: A \rightarrow B$ is the onto function

$$f = \{(1, x), (2, z), (3, x), (4, y), (5, z), (6, y), (7, x)\}.$$

The relation $\mathcal{R}$ defined on A by $a \mathcal{R} b$ if $f(a) = f(b)$ was shown to be an equivalence relation. Here

$$f^{-1}(x) = \{1, 3, 7\} = [1] \, (= [3] = [7]),$$
$$f^{-1}(y) = \{4, 6\} = [4] \, (= [6]), \quad \text{and}$$
$$f^{-1}(z) = \{2, 5\} = [2] \, (= [5]).$$

With $A = [1] \cup [4] \cup [2] = f^{-1}(x) \cup f^{-1}(y) \cup f^{-1}(z)$, we see that $\{f^{-1}(x), f^{-1}(y), f^{-1}(z)\}$ determines a partition of A.

   In fact, for any nonempty sets A, B, if $f: A \rightarrow B$ is an onto function, then $A = \bigcup_{b \in B} f^{-1}(b)$ and $\{f^{-1}(b) | b \in B\}$ provides us with a partition of A.

EXAMPLE 7.57

In the programming language C++ a nonexecutable specification statement called the *union construct* allows two or more variables in a given program to refer to the same memory location.

For example, within a program the statements

```
union
{
  int a;
  int c;
  int p;
};
union
{
  int up;
  int down;
};
```

inform the C++ compiler that the integer variables a, c, and p will share one memory location while the integer variables *up* and *down* will share another. Here the set of all program variables is partitioned by the equivalence relation $\mathcal{R}$, where $v_1 \mathcal{R} v_2$ if v_1 and v_2 are program variables that share the same memory location.

EXAMPLE 7.58

Having seen examples of how an equivalence relation induces a partition of a set, we now go backward. If an equivalence relation $\mathcal{R}$ on $A = \{1, 2, 3, 4, 5, 6, 7\}$ induces the partition $A = \{1, 2\} \cup \{3\} \cup \{4, 5, 7\} \cup \{6\}$, what is $\mathcal{R}$?

Consider the cell $\{1, 2\}$ of the partition. This subset implies that $[1] = \{1, 2\} = [2]$, and so $(1, 1), (2, 2), (1, 2), (2, 1) \in \mathcal{R}$. (The first two ordered pairs are necessary for the reflexive property of $\mathcal{R}$; the others preserve symmetry.)

In like manner, the cell $\{4, 5, 7\}$ implies that under $\mathcal{R}$, $[4] = [5] = [7] = \{4, 5, 7\}$ and that, as an equivalence relation, $\mathcal{R}$ must contain $\{4, 5, 7\} \times \{4, 5, 7\}$. In fact,

$$\mathcal{R} = (\{1, 2\} \times \{1, 2\}) \cup (\{3\} \times \{3\}) \cup (\{4, 5, 7\} \times \{4, 5, 7\}) \cup (\{6\} \times \{6\}),$$

and

$$|\mathcal{R}| = 2^2 + 1^2 + 3^2 + 1^2 = 15.$$

The results in Examples 7.54, 7.55, 7.56, and 7.58 lead us to the following.

THEOREM 7.7

If A is a set, then

a) any equivalence relation $\mathcal{R}$ on A induces a partition of A, and

b) any partition of A gives rise to an equivalence relation $\mathcal{R}$ on A.

Proof: Part (a) follows from parts (a) and (c) of Theorem 7.6. For part (b), given a partition $\{A_i\}_{i \in I}$ of A, define relation $\mathcal{R}$ on A by $x \mathcal{R} y$, if x and y are in the same cell of the partition. We leave to the reader the details of verifying that $\mathcal{R}$ is an equivalence relation.

On the basis of this theorem and the examples we have examined, we state the next result. A proof for it is outlined in Exercise 16 at the end of the section.

THEOREM 7.8   For any set A, there is a one-to-one correspondence between the set of equivalence relations on A and the set of partitions of A.

We are primarily concerned with using this result for finite sets.

EXAMPLE 7.59

a) If $A = \{1, 2, 3, 4, 5, 6\}$, how many relations on A are equivalence relations?

We solve this problem by counting the partitions of A, realizing that a partition of A is a distribution of the (distinct) elements of A into identical containers, with no container left empty. From Section 5.3 we know, for example, that there are $S(6, 2)$ partitions of A into two identical nonempty containers. Using the Stirling numbers of the second kind, as the number of containers varies from 1 to 6, we have $\sum_{i=1}^{6} S(6, i) = 203$ different partitions of A. Consequently, there are 203 equivalence relations on A.

b) How many of the equivalence relations in part (a) satisfy $1, 2 \in [4]$?

Identifying 1, 2, and 4 as the "same" element under these equivalence relations, we count as in part (a) for the set $B = \{1, 3, 5, 6\}$ and find that there are $\sum_{i=1}^{4} S(4, i) = 15$ equivalence relations on A for which $[1] = [2] = [4]$.

We close by noting that if A is a finite set with $|A| = n$, then for all $n \le r \le n^2$, there is an equivalence relation $\mathcal{R}$ on A with $|\mathcal{R}| = r$ if and only if there exist $n_1, n_2, \ldots, n_k \in \mathbf{Z}^+$ with $\sum_{i=1}^{k} n_i = n$ and $\sum_{i=1}^{k} n_i^2 = r$.

EXERCISES 7.4

1. Determine whether each of the following collections of sets is a partition for the given set A. If the collection is not a partition, explain why it fails to be.

a) $A = \{1, 2, 3, 4, 5, 6, 7, 8\}$; $A_1 = \{4, 5, 6\}$, $A_2 = \{1, 8\}$, $A_3 = \{2, 3, 7\}$.

b) $A = \{a, b, c, d, e, f, g, h\}$; $A_1 = \{d, e\}$, $A_2 = \{a, c, d\}$, $A_3 = \{f, h\}$, $A_4 = \{b, g\}$.

2. Let $A = \{1, 2, 3, 4, 5, 6, 7, 8\}$. In how many ways can we partition A as $A_1 \cup A_2 \cup A_3$ with

a) $1, 2 \in A_1$, $3, 4 \in A_2$,   and $5, 6, 7 \in A_3$?

b) $1, 2 \in A_1$, $3, 4 \in A_2$, $5, 6 \in A_3$,   and $|A_1| = 3$?

c) $1, 2 \in A_1$, $3, 4 \in A_2$,   and $5, 6 \in A_3$?

3. If $A = \{1, 2, 3, 4, 5\}$ and $\mathcal{R}$ is the equivalence relation on A that induces the partition $A = \{1, 2\} \cup \{3, 4\} \cup \{5\}$, what is $\mathcal{R}$?

4. For $A = \{1, 2, 3, 4, 5, 6\}$, $\mathcal{R} = \{(1, 1), (1, 2), (2, 1), (2, 2), (3, 3), (4, 4), (4, 5), (5, 4), (5, 5), (6, 6)\}$ is an equivalence relation on A. (a) What are $[1]$, $[2]$, and $[3]$ under this equivalence relation? (b) What partition of A does $\mathcal{R}$ induce?

5. If $A = A_1 \cup A_2 \cup A_3$, where $A_1 = \{1, 2\}$, $A_2 = \{2, 3, 4\}$, and $A_3 = \{5\}$, define relation $\mathcal{R}$ on A by $x \mathcal{R} y$ if x and y are in the same subset A_i, for $1 \le i \le 3$. Is $\mathcal{R}$ an equivalence relation?

6. For $A = \mathbf{R}^2$, define $\mathcal{R}$ on A by $(x_1, y_1) \mathcal{R} (x_2, y_2)$ if $x_1 = x_2$.

a) Verify that $\mathcal{R}$ is an equivalence relation on A.

b) Describe geometrically the equivalence classes and partition of A induced by $\mathcal{R}$.

7. Let $A = \{1, 2, 3, 4, 5\} \times \{1, 2, 3, 4, 5\}$, and define $\mathcal{R}$ on A by $(x_1, y_1) \mathcal{R} (x_2, y_2)$ if $x_1 + y_1 = x_2 + y_2$.

a) Verify that $\mathcal{R}$ is an equivalence relation on A.

b) Determine the equivalence classes $[(1, 3)]$, $[(2, 4)]$, and $[(1, 1)]$.

c) Determine the partition of A induced by $\mathcal{R}$.

8. If $A = \{1, 2, 3, 4, 5, 6, 7\}$, define $\mathcal{R}$ on A by $(x, y) \in \mathcal{R}$ if $x - y$ is a multiple of 3.

a) Show that $\mathcal{R}$ is an equivalence relation on A.

b) Determine the equivalence classes and partition of A induced by $\mathcal{R}$.

9. For $A = \{(-4, -20), (-3, -9), (-2, -4), (-1, -11), (-1, -3), (1, 2), (1, 5), (2, 10), (2, 14), (3, 6), (4, 8), (4, 12)\}$, define the relation $\mathcal{R}$ on A by $(a, b) \mathcal{R} (c, d)$ if $ad = bc$.

a) Verify that $\mathcal{R}$ is an equivalence relation on A.

b) Find the equivalence classes $[(2, 14)]$, $[(-3, -9)]$, and $[(4, 8)]$.

c) How many cells are there in the partition of A induced by $\mathcal{R}$?

10. Let A be a nonempty set and fix the set B, where $B \subseteq A$. Define the relation $\mathcal{R}$ on $\mathcal{P}(A)$ by $X \mathcal{R} Y$, for $X, Y \subseteq A$, if $B \cap X = B \cap Y$.

    a) Verify that $\mathcal{R}$ is an equivalence relation on $\mathcal{P}(A)$.

    b) If $A = \{1, 2, 3\}$ and $B = \{1, 2\}$, find the partition of $\mathcal{P}(A)$ induced by $\mathcal{R}$.

    c) If $A = \{1, 2, 3, 4, 5\}$ and $B = \{1, 2, 3\}$, find $[X]$ if $X = \{1, 3, 5\}$.

    d) For $A = \{1, 2, 3, 4, 5\}$ and $B = \{1, 2, 3\}$, how many equivalence classes are in the partition induced by $\mathcal{R}$?

11. How many of the equivalence relations on $A = \{a, b, c, d, e, f\}$ have (a) exactly two equivalence classes of size 3? (b) exactly one equivalence class of size 3? (c) one equivalence class of size 4? (d) at least one equivalence class with three or more elements?

12. Let $A = \{v, w, x, y, z\}$. Determine the number of relations on A that are (a) reflexive and symmetric; (b) equivalence relations; (c) reflexive and symmetric but not transitive; (d) equivalence relations that determine exactly two equivalence classes; (e) equivalence relations where $w \in [x]$; (f) equivalence relations where $v, w \in [x]$; (g) equivalence relations where $w \in [x]$ and $y \in [z]$; and (h) equivalence relations where $w \in [x]$, $y \in [z]$, and $[x] \neq [z]$.

13. If $|A| = 30$ and the equivalence relation $\mathcal{R}$ on A partitions A into (disjoint) equivalence classes A_1, A_2, and A_3, where $|A_1| = |A_2| = |A_3|$, what is $|\mathcal{R}|$?

14. Let $A = \{1, 2, 3, 4, 5, 6, 7\}$. For each of the following values of r, determine an equivalence relation $\mathcal{R}$ on A with $|\mathcal{R}| = r$, or explain why no such relation exists. (a) $r = 6$; (b) $r = 7$; (c) $r = 8$; (d) $r = 9$; (e) $r = 11$; (f) $r = 22$; (g) $r = 23$; (h) $r = 30$; (i) $r = 31$.

15. Provide the details for the proof of part (b) of Theorem 7.7.

16. For any set $A \neq \emptyset$, let $P(A)$ denote the set of all partitions of A, and let $E(A)$ denote the set of all equivalence relations on A. Define the function $f: E(A) \rightarrow P(A)$ as follows: If $\mathcal{R}$ is an equivalence relation on A, then $f(\mathcal{R})$ is the partition of A induced by $\mathcal{R}$. Prove that f is one-to-one and onto, thus establishing Theorem 7.8.

17. Let $f: A \rightarrow B$. If $\{B_1, B_2, B_3, \ldots, B_n\}$ is a partition of B, prove that $\{f^{-1}(B_i) | 1 \leq i \leq n, f^{-1}(B_i) \neq \emptyset\}$ is a partition of A.

7.5
Finite State Machines: The Minimization Process

In Section 6.3 we encountered two finite state machines that performed the same task but had different numbers of internal states. (See Figs. 6.9 and 6.10.) The machine with the larger number of internal states contains *redundant* states — states that can be eliminated because other states will perform their functions. Since minimization of the number of states in a machine reduces its complexity and cost, we seek a process for transforming a given machine into one that has no redundant internal states. This process is known as the *minimization process*, and its development relies on the concepts of equivalence relation and partition.

    Starting with a given finite state machine $M = (S, \mathcal{I}, \mathcal{O}, \nu, \omega)$, we define the relation E_1 on S by $s_1 E_1 s_2$ if $\omega(s_1, x) = \omega(s_2, x)$, for all $x \in \mathcal{I}$. This relation E_1 is an equivalence relation on S, and it partitions S into subsets such that two states are in the same subset if they produce the same output for each $x \in \mathcal{I}$. Here the states s_1, s_2 are called *1-equivalent*.

    For each $k \in \mathbf{Z}^+$, we say that the states s_1, s_2 are *k-equivalent* if $\omega(s_1, x) = \omega(s_2, x)$ for all $x \in \mathcal{I}^k$. Here ω is the extension of the given output function to $S \times \mathcal{I}^*$. The relation of k-equivalence is also an equivalence relation on S; it partitions S into subsets of k-equivalent states. We write $s_1 E_k s_2$ to denote that s_1 and s_2 are k-equivalent.

    Finally, if $s_1, s_2 \in S$ and s_1, s_2 are k-equivalent for all $k \geq 1$, then we call s_1 and s_2 *equivalent* and write $s_1 E s_2$. When this happens, we find that if we keep s_1 in our machine, then s_2 will be redundant and can be removed. Hence our objective is to determine the partition of S induced by E and to select one state for each equivalence class. Then we shall have a minimal realization of the given machine.

To accomplish this, let us start with the following observations.

a) If two states in a machine are not 2-equivalent, could they possibly be 3-equivalent? (or k-equivalent, for $k \geq 4$?)

The answer is no. If $s_1, s_2 \in S$ and $s_1 \not\mathrel{E_2} s_2$ (that is, s_1 and s_2 are not 2-equivalent), then there is at least one string $xy \in \mathcal{I}^2$ such that $\omega(s_1, xy) = v_1 v_2 \neq w_1 w_2 = \omega(s_2, xy)$, where $v_1, v_2, w_1, w_2 \in \mathbb{O}$. So with regard to E_3, we find that $s_1 \not\mathrel{E_3} s_2$ because for any $z \in \mathcal{I}$, $\omega(s_1, xyz) = v_1 v_2 v_3 \neq w_1 w_2 w_3 = \omega(s_2, xyz)$.

In general, to find states that are $(k+1)$-equivalent, we look at states that are k-equivalent.

b) Now suppose that $s_1, s_2 \in S$ and $s_1 \mathrel{E_2} s_2$. We wish to determine whether $s_1 \mathrel{E_3} s_2$. That is, does $\omega(s_1, x_1 x_2 x_3) = \omega(s_2, x_1 x_2 x_3)$ for all strings $x_1 x_2 x_3 \in \mathcal{I}^3$? Consider what happens. First we get $\omega(s_1, x_1) = \omega(s_2, x_1)$, because $s_1 \mathrel{E_2} s_2 \Rightarrow s_1 \mathrel{E_1} s_2$. Then there is a transition to the states $\nu(s_1, x_1)$ and $\nu(s_2, x_1)$. Consequently, $\omega(s_1, x_1 x_2 x_3) = \omega(s_2, x_1 x_2 x_3)$ if $\omega(\nu(s_1, x_1), x_2 x_3) = \omega(\nu(s_2, x_1), x_2 x_3)$ [that is, if $\nu(s_1, x_1) \mathrel{E_2} \nu(s_2, x_1)$].

In general, for $s_1, s_2, \in S$, where $s_1 \mathrel{E_k} s_2$, we find that $s_1 \mathrel{E_{k+1}} s_2$ if (and only if) $\nu(s_1, x) \mathrel{E_k} \nu(s_2, x)$ for all $x \in \mathcal{I}$.

With these observations to guide us, we now present an algorithm for the minimization of a finite state machine M.

Step 1: Set $k = 1$. We determine the states that are 1-equivalent by examining the rows in the state table for M. For $s_1, s_2 \in S$ it follows that $s_1 \mathrel{E_1} s_2$ when s_1, s_2 have the same output rows.

Let P_1 be the partition of S induced by E_1.

Step 2: Having determined P_k, we obtain P_{k+1} by noting that if $s_1 \mathrel{E_k} s_2$, then $s_1 \mathrel{E_{k+1}} s_2$ when $\nu(s_1, x) \mathrel{E_k} \nu(s_2, x)$ for all $x \in \mathcal{I}$. We have $s_1 \mathrel{E_k} s_2$ if s_1, s_2 are in the same cell of the partition P_k. Likewise, $\nu(s_1, x) \mathrel{E_k} \nu(s_2, x)$ for each $x \in \mathcal{I}$, if $\nu(s_1, x)$ and $\nu(s_2, x)$ are in the same cell of the partition P_k. In this way P_{k+1} is obtained from P_k.

Step 3: If $P_{k+1} = P_k$, the process is complete. We select one state from each equivalence class and these states yield a minimal realization of M.

If $P_{k+1} \neq P_k$, we increase k by 1 and return to step (2).

We illustrate the algorithm in the following example.

EXAMPLE 7.60

With $\mathcal{I} = \mathbb{O} = \{0, 1\}$, let M be given by the state table shown in Table 7.1. Looking at the output rows, we see that s_3 and s_4 are 1-equivalent, as are s_2, s_5, and s_6. Here E_1 partitions S as follows:

$$P_1: \{s_1\}, \{s_2, s_5, s_6\}, \{s_3, s_4\}.$$

For each $s \in S$ and each $k \in \mathbf{Z}^+$, $s \mathrel{E_k} s$, so as we continue this process to determine P_2, we shall not concern ourselves with equivalence classes of only one state.

Since $s_3 \mathrel{E_1} s_4$, there is a chance that we could have $s_3 \mathrel{E_2} s_4$. Here $\nu(s_3, 0) = s_2$, $\nu(s_4, 0) = s_5$ with $s_2 \mathrel{E_1} s_5$, and $\nu(s_3, 1) = s_4$, $\nu(s_4, 1) = s_3$ with $s_4 \mathrel{E_1} s_3$. Hence $\nu(s_3, x) \mathrel{E_1} \nu(s_4, x)$, for all $x \in \mathcal{I}$, and $s_3 \mathrel{E_2} s_4$. Similarly, $\nu(s_2, 0) = s_5$, $\nu(s_5, 0) = s_2$ with $s_5 \mathrel{E_1} s_2$, and $\nu(s_2, 1) = s_2$, $\nu(s_5, 1) = s_5$ with $s_2 \mathrel{E_1} s_5$. Thus $s_2 \mathrel{E_2} s_5$. Finally, $\nu(s_5, 0) = s_2$ and

$\nu(s_6, 0) = s_1$, but $s_2 \not{E}_1 s_1$, so $s_5 \not{E}_2 s_6$. (Why don't we investigate the possibility of $s_2 E_2 s_6$?) Equivalence relation E_2 partitions S as follows:

$$P_2: \{s_1\}, \{s_2, s_5\}, \{s_3, s_4\}, \{s_6\}.$$

Since $P_2 \neq P_1$, we continue the process to get P_3. In determining whether $s_2 E_3 s_5$, we see that $\nu(s_2, 0) = s_5$, $\nu(s_5, 0) = s_2$, and $s_5 E_2 s_2$. Also, $\nu(s_2, 1) = s_2$, $\nu(s_5, 1) = s_5$, and $s_2 E_2 s_5$. With $\nu(s_2, x) E_2 \nu(s_5, x)$ for all $x \in \mathcal{I}$, we have $s_2 E_3 s_5$. For s_3, s_4, $(\nu(s_3, 0) = s_2)$ E_2 $(s_5 = \nu(s_4, 0))$ and $(\nu(s_3, 1) = s_4) E_2 (s_3 = \nu(s_4, 1))$, so $s_3 E_3 s_4$ and E_3 induces the partition $P_3: \{s_1\}, \{s_2, s_5\}, \{s_3, s_4\}, \{s_6\}$.

Table 7.1

	ν		ω	
	0	1	0	1
s_1	s_4	s_3	0	1
s_2	s_5	s_2	1	0
s_3	s_2	s_4	0	0
s_4	s_5	s_3	0	0
s_5	s_2	s_5	1	0
s_6	s_1	s_6	1	0

Table 7.2

	ν		ω	
	0	1	0	1
s_1	s_3	s_3	0	1
s_2	s_2	s_2	1	0
s_3	s_2	s_3	0	0
s_6	s_1	s_6	1	0

Now $P_3 = P_2$ so the process is completed, as indicated in step (3) of the algorithm. We find that s_5 and s_4 may be regarded as redundant states. Removing them from the table, and replacing all further occurrences of them by s_2 and s_3, respectively, we arrive at Table 7.2. This is a minimal machine that performs the same tasks as the machine given in Table 7.1.

If we do not want states that skip a subscript, we can always relabel the states in this minimal machine. Here we would have $s_1, s_2, s_3, s_4 \ (= s_6)$, but this s_4 is not the same s_4 we started with in Table 7.1.

You may be wondering how we knew that we could stop the process when $P_3 = P_2$. For after all, couldn't it happen that perhaps $P_4 \neq P_3$, or that $P_4 = P_3$ but $P_5 \neq P_4$? To prove that this never occurs, we define the following idea.

Definition 7.23

If P_1, P_2 are partitions of a set A, then P_2 is called a *refinement* of P_1, and we write $P_2 \leq P_1$, if every cell of P_2 is contained in a cell of P_1. When $P_2 \leq P_1$ and $P_2 \neq P_1$ we write $P_2 < P_1$. This occurs when at least one cell in P_2 is properly contained in a cell in P_1.

In the minimization process of Example 7.60, we had $P_3 = P_2 < P_1$. Whenever we apply the algorithm, as we get P_{k+1} from P_k, we always find that $P_{k+1} \leq P_k$, because $(k + 1)$-equivalence implies k-equivalence. So each successive partition refines the preceding partition.

THEOREM 7.9

In applying the minimization process, if $k \geq 1$ and P_k and P_{k+1} are partitions with $P_{k+1} = P_k$, then $P_{r+1} = P_r$ for all $r \geq k + 1$.

Proof: If not, let $r \ (\geq k + 1)$ be the smallest subscript such that $P_{r+1} \neq P_r$. Then $P_{r+1} < P_r$, so there exist $s_1, s_2 \in S$ with $s_1 E_r s_2$ but $s_1 \not{E}_{r+1} s_2$. But $s_1 E_r s_2 \Rightarrow \nu(s_1, x) E_{r-1} \nu(s_2, x)$, for all $x \in \mathscr{I}$, and with $P_r = P_{r-1}$, we then find that $\nu(s_1, x) \; E_r \; \nu(s_2, x)$, for all $x \in \mathscr{I}$, so $s_1 \; E_{r+1} \; s_2$. Consequently, $P_{r+1} = P_r$.

We close this section with the following related idea. Let M be a finite state machine with $s_1, s_2 \in S$, and s_1, s_2 not equivalent. If $s_1 \; \not{E}_1 \; s_2$, then these states produce different output rows in the state table for M. In this case it is easy to find an $x \in \mathscr{I}$ such that $\omega(s_1, x) \neq \omega(s_2, x)$, and this distinguishes these nonequivalent states. Otherwise, s_1 and s_2 produce the same output rows in the table but there is a smallest integer $k \geq 1$ such that $s_1 \; E_k \; s_2$ but $s_1 \; \not{E}_{k+1} \; s_2$. Now if we are to distinguish these states, we need to find a string $x = x_1 x_2 \cdots x_k x_{k+1} \in \mathscr{I}^{k+1}$ such that $\omega(s_1, x) \neq \omega(s_2, x)$, even though $\omega(s_1, x_1 x_2 \cdots x_k) = \omega(s_2, x_1 x_2 \cdots x_k)$. Such a string x is called a *distinguishing string* for the states s_1 and s_2. There may be more than one such string, but each has the same (minimal) length $k + 1$.

Before we try to find a distinguishing string for two nonequivalent states in a specific finite state machine, let us examine the major idea at play here. So suppose that $s_1, s_2 \in S$ and that for some (fixed) $k \in \mathbf{Z}^+$ we have $s_1 \; E_k \; s_2$ but $s_1 \; \not{E}_{k+1} \; s_2$. What can we conclude?

We find that

$$s_1 \; \not{E}_{k+1} \; s_2 \Rightarrow \exists x_1 \in \mathscr{I} \; [\nu(s_1, x_1) \; \not{E}_k \; \nu(s_2, x_1)]$$

$$\Rightarrow \exists x_1 \in \mathscr{I} \; \exists x_2 \in \mathscr{I} \; [\nu(\nu(s_1, x_1), x_2) \; \not{E}_{k-1} \; \nu(\nu(s_2, x_1), x_2)],$$

$$\text{or} \quad \exists x_1 \in \mathscr{I} \; \exists x_2 \in \mathscr{I} \; [\nu(s_1, x_1 x_2) \; \not{E}_{k-1} \; \nu(s_2, x_1 x_2)]$$

$$\Rightarrow \exists x_1, x_2, x_3 \in \mathscr{I} \; [\nu(s_1, x_1 x_2 x_3) \; \not{E}_{k-2} \; \nu(s_2, x_1 x_2 x_3)]$$

$$\Rightarrow \cdots$$

$$\Rightarrow \exists x_1, x_2, \ldots, x_i \in \mathscr{I} \; [\nu(s_1, x_1 x_2 \cdots x_i) \; \not{E}_{k+1-i} \; \nu(s_2, x_1 x_2 \cdots x_i)]$$

$$\Rightarrow \cdots$$

$$\Rightarrow \exists x_1, x_2, \ldots, x_k \in \mathscr{I} \; [\nu(s_1, x_1 x_2 \cdots x_k) \; \not{E}_1 \; \nu(s_2, x_1 x_2 \cdots x_k)].$$

This last statement about the states $\nu(s_1, x_1 x_2 \cdots x_k)$, $\nu(s_2, x_1 x_2 \cdots x_k)$ not being 1-equivalent implies that we can find $x_{k+1} \in \mathscr{I}$ where

$$\omega(\nu(s_1, x_1 x_2 \cdots x_k), x_{k+1}) \neq \omega(\nu(s_2, x_1 x_2 \cdots x_k), x_{k+1}). \tag{1}$$

That is, these *single* output symbols from $\mathbb{O}$ are different.

The result denoted by Eq. (1) also implies that

$$\omega(s_1, x) = \omega(s_1, x_1 x_2 \cdots x_k x_{k+1}) \neq \omega(s_2, x_1 x_2 \cdots x_k x_{k+1}) = \omega(s_2, x).$$

In this case we have two output strings of length $k + 1$ that agree for the first k symbols and differ in the $(k + 1)$st symbol.

We shall use the preceding observations, together with the partitions $P_1, P_2, \ldots, P_k$, P_{k+1} of the minimization process, in order to deal with the following example.

EXAMPLE 7.61

From Example 7.60 we have the partitions shown below. Here $s_2 \; E_1 \; s_6$, but $s_2 \; \not{E}_2 \; s_6$. So we seek an input string x of length 2 such that $\omega(s_2, x) \neq \omega(s_6, x)$.

1) We start at P_2, where for s_2, s_6, we find that $\nu(s_2, 0) = s_5$ and $\nu(s_6, 0) = s_1$ are in different cells of P_1 — that is,

$$s_5 = \nu(s_2, 0) \; \not{E}_1 \; \nu(s_6, 0) = s_1.$$

[The input 0 and output 1 (for $\omega(s_2, 0) = 1 = \omega(s_6, 0)$) provide the labels for the arrows going from the cells of P_2 to those of P_1.]

$$P_2: \{s_1\}, \{s_2, s_5\}, \{s_3, s_4\}, \{s_6\}$$

$$0, 1 \qquad\qquad\qquad 0, 1$$

$$P_1: \{s_1\}, \{s_2, s_5, s_6\}, \{s_3, s_4\}$$

$$0, 0 \qquad 0, 1$$

2) Working with s_1 and s_5 in the partition P_1 we see that

$$\omega(v(s_2, 0), 0) = \omega(s_5, 0) = 1 \neq 0 = \omega(s_1, 0) = \omega(v(s_6, 0), 0).$$

3) Hence $x = 00$ is a minimal distinguishing string for s_2 and s_6 because $\omega(s_2, 00) = 11 \neq 10 = \omega(s_6, 00)$.

EXAMPLE 7.62

Applying the minimization process to the machine given by the state table in part (a) of Table 7.3, we obtain the partitions in part (b) of the table. (Here $P_4 = P_3$.) We find that the states s_1 and s_4 are 2-equivalent but not 3-equivalent. To construct a minimal distinguishing string for these two states, we proceed as follows:

1) Since $s_1 \not\equiv_3 s_4$, we use partitions P_3 and P_2 to find $x_1 \in \mathscr{I}$ (namely, $x_1 = 1$) so that

$$(v(s_1, 1) = s_2) \not\equiv_2 (s_5 = v(s_4, 1)).$$

2) Then $v(s_1, 1) \not\equiv_2 v(s_4, 1) \Rightarrow \exists x_2 \in \mathscr{I}$ (here $x_2 = 1$) with $(v(s_1, 1), 1) \not\equiv_1 (v(s_4, 1), 1)$, or $v(s_1, 11) \not\equiv_1 v(s_4, 11)$. We used the partitions P_2 and P_1 to obtain $x_2 = 1$.

3) Now we use the partition P_1 where we find that for $x_3 = 1 \in \mathscr{I}$,

$$\omega(v(s_1, 11), 1) = 0 \neq 1 = \omega(v(s_4, 11), 1) \quad \text{or}$$
$$\omega(s_1, 111) = 100 \neq 101 = \omega(s_4, 111).$$

In part (b) of Table 7.3, we see how we arrived at the minimal distinguishing string $x = 111$ for these states. (Also note how this part of the table indicates that 11 is a minimal distinguishing string for the states s_2 and s_5, which are 1-equivalent but not 2-equivalent.)

Table 7.3

	v		ω		$P_3: \{s_1, s_3\}, \{s_2\}, \{s_4\}, \{s_5\}$
	0	1	0	1	$1, 1 \qquad\qquad 1, 1$
s_1	s_4	s_2	0	1	$P_2: \{s_1, s_3, s_4\}, \{s_2\}, \{s_5\}$
s_2	s_5	s_2	0	0	$1, 0 \quad 1, 0$
s_3	s_4	s_2	0	1	$P_1: \{s_1, s_3, s_4\}, \{s_2, s_5\}$
s_4	s_3	s_5	0	1	$1, 1 \downarrow \qquad \downarrow 1, 0$
s_5	s_2	s_3	0	0	

(a)	(b)

A great deal more can be done with finite state machines. Among other omissions, we have avoided offering any rigorous explanation or proof of why the minimization process works. The interested reader should consult the chapter references for more on this topic.

1. Apply the minimization process to each machine in Table 7.4.

Table 7.4

	v		ω	
	0	1	0	1
s_1	s_4	s_1	0	1
s_2	s_3	s_3	1	0
s_3	s_1	s_4	1	0
s_4	s_1	s_3	0	1
s_5	s_3	s_3	1	0

(a)

	v		ω	
	0	1	0	1
s_1	s_6	s_3	0	0
s_2	s_5	s_4	0	1
s_3	s_6	s_2	1	1
s_4	s_4	s_3	1	0
s_5	s_2	s_4	0	1
s_6	s_4	s_6	0	0

(b)

	v		ω	
	0	1	0	1
s_1	s_6	s_3	0	0
s_2	s_3	s_1	0	0
s_3	s_2	s_4	0	0
s_4	s_7	s_4	0	0
s_5	s_6	s_7	0	0
s_6	s_5	s_2	1	0
s_7	s_4	s_1	0	0

(c)

2. For the machine in Table 7.4(c), find a (minimal) distinguishing string for each given pair of states: (a) s_1, s_5; (b) s_2, s_3; (c) s_5, s_7.

3. Let M be the finite state machine given in the state diagram shown in Fig. 7.26.

a) Minimize machine M.

b) Find a (minimal) distinguishing string for each given pair of states: (i) s_3, s_6; (ii) s_3, s_4; and (iii) s_1, s_2.

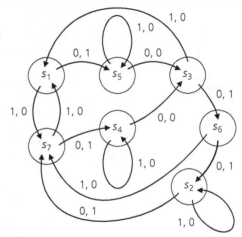

Figure 7.26

7.6
Summary and Historical Review

Once again the relation concept surfaces. In Chapter 5 this idea was introduced as a generalization of the function. Here in Chapter 7 we concentrated on relations and the special properties: reflexive, symmetric, antisymmetric, and transitive. As a result we focused on two special kinds of relations: partial orders and equivalence relations.

A relation $\mathscr{R}$ on a set A is a partial order, making A into a poset, if $\mathscr{R}$ is reflexive, antisymmetric, and transitive. Such a relation generalizes the familiar "less than or equal to" relation on the real numbers. Try to imagine calculus, or even elementary algebra, without it! Or take a simple computer program and see what happens if the program is entered into the computer haphazardly, permuting the order of the statements. Order is with us wherever we turn. We have grown so accustomed to it that we sometimes take it for granted. The origins of the subject of partially ordered sets (and lattices) came about during the nineteenth century in the work of George Boole (1815–1864), Richard Dedekind (1831–1916), Charles Sanders Peirce (1839–1914), and Ernst Schröder (1841–1902). The work of Garrett Birkhoff (1911–1996) in the 1930s, however, is where the initial work on partially ordered sets and lattices was developed to the point where these areas emerged as subjects in their own right.

For a finite poset, the Hasse diagram, a special type of directed graph, provides a pictorial representation of the order defined by the poset; it also proves useful when a total order, including the given partial order, is needed. These diagrams are named for the German number theorist Helmut Hasse (1898–1979). He introduced them in his textbook *Höhere Algebra* (published in 1926) as an aid in the study of the solutions of polynomial equations. The method we employed to derive a total order from a partial order is called topological sorting and it is used in the solution of PERT (Program Evaluation and Review Technique) networks. As mentioned earlier, this method was developed and first used by the U.S. Navy.

Although the equivalence relation differs from the partial order in only one property, it is quite different in structure and application. We make no attempt to trace the origin of the equivalence relation, but the ideas behind the reflexive, symmetric, and transitive properties can be found in *I Principii di Geometria* (1889), the work of the Italian mathematician Giuseppe Peano (1858–1932). The work of Carl Friedrich Gauss (1777–1855) on *congruence*, which he developed in the 1790s, also utilizes these ideas in spirit, if not in name.

Giuseppe Peano (1858–1932)

Carl Friedrich Gauss (1777–1855)

Basically, an equivalence relation $\mathscr{R}$ on a set A generalizes equality; it induces a characteristic of "sameness" among the elements of A. This "sameness" notion then causes the set A to be partitioned into subsets called *equivalence classes*. Conversely, we find that a partition of a set A induces an equivalence relation on A. The partition of a set arises in many places in mathematics and computer science. In computer science many searching algorithms rely on a technique that successively reduces the size of a given set A that is being searched. By partitioning A into smaller and smaller subsets, we apply the searching procedure in a more efficient manner. Each successive partition refines its predecessor, the key needed, for example, in the minimization process for finite state machines.

Throughout the chapter we emphasized the interplay between relations, directed graphs, and (0, 1)-matrices. These matrices provide a rectangular array of information about a relation, or graph, and prove useful in certain calculations. Storing information like this, in rectangular arrays and in consecutive memory locations, has been practiced in computer science since the late 1940s and early 1950s. For more on the historical background of such considerations, consult pages 456–462 of D. E. Knuth [3]. Another way to store information about a graph is the *adjacency list representation*. (See Supplementary Exercise 11.) In the study of data structures, *linked lists* and *doubly linked lists* are prominent in implementing such a representation. For more on this, consult the text by A. V. Aho, J. E. Hopcroft, and J. D. Ullman [1].

With regard to graph theory, we are in an area of mathematics that dates back to 1736 when the Swiss mathematician Leonhard Euler (1707–1783) solved the problem of the seven bridges of Königsberg. Since then, much more has evolved in this area, especially in conjunction with data structures in computer science.

For similar coverage of some of the topics in this chapter, see Chapter 3 of D. F. Stanat and D. F. McAllister [6]. An interesting presentation of the "Equivalence Problem" can be found on pages 353–355 of D. E. Knuth [3] for those wanting more information on the role of the computer in conjunction with the concept of the equivalence relation.

The early work on the development of the minimization process can be found in the paper by E. F. Moore [5], which builds upon prior ideas of D. A. Huffman [2]. Chapter 10 of Z. Kohavi [4] covers the minimization process for different types of finite state machines and includes some hardware considerations in their design.

REFERENCES

1. Aho, Alfred V., Hopcroft, John E., and Ullman, Jeffrey D. *Data Structures and Algorithms.* Reading, Mass.: Addison-Wesley, 1983.
2. Huffman, David A. "The Synthesis of Sequential Switching Circuits." *Journal of the Franklin Institute* 257, no. 3: pp. 161–190; no. 4: pp. 275–303, 1954.
3. Knuth, Donald E. *The Art of Computer Programming*, 2nd ed., Volume 1, *Fundamental Algorithms*. Reading, Mass.: Addison-Wesley, 1973.
4. Kohavi, Zvi. *Switching and Finite Automata Theory*, 2nd ed. New York: McGraw-Hill, 1978.
5. Moore, E. F. "Gedanken-experiments on Sequential Machines." *Automata Studies, Annals of Mathematical Studies*, no. 34: pp. 129–153. Princeton, N.J.: Princeton University Press, 1956.
6. Stanat, Donald F., and McAllister, David F. *Discrete Mathematics in Computer Science.* Englewood Cliffs, N.J.: Prentice-Hall, 1977.

SUPPLEMENTARY EXERCISES

1. Let A be a set and I an index set where, for each $i \in I$, $\mathcal{R}_i$ is a relation on A. Prove or disprove each of the following.

a) $\displaystyle\bigcup_{i \in I} \mathcal{R}_i$ is reflexive on A if and only if each $\mathcal{R}_i$ is reflexive on A.

b) $\displaystyle\bigcap_{i \in I} \mathcal{R}_i$ is reflexive on A if and only if each $\mathcal{R}_i$ is reflexive on A.

2. Repeat Exercise 1 with "reflexive" replaced by (i) symmetric; (ii) antisymmetric; (iii) transitive.

3. For a set A, let $\mathcal{R}_1$ and $\mathcal{R}_2$ be symmetric relations on A. If $\mathcal{R}_1 \circ \mathcal{R}_2 \subseteq \mathcal{R}_2 \circ \mathcal{R}_1$, prove that $\mathcal{R}_1 \circ \mathcal{R}_2 = \mathcal{R}_2 \circ \mathcal{R}_1$.

4. For each of the following relations on the set specified, determine whether the relation is reflexive, symmetric, antisymmetric, or transitive. Also determine whether it is a partial order or an equivalence relation, and, if the latter, describe the partition induced by the relation.

   a) $\mathcal{R}$ is the relation on **Q** where $a \mathcal{R} b$ if $|a - b| < 1$.

   b) Let T be the set of all triangles in the plane. For $t_1, t_2 \in T$, define $t_1 \mathcal{R} t_2$ if t_1, t_2 have the same area.

   c) For T as in part (b), define $\mathcal{R}$ by $t_1 \mathcal{R} t_2$ if at least two sides of t_1 are contained within the perimeter of t_2.

   d) Let $A = \{1, 2, 3, 4, 5, 6, 7\}$. Define $\mathcal{R}$ on A by $x \mathcal{R} y$ if $xy \geq 10$.

5. For sets A, B, and C with relations $\mathcal{R}_1 \subseteq A \times B$ and $\mathcal{R}_2 \subseteq B \times C$, prove or disprove that $(\mathcal{R}_1 \circ \mathcal{R}_2)^c = \mathcal{R}_2^c \circ \mathcal{R}_1^c$.

6. For a set A, let $C = \{P_i | P_i \text{ is a partition of } A\}$. Define relation $\mathcal{R}$ on C by $P_i \mathcal{R} P_j$ if $P_i \leq P_j$ — that is, P_i is a refinement of P_j.

   a) Verify that $\mathcal{R}$ is a partial order on C.

   b) For $A = \{1, 2, 3, 4, 5\}$, let P_i, $1 \leq i \leq 4$, be the following partitions: P_1: $\{1, 2\}, \{3, 4, 5\}$; P_2: $\{1, 2\}, \{3, 4\}, \{5\}$; P_3: $\{1\}, \{2\}, \{3, 4, 5\}$; P_4: $\{1, 2\}, \{3\}, \{4\}, \{5\}$. Draw the Hasse diagram for $C = \{P_i | 1 \leq i \leq 4\}$, where C is partially ordered by refinement.

7. Give an example of a poset with 5 minimal (maximal) elements but no least (greatest) element.

8. Let $A = \{1, 2, 3, 4, 5, 6\} \times \{1, 2, 3, 4, 5, 6\}$. Define $\mathcal{R}$ on A by $(x_1, y_1) \mathcal{R} (x_2, y_2)$, if $x_1 y_1 = x_2 y_2$.

   a) Verify that $\mathcal{R}$ is an equivalence relation on A.

   b) Determine the equivalence classes $[(1, 1)]$, $[(2, 2)]$, $[(3, 2)]$, and $[(4, 3)]$.

9. If the complete graph K_n has 45 edges, what is n?

10. Let $\mathcal{F} = \{f : \mathbf{Z}^+ \to \mathbf{R}\}$ — that is, $\mathcal{F}$ is the set of all functions with domain $\mathbf{Z}^+$ and codomain $\mathbf{R}$.

   a) Define the relation $\mathcal{R}$ on $\mathcal{F}$ by $g \mathcal{R} h$, for $g, h \in \mathcal{F}$, if g is dominated by h and h is dominated by g — that is, $g \in \Theta(h)$. (See Exercises 14, 15 for Section 5.7.) Prove that $\mathcal{R}$ is an equivalence relation on $\mathcal{F}$.

   b) For $f \in \mathcal{F}$, let $[f]$ denote the equivalence class of f for the relation $\mathcal{R}$ of part (a). Let $\mathcal{F}'$ be the set of equivalence classes induced by $\mathcal{R}$. Define the relation $\mathcal{S}$ on $\mathcal{F}'$ by $[g] \mathcal{S} [h]$, for $[g], [h] \in \mathcal{F}'$, if g is dominated by h. Verify that $\mathcal{S}$ is a partial order.

   c) For $\mathcal{R}$ in part (a), let $f, f_1, f_2 \in \mathcal{F}$ with $f_1, f_2 \in [f]$. If $f_1 + f_2 : \mathbf{Z}^+ \to \mathbf{R}$ is defined by $(f_1 + f_2)(n) = f_1(n) + f_2(n)$, for $n \in \mathbf{Z}^+$, prove or disprove that $f_1 + f_2 \in [f]$.

11. We have seen that the adjacency matrix can be used to represent a graph. However, this method proves to be rather inefficient when there are many 0's (that is, few edges) present. A better method uses the *adjacency list representation*, which is made up of an *adjacency list* for each vertex v and an *index list*. For the graph shown in Fig. 7.27, the representation is given by the two lists in Table 7.5.

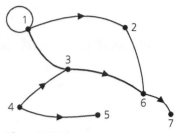

Figure 7.27

Table 7.5

Adjacency List		Index List	
1	1	1	1
2	2	2	4
3	3	3	5
4	6	4	7
5	1	5	9
6	6	6	9
7	3	7	11
8	5	8	11
9	2		
10	7		

For each vertex v in the graph, we list, preferably in numerical order, each vertex w that is adjacent from v. Hence for 1, we list 1, 2, 3 as the first three adjacencies in our adjacency list. Next to 2 in the index list we place a 4, which tells us where to start looking in the adjacency list for the adjacencies from 2. Since there is a 5 to the right of 3 in the index list, we know that the only adjacency from 2 is 6. Likewise, the 7 to the right of 4 in the index list directs us to the seventh entry in the adjacency list — namely, 3 — and we find that vertex 4 is adjacent to vertices 3 (the seventh vertex in the adjacency list) and 5 (the eighth vertex in the adjacency list). We stop at vertex 5 because of the 9 to the right of vertex 5 in the index list. The 9's in the index list next to 5 and 6 indicate that no vertex is adjacent from vertex 5. In a similar way, the 11's next to 7 and 8 in the index list tell us that vertex 7 is not adjacent to any vertex in the given directed graph.

In general, this method provides an easy way to determine the vertices adjacent from a vertex v. They are listed in the positions index(v), index(v) + 1, ..., index($v + 1$) − 1 of the adjacency list.

Finally, the last pair of entries in the index list — namely, 8 and 11 — is a "phantom" that indicates where the adjacency list would pick up from *if* there were an eighth vertex in the graph.

Represent each of the graphs in Fig. 7.28 in this manner.

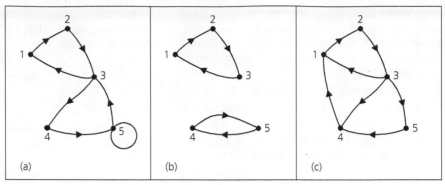

Figure 7.28

12. The adjacency list representation of a directed graph G is given by the lists in Table 7.6. Construct G from this representation.

Table 7.6

Adjacency List		Index List	
1	2	1	1
2	3	2	4
3	6	3	5
4	3	4	5
5	3	5	8
6	4	6	10
7	5	7	10
8	3	8	10
9	6		

13. Let G be an undirected graph with vertex set V. Define the relation $\mathcal{R}$ on V by $v\,\mathcal{R}\,w$ if $v = w$ or if there is a path from v to w (or from w to v since G is undirected). (a) Prove that $\mathcal{R}$ is an equivalence relation on V. (b) What can we say about the associated partition?

14. a) For the finite state machine given in Table 7.7, determine a minimal machine that is equivalent to it.

b) Find a minimal string that distinguishes states s_4 and s_6.

15. At the computer center Maria is faced with running 10 computer programs which, because of priorities, are restricted by the following conditions: (a) $10 > 8, 3$; (b) $8 > 7$; (c) $7 > 5$; (d) $3 > 9, 6$; (e) $6 > 4, 1$; (f) $9 > 4, 5$; (g) $4, 5, 1 > 2$; where, for example, $10 > 8, 3$ means that program number 10 must be run before programs 8 and 3. Determine an order for running these programs so that the priorities are satisfied.

16. a) Draw the Hasse diagram for the set of positive integer divisors of (i) 2; (ii) 4; (iii) 6; (iv) 8; (v) 12; (vi) 16; (vii) 24; (viii) 30; (ix) 32.

Table 7.7

	v		ω	
	0	1	0	1
s_1	s_7	s_6	1	0
s_2	s_7	s_7	0	0
s_3	s_7	s_2	1	0
s_4	s_2	s_3	0	0
s_5	s_3	s_7	0	0
s_6	s_4	s_1	0	0
s_7	s_3	s_5	1	0
s_8	s_7	s_3	0	0

b) For all $2 \le n \le 35$, show that the Hasse diagram for the set of positive-integer divisors of n looks like one of the nine diagrams in part (a). (Ignore the numbers at the vertices and concentrate on the structure given by the vertices and edges.) What happens for $n = 36$?

c) For $n \in \mathbf{Z}^+$, $\tau(n) =$ the number of positive-integer divisors of n. (See Supplementary Exercise 32 in Chapter 5.) Let $m, n \in \mathbf{Z}^+$ and S, T be the sets of all positive-integer divisors of m, n, respectively. The results of parts (a) and (b) imply that if the Hasse diagrams of S, T are structurally the same, then $\tau(m) = \tau(n)$. But is the converse true?

d) Show that each Hasse diagram in part (a) is a lattice if we define glb$\{x, y\} = \gcd(x, y)$ and lub$\{x, y\} = \operatorname{lcm}(x, y)$.

17. Let U denote the set of all points in and on the unit square shown in Fig. 7.29. That is, $U = \{(x, y)\,|\,0 \le x \le 1, 0 \le y \le 1\}$. Define the relation $\mathcal{R}$ on U by $(a, b)\,\mathcal{R}\,(c, d)$ if (1) $(a, b) = (c, d)$, or (2) $b = d$ and $a = 0$ and $c = 1$, or (3) $b = d$ and $a = 1$ and $c = 0$.

a) Verify that $\mathcal{R}$ is an equivalence relation on U.

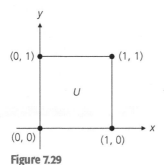

U

(0, 1) ●   ● (1, 1)

(0, 0) ●   ● (1, 0)

Figure 7.29

b) List the ordered pairs in the equivalence classes $[(0.3, 0.7)], [(0.5, 0)], [(0.4, 1)], [(0, 0.6)], [(1, 0.2)]$. For $0 \le a \le 1, 0 \le b \le 1$, how many ordered pairs are in $[(a, b)]$?

c) If we "glue together" the ordered pairs in each equivalence class, what type of surface comes about?

18. a) For $\mathcal{U} = \{1, 2, 3\}$, let $A = \mathcal{P}(\mathcal{U})$. Define the relation $\mathcal{R}$ on A by $B \mathcal{R} C$ if $B \subseteq C$. How many ordered pairs are there in the relation $\mathcal{R}$?

b) Answer part (a) for $\mathcal{U} = \{1, 2, 3, 4\}$.

c) Generalize the results of parts (a) and (b).

19. For $n \in \mathbf{Z}^+$, let $\mathcal{U} = \{1, 2, 3, \ldots, n\}$. Define the relation $\mathcal{R}$ on $\mathcal{P}(\mathcal{U})$ by $A \mathcal{R} B$ if $A \nsubseteq B$ and $B \nsubseteq A$. How many ordered pairs are there in this relation?

20. Let A be a finite nonempty set with $B \subseteq A$ (B fixed), and $|A| = n, |B| = m$. Define the relation $\mathcal{R}$ on $\mathcal{P}(A)$ by $X \mathcal{R} Y$, for $X, Y \subseteq A$, if $X \cap B = Y \cap B$. Then $\mathcal{R}$ is an equivalence relation, as verified in Exercise 10 of Section 7.4. (a) How many equivalence classes are in the partition of $\mathcal{P}(A)$ induced by $\mathcal{R}$? (b) How many subsets of A are in each equivalence class of the partition induced by $\mathcal{R}$?

21. For $A \ne \emptyset$, let $(A, \mathcal{R})$ be a poset, and let $\emptyset \ne B \subseteq A$ such that $\mathcal{R}' = (B \times B) \cap \mathcal{R}$. If $(B, \mathcal{R}')$ is totally ordered, we call $(B, \mathcal{R}')$ a *chain* in $(A, \mathcal{R})$. In the case where B is finite, we may order the elements of B by $b_1 \mathcal{R}' b_2 \mathcal{R}' b_3 \mathcal{R}' \cdots \mathcal{R}' b_{n-1} \mathcal{R}' b_n$ and say that the chain has *length n*. A chain (of length n) is called *maximal* if there is no element $a \in A$ where $a \notin \{b_1, b_2, b_3, \ldots, b_n\}$ and $a \mathcal{R} b_1, b_n \mathcal{R} a$, or $b_i \mathcal{R} a \mathcal{R} b_{i+1}$, for some $1 \le i \le n - 1$.

a) Find two chains of length 3 for the poset given by the Hasse diagram in Fig. 7.20. Find a maximal chain for this poset. How many such maximal chains does it have?

b) For the poset given by the Hasse diagram in Fig. 7.18(d), find two maximal chains of different lengths. What is the length of a longest (maximal) chain for this poset?

c) Let $\mathcal{U} = \{1, 2, 3, 4\}$ and $A = \mathcal{P}(\mathcal{U})$. For the poset

$(A, \subseteq)$, find two maximal chains. How many such maximal chains are there for this poset?

d) If $\mathcal{U} = \{1, 2, 3, \ldots, n\}$, how many maximal chains are there in the poset $(\mathcal{P}(\mathcal{U}), \subseteq)$?

22. For $\emptyset \ne C \subseteq A$, let $(C, \mathcal{R}')$ be a maximal chain in the poset $(A, \mathcal{R})$, where $\mathcal{R}' = (C \times C) \cap \mathcal{R}$. If the elements of C are ordered as $c_1 \mathcal{R}' c_2 \mathcal{R}' \cdots \mathcal{R}' c_n$, prove that c_1 is a minimal element in $(A, \mathcal{R})$ and that c_n is maximal in $(A, \mathcal{R})$.

23. Let $(A, \mathcal{R})$ be a poset in which the length of a longest (maximal) chain is $n \ge 2$. Let M be the set of all maximal elements in $(A, \mathcal{R})$, and let $B = A - M$. If $\mathcal{R}' = (B \times B) \cap \mathcal{R}$, prove that the length of a longest chain in $(B, \mathcal{R}')$ is $n - 1$.

24. Let $(A, \mathcal{R})$ be a poset, and let $\emptyset \ne C \subseteq A$. If $(C \times C) \cap \mathcal{R} = \emptyset$, then for all distinct $x, y \in C$ we have $x \not\mathcal{R} y$ and $y \not\mathcal{R} x$. The elements of C are said to form an *antichain* in the poset $(A, \mathcal{R})$.

a) Find an antichain with three elements for the poset given in the Hasse diagram of Fig. 7.18(d). Determine a largest antichain containing the element 6. Determine a largest antichain for this poset.

b) If $\mathcal{U} = \{1, 2, 3, 4\}$, let $A = \mathcal{P}(\mathcal{U})$. Find two different antichains for the poset $(A, \subseteq)$. How many elements occur in a largest antichain for this poset?

c) Prove that in any poset $(A, \mathcal{R})$, the set of all maximal elements and the set of all minimal elements are antichains.

25. Let $(A, \mathcal{R})$ be a poset in which the length of a longest chain is n. Use mathematical induction to prove that the elements of A can be partitioned into n antichains $C_1, C_2, \ldots, C_n$ (where $C_i \cap C_j = \emptyset$, for $1 \le i < j \le n$).

26. a) In how many ways can one totally order the partial order of positive-integer divisors of 96?

b) How many of the total orders in part (a) start with $96 > 32$?

c) How many of the total orders in part (a) end with $3 > 1$?

d) How many of the total orders in part (a) start with $96 > 32$ and end with $3 > 1$?

e) How many of the total orders in part (a) start with $96 > 48 > 32 > 16$?

27. Let n be a fixed positive integer and let $A_n = \{0, 1, \ldots, n\} \subseteq \mathbf{N}$. (a) How many edges are there in the Hasse diagram for the total order $(A_n, \le)$, where "$\le$" is the ordinary "less than or equal to" relation? (b) In how many ways can the edges in the Hasse diagram of part (a) be partitioned so that the edges in each cell (of the partition) provide a path (of one or more edges)? (c) In how many ways can the edges in the Hasse diagram for $(A_{12}, \le)$ be partitioned so that the edges in each cell (of the partition) provide a path (of one or more edges) and one of the cells is $\{(3, 4), (4, 5), (5, 6), (6, 7)\}$?

FURTHER TOPICS IN ENUMERATION

8

The Principle of Inclusion and Exclusion

We now return to the topic of enumeration as we investigate the *Principle of Inclusion and Exclusion*. Extending the ideas in the counting problems on Venn diagrams in Chapter 3, this principle will assist us in establishing the formula we conjectured in Section 5.3 for the number of onto functions $f: A \to B$, where A, B are finite (nonempty) sets. Other applications of this principle will demonstrate its versatile nature in combinatorial mathematics.

8.1
The Principle of Inclusion and Exclusion

In this section we develop some notation for stating this new counting principle. Then we establish the principle by a combinatorial argument. Following this, a wide range of examples demonstrate how this principle may be applied.

We shall motivate the Principle of Inclusion and Exclusion with a series of three examples, the first two of which will be reminiscent of the work we did with counting and Venn diagrams in Section 3.3.

EXAMPLE 8.1

Let S represent the set of 100 students enrolled in the freshman engineering program at Central College. Then $|S| = 100$. Now let c_1, c_2 denote the following conditions (or properties) satisfied by some of the elements of S:

c_1: A student at Central College is among the 100 students in the freshman engineering program and is enrolled in Freshman Composition.

c_2: A student at Central College is among the 100 students in the freshman engineering program and is enrolled in Introduction to Economics.

Suppose that 35 of these 100 students are enrolled in Freshman Composition and that 30 of them are enrolled in Introduction to Economics. We shall denote this by

$$N(c_1) = 35 \quad \text{and} \quad N(c_2) = 30.$$

If nine of these 100 students are enrolled in both Freshman Composition and Introduction to Economics then we write $N(c_1 c_2) = 9$.

Further, of these 100 students, there are $100 - 35 = 65$ who are *not* taking Freshman Composition. Denoting $|S|$ by N, we can designate this by writing $N(\overline{c}_1) = N - N(c_1)$. In a similar way we designate that there are $N(\overline{c}_2) = N - N(c_2) = 100 - 30 = 70$ of these students who are not taking Introduction to Economics. The number who *are* taking Freshman Composition and who are *not* taking Introduction to Economics is $N(c_1\overline{c}_2) = N(c_1) - N(c_1c_2) = 35 - 9 = 26$. Likewise, of these 100 students, there are $N(\overline{c}_1c_2) = N(c_2) - N(c_1c_2) = 30 - 9 = 21$ who are enrolled in Introduction to Economics but not in Freshman Composition. Of particular interest are those students (from among these 100 freshmen) who are taking neither Freshman Composition nor Introduction to Economics — that is, they are *not* taking Freshman Composition and they are also *not* taking Introduction to Economics. Their number is $N(\overline{c}_1\overline{c}_2)$. And since $N(\overline{c}_1) = N(\overline{c}_1c_2) + N(\overline{c}_1\overline{c}_2)$, we learn that $N(\overline{c}_1\overline{c}_2) = N(\overline{c}_1) - N(\overline{c}_1c_2) = 65 - 21 = 44$.

The preceding observations also demonstrate that

$$N(\overline{c}_1\overline{c}_2) = N(\overline{c}_1) - N(\overline{c}_1c_2) = [N - N(c_1)] - [N(c_2) - N(c_1c_2)]$$
$$= N - N(c_1) - N(c_2) + N(c_1c_2) = N - [N(c_1) + N(c_2)] + N(c_1c_2)$$
$$= 100 - [35 + 30] + 9 = 44, \text{ as we saw above.}$$

From the Venn diagram in Fig. 8.1, we see that if $N(c_1)$ denotes the number of elements of S in the left-hand circle and $N(c_2)$ denotes the number in the right-hand circle, then $N(c_1c_2)$ is the number of these elements from S in the overlap, while $N(\overline{c}_1\overline{c}_2)$ counts those elements of S that are outside the union of these two circles. Consequently, we see once again — this time from the figure — that

$$N(\overline{c}_1\overline{c}_2) = N - [N(c_1) + N(c_2)] + N(c_1c_2),$$

where the last term is added on because it was eliminated twice in the term $[N(c_1) + N(c_2)]$. (Also, at this point, the reader may wish to look back at the second formula following Example 3.25 to find the same result presented with a different notation.)

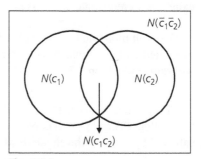

Figure 8.1

[Before we advance to our next example where we will introduce a third condition, let us note that $N(\overline{c}_1\overline{c}_2)$ is *not* the same as $N(\overline{c_1c_2})$. For $N(\overline{c_1c_2}) = N - N(c_1c_2) = 100 - 9 = 91$, in this example, while $N(\overline{c}_1\overline{c}_2) = 44$, as we learned earlier. However, $N(\overline{c}_1 \text{ or } \overline{c}_2) = N(\overline{c_1c_2}) = 91 = 65 + 70 - 44 = N(\overline{c}_1) + N(\overline{c}_2) - N(\overline{c}_1\overline{c}_2)$.]

EXAMPLE 8.2

We start with the same 100 students as in Example 8.1 and the same conditions c_1, c_2, but now we consider a third condition, given as follows:

c_3: A student at Central College is among the 100 students in the freshman engineering program and is enrolled in Fundamentals of Computer Programming.

It is still the case that $N(c_1) = 35$, $N(c_2) = 30$, and $N(c_1c_2) = 9$, but now we are also given that $N(c_3) = 30$, $N(c_1c_3) = 11$, $N(c_2c_3) = 10$, and $N(c_1c_2c_3) = 5$ (that is, there are five of these 100 freshmen who are taking Freshman Composition, Introduction to Economics, and Fundamentals of Computer Programming). Looking to Fig. 8.2, we learn that

$$N(\overline{c}_1\overline{c}_2\overline{c}_3) = N - [N(c_1) + N(c_2) + N(c_3)] + [N(c_1c_2) + N(c_1c_3) + N(c_2c_3)]$$
$$- N(c_1c_2c_3).$$

So here we have $N(\overline{c}_1\overline{c}_2\overline{c}_3) = 100 - [35 + 30 + 30] + [9 + 11 + 10] - 5 = 30$. That is, out of these 100 students there are 30 who are *not* enrolled in any of the courses: (i) Freshman Composition; (ii) Introduction to Economics; or (iii) Fundamentals of Computer Programming.

[We also learn here that $N(\overline{c}_3) = 70 = 100 - 30 = N - N(c_3)$, $N(\overline{c}_1\overline{c}_3) = 46 = 100 - [35 + 30] + 11 = N - [N(c_1) + N(c_3)] + N(c_1c_3)$, and $N(\overline{c}_2\overline{c}_3) = 50 = 100 - [30 + 30] + 10 = N - [N(c_2) + N(c_3)] + N(c_2c_3)$. Furthermore, we note the similarity here with the result for $|\overline{A} \cap \overline{B} \cap \overline{C}|$ given in the second formula following Example 3.26.]

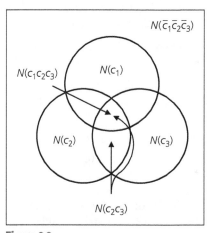

Figure 8.2

EXAMPLE 8.3

Based on the results in the previous two examples we may now feel that for a given finite set S (with $|S| = N$) and four conditions c_1, c_2, c_3, c_4 we should have

$$N(\overline{c}_1\overline{c}_2\overline{c}_3\overline{c}_4) = N - [N(c_1) + N(c_2) + N(c_3) + N(c_4)] \qquad (*)$$
$$+ [N(c_1c_2) + N(c_1c_3) + N(c_1c_4) + N(c_2c_3) + N(c_2c_4) + N(c_3c_4)]$$
$$- [N(c_1c_2c_3) + N(c_1c_2c_4) + N(c_1c_3c_4) + N(c_2c_3c_4)]$$
$$+ N(c_1c_2c_3c_4).$$

To show that this is the case we consider an arbitrary element x from S and show that it is counted the same number of times on both sides of the above equation.

0) If x satisfies none of the four conditions, then it is counted once on the left side of Eq. (*) [in $N(\overline{c}_1\overline{c}_2\overline{c}_3\overline{c}_4)$], and once on the right side of Eq. (*) [in N].

1) If x satisfies only one of the conditions, say c_1, then it is not counted at all on the left side of Eq. (*). But on the right side of Eq. (*), x is counted once in N and once in $N(c_1)$, for a total of $1 - 1 = 0$ times.

2) Now suppose that x satisfies conditions c_2, c_4 but does *not* satisfy conditions c_1, c_3. Once again x is not counted on the left side of Eq. (*). For the right side of Eq. (*), x is counted once in N, once in each of $N(c_2)$ and $N(c_4)$, and then once in $N(c_2 c_4)$, totaling $1 - [1 + 1] + 1 = 1 - \binom{2}{1} + \binom{2}{2} = 0$ times.

3) Continuing with the case for three conditions, we'll suppose here that x satisfies conditions c_1, c_2, and c_4, but *not* c_3. As in the previous two cases, x is not counted on the left side of Eq. (*). On the right side of Eq. (*), x is counted once in N, once in each of $N(c_1)$, $N(c_2)$, and $N(c_4)$, once in each of $N(c_1 c_2)$, $N(c_1 c_4)$, and $N(c_2 c_4)$, and, finally, once in $N(c_1 c_2 c_4)$. So on the right side of Eq. (*), x is counted $1 - [1 + 1 + 1] + [1 + 1 + 1] - 1 = 1 - \binom{3}{1} + \binom{3}{2} - \binom{3}{3} = 0$ times, in total.

4) Finally, if x satisfies all four of the conditions c_1, c_2, c_3, c_4, then once again it is not counted on the left side of Eq. (*). On the right side of Eq. (*), x is counted once for each of the 16 terms on the right side of this equation — for a total of $1 - [1 + 1 + 1 + 1] + [1 + 1 + 1 + 1 + 1 + 1] - [1 + 1 + 1 + 1] + 1 = 1 - \binom{4}{1} + \binom{4}{2} - \binom{4}{3} + \binom{4}{4} = 0$ times.

Consequently, from these preceding five cases we have shown that the two sides of Eq. (*) count the same elements from S, and this provides a combinatorial proof for the formula for $N(\overline{c}_1 \overline{c}_2 \overline{c}_3 \overline{c}_4)$.

So now we shall reconsider the situation in Example 8.2 and introduce a fourth condition as follows:

c_4:   A student at Central College is among the 100 students in the freshman engineering program and is enrolled in Introduction to Design.

We already know that $N(c_1) = 35$, $N(c_2) = 30$, $N(c_3) = 30$, $N(c_1 c_2) = 9$, $N(c_1 c_3) = 11$, $N(c_2 c_3) = 10$, and $N(c_1 c_2 c_3) = 5$. If $N(c_4) = 41$, $N(c_1 c_4) = 13$, $N(c_2 c_4) = 14$, $N(c_3 c_4) = 10$, $N(c_1 c_2 c_4) = 6$, $N(c_1 c_3 c_4) = 6$, $N(c_2 c_3 c_4) = 6$, and $N(c_1 c_2 c_3 c_4) = 4$, then, using the equation we derived above, it follows that $N(\overline{c}_1 \overline{c}_2 \overline{c}_3 \overline{c}_4) = 100 - [35 + 30 + 30 + 41] + [9 + 11 + 13 + 10 + 14 + 10] - [5 + 6 + 6 + 6] + 4 = 100 - 136 + 67 - 23 + 4 = 12$. Thus, of the 100 students in the freshman engineering program at Central College, there are 12 who are not taking any of the four courses: Freshman Composition, Introduction to Economics, Fundamentals of Computer Programming, or Introduction to Design.

If we are interested in the number (from these 100 students) who are taking Freshman Composition, but none of the other three courses, then we should want to compute $N(c_1 \overline{c}_2 \overline{c}_3 \overline{c}_4)$. To do so we start by observing that

$$N(\overline{c}_2 \overline{c}_3 \overline{c}_4) = N(c_1 \overline{c}_2 \overline{c}_3 \overline{c}_4) + N(\overline{c}_1 \overline{c}_2 \overline{c}_3 \overline{c}_4),$$

which can be established by an argument similar to the one above for $N(\overline{c}_1 \overline{c}_2 \overline{c}_3 \overline{c}_4)$. This then leads us to

$$N(c_1 \overline{c}_2 \overline{c}_3 \overline{c}_4) = N(\overline{c}_2 \overline{c}_3 \overline{c}_4) - N(\overline{c}_1 \overline{c}_2 \overline{c}_3 \overline{c}_4).$$

Using the result in Example 8.2 we find that

$$N(\overline{c}_2 \overline{c}_3 \overline{c}_4) = N - [N(c_2) + N(c_3) + N(c_4)] + [N(c_2 c_3) + N(c_2 c_4) + N(c_3 c_4)]$$
$$- N(c_2 c_3 c_4)$$
$$= 100 - [30 + 30 + 41] + [10 + 14 + 10] - 6 = 27, \text{ and}$$
$$N(c_1 \overline{c}_2 \overline{c}_3 \overline{c}_4) = N(\overline{c}_2 \overline{c}_3 \overline{c}_4) - N(\overline{c}_1 \overline{c}_2 \overline{c}_3 \overline{c}_4) = 27 - 12 = 15.$$

So there are 15 students in this set of 100 who are taking Freshman Composition, but none of the other courses: Introduction to Economics, Fundamentals of Computer Programming, or Introduction to Design.

Further, we also observe that

$$N(c_1\bar{c}_2\bar{c}_3\bar{c}_4) = N(\bar{c}_2\bar{c}_3\bar{c}_4) - N(\bar{c}_1\bar{c}_2\bar{c}_3\bar{c}_4)$$

$$= \big\{N - [N(c_2) + N(c_3) + N(c_4)] + [N(c_2c_3) + N(c_2c_4) + N(c_3c_4)]$$

$$- N(c_2c_3c_4)\big\} - \big\{N - [N(c_1) + N(c_2) + N(c_3) + N(c_4)]$$

$$+ [N(c_1c_2) + N(c_1c_3) + N(c_1c_4) + N(c_2c_3) + N(c_2c_4) + N(c_3c_4)]$$

$$- [N(c_1c_2c_3) + N(c_1c_2c_4) + N(c_1c_3c_4) + N(c_2c_3c_4)] + N(c_1c_2c_3c_4)\big\}, \text{ or}$$

$$N(c_1\bar{c}_2\bar{c}_3\bar{c}_4) = N(c_1) - [N(c_1c_2) + N(c_1c_3) + N(c_1c_4)]$$

$$+ [N(c_1c_2c_3) + N(c_1c_2c_4) + N(c_1c_3c_4)] - N(c_1c_2c_3c_4).$$

So here $N(c_1\bar{c}_2\bar{c}_3\bar{c}_4) = 35 - [9 + 11 + 13] + [5 + 6 + 6] - 4 = 35 - 33 + 17 - 4 = 15$, as we found above.

Having seen the results in Examples 8.1, 8.2, and 8.3, now it is time for us to generalize these results and establish the Principle of Inclusion and Exclusion. To do so we once again let S be a set with $|S| = N$, and we let $c_1, c_2, \ldots, c_t$ be a collection of t conditions or properties — each of which may be satisfied by some of the elements of S. Some elements of S may satisfy more than one of the conditions, whereas others may not satisfy any of them. For all $1 \le i \le t$, $N(c_i)$ will denote the number of elements in S that satisfy condition c_i. (Elements of S are counted here when they satisfy only condition c_i, as well as when they satisfy c_i and other conditions c_j, for $j \ne i$.) For all $i, j \in \{1, 2, 3, \ldots, t\}$ where $i \ne j$, $N(c_ic_j)$ will denote the number of elements in S that satisfy both of the conditions c_i, c_j, and perhaps some others. [$N(c_ic_j)$ does *not* count the elements of S that satisfy *only* c_i, c_j.] Continuing, if $1 \le i, j, k \le t$ are three distinct integers, then $N(c_ic_jc_k)$ denotes the number of elements in S satisfying, perhaps among others, each of the conditions c_i, c_j, and c_k.

For each $1 \le i \le t$, $N(\bar{c}_i) = N - N(c_i)$ denotes the number of elements in S that do not satisfy condition c_i. If $1 \le i, j \le t$ with $i \ne j$, $N(\bar{c}_i\bar{c}_j) = $ the number of elements in S that do not satisfy either of the conditions c_i or c_j. [This is *not* the same as $N(\overline{c_ic_j})$, as we observed at the end of Example 8.1.]

With the necessary preliminaries now in hand we state the following theorem.

THEOREM 8.1

The Principle of Inclusion and Exclusion. Consider a set S, with $|S| = N$, and conditions c_i, $1 \le i \le t$, each of which may be satisfied by some of the elements of S. The number of elements of S that satisfy *none* of the conditions c_i, $1 \le i \le t$, is denoted by $\overline{N} = N(\bar{c}_1\bar{c}_2\bar{c}_3 \cdots \bar{c}_t)$ where

$$\overline{N} = N - [N(c_1) + N(c_2) + N(c_3) + \cdots + N(c_t)] \tag{1}$$

$$+ [N(c_1c_2) + N(c_1c_3) + \cdots + N(c_1c_t) + N(c_2c_3) + \cdots + N(c_{t-1}c_t)]$$

$$- [N(c_1c_2c_3) + N(c_1c_2c_4) + \cdots + N(c_1c_2c_t) + N(c_1c_3c_4) + \cdots$$

$$+ N(c_1c_3c_t) + \cdots + N(c_{t-2}c_{t-1}c_t)] + \cdots + (-1)^t N(c_1c_2c_3 \cdots c_t),$$

or

$$\overline{N} = N - \sum_{1 \le i \le t} N(c_i) + \sum_{1 \le i < j \le t} N(c_i c_j) - \sum_{1 \le i < j < k \le t} N(c_i c_j c_k) + \cdots \qquad (2)$$

$$+ (-1)^t N(c_1 c_2 c_3 \cdots c_t).$$

Proof: Although this result can be established by applying the Principle of Mathematical Induction to the number t of conditions, we shall give a combinatorial proof. The argument will be reminiscent of the ideas we saw in Example 8.3 in establishing the formula for $N(\overline{c}_1 \overline{c}_2 \overline{c}_3 \overline{c}_4)$.

For each $x \in S$ we show that x contributes the same count, either 0 or 1, to each side of Eq. (2).

If x satisfies none of the conditions, then x is counted once in $\overline{N}$ and once in N, but not in any of the other terms in Eq. (2). Consequently, x contributes a count of 1 to each side of the equation.

The other possibility is that x satisfies *exactly* r of the conditions where $1 \le r \le t$. In this case x contributes nothing to $\overline{N}$. But on the right-hand side of Eq. (2), x is counted

(1) One time in N.

(2) r times in $\displaystyle\sum_{1 \le i \le t} N(c_i)$. (Once for each of the r conditions.)

(3) $\dbinom{r}{2}$ times in $\displaystyle\sum_{1 \le i < j \le t} N(c_i c_j)$. (Once for each pair of conditions selected from the r conditions it satisfies.)

(4) $\dbinom{r}{3}$ times in $\displaystyle\sum_{1 \le i < j < k \le t} N(c_i c_j c_k)$. (Why?)

. . . .

(r + 1) $\dbinom{r}{r} = 1$ time in $\displaystyle\sum N(c_{i_1} c_{i_2} \cdots c_{i_r})$, where the summation is taken over all selections of size r from the t conditions.

Consequently, on the right-hand side of Eq. (2), x is counted

$$1 - r + \binom{r}{2} - \binom{r}{3} + \cdots + (-1)^r \binom{r}{r} = [1 + (-1)]^r = 0^r = 0 \text{ times,}$$

by the binomial theorem. Therefore, the two sides of Eq. (2) count the same elements from S, and the equality is verified.

An immediate corollary of this principle is given as follows:

COROLLARY 8.1   Under the hypotheses of Theorem 8.1, the number of elements in S that satisfy at least one of the conditions c_i, where $1 \le i \le t$, is given by $N(c_1 \text{ or } c_2 \text{ or } \ldots \text{ or } c_t) = N - \overline{N}$.

Before solving some examples, we examine some further notation for simplifying the statement of Theorem 8.1.

We write

$$S_0 = N,$$
$$S_1 = [N(c_1) + N(c_2) + \cdots + N(c_t)],$$
$$S_2 = [N(c_1 c_2) + N(c_1 c_3) + \cdots + N(c_1 c_t) + N(c_2 c_3) + \cdots + N(c_{t-1} c_t)],$$

and, in general,

$$S_k = \sum N(c_{i_1} c_{i_2} \cdots c_{i_k}), \ 1 \le k \le t,$$

where the summation is taken over all selections of size k from the collection of t conditions. Hence S_k has $\binom{t}{k}$ summands in it.

Using this notation we can rewrite the result in Eq. (2) as

$$\overline{N} = S_0 - S_1 + S_2 - S_3 + \cdots + (-1)^t S_t.$$

Now let us look at how this principle is used to solve certain enumeration problems.

EXAMPLE 8.4

Determine the number of positive integers n where $1 \le n \le 100$ and n is *not* divisible by 2, 3, or 5.

Here $S = \{1, 2, 3, \ldots, 100\}$ and $N = 100$. For $n \in S$, n satisfies

a) condition c_1 if n is divisible by 2,

b) condition c_2 if n is divisible by 3, and

c) condition c_3 if n is divisible by 5.

Then the answer to this problem is $N(\overline{c}_1 \overline{c}_2 \overline{c}_3)$.

As in Section 5.2 we use the notation $\lfloor r \rfloor$ to denote the greatest integer less than or equal to r, for any real number r. This function proves to be helpful in this problem as we find that

$N(c_1) = \lfloor 100/2 \rfloor = 50$ [since the 50 ($= \lfloor 100/2 \rfloor$) positive integers 2, 4, 6, 8, ..., 96, 98 ($= 2 \cdot 49$), 100 ($= 2 \cdot 50$) are divisible by 2];

$N(c_2) = \lfloor 100/3 \rfloor = \lfloor 33\,1/3 \rfloor = 33$ [since the 33 ($= \lfloor 100/3 \rfloor$) positive integers 3, 6, 9, 12, ..., 96 ($= 3 \cdot 32$), 99 ($= 3 \cdot 33$) are divisible by 3];

$N(c_3) = \lfloor 100/5 \rfloor = 20$;

$N(c_1 c_2) = \lfloor 100/6 \rfloor = 16$ [since there are 16 ($= \lfloor 100/6 \rfloor$) elements in S that are divisible by both 2 and 3 — hence divisible by lcm(2, 3) $= 2 \cdot 3 = 6$];

$N(c_1 c_3) = \lfloor 100/10 \rfloor = 10$;

$N(c_2 c_3) = \lfloor 100/15 \rfloor = 6$; and

$N(c_1 c_2 c_3) = \lfloor 100/30 \rfloor = 3$.

Applying the Principle of Inclusion and Exclusion, we find that

$$N(\overline{c}_1 \overline{c}_2 \overline{c}_3) = S_0 - S_1 + S_2 - S_3 = N - [N(c_1) + N(c_2) + N(c_3)]$$
$$+ [N(c_1 c_2) + N(c_1 c_3) + N(c_2 c_3)] - N(c_1 c_2 c_3)$$
$$= 100 - [50 + 33 + 20] + [16 + 10 + 6] - 3 = 26.$$

(These 26 numbers are 1, 7, 11, 13, 17, 19, 23, 29, 31, 37, 41, 43, 47, 49, 53, 59, 61, 67, 71, 73, 77, 79, 83, 89, 91, and 97.)

EXAMPLE 8.5

In Chapter 1 we found the number of nonnegative integer solutions to the equation $x_1 + x_2 + x_3 + x_4 = 18$. We now answer the same question with the extra restriction that $x_i \leq 7$, for all $1 \leq i \leq 4$.

Here S is the set of solutions of $x_1 + x_2 + x_3 + x_4 = 18$, with $0 \leq x_i$ for all $1 \leq i \leq 4$. So $|S| = N = S_0 = \binom{4 + 18 - 1}{18} = \binom{21}{18}$.

We say that a solution x_1, x_2, x_3, x_4 satisfies condition c_i, where $1 \leq i \leq 4$, if $x_i > 7$ (or $x_i \geq 8$). The answer to the problem is then $N(\bar{c}_1 \bar{c}_2 \bar{c}_3 \bar{c}_4)$.

Here by symmetry $N(c_1) = N(c_2) = N(c_3) = N(c_4)$. To compute $N(c_1)$, we consider the integer solutions for $x_1 + x_2 + x_3 + x_4 = 10$, with each $x_i \geq 0$ for all $1 \leq i \leq 4$. Then we add 8 to the value of x_1 and get the solutions of $x_1 + x_2 + x_3 + x_4 = 18$ that satisfy condition c_1. Hence $N(c_i) = \binom{4 + 10 - 1}{10} = \binom{13}{10}$, for each $1 \leq i \leq 4$, and $S_1 = \binom{4}{1}\binom{13}{10}$.

Likewise, $N(c_1 c_2)$ is the number of integer solutions of $x_1 + x_2 + x_3 + x_4 = 2$, where $x_i \geq 0$ for all $1 \leq i \leq 4$. So $N(c_1 c_2) = \binom{4 + 2 - 1}{2} = \binom{5}{2}$, and $S_2 = \binom{4}{2}\binom{5}{2}$.

Since $N(c_i c_j c_k) = 0$ for every selection of three conditions, and $N(c_1 c_2 c_3 c_4) = 0$, we have

$$N(\bar{c}_1 \bar{c}_2 \bar{c}_3 \bar{c}_4) = S_0 - S_1 + S_2 - S_3 + S_4 = \binom{21}{18} - \binom{4}{1}\binom{13}{10} + \binom{4}{2}\binom{5}{2} - 0 + 0 = 246.$$

So of the 1330 nonnegative integer solutions of $x_1 + x_2 + x_3 + x_4 = 18$, only 246 of them satisfy $x_i \leq 7$ for each $1 \leq i \leq 4$.

Our next example establishes the formula conjectured in Section 5.3 for counting onto functions.

EXAMPLE 8.6

For finite sets A, B, where $|A| = m \geq n = |B|$, let $A = \{a_1, a_2, \ldots, a_m\}$, $B = \{b_1, b_2, \ldots, b_n\}$, and S = the set of all functions $f : A \to B$. Then $N = S_0 = |S| = n^m$.

For all $1 \leq i \leq n$, let c_i denote the condition on S where a function $f : A \to B$ satisfies c_i if b_i is *not* in the range of f. (Note the difference between c_i here and c_i in Examples 8.4 and 8.5.) Then $N(\bar{c}_i)$ is the number of functions in S that have b_i in their range, and $N(\bar{c}_1 \bar{c}_2 \cdots \bar{c}_n)$ counts the number of onto functions $f : A \to B$.

For all $1 \leq i \leq n$, $N(c_i) = (n - 1)^m$, because each element of B, except b_i, can be used as the second component of an ordered pair for a function $f : A \to B$, whose range does not include b_i. Likewise, for all $1 \leq i < j \leq n$, there are $(n - 2)^m$ functions $f : A \to B$ whose range contains neither b_i nor b_j. From these observations we have $S_1 = [N(c_1) + N(c_2) + \cdots + N(c_n)] = n(n - 1)^m = \binom{n}{1}(n - 1)^m$, and $S_2 = [N(c_1 c_2) + N(c_1 c_3) + \cdots + N(c_1 c_n) + N(c_2 c_3) + \cdots + N(c_2 c_n) + \cdots + N(c_{n-1} c_n)] = \binom{n}{2}(n - 2)^m$. In general, for each $1 \leq k \leq n$,

$$S_k = \sum_{1 \leq i_1 < i_2 < \cdots < i_k \leq n} N(c_{i_1} c_{i_2} \cdots c_{i_k}) = \binom{n}{k}(n - k)^m.$$

It then follows by the Principle of Inclusion and Exclusion that the number of onto functions from A to B is

$$N(\bar{c}_1\bar{c}_2\bar{c}_3 \cdots \bar{c}_n) = S_0 - S_1 + S_2 - S_3 + \cdots + (-1)^n S_n$$

$$= n^m - \binom{n}{1}(n-1)^m + \binom{n}{2}(n-2)^m - \binom{n}{3}(n-3)^m$$

$$+ \cdots + (-1)^n(n-n)^m = \sum_{i=0}^{n}(-1)^i\binom{n}{i}(n-i)^m$$

$$= \sum_{i=0}^{n}(-1)^i\binom{n}{n-i}(n-i)^m.$$

Before we finish discussing this example, let us note that

$$\sum_{i=0}^{n}(-1)^i\binom{n}{n-i}(n-i)^m$$

can also be evaluated even if $m < n$. Furthermore, for $m < n$, the expression

$$N(\bar{c}_1\bar{c}_2\bar{c}_3 \cdots \bar{c}_n)$$

still counts the number of functions $f: A \to B$, where $|A| = m$, $|B| = n$, and each element of B is in the range of f. But now this number is 0.

For example, suppose that $m = 3 < 7 = n$. Then $N(\bar{c}_1\bar{c}_2\bar{c}_3 \cdots \bar{c}_7)$ counts the number of onto functions $f: A \to B$ for $|A| = 3$ and $|B| = 7$. We know this number is 0, and we also find that

$$\sum_{i=0}^{7}(-1)^i\binom{7}{7-i}(7-i)^3 = \binom{7}{7}7^3 - \binom{7}{6}6^3 + \binom{7}{5}5^3 - \binom{7}{4}4^3 + \binom{7}{3}3^3 - \binom{7}{2}2^3 + \binom{7}{1}1^3 - \binom{7}{0}0^3$$

$$= 343 - 1512 + 2625 - 2240 + 945 - 168 + 7 - 0 = 0.$$

Hence, for all $m, n \in \mathbf{Z}^+$, if $m < n$, then

$$\sum_{i=0}^{n}(-1)^i\binom{n}{n-i}(n-i)^m = 0.$$

We now solve a problem similar to those in Chapter 3 that dealt with Venn diagrams.

EXAMPLE 8.7

In how many ways can the 26 letters of the alphabet be permuted so that none of the patterns *car*, *dog*, *pun*, or *byte* occurs?

Let S denote the set of all permutations of the 26 letters. Then $|S| = 26!$ For each $1 \le i \le 4$, a permutation in S is said to satisfy condition c_i if the permutation contains the pattern *car*, *dog*, *pun*, or *byte*, respectively.

In order to compute $N(c_1)$, for example, we count the number of ways the 24 symbols *car*, $b, d, e, f, \ldots, p, q, s, t, \ldots, x, y, z$ can be permuted. So $N(c_1) = 24!$, and in a similar way we obtain

$$N(c_2) = N(c_3) = 24!, \quad \text{while } N(c_4) = 23!$$

For $N(c_1c_2)$ we deal with the 22 symbols *car*, *dog*, $b, e, f, h, i, \ldots, m, n, p, q, s, t, \ldots, x, y, z$, which can be permuted in 22! ways. Hence $N(c_1c_2) = 22!$, and comparable calculations give

$$N(c_1c_3) = N(c_2c_3) = 22!, \qquad N(c_ic_4) = 21!, \quad i \ne 4.$$

Furthermore,

$$N(c_1 c_2 c_3) = 20!, \qquad N(c_i c_j c_4) = 19!, \qquad 1 \le i < j \le 3,$$
$$N(c_1 c_2 c_3 c_4) = 17!$$

So the number of permutations in S that contain none of the given patterns is

$$N(\bar{c}_1 \bar{c}_2 \bar{c}_3 \bar{c}_4) = 26! - [3(24!) + 23!] + [3(22!) + 3(21!)] - [20! + 3(19!)] + 17!$$

Our next example deals with a number theory problem.

EXAMPLE 8.8

For $n \in \mathbf{Z}^+$, $n \ge 2$, let $\phi(n)$ be the number of positive integers m, where $1 \le m < n$ and $\gcd(m, n) = 1$ — that is, m, n are relatively prime. This function is known as *Euler's phi function*, and it arises in several situations in abstract algebra involving enumeration. We find that $\phi(2) = 1$, $\phi(3) = 2$, $\phi(4) = 2$, $\phi(5) = 4$, and $\phi(6) = 2$. For each prime p, $\phi(p) = p - 1$. We would like to derive a formula for $\phi(n)$ that is related to n so that we need not make a case-by-case comparison for each m, $1 \le m < n$, against the integer n.

The derivation of our formula will use the Principle of Inclusion and Exclusion as in Example 8.4. We proceed as follows: For $n \ge 2$, use the Fundamental Theorem of Arithmetic to write $n = p_1^{e_1} p_2^{e_2} \cdots p_t^{e_t}$, where $p_1, p_2, \ldots, p_t$ are distinct primes and $e_i \ge 1$, for all $1 \le i \le t$. We consider the case where $t = 4$. This will be enough to demonstrate the general idea.

With $S = \{1, 2, 3, \ldots, n\}$, we have $N = S_0 = |S| = n$, and for each $1 \le i \le 4$ we say that $k \in S$ satisfies condition c_i if k is divisible by p_i. For $1 \le k < n$, $\gcd(k, n) = 1$ if k is not divisible by any of the primes p_i, where $1 \le i \le 4$. Hence $\phi(n) = N(\bar{c}_1 \bar{c}_2 \bar{c}_3 \bar{c}_4)$.

For each $1 \le i \le 4$, we have $N(c_i) = n/p_i$; $N(c_i c_j) = n/(p_i p_j)$, for all $1 \le i < j \le 4$. Also, $N(c_i c_j c_\ell) = n/(p_i p_j p_\ell)$, for all $1 \le i < j < \ell \le 4$, and $N(c_1 c_2 c_3 c_4) = n/(p_1 p_2 p_3 p_4)$. So

$$\phi(n) = S_0 - S_1 + S_2 - S_3 + S_4$$

$$= n - \left[\frac{n}{p_1} + \cdots + \frac{n}{p_4} \right] + \left[\frac{n}{p_1 p_2} + \frac{n}{p_1 p_3} + \cdots + \frac{n}{p_3 p_4} \right]$$

$$- \left[\frac{n}{p_1 p_2 p_3} + \cdots + \frac{n}{p_2 p_3 p_4} \right] + \frac{n}{p_1 p_2 p_3 p_4}$$

$$= n \left[1 - \left(\frac{1}{p_1} + \cdots + \frac{1}{p_4} \right) + \left(\frac{1}{p_1 p_2} + \frac{1}{p_1 p_3} + \cdots + \frac{1}{p_3 p_4} \right) \right.$$

$$\left. - \left(\frac{1}{p_1 p_2 p_3} + \cdots + \frac{1}{p_2 p_3 p_4} \right) + \frac{1}{p_1 p_2 p_3 p_4} \right]$$

$$= \frac{n}{p_1 p_2 p_3 p_4} \left[p_1 p_2 p_3 p_4 - (p_2 p_3 p_4 + p_1 p_3 p_4 + p_1 p_2 p_4 + p_1 p_2 p_3) \right.$$

$$+ (p_3 p_4 + p_2 p_4 + p_2 p_3 + p_1 p_4 + p_1 p_3 + p_1 p_2)$$

$$\left. - (p_4 + p_3 + p_2 + p_1) + 1 \right]$$

$$= \frac{n}{p_1 p_2 p_3 p_4} \left[(p_1 - 1)(p_2 - 1)(p_3 - 1)(p_4 - 1) \right]$$

$$= n \left[\frac{p_1 - 1}{p_1} \cdot \frac{p_2 - 1}{p_2} \cdot \frac{p_3 - 1}{p_3} \cdot \frac{p_4 - 1}{p_4} \right] = n \prod_{i=1}^{4} \left(1 - \frac{1}{p_i} \right).$$

In general, $\phi(n) = n \prod_{p|n}(1 - (1/p))$, where the product is taken over all primes p dividing n. When $n = p$, a prime, $\phi(n) = \phi(p) = p[1 - (1/p)] = p - 1$, as we observed earlier. If $n = 23{,}100$, for example, we find that

$$\phi(23{,}100) = \phi(2^2 \cdot 3 \cdot 5^2 \cdot 7 \cdot 11)$$
$$= (23{,}100)(1 - (1/2))(1 - (1/3))(1 - (1/5))(1 - (1/7))(1 - (1/11))$$
$$= 4800.$$

The Euler phi function has many interesting properties. We shall investigate some of them in the exercises for this section and in the Supplementary Exercises.

The next example provides another encounter with the circular arrangements introduced in Chapter 1.

| EXAMPLE 8.9 |

Six married couples are to be seated at a circular table. In how many ways can they arrange themselves so that no wife sits next to her husband? (Here, as in Example 1.16, two seating arrangements are considered the same if one is a rotation of the other.)

For $1 \le i \le 6$, we let c_i denote the condition where a seating arrangement has couple i seated next to each other.

To determine $N(c_1)$, for instance, we consider arranging 11 distinct objects — namely, couple 1 (considered as one object) and the other 10 people. Eleven distinct objects can be arranged around a circular table in $(11 - 1)! = 10!$ ways. However, here $N(c_1) = 2(10!)$, where the 2 takes into account whether the wife in couple 1 is seated to the left or right of her husband. Similarly, $N(c_i) = 2(10!)$, for $2 \le i \le 6$, and $S_1 = \binom{6}{1}2(10!)$.

Continuing, let us now compute $N(c_i c_j)$, for $1 \le i < j \le 6$. Here we are arranging 10 distinct objects — couple i (considered as one object), couple j (likewise considered as one object), and the other eight people. Ten distinct objects can be arranged around a circular table in $(10 - 1)! = 9!$ ways. So here $N(c_i c_j) = 2^2(9!)$ because there are two ways for the wife in couple i to be seated next to her husband, and two ways for the wife in couple j to be seated next to her husband. Consequently, $S_2 = \binom{6}{2}2^2(9!)$.

Similar reasoning shows us that

$$N(c_1 c_2 c_3) = 2^3(8!), \quad S_3 = \binom{6}{3}2^3(8!) \qquad N(c_1 c_2 c_3 c_4) = 2^4(7!), \quad S_4 = \binom{6}{4}2^4(7!)$$
$$N(c_1 c_2 c_3 c_4 c_5) = 2^5(6!), \quad S_5 = \binom{6}{5}2^5(6!) \qquad N(c_1 c_2 c_3 c_4 c_5 c_6) = 2^6(5!), \quad S_6 = \binom{6}{6}2^6(5!).$$

With S_0 (the total number of arrangements of the 12 people) $= (12 - 1)! = 11!$, we find that the number of arrangements where no couple is seated side by side is

$$N(\overline{c_1}\overline{c_2}\cdots\overline{c_6}) = \sum_{i=0}^{6}(-1)^i S_i = \sum_{i=0}^{6}(-1)^i \binom{6}{i}2^i(11-i)!$$
$$= 39{,}916{,}800 - 43{,}545{,}600 + 21{,}772{,}800 - 6{,}451{,}200$$
$$+ 1{,}209{,}600 - 138{,}240 + 7680$$
$$= 12{,}771{,}840.$$

Our final example recalls some of the graph theory we studied in Chapter 7.

| EXAMPLE 8.10 |

In a certain area of the countryside are five villages. An engineer is to devise a system of two-way roads so that after the system is completed, no village will be isolated. In how many ways can he do this?

Calling the villages a, b, c, d, and e, we seek the number of loop-free undirected graphs on these vertices, where no vertex is isolated. Consequently, we want to count situations such as those illustrated in parts (a) and (b) of Fig. 8.3, but not situations such as those shown in parts (c) and (d).

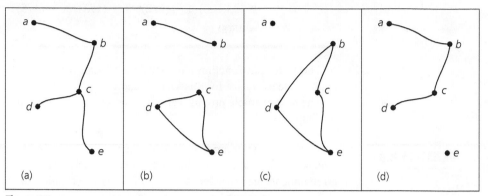

Figure 8.3

Let S be the set of loop-free undirected graphs G on $V = \{a, b, c, d, e\}$. Then $N = S_0 = |S| = 2^{10}$ because there are $\binom{5}{2} = 10$ possible two-way roads for these five villages, and each road can be either included or excluded.

For each $1 \leq i \leq 5$, let c_i be the condition that a system of these roads isolates village a, b, c, d, and e, respectively. Then the answer to the problem is $N(\overline{c}_1\overline{c}_2\overline{c}_3\overline{c}_4\overline{c}_5)$.

For condition c_1 village a is isolated, so we consider the six edges (roads) $\{b, c\}$, $\{b, d\}$, $\{b, e\}$, $\{c, d\}$, $\{c, e\}$, $\{d, e\}$. With two choices for each edge — namely, put the edge in the graph or leave the edge out — we find that $N(c_1) = 2^6$. Then by symmetry $N(c_i) = 2^6$ for all $2 \leq i \leq 5$, so $S_1 = \binom{5}{1}2^6$.

When villages a and b are to be isolated, each of the edges $\{c, d\}$, $\{d, e\}$, $\{c, e\}$ may be put in or left out of our graph. This results in 2^3 possibilities, so $N(c_1c_2) = 2^3$, and $S_2 = \binom{5}{2}2^3$.

Similar arguments tell us that $N(c_1c_2c_3) = 2^1$ and $S_3 = \binom{5}{3}2^1$; $N(c_1c_2c_3c_4) = 2^0$ and $S_4 = \binom{5}{4}2^0$; and $N(c_1c_2c_3c_4c_5) = 2^0$ and $S_5 = \binom{5}{5}2^0$.

Consequently,

$$N(\overline{c}_1\overline{c}_2\overline{c}_3\overline{c}_4\overline{c}_5) = 2^{10} - \binom{5}{1}2^6 + \binom{5}{2}2^3 - \binom{5}{3}2^1 + \binom{5}{4}2^0 - \binom{5}{5}2^0 = 768.$$

EXERCISES 8.1

1. Let S be a finite set with $|S| = N$ and let c_1, c_2, c_3, c_4 be four conditions, each of which may be satisfied by one or more of the elements of S. Prove that $N(\overline{c}_2\overline{c}_3\overline{c}_4) = N(c_1\overline{c}_2\overline{c}_3\overline{c}_4) + N(\overline{c}_1\overline{c}_2\overline{c}_3\overline{c}_4)$.

2. Establish the Principle of Inclusion and Exclusion by applying the Principle of Mathematical Induction to the number t of conditions.

3. Of the 100 students in Example 8.3, how many are taking (a) Fundamentals of Computer Programming but none of the other three courses; (b) Fundamentals of Computer Programming and Introduction to Economics but neither of the other two courses?

4. Annually, the 65 members of the maintenance staff sponsor a "Christmas in July" picnic for the 400 summer employees at their company. For these 65 people, 21 bring hot dogs, 35 bring fried chicken, 28 bring salads, 32 bring desserts, 13 bring hot dogs and fried chicken, 10 bring hot dogs and salads, 9 bring hot dogs and desserts, 12 bring fried chicken and salads, 17 bring fried chicken and desserts, 14 bring salads and desserts, 4 bring hot dogs, fried chicken, and salads, 6 bring hot dogs, fried chicken, and desserts, 5 bring hot dogs, salads, and desserts, 7 bring fried chicken, salads, and desserts, and 2 bring all four food items. Those (of the 65) who do not bring any of these four food items are responsible for setting up and cleaning up for the picnic. How many of the 65 maintenance staff will (a) help to set up and clean up for the picnic? (b) bring only hot dogs? (c) bring exactly one food item?

5. Determine the number of positive integers n, $1 \leq n \leq 2000$, that are

   a) not divisible by 2, 3, or 5

   b) not divisible by 2, 3, 5, or 7

   c) not divisible by 2, 3, or 5, but are divisible by 7

6. Determine how many integer solutions there are to $x_1 + x_2 + x_3 + x_4 = 19$, if

   a) $0 \leq x_i$ for all $1 \leq i \leq 4$

   b) $0 \leq x_i < 8$ for all $1 \leq i \leq 4$

   c) $0 \leq x_1 \leq 5, 0 \leq x_2 \leq 6, 3 \leq x_3 \leq 7, 3 \leq x_4 \leq 8$

7. In how many ways can one arrange all of the letters in the word INFORMATION so that no pair of consecutive letters occurs more than once? [Here we want to count arrangements such as IINNOOFRMTA and FORTMAIINON but not INFORIN-MOTA (where "IN" occurs twice) or NORTFNOIAMI (where "NO" occurs twice).]

8. Determine the number of integer solutions to $x_1 + x_2 + x_3 + x_4 = 19$ where $-5 \leq x_i \leq 10$ for all $1 \leq i \leq 4$.

9. Determine the number of positive integers x where $x \leq 9{,}999{,}999$ and the sum of the digits in x equals 31.

10. Professor Bailey has just completed writing the final examination for his course in advanced engineering mathematics. This examination has 12 questions, whose total value is to be 200 points. In how many ways can Professor Bailey assign the 200 points if each question must count for at least 10, but not more than 25, points and the point value for each question is to be a multiple of 5?

11. At Flo's Flower Shop, Flo wants to arrange 15 different plants on five shelves for a window display. In how many ways can she arrange them so that each shelf has at least one, but no more than four, plants?

12. In how many ways can Troy select nine marbles from a bag of twelve (identical except for color), where three are red, three blue, three white, and three green?

13. Find the number of permutations of a, b, c, ..., x, y, z, in which none of the patterns *spin*, *game*, *path*, or *net* occurs.

14. Answer the question in Example 8.10 for the case of six villages.

15. If eight distinct dice are rolled, what is the probability that all six numbers appear?

16. How many social security numbers (nine-digit sequences) have each of the digits 1, 3, and 7 appearing at least once?

17. In how many ways can three x's, three y's, and three z's be arranged so that no consecutive triple of the same letter appears?

18. Frostburg township sponsors four Boy Scout troops, each with 20 boys. If the head scoutmaster selects 50 of these boys to represent this township at the state jamboree, what is the probability that his selection will include at least one boy from each of the four troops?

19. If Zachary rolls a fair die five times, what is the probability that the sum of his five rolls is 20?

20. At a 12-week conference in mathematics, Sharon met seven of her friends from college. During the conference she met each friend at lunch 35 times, every pair of them 16 times, every trio eight times, every foursome four times, each set of five twice, and each set of six once, but never all seven at once. If she had lunch every day during the 84 days of the conference, did she ever have lunch alone?

21. Compute $\phi(n)$ for n equal to (a) 51; (b) 420; (c) 12300.

22. Compute $\phi(n)$ for n equal to (a) 5186; (b) 5187; (c) 5188.

23. Let $n \in \mathbf{Z}^+$. (a) Determine $\phi(2^n)$. (b) Determine $\phi(2^n p)$, where p is an odd prime.

24. For which $n \in \mathbf{Z}^+$ is $\phi(n)$ odd?

25. How many positive integers n less than 6000 (a) satisfy $\gcd(n, 6000) = 1$? (b) share a common prime divisor with 6000?

26. If $m, n \in \mathbf{Z}^+$, prove that $\phi(n^m) = n^{m-1}\phi(n)$.

27. Find three values for $n \in \mathbf{Z}^+$ where $\phi(n) = 16$.

28. For which positive integers n is $\phi(n)$ a power of 2?

29. For which positive integers n does 4 divide $\phi(n)$?

30. At an upcoming family reunion, five families — each consisting of a husband, wife, and one child — are to be seated around a circular table. In how many ways can these 15 people be arranged around the table so that no family is seated all together? (Here, as in Example 8.9, two seating arrangements are considered the same if one is a rotation of the other.)

8.2
Generalizations of the Principle

Consider a set S with $|S| = N$, and conditions $c_1, c_2, \ldots, c_t$ satisfied by some of the elements of S. In Section 8.1 we saw how the Principle of Inclusion and Exclusion provides a way to determine $N(\overline{c}_1\overline{c}_2 \cdots \overline{c}_t)$, the number of elements in S that satisfy none of the t conditions. If $m \in \mathbf{Z}^+$ and $1 \leq m \leq t$, we now want to determine E_m, which denotes the number of elements in S that satisfy *exactly m* of the t conditions. (At present we can obtain E_0.)

We can write formulas such as

$$E_1 = N(c_1\bar{c}_2\bar{c}_3 \cdots \bar{c}_t) + N(\bar{c}_1 c_2 \bar{c}_3 \cdots \bar{c}_t) + \cdots + N(\bar{c}_1 \bar{c}_2 \bar{c}_3 \cdots \bar{c}_{t-1} c_t),$$

and

$$E_2 = N(c_1 c_2 \bar{c}_3 \cdots \bar{c}_t) + N(c_1 \bar{c}_2 c_3 \cdots \bar{c}_t) + \cdots + N(\bar{c}_1 \bar{c}_2 \bar{c}_3 \cdots \bar{c}_{t-2} c_{t-1} c_t),$$

and although these results do not assist us as much as we should like, they will be a useful starting place as we examine the Venn diagrams for the cases where $t = 3$ and 4.

For Fig. 8.4, where $t = 3$, we place a numbered condition beside the circle representing those elements of S satisfying that particular condition and we also number each of the individual regions shown. Then E_1 equals the number of elements in regions 2, 3, and 4. But we can also write

$$E_1 = N(c_1) + N(c_2) + N(c_3) - 2\left[N(c_1 c_2) + N(c_1 c_3) + N(c_2 c_3)\right] + 3N(c_1 c_2 c_3).$$

In $N(c_1) + N(c_2) + N(c_3)$ we count the elements in regions 5, 6, and 7 twice and those in region 8 three times. In the next term, the elements in regions 5, 6, and 7 are deleted twice. We remove the elements in region 8 six times in $2\left[N(c_1 c_2) + N(c_1 c_3) + N(c_2 c_3)\right]$, so we then add on the term $3N(c_1 c_2 c_3)$ and end up not counting the elements in region 8 at all. Hence we have $E_1 = S_1 - 2S_2 + 3S_3 = S_1 - \binom{2}{1}S_2 + \binom{3}{2}S_3$.

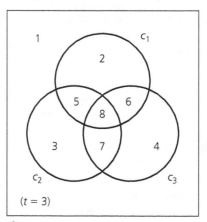

Figure 8.4

When we turn to E_2, our earlier formula indicates that we want to count the elements of S in regions 5, 6, and 7. From the Venn diagram,

$$E_2 = N(c_1 c_2) + N(c_1 c_3) + N(c_2 c_3) - 3N(c_1 c_2 c_3) = S_2 - 3S_3 = S_2 - \binom{3}{1}S_3,$$

and

$$E_3 = N(c_1 c_2 c_3) = S_3.$$

In Fig. 8.5, the conditions c_1, c_2, c_3 are associated with circular subsets of S, whereas c_4 is paired with the rather irregularly shaped area made up of regions 4, 8, 9, 11, 12, 13, 14, and 16. For each $1 \leq i \leq 4$, E_i is determined as follows:

E_1 [regions 2, 3, 4, 5]:

$$E_1 = [N(c_1) + N(c_2) + N(c_3) + N(c_4)]$$
$$- 2\,[N(c_1c_2) + N(c_1c_3) + N(c_1c_4) + N(c_2c_3) + N(c_2c_4) + N(c_3c_4)]$$
$$+ 3\,[N(c_1c_2c_3) + N(c_1c_2c_4) + N(c_1c_3c_4) + N(c_2c_3c_4)]$$
$$- 4N(c_1c_2c_3c_4)$$
$$= S_1 - 2S_2 + 3S_3 - 4S_4 = S_1 - \binom{2}{1}S_2 + \binom{3}{2}S_3 - \binom{4}{3}S_4.$$

Note: Taking an element in region 3, we find that it is counted once in E_1 and once in S_1 [in $N(c_3)$]. Taking an element in region 6, we find that it is not counted in E_1; it is counted twice in S_1 [in both $N(c_2)$ and $N(c_3)$] but removed twice in $2S_2$ [for it is counted once in S_2 in $N(c_2c_3)$], so overall it is not counted. The reader should now consider an element from region 12 and one from region 16 and show that each contributes a count of 0 to both sides of the formula for E_1.

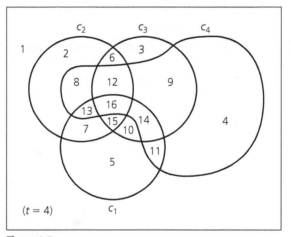

Figure 8.5

E_2 [regions 6–11]:

From Fig. 8.5, $E_2 = S_2 - 3S_3 + 6S_4 = S_2 - \binom{3}{1}S_3 + \binom{4}{2}S_4$. For details on this formula we examine the results in Table 8.1, where next to each summand of S_2, S_3, and S_4 we list the regions whose elements are counted in determining that particular summand. In calculating $S_2 - 3S_3 + 6S_4$ we find the elements in regions 6–11, which are precisely those that are to be counted for E_2.

Table 8.1

S_2	S_3	S_4
$N(c_1c_2)$: 7, 13, 15, 16	$N(c_1c_2c_3)$: 15, 16	$N(c_1c_2c_3c_4)$: 16
$N(c_1c_3)$: 10, 14, 15, 16	$N(c_1c_2c_4)$: 13, 16	
$N(c_1c_4)$: 11, 13, 14, 16	$N(c_1c_3c_4)$: 14, 16	
$N(c_2c_3)$: 6, 12, 15, 16	$N(c_2c_3c_4)$: 12, 16	
$N(c_2c_4)$: 8, 12, 13, 16		
$N(c_3c_4)$: 9, 12, 14, 16		

Finally, the elements for E_3 are found in regions 12–15, and $E_3 = S_3 - 4S_4 = S_3 - \binom{4}{1} S_4$; the elements for E_4 are those in region 16, and $E_4 = S_4$.

These results suggest the following theorem.

THEOREM 8.2

Under the hypotheses of Theorem 8.1, for each $1 \leq m \leq t$, the number of elements in S that satisfy *exactly m* of the conditions $c_1, c_2, \ldots, c_t$ is given by

$$E_m = S_m - \binom{m+1}{1} S_{m+1} + \binom{m+2}{2} S_{m+2} - \cdots + (-1)^{t-m} \binom{t}{t-m} S_t. \qquad (1)$$

(If $m = 0$, we obtain Theorem 8.1.)

Proof: Arguing as in Theorem 8.1, let $x \in S$ and consider the following three cases.

    a) When x satisfies fewer than m conditions, it contributes a count of 0 to each of the terms $E_m, S_m, S_{m+1}, \ldots, S_t$, so it is not counted on either side of the equation.

    b) When x satisfies exactly m of the conditions, it is counted once in E_m and once in S_m, but not in $S_{m+1}, \ldots, S_t$. Consequently, it is included once in the count for either side of the equation.

    c) Suppose x satisfies r of the conditions, where $m < r \leq t$. Then x contributes nothing to E_m. Yet it is counted $\binom{r}{m}$ times in S_m, $\binom{r}{m+1}$ times in $S_{m+1}, \ldots$, and $\binom{r}{r}$ times in S_r, but 0 times for any term beyond S_r. So on the right-hand side of the equation, x is counted $\binom{r}{m} - \binom{m+1}{1}\binom{r}{m+1} + \binom{m+2}{2}\binom{r}{m+2} - \cdots + (-1)^{r-m}\binom{r}{r-m}\binom{r}{r}$ times. For $0 \leq k \leq r - m$,

$$\binom{m+k}{k}\binom{r}{m+k} = \frac{(m+k)!}{k!\,m!} \cdot \frac{r!}{(m+k)!(r-m-k)!}$$

$$= \frac{r!}{m!} \cdot \frac{1}{k!(r-m-k)!} = \frac{r!}{m!(r-m)!} \cdot \frac{(r-m)!}{k!(r-m-k)!}$$

$$= \binom{r}{m}\binom{r-m}{k}.$$

Consequently, on the right-hand side of Eq. (1), x is counted

$$\binom{r}{m}\binom{r-m}{0} - \binom{r}{m}\binom{r-m}{1} + \binom{r}{m}\binom{r-m}{2} - \cdots + (-1)^{r-m}\binom{r}{m}\binom{r-m}{r-m}$$

$$= \binom{r}{m}\left[\binom{r-m}{0} - \binom{r-m}{1} + \binom{r-m}{2} - \cdots + (-1)^{r-m}\binom{r-m}{r-m}\right]$$

$$= \binom{r}{m}[1-1]^{r-m} = \binom{r}{m} \cdot 0 = 0 \text{ times,}$$

and the formula is verified.

Based on this result, if L_m denotes the number of elements of S (under the hypotheses of Theorem 8.1) that satisfy *at least m* of the t conditions, then we have the following formula.

COROLLARY 8.2

$$L_m = S_m - \binom{m}{m-1} S_{m+1} + \binom{m+1}{m-1} S_{m+2} - \cdots + (-1)^{t-m}\binom{t-1}{m-1} S_t.$$

Proof: A proof is outlined in the exercises at the end of this section.

When $m = 1$, the result in Corollary 8.2 becomes

$$L_1 = S_1 - \binom{1}{0}S_2 + \binom{2}{0}S_3 - \cdots + (-1)^{t-1}\binom{t-1}{0}S_t$$
$$= S_1 - S_2 + S_3 - \cdots + (-1)^{t-1}S_t.$$

Comparing this with the result in Theorem 8.1, we find that

$$L_1 = N - \overline{N} = |S| - \overline{N}.$$

This result is not much of a surprise, because an element x of S is counted in L_1 if it satisfies at least one of the conditions $c_1, c_2, c_3, \ldots, c_t$ — that is, if $x \in S$ and x is not counted in $\overline{N} = N(\overline{c}_1\overline{c}_2\overline{c}_3 \cdots \overline{c}_t)$.

EXAMPLE 8.11	Looking back to Example 8.10, we shall find the numbers of systems of two-way roads so that exactly (E_2) and at least (L_2) two of the villages remain isolated.

The previously calculated results for this example show

$$E_2 = S_2 - \binom{3}{1}S_3 + \binom{4}{2}S_4 - \binom{5}{3}S_5 = 80 - 3(20) + 6(5) - 10(1) = 40,$$
$$L_2 = S_2 - \binom{2}{1}S_3 + \binom{3}{1}S_4 - \binom{4}{1}S_5 = 80 - 2(20) + 3(5) - 4(1) = 51.$$

EXERCISES 8.2

1. For the situation in Examples 8.10 and 8.11 compute E_i for $0 \le i \le 5$ and show that $\sum_{i=0}^{5} E_i = N = |S|$.

2. a) In how many ways can the letters in ARRANGEMENT be arranged so that there are exactly two pairs of consecutive identical letters? at least two pairs of consecutive identical letters?

b) Answer part (a), replacing two with three.

3. In how many ways can one arrange the letters in CORRESPONDENTS so that (a) there is no pair of consecutive identical letters? (b) there are exactly two pairs of consecutive identical letters? (c) there are at least three pairs of consecutive identical letters?

4. Let $A = \{1, 2, 3, \ldots, 10\}$, and $B = \{1, 2, 3, \ldots, 7\}$. How many functions $f: A \to B$ satisfy $|f(A)| = 4$? How many have $|f(A)| \le 4$?

5. In how many ways can one distribute ten distinct prizes among four students with exactly two students getting nothing? How many ways have at least two students getting nothing?

6. Zelma is having a luncheon for herself and nine of the women in her tennis league. On the morning of the luncheon she places name cards at the ten places at her table and then leaves to run a last-minute errand. Her husband, Herbert, comes home from his morning tennis match and unfortunately leaves the back door open. A gust of wind scatters the ten name cards. In how many ways can Herbert replace the ten cards at the places at the table so that exactly four of the ten women will be seated where Zelma had wanted them? In how many ways will at least four of them be seated where they were supposed to be?

7. If 13 cards are dealt from a standard deck of 52, what is the probability that these 13 cards include (a) at least one card from each suit? (b) exactly one void (for example, no clubs)? (c) exactly two voids?

8. The following provides an outline for proving Corollary 8.2. Fill in the needed details.

a) First note that $E_t = L_t = S_t$.

b) What is E_{t-1}, and how are L_t and L_{t-1} related?

c) Show that $L_{t-1} = S_{t-1} - \binom{t-1}{t-2}S_t$.

d) For all $1 \le m \le t-1$, how are L_m, L_{m+1}, and E_m related?

e) Using the results in steps (a) through (d), establish the corollary by a backward type of induction.

8.3
Derangements: Nothing Is in Its Right Place

In elementary calculus the Maclaurin series for the exponential function is given by

$$e^x = 1 + x + \frac{x^2}{2!} + \frac{x^3}{3!} + \cdots = \sum_{n=0}^{\infty} \frac{x^n}{n!},$$

so

$$e^{-1} = \sum_{n=0}^{\infty} \frac{(-1)^n}{n!} = 1 - 1 + \frac{1}{2!} - \frac{1}{3!} + \cdots.$$

To five places, $e^{-1} = 0.36788$ and $1 - 1 + (1/2!) - (1/3!) + \cdots - (1/7!) \doteq 0.36786$. Consequently, for all $k \in \mathbf{Z}^+$, if $k \geq 7$, then $\sum_{n=0}^{k}((-1)^n)/n!$ is a very good approximation to e^{-1}.

We find these ideas helpful in working some of the following examples.

EXAMPLE 8.12

While at the racetrack, Ralph bets on each of the ten horses in a race to come in according to how they are favored. In how many ways can they reach the finish line so that he loses all of his bets?

Removing the words *horses* and *racetrack* from the problem, we really want to know in how many ways we can arrange the numbers 1, 2, 3, ..., 10 so that 1 is not in first place (its natural position), 2 is not in second place (its natural position), ..., and 10 is not in tenth place (its natural position). These arrangements are called the *derangements* of 1, 2, 3, ..., 10.

The Principle of Inclusion and Exclusion provides the key to calculating the number of derangements. For each $1 \leq i \leq 10$, an arrangement of 1, 2, 3, ..., 10 is said to satisfy condition c_i if integer i is in the ith place. We obtain the number of derangements, denoted by d_{10}, as follows:

$$d_{10} = N(\bar{c}_1 \bar{c}_2 \bar{c}_3 \cdots \bar{c}_{10}) = 10! - \binom{10}{1}9! + \binom{10}{2}8! - \binom{10}{3}7! + \cdots + \binom{10}{10}0!$$

$$= 10! \left[1 - \binom{10}{1}(9!/10!) + \binom{10}{2}(8!/10!) - \binom{10}{3}(7!/10!) + \cdots + \binom{10}{10}(0!/10!) \right]$$

$$= 10! [1 - 1 + (1/2!) - (1/3!) + \cdots + (1/10!)] \doteq (10!)(e^{-1}).$$

The sample space here consists of the 10! ways the horses can finish. So the *probability* that Ralph will lose every bet is approximately $(10!)(e^{-1})/(10!) = e^{-1}$. This probability remains (more or less) the same if the number of horses in the race is 11, 12, ... On the other hand, for n horses, where $n \geq 10$, the probability that our gambler wins at least one of his bets is approximately $1 - e^{-1} \doteq 0.63212$.

EXAMPLE 8.13

The number of derangements of 1, 2, 3, 4 is

$$d_4 = 4![1 - 1 + (1/2!) - (1/3!) + (1/4!)]$$
$$= 4![(1/2!) - (1/3!) + (1/4!)] = (4)(3) - 4 + 1 = 9.$$

These nine derangements are

2143	3142	4123
2341	3412	4312
2413	3421	4321.

Among the $24 - 9 = 15$ permutations of $1, 2, 3, 4$ that are *not* derangements one finds 1234, 2314, 3241, 1342, 2431, and 2314.

EXAMPLE 8.14	Peggy has seven books to review for the C–H Company, so she hires seven people to review them. She wants two reviews per book, so the first week she gives each person one book to read and then redistributes the books at the start of the second week. In how many ways can she make these two distributions so that she gets two reviews (by different people) of each book?

She can distribute the books in 7! ways the first week. Numbering both the books and the reviewers (for the first week) as $1, 2, \ldots, 7$, for the second distribution she must arrange these numbers so that none of them is in its natural position. This she can do in d_7 ways. By the rule of product, she can make the two distributions in $(7!)d_7 \doteq (7!)^2(e^{-1})$ ways.

EXERCISES 8.3

1. In how many ways can the integers $1, 2, 3, \ldots, 10$ be arranged in a line so that no even integer is in its natural position?

2. a) List all the derangements of $1, 2, 3, 4, 5$ where the first three numbers are 1, 2, and 3, in some order.

b) List all the derangements of $1, 2, 3, 4, 5, 6$ where the first three numbers are 1, 2, and 3, in some order.

3. How many derangements are there for $1, 2, 3, 4, 5$?

4. How many permutations of $1, 2, 3, 4, 5, 6, 7$ are not derangements?

5. a) Let $A = \{1, 2, 3, \ldots, 7\}$. A function $f: A \to A$ is said to have a *fixed point* if for some $x \in A$, $f(x) = x$. How many one-to-one functions $f: A \to A$ have at least one fixed point?

b) In how many ways can we devise a secret code by assigning to each letter of the alphabet a different letter to represent it?

6. How many derangements of $1, 2, 3, 4, 5, 6, 7, 8$ start with (a) 1, 2, 3, and 4, in some order? (b) 5, 6, 7, and 8, in some order?

7. For the positive integers $1, 2, 3, \ldots, n - 1, n$, there are 11,660 derangements where 1, 2, 3, 4, and 5 appear in the first five positions. What is the value of n?

8. Four applicants for a job are to be interviewed for 30 minutes each: 15 minutes with each of supervisors Nancy and Yolanda. (The interviews are in separate rooms, and interviewing starts at 9:00 A.M.) (a) In how many ways can these interviews be scheduled during a one-hour period? (b) One applicant, named Josephine, arrives at 9:00 A.M. What is the probability that she will have her two interviews one after the other? (c) Regina, another applicant, arrives at 9:00 A.M. and

hopes to be finished in time to leave by 9:50 A.M. for another appointment. What is the probability that Regina will be able to leave on time?

9. In how many ways can Mrs. Ford distribute ten distinct books to her ten children (one book to each child) and then collect and redistribute the books so that each child has the opportunity to peruse two different books?

10. a) When n balls, numbered $1, 2, 3, \ldots, n$ are taken in succession from a container, a *rencontre* occurs if the mth ball withdrawn is numbered m, for some $1 \le m \le n$. Find the probability of getting (i) no rencontres; (ii) (exactly) one rencontre, (iii) at least one rencontre; and (iv) r rencontres, where $1 \le r \le n$.

b) Approximate the answers to the questions in part (a).

11. Ten women attend a business luncheon. Each woman checks her coat and attaché case. Upon leaving, each woman is given a coat and case at random. (a) In how many ways can the coats and cases be distributed so that no woman gets either of her possessions? (b) In how many ways can they be distributed so that no woman gets back both of her possessions?

12. Ms. Pezzulo teaches geometry and then biology to a class of 12 advanced students in a classroom that has only 12 desks. In how many ways can she assign the students to these desks so that (a) no student is seated at the same desk for both classes? (b) there are exactly six students each of whom occupies the same desk for both classes?

13. Give a combinatorial argument to verify that for all $n \in \mathbf{Z}^+$,

$$n! = \binom{n}{0}d_0 + \binom{n}{1}d_1 + \binom{n}{2}d_2 + \cdots + \binom{n}{n}d_n = \sum_{k=0}^{n} \binom{n}{k}d_k.$$

(For each $1 \le k \le n$, $d_k =$ the number of derangements of $1, 2, 3, \ldots, k$; $d_0 = 1$.)

14. a) In how many ways can the integers $1, 2, 3, \ldots, n$ be arranged in a line so that none of the patterns $12, 23, 34, \ldots, (n-1)n$ occurs?

b) Show that the result in part (a) equals $d_{n-1} + d_n$. (d_n = the number of derangements of $1, 2, 3, \ldots, n$.)

15. Answer part (a) of Exercise 14 if the numbers are arranged in a circle, and, as we count clockwise about the circle, none of the patterns $12, 23, 34, \ldots, (n-1)n, n1$ occurs.

16. What is the probability that the gambler in Example 8.12 wins (a) (exactly) five of his bets? (b) at least five of his bets?

8.4
Rook Polynomials

Consider the six-square "chessboard" shown in Fig. 8.6 (*Note:* The shaded squares are not part of the chessboard.). In chess a piece called a *rook* or *castle* is allowed at one turn to be moved horizontally or vertically over as many unoccupied spaces as one wishes. Here a rook in square 3 of the figure could be moved in one turn to squares 1, 2, or 4. A rook at square 5 could be moved to square 6 or square 2 (even though there is no square between squares 5 and 2).

For $k \in \mathbf{Z}^+$ we want to determine the number of ways in which k rooks can be placed on the unshaded squares of this chessboard so that no two of them can take each other — that is, no two of them are in the same row or column of the chessboard. This number is denoted by r_k, or by $r_k(C)$ if we wish to stress that we are working on a particular chessboard C.

For any chessboard, r_1 is the number of squares on the board. Here $r_1 = 6$. Two nontaking rooks can be placed at the following pairs of positions: $\{1, 4\}, \{1, 5\}, \{2, 4\}, \{2, 6\}, \{3, 5\}, \{3, 6\}, \{4, 5\}$, and $\{4, 6\}$, so $r_2 = 8$. Continuing, we find that $r_3 = 2$, using the locations $\{1, 4, 5\}$ and $\{2, 4, 6\}$; $r_k = 0$, for $k \geq 4$.

With $r_0 = 1$, the *rook polynomial*, $r(C, x)$, for the chessboard in Fig. 8.6 is defined as $r(C, x) = 1 + 6x + 8x^2 + 2x^3$. For each $k \geq 0$, the coefficient of x^k is the number of ways we can place k nontaking rooks on chessboard C.

What we have done here (using a case-by-case analysis) soon proves tedious. As the size of the board increases, we have to consider cases wherein numbers such as r_4 and r_5 are nonzero. Consequently, we shall now make some observations that will allow us to make use of small boards and somehow break up a large board into smaller *subboards*.

The chessboard C in Fig. 8.7 is made up of 11 unshaded squares. We note that C consists of a 2×2 subboard C_1 located in the upper left corner and a seven-square subboard C_2 located in the lower right corner. These subboards are *disjoint* because they have no squares in the same row or column of C.

Calculating as we did for our first chessboard, here we find

$$r(C_1, x) = 1 + 4x + 2x^2, \qquad r(C_2, x) = 1 + 7x + 10x^2 + 2x^3,$$
$$r(C, x) = 1 + 11x + 40x^2 + 56x^3 + 28x^4 + 4x^5 = r(C_1, x) \cdot r(C_2, x).$$

Figure 8.6

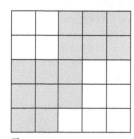

Figure 8.7

Hence $r(C, x) = r(C_1, x) \cdot r(C_2, x)$. But did this occur by luck or is something happening here that we should examine more closely? For example, to obtain r_3 for C, we need to know in how many ways three nontaking rooks can be placed on board C. These fall into three cases:

a) All three rooks are on subboard C_2 (and none is on C_1): $(2)(1) = 2$ ways.

b) Two rooks are on subboard C_2 and one is on C_1: $(10)(4) = 40$ ways.

c) One rook is on subboard C_2 and two are on C_1: $(7)(2) = 14$ ways.

Consequently, three nontaking rooks can be placed on board C in $(2)(1) + (10)(4) + (7)(2) = 56$ ways. Here we see that 56 arises just as the coefficient of x^3 does in the product $r(C_1, x) \cdot r(C_2, x)$.

In general, if C is a chessboard made up of *pairwise* disjoint subboards $C_1, C_2, \ldots, C_n$, then $r(C, x) = r(C_1, x)r(C_2, x) \cdots r(C_n, x)$.

The last result for this section demonstrates the type of principle we have seen in other results in combinatorial and discrete mathematics: Given a large chessboard, break it into smaller subboards whose rook polynomials can be determined by inspection.

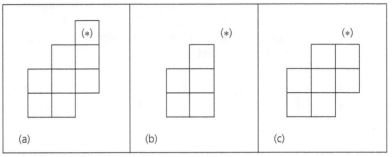

(a)   (b)   (c)

Figure 8.8

Consider chessboard C in Fig. 8.8(a). For $k \geq 1$, suppose we wish to place k nontaking rooks on C. For each square of C, such as the one designated by $(*)$, there are two possibilities to examine.

a) Place one rook on the designated square. Then we remove, as possible locations for the other $k - 1$ rooks, all other squares of C in the same row or column as the designated square. We use C_s to denote the remaining smaller subboard [seen in Fig. 8.8(b)].

b) We do not use the designated square at all. The k rooks are placed on the subboard C_e [C with the one designated square eliminated — as shown in Fig. 8.8(c)].

Since these two cases are all-inclusive and mutually disjoint,

$$r_k(C) = r_{k-1}(C_s) + r_k(C_e).$$

From this we see that

$$r_k(C)x^k = r_{k-1}(C_s)x^k + r_k(C_e)x^k. \tag{1}$$

If n is the number of squares in the chessboard (here n is 8), then Eq. (1) is valid for all $1 \leq k \leq n$, and we write

$$\sum_{k=1}^{n} r_k(C)x^k = \sum_{k=1}^{n} r_{k-1}(C_s)x^k + \sum_{k=1}^{n} r_k(C_e)x^k. \tag{2}$$

For Eq. (2) we realize that the summations may stop before $k = n$. We have seen cases, as in Fig. 8.6, where r_n and some prior r_k's are 0. The summations start at $k = 1$, for otherwise we could find ourselves with the term $r_{-1}(C_s)x^0$ in the first summand on the right-hand side of Eq. (2).

Equation (2) may be rewritten as

$$\sum_{k=1}^{n} r_k(C)x^k = x \sum_{k=1}^{n} r_{k-1}(C_s)x^{k-1} + \sum_{k=1}^{n} r_k(C_e)x^k \tag{3}$$

or

$$1 + \sum_{k=1}^{n} r_k(C)x^k = x \cdot r(C_s, x) + \sum_{k=1}^{n} r_k(C_e)x^k + 1,$$

from which it follows that

$$r(C, x) = x \cdot r(C_s, x) + r(C_e, x). \tag{4}$$

We now use this final equation to determine the rook polynomial for the chessboard shown in part (a) of Fig. 8.8. Each time the idea in Eq. (4) is used, we mark the special square we are using with $(*)$. Parentheses are placed about each chessboard to denote the rook polynomial of the board.

$$= x^2(1 + 2x) + 2x(1 + 4x + 2x^2) + x(1 + 3x + x^2)$$

$$= 3x + 12x^2 + 7x^3 + x(1 + 2x) + (1 + 4x + 2x^2) = 1 + 8x + 16x^2 + 7x^3.$$

8.5
Arrangements with Forbidden Positions

The rook polynomials of the previous section seem interesting on their own. Now we shall find them useful in solving the following problems.

EXAMPLE 8.15

In making seating arrangements for their son's wedding reception, Grace and Nick are down to four relatives, denoted R_i, for $1 \le i \le 4$, who do not get along with one another. There is a single open seat at each of the five tables T_j, where $1 \le j \le 5$. Because of family differences,

a) R_1 will not sit at T_1 or T_2.

b) R_2 will not sit at T_2.

c) R_3 will not sit at T_3 or T_4.

d) R_4 will not sit at T_4 or T_5.

This situation is represented in Fig. 8.9. The number of ways we can seat these four people at four different tables, and satisfy conditions (a) through (d), is the number of ways four nontaking rooks can be placed on the chessboard made up of the *unshaded* squares. However, since there are only seven shaded squares, as opposed to thirteen unshaded ones, it would be easier to work with the shaded chessboard.

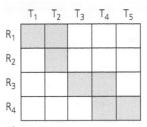

Figure 8.9

We start with the conditions that are required for us to apply the Principle of Inclusion and Exclusion: For each $1 \leq i \leq 4$, let c_i be the condition where a seating assignment of these four people (at different tables) is made with relative R_i in a forbidden (shaded) position. As usual, $|S|$ denotes the total number of ways we can place the four relatives, one to a table. Then $|S| = N = S_0 = 5!$

To determine S_1 we consider each of the following:

- $N(c_1) = 4! + 4!$, for there are $4!$ ways to seat R_2, R_3, and R_4 if R_1 is in forbidden position T_1 and another $4!$ ways if R_1 is at table T_2, his or her other forbidden position.

- $N(c_2) = 4!$, for after placing R_2 at forbidden table T_2, we must place R_1, R_3, and R_4 at T_1, T_3, T_4, and T_5, one person to a table.

- $N(c_3) = 4! + 4!$, one summand for R_3 being in forbidden position T_3, and the other summand for R_3 being in the forbidden position T_4.

- $N(c_4) = 4! + 4!$, each of the two summands arising when R_4 is placed at each of the two forbidden positions T_4 and T_5.

Hence $S_1 = 7(4!)$.

Turning to S_2 we have these considerations:

- $N(c_1 c_2) = 3!$, because after we place R_1 at T_1 and R_2 at T_2, there are three tables (T_3, T_4, and T_5) where R_3 and R_4 can be seated.

- $N(c_1 c_3) = 3! + 3! + 3! + 3!$, because there are four cases where R_1 and R_3 are located at forbidden positions:

  i) R_1 at T_1; R_3 at T_3 **ii)** R_1 at T_2; R_3 at T_3

  iii) R_1 at T_1; R_3 at T_4 **iv)** R_1 at T_2; R_3 at T_4.

In a similar manner we find that $N(c_1 c_4) = 4(3!)$, $N(c_2 c_3) = 2(3!)$, $N(c_2 c_4) = 2(3!)$, and $N(c_3 c_4) = 3(3!)$. Consequently, $S_2 = 16(3!)$.

Before continuing, we make a few observations about S_1 and S_2. For S_1 we have $7(4!) = 7(5 - 1)!$, where 7 is the number of shaded squares in Fig. 8.9. Also, $S_2 = 16(3!) = 16(5 - 2)!$, where 16 is the number of ways two nontaking rooks can be placed on the shaded chessboard.

In general, for all $0 \leq i \leq 4$, $S_i = r_i (5 - i)!$, where r_i is the number of ways in which it is possible to place i nontaking rooks on the shaded chessboard shown in Fig. 8.9.

Consequently, to expedite the solution of this problem, we turn to $r(C, x)$, the rook polynomial of this shaded chessboard. Using the decomposition of C into the disjoint subboards in the upper left and lower right corners, we find that

$$r(C, x) = (1 + 3x + x^2)(1 + 4x + 3x^2) = 1 + 7x + 16x^2 + 13x^3 + 3x^4,$$

so

$$N(\bar{c}_1\bar{c}_2\bar{c}_3\bar{c}_4) = S_0 - S_1 + S_2 - S_3 + S_4 = 5! - 7(4!) + 16(3!) - 13(2!) + 3(1!)$$

$$= \sum_{i=0}^{4}(-1)^i r_i (5 - i)! = 25.$$

Grace and Nick can breathe a sigh of relief. There are 25 ways in which they can seat these last four relatives at the reception and avoid any squabbling.

The next example demonstrates how a bit of rearranging of our chessboard can help in our calculations.

EXAMPLE 8.16

We have a pair of dice; one is red, the other green. We roll these dice six times. What is the probability that we obtain all six values on both the red die and the green die if we know that the ordered pairs $(1, 2)$, $(2, 1)$, $(2, 5)$, $(3, 4)$, $(4, 1)$, $(4, 5)$, and $(6, 6)$ did not occur? [Here an ordered pair (a, b) indicates a on the red die and b on the green.]

Recognizing this problem as one dealing with permutations and forbidden positions, we construct the chessboard shown in Fig. 8.10(a), where the row labels represent the outcome on the red die, the column labels the outcome on the green die, and the shaded squares constitute the forbidden positions. In this figure the shaded squares are scattered. Relabeling the rows and columns, we can redraw the chessboard as shown in Fig. 8.10(b), where we have taken shaded squares in the same row (or column) of the board shown in part (a) and made them adjacent. In Fig. 8.10(b), the chessboard C (of seven shaded squares) is the union of four pairwise disjoint subboards, and so

$$r(C, x) = (1 + 4x + 2x^2)(1 + x)^3 = 1 + 7x + 17x^2 + 19x^3 + 10x^4 + 2x^5.$$

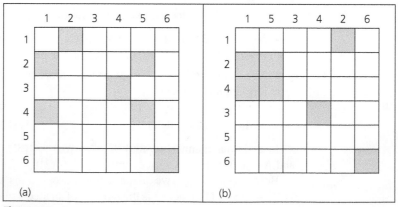

Figure 8.10

For each $1 \le i \le 6$, define c_i as the condition where, having rolled the dice six times, we find that all six values occur on both the red die and the green die, but i on the red die is paired with one of the forbidden numbers on the green die. [Note that $N(c_5) = 0$.] Then the number of (ordered) sequences of the six rolls of the dice for the event we are interested in is

$$(6!)N(\overline{c}_1\overline{c}_2\overline{c}_3\overline{c}_4\overline{c}_5\overline{c}_6) = (6!)\sum_{i=0}^{6}(-1)^i S_i = (6!)\sum_{i=0}^{6}(-1)^i r_i(6-i)!$$

$$= 6![6! - 7(5!) + 17(4!) - 19(3!) + 10(2!) - 2(1!) + 0(0!)]$$

$$= 6![192] = 138,240.$$

Since the sample space consists of all sequences of six ordered pairs selected with repetition from the 29 unshaded squares of the chessboard, the probability of this event is $138,240/(29)^6 \doteq 0.00023$.

Our last example provides a unifying idea for what we have done in this section.

EXAMPLE 8.17

Let $A = \{1, 2, 3, 4\}$ and $B = \{u, v, w, x, y, z\}$. How many one-to-one functions $f: A \to B$ satisfy none of the following conditions:

$$c_1: f(1) = u \text{ or } v \qquad c_2: f(2) = w \qquad c_3: f(3) = w \text{ or } x \qquad c_4: f(4) = x, y, \text{ or } z$$

As in our two prior examples, we construct a chessboard, as shown in Fig. 8.11. Here we are really interested in the chessboard C made up of the eight shaded squares (which comprise two disjoint subboards). Now

$$r(C, x) = (1 + 2x)(1 + 6x + 9x^2 + 2x^3) = 1 + 8x + 21x^2 + 20x^3 + 4x^4.$$

So

$$N(\overline{c}_1\overline{c}_2\overline{c}_3\overline{c}_4) = S_0 - S_1 + S_2 - S_3 + S_4$$

$$= (6!/2!) - 8(5!/2!) + 21(4!/2!) - 20(3!/2!) + 4(2!/2!)$$

$$= \sum_{i=0}^{4}(-1)^i r_i(6-i)!/2! = 76$$

and there are 76 one-to-one functions $f: A \to B$ where none of the conditions c_1, c_2, c_3, c_4 is satisfied.

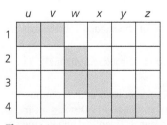

Figure 8.11

Even more so, look back at $N(\overline{c}_1\overline{c}_2\overline{c}_3\overline{c}_4)$ in Example 8.15. Disregarding the vocabulary of the "relatives" and "tables," we realize that we are counting the number of one-to-one functions $g: \{R_1, R_2, R_3, R_4\} \to \{T_1, T_2, T_3, T_4, T_5\}$ where none of the conditions c_1, c_2, c_3, c_4 is satisfied. (The situation is similar for $N(\overline{c}_1\overline{c}_2\overline{c}_3\overline{c}_4\overline{c}_5\overline{c}_6)$ in Example 8.16.)

Finally, for $A = \{1, 2, 3, 4, 5, 6, 7, 8\}$, suppose we want to count the number of one-to-one functions $h: A \to A$ where $h(i) \neq i$ for all $i \in A$. Here the rook polynomial would be

$$r(C, x) = (1 + x)^8 = \sum_{k=0}^{8} \binom{8}{k} x^k$$

and we find that the number of such one-to-one functions h is

$$\binom{8}{0} 8! - \binom{8}{1} 7! + \binom{8}{2} 6! - \binom{8}{3} 5! + \cdots + \binom{8}{8} 0!$$

$$= 8! \left[1 - 1 + \frac{1}{2!} - \frac{1}{3!} + \cdots + \frac{1}{8!} \right]$$

$$= d_8, \text{ the number of derangements of } 1, 2, 3, \ldots, 8.$$

EXERCISES 8.4 AND 8.5

1. Verify directly the rook polynomials for (a) the unshaded chessboards in Figs. 8.7 and 8.8(a), and (b) the shaded chessboards in Figs. 8.9 and 8.10(b).

2. Construct or describe a smallest (least number of squares) chessboard for which $r_{10} \neq 0$.

3. a) Find the rook polynomial for the standard 8×8 chessboard.

b) Answer part (a) with 8 replaced by n, for $n \in \mathbf{Z}^+$.

4. Find the rook polynomials for the shaded chessboards in Fig. 8.12.

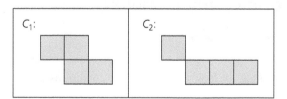

Figure 8.12

5. a) Find the rook polynomials for the shaded chessboards in Fig. 8.13.

b) Generalize the chessboard (and rook polynomial) for Fig. 8.13(i).

6. a) Let C be a chessboard that has m rows and n columns, with $m \leq n$ (for a total of mn squares). For $0 \leq k \leq m$, in how many ways can we arrange k (identical) nontaking rooks on C?

b) For the chessboard C in part (a), determine the rook polynomial $r(C, x)$.

7. Professor Ruth has five graders to correct programs in her courses in Java, C++, SQL, Perl, and VHDL. Graders Jeanne

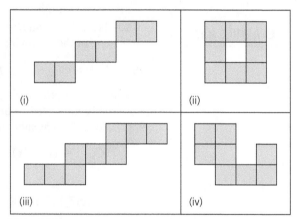

Figure 8.13

and Charles both dislike SQL, Sandra wants to avoid C++ and VHDL. Paul detests Java and C++, and Todd refuses to work in SQL and Perl. In how many ways can Professor Ruth assign each grader to correct programs in one language, cover all five languages, and keep everyone content?

8. Why do we have 6! in the term $(6!) N(\bar{c}_1 \bar{c}_2 \cdots \bar{c}_6)$ for the solution of Example 8.16?

9. Five professors named Al, Violet, Lynn, Jack, and Mary Lou are to be assigned to teach one class each from among calculus I, calculus II, calculus III, statistics, and combinatorics. Al will not teach calculus II or combinatorics, Lynn cannot stand statistics, Violet and Mary Lou both refuse to teach calculus I or calculus III, and Jack detests calculus II.

a) In how many ways can the head of the mathematics department assign each of these professors one of these five courses and still keep peace in the department?

b) For the assignments in part (a), what is the probability that Violet will get to teach combinatorics?

10. A pair of dice, one red and the other green, is rolled six times. We know that the ordered pairs $(1, 1)$, $(1, 5)$, $(2, 4)$, $(3, 6)$, $(4, 2)$, $(4, 4)$, $(5, 1)$, and $(5, 5)$ did not come up. What is the probability that every value came up on both the red die and the green one?

11. A computer dating service wants to match each of four women with one of six men. According to the information these applicants provided when they joined the service, we can draw the following conclusions.

- Woman 1 would not be compatible with man 1, 3, or 6.

- Woman 2 would not be compatible with man 2 or 4.

- Woman 3 would not be compatible with man 3 or 6.

- Woman 4 would not be compatible with man 4 or 5.

In how many ways can the service successfully match each of the four women with a compatible partner?

12. For $A = \{1, 2, 3, 4, 5\}$ and $B = \{u, v, w, x, y, z\}$, determine the number of one-to-one functions $f: A \rightarrow B$ where $f(1) \neq v, w$; $f(2) \neq u, w$; $f(3) \neq x$; and $f(4) \neq v, x, y$.

8.6
Summary and Historical Review

In the first and third chapters of this text we were concerned with enumeration problems in which we had to be careful of situations wherein arrangements or selections were overcounted. This situation became even more involved in Chapter 5 when we tried to count the number of onto functions for two finite sets.

With Venn diagrams to lead the way, in this chapter we obtained a pattern called the Principle of Inclusion and Exclusion. Using this principle, we restated each problem in terms of conditions and subsets. Using enumeration formulas on permutations and combinations that were developed earlier, we solved some simpler subproblems and let the principle manage our concern about overcounting. As a result, we were able to solve a variety of problems, some dealing with number theory and one with graph theory. We also proved the formula conjectured earlier in Section 5.3 for the number of onto functions for two finite sets.

This principle has an interesting history, being found in different manuscripts under such names as the "Sieve Method" or the "Principle of Cross Classification." A set-theoretic version of the principle, which concerned itself with set unions and intersections, is found in *Doctrine of Chances* (1718), a text on probability theory by Abraham DeMoivre (1667–1754). Somewhat earlier, in 1708, Pierre Rémond de Montmort (1678–1719) used the idea behind the principle in his solution of the problem generally known as *le problème des rencontres* (matches). (In this old French card game the 52 cards in a first deck are arranged face up in a row — perhaps on a table. Then the 52 cards of a second deck are dealt, with one new card being placed on each of the 52 cards previously arranged on the table top. The score for the game is determined by counting the resulting matches, where both the suit and the face value for each of the two cards must match.)

Credit for the way we developed and dealt with the Principle of Inclusion and Exclusion belongs to James Joseph Sylvester (1814–1897). (This colorful English-born mathematician also made major contributions in the theory of equations; the theory of matrices and determinants; and invariant theory, which he founded with Arthur Cayley (1821–1895). In addition Sylvester founded the *American Journal of Mathematics*, the first American journal established for mathematical research.) The importance of the inclusion-exclusion technique was not generally appreciated, however, until somewhat later, when the publication *Choice and Chance* by W. A. Whitworth [10] made mathematicians more aware of its potential and use.

James Joseph Sylvester (1814–1897)

For more on the application of this principle, examine Chapter 4 of C. L. Liu [4], Chapter 2 of H. J. Ryser [8], or Chapter 8 of A. Tucker [9]. More number-theoretic results related to the principle, including the Möbius inversion formula, can be found in Chapter 2 of M. Hall [1], Chapter X of C. L. Liu [5], and Chapter 16 of G. H. Hardy and E. M. Wright [3]. An extension of this formula is given in the article by G. C. Rota [7].

The article by D. Hanson, K. Seyffarth, and J. H. Weston [2] provides an interesting generalization of the derangement problem discussed in Section 8.3. The ideas behind the rook polynomials and their applications were developed in the late 1930s and during the 1940s and 1950s. Additional material on this topic is found in Chapters 7 and 8 of J. Riordan [6].

REFERENCES

1. Hall, Marshall, Jr. *Combinatorial Theory*. Waltham, Mass.: Blaisdell, 1967.
2. Hanson, Denis, Seyffarth, Karen, and Weston, J. Harley. "Matchings, Derangements, Rencontres." *Mathematics Magazine* 56, no. 4 (September 1983): pp. 224–229.
3. Hardy, Godfrey Harold, and Wright, Edward Maitland. *An Introduction to the Theory of Numbers*, 5th ed. Oxford: Oxford University Press, 1979.
4. Liu, C. L. *Introduction to Combinatorial Mathematics*. New York: McGraw-Hill, 1968.
5. Liu, C. L. *Topics in Combinatorial Mathematics*. Mathematical Association of America, 1972.
6. Riordan, John. *An Introduction to Combinatorial Analysis*. Princeton, N.J.: Princeton University Press, 1980. (Originally published in 1958 by John Wiley & Sons.)
7. Rota, Gian Carlo. "On the Foundations of Combinatorial Theory, I. Theory of Möbius Functions." *Zeitschrift für Wahrscheinlichkeits Theorie* 2 (1964): pp. 340–368.
8. Ryser, Herbert J. *Combinatorial Mathematics*. Carus Mathematical Monograph, No. 14. Published by the Mathematical Association of America, distributed by John Wiley & Sons, New York, 1963.
9. Tucker, Alan. *Applied Combinatorics*, 4th ed. New York: Wiley, 2002.
10. Whitworth, William Allen. *Choice and Chance*. Originally published at Cambridge in 1867. Reprint of the 5th ed. (1901), Hafner, New York, 1965.

SUPPLEMENTARY EXERCISES

1. Determine how many $n \in \mathbf{Z}^+$ satisfy $n \le 500$ and are not divisible by 2, 3, 5, 6, 8, or 10.

2. How many integers n are such that $0 \le n < 1,000,000$ and the sum of the digits in n is less than or equal to 37?

3. At next week's church bazaar, Joseph and his cousin Jeffrey must arrange six baseballs, six footballs, six soccer balls, and six volleyballs on the four shelves in the sports booth sponsored by their Boy Scout troop. In how many ways can they do this so that there are at least two, but no more than seven, balls on each shelf? (Here all six balls for any one of the four sports are identical in appearance.)

4. Find the number of positive integers n where $1 \le n \le 1000$ and n is *not* a perfect square, cube, or fourth power.

5. In how many ways can we arrange the integers 1, 2, 3, ..., 8 in a line so that there are no occurrences of the patterns 12, 23, ..., 78, 81?

6. a) If we have k different colors available, in how many ways can we paint the walls of a pentagonal room if adjacent walls are to be painted with different colors?

b) What is the smallest value of k for which such a coloring is possible?

7. Ten students take a physics test in a certain room. When the test is over the students take a break and then return to the room to discuss their answers to the test questions. If there are 14 chairs in this room, in how many ways can the students seat themselves after the break so that no one is in the same chair he, or she, occupied during the test?

8. Using the result of Theorem 8.2, prove that the number of ways we can place s different objects in n distinct containers with m containers each containing exactly r of the objects is

$$\frac{(-1)^m n! \, s!}{m!} \sum_{i=m}^{n} \frac{(-1)^i (n-i)^{s-ir}}{(i-m)!(n-i)!(s-ir)!(r!)^i}.$$

9. If an arrangement of the letters in SURREPTITIOUS is selected at random, what is the probability that it contains (a) (exactly) three pairs of consecutive identical letters? (b) at most three pairs of consecutive identical letters?

10. In how many ways can four w's, four x's, four y's, and four z's be arranged so that there is no consecutive quadruple of the same letter?

11. a) Given n distinct objects, in how many ways can we select r of these objects so that each selection includes some particular m of the n objects? (Here $m \le r \le n$.)

b) Using the Principle of Inclusion and Exclusion, prove that for $m \le r \le n$,

$$\binom{n-m}{n-r} = \sum_{i=0}^{m} (-1)^i \binom{m}{i} \binom{n-i}{r}.$$

12. a) Let $\lambda \in \mathbf{Z}^+$. If we have λ different colors available, in how many ways can we color the vertices of the graph shown in Fig. 8.14(a) so that no adjacent vertices share the same color? This result in λ is called the *chromatic polynomial* of the graph, and the smallest value of λ for which the value of this polynomial is positive is called the *chromatic number* of the graph. What is the chromatic number of this graph? (We shall pursue this idea further in Chapter 11.)

b) If there are six colors available, in how many ways can the rooms R_i, $1 \le i \le 5$, shown in Fig. 8.14(b) be painted so that rooms with a common doorway, D_j, $1 \le j \le 5$, are painted with different colors?

13. Find the number of ways to arrange the letters in LAPTOP so that none of the letters L, A, T, O is in its original position and the letter P is not in the third or sixth position.

14. For $n \in \mathbf{Z}^+$ prove that if $\phi(n) = n - 1$ then n is prime.

15. Let D_{18} denote the set of positive divisors of 18. For $d \in D_{18}$ let $S_d = \{n \mid 0 < n \le 18 \text{ and } \gcd(n, 18) = d\}$. (a) Show that the collection S_d, $d \in D_{18}$, provides a partition of $\{1, 2, 3, 4, \ldots, 17, 18\}$. (b) Note that $|S_1| = 6 = \phi(18)$ and $|S_2| = 6 = \phi(9)$. For each $d \in D_{18}$, express $|S_d|$ in terms of Euler's phi function.

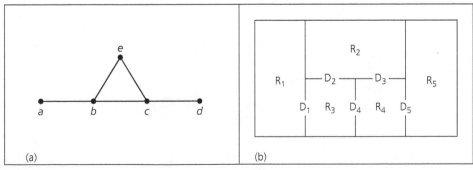

Figure 8.14

16. For $m \in \mathbf{Z}^+$ let $D_m = \{d \in \mathbf{Z}^+ | d$ divides $m\}$. For $d \in D_m$ let $S_d = \{n | 0 < n \leq m$ and $\gcd(n, m) = d\}$. (a) Show that the collection $S_d, d \in D_m$, provides a partition of $\{1, 2, 3, 4, \ldots, m - 1, m\}$. (b) Determine $|S_d|$ for each $d \in D_m$.

17. If $n \in \mathbf{Z}^+$, prove that (a) $\phi(2n) = 2\phi(n)$ when n is even; and (b) $\phi(2n) = \phi(n)$ when n is odd.

18. Let $a, b, c \in \mathbf{Z}^+$ with $c = \gcd(a, b)$. Prove that

$$\phi(ab)\phi(c) = \phi(a)\phi(b)c.$$

19. Caitlyn has 48 different books: 12 each in mathematics, chemistry, physics, and computer science. These books are arranged on four shelves in her office with all books on any one subject on its own shelf. When her office is cleaned, the 48 books are taken down and then replaced on the shelves — once again with all 12 books on any one subject on its own shelf. In how many ways can this be done so that (a) no subject is on its original shelf? (b) one subject is on its original shelf? (c) no subject is on its original shelf and no book is in its original position? [For example, the book originally in the third (from the left) position on the first shelf must not be replaced on the first shelf and must not be in the third (from the left) position on the shelf where it is placed.]

9

Generating Functions

In this chapter and the next, we continue our study of enumeration, introducing at this time the important concept of the *generating function*.

The problem of making selections, with repetitions allowed, was studied in Chapter 1. There we sought, for example, the number of integer solutions to the equation $c_1 + c_2 + c_3 + c_4 = 25$ where $c_i \geq 0$ for all $1 \leq i \leq 4$. With the Principle of Inclusion and Exclusion, in Chapter 8, we were able to solve a more restricted version of the problem, such as $c_1 + c_2 + c_3 + c_4 = 25$ with $0 \leq c_i < 10$ for all $1 \leq i \leq 4$. If, in addition, we wanted c_2 to be even and c_3 to be a multiple of 3, we could apply the results of Chapters 1 and 8 to several subcases.

The power of the generating function rests upon its ability not only to solve the kinds of problems we have considered so far but also to aid us in new situations where additional restrictions may be involved.

9.1
Introductory Examples

Instead of defining a generating function at this point, we shall examine some examples that motivate the idea.

EXAMPLE 9.1   While shopping one Saturday, Mildred buys 12 oranges for her children, Grace, Mary, and Frank. In how many ways can she distribute the oranges so that Grace gets at least four, and Mary and Frank get at least two, but Frank gets no more than five? Table 9.1 lists

Table 9.1

G	M	F	G	M	F
4	3	5	6	2	4
4	4	4	6	3	3
4	5	3	6	4	2
4	6	2	7	2	3
5	2	5	7	3	2
5	3	4	8	2	2
5	4	3			
5	5	2			

all the possible distributions. We see that we have all the integer solutions to the equation $c_1 + c_2 + c_3 = 12$ where $4 \leq c_1$, $2 \leq c_2$, and $2 \leq c_3 \leq 5$.

Considering the first two cases in this table, we find the solutions $4 + 3 + 5 = 12$ and $4 + 4 + 4 = 12$. Now where in our prior algebraic experiences did anything like this happen? When multiplying polynomials we add the powers of the variable, and here, when we multiply the three polynomials,

$$(x^4 + x^5 + x^6 + x^7 + x^8)(x^2 + x^3 + x^4 + x^5 + x^6)(x^2 + x^3 + x^4 + x^5),$$

two of the ways to obtain x^{12} are as follows:

1) From the product $x^4 x^3 x^5$, where x^4 is taken from $(x^4 + x^5 + x^6 + x^7 + x^8)$, x^3 from $(x^2 + x^3 + x^4 + x^5 + x^6)$, and x^5 from $(x^2 + x^3 + x^4 + x^5)$.

2) From the product $x^4 x^4 x^4$, where the first x^4 is found in the first polynomial, the second x^4 in the second polynomial, and the third x^4 in the third polynomial.

Examining the product

$$(x^4 + x^5 + x^6 + x^7 + x^8)(x^2 + x^3 + x^4 + x^5 + x^6)(x^2 + x^3 + x^4 + x^5)$$

more closely, we realize that we obtain the product $x^i x^j x^k$ for every triple (i, j, k) that appears in Table 9.1. Consequently, the coefficient of x^{12} in

$$f(x) = (x^4 + x^5 + x^6 + x^7 + x^8)(x^2 + x^3 + x^4 + x^5 + x^6)(x^2 + x^3 + x^4 + x^5)$$

counts the number of distributions—namely, 14—that we seek. The function $f(x)$ is called a *generating function* for the distributions.

But where did the factors in this product come from?

The factor $x^4 + x^5 + x^6 + x^7 + x^8$, for example, indicates that we can give Grace 4 *or* 5 *or* 6 *or* 7 *or* 8 of the oranges. Once again we make use of the interplay between the exclusive *or* and ordinary addition. The coefficient of each power of x is 1 because, considering the oranges as identical objects, there is only one way to give Grace four oranges, one way to give her five oranges, and so on. Since Mary and Frank must each receive at least two oranges, the other terms $(x^2 + x^3 + x^4 + x^5 + x^6)$ and $(x^2 + x^3 + x^4 + x^5)$ start with x^2, and for Frank we stop at x^5 so that he doesn't receive more than five oranges. (Why does the term for Mary stop at x^6?)

Most of us are reasonably convinced now that the coefficient of x^{12} in $f(x)$ yields the answer. Some, however, may be a bit skeptical about this new idea. It seems that we could list the cases in Table 9.1 faster than we could multiply out the three factors in $f(x)$ or calculate the coefficient x^{12} in $f(x)$. At present that may seem true. But as we progress to problems with more unknowns and larger quantities to distribute, the generating function will more than demonstrate its worth. (The reader may realize that the rook polynomials of Chapter 8 are examples of generating functions.) For now we consider two more examples.

EXAMPLE 9.2

If there is an unlimited number (or at least 24 of each color) of red, green, white, and black jelly beans, in how many ways can Douglas select 24 of these candies so that he has an even number of white beans and at least six black ones?

The polynomials associated with the jelly bean colors are as follows:

- red (green): $1 + x + x^2 + \cdots + x^{24}$, where the leading 1 is for $1x^0$, because one possibility for the red (and green) jelly beans is that none of that color is selected
- white: $(1 + x^2 + x^4 + x^6 + \cdots + x^{24})$
- black: $(x^6 + x^7 + x^8 + \cdots + x^{24})$

So the answer to the problem is the coefficient of x^{24} in the generating function

$$f(x) = (1 + x + x^2 + \cdots + x^{24})^2(1 + x^2 + x^4 + \cdots + x^{24})(x^6 + x^7 + \cdots + x^{24}).$$

One such selection is five red, three green, eight white, and eight black jelly beans. This arises from x^5 in the first factor, x^3 in the second factor, and x^8 in the last two factors.

One more example before closing this section!

EXAMPLE 9.3

How many integer solutions are there for the equation $c_1 + c_2 + c_3 + c_4 = 25$ if $0 \le c_i$ for all $1 \le i \le 4$?

We can alternatively ask in how many ways 25 (identical) pennies can be distributed among four children.

For each child the possibilities can be described by the polynomial $1 + x + x^2 + x^3 + \cdots + x^{25}$. Then the answer to this problem is the coefficient of x^{25} in the generating function

$$f(x) = (1 + x + x^2 + \cdots + x^{25})^4.$$

The answer can also be obtained as the coefficient of x^{25} in the generating function

$$g(x) = (1 + x + x^2 + x^3 + \cdots + x^{25} + x^{26} + \cdots)^4,$$

if we rephrase the question in terms of distributing, from a large (or unlimited) number of pennies, 25 pennies among four children. [Whereas $f(x)$ is a polynomial, $g(x)$ is a *power series* in x.] Note that the terms x^k, for all $k \ge 26$, are never used. So why bother with them? Because there will be times when it is easier to compute with a power series than with a polynomial.

EXERCISES 9.1

1. For each of the following, determine a generating function and indicate the coefficient in the function that is needed to solve the problem. (Give both the polynomial and power series forms of the generating function, wherever appropriate.)

Find the number of integer solutions for the following equations:

a) $c_1 + c_2 + c_3 + c_4 = 20$, $0 \le c_i \le 7$ for all $1 \le i \le 4$

b) $c_1 + c_2 + c_3 + c_4 = 20$, $0 \le c_i$ for all $1 \le i \le 4$, with c_2 and c_3 even

c) $c_1 + c_2 + c_3 + c_4 + c_5 = 30$, $2 \le c_1 \le 4$ and $3 \le c_i \le 8$ for all $2 \le i \le 5$

d) $c_1 + c_2 + c_3 + c_4 + c_5 = 30$, $0 \le c_i$ for all $1 \le i \le 5$, with c_2 even and c_3 odd

2. Determine the generating function for the number of ways to distribute 35 pennies (from an unlimited supply) among five children if (a) there are no restrictions; (b) each child gets at least 1¢; (c) each child gets at least 2¢; (d) the oldest child gets at least 10¢; and (e) the two youngest children must each get at least 10¢.

3. a) Find the generating function for the number of ways to select 10 candy bars from large supplies of six different kinds.

b) Find the generating function for the number of ways to select, with repetitions allowed, r objects from a collection of n distinct objects.

4. a) Explain why the generating function for the number of ways to have n cents in pennies and nickels is

$$(1 + x + x^2 + x^3 + \cdots)(1 + x^5 + x^{10} + \cdots).$$

b) Find the generating function for the number of ways to have n cents in pennies, nickels, and dimes.

5. Find the generating function for the number of integer solutions to the equation $c_1 + c_2 + c_3 + c_4 = 20$ where $-3 \le c_1$, $-3 \le c_2$, $-5 \le c_3 \le 5$, and $0 \le c_4$.

6. For $S = \{a, b, c\}$, consider the function

$$f(x) = (1 + ax)(1 + bx)(1 + cx)$$
$$= 1 + ax + bx + cx + abx^2 + acx^2$$
$$+ bcx^2 + abcx^3.$$

Here, in $f(x)$

- The coefficient of x^0 is 1 — for the subset $\emptyset$ of S.
- The coefficient of x^1 is $a + b + c$ — for the subsets $\{a\}$, $\{b\}$, and $\{c\}$ of S.
- The coefficient of x^2 is $ab + ac + bc$ — for the subsets $\{a, b\}$, $\{a, c\}$, and $\{b, c\}$ of S.

• The coefficient of x^3 is abc — for the subset $\{a, b, c\} = S$.

Consequently, $f(x)$ is the generating function for the subsets of S. For when we calculate $f(1)$, we obtain a sum wherein each of the eight summands corresponds with a subset of S; the summand 1 corresponds with $\emptyset$. [If we go one step further and set $a = b = c = 1$ in $f(x)$, then $f(1) = 8$, the number of subsets of S.]

a) Give the generating function for the subsets of

$$S = \{a, b, c, \ldots, r, s, t\}.$$

b) Answer part (a) for selections wherein each element can be rejected or selected as many as three times.

9.2
Definition and Examples:
Calculational Techniques

In this section we shall examine a number of formulas and examples dealing with power series. These will be used to obtain the coefficients of particular terms in a generating function.

We start with the following concept.

Definition 9.1

Let $a_0, a_1, a_2, \ldots$ be a sequence of real numbers. The function

$$f(x) = a_0 + a_1 x + a_2 x^2 + \cdots = \sum_{i=0}^{\infty} a_i x^i$$

is called the *generating function* for the given sequence.

Where could this idea have come from?

EXAMPLE 9.4

For any $n \in \mathbf{Z}^+$,

$$(1 + x)^n = \binom{n}{0} + \binom{n}{1} x + \binom{n}{2} x^2 + \cdots + \binom{n}{n} x^n,$$

so $(1 + x)^n$ is the generating function for the sequence

$$\binom{n}{0}, \binom{n}{1}, \binom{n}{2}, \ldots, \binom{n}{n}, 0, 0, 0, \ldots.$$

EXAMPLE 9.5

a) For $n \in \mathbf{Z}^+$,

$$(1 - x^{n+1}) = (1 - x)(1 + x + x^2 + x^3 + \cdots + x^n).$$

So

$$\frac{1 - x^{n+1}}{1 - x} = 1 + x + x^2 + \cdots + x^n,$$

and $(1 - x^{n+1})/(1 - x)$ is the generating function for the sequence $1, 1, 1, \ldots, 1, 0, 0, 0, \ldots$, where the first $n + 1$ terms are 1.

b) Extending the idea in part (a), we find that

$$1 = (1 - x)(1 + x + x^2 + x^3 + x^4 + \cdots),$$

so

$$\frac{1}{1 - x}$$

is the generating function for the sequence 1, 1, 1, 1, ... [Note that $1/(1 - x) = 1 + x + x^2 + x^3 + \cdots$ is valid for all real x where $|x| < 1$; it is for this set of values that the *geometric series* $1 + x + x^2 + x^3 + \cdots$ *converges*. In our work with generating functions we shall be primarily concerned with the coefficients of the powers of x. However, later in Example 9.18, we shall use this and two other related series to evaluate infinite sums for values within the set of values for which each such infinite series converges.]

c) With

$$\frac{1}{1 - x} = 1 + x + x^2 + x^3 + \cdots = \sum_{i=0}^{\infty} x^i,$$

taking the derivative yields

$$\frac{d}{dx} \frac{1}{1 - x} = \frac{d}{dx}(1 - x)^{-1} = (-1)(1 - x)^{-2}(-1) = \frac{1}{(1 - x)^2}$$

$$= \frac{d}{dx}(1 + x + x^2 + x^3 + \cdots) = 1 + 2x + 3x^2 + 4x^3 + \cdots$$

Consequently,

$$\frac{1}{(1 - x)^2}$$

is the generating function for the sequence 1, 2, 3, 4, ..., while

$$\frac{x}{(1 - x)^2} = 0 + 1x + 2x^2 + 3x^3 + 4x^4 + \cdots$$

is the generating function for the sequence 0, 1, 2, 3, ...

d) Continuing from part (c),

$$\frac{d}{dx} \frac{x}{(1 - x)^2} = \frac{d}{dx}(0 + x + 2x^2 + 3x^3 + \cdots),$$

or

$$\frac{x + 1}{(1 - x)^3} = 1 + 2^2 x + 3^2 x^2 + 4^2 x^3 + \cdots.$$

Hence,

$$\frac{x + 1}{(1 - x)^3}$$

generates $1^2, 2^2, 3^2, \ldots$, and

$$\frac{x(x + 1)}{(1 - x)^3}$$

generates $0^2, 1^2, 2^2, 3^2, \ldots$.

e) Now let us take one more look at the results in parts (b), (c), (d) — along with some extensions. But this time we have a change in the notation:

$$f_0(x) = \frac{1}{1-x} = 1 + x + x^2 + x^3 + \cdots$$

$$f_1(x) = x\frac{d}{dx}f_0(x) = \frac{x}{(1-x)^2}$$

$$= 0 + x + 2x^2 + 3x^3 + \cdots$$

$$f_2(x) = x\frac{d}{dx}f_1(x) = \frac{x^2 + x}{(1-x)^3}$$

$$= 0^2 + 1^2x + 2^2x^2 + 3^2x^3 + \cdots$$

$$f_3(x) = x\frac{d}{dx}f_2(x) = \frac{x^3 + 4x^2 + x}{(1-x)^4}$$

$$= 0^3 + 1^3x + 2^3x^2 + 3^3x^3 + \cdots$$

$$f_4(x) = x\frac{d}{dx}f_3(x) = \frac{x^4 + 11x^3 + 11x^2 + x}{(1-x)^5}$$

$$= 0^4 + 1^4x + 2^4x^2 + 3^4x^3 + \cdots$$

Now look at the output for the Maple code in Fig. 9.1. Here we find the numerators for $f_0(x)$, $f_1(x)$, ..., $f_4(x)$, along with those for $f_5(x)$ and $f_6(x)$ [where the denominators are $(1-x)^6$ and $(1-x)^7$, respectively]. The coefficients for these numerators are exactly the Eulerian numbers we introduced in Example 4.21. We choose not to pursue this here, but the interested reader, who wants to examine this connection further, should look into reference [4].

```
> f||0(x) := 1/(1-x);
```
$$f0(x) := \frac{1}{1-x}$$
```
> for i from 1 to 6 do
>     f||i(x) := simplify(x*diff(f||(i-1)(x),x)):
>     print(sort(expand((-1)^(i+1)*numer(f||i(x)))));
> od:
```
$$x$$
$$x^2 + x$$
$$x^3 + 4x^2 + x$$
$$x^4 + 11x^3 + 11x^2 + x$$
$$x^5 + 26x^4 + 66x^3 + 26x^2 + x$$
$$x^6 + 57x^5 + 302x^4 + 302x^3 + 57x^2 + x$$

Figure 9.1

EXAMPLE 9.6

a) Rewriting the result in part (b) of Example 9.5, we have

$$\frac{1}{1-y} = 1 + y + y^2 + y^3 + \cdots.$$

Upon substituting $2x$ for y, we then learn that

$$\frac{1}{1-2x} = 1 + (2x) + (2x)^2 + (2x)^3 + \cdots = 1 + 2x + 2^2 x^2 + 2^3 x^3 + \cdots,$$

so $1/(1-2x)$ is the generating function for the sequence $1\ (= 2^0)$, $2\ (= 2^1)$, 2^2, $2^3, \ldots$. In fact, for each $a \in \mathbf{R}$, it follows that $1/(1-ax) = 1 + (ax) + (ax)^2 + (ax)^3 + \cdots = 1 + ax + a^2 x^2 + a^3 x^3 + \cdots$, so $1/(1-ax)$ is the generating function for the sequence $1\ (= a^0)$, $a\ (= a^1)$, a^2, $a^3, \ldots$. [Here we want $0^0 = 1$ for the case where $a = 0$.]

b) Again, from part (b) of Example 9.5, we know that the generating function for the sequence $1, 1, 1, 1, \ldots$ is $f(x) = 1/(1-x)$. Therefore the function

$$g(x) = f(x) - x^2 = \frac{1}{1-x} - x^2$$

is the generating function for the sequence $1, 1, 0, 1, 1, 1, \ldots$, while the function

$$h(x) = f(x) + 2x^3 = \frac{1}{1-x} + 2x^3$$

generates the sequence $1, 1, 1, 3, 1, 1, \ldots$.

c) Finally, can we use the results of Example 9.5 to find a generating function for the sequence $0, 2, 6, 12, 20, 30, 42, \ldots$?

Here we observe that

$$\begin{aligned}
a_0 &= 0 = 0^2 + 0, & a_1 &= 2 = 1^2 + 1, \\
a_2 &= 6 = 2^2 + 2, & a_3 &= 12 = 3^2 + 3, \\
a_4 &= 20 = 4^2 + 4, \ldots.
\end{aligned}$$

In general, we have $a_n = n^2 + n$, for each $n \geq 0$.

Using the results from parts (c) and (d) of Example 9.5, we now find that

$$\frac{x(x+1)}{(1-x)^3} + \frac{x}{(1-x)^2} = \frac{x(x+1) + x(1-x)}{(1-x)^3} = \frac{2x}{(1-x)^3}$$

is the generating function for the given sequence. (The solution here depends upon our ability to recognize each a_n as the sum of n^2 and n. If we do not see this, we may be unable to answer the given question. Consequently, in Example 10.6 of the next chapter, we shall examine another technique to help us recognize the formula for a_n.)

For each $n \in \mathbf{Z}^+$, the binomial theorem tells us that $(1+x)^n = \binom{n}{0} + \binom{n}{1}x + \binom{n}{2}x^2 + \cdots + \binom{n}{n}x^n$. We want to extend this idea to cases where (a) $n < 0$ and (b) n is not necessarily an integer.

With $n, r \in \mathbf{Z}^+$ and $n \geq r > 0$, we have

$$\binom{n}{r} = \frac{n!}{r!(n-r)!} = \frac{n(n-1)(n-2)\cdots(n-r+1)}{r!}.$$

If $n \in \mathbf{R}$, we use

$$\frac{n(n-1)(n-2)\cdots(n-r+1)}{r!}$$

as the definition of $\binom{n}{r}$.

Then, for example, if $n \in \mathbf{Z}^+$, we have

$$\binom{-n}{r} = \frac{(-n)(-n-1)(-n-2)\cdots(-n-r+1)}{r!}$$

$$= \frac{(-1)^r(n)(n+1)(n+2)\cdots(n+r-1)}{r!}$$

$$= \frac{(-1)^r(n+r-1)!}{(n-1)!\,r!} = (-1)^r\binom{n+r-1}{r}.$$

Finally, for each *real* number n, we define $\binom{n}{0} = 1$.

EXAMPLE 9.7

For $n \in \mathbf{Z}^+$, the Maclaurin series expansion for $(1+x)^{-n}$ is given by

$$(1+x)^{-n} = 1 + (-n)x + (-n)(-n-1)x^2/2!$$
$$+ (-n)(-n-1)(-n-2)x^3/3! + \cdots$$
$$= 1 + \sum_{r=1}^{\infty} \frac{(-n)(-n-1)(-n-2)\cdots(-n-r+1)}{r!}x^r$$
$$= \sum_{r=0}^{\infty}(-1)^r\binom{n+r-1}{r}x^r.$$

Hence $(1+x)^{-n} = \binom{-n}{0} + \binom{-n}{1}x + \binom{-n}{2}x^2 + \cdots = \sum_{r=0}^{\infty}\binom{-n}{r}x^r$. This generalizes the binomial theorem of Chapter 1 and shows us that $(1+x)^{-n}$ is the generating function for the sequence $\binom{-n}{0}, \binom{-n}{1}, \binom{-n}{2}, \binom{-n}{3}, \ldots$.

EXAMPLE 9.8

Find the coefficient of x^5 in $(1-2x)^{-7}$.

With $y = -2x$, use the result in Example 9.7 to write $(1-2x)^{-7} = (1+y)^{-7} = \sum_{r=0}^{\infty}\binom{-7}{r}y^r = \sum_{r=0}^{\infty}\binom{-7}{r}(-2x)^r$. Consequently, the coefficient of x^5 is $\binom{-7}{5}(-2)^5 = (-1)^5\binom{7+5-1}{5}(-32) = (32)\binom{11}{5} = 14{,}784$.

EXAMPLE 9.9

For each real number n, the Maclaurin series expansion for $(1+x)^n$ is

$$1 + nx + n(n-1)x^2/2! + n(n-1)(n-2)x^3/3! + \cdots$$
$$= 1 + \sum_{r=1}^{\infty}\frac{n(n-1)(n-2)\cdots(n-r+1)}{r!}x^r.$$

Therefore,

$$(1 + 3x)^{-1/3} = 1 + \sum_{r=1}^{\infty} \frac{(-1/3)(-4/3)(-7/3) \cdots ((-3r+2)/3)}{r!}(3x)^r$$

$$= 1 + \sum_{r=1}^{\infty} \frac{(-1)(-4)(-7) \cdots (-3r+2)}{r!} x^r,$$

and $(1 + 3x)^{-1/3}$ generates the sequence $1, -1, (-1)(-4)/2!, (-1)(-4)(-7)/3!, \ldots,$ $(-1)(-4)(-7) \cdots (-3r+2)/r!, \ldots$.

EXAMPLE 9.10

Determine the coefficient of x^{15} in $f(x) = (x^2 + x^3 + x^4 + \cdots)^4$.

Since $(x^2 + x^3 + x^4 + \cdots) = x^2(1 + x + x^2 + \cdots) = x^2/(1-x)$, the coefficient of x^{15} in $f(x)$ is the coefficient of x^{15} in $(x^2/(1-x))^4 = x^8/(1-x)^4$. Hence the coefficient sought is that of x^7 in $(1-x)^{-4}$, namely, $\binom{-4}{7}(-1)^7 = (-1)^7\binom{4+7-1}{7}(-1)^7 = \binom{10}{7} = 120$.

In general, for $n \in \mathbf{Z}^+$, the coefficient of x^n in $f(x)$ is 0, when $0 \leq n \leq 7$. For all $n \geq 8$, the coefficient of x^n in $f(x)$ is the coefficient of x^{n-8} in $(1-x)^{-4}$, which is $\binom{-4}{n-8} \cdot (-1)^{n-8} = \binom{n-5}{n-8}$.

Before continuing, we collect the identities shown in Table 9.2 (on page 424) for future reference.

The next two examples show how generating functions can be applied to derive some of our earlier results.

EXAMPLE 9.11

In how many ways can we select, with repetitions allowed, r objects from n distinct objects?

For each of the n distinct objects, the geometric series $1 + x + x^2 + x^3 + \cdots$ represents the possible choices for that object (namely none, one, two, $\ldots$). Considering all of the n distinct objects, the generating function is

$$f(x) = (1 + x + x^2 + x^3 + \cdots)^n,$$

and the required answer is the coefficient of x^r in $f(x)$. Now from identities 5 and 8 in Table 9.2 we have

$$(1 + x + x^2 + x^3 + \cdots)^n = \left(\frac{1}{1-x}\right)^n = \frac{1}{(1-x)^n} = \sum_{i=0}^{\infty} \binom{n+i-1}{i} x^i,$$

so the coefficient of x^r is

$$\binom{n+r-1}{r},$$

the result we found in Chapter 1.

EXAMPLE 9.12

Once again we consider the problem of counting the compositions of a positive integer n — this time using generating functions.

Start with

$$\frac{x}{1-x} = x + x^2 + x^3 + x^4 + \cdots$$

Table 9.2

For all $m, n \in \mathbf{Z}^+$, $a \in \mathbf{R}$,

1) $(1 + x)^n = \binom{n}{0} + \binom{n}{1}x + \binom{n}{2}x^2 + \cdots + \binom{n}{n}x^n$

2) $(1 + ax)^n = \binom{n}{0} + \binom{n}{1}ax + \binom{n}{2}a^2x^2 + \cdots + \binom{n}{n}a^nx^n$

3) $(1 + x^m)^n = \binom{n}{0} + \binom{n}{1}x^m + \binom{n}{2}x^{2m} + \cdots + \binom{n}{n}x^{nm}$

4) $(1 - x^{n+1})/(1 - x) = 1 + x + x^2 + \cdots + x^n$

5) $1/(1 - x) = 1 + x + x^2 + x^3 + \cdots = \sum_{i=0}^{\infty} x^i$

6) $1/(1 - ax) = 1 + (ax) + (ax)^2 + (ax)^3 + \cdots$

$\qquad = \sum_{i=0}^{\infty}(ax)^i = \sum_{i=0}^{\infty} a^i x^i$

$\qquad = 1 + ax + a^2x^2 + a^3x^3 + \cdots$

7) $1/(1 + x)^n = \binom{-n}{0} + \binom{-n}{1}x + \binom{-n}{2}x^2 + \cdots$

$\qquad = \sum_{i=0}^{\infty} \binom{-n}{i}x^i$

$\qquad = 1 + (-1)\binom{n+1-1}{1}x + (-1)^2\binom{n+2-1}{2}x^2 + \cdots$

$\qquad = \sum_{i=0}^{\infty}(-1)^i\binom{n+i-1}{i}x^i$

8) $1/(1 - x)^n = \binom{-n}{0} + \binom{-n}{1}(-x) + \binom{-n}{2}(-x)^2 + \cdots$

$\qquad = \sum_{i=0}^{\infty} \binom{-n}{i}(-x)^i$

$\qquad = 1 + (-1)\binom{n+1-1}{1}(-x) + (-1)^2\binom{n+2-1}{2}(-x)^2 + \cdots$

$\qquad = \sum_{i=0}^{\infty} \binom{n+i-1}{i}x^i$

If $f(x) = \sum_{i=0}^{\infty} a_i x^i$, $g(x) = \sum_{i=0}^{\infty} b_i x^i$, and $h(x) = f(x)g(x)$, then $h(x) = \sum_{i=0}^{\infty} c_i x^i$, where for all $k \geq 0$,

$$c_k = a_0 b_k + a_1 b_{k-1} + \cdots + a_{k-1}b_1 + a_k b_0 = \sum_{j=0}^{k} a_j b_{k-j}.$$

where, for example, the coefficient of x^4 is 1, for the one-summand composition of 4 — namely, 4. To obtain the number of compositions of n where there are two summands, we need the coefficient of x^n in $(x + x^2 + x^3 + x^4 + \cdots)^2 = [x/(1 - x)]^2 = x^2/(1 - x)^2$. Here, for instance, we obtain x^4 in $(x + x^2 + x^3 + x^4 + \cdots)^2$ from the products $x^1 \cdot x^3$, $x^2 \cdot x^2$, and $x^3 \cdot x^1$. So the coefficient of x^4 in $x^2/(1 - x)^2$ is 3 — for the three two-summand compositions $1 + 3$, $2 + 2$, and $3 + 1$ (of 4). Continuing with the three-summand compositions we now examine $(x + x^2 + x^3 + x^4 + \cdots)^3 = [x/(1 - x)]^3 = x^3/(1 - x)^3$. Once again we look at the ways x^4 comes about — namely, from the products $x^1 \cdot x^1 \cdot x^2$, $x^1 \cdot x^2 \cdot x^1$, $x^2 \cdot x^1 \cdot x^1$. So here the coefficient of x^4 is 3, which accounts for the compositions $1 + 1 + 2$, $1 + 2 + 1$, and $2 + 1 + 1$ (of 4). Finally, the coefficient of x^4 in $(x + x^2 + x^3 + x^4 + \cdots)^4 = [x/(1 - x)]^4 = x^4/(1 - x)^4$ is 1 — for the one four-summand composition $1 + 1 + 1 + 1$ (of 4).

The results in the previous paragraph tell us that the coefficient of x^4 in $\sum_{i=1}^{4}[x/(1 - x)]^i$ is $1 + 3 + 3 + 1 = 8 \, (= 2^3)$, the number of compositions of 4. In fact, this is also the coefficient of x^4 in $\sum_{i=1}^{\infty}[x/(1 - x)]^i$. Generalizing the situation we find that the number of compositions of a positive integer n is the coefficient of x^n in the generating function

$f(x) = \sum_{i=1}^{\infty} [x/(1-x)]^i$. But if we set $y = x/(1-x)$, it then follows that

$$f(x) = \sum_{i=1}^{\infty} y^i = y \sum_{i=0}^{\infty} y^i = y \left(\frac{1}{1-y} \right) = \left(\frac{x}{1-x} \right) \left[\frac{1}{1 - \left(\frac{x}{1-x} \right)} \right] = \left(\frac{x}{1-x} \right) \left[\frac{1}{\frac{1-x-x}{1-x}} \right]$$

$$= x/(1-2x) = x[1 + (2x) + (2x)^2 + (2x)^3 + \cdots]$$

$$= 2^0 x + 2^1 x^2 + 2^2 x^3 + 2^3 x^4 + \cdots.$$

So the number of compositions of a positive integer n is the coefficient of x^n in $f(x)$ — and this is 2^{n-1} (as we found earlier in Examples 1.37, 3.11, and 4.12.)

EXAMPLE 9.13

Before we look at any specific compositions, let us start by examining identity 4 in Table 9.2. When x is replaced by 2 in this identity, the result tells us that for all $n \in \mathbf{Z}^+$, $1 + 2 + 2^2 + \cdots + 2^n = (1 - 2^{n+1})/(1 - 2) = 2^{n+1} - 1$. [This result was also established by the Principle of Mathematical Induction — in part (a) of Exercise 2 for Section 4.1.] All well and good — but where would one ever use such a formula? In Table 9.3 we find the special compositions of 6 and 7 that read the same left to right as right to left. These are the *palindromes* of 6 and 7. We find that for 7 there are $1 + (1 + 2 + 4) = 1 + (1 + 2^1 + 2^2) = 1 + (2^3 - 1) = 2^3$ palindromes. There is one palindrome with one summand — namely, 7. There is also one palindrome where the center summand is 5 and where we place the one composition of 1 on either side of this summand.

Table 9.3

1)	6	(1)	1)	7	(1)
2)	$1 + 4 + 1$	(1)	2)	$1 + 5 + 1$	(1)
3)	$2 + 2 + 2$	(2)	3)	$2 + 3 + 2$	(2)
4)	$1 + 1 + 2 + 1 + 1$		4)	$1 + 1 + 3 + 1 + 1$	
5)	$3 + 3$		5)	$3 + 1 + 3$	
6)	$1 + 2 + 2 + 1$	(4)	6)	$1 + 2 + 1 + 2 + 1$	(4)
7)	$2 + 1 + 1 + 2$		7)	$2 + 1 + 1 + 1 + 2$	
8)	$1 + 1 + 1 + 1 + 1 + 1$		8)	$1 + 1 + 1 + 1 + 1 + 1 + 1$	

For the center summand 3 we place one of the two compositions of 2 on the right (of 3) and then match it on the left, with the same composition, in reverse order. This procedure provides the third and fourth palindromes of 7 in the table. Finally, when the center summand is 1, we put a given composition of 3 on the right of this 1 and match it on the left with the same composition, in reverse order. There are $2^{3-1} = 4$ compositions of 3, so this procedure results in the last four palindromes of 7 in the table.

The situation is similar for the palindromes of 6 except for the case where, instead of 0 as the center summand, a plus sign appears in the center. Here we obtain the last $2^{3-1} = 4$ palindromes of 6 in the table — one for each composition of 3. Summarizing for $n = 6$ we have

    i) Center summand 6   1 palindrome

    ii) Center summand 4   1 ($= 2^{1-1}$) palindrome

    iii) Center summand 2   2 ($= 2^{2-1}$) palindromes

    iv) Plus sign at the center   4 ($= 2^{3-1}$) palindromes

So there are $1 + (1 + 2^1 + 2^2) = 1 + (2^3 - 1) = 2^3$ palindromes for 6.

Now we look at the general situation. For $n = 1$ there is one palindrome. If $n = 2k + 1$, for $k \in \mathbf{Z}^+$, then there is one palindrome with center summand n. For $1 \le t \le k$, there are 2^{t-1} palindromes of n with center summand $n - 2t$. (One palindrome for each of the 2^{t-1} compositions of t.) Hence the total number of palindromes of n is $1 + (1 + 2^1 + 2^2 + \cdots + 2^{k-1}) = 1 + (2^k - 1) = 2^k = 2^{(n-1)/2}$. Now consider n even, say $n = 2k$, for $k \in \mathbf{Z}^+$. Here there is also one palindrome with center summand n and, for $1 \le s \le k - 1$, there are 2^{s-1} palindromes of n with center summand $n - 2s$. (One palindrome for each of the 2^{s-1} compositions of s.) In addition, there are 2^{k-1} palindromes where a plus sign is at the center. (One palindrome for each of the 2^{k-1} compositions of k.) In total, n has $1 + (1 + 2^1 + 2^2 + \cdots + 2^{k-2} + 2^{k-1}) = 1 + (2^k - 1) = 2^k = 2^{n/2}$ palindromes.

The preceding results can be simplified. Observe that for $n \in \mathbf{Z}^+$, n has $2^{\lfloor n/2 \rfloor}$ palindromes.

Having dealt with compositions (once again) and palindromes, we continue at this point with some additional examples dealing with generating functions.

EXAMPLE 9.14

In how many ways can a police captain distribute 24 rifle shells to four police officers so that each officer gets at least three shells, but not more than eight?

The choices for the number of shells each officer receives are given by $x^3 + x^4 + \cdots + x^8$. There are four officers, so the resulting generating function is

$$f(x) = (x^3 + x^4 + \cdots + x^8)^4.$$

We seek the coefficient of x^{24} in $f(x)$. With $(x^3 + x^4 + \cdots + x^8)^4 = x^{12}(1 + x + x^2 + \cdots + x^5)^4 = x^{12}((1 - x^6)/(1 - x))^4$, the answer is the coefficient of x^{12} in $(1 - x^6)^4 \cdot (1 - x)^{-4} = [1 - \binom{4}{1}x^6 + \binom{4}{2}x^{12} - \binom{4}{3}x^{18} + x^{24}][\binom{-4}{0} + \binom{-4}{1}(-x) + \binom{-4}{2}(-x)^2 + \cdots]$, which is $[\binom{-4}{12}(-1)^{12} - \binom{4}{1}\binom{-4}{6}(-1)^6 + \binom{4}{2}\binom{-4}{0}] = [\binom{15}{12} - \binom{4}{1}\binom{9}{6} + \binom{4}{2}] = 125$.

EXAMPLE 9.15

Verify that for all $n \in \mathbf{Z}^+$, $\binom{2n}{n} = \sum_{i=0}^{n} \binom{n}{i}^2$.

Since $(1 + x)^{2n} = [(1 + x)^n]^2$, by comparison of coefficients (of like powers of x), the coefficient of x^n in $(1 + x)^{2n}$, which is $\binom{2n}{n}$, must equal the coefficient of x^n in $[\binom{n}{0} + \binom{n}{1}x + \binom{n}{2}x^2 + \cdots + \binom{n}{n}x^n]^2$, and this is $\binom{n}{0}\binom{n}{n} + \binom{n}{1}\binom{n}{n-1} + \binom{n}{2}\binom{n}{n-2} + \cdots + \binom{n}{n}\binom{n}{0}$. With $\binom{n}{r} = \binom{n}{n-r}$, for all $0 \le r \le n$, the result follows.

EXAMPLE 9.16

Determine the coefficient of x^8 in $\dfrac{1}{(x - 3)(x - 2)^2}$.

Since $1/(x - a) = (-1/a)(1/(1 - (x/a))) = (-1/a)[1 + (x/a) + (x/a)^2 + \cdots]$ for any $a \ne 0$, we could solve this problem by finding the coefficient of x^8 in $1/[(x - 3)(x - 2)^2]$ expressed as $(-1/3)[1 + (x/3) + (x/3)^2 + \cdots](1/4)[\binom{-2}{0} + \binom{-2}{1}(-x/2) + \binom{-2}{2}(-x/2)^2 + \cdots]$.

An alternative technique uses the partial fraction decomposition:

$$\frac{1}{(x - 3)(x - 2)^2} = \frac{A}{x - 3} + \frac{B}{x - 2} + \frac{C}{(x - 2)^2}.$$

This decomposition implies that

$$1 = A(x - 2)^2 + B(x - 2)(x - 3) + C(x - 3),$$

or

$$0 \cdot x^2 + 0 \cdot x + 1 = 1 = (A + B)x^2 + (-4A - 5B + C)x + (4A + 6B - 3C).$$

By comparing coefficients (for x^2, x, and 1, respectively), we find that $A + B = 0$, $-4A - 5B + C = 0$, and $4A + 6B - 3C = 1$. Solving these equations yields $A = 1$, $B = -1$, and $C = -1$. Hence

$$\frac{1}{(x-3)(x-2)^2} = \frac{1}{x-3} - \frac{1}{x-2} - \frac{1}{(x-2)^2}$$

$$= \left(\frac{-1}{3}\right)\frac{1}{1-(x/3)} + \left(\frac{1}{2}\right)\frac{1}{1-(x/2)} + \left(\frac{-1}{4}\right)\frac{1}{(1-(x/2))^2}$$

$$= \left(\frac{-1}{3}\right)\sum_{i=0}^{\infty}\left(\frac{x}{3}\right)^i + \left(\frac{1}{2}\right)\sum_{i=0}^{\infty}\left(\frac{x}{2}\right)^i$$

$$+ \left(\frac{-1}{4}\right)\left[\binom{-2}{0} + \binom{-2}{1}\left(\frac{-x}{2}\right) + \binom{-2}{2}\left(\frac{-x}{2}\right)^2 + \cdots\right].$$

The coefficient of x^8 is $(-1/3)(1/3)^8 + (1/2)(1/2)^8 + (-1/4)\binom{-2}{8}(-1/2)^8 = -[(1/3)^9 + 7(1/2)^{10}]$.

EXAMPLE 9.17

Use generating functions to determine how many four-element subsets of $S = \{1, 2, 3, \ldots, 15\}$ contain no consecutive integers.

a) Consider one such subset (say $\{1, 3, 7, 10\}$), and write $1 \le 1 < 3 < 7 < 10 \le 15$. We see that this set of inequalities determines the differences $1 - 1 = 0$, $3 - 1 = 2$, $7 - 3 = 4$, $10 - 7 = 3$, and $15 - 10 = 5$, and these differences sum to 14. Considering another such subset — say $\{2, 5, 11, 15\}$, we write $1 \le 2 < 5 < 11 < 15 \le 15$; these inequalities yield the differences 1, 3, 6, 4, and 0, which also sum to 14.

Turning things around, we find that the nonnegative integers 0, 2, 3, 2, and 7 sum to 14 and they are the differences that arise from the inequalities $1 \le 1 < 3 < 6 < 8 \le 15$ (for the subset $\{1, 3, 6, 8\}$).

These examples suggest a one-to-one correspondence between the four-element subsets to be counted and the integer solutions to $c_1 + c_2 + c_3 + c_4 + c_5 = 14$ where $0 \le c_1, c_5$, and $2 \le c_2, c_3, c_4$. (*Note:* Here $c_2, c_3, c_4 \ge 2$ guarantee that there are no consecutive integers in the subset.) The answer is the coefficient of x^{14} in

$$f(x) = (1 + x + x^2 + x^3 + \cdots)(x^2 + x^3 + x^4 + \cdots)^3(1 + x + x^2 + x^3 + \cdots)$$
$$= x^6(1 - x)^{-5}.$$

This then is the coefficient of x^8 in $(1 - x)^{-5}$, which is $\binom{-5}{8}(-1)^8 = \binom{5+8-1}{8} = \binom{12}{8} = 495$.

b) Another way to look at the problem is as follows.

For the subset $\{1, 3, 7, 10\}$, we examine the strict inequalities $0 < 1 < 3 < 7 < 10 < 16$ and consider how many integers there are strictly between each successive pair of these numbers. Here we get 0, 1, 3, 2, and 5: 0 because there is no integer between 0 and 1, 1 for the integer 2 between 1 and 3, 3 for the integers 4, 5, 6 between 3 and 7, and so on. These five integers sum to 11. When we do the same thing for the subset $\{2, 5, 11, 15\}$, the strict inequalities $0 < 2 < 5 < 11 < 15 < 16$ yield the results 1, 2, 5, 3, and 0, which also sum to 11.

On the other hand, we find that the nonnegative integers 0, 1, 2, 1, and 7 add up to 11 and they arise as the numbers of distinct integers between the integers in the five successive strict inequalities $0 < 1 < 3 < 6 < 8 < 16$. These correspond to the subset $\{1, 3, 6, 8\}$.

These results suggest a one-to-one correspondence between the desired subsets and the integer solutions to $b_1 + b_2 + b_3 + b_4 + b_5 = 11$, where $0 \leq b_1, b_5$ and $1 \leq b_2, b_3, b_4$. (*Note*: In this case, $b_1, b_2, b_3 \geq 1$ guarantee that there are no consecutive integers in the subset.) The number of these solutions is the coefficient of x^{11} in

$$g(x) = (1 + x + x^2 + \cdots)(x + x^2 + x^3 + \cdots)^3(1 + x + x^2 + \cdots)$$
$$= x^3(1 - x)^{-5}.$$

The answer is $\binom{-5}{8}(-1)^8 = 495$, as above. (The reader may now wish to look back at Supplementary Exercise 15 in Chapter 3.)

Our next example takes us back to the optional material in Chapter 3 where we first encountered the idea of the sample space. But now that we know about generating functions we will be able to deal with a sample space that is discrete but *not* finite — that is, a countably infinite[†] sample space.

EXAMPLE 9.18[‡]

a) Suppose that Brianna takes an actuarial examination until she passes it. Further, suppose the probability that Brianna passes the examination on any given attempt is 0.8 and that the result of each attempt, after the first, is independent of any previous attempt. If we let P denote "pass" and F denote "fail", for any given attempt, then here our sample space may be expressed as $\mathscr{S} = \{P, FP, FFP, FFFP, \ldots\}$, where, for example, $Pr(FFP)$ — the probability Brianna fails the exam twice before she passes it — is given by $(0.2)^2(0.8)$. In addition, the sum of the probabilities for the outcomes in $\mathscr{S}$ is $(0.8) + (0.2)(0.8) + (0.2)^2(0.8) + (0.2)^3(0.8) + \cdots = \sum_{i=0}^{\infty}(0.2)^i(0.8) = (0.8)\sum_{i=0}^{\infty}(0.2)^i = (0.8)\left(\frac{1}{1-0.2}\right) = (0.8)\left(\frac{1}{0.8}\right) = 1$, as it should be — for according to the second axiom of probability (in Section 3.5) we expect $Pr(\mathscr{S}) = 1$. [Note that $\sum_{i=0}^{\infty}(0.2)^i = \frac{1}{1-0.2}$ follows from the result in part (b) of Example 9.5. The given geometric series converges to $\frac{1}{1-0.2}$ because $|0.2| < 1$.]

b) Now suppose we want to know the probability Brianna passes the exam on an even-numbered attempt. That is, we want $Pr(A)$ where A is the event $\{FP, FFFP, \ldots\}$.

At this point let us introduce the discrete random variable Y where Y counts the number of attempts up to and including the one where Brianna passes the exam. Then the probability distribution for Y is given by $Pr(Y = y) = (0.2)^{y-1}(0.8)$, $y \geq 1$. So $Pr(A)$ can be determined as follows: $Pr(A) = \sum_{i=1}^{\infty} Pr(Y = 2i) = \sum_{i=1}^{\infty}(0.2)^{2i-1}(0.8) = (0.8)\sum_{i=1}^{\infty}(0.2)^{2i-1} = 0.8[(0.2) + (0.2)^3 + (0.2)^5 + \cdots] = (0.8)(0.2)[1 + (0.2)^2 + (0.2)^4 + \cdots] = (0.8)(0.2)\frac{1}{1-(0.2)^2} = \frac{(0.8)(0.2)}{0.96} = \frac{1}{6}$. And once again we have used the result in part (b) of Example 9.5, this time with $x = (0.2)^2$, where $|(0.2)^2| = |0.04| < 1$.

[†] The reader can learn more about countably infinite sets from the material in Appendix 3.

[‡] This example uses material from the optional sections of Chapter 3. It may be skipped without any loss of continuity.

c) Continuing with Y, now we'd like to find $E(Y)$, the number of times Brianna expects to take the actuarial exam before she passes it. To determine $E(Y)$ we'll start with the formula $1/(1-t) = 1 + t + t^2 + t^3 + \cdots$ and go one step further. Taking the derivative of both sides, we find [as in Example 9.5(c)] that

$$(-1)(1-t)^{-2}(-1) = \frac{1}{(1-t)^2} = \frac{d}{dt}\left[\frac{1}{1-t}\right] = 1 + 2t + 3t^2 + 4t^3 + \cdots,$$

where this series likewise converges† for $|t| < 1$. Therefore,

$$E(Y) = \sum_{y=1}^{\infty} yPr(Y = y) = \sum_{y=1}^{\infty} y(0.2)^{y-1}(0.8)$$

$$= (0.8)\sum_{y=1}^{\infty} y(0.2)^{y-1} = (0.8)[1 + 2(0.2) + 3(0.2)^2 + 4(0.2)^3 + \cdots]$$

$$= (0.8)\frac{1}{(1-0.2)^2} = (0.8)\frac{1}{(0.8)^2} = \frac{1}{0.8} = \frac{5}{4} = 1.25.$$

So Brianna expects to take the exam 1.25 times before she passes it.

d) Finally, to determine $\text{Var}(Y)$ we first want to find $E(Y^2)$. To do so we first multiply the result in part (c) by t and find [as in Example 9.5(c)] that

$$\frac{t}{(1-t)^2} = t + 2t^2 + 3t^3 + 4t^4 + \cdots.$$

Differentiating both sides of this equation now gives us

$$\frac{(1-t)^2(1) - t(2)(1-t)(-1)}{(1-t)^4} = \frac{1+t}{(1-t)^3} = \frac{d}{dt}\left[\frac{t}{(1-t)^2}\right]$$

$$= 1^2 + 2^2t + 3^2t^2 + 4^2t^3 + \cdots,$$

and this series is also convergent‡ for $|t| < 1$. So now we have

$$E(Y^2) = \sum_{y=1}^{\infty} y^2 Pr(Y = y) = \sum_{y=1}^{\infty} y^2(0.2)^{y-1}(0.8)$$

$$= (0.8)\sum_{y=1}^{\infty} y^2(0.2)^{y-1} = (0.8)[1^2 + 2^2(0.2) + 3^2(0.2)^2 + 4^2(0.2)^3 + \cdots]$$

$$= (0.8)\left[\frac{1+0.2}{(1-0.2)^3}\right] = \frac{1.2}{(0.8)^2} = \frac{15}{8}.$$

---

†Using the Ratio Test from calculus, one finds that

$$\lim_{n\to\infty}\left|\frac{(n+1)t^n}{nt^{n-1}}\right| = |t| \lim_{n\to\infty}\frac{n+1}{n} = |t| \lim_{n\to\infty}\left(1 + \frac{1}{n}\right) = |t|(1) = |t|.$$

When $t = \pm 1$, $\lim_{n\to\infty} nt^{n-1} \neq 0$, so the series does not converge for $t = \pm 1$. Consequently, this infinite series converges for $|t| < 1$.

‡Once again we use the Ratio Test from calculus. Here

$$\lim_{n\to\infty}\left|\frac{(n+1)^2 t^n}{n^2 t^{n-1}}\right| = |t| \lim_{n\to\infty}\frac{(n+1)^2}{n^2} = |t| \lim_{n\to\infty}\left(1 + \frac{1}{n}\right)^2 = |t|(1)^2 = |t|.$$

When $t = \pm 1$, $\lim_{n\to\infty} n^2 t^{n-1} \neq 0$, so the series does not converge for $t = \pm 1$. Consequently, this infinite series converges for $|t| < 1$.

Consequently,

$$\text{Var}(Y) = E(Y^2) - [E(Y)]^2 = \frac{15}{8} - \left(\frac{5}{4}\right)^2 = \frac{30-25}{16} = \frac{5}{16}.$$

The preceding example introduced us to a new discrete random variable — namely, the *geometric random variable*. In this situation we perform a Bernoulli trial until we are successful (for the first time). As with the binomial random variable the outcome of each trial, after the first, is independent of the outcome for any previous trial. Further, the probability of success for each Bernoulli trial is p, and the probability of failure is $q = 1 - p$.

If we let the random variable Y count the number of trials until we are finally successful, then Y is a discrete random variable with probability distribution given by

$$Pr(Y = y) = q^{y-1}p, \quad y = 1, 2, 3, \ldots .$$

In addition, we find that

$$E(Y) = \frac{1}{p} \quad \text{and} \quad \text{Var}(Y) = \frac{q}{p^2}.$$

The following example uses the last identity in Table 9.2. (This identity was used earlier in Examples 9.14 and 9.15 — but rather implicitly.)

EXAMPLE 9.19

Let $f(x) = x/(1-x)^2$. This is the generating function for the sequence $a_0, a_1, a_2, \ldots$, where $a_k = k$ for all $k \in \mathbf{N}$. The function $g(x) = x(x+1)/(1-x)^3$ generates the sequence $b_0, b_1, b_2, \ldots$, for $b_k = k^2$, $k \in \mathbf{N}$.

The function $h(x) = f(x)g(x)$ consequently gives us $a_0b_0 + (a_0b_1 + a_1b_0)x + (a_0b_2 + a_1b_1 + a_2b_0)x^2 + \cdots$, so $h(x)$ is the generating function for the sequence $c_0, c_1, c_2, \ldots$, where for each $k \in \mathbf{N}$,

$$c_k = a_0b_k + a_1b_{k-1} + a_2b_{k-2} + \cdots + a_{k-2}b_2 + a_{k-1}b_1 + a_kb_0.$$

Here, for example, we find that

$$c_0 = 0 \cdot 0^2 = 0$$
$$c_1 = 0 \cdot 1^2 + 1 \cdot 0^2 = 0$$
$$c_2 = 0 \cdot 2^2 + 1 \cdot 1^2 + 2 \cdot 0^2 = 1$$
$$c_3 = 0 \cdot 3^2 + 1 \cdot 2^2 + 2 \cdot 1^2 + 3 \cdot 0^2 = 6$$

and, in general, $c_k = \sum_{i=0}^{k} i(k-i)^2$. (We shall simplify this summation formula in the Section Exercises.)

Whenever a sequence $c_0, c_1, c_2, \ldots$ arises from two generating functions $f(x)$ [for $a_0, a_1, a_2, \ldots$] and $g(x)$ [for $b_0, b_1, b_2, \ldots$], as in this example, the sequence $c_0, c_1, c_2, \ldots$ is called the *convolution* of the sequences $a_0, a_1, a_2, \ldots$ and $b_0, b_1, b_2, \ldots$.

Our last example provides one more instance of the convolution of sequences.

EXAMPLE 9.20

For $f(x) = 1/(1-x) = 1 + x + x^2 + x^3 + \cdots$ and $g(x) = 1/(1+x) = 1 - x + x^2 - x^3 + \cdots$, we find that

$$f(x)g(x) = 1/[(1-x)(1+x)] = 1/(1-x^2) = 1 + x^2 + x^4 + x^6 + \cdots .$$

Consequently, the sequence 1, 0, 1, 0, 1, 0, . . . is the convolution of the sequences 1, 1, 1, 1, 1, 1, . . . and 1, −1, 1, −1, 1, −1, ...

EXERCISES 9.2

1. Find generating functions for the following sequences. [For example, in the case of the sequence 0, 1, 3, 9, 27, . . . , the answer required is $x/(1 − 3x)$, not $\sum_{i=0}^{\infty} 3^i x^{i+1}$ or simply $0 + x + 3x^2 + 9x^3 + \cdots$.]

a) $\binom{8}{0}, \binom{8}{1}, \binom{8}{2}, \ldots, \binom{8}{8}$

b) $\binom{8}{1}, 2\binom{8}{2}, 3\binom{8}{3}, \ldots, 8\binom{8}{8}$

c) 1, −1, 1, −1, 1, −1, . . .

d) 0, 0, 0, 6, −6, 6, −6, 6, . . .

e) 1, 0, 1, 0, 1, 0, 1, . . .

f) $0, 0, 1, a, a^2, a^3, \ldots, a \neq 0$

2. Determine the sequence generated by each of the following generating functions.

a) $f(x) = (2x − 3)^3$ **b)** $f(x) = x^4/(1 − x)$

c) $f(x) = x^3/(1 − x^2)$ **d)** $f(x) = 1/(1 + 3x)$

e) $f(x) = 1/(3 − x)$

f) $f(x) = 1/(1 − x) + 3x^7 − 11$

3. In each of the following, the function $f(x)$ is the generating function for the sequence $a_0, a_1, a_2, \ldots$, whereas the sequence $b_0, b_1, b_2, \ldots$ is generated by the function $g(x)$. Express $g(x)$ in terms of $f(x)$.

a) $b_3 = 3$
$b_n = a_n, n \in \mathbf{N}, n \neq 3$

b) $b_3 = 3$
$b_7 = 7$
$b_n = a_n, n \in \mathbf{N}, n \neq 3, 7$

c) $b_1 = 1$
$b_3 = 3$
$b_n = 2a_n, n \in \mathbf{N}, n \neq 1, 3$

d) $b_1 = 1$
$b_3 = 3$
$b_7 = 7$
$b_n = 2a_n + 5, n \in \mathbf{N}, n \neq 1, 3, 7$

4. Determine the constant (that is, the coefficient of x^0) in $(3x^2 − (2/x))^{15}$.

5. a) Find the coefficient of x^7 in
$$(1 + x + x^2 + x^3 + \cdots)^{15}.$$

b) Find the coefficient of x^7 in
$$(1 + x + x^2 + x^3 + \cdots)^n \text{ for } n \in \mathbf{Z}^+.$$

6. Find the coefficient of x^{50} in $(x^7 + x^8 + x^9 + \cdots)^6$.

7. Find the coefficient of x^{20} in $(x^2 + x^3 + x^4 + x^5 + x^6)^5$.

8. For $n \in \mathbf{Z}^+$, find in $(1 + x + x^2)(1 + x)^n$ the coefficient of (a) x^7; (b) x^8; and (c) x^r for $0 \leq r \leq n + 2, r \in \mathbf{Z}$.

9. Find the coefficient of x^{15} in each of the following.

a) $x^3(1 − 2x)^{10}$

b) $(x^3 − 5x)/(1 − x)^3$

c) $(1 + x)^4/(1 − x)^4$

10. In how many ways can two dozen identical robots be assigned to four assembly lines with (a) at least three robots assigned to each line? (b) at least three, but no more than nine, robots assigned to each line?

11. In how many ways can 3000 identical envelopes be divided, in packages of 25, among four student groups so that each group gets at least 150, but not more than 1000, of the envelopes?

12. Two cases of soft drinks, 24 bottles of one type and 24 of another, are distributed among five surveyors who are conducting taste tests. In how many ways can the 48 bottles be distributed so that each surveyor gets (a) at least two bottles of each type? (b) at least two bottles of one particular type and at least three of the other?

13. If a fair die is rolled 12 times, what is the probability that the sum of the rolls is 30?

14. Carol is collecting money from her cousins to have a party for her aunt. If eight of the cousins promise to give $2, $3, $4, or $5 each, and two others each give $5 or $10, what is the probability that Carol will collect exactly $40?

15. In how many ways can Traci select n marbles from a large supply of blue, red, and yellow marbles (all of the same size) if the selection must include an even number of blue ones?

16. How can Mary split up 12 hamburgers and 16 hot dogs among her sons Richard, Peter, Christopher, and James in such a way that James gets at least one hamburger and three hot dogs, and each of his brothers gets at least two hamburgers but at most five hot dogs?

17. Verify that $(1 − x − x^2 − x^3 − x^4 − x^5 − x^6)^{-1}$ is the generating function for the number of ways the sum n, where $n \in \mathbf{N}$, can be obtained when a single die is rolled an arbitrary number of times.

18. Show that $(1 − 4x)^{-1/2}$ generates the sequence $\binom{2n}{n}$, $n \in \mathbf{N}$.

19. a) If a computer generates a random composition of 8, what is the probability the composition is a palindrome?

b) Answer the question in part (a) after replacing 8 by n, a fixed positive integer.

20. a) How many palindromes of 11 start with 1? with 2? with 3? with 4?

b) How many palindromes of 12 start with 1? with 2? with 3? with 4?

21. Let n be a (fixed) positive integer, with $n \geq 2$. If $1 \leq t \leq \lfloor n/2 \rfloor$, how many palindromes of n start with t?

22. Let $n \in \mathbf{Z}^+$, n odd. Can a palindrome of n have an even number of summands?

23. Let $n \in \mathbf{Z}^+$, n even. How many palindromes of n have an even number of summands? How many have an odd number of summands?

24. Determine the number of palindromes of n, where all summands are even, for (a) $n = 10$; (b) $n = 12$; and (c) n even.

25. Shay rolls a fair die until she gets a 6. If the random variable Y counts the number of times Shay rolls the die until she gets her first 6, determine (a) the probability distribution for Y; (b) $E(Y)$; and (c) σ_Y.

26. Referring back to the preceding exercise, what is the probability Shay rolls her first 6 on an even-numbered roll?

27. Leroy has a biased coin where $Pr(\text{H}) = \frac{2}{3}$ and $Pr(\text{T}) = \frac{1}{3}$. Assuming that each toss, after the first, is independent of any previous outcome, if Leroy tosses the coin until he gets a tail, what is the probability he tosses it an odd number of times?

28. If Y is a geometric random variable with $E(Y) = \frac{7}{3}$, determine (a) $Pr(Y = 3)$; (b) $Pr(Y \geq 3)$; (c) $Pr(Y \geq 5)$; (d) $Pr(Y \geq 5 | Y \geq 3)$; (e) $Pr(Y \geq 6 | Y \geq 4)$; and (f) σ_Y.

29. Consider part (a) of Example 9.17.

a) Determine the differences for the inequalities that result from the subset $\{3, 6, 8, 15\}$ of S, and verify that those differences add to the correct sum.

b) Find the subset of S that determines the differences 2, 2, 3, 7, and 0.

c) Find the subset of S that determines the differences $a, b, c, d,$ and e, where $0 \leq a, e,$ and $2 \leq b, c, d$.

30. In how many ways can we select seven nonconsecutive integers from $\{1, 2, 3, \ldots, 50\}$?

31. Use the following summation formulas to simplify the expression for c_k in Example 9.19:

$$\sum_{i=0}^{k} i = \sum_{i=1}^{k} i = \frac{k(k+1)}{2},$$

$$\sum_{i=0}^{k} i^2 = \sum_{i=1}^{k} i^2 = \frac{k(k+1)(2k+1)}{6}, \quad \text{and}$$

$$\sum_{i=0}^{k} i^3 = \sum_{i=1}^{k} i^3 = \frac{k^2(k+1)^2}{4}.$$

32. a) Find the first four terms $c_0, c_1, c_2,$ and c_3 of the convolutions for each of the following pairs of sequences.

    i) $a_n = 1, b_n = 1,$ for all $n \in \mathbf{N}$
    ii) $a_n = 1, b_n = 2^n,$ for all $n \in \mathbf{N}$
    iii) $a_0 = a_1 = a_2 = a_3 = 1; \quad a_n = 0, \quad n \in \mathbf{N},$
    $n \neq 0, 1, 2, 3; b_n = 1,$ for all $n \in \mathbf{N}$

b) Find a general formula for c_n in each of the results of part (a).

33. Find a formula for the convolution of each of the following pairs of sequences.

a) $a_n = 1, 0 \leq n \leq 4, a_n = 0,$ for all $n \geq 5$; $b_n = n,$ for all $n \in \mathbf{N}$

b) $a_n = (-1)^n, b_n = (-1)^n,$ for all $n \in \mathbf{N}$

9.3
Partitions of Integers

In number theory, we are confronted with partitioning a positive integer n into positive summands and seeking the number of such partitions, without regard to order. This number is denoted by $p(n)$. For example,

$$p(1) = 1: \quad 1$$
$$p(2) = 2: \quad 2 = 1 + 1$$
$$p(3) = 3: \quad 3 = 2 + 1 = 1 + 1 + 1$$
$$p(4) = 5: \quad 4 = 3 + 1 = 2 + 2 = 2 + 1 + 1 = 1 + 1 + 1 + 1$$
$$p(5) = 7: \quad 5 = 4 + 1 = 3 + 2 = 3 + 1 + 1 = 2 + 2 + 1$$
$$= 2 + 1 + 1 + 1 = 1 + 1 + 1 + 1 + 1$$

We should like to obtain $p(n)$ for a given n without having to list all the partitions. We need a tool to keep track of the numbers of 1's, 2's, $\ldots$, n's that are used as summands for n.

If $n \in \mathbf{Z}^+$, the number of 1's we can use is 0 or 1 or 2 or.... The power series $1 + x + x^2 + x^3 + x^4 + \cdots$ keeps account of this for us. In like manner, $1 + x^2 + x^4 + x^6 + \cdots$ keeps track of the number of 2's in the partition of n, while $1 + x^3 + x^6 + x^9 + \cdots$ accounts for the number of 3's. Therefore, in order to determine $p(10)$, for instance, we want the coefficient of x^{10} in $f(x) = (1 + x + x^2 + x^3 + \cdots)(1 + x^2 + x^4 + x^6 + \cdots)(1 + x^3 + x^6 + x^9 + \cdots) \cdots (1 + x^{10} + x^{20} + \cdots)$ or in $g(x) = (1 + x + x^2 + x^3 + \cdots + x^{10})(1 + x^2 + x^4 + \cdots + x^{10})(1 + x^3 + x^6 + x^9) \cdots (1 + x^{10})$.

We prefer to work with $f(x)$, because it can be written in the more compact form

$$f(x) = \frac{1}{(1-x)} \frac{1}{(1-x^2)} \frac{1}{(1-x^3)} \cdots \frac{1}{(1-x^{10})} = \prod_{i=1}^{10} \frac{1}{(1-x^i)}.$$

If this product is extended beyond $i = 10$, we get $P(x) = \prod_{i=1}^{\infty}[1/(1-x^i)]$, which generates the sequence $p(0), p(1), p(2), p(3), \ldots$, where we define $p(0) = 1$.

Unfortunately, it is impossible to actually calculate the infinite number of terms in the product $P(x)$. If we consider only $\prod_{i=1}^{r}[1/(1-x^i)]$ for some fixed r, then the coefficient of x^n here is the number of partitions of n into summands that do not exceed r.

Despite the difficulty in calculating $p(n)$ from $P(x)$ for large values of n, the idea of the generating function will be useful in studying certain kinds of partitions.

EXAMPLE 9.21

Find the generating function for the number of ways an advertising agent can purchase n minutes ($n \in \mathbf{Z}^+$) of air time if time slots for commercials come in blocks of 30, 60, or 120 seconds.

Let 30 seconds represent one time unit. Then the answer is the number of integer solutions to the equation $a + 2b + 4c = 2n$ with $0 \le a, b, c$.

The associated generating function is

$$f(x) = (1 + x + x^2 + \cdots)(1 + x^2 + x^4 + \cdots)(1 + x^4 + x^8 + \cdots)$$

$$= \frac{1}{1-x} \frac{1}{1-x^2} \frac{1}{1-x^4},$$

and the coefficient of x^{2n} is the number of partitions of $2n$ into 1's, 2's, and 4's, the answer to the problem.

EXAMPLE 9.22

Find the generating function for $p_d(n)$, the number of partitions of a positive integer n into *distinct* summands.

Before we start, let us consider the 11 partitions of 6:

1) $1 + 1 + 1 + 1 + 1 + 1$ 2) $1 + 1 + 1 + 1 + 2$
3) $1 + 1 + 1 + 3$ 4) $1 + 1 + 4$
5) $1 + 1 + 2 + 2$ 6) $1 + 5$
7) $1 + 2 + 3$ 8) $2 + 2 + 2$
9) $2 + 4$ 10) $3 + 3$
11) 6

Partitions (6), (7), (9), and (11) have distinct summands, so $p_d(6) = 4$.

In calculating $p_d(n)$, for each $k \in \mathbf{Z}^+$ there are two choices: Either k is not used as one of the summands of n, or it is. This can be accounted for by the polynomial $1 + x^k$, and consequently, the generating function for these partitions is

$$P_d(x) = (1 + x)(1 + x^2)(1 + x^3) \cdots = \prod_{i=1}^{\infty} (1 + x^i).$$

For each $n \in \mathbf{Z}^+$, $p_d(n)$ is the coefficient of x^n in $(1 + x)(1 + x^2) \cdots (1 + x^n)$. [We define $p_d(0) = 1$.] When $n = 6$, the coefficient of x^6 in $(1 + x)(1 + x^2) \cdots (1 + x^6)$ is 4.

EXAMPLE 9.23

Considering the partitions in Example 9.22, we see that there are four partitions of 6 into odd summands: namely, (1), (3), (6), and (10). We also have $p_d(6) = 4$. Is this a coincidence?

Let $p_o(n)$ denote the number of partitions of n into odd summands, when $n \geq 1$. We define $p_o(0) = 1$. The generating function for the sequence $p_o(0)$, $p_o(1)$, $p_o(2)$, ... is given by

$$P_o(x) = (1 + x + x^2 + x^3 + \cdots)(1 + x^3 + x^6 + \cdots)(1 + x^5 + x^{10} + \cdots) \cdot$$

$$(1 + x^7 + \cdots) \cdots = \frac{1}{1 - x} \frac{1}{1 - x^3} \frac{1}{1 - x^5} \frac{1}{1 - x^7} \cdots.$$

Now because

$$1 + x = \frac{1 - x^2}{1 - x}, \qquad 1 + x^2 = \frac{1 - x^4}{1 - x^2}, \qquad 1 + x^3 = \frac{1 - x^6}{1 - x^3}, \qquad \cdots,$$

we have

$$P_d(x) = (1 + x)(1 + x^2)(1 + x^3)(1 + x^4) \cdots$$

$$= \frac{1 - x^2}{1 - x} \frac{1 - x^4}{1 - x^2} \frac{1 - x^6}{1 - x^3} \frac{1 - x^8}{1 - x^4} \cdots = \frac{1}{1 - x} \frac{1}{1 - x^3} \cdots = P_o(x).$$

From the equality of the generating functions, $p_d(n) = p_o(n)$, for all $n \geq 0$.

EXAMPLE 9.24

Once again we shall permit only odd summands, but in this example each such (odd) summand must occur an odd number of times — or not at all. Here, for example, there is one such partition of the integer 1 — namely, 1 — but there are *no* such partitions of the integer 2. For the integer 3 we have two of these partitions: 3 and $1 + 1 + 1$. When we examine the possibilities for the integer 4, we find the one partition $3 + 1$.

The generating function for the partitions described here is given by

$$f(x) = (1 + x + x^3 + x^5 + \cdots)(1 + x^3 + x^9 + x^{15} + \cdots)(1 + x^5 + x^{15} + x^{25} + \cdots) \cdots$$

$$= \prod_{k=0}^{\infty} \left(1 + \sum_{i=0}^{\infty} x^{(2k+1)(2i+1)} \right).$$

The generating function is *not* given by

$$(x + x^3 + x^5 + \cdots)(x^3 + x^9 + x^{15} + \cdots)(x^5 + x^{15} + x^{25} + \cdots) \cdots \qquad (*)$$

If it were, then the product could not contain any terms where x would appear to a finite power. The situation given by equation $(*)$ would occur if we were to believe that *every* odd positive integer must appear as a summand at least once. And in such a "partition" the number of summands and the sum itself would both be infinite. Consequently, whether or not it is stated, we must realize that each odd summand may not appear at all — and this condition is accounted for by the (first) summand, $1 = x^0$, that appears in each factor of

$f(x)$. In fact, for all but a finite number of odd summands, this is the case. Of course, when an odd summand does appear in a partition, it does so an odd number of times.

We close this section with an idea called the *Ferrers graph*. This graph uses rows of dots to represent a partition of an integer where the number of dots per row does not increase as we go from any row to the one below it.

In Fig. 9.2 we find the Ferrers graphs for two partitions of 14: (a) $4 + 3 + 3 + 2 + 1 + 1$ and (b) $6 + 4 + 3 + 1$. The graph in part (b) is said to be the *transposition* of the graph in part (a), and vice versa, because one graph can be obtained from the other by interchanging rows and columns.

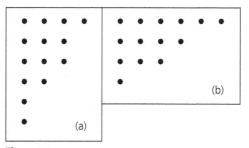

Figure 9.2

These graphs often suggest results about partitions. Here we see a partition of 14 into summands, where 4 is the largest summand, and a second partition of 14 into exactly four summands. There is a one-to-one correspondence between a Ferrers graph and its transposition, so this example demonstrates a particular instance of the general result: The number of partitions of an integer n into m summands is equal to the number of partitions of n into summands where m is the largest summand.

EXERCISES 9.3

1. Find all partitions of 7.

2. Determine the generating function for the sequence a_0, a_1, a_2, ..., where a_n is the number of partitions of the nonnegative integer n into (a) even summands; (b) distinct even summands; and (c) distinct odd summands.

3. In $f(x) = [1/(1-x)][1/(1-x^2)][1/(1-x^3)]$, the coefficient of x^6 is 7. Interpret this result in terms of partitions of 6.

4. Find the generating function for the number of integer solutions of

  a) $2w + 3x + 5y + 7z = n$, $0 \leq w, x, y, z$

  b) $2w + 3x + 5y + 7z = n$, $0 \leq w$, $4 \leq x, y$, $5 \leq z$

5. Find the generating function for the number of partitions of the nonnegative integer n into summands where (a) each summand must appear an even number of times; and (b) each summand must be even.

6. What is the generating function for the number of partitions of $n \in \mathbf{N}$ into summands that (a) cannot occur more than five times; and (b) cannot exceed 12 and cannot occur more than five times?

7. Show that the number of partitions of a positive integer n where no summand appears more than twice equals the number of partitions of n where no summand is divisible by 3.

8. Show that the number of partitions of $n \in \mathbf{Z}^+$ where no summand is divisible by 4 equals the number of partitions of n where no even summand is repeated (although odd summands may or may not be repeated).

9. Using a Ferrers graph, show that the number of partitions of an integer n into summands not exceeding m is equal to the number of partitions of n into at most m summands.

10. Using a Ferrers graph, show that the number of partitions of n is equal to the number of partitions of $2n$ into n summands.

9.4
The Exponential Generating Function

The type of generating function we have been dealing with is often referred to as the *ordinary* generating function for a given sequence. This function arose in selection problems, where order was not relevant. However, turning now to problems of arrangement, where order is crucial, we seek a comparable tool. To find such a tool, we return to the binomial theorem.

For each $n \in \mathbf{Z}^+$, $(1 + x)^n = \binom{n}{0} + \binom{n}{1}x + \binom{n}{2}x^2 + \cdots + \binom{n}{n}x^n$, so $(1 + x)^n$ is the (ordinary) generating function for the sequence $\binom{n}{0}$, $\binom{n}{1}$, $\binom{n}{2}$, ..., $\binom{n}{n}$, 0, 0, ... When dealing with this idea in Chapter 1, we also wrote $\binom{n}{r} = C(n, r)$ when we wanted to emphasize that $\binom{n}{r}$ represented the number of combinations of n objects taken r at a time, with $0 \le r \le n$. Consequently, $(1 + x)^n$ generates the sequence $C(n, 0)$, $C(n, 1)$, $C(n, 2)$, ..., $C(n, n)$, 0, 0, ...

Now for all $0 \le r \le n$,

$$C(n, r) = \frac{n!}{r!(n - r)!} = \left(\frac{1}{r!}\right) P(n, r),$$

where $P(n, r)$ denotes the number of permutations of n objects taken r at a time. So

$$(1 + x)^n = C(n, 0) + C(n, 1)x + C(n, 2)x^2 + C(n, 3)x^3 + \cdots + C(n, n)x^n$$

$$= P(n, 0) + P(n, 1)x + P(n, 2)\frac{x^2}{2!} + P(n, 3)\frac{x^3}{3!} + \cdots + P(n, n)\frac{x^n}{n!}.$$

Hence, if in $(1 + x)^n$ we consider the coefficient of $x^r/r!$, with $0 \le r \le n$, we obtain $P(n, r)$. On the basis of this observation, we have the following definition.

Definition 9.2   For a sequence $a_0, a_1, a_2, a_3, \ldots$ of real numbers,

$$f(x) = a_0 + a_1 x + a_2\frac{x^2}{2!} + a_3\frac{x^3}{3!} + \cdots = \sum_{i=0}^{\infty} a_i \frac{x^i}{i!},$$

is called the *exponential generating function* for the given sequence.

EXAMPLE 9.25   Examining the Maclaurin series expansion for e^x, we find

$$e^x = 1 + x + \frac{x^2}{2!} + \frac{x^3}{3!} + \frac{x^4}{4!} + \cdots = \sum_{i=0}^{\infty} \frac{x^i}{i!},$$

so e^x is the exponential generating function for the sequence 1, 1, 1, ... (The function e^x is the ordinary generating function for the sequence 1, 1, 1/2!, 1/3!, 1/4!, ...)

Our next example shows how this idea can help us count certain types of arrangements.

EXAMPLE 9.26   In how many ways can four of the letters in ENGINE be arranged?

In Table 9.4 we list the possible selections of size 4 from the letters E, N, G, I, N, E, along with the number of arrangements those four letters determine.

We now obtain the answer by means of an exponential generating function. For the letter E we use $[1 + x + (x^2/2!)]$ because there are 0, 1, or 2 E's to arrange. Note that the coefficient of $x^2/2!$ is 1, the number of distinct ways to arrange (only) two E's. In like

Table 9.4

E	E	N	N	4!/(2! 2!)	E	G	N	N	4!/2!
E	E	G	N	4!/2!	E	I	N	N	4!/2!
E	E	I	N	4!/2!	G	I	N	N	4!/2!
E	E	G	I	4!/2!	E	I	G	N	4!

manner, we have $[1 + x + (x^2/2!)]$ for the arrangements of 0, 1, or 2 N's. The arrangements for each of the letters G and I are represented by $(1 + x)$.

Consequently, we find here that the exponential generating function is

$$f(x) = [1 + x + (x^2/2!)]^2(1 + x)^2,$$

and we claim that the required answer is the coefficient of $x^4/4!$ in $f(x)$.

In order to motivate our claim, let us consider two of the eight ways in which the term $x^4/4!$ arises in the expansion of

$$f(x) = [1 + x + (x^2/2!)][1 + x + (x^2/2!)](1 + x)(1 + x).$$

1) From the product $(x^2/2!)(x^2/2!)(1)(1)$, where $(x^2/2!)$ is taken from each of the first two factors (namely, $[1 + x + (x^2/2!)]$) and 1 is taken from each of the last two factors [namely, $(1 + x)$]. Then $(x^2/2!)(x^2/2!)(1)(1) = x^4/(2! 2!) = (4!/(2! 2!)) \cdot (x^4/4!)$, and the coefficient of $x^4/4!$ is $4!/(2! 2!)$ — the number of ways one can arrange the four letters E, E, N, N.

2) From the product $(x^2/2!)(1)(x)(x)$, where $(x^2/2!)$ is taken from the first factor (namely, $[1 + x + (x^2/2!)]$), 1 is taken from the second factor (again, $[1 + x + (x^2/2!)]$), and x is taken from each of the last two factors [namely, $(1 + x)$]. Here $(x^2/2!)(1)(x)(x) = x^4/2! = (4!/2!)(x^4/4!)$, so the coefficient of $x^4/4!$ is $4!/2!$ — the number of ways the four letters E, E, G, I can be arranged.

In the complete expansion of $f(x)$, the term involving x^4 [and, consequently, $x^4/4!$] is

$$\left(\frac{x^4}{2! \, 2!} + \frac{x^4}{2!} + \frac{x^4}{2!} + \frac{x^4}{2!} + \frac{x^4}{2!} + \frac{x^4}{2!} + \frac{x^4}{2!} + x^4 \right)$$

$$= \left[\left(\frac{4!}{2! \, 2!} \right) + \left(\frac{4!}{2!} \right) + \left(\frac{4!}{2!} \right) + \left(\frac{4!}{2!} \right) + \left(\frac{4!}{2!} \right) + \left(\frac{4!}{2!} \right) + \left(\frac{4!}{2!} \right) + 4! \right] \left(\frac{x^4}{4!} \right),$$

where the coefficient of $x^4/4!$ is the answer (102 arrangements) produced by the eight results in the table.

EXAMPLE 9.27

Consider the Maclaurin series expansions of e^x and e^{-x}.

$$e^x = 1 + x + \frac{x^2}{2!} + \frac{x^3}{3!} + \frac{x^4}{4!} + \cdots \qquad e^{-x} = 1 - x + \frac{x^2}{2!} - \frac{x^3}{3!} + \frac{x^4}{4!} - \cdots$$

Adding these series together, we find that

$$e^x + e^{-x} = 2 \left(1 + \frac{x^2}{2!} + \frac{x^4}{4!} + \cdots \right),$$

or

$$\frac{e^x + e^{-x}}{2} = 1 + \frac{x^2}{2!} + \frac{x^4}{4!} + \cdots.$$

Subtracting e^{-x} from e^x yields

$$\frac{e^x - e^{-x}}{2} = x + \frac{x^3}{3!} + \frac{x^5}{5!} + \cdots.$$

These results now help us in the following.

EXAMPLE 9.28

A ship carries 48 flags, 12 each of the colors red, white, blue, and black. Twelve of these flags are placed on a vertical pole in order to communicate a signal to other ships.

a) How many of these signals use an even number of blue flags and an odd number of black flags?

The exponential generating function

$$f(x) = \left(1 + x + \frac{x^2}{2!} + \frac{x^3}{3!} + \cdots\right)^2 \left(1 + \frac{x^2}{2!} + \frac{x^4}{4!} + \cdots\right) \left(x + \frac{x^3}{3!} + \frac{x^5}{5!} + \cdots\right)$$

considers all such signals made up of n flags, where $n \geq 1$. The last two factors in $f(x)$ restrict the signals to an even number of blue flags and an odd number of black flags, respectively.

Since

$$f(x) = (e^x)^2 \left(\frac{e^x + e^{-x}}{2}\right) \left(\frac{e^x - e^{-x}}{2}\right) = \left(\frac{1}{4}\right)(e^{2x})(e^{2x} - e^{-2x}) = \frac{1}{4}(e^{4x} - 1)$$

$$= \frac{1}{4}\left(\sum_{i=0}^{\infty} \frac{(4x)^i}{i!} - 1\right) = \left(\frac{1}{4}\right)\sum_{i=1}^{\infty} \frac{(4x)^i}{i!},$$

the coefficient of $x^{12}/12!$ in $f(x)$ yields $(1/4)(4^{12}) = 4^{11}$ signals made up of 12 flags with an even number of blue flags and an odd number of black flags.

b) How many of the signals have at least three white flags or no white flags at all? In this situation we use the exponential generating function

$$g(x) = \left(1 + x + \frac{x^2}{2!} + \frac{x^3}{3!} + \cdots\right) \left(1 + \frac{x^3}{3!} + \frac{x^4}{4!} + \cdots\right) \left(1 + x + \frac{x^2}{2!} + \frac{x^3}{3!} + \cdots\right)^2$$

$$= e^x \left(e^x - x - \frac{x^2}{2!}\right)(e^x)^2 = e^{3x}\left(e^x - x - \frac{x^2}{2!}\right) = e^{4x} - xe^{3x} - \left(\frac{1}{2}\right)x^2 e^{3x}$$

$$= \sum_{i=0}^{\infty} \frac{(4x)^i}{i!} - x \sum_{i=0}^{\infty} \frac{(3x)^i}{i!} - \left(\frac{x^2}{2}\right)\left(\sum_{i=0}^{\infty} \frac{(3x)^i}{i!}\right).$$

Here the factor $(1 + \frac{x^3}{3!} + \frac{x^4}{4!} + \cdots) = e^x - x - \frac{x^2}{2!}$ in $g(x)$ restricts the signals to those that contain three or more of the 12 white flags, or none at all. The answer for the number of signals sought here is the coefficient of $x^{12}/12!$ in $g(x)$. As we consider each summand (involving an infinite summation), we find:

i) $\sum_{i=0}^{\infty} \frac{(4x)^i}{i!}$ — Here we have the term $\frac{(4x)^{12}}{12!} = 4^{12}\left(\frac{x^{12}}{12!}\right)$, so the coefficient of $x^{12}/12!$ is 4^{12};

ii) $x \left(\sum_{i=0}^{\infty} \frac{(3x)^i}{i!} \right)$ —Now we see that in order to get $x^{12}/12!$ we need to consider the term $x[(3x)^{11}/11!] = 3^{11}(x^{12}/11!) = (12)(3^{11})(x^{12}/12!)$, and here the coefficient of $x^{12}/12!$ is $(12)(3^{11})$; and

iii) $(x^2/2) \left(\sum_{i=0}^{\infty} \frac{(3x)^i}{i!} \right)$ —For this last summand we observe that

$(x^2/2)[(3x)^{10}/10!] = (1/2)(3^{10})(x^{12}/10!) = (1/2)(12)(11)(3^{10})(x^{12}/12!),$

where this time the coefficient of $x^{12}/12!$ is $(1/2)(12)(11)(3^{10})$.

Consequently, the number of 12 flag signals with at least three white flags, or none at all, is

$$4^{12} - 12(3^{11}) - (1/2)(12)(11)(3^{10}) = 10,754,218.$$

Our final example is reminiscent of past results.

EXAMPLE 9.29

A company hires 11 new employees, each of whom is to be assigned to one of four subdivisions. Each subdivision will get at least one new employee. In how many ways can these assignments be made?

Calling the subdivisions A, B, C, and D, we can equivalently count the number of 11-letter sequences in which there is at least one occurrence of each of the letters A, B, C, and D. The exponential generating function for these arrangements is

$$f(x) = \left(x + \frac{x^2}{2!} + \frac{x^3}{3!} + \frac{x^4}{4!} + \cdots \right)^4 = (e^x - 1)^4 = e^{4x} - 4e^{3x} + 6e^{2x} - 4e^x + 1.$$

The answer then is the coefficient of $x^{11}/11!$ in $f(x)$:

$$4^{11} - 4(3^{11}) + 6(2^{11}) - 4(1^{11}) = \sum_{i=0}^{4} (-1)^i \binom{4}{i} (4 - i)^{11}.$$

This form of the answer should bring to mind some of the enumeration problems in Chapter 5. Once the vocabulary is set aside, we are counting the number of onto functions $g: X \rightarrow Y$ where $|X| = 11$, $|Y| = 4$.

EXERCISES 9.4

1. Find the exponential generating function for each of the following sequences.

a) $1, -1, 1, -1, 1, -1, \ldots$

b) $1, 2, 2^2, 2^3, 2^4, \ldots$

c) $1, -a, a^2, -a^3, a^4, \ldots, \quad a \in \mathbf{R}$

d) $1, a^2, a^4, a^6, \ldots, \quad a \in \mathbf{R}$

e) $a, a^3, a^5, a^7, \ldots, \quad a \in \mathbf{R}$

f) $0, 1, 2(2), 3(2^2), 4(2^3), \ldots$

2. Determine the sequence generated by each of the following exponential generating functions.

a) $f(x) = 3e^{3x}$

b) $f(x) = 6e^{5x} - 3e^{2x}$

c) $f(x) = e^x + x^2$

d) $f(x) = e^{2x} - 3x^3 + 5x^2 + 7x$

e) $f(x) = 1/(1 - x)$

f) $f(x) = 3/(1 - 2x) + e^x$

3. In each of the following, the function $f(x)$ is the exponential generating function for the sequence $a_0, a_1, a_2, \ldots,$ whereas the function $g(x)$ is the exponential generating function for the sequence $b_0, b_1, b_2, \ldots .$ Express $g(x)$ in terms of $f(x)$ if

a) $b_3 = 3$
$b_n = a_n, n \in \mathbf{N}, n \neq 3$

b) $a_n = 5^n, n \in \mathbf{N}$
$b_3 = -1$
$b_n = a_n, n \in \mathbf{N}, n \neq 3$

c) $b_1 = 2$
$b_2 = 4$
$b_n = 2a_n, n \in \mathbf{N}, n \neq 1, 2$

d) $b_1 = 2$
$b_2 = 4$
$b_3 = 8$
$b_n = 2a_n + 3, n \in \mathbf{N}, n \neq 1, 2, 3$

4. a) For the ship in Example 9.28, how many signals use at least one flag of each color? (Solve this with an exponential generating function.)

b) Restate part (a) in an alternative way that uses the concept of an onto function.

c) How many signals are there in Example 9.28, where the total number of blue and black flags is even?

5. Find the exponential generating function for the sequence $0!, 1!, 2!, 3!, \ldots$.

6. a) Find the exponential generating function for the number of ways to arrange n letters, $n \geq 0$, selected from each of the following words.

   i) HAWAII

   ii) MISSISSIPPI

   iii) ISOMORPHISM

b) For section (ii) of part (a), what is the exponential generating function if the arrangement must contain at least two I's?

7. Say the company in Example 9.29 hires 25 new employees. Give the exponential generating function for the number of ways to assign these people to the four subdivisions so that each subdivision receives at least 3, but no more than 10, new people.

8. Given the sequences $a_0, a_1, a_2, \ldots$ and $b_0, b_1, b_2, \ldots$, with exponential generating functions $f(x), g(x)$, respectively, show that if $h(x) = f(x)g(x)$, then $h(x)$ is the exponential generating function of the sequence $c_0, c_1, c_2, \ldots$, where $c_n = \sum_{i=0}^{n} \binom{n}{i} a_i b_{n-i}$, for each $n \geq 0$.

9. If a 20-digit ternary $(0, 1, 2)$ sequence is randomly generated, what is the probability that: (a) It has an even number of 1's? (b) It has an even number of 1's and an even number of 2's? (c) It has an odd number of 0's? (d) The total number of 0's and 1's is odd? (e) The total number of 0's and 1's is even?

10. How many 20-digit quaternary $(0, 1, 2, 3)$ sequences are there where: (a) There is at least one 2 and an odd number of 0's? (b) No symbol occurs exactly twice? (c) No symbol occurs exactly three times? (d) There are exactly two 3's or none at all?

9.5
The Summation Operator

This final section introduces a technique that helps us go from the (ordinary) generating function for the sequence $a_0, a_1, a_2, \ldots$ to the generating function for the sequence $a_0, a_0 + a_1, a_0 + a_1 + a_2, \ldots$.

For $f(x) = a_0 + a_1 x + a_2 x^2 + a_3 x^3 + \cdots$, consider the function $f(x)/(1-x)$.

$$\frac{f(x)}{1-x} = f(x) \cdot \frac{1}{1-x} = [a_0 + a_1 x + a_2 x^2 + a_3 x^3 + \cdots][1 + x + x^2 + x^3 + \cdots]$$

$$= a_0 + (a_0 + a_1)x + (a_0 + a_1 + a_2)x^2 + (a_0 + a_1 + a_2 + a_3)x^3 + \cdots,$$

so $f(x)/(1-x)$ generates the sequence of sums $a_0, a_0 + a_1, a_0 + a_1 + a_2, a_0 + a_1 + a_2 + a_3, \ldots$. This is why we refer to $1/(1-x)$ as the *summation operator*. Furthermore we see that the sequence $a_0, a_0 + a_1, a_0 + a_1 + a_2, a_0 + a_1 + a_2 + a_3, \ldots$ is the convolution of the sequence $a_0, a_1, a_2, \ldots$ and the sequence $b_0, b_1, b_2, \ldots$, where $b_n = 1$ for all $n \in \mathbf{N}$.

We find this technique handy in the following examples.

EXAMPLE 9.30

a) We know from part (b) of Example 9.5 that $1/(1-x)$ is the generating function for the sequence $1, 1, 1, \ldots$. Consequently, upon applying the summation operator, $1/(1-x)$, we see that $(1/(1-x))(1/(1-x))$ is the generating function for the sequence $1, 1 + 1, 1 + 1 + 1, \ldots$—that is, $1/(1-x)^2$ is the generating function for the sequence $1, 2, 3, \ldots$, as we found in part (c) of Example 9.5.

b) Now let us start with the polynomial $x + x^2$, the generating function for the sequence $0, 1, 1, 0, 0, 0, \ldots$. Applying the summation operator, we have $(x + x^2)(1/(1 - x)) = (x + x^2)/(1 - x)$, the generating function for the sequence $0, 0 + 1, 0 + 1 + 1, 0 + 1 + 1 + 0, \ldots$,—that is, the sequence $0, 1, 2, 2, \ldots$. A second application of the summation operator tells us that $(x + x^2)/(1 - x)^2$ is the generating function for the sequence $0, 0 + 1, 0 + 1 + 2, 0 + 1 + 2 + 2, \ldots$,— that is, the sequence $0, 1, 3, 5, \ldots$. A final application of the summation operator tells us that $(x + x^2)/(1 - x)^3$ is the generating function for the sequence $0, 0 + 1, 0 + 1 + 3, 0 + 1 + 3 + 5, \ldots$,—that is, the sequence $0, 1, 4, 9, \ldots$. This *suggests* that, for $n \geq 1$, $\sum_{k=1}^{n}(2k - 1) = n^2$. To verify this suggestion, we look at the coefficient of x^n in $(x + x^2)/(1 - x)^3 = x(1 - x)^{-3} + x^2(1 - x)^{-3}$. The coefficient of x^{n-1} in $(1 - x)^{-3}$ [which is the coefficient of x^n in $x(1 - x)^{-3}$] is

$$\binom{-3}{n - 1}(-1)^{n-1} = (-1)^{n-1}\binom{3 + (n - 1) - 1}{n - 1}(-1)^{n-1} = \binom{n + 1}{n - 1} = \frac{1}{2}(n + 1)(n).$$

The coefficient of x^{n-2} in $(1 - x)^{-3}$ [which is the coefficient of x^n in $x^2(1 - x)^{-3}$] is $\binom{-3}{n - 2}(-1)^{n-2} = (-1)^{n-2}\binom{3 + (n - 2) - 1}{n - 2}(-1)^{n-2} = \binom{n}{n - 2} = \frac{1}{2}(n)(n - 1)$. Consequently, for $n \geq 1$, $\sum_{k=1}^{n}(2k - 1) = $ the coefficient of x^n in $(x + x^2)/(1 - x)^3 = \frac{1}{2}(n + 1)(n) + \frac{1}{2}(n)(n - 1) = \frac{1}{2}(n)[(n + 1) + (n - 1)] = n^2$, as we learned earlier in Example 4.7, using the Principle of Mathematical Induction.

Our last example provides us with a method for deriving some of the summation formulas we encountered in earlier chapters.

EXAMPLE 9.31

Find a formula to express $0^2 + 1^2 + 2^2 + \cdots + n^2$ as a function of n.

As in Section 9.2, we start with $g(x) = 1/(1 - x) = 1 + x + x^2 + \cdots$. Then

$$(-1)(1 - x)^{-2}(-1) = \frac{1}{(1 - x)^2} = \frac{dg(x)}{dx} = 1 + 2x + 3x^2 + 4x^3 + \cdots,$$

so $x/(1 - x)^2$ is the generating function for $0, 1, 2, 3, 4, \ldots$. Repeating this technique, we find that

$$x\frac{d}{dx}\left[x\left(\frac{dg(x)}{dx}\right)\right] = \frac{x(1 + x)}{(1 - x)^3} = x + 2^2x^2 + 3^2x^3 + \cdots,$$

so $x(1 + x)/(1 - x)^3$ generates $0^2, 1^2, 2^2, 3^2, \ldots$. As a consequence of our earlier observations about the summation operator, we find that

$$\frac{x(1 + x)}{(1 - x)^3}\frac{1}{(1 - x)} = \frac{x(1 + x)}{(1 - x)^4}$$

is the generating function for $0^2, 0^2 + 1^2, 0^2 + 1^2 + 2^2, 0^2 + 1^2 + 2^2 + 3^2, \ldots$. Hence the coefficient of x^n in $[x(1 + x)]/(1 - x)^4$ is $\sum_{i=0}^{n} i^2$. But the coefficient of x^n in $[x(1 + x)]/(1 - x)^4$ can also be calculated as follows:

$$\frac{x(1 + x)}{(1 - x)^4} = (x + x^2)(1 - x)^{-4} = (x + x^2)\left[\binom{-4}{0} + \binom{-4}{1}(-x) + \binom{-4}{2}(-x)^2 + \cdots\right],$$

so the coefficient of x^n is

$$\binom{-4}{n-1}(-1)^{n-1} + \binom{-4}{n-2}(-1)^{n-2}$$

$$= (-1)^{n-1}\binom{4+(n-1)-1}{n-1}(-1)^{n-1} + (-1)^{n-2}\binom{4+(n-2)-1}{n-2}(-1)^{n-2}$$

$$= \binom{n+2}{n-1} + \binom{n+1}{n-2} = \frac{(n+2)!}{3!(n-1)!} + \frac{(n+1)!}{3!(n-2)!}$$

$$= \frac{1}{6}[(n+2)(n+1)(n) + (n+1)(n)(n-1)]$$

$$= \frac{1}{6}(n)(n+1)[(n+2)+(n-1)] = \frac{n(n+1)(2n+1)}{6}.$$

EXERCISES 9.5

1. Find the generating function for the sequences (a) 1, 2, 3, 3, 3, . . . ; (b) 1, 2, 3, 4, 4, 4, . . . ; (c) 1, 4, 7, 10, 13, ...

2. a) Find the generating function for the sequences (i) 0, 1, 0, 0, 0, . . . ; (ii) 0, 1, 1, 1, 1, . . . ; (iii) 0, 1, 2, 3, 4, . . . ; (iv) 0, 1, 3, 6, 10, ...

  b) Use result (iv) from part (a) to find a formula for $\sum_{k=1}^{n} k$.

3. Continue the development of the ideas set forth in Example 9.31 and derive the formula $\sum_{i=0}^{n} i^3 = [n(n+1)/2]^2$.

4. If $f(x) = \sum_{n=0}^{\infty} a_n x^n$, what is the generating function for the sequence $a_0, a_0 + a_1, a_1 + a_2, a_2 + a_3, \ldots$? What is the generating function for the sequence $a_0, a_0 + a_1, a_0 + a_1 + a_2, a_1 + a_2 + a_3, a_2 + a_3 + a_4, \ldots$? What is the generating function for the sequence $\frac{a_0}{4}, \frac{a_0}{2} + \frac{a_1}{4}, \frac{a_0}{4} + \frac{a_1}{2} + \frac{a_2}{4}, \frac{a_1}{4} + \frac{a_2}{2} + \frac{a_3}{4}, \ldots$?

5. Let $f(x)$ be the generating function for the sequence $a_0, a_1, a_2, \ldots$. For what sequence is $(1-x)f(x)$ the generating function?

6. Let $f(x) = \sum_{i=0}^{\infty} a_i x^i$ with $f(1) = \sum_{i=0}^{\infty} a_i$, a finite number. Verify that the quotient $[f(x) - f(1)]/(x-1)$ is the generating function for the sequence $s_0, s_1, s_2, \ldots$, where $s_n = \sum_{i=n+1}^{\infty} a_i$, $n \in \mathbf{N}$.

7. Find the generating function for the sequence $a_0, a_1, a_2, \ldots$, where $a_n = \sum_{i=0}^{n}(1/i!)$, $n \in \mathbf{N}$.

8. a) Find the generating function for the sequence 0, 1, 3, 6, 10, 15, . . . (where 1, 3, 6, 10, 15, . . . are the triangular numbers of Example 4.5).

  b) For $n \in \mathbf{Z}^+$, determine a formula for the sum of the first n triangular numbers.

9.6
Summary and Historical Review

In the early thirteenth century the Italian mathematician Leonardo of Pisa (c. 1175–1250), in his *Liber Abaci*, introduced the European world to the Hindu-Arabic notation for numerals and algorithms for arithmetic. In this text he also originated the study of the sequence 0, 1, 1, 2, 3, 5, 8, 13, 21, . . . , which can be given recursively by $F_0 = 0$, $F_1 = 1$, and $F_{n+2} = F_{n+1} + F_n$, $n \geq 0$. Since Leonardo was the son of Bonaccio, the sequence has come to be called the Fibonacci numbers. (Filius Bonaccii is the Latin form for "son of Bonaccio.")

If we consider the formula

$$F_n = \frac{1}{\sqrt{5}}\left[\left(\frac{1+\sqrt{5}}{2}\right)^n - \left(\frac{1-\sqrt{5}}{2}\right)^n\right], \qquad n \geq 0,$$

we find $F_0 = 0$, $F_1 = 1$, $F_2 = 1$, $F_3 = 2$, $F_4 = 3$, ... Yes, this formula determines each Fibonacci number as a function of n. (Here we have the solution for the recursive Fibonacci relation. We shall learn more about this in the next chapter.) This formula was not derived, however, until 1718, when Abraham DeMoivre (1667–1754) obtained the result from the generating function

$$f(x) = \frac{x}{1 - x - x^2} = \frac{1}{\sqrt{5}} \left[\frac{1}{1 - \left(\frac{1 + \sqrt{5}}{2}\right) x} - \frac{1}{1 - \left(\frac{1 - \sqrt{5}}{2}\right) x} \right].$$

Extending the existing techniques of the generating function, Leonhard Euler (1707–1783) advanced the study of the partitions of integers in his 1748 two-volume opus, *Introductio in Analysin Infinitorum*. With

$$P(x) = \frac{1}{1 - x} \frac{1}{1 - x^2} \frac{1}{1 - x^3} \cdots = \prod_{i=1}^{\infty} \frac{1}{1 - x^i},$$

we have the generating function for $p(0)$, $p(1)$, $p(2)$, ..., where $p(n)$ is the number of partitions of n into positive summands and $p(0)$ is defined to be 1.

Leonhard Euler (1707–1783)

In the latter part of the eighteenth century, further developments on generating functions arose in conjunction with ideas in probability theory, especially with what is now called the "moment generating function." These related notions were presented in their first complete treatment by the great scholar Pierre-Simon de Laplace (1749–1827) in his 1812 publication *Théorie Analytique des Probabilités*.

Finally, we mention Norman Macleod Ferrers (1829–1903), after whom the diagram we called the Ferrers graph is named.

For us the study of the ordinary and exponential generating functions provided a powerful technique that unified ideas found in Chapters 1, 5, and 8. Extending our prior experience with polynomials to power series, and extending the binomial theorem to $(1 + x)^n$ for the cases where n need not be positive or even an integer, we found the necessary tools to compute the coefficients in these generating functions. This was more than worth the effort because the algebraic calculations we performed took into account all of the selection processes we were trying to consider. We also found that we had seen some generating functions in a prior chapter and saw how they arose in the study of partitions.

The concept of a partition of a positive integer now enables us to complete the summaries of our earlier discussions on distributions, as given in Tables 1.11 and 5.13. Here we can now deal with the distributions of m objects into n ($\leq m$) containers for the cases where neither the objects nor the containers are distinct. These are covered by the entries in the second and fourth rows of Table 9.5. The notation $p(m, n)$, which appears in the last column for these entries, is used to denote the number of partitions of the positive integer m into exactly n (positive) summands. (This idea will be examined further in Supplementary Exercise 3 of the next chapter.) The types of distributions in the first and third rows of this table were also listed in Table 5.13. We include them here a second time for the sake of comparison and completeness.

Table 9.5

Objects Are Distinct	Containers Are Distinct	Some Container(s) May Be Empty	Number of Distributions
No	Yes	Yes	$\binom{n+m-1}{m}$
No	No	Yes	(1) $p(m)$, for $n = m$ (2) $p(m, 1) + p(m, 2) + \cdots + p(m, n)$, for $n < m$
No	Yes	No	$\binom{n+(m-n)-1}{(m-n)} = \binom{m-1}{m-n} = \binom{m-1}{n-1}$
No	No	No	$p(m, n)$

For comparable coverage of the material presented in this chapter, the interested reader should consult Chapter 2 of C. L. Liu [3] and Chapter 6 of A. Tucker [8]. The text by J. Riordan [6] has extensive coverage of ordinary and exponential generating functions. An interesting survey article on generating functions, written by Richard P. Stanley, can be found in the text edited by G-C. Rota [7]. The text by H. S. Wilf [9] deals with generating functions and some of the ways they are applied in discrete mathematics. This work also demonstrates how these functions provide a bridge between discrete mathematics and continuous analysis (in particular, the theory of functions of a complex variable).

The reader interested in learning more about the theory of partitions should consult Chapter 10 of I. Niven, H. Zuckerman, and H. Montgomery [5].

Finally, a great deal about the moment generating function and its use in probability theory can be found in Chapter 3 of H. J. Larson [2] and in Chapter XI of the comprehensive work by W. Feller [1].

REFERENCES

1. Feller, William. *An Introduction to Probability Theory and Its Applications*, Vol. I, 3rd ed. New York: Wiley, 1968.
2. Larson, Harold J. *Introduction to Probability Theory and Statistical Inference*, 2nd ed. New York: Wiley, 1969.
3. Liu, C. L. *Introduction to Combinatorial Mathematics*. New York: McGraw-Hill, 1968.
4. Neal, David. "The Series $\sum_{n=1}^{\infty} n^m x^n$ and a Pascal-like Triangle." *The College Mathematics Journal* 25, No. 2 (March 1994): pp. 99–101.

5. Niven, Ivan, Zuckerman, Herbert, and Montgomery, Hugh. *An Introduction to the Theory of Numbers*, 5th ed. New York: Wiley, 1991.

6. Riordan, John. *An Introduction to Combinatorial Analysis*. Princeton, N.J.: Princeton University Press, 1980. (Originally published in 1958 by John Wiley & Sons.)

7. Rota, Gian-Carlo, ed. *Studies in Combinatorics*, Studies in Mathematics, Vol. 17. Washington, D.C.: The Mathematical Association of America, 1978.

8. Tucker, Alan. *Applied Combinatorics*, 4th ed. New York: Wiley, 2002.

9. Wilf, Herbert S. *Generatingfunctionology*, 2nd ed. San Diego, Calif.: Academic Press, 1994.

SUPPLEMENTARY EXERCISES

1. Find the generating function for each of the following sequences.

   a) $7, 8, 9, 10, \ldots$

   b) $1, a, a^2, a^3, a^4, \ldots, \quad a \in \mathbf{R}$

   c) $1, (1+a), (1+a)^2, (1+a)^3, \ldots, \quad a \in \mathbf{R}$

   d) $2, 1+a, 1+a^2, 1+a^3, \ldots, \quad a \in \mathbf{R}$

2. Find the coefficient of x^{83} in

$$f(x) = (x^5 + x^8 + x^{11} + x^{14} + x^{17})^{10}.$$

3. Sergeant Bueti must distribute 40 bullets (20 for rifles and 20 for handguns) among four police officers so that each officer gets at least two, but no more than seven, bullets of each type. In how many ways can he do this?

4. Find a generating function for the number of ways to partition a positive integer n into positive-integer summands, where each summand appears an odd number of times or not at all.

5. For $n \in \mathbf{Z}^+$, show that the number of partitions of n in which no even summand is repeated (an odd summand may or may not be repeated) is the same as the number of partitions of n where no summand occurs more than three times.

6. How many 10-digit telephone numbers use only the digits 1, 3, 5 and 7, with each digit appearing at least twice or not at all?

7. a) For what sequence of numbers is $g(x) = (1 - 2x)^{-5/2}$ the exponential generating function?

   b) Find a and b so that $(1 - ax)^b$ is the exponential generating function for the sequence $1, 7, 7 \cdot 11, 7 \cdot 11 \cdot 15, \ldots$.

8. For integers $n, k \geq 0$ let

- P_1 be the number of partitions of n.

- P_2 be the number of partitions of $2n + k$, where $n + k$ is the greatest summand.

- P_3 be the number of partitions of $2n + k$ into precisely $n + k$ summands.

Using the concept of the Ferrers graph, prove that $P_1 = P_2$ and $P_2 = P_3$, thus concluding that the number of partitions of $2n + k$ into precisely $n + k$ summands is the same for all k.

9. Simplify the following sum where $n \in \mathbf{Z}^+$: $\binom{n}{1} + 2\binom{n}{2} + 3\binom{n}{3} + \cdots + n\binom{n}{n}$. (*Hint*: You may wish to start with the binomial theorem.)

10. Determine the generating function for the number of partitions of $n \in \mathbf{N}$ where 1 occurs at most once, 2 occurs at most twice, 3 at most thrice, and, in general, k occurs at most k times, for every $k \in \mathbf{Z}^+$.

11. In a rural area 12 mailboxes are located at a general store.

   a) If a newscarrier has 20 identical fliers, in how many ways can she distribute the fliers so that each mailbox gets at least one flier?

   b) If the mailboxes are in two rows of six each, what is the probability that a distribution from part (a) will have 10 fliers distributed to the top six boxes and 10 to the bottom six?

12. Let S be a set containing n distinct objects. Verify that $e^x/(1 - x)^k$ is the exponential generating function for the number of ways to choose m of the objects in S, for $0 \leq m \leq n$, and distribute these objects among k distinct containers, with the order of the objects in any container relevant for the distribution.

13. a) For $a, d \in \mathbf{R}$, find the generating function for the sequence $a, a + d, a + 2d, a + 3d, \ldots$.

   b) For $n \in \mathbf{Z}^+$, use the result from part (a) to find a formula for the sum of the first n terms of the arithmetic progression $a, a + d, a + 2d, a + 3d, \ldots$.

14. a) For the alphabet $\Sigma = \{0, 1\}$, let a_n count the number of strings of length n in Σ^*—that is, for $n \in \mathbf{N}$, $a_n = |\Sigma^n|$. Determine the generating function for the sequence $a_0, a_1, a_2, \ldots$.

   b) Answer the question posed in part (a) when $|\Sigma| = k$, a fixed positive integer.

15. Let $f(x) = a_0 + a_1 x + a_2 x^2 + a_3 x^3 + \ldots$, the generating function for the sequence $a_0, a_1, a_2, a_3, \ldots$. Now let $n \in \mathbf{Z}^+$, n fixed.

   a) Find the generating function for the sequence $0, 0, 0, \ldots 0, a_0, a_1, a_2, a_3, \ldots$, where there are n leading zeros.

   b) Find the generating function for the sequence $a_n, a_{n+1}, a_{n+2}, \ldots$.

16. Suppose that X is a discrete random variable with probability distribution given by

$$Pr(X = x) = \begin{cases} k \left(\frac{1}{4}\right)^x, & x = 0, 1, 2, 3, \ldots \\ 0, & \text{otherwise,} \end{cases}$$

where k is a constant. Determine (a) the value of k; (b) $Pr(X = 3)$, $Pr(X \le 3)$, $Pr(X > 3)$, $Pr(X \ge 2)$; and (c) $Pr(X \ge 4 | X \ge 2)$, $Pr(X \ge 104 | X \ge 102)$.

17. Suppose that Y is a geometric random variable where the probability of success for each Bernoulli trial is p. If $m, n \in \mathbf{Z}^+$ with $m > n$, determine $Pr(Y \ge m | Y \ge n)$.

18. A test car is driven a fixed distance of n miles along a straight highway. (Here $n \in \mathbf{Z}^+$.) The car travels at one mile per hour for the first mile, two miles per hour for the second mile, four miles per hour for the third mile, $\ldots$, and 2^{n-1} miles per hour for the nth mile.

a) What is the car's average velocity for the first four miles?

b) For a given value of n, what is the car's average velocity for the first n miles?

c) Find the smallest value of n for which the car's average velocity for the first n miles exceeds 10 miles per hour.

10

Recurrence Relations

In earlier sections of the text we saw some recursive definitions and constructions. In Definitions 5.19, 6.7, 6.12, and 7.9, we obtained concepts at level $n + 1$ (or of size $n + 1$) from comparable concepts at level n (or of size n), after establishing the concept at a first value of n, such as 0 or 1. When we dealt with the Fibonacci and Lucas numbers in Section 4.2, the results at level $n + 1$ turned out to depend on those at levels n and $n - 1$; and for each of these sequences of integers the basis consisted of the first two integers (of the sequence). Now we shall find ourselves in a somewhat similar situation. We shall investigate functions $a(n)$, preferably written as a_n (for $n \geq 0$), where a_n depends on some of the prior terms $a_{n-1}, a_{n-2}, \ldots, a_1, a_0$. This study of what are called either *recurrence relations* or *difference equations* is the discrete counterpart to ideas applied in ordinary differential equations.

Our development will not employ any ideas from differential equations but will start with the notion of a geometric progression. As further ideas are developed, we shall see some of the many applications that make this topic so important.

10.1
The First-Order Linear Recurrence Relation

A *geometric progression* is an infinite sequence of numbers, such as 5, 15, 45, 135, ..., where the division of each term, other than the first, by its immediate predecessor is a constant, called the *common ratio*. For our sequence this common ratio is 3: $15 = 3(5)$, $45 = 3(15)$, and so on. If $a_0, a_1, a_2, \ldots$ is a geometric progression, then $a_1/a_0 = a_2/a_1 = \cdots = a_{n+1}/a_n = \cdots = r$, the common ratio. In our particular geometric progression we have $a_{n+1} = 3a_n$, $n \geq 0$.

The *recurrence relation* $a_{n+1} = 3a_n$, $n \geq 0$, does not define a unique geometric progression. The sequence 7, 21, 63, 189, ... also satisfies the relation. To pinpoint a particular sequence described by $a_{n+1} = 3a_n$, we need to know one of the terms of that sequence. Hence

$$a_{n+1} = 3a_n, \qquad n \geq 0, \qquad a_0 = 5,$$

uniquely defines the sequence 5, 15, 45, ..., whereas

$$a_{n+1} = 3a_n, \qquad n \geq 0, \qquad a_1 = 21,$$

identifies 7, 21, 63, ... as the geometric progression under study.

The equation $a_{n+1} = 3a_n$, $n \geq 0$ is a recurrence relation because the value of a_{n+1} (the present consideration) is dependent on a_n (a prior consideration). Since a_{n+1} depends only on its immediate predecessor, the relation is said to be of *first order*. In particular, this is a *first-order linear homogeneous* recurrence relation *with constant coefficients*. (We'll say more about these ideas later.) The general form of such an equation can be written $a_{n+1} = da_n$, $n \geq 0$, where d is a constant.

Values such as a_0 or a_1, given in addition to the recurrence relations, are called *boundary conditions*. The expression $a_0 = A$, where A is a constant, is also referred to as an *initial condition*. Our examples show the importance of the boundary condition in determining the unique solution.

Let us return now to the recurrence relation

$$a_{n+1} = 3a_n, \qquad n \geq 0, \qquad a_0 = 5.$$

The first four terms of this sequence are

$$a_0 = 5,$$
$$a_1 = 3a_0 = 3(5),$$
$$a_2 = 3a_1 = 3(3a_0) = 3^2(5), \quad \text{and}$$
$$a_3 = 3a_2 = 3(3^2(5)) = 3^3(5).$$

These results suggest that for each $n \geq 0$, $a_n = 5(3^n)$. This is the *unique solution* of the given recurrence relation. In this solution, the value of a_n is a function of n and there is no longer any dependence on prior terms of the sequence, once we define a_0. To compute a_{10}, for example, we simply calculate $5(3^{10}) = 295{,}245$; there is no need to start at a_0 and build up to a_9 in order to obtain a_{10}.

From this example we are directed to the following. (This result can be established by the Principle of Mathematical Induction.)

The unique solution of the recurrence relation

$$a_{n+1} = da_n, \qquad \text{where } n \geq 0, \quad d \text{ is a constant, and} \quad a_0 = A,$$

is given by

$$a_n = Ad^n, \qquad n \geq 0.$$

Thus the solution $a_n = Ad^n$, $n \geq 0$, defines a discrete function whose domain is the set **N** of all nonnegative integers.

EXAMPLE 10.1

Solve the recurrence relation $a_n = 7a_{n-1}$, where $n \geq 1$ and $a_2 = 98$.

This is just an alternative form of the relation $a_{n+1} = 7a_n$ for $n \geq 0$ and $a_2 = 98$. Hence the solution has the form $a_n = a_0(7^n)$. Since $a_2 = 98 = a_0(7^2)$, it follows that $a_0 = 2$, and $a_n = 2(7^n)$, $n \geq 0$, is the unique solution.

EXAMPLE 10.2

A bank pays 6% (annual) interest on savings, compounding the interest monthly. If Bonnie deposits $1000 on the first day of May, how much will this deposit be worth a year later?

The annual interest rate is 6%, so the monthly rate is $6\%/12 = 0.5\% = 0.005$. For $0 \leq n \leq 12$, let p_n denote the value of Bonnie's deposit at the end of n months. Then $p_{n+1} = p_n + 0.005p_n$, where $0.005p_n$ is the interest earned on p_n during month $n + 1$, for $0 \leq n \leq 11$, and $p_0 = \$1000$.

The relation $p_{n+1} = (1.005)p_n$, $p_0 = \$1000$, has the solution $p_n = p_0(1.005)^n = \$1000(1.005)^n$. Consequently, at the end of one year, Bonnie's deposit is worth $\$1000(1.005)^{12} = \$1061.68$.

In the next example we find a fifth way to count the number of compositions of a positive integer. The reader may recall that this situation was examined earlier in Examples 1.37, 3.11, 4.12, and 9.12.

EXAMPLE 10.3

Figure 10.1 provides the compositions of 3 and 4. Here we see that compositions $(1')$–$(4')$ of 4 arise from the corresponding compositions of 3 by increasing the last summand (in each corresponding composition of 3) by 1. The other four compositions of 4, namely, $(1'')$–$(4'')$, are obtained from the compositions of 3 by appending "+1" to each of the corresponding compositions of 3. (The reader may recall seeing such results in Fig. 4.7.)

		$(1')$	4
		$(2')$	$1+3$
(1)	3	$(3')$	$2+2$
(2)	$1+2$	$(4')$	$1+1+2$
(3)	$2+1$		
(4)	$1+1+1$	$(1'')$	$3+1$
		$(2'')$	$1+2+1$
		$(3'')$	$2+1+1$
		$(4'')$	$1+1+1+1$

Figure 10.1

What happens in Fig. 10.1 exemplifies the general situation. So if we let a_n count the number of compositions of n, for $n \in \mathbf{Z}^+$, we find that

$$a_{n+1} = 2a_n, \qquad n \geq 1, \qquad a_1 = 1.$$

However, in order to apply the formula for the unique solution (where $n \geq 0$) to this recurrence relation, we let $b_n = a_{n+1}$. Then we have

$$b_{n+1} = 2b_n, \qquad n \geq 0, \qquad b_0 = 1,$$

so $b_n = b_0(2^n) = 2^n$, and $a_n = b_{n-1} = 2^{n-1}$, $n \geq 1$.

The recurrence relation $a_{n+1} - da_n = 0$ is called *linear* because each subscripted term appears to the first power (as do the variables x and y in the equation of a line in the plane). In a linear relation there are no products such as $a_n a_{n-1}$, which appears in the nonlinear recurrence relation $a_{n+1} - 3a_n a_{n-1} = 0$. However, there are times when a nonlinear recurrence relation can be transformed into a linear one by a suitable algebraic substitution.

EXAMPLE 10.4

Find a_{12} if $a_{n+1}^2 = 5a_n^2$, where $a_n > 0$ for $n \geq 0$, and $a_0 = 2$.

Although this recurrence relation is not linear in a_n, if we let $b_n = a_n^2$, then the new relation $b_{n+1} = 5b_n$ for $n \geq 0$, and $b_0 = 4$, is a linear relation whose solution is $b_n = 4 \cdot 5^n$. Therefore, $a_n = 2(\sqrt{5})^n$ for $n \geq 0$, and $a_{12} = 2(\sqrt{5})^{12} = 31,250$.

The general first-order linear recurrence relation with constant coefficients has the form $a_{n+1} + ca_n = f(n)$, $n \geq 0$, where c is a constant and $f(n)$ is a function on the set **N** of nonnegative integers.

When $f(n) = 0$ for all $n \in \mathbf{N}$, the relation is called *homogeneous*; otherwise it is called *nonhomogeneous*. So far we have only dealt with homogeneous relations. Now we shall solve a nonhomogeneous relation. We shall develop specific techniques that work for all linear homogeneous recurrence relations with constant coefficients. However, many different techniques prove useful when we deal with a nonhomogeneous problem, although none allows us to solve everything that can arise.

EXAMPLE 10.5

Perhaps the most popular, though not the most efficient, method of sorting numeric data is a technique called the *bubble sort*. Here the input is a positive integer n and an array $x_1, x_2, x_3, \ldots, x_n$ of real numbers that are to be sorted into ascending order.

The pseudocode procedure in Fig. 10.2 provides an implementation for an algorithm to carry out this sorting process. Here the integer variable i is the counter for the outer **for** loop, whereas the integer variable j is the counter for the inner **for** loop. Finally, the real variable *temp* is used for storage that is needed when an exchange takes place.

```
procedure BubbleSort(n: positive integer; x₁,x₂,x₃,...,xₙ: real numbers)
begin
  for i := 1 to n − 1 do
    for j := n downto i + 1 do
      if xⱼ < xⱼ₋₁ then
        begin           {interchange}
          temp := xⱼ₋₁
          xⱼ₋₁ := xⱼ
          xⱼ := temp
        end
end
```

Figure 10.2

We compare the last entry, x_n, in the given array with its immediate predecessor, x_{n-1}. If $x_n < x_{n-1}$, we interchange the values stored in x_{n-1} and x_n. In any event we will now have $x_{n-1} \leq x_n$. Then we compare x_{n-1} with its immediate predecessor, x_{n-2}. If $x_{n-1} < x_{n-2}$, we interchange them. We continue the process. After $n - 1$ such comparisons, the smallest number in the list is stored in x_1. We then repeat this process for the $n - 1$ numbers now stored in the (smaller) array $x_2, x_3, \ldots, x_n$. In this way, each time (counted by i) this process is carried out, the smallest number in the remaining sublist "bubbles up" to the front of that sublist.

A small example wherein $n = 5$ and $x_1 = 7$, $x_2 = 9$, $x_3 = 2$, $x_4 = 5$, and $x_5 = 8$ is given in Fig. 10.3 to show how the bubble sort of Fig. 10.2 places a given sequence in ascending order. In this figure each comparison that leads to an interchange is denoted by the symbol $\supsetneq$; the symbol } indicates a comparison that results in no interchange.

To determine the time-complexity function $h(n)$ when this algorithm is used on an input (array) of size $n \geq 1$, we count the total number of *comparisons* made in order to sort the n given numbers into ascending order.

If a_n denotes the number of comparisons needed to sort n numbers in this way, then we get the following recurrence relation:

$$a_n = a_{n-1} + (n - 1), \qquad n \geq 2, \qquad a_1 = 0.$$

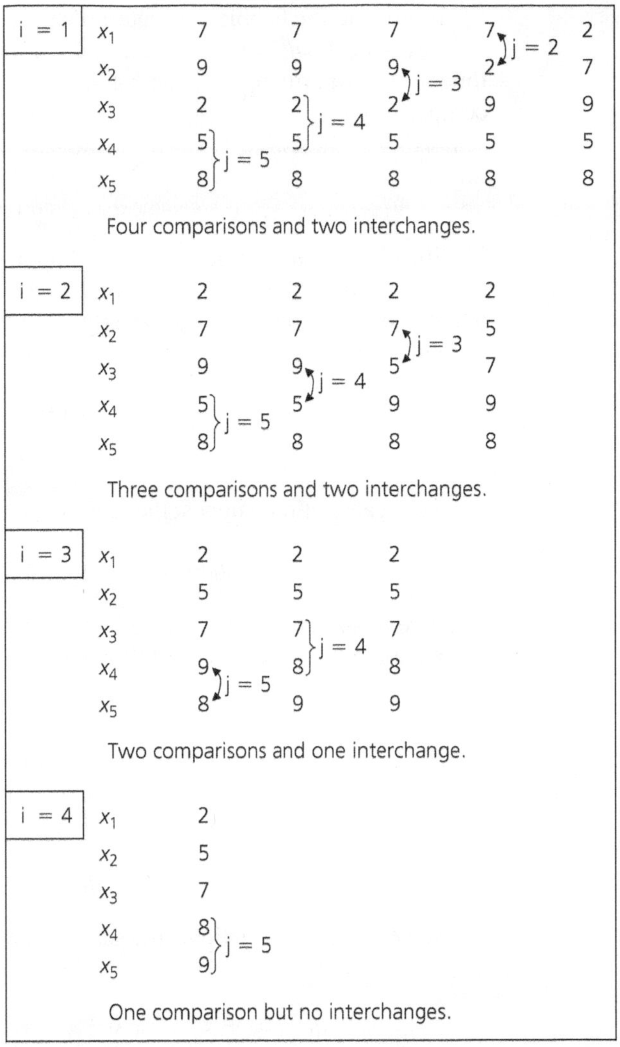

i = 1	x_1	7	7	7	7 ⌐ j = 2	2
	x_2	9	9	9 ⌐ j = 3	2 ⌐	7
	x_3	2	2 ⌐ j = 4	2 ⌐	9	9
	x_4	5 ⌐ j = 5	5 ⌐	5	5	5
	x_5	8 ⌐	8	8	8	8

Four comparisons and two interchanges.

i = 2	x_1	2	2	2	2
	x_2	7	7	7 ⌐ j = 3	5
	x_3	9	9 ⌐ j = 4	5 ⌐	7
	x_4	5 ⌐ j = 5	5 ⌐	9	9
	x_5	8 ⌐	8	8	8

Three comparisons and two interchanges.

i = 3	x_1	2	2	2
	x_2	5	5	5
	x_3	7	7 ⌐ j = 4	7
	x_4	9 ⌐ j = 5	8 ⌐	8
	x_5	8 ⌐	9	9

Two comparisons and one interchange.

i = 4	x_1	2
	x_2	5
	x_3	7
	x_4	8 ⌐ j = 5
	x_5	9 ⌐

One comparison but no interchanges.

Figure 10.3

This arises as follows. Given a list of n numbers, we make $n - 1$ comparisons to bubble the smallest number up to the start of the list. The remaining sublist of $n - 1$ numbers then requires a_{n-1} comparisons in order to be completely sorted.

This relation is a linear first-order relation with constant coefficients, but the term $n - 1$ makes it nonhomogeneous. Since we have no technique for attacking such a relation, let us list some terms and see whether there is a recognizable pattern.

$$a_1 = 0$$
$$a_2 = a_1 + (2 - 1) = 1$$
$$a_3 = a_2 + (3 - 1) = 1 + 2$$
$$a_4 = a_3 + (4 - 1) = 1 + 2 + 3$$
$$\cdots \quad \cdots \quad \cdots \quad \cdots$$

In general, $a_n = 1 + 2 + \cdots + (n - 1) = [(n - 1)n]/2 = (n^2 - n)/2$.

As a result, the bubble sort determines the time-complexity function $h: \mathbf{Z}^+ \to \mathbf{R}$ given by $h(n) = a_n = (n^2 - n)/2$. [Here $h(\mathbf{Z}^+) \subset \mathbf{N}$.] Consequently, as a measure of the running time for the algorithm, we write $h \in O(n^2)$. Hence the bubble sort is said to require $O(n^2)$ comparisons.

EXAMPLE 10.6

In part (c) of Example 9.6 we sought the generating function for the sequence 0, 2, 6, 12, 20, 30, 42, . . . , and the solution rested upon our ability to recognize that $a_n = n^2 + n$ for each $n \in \mathbf{N}$. If we fail to see this, perhaps we can examine the given sequence and determine whether there is some other pattern that will help us.

Here $a_0 = 0, a_1 = 2, a_2 = 6, a_3 = 12, a_4 = 20, a_5 = 30, a_6 = 42$, and

$$a_1 - a_0 = 2 \qquad a_3 - a_2 = 6 \qquad a_5 - a_4 = 10$$
$$a_2 - a_1 = 4 \qquad a_4 - a_3 = 8 \qquad a_6 - a_5 = 12.$$

These calculations suggest the recurrence relation

$$a_n - a_{n-1} = 2n, \qquad n \geq 1, \qquad a_0 = 0.$$

To solve this relation, we proceed in a slightly different manner from the method we used in Example 10.5. Consider the following n equations:

$$a_1 - a_0 = 2$$
$$a_2 - a_1 = 4$$
$$a_3 - a_2 = 6$$
$$\vdots \quad \vdots \quad \vdots$$
$$a_n - a_{n-1} = 2n.$$

When we add these equations, the sum for the left-hand side will contain a_i and $-a_i$ for all $1 \leq i \leq n - 1$. So we obtain

$$a_n - a_0 = 2 + 4 + 6 + \cdots + 2n = 2(1 + 2 + 3 + \cdots + n)$$
$$= 2[n(n+1)/2] = n^2 + n.$$

Since $a_0 = 0$, it follows that $a_n = n^2 + n$ for all $n \in \mathbf{N}$, as we found earlier in part (c) of Example 9.6.

At this point we shall examine a recurrence relation with a variable coefficient.

EXAMPLE 10.7

Solve the relation $a_n = n \cdot a_{n-1}$, where $n \geq 1$ and $a_0 = 1$.

Writing the first five terms defined by the relation, we have

$$a_0 = 1 \qquad\qquad a_2 = 2 \cdot a_1 = 2 \cdot 1 \qquad\qquad a_4 = 4 \cdot a_3 = 4 \cdot 3 \cdot 2 \cdot 1$$
$$a_1 = 1 \cdot a_0 = 1 \qquad a_3 = 3 \cdot a_2 = 3 \cdot 2 \cdot 1$$

Therefore, $a_n = n!$ and the solution is the discrete function a_n, which counts the number of permutations of n objects, $n \geq 0$.

While on the subject of permutations, we shall examine a recursive algorithm for generating the permutations of $\{1, 2, 3, \ldots, n - 1, n\}$ from those for $\{1, 2, 3, \ldots, n - 1\}$.[†] There is only one permutation of $\{1\}$. Examining the permutations of $\{1, 2\}$,

$$
\begin{array}{cc}
1 & 2 \\
2 & 1
\end{array}
$$

we see that after writing the permutation 1 twice, we intertwine the number 2 about 1 to get the permutations listed. Writing each of these two permutations three times, we intertwine the number 3 and obtain

$$
\begin{array}{ccc}
1 & & 2 \quad 3 \\
1 & 3 & 2 \\
3 \quad 1 & & 2 \\
3 \quad 2 & & 1 \\
2 & 3 & 1 \\
2 & & 1 \quad 3
\end{array}
$$

We see here that the first permutation is 123 and that we obtain each of the next two permutations from its immediate predecessor by interchanging two numbers: 3 and the integer to its left. When 3 reaches the left side of the permutation, we examine the remaining numbers and permute them according to the list of permutations we generated for $\{1, 2\}$. (This makes the procedure recursive.) After that we interchange 3 with the integer on its right until 3 is on the right side of the permutation. We note that if we interchange 1 and 2 in the last permutation, we get 123, the first permutation listed.

Continuing for $S = \{1, 2, 3, 4\}$, we first list each of the six permutations of $\{1, 2, 3\}$ four times. Starting with the permutation 1234, we intertwine the 4 throughout the remaining 23 permutations as indicated in Table 10.1 (on page 454). The only new idea here develops as follows. When progressing from permutation (5) to (6) to (7) to (8), we interchange 4 with the integer to its right. At permutation (8), where 4 has reached the right side, we obtain permutation (9) by keeping the location of 4 fixed and replacing the permutation 132 by 312 from the list of permutations of $\{1, 2, 3\}$. After that we continue as for the first eight permutations until we reach permutation (16), where 4 is again on the right. We then permute 321 to obtain 231 and continue intertwining 4 until all 24 permutations have been generated. Once again, if 1 and 2 are interchanged in the last permutation, we obtain the first permutation in our list.

The chapter references provide more information on recursive procedures for generating permutations and combinations.

We shall close this first section by returning to an earlier idea — the greatest common divisor of two positive integers.

EXAMPLE 10.8

Recursive methods are fundamental in the areas of discrete mathematics and the analysis of algorithms. Such methods arise when we want to solve a given problem by breaking it down, or referring it, to smaller similar problems. In many programming languages this can be implemented by the use of recursive functions and procedures, which are permitted to invoke themselves. This example will provide one such procedure.

[†]The material from here to the end of this section is a digression that uses the idea of recursion. It does not deal with methods for solving recurrence relations and may be omitted with no loss of continuity.

Table 10.1

(1)		1		2		3	4
(2)		1		2	4	3	
(3)		1	4	2		3	
(4)	4	1		2		3	
(5)	4	1		3		2	
(6)		1	4	3		2	
(7)		1		3	4	2	
(8)		1		3		2	4
(9)		3		1		2	4
(10)		3		1	4	2	
(11)		3	4	1		2	
$\cdots$		$\cdots$		$\cdots$		$\cdots$	
(15)		3		2	4	1	
(16)		3		2		1	4
(17)		2		3		1	4
$\cdots$		$\cdots$		$\cdots$		$\cdots$	
(22)		2	4	1		3	
(23)		2		1	4	3	
(24)		2		1		3	4

In computing gcd(333, 84) we obtain the following calculations when we use the Euclidean algorithm (presented in Section 4.4).

$$333 = 3(84) + 81 \qquad 0 < 81 < 84 \tag{1}$$

$$84 = 1(81) + 3 \qquad 0 < 3 < 81 \tag{2}$$

$$81 = 27(3) + 0. \tag{3}$$

Since 3 is the last nonzero remainder, the Euclidean algorithm tells us that gcd(333, 84) = 3. However, if we use only the calculations in Eqs. (2) and (3), then we find that gcd(84, 81) = 3. And Eq. (3) alone implies that gcd(81, 3) = 3 because 3 divides 81. Consequently,

$$\gcd(333, 84) = \gcd(84, 81) = \gcd(81, 3) = 3,$$

where the integers involved in the successive calculations get smaller as we go from Eq. (1) to Eq. (2) to Eq. (3).

We also observe that

$$81 = 333 \bmod 84 \qquad \text{and} \qquad 3 = 84 \bmod 81.$$

Therefore it follows that

$$\gcd(333, 84) = \gcd(84, 333 \bmod 84) = \gcd(333 \bmod 84, 84 \bmod (333 \bmod 84)).$$

These results suggest the following recursive method for computing gcd(a, b), where $a, b \in \mathbf{Z}^+$.

Say we have the input $a, b \in \mathbf{Z}^+$.

Step 1: If $b \mid a$ (or $a \bmod b = 0$), then gcd(a, b) = b.

Step 2: If $b \nmid a$, then perform the following tasks in the order specified.

    i) Set $a = b$.

ii) Set $b = a \bmod b$, where the value of a for this assignment is the *old* value of a.

iii) Return to step (1).

These ideas are used in the pseudocode procedure in Fig. 10.4. (The reader may wish to compare this procedure with the one given in Fig. 4.11.)

```
procedure gcd2(a, b: positive integers)
begin
    if a mod b = 0 then
        gcd = b
    else gcd = gcd2(b, a mod b)
end
```

Figure 10.4

EXERCISES 10.1

1. Find a recurrence relation, with initial condition, that uniquely determines each of the following geometric progressions.

    a) 2, 10, 50, 250, . . .

    b) 6, −18, 54, −162, . . .

    c) 7, 14/5, 28/25, 56/125, . . .

2. Find the unique solution for each of the following recurrence relations.

    a) $a_{n+1} - 1.5a_n = 0, \quad n \geq 0$

    b) $4a_n - 5a_{n-1} = 0, \quad n \geq 1$

    c) $3a_{n+1} - 4a_n = 0, \quad n \geq 0, \quad a_1 = 5$

    d) $2a_n - 3a_{n-1} = 0, \quad n \geq 1, \quad a_4 = 81$

3. If a_n, $n \geq 0$, is the unique solution of the recurrence relation $a_{n+1} - da_n = 0$, and $a_3 = 153/49$, $a_5 = 1377/2401$, what is d?

4. The number of bacteria in a culture is 1000 (approximately), and this number increases 250% every two hours. Use a recurrence relation to determine the number of bacteria present after one day.

5. If Laura invests $100 at 6% interest compounded quarterly, how many months must she wait for her money to double? (She cannot withdraw the money before the quarter is up.)

6. Paul invested the stock profits he received 15 years ago in an account that paid 8% interest compounded quarterly. If his account now has $7218.27 in it, what was his initial investment?

7. Let $x_1, x_2, \ldots, x_{20}$ be a list of distinct real numbers to be sorted by the bubble-sort technique of Example 10.5. (a) After how many comparisons will the 10 smallest numbers of the original list be arranged in ascending order? (b) How many more comparisons are needed to finish this sorting job?

8. For the implementation of the bubble sort given in Fig. 10.2, the outer **for** loop is executed $n - 1$ times. This occurs regardless of whether any interchanges take place during the execution of the inner **for** loop. Consequently, for $i = k$, where $1 \leq k \leq n - 2$, if the execution of the inner **for** loop results in no interchanges, then the list is in ascending order. So the execution of the outer **for** loop for $k + 1 \leq i \leq n - 1$ is not needed.

    a) For the situation described here, how many unnecessary comparisons are made if the execution of the inner **for** loop for $i = k$ ($1 \leq k \leq n - 2$) results in no interchanges?

    b) Write an improved version of the bubble sort shown in Fig. 10.2. (Your result should eliminate the unnecessary comparisons discussed at the start of this exercise.)

    c) Using the number of comparisons as a measure of its running time, determine the best-case and the worst-case time complexities for the algorithm implemented in part (b).

9. Say the permutations of $\{1, 2, 3, 4, 5\}$ are generated by the procedure developed after Example 10.7. (a) What is the last permutation in the list? (b) What two permutations precede 25134? (c) What three permutations follow 25134?

10. For $n > 1$, a permutation $p_1, p_2, p_3, \ldots, p_n$ of the integers $1, 2, 3, \ldots, n$ is called *orderly* if, for each $i = 1, 2, 3, \ldots, n - 1$, there exists a $j > i$ such that $|p_j - p_i| = 1$. [If $n = 2$, the permutations 1, 2 and 2, 1 are both orderly. When $n = 3$ we find that 3, 1, 2 is an orderly permutation, while 2, 3, 1 is not. (Why not?)] (a) List all the orderly permutations for 1, 2, 3. (b) List all the orderly permutations for 1, 2, 3, 4. (c) If p_1, p_2, p_3, p_4, p_5 is an orderly permutation of 1, 2, 3, 4, 5, what value(s) can p_1 be? (d) For $n > 1$, let a_n count the number of orderly permutations for 1, 2, 3, . . . , n. Find and solve a recurrence relation for a_n.

10.2
The Second-Order Linear
Homogeneous Recurrence
Relation with Constant Coefficients

Let $k \in \mathbf{Z}^+$ and $C_0 \ (\neq 0), C_1, C_2, \ldots, C_k \ (\neq 0)$ be real numbers. If a_n, for $n \geq 0$, is a discrete function, then

$$C_0 a_n + C_1 a_{n-1} + C_2 a_{n-2} + \cdots + C_k a_{n-k} = f(n), \qquad n \geq k,$$

is a linear recurrence relation (with constant coefficients) of *order k*. When $f(n) = 0$ for all $n \geq 0$, the relation is called *homogeneous*; otherwise, it is called *nonhomogeneous*.

In this section we shall concentrate on the homogeneous relation of order two:

$$C_0 a_n + C_1 a_{n-1} + C_2 a_{n-2} = 0, \qquad n \geq 2.$$

On the basis of our work in Section 10.1, we seek a solution of the form $a_n = cr^n$, where $c \neq 0$ and $r \neq 0$.

Substituting $a_n = cr^n$ into $C_0 a_n + C_1 a_{n-1} + C_2 a_{n-2} = 0$, we obtain

$$C_0 cr^n + C_1 cr^{n-1} + C_2 cr^{n-2} = 0.$$

With $c, r \neq 0$, this becomes $C_0 r^2 + C_1 r + C_2 = 0$, a quadratic equation which is called the *characteristic equation*. The roots r_1, r_2 of this equation determine the following three cases: (a) r_1, r_2 are distinct real numbers; (b) r_1, r_2 form a complex conjugate pair; or (c) r_1, r_2 are real, but $r_1 = r_2$. In all cases, r_1 and r_2 are called the *characteristic roots*.

Case (A): (Distinct Real Roots)

EXAMPLE 10.9

Solve the recurrence relation $a_n + a_{n-1} - 6a_{n-2} = 0$, where $n \geq 2$ and $a_0 = -1, a_1 = 8$.

If $a_n = cr^n$ with $c, r \neq 0$, we obtain $cr^n + cr^{n-1} - 6cr^{n-2} = 0$ from which the characteristic equation $r^2 + r - 6 = 0$ follows:

$$0 = r^2 + r - 6 = (r + 3)(r - 2) \Rightarrow r = 2, -3.$$

Since we have two distinct real roots, $a_n = 2^n$ and $a_n = (-3)^n$ are both solutions [as are $b(2^n)$ and $d(-3)^n$, for arbitrary constants b, d]. They are *linearly independent solutions* because one is not a multiple of the other; that is, there is no real constant k such that $(-3)^n = k(2^n)$ for *all* $n \in \mathbf{N}$.[†] We write $a_n = c_1(2^n) + c_2(-3)^n$ for the *general* solution, where c_1, c_2 are arbitrary constants.

With $a_0 = -1$ and $a_1 = 8$, c_1 and c_2 are determined as follows:

$$-1 = a_0 = c_1(2^0) + c_2(-3)^0 = c_1 + c_2$$
$$8 = a_1 = c_1(2^1) + c_2(-3)^1 = 2c_1 - 3c_2.$$

Solving this system of equations, one finds $c_1 = 1, c_2 = -2$. Therefore, $a_n = 2^n - 2(-3)^n$, $n \geq 0$, is the *unique* solution of the given recurrence relation.

The reader should realize that to determine the unique solution of a second-order linear homogeneous recurrence relation with constant coefficients one needs two initial conditions

---

[†]We can also call the solutions $a_n = 2^n$ and $a_n = (-3)^n$ *linearly independent* when the following condition is satisfied: For $k_1, k_2 \in \mathbf{R}$, if $k_1(2^n) + k_2(-3)^n = 0$ for *all* $n \in \mathbf{N}$, then $k_1 = k_2 = 0$.

(values) — that is, the value of a_n for two values of n, very often $n = 0$ and $n = 1$, or $n = 1$ and $n = 2$.

An interesting second-order homogeneous recurrence relation is the *Fibonacci relation*. (This was mentioned earlier in Sections 4.2 and 9.6.)

EXAMPLE 10.10

Solve the recurrence relation $F_{n+2} = F_{n+1} + F_n$, where $n \geq 0$ and $F_0 = 0$, $F_1 = 1$.

As in the previous example, let $F_n = cr^n$, for $c, r \neq 0, n \geq 0$. Upon substitution we get $cr^{n+2} = cr^{n+1} + cr^n$. This gives the characteristic equation $r^2 - r - 1 = 0$. The characteristic roots are $r = (1 \pm \sqrt{5})/2$, so the general solution is

$$F_n = c_1 \left(\frac{1 + \sqrt{5}}{2} \right)^n + c_2 \left(\frac{1 - \sqrt{5}}{2} \right)^n.$$

To solve for c_1, c_2, we use the given initial values and write $0 = F_0 = c_1 + c_2$, $1 = F_1 = c_1[(1 + \sqrt{5})/2] + c_2[(1 - \sqrt{5})/2]$. Since $-c_1 = c_2$, we have $2 = c_1(1 + \sqrt{5}) - c_1(1 - \sqrt{5})$ and $c_1 = 1/\sqrt{5}$. The general solution is given by

$$F_n = \frac{1}{\sqrt{5}} \left[\left(\frac{1 + \sqrt{5}}{2} \right)^n - \left(\frac{1 - \sqrt{5}}{2} \right)^n \right], \qquad n \geq 0.$$

When dealing with the Fibonacci numbers one often finds the assignments $\alpha = (1 + \sqrt{5})/2$ and $\beta = (1 - \sqrt{5})/2$, where α is known as the *golden ratio*. As a result, we find that

$$F_n = \frac{1}{\sqrt{5}}(\alpha^n - \beta^n) = \frac{\alpha^n - \beta^n}{\alpha - \beta}, \qquad n \geq 0.$$

[This representation is referred to as the *Binet form* for F_n, as it was first published in 1843 by Jacques Philippe Marie Binet (1786–1856).]

EXAMPLE 10.11

For $n \geq 0$, let $S = \{1, 2, 3, \ldots, n\}$ (when $n = 0$, $S = \emptyset$), and let a_n denote the number of subsets of S that contain no consecutive integers. Find and solve a recurrence relation for a_n.

For $0 \leq n \leq 4$, we have $a_0 = 1$, $a_1 = 2$, $a_2 = 3$, $a_3 = 5$, and $a_4 = 8$. [For example, $a_3 = 5$ because $S = \{1, 2, 3\}$ has $\emptyset$, $\{1\}$, $\{2\}$, $\{3\}$, and $\{1, 3\}$ as subsets with no consecutive integers (and no other such subsets).] These first five terms are reminiscent of the Fibonacci sequence. But do things change as we continue?

Let $n \geq 2$ and $S = \{1, 2, 3, \ldots, n - 2, n - 1, n\}$. If $A \subseteq S$ and A is to be counted in a_n, there are two possibilities:

a) $n \in A$: When this happens $(n - 1) \notin A$, and $A - \{n\}$ would be counted in a_{n-2}.

b) $n \notin A$: For this case A would be counted in a_{n-1}.

These two cases are exhaustive and mutually disjoint, so we conclude that $a_n = a_{n-1} + a_{n-2}$, where $n \geq 2$ and $a_0 = 1$, $a_1 = 2$, is the recurrence relation for the problem. Now we could solve for a_n, but if we notice that $a_n = F_{n+2}, n \geq 0$, then the result of Example 10.10 implies that

$$a_n = \frac{1}{\sqrt{5}} \left[\left(\frac{1 + \sqrt{5}}{2} \right)^{n+2} - \left(\frac{1 - \sqrt{5}}{2} \right)^{n+2} \right], \qquad n \geq 0.$$

EXAMPLE 10.12

Suppose we have a $2 \times n$ chessboard, for $n \in \mathbf{Z}^+$. The case for $n = 4$ is shown in part (a) of Fig. 10.5. We wish to cover such a chessboard using 2×1 (vertical) dominoes, which can also be used as 1×2 (horizontal) dominoes. Such dominoes (or tiles) are shown in part (b) of Fig. 10.5.

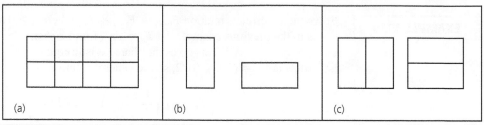

(a)   (b)   (c)

Figure 10.5

For $n \in \mathbf{Z}^+$ we let b_n count the number of ways we can cover (or tile) a $2 \times n$ chessboard using our 2×1 and 1×2 dominoes. Here $b_1 = 1$, for a 2×1 chessboard necessitates one 2×1 (vertical) domino. A 2×2 chessboard can be covered in two ways — using two 2×1 (vertical) dominoes or two 1×2 (horizontal) dominoes, as shown in part (c) of the figure. Hence $b_2 = 2$. For $n \geq 3$, consider the last (nth) column of a $2 \times n$ chessboard. This column can be covered in two ways.

i) By one 2×1 (vertical) domino: Here the remaining $2 \times (n - 1)$ subboard can be covered in b_{n-1} ways.

ii) By the right squares of two 1×2 (horizontal) dominoes placed one above the other: Now the remaining $2 \times (n - 2)$ subboard can be covered in b_{n-2} ways.

Since these two ways have nothing in common and deal with all possibilities, we may write

$$b_n = b_{n-1} + b_{n-2}, \qquad n \geq 3, \qquad b_1 = 1, \qquad b_2 = 2.$$

We find that $b_n = F_{n+1}$, so here is another situation where the Fibonacci numbers arise. The result from Example 10.10 gives us $b_n = (1/\sqrt{5})[((1 + \sqrt{5})/2)^{n+1} - ((1 - \sqrt{5})/2)^{n+1}]$, $n \geq 1$.

EXAMPLE 10.13

At this point we examine an interesting application where the number $\alpha = (1 + \sqrt{5})/2$ plays a major role. This application deals with Gabriel Lamé's work in estimating the number of divisions used in the Euclidean algorithm to find $\gcd(a, b)$, where $a, b \in \mathbf{Z}^+$ with $a \geq b \geq 2$. To find this estimate we need the following property of the Fibonacci numbers, which can be established by the alternative form of the Principle of Mathematical Induction. (A proof is requested in the Section Exercises.)

Property: For $n \geq 3$, $F_n > \alpha^{n-2}$.

Addressing the problem at hand — namely, estimating the number of divisions when the Euclidean algorithm is used to find $\gcd(a, b)$ — we recall the following steps from Theorem 4.7.

Letting $r_0 = a$ and $r_1 = b$, we have

$$r_0 = q_1 r_1 + r_2, \qquad 0 < r_2 < r_1$$
$$r_1 = q_2 r_2 + r_3, \qquad 0 < r_3 < r_2$$
$$r_2 = q_3 r_3 + r_4, \qquad 0 < r_4 < r_3$$
$$\cdots\cdots\cdots\cdots$$
$$r_{n-2} = q_{n-1} r_{n-1} + r_n, \qquad 0 < r_n < r_{n-1}$$
$$r_{n-1} = q_n r_n.$$

So r_n, the last nonzero remainder, is $\gcd(a, b)$.

From the subscripts on r we see that n divisions have been performed in determining $r_n = \gcd(a, b)$. In addition, $q_i \geq 1$, for all $1 \leq i \leq n - 1$, and $q_n \geq 2$ because $r_n < r_{n-1}$. Examining the n nonzero remainders $r_n, r_{n-1}, r_{n-2}, \ldots, r_2$, and $r_1 (= b)$, we learn that

$$r_n > 0, \text{ so } r_n \geq 1 = F_2.$$
$$[(q_n \geq 2) \wedge (r_n \geq 1)] \Rightarrow r_{n-1} = q_n r_n \geq 2 \cdot 1 = 2 = F_3$$
$$r_{n-2} = q_{n-1} r_{n-1} + r_n \geq 1 \cdot r_{n-1} + r_n \geq F_3 + F_2 = F_4$$
$$\cdots\cdots\cdots\cdots$$
$$r_2 = q_3 r_3 + r_4 \geq 1 \cdot r_3 + r_4 \geq F_{n-1} + F_{n-2} = F_n$$
$$b = r_1 = q_2 r_2 + r_3 \geq 1 \cdot r_2 + r_3 \geq F_n + F_{n-1} = F_{n+1}.$$

Therefore, if n divisions are performed by the Euclidean algorithm to determine $\gcd(a, b)$, with $a \geq b \geq 2$, then $b \geq F_{n+1}$. So by virtue of the property introduced earlier, we may write $b > \alpha^{(n+1)-2} = \alpha^{n-1} = [(1 + \sqrt{5})/2]^{n-1}$. Consequently, we find now that

$$b > \alpha^{n-1} \Rightarrow \log_{10} b > \log_{10}(\alpha^{n-1}) = (n - 1) \log_{10} \alpha > \frac{n - 1}{5},$$

since $\log_{10} \alpha = \log_{10}[(1 + \sqrt{5})/2] \doteq 0.208988 > 0.2 = \frac{1}{5}$.

At this point suppose that $10^{k-1} \leq b < 10^k$, so that the decimal (base 10) representation of b has k digits. Then

$$k = \log_{10} 10^k > \log_{10} b > \frac{n - 1}{5}, \quad \text{and} \quad n < 5k + 1.$$

With $n, k \in \mathbf{Z}^+$ we have $n < 5k + 1 \Rightarrow n \leq 5k$, and this last inequality now completes a proof for the following.

Lamé's Theorem: Let $a, b \in \mathbf{Z}^+$ with $a \geq b \geq 2$. Then the number of divisions needed, in the Euclidean algorithm, to determine $\gcd(a, b)$ is at most 5 times the number of decimal digits in b.

Before closing this example, we learn one more fact from Lamé's Theorem. Since $b \geq 2$, it follows that $\log_{10} b \geq \log_{10} 2$, so $5 \log_{10} b \geq 5 \log_{10} 2 = \log_{10} 2^5 = \log_{10} 32 > 1$. From above we know that $n - 1 < 5 \log_{10} b$, so

$$n < 1 + 5 \log_{10} b < 5 \log_{10} b + 5 \log_{10} b = 10 \log_{10} b$$

and $n \in O(\log_{10} b)$. [Hence, the number of divisions needed, in the Euclidean algorithm, to determine $\gcd(a, b)$, for $a, b \in \mathbf{Z}^+$ with $a \geq b \geq 2$, is $O(\log_{10} b)$ — that is, on the order of the number of decimal digits in b.]

Returning to the theme of the section we now examine a recurrence relation in a computer science application.

EXAMPLE 10.14

In many programming languages one may consider those legal arithmetic expressions, *without parentheses*, that are made up of the digits 0, 1, 2, . . . , 9 and the binary operation symbols $+$, $*$, $/$. For example, $3 + 4$ and $2 + 3 * 5$ are legal arithmetic expressions; $8 + * 9$ is not. Here $2 + 3 * 5 = 17$, since there is a hierarchy of operations: Multiplication and division are performed before addition. Operations at the same level are performed in their order of appearance as the expression is scanned from left to right.

For $n \in \mathbf{Z}^+$, let a_n be the number of these (legal) arithmetic expressions that are made up of n symbols. Then $a_1 = 10$, since the arithmetic expressions of one symbol are the 10 digits. Next $a_2 = 100$. This accounts for the expressions 00, 01, . . . , 09, 10, 11, . . . , 99. (There are no unnecessary leading plus signs.) When $n \geq 3$, we consider two cases in order to derive a recurrence relation for a_n:

1) If x is an arithmetic expression of $n - 1$ symbols, the last symbol must be a digit. Adding one more digit to the right of x, we get $10a_{n-1}$ arithmetic expressions of n symbols where the last two symbols are digits.

2) Now let y be an arithmetic expression of $n - 2$ symbols. To obtain an arithmetic expression with n symbols (that is not counted in case 1), we adjoin to the right of y one of the 29 two-symbol expressions $+1, \ldots, +9, +0, *1, \ldots, *9, *0, /1, \ldots, /9$.

From these two cases we have $a_n = 10a_{n-1} + 29a_{n-2}$, where $n \geq 3$ and $a_1 = 10$, $a_2 = 100$. Here the characteristic roots are $5 \pm 3\sqrt{6}$ and the solution is $a_n = (5/(3\sqrt{6})) \cdot [(5 + 3\sqrt{6})^n - (5 - 3\sqrt{6})^n]$ for $n \geq 1$. (Verify this result.)

Another way to complete the solution of this problem is to use the recurrence relation $a_n = 10a_{n-1} + 29a_{n-2}$, with $a_2 = 100$ and $a_1 = 10$, to calculate a value for a_0 — namely, $a_0 = (a_2 - 10a_1)/29 = 0$. The solution for the recurrence relation

$$a_n = 10a_{n-1} + 29a_{n-2}, \qquad n \geq 2, \qquad a_0 = 0, \qquad a_1 = 10$$

is

$$a_n = (5/(3\sqrt{6}))[(5 + 3\sqrt{6})^n - (5 - 3\sqrt{6})^n], \qquad n \geq 0.$$

A second method for counting palindromes arises in our next example.

EXAMPLE 10.15

In Fig. 10.6 we find the palindromes of 3, 4, 5, and 6 — that is, the compositions of 3, 4, 5, and 6 that read the same left to right as right to left. (We saw this concept earlier in Example 9.13.) Consider first the palindromes of 3 and 5. To build the palindromes of 5 from those of 3 we do the following:

i) Add 1 to the first and last summands in a palindrome of 3. This is how we get palindromes (1') and (2') for 5 from the respective palindromes (1) and (2) for 3. [*Note:* When we have a one summand palindrome n we get the one summand palindrome $n + 2$. That is how we build palindrome (1') for 5 from palindrome (1) for 3.]

ii) Append "$1+$" to the start and "$+1$" to the end of each palindrome of 3. This technique generates the palindromes (1") and (2") for 5 from the respective palindromes (1) and (2) for 3.

(1)	3	(1′)	5	(1)	4	(1′)	6
(2)	$1+1+1$	(2′)	$2+1+2$	(2)	$1+2+1$	(2′)	$2+2+2$
		(1″)	$1+3+1$	(3)	$2+2$	(3′)	$3+3$
		(2″)	$1+1+1+1+1$	(4)	$1+1+1+1$	(4′)	$2+1+1+2$
						(1″)	$1+4+1$
						(2″)	$1+1+2+1+1$
						(3″)	$1+2+2+1$
						(4″)	$1+1+1+1+1+1$

Figure 10.6

The situation is similar for building the palindromes of 6 from those of 4.

The preceding observations lead us to the following. For $n \in \mathbf{Z}^+$, let p_n count the number of palindromes of n. Then

$$p_n = 2p_{n-2}, \qquad n \geq 3, \qquad p_1 = 1, \qquad p_2 = 2.$$

Substituting $p_n = cr^n$, for $c, r \neq 0$, $n \geq 1$, into this recurrence relation, the resulting characteristic equation is $r^2 - 2 = 0$. The characteristic roots are $r = \pm\sqrt{2}$, so $p_n = c_1(\sqrt{2})^n + c_2(-\sqrt{2})^n$. From

$$1 = p_1 = c_1(\sqrt{2}) + c_2(-\sqrt{2})$$
$$2 = p_2 = c_1(\sqrt{2})^2 + c_2(-\sqrt{2})^2$$

we find that $c_1 = \left(\frac{1}{2} + \frac{1}{2\sqrt{2}}\right)$, $c_2 = \left(\frac{1}{2} - \frac{1}{2\sqrt{2}}\right)$, so

$$p_n = \left(\frac{1}{2} + \frac{1}{2\sqrt{2}}\right)(\sqrt{2})^n + \left(\frac{1}{2} - \frac{1}{2\sqrt{2}}\right)(-\sqrt{2})^n, \qquad n \geq 1.$$

Unfortunately, this does not look like the result found in Example 9.13. After all, that answer contained no radical terms. However, suppose we consider n even, say $n = 2k$. Then

$$p_n = \left(\frac{1}{2} + \frac{1}{2\sqrt{2}}\right)(\sqrt{2})^{2k} + \left(\frac{1}{2} - \frac{1}{2\sqrt{2}}\right)(-\sqrt{2})^{2k}$$

$$= \left(\frac{1}{2} + \frac{1}{2\sqrt{2}}\right)2^k + \left(\frac{1}{2} - \frac{1}{2\sqrt{2}}\right)2^k = 2^k = 2^{n/2}.$$

For n odd, say $n = 2k - 1$, $k \in \mathbf{Z}^+$, we leave it for the reader to show that $p_n = 2^{k-1} = 2^{(n-1)/2}$.

The preceding results can be expressed by $p_n = 2^{\lfloor n/2 \rfloor}$, $n \geq 1$, as we found in Example 9.13.

The recurrence relation for the next example will be set up in two ways. In the first part we shall see how auxiliary variables may be helpful.

EXAMPLE 10.16   Find a recurrence relation for the number of binary sequences of length n that have no consecutive 0's.

a) For $n \geq 1$, let a_n be the number of such sequences of length n. Let $a_n^{(0)}$ count those that end in 0, and $a_n^{(1)}$ those that end in 1. Then $a_n = a_n^{(0)} + a_n^{(1)}$.

We derive a recurrence relation for a_n, $n \geq 1$, by computing $a_1 = 2$ and then considering each sequence x of length $n - 1$ (> 0) where x contains no consecutive 0's. If x ends in 1, then we can append a 0 or a 1 to it, giving us $2a_{n-1}^{(1)}$ of the sequences counted by a_n. If the sequence x ends in 0, then only 1 can be appended, resulting in $a_{n-1}^{(0)}$ sequences counted by a_n. Since these two cases exhaust all possibilities and have nothing in common, we have

$$a_n = 2 \cdot a_{n-1}^{(1)} + 1 \cdot a_{n-1}^{(0)}$$

The *n*th position can be 0 or 1.   **The *n*th position can only be 1.**

If we consider any sequence y counted in a_{n-2} we find that the sequence $y1$ is counted in $a_{n-1}^{(1)}$. Likewise, if the sequence $z1$ is counted in $a_{n-1}^{(1)}$, then z is counted in a_{n-2}. Consequently, $a_{n-2} = a_{n-1}^{(1)}$ and

$$a_n = a_{n-1}^{(1)} + \left[a_{n-1}^{(1)} + a_{n-1}^{(0)} \right] = a_{n-1}^{(1)} + a_{n-1} = a_{n-1} + a_{n-2}.$$

Therefore the recurrence relation for this problem is $a_n = a_{n-1} + a_{n-2}$, where $n \geq 3$ and $a_1 = 2$, $a_2 = 3$. (We leave the details of the solution for the reader.)

b) Alternatively, if $n \geq 1$ and a_n counts the number of binary sequences with no consecutive 0's, then $a_1 = 2$ and $a_2 = 3$, and for $n \geq 3$ we consider the binary sequences counted by a_n. There are two possibilities for these sequences:

(Case 1: The nth symbol is 1) Here we find that the preceding $n - 1$ symbols form a binary sequence with no consecutive 0's. There are a_{n-1} such sequences.

(Case 2: The nth symbol is 0) Here each such sequence actually ends in 10 and the first $n - 2$ symbols provide a binary sequence with no consecutive 0's. In this case there are a_{n-2} such sequences.

Since these two cases cover all the possibilities and have no such sequence in common, we may write

$$a_n = a_{n-1} + a_{n-2}, \qquad n \geq 3, \qquad a_1 = 2, \qquad a_2 = 3,$$

as we found in part (a).

In both part (a) and part (b) we can use the recurrence relation and $a_1 = 2$, $a_2 = 3$ to go back and determine a value for a_0 — namely, $a_0 = a_2 - a_1 = 3 - 2 = 1$. Then we can solve the recurrence relation

$$a_n = a_{n-1} + a_{n-2}, \qquad n \geq 2, \qquad a_0 = 1, \qquad a_1 = 2.$$

Before going any further we want to be sure that the reader understands why a general argument is needed when we develop our recurrence relations. When we are proving a theorem we do *not* draw any general conclusions from a few (or even, perhaps, many) particular instances. The same is true here. The following example should serve to drive this point home.

EXAMPLE 10.17

We start with n identical pennies and let a_n count the number of ways we can arrange these pennies — *contiguous* in each row where each penny above the bottom row touches two pennies in the row below it. (In these arrangements we are *not* concerned with whether any given penny is heads up or heads down.) In Fig. 10.7 we have the possible arrangements for $1 \leq n \leq 6$. From this it follows that

$$a_1 = 1, \qquad a_2 = 1, \qquad a_3 = 2, \qquad a_4 = 3, \qquad a_5 = 5, \quad \text{and} \quad a_6 = 8.$$

Consequently, these results might *suggest* that, in general, $a_n = F_n$, the nth Fibonacci number. Unfortunately, we have been led astray, as one finds, for example, that

$$a_7 = 12 \neq 13 = F_7, \qquad a_8 = 18 \neq 21 = F_8, \quad \text{and} \quad a_9 = 26 \neq 34 = F_9.$$

(The arrangements in this example were studied by F. C. Auluck in reference [2].)

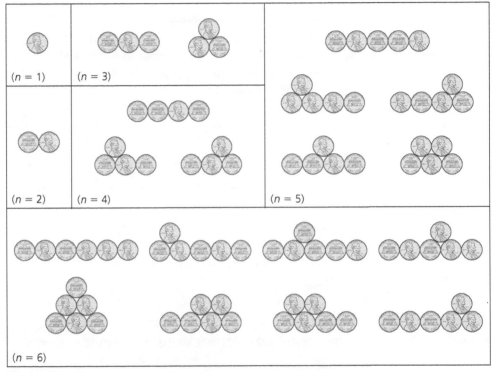

Figure 10.7

The last two examples for case (A) show us how to extend the results for second-order recurrence relations to those of higher order.

EXAMPLE 10.18

Solve the recurrence relation

$$2a_{n+3} = a_{n+2} + 2a_{n+1} - a_n, \qquad n \geq 0, \qquad a_0 = 0, \qquad a_1 = 1, \qquad a_2 = 2.$$

Letting $a_n = cr^n$ for $c, r \neq 0$ and $n \geq 0$, we obtain the characteristic equation $2r^3 - r^2 - 2r + 1 = 0 = (2r - 1)(r - 1)(r + 1)$. The characteristic roots are $1/2$, 1, and -1, so the solution is $a_n = c_1(1)^n + c_2(-1)^n + c_3(1/2)^n = c_1 + c_2(-1)^n + c_3(1/2)^n$. [The solutions 1, $(-1)^n$, and $(1/2)^n$ are called linearly independent because it is impossible to express any one of them as a linear combination of the other two.[†]] From $0 = a_0$, $1 = a_1$, and $2 = a_2$, we derive $c_1 = 5/2$, $c_2 = 1/6$, $c_3 = -8/3$. Consequently, $a_n = (5/2) + (1/6)(-1)^n + (-8/3)(1/2)^n$, $n \geq 0$.

EXAMPLE 10.19

For $n \geq 1$ we want to tile a $2 \times n$ chessboard using the two types of tiles shown in part (a) of Fig. 10.8. Letting a_n count the number of such tilings, we find that $a_1 = 1$, since we can tile a 2×1 chessboard (of one column) in only one way—using two 1×1 square tiles. Part (b) of the figure shows us that $a_2 = 5$. Finally, for the 2×3 chessboard there are 11 possible tilings: (i) one that uses six 1×1 square tiles; (ii) eight that use three 1×1 square tiles and one of the larger tiles; and (iii) two that use two of the larger tiles. When $n \geq 4$ we consider the nth column of the $2 \times n$ chessboard. There are three cases to examine:

1) the nth column is covered by two 1×1 square tiles—this case provides a_{n-1} tilings;

2) the $(n-1)$st and nth columns are tiled with one 1×1 square tile and one larger tile—this case accounts for $4a_{n-2}$ tilings; and

3) the $(n-2)$nd, $(n-1)$st, and nth columns are tiled with two of the larger tiles—this results in $2a_{n-3}$ tilings.

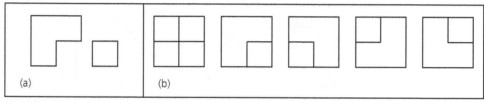

(a)   (b)

Figure 10.8

These three cases cover all possibilities and no two of the cases have anything in common, so

$$a_n = a_{n-1} + 4a_{n-2} + 2a_{n-3}, \qquad n \geq 4, \qquad a_1 = 1, \qquad a_2 = 5, \qquad a_3 = 11.$$

The characteristic equation $x^3 - x^2 - 4x - 2 = 0$ can be written as $(x + 1)(x^2 - 2x - 2) = 0$, so the characteristic roots are -1, $1 + \sqrt{3}$, and $1 - \sqrt{3}$. Consequently, $a_n = c_1(-1)^n + c_2(1 + \sqrt{3})^n + c_3(1 - \sqrt{3})^n$, $n \geq 1$. From $1 = a_1 = -c_1 + c_2(1 + \sqrt{3}) + c_3(1 - \sqrt{3})$, $5 = a_2 = c_1 + c_2(1 + \sqrt{3})^2 + c_3(1 - \sqrt{3})^2$, and $11 = a_3 = -c_1 + c_2(1 + \sqrt{3})^3 + c_3(1 - \sqrt{3})^3$, we have $c_1 = 1$, $c_2 = 1/\sqrt{3}$, and $c_3 = -1/\sqrt{3}$. So

$$a_n = (-1)^n + (1/\sqrt{3})(1 + \sqrt{3})^n + (-1/\sqrt{3})(1 - \sqrt{3})^n, \qquad n \geq 1.$$

Case (B): (Complex Roots)

Before getting into the case of complex roots, we recall DeMoivre's Theorem:

$$(\cos \theta + i \sin \theta)^n = \cos n\theta + i \sin n\theta, \qquad n \geq 0.$$

[This is part (b) of Exercise 12 of Section 4.1.]

[†]Alternatively, the solutions 1, $(-1)^n$, and $(1/2)^n$ are linearly independent, because if k_1, k_2, k_3 are real numbers, and $k_1(1) + k_2(-1)^n + k_3(1/2)^n = 0$ for *all* $n \in \mathbf{N}$, then $k_1 = k_2 = k_3 = 0$.

If $z = x + iy \in \mathbf{C}$, $z \neq 0$, we can write $z = r(\cos\theta + i\sin\theta)$, where $r = \sqrt{x^2 + y^2}$ and $(y/x) = \tan\theta$, for $x \neq 0$. If $x = 0$, then for $y > 0$,

$$z = yi = yi\sin(\pi/2) = y(\cos(\pi/2) + i\sin(\pi/2)),$$

and for $y < 0$,

$$z = yi = |y|i\sin(3\pi/2) = |y|(\cos(3\pi/2) + i\sin(3\pi/2)).$$

In all cases, $z^n = r^n(\cos n\theta + i\sin n\theta)$, for $n \geq 0$, by DeMoivre's Theorem.

EXAMPLE 10.20

Determine $(1 + \sqrt{3}\,i)^{10}$.

Figure 10.9 shows a geometric way to represent the complex number $1 + \sqrt{3}\,i$ as the point $(1, \sqrt{3})$ in the xy-plane. Here $r = \sqrt{1^2 + (\sqrt{3})^2} = 2$, and $\theta = \pi/3$.

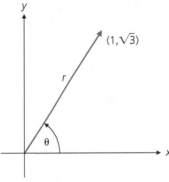

Figure 10.9

So $1 + \sqrt{3}\,i = 2(\cos(\pi/3) + i\sin(\pi/3))$, and

$$(1 + \sqrt{3}\,i)^{10} = 2^{10}(\cos(10\pi/3) + i\sin(10\pi/3)) = 2^{10}(\cos(4\pi/3) + i\sin(4\pi/3))$$
$$= 2^{10}((-1/2) - (\sqrt{3}/2)i) = (-2^9)(1 + \sqrt{3}\,i).$$

We'll use such results in the following examples.

EXAMPLE 10.21

Solve the recurrence relation $a_n = 2(a_{n-1} - a_{n-2})$, where $n \geq 2$ and $a_0 = 1$, $a_1 = 2$.

Letting $a_n = cr^n$, for $c, r \neq 0$, we obtain the characteristic equation $r^2 - 2r + 2 = 0$, whose roots are $1 \pm i$. Consequently, the general solution has the form $c_1(1 + i)^n + c_2(1 - i)^n$, where c_1 and c_2 presently denote arbitrary *complex* constants. [As in case (A), there are two independent solutions: $(1 + i)^n$ and $(1 - i)^n$.]

$$1 + i = \sqrt{2}(\cos(\pi/4) + i\sin(\pi/4))$$

and

$$1 - i = \sqrt{2}(\cos(-\pi/4) + i\sin(-\pi/4)) = \sqrt{2}(\cos(\pi/4) - i\sin(\pi/4)).$$

This yields

$$
\begin{aligned}
a_n &= c_1(1+i)^n + c_2(1-i)^n \\
&= c_1[\sqrt{2}(\cos(\pi/4) + i\,\sin(\pi/4))]^n + c_2[\sqrt{2}(\cos(-\pi/4) + i\,\sin(-\pi/4))]^n \\
&= c_1(\sqrt{2})^n(\cos(n\pi/4) + i\,\sin(n\pi/4)) + c_2(\sqrt{2})^n(\cos(-n\pi/4) + i\,\sin(-n\pi/4)) \\
&= c_1(\sqrt{2})^n(\cos(n\pi/4) + i\,\sin(n\pi/4)) + c_2(\sqrt{2})^n(\cos(n\pi/4) - i\,\sin(n\pi/4)) \\
&= (\sqrt{2})^n[k_1\cos(n\pi/4) + k_2\sin(n\pi/4)],
\end{aligned}
$$

where $k_1 = c_1 + c_2$ and $k_2 = (c_1 - c_2)i$.

$$
1 = a_0 = [k_1\cos 0 + k_2\sin 0] = k_1
$$
$$
2 = a_1 = \sqrt{2}[1 \cdot \cos(\pi/4) + k_2\sin(\pi/4)], \ \text{ or } 2 = 1 + k_2, \ \text{ and } k_2 = 1.
$$

The solution for the given initial conditions is then given by

$$
a_n = (\sqrt{2})^n[\cos(n\pi/4) + \sin(n\pi/4)], \qquad n \geq 0.
$$

[*Note:* This solution contains no complex numbers. A small point may bother the reader here. How did we start with c_1, c_2 complex and end up with $k_1 = c_1 + c_2$ and $k_2 = (c_1 - c_2)i$ real? This happens if c_1, c_2 are complex conjugates.]

Let us now examine an application from linear algebra.

EXAMPLE 10.22

For $b \in \mathbf{R}^+$, consider the $n \times n$ determinant[†] D_n given by

$$
\begin{vmatrix}
b & b & 0 & 0 & 0 & \cdots\cdots & 0 & 0 & 0 & 0 & 0 \\
b & b & b & 0 & 0 & \cdots\cdots & 0 & 0 & 0 & 0 & 0 \\
0 & b & b & b & 0 & \cdots\cdots & 0 & 0 & 0 & 0 & 0 \\
0 & 0 & b & b & b & \cdots\cdots & 0 & 0 & 0 & 0 & 0 \\
\cdot & \cdot & \cdot & \cdot & \cdot & \cdots\cdots & \cdot & \cdot & \cdot & \cdot & \cdot \\
0 & 0 & 0 & 0 & 0 & \cdots\cdots & b & b & b & 0 & 0 \\
0 & 0 & 0 & 0 & 0 & \cdots\cdots & 0 & b & b & b & 0 \\
0 & 0 & 0 & 0 & 0 & \cdots\cdots & 0 & 0 & b & b & b \\
0 & 0 & 0 & 0 & 0 & \cdots\cdots & 0 & 0 & 0 & b & b
\end{vmatrix}
$$

Find the value of D_n as a function of n.

Let a_n, $n \geq 1$, denote the value of the $n \times n$ determinant D_n. Then

$$
a_1 = |b| = b \quad \text{and} \quad a_2 = \begin{vmatrix} b & b \\ b & b \end{vmatrix} = 0 \quad \left(\text{and} \quad a_3 = \begin{vmatrix} b & b & 0 \\ b & b & b \\ 0 & b & b \end{vmatrix} = -b^3\right).
$$

[†]The expansion of determinants is discussed in Appendix 2.

Expanding D_n by its first row, we have $D_n =$

$$b \begin{vmatrix} b & b & 0 & 0 & \cdots & 0 & 0 & 0 & 0 \\ b & b & b & 0 & \cdots & 0 & 0 & 0 & 0 \\ 0 & b & b & b & \cdots & 0 & 0 & 0 & 0 \\ \cdot & \cdot & \cdot & \cdot & \cdots & \cdot & \cdot & \cdot & \cdot \\ 0 & 0 & 0 & 0 & \cdots & b & b & b & 0 \\ 0 & 0 & 0 & 0 & \cdots & 0 & b & b & b \\ 0 & 0 & 0 & 0 & \cdots & 0 & 0 & b & b \end{vmatrix} - b \begin{vmatrix} b & b & 0 & 0 & \cdots & 0 & 0 & 0 & 0 \\ 0 & b & b & 0 & \cdots & 0 & 0 & 0 & 0 \\ 0 & b & b & b & \cdots & 0 & 0 & 0 & 0 \\ \cdot & \cdot & \cdot & \cdot & \cdots & \cdot & \cdot & \cdot & \cdot \\ 0 & 0 & 0 & 0 & \cdots & b & b & b & 0 \\ 0 & 0 & 0 & 0 & \cdots & 0 & b & b & b \\ 0 & 0 & 0 & 0 & \cdots & 0 & 0 & b & b \end{vmatrix}$$

$\underbrace{\qquad\qquad\qquad\qquad\qquad}_{\text{(This is } D_{n-1}.)}$

When we expand the second determinant by its first column, we find that $D_n = bD_{n-1} - (b)(b)D_{n-2} = bD_{n-1} - b^2 D_{n-2}$. This translates into the relation $a_n = ba_{n-1} - b^2 a_{n-2}$, for $n \geq 3$, $a_1 = b$, $a_2 = 0$.

If we let $a_n = cr^n$ for $c, r \neq 0$ and $n \geq 1$, the characteristic equation produces the roots $b[(1/2) \pm i\sqrt{3}/2]$.

Hence

$$a_n = c_1[b((1/2) + i\sqrt{3}/2)]^n + c_2[b((1/2) - i\sqrt{3}/2)]^n$$
$$= b^n[c_1(\cos(\pi/3) + i\sin(\pi/3))^n + c_2(\cos(\pi/3) - i\sin(\pi/3))^n]$$
$$= b^n[k_1\cos(n\pi/3) + k_2\sin(n\pi/3)].$$

$b = a_1 = b[k_1\cos(\pi/3) + k_2\sin(\pi/3)]$, so $1 = k_1(1/2) + k_2(\sqrt{3}/2)$, or $k_1 + \sqrt{3}\,k_2 = 2$.

$0 = a_2 = b^2[k_1\cos(2\pi/3) + k_2\sin(2\pi/3)]$, so $0 = (k_1)(-1/2) + k_2(\sqrt{3}/2)$, or

$$k_1 = \sqrt{3}\,k_2.$$

Hence $k_1 = 1$, $k_2 = 1/\sqrt{3}$ and the value of D_n is

$$b^n[\cos(n\pi/3) + (1/\sqrt{3})\sin(n\pi/3)].$$

Case (C): (Repeated Real Roots)

EXAMPLE 10.23

Solve the recurrence relation $a_{n+2} = 4a_{n+1} - 4a_n$, where $n \geq 0$ and $a_0 = 1$, $a_1 = 3$.

As in the other two cases, we let $a_n = cr^n$, where $c, r \neq 0$ and $n \geq 0$. Then the characteristic equation is $r^2 - 4r + 4 = 0$ and the characteristic roots are both $r = 2$. (So $r = 2$ is called "a root of multiplicity 2.") Unfortunately, we now lack two independent solutions: 2^n and 2^n are definitely multiples of each other. We need one more independent solution. Let us try $g(n)2^n$ where $g(n)$ is not a constant. Substituting this into the given relation yields

$$g(n+2)2^{n+2} = 4g(n+1)2^{n+1} - 4g(n)2^n$$

or

$$g(n+2) = 2g(n+1) - g(n). \tag{1}$$

One finds that $g(n) = n$ satisfies Eq. (1).[†] So $n2^n$ is a second independent solution. (It is independent because it is impossible to have $n2^n = k2^n$ for *all* $n \geq 0$ if k is a constant.)

[†]Actually, the general solution is $g(n) = an + b$, for arbitrary constants a, b, with $a \neq 0$. Here we chose $a = 1$ and $b = 0$ to make $g(n)$ as simple as possible.

The general solution is of the form $a_n = c_1(2^n) + c_2 n(2^n)$. With $a_0 = 1$, $a_1 = 3$ we find $a_n = 2^n + (1/2)n(2^n) = 2^n + n(2^{n-1})$, $n \geq 0$.

In general, if $C_0 a_n + C_1 a_{n-1} + C_2 a_{n-2} + \cdots + C_k a_{n-k} = 0$, with $C_0\ (\neq 0)$, C_1, C_2, $\ldots$, $C_k\ (\neq 0)$ real constants, and r a characteristic root of multiplicity m, where $2 \leq m \leq k$, then the part of the general solution that involves the root r has the form

$$A_0 r^n + A_1 n r^n + A_2 n^2 r^n + \cdots + A_{m-1} n^{m-1} r^n$$
$$= (A_0 + A_1 n + A_2 n^2 + \cdots + A_{m-1} n^{m-1}) r^n,$$

where $A_0, A_1, A_2, \ldots, A_{m-1}$ are arbitrary constants.

Our last example involves a little probability.

EXAMPLE 10.24

If a first case of measles is recorded in a certain school system, let p_n denote the probability that at least one case is reported during the nth week after the first recorded case. School records provide evidence that $p_n = p_{n-1} - (0.25)p_{n-2}$, where $n \geq 2$. Since $p_0 = 0$ and $p_1 = 1$, if the first case (of a new outbreak) is recorded on Monday, March 3, 2003, when did the probability for the occurrence of a new case decrease to less than 0.01 for the first time?

With $p_n = cr^n$ for c, $r \neq 0$, the characteristic equation for the recurrence relation is $r^2 - r + (1/4) = 0 = (r - (1/2))^2$. The general solution has the form $p_n = (c_1 + c_2 n)(1/2)^n$, $n \geq 0$. For $p_0 = 0$, $p_1 = 1$, we get $c_1 = 0$, $c_2 = 2$, so $p_n = n2^{-n+1}$, $n \geq 0$.

The first integer n for which $p_n < 0.01$ is 12. Hence, it was not until the week of May 19, 2003, that the probability of another new case occurring was less than 0.01.

EXERCISES 10.2

1. Solve the following recurrence relations. (No final answer should involve complex numbers.)

a) $a_n = 5a_{n-1} + 6a_{n-2}$, $n \geq 2$, $a_0 = 1$, $a_1 = 3$

b) $2a_{n+2} - 11a_{n+1} + 5a_n = 0$, $n \geq 0$, $a_0 = 2$, $a_1 = -8$

c) $a_{n+2} + a_n = 0$, $n \geq 0$, $a_0 = 0$, $a_1 = 3$

d) $a_n - 6a_{n-1} + 9a_{n-2} = 0$, $n \geq 2$, $a_0 = 5$, $a_1 = 12$

e) $a_n + 2a_{n-1} + 2a_{n-2} = 0$, $n \geq 2$, $a_0 = 1$, $a_1 = 3$

2. a) Verify the final solutions in Examples 10.14 and 10.23.

   b) Solve the recurrence relation in Example 10.16.

3. If $a_0 = 0$, $a_1 = 1$, $a_2 = 4$, and $a_3 = 37$ satisfy the recurrence relation $a_{n+2} + ba_{n+1} + ca_n = 0$, where $n \geq 0$ and b, c are constants, determine b, c and solve for a_n.

4. Find and solve a recurrence relation for the number of ways to park motorcycles and compact cars in a row of n spaces if each cycle requires one space and each compact needs two. (All cycles are identical in appearance, as are the cars, and we want to use up all the n spaces.)

5. Answer the question posed in Exercise 4 if (a) the motorcycles come in two distinct models; (b) the compact cars come in three different colors; and (c) the motorcycles come in two distinct models and the compact cars come in three different colors.

6. Answer the questions posed in Exercise 5 if empty spaces are allowed.

7. In Exercise 12 of Section 4.2 we learned that $F_0 + F_1 + F_2 + \cdots + F_n = \sum_{i=0}^{n} F_i = F_{n+2} - 1$. This is one of many such properties of the Fibonacci numbers that were discovered by the French mathematician François Lucas (1842–1891). Although we established the result by the Principle of Mathematical Induction, we see that it is easy to develop this formula by adding the system of $n + 1$ equations

$$F_0 = F_2 - F_1$$
$$F_1 = F_3 - F_2$$
$$\cdots \quad \cdots \quad \cdots$$
$$F_{n-1} = F_{n+1} - F_n$$
$$F_n = F_{n+2} - F_{n+1}.$$

Develop formulas for each of the following sums, and then check the general result by the Principle of Mathematical Induction.

  a) $F_1 + F_3 + F_5 + \cdots + F_{2n-1}$, where $n \in \mathbf{Z}^+$

  b) $F_0 + F_2 + F_4 + \cdots + F_{2n}$, where $n \in \mathbf{Z}^+$

8. a) Prove that

$$\lim_{n \to \infty} \frac{F_{n+1}}{F_n} = \frac{1 + \sqrt{5}}{2}.$$

(This limit has come to be known as the *golden ratio* and is often designated by α, as we mentioned in Example 10.10.)

  b) Consider a regular pentagon *ABCDE* inscribed in a circle, as shown in Fig. 10.10.

  i) Use the law of sines and the double angle formula for the sine to show that $AC/AX = 2 \cos 36°$.

  ii) As $\cos 18° = \sin 72°$
  $= 4 \sin 18° \cos 18° (1 - 2 \sin^2 18°)$ (Why?), show that $\sin 18°$ is a root of the polynomial equation $8x^3 - 4x + 1 = 0$, and deduce that $\sin 18° = (\sqrt{5} - 1)/4$.

  c) Verify that $AC/AX = (1 + \sqrt{5})/2$.

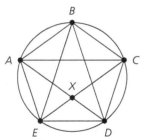

Figure 10.10

9. For $n \geq 0$, let a_n count the number of ways a sequence of 1's and 2's will sum to n. For example, $a_3 = 3$ because (1) 1, 1, 1; (2) 1, 2; and (3) 2, 1 sum to 3. Find and solve a recurrence relation for a_n.

10. For $\Sigma = \{0, 1\}$, let $A \subseteq \Sigma^*$, where $A = \{00, 1\}$. For $n \geq 1$, let a_n count the number of strings in A^* of length n. Find and solve a recurrence relation for a_n. (The reader may wish to refer to Exercise 25 for Section 6.1.)

11. a) For $n \geq 1$, let a_n count the number of binary strings of length n, where there are no consecutive 1's. Find and solve a recurrence relation for a_n.

  b) For $n \geq 1$, let b_n count the number of binary strings of length n, where there are no consecutive 1's and the first and last bit of the string are not both 1. Find and solve a recurrence relation for b_n.

12. Suppose that poker chips come in four colors — red, white, green, and blue. Find and solve a recurrence relation for the number of ways to stack n of these poker chips so that there are no consecutive blue chips.

13. An alphabet Σ consists of the four numeric characters 1, 2, 3, 4, and the seven alphabetic characters a, b, c, d, e, f, g. Find and solve a recurrence relation for the number of words of length n (in Σ^*), where there are no consecutive (identical or distinct) alphabetic characters.

14. An alphabet Σ consists of seven numeric characters and k alphabetic characters. For $n \geq 0$, a_n counts the number of strings (in Σ^*) of length n that contain no consecutive (identical or distinct) alphabetic characters. If $a_{n+2} = 7a_{n+1} + 63a_n$, $n \geq 0$, what is the value of k?

15. Solve the recurrence relation $a_{n+2} = a_{n+1}a_n$, $n \geq 0$, $a_0 = 1$, $a_1 = 2$.

16. For $n \geq 1$, let a_n be the number of ways to write n as an ordered sum of positive integers, where each summand is at least 2. (For example, $a_5 = 3$ because here we may represent 5 by 5, by $2 + 3$, and by $3 + 2$.) Find and solve a recurrence relation for a_n.

17. a) For a fixed nonnegative integer n, how many compositions of $n + 3$ have no 1 as a summand?

  b) For the compositions in part (a), how many start with (i) 2; (ii) 3; (iii) k, where $2 \leq k \leq n + 1$?

  c) How many of the compositions in part (a) start with $n + 2$ or $n + 3$?

  d) How are the results in parts (a)–(c) related to the formula derived at the start of Exercise 7?

18. Determine the points of intersection of the parabola $y = x^2 - 1$ and the line $y = x$.

19. Find the points of intersection of the hyperbola $y = 1 + \frac{1}{x}$ and the line $y = x$.

20. a) For $\alpha = (1 + \sqrt{5})/2$, show that $\alpha^2 = \alpha + 1$.

  b) If $n \in \mathbf{Z}^+$, prove that $\alpha^n = \alpha F_n + F_{n-1}$.

21. Let F_n denote the nth Fibonacci number, for $n \geq 0$, and let $\alpha = (1 + \sqrt{5})/2$. For $n \geq 3$, prove that (a) $F_n > \alpha^{n-2}$ and (b) $F_n < \alpha^{n-1}$.

22. a) For $n \in \mathbf{Z}^+$, let a_n count the number of palindromes of $2n$. Then $a_{n+1} = 2a_n$, $n \geq 1$, $a_1 = 2$. Solve this first-order recurrence relation for a_n.

  b) For $n \in \mathbf{Z}^+$, let b_n count the number of palindromes of $2n - 1$. Set up and solve a first-order recurrence relation for b_n.

(You may want to compare your solutions here with those given in Examples 9.13 and 10.15.)

23. Consider ternary strings — that is, strings where 0, 1, 2 are the only symbols used. For $n \geq 1$, let a_n count the number of ternary strings of length n where there are no consecutive 1's and no consecutive 2's. Find and solve a recurrence relation for a_n.

24. For $n \geq 1$, let a_n count the number of ways to tile a $2 \times n$ chessboard using horizontal (1×2) dominoes [which can also be used as vertical (2×1) dominoes] and square (2×2) tiles. Find and solve a recurrence relation for a_n.

25. In how many ways can one tile a 2×10 chessboard using dominoes and square tiles (as in Exercise 24) if the dominoes come in four colors and the square tiles come in five colors?

26. Let $\Sigma = \{0, 1\}$ and $A = \{0, 01, 11\} \subseteq \Sigma^*$. For $n \geq 1$, let a_n count the number of strings in A^* of length n. Find and solve a recurrence relation for a_n.

27. Let $\Sigma = \{0, 1\}$ and $A = \{0, 01, 011, 111\} \subseteq \Sigma^*$. For $n \geq 1$, let a_n count the number of strings in A^* of length n. Find and solve a recurrence relation for a_n.

28. Let $\Sigma = \{0, 1\}$ and $A = \{0, 01, 011, 0111, 1111\} \subseteq \Sigma^*$. For $n \geq 1$, let a_n count the number of strings in A^* of length n. Find and solve a recurrence relation for a_n.

29. A particle moves horizontally to the right. For $n \in \mathbf{Z}^+$, the distance the particle travels in the $(n + 1)$st second is equal to twice the distance it travels during the nth second. If $x_n, n \geq 0$, denotes the position of the particle at the start of the $(n + 1)$st second, find and solve a recurrence relation for x_n, where $x_0 = 1$ and $x_1 = 5$.

30. For $n \geq 1$, let D_n be the following $n \times n$ determinant.

$$
\begin{vmatrix}
2 & 1 & 0 & 0 & 0 & \cdots & 0 & 0 & 0 & 0 \\
1 & 2 & 1 & 0 & 0 & \cdots & 0 & 0 & 0 & 0 \\
0 & 1 & 2 & 1 & 0 & \cdots & 0 & 0 & 0 & 0 \\
\cdot & \cdot & \cdot & \cdot & \cdot & \cdots & \cdot & \cdot & \cdot & \cdot \\
0 & 0 & 0 & 0 & 0 & \cdots & 1 & 2 & 1 & 0 \\
0 & 0 & 0 & 0 & 0 & \cdots & 0 & 1 & 2 & 1 \\
0 & 0 & 0 & 0 & 0 & \cdots & 0 & 0 & 1 & 2
\end{vmatrix}
$$

Find and solve a recurrence relation for the value of D_n.

31. Solve the recurrence relation $a_{n+2}^2 - 5a_{n+1}^2 + 4a_n^2 = 0$, where $n \geq 0$ and $a_0 = 4$, $a_1 = 13$.

32. Determine the constants b and c if $a_n = c_1 + c_2(7^n), n \geq 0$, is the general solution of the relation $a_{n+2} + ba_{n+1} + ca_n = 0, n \geq 0$.

33. Prove that any two consecutive Fibonacci numbers are relatively prime.

34. Write a computer program (or develop an algorithm) to determine whether a given nonnegative integer is a Fibonacci number.

10.3
The Nonhomogeneous
Recurrence Relation

We now turn to the recurrence relations

$$a_n + C_1 a_{n-1} = f(n), \qquad n \geq 1, \tag{1}$$

$$a_n + C_1 a_{n-1} + C_2 a_{n-2} = f(n), \qquad n \geq 2, \tag{2}$$

where C_1 and C_2 are constants, $C_1 \neq 0$ in Eq. (1), $C_2 \neq 0$, and $f(n)$ is not identically 0. Although there is no general method for solving all nonhomogeneous relations, for certain functions $f(n)$ we shall find a successful technique.

We start with the special case for Eq. (1), when $C_1 = -1$. For the nonhomogeneous relation $a_n - a_{n-1} = f(n)$, we have

$$a_1 = a_0 + f(1)$$
$$a_2 = a_1 + f(2) = a_0 + f(1) + f(2)$$
$$a_3 = a_2 + f(3) = a_0 + f(1) + f(2) + f(3)$$
$$\vdots$$
$$a_n = a_{n-1} + f(n) = a_0 + f(1) + \cdots + f(n) = a_0 + \sum_{i=1}^{n} f(i).$$

We can solve this type of relation in terms of n, if we can find a suitable summation formula for $\sum_{i=1}^{n} f(i)$.

EXAMPLE 10.25	Solve the recurrence relation $a_n - a_{n-1} = 3n^2$, where $n \geq 1$ and $a_0 = 7$.

Here $f(n) = 3n^2$, so the unique solution is

$$a_n = a_0 + \sum_{i=1}^{n} f(i) = 7 + 3 \sum_{i=1}^{n} i^2 = 7 + \frac{1}{2}(n)(n+1)(2n+1).$$

When a formula for the summation is not known, the following procedure will handle Eq. (1) for certain functions $f(n)$, regardless of the value of $C_1 \, (\neq 0)$. It also works for the second-order nonhomogeneous relation in Eq. (2) — again, for certain functions $f(n)$. Known as the *method of undetermined coefficients*, it relies on the associated homogeneous relation obtained when $f(n)$ is replaced by 0.

For either of Eq. (1) or Eq. (2), we let $a_n^{(h)}$ denote the general solution of the associated homogeneous relation, and we let $a_n^{(p)}$ be a solution of the given nonhomogeneous relation. The term $a_n^{(p)}$ is called a *particular* solution. Then $a_n = a_n^{(h)} + a_n^{(p)}$ is the general solution of the given relation. To determine $a_n^{(p)}$ we use the form of $f(n)$ to suggest a form for $a_n^{(p)}$.

EXAMPLE 10.26	Solve the recurrence relation $a_n - 3a_{n-1} = 5(7^n)$, where $n \geq 1$ and $a_0 = 2$.

The solution of the associated homogeneous relation is $a_n^{(h)} = c(3^n)$. Since $f(n) = 5(7^n)$, we seek a particular solution $a_n^{(p)}$ of the form $A(7^n)$. As $a_n^{(p)}$ is to be a solution of the given nonhomogeneous relation, we place $a_n^{(p)} = A(7^n)$ into the given relation and find that $A(7^n) - 3A(7^{n-1}) = 5(7^n)$, $n \geq 1$. Dividing by 7^{n-1}, we find that $7A - 3A = 5(7)$, so $A = 35/4$, and $a_n^{(p)} = (35/4)7^n = (5/4)7^{n+1}$, $n \geq 0$. The general solution is $a_n = c(3^n) + (5/4)7^{n+1}$. With $2 = a_0 = c + (5/4)(7)$, it follows that $c = -27/4$ and $a_n = (5/4)(7^{n+1}) - (1/4)(3^{n+3})$, $n \geq 0$.

EXAMPLE 10.27	Solve the recurrence relation $a_n - 3a_{n-1} = 5(3^n)$, where $n \geq 1$ and $a_0 = 2$.

As in Example 10.26, $a_n^{(h)} = c(3^n)$, but here $a_n^{(h)}$ and $f(n)$ are not linearly independent. As a result we consider a particular solution $a_n^{(p)}$ of the form $Bn(3^n)$. (What happens if we substitute $a_n^{(p)} = B(3^n)$ into the given relation?)

Substituting $a_n^{(p)} = Bn3^n$ into the given relation yields

$$Bn(3^n) - 3B(n-1)(3^{n-1}) = 5(3^n), \quad \text{or} \quad Bn - B(n-1) = 5, \quad \text{so} \quad B = 5.$$

Hence $a_n = a_n^{(h)} + a_n^{(p)} = (c + 5n)3^n$, $n \geq 0$. With $a_0 = 2$, the unique solution is $a_n = (2 + 5n)(3^n)$, $n \geq 0$.

From the two preceding examples we generalize as follows.

Consider the nonhomogeneous first-order relation

$$a_n + C_1 a_{n-1} = kr^n,$$

where k is a constant and $n \in \mathbf{Z}^+$. If r^n is *not* a solution of the associated homogeneous relation

$$a_n + C_1 a_{n-1} = 0,$$

then $a_n^{(p)} = Ar^n$, where A is a constant. When r^n is a solution of the associated homogeneous relation, then $a_n^{(p)} = Bnr^n$, for B a constant.

Now consider the case of the nonhomogeneous second-order relation

$$a_n + C_1 a_{n-1} + C_2 a_{n-2} = kr^n,$$

where k is a constant. Here we find that

a) $a_n^{(p)} = Ar^n$, for A a constant, if r^n is not a solution of the associated homogeneous relation;

b) $a_n^{(p)} = Bnr^n$, where B is a constant, if $a_n^{(h)} = c_1 r^n + c_2 r_1^n$, where $r_1 \neq r$; and

c) $a_n^{(p)} = Cn^2 r^n$, for C a constant, when $a_n^{(h)} = (c_1 + c_2 n)r^n$.

EXAMPLE 10.28

The Towers of Hanoi. Consider n circular disks (having different diameters) with holes in their centers. These disks can be stacked on any of the pegs shown in Fig. 10.11. In the figure, $n = 5$ and the disks are stacked on peg 1 with no disk resting upon a smaller one. The objective is to transfer the disks one at a time so that we end up with the original stack on peg 3. Each of pegs 1, 2, and 3 may be used as a temporary location for any disk(s), but at no time are we allowed to have a larger disk on top of a smaller one on any peg. What is the minimum number of moves needed to do this for n disks?

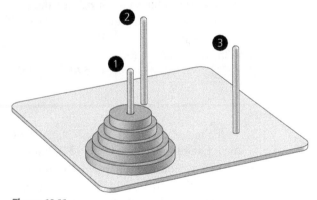

Figure 10.11

For $n \geq 0$, let a_n count the *minimum* number of moves it takes to transfer n disks from peg 1 to peg 3 in the manner described. Then, for $n + 1$ disks we can do the following:

a) Transfer the top n disks from peg 1 to peg 2 according to the directions that are given. This takes at least a_n moves.

b) Transfer the largest disk from peg 1 to peg 3. This takes one move.

c) Finally, transfer the n disks on peg 2 onto the largest disk, now on peg 3 — once again following the specified directions. This also requires at least a_n moves.

Consequently, at this point we know that a_{n+1} is no more than $2a_n + 1$ — that is, $a_{n+1} \leq 2a_n + 1$. But could there be a method where we actually have $a_{n+1} < 2a_n + 1$? Alas, no! For at some point the largest disk (the one at the bottom of the original stack — on peg 1) must be moved to peg 3. This move requires that peg 3 has no disks on it. So this largest disk may only be moved to peg 3 after the n smaller disks have moved to peg 2 [where they are stacked in increasing size from the smallest (on the top) to the largest (on the bottom)]. Getting these n smaller disks moved, accordingly, requires at least a_n moves. The largest disk must be moved at least once to get it to peg 3. Then, to get the n smaller disks on top of the largest disk (all on peg 3), according to the requirements, requires at least a_n more steps. So $a_{n+1} \geq a_n + 1 + a_n = 2a_n + 1$.

With $2a_n + 1 \leq a_{n+1} \leq 2a_n + 1$, we now obtain the relation $a_{n+1} = 2a_n + 1$, where $n \geq 0$ and $a_0 = 0$.

For $a_{n+1} - 2a_n = 1$, we know that $a_n^{(h)} = c(2^n)$. Since $f(n) = 1 = (1)^n$ is not a solution of $a_{n+1} - 2a_n = 0$, we set $a_n^{(p)} = A(1)^n = A$ and find from the given relation that $A = 2A + 1$, so $A = -1$ and $a_n = c(2^n) - 1$. From $a_0 = 0 = c - 1$ it then follows that $c = 1$, so $a_n = 2^n - 1, n \geq 0$.

The next example arises from the mathematics of finance.

EXAMPLE 10.29

Pauline takes out a loan of S dollars that is to be paid back in T periods of time. If r is the interest rate per period for the loan, what (constant) payment P must she make at the end of each period?

We let a_n denote the amount still owed on the loan at the end of the nth period (following the nth payment). Then at the end of the $(n + 1)$st period, the amount Pauline still owes on her loan is a_n (the amount she owed at the end of the nth period) $+ ra_n$ (the interest that accrued during the $(n + 1)$st period) $- P$ (the payment she made at the end of the $(n + 1)$st period). This gives us the recurrence relation

$$a_{n+1} = a_n + ra_n - P, \qquad 0 \leq n \leq T - 1, \qquad a_0 = S, \qquad a_T = 0.$$

For this relation $a_n^{(h)} = c(1 + r)^n$, while $a_n^{(p)} = A$ since no constant is a solution of the associated homogeneous relation. With $a_n^{(p)} = A$ we find $A - (1 + r)A = -P$, so $A = P/r$. From $a_0 = S$, we obtain $a_n = (S - (P/r))(1 + r)^n + (P/r), 0 \leq n \leq T$.

Since $0 = a_T = (S - (P/r))(1 + r)^T + (P/r)$, it follows that

$$(P/r) = ((P/r) - S)(1 + r)^T \quad \text{and} \quad P = (Sr)[1 - (1 + r)^{-T}]^{-1}.$$

We now consider a problem in the analysis of algorithms.

EXAMPLE 10.30

For $n \geq 1$, let S be a set containing 2^n real numbers.

The following procedure is used to determine the maximum and minimum elements of S. We wish to determine the number of comparisons made between pairs of elements in S during the execution of this procedure.

If a_n denotes the number of needed comparisons, then $a_1 = 1$. When $n = 2, |S| = 2^2 = 4$, so $S = \{x_1, x_2, y_1, y_2\} = S_1 \cup S_2$ where $S_1 = \{x_1, x_2\}, S_2 = \{y_1, y_2\}$. Since $a_1 = 1$, it takes one comparison to determine the maximum and minimum elements in each of S_1, S_2. Comparing the minimum elements of S_1 and S_2 and then their maximum elements, we learn the maximum and minimum elements in S and find that $a_2 = 4 = 2a_1 + 2$. In general, if $|S| = 2^{n+1}$, we write $S = S_1 \cup S_2$ where $|S_1| = |S_2| = 2^n$. To determine the maximum and minimum elements in each of S_1 and S_2 requires a_n comparisons. Comparing the maximum (minimum) elements of S_1 and S_2 requires one more comparison; consequently, $a_{n+1} = 2a_n + 2, n \geq 1$.

Here $a_n^{(h)} = c(2^n)$ and $a_n^{(p)} = A$, a constant. Substituting $a_n^{(p)}$ into the relation, we find that $A = 2A + 2$, or $A = -2$. So $a_n = c2^n - 2$, and with $a_1 = 1 = 2c - 2$, we obtain $c = 3/2$. Therefore $a_n = (3/2)(2^n) - 2$.

A note of caution! The existence of this procedure, which requires $(3/2)(2^n) - 2$ comparisons, does *not* exclude the possibility that we could achieve the same results via another remarkably clever method that requires fewer comparisons.

An example on counting certain strings of length 10, for the quaternary alphabet $\Sigma = \{0, 1, 2, 3\}$, provides a slight twist to what we've been doing so far.

EXAMPLE 10.31

For the alphabet $\Sigma = \{0, 1, 2, 3\}$, there are $4^{10} = 1{,}048{,}576$ strings of length 10 (in Σ^{10}, or Σ^*). Now we want to know how many of these more than 1 million strings contain an even number of 1's.

Instead of being so specific about the length of the strings, we will start by letting a_n count those strings among the 4^n strings in Σ^n where there are an even number of 1's. To determine how the strings counted by a_n, for $n \geq 2$, are related to those counted by a_{n-1}, consider the nth symbol of one of these strings of length n (where there is an even number of 1's). Two cases arise:

1) The nth symbol is 0, 2, or 3: Here the preceding $n - 1$ symbols provide one of the strings counted by a_{n-1}. So this case provides $3a_{n-1}$ of the strings counted by a_n.

2) The nth symbol is 1: In this case, there must be an odd number of 1's among the first $n - 1$ symbols. There are 4^{n-1} strings of length $n - 1$ and we want to avoid those that have an even number of 1's — there are $4^{n-1} - a_{n-1}$ such strings. Consequently, this second case gives us $4^{n-1} - a_{n-1}$ of the strings counted by a_n.

These two cases are exhaustive and mutually disjoint, so we may write

$$a_n = 3a_{n-1} + (4^{n-1} - a_{n-1}) = 2a_{n-1} + 4^{n-1}, \qquad n \geq 2.$$

Here $a_1 = 3$ (for the strings 0, 2, and 3). We find that $a_n^{(h)} = c(2^n)$ and $a_n^{(p)} = A(4^{n-1})$. Upon substituting $a_n^{(p)}$ into the above relation we have $A(4^{n-1}) = 2A(4^{n-2}) + 4^{n-1}$, so $4A = 2A + 4$ and $A = 2$. Hence, $a_n = c(2^n) + 2(4^{n-1})$, $n \geq 2$. From $3 = a_1 = 2c + 2$ it follows that $c = 1/2$, so $a_n = 2^{n-1} + 2(4^{n-1})$, $n \geq 1$.

When $n = 10$, we learn that of the $4^{10} = 1{,}048{,}576$ strings in Σ^{10}, there are $2^9 + 2(4^9) = 524{,}800$ that contain an even number of 1's.

Before continuing we realize that the answer here for a_n can be checked by using the exponential generating function $f(x) = \sum_{n=0}^{\infty} a_n \frac{x^n}{n!}$ (where $a_0 = 1$). From the techniques developed in Section 9.4 we have

$$f(x) = \left(1 + x + \frac{x^2}{2!} + \cdots\right)\left(1 + \frac{x^2}{2!} + \frac{x^4}{4!} + \cdots\right)\left(1 + x + \frac{x^2}{2!} + \cdots\right)\left(1 + x + \frac{x^2}{2!} + \cdots\right)$$

$$= e^x \cdot \left(\frac{e^x + e^{-x}}{2}\right) \cdot e^x \cdot e^x$$

$$= \left(\frac{1}{2}\right) e^{4x} + \left(\frac{1}{2}\right) e^{2x}$$

$$= \left(\frac{1}{2}\right) \sum_{n=0}^{\infty} \frac{(4x)^n}{n!} + \left(\frac{1}{2}\right) \sum_{n=0}^{\infty} \frac{(2x)^n}{n!}.$$

Here a_n = the coefficient of $\frac{x^n}{n!}$ in $f(x) = \left(\frac{1}{2}\right) 4^n + \left(\frac{1}{2}\right) 2^n = 2^{n-1} + 2(4^{n-1})$, as above.

EXAMPLE 10.32

In 1904, the Swedish mathematician Helge von Koch (1870–1924) created the intriguing curve now known as the Koch "snowflake" curve. The construction of this curve starts with an equilateral triangle, as shown in part (a) of Fig. 10.12, where the triangle has side 1, perimeter 3, and area $\sqrt{3}/4$. (Recall that an equilateral triangle of side s has perimeter $3s$ and area $s^2\sqrt{3}/4$.) The triangle is then transformed into the Star of David in Fig. 10.12(b) by removing the middle one-third of each side (of the original equilateral triangle) and attaching a new equilateral triangle whose side has length 1/3. So as we go from part (a) to part (b) in the figure, each side of length 1 is transformed into 4 sides of length 1/3, and we get a 12-sided polygon of area $(\sqrt{3}/4) + (3)(\sqrt{3}/4)(1/3)^2 = \sqrt{3}/3$. Continuing the process, we transform the figure of part (b) into that of part (c) by removing the middle one-third of each of the 12 sides in the Star of David and attaching an equilateral triangle of side $1/9 \, (= (1/3)^2)$. Now we have [in Fig. 10.12(c)] a $4^2(3)$-sided polygon whose area is

$$(\sqrt{3}/3) + (4)(3)(\sqrt{3}/4)[(1/3)^2]^2 = 10\sqrt{3}/27.$$

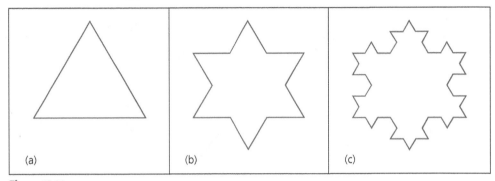

Figure 10.12

For $n \geq 0$, let a_n denote the area of the polygon P_n obtained from the original equilateral triangle after we apply n transformations of the type described above [the first from P_0 in Fig. 10.12(a) to P_1 in Fig. 10.12(b) and the second from P_1 in Fig. 10.12(b) to P_2 in Fig. 10.12(c)]. As we go from P_n (with $4^n(3)$ sides) to P_{n+1} (with $4^{n+1}(3)$ sides), we find that

$$a_{n+1} = a_n + (4^n(3))(\sqrt{3}/4)(1/3^{n+1})^2 = a_n + (1/(4\sqrt{3}))(4/9)^n$$

because in transforming P_n into P_{n+1} we remove the middle one-third of each of the $4^n(3)$ sides of P_n and attach an equilateral triangle of side $(1/3^{n+1})$.

The homogeneous part of the solution for this first-order nonhomogeneous recurrence relation is $a_n^{(h)} = A(1)^n = A$. Since $(4/9)^n$ is not a solution of the associated homogeneous relation, the particular solution is given by $a_n^{(p)} = B(4/9)^n$, where B is a constant. Substituting this into the recurrence relation $a_{n+1} = a_n + (1/(4\sqrt{3}))(4/9)^n$, we find that $B = (-9/5)(1/(4\sqrt{3}))$. Consequently,

$$a_n = A + (-9/5)(1/(4\sqrt{3}))(4/9)^n = A - (1/(5\sqrt{3}))(4/9)^{n-1}, \qquad n \geq 0.$$

Since $\sqrt{3}/4 = a_0 = A - (1/(5\sqrt{3}))(4/9)^{-1}$, it follows that $A = 6/(5\sqrt{3})$ and

$$a_n = (6/(5\sqrt{3})) - (1/(5\sqrt{3}))(4/9)^{n-1} = (1/(5\sqrt{3}))[6 - (4/9)^{n-1}], \qquad n \geq 0.$$

[As n grows larger, we find that $(4/9)^{n-1}$ tends to 0 and a_n approaches $6/(5\sqrt{3})$. We can also obtain this value by continuing the calculations we had before we introduced our recurrence relation, thus noting that this limiting area is also given by

$$(\sqrt{3}/4) + (\sqrt{3}/4)(3)(1/3)^2 + (\sqrt{3}/4)(4)(3)(1/3^2)^2 + (\sqrt{3}/4)(4^2)(3)(1/3^3)^2 + \cdots$$

$$= (\sqrt{3}/4) + (\sqrt{3}/4)(3)\sum_{n=0}^{\infty} 4^n (1/3^{n+1})^2 = (\sqrt{3}/4) + (1/(4\sqrt{3}))\sum_{n=0}^{\infty}(4/9)^n$$

$$= (\sqrt{3}/4) + (1/(4\sqrt{3}))[1/(1 - (4/9))] = (\sqrt{3}/4) + (1/(4\sqrt{3}))(9/5) = 6/(5\sqrt{3}),$$

by using the result for the sum of a geometric series from part (b) of Example 9.5.]

EXAMPLE 10.33

For $n \geq 1$, let $X_n = \{1, 2, 3, \ldots, n\}$; $\mathcal{P}(X_n)$ denotes the power set of X_n. We want to determine a_n, the number of edges in the Hasse diagram for the partial order $(\mathcal{P}(X_n), \subseteq)$. Here $a_1 = 1$ and $a_2 = 4$, and from Fig. 10.13 it follows that

$$a_3 = 2a_2 + 2^2.$$

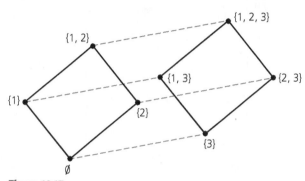

Figure 10.13

This is because the Hasse diagram for $(\mathcal{P}(X_3), \subseteq)$ contains the a_2 edges in the Hasse diagram for $(\mathcal{P}(X_2), \subseteq)$ as well as the a_2 edges in the Hasse diagram for the partial order $(\{\{3\}, \{1, 3\}, \{2, 3\}, \{1, 2, 3\}\}, \subseteq)$. [Note the identical structure shared by the partial orders $(\mathcal{P}(\{1, 2\}), \subseteq)$ and $(\{\{3\}, \{1, 3\}, \{2, 3\}, \{1, 2, 3\}\}, \subseteq)$.] In addition, there are 2^2 other (dashed) edges — one for each subset of $\{1, 2\}$. Now for $n \geq 1$, consider the Hasse diagrams for the partial orders $(\mathcal{P}(X_n), \subseteq)$ and $(\{T \cup \{n + 1\}|T \in \mathcal{P}(X_n)\}, \subseteq)$. For each $S \in \mathcal{P}(X_n)$, draw an edge from S in $(\mathcal{P}(X_n), \subseteq)$ to $S \cup \{n + 1\}$ in $(\{T \cup \{n + 1\}|T \in \mathcal{P}(X_n)\}, \subseteq)$. The result is the Hasse diagram for $(\mathcal{P}(X_{n+1}), \subseteq)$. From the construction we see that

$$a_{n+1} = 2a_n + 2^n, \qquad n \geq 1, \qquad a_1 = 1.$$

The solution to this recurrence relation, with the given condition $a_1 = 1$, is $a_n = n2^{n-1}$, $n \geq 1$.

Each of our next two examples deals with a second-order relation.

EXAMPLE 10.34

Solve the recurrence relation

$$a_{n+2} - 4a_{n+1} + 3a_n = -200, \qquad n \geq 0, \qquad a_0 = 3000, \qquad a_1 = 3300.$$

Here $a_n^{(h)} = c_1(3^n) + c_2(1^n) = c_1(3^n) + c_2$. Since $f(n) = -200 = -200(1^n)$ is a solution of the associated homogeneous relation, here $a_n^{(p)} = An$ for some constant A. This leads us to

$$A(n+2) - 4A(n+1) + 3An = -200, \quad \text{so} \quad -2A = -200, \quad A = 100.$$

Hence $a_n = c_1(3^n) + c_2 + 100n$. With $a_0 = 3000$ and $a_1 = 3300$, we have $a_n = 100(3^n) + 2900 + 100n, n \geq 0$.

Before proceeding any further, a point needs to be made about the role of technology in solving recurrence relations. When a computer algebra system is available, we are spared much of the drudgery of computation. Consequently, all our effort can be directed to analyzing the situation at hand and setting up the recurrence relation with its initial condition(s). Once this is done our job is just about finished. A line or two of code will often do the trick! For example, the Maple code in Fig. 10.14 shows how one can readily solve the recurrence relations of Examples 10.33 and 10.34.

```
> rsolve({a(n+1)=2*a(n)+2^n,a(1)=1},a(n));
```
$$-\frac{2^n}{2} + \left(\frac{n}{2} + \frac{1}{2}\right) 2^n$$
```
> simplify(%);
```
$$2^{(n-1)} n$$
```
> rsolve({a(n+2)=4*a(n+1)+3*a(n)=-200,a(0)=3000,a(1)=3300},a(n));
```
$$100 \cdot 3^n + 2900 + 100 \, n$$

Figure 10.14

EXAMPLE 10.35

In part (a) of Fig. 10.15 we have an *iterative* algorithm (written as a pseudocode procedure) for computing the nth Fibonacci number, for $n \geq 0$. Here the input is a nonnegative integer n and the output is the Fibonacci number F_n. The variables i, *fib*, *last*, *next_to_last*, and *temp* are integer variables. In this algorithm we calculate F_n (in this case for $n \geq 0$) by first assigning or computing all of the previous values $F_0, F_1, F_2, \ldots, F_{n-1}$. Here the number of additions needed to determine F_n is 0 for $n = 0, 1$ and $n - 1$ (within the **for** loop) for $n \geq 2$.

Part (b) of Fig. 10.15 provides a pseudocode procedure to implement a *recursive* algorithm for calculating F_n for $n \in \mathbf{N}$. Here the variable *fib* is likewise an integer variable. For this procedure we wish to determine a_n, the number of additions performed in computing $F_n, n \geq 0$. We find that $a_0 = 0, a_1 = 0$, and from the shaded line in the procedure — namely,

$$fib := FibNum2(n - 1) + FibNum2(n - 2) \tag{$*$}$$

we obtain the nonhomogeneous recurrence relation

$$a_n = a_{n-1} + a_{n-2} + 1, \quad n \geq 2,$$

where the summand of 1 is due to the addition in Eq. (*).

```
procedure FibNum1(n: nonnegative integer)
begin
  if n = 0 then
    fib := 0
  else if n = 1 then
    fib := 1
  else
    begin
      last := 1
      next_to_last := 0
      for i := 2 to n do
        begin
          temp := last
          last := last + next_to_last
          next_to_last := temp
        end
      fib := last
    end
end                                          (a)
```

```
procedure FibNum2(n: nonnegative integer)
begin
  if n = 0 then
    fib := 0
  else if n = 1 then
    fib := 1
  else
    fib := FibNum2(n - 1) + FibNum2(n - 2)
end                                          (b)
```

Figure 10.15

Here we find that $a_n^{(h)} = c_1\left(\frac{1+\sqrt{5}}{2}\right)^n + c_2\left(\frac{1-\sqrt{5}}{2}\right)^n$ and that $a_n^{(p)} = A$, a constant. Upon substituting $a_n^{(p)}$ into the nonhomogeneous recurrence relation we find that

$$A = A + A + 1,$$

so $A = -1$ and $a_n = c_1\left(\frac{1+\sqrt{5}}{2}\right)^n + c_2\left(\frac{1-\sqrt{5}}{2}\right)^n - 1$.

Since $a_0 = 0$ and $a_1 = 0$ it follows that

$$c_1 + c_2 = 1 \quad \text{and} \quad c_1\left(\frac{1+\sqrt{5}}{2}\right) + c_2\left(\frac{1-\sqrt{5}}{2}\right) = 1.$$

From these equations we learn that $c_1 = (1 + \sqrt{5})/(2\sqrt{5})$, $c_2 = (\sqrt{5} - 1)/(2\sqrt{5})$. Therefore,

$$a_n = \left(\frac{1+\sqrt{5}}{2\sqrt{5}}\right)\left(\frac{1+\sqrt{5}}{2}\right)^n - \left(\frac{1-\sqrt{5}}{2\sqrt{5}}\right)\left(\frac{1-\sqrt{5}}{2}\right)^n - 1$$

$$= \frac{1}{\sqrt{5}}\left(\frac{1+\sqrt{5}}{2}\right)^{n+1} - \frac{1}{\sqrt{5}}\left(\frac{1-\sqrt{5}}{2}\right)^{n+1} - 1.$$

As n gets larger $[(1 - \sqrt{5})/2]^{n+1}$ approaches 0 since $|(1 - \sqrt{5})/2| < 1$, and $a_n \doteq (1/\sqrt{5})[(1 + \sqrt{5})/2]^{n+1} = ((1 + \sqrt{5})/(2\sqrt{5}))((1 + \sqrt{5})/2)^n$.

Consequently, we can see that, as the value of n increases, the first procedure requires far less computation than the second one does.

We now summarize and extend the solution techniques already discussed in Examples 10.26 through 10.35.

Given a linear nonhomogeneous recurrence relation (with constant coefficients) of the form $C_0 a_n + C_1 a_{n-1} + C_2 a_{n-2} + \cdots + C_k a_{n-k} = f(n)$, where $C_0 \neq 0$ and $C_k \neq 0$, let $a_n^{(h)}$ denote the homogeneous part of the solution a_n.

1) If $f(n)$ is a constant multiple of one of the forms in the first column of Table 10.2 and is not a solution of the associated homogeneous relation, then $a_n^{(p)}$ has the form shown in the second column of Table 10.2. (Here A, B, A_0, A_1, A_2, $\ldots$, A_{t-1}, A_t are constants determined by substituting $a_n^{(p)}$ into the given relation; t, r, and θ are also constants.)

Table 10.2

	$a_n^{(p)}$
c, a constant	A, a constant
n	$A_1 n + A_0$
n^2	$A_2 n^2 + A_1 n + A_0$
$n^t, t \in \mathbf{Z}^+$	$A_t n^t + A_{t-1} n^{t-1} + \cdots + A_1 n + A_0$
$r^n, r \in \mathbf{R}$	$A r^n$
$\sin \theta n$	$A \sin \theta n + B \cos \theta n$
$\cos \theta n$	$A \sin \theta n + B \cos \theta n$
$n^t r^n$	$r^n (A_t n^t + A_{t-1} n^{t-1} + \cdots + A_1 n + A_0)$
$r^n \sin \theta n$	$A r^n \sin \theta n + B r^n \cos \theta n$
$r^n \cos \theta n$	$A r^n \sin \theta n + B r^n \cos \theta n$

2) When $f(n)$ comprises a sum of constant multiples of terms such as those in the first column of the table for item (1), and none of these terms is a solution of the associated homogeneous relation, then $a_n^{(p)}$ is made up of the sum of the corresponding terms in the column headed by $a_n^{(p)}$. For example, if $f(n) = n^2 + 3 \sin 2n$ and no summand of $f(n)$ is a solution of the associated homogeneous relation, then $a_n^{(p)} = (A_2 n^2 + A_1 n + A_0) + (A \sin 2n + B \cos 2n)$.

3) Things get trickier if a summand $f_1(n)$ of $f(n)$ is a solution of the associated homogeneous relation. This happens, for example, when $f(n)$ contains summands such as cr^n or $(c_1 + c_2 n)r^n$ and r is a characteristic root. If $f_1(n)$ causes this problem, we multiply the trial solution $(a_n^{(p)})_1$ corresponding to $f_1(n)$ by the smallest power of n, say n^s, for which no summand of $n^s f_1(n)$ is a solution of the associated homogeneous relation. Then $n^s (a_n^{(p)})_1$ is the corresponding part of $a_n^{(p)}$.

In order to check some of our preceding remarks on particular solutions for nonhomogeneous recurrence relations, the next application provides us with a situation that can be solved in more than one way.

EXAMPLE 10.36

For $n \geq 2$, suppose that there are n people at a party and that each of these people shakes hands (exactly one time) with all of the other people there (and no one shakes hands with himself or herself). If a_n counts the total number of handshakes, then

$$a_{n+1} = a_n + n, \qquad n \geq 2, \qquad a_2 = 1, \tag{3}$$

because when the $(n + 1)$st person arrives, he or she will shake hands with the n other people who have already arrived.

According to the results in Table 10.2, we might think that the trial (particular) solution for Eq. (3) is $A_1 n + A_0$, for constants A_0 and A_1. But here the associated homogeneous relation is $a_{n+1} = a_n$, or $a_{n+1} - a_n = 0$, for which $a_n^{(h)} = c(1^n) = c$, where c denotes an arbitrary constant. Therefore, the summand A_0 (in $A_1 n + A_0$) is a solution of the associated homogeneous relation. Consequently, the third remark (given with Table 10.2) tells us that we must multiply $A_1 n + A_0$ by the smallest power of n for which we no longer have any constant summand. This is accomplished by multiplying $A_1 n + A_0$ by n^1, and so we find here that

$$a_n^{(p)} = A_1 n^2 + A_0 n.$$

When we substitute this result into Eq. (3) we have

$$A_1 (n + 1)^2 + A_0 (n + 1) = A_1 n^2 + A_0 n + n,$$

or $A_1 n^2 + (2A_1 + A_0)n + (A_1 + A_0) = A_1 n^2 + (A_0 + 1)n$.

By comparing the coefficients on like powers of n we find that

(n^2): $A_1 = A_1$;

(n): $2A_1 + A_0 = A_0 + 1$; and

(n^0): $A_1 + A_0 = 0$.

Consequently, $A_1 = 1/2$ and $A_0 = -1/2$, so $a_n^{(p)} = (1/2)n^2 + (-1/2)n$ and $a_n = a_n^{(h)} + a_n^{(p)} = c + (1/2)(n)(n - 1)$. Since $a_2 = 1$, it follows from $1 = a_2 = c + (1/2)(2)(1)$ that $c = 0$, and $a_n = (1/2)(n)(n - 1)$, for $n \geq 2$.

We can also obtain this result by considering the n people in the room and realizing that each possible handshake corresponds with a selection of size 2 from this set of size n — and there are $\binom{n}{2} = (n!)/(2!(n - 2)!) = (1/2)(n)(n - 1)$ such selections. [Or we can consider the n people as vertices of an undirected graph (with no loops) where an edge corresponds with a handshake. Our answer is then the number of edges in the complete graph K_n, and there are $\binom{n}{2} = (1/2)(n)(n - 1)$ such edges.]

Our last example further demonstrates how we may use the results in Table 10.2.

EXAMPLE 10.37

a) Consider the nonhomogeneous recurrence relation

$$a_{n+2} - 10a_{n+1} + 21a_n = f(n), \qquad n \geq 0.$$

Here the homogeneous part of the solution is

$$a_n^{(h)} = c_1(3^n) + c_2(7^n),$$

for arbitrary constants c_1, c_2.

In Table 10.3 we list the form for the particular solution for certain choices of $f(n)$. Here the values of the 11 constants A_i, for $0 \leq i \leq 10$, are determined by substituting $a_n^{(p)}$ into the given nonhomogeneous recurrence relation.

Table 10.3

$f(n)$	$a_n^{(p)}$
5	A_0
$3n^2 - 2$	$A_3 n^2 + A_2 n + A_1$
$7(11^n)$	$A_4(11^n)$
$31(r^n), r \neq 3, 7$	$A_5(r^n)$
$6(3^n)$	$A_6 n 3^n$
$2(3^n) - 8(9^n)$	$A_7 n 3^n + A_8(9^n)$
$4(3^n) + 3(7^n)$	$A_9 n 3^n + A_{10} n 7^n$

b) The homogeneous component of the solution for

$$a_n + 4a_{n-1} + 4a_{n-2} = f(n), \qquad n \geq 2,$$

is

$$a_n^{(h)} = c_1(-2)^n + c_2 n(-2)^n,$$

where c_1, c_2 denote arbitrary constants. Consequently,

1) if $f(n) = 5(-2)^n$, then $a_n^{(p)} = An^2(-2)^n$;

2) if $f(n) = 7n(-2)^n$, then $a_n^{(p)} = n^2(-2)^n(A_1 n + A_0)$; and

3) if $f(n) = -11n^2(-2)^n$, then $a_n^{(p)} = n^2(-2)^n(B_2 n^2 + B_1 n + B_0)$.

(Here, the constants A, A_0, A_1, B_0, B_1, and B_2 are determined by substituting $a_n^{(p)}$ into the given nonhomogeneous recurrence relation.)

EXERCISES 10.3

1. Solve each of the following recurrence relations.

a) $a_{n+1} - a_n = 2n + 3, \quad n \geq 0, \quad a_0 = 1$

b) $a_{n+1} - a_n = 3n^2 - n, \quad n \geq 0, \quad a_0 = 3$

c) $a_{n+1} - 2a_n = 5, \quad n \geq 0, \quad a_0 = 1$

d) $a_{n+1} - 2a_n = 2^n, \quad n \geq 0, \quad a_0 = 1$

2. Use a recurrence relation to derive the formula for $\sum_{i=0}^{n} i^2$.

3. a) Let n lines be drawn in the plane such that each line intersects every other line but no three lines are ever co-incident. For $n \geq 0$, let a_n count the number of regions into which the plane is separated by the n lines. Find and solve a recurrence relation for a_n.

b) For the situation in part (a), let b_n count the number of infinite regions that result. Find and solve a recurrence relation for b_n.

4. On the first day of a new year, Joseph deposits $1000 in an account that pays 6% interest compounded monthly. At the beginning of each month he adds $200 to his account. If he continues to do this for the next four years (so that he makes 47 additional deposits of $200), how much will his account be worth exactly four years after he opened it?

5. Solve the following recurrence relations.

a) $a_{n+2} + 3a_{n+1} + 2a_n = 3^n, \quad n \geq 0, \quad a_0 = 0, \quad a_1 = 1$

b) $a_{n+2} + 4a_{n+1} + 4a_n = 7, \quad n \geq 0, \quad a_0 = 1, \quad a_1 = 2$

6. Solve the recurrence relation $a_{n+2} - 6a_{n+1} + 9a_n = 3(2^n) + 7(3^n)$, where $n \geq 0$ and $a_0 = 1, a_1 = 4$.

7. Find the general solution for the recurrence relation $a_{n+3} - 3a_{n+2} + 3a_{n+1} - a_n = 3 + 5n, n \geq 0$.

8. Determine the number of n-digit quaternary $(0, 1, 2, 3)$ sequences in which there is never a 3 anywhere to the right of a 0.

9. Meredith borrows $2500, at 12% compounded monthly, to buy a computer. If the loan is to be paid back over two years, what is his monthly payment?

10. The general solution of the recurrence relation $a_{n+2} + b_1 a_{n+1} + b_2 a_n = b_3 n + b_4, n \geq 0$, with b_i constant for $1 \leq i \leq 4$, is $c_1 2^n + c_2 3^n + n - 7$. Find b_i for each $1 \leq i \leq 4$.

11. Solve the following recurrence relations.

a) $a_{n+2}^2 - 5a_{n+1}^2 + 6a_n^2 = 7n, \quad n \geq 0, \quad a_0 = a_1 = 1$

b) $a_n^2 - 2a_{n-1} = 0, \quad n \geq 1, \quad a_0 = 2$ (Let $b_n = \log_2 a_n, n \geq 0$.)

12. Let $\Sigma = \{0, 1, 2, 3\}$. For $n \geq 1$, let a_n count the number of strings in Σ^n containing an odd number of 1's. Find and solve a recurrence relation for a_n.

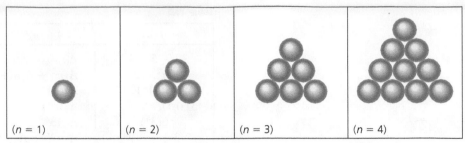

Figure 10.16

13. a) For the binary string 001110, there are three runs: 00, 111, and 0. Meanwhile, the string 000111 has only two runs: 000 and 111; while the string 010101 determines the six runs: 0, 1, 0, 1, 0, 1. For $n = 1$, we consider two binary strings, namely, 0 and 1 — these two strings (of length 1) determine a total of two runs. There are four binary strings of length $n = 2$ and these strings determine 1 (for 00) + 2 (for 01) + 2 (for 10) + 1 (for 11) = 6 runs. Find and solve a recurrence relation for t_n, the total number of runs determined by the 2^n binary strings of length n, where $n \geq 1$.

b) Answer the question posed in part (a) for quaternary strings of length n. (Here the alphabet comprises 0, 1, 2, 3.)

c) Generalize the results of parts (a) and (b).

14. a) For $n \geq 1$, the nth *triangular number* t_n is defined by $t_n = 1 + 2 + \cdots + n = n(n+1)/2$. Find and solve a recurrence relation for s_n, $n \geq 1$, where $s_n = t_1 + t_2 + \cdots + t_n$, the sum of the first n triangular numbers. [The reader may wish to compare the result obtained here with the formula given in Example 4.5 or with the result requested in part (b) of Exercise 8 of Section 9.5.]

b) In an organic laboratory, Kelsey synthesizes a crystalline structure that is made up of 10,000,000 triangular layers of atoms. The first layer of the structure has one atom, the second layer has three atoms, and, in general, the nth layer has $1 + 2 + \cdots + n = t_n$ atoms. (Consider each layer, other than the last, as if it were placed upon the spaces that result among the neighboring atoms of the succeeding layer. See Fig. 10.16.) (i) How many atoms are there in one of these crystalline structures? (ii) How many atoms are packed (strictly) between the 10,000th and 100,000th layer?

15. Write a computer program (or develop an algorithm) to solve the problem of the Towers of Hanoi. For $n \in \mathbf{Z}^+$, the program should provide the necessary steps for transferring the n disks from peg 1 to peg 3 under the restrictions specified in Example 10.28.

10.4
The Method of Generating Functions

With all the different cases we had to consider for the nonhomogeneous linear recurrence relation, we now get some assistance from the generating function. This technique will find both the homogeneous and the particular solutions for a_n, and it will incorporate the given initial conditions as well. Furthermore, we'll be able to do even more with this method.

We demonstrate the method in the following examples.

EXAMPLE 10.38

Solve the relation $a_n - 3a_{n-1} = n$, $n \geq 1$, $a_0 = 1$.

This relation represents an infinite set of equations:

$$(n = 1) \quad a_1 - 3a_0 = 1$$
$$(n = 2) \quad a_2 - 3a_1 = 2$$
$$\vdots \qquad \vdots \qquad \vdots$$

Multiplying the first of these equations by x, the second by x^2, and so on, we obtain

$$(n = 1) \qquad a_1 x^1 - 3a_0 x^1 = 1x^1$$
$$(n = 2) \qquad a_2 x^2 - 3a_1 x^2 = 2x^2$$
$$\vdots \qquad\qquad \vdots \qquad \vdots$$

Adding this second set of equations, we find that

$$\sum_{n=1}^{\infty} a_n x^n - 3 \sum_{n=1}^{\infty} a_{n-1} x^n = \sum_{n=1}^{\infty} n x^n. \qquad (1)$$

We want to solve for a_n in terms of n. To accomplish this, let $f(x) = \sum_{n=0}^{\infty} a_n x^n$ be the (ordinary) generating function for the sequence $a_0, a_1, a_2 \ldots$. Then Eq. (1) can be re-written as

$$(f(x) - a_0) - 3x \sum_{n=1}^{\infty} a_{n-1} x^{n-1} = \sum_{n=1}^{\infty} n x^n \left(= \sum_{n=0}^{\infty} n x^n \right). \qquad (2)$$

Since $\sum_{n=1}^{\infty} a_{n-1} x^{n-1} = \sum_{n=0}^{\infty} a_n x^n = f(x)$ and $a_0 = 1$, the left-hand side of Eq. (2) becomes $(f(x) - 1) - 3xf(x)$.

Before we can proceed, we need the generating function for the sequence $0, 1, 2, 3, \ldots$. Recall from part (c) of Example 9.5 that

$$\frac{x}{(1-x)^2} = x + 2x^2 + 3x^3 + \cdots, \quad \text{so}$$

$$(f(x) - 1) - 3xf(x) = \frac{x}{(1-x)^2}, \quad \text{and} \quad f(x) = \frac{1}{(1-3x)} + \frac{x}{(1-x)^2(1-3x)}.$$

Using a partial fraction decomposition, we find that

$$\frac{x}{(1-x)^2(1-3x)} = \frac{A}{(1-x)} + \frac{B}{(1-x)^2} + \frac{C}{(1-3x)},$$

or

$$x = A(1-x)(1-3x) + B(1-3x) + C(1-x)^2.$$

From the following assignments for x, we get

$$(x = 1): \quad 1 = B(-2), \qquad B = -\frac{1}{2}.$$

$$\left(x = \frac{1}{3}\right): \quad \frac{1}{3} = C\left(\frac{2}{3}\right)^2, \qquad C = \frac{3}{4}.$$

$$(x = 0): \quad 0 = A + B + C, \qquad A = -(B + C) = -\frac{1}{4}.$$

Therefore,

$$f(x) = \frac{1}{1-3x} + \frac{(-1/4)}{(1-x)} + \frac{(-1/2)}{(1-x)^2} + \frac{(3/4)}{(1-3x)}$$

$$= \frac{(7/4)}{(1-3x)} + \frac{(-1/4)}{(1-x)} + \frac{(-1/2)}{(1-x)^2}.$$

We find a_n by determining the coefficient of x^n in each of the three summands.

a) $(7/4)/(1 - 3x) = (7/4)[1/(1 - 3x)]$
$$= (7/4)[1 + (3x) + (3x)^2 + (3x)^3 + \cdots], \text{ and the coefficient of}$$
x^n is $(7/4)3^n$.

b) $(-1/4)/(1 - x) = (-1/4)[1 + x + x^2 + \cdots]$, and the coefficient of x^n here is $(-1/4)$.

c) $(-1/2)/(1 - x)^2 = (-1/2)(1 - x)^{-2}$
$$= (-1/2) \left[\binom{-2}{0} + \binom{-2}{1}(-x) + \binom{-2}{2}(-x)^2 + \binom{-2}{3}(-x)^3 + \cdots \right]$$
and the coefficient of x^n is given by $(-1/2)\binom{-2}{n}(-1)^n = (-1/2)(-1)^n \binom{2+n-1}{n} \cdot$
$(-1)^n = (-1/2)(n + 1)$.

Therefore $a_n = (7/4)3^n - (1/2)n - (3/4), n \geq 0$. (Note that there is no special concern here with $a_n^{(p)}$. Also, the same answer is obtained by using the techniques of Section 10.3.)

In our next example we extend what we learned in Example 10.38 to a second-order relation. This time we present the solution within a list of instructions one can follow in order to apply the generating-function method.

EXAMPLE 10.39   Consider the recurrence relation

$$a_{n+2} - 5a_{n+1} + 6a_n = 2, \qquad n \geq 0, \qquad a_0 = 3, \qquad a_1 = 7.$$

1) We first multiply this given relation by x^{n+2} because $n + 2$ is the largest subscript that appears. This gives us

$$a_{n+2}x^{n+2} - 5a_{n+1}x^{n+2} + 6a_n x^{n+2} = 2x^{n+2}.$$

2) Then we sum all of the equations represented by the result in step (1) and obtain

$$\sum_{n=0}^{\infty} a_{n+2}x^{n+2} - 5\sum_{n=0}^{\infty} a_{n+1}x^{n+2} + 6\sum_{n=0}^{\infty} a_n x^{n+2} = 2\sum_{n=0}^{\infty} x^{n+2}.$$

3) In order to have each of the subscripts on a match the corresponding exponent on x, we rewrite the equation in step (2) as

$$\sum_{n=0}^{\infty} a_{n+2}x^{n+2} - 5x\sum_{n=0}^{\infty} a_{n+1}x^{n+1} + 6x^2\sum_{n=0}^{\infty} a_n x^n = 2x^2\sum_{n=0}^{\infty} x^n.$$

Here we also rewrite the power series on the right-hand side of the equation in a form that will permit us to use what we learned in Section 2 of Chapter 9.

4) Let $f(x) = \sum_{n=0}^{\infty} a_n x^n$ be the generating function for the solution. The equation in step (3) now takes the form

$$(f(x) - a_0 - a_1 x) - 5x(f(x) - a_0) + 6x^2 f(x) = \frac{2x^2}{1 - x},$$

or

$$(f(x) - 3 - 7x) - 5x(f(x) - 3) + 6x^2 f(x) = \frac{2x^2}{1 - x}.$$

5) Solving for $f(x)$ we have

$$(1 - 5x + 6x^2)f(x) = 3 - 8x + \frac{2x^2}{1-x} = \frac{3 - 11x + 10x^2}{1-x},$$

from which it follows that

$$f(x) = \frac{3 - 11x + 10x^2}{(1 - 5x + 6x^2)(1-x)} = \frac{(3 - 5x)(1 - 2x)}{(1 - 3x)(1 - 2x)(1-x)} = \frac{3 - 5x}{(1 - 3x)(1-x)}.$$

A partial-fraction decomposition (by hand, or via a computer algebra system) gives us

$$f(x) = \frac{2}{1 - 3x} + \frac{1}{1-x} = 2\sum_{n=0}^{\infty}(3x)^n + \sum_{n=0}^{\infty} x^n.$$

Consequently, $a_n = 2(3^n) + 1, n \geq 0$.

We consider a third example, which has a familiar result.

EXAMPLE 10.40

Let $n \in \mathbf{N}$. For $r \geq 0$, let $a(n, r) =$ the number of ways we can select, with repetitions allowed, r objects from a set of n distinct objects.

For $n \geq 1$, let $\{b_1, b_2, \ldots, b_n\}$ be the set of these objects, and consider object b_1. Exactly one of two things can happen.

a) The object b_1 is never selected. Hence the r objects are selected from $\{b_2, \ldots, b_n\}$. This we can do in $a(n - 1, r)$ ways.

b) The object b_1 is selected at least once. Then we must select $r - 1$ objects from $\{b_1, b_2, \ldots, b_n\}$, so we can continue to select b_1 in addition to the one selection of it we've already made. There are $a(n, r - 1)$ ways to accomplish this.

Then $a(n, r) = a(n - 1, r) + a(n, r - 1)$ because these two cases cover all possibilities and are mutually disjoint.

Let $f_n = \sum_{r=0}^{\infty} a(n, r)x^r$ be the generating function for the sequence $a(n, 0), a(n, 1),$ $a(n, 2), \ldots$ [Here f_n is an abbreviation for $f_n(x)$.] From $a(n, r) = a(n - 1, r) + a(n, r - 1)$, where $n \geq 1$ and $r \geq 1$, it follows that

$$a(n, r)x^r = a(n - 1, r)x^r + a(n, r - 1)x^r \quad \text{and}$$

$$\sum_{r=1}^{\infty} a(n, r)x^r = \sum_{r=1}^{\infty} a(n - 1, r)x^r + \sum_{r=1}^{\infty} a(n, r - 1)x^r.$$

Realizing that $a(n, 0) = 1$ for $n \geq 0$ and $a(0, r) = 0$ for $r > 0$, we write

$$f_n - a(n, 0) = f_{n-1} - a(n - 1, 0) + x\sum_{r=1}^{\infty} a(n, r - 1)x^{r-1},$$

so $f_n - 1 = f_{n-1} - 1 + xf_n$. Therefore, $f_n - xf_n = f_{n-1}$, or $f_n = f_{n-1}/(1 - x)$.

If $n = 5$, for example, then

$$f_5 = \frac{f_4}{(1-x)} = \frac{1}{(1-x)} \cdot \frac{f_3}{(1-x)} = \frac{f_3}{(1-x)^2} = \frac{f_2}{(1-x)^3} = \frac{f_1}{(1-x)^4}$$

$$= \frac{f_0}{(1-x)^5} = \frac{1}{(1-x)^5},$$

since $f_0 = a(0, 0) + a(0, 1)x + a(0, 2)x^2 + \cdots = 1 + 0 + 0 + \cdots.$

In general, $f_n = 1/(1-x)^n = (1-x)^{-n}$, so $a(n, r)$ is the coefficient of x^r in $(1-x)^{-n}$, which is $\binom{-n}{r}(-1)^r = \binom{n+r-1}{r}$.

[Here we dealt with a recurrence relation for $a(n, r)$, a discrete function of the two (integer) variables $n, r \geq 0$.]

Our last example shows how generating functions may be used to solve a system of recurrence relations.

EXAMPLE 10.41

This example provides an approximate model for the propagation of high- and low-energy neutrons as they strike the nuclei of fissionable material (such as uranium) and are absorbed. Here we deal with a fast reactor where there is no moderator (such as water). (In reality, all the neutrons have fairly high energy and there are not just two energy levels. There is a continuous spectrum of energy levels, and these neutrons at the upper end of the spectrum are called the high-energy neutrons. The higher-energy neutrons tend to produce more new neutrons than the lower-energy ones.)

Consider the reactor at time 0 and suppose one high-energy neutron is injected into the system. During each time interval thereafter (about 1 microsecond, or 10^{-6} second) the following events occur.

a) When a high-energy neutron interacts with a nucleus (of fissionable material), upon absorption this results (one microsecond later) in two new high-energy neutrons and one low-energy one.

b) For interactions involving a low-energy neutron, only one neutron of each energy level is produced.

Assuming that all free neutrons interact with nuclei one microsecond after their creation, find functions of n such that

$$a_n = \text{the number of high-energy neutrons,}$$

$$b_n = \text{the number of low-energy neutrons,}$$

in the reactor after n microseconds, $n \geq 0$.

Here we have $a_0 = 1$, $b_0 = 0$ and the *system* of recurrence relations

$$a_{n+1} = 2a_n + b_n \tag{3}$$

$$b_{n+1} = a_n + b_n. \tag{4}$$

Let $f(x) = \sum_{n=0}^{\infty} a_n x^n$, $g(x) = \sum_{n=0}^{\infty} b_n x^n$ be the generating functions for the sequences $\{a_n | n \geq 0\}$, $\{b_n | n \geq 0\}$, respectively. From Eqs. (3) and (4), when $n \geq 0$

$$a_{n+1}x^{n+1} = 2a_n x^{n+1} + b_n x^{n+1} \tag{3$'$}$$

$$b_{n+1}x^{n+1} = a_n x^{n+1} + b_n x^{n+1}. \tag{4$'$}$$

Summing Eq. (3)$'$ over all $n \geq 0$, we have

$$\sum_{n=0}^{\infty} a_{n+1}x^{n+1} = 2x \sum_{n=0}^{\infty} a_n x^n + x \sum_{n=0}^{\infty} b_n x^n. \tag{3$''$}$$

In similar fashion, Eq. (4)$'$ yields

$$\sum_{n=0}^{\infty} b_{n+1}x^{n+1} = x \sum_{n=0}^{\infty} a_n x^n + x \sum_{n=0}^{\infty} b_n x^n. \tag{4$''$}$$

Introducing the generating functions at this point, we get

$$f(x) - a_0 = 2xf(x) + xg(x) \tag{3}''$$

$$g(x) - b_0 = xf(x) + xg(x), \tag{4}''$$

a system of equations relating the generating functions. Solving this system, we find that

$$f(x) = \frac{1-x}{x^2 - 3x + 1} = \left(\frac{5+\sqrt{5}}{10}\right)\left(\frac{1}{\gamma - x}\right) + \left(\frac{5-\sqrt{5}}{10}\right)\left(\frac{1}{\delta - x}\right) \quad \text{and}$$

$$g(x) = \frac{x}{x^2 - 3x + 1} = \left(\frac{-5 - 3\sqrt{5}}{10}\right)\left(\frac{1}{\gamma - x}\right) + \left(\frac{-5 + 3\sqrt{5}}{10}\right)\left(\frac{1}{\delta - x}\right),$$

where

$$\gamma = \frac{3+\sqrt{5}}{2}, \qquad \delta = \frac{3-\sqrt{5}}{2}.$$

Consequently,

$$a_n = \left(\frac{5+\sqrt{5}}{10}\right)\left(\frac{3-\sqrt{5}}{2}\right)^{n+1} + \left(\frac{5-\sqrt{5}}{10}\right)\left(\frac{3+\sqrt{5}}{2}\right)^{n+1} \quad \text{and}$$

$$b_n = \left(\frac{-5 - 3\sqrt{5}}{10}\right)\left(\frac{3-\sqrt{5}}{2}\right)^{n+1} + \left(\frac{-5 + 3\sqrt{5}}{10}\right)\left(\frac{3+\sqrt{5}}{2}\right)^{n+1}, \qquad n \geq 0.$$

EXERCISES 10.4

1. Solve the following recurrence relations by the method of generating functions.

a) $a_{n+1} - a_n = 3^n$, $n \geq 0$, $a_0 = 1$

b) $a_{n+1} - a_n = n^2$, $n \geq 0$, $a_0 = 1$

c) $a_{n+2} - 3a_{n+1} + 2a_n = 0$, $n \geq 0$, $a_0 = 1$, $a_1 = 6$

d) $a_{n+2} - 2a_{n+1} + a_n = 2^n$, $n \geq 0$, $a_0 = 1$, $a_1 = 2$

2. For n distinct objects, let $a(n, r)$ denote the number of ways we can select, without repetition, r of the n objects when

$0 \leq r \leq n$. Here $a(n, r) = 0$ when $r > n$. Use the recurrence relation $a(n, r) = a(n - 1, r - 1) + a(n - 1, r)$, where $n \geq 1$ and $r \geq 1$, to show that $f(x) = (1 + x)^n$ generates $a(n, r)$, $r \geq 0$.

3. Solve the following systems of recurrence relations.

a) $a_{n+1} = -2a_n - 4b_n$
$b_{n+1} = 4a_n + 6b_n$
$n \geq 0$, $a_0 = 1$, $b_0 = 0$

b) $a_{n+1} = 2a_n - b_n + 2$
$b_{n+1} = -a_n + 2b_n - 1$
$n \geq 0$, $a_0 = 0$, $b_0 = 1$

10.5
A Special Kind of Nonlinear
Recurrence Relation (Optional)

Thus far our study of recurrence relations has dealt with linear relations with constant coefficients. The study of nonlinear recurrence relations and of relations with variable coefficients is not a topic we shall pursue except for one special nonlinear relation that lends itself to the method of generating functions.

We shall develop the method in a counting problem on data structures. Before doing so, however, we first observe that if $f(x) = \sum_{i=0}^{\infty} a_i x^i$ is the generating function for $a_0, a_1, a_2, \ldots$, then $[f(x)]^2$ generates $a_0 a_0, a_0 a_1 + a_1 a_0, a_0 a_2 + a_1 a_1 + a_2 a_0, \ldots$,

$a_0 a_n + a_1 a_{n-1} + a_2 a_{n-2} + \cdots + a_{n-1} a_1 + a_n a_0, \ldots,$ the convolution of the sequence $a_0, a_1, a_2, \ldots,$ with itself.

EXAMPLE 10.42

In Sections 3.4 and 5.1, we encountered the idea of a tree diagram. In general, a *tree* is an undirected graph that is connected and has no loops or cycles. Here we examine rooted binary trees.

In Fig. 10.17 we see two such trees, where the circled vertex denotes the *root*. These trees are called *binary* because from each vertex there are at most two edges (called *branches*) descending (since a rooted tree is a directed graph) from that vertex.

In particular, these rooted binary trees are *ordered* in the sense that a left branch descending from a vertex is considered different from a right branch descending from that vertex. For the case of three vertices, the five possible ordered rooted binary trees are shown in Fig. 10.18. (If no attention were paid to order, then the last four rooted trees would be the same structure.)

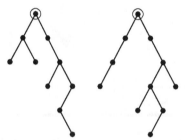

Figure 10.17

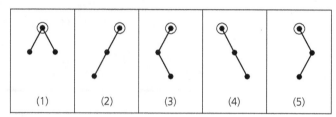

Figure 10.18

Our objective is to count, for $n \geq 0$, the number b_n of rooted ordered binary trees on n vertices. Assuming that we know the values of b_i for $0 \leq i \leq n$, in order to obtain b_{n+1} we select one vertex as the root and note, as in Fig. 10.19, that the substructures descending on the left and right of the root are smaller (rooted ordered binary) trees whose total number of vertices is n. These smaller trees are called *subtrees* of the given tree. Among these possible subtrees is the empty subtree, of which there is only 1 ($= b_0$).

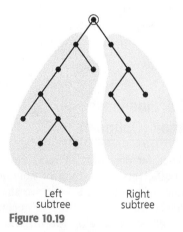

Left   Right
subtree   subtree

Figure 10.19

Now consider how the n vertices in these two subtrees can be divided up.

(1) 0 vertices on the left, n vertices on the right. This results in $b_0 b_n$ overall substructures to be counted in b_{n+1}.

(2) 1 vertex on the left, $n - 1$ vertices on the right, yielding $b_1 b_{n-1}$ rooted ordered binary trees on $n + 1$ vertices.

. . . . . .

($i + 1$) i vertices on the left, $n - i$ on the right, for a count of $b_i b_{n-i}$ toward b_{n+1}.

. . . . . .

($n + 1$) n vertices on the left and none on the right, contributing $b_n b_0$ of the trees.

Hence, for all $n \geq 0$,

$$b_{n+1} = b_0 b_n + b_1 b_{n-1} + b_2 b_{n-2} + \cdots + b_{n-1} b_1 + b_n b_0,$$

and

$$\sum_{n=0}^{\infty} b_{n+1} x^{n+1} = \sum_{n=0}^{\infty} (b_0 b_n + b_1 b_{n-1} + \cdots + b_{n-1} b_1 + b_n b_0) x^{n+1}. \tag{1}$$

Now let $f(x) = \sum_{n=0}^{\infty} b_n x^n$ be the generating function for $b_0, b_1, b_2, \ldots$. We rewrite Eq. (1) as

$$(f(x) - b_0) = x \sum_{n=0}^{\infty} (b_0 b_n + b_1 b_{n-1} + \cdots + b_n b_0) x^n = x[f(x)]^2.$$

This brings us to the quadratic [in $f(x)$]

$$x[f(x)]^2 - f(x) + 1 = 0, \quad \text{so} \quad f(x) = [1 \pm \sqrt{1 - 4x}]/(2x).$$

But $\sqrt{1 - 4x} = (1 - 4x)^{1/2} = \binom{1/2}{0} + \binom{1/2}{1}(-4x) + \binom{1/2}{2}(-4x)^2 + \cdots$, where the coefficient of x^n, $n \geq 1$, is

$$\binom{1/2}{n}(-4)^n = \frac{(1/2)((1/2) - 1)((1/2) - 2) \cdots ((1/2) - n + 1)}{n!}(-4)^n$$

$$= (-1)^{n-1} \frac{(1/2)(1/2)(3/2) \cdots ((2n - 3)/2)}{n!}(-4)^n$$

$$= \frac{(-1)2^n (1)(3) \cdots (2n - 3)}{n!}$$

$$= \frac{(-1)2^n (n!)(1)(3) \cdots (2n - 3)(2n - 1)}{(n!)(n!)(2n - 1)}$$

$$= \frac{(-1)(2)(4) \cdots (2n)(1)(3) \cdots (2n - 1)}{(2n - 1)(n!)(n!)} = \frac{(-1)}{(2n - 1)} \binom{2n}{n}.$$

In $f(x)$ we select the negative radical; otherwise, we would have negative values for the b_n's. Then

$$f(x) = \frac{1}{2x} \left[1 - \left[1 - \sum_{n=1}^{\infty} \frac{1}{(2n - 1)} \binom{2n}{n} x^n \right] \right],$$

and b_n, the coefficient of x^n in $f(x)$, is half the coefficient of x^{n+1} in

$$\sum_{n=1}^{\infty} \frac{1}{(2n - 1)} \binom{2n}{n} x^n.$$

So

$$b_n = \frac{1}{2}\left[\frac{1}{2(n+1)-1}\right]\binom{2(n+1)}{n+1} = \frac{(2n)!}{(n+1)!(n!)} = \frac{1}{(n+1)}\binom{2n}{n}.$$

The numbers b_n are called the *Catalan numbers*—the same sequence of numbers we encountered in Section 1.5. As we mentioned earlier (following Example 1.42), these numbers are named after the Belgian mathematician Eugène Charles Catalan (1814–1894), who used them in determining the number of ways to parenthesize the expression $x_1 x_2 x_3 \cdots x_n$. The first nine Catalan numbers are $b_0 = 1$, $b_1 = 1$, $b_2 = 2$, $b_3 = 5$, $b_4 = 14$, $b_5 = 42$, $b_6 = 132$, $b_7 = 429$, and $b_8 = 1430$.

We continue now with a second application of the Catalan numbers. This is based on an example given by Shimon Even. (See page 86 of reference [6].)

EXAMPLE 10.43

An important data structure that arises in computer science is the *stack*. This structure allows the storage of data items according to the following restrictions.

1) All insertions take place at *one* end of the structure. This is called the *top* of the stack, and the insertion process is referred to as the *push* procedure.

2) All deletions from the (nonempty) stack also take place from the top. We call the deletion process the *pop* procedure.

Since the *last* item inserted *in* this structure is the *first* item that can then be popped *out* of it, the stack is often referred to as a "last-in-first-out" (LIFO) structure.

Intuitive models for this data structure include a pile of poker chips on a table, a stack of trays in a cafeteria, and the discard pile used in playing certain card games. In all three of these cases, we can only (1) insert a new entry at the top of the pile or stack or (2) take (delete) the entry at the top of the (nonempty) pile or stack.

Here we shall use this data structure, with its push and pop procedures, to help us permute the (ordered) list $1, 2, 3, \ldots, n$, for $n \in \mathbf{Z}^+$. The diagram in Fig. 10.20 shows how each integer of the input $1, 2, 3, \ldots, n$ must be pushed onto the top of the stack in the order given. However, we may pop an entry from the top of the (nonempty) stack at any time. But once an entry is popped from the stack, it may not be returned to either the top of the stack or the input left to be pushed onto the stack. The process continues until no entry is left in the stack. Thus the ordered sequence of elements popped from the stack determines a permutation of $1, 2, 3, \ldots, n$.

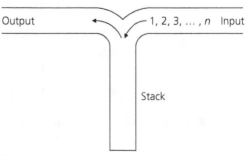

Figure 10.20

If $n = 1$, our input list consists of only the integer 1. We insert 1 at the top of the (empty) stack and then pop it out. This results in the permutation 1.

For $n = 2$, there are two permutations possible for 1, 2, and we can get both of them using the stack.

1) To get 1, 2 we place 1 at the top of the (empty) stack and then pop it. Then 2 is placed at the top of the (empty) stack and it is popped.

2) The permutation 2, 1 is obtained when 1 is placed at the top of the (empty) stack and 2 is then pushed onto the top of this (nonempty) stack. Upon popping first 2 from the top of the stack, and then 1, we obtain 2, 1.

Turning to the case where $n = 3$, we find that we can obtain only five of the $3! = 6$ possible permutations of 1, 2, 3 in this situation. For example, the permutation 2, 3, 1 results when we take the following steps.

- Place 1 at the top of the (empty) stack.
- Push 2 onto the top of the stack (on top of 1).
- Pop 2 from the stack.
- Push 3 onto the top of the stack (on top of 1).
- Pop 3 from the stack.
- Pop 1 from the stack, leaving it empty.

The reason we fail to obtain all six permutations of 1, 2, 3 is that we cannot generate the permutation 3, 1, 2 using the stack. For in order to have 3 in the first position of the permutation, we must build the stack by first pushing 1 onto the (empty) stack, then pushing 2 onto the top of the stack (on top of 1), and finally pushing 3 onto the stack (on top of 2). After 3 is popped from the top of the stack, we get 3 as the first number in our permutation. But with 2 now at the top of the stack, we cannot pop 1 until after 2 has been popped, so the permutation 3, 1, 2 cannot be generated.

When $n = 4$, there are 14 permutations of the (ordered) list 1, 2, 3, 4 that can be generated by the stack method. We list them in the four columns of Table 10.4 according to the location of 1 in the permutation.

Table 10.4

1, 2, 3, 4	2, 1, 3, 4	2, 3, 1, 4	2, 3, 4, 1
1, 2, 4, 3	2, 1, 4, 3	3, 2, 1, 4	2, 4, 3, 1
1, 3, 2, 4			3, 2, 4, 1
1, 3, 4, 2			3, 4, 2, 1
1, 4, 3, 2			4, 3, 2, 1

1) There are five permutations with 1 in the first position, because after 1 is pushed onto and popped from the stack, there are five ways to permute 2, 3, 4 using the stack.

2) When 1 is in the second position, 2 must be in the first position. This is because we pushed 1 onto the (empty) stack, then pushed 2 on top of it and then popped 2 and then 1. There are two permutations in column 2, because 3, 4 can be permuted in two ways on the stack.

3) For column 3 we have 1 in position three. We note that the only numbers that can precede it are 2 and 3, which can be permuted on the stack (with 1 on the bottom) in two ways. Then 1 is popped, and we push 4 onto the (empty) stack and then pop it.

4) In the last column we obtain five permutations: After we push 1 onto the top of the (empty) stack, there are five ways to permute 2, 3, 4 using the stack (with 1 on the bottom). Then 1 is popped from the stack to complete the permutation.

On the basis of these observations, for $1 \leq i \leq 4$, let a_i count the number of ways to permute the integers 1, 2, 3, . . . , i (or any list of i consecutive integers) using the stack. Also, we define $a_0 = 1$ since there is only one way to permute nothing, using the stack. Then

$$a_4 = a_0 a_3 + a_1 a_2 + a_2 a_1 + a_3 a_0,$$

where

a) Each summand $a_j a_k$ satisfies $j + k = 3$.

b) The subscript j tells us that there are j integers to the left of 1 in the permutation — in particular, for $j \geq 1$, these are the integers from 2 to $j + 1$, inclusive.

c) The subscript k indicates that there are k integers to the right of 1 in the permutation — for $k \geq 1$, these are the integers from $4 - (k - 1)$ to 4.

This permutation problem can now be generalized to any $n \in \mathbf{N}$, so that

$$a_{n+1} = a_0 a_n + a_1 a_{n-1} + a_2 a_{n-2} + \cdots + a_{n-1} a_1 + a_n a_0,$$

with $a_0 = 1$. From the result in Example 10.42 we know that

$$a_n = \frac{1}{(n + 1)} \binom{2n}{n}.$$

Now let us make one final observation about the permutations in Table 10.4. Consider, for example, the permutation 3, 2, 4, 1. How did this permutation come about? First 1 is pushed onto the empty stack. This is then followed by pushing 2 on top of 1 and then pushing 3 on top of 2. Now 3 is popped from the top of the stack, leaving 2 and 1; then 2 is popped from the top of the stack, leaving just 1. At this point 4 is pushed on top of 1 and then popped, leaving 1 on the stack. Finally, 1 is popped from the (top of the) stack, leaving the stack empty. So the permutation 3, 2, 4, 1 comes about from the following sequence of four pushes and four pops:

push, push, push, pop, pop, push, pop, pop.

Now replace each "push" with a "1" and each "pop" with a "0". The result is the sequence

1   1   1   0   0   1   0   0.

Similarly, the permutation 1, 3, 4, 2 is determined by the sequence

push, pop, push, push, pop, push, pop, pop

and this corresponds with the sequence

1   0   1   1   0   1   0   0.

In fact, each permutation in Table 10.4 gives rise to a sequence of four 1's and four 0's. But there are $8!/(4!\ 4!) = 70$ ways to list four 1's and four 0's. Do these 14 sequences have some special property? Yes! As we go from left to right in each of these sequences, the number of 1's (pushes) is never exceeded by the number of 0's (pops) [just like in part (b) of Example 1.43 — another situation counted by the Catalan numbers].

Our last example for this section is comparable to Example 10.17. Once again we see that we must guard against trying to obtain a general result without a general argument — no matter what a few special cases might suggest.

EXAMPLE 10.44

Here we start with *n distinct objects* and, for $n \geq 1$, we distribute them among at most *n identical containers*, but we do not allow more than three objects in any container, and we are not concerned about how the objects are arranged within any one container. We let a_n count the number of these distributions, and from Fig. 10.21 we see that

$$a_0 = 1, \qquad a_1 = 1, \qquad a_2 = 2, \qquad a_3 = 5, \quad \text{and} \quad a_4 = 14.$$

It appears that we might have the first five terms in the sequence of Catalan numbers. Unfortunately, the pattern breaks down and we find, for example, that

$$a_5 = 46 \neq 42 \text{ (the sixth Catalan number)} \quad \text{and}$$

$$a_6 = 166 \neq 132 \text{ (the seventh Catalan number)}.$$

(The distributions in this example were studied by F. L. Miksa, L. Moser, and M. Wyman in reference [22].)

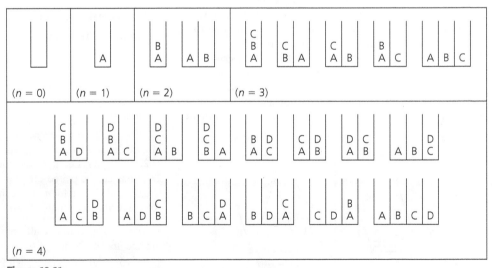

Figure 10.21

Other examples that involve the Catalan numbers can be found in the chapter references.

<hr/>

EXERCISES 10.5

1. For the rooted ordered binary trees of Example 10.42, calculate b_4 and draw all of these four-vertex structures.

2. Verify that for all $n \geq 0$,

$$\frac{1}{2}\left(\frac{1}{2n+1}\right)\binom{2n+2}{n+1} = \left(\frac{1}{n+1}\right)\binom{2n}{n}.$$

3. Show that for all $n \geq 2$,

$$\binom{2n-1}{n} - \binom{2n-1}{n-2} = \frac{1}{(n+1)}\binom{2n}{n}.$$

4. Which of the following permutations of $1, 2, 3, 4, 5, 6, 7, 8$ can be obtained using the stack (of Example 10.43)?

a) $4, 2, 3, 1, 5, 6, 7, 8$ **b)** $5, 4, 3, 6, 2, 1, 8, 7$

c) $4, 5, 3, 2, 1, 8, 6, 7$ **d)** $3, 4, 2, 1, 7, 6, 8, 5$

5. Suppose that the integers 1, 2, 3, 4, 5, 6, 7, 8 are permuted using the stack (of Example 10.43). (a) How many permutations are possible? (b) How many permutations have 1 in position 4 and 5 in position 8? (c) How many permutations have 1 in position 6? (d) How many permutations start with 321?

6. This exercise deals with a problem that was first proposed by Leonard Euler. The problem examines a given convex polygon of n (≥ 3) sides — that is, a polygon of n sides that satisfies the property: For all points P_1, P_2 within the interior of the polygon, the line segment joining P_1 and P_2 also lies within the interior of the polygon. Given a convex polygon of n sides, Euler wanted to count the number of ways the interior of the polygon could be triangulated (subdivided into triangles) by drawing diagonals that do not intersect.

For a convex polygon of $n \geq 3$ sides, let t_n count the number of ways the interior of the polygon can be triangulated by drawing nonintersecting diagonals.

a) Define $t_2 = 1$ and verify that

$$t_{n+1} = t_2 t_n + t_3 t_{n-1} + \cdots + t_{n-1} t_3 + t_n t_2.$$

b) Express t_n as a function of n.

7. In Fig. 10.22 we have two of the five ways in which we can triangulate the interior of a convex pentagon with no intersecting diagonals. Here we have labeled four of the sides — with the letters a, b, c, d — as well as the five vertices. In part (i) we use the labels on sides a and b to give us the label ab on the diagonal connecting vertices 2 and 4. This is because this diagonal (labeled ab), together with the sides a and b, provides us with one of the interior triangles for this triangulation of the convex pentagon. Then the diagonal ab and the side c give rise to the label $(ab)c$ on the diagonal determined by vertices

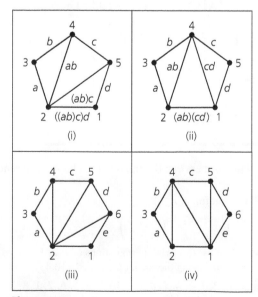

Figure 10.22

2 and 5 — and the sides labeled ab, c and $(ab)c$ provide a second interior triangle for this triangulation. Continuing in this way, we label the base edge connecting vertices 1 and 2 with $((ab)c)d$ — one of the five ways we can introduce parentheses in order to obtain the three products (of two numbers at a time) needed to compute $abcd$. The triangulation in part (ii) of the figure corresponds with the parenthesized product $(ab)(cd)$.

a) Determine the parenthesized product involving a, b, c, d for the other three triangulations of the convex pentagon.

b) Find the parenthesized product for each of the triangulated convex hexagons in parts (iii) and (iv) of Fig. 10.22

[From part (a) we learn that there are five ways to parenthesize the expression $abcd$ (and five ways to triangulate a convex pentagon). Part (b) shows us two of the 14 ways one can introduce parentheses for the expression $abcde$ (and triangulate a convex hexagon). In general, there are $\frac{1}{n+1}\binom{2n}{n}$ ways to parenthesize the expression $x_1 x_2 x_3 \cdots x_{n-1} x_n x_{n+1}$. It was in solving this problem that Eugène Charles Catalan discovered the sequence that now bears his name.]

8. For $n \geq 0$,

$$b_n = \left(\frac{1}{n+1}\right)\binom{2n}{n}$$

is the nth Catalan number.

a) Show that for all $n \geq 0$,

$$b_{n+1} = \frac{2(2n+1)}{(n+2)} b_n.$$

b) Use the result of part (a) to write a computer program (or develop an algorithm) that calculates the first 15 Catalan numbers.

9. For $n \geq 0$, evenly distribute $2n$ points on the circumference of a circle, and label these points cyclically with the integers 1, 2, 3, $\ldots$, $2n$. Let a_n be the number of ways in which these $2n$ points can be paired off as n chords where no two chords intersect. (The case for $n = 3$ is shown in Fig. 10.23.) Find and solve a recurrence relation for a_n, $n \geq 0$.

10. For $n \in \mathbf{N}$, consider all paths from $(0, 0)$ to $(2n, 0)$ using the moves N: $(x, y) \rightarrow (x+1, y+1)$ and S: $(x, y) \rightarrow (x+1, y-1)$, where any such path can never fall below the x-axis. The five paths (generally called *mountain ranges*) for $n = 3$ are shown in Fig. 10.24. How many mountain ranges are there for each $n \in \mathbf{N}$? (Verify your claim!)

11. For $n \in \mathbf{Z}^+$, let $f: \{1, 2, \ldots, n\} \rightarrow \{1, 2, \ldots, n\}$, where f is monotone increasing [that is, $1 \leq i < j \leq n \Rightarrow f(i) \leq f(j)$] and $f(i) \geq i$ for all $1 \leq i \leq n$. (a) Determine the five monotone increasing functions $f: \{1, 2, 3\} \rightarrow \{1, 2, 3\}$, where $f(i) \geq i$ for all $1 \leq i \leq 3$. (b) Use the graphs of the functions from part (a) to set up a one-to-one correspondence with the paths from $(0, 0)$ to $(3, 3)$ using the moves R: $(x, y) \rightarrow (x+1, y)$, U: $(x, y) \rightarrow (x, y+1)$, where each such path never falls below the line $y = x$. (The reader may

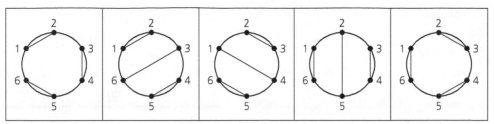

Figure 10.23

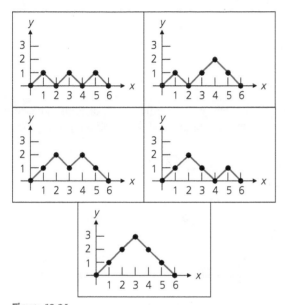

Figure 10.24

wish to check Exercise 3 for Section 1.5.) (c) If the paths in part (b) are rotated clockwise through $45°$, what results do we find? (d) How many monotone increasing functions f have domain and codomain equal to $\{1, 2, 3, \ldots, n\}$, for $n \in \mathbf{Z}^+$, and satisfy $f(i) \geq i$ for all $1 \leq i \leq n$?

12. For $n \in \mathbf{Z}^+$, let $g: \{1, 2, \ldots, n\} \to \{1, 2, \ldots, n\}$, where $g(i) \leq i$ for all $1 \leq i \leq n$. (a) Determine the five functions $g: \{1, 2, 3\} \to \{1, 2, 3\}$ where $g(i) \leq i$ for all $1 \leq i \leq 3$. (b) Set up a one-to-one correspondence between the functions in part (a) here and those in part (a) of the previous exercise. [You want a one-to-one correspondence that will generalize when you examine the functions $f, g: \{1, 2, \ldots, n\} \to \{1, 2, \ldots, n\}, n \in \mathbf{Z}^+$, where $f(i) \geq i$ and $g(i) \leq i$ for all $1 \leq i$

$\leq n$.] (c) How many functions g have domain and codomain equal to $\{1, 2, 3, \ldots, n\}$, for $n \in \mathbf{Z}^+$, and satisfy $g(i) \leq i$ for all $1 \leq i \leq n$?

13. For $n \in \mathbf{N}$, consider the arrangements of pennies built on a contiguous row of n pennies. Each penny that is not in the bottom row (of n pennies) rests upon the two pennies below it, and there is no concern about whether heads or tails appears. The situation for $n = 3$ is shown in Fig. 10.25. How many such arrangements are there for a contiguous row of n pennies, $n \in \mathbf{N}$?

14. For $n \in \mathbf{N}$, let s_n count the number of ways one can travel from $(0, 0)$ to (n, n) using the moves R: $(x, y) \to (x + 1, y)$, U: $(x, y) \to (x, y + 1)$, D: $(x, y) \to (x + 1, y + 1)$, where the path can never rise above the line $y = x$. (a) Determine s_2. (b) How is s_2 related to the Catalan numbers b_0, b_1, b_2? (c) How is s_3 related to b_0, b_1, b_2, b_3? What is s_3? (d) For $n \in \mathbf{N}$, how is s_n related to $b_0, b_1, b_2, \ldots, b_n$? (The numbers $s_0, s_1, s_2, \ldots$ are known as the *Schröder numbers*.)

15. A one-to-one function $f: \{1, 2, 3, \ldots, n\} \to \{1, 2, 3, \ldots, n\}$ is often called a *permutation*. Such a permutation is termed a *rise/fall* permutation when $f(1) < f(2), f(2) > f(3)$, $f(3) < f(4), \ldots$. For example, if $n = 4$ the five permutations 1324 (where $f(1) = 1$, $f(2) = 3$, $f(3) = 2$, $f(4) = 4$), 1423, 2314, 2413, and 3412 are the rise/fall permutations (for 1, 2, 3, 4). This we denote by writing $E_4 = 5$, where, in general, E_n counts the number of rise/fall permutations for 1, 2, 3, $\ldots$, n. The numbers $E_0, E_1, E_2, E_3, \ldots$ are called the *Euler numbers* (not to be confused with the Eulerian numbers in Example 4.21). We define $E_0 = 1$ and find that $E_1 = 1$, $E_2 = 1$.

a) Find the rise/fall permutations for 1, 2, 3. What is E_3?

b) Find the rise/fall permutations for 1, 2, 3, 4, 5. What is E_5?

c) Explain why in each rise/fall permutation of 1, 2, 3, $\ldots n$, we find n at position $2i$ for some $1 \leq i \leq \lfloor n/2 \rfloor$, if $n > 1$.

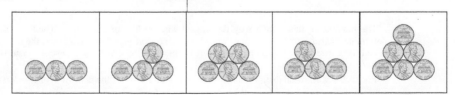

Figure 10.25

d) For $n \geq 2$, show that

$$E_n = \sum_{i=1}^{\lfloor n/2 \rfloor} \binom{n-1}{2i-1} E_{2i-1} E_{n-2i}, \quad E_0 = E_1 = 1.$$

e) Where do we find 1 in a rise/fall permutation of $1, 2, 3, \ldots, n$?

f) For $n \geq 1$, show that

$$E_n = \sum_{i=0}^{\lfloor (n-1)/2 \rfloor} \binom{n-1}{2i} E_{2i} E_{n-2i-1}, \quad E_0 = 1.$$

g) Prove that for $n \geq 2$,

$$E_n = \left(\frac{1}{2}\right) \sum_{i=0}^{n-1} \binom{n-1}{i} E_i E_{n-i-1}, \quad E_0 = E_1 = 1.$$

h) Use the result in part (g) to find E_6 and E_7.

i) Find the Maclaurin series expansion for $f(x) = \sec x + \tan x$. Conjecture (no proof required) the sequence for which this is the exponential generating function.

10.6
Divide-and-Conquer
Algorithms (Optional)[†]

One of the most important and widely applicable types of efficient algorithms is based on a *divide-and-conquer* approach. Here the strategy, in general, is to solve a given problem of size n $(n \in \mathbf{Z}^+)$ by

1) Solving the problem for a small value of n directly (this provides an initial condition for the resulting recurrence relation).

2) Breaking the general problem of size n into a smaller problems of the same type and (approximately) the same size — either $\lceil n/b \rceil$ or $\lfloor n/b \rfloor$,[‡] where $a, b \in \mathbf{Z}^+$ with $1 \leq a < n$ and $1 < b < n$.

Then we solve the a smaller problems and use their solutions to construct a solution for the original problem of size n. We shall be especially interested in cases where n is a power of b, and $b = 2$.

We shall study those divide-and-conquer algorithms where

1) The time to solve the initial problem of size $n = 1$ is a constant $c \geq 0$, and

2) The time to break the given problem of size n into a smaller (similar) problems, together with the time to combine the solutions of these smaller problems to get a solution for the given problem, is $h(n)$, a function of n.

Our concern here will actually be with the time-complexity function $f(n)$ for these algorithms. Consequently, we shall use the notation $f(n)$ here, instead of the subscripted notation a_n that we used in the earlier sections of this chapter.

The conditions that have now been stated lead to the following recurrence relation.

$$f(1) = c,$$
$$f(n) = af(n/b) + h(n), \quad \text{for } n = b^k, \quad k \geq 1.$$

We note that the domain of f is $\{1, b, b^2, b^3, \ldots\} = \{b^i \mid i \in \mathbf{N}\} \subset \mathbf{Z}^+$.

[†]The material in this section may be skipped with no loss of continuity. It will be used in Section 12.3 to determine the time-complexity function for the merge sort algorithm. However, the result there will also be obtained for a special case of the merge sort by another method that does not use the material developed in this section.

[‡]For each $x \in \mathbf{R}$, recall that $\lceil x \rceil$ denotes the *ceiling* of x and $\lfloor x \rfloor$ the *floor* of x, or *greatest integer* in x, where

    a) $\lfloor x \rfloor = \lceil x \rceil = x$, for $x \in \mathbf{Z}$.
    b) $\lfloor x \rfloor =$ the integer directly to the left of x, for $x \in \mathbf{R} - \mathbf{Z}$.
    c) $\lceil x \rceil =$ the integer directly to the right of x, for $x \in \mathbf{R} - \mathbf{Z}$.

In our first result, the solution of this recurrence relation is derived for the case where $h(n)$ is the constant c.

THEOREM 10.1   Let $a, b, c \in \mathbf{Z}^+$ with $b \geq 2$, and let $f: \mathbf{Z}^+ \to \mathbf{R}$. If

$$f(1) = c, \quad \text{and}$$
$$f(n) = af(n/b) + c, \qquad \text{for } n = b^k, \qquad k \geq 1,$$

then for all $n = 1, b, b^2, b^3, \ldots$,

1) $f(n) = c(\log_b n + 1)$, when $a = 1$, and

2) $f(n) = \dfrac{c(an^{\log_b a} - 1)}{a - 1}$, when $a \geq 2$.

Proof: For $k \geq 1$ and $n = b^k$, we write the following system of k equations. [Starting with the second equation, we obtain each of these equations from its immediate predecessor by (i) replacing each occurrence of n in the prior equation by n/b and (ii) multiplying the resulting equation in (i) by a.]

$$f(n) = af(n/b) + c$$
$$af(n/b) = a^2 f(n/b^2) + ac$$
$$a^2 f(n/b^2) = a^3 f(n/b^3) + a^2 c$$
$$\vdots \qquad\quad \vdots \qquad\quad \vdots$$
$$a^{k-2} f(n/b^{k-2}) = a^{k-1} f(n/b^{k-1}) + a^{k-2} c$$
$$a^{k-1} f(n/b^{k-1}) = a^k f(n/b^k) + a^{k-1} c$$

We see that each of the terms $af(n/b), a^2 f(n/b^2), \ldots, a^{k-1} f(n/b^{k-1})$ occurs one time as a summand on both the left-hand and right-hand sides of these equations. Therefore, upon adding both sides of the k equations and canceling these common summands, we obtain

$$f(n) = a^k f(n/b^k) + [c + ac + a^2 c + \cdots + a^{k-1} c].$$

Since $n = b^k$ and $f(1) = c$, we have

$$f(n) = a^k f(1) + c[1 + a + a^2 + \cdots + a^{k-1}]$$
$$= c[1 + a + a^2 + \cdots + a^{k-1} + a^k].$$

1) If $a = 1$, then $f(n) = c(k + 1)$. But $n = b^k \iff \log_b n = k$, so $f(n) = c(\log_b n + 1)$, for $n \in \{b^i \,|\, i \in \mathbf{N}\}$.

2) When $a \geq 2$, then $f(n) = \dfrac{c(1 - a^{k+1})}{1 - a} = \dfrac{c(a^{k+1} - 1)}{a - 1}$, from identity 4 of Table 9.2. Now $n = b^k \iff \log_b n = k$, so

$$a^k = a^{\log_b n} = (b^{\log_b a})^{\log_b n} = (b^{\log_b n})^{\log_b a} = n^{\log_b a},$$

and

$$f(n) = \frac{c(an^{\log_b a} - 1)}{(a - 1)}, \qquad \text{for } n \in \{b^i \,|\, i \in \mathbf{N}\}.$$

EXAMPLE 10.45

a) Let $f: \mathbf{Z}^+ \to \mathbf{R}$, where

$$f(1) = 3, \quad \text{and}$$
$$f(n) = f(n/2) + 3, \qquad \text{for } n = 2^k, \qquad k \in \mathbf{Z}^+.$$

So by part (1) of Theorem 10.1, with $c = 3$, $b = 2$, and $a = 1$, it follows that $f(n) = 3(\log_2 n + 1)$ for $n \in \{1, 2, 4, 8, 16, \ldots\}$.

b) Suppose that $g: \mathbf{Z}^+ \to \mathbf{R}$ with

$$g(1) = 7, \quad \text{and}$$
$$g(n) = 4g(n/3) + 7, \qquad \text{for } n = 3^k, \qquad k \in \mathbf{Z}^+.$$

Then with $c = 7$, $b = 3$, and $a = 4$, part (2) of Theorem 10.1 implies that $g(n) = (7/3)(4n^{\log_3 4} - 1)$, when $n \in \{1, 3, 9, 27, 81, \ldots\}$.

c) Finally, consider $h: \mathbf{Z}^+ \to \mathbf{R}$, where

$$h(1) = 5, \quad \text{and}$$
$$h(n) = 7h(n/7) + 5, \qquad \text{for } n = 7^k, \qquad k \in \mathbf{Z}^+.$$

Once again we use part (2) of Theorem 10.1, this time with $a = b = 7$ and $c = 5$. Here we learn that $h(n) = (5/6)(7n^{\log_7 7} - 1) = (5/6)(7n - 1)$ for $n \in \{1, 7, 49, 343, \ldots\}$.

Considering Theorem 10.1, we must unfortunately realize that although we know about f for $n \in \{1, b, b^2, \ldots\}$, we cannot say anything about the value of f for the integers in $\mathbf{Z}^+ - \{1, b, b^2, \ldots\}$. So at this time we are unable to deal with the concept of f as a time-complexity function. To overcome this, we now generalize Definition 5.23, wherein the idea of function dominance was first introduced.

Definition 10.1

Let $f, g: \mathbf{Z}^+ \to \mathbf{R}$ with S an infinite subset of $\mathbf{Z}^+$. We say that g *dominates f on S* (or f is *dominated by g on S*) if there exist constants $m \in \mathbf{R}^+$ and $k \in \mathbf{Z}^+$ such that $|f(n)| \leq m|g(n)|$ for all $n \in S$, where $n \geq k$.

Under these conditions we also say that $f \in O(g)$ *on S*.

EXAMPLE 10.46

Let $f: \mathbf{Z}^+ \to \mathbf{R}$ be defined so that

$$f(n) = n, \qquad \text{for } n \in \{1, 3, 5, 7, \ldots\} = S_1,$$
$$f(n) = n^2, \qquad \text{for } n \in \{2, 4, 6, 8, \ldots\} = S_2.$$

Then $f \in O(n)$ on S_1 and $f \in O(n^2)$ on S_2. However, we *cannot* conclude that $f \in O(n)$.

EXAMPLE 10.47

From Example 10.45, it now follows from Definition 10.1 that

a) $f \in O(\log_2 n)$ on $\{2^k | k \in \mathbf{N}\}$ b) $g \in O(n^{\log_3 4})$ on $\{3^k | k \in \mathbf{N}\}$

c) $h \in O(n)$ on $\{7^k | k \in \mathbf{N}\}$.

Using Definition 10.1, we now consider the following corollaries for Theorem 10.1. The first is a generalization of the first two results given in Example 10.47.

COROLLARY 10.1   Let a, b, $c \in \mathbf{Z}^+$ with $b \geq 2$, and let $f: \mathbf{Z}^+ \to \mathbf{R}$. If

$$f(1) = c, \quad \text{and}$$
$$f(n) = af(n/b) + c, \qquad \text{for } n = b^k, \qquad k \geq 1,$$

then

1) $f \in O(\log_b n)$ on $\{b^k | k \in \mathbf{N}\}$, when $a = 1$, and
2) $f \in O(n^{\log_b a})$ on $\{b^k | k \in \mathbf{N}\}$, when $a \geq 2$.

Proof: This proof is left as an exercise for the reader.

This second corollary changes the equal signs of Theorem 10.1 to inequalities. As a result, the codomain of f must be restricted from $\mathbf{R}$ to $\mathbf{R}^+ \cup \{0\}$.

COROLLARY 10.2   For a, b, $c \in \mathbf{Z}^+$ with $b \geq 2$, let $f: \mathbf{Z}^+ \to \mathbf{R}^+ \cup \{0\}$. If

$$f(1) \leq c, \quad \text{and}$$
$$f(n) \leq af(n/b) + c, \qquad \text{for } n = b^k, \qquad k \geq 1,$$

then for all $n = 1, b, b^2, b^3, \ldots$,

1) $f \in O(\log_b n)$, when $a = 1$, and
2) $f \in O(n^{\log_b a})$, when $a \geq 2$.

Proof: Consider the function $g: \mathbf{Z}^+ \to \mathbf{R}^+ \cup \{0\}$, where

$$g(1) = c, \quad \text{and}$$
$$g(n) = ag(n/b) + c, \qquad \text{for } n \in \{1, b, b^2, \ldots\}.$$

By Corollary 10.1,

$$g \in O(\log_b n) \quad \text{on} \quad \{b^k | k \in \mathbf{N}\}, \quad \text{when } a = 1, \quad \text{and}$$
$$g \in O(n^{\log_b a}) \quad \text{on} \quad \{b^k | k \in \mathbf{N}\}, \quad \text{when } a \geq 2.$$

We claim that $f(n) \leq g(n)$ for all $n \in \{1, b, b^2, \ldots\}$. To prove our claim, we induct on k where $n = b^k$. If $k = 0$, then $n = b^0 = 1$ and $f(1) \leq c = g(1)$ — so the result is true for this first case. Assuming the result is true for some $t \in \mathbf{N}$, we have $f(n) = f(b^t) \leq g(b^t) = g(n)$, for $n = b^t$. Then for $k = t + 1$ and $n = b^k = b^{t+1}$, we find that

$$f(n) = f(b^{t+1}) \leq af(b^{t+1}/b) + c = af(b^t) + c \leq ag(b^t) + c = g(b^{t+1}) = g(n).$$

Therefore, it follows by the Principle of Mathematical Induction that $f(n) \leq g(n)$ for all $n \in \{1, b, b^2, \ldots\}$. Consequently, $f \in O(g)$ on $\{b^k | k \in \mathbf{N}\}$, and the corollary follows because of our earlier statement about g.

Up to this point, our study of divide-and-conquer algorithms has been predominantly theoretical. It is high time we gave an example in which these ideas can be applied. The following result will confirm one of our earlier examples.

EXAMPLE 10.48

For $n = 1, 2, 4, 8, 16, \ldots$, let $f(n)$ count the number of comparisons needed to find the maximum and minimum elements in a set $S \subset \mathbf{R}$, where $|S| = n$ and the procedure in Example 10.30 is used.

If $n = 1$, then the maximum and minimum elements are the same element. Therefore, no comparisons are necessary and $f(1) = 0$.

If $n > 1$, then $n = 2^k$ for some $k \in \mathbf{Z}^+$, and we partition S as $S_1 \cup S_2$ where $|S_1| = |S_2| = n/2 = 2^{k-1}$. It takes $f(n/2)$ comparisons to find the maximum M_i and the minimum m_i for each set S_i, $i = 1, 2$. For $n \geq 4$, knowing m_1, M_1, m_2, and M_2, we then compare m_1 with m_2 and M_1 and M_2 to determine the minimum and maximum elements in S. Therefore,

$$f(n) = 2f(n/2) + 1, \qquad \text{when } n = 2, \quad \text{and}$$
$$f(n) = 2f(n/2) + 2, \qquad \text{when } n = 4, 8, 16, \ldots.$$

Unfortunately, these results do not provide the hypotheses of Theorem 10.1. However, if we change our equations into the inequalities

$$f(1) \leq 2$$
$$f(n) \leq 2f(n/2) + 2, \qquad \text{for } n = 2^k, \qquad k \geq 1,$$

then by Corollary 10.2 the time-complexity function $f(n)$, measured by the number of comparisons made in this recursive procedure, satisfies $f \in O(n^{\log_2 2}) = O(n)$, for all $n = 1, 2, 4, 8, \ldots$.

We can examine the relationship between this example and Example 10.30 even further. From that earlier result, we know that if $|S| = n = 2^k$, $k \geq 1$, then the number of comparisons $f(n)$ we need (in the given procedure) to find the maximum and minimum elements in S is $(3/2)(2^k) - 2$. (*Note*: Our statement here replaces the variable n of Example 10.30 by the variable k.)

Since $n = 2^k$, we find that we can now write

$$f(1) = 0$$
$$f(n) = f(2^k) = (3/2)(2^k) - 2 = (3/2)n - 2, \qquad \text{for } n = 2, 4, 8, 16, \ldots.$$

Hence $f \in O(n)$ for $n \in \{2^k | k \in \mathbf{N}\}$, just as we obtained above using Corollary 10.2.

All of our results have required that $n = b^k$, for some $k \in \mathbf{N}$, so it is only natural to ask whether we can do anything in the case where n is allowed to be an arbitrary positive integer. To find out, we introduce the following idea.

Definition 10.2

A function $f \colon \mathbf{Z}^+ \to \mathbf{R}^+ \cup \{0\}$ is called *monotone increasing* if for all $m, n \in \mathbf{Z}^+$, $m < n \Rightarrow f(m) \leq f(n)$.

This permits us to consider results for all $n \in \mathbf{Z}^+$ — under certain circumstances.

THEOREM 10.2

Let $f \colon \mathbf{Z}^+ \to \mathbf{R}^+ \cup \{0\}$ be monotone increasing, and let $g \colon \mathbf{Z}^+ \to \mathbf{R}$. For $b \in \mathbf{Z}^+$, $b \geq 2$, suppose that $f \in O(g)$ for all $n \in S = \{b^k | k \in \mathbf{N}\}$. Under these conditions,

  a) If $g \in O(\log n)$, then $f \in O(\log n)$.

  b) If $g \in O(n \log n)$, then $f \in O(n \log n)$.

  c) If $g \in O(n^r)$, then $f \in O(n^r)$, for $r \in \mathbf{R}^+ \cup \{0\}$.

Proof: We shall prove part (a) and leave parts (b) and (c) for the Section Exercises. Before starting, we should note that the base for the logarithms in parts (a) and (b) is any positive real number greater than 1.

Since $f \in O(g)$ on S, and $g \in O(\log n)$, we at least have $f \in O(\log n)$ on S. Therefore, by Definition 10.1, there exist constants $m \in \mathbf{R}^+$ and $s \in \mathbf{Z}^+$ such that $f(n) = |f(n)| \le m|\log n| = m \log n$ for all $n \in S$, $n \ge s$. We need to find a constant $M \in \mathbf{R}^+$ such that $f(n) \le M \log n$ for *all* $n \ge s$, not just those $n \in S$.

First let us agree to choose s large enough so that $\log s \ge 1$. Now let $n \in \mathbf{Z}^+$, where $n \ge s$ but $n \notin S$. Then there exists $k \in \mathbf{Z}^+$ such that $s \le b^k < n < b^{k+1}$. Since f is monotone increasing and positive,

$$
\begin{aligned}
f(n) \le f(b^{k+1}) &\le m \log(b^{k+1}) = m[\log(b^k) + \log b] \\
&= m \log(b^k) + m \log b \\
&< m \log(b^k) + m \log b \log(b^k) \\
&= m(1 + \log b) \log(b^k) \\
&< m(1 + \log b) \log n.
\end{aligned}
$$

So with $M = m(1 + \log b)$ we find that for all $n \in \mathbf{Z}^+ - S$, if $n \ge s$ then $f(n) < M \log n$. Hence $f(n) \le M \log n$ for *all* $n \in \mathbf{Z}^+$, where $n \ge s$, and $f \in O(\log n)$.

We shall now use the result of Theorem 10.2 in determining the time-complexity function $f(n)$ for a searching algorithm known as *binary search*.

In Example 5.70 we analyzed an algorithm wherein an array $a_1, a_2, a_3, \ldots, a_n$ of integers was searched for the presence of a particular integer called *key*. At that time the array entries were not given in any particular order, so we simply compared the value of *key* with those of the array elements $a_1, a_2, a_3, \ldots, a_n$. This would not be very efficient, however, if we knew that $a_1 < a_2 < a_3 < \cdots < a_n$. (After all, one does not search a telephone book for the telephone number of a particular person by starting at page 1 and examining every name in succession. The alphabetical ordering of the last names is used to speed up the searching process.) Let us look at a particular example.

EXAMPLE 10.49

Consider the array $a_1, a_2, a_3, \ldots, a_7$ of integers, where $a_1 = 2$, $a_2 = 4$, $a_3 = 5$, $a_4 = 7$, $a_5 = 10$, $a_6 = 17$, and $a_7 = 20$, and let *key* $= 9$. We search this array as follows:

1) Compare *key* with the entry at the center of the array; here it is $a_4 = 7$. Since *key* $> a_4$, we now concentrate on the remaining elements in the subarray a_5, a_6, a_7.

2) Now compare *key* with the center element a_6. Since *key* $= 9 < 17 = a_6$, we now turn to the subarray (of a_5, a_6, a_7) that consists of those elements smaller than a_6. Here this is only the element a_5.

3) Comparing *key* with a_5, we find that *key* $\ne a_5$, so *key* is not present in the given array $a_1, a_2, a_3, \ldots, a_7$.

From the results of Example 10.49, we make the following observations for a general (ordered) array of integers (or real numbers). Let $a_1, a_2, a_3, \ldots, a_n$ denote the given array, and let *key* denote the integer (or real number) for which we are searching. Unlike our array in Example 5.70, here

$$a_1 < a_2 < a_3 < \cdots < a_n.$$

1) First we compare the value of *key* with the array entry at or near the center. This entry is $a_{(n+1)/2}$ for n odd or $a_{n/2}$ for n even.

   Whether n is even or odd, the array element subscripted by $c = \lfloor (n+1)/2 \rfloor$ is the center, or near center, element. Note that at this point 1 is the value of the smallest subscript for the array subscripts, whereas n is the value of the largest subscript.

2) If *key* is a_c, we are finished. If not, then

   a) If *key* exceeds a_c, we search (with this dividing process) the subarray a_{c+1}, $a_{c+2}, \ldots, a_n$.

   b) If *key* is smaller than a_c, then the dividing process is applied in searching the subarray $a_1, a_2, \ldots, a_{c-1}$.

The preceding observations have been used in developing the pseudocode procedure in Fig. 10.26. Here the input is an ordered array $a_1, a_2, a_3, \ldots, a_n$ of integers, or real numbers, in ascending order, the positive integer n (for the number of entries in the given array), and the value of the integer variable *key*. If the array elements are integers (real numbers), then *key* should be an integer (real number). The variables s and l are integer variables used for storing the smallest and largest subscripts for the subscripts of the array or subarray being searched. The integer variable c stores the index for the array (subarray) element at, or near, the center of the array (subarray). In general, $c = \lfloor (s+l)/2 \rfloor$. The integer variable *location* stores the subscript of the array entry where *key* is located; the value of *location* is 0 when *key* is not present in the given array.

```
procedure BinarySearch(n: positive integer; key, a₁, a₂, a₃, ..., aₙ: integers)
begin
s := 1    {s is the smallest subscript of the subarray being searched}
l := n    {l is the largest subscript of the subarray being searched}
location := 0
while s ≤ l do
  begin
    c := ⌊(s + l)/2⌋
    if key = a_c then
      begin
        location := c
        s := l + 1
      end
    else if key < a_c then
      l := c - 1
    else s := c + 1
  end
end
```

Figure 10.26

We want to measure the (worst-case) time complexity for the algorithm implemented in Fig. 10.26. Here $f(n)$ will count the maximum number of comparisons (between *key* and a_c) needed to determine whether the given number *key* appears in the ordered array $a_1, a_2, a_3, \ldots, a_n$.

- For $n = 1$, *key* is compared to a_1 and $f(1) = 1$.
- When $n = 2$, in the worst case *key* is compared to a_1 and then to a_2, so $f(2) = 2$.
- In the case of $n = 3$, $f(3) = 2$ (in the worst case).
- When $n = 4$, the worst case occurs when *key* is first compared to a_2 and then a binary search of a_3, a_4 follows. Searching a_3, a_4 requires (in the worst case) $f(2)$ comparisons. So $f(4) = 1 + f(2) = 3$.

At this point we see that $f(1) \le f(2) \le f(3) \le f(4)$, and we conjecture that f is a monotone increasing function. To verify this, we shall use the Principle of Mathematical Induction in its alternative form. Here we assume that for all $i, j \in \{1, 2, 3, \ldots, n\}$, $i < j \Rightarrow f(i) \le f(j)$. Now consider the integer $n + 1$. We have two cases to examine.

1) $n + 1$ is odd: Here we write $n = 2k$ and $n + 1 = 2k + 1$, for some $k \in \mathbf{Z}^+$. In the worst case, $f(n + 1) = f(2k + 1) = 1 + f(k)$, where 1 counts the comparison of *key* with a_{k+1}, and $f(k)$ counts the (maximum) number of comparisons needed in a binary search of the subarray $a_1, a_2, \ldots, a_k$ or the subarray $a_{k+2}, a_{k+3}, \ldots, a_{2k+1}$.

   Now $f(n) = f(2k) = 1 + \max\{f(k-1), f(k)\}$. Since $k - 1, k < n$, by the induction hypothesis we have $f(k-1) \le f(k)$, so $f(n) = 1 + f(k) = f(n+1)$.

2) $n + 1$ is even: At this time we have $n + 1 = 2r$, for some $r \in \mathbf{Z}^+$, and in the worst case, $f(n + 1) = 1 + \max\{f(r-1), f(r)\} = 1 + f(r)$, by the induction hypothesis. Therefore,

$$f(n) = f(2r - 1) = 1 + f(r - 1) \le 1 + f(r) = f(n + 1).$$

Consequently, the function f is monotone increasing.

Now it is time to determine the worst-case time complexity for the binary search algorithm, using the function $f(n)$. Since

$$f(1) = 1, \quad \text{and}$$
$$f(n) = f(n/2) + 1, \qquad \text{for } n = 2^k, \qquad k \ge 1,$$

it follows from Theorem 10.1 (with $a = 1$, $b = 2$, and $c = 1$) that

$$f(n) = \log_2 n + 1, \quad \text{and} \quad f \in O(\log_2 n) \qquad \text{for } n \in \{1, 2, 4, 8, \ldots\}.$$

But with f monotone increasing, from Theorem 10.2 it now follows that $f \in O(\log_2 n)$ (for all $n \in \mathbf{Z}^+$). Consequently, binary search is an $O(\log_2 n)$ algorithm, whereas the searching algorithm of Example 5.70 is $O(n)$. Therefore, as the value of n increases, binary search is the more efficient algorithm — but then it requires the additional condition that the array be ordered.

This section has introduced some of the basic ideas in the study of divide-and-conquer algorithms. It also extends the material first introduced on computational complexity and the analysis of algorithms in Sections 5.7 and 5.8.

The Section Exercises include some extensions of the results developed in this section. The reader who wants to pursue this topic further should find the chapter references both helpful and interesting.

1. In each of the following, $f: \mathbf{Z}^+ \to \mathbf{R}$. Solve for $f(n)$ relative to the given set S, and determine the appropriate "big-Oh" form for f on S.

a) $f(1) = 5$
$f(n) = 4f(n/3) + 5, \quad n = 3, 9, 27, \ldots$
$S = \{3^i | i \in \mathbf{N}\}$

b) $f(1) = 7$
$f(n) = f(n/5) + 7, \quad n = 5, 25, 125, \ldots$
$S = \{5^i | i \in \mathbf{N}\}$

2. Let $a, b, c \in \mathbf{Z}^+$ with $b \geq 2$, and let $d \in \mathbf{N}$. Prove that the solution for the recurrence relation

$f(1) = d$

$f(n) = af(n/b) + c, \qquad n = b^k, \qquad k \geq 1$

satisfies

a) $f(n) = d + c\log_b n$, for $n = b^k$, $k \in \mathbf{N}$, when $a = 1$.

b) $f(n) = dn^{\log_b a} + (c/(a-1))[n^{\log_b a} - 1]$, for $n = b^k$, $k \in \mathbf{N}$, when $a \geq 2$.

3. Determine the appropriate "big-Oh" forms for f on $\{b^k | k \in \mathbf{N}\}$ in parts (a) and (b) of Exercise 2.

4. In each of the following, $f: \mathbf{Z}^+ \to \mathbf{R}$. Solve for $f(n)$ relative to the given set S, and determine the appropriate "big-Oh" form for f on S.

a) $f(1) = 0$
$f(n) = 2f(n/5) + 3, \quad n = 5, 25, 125, \ldots$
$S = \{5^i | i \in \mathbf{N}\}$

b) $f(1) = 1$
$f(n) = f(n/2) + 2, \quad n = 2, 4, 8, \ldots$
$S = \{2^i | i \in \mathbf{N}\}$

5. Consider a tennis tournament for n players, where $n = 2^k$, $k \in \mathbf{Z}^+$. In the first round $n/2$ matches are played, and the $n/2$ winners advance to round 2, where $n/4$ matches are played. This halving process continues until a winner is determined.

a) For $n = 2^k$, $k \in \mathbf{Z}^+$, let $f(n)$ count the total number of matches played in the tournament. Find and solve a recurrence relation for $f(n)$ of the form

$f(1) = d$

$f(n) = af(n/2) + c, \qquad n = 2, 4, 8, \ldots,$

where a, c, and d are constants.

b) Show that your answer in part (a) also solves the recurrence relation

$f(1) = d$

$f(n) = f(n/2) + (n/2), \qquad n = 2, 4, 8, \ldots.$

6. Complete the proofs for Corollary 10.1 and parts (b) and (c) of Theorem 10.2.

7. What is the best-case time-complexity function for binary search?

8. a) Modify the procedure in Example 10.48 as follows: For any $S \subset \mathbf{R}$, where $|S| = n$, partition S as $S_1 \cup S_2$, where $|S_1| = |S_2|$, for n even, and $|S_1| = 1 + |S_2|$, for n odd. Show that if $f(n)$ counts the number of comparisons needed (in this procedure) to find the maximum and minimum elements of S, then f is a monotone increasing function.

b) What is the appropriate "big-Oh" form for the function f of part (a)?

9. In Corollary 10.2 we were concerned with finding the appropriate "big-Oh" form for a function $f: \mathbf{Z}^+ \to \mathbf{R}^+ \cup \{0\}$ where

$f(1) \leq c, \qquad$ for $c \in \mathbf{Z}^+$

$f(n) \leq af(n/b) + c,$

$\qquad$ for $a, b \in \mathbf{Z}^+$ with $b \geq 2$, and $n = b^k$, $k \in \mathbf{Z}^+$.

Here the constant c in the second inequality is interpreted as the amount of time needed to break down the given problem of size n into a smaller (similar) problems of size n/b and to combine the a solutions of these smaller problems in order to get a solution for the original problem of size n. Now we shall examine a situation wherein this amount of time is no longer constant but depends on n.

a) Let $a, b, c \in \mathbf{Z}^+$, with $b \geq 2$. Let $f: \mathbf{Z}^+ \to \mathbf{R}^+ \cup \{0\}$ be a monotone increasing function, where

$f(1) \leq c$

$f(n) \leq af(n/b) + cn, \qquad$ for $n = b^k$, $\qquad k \in \mathbf{Z}^+$.

Use an argument similar to the one given (for equalities) in Theorem 10.1 to show that for all $n = 1, b, b^2, b^3, \ldots$,

$$f(n) \leq cn \sum_{i=0}^{k} (a/b)^i.$$

b) Use the result of part (a) to show that $f \in O(n \log n)$, where $a = b$. (The base for the log function here is any real number greater than 1.)

c) When $a \neq b$, show that part (a) implies that

$$f(n) \leq \left(\frac{c}{a-b}\right)(a^{k+1} - b^{k+1}).$$

d) From part (c), prove that (i) $f \in O(n)$, when $a < b$; and (ii) $f \in O(n^{\log_b a})$, when $a > b$. [*Note*: The "big-Oh" form for f here and in part (b) is for f on $\mathbf{Z}^+$, not just $\{b^k | k \in \mathbf{N}\}$.]

10. In this exercise we briefly introduce the *Master Theorem*. (For more on this result, including a proof, we refer the reader to pp. 73–84 of reference [5] by T. H. Cormen, C. E. Leiserson, R. L. Rivest, and C. Stein.)

Consider the recurrence relation

$f(1) = 1,$

$f(n) = af(n/b) + h(n),$

where $n \in \mathbf{Z}^+$, $n > 1$, $a \in \mathbf{Z}^+$, $a < n$, and $b \in \mathbf{R}^+$, $1 < b < n$. The function h accounts for the time (or cost) of dividing the given problem of size n into a smaller (similar) problems of size approximately n/b and then combining the results from the a smaller problems. Further, there exists $k \in \mathbf{Z}^+$ such that $h(n) > 0$ for all $n \geq k$. (Since n/b need not be an integer, the recurrence relation is not properly defined. To get around this we need to replace n/b by either $\lfloor n/b \rfloor$ or $\lceil n/b \rceil$. But as this does not affect the outcome of the result, for large values of n, we shall not concern ourselves with such details.)

Under the above hypothesis we find the following [where Θ (big theta) and Ω (big omega) are as given in Exercises 11–16 for Section 5.7]:

i) If $h \in O(n^{\log_b a - \epsilon})$, for some fixed $\epsilon > 0$, then $f \in \Theta(n^{\log_b a})$;

ii) If $h \in \Theta(n^{\log_b a})$, then $f \in \Theta(n^{\log_b a} \log_2 n)$; and

iii) If $h \in \Omega(n^{\log_b a + \epsilon})$ for some fixed $\epsilon > 0$, and if $a h(n/b) \leq c h(n)$, for some fixed c, where $0 < c < 1$, and for all sufficiently large n, then $f \in \Theta(h)$.

In all three cases, the function h is compared with $n^{\log_b a}$ and, roughly speaking, the Master Theorem then determines the complexity of the solution $f(n)$ as the larger of the two functions in cases (i) and (iii), while in case (ii) we find the added factor $\log_2 n$. However, it is important to realize that there are some recurrence relations of this type that do not fall under any of these three cases.

For now we consider the following, where $f(1) = 1$ for all three examples.

1) $f(n) = 16 f(n/4) + n$

Here $a = 16$, $b = 4$, $n^{\log_b a} = n^{\log_4 16} = n^2$, and $h(n) = n$. So $h \in O(n^{\log_4 16 - \epsilon})$ with $\epsilon = 1$. Consequently, h falls under the hypothesis for case (i) and it follows that $f \in \Theta(n^2)$.

2) $f(n) = f(3n/4) + 5$

Now we have $a = 1$, $b = 4/3$, $n^{\log_b a} = n^{\log_{4/3} 1} = n^0 = 1$, and $h(n) = 5$. Consequently, $h \in \Theta(n^{\log_{4/3} 1})$ and from case (ii) we learn that $f \in \Theta(n^{\log_{4/3} 1} \log_2 n) = \Theta(\log_2 n)$.

3) $f(n) = 7 f(n/8) + n \log_2 n$

For this recurrence relation we have $a = 7$, $b = 8$, $n^{\log_b a} = n^{\log_8 7} \doteq n^{0.936}$, and $h(n) = n \log_2 n$. So $h \in \Omega(n^{\log_8 7 + \epsilon})$, where $\epsilon \doteq 0.064 > 0$. Further, for all sufficiently large n, $a h(n/b) = 7(n/8) \log_2(n/8) = (7/8)n[\log_2 n - \log_2 8] \leq (7/8)n \log_2 n = c h(n)$, for $0 < c = 7/8 < 1$. Thus, h satisfies the hypotheses for case (iii) and we have $f \in \Theta(n \log_2 n)$.

Use the Master Theorem to determine the complexity of f in each of the following, where $f(1) = 1$:

a) $f(n) = 9 f(n/3) + n$ **b)** $f(n) = 2 f(n/2) + 1$

c) $f(n) = f(2n/3) + 1$ **d)** $f(n) = 2 f(n/3) + n$

e) $f(n) = 4 f(n/2) + n^2$

10.7
Summary and Historical Review

In this chapter the recurrence relation has emerged as another tool for solving combinatorial problems. In these problems we analyze a given situation and then express the result a_n in terms of the results for certain smaller nonnegative integers. Once the recurrence relation is determined, we can solve for any value of a_n (within reason). When we have access to a computer, such relations are particularly valuable, especially if they cannot be solved explicitly.

The study of recurrence relations can be traced back to the Fibonacci relation $F_{n+2} = F_{n+1} + F_n$, $n \geq 0$, $F_0 = 0$, $F_1 = 1$, which was given by Leonardo of Pisa (c. 1175–1250) in 1202. In his *Liber Abaci*, he deals with a problem concerning the number of pairs of rabbits that result in one year if one starts with a single pair that produces another pair at the end of each month. Each new pair starts to breed similarly one month after its birth, and we assume that no rabbits die during the given year. Hence, at the end of the first month there are two pairs of rabbits; three pairs after two months; five pairs after three months; and so on. [As mentioned in the summary of Chapter 9, Abraham DeMoivre (1667–1754) obtained this result by the method of generating functions in 1718.] This same sequence appears in the work of the German mathematician Johannes Kepler (1571–1630), who used it in his studies on how the leaves of a plant or flower are arranged about its stem. In 1844 the French mathematician Gabriel Lamé (1795–1870) used the sequence in his analysis of the efficiency of the Euclidean algorithm. Later, François Èdouard Anatole Lucas (1842–1891), who popularized the Towers of Hanoi puzzle, derived many properties of this sequence and was the first to call these numbers the Fibonacci sequence.

Leonardo Fibonacci (c. 1175–1250)
Reproduced courtesy of The Granger Collection, New York

For an elementary coverage of examples and properties for the Fibonacci numbers one should examine the book by T. H. Garland [10]. Even more can be learned from the texts by V. E. Hoggatt, Jr. [14] and S. Vajda [29]. The UMAP article by R.V. Jean [16] gives many applications of this sequence. Chapter 8 of the mathematical exposition by R. Honsberger [15] provides an interesting account of the Fibonacci numbers and of the related sequence called the Lucas numbers. The text by R. L. Graham, D. E. Knuth, and O. Patashnik [12] also includes many interesting examples and properties of both the Fibonacci numbers and the Catalan numbers. More counterexamples for the Fibonacci and Catalan numbers, like those found in Examples 10.17 and 10.44, respectively, can be found in the article by R. K. Guy [13]. Additional material on the role of the golden ratio in such areas as geometry, probability, and fractals is given in the book by H. Walser [30]. The book by T. Koshy [19] provides a definitive history and extensive analysis of the Fibonacci and Lucas numbers, together with a wide variety of applications, examples, and exercises.

Comparable coverage of the material presented in this chapter can be found in Chapter 3 of C. L. Liu [21]. For more on the theoretical development of linear recurrence relations with constant coefficients, examine Chapter 9 of N. Finizio and G. Ladas [8].

Applications in probability theory dealing with recurrent events, random walks, and ruin problems can be found in Chapters XIII and XIV of the classic text by W. Feller [7]. The UMAP module by D. R. Sherbert [24] introduces difference equations and includes an application in economics known as the *Cobweb Theorem*. The text by S. Goldberg [11] has more on applications in the social sciences.

Recursive techniques in the generation of permutations and combinations are developed in Chapter 4 of R. A. Brualdi [3]. The algorithm presented in Section 10.1 for the permutations of $\{1, 2, 3, \ldots, n\}$ first appeared in the work of H. D. Steinhaus [27] and is often referred to as the *adjacent mark ordering algorithm*. This result was rediscovered later, independently by H. F. Trotter [28] and S. M. Johnson [17]. Efficient sorting methods for permutations and other combinatorial structures are analyzed in the text by D. E. Knuth [18]. The work of E. M. Reingold, J. Nievergelt, and N. Deo [23] also deals with such algorithms.

For those who enjoyed the rooted ordered binary trees in Section 10.5, Chapter 3 of A. V. Aho, J. E. Hopcroft, and J. D. Ullman [1] should prove interesting. The basis for the example on stacks is given on page 86 of the text by S. Even [6]. The article by M. Gardner [9] provides other examples where the Catalan numbers arise. Computational considerations in determining Catalan numbers are examined in the article by D. M. Campbell [4]. Much more about the Catalan numbers can be found in the text by R. P. Stanley [26] — in particular, 66 situations, where these numbers arise, are provided on pp. 219–229.

Finally, the coverage on divide-and-conquer algorithms in Section 10.6 is modeled after D. F. Stanat and D. F. McAllister's presentation in Section 5.3 of [25]. Chapter 10 of the text by A. V. Aho, J. E. Hopcroft, and J. D. Ullman [1] provides some further information on this topic. An application of this method in a matrix multiplication algorithm appears in Chapter 10 of the text by C. L. Liu [20]. Additional coverage and a proof for the Master Theorem are given in Chapter 4 of the text by T. H. Cormen, C. E. Leiserson, R. L. Rivest, and C. Stein [5].

REFERENCES

1. Aho, Alfred V., Hopcroft, John E., and Ullman, Jeffery D. *Data Structures and Algorithms*. Reading, Mass.: Addison-Wesley, 1983.
2. Auluck, F. C. "On Some New Types of Partitions Associated with Generalized Ferrers Graphs." *Proceedings of the Cambridge Philosophical Society* 47 (1951): pp. 679–685.
3. Brualdi, Richard A. *Introductory Combinatorics*, 3rd ed. Upper Saddle River, N.J.: Prentice-Hall, 1999.
4. Campbell, Douglas M. "The Computation of Catalan Numbers." *Mathematics Magazine* 57, no. 4 (September 1984): pp. 195–208.
5. Cormen, Thomas·H., Leiserson, Charles E., Rivest, Ronald L., and Stein, Clifford. *Introduction to Algorithms*, 2nd ed. Boston, Mass.: McGraw-Hill, 2001.
6. Even, Shimon. *Graph Algorithms*. Rockville, Md.: Computer Science Press, 1979.
7. Feller, William. *An Introduction to Probability Theory and Its Applications*, Vol. I, 3rd ed. New York: Wiley, 1968.
8. Finizio, N., and Ladas, G. *An Introduction to Differential Equations*. Belmont, Calif.: Wadsworth Publishing Company, 1982.
9. Gardner, Martin. "Mathematical Games, Catalan Numbers: An Integer Sequence that Materializes in Unexpected Places." *Scientific American* 234, no. 6 (June 1976): pp. 120–125.
10. Garland, Trudi Hammel. *Fascinating Fibonaccis*. Palo Alto, Calif.: Dale Seymour Publications, 1987.
11. Goldberg, Samuel. *Introduction to Difference Equations*. New York: Wiley, 1958.
12. Graham, Ronald Lewis, Knuth, Donald Ervin, and Patashnik, Oren. *Concrete Mathematics*, 2nd ed. Reading, Mass.: Addison-Wesley, 1994.
13. Guy, Richard K. "The Second Strong Law of Small Numbers." *Mathematics Magazine* 63, no. 1 (February 1990): pp. 3–20.
14. Hoggatt, Verner E., Jr. *Fibonacci and Lucas Numbers*. Boston, Mass.; Houghton Mifflin, 1969.
15. Honsberger, Ross. *Mathematical Gems III* (The Dolciani Mathematical Expositions, Number Nine). Washington, D.C.: The Mathematical Association of America, 1985.
16. Jean, Roger V. "The Fibonacci Sequence." *The UMAP Journal* 5, no. 1 (1984): pp. 23–47.
17. Johnson, Selmer M. "Generation of Permutations by Adjacent Transposition." *Mathematics of Computation* 17 (1963): pp. 282–285.
18. Knuth, Donald E. *The Art of Computer Programming/Volume 3 Sorting and Searching*. Reading, Mass: Addison-Wesley, 1973.
19. Koshy, Thomas. *Fibonacci and Lucas Numbers with Applications*. New York: Wiley, 2001.
20. Liu, C. L. *Elements of Discrete Mathematics*, 2nd ed. New York: McGraw-Hill, 1985.
21. Liu, C. L. *Introduction to Combinatorial Mathematics*. New York: McGraw-Hill, 1968.
22. Miksa, F. L., Moser, L., and Wyman, M. "Restricted Partitions of Finite Sets." *Canadian Mathematics Bulletin* 1 (1958): pp. 87–96.

23. Reingold, E. M., Nievergelt, J., and Deo, N. *Combinatorial Algorithms: Theory and Practice.* Englewood Cliffs, N.J.: Prentice-Hall, 1977.

24. Sherbert, Donald R. *Difference Equations with Applications*, UMAP Module 322. Cambridge, Mass.: Birkhauser Boston, 1980.

25. Stanat, Donald F., and McAllister, David F. *Discrete Mathematics in Computer Science*. Englewood Cliffs, N.J.: Prentice-Hall, 1977.

26. Stanley, Richard P. *Enumerative Combinatorics*, Vol. 2. New York: Cambridge University Press, 1999.

27. Steinhaus, Hugo D. *One Hundred Problems in Elementary Mathematics*. New York: Basic Books, 1964.

28. Trotter, H. F. "ACM Algorithm 115 — Permutations." *Communications of the ACM* 5 (1962): pp. 434–435.

29. Vajda, S. *Fibonacci & Lucas Numbers, and the Golden Section*. New York: Halsted Press (a division of John Wiley & Sons), 1989.

30. Walser, Hans. *The Golden Section*. Washington, D.C.: The Mathematical Association of America, 2001.

SUPPLEMENTARY EXERCISES

1. For $n \in \mathbf{Z}^+$ and $n \geq k + 1 \geq 1$, verify algebraically the recursion formula

$$\binom{n}{k+1} = \left(\frac{n-k}{k+1}\right)\binom{n}{k}.$$

2. a) For $n \geq 0$, let B_n denote the number of partitions of $\{1, 2, 3, \ldots, n\}$. Set $B_0 = 1$ for the partitions of $\emptyset$. Verify that for all $n \geq 0$,

$$B_{n+1} = \sum_{i=0}^{n}\binom{n}{n-i}B_i = \sum_{i=0}^{n}\binom{n}{i}B_i.$$

[The numbers $B_i, i \geq 0$, are referred to as the *Bell numbers* after Eric Temple Bell (1883–1960).]

b) How are the Bell numbers related to the Stirling numbers of the second kind?

3. Let $n, k \in \mathbf{Z}^+$, and define $p(n, k)$ to be the number of partitions of n into exactly k (positive-integer) summands. Prove that $p(n, k) = p(n - 1, k - 1) + p(n - k, k)$.

4. For $n \geq 1$, let a_n count the number of ways to write n as an ordered sum of odd positive integers. (For example, $a_4 = 3$ since $4 = 3 + 1 = 1 + 3 = 1 + 1 + 1 + 1$.) Find and solve a recurrence relation for a_n.

5. Let $A = \begin{bmatrix} 1 & 1 \\ 1 & 0 \end{bmatrix}$.

a) Compute A^2, A^3, and A^4.

b) Conjecture a general formula for $A^n, n \in \mathbf{Z}^+$, and establish your conjecture by the Principle of Mathematical Induction.

6. Let $M = \begin{bmatrix} 1 & 1 \\ 1 & 2 \end{bmatrix}$.

a) Compute M^2, M^3, and M^4.

b) Conjecture a general formula for $M^n, n \in \mathbf{Z}^+$, and establish your conjecture by the Principle of Mathematical Induction.

7. Determine the points of intersection of the parabola $y = x^2 - 1$ and the hyperbola $y = 1 + \frac{1}{x}$.

8. Let $\alpha = (1 + \sqrt{5})/2$ and $\beta = (1 - \sqrt{5})/2$.

a) Verify that $\alpha^2 = \alpha + 1$ and $\beta^2 = \beta + 1$.

b) Prove that for all $n \geq 0$, $\sum_{k=0}^{n}\binom{n}{k}F_k = F_{2n}$.

c) Show that $\alpha^3 = 1 + 2\alpha$ and $\beta^3 = 1 + 2\beta$.

d) Prove that for all $n \geq 0$, $\sum_{k=0}^{n}\binom{n}{k}2^k F_k = F_{3n}$.

9. a) For $\alpha = (1 + \sqrt{5})/2$, verify that $\alpha^2 + 1 = 2 + \alpha$ and $(2 + \alpha)^2 = 5\alpha^2$.

b) Show that for $\beta = (1 - \sqrt{5})/2$, $\beta^2 + 1 = 2 + \beta$ and $(2 + \beta)^2 = 5\beta^2$.

c) If $n, m \in \mathbf{N}$ prove that

$$\sum_{k=0}^{2n}\binom{2n}{k}F_{2k+m} = 5^n F_{2n+m}.$$

10. Renu wants to sell her laptop for \$4000. Narmada offers to buy it for \$3000. Renu then splits the difference and asks for \$3500. Narmada likewise splits the difference and makes a new offer of \$3250. (a) If the women continue this process (of asking prices and counteroffers), what will Narmada be willing to pay on her 5th offer? 10th offer? kth offer, $k \geq 1$? (b) If the women continue this process (providing many, many new asking prices and counteroffers), what price will they approach? (c) Suppose that Narmada was willing to buy the laptop for \$3200. What should she have offered to pay Renu the first time?

11. Parts (a) and (b) of Fig. 10.27 provide the Hasse diagrams for two partial orders referred to as the *fences* $\mathscr{F}_5$, $\mathscr{F}_6$ [on 5, 6 (distinct) elements, respectively]. If, for instance, $\mathscr{R}$ denotes the partial order for the fence $\mathscr{F}_5$, then $a_1 \, \mathscr{R} \, a_2$, $a_3 \, \mathscr{R} \, a_2$, $a_3 \, \mathscr{R} \, a_4$, and $a_5 \, \mathscr{R} \, a_4$. For each such fence $\mathscr{F}_n$, $n \geq 1$, we follow the convention that an element with an odd subscript is minimal and one with an even subscript is maximal. Let $(\{1, 2\}, \leq)$ denote the partial order where $\leq$ denotes the usual "less than or equal to" relation. As in Exercise 26 of Section 7.3, a function $f : \mathscr{F}_n \to \{1, 2\}$ is called *order-preserving* when for all $x, y \in \mathscr{F}_n$, $x \, \mathscr{R} \, y \Rightarrow f(x) \leq f(y)$. Let c_n count the number of such order-preserving functions. Find and solve a recurrence relation for c_n.

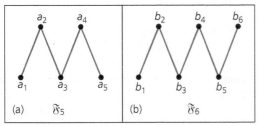

Figure 10.27

12. For $n \geq 0$, let $m = \lfloor (n+1)/2 \rfloor$. Prove that $F_{n+2} = \sum_{k=0}^{m} \binom{n-k+1}{k}$. (You may want to look back at Examples 9.17 and 10.11.)

13. a) For $n \in \mathbf{Z}^+$, determine the number of ways one can tile a $1 \times n$ chessboard using 1×1 white (square) tiles and 1×2 blue (rectangular) tiles.

b) How many of the tilings in part (a) use (i) no blue tiles; (ii) exactly one blue tile; (iii) exactly two blue tiles; (iv) exactly three blue tiles; and (v) exactly k blue tiles, where $0 \leq k \leq \lfloor n/2 \rfloor$?

c) How are the results in parts (a) and (b) related?

14. Let $c = \sqrt{1 + \sqrt{1 + \sqrt{1 + \sqrt{1 + \cdots}}}}$. How is c^2 related to c? What is the value of c?

15. For $n \in \mathbf{Z}^+$, d_n denotes the number of derangements of $\{1, 2, 3, \ldots, n\}$, as discussed in Section 8.3.

a) If $n > 2$, show that d_n satisfies the recurrence relation

$$d_n = (n-1)(d_{n-1} + d_{n-2}), \qquad d_2 = 1, \qquad d_1 = 0.$$

b) How can we define d_0 so that the result in part (a) is valid for $n \geq 2$?

c) Rewrite the result in part (a) as

$$d_n - nd_{n-1} = -[d_{n-1} - (n-1)d_{n-2}].$$

How can $d_n - nd_{n-1}$ be expressed in terms of d_{n-2}, d_{n-3}?

d) Show that $d_n - nd_{n-1} = (-1)^n$.

e) Let $f(x) = \sum_{n=0}^{\infty} (d_n x^n)/n!$. After multiplying both sides of the equation in part (d) by $x^n/n!$ and summing

for $n \geq 2$, verify that $f(x) = (e^{-x})/(1-x)$. Hence

$$d_n = n! \left[1 - \frac{1}{1!} + \frac{1}{2!} - \frac{1}{3!} + \cdots + \frac{(-1)^n}{n!} \right].$$

16. For $n \geq 0$, draw n ovals in the plane so that each oval intersects each of the others in exactly two points and no three ovals are coincident. If a_n denotes the number of regions in the plane that results from these n ovals, find and solve a recurrence relation for a_n.

17. For $n \geq 0$, let us toss a coin $2n$ times.

a) If a_n is the number of sequences of $2n$ tosses where n heads and n tails occur, find a_n in terms of n.

b) Find constants r, s, and t so that $(r + sx)^t = f(x) = \sum_{n=0}^{\infty} a_n x^n$.

c) Let b_n denote the number of sequences of $2n$ tosses where the numbers of heads and tails are equal for the first time only after all $2n$ tosses have been made. (For example, if $n = 3$, then HHHTTT and HHTHTT are counted in b_n, but HTHHTT and HHTTHT are not.)

Define $b_0 = 0$ and show that for all $n \geq 1$,

$$a_n = a_0 b_n + a_1 b_{n-1} + \cdots + a_{n-1} b_1 + a_n b_0.$$

d) Let $g(x) = \sum_{n=0}^{\infty} b_n x^n$. Show that $g(x) = 1 - 1/f(x)$, and then solve for b_n, $n \geq 1$.

18. For $\alpha = (1 + \sqrt{5})/2$ and $\beta = (1 - \sqrt{5})/2$, show that $\sum_{k=0}^{\infty} \beta^k = -\beta = \alpha - 1$ and that $\sum_{k=0}^{\infty} |\beta|^k = \alpha^2$.

19. Let a, b, c be fixed real numbers with $ab = 1$ and let $f : \mathbf{R} \times \mathbf{R} \to \mathbf{R}$ be the binary operation, where $f(x, y) = a + bxy + c(x + y)$. Determine the value(s) of c for which f will be associative.

20. a) For $\alpha = (1 + \sqrt{5})/2$ and $\beta = (1 - \sqrt{5})/2$, verify that $\alpha^2 - \alpha^{-2} = \alpha - \beta = \beta^{-2} - \beta^2$.

b) Prove that $F_{2n} = F_{n+1}^2 - F_{n-1}^2$, $n \geq 1$.

c) For $n \geq 1$, let T be an isosceles trapezoid with bases of length F_{n-1} and F_{n+1}, and sides of length F_n. Prove that the area of T is $(\sqrt{3}/4) F_{2n}$. [Note that, when $n = 1$, the trapezoid degenerates into a triangle. However, the formula is still correct.]

21. Let $\mathscr{S}$ be the sample space for an experiment $\mathscr{E}$. If A, B are events from $\mathscr{S}$ with $A \cup B = \mathscr{S}$, $A \cap B = \emptyset$, $Pr(A) = p$, and $Pr(B) = p^2$, determine p.

22. De'Jzaun and Sandra toss a loaded coin, where $Pr(H) = p > 0$. The first to obtain a head is the winner. Sandra goes first but, if she tosses a tail, then De'Jzaun gets two chances. If he tosses two tails, then Sandra again tosses the coin and, if her toss is a tail, then De'Jzaun again goes twice (if his first toss is a tail). This continues until someone tosses a head. What value of p makes this a fair game (that is, a game where both Sandra and De'Jzaun have probability $\frac{1}{2}$ of winning)?

23. For $n \geq 1$, let a_n count the number of binary strings of length n, where there is no run of 1's of odd length. Consequently, when $n = 6$, for instance, we want to include the strings 110000 (which has a run of two 1's and a run of four 0's) and 011110 (which has two runs of one 0 and one run of four 1's), but we do not include either 100011 (which starts with a run of one 1) or 110111 (which ends with a run of three 1's). Find and solve a recurrence relation for a_n.

24. Let a, b be fixed nonzero real numbers. Determine x_n if $x_n = x_{n-1} x_{n-2}, n \geq 2, x_0 = a, x_1 = b$.

25. a) Evaluate $F_{n+1}^2 - F_n F_{n+1} - F_n^2$ for $n = 0, 1, 2, 3$.

b) From the results in part (a), conjecture a formula for $F_{n+1}^2 - F_n F_{n+1} - F_n^2$ for $n \in \mathbf{N}$.

c) Establish the conjecture in part (b) using the Principle of Mathematical Induction.

26. Let $n \in \mathbf{Z}^+$. On a $1 \times n$ chessboard two kings are called *nontaking*, if they do not occupy adjacent squares. In how many ways can one place 0 or more nontaking kings on a $1 \times n$ chessboard?

27. a) For $1 \leq i \leq 6$, determine the rook polynomial $r(C_i, x)$ for the chessboard C_i shown in Fig. 10.28.

b) For each rook polynomial in part (a), find the sum of the coefficients of the powers of x — that is, determine $r(C_i, 1)$ for $1 \leq i \leq 6$.

28. (Gambler's Ruin) When Cathy and Jill play checkers, each has probability $\frac{1}{2}$ of winning. There is never a tie, and the games are independent in the sense that no matter how many games the girls have played, each girl still has probability $\frac{1}{2}$ of winning the next game. After each game the loser gives the winner a quarter. If Cathy has $2.00 to play with and Jill has $2.50 and

they play until one of them is broke, what is the probability that Cathy gets wiped out?

29. For $n, m \in \mathbf{Z}^+$, let $f(n, m)$ count the number of partitions of n where the summands form a nonincreasing sequence of positive integers and no summand exceeds m. With $n = 4$ and $m = 2$, for example, we find that $f(4, 2) = 3$ because here we are concerned with the three partitions

$$4 = 2 + 2, \qquad 4 = 2 + 1 + 1, \qquad 4 = 1 + 1 + 1 + 1.$$

a) Verify that for all $n, m \in \mathbf{Z}^+$,

$$f(n, m) = f(n - m, m) + f(n, m - 1).$$

b) Write a computer program (or develop an algorithm) to compute $f(n, m)$ for $n, m \in \mathbf{Z}^+$.

c) Write a computer program (or develop an algorithm) to compute $p(n)$, the number of partitions of a given positive integer n.

30. Let A, B be sets with $|A| = m \geq n = |B|$, and let $a(m, n)$ count the number of *onto* functions from A to B. Show that

$$a(m, 1) = 1$$

$$a(m, n) = n^m - \sum_{i=1}^{n-1} \binom{n}{i} a(m, i), \qquad \text{when } m \geq n > 1.$$

31. When one examines the units digit of each Fibonacci number F_n, $n \geq 0$, one finds that these digits form a sequence that repeats after 60 terms. [This was first proved by Joseph-Louis Lagrange (1736–1813).] Write a computer program (or develop an algorithm) to calculate this sequence of 60 digits.

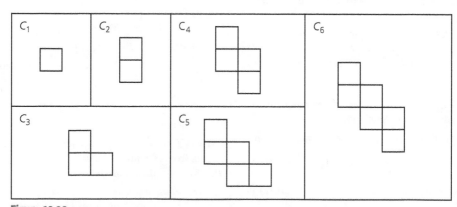

Figure 10.28

GRAPH THEORY AND APPLICATIONS

11

An Introduction to Graph Theory

With this chapter we start to develop another major topic of this text. Unlike other areas in mathematics, the theory of graphs has a definite starting place, a paper published in 1736 by the Swiss mathematician Leonhard Euler (1707–1783). The main idea behind this work grew out of a now-popular problem known as the seven bridges of Königsberg. We shall examine the solution of this problem, from which Euler developed some of the fundamental concepts for the theory of graphs.

Unlike the continuous graphs of early algebra courses, the graphs we examine here are finite in structure and can be used to analyze relationships and applications in many different settings. We have seen some examples of applications of graph theory in earlier chapters (3, 5–8, and 10). However, the development here is independent of these prior discussions.

11.1
Definitions and Examples

When we use a road map, we are often concerned with seeing how to get from one town to another by means of the roads indicated on the map. Consequently, we are dealing with two distinct sets of objects: towns and roads. As we have seen many times before, such sets of objects can be used to define a relation. If V denotes the set of towns and E the set of roads, we can define a relation $\mathcal{R}$ on V by $a \, \mathcal{R} \, b$ if we can travel from a to b using only the roads in E. If the roads in E that take us from a to b are all two-way roads, then we also have $b \, \mathcal{R} \, a$. Should all the roads under consideration be two-way, we have a symmetric relation.

One way to represent a relation is by listing the ordered pairs that are its elements. Here, however, it is more convenient to use a picture, as shown in Fig. 11.1. This figure demonstrates the possible ways of traveling among six towns using the eight roads indicated. It shows that there is at least one set of roads connecting any two towns (identical or distinct). This pictorial representation is a lot easier to work with than the 36 ordered pairs of the relation $\mathcal{R}$.

At the same time, Fig. 11.1 would be appropriate for representing six communication centers, with the eight "roads" interpreted as communication links. If each link provides two-way communication, we should be quite concerned about the vulnerability of center a to such hazards as equipment breakdown or enemy attack. Without center a, neither b nor c can communicate with any of d, e, or f.

From these observations we consider the following concepts.

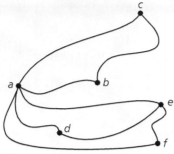

Figure 11.1

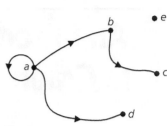

Figure 11.2

Definition 11.1

Let V be a finite nonempty set, and let $E \subseteq V \times V$. The pair (V, E) is then called a *directed graph* (on V), or *digraph*[†] (on V), where V is the set of *vertices*, or *nodes*, and E is its set of *(directed) edges* or *arcs*. We write $G = (V, E)$ to denote such a graph.

When there is no concern about the direction of any edge, we still write $G = (V, E)$. But now E is a set of unordered pairs of elements taken from V, and G is called an *undirected graph*.

Whether $G = (V, E)$ is directed or undirected, we often call V the *vertex set* of G and E the *edge set* of G.

Figure 11.2 provides an example of a directed graph on $V = \{a, b, c, d, e\}$ with $E = \{(a, a), (a, b), (a, d), (b, c)\}$. The direction of an edge is indicated by placing a directed arrow on the edge, as shown here. For any edge, such as (b, c), we say that the edge is *incident* with the vertices b, c; b is said to be *adjacent to c*, whereas c is *adjacent from b*. In addition, vertex b is called the *origin*, or *source*, of the edge (b, c), and vertex c is the *terminus*, or *terminating vertex*. The edge (a, a) is an example of a *loop*, and the vertex e that has no incident edges is called an *isolated* vertex.

An undirected graph is shown in Fig. 11.3(a). This graph is a more compact way of describing the directed graph given in Fig. 11.3(b). In an undirected graph, there are undirected edges such as $\{a, b\}, \{b, c\}, \{a, c\}, \{c, d\}$ in Fig. 11.3(a). An edge such as $\{a, b\}$ stands for $\{(a, b), (b, a)\}$. Although $(a, b) = (b, a)$ only when $a = b$, we do have $\{a, b\} = \{b, a\}$

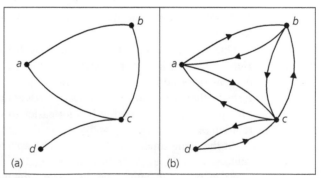

(a)   (b)

Figure 11.3

[†] Since the terminology of graph theory is not standard, the reader may find some differences between terms used here and in other texts.

for any a, b. We can write $\{a, a\}$ to denote a loop in an undirected graph, but $\{a, a\}$ is considered the same as (a, a).

In general, if a graph G is not specified as directed or undirected, it is assumed to be undirected. When it contains no loops it is called *loop-free*.

In the next two definitions we shall not concern ourselves with any loops that may be present in the undirected graph G.

Definition 11.2

Let x, y be (not necessarily distinct) vertices in an undirected graph $G = (V, E)$. An x-y *walk* in G is a (loop-free) finite alternating sequence

$$x = x_0, e_1, x_1, e_2, x_2, e_3, \ldots, e_{n-1}, x_{n-1}, e_n, x_n = y$$

of vertices and edges from G, starting at vertex x and ending at vertex y and involving the n edges $e_i = \{x_{i-1}, x_i\}$, where $1 \le i \le n$.

The *length* of this walk is n, the number of edges in the walk. (When $n = 0$, there are no edges, $x = y$, and the walk is called *trivial*. These walks are not considered very much in our work.)

Any x-y walk where $x = y$ (and $n > 1$) is called a *closed walk*. Otherwise the walk is called *open*.

Note that a walk may repeat both vertices and edges.

EXAMPLE 11.1

For the graph in Fig. 11.4 we find, for example, the following three open walks. We can list the edges only or the vertices only (if the other is clearly implied).

1) $\{a, b\}, \{b, d\}, \{d, c\}, \{c, e\}, \{e, d\}, \{d, b\}$: This is an a-b walk of length 6 in which we find the vertices d and b repeated, as well as the edge $\{b, d\} (= \{d, b\})$.

2) $b \to c \to d \to e \to c \to f$: Here we have a b-f walk where the length is 5 and the vertex c is repeated, but no edge appears more than once.

3) $\{f, c\}, \{c, e\}, \{e, d\}, \{d, a\}$: In this case the given f-a walk has length 4 with no repetition of either vertices or edges.

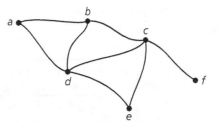

Figure 11.4

Since the graph of Fig. 11.4 is undirected, the a-b walk in part (1) is also a b-a walk (we read the edges, if necessary, as $\{b, d\}, \{d, e\}, \{e, c\}, \{c, d\}, \{d, b\}$, and $\{b, a\}$). Similar remarks hold for the walks in parts (2) and (3).

Finally, the edges $\{b, c\}, \{c, d\}$, and $\{d, b\}$ provide a b-b (closed) walk. These edges (ordered appropriately) also define (closed) c-c and d-d walks.

Now let us examine special types of walks.

Definition 11.3   Consider any x-y walk in an undirected graph $G = (V, E)$.

  a) If no edge in the x-y walk is repeated, then the walk is called an x-y *trail*. A closed x-x trail is called a *circuit*.

  b) If no vertex of the x-y walk occurs more than once, then the walk is called an x-y *path*. When $x = y$, the term *cycle* is used to describe such a closed path.

Convention: In dealing with circuits, we shall always understand the presence of at least one edge. When there is only one edge, then the circuit is a loop (and the graph is no longer loop-free). Circuits with two edges arise in multigraphs, a concept we shall define shortly.

The term *cycle* will always imply the presence of at least three distinct edges (from the graph).

EXAMPLE 11.2   a) The b-f walk in part (2) of Example 11.1 is a b-f trail, but it is not a b-f path because of the repetition of vertex c. However, the f-a walk in part (3) of that example is both an f-a trail (of length 4) and an f-a path (of length 4).

  b) In Fig. 11.4, the edges $\{a, b\}$, $\{b, d\}$, $\{d, c\}$, $\{c, e\}$, $\{e, d\}$, and $\{d, a\}$ provide an a-a circuit. The vertex d is repeated, so the edges do *not* give us an a-a cycle.

  c) The edges $\{a, b\}$, $\{b, c\}$, $\{c, d\}$, and $\{d, a\}$ provide an a-a cycle (of length 4) in Fig. 11.4. When ordered appropriately these same edges may also define a b-b, c-c, or d-d cycle. Each of these cycles is also a circuit.

For a directed graph we shall use the adjective *directed*, as in, for example, *directed walks, directed paths*, and *directed cycles*.

Before continuing, we summarize (in Table 11.1) for future reference the results of Definitions 11.2 and 11.3. Each occurrence of "Yes" in the first two columns here should be interpreted as "Yes, possibly." Table 11.1 reflects the fact that a path is a trail, which in turn is an open walk. Furthermore, every cycle is a circuit, and every circuit (with at least two edges) is a closed walk.

Table 11.1

Repeated Vertex (Vertices)	Repeated Edge(s)	Open	Closed	Name
Yes	Yes	Yes		Walk (open)
Yes	Yes		Yes	Walk (closed)
Yes	No	Yes		Trail
Yes	No		Yes	Circuit
No	No	Yes		Path
No	No		Yes	Cycle

Considering how many concepts we have introduced, it is time to prove a first result in this new theory.

THEOREM 11.1   Let $G = (V, E)$ be an undirected graph, with $a, b \in V$, $a \neq b$. If there exists a trail (in G) from a to b, then there is a path (in G) from a to b.

Proof: Since there is a trail from a to b, we select one of shortest length, say $\{a, x_1\}$, $\{x_1, x_2\}, \ldots, \{x_n, b\}$. If this trail is not a path, we have the situation $\{a, x_1\}, \{x_1, x_2\}, \ldots, \{x_{k-1}, x_k\}, \{x_k, x_{k+1}\}, \{x_{k+1}, x_{k+2}\}, \ldots, \{x_{m-1}, x_m\}, \{x_m, x_{m+1}\}, \ldots, \{x_n, b\}$, where $k < m$ and $x_k = x_m$, possibly with $k = 0$ and $a \ (= x_0) = x_m$, or $m = n + 1$ and $x_k = b \ (= x_{n+1})$. But then we have a contradiction because $\{a, x_1\}, \{x_1, x_2\}, \ldots, \{x_{k-1}, x_k\}, \{x_m, x_{m+1}\}, \ldots, \{x_n, b\}$ is a shorter trail from a to b.

The notion of a path is needed in the following graph property.

Definition 11.4   Let $G = (V, E)$ be an undirected graph. We call G *connected* if there is a path between any two distinct vertices of G.

Let $G = (V, E)$ be a directed graph. Its associated undirected graph is the graph obtained from G by ignoring the directions on the edges. If more than one undirected edge results for a pair of distinct vertices in G, then only one of these edges is drawn in the associated undirected graph. When this associated graph is connected, we consider G connected.

A graph that is not connected is called *disconnected*.

The graphs in Figs. 11.1, 11.3, and 11.4 are connected. In Fig. 11.2 the graph is not connected because, for example, there is no path from a to e.

EXAMPLE 11.3   In Fig. 11.5 we have an undirected graph on $V = \{a, b, c, d, e, f, g\}$. This graph is not connected because, for example, there is no path from a to e. However, the graph is composed of pieces (with vertex sets $V_1 = \{a, b, c, d\}$, $V_2 = \{e, f, g\}$, and edge sets $E_1 = \{\{a, b\}, \{a, c\}, \{a, d\}, \{b, d\}\}$, $E_2 = \{\{e, f\}, \{f, g\}\}$) that are themselves connected, and these pieces are called the (*connected*) *components* of the graph. Hence an undirected graph $G = (V, E)$ is disconnected if and only if V can be partitioned into at least two subsets V_1, V_2 such that there is no edge in E of the form $\{x, y\}$, where $x \in V_1$ and $y \in V_2$. A graph is connected if and only if it has only one component.

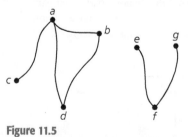

Figure 11.5

Definition 11.5   For any graph $G = (V, E)$, the number of components of G is denoted by $\kappa(G)$.

EXAMPLE 11.4   For the graphs in Figs. 11.1, 11.3, and 11.4, $\kappa(G) = 1$ because these graphs are connected; $\kappa(G) = 2$ for the graphs in Figs. 11.2 and 11.5.

Before closing this first section, we extend our concept of a graph. Thus far we have allowed at most one edge between two vertices; we now consider an extension.

Definition 11.6

Let V be a finite nonempty set. We say that the pair (V, E) determines a *multigraph* G with vertex set V and edge set $E^\dagger$ if, for some $x, y \in V$, there are two or more edges in E of the form (a) (x, y) (for a directed multigraph), or (b) $\{x, y\}$ (for an undirected multigraph). In either case, we write $G = (V, E)$ to designate the multigraph, just as we did for graphs.

Figure 11.6 shows an example of a directed multigraph. There are three edges from a to b, so we say that the edge (a, b) has *multiplicity* 3. The edges (b, c) and (d, e) both have multiplicity 2. Also, the edge (e, d) and either one of the edges (d, e) form a (directed) circuit of length 2 in the multigraph.

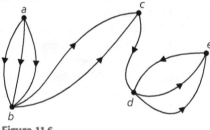

Figure 11.6

We shall need the idea of a multigraph later in the chapter when we solve the problem of the seven bridges of Königsberg. (*Note*: Whenever we are dealing with a multigraph G, we shall state explicitly that G is a multigraph.)

EXERCISES 11.1

1. List three situations, different from those in this section, where a graph could prove useful.

2. For the graph in Fig. 11.7, determine (a) a walk from b to d that is not a trail; (b) a b-d trail that is not a path; (c) a path from b to d; (d) a closed walk from b to b that is not a circuit; (e) a circuit from b to b that is not a cycle; and (f) a cycle from b to b.

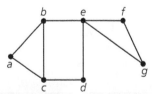

Figure 11.7

3. For the graph in Fig. 11.7, how many paths are there from b to f?

4. For $n \geq 2$, let $G = (V, E)$ be the loop-free undirected graph, where V is the set of binary n-tuples (of 0's and 1's) and $E = \{\{v, w\} | v, w \in V$ and v, w differ in (exactly) two positions$\}$. Find $\kappa(G)$.

5. Let $G = (V, E)$ be the undirected graph in Fig. 11.8. How many paths are there in G from a to h? How many of these paths have length 5?

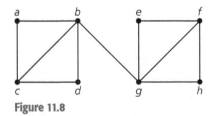

Figure 11.8

6. If a, b are distinct vertices in a connected undirected graph G, the *distance* from a to b is defined to be the length of a shortest path from a to b (when $a = b$ the *distance* is defined to be

---

$\dagger$We now allow a set to have repeated elements in order to account for multiple edges. We realize that this is a change from the way we dealt with sets in Chapter 3. To overcome this the term *multiset* is often used to describe E in this case.

0). For the graph in Fig. 11.9, find the distances from d to (each of) the other vertices in G.

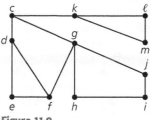

Figure 11.9

7. Seven towns a, b, c, d, e, f, and g are connected by a system of highways as follows: (1) I-22 goes from a to c, passing through b; (2) I-33 goes from c to d and then passes through b as it continues to f; (3) I-44 goes from d through e to a; (4) I-55 goes from f to b, passing through g; and (5) I-66 goes from g to d.

a) Using vertices for towns and directed edges for segments of highways between towns, draw a directed graph that models this situation.

b) List the paths from g to a.

c) What is the smallest number of highway segments that would have to be closed down in order for travel from b to d to be disrupted?

d) Is it possible to leave town c and return there, visiting each of the other towns only once?

e) What is the answer to part (d) if we are not required to return to c?

f) Is it possible to start at some town and drive over each of these highways exactly once? (You are allowed to visit a town more than once, and you need not return to the town from which you started.)

8. Figure 11.10 shows an undirected graph representing a section of a department store. The vertices indicate where cashiers are located; the edges denote unblocked aisles between cashiers. The department store wants to set up a security system where (plainclothes) guards are placed at certain cashier locations so that each cashier either has a guard at his or her location or is only one aisle away from a cashier who has a guard. What is the smallest number of guards needed?

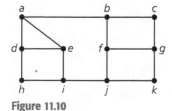

Figure 11.10

9. Let $G = (V, E)$ be a loop-free connected undirected graph, and let $\{a, b\}$ be an edge of G. Prove that $\{a, b\}$ is part of a cycle

if and only if its removal (the vertices a and b are left) does not disconnect G.

10. Give an example of a connected graph G where removing any edge of G results in a disconnected graph.

11. Let G be a graph that satisfies the condition in Exercise 10. (a) Must G be loop-free? (b) Could G be a multigraph? (c) If G has n vertices, can we determine how many edges it has?

12. a) If $G = (V, E)$ is an undirected graph with $|V| = v$, $|E| = e$, and no loops, prove that $2e \leq v^2 - v$.

b) State the corresponding inequality for the case when G is directed.

13. Let $G = (V, E)$ be an undirected graph. Define a relation $\mathcal{R}$ on V by $a \mathcal{R} b$ if $a = b$ or if there is a path in G from a to b. Prove that $\mathcal{R}$ is an equivalence relation. Describe the partition of V induced by $\mathcal{R}$.

14. a) Consider the three connected undirected graphs in Fig. 11.11. The graph in part (a) of the figure consists of a cycle (on the vertices u_1, u_2, u_3) and a vertex u_4 with edges (spokes) drawn from u_4 to the other three vertices. This graph is called the *wheel with three spokes* and is denoted by W_3. In part (b) of the figure we find the graph

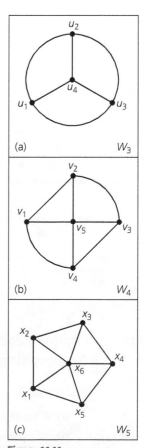

Figure 11.11

W_4 —the wheel with four spokes. The wheel W_5 with five spokes appears in Fig. 11.11(c). Determine how many cycles of length 4 there are in each of these graphs.

b) In general, if $n \in \mathbf{Z}^+$ and $n \geq 3$, then the *wheel with n spokes* is the graph made up of a cycle of length n together with an additional vertex that is adjacent to the n vertices of the cycle. The graph is denoted by W_n. (i) How many cycles of length 4 are there in W_n? (ii) How many cycles in W_n have length n?

15. For the undirected graph in Fig. 11.12, find and solve a recurrence relation for the number of closed v-v walks of length $n \geq 1$, if we allow such a walk, in this case, to contain or consist of one or more loops.

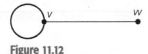

Figure 11.12

16. *Unit-Interval Graphs.* For $n \geq 1$, we start with n closed intervals of unit length and draw the corresponding unit-interval graph on n vertices, as shown in Fig. 11.13. In part (a) of the figure we have one unit interval. This corresponds to the single vertex u; both the interval and the unit-interval graph can be represented by the binary sequence 01. In parts (b), (c) of the figure we have the two unit-interval graphs determined by two unit intervals. When two unit intervals overlap [as in part (c)] an edge is drawn in the unit-interval graph joining the vertices corresponding to these unit intervals. Hence the unit-interval graph in part (b) consists of the two isolated vertices v_1, v_2 that correspond with the nonoverlapping unit intervals. In part (c) the unit intervals overlap so the corresponding unit-interval graph consists of a single edge joining the vertices v_1, v_2 (that correspond to the given unit intervals). A closer look at the unit intervals in part (c) reveals how we can represent the positioning of these intervals and the corresponding unit-interval graph by the binary sequence 0011. In parts (d)–(f) of the figure we have three of the unit-interval graphs for three unit intervals — together with their corresponding binary sequences.

a) How many other unit-interval graphs are there for three unit intervals? What are the corresponding binary sequences for these graphs?

b) How many unit-interval graphs are there for four unit intervals?

c) For $n \geq 1$, how many unit-interval graphs are there for n unit intervals?

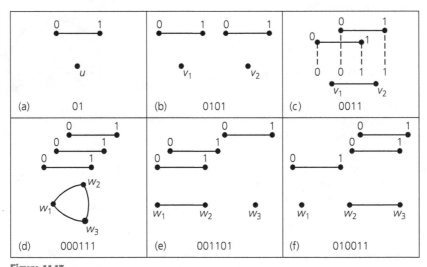

Figure 11.13

11.2
Subgraphs, Complements, and Graph Isomorphism

In this section we shall focus on the following two ideas:

a) What types of substructures are present in a graph?

b) Is it possible to draw two graphs that appear distinct but have the same underlying structure?

To answer the question in part (a) we introduce the following definition.

Definition 11.7   If $G = (V, E)$ is a graph (directed or undirected), then $G_1 = (V_1, E_1)$ is called a *subgraph* of G if $\emptyset \neq V_1 \subseteq V$ and $E_1 \subseteq E$, where each edge in E_1 is incident with vertices in V_1.

Figure 11.14(a) provides us with an undirected graph G and two of its subgraphs, G_1 and G_2. The vertices a, b are isolated in subgraph G_1. Part (b) of the figure provides a directed example. Here vertex w is isolated in the subgraph G'.

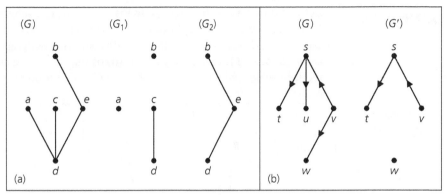

Figure 11.14

Certain special types of subgraphs arise as follows:

Definition 11.8   Given a (directed or undirected) graph $G = (V, E)$, let $G_1 = (V_1, E_1)$ be a subgraph of G. If $V_1 = V$, then G_1 is called a *spanning subgraph* of G.

In part (a) of Fig. 11.14 neither G_1 nor G_2 is a spanning subgraph of G. The subgraphs G_3 and G_4 — shown in part (a) of Fig. 11.15 — are both spanning subgraphs of G. The directed graph G' in part (b) of Fig. 11.14 is a subgraph, but *not* a spanning subgraph, of the directed graph G given in that part of the figure. In part (b) of Fig. 11.15 the directed graphs G'' and G''' are two of the $2^4 = 16$ possible spanning subgraphs.

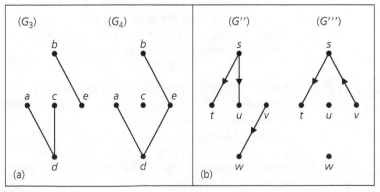

Figure 11.15

Definition 11.9

Let $G = (V, E)$ be a graph (directed or undirected). If $\emptyset \neq U \subseteq V$, the *subgraph of G induced by U* is the subgraph whose vertex set is U and which contains all edges (from G) of either the form (a) (x, y), for $x, y \in U$ (when G is directed), or (b) $\{x, y\}$, for $x, y \in U$ (when G is undirected). We denote this subgraph by $\langle U \rangle$.

A subgraph G' of a graph $G = (V, E)$ is called an *induced subgraph* if there exists $\emptyset \neq U \subseteq V$, where $G' = \langle U \rangle$.

For the subgraphs in Fig. 11.14(a), we find that G_2 is an induced subgraph of G but the subgraph G_1 is not an induced subgraph because edge $\{a, d\}$ is missing.

EXAMPLE 11.5

Let $G = (V, E)$ denote the graph in Fig. 11.16(a). The subgraphs in parts (b) and (c) of the figure are induced subgraphs of G. For the connected subgraph in part (b), $G_1 = \langle U_1 \rangle$ for $U_1 = \{b, c, d, e\}$. In like manner, the disconnected subgraph in part (c) is $G_2 = \langle U_2 \rangle$ for $U_2 = \{a, b, e, f\}$. Finally, G_3 in part (d) of Fig. 11.16 is a subgraph of G. But it is not an induced subgraph; the vertices c, e are in G_3, but the edge $\{c, e\}$ (of G) is not present.

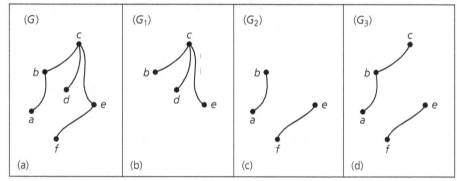

Figure 11.16

Another special type of subgraph comes about when a certain vertex or edge is deleted from the given graph. We formalize these ideas in the following definition.

Definition 11.10

Let v be a vertex in a directed or an undirected graph $G = (V, E)$. The subgraph of G denoted by $G - v$ has the vertex set $V_1 = V - \{v\}$ and the edge set $E_1 \subseteq E$, where E_1 contains all the edges in E except for those that are incident with the vertex v. (Hence $G - v$ is the subgraph of G induced by V_1.)

In a similar way, if e is an edge of a directed or an undirected graph $G = (V, E)$, we obtain the subgraph $G - e = (V_1, E_1)$ of G, where the set of edges $E_1 = E - \{e\}$, and the vertex set is unchanged (that is, $V_1 = V$).

EXAMPLE 11.6

Let $G = (V, E)$ be the undirected graph in Fig. 11.17(a). Part (b) of this figure is the subgraph G_1 (of G), where $G_1 = G - c$. It is also the subgraph of G induced by the set of vertices $U_1 = \{a, b, d, f, g, h\}$, so $G_1 = \langle V - \{c\} \rangle = \langle U_1 \rangle$. In part (c) of Fig. 11.17 we find the subgraph G_2 of G, where $G_2 = G - e$ for e the edge $\{c, d\}$. The result in Fig. 11.17(d) shows how the ideas in Definition 11.10 can be extended to the deletion of more than one vertex (edge). We may represent this subgraph of G as $G_3 = (G - b) - f = (G - f) - b = G - \{b, f\} = \langle U_3 \rangle$, for $U_3 = \{a, c, d, g, h\}$.

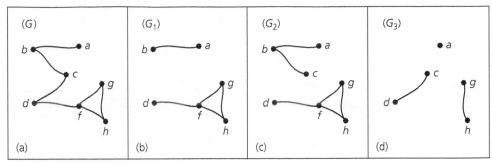

Figure 11.17

The idea of a subgraph gives us a way to develop the complement of an undirected loop-free graph. Before doing so, however, we define a type of graph that is maximal in size for a given number of vertices.

Definition 11.11   Let V be a set of n vertices. The *complete graph* on V, denoted K_n, is a loop-free undirected graph, where for all $a, b \in V$, $a \neq b$, there is an edge $\{a, b\}$.

Figure 11.18 provides the complete graphs K_n, for $1 \leq n \leq 4$. We shall realize, when we examine the idea of graph isomorphism, that these are the only possible complete graphs for the given number of vertices.

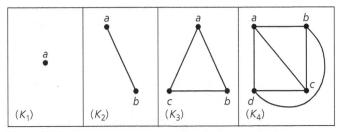

Figure 11.18

In determining the complement of a set in Chapter 3, we needed to know the universal set under consideration. The complete graph plays a role similar to a universal set.

Definition 11.12   Let G be a loop-free undirected graph on n vertices. The *complement of G*, denoted $\overline{G}$, is the subgraph of K_n consisting of the n vertices in G and all edges that are not in G. (If $G = K_n$, $\overline{G}$ is a graph consisting of n vertices and no edges. Such a graph is called a *null* graph.)

Figure 11.19(a) shows an undirected graph on four vertices. Its complement is shown in part (b) of the figure. In the complement, vertex a is isolated.

Once again we have reached a point where many new ideas have been defined. To demonstrate why some of these ideas are important, we apply them now to the solution of an interesting puzzle.

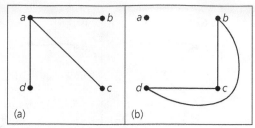

Figure 11.19

| EXAMPLE 11.7 |

Instant Insanity. The game of Instant Insanity is played with four cubes. Each of the six faces on a cube is painted with one of the colors red (R), white (W), blue (B), or yellow (Y). The object of the game is to place the cubes in a column of four such that all four (different) colors appear on each of the four sides of the column.

Consider the cubes in Fig. 11.20 and number them as shown. (These cubes are only one example of this game. Many others exist.) First we shall estimate the number of arrangements that are possible here. If we wish to place cube 1 at the bottom of the column, there are at most three different ways in which we can do this. In Fig. 11.20 cube 1 is unfolded, and we see that it makes no difference whether we place the red face on the table or the opposite white face on the table. We are concerned only with the other four faces at the base of our column. With three pairs of opposite faces there will be at most three ways to place the *first* cube for the base of the column. Now consider cube 2. Although some colors are repeated, no pair of opposite faces has the same color. Hence we have six ways to place the second cube on top of the first. We can then rotate the second cube without changing either the face on the top of the first cube or the face on the bottom of the second cube. With four possible rotations we may place the second cube on top of the first in as many as 24 different ways. Continuing the argument, we find that there can be as many as $(3)(24)(24)(24) = 41,472$ possibilities to consider. And there may not even be a solution!

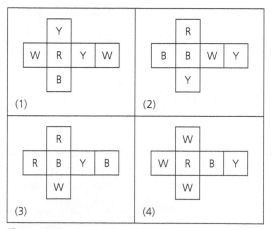

Figure 11.20

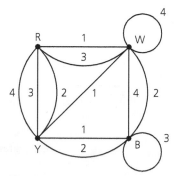

Figure 11.21

In solving this puzzle we realize that it is difficult to keep track of (1) colors on opposite faces of cubes and (2) columns of colors. A graph (actually a labeled multigraph) helps us to visualize the situation. In Fig. 11.21 we have a graph on four vertices R, W, B, and Y. As we consider each cube, we examine its three pairs of opposite faces. For example, cube

1 has a pair of opposite faces painted yellow and blue, so we draw an edge connecting Y and B and label it 1 (for cube 1). The other two edges in the figure that are labeled with 1 account for the pairs of opposite faces that are white and yellow, and red and white. Doing likewise for the other cubes, we arrive at the graph in the figure. A loop, such as the one at B, with label 3, indicates a pair of opposite faces with the same color (for cube 3).

In the graph we see a total of 12 edges falling into four sets of 3, according to the labels for the cubes. At each vertex the number of edges incident to (or from) the vertex counts the number of faces on the four cubes that have that color. (We count a loop twice.) Hence Fig. 11.21 tells us that for our four cubes we have five red faces, seven white ones, six blue ones, and six that are yellow.

With the four cubes stacked in a column, we examine two opposite sides of the column. This arrangement gives us four edges in the graph of Fig. 11.21, where each label appears once. Since each color is to appear only once on a side of the column, each color must appear twice as an endpoint of these four edges. If we can accomplish the same result for the other two sides of the column, we have solved the puzzle. In Fig. 11.22(a) we see that each side in one pair of opposite sides of our column has the four colors if the cubes are arranged according to the information provided by the subgraph shown there. However, to accomplish this for the other two sides of the column also, we need a second such subgraph that doesn't use any edge in part (a). In this case a second such subgraph does exist, as shown in part (b) of the figure.

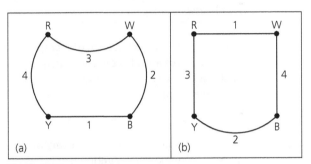

Figure 11.22

Figure 11.23 shows how to arrange the cubes as indicated by the subgraphs in Fig. 11.22.

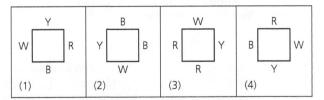

Figure 11.23

In general, for any four cubes we construct a labeled multigraph and try to find two subgraphs where (1) each subgraph contains all four vertices, and four edges, one for each label; (2) in each subgraph, each vertex is incident with exactly two edges (a loop is counted twice); and (3) no (labeled) edge of the labeled multigraph appears in both subgraphs.

Now we turn to the second question posed at the start of the section.

Parts (a) and (b) of Fig. 11.24 show two undirected graphs on four vertices. Since straight edges and curved edges are considered the same here, each graph represents six adjacent pairs of vertices. In fact, we probably feel that these graphs are both examples of the graph K_4. We make this feeling mathematically rigorous in the following definition.

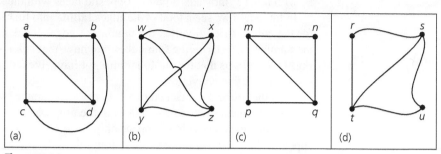

Figure 11.24

Definition 11.13

Let $G_1 = (V_1, E_1)$ and $G_2 = (V_2, E_2)$ be two undirected graphs. A function $f\colon V_1 \to V_2$ is called a *graph isomorphism* if (a) f is one-to-one and onto, and (b) for all $a, b \in V_1$, $\{a, b\} \in E_1$ if and only if $\{f(a), f(b)\} \in E_2$. When such a function exists, G_1 and G_2 are called *isomorphic graphs*.

The vertex correspondence of a graph isomorphism preserves adjacencies. Since which pairs of vertices are adjacent and which are not is the only essential property of an undirected graph, in this way the structure of the graphs is preserved.

For the graphs in parts (a) and (b) of Fig. 11.24 the function f defined by

$$f(a) = w, \qquad f(b) = x, \qquad f(c) = y, \qquad f(d) = z$$

provides an isomorphism. [In fact, any one-to-one correspondence between $\{a, b, c, d\}$ and $\{w, x, y, z\}$ will be an isomorphism because both of the given graphs are complete graphs. This would also be true if each of the given graphs had only four isolated vertices (and no edges).] Consequently, as far as (graph) structure is concerned, these graphs are considered the same — each is (isomorphic to) the complete graph K_4.

For the graphs in parts (c) and (d) of Fig. 11.24 we need to be a little more careful. The function g defined by

$$g(m) = r, \qquad g(n) = s, \qquad g(p) = t, \qquad g(q) = u$$

is one-to-one and onto (for the given vertex sets). However, although $\{m, q\}$ is an edge in the graph of part (c), $\{g(m), g(q)\} = \{r, u\}$ is not an edge in the graph of part (d). Consequently, the function g does *not* define a graph isomorphism. To maintain the correspondence of edges, we consider the one-to-one onto function h where

$$h(m) = s, \qquad h(n) = r, \qquad h(p) = u, \qquad h(q) = t.$$

In this case we have the edge correspondences

$$\{m, n\} \leftrightarrow \{h(m), h(n)\} = \{s, r\}, \qquad \{n, q\} \leftrightarrow \{h(n), h(q)\} = \{r, t\},$$
$$\{m, p\} \leftrightarrow \{h(m), h(p)\} = \{s, u\}, \qquad \{p, q\} \leftrightarrow \{h(p), h(q)\} = \{u, t\},$$
$$\{m, q\} \leftrightarrow \{h(m), h(q)\} = \{s, t\},$$

so h is a graph isomorphism. [We also notice how, for example, the cycle $m \to n \to q \to m$ corresponds with the cycle $s \, (= h(m)) \to r \, (= h(n)) \to t \, (= h(q)) \to s \, (= h(m))$.]

Finally, since the graph in part (a) of Fig. 11.24 has six edges and that in part (c) has only five edges, these two graphs cannot be isomorphic.

Now let us examine the idea of graph isomorphism in a more difficult situation.

EXAMPLE 11.8

In Fig. 11.25 we have two graphs, each on ten vertices. Unlike the graphs in Fig. 11.24, it is not immediately apparent whether or not these graphs are isomorphic.

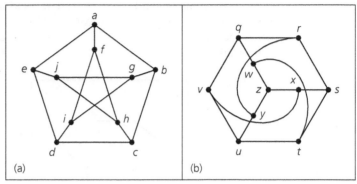

Figure 11.25

One finds that the correspondence given by

$$
\begin{array}{lllll}
a \to q & c \to u & e \to r & g \to x & i \to z \\
b \to v & d \to y & f \to w & h \to t & j \to s
\end{array}
$$

preserves all adjacencies. For example, $\{f, h\}$ is an edge in graph (a) with $\{w, t\}$ the corresponding edge in graph (b). But how did we come up with the correspondence? The following discussion provides some clues.

We note that because an isomorphism preserves adjacencies, it preserves graph substructures such as paths and cycles. In graph (a) the edges $\{a, f\}$, $\{f, i\}$, $\{i, d\}$, $\{d, e\}$, and $\{e, a\}$ constitute a cycle of length 5. Hence we must preserve this as we try to find an isomorphism. One possibility for the corresponding edges in graph (b) is $\{q, w\}$, $\{w, z\}$, $\{z, y\}$, $\{y, r\}$, and $\{r, q\}$, which also provides a cycle of length 5. (A second possible choice is given by the edges in the cycle $y \to r \to s \to t \to u \to y$.) In addition, starting at vertex a in graph (a), we find a path that will "visit" each vertex only once. We express this path by $a \to f \to h \to c \to b \to g \to j \to e \to d \to i$. For the graphs to be isomorphic there must be a corresponding path in graph (b). Here the path described by $q \to w \to t \to u \to v \to x \to s \to r \to y \to z$ is the counterpart.

These are some of the ideas we can use to try to develop an isomorphism and determine whether two graphs are isomorphic. Other considerations will be discussed throughout the chapter. However, there is no simple, foolproof method — especially when we are confronted with larger graphs $G_1 = (V_1, E_1)$ and $G_2 = (V_2, E_2)$, where $|V_1| = |V_2|$ and $|E_1| = |E_2|$.

We close this section with one more example involving graph isomorphism.

EXAMPLE 11.9

Each of the two graphs in Fig. 11.26 has six vertices and nine edges. Therefore it is reasonable to ask whether they are isomorphic.

In graph (a), vertex a is adjacent to two other vertices of the graph. Consequently, if we try to construct an isomorphism between these graphs, we should associate vertex a with a comparable vertex in graph (b), say vertex u. A similar situation exists for vertex d and either vertex x or vertex z. But no matter which of the vertices x or z we use, there remains one vertex in graph (b) that is adjacent to two other vertices. And there is no other such vertex in graph (a) to continue our one-to-one structure preserving correspondence. Consequently, these graphs are not isomorphic.

Furthermore, in graph (b) it is possible to start at any vertex and find a circuit that includes every edge of the graph. For example, if we start at vertex u, the circuit $u \rightarrow w \rightarrow v \rightarrow y \rightarrow w \rightarrow z \rightarrow y \rightarrow x \rightarrow v \rightarrow u$ exhibits this property. This does not happen in graph (a) where the only trails that include each edge start at either b or f and then terminate at f or b, respectively.

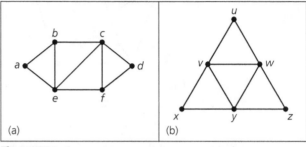

Figure 11.26

EXERCISES 11.2

1. Let G be the undirected graph in Fig. 11.27(a).

a) How many connected subgraphs of G have four vertices and include a cycle?

b) Describe the subgraph G_1 (of G) in part (b) of the figure first, as an induced subgraph and second, in terms of deleting a vertex of G.

c) Describe the subgraph G_2 (of G) in part (c) of the figure first, as an induced subgraph and second, in terms of the deletion of vertices of G.

d) Draw the subgraph of G induced by the set of vertices $U = \{b, c, d, f, i, j\}$.

e) For the graph G, let the edge $e = \{c, f\}$. Draw the subgraph $G - e$.

2. a) Let $G = (V, E)$ be an undirected graph, with $G_1 = (V_1, E_1)$ a subgraph of G. Under what condition(s) is G_1 *not* an induced subgraph of G?

b) For the graph G in Fig. 11.27(a), find a subgraph that is not an induced subgraph.

3. a) How many spanning subgraphs are there for the graph G in Fig. 11.27(a)?

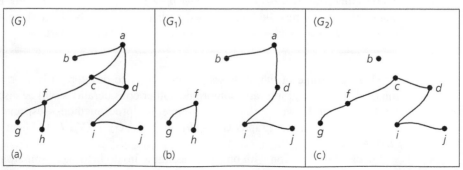

Figure 11.27

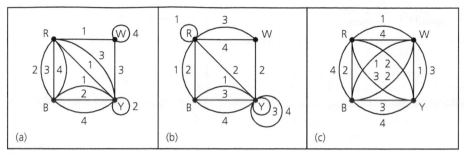

Figure 11.28

b) How many connected spanning subgraphs are there in part (a)?

c) How many of the spanning subgraphs in part (a) have vertex a as an isolated vertex?

4. If $G = (V, E)$ is an undirected graph, how many spanning subgraphs of G are also induced subgraphs?

5. Let $G = (V, E)$ be an undirected graph, where $|V| \geq 2$. If every induced subgraph of G is connected, can we identify the graph G?

6. Find all (loop-free) nonisomorphic undirected graphs with four vertices. How many of these graphs are connected?

7. Each of the labeled multigraphs in Fig. 11.28 arises in the analysis of a set of four blocks for the game of Instant Insanity. In each case determine a solution to the puzzle, if possible.

8. a) How many paths of length 4 are there in the complete graph K_7? (Remember that a path such as $v_1 \rightarrow v_2 \rightarrow v_3 \rightarrow v_4 \rightarrow v_5$ is considered to be the same as the path $v_5 \rightarrow v_4 \rightarrow v_3 \rightarrow v_2 \rightarrow v_1$.)

b) Let $m, n \in \mathbf{Z}^+$ with $m < n$. How many paths of length m are there in the complete graph K_n?

9. For each pair of graphs in Fig. 11.29, determine whether or not the graphs are isomorphic.

10. Let G be an undirected (loop-free) graph with v vertices and e edges. How many edges are there in $\overline{G}$?

11. a) If G_1, G_2 are (loop-free) undirected graphs, prove that G_1, G_2 are isomorphic if and only if $\overline{G}_1, \overline{G}_2$ are isomorphic.

b) Determine whether the graphs in Fig. 11.30 are isomorphic.

12. a) Let G be an undirected graph with n vertices. If G is isomorphic to its own complement $\overline{G}$, how many edges must G have? (Such a graph is called *self-complementary*.)

b) Find an example of a self-complementary graph on four vertices and one on five vertices.

c) If G is a self-complementary graph on n vertices, where $n > 1$, prove that $n = 4k$ or $n = 4k + 1$, for some $k \in \mathbf{Z}^+$.

13. Let G be a cycle on n vertices. Prove that G is self-complementary if and only if $n = 5$.

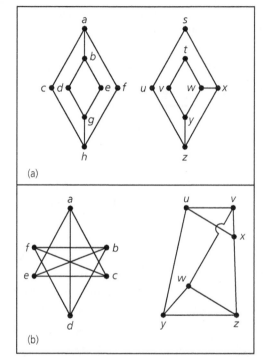

Figure 11.29

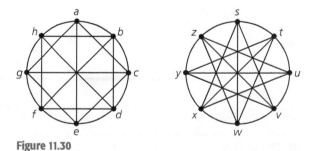

Figure 11.30

14. a) Find a graph G where both G and $\overline{G}$ are connected.

b) If G is a graph on n vertices, for $n \geq 2$, and G is not connected, prove that $\overline{G}$ is connected.

15. a) Extend Definition 11.13 to directed graphs.

b) Determine whether the directed graphs in Fig. 11.31 are isomorphic.

16. a) How many subgraphs $H = (V, E)$ of K_6 satisfy $|V| = 3$? (If two subgraphs are isomorphic but have different vertex sets, consider them distinct.)

b) How many subgraphs $H = (V, E)$ of K_6 satisfy $|V| = 4$?

c) How many subgraphs does K_6 have?

d) For $n \geq 3$, how many subgraphs does K_n have?

17. Let v, w be two vertices in K_n, $n \geq 3$. How many walks of length 3 are there from v to w?

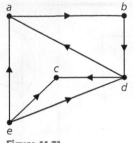
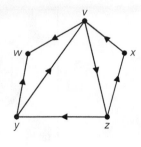

Figure 11.31

11.3
Vertex Degree: Euler Trails and Circuits

In Example 11.9 the number of edges incident with a vertex was used to show that two undirected graphs were not isomorphic. We now find this idea even more helpful.

Definition 11.14

Let G be an undirected graph or multigraph. For each vertex v of G, the *degree of v*, written $\deg(v)$, is the number of edges in G that are incident with v. Here a loop at a vertex v is considered as two incident edges for v.

EXAMPLE 11.10

For the graph in Fig. 11.32, $\deg(b) = \deg(d) = \deg(f) = \deg(g) = 2$, $\deg(c) = 4$, $\deg(e) = 0$, and $\deg(h) = 1$. For vertex a we have $\deg(a) = 3$ because we count a loop twice. Since h has degree 1, it is called a *pendant* vertex.

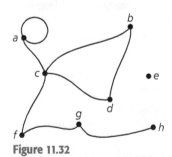

Figure 11.32

Using the idea of vertex degree, we have the following result.

THEOREM 11.2

If $G = (V, E)$ is an undirected graph or multigraph, then $\sum_{v \in V} \deg(v) = 2|E|$.

Proof: As we consider each edge $\{a, b\}$ in graph G, we find that the edge contributes a count of 1 to each of $\deg(a)$, $\deg(b)$, and consequently a count of 2 to $\sum_{v \in V} \deg(v)$. Thus $2|E|$ accounts for $\deg(v)$, for all $v \in V$, and $\sum_{v \in V} \deg(v) = 2|E|$.

This theorem provides some insight into the number of odd-degree vertices that can exist in a graph.

COROLLARY 11.1

For any undirected graph or multigraph, the number of vertices of odd degree must be even.

Proof: We leave the proof for the reader.

We apply Theorem 11.2 in the following example.

EXAMPLE 11.11

An undirected graph (or multigraph) where each vertex has the same degree is called a *regular* graph. If $\deg(v) = k$ for all vertices v, then the graph is called *k-regular*. Is it possible to have a 4-regular graph with 10 edges?

From Theorem 11.2, $2|E| = 20 = 4|V|$, so we have five vertices of degree 4. Figure 11.33 provides two nonisomorphic examples that satisfy the requirements.

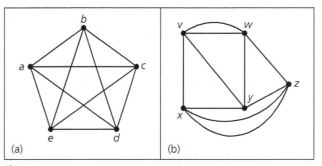

Figure 11.33

If we want each vertex to have degree 4, with 15 edges in the graph, we find that $2|E| = 30 = 4|V|$, from which it follows that no such graph is possible.

Our next example introduces a regular graph that arises in the study of computer architecture.

EXAMPLE 11.12

The Hypercube. In order to build a parallel computer one needs to have multiple CPUs (central processing units), where each such processor works on part of a problem. But often we cannot actually decompose a problem completely, so at some point the processors (each with its own memory) have to be able to communicate with one another.

We envisage this situation as follows. The accumulated data for a given problem are taken from a central storage location and divided up among the processors. The processors go through a phase where each computes on its own for a certain period of time and then some intercommunication takes place. Then the processors return to computing on their own and continue back and forth between operating individually and communicating with one another. This situation adequately describes how parallel algorithms work in practice.

To model the communication between the processors we use a loop-free connected undirected graph where each processor is assigned a vertex. When two processors, say p_1, p_2, are able to communicate directly with one another we draw the edge $\{p_1, p_2\}$ to represent this (line of) possible communication. How can we decide on a model (that is, a graph) to speed up the processing time? The complete graph (on all of our processors as vertices) would be ideal — but prohibitively expensive because of all the necessary connections. On the other hand, one can connect n processors along a path with $n - 1$ edges or on a cycle with n edges. Another possible model is a *grid* (or, *mesh*) graph, examples of which are shown in Fig. 11.34.

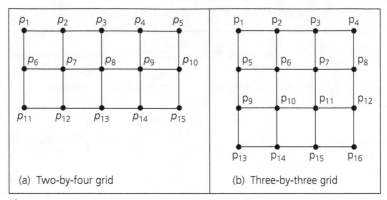

(a) Two-by-four grid   (b) Three-by-three grid

Figure 11.34

But in these last three models the distances (as measured by the number of edges in the shortest paths) between pairs of processors get longer and longer as the number of processors increases. A compromise that weighs the number of edges (direct connections) against the distance between pairs of vertices (processors) is embodied in the regular graph called the *hypercube*.

For $n \in \mathbf{N}$, the *n*-dimensional hypercube (or *n-cube*) is denoted by Q_n. It is a loop-free connected undirected graph with 2^n vertices. For $n \geq 1$, these vertices are labeled by the 2^n *n*-bit sequences representing $0, 1, 2, \ldots, 2^n - 1$. For instance, Q_3 has eight vertices — labeled 000, 001, 010, 011, 100, 101, 110, and 111. Two vertices v_1, v_2 of Q_n are joined by the edge $\{v_1, v_2\}$ when the binary labels for v_1, v_2 differ in exactly one position. Then for any vertices u, w in Q_n there is a shortest path of length d, when d is the number of positions where the binary labels for u, w differ. [This insures that Q_n is connected.]

Figure 11.35 shows Q_n for $n = 0, 1, 2, 3$. In general, for $n \geq 0$, Q_{n+1} can be constructed recursively from two copies of Q_n as follows. Prefix the vertex labels of one copy of Q_n with 0 (call the result $Q_{0,n}$) and those of the other copy with 1 (call this result $Q_{1,n}$). For x in $Q_{0,n}$ and y in $Q_{1,n}$ draw the edge $\{x, y\}$ if the (newly prefixed) binary labels for x, y differ only in the first (newly prefixed) position. The case for $n = 3$ (so $n + 1 = 4$) is demonstrated in Fig. 11.36. The blue edges are the new edges described above for constructing Q_4 from two copies of Q_3.

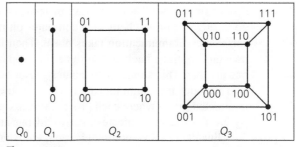

Figure 11.35

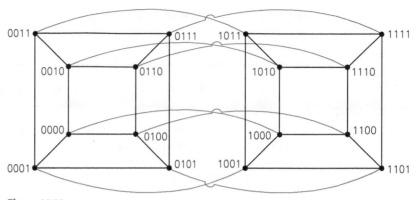

Figure 11.36

In summary, we reiterate that for $n \in \mathbf{N}$, the hypercube Q_n is an n-regular loop-free undirected graph with 2^n vertices. Further, it is connected with the distance between any two vertices at most n. From Theorem 11.2 it follows that Q_n has $(1/2)n2^n = n2^{n-1}$ edges. [Referring back to Example 10.33, we find that $n2^{n-1}$ is likewise the number of edges for the Hasse diagram of the partial order $(\mathscr{P}(X_n), \subseteq)$, where $X_n = \{1, 2, 3, \ldots, n\}$ and $\mathscr{P}(X_n)$ is the power set of X_n. This is no mere coincidence! If we use the Gray code of Example 3.9 to label the vertices of this Hasse diagram, we find we have the hypercube Q_n.]

Finally, note that in Q_4 there are 16 vertices (processors) and the longest distance between vertices is 4. Contrast this with the grids in Fig. 11.34, where there are 15 vertices in part (a) and 16 in part (b) — yet the longest distance is 6 in both grids.

We turn now to the reason why Euler developed the idea of the degree of a vertex: to solve the problem dealing with the seven bridges of Königsberg.

EXAMPLE 11.13

The Seven Bridges of Königsberg. During the eighteenth century, the city of Königsberg (in East Prussia) was divided into four sections (including the island of Kneiphof) by the Pregel River. Seven bridges connected these regions, as shown in Fig. 11.37(a). It was said that residents spent their Sunday walks trying to find a way to walk about the city so as to cross each bridge exactly once and then return to the starting point.

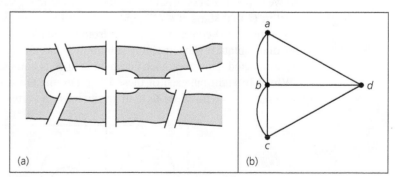

(a)   (b)

Figure 11.37

In order to determine whether or not such a circuit existed, Euler represented the four sections of the city and the seven bridges by the multigraph shown in Fig. 11.37(b). Here he found four vertices with $\deg(a) = \deg(c) = \deg(d) = 3$ and $\deg(b) = 5$. He also found that the existence of such a circuit depended on the number of vertices of odd degree in the graph.

Before proving the general result, we give the following definition.

Definition 11.15

Let $G = (V, E)$ be an undirected graph or multigraph with no isolated vertices. Then G is said to have an *Euler circuit* if there is a circuit in G that traverses every edge of the graph exactly once. If there is an open trail from a to b in G and this trail traverses each edge in G exactly once, the trail is called an *Euler trail*.

The problem of the seven bridges is now settled as we characterize the graphs that have an Euler circuit.

THEOREM 11.3

Let $G = (V, E)$ be an undirected graph or multigraph with no isolated vertices. Then G has an Euler circuit if and only if G is connected and every vertex in G has even degree.

Proof: If G has an Euler circuit, then for all a, $b \in V$ there is a trail from a to b — namely, that part of the circuit that starts at a and terminates at b. Therefore, it follows from Theorem 11.1 that G is connected.

Let s be the starting vertex of the Euler circuit. For any other vertex v of G, each time the circuit comes to v it then departs from the vertex. Thus the circuit has traversed either two (new) edges that are incident with v or a (new) loop at v. In either case a count of 2 is contributed to $\deg(v)$. Since v is not the starting point and each edge incident to v is traversed only once, a count of 2 is obtained each time the circuit passes through v, so $\deg(v)$ is even. As for the starting vertex s, the first edge of the circuit must be distinct from the last edge, and because any other visit to s results in a count of 2 for $\deg(s)$, we have $\deg(s)$ even.

Conversely, let G be connected with every vertex of even degree. If the number of edges in G is 1 or 2, then G must be as shown in Fig. 11.38. Euler circuits are immediate in these cases. We proceed now by induction and assume the result true for all situations where there are fewer than n edges. If G has n edges, select a vertex s in G as a starting point to build an Euler circuit. The graph (or multigraph) G is connected and each vertex has even degree, so we can at least construct a circuit C containing s. (Verify this by considering the longest trail in G that starts at s.) Should the circuit contain every edge of G, we are finished. If not, remove the edges of the circuit from G, making sure to remove any vertex that would become isolated. The remaining subgraph K has all vertices of even degree, but it may not be connected. However, each component of K is connected and will have an Euler circuit. (Why?) In addition, each of these Euler circuits has a vertex that is on C. Consequently, starting at s we travel on C until we arrive at a vertex s_1 that is on the Euler circuit of a

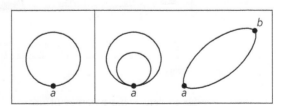

Figure 11.38

component C_1 of K. Then we traverse this Euler circuit and, returning to s_1, continue on C until we reach a vertex s_2 that is on the Euler circuit of component C_2 of K. Since the graph G is finite, as we continue this process we construct an Euler circuit for G.

Should G be connected and not have too many vertices of odd degree, we can at least find an Euler trail in G.

COROLLARY 11.2

If G is an undirected graph or multigraph with no isolated vertices, then we can construct an Euler trail in G if and only if G is connected and has exactly two vertices of odd degree.

Proof: If G is connected and a and b are the vertices of G that have odd degree, add an additional edge $\{a, b\}$ to G. We now have a graph G_1 that is connected and has every vertex of even degree. Hence G_1 has an Euler circuit C, and when the edge $\{a, b\}$ is removed from C, we obtain an Euler trail for G. (Thus the Euler trail starts at one of the vertices of odd degree and terminates at the other odd vertex.) We leave the details of the converse for the reader.

Returning now to the seven bridges of Königsberg, we realize that Fig. 11.37(b) is a connected multigraph, but it has four vertices of odd degree. Consequently, it has no Euler trail or Euler circuit.

Now that we have seen how the solution of an eighteenth-century problem led to the start of graph theory, is there a somewhat more contemporary context in which we might be able to apply what we have learned?

To answer this question (in the affirmative), we shall state the directed version of Theorem 11.3. But first we need to refine the concept of the degree of a vertex.

Definition 11.16

Let $G = (V, E)$ be a directed graph or multigraph. For each $v \in V$,

  a) The *incoming*, or *in*, *degree* of v is the number of edges in G that are incident into v, and this is denoted by $id(v)$.

  b) The *outgoing*, or *out*, *degree* of v is the number of edges in G that are incident from v, and this is denoted by $od(v)$.

For the case where the directed graph or multigraph contains one or more loops, each loop at a given vertex v contributes a count of 1 to each of $id(v)$ and $od(v)$.

The concepts of the in degree and the out degree for vertices now lead us to the following theorem.

THEOREM 11.4

Let $G = (V, E)$ be a directed graph or multigraph with no isolated vertices. The graph G has a directed Euler circuit if and only if G is connected and $id(v) = od(v)$ for all $v \in V$.

Proof: The proof of this theorem is left for the reader.

At this time we consider an application of Theorem 11.4. This example is based on a telecommunication problem given by C. L. Liu on pages 176–178 of reference [23].

In Fig. 11.39(a) we have the surface of a rotating drum that is divided into eight sectors of equal area. In part (b) of the figure we have placed conducting (shaded sectors and inner circle) and nonconducting (unshaded sectors) material on the drum. When the three terminals (shown in the figure) make contact with the three designated sectors, the nonconducting material results in no flow of current and a 1 appears on the display of a digital device. For the sectors with the conducting material, a flow of current takes place and a 0 appears on the display in each case. If the drum were rotated 45 degrees (clockwise), the screen would read 110 (from top to bottom). So we can obtain at least two (namely, 100 and 110) of the eight binary representations from 000 (for 0) to 111 (for 7). But can we represent all eight of them as the drum continues to rotate? And could we extend the problem to the 16 four-bit binary representations from 0000 through 1111, and perhaps generalize the results even further?

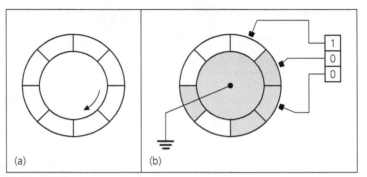

(a)   (b)

Figure 11.39

To answer the question for the problem in the figure, we construct a directed graph $G = (V, E)$, where $V = \{00, 01, 10, 11\}$ and E is constructed as follows: If $b_1b_2, b_2b_3 \in V$, draw the edge (b_1b_2, b_2b_3). This results in the directed graph of Fig. 11.40(a), where $|E| = 8$. We see that this graph is connected and that for all $v \in V$, $id(v) = od(v)$. Consequently, by Theorem 11.4, it has a directed Euler circuit. One such circuit is given by

$$
\begin{array}{ccccccc}
100 & 000 & 001 & 010 & 101 & 011 & 111 \\
\end{array}
$$

$$
\overset{\curvearrowright}{} 10 \longrightarrow 00 \longrightarrow 00 \longrightarrow 01 \longrightarrow 10 \longrightarrow 01 \longrightarrow 11 \longrightarrow 11 \overset{\curvearrowleft}{}
$$

$$
110
$$

Here the label on each edge $e = (a, c)$, as shown in part (b) of Fig. 11.40, is the three-bit sequence $x_1x_2x_3$, where $a = x_1x_2$ and $c = x_2x_3$. Since the vertices of G are the four distinct two-bit sequences 00, 01, 10, and 11, the labels on the eight edges of G determine the eight distinct three-bit sequences. Also, any two consecutive edge labels in the Euler circuit are of the form $y_1y_2y_3$ and $y_2y_3y_4$.

Starting with the edge label 100, in order to get the next label, 000, we concatenate the last bit in 000, namely 0, to the string 100. The resulting string 1000 then provides 100 (1000) and 000 (1000). The next edge label is 001, so we concatenate the 1 (the last bit in 001) to our present string 1000 and get 10001, which provides the three distinct three-bit sequences 100 (10001), 000 (10001), and 001 (10001). Continuing in this way, we arrive at the eight-bit sequence 10001011 (where the last 1 is *wrapped around*), and these eight bits are then arranged in the sectors of the rotating drum as in Fig. 11.41. It is from this figure that the result in Fig. 11.39(b) is obtained. And as the drum in Fig. 11.39(b) rotates, all of the eight three-bit sequences 100, 110, 111, 011, 101, 010, 001, and 000 are obtained.

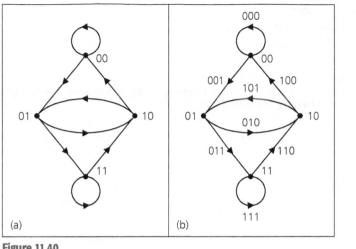

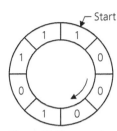

Figure 11.40

Figure 11.41

In closing this section, we wish to call the reader's attention to reference [24] by Anthony Ralston. This article is a good source for more ideas and generalizations related to the problem discussed in Example 11.14.

EXERCISES 11.3

1. Determine $|V|$ for the following graphs or multigraphs G.

a) G has nine edges and all vertices have degree 3.

b) G is regular with 15 edges.

c) G has 10 edges with two vertices of degree 4 and all others of degree 3.

2. If $G = (V, E)$ is a connected graph with $|E| = 17$ and $\deg(v) \geq 3$ for all $v \in V$, what is the maximum value for $|V|$?

3. Let $G = (V, E)$ be a connected undirected graph.

a) What is the largest possible value for $|V|$ if $|E| = 19$ and $\deg(v) \geq 4$ for all $v \in V$?

b) Draw a graph to demonstrate each possible case in part (a).

4. a) Let $G = (V, E)$ be a loop-free undirected graph, where $|V| = 6$ and $\deg(v) = 2$ for all $v \in V$, Up to isomorphism how many such graphs G are there?

b) Answer part (a) for $|V| = 7$.

c) Let $G_1 = (V_1, E_1)$ be a loop-free undirected 3-regular graph with $|V_1| = 6$. Up to isomorphism how many such graphs G_1 are there?

d) Answer part (c) for $|V_1| = 7$ and G_1 4-regular.

e) Generalize the results in parts (c) and (d).

5. Let $G_1 = (V_1, E_1)$ and $G_2 = (V_2, E_2)$ be the loop-free undirected connected graphs in Fig. 11.42.

a) Determine $|V_1|$, $|E_1|$, $|V_2|$, and $|E_2|$.

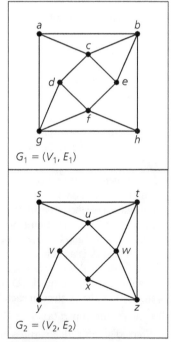

$G_1 = (V_1, E_1)$

$G_2 = (V_2, E_2)$

Figure 11.42

b) Find the degree of each vertex in V_1. Do likewise for each vertex in V_2.

c) Are the graphs G_1 and G_2 isomorphic?

6. Let $V = \{a, b, c, d, e, f\}$. Draw three nonisomorphic loop-free undirected graphs $G_1 = (V, E_1)$, $G_2 = (V, E_2)$, and

$G_3 = (V, E_3)$, where, in all three graphs, we have $\deg(a) = 3$, $\deg(b) = \deg(c) = 2$, and $\deg(d) = \deg(e) = \deg(f) = 1$.

7. a) How many different paths of length 2 are there in the undirected graph G in Fig. 11.43?

   b) Let $G = (V, E)$ be a loop-free undirected graph, where $V = \{v_1, v_2, \ldots, v_n\}$ and $\deg(v_i) = d_i$, for all $1 \le i \le n$. How many different paths of length 2 are there in G?

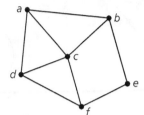

Figure 11.43

8. a) Find the number of edges in Q_8.

   b) Find the maximum distance between pairs of vertices in Q_8. Give an example of one such pair that achieves this distance.

   c) Find the length of a longest path in Q_8.

9. a) What is the dimension of the hypercube with 524,288 edges?

   b) How many vertices are there for a hypercube with 4,980,736 edges?

10. For $n \in \mathbf{Z}^+$, how many distinct (though isomorphic) paths of length 2 are there in the n-dimensional hypercube Q_n?

11. Let $n \in \mathbf{Z}^+$, with $n \ge 9$. Prove that if the edges of K_n can be partitioned into subgraphs isomorphic to cycles of length 4 (where any two such cycles share no common edge), then $n = 8k + 1$ for some $k \in \mathbf{Z}^+$.

12. a) For $n \ge 2$, let V denote the vertices in Q_n. For $1 \le k < \ell \le n$, define the relation $\mathcal{R}$ on V as follows: If $w, x \in V$, then $w \mathcal{R} x$ if w and x have the same bit (0, or 1) in position k and the same bit (0, or 1) in position ℓ of their binary labels. [For example, if $n = 7$ and $k = 3, \ell = 6$, then 1100010 $\mathcal{R}$ 0000011.] Show that $\mathcal{R}$ is an equivalence relation. How many blocks are there for this equivalence relation? How many vertices are there in each block? Describe the subgraph of Q_n induced by the vertices in each block.

   b) Generalize the results of part (a).

13. If G is an undirected graph with n vertices and e edges, let $\delta = \min_{v \in V}\{\deg(v)\}$ and let $\Delta = \max_{v \in V}\{\deg(v)\}$. Prove that $\delta \le 2(e/n) \le \Delta$.

14. Let $G = (V, E)$, $H = (V', E')$ be undirected graphs with $f: V \to V'$ establishing an isomorphism between the graphs. (a) Prove that $f^{-1}: V' \to V$ is also an isomorphism for G and H. (b) If $a \in V$, prove that $\deg(a)$ (in G) $= \deg(f(a))$ (in H).

15. For all $k \in \mathbf{Z}^+$ where $k \ge 2$, prove that there exists a loop-free connected undirected graph $G = (V, E)$, where $|V| = 2k$ and $\deg(v) = 3$ for all $v \in V$.

16. Prove that for each $n \in \mathbf{Z}^+$ there exists a loop-free connected undirected graph $G = (V, E)$, where $|V| = 2n$ and which has two vertices of degree i for every $1 \le i \le n$.

17. Complete the proofs of Corollaries 11.1 and 11.2.

18. Let k be a fixed positive integer and let $G = (V, E)$ be a loop-free undirected graph, where $\deg(v) \ge k$ for all $v \in V$. Prove that G contains a path of length k.

19. a) Explain why it is not possible to draw a loop-free connected undirected graph with eight vertices, where the degrees of the vertices are 1, 1, 1, 2, 3, 4, 5, and 7.

   b) Give an example of a loop-free connected undirected multigraph with eight vertices, where the degrees of the vertices are 1, 1, 1, 2, 3, 4, 5, and 7.

20. a) Find an Euler circuit for the graph in Fig. 11.44.

   b) If the edge $\{d, e\}$ is removed from this graph, find an Euler trail for the resulting subgraph.

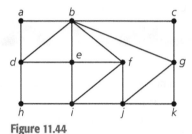

Figure 11.44

21. Determine the value(s) of n for which the complete graph K_n has an Euler circuit. For which n does K_n have an Euler trail but not an Euler circuit?

22. For the graph in Fig. 11.37(b), what is the smallest number of bridges that must be removed so that the resulting subgraph has an Euler trail but not an Euler circuit? Which bridge(s) should we remove?

23. When visiting a chamber of horrors, Paul and David try to figure out whether they can travel through the seven rooms and surrounding corridor of the attraction without passing through any door more than once. If they must start from the starred position in the corridor shown in Fig. 11.45, can they accomplish their goal?

24. Let $G = (V, E)$ be a directed graph, where $|V| = n$ and $|E| = e$. What are the values for $\sum_{v \in V} id(v)$ and $\sum_{v \in V} od(v)$?

25. a) Find the maximum length of a trail in

   i) K_6 **ii)** K_8

   iii) K_{10} **iv)** $K_{2n}, n \in \mathbf{Z}^+$

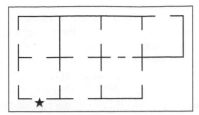

Figure 11.45

b) Find the maximum length of a circuit in

i) K_6 **ii)** K_8

iii) K_{10} **iv)** $K_{2n}, n \in \mathbf{Z}^+$

26. a) Let $G = (V, E)$ be a directed graph or multigraph with no isolated vertices. Prove that G has a directed Euler circuit if and only if G is connected and $od(v) = id(v)$ for all $v \in V$.

b) A directed graph is called *strongly connected* if there is a directed path from a to b for all vertices a, b, where $a \neq b$. Prove that if a directed graph has a directed Euler circuit, then it is strongly connected. Is the converse true?

27. Let G be a directed graph on n vertices. If the associated undirected graph for G is K_n, prove that $\sum_{v \in V} [od(v)]^2 = \sum_{v \in V} [id(v)]^2$.

28. If $G = (V, E)$ is a directed graph or multigraph with no isolated vertices, prove that G has a directed Euler trail if and only if (i) G is connected; (ii) $od(v) = id(v)$ for all but two vertices x, y in V; and (iii) $od(x) = id(x) + 1, id(y) = od(y) + 1$.

29. Let $V = \{000, 001, 010, \ldots, 110, 111\}$. For each four-bit sequence $b_1 b_2 b_3 b_4$ draw an edge from the element $b_1 b_2 b_3$ to the element $b_2 b_3 b_4$ in V. (a) Draw the graph $G = (V, E)$ as described. (b) Find a directed Euler circuit for G. (c) Equally space eight 0's and eight 1's around the edge of a rotating (clockwise) drum so that these 16 bits form a circular sequence where the (consecutive) subsequences of length 4 provide the binary representations of $0, 1, 2, \ldots, 14, 15$ in some order.

30. Carolyn and Richard attended a party with three other married couples. At this party a good deal of handshaking took place, but (1) no one shook hands with her or his spouse; (2) no one shook hands with herself or himself; and (3) no one shook hands with anyone more than once. Before leaving the party, Carolyn asked the other seven people how many hands she or he had shaken. She received a different answer from each of the seven. How many times did Carolyn shake hands at this party? How many times did Richard?

31. Let $G = (V, E)$ be a loop-free connected undirected graph with $|V| \geq 2$. Prove that G contains two vertices v, w, where $\deg(v) = \deg(w)$.

32. If $G = (V, E)$ is an undirected graph with $|V| = n$ and $|E| = k$, the following matrices are used to represent G.

Let $V = \{v_1, v_2, \ldots, v_n\}$. Define the *adjacency matrix* $A = (a_{ij})_{n \times n}$ where $a_{ij} = 1$ if $\{v_i, v_j\} \in E$, otherwise $a_{ij} = 0$.

If $E = \{e_1, e_2, \ldots, e_k\}$, the *incidence matrix* I is the $n \times k$ matrix $(b_{ij})_{n \times k}$ where $b_{ij} = 1$ if v_i is a vertex on the edge e_j, otherwise $b_{ij} = 0$.

a) Find the adjacency and incidence matrices associated with the graph in Fig. 11.46.

b) Calculating A^2 and using the Boolean operations where $0 + 0 = 0, 0 + 1 = 1 + 0 = 1 + 1 = 1$, and $0 \cdot 0 = 0 \cdot 1 = 1 \cdot 0 = 0, 1 \cdot 1 = 1$, prove that the entry in row i and column j of A^2 is 1 if and only if there is a walk of length 2 between the ith and jth vertices of V.

c) If we calculate A^2 using ordinary addition and multiplication, what do the entries in the matrix reveal about G?

d) What is the column sum for each column of A? Why?

e) What is the column sum for each column of I? Why?

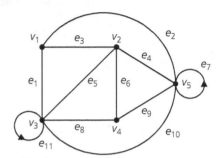

Figure 11.46

33. Determine whether or not the loop-free undirected graphs with the following adjacency matrices are isomorphic.

a) $\begin{bmatrix} 0 & 0 & 1 \\ 0 & 0 & 1 \\ 1 & 1 & 0 \end{bmatrix}, \begin{bmatrix} 0 & 1 & 1 \\ 1 & 0 & 0 \\ 1 & 0 & 0 \end{bmatrix}$

b) $\begin{bmatrix} 0 & 1 & 0 & 1 \\ 1 & 0 & 1 & 1 \\ 0 & 1 & 0 & 1 \\ 1 & 1 & 1 & 0 \end{bmatrix}, \begin{bmatrix} 0 & 1 & 1 & 1 \\ 1 & 0 & 1 & 0 \\ 1 & 1 & 0 & 1 \\ 1 & 0 & 1 & 0 \end{bmatrix}$

c) $\begin{bmatrix} 0 & 1 & 1 & 1 \\ 1 & 0 & 1 & 0 \\ 1 & 1 & 0 & 0 \\ 1 & 0 & 0 & 0 \end{bmatrix}, \begin{bmatrix} 0 & 1 & 0 & 1 \\ 1 & 0 & 1 & 0 \\ 0 & 1 & 0 & 1 \\ 1 & 0 & 1 & 0 \end{bmatrix}$

34. Determine whether or not the loop-free undirected graphs with the following incidence matrices are isomorphic.

a) $\begin{bmatrix} 1 & 0 & 1 \\ 0 & 1 & 1 \\ 1 & 1 & 0 \end{bmatrix}, \begin{bmatrix} 0 & 1 & 1 \\ 1 & 1 & 0 \\ 1 & 0 & 1 \end{bmatrix}$

b) $\begin{bmatrix} 1 & 0 & 1 & 1 \\ 1 & 1 & 0 & 0 \\ 0 & 1 & 1 & 0 \\ 0 & 0 & 0 & 1 \end{bmatrix}, \begin{bmatrix} 1 & 0 & 0 & 1 \\ 1 & 1 & 0 & 0 \\ 0 & 1 & 1 & 0 \\ 0 & 0 & 1 & 1 \end{bmatrix}$

$$
\textbf{c)} \quad
\begin{bmatrix}
1 & 0 & 0 & 0 & 1 \\
0 & 1 & 1 & 0 & 1 \\
0 & 1 & 0 & 1 & 0 \\
1 & 0 & 1 & 1 & 0
\end{bmatrix},
\begin{bmatrix}
1 & 1 & 0 & 0 & 0 \\
0 & 1 & 1 & 0 & 1 \\
0 & 0 & 0 & 1 & 1 \\
1 & 0 & 1 & 1 & 0
\end{bmatrix}
$$

35. There are 15 people at a party. Is it possible for each of these people to shake hands with (exactly) three others?

36. Consider the two-by-four grid in Fig. 11.34. Assign the partial Gray code $A = \{00, 01, 11\}$ to the three horizontal levels: top (00), middle (01), and bottom (11). Now assign the partial Gray code $B = \{000, 001, 011, 010, 110\}$ to the five vertical levels: left, or first (000), second (001), third (011), fourth (010), and right, or fifth (110). Use the elements of $A \times B$ to label the 15 processors of this grid; for example, p_1 is labeled (00,000), p_2 is labeled (00, 001), p_8 is labeled (01, 011), p_{14} is labeled (11, 010), and p_{15} is labeled (11, 110). Show that the two-by-four grid is isomorphic to a subgraph of the hypercube Q_5. (Thus we can consider the two-by-four grid to be embedded in the hypercube Q_5.)

37. Prove that the three-by-three grid of Fig. 11.34 is isomorphic to a subgraph of the hypercube Q_4.

11.4
Planar Graphs

On a road map the lines indicating the roads and highways usually intersect only at junctions or towns. But sometimes roads seem to intersect when one road is located above another, as in the case of an overpass. In this case the two roads are at different levels, or planes. This type of situation leads us to the following definition.

Definition 11.17

A graph (or multigraph) G is called *planar* if G can be drawn in the plane with its edges intersecting only at vertices of G. Such a drawing of G is called an *embedding* of G in the plane.

EXAMPLE 11.15

The graphs in Fig. 11.47 are planar. The first is a 3-regular graph, because each vertex has degree 3; it is planar because no edges intersect except at the vertices. In graph (b) it appears that we have a *nonplanar* graph; the edges $\{x, z\}$ and $\{w, y\}$ overlap at a point other than a vertex. However, we can redraw this graph as shown in part (c) of the figure. Consequently, K_4 is planar.

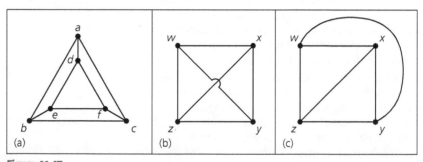

Figure 11.47

EXAMPLE 11.16

Just as K_4 is planar, so are the graphs K_1, K_2, and K_3.

An attempt to embed K_5 in the plane is shown in Fig. 11.48. If K_5 were planar, then any embedding would have to contain the pentagon in part (a) of the figure. Since a complete graph contains an edge for every pair of distinct vertices, we add edge $\{a, c\}$ as shown in part (b). This edge is contained entirely within the interior of the pentagon in part (a). (We could have drawn the edge in the exterior region determined by the pentagon. The reader will be asked in the exercises to show that the same conclusion arises in this case.) Moving

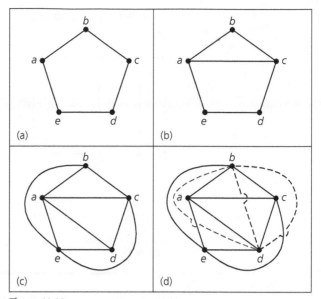

Figure 11.48

to part (c), we add in the edges $\{a, d\}$, $\{c, e\}$, and $\{b, e\}$. Now we consider the vertices b and d. We need the edge $\{b, d\}$ in order to have K_5. Vertex d is inside the region formed by the cycle edges $\{a, c\}$, $\{c, e\}$, and $\{e, a\}$, whereas b is outside the region. Thus in drawing the edge $\{b, d\}$, we must intersect one of the existing edges at least once, as shown by the dotted edges in part (d). Consequently, K_5 is nonplanar. (Since this proof appeals to a diagram, it definitely lacks rigor. However, later in the section we shall prove that K_5 is nonplanar by another method.)

Before we can characterize all nonplanar graphs we need to examine another class of graphs.

Definition 11.18

A graph $G = (V, E)$ is called *bipartite* if $V = V_1 \cup V_2$ with $V_1 \cap V_2 = \emptyset$, and every edge of G is of the form $\{a, b\}$ with $a \in V_1$ and $b \in V_2$. If each vertex in V_1 is joined with every vertex in V_2, we have a *complete bipartite* graph. In this case, if $|V_1| = m$, $|V_2| = n$, the graph is denoted by $K_{m,n}$.

EXAMPLE 11.17

Figure 11.49 indicates how we may partition the vertices of the hypercubes Q_1, Q_2, Q_3 to demonstrate that these graphs are bipartite. In general, for each $n \geq 1$, partition the vertices of Q_n as $V_1 \cup V_2$, where V_1 consists of all vertices whose binary labels have an even number of 1's, while V_2 consists of those whose binary labels have an odd number of 1's. Could there exist an edge $\{x, y\}$ in Q_n where $x, y \in V_1$? Recall that edges in Q_n connect vertices that differ in exactly one of the n positions in their binary labels. Suppose that the binary labels of x, y differ only in position i, for some $1 \leq i \leq n$. Then the total number of 1's in the binary labels for x, y is $2 \cdot$ [the number of 1's in x (or y) in all positions other than position i] $+ 1$, an odd total. But with $x, y \in V_1$, their binary labels each contain an even number of 1's — so the total number of 1's in these binary labels is even! This contradiction tells us that there is no edge $\{x, y\}$ in Q_n where $x, y \in V_1$. A similar argument can be given to rule out the possibility of an edge $\{u, w\}$, where $u, w \in V_2$. Consequently, Q_n is bipartite for all $n \geq 1$.

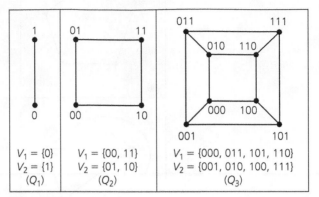

$V_1 = \{0\}$
$V_2 = \{1\}$
(Q_1)

$V_1 = \{00, 11\}$
$V_2 = \{01, 10\}$
(Q_2)

$V_1 = \{000, 011, 101, 110\}$
$V_2 = \{001, 010, 100, 111\}$
(Q_3)

Figure 11.49

Figure 11.50 shows two bipartite graphs. The graph in part (a) satisfies the definition for $V_1 = \{a, b\}$ and $V_2 = \{c, d, e\}$. If we add the edges $\{b, d\}$ and $\{b, c\}$, the result is the complete bipartite graph $K_{2,3}$, which is planar. Graph (b) of the figure is $K_{3,3}$. Let $V_1 = \{h_1, h_2, h_3\}$ and $V_2 = \{u_1, u_2, u_3\}$, and interpret V_1 as a set of houses and V_2 as a set of utilities. Then $K_{3,3}$ is called the *utility graph*. Can we hook up each of the houses with each of the utilities and avoid having overlapping utility lines? In Fig. 11.50(b) it appears that this is not possible and that $K_{3,3}$ is nonplanar. (Once again we deduce the nonplanarity of a graph from a diagram. However, we shall verify that $K_{3,3}$ is nonplanar by another method, later in Example 11.21 of this section.)

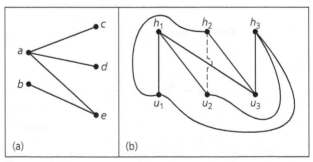

(a)   (b)

Figure 11.50

We shall see that when we are dealing with nonplanar graphs, either K_5 or $K_{3,3}$ will be the source of the problem. Before stating the general result, however, we need to develop one final new idea.

Definition 11.19

Let $G = (V, E)$ be a loop-free undirected graph, where $E \neq \emptyset$. An *elementary subdivision* of G results when an edge $e = \{u, w\}$ is removed from G and then the edges $\{u, v\}, \{v, w\}$ are added to $G - e$, where $v \notin V$.

The loop-free undirected graphs $G_1 = (V_1, E_1)$ and $G_2 = (V_2, E_2)$ are called *homeomorphic* if they are isomorphic or if they can both be obtained from the same loop-free undirected graph H by a sequence of elementary subdivisions.

EXAMPLE 11.18

a) Let $G = (V, E)$ be a loop-free undirected graph with $|E| \geq 1$. If G' is obtained from G by an elementary subdivision, then the graph $G' = (V', E')$ satisfies $|V'| = |V| + 1$ and $|E'| = |E| + 1$.

b) Consider the graphs G, G_1, G_2, and G_3 in Fig. 11.51. Here G_1 is obtained from G by means of one elementary subdivision: Delete edge $\{a, b\}$ from G and then add the edges $\{a, w\}$ and $\{w, b\}$. The graph G_2 is obtained from G by two elementary subdivisions. Hence G_1 and G_2 are homeomorphic. Also, G_3 can be obtained from G by four elementary subdivisions, so G_3 is homeomorphic to both G_1 and G_2.

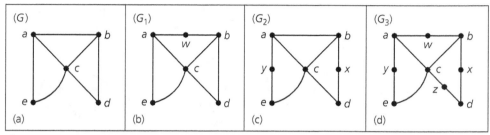

Figure 11.51

However, we cannot obtain G_1 from G_2 (or G_2 from G_1) by a sequence of elementary subdivisions. Furthermore, the graph G_3 can be obtained from either G_1 or G_2 by a sequence of elementary subdivisions: six (such sequences of three elementary subdivisions) for G_1 and two for G_2. But neither G_1 nor G_2 can be obtained from G_3 by a sequence of elementary subdivisions.

One may think of homeomorphic graphs as being isomorphic except, possibly, for vertices of degree 2. In particular, if two graphs are homeomorphic, they are either both planar or they are both nonplanar.

These preliminaries lead us to the following result.

THEOREM 11.5

Kuratowski's Theorem. A graph is nonplanar if and only if it contains a subgraph that is homeomorphic to either K_5 or $K_{3,3}$.

Proof: (This theorem was first proved by the Polish mathematician Kasimir Kuratowski in 1930.) If a graph G has a subgraph homeomorphic to either K_5 or $K_{3,3}$, it is clear that G is nonplanar. The converse of this theorem, however, is much more difficult to prove. (A proof can be found in Chapter 8 of C. L. Liu [23] or Chapter 6 of D. B. West [32].)

We demonstrate the use of Kuratowski's Theorem in the following example.

EXAMPLE 11.19

a) Figure 11.52(a) is a familiar graph called the *Petersen graph*. Part (b) of the figure provides a subgraph of the Petersen graph that is homeomorphic to $K_{3,3}$. (Figure 11.53 shows how the subgraph is obtained from $K_{3,3}$ by a sequence of four elementary subdivisions.) Hence the Petersen graph is nonplanar.

b) In part (a) of Fig. 11.54 we find the 3-regular graph G, which is isomorphic to the 3-dimensional hypercube Q_3. The 4-regular complement of G is shown in Fig. 11.54(b), where the edges $\{a, g\}$ and $\{d, f\}$ suggest that G *may be* nonplanar. Figure 11.54(c) depicts a subgraph H of $\overline{G}$ that is homeomorphic to K_5, so by Kuratowski's Theorem it follows that $\overline{G}$ *is* nonplanar.

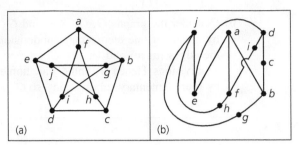

Figure 11.52

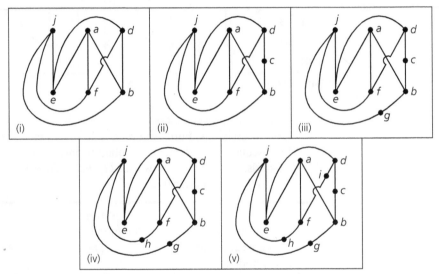

Figure 11.53

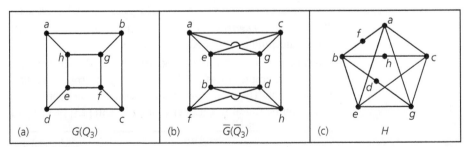

Figure 11.54

When a graph or multigraph is planar and connected, we find the following relation, which was discovered by Euler. For this relation we need to be able to count the number of regions determined by a planar connected graph or multigraph — the number (of these regions) being defined only when we have a planar embedding of the graph. For instance, the planar embedding of K_4 in part (a) of Fig. 11.55 demonstrates how this depiction of K_4 determines four regions in the plane: three of finite area — namely, R_1, R_2, and R_3 — and the infinite region R_4. When we look at Fig. 11.55(b) we might think that here K_4 determines five regions, but this depiction does *not* present a planar embedding of K_4. So the result in Fig. 11.55(a) is the only one we actually want to deal with here.

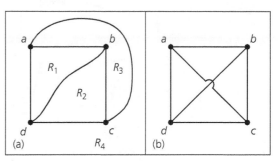

Figure 11.55

THEOREM 11.6   Let $G = (V, E)$ be a connected planar graph or multigraph with $|V| = v$ and $|E| = e$. Let r be the number of regions in the plane determined by a planar embedding (or, depiction) of G; one of these regions has infinite area and is called *the infinite region*. Then $v - e + r = 2$.

Proof: The proof is by induction on e. If $e = 0$ or 1, then G is isomorphic to one of the graphs in Fig. 11.56. The graph in part (a) has $v = 1, e = 0$, and $r = 1$; so, $v - e + r = 1 - 0 + 1 = 2$. For graph (b), $v = 1, e = 1$, and $r = 2$. Graph (c) has $v = 2, e = 1$, and $r = 1$. In both cases, $v - e + r = 2$.

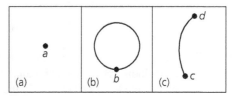

Figure 11.56

Now let $k \in \mathbf{N}$ and assume that the result is true for every connected planar graph or multigraph with e edges, where $0 \le e \le k$. If $G = (V, E)$ is a connected planar graph or multigraph with v vertices, r regions, and $e = k + 1$ edges, let $a, b \in V$ with $\{a, b\} \in E$. Consider the subgraph H of G obtained by deleting the edge $\{a, b\}$ from G. (If G is a multigraph and $\{a, b\}$ is one of a set of edges between a and b, then we remove it only once.) Consequently, we may write $H = G - \{a, b\}$ or $G = H + \{a, b\}$. We consider the following two cases, depending on whether H is connected or disconnected.

Case 1: The results in parts (a), (b), (c), and (d) of Fig. 11.57 show us how a graph G may be obtained from a connected graph H when the (new) loop $\{a, a\}$ is drawn as in parts (a) and (b) or when the (new) edge $\{a, b\}$ joins two distinct vertices in H as in parts (c) and (d). In all of these situations, H has v vertices, k edges, and $r - 1$ regions because one of the regions for H is split into two regions for G. The induction hypothesis applied to graph H tells us that $v - k + (r - 1) = 2$, and from this it follows that $2 = v - (k + 1) + r = v - e + r$. So Euler's Theorem is true for G in this case.

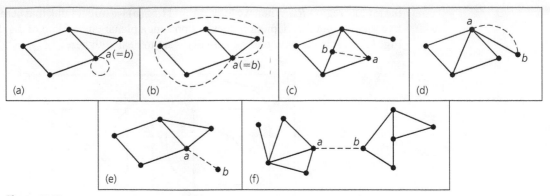

Figure 11.57

Case 2: Now we consider the case where $G - \{a, b\} = H$ is a disconnected graph [as demonstrated in Fig. 11.57(e) and (f)]. Here H has v vertices, k edges, and r regions. Also, H has two components H_1 and H_2, where H_i has v_i vertices, e_i edges, and r_i regions, for $i = 1, 2$. [Part (e) of Fig. 11.57 indicates that one component could consist of just an isolated vertex.] Furthermore, $v_1 + v_2 = v$, $e_1 + e_2 = k \, (= e - 1)$, and $r_1 + r_2 = r + 1$ because each of H_1 and H_2 determines an infinite region. When we apply the induction hypothesis to each of H_1 and H_2 we learn that

$$v_1 - e_1 + r_1 = 2 \quad \text{and} \quad v_2 - e_2 + r_2 = 2.$$

Consequently, $(v_1 + v_2) - (e_1 + e_2) + (r_1 + r_2) = v - (e - 1) + (r + 1) = 4$, and from this it follows that $v - e + r = 2$, thus establishing Euler's Theorem for G in this case.

The following corollary for Theorem 11.6 provides two inequalities relating the number of edges in a loop-free connected planar graph G with (1) the number of regions determined by a planar embedding of G; and (2) the number of vertices in G. Before we examine this corollary, however, let us look at the following helpful idea. For each region R in a planar embedding of a (planar) graph or multigraph, the *degree of R*, denoted $\deg(R)$, is the number of edges traversed in a (shortest) closed walk about (the edges in) the boundary of R. If $G = (V, E)$ is the graph of Fig. 11.58(a), then this planar embedding of G has four regions where

$$\deg(R_1) = 5, \qquad \deg(R_2) = 3, \qquad \deg(R_3) = 3, \qquad \deg(R_4) = 7.$$

[Here $\deg(R_4) = 7$, as determined by the closed walk: $a \to b \to g \to h \to g \to f \to d \to a$.] Part (b) of the figure shows a second planar embedding of G — again with four regions — and here

$$\deg(R_5) = 4, \qquad \deg(R_6) = 3, \qquad \deg(R_7) = 5, \qquad \deg(R_8) = 6.$$

[The closed walk $b \to g \to h \to g \to f \to b$ gives us $\deg(R_7) = 5$.)]

We see that $\sum_{i=1}^{4} \deg(R_i) = 18 = \sum_{i=5}^{8} \deg(R_i) = 2 \cdot 9 = 2|E|$. This is true in general because each edge of the planar embedding is either part of the boundary of two regions [like $\{b, c\}$ in parts (a) and (b)] or occurs twice in the closed walk about the edges in the boundary for one region [like $\{g, h\}$ in parts (a) and (b)].

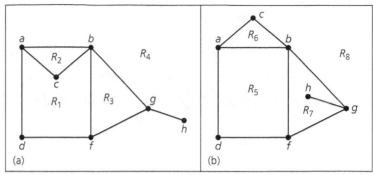

Figure 11.58

Now let us consider the following.

COROLLARY 11.3

Let $G = (V, E)$ be a loop-free connected planar graph with $|V| = v$, $|E| = e > 2$, and r regions. Then $3r \leq 2e$ and $e \leq 3v - 6$.

Proof: Since G is loop-free and is not a multigraph, the boundary of each region (including the infinite region) contains at least three edges — hence, each region has degree ≥ 3. Consequently, $2e = 2|E| =$ the sum of the degrees of the r regions determined by G and $2e \geq 3r$. From Euler's Theorem, $2 = v - e + r \leq v - e + (2/3)e = v - (1/3)e$, so $6 \leq 3v - e$, or $e \leq 3v - 6$.

We now consider what this corollary does and does not imply. If $G = (V, E)$ is a loop-free connected graph with $|E| > 2$, then if $e > 3v - 6$, it follows that G is not planar. However, if $e \leq 3v - 6$, we cannot conclude that G is planar.

EXAMPLE 11.20

The graph K_5 is loop-free and connected with ten edges and five vertices. Consequently, $3v - 6 = 15 - 6 = 9 < 10 = e$. Therefore, by Corollary 11.3, we find that K_5 is nonplanar.

EXAMPLE 11.21

The graph $K_{3,3}$ is loop-free and connected with nine edges and six vertices. Here $3v - 6 = 18 - 6 = 12 \geq 9 = e$. It would be a mistake to conclude from this that $K_{3,3}$ is planar. It would be the mistake of arguing by the converse.

However, $K_{3,3}$ is nonplanar. If $K_{3,3}$ were planar, then since each region in the graph is bounded by at least four edges, we have $4r \leq 2e$. (We found a similar situation in the proof of Corollary 11.3.) From Euler's Theorem, $v - e + r = 2$, or $r = e - v + 2 = 9 - 6 + 2 = 5$, so $20 = 4r \leq 2e = 18$. From this contradiction we have $K_{3,3}$ being nonplanar.

EXAMPLE 11.22

We use Euler's Theorem to characterize the *Platonic solids*. [For these solids all faces are congruent and all (interior) solid angles are equal.] In Fig. 11.59 we have two of these solids. Part (a) of the figure shows the regular tetrahedron, which has four faces, each an equilateral triangle. Concentrating on the edges of the tetrahedron, we focus on its underlying framework. As we view this framework from a point directly above the center of one of the faces, we picture the planar representation in part (b). This planar graph determines four regions (corresponding to the four faces); three regions meet at each of the four vertices. Part (c) of the figure provides another Platonic solid, the cube. Its associated planar graph is given in part (d). In this graph there are six regions with three regions meeting at each vertex.

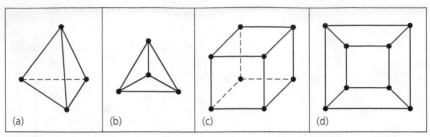

Figure 11.59

On the basis of our observations for the regular tetrahedron and the cube, we shall determine the other Platonic solids by means of their associated planar graphs. In these graphs $G = (V, E)$ we have $v = |V|$; $e = |E|$; $r =$ the number of planar regions determined by G; $m =$ the number of edges in the boundary of each region; and $n =$ the number of regions that meet at each vertex. Thus the constants $m, n \geq 3$. Since each edge is used in the boundary of two regions and there are r regions, each with m edges, it follows that $2e = mr$. Counting the endpoints of the edges, we get $2e$. But all these endpoints can also be counted by considering what happens at each vertex. Since n regions meet at each vertex, n edges meet there, so there are n endpoints of edges to count at each of the v vertices. This totals nv endpoints of edges, so $2e = nv$. From Euler's Theorem we have

$$0 < 2 = v - e + r = \frac{2e}{n} - e + \frac{2e}{m} = e\left(\frac{2m - mn + 2n}{mn}\right).$$

With $e, m, n > 0$, we find that

$$2m - mn + 2n > 0 \Rightarrow mn - 2m - 2n < 0$$
$$\Rightarrow mn - 2m - 2n + 4 < 4 \Rightarrow (m - 2)(n - 2) < 4.$$

Since $m, n \geq 3$, we have $(m - 2), (n - 2) \in \mathbf{Z}^+$, and there are only five cases to consider:

1) $(m - 2) = (n - 2) = 1$; $m = n = 3$ (The regular tetrahedron)

2) $(m - 2) = 2, (n - 2) = 1$; $m = 4, n = 3$ (The cube)

3) $(m - 2) = 1, (n - 2) = 2$; $m = 3, n = 4$ (The octahedron)

4) $(m - 2) = 3, (n - 2) = 1$; $m = 5, n = 3$ (The dodecahedron)

5) $(m - 2) = 1, (n - 2) = 3$; $m = 3, n = 5$ (The icosahedron)

The planar graphs for cases 3–5 are shown in Fig. 11.60.

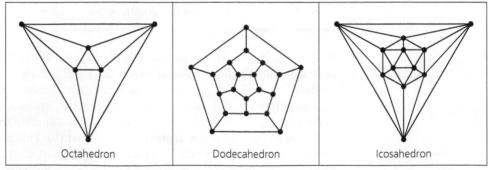

Octahedron   Dodecahedron   Icosahedron

Figure 11.60

The last idea we shall discuss for planar graphs is the notion of a *dual* graph. This concept is also valid for planar graphs with loops and for planar multigraphs. To construct a dual (relative to a particular embedding) for a planar graph or multigraph G with $V = \{a, b, c, d, e, f\}$, place a point (vertex) inside each region, including the infinite region, determined by the graph, as in Fig. 11.61(a). For each edge shared by two regions, draw an edge connecting the vertices inside these regions. For an edge that is traversed twice in the closed walk about the edges of one region, draw a loop at the vertex for this region. In Fig. 11.61(b), G^d is a dual for the graph $G = (V, E)$. From this example we make the following observations:

1) An edge in G corresponds with an edge in G^d, and conversely.

2) A vertex of degree 2 in G yields a pair of edges in G^d that connect the same two vertices. Hence G^d may be a multigraph. (Here vertex e provides the edges $\{a, e\}$, $\{e, f\}$ in G that brought about the two edges connecting v and z in G^d.)

3) Given a loop in G, if the interior of the (finite area) region determined by the loop contains no other vertex or edge of G, then the loop yields a pendant vertex in G^d. (It is also true that a pendant vertex in G yields a loop in G^d.)

4) The degree of a vertex in G^d is the number of edges in the boundary of the closed walk about the region in G that contains that vertex.

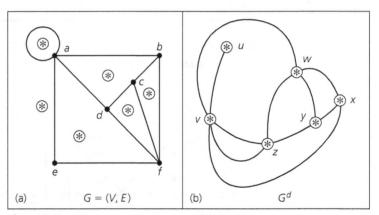

(a) $G = (V, E)$ (b) G^d

Figure 11.61

(Why is G^d called *a* dual of G instead of *the* dual of G? The Section Exercises will show that it is possible to have isomorphic graphs G_1 and G_2 with respective duals G_1^d, G_2^d that are not isomorphic.)

In order to examine further the relationship between a graph G and a dual G^d of G, we introduce the following idea. [Here we recall (from Definition 11.5) that $\kappa(G)$ counts the number of components of G.]

Definition 11.20

Let $G = (V, E)$ be an undirected graph or multigraph. A subset E' of E is called a *cut-set* of G if by removing the edges (but not the vertices) in E' from G, we have $\kappa(G) < \kappa(G')$, where $G' = (V, E - E')$; but when we remove (from E) any proper subset E'' of E', we have $\kappa(G) = \kappa(G'')$, for $G'' = (V, E - E'')$.

EXAMPLE 11.23

For a given connected graph, a cut-set is a *minimal* disconnecting set of edges. In the graph in Fig. 11.62(a), note that each of the sets $\{\{a, b\}, \{a, c\}\}$, $\{\{a, b\}, \{c, d\}\}$, $\{\{e, h\}, \{f, h\}, \{g, h\}\}$, and $\{\{d, f\}\}$ is a cut-set. For the graph in part (b) of the figure, the edge set $\{\{n, p\}, \{r, p\}, \{r, s\}\}$ is a cut-set. Note that the edges in this cut-set are *not* all incident to some single vertex. Here the cut-set separates the vertices m, n, r from the vertices p, s, t. The edge set $\{\{s, t\}\}$ is also a cut-set for this graph — the removal of the edge $\{s, t\}$ from this connected graph results in a subgraph with two components, one of which is the isolated vertex t.

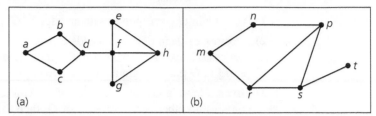

Figure 11.62

Whenever a cut-set for a connected graph consists of only one edge, that edge is called a *bridge* for the graph. For the graph in Fig. 11.62(a), the edge $\{d, f\}$ is the only bridge; the edge $\{s, t\}$ is the only bridge in part (b) of the figure.

We return now to the graphs in Fig. 11.61, redrawing them as shown in Fig. 11.63 in order to emphasize the correspondence between their edges.

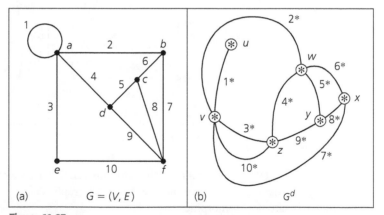

Figure 11.63

Here the edges in G are labeled $1, 2, \ldots, 10$. The numbering scheme for G^d is obtained as follows: The edge labeled 4^*, for example, connects the vertices w and z in G^d. We drew this edge because edge 4 in G was a common edge of the regions containing these vertices. Likewise, edge 7 is common to the region containing x and the infinite region containing v. Hence we label the edge in G^d that connects x and v with 7^*.

In graph G the set of edges labeled $6, 7, 8$ constitutes a cycle. What about the edges labeled $6^*, 7^*, 8^*$ in G^d? If they are removed from G^d, then vertex x becomes isolated and G^d is disconnected. Since we cannot disconnect G^d by removing any proper subset of $\{6^*, 7^*, 8^*\}$, these edges form a cut-set in G^d. In similar fashion, edges 2, 4, 10 form a cut-set in G, whereas in G^d the edges $2^*, 4^*, 10^*$ yield a cycle.

We also have the two-edge cut-set $\{3, 10\}$ in G, and we find that the edges $3^*, 10^*$ provide a two-edge circuit in G^d. Another observation: The one-edge cut set $\{1^*\}$ in G^d comes about from edge 1, a loop in G.

In general, there is a one-to-one correspondence between the following sets of edges in a planar graph G and a dual G^d of G.

1) Cycles (cut-sets) of n (≥ 3) edges in G correspond with cut-sets (cycles) of n edges in G^d.

2) A loop in G corresponds with a one-edge cut-set in G^d.

3) A one-edge cut-set in G corresponds with a loop in G^d.

4) A two-edge cut-set in G corresponds with a two-edge circuit in G^d.

5) If G is a planar multigraph, then each two-edge circuit in G determines a two-edge cut-set of G^d.

All these theoretical observations are interesting, but let us stop here and see how we might apply the idea of a dual.

EXAMPLE 11.24

If we consider the five finite regions in Fig. 11.64(a) as countries on a map, and we construct the subgraph (because we do not use the infinite region) of a dual as shown in part (b), then we find the following relationship.

Suppose we are confronted with the "mapmaker's problem" whereby we want to color the five regions of the map in part (a) so that two countries that share a common border are colored with different colors. This type of coloring can be translated into the dual notion of coloring the vertices in part (b) so that adjacent vertices are colored with different colors. (Such coloring problems will be examined further in Section 11.6.)

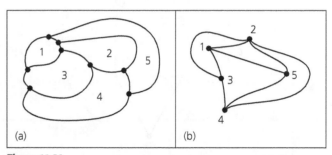

(a)   (b)

Figure 11.64

The final result for this section provides us with an application for an electrical network. This material is based on Example 8.6 on pp. 227–230 of the text by C. L. Liu [23].

EXAMPLE 11.25

In Fig. 11.65 we see an electrical network with nine contacts (switches) that control the excitation of a light. We want to construct a dual network where a second light will go on (off) whenever the light in our given network is off (on).

The contacts (switches) are of two types: normally open (as shown in Fig. 11.65) and normally closed. We use a and a' as in Fig. 11.66 to represent the normally open and normally closed contacts, respectively.

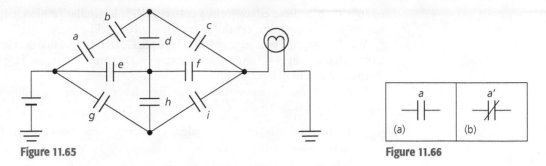

Figure 11.65

Figure 11.66

In Fig. 11.67(a) a *one-terminal-pair-graph* represents the network in Fig. 11.65. Here the special pair of vertices is labeled 1 and 2. These vertices are called the *terminals* of the graph. Also each edge is labeled according to its corresponding contact in Fig. 11.65.

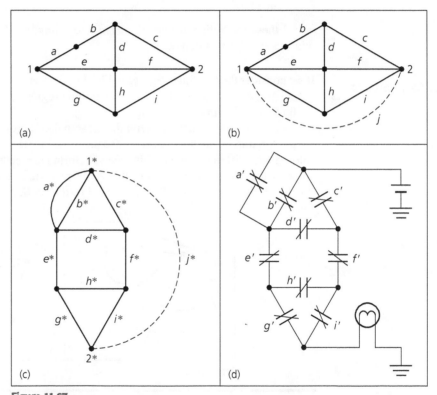

Figure 11.67

A one-terminal-pair-graph G is called a *planar-one-terminal-pair-graph* if G is planar, and the resulting graph is also planar when an edge connecting the terminals is added to G. Figure 11.67(b) shows this situation. Constructing a dual of part (b), we obtain the graph in part (c) of the figure. Removal of the dotted edge results in the terminals 1*, 2* for this dual, which is a one-terminal-pair-graph. This graph provides the dual network in Fig. 11.67(d). We make two observations in closing.

1) When the contacts at a, b, c are closed in the original network (Fig. 11.65), the light is on. In Fig. 11.67(b) the edges a, b, c, j form a cycle that includes the terminals.

In part (c) of the figure, the edges a^*, b^*, c^*, j^* form a cut-set disconnecting the terminals 1^*, 2^*. Finally, with a', b', c' open in part (d) of the figure, no current gets past the first level of contacts (switches) and the light is off.

2) In like manner, the edges c, d, e, g, j form a cut-set that separates the terminals in Fig. 11.67(b). (When the contacts at c, d, e, g are open in Fig. 11.65, the light is off.) Figure 11.67(c) shows how c^*, d^*, e^*, g^*, j^* form a cycle that includes 1^*, 2^*. If c', d', e', g' are closed in part (d), current flows through the dual network and the light is on.

EXERCISES 11.4

1. Verify that the conclusion in Example 11.16 is unchanged if Fig. 11.48(b) has edge $\{a, c\}$ drawn in the exterior of the pentagon.

2. Show that when any edge is removed from K_5, the resulting subgraph is planar. Is this true for the graph $K_{3,3}$?

3. a) How many vertices and how many edges are there in the complete bipartite graphs $K_{4,7}$, $K_{7,11}$, and $K_{m,n}$, where $m, n, \in \mathbf{Z}^+$?

b) If the graph $K_{m,12}$ has 72 edges, what is m?

4. Prove that any subgraph of a bipartite graph is bipartite.

5. For each graph in Fig. 11.68 determine whether or not the graph is bipartite.

6. Let $n \in \mathbf{Z}^+$ with $n \geq 4$. How many subgraphs of K_n are isomorphic to the complete bipartite graph $K_{1,3}$?

7. Let $m, n \in \mathbf{Z}^+$ with $m \geq n \geq 2$. (a) Determine how many distinct cycles of length 4 there are in $K_{m,n}$. (b) How many different paths of length 2 are there in $K_{m,n}$? (c) How many different paths of length 3 are there in $K_{m,n}$?

8. What is the length of a longest path in each of the following graphs?

a) $K_{1,4}$ **b)** $K_{3,7}$ **c)** $K_{7,12}$

d) $K_{m,n}$, where $m, n \in \mathbf{Z}^+$ with $m < n$.

9. How many paths of longest length are there in each of the following graphs? (Remember that a path such as $v_1 \to v_2 \to v_3$ is considered to be the same as the path $v_3 \to v_2 \to v_1$.)

a) $K_{1,4}$ **b)** $K_{3,7}$ **c)** $K_{7,12}$

d) $K_{m,n}$, where $m, n \in \mathbf{Z}^+$ with $m < n$.

10. Can a bipartite graph contain a cycle of odd length? Explain.

11. Let $G = (V, E)$ be a loop-free connected graph with $|V| = v$. If $|E| > (v/2)^2$, prove that G cannot be bipartite.

12. a) Find all the nonisomorphic complete bipartite graphs $G = (V, E)$, where $|V| = 6$.

b) How many nonisomorphic complete bipartite graphs $G = (V, E)$ satisfy $|V| = n \geq 2$?

13. a) Let $X = \{1, 2, 3, 4, 5\}$. Construct the loop-free undirected graph $G = (V, E)$ as follows:

- (V): Let each two-element subset of X represent a vertex in G.
- (E): If $v_1, v_2 \in V$ correspond to subsets $\{a, b\}$ and $\{c, d\}$, respectively, of X, then draw the edge $\{v_1, v_2\}$ in G if $\{a, b\} \cap \{c, d\} = \emptyset$.

b) To what graph is G isomorphic?

14. Determine which of the graphs in Fig. 11.69 are planar. If a graph is planar, redraw it with no edges overlapping. If it is nonplanar, find a subgraph homeomorphic to either K_5 or $K_{3,3}$.

15. Let $m, n \in \mathbf{Z}^+$ with $m \leq n$. Under what condition(s) on m, n will every edge in $K_{m,n}$ be in exactly one of two isomorphic subgraphs of $K_{m,n}$?

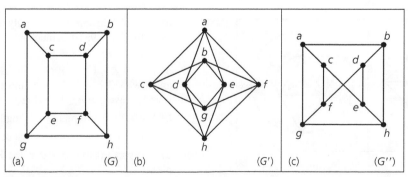

Figure 11.68

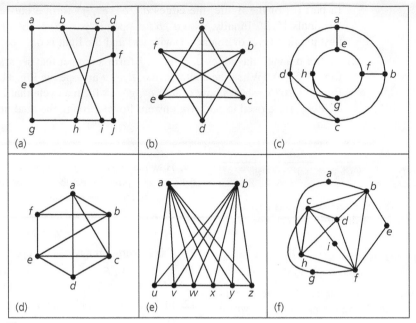

Figure 11.69

16. Prove that the Petersen graph is isomorphic to the graph in Fig. 11.70

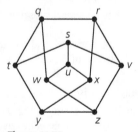

Figure 11.70

17. Determine the number of vertices, the number of edges, and the number of regions for each of the planar graphs in Fig. 11.71. Then show that your answers satisfy Euler's Theorem for connected planar graphs.

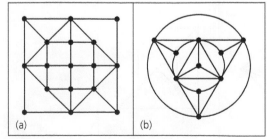

Figure 11.71

18. Let $G = (V, E)$ be an undirected connected loop-free graph. Suppose further that G is planar and determines 53 re-

gions. If, for some planar embedding of G, each region has at least five edges in its boundary, prove that $|V| \geq 82$.

19. Let $G = (V, E)$ be a loop-free connected 4-regular planar graph. If $|E| = 16$, how many regions are there in a planar depiction of G?

20. Suppose that $G = (V, E)$ is a loop-free planar graph with $|V| = v$, $|E| = e$, and $\kappa(G) = $ the number of components of G. (a) State and prove an extension of Euler's Theorem for such a graph. (b) Prove that Corollary 11.3 remains valid if G is loop-free and planar but not connected.

21. Prove that every loop-free connected planar graph has a vertex v with $\deg(v) < 6$.

22. a) Let $G = (V, E)$ be a loop-free connected graph with $|V| \geq 11$. Prove that either G or its complement $\overline{G}$ must be nonplanar.

b) The result in part (a) is actually true for $|V| \geq 9$, but the proof for $|V| = 9$, 10, is much harder. Find a counterexample to part (a) for $|V| = 8$.

23. a) Let $k \in \mathbf{Z}^+$, $k \geq 3$. If $G = (V, E)$ is a connected planar graph with $|V| = v$, $|E| = e$, and each cycle of length at least k, prove that $e \leq \left(\frac{k}{k-2}\right)(v - 2)$.

b) What is the minimal cycle length in $K_{3,3}$?

c) Use parts (a) and (b) to conclude that $K_{3,3}$ is nonplanar.

d) Use part (a) to prove that the Petersen graph is nonplanar.

24. a) Find a dual graph for each of the two planar graphs and the one planar multigraph in Fig. 11.72.

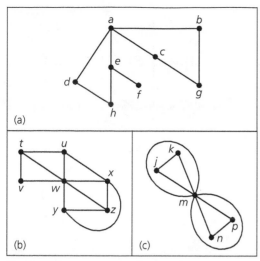

Figure 11.72

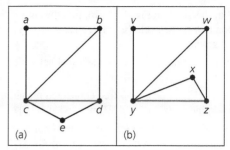

Figure 11.73

1) In Fig. 11.74 we split a vertex, namely r, of G and obtain the graph H, which is disconnected.

2) In Fig. 11.75 we obtain graph (d) from graph (a) by

   i) first splitting the two distinct vertices j and q — disconnecting the graph,

   ii) then reflecting one subgraph about the horizontal axis, and

   iii) then identifying vertex $j(q)$ in one subgraph with vertex $q(j)$ in the other subgraph.

Prove that the dual graphs obtained in part (c) are 2-isomorphic.

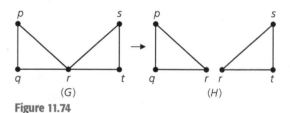

Figure 11.74

e) For the cut-set $\{\{a, b\}, \{c, b\}, \{d, b\}\}$ in part (a) of Fig. 11.73, find the corresponding cycle in its dual. In the

b) Does the dual for the multigraph in part (c) have any pendant vertices? If not, does this contradict the third observation made prior to Definition 11.20?

25. a) Find duals for the planar graphs that correspond with the five Platonic solids.

  b) Find the dual of the graph W_n, the wheel with n spokes (as defined in Exercise 14 of Section 11.1).

26. a) Show that the graphs in Fig. 11.73 are isomorphic.

  b) Draw a dual for each graph.

  c) Show that the duals obtained in part (b) are not isomorphic.

  d) Two graphs G and H are called *2-isomorphic* if one can be obtained from the other by applying either or both of the following procedures a finite number of times.

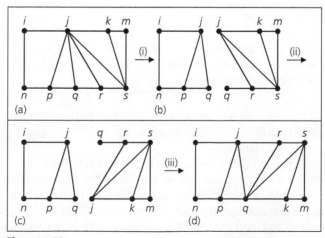

Figure 11.75

dual of the graph in Fig. 11.73(b), find the cut-set that corresponds with the cycle $\{w, z\}, \{z, x\}, \{x, y\}, \{y, w\}$ in the given graph.

27. Find the dual network for the electrical network shown in Fig. 11.76.

28. Let $G = (V, E)$ be a loop-free connected planar graph. If G is isomorphic to its dual and $|V| = n$, what is $|E|$?

29. Let G_1, G_2 be two loop-free connected undirected graphs. If G_1, G_2 are homeomorphic, prove that (a) G_1, G_2 have the same number of vertices of odd degree; (b) G_1 has an Euler trail if and only if G_2 has an Euler trail; and (c) G_1 has an Euler circuit if and only if G_2 has an Euler circuit.

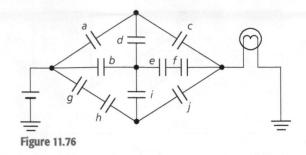

Figure 11.76

11.5
Hamilton Paths and Cycles

In 1859 the Irish mathematician Sir William Rowan Hamilton (1805–1865) developed a game that he sold to a Dublin toy manufacturer. The game consisted of a wooden regular dodecahedron with the 20 corner points (vertices) labeled with the names of prominent cities. The objective of the game was to find a cycle along the edges of the solid so that each city was on the cycle (exactly once). Figure 11.77 is the planar graph for this Platonic solid; such a cycle is designated by the darkened edges. This illustration leads us to the following definition.

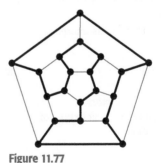

Figure 11.77

Definition 11.21

If $G = (V, E)$ is a graph or multigraph with $|V| \geq 3$, we say that G has a *Hamilton cycle* if there is a cycle in G that contains every vertex in V. A *Hamilton path* is a path (and not a cycle) in G that contains each vertex.

Given a graph with a Hamilton cycle, we find that the deletion of any edge in the cycle results in a Hamilton path. It is possible, however, for a graph to have a Hamilton path without having a Hamilton cycle.

It may seem that the existence of a Hamilton cycle (path) and the existence of an Euler circuit (trail) for a graph are similar problems. The Hamilton cycle (path) is designed to visit each vertex in a graph only once; the Euler circuit (trail) traverses the graph so that each edge is traveled exactly once. Unfortunately, there is no helpful connection between the two ideas, and unlike the situation for Euler circuits (trails), there do not exist necessary and sufficient conditions on a graph G that guarantee the existence of a Hamilton cycle (path). If a graph has a Hamilton cycle, then it will at least be connected. Many theorems exist that establish either necessary or sufficient conditions for a connected graph to have a Hamilton cycle or path. We shall investigate several of these results later. When confronted with particular graphs, however, we shall often resort to trial and error, with a few helpful observations.

EXAMPLE 11.26

Referring back to the hypercubes in Fig. 11.35 we find in Q_2 the cycle

$$00 \longrightarrow 10 \longrightarrow 11 \longrightarrow 01 \longrightarrow 00$$

and in Q_3 the cycle

$$000 \longrightarrow 100 \longrightarrow 110 \longrightarrow 010 \longrightarrow 011 \longrightarrow 111 \longrightarrow 101 \longrightarrow 001 \longrightarrow 000.$$

Hence Q_2 and Q_3 have Hamilton cycles (and paths). In fact, for all $n \geq 2$, we find that Q_n has a Hamilton cycle. (The reader is asked to establish this in the Section Exercises.) [Note, in addition, that the listings: 00, 10, 11, 01 and 000, 100, 110, 010, 011, 111, 101, 001 are examples of Gray codes (which were introduced in Example 3.9).]

EXAMPLE 11.27

If G is the graph in Fig. 11.78, the edges $\{a, b\}, \{b, c\}, \{c, f\}, \{f, e\}, \{e, d\}, \{d, g\}, \{g, h\},$ $\{h, i\}$ yield a Hamilton path for G. But does G have a Hamilton cycle?

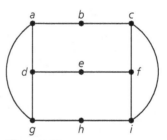

Figure 11.78

Since G has nine vertices, if there is a Hamilton cycle in G it must contain nine edges. Let us start at vertex b and try to build a Hamilton cycle. Because of the symmetry in the graph, it doesn't matter whether we go from b to c or to a. We'll go to c. At c we can go either to f or to i. Using symmetry again, we go to f. Then we delete edge $\{c, i\}$ from further consideration because we cannot return to vertex c. In order to include vertex i in our cycle, we must now go from f to i (to h to g). With edges $\{c, f\}$ and $\{f, i\}$ in the cycle, we cannot have edge $\{e, f\}$ in the cycle. [Otherwise, in the cycle we would have $\deg(f) > 2$.] But then once we get to e we are stuck. Hence there is no Hamilton cycle for the graph.

Example 11.27 indicates a few helpful hints for trying to find a Hamilton cycle in a graph $G = (V, E)$.

1) If G has a Hamilton cycle, then for all $v \in V$, $\deg(v) \geq 2$.

2) If $a \in V$ and $\deg(a) = 2$, then the two edges incident with vertex a must appear in every Hamilton cycle for G.

3) If $a \in V$ and $\deg(a) > 2$, then as we try to build a Hamilton cycle, once we pass through vertex a, any unused edges incident with a are deleted from further consideration.

4) In building a Hamilton cycle for G, we cannot obtain a cycle for a subgraph of G unless it contains all the vertices of G.

Our next example provides an interesting technique for showing that a special type of graph has no Hamilton path.

EXAMPLE 11.28

In Fig. 11.79(a) we have a connected graph G, and we wish to know whether G contains a Hamilton path. Part (b) of the figure provides the same graph with a set of labels x, y. This labeling is accomplished as follows: First we label vertex a with the letter x. Those vertices adjacent to a (namely, b, c, and d) are then labeled with the letter y. Then we label the unlabeled vertices adjacent to b, c, or d with x. This results in the label x on the vertices e, g, and i. Finally, we label the unlabeled vertices adjacent to e, g, or i with the label y. At this point, all the vertices in G are labeled. Now, since $|V| = 10$, if G is to have a Hamilton path there must be an alternating sequence of five x's and five y's. Only four vertices are labeled with x, so this is impossible. Hence G has no Hamilton path (or cycle).

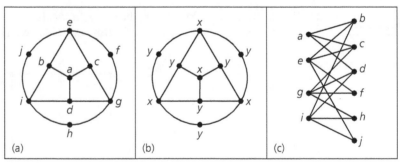

Figure 11.79

But why does this argument work here? In part (c) of Fig. 11.79 we have redrawn the given graph, and we see that it is bipartite. From Exercise 10 in the previous section we know that a bipartite graph cannot have a cycle of odd length. It is also true that if a graph has no cycle of odd length, then it is bipartite. (The proof is requested of the reader in Exercise 9 of this section.) Consequently, whenever a connected graph has no odd cycle (and is bipartite), the method described above may be helpful in determining when the graph does *not* have a Hamilton path. (Exercise 10 in this section examines this idea further.)

Our next example provides an application that calls for Hamilton cycles in a complete graph.

EXAMPLE 11.29

At Professor Alfred's science camp, 17 students have lunch together each day at a circular table. They are trying to get to know one another better, so they make an effort to sit next to two different colleagues each afternoon. For how many afternoons can they do this? How can they arrange themselves on these occasions?

To solve this problem we consider the graph K_n, where $n \geq 3$ and is odd. This graph has n vertices (one for each student) and $\binom{n}{2} = n(n - 1)/2$ edges. A Hamilton cycle in K_n

corresponds to a seating arrangement. Each of these cycles has n edges, so we can have at most $(1/n)\binom{n}{2} = (n-1)/2$ Hamilton cycles with no two having an edge in common.

Consider the circle in Fig. 11.80 and the subgraph of K_n consisting of the n vertices and the n edges $\{1, 2\}, \{2, 3\}, \ldots, \{n-1, n\}, \{n, 1\}$. Keep the vertices on the circumference fixed and rotate this Hamilton cycle clockwise through the angle $[1/(n-1)](2\pi)$. This gives us the Hamilton cycle (Fig. 11.81) made up of edges $\{1, 3\}, \{3, 5\}, \{5, 2\}, \{2, 7\}, \ldots,$ $\{n, n-3\}, \{n-3, n-1\}, \{n-1, 1\}$. This Hamilton cycle has no edge in common with the first cycle. When $n \geq 7$ and we continue to rotate the cycle in Fig. 11.80 in this way through angles $[k/(n-1)](2\pi)$, where $2 \leq k \leq (n-3)/2$, we obtain a total of $(n-1)/2$ Hamilton cycles, no two of which have an edge in common.

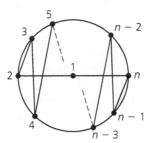

Figure 11.80   Figure 11.81

Therefore the 17 students at the science camp can dine for $[(17-1)/2] = 8$ days before some student will have to sit next to another student for a second time. Using Fig. 11.80 with $n = 17$, we can obtain eight such possible arrangements.

We turn now to some further results on Hamilton paths and cycles. Our first result was established in 1934 by L. Redei.

THEOREM 11.7   Let K_n^* be a complete directed graph — that is, K_n^* has n vertices and for each distinct pair x, y of vertices, exactly one of the edges (x, y) or (y, x) is in K_n^*. Such a graph (called a *tournament*) always contains a (directed) Hamilton path.

Proof: Let $m \geq 2$ with p_m a path containing the $m-1$ edges $(v_1, v_2), (v_2, v_3), \ldots,$ (v_{m-1}, v_m). If $m = n$, we're finished. If not, let v be a vertex that doesn't appear in p_m.

If (v, v_1) is an edge in K_n^*, we can extend p_m by adjoining this edge. If not, then (v_1, v) must be an edge. Now suppose that (v, v_2) is in the graph. Then we have the larger path: $(v_1, v), (v, v_2), (v_2, v_3), \ldots, (v_{m-1}, v_m)$. If (v, v_2) is not an edge in K_n^*, then (v_2, v) must be. As we continue this process there are only two possibilities: (a) For some $1 \leq k \leq m-1$ the edges $(v_k, v), (v, v_{k+1})$ are in K_n^* and we replace (v_k, v_{k+1}) with this pair of edges; or (b) (v_m, v) is in K_n^* and we add this edge to p_m. Either case results in a path p_{m+1} that includes $m + 1$ vertices and has m edges. This process can be repeated until we have such a path p_n on n vertices.

EXAMPLE 11.30   In a round-robin tournament each player plays every other player exactly once. We want to somehow rank the players according to the results of the tournament. Since we could have players a, b, and c where a beats b and b beats c, but c beats a, it is not always possible to have a ranking where a player in a certain position has beaten all of the opponents in later positions. Representing the players by vertices, construct a directed graph G on these vertices by drawing edge (x, y) if x beats y. Then by Theorem 11.7, it is possible to list the players such that each has beaten the next player on the list.

THEOREM 11.8

Let $G = (V, E)$ be a loop-free graph with $|V| = n \geq 2$. If $\deg(x) + \deg(y) \geq n - 1$ for all $x, y \in V$, $x \neq y$, then G has a Hamilton path.

Proof: First we prove that G is connected. If not, let C_1, C_2 be two components of G and let $x, y \in V$ with x a vertex in C_1 and y a vertex in C_2. Let C_i have n_i vertices, $i = 1, 2$. Then $\deg(x) \leq n_1 - 1$, $\deg(y) \leq n_2 - 1$, and $\deg(x) + \deg(y) \leq (n_1 + n_2) - 2 \leq n - 2$, contradicting the condition given in the theorem. Consequently, G is connected.

Now we build a Hamilton path for G. For $m \geq 2$, let p_m be the path $\{v_1, v_2\}, \{v_2, v_3\}, \ldots, \{v_{m-1}, v_m\}$ of length $m - 1$. (We relabel vertices if necessary.) Such a path exists, because for $m = 2$ all that is needed is one edge. If v_1 is adjacent to any vertex v other than $v_2, v_3, \ldots, v_m$, we add the edge $\{v, v_1\}$ to p_m to get p_{m+1}. The same type of procedure is carried out if v_m is adjacent to a vertex other than $v_1, v_2, \ldots, v_{m-1}$. If we are able to enlarge p_m to p_n in this way, we get a Hamilton path. Otherwise the path $p_m: \{v_1, v_2\}, \ldots, \{v_{m-1}, v_m\}$ has v_1, v_m adjacent only to vertices in p_m, and $m < n$. When this happens we claim that G contains a cycle on these vertices. If v_1 and v_m are adjacent, then the cycle is $\{v_1, v_2\}, \{v_2, v_3\}, \ldots, \{v_{m-1}, v_m\}, \{v_m, v_1\}$. If v_1 and v_m are not adjacent, then v_1 is adjacent to a subset S of the vertices in $\{v_2, v_3, \ldots, v_{m-1}\}$. If there is a vertex $v_t \in S$ such that v_m is adjacent to v_{t-1}, then we can get the cycle by adding $\{v_1, v_t\}, \{v_{t-1}, v_m\}$ to p_m and deleting $\{v_{t-1}, v_t\}$ as shown in Fig. 11.82. If not, let $|S| = k < m - 1$. Then $\deg(v_1) = k$ and $\deg(v_m) \leq (m - 1) - k$, and we have the contradiction $\deg(v_1) + \deg(v_m) \leq m - 1 < n - 1$. Hence there is a cycle connecting $v_1, v_2, \ldots, v_m$.

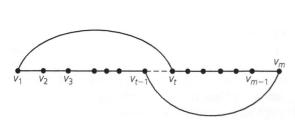

Figure 11.82

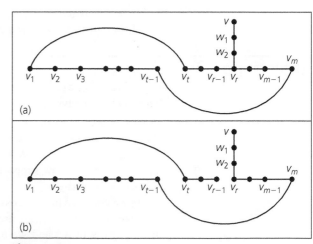

Figure 11.83

Now consider a vertex $v \in V$ that is not found on this cycle. The graph G is connected, so there is a path from v to a first vertex v_r in the cycle, as shown in Fig. 11.83(a). Removing the edge $\{v_{r-1}, v_r\}$ (or $\{v_1, v_t\}$ if $r = t$), we get the path (longer than the original p_m) shown in Fig. 11.83(b). Repeating this process (applied to p_m) for the path in Fig. 11.83(b), we continue to increase the length of the path until it includes every vertex of G.

COROLLARY 11.4   Let $G = (V, E)$ be a loop-free graph with n (≥ 2) vertices. If $\deg(v) \geq (n - 1)/2$ for all $v \in V$, then G has a Hamilton path.

Proof: The proof is left as an exercise for the reader.

Our last theorem for this section provides a sufficient condition for the existence of a Hamilton cycle in a loop-free graph. This was first proved by Oystein Ore in 1960.

THEOREM 11.9   Let $G = (V, E)$ be a loop-free undirected graph with $|V| = n \geq 3$. If $\deg(x) + \deg(y) \geq n$ for all nonadjacent $x, y \in V$, then G contains a Hamilton cycle.

Proof: Assume that G does not contain a Hamilton cycle. We add edges to G until we arrive at a subgraph H of K_n, where H has no Hamilton cycle, but, for any edge e (of K_n) not in H, $H + e$ does have a Hamilton cycle.

Since $H \neq K_n$, there are vertices $a, b \in V$, where $\{a, b\}$ is not an edge of H but $H + \{a, b\}$ has a Hamilton cycle C. The graph H has no such cycle, so the edge $\{a, b\}$ is a part of cycle C. Let us list the vertices of H (and G) on cycle C as follows:

$$a\,(= v_1) \rightarrow b\,(= v_2) \rightarrow v_3 \rightarrow v_4 \rightarrow \cdots \rightarrow v_{n-1} \rightarrow v_n$$

For each $3 \leq i \leq n$, if the edge $\{b, v_i\}$ is in the graph H, then we claim that the edge $\{a, v_{i-1}\}$ cannot be an edge of H. For if both of these edges are in H, for some $3 \leq i \leq n$, then we get the Hamilton cycle

$$b \rightarrow v_i \rightarrow v_{i+1} \rightarrow \cdots \rightarrow v_{n-1} \rightarrow v_n \rightarrow a \rightarrow v_{i-1} \rightarrow v_{i-2} \rightarrow \cdots v_4 \rightarrow v_3$$

for the graph H (which has no Hamilton cycle). Therefore, for each $3 \leq i \leq n$, at most one of the edges $\{b, v_i\}$, $\{a, v_{i-1}\}$ is in H. Consequently,

$$\deg_H(a) + \deg_H(b) < n,$$

where $\deg_H(v)$ denotes the degree of vertex v in graph H. For all $v \in V$, $\deg_H(v) \geq \deg_G(v) = \deg(v)$, so we have nonadjacent (in G) vertices a, b, where

$$\deg(a) + \deg(b) < n.$$

This contradicts the hypothesis that $\deg(x) + \deg(y) \geq n$ for all nonadjacent $x, y \in V$, so we reject our assumption and find that G contains a Hamilton cycle.

Now we shall obtain the following two results from Theorem 11.9. Each will give us a sufficient condition for a loop-free undirected graph $G = (V, E)$ to have a Hamilton cycle. The first result is similar to Corollary 11.4 and is concerned with the degree of each vertex v in V. The second result examines the size of the edge set E.

COROLLARY 11.5   If $G = (V, E)$ is a loop-free undirected graph with $|V| = n \geq 3$, and if $\deg(v) \geq n/2$ for all $v \in V$, then G has a Hamilton cycle.

Proof: We shall leave the proof of this result for the Section Exercises.

COROLLARY 11.6   If $G = (V, E)$ is a loop-free undirected graph with $|V| = n \geq 3$, and if $|E| \geq \binom{n-1}{2} + 2$, then G has a Hamilton cycle.

Proof: Let $a, b \in V$, where $\{a, b\} \notin E$. [Since a, b are nonadjacent, we want to show that $\deg(a) + \deg(b) \geq n$.] Remove the following from the graph G: (i) all edges of the form $\{a, x\}$, where $x \in V$; (ii) all edges of the form $\{y, b\}$, where $y \in V$; and (iii) the vertices a and b. Let $H = (V', E')$ denote the resulting subgraph. Then $|E| = |E'| + \deg(a) + \deg(b)$ because $\{a, b\} \notin E$.

Since $|V'| = n - 2$, H is a subgraph of the complete graph K_{n-2}, so $|E'| \leq \binom{n-2}{2}$. Consequently, $\binom{n-1}{2} + 2 \leq |E| = |E'| + \deg(a) + \deg(b) \leq \binom{n-2}{2} + \deg(a) + \deg(b)$, and we find that

$$\deg(a) + \deg(b) \geq \binom{n-1}{2} + 2 - \binom{n-2}{2}$$

$$= \left(\frac{1}{2}\right)(n-1)(n-2) + 2 - \left(\frac{1}{2}\right)(n-2)(n-3)$$

$$= \left(\frac{1}{2}\right)(n-2)[(n-1) - (n-3)] + 2$$

$$= \left(\frac{1}{2}\right)(n-2)(2) + 2 = (n-2) + 2 = n.$$

Therefore it follows from Theorem 11.9 that the given graph G has a Hamilton cycle.

A problem that is related to the search for Hamilton cycles in a graph is the *traveling salesman problem*. (An article dealing with this problem was published by Thomas P. Kirkman in 1855.) Here a traveling salesperson leaves his or her home and must visit certain locations before returning. The objective is to find an order in which to visit the locations that is most efficient (perhaps in terms of total distance traveled or total cost). The problem can be modeled with a labeled (edges have distances or costs associated with them) graph where the most efficient Hamilton cycle is sought.

The references by R. Bellman, K. L. Cooke, and J. A. Lockett [7]; M. Bellmore and G. L. Nemhauser [8]; E. A. Elsayed [15]; E. A. Elsayed and R. G. Stern [16]; and L. R. Foulds [17] should prove interesting to the reader who wants to learn more about this important optimization problem. Also, the text edited by E. L. Lawler, J. K. Lenstra, A. H. G. Rinnooy Kan, and D. B. Shmoys [22] presents 12 papers on various facets of this problem.

Even more on the traveling salesman problem and its applications can be found in the handbooks edited by M. O. Ball, T. L. Magnanti, C. L. Monma, and G. L. Nemhauser — in particular, the articles by R. K. Ahuja, T. L. Magnanti, J. B. Orlin, and M. R. Reddy [2], and by M. Jünger, G. Reinelt, and G. Rinaldi [21].

EXERCISES 11.5

1. Give an example of a connected graph that has (a) Neither an Euler circuit nor a Hamilton cycle. (b) An Euler circuit but no Hamilton cycle. (c) A Hamilton cycle but no Euler circuit. (d) Both a Hamilton cycle and an Euler circuit.

2. Characterize the type of graph in which an Euler trail (circuit) is also a Hamilton path (cycle).

3. Find a Hamilton cycle, if one exists, for each of the graphs or multigraphs in Fig. 11.84. If the graph has no Hamilton cycle, determine whether it has a Hamilton path.

4. a) Show that the Petersen graph [Fig. 11.52(a)] has no Hamilton cycle but that it has a Hamilton path.

b) Show that if any vertex (and the edges incident to it) is removed from the Petersen graph, then the resulting subgraph has a Hamilton cycle.

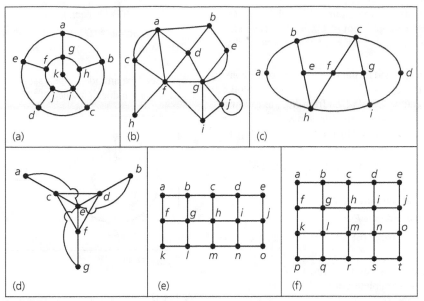

Figure 11.84

5. Consider the graphs in parts (d) and (e) of Fig. 11.84. Is it possible to remove one vertex from each of these graphs so that each of the resulting subgraphs has a Hamilton cycle?

6. If $n \geq 3$, how many different Hamilton cycles are there in the wheel graph W_n? (The graph W_n was defined in Exercise 14 of Section 11.1.)

7. a) For $n \geq 3$, how many different Hamilton cycles are there in the complete graph K_n?

b) How many edge-disjoint Hamilton cycles are there in K_{21}?

c) Nineteen students in a nursery school play a game each day where they hold hands to form a circle. For how many days can they do this with no student holding hands with the same playmate twice?

8. a) For $n \in \mathbf{Z}^+$, $n \geq 2$, show that the number of distinct Hamilton cycles in the graph $K_{n,n}$ is $(1/2)(n-1)! \, n!$.

b) How many different Hamilton paths are there for $K_{n,n}$, $n \geq 1$?

9. Let $G = (V, E)$ be a loop-free undirected graph. Prove that if G contains no cycle of odd length, then G is bipartite.

10. a) Let $G = (V, E)$ be a connected bipartite undirected graph with V partitioned as $V_1 \cup V_2$. Prove that if $|V_1| \neq |V_2|$, then G cannot have a Hamilton cycle.

b) Prove that if the graph G in part (a) has a Hamilton path, then $|V_1| - |V_2| = \pm 1$.

c) Give an example of a connected bipartite undirected graph $G = (V, E)$, where V is partitioned as $V_1 \cup V_2$ and $|V_1| = |V_2| - 1$, but G has no Hamilton path.

11. a) Determine all nonisomorphic tournaments with three vertices.

b) Find all of the nonisomorphic tournaments with four vertices. List the in degree and the out degree for each vertex, in each of these tournaments.

12. Prove that for $n \geq 2$, the hypercube Q_n has a Hamilton cycle.

13. Let $T = (V, E)$ be a tournament with $v \in V$ of maximum out degree. If $w \in V$ and $w \neq v$, prove that either $(v, w) \in E$ or there is a vertex y in V where $y \neq v, w$, and $(v, y), (y, w) \in E$. (Such a vertex v is called a *king* for the tournament.)

14. Find a counterexample to the converse of Theorem 11.8.

15. Give an example of a loop-free connected undirected multigraph $G = (V, E)$ such that $|V| = n$ and $\deg(x) + \deg(y) \geq n - 1$ for all $x, y \in V$, but G has no Hamilton path.

16. Prove Corollaries 11.4 and 11.5.

17. Give an example to show that the converse of Corollary 11.5 need not be true.

18. Helen and Dominic invite 10 friends to dinner. In this group of 12 people everyone knows at least 6 others. Prove that the 12 can be seated around a circular table in such a way that each person is acquainted with the persons sitting on either side.

19. Let $G = (V, E)$ be a loop-free undirected graph that is 6-regular. Prove that if $|V| = 11$, then G contains a Hamilton cycle.

20. Let $G = (V, E)$ be a loop-free undirected n-regular graph with $|V| \geq 2n + 2$. Prove that $\overline{G}$ (the complement of G) has a Hamilton cycle.

21. For $n \geq 3$, let C_n denote the undirected cycle on n vertices. The graph $\overline{C}_n$, the complement of C_n, is often called the *cocycle* on n vertices. Prove that for $n \geq 5$ the cocycle $\overline{C}_n$ has a Hamilton cycle.

22. Let $n \in \mathbf{Z}^+$ with $n \geq 4$, and let the vertex set V' for the complete graph K_{n-1} be $\{v_1, v_2, v_3, \ldots, v_{n-1}\}$. Now construct the loop-free undirected graph $G_n = (V, E)$ from K_{n-1} as follows: $V = V' \cup \{v\}$, and E consists of all the edges in K_{n-1} except for the edge $\{v_1, v_2\}$, which is replaced by the pair of edges $\{v_1, v\}$ and $\{v, v_2\}$.

   a) Determine $\deg(x) + \deg(y)$ for all nonadjacent vertices x and y in V.

   b) Does G_n have a Hamilton cycle?

   c) How large is the edge set E?

   d) Do the results in parts (b) and (c) contradict Corollary 11.6?

23. For $n \in \mathbf{Z}^+$ where $n \geq 4$, let $V' = \{v_1, v_2, v_3, \ldots, v_{n-1}\}$ be the vertex set for the complete graph K_{n-1}. Construct the loop-free undirected graph $H_n = (V, E)$ from K_{n-1} as follows: $V = V' \cup \{v\}$, and E consists of all the edges in K_{n-1} together with the new edge $\{v, v_1\}$.

   a) Show that H_n has a Hamilton path but no Hamilton cycle.

   b) How large is the edge set E?

24. Let $n = 2^k$ for $k \in \mathbf{Z}^+$. We use the n k-bit sequences (of 0's and 1's) to represent $1, 2, 3, \ldots, n$, so that for two consecutive integers $i, i + 1$, the corresponding k-bit sequences differ in exactly one component. This representation is called a *Gray code* (comparable to what we saw in Example 3.9).

   a) For $k = 3$, use a graph model with $V = \{000, 001, 010, \ldots, 111\}$ to find such a code for $1, 2, 3, \ldots, 8$. How is this related to the concept of a Hamilton path?

   b) Answer part (a) for $k = 4$.

25. If $G = (V, E)$ is an undirected graph, a subset I of V is called *independent* if no two vertices in I are adjacent. An independent set I is called *maximal* if no vertex v can be added to I with $I \cup \{v\}$ independent. The *independence number* of G, denoted $\beta(G)$, is the size of a largest independent set in G.

   a) For each graph in Fig. 11.85 find two maximal independent sets with different sizes.

   b) Find $\beta(G)$ for each graph in part (a).

   c) Determine $\beta(G)$ for each of the following graphs: (i) $K_{1,3}$; (ii) $K_{2,3}$; (iii) $K_{3,2}$; (iv) $K_{4,4}$; (v) $K_{4,6}$; (vi) $K_{m,n}$, $m, n \in \mathbf{Z}^+$.

   d) Let I be an independent set in $G = (V, E)$. What type of subgraph does I induce in $\overline{G}$?

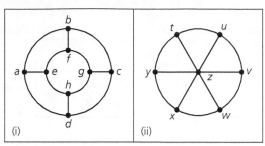

Figure 11.85

26. Let $G = (V, E)$ be an undirected graph with subset I of V an independent set. For each $a \in I$ and each Hamilton cycle C for G, there will be $\deg(a) - 2$ edges in E that are incident with a and not in C. Therefore there are at least $\sum_{a \in I} [\deg(a) - 2] = \sum_{a \in I} \deg(a) - 2|I|$ edges in E that do not appear in C.

   a) Why are these $\sum_{a \in I} \deg(a) - 2|I|$ edges distinct?

   b) Let $v = |V|$, $e = |E|$. Prove that if

   $$e - \sum_{a \in I} \deg(a) + 2|I| < v,$$

   then G has no Hamilton cycle.

   c) Select a suitable independent set I and use part (b) to show that the graph in Fig. 11.86 (known as the Herschel graph) has no Hamilton cycle.

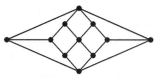

Figure 11.86

11.6
Graph Coloring
and Chromatic Polynomials

At the J. & J. Chemical Company, Jeannette is in charge of the storage of chemical compounds in the company warehouse. Since certain types of compounds (such as acids and bases) should not be kept in the same vicinity, she decides to have her partner Jack partition the warehouse into separate storage areas so that incompatible chemical reagents can be stored in separate compartments. How can she determine the number of storage compartments that Jack will have to build?

If this company sells 25 chemical compounds, let $\{c_1, c_2, \ldots, c_{25}\} = V$, a set of vertices. For all $1 \le i < j \le 25$, we draw the edge $\{c_i, c_j\}$ if c_i and c_j must be stored in separate compartments. This gives us an undirected graph $G = (V, E)$.

We now introduce the following concept.

Definition 11.22

If $G = (V, E)$ is an undirected graph, a *proper coloring* of G occurs when we color the vertices of G so that if $\{a, b\}$ is an edge in G, then a and b are colored with different colors. (Hence adjacent vertices have different colors.) The minimum number of colors needed to properly color G is called the *chromatic number* of G and is written $\chi(G)$.

Returning to assist Jeannette at the warehouse, we find that the number of storage compartments Jack must build is equal to $\chi(G)$ for the graph we constructed on $V = \{c_1, c_2, \ldots, c_{25}\}$. But how do we compute $\chi(G)$? Before we present any work on how to determine the chromatic number of a graph, we turn to the following related idea.

In Example 11.24 we mentioned the connection between coloring the regions in a planar map (with neighboring regions having different colors) and properly coloring the vertices in an associated graph. Determining the smallest number of colors needed to color planar maps in this way has been a problem of interest for over a century.

In about 1850, Francis Guthrie (1831–1899) became interested in the general problem after showing how to color the counties on a map of England with only four colors. Shortly thereafter, he showed the "Four-color Problem" to his younger brother Frederick (1833–1866), who was then a student of Augustus DeMorgan (1806–1871). DeMorgan communicated the problem (in 1852) to William Hamilton (1805–1865). The problem did not interest Hamilton and lay dormant for about 25 years. Then, in 1878, the scientific community was made aware of the problem through an announcement by Arthur Cayley (1821–1895) at a meeting of the London Mathematical Society. In 1879 Cayley stated the problem in the first volume of the *Proceedings of the Royal Geographical Society*. Shortly thereafter, the British barrister (and keen amateur mathematician) Sir Alfred Kempe (1849–1922) devised a proof that remained unquestioned for over a decade. In 1890, however, the British mathematician Percy John Heawood (1861–1955) found a mistake in Kempe's work.

The problem remained unsolved until 1976, when it was finally settled by Kenneth Appel and Wolfgang Haken. Their proof employs a very intricate computer analysis of 1936 (reducible) configurations.

Although only four colors are needed to properly color the regions in a planar map, we need more than four colors to properly color the vertices of some nonplanar graphs.

We start with some small examples. Then we shall find a way to determine $\chi(G)$ from smaller subgraphs of G—in certain situations. [In general, computing $\chi(G)$ is a very difficult problem.] We shall also obtain what is called the chromatic polynomial for G and see how it can be used in computing $\chi(G)$.

EXAMPLE 11.31

For the graph G in Fig. 11.87, we start at vertex a and next to each vertex write the number of a color needed to properly color the vertices of G that have been considered up to that point. Going to vertex b, the 2 indicates the need for a second color because vertices a and b are adjacent. Proceeding alphabetically to f, we find that two colors are needed to properly color $\{a, b, c, d, e, f\}$. For vertex g a third color is needed; this third color can also be used for vertex h because $\{g, h\}$ is not an edge in G. Thus this sequential coloring (labeling) method gives us a proper coloring for G, so $\chi(G) \leq 3$. Since K_3 is a subgraph of G [for example, the subgraph induced by a, b and g is (isomorphic to) K_3], we have $\chi(G) \geq 3$, so $\chi(G) = 3$.

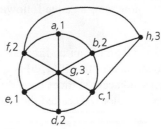

Figure 11.87

EXAMPLE 11.32

a) For all $n \geq 1$, $\chi(K_n) = n$.

b) The chromatic number of the Herschel graph (Fig. 11.86) is 2.

c) If G is the Petersen graph [see Fig. 11.52 (a)], then $\chi(G) = 3$.

EXAMPLE 11.33

Let G be the graph shown in Fig. 11.88. For $U = \{b, f, h, i\}$, the induced subgraph $\langle U \rangle$ of G is isomorphic to K_4, so $\chi(G) \geq \chi(K_4) = 4$. Therefore, if we can determine a way to properly color the vertices of G with four colors, then we shall know that $\chi(G) = 4$. One way to accomplish this is to color the vertices e, f, g blue; the vertices b, j red; the vertices c, h white; and the vertices a, d, i green.

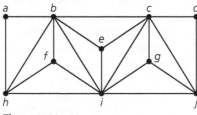

Figure 11.88

We turn now to a method for determining $\chi(G)$. Our coverage follows the development in the survey article [25] by R. C. Read.

Let G be an undirected graph, and let λ be the number of colors that we have available for properly coloring the vertices of G. Our objective is to find a polynomial function $P(G, \lambda)$, in the variable λ, called the *chromatic polynomial* of G, that will tell us in how many different ways we can properly color the vertices of G, using at most λ colors.

Throughout this discussion, the vertices in an undirected graph $G = (V, E)$ are distinguished by labels. Consequently, two proper colorings of such a graph will be considered different in the following sense: A proper coloring (of the vertices of G) that uses at most λ colors is a function f, with domain V and codomain $\{1, 2, 3, \ldots, \lambda\}$, where $f(u) \neq f(v)$, for adjacent vertices $u, v \in V$. Proper colorings are then different in the same way that these functions are different.

EXAMPLE 11.34

a) If $G = (V, E)$ with $|V| = n$ and $E = \emptyset$, then G consists of n isolated points, and by the rule of product, $P(G, \lambda) = \lambda^n$.

b) If $G = K_n$, then at least n colors must be available for us to color G properly. Here, by the rule of product, $P(G, \lambda) = \lambda(\lambda - 1)(\lambda - 2) \cdots (\lambda - n + 1)$, which we denote by $\lambda^{(n)}$. For $\lambda < n$, $P(G, \lambda) = 0$ and there are no ways to properly color K_n. $P(G, \lambda) > 0$ for the first time when $\lambda = n = \chi(G)$.

c) For each path in Fig. 11.89, we consider the number of choices (of the λ colors) at each successive vertex. Proceeding alphabetically, we find that $P(G_1, \lambda) = \lambda(\lambda - 1)^3$ and $P(G_2, \lambda) = \lambda(\lambda - 1)^4$. Since $P(G_1, 1) = 0 = P(G_2, 1)$, but $P(G_1, 2) = 2 = P(G_2, 2)$, it follows that $\chi(G_1) = \chi(G_2) = 2$. If five colors are available we can properly color G_1 in $5(4)^3 = 320$ ways; G_2 can be so colored in $5(4)^4 = 1280$ ways.

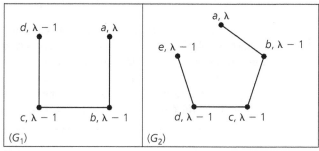

Figure 11.89

In general, if G is a path on n vertices, then $P(G, \lambda) = \lambda(\lambda - 1)^{n-1}$.

d) If G is made up of components $G_1, G_2, \ldots, G_k$, then again by the rule of product, it follows that $P(G, \lambda) = P(G_1, \lambda) \cdot P(G_2, \lambda) \cdots P(G_k, \lambda)$.

As a result of Example 11.34(d), we shall concentrate on connected graphs. In many instances in discrete mathematics, methods have been employed to solve problems in large cases by breaking these down into two or more smaller cases. Once again we use this method of attack. To do so, we need the following ideas and notation.

Let $G = (V, E)$ be an undirected graph. For $e = \{a, b\} \in E$, let G_e denote the subgraph of G obtained by deleting e from G, without removing vertices a and b; that is, $G_e = G - e$ as defined in Section 11.2. From G_e a second subgraph of G is obtained by coalescing (or, identifying) the vertices a and b. This second subgraph is denoted by G'_e.

EXAMPLE 11.35

Figure 11.90 shows G_e and G'_e for graph G with the edge e as specified. Note how the coalescing of a and b in G'_e results in the coalescing of the two pairs of edges $\{d, b\}$, $\{d, a\}$ and $\{a, c\}$, $\{b, c\}$.

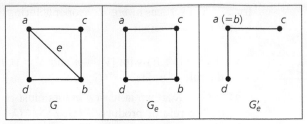

Figure 11.90

Using these special subgraphs, we turn now to the main result.

THEOREM 11.10

Decomposition Theorem for Chromatic Polynomials. If $G = (V, E)$ is a connected graph and $e \in E$, then

$$P(G_e, \lambda) = P(G, \lambda) + P(G'_e, \lambda).$$

Proof: Let $e = \{a, b\}$. The number of ways to properly color the vertices in G_e with (at most) λ colors is $P(G_e, \lambda)$. Those colorings where a and b have different colors are proper colorings of G. The colorings of G_e that are not proper colorings of G occur when a and b have the same color. But each of these colorings corresponds with a proper coloring for G'_e. This partition of the $P(G_e, \lambda)$ proper colorings of G_e into two disjoint subsets results in the formula $P(G_e, \lambda) = P(G, \lambda) + P(G'_e, \lambda)$.

When calculating chromatic polynomials, we shall place brackets about a graph to indicate its chromatic polynomial.

EXAMPLE 11.36

The following calculations yield $P(G, \lambda)$ for G a cycle of length 4.

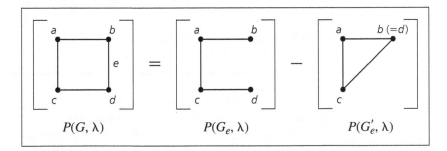

From Example 11.34(c) it follows that $P(G_e, \lambda) = \lambda(\lambda - 1)^3$. With $G'_e = K_3$ we have $P(G'_e, \lambda) = \lambda^{(3)}$. Therefore,

$$P(G, \lambda) = \lambda(\lambda - 1)^3 - \lambda(\lambda - 1)(\lambda - 2) = \lambda(\lambda - 1)[(\lambda - 1)^2 - (\lambda - 2)]$$
$$= \lambda(\lambda - 1)[\lambda^2 - 3\lambda + 3] = \lambda^4 - 4\lambda^3 + 6\lambda^2 - 3\lambda.$$

Since $P(G, 1) = 0$ while $P(G, 2) = 2 > 0$, we know that $\chi(G) = 2$.

EXAMPLE 11.37

Here we find a second application of Theorem 11.10.

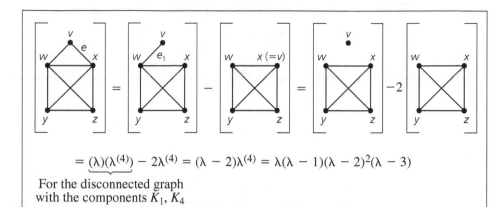

$$= \underbrace{(\lambda)(\lambda^{(4)})}_{\text{For the disconnected graph}} - 2\lambda^{(4)} = (\lambda - 2)\lambda^{(4)} = \lambda(\lambda - 1)(\lambda - 2)^2(\lambda - 3)$$

For the disconnected graph
with the components K_1, K_4

For each $1 \leq \lambda \leq 3$, $P(G, \lambda) = 0$, but $P(G, \lambda) > 0$ for all $\lambda \geq 4$. Consequently, the given graph has chromatic number 4.

The chromatic polynomials given in Examples 11.36 and 11.37 suggest the following results.

THEOREM 11.11

For each graph G, the constant term in $P(G, \lambda)$ is 0.

Proof: For each graph G, $\chi(G) > 0$ because $V \neq \emptyset$. If $P(G, \lambda)$ has constant term a, then $P(G, 0) = a \neq 0$. This implies that there are a ways to color G properly with 0 colors, a contradiction.

THEOREM 11.12

Let $G = (V, E)$ with $|E| > 0$. Then the sum of the coefficients in $P(G, \lambda)$ is 0.

Proof: Since $|E| \geq 1$, we have $\chi(G) \geq 2$, so we cannot properly color G with only one color. Consequently, $P(G, 1) = 0 =$ the sum of the coefficients in $P(G, \lambda)$.

Since the chromatic polynomial of a complete graph is easy to determine, an alternative method for finding $P(G, \lambda)$ can be obtained. Theorem 11.10 reduced the problem to smaller graphs. Here we add edges to a given graph until we reach complete graphs.

THEOREM 11.13

Let $G = (V, E)$, with $a, b \in V$ but $\{a, b\} = e \notin E$. We write G_e^+ for the graph we obtain from G by adding the edge $e = \{a, b\}$. Coalescing the vertices a and b in G gives us the subgraph G_e^{++} of G. Under these circumstances $P(G, \lambda) = P(G_e^+, \lambda) + P(G_e^{++}, \lambda)$.

Proof: This result follows as in Theorem 11.10 because $P(G_e^+, \lambda) = P(G, \lambda) - P(G_e^{++}, \lambda)$.

EXAMPLE 11.38

Let us now apply Theorem 11.13.

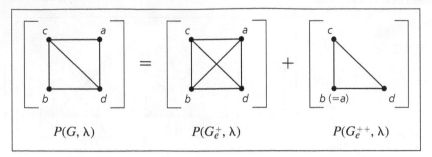

$$P(G, \lambda) \qquad\qquad P(G_e^+, \lambda) \qquad\qquad P(G_e^{++}, \lambda)$$

Here $P(G, \lambda) = \lambda^{(4)} + \lambda^{(3)} = \lambda(\lambda - 1)(\lambda - 2)^2$, so $\chi(G) = 3$. In addition, if six colors are available, the vertices in G can be properly colored in $6(5)(4)^2 = 480$ ways.

Our next result again uses complete graphs — along with the following concepts. For all graphs $G_1 = (V_1, E_1)$ and $G_2 = (V_2, E_2)$.

i) the *union* of G_1 and G_2, denoted $G_1 \cup G_2$, is the graph with vertex set $V_1 \cup V_2$ and edge set $E_1 \cup E_2$; and

ii) when $V_1 \cap V_2 \neq \emptyset$, the *intersection* of G_1 and G_2, denoted $G_1 \cap G_2$, is the graph with vertex set $V_1 \cap V_2$ and edge set $E_1 \cap E_2$.

THEOREM 11.14

Let G be an undirected graph with subgraphs G_1, G_2. If $G = G_1 \cup G_2$ and $G_1 \cap G_2 = K_n$, for some $n \in \mathbf{Z}^+$, then

$$P(G, \lambda) = \frac{P(G_1, \lambda) \cdot P(G_2, \lambda)}{\lambda^{(n)}}.$$

Proof: Since $G_1 \cap G_2 = K_n$, it follows that K_n is a subgraph of both G_1 and G_2 and that $\chi(G_1), \chi(G_2) \geq n$. Given λ colors, there are $\lambda^{(n)}$ proper colorings of K_n. For each of these $\lambda^{(n)}$ colorings there are $P(G_1, \lambda)/\lambda^{(n)}$ ways to properly color the remaining vertices in G_1. Likewise, there are $P(G_2, \lambda)/\lambda^{(n)}$ ways to properly color the remaining vertices in G_2. By the rule of product,

$$P(G, \lambda) = P(K_n, \lambda) \cdot \frac{P(G_1, \lambda)}{\lambda^{(n)}} \cdot \frac{P(G_2, \lambda)}{\lambda^{(n)}} = \frac{P(G_1, \lambda) \cdot P(G_2, \lambda)}{\lambda^{(n)}}.$$

EXAMPLE 11.39

Consider the graph in Example 11.37. Let G_1 be the subgraph induced by the vertices w, x, y, z. Let G_2 be the complete graph K_3 — with vertices $v, w,$ and x. Then $G_1 \cap G_2$ is the edge $\{w, x\}$, so $G_1 \cap G_2 = K_2$.

Therefore

$$P(G, \lambda) = \frac{P(G_1, \lambda) \cdot P(G_2, \lambda)}{\lambda^{(2)}} = \frac{\lambda^{(4)} \cdot \lambda^{(3)}}{\lambda^{(2)}}$$

$$= \frac{\lambda^2(\lambda - 1)^2(\lambda - 2)^2(\lambda - 3)}{\lambda(\lambda - 1)}$$

$$= \lambda(\lambda - 1)(\lambda - 2)^2(\lambda - 3),$$

agreeing with the answer obtained in Example 11.37.

Much more can be said about chromatic polynomials—in particular, there are many unanswered questions. For example, no one has found a set of conditions that indicate whether a given polynomial in λ is the chromatic polynomial for some graph. More about this topic is introduced in the article by R. C. Read [25].

EXERCISES 11.6

1. A pet-shop owner receives a shipment of tropical fish. Among the different species in the shipment are certain pairs where one species feeds on the other. These pairs must consequently be kept in different aquaria. Model this problem as a graph-coloring problem, and tell how to determine the smallest number of aquaria needed to preserve all the fish in the shipment.

2. As the chair for church committees, Mrs. Blasi is faced with scheduling the meeting times for 15 committees. Each committee meets for one hour each week. Two committees having a common member must be scheduled at different times. Model this problem as a graph-coloring problem, and tell how to determine the least number of meeting times Mrs. Blasi has to consider for scheduling the 15 committee meetings.

3. a) At the J. & J. Chemical Company, Jeannette has received three shipments that contain a total of seven different chemicals. Furthermore, the nature of these chemicals is such that for all $1 \leq i \leq 5$, chemical i cannot be stored in the same storage compartment as chemical $i + 1$ or chemical $i + 2$. Determine the smallest number of separate storage compartments that Jeannette will need to safely store these seven chemicals.

b) Suppose that in addition to the conditions in part (a), the following four pairs of these same seven chemicals also require separate storage compartments: 1 and 4, 2 and 5, 2 and 6, and 3 and 6. What is the smallest number of storage compartments that Jeannette now needs to safely store the seven chemicals?

4. Give an example of an undirected graph $G = (V, E)$, where $\chi(G) = 3$ but no subgraph of G is isomorphic to K_3.

5. a) Determine $P(G, \lambda)$ for $G = K_{1,3}$.

b) For $n \in \mathbf{Z}^+$, what is the chromatic polynomial for $K_{1,n}$? What is its chromatic number?

6. a) Consider the graph $K_{2,3}$ shown in Fig. 11.91, and let $\lambda \in \mathbf{Z}^+$ denote the number of colors available to properly color the vertices of $K_{2,3}$. (i) How many proper colorings of $K_{2,3}$ have vertices a, b colored the same? (ii) How many proper colorings of $K_{2,3}$ have vertices a, b colored with different colors?

b) What is the chromatic polynomial for $K_{2,3}$? What is $\chi(K_{2,3})$?

c) For $n \in \mathbf{Z}^+$, what is the chromatic polynomial for $K_{2,n}$? What is $\chi(K_{2,n})$?

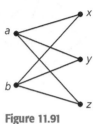

Figure 11.91

7. Find the chromatic number of the following graphs.

a) The complete bipartite graphs $K_{m,n}$.

b) A cycle on n vertices, $n \geq 3$.

c) The graphs in Figs. 11.59(d), 11.62(a), and 11.85.

d) The n-cube Q_n, $n \geq 1$.

8. If G is a loop-free undirected graph with at least one edge, prove that G is bipartite if an only if $\chi(G) = 2$.

9. a) Determine the chromatic polynomials for the graphs in Fig. 11.92

b) Find $\chi(G)$ for each graph.

c) If five colors are available, in how many ways can the vertices of each graph be properly colored?

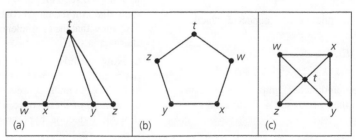

Figure 11.92

10. a) Determine whether the graphs in Fig. 11.93 are isomorphic.

b) Find $P(G, \lambda)$ for each graph.

c) Comment on the results found in parts (a) and (b).

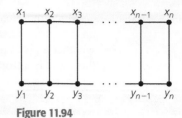

Figure 11.94

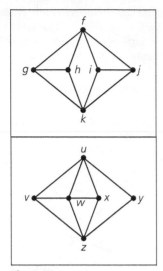

Figure 11.93

11. For $n \geq 3$, let $G_n = (V, E)$ be the undirected graph obtained from the complete graph K_n upon deletion of one edge. Determine $P(G_n, \lambda)$ and $\chi(G_n)$.

12. Consider the complete graph K_n for $n \geq 3$. Color r of the vertices in K_n red and the remaining $n - r$ ($= g$) vertices green. For any two vertices v, w in K_n color the edge $\{v, w\}$ (1) red if v, w are both red; (2) green if v, w are both green; or (3) blue if v, w have different colors. Assume that $r \geq g$.

a) Show that for $r = 6$ and $g = 3$ (and $n = 9$) the total number of red and green edges in K_9 equals the number of blue edges in K_9.

b) Show that the total number of red and green edges in K_n equals the number of blue edges in K_n if and only if $n = r + g$, where g, r are consecutive triangular numbers. [The triangular numbers are defined recursively by $t_1 = 1$, $t_{n+1} = t_n + (n + 1)$, $n \geq 1$; so $t_n = n(n+1)/2$. Hence $t_1 = 1, t_2 = 3, t_3 = 6, \ldots$.]

13. Let $G = (V, E)$ be the undirected connected "ladder graph" shown in Fig. 11.94.

a) Determine $|V|$ and $|E|$.

b) Prove that $P(G, \lambda) = \lambda(\lambda - 1)(\lambda^2 - 3\lambda + 3)^{n-1}$.

14. Let G be a loop-free undirected graph, where $\Delta = \max_{v \in V}\{\deg(v)\}$. (a) Prove that $\chi(G) \leq \Delta + 1$. (b) Find two types of graphs G, where $\chi(G) = \Delta + 1$.

15. For $n \geq 3$, let C_n denote the cycle of length n.

a) What is $P(C_3, \lambda)$?

b) If $n \geq 4$, show that
$$P(C_n, \lambda) = P(P_{n-1}, \lambda) - P(C_{n-1}, \lambda),$$
where P_{n-1} denotes the path of length $n - 1$.

c) Verify that $P(P_{n-1}, \lambda) = \lambda(\lambda - 1)^{n-1}$, for all $n \geq 2$.

d) Establish the relations
$$P(C_n, \lambda) - (\lambda - 1)^n = (\lambda - 1)^{n-1} - P(C_{n-1}, \lambda), \qquad n \geq 4,$$
$$P(C_n, \lambda) - (\lambda - 1)^n = P(C_{n-2}, \lambda) - (\lambda - 1)^{n-2}, \qquad n \geq 5.$$

e) Prove that for all $n \geq 3$,
$$P(C_n, \lambda) = (\lambda - 1)^n + (-1)^n(\lambda - 1).$$

16. For $n \geq 3$, recall that the wheel graph, W_n, is obtained from a cycle of length n by placing a new vertex within the cycle and adding edges (spokes) from this new vertex to each vertex of the cycle.

a) What relationship is there between $\chi(C_n)$ and $\chi(W_n)$?

b) Use part (e) of Exercise 15 to show that
$$P(W_n, \lambda) = \lambda(\lambda - 2)^n + (-1)^n\lambda(\lambda - 2).$$

c) i) If we have k different colors available, in how many ways can we paint the walls and ceiling of a pentagonal room if adjacent walls, and any wall and the ceiling, are to be painted with different colors?

ii) What is the smallest value of k for which such a coloring is possible?

17. Let $G = (V, E)$ be a loop-free undirected graph with chromatic polynomial $P(G, \lambda)$ and $|V| = n$. Use Theorem 11.13 to prove that $P(G, \lambda)$ has degree n and leading coefficient 1 (that is, the coefficient of λ^n is 1).

18. Let $G = (V, E)$ be a loop-free undirected graph.

a) For each such graph, where $|V| \leq 3$, find $P(G, \lambda)$ and show that in it the terms contain consecutive powers of λ. Also show that the coefficients of these consecutive powers alternate in sign.

b) Now consider $G = (V, E)$, where $|V| = n \geq 4$ and $|E| = k$. Prove by mathematical induction that the terms in $P(G, \lambda)$ contain consecutive powers of λ and that the coefficients of these consecutive powers alternate in sign. [For the induction hypothesis, assume that the result is true for all loop-free undirected graphs $G = (V, E)$, where either (i) $|V| = n - 1$ or (ii) $|V| = n$, but $|E| = k - 1$.]

c) Prove that if $|V| = n$, then the coefficient of λ^{n-1} in $P(G, \lambda)$ is the negative of $|E|$.

19. Let $G = (V, E)$ be a loop-free undirected graph. We call G *color-critical* if $\chi(G) > \chi(G - v)$ for all $v \in V$.

a) Explain why cycles with an odd number of vertices are color-critical while cycles with an even number of vertices are not color-critical.

b) For $n \in \mathbf{Z}^+$, $n \geq 2$, which of the complete graph K_n are color-critical?

c) Prove that a color-critical graph must be connected.

d) Prove that if G is color-critical with $\chi(G) = k$, then $\deg(v) \geq k - 1$ for all $v \in V$.

11.7
Summary and Historical Review

Unlike other areas in mathematics, graph theory traces its beginnings to a definite time and place: the problem of the seven bridges of Königsberg, which was solved in 1736 by Leonhard Euler (1707–1783). And in 1752 we find Euler's Theorem for planar graphs. (This result was originally presented in terms of polyhedra.) However, after these developments, little was accomplished in this area for almost a century.

Then, in 1847, Gustav Kirchhoff (1824–1887) examined a special type of graph called a tree. (A *tree* is a loop-free undirected graph that is connected but contains no cycles.) Kirchhoff used this concept in applications dealing with electrical networks in his extension of Ohm's laws for electrical flow. Ten years later Arthur Cayley (1821–1895) developed this same type of graph in order to count the distinct isomers of the saturated hydrocarbons $C_n H_{2n+2}, n \in \mathbf{Z}^+$.

This period also saw two other major ideas come to light. The *four-color conjecture* was first investigated by Francis Guthrie (1831–1899) in about 1850. In Section 11.6 we related some of the history of this problem, which was solved via an intricate computer analysis in 1976 by Kenneth Appel and Wolfgang Haken.

The second major idea was the Hamilton cycle. This cycle is named for Sir William Rowan Hamilton (1805–1865), who used the idea in 1859 for an intriguing puzzle that used the edges on a regular dodecahedron. A solution to this puzzle is not very difficult to find, but mathematicians still search for necessary and sufficient conditions to characterize those undirected graphs that possess a Hamilton path or cycle.

Following these developments, we find little activity until after 1920. The characterization of planar graphs was solved by the Polish mathematician Kasimir Kuratowski (1896–1980) in 1930. In 1936 we find the publication of the first book on graph theory, written by the Hungarian mathematician Dénes König (1884–1944), a prominent researcher in the field. Since then there has been a great deal of activity in the area, the computer providing assistance in the last five decades. Among the many contemporary researchers (not mentioned in the chapter references) in this and related fields one finds the names Claude Berge, V. Chvátal, Paul Erdös, Laszlo Lovász, W. T. Tutte, and Hassler Whitney.

Comparable coverage of the material presented in this chapter is contained in Chapters 6, 8, and 9 of C. L. Liu [23]. More advanced work is found in the works by J. A. Bondy and U. S. R. Murty [10], N. Hartsfield and G. Ringel [20], and D. B. West [32]. The book by F. Buckley and F. Harary [11] revises the classic work of F. Harary [18] and brings the reader up to date on the topics covered in the original 1969 work. The text by G. Chartrand and L. Lesniak [12] provides a more algorithmic approach in its presentation. A proof of

William Rowan Hamilton (1805–1865)
Reproduced courtesy of The Granger Collection, New York

Paul Erdös (1913–1996)
Reproduced courtesy of Christopher Barker

Kuratowski's Theorem appears in Chapter 8 of C. L. Liu [23] and Chapter 6 of D. B. West [32]. The article by G. Chartrand and R. J. Wilson [13] develops many important concepts in graph theory by focusing on one particular graph — the Petersen graph. This graph (which we mentioned in Section 11.4) is named for the Danish mathematician Julius Peter Christian Petersen (1839–1910), who discussed the graph in a paper in 1898.

Applications of graph theory in electrical networks can be found in S. Seshu and M. B. Reed [30]. In the text by N. Deo [14], applications in coding theory, electrical networks, operations research, computer programming, and chemistry occupy Chapters 12–15. The text by F. S. Roberts [26] applies the methods of graph theory to the social sciences. Applications of graph theory in chemistry are given in the article by D. H. Rouvray [29].

More on chromatic polynomials can be found in the survey article by R. C. Read [25]. The role of Polya's theory[†] in graphical enumeration is examined in Chapter 10 of N. Deo [14]. A thorough coverage of this topic is found in the text by F. Harary and E. M. Palmer [19].

Additional coverage on the historical development of graph theory is given in N. Biggs, E. K. Lloyd, and R. J. Wilson [9].

Many applications in graph theory involve large graphs that require the computationally intensive talents of a computer in conjunction with the ingenuity of mathematical methods. Chapter 11 of N. Deo [14] presents computer algorithms dealing with several of the graph-theoretic properties we have studied here. Along the same line, the text by A. V. Aho, J. E. Hopcroft, and J. D. Ullman [1] provides even more for the reader interested in computer science.

As mentioned at the end of Section 11.5, the traveling salesman problem is closely related to the search for a Hamilton cycle in a graph. This is a graph-theoretic problem of interest in both operations research and computer science. The article by M. Bellmore and G. L.

[†]We shall introduce the basic ideas behind this method of enumeration in Chapter 16.

Nemhauser [8] provides a good introductory survey of results on this problem. The text by R. Bellman, K. L. Cooke, and J. A. Lockett [7] includes an algorithmic treatment of this problem along with other graph problems. A number of heuristics for obtaining an approximate solution to the problem are given in Chapter 4 of the text by L. R. Foulds [17]. The text edited by E. L. Lawler, J. K. Lenstra, A. H. G. Rinnooy Kan, and D. B. Shmoys [22] contains 12 papers dealing with various aspects of this problem, including historical considerations as well as some results on computational complexity. Applications, where a robot visits different locations in an automated warehouse in order to fill a given order, are examined in the articles by E. A. Elsayed [15] and by E. A. Elsayed and R. G. Stern [16].

The solution of the four-color problem can be examined further by starting with the paper by K. Appel and W. Haken [3]. The problem, together with its history and solution, is examined in the text by D. Barnette [6] and in the *Scientific American* article by K. Appel and W. Haken [4]. The proof uses a computer analysis to handle a large number of cases; the article by T. Tymoczko [31] examines the role of such techniques in pure mathematics. In [5] K. Appel and W. Haken further examine their proof in the light of the computer analysis that was used. The articles by N. Robertson, D. P. Sanders, P. D. Seymour, and R. Thomas [27, 28] provide a simplified proof. In 1997 their computer code was made available on the Internet. This code could prove the four-color problem on a desktop workstation in roughly three hours.

Finally, the article by A. Ralston [24] demonstrates some of the connections among coding theory, combinatorics, graph theory, and computer science.

REFERENCES

1. Aho, Alfred V., Hopcroft, John E., and Ullman, Jeffrey D. *Data Structures and Algorithms.* Reading, Mass.: Addison-Wesley, 1983.
2. Ahuja, Ravindra K., Magnanti, Thomas L., Orlin, James B., and Reddy, M. R. "Applications of Network Optimization." In M. O. Ball, Thomas L. Magnanti, C. L. Monma, and G. L. Nemhauser, eds., *Handbooks in Operations Research and Management Science,* Vol. 7, *Network Models.* Amsterdam, Holland: Elsevier, 1995, pp. 1–83.
3. Appel, Kenneth, and Haken, Wolfgang. "Every Planar Map Is Four Colorable." *Bulletin of the American Mathematical Society* 82 (1976): pp. 711–712.
4. Appel, Kenneth, and Haken, Wolfgang. "The Solution of the Four-Color-Map Problem. " *Scientific American* 237 (October 1977): pp. 108–121.
5. Appel, Kenneth, and Haken, Wolfgang. "The Four Color Proof Suffices." *Mathematical Intelligencer* 8, no. 1 (1986): pp. 10– 20.
6. Barnette, David. *Map Coloring, Polyhedra, and the Four-Color Problem.* Washington, D.C.: The Mathematical Association of America, 1983.
7. Bellman, R., Cooke, K. L., and Lockett, J. A. *Algorithms, Graphs, and Computers.* New York: Academic Press, 1970.
8. Bellmore, M., and Nemhauser, G. L. "The Traveling Salesman Problem: A Survey." *Operations Research* 16 (1968): pp. 538–558.
9. Biggs, N., Lloyd, E. K., and Wilson, R. J. *Graph Theory (1736–1936).* Oxford, England: Clarendon Press, 1976.
10. Bondy, J. A., and Murty, U. S. R. *Graph Theory with Applications.* New York: Elsevier North-Holland, 1976.
11. Buckley, Fred, and Harary, Frank. *Distance in Graphs.* Reading, Mass.: Addison-Wesley, 1990.
12. Chartrand, Gary, and Lesniak, Linda. *Graphs and Digraphs,* 3rd ed. Boca Raton, Fla.: CRC Press, 1996.
13. Chartrand, Gary, and Wilson, Robin J. "The Petersen Graph." In Frank Harary and John S. Maybee, eds., *Graphs and Applications.* New York: Wiley, 1985.

14. Deo, Narsingh. *Graph Theory with Applications to Engineering and Computer Science.* Englewood Cliffs, N. J.: Prentice-Hall, 1974.

15. Elsayed, E. A."Algorithms for Optimal Material Handling in Automatic Warehousing Systems." *Int. J. Prod. Res.* 19 (1981): pp. 525–535.

16. Elsayed, E. A., and Stern, R. G."Computerized Algorithms for Order Processing in Automated Warehousing Systems." *Int. J. Prod. Res.* 21 (1983): pp. 579–586.

17. Foulds, L. R. *Combinatorial Optimization for Undergraduates.* New York: Springer-Verlag, 1984.

18. Harary, Frank. *Graph Theory.* Reading, Mass.: Addison-Wesley, 1969.

19. Harary, Frank, and Palmer, Edgar M. *Graphical Enumeration.* New York: Academic Press, 1973.

20. Hartsfield, Nora, and Ringel, Gerhard. *Pearls in Graph Theory: A Comprehensive Introduction.* Boston, Mass.: Harcourt/Academic Press, 1994.

21. Jünger, M., Reinelt, G., and Rinaldi, G."The Traveling Salesman Problem." In M. O. Ball, Thomas L. Magnanti, C. L. Monma, and G. L. Nemhauser, eds., *Handbooks in Operations Research and Management Science*, Vol. 7, *Network Models.* Amsterdam, Holland: Elsevier, 1995, pp. 225–330.

22. Lawler, E. L., Lenstra, J. K., Rinnooy Kan, A. H. G., and Shmoys, D. B., eds. *The Traveling Salesman Problem.* New York: Wiley, 1986.

23. Liu, C. L. *Introduction to Combinatorial Mathematics.* New York: McGraw-Hill, 1968.

24. Ralston, Anthony. "De Bruijn Sequences — A Model Example of the Interaction of Discrete Mathematics and Computer Science." *Mathematics Magazine* 55, no. 3 (May 1982): pp. 131–143.

25. Read, R. C."An Introduction to Chromatic Polynomials." *Journal of Combinatorial Theory* 4 (1968): pp. 52–71.

26. Roberts, Fred S. *Discrete Mathematical Models.* Englewood Cliffs, N. J.: Prentice-Hall, 1976.

27. Robertson, N., Sanders, D. P., Seymour, P. D., and Thomas, R. "Efficiently Four-coloring Planar Graphs." *Proceedings of the 28th ACM Symposium on the Theory of Computation.* ACM Press (1996): pp. 571–575.

28. Robertson, N., Sanders, D. P., Seymour, P. D., and Thomas, R. "The Four-color Theorem." *Journal of Combinatorial Theory Series B* 70 (1997): pp. 166–183.

29. Rouvray, Dennis H. "Predicting Chemistry from Topology." *Scientific American* 255, no. 3 (September 1986): pp. 40–47.

30. Seshu, S., and Reed, M. B. *Linear Graphs and Electrical Networks.* Reading, Mass.: Addison-Wesley, 1961.

31. Tymoczko, Thomas. "Computers, Proofs and Mathematicians: A Philosophical Investigation of the Four-Color Proof." *Mathematics Magazine* 53, no. 3 (May 1980): pp. 131–138.

32. West, Douglas B. *Introduction to Graph Theory*, 2nd ed. Upper Saddle River, N.J.: Prentice-Hall, 2001.

SUPPLEMENTARY EXERCISES

1. Let G be a loop-free undirected graph on n vertices. If G has 56 edges and $\overline{G}$ has 80 edges, what is n?

2. Determine the number of cycles of length 4 in the hypercube Q_n.

3. a) If the edges of K_6 are painted either red or blue, prove that there is a red triangle or a blue triangle that is a subgraph.

b) Prove that in any group of six people there must be three who are total strangers to one another or three who are mutual friends.

4. a) Let $G = (V, E)$ be a loop-free undirected graph. Recall that G is called self-complementary if G and $\overline{G}$ are isomorphic. If G is self-complementary (i) determine $|E|$ if $|V| = n$; (ii) prove that G is connected.

b) Let $n \in \mathbf{Z}^+$, where $n = 4k$ $(k \in \mathbf{Z}^+)$ or $n = 4k + 1$ $(k \in \mathbf{N})$. Prove that there exists a self-complementary graph $G = (V, E)$, where $|V| = n$.

5. a) Show that the graphs G_1 and G_2, in Fig. 11.95, are isomorphic.

b) How many different isomorphisms $f: G_1 \to G_2$ are possible here?

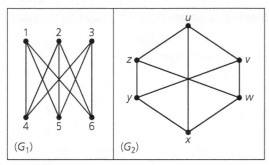

Figure 11.95

6. Are any of the planar graphs for the five Platonic solids bipartite?

7. a) How many paths of length 5 are there in the complete bipartite graph $K_{3,7}$? (Remember that a path such as $v_1 \to v_2 \to v_3 \to v_4 \to v_5 \to v_6$ is considered to be the same as the path $v_6 \to v_5 \to v_4 \to v_3 \to v_2 \to v_1$.)

b) How many paths of length 4 are there in $K_{3,7}$?

c) Let $m, n, p \in \mathbf{Z}^+$ with $2m < n$ and $1 \le p \le 2m$. How many paths of length p are there in the complete bipartite graph $K_{m,n}$?

8. Let $X = \{1, 2, 3, \ldots, n\}$, where $n \ge 2$. Construct the loop-free undirected graph $G = (V, E)$ as follows:

- (V): Each two-element subset of X determines a vertex of G.
- (E): If $v_1, v_2 \in V$ correspond to subsets $\{a, b\}$ and $\{c, d\}$, respectively, of X, draw the edge $\{v_1, v_2\}$ in G when $\{a, b\} \cap \{c, d\} = \emptyset$.

a) Show that G is an isolated vertex when $n = 2$ and that G is disconnected for $n = 3, 4$.

b) Show that for $n \ge 5$, G is connected. (In fact, for all $v_1, v_2 \in V$, either $\{v_1, v_2\} \in E$ or there is a path of length 2 connecting v_1 and v_2.)

c) Prove that G is nonplanar for $n \ge 5$.

d) Prove that for $n \ge 8$, G has a Hamilton cycle.

9. If $G = (V, E)$ is an undirected graph, a subset K of V is called a *covering* of G if for every edge $\{a, b\}$ of G either a or b is in K. The set K is a *minimal covering* if $K - \{x\}$ fails to cover G for each $x \in K$. The number of vertices in a smallest covering is called the *covering number* of G.

a) Prove that if $I \subseteq V$, then I is an independent set in G if and only if $V - I$ is a covering of G.

b) Verify that $|V|$ is the sum of the independence number of G (as defined in Exercise 25 for Section 11.5) and its covering number.

10. If $G = (V, E)$ is an undirected graph, a subset D of V is called a *dominating set* if for all $v \in V$, either $v \in D$ or v is adjacent to a vertex in D. If D is a dominating set and no proper subset of D has this property, then D is called *minimal*. The size of any smallest dominating set in G is denoted by $\gamma(G)$ and is called the *domination number* of G.

a) If G has no isolated vertices, prove that if D is a minimal dominating set, then $V - D$ is a dominating set.

b) If $I \subseteq V$ is independent, prove that I is a dominating set if and only if I is maximal independent.

c) Show that $\gamma(G) \le \beta(G)$, and that $|V| \le \beta(G)\chi(G)$. [Here $\beta(G)$ is the independence number of G — first given in Exercise 25 of Section 11.5.]

11. Let $G = (V, E)$ be the undirected connected "ladder graph" shown in Fig. 11.94. For $n \ge 0$, let a_n denote the number of ways one can select n of the edges in G so that no two edges share a common vertex. Find and solve a recurrence relation for a_n.

12. Consider the four *comb* graphs in parts (i), (ii), (iii), and (iv) of Fig. 11.96. These graphs have 1 tooth, 2 teeth, 3 teeth, and n teeth, respectively. For $n \ge 1$, let a_n count the number of independent subsets in $\{x_1, x_2, \ldots, x_n, y_1, y_2, \ldots, y_n\}$. Find and solve a recurrence relation for a_n.

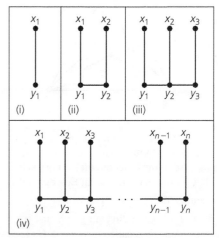

Figure 11.96

13. Consider the four graphs in parts (i), (ii), (iii), and (iv) of Fig. 11.97. If a_n counts the number of independent subsets of $\{x_1, x_2, \ldots, x_n, y_1, y_2, \ldots, y_n\}$, where $n \ge 1$, find and solve a recurrence relation for a_n.

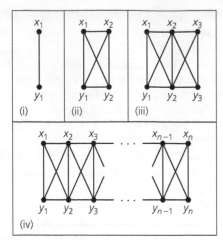

Figure 11.97

14. For $n \geq 1$, let $a_n = \binom{n}{2}$, the number of edges in K_n, and let $a_0 = 0$. Find the generating function $f(x) = \sum_{n=0}^{\infty} a_n x^n$.

15. For the graph G in Fig. 11.98, answer the following questions.

  a) What are $\gamma(G)$, $\beta(G)$, and $\chi(G)$?

  b) Does G have an Euler circuit or a Hamilton cycle?

  c) Is G bipartite? Is it planar?

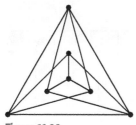

Figure 11.98

16. a) Suppose that the complete bipartite graph $K_{m,n}$ contains 16 edges and satisfies $m \leq n$. Determine m, n so that $K_{m,n}$ possesses (i) an Euler circuit but not a Hamilton cycle; (ii) both a Hamilton cycle and an Euler circuit.

  b) Generalize the results of part (a).

17. If $G = (V, E)$ is an undirected graph, any subgraph of G that is a complete graph is called a *clique* in G. The number of vertices in a largest clique in G is called the *clique number* for G and is denoted by $\omega(G)$.

  a) How are $\chi(G)$ and $\omega(G)$ related?

  b) Is there any relationship between $\omega(G)$ and $\beta(\overline{G})$?

18. If $G = (V, E)$ is an undirected loop-free graph, the *line graph* of G, denoted $L(G)$, is a graph with the set E as vertices,

where we join two vertices e_1, e_2 in $L(G)$ if and only if e_1, e_2 are adjacent edges in G.

  a) Find $L(G)$ for each of the graphs in Fig. 11.99.

  b) Assuming that $|V| = n$ and $|E| = e$, show that $L(G)$ has e vertices and $(1/2) \sum_{v \in V} \deg(v)[\deg(v) - 1] = [(1/2) \sum_{v \in V} [\deg(v)]^2] - e = \sum_{v \in V} \binom{\deg(v)}{2}$ edges.

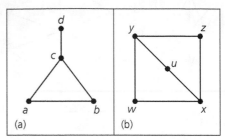

Figure 11.99

  c) Prove that if G has an Euler circuit, then $L(G)$ has both an Euler circuit and a Hamilton cycle.

  d) If $G = K_4$, examine $L(G)$ to show that the converse of part (c) is false.

  e) Prove that if G has a Hamilton cycle, then so does $L(G)$.

  f) Examine $L(G)$ for the graph in Fig. 11.99(b) to show that the converse of part (e) is false.

  g) Verify that $L(G)$ is nonplanar for $G = K_5$ and $G = K_{3,3}$.

  h) Give an example of a graph G, where G is planar but $L(G)$ is not.

19. Explain why each of the following polynomials in λ cannot be a chromatic polynomial.

  a) $\lambda^4 - 5\lambda^3 + 7\lambda^2 - 6\lambda + 3$

  b) $3\lambda^3 - 4\lambda^2 + \lambda$

  c) $\lambda^4 - 3\lambda^3 + 5\lambda^2 - 4\lambda$

20. a) For all $x, y \in \mathbf{Z}^+$, prove that $x^3 y - x y^3$ is even.

  b) Let $V = \{1, 2, 3, \ldots, 8, 9\}$. Construct the loop-free undirected graph $G = (V, E)$ as follows: For $m, n \in V$, $m \neq n$, draw the edge $\{m, n\}$ in G if 5 divides $m + n$ or $m - n$.

  c) Given any three distinct positive integers, prove that there are two of these, say x and y, where 10 divides $x^3 y - x y^3$.

21. a) For $n \geq 1$, let P_{n-1} denote the path made up of n vertices and $n - 1$ edges. Let a_n be the number of independent subsets of vertices in P_{n-1}. (The empty subset is considered one of these independent subsets.) Find and solve a recurrence relation for a_n.

b) Determine the number of independent subsets (of vertices) in each of the graphs G_1, G_2, and G_3, of Fig. 11.100.

c) For each of the graphs H_1, H_2, and H_3, of Fig. 11.101, find the number of independent subsets of vertices.

d) Let $G = (V, E)$ be a loop-free undirected graph with $V = \{v_1, v_2, \ldots, v_r\}$ and where there are m independent subsets of vertices. The graph $G' = (V', E')$ is constructed from G as follows: $V' = V \cup \{x_1, x_2, \ldots, x_s\}$, with no x_i in V, for all $1 \le i \le s$; and $E' = E \cup \{\{x_i, v_j\} | 1 \le i \le s, 1 \le j \le r\}$. How many subsets of V' are independent?

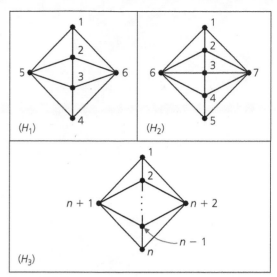

Figure 11.101

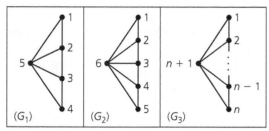

Figure 11.100

22. Suppose that $G = (V, E)$ is a loop-free undirected graph. If G is 5-regular and $|V| = 10$, prove that G is nonplanar.

12

Trees

Continuing our study of graph theory, we shall now focus on a special type of graph called a tree. First used in 1847 by Gustav Kirchhoff (1824–1887) in his work on electrical networks, trees were later redeveloped and named by Arthur Cayley (1821–1895). In 1857 Cayley used these special graphs in order to enumerate the different isomers of the saturated hydrocarbons C_nH_{2n+2}, $n \in \mathbf{Z}^+$.

With the advent of digital computers, many new applications were found for trees. Special types of trees are prominent in the study of data structures, sorting, and coding theory, and in the solution of certain optimization problems.

12.1
Definitions, Properties, and Examples

Definition 12.1   Let $G = (V, E)$ be a loop-free undirected graph. The graph G is called a *tree*[†] if G is connected and contains no cycles.

In Fig. 12.1 the graph G_1 is a tree, but the graph G_2 is not a tree because it contains the cycle $\{a, b\}$, $\{b, c\}$, $\{c, a\}$. The graph G_3 is not connected, so it cannot be a tree. However, each component of G_3 is a tree, and in this case we call G_3 a *forest*.

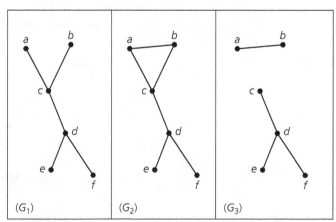

Figure 12.1

[†] As in the case of graphs, the terminology in the study of trees is not standard and the reader may find some differences from one textbook to another.

When a graph is a tree we write T instead of G to emphasize this structure.

In Fig. 12.1 we see that G_1 is a subgraph of G_2 where G_1 contains all the vertices of G_2 and G_1 is a tree. In this situation G_1 is a spanning tree for G_2. Hence a *spanning tree* for a connected graph is a spanning subgraph that is also a tree. We may think of a spanning tree as providing minimal connectivity for the graph and as a minimal skeletal framework holding the vertices together. The graph G_3 provides a *spanning forest* for the graph G_2.

We now examine some properties of trees.

THEOREM 12.1   If a, b are distinct vertices in a tree $T = (V, E)$, then there is a unique path that connects these vertices.

Proof: Since T is connected, there is at least one path in T that connects a and b. If there were more, then from two such paths some of the edges would form a cycle. But T has no cycles.

THEOREM 12.2   If $G = (V, E)$ is an undirected graph, then G is connected if and only if G has a spanning tree.

Proof: If G has a spanning tree T, then for every pair a, b of distinct vertices in V a subset of the edges in T provides a (unique) path between a and b, and so G is connected. Conversely, if G is connected and G is not a tree, remove all loops from G. If the resulting subgraph G_1 is not a tree, then G_1 must contain a cycle C_1. Remove an edge e_1 from C_1 and let $G_2 = G_1 - e_1$. If G_2 contains no cycles, then G_2 is a spanning tree for G because G_2 contains all the vertices in G, is loop-free, and is connected. If G_2 does contain a cycle — say, C_2 — then remove an edge e_2 from C_2 and consider the subgraph $G_3 = G_2 - e_2 = G_1 - \{e_1, e_2\}$. Once again, if G_3 contains no cycles, then we have a spanning tree for G. Otherwise we continue this procedure a finite number of additional times until we arrive at a spanning subgraph of G that is loop-free and connected and contains no cycles (and, consequently, is a spanning tree for G).

Figure 12.2 shows three nonisomorphic trees that exist for five vertices. Although they are not isomorphic, they all have the same number of edges, namely, four. This leads us to the following general result.

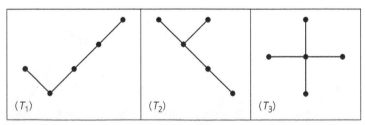

Figure 12.2

THEOREM 12.3   In every tree $T = (V, E)$, $|V| = |E| + 1$.

Proof: The proof is obtained by applying the alternative form of the Principle of Mathematical Induction to $|E|$. If $|E| = 0$, then the tree consists of a single isolated vertex, as in

Fig. 12.3(a). Here $|V| = 1 = |E| + 1$. Parts (b) and (c) of the figure verify the result for the cases where $|E| = 1$ or 2.

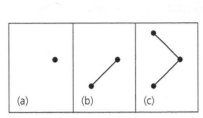

Figure 12.3

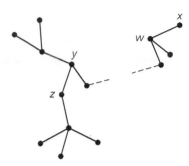

Figure 12.4

Assume the theorem is true for every tree that contains at most k edges, where $k \geq 0$. Now consider a tree $T = (V, E)$, as in Fig. 12.4, where $|E| = k + 1$. [The dotted edge(s) indicates that some of the tree doesn't appear in the figure.] If, for instance, the edge with endpoints y, z is removed from T, we obtain two *subtrees*, $T_1 = (V_1, E_1)$ and $T_2 = (V_2, E_2)$, where $|V| = |V_1| + |V_2|$ and $|E_1| + |E_2| + 1 = |E|$. (One of these subtrees could consist of just a single vertex if, for example, the edge with endpoints w, x were removed.) Since $0 \leq |E_1| \leq k$ and $0 \leq |E_2| \leq k$, it follows, by the induction hypothesis, that $|E_i| + 1 = |V_i|$, for $i = 1, 2$. Consequently, $|V| = |V_1| + |V_2| = (|E_1| + 1) + (|E_2| + 1) = (|E_1| + |E_2| + 1) + 1 = |E| + 1$, and the theorem follows by the alternative form of the Principle of Mathematical Induction.

As we examine the trees in Fig. 12.2 we also see that each tree has at least two pendant vertices — that is, vertices of degree 1. This is also true in general.

THEOREM 12.4

For every tree $T = (V, E)$, if $|V| \geq 2$, then T has at least two pendant vertices.

Proof: Let $|V| = n \geq 2$. From Theorem 12.3 we know that $|E| = n - 1$, so by Theorem 11.2 it follows that $2(n - 1) = 2|E| = \sum_{v \in V} \deg(v)$. Since T is connected, we have $\deg(v) \geq 1$ for all $v \in V$. If there are k pendant vertices in T, then each of the other $n - k$ vertices has degree at least 2 and

$$2(n - 1) = 2|E| = \sum_{v \in V} \deg(v) \geq k + 2(n - k).$$

From this we see that $[2(n - 1) \geq k + 2(n - k)] \Rightarrow [(2n - 2) \geq (k + 2n - 2k)] \Rightarrow [-2 \geq -k] \Rightarrow [k \geq 2]$, and the result is consequently established.

EXAMPLE 12.1

In Fig. 12.5 we have two trees, each with 14 vertices (labeled with C's and H's) and 13 edges. Each vertex has degree 4 (C, carbon atom) or degree 1 (H, hydrogen atom). Part (b) of the figure has a carbon atom (C) at the center of the tree. This carbon atom is adjacent to four vertices, three of which have degree 4. There is no vertex (C atom) in part (a) that possesses this property, so the two trees are not isomorphic. They serve as models for the two chemical isomers that correspond with the saturated[†] hydrocarbon C_4H_{10}. Part (a) represents n-butane (formerly called butane); part (b) represents 2-methyl propane (formerly called isobutane).

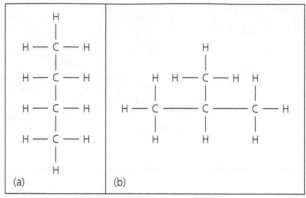

(a)   (b)

Figure 12.5

A second result from chemistry is given in the following example.

EXAMPLE 12.2

If a saturated hydrocarbon [in particular, an acyclic (no cycles), single-bond hydrocarbon — called an *alkane*] has n carbon atoms, show that it has $2n + 2$ hydrogen atoms.

Considering the saturated hydrocarbon as a tree $T = (V, E)$, let k equal the number of pendant vertices, or hydrogen atoms, in the tree. Then with a total of $n + k$ vertices, where each of the n carbon atoms has degree 4, we find that

$$4n + k = \sum_{v \in V} \deg(v) = 2|E| = 2(|V| - 1) = 2(n + k - 1),$$

and

$$4n + k = 2(n + k - 1) \Rightarrow k = 2n + 2.$$

We close this section with a theorem that provides several different ways to characterize trees.

THEOREM 12.5

The following statements are equivalent for a loop-free undirected graph $G = (V, E)$.

a) G is a tree.

b) G is connected, but the removal of any edge from G disconnects G into two subgraphs that are trees.

c) G contains no cycles, and $|V| = |E| + 1$.

d) G is connected, and $|V| = |E| + 1$.

[†]The adjective *saturated* is used here to indicate that for the number of carbon atoms present in the molecule, we have the maximum number of hydrogen atoms.

e) G contains no cycles, and if $a, b \in V$ with $\{a, b\} \notin E$, then the graph obtained by adding edge $\{a, b\}$ to G has precisely one cycle.

Proof: We shall prove that (a) $\Rightarrow$ (b), (b) $\Rightarrow$ (c), and (c) $\Rightarrow$ (d), leaving to the reader the proofs for (d) $\Rightarrow$ (e) and (e) $\Rightarrow$ (a).

[(a) $\Rightarrow$ (b)]: If G is a tree, then G is connected. So let $e = \{a, b\}$ be any edge of G. Then if $G - e$ is connected, there are at least two paths in G from a to b. But this contradicts Theorem 12.1. Hence $G - e$ is disconnected and so the vertices in $G - e$ may be partitioned into two subsets: (1) vertex a and those vertices that can be reached from a by a path in $G - e$; and (2) vertex b and those vertices that can be reached from b by a path in $G - e$. These two connected components are trees because a loop or cycle in either component would also be in G.

[(b) $\Rightarrow$ (c)]: If G contains a cycle, then let $e = \{a, b\}$ be an edge of the cycle. But then $G - e$ is connected, contradicting the hypothesis in part (b). So G contains no cycles, and since G is a loop-free connected undirected graph, we know that G is a tree. Consequently, it follows from Theorem 12.3 that $|V| = |E| + 1$.

[(c) $\Rightarrow$ (d)]: Let $\kappa(G) = r$ and let $G_1, G_2, \ldots, G_r$ be the components of G. For $1 \leq i \leq r$, select a vertex $v_i \in G_i$ and add the $r - 1$ edges $\{v_1, v_2\}, \{v_2, v_3\}, \ldots, \{v_{r-1}, v_r\}$ to G to form the graph $G' = (V, E')$, which is a tree. Since G' is a tree, we know that $|V| = |E'| + 1$ because of Theorem 12.3. But from part (c), $|V| = |E| + 1$, so $|E| = |E'|$ and $r - 1 = 0$. With $r = 1$, it follows that G is connected.

EXERCISES 12.1

1. a) Draw the graphs of all nonisomorphic trees on six vertices.

b) How many isomers does hexane (C_6H_{14}) have?

2. Let $T_1 = (V_1, E_1)$, $T_2 = (V_2, E_2)$ be two trees where $|E_1| = 17$ and $|V_2| = 2|V_1|$. Determine $|V_1|$, $|V_2|$, and $|E_2|$.

3. a) Let $F_1 = (V_1, E_1)$ be a forest of seven trees where $|E_1| = 40$. What is $|V_1|$?

b) If $F_2 = (V_2, E_2)$ is a forest with $|V_2| = 62$ and $|E_2| = 51$, how many trees determine F_2?

4. If $G = (V, E)$ is a forest with $|V| = v$, $|E| = e$, and κ components (trees), what relationship exists among v, e, and κ?

5. What kind of trees have exactly two pendant vertices?

6. a) Verify that all trees are planar.

b) Derive Theorem 12.3 from part (a) and Euler's Theorem for planar graphs.

7. Give an example of an undirected graph $G = (V, E)$ where $|V| = |E| + 1$ but G is not a tree.

8. a) If a tree has four vertices of degree 2, one vertex of degree 3, two of degree 4, and one of degree 5, how many pendant vertices does it have?

b) If a tree $T = (V, E)$ has v_2 vertices of degree 2, v_3 vertices of degree 3, ..., and v_m vertices of degree m, what are $|V|$ and $|E|$?

9. If $G = (V, E)$ is a loop-free undirected graph, prove that G is a tree if there is a unique path between any two vertices of G.

10. The connected undirected graph $G = (V, E)$ has 30 edges. What is the maximum value that $|V|$ can have?

11. Let $T = (V, E)$ be a tree with $|V| = n \geq 2$. How many distinct paths are there (as subgraphs) in T?

12. Let $G = (V, E)$ be a loop-free connected undirected graph where $V = \{v_1, v_2, v_3, \ldots, v_n\}, n \geq 2, \deg(v_1) = 1$, and $\deg(v_i) \geq 2$ for $2 \leq i \leq n$. Prove that G must have a cycle.

13. Find two nonisomorphic spanning trees for the complete bipartite graph $K_{2,3}$. How many nonisomorphic spanning trees are there for $K_{2,3}$?

14. For $n \in \mathbf{Z}^+$, how many nonisomorphic spanning trees are there for $K_{2,n}$?

15. Determine the number of nonidentical (though some may be isomorphic) spanning trees that exist for each of the graphs shown in Fig. 12.6.

16. For each graph in Fig. 12.7, determine how many nonidentical (though some may be isomorphic) spanning trees exist.

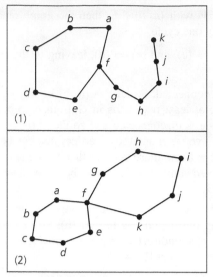

(1)

(2)

Figure 12.6

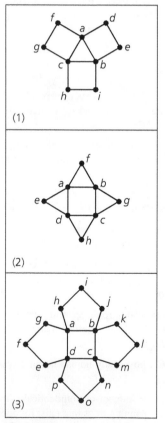

(1)

(2)

(3)

Figure 12.7

17. Let $T = (V, E)$ be a tree where $|V| = n$. Suppose that for each $v \in V$, $\deg(v) = 1$ or $\deg(v) \geq m$, where m is a fixed positive integer and $m \geq 2$.

a) What is the smallest value possible for n?

b) Prove that T has at least m pendant vertices.

18. Suppose that $T = (V, E)$ is a tree with $|V| = 1000$. What is the sum of the degrees of all the vertices in T?

19. Let $G = (V, E)$ be a loop-free connected undirected graph. Let H be a subgraph of G. The *complement of H in G* is the subgraph of G made up of those edges in G that are not in H (along with the vertices incident to these edges).

a) If T is a spanning tree of G, prove that the complement of T in G does not contain a cut-set of G.

b) If C is a cut-set of G, prove that the complement of C in G does not contain a spanning tree of G.

20. Complete the proof of Theorem 12.5.

21. A labeled tree is one wherein the vertices are labeled. If the tree has n vertices, then $\{1, 2, 3, \ldots, n\}$ is used as the set of labels. We find that two trees that are isomorphic without labels may become nonisomorphic when labeled. In Fig. 12.8, the first two trees are isomorphic as labeled trees. The third tree is isomorphic to the other two if we ignore the labels; as a labeled tree, however, it is not isomorphic to either of the other two.

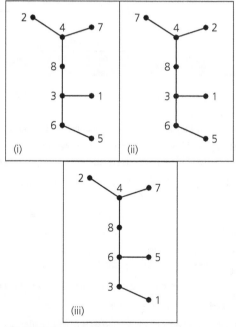

Figure 12.8

The number of nonisomorphic trees with n labeled vertices can be counted by setting up a one-to-one correspondence between these trees and the n^{n-2} sequences (with repetitions allowed) $x_1, x_2, \ldots, x_{n-2}$ whose entries are taken from $\{1, 2, 3, \ldots, n\}$. If T is one such labeled tree, we use the following algorithm to find its corresponding sequence — called the *Prüfer code* for the tree. (Here T has at least one edge.)

Step 1: Set the counter i to 1.

Step 2: Set $T(i) = T$.

Step 3: Since a tree has at least two pendant vertices, select the pendant vertex in $T(i)$ with the smallest label y_i. Now remove the edge $\{x_i, y_i\}$ from $T(i)$ and use x_i for the ith component of the sequence.

Step 4: If $i = n - 2$, we have the sequence corresponding to the given labeled tree $T(1)$. If $i \neq n - 2$, increase i by 1, set $T(i)$ equal to the resulting subtree obtained in step (3), and return to step (3).

a) Find the six-digit sequence (Prüfer code) for trees (i) and (iii) in Fig. 12.8.

b) If v is a vertex in T, show that the number of times the label on v appears in the Prüfer code $x_1, x_2, \ldots, x_{n-2}$ is $\deg(v) - 1$.

c) Reconstruct the labeled tree on eight vertices that is associated with the Prüfer code 2, 6, 5, 5, 5, 5.

d) Develop an algorithm for reconstructing a tree from a given Prüfer code $x_1, x_2, \ldots, x_{n-2}$.

22. Let $n \in \mathbf{Z}^+$, $n \geq 3$. If v is a vertex in K_n, how many of the n^{n-2} spanning trees of K_n have v as a pendant vertex?

23. Characterize the trees whose Prüfer codes

    a) contain only one integer, or

    b) have distinct integers in all positions.

24. Show that the number of labeled trees with n vertices, k of which are pendant vertices, is $\binom{n}{k}(n-k)!S(n-2, n-k) = (n!/k!)S(n-2, n-k)$, where $S(n-2, n-k)$ is a Stirling number of the second kind. (This result was first established in 1959 by A. Rényi.)

25. Let $G = (V, E)$ be the undirected graph in Fig. 12.9. Show that the edge set E can be partitioned as $E_1 \cup E_2$ so that the subgraphs $G_1 = (V, E_1)$, $G_2 = (V, E_2)$ are isomorphic spanning trees of G.

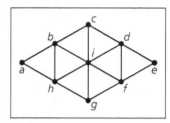

Figure 12.9

12.2
Rooted Trees

We turn now to directed trees. We find a variety of applications for a special type of directed tree called a rooted tree.

Definition 12.2   If G is a directed graph, then G is called a *directed tree* if the undirected graph associated with G is a tree. When G is a directed tree, G is called a *rooted tree* if there is a unique vertex r, called the *root*, in G with the in degree of $r = id(r) = 0$, and for all other vertices v, the in degree of $v = id(v) = 1$.

The tree in part (a) of Fig. 12.10 is directed but not rooted; the tree in part (b) is rooted with root r.

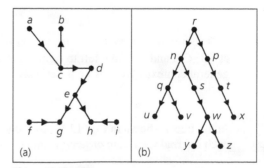

Figure 12.10

We draw rooted trees as in Fig. 12.10(b) but with the directions understood as going from the upper level to the lower level, so that the arrows aren't needed. In a rooted tree, a vertex with out degree 0 is called a *leaf* (or *terminal vertex.*) Vertices u, v, x, y, z are leaves in Fig. 12.10(b). All other vertices are called *branch nodes* (or *internal vertices*).

Consider the vertex s in this rooted tree [Fig. 12.10(b)]. The path from the root, r, to s is of length 2, so we say that s is at *level* 2 in the tree, or that s has *level number* 2. Similarly, x is at level 3, whereas y has level number 4. We call s a *child* of n, and we call n the *parent* of s. Vertices w, y, and z are considered *descendants* of s, n, and r, while s, n, and r are called *ancestors* of w, y, and z. In general, if v_1 and v_2 are vertices in a rooted tree and v_1 has the smaller level number, then v_1 is an ancestor of v_2 (or v_2 is a descendant of v_1) if there is a (directed) path from v_1 to v_2. Two vertices with a common parent are referred to as *siblings*. Such is the case for vertices q and s, whose common parent is vertex n. Finally, if v_1 is any vertex of the tree, the *subtree at* v_1 is the subgraph induced by the root v_1 and all of its descendants (there may be none).

EXAMPLE 12.3

In Fig. 12.11(a) a rooted tree is used to represent the table of contents of a three-chapter ($C1$, $C2$, $C3$) book. Vertices with level number 2 are for sections within a chapter; those at level 3 represent subsections within a section. Part (b) of the figure displays the natural order for the table of contents of this book.

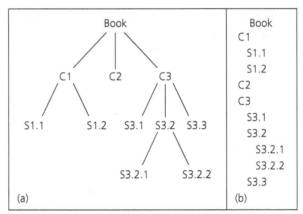

Figure 12.11

The tree in Fig. 12.11(a) suggests an order for the vertices if we examine the subtrees at $C1$, $C2$, and $C3$ from left to right. (This order will recur again in this section, in a more general context.) We now consider a second example that provides such an order.

EXAMPLE 12.4

In the tree T shown in Fig. 12.12, the edges (or branches, as they are often called) leaving each internal vertex are *ordered* from left to right. Hence T is called an *ordered rooted tree*.

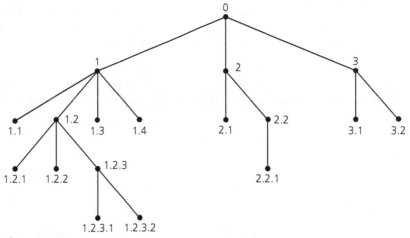

Figure 12.12

We label the vertices for this tree by the following algorithm.

Step 1: First assign the root the label (or *address*) 0.

Step 2: Next assign the positive integers 1, 2, 3, . . . to the vertices at level 1, going from left to right.

Step 3: Now let v be an internal vertex at level $n \geq 1$, and let $v_1, v_2, \ldots, v_k$ denote the children of v (going from left to right). If a is the label assigned to vertex v, assign the labels $a.1, a.2, \ldots, a.k$ to the children $v_1, v_2, \ldots, v_k$, respectively.

Consequently, each vertex in T, other than the root, has a label of the form $a_1.a_2.a_3 . . . . . a_n$ if and only if that vertex has level number n. This is known as the *universal address system*.

This system provides a way to *order* all vertices in T. If u and v are two vertices in T with addresses b and c, respectively, we define $b < c$ if (a) $b = a_1.a_2 . . . . . a_m$ and $c = a_1.a_2 . . . . . a_m.a_{m+1} . . . . . a_n$, with $m < n$; or (b) $b = a_1.a_2 . . . . . a_m.x_1 . . . . . y$ and $c = a_1.a_2 . . . . . a_m.x_2 . . . . . z$, where $x_1, x_2 \in \mathbf{Z}^+$ and $x_1 < x_2$.

For the tree under consideration, this ordering yields

$$
\begin{array}{cccccc}
0 & 1.2 & 1.2.3 & 1.3 & 2.1 & 3 \\
1 & 1.2.1 & 1.2.3.1 & 1.4 & 2.2 & 3.1 \\
1.1 & 1.2.2 & 1.2.3.2 & 2 & 2.2.1 & 3.2
\end{array}
$$

Since this resembles the alphabetical ordering in a dictionary, the order is called the *lexicographic*, or *dictionary, order*.

We now consider an application of a rooted tree in the study of computer science.

EXAMPLE 12.5

a) A rooted tree is a *binary* rooted tree if for each vertex v, $od(v) = 0$, 1, or 2 — that is, if v has at most two children. If $od(v) = 0$ or 2 for all $v \in V$, then the rooted tree is called a *complete* binary tree. Such a tree can represent a binary operation, as in parts

(a) and (b) of Fig. 12.13. To avoid confusion when dealing with a noncommutative operation ∘, we label the root as ∘ and require the result to be $a \circ b$, where a is the left child, and b the right child, of the root.

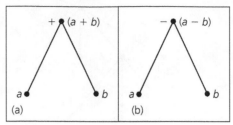

Figure 12.13

b) In Fig. 12.14 we extend the ideas presented in Fig. 12.13 in order to construct the binary rooted tree for the algebraic expression

$$((7 - a)/5) * ((a + b) \uparrow 3),$$

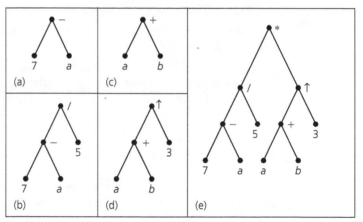

Figure 12.14

where $*$ denotes multiplication and $\uparrow$ denotes exponentiation. Here we construct this tree, as shown in part (e) of the figure, from the bottom up. First, a subtree for the expression $7 - a$ is constructed in part (a) of Fig. 12.14. This is then incorporated (as the left subtree for $/$) in the binary rooted tree for the expression $(7 - a)/5$ in Fig. 12.14 (b). Then, in a similar way, the binary rooted trees in parts (c) and (d) of the figure are constructed for the expressions $a + b$ and $(a + b) \uparrow 3$, respectively. Finally, the two subtrees in parts (b) and (d) are used as the left and right subtrees, respectively, for $*$ and give us the binary rooted tree [in Fig. 12.14(e)] for $((7 - a)/5) * ((a + b) \uparrow 3)$.

The same ideas are used in Fig. 12.15, where we find the binary rooted trees for the algebraic expressions

$$(a - (3/b)) + 5 \text{ [in part (a)]} \qquad \text{and} \qquad a - (3/(b + 5)) \text{ [in part (b)]}.$$

c) In evaluating $t + (uv)/(w + x - y^z)$ in certain procedural languages, we write the expression in the form $t + (u * v)/(w + x - y \uparrow z)$. When the computer evaluates this expression, it performs the binary operations (within each parenthesized part) according to a hierarchy of operations whereby exponentiation precedes multiplication

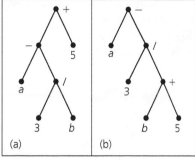

Figure 12.15

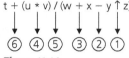

Figure 12.16

and division, which in turn precede addition and subtraction. In Fig. 12.16 we number the operations in the order in which they are performed by the computer. For the computer to evaluate this expression, it must somehow scan the expression in order to perform the operations in the order specified.

Instead of scanning back and forth continuously, however, the machine converts the expression into a notation that is independent of parentheses. This is known as Polish notation, in honor of the Polish (actually Ukrainian) logician Jan Lukasiewicz (1878–1956). Here the *infix* notation $a \circ b$ for a binary operation $\circ$ becomes $\circ ab$, the *prefix* (or Polish) notation. The advantage is that the expression in Fig. 12.16 can be rewritten without parentheses as

$$+\,t\,/\ast uv + w - x \uparrow yz,$$

where the evaluation proceeds from right to left. When a binary operation is encountered, it is performed on the two operands to its right. The result is then treated as one of the operands for the next binary operation encountered as we continue to the left. For instance, given the assignments $t = 4$, $u = 2$, $v = 3$, $w = 1$, $x = 9$, $y = 2$, $z = 3$, the following steps take place in the evaluation of the expression

$$+\,t\,/\ast uv + w - x \uparrow yz.$$

1) $+\,4\,/\ast 2\ 3 + 1 - 9 \underbrace{\uparrow 2\ 3}$
$\qquad\qquad\qquad\qquad\quad 2 \uparrow 3 = 8$

2) $+\,4\,/\ast 2\ 3 + 1 \underbrace{- 9\ 8}$
$\qquad\qquad\qquad\qquad 9 - 8 = 1$

3) $+\,4\,/\ast 2\ 3 \underbrace{+ 1\ 1}$
$\qquad\qquad\qquad\quad 1 + 1 = 2$

4) $+\,4\,/ \underbrace{\ast 2\ 3}\ 2$
$\qquad\qquad\quad 2 \ast 3 = 6$

5) $+\,4\, \underbrace{/\ 6\ 2}$
$\qquad\qquad\quad 6\ /\ 2 = 3$

6) $\underbrace{+\,4\ 3}$
$\quad 4 + 3 = 7$

So the value of the given expression for the preceding assignments is 7.

The use of Polish notation is important for the compilation of computer programs and can be obtained by representing a given expression by a rooted tree, as shown in Fig. 12.17. Here each variable (or constant) is used to label a leaf of the tree. Each internal vertex is labeled by a binary operation whose left and right operands are the left and right subtrees it determines. Starting at the root, as we transverse the tree from top to bottom and left to right, as shown in Fig. 12.17, we find the Polish notation by writing down the labels of the vertices in the order in which they are visited.

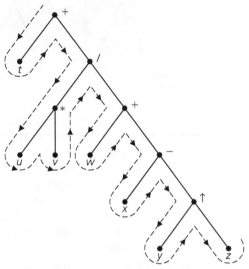

Figure 12.17

The last two examples illustrate the importance of order. Several methods exist for systematically ordering the vertices in a tree. Two of the most prevalent in the study of data structures are the preorder and postorder. These are defined recursively in the following definition.

Definition 12.3 Let $T = (V, E)$ be a rooted tree with root r. If T has no other vertices, then the root by itself constitutes the *preorder* and *postorder traversals* of T. If $|V| > 1$, let $T_1, T_2, T_3, \ldots, T_k$ denote the subtrees of T as we go from left to right (as in Fig. 12.18).

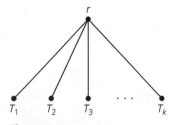

Figure 12.18

a) The *preorder traversal* of T first visits r and then traverses the vertices of T_1 in preorder, then the vertices of T_2 in preorder, and so on until the vertices of T_k are traversed in preorder.

b) The *postorder traversal* of T traverses in postorder the vertices of the subtrees T_1, $T_2, \ldots, T_k$ and then visits the root.

We demonstrate these ideas in the following example.

EXAMPLE 12.6

Consider the rooted tree shown in Fig. 12.19.

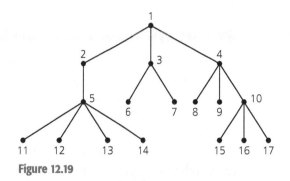

Figure 12.19

a) *Preorder*: After visiting vertex 1 we visit the subtree T_1 rooted at vertex 2. After visiting vertex 2 we proceed to the subtree rooted at vertex 5, and after visiting vertex 5 we go to the subtree rooted at vertex 11. This subtree has no other vertices, so we visit vertex 11 and then return to vertex 5 from which we visit, in succession, vertices 12, 13, and 14. Following this we *backtrack* (14 to 5 to 2 to 1) to the root and then visit the vertices in the subtree T_2 in the preorder 3, 6, 7. Finally, after returning to the root for the last time, we traverse the subtree T_3 in the preorder 4, 8, 9, 10, 15, 16, 17. Hence the preorder listing of the vertices in this tree is 1, 2, 5, 11, 12, 13, 14, 3, 6, 7, 4, 8, 9, 10, 15, 16, 17.

In this ordering we start at the root and build a path as far as we can. At each level we go to the leftmost vertex (not previously visited) at the next level, until we reach a leaf ℓ. Then we backtrack to the parent p of this leaf ℓ and visit ℓ's sibling s (and the subtree that s determines) directly on its right. If no such sibling s exists, we backtrack to the grandparent g of the leaf ℓ and visit, if it exists, a vertex u that is the sibling of p directly to its right in the tree. Continuing in this manner, we eventually visit (the first time each one is encountered) all of the vertices in the tree.

The vertices in Figs. 12.11(a), 12.12, and 12.17 are visited in preorder. The preorder traversal for the tree in Fig. 12.11(a) provides the ordering in Fig. 12.11(b). The lexicographic order in Example 12.4 arises from the preorder traversal of the tree in Fig. 12.12.

b) *Postorder*: For the postorder traversal of a tree, we start at the root r and build the longest path, going to the leftmost child of each internal vertex whenever we can. When we arrive at a leaf ℓ we visit this vertex and then backtrack to its parent p. However, we do not visit p until after all of its descendants are visited. The next vertex we visit is found by applying the same procedure at p that was originally applied at r in obtaining ℓ — except that now we first go from p to the sibling of ℓ directly to the right (of ℓ). And at no time is any vertex visited more than once or before any of its descendants.

For the tree given in Fig. 12.19, the postorder traversal starts with a postorder traversal of the subtree T_1 rooted at vertex 2. This yields the listing 11, 12, 13, 14, 5, 2. We proceed to the subtree T_2, and the postorder listing continues with 6, 7, 3. Then for T_3 we find 8, 9, 15, 16, 17, 10, 4 as the postorder listing. Finally, vertex 1 is visited. Consequently, for this tree, the postorder traversal visits the vertices in the order 11, 12, 13, 14, 5, 2, 6, 7, 3, 8, 9, 15, 16, 17, 10, 4, 1.

In the case of binary rooted trees, a third type of tree traversal called the *inorder* traversal may be used. Here we do *not* consider subtrees as first and second, but rather in terms of left and right. The formal definition is recursive, as were the definitions of preorder and postorder traversals.

Definition 12.4   Let $T = (V, E)$ be a binary rooted tree with vertex r the root.

1) If $|V| = 1$, then the vertex r constitutes the *inorder traversal* of T.

2) When $|V| > 1$, let T_L and T_R denote the left and right subtrees of T. The *inorder traversal* of T first traverses the vertices T_L in inorder, then it visits the root r, and then it traverses, in inorder, the vertices of T_R.

We realize that here a left or right subtree may be empty. Also, if v is a vertex in such a tree and $od(v) = 1$, then if w is the child of v, we must distinguish between w's being the left child and its being the right child.

EXAMPLE 12.7   As a result of the previous comments, the two binary rooted trees shown in Fig. 12.20 are not considered the same, when viewed as *ordered trees*. As rooted binary trees they are the same. (Each tree has the same set of vertices and the same set of directed edges.) However, when we consider the additional concept of left and right children, we see that in part (a) of the figure vertex v has right child a, whereas in part (b) vertex a is the left child of v. Consequently, when the difference between left and right children is taken into consideration, these trees are no longer viewed as the same tree.

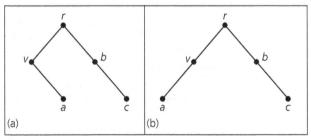

Figure 12.20

In visiting the vertices for the tree in part (a) of Fig. 12.20, we first visit in inorder the left subtree of the root r. This subtree consists of the root v and its *right child a*. (Here the left child is *null*, or nonexistent.) Since v has no left subtree, we visit in inorder vertex v and then its right subtree, namely, a. Having traversed the left subtree of r, we now visit vertex r and then traverse, in inorder, the vertices in the right subtree of r. This results in our visiting first vertex b (because b has no left subtree) and then vertex c. Hence the inorder listing for the tree shown in Fig. 12.20(a) is v, a, r, b, c.

When we consider the tree in part (b) of the figure, once again we start by visiting, in inorder, the vertices in the left subtree of the root r. Here, however, this left subtree consists of vertex v (the root of the subtree) and its *left child a*. (In this case, the right child of v is null, or nonexistent.) Therefore this inorder traversal first visits vertex a (the left subtree of v), and then vertex v. Since v has no right subtree, we are now finished visiting the left subtree of r, in inorder. So next the root r is visited, and then the vertices of the right subtree of r are traversed, in inorder. This results in the inorder listing a, v, r, b, c for the tree shown in Fig. 12.20(b).

We should note, however, that for the preorder traversal *in this particular example*,[†] the same result is obtained for both trees:

Preorder listing: r, v, a, b, c.

Likewise, this particular example is such that the postorder traversal for either tree gives us the following:

Postorder listing: a, v, c, b, r.

It is only for the inorder traversal, with its distinctions between left and right children and between left and right subtrees, that a difference occurs. For the trees in parts (a) and (b) of Fig. 12.20 we found the respective inorder listings to be

(a) v, a, r, b, c and   (b) a, v, r, b, c.

EXAMPLE 12.8

If we apply the inorder traversal to the binary rooted tree shown in Fig. 12.21, we find that the inorder listing for the vertices is $p, j, q, f, c, k, g, a, d, r, b, h, s, m, e, i, t, n, u$.

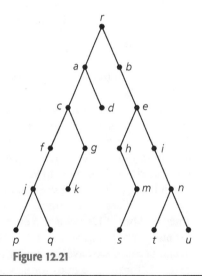

Figure 12.21

Our next example shows how the preorder traversal can be used in a counting problem dealing with binary trees.

EXAMPLE 12.9[‡]

For $n \geq 0$, consider the complete binary trees on $2n + 1$ vertices. The cases for $0 \leq n \leq 3$ are shown in Fig. 12.22. Here we distinguish left from right. So, for example, the two

---

[†]A note of caution! If we interchange the order of the two existing children (of a certain parent) in a binary rooted tree, then a change results in the preorder, postorder, and inorder traversals. If one child is "null," however, then only the inorder traversal changes.

[‡]This example uses material developed in the optional Sections 1.5 and 10.5. It may be omitted with no loss of continuity.

complete binary trees for $n = 2$ are considered distinct. [If we do not distinguish left from right, these trees are (isomorphic and) no longer counted as two different trees.]

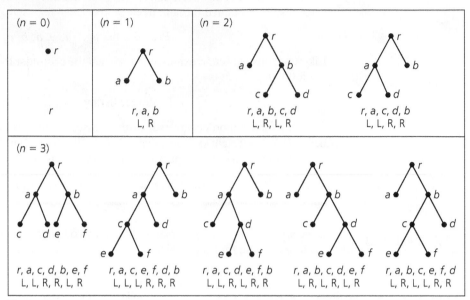

Figure 12.22

Below each tree in the figure we list the vertices for a preorder traversal. In addition, for $1 \le n \le 3$, we find a list of n L's and n R's under each preorder traversal. These lists are determined as follows. The first tree for $n = 2$, for instance, has the list L, R, L, R because, after visiting the root r, we go to the left (L) subtree rooted at a and visit vertex a. Then we backtrack to r and go to the right (R) subtree rooted at b. After visiting vertex b we go to the left (L) subtree of b rooted at c and visit vertex c. Then, lastly, we backtrack to b and go to its right (R) subtree to visit vertex d. This generates the list L, R, L, R and the other seven lists of L's and R's are obtained in the same way.

Since we are traversing these trees in preorder, each list starts with an L. There is an equal number of L's and R's in each list because the trees are complete binary trees. Finally, the number of R's never exceeds the number of L's as a given list is read from left to right — again, because we have a preorder traversal. Should we replace each L by a 1 and each R by a -1, for the five trees for $n = 3$, we find ourselves back in part (a) of Example 1.43, where we have one of our early examples of the Catalan numbers. Hence, for $n \ge 0$, we see that the number of complete binary trees on $2n + 1$ vertices is $\frac{1}{n+1}\binom{2n}{n}$, the nth Catalan number. [Note that if we *prune* the five trees for $n = 3$ by removing the four leaves for each tree, we obtain the five rooted ordered binary trees in Fig. 10.18.]

The notion of preorder now arises in the following procedure for finding a spanning tree for a connected graph.

Let $G = (V, E)$ be a loop-free connected undirected graph with $r \in V$. Starting from r, we construct a path in G that is as long as possible. If this path includes every vertex in V, then the path is a spanning tree T for G and we are finished. If not, let x and y be the last two vertices visited along this path, with y the last vertex. We then return, or *backtrack*, to the vertex x and construct a second path in G that is as long as possible, starts at x, and doesn't include any vertex already visited. If no such path exists, backtrack to the parent p of x and see how far it is possible to branch off from p, building a path (that is as long as possible and has no previously visited vertices) to a new vertex y_1 (which will be a new leaf for T). Should all edges from the vertex p lead to vertices already encountered, backtrack one level higher and continue the process. Since the graph is finite and connected, this technique, which is called *backtracking*, or *depth-first search*, eventually determines a spanning tree T for G, where r is regarded as the root of T. Using T, we then order the vertices of G in a preorder listing.

The depth-first search serves as a framework around which many algorithms can be designed to test for certain graph properties. One such algorithm will be examined in detail in Section 12.5.

One way to help implement the depth-first search in a computer program is to assign a fixed order to the vertices of the given graph $G = (V, E)$. Then if there are two or more vertices adjacent to a vertex v and none of these vertices has already been visited, we shall know exactly which vertex to visit first. This order now helps us to develop the foregoing description of the depth-first search as an algorithm.

Let $G = (V, E)$ be a loop-free connected undirected graph where $|V| = n$ and the vertices are ordered as $v_1, v_2, v_3, \ldots, v_n$. To find the rooted ordered depth-first spanning tree for the prescribed order, we apply the following algorithm, wherein the variable v is used to store the vertex presently being examined.

Depth-First Search Algorithm

Step 1: Assign v_1 to the variable v and initialize T as the tree consisting of just this one vertex. (The vertex v_1 will be the root of the spanning tree that develops.) Visit v_1.

Step 2: Select the smallest subscript i, for $2 \leq i \leq n$, such that $\{v, v_i\} \in E$ and v_i has not already been visited.

   If no such subscript is found, then go to step (3). Otherwise, perform the following: (1) Attach the edge $\{v, v_i\}$ to the tree T and visit v_i; (2) Assign v_i to v; and (3) Return to step (2).

Step 3: If $v = v_1$, the tree T is the (rooted ordered) spanning tree for the order specified.

Step 4: For $v \neq v_1$, backtrack from v to its parent u in T. Then assign u to v and return to step (2).

EXAMPLE 12.10

We now apply this algorithm to the graph $G = (V, E)$ shown in Fig. 12.23(a). Here the order for the vertices is alphabetic: $a, b, c, d, e, f, g, h, i, j$.

First we assign the vertex a to the variable v and initialize T as just the vertex a (the root). We visit vertex a. Then, going to step (2), we find that the vertex b is the first vertex w such that $\{a, w\} \in E$ and w has not been visited earlier. So we attach edge $\{a, b\}$ to T and visit b, assign b to v, and then return to step (2).

At $v = b$ we find that the first vertex (not visited earlier) that provides an edge for the spanning tree is d. Consequently, the edge $\{b, d\}$ is attached to T and d is visited, then d is assigned to v, and we again return to step (2).

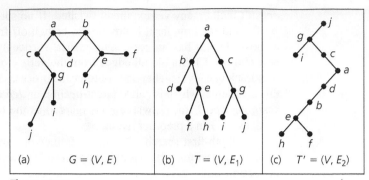

Figure 12.23

This time, however, there is no new vertex that we can obtain from d, because vertices a and b have already been visited. So we go to step (3). But here the value of v is d, not a, and we go to step (4). Now we backtrack from d, assigning the vertex b to v, and then we return to step (2). At this time we add the edge $\{b, e\}$ to T and visit e.

Continuing the process, we attach the edge $\{e, f\}$ (and visit f) and then the edge $\{e, h\}$ (and visit h). But now the vertex h has been assigned to v, and we must backtrack from h to e to b to a. When v is assigned the vertex a this (second) time, the new edge $\{a, c\}$ is obtained and vertex c is visited. Then we proceed to attach the edges $\{c, g\}$, $\{g, i\}$, and $\{g, j\}$ (visiting the vertices g, i, and j, respectively). At this point all of the vertices in G have been visited, and we backtrack from j to g to c to a. With $v = a$ once again we return to step (2) and from there to step (3), where the process terminates.

The resulting tree $T = (V, E_1)$ is shown in part (b) of Fig. 12.23. Part (c) of the figure shows the tree T' that results for the vertex ordering: $j, i, h, g, f, e, d, c, b, a$.

A second method for searching the vertices of a loop-free connected undirected graph is the *breadth-first search*. Here we designate one vertex as the root and fan out to all vertices adjacent to the root. From each child of the root we then fan out to those vertices (not previously visited) that are adjacent to one of these children. As we continue this process, we never list a vertex twice, so no cycle is constructed, and with G finite the process eventually terminates.

We actually used this technique earlier in Example 11.28 of Section 11.5.

A certain data structure proves useful in developing an algorithm for this second searching method. A *queue* is an ordered list wherein items are inserted at one end (called the *rear*) of the list and deleted at the other end (called the *front*). The *first* item inserted *in* the queue is the *first* item that can be taken out of it. Consequently, a queue is referred to as a "first-in, first-out," or FIFO, structure.

As in the depth-first search, we again assign an order to the vertices of our graph.

We start with a loop-free connected undirected graph $G = (V, E)$, where $|V| = n$ and the vertices are ordered as $v_1, v_2, v_3, \ldots, v_n$. The following algorithm generates the (rooted ordered) breadth-first spanning tree T of G for the given order.

Breadth-First Search Algorithm

Step 1: Insert vertex v_1 at the rear of the (initially empty) queue Q and initialize T as the tree made up of this one vertex v_1 (the root of the final version of T). Visit v_1.

Step 2: While the queue Q is not empty, delete the vertex v from the front of Q. Now examine the vertices v_i (for $2 \le i \le n$) that are adjacent to v — in the specified order. If v_i has not been visited, perform the following: (1) Insert v_i at the rear of Q; (2) Attach the edge $\{v, v_i\}$ to T; and (3) Visit vertex v_i. [If we examine *all* of the vertices previously in the queue Q and obtain no new edges, then the tree T (generated to this point) is the (rooted ordered) spanning tree for the given order.]

EXAMPLE 12.11

We shall employ the graph of Fig. 12.23(a) with the prescribed order $a, b, c, d, e, f, g, h, i, j$ to illustrate the use of the algorithm for the breadth-first search.

Start with vertex a. Insert a at the rear of (the presently empty) queue Q, initialize T as this one vertex (the root of the resulting tree), and visit vertex a.

In step (2) we now delete a from (the front of) Q and examine the vertices adjacent to a — namely, the vertices b, c, d. (These vertices have not been previously visited.) This results in our (i) inserting vertex b at the rear of Q, attaching the edge $\{a, b\}$ to T, and visiting vertex b; (ii) inserting vertex c at the rear of Q (after b), attaching the edge $\{a, c\}$ to T, and visiting vertex c; and (iii) inserting vertex d at the rear of Q (after c), attaching the edge $\{a, d\}$ to T, and visiting vertex d.

Since the queue Q is not empty, we execute step (2) again. Upon deleting vertex b from the front of Q, we now find that the only vertex adjacent to b (that has not been previously visited) is e. So we insert vertex e at the rear of Q (after d), attach the edge $\{b, e\}$ to T, and visit vertex e. Continuing with vertex c we obtain the new (unvisited) vertex g. So we insert vertex g at the rear of Q (after e), attach the edge $\{c, g\}$ to T, and visit vertex g. And now we delete vertex d from the front of Q. But at this point there are no unvisited vertices adjacent to d, so we then delete vertex e from the front of Q. This vertex leads to the following: inserting vertex f at the rear of Q (after g), attaching the edge $\{e, f\}$ to T, and visiting vertex f. This is followed by: inserting vertex h at the rear of Q (after f), attaching edge $\{e, h\}$ to T, and visiting vertex h. Continuing with vertex g, we insert vertex i at the rear of Q (after h), attach edge $\{g, i\}$ to T, and visit vertex i, and then we insert vertex j at the rear of Q (after i), attach edge $\{g, j\}$ to T, and visit vertex j.

Once again we return to the beginning of step (2). But now when we delete (from the front of Q) and examine each of the vertices f, h, i, and j (in this order), we find no unvisited vertices for any of these four vertices. Consequently, the queue Q now remains empty and the tree T in Fig. 12.24(a) is the breadth-first spanning tree for G, for the order

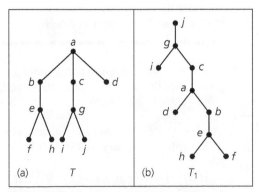

(a) T (b) T_1

Figure 12.24

prescribed. (The tree T_1, shown in part (b) of the figure, arises for the order $j, i, h, g, f, e,$ d, c, b, a.)

Let us apply these ideas on graph searching to one more example.

EXAMPLE 12.12

Let $G = (V, E)$ be an undirected graph (with loops) where the vertices are ordered as $v_1, v_2, \ldots, v_7$. If Fig. 12.25(a) is the adjacency matrix $A(G)$ for G, how can we use this representation of G to determine whether G is connected, without drawing the graph?

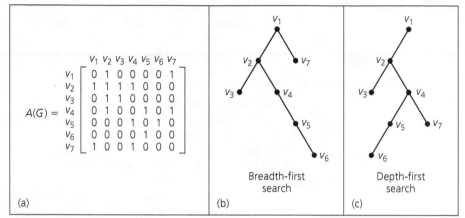

Figure 12.25

Using v_1 as the root, in part (b) of the figure we search the graph by means of its adjacency matrix, using a breadth-first search. [Here we ignore the loops by ignoring any 1's on the main diagonal (extending from the upper left to the lower right).] First we visit the vertices adjacent to v_1, listing them in ascending order according to the subscripts on the v's in $A(G)$. The search continues, and as all vertices in G are reached, G is shown to be connected.

The same conclusion follows from the depth-first search in part (c). The tree here also has v_1 as its root. As the tree branches out to search the graph, it does so by listing the first vertex found adjacent to v_1 according to the row in $A(G)$ for v_1. Likewise, from v_2 the first new vertex in this search is found from $A(G)$ to be v_3. The vertex v_3 is a leaf in this tree because no new vertex can be visited from v_3. As we backtrack to v_2, row 2 of $A(G)$ indicates that v_4 can now be visited from v_2. As this process continues, the connectedness of G follows from part (c) of the figure.

It is time now to return to our main discussion on rooted trees. The following definition generalizes the ideas that were introduced for Example 12.5.

Definition 12.5

Let $T = (V, E)$ be a rooted tree, and let $m \in \mathbf{Z}^+$.

We call T an *m-ary tree* if $od(v) \leq m$ for all $v \in V$. When $m = 2$, the tree is called a *binary tree*.

If $od(v) = 0$ or m, for all $v \in V$, then T is called a *complete m-ary tree*. The special case of $m = 2$ results in a *complete binary tree*.

In a complete m-ary tree, each internal vertex has exactly m children. (Each leaf of this tree still has no children.)

Some properties of these trees are considered in the following theorem.

THEOREM 12.6

Let $T = (V, E)$ be a complete m-ary tree with $|V| = n$. If T has ℓ leaves and i internal vertices, then (a) $n = mi + 1$; (b) $\ell = (m - 1)i + 1$; and (c) $i = (\ell - 1)/(m - 1) = (n - 1)/m$.

Proof: This proof is left for the Section Exercises.

EXAMPLE 12.13

The Wimbledon tennis championship is a single-elimination tournament wherein a player (or doubles team) is eliminated after a single loss. If 27 women compete in the singles championship, how many matches must be played to determine the number-one female player?

Consider the tree shown in Fig. 12.26. With 27 women competing, there are 27 leaves in this complete binary tree, so from Theorem 12.6(c) the number of internal vertices (which is the number of matches) is $i = (\ell - 1)/(m - 1) = (27 - 1)/(2 - 1) = 26$.

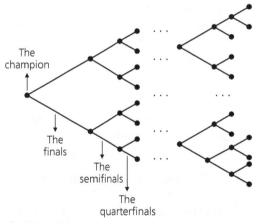

Figure 12.26

EXAMPLE 12.14

A classroom contains 25 microcomputers that must be connected to a wall socket that has four outlets. Connections are made by using extension cords that have four outlets each. What is the least number of cords needed to get these computers set up for class use?

The wall socket is considered the root of a complete m-ary tree for $m = 4$. The microcomputers are the leaves of this tree, so $\ell = 25$. Each internal vertex, except the root, corresponds with an extension cord. So by part (c) of Theorem 12.6, there are $(\ell - 1)/(m - 1) = (25 - 1)/(4 - 1) = 8$ internal vertices. Hence we need $8 - 1$ (where the 1 is subtracted for the root) = 7 extension cords.

Definition 12.6

If $T = (V, E)$ is a rooted tree and h is the largest level number achieved by a leaf of T, then T is said to have *height* h. A rooted tree T of height h is said to be *balanced* if the level number of every leaf in T is $h - 1$ or h.

The rooted tree shown in Fig. 12.19 is a balanced tree of height 3. Tree T' in Fig. 12.23(c) has height 7 but is not balanced. (Why?)

The tree for the tournament in Example 12.13 must be balanced so that the tournament will be as fair as possible. If it is not balanced, some competitor will receive more than one bye (an opportunity to advance without playing a match).

Before stating our next theorem, let us recall that for all $x \in \mathbf{R}$, $\lfloor x \rfloor$ denotes the greatest integer in x, or floor of x, whereas $\lceil x \rceil$ designates the ceiling of x.

THEOREM 12.7

Let $T = (V, E)$ be a complete m-ary tree of height h with ℓ leaves. Then $\ell \leq m^h$ and $h \geq \lceil \log_m \ell \rceil$.

Proof: The proof that $\ell \leq m^h$ will be established by induction on h. When $h = 1$, T is a tree with a root and m children. In this case $\ell = m = m^h$, and the result is true. Assume the result true for all trees of height $< h$, and consider a tree T with height h and ℓ leaves. (The level numbers that are possible for these leaves are $1, 2, \ldots, h$, with at least m of the leaves at level h.) The ℓ leaves of T are also the ℓ leaves (total) for the m subtrees T_i, $1 \leq i \leq m$, of T rooted at each of the children of the root. For $1 \leq i \leq m$, let ℓ_i be the number of leaves in subtree T_i. (In the case where leaf and root coincide, $\ell_i = 1$. But since $m \geq 1$ and $h - 1 \geq 0$, we have $m^{h-1} \geq 1 = \ell_i$.) By the induction hypothesis, $\ell_i \leq m^{h(T_i)} \leq m^{h-1}$, where $h(T_i)$ denotes the height of the subtree T_i, and so $\ell = \ell_1 + \ell_2 + \cdots + \ell_m \leq m(m^{h-1}) = m^h$.

With $\ell \leq m^h$, we find that $\log_m \ell \leq \log_m(m^h) = h$, and since $h \in \mathbf{Z}^+$, it follows that $h \geq \lceil \log_m \ell \rceil$.

COROLLARY 12.1

Let T be a balanced complete m-ary tree with ℓ leaves. Then the height of T is $\lceil \log_m \ell \rceil$.

Proof: This proof is left as an exercise.

We close this section with an application that uses a complete ternary ($m = 3$) tree.

EXAMPLE 12.15

Decision Trees. There are eight coins (identical in appearance) and a pan balance. If exactly one of these coins is counterfeit and heavier than the other seven, find the counterfeit coin.

Let the coins be labeled $1, 2, 3, \ldots, 8$. In using the pan balance to compare sets of coins there are three outcomes to consider: (a) the two sides balance to indicate that the coins in the two pans are not counterfeit; (b) the left pan of the balance goes down, indicating that the counterfeit coin is in the left pan; or (c) the right pan goes down, indicating that it holds the counterfeit coin.

In Fig. 12.27(a), we search for the counterfeit coin by first balancing coins 1, 2, 3, 4 against 5, 6, 7, 8. If the balance tips to the right, we follow the right branch from the root to then analyze coins 5, 6 against 7, 8. If the balance tips to the left, we test coins 1, 2 against 3, 4. At each successive level, we have half as many coins to test, so at level 3 (after three weighings) the heavier counterfeit coin has been identified.

The tree in part (b) of the figure finds the heavier coin in two weighings. The first weighing balances coins 1, 2, 3 against 6, 7, 8. Three possible outcomes can occur: (i) the balance tips to the right, indicating that the heavier coin is 6, 7, or 8, and we follow the right branch from the root; (ii) the balance tips to the left and we follow the left branch to find which of 1, 2, 3 is the heavier; or (iii) the pans balance and we follow the center branch to find which of 4, 5 is heavier. At each internal vertex the label indicates which coins are being compared.

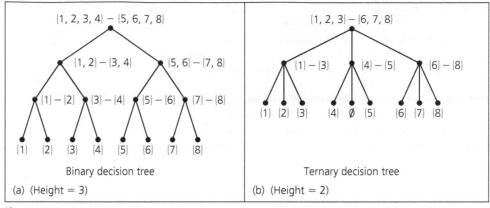

Figure 12.27

Unlike part (a), a conclusion may be deduced in part (b) when a coin is not included in a weighing. Finally, when comparing coins 4 and 5, because equality cannot take place we label the center leaf with Ø.

In this particular problem, we claim that the height of the complete ternary tree used must be at least 2. With eight coins involved, the tree will have at least eight leaves. Consequently, with $\ell \geq 8$, it follows from Theorem 12.7 that $h \geq \lceil \log_3 \ell \rceil \geq \lceil \log_3 8 \rceil = 2$, so at least two weighing are needed. If n coins are involved, the complete ternary tree will have ℓ leaves where $\ell \geq n$, and its height h satisfies $h \geq \lceil \log_3 n \rceil$.

EXERCISES 12.2

1. Answer the following questions for the tree shown in Fig. 12.28.

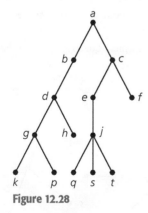

Figure 12.28

a) Which vertices are the leaves?

b) Which vertex is the root?

c) Which vertex is the parent of g?

d) Which vertices are the descendants of c?

e) Which vertices are the siblings of s?

f) What is the level number of vertex f?

g) Which vertices have level number 4?

2. Let $T = (V, E)$ be a binary tree. In Fig. 12.29 we find the subtree of T rooted at vertex p. (The dashed line coming into vertex p indicates that there is more to the tree T than what appears in the figure.) If the level number for vertex u is 37, (a) what are the level numbers for vertices p, s, t, v, w, x, y, and z? (b) how many ancestors does vertex u have? (c) how many ancestors does vertex y have?

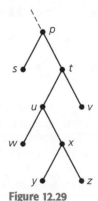

Figure 12.29

3. a) Write the expression $(w + x - y)/(\pi * z^3)$ in Polish notation, using a rooted tree.

b) What is the value of the expression (in Polish notation) $/ \uparrow a - bc + d * ef$, if $a = c = d = e = 2$, $b = f = 4$?

4. Let $T = (V, E)$ be a rooted tree ordered by a universal address system. (a) If vertex v in T has address 2.1.3.6, what is the smallest number of siblings that v must have? (b) For the vertex v in part (a), find the address of its parent. (c) How many ancestors does the vertex v in part (a) have? (d) With the presence of v in T, what other addresses must there be in the system?

5. For the tree shown in Fig. 12.30, list the vertices according to a preorder traversal, an inorder traversal, and a postorder traversal.

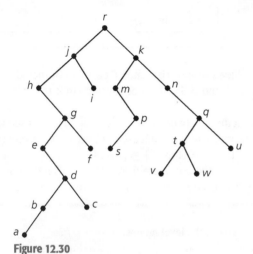

Figure 12.30

6. List the vertices in the tree shown in Fig. 12.31 when they are visited in a preorder traversal and in a postorder traversal.

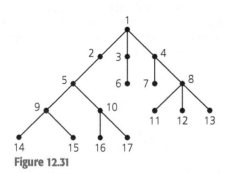

Figure 12.31

7. a) Find the depth-first spanning tree for the graph shown in Fig. 11.72(a) if the order of the vertices is given as (i) a, b, c, d, e, f, g, h; (ii) h, g, f, e, d, c, b, a; (iii) a, b, c, d, h, g, f, e.

b) Repeat part (a) for the graph shown in Fig. 11.85(i).

8. Find the breadth-first spanning trees for the graphs and prescribed orders given in Exercise 7.

9. Let $G = (V, E)$ be an undirected graph with adjacency matrix $A(G)$ as shown here.

	v_1	v_2	v_3	v_4	v_5	v_6	v_7	v_8
v_1	0	1	0	0	0	0	1	0
v_2	1	1	0	1	1	0	1	0
v_3	0	0	0	1	0	1	0	1
v_4	0	1	1	0	0	0	0	0
v_5	0	1	0	0	0	0	1	0
v_6	0	0	1	0	0	1	0	0
v_7	1	1	0	0	1	0	0	0
v_8	0	0	1	0	0	0	0	0

Use a breadth-first search based on $A(G)$ to determine whether G is connected.

10. a) Let $T = (V, E)$ be a binary tree. If $|V| = n$, what is the maximum height that T can attain?

b) If $T = (V, E)$ is a complete binary tree and $|V| = n$, what is the maximum height that T can reach in this case?

11. Prove Theorem 12.6 and Corollary 12.1.

12. With m, n, i, ℓ as in Theorem 12.6, prove that

a) $n = (m\ell - 1)/(m - 1)$.   **b)** $\ell = [(m - 1)n + 1]/m$.

13. a) A complete ternary (or 3-ary) tree $T = (V, E)$ has 34 internal vertices. How many edges does T have? How many leaves?

b) How many internal vertices does a complete 5-ary tree with 817 leaves have?

14. The complete binary tree $T = (V, E)$ has $V = \{a, b, c, \ldots, i, j, k\}$. The postorder listing of V yields $d, e, b, h, i, f, j, k, g, c, a$. From this information draw T if (a) the height of T is 3; (b) the height of the left subtree of T is 3.

15. For $m \geq 3$, a complete m-ary tree can be transformed into a complete binary tree by applying the idea shown in Fig. 12.32.

a) Use this technique to transform the complete ternary decision tree shown in Fig. 12.27(b).

b) If T is a complete quaternary tree of height 3, what is the maximum height that T can have after it is transformed into a complete binary tree? What is the minimum height?

c) Answer part (b) if T is a complete m-ary tree of height h.

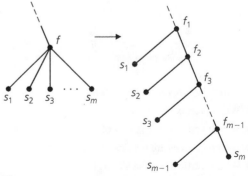

Figure 12.32

16. a) At a men's singles tennis tournament, each of 25 players brings a can of tennis balls. When a match is played, one can of balls is opened and used, then kept by the loser. The winner takes the unopened can on to his next match. How many cans of tennis balls will be opened during this tournament? How many matches are played in the tournament?

b) In how many matches did the tournament champion play?

17. What is the maximum number of internal vertices that a complete quaternary tree of height 8 can have? What is the number for a complete m-ary tree of height h?

18. On the first Sunday of 2003 Rizzo and Frenchie start a chain letter, each of them sending five letters (to ten different friends between them). Each person receiving the letter is to send five copies to five new people on the Sunday following the letter's arrival. After the first seven Sundays have passed, what is the total number of chain letters that have been mailed? How many were mailed on the last three Sundays?

19. Use a complete ternary decision tree to repeat Example 12.15 for a set of 12 coins, exactly one of which is heavier (and counterfeit).

20. Let $T = (V, E)$ be a balanced complete m-ary tree of height $h \geq 2$. If T has ℓ leaves and b_{h-1} internal vertices at level $h - 1$, explain why $\ell = m^{h-1} + (m - 1)b_{h-1}$.

21. Consider the complete binary trees on 31 vertices. (Here we distinguish left from right as in Example 12.9.) How many of these trees have 11 vertices in the left subtree of the root? How many have 21 vertices in the right subtree of the root?

22. For $n \geq 0$, let a_n count the number of complete binary trees on $2n + 1$ vertices. (Here we distinguish left from right as in Example 12.9.) How is a_{n+1} related to $a_0, a_1, a_2, \ldots, a_{n-1}, a_n$?

23. Consider the following algorithm where the input is a rooted tree with root r.

Step 1: Push r onto the (empty) stack

Step 2: While the stack is not empty
Pop the vertex at the top of
the stack and record its label
Push the children — going from
right to left — of this vertex
onto the stack

(The stack data structure was explained in Example 10.43).

What is the output when this algorithm is applied to (a) the tree in Fig. 12.19? (b) any rooted tree?

24. Consider the following algorithm where the input is a rooted tree with root r.

Step 1: Push r onto the (empty) stack

Step 2: While the stack is not empty
If the entry at the top of the stack is
not marked
Then mark it and push its
children — right to left — onto
the stack
Else
Pop the vertex at the top of the
stack and record its label

What is the output when the algorithm is applied to (a) the tree in Fig. 12.19? (b) any rooted tree?

12.3
Trees and Sorting

In Example 10.5, the bubble sort was introduced. There we found that the number of comparisons needed to sort a list of n items is $n(n - 1)/2$. Consequently, this algorithm determines a function $h: \mathbf{Z}^+ \to \mathbf{R}$ defined by $h(n) = n(n - 1)/2$. This is the (worst-case) time-complexity function for the algorithm, and we often express this by writing $h \in O(n^2)$. Consequently, the bubble sort is said to require $O(n^2)$ comparisons. We interpret this to mean that for large n, the number of comparisons is bounded above by cn^2, where c is a constant that is generally not specified because it depends on such factors as the compiler and the computer that are used.

In this section we shall study a second method for sorting a given list of n items into ascending order. The method is called the *merge sort*, and we shall find that the order of its worst-case time-complexity function is $O(n \log_2 n)$. This will be accomplished in the following manner:

1) First we shall measure the number of comparisons needed when n is a power of 2. Our method will employ a pair of balanced complete binary trees.

2) Then we shall cover the case for general n by using the optional material on divide-and-conquer algorithms in Section 10.6.

For the case where n is an arbitrary positive integer, we start by considering the following procedure.

Given a list of n items to sort into ascending order, the *merge sort* recursively splits the given list and all subsequent sublists in half (or as close as possible to half) until each sublist contains a single element. Then the procedure merges these sublists in ascending order until the original n items have been so sorted. The splitting and merging processes can best be described by a pair of balanced complete binary trees, as in the next example.

EXAMPLE 12.16

Merge Sort. Using the merge sort, Fig. 12.33 sorts the list 6, 2, 7, 3, 4, 9, 5, 1, 8. The tree at the top of the figure shows how the process first splits the given list into sublists of size 1. The merging process is then outlined by the tree at the bottom of the figure.

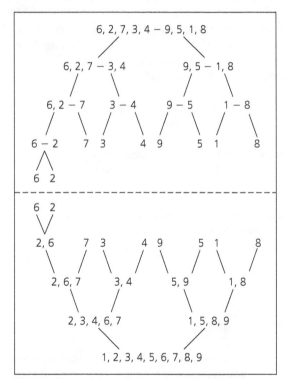

Figure 12.33

To compare the merge sort to the bubble sort, we want to determine its (worst-case) time-complexity function. The following lemma will be needed for this task.

LEMMA 12.1

Let L_1 and L_2 be two sorted lists of ascending numbers, where L_i contains n_i elements, for $i = 1, 2$. Then L_1 and L_2 can be merged into one ascending list L using at most $n_1 + n_2 - 1$ comparisons.

Proof: To merge L_1, L_2 into list L, we perform the following algorithm.

Step 1: Set L equal to the empty list $\emptyset$.

Step 2: Compare the first elements in L_1, L_2. Remove the smaller of the two from the list it is in and place it at the end of L.

Step 3: For the *present* lists L_1, L_2 [one change is made in one of these lists each time step (2) is executed], there are two considerations.

    a) If either of L_1, L_2 is empty, then the other list is concatenated to the end of L. This completes the merging process.

    b) If not, return to step (2).

Each comparison of a number from L_1 with one from L_2 results in the placement of an element at the end of list L, so there cannot be more than $n_1 + n_2$ comparisons. When one of the lists L_1 or L_2 becomes empty no further comparisons are needed, so the maximum number of comparisons needed is $n_1 + n_2 - 1$.

To determine the (worst-case) time-complexity function of the merge sort, consider a list of n elements. For the moment, we do not treat the general problem, assuming here that $n = 2^h$.[†] In the splitting process, the list of 2^h elements is first split into two sublists of size 2^{h-1}. (These are the level 1 vertices in the tree representing the splitting process.) As the process continues, each successive list of size 2^{h-k}, $h > k$, is at level k and splits into two sublists of size $(1/2)(2^{h-k}) = 2^{h-k-1}$. At level h the sublists each contain $2^{h-h} = 1$ element.

Reversing the process, we first merge the $n = 2^h$ leaves into 2^{h-1} ordered sublists of size 2. These sublists are at level $h - 1$ and require $(1/2)(2^h) = 2^{h-1}$ comparisons (one per pair). As this merging process continues, at each of the 2^k vertices at level k, $1 \leq k < h$, there is a sublist of size 2^{h-k}, obtained from merging the two sublists of size 2^{h-k-1} at its children (on level $k + 1$). From Lemma 12.1, this merging requires at most $2^{h-k-1} + 2^{h-k-1} - 1 = 2^{h-k} - 1$ comparisons. When the children of the root are reached, there are two sublists of size 2^{h-1} (at level 1). To merge these sublists into the final list requires at most $2^{h-1} + 2^{h-1} - 1 = 2^h - 1$ comparisons.

Consequently, for $1 \leq k \leq h$, at level k there are 2^{k-1} pairs of vertices. At each of these vertices is a sublist of size 2^{h-k}, so it takes at most $2^{h-k+1} - 1$ comparisons to merge each pair of sublists. With 2^{k-1} pairs of vertices at level k, the total number of comparisons at level k is at most $2^{k-1}(2^{h-k+1} - 1)$. When we sum over all levels k, where $1 \leq k \leq h$, we find that the total number of comparisons is at most

$$\sum_{k=1}^{h} 2^{k-1}(2^{h-k+1} - 1) = \sum_{k=0}^{h-1} 2^k(2^{h-k} - 1) = \sum_{k=0}^{h-1} 2^h - \sum_{k=0}^{h-1} 2^k = h \cdot 2^h - (2^h - 1).$$

With $n = 2^h$, we have $h = \log_2 n$ and

$$h \cdot 2^h - (2^h - 1) = n \log_2 n - (n - 1) = n \log_2 n - n + 1,$$

[†]The result obtained here for $n = 2^h$, $h \in \mathbf{N}$, is actually true for all $n \in \mathbf{Z}^+$. However, the derivation for general n requires the optional material in Section 10.6. That is why this counting argument is included here — for the benefit of those readers who did not cover Section 10.6.

where $n \log_2 n$ is the dominating term for large n. Thus the (worst-case) time-complexity function for this sorting procedure is $g(n) = n \log_2 n - n + 1$ and $g \in O(n \log_2 n)$, for $n = 2^h$, $h \in \mathbf{Z}^+$. Hence the number of comparisons needed to merge sort a list of n items is bounded above by $dn \log_2 n$ for some constant d, and for all $n \geq n_0$, where n_0 is some particular (large) positive integer.

To show that the order of the merge sort is $O(n \log_2 n)$ for all $n \in \mathbf{Z}^+$, our second approach will use the result of Exercise 9 from Section 10.6. We state that now:

Let a, b, $c \in \mathbf{Z}^+$, with $b \geq 2$. If $g: \mathbf{Z}^+ \to \mathbf{R}^+ \cup \{0\}$ is a monotone increasing function, where

$$g(1) \leq c,$$
$$g(n) \leq ag(n/b) + cn, \quad \text{for } n = b^h, h \in \mathbf{Z}^+,$$

then for the case where $a = b$, we have $g \in O(n \log n)$, for *all* $n \in \mathbf{Z}^+$. (The base for the log function may be any real number greater than 1. Here we shall use the base 2.)

Before we can apply this result to the merge sort, we wish to formulate this sorting process (illustrated in Fig. 12.33) as a precise algorithm. To do so, we call the procedure outlined in Lemma 12.1 the "merge" algorithm. Then we shall write "merge (L_1, L_2)" in order to represent the application of that procedure to the lists L_1, L_2, which are in ascending order.

The algorithm for merge sort is a recursive procedure because it may invoke itself. Here the input is an array (called List) of n items, such as real numbers.

The MergeSort Algorithm

Step 1: If $n = 1$, then List is already sorted and the process terminates. If $n > 1$, then go to step (2).

Step 2: (Divide the array and sort the subarrays.) Perform the following:
  1) Assign m the value $\lfloor n/2 \rfloor$.
  2) Assign to List 1 the subarray

$$\text{List}[1], \text{List}[2], \ldots, \text{List}[m].$$

  3) Assign to List 2 the subarray

$$\text{List}[m+1], \text{List}[m+2], \ldots, \text{List}[n].$$

  4) Apply MergeSort to List 1 (of size m) and to List 2 (of size $n - m$).

Step 3: Merge (List 1, List 2).

The function $g: \mathbf{Z}^+ \to \mathbf{R}^+ \cup \{0\}$ will measure the (worst-case) time-complexity for this algorithm by counting the maximum number of comparisons needed to merge sort an array of n items. For $n = 2^h$, $h \in \mathbf{Z}^+$, we have

$$g(n) = 2g(n/2) + [(n/2) + (n/2) - 1].$$

The term $2g(n/2)$ results from step (2) of the MergeSort algorithm, and the summand $[(n/2) + (n/2) - 1]$ follows from step (3) of the algorithm and Lemma 12.1.

With $g(1) = 0$, the preceding equation provides the inequalities

$$g(1) = 0 \le 1,$$
$$g(n) = 2g(n/2) + (n - 1) \le 2g(n/2) + n, \quad \text{for } n = 2^h, h \in \mathbf{Z}^+.$$

We also observe that $g(1) = 0$, $g(2) = 1$, $g(3) = 3$, and $g(4) = 5$, so $g(1) \le g(2) \le g(3) \le g(4)$. Consequently, it appears that g may be a monotone increasing function. The proof that it is monotone increasing is similar to that given for the time-complexity function of binary search. This follows Example 10.49 in Section 10.6, so we leave the details showing that g is monotone increasing to the Section Exercises.

Now with $a = b = 2$ and $c = 1$, the result stated earlier implies that $g \in O(n \log_2 n)$ for *all* $n \in \mathbf{Z}^+$.

Although $n \log_2 n \le n^2$ for all $n \in \mathbf{Z}^+$, it does *not* follow that because the bubble sort is $O(n^2)$ and the merge sort is $O(n \log_2 n)$, the merge sort is more efficient than the bubble sort for all $n \in \mathbf{Z}^+$. The bubble sort requires less programming effort and generally takes less time than the merge sort for small values of n (depending on factors such as the programming language, the compiler, and the computer). However, as n increases, the ratio of the worst-case running times, as measured by $(cn^2)/(dn \log_2 n) = (c/d)(n/\log_2 n)$, gets arbitrarily large. Consequently, as the input list increases in size, the $O(n^2)$ algorithm (bubble sort) takes significantly more time than the $O(n \log_2 n)$ algorithm (merge sort).

For more on sorting algorithms and their time-complexity functions, the reader should examine [1], [3], [4], [7], and [8] in the chapter references.

EXERCISES 12.3

1. a) Give an example of two lists L_1, L_2, each of which is in ascending order and contains five elements, and where nine comparisons are needed to merge L_1, L_2 by the algorithm given in Lemma 12.1.

b) Let $m, n \in \mathbf{Z}^+$ with $m < n$. Give an example of two lists L_1, L_2, each of which is in ascending order, where L_1 has m elements, L_2 has n elements, and $m + n - 1$ comparisons are needed to merge L_1, L_2 by the algorithm given in Lemma 12.1.

2. Apply the merge sort to each of the following lists. Draw the splitting and merging trees for each application of the procedure.

  a) $-1, 0, 2, -2, 3, 6, -3, 5, 1, 4$

  b) $-1, 7, 4, 11, 5, -8, 15, -3, -2, 6, 10, 3$

3. Related to the merge sort is a somewhat more efficient procedure called the *quick sort*. Here we start with a list $L: a_1, a_2, \ldots, a_n$, and use a_1 as a pivot to develop two sublists L_1 and L_2 as follows. For $i > 1$, if $a_i < a_1$, place a_i at the end of the first list being developed (this is L_1 at the end of the process); otherwise, place a_i at the end of the second list L_2.

After all a_i, $i > 1$, have been processed, place a_1 at the end of the first list. Now apply quick sort recursively to each of the lists L_1 and L_2 to obtain sublists L_{11}, L_{12}, L_{21}, and L_{22}. Continue the process until each of the resulting sublists contains one element. The sublists are then ordered, and their concatenation gives the ordering sought for the original list L.

Apply quick sort to each list in Exercise 2.

4. Prove that the function g used in the second method to analyze the (worst-case) time-complexity of the merge sort is monotone increasing.

12.4
Weighted Trees and Prefix Codes

Among the topics to which discrete mathematics is applied, coding theory is one wherein different finite structures play a major role. These structures enable us to represent and transmit information that is coded in terms of the symbols in a given alphabet. For instance, the way we most often code, or represent, characters internally in a computer is by means of strings of fixed length, using the symbols 0 and 1.

The codes developed in this section, however, will use strings of different lengths. Why a person should want to develop such a coding scheme and how the scheme can be constructed will be our major concerns in this section.

Suppose we wish to develop a way to represent the letters of the alphabet using strings of 0's and 1's. Since there are 26 letters, we should be able to encode these symbols in terms of sequences of five bits, given that $2^4 < 26 < 2^5$. However, in the English (or any other) language, not all letters occur with the same frequency. Consequently, it would be more efficient to use binary sequences of different lengths, with the most frequently occurring letters (such as e, i, t) represented by the shortest possible sequences. For example, consider $S = \{a, e, n, r, t\}$, a subset of the alphabet. Represent the elements of S by the binary sequences

$$a: 01 \qquad e: 0 \qquad n: 101 \qquad r: 10 \qquad t: 1.$$

If the message "*ata*" is to be transmitted, the binary sequence 01101 is sent. Unfortunately, this sequence is also transmitted for the messages "*etn*", "*atet*", and "*an*".

Consider a second encoding scheme, one given by

$$a: 111 \qquad e: 0 \qquad n: 1100 \qquad r: 1101 \qquad t: 10.$$

Here the message "*ata*" is represented by the sequence 11110111 and there are no other possibilities to confuse the situation. What's more, the labeled complete binary tree shown in Fig. 12.34 can be used to decode the sequence 11110111. Starting at the root, traverse the edge labeled 1 to the right child (of the root). Continuing along the next two edges labeled with 1, we arrive at the leaf labeled a. Hence the unique path from the root to the vertex at a is unambiguously determined by the first three 1's in the sequence 11110111. After we return to the root, the next two symbols in the sequence — namely, 10 — determine the unique path along the edge from the root to its right child, followed by the edge from that child to its left child. This terminates at the vertex labeled t. Again returning to the root, the final three bits of the sequence determine the letter a for a second time. Hence the tree "decodes" 11110111 as *ata*.

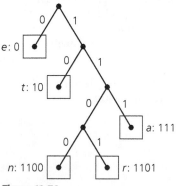

Figure 12.34

Why did the second encoding scheme work out so readily when the first led to ambiguities? In the first scheme, r is represented as 10 and n as 101. If we encounter the symbols 10, how can we determine whether the symbols represent r or the first two symbols of 101, which represent n? The problem is that the sequence for r is a prefix of the sequence for

n. This ambiguity does not occur in the second encoding scheme, suggesting the following definition.

Definition 12.7

A set P of binary sequences (representing a set of symbols) is called a *prefix code* if no sequence in P is the prefix of any other sequence in P.

Consequently, the binary sequences 111, 0, 1100, 1101, 10 constitute a prefix code for the letters a, e, n, r, t, respectively. But how did the complete binary tree of Fig. 12.34 come about? To deal with this problem, we need the following concept.

Definition 12.8

If T is a complete binary tree of height h, then T is called a *full* binary tree if all the leaves in T are at level h.

EXAMPLE 12.17

For the prefix code $P = \{111, 0, 1100, 1101, 10\}$, the longest binary sequence has length 4. Draw the labeled full binary tree of height 4, as shown in Fig. 12.35. The elements of P are assigned to the vertices of this tree as follows. For example, the sequence 10 traces the path from the root r to its right child c_R. Then it continues to the left child of c_R, where the box (marked with the asterisk) indicates completion of the sequence. Returning to the root, the other four sequences are traced out in similar fashion, resulting in the other four boxed vertices. For each boxed vertex remove the subtree (except for the root) that it determines. The resulting pruned tree is the complete binary tree of Fig. 12.34, where no "box" is an ancestor of another "box."

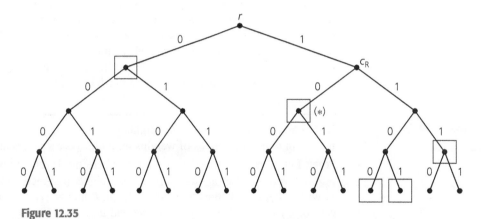

Figure 12.35

We turn now to a method for determining a labeled tree that models a prefix code, where the frequency of occurrence of each symbol in the average text is taken into account — in other words, a prefix code wherein the shorter sequences are used for the more frequently occurring symbols. If there are many symbols, such as all 26 letters of the alphabet, a trial-and-error method for constructing such a tree is not efficient. An elegant construction developed by David A. Huffman (1925–1999) provides a technique for constructing such trees.

The general problem of constructing an efficient tree can be described as follows.

Let $w_1, w_2, \ldots, w_n$ be a set of positive numbers called *weights*, where $w_1 \leq w_2 \leq \cdots \leq w_n$. If $T = (V, E)$ is a complete binary tree with n leaves, assign these weights (in any one-to-one manner) to the n leaves. The result is called a *complete binary tree for the weights* $w_1, w_2, \ldots, w_n$. The *weight of the tree*, denoted $W(T)$, is defined as $\sum_{i=1}^{n} w_i \ell(w_i)$ where, for each $1 \leq i \leq n$, $\ell(w_i)$ is the level number of the leaf assigned the weight w_i. The objective is to assign the weights so that $W(T)$ is as small as possible. A complete binary tree T' for these weights is said to be an *optimal* tree if $W(T') \leq W(T)$ for any other complete binary tree T for the weights.

Figure 12.36 shows two complete binary trees for the weights 3, 5, 6, and 9. For tree T_1, $W(T_1) = \sum_{i=1}^{4} w_i \ell(w_i) = (3 + 9 + 5 + 6) \cdot 2 = 46$ because each leaf has level number 2. In the case of T_2, $W(T_2) = 3 \cdot 3 + 5 \cdot 3 + 6 \cdot 2 + 9 \cdot 1 = 45$, which we shall find is optimal.

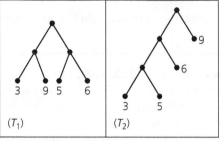

Figure 12.36

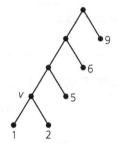

Figure 12.37

The major idea behind Huffman's construction is that in order to obtain an optimal tree T for the n weights $w_1, w_2, w_3, \ldots, w_n$, one considers an optimal tree T' for the $n - 1$ weights $w_1 + w_2, w_3, \ldots, w_n$. (It cannot be assumed that $w_1 + w_2 \leq w_3$.) In particular, the tree T' is transformed into T by replacing the leaf v having weight $w_1 + w_2$ by a tree rooted at v of height 1 with left child of weight w_1 and right child of weight w_2. To illustrate, if the tree T_2 in Fig. 12.36 is optimal for the four weights $1 + 2$, 5, 6, 9, then the tree in Fig. 12.37 will be optimal for the five weights 1, 2, 5, 6, 9.

We need the following lemma to establish these claims.

LEMMA 12.2   If T is an optimal tree for the n weights $w_1 \leq w_2 \leq \cdots \leq w_n$, then there exists an optimal tree T' in which the leaves of weights w_1 and w_2 are siblings at the maximal level (in T').

Proof: Let v be an internal vertex of T where the level number of v is maximal for all internal vertices. Let w_x and w_y be the weights assigned to the children x, y of vertex v, with $w_x \leq w_y$. By the choice of vertex v, $\ell(w_x) = \ell(w_y) \geq \ell(w_1)$, $\ell(w_2)$. Consider the case of $w_1 < w_x$. (If $w_1 = w_x$, then w_1 and w_x can be interchanged and we would consider the case of $w_2 < w_y$. Applying the following proof to this case, we would find that w_y and w_2 can be interchanged.)

If $\ell(w_x) > \ell(w_1)$, let $\ell(w_x) = \ell(w_1) + j$, for some $j \in \mathbf{Z}^+$. Then $w_1 \ell(w_1) + w_x \ell(w_x) = w_1 \ell(w_1) + w_x [\ell(w_1) + j] = w_1 \ell(w_1) + w_x j + w_x \ell(w_1) > w_1 \ell(w_1) + w_1 j + w_x \ell(w_1) = w_1 \ell(w_x) + w_x \ell(w_1)$. So $W(T) = w_1 \ell(w_1) + w_x \ell(w_x) + \sum_{i \neq 1, x} w_i \ell(w_i) > w_1 \ell(w_x) + w_x \ell(w_1) + \sum_{i \neq 1, x} w_i \ell(w_i)$. Consequently, by interchanging the locations of the weights w_1 and w_x, we obtain a tree of smaller weight. But this contradicts the choice of T as an optimal tree. Therefore $\ell(w_x) = \ell(w_1) = \ell(w_y)$. In a similar manner, it can be shown that $\ell(w_y) = \ell(w_2)$, so $\ell(w_x) = \ell(w_y) = \ell(w_1) = \ell(w_2)$. Interchanging the locations of the pair w_1, w_x, and the pair w_2, w_y, we obtain an optimal tree T', where w_1, w_2 are siblings.

From this lemma we see that smaller weights will appear at the higher levels (and thus have higher level numbers) in an optimal tree.

THEOREM 12.8   Let T be an optimal tree for the weights $w_1 + w_2, w_3, \ldots, w_n$, where $w_1 \leq w_2 \leq w_3 \leq \cdots \leq w_n$. At the leaf with weight $w_1 + w_2$ place a (complete) binary tree of height 1 and assign the weights w_1, w_2 to the children (leaves) of this former leaf. The new binary tree T_1 so constructed is then optimal for the weights $w_1, w_2, w_3, \ldots, w_n$.

Proof: Let T_2 be an optimal tree for the weights $w_1, w_2, \ldots, w_n$, where the leaves for weights w_1, w_2 are siblings. Remove the leaves of weights w_1, w_2 and assign the weight $w_1 + w_2$ to their parent (now a leaf). This complete binary tree is denoted T_3 and $W(T_2) = W(T_3) + w_1 + w_2$. Also, $W(T_1) = W(T) + w_1 + w_2$. Since T is optimal, $W(T) \leq W(T_3)$. If $W(T) < W(T_3)$, then $W(T_1) < W(T_2)$, contradicting the choice of T_2 as optimal. Hence $W(T) = W(T_3)$ and, consequently, $W(T_1) = W(T_2)$. So T_1 is optimal for the weights $w_1, w_2, \ldots, w_n$.

Remark. The preceding proof started with an optimal tree T_2 whose existence rests on the fact that there is only a finite number of ways in which we can assign n weights to a complete binary tree with n leaves. Consequently, with a finite number of assignments there is at least one where $W(T)$ is minimal. But finite numbers can be large. This proof establishes the existence of an optimal tree for a set of weights and develops a way for constructing such a tree. To construct such a (Huffman) tree we consider the following algorithm.

Given the n (≥ 2) weights $w_1, w_2, \ldots, w_n$, proceed as follows:

Step 1: Assign the given weights, one each to a set S of n isolated vertices. [Each vertex is the root of a complete binary tree (of height 0) with a weight assigned to it.]

Step 2: While $|S| > 1$ perform the following:
  a) Find two trees T, T' in S with the smallest two root weights w, w', respectively.
  b) Create the new (complete binary) tree T^* with root weight $w^* = w + w'$ and having T, T' as its left and right subtrees, respectively.
  c) Place T^* in S and delete T and T'. [Where $|S| = 1$, the one complete binary tree in S is a Huffman tree.]

We now use this algorithm in the following example.

EXAMPLE 12.18'   Construct an optimal prefix code for the symbols a, o, q, u, y, z that occur (in a given sample) with frequencies 20, 28, 4, 17, 12, 7, respectively.

Figure 12.38 shows the construction that follows Huffman's procedure. In part (b) weights 4 and 7 are combined so that we then consider the construction for the weights 11, 12, 17, 20, 28. At each step [in parts (c)–(f) of Fig. 12.38] we create a tree with subtrees rooted at the two smallest weights. These two smallest weights belong to vertices each of which is originally either isolated (a tree with just a root) or the root of a tree obtained earlier in the construction. From the last result, a prefix code is determined as

$$a: 01 \qquad o: 11 \qquad q: 1000 \qquad u: 00 \qquad y: 101 \qquad z: 1001.$$

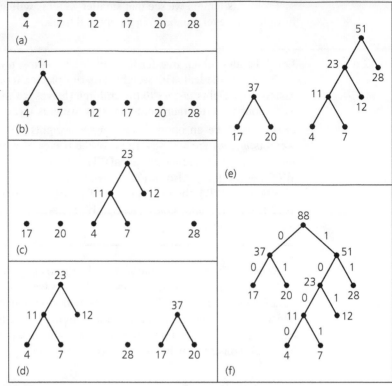

Figure 12.38

Different prefix codes may result from the way the trees T, T' are selected and assigned as the left or right subtree in steps 2(a) and 2(b) in our algorithm and from the assignment of 0 or 1 to the branches (edges) of our final (Huffman) tree.

EXERCISES 12.4

1. For the prefix code given in Fig. 12.34, decode the sequences (a) 1001111101; (b) 10111100110001101; (c) 1101111110010.

2. A code for $\{a, b, c, d, e\}$ is given by $a: 00$ $b: 01$ $c: 101$ $d: x10$ $e: yz1$, where $x, y, z \in \{0, 1\}$. Determine x, y, and z so that the given code is a prefix code.

3. Construct an optimal prefix code for the symbols $a, b, c, \ldots, i, j$ that occur (in a given sample) with respective frequencies 78, 16, 30, 35, 125, 31, 20, 50, 80, 3.

4. How many leaves does a full binary tree have if its height is (a) 3? (b) 7? (c) 12? (d) h?

5. Let $T = (V, E)$ be a complete m-ary tree of height h. This tree is called a *full* m-ary tree if all of its leaves are at level h. If T is a full m-ary tree with height 7 and 279,936 leaves, how many internal vertices are there in T?

6. Let T be a full m-ary tree with height h and v vertices. Determine h in terms of m and v.

7. Using the weights 2, 3, 5, 10, 10, show that the height of a Huffman tree for a given set of weights is not unique. How would you modify the algorithm so as to always produce a Huffman tree of minimal height for the given weights?

8. Let L_i, for $1 \le i \le 4$, be four lists of numbers, each sorted in ascending order. The numbers of entries in these lists are 75, 40, 110, and 50, respectively.

  a) How many comparisons are needed to merge these four lists by merging L_1 and L_2, merging L_3 and L_4, and then merging the two resulting lists?

  b) How many comparisons are needed if we first merge L_1 and L_2, then merge the result with L_3, and finally merge this result with L_4?

  c) In order to minimize the total number of comparisons in this merging of the four lists, what order should the merging follow?

  d) Extend the result in part (c) to n sorted lists $L_1, L_2, \ldots, L_n$.

12.5
Biconnected Components
and Articulation Points

Let $G = (V, E)$ be the loop-free connected undirected graph shown in Fig. 12.39(a), where each vertex represents a communication center. Here an edge $\{x, y\}$ indicates the existence of a communication link between the centers at x and y.

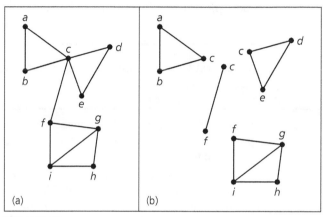

Figure 12.39

By splitting the vertices at c and f, in the suggested fashion, we obtain the collection of subgraphs in part (b) of the figure. These vertices are examples of the following.

Definition 12.9

A vertex v in a loop-free undirected graph $G = (V, E)$ is called an *articulation point* if $\kappa(G - v) > \kappa(G)$; that is, the subgraph $G - v$ has more components than the given graph G.

A loop-free connected undirected graph with no articulation points is called *biconnected*.

A *biconnected component* of a graph is a maximal biconnected subgraph — a biconnected subgraph that is not properly contained in a larger biconnected subgraph.

The graph shown in Fig. 12.39 has the two articulation points, c and f, and its four biconnected components are shown in part (b) of the figure.

In terms of communication centers and links, the articulation points of the graph indicate where the system is most vulnerable. Without articulation points, such a system is more likely to survive disruptions at a communication center, regardless of whether these disruptions are caused by the breakdown of a technical device or by external forces.

The problem of finding the articulation points in a connected graph provides an application for the depth-first spanning tree. The objective here is the development of an algorithm that determines the articulation points of a loop-free connected undirected graph. If no such points exist, then the graph is biconnected. Should such vertices exist, the resulting biconnected components can be used to provide information about such properties as the planarity and chromatic number of the given graph.

The following preliminaries are needed for developing this algorithm.

Returning to Fig. 12.39(a), we see that there are four paths from a to e—namely, (1) $a \to c \to e$; (2) $a \to c \to d \to e$; (3) $a \to b \to c \to e$; and (4) $a \to b \to c \to d \to e$. Now what do these four paths have in common? They all pass through the vertex c, one of the articulation points of G. This observation now motivates our first preliminary result.

LEMMA 12.3   Let $G = (V, E)$ be a loop-free connected undirected graph with $z \in V$. The vertex z is an articulation point of G if and only if there exist distinct $x, y \in V$ with $x \neq z$, $y \neq z$, and such that every path in G connecting x and y contains the vertex z.

Proof: This result follows from Definition 12.9. A proof is requested of the reader in the Section Exercises.

Our next lemma provides an important and useful property of the depth-first spanning tree.

LEMMA 12.4   Let $G = (V, E)$ be a loop-free connected undirected graph with $T = (V, E')$ a depth-first spanning tree of G. If $\{a, b\} \in E$ but $\{a, b\} \notin E'$, then a is either an ancestor or a descendant of b in the tree T.

Proof: From the depth-first spanning tree T, we obtain a preorder listing for the vertices in V. For all $v \in V$, let dfi(v) denote the depth-first index of vertex v—that is, the position of v in the preorder listing. Assume that dfi(a) < dfi(b). Consequently, a is encountered before b in the preorder traversal of T, so a cannot be a descendant of b. If, in addition, vertex a is not an ancestor of b, then b is not in the subtree T_a of T rooted at a. But when we backtrack (through T_a) to a, we find that because $\{a, b\} \in E$, it should have been possible for the depth-first search to go from a to b and to use the edge $\{a, b\}$ in T. This contradiction shows that b is in T_a, so a is an ancestor of b.

If $G = (V, E)$ is a loop-free connected undirected graph, let $T = (V, E')$ be a depth-first spanning tree for G, as shown in Fig. 12.40. By Lemma 12.4, the dotted edge $\{a, b\}$, which is not part of T, indicates an edge that could exist in G. Such an edge is called a *back edge* (relative to T), and here a is an ancestor of b. [Here dfi(a) = 3, whereas dfi(b) = 6.] The dotted edge $\{b, d\}$ in the figure cannot exist in G, also because of Lemma 12.4. Thus all edges of G are either edges in T or back edges (relative to T).

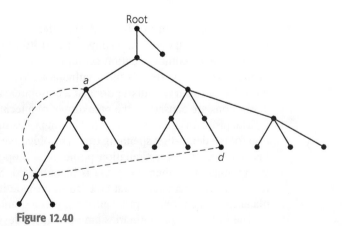

Figure 12.40

Our next example provides further insight into the relationship between the articulation points of a graph G and a depth-first spanning tree of G.

EXAMPLE 12.19

In part (1) of Fig. 12.41 we have a loop-free connected undirected graph $G = (V, E)$. Applying Lemma 12.3 to vertex a, for example, we find that the only path in G from b to i passes through a. In the case of vertex d, we apply the same lemma and consider the vertices a and h. Now we find that although there are four paths from a to h, all four pass through vertex d. Consequently, vertices a and d are two of the articulation points in G. The vertex h is the only other articulation point. Can you find two vertices in G for which all connecting paths (for these vertices) in G pass through h?

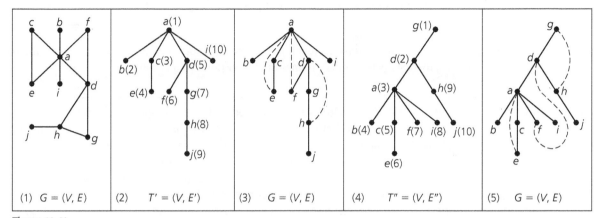

Figure 12.41

Applying the depth-first search algorithm, with the vertices of G ordered alphabetically, in part (2) of Fig. 12.41, we find the depth-first spanning tree $T' = (V, E')$ for G, where a has been chosen as the root. The parenthesized integer next to each vertex indicates the order in which that vertex is visited during the prescribed depth-first search. Part (3) of the figure incorporates the three back edges (relative to T, in G) that are missing from part (2).

For the tree T', the root a, which is an articulation point in G, has more than one child. The articulation point d has a child — namely, g — with no back edge from g or any of its descendants (h and j) to an ancestor of d [as we see in part (3) of Fig. 12.41]. The same is true for the articulation point h. Its child j has (no children and) no back edge to an ancestor of h.

In part (4) of the figure, $T'' = (V, E'')$ is the depth-first spanning tree for the vertices ordered alphabetically once again, but this time vertex g has been chosen as the root. As in part (2) of the figure, the parenthesized integer next to each vertex indicates the order in which that vertex is visited during this depth-first search. The three back edges (relative to T'', in G) that are missing from T'' are shown in part (5) of the figure.

The root g of T'' has only one child and g is not an articulation point in G. Further, for each of the articulation points there is at least one child with no back edge from that child or one of its descendants to an ancestor of the articulation point. To be more specific, from part (5) of Fig. 12.41 we find that for the articulation point a we may use any of the children b, c or i, but not f; for d that child is a; and for h the child is j.

The observations made in Example 12.19 now lead us to the following.

LEMMA 12.5

Let $G = (V, E)$ be a loop-free connected undirected graph with $T = (V, E')$ a depth-first spanning tree of G. If r is the root of T, then r is an articulation point of G if and only if r has at least two children in T.

Proof: If r has only one child — say, c — then all the other vertices of G are descendants of c (and r) in T. So if x, y are two distinct vertices of T, neither of which is r, then in the subtree T_c, rooted at c, there is a path from x to y. Since r is not a vertex in T_c, r is not on this path. Consequently, r is not an articulation point in G — by virtue of Lemma 12.3. Conversely, let r be the root of the depth-first spanning tree T and let c_1, c_2 be children of r. Let x be a vertex in T_{c_1}, the subtree of T rooted at c_1. Similarly, let y be a vertex in T_{c_2}, the subtree of T rooted at c_2. Could there be a path from x to y in G that avoids r? If so, there is an edge $\{v_1, v_2\}$ in G with v_1 in T_{c_1} and v_2 in T_{c_2}. But this contradicts Lemma 12.4.

Our final preliminary result settles the issue of when a vertex, that is not the root of a depth-first spanning tree, is an articulation point of a graph.

LEMMA 12.6

Let $G = (V, E)$ be a loop-free connected undirected graph with $T = (V, E')$ a depth-first spanning tree for G. Let r be the root of T and let $v \in V$, $v \neq r$. Then v is an articulation point of G if and only if there exists a child c of v with no back edge (relative to T, in G) from a vertex in T_c, the subtree rooted at c, to an ancestor of v.

Proof: Suppose that vertex v has a child c such that there is no back edge (relative to T, in G) from a vertex in T_c to an ancestor of v. Then every path (in G) from r to c passes through v. From Lemma 12.3 it then follows that v is an articulation point of G.

To establish the converse, let the nonroot vertex v of T satisfy the following: For each child c of v there is a back edge (relative to T, in G) from a vertex in T_c, the subtree rooted at c, to an ancestor of v. Now let $x, y \in V$ with $x \neq v$, $y \neq v$. We consider the following three possibilities:

1) If neither x nor y is a descendant of v, as in part (1) of Fig. 12.42, delete from T the subtree T_v rooted at v. The resulting subtree (of T) contains x, y and a path from x to y that does not pass through v, so v is not an articulation point of G.

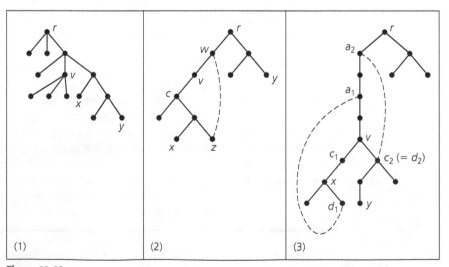

Figure 12.42

2) If one of x, y — say, x — is a descendant of v but y is not, then x is a child of v or a descendant of a child c of v [as in part (2) of Fig. 12.42]. From the hypothesis there is a back edge (relative to T, in G) from some $z \in T_c$ to an ancestor w of v. Since x, $z \in T_c$, there is a path p_1 from x to z (that does not pass through v). Then, as neither w nor y is a descendant of v, from part (1) there is a path p_2 from w to y that does not pass through v. The edges in p_1, p_2 together with the edge $\{z, w\}$ provide a path from x to y that does not pass through v — and once again, v is not an articulation point.

3) Finally, suppose that both x, y are descendants of v, as in part (3) of Fig. 12.42. Here c_1, c_2 are children of v — perhaps, with $c_1 = c_2$ — and x is a vertex in T_{c_1}, the subtree rooted at c_1, while y is a vertex in T_{c_2}, the subtree rooted at c_2. From the hypothesis, there exist back edges $\{d_1, a_1\}$ and $\{d_2, a_2\}$ (relative to T, in G), where d_1, d_2 are descendants of v and a_1, a_2 are ancestors of v. Further, there is a path p_1 from x to d_1 in T_{c_1} and a path p_2 from y to d_2 in T_{c_2}. As neither a_1 nor a_2 is a descendant of v, from part (1) we have a path p (in T) from a_1 to a_2, where p avoids v. Now we can do the following: (i) Go from x to d_1 using path p_1; (ii) Go from d_1 to a_1 on the edge $\{d_1, a_1\}$; (iii) Continue to a_2 using path p; (iv) Go from a_2 to d_2 on the edge $\{a_2, d_2\}$; and (v) Finish at y using the path p_2 from d_2 to y. This provides a path from x to y that avoids v so v is not an articulation point of G and this completes the proof.

Using the results from the preceding four lemmas, we once again start with a loop-free connected undirected graph $G = (V, E)$ with depth-first spanning tree T. For $v \in V$, where v is not the root of T, we let $T_{v,c}$ be the subtree consisting of edge $\{v, c\}$ (c a child of v) together with the tree T_c rooted at c. If there is no back edge from a descendant of v in $T_{v,c}$ to an ancestor of v (and v has at least one ancestor — the root of T), then the splitting of vertex v results in the separation of $T_{v,c}$ from G, and v is an articulation point. If no other articulation points of G occur in $T_{v,c}$, then the addition to $T_{v,c}$ of all other edges in G determined by the vertices in $T_{v,c}$ (the subgraph of G induced by the vertices in $T_{v,c}$) results in a biconnected component of G. A root has no ancestors, and it is an articulation point if and only if it has more than one child.

The depth-first spanning tree preorders the vertices of G. For $x \in V$ let dfi(x) denote the depth-first index of x in that preorder. If y is a descendant of x, then dfi(x) < dfi(y). For y an ancestor of x, dfi(x) > dfi(y). Define low(x) = min$\{$dfi(y)$|y$ is adjacent in G to either x or a descendant of $x\}$. If z is the parent of x (in T), then there are two possibilities to consider:

1) low(x) = dfi(z): In this case T_x, the subtree rooted at x, contains no vertex that is adjacent to an ancestor of z by means of a back edge of T. Hence z is an articulation point of G. If T_x contains no articulation points, then T_x together with edge $\{z, x\}$ spans a biconnected component of G (that is, the subgraph of G induced by vertex z and the vertices in T_x is a biconnected component of G). Now remove T_x and the edge $\{z, x\}$ from T, and apply this idea to the remaining subtree of T.

2) low(x) < dfi(z): Here there is a descendant of z in T_x that is joined [by a back edge (relative to T, in G)] to an ancestor of z.

To deal in an efficient manner with these ideas, we develop the following algorithm. Let $G = (V, E)$ be a loop-free connected undirected graph.

Step 1: Find the depth-first spanning tree T for G according to a prescribed order. Let $x_1, x_2, \ldots, x_n$ be the vertices of G preordered by T. Then $\mathrm{dfi}(x_j) = j$ for all $1 \le j \le n$.

Step 2: Start with x_n and continue back to $x_{n-1}, x_{n-2}, \ldots, x_3, x_2, x_1$, determining $\mathrm{low}(x_j)$, for $j = n, n-1, n-2, \ldots, 3, 2, 1$, recursively, as follows:

  a) $\mathrm{low}'(x_j) = \min\{\mathrm{dfi}(z) | z$ is adjacent in G to $x_j\}$.

  b) If $c_1, c_2, \ldots, c_m$ are the children of x_j, then $\mathrm{low}(x_j) = \min\{\mathrm{low}'(x_j), \mathrm{low}(c_1), \mathrm{low}(c_2), \ldots, \mathrm{low}(c_m)\}$. [No problem arises here, for the vertices are examined in the reverse order to the given preorder. Consequently, if c is a child of p, then $\mathrm{low}(c)$ is determined before $\mathrm{low}(p)$.]

Step 3: Let w_j be the parent of x_j in T. If $\mathrm{low}(x_j) = \mathrm{dfi}(w_j)$, then w_j is an articulation point of G, unless w_j is the root of T and w_j has no child in T other than x_j. Moreover, in either situation the subtree rooted at x_j together with the edge $\{w_j, x_j\}$ is part of a biconnected component of G.

EXAMPLE 12.20

We apply this algorithm to the graph $G = (V, E)$ shown in part (i) of Fig. 12.43.

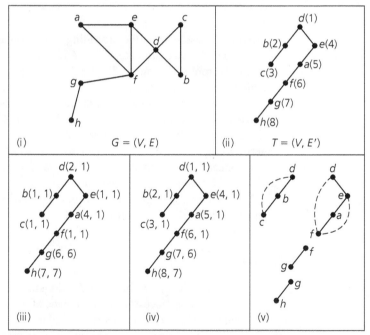

Figure 12.43

In part (ii) of the figure we have the depth-first spanning tree $T = (V, E')$ for G with d as the root. (Here the order followed for the vertices of G is alphabetic.) Next to each vertex v of T [in part (ii)] is the $\mathrm{dfi}(v)$. These labels tell us the order in which the vertices of G are first visited.

For step (2) of the algorithm we go in the reverse order from the depth-first search and start with vertex $h(= x_8)$. Since $\{g, h\} \in E$ and h is not adjacent to any other vertex of G we have $\mathrm{low}'(h) = \mathrm{dfi}(g) [= \mathrm{dfi}(x_7)] = 7$. Further, as h has no children, it follows that $\mathrm{low}(h) = \mathrm{low}'(h) = 7$. This accounts for the label $(7, 7) [= (\mathrm{low}'(h), \mathrm{low}(h))]$ next to h in part (iii) of Fig. 12.43. Continuing next with g, and then f, we obtain the labels (6, 6) for g, and (1, 1) for f, since $\text{low}'(g) = \text{low}(g) = 6$ and $\text{low}'(f) = \text{low}(f) = 1$. Since $\{a, e\}, \{a, f\} \in E$ with $\text{dfi}(e) = 4$ and $\text{dfi}(f) = 6$, for vertex a we have $\text{low}'(a) = \min\{4, 6\} = 4$. Then we find that $\text{low}(a) = \min\{4, \text{low}(f)\} = \min\{4, 1\} = 1$. Hence the label (4, 1) for vertex a. Continuing back through e, c, b, and d, we obtain the labels $(\text{low}'(x_i), \text{low}(x_i))$ for $i = 4, 3, 2, 1$. Consequently, by applying step (2) of the algorithm we arrive at the tree in Fig. 12.43 (iii).

In part (iv) of Fig. 12.43 the ordered pair next to each vertex v is $(\text{dfi}(v), \text{low}(v))$. Applying step (3) of the algorithm to the tree in part (iv), at this point we go in reverse order once again. First we deal with vertex h ($= x_8$). Since g is the parent of h (in T) and $\text{low}(h) = 7 = \text{dfi}(g)$, g is an articulation point of G and the edge $\{h, g\}$ is a biconnected component of G. Deleting the subtree rooted at g from T, we continue with vertex g ($= x_7$). Here f is the parent of g (in the tree $T - h$) and $\text{low}(g) = 6 = \text{dfi}(f)$, so f is another articulation point—with edge $\{g, f\}$ the corresponding biconnected component.

Continuing now with the tree $(T - h) - g$, as we go from f to a to e, and then from c to b, we find no new articulation points among the four vertices a, e, c, and b. Since vertex d is the root of T and d has two children—namely, the vertices b and e, it then follows from Lemma 12.5 that d is an articulation point of G. The vertices d, e, a, f induce the biconnected component consisting of the tree edges $\{f, a\}, \{a, e\}, \{e, d\}$ and the back edges (relative to T, in G) $\{f, e\}$ and $\{f, d\}$. Finally, the cycle induced (in G) by the vertices b, c and d provides the fourth biconnected component.

Part (v) of Fig. 12.43 shows the three articulation points g, f, and d, and the four biconnected components of G.

---

1. Find the articulation points and biconnected components for the graph shown in Fig. 12.44.

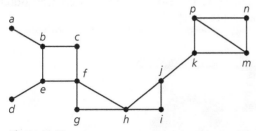

Figure 12.44

2. Prove Lemma 12.3.

3. Let $T = (V, E)$ be a tree with $|V| = n \geq 3$.

a) What are the smallest and the largest numbers of articulation points that T can have? Describe the trees for each of these cases.

b) How many biconnected components does T have in each of the cases in part (a)?

4. a) Let $T = (V, E)$ be a tree. If $v \in V$, prove that v is an articulation point of T if and only if $\deg(v) > 1$.

b) Let $G = (V, E)$ be a loop-free connected undirected graph with $|E| \geq 1$. Prove that G has at least two vertices that are not articulation points.

5. If $B_1, B_2, \ldots, B_k$ are the biconnected components of a loop-free connected undirected graph G, how is $\chi(G)$ related to $\chi(B_i)$, $1 \leq i \leq k$? [Recall that $\chi(G)$ denotes the chromatic number of G, as defined in Section 11.6.]

6. Let $G = (V, E)$ be a loop-free connected undirected graph with biconnected components $B_1, B_2, \ldots, B_8$. For $1 \leq i \leq 8$, the number of distinct spanning trees for B_i is n_i. How many distinct spanning trees exist for G?

7. Let $G = (V, E)$ be a loop-free connected undirected graph with $|V| \geq 3$. If G has no articulation points, prove that G has no pendant vertices.

8. For the loop-free connected undirected graph G in Fig. 12.43(i), order the vertices alphabetically.

a) Determine the depth-first spanning tree T for G with e as the root.

b) Apply the algorithm developed in this section to the tree T in part (a) to find the articulation points and biconnected components of G.

9. Answer the questions posed in the previous exercise but this time order the vertices as h, g, f, e, d, c, b, a and let c be the root of T.

10. Let $G = (V, E)$ be a loop-free connected undirected graph, where $V = \{a, b, c, \ldots, h, i, j\}$. Ordering the vertices alphabetically, the depth-first spanning tree T for G — with a as the root — is given in Fig. 12.45(i). In part (ii) of the figure the ordered pair next to each vertex v provides $(\text{low}'(v), \text{low}(v))$. Determine the articulation points and the spanning trees for the biconnected components of G.

11. In step (2) of the algorithm for articulation points, is it really necessary to compute $\text{low}(x_1)$ and $\text{low}(x_2)$?

12. Let $G = (V, E)$ be a loop-free connected undirected graph with $v \in V$.

    a) Prove that $\overline{G - v} = \overline{G} - v$.

    b) If v is an articulation point of G, prove that v cannot be an articulation point of $\overline{G}$.

13. If $G = (V, E)$ is a loop-free undirected graph, we call G *color-critical* if $\chi(G - v) < \chi(G)$ for all $v \in V$. (We examined such graphs earlier, in Exercise 19 of Section 11.6.) Prove that a color-critical graph has no articulation points.

14. Does the result in Lemma 12.4 remain true if $T = (V, E')$ is a breadth-first spanning tree for $G = (V, E)$?

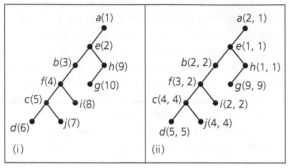

Figure 12.45

12.6
Summary and Historical Review

The structure now called a tree first appeared in 1847 in the work of Gustav Kirchhoff (1824–1887) on electrical networks. The concept also appeared at this time in *Geometrie die Lage*, by Karl von Staudt (1798–1867). In 1857 trees were rediscovered by Arthur Cayley (1821–1895), who was unaware of these earlier developments. The first to call the structure a "tree," Cayley used it in applications dealing with chemical isomers. He also investigated the enumeration of certain classes of trees. In his first work on trees, Cayley enumerated unlabeled rooted trees. This was then followed by the enumeration of unlabeled ordered trees. Two of Cayley's contemporaries who also studied trees were Carl Borchardt (1817–1880) and Marie Ennemond Jordan (1838–1922).

Arthur Cayley (1821–1895)

The formula n^{n-2} for the number of labeled trees on n vertices (Exercise 21 at the end of Section 12.1) was discovered in 1860 by Carl Borchardt. Cayley later gave an independent development of the formula, in 1889. Since then, there have been other derivations. These are surveyed in the book by J. W. Moon [10].

The paper by G. Polya [11] is a pioneering work on the enumeration of trees and other combinatorial structures. Polya's theory of enumeration, which we shall see in Chapter 16, was developed in this work. For more on the enumeration of trees, the reader should see Chapter 15 of F. Harary [5]. The article by D. R. Shier [12] provides a labyrinth of several different techniques for calculating the number of spanning trees for $K_{2,n}$.

The high-speed digital computer has proved to be a constant impetus for the discovery of new applications of trees. The first application of these structures was in the manipulation of algebraic formulae. This dates back to 1951 in the work of Grace Murray Hopper. Since then, computer applications of trees have been widely investigated. In the beginning, particular results appeared only in the documentation of specific algorithms. The first general survey of the applications of trees was made in 1961 by Kenneth Iverson as part of a broader survey on data structures. Such ideas as preorder and postorder can be traced to the early 1960s, as evidenced in the work of Zdzislaw Pawlak, Lyle Johnson, and Kenneth Iverson. At this time Kenneth Iverson also introduced the name and the notation, namely $\lceil x \rceil$, for the ceiling of a real number x. Additional material on these orders and the procedures for their implementation on a computer can be found in Chapter 3 of the text by A. V. Aho, J. E. Hopcroft, and J. D. Ullman [1]. In the article by J. E. Atkins, J. S. Dierckman, and K. O'Bryant [2], the notion of preorder is used to develop an optimal route for snow removal.

Rear Admiral Grace Murray Hopper (1906–1992) salutes as she and Navy Secretary John Lehman leave the U.S.S *Constitution*.

AP/World Wide Photos

If $G = (V, E)$ is a loop-free undirected graph, then the depth-first search and the breadth-first search (given in Section 12.2) provide ways to determine whether the given graph is connected. The algorithms developed for these searching procedures are also important in developing other algorithms. For example, the depth-first search arises in the algorithm for finding the articulation points and biconnected components of a loop-free connected undirected graph. If $|V| = n$ and $|E| = e$, then it can be shown that both the depth-first search and the breadth-first search have time-complexity $O(\max\{n, e\})$. For most graphs $e > n$, so the algorithms are generally considered to have time-complexity $O(e)$. These ideas are developed in great detail in Chapter 7 of S. Baase and A. Van Gelder [3], where the coverage also includes an analysis of the time-complexity function for the algorithm (of Section 12.5) that determines articulation points (and biconnected components). Chapter 6 of the text by A. V. Aho, J. E. Hopcroft, and J. D. Ullman [1] also deals with the depth-first search, whereas Chapter 7 covers the breadth-first search and the algorithm for articulation points.

More on the properties and computer applications of trees is given in Section 3 of Chapter 2 in the work by D. E. Knuth [7]. Sorting techniques and their use of trees can be further studied in Chapter 11 of A. V. Aho, J. E. Hopcroft, and J. D. Ullman [1] and in Chapter 7 of T. H. Cormen, C. E. Leiserson, R. L. Rivest, and C. Stein [4]. An extensive investigation will warrant the coverage found in the text by D. E. Knuth [8].

The technique in Section 12.4 for designing prefix codes is based on methods developed by D. A. Huffman [6].

David A. Huffman

University of Florida, Department of Computer and Information Science and Engineering

Finally, Chapter 7 of C. L. Liu [9] deals with trees, cycles, cut-sets, and the vector spaces associated with these ideas. The reader with a background in linear or abstract algebra should find this material of interest.

REFERENCES

1. Aho, Alfred V., Hopcroft, John E., and Ullman, Jeffrey D. *Data Structures and Algorithms.* Reading, Mass.: Addison-Wesley, 1983.
2. Atkins, Joel E., Dierckman, Jeffrey S., and O'Bryant, Kevin. "A Real Snow Job." *The UMAP Journal*, Fall no. 3 (1990): pp. 231–239.

3. Baase, Sara, and Van Gelder, Allen. *Computer Algorithms: Introduction to Design and Analysis*, 3rd ed. Reading, Mass.: Addison-Wesley, 2000.

4. Cormen, Thomas H., Leiserson, Charles E., Rivest, Ronald L., and Stein, Clifford. *Introduction to Algorithms*, 2nd ed. Boston, Mass.: McGraw-Hill, 2001.

5. Harary, Frank. *Graph Theory*. Reading, Mass.: Addison-Wesley, 1969.

6. Huffman, David A. "A Method for the Construction of Minimum Redundancy Codes." *Proceedings of the IRE* 40 (1952): pp. 1098–1101.

7. Knuth, Donald E. *The Art of Computer Programming*, Vol. 1, 2nd ed. Reading, Mass.: Addison-Wesley, 1973.

8. Knuth, Donald E. *The Art of Computer Programming*, Vol. 3. Reading, Mass.: Addison-Wesley, 1973.

9. Liu, C. L. *Introduction to Combinatorial Mathematics*. New York: McGraw-Hill, 1968.

10. Moon, John Wesley. *Counting Labelled Trees*. Canadian Mathematical Congress, Montreal, Canada, 1970.

11. Polya, George. "Kombinatorische Anzahlbestimmungen für Gruppen, Graphen und Chemische Verbindungen." *Acta Mathematica* 68 (1937): pp. 145–234.

12. Shier, Douglas R. "Spanning Trees: Let Me Count the Ways." *Mathematics Magazine* 73 (2000): pp. 376–381.

SUPPLEMENTARY EXERCISES

1. Let $G = (V, E)$ be a loop-free undirected graph with $|V| = n$. Prove that G is a tree if and only if $P(G, \lambda) = \lambda(\lambda - 1)^{n-1}$.

2. A telephone communication system is set up at a company where 125 executives are employed. The system is initialized by the president, who calls her four vice presidents. Each vice president then calls four other executives, some of whom in turn call four others, and so on. (Each executive who does make a call will actually make four calls.)

   a) How many calls are made in reaching all 125 executives?

   b) How many executives, aside from the president, are required to make calls?

3. Let T be a complete binary tree with the vertices of T ordered by a preorder traversal. This traversal assigns the label 1 to all internal vertices of T and the label 0 to each leaf. The sequence of 0's and 1's that results from the preorder traversal of T is called the tree's *characteristic sequence*.

   a) Find the characteristic sequence for the complete binary tree shown in Fig. 12.17.

   b) Determine the complete binary trees for the characteristic sequences
   i) 1011001010100 and
   ii) 1011110000101011000.

   c) What are the last two symbols in the characteristic sequence for all complete binary trees? Why?

4. For $k \in \mathbf{Z}^+$, let $n = 2^k$, and consider the list $L: a_1, a_2, a_3, \ldots, a_n$. To sort L in ascending order, first compare the entries a_i and $a_{i+(n/2)}$, for each $1 \le i \le n/2$. For the resulting 2^{k-1} ordered pairs, merge sort the ith and $(i + (n/4))$-th ordered pairs, for each $1 \le i \le n/4$. Now do a merge sort on the ith and $(i + (n/8))$-th ordered quadruples, for each $1 \le i \le n/8$. Continue the process until the elements of L are in ascending order.

   a) Apply this sorting procedure to the list

$$L: 11, 3, 4, 6, -5, 7, 35,$$
$$-2, 1, 23, 9, 15, 18, 2, -10, 5.$$

   b) If $n = 2^k$, how many comparisons at most does this procedure require?

5. Let $G = (V, E)$ be a loop-free undirected graph. If $\deg(v) \ge 2$ for all $v \in V$, prove that G contains a cycle.

6. Let $T = (V, E)$ be a rooted tree with root r. Define the relation $\mathcal{R}$ on V by $x \mathcal{R} y$, for $x, y \in V$, if $x = y$ or if x is on the path from r to y. Prove that $\mathcal{R}$ is a partial order.

7. Let $T = (V, E)$ be a tree with $V = \{v_1, v_2, \ldots, v_n\}$, for $n \ge 2$. Prove that the number of pendant vertices in T is equal to

$$2 + \sum_{\deg(v_i) \ge 3} (\deg(v_i) - 2).$$

8. Let $G = (V, E)$ be a loop-free undirected graph. Define the relation $\mathcal{R}$ on E as follows: If $e_1, e_2 \in E$, then $e_1 \mathcal{R} e_2$ if $e_1 = e_2$ or if e_1 and e_2 are edges of a cycle C in G.

   a) Verify that $\mathcal{R}$ is an equivalence relation on E.

   b) Describe the partition of E induced by $\mathcal{R}$.

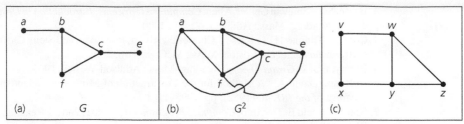

Figure 12.46

9. If $G = (V, E)$ is a loop-free connected undirected graph and $a, b \in V$, then we define the *distance* from a to b (or from b to a), denoted $d(a, b)$, as the length of a shortest path (in G) connecting a and b. (This is the number of edges in a shortest path connecting a and b and is 0 when $a = b$.)

For any loop-free connected undirected graph $G = (V, E)$, the *square* of G, denoted G^2, is the graph with vertex set V (the same as G) and edge set defined as follows: For distinct $a, b \in V$, $\{a, b\}$ is an edge in G^2 if $d(a, b) \le 2$ (in G). In parts (a) and (b) of Fig. 12.46, we have a graph G and its square.

a) Find the square of the graph in part (c) of the figure.

b) Find G^2 if G is the graph $K_{1,3}$.

c) If G is the graph $K_{1,n}$, for $n \ge 4$, how many edges are added to G in order to construct G^2?

d) For any loop-free connected undirected graph G, prove that G^2 has no articulation points.

10. a) Let $T = (V, E)$ be a complete 6-ary tree of height 8. If T is balanced, but not full, determine the minimum and maximum values for $|V|$.

b) Answer part (a) if $T = (V, E)$ is a complete m-ary tree of height h.

11. The rooted *Fibonacci trees* T_n, $n \ge 1$, are defined recursively as follows:

1) T_1 is the rooted tree consisting of only the root;

2) T_2 is the same as T_1 — it too is a rooted tree that consists of a single vertex; and

3) For $n \ge 3$, T_n is the rooted binary tree with T_{n-1} as its left subtree and T_{n-2} as its right subtree.

The first six rooted Fibonacci trees are shown in Fig. 12.47:

a) For $n \ge 1$, let ℓ_n count the number of leaves in T_n. Find and solve a recurrence relation for ℓ_n.

b) Let i_n count the number of internal vertices for the tree T_n, where $n \ge 1$. Find and solve a recurrence relation for i_n.

c) Determine a formula for v_n, the total number of vertices in T_n, where $n \ge 1$.

12. a) The graph in part (a) of Fig. 12.48 has exactly one spanning tree — namely, the graph itself. The graph in Fig. 12.48(b) has four nonidentical, though isomorphic, spanning trees. In part (c) of the figure we find three of the nonidentical spanning trees for the graph in part (d). Note that T_2 and T_3 are isomorphic, but T_1 is not isomorphic to T_2 (or T_3). How many nonidentical spanning trees exist for the graph in Fig. 12.48(d)?

b) In Fig. 12.48(e) we generalize the graphs in parts (a), (b), and (d) of the figure. For each $n \in \mathbf{Z}^+$, the graph G_n is $K_{2,n}$.

If t_n counts the number of nonidentical spanning trees for G_n, find and solve a recurrence relation for t_n.

13. Let $G = (V, E)$ be the undirected connected "ladder graph" shown in Fig. 12.49. For $n \ge 0$, let a_n count the number of spanning trees of G, whereas b_n counts the number of these spanning trees that contain the edge $\{x_1, y_1\}$.

a) Explain why $a_n = a_{n-1} + b_n$.

b) Find an equation that expresses b_n in terms of a_{n-1} and b_{n-1}.

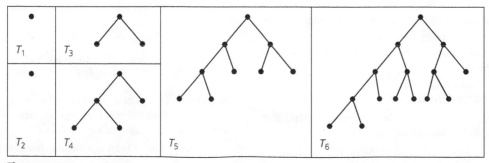

Figure 12.47

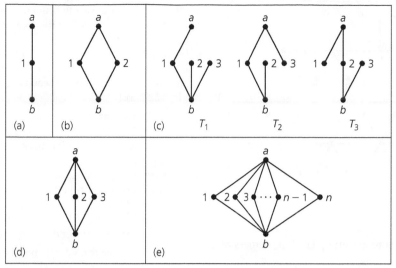

Figure 12.48

c) Use the results in parts (a) and (b) to set up and solve a recurrence relation for a_n.

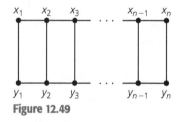

Figure 12.49

14. Let $T = (V, E)$ be a tree where $|V| = v$ and $|E| = e$. The tree T is called *graceful* if it is possible to assign the labels $\{1, 2, 3, \ldots, v\}$ to the vertices of T in such a manner that the induced edge labeling — where each edge $\{i, j\}$ is assigned the label $|i - j|$, for $i, j \in \{1, 2, 3, \ldots, v\}$, $i \neq j$ — results in the e edges being labeled by $1, 2, 3, \ldots, e$.

a) Prove that every path on n vertices, $n \geq 2$, is graceful.

b) For $n \in \mathbf{Z}^+$, $n \geq 2$, show that $K_{1,n}$ is graceful.

c) If $T = (V, E)$ is a tree with $4 \leq |V| \leq 6$, show that T is graceful. (It has been conjectured that every tree is graceful.)

15. For an undirected graph $G = (V, E)$ a subset of I of V is called *independent* when no two vertices in I are adjacent. If, in addition, $I \cup \{x\}$ is not independent for each $x \in V - I$, then we say that I is a *maximal independent* set (of vertices).

The two graphs in Fig. 12.50 are examples of special kinds of trees called caterpillars. In general, a tree $T = (V, E)$ is a *caterpillar* when there is a (maximal) path p such that, for all $v \in V$, either v is on the path p or v is adjacent to a vertex on the path p. This path p is called the *spine* of the caterpillar.

a) How many maximal independent sets of vertices are there for each of the caterpillars in parts (i) and (ii) of Fig. 12.50?

b) For $n \in \mathbf{Z}^+$, with $n \geq 3$, let a_n count the number of maximal independent sets in a caterpillar T whose spine contains n vertices. Find and solve a recurrence relation for a_n. [The reader may wish to reexamine part (a) of Supplementary Exercise 21 in Chapter 11.]

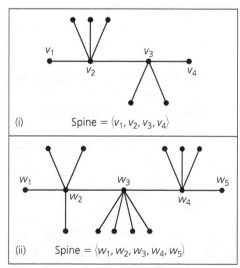

Figure 12.50

16. In part (i) of Fig. 12.51 we find a graceful labeling of the caterpillar shown in part (i) of Fig. 12.50. Find a graceful labeling for the caterpillars in part (ii) of Figs. 12.50 and 12.51.

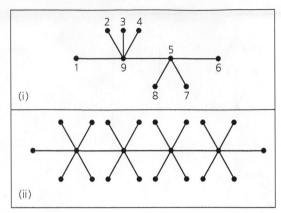

(i)

(ii)

Figure 12.51

17. Develop an algorithm to gracefully label the vertices of a caterpillar with at least two edges.

18. Consider the caterpillar in part (i) of Fig. 12.50. If we label each edge of the spine with a 1 and each of the other edges with a 0, the caterpillar can be represented by a binary string. Here that binary string is 10001001 where the first 1 is for the first (left-most) edge of the spine, the next three 0's are for the (nonspine) edges at v_2, the second 1 is for edge $\{v_2, v_3\}$, the two 0's are for the (nonspine) leaves at v_3, and the final 1 accounts for the third (right-most) edge of the spine.

We also note that the reversal of the binary string 10001001 — namely, 10010001 — corresponds with a second caterpillar that is isomorphic to the one in part (i) of Fig. 12.50.

a) Find the binary strings for each of the caterpillars in part (ii) of Figs. 12.50 and 12.51.

b) Can a caterpillar have a binary string of all 1's?

c) Can the binary string for a caterpillar have only two 1's?

d) Draw all the nonisomorphic caterpillars on five vertices. For each caterpillar determine its binary string. How many of these binary strings are palindromes?

e) Answer the question posed in part (d) upon replacing "five" by "six."

f) For $n \geq 3$, prove that the number of nonisomorphic caterpillars on n vertices is $(1/2)(2^{n-3} + 2^{\lceil (n-3)/2 \rceil}) = 2^{n-4} + 2^{\lfloor (n-4)/2 \rfloor} = 2^{n-4} + 2^{\lfloor n/2 \rfloor - 2}$. (This was first established in 1973 by F. Harary and A. J. Schwenk.)

19. For $n \geq 0$, we want to count the number of ordered rooted trees on $n + 1$ vertices. The five trees in Fig. 12.52(a) cover the case for $n = 3$.

[*Note:* Although the two trees in Fig. 12.52(b) are distinct as binary rooted trees, as ordered rooted trees they are considered the same tree and each is accounted for by the fourth tree in Fig. 12.52(a).]

a) Performing a postorder traversal of each tree in Fig. 12.52(a), we traverse each edge twice — once going down and once coming back up. When we traverse an edge going down, we shall write "1" and when we traverse one coming back up, we shall write "−1." Hence the postorder traversal for the first tree in Fig. 12.52(a) generates the list 1, 1, 1, −1, −1, −1. The list 1, 1, −1, −1, 1, −1 arises for the second tree in part (a) of the figure. Find the corresponding lists for the other three trees in Fig. 12.52(a).

b) Determine the ordered rooted trees on five vertices that generate the lists: (i) 1, −1, 1, 1, −1, 1, −1, −1; (ii) 1, 1, −1, −1, 1, 1, −1, −1; and (iii) 1, −1, 1, −1, 1, 1, −1, −1. How many such trees are there on five vertices?

c) For $n \geq 0$, how many ordered rooted trees are there for $n + 1$ vertices?

20. For $n \geq 1$, let t_n count the number of spanning trees for the *fan* on $n + 1$ vertices. The fan for $n = 4$ is shown in Fig. 12.53.

a) Show that $t_{n+1} = t_n + \sum_{i=0}^{n} t_i$, where $n \geq 1$ and $t_0 = 1$.

b) For $n \geq 2$, show that $t_{n+1} = 3t_n - t_{n-1}$.

c) Solve the recurrence relation in part (b) and show that for $n \geq 1$, $t_n = F_{2n}$, the $2n$th Fibonacci number.

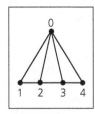

Figure 12.53

21. a) Consider the subgraph of G (in Fig. 12.54) induced by the vertices a, b, c, d. This graph is called a *kite*. How many nonidentical (though some may be isomorphic) spanning trees are there for this kite?

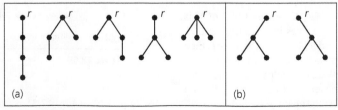

(a)   (b)

Figure 12.52

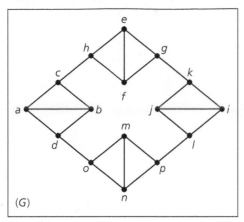

Figure 12.54

b) How many nonidentical (though some may be isomorphic) spanning trees of G do not contain the edge $\{c, h\}$?

c) How many nonidentical (though some may be isomorphic) spanning trees of G contain all four of the edges $\{c, h\}$, $\{g, k\}$, $\{l, p\}$, and $\{d, o\}$?

d) How many nonidentical (though some may be isomorphic) spanning trees exist for G?

e) We generalize the graph G as follows. For $n \geq 2$, start with a cycle on the $2n$ vertices $v_1, v_2, \ldots, v_{2n-1}, v_{2n}$. Replace each of the n edges $\{v_1, v_2\}$, $\{v_3, v_4\}, \ldots,$ $\{v_{2n-1}, v_{2n}\}$ with a (labeled) kite so that the resulting graph is 3-regular. (The case for $n = 4$ appears in Fig. 12.54.) How many nonidentical (though some may be isomorphic) spanning trees are there for this graph?

13

Optimization and Matching

Using the structures of trees and graphs, the final chapter for this part of the text introduces techniques that arise in the area of mathematics called *operations research*. These optimization techniques can be applied to graphs and multigraphs that have a positive real number (in Sections 13.1 and 13.2) or a nonnegative integer (in Section 13.3), called a weight, associated with each edge of the graph or multigraph. These numbers relate information such as the distance between the vertices that are the endpoints of the edge, or perhaps the amount of material that can be shipped from one vertex to another along an edge that represents a highway or air route. With the graphs providing the framework, the optimization methods are developed in an algorithmic manner to facilitate their implementation on a computer. Among the problems we analyze are the determinations of:

1) The shortest distance between a designated vertex v_0 and each of the other vertices in a loop-free connected directed graph.

2) A spanning tree for a given graph or multigraph, where the sum of the weights of the edges in the tree is minimal.

3) The maximum amount of material that can be transported from a starting point (the source) to a terminating point (the sink), where the weight of an edge indicates its capacity for handling the material being transported.

13.1
Dijkstra's Shortest-Path Algorithm

We start with a loop-free connected directed graph $G = (V, E)$. Now to each edge $e = (a, b)$ of this graph, we assign a positive real number called the *weight* of e. This is denoted by $\text{wt}(e)$, or $\text{wt}(a, b)$. If $x, y \in V$ but $(x, y) \notin E$, we define $\text{wt}(x, y) = \infty$.

For each $e = (a, b) \in E$, $\text{wt}(e)$ may represent (1) the length of a road from a to b, (2) the time it takes to travel on this road from a to b, or (3) the cost of traveling from a to b on this road.

Whenever such a graph $G = (V, E)$ is given with the weight assignments described here, the graph is referred to as a *weighted* graph.

EXAMPLE 13.1   In Fig. 13.1 the weighted graph $G = (V, E)$ represents travel routes between certain pairs of cities. Here the weight of each edge (x, y) indicates the approximate flying time for a direct flight from city x to city y.

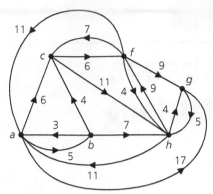

Figure 13.1

In this directed graph there are situations where $\text{wt}(x, y) \neq \text{wt}(y, x)$ for certain edges (x, y) and (y, x) in G. For example, $\text{wt}(c, f) = 6 \neq 7 = \text{wt}(f, c)$. Perhaps this is due to tailwinds. As a plane flies from c to f, the plane may be assisted by tailwinds that, in turn, slow it down when it is flying in the opposite direction (from f to c).

We see that $c, g \in V$ but $(c, g), (g, c) \notin E$, so $\text{wt}(g, c) = \text{wt}(c, g) = \infty$. This is also true for other pairs of vertices. On the other hand, for certain pairs of vertices such as a, f, we have $\text{wt}(a, f) = \infty$ whereas $\text{wt}(f, a) = 11$, a finite number.

Our objective in this section has two parts. Given a weighted graph $G = (V, E)$, for each $e = (x, y) \in E$, we shall interpret $\text{wt}(e)$ as the length of a direct route (whether by automobile, plane, or boat) from x to y. For $a, b \in V$, suppose that $v_1, v_2, \ldots, v_n \in V$ and that the edges $(a, v_1), (v_1, v_2), \ldots, (v_n, b)$ provide a directed path (in G) from a to b. The *length* of this path is defined as $\text{wt}(a, v_1) + \text{wt}(v_1, v_2) + \cdots + \text{wt}(v_n, b)$. We write $d(a, b)$ for the (shortest) distance from a to b — that is, the length of a shortest directed path (in G) from a to b. If no such path exists (in G) from a to b, then we define $d(a, b) = \infty$. And for all $a \in V$, $d(a, a) = 0$. Consequently, we have the distance function $d: V \times V \to \mathbf{R}^+ \cup \{0, \infty\}$.

Now fix $v_0 \in V$. Then for all $v \in V$, we shall determine

1) $d(v_0, v)$; and

2) a directed path from v_0 to v [of length $d(v_0, v)$] if $d(v_0, v)$ is finite.

To accomplish these objectives, we shall introduce a version of the algorithm that was developed by Edsger Wybe Dijkstra (1930–2002) in 1959. This procedure is an example of a *greedy* algorithm, for what we do to obtain the best result *locally* (for vertices "close" to v_0) turns out to be the best result *globally* (for all vertices of the graph).

Before we state the algorithm, we wish to examine some properties of the distance function d. These properties will help us understand why the algorithm works.

With $v_0 \in V$ fixed (as it was earlier), let $S \subset V$ with $v_0 \in S$, and $\overline{S} = V - S$. Then we define the *distance from v_0 to $\overline{S}$* by

$$d(v_0, \overline{S}) = \min_{v \in \overline{S}}\{d(v_0, v)\}.$$

When $d(v_0, \overline{S}) < \infty$, then $d(v_0, \overline{S})$ is the length of a shortest directed path from v_0 to a vertex in $\overline{S}$. In this case there will exist at least one vertex v_{m+1} in $\overline{S}$ with $d(v_0, \overline{S}) = d(v_0, v_{m+1})$.

Here $P: (v_0, v_1), (v_1, v_2), \ldots, (v_{m-1}, v_m), (v_m, v_{m+1})$ is a shortest directed path (in G) from v_0 to v_{m+1}. So, at this point, we claim that

1) $v_0, v_1, v_2, \ldots, v_m \in S$; and

2) $P': (v_0, v_1), (v_1, v_2), \ldots, (v_{k-1}, v_k)$ is a shortest directed path (in G) from v_0 to v_k, for each $1 \leq k \leq m$.

(The proofs for these two results are requested in the first exercise at the end of this section.)
From these observations it follows that

$$d(v_0, \overline{S}) = \min\{d(v_0, u) + \mathrm{wt}(u, w)\},$$

where the minimum is evaluated over all $u \in S$, $w \in \overline{S}$. If a minimum occurs for $u = x$ and $w = y$, then

$$d(v_0, y) = d(v_0, x) + \mathrm{wt}(x, y)$$

is the (shortest) distance from v_0 to y.

The formula for $d(v_0, \overline{S})$ is the cornerstone of the algorithm. We start with the set $S_0 = \{v_0\}$ and then determine

$$d(v_0, \overline{S}_0) = \min_{\substack{u \in S_0 \\ w \in \overline{S}_0}} \{d(v_0, u) + \mathrm{wt}(u, w)\}.$$

This gives us $d(v_0, \overline{S}_0) = \min_{w \in \overline{S}_0}\{\mathrm{wt}(v_0, w)\}$ since $S_0 = \{v_0\}$ and $d(v_0, v_0) = 0$. If $v_1 \in \overline{S}_0$ and $d(v_0, \overline{S}_0) = \mathrm{wt}(v_0, v_1)$, then we enlarge S_0 to $S_1 = S_0 \cup \{v_1\}$ and determine

$$d(v_0, \overline{S}_1) = \min_{\substack{u \in S_1 \\ w \in \overline{S}_1}} \{d(v_0, u) + \mathrm{wt}(u, w)\}.$$

This leads us to a vertex v_2 in $\overline{S}_1$ with $d(v_0, \overline{S}_1) = d(v_0, v_2)$. Continuing the process, if $S_i = \{v_0, v_1, v_2, \ldots, v_i\}$ has been determined and $v_{i+1} \in \overline{S}_i$ with $d(v_0, v_{i+1}) = d(v_0, \overline{S}_i)$, then we enlarge S_i to $S_{i+1} = S_i \cup \{v_{i+1}\}$. We stop when we reach $\overline{S}_{n-1} = \emptyset$ (where $n = |V|$) or when $d(v_0, \overline{S}_i) = \infty$ for some $0 \leq i \leq n - 2$.

Throughout this process, various labels will be placed on each vertex $v \in V$. The final set of labels appearing on the vertices will have the form $(L(v), u)$, where $L(v) = d(v_0, v)$, the distance from v_0 to v, and u is the vertex (if one exists) that precedes v along a shortest path from v_0 to v. That is, (u, v) is the last edge in a directed path from v_0 to v, and this path determines $d(v_0, v)$. At first we label v_0 with $(0, -)$ and all of the other vertices v with the label $(\infty, -)$. As we apply the algorithm, the label on each $v \neq v_0$ will change (sometimes more than once) from $(\infty, -)$ to the final label $(L(v), u) = (d(v_0, v), u)$, unless $d(v_0, v) = \infty$.

Now that these preliminaries are behind us, it is time to formally state the algorithm.

Let $G = (V, E)$ be a weighted graph, with $|V| = n$. To find the shortest distance from a fixed vertex v_0 to all other vertices in G, as well as a shortest directed path for each of these vertices, we apply the following algorithm.

Dijkstra's Shortest-Path Algorithm

Step 1: Set the counter $i = 0$ and $S_0 = \{v_0\}$. Label v_0 with $(0, -)$ and each $v \neq v_0$ with $(\infty, -)$.

If $n = 1$, then $V = \{v_0\}$ and the problem is solved.
If $n > 1$, continue to step (2).

Step 2: For each $v \in \overline{S}_i$ replace, when possible, the label on v by the new label $(L(v), y)$ where

$$L(v) = \min_{u \in S_i}\{L(v), L(u) + \text{wt}(u, v)\},$$

and y is a vertex in S_i that produces the minimum $L(v)$. [When a replacement does take place, it is due to the fact that we can go from v_0 to v and travel a shorter distance by going along a path that includes the edge (y, v).]

Step 3: If every vertex in $\overline{S}_i$ (for some $0 \leq i \leq n - 2$) has the label $(\infty, -)$, then the labeled graph contains the information we are seeking.

If not, then there is at least one vertex $v \in \overline{S}_i$ that is not labeled by $(\infty, -)$, and we perform the following tasks:

1) Select a vertex v_{i+1} where $L(v_{i+1})$ is a minimum (for all such v). There may be more than one such vertex, in which case we are free to choose among the possible candidates. The vertex v_{i+1} is an element of $\overline{S}_i$ that is closest to v_0.
2) Assign $S_i \cup \{v_{i+1}\}$ to S_{i+1}.
3) Increase the counter i by 1.
   If $i = n - 1$, the labeled graph contains the information we want.
   If $i < n - 1$, return to step (2).

We now apply this algorithm in the following example.

EXAMPLE 13.2   Apply Dijkstra's algorithm to the weighted graph $G = (V, E)$ shown in Fig. 13.1 in order to find the shortest distance from vertex c $(= v_0)$ to each of the other five vertices in G.

Initialization: $(i = 0)$. Set $S_0 = \{c\}$. Label c with $(0, -)$ and all other vertices in G with $(\infty, -)$.

First Iteration: $(\overline{S}_0 = \{a, b, f, g, h\})$. Here $i = 0$ in step (2) and we find, for example, that

$$L(a) = \min\{L(a), L(c) + \text{wt}(c, a)\}$$
$$= \min\{\infty, 0 + \infty\} = \infty,$$

whereas

$$L(f) = \min\{L(f), L(c) + \text{wt}(c, f)\}$$
$$= \min\{\infty, 0 + 6\} = 6.$$

Similar calculations yield $L(b) = L(g) = \infty$ and $L(h) = 11$. So we label the vertex f with $(6, c)$ and the vertex h with $(11, c)$. The other vertices in $\overline{S}_0$ remain labeled by $(\infty, -)$. [See Fig. 13.2(a).] In step (3) we see that f is the vertex v_1 in $\overline{S}_0$ closest to v_0, so we assign to S_1 the set $S_0 \cup \{f\} = \{c, f\}$ and increase the counter i to 1. Since $i = 1 < 5$ $(= 6 - 1)$, we return to step (2).

Second Iteration: $(\overline{S}_1 = \{a, b, g, h\})$. Now $i = 1$ in step (2), and for each $v \in \overline{S}_1$ we set

$$L(v) = \min_{u \in S_1}\{L(v), L(u) + \text{wt}(u, v)\}.$$

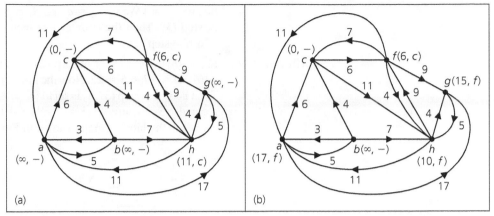

Figure 13.2

This yields

$$L(a) = \min\{L(a), L(c) + \text{wt}(c, a), L(f) + \text{wt}(f, a)\}$$
$$= \min\{\infty, 0 + \infty, 6 + 11\} = 17,$$

so vertex a is labeled $(17, f)$. In a similar manner, we find

$$L(b) = \min\{\infty, 0 + \infty, 6 + \infty\} = \infty,$$
$$L(g) = \min\{\infty, 0 + \infty, 6 + 9\} = 15,$$
$$L(h) = \min\{11, 0 + 11, 6 + 4\} = 10.$$

[These results provide the labeling in Fig. 13.2(b).] In step (3) we find that the vertex v_2 is h because $h \in \overline{S}_1$ and $L(h)$ is a minimum. Then S_2 is assigned $S_1 \cup \{h\} = \{c, f, h\}$, the counter is increased to 2, and since $2 < 5$, the algorithm directs us back to step (2).

Third Iteration: $(\overline{S}_2 = \{a, b, g\})$. With $i = 2$ in step (2) the following are now computed:

$$L(a) = \min_{u \in S_2}\{L(a), L(u) + \text{wt}(u, a)\}$$
$$= \min\{17, 0 + \infty, 6 + 11, 10 + 11\} = 17$$

(so the label on a is not changed);

$$L(b) = \min\{\infty, 0 + \infty, 6 + \infty, 10 + \infty\} = \infty$$

(so the label on b remains ∞); and

$$L(g) = \min\{15, 0 + \infty, 6 + 9, 10 + 4\} = 14 < 15,$$

so the label on g is changed to $(14, h)$ because $14 = L(h) + \text{wt}(h, g)$. Among the vertices in $\overline{S}_2$, g is the closest to v_0 since $L(g)$ is a minimum. In step (3), vertex v_3 is defined as g and $S_3 = S_2 \cup \{g\} = \{c, f, h, g\}$. Then the counter i is increased to $3 < 5$, and we return to step (2).

Fourth Iteration: $(\overline{S}_3 = \{a, b\})$. With $i = 3$, the following are determined in step (2): $L(a) = 17$; $L(b) = \infty$. (Thus no labels are changed during this iteration.) We set $v_4 = a$ and $S_4 = S_3 \cup \{a\} = \{c, f, h, g, a\}$ in step (3). Then the counter i is increased to 4 (< 5), and we return to step (2).

Fifth Iteration: ($\overline{S}_4 = \{b\}$). Here $i = 4$ in step (2), and we find $L(b) = L(a) +$ wt$(a, b) = 17 + 5 = 22$. Now the label on b is changed to $(22, a)$. Then $v_5 = b$ in step (3), S_5 is set to $\{c, f, h, g, a, b\}$, and i is incremented to 5. But now that $i = 5 = |V| - 1$, the process terminates. We reach the labeled graph shown in Fig. 13.3.

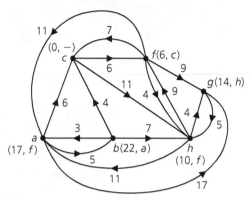

Figure 13.3

From the labels in Fig. 13.3 we have the following shortest distances from c to the other five vertices in G:

1) $d(c, f) = L(f) = 6$. **2)** $d(c, h) = L(h) = 10$.

3) $d(c, g) = L(g) = 14$. **4)** $d(c, a) = L(a) = 17$.

5) $d(c, b) = L(b) = 22$.

To determine, for example, a shortest directed path from c to b, we start at vertex b, which is labeled $(22, a)$. Hence a is the predecessor of b on this shortest path. The label on a is $(17, f)$, so f precedes a on the path. Finally, the label on f is $(6, c)$, so we are back at vertex c, and the shortest directed path from c to b determined by the algorithm is given by the edges (c, f), (f, a), and (a, b).

Now that we have demonstrated one application of this algorithm, our next concern is the order of its worst-case time-complexity function $f(n)$, where $n = |V|$ in the weighted graph $G = (V, E)$. We shall estimate the worst-case complexity in terms of the number of additions and comparisons that are made in steps (2) and (3) during execution of the algorithm.

Following the initialization process in step (1), there are at most $n - 1$ iterations because each iteration determines the next closest vertex to v_0 and $n - 1 = |V - \{v_0\}|$.

If $0 \le i \le n - 2$, then in step (2) for that iteration [the $(i + 1)$st], we find that the following takes place for each $v \in \overline{S}_i$:

1) When $0 \le i \le n - 2$, we perform at most $n - 1$ additions to calculate

$$L(v) = \min_{u \in S_i}\{L(v), L(u) + \text{wt}(u, v)\}$$

—one addition for each $u \in S_i$.

2) We compare the present value of $L(v)$ with each of the (possibly infinite) numbers $L(v) + \text{wt}(u, v)$ — one for each $u \in S_i$, where $|S_i| \leq n - 1$ — in order to determine the updated value of $L(v)$. This requires at most $n - 1$ comparisons. Therefore, before we get to step (3) we have performed at most $2(n - 1)$ steps for *each* $v \in \overline{S}_i$ — a total of at most $2(n - 1)^2$ steps for *all* $v \in \overline{S}_i$.

Continuing to step (3), we now must select the minimum from among at most $n - 1$ numbers $L(v)$, where $v \in \overline{S}_i$. This requires $n - 2$ additional comparisons — in the worst case.

Consequently, each iteration needs no more than $2(n - 1)^2 + (n - 2)$ steps in all. It is possible to have as many as $n - 1$ iterations, so it follows that

$$f(n) \leq (n - 1)[2(n - 1)^2 + (n - 2)] \in O(n^3).$$

We shall close this section with some observations that can be used to improve the worst-case time-complexity of this algorithm. First we should observe that for $0 \leq i \leq n - 2$, the $(i + 1)$st iteration of our present algorithm generated the $(i + 1)$st closest vertex to v_0. This was the vertex v_{i+1}. In our example we found $v_1 = f$, $v_2 = h$, $v_3 = g$, $v_4 = a$, and $v_5 = b$.

Second, note how much duplication we had when computing $L(v)$. This is seen quite readily in the second and third iterations of Example 13.2. We should like to cut back on such unnecessary calculations, so let us try a slightly different approach to our shortest-path problem. Once again we start with a weighted graph $G = (V, E)$ with $|V| = n$ and $v_0 \in V$. We shall now let v_i denote the ith closest vertex to v_0, where $0 \leq i \leq n - 1$, $S_i = \{v_0, v_1, \ldots, v_i\}$, and $\overline{S}_i = V - S_i$. At the start we assign to each $v \in V$ the number $L_0(v)$ as follows:

$$L_0(v_0) = 0 \qquad \text{because } d(v_0, v_0) = 0 \quad \text{and}$$
$$L_0(v) = \infty, \quad \text{for } v \neq v_0.$$

Then for $i \geq 0$ and $v \in \overline{S}_i$, we define

$$L_{i+1}(v) = \min\{L_i(v), L_i(v_i) + \text{wt}(v_i, v)\},$$

where v_i is a vertex for which $L_i(v_i)$ is minimal: a vertex that is ith closest to v_0. We find that

$$L_{i+1}(v) = \min_{1 \leq j \leq i}\{d(v_0, v_j) + \text{wt}(v_j, v)\}.$$

Now let us see what happens at each of the (at most) $n - 1$ iterations when we employ the definition of $L_{i+1}(v)$ that uses the vertex v_i.

For each $v \in \overline{S}_i$ we need only one addition [namely, $L_i(v_i) + \text{wt}(v_i, v)$] and one comparison [between $L_i(v)$ and $L_i(v_i) + \text{wt}(v_i, v)$] in order to compute $L_{i+1}(v)$. Since there are at most $n - 1$ vertices in $\overline{S}_i$, this necessitates at most $2(n - 1)$ steps to obtain $L_{i+1}(v)$ for *all* $v \in \overline{S}_i$. Finding the minimum of $\{L_{i+1}(v) | v \in \overline{S}_i\}$ requires at most $n - 2$ comparisons, so at each iteration we can obtain v_{i+1} — a vertex $v \in \overline{S}_i$ where $L_{i+1}(v)$ is a minimum — in at most $2(n - 1) + (n - 2) = 3n - 4$ steps. We perform at most $n - 1$ iterations, so we find for this version of Dijkstra's algorithm, that the worst-case time-complexity is $O(n^2)$.

In order to find a shortest path from v_0 to each $v \in V$, $v \neq v_0$, we see that whenever $L_{i+1}(v) < L_i(v)$, for any $0 \leq i \leq n - 2$, we need to keep track of the vertex $y \in S_i$ for which $L_{i+1}(v) = d(v_0, y) + \text{wt}(y, v)$.

Other implementations of Dijkstra's algorithm use a data structure called a *heap*. For a weighted graph $G = (V, E)$, where $|V| = n$ and $|E| = m$, we find, for example, that the *binary heap* implementation of this algorithm has worst-case time-complexity $O(m \log_2 n)$. (This, and much more, is discussed on pp. 108–122 of the text by R. K. Ahuja, T. L. Magnanti, and J. B. Orlin [2]. The reader can also find more about various kinds of heaps on pp. 773–787 of this text. Another source for the implementation and running-time of Dijkstra's algorithm is Section 24.3 (pp. 595–601) of the text by T. H. Cormen, C. E. Leiserson, R. L. Rivest, and C. Stein [7].)

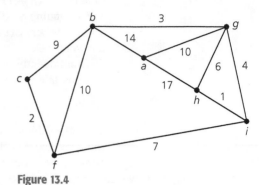

Figure 13.4

EXERCISES 13.1

1. Let $G = (V, E)$ be a weighted graph, where for each edge $e = (a, b)$ in E, wt(a, b) equals the distance from a to b along edge e. If $(a, b) \notin E$, then wt$(a, b) = \infty$.

Fix $v_0 \in V$ and let $S \subset V$, with $v_0 \in S$. Then for $\overline{S} = V - S$ we define $d(v_0, \overline{S}) = \min_{v \in \overline{S}}\{d(v_0, v)\}$. If $v_{m+1} \in \overline{S}$ and $d(v_0, \overline{S}) = d(v_0, v_{m+1})$, then $P: (v_0, v_1), (v_1, v_2), \ldots,$ $(v_{m-1}, v_m), (v_m, v_{m+1})$ is a shortest directed path (in G) from v_0 to v_{m+1}. Prove that

a) $v_0, v_1, v_2, \ldots, v_{m-1}, v_m \in S.$

b) $P': (v_0, v_1), (v_1, v_2), \ldots, (v_{k-1}, v_k)$ is a shortest directed path (in G) from v_0 to v_k, for each $1 \leq k \leq m$.

2. a) Apply Dijkstra's algorithm to the weighted graph $G = (V, E)$ in Fig. 13.4, and determine the shortest distance from vertex a to each of the other six vertices in G. Here wt$(e) = $ wt$(x, y) = $ wt(y, x) for each edge $e = \{x, y\}$ in E.

b) Determine a shortest path from vertex a to each of the vertices c, f, and i.

3. a) Apply Dijkstra's algorithm to the graph shown in Fig. 13.1 and determine the shortest distance from vertex a to each of the other vertices in the graph.

b) Find a shortest path from vertex a to each of the vertices f, g, and h.

4. Use the ideas developed at the end of the section to confirm the result obtained in (a) Example 13.2; and (b) part (a) of Exercise 2.

5. Prove or disprove the following for a weighted graph $G = (V, E)$, where $V = \{v_0, v_1, v_2, \ldots, v_n\}$ and $e_1 \in E$ with wt$(e_1) < $ wt(e) for all $e \in E$, $e \neq e_1$. If Dijkstra's algorithm is applied to G, and the shortest distance $d(v_0, v_i)$ is computed for each vertex v_i, $1 \leq i \leq n$, then there exists a vertex v_j, for some $1 \leq j \leq n$, where the edge e_1 is used in the shortest path from v_0 to v_j.

13.2
Minimal Spanning Trees:
The Algorithms of Kruskal and Prim

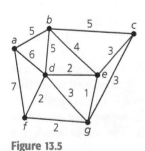

Figure 13.5

A loosely coupled computer network is to be set up for a system of seven computers. The graph G in Fig. 13.5 models the situation. The computers are represented by the vertices in the graph; the edges represent transmission lines that are being considered for linking certain pairs of computers. Associated with each edge e in G is a positive real number wt(e), the weight of e. Here the weight of an edge indicates the projected cost for constructing that particular transmission line. The objective is to link all the computers while minimizing the total cost of construction. To do so requires a spanning tree T, where the sum of the weights of the edges in T is minimal. The construction of such an *optimal* spanning tree can be accomplished by using the algorithms that were developed by Joseph Bernard Kruskal (1928–) and Robert Clay Prim (1921–).

Like Dijkstra's algorithm, these algorithms are *greedy*; when each is used, at each step of the process an optimal (here minimal) choice is made from the remaining available data. Once again, if what appears to be the best choice *locally* (for example, for a vertex c and the vertices near c) turns out to be the best choice *globally* (for all vertices of the graph), then the greedy algorithm will lead to an optimal solution.

We first consider Kruskal's algorithm. This algorithm is given as follows.

Let $G = (V, E)$ be a loop-free undirected connected graph, where $|V| = n$ and each edge e is assigned a positive real number wt(e). To find an optimal (minimal) spanning tree for G, apply the following algorithm.

Kruskal's Algorithm

Step 1: Set the counter $i = 1$ and select an edge e_1 in G, where wt(e_1) is as small as possible.

Step 2: For $1 \leq i \leq n - 2$, if edges $e_1, e_2, \ldots, e_i$ have been selected, then select edge e_{i+1} from the remaining edges in G so that (a) wt(e_{i+1}) is as small as possible and (b) the subgraph of G determined by the edges $e_1, e_2, \ldots, e_i, e_{i+1}$ (and the vertices they are incident with) contains no cycles.

Step 3: Replace i by $i + 1$.

If $i = n - 1$, the subgraph of G determined by edges $e_1, e_2, \ldots, e_{n-1}$ is connected with n vertices and $n - 1$ edges, and is an optimal spanning tree for G.

If $i < n - 1$, return to step (2).

Before establishing the validity of the algorithm, we consider the following example.

EXAMPLE 13.3

Apply Kruskal's algorithm to the graph shown in Fig. 13.5.

Initialization: $(i = 1)$. Since there is a unique edge — namely, $\{e, g\}$ — of smallest weight 1, start with $T = \{\{e, g\}\}$. (T starts as a tree with one edge, and after each iteration it grows into a larger tree or forest. After the last iteration the subgraph T is an optimal spanning tree for the given graph G.)

First Iteration: Among the remaining edges in G, three have the next smallest weight 2. Select $\{d, f\}$, which satisfies the conditions in step (2). Now T is the forest $\{\{e, g\}, \{d, f\}\}$, and i is increased to 2. With $i = 2 < 6$, return to step (2).

Second Iteration: Two remaining edges have weight 2. Select $\{d, e\}$. Now T is the tree $\{\{e, g\}, \{d, f\}, \{d, e\}\}$, and i increases to 3. But because $3 < 6$, the algorithm directs us back to step (2).

Third Iteration: Among the edges of G that are not in T, edge $\{f, g\}$ has minimal weight 2. However, if this edge is added to T, the result contains a cycle, which destroys the tree structure being sought. Consequently, the edges $\{c, e\}$, $\{c, g\}$, and $\{d, g\}$ are considered. Edge $\{d, g\}$ brings about a cycle, but either $\{c, e\}$ or $\{c, g\}$ satisfies the conditions in step (2). Select $\{c, e\}$. T grows to $\{\{e, g\}, \{d, f\}, \{d, e\}, \{c, e\}\}$ and i is increased to 4. Returning to step (2), we find that the fourth and fifth iterations provide the following.

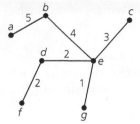

Figure 13.6

Fourth Iteration: $T = \{\{e, g\}, \{d, f\}, \{d, e\}, \{c, e\}, \{b, e\}\}$; i increases to 5.

Fifth Iteration: $T = \{\{e, g\}, \{d, f\}, \{d, e\}, \{c, e\}, \{b, e\}, \{a, b\}\}$. The counter i now becomes $6 = $ (number of vertices in G) $- 1$. So T is an optimal tree for graph G and has weight $1 + 2 + 2 + 3 + 4 + 5 = 17$.

Figure 13.6 shows this spanning tree of minimal weight.

Example 13.3 demonstrates that Kruskal's algorithm does generate a spanning tree. This follows from parts (a) and (d) of Theorem 12.5 since the resulting subgraph has n ($= |V|$) vertices and $n - 1$ edges and is connected. In general, if $G = (V, E)$ is a loop-free weighted connected undirected graph and T is the subgraph of G that is generated by Kruskal's algorithm, then T has no cycles. Furthermore, T is a spanning subgraph of G. For if $v \in V$ and v is not in T, then we can add an edge e of G to T where e is incident with v — and the resulting subgraph of G still contains no cycles. Finally, T is connected. Otherwise T has at least two components, say T_1 and T_2, and since G is connected we could add to T an edge $\{x, y\}$ from G where x is in T_1 and y is in T_2 — and no cycle would be present in this subgraph. Consequently, the subgraph T of G is a connected spanning subgraph of G with no cycles (or loops), so T is a spanning tree of G.

The algorithm is greedy; it selects from the remaining edges an edge of minimal weight that doesn't create a cycle. The following result guarantees that the spanning tree obtained is optimal.

THEOREM 13.1   Let $G = (V, E)$ be a loop-free weighted connected undirected graph. Any spanning tree for G that is obtained by Kruskal's algorithm is optimal.

Proof: Let $|V| = n$, and let T be a spanning tree for G obtained by Kruskal's algorithm. The edges in T are labeled $e_1, e_2, \ldots, e_{n-1}$, according to the order in which they are generated by the algorithm. For each optimal tree T' of G, define $d(T') = k$ if k is the smallest positive integer such that T and T' both contain $e_1, e_2, \ldots, e_{k-1}$, but $e_k \notin T'$.

Let T_1 be an optimal tree for which $d(T_1) = r$ is maximal. If $r = n$, then $T = T_1$ and the result follows. Otherwise, $r \leq n - 1$ and adding edge e_r (of T) to T_1 produces the cycle C, where there exists an edge e'_r of C that is in T_1 but not in T.

Start with tree T_1. Adding e_r to T_1 and deleting e'_r, we obtain a connected graph with n vertices and $n - 1$ edges. This graph is a spanning tree, T_2. The weights of T_1 and T_2 satisfy $\mathrm{wt}(T_2) = \mathrm{wt}(T_1) + \mathrm{wt}(e_r) - \mathrm{wt}(e'_r)$.

Following the selection of $e_1, e_2, \ldots, e_{r-1}$ in Kruskal's algorithm, the edge e_r is chosen so that $\mathrm{wt}(e_r)$ is minimal and no cycle results when e_r is added to the subgraph H of G determined by $e_1, e_2, \ldots, e_{r-1}$. Since e'_r produces no cycle when added to the subgraph H, by the minimality of $\mathrm{wt}(e_r)$ it follows that $\mathrm{wt}(e'_r) \geq \mathrm{wt}(e_r)$. Hence $\mathrm{wt}(e_r) - \mathrm{wt}(e'_r) \leq 0$, so $\mathrm{wt}(T_2) \leq \mathrm{wt}(T_1)$. But with T_1 optimal, we must have $\mathrm{wt}(T_2) = \mathrm{wt}(T_1)$, so T_2 is optimal.

The tree T_2 is optimal and has the edges $e_1, e_2, \ldots, e_{r-1}, e_r$ in common with T, so $d(T_2) \geq r + 1 > r = d(T_1)$, contradicting the choice of T_1. Consequently, $T_1 = T$ and the tree T produced by Kruskal's algorithm is optimal.

We measure the worst-case time-complexity for Kruskal's algorithm by making the following observations. Given a loop-free weighted connected undirected graph $G = (V, E)$, where $|V| = n$ and $|E| = m \geq 2$, we can use the merge sort of Section 12.3 to list (and relabel, if necessary) the edges in E as $e_1, e_2, \ldots, e_m$, where $\text{wt}(e_1) \leq \text{wt}(e_2) \leq \cdots \leq \text{wt}(e_m)$. The number of comparisons needed to do this is $O(m \log_2 m)$. Then once we have the edges of G listed in this order (of nondecreasing weights), step (2) of the algorithm is carried out at most $m - 1$ times — once for each of the edges $e_2, e_3, \ldots, e_m$.

For each edge e_i, $2 \leq i \leq m$, we must determine whether e_i causes the formation of a cycle in the tree, or forest, that we have developed (after considering the edges $e_1, e_2, \ldots, e_{i-1}$). This can be done for each edge in a constant [that is, $O(1)$] amount of time, if we use additional data structures, such as the *component flag* data structure. Unfortunately, the updating of this data structure cannot be performed in a constant amount of time. However, it does turn out that *all* of the work needed for cycle detection can be carried out in at most $O(n \log_2 n)$ steps.[†]

Consequently, we shall define the worst-case time-complexity function f, for $m \geq 2$, as the sum of the following:

1) The total number of comparisons needed to sort the edges of G into nondecreasing order, and

2) The total number of steps that are carried out in step (2) in order to detect the formation of a cycle.

Unless G is a tree, it follows that $|V| = n \leq m = |E|$ because G is connected. As a result, $n \log_2 n \leq m \log_2 m$ and $f \in O(m \log_2 m)$.

A measure in terms of n, the number of vertices in G, can also be given. Here $n - 1 \leq m$ because the graph is connected, and $m \leq \binom{n}{2} = (1/2)(n)(n - 1)$, the number of edges in K_n. Consequently $m \log_2 m \leq n^2 \log_2 n^2 = 2n^2 \log_2 n$, and we can express the worst-case time-complexity of Kruskal's algorithm as $O(n^2 \log_2 n)$, although this is less precise than $O(m \log_2 m)$.

A second technique for constructing an optimal tree was developed by Robert Clay Prim. In this greedy algorithm, the vertices in the graph are partitioned into two sets: processed and not processed. At first only one vertex is in the set P of processed vertices, and all other vertices are in the set N of vertices to be processed. Each iteration of the algorithm increases the set P by one vertex while the size of set N decreases by one. The algorithm is summarized as follows.

Let $G = (V, E)$ be a loop-free weighted connected undirected graph. To obtain an optimal tree T for G, apply the following procedure.

Prim's Algorithm

Step 1: Set the counter $i = 1$ and place an arbitrary vertex $v_1 \in V$ into set P. Define $N = V - \{v_1\}$ and $T = \emptyset$.

Step 2: For $1 \leq i \leq n - 1$, where $|V| = n$, let $P = \{v_1, v_2, \ldots, v_i\}$, $T = \{e_1, e_2, \ldots, e_{i-1}\}$, and $N = V - P$. Add to T a shortest edge (an edge of minimal weight) in G that connects a vertex x in P with a vertex $y (= v_{i+1})$ in N. Place y in P and delete it from N.

[†] For more on the analysis of the segment dealing with cycle detection, we refer the reader to Chapter 8 of the text by S. Baase and A. Van Gelder [3] and to Chapter 4 of the text by E. Horowitz and S. Sahni [17].

Step 3: Increase the counter by 1.
  If $i = n$, the subgraph of G determined by the edges $e_1, e_2, \ldots, e_{n-1}$ is connected with n vertices and $n - 1$ edges and is an optimal tree for G.
  If $i < n$, return to step (2).

We use this algorithm to find an optimal tree for the graph in Fig. 13.5.

EXAMPLE 13.4

Prim's algorithm generates an optimal tree as follows.

Initialization:	$i = 1$; $P = \{a\}$; $N = \{b, c, d, e, f, g\}$; $T = \emptyset$.
First Iteration:	$T = \{\{a, b\}\}$; $P = \{a, b\}$; $N = \{c, d, e, f, g\}$; $i = 2$.
Second Iteration:	$T = \{\{a, b\}, \{b, e\}\}$; $P = \{a, b, e\}$; $N = \{c, d, f, g\}$; $i = 3$.
Third Iteration:	$T = \{\{a, b\}, \{b, e\}, \{e, g\}\}$; $P = \{a, b, e, g\}$; $N = \{c, d, f\}$; $i = 4$.
Fourth Iteration:	$T = \{\{a, b\}, \{b, e\}, \{e, g\}, \{d, e\}\}$; $P = \{a, b, e, g, d\}$; $N = \{c, f\}$; $i = 5$.
Fifth Iteration:	$T = \{\{a, b\}, \{b, e\}, \{e, g\}, \{d, e\}, \{f, g\}\}$; $P = \{a, b, e, g, d, f\}$; $N = \{c\}$; $i = 6$.
Sixth Iteration:	$T = \{\{a, b\}, \{b, e\}, \{e, g\}, \{d, e\}, \{f, g\}, \{c, g\}\}$; $P = \{a, b, e, g, d, f, c\} = V$; $N = \emptyset$; $i = 7 = |V|$. Hence T is an optimal spanning tree of weight 17 for G, as seen in Fig. 13.7.

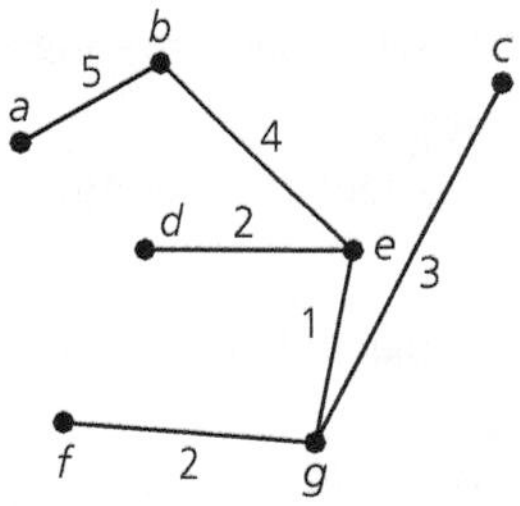

Figure 13.7

Note that the minimal spanning tree obtained here differs from that in Fig. 13.6. So this type of spanning tree need not be unique.

We shall only state the following theorem, which establishes the validity of Prim's algorithm. The proof is left for the reader.

THEOREM 13.2   Let $G = (V, E)$ be a loop-free weighted connected undirected graph. Any spanning tree for G that is obtained by Prim's algorithm is optimal.

Note that at each iteration Prim's algorithm always grows a tree. Some iteration(s) of Kruskal's algorithm may grow a forest (which is not a tree). Also observe that Prim's algorithm can be started at *any* vertex in the graph.

We conclude this section with a few words and references about the worst-case time-complexity for Prim's algorithm. When the algorithm is applied to a loop-free weighted connected undirected graph $G = (V, E)$, where $|V| = n$ and $|E| = m$, the typical implementations require $O(n^2)$ steps. (This can be found in Chapter 7 of A. V. Aho, J. E. Hopcroft, and J. D. Ullman [1]; in Chapter 8 of S. Baase and A. Van Gelder [3]; and in Chapter 4 of E. Horowitz and S. Sahni [17].) Other implementations of the algorithm have improved the situation so that it requires $O(m \log_2 n)$ steps. (This is discussed in the articles by R. L. Graham and P. Hell [16]; by D. B. Johnson [18]; and by A. Kershenbaum and R. Van Slyke [19].) The worst-case time-complexities for various heap implementations are discussed in

Section 13.5 of R. V. Ahuja, T. L. Magnanti, and J. B. Orlin [2] and Section 23.2 of T. H. Cormen, C. E. Leiserson, R. L. Rivest, and C. Stein [7].

EXERCISES 13.2

1. Apply Kruskal's and Prim's algorithms to determine minimal spanning trees for the graph shown in Fig. 13.8.

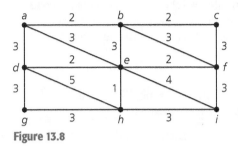

Figure 13.8

2. Let $G = W_4$, the wheel on four spokes. Assign the weights 1, 1, 2, 2, 3, 3, 4, 4 to the edges of G so that (a) G has a unique minimal spanning tree; (b) G has more than one minimal spanning tree.

3. Let $G = (V, E)$ be a loop-free weighted connected undirected graph with $T = (V, E')$, a minimal spanning tree for G. For $v, w \in V$, is the path from v to w in T a path of minimum weight in G?

4. Table 13.1 provides information on the distance (in miles) between pairs of cities in the state of Indiana.

A system of highways connecting these seven cities is to be constructed. Determine which highways should be constructed so that the cost of construction is minimal. (Assume that the cost of construction of a mile of highway is the same between every pair of cities.)

5. a) Answer Exercise 4 under the additional requirement that the system includes a highway directly linking Evansville and Indianapolis.

b) If there must be a direct link between Fort Wayne and Gary in addition to the one connecting Evansville and Indianapolis, find the minimum number of miles of highway that must be constructed.

6. Let $G = (V, E)$ be a loop-free weighted connected undirected graph. For $n \in \mathbf{Z}^+$, let $\{e_1, e_2, \ldots, e_n\}$ be a set of edges (from E) that includes no cycle in G. Modify Kruskal's algorithm in order to obtain a spanning tree of G that is minimal among all the spanning trees of G that include the edges $e_1, e_2, \ldots, e_n$.

7. a) Modify Kruskal's algorithm to determine an optimal tree of *maximal* weight.

b) Interpret the information of Exercise 4 in terms of the number of calls that can be placed between pairs of cities via the adoption of certain new telephone transmission lines. (Cities that are not directly linked must communicate through one or more intermediate cities.) How can the seven cities be minimally connected and allow a maximum number of calls to be placed?

8. Prove Theorem 13.2.

9. Let $G = (V, E)$ be a loop-free weighted connected undirected graph, where for each pair of distinct edges $e_1, e_2 \in E$, $wt(e_1) \neq wt(e_2)$. Prove that G has only one minimal spanning tree.

Table 13.1

	Bloomington	Evansville	Fort Wayne	Gary	Indianapolis	South Bend
Evansville	119	—	—	—	—	—
Fort Wayne	174	290	—	—	—	—
Gary	198	277	132	—	—	—
Indianapolis	51	168	121	153	—	—
South Bend	198	303	79	58	140	—
Terre Haute	58	113	201	164	71	196

13.3
Transport Networks:
The Max-Flow Min-Cut Theorem

This section provides an application for weighted directed graphs to the flow of a commodity from a source to a prescribed destination. Such commodities may be gallons of oil that flow through pipelines or numbers of telephone calls transmitted in a communication system. In modeling such situations, we interpret the weight of an edge in the directed graph as a capacity that places an upper limit on, for example, the amount of oil that can flow through a certain part of a system of pipelines. These ideas are expressed formally in the following definition.

Definition 13.1

Let $N = (V, E)$ be a loop-free connected directed graph. Then N is called a *network*, or *transport network*, if the following conditions are satisfied:

a) There exists a unique vertex $a \in V$ with $id(a)$, the in degree of a, equal to 0. This vertex a is called the *source*.

b) There is a unique vertex $z \in V$, called the *sink*, where $od(z)$, the out degree of z, equals 0.

c) The graph N is weighted, so there is a function from E to the set of nonnegative integers that assigns to each edge $e = (v, w) \in E$ a *capacity*, denoted by $c(e) = c(v, w)$.

EXAMPLE 13.5

The graph in Fig. 13.9 is a transport network. Here vertex a is the source, the sink is at vertex z, and capacities are shown beside each edge. Since $c(a, b) + c(a, g) = 5 + 7 = 12$, the amount of the commodity being transported from a to z cannot exceed 12. With $c(d, z) + c(h, z) = 5 + 6 = 11$, the amount is further restricted to be no greater than 11. To determine the maximum amount that can be transported from a to z, we must consider the capacities of all edges in the network.

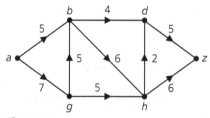

Figure 13.9

The following definition is introduced to assist us in solving this problem.

Definition 13.2

If $N = (V, E)$ is a transport network, a function f from E to the nonnegative integers is called a *flow* for N if

a) $f(e) \leq c(e)$ for each edge $e \in E$; and

b) for each $v \in V$, other than the source a or the sink z, $\sum_{w \in V} f(w, v) = \sum_{w \in V} f(v, w)$. (If there is no edge (v, w), then $f(v, w) = 0$.)

The first property specifies that the amount of material transported along a given edge cannot exceed the capacity of that edge. Property (b) enforces a conservation condition: The amount of material flowing into a vertex v must equal the amount that flows out from this vertex. This is so for all vertices except the source and the sink.

EXAMPLE 13.6

For the networks in Fig. 13.10, the label x, y on each edge e is determined so that $x = c(e)$ and y is the value assigned for a possible flow f. The label on each edge e satisfies $f(e) \leq c(e)$. In part (a) of the figure, the "flow" into vertex g is 5, but the "flow" out from that vertex is $2 + 2 = 4$. Hence the function f is not a flow in this case. The function f for part (b) does satisfy both properties, so it is a flow for the given network.

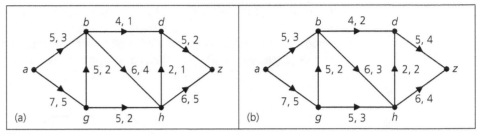

Figure 13.10

Definition 13.3

Let f be a flow for a transport network $N = (V, E)$.

a) An edge e of the network is called *saturated* if $f(e) = c(e)$. When $f(e) < c(e)$, the edge is called *unsaturated*.

b) If a is the source of N, then $\mathrm{val}(f) = \sum_{v \in V} f(a, v)$ is called the *value of the flow*.

EXAMPLE 13.7

For the network in Fig. 13.10(b), only the edge (h, d) is saturated. All other edges are unsaturated. The value of the flow in this network is

$$\mathrm{val}(f) = \sum_{v \in V} f(a, v) = f(a, b) + f(a, g) = 3 + 5 = 8.$$

But is there another flow f_1 such that $\mathrm{val}(f_1) > 8$? The determination of a *maximal flow* (a flow that achieves the greatest possible value) is the objective of the remainder of this section. To accomplish this, we observe that in the network of Fig. 13.10(b),

$$\sum_{v \in V} f(a, v) = 3 + 5 = 8 = 4 + 4 = f(d, z) + f(h, z) = \sum_{v \in V} f(v, z).$$

Consequently, the total flow leaving the source a equals the total flow into the sink z.

The last remark in Example 13.7 seems like a reasonable circumstance, but will it occur in general? To prove the result for every network, we need the following special type of cut-set.

Definition 13.4

If $N = (V, E)$ is a transport network and C is a cut-set for the undirected graph associated with N, then C is called a *cut*, or an *a-z cut*, if the removal of the edges in C from the network results in the separation of a and z.

<table>
<tr><td>**EXAMPLE 13.8**</td><td>Each of the dotted curves in Fig. 13.11 indicates a cut for the given network. The cut C_1 consists of the undirected edges $\{a, g\}$, $\{b, d\}$, $\{b, g\}$, and $\{b, h\}$. This cut partitions the vertices of the network into the two sets $P = \{a, b\}$ and its complement $\overline{P} = \{d, g, h, z\}$, so C_1 is denoted as $(P, \overline{P})$. The *capacity of a cut*, denoted $c(P, \overline{P})$, is defined by</td></tr>
</table>

$$c(P, \overline{P}) = \sum_{\substack{v \in P \\ w \in \overline{P}}} c(v, w),$$

the sum of the capacities of all edges (v, w), where $v \in P$ and $w \in \overline{P}$. In this example, $c(P, \overline{P}) = c(a, g) + c(b, d) + c(b, h) = 7 + 4 + 6 = 17$. [Considering the *directed* edges (from P to $\overline{P}$) in the cut $C_1 = (P, \overline{P})$ — namely, (a, g), (b, d), (b, h) — we find that the removal of these edges does not result in a subgraph with two components. However, the removal of these three edges eliminates all possible directed paths from a to z and no proper subset of $\{(a, g), (b, d), (b, h)\}$ has this separating property.]

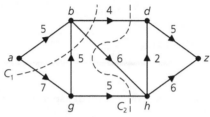

Figure 13.11

The cut C_2 induces the vertex partition $Q = \{a, b, g\}$, $\overline{Q} = \{d, h, z\}$ and has capacity $c(Q, \overline{Q}) = c(b, d) + c(b, h) + c(g, h) = 4 + 6 + 5 = 15$.

A third cut of interest is the one that induces the vertex partition $S = \{a, b, d, g, h\}$, $\overline{S} = \{z\}$. (What are the edges in this cut?) Its capacity is 11.

Using the idea of the capacity of a cut, this next result provides an upper bound for the value of a flow in a network.

<table>
<tr><td>**THEOREM 13.3**</td><td>Let f be a flow in a network $N = (V, E)$. If $C = (P, \overline{P})$ is any cut in N, then val(f) cannot exceed $c(P, \overline{P})$.

Proof: Let vertex a be the source in N and vertex z the sink. Since $id(a) = 0$, it follows that for all $w \in V$, $f(w, a) = 0$. Consequently,</td></tr>
</table>

$$\text{val}(f) = \sum_{v \in V} f(a, v) = \sum_{v \in V} f(a, v) - \sum_{w \in V} f(w, a).$$

By property (b) in the definition of a flow, for all $x \in P$, $x \neq a$, $\sum_{v \in V} f(x, v) - \sum_{w \in V} f(w, x) = 0$.

Adding the results in the above equations yields

$$\mathrm{val}(f) = \left[\sum_{v \in V} f(a, v) - \sum_{w \in V} f(w, a) \right] + \sum_{\substack{x \in P \\ x \neq a}} \left[\sum_{v \in V} f(x, v) - \sum_{w \in V} f(w, x) \right]$$

$$= \sum_{\substack{x \in P \\ v \in V}} f(x, v) - \sum_{\substack{x \in P \\ w \in V}} f(w, x)$$

$$= \left[\sum_{\substack{x \in P \\ v \in P}} f(x, v) + \sum_{\substack{x \in P \\ v \in \overline{P}}} f(x, v) \right] - \left[\sum_{\substack{x \in P \\ w \in P}} f(w, x) + \sum_{\substack{x \in P \\ w \in \overline{P}}} f(w, x) \right].$$

Since

$$\sum_{\substack{x \in P \\ v \in P}} f(x, v) \quad \text{and} \quad \sum_{\substack{x \in P \\ w \in P}} f(w, x)$$

are summed over the same set of all ordered pairs in $P \times P$, these summations are equal. Consequently,

$$\mathrm{val}(f) = \sum_{\substack{x \in P \\ v \in \overline{P}}} f(x, v) - \sum_{\substack{x \in P \\ w \in \overline{P}}} f(w, x).$$

For all $x, w \in V, f(w, x) \geq 0$, so

$$\sum_{\substack{x \in P \\ w \in \overline{P}}} f(w, x) \geq 0 \quad \text{and} \quad \mathrm{val}(f) \leq \sum_{\substack{x \in P \\ v \in \overline{P}}} f(x, v) \leq \sum_{\substack{x \in P \\ v \in \overline{P}}} c(x, v) = c(P, \overline{P}).$$

From Theorem 13.3 we find that in a network N, the value for *any* flow is less than or equal to the capacity of *any* cut in that network. Hence the value of the maximum flow cannot exceed the minimum capacity over all cuts in a network. For the network in Fig. 13.11, it can be shown that the cut consisting of edges (d, z) and (h, z) has minimum capacity 11. Consequently, the maximum flow f for the network satisfies $\mathrm{val}(f) \leq 11$. It will turn out that the value of the maximum flow is 11. How to construct such a flow and why its value equals the minimum capacity among all cuts will be dealt with in this section.

However, before we deal with this construction, let us note that in the proof of Theorem 13.3, the value of a flow is given by

$$\mathrm{val}(f) = \sum_{\substack{x \in P \\ v \in \overline{P}}} f(x, v) - \sum_{\substack{x \in P \\ w \in \overline{P}}} f(w, x),$$

where $(P, \overline{P})$ is *any* cut in N. Therefore, once a flow is constructed in a network, then for any cut $(P, \overline{P})$ in the network, the value of the flow equals the sum of the flows in the edges directed from the vertices in P to those in $\overline{P}$ minus the sum of the flows in the edges directed from the vertices in $\overline{P}$ to those in P.

This observation leads to the following result.

COROLLARY 13.1

If f is a flow in a transport network $N = (V, E)$, then the value of the flow from the source a is equal to the value of the flow into the sink z.

Proof: Let $P = \{a\}$, $\overline{P} = V - \{a\}$, and $Q = V - \{z\}$, $\overline{Q} = \{z\}$. From the above observation,

$$\sum_{\substack{x \in P \\ v \in \overline{P}}} f(x, v) - \sum_{\substack{x \in P \\ w \in \overline{P}}} f(w, x) = \text{val}(f) = \sum_{\substack{y \in Q \\ v \in \overline{Q}}} f(y, v) - \sum_{\substack{y \in Q \\ w \in \overline{Q}}} f(w, y).$$

With $P = \{a\}$ and $id(a) = 0$, we find that $\sum_{x \in P, w \in \overline{P}} f(w, x) = \sum_{w \in \overline{P}} f(w, a) = 0$. Similarly, for $\overline{Q} = \{z\}$ and $od(z) = 0$, it follows that $\sum_{y \in Q, w \in \overline{Q}} f(w, y) = \sum_{y \in Q} f(z, y) = 0$. Consequently,

$$\sum_{\substack{x \in P \\ v \in \overline{P}}} f(x, v) = \sum_{v \in \overline{P}} f(a, v) = \text{val}(f) = \sum_{\substack{y \in Q \\ v \in \overline{Q}}} f(y, v) = \sum_{y \in Q} f(y, z),$$

and this establishes the corollary.

Additional properties of flows and cuts in a network are given in the following corollaries.

COROLLARY 13.2

Let f be a flow in a transport network $N = (V, E)$ and let $(P, \overline{P})$ be a cut, where $\text{val}(f) = c(P, \overline{P})$. Then f is a maximum flow for the network N and $(P, \overline{P})$ is a minimum cut [that is, $(P, \overline{P})$ has minimum capacity in N].

Proof: If f_1 is any flow in N, then from Theorem 13.3 it follows that

$$\text{val}(f_1) \leq c(P, \overline{P}) = \text{val}(f),$$

so f is a maximum flow. Likewise, for any cut $(Q, \overline{Q})$ in N we have

$$c(P, \overline{P}) = \text{val}(f) \leq c(Q, \overline{Q}),$$

so $(P, \overline{P})$ is a minimum cut — again, by Theorem 13.3.

COROLLARY 13.3

If f is a maximum flow in a transport network $N = (V, E)$ and $(P, \overline{P})$ is a minimum cut, then $\text{val}(f) \leq c(P, \overline{P})$.

Proof: The proof of this corollary is requested in the Section Exercises.

COROLLARY 13.4

For a transport network $N = (V, E)$, let f be a flow in N and let $(P, \overline{P})$ be a cut. Then $\text{val}(f) = c(P, \overline{P})$ if and only if

a) $f(e) = c(e)$ for each edge $e = (x, y)$, where $x \in P$ and $y \in \overline{P}$, and

b) $f(e) = 0$ for each edge $e = (v, w)$, where $v \in \overline{P}$ and $w \in P$.

Furthermore, under these circumstances, f is a maximum flow and $(P, \overline{P})$ is a minimum cut.

Proof: The proof of this corollary is requested in the Section Exercises.

We turn now to the main results of the section—namely, (1) developing an efficient algorithm to solve the Maximum Flow–Minimum Cut (Max-Flow Min-Cut) problem, and (2) establishing the Max-Flow Min-Cut Theorem. The algorithm we introduce was initially presented in the work of Lester R. Ford, Jr., and Delbert Ray Fulkerson. Basically, it is designed to increase the flow in a transport network N iteratively, until no further increase is possible.

In order to motivate the concepts we shall need here, we start by considering the following example.

EXAMPLE 13.9
Let $N = (V, E)$ be the transport network shown in part (i) of Fig. 13.12. Examining the edges (b, z) and (g, z), we see that the value of the flow is $6 + 2 = 8$. But neither of these two edges is saturated, nor is any other edge in N, so we shall try to increase the present flow. To do so, consider a directed path from a to z—for example, the path p made up of the edges (a, b) and (b, z) [as in part (ii) of the figure]. For this path we define $\Delta_p = \min_{e \in p}\{c(e) - f(e)\} = \min\{8 - 4, 8 - 6\} = \min\{4, 2\} = 2$. This tells us that the flow in each of these two edges can be increased by 2, with the conservation of flow still maintained. The resulting network, in part (iii) of the figure, now has flow value $8 + 2 = 10$.

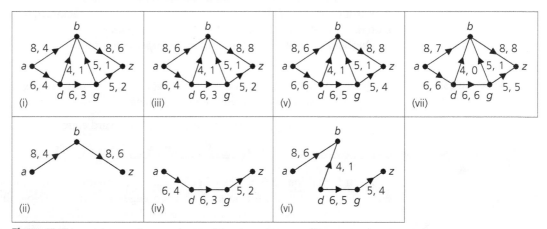

Figure 13.12

So far, so good. Now let us try to increase the flow again. This time we use the directed path p_1 from a to z as shown in part (iv) of Fig. 13.12. This path comprises the edges (a, d), (d, g), and (g, z) and here $\Delta_{p_1} = \min_{e \in p_1}\{c(e) - f(e)\} = \min\{6 - 4, 6 - 3, 5 - 2\} = \min\{2, 3, 3\} = 2$. The resulting network, with the adjustment $\Delta_{p_1} = 2$, is shown in Fig. 13.12(v) and it has flow value 12.

Now, at this point, any possible directed a-z path in N [of Fig. 13.12(v)] must use either edge (a, d) or edge (b, z), both of which are saturated—that is, $c(e) = f(e)$. Consequently, it may seem that the current flow of 12 is the maximum flow possible.

If, however, we disregard the directions on the edges of the network, it is possible to find other paths from a to z. Consider one such path—the path p_2 shown in part (vi) of the figure. This undirected path comprises the edges $\{a, b\}$, $\{b, d\}$, $\{d, g\}$, and $\{g, z\}$. Here we define

$\Delta_{p_2} = \min_{e \in p_2}\{\Delta_e\}$, where $\Delta_e = c(e) - f(e)$ for the forward edges (a, b), (d, g), (g, z), and $\Delta_e = f(e)$ for the backward edge going from b to d [the opposite of the direction for edge (d, b) in N]. So $\Delta_{p_2} = \min[\{8 - 6, 6 - 5, 5 - 4\} \cup \{1\}] = 1$. This increase of one unit of flow is added to the flow for each of the three forward edges and subtracted from the flow for the one backward edge. The resulting final network appears in part (vii) of Fig. 13.12, where we see that by decreasing the flow from d to b by one unit (of flow) we have been able to redirect this one unit from d to g and then from g to z. So now the flow value for N is $12 + 1 = 13$ and this is the maximum flow value possible — for the edges (b, z) and (g, z) are saturated.

What has taken place in Example 13.9 now leads us to the following.

Definition 13.5   Let $N = (V, E)$ be a transport network and let

$$a = v_0, e_1, v_1, e_2, v_2, \ldots, v_{n-1}, e_n, v_n = z$$

be an alternating sequence of vertices and edges, where the edges are taken from the undirected graph associated with N. This sequence is called a *semipath*.[†]
For $2 \le i \le n - 1$, if $e_i = (v_{i-1}, v_i)$ — that is, e_i is the directed edge in N from v_{i-1} to v_i — then e_i is called a *forward edge*. In the case where $2 \le j \le n - 1$ and $e_j = (v_j, v_{j-1})$ — that is, (v_{j-1}, v_j) is the actual directed edge in N — then e_j is called a *backward edge*.

When all of the edges in a semipath are forward edges (in N), then we have a directed path from a to z in N. It is only when there is at least one backward edge (from N) that the path in the associated undirected graph is a semipath.
Our next idea takes the notion of the semipath one step further.

Definition 13.6   Let f be a flow in a transport network $N = (V, E)$. An *f-augmenting path* p is a semipath (from a to z) where for each edge e on p we have

$$f(e) < c(e), \quad \text{for } e \text{ a forward edge}$$
$$f(e) > 0, \qquad \text{for } e \text{ a backward edge.}$$

From Definition 13.6 we see that along an f-augmenting path p the flow on a forward edge can be increased, for no such forward edge is saturated. [Note that here we could have $f(e) = 0$.] For each backward edge the flow is positive, so it can be decreased (and redirected elsewhere). The maximum possible increase or decrease is given in terms of Δ_e, the tolerance on an edge e, as we learn in the following.

Definition 13.7   Let p be an f-augmenting path in a transport network $N = (V, E)$. For each edge e on the semipath p,

$$\Delta_e = \begin{cases} c(e) - f(e), & \text{for } e \text{ a forward edge} \\ f(e), & \text{for } e \text{ a backward edge.} \end{cases}$$

The quantity Δ_e is often called the *tolerance* on edge e.

[†] Some authors use the term *chain* or *quasi-path* in place of semipath.

Note that in Definition 13.7 we have $\Delta_e > 0$ for each edge e on p. Further, we find that $\Delta_p = \min_{e \in p}\{\Delta_e\}$ is the maximum increase (for the forward edges) and maximum decrease (for the backward edges) that we can have and still maintain the conservation condition in part (b) of Definition 13.2.

Our next result formally establishes what was described in Definition 13.7 and the paragraph that followed.

THEOREM 13.4

Let f be a flow in a transport network $N = (V, E)$ and let p be an f-augmenting path in N with $\Delta_p = \min_{e \in p}\{\Delta_e\}$. Define $f_1: E \to \mathbf{N}$ by

$$f_1(e) = \begin{cases} f(e) + \Delta_p, & e \in p, e \text{ a forward edge} \\ f(e) - \Delta_p, & e \in p, e \text{ a backward edge} \\ f(e), & e \in E, e \notin p \end{cases}$$

Then f_1 is a flow in N with $\text{val}(f_1) = \text{val}(f) + \Delta_p$.

Proof: From the definition of Δ_p we have $0 \le f_1(e) \le c(e)$, for each $e \in E$. So f_1 satisfies condition (a) of Definition 13.2. To establish condition (b) of Definition 13.2 for f_1, we only need to consider those $v \in V$ where v is on the semipath p and $v \ne a, z$. So let $\{v_i, v\}$ and $\{v, v_{i+2}\}$ be the two edges in p that are incident with v. When we consider the *net* change at v, we see in the four cases of Fig. 13.13 that this change is 0. Consequently, f_1 satisfies condition (b) and is a flow.

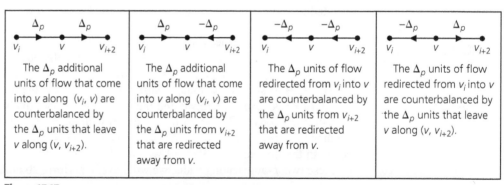

Figure 13.13

To determine $\text{val}(f_1)$ we consider $e_1 = (v_0, v_1) = (a, v_1)$, the first edge on the f-augmenting path p. Then e_1 is adjacent from the source a and it follows from part (b) of Definition 13.3 that $\text{val}(f_1) = \sum_{v \in V} f_1(a, v) = \sum_{v \in V - \{v_1\}} f_1(a, v) + f_1(a, v_1) = \sum_{v \in V - \{v_1\}} f(a, v) + f(a, v_1) + \Delta_p = \sum_{v \in V} f(a, v) + \Delta_p = \text{val}(f) + \Delta_p$.

The result of Theorem 13.4 now helps us in characterizing a maximum flow in a transport network.

THEOREM 13.5

Let $N = (V, E)$ be a transport network with flow f. The flow f is a maximum flow in N if and only if there exists no f-augmenting path in N.

Proof: If f is a maximum flow in N, then it follows from Theorem 13.4 that there is no f-augmenting path in N.

Conversely, if there is no f-augmenting path in N, consider the set of all partial semipaths in N that start at a. We call each of these edge sets a partial semipath because it cannot reach z, without contradicting the hypothesis. Let P be the union of the vertices in these partial semipaths. Then $a \in P$, and $\overline{P} \neq \emptyset$ as $z \in \overline{P}$. Further, $(P, \overline{P})$ is a cut for N and,

i) if $e = (u, w) \in E$ with $u \in P$, $w \in \overline{P}$, then $f(e) = c(e)$ — otherwise, $w \in P$;

ii) if $e = (u, w) \in E$ with $w \in P$, $u \in \overline{P}$, then $f(e) = 0$ — otherwise, $f(e) > 0$ and $u \in P$.

Consequently, from Corollary 13.4, it follows that f is a maximum flow.

We now turn to the main result of the section.

THEOREM 13.6

The Max-Flow Min-Cut Theorem. For a transport network $N = (V, E)$, the maximum flow value that can be attained in N is equal to the minimum capacity over all cuts in the network.

Proof: Let f be a flow for which $\text{val}(f)$ is a maximum. Then let $(P, \overline{P})$ be the cut constructed as in Theorem 13.5. We know from Corollary 13.4 that $\text{val}(f) = c(P, \overline{P})$. And then Corollary 13.2 shows us that $(P, \overline{P})$ is a minimum cut.

Now that we have dispensed with the necessary theory it is time to develop an efficient way of determining a maximum flow and minimum cut for a given transport network N. The discussion in Example 13.9 might suggest that we should simply find f-augmenting paths and use them to continue increasing the existing flow in N. However, this may prove to be tedious and inefficient as our next example demonstrates.

EXAMPLE 13.10

Consider the transport network $N = (V, E)$ in Fig. 13.14(i), where the initial flow is given as $f(e) = 0$ for each $e \in E$. The capacities for the edges are $c(a, b) = c(b, z) = c(a, d) = c(d, z) = 10$ and $c(d, b) = 1$. If we use the directed paths $(a, b), (b, z)$ and then $(a, d), (d, z)$ as successive f-augmenting paths, we attain the flow in part (ii) of the figure after two iterations. Here we find that $\text{val}(f) = 20$ and this is a maximum flow since $20 = c(P, \overline{P})$ for $P = \{a\}$. If, instead, we start with the directed path $(a, d), (d, b), (b, z)$ and then the semipath $\{a, b\}, \{b, d\}, \{d, z\}$ as our first two successive f-augmenting paths, we attain the flow in Fig. 13.14(iii) where $\text{val}(f) = 2$. Should we continue to alternately use these two f-augmenting paths, we will have to perform 20 iterations in total before we attain the flow in part (ii) of the figure.

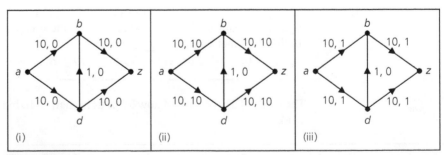

Figure 13.14

What do we observe here? The directed paths $(a, b), (b, z)$ and $(a, d), (d, z)$ each have two edges, while the directed path $(a, d), (d, b), (b, z)$ and the semipath $\{a, b\}, \{b, d\}, \{d, z\}$ each have three edges. Further, note how the first iteration in Example 13.9 used a directed path with two edges, the second iteration a directed path with three edges, and the third iteration a semipath of four edges.

The observations made in Example 13.10 suggest that for each iteration it is more efficient to use an f-augmenting path with the least number of edges. This idea was used by Jack Edmonds and Richard M. Karp in the development of an algorithm to find such f-augmenting paths. Their approach uses a breadth-first search and, as in Prim's algorithm, the vertex set V is partitioned as $P \cup \overline{P}$, where P accounts for the processed vertices. However, before we can deal with this algorithm we need one additional idea.

Definition 13.8

Let $N = (V, E)$ be a transport network with flow f. Start to construct a breadth-first spanning tree T for N (as an undirected graph) using the source a as the root, and a prescribed order for the other vertices in V. While the sink z is not a vertex in T, let $e = \{v, w\}$ be the newest edge appended in the construction of T, with v in the present tree and w the new vertex. The edge e is called *usable* if

$$e = (v, w) \text{ with } f(e) < c(e), \quad \text{or}$$
$$e = (w, v) \text{ with } f(e) > 0.$$

Now we are ready to deal with the following algorithm. Here the input is a transport network $N = (V, E)$ with flow f. The output is an f-augmenting path p, with a minimum number of edges, if one exists; otherwise, the output is a minimum cut $(P, \overline{P})$ with $c(P, \overline{P}) = \text{val}(f)$.

The Edmonds-Karp Algorithm

Step 1: Place the source a into set P (thus initializing the set of processed vertices.) Assign the label $(\ , 1)$ to a and set the counter $i = 2$.

Step 2: While the sink z is not in P
      If there is a usable edge in N
            Let $e = \{v, w\}$ be usable with labeled vertex v having
                  the smallest counter assignment
            If w is unlabeled
                Label w with (v, i)
                Place w in P
                Increase the counter i by 1.
      Else
            Return the minimum cut $(P, \overline{P})$.

Step 3: If z is in P, start with z and backtrack to a using the first component of the vertex labels. (This provides an f-augmenting path p with the smallest number of edges.)

At this point we have finally arrived at the algorithm for determining a maximum flow and minimum cut for a transport network $N = (V, E)$. The original version of this algorithm was developed by Lester R. Ford, Jr., and Delbert Ray Fulkerson. Here we shall incorporate the previous algorithm by Jack Edmonds and Richard M. Karp in order to improve the efficiency of the original algorithm.

As with the preceding algorithm, the input is again a transport network $N = (V, E)$. The output is a maximum flow and minimum cut for N.

The Ford-Fulkerson Algorithm

Step 1: Define the initial flow f on the edges of N by $f(e) = 0$ for each $e \in E$.

Step 2: Repeat

      Apply the Edmonds-Karp algorithm to determine
      an f-augmenting path p.

      Let $\Delta_p = \min_{e \in p}\{\Delta_e\}$.

      For each $e \in p$

        If e is a forward edge

          $f(e) := f(e) + \Delta_p$

        Else (e is a backward edge)

          $f(e) := f(e) - \Delta_p$

    Until no f-augmenting path p can be found in N.

    Return the maximum flow f.

Step 3: Return the minimum cut $(P, \overline{P})$ (from the last application of the Edmonds-Karp algorithm, where no further f-augmenting path could be constructed).

Before demonstrating the use of the Ford-Fulkerson and Edmonds-Karp algorithms we state one last corollary and some related comments. The proof of the corollary is left as an exercise.

COROLLARY 13.5
Let $N = (V, E)$ be a transport network where for each $e \in E$, $c(e)$ is a positive integer. Then there is a maximum flow f for N, where $f(e)$ is a nonnegative integer for each edge e.

The definition of transport network and flow (in a transport network) may be modified to allow nonnegative real-valued capacity and flow functions. If the capacities in a transport network are rational numbers, then the Ford-Fulkerson algorithm will terminate and attain a maximum flow and minimum cut. When some capacities are irrational, however, the original algorithm developed by L. R. Ford, Jr., and D. R. Fulkerson may not terminate correctly. Furthermore, Ford and Fulkerson [14] showed that their algorithm could result in a flow — but that the flow need not be a maximum flow. When irrational capacities do arise, the modification given by Edmonds and Karp [11] terminates and attains a maximum flow. Further, the Edmonds-Karp algorithm can be implemented so that its worst-case time-complexity is $O(nm^2)$, where $n = |V|$, $m = |E|$, for $N = (V, E)$. (For more on the time-complexity of this algorithm one should examine Section 6.5 of Ahuja, Magnanti, and Orlin [2] and Chapter 26 of Cormen, Leiserson, Rivest, and Stein [7].)

EXAMPLE 13.11
Use the Ford-Fulkerson and Edmonds-Karp algorithms to find a maximum flow for the transport network in Fig. 13.15(i).

In the transport network $N = (V, E)$ [of Fig. 13.15(i)], each edge is labeled with a pair of nonnegative integers x, y, where x is the capacity of the edge and $y = 0$ indicates an initial flow. This follows from step (1) of the Ford-Fulkerson algorithm.

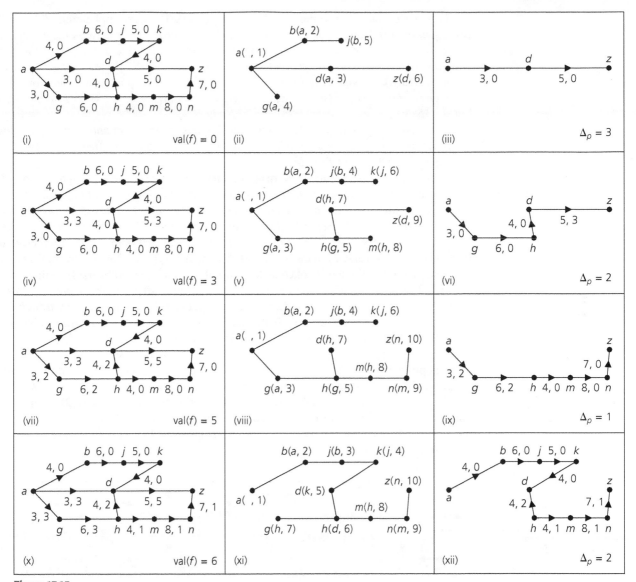

Figure 13.15

When applying the Edmonds-Karp algorithm the prescribed order for the vertices $V -$ $\{a\}$ will be alphabetic. Applying this algorithm for the first time, in step (1) we label a with (,1), place a in P, and set the counter i to 2. In step (2) we find there are three usable (forward) edges: (a, b), (a, d), and (a, g). Following the prescribed order, we select (a, b), label b with $(a, 2)$, place b in P, and increase the counter to 3. Executing step (2) a second time, we select (a, d), label d with $(a, 3)$, place d in P, and increase the counter to 4. At this point, step (2) is executed a third time, for edge (a, g). So we label g with $(a, 4)$, place g in P, and increase the counter to 5.

The edge (b, j) is usable with b having the smallest counter label. [None of the edges (a, b), (a, d), (a, g) is usable at this stage.] Now in step (2) the vertex j is labeled with $(b, 5)$, b is placed in P, and the counter is increased to 6. For the vertex d in P, the edge

(k, d) is not usable because the flow in this edge is 0. The next application of step (2), consequently, results in the label $(d, 6)$ on z, places z in P, and increases the counter to 7. But with z in P we are finished with step (2), and so we arrive at the partial breadth-first spanning tree (for the undirected graph associated with N) rooted at a — as shown in Fig. 13.15(ii). Backtracking in step (3) of the Edmonds-Karp algorithm now provides the f-augmenting path p: (a, d), (d, z), where $\Delta_p = \min\{3 - 0, 5 - 0\} = 3$, as shown in Fig. 13.15(iii).

At this point, we go to step (2) of the Ford-Fulkerson algorithm and increase the flow on (a, d) from 0 to 3 and that on (d, z) from 0 to 3. The result is the transport network in Fig. 13.15(iv), where $val(f) = 3$.

We now return to the Edmonds-Karp algorithm to determine the next f-augmenting path. The resulting partial breadth-first spanning tree for this is shown in part (v) of the figure. The corresponding f-augmenting path p in Fig. 13.15(vi) has tolerance $\Delta_p = \min\{3 - 0, 6 - 0, 4 - 0, 5 - 3\} = 2$. Step (2) of the Ford-Fulkerson algorithm then provides the network in Fig. 13.15(vii), where $val(f) = 3 + \Delta_p = 5$. The next (similar) iteration takes us from this transport network to the one in Fig. 13.15(x), where the flow is now 6. When the Edmonds-Karp algorithm is invoked at this stage, the resulting breadth-first spanning tree is shown in Fig. 13.15(xi). In this application of the algorithm, after we label d with $(k, 5)$, we next label h because we now have the usable (back) edge (h, d) — for the flow from h to d is $2 (> 0)$. Backtracking from z to a in the tree in part (xi) results in the f-augmenting path p in part (xii) with $\Delta_p = \min\{4 - 0, 6 - 0, 5 - 0, 4 - 0, 2, 4 - 1, 8 - 1, 7 - 1\} = 2$.

This now brings us to the transport network in Fig. 13.16(i), where $val(f) = 8$. If we try to apply the Edmonds-Karp algorithm to find the next f-augmenting path, we obtain the partial breadth-first spanning tree in Fig. 13.16(ii). At this point, $P = \{a, b, j, k, d\}$ so $z \notin P$, and there are no other usable edges. Consequently, the last line of step (2) provides the minimum cut $(P, \overline{P})$, where $\overline{P} = \{g, h, m, n, z\}$, as shown in Fig. 13.16(iii). Further, from the edges that are crossed by the dotted curve, we have $val(f) = f((a, g)) + f((d, z)) - f((h, d)) = 3 + 5 - 0 = 8 = c(P, \overline{P})$.

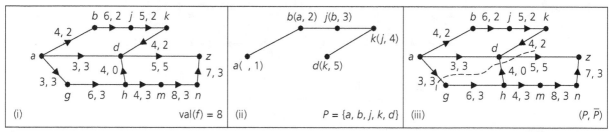

Figure 13.16

We close this section with three examples that are modeled with the concept of the transport network. After setting up the models, the final solution of each example is left to the Section Exercises.

EXAMPLE 13.12

Computer chips are manufactured (in units of a thousand) at three companies, c_1, c_2, and c_3. These chips are then distributed to two computer manufacturers, m_1 and m_2, through the "transport network" in Fig. 13.17(a), where there are the three sources — c_1, c_2, and c_3 — and the two sinks, m_1 and m_2. Company c_1 can produce up to 15 units, company c_2 up to 20 units, and company c_3 up to 25 units. If each manufacturer needs 25 units, how many units should each company produce so that together they can meet the demand of each manufacturer or at least supply them with as many units as the network will allow?

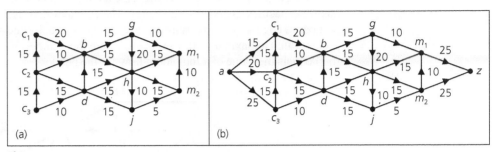

Figure 13.17

In order to model this example with a transport network, we introduce a source a and a sink z, as shown in Fig. 13.17(b). The manufacturing capabilities of the three companies are then used to define capacities for the edges (a, c_1), (a, c_2), and (a, c_3). For the edges (m_1, z) and (m_2, z) the demands are used as capacities. To answer the question posed here, one applies the Edmonds-Karp and Ford-Fulkerson algorithms to this network to find the value of a maximum flow.

EXAMPLE 13.13

The transport network shown in Fig. 13.18(a) has an added restriction, for now there are capacities assigned to vertices other than the source and sink. Such a capacity places an upper limit on the amount of the commodity in question that may *pass through* a given vertex. Part (b) of the figure shows how to redraw the network in order to obtain one where the Edmonds-Karp and Ford-Fulkerson algorithms can be applied. For each vertex v other than a or z, split v into vertices v_1 and v_2. Draw an edge from v_1 to v_2 and label it with the capacity originally assigned to v. An edge of the form (v, w), where $v \neq a$, $w \neq z$, then becomes the edge (v_2, w_1), maintaining the capacity of (v, w). Edges of the form (a, v) become (a, v_1) with capacity $c(a, v)$. An edge such as (w, z) is replaced by the edge (w_2, z), with capacity $c(w, z)$.

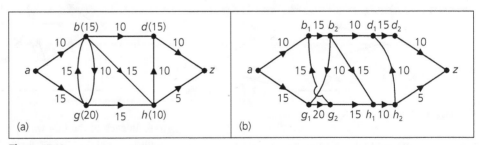

Figure 13.18

The maximum flow for the given network is now determined by applying the Edmonds-Karp and Ford-Fulkerson algorithms to the network shown in Fig. 13.18(b).

EXAMPLE 13.14

During the practice of war games, messengers must deliver information from headquarters (vertex a) to a field command station (vertex z). Since certain roads may be blocked or destroyed, how many messengers should be sent out so that each travels along a path that has no edge in common with any other path taken?

Since the distances between vertices are not relevant here, the graph shown in Fig. 13.19 has no capacities assigned to its edges. The problem here is to determine the maximum number of edge-disjoint paths from a to z. Assigning each edge a capacity of 1 converts the problem into a maximum-flow problem, where the number of edge-disjoint paths (from a to z) equals the value of a maximum flow for the network.

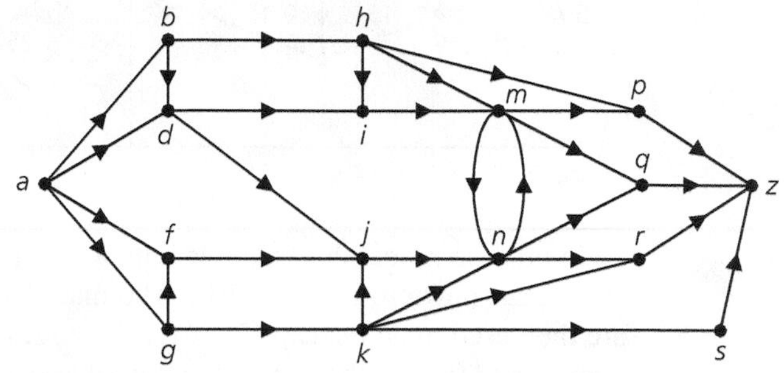

Figure 13.19

EXERCISES 13.3

1. a) For the network shown in Fig. 13.20, let the capacity of each edge be 10. If each edge e in the figure is labeled by a function f, as shown, determine the values of $s, t, w, x,$ and y so that f is a flow in the network.

b) What is the value of this flow?

c) Find three cuts $(P, \overline{P})$ in this network that have capacity 30.

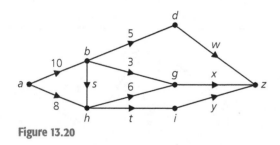

Figure 13.20

2. Prove Corollaries 13.3 and 13.4.

3. Find a maximum flow and the corresponding minimum cut for each transport network shown in Fig. 13.21.

4. Apply the Edmonds-Karp and Ford-Fulkerson algorithms to find a maximum flow in Examples 13.12, 13.13, and 13.14.

5. Prove Corollary 13.5.

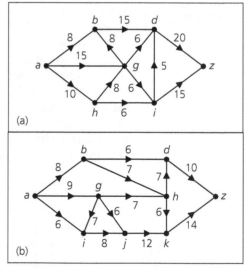

(a)

(b)

Figure 13.21

6. In each of the following "transport networks" two companies, c_1 and c_2, produce a certain product that is used by two manufacturers, m_1 and m_2. For the network shown in part (a) of Fig. 13.22, company c_1 can produce 8 units and company c_2 can produce 7 units; manufacturer m_1 requires 7 units and manufacturer m_2 needs 6 units. In the network shown in Fig. 13.22(b), each company can produce 7 units and each manufacturer needs 6 units. In which situation(s) can the producers meet the manufacturers' demands?

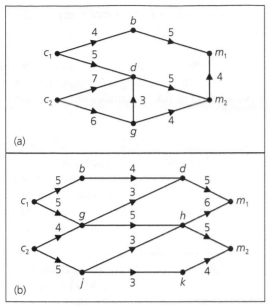

(a)

(b)

Figure 13.22

7. Find a maximum flow for the network shown in Fig. 13.23. The capacities on the undirected edges indicate that the capacity is the same in either direction. [However, for an undirected edge a flow can go in only one direction at a time as opposed to the situation for vertices b, g in Fig. 13.18(a).]

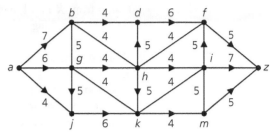

Figure 13.23

13.4
Matching Theory

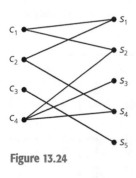

Figure 13.24

The Villa school district must hire four teachers to teach classes in the following subjects: mathematics (s_1), computer science (s_2), chemistry (s_3), physics (s_4), and biology (s_5). Four candidates who are interested in teaching in this district are Miss Carelli (c_1), Mr. Ritter (c_2), Ms. Camille (c_3), and Mrs. Lewis (c_4). Miss Carelli is certified in mathematics and computer science; Mr. Ritter in mathematics and physics; Ms. Camille in biology; and Mrs. Lewis in chemistry, physics, and computer science. If the district hires all four candidates, can each teacher be assigned to teach a (different) subject in which he or she is certified?

This problem is an example of a general situation called the *assignment problem*. Using the Principle of Inclusion and Exclusion in conjunction with the rook polynomial (see Sections 8.4 and 8.5), one can determine in how many ways, if any, the four teachers may be assigned so that each teaches a different subject for which he or she is qualified. However, these techniques do not provide a means of setting up any of these assignments. In Fig. 13.24 the problem is modeled by means of a bipartite graph $G = (V, E)$, where V is partitioned as $X \cup Y$ with $X = \{c_1, c_2, c_3, c_4\}$ and $Y = \{s_1, s_2, s_3, s_4, s_5\}$, and the edges of G represent the qualifications for the individual teachers. The edges $\{c_1, s_2\}, \{c_2, s_4\}, \{c_3, s_5\}, \{c_4, s_3\}$ demonstrate such an assignment of X into Y.

To examine this idea further, the following concepts are introduced.

Definition 13.9   Let $G = (V, E)$ be a bipartite graph with V partitioned as $X \cup Y$. (Each edge of E has the form $\{x, y\}$ with $x \in X$ and $y \in Y$.)

    a) A *matching* in G is a subset of E such that no two edges share a common vertex in X or Y.

b) A *complete matching* of X into Y is a matching in G such that every $x \in X$ is the endpoint of an edge.

In terms of functions, a matching is a function that establishes a one-to-one correspondence between a subset of X and a subset of Y. When the matching is complete, a one-to-one function from X into Y is defined. The example in Fig. 13.24 contains such a function and a complete matching.

For a bipartite graph $G = (V, E)$ with V partitioned as $X \cup Y$, a complete matching of X into Y requires $|X| \le |Y|$. If $|X|$ is large, then the construction of such a matching cannot be accomplished just by observation or trial and error. The following theorem, due to the English mathematician Philip Hall (1935), provides a necessary and sufficient condition for the existence of such a matching. The proof of the theorem, however, is not that given by Hall. A constructive proof that uses the material developed on transport networks is given.

THEOREM 13.7   Let $G = (V, E)$ be bipartite with V partitioned as $X \cup Y$. A complete matching of X into Y exists if and only if for every subset A of X, $|A| \le |R(A)|$, where $R(A)$ is the subset of Y consisting of those vertices each of which is adjacent to at least one vertex in A.

Before proving the theorem, we illustrate its use in the following example.

EXAMPLE 13.15

a) The bipartite graph shown in Fig. 13.25(a) has no complete matching. Any attempt to construct such a matching must include $\{x_1, y_1\}$ and either $\{x_2, y_3\}$ or $\{x_3, y_3\}$. If $\{x_2, y_3\}$ is included, there is no match for x_3. Likewise, if $\{x_3, y_3\}$ is included, we are not able to match x_2. If $A = \{x_1, x_2, x_3\} \subseteq X$, then $R(A) = \{y_1, y_3\}$. With $|A| = 3 > 2 = |R(A)|$, it follows from Theorem 13.7 that no complete matching can exist.

Table 13.2

A	$R(A)$	$|A|$	$|R(A)|$
$\emptyset$	$\emptyset$	0	0
$\{x_1\}$	$\{y_1, y_2, y_3\}$	1	3
$\{x_2\}$	$\{y_2\}$	1	1
$\{x_3\}$	$\{y_2, y_3, y_5\}$	1	3
$\{x_4\}$	$\{y_4, y_5\}$	1	2
$\{x_1, x_2\}$	$\{y_1, y_2, y_3\}$	2	3
$\{x_1, x_3\}$	$\{y_1, y_2, y_3, y_4\}$	2	4
$\{x_1, x_4\}$	Y	2	5
$\{x_2, x_3\}$	$\{y_2, y_3, y_5\}$	2	3
$\{x_2, x_4\}$	$\{y_2, y_4, y_5\}$	2	3
$\{x_3, x_4\}$	$\{y_2, y_3, y_4, y_5\}$	2	4
$\{x_1, x_2, x_3\}$	$\{y_1, y_2, y_3, y_5\}$	3	4
$\{x_1, x_2, x_4\}$	Y	3	5
$\{x_1, x_3, x_4\}$	Y	3	5
$\{x_2, x_3, x_4\}$	$\{y_2, y_3, y_4, y_5\}$	3	4
X	Y	4	5

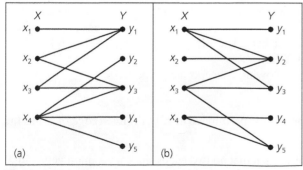

Figure 13.25

b) For the graph in part (b) of the figure, consider the exhaustive listing in Table 13.2. Assuming the validity of Theorem 13.7, this listing indicates that the graph contains a complete matching.

We turn now to a proof of the theorem.

Proof: With V partitioned as $X \cup Y$, let $X = \{x_1, x_2, \ldots, x_m\}$ and $Y = \{y_1, y_2, \ldots, y_n\}$. Construct a transport network N that extends graph G by introducing two new vertices a (the source) and z (the sink). For each vertex x_i, $1 \le i \le m$, draw edge (a, x_i); for each vertex y_j, $1 \le j \le n$, draw edge (y_j, z). Each new edge is given a capacity of 1. Let M be any positive integer that exceeds $|X|$. Assign each edge in G the capacity M. The original graph G and its associated network N appear as shown in Fig. 13.26. It follows that a complete matching exists in G if and only if there is a maximum flow in N that uses all edges (a, x_i), $1 \le i \le m$. Then the value of such a maximum flow is $m = |X|$.

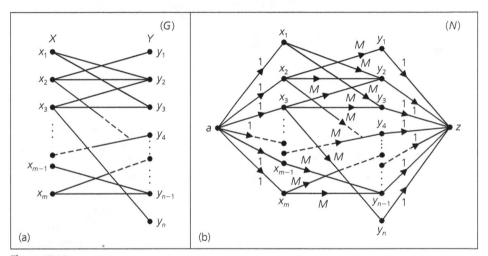

Figure 13.26

We shall prove that there is a complete matching in G by showing that $c(P, \overline{P}) \ge |X|$ for each cut $(P, \overline{P})$ in N. So if $(P, \overline{P})$ is an arbitrary cut in the transport network N, let us define $A = X \cap P$ and $B = Y \cap P$. Then $A \subseteq X$ where we shall write $A = \{x_1, x_2, \ldots, x_i\}$ for some $0 \le i \le m$. (The elements of X are relabeled, if necessary, so that the subscripts on the elements of A are consecutive. When $i = 0$, $A = \emptyset$.) Now P consists of the source a together with the vertices in A and the set $B \subseteq Y$, as shown in Fig. 13.27(a). (Elements of Y are also relabeled if necessary.) In addition, $\overline{P} = (X - A) \cup (Y - B) \cup \{z\}$. If there is an edge $\{x, y\}$ with $x \in A$ and $y \in (Y - B)$, then the capacity of that edge is a summand in $c(P, \overline{P})$ and $c(P, \overline{P}) \ge M > |X|$. Should no such edge exist, then $c(P, \overline{P})$ is determined by the capacities of (1) the edges from the source a to the vertices in $X - A$ and (2) the edges from the vertices in B to the sink z. Since each of these edges has capacity 1, $c(P, \overline{P}) = |X - A| + |B| = |X| - |A| + |B|$. With $B \supseteq R(A)$, we have $|B| \ge |R(A)|$, and since $|R(A)| \ge |A|$, it follows that $|B| \ge |A|$. Consequently, $c(P, \overline{P}) = |X| + (|B| - |A|) \ge |X|$. Therefore, since every cut in network N has capacity at least $|X|$ and the cut $(\{a\}, V - \{a\})$ achieves a capacity of $|X|$, by Theorem 13.6 any maximum flow for N has value $|X|$. Such a flow will result in exactly $|X|$ edges from X to Y having flow 1, and this flow provides a complete matching of X into Y.

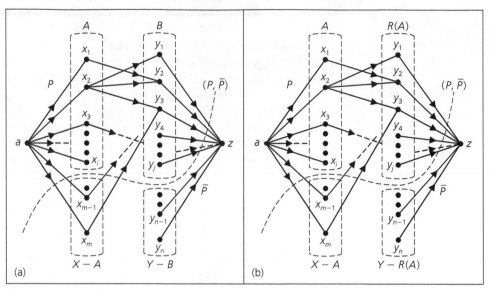

Figure 13.27

Conversely, suppose that there exists a subset A of X where $|A| > |R(A)|$. Let $(P, \overline{P})$ be the cut shown for the network in Fig. 13.27(b), with $P = \{a\} \cup A \cup R(A)$ and $\overline{P} = (X - A) \cup (Y - R(A)) \cup \{z\}$. Then $c(P, \overline{P})$ is determined by (1) the edges from the source a to the vertices in $X - A$ and (2) the edges from the vertices in $R(A)$ to the sink z. Hence $c(P, \overline{P}) = |X - A| + |R(A)| = |X| - (|A| - |R(A)|) < |X|$, since $|A| > |R(A)|$. The network has a cut of capacity less than $|X|$, so once again by Theorem 13.6 it follows that any maximum flow in the network has value smaller than $|X|$. Therefore there is no complete matching from X into Y for the given bipartite graph G.

EXAMPLE 13.16

Five students, s_1, s_2, s_3, s_4, and s_5, are members of three committees, c_1, c_2, and c_3. The bipartite graph shown in Fig. 13.28(a) indicates the committee memberships. Each committee is to select a student representative to meet with the school president. Can a selection be made in such a way that each committee has a distinct representative?

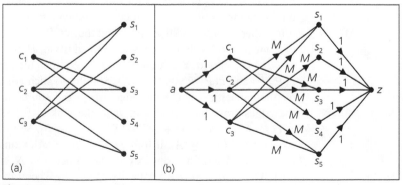

Figure 13.28

Although this problem is small enough to solve by inspection, we use the ideas developed in Section 13.3. Figure 13.28(b) provides the network for the given bipartite graph. Here we consider the vertices, other than the source a, ordered as $c_1, c_2, c_3, s_1, s_2, s_3, s_4, s_5, z$. In Fig. 13.29(a), the Edmonds-Karp algorithm is applied for the first time and provides the f-augmenting path $p: (a, c_1), (c_1, s_3), (s_3, z)$ with $\Delta_p = 1$. Applying the Ford-Fulkerson algorithm results in the network in part (b) of the figure, and this network indicates the edge (c_1, s_3) as the start for a possible complete matching. [Many edge labels are omitted in parts (b) and (c) of the figure in order to simplify the diagrams. Every unlabeled edge that starts at a or terminates at z should have the label 1, 0 to indicate a capacity of 1 and a flow of 0; all other unlabeled edges should bear the label M, 0.] The next application of these two algorithms provides the f-augmenting path $(a, c_2), (c_2, s_1), (s_1, z)$ and the edge (c_2, s_1) to extend the matching. Finally, the last application of the Edmonds-Karp and Ford-Fulkerson algorithms gives us the f-augmenting path $(a, c_3), (c_3, s_2), (s_2, z)$ and the final edge — namely, (c_3, s_2) — for the complete matching. This is indicated by the maximum flow in part (c) of Fig. 13.29.

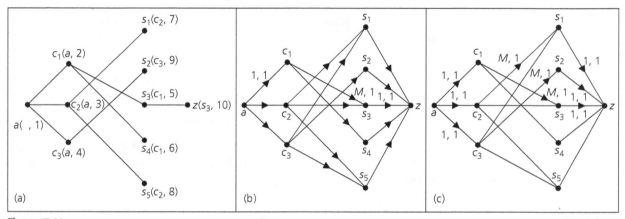

Figure 13.29

This example is a particular instance of a problem studied by Philip Hall. He considered a collection of sets $A_1, A_2, \ldots, A_n$, where the elements $a_1, a_2, \ldots, a_n$ were called a *system of distinct representatives* for the collection if (a) $a_i \in A_i$, for all $1 \le i \le n$; and (b) $a_i \ne a_j$, whenever $1 \le i < j \le n$. Rewording Theorem 13.7 in this context, we find that the collection $A_1, A_2, \ldots, A_n$ has a system of distinct representatives if and only if, for all $1 \le i \le n$, the union of any i of the sets $A_1, A_2, \ldots, A_n$ contains at least i elements.

Although the condition in Theorem 13.7 may be very tedious to check, the following corollary provides a sufficient condition for the existence of a complete matching.

COROLLARY 13.6

Let $G = (V, E)$ be a bipartite graph with V partitioned as $X \cup Y$. There is a complete matching of X into Y if, for some $k \in \mathbf{Z}^+$, $\deg(x) \ge k \ge \deg(y)$ for all vertices $x \in X$ and $y \in Y$.

Proof: This proof is left for the Section Exercises.

EXAMPLE 13.17

a) Corollary 13.6 is applicable to the graph shown in Fig. 13.28(a). Here the appropriate value of k is 2.

b) There are 50 students (25 females and 25 males) in the senior class at Bell High School. If each female in the class is appreciated by exactly five of the males, and each male enjoys the company of exactly five of the females in the class, then it is possible for each male to go to the class party with a female he likes and each female will attend with a male who likes her. (As a result of problems of this type, the condition in Theorem 13.7 has often been referred to in the literature as *Hall's Marriage Condition*.)

For problems such as the one in Example 13.15(a), where a complete matching does not exist, the following type of matching is often of interest.

Definition 13.10

If $G = (V, E)$ is a bipartite graph with V partitioned as $X \cup Y$, a *maximal matching* in G is one that matches as many vertices in X as possible with the vertices in Y.

To investigate the existence and construction of a maximal matching, the following new idea is presented.

Definition 13.11

Let $G = (V, E)$ be a bipartite graph, where V is partitioned as $X \cup Y$. If $A \subseteq X$, then $\delta(A) = |A| - |R(A)|$ is called the *deficiency of* A. The *deficiency of graph* G, denoted $\delta(G)$, is given by $\delta(G) = \max\{\delta(A)|A \subseteq X\}$.

For $\emptyset \subseteq X$, we have $R(\emptyset) = \emptyset$, so $\delta(\emptyset) = 0$ and $\delta(G) \geq 0$. If $\delta(G) > 0$, there is a subset A of X with $|A| - |R(A)| > 0$, so $|A| > |R(A)|$ and from Theorem 13.7 we know that there is no complete matching of X into Y.

EXAMPLE 13.18

The graph in Fig. 13.30(a) has no complete matching. [See Example 13.15(a).] For $A = \{x_1, x_2, x_3\}$, we find that $R(A) = \{y_1, y_3\}$ and $\delta(A) = 3 - 2 = 1$. As a result of this subset A we find that $\delta(G) = 1$. Removing one of the vertices from A (and the edges incident with it), we obtain the subgraph shown in part (b) of the figure. This (bipartite) subgraph contains a complete matching from $X_1 = \{x_2, x_3, x_4\}$ into Y. The edges $\{x_2, y_1\}$, $\{x_3, y_3\}$, and $\{x_4, y_4\}$ indicate one such matching that is also a maximal matching of X into Y.

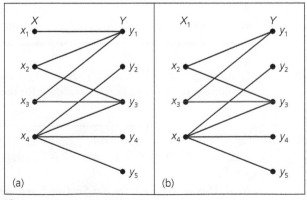

(a)   (b)

Figure 13.30

The ideas developed in Example 13.18 lead to the following theorem.

THEOREM 13.8

Let $G = (V, E)$ be bipartite with V partitioned as $X \cup Y$. The maximum number of vertices in X that can be matched with those in Y is $|X| - \delta(G)$.

Proof: We provide a constructive proof, using transport networks as in the proof of Theorem 13.7. As in Figure 13.26, let N be the network associated with the bipartite graph G. The result will follow when we show that (a) the capacity of every cut $(P, \overline{P})$ in N is greater than or equal to $|X| - \delta(G)$, and (b) there exists a cut with capacity $|X| - \delta(G)$.

Let $(P, \overline{P})$ be a cut in N, where P is made up of the source a, the vertices in $A = P \cap X \subseteq X$, and the vertices in $B = P \cap Y \subseteq Y$. [See Fig. 13.27(a).] As in the proof of Theorem 13.7, the subsets A, B may be $\emptyset$.

1) If edge (x, y) is in N with $x \in A$ and $y \in Y - B$, then $c(x, y)$ is a summand in $c(P, \overline{P})$. Since $c(x, y) = M > |X|$, it follows that $c(P, \overline{P}) > |X| \geq |X| - \delta(G)$.

2) If no such edge as in (1) exists, then $c(P, \overline{P})$ is determined by the $|X - A|$ edges from a to $X - A$ and the $|B|$ edges from B to z. Since each of these edges has capacity 1, we find that $c(P, \overline{P}) = |X - A| + |B| = |X| - |A| + |B|$. No edge connects a vertex in A with a vertex in $Y - B$, so $R(A) \subseteq B$ and $|R(A)| \leq |B|$. Consequently, $c(P, \overline{P}) = (|X| - |A|) + |B| \geq (|X| - |A|) + |R(A)| = |X| - (|A| - |R(A)|) = |X| - \delta(A) \geq |X| - \delta(G)$.

Therefore, in either case, $c(P, \overline{P}) \geq |X| - \delta(G)$ for every cut $(P, \overline{P})$ in N.

To complete the proof, we must establish the existence of a cut with capacity $|X| - \delta(G)$. Since $\delta(G) = \max\{\delta(A) | A \subseteq X\}$, we can select a subset A of X with $\delta(G) = \delta(A)$. Examining Fig. 13.27(b), we let $P = \{a\} \cup A \cup R(A)$. Then $\overline{P} = (X - A) \cup (Y - R(A)) \cup \{z\}$. There is no edge between the vertices in A and those in $Y - R(A)$, so $c(P, \overline{P}) = |X - A| + |R(A)| = |X| - (|A| - |R(A)|) = |X| - \delta(A) = |X| - \delta(G)$.

We close this section with an example that deals with these concepts.

EXAMPLE 13.19

Let $G = (V, E)$ be bipartite with V partitioned as $X \cup Y$. For each $x \in X$, $\deg(x) \geq 4$ and, for each $y \in Y$, $\deg(y) \leq 5$. If $|X| \leq 15$, find an upper bound (as small as possible) for $\delta(G)$.

Let $\emptyset \neq A \subseteq X$ and let $E_1 \subseteq E$, where $E_1 = \{\{a, b\} | a \in A, b \in R(A)\}$. Since $\deg(a) \geq 4$ for all $a \in A$, $|E_1| \geq 4|A|$. With $\deg(b) \leq 5$ for all $b \in R(A)$, $|E_1| \leq 5|R(A)|$. Hence $4|A| \leq 5|R(A)|$ and $\delta(A) = |A| - |R(A)| \leq |A| - (4/5)|A| = (1/5)|A|$. Since $A \subseteq X$, we have $|A| \leq 15$, so $\delta(A) \leq (1/5)(15) = 3$. Consequently, $\delta(G) = \max\{\delta(A) | A \subseteq X\} \leq 3$, so there exists a maximal matching M of X into Y such that $|M| \geq |X| - 3$.

EXERCISES 13.4

1. For the graph shown in Fig. 13.24, if four edges are selected at random, what is the probability that they provide a complete matching of X into Y?

2. Cathy is liked by Albert, Joseph, and Robert; Janice by Joseph and Dennis; Theresa by Albert and Joseph; Nettie by Dennis, Joseph, and Frank; and Karen by Albert, Joseph, and Robert. (a) Set up a bipartite graph to model the matching problem where each man is paired with a woman he likes. (b) Draw

the associated network for the graph in part (a) and determine a maximum flow for this network. What complete matching does this determine? (c) Is there a complete matching that pairs Janice with Dennis and Nettie with Frank? (d) Is it possible to determine two complete matchings where each man is paired with two different women?

3. At Rydell High School the senior class is represented on six school committees by Annemarie (A), Gary (G), Jill (J), Kenneth (K), Michael (M), Norma (N), Paul (P), and Rosemary (R). The senior members of these committees are $\{A, G, J, P\}$,

{G, J, K, R}, {A, M, N, P}, {A, G, M, N, P}, {A, G, K, N, R}, and {G, K, N, R}. **(a)** The student government calls a meeting that requires the presence of exactly one senior member from each committee. Find a selection that maximizes the number of seniors involved. **(b)** Before the meeting, the finances of each committee are to be reviewed by a senior who is not on that committee. Can this be accomplished so that six different seniors are involved in this review process? If so, how?

4. Let $G = (V, E)$ be a bipartite graph with V partitioned as $X \cup Y$, where $X = \{x_1, x_2, \ldots, x_m\}$ and $Y = \{y_1, y_2, \ldots, y_n\}$. How many complete matchings of X into Y are there if

  a) $m = 2$, $n = 4$, and $G = K_{m,n}$?

  b) $m = 4$, $n = 4$, and $G = K_{m,n}$?

  c) $m = 5$, $n = 9$, and $G = K_{m,n}$?

  d) $m \le n$ and $G = K_{m,n}$?

5. If $G = (V, E)$ is an undirected graph, a spanning subgraph H of G in which each vertex has degree 1 is called a *one-factor* (or *perfect matching*) for G.

  a) If G has a one-factor, prove that $|V|$ is even.

  b) Does the Petersen graph have a one-factor? (The Petersen graph was first introduced in Example 11.19.)

  c) In Fig. 13.31 we find the graph K_4 in part (a), while part (b) provides the three possible one-factors for K_4. How many one-factors are there for the graph K_6?

  d) For $n \in \mathbf{Z}^+$, let a_n count the number of one-factors that exist for the graph K_{2n}. Find and solve a recurrence relation for a_n.

6. Prove Corollary 13.6.

7. Fritz is in charge of assigning students to part-time jobs at the college where he works. He has 25 student applications, and there are 25 different part-time jobs available on the campus. Each applicant is qualified for at least four of the jobs, but each job can be performed by at most four of the applicants. Can Fritz assign all the students to jobs for which they are qualified? Explain.

8. For each of the following collections of sets, determine, if possible, a system of distinct representatives. If no such system exists, explain why.

  a) $A_1 = \{2, 3, 4\}$, $A_2 = \{3, 4\}$, $A_3 = \{1\}$, $A_4 = \{2, 3\}$

  b) $A_1 = A_2 = A_3 = \{2, 4, 5\}$, $A_4 = A_5 = \{1, 2, 3, 4, 5\}$

  c) $A_1 = \{1, 2\}$, $A_2 = \{2, 3, 4\}$, $A_3 = \{2, 3\}$, $A_4 = \{1, 3\}$, $A_5 = \{2, 4\}$

9. a) Determine all systems of distinct representatives for the collection of sets $A_1 = \{1, 2\}$, $A_2 = \{2, 3\}$, $A_3 = \{3, 4\}$, $A_4 = \{4, 1\}$.

  b) Given the collection of sets $A_1 = \{1, 2\}$, $A_2 = \{2, 3\}, \ldots, A_n = \{n, 1\}$, determine how many different systems of distinct representatives exist for the collection.

10. Let $A_1, A_2, \ldots, A_n$ be a collection of sets, where $A_1 = A_2 = \cdots = A_n$ and $|A_i| = k > 0$ for all $1 \le i \le n$. **(a)** Prove that the given collection has a system of distinct representatives if and only if $n \le k$. **(b)** When $n \le k$, how many different systems exist for the collection?

11. Let $G = (V, E)$ be a bipartite graph, where V is partitioned as $X \cup Y$. If $\deg(x) \ge 4$ for all $x \in X$ and $\deg(y) \le 5$ for all $y \in Y$, prove that if $|X| \le 10$ then $\delta(G) \le 2$.

12. Let $G = (V, E)$ be bipartite with V partitioned as $X \cup Y$. For all $x \in X$, $\deg(x) \ge 3$, and for all $y \in Y$, $\deg(y) \le 7$. If $|X| \le 50$, find an upper bound (that is as small as possible) on $\delta(G)$.

13. a) Let $G = (V, E)$ be the bipartite graph shown in Fig. 13.32, with V partitioned as $X \cup Y$. Determine $\delta(G)$ and a maximal matching of X into Y.

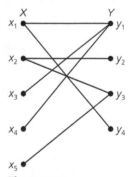

Figure 13.32

  b) For any bipartite graph $G = (V, E)$, with V partitioned as $X \cup Y$, if $\beta(G)$ denotes the independence number of G, show that $|Y| = \beta(G) - \delta(G)$. (The independence number

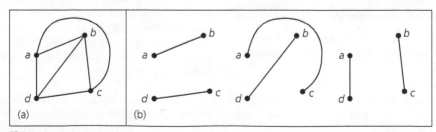

(a)   (b)

Figure 13.31

of an undirected graph is defined in Exercise 25 for Section 11.5.)

c) Determine a largest maximal independent set of vertices for the graphs shown in Fig. 13.30(a) and Fig. 13.32.

14. For $n \geq 2$, prove that the hypercube Q_n has at least $2^{(2^{n-2})}$ perfect matchings (as defined above in Exercise 5).

13.5
Summary and Historical Review

This chapter has provided us with a sample of the ways in which graph theory enters into an area of mathematics called operations research. Each topic was presented in an algorithmic manner that can be used in the computer implementation needed for solving each type of problem. Comparable coverage of this material can be found in Chapters 10 and 11 of the text by C. L. Liu [22]. Chapters 4 and 5 of E. Lawler [21] offer an extensive coverage of many other developments on networks and matching. This text provides a wide variety of applications and includes references for additional reading.

In Section 13.1 we examined a shortest-path algorithm for weighted graphs. The full development of the algorithm is given in the article by E. W. Dijkstra [10].

Edsger W. Dijkstra (1930–2002)

Joseph B. Kruskal (1928–)

Section 13.2 provided two techniques for finding a minimal spanning tree in a weighted loop-free connected undirected graph. These techniques were developed in the late 1950s by J. B. Kruskal [20] and R. C. Prim [25]. Actually, however, methods for constructing minimal spanning trees can be traced back to 1926, to the work of Otakar Borůvka dealing with the construction of an electric power network. Even before this (1909–1911) the anthropologist Jan Czekanowski, in his work on various classification schemes, was very close to recognizing the minimal spanning tree problem and to providing a greedy algorithm for its solution. The survey paper by R. L. Graham and P. Hell [16] mentions the contributions made by Borůvka and Czekanowski and gives more information on the history and applications of this structure.

The computer implementation of all the techniques given in the first two sections can be found in Chapters 6 and 7 of A. V. Aho, J. E. Hopcroft, and J. D. Ullman [1]; in Chapter 8 of S. Baase and A. Van Gelder [3]; in Chapters 23 and 24 of T. H. Cormen, C. E. Leiserson,

R. L. Rivest, and C. Stein [7]; and in Chapter 4 of E. Horowitz and S. Sahni [17]. These references also discuss the efficiency and speed of these algorithms. Sections 4.5–4.9 of the text by R. K. Ahuja, T. L. Magnanti, and J. B. Orlin [2] provide more on different implementations of Dijkstra's algorithm, along with discussions on their features and worst-case time-complexities. Six applications of the algorithm are described in Section 4.2 of this text. As we mentioned at the end of Section 13.2, the articles by R. L. Graham and P. Hell [16], by D. B. Johnson [18], and by A. Kershenbaum and R. Van Slyke [19] discuss other implementations of Prim's algorithm. An interesting application of the concept of the minimal spanning tree in a physical science setting is provided in the article by D. R. Shier [27]. Other applications are discussed in Section 13.2 of R. K. Ahuja, T. L. Magnanti, and J. B. Orlin [2].

As we noted in Section 13.3, problems dealing with the allocation of resources or the shipment of goods can be modeled by means of transport networks. The fundamental work by G. B. Dantzig, L. R. Ford, and D. R. Fulkerson can be found in their pioneering articles [8, 9, 12, 13]. The classic text by L. R. Ford and D. R. Fulkerson [14] provides excellent coverage of this topic. In addition, the reader may wish to examine Chapter 6 of R. K. Ahuja, T. L. Magnanti, and J. B. Orlin [2], Chapter 8 of the text by C. Berge [4], Chapter 7 of the book by R. G. Busacker and T. L. Saaty [6], or Chapter 26 of T. H. Cormen, C. E. Leiserson, R. L. Rivest and C. Stein [7]. Chapter 10 in C. L. Liu [22] includes coverage on an extension to networks wherein the flow in each edge is restricted by a lower as well as an upper capacity. For more applications the reader should examine the article by D. R. Fulkerson on pages 139–171 of [15]. Section 6.2 of R. K. Ahuja, T. L. Magnanti, and J. B. Orlin [2] contains six additional applications.

The last topic discussed here dealt with matching in a bipartite graph. The theory behind this was first developed by Philip Hall in 1935, but here the ideas on transport networks were used to provide an algorithm for a solution. Chapter 7 of the text by O. Ore [24] provides a very readable introduction to this topic, along with some applications. For more on systems of representatives, the reader should examine Chapter 5 of the monograph by H. J. Ryser [26]. A second method for finding a maximal matching in a bipartite graph is called the *Hungarian method*. This is given in Chapter 5 of the text by J. A. Bondy and U. S. R. Murty [5] and in Chapter 10 of the book by C. Berge [4]. In addition to its application in solving the assignment problem, matching theory has many interesting combinatorial implications. One may learn more about these in the survey article by L. Mirsky and H. Perfect [23].

REFERENCES

1. Aho, Alfred V., Hopcroft, John E., and Ullman, Jeffrey D. *Data Structures and Algorithms*. Reading, Mass.: Addison-Wesley, 1983.

2. Ahuja, Ravindra K., Magnanti, Thomas L., and Orlin, James B. *Network Flows*. Englewood Cliffs, N.J.: Prentice Hall, 1993.

3. Baase, Sara, and Van Gelder, Allen. *Computer Algorithms, Introduction to Design and Analysis*, 3rd ed. Reading Mass.: Addison-Wesley, 2000.

4. Berge, Claude. *The Theory of Graphs and Its Applications*. New York: Wiley, 1962.

5. Bondy, J. A., and Murty, U. S. R. *Graph Theory with Applications*. New York: Elsevier North Holland, 1976.

6. Busacker, Robert G., and Saaty, Thomas L. *Finite Graphs and Networks*. New York: McGraw-Hill, 1965.

7. Cormen, Thomas H., Leiserson, Charles E., Rivest, Ronald L., and Stein, Clifford. *Introduction to Algorithms*, 2nd ed. New York: McGraw-Hill, 2001.

8. Dantzig, George B., and Fulkerson, Delbert Ray. *Computation of Maximal Flows in Networks*. The RAND Corporation, P-677, 1955.

9. Dantzig, George B., and Fulkerson, Delbert Ray. *On the Max Flow Min Cut Theorem.* The RAND Corporation, RM-1418-1, 1955.

10. Dijkstra, Edsger W. "A Note on Two Problems in Connexion with Graphs." *Numerische Mathematik* 1 (1959): pp. 269–271.

11. Edmonds, Jack, and Karp, Richard M. "Theoretical Improvements in Algorithmic Efficiency for Network Flow Problems." *J. Assoc. Comput. Mach.* 19 (1972): pp. 248–264.

12. Ford, Lester R., Jr. *Network Flow Theory.* The RAND Corporation, P-923, 1956.

13. Ford, Lester R., Jr., and Fulkerson, Delbert Ray. "Maximal Flow Through a Network." *Canadian Journal of Mathematics* 8 (1956): pp. 399–404.

14. Ford, Lester R., Jr., and Fulkerson, Delbert Ray. *Flows in Networks.* Princeton, N.J.: Princeton University Press, 1962.

15. Fulkerson, Delbert Ray, ed. *Studies in Graph Theory*, Part I. *MAA Studies in Mathematics*, Vol. 11, The Mathematical Association of America, 1975.

16. Graham, Ronald L., and Hell, Pavol. "On the History of the Minimum Spanning Tree Problem." *Annals of the History of Computing* 7, no. 1 (January 1985): pp. 43–57.

17. Horowitz, Ellis, and Sahni, Sartaj. *Fundamentals of Computer Algorithms.* Potomac, Md.: Computer Science Press, 1978.

18. Johnson, D. B. "Priority Queues with Update and Minimum Spanning Trees." *Information Processing Letters* 4 (1975): pp. 53–57.

19. Kershenbaum, A., and Van Slyke, R. "Computing Minimum Spanning Trees Efficiently." *Proceedings of the Annual ACM Conference*, 1972, pp. 518–527.

20. Kruskal, Joseph B. "On the Shortest Spanning Subtree of a Graph and the Traveling Salesman Problem." *Proceedings of the AMS* 1, no. 1 (1956): pp. 48–50.

21. Lawler, Eugene. *Combinatorial Optimization: Networks and Matroids.* New York: Holt, 1976.

22. Liu, C. L. *Introduction to Combinatorial Mathematics.* New York: McGraw-Hill, 1968.

23. Mirsky, L., and Perfect, H. "Systems of Representatives." *Journal of Mathematical Analysis and Applications* 3 (1966): pp. 520–568.

24. Ore, Oystein. *Theory of Graphs.* Providence, R.I.: American Mathematical Society, 1962.

25. Prim, Robert C. "Shortest Connection Networks and Some Generalizations." *Bell System Technical Journal* 36 (1957): pp. 1389–1401.

26. Ryser, Herbert J. *Combinatorial Mathematics.* Carus Mathematical Monographs, Number 14, Mathematical Association of America, 1963.

27. Shier, Douglas R. "Testing for Homogeneity Using Minimum Spanning Trees." *The UMAP Journal* 3, no. 3 (1982): pp. 273–283.

SUPPLEMENTARY EXERCISES

1. Apply Dijkstra's algorithm to the weighted directed multigraph shown in Fig. 13.33, and find the shortest distance from vertex a to the other seven vertices in the graph.

2. For her class in the analysis of algorithms, Stacy writes the following algorithm to determine the shortest distance from a vertex a to a vertex b in a weighted directed graph $G = (V, E)$.

Step 1: Set Distance equal to 0, assign vertex a to the variable v, and let $T = V$.

Step 2: If $v = b$, the value of Distance is the answer to the problem. If $v \neq b$, then

    1) Replace T by $T - \{v\}$ and select $w \in T$ with wt(v, w) minimal.

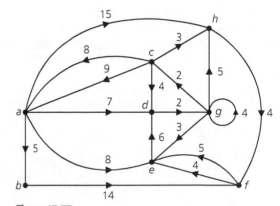

Figure 13.33

    2) Set Distance equal to Distance + wt(v, w).

    3) Assign vertex w to the variable v and then return to step (2).

Is Stacy's algorithm correct? If so, prove it. If not, provide a counterexample.

3. a) Let $G = (V, E)$ be a loop-free weighted connected undirected graph. If $e_1 \in E$ with $\text{wt}(e_1) < \text{wt}(e)$ for all other edges $e \in E$, prove that edge e_1 is part of every minimal spanning tree for G.

b) With G as in part (a), suppose that there are edges $e_1, e_2 \in E$ with $\text{wt}(e_1) < \text{wt}(e_2) < \text{wt}(e)$ for all other edges $e \in E$. Prove or disprove: Edge e_2 is part of every minimal spanning tree for G.

4. a) Let $G = (V, E)$ be a loop-free weighted connected undirected graph where each edge e of G is part of a cycle. Prove that if $e_1 \in E$ with $\text{wt}(e_1) > \text{wt}(e)$ for all other edges $e \in E$, then no spanning tree for G that contains e_1 can be minimal.

b) With G as in part (a), suppose that $e_1, e_2 \in E$ with $\text{wt}(e_1) > \text{wt}(e_2) > \text{wt}(e)$ for all other edges $e \in E$. Prove or disprove: Edge e_2 is not part of any minimal spanning tree for G.

5. Using the concept of flow in a transport network, construct a directed multigraph $G = (V, E)$, with $V = \{u, v, w, x, y\}$ and $id(u) = 1$, $od(u) = 3$; $id(v) = 3$, $od(v) = 3$; $id(w) = 3$, $od(w) = 4$; $id(x) = 5$, $od(x) = 4$; and $id(y) = 4$, $od(y) = 2$.

6. A set of words $\{qs, tq, ut, pqr, srt\}$ is to be transmitted using a binary code for each letter. (a) Show that it is possible to select one letter from each word as a system of distinct representatives for these words. (b) If a letter is selected at random from each of the five words, what is the probability that the selection is a system of distinct representatives for the words?

7. For $n \in \mathbf{Z}^+$ and for each $1 \le i \le n$, let $A_i = \{1, 2, 3, \ldots, n\} - \{i\}$. How many different systems of distinct representatives exist for the collection $A_1, A_2, A_3, \ldots, A_n$?

8. This exercise outlines a proof of the Birkhoff-von Neumann Theorem.

a) For $n \in \mathbf{Z}^+$, an $n \times n$ matrix is called a *permutation* matrix if there is exactly one 1 in each row and column, and all other entries are 0. How many 5×5 permutation matrices are there? How many $n \times n$?

b) An $n \times n$ matrix B is called *doubly stochastic* if $b_{ij} \ge 0$ for all $1 \le i \le n$, $1 \le j \le n$, and the sum of the entries in each row or column is 1. If

$$B = \begin{bmatrix} 0.2 & 0.1 & 0.7 \\ 0.4 & 0.5 & 0.1 \\ 0.4 & 0.4 & 0.2 \end{bmatrix},$$

verify that B is doubly stochastic.

c) Find four positive real numbers c_1, c_2, c_3, and c_4, and four permutation matrices P_1, P_2, P_3, and P_4, such that $c_1 + c_2 + c_3 + c_4 = 1$ and $B = c_1 P_1 + c_2 P_2 + c_3 P_3 + c_4 P_4$.

d) Part (c) is a special case of the Birkhoff-von Neumann Theorem: If B is an $n \times n$ doubly stochastic matrix, then there exist positive real numbers $c_1, c_2, \ldots, c_k$ and permutation matrices $P_1, P_2, \ldots, P_k$ such that $\sum_{i=1}^{k} c_i = 1$ and $\sum_{i=1}^{k} c_i P_i = B$. To prove this result, proceed as follows: Construct a bipartite graph $G = (V, E)$ with V partitioned as $X \cup Y$, where $X = \{x_1, x_2, \ldots, x_n\}$ and $Y = \{y_1, y_2, \ldots, y_n\}$. The vertex x_i, for all $1 \le i \le n$, corresponds with the ith row of B; the vertex y_j, for all $1 \le j \le n$, corresponds with the jth column of B. The edges of G are of the form $\{x_i, y_j\}$ if and only if $b_{ij} > 0$. We claim that there is a complete matching of X into Y.

If not, there is a subset A of X with $|A| > |R(A)|$. That is, there is a set of r rows of B having positive entries in s columns and $r > s$. What is the sum of these r rows of B? Yet the sum of these same entries, when added column by column, is less than or equal to s. (Why?) Consequently, we have a contradiction.

As a result of the complete matching of X into Y, there are n positive entries in B that occur so that no two are in the same row or column. (Why?) If c_1 is the smallest of these entries, then we may write $B = c_1 P_1 + B_1$, where P_1 is an $n \times n$ permutation matrix wherein the 1's are located according to the positive entries in B that came about from the complete matching. What are the sums of the entries in the rows and columns of B_1?

e) How is the proof completed?

9. Let $G = (V, E)$ be a bipartite graph with V partitioned as $X \cup Y$. If $E' \subseteq E$, and E' determines a complete matching of X into Y, what property do the vertices determined by E' in the line graph $L(G)$ have? [The line graph $L(G)$ for a loop-free undirected graph G is defined in Supplementary Exercise 18 of Chapter 11.]

MODERN APPLIED ALGEBRA

14

Rings and Modular Arithmetic

In this fourth and final part of the text, the emphasis will be on structure again as we begin the investigation of sets of elements that are closed under two binary operations. The concepts of structure and enumeration often reinforce each other. Here we will see this occur as ideas seen in Chapters 1, 4, 5, and 8 come to the forefront again.

When we examined the set **Z** in Chapter 4, it was in conjunction with the closed binary operations of addition and multiplication. In this chapter we emphasize these operations by writing $(\mathbf{Z}, +, \cdot)$, instead of just **Z**. Patterned after some of the properties of $(\mathbf{Z}, +, \cdot)$, the algebraic structure called a *ring* will be defined. Without knowing it, we have been dealing with rings in many mathematical settings. Now we shall be concerned with finite rings that arise in number theory and computer science applications. Of particular interest in the study of computer science is the *hashing function*, which we find provides a means of identifying records stored in a table.

14.1
The Ring Structure:
Definition and Examples

We start by defining the ring structure, realizing as we do that most abstract definitions, like theorems, come about from a study of many examples where one recognizes the common idea or ideas present in what may seem to be a collection of unrelated objects.

Definition 14.1 Let R be a nonempty set on which we have two closed binary operations, denoted by $+$ and $\cdot$ (which may be quite different from the ordinary addition and multiplication to which we are accustomed). Then $(R, +, \cdot)$ is a *ring* if for all a, b, $c \in R$, the following conditions are satisfied:

a) $a + b = b + a$ Commutative Law of $+$

b) $a + (b + c) = (a + b) + c$ Associative Law of $+$

c) There exists $z \in R$ such that   Existence of an identity for $+$
     $a + z = z + a = a$ for every $a \in R$.

d) For each $a \in R$ there is an element   Existence of inverses under $+$
     $b \in R$ with $a + b = b + a = z$.

e) $a \cdot (b \cdot c) = (a \cdot b) \cdot c$ Associative Law of $\cdot$

f) $a \cdot (b + c) = a \cdot b + a \cdot c$ Distributive Laws of $\cdot$ over $+$
$(b + c) \cdot a = b \cdot a + c \cdot a$

Since the closed binary operations of $+$ (ring addition) and $\cdot$ (ring multiplication) are both associative, no ambiguity will arise if we write $a + b + c$ for either $(a + b) + c$ or $a + (b + c)$, or $a \cdot b \cdot c$ for either $(a \cdot b) \cdot c$ or $a \cdot (b \cdot c)$. When dealing with the (closed) binary operation of ring multiplication, we shall often write ab for $a \cdot b$. In addition, we can extend the associative laws (given in the definition of a ring) as we did in Exercises 8 and 9 of Section 4.2. Using mathematical induction, it can be verified that for all $r, n \in \mathbf{Z}^+$, with $n \geq 3$ and $1 \leq r < n$,

$$(a_1 + a_2 + \cdots + a_r) + (a_{r+1} + \cdots + a_n) = a_1 + a_2 + \cdots + a_r + a_{r+1} + \cdots + a_n,$$

and

$$(a_1 a_2 \cdots a_r)(a_{r+1} \cdots a_n) = a_1 a_2 \cdots a_r a_{r+1} \cdots a_n,$$

where $a_1, a_2, \ldots, a_r, a_{r+1}, \ldots, a_n$ are elements of a given ring $(R, +, \cdot)$. In a corresponding way, the distributive laws generalize as follows:

$$a(b_1 + b_2 + \cdots + b_n) = ab_1 + ab_2 + \cdots + ab_n,$$

$$(b_1 + b_2 + \cdots + b_n)a = b_1 a + b_2 a + \cdots + b_n a,$$

for arbitrary ring elements $a, b_1, b_2, \ldots, b_n$ and all $n \in \mathbf{Z}^+$ where $n \geq 3$.

In the next section we shall learn that the additive identity (or zero element) is unique, as is the additive inverse of each ring element. For now, let us consider some examples of rings.

EXAMPLE 14.1

Under the (closed) binary operations of ordinary addition and multiplication, we find that $\mathbf{Z}$, $\mathbf{Q}$, $\mathbf{R}$, and $\mathbf{C}$ are rings. In all of these rings the additive identity z is the integer 0, and the additive inverse of each number x is the familiar $-x$.

EXAMPLE 14.2

Let $M_2(\mathbf{Z})$ denote the set of all 2×2 matrices with integer entries. [The sets $M_2(\mathbf{Q})$, $M_2(\mathbf{R})$, and $M_2(\mathbf{C})$ are defined similarly.] In $M_2(\mathbf{Z})$ two matrices are equal if their corresponding entries are equal in $\mathbf{Z}$.

Here we define $+$ and $\cdot$ by

$$\begin{bmatrix} a & b \\ c & d \end{bmatrix} + \begin{bmatrix} e & f \\ g & h \end{bmatrix} = \begin{bmatrix} a + e & b + f \\ c + g & d + h \end{bmatrix}, \qquad \begin{bmatrix} a & b \\ c & d \end{bmatrix} \begin{bmatrix} e & f \\ g & h \end{bmatrix} = \begin{bmatrix} ae + bg & af + bh \\ ce + dg & cf + dh \end{bmatrix}.$$

Under these (closed) binary operations, $M_2(\mathbf{Z})$ is a ring. Here $z = \begin{bmatrix} 0 & 0 \\ 0 & 0 \end{bmatrix}$ and the additive inverse of $\begin{bmatrix} a & b \\ c & d \end{bmatrix}$ is $\begin{bmatrix} -a & -b \\ -c & -d \end{bmatrix}$.

A few things happen here, however, that do not occur in the rings of Example 14.1. For example,

$$\begin{bmatrix} 1 & 2 \\ 1 & 1 \end{bmatrix} \begin{bmatrix} 3 & 7 \\ 1 & 0 \end{bmatrix} = \begin{bmatrix} 5 & 7 \\ 4 & 7 \end{bmatrix} \neq \begin{bmatrix} 10 & 13 \\ 1 & 2 \end{bmatrix} = \begin{bmatrix} 3 & 7 \\ 1 & 0 \end{bmatrix} \begin{bmatrix} 1 & 2 \\ 1 & 1 \end{bmatrix}$$

shows that multiplication need not be commutative in a ring. That is why there are two distributive laws. Also,

$$\begin{bmatrix} 1 & -1 \\ -1 & 1 \end{bmatrix} \begin{bmatrix} 2 & 1 \\ 2 & 1 \end{bmatrix} = \begin{bmatrix} 0 & 0 \\ 0 & 0 \end{bmatrix},$$

even though neither $\begin{bmatrix} 1 & -1 \\ -1 & 1 \end{bmatrix}$ nor $\begin{bmatrix} 2 & 1 \\ 2 & 1 \end{bmatrix}$ is the additive identity. Hence a ring may contain what are called *proper divisors of zero* — that is, nonzero elements whose product is the zero element of the ring.

We extend our study of the ring structure in the following.

Definition 14.2   Let $(R, +, \cdot)$ be a ring.

   a) If $ab = ba$ for all $a, b \in R$, then R is called a *commutative* ring.

   b) The ring R is said to have no *proper divisors of zero* if for all $a, b \in R$, $ab = z \Rightarrow a = z$ or $b = z$.

   c) If an element $u \in R$ is such that $u \neq z$ and $au = ua = a$ for all $a \in R$, we call u a *unity*, or *multiplicative identity*, of R. Here R is called a *ring with unity*.

It follows from part (c) of Definition 14.2 that whenever we have a ring R with unity, then R contains at least two elements. Furthermore, if a ring has a unity we shall learn in the next section that it is unique.

The rings in Example 14.1 are all commutative rings whose unity is the integer 1. None of these rings has any proper divisors of zero. Meanwhile, the ring $M_2(\mathbf{Z})$ is a noncommutative ring whose unity is the matrix $\begin{bmatrix} 1 & 0 \\ 0 & 1 \end{bmatrix}$. This ring does contain proper divisors of zero.

Also, whenever we want to verify that a particular structure $(R, +, \cdot)$ is a ring, we can start by showing that R is closed under both binary operations. Then we can continue and verify conditions (a)–(e) of Definition 14.1. Before we try to establish the distributive laws, however, we might want to first determine if the multiplication operation is commutative. Should we find this operation to be commutative, then we need only establish one of the distributive laws (for the other will follow automatically). Further, if we are able to verify all of the preceding conditions, then we'll know that $(R, +, \cdot)$ is not just a ring, but a commutative ring.

Now let us study another example as we further investigate the ideas set down in Definitions 14.1 and 14.2.

EXAMPLE 14.3   Consider the set $\mathbf{Z}$ together with the binary operations of $\oplus$ and $\odot$, which are defined by

$$x \oplus y = x + y - 1, \qquad x \odot y = x + y - xy.$$

Consequently, here we find, for instance, that $3 \oplus 7 = 3 + 7 - 1 = 9$ and $3 \odot 7 = 3 + 7 - 3 \cdot 7 = -11$.

Since ordinary addition, subtraction, and multiplication are closed binary operations for $\mathbf{Z}$, these new binary operations — namely, $\oplus$ and $\odot$ — are also closed for $\mathbf{Z}$. In fact, we shall find that $(\mathbf{Z}, \oplus, \odot)$ is a ring.

a) In order to verify that $(\mathbf{Z}, \oplus, \odot)$ is a ring we must establish the six conditions given in Definition 14.1. We shall examine three of these conditions and leave the other three for the Section Exercises.

1) First, since ordinary addition is a commutative binary operation for $\mathbf{Z}$, we find that for all $x, y \in \mathbf{Z}$,

$$x \oplus y = x + y - 1 = y + x - 1 = y \oplus x.$$

So the binary operation $\oplus$ is also commutative for $\mathbf{Z}$.

2) When we examine condition (c) we realize that we need to find an integer z such that $a \oplus z = z \oplus a = a$, for every a in $\mathbf{Z}$. Therefore, we must solve the equation $a + z - 1 = a$, which leads us to $z = 1$. Hence the *nonzero* integer 1 is the *zero* element (or additive identity) for $\oplus$.

3) What about additive inverses? At this point if we are given an (arbitrary) integer a, we want to know if there is an integer b such that $a \oplus b = b \oplus a = z$. From part (2) above and the definition of $\oplus$ this says that the integer b must satisfy $a + b - 1 = 1$, and it follows that $b = 2 - a$. So, for instance, the additive inverse of 7 is $2 - 7 = -5$ and the additive inverse for -42 is $2 - (-42) = 44$. After all, in the case of 7 we find that $7 \oplus (-5) = 7 + (-5) - 1 = 7 - 5 - 1 = 1$, where 1 is the additive identity. [*Note*: Since we showed in part (1) that $\oplus$ is commutative, we also know that $(-5) \oplus 7 = 1$.]

b) Furthermore, the ring $(\mathbf{Z}, \oplus, \odot)$ also possesses the additional properties given in Definition 14.2. In particular this ring has a unity (that is, a multiplicative identity). To determine the unity, let a be any integer and consider the element $u \ (\neq z = 1)$ where

$$a \odot u = u \odot a = a.$$

Since $a \odot u = a + u - au$, we solve $a + u - au = a$ to find that $u(1 - a) = 0$. Since a is arbitrary, this must hold even when $a \neq 1$. Consequently, the integer $u = 0$ is the unity for the ring $(\mathbf{Z}, \oplus, \odot)$.

After these examples of infinite rings, we turn now to rings with finitely many elements.

EXAMPLE 14.4

Let $\mathcal{U} = \{1, 2\}$ and $R = \mathscr{P}(\mathcal{U})$. Define $+$ and $\cdot$ on the elements of R by

$$A + B = A \triangle B = \{x \mid x \in A \text{ or } x \in B, \text{ but not both}\}$$
$$A \cdot B = A \cap B = \text{the intersection of sets } A, B \subseteq \mathcal{U}.$$

We form Tables 14.1(a) and (b) for these operations.

From results in Chapter 3, one finds that R satisfies conditions (a), (b), (e), and (f) of Definition 14.1 for these (closed) binary operations of "addition" and "multiplication." The table for "addition" shows that $\varnothing$ is the additive identity. For each $x \in R$, the additive inverse of x is x itself. The multiplication table is symmetric about the diagonal from the upper left to the lower right, so the operation described by the table is commutative. The table also indicates that R has *unity* $\mathcal{U}$. So R is a finite commutative ring with unity. The elements $\{1\}, \{2\}$ provide an example of proper divisors of zero.

Table 14.1

+ (△)	∅	{1}	{2}	𝒰
∅	∅	{1}	{2}	𝒰
{1}	{1}	∅	𝒰	{2}
{2}	{2}	𝒰	∅	{1}
𝒰	𝒰	{2}	{1}	∅

(a)

· (∩)	∅	{1}	{2}	𝒰
∅	∅	∅	∅	∅
{1}	∅	{1}	∅	{1}
{2}	∅	∅	{2}	{2}
𝒰	∅	{1}	{2}	𝒰

(b)

EXAMPLE 14.5

For $R = \{a, b, c, d, e\}$ we define $+$ and $\cdot$ by Tables 14.2(a) and (b).

Table 14.2

+	a	b	c	d	e
a	a	b	c	d	e
b	b	c	d	e	a
c	c	d	e	a	b
d	d	e	a	b	c
e	e	a	b	c	d

(a)

·	a	b	c	d	e
a	a	a	a	a	a
b	a	b	c	d	e
c	a	c	e	b	d
d	a	d	b	e	c
e	a	e	d	c	b

(b)

Although we do not verify them here, the 125 equalities needed to establish each of the associative laws and the distributive laws all hold, so $(R, +, \cdot)$ turns out to be a finite commutative ring with unity, and it has no proper divisors of zero. The element a is the zero (that is, the additive identity) of R, whereas b is the unity. Here every nonzero element x has a *multiplicative inverse* y, where $xy = yx = b$, the unity. Elements c and d are multiplicative inverses of each other; b is its own inverse, as is e.

We now consider the concept of a multiplicative inverse for a ring element in general.

Definition 14.3

Let R be a ring with unity u. If $a \in R$ and there exists $b \in R$ such that $ab = ba = u$, then b is called a *multiplicative inverse* of a and a is called a *unit* of R. (The element b is also a unit of R.)

In Section 14.2 we shall see that if a ring element does have a multiplicative inverse, then it has only one such inverse. In the meantime, we'll examine two special kinds of ring structures.

Definition 14.4

Let R be a commutative ring with unity. Then

a) R is called an *integral domain* if R has no proper divisors of zero.

b) R is called a *field* if every nonzero element of R is a unit.

The ring $(\mathbf{Z}, +, \cdot)$ is an integral domain but not a field, while $\mathbf{Q}$, $\mathbf{R}$, $\mathbf{C}$, under ordinary addition and multiplication, are both integral domains and fields. The ring in Example 14.5 is both an integral domain and a field.

It follows from part (c) of Definition 14.2 that if R is an integral domain or a field, then $|R| \geq 2$.

EXAMPLE 14.6   For our last ring of this section we let $R = \{s, t, v, w, x, y\}$ and $+$ and $\cdot$ are given by Tables 14.3(a) and(b).

Table 14.3

+	s	t	v	w	x	y
s	s	t	v	w	x	y
t	t	v	w	x	y	s
v	v	w	x	y	s	t
w	w	x	y	s	t	v
x	x	y	s	t	v	w
y	y	s	t	v	w	x

·	s	t	v	w	x	y
s	s	s	s	s	s	s
t	s	t	v	w	x	y
v	s	v	x	s	v	x
w	s	w	s	w	s	w
x	s	x	v	s	x	v
y	s	y	x	w	v	t

(a)   (b)

From these tables we see that $(R, +, \cdot)$ is a commutative ring with unity, but it is neither an integral domain nor a field. The element t is the unity, and t and y are the units of R.

We also note that $vv = vy$, and even though v is not the zero element of R, we cannot cancel and say that $v = y$. So a general ring does *not* satisfy the cancellation law of multiplication that we may sometimes take for granted. We shall look at this idea again in the next section.

EXERCISES 14.1

1. Find the additive inverse for each element in the rings of Examples 14.5 and 14.6.

2. Determine whether or not each of the following sets of numbers is a ring under ordinary addition and multiplication.

  a) $R =$ the set of positive integers and zero

  b) $R = \{kn \mid n \in \mathbf{Z}, k \text{ a fixed positive integer}\}$

  c) $R = \{a + b\sqrt{2} \mid a, b \in \mathbf{Z}\}$

  d) $R = \{a + b\sqrt{2} + c\sqrt{3} \mid a \in \mathbf{Z}, b, c \in \mathbf{Q}\}$

3. Let $(R, +, \cdot)$ be a ring with a, b, c, d elements of R. State the conditions (from the definition of a ring) that are needed to prove each of the following results.

  a) $(a + b) + c = b + (c + a)$

  b) $d + a(b + c) = ab + (d + ac)$

  c) $c(d + b) + ab = (a + c)b + cd$

  d) $a(bc) + (ab)d = (ab)(d + c)$

4. For the set R in Example 14.4, keep $A \cdot B = A \cap B$, but define $A + B = A \cup B$. Is $(R, \cup, \cap)$ a ring?

5. Consider the set $\mathbf{Z}$ together with the binary operations $\oplus$ and $\odot$ given in Example 14.3. (a) Verify the associative laws for $\oplus$ and $\odot$ and the distributive laws in order to complete the work started in part (a) of Example 14.3. [This now establishes that $(\mathbf{Z}, \oplus, \odot)$ is a ring.] (b) Is this ring commutative? (c) In part (b) of Example 14.3 we showed that 0 is the unity for $(\mathbf{Z}, \oplus, \odot)$. What are the units for this ring? (d) Is this ring an integral domain? a field?

6. Define the binary operations $\oplus$ and $\odot$ on $\mathbf{Z}$ by $x \oplus y = x + y - 7, x \odot y = x + y - 3xy$, for all x, $y \in \mathbf{Z}$. Explain why $(\mathbf{Z}, \oplus, \odot)$ is *not* a ring.

7. Let k, m be fixed integers. Find all values for k, m for which $(\mathbf{Z}, \oplus, \odot)$ is a ring under the binary operations $x \oplus y = x + y - k$, $x \odot y = x + y - mxy$, where x, $y \in \mathbf{Z}$.

8. Tables 14.4(a) and (b) make $(R, +, \cdot)$ into a ring, where $R = \{s, t, x, y\}$. (a) What is the zero for this ring? (b) What is the additive inverse of each element? (c) What is $t(s + xy)$? (d) Is R a commutative ring? (e) Does R have a unity? (f) Find a pair of zero divisors.

9. Define addition and multiplication, denoted by $\oplus$ and $\odot$, respectively, on the set $\mathbf{Q}$ as follows. For $a, b \in \mathbf{Q}$, $a \oplus b =$

Table 14.4

+	s	t	x	y
s	y	x	s	t
t	x	y	t	s
x	s	t	x	y
y	t	s	y	x

(a)

$\cdot$	s	t	x	y
s	y	y	x	x
t	y	y	x	x
x	x	x	x	x
y	x	x	x	x

(b)

$a + b + 7$, $a \odot b = a + b + (ab/7)$. (a) Prove that $(\mathbf{Q}, \oplus, \odot)$ is a ring. (b) Is this ring commutative? (c) Does the ring have a unity? What about units? (d) Is this ring an integral domain? a field?

10. Let $(\mathbf{Q}, \oplus, \odot)$ denote the field where $\oplus$ and $\odot$ are defined by

$$a \oplus b = a + b - k, \qquad a \odot b = a + b + (ab/m),$$

for fixed elements k, m ($\neq 0$) of $\mathbf{Q}$. Determine the value for k and the value for m in each of the following.

    a) The zero element for the field is 3.

    b) The additive inverse of the element 6 is -9.

    c) The multiplicative inverse of 2 is $1/8$.

11. Let $R = \{a + bi \,|\, a, b \in \mathbf{Z}, \ i^2 = -1\}$, with addition and multiplication defined by $(a + bi) + (c + di) = (a + c) + (b + d)i$ and $(a + bi)(c + di) = (ac - bd) + (bc + ad)i$, respectively. (a) Verify that R is an integral domain. (b) Determine all units in R.

12. a) Determine the multiplicative inverse of the matrix $\begin{bmatrix} 1 & 2 \\ 3 & 7 \end{bmatrix}$ in the ring $M_2(\mathbf{Z})$ — that is, find a, b, c, d so that

$$\begin{bmatrix} 1 & 2 \\ 3 & 7 \end{bmatrix} \begin{bmatrix} a & b \\ c & d \end{bmatrix} = \begin{bmatrix} a & b \\ c & d \end{bmatrix} \begin{bmatrix} 1 & 2 \\ 3 & 7 \end{bmatrix} = \begin{bmatrix} 1 & 0 \\ 0 & 1 \end{bmatrix}.$$

    b) Show that $\begin{bmatrix} 1 & 2 \\ 3 & 8 \end{bmatrix}$ is a unit in the ring $M_2(\mathbf{Q})$ but not a unit in $M_2(\mathbf{Z})$.

13. If $\begin{bmatrix} a & b \\ c & d \end{bmatrix} \in M_2(\mathbf{R})$, prove that $\begin{bmatrix} a & b \\ c & d \end{bmatrix}$ is a unit of this ring if and only if $ad - bc \neq 0$.

14. Give an example of a ring with eight elements. How about one with 16 elements? Generalize.

15. For $R = \{s, t, x, y\}$, define $+$ and $\cdot$, making R into a ring, by Table 14.5(a) for $+$ and by the partial table for $\cdot$ in Table 14.5(b).

Table 14.5

+	s	t	x	y
s	s	t	x	y
t	t	s	y	x
x	x	y	s	t
y	y	x	t	s

(a)

$\cdot$	s	t	x	y
s	s	s	s	s
t	s	t	$?$	$?$
x	s	t	$?$	y
y	s	$?$	s	$?$

(b)

    a) Using the associative and distributive laws, determine the entries for the missing spaces in the multiplication table.

    b) Is this ring commutative?

    c) Does it have a unity? How about units?

    d) Is the ring an integral domain or a field?

14.2
Ring Properties and Substructures

In each ring of Section 14.1 we were concerned with *the* zero element of the ring and *the* additive inverse of each ring element. It is time now to show, along with other properties, that these elements are truly unique.

THEOREM 14.1   In any ring $(R, +, \cdot)$,

    a) the zero element z is unique, and

    b) the additive inverse of each ring element is unique.

    Proof:

    a) If R has more than one additive identity, let z_1, z_2 denote two such elements. Then

$$z_1 = z_1 + z_2 = z_2.$$

              ↗   ↖

      Since z_2 is an   Since z_1 is an
     additive identity   additive identity

b) For $a \in R$, suppose there are two elements $b, c \in R$ where $a + b = b + a = z$ and $a + c = c + a = z$. Then $b = b + z = b + (a + c) = (b + a) + c = z + c = c$. (The reader should supply the condition that establishes each equality.)

As a result of the uniqueness in part (b), from this point on we shall denote *the* additive inverse of $a \in R$ by $-a$. Further, we may now speak of *subtraction* in the ring, where we understand that $a - b = a + (-b)$.

From Theorem 14.1(b) we also obtain the following for any ring R.

THEOREM 14.2

The Cancellation Laws of Addition. For all $a, b, c \in R$,

    a) $a + b = a + c \Rightarrow b = c$, and

    b) $b + a = c + a \Rightarrow b = c$.

Proof:

    a) Since $a \in R$, it follows that $-a \in R$ and we have

$$a + b = a + c \Rightarrow (-a) + (a + b) = (-a) + (a + c)$$
$$\Rightarrow [(-a) + a] + b = [(-a) + a] + c$$
$$\Rightarrow z + b = z + c \Rightarrow b = c.$$

    b) We leave this similar proof for the reader.

Note that when we examine the addition table for a finite ring we find that each element of the ring occurs exactly one time in each row and column of the table. This is a direct consequence of Theorem 14.2 — where part (a) handles the rows and part (b) the columns.

THEOREM 14.3

For any ring $(R, +, \cdot)$ and any $a \in R$, we have $az = za = z$.

Proof: If $a \in R$, then $az = a(z + z)$ because $z + z = z$. Hence $z + az = az = az + az$. (Why?) Using the cancellation law of addition, we have $z = az$.

The proof that $za = z$ is done similarly.

The reader may feel that the result of Theorem 14.3 is obvious. But we are not dealing with just $\mathbf{Z}$ or $\mathbf{Q}$ or $M_2(\mathbf{Z})$. Our objective is to show that *any* ring satisfies such a result, and to get the result we may only use the conditions in the definition of a ring and whatever properties we've derived for arbitrary rings up to this point.

The uniqueness of additive inverses [from part (b) of Theorem 14.1] now implies the following result.

THEOREM 14.4

Given a ring $(R, +, \cdot)$, for all $a, b \in R$,

    a) $-(-a) = a$,

    b) $a(-b) = (-a)b = -(ab)$, and

    c) $(-a)(-b) = ab$.

Proof:

a) By the convention stated after Theorem 14.1, $-(-a)$ denotes *the* additive inverse of $-a$. Since $(-a) + a = z$, a is also an additive inverse for $-a$. Consequently, by the uniqueness of such inverses, $-(-a) = a$.

b) We shall prove that $a(-b) = -(ab)$ and shall leave the other part for the reader. We know that $-(ab)$ denotes *the* additive inverse of ab. However, $ab + a(-b) = a[b + (-b)] = az = z$, by Theorem 14.3, so by the uniqueness of additive inverses, $a(-b) = -(ab)$.

c) Here we establish an idea we have used in algebra since our first encounter with *signed numbers*. "Minus times minus does indeed equal plus," and the proof follows from the properties and definition of a ring. From part (b) we have $(-a)(-b) = -[a(-b)] = -[-(ab)]$, and the result then follows from part (a).

For the operation of multiplication one also finds the following, which is comparable to Theorem 14.1.

THEOREM 14.5   For a ring $(R, +, \cdot)$,

a) if R has a unity, then it is unique, and

b) if R has a unity, and x is a unit of R, then the multiplicative inverse of x is unique.

Proof: The proofs of these results are left to the reader.

As a result of this theorem, when $(R, +, \cdot)$ is a ring with unity, we shall denote *the* unity by u. Furthermore, in such a ring *the* multiplicative inverse of each unit x will be denoted by x^{-1}. Also, one may now restate the definition of a field as a commutative ring F with unity, such that for all $x \in F$, $x \neq z \Rightarrow x^{-1} \in F$.

With this notion to assist us, we examine some further properties and relations between fields and integral domains.

THEOREM 14.6   Let $(R, +, \cdot)$ be a commutative ring with unity. Then R is an integral domain if and only if, for all $a, b, c \in R$ where $a \neq z$, $ab = ac \Rightarrow b = c$. (Hence, a commutative ring with unity that satisfies the *cancellation law of multiplication* is an integral domain.)

Proof: If R is an integral domain and $x, y \in R$, then $xy = z \Rightarrow x = z$ or $y = z$. Now if $ab = ac$, then $ab - ac = a(b - c) = z$, and because $a \neq z$, it follows that $b - c = z$ or $b = c$. Conversely, if R is commutative with unity and R satisfies multiplicative cancellation, then let $a, b \in R$ with $ab = z$. If $a = z$, we are finished. If not, as $az = z$, we can write $ab = az$ and conclude that $b = z$. So there are no proper divisors of zero and R is an integral domain.

Before going on, let us realize that the cancellation law of multiplication does *not* imply the existence of multiplicative inverses. The integral domain $(\mathbf{Z}, +, \cdot)$ satisfies multiplicative cancellation, but it contains only two elements — namely, 1 and -1 — that are units. Hence, an integral domain need not be a field. But what about a field? Is it necessarily an integral domain?

THEOREM 14.7

If $(F, +, \cdot)$ is a field, then it is an integral domain.

Proof: Let $a, b \in F$ with $ab = z$. If $a = z$, we are finished. If not, a has a multiplicative inverse a^{-1} because F is a field. Then

$$ab = z \Rightarrow a^{-1}(ab) = a^{-1}z \Rightarrow (a^{-1}a)b = a^{-1}z \Rightarrow ub = z \Rightarrow b = z.$$

Hence F has no proper divisors of zero and is an integral domain.

In Chapter 5 we found that functions $f: A \to A$ could be one-to-one (or onto) without being onto (or one-to-one). However, if A were *finite*, such a function f was one-to-one if and only if it was onto. (See Theorem 5.11.) The same situation occurs with *finite* integral domains. An integral domain need not be a field, but when it is finite we find that this structure is a field.

THEOREM 14.8

A *finite* integral domain $(D, +, \cdot)$ is a field.

Proof: Since D is finite, we may list the elements of D as $\{d_1, d_2, \ldots, d_n\}$. For $d \in D$, where $d \neq z$, we have $dD = \{dd_1, dd_2, \ldots, dd_n\} \subseteq D$ because D is closed under multiplication. Now $|D| = n$ and $dD \subseteq D$, so if we could show that dD contains n elements, we would have $dD = D$. If $|dD| < n$, then $dd_i = dd_j$, for some $1 \leq i < j \leq n$. But since D is an integral domain and $d \neq z$, we have $d_i = d_j$, when they are supposed to be distinct. So $dD = D$ and for some $1 \leq k \leq n$, $dd_k = u$, the unity of D. Then $dd_k = u \Rightarrow d$ is a unit of D, and since d was chosen arbitrarily, it follows that $(D, +, \cdot)$ is a field.

From the proof of Theorem 14.8 we also realize that when we are dealing with the *non-zero* elements of a finite field, the multiplication table for these elements is such that each element of the field occurs exactly once in each of the rows and columns.

In the next section we shall look at finite fields that are useful in discrete mathematics. Before closing this section, however, let us examine some special subsets of a ring.

When we were dealing with finite state machines in Chapter 6, we saw instances where subsets of the set of internal states gave rise to machines on their own (when the next state and output functions of the original machines were suitably restricted). These were called submachines. Since closed binary operations are special kinds of functions, we encounter a similar idea in the following definition.

Definition 14.5

For a ring $(R, +, \cdot)$, a nonempty subset S of R is called a *subring* of R if $(S, +, \cdot)$ — that is, S under the addition and multiplication of R, restricted to S — is a ring.

EXAMPLE 14.7

For every ring R, the subsets $\{z\}$ and R are always subrings of R.

EXAMPLE 14.8

a) The set of all even integers is a subring of $(\mathbf{Z}, +, \cdot)$. In fact, for each $n \in \mathbf{Z}^+$, $n\mathbf{Z} = \{nx \mid x \in \mathbf{Z}\}$ is a subring of $(\mathbf{Z}, +, \cdot)$.

b) $(\mathbf{Z}, +, \cdot)$ is a subring of $(\mathbf{Q}, +, \cdot)$, which is a subring of $(\mathbf{R}, +, \cdot)$, which is a subring of $(\mathbf{C}, +, \cdot)$.

EXAMPLE 14.9

In Example 14.6, the subsets $S = \{s, w\}$ and $T = \{s, v, x\}$ are subrings of R.

The next result characterizes those subsets of a ring that are subrings.

THEOREM 14.9

Given a ring $(R, +, \cdot)$, a nonempty subset S of R is a subring of R if and only if

1) for all $a, b \in S$, we have $a + b, ab \in S$ (that is, S is closed under the binary operations of addition and multiplication defined on R), and

2) for all $a \in S$, we have $-a \in S$.

Proof: If $(S, +, \cdot)$ is a subring of R, then in its own right it satisfies all the conditions of a ring. Hence it satisfies conditions 1 and 2 of the theorem. Conversely, let S be a nonempty subset of R that satisfies conditions 1 and 2. Conditions (a), (b), (e), and (f) of the definition of a ring are inherited by the elements of S, because they are also elements of R. Thus, all we need to verify here is that S has an additive identity. Now $S \neq \emptyset$, so there is an element $a \in S$, and by condition 2, $-a \in S$. Then by condition 1, $z = a + (-a) \in S$.

EXAMPLE 14.10

Consider the ring $(\mathbf{Z}, \oplus, \odot)$ that we examined in Example 14.3 and Exercise 5 of Section 14.1. Here we have $x \oplus y = x + y - 1$ and $x \odot y = x + y - xy$. Now consider the subset $S = \{\ldots, -5, -3, -1, 1, 3, 5, \ldots\}$ of all odd integers. Since, for example, 3 and 5 are in S but the ordinary sum $3 + 5 = 8 \notin S$, this set S is *not* a subring of $(\mathbf{Z}, +, \cdot)$. However, $3 \oplus 5 = 3 + 5 - 1 = 7 \in S$. In fact, for all $a, b \in S$ we have $a \oplus b = a + b - 1$, where $a + b$ is even, and $a + b - 1$ is odd — so $a \oplus b \in S$. Also, $a \odot b = a + b - ab$, where $a + b$ is even and ab is odd — so $a \odot b \in S$. Finally, $-a$ [the additive inverse of a in the ring $(\mathbf{Z}, \oplus, \odot)$] is equal to $2 - a$, which is odd whenever a is odd. Consequently, if $a \in S$ then $-a \in S$, and it follows from Theorem 14.9 that S is a subring of $(\mathbf{Z}, \oplus, \odot)$.

Note that $(\mathbf{Z}^+, +, \cdot)$ satisfies condition 1 in Theorem 14.9, but not condition 2, so it is not a subring of $(\mathbf{Z}, +, \cdot)$.

The result in Theorem 14.9 can also be given as follows.

THEOREM 14.10

For any ring $(R, +, \cdot)$, if $\emptyset \neq S \subseteq R$,

a) then $(S, +, \cdot)$ is a subring of R if and only if for all $a, b \in S$, we have $a - b \in S$ and $ab \in S$;

b) and if S is finite, then $(S, +, \cdot)$ is a subring of R if and only if for all $a, b \in S$, we have $a + b, ab \in S$. (Once again, additional help comes from a finiteness condition.)

Proof: These proofs we leave for the reader.

The next example demonstrates how one might use the first part of the preceding theorem.

EXAMPLE 14.11

Let us consider the ring $R = M_2(\mathbf{Z})$ and the subset

$$S = \left\{ \begin{bmatrix} x & x+y \\ x+y & x \end{bmatrix} \middle| x, y \in \mathbf{Z} \right\}$$

of R. When $x = y = 0$ it follows that $\begin{bmatrix} 0 & 0 \\ 0 & 0 \end{bmatrix} \in S$, and $S \neq \emptyset$. So now we examine any two elements of S — namely, two matrices of the form

$$\begin{bmatrix} x & x+y \\ x+y & x \end{bmatrix} \quad \text{and} \quad \begin{bmatrix} v & v+w \\ v+w & v \end{bmatrix},$$

where $x, y, v, w \in \mathbf{Z}$. We find that

$$\begin{bmatrix} x & x+y \\ x+y & x \end{bmatrix} - \begin{bmatrix} v & v+w \\ v+w & v \end{bmatrix} = \begin{bmatrix} x-v & (x-v)+(y-w) \\ (x-v)+(y-w) & x-v \end{bmatrix},$$

so S is closed under subtraction. Turning to multiplication we have

$$\begin{bmatrix} x & x+y \\ x+y & x \end{bmatrix} \begin{bmatrix} v & v+w \\ v+w & v \end{bmatrix}$$

$$= \begin{bmatrix} xv + (x+y)(v+w) & x(v+w) + (x+y)v \\ (x+y)v + x(v+w) & (x+y)(v+w) + xv \end{bmatrix}$$

$$= \begin{bmatrix} xv + xv + yv + xw + yw & xv + xw + xv + yv \\ xv + yv + xv + xw & xv + yv + xw + yw + xv \end{bmatrix}$$

$$= \begin{bmatrix} xv + xv + yv + xw + yw & (xv + xv + yv + xw + yw) + (-yw) \\ (xv + xv + yv + xw + yw) + (-yw) & xv + xv + yv + xw + yw \end{bmatrix},$$

so S is also closed under multiplication.

Appealing now to part (a) of Theorem 14.10, one finds that S is a subring of R.

We shall now single out an important type of subring.

Definition 14.6   A nonempty subset I of a ring R is called an *ideal* of R if for all $a, b \in I$ and all $r \in R$, we have (a) $a - b \in I$ and (b) $ar, ra \in I$.

An ideal is a subring, but the converse does not necessarily hold: $(\mathbf{Z}, +, \cdot)$ is a subring of $(\mathbf{Q}, +, \cdot)$ but not an ideal because, for example, $(1/2)9 \notin \mathbf{Z}$ for $(1/2) \in \mathbf{Q}, 9 \in \mathbf{Z}$. Meanwhile, all the subrings in Example 14.8(a) are ideals of $(\mathbf{Z}, +, \cdot)$.

Looking back to Example 14.10 we see that if $a \in S, x \in \mathbf{Z}$, then $a \odot x = a + x - ax$ $(= x \odot a)$, and if x is even (because the case for x odd has already been covered within Example 14.10), then $a + x$ is odd and ax is even, making $a + x - ax$ odd. Consequently, for all $a \in S$ and all $x \in \mathbf{Z}$, $a \odot x$ and $x \odot a$ are in S, so S is an ideal of the ring $(\mathbf{Z}, \oplus, \odot)$.

EXERCISES 14.2

1. Complete the proofs of Theorems 14.2, 14.4, 14.5, and 14.10.

2. If a, b, and c are any elements in a ring $(R, +, \cdot)$, prove that (a) $a(b - c) = ab - (ac) = ab - ac$ and (b) $(b - c)a = ba - (ca) = ba - ca$.

3. a) If R is a ring with unity and a, b are units of R, prove that ab is a unit of R and that $(ab)^{-1} = b^{-1}a^{-1}$.

b) For the ring $M_2(\mathbf{Z})$, find $A^{-1}, B^{-1}, (AB)^{-1}, (BA)^{-1}$,

and $B^{-1}A^{-1}$ if

$$A = \begin{bmatrix} 4 & 7 \\ 1 & 2 \end{bmatrix}, \qquad B = \begin{bmatrix} 5 & 2 \\ 2 & 1 \end{bmatrix}.$$

4. Prove that a unit in a ring R cannot be a proper divisor of zero.

5. If a is a unit in ring R, prove that $-a$ is also a unit in R.

6. a) Verify that the subsets $S = \{s, w\}$ and $T = \{s, v, x\}$ are subrings of the ring R in Example 14.6. (The binary operations for the elements of S, T are those given in Table 14.3.)

b) Are the subrings in part (a) ideals of R?

7. Let S and T be subrings of a ring R. Prove that $S \cap T$ is a subring of R.

8. Let $R = M_2(\mathbf{Z})$ and let S be the subset of R where

$$S = \left\{ \begin{bmatrix} x & x - y \\ x - y & y \end{bmatrix} \middle| x, y \in \mathbf{Z} \right\}.$$

Prove that S is a subring of R.

9. Let $(R, +, \cdot)$ be a ring. If S, T_1, and T_2 are subrings of R, and $S \subseteq T_1 \cup T_2$, prove that $S \subseteq T_1$ or $S \subseteq T_2$.

10. a) Let $(R, +, \cdot)$ be a finite commutative ring with unity u. If $r \in R$ and r is not the zero element of R, prove that r is either a unit or a proper divisor of zero.

b) Does the result in part (a) remain valid when R is infinite?

11. a) For $R = M_2(\mathbf{Z})$, prove that

$$S = \left\{ \begin{bmatrix} a & 0 \\ 0 & 0 \end{bmatrix} \middle| a \in \mathbf{Z} \right\}$$

is a subring of R.

b) What is the unity of R?

c) Does S have a unity?

d) Does S have any properties that R does not have?

e) Is S an ideal of R?

12. Let S and T be the following subsets of the ring $R = M_2(\mathbf{Z})$:

$$S = \left\{ \begin{bmatrix} a & 0 \\ b & c \end{bmatrix} \middle| a, b, c \in \mathbf{Z} \right\},$$

$$T = \left\{ \begin{bmatrix} 2a & 2b \\ 2c & 2d \end{bmatrix} \middle| a, b, c, d \in \mathbf{Z} \right\}.$$

a) Verify that S is a subring of R. Is it an ideal?

b) Verify that T is a subring of R. Is it an ideal?

13. Let $(R, +, \cdot)$ be a commutative ring, and let z denote the zero element of R. For a fixed element $a \in R$, define $N(a) = \{r \in R | ra = z\}$. Prove that $N(a)$ is an ideal of R.

14. Let R be a commutative ring with unity u, and let I be an ideal of R. (a) If $u \in I$, prove that $I = R$. (b) If I contains a unit of R, prove that $I = R$.

15. If R is a field, how many ideals does R have?

16. Let $(R, +, \cdot)$ be the (finite) commutative ring with unity given by Tables 14.6(a) and (b).

Table 14.6

+	z	u	a	b
z	z	u	a	b
u	u	z	b	a
a	a	b	z	u
b	b	a	u	z

(a)

·	z	u	a	b
z	z	z	z	z
u	z	u	a	b
a	z	a	b	u
b	z	b	u	a

(b)

a) Verify that R is a field.

b) Find a subring of R that is not an ideal.

c) Let x and y be unknowns. Solve the following system of linear equations in R: $bx + y = u$; $x + by = z$.

17. Let R be a commutative ring with unity u.

a) For any (fixed) $a \in R$, prove that $aR = \{ar | r \in R\}$ is an ideal of R.

b) If the only ideals of R are $\{z\}$ and R, prove that R is a field.

18. Let $(S, +, \cdot)$ and $(T, +', \cdot')$ be two rings. For $R = S \times T$, define addition "$\oplus$" and multiplication "$\odot$" by

$$(s_1, t_1) \oplus (s_2, t_2) = (s_1 + s_2, t_1 +' t_2),$$
$$(s_1, t_1) \odot (s_2, t_2) = (s_1 \cdot s_2, t_1 \cdot' t_2).$$

a) Prove that under these closed binary operations, R is a ring.

b) If both S and T are commutative, prove that R is commutative.

c) If S has unity u_S and T has unity u_T, what is the unity of R?

d) If S and T are fields, is R also a field?

19. Let $(R, +, \cdot)$ be a ring with unity u, and $|R| = 8$. On $R^4 = R \times R \times R \times R$, define $+$ and $\cdot$ as suggested by Exercise 18. In the ring R^4, (a) how many elements have exactly two nonzero components? (b) how many elements have all nonzero components? (c) is there a unity? (d) how many units are there if R has four units?

20. Let $(R, +, \cdot)$ be a ring, with $a \in R$. Define $0a = z$, $1a = a$, and $(n + 1)a = na + a$, for all $n \in \mathbf{Z}^+$. (Here we are multiplying elements of R by elements of $\mathbf{Z}$, so we have yet another operation that is different from the multiplications in either of $\mathbf{Z}$ or R.) For $n > 0$, we define $(-n)a = n(-a)$, so, for example, $(-3)a = 3(-a) = 2(-a) + (-a) = [(-a) + (-a)] + (-a) = [-(a + a)] + (-a) = -[(a + a) + a] = -[2a + a] = -(3a)$.

For all $a, b \in R$, and all $m, n \in \mathbf{Z}$, prove that

a) $ma + na = (m + n)a$ **b)** $m(na) = (mn)a$

c) $n(a + b) = na + nb$ **d)** $n(ab) = (na)b = a(nb)$

e) $(ma)(nb) = (mn)(ab) = (na)(mb)$

21. a) For ring $(R, +, \cdot)$ and each $a \in R$, we define $a^1 = a$, and $a^{n+1} = a^n a$, for all $n \in \mathbf{Z}^+$. Prove that for all $m, n \in \mathbf{Z}^+$, $(a^m)(a^n) = a^{m+n}$ and $(a^m)^n = a^{mn}$.

b) Can you suggest how we might define a^0 or a^{-n}, $n \in \mathbf{Z}^+$, including any necessary conditions that R must satisfy for these definitions to make sense?

14.3
The Integers Modulo n

Enough abstraction for a while! We shall now concentrate on the construction and use of special finite rings and fields.

Definition 14.7

Let $n \in \mathbf{Z}^+$, $n > 1$. For $a, b \in \mathbf{Z}$, we say that a *is congruent to b modulo n*, and we write $a \equiv b \pmod{n}$, if $n | (a - b)$, or, equivalently, $a = b + kn$ for some $k \in \mathbf{Z}$.

EXAMPLE 14.12

i) We find that $17 \equiv 2 \pmod{5}$, since $17 - 2 = 15 = 3(5)$, or $17 = 2 + 3(5)$.

ii) As $-7 + 49 = -7 - (-49) = 42 = 7(6)$ [or, $-7 = -49 + 7(6)$], we have $-7 \equiv -49 \pmod{6}$.

iii) Since $11 - (-5) = 16 = 2(8)$ [or, $11 = -5 + 2(8)$], it follows that $11 \equiv -5 \pmod{8}$.

Before we examine our first theorem, let us make three observations about this new concept of congruence modulo n. Here, as above, we have $a, b, n \in \mathbf{Z}$, with $n > 1$.

i) Using the division algorithm, we can write $a = q_1 n + r_1$ and $b = q_2 n + r_2$, with $0 \le r_1 < n, 0 \le r_2 < n$. So $a - b = (q_1 - q_2)n + (r_1 - r_2)$. Then, if $a \equiv b \pmod{n}$, it follows that $n | (a - b)$, and, consequently, $n | (r_1 - r_2)$. But with $0 \le |r_1 - r_2| < n$, we now find that $r_1 = r_2$.

Hence, if $a \equiv b \pmod{n}$, then a, b have the same remainder upon division by n.

ii) The converse of the result in (i) is also true. That is, if $a = q_1 n + r_1$ and $b = q_2 n + r_2$, with $r_1 = r_2$, then $a - b = (q_1 - q_2)n$ and $a \equiv b \pmod{n}$.

iii) Although $a = b \Rightarrow a \equiv b \pmod{n}$, we can*not* expect $a \equiv b \pmod{n} \Rightarrow a = b$. However, if $a \equiv b \pmod{n}$ and $a, b \in \{0, 1, 2, \ldots, n - 1\}$, then $a = b$.

THEOREM 14.11

Congruence modulo n is an equivalence relation on $\mathbf{Z}$.

Proof: The proof is left for the reader.

Since an equivalence relation on a set induces a partition of that set, for $n \ge 2$, congruence modulo n partitions $\mathbf{Z}$ into the n equivalence classes

$$[0] = \{\ldots, -2n, -n, 0, n, 2n, \ldots\} = \{0 + nx | x \in \mathbf{Z}\}$$
$$[1] = \{\ldots, -2n + 1, -n + 1, 1, n + 1, 2n + 1, \ldots\} = \{1 + nx | x \in \mathbf{Z}\}$$
$$[2] = \{\ldots, -2n + 2, -n + 2, 2, n + 2, 2n + 2, \ldots\} = \{2 + nx | x \in \mathbf{Z}\}$$
$$\vdots$$
$$[n - 1] = \{\ldots, -n - 1, -1, n - 1, 2n - 1, 3n - 1, \ldots\}$$
$$= \{(n - 1) + nx | x \in \mathbf{Z}\}.$$

For all $t \in \mathbf{Z}$, by the division algorithm (of Section 4.3) we can write $t = qn + r$, where $0 \le r < n$, so $t \in [r]$, or $[t] = [r]$. We use the notation $\mathbf{Z}_n$ to denote $\{[0], [1], [2], \ldots, [n - 1]\}$. (When there is no danger of ambiguity, we often replace $[a]$ by a and write $\mathbf{Z}_n = \{0, 1, 2, \ldots, n - 1\}$.) Our objective now is to define closed binary operations of addition and multiplication on the set $\mathbf{Z}_n$ of equivalence classes so that we obtain a ring.

For $[a]$, $[b] \in \mathbf{Z}_n$, define $+$ and $\cdot$ by

$$[a] + [b] = [a + b] \quad \text{and} \quad [a] \cdot [b] = [a][b] = [ab].$$

For example, if $n = 7$, then $[2] + [6] = [2 + 6] = [8] = [1]$, and $[2][6] = [12] = [5]$.

Before these definitions are so readily accepted, we must investigate whether or not these (closed binary) operations are *well-defined* in the sense that if $[a] = [c]$, $[b] = [d]$, then $[a] + [b] = [c] + [d]$ and $[a][b] = [c][d]$. Since $[a] = [c]$ can occur with $a \neq c$, do the results of our addition and multiplication depend on which representatives are chosen from the equivalence classes? We shall prove that the results of the two operations are independent of the choice of class representatives and that the operations are very definitely well-defined.

First, we observe that $[a] = [c] \Rightarrow a = c + sn$, for some $s \in \mathbf{Z}$, and $[b] = [d] \Rightarrow b = d + tn$, for some $t \in \mathbf{Z}$. Hence

$$a + b = (c + sn) + (d + tn) = c + d + (s + t)n,$$

so $(a + b) \equiv (c + d) \pmod{n}$ and $[a + b] = [c + d]$. Also,

$$ab = (c + sn)(d + tn) = cd + (sd + ct + stn)n$$

and $ab \equiv cd \pmod{n}$, or $[ab] = [cd]$.

This result now leads us to the following.

THEOREM 14.12

For $n \in \mathbf{Z}^+$, $n > 1$, under the closed binary operations defined above, $\mathbf{Z}_n$ is a commutative ring with unity $[1]$ (and additive identity $[0]$).

Proof: The proof is left to the reader. Verification of the ring properties follows from the definitions of addition and multiplication in $\mathbf{Z}_n$ and from the corresponding properties for the ring $(\mathbf{Z}, +, \cdot)$.

Before stating any further results, let us examine two particular examples, $\mathbf{Z}_5$ and $\mathbf{Z}_6$. In Tables 14.7(a) and (b) and 14.8(a) and (b), we simplify $[a]$ by writing a.

Table 14.7

$\mathbf{Z}_5$ $+$	0	1	2	3	4		$\cdot$	0	1	2	3	4
0	0	1	2	3	4		0	0	0	0	0	0
1	1	2	3	4	0		1	0	1	2	3	4
2	2	3	4	0	1		2	0	2	4	1	3
3	3	4	0	1	2		3	0	3	1	4	2
4	4	0	1	2	3		4	0	4	3	2	1

   (a)   (b)

In $\mathbf{Z}_5$ every nonzero element has a multiplicative inverse, so $\mathbf{Z}_5$ is a field. For $\mathbf{Z}_6$, however, 1 and 5 are the only units and 2, 3, 4 are proper divisors of zero. Meanwhile, in $\mathbf{Z}_9$, $3 \cdot 3 = 3 \cdot 6 = 0$, so 3 and 6 are proper divisors of zero. Consequently, for $\mathbf{Z}_n$, $n > 2$, to be a field, we need more than just an odd modulus.

Table 14.8

$\mathbf{Z}_6$	+	0	1	2	3	4	5
	0	0	1	2	3	4	5
	1	1	2	3	4	5	0
	2	2	3	4	5	0	1
	3	3	4	5	0	1	2
	4	4	5	0	1	2	3
	5	5	0	1	2	3	4

(a)

·	0	1	2	3	4	5
0	0	0	0	0	0	0
1	0	1	2	3	4	5
2	0	2	4	0	2	4
3	0	3	0	3	0	3
4	0	4	2	0	4	2
5	0	5	4	3	2	1

(b)

THEOREM 14.13

$\mathbf{Z}_n$ is a field if and only if n is a prime.

Proof: Let n be a prime, and suppose that $0 < a < n$. Then $\gcd(a, n) = 1$, so as we learned in Section 4.4 there are integers s, t with $as + tn = 1$. Thus $as \equiv 1 \pmod{n}$, or $[a][s] = [1]$, and $[a]$ is a unit of $\mathbf{Z}_n$, which is consequently a field.

Conversely, if n is not a prime, then $n = n_1 n_2$, where $1 < n_1, n_2 < n$. So $[n_1] \neq [0]$ and $[n_2] \neq [0]$ but $[n_1][n_2] = [n_1 n_2] = [0]$, and $\mathbf{Z}_n$ is not even an integral domain, so it cannot be a field.

In $\mathbf{Z}_6$, $[5]$ is a unit and $[3]$ is a zero divisor. We seek a way to recognize when $[a]$ is a unit in $\mathbf{Z}_n$, for n composite.

THEOREM 14.14

In $\mathbf{Z}_n$, $[a]$ is a unit if and only if $\gcd(a, n) = 1$.

Proof: If $\gcd(a, n) = 1$, the result follows as in the proof of Theorem 14.13. For the converse, let $[a] \in \mathbf{Z}_n$ and $[a]^{-1} = [s]$. Then $[as] = [a][s] = [1]$, so $as \equiv 1 \pmod{n}$ and $as = 1 + tn$, for some $t \in \mathbf{Z}$. But $1 = as + n(-t) \Rightarrow \gcd(a, n) = 1$.

EXAMPLE 14.13

Find $[25]^{-1}$ in $\mathbf{Z}_{72}$.

Since $\gcd(25, 72) = 1$, the Euclidean algorithm leads us to

$$72 = 2(25) + 22, \qquad 0 < 22 < 25$$
$$25 = 1(22) + 3, \qquad 0 < 3 < 22$$
$$22 = 7(3) + 1, \qquad 0 < 1 < 3.$$

As 1 is the last nonzero remainder, we have

$$1 = 22 - 7(3) = 22 - 7[25 - 22] = (-7)(25) + (8)(22)$$
$$= (-7)(25) + 8[72 - 2(25)] = 8(72) - 23(25).$$

But

$$1 = 8(72) - 23(25) \Rightarrow 1 \equiv (-23)(25) \equiv (-23 + 72)(25) \pmod{72},$$

so $[1] = [49][25]$ and $[25]^{-1} = [49]$ in $\mathbf{Z}_{72}$.

In addition, from this result we are now able to solve the following linear congruences for x:

1) If $25x \equiv 1 \pmod{72}$, then $x \equiv 49 \pmod{72}$.

2) If $25x \equiv 3 \pmod{72}$, then $x \equiv 49 \cdot 3 \pmod{72} \equiv 3 \pmod{72}$.

Now [25] is a unit in $\mathbf{Z}_{72}$, but is there any way of knowing how many units this ring has? From Theorem 14.14, if $1 \leq a < 72$, then $[a]^{-1}$ exists if and only if $\gcd(a, 72) = 1$. Consequently, the number of units in $\mathbf{Z}_{72}$ is the number of integers a, such that $1 \leq a < 72$ and $\gcd(a, 72) = 1$. Using Euler's phi function (Example 8.8), we find that this is

$$\phi(72) = \phi(2^3 3^2) = (72)[1 - (1/2)][1 - (1/3)] = (72)(1/2)(2/3) = 24.$$

In general, for any $n \in \mathbf{Z}^+, n > 1$, there are $\phi(n)$ units and $n - 1 - \phi(n)$ proper divisors of zero in $\mathbf{Z}_n$.

Before we continue with some examples where congruence plays a role, we want to look back at the binary operation **mod** that was introduced earlier in Examples 4.36 and 10.8. In those examples we considered $x, y \in \mathbf{Z}^+$ and defined x **mod** y as the remainder obtained when we divide x by y. At this point we shall extend this concept to include the case where $x \leq 0$. Hence, for $x \in \mathbf{Z}$ and $y \in \mathbf{Z}^+$, x **mod** y is the remainder that results upon division of x by y.

But, now, how is **mod** related to the mod of Definition 14.7? Here we find that if $a, b, n \in \mathbf{Z}$, with $n > 1$, then $a \equiv b \pmod{n}$ if and only if a **mod** $n = b$ **mod** n. (This follows from the observations we made prior to Theorem 14.11.)

And now the time has arrived for some additional examples.

EXAMPLE 14.14

Randomly generated numbers arise in many applications. In particular, they are often used for the computer simulation of experiments that are too expensive, too dangerous, or just plain impossible to conduct in the real world.

The idea of using a computer to generate random numbers was first developed by John von Neumann (1903–1957) in 1946. However, although these numbers may *appear* to be random, they are not — hence the title *pseudorandom* numbers.

Proposed in 1949 by Derrick H. Lehmer (1905–1991), the most commonly used technique for generating such pseudorandom numbers employs the notion of congruence. For the *linear congruential generator*, one starts with the four integers: the multiplier a, the increment c, the modulus m, and the seed x_0, where

$$2 \leq a < m, \qquad 0 \leq c < m, \qquad \text{and} \qquad 0 \leq x_0 < m.$$

These nonnegative integers are used to generate a sequence of pseudorandom numbers, $x_1, x_2, x_3, \ldots$, recursively, by

$$x_{n+1} = (ax_n + c) \textbf{ mod } m.$$

So $0 \leq x_{n+1} < m$, for $n \geq 0$. For example, if $a = 3$, $c = 2$, $m = 11$, and $x_0 = 1$, then

$$x_1 = (ax_0 + c) \textbf{ mod } m = [3(1) + 2] \textbf{ mod } 11 = 5, \text{ so } x_1 = 5.$$

Similarly, $x_2 = (ax_1 + c) \textbf{ mod } m = [3(5) + 2] \textbf{ mod } 11 = 17 \textbf{ mod } 11 = 6$, so $x_2 = 6$.

Continuing in this manner, one finds that $x_3 = 9$, $x_4 = 7$, and $x_5 = 1$, the seed. Consequently, this linear congruential generator produces five distinct integers before repeating. The sequence of pseudorandom numbers thus obtained is 1, 5, 6, 9, 7, 1, 5, 6, ...

With $a = 3$, $c = 5$, $m = 12$, and $x_0 = 6$, we first learn that $x_1 = [3(6) + 5]$ **mod** $12 = 11$, so $x_1 = 11$. Next, $x_2 = [3(11) + 5]$ **mod** $12 = 38$ **mod** $12 = 2$, so $x_2 = 2$. Further computation yields $x_3 = 11$. This time the linear congruential generator yields only three distinct integers before repeating. The sequence of pseudorandom numbers generated here is 6, 11, 2, 11, 2, 11, 2, . . . , where the seed is not repeated.

In practice large values for a and m are used — especially for critical simulations. For $a = 16,807$ ($= 7^5$), $c = 0$, $m = 2,147,483,647$ ($= 2^{31} - 1$, a prime), and $x_0 = 1$, one obtains a sequence of 2,147,483,647 pseudorandom numbers before a repeated integer appears.

EXAMPLE 14.15

a) Whether it's youngsters using decoder rings or military leaders sending battle plans to troops, throughout history, various people have wanted to keep certain information unintelligible, should it fall into the wrong hands.

As early as the first century B.C., the Roman general Gaius Julius Caesar (100 B.C.– 44 B.C.) used a *cipher shift* to make the contents of certain messages understandable only for those he intended the messages to reach. To describe this early form of *cryptosystem* — often termed the *Caesar cipher* — we shall make certain conventions to simplify the presentation. First, we shall write the original message, the *plaintext*, using only lowercase letters, with no punctuation or spaces. Then to encrypt the plaintext, each lowercase letter, from a to w, is *shifted* to the letter three places forward in the alphabet, and the last three letters — namely, x, y, and z — are shifted to the first three letters, respectively. We use the uppercase letters for the resulting *ciphertext*. Consequently, a is encrypted as D, b as E, c as F, . . . , j as M, . . . , m as P, . . . , y as B, and z as C.

If Caesar wanted to inform a senator in Rome of a recent victory, he might have sent the message "I came, I saw, I conquered." Encryption of this message takes place as follows:

Plaintext   *i c a m e i s a w i c o n q u e r e d*

Ciphertext   *L F D P H L V D Z L F R Q T X H U H G*

Upon receiving the ciphertext, as long as this senator knows the size and direction of the shift, he can reverse the process. Decryption then results by replacing each uppercase letter, from D to Z, in the ciphertext by the lowercase letter three places back in the alphabet, and A by x, B by y, and C by z. After decrypting, one then inserts the appropriate spaces and punctuation in the plaintext. (Note that by removing spaces in the plaintext, the resulting absence of spaces in the ciphertext helps make the message more unintelligible. If one does not know the size and direction of the shift for decryption, the presence of spaces may suggest certain information about the structure of the original message.)

b) The idea of Caesar's cipher can be generalized and modeled mathematically by using the concept of congruence. Start by assigning each of the 26 letters of the plaintext a nonnegative integer as shown:

a	b	c	d	$\cdots$	k	ℓ	m	n	$\cdots$	w	x	y	z
0	1	2	3	$\cdots$	10	11	12	13	$\cdots$	22	23	24	25

The 26 letters for the ciphertext are assigned the same integers — that is, A is assigned 0, B is assigned 1, ..., Y is assigned 24, and Z is assigned 25.

Now select a nonnegative integer κ, where $0 \leq \kappa \leq 25$. For instance, Caesar chose $\kappa = 3$. This integer κ is called the *key* and helps us define the encrypting function $E: \mathbf{Z}_{26} \to \mathbf{Z}_{26}$ as follows. Given a letter of the plaintext, let θ be the nonnegative integer to which it corresponds. Then $E(\theta) = (\theta + \kappa) \bmod 26$ and this result determines the corresponding ciphertext letter for the plaintext letter assigned the nonnegative integer θ. To decrypt we apply the inverse function $D: \mathbf{Z}_{26} \to \mathbf{Z}_{26}$ where we write $D(\theta) = (\theta - \kappa) \bmod 26$. Replacing each nonnegative integer with its corresponding plaintext letter, one captures the plaintext version of the original message.

If we do not know the key, a trial-and-error approach can be used. There are 26 possibilities — one for each of the 26 possible values of κ. A more efficient method of attack takes into account the most frequently occurring letters in the alphabet and the most frequently occurring letters in the ciphertext. In the English language, the letter e occurs most often, with t, a, o and i the next four most frequently occurring letters.

Now if a parent receives the ciphertext $Z \ L \ U \ K \ T \ V \ Y \ L \ T \ V \ U \ L \ F$ from a college student, and does not know the key, what can the parent do? Since the most frequently occurring letter in the ciphertext is L, the parent can correspond e with L under the encryption. This suggests that $E: \mathbf{Z}_{26} \to \mathbf{Z}_{26}$ be defined by $E(\theta) = (\theta + 7) \bmod 26$, since L is seven places after e in the alphabet. So here the key, κ, is 7 and the decryption function is $D: \mathbf{Z}_{26} \to \mathbf{Z}_{26}$ with $D(\theta) = (\theta - 7) \bmod 26$.

Decoding the ciphertext message received by the parent can be analyzed as follows:

(1)	Z	L	U	K	T	V	Y	L	T	V	U	L	F
(2)	25	11	20	10	19	21	24	11	19	21	20	11	5
(3)	18	4	13	3	12	14	17	4	12	14	13	4	24
(4)	s	e	n	d	m	o	r	e	m	o	n	e	y

Here (1) provides the given (encrypted) ciphertext. In (2) each ciphertext letter is replaced by the nonnegative integer assigned to it. Upon applying the decryption function D, the results in (2) provide the assignments in (3). Replacing each nonnegative integer in (3) by its corresponding plaintext letter yields the original message

"Send more money."

c) The security of the shift cipher in part (b) can be slightly enhanced by means of the *affine cipher*. The letters of the plaintext and ciphertext are assigned nonnegative integers, as in part (b). Here, however, the encryption function E is given by $E(\theta) = (\alpha\theta + \kappa) \bmod 26$, where $0 \leq \alpha, \kappa \leq 25$, and $\gcd(\alpha, 26) = 1$.

If $\theta_1, \theta_2 \in \mathbf{Z}_{26}$, then $E(\theta_1) = E(\theta_2) \Rightarrow (\alpha\theta_1 + \kappa) \bmod 26 = (\alpha\theta_2 + \kappa) \bmod 26 \Rightarrow \alpha\theta_1 \bmod 26 = \alpha\theta_2 \bmod 26 \Rightarrow \theta_1 = \theta_2$, by Theorem 14.14. So E is one-to-one. Further, E is also onto and invertible, by Theorem 5.11, because $\mathbf{Z}_{26}$ is finite.

Let us consider a specific example. Suppose $\alpha = 11$ and $\kappa = 7$. Then the encryption of the plaintext letter g proceeds as follows:

i) g is assigned the nonnegative integer 6;

ii) applying E, we have $E(6) = (11 \cdot 6 + 7) \bmod 26 = 73 \bmod 26 = 21$; and

iii) the nonnegative integer 21 determines the ciphertext letter V.

[So using this affine cipher, where $E(\theta) = (11\theta + 7) \bmod 26$, the plaintext letter g is encrypted as the ciphertext letter V.]

Now suppose we have the following ciphertext for a message encrypted by an affine cipher:

$$QYYFGCULBLKYZVOSTCOY PURGCULYZYWKYOSTCOYL$$

With no knowledge of α or κ, one might have to examine as many as $[\phi(26)](26) = \left[26\left(1 - \frac{1}{2}\right)\left(1 - \frac{1}{13}\right)\right](26) = \left[26\left(\frac{1}{2}\right)\left(\frac{12}{13}\right)\right](26) = (12)(26) = 312$ cases for the key α, κ. However, let's say that by some means — perhaps by considering the frequencies of occurrence for the letters in the plaintext and ciphertext — we deduce two correspondences. Specifically, we know that e and Y correspond, as do t and R. In addition, the nonnegative integers 4 and 19 are the replacements for the plaintext letters e and t, respectively, while 24 and 17 are the respective replacements for Y and R, in the ciphertext, so the encryption function E is determined as follows:

1) The correspondence of $e(4)$ and $Y(24)$ tells us that $E(4) = (4\alpha + \kappa) \bmod 26 = 24$.

2) The correspondence of $t(19)$ and $R(17)$ tells us that $E(19) = (19\alpha + \kappa) \bmod 26 = 17$.

Consequently, $E(19) - E(4) = [(19\alpha + \kappa) - (4\alpha + \kappa)] \bmod 26 = 15\alpha \bmod 26 = (17 - 24) \bmod 26 = -7 \bmod 26 = 19$. Since $15 \cdot 7 = 105 = 1 + 104 = 1 + 4(26)$, we have $15 \cdot 7 = 1 \bmod 26$, so $15^{-1} = 7$ (in $\mathbf{Z}_{26}$). Then $15\alpha = 19 \bmod 26 \Rightarrow \alpha = 15^{-1} \cdot 19 \bmod 26 = 7 \cdot 19 \bmod 26 = 133 \bmod 26 = 3$, as $133 = 3 + 5(26)$. With $\alpha = 3 \bmod 26$ it now follows from (1) that $\kappa = (24 - 4\alpha) \bmod 26 = (24 - 12) \bmod 26 = 12$. [Or, from (2), $\kappa = (17 - 19\alpha) \bmod 26 = (17 - 57) \bmod 26 = -40 \bmod 26 = 12$.]

Consequently, $E: \mathbf{Z}_{26} \to \mathbf{Z}_{26}$ is defined by $E(\theta) = (3\theta + 12) \bmod 26$ and the decryption function $D: \mathbf{Z}_{26} \to \mathbf{Z}_{26}$ is given by $D(\theta) = (9\theta + 22) \bmod 26$, since $E^{-1}(\theta) = 3^{-1}(\theta - 12) \bmod 26 = 9(\theta - 12) \bmod 26 = (9\theta - 108) \bmod 26 = (9\theta + 22) \bmod 26$. This function D is used in the following to obtain the results in row 3 from the nonnegative integers (that replace the ciphertext letters) in row 2.

(1) Ciphertext	Q	Y	Y	F	G	C	U	L	B	L	K	Y	Z	V	O	S	T	C	O	Y
(2)	16	24	24	5	6	2	20	11	1	11	10	24	25	21	14	18	19	2	14	24
(3)	10	4	4	15	24	14	20	17	5	17	8	4	13	3	18	2	11	14	18	4
(4) Plaintext	k	e	e	p	y	o	u	r	f	r	i	e	n	d	s	c	ℓ	o	s	e

(1) Ciphertext	P	U	R	G	C	U	L	Y	Z	Y	W	K	Y	O	S	T	C	O	Y	L
(2)	15	20	17	6	2	20	11	24	25	24	22	10	24	14	18	19	2	14	24	11
(3)	1	20	19	24	14	20	17	4	13	4	12	8	4	18	2	11	14	18	4	17
(4) Plaintext	b	u	t	y	o	u	r	e	n	e	m	i	e	s	c	ℓ	o	s	e	r

Here, for example, the ciphertext letter Q is replaced by the nonnegative integer 16. Applying the decryption function D to 16 we have $D(16) = (9 \cdot 16 + 22) \bmod 26 = 166 \bmod 26 = 10$, and 10 is the nonnegative integer that corresponds to the plaintext letter k.

The decrypted message now reveals the sage advice given by Don Vito Corleone (of Mario Puzo's *The Godfather*) to his youngest son, Michael — namely, "Keep your friends close but your enemies closer."

The security of each of the cryptosystems in Example 14.15 depends on the key [$\kappa = 3$ in part (a), κ in part (b), and α, κ in part (c)]. For such *private key cryptosystems*, the two people wishing to use the system need to securely exchange the key. Should any unauthorized person discover the key, then that person could readily encrypt or decrypt messages.

Our next example deals with modular exponentiation.

EXAMPLE 14.16

In the study of cryptology[†] one often needs to perform modular exponentiation to compute a result such as $b^e \bmod n$, where b, e, and n are large integers. To demonstrate this — on a somewhat smaller scale — let us determine $5^{143} \bmod 222$. We realize that it is rather inefficient to actually compute 5^{143} (a very large integer) and then find the remainder upon dividing the result (for 5^{143}) by 222. A more efficient approach starts with the binary representation for the exponent — here, 143. With

$$143 = 1(128) + 0(64) + 0(32) + 0(16) + 1(8) + 1(4) + 1(2) + 1(1)$$
$$= 1(2^7) + 0(2^6) + 0(2^5) + 0(2^4) + 1(2^3) + 1(2^2) + 1(2^1) + 1(2^0)$$
$$= (10001111)_2,$$

we compute $5^{143} \bmod 222$ by using the binary representation (of 143) in reverse order — that is, going from the right to the left. The pseudocode procedure in Fig. 14.1 provides the necessary steps for this computation. Here the input is an integer b, the positive integer n (the modulus), and the binary representation $(a_m a_{m-1} \cdots a_2 a_1 a_0)_2$ for the exponent e, another positive integer. The output x equals $b^e \bmod n$.

```
procedure ModularExponentiation(b: integer;
                  n, e = (a_m a_{m-1} ··· a_2 a_1 a_0)_2: positive integers)
begin
  x := 1
  power := b mod n
  for i = 0 to m do
    begin
      if a_i = 1 then x := (x * power) mod n
      power := (power * power) mod n
    end
end
```

Figure 14.1

For our example, $b = 5$, $e = 143 = (10001111)_2 = (a_7 a_6 a_5 \cdots a_2 a_1 a_0)_2$ [so $m = 7$], and $n = 222$. The results in Table 14.9 show us the steps that are followed in the execution of the **for** loop. This is after the initial assignments are made: x is 1 and *power* is $b \bmod n$ — that is, $5 \bmod 222 = 5$.

Following the execution of this procedure, the last entry in the column for x tells us that $5^{143} \bmod 222$ is 89.

[†]For more on cryptology (and related topics), the reader should find the references by T. H. Barr [3], P. Garrett [6], and W. Trappe and L. C. Washington [13] of interest.

Table 14.9

i	a_i	x	power
0	1	$1 * 5 = 5$	$5^2 \, (= 25) \, \textbf{mod} \, 222 = 25$
1	1	$5 * 25 \, \textbf{mod} \, 222 = 125$	$25^2 \, (= 625) \, \textbf{mod} \, 222 = 181$
2	1	$125 * 181 \, \textbf{mod} \, 222 = 203$	$181^2 \, (= 32761) \, \textbf{mod} \, 222 = 127$
3	1	$203 * 127 \, \textbf{mod} \, 222 = 29$	$127^2 \, (= 16129) \, \textbf{mod} \, 222 = 145$
4	0	29	$145^2 \, (= 21025) \, \textbf{mod} \, 222 = 157$
5	0	29	$157^2 \, (= 24649) \, \textbf{mod} \, 222 = 7$
6	0	29	$7^2 \, (= 49) \, \textbf{mod} \, 222 = 49$
7	1	$29 * 49 \, \textbf{mod} \, 222 = 89$	$49^2 \, (= 2401) \, \textbf{mod} \, 222 = 181$

The next example provides an application of modular congruence in information retrieval.

EXAMPLE 14.17

When searching a table of records stored in a computer, each record is assigned a memory location or *address* in the computer's memory. The record itself is often made up of fields (this has nothing to do with ring structures). For instance, a college registrar keeps a record on each student, with the record containing information on the student's social security number, name, and major, for a total of three fields.

In searching for a particular student's record, we can use his or her social security number as the *key* to the record because it uniquely identifies that record. As a result, we develop a function from the set of keys to the set of addresses in the table.

If the college is small enough, we may find that the first four digits of the social security number are enough for identification. We develop a *hashing* (or *scattering*) function h from the set of keys (still social security numbers) to the set of addresses, determined now by the first four digits of the key. For example, $h(081\text{-}37\text{-}6495)$ identifies the record at the address associated with 0813. In this way we can store the table using at most 10,000 addresses. All is well as long as h is one-to-one. Should a second student have social security number 081-39-0207, then h would no longer uniquely identify a student's record. When this happens, a *collision* is said to occur. Since increasing the size of the stored table often results in more unused storage, we must balance the cost of this storage against the cost of handling such collisions. Techniques for resolving collisions have been devised. They depend on the data structures (such as vectors or linear linked lists) that are used to store the records.

Different kinds of hashing functions that have been developed include the following.

a) The *division* method: Here we restrict the number of addresses we want to use to a fixed integer n. For any key k (a positive integer), we define $h(k) = r$, where $r = k \, \textbf{mod} \, n$ — that is, $r \equiv k \pmod{n}$ and $0 \leq r < n$.

b) Often implemented is the *folding* method, where the key is split into parts and the parts are added together to give $h(\text{key})$. For example, $h(081\text{-}37\text{-}6495) = 081 + 37 + 6495 = 6613$ utilizes folding, and if we want only three-digit addresses, suppressing the first digit 6, we can have $h(081\text{-}37\text{-}6495) = 613$.

The importance of choosing a pertinent hashing function cannot be emphasized enough as we try to improve efficiency in terms of greater speed and less unused storage.

Using the modular concept, we can develop a hashing function h, using the same keys as above, where

$$h(x_1x_2x_3\text{-}x_4x_5\text{-}x_6x_7x_8x_9) = y_1y_2y_3,$$

with

$$y_1 = (x_1 + x_2 + x_3) \textbf{ mod } 5$$
$$y_2 = (x_4 + x_5) \textbf{ mod } 3$$
$$y_3 = (x_6 + x_7 + x_8 + x_9) \textbf{ mod } 7.$$

Here, for example, $h(081\text{-}37\text{-}6495) = 413$.

Our last example for this section provides one more encounter with the Catalan numbers (of Sections 1.5 and 10.5).

EXAMPLE 14.18

In how many ways can we select three elements a, b, c from $\{0, 1, 2, 3\}$, if *repetitions are allowed* and we want $a + b + c \equiv 0 \pmod 4$? The selections are listed in column 1 of Table 14.10. (Here each selection sums to 0, 4, or 8, and order is *not* relevant. For instance, $a = 0$, $b = 1$, $c = 3$ is considered the same selection as $a = 1$, $b = 0$, $c = 3$.) We see that there are five such selections and we recall that $5 = \left(\frac{1}{3+1}\right)\binom{2 \cdot 3}{3}$, the third Catalan number. Furthermore, by adding 1 to each entry of the selection 0, 0, 0 (in row 1 and column 1) we obtain the selection 1, 1, 1 (in row 1 and column 2). Likewise, the selection 2, 3, 1 (in row 2 and column 3) arises by adding 2 to each entry of the selection 0, 1, 3 (in row 2 and column 1) and reducing each sum modulo 4. Similar computations provide the other 13 selections in columns 2, 3, 4.

Table 14.10

Sum Is 0 (mod 4)	Sum Is 3 (mod 4)	Sum Is 2 (mod 4)	Sum Is 1 (mod 4)
0, 0, 0	1, 1, 1	2, 2, 2	3, 3, 3
0, 1, 3	1, 2, 0	2, 3, 1	3, 0, 2
0, 2, 2	1, 3, 3	2, 0, 0	3, 1, 1
1, 1, 2	2, 2, 3	3, 3, 0	0, 0, 1
2, 3, 3	3, 0, 0	0, 1, 1	1, 2, 2

To generalize this result, we count the number of selections x_1, x_2, ..., x_n, from $\{0, 1, 2, 3, \ldots, n\}$, where repetitions are allowed and $x_1 + x_2 + \cdots + x_n \equiv 0 \pmod{n + 1}$. From Section 1.4 we know there are $\binom{(n+1)+n-1}{n} = \binom{2n}{n}$ ways to select n objects from $n + 1$ distinct objects, with repetitions allowed. Let Sel_n denote the set of these $\binom{2n}{n}$ selections. (The 20 selections in Table 14.10 illustrate Sel_3.) Define the relation $\mathcal{R}$ on Sel_n by $s_1 \mathcal{R} s_2$, if the sum of the entries in selection s_1 is the same, *modulo* $n + 1$, as the sum of the entries in selection s_2. Then $\mathcal{R}$ is an equivalence relation, so Sel_n can be partitioned into $n + 1$ equivalence classes (one for each of the selection sums 0, 1, 2, ..., n — taken *modulo* $n + 1$). [*Note:* We get all $n + 1$ possible selection sums, for if $0 \le k_1 \le n$, $0 \le k_2 \le n$, and $nk_1 \equiv nk_2 \pmod{n + 1}$, then $k_1 \equiv k_2 \pmod{n + 1}$. This is due to Theorem 14.14 since $\gcd(n, n + 1) = 1$. With $k_1, k_2 \in \{0, 1, \ldots, n\}$ it then follows that $k_1 = k_2$.]

For $0 \le s \le n$, let Sel_n^s denote the selections that sum to s, *modulo* $n + 1$. When $1 \le s \le n$, write $s = nk$ (for $k = n^{-1}s$). Define $f: Sel_n^0 \to Sel_n^s$ as follows. For

$\{x_1, x_2, \ldots, x_n\} \in Sel_n^0$, $f(\{x_1, x_2, \ldots, x_n\}) = \{x_1 + k, x_2 + k, \ldots, x_n + k\}$, where $x_i + k$ is reduced *modulo* $n + 1$. Now consider $\{y_1, y_2, \ldots, y_n\} \in Sel_n^s$ and define $g: Sel_n^s \to Sel_n^0$ by $g(\{y_1, y_2, \ldots, y_n\}) = \{y_1 + (n + 1 - k), y_2 + (n + 1 - k), \ldots, y_n + (n + 1 - k)\}$. One finds that $g = f^{-1}$ so $\left|Sel_n^0\right| = \left|Sel_n^1\right| = \cdots = \left|Sel_n^n\right|$. Consequently, each equivalence class has the same size, namely, $\left(\frac{1}{n+1}\right) \binom{2n}{n}$, the nth Catalan number.

EXERCISES 14.3

1. a) Determine whether each of the following pairs of integers is congruent modulo 8.

    i) 62, 118   **ii)** −43, −237   **iii)** −90, 230

b) Determine whether each of the following pairs of integers is congruent modulo 9.

    i) 76, 243   **ii)** −137, 700   **iii)** −56, −1199

2. For each of the following determine the value(s) of the integer $n > 1$ for which the given congruence is true.

  a) $28 \equiv 6 \pmod{n}$ **b)** $68 \equiv 37 \pmod{n}$

  c) $301 \equiv 233 \pmod{n}$ **d)** $49 \equiv 2 \pmod{n}$

3. List four elements in each of the following equivalence classes.

  a) $[1]$ in $\mathbf{Z}_7$ **b)** $[2]$ in $\mathbf{Z}_{11}$ **c)** $[10]$ in $\mathbf{Z}_{17}$

4. Prove that if $a, b, c, n \in \mathbf{Z}$ with $a, n > 0$, and $b \equiv c \pmod{n}$, then $ab \equiv ac \pmod{an}$.

5. Let $a, b, m, n \in \mathbf{Z}$ with $m, n > 0$. Prove that if $a \equiv b \pmod{n}$ and $m|n$, then $a \equiv b \pmod{m}$.

6. Let $m, n \in \mathbf{Z}^+$ with $\gcd(m, n) = 1$ and let $a, b \in \mathbf{Z}$. Prove that $a \equiv b \pmod{m}$ and $a \equiv b \pmod{n}$ if and only if $a \equiv b \pmod{mn}$.

7. Provide a counterexample to show that the result in the preceding exercise is false if $\gcd(m, n) > 1$.

8. Prove that for all integers n exactly one of n, $2n - 1$, and $2n + 1$ is divisible by 3.

9. If $n \in \mathbf{Z}^+$ and $n > 2$ prove that

$$\sum_{i=1}^{n-1} i \equiv \begin{cases} 0 & \pmod{n}, \quad n \text{ odd} \\ \frac{n}{2} & \pmod{n}, \quad n \text{ even.} \end{cases}$$

10. Complete the proofs of Theorems 14.11 and 14.12.

11. Define relation $\mathcal{R}$ on $\mathbf{Z}^+$ by $a \mathrel{\mathcal{R}} b$, if $\tau(a) = \tau(b)$, where $\tau(a) =$ the number of positive (integer) divisors of a. For example, $2 \mathrel{\mathcal{R}} 3$ and $4 \mathrel{\mathcal{R}} 25$ but $5 \not\mathrel{\mathcal{R}} 9$.

a) Verify that $\mathcal{R}$ is an equivalence relation on $\mathbf{Z}^+$.

b) For the equivalence classes $[a]$ and $[b]$ induced by $\mathcal{R}$, define operations of addition and multiplication by $[a] + [b] = [a + b]$ and $[a][b] = [ab]$. Are these operations well-defined [that is, does $a \mathrel{\mathcal{R}} c$, $b \mathrel{\mathcal{R}} d \Rightarrow (a + b) \mathrel{\mathcal{R}} (c + d)$, $(ab) \mathrel{\mathcal{R}} (cd)$]?

12. Find the multiplicative inverse of each element in $\mathbf{Z}_{11}, \mathbf{Z}_{13}$, and $\mathbf{Z}_{17}$.

13. Find $[a]^{-1}$ in $\mathbf{Z}_{1009}$ for (a) $a = 17$, (b) $a = 100$, and (c) $a = 777$.

14. a) Find all subrings of $\mathbf{Z}_{12}, \mathbf{Z}_{18}$, and $\mathbf{Z}_{24}$.

b) Construct the Hasse diagram for each of these collections of subrings, where the partial order arises from set inclusion. Compare these diagrams with those for the set of positive divisors of n ($n = 12$; 18; 24), where the partial order now comes from the divisibility relation.

c) Find the formula for the number of subrings in $\mathbf{Z}_n, n > 1$.

15. How many units and how many (proper) zero divisors are there in (a) $\mathbf{Z}_{17}$? (b) $\mathbf{Z}_{117}$? (c) $\mathbf{Z}_{1117}$?

16. Prove that in any list of n consecutive integers, one of the integers is divisible by n.

17. If three distinct integers are randomly selected from the set $\{1, 2, 3, \ldots, 1000\}$, what is the probability that their sum is divisible by 3?

18. a) For $c, d, n, m \in \mathbf{Z}$, with $n > 1$ and $m > 0$, prove that if $c \equiv d \pmod{n}$, then $mc \equiv md \pmod{n}$ and $c^m \equiv d^m \pmod{n}$.

b) If $x_n x_{n-1} \cdots x_1 x_0 = x_n \cdot 10^n + \cdots + x_1 \cdot 10 + x_0$ denotes an $(n + 1)$-digit integer, then prove that

$$x_n x_{n-1} \cdots x_1 x_0 \equiv x_n + x_{n-1} + \cdots + x_1 + x_0 \pmod{9}.$$

19. a) Prove that for all $n \in \mathbf{N}$, $10^n \equiv (-1)^n \pmod{11}$.

b) Consider the result for mod 9 in part (b) of Exercise 18. State and prove a comparable result for mod 11.

20. For p a prime determine all elements $a \in \mathbf{Z}_p$ where $a^2 = a$.

21. For $a, b, n \in \mathbf{Z}^+$ and $n > 1$, prove that $a \equiv b \pmod{n} \Rightarrow \gcd(a, n) = \gcd(b, n)$.

22. a) Show that for all $[a] \in \mathbf{Z}_7$, if $[a] \neq [0]$, then

$$[a]^6 = [1].$$

b) Let $n \in \mathbf{Z}^+$ with $\gcd(n, 7) = 1$. Prove that

$$7 | (n^6 - 1).$$

23. Use the Caesar cipher to encrypt the plaintext: "All Gaul is divided into three parts."

24. The ciphertext $FTQIMKIQIQDQ$ was encrypted using the encryption function $E: \mathbf{Z}_{26} \to \mathbf{Z}_{26}$ where $E(\theta) = (\theta + \kappa) \bmod 26$. Considering the frequencies of occurrence for the letters in the ciphertext, determine (a) the key κ for this cipher shift; (b) the decryption function D; and (c) the original (plaintext) message.

25. Determine the total number of affine ciphers for an alphabet of (a) 24 letters; (b) 25 letters; (c) 27 letters; and (d) 30 letters.

26. The ciphertext

$$RWJWQTOOMYHKUXGOEMYP$$

was encrypted with an affine cipher. Given that the plaintext letters e, t are encrypted as the ciphertext letters W, X, respectively, determine (a) the encryption function E; (b) the decryption function D; and (c) the original (plaintext) message.

27. (a) How many distinct terms does the linear congruential generator with $a = 5$, $c = 3$, $m = 19$, and $x_0 = 10$, produce? (b) What is the sequence of pseudorandom members generated?

28. Given the modulus m and the two seeds x_0, x_1, with $0 < x_0, x_1 < m$, a sequence of pseudorandom numbers can be generated recursively from $x_n = (x_{n-1} + x_{n-2}) \bmod m$, $n \geq 2$. This generator is called the *Fibonacci generator.*

Find the first ten pseudorandom numbers generated when $m = 37$ and $x_0 = 1$, $x_1 = 28$.

29. Let $x_{n+1} = (ax_n + c) \bmod m$, where $2 \leq a < m$, $0 \leq c < m$, $0 \leq x_0 < m$, $0 \leq x_{n+1} < m$, and $n \geq 0$. Prove that

$$x_n = (a^n x_0 + c[(a^n - 1)/(a - 1)]) \bmod m, 0 \leq x_n < m.$$

30. Consider the linear congruential generator with $a = 7$, $c = 4$, and $m = 9$. If $x_4 = 1$, determine the seed x_0.

31. Prove that the sum of the cubes of three consecutive integers is divisible by 9.

32. Determine the last digit in 3^{55}.

33. For $m, n, r \in \mathbf{Z}^+$, let $p(m, n, r)$ count the number of partitions of m into at most n (positive) summands each no larger than r. Evaluate $\sum_{k=0}^{n-1} p(k(n+1), n, n)$, $n \in \mathbf{Z}^+$.

34. Given a ring $(R, +, \cdot)$, an element $r \in R$ is called *idempotent* when $r^2 = r$. If $n \in \mathbf{Z}^+$ with $n \geq 2$, prove that if $k \in \mathbf{Z}_n$ and k is idempotent, then $n - k + 1$ is idempotent.

35. For the hashing function at the end of Example 14.17, find (a) $h(123\text{-}04\text{-}2275)$; (b) a social security number n such that $h(n) = 413$, thus causing a collision with the number 081-37-6495 of the example.

36. Write a computer program (or develop an algorithm) that implements the hashing function of Exercise 35.

37. The parking lot for a local restaurant has 41 parking spaces, numbered consecutively from 0 to 40. Upon driving into this lot, a patron is assigned a parking space by the parking attendant who uses the hashing function $h(k) = k \bmod 41$, where k is the integer obtained from the last three digits on the patron's license plate. Further, to avoid a *collision* (where an occupied space might be assigned), when such a situation arises, the patron is directed to park in the next (consecutive) available space — where 0 is assumed to follow 40.

a) Suppose that eight automobiles arrive as the restaurant opens. If the last three digits in the license plates for these eight patrons (in their order of arrival) are

206, 807, 137, 444, 617, 330, 465, 905,

respectively, which spaces are assigned to the drivers of these eight automobiles by the parking attendant?

b) Following the arrival of the eight patrons in part (a), and before any of the eight could leave, a ninth patron arrives with a license plate where the last three digits are $00x$. If this patron is assigned to space 5, what is (are) the possible value(s) of x?

38. Solve the following linear congruences for x.

a) $3x \equiv 7 \pmod{31}$ b) $5x \equiv 8 \pmod{37}$

c) $6x \equiv 97 \pmod{125}$

14.4
Ring Homomorphisms and Isomorphisms

In this final section we shall examine functions (between rings) that obey special properties which depend on the closed binary operations in the rings.

EXAMPLE 14.19

Consider the rings $(\mathbf{Z}, +, \cdot)$ and $(\mathbf{Z}_6, +, \cdot)$, where addition and multiplication in $\mathbf{Z}_6$ are as defined in Section 14.3.

Define $f: \mathbf{Z} \to \mathbf{Z}_6$ by $f(x) = [x]$. For example, $f(1) = [1] = [7] = f(7)$ and $f(2) = f(8) = f(2 + 6k) = [2]$, for all $k \in \mathbf{Z}$. (So f is onto though not one-to-one.)

For $2, 3 \in \mathbf{Z}$, $f(2) = [2]$, $f(3) = [3]$ and we have $f(2 + 3) = f(5) = [5] = [2] + [3] = f(2) + f(3)$, and $f(2 \cdot 3) = f(6) = [0] = [2][3] = f(2) \cdot f(3)$.

In fact, for all $x, y \in \mathbf{Z}$,

$$f(x + y) = [x + y] = [x] + [y] = f(x) + f(y),$$

↑   ↑
Addition in Z   **Addition in $\mathbf{Z}_6$**

and

$$f(x \cdot y) = [xy] = [x][y] = f(x) \cdot f(y).$$

↑   ↑
Multiplication in Z   **Multiplication in $\mathbf{Z}_6$**

This example suggests the following definition.

Definition 14.8

Let $(R, +, \cdot)$ and $(S, \oplus, \odot)$ be rings. A function $f: R \rightarrow S$ is called a *ring homomorphism* if for all $a, b \in R$,

a) $f(a + b) = f(a) \oplus f(b)$, and

b) $f(a \cdot b) = f(a) \odot f(b)$.

When the function f is onto we say that S is a *homomorphic image* of R.

This function is said to *preserve* the ring operations for the following reasons: Consider $f(a + b) = f(a) \oplus f(b)$. *Adding a, b in R* first and then finding the image (under f) in S of this sum, we get the same result as when we first determine the images (under f) in S of a, b, and then *add* these images *in S*. (Hence we have the function operation and the additive operations commuting with each other.) Similar remarks can be made about the multiplicative operations in the rings.

For the rings $\mathbf{Z}_4$ and $\mathbf{Z}_8$, define the function $f: \mathbf{Z}_4 \rightarrow \mathbf{Z}_8$ by $f([a]) = [a]^2 (= [a^2])$. Then for all $[a], [b] \in \mathbf{Z}_4$, we have

$$f([a][b]) = f([ab]) = [ab]^2 = ([a][b])^2 = [a]^2[b]^2 = f([a])f([b]).$$

↑   ↑
Multiplication in $\mathbf{Z}_4$ **Multiplication in $\mathbf{Z}_8$**

Consequently, this function f preserves the multiplicative operations in the rings. However, for $[1], [2] \in \mathbf{Z}_4$, we find that $f([1] + [2]) = f([3]) = [3]^2 = [1]$, while $f([1]) + f([2]) = [1]^2 + [2]^2 = [1] + [4] = [5] (\neq [1]$ in $\mathbf{Z}_8$). So f does *not* preserve the additive operations in the rings — hence, f is *not* a ring homomorphism.

The function $g: \mathbf{Z}_4 \rightarrow \mathbf{Z}_8$, defined by $g([a]) = 3[a]$, preserves the additive operations, but not the multiplicative operations, in the rings.

Definition 14.9

Let $f: (R, +, \cdot) \rightarrow (S, \oplus, \odot)$ be a ring homomorphism. If f is one-to-one and onto, then f is called a *ring isomorphism* and we say that R and S are *isomorphic rings*.

We can think of isomorphic rings arising when the "same" ring is dealt with in two different languages. The function f then provides a dictionary for unambiguously translating from one language into the other.

The terms "homomorphism" and "isomorphism" come from the Greek, where *morphe* refers to shape or structure, *homo* means similar, and *iso* means identical or same. Hence homomorphic rings (that is, rings where one is a homomorphic image of the other) may be thought of as similar in structure, while isomorphic rings are (abstractly) replicas of the same structure.

In Definition 11.13 we defined the concept of *graph* isomorphism. There we called the undirected graphs $G_1 = (V_1, E_1)$ and $G_2 = (V_2, E_2)$ isomorphic when we could find a function $f: V_1 \rightarrow V_2$ such that

a) f is one-to-one and onto, and

b) $\{a, b\} \in E_1$ if and only if $\{f(a), f(b)\} \in E_2$.

In light of our statements about ring isomorphisms, another way to think about condition (b) here is in terms of the function f preserving the structures of the undirected graphs G_1 and G_2. When $|V_1| = |V_2|$, it is not difficult to find a function $f: V_1 \rightarrow V_2$ that is one-to-one and onto. However, for a given set V of vertices, what determines the structure of an undirected graph $G = (V, E)$ is its set of edges (where the vertex adjacencies are defined). Therefore a one-to-one correspondence $f: V_1 \rightarrow V_2$ is a graph isomorphism when it preserves the structures of G_1 and G_2 by preserving these vertex adjacencies.

EXAMPLE 14.20

For the ring R in Example 14.5 and the ring $\mathbf{Z}_5$, the function $f: R \rightarrow \mathbf{Z}_5$ given by

$$f(a) = [0], \qquad f(b) = [1], \qquad f(c) = [2], \qquad f(d) = [3], \qquad f(e) = [4]$$

provides us with a ring isomorphism.

For example, $f(c + d) = f(a) = [0] = [2] + [3] = f(c) + f(d)$, while $f(be) = f(e) = [4] = [1][4] = f(b)f(e)$. (In the absence of other methods and theorems, there are 25 such equalities that must be verified for the preservation of each of the binary operations.)

Inasmuch as there are $5! = 120$ one-to-one functions from R onto $\mathbf{Z}_5$, is there any assistance we can call upon in attempting to determine when one of these functions is an isomorphism? Suggested by Example 14.20, the following theorem provides ways of at least starting to determine when functions between rings can be homomorphisms and isomorphisms. [Parts (c) and (d) of this theorem rely on the results of Exercises 20 and 21 in Section 14.2.]

THEOREM 14.15

If $f: (R, +, \cdot) \rightarrow (S, \oplus, \odot)$ is a ring homomorphism, then

a) $f(z_R) = z_S$, where z_R, z_S are the zero elements of R, S, respectively;

b) $f(-a) = -f(a)$, for all $a \in R$;

c) $f(na) = nf(a)$, for all $a \in R$, $n \in \mathbf{Z}$;

d) $f(a^n) = [f(a)]^n$, for all $a \in R$, $n \in \mathbf{Z}^+$; and

e) if A is a subring of R, it follows that $f(A)$ is a subring of S.

Proof:

a) $z_S \oplus f(z_R) = f(z_R) = f(z_R + z_R) = f(z_R) \oplus f(z_R)$. (Why?) So by the cancellation law of addition in S, we have $f(z_R) = z_S$.

b) $z_S = f(z_R) = f(a + (-a)) = f(a) \oplus f(-a)$. Since additive inverses in S are unique and $f(-a)$ is an additive inverse of $f(a)$, it follows that $f(-a) = -f(a)$.

c) If $n = 0$, then $f(na) = f(z_R) = z_S = nf(a)$. The result is also true for $n = 1$, so we assume the truth for $n = k$ (≥ 1). Proceeding by mathematical induction, we examine the case where $n = k + 1$. By the results of Exercise 20 of Section 14.2, we get $f((k + 1)a) = f(ka + a) = f(ka) \oplus f(a) = kf(a) \oplus f(a)$ (Why?) $= (k + 1)(f(a))$ (Why?). (*Note:* There are three different kinds of addition here.)

When $n > 0$, $f(-na) = -nf(a)$. This follows from our prior proof by induction, part (b) of this proof, and part (b) of Theorem 14.1, because $f(-na) + f(na) = f(n(-a)) + f(na) = nf(-a) + nf(a) = n[f(-a) + f(a)] = n[-f(a) + f(a)] = nz_S = z_S$. Hence the result follows for all $n \in \mathbf{Z}$.

d) We leave this result for the reader to prove.

e) Since $A \neq \emptyset$, $f(A) \neq \emptyset$. If $x, y \in f(A)$, then $x = f(a), y = f(b)$ for some $a, b \in A$. Then $x \oplus y = f(a) \oplus f(b) = f(a + b)$, and $x \odot y = f(a) \odot f(b) = f(ab)$, with $a + b, ab \in A$ (Why?), so $x \oplus y, x \odot y \in f(A)$. Also, if $x \in f(A)$ then $x = f(a)$ for some $a \in A$. So we have $f(-a) = -f(a) = -x$, and because $-a \in A$ (Why?), we have $-x \in f(A)$. Therefore $f(A)$ is a subring of S.

When the homomorphism is onto, we obtain the following theorem.

THEOREM 14.16

If $f: (R, +, \cdot) \to (S, \oplus, \odot)$ is a ring homomorphism from R *onto* S, where $|S| > 1$, then

a) if R has unity u_R, then $f(u_R)$ is the unity of S;

b) if R has unity u_R and a is a unit in R, then $f(a)$ is a unit in S and $f(a^{-1}) = [f(a)]^{-1}$;

c) if R is commutative, then S is commutative; and

d) if I is an ideal of R, then $f(I)$ is an ideal of S.

Proof: We shall prove part (d) and leave the other parts to the reader. Since I is a subring of R, it follows that $f(I)$ is a subring of S by part (e) of Theorem 14.15. To verify that $f(I)$ is an ideal, let $x \in f(I)$ and $s \in S$. Then $x = f(a)$ and $s = f(r)$, for some $a \in I, r \in R$. So $s \odot x = f(r) \odot f(a) = f(ra)$, with $ra \in I$, and we have $s \odot x \in f(I)$. Similarly, $x \odot s \in f(I)$, so $f(I)$ is an ideal of S.

These theorems reinforce the way in which homomorphisms and isomorphisms preserve structure. But can we find any use for these functions, aside from using them to prove more theorems? To help answer this, we start by considering the following example.

EXAMPLE 14.21

Extending the idea developed in Exercise 18 of Section 14.2, let R be the ring $\mathbf{Z}_2 \times \mathbf{Z}_3 \times \mathbf{Z}_5$. Then $|R| = |\mathbf{Z}_2| \cdot |\mathbf{Z}_3| \cdot |\mathbf{Z}_5| = 30$, and the operations of addition and multiplication are defined in R as follows:

For all $(a_1, a_2, a_3), (b_1, b_2, b_3) \in R$ where $a_1, b_1 \in \mathbf{Z}_2$, $a_2, b_2 \in \mathbf{Z}_3$, and $a_3, b_3 \in \mathbf{Z}_5$,

$$(a_1, a_2, a_3) + (b_1, b_2, b_3) = (a_1 + b_1, a_2 + b_2, a_3 + b_3)$$

$$\begin{array}{cccc} \uparrow & \uparrow & \uparrow & \uparrow \\ \textbf{Addition} & \textbf{Addition} & \textbf{Addition} & \textbf{Addition} \\ \textbf{in R} & \textbf{in } \mathbf{Z_2} & \textbf{in } \mathbf{Z_3} & \textbf{in } \mathbf{Z_5} \end{array}$$

and

$$(a_1, a_2, a_3) \cdot (b_1, b_2, b_3) = (a_1 \cdot b_1, a_2 \cdot b_2, a_3 \cdot b_3).$$

| Multiplication in R | Multiplication in $\mathbf{Z}_2$ | Multiplication in $\mathbf{Z}_3$ | Multiplication in $\mathbf{Z}_5$ |

Define the function $f: \mathbf{Z}_{30} \to R$ by $f(x) = (x_1, x_2, x_3)$, where

$$x_1 = x \bmod 2$$
$$x_2 = x \bmod 3$$
$$x_3 = x \bmod 5.$$

In other words, x_1, x_2, and x_3 are the remainders that result when x is divided by 2, 3, and 5, respectively.

The results in Table 14.11 show that f is a function that is one-to-one and onto.

Table 14.11

x (in $\mathbf{Z}_{30}$)	$f(x)$ (in R)	x (in $\mathbf{Z}_{30}$)	$f(x)$ (in R)	x (in $\mathbf{Z}_{30}$)	$f(x)$ (in R)
0	(0, 0, 0)	10	(0, 1, 0)	20	(0, 2, 0)
1	(1, 1, 1)	11	(1, 2, 1)	21	(1, 0, 1)
2	(0, 2, 2)	12	(0, 0, 2)	22	(0, 1, 2)
3	(1, 0, 3)	13	(1, 1, 3)	23	(1, 2, 3)
4	(0, 1, 4)	14	(0, 2, 4)	24	(0, 0, 4)
5	(1, 2, 0)	15	(1, 0, 0)	25	(1, 1, 0)
6	(0, 0, 1)	16	(0, 1, 1)	26	(0, 2, 1)
7	(1, 1, 2)	17	(1, 2, 2)	27	(1, 0, 2)
8	(0, 2, 3)	18	(0, 0, 3)	28	(0, 1, 3)
9	(1, 0, 4)	19	(1, 1, 4)	29	(1, 2, 4)

To verify that f is an isomorphism, let $x, y \in \mathbf{Z}_{30}$. Then

$$f(x + y) = ((x + y) \bmod 2, (x + y) \bmod 3, (x + y) \bmod 5)$$
$$= (x \bmod 2, x \bmod 3, x \bmod 5) + (y \bmod 2, y \bmod 3, y \bmod 5)$$
$$= f(x) + f(y),$$

and

$$f(xy) = (xy \bmod 2, xy \bmod 3, xy \bmod 5)$$
$$= (x \bmod 2, x \bmod 3, x \bmod 5) \cdot (y \bmod 2, y \bmod 3, y \bmod 5)$$
$$= f(x)f(y),$$

so f is an isomorphism.

In examining Table 14.11 we find, for example, that

1) $f(0) = (0, 0, 0)$, where 0 is the zero element of $\mathbf{Z}_{30}$ and $(0, 0, 0)$ is the zero element of $\mathbf{Z}_2 \times \mathbf{Z}_3 \times \mathbf{Z}_5$.

2) $f(2 + 4) = f(6) = (0, 0, 1) = (0, 2, 2) + (0, 1, 4) = f(2) + f(4)$.

3) The element 21 is the additive inverse of 9 in $\mathbf{Z}_{30}$, whereas $f(21) = (1, 0, 1)$ is the additive inverse of $(1, 0, 4) = f(9)$ in $\mathbf{Z}_2 \times \mathbf{Z}_3 \times \mathbf{Z}_5$.

4) $\{0, 5, 10, 15, 20, 25\}$ is a subring of $\mathbf{Z}_{30}$ with $\{(0, 0, 0) \,(= f(0)), (1, 2, 0) \,(= f(5)),$
$(0, 1, 0) \,(= f(10)), (1, 0, 0) \,(= f(15)), (0, 2, 0) \,(= f(20)), (1, 1, 0) \,(= f(25))\}$
the corresponding subring in $\mathbf{Z}_2 \times \mathbf{Z}_3 \times \mathbf{Z}_5$.

But what else can we do with this isomorphism between $\mathbf{Z}_{30}$ and $\mathbf{Z}_2 \times \mathbf{Z}_3 \times \mathbf{Z}_5$? Suppose, for example, that we need to calculate $28 \cdot 17$ in $\mathbf{Z}_{30}$. We can transfer the problem to $\mathbf{Z}_2 \times \mathbf{Z}_3 \times \mathbf{Z}_5$ and compute $f(28) \cdot f(17) = (0, 1, 3) \cdot (1, 2, 2)$, where the moduli 2, 3, and 5 are smaller than 30 and easier to work with. Since $(0, 1, 3) \cdot (1, 2, 2) = (0 \cdot 1, 1 \cdot 2, 3 \cdot 2) = (0, 2, 1)$ and $f^{-1}(0, 2, 1) = 26$, it follows that $28 \cdot 17$ (in $\mathbf{Z}_{30}$) is 26.

In Example 14.21 we see that if we are given an element (x_1, x_2, x_3) in $\mathbf{Z}_2 \times \mathbf{Z}_3 \times \mathbf{Z}_5$, then we can use Table 14.11 to find the unique element x in $\mathbf{Z}_{30}$ so that $f(x) = (x_1, x_2, x_3)$. But what would we do if we did not have such a table—especially, if we found ourselves working with larger rings, such as $\mathbf{Z}_{32736}$ and $\mathbf{Z}_{31} \times \mathbf{Z}_{32} \times \mathbf{Z}_{33}$, and the isomorphism $g: \mathbf{Z}_{32736} \to \mathbf{Z}_{31} \times \mathbf{Z}_{32} \times \mathbf{Z}_{33}$ where $g(x) = (x \bmod 31, x \bmod 32, x \bmod 33)$ for $x \in \mathbf{Z}_{32736}$? The following result provides a technique for determining the unique preimage for a given element of the codomain for such an isomorphism g.

THEOREM 14.17

The Chinese Remainder Theorem. Let $m_1, m_2, \ldots, m_k \in \mathbf{Z}^+ - \{1\}$ with $k \geq 2$, and with $\gcd(m_i, m_j) = 1$ for all $1 \leq i < j \leq k$. Then the system of k congruences

$$x \equiv a_1 \pmod{m_1}$$

$$x \equiv a_2 \pmod{m_2}$$

$$\vdots$$

$$x \equiv a_k \pmod{m_k}$$

has a simultaneous solution. Further, any two such solutions of the system are congruent modulo $m_1 m_2 \cdots m_k$.

Proof: We start by showing how to construct a simultaneous solution of the system of k congruences.

Let $m = m_1 m_2 \cdots m_k$ and, for $1 \leq j \leq k$, let $M_j = m/m_j$. [So, for example, $M_1 = m_2 m_3 m_4 \cdots m_k$ and $M_2 = m_1 m_3 m_4 \cdots m_k$.] We find that for all $1 \leq j \leq k$, $\gcd(m_j, M_j) = 1$. If not, then for some (fixed) j, with $1 \leq j \leq k$, there exists a prime p such that $p | m_j$ and $p | M_j$. But from Lemma 4.3 it follows that if $p | M_j$ then $p | m_i$ for some $1 \leq i \leq k$, where $i \neq j$. Consequently, we find that $p | m_j$ and $p | m_i$ for $i \neq j$, and this contradicts $\gcd(m_j, m_j) = 1$.

For each $1 \leq j \leq k$, $\gcd(m_j, M_j) = 1$. Consequently, from Theorem 14.14 we know that M_j is a unit in $\mathbf{Z}_{m_j}$. So there exists $x_j \in \mathbf{Z}_{m_j}$ such that $M_j x_j \equiv 1 \pmod{m_j}$. Now consider the sum

$$x = a_1 M_1 x_1 + a_2 M_2 x_2 + \cdots + a_k M_k x_k.$$

We claim that x is a simultaneous solution of the system of k congruences. Note that for $1 \leq j \leq k$ and $1 \leq i \leq k$, if $i \neq j$ then $M_i \equiv 0 \pmod{m_j}$ because $m_j | M_i$. Hence $M_i x_i \equiv 0 \pmod{m_j}$. Since $M_j x_j \equiv 1 \pmod{m_j}$ we find that

$$x \equiv a_j M_j x_j \equiv a_j \pmod{m_j},$$

for each $1 \leq j \leq k$.

Now suppose that x, y are both simultaneous solutions of the system of k congruences. Then $x \equiv y \pmod{m_j}$ for all $1 \leq j \leq k$. Consider the prime factorization of $m = m_1 m_2 \cdots m_k$. Let p be a prime such that $p^t | m$ but $p^{t+1} \nmid m$, for some $t \in \mathbf{Z}^+$. Since $\gcd(m_i, m_j) = 1$ for all $1 \leq i < j \leq k$, it follows that $p^t | m_j$ for one (and only one) modulus m_j. Consequently, we see that $p^t | (x - y)$, and so it follows from the Fundamental Theorem of Arithmetic that $m | (x - y)$, or $x \equiv y \pmod{m}$.

Now let us see how one can apply the Chinese Remainder Theorem.

EXAMPLE 14.22

In Marjorie's fourth-grade arithmetic class, three students — namely, Megan, Avery, and Elizabeth — enjoy doing long-division problems (without a calculator). So Marjorie selects a positive integer n and asks for the remainder upon division by three different divisors. Upon dividing by 31 Megan learns that the remainder is 14. Avery divides n by 32 and finds the remainder is 16. Meanwhile, Elizabeth obtains the remainder of 18 when she divides n by 33. What is the smallest value of n that Marjorie could have selected?

Here we seek a simultaneous solution for the three congruences

$$x \equiv 14 \pmod{31}, \qquad x \equiv 16 \pmod{32}, \qquad x \equiv 18 \pmod{33}.$$

So $a_1 = 14$, $a_2 = 16$, $a_3 = 18$, $m_1 = 31$, $m_2 = 32$, $m_3 = 33$, and $m = m_1 m_2 m_3 = 32736$. Further, $M_1 = m/m_1 = 1056$, $M_2 = m/m_2 = 1023$, and $M_3 = m/m_3 = 992$. Using the Euclidean algorithm (when necessary), as in Example 14.13, we learn that

$$[x_1] = [M_1]^{-1} = [1056]^{-1} = [34(31) + 2]^{-1} = [2]^{-1} = [16] \text{ in } \mathbf{Z}_{m_1} = \mathbf{Z}_{31},$$
$$[x_2] = [M_2]^{-1} = [1023]^{-1} = [31(32) + 31]^{-1} = [31]^{-1} = [31] \text{ in } \mathbf{Z}_{m_2} = \mathbf{Z}_{32}, \quad \text{and}$$
$$[x_3] = [M_3]^{-1} = [992]^{-1} = [30(33) + 2]^{-1} = [2]^{-1} = [17] \text{ in } \mathbf{Z}_{m_3} = \mathbf{Z}_{33}.$$

Hence,

$$x \equiv (14)(1056)(16) + (16)(1023)(31) + (18)(992)(17) \pmod{32736}$$
$$\equiv 1047504 \pmod{32736}$$
$$\equiv 31(32736) + 32688 \pmod{32736}$$
$$\equiv 32688 \pmod{32736}.$$

So the (smallest) positive integer n that Marjorie could have selected is 32688.

(As a check we find that $32688 = 1054(31) + 14 = 1021(32) + 16 = 990(33) + 18$, so x satisfies the given system of three congruences and is the smallest positive integer that does so.)

Now if we look back at the isomorphism $g: \mathbf{Z}_{32736} \to \mathbf{Z}_{31} \times \mathbf{Z}_{32} \times \mathbf{Z}_{33}$ (that we mentioned prior to stating the Chinese Remainder Theorem) we see that for the codomain element $(14, 16, 18)$ in $\mathbf{Z}_{31} \times \mathbf{Z}_{32} \times \mathbf{Z}_{33}$, the element 32688 in the domain $\mathbf{Z}_{32736}$ is the (unique) preimage. That is, $g(32688) = (14, 16, 18)$ and for any other integer y, if $g(y) = (14, 16, 18)$, then $y \equiv 32688 \pmod{32736}$ — so 32688 is the only solution in $\{0, 1, 2, 3, \dots, 32735\}$.

The isomorphisms f (of Example 14.21) and g (of Example 14.22) are special cases of a more general result† that we shall now state. If $n = n_1 n_2 \cdots n_k$, where $n_i > 1$ for all $1 \leq i \leq k$ and $\gcd(n_i, n_j) = 1$ for all $1 \leq i < j \leq k$, then the rings $\mathbf{Z}_n$ and $\mathbf{Z}_{n_1} \times \mathbf{Z}_{n_2} \times \cdots \times \mathbf{Z}_{n_k}$ are isomorphic. In particular, we know from the Fundamental Theorem of Arithmetic that for each $n \in \mathbf{Z}^+ - \{1\}$, we can factor n as $p_1^{e_1} p_2^{e_2} \cdots p_t^{e_t}$, where $p_1, p_2, \ldots, p_t$ are t distinct primes, $t \geq 1$, and $e_1, e_2, \ldots, e_t \in \mathbf{Z}^+$. It then follows that the rings $\mathbf{Z}_n$ and $\mathbf{Z}_{m_1} \times \mathbf{Z}_{m_2} \times \cdots \times \mathbf{Z}_{m_t}$ are isomorphic for $m_1 = p_1^{e_1}$, $m_2 = p_2^{e_2}, \ldots, m_t = p_t^{e_t}$.

As a result of this isomorphism, arithmetic involving large integers (that exceed the word size of a given computer) can be performed using the smaller different moduli. Further, the computation for these smaller moduli can be carried out in parallel—thus, reducing computation time. [For more on the Chinese Remainder Theorem in conjunction with applications of residue arithmetic in computers, we direct the interested reader to pages 146–149 of the text by K. H. Rosen [12], pages 344–359 of the text by J. P. Tremblay and R. Manohar [14], as well as the text by D. E. Knuth [8].

EXERCISES 14.4

1. If R is the ring of Example 14.6, construct an isomorphism $f: R \to \mathbf{Z}_6$.

2. Complete the proofs of Theorems 14.15 and 14.16.

3. If R, S, and T are rings and $f: R \to S$, $g: S \to T$ are ring homomorphisms, prove that the composite function $g \circ f: R \to T$ is a ring homomorphism.

4. If $S = \left\{ \begin{bmatrix} a & 0 \\ 0 & a \end{bmatrix} \middle| a \in \mathbf{R} \right\}$, then S is a ring under matrix addition and multiplication. Prove that $\mathbf{R}$ is isomorphic to S.

5. a) Let $(R, +, \cdot)$ and $(S, \oplus, \odot)$ be rings with zero elements z_R and z_S, respectively. If $f: R \to S$ is a ring homomorphism, let $K = \{a \in R \mid f(a) = z_S\}$. Prove that K is an ideal of R. (K is called the *kernel* of the homomorphism f.)

b) Find the kernel of the homomorphism in Example 14.19.

c) Let f, $(R, +, \cdot)$, and $(S, \oplus, \odot)$ be as in part (a). Prove that f is one-to-one if and only if the kernel of f is $\{z_R\}$.

6. Use the information in Table 14.11 to compute each of the following in $\mathbf{Z}_{30}$.

a) $(13)(23) + 18$ **b)** $(11)(21) - 20$

c) $(13 + 19)(27)$ **d)** $(13)(29) + (24)(8)$

7. a) Construct a table (as in Example 14.21) for the isomorphism $f: \mathbf{Z}_{20} \to \mathbf{Z}_4 \times \mathbf{Z}_5$.

b) Use the table from part (a) to compute the following in $\mathbf{Z}_{20}$.

i) $(17)(19) + (12)(14)$
ii) $(18)(11) - (9)(15)$

8. Let $n, r, s \in \mathbf{Z}^+$ with $n, r, s \geq 2$, $n = rs$, and $\gcd(r, s) = 1$. If $f: \mathbf{Z}_n \to \mathbf{Z}_r \times \mathbf{Z}_s$ is a ring isomorphism with $f(a) = (1, 0)$ and $f(b) = (0, 1)$, prove that if $(m, t) \in \mathbf{Z}_r \times \mathbf{Z}_s$, then $f^{-1}(m, t) \equiv ma + tb \pmod{n}$.

9. a) How many units are there in the ring $\mathbf{Z}_8$?

b) How many units are there in the ring $\mathbf{Z}_2 \times \mathbf{Z}_2 \times \mathbf{Z}_2$?

c) Are $\mathbf{Z}_8$ and $\mathbf{Z}_2 \times \mathbf{Z}_2 \times \mathbf{Z}_2$ isomorphic rings?

10. a) How many units are there in $\mathbf{Z}_{15}$? How many in $\mathbf{Z}_3 \times \mathbf{Z}_5$?

b) Are $\mathbf{Z}_{15}$ and $\mathbf{Z}_3 \times \mathbf{Z}_5$ isomorphic?

11. Are $\mathbf{Z}_4$ and the ring in Example 14.4 isomorphic?

12. If $f: R \to S$ is a ring homomorphism and J is an ideal of S, prove that $f^{-1}(J) = \{a \in R \mid f(a) \in J\}$ is an ideal of R.

13. Find a simultaneous solution for the system of two congruences:

$$x \equiv 5 \pmod{8}$$
$$x \equiv 73 \pmod{81}.$$

14. A band of 17 pirates captures a treasure chest full of (identical) gold coins. When the coins are divided up into equal numbers, three coins remain. One pirate accuses the distributor of miscounting and kills him in a duel. As a result, the second time the coins are distributed, in equal numbers, among the 16 surviving pirates, there are 10 coins remaining. An argument erupts and leads to gun play, resulting in the demise of another pirate. Now when the coins are divided up, in 15 equal piles, there are no remaining coins. What is the smallest number of coins that could have been in the chest?

15. Find a simultaneous solution for the system of four congruences:

$$x \equiv 1 \pmod{2}$$
$$x \equiv 2 \pmod{3}$$
$$x \equiv 3 \pmod{5}$$
$$x \equiv 5 \pmod{7}.$$

---

†In some textbooks this result is referred to as the Chinese Remainder Theorem.

14.5
Summary and Historical Review

Emphasizing structure induced by two closed binary operations, this chapter has introduced us to the mathematical system called a ring. Throughout the development of mathematics, the ring of integers has played a key role. In the branch of mathematics called number theory, we examine the basic properties of $(\mathbf{Z}, +, \cdot)$, as well as the finite rings $(\mathbf{Z}_n, +, \cdot)$. The matrix rings provide familiar examples of noncommutative rings.

Pierre de Fermat (1601–1665)

Sophie Germain (1776–1831)

This chapter contains the development of an abstract theory. On the basis of the definition of a ring, we established principles of elementary algebra that we have been using since our early encounters with arithmetic, signed numbers, and the manipulation of unknowns. The reader may have found some of the proofs tedious, as we justified all the steps in the derivations. Faced with the challenge of trying to prove a result in abstract mathematics, one should follow the advice given by the Roman rhetorician Marcus Fabius Quintilianus (first century A.D.), when he said, "One should not aim at being possible to understand (or follow), but at being impossible to be misunderstood."

A famous problem in number theory, known as *Fermat's Last Theorem*, claims that the equation $x^n + y^n = z^n$, $n \in \mathbf{Z}^+$, $n > 1$, has no solutions in $\mathbf{Z}^+$ when $n > 2$. In 1637 the French mathematician Pierre de Fermat (1601–1665) wrote that he had proved this result but that the proof was too long to be included in the margin of his manuscript. Many renowned mathematicians of the eighteenth and nineteenth centuries tried to prove this result — among them Leonhard Euler (1707–1783), Peter Gustav Lejeune Dirichlet (1805–1859), Carl Friedrich Gauss (1777–1855), Sophie Germain (1776–1831), Adrien-Marie Legendre (1752–1833), Niels Henrik Abel (1802–1829), Gabriel Lamé (1795–1870), and Leopold Kronecker (1823–1891). Although unsuccessful, attempts to resolve Fermat's claim did result in new mathematical ideas and theories. The twentieth century also produced scholars who expended tremendous efforts on this problem. One such scholar was born in Cambridge, England, in 1953. There, at the age of 10, he went to the public library in his town and looked into a book on mathematics. As he read about Fermat's Last Theorem, it seemed so simple — and he wanted to prove it. In the 1970s Andrew Wiles went to Cambridge University, and after he finished his degree, he became a research student there, working in number theory — in an area called Iwasawa theory. For at this time Fermat's Last Theorem was not in fashion. When Wiles completed his doctorate, he moved to the United States, to a position at Princeton University. In the 1980s his enthusiasm for his childhood dream was rekindled and he spent close to seven years working alone — locked up in his attic office. He finally confided in his colleague Nick Katz — in January 1993. Then in June 1993 Professor Wiles returned to Cambridge to deliver a series of three lectures at a number-theory conference. The last lecture ended in grand applause, accompanied by flashing cameras and reporters' questions. It appeared that he had solved Fermat's Last Theorem. Unfortunately, when his 200-page write-up was peer-reviewed, by experts such as Nick Katz, problems started to arise, and a hole in the proof caused everything to collapse like a house of cards. The fall of 1993 found Wiles back at Princeton — now crestfallen, angry, and humiliated. But then, after renewed effort, on September 19, 1994, he took one last look at his proposed proof. The next morning he wrote up a new proof, as everything fell into place. This time no one could find any flaws. The May 1995 issue of the journal *Annals of Mathematics* contains the original Cambridge paper by Andrew Wiles and the correction by Wiles and his friend and former student Richard Taylor. At last Fermat's Last Theorem was laid to rest. (Although Wiles gets much of the praise, other mathematicians deserve accolades as well — among them, Kenneth Ribet, Barry Mazur, Goro Shimura, Yutaka Taniyama, Gerhard Frey, Matthias Flach, and Richard Taylor.) For more on the history and development of the proof of this famous theorem, the reader is directed to the very readable account given by A. D. Aczel [1].

Andrew John Wiles (1953–)

AP/Wide World Photos

In trying to prove Fermat's Last Theorem, the German mathematician Ernst Kummer (1810–1893) developed the foundations for the concept of the *ideal*. This concept was later formulated, named, and utilized by his countryman Richard Dedekind (1831–1916) in his research on what are now called Dedekind domains. Use of the term "ring," however, seems to be attributable to the German mathematician David Hilbert (1862–1943).

Ring homomorphisms and their interplay with ideals were extensively developed by the German mathematician Emmy Noether (1882–1935). This great genius received little remuneration, financial or otherwise, from the governing bodies of her native land because of the sexual bias that was prevalent in the universities at that time. Emmy Noether's talents were nonetheless recognized by her colleagues, and she was eulogized in the *New York Times* on May 3, 1935, by Albert Einstein (1879–1955), who acknowledged the influence and importance of her work for the development of relativity theory. In addition to enduring sexual bias, as a Jew she was forced to flee her homeland in 1933, when the Nazis came to power. She spent the last two years of her life guiding young mathematicians in the United States. For more on the life of this fascinating person, examine the biography by A. Dick [4] and the article by C. Kimberling [7].

The special rings called *fields* arise in the rational, real, and complex number systems. But we also saw some interesting finite fields. These structures will be examined again in Chapter 17 in connection with combinatorial designs. The field theory developed by the French genius Evariste Galois (1811–1832) answered questions about the solutions of polynomial equations of degree > 4. These questions had baffled mathematicians for centuries, and his ideas, now known as Galois theory, still comprise one of the most elegant mathematical theories ever developed. More on Galois theory appears in the text by O. Zariski and P. Samuel [16].

Emmy Noether (1882–1935)

For supplemental reading on ring theory at the introductory level, the interested reader should examine Chapters 12–18 of J. A. Gallian [5], Chapter 6 of V. H. Larney [9], and Chapters 6, 7, and 12 of N. H. McCoy and T. R. Berger [10]. A somewhat more advanced coverage can be found in Chapter 4 of the text by E. A. Walker [15].

The development of modular congruence, along with many related ideas, we owe primarily to Carl Friedrich Gauss. Problems involving systems of congruences date back to the late first century where they appear in the work of the Greek mathematician Nicomachus of Gerasa. Systems of two congruences can also be found in the writings of the seventh-century mathematician Brahmagupta (born in 1598 in northwestern India). However, it was not until 1247 that we find the publication of a general method for solving systems of linear congruences. In his *Shushu jiuzhang (Mathematical Treatise in Nine Sections)*, the method now called the Chinese Remainder Theorem is presented by the Chinese mathematician Qin Jiushao (c. 1202–1261). Born in the province of Sichuan during the time of Genghis Khan, this remarkable mathematical talent was also an accomplished architect, musician, and poet, as well as being quite the sportsman — in archery, fencing, and horsemanship.

More on the solution of congruences and the Chinese Remainder Theorem can be found in the texts by I. Niven, H. S. Zuckerman, and H. L. Montgomery [11] and K. H. Rosen [12].

As mentioned earlier (in the footnote in Example 14.16), more on the history, development, and applications of cryptology can be found in the texts by T. H. Barr [3], P. Garrett [6], and W. Trappe and L. C. Washington [13].

Finally, the topic of hashing, or scattering, can be further investigated in Chapter 2 of J. P. Tremblay and R. Manohar [14]. Chapter 4 of A. V. Aho, J. E. Hopcroft, and J. D. Ullman [2] includes a discussion on the efficiency of hashing functions and a probabilistic investigation of the collision problem that arises for these functions.

REFERENCES

1. Aczel, Amir D. *Fermat's Last Theorem: Unlocking the Secret of an Ancient Mathematical Problem.* New York: Four Walls Eight Windows, 1996.

2. Aho, Alfred V., Hopcroft, John E., and Ullman, Jeffrey D. *Data Structures and Algorithms.* Reading, Mass.: Addison-Wesley, 1983.

3. Barr, Thomas H. *Invitation to Cryptology.* Upper Saddle River, N.J.: Prentice-Hall, 2002.

4. Dick, Auguste. *Emmy Noether (1882–1935)*, trans. Heidi Blocher. Boston: Birkhäuser-Boston, 1981.

5. Gallian, Joseph A. *Contemporary Abstract Algebra*, 5th ed. Boston: Houghton Mifflin, 2002.

6. Garrett, Paul. *Making, Breaking Codes: An Introduction to Cryptology.* Upper Saddle River, N.J.: Prentice-Hall, 2001.

7. Kimberling, Clark. "Emmy Noether, Greatest Woman Mathematician." *Mathematics Teacher* (March 1982): pp. 246–249.

8. Knuth, Donald Ervin. *The Art of Computer Programming*, 3rd ed., Volume 2, *Semi-Numerical Algorithms.* Reading, Mass.: Addison-Wesley, 1997.

9. Larney, Violet Hachmeister. *Abstract Algebra: A First Course.* Boston: Prindle, Weber & Schmidt, 1975.

10. McCoy, Neal H., and Berger, Thomas R. *Algebra: Groups, Rings and Other Topics.* Boston: Allyn and Bacon, 1977.

11. Niven, Ivan, Zuckerman, Herbert S., and Montgomery, Hugh L. *An Introduction to the Theory of Numbers*, 5th ed. New York: Wiley, 1991.

12. Rosen, Kenneth H. *Elementary Number Theory*, 4th ed. Reading, Mass.: Addison-Wesley, 1999.

13. Trappe, Wade, and Washington, Lawrence C. *Introduction to Cryptography with Coding Theory.* Upper Saddle River, N.J.: Prentice-Hall, 2002.

14. Tremblay, Jean-Paul, and Manohar, R. *Discrete Mathematical Structures with Applications to Computer Science.* New York: McGraw-Hill, 1975.

15. Walker, Elbert A. *Introduction to Abstract Algebra.* New York: Random House/Birkhäuser, 1987.

16. Zariski, Oscar, and Samuel, Pierre. *Commutative Algebra*, Vol. I. Princeton, N.J.: Van Nostrand, 1958.

SUPPLEMENTARY EXERCISES

1. Determine whether each of the following statements is true or false. For each false statement give a counterexample.

a) If $(R, +, \cdot)$ is a ring, and $\emptyset \neq S \subseteq R$ with S closed under $+$ and $\cdot$, then S is a subring of R.

b) If $(R, +, \cdot)$ is a ring with unity, and S is a subring of R, then S has a unity.

c) If $(R, +, \cdot)$ is a ring with unity u_R, and S is a subring of R with unity u_S, then $u_R = u_S$.

d) Every field is an integral domain.

e) Every subring of a field is a field.

f) A field can have only two subrings.

g) Every finite field has a prime number of elements.

h) The field $(\mathbf{Q}, +, \cdot)$ has an infinite number of subrings.

2. Prove that a ring R is commutative if and only if $(a + b)^2 = a^2 + 2ab + b^2$, for all $a, b \in R$.

3. A ring R is called *Boolean* if $a^2 = a$ for all $a \in R$. If R is Boolean, prove that (a) $a + a = 2a = z$, for all $a \in R$; and (b) R is commutative.

4. With $\mathbf{C}$ the field of complex numbers and S the ring of 2×2 real matrices of the form $\begin{bmatrix} a & b \\ -b & a \end{bmatrix}$, define $f: \mathbf{C} \to S$ by $f(a + bi) = \begin{bmatrix} a & b \\ -b & a \end{bmatrix}$, for $a + bi \in \mathbf{C}$. Prove that f is a ring isomorphism.

5. If $(R, +, \cdot)$ is a ring, prove that $C = \{r \in R | ar = ra$, for all $a \in R\}$ is a subring of R. (The subring C is called the *center* of R.)

6. Given a finite field F, let $M_2(F)$ denote the set of all 2×2 matrices with entries from F. As in Example 14.2, $(M_2(F), +, \cdot)$ becomes a noncommutative ring with unity.

  a) Determine the number of elements in $M_2(F)$ if F is

  i) $\mathbf{Z}_2$ **ii)** $\mathbf{Z}_3$ **iii)** $\mathbf{Z}_p$, p a prime

  b) As in Exercise 13 of Section 14.1, $A = \begin{bmatrix} a & b \\ c & d \end{bmatrix} \in M_2(\mathbf{Z}_p)$ is a unit if and only if $ad - bc \neq z$. This occurs if the first row of A does not contain all zeros (that is, z's) and the second row is not a multiple (by an element of $\mathbf{Z}_p$) of the first. Use this observation to determine the number of units in

  i) $M_2(\mathbf{Z}_2)$ **ii)** $M_2(\mathbf{Z}_3)$ **iii)** $M_2(\mathbf{Z}_p)$, p a prime

7. Given an integral domain $(D, +, \cdot)$ with zero element z, let $a, b \in D$ with $ab \neq z$. (a) If $a^3 = b^3$ and $a^5 = b^5$, prove that $a = b$. (b) Let $m, n \in \mathbf{Z}^+$ with $\gcd(m, n) = 1$. If $a^m = b^m$ and $a^n = b^n$, prove that $a = b$.

8. Let $A = \mathbf{R}^+$. Define $\oplus$ and $\odot$ on A by $a \oplus b = ab$, the ordinary product of a, b; and $a \odot b = a^{\log_2 b}$.

  a) Verify that $(A, \oplus, \odot)$ is a commutative ring with unity.

  b) Is this ring an integral domain or field?

9. Let R be a ring with ideals A and B. Define $A + B = \{a + b | a \in A, b \in B\}$. Prove that $A + B$ is an ideal of R. (For any ring R, the ideals of R form a poset under set inclusion. If A and B are ideals of R, with $\text{glb}\{A, B\} = A \cap B$ and $\text{lub}\{A, B\} = A + B$, the poset is a lattice.)

10. a) If p is a prime, prove that p divides $\binom{p}{k}$, for all $0 < k < p$.

  b) If $a, b \in \mathbf{Z}$, prove that $(a + b)^p \equiv a^p + b^p \pmod{p}$.

11. Given n positive integers $x_1, x_2, \ldots, x_n$, not necessarily distinct, prove that either $n | (x_1 + x_2 + \cdots + x_i)$,

for some $1 \leq i \leq n$, or there exist $1 \leq i < j \leq n$ such that $n | (x_{i+1} + \cdots + x_{j-1} + x_j)$.

12. Consider the ring $(\mathbf{Z}^3, \oplus, \odot)$ where addition and multiplication are defined by $(a, b, c) \oplus (d, e, f) = (a + d, b + e, c + f)$ and $(a, b, c) \odot (d, e, f) = (ad, be, cf)$. (Here, for example, $a + d$ and ad are computed by using the standard binary operations of addition and multiplication in $\mathbf{Z}$.) Let S be the subset of $\mathbf{Z}^3$ where $S = \{(a, b, c) | a = b + c\}$. Prove that S is *not* a subring of $(\mathbf{Z}^3, \oplus, \odot)$.

13. a) In how many ways can one select two positive integers m, n, not necessarily distinct, so that $1 \leq m \leq 100$, $1 \leq n \leq 100$ and the last digit of $7^m + 3^n$ is 8?

  b) Answer part (a) for the case where $1 \leq m \leq 125$, $1 \leq n \leq 125$.

  c) If one randomly selects m, n [as in part (a)], what is the probability that 2 is now the last digit of $7^m + 3^n$?

14. Let $n \in \mathbf{Z}^+$ with $n > 1$.

  a) If $n = 2k$ where k is an odd integer, prove that
$$k^3 \equiv k \pmod{n}.$$

  b) If $n = 4k$ for some $k \in \mathbf{Z}^+$, prove that
$$(2k)^2 \equiv 0 \pmod{n}.$$

  c) Prove that
$$\sum_{i=1}^{n-1} i^3 \equiv \begin{cases} \frac{n}{2} \pmod{n}, & \text{for } n \text{ even with } \frac{n}{2} \text{ odd,} \\ 0 \pmod{n}, & \text{otherwise.} \end{cases}$$

15. Suppose that $a, b, c \in \mathbf{Z}$ and $5 | (a^2 + b^2 + c^2)$. Prove that $5 | a$ or $5 | b$ or $5 | c$.

16. Write a computer program (or develop an algorithm) that reverses the order of the digits in a given positive integer. For example, the input 1374 should result in the output 4731.

17. Suppose that $a, b, k \in \mathbf{Z}^+$ with $a - b = p_1^{e_1} p_2^{e_2} \cdots p_k^{e_k}$, for $p_1, p_2, \ldots, p_k$ prime and $e_1, e_2, \ldots, e_k \in \mathbf{Z}^+$. For how many values of n (> 1) is $a \equiv b \pmod{n}$ true?

18. As the co-chairs of the Homecoming Parade Committee, Jerina and Noor must organize the freshmen for a pregame presentation. When they arrange these students in rows of 8, there are three students remaining. When rows of 11 are tried, four students remain. Finally, rows of 15 leave five students remaining. So the co-chairs use the rows of 15 and place the remaining five students at the center (in positions 6–10) of the first row. What is the smallest number of freshmen Jerina and Noor are trying to organize?

15

Boolean Algebra and Switching Functions

Again we encounter an algebraic system in which the structure depends primarily on two closed binary operations. Unlike the situation for rings, in dealing with Boolean algebras we shall stress applications more than the abstract nature of the system. Nonetheless, we shall carefully examine the structure of a Boolean algebra, and in our study we shall find results that are quite different from those for rings. Among other things, a finite Boolean algebra must have 2^n elements, for some $n \in \mathbf{Z}^+$. Yet we know of at least one ring for each $m \in \mathbf{Z}^+$, $m > 1$ — namely, the ring $(\mathbf{Z}_m, +, \cdot)$.

In 1854 the English mathematician George Boole published his monumental work *An Investigation of the Laws of Thought*. Within this work Boole created a system of mathematical logic that he developed in terms of what is now called a Boolean algebra.

In 1938 Claude Elwood Shannon developed the algebra of switching functions and showed how its structure was related to the ideas established by Boole. As a result of this work, an example of abstract mathematics in the nineteenth century became an applied mathematical discipline in the twentieth century.

15.1
Switching Functions: Disjunctive and Conjunctive Normal Forms

An electric switch can be turned on (allowing the flow of current) or off (preventing the flow of current). Similarly, in a transistor, current is either passing (conducting) or not passing (nonconducting). These are two examples of *two-state devices*. (In Section 2.2 we saw how the electric switch was related to the two-valued logic.)

In order to investigate such two-state devices, we abstract these notions of "true" and "false," "on" and "off," as follows.

Let $B = \{0, 1\}$. We define addition, multiplication, and complements for the elements of B by

a)	$0 + 0 = 0; \quad 0 + 1 = 1 + 0 = 1 + 1 = 1.$
b)	$0 \cdot 0 = 0 = 1 \cdot 0 = 0 \cdot 1; \quad 1 \cdot 1 = 1.$
c)	$\overline{0} = 1; \quad \overline{1} = 0.$

A variable x is called a *Boolean variable* if x takes on only values in B. Consequently, $x + x = x$ and $x^2 = x \cdot x = xx = x$ for every Boolean variable x.

If x, y are Boolean variables, then

1) $x + y = 0$ if and only if $x = y = 0$, and

2) $xy = 1$ if and only if $x = y = 1$.

If $n \in \mathbf{Z}^+$, $B^n = \{(b_1, b_2, \ldots, b_n) | b_i \in \{0, 1\}, 1 \leq i \leq n\}$. A function $f \colon B^n \to B$ is called a *Boolean*, or *switching, function* of n variables. The n variables are emphasized by writing $f(x_1, x_2, \ldots, x_n)$, where each x_i, for $1 \leq i \leq n$, is a Boolean variable.

EXAMPLE 15.1

Let $f \colon B^3 \to B$, where $f(x, y, z) = xy + z$.[†] (We write xy for $x \cdot y$.) This Boolean function is determined by evaluating f for each of the eight possible assignments to the variables x, y, z, as Table 15.1 demonstrates.

Table 15.1

x	y	z	xy	$f(x, y, z) = xy + z$
0	0	0	0	0
0	0	1	0	1
0	1	0	0	0
0	1	1	0	1
1	0	0	0	0
1	0	1	0	1
1	1	0	1	1
1	1	1	1	1

Definition 15.1

For $n \in \mathbf{Z}^+$, $n \geq 2$, let $f, g \colon B^n \to B$ be two Boolean functions of the n Boolean variables $x_1, x_2, \ldots, x_n$. We say that f and g are *equal* and write $f = g$ if the columns for f, g [in their respective (function) tables] are exactly the same. [The tables show that $f(b_1, b_2, \ldots, b_n) = g(b_1, b_2, \ldots, b_n)$ for each of the 2^n possible assignments of either 0 or 1 to each of the n Boolean variables $x_1, x_2, \ldots, x_n$.]

Definition 15.2

If $f \colon B^n \to B$, then the *complement* of f, denoted $\overline{f}$, is the Boolean function defined on B^n by

$$\overline{f}(b_1, b_2, \ldots, b_n) = \overline{f(b_1, b_2, \ldots, b_n)}.$$

If $g \colon B^n \to B$, we define $f + g$, $f \cdot g \colon B^n \to B$, the *sum* and *product* of f, g, respectively, by

$$(f + g)(b_1, b_2, \ldots, b_n) = f(b_1, b_2, \ldots, b_n) + g(b_1, b_2, \ldots, b_n)$$

and

$$(f \cdot g)(b_1, b_2, \ldots, b_n) = f(b_1, b_2, \ldots, b_n) \cdot g(b_1, b_2, \ldots, b_n).$$

[†]When dealing with Boolean variables multiplication is performed before addition. Hence $xy + z$ represents $(xy) + z$, not $x(y + z)$.

Ten laws—important consequences of these definitions—are summarized in Table 15.2.

Table 15.2

1) $\overline{\overline{f}} = f$	$\overline{\overline{x}} = x$	Law of the *Double Complement*
2) $\overline{f + g} = \overline{f}\,\overline{g}$ $\overline{fg} = \overline{f} + \overline{g}$	$\overline{x + y} = \overline{x}\,\overline{y}$ $\overline{xy} = \overline{x} + \overline{y}$	*DeMorgan's* Laws
3) $f + g = g + f$ $fg = gf$	$x + y = y + x$ $xy = yx$	*Commutative* Laws
4) $f + (g + h) = (f + g) + h$ $f(gh) = (fg)h$	$x + (y + z) = (x + y) + z$ $x(yz) = (xy)z$	*Associative* Laws
5) $f + gh = (f + g)(f + h)$ $f(g + h) = fg + fh$	$x + yz = (x + y)(x + z)$ $x(y + z) = xy + xz$	*Distributive* Laws
6) $f + f = f$ $ff = f$	$x + x = x$ $xx = x$	*Idempotent* Laws
7) $f + \mathbf{0} = f$ $f \cdot \mathbf{1} = f$	$x + 0 = x$ $x \cdot 1 = x$	*Identity* Laws
8) $f + \overline{f} = \mathbf{1}$ $f\overline{f} = \mathbf{0}$	$x + \overline{x} = 1$ $x\overline{x} = 0$	*Inverse* Laws
9) $f + \mathbf{1} = \mathbf{1}$ $f \cdot \mathbf{0} = \mathbf{0}$	$x + 1 = 1$ $x \cdot 0 = 0$	*Dominance* Laws
10) $f + fg = f$ $f(f + g) = f$	$x + xy = x$ $x(x + y) = x$	*Absorption* Laws

As with the laws of logic (in Chapter 2) and the laws of set theory (in Chapter 3), the properties shown in Table 15.2 are satisfied by all Boolean functions $f, g, h: B^n \to B$ and by all Boolean variables x, y, z. (We write fg for $f \cdot g$.)

The symbol $\mathbf{0}$ denotes the constant Boolean function whose value is always 0, and $\mathbf{1}$ is the function whose only value is 1. (*Note*: $\mathbf{0}, \mathbf{1} \notin B$.)

Once again the idea of duality appears in properties 2–10. If s stands for a theorem about the equality of Boolean functions, then s^d, the *dual* of s, is obtained by replacing in s all occurrences of $+ (\cdot)$ by $\cdot (+)$ and all occurrences of $\mathbf{0}$ ($\mathbf{1}$) by $\mathbf{1}$ ($\mathbf{0}$). By the principle of duality (which we shall examine in Section 15.4) the statement s^d is also a theorem. The same is true for a theorem dealing with the equality of Boolean variables, except here it is the Boolean values 0 and 1 that are replaced, not the constant functions $\mathbf{0}$ and $\mathbf{1}$.

The principle of duality is handy for establishing property 5 of Table 15.2 for Boolean functions and Boolean variables.

EXAMPLE 15.2

The Distributive Law of $+$ over $\cdot$. The last two columns of Table 15.3 show that $f + gh = (f + g)(f + h)$. We also see that $x + yz = (x + y)(x + z)$ is a special case of this property for the situation where $f, g, h: B^3 \to B$, with $f(x, y, z) = x$, $g(x, y, z) = y$, and $h(x, y, z) = z$. Hence no additional tables are needed to establish this property for Boolean variables.

Table 15.3

f	g	h	gh	$f+g$	$f+h$	$f+gh$	$(f+g)(f+h)$
0	0	0	0	0	0	0	0
0	0	1	0	0	1	0	0
0	1	0	0	1	0	0	0
0	1	1	1	1	1	1	1
1	0	0	0	1	1	1	1
1	0	1	0	1	1	1	1
1	1	0	0	1	1	1	1
1	1	1	1	1	1	1	1

By the principle of duality, we obtain $f(g+h) = fg + fh$.

EXAMPLE 15.3

a) To establish the first absorption property for Boolean variables, instead of relying on table construction we argue as follows:

$$
\begin{aligned}
x + xy &= x1 + xy && \text{Identity Law} \\
&= x(1+y) && \text{Distributive Law of } \cdot \text{ over } + \\
&= x1 && \text{Dominance Law (and Commutative Law of } +) \\
&= x && \text{Identity Law}
\end{aligned}
$$

This result indicates that some of our laws can be derived from others. The question then is which properties we must establish with tables so that we can derive the other properties as we did here. We shall consider this later in Section 15.4 when we study the structure of a Boolean algebra.

In the meantime, let us demonstrate how the results of Table 15.2 can be used to simplify another Boolean expression.

b) Simplify the expression $wx + \overline{\overline{x}\,\overline{z}} + (y + \overline{z})$, where w, x, y, and z are Boolean variables.

$$
\begin{aligned}
wx + \overline{\overline{x}\,\overline{z}} + (y+\overline{z}) &= wx + (\overline{\overline{x}} + \overline{z}) + (y+\overline{z}) && \text{DeMorgan's Law} \\
&= wx + (x+\overline{z}) + (y+\overline{z}) && \text{Law of the Double Complement} \\
&= [(wx+x)+\overline{z}] + (y+\overline{z}) && \text{Associative Law of } + \\
&= (x+\overline{z}) + (y+\overline{z}) && \text{Absorption Law (and the Commutative Laws of } + \text{ and } \cdot) \\
&= x + (\overline{z}+\overline{z}) + y && \text{Commutative and Associative Laws of } + \\
&= x + \overline{z} + y && \text{Idempotent Law of } +
\end{aligned}
$$

Up to this point we have repeated for Boolean functions what we did in Chapter 2 for statements. When given a Boolean function (in algebraic terms), we construct its table of values. Now we consider the reverse process: Given a table of values, we shall find a Boolean function (described in algebraic terms) for which it is the correct table.

EXAMPLE 15.4	Given three Boolean variables x, y, z, find formulas for functions f, g, h: $B^3 \to B$ for the columns specified in Table 15.4.

For the column under f we want a result that has the value 1 only in the case where $x = y = 0$ and $z = 1$. The function $f(x, y, z) = \bar{x}\,\bar{y}z$ is one such function. In the same way, $g(x, y, z) = x\bar{y}\,\bar{z}$ yields the value 1 for $x = 1$, $y = z = 0$, and is 0 in all other cases. As each of f and g has the value 1 in only one case and these cases are distinct from each other, their sum $f + g$ has the value 1 in exactly these two cases. So $h(x, y, z) = f(x, y, z) + g(x, y, z) = \bar{x}\,\bar{y}z + x\bar{y}\,\bar{z}$ has the column of values given under h.

Table 15.4

x	y	z	f	g	h
0	0	0	0	0	0
0	0	1	1	0	1
0	1	0	0	0	0
0	1	1	0	0	0
1	0	0	0	1	1
1	0	1	0	0	0
1	1	0	0	0	0
1	1	1	0	0	0

This example leads us to the following definition.

Definition 15.3	For all $n \in \mathbf{Z}^+$, if f is a Boolean function on the n variables $x_1, x_2, \ldots, x_n$, we call

    a) each term x_i or its complement $\bar{x}_i$, for $1 \le i \le n$, a *literal*;

    b) a term of the form $y_1 y_2 \cdots y_n$, where each $y_i = x_i$ or $\bar{x}_i$, for $1 \le i \le n$, a *fundamental conjunction*; and

    c) a representation of f as a sum of fundamental conjunctions a *disjunctive normal form* (d.n.f.) of f.

Although no formal proof is given here, the following examples suggest that each $f: B^n \to B$, $f \neq \mathbf{0}$, has a unique (up to the order of fundamental conjunctions) representation as a d.n.f.

EXAMPLE 15.5	Find the d.n.f. for $f: B^3 \to B$, where $f(x, y, z) = xy + \bar{x}z$.

From Table 15.5, we see that the column for f contains four 1's. They indicate the four fundamental conjunctions needed in the d.n.f. of f, so $f(x, y, z) = \bar{x}\,\bar{y}z + \bar{x}yz + xy\bar{z} + xyz$.

Another way to solve this problem is to take each product term appearing in f — namely, xy and $\bar{x}z$ — and somehow involve whichever variables are missing. Using the properties of these variables, we have $xy + \bar{x}z = xy(z + \bar{z}) + \bar{x}(y + \bar{y})z$ (Why?) $= xyz + xy\bar{z} + \bar{x}yz + \bar{x}\,\bar{y}z$.

Table 15.5

x	y	z	xy	$\overline{x}z$	f
0	0	0	0	0	0
0	0	1	0	1	1
0	1	0	0	0	0
0	1	1	0	1	1
1	0	0	0	0	0
1	0	1	0	0	0
1	1	0	1	0	1
1	1	1	1	0	1

EXAMPLE 15.6

Find the d.n.f. for $g(w, x, y, z) = wx\overline{y} + wy\overline{z} + xy$.
We examine each term, as follows:

a) $wx\overline{y} = wx\overline{y}(z + \overline{z}) = wx\overline{y}z + wx\overline{y}\,\overline{z}$

b) $wy\overline{z} = w(x + \overline{x})y\overline{z} = wxy\overline{z} + w\overline{x}y\overline{z}$

c) $xy = (w + \overline{w})xy(z + \overline{z}) = wxyz + wxy\overline{z} + \overline{w}xyz + \overline{w}xy\overline{z}$

It follows from the idempotent property of $+$ that the d.n.f. of g is

$$g(w, x, y, z) = wx\overline{y}z + wx\overline{y}\,\overline{z} + wxy\overline{z} + w\overline{x}y\overline{z} + wxyz + \overline{w}xyz + \overline{w}xy\overline{z}.$$

Consider the first three columns in Table 15.6. If we agree to list the Boolean variables in alphabetical order, we see that the values for x, y, z in any row determine a binary label. These binary labels for $0, 1, 2, \ldots, 7$ arise for rows $1, 2, \ldots, 8$, respectively, as shown in columns 4 and 5 of Table 15.6. [We note, for instance, that the first row has row number 1 but binary label 000 ($= 0$). Likewise, the seventh row — where $x = 1$, $y = 1$, $z = 0$ — has row number 7 but binary label 110 ($= 6$).] As a result, the d.n.f. of a nonzero Boolean function can be expressed more compactly. For instance, the function f in Example 15.5 can be given by $f = \sum m(1, 3, 6, 7)$, where m indicates the *minterms* (that is, fundamental conjunctions — each here on three literals) at rows 2, 4, 7, 8, with the respective binary labels 1, 3, 6, 7. The word *minterm* is used here to emphasize that the fundamental conjunction has the value 1 a minimal number of times — namely, one time — without being identically **0**. For example, $m(1)$ denotes the minterm for the row with binary label 001 ($= 1$) where

Table 15.6

x	y	z	Binary Label	Row Number
0	0	0	000 ($= 0$)	1
0	0	1	001 ($= 1$)	2
0	1	0	010 ($= 2$)	3
0	1	1	011 ($= 3$)	4
1	0	0	100 ($= 4$)	5
1	0	1	101 ($= 5$)	6
1	1	0	110 ($= 6$)	7
1	1	1	111 ($= 7$)	8

$x = y = 0$ and $z = 1$; this corresponds with the fundamental conjunction $\overline{x}\,\overline{y}z$, which has the value 1 for exactly one assignment (where $x = y = 0$ and $z = 1$).

Lacking a table, we can still represent the d.n.f. of the function g of Example 15.6, for instance, as a *sum of minterms*. For each fundamental conjunction $c_1 c_2 c_3 c_4$, where $c_1 = w$ or $\overline{w}, \ldots, c_4 = z$ or $\overline{z}$, we replace each c_i, $1 \le i \le 4$, by 0 if c_i is a complemented variable, and by 1 otherwise. In this way the binary label associated with that fundamental conjunction is obtained. As a sum of minterms, we find that $g = \sum m(6, 7, 10, 12, 13, 14, 15)$.

Dual to the disjunctive normal form is the conjunctive normal form, which we discuss before closing this section.

EXAMPLE 15.7

Table 15.7

x	y	z	f
0	0	0	0
0	0	1	1
0	1	0	0
0	1	1	1
1	0	0	1
1	0	1	1
1	1	0	0
1	1	1	1

Let $f: B^3 \to B$ be given by Table 15.7. A term of the form $c_1 + c_2 + c_3$, where $c_1 = x$ or $\overline{x}$, $c_2 = y$ or $\overline{y}$, and $c_3 = z$ or $\overline{z}$, is called a *fundamental disjunction*. The fundamental disjunction $x + y + z$ has value 1 in all cases except where the value for each of x, y, z is 0. Similarly, $x + \overline{y} + z$ has value 1 except when $x = z = 0$ and $y = 1$. Since each of these fundamental disjunctions has the value 0 in only one case, and these cases do not occur simultaneously, the product $(x + y + z)(x + \overline{y} + z)$ has the value 0 in precisely the two cases just given. Continuing in this manner, we may represent the function f as

$$f = (x + y + z)(x + \overline{y} + z)(\overline{x} + \overline{y} + z)$$

and we call this the *conjunctive normal form* (c.n.f.) for f.

Since the fundamental disjunction $x + y + z$ has the value 1 a maximum number of times (without being identically **1**), it is called a *maxterm*, especially when we use a binary row label to represent it. Using the binary labels to index the rows of the table, we may write $f = \prod M(0, 2, 6)$, a *product of maxterms*.

Such a representation exists for each $f \ne \mathbf{1}$, and it is unique up to the order of the fundamental disjunctions (or maxterms).

EXAMPLE 15.8

Let $g: B^4 \to B$, where $g(w, x, y, z) = (w + x + y)(x + \overline{y} + z)(w + \overline{y})$. To obtain the c.n.f. for g, we rewrite each disjunction in the product as follows:

a) $w + x + y = w + x + y + 0 = w + x + y + z\overline{z}$
$\qquad = (w + x + y + z)(w + x + y + \overline{z})$

b) $x + \overline{y} + z = w\overline{w} + x + \overline{y} + z = (w + x + \overline{y} + z)(\overline{w} + x + \overline{y} + z)$

c) $w + \overline{y} = w + x\overline{x} + \overline{y} = (w + x + \overline{y})(w + \overline{x} + \overline{y})$
$\qquad = (w + x + \overline{y} + z\overline{z})(w + \overline{x} + \overline{y} + z\overline{z})$
$\qquad = (w + x + \overline{y} + z)(w + x + \overline{y} + \overline{z})(w + \overline{x} + \overline{y} + z)(w + \overline{x} + \overline{y} + \overline{z})$

Consequently, using the idempotent law of $\cdot$, we have $g(w, x, y, z) = (w + x + y + z) \cdot (w + x + y + \overline{z})(w + x + \overline{y} + z)(\overline{w} + x + \overline{y} + z)(w + x + \overline{y} + \overline{z})(w + \overline{x} + \overline{y} + z) \cdot (w + \overline{x} + \overline{y} + \overline{z})$.

To obtain g as a product of maxterms, we associate with each fundamental disjunction $d_1 + d_2 + d_3 + d_4$ the binary number $b_1 b_2 b_3 b_4$, where $b_1 = 0$ if $d_1 = w$; $b_1 = 1$ if $d_1 = \overline{w}; \ldots; b_4 = 0$ if $d_4 = z$; $b_4 = 1$ if $d_4 = \overline{z}$. As a result, $g = \prod M(0, 1, 2, 3, 6, 7, 10)$.

Our last example in this section reviews what we have learned about the ways to represent a nonconstant Boolean function f (that is, $f \ne \mathbf{0}$ and $f \ne \mathbf{1}$).

EXAMPLE 15.9

If $h(w, x, y, z) = wx + \overline{w}y + \overline{x}yz$, then we may rewrite each summand in h as follows:

i) $wx = wx(y + \overline{y})(z + \overline{z}) = wxyz + wxy\overline{z} + wx\overline{y}z + wx\overline{y}\overline{z}$

ii) $\overline{w}y = \overline{w}(x + \overline{x})y(z + \overline{z}) = \overline{w}xyz + \overline{w}xy\overline{z} + \overline{w}\overline{x}yz + \overline{w}\overline{x}y\overline{z}$

iii) $\overline{x}yz = (w + \overline{w})\overline{x}yz = w\overline{x}yz + \overline{w}\overline{x}yz$

Using the idempotent law of $+$, we find that the d.n.f. for h is

$$wxyz + wxy\overline{z} + wx\overline{y}z + wx\overline{y}\overline{z} + \overline{w}xyz + \overline{w}xy\overline{z} + \overline{w}\overline{x}yz + \overline{w}\overline{x}y\overline{z} + w\overline{x}yz.$$

Considering each fundamental conjunction in the d.n.f. for h, we obtain the following binary labels and minterm numbers:

$wxyz$:	1111 ($= 15$)	$wx\overline{y}\overline{z}$:	1100 ($= 12$)	$\overline{w}\overline{x}yz$:	0011 ($= 3$)
$wxy\overline{z}$:	1110 ($= 14$)	$\overline{w}xyz$:	0111 ($= 7$)	$\overline{w}\overline{x}y\overline{z}$:	0010 ($= 2$)
$wx\overline{y}z$:	1101 ($= 13$)	$\overline{w}xy\overline{z}$:	0110 ($= 6$)	$w\overline{x}yz$:	1011 ($= 11$)

So we may write $h = \sum m(2, 3, 6, 7, 11, 12, 13, 14, 15)$. And from this representation using minterms we have $h = \prod M(0, 1, 4, 5, 8, 9, 10)$, a product of maxterms.

Finally, we take the binary label for each maxterm and determine its corresponding fundamental disjunction:

$0 = 0000$:	$w + x + y + z$	$8 = 1000$:	$\overline{w} + x + y + z$
$1 = 0001$:	$w + x + y + \overline{z}$	$9 = 1001$:	$\overline{w} + x + y + \overline{z}$
$4 = 0100$:	$w + \overline{x} + y + z$	$10 = 1010$:	$\overline{w} + x + \overline{y} + z$
$5 = 0101$:	$w + \overline{x} + y + \overline{z}$		

This tells us that the c.n.f. for h is

$$(w + x + y + z)(w + x + y + \overline{z})(w + \overline{x} + y + z)(w + \overline{x} + y + \overline{z}) \cdot$$
$$(\overline{w} + x + y + z)(\overline{w} + x + y + \overline{z})(\overline{w} + x + \overline{y} + z).$$

Hence,

$$wxyz + wxy\overline{z} + wx\overline{y}z + wx\overline{y}\overline{z} + \overline{w}xyz + \overline{w}xy\overline{z} + \overline{w}\overline{x}yz + \overline{w}\overline{x}y\overline{z} + w\overline{x}yz =$$
$$\sum m(2, 3, 6, 7, 11, 12, 13, 14, 15) = \prod M(0, 1, 4, 5, 8, 9, 10) =$$
$$(w + x + y + z)(w + x + y + \overline{z})(w + \overline{x} + y + z)(w + \overline{x} + y + \overline{z}) \cdot$$
$$(\overline{w} + x + y + z)(\overline{w} + x + y + \overline{z})(\overline{w} + x + \overline{y} + z).$$

EXERCISES 15.1

1. Find the value of each of the following Boolean expressions if the values of the Boolean variables $w, x, y,$ and z are 1, 1, 0, and 0, respectively.

a) $\overline{xy} + \overline{x}\,\overline{y}$ **b)** $w + \overline{x}y$ **c)** $wx + \overline{y} + yz$

d) $(wx + y\overline{z}) + w\overline{y} + \overline{(w + y)(\overline{x} + y)}$

2. Let $w, x,$ and y be Boolean variables where the value of x is 1. For each of the following Boolean expressions, determine, if possible, the value of the expression. If you cannot determine the value of the expression, then find the number of assign-

ments of values for w and y that will result in the value 1 for the expression.

a) $x + xy + w$ **b)** $xy + w$

c) $\overline{x}y + xw$ **d)** $\overline{x}y + w$

3. a) How many rows are needed to construct the (function) table for a Boolean function of n variables?

b) How many different Boolean functions of n variables are there?

4. a) Find the fundamental conjunction made up from the variables $w, x, y, z,$ or their complements, where the value of the conjunction is 1 precisely when

i) $w = x = 0$, $y = z = 1$.

ii) $w = 0$, $x = 1$, $y = 1$, $z = 0$.

iii) $w = 0$, $x = y = z = 1$.

iv) $w = x = y = z = 0$.

b) Answer part (a) this time for fundamental disjunctions, instead of fundamental conjunctions, where the value of each fundamental disjunction is 0 precisely for the stated values of w, x, y, z.

5. Suppose that $f: B^3 \rightarrow B$ is defined by

$$f(x, y, z) = \overline{\overline{(x + y)} + (\overline{x}z)}.$$

a) Determine the d.n.f. and c.n.f. for f.

b) Write f as a sum of minterms and as a product of maxterms (utilizing binary labels).

6. Let $g: B^4 \rightarrow B$ be defined by

$$g(w, x, y, z) = (wz + xyz)(x + \overline{x}\,\overline{y}z).$$

a) Find the d.n.f. and c.n.f. for g.

b) Write g as a sum of minterms and as a product of maxterms (utilizing binary labels).

7. Let F_6 denote the set of all Boolean functions $f: B^6 \rightarrow B$. (a) What is $|F_6|$? (b) How many fundamental conjunctions (disjunctions) are there in F_6? (c) How many minterms (maxterms) are there in F_6?

8. Let $f: B^4 \rightarrow B$. Find the disjunctive normal form for f if

a) $f^{-1}(1) = \{0101$ (that is, $w = 0$, $x = 1$, $y = 0$, $z = 1$), 0110, 1000, 1011$\}$.

b) $f^{-1}(0) = \{0000, 0001, 0010, 0100, 1000, 1001, 0110\}$.

9. Let $B^n \rightarrow B$. If the d.n.f. of f has m fundamental conjunctions and its c.n.f. has k fundamental disjunctions, how are m, n, and k related?

10. If x, y, and z are Boolean variables and $x + y + z = xyz$, prove that x, y, z all have the same value.

11. Simplify the following Boolean expressions.

a) $xy + (x + y)\overline{z} + y$

b) $x + y + \overline{(\overline{x} + y + z)}$

c) $yz + wx + z + [wz(xy + wz)]$

12. Find the values of the Boolean variables w, x, y, z that satisfy the following system of simultaneous (Boolean) equations.

$$x + \overline{x}y = 0 \qquad \overline{x}y = \overline{x}z \qquad \overline{x}y + \overline{x}\,\overline{z} + zw = \overline{z}w$$

13. a) For $f, g, h: B^n \rightarrow B$, prove that $fg + \overline{f}h + gh = fg + \overline{f}h$ and that $fg + f\overline{g} + \overline{f}g + \overline{f}\,\overline{g} = \mathbf{1}$.

b) State the dual of each result in part (a).

14. Let $f, g: B^n \rightarrow B$. Define the relation "$\leq$" on F_n, the set of all Boolean functions of n variables, by $f \leq g$ if the value of g is 1 at least whenever the value of f is 1.

a) Prove that this relation is a partial order on F_n.

b) Prove that $fg \leq f$ and $f \leq f + g$.

c) For $n = 2$, draw the Hasse diagram for the 16 functions in F_2. Where are the minterms and maxterms located in the diagram? Compare this diagram with that for the power set of $\{a, b, c, d\}$ partially ordered under the subset relation.

15. Define the closed binary operation $\oplus$ (Exclusive Or) on F_n, the set of all Boolean functions on n variables, by $f \oplus g = f\overline{g} + \overline{f}g$, where $f, g: B^n \rightarrow B$.

a) Determine $f \oplus f$, $f \oplus \overline{f}$, $f \oplus \mathbf{1}$, and $f \oplus \mathbf{0}$.

b) Prove or disprove each of the following.

i) $f \oplus g = \mathbf{0} \Rightarrow f = g$

ii) $f \oplus (g \oplus h) = (f \oplus g) \oplus h$

iii) $f \oplus g = \overline{f} \oplus \overline{g}$

iv) $f \oplus gh = (f \oplus g)(f \oplus h)$

v) $f(g \oplus h) = fg \oplus fh$

vi) $\overline{(f \oplus g)} = \overline{f} \oplus g = f \oplus \overline{g}$

vii) $f \oplus g = f \oplus h \Rightarrow g = h$

15.2
Gating Networks: Minimal Sums of Products: Karnaugh Maps

The switching functions of Section 15.1 present an interesting mathematical theory. Their importance lies in their implementation by means of *logic gates* (devices in a digital computer that perform specified tasks in the processing of data). The electrical and mechanical components of such gates depend on the state of the art; we shall not concern ourselves here with questions relating to hardware.

Figure 15.1 contains the logic gates for negation (complement), conjunction, and disjunction in parts (a), (b), and (c), respectively. Since the Boolean operations of $+$ and $\cdot$ are associative, we may have more than two inputs for an AND gate or an OR gate.

Figure 15.2 shows the *logic*, or *gating, network* for the expression $(w + \overline{x})(y + xz)$. Symbols on a line to the left of a gate (or inverter) are *inputs*. When they are on line

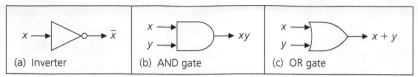

Figure 15.1

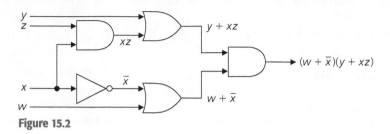

Figure 15.2

segments to the right of a gate, they are *outputs*. We have *split* the input line for x, so that x may serve as input for both an AND gate and an inverter.

The exercises will provide practice in drawing the logic network for a Boolean expression and in going from the network to the expression. Meanwhile certain features of these networks need to be emphasized.

1) An input line may be split to provide that input to more than one gate.

2) Input and output lines come together only at gates.

3) There is no doubling back; that is, the output from a gate g cannot be used as an input for the same gate g or for any gate (directly or indirectly) leading into g.

4) We assume that the output of a gating network is an instantaneous function of the present inputs. There is no time dependence and we attach no importance to prior inputs, as we do with finite state machines.

With these ideas in mind, let us analyze the computer addition of binary numbers.

EXAMPLE 15.10

When we add two bits (binary digits), the result consists of a sum s and a carry c. In three of four cases the carry is 0, so we shall concentrate on the computation of $1 + 1$. Examining parts (b) and (c) of Table 15.8, we consider the sum s and the carry c as Boolean functions of the variables x and y. Then $c = xy$ and $s = \overline{x}y + x\overline{y} = x \oplus y = (x + y)(\overline{xy})$. (Recall that $\oplus$ denotes exclusive OR.)

Table 15.8

x	y	Binary Sum
0	0	$0 + 0 = 0$
0	1	$0 + 1 = 1$
1	0	$1 + 0 = 1$
1	1	$1 + 1 = 10$

(a)

x	y	Sum
0	0	0
0	1	1
1	0	1
1	1	0

(b)

x	y	Carry
0	0	0
0	1	0
1	0	0
1	1	1

(c)

Figure 15.3 is a gating network with two outputs. It is referred to as a *multiple output network*. This device, called a *half-adder*, implements the results in parts (b) and (c) of

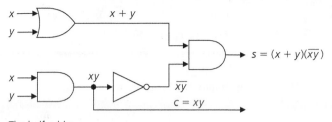

The half-adder

Figure 15.3

Table 15.8. Using two half-adders and an OR gate, we construct the *full-adder* shown in Fig. 15.4(a). If $x = x_n x_{n-1} \ldots x_2 x_1 x_0$ and $y = y_n y_{n-1} \ldots y_2 y_1 y_0$, consider the process of adding the bits x_i and y_i in finding the sum $x + y$. Here c_{i-1} is the carry from the addition of x_{i-1} and y_{i-1} (and a possible carry c_{i-2}). The input c_{i-1}, together with the inputs x_i and y_i, produce the sum s_i and the carry c_i as shown in the figure. Finally, in Fig. 15.4(b) two full-adders and a half-adder are combined to produce the sum of the two binary numbers $x_2 x_1 x_0$ and $y_2 y_1 y_0$, whose sum is $c_2 s_2 s_1 s_0$.

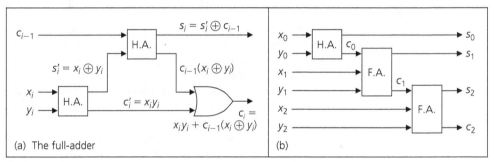

(a) The full-adder   (b)

Figure 15.4

The next example introduces the main theme of this section — the minimal-sum-of-products representation of a Boolean function.

EXAMPLE 15.11

Find a gating network for the Boolean function

$$f(w, x, y, z) = \sum m(4, 5, 7, 8, 9, 11).$$

Consider the order of the variables as w, x, y, z. We can determine the d.n.f. of f by writing each minterm number in binary notation and then finding its corresponding fundamental conjunction. For example, (a) $5 = 0101$, indicating the fundamental conjunction $\overline{w}x\overline{y}z$; and (b) $7 = 0111$, indicating $\overline{w}xyz$. Continuing in this way, we have $f(w, x, y, z) = \overline{w}x\overline{y}z + \overline{w}xyz + w\overline{x}\,\overline{y}\,\overline{z} + w\overline{x}\,\overline{y}z + w\overline{x}yz + \overline{w}x\overline{y}\,\overline{z}$.

Using properties of Boolean variables, we find that

$$f = \overline{w}xz(\overline{y} + y) + w\overline{x}\,\overline{y}(\overline{z} + z) + w\overline{x}yz + \overline{w}x\overline{y}\,\overline{z}$$
$$= \overline{w}xz + w\overline{x}\,\overline{y} + w\overline{x}yz + \overline{w}x\overline{y}\,\overline{z} = \overline{w}x(z + \overline{y}\,\overline{z}) + w\overline{x}(\overline{y} + yz)$$
$$= \overline{w}x(z + \overline{y}) + w\overline{x}(\overline{y} + z) \text{ (Why?)} = \overline{w}x(\overline{y} + z) + w\overline{x}(\overline{y} + z),$$

so

a) $f(w, x, y, z) = \overline{w}xz + \overline{w}x\overline{y} + w\overline{x}\,\overline{y} + w\overline{x}z$; or

b) $f(w, x, y, z) = \overline{w}x(\overline{y} + z) + w\overline{x}(\overline{y} + z)$.

In Example 15.11, the result

$$f(w, x, y, z) = \overline{w}xz + \overline{w}x\overline{y} + w\overline{x}\,\overline{y} + w\overline{x}z$$

is often referred to as a *minimal-sum-of-products* representation for the function $f(w, x, y, z) = \sum m(4, 5, 7, 8, 9, 11)$. We see that this representation is a sum of four products — where each product is made up of three literals. When we call such a representation minimal we mean two things:

1) Any possible further modification will result in a representation that is *not* a sum of such products; and

2) If f can be represented in a second way as a sum of products (of literals), then we will have at least four product terms — each with at least three literals.
   [*Note*: A minimal sum of products for a given Boolean function f ($\neq \mathbf{0}$) need not be unique — as we shall find in Example 15.15.]

In this text our discussion of this idea will be somewhat informal. We shall not attempt to prove that each nonzero Boolean function has such a minimal-sum-of-products representation. Instead we shall assume the existence of this representation and simply continue our study of how to obtain such a result.

From this point on we shall consider an input of the form $\overline{w}$ as an exact input, which has not passed through any gates, instead of regarding it as the result obtained from inputting w and passing it through an inverter.

In Fig. 15.5(a), we have a gating network implementing the d.n.f. of the function f in Example 15.11. Part (b) of the figure is the gating network for f as a minimal sum of products. Figure 15.5(c) has a gating network for $f = \overline{w}x(\overline{y} + z) + w\overline{x}(\overline{y} + z)$.

The network in part (c) has only four logic gates, whereas that in part (b) has five such devices. Consequently, we may feel that the network in part (c) is better with regard to minimizing cost because each extra gate increases the cost of production. However, even though there are fewer inputs and fewer gates for the implementation in part (c), some of the inputs (namely, $\overline{y}$ and z) must pass through three *levels of gating* before providing the output f. For the minimal sum of products in part (b), there are only two levels of gating. In the study of gating networks, outputs are considered instantaneous functions of the input. In practice, however, each level of gating adds a delay in the development of the function f. For high-speed digital equipment we want to minimize delay, so we opt for more speed at the price of increased manufacturing cost.

It is this need to maximize speed that makes us want to represent a Boolean function as a minimal sum of products. In order to accomplish this for functions of not more than six variables, we use a pictorial method called the *Karnaugh map*, developed in 1953 by Maurice Karnaugh (1924 –). Karnaugh maps always produce forms with at most two levels of gating, and we shall find that the d.n.f. of a Boolean function is a major key behind this technique.

In simplifying the d.n.f. of f in Example 15.11, we combined the two fundamental conjunctions $\overline{w}x\overline{y}z$ and $\overline{w}xyz$ into the product term $\overline{w}xz$ because $\overline{w}x\overline{y}z + \overline{w}xyz = \overline{w}xz(\overline{y} + y) = \overline{w}xz(1) = \overline{w}xz$. This indicates that if two fundamental conjunctions differ in exactly one literal, then they can be combined into a product term with that literal missing.

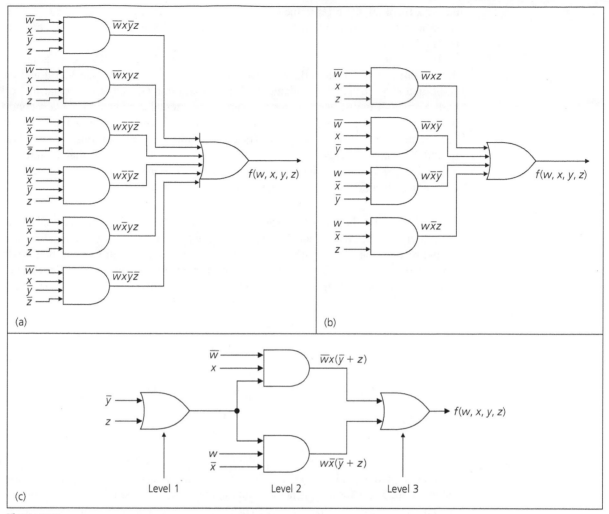

Figure 15.5

For $g: B^4 \to B$, where $g(w, x, y, z) = wx\overline{y}\,\overline{z} + wx\overline{y}z + wxyz + wxy\overline{z}$, each fundamental conjunction (except the first) differs from its predecessor in exactly one literal. Here we can simplify g as $g = wx\overline{y}(\overline{z} + z) + wxy(z + \overline{z}) = wx\overline{y} + wxy = wx(\overline{y} + y) = wx$. We could have also written

$$g = wx(\overline{y}\,\overline{z} + \overline{y}z + yz + y\overline{z}) = wx(y + \overline{y})(z + \overline{z}) = wx.$$

The key to this reduction process is the recognition of pairs (quadruples, ..., 2^n-tuples) of fundamental conjunctions where any two adjacent terms differ in exactly one literal. If $h: B^4 \to B$, and the d.n.f. of h has 12 terms, can we move these terms around to recognize the best reductions? The Karnaugh map organizes these terms for us.

We start with the case of two variables, w and x. Table 15.9 shows the Karnaugh maps for the functions $f(w, x) = wx$ and $g(w, x) = w + x$. (The 0's are suppressed in the tables for these maps.)

In part (a), the 1 interior to the table indicates the fundamental conjunction wx. This occurs in the row for $w = 1$ and the column for $x = 1$, the one case when $wx = 1$. In

Table 15.9

$w \backslash x$	0	1
0		
1		1

(a) wx

$w \backslash x$	0	1
0		1
1	1	1

(b) $w + x$

part (b), there are three 1's in the table. The top 1 is for $\overline{w}x$, which has the value 1 exactly when $w = 0$, $x = 1$. The bottom two 1's are for $w\overline{x}$ and wx, as we read the bottom row from left to right.

Table 15.9(b) represents the d.n.f. $\overline{w}x + w\overline{x} + wx$. As a result of their adjacency in the bottom row, the table indicates that $w\overline{x}$ and wx differ in only one literal and can be combined to yield w. By the idempotent law of addition (which is so crucial in working with Karnaugh maps), we can use the same fundamental conjunction wx a second time in this reduction process. The adjacency in the second column of the table indicates the combining of $\overline{w}x$ and wx to get x. (In the x column all possibilities for w—namely, w and $\overline{w}$—appear. This is a way to recognize x as the result for that column.) Thus Table 15.9(b) illustrates that $\overline{w}x + w\overline{x} + wx = \overline{w}x + w\overline{x} + wx + wx = (w\overline{x} + wx) + (\overline{w}x + wx) = w(\overline{x} + x) + (\overline{w} + w)x = w(1) + (1)x = w + x$.

<table>
<tr><td>**EXAMPLE 15.12**</td></tr>
</table>

We now consider three Boolean variables w, x, y. In Table 15.10, the first new idea we encounter is in the column headings for xy. These are not the same as the headings we had for the rows in the function tables. We see here, in going from left to right, that 00 differs from 01 in exactly one place, 01 differs from 11 in exactly one place, 11 differs from 10 in exactly one place, and, upon wrapping around, 10 differs from 00 in exactly one place.

Table 15.10

$w \backslash xy$	00	01	11	10
0	①			①
1	①		1	

If $f(w, x, y) = \sum m(0, 2, 4, 7)$, then because $0 = 000(\overline{w}\,\overline{x}\,\overline{y})$, $2 = 010(\overline{w}x\overline{y})$, $4 = 100(w\overline{x}\,\overline{y})$, and $7 = 111(wxy)$, we can represent these terms by placing 1's as shown in Table 15.10. The 1 for wxy is not adjacent to any other 1 in the table, so it is *isolated*; we shall have wxy as one of the summands in the minimal sum of products representing f. The 1 for $\overline{w}x\overline{y}$ (at the right end of the first row) is not isolated, for once again we consider the table as wrapping around, making this 1 adjacent to the 1 for $\overline{w}\,\overline{x}\,\overline{y}$ (at the left end of the first row). These combine (under addition) to give us $\overline{w}x\overline{y} + \overline{w}\,\overline{x}\,\overline{y} = \overline{w}\,\overline{y}(x + \overline{x}) = \overline{w}\,\overline{y}(1) = \overline{w}\,\overline{y}$. Finally, the 1's in the column for $x = y = 0$ indicate a reduction of $\overline{w}\,\overline{x}\,\overline{y} + w\overline{x}\,\overline{y}$ to $(\overline{w} + w)\overline{x}\,\overline{y} = (1)\overline{x}\,\overline{y} = \overline{x}\,\overline{y}$. Hence, as a minimal sum of products, $f = wxy + \overline{w}\,\overline{y} + \overline{x}\,\overline{y}$.

<table>
<tr><td>**EXAMPLE 15.13**</td></tr>
</table>

From the respective parts of Table 15.11 we have

a) $f(w, x, y) = \sum m(0, 2, 4, 6) = \sum m(0, 4) + \sum m(2, 6) = (\overline{w}\,\overline{x}\,\overline{y} + w\overline{x}\,\overline{y}) + (\overline{w}x\overline{y} + wx\overline{y}) = (\overline{w} + w)\overline{x}\,\overline{y} + (\overline{w} + w)x\overline{y} = (1)\overline{x}\,\overline{y} + (1)x\overline{y} = \overline{x}\,\overline{y} + x\overline{y} = (\overline{x} + x)\overline{y} = (1)\overline{y} = \overline{y}$, the only variable whose value does not change when the four terms designated by the 1's are considered. [The value of y is 0 here, so $f(w, x, y) = \overline{y}$.]

b) $f(w, x, y) = \sum m(0, 1, 2, 3) = \overline{w}\,\overline{x}\,\overline{y} + \overline{w}\,\overline{x}y + \overline{w}x\overline{y} + \overline{w}xy = \overline{w}(\overline{x}\,\overline{y} + \overline{x}y + x\overline{y} + xy) = \overline{w}(\overline{x} + x)(\overline{y} + y) = \overline{w}(1)(1) = \overline{w}$.

c) $f(w, x, y) = \sum m(1, 2, 3, 5, 6, 7) = \sum m(1, 3, 5, 7) + \sum m(2, 3, 6, 7) = y + x$.

Table 15.11

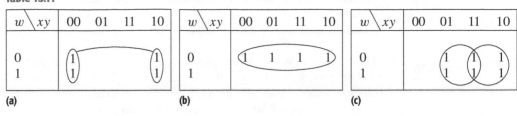

(a) (b) (c)

Advancing to four variables, we consider the following example.

EXAMPLE 15.14

Find a minimal-sum-of-products representation for the function

$$f(w, x, y, z) = \sum m(0, 1, 2, 3, 8, 9, 10).$$

The Karnaugh map for f in Table 15.12 combines the 1's in the four (adjacent) corners to give the term $\overline{w}\,\overline{x}\,\overline{y}\,\overline{z} + \overline{w}\,\overline{x}y\overline{z} + w\overline{x}\,\overline{y}\,\overline{z} + w\overline{x}y\overline{z} = \overline{x}\,\overline{z}(\overline{w}\,\overline{y} + \overline{w}y + w\overline{y} + wy) = \overline{x}\,\overline{z}$. The four 1's in the top row combine to give $\overline{w}\,\overline{x}$. (Using only the middle two 1's, we do not make use of all the available adjacencies and get the term $\overline{w}\,\overline{x}z$, which has one more literal than $\overline{w}\,\overline{x}$.) Finally, the 1 in the row ($w = 1$, $x = 0$) and the column ($y = 0$, $z = 1$) can be combined with the 1 on its left, *and* these can then be combined with the first two 1's in the top row to give $\overline{w}\,\overline{x}\,\overline{y}\,\overline{z} + \overline{w}\,\overline{x}\,\overline{y}z + w\overline{x}\,\overline{y}\,\overline{z} + w\overline{x}\,\overline{y}z = \overline{x}\,\overline{y}$. Hence, as a minimal sum of products, $f(w, x, y, z) = \overline{x}\,\overline{z} + \overline{w}\,\overline{x} + \overline{x}\,\overline{y}$.

Table 15.12

$wx \backslash yz$	00	01	11	10
00	1	1	1	1
01				
11				
10	1	1		1

EXAMPLE 15.15

The map for $f(w, x, y, z) = \sum m(9, 10, 11, 12, 13)$ appears in Table 15.13. The only 1 in the table that has not been combined with another term is adjacent to a 1 on its right (this combination yields $w\overline{x}z$) and to a 1 above it (this combination yields $w\overline{y}z$). Consequently, we can represent f as a minimal sum of products in two ways: $wx\overline{y} + w\overline{x}y + w\overline{x}z$ and $wx\overline{y} + w\overline{x}y + w\overline{y}z$. This type of representation, then, is not unique. However, we should

Table 15.13

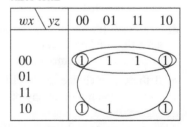

$wx \backslash yz$	00	01	11	10
00				
01				
11	1	1		
10		1	1	1

observe that the same number of product terms and the same total number of literals appear in each case.

EXAMPLE 15.16

There is a right way and there is a wrong way to use a Karnaugh map.

Let $f(w, x, y, z) = \sum m(3, 4, 5, 7, 9, 13, 14, 15)$. In Table 15.14(a) we combine a block of four 1's into the term xz. But when we account for the other four 1's, we do what is shown in part (b). So the result in part (b) will yield f as a sum of four terms (each with three literals), whereas the method suggested in part (a) adds the extra (unneeded) term xz.

Table 15.14

$wx \backslash yz$	00	01	11	10
00			1	
01	1	1	1	
11		1	1	1
10		1		

$wx \backslash yz$	00	01	11	10
00			1	
01	1	1	1	
11			1	1
10			1	

(a)　　　　　　　　　**(b)**

The following suggestions on the use of Karnaugh maps are based on what we have done so far. We state them now so that they may be used for larger maps.

1) Start by combining those terms in the table where there is at most one possibility for simplification.

2) Check the four corners of a table. They may contain adjacent 1's even though the 1's appear isolated.

3) In all simplifications, try to obtain the largest possible block of adjacent 1's in order to get a minimal product term. (Recall that 1's can be used more than once, if necessary, because of the idempotent law of +.)

4) If there is a choice in simplifying an entry in the table, try to use adjacent 1's that have not been used in any prior simplification.

EXAMPLE 15.17

If $f(v, w, x, y, z) = \sum m(1, 5, 10, 11, 14, 15, 18, 26, 27, 30, 31)$, we construct two 4×4 tables, one for $v = 0$, the other for $v = 1$. (See Table 15.15.)

Table 15.15

$wx \backslash yz$	00	01	11	10
00		1		
01		1		
11			1	1
10			1	1

$wx \backslash yz$	00	01	11	10
00				1
01				
11			1	1
10			1	1

$(v = 0)$　　　　　　　　**$(v = 1)$**

Following the order of the variables, we write, for example, $5 = 00101$ in order to indicate the need for a 1 in the second row and second column of the table for $v = 0$. The other five 1's in the table where $v = 0$ are for the minterms for 1, 10, 11, 14, 15. The minterms for

18, 26, 27, 30, 31 are represented by the five 1's in the table where $v = 1$. After filling in all the 1's, we see that the 1 in the first row, fourth column of the table for $v = 1$ can be combined with another term in only one way — with $vw\overline{x}y\overline{z}$ — yielding the product $v\overline{x}y\overline{z}$. This is also true for the two 1's in the second column of the ($v = 0$) table. These give the product $\overline{v}\,\overline{w}\,\overline{y}z$. The block of eight 1's yields wy, and we have $f(v, w, x, y, z) = wy + \overline{v}\,\overline{w}\,\overline{y}z + v\overline{x}y\overline{z}$.

A function f of the six variables t, v, w, x, y, and z requires four tables — one for each of the cases (a) $t = 0$, $v = 0$; (b) $t = 0$, $v = 1$; (c) $t = 1$, $v = 1$; and (d) $t = 1$, $v = 0$. Beyond six variables, this method becomes overly complicated. Another procedure, the *Quine-McCluskey Method*, can be used. For a large number of variables the method is tedious to perform by hand, but it is a systematic procedure suitable for computer implementation, particularly for computers possessing some type of "binary compare" command. (More about this technique is given in Chapter 7 of Reference [3].)

We close this section with an example involving the dual concept — namely, a *minimal product of sums*.

EXAMPLE 15.18

For $g(w, x, y, z) = \prod M(1, 5, 7, 9, 10, 13, 14, 15)$, this time we place a 0 in each of the positions for the binary equivalents of the maxterms listed. This yields the results shown in Table 15.16 (where the 1's are suppressed).

Table 15.16

wx \ yz	00	01	11	10
00		0		
01		0	0	
11		0	0	0
10		0		0

The 0 in the lower right-hand corner can only be combined with the 0 above it, and so we have $(\overline{w} + x + \overline{y} + z)(\overline{w} + \overline{x} + \overline{y} + z) = (\overline{w} + \overline{y} + z) + x\overline{x} = (\overline{w} + \overline{y} + z) + 0 = \overline{w} + \overline{y} + z$. The block of four 0's (for the maxterms for 5, 7, 13, 15) simplifies to $\overline{x} + \overline{z}$, whereas the four 0's (for the maxterms for 1, 5, 9, 13) in the second column yield $y + \overline{z}$. So $g(w, x, y, z) = (\overline{w} + \overline{y} + z)(\overline{x} + \overline{z})(y + \overline{z})$, a minimal product of sums.

EXERCISES 15.2

1. Using inverters, AND gates, and OR gates, construct the gates shown in Fig. 15.6.

2. Using only NAND[†] gates (see Fig. 15.6), construct the inverter, AND gate, and OR gate.

3. Answer Exercise 2, replacing NAND by NOR.

4. Using inverters, AND gates, and OR gates, construct gating networks for

a) $f(x, y, z) = x\overline{z} + y\overline{z} + x$

b) $g(x, y, z) = (x + \overline{z})(y + \overline{z})\overline{x}$

c) $h(x, y, z) = \overline{(xy \oplus yz)}$

[†]The NAND gate is constructed in a very simple manner from transistors — both in the old-fashioned technology of semiconductors as well as in the more recent techniques of silicon chip fabrication. Furthermore, most of the gating networks that represent what is actually happening inside of today's computers contain large numbers of these NAND gates.

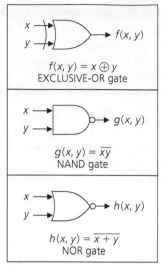

Figure 15.6

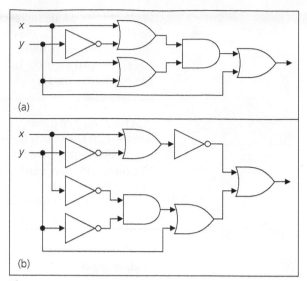

(a)

(b)

Figure 15.8

5. For the network in Fig. 15.7, express f as a function of w, x, y, z.

6. Implement the half-adder of Fig. 15.3 using only (a) NAND gates; (b) NOR gates.

7. For each of the networks in Fig. 15.8 express the output in terms of the Boolean variables x, y or their complements. Then use the expression for the output to simplify the given network.

8. For each of the following Boolean functions f, design a two-level gating network for f as a minimal sum of products.

   a) $f: B^3 \to B$, where $f(x, y, z) = 1$ if and only if exactly two of the variables have the value 1.

   b) $f: B^4 \to B$, where $f(w, x, y, z) = 1$ if and only if an odd number of variables have the value 1.

9. Find a minimal-sum-of-products representation for

   a) $f(w, x, y) = \sum m(1, 2, 5, 6)$

   b) $f(w, x, y) = \prod M(0, 1, 4, 5)$

   c) $f(w, x, y, z) = \sum m(0, 2, 5, 7, 8, 10, 13, 15)$

   d) $f(w, x, y, z) = \sum m(5, 6, 8, 11, 12, 13, 14, 15)$

   e) $f(w, x, y, z) = \sum m(7, 9, 10, 11, 14, 15)$

   f) $f(v, w, x, y, z) =$
   $\sum m(1, 2, 3, 4, 10, 17, 18, 19, 22, 23, 27, 28, 30, 31)$

10. Obtain a minimal-product-of-sums representation for $f(w, x, y, z) = \prod M(0, 1, 2, 4, 5, 10, 12, 13, 14)$.

11. Let $f: B^n \to B$ be a function of the Boolean variables $x_1, x_2, \ldots, x_n$. Determine n if the number of 1's needed to express x_1 in the Karnaugh map for f is (a) 2; (b) 4; (c) 8; (d) 2^k, for $k \in \mathbf{Z}^+$ with $1 \le k \le n - 1$.

12. If $g: B^7 \to B$ is a Boolean function of the Boolean variables $x_1, x_2, \ldots, x_7$, how many 1's are needed in the Karnaugh map of g in order to represent the product term (a) x_1; (b) $x_1 x_2$; (c) $x_1 \overline{x}_2 x_3$; (d) $x_1 x_3 x_5 x_7$?

13. In each of the following, $f: B^4 \to B$, where the Boolean variables (in order) are w, x, y, and z. Determine $|f^{-1}(0)|$ and $|f^{-1}(1)|$ if, as a minimal sum of products, f reduces to

   a) $\overline{x}$ **b)** wy **c)** $w\overline{y}z$

   d) $x + y$ **e)** $xy + z$ **f)** $xy\overline{z} + w$

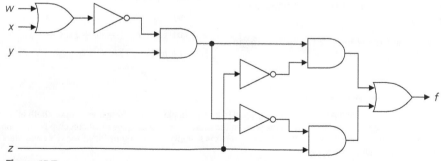

Figure 15.7

15.3
Further Applications:
Don't-Care Conditions

Our objective now is to use the ideas we have developed in the first two sections in a variety of applications.

EXAMPLE 15.19

As head of the church bazaar, Paula has volunteered to leave her automobile dealership early one evening in order to bake a cake that will be sold at the bazaar. Members of the bazaar committee volunteer to donate the needed ingredients as shown in Table 15.17.

Table 15.17

	Flour	Milk	Butter	Pecans	Eggs
Sue	×		×		
Dorothy			×	×	
Bettie	×	×			
Theresa		×			×
Ruthanne		×	×	×	

Paula sends her daughter Amy to pick up the ingredients. Write a Boolean expression to help Paula determine which (minimal) sets of volunteers she should consider so that Amy can collect all of the necessary ingredients.

Let s, d, b, t, and r denote five Boolean variables corresponding, respectively, to the women listed in the first column of the table. To get the flour, Amy must visit Sue or Bettie. In Boolean terminology, we can say that flour determines the sum $s + b$. This term will be part of a product of sums. For the other ingredients, the following sums denote the choices.

$$\text{milk: } b + t + r \qquad \text{butter: } s + d + r \qquad \text{pecans: } d + r \qquad \text{eggs: } t$$

To answer the question posed here, we seek a minimal sum of products for the function $f(s, d, b, t, r) = (s + b)(b + t + r)(s + d + r)(d + r)t$. The answer can be obtained by multiplying everything out and then simplifying the result, or by using a Karnaugh map. This time we'll use the map (in Table 15.18).

Table 15.18

$db \setminus tr$	00	01	11	10
00	0	0	0	0
01	0	0	1	0
11	0	0	1	1
10	0	0	0	0

($s = 0$)

$db \setminus tr$	00	01	11	10
00	0	0	1	0
01	0	0	1	0
11	0	0	1	1
10	0	0	1	1

($s = 1$)

We are starting with f as a product (*not* minimal) of sums. Consequently, we first fill in the 0's of the table as follows: Here $s + b$, for example, is represented by the eight 0's in the first and fourth rows of the table for $s = 0$ — these are the eight assignments for s, d, b,

t, r where $s + b$ has the value 0; for t we need the 16 0's in the first two columns of both tables. After filling in the 0's for the other three sums in the product, we then place a 1 in the nine remaining spaces and arrive at the table shown. Now we need a minimal sum of products for the nine 1's in the table. We find the result is $srt + sdt + brt + dbt$. (Verify this.) Therefore, Amy can be sent to collect the ingredients in one of four ways. She may call upon Sue, Ruthanne, and Theresa — or, perhaps, Dorothy, Bettie, and Theresa — or she may follow through with one of her other two options.

In our next application, we examine a certain property of graphs. This property was introduced earlier in Supplementary Exercise 10 of Chapter 11. The development here, however, does not rely on that prior presentation.

Definition 15.4

Let $G = (V, E)$ denote a graph (undirected) with vertex set V and edge set E. A subset D of V is called a *dominating set* for G if for every $v \in V$, either $v \in D$ or v is adjacent to a vertex in D.

For the graph shown in Fig. 15.9, the sets $\{a, d\}$, $\{a, c, e\}$ and $\{b, d, e, f\}$ are examples of dominating sets. The set $\{a, c, e\}$ is a *minimal dominating set*, for if any of the three vertices a, c, or e is removed, the remaining two no longer dominate the graph. The set $\{a, d\}$ is also minimal, but $\{b, d, e, f\}$ is not because $\{b, d, e\}$ already dominates G.

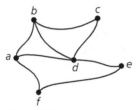

Figure 15.9

EXAMPLE 15.20

For the graph shown in Fig. 15.9, let the vertices represent cities and the edges highways. We wish to build hospitals in some of these cities so that each city either has a hospital or is adjacent to a city that does. In how many ways can this be accomplished by building a minimal number of hospitals in each case?

To answer this question, we need the minimal dominating sets for G. Consider vertex a. To guarantee that a will satisfy our objective, we must build a hospital in a, or b, or d, or f (since b, d, and f are all adjacent to a). Hence we have the term $a + b + d + f$. For b to satisfy our objective, we generate the term $a + b + c + d$. Continuing with the other four locations, we find that the answer is then a minimal-sum-of-products representation for the Boolean function $g(a, b, c, d, e, f) = (a + b + d + f)(a + b + c + d)(b + c + d) \cdot (a + b + c + d + e)(d + e + f)(a + e + f)$. Using the properties of Boolean variables, we have

$$g = (a + b + d + f)(b + c + d) \cdot \qquad \text{Absorption Law}$$
$$(d + e + f)(a + e + f)$$
$$= [(a + f)c + (b + d)][da + (e + f)] \qquad \text{Distributive Law of } + \text{ over } \cdot,$$
$$\text{and the Commutative Law of } +$$

$= [ac + fc + b + d][da + e + f]$	Distributive Law of $\cdot$ over $+$
$= acda + ace + acf + fcda + fce + fcf$ $+ bda + be + bf + dda + de + df$	Distributive Law of $\cdot$ over $+$
$= ace + (acf + acdf + cef + cf)$ $+ (acd + abd + ad) + be + bf$ $+ de + df$	Commutative and Associative Laws of $+$ and $\cdot$, and the Idempotent Law of $\cdot$
$= ace + cf + ad + be + bf + de + df$	Absorption Law

Consequently, in six of the cases the objective can be achieved by building only two hospitals. If a and c have the largest populations and we want to locate hospitals in each of these cities, then we would also have to construct a hospital at e.

The next application we shall examine introduces the notion of *"don't-care" conditions*.

EXAMPLE 15.21

The four input lines for the gating network shown in Fig. 15.10 provide the binary equivalents of the digits 0, 1, 2, ..., 9, with each number represented as $abce$ (e is least significant). Construct a gating network with two levels of gating such that the output function f equals 1 for the input that represents the digits 0, 3, 6, 9 (that is, f detects digits divisible by 3).

Table 15.19

a	b	c	e	f	a	b	c	e	f
0	0	0	0	1	1	0	0	0	0
0	0	0	1	0	1	0	0	1	1
0	0	1	0	0	1	0	1	0	$\times$
0	0	1	1	1	1	0	1	1	$\times$
0	1	0	0	0	1	1	0	0	$\times$
0	1	0	1	0	1	1	0	1	$\times$
0	1	1	0	1	1	1	1	0	$\times$
0	1	1	1	0	1	1	1	1	$\times$

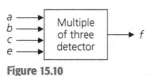

a
b
c
e → Multiple of three detector → f

Figure 15.10

Before concluding that $f = 0$ for the other 12 cases, we examine Table 15.19, where an "$\times$" appears for the value of f in the last six cases. These input combinations do not occur (because of certain external constraints), so we *don't care* what the value of f is in these situations. For such occurrences, the outputs are referred to as *unspecified* and f is called *incompletely specified*. Therefore, we write $f = \sum m(0, 3, 6, 9) + d(10, 11, 12, 13, 14, 15)$, where $d(10, 11, 12, 13, 14, 15)$ denotes the six don't-care conditions for the rows with the binary labels for 10, 11, 12, 13, 14, 15. When seeking a minimal-sum-of-products representation for f, we can use any or all of these don't-care conditions in the simplification process.

From the Karnaugh map in Table 15.20, we write f as a minimal sum of products, obtaining

$$f = \overline{a}\overline{b}\overline{c}\overline{e} + \overline{b}ce + bc\overline{e} + ae.$$

The first summand in f is for recognition of 0; $\overline{b}ce$ provides recognition for 3 because it stands for 0011 ($\overline{a}\overline{b}ce$), since 1011 ($a\overline{b}ce$) does not occur. Likewise $bc\overline{e}$ is needed to recognize 6, whereas ae takes care of 9. Figure 15.11 provides the interior details (minus the inverters) of Fig. 15.10. (Note that in Table 15.20 there are some don't-care conditions that were not used.)

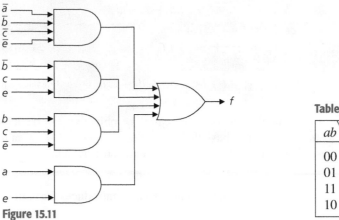

Figure 15.11

Table 15.20

$ab \backslash ce$	00	01	11	10
00	①		①	
01				1
11	×	×	×	×
10		1	⊗	×

We close this section with one more example on how to use don't-care conditions.

EXAMPLE 15.22

Find a minimal-sum-of-products representation for the incompletely specified Boolean function

$$f(w, x, y, z) = \sum m(0, 1, 2, 8, 15) + d(9, 11, 12).$$

Consider the Karnaugh map in Table 15.21. As in the previous examples each minterm is represented by a 1 in the table; each don't-care condition is designated by an ×. The 1 representing $\overline{w}\overline{x}y\overline{z}$ (at the right end of the first row) can be simplified in only one way — using the "adjacent" 1 for $\overline{w}\overline{x}\overline{y}\overline{z}$. This gives us $\overline{w}\overline{x}y\overline{z} + \overline{w}\overline{x}\overline{y}\overline{z} = \overline{w}\overline{x}\overline{z}(y + \overline{y}) = \overline{w}\overline{x}\overline{z}$. Likewise the 1 for the fundamental conjunction $wxyz$ is only adjacent to an × — for the don't-care condition $w\overline{x}yz$. This adjacency simplifies to $wxyz + w\overline{x}yz = wyz$. Finally, the remaining 1's for the fundamental conjunctions $\overline{w}\overline{x}\overline{y}z$ and $w\overline{x}\overline{y}\overline{z}$ can be used with the minterm for 0 — namely, $\overline{w}\overline{x}\overline{y}\overline{z}$ — and the don't-care condition $w\overline{x}\overline{y}z$. This gives us $\overline{w}\overline{x}\overline{y}\overline{z} + w\overline{x}\overline{y}z + \overline{w}\overline{x}\overline{y}z + w\overline{x}\overline{y}\overline{z} = (\overline{w}z + w\overline{z} + \overline{w}\overline{z} + wz)\overline{x}\overline{y} = (w + \overline{w})(z + \overline{z})\overline{x}\overline{y} = \overline{x}\overline{y}$.

Table 15.21

$wx \backslash yz$	00	01	11	10
00	①	①		①
01				
11		×		1
10	1	×		×

[Note the following:

1) In the third simplification we used the fundamental conjunction $\overline{w}\overline{x}\overline{y}\overline{z}$ a second time. It was also used in the first simplification since it is adjacent to the fundamental conjunction $\overline{w}\overline{x}y\overline{z}$. However, this does not present a problem here because of the idempotent law of $+$.

2) The don't-care condition for $wx\overline{y}\overline{z}$ was not used.]

Consequently, as a minimal sum of products, $f(w, x, y, z) = \sum m(0, 1, 2, 8, 15) + d(9, 11, 12) = \sum m(0, 1, 2, 8, 15) + d(9, 11) = \overline{w}\,\overline{x}\,\overline{z} + wyz + \overline{x}\,\overline{y}$.

---

1. For his tenth birthday, Mona wants to buy her son Jason some stamps for his collection. At the hobby shop she finds six different packages (which we shall call u, v, w, x, y, z). The kinds of stamps in each of these packages are shown in Table 15.22.

Determine all minimal combinations of packages Mona can buy so that Jason will get some stamps from all four geographical locations.

Table 15.22

	United States	European	Asian	African
u		✓		✓
v	✓		✓	
w	✓	✓		
x	✓			
y	✓			✓
z			✓	✓

2. Rework Example 15.20 using a Karnaugh map on six variables.

3. Find a minimal-sum-of-products representation for

a) $f(w, x, y, z) = \sum m(1, 3, 5, 7, 9) + d(10, 11, 12, 13, 14, 15)$

b) $f(w, x, y, z) = \sum m(0, 5, 6, 8, 13, 14) + d(4, 9, 11)$

c) $f(v, w, x, y, z) = \sum m(0, 2, 3, 4, 5, 6, 12, 19, 20, 24, 28) + d(1, 13, 16, 29, 31)$

4. The four input lines for the gating network shown in Fig. 15.12 provide the binary equivalents of the numbers $0, 1, 2, \ldots, 15$, where each number is represented as $abce$, with e the least significant bit.

a) Determine the d.n.f. of f, whose value is 1 for $abce$ prime, and 0 otherwise.

b) Draw the two-level gating network for f as a minimal sum of products.

c) We are informed that the given network is part of a larger network and that, as a result, the binary equivalents of the numbers 10 through 15 are never provided as input. Design a two-level gating network for f under these circumstances.

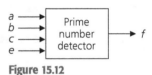

Figure 15.12

5. Determine all minimal dominating sets for the graph G shown in Fig. 15.13.

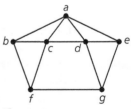

Figure 15.13

15.4
The Structure of a Boolean Algebra (Optional)

In this last section we analyze the structure of a Boolean algebra and determine those $m \in \mathbf{Z}^+$ for which there is a Boolean algebra of m elements.

Definition 15.5   Let $\mathcal{B}$ be a nonempty set that contains two special elements 0 (the zero element) and 1 (the unity, or one, element) and on which we define closed binary operations $+$, $\cdot$, and a monary (or unary) operation $^-$. Then $(\mathcal{B}, +, \cdot, ^-, 0, 1)$ is called a *Boolean algebra* if the following conditions are satisfied for all $x, y, z \in \mathcal{B}$.

a) $x + y = y + x$	a)' $xy = yx$	Commutative Laws
b) $x(y + z) = xy + xz$	b)' $x + yz = (x + y)(x + z)$	Distributive Laws
c) $x + 0 = x$	c)' $x1 = x \cdot 1 = x$	Identity Laws
d) $x + \bar{x} = 1$	d)' $x\bar{x} = x \cdot \bar{x} = 0$	Inverse Laws
e) $0 \neq 1$		

As seen in Definition 15.5, we often write xy for $x \cdot y$. When the operations and identity elements are known, we write $\mathcal{B}$ instead of $(\mathcal{B}, +, \cdot, {}^-, 0, 1)$.

From our past experience we have the following examples.

EXAMPLE 15.23

If $\mathcal{U}$ is a (finite) set, then $\mathcal{B} = \mathcal{P}(\mathcal{U})$ is a Boolean algebra where for $A, B \subseteq \mathcal{U}$, we have $A + B = A \cup B$, $AB = A \cap B$, $\bar{A}$ = the complement of A (in $\mathcal{U}$), and where $\emptyset$ is the zero element and $\mathcal{U}$ is the unity.

EXAMPLE 15.24

For $n \in \mathbf{Z}^+$, $F_n = \{f: B^n \to B\}$, the set of Boolean functions on n Boolean variables, is a Boolean algebra where $+$, $\cdot$, and $^-$ are as defined in Definition 15.2, and where the zero element is the constant function **0**, while the constant function **1** is the one element.

Let us now examine a new type of Boolean algebra.

EXAMPLE 15.25

Let $\mathcal{B}$ be the set of all positive integer divisors of 30: $\mathcal{B} = \{1, 2, 3, 5, 6, 10, 15, 30\}$. For all $x, y \in \mathcal{B}$, define $x + y = \text{lcm}(x, y)$; $xy = \gcd(x, y)$; and $\bar{x} = 30/x$. Then with 1 as the zero element and 30 as the unity element, one can verify that $(\mathcal{B}, +, \cdot, {}^-, 1, 30)$ is a Boolean algebra. We shall establish one of the distributive laws for this Boolean algebra and leave the other conditions for the reader to check.

For the first distributive law we want to show that

$$\gcd(x, \text{lcm}(y, z)) = \text{lcm}(\gcd(x, y), \gcd(x, z)),$$

for all $x, y, z \in \mathcal{B}$. In order to do so we write

$$x = 2^{k_1} 3^{k_2} 5^{k_3}, \qquad y = 2^{m_1} 3^{m_2} 5^{m_3}, \qquad \text{and} \quad z = 2^{n_1} 3^{n_2} 5^{n_3},$$

where $0 \leq k_i, m_i, n_i \leq 1$ for all $1 \leq i \leq 3$.

Then $\text{lcm}(y, z) = 2^{s_1} 3^{s_2} 5^{s_3}$, where $s_i = \max\{m_i, n_i\}$, for all $1 \leq i \leq 3$, and so $\gcd(x, \text{lcm}(y, z)) = 2^{t_1} 3^{t_2} 5^{t_3}$, where $t_i = \min\{k_i, \max\{m_i, n_i\}\}$, for all $1 \leq i \leq 3$. Also, $\gcd(x, y) = 2^{u_1} 3^{u_2} 5^{u_3}$, where $u_i = \min\{k_i, m_i\}$, when $1 \leq i \leq 3$, and $\gcd(x, z) = 2^{v_1} 3^{v_2} 5^{v_3}$ with $v_i = \min\{k_i, n_i\}$ for all $1 \leq i \leq 3$. So $\text{lcm}(\gcd(x, y), \gcd(x, z)) = 2^{w_1} 3^{w_2} 5^{w_3}$, where $w_i = \max\{u_i, v_i\}$, for all $1 \leq i \leq 3$.

Therefore, for each $i \in \{1, 2, 3\}$, $w_i = \max\{u_i, v_i\} = \max\{\min\{k_i, m_i\}, \min\{k_i, n_i\}\}$, and $t_i = \min\{k_i, \max\{m_i, n_i\}\}$. To verify the result, we need to show that $w_i = t_i$ for all $1 \leq i \leq 3$. If $k_i = 0$, then $w_i = 0 = t_i$. If $k_i = 1$, then $w_i = \max\{m_i, n_i\} = t_i$. This exhausts all possibilities, so $w_i = t_i$ for $1 \leq i \leq 3$ and

$$\gcd(x, \text{lcm}(y, z)) = \text{lcm}(\gcd(x, y), \gcd(x, z)).$$

If we analyze this result further, we find that 30 can be replaced by any number $m = p_1 p_2 p_3$, where p_1, p_2, p_3 are distinct primes. In fact, the result follows for the set of all divisors of $p_1 p_2 \cdots p_n$, a product of n distinct primes. (Note that such a product is square-free; that is, there is no $k \in \mathbf{Z}^+$, $k > 1$, with k^2 dividing it.)

EXAMPLE 15.26

A word about the propositional calculus. If p, q are two primitive propositions, we may feel that the collection of all propositions obtained from p, q, using $\vee$, $\wedge$, and $\neg$, should be a Boolean algebra. After all, just look at the laws of logic and the way they compare with the comparable results for set theory and Boolean functions. There is one main difference. In our study of logic we found, for example, that $p \wedge q \Longleftrightarrow q \wedge p$, not that $p \wedge q = q \wedge p$. To get around this we define a relation $\mathcal{R}$ on the set S of all propositions so obtained from p, q, where $s_1 \mathcal{R} s_2$ if $s_1 \Longleftrightarrow s_2$. Then $\mathcal{R}$ is an equivalence relation on S and partitions S, in this case, into 16 equivalence classes. If we define $+$, $\cdot$, and $^-$ on these equivalence classes by $[s_1] + [s_2] = [s_1 \vee s_2]$, $[s_1][s_2] = [s_1 \wedge s_2]$, and $\overline{[s_1]} = [\neg s_1]$, and if we recognize $[T_0]$ as the one element and $[F_0]$ as the zero element, then we get a Boolean algebra.

In the definition of a Boolean algebra, there are nine conditions. Yet in the lists of properties we examined for set theory, logic, and Boolean functions, we listed 19 properties. And there were even more! Undoubtedly, there is a way to get the remaining properties, and others not listed among the 19, from the ones given in the definition.

THEOREM 15.1

The Idempotent Laws. For all $x \in \mathcal{B}$, a Boolean algebra, (i) $x + x = x$; and (ii) $xx = x$.

Proof: (To the right of each equality appearing in this proof, we list the letter of the condition from Definition 15.5 that justifies it.)

i) $x = x + 0$	**c)**	**ii)** $x = x \cdot 1$	**c)**$'$
$= x + x\overline{x}$	**d)**$'$	$= x(x + \overline{x})$	**d)**
$= (x + x)(x + \overline{x})$	**b)**$'$	$= xx + x\overline{x}$	**b)**
$= (x + x) \cdot 1$	**d)**	$= xx + 0$	**d)**$'$
$= x + x$	**c)**$'$	$= xx$	**c)**

In proving this theorem we can obtain the proof of part (ii) from that of part (i) by changing all occurrences of $+$ to $\cdot$, and vice versa, and all occurrences of 0 to 1, and vice versa. Also, the justifications for the corresponding steps constitute a pair of conditions in Definition 15.5. As in the past, these pairs are said to be *duals* of each other; condition (e) is called *self-dual*. This now leads us to the following result.

THEOREM 15.2

The Principle of Duality. If s is a theorem about a Boolean algebra, and s can be proved from the conditions in Definition 15.5 and properties derived from these same conditions, then its dual s^d is likewise a theorem.

Proof: Let s be such a theorem. Dualizing all the steps and reasons in the proof of s (as in the proof of Theorem 15.1), we obtain a proof for s^d.

We now list some further properties for a Boolean algebra. We shall prove some of these properties and leave the remaining proofs for the reader.

THEOREM 15.3

For every Boolean algebra $\mathcal{B}$, if $x, y, z \in \mathcal{B}$, then

a) $x \cdot 0 = 0$	**a)**$'$ $x + 1 = 1$	Dominance Laws
b) $x(x + y) = x$	**b)**$'$ $x + xy = x$	Absorption Laws

c) $[xy = xz$ and $\overline{x}y = \overline{x}z] \Rightarrow y = z$ Cancellation Laws

c)′ $[x + y = x + z$ and $\overline{x} + y = \overline{x} + z] \Rightarrow y = z$

d) $x(yz) = (xy)z$ d)′ $x + (y + z) = (x + y) + z$ Associative Laws

e) $[x + y = 1$ and $xy = 0] \Rightarrow y = \overline{x}$ Uniqueness of Complements
(Inverses)

f) $\overline{\overline{x}} = x$ Law of the Double
Complement

g) $\overline{xy} = \overline{x} + \overline{y}$ g)′ $\overline{x + y} = \overline{x}\,\overline{y}$ DeMorgan's Laws

h) $\overline{0} = 1$ h)′ $\overline{1} = 0$

i) $x\overline{y} = 0$ if and only if   i)′ $x + \overline{y} = 1$ if and only if
$xy = x$ $x + y = x$

Proof:

a) $x \cdot 0 = 0 + x \cdot 0,$ by Definition 15.5(c), (a)
$= x \cdot \overline{x} + x \cdot 0,$ by Definition 15.5(d)′
$= x \cdot (\overline{x} + 0),$ by Definition 15.5(b)
$= x \cdot \overline{x},$ by Definition 15.5(c)
$= 0,$ by Definition 15.5(d)′

a)′ This follows from part (a) by the Principle of Duality.

c) Here $y = 1 \cdot y = (x + \overline{x})y = xy + \overline{x}y = xz + \overline{x}z = (x + \overline{x})z = 1 \cdot z = z.$ (Verify all equalities.)

c)′ This is the dual of part (c).

d) To establish this result, we use result (c)′ and arrive at the conclusion by showing that $x + [x(yz)] = x + [(xy)z]$ and $\overline{x} + [x(yz)] = \overline{x} + [(xy)z]$. Using the absorption law, we find that $x + [x(yz)] = x$. Likewise $x + [(xy)z] = [x + (xy)](x + z) = x(x + z) = x$. Then $\overline{x} + [x(yz)] = (\overline{x} + x)(\overline{x} + yz) = 1 \cdot (\overline{x} + yz) = \overline{x} + yz$, whereas $\overline{x} + [(xy)z] = (\overline{x} + xy)(\overline{x} + z) = ((\overline{x} + x)(\overline{x} + y))(\overline{x} + z) = (1 \cdot (\overline{x} + y))(\overline{x} + z) = (\overline{x} + y)(\overline{x} + z) = \overline{x} + yz$. (Verify all equalities.)
The result now follows by the cancellation law in part (c)′.

d)′ Fortunately, this is the dual of part (d).

e) We find here that $\overline{x} = \overline{x} + 0 = \overline{x} + xy = (\overline{x} + x)(\overline{x} + y) = 1 \cdot (\overline{x} + y) = (\overline{x} + y) \cdot 1 = (\overline{x} + y)(x + y) = \overline{x}x + y = 0 + y = y$. (Verify all equalities.)
We note that statement (e) is self-dual. Statement (f) is a corollary of (e) because $\overline{\overline{x}}$ and x are both complements (inverses) of $\overline{x}$.

g) This result will follow from part (e) if we can show that $\overline{x} + \overline{y}$ is a complement of xy.

$$xy + (\overline{x} + \overline{y}) = (xy + \overline{x}) + \overline{y} = (x + \overline{x})(y + \overline{x}) + \overline{y}$$
$$= 1 \cdot (y + \overline{x}) + \overline{y} = (y + \overline{y}) + \overline{x} = 1 + \overline{x} = 1.$$

Also, $xy(\overline{x} + \overline{y}) = (xy\overline{x}) + (xy\overline{y}) = ((x\overline{x})y) + (x(y\overline{y})) = 0 \cdot y + x \cdot 0 = 0 + 0 = 0$.
Consequently, $\overline{x} + \overline{y}$ is a complement of xy, and by uniqueness of complements, it follows that $\overline{xy} = \overline{x} + \overline{y}$.

Enough proving for a while! Now we are going to investigate how to impose an order on the elements of a Boolean algebra. In fact, we shall want a partial order, and for this reason we turn now to the Hasse diagram.

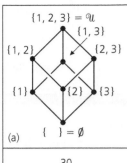

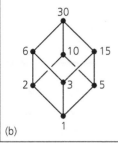

Figure 15.14

Let us start by considering the Hasse diagrams for the following two Boolean algebras.

a) $(\mathscr{P}(\mathscr{U}), \cup, \cap, {}^-, \emptyset, \mathscr{U})$, where $\mathscr{U} = \{1, 2, 3\}$, and the partial order is induced by the subset relation.

b) $(\mathscr{S}, +, \cdot, {}^-, 1, 30)$, where $\mathscr{S} = \{1, 2, 3, 5, 6, 10, 15, 30\}$, $x + y = \text{lcm}(x, y)$, $xy = \gcd(x, y)$, and $\overline{x} = 30/x$. Hence the zero element is the divisor 1 and the one element is the divisor 30. The relation $\mathscr{R}$ on $\mathscr{S}$, defined by $x \,\mathscr{R}\, y$ if x divides y, makes $\mathscr{S}$ into a poset.

Figure 15.14 shows the Hasse diagrams for these two Boolean algebras. Ignoring the labels at the vertices in each diagram, we see that the underlying structures are the same. This suggests how we should define the concept of *isomorphism for Boolean algebras*.

These examples also suggest two other ideas.

1) Can we partially order any finite Boolean algebra?

2) Looking at Fig. 15.14(a), we see that the nonzero elements just above $\emptyset$ are such that every element other than $\emptyset$ can be obtained as a Boolean sum of these three. For example, $\{1, 3\} = \{1\} \cup \{3\}$ and $\{1, 2, 3\} = \{1\} \cup \{2\} \cup \{3\}$. For part (b), the numbers 2, 3, and 5 are such that every divisor other than 1 is realized as the Boolean sum of these three. For example, $6 = \text{lcm}(2, 3)$ and $30 = \text{lcm}(2, \text{lcm}(3, 5)) = \text{lcm}(2, 3, 5)$.

We now start to deal formally with these suggestions.

When dealing with sets in Chapter 3 we related the operations of $\cup, \cap$, and $^-$ to the subset relation by the equivalence of the statements: (a) $A \subseteq B$; (b) $A \cap B = A$; (c) $A \cup B = B$; and (d) $\overline{B} \subseteq \overline{A}$, where $A, B \subseteq \mathscr{U}$. We now use parts (a) and (b) in an attempt to partially order any Boolean algebra $\mathscr{B}$.

Definition 15.6   If $x, y \in \mathscr{B}$, define $x \leq y$ if $xy = x$.

Hence we define a new concept — namely, "$\leq$" — in terms of notions we have in $\mathscr{B}$ — namely, $\cdot$ and the notion of equality. We can make up definitions! But does this one lead us anywhere?

THEOREM 15.4   The relation "$\leq$", just defined, is a partial order.

Proof: Since $xx = x$ for all $x \in \mathscr{B}$, we have $x \leq x$ and the relation is reflexive. To establish antisymmetry, suppose that $x, y \in \mathscr{B}$ with $x \leq y$ and $y \leq x$. Then $xy = x$ and $yx = y$. By the commutative property, $xy = yx$, so $x = y$. Finally, if $x \leq y$ and $y \leq z$, then $xy = x$ and $yz = y$, so $x = xy = x(yz) = (xy)z = xz$, and with $x = xz$, we have $x \leq z$, so the relation is transitive.

Now we can partially order any Boolean algebra, and we note that for all x in a Boolean algebra, $0 \leq x$ and $x \leq 1$. (Why?) Before going on, however, let us consider the Boolean algebra consisting of the divisors of 30. How do we apply Theorem 15.4 in this example? Here the partial order is given by $x \leq y$ if $xy = x$. Since xy is $\gcd(x, y)$, if $\gcd(x, y) = x$, then x divides y. But this was precisely the partial order we had on this Boolean algebra when we started.

Armed with this concept of partial order, we return to the observations we made earlier about the elements in the Hasse diagrams of Fig. 15.14.

Definition 15.7	Let 0 denote the zero element of a Boolean algebra $\mathcal{B}$. An element $x \in \mathcal{B}$, $x \neq 0$, is called an *atom* of $\mathcal{B}$ if for all $y \in \mathcal{B}$, $y \leq x \Rightarrow y = 0$ or $y = x$.

EXAMPLE 15.27	**a)** For the Boolean algebra of all subsets of $\mathcal{U} = \{1, 2, 3\}$, the atoms are $\{1\}$, $\{2\}$, and $\{3\}$.
	b) When we are dealing with the positive integer divisors of 30, the atoms of this Boolean algebra are 2, 3, and 5.
	c) The atoms in the Boolean algebra $F_n = \{f : B^n \to B\}$ are the minterms (or fundamental conjunctions).

The atoms of a finite Boolean algebra satisfy the following properties.

THEOREM 15.5	**a)** If x is an atom in a Boolean algebra $\mathcal{B}$, then for all $y \in \mathcal{B}$, $xy = 0$ or $xy = x$.
	b) If x_1, x_2 are atoms of $\mathcal{B}$ and $x_1 \neq x_2$, then $x_1 x_2 = 0$.

Proof:

a) For all $x, y \in \mathcal{B}$, $xy \leq x$, because $(xy)x = x(yx) = x(xy) = (xx)y = xy$. For x an atom, $xy \leq x \Rightarrow xy = 0$ or $xy = x$.

b) This follows from part (a). The reader should supply the details.

THEOREM 15.6	If $x_1, x_2, \ldots, x_n$ are all the atoms in a finite Boolean algebra $\mathcal{B}$ and $x \in \mathcal{B}$ with $xx_i = 0$ for all $1 \leq i \leq n$, then $x = 0$.

Proof: If $x \neq 0$, let $S = \{y \in \mathcal{B} \mid 0 < y \leq x\}$. ($0 < y$ denotes $0 \leq y$ and $0 \neq y$.) With $x \in S$, we have $S \neq \emptyset$. Since S is finite, we can find an element z in $\mathcal{B}$ where $0 < z \leq x$ and no element of $\mathcal{B}$ is between 0 and z. Then z is an atom and $0 = xz = z > 0$. This possibility has led us to a contradiction, so we cannot have $x \neq 0$; that is, $x = 0$.

This leads us to the following result on representation.

THEOREM 15.7	Given a finite Boolean algebra $\mathcal{B}$ with atoms $x_1, x_2, \ldots, x_n$, each $x \in \mathcal{B}$, $x \neq 0$, can be written as a sum of atoms uniquely, up to order.

Proof: Since $x \neq 0$, by Theorem 15.6, $S = \{x_i \mid xx_i \neq 0\} \neq \emptyset$. Let $S = \{x_{i_1}, x_{i_2}, \ldots, x_{i_k}\}$, and $y = x_{i_1} + x_{i_2} + \cdots + x_{i_k}$. Then $xy = x(x_{i_1} + x_{i_2} + \cdots + x_{i_k}) = xx_{i_1} + xx_{i_2} + \cdots + xx_{i_k} = x_{i_1} + x_{i_2} + \cdots + x_{i_k}$, by Theorem 15.5(a). So $xy = y$.

Now consider $(x\overline{y})x_i$ for each $1 \leq i \leq n$. If $x_i \notin S$, then $xx_i = 0$, and $(x\overline{y})x_i = 0$. For $x_i \in S$, we have $(x\overline{y})x_i = xx_i(\overline{x_{i_1} + x_{i_2} + \cdots + x_{i_k}}) = xx_i(\overline{x}_{i_1}\overline{x}_{i_2} \cdots \overline{x}_{i_k}) = x(x_i\overline{x}_i)(z)$, where z is the product of the complements of all elements in $S - \{x_i\}$. As $x_i\overline{x}_i = 0$, it follows that $(x\overline{y})x_i = 0$. So $(x\overline{y})x_i = 0$ for all x_i, where $1 \leq i \leq n$. By Theorem 15.6, we have $x\overline{y} = 0$.

With $xy = y$ and $x\overline{y} = 0$, it follows that $x = x \cdot 1 = x(y + \overline{y}) = xy + x\overline{y} = xy + 0 = y = x_{i_1} + x_{i_2} + \cdots + x_{i_k}$, a sum of atoms.

To show that this representation for x is unique, up to order, suppose $x = x_{j_1} + x_{j_2} + \cdots + x_{j_\ell}$.

If x_{j_1} does not appear as a summand in $x_{i_1} + x_{i_2} + \cdots + x_{i_k}$, then $x_{j_1} = x_{j_1} \cdot x_{j_1} = x_{j_1}(x_{j_1} + x_{j_2} + \cdots + x_{j_\ell})$ [by Theorem 15.5(b)] $= x_{j_1}x = x_{j_1}(x_{i_1} + x_{i_2} + \cdots + x_{i_k}) = 0$ [again, by Theorem 15.5(b)]. Hence x_{j_1} must appear as a summand in $x_{i_1} + x_{i_2} + \cdots + x_{i_k}$, as must $x_{j_2}, \ldots, x_{j_\ell}$. So $\ell \le k$. By the same reasoning, we get $k \le \ell$ and find the representations identical, except for order.

From this result we see that if $\mathcal{B}$ is a finite Boolean algebra with atoms $x_1, x_2, \ldots, x_n$, then each $x \in \mathcal{B}$ can be uniquely written as $\sum_{i=1}^{n} c_i x_i$, with each $c_i \in \{0, 1\}$ (and because $\mathcal{B}$ is closed under $+$, each such *linear combination* of atoms is in $\mathcal{B}$). If $c_i = 0$, this indicates that x_i is not in the representation of x; $c_i = 1$ indicates that it is. Consequently, each $x \in \mathcal{B}$ is associated with a unique n-tuple $(c_1, c_2, \ldots, c_n)$, and there are 2^n such n-tuples. Therefore we have proved the following result.

THEOREM 15.8 If $\mathcal{B}$ is a finite Boolean algebra with n atoms, then $|\mathcal{B}| = 2^n$.

There is one final question to resolve. If $n \in \mathbf{Z}^+$, how many different Boolean algebras of size 2^n are there? Looking at the Hasse diagrams in Fig. 15.14, we see two different pictures. But if we ignore the labels on the vertices, the underlying structures emerge as exactly the same. Hence these two Boolean algebras are said to be abstractly identical or isomorphic.

Definition 15.8 Suppose $(\mathcal{B}_1, +, \cdot, {}^-, 0, 1)$ and $(\mathcal{B}_2, +, \cdot, {}^-, 0, 1)$ are Boolean algebras. Then $\mathcal{B}_1, \mathcal{B}_2$ are called *isomorphic* if there is a function $f: \mathcal{B}_1 \to \mathcal{B}_2$ such that f is one-to-one and onto, and for all $x_1, y_1 \in \mathcal{B}_1$,

a) $f(x_1 + y_1) = f(x_1) + f(y_1)$
$$\underset{\text{(in } \mathcal{B}_1)}{\uparrow} \qquad\qquad \underset{\text{(in } \mathcal{B}_2)}{\uparrow}$$

b) $f(x_1 \cdot y_1) = f(x_1) \cdot f(y_1)$
$$\underset{\text{(in } \mathcal{B}_1)}{\uparrow} \qquad\qquad \underset{\text{(in } \mathcal{B}_2)}{\uparrow}$$

c) $f(\overline{x_1}) = \overline{f(x_1)}$ [In $f(\overline{x_1})$ we take the complement in $\mathcal{B}_1$, while for $\overline{f(x_1)}$ the complement is taken in $\mathcal{B}_2$.]

Such a function f *preserves* the operations of the algebraic structures.

EXAMPLE 15.28 For the two Boolean algebras in Fig. 15.14, define f by

$$f: \emptyset \to 1 \qquad f: \{2\} \to 3 \qquad f: \{1, 2\} \to 6 \qquad f: \{2, 3\} \to 15$$
$$f: \{1\} \to 2 \qquad f: \{3\} \to 5 \qquad f: \{1, 3\} \to 10 \qquad f: \{1, 2, 3\} \to 30$$

Note the following:

a) The zero elements correspond under f, as do the unity elements.

b) $f(\{1\} \cup \{2\}) = f(\{1, 2\}) = 6 = \mathrm{lcm}(2, 3) = \mathrm{lcm}(f(\{1\}), f(\{2\}))$

c) $f(\{1, 2\} \cap \{2, 3\}) = f(\{2\}) = 3 = \gcd(6, 15) = \gcd(f(\{1, 2\}), f(\{2, 3\}))$

d) $f(\overline{\{2\}}) = f(\{1, 3\}) = 10 = 30/3 = \overline{3} = \overline{f(\{2\})}$

e) The image of each atom ($\{1\}, \{2\}, \{3\}$) is an atom (2, 3, 5, respectively).

This function is an isomorphism. Once we establish a correspondence between the respective zero elements and between the respective atoms, the remaining correspondences are determined from these by Theorem 15.7 and the preservation of the operations under f.

From this example we have our final result.

THEOREM 15.9

Every finite Boolean algebra $\mathcal{B}$ is isomorphic to a Boolean algebra of sets.

Proof: Since $\mathcal{B}$ is finite, $\mathcal{B}$ has n atoms x_i, $1 \leq i \leq n$, and $|\mathcal{B}| = 2^n$. Let $\mathcal{U} = \{1, 2, \ldots, n\}$ and $\mathcal{P}(\mathcal{U})$ be the Boolean algebra of subsets of $\mathcal{U}$.

We define $f: \mathcal{B} \to \mathcal{P}(\mathcal{U})$ as follows. For each $x \in \mathcal{B}$, it follows from Theorem 15.7 that we can write $x = \sum_{i=1}^{n} c_i x_i$, where each c_i is 0 or 1. [Here $c_i \in \{0, 1\}$ ($= B$) and for each atom a in $\mathcal{B}$, $c_i a = 0$ (the zero element in $\mathcal{B}$) if $c_i = 0$, while $c_i a = a$ when $c_i = 1$.] Then we define

$$f(x) = \{i \,|\, 1 \leq i \leq n \quad \text{and} \quad c_i = 1\}.$$

[For example, (1) $f(0) = \emptyset$; (2) $f(x_i) = \{i\}$ for each atom x_i, where $1 \leq i \leq n$; (3) $f(x_1 + x_2) = \{1, 2\}$; and (4) $f(x_2 + x_4 + x_7) = \{2, 4, 7\}$.] Now consider $x, y \in \mathcal{B}$, with $x = \sum_{i=1}^{n} c_i x_i$ and $y = \sum_{i=1}^{n} d_i x_i$, where $c_i, d_i \in \{0, 1\}$ for all $1 \leq i \leq n$.

First we find that $x + y = \sum_{i=1}^{n} s_i x_i$, where $s_i = c_i + d_i$ for each $1 \leq i \leq n$. (Remember that here $1 + 1 = 1$.) Consequently,

$$\begin{aligned}
f(x + y) &= \{i \,|\, 1 \leq i \leq n \text{ and } s_i = 1\} \\
&= \{i \,|\, 1 \leq i \leq n \text{ and } (c_i = 1 \text{ or } d_i = 1)\} \\
&= \{i \,|\, 1 \leq i \leq n \text{ and } c_i = 1\} \cup \{i \,|\, 1 \leq i \leq n \text{ and } d_i = 1\} \\
&= f(x) \cup f(y).
\end{aligned}$$

Theorem 15.5(b) tells us that

$$x \cdot y = \sum_{i=1}^{n} t_i x_i,$$

where $t_i = c_i d_i$ for all $1 \leq i \leq n$, and so, in a similar way, we get

$$\begin{aligned}
f(x \cdot y) &= \{i \,|\, 1 \leq i \leq n \text{ and } t_i = 1\} \\
&= \{i \,|\, 1 \leq i \leq n \text{ and } (c_i = 1 \text{ and } d_i = 1)\} \\
&= \{i \,|\, 1 \leq i \leq n \text{ and } c_i = 1\} \cap \{i \,|\, 1 \leq i \leq n \text{ and } d_i = 1\} \\
&= f(x) \cap f(y).
\end{aligned}$$

To complete the proof that f is an isomorphism, we should first observe that if $x = \sum_{i=1}^{n} c_i x_i$, then $\overline{x} = \sum_{i=1}^{n} \overline{c}_i x_i$. This follows from Theorems 15.3(e) and 15.5(b) because

$$\sum_{i=1}^{n} c_i x_i + \sum_{i=1}^{n} \overline{c}_i x_i = \sum_{i=1}^{n} (c_i + \overline{c}_i) x_i = \sum_{i=1}^{n} x_i = 1$$

(Why is this true? See Exercise 15 for this section.) and

$$\left(\sum_{i=1}^{n} c_i x_i\right)\left(\sum_{i=1}^{n} \overline{c}_i x_i\right) = \sum_{i=1}^{n} c_i \overline{c}_i x_i = \sum_{i=1}^{n} 0 x_i = 0.$$

Now we find that

$$f(\overline{x}) = \{i \mid 1 \leq i \leq n \text{ and } \overline{c}_i = 1\}$$
$$= \{i \mid 1 \leq i \leq n \text{ and } c_i = 0\}$$
$$= \overline{\{i \mid 1 \leq i \leq n \text{ and } c_i = 1\}}$$
$$= \overline{f(x)},$$

so the function f preserves the operations in the Boolean algebras $\mathcal{B}$ and $\mathcal{P}(\mathcal{U})$. We leave to the reader the details showing that f is one-to-one and onto.

EXERCISES 15.4

1. Verify the second distributive law and the identity and inverse laws for Example 15.25.

2. Complete the proof of Theorem 15.3.

3. Let $\mathcal{B}$ be the set of positive integer divisors of 210, and define $+$, $\cdot$, and $^-$ for $\mathcal{B}$ by $x + y = \text{lcm}(x, y)$, $x \cdot y = xy = \gcd(x, y)$, and $\overline{x} = 210/x$. Determine each of the following:

a) $30 + 5 \cdot 7$
b) $(30 + 5) \cdot (30 + 7)$
c) $\overline{(14 + 15)}$
d) $21(2 + \overline{10})$
e) $(2 + 3) + 5$
f) $(6 + 35)(7 + 10)$

4. For a Boolean algebra $\mathcal{B}$ the relation "$\leq$" on $\mathcal{B}$, defined by $x \leq y$ if $xy = x$, was shown to be a partial order. Prove that: (a) if $x \leq y$ then $x + y = y$; and (b) if $x \leq y$ then $\overline{y} \leq \overline{x}$.

5. Let $(\mathcal{B}, +, \cdot, ^-, 0, 1)$ be a Boolean algebra that is partially ordered by $\leq$.

a) If $w \in \mathcal{B}$ and $w \leq 0$, prove that $w = 0$.
b) If $x \in \mathcal{B}$ and $1 \leq x$, prove that $x = 1$.
c) If $y, z \in \mathcal{B}$ with $y \leq z$ and $y \leq \overline{z}$, prove that $y = 0$.

6. Let $(\mathcal{B}, +, \cdot, ^-, 0, 1)$ be a Boolean algebra that is partially ordered by $\leq$. If $w, x, y, z \in \mathcal{B}$ with $w \leq x$ and $y \leq z$, prove that (a) $wy \leq xy$; and (b) $w + y \leq x + z$.

7. If $\mathcal{B}$ is a Boolean algebra, partially ordered by $\leq$, and $x, y \in \mathcal{B}$, what is the dual of the statement "$x \leq y$"?

8. Let $F_n = \{f: B^n \to B\}$ be the Boolean algebra of all Boolean functions on n Boolean variables. How many atoms does F_n have?

9. Verify Theorem 15.5(b).

10. If $\mathcal{B}$ is a Boolean algebra, prove that the zero element and the one element of $\mathcal{B}$ are unique.

11. Let $f: \mathcal{B}_1 \to \mathcal{B}_2$ be an isomorphism of Boolean algebras. Prove each of the following:

a) $f(0) = 0$.
b) $f(1) = 1$.
c) If $x, y \in \mathcal{B}_1$ with $x \leq y$, then in $\mathcal{B}_2$, $f(x) \leq f(y)$.
d) If x is an atom of $\mathcal{B}_1$, then $f(x)$ is an atom in $\mathcal{B}_2$.

12. Let $\mathcal{B}_1$ be the Boolean algebra of all positive integer divisors of 2310, with $\mathcal{B}_2$ the Boolean algebra of all subsets of $\{a, b, c, d, e\}$.

a) Define $f: \mathcal{B}_1 \to \mathcal{B}_2$ so that $f(2) = \{a\}$, $f(3) = \{b\}$, $f(5) = \{c\}$, $f(7) = \{d\}$, $f(11) = \{e\}$. For f to be an isomorphism, what must the images of 35, 110, 210, and 330 be?

b) How many different isomorphisms can one define between $\mathcal{B}_1$ and $\mathcal{B}_2$?

13. a) If $\mathcal{B}_1$, $\mathcal{B}_2$ are Boolean algebras and $f: \mathcal{B}_1 \to \mathcal{B}_2$ is one-to-one, onto, and such that $f(x + y) = f(x) + f(y)$ and $f(\overline{x}) = \overline{f(x)}$, for all $x, y \in \mathcal{B}_1$, prove that f is an isomorphism.

b) State and prove another result comparable to that in part (a). (What principle is used here?)

14. Prove that the function f in Theorem 15.9 is one-to-one and onto.

15. Let $\mathcal{B}$ be a finite Boolean algebra with the n atoms $x_1, x_2, \ldots, x_n$. (So $|\mathcal{B}| = 2^n$.) Prove that

$$1 = x_1 + x_2 + \cdots + x_n.$$

15.5
Summary and Historical Review

The modern concept of abstract algebra was developed by George Boole in his study of general abstract systems, as opposed to particular examples of such systems. In his 1854 publication *An Investigation of the Laws of Thought*, he formulated the mathematical structure now called a Boolean algebra. Although abstract in nature during the nineteenth century, the study of Boolean algebra was investigated in the twentieth century for its applicative value.

Starting in 1938, Claude Elwood Shannon (1916–2001) made the first major contribution in applied Boolean algebra in [8]. He devised the algebra of switching circuits and showed its relation to the algebra of logic. Additional developments that were made in this area during the 1940s and 1950s are noted in the paper by C. E. Shannon [9] and in the report of the Harvard University Computation Laboratory [10]. (The computer term *bit* was coined by Claude E. Shannon, who was also one of the first to represent information in terms of bits.)

Claude Elwood Shannon (1916–2001)

We found that switching functions can be represented by their disjunctive and conjunctive normal forms. These forms allowed us to write such functions in a compact way using binary labels. The minimization process showed us how to represent a given Boolean function as a minimal sum of products or as a minimal product of sums. Based on the map method by E. W. Veitch [11], Maurice Karnaugh's modification [4] was developed here as a pictorial method for the simplification of Boolean functions. Another technique that we mentioned in the text is the tabulation algorithm known as the Quine-McCluskey method. Originally developed by Willard Van Orman Quine (1908–2000) [6, 7], this technique was modified by Edward J. McCluskey, Jr. (1929–) [5]. It is very useful for functions with more than six variables and lends itself to computer implementation. The interested reader can find more about Karnaugh maps in Chapter 6 of F. J. Hill and G. R. Peterson [3]. Chapter 7 of [3] provides an excellent coverage of the Quine-McCluskey method. J. F. Wakerly [12] examines digital circuits in the light of contemporary technology, whereas T. L. Booth [1] investigates some specific applications of logic design in the study of computers. A more advanced coverage of the applications given in this chapter (along with many other related concepts) is given in the text by K. G. Gopolan [2].

Although the major part of this chapter was applied in nature, Section 15.4 found us investigating the structure of a Boolean algebra. Unlike commutative rings with unity, which come in all possible sizes, we found that a Boolean algebra can contain only 2^n elements, where $n \in \mathbf{Z}^+$. Uniqueness of representation appeared as we found the atoms of a Boolean algebra used to build the rest of the algebra (except for the zero element). The Boolean algebra of sets that we studied in Chapter 3 was found to represent all finite Boolean algebras in the sense that a finite Boolean algebra with n atoms is isomorphic to the Boolean algebra of all subsets of $\{1, 2, 3, \ldots, n\}$.

REFERENCES

1. Booth, Taylor L. *Digital Networks and Computer Systems*, 2nd ed. New York: Wiley, 1978.
2. Gopolan, K. Gopal. *Introduction to Digital Microelectronic Circuits*. Chicago: Irwin, 1996.
3. Hill, Frederick J., and Peterson, Gerald R. *Introduction to Switching Theory and Logical Design*, 3rd ed. New York: Wiley, 1981.
4. Karnaugh, Maurice. "The Map Method for Synthesis of Combinational Logic Circuits." *Transactions of the AIEE*, part I, vol. 72, no. 9 (1953): pp. 593–599.
5. McCluskey, Edward J., Jr. "Minimization of Boolean Functions." *Bell System Technical Journal* 35, no. 6 (November 1956): pp. 1417–1444.
6. Quine, Willard V. "The Problem of Simplifying Truth Functions." *American Mathematical Monthly* 59, no. 8 (October 1952): pp. 521–531.
7. Quine, Willard V. "A Way to Simplify Truth Functions." *American Mathematical Monthly* 62, no. 9 (November 1955): pp. 627–631.
8. Shannon, Claude E. "A Symbolic Analysis of Relay and Switching Circuits." *Transactions of the AIEE*, vol. 57 (1938): pp: 713–723.
9. Shannon, Claude E. "The Synthesis of Two-terminal Switching Circuits." *Bell System Technical Journal*, vol. 28 (1949): pp. 59–98.
10. Staff of the Computation Laboratory. *Synthesis of Electronic Computing and Control Circuits*, Annals 27. Cambridge, Mass.: Harvard University Press, 1951.
11. Veitch, E. W. "A Chart Method for Simplifying Truth Functions." *Proceedings of the ACM*. Pittsburgh, Penn. (May 1952): pp. 127–133.
12. Wakerly, John F. *Digital Design: Principles and Practices*, 2nd ed. Englewood Cliffs, N.J.: Prentice-Hall, 1994.

SUPPLEMENTARY EXERCISES

1. Let $n \geq 2$. If x_i is a Boolean variable for all $1 \leq i \leq n$, prove that

a) $\overline{(x_1 + x_2 + \cdots + x_n)} = \overline{x}_1 \overline{x}_2 \cdots \overline{x}_n$

b) $\overline{(x_1 x_2 \cdots x_n)} = \overline{x}_1 + \overline{x}_2 + \cdots + \overline{x}_n$

2. Let $f, g: B^5 \to B$ be Boolean functions, where $f = \sum m(1, 2, 4, 7, x)$ and $g = \sum m(0, 1, 2, 3, y, z, 16, 25)$. If $f \leq g$, what are x, y, z?

3. Eileen is having a party and finds herself confronted with decisions about inviting five of her friends.

a) If she invites Margaret, she must also invite Joan.

b) If Kathleen is invited, Nettie and Margaret must also be invited.

c) She can invite Cathy or Joan, but definitely not both of them.

d) Neither Cathy nor Nettie will show up if the other is not invited.

e) Either Kathleen or Nettie or both must be invited.

Determine which subsets of these five friends Eileen can invite to her party and still satisfy conditions (a) through (e).

4. Let $f, g: B^4 \to B$, where $f = \sum m(2, 4, 6, 8)$, and $g = \sum m(1, 2, 3, 4, 5, 6, 7, 8, 9, 11, 13, 15)$. Find a function $h: B^4 \to B$ such that $f = gh$.

5. Let $\mathcal{B}$ be a Boolean algebra that is partially ordered by $\leq$. If $x, y, z \in \mathcal{B}$, prove that $x + y \leq z$ if and only if $x \leq z$ and $y \leq z$.

6. State and prove the dual of the result in Exercise 5.

7. Let $\mathscr{B}$ be a Boolean algebra that is partially ordered by $\leq$. For all $x, y \in \mathscr{B}$ prove that

a) $x \leq y$ if and only if $\overline{x} + y = 1$; and

b) $x \leq \overline{y}$ if and only if $xy = 0$.

8. Let x, y be elements in the Boolean algebra $\mathscr{B}$. Prove that $x = y$ if and only if $x\overline{y} + \overline{x}y = 0$.

9. Use a Karnaugh map to find a minimal-sum-of products representation for

a) $f(w, x, y, z) = \sum m(0, 1, 2, 3, 6, 7, 14, 15)$

b) $g(v, w, x, y, z) = \prod M(1, 2, 4, 6, 9, 10, 11, 14, 17,$ $18, 19, 20, 22, 25, 26, 27, 30)$

10. The four input lines for the network in Fig. 15.15 provide the binary equivalents of the numbers $0, 1, 2, \ldots, 15$, where each number is represented as $abce$, with e the least significant bit.

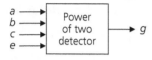

Figure 15.15

a) Find the d.n.f. of g, whose value is 1 exactly when $abce$ is the binary equivalent of 1, 2, 4, or 8.

b) Draw the two-level gating network for g as a minimal sum of products.

c) If this network is part of a larger network and, consequently, the binary equivalents of the numbers 10 through 15 never occur as inputs, design a two-level gating network for g in this case.

11. For n Boolean variables there are 2^{2^n} Boolean functions, each of which can be represented by a function table.

a) A Boolean function f on the n variables $x_1, x_2, \ldots, x_n$ is called *self-dual* if

$$f(x_1, x_2, \ldots, x_n) = \overline{f(\overline{x}_1, \overline{x}_2, \ldots, \overline{x}_n)}.$$

How many Boolean functions on n variables are self-dual?

b) Let $f: B^3 \to B$. Then f is called *symmetric* if

$$f(x, y, z) = f(x, z, y) = f(y, x, z)$$
$$= f(y, z, x) = f(z, x, y) = f(z, y, x).$$

So the value of f is unchanged when we rearrange the three columns of values listed under x, y, and z in the table for f. How many such functions are there on three Boolean variables? How many are there on n Boolean variables?

12. Let $\mathscr{B}_1$ be the Boolean algebra of all positive integer divisors of 30030, and let $\mathscr{B}_2$ be the Boolean algebra of all subsets of $\{u, v, w, x, y, z\}$. How many isomorphisms $f: \mathscr{B}_1 \to \mathscr{B}_2$ satisfy $f(2) = \{u\}$ and $f(6) = \{u, v\}$?

13. For (a) $n = 60$, and (b) $n = 120$, explain why the positive integer divisors of n do not yield a Boolean algebra. (Here $x + y = \text{lcm}(x, y)$, $xy = \gcd(x, y)$, $\overline{x} = n/x$, 1 is the zero element, and n is the one element.)

14. Let $a, b, c \in \mathscr{B}$, a Boolean algebra. Prove that $ab + c = a(b + c)$ if and only if $c \leq a$.

16

Groups, Coding Theory, and Polya's Method of Enumeration

In our study of algebraic structures we examine properties shared by particular mathematical systems. Then we generalize our findings in order to study the underlying structure common to these particular examples.

In Chapter 14 we did this with the ring structure, which depended on two closed binary operations. Now we turn to a structure involving one closed binary operation. This structure is called a *group*.

Our study of groups will examine many ideas comparable to those for rings. However, here we shall dwell primarily on those aspects of the structure that are needed for applications in cryptology, coding theory, and a counting method developed by George Polya.

16.1
Definition, Examples, and Elementary Properties

Definition 16.1   If G is a nonempty set and $\circ$ is a binary operation on G, then $(G, \circ)$ is called a *group* if the following conditions are satisfied.

1) For all $a, b \in G$, $a \circ b \in G$. (Closure of G under $\circ$)

2) For all $a, b, c \in G$, $a \circ (b \circ c) = (a \circ b) \circ c$. (The Associative Property)

3) There exists $e \in G$ with $a \circ e = e \circ a = a$, for all $a \in G$. (The Existence of an Identity)

4) For each $a \in G$ there is an element $b \in G$ such that $a \circ b = b \circ a = e$. (Existence of Inverses)

Furthermore, if $a \circ b = b \circ a$ for all $a, b \in G$, then G is called a *commutative*, or *abelian*, group. The adjective *abelian* honors the Norwegian mathematician Niels Henrik Abel (1802–1829).

We realize that the first condition in Definition 16.1 could have been omitted if we simply required the binary operation for G to be a *closed* binary operation.

Following Definition 14.1 (for a ring) we mentioned how the associative laws for the closed binary operations of $+$ (ring addition) and $\cdot$ (ring multiplication) could be extended by mathematical induction. The same type of situation arises for groups. If $(G, \circ)$ is any group, and $r, n \in \mathbf{Z}^+$ with $n \geq 3$ and $1 \leq r < n$, then

$$(a_1 \circ a_2 \circ \cdots \circ a_r) \circ (a_{r+1} \circ \cdots \circ a_n) = a_1 \circ a_2 \circ \cdots \circ a_r \circ a_{r+1} \circ \cdots \circ a_n,$$

where $a_1, a_2, \ldots, a_r, a_{r+1}, \ldots, a_n$ are all elements from G.

EXAMPLE 16.1

Under ordinary addition, each of $\mathbf{Z}, \mathbf{Q}, \mathbf{R}, \mathbf{C}$ is an abelian group. None of these is a group under multiplication because 0 has no multiplicative inverse. However, $\mathbf{Q}^*, \mathbf{R}^*$, and $\mathbf{C}^*$ (the nonzero elements of $\mathbf{Q}, \mathbf{R}$, and $\mathbf{C}$, respectively) are abelian groups under ordinary multiplication.

If $(R, +, \cdot)$ is a ring, then $(R, +)$ is an abelian group; the nonzero elements of a *field* $(F, +, \cdot)$ form the abelian group $(F^*, \cdot)$.

EXAMPLE 16.2

For $n \in \mathbf{Z}^+, n > 1$, we find that $(\mathbf{Z}_n, +)$ is an abelian group. When p is a prime, $(\mathbf{Z}_p^*, \cdot)$ is an abelian group. Tables 16.1 and 16.2 demonstrate this for $n = 6$ and $p = 7$, respectively. (Recall that in $\mathbf{Z}_n$ we often write a for $[a] = \{a + kn | k \in \mathbf{Z}\}$. The same notation is used in $\mathbf{Z}_p^*$.)

Table 16.1

+	0	1	2	3	4	5
0	0	1	2	3	4	5
1	1	2	3	4	5	0
2	2	3	4	5	0	1
3	3	4	5	0	1	2
4	4	5	0	1	2	3
5	5	0	1	2	3	4

Table 16.2

$\cdot$	1	2	3	4	5	6
1	1	2	3	4	5	6
2	2	4	6	1	3	5
3	3	6	2	5	1	4
4	4	1	5	2	6	3
5	5	3	1	6	4	2
6	6	5	4	3	2	1

Definition 16.2

For every group G the number of elements in G is called the *order* of G and this is denoted by $|G|$. When the number of elements in a group is not finite we say that G has infinite order.

EXAMPLE 16.3

For all $n \in \mathbf{Z}^+, |(\mathbf{Z}_n, +)| = n$, while $|(\mathbf{Z}_p^*, \cdot)| = p - 1$ for each prime p.

EXAMPLE 16.4

Let us start with the ring $(\mathbf{Z}_9, +, \cdot)$ and consider the subset $U_9 = \{a \in \mathbf{Z}_9 | a$ is a unit in $\mathbf{Z}_9\} = \{a \in \mathbf{Z}_9 | a^{-1}$ exists$\} = \{1, 2, 4, 5, 7, 8\} = \{a \in \mathbf{Z}^+ | 1 \leq a \leq 8$ and $\gcd(a, 9) = 1\}$. The results in Table 16.3 show us that U_9 is closed under the multiplication for the ring $(\mathbf{Z}_9, +, \cdot)$ — namely, multiplication modulo 9. Furthermore, we also see that 1 is the identity element and that each element has an inverse (in U_9). For instance, 5 is the inverse for 2, and 7 is the inverse for 4. Finally, since every ring is associative under the operation of (ring) multiplication, it follows that $a \cdot (b \cdot c) = (a \cdot b) \cdot c$ for all $a, b, c \in U_9$. Consequently, $(U_9, \cdot)$ is a group of order 6 — in fact, it is an abelian group of order 6.

Table 16.3

·	1	2	4	5	7	8
1	1	2	4	5	7	8
2	2	4	8	1	5	7
4	4	8	7	2	1	5
5	5	1	2	7	8	4
7	7	5	1	8	4	2
8	8	7	5	4	2	1

In general, for each $n \in \mathbf{Z}^+$, where $n > 1$, if $U_n = \{a \in (\mathbf{Z}_n, +, \cdot) \mid a$ is a unit$\} = \{a \in \mathbf{Z}^+ \mid 1 \le a \le n - 1$ and $\gcd(a, n) = 1\}$, then $(U_n, \cdot)$ is an abelian group under the (closed) binary operation of multiplication modulo n. The group $(U_n, \cdot)$ is called the *group of units* for the ring $(\mathbf{Z}_n, +, \cdot)$ and it has order $\phi(n)$, where ϕ denotes the Euler phi function of Section 8.1.

From here on the group operation will be written multiplicatively, unless it is given otherwise. So $a \circ b$ now becomes ab.

The following theorem provides several properties shared by all groups.

THEOREM 16.1   For every group G,

   a) the identity of G is unique.

   b) the inverse of each element of G is unique.

   c) if $a, b, c \in G$ and $ab = ac$, then $b = c$. (Left-cancellation property)

   d) if $a, b, c \in G$ and $ba = ca$, then $b = c$. (Right-cancellation property)

Proof:

   a) If e_1, e_2 are both identities in G, then $e_1 = e_1 e_2 = e_2$. (Justify each equality.)

   b) Let $a \in G$ and suppose that b, c are both inverses of a. Then $b = be = b(ac) = (ba)c = ec = c$. (Justify each equality.)

The proofs of properties (c) and (d) are left for the reader. (It is because of these properties that we find each group element appearing exactly once in each row and each column of the table for a finite group.)

On the basis of the result in Theorem 16.1(b) the unique inverse of a will be designated by a^{-1}. When the group is written additively, $-a$ is used to denote the (additive) inverse of a.

As in the case of multiplication in a ring, we have powers of elements in a group. We define $a^0 = e$, $a^1 = a$, $a^2 = a \cdot a$, and in general $a^{n+1} = a^n \cdot a$, for all $n \in \mathbf{N}$. Since each group element has an inverse, for $n \in \mathbf{Z}^+$, we define $a^{-n} = (a^{-1})^n$. Then a^n is defined for all $n \in \mathbf{Z}$, and it can be shown that for all $m, n \in \mathbf{Z}$, $a^m \cdot a^n = a^{m+n}$ and $(a^m)^n = a^{mn}$.

If the group operation is addition, then multiples replace powers and for all $m, n \in \mathbf{Z}$, and all $a \in G$, we find that

$$ma + na = (m + n)a \qquad m(na) = (mn)a.$$

In this case the identity is written as 0, rather than e. And here, for all $a \in G$, we have $0a = 0$, where the "0" in front of a is the integer 0 (in $\mathbf{Z}$) while the "0" on the right side of the equation is the identity 0 (in G). [So these two "0"'s are different.]

For an abelian group G we also find that for all $n \in \mathbf{Z}$ and all $a, b \in G$, (1) $(ab)^n = a^n b^n$, when G is written multiplicatively; and (2) $n(a + b) = na + nb$, when the additive notation is used for G.

We now take a look at a special subset of a group.

EXAMPLE 16.5

Let $G = (\mathbf{Z}_6, +)$. If $H = \{0, 2, 4\}$, then H is a nonempty subset of G. Table 16.4 shows that $(H, +)$ is also a group under the binary operation of G.

Table 16.4

+	0	2	4
0	0	2	4
2	2	4	0
4	4	0	2

This situation motivates the following definition.

Definition 16.3

Let G be a group and $\emptyset \neq H \subseteq G$. If H is a group under the binary operation of G, then we call H a *subgroup* of G.

EXAMPLE 16.6

a) Every group G has $\{e\}$ and G as subgroups. These are the *trivial* subgroups of G. All others are termed *nontrivial*, or *proper*.

b) In addition to $H = \{0, 2, 4\}$, the subset $K = \{0, 3\}$ is also a (proper) subgroup of $G = (\mathbf{Z}_6, +)$.

c) Each of the nonempty subsets $\{1, 8\}$ and $\{1, 4, 7\}$ is a subgroup of $(U_9, \cdot)$.

d) The group $(\mathbf{Z}, +)$ is a subgroup of $(\mathbf{Q}, +)$, which is a subgroup of $(\mathbf{R}, +)$. Yet $\mathbf{Z}^*$ under multiplication is not a subgroup of $(\mathbf{Q}^*, \cdot)$. (Why not?)

For a group G and $\emptyset \neq H \subseteq G$, the following tells us when H is a subgroup of G.

THEOREM 16.2

If H is a nonempty subset of a group G, then H is a subgroup of G if and only if (a) for all $a, b \in H$, $ab \in H$, and (b) for all $a \in H$, $a^{-1} \in H$.

Proof: If H is a subgroup of G, then by Definition 16.3 H is a group under the same binary operation. Hence it satisfies all the group conditions, including the two mentioned here. Conversely, let $\emptyset \neq H \subseteq G$ with H satisfying conditions (a) and (b). For all $a, b, c \in H$, $(ab)c = a(bc)$ in G, so $(ab)c = a(bc)$ in H. (We say that H "inherits" the associative property from G.) Finally, as $H \neq \emptyset$, let $a \in H$. By condition (b), $a^{-1} \in H$ and by condition (a), $aa^{-1} = e \in H$, so H contains the identity element and is a group.

A finiteness condition modifies the situation.

THEOREM 16.3

If G is a group and $\emptyset \neq H \subseteq G$, with H *finite*, then H is a subgroup of G if and only if H is closed under the binary operation of G.

Proof: As in the proof of Theorem 16.2, if H is a subgroup of G, then H is closed under the binary operation of G. Conversely, let H be a finite nonempty subset of G that is so closed. If $a \in H$, then $aH = \{ah | h \in H\} \subseteq H$ because of the closure condition. By left-cancellation in G, $ah_1 = ah_2 \Rightarrow h_1 = h_2$, so $|aH| = |H|$. With $aH \subseteq H$ and $|aH| = |H|$, it follows from H being *finite* that $aH = H$. As $a \in H$, there exists $b \in H$ with $ab = a$. But (in G) $ab = a = ae$, so $b = e$ and H contains the identity. Since $e \in H = aH$, there is an element $c \in H$ such that $ac = e$. Then $(ca)^2 = (ca)(ca) = (c(ac))a = (ce)a = ca = (ca)e$, so $ca = e$, and $c = a^{-1} \in H$. Consequently, by Theorem 16.2, H is a subgroup of G.

The finiteness condition in Theorem 16.3 is crucial. Both $\mathbf{Z}^+$ and $\mathbf{N}$ are nonempty closed subsets of the group $(\mathbf{Z}, +)$, yet neither has the additive inverses needed for the group structure.

The next example provides a nonabelian group.

EXAMPLE 16.7

Consider the first equilateral triangle shown in Fig. 16.1(a). When we rotate this triangle counterclockwise (within its plane) through 120° about an axis perpendicular to its plane and passing through its center C, we obtain the second triangle shown in Fig. 16.1(a). As a result, the vertex originally labeled 1 in Fig. 16.1(a) is now in the position that was originally labeled 3. Likewise, 2 is now in the position originally occupied by 1, and 3 has moved to where 2 was. This can be described by the function $\pi_1 : \{1, 2, 3\} \to \{1, 2, 3\}$, where $\pi_1(1) = 3$, $\pi_1(2) = 1$, $\pi_1(3) = 2$. A more compact notation, $\left(\begin{smallmatrix} 1 & 2 & 3 \\ 3 & 1 & 2 \end{smallmatrix}\right)$, where we write $\pi_1(i)$ below i for each $1 \leq i \leq 3$, emphasizes that π_1 is a permutation of $\{1, 2, 3\}$. If π_2 denotes the counterclockwise rotation through 240°, then $\pi_2 = \left(\begin{smallmatrix} 1 & 2 & 3 \\ 2 & 3 & 1 \end{smallmatrix}\right)$. For the identity π_0 — that is, the rotation through $n(360°)$ for $n \in \mathbf{Z}$ — we write $\pi_0 = \left(\begin{smallmatrix} 1 & 2 & 3 \\ 1 & 2 & 3 \end{smallmatrix}\right)$. These rotations are called *rigid motions* of the triangle. They are two-dimensional motions that keep the center C fixed and preserve the shape of the triangle. Hence the triangle looks the same as when we started, except for a possible rearrangement of the labels on some of its vertices.

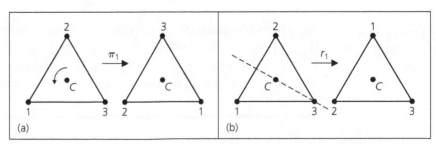

(a)   (b)

Figure 16.1

In addition to these rotations, the triangle can be reflected along an axis passing through a vertex and the midpoint of the opposite side. For the diagonal axis that bisects the base angle on the right, the reflection gives the result in Fig. 16.1(b). This we represent by $r_1 = \begin{pmatrix} 1 & 2 & 3 \\ 2 & 1 & 3 \end{pmatrix}$. A similar reflection about the axis bisecting the left base angle yields the permutation $r_2 = \begin{pmatrix} 1 & 2 & 3 \\ 1 & 3 & 2 \end{pmatrix}$. When the triangle is reflected about its vertical axis, we have $r_3 = \begin{pmatrix} 1 & 2 & 3 \\ 3 & 2 & 1 \end{pmatrix}$. Each r_i, for $1 \le i \le 3$, is a three-dimensional rigid motion.

Let $G = \{\pi_0, \pi_1, \pi_2, r_1, r_2, r_3\}$, the set of rigid motions (in space) of the equilateral triangle. We make G into a group by defining the rigid motion $\alpha\beta$, for $\alpha, \beta \in G$, as that motion obtained by applying first α and then following up with β. Hence, for example, $\pi_1 r_1 = r_3$. We can see this geometrically, but it will be handy to consider the permutations as follows: $\pi_1 r_1 = \begin{pmatrix} 1 & 2 & 3 \\ 3 & 1 & 2 \end{pmatrix}\begin{pmatrix} 1 & 2 & 3 \\ 2 & 1 & 3 \end{pmatrix}$, where, for example, $\pi_1(1) = 3$ and $r_1(3) = 3$ and we write $1 \xrightarrow{\pi_1} 3 \xrightarrow{r_1} 3$. So $1 \xrightarrow{\pi_1 r_1} 3$ in the product $\pi_1 r_1$. (Note that the order in which we write the product $\pi_1 r_1$ here is the opposite of the order for their composite function as defined in Section 5.6. The notation of Section 5.6 occurs in analysis, whereas in algebra there is a tendency to employ this opposite order.) Also, since $2 \xrightarrow{\pi_1} 1 \xrightarrow{r_1} 2$ and $3 \xrightarrow{\pi_1} 2 \xrightarrow{r_1} 1$, it follows that $\pi_1 r_1 = \begin{pmatrix} 1 & 2 & 3 \\ 3 & 2 & 1 \end{pmatrix} = r_3$.

Table 16.5 verifies that under this binary operation G is closed, with identity π_0. Also $\pi_1^{-1} = \pi_2$, $\pi_2^{-1} = \pi_1$, and every other element is its own inverse. Since the elements of G are actually functions, the associative property follows from Theorem 5.6 (although in reverse order).

Table 16.5

·	π_0	π_1	π_2	r_1	r_2	r_3
π_0	π_0	π_1	π_2	r_1	r_2	r_3
π_1	π_1	π_2	π_0	r_3	r_1	r_2
π_2	π_2	π_0	π_1	r_2	r_3	r_1
r_1	r_1	r_2	r_3	π_0	π_1	π_2
r_2	r_2	r_3	r_1	π_2	π_0	π_1
r_3	r_3	r_1	r_2	π_1	π_2	π_0

We computed $\pi_1 r_1$ as r_3, but from Table 16.5 we see that $r_1 \pi_1 = r_2$. With $\pi_1 r_1 = r_3 \ne r_2 = r_1 \pi_1$, it follows that G is nonabelian.

This group can also be obtained as the group of all permutations of the set $\{1, 2, 3\}$ under the binary operation of function composition. It is denoted by S_3 (the *symmetric* group on three symbols).

EXAMPLE 16.8

The symmetric group S_4 consists of the 24 permutations of $\{1, 2, 3, 4\}$. Here $\pi_0 = \begin{pmatrix} 1 & 2 & 3 & 4 \\ 1 & 2 & 3 & 4 \end{pmatrix}$ is the identity. If $\alpha = \begin{pmatrix} 1 & 2 & 3 & 4 \\ 2 & 1 & 4 & 3 \end{pmatrix}$, $\beta = \begin{pmatrix} 1 & 2 & 3 & 4 \\ 3 & 1 & 2 & 4 \end{pmatrix}$, then $\alpha\beta = \begin{pmatrix} 1 & 2 & 3 & 4 \\ 1 & 3 & 4 & 2 \end{pmatrix}$ but $\beta\alpha = \begin{pmatrix} 1 & 2 & 3 & 4 \\ 4 & 2 & 1 & 3 \end{pmatrix}$, so S_4 is nonabelian. Also, $\beta^{-1} = \begin{pmatrix} 1 & 2 & 3 & 4 \\ 2 & 3 & 1 & 4 \end{pmatrix}$ and $\alpha^2 = \pi_0 = \beta^3$. Within S_4 there is a subgroup of order 8 that represents the group of rigid motions for a square.

We turn now to a construction for making larger groups out of smaller ones.

THEOREM 16.4 Let $(G, \circ)$ and $(H, *)$ be groups. Define the binary operation $\cdot$ on $G \times H$ by $(g_1, h_1) \cdot (g_2, h_2) = (g_1 \circ g_2, h_1 * h_2)$. Then $(G \times H, \cdot)$ is a group and is called the *direct product* of G and H.

Proof: The verification of the group properties for $(G \times H, \cdot)$ is left to the reader.

EXAMPLE 16.9 Consider the groups $(\mathbf{Z}_2, +)$, $(\mathbf{Z}_3, +)$. On $G = \mathbf{Z}_2 \times \mathbf{Z}_3$, define $(a_1, b_1) \cdot (a_2, b_2) = (a_1 + a_2, b_1 + b_2)$. Then G is a group of order 6 where the identity is $(0, 0)$, and the inverse, for example, of the element $(1, 2)$ is $(1, 1)$.

EXERCISES 16.1

1. For each of the following sets, determine whether or not the set is a group under the stated binary operation. If so, determine its identity and the inverse of each of its elements. If it is not a group, state the condition(s) of the definition that it violates.

  a) $\{-1, 1\}$ under multiplication

  b) $\{-1, 1\}$ under addition

  c) $\{-1, 0, 1\}$ under addition

  d) $\{10n \,|\, n \in \mathbf{Z}\}$ under addition

  e) The set of all one-to-one functions $g: A \rightarrow A$, where $A = \{1, 2, 3, 4\}$, under function composition

  f) $\{a/2^n \,|\, a, n \in \mathbf{Z}, n \geq 0\}$ under addition

2. Prove parts (c) and (d) of Theorem 16.1.

3. Why is the set $\mathbf{Z}$ *not* a group under subtraction?

4. Let $G = \{q \in \mathbf{Q} \,|\, q \neq -1\}$. Define the binary operation $\circ$ on G by $x \circ y = x + y + xy$. Prove that $(G, \circ)$ is an abelian group.

5. Define the binary operation $\circ$ on $\mathbf{Z}$ by $x \circ y = x + y + 1$. Verify that $(\mathbf{Z}, \circ)$ is an abelian group.

6. Let $S = \mathbf{R}^* \times \mathbf{R}$. Define the binary operation $\circ$ on S by $(u, v) \circ (x, y) = (ux, vx + y)$. Prove that $(S, \circ)$ is a nonabelian group.

7. Find the elements in the groups U_{20} and U_{24} — the groups of units for the rings $(\mathbf{Z}_{20}, +, \cdot)$ and $(\mathbf{Z}_{24}, +, \cdot)$, respectively.

8. For any group G prove that G is abelian if and only if $(ab)^2 = a^2 b^2$ for all $a, b \in G$.

9. If G is a group, prove that for all $a, b \in G$,

  a) $(a^{-1})^{-1} = a$ **b)** $(ab)^{-1} = b^{-1} a^{-1}$

10. Prove that a group G is abelian if and only if for all $a, b \in G$, $(ab)^{-1} = a^{-1} b^{-1}$.

11. Find all subgroups in each of the following groups.

  a) $(\mathbf{Z}_{12}, +)$ **b)** $(\mathbf{Z}_{11}^*, \cdot)$ **c)** S_3

12. a) How many rigid motions (in two or three dimensions) are there for a square?

b) Make a group table for these rigid motions like the one in Table 16.5 for the equilateral triangle. What is the identity for this group? Describe the inverse of each element geometrically.

13. a) How many rigid motions (in two or three dimensions) are there for a regular pentagon? Describe them geometrically.

  b) Answer part (a) for a regular n-gon, $n \geq 3$.

14. In the group S_5, let

$$\alpha = \begin{pmatrix} 1 & 2 & 3 & 4 & 5 \\ 2 & 3 & 1 & 4 & 5 \end{pmatrix} \quad \text{and} \quad \beta = \begin{pmatrix} 1 & 2 & 3 & 4 & 5 \\ 2 & 1 & 5 & 3 & 4 \end{pmatrix}.$$

Determine $\alpha\beta$, $\beta\alpha$, α^3, β^4, α^{-1}, β^{-1}, $(\alpha\beta)^{-1}$, $(\beta\alpha)^{-1}$, and $\beta^{-1}\alpha^{-1}$.

15. If G is a group, let $H = \{a \in G \,|\, ag = ga \text{ for all } g \in G\}$. Prove that H is a subgroup of G. (The subgroup H is called the *center* of G.)

16. Let ω be the complex number $(1/\sqrt{2})(1 + i)$.

  a) Show that $\omega^8 = 1$ but $\omega^n \neq 1$ for $n \in \mathbf{Z}^+$, $1 \leq n \leq 7$.

  b) Verify that $\{\omega^n \,|\, n \in \mathbf{Z}^+, 1 \leq n \leq 8\}$ is an abelian group under multiplication.

17. a) Prove Theorem 16.4.

  b) Extending the idea developed in Theorem 16.4 and Example 16.9 to the group $\mathbf{Z}_6 \times \mathbf{Z}_6 \times \mathbf{Z}_6 = \mathbf{Z}_6^3$, answer the following.

   i) What is the order of this group?

   ii) Find a subgroup of $\mathbf{Z}_6^3$ of order 6, one of order 12, and one of order 36.

   iii) Determine the inverse of each of the elements $(2, 3, 4)$, $(4, 0, 2)$, $(5, 1, 2)$.

18. a) If H, K are subgroups of a group G, prove that $H \cap K$ is also a subgroup of G.

  b) Give an example of a group G with subgroups H, K such that $H \cup K$ is *not* a subgroup of G.

19. a) Find all x in $(\mathbf{Z}_5^*, \cdot)$ such that $x = x^{-1}$.

  b) Find all x in $(\mathbf{Z}_{11}^*, \cdot)$ such that $x = x^{-1}$.

  c) Let p be a prime. Find all x in $(\mathbf{Z}_p^*, \cdot)$ such that $x = x^{-1}$.

d) Prove that $(p - 1)! \equiv -1 \pmod{p}$, for p a prime. [This result is known as Wilson's Theorem, although it was only conjectured by John Wilson (1741–1793). The first proof was given in 1770 by Joseph Louis Lagrange (1736–1813).]

20. a) Find x in $(U_g, \cdot)$ where $x \neq 1$, $x \neq 7$ but $x = x^{-1}$.

b) Find x in $(U_{16}, \cdot)$ where $x \neq 1$, $x \neq 15$ but $x = x^{-1}$.

c) Let $k \in \mathbf{Z}^+$, $k \geq 3$. Find x in $(U_{2^k}, \cdot)$ where $x \neq 1$, $x \neq 2^k - 1$ but $x = x^{-1}$.

16.2
Homomorphisms, Isomorphisms, and Cyclic Groups

We turn our attention once again to functions that preserve structure.

EXAMPLE 16.10

Let $G = (\mathbf{Z}, +)$ and $H = (\mathbf{Z}_4, +)$. Define $f: G \to H$ by

$$f(x) = [x] = \{x + 4k | \, k \in \mathbf{Z}\}.$$

For all $x, y \in G$,

$$f(x + y) = [x + y] = [x] + [y] = f(x) + f(y),$$

               ↑   ↑

     The operation in **G**   The operation in **H**

where the second equality follows from the way the addition of equivalence classes was developed in Section 14.3. Consequently, here f preserves the group operations and is an example of a special type of function that we shall now define.

Definition 16.4

If $(G, \circ)$ and $(H, *)$ are groups and $f: G \to H$, then f is called a *group homomorphism* if for all $a, b \in G$, $f(a \circ b) = f(a) * f(b)$.

When we know that the given structures are groups, the function f is simply called a homomorphism.

Some properties of homomorphisms are given in the following theorem.

THEOREM 16.5

Let $(G, \circ)$, $(H, *)$ be groups with respective identities e_G, e_H. If $f: G \to H$ is a homomorphism, then

a) $f(e_G) = e_H$.   **b)** $f(a^{-1}) = [f(a)]^{-1}$ for all $a \in G$.

c) $f(a^n) = [f(a)]^n$ for all $a \in G$ and all $n \in \mathbf{Z}$.

d) $f(S)$ is a subgroup of H for each subgroup S of G.

Proof:

    a) $e_H * f(e_G) = f(e_G) = f(e_G \circ e_G) = f(e_G) * f(e_G)$, so by right-cancellation [Theorem 16.1(d)], it follows that $f(e_G) = e_H$.

b) & c) The proofs of these parts are left for the reader.

d) If S is a subgroup of G, then $S \neq \emptyset$, so $f(S) \neq \emptyset$. Let $x, y \in f(S)$. Then $x = f(a)$, $y = f(b)$, for some $a, b \in S$. Since S is a subgroup of G, it follows that $a \circ b \in S$, so $x * y = f(a) * f(b) = f(a \circ b) \in f(S)$. Finally, $x^{-1} = [f(a)]^{-1} = f(a^{-1}) \in f(S)$ because $a^{-1} \in S$ when $a \in S$. Consequently, by Theorem 16.2, $f(S)$ is a subgroup of H.

Definition 16.5   If $f: (G, \circ) \to (H, *)$ is a homomorphism, we call f an *isomorphism* if it is one-to-one and onto. In this case G, H are said to be *isomorphic groups*.

EXAMPLE 16.11   Let $f: (\mathbf{R}^+, \cdot) \to (\mathbf{R}, +)$ where $f(x) = \log_{10}(x)$. This function is both one-to-one and onto. (Verify these properties.) For all $a, b \in \mathbf{R}^+$, $f(ab) = \log_{10}(ab) = \log_{10} a + \log_{10} b = f(a) + f(b)$. Therefore, f is an isomorphism and the group of positive real numbers under multiplication is abstractly the same as the group of all real numbers under addition. Here the function f translates a problem in the multiplication of real numbers (a somewhat difficult problem without a calculator) into a problem dealing with the addition of real numbers (an easier arithmetic consideration). This was a major reason behind the use of logarithms before the advent of calculators.

EXAMPLE 16.12   Let G be the group of complex numbers $\{1, -1, i, -i\}$ under multiplication. Table 16.6 shows the multiplication table for this group. With $H = (\mathbf{Z}_4, +)$, consider $f: G \to H$ defined by

$$f(1) = [0] \qquad f(-1) = [2] \qquad f(i) = [1] \qquad f(-i) = [3].$$

Then $f((i)(-i)) = f(1) = [0] = [1] + [3] = f(i) + f(-i)$, and $f((-1)(-i)) = f(i) = [1] = [2] + [3] = f(-1) + f(-i)$.

Although we have not checked all possible cases, the function is an isomorphism. Note that the image under f of the subgroup $\{1, -1\}$ of G is $\{[0], [2]\}$, a subgroup of H.

Table 16.6

$\cdot$	1	-1	i	$-i$
1	1	-1	i	$-i$
-1	-1	1	$-i$	i
i	i	$-i$	-1	1
$-i$	$-i$	i	1	-1

Let us take a closer look at this group G. Here $i^1 = i$, $i^2 = -1$, $i^3 = -i$, and $i^4 = 1$, so every element of G is a power of i, and we say that i *generates* G. This is denoted by $G = \langle i \rangle$. (It is also true that $G = \langle -i \rangle$. Verify this.)

The last part of the preceding example leads us to the following definition.

Definition 16.6   A group G is called *cyclic* if there is an element $x \in G$ such that for each $a \in G$, $a = x^n$ for some $n \in \mathbf{Z}$.

EXAMPLE 16.13

a) The group $H = (\mathbf{Z}_4, +)$ is cyclic. Here the operation is addition, so we have multiples instead of powers. We find that both [1] and [3] generate H. For the case of [3], we have $1 \cdot [3] = [3]$, $2 \cdot [3] (= [3] + [3]) = [2]$, $3 \cdot [3] = [1]$, and $4 \cdot [3] = [0]$. Hence $H = \langle [3] \rangle = \langle [1] \rangle$.

b) Consider the multiplicative group $U_9 = \{1, 2, 4, 5, 7, 8\}$ that we examined in Example 16.4. Here we find that $2^1 = 2$, $2^2 = 4$, $2^3 = 8$, $2^4 = 7$, $2^5 = 5$, $2^6 = 1$, so U_9 is a cyclic group of order 6 and $U_9 = \langle 2 \rangle$. It is also true that $U_9 = \langle 5 \rangle$ because $5^1 = 5$, $5^2 = 7$, $5^3 = 8$, $5^4 = 4$, $5^5 = 2$, $5^6 = 1$.

The concept of a cyclic group leads to a related idea. Given a group G, if $a \in G$ consider the set $S = \{a^k | k \in \mathbf{Z}\}$. From Theorem 16.2 it follows that S is a subgroup of G. This subgroup is called the *subgroup generated by* a and is designated by $\langle a \rangle$. In Example 16.12 $\langle i \rangle = \langle -i \rangle = G$; also, $\langle -1 \rangle = \{-1, 1\}$ and $\langle 1 \rangle = \{1\}$. For part (a) of Example 16.13 we consider multiples instead of powers and find that $H = \langle [1] \rangle = \langle [3] \rangle$, $\langle [2] \rangle = \{[0], [2]\}$, and $\langle [0] \rangle = \{[0]\}$. When we examine the group U_9 in part (b) of that example we see that $U_9 = \langle 2 \rangle$ (or $\langle [2] \rangle$) $= \langle 5 \rangle$, $\langle 4 \rangle = \{1, 4, 7\} = \langle 7 \rangle$, $\langle 8 \rangle = \{1, 8\}$, and $\langle 1 \rangle = \{1\}$.

Definition 16.7

If G is a group and $a \in G$, the *order of* a, denoted $\circ(a)$, is $|\langle a \rangle|$. (If $|\langle a \rangle|$ is infinite, we say that a has infinite order.)

In Example 16.12, $\circ(1) = 1$, $\circ(-1) = 2$, whereas both i and $-i$ have order 4.

Let us take a second look at the idea of order for the case where $|\langle a \rangle|$ is finite. When $|\langle a \rangle| = 1$ then $a = e$ because $a = a^1 \in \langle a \rangle$ and $e = a^0 \in \langle a \rangle$. If $|\langle a \rangle|$ is finite but $a \neq e$, then $\langle a \rangle = \{a^m | m \in \mathbf{Z}\}$ is finite, so $\{a, a^2, a^3, \ldots\} = \{a^m | m \in \mathbf{Z}^+\}$ is also finite. Consequently, there exist $s, t \in \mathbf{Z}^+$, where $1 \leq s < t$ and $a^s = a^t$ — from which it follows that $a^{t-s} = e$, with $t - s \in \mathbf{Z}^+$. Since $e \in \{a^m | m \in \mathbf{Z}^+\}$, let n be the smallest positive integer such that $a^n = e$. We claim that $\langle a \rangle = \{a, a^2, a^3, \ldots, a^{n-1}, a^n (= e)\}$.

First we observe that $|\{a, a^2, a^3, \ldots, a^{n-1}, a^n (= e)\}| = n$. Otherwise, we have $a^u = a^v$ for positive integers u, v where $1 \leq u < v \leq n$, and then $a^{v-u} = e$ with $0 < v - u < n$. This, however, contradicts the minimality of n. So now we know that $|\langle a \rangle| \geq n$. But for each $k \in \mathbf{Z}$, it follows from the division algorithm that $k = qn + r$, where $0 \leq r < n$, and so $a^k = a^{qn+r} = (a^n)^q (a^r) = (e^q)(a^r) = a^r \in \{a, a^2, a^3, \ldots, a^{n-1}, a^n (= e = a^0)\}$. Therefore, $\langle a \rangle = \{a, a^2, a^3, \ldots, a^{n-1}, a^n (= e)\}$ and we can also define $\circ(a)$ as the *smallest positive integer* n for which $a^n = e$. This alternative definition for the order of a group element (of finite order) proves to be of value in the following theorem.

THEOREM 16.6

Let $a \in G$ with $\circ(a) = n$. If $k \in \mathbf{Z}$ and $a^k = e$, then $n | k$.

Proof: By the division algorithm (again), we have $k = qn + r$, for $0 \leq r < n$, and so it follows that $e = a^k = a^{qn+r} = (a^n)^q (a^r) = (e^q)(a^r) = a^r$. If $0 < r < n$, we contradict the definition of n as $\circ(a)$. Hence $r = 0$ and $k = qn$.

We now examine some further results on cyclic groups. The next example helps us to motivate part (b) of Theorem 16.7.

EXAMPLE 16.14

It is known from part (b) of Example 16.13 that $U_9 = \{1, 2, 4, 5, 7, 8\} = \langle 2 \rangle$. We use this fact to define the function $f: U_9 \to (\mathbf{Z}_6, +)$ as follows:

$$f(1) = [0] \qquad f(2) = [1] \qquad f(4) = [2]$$
$$f(5) = f(2^5) = [5] \qquad f(7) = f(2^4) = [4] \qquad f(8) = f(2^3) = [3].$$

So, in general, for each $a \in U_9$ we write $a = 2^k$, for some $0 \le k \le 5$, and have $f(a) = f(2^k) = [k]$. This function f is one-to-one and onto and we find, for example, that $f(2 \cdot 5) = f(1) = [0] = [1] + [5] = f(2) + f(5)$, and $f(7 \cdot 8) = f(2) = [1] = [4] + [3] = f(7) + f(8)$.

In general, for a, b in U_9 we may write $a = 2^m$ and $b = 2^n$, where $0 \le m \le 5$ and $0 \le n \le 5$. It then follows that

$$f(a \cdot b) = f(2^m \cdot 2^n) = f(2^{m+n}) = [m + n] = [m] + [n] = f(a) + f(b).$$

Consequently, the function f is an isomorphism and the groups U_9 and $(\mathbf{Z}_6, +)$ are isomorphic.

[Note how the function f links the generators of the two cyclic groups. Also note that the function $g: U_9 \to (\mathbf{Z}_6, +)$ where

$$g(1) = [0] \qquad g(5) = [1] \qquad g(7) = g(5^2) = [2]$$
$$g(8) = g(5^3) = [3] \qquad g(4) = g(5^4) = [4] \qquad g(2) = g(5^5) = [5]$$

is another isomorphism between these two cyclic groups.]

THEOREM 16.7   Let G be a cyclic group.

a) If $|G|$ is infinite, then G is isomorphic to $(\mathbf{Z}, +)$.

b) If $|G| = n$, where $n > 1$, then G is isomorphic to $(\mathbf{Z}_n, +)$.

Proof:

a) For $G = \langle a \rangle = \{a^k | k \in \mathbf{Z}\}$, let $f: G \to \mathbf{Z}$ be defined by $f(a^k) = k$. (Could we have $a^k = a^t$ with $k \ne t$? If so, f would not be a function.) For $a^m, a^n \in G$, $f(a^m \cdot a^n) = f(a^{m+n}) = m + n = f(a^m) + f(a^n)$, so f is a homomorphism. We leave to the reader the verification that f is one-to-one and onto.

b) If $G = \langle a \rangle = \{a, a^2, \dots, a^{n-1}, a^n = e\}$, then the function $f: G \to \mathbf{Z}_n$ defined by $f(a^k) = [k]$ is an isomorphism. (Verify this.)

EXAMPLE 16.15   If $G = \langle g \rangle$, G is abelian because $g^m \cdot g^n = g^{m+n} = g^{n+m} = g^n \cdot g^m$ for all $m, n \in \mathbf{Z}$. The converse, however, is false. The group H of Table 16.7 is abelian, and $\circ(e) = 1$, $\circ(a) = \circ(b) = \circ(c) = 2$. Since no element of H has order 4, H cannot be cyclic. (The group H is the smallest noncyclic group and is known as the *Klein Four* group.)

Table 16.7

·	e	a	b	c
e	e	a	b	c
a	a	e	c	b
b	b	c	e	a
c	c	b	a	e

Our last result concerns the structure of subgroups in a cyclic group.

THEOREM 16.8   Every subgroup of a cyclic group is cyclic.

Proof: Let $G = \langle a \rangle$. If H is a subgroup of G, each element of H has the form a^k, for some $k \in \mathbf{Z}$. For $H \neq \{e\}$, let t be the smallest positive integer such that $a^t \in H$. (How do we know such an integer t exists?) We claim that $H = \langle a^t \rangle$. Since $a^t \in H$, by the closure property for the subgroup H, $\langle a^t \rangle \subseteq H$. For the opposite inclusion, let $b \in H$, with $b = a^s$, for some $s \in \mathbf{Z}$. By the division algorithm, $s = qt + r$, where $q, r \in \mathbf{Z}$ and $0 \leq r < t$. Consequently, $a^s = a^{qt+r}$ and so $a^r = a^{-qt}a^s = (a^t)^{-q}b$. H is a subgroup of G, so $a^t \in H \Rightarrow (a^t)^{-q} \in H$. Then with $(a^t)^{-q}, b \in H$, it follows that $a^r = (a^t)^{-q}b \in H$. But if $a^r \in H$ with $r > 0$, then we contradict the minimality of t. Hence $r = 0$ and $b = a^{qt} = (a^t)^q \in \langle a^t \rangle$, so $H = \langle a^t \rangle$, a cyclic group.

EXERCISES 16.2

1. Prove parts (b) and (c) of Theorem 16.5.

2. Let $A = \begin{bmatrix} 0 & 1 \\ -1 & 0 \end{bmatrix}$.

   a) Determine A^2, A^3, and A^4.

   b) Verify that $\{A, A^2, A^3, A^4\}$ is an abelian group under ordinary matrix multiplication.

   c) Prove that the group in part (b) is isomorphic to the group shown in Table 16.6.

3. If $G = (\mathbf{Z}_6, +)$, $H = (\mathbf{Z}_3, +)$, and $K = (\mathbf{Z}_2, +)$, find an isomorphism for the groups $H \times K$ and G.

4. Let $f \colon G \to H$ be a group homomorphism onto H. If G is abelian, prove that H is abelian.

5. Let $(\mathbf{Z} \times \mathbf{Z}, \oplus)$ be the abelian group where $(a, b) \oplus (c, d) = (a + c, b + d)$ — here $a + c$ and $b + d$ are computed using ordinary addition in $\mathbf{Z}$ — and let $(G, +)$ be an additive group. If $f \colon \mathbf{Z} \times \mathbf{Z} \to G$ is a group homomorphism where $f(1, 3) = g_1$ and $f(3, 7) = g_2$, express $f(4, 6)$ in terms of g_1 and g_2.

6. Let $f \colon (\mathbf{Z} \times \mathbf{Z}, \oplus) \to (\mathbf{Z}, +)$ be the function defined by $f(x, y) = x - y$. [Here $(\mathbf{Z} \times \mathbf{Z}, \oplus)$ is the same group as in Exercise 5, and $(\mathbf{Z}, +)$ is the group of integers under ordinary addition.]

   a) Prove that f is a homomorphism onto $\mathbf{Z}$.

   b) Determine all $(a, b) \in \mathbf{Z} \times \mathbf{Z}$ with $f(a, b) = 0$.

   c) Find $f^{-1}(7)$.

   d) If $E = \{2n \mid n \in \mathbf{Z}\}$, what is $f^{-1}(E)$?

7. Find the order of each element in the group of rigid motions of (a) the equilateral triangle; and (b) the square.

8. In S_5 find an element of order n, for all $2 \leq n \leq 5$. Also determine the (cyclic) subgroup of S_5 that each of these elements generates.

9. a) Find all the elements of order 10 in $(\mathbf{Z}_{40}, +)$.

   b) Let $G = \langle a \rangle$ be a cyclic group of order 40. Which elements of G have order 10?

10. a) Determine U_{14}, the group of units for the ring $(\mathbf{Z}_{14}, +, \cdot)$.

   b) Show that U_{14} is cyclic and find all of its generators.

11. Verify that $(\mathbf{Z}_p^*, \cdot)$ is cyclic for the primes 5, 7, and 11.

12. For a group G, prove that the function $f \colon G \to G$ defined by $f(a) = a^{-1}$ is an isomorphism if and only if G is abelian.

13. If $f \colon G \to H$, $g \colon H \to K$ are homomorphisms, prove that the composite function $g \circ f \colon G \to K$, where $(g \circ f)(x) = g(f(x))$, is a homomorphism.

14. For $\omega = (1/\sqrt{2})(1 + i)$, let G be the multiplicative group $\{\omega^n \mid n \in \mathbf{Z}^+, 1 \leq n \leq 8\}$.

   a) Show that G is cyclic and find each element $x \in G$ such that $\langle x \rangle = G$.

   b) Prove that G is isomorphic to the group $(\mathbf{Z}_8, +)$.

15. a) Find all generators of the cyclic groups $(\mathbf{Z}_{12}, +)$, $(\mathbf{Z}_{16}, +)$, and $(\mathbf{Z}_{24}, +)$.

   b) Let $G = \langle a \rangle$ with $o(a) = n$. Prove that $a^k, k \in \mathbf{Z}^+$, generates G if and only if k and n are relatively prime.

   c) If G is a cyclic group of order n, how many distinct generators does it have?

16. Let $f \colon G \to H$ be a group homomorphism. If $a \in G$ with $o(a) = n$, and $o(f(a)) = k$ (in H), prove that $k \mid n$.

16.3
Cosets and Lagrange's Theorem

In the last two sections, for all finite groups G and subgroups H of G, we had $|H|$ dividing $|G|$. In this section we'll see that this was not mere chance but is true in general. To prove this we need one new idea.

Definition 16.8

If H is a subgroup of G, then for each $a \in G$, the set $aH = \{ah | h \in H\}$ is called a *left coset* of H in G. The set $Ha = \{ha | h \in H\}$ is a *right coset* of H in G.

If the operation in G is addition, we write $a + H$ in place of aH, where $a + H = \{a + h | h \in H\}$.

When the term *coset* is used in this chapter, it will refer to a left coset. For abelian groups there is no need to distinguish between left and right cosets. However, at the end of the next example we'll see that this is not the case for nonabelian groups.

EXAMPLE 16.16

If G is the group of Example 16.7 and $H = \{\pi_0, \pi_1, \pi_2\}$, the coset $r_1 H = \{r_1 \pi_0, r_1 \pi_1, r_1 \pi_2\}$ $= \{r_1, r_2, r_3\}$. Likewise we have $r_2 H = r_3 H = \{r_1, r_2, r_3\}$, whereas $\pi_0 H = \pi_1 H = \pi_2 H = H$.

We see that $|\alpha H| = |H|$ for each $\alpha \in G$ and that $G = H \cup r_1 H$ is a partition of G.

For the subgroup $K = \{\pi_0, r_1\}$, we find $r_2 K = \{r_2, \pi_2\}$ and $r_3 K = \{r_3, \pi_1\}$. Again a partition of G arises: $G = K \cup r_2 K \cup r_3 K$. (*Note:* $Kr_2 = \{\pi_0 r_2, r_1 r_2\} = \{r_2, \pi_1\} \neq r_2 K$.)

EXAMPLE 16.17

For $G = (\mathbf{Z}_{12}, +)$ and $H = \{[0], [4], [8]\}$, we find that

$$[0] + H = \{[0], [4], [8]\} = [4] + H = [8] + H = H$$
$$[1] + H = \{[1], [5], [9]\} = [5] + H = [9] + H$$
$$[2] + H = \{[2], [6], [10]\} = [6] + H = [10] + H$$
$$[3] + H = \{[3], [7], [11]\} = [7] + H = [11] + H,$$

and $H \cup ([1] + H) \cup ([2] + H) \cup ([3] + H)$ is a partition of G.

These examples now prepare us for the following results.

LEMMA 16.1

If H is a subgroup of the finite group G, then for all $a, b \in G$, (a) $|aH| = |H|$; and (b) either $aH = bH$ or $aH \cap bH = \emptyset$.

Proof:

a) Since $aH = \{ah | h \in H\}$, it follows that $|aH| \leq |H|$. If $|aH| < |H|$, we have $ah_i = ah_j$ with h_i, h_j distinct elements of H. By left-cancellation in G we then get the contradiction $h_i = h_j$, so $|aH| = |H|$.

b) If $aH \cap bH \neq \emptyset$, let $c = ah_1 = bh_2$, for some $h_1, h_2 \in H$. If $x \in aH$, then $x = ah$ for some $h \in H$, and so $x = (bh_2 h_1^{-1})h = b(h_2 h_1^{-1} h) \in bH$, and $aH \subseteq bH$. Similarly, $y \in bH \Rightarrow y = bh_3$, for some $h_3 \in H \Rightarrow y = (ah_1 h_2^{-1})h_3 = a(h_1 h_2^{-1} h_3) \in aH$, so $bH \subseteq aH$. Therefore aH and bH are either disjoint or identical.

We observe that if $g \in G$, then $g \in gH$ because $e \in H$. Also, by part (b) of Lemma 16.1, G can be partitioned into mutually disjoint cosets.

At this point we are ready to prove the main result of this section.

THEOREM 16.9   *Lagrange's Theorem.* If G is a finite group of order n with H a subgroup of order m, then m divides n.

Proof: If $H = G$ the result follows. Otherwise $m < n$ and there exists an element $a \in G - H$. Since $a \notin H$, it follows that $aH \neq H$, so $aH \cap H = \emptyset$. If $G = aH \cup H$, then $|G| = |aH| + |H| = 2|H|$ and the theorem follows. If not, there is an element $b \in G - (H \cup aH)$, with $bH \cap H = \emptyset = bH \cap aH$ and $|bH| = |H|$. If $G = bH \cup aH \cup H$, we have $|G| = 3|H|$. Otherwise we're back to an element $c \in G$ with $c \notin bH \cup aH \cup H$. The group G is finite, so this process terminates and we find that $G = a_1 H \cup a_2 H \cup \cdots \cup a_k H$. Therefore, $|G| = k|H|$ and m divides n.

An alternative method for proving this theorem is given in Exercise 12 for this section. We close with the statements of two corollaries. Their proofs are requested in the Section Exercises.

COROLLARY 16.1   If G is a finite group and $a \in G$, then $\mathfrak{o}(a)$ divides $|G|$.

COROLLARY 16.2   Every group of prime order is cyclic.

EXERCISES 16.3

1. Let $G = S_4$. (a) For $\alpha = \begin{pmatrix} 1 & 2 & 3 & 4 \\ 2 & 3 & 4 & 1 \end{pmatrix}$, find the subgroup $H = \langle \alpha \rangle$. (b) Determine the left cosets of H in G.

2. Answer Exercise 1 for the case where α is replaced by $\beta = \begin{pmatrix} 1 & 2 & 3 & 4 \\ 2 & 3 & 1 & 4 \end{pmatrix}$.

3. If $\gamma = \begin{pmatrix} 1 & 2 & 3 & 4 \\ 2 & 1 & 4 & 3 \end{pmatrix} \in S_4$, how many cosets does $\langle \gamma \rangle$ determine?

4. For $G = (\mathbf{Z}_{24}, +)$, find the cosets determined by the subgroup $H = \langle [3] \rangle$. Do likewise for the subgroup $K = \langle [4] \rangle$.

5. Let G be a group with subgroups H and K. If $|G| = 660$, $|K| = 66$, and $K \subset H \subset G$, what are the possible values for $|H|$?

6. Let R be a ring with unity u. Prove that the units of R form a group under the multiplication of the ring.

7. Let $G = S_4$, the symmetric group on four symbols, and let H be the subset of G where

$$H = \left\{ \begin{pmatrix} 1 & 2 & 3 & 4 \\ 1 & 2 & 3 & 4 \end{pmatrix}, \begin{pmatrix} 1 & 2 & 3 & 4 \\ 2 & 1 & 4 & 3 \end{pmatrix}, \begin{pmatrix} 1 & 2 & 3 & 4 \\ 3 & 4 & 1 & 2 \end{pmatrix}, \begin{pmatrix} 1 & 2 & 3 & 4 \\ 4 & 3 & 2 & 1 \end{pmatrix} \right\}.$$

a) Construct a table to show that H is an abelian subgroup of G.

b) How many left cosets of H are there in G?

c) Consider the group $(\mathbf{Z}_2 \times \mathbf{Z}_2, \oplus)$ where $(a, b) \oplus (c, d) = (a + c, b + d)$ — and the sums $a + c, b + d$ are computed using addition modulo 2. Prove that H is isomorphic to this group.

8. If G is a group of order n and $a \in G$, prove that $a^n = e$.

9. Let p be a prime. (a) If G has order $2p$, prove that every proper subgroup of G is cyclic. (b) If G has order p^2, prove that G has a subgroup of order p.

10. Prove Corollaries 16.1 and 16.2.

11. Let H and K be subgroups of a group G, where e is the identity of G.

a) Prove that if $|H| = 10$ and $|K| = 21$, then $H \cap K = \{e\}$.

b) If $|H| = m$ and $|K| = n$, with $\gcd(m, n) = 1$, prove that $H \cap K = \{e\}$.

12. The following provides an alternative way to establish Lagrange's Theorem. Let G be a group of order n, and let H be a subgroup of G of order m.

a) Define the relation $\mathcal{R}$ on G as follows: If $a, b \in G$, then $a \mathcal{R} b$ if $a^{-1}b \in H$. Prove that $\mathcal{R}$ is an equivalence relation on G.

b) For $a, b \in G$, prove that $a \mathcal{R} b$ if and only if $aH = bH$.

c) If $a \in G$, prove that $[a]$, the equivalence class of a under $\mathcal{R}$, satisfies $[a] = aH$.

d) For each $a \in G$, prove that $|aH| = |H|$.

e) Now establish the conclusion of Lagrange's Theorem, namely that $|H|$ divides $|G|$.

13. a) *Fermat's Theorem.* If p is a prime, prove that $a^p \equiv a$ (mod p) for each $a \in \mathbf{Z}$. [How is this related to Exercise 22(a) of Section 14.3?]

b) *Euler's Theorem.* For each $n \in \mathbf{Z}^+$, $n > 1$, and each $a \in \mathbf{Z}$, prove that if $\gcd(a, n) = 1$, then $a^{\phi(n)} \equiv 1 \pmod{n}$.

c) How are the theorems in parts (a) and (b) related?

d) Is there any connection between these two theorems and the results in Exercises 6 and 8?

16.4
The RSA Cryptosystem (Optional)

This section provides us with an opportunity to use some of the theoretical ideas we encountered in Sections 14.3 and 16.3 in a more contemporary application.

In Example 14.15 of Section 14.3 we introduced two private-key cryptosystems: the cipher shift and the affine cipher. For an alphabet of m characters, the encryption function $E: \mathbf{Z}_m \to \mathbf{Z}_m$, for the cipher-shift cryptosystem, is given by $E(\theta) = (\theta + \kappa) \bmod m$, where $\theta, \kappa \in \mathbf{Z}_m$, for κ ($\neq 0$) fixed. (Using $\kappa = 0$ would not alter any of the characters in a message.) Consequently, there are $m - 1$ possibilities to examine in an attempt to discover the value of the key κ. Further, once we know the value of κ, we also know the decryption function $D: \mathbf{Z}_m \to \mathbf{Z}_m$, for $D(\theta) = (\theta - \kappa) \bmod m$. In the case of the affine-cipher cryptosystem (also with an alphabet of m characters) the encryption function $E: \mathbf{Z}_m \to \mathbf{Z}_m$ is now given by $E(\theta) = (\alpha\theta + \kappa) \bmod m$, where $\theta, \alpha, \kappa \in \mathbf{Z}_m$, for fixed α, κ, with α invertible in $\mathbf{Z}_m$ [or, equivalently, with $\gcd(\alpha, m) = 1$]. Here the decryption function $D: \mathbf{Z}_m \to \mathbf{Z}_m$ is given by $D(\theta) = [\alpha^{-1}(\theta - \kappa)] \bmod m$. Without prior knowledge of the key (α, κ), now one would have to check $m\phi(m)$ possibilities to discover the appropriate values of α and κ for this private-key cryptosystem.

The security of either of the above cryptosystems depends on having the key [be it κ or (α, κ)] known only to the sender and the recipient of the messages.

The RSA cryptosystem is an example of a *public-key* cryptosystem. This cryptosystem was developed in the 1970s (and patented in 1983) by Ronald Rivest (1948–), Adi Shamir (1952–), and Leonard Adleman (1945–). (Taking the first letter from the surname of each of these three men provides the adjective RSA.)

We shall describe how this cryptosystem works and provide an example for encryption and decryption. In so doing, we shall find ourselves using some of the results from Sections 14.3 and 16.3.

EXAMPLE 16.18

As with the two private-key cryptosystems, once again we have an alphabet of m characters. We start with two distinct primes p, q. In practice, these should be large primes — each with 100 or more digits. (However, for our example we shall use much smaller primes.) After selecting the primes p, q, we then consider the integers $n = pq$ and $r = (p - 1) \cdot (q - 1) = \phi(p)\phi(q) = \phi(pq) = \phi(n)$, and, at this point, we choose an invertible element e in $\mathbf{Z}_r = (\mathbf{Z}_{\phi(n)})$.

[Here, if the element e is chosen at random, then the only time we fail to obtain an invertible element is when the element chosen is a multiple of p (there are q possibilities) or a multiple of q (there are p possibilities). In this count of $p + q$ elements we have accounted for pq twice, so there are only $p + q - 1$ possibilities for failure. Hence, the probability for failure is $(p + q - 1)/(pq) = (1/q) + (1/p) - (1/(pq))$, a very small number if p and q each have 100 or more digits.]

For instance, consider $p = 61$, $q = 127$, with $n = (61)(127) = 7747$ and $r = \phi(61) \cdot \phi(127) = (60)(126) = 7560$. Now suppose we select e as 17.

Consider the following message that we wish to encrypt.

INVEST IN BONDS

Using the same plaintext assignments as in part (b) of Example 14.15, here we would replace the letter "I" by 08 (not merely 8). Then we replace "N" by 13. This provides us with the first block of four digits — namely, 0813 — for the first two letters "IN". The assignment for the complete message is as follows [where we have appended the letter "X" to the right end, in order for the final block to have two letters (or, four digits)]:

I	N	V	E	S	T	I	N	B	O	N	D	S	X
08	13	21	04	18	19	08	13	01	14	13	03	18	23

We now encrypt each block B of four digits by the encryption function E, where $E(B) = B^e \bmod n$. (This modular exponentiation can be carried out efficiently by using the procedure in Example 14.16.) So here the domain of E is the concatenation of $\mathbf{Z}_{26}$ with itself, and we find that

$0813^{17} \bmod 7747 = 2169 \qquad 2104^{17} \bmod 7747 = 0628 \qquad 1819^{17} \bmod 7747 = 5540$

$0813^{17} \bmod 7747 = 2169 \qquad 0114^{17} \bmod 7747 = 6560 \qquad 1303^{17} \bmod 7747 = 6401$

$1823^{17} \bmod 7747 = 4829.$

Consequently, the recipient of the encrypted assignment (for the given plaintext message) receives the ciphertext

2169   0628   5540   2169   6560   6401   4829.

Now the question is: "How does the recipient decrypt the ciphertext received?"

Since e is a unit in $\mathbf{Z}_r$ $(= \mathbf{Z}_{\phi(n)})$, we can use the Euclidean algorithm (as in Example 14.13) to compute $e^{-1} = d$. Then we define the decryption function D, where $D(C) = C^d \bmod n$, for a block C of four digits. Since $e^{-1} = d$, it follows that $ed \equiv 1 \bmod \phi(n)$ — that is, $ed \bmod \phi(n) = 1$. Therefore, $ed = k\phi(n) + 1$, for some $k \in \mathbf{Z}$. Now recall the argument given earlier for the probability that a randomly selected element e from $\mathbf{Z}_n$ is invertible (or a unit in $\mathbf{Z}_n$). For any block B of four digits, we consider B as an element of $\mathbf{Z}_n$ — in fact, we consider B as a unit in $\mathbf{Z}_n$. Since the units in the ring $(\mathbf{Z}_n, +, \cdot)$ form a group of order $\phi(n)$ under multiplication, it follows from the result in Exercise 8 of Section 16.3 that $B^{ed} = B^{k\phi(n)+1} = (B^{\phi(n)})^k B^1 \equiv B \pmod{n}$, or $B^{ed} \bmod n = B$. [This is also a consequence of Euler's Theorem, as stated in part (b) of Exercise 13 in Section 16.3.]

Applying the result from the previous paragraph in our example we have $p = 61$, $q = 127$, $n = pq = 7747$, $r = \phi(n) = (p - 1)(q - 1) = (60)(126) = 7560$, and $e = 17$. From the Euclidean algorithm we calculate $d = e^{-1} = 3113$. Now we find, for instance, that $2169^{3113} \bmod 7747 = 0813$ and that $0628^{3113} \bmod 7747 = 2104$. Continuing, the recipient determines the numeric assignment for the original plaintext and then the plaintext.

Now what makes the RSA cryptosystem more secure than the private-key cryptosystems we studied? First, we should relate that the RSA cryptosystem is *not* a private-key cryptosystem. This system is an example of a *public-key* cryptosystem, where the key (n, e) is made public. So it seems that all one needs to do to decrypt the encrypted assignment is to determine $d = e^{-1}$ in $\mathbf{Z}_r$ $(= \mathbf{Z}_{\phi(n)})$. Now it is time to realize that by knowing n we do not immediately know r. For to be able to determine $r = (p-1)(q-1)$, we need to know p, q, the prime factors of n. And this is what makes this system so much more secure than the other cryptosystems we mentioned. Determining the primes p, q, when they are 100 or more digits long, is not a feasible problem. However, as computer power continues to improve, to keep the RSA cryptosystem secure, one may need to redefine the key using primes with more and more digits.

In closing, we show how the problem of factoring the modulus n as pq is related to the problem of determining $r = (p-1)(q-1)$. We start by observing that

$$p + q = pq - (p-1)(q-1) + 1 = n - \phi(n) + 1 = n - r + 1,$$

while

$$p - q = \sqrt{(p-q)^2} = \sqrt{(p-q)^2 + 4pq - 4pq} = \sqrt{(p+q)^2 - 4pq}$$
$$= \sqrt{(p+q)^2 - 4n} = \sqrt{(n-r+1)^2 - 4n}.$$

Then, from these two equations, we learn that

$$p = (1/2)[(p+q) + (p-q)] = (1/2)[(n-r+1) + \sqrt{(n-r+1)^2 - 4n}]$$

and

$$q = (1/2)[(p+q) - (p-q)] = (1/2)[(n-r+1) - \sqrt{(n-r+1)^2 - 4n}].$$

Consequently, when we know n and r, then we can readily determine the primes p, q such that $n = pq$.

EXERCISES 16.4

The use of a computer algebra system is strongly recommended for the first four exercises.

1. Determine the ciphertext for the plaintext INVEST IN STOCKS, when using RSA encryption with $e = 7$ and $n = 2573$.

2. Determine the ciphertext for the plaintext ORDER A PIZZA, when using RSA encryption with $e = 5$ and $n = 1459$.

3. Determine the plaintext for the RSA ciphertext 1418 1436 2370 1102 1805 0250, if $e = 11$ and $n = 2501$.

4. Determine the plaintext for the RSA ciphertext 0986 3029 1134 1105 1232 2281 2967 0272 1818 2398 1153, if $e = 17$ and $n = 3053$.

5. Find the primes p, q if $n = pq = 121{,}361$ and $\phi(n) = 120{,}432$.

6. Find the primes p, q if $n = pq = 5{,}446{,}367$ and $\phi(n) = 5{,}441{,}640$.

16.5
Elements of Coding Theory

In this and the next four sections we introduce an area of applied mathematics called *algebraic coding theory*. This theory was inspired by the fundamental paper of Claude Shannon (1948) along with results by Marcel Golay (1949) and Richard Hamming (1950). Since that time it has become an area of great interest where algebraic structures, probability, and combinatorics all play a role.

Our coverage will be held to an introductory level as we seek to model the transmission of information represented by strings of the signals 0 and 1.

In digital communications, when information is transmitted in the form of strings of 0's and 1's, certain problems arise. As a result of "noise" in the channel, when a certain signal is transmitted a different signal may be received, thus causing the receiver to make a wrong decision. Hence we want to develop techniques to help us detect, and perhaps even correct, transmission errors. However, we can only improve the chances of correct transmission; there are no guarantees.

Our model uses a *binary symmetric channel*, as shown in Fig. 16.2. The adjective *binary* appears because an individual signal is represented by one of the bits 0 or 1. When a transmitter sends the signal 0 or 1 in such a channel, associated with either signal is a (constant) probability p for incorrect transmission. When that probability p is the same for both signals, the channel is called *symmetric*. Here, for example, we have probability p of sending 0 and having 1 received. The probability of sending signal 0 and having it received correctly is then $1 - p$. All possibilities are illustrated in Fig. 16.2.

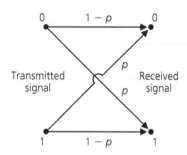

The Binary Symmetric Channel

Figure 16.2

EXAMPLE 16.19

Consider the string $c = 10110$. We regard c as an element of the group $\mathbf{Z}_2^5$, formed from the direct product of five copies of $(\mathbf{Z}_2, +)$. To shorten notation we write 10110 instead of $(1, 0, 1, 1, 0)$. When sending each bit (individual signal) of c through the binary symmetric channel, we assume that the probability of incorrect transmission is $p = 0.05$, so that the probability of transmitting c with no errors is $(0.95)^5 \doteq 0.77$.

Here, and throughout our discussion of coding theory, we assume that the transmission of each signal does not depend in any way on the transmissions of prior signals. Consequently, the probability of the occurrence of all of these *independent* events (in their prescribed order) is given by the product of their individual probabilities.

What is the probability that the party receiving the five-bit message receives the string $r = 00110$ — that is, the original message with an error in the first position? The probability of incorrect transmission for the first bit is 0.05, so with the assumption of independent events, $(0.05)(0.95)^4 \doteq 0.041$ is the probability of sending $c = 10110$ and receiving $r = 00110$. With $e = 10000$, we can write $c + e = r$ and interpret r as the result of the sum of the original message c and the particular *error pattern* $e = 10000$. Since $c, r, e \in \mathbf{Z}_2^5$ and $-1 = 1$ in $\mathbf{Z}_2$, we also have $c + r = e$ and $r + e = c$.

In transmitting $c = 10110$, the probability of receiving $r = 00100$ is

$$(0.05)(0.95)^2(0.05)(0.95) \doteq 0.002,$$

so this multiple error is not very likely to occur.

Finally if we transmit $c = 10110$, what is the probability that r differs from c in exactly two places? To answer this we sum the probabilities for each error pattern consisting of two 1's and three 0's. Each such pattern has probability 0.002. There are $\binom{5}{2}$ such patterns, so the probability of two errors in transmission is given by

$$\binom{5}{2}(0.05)^2(0.95)^3 \doteq 0.021.$$

These results lead us to the following theorem.

THEOREM 16.10   Let $c \in \mathbf{Z}_2^n$. For the transmission of c through a binary symmetric channel with probability p of incorrect transmission,

   a) the probability of receiving $r = c + e$, where e is a *particular* error pattern consisting of k 1's and $(n - k)$ 0's, is $p^k(1 - p)^{n-k}$.

   b) the probability that (exactly) k errors are made in the transmission is

$$\binom{n}{k}p^k(1 - p)^{n-k}.^{\dagger}$$

In Example 16.19, the probability of making at most one error in the transmission of $c = 10110$ is $(0.95)^5 + \binom{5}{1}(0.05)(0.95)^4 \doteq 0.977$. Thus the chance for multiple errors in transmission will be considered negligible throughout the discussion in this chapter. Such an assumption is valid when p is small. In actuality, a binary symmetric channel is considered "good" when $p < 10^{-5}$. However, no matter what else we stipulate, we always want $p < 1/2$.

To improve the accuracy of transmission in a binary symmetric channel, certain types of coding schemes can be used where extra bits are provided.

For $m, n \in \mathbf{Z}^+$, let $n > m$. Consider $\emptyset \neq W \subseteq \mathbf{Z}_2^m$. The set W consists of the *messages* to be transmitted. To each $w \in W$ are appended $n - m$ extra bits to form the *code word* c, where $c \in \mathbf{Z}_2^n$. This process is called *encoding* and is represented by the function $E: W \to \mathbf{Z}_2^n$. Then $E(w) = c$ and $E(W) = C \subseteq \mathbf{Z}_2^n$. Since the function E simply appends extra bits to the (distinct) messages, the encoding process is one-to-one. Upon transmission, c is received as $T(c)$, where $T(c) \in \mathbf{Z}_2^n$. Unfortunately, T is not a function because $T(c)$ may be different at different transmission times (for the noise in the channel changes with time). (See Fig. 16.3.)

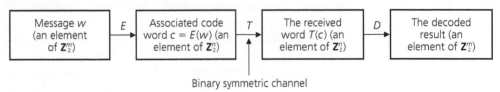

Figure 16.3

Upon receiving $T(c)$, we want to apply a decoding function $D: \mathbf{Z}_2^n \to \mathbf{Z}_2^m$ to remove the extra bits and, we hope, obtain the original message w. Ideally $D \circ T \circ E$ should be the identity function on W, with $D: C \to W$. Since this cannot be expected, we seek functions E and D such that there is a high probability of correctly decoding the received word $T(c)$ and recapturing the original message w. In addition, we want the ratio m/n to be as large as possible so that an excessive number of bits are not appended to w in getting the code

---

†This is the binomial probability distribution that was developed in (optional) Sections 3.5 and 3.7.

word $c = E(w)$. This ratio m/n measures the *efficiency* of our scheme and is called the *rate* of the code. Finally, the functions E and D should be more than theoretical results; they must be practical in the sense that they can be implemented electronically.

In such a scheme, the functions E and D are called the *encoding* and *decoding* functions, respectively, of an (n, m) *block code*.

We illustrate these ideas in the following two examples.

EXAMPLE 16.20

Consider the $(m + 1, m)$ block code for $m = 8$. Let $W = \mathbf{Z}_2^8$. For each $w = w_1 w_2 \cdots w_8 \in W$, define $E: \mathbf{Z}_2^8 \to \mathbf{Z}_2^9$ by $E(w) = w_1 w_2 \cdots w_8 w_9$, where $w_9 = \sum_{i=1}^8 w_i$, with the addition performed modulo 2. For example, $E(11001101) = 110011011$, and $E(00110011) = 001100110$.

For all $w \in \mathbf{Z}_2^8$, $E(w)$ contains an even number of 1's. So for $w = 11010110$ and $E(w) = 110101101$, if we receive $T(c) = T(E(w))$ as 100101101, from the odd number of 1's in $T(c)$ we know that a mistake has occurred in transmission. Hence we are able to *detect* single errors in transmission. But we seem to have no way to correct such errors.

The probability of sending the code word 110101101 and making at most one error in transmission is

$$\underbrace{(1 - p)^9}_{\substack{\text{All nine bits are} \\ \text{correctly transmitted.}}} + \underbrace{\tbinom{9}{1} p (1 - p)^8}_{\substack{\text{One bit is changed in} \\ \text{transmission and an error is detected.}}}.$$

For $p = 0.001$ this gives $(0.999)^9 + \binom{9}{1}(0.001)(0.999)^8 \doteq 0.99996417$.

If we detect an error and we are able to relay a signal back to the transmitter to repeat the transmission of the code word, and continue this process until the received word has an even number of 1's, then the probability of sending the code word 110101101 and receiving the correct transmission is approximately 0.99996393.[†]

Should an even positive number of errors occur in transmission, $T(c)$ is unfortunately accepted as the correct code word and we interpret its first eight components as the original message. This scheme is called the $(m + 1, m)$ *parity-check code* and is appropriate only when multiple errors are not likely to occur.

If we send the message 11010110 through the channel, we have probability $(0.999)^8 = 0.99202794$ of correct transmission. By using this parity-check code, we increase our chances of getting the correct message to (approximately) 0.99996393. However, an extra signal is sent (and perhaps additional transmissions are needed) and the rate of the code has decreased from 1 to 8/9.

But suppose that instead of sending eight bits we sent 160 bits, in successive strings of length 8. The chances of receiving the correct message without any coding scheme would be

---

[†]For $p = 0.001$ the probability that an odd number of errors occurs in the transmission of the code word 110101101 is

$$P_{\text{odd}} = \tbinom{9}{1}(0.999)^8(0.001) + \tbinom{9}{3}(0.999)^6(0.001)^3 + \tbinom{9}{5}(0.999)^4(0.001)^5 + \tbinom{9}{7}(0.999)^2(0.001)^7 + \tbinom{9}{9}(0.001)^9$$
$$\doteq 0.008928251 + 0.000000083 + 0.000000000 + 0.000000000 + 0.000000000 = 0.008928334.$$

With $q =$ the probability of the correct transmission of 110101101 $= (0.999)^9$, the probability that this code word is transmitted and correctly received under these conditions (of retransmission) is then given by

$$q + p_{\text{odd}} \cdot q + (p_{\text{odd}})^2 q + (p_{\text{odd}})^3 q + \cdots = q/(1 - p_{\text{odd}}) \doteq 0.99996393 \text{ (to eight decimal places)}.$$

$(0.999)^{160} \doteq 0.85207557$. With the parity-check method we send 180 bits, but the chances for correct transmission now increase to $(0.999964)^{20} \doteq 0.99928025$.

EXAMPLE 16.21

The $(3m, m)$ *triple repetition code* is one where we can both *detect* and *correct* single errors in transmission. With $m = 8$ and $W = \mathbf{Z}_2^8$, we define $E: \mathbf{Z}_2^8 \to \mathbf{Z}_2^{24}$ by $E(w_1 w_2 \cdots w_7 w_8) = w_1 w_2 \cdots w_8 w_1 w_2 \cdots w_8 w_1 w_2 \cdots w_8$.

Hence if $w = 10110111$, then $c = E(w) = 101101111011011110110111$.

The decoding function $D: \mathbf{Z}_2^{24} \to \mathbf{Z}_2^8$ is carried out by the majority rule. For example, if $T(c) = 101001110011011110110110$, then we have three errors occurring in positions 4, 9, and 24. We decode $T(c)$, by examining the first, ninth, and seventeenth positions to see which signal appears more times. Here it is 1 (which occurs twice), so we decode the first entry in the decoded message as 1. Continuing with the entries in the second, tenth, and eighteenth positions, the result for the second entry of the decoded message is 0 (which occurs all three times). As we proceed, we recapture the correct message, 10110111.

Although we have more than one transmission error here, all is well unless two (or more) errors occur with the second error eight or sixteen spaces after the first — that is, if two (or more) incorrect transmissions occur for the same bit of the original message.

Now how does this scheme compare with the other methods we have? With $p = 0.001$, the probability of correctly decoding a single bit is $(0.999)^3 + \binom{3}{1}(0.001)(0.999)^2 \doteq 0.99999700$. So the probability of receiving and correctly decoding the eight-bit message is $(0.99999700)^8 = 0.99997600$, just slightly better than the result from the parity-check method (where we may have to retransmit, thus increasing the overall transmission time). Here we transmit 24 signals for this message, so our rate is now $1/3$. For this increased accuracy and the ability to detect and now *correct* single errors (which we could not do in any previous schemes), we may pay with an increase in transmission time. But we do not waste time with retransmissions.

EXERCISES 16.5

1. Let C be a set of code words, where $C \subseteq \mathbf{Z}_2^7$. In each of the following, two of e (error pattern), r (received word) and c (code word) are given, with $r = c + e$. Determine the third term.

a) $c = 1010110$, $r = 1011111$

b) $c = 1010110$, $e = 0101101$

c) $e = 0101111$, $r = 0000111$

2. A binary symmetric channel has probability $p = 0.05$ of incorrect transmission. If the code word $c = 011011101$ is transmitted, what is the probability that (a) we receive $r = 011111101$? (b) we receive $r = 111011100$? (c) a single error occurs? (d) a double error occurs? (e) a triple error occurs? (f) three errors occur, no two of them consecutive?

3. Let $E: \mathbf{Z}_2^3 \to \mathbf{Z}_2^9$ be the encoding function for the (9, 3) triple repetition code.

a) If $D: \mathbf{Z}_2^9 \to \mathbf{Z}_2^3$ is the corresponding decoding function, apply D to decode the received words (i) 111101100;

(ii) 000100011; (iii) 010011111.

b) Find three different received words r for which $D(r) = 000$.

c) For each $w \in \mathbf{Z}_2^3$, what is $|D^{-1}(w)|$?

4. The $(5m, m)$ five-times repetition code has encoding function $E: \mathbf{Z}_2^m \to \mathbf{Z}_2^{5m}$, where $E(w) = wwwww$. Decoding with $D: \mathbf{Z}_2^{5m} \to \mathbf{Z}_2^m$ is accomplished by the majority rule. (Here we are able to correct single and double errors made in transmission.)

a) With $p = 0.05$, what is the probability for the transmission and correct decoding of the signal 0?

b) Answer part (a) for the message 110 in place of the signal 0.

c) For $m = 2$, decode the received word
$$r = 0111001001.$$

d) If $m = 2$, find three received words r where $D(r) = 00$.

e) For $m = 2$ and $D: \mathbf{Z}_2^{10} \to \mathbf{Z}_2^2$, what is $|D^{-1}(w)|$ for each $w \in \mathbf{Z}_2^2$?

16.6
The Hamming Metric

In this section we develop the general principles for discussing the error-detecting and error-correcting capabilities of a coding scheme. These ideas were developed by Richard Wesley Hamming (1915–1998).

We start by considering a code $C \subseteq \mathbf{Z}_2^4$, where $c_1 = 0111$, $c_2 = 1111 \in C$. Now both the transmitter and the receiver know the elements of C. So if the transmitter sends c_1 but the person receiving the code word receives $T(c_1)$ as 1111, then he or she feels that c_2 was transmitted and makes whatever decision (a wrong one) c_2 implies. Consequently, although only one transmission error was made, the results could be unpleasant. Why is this? Unfortunately we have two code words that are almost the same. They are rather *close* to each other, for they differ in only one component.

We describe this notion of closeness more precisely as follows.

Definition 16.9

For each element $x = x_1 x_2 \cdots x_n \in \mathbf{Z}_2^n$, where $n \in \mathbf{Z}^+$, the *weight of* x, denoted $\mathrm{wt}(x)$, is the number of components x_i of x, for $1 \leq i \leq n$, where $x_i = 1$. If $y \in \mathbf{Z}_2^n$, the *distance between* x *and* y, denoted $d(x, y)$, is the number of components where $x_i \neq y_i$, for $1 \leq i \leq n$.

EXAMPLE 16.22

For $n = 5$, let $x = 01001$ and $y = 11101$. Then $\mathrm{wt}(x) = 2$, $\mathrm{wt}(y) = 4$, and $d(x, y) = 2$. In addition, $x + y = 10100$, so $\mathrm{wt}(x + y) = 2$. Is it just by chance that $d(x, y) = \mathrm{wt}(x + y)$? For each $1 \leq i \leq 5$, $x_i + y_i$ contributes a count of 1 to $\mathrm{wt}(x + y) \Longleftrightarrow x_i \neq y_i \Longleftrightarrow x_i, y_i$ contribute a count of 1 to $d(x, y)$. [This is actually true for all $n \in \mathbf{Z}^+$, so $\mathrm{wt}(x + y) = d(x, y)$ for all $x, y \in \mathbf{Z}_2^n$.]

When $x, y \in \mathbf{Z}_2^n$, we write $d(x, y) = \sum_{i=1}^{n} d(x_i, y_i)$ where,

$$\text{for each } 1 \leq i \leq n, \qquad d(x_i, y_i) = \begin{cases} 0 & \text{if } x_i = y_i \\ 1 & \text{if } x_i \neq y_i. \end{cases}$$

LEMMA 16.2

For all $x, y \in \mathbf{Z}_2^n$, $\mathrm{wt}(x + y) \leq \mathrm{wt}(x) + \mathrm{wt}(y)$.

Proof: We prove this lemma by examining, for each $1 \leq i \leq n$, the components $x_i, y_i, x_i + y_i$, of $x, y, x + y$, respectively. Only one situation would cause this inequality to be false: if $x_i + y_i = 1$ while $x_i = 0$ and $y_i = 0$, for some $1 \leq i \leq n$. But this never occurs because $x_i + y_i = 1$ implies that exactly one of x_i and y_i is 1.

In Example 16.22 we found that

$$\mathrm{wt}(x + y) = \mathrm{wt}(10100) = 2 \leq 2 + 4 = \mathrm{wt}(01001) + \mathrm{wt}(11101) = \mathrm{wt}(x) + \mathrm{wt}(y).$$

THEOREM 16.11

The distance function d defined on $\mathbf{Z}_2^n \times \mathbf{Z}_2^n$ satisfies the following for all $x, y, z \in \mathbf{Z}_2^n$.

a) $d(x, y) \geq 0$ **b)** $d(x, y) = 0 \Longleftrightarrow x = y$

c) $d(x, y) = d(y, x)$ **d)** $d(x, z) \leq d(x, y) + d(y, z)$

Proof: We leave the first three parts for the reader and prove part (d).

In $\mathbf{Z}_2^n$, $y + y = 0$, so $d(x, z) = \text{wt}(x + z) = \text{wt}(x + (y + y) + z) = \text{wt}((x + y) + (y + z)) \leq \text{wt}(x + y) + \text{wt}(y + z)$, by Lemma 16.2. With $\text{wt}(x + y) = d(x, y)$ and $\text{wt}(y + z) = d(y, z)$, the result follows. (This property is generally called the *Triangle Inequality*.)

When a function satisfies the four properties listed in Theorem 16.11, it is called a *distance function* or *metric*, and we call $(\mathbf{Z}_2^n, d)$ a *metric space*. Hence d (as given above) is often referred to as the *Hamming metric*. This metric is used in the following.

Definition 16.10	For $n, k \in \mathbf{Z}^+$ and $x \in \mathbf{Z}_2^n$, the *sphere* of radius k centered at x is defined as $S(x, k) = \{y \in \mathbf{Z}_2^n \mid d(x, y) \leq k\}$.

EXAMPLE 16.23	For $n = 3$ and $x = 110 \in \mathbf{Z}_2^3$, $S(x, 1) = \{110, 010, 100, 111\}$ and $S(x, 2) = \{110, 010, 100, 111, 000, 101, 011\}$.

With these preliminaries in hand we turn now to the two major results of this section.

THEOREM 16.12   Let $E: W \to C$ be an encoding function with the set of messages $W \subseteq \mathbf{Z}_2^m$ and the set of code words $E(W) = C \subseteq \mathbf{Z}_2^n$, where $m < n$. If our objective is error detection, then for $k \in \mathbf{Z}^+$, we can detect all transmission errors of weight $\leq k$ if and only if the minimum distance between code words is at least $k + 1$.

Proof: The set C is known to both the transmitter and the receiver, so if $w \in W$ is the message and $c = E(w)$ is transmitted, let $c \neq T(c) = r$. If the minimum distance between code words is at least $k + 1$, then the transmission of c can result in as many as k errors and r will not be listed in C. Hence we can detect all errors e where $\text{wt}(e) \leq k$. Conversely, let c_1, c_2 be code words with $d(c_1, c_2) < k + 1$. Then $c_2 = c_1 + e$ where $\text{wt}(e) \leq k$. If we send c_1 and $T(c_1) = c_2$, then we would feel that c_2 had been sent, thus failing to detect an error of weight $\leq k$.

What can we say about error-correcting capability?

THEOREM 16.13   Let E, W, and C be as in Theorem 16.12. If our objective is error correction, then for $k \in \mathbf{Z}^+$, we can construct a decoding function $D: \mathbf{Z}_2^n \to W$ that corrects all transmission errors of weight $\leq k$ if and only if the minimum distance between code words is at least $2k + 1$.

Proof: For $c \in C$, consider $S(c, k) = \{x \in \mathbf{Z}_2^n \mid d(c, x) \leq k\}$. Define $D: \mathbf{Z}_2^n \to W$ as follows. If $r \in \mathbf{Z}_2^n$ and $r \in S(c, k)$ for some code word c, then $D(r) = w$ where $E(w) = c$. [Here c is the (unique) code word *nearest* to r.] If $r \notin S(c, k)$ for any $c \in C$, then we define $D(r) = w_0$, where w_0 is some arbitrary message that remains fixed once it is chosen. The only problem we could face here is that D might not be a function. This will happen if there is an element r in $\mathbf{Z}_2^n$ with r in both $S(c_1, k)$ and $S(c_2, k)$ for distinct code words c_1, c_2. But $r \in S(c_1, k) \Rightarrow d(c_1, r) \leq k$, and $r \in S(c_2, k) \Rightarrow d(c_2, r) \leq k$, so $d(c_1, c_2) \leq d(c_1, r) + d(r, c_2) \leq k + k < 2k + 1$. Consequently, if the minimum distance between code words is at least $2k + 1$, then D is a function, and it will decode all possible received words, correcting any transmission error of weight $\leq k$. Conversely, if $c_1, c_2 \in C$ and $d(c_1, c_2) \leq 2k$, then c_2 can be obtained from c_1 by making at most $2k$ changes. Starting at code word c_1 we make approximately half (exactly, $\lfloor d(c_1, c_2)/2 \rfloor$) of these changes. This brings us to $r = c_1 + e_1$ with $\text{wt}(e_1) \leq k$. Continuing from r, we make the remaining changes to get to c_2 and find $r + e_2 = c_2$ with $\text{wt}(e_2) \leq k$. But then $r = c_2 + e_2$. Now with $c_1 + e_1 = r = c_2 + e_2$ and $\text{wt}(e_1), \text{wt}(e_2) \leq k$, how can one decide on the code word from which r arises? This ambiguity results in a possible error of weight $\leq k$ that cannot be corrected.

EXAMPLE 16.24

With $W = \mathbf{Z}_2^2$ let $E: W \to \mathbf{Z}_2^6$ be given by

$$E(00) = 000000 \qquad E(10) = 101010 \qquad E(01) = 010101 \qquad E(11) = 111111.$$

Then the minimum distance between code words is 3, so we can correct all single errors.
With

$$S(000000, 1) = \{x \in \mathbf{Z}_2^6 | d(000000, x) \leq 1\}$$
$$= \{000000, 100000, 010000, 001000, 000100, 000010, 000001\},$$

the decoding function $D: \mathbf{Z}_2^6 \to W$ gives $D(x) = 00$ for all $x \in S(000000, 1)$.
Similarly,

$$S(010101, 1) = \{x \in \mathbf{Z}_2^6 | d(010101, x) \leq 1\}$$
$$= \{010101, 110101, 000101, 011101, 010001, 010111, 010100\},$$

and here $D(x) = 01$ for each $x \in S(010101, 1)$. At this point our definition of D accounts for 14 of the elements in $\mathbf{Z}_2^6$. Continuing to define D for the 14 elements in $S(101010, 1)$ and $S(111111, 1)$ there remain 36 other elements to account for. We define $D(x) = 00$ (or any other message) for these 36 other elements and have a decoding function that will correct single errors.

Beware! There is a subtle point that needs to be made about Theorems 16.12 and 16.13. For example, if the minimum distance between code words is $2k + 1$ one may feel that we can detect all errors of weight $\leq 2k$ *and* correct all errors of weight $\leq k$. This is not necessarily true. That is, error detection and error correction need not take place at the same time and at the maximum levels. To see this, reconsider the (6, 2)-triple repetition code of Example 16.24. Here the encoding function $E: W(= \mathbf{Z}_2^2) \to \mathbf{Z}_2^6$ is given by $E(w_1 w_2) = w_1 w_2 w_1 w_2 w_1 w_2$ and the code comprises the four elements of $\mathbf{Z}_2^6$ in the range of E. Since the minimum distance between any two elements of $\mathbf{Z}_2^2$ is 1, it follows that the minimum distance between code words is 3 (as observed earlier in Example 16.24).

Now suppose that our major objective is error correction and that $r = 100000 [\notin E(W)]$ is received. We see that $d(000000, r) = 1$, $d(101010, r) = 2$, $d(010101, r) = 4$, and $d(111111, r) = 5$. Consequently, we should choose to decode r as 000000, the unique code word nearest to r. Unfortunately, suppose that the actual message were 10 (with corresponding code word 101010), but we received $r = 100000$. Upon correcting r as 000000, we should then decode 000000 to get the incorrect message 00. And, in so doing, we have failed to detect an error of weight 2.

In this type of situation one can develop a scheme where a mixed strategy is used. Here both error correction *and* error detection may be carried out at some levels.

For $t \in \mathbf{N}$, if the received word is r and there is a unique code word c_1 such that $d(c_1, r) \leq t$, then we decode r as c_1. (*Note:* The case where $r = c_1$ is covered when $t = 0$.) If there exists a second code word c_2 such that $d(c_2, r) = d(c_1, r)$, or if $d(c, r) > t$ for all code words c, then an error is declared (and retransmission is generally requested). Using this scheme, if the minimum distance between code words is at least $2t + s + 1$, for $s \in \mathbf{N}$, then we can correct all errors of weight $\leq t$ and detect all errors with weights between $t + 1$ and $t + s$, inclusive.

When using this scheme for the $(6, 2)$-triple repetition code, our options include:

1) $t = 0$; $s = 2$: Here we can detect all errors of weight ≤ 2 but we have no error-correction capability.

2) $t = 1$; $s = 0$: Single errors are corrected here but there is no error-detecting capability.

If we use the $(10, 2)$-five-times repetition code, then the minimum distance is 5. Applying the above scheme in this case, our options now include:

1) $t = 0$; $s = 4$: Here we can detect all errors of weight ≤ 4 but we have no error-correction capability.

2) $t = 1$; $s = 2$: Now single errors are corrected and we can also detect all errors e, where $2 \leq \text{wt}(e) \leq 3$.

3) $t = 2$; $s = 0$: All errors of weight ≤ 2 are corrected but there is no error-detecting capability.

[For more on this, the interested reader should examine Chapter 4 of the text by S. Roman [24].]

16.7
The Parity-Check and Generator Matrices

In this section we introduce an example where the encoding and decoding functions are given by matrices over $\mathbf{Z}_2$. One of these matrices will help us to locate the *nearest* code word for a given received word. This will be especially helpful as the set C of code words grows larger.

EXAMPLE 16.25

Let

$$G = \begin{bmatrix} 1 & 0 & 0 & 1 & 1 & 0 \\ 0 & 1 & 0 & 0 & 1 & 1 \\ 0 & 0 & 1 & 1 & 0 & 1 \end{bmatrix}$$

be a 3×6 matrix over $\mathbf{Z}_2$. The first three columns of G form the 3×3 identity matrix I_3. Letting A denote the matrix formed from the last three columns of G, we write $G = [I_3 | A]$ to denote its structure. The (partitioned) matrix G is called a *generator matrix*.

We use G to define an encoding function $E: \mathbf{Z}_2^3 \to \mathbf{Z}_2^6$ as follows. For $w \in \mathbf{Z}_2^3$, $E(w) = wG$ is the element in $\mathbf{Z}_2^6$ obtained by multiplying w, considered as a three-dimensional row vector, by the matrix G on its right. Unlike the results on matrix multiplication in Chapter 7, in the calculations here we have $1 + 1 = 0$, not $1 + 1 = 1$.

(Even if the set W of messages is not all of $\mathbf{Z}_2^3$, we'll assume that all of $\mathbf{Z}_2^3$ is encoded and that the transmitter and receiver will both know the real messages of importance and their corresponding code words.)

We find here, for example, that

$$E(110) = (110)G = [110] \begin{bmatrix} 1 & 0 & 0 & 1 & 1 & 0 \\ 0 & 1 & 0 & 0 & 1 & 1 \\ 0 & 0 & 1 & 1 & 0 & 1 \end{bmatrix} = [110101],$$

and

$$E(010) = (010)G = [010] \begin{bmatrix} 1 & 0 & 0 & 1 & 1 & 0 \\ 0 & 1 & 0 & 0 & 1 & 1 \\ 0 & 0 & 1 & 1 & 0 & 1 \end{bmatrix} = [010011].$$

Note that $E(110)$ can be obtained by adding the first two rows of G, whereas $E(010)$ is simply the second row of G.

The set of code words obtained by this method is

$$C = \{000000, 100110, 010011, 001101, 110101, 101011, 011110, 111000\} \subseteq \mathbf{Z}_2^6,$$

and one can recapture the corresponding message by simply dropping the last three components of the code word. In addition, the minimum distance between code words is 3, so we can detect errors of weight ≤ 2 or correct single errors. (We shall assume that multiple errors are rare and concentrate on error correction.)

For all $w = w_1 w_2 w_3 \in \mathbf{Z}_2^3$, $E(w) = w_1 w_2 w_3 w_4 w_5 w_6 \in \mathbf{Z}_2^6$. Since

$$E(w) = [w_1 w_2 w_3] \begin{bmatrix} 1 & 0 & 0 & 1 & 1 & 0 \\ 0 & 1 & 0 & 0 & 1 & 1 \\ 0 & 0 & 1 & 1 & 0 & 1 \end{bmatrix}$$

$$= [w_1 w_2 w_3 (w_1 + w_3)(w_1 + w_2)(w_2 + w_3)],$$

we have $w_4 = w_1 + w_3$, $w_5 = w_1 + w_2$, $w_6 = w_2 + w_3$, and these equations are called the *parity-check equations*. Since $w_i \in \mathbf{Z}_2$ for each $1 \leq i \leq 6$, it follows that $w_i = -w_i$ and so the equations can be rewritten as

$$
\begin{aligned}
w_1 \phantom{{}+w_2} + w_3 + w_4 \phantom{{}+w_5 +w_6} &= 0 \\
w_1 + w_2 \phantom{{}+w_3+w_4} + w_5 \phantom{{}+w_6} &= 0 \\
\phantom{w_1 +{}} w_2 + w_3 \phantom{{}+w_4+w_5} + w_6 &= 0.
\end{aligned}
$$

Thus we find that

$$\begin{bmatrix} 1 & 0 & 1 & 1 & 0 & 0 \\ 1 & 1 & 0 & 0 & 1 & 0 \\ 0 & 1 & 1 & 0 & 0 & 1 \end{bmatrix} \begin{bmatrix} w_1 \\ w_2 \\ w_3 \\ w_4 \\ w_5 \\ w_6 \end{bmatrix} = H \cdot (E(w))^{\text{tr}} = \begin{bmatrix} 0 \\ 0 \\ 0 \end{bmatrix},$$

where $(E(w))^{\text{tr}}$ denotes the transpose of $E(w)$. Consequently, if $r = r_1 r_2 \cdots r_6 \in \mathbf{Z}_2^6$, we can identify r as a code word if and only if

$$H \cdot r^{\text{tr}} = \begin{bmatrix} 0 \\ 0 \\ 0 \end{bmatrix}.$$

Writing $H = [B | I_3]$, we notice that if the rows and columns of B are interchanged, then we get A. Hence $B = A^{\text{tr}}$.

From the theory developed earlier on error correction, because the minimum distance between the code words of this example is 3, we should be able to develop a decoding function that corrects single errors.

Suppose we receive $r = 110110$. We want to find the code word c that is the *nearest neighbor* of r. If there is a long list of code words against which to check r, we would be better off to first examine $H \cdot r^{tr}$, which is called the *syndrome* of r. Here

$$H \cdot r^{tr} = \begin{bmatrix} 1 & 0 & 1 & 1 & 0 & 0 \\ 1 & 1 & 0 & 0 & 1 & 0 \\ 0 & 1 & 1 & 0 & 0 & 1 \end{bmatrix} \begin{bmatrix} 1 \\ 1 \\ 0 \\ 1 \\ 1 \\ 0 \end{bmatrix} = \begin{bmatrix} 0 \\ 1 \\ 1 \end{bmatrix},$$

so r is not a code word. Hence we at least detect an error. Looking back at the list of code words, we see that $d(100110, r) = 1$. For all other $c \in C$, $d(r, c) \geq 2$. Writing $r = c + e = 100110 + 010000$, we find that the transmission error (of weight 1) occurs in the second component of r. Is it just a coincidence that the syndrome $H \cdot r^{tr}$ produced the second column of H? If not, then we can use this result in order to realize that if a single transmission error occurred, it took place at the second component. Changing the second component of r, we get c; the message w comprises the first three components of c.

Let $r = c + e$, where c is a code word and e is an error pattern of weight 1. Suppose that 1 is in the ith component of e, where $1 \leq i \leq 6$. Then

$$H \cdot r^{tr} = H \cdot (c + e)^{tr} = H \cdot (c^{tr} + e^{tr}) = H \cdot c^{tr} + H \cdot e^{tr}.$$

With c a code word, it follows that $H \cdot c^{tr} = \mathbf{0}$, so $H \cdot r^{tr} = H \cdot e^{tr} = i$th column of matrix H. Thus c and r differ only in the ith component, and we can determine c by simply changing the ith component of r.

Since we are primarily concerned with transmissions where multiple errors are rare, this technique is of definite value. If we ask for more, however, we find ourselves expecting too much.

Suppose that we receive $r = 000111$. Computing the syndrome

$$H \cdot r^{tr} = \begin{bmatrix} 1 & 0 & 1 & 1 & 0 & 0 \\ 1 & 1 & 0 & 0 & 1 & 0 \\ 0 & 1 & 1 & 0 & 0 & 1 \end{bmatrix} \begin{bmatrix} 0 \\ 0 \\ 0 \\ 1 \\ 1 \\ 1 \end{bmatrix} = \begin{bmatrix} 1 \\ 1 \\ 1 \end{bmatrix},$$

we obtain a result that is not one of the columns of H. Yet $H \cdot r^{tr}$ can be obtained as the sum of two columns from H. If $H \cdot r^{tr}$ came from the first and sixth columns of H, correcting these components in r results in the code word 100110. If we sum the third and fifth columns of H to get this syndrome, upon changing the third and fifth components of r we get a second code word, 001101. So we cannot expect H to correct multiple errors. This is no surprise since the minimum distance between code words is 3.

We summarize the results of Example 16.25 for the general situation. For $m, n \in \mathbf{Z}^+$ with $m < n$, the encoding function $E: \mathbf{Z}_2^m \to \mathbf{Z}_2^n$ is given by an $m \times n$ matrix G over $\mathbf{Z}_2$. This matrix G is called the generator matrix for the code and has the form $[I_m | A]$, where

A is an $m \times (n - m)$ matrix. Here $E(w) = wG$ for each message $w \in \mathbf{Z}_2^m$, and the code $C = E(\mathbf{Z}_2^m) \subset \mathbf{Z}_2^n$.

The associated *parity-check matrix H* is an $(n - m) \times n$ matrix of the form $[A^{\mathrm{tr}} | I_{n-m}]$. This matrix can also be used to define the encoding function E, because if $w = w_1 w_2 \cdots w_m \in \mathbf{Z}_2^m$, then $E(w) = w_1 w_2 \cdots w_m w_{m+1} \cdots w_n$, where $w_{m+1}, \ldots, w_n$ can be determined from the set of $n - m$ (parity-check) equations that arise from $H \cdot (E(w))^{\mathrm{tr}} = \mathbf{0}$, the column vector of $n - m$ 0's.

This unique parity-check matrix H also provides a decoding scheme that corrects single errors in transmission if:

a) H does not contain a column of 0's. (If the ith column of H had all 0's and $H \cdot r^{\mathrm{tr}} = \mathbf{0}$ for a received word r, we couldn't decide whether r was a code word or a received word whose ith component was incorrectly transmitted. We do not want to compare r with all code words when C is large.)

b) No two columns of H are the same. (If the ith and jth columns of H are the same and $H \cdot r^{\mathrm{tr}}$ equals this repeated column, how would we decide which component of r to change?)

When H satisfies these two conditions, we get the following decoding algorithm. For each $r \in \mathbf{Z}_2^n$, if $T(c) = r$, then:

1) With $H \cdot r^{\mathrm{tr}} = \mathbf{0}$, we feel that the transmission was correct and that r is the code word that was transmitted. The decoded message then consists of the first m components of r.

2) With $H \cdot r^{\mathrm{tr}}$ equal to the ith column of H, we feel that there has been a single error in transmission and change the ith component of r in order to get the code word c. Here the first m components of c yield the original message.

3) If neither case 1 nor case 2 occurs, we feel that there has been more than one transmission error and we cannot provide a reliable way to decode in this situation.

We close with one final comment on the matrix H. If we start with a parity-check matrix $H = [B | I_{n-m}]$ and use it, as described above, to define the function E, then we obtain the same set of code words that is generated by the unique associated generator matrix $G = [I_m | B^{\mathrm{tr}}]$.

EXERCISES 16.6 and 16.7

1. For Example 16.24, list the elements in $S(101010, 1)$ and $S(111111, 1)$.

2. Decode each of the following received words for Example 16.24.

    a) 110101   **b)** 101011

    c) 001111   **d)** 110000

3. a) If $x \in \mathbf{Z}_2^{10}$, determine $|S(x, 1)|, |S(x, 2)|, |S(x, 3)|$.

    b) For $n, k \in \mathbf{Z}^+$ with $1 \le k \le n$, if $x \in \mathbf{Z}_2^n$, what is $|S(x, k)|$?

4. Let $E: \mathbf{Z}_2^5 \to \mathbf{Z}_2^{25}$ be an encoding function where the minimum distance between code words is 9. What is the largest value of k such that we can detect errors of weight $\le k$? If we wish to correct errors of weight $\le n$, what is the maximum value for n?

5. For each of the following encoding functions, find the minimum distance between the code words. Discuss the error-detecting and error-correcting capabilities of each code.

    a) $E: \mathbf{Z}_2^2 \to \mathbf{Z}_2^5$

        $00 \to 00001$ $01 \to 01010$

        $10 \to 10100$ $11 \to 11111$

    b) $E: \mathbf{Z}_2^2 \to \mathbf{Z}_2^{10}$

        $00 \to 0000000000$ $01 \to 0000011111$

        $10 \to 1111100000$ $11 \to 1111111111$

c) $E: \mathbf{Z}_2^3 \to \mathbf{Z}_2^6$

$000 \to 000111$ $001 \to 001001$
$010 \to 010010$ $011 \to 011100$
$100 \to 100100$ $101 \to 101010$
$110 \to 110001$ $111 \to 111000$

d) $E: \mathbf{Z}_2^3 \to \mathbf{Z}_2^8$

$000 \to 00011111$ $001 \to 00111010$
$010 \to 01010101$ $011 \to 01110000$
$100 \to 10001101$ $101 \to 10101000$
$110 \to 11000100$ $111 \to 11100011$

6. a) Use the parity-check matrix H of Example 16.25 to decode the following received words.

    i) 111101   **ii)** 110101
    iii) 001111   **iv)** 100100
    v) 110001   **vi)** 111111
    vii) 111100   **viii)** 010100

b) Are all the results in part (a) uniquely determined?

7. The encoding function $E: \mathbf{Z}_2^2 \to \mathbf{Z}_2^5$ is given by the generator matrix

$$G = \begin{bmatrix} 1 & 0 & 1 & 1 & 0 \\ 0 & 1 & 0 & 1 & 1 \end{bmatrix}.$$

a) Determine all code words. What can we say about the error-detection capability of this code? What about its error-correction capability?

b) Find the associated parity-check matrix H.

c) Use H to decode each of the following received words.

    i) 11011   **ii)** 10101   **iii)** 11010
    iv) 00111   **v)** 11101   **vi)** 00110

8. Define the encoding function $E: \mathbf{Z}_2^3 \to \mathbf{Z}_2^6$ by means of the parity-check matrix

$$H = \begin{bmatrix} 1 & 0 & 1 & 1 & 0 & 0 \\ 1 & 1 & 0 & 0 & 1 & 0 \\ 1 & 0 & 1 & 0 & 0 & 1 \end{bmatrix}.$$

a) Determine all code words.

b) Does this code correct all single errors in transmission?

9. Find the generator and parity-check matrices for the $(9, 8)$ single parity-check coding scheme of Example 16.20.

10. a) Show that the 1×9 matrix $G = [1 \quad 1 \quad 1 \dots 1]$ is the generator matrix for the $(9, 1)$ nine-times repetition code.

b) What is the associated parity-check matrix H in this case?

11. For an (n, m) code C with generator matrix $G = [I_m | A]$ and parity-check matrix $H = [A^{tr} | I_{n-m}]$, the $(n, n - m)$ code C^d with generator matrix $[I_{n-m} | A^{tr}]$ and parity-check matrix $[A | I_m]$ is called the *dual code* of C. Show that the codes in each of Exercises 9 and 10 constitute a pair of dual codes.

12. Given $n \in \mathbf{Z}^+$, let the set $M(n, k) \subseteq \mathbf{Z}_2^n$ contain the maximum number of code words of length n, where the minimum distance between code words is $2k + 1$. Prove that

$$\frac{2^n}{\sum_{i=0}^{2k} \binom{n}{i}} \leq |M(n, k)| \leq \frac{2^n}{\sum_{i=0}^{k} \binom{n}{i}}.$$

(The upper bound on $|M(n, k)|$ is called the *Hamming bound*; the lower bound is referred to as the *Gilbert bound*.)

16.8
Group Codes:
Decoding with Coset Leaders

Now that we've examined some introductory material on coding theory, it is time to see how the group structure enters the picture.

Definition 16.11   Let $E: \mathbf{Z}_2^m \to \mathbf{Z}_2^n$, for $n > m$, be an encoding function. The code $C = E(\mathbf{Z}_2^m)$ is called a *group code* if C is a subgroup of $\mathbf{Z}_2^n$.

Recall the encoding function $E: \mathbf{Z}_2^2 \to \mathbf{Z}_2^6$ (of Example 16.24) where

$$E(00) = 000000 \qquad E(10) = 101010 \qquad E(01) = 010101 \qquad E(11) = 111111.$$

Here $\mathbf{Z}_2^2$ and $\mathbf{Z}_2^6$ are groups under componentwise addition modulo 2; the subset $C = E(\mathbf{Z}_2^2) = \{000000, 101010, 010101, 111111\}$ is a subgroup of $\mathbf{Z}_2^6$, and an example of a group code. (Note that C contains 000000, the zero element of $\mathbf{Z}_2^6$.)

In general when the code words form a group, we find that it is easier to compute the minimum distance between code words.

THEOREM 16.14

In a group code, the minimum distance between distinct code words is the minimum of the weights of the nonzero elements of the code.

Proof: Let $a, b, c \in C$ where $a \neq b$, $d(a, b)$ is minimum, and c is nonzero with minimum weight. By closure in the group C, $a + b$ is a code word. Since $d(a, b) = \text{wt}(a + b)$, by the choice of c we have $d(a, b) \geq \text{wt}(c)$. Also, $\text{wt}(c) = d(c, \mathbf{0})$, where $\mathbf{0}$ is a code word because C is a group. Then $d(c, \mathbf{0}) \geq d(a, b)$ by the choice of a, b, so $\text{wt}(c) \geq d(a, b)$. Consequently, $d(a, b) = \text{wt}(c)$.

If C is a set of code words and $|C| = 1024$, we have to compute $\binom{1024}{2} = 523{,}776$ distances to find the minimum distance between code words. But if we can recognize that C possesses a group structure, we need only compute the weights of the 1023 nonzero elements of C.

Is there some way to guarantee that the code words form a group? By Theorem 16.5(d), the homomorphic image of a subgroup is a subgroup, so if $E: \mathbf{Z}_2^m \rightarrow \mathbf{Z}_2^n$ is a group homomorphism, then $C = E(\mathbf{Z}_2^m)$ will be a subgroup of $\mathbf{Z}_2^n$. Our next result will use this fact to show that the codes we obtain when using a generator matrix G or a parity-check matrix H are group codes. Furthermore, the proof of this result reconfirms the observation we made (at the end of the previous section) about *the* code that arises from a generator matrix G or its associated parity-check matrix H.

THEOREM 16.15

Let $E: \mathbf{Z}_2^m \rightarrow \mathbf{Z}_2^n$ be an encoding function given by a generator matrix G or the associated parity-check matrix H. Then $C = E(\mathbf{Z}_2^m)$ is a group code.

Proof: We establish these results by proving that the function E arising from G or H is a group homomorphism.

If $x, y \in \mathbf{Z}_2^m$, then $E(x + y) = (x + y)G = xG + yG = E(x) + E(y)$. Hence E is a homomorphism and $C = E(\mathbf{Z}_2^m)$ is a group code [by virtue of part (d) of Theorem 16.5].

For the case of H, if x is a message, then $E(x) = x_1 x_2 \cdots x_m x_{m+1} \cdots x_n$, where $x = x_1 x_2 \cdots x_m \in \mathbf{Z}_2^m$ and $H \cdot (E(x))^{\text{tr}} = \mathbf{0}$. In particular, $E(x)$ is uniquely determined by these two properties. If y is also a message, then $x + y$ is likewise, and $E(x + y)$ has $(x_1 + y_1), (x_2 + y_2), \ldots, (x_m + y_m)$ as its first m components, as does $E(x) + E(y)$. Further, $H \cdot (E(x) + E(y))^{\text{tr}} = H \cdot (E(x)^{\text{tr}} + E(y)^{\text{tr}}) = H \cdot E(x)^{\text{tr}} + H \cdot E(y)^{\text{tr}} = \mathbf{0} + \mathbf{0} = \mathbf{0}$. Since $E(x + y)$ is the unique element of $\mathbf{Z}_2^n$ with $(x_1 + y_1), (x_2 + y_2), \ldots, (x_m + y_m)$ as its first m components and with $H \cdot (E(x + y))^{\text{tr}} = \mathbf{0}$, it follows that $E(x + y) = E(x) + E(y)$. So E is a group homomorphism and, consequently, $C = \{c \in \mathbf{Z}_2^n | H \cdot c^{\text{tr}} = \mathbf{0}\}$ is a group code.

Now we use the group structure of C, together with its cosets in $\mathbf{Z}_2^n$, to develop a scheme for decoding. Our example uses the code developed in Example 16.25, but the procedure applies for every group code.

EXAMPLE 16.26

We develop a table for decoding as follows.

1) First list in a row the elements of the group code C, starting with the identity.

000000   100110   010011   001101   110101   101011   011110   111000.

2) Next select an element x of $\mathbf{Z}_2^6$ ($\mathbf{Z}_2^n$, in general) where x does not appear anywhere in the table developed so far and has minimum weight. Then list the elements of the coset $x + C$, with $x + c$ directly below c for each $c \in C$. For $x = 100000$ we have

000000 100110 010011 001101 110101 101011 011110 111000
100000 000110 110011 101101 010101 001011 111110 011000.

3) Repeat step (2) until the cosets provide a partition of $\mathbf{Z}_2^6$ ($\mathbf{Z}_2^n$, in general). This results in the *decoding table* shown in Table 16.8.

4) Once the decoding table is constructed, for each received word r we find the column containing r and use the first three components of the code word c at the top of the column to decode r.

Table 16.8 Decoding Table for the Code of Example 16.25

000000	100110	010011	001101	110101	101011	011110	111000
100000	000110	110011	101101	010101	001011	111110	011000
010000	110110	000011	011101	100101	111011	001110	101000
001000	101110	011011	000101	111101	100011	010110	110000
000100	100010	010111	001001	110001	101111	011010	111100
000010	100100	010001	001111	110111	101001	011100	111010
000001	100111	010010	001100	110100	101010	011111	111001
010100	110010	000111	011001	100001	111111	001010	101100

From the table we find that the code words for the received words

$$r_1 = 101001 \qquad r_2 = 111010 \qquad r_3 = 001001 \qquad r_4 = 111011$$

are

$$c_1 = 101011 \qquad c_2 = 111000 \qquad c_3 = 001101 \qquad c_4 = 101011,$$

respectively. From these results the respective messages are

$$w_1 = 101 \qquad w_2 = 111 \qquad w_3 = 001 \qquad w_4 = 101.$$

The entries in the first column of Table 16.8 are called the *coset leaders*. For the first seven rows, the coset leaders are the same in all tables, with some permutations of rows possible. However, for the last row, either 100001 or 001010 could have been used in place of 010100 because they also have minimum weight 2. So the table need not be unique. [As a result, not all double errors can be corrected because there may not be a unique code word at a minimum distance for each r in the last coset (the one with coset leader 010100). For example, $r = 001010$ has three closest code words (at distance 2)—namely, 000000, 101011, and 011110.]

How do the coset leaders really help us? It seems that the code words in the first row are what we used to decode r_1, r_2, r_3, and r_4 above.

Consider the received words $r_1 = 101001$ and $r_2 = 111010$ in the sixth row, where the coset leader is $x = 000010$. Computing syndromes, we find that

$$H \cdot (r_1)^{\mathrm{tr}} = \begin{bmatrix} 0 \\ 1 \\ 0 \end{bmatrix} = H \cdot (r_2)^{\mathrm{tr}} = H \cdot x^{\mathrm{tr}}.$$

This is not just a coincidence.

THEOREM 16.16

Let $C \subseteq \mathbf{Z}_2^n$ be a group code for a parity-check matrix H, and let $r_1, r_2 \in \mathbf{Z}_2^n$. For the table of cosets of C in $\mathbf{Z}_2^n$, r_1 and r_2 are in the same coset of C if and only if $H \cdot (r_1)^{\mathrm{tr}} = H \cdot (r_2)^{\mathrm{tr}}$.

Proof: If r_1 and r_2 are in the same coset, then $r_1 = x + c_1$ and $r_2 = x + c_2$, where x is the coset leader, and c_1 and c_2 are the code words at the tops of the respective columns for r_1 and r_2. Then $H \cdot (r_1)^{\mathrm{tr}} = H \cdot (x + c_1)^{\mathrm{tr}} = H \cdot x^{\mathrm{tr}} + H \cdot c_1^{\mathrm{tr}} = H \cdot x^{\mathrm{tr}} + \mathbf{0} = H \cdot x^{\mathrm{tr}}$ because c_1 is a code word. Likewise, $H \cdot (r_2)^{\mathrm{tr}} = H \cdot x^{\mathrm{tr}}$, so r_1, r_2 have the same syndrome. Conversely, $H \cdot (r_1)^{\mathrm{tr}} = H \cdot (r_2)^{\mathrm{tr}} \Rightarrow H \cdot (r_1 + r_2)^{\mathrm{tr}} = \mathbf{0} \Rightarrow r_1 + r_2$ is a code word c. Hence $r_1 + r_2 = c$, so $r_1 = r_2 + c$ and $r_1 \in r_2 + C$. Since $r_2 \in r_2 + C$, we have r_1, r_2 in the same coset.

In decoding received words, when Table 16.8 is used we must search through 64 elements to find a given received word. For $C \subseteq \mathbf{Z}_2^{12}$ there are 4096 strings, each with 12 bits. Such a searching process is tedious, so perhaps we should be thinking about having a computer do the searching. Presently it appears that this means storing the entire table: $6 \times 64 = 384$ bits of storage for Table 16.8; $12 \times 4096 = 49,152$ bits for $C \subseteq \mathbf{Z}_2^{12}$. We should like to improve this situation. Before things get better, however, they'll look worse as we enlarge Table 16.8, as shown in Table 16.9. This new table includes to the left of the coset leaders (the transposes of) the syndromes for each row.

Table 16.9 Decoding Table 16.8 with Syndromes

000	000000	100110	010011	001101	110101	101011	011110	111000
110	100000	000110	110011	101101	010101	001011	111110	011000
011	010000	110110	000011	011101	100101	111011	001110	101000
101	001000	101110	011011	000101	111101	100011	010110	110000
100	000100	100010	010111	001001	110001	101111	011010	111100
010	000010	100100	010001	001111	110111	101001	011100	111010
001	000001	100111	010010	001100	110100	101010	011111	111001
111	010100	110010	000111	011001	100001	111111	001010	101100

Now we can decode a received word r by the following procedure.

1) Compute the syndrome $H \cdot r^{\mathrm{tr}}$.

2) Find the coset leader x to the right of $H \cdot r^{\mathrm{tr}}$.

3) Add x to r to get c. (The code word c that we are seeking at the top of the column containing r satisfies $c + x = r$, or $c = x + r$.)

Consequently, all that is needed from Table 16.9 are the first two columns, which will require $(3)(8) + (6)(8) = 72$ storage bits. With 18 more storage bits for H we can store what we need for this decoding process, called *decoding by coset leaders*, in 90 storage bits, as opposed to the original estimate of 384 bits.

Applying this procedure to $r = 110110$, we find the syndrome

$$H \cdot r^{\mathrm{tr}} = \begin{bmatrix} 0 \\ 1 \\ 1 \end{bmatrix}.$$

Since 011 is to the left of the coset leader $x = 010000$, the code word $c = x + r = 010000 + 110110 = 100110$, from which we recapture the original message, 100.

The code here is a group code where the minimum weight of the nonzero code words is 3, so we expected to be able to find a decoding scheme that corrected single errors. Here this is accomplished because the error patterns of weight 1 are all coset leaders. We cannot correct all double errors; only one error pattern of weight 2 is a coset leader. All error patterns of weight 1 or 2 would have to be coset leaders before our decoding scheme could correct both single and double errors in transmission.

Unlike the situation in Example 16.25, where syndromes were also used for decoding, things here are a bit different. Once we have a complete table listing all of the cosets of C in $\mathbf{Z}_2^6$, the process of decoding by coset leaders will give us an answer for *all* received words, not just for those that are code words or have syndromes that appear among the columns of the parity-check matrix H. However, we do realize that there is still a problem here because the last row of our table is not unique. Nonetheless, as our last result will affirm, this method provides a decoding scheme that is as good as any other.

THEOREM 16.17   When we are decoding by coset leaders, if $r \in \mathbf{Z}_2^n$ is a received word and r is decoded as the code word c^* (which we then decode to recapture the message), then $d(c^*, r) \leq d(c, r)$ for all code words c.

Proof: Let x be the coset leader for the coset containing r. Then $r = c^* + x$, or $r + c^* = x$, so $d(c^*, r) = \text{wt}(r + c^*) = \text{wt}(x)$. If c is any code word, then $d(c, r) = \text{wt}(c + r)$, and we have $c + r = c + (c^* + x) = (c + c^*) + x$. Since C is a group code, it follows that $c + c^* \in C$ and so $c + r$ is in the coset $x + C$. Among the elements in the coset $x + C$, the coset leader x is chosen to have minimum weight, so $\text{wt}(c + r) \geq \text{wt}(x)$. Consequently, $d(c^*, r) = \text{wt}(x) \leq \text{wt}(c + r) = d(c, r)$.

16.9
Hamming Matrices

We found the parity-check matrix H helpful in correcting single errors in transmission when (a) H had no column of 0's and (b) no two columns of H were the same. For the matrix

$$H = \begin{bmatrix} 1 & 1 & 0 & 1 & 1 & 0 & 0 \\ 1 & 0 & 1 & 1 & 0 & 1 & 0 \\ 0 & 1 & 1 & 1 & 0 & 0 & 1 \end{bmatrix}$$

we find that H satisfies these two conditions and that for the number of rows ($r = 3$) in H we have the maximum number of columns possible. If an additional column is added, H will no longer be useful for correcting single errors.

The generator matrix G associated with H is

$$G = \begin{bmatrix} 1 & 0 & 0 & 0 & 1 & 1 & 0 \\ 0 & 1 & 0 & 0 & 1 & 0 & 1 \\ 0 & 0 & 1 & 0 & 0 & 1 & 1 \\ 0 & 0 & 0 & 1 & 1 & 1 & 1 \end{bmatrix}.$$

Consequently we have a (7, 4) group code. The encoding function $E: \mathbf{Z}_2^4 \to \mathbf{Z}_2^7$ encodes four-bit messages into seven-bit code words. We realize that because H is determined by three parity-check equations, we have now maximized the number of bits we can have in the messages (under our present coding scheme). In addition, the columns of H, read from top to bottom, are the binary equivalents of the integers from 1 to 7.

In general, if we start with r parity-check equations, then the parity-check matrix H can have as many as $2^r - 1$ columns and still be used to correct single errors. Under these circumstances $H = [B \mid I_r]$, where B is an $r \times (2^r - 1 - r)$ matrix, and $G = [I_m \mid B^{tr}]$ with $m = 2^r - 1 - r$. The parity-check matrix H associated with a $(2^r - 1, 2^r - 1 - r)$ group code in this way is called a *Hamming matrix*, and the code is referred to as a *Hamming code*.

EXAMPLE 16.27

If $r = 4$, then $2^r - 1 = 15$ and $2^r - 1 - r = 11$. The one (up to a permutation of the columns) possible Hamming matrix H for $r = 4$ is

$$\begin{bmatrix} 1 & 1 & 1 & 1 & 1 & 1 & 1 & 0 & 0 & 0 & 0 & 1 & 0 & 0 & 0 \\ 1 & 1 & 1 & 1 & 0 & 0 & 0 & 1 & 1 & 1 & 0 & 0 & 1 & 0 & 0 \\ 1 & 1 & 0 & 0 & 1 & 1 & 0 & 1 & 1 & 0 & 1 & 0 & 0 & 1 & 0 \\ 1 & 0 & 1 & 0 & 1 & 0 & 1 & 1 & 0 & 1 & 1 & 0 & 0 & 0 & 1 \end{bmatrix}.$$

Once again, the columns of H contain the binary equivalents of the integers from 1 to 15 $(= 2^4 - 1)$.

This matrix H is the parity-check matrix of a Hamming $(15, 11)$ code whose rate is $11/15$.

With regard to the rate of these Hamming codes, for all $r \geq 2$, the rate m/n of such a code is given by $m/n = (2^r - 1 - r)/(2^r - 1) = 1 - [r/(2^r - 1)]$. As r increases, $r/(2^r - 1)$ goes to 0 and the rate approaches 1.

We close our discussion on coding theory with one final observation. In Section 16.7 we presented G (and H) in what is called the *systematic form*. Other arrangements of the rows and columns of these matrices are also possible, and these yield *equivalent codes*. (More on this can be found in the text by L. L. Dornhoff and F. E. Hohn [4].) We mention this here because it is often common practice to list the columns in a Hamming matrix of r rows so that the binary representations of the integers from 1 to $2^r - 1$ appear as the columns of H are read from left to right. For the Hamming $(7, 4)$ code, the matrix H mentioned at the start of this section would take the (equivalent) form

$$H_1 = \begin{bmatrix} 0 & 0 & 0 & 1 & 1 & 1 & 1 \\ 0 & 1 & 1 & 0 & 0 & 1 & 1 \\ 1 & 0 & 1 & 0 & 1 & 0 & 1 \end{bmatrix}.$$

Here the identity appears in the first, second, and fourth columns instead of in the last three. Consequently, we would use these components for the parity checks and find that if we send the message $w = w_1 w_2 w_3 w_4$, then the corresponding code word $E(w)$ is $c_1 c_2 w_1 c_3 w_2 w_3 w_4$, where

$$c_1 = w_1 + w_2 \qquad + w_4$$
$$c_2 = w_1 \qquad + w_3 + w_4$$
$$c_3 = \qquad w_2 + w_3 + w_4,$$

so that $H_1 \cdot (E(w))^{tr} = \mathbf{0}$.

In particular, if we send the message $w = w_1 w_2 w_3 w_4 = 1010$, the corresponding code word would be $E(w) = c = c_1 c_2 w_1 c_3 w_2 w_3 w_4 = c_1 c_2 1 c_3 010$, where $c_1 = w_1 + w_2 +$

$w_4 = 1 + 0 + 0 = 1$, $c_2 = w_1 + w_3 + w_4 = 1 + 1 + 0 = 0$, and $c_3 = w_2 + w_3 + w_4 = 0 + 1 + 0 = 1$. Then $c = 1011010$ and $H_1 \cdot (E(w))^{tr} = H_1 \cdot (E(1010))^{tr} = H_1 \cdot (1011010)^{tr} = \mathbf{0}$. (Verify this!) So if $c = 1011010$ is sent but $r = 1001010$ is received, we have $H_1 \cdot r^{tr} = H_1 \cdot (1001010)^{tr} = (011)^{tr}$. (Verify this as well!) Since 011 is the binary representation for 3 we know that the error is in position 3 — and this time we did *not* have to examine the columns of H_1. So using a parity-check matrix of the form H_1 simplifies syndrome decoding. In general, for $c = c_1 c_2 w_1 c_3 w_2 w_3 w_4$, let $r = c + e$, where e is an error pattern of weight 1. And suppose that the 1 in e is in position i, where $1 \le i \le 7$. Then the syndrome $H_1 \cdot r^{tr}$ provides the binary representation for i and we can determine c without examining the columns of H_1. From the third, fifth, sixth, and seventh components of c we can then recapture the original message w.

EXERCISES 16.8 and 16.9

1. Let $E: \mathbf{Z}_2^8 \to \mathbf{Z}_2^{12}$ be the encoding function for a code C. How many calculations are needed to find the minimum distance between code words? How many calculations are needed if E is a group homomorphism?

2. a) Use Table 16.9 to decode the following received words.

000011	100011	111110	100001
001100	011110	001111	111100

b) Do any of the results in part (a) change if a different set of coset leaders is used?

3. a) Construct a decoding table (with syndromes) for the group code given by the generator matrix

$$G = \begin{bmatrix} 1 & 0 & 1 & 1 & 0 \\ 0 & 1 & 0 & 1 & 1 \end{bmatrix}.$$

b) Use the table from part (a) to decode the following received words.

11110	11101	11011	10100
10011	10101	11111	01100

c) Does this code correct single errors in transmission?

4. Let

$$H = \begin{bmatrix} 1 & 1 & 0 & 1 & 1 & 0 & 0 \\ 1 & 0 & 1 & 1 & 0 & 1 & 0 \\ 0 & 1 & 1 & 1 & 0 & 0 & 1 \end{bmatrix}$$

be the parity-check matrix for a Hamming (7, 4) code.

a) Encode the following messages:

1000	1100	1011	1110	1001	1111.

b) Decode the following received words:

1100001	1110111	0010001	0011100.

c) Construct a decoding table consisting of the syndromes and coset leaders for this code.

d) Use the result in part (c) to decode the received words given in part (b).

5. a) What are the dimensions of the generator matrix for the Hamming (63, 57) code? What are the dimensions for the associated parity-check matrix H?

b) What is the rate of this code?

6. Compare the rates of the Hamming (7, 4) code and the (3, 1) triple-repetition code.

7. a) Let $p = 0.01$ be the probability of incorrect transmission for a binary symmetric channel. If the message 1011 is sent via the Hamming (7, 4) code, what is the probability of correct decoding?

b) Answer part (a) for a 20-bit message sent in five blocks of length 4.

16.10
Counting and Equivalence: Burnside's Theorem

In this section and the next two we shall develop a counting technique known as Polya's Method of Enumeration. Our development will not be very rigorous. Often we shall only state the general results of the theory as seen in the solution of a specific problem. Our first encounter with the type of problem to which this counting technique applies is presented in the following example.

EXAMPLE 16.28

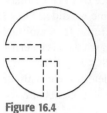

Figure 16.4

We have a set of sticks, all of the same length and color, and a second set of round plastic disks. Each disk contains two holes, as shown in Fig. 16.4, into which the sticks can be inserted in order to form different shapes, such as a square. (See Fig. 16.5.) If each disk is either red or white, how many distinct squares can we form?

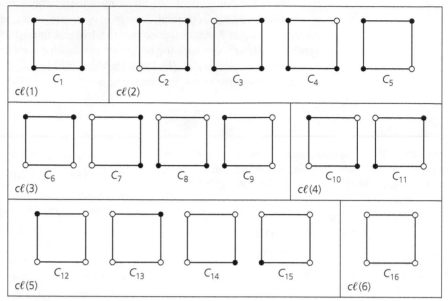

Figure 16.5

If the square is considered stationary, then the four disks are located at four distinct locations; a red or white disk is used at each location. Thus there are $2^4 = 16$ different configurations, as shown in Fig. 16.5, where a dark circle indicates a red disk. The configurations have been split into six classes, $c\ell(1)$, $c\ell(2)$, ..., $c\ell(6)$, according to the number and relative location of the red disks.

Now suppose that the square is not fixed but can be moved about in space. Unless the vertices (disks) are marked somehow, certain configurations in Fig. 16.5 are indistinguishable when we move them about.

To place these notions in a more mathematical setting, we use the nonabelian group of three-dimensional rigid motions of a square to define an equivalence relation on the configurations in Fig. 16.5. Since this group will be used throughout this section and the next two sections, we now give a detailed description of its elements.

In Fig. 16.6 we have the group $G = \{\pi_0, \pi_1, \pi_2, \pi_3, r_1, r_2, r_3, r_4\}$ for the rigid motions of the square in part (a), where we have labeled the vertices with 1, 2, 3, and 4. Parts (b) through (i) of the figure show how each element of G is applied. We have expressed each group element as a permutation of $\{1, 2, 3, 4\}$ and in a new form called a *product of disjoint cycles*. For example, in part (b) we find $\pi_1 = (1234)$. The cycle (1234) indicates that if we start with the square in part (a), after applying π_1, we find that 1 has moved to the position originally occupied by 2, 2 to that of 3, 3 to that of 4, and 4 to that of 1. In general, if xy appears in a cycle, then x moves to the position originally occupied by y. Also, for a cycle where x and y appear as $(x \ldots y)$, y moves to the position originally occupied by x when the motion described by this cycle is applied. Note that $(1234) = (2341) = (3412) = (4123)$. We say that each of these cycles has *length* 4, the number of elements in the cycle. In the case of r_1 in part (f) of the figure, starting with 1 we find that r_1 sends 1 to 4, so we have

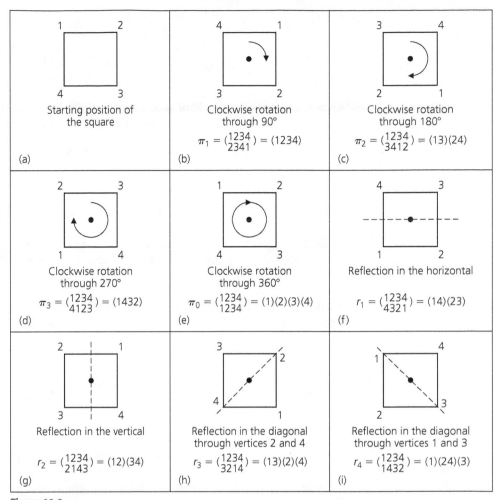

Figure 16.6

$(14 \ldots)$ as the start of our first cycle in this *decomposition* of r_1. However, here r_1 sends 4 to 1, so we have completed a portion—namely, (14)—of the complete decomposition. We then select a vertex that has not yet appeared—for example, vertex 2. Since r_1 sends 2 to 3 and 3, in turn, to 2, we get a second cycle (23). This exhausts all vertices and so $(14)(23) = r_1$, where the cycles (14) and (23) have no vertex in common. Here $(14)(23) = (23)(14) = (23)(41) = (32)(41)$ all provide a representation of r_1 as a product of disjoint cycles, each of length 2. Last, for the group element $r_3 = (13)(2)(4)$, the cycle (2) indicates that 2 is fixed, or *invariant*, under the permutation r_3. When the number of vertices involved is known, the permutation r_3 may also be written as $r_3 = (13)$, where the missing elements are understood to be fixed. However, we shall write all of the cycles in our decompositions, for this will be useful later in our discussion.

Before continuing with the main discussion concerning the disks and sticks, let us examine some further results on disjoint cycles.

In the group S_6 of all permutations of $\{1, 2, 3, 4, 5, 6\}$, let $\pi = \begin{pmatrix} 1 & 2 & 3 & 4 & 5 & 6 \\ 2 & 3 & 1 & 4 & 6 & 5 \end{pmatrix}$. As a product of disjoint cycles,

$$\pi = (123)(4)(56) = (56)(4)(123) = (4)(231)(65).$$

If $\sigma \in S_6$, with $\sigma = \begin{pmatrix} 1 & 2 & 3 & 4 & 5 & 6 \\ 2 & 4 & 5 & 1 & 6 & 3 \end{pmatrix}$, then

$$\sigma = (124)(356) = \begin{pmatrix} 1 & 2 & 3 & 4 & 5 & 6 \\ 2 & 4 & 3 & 1 & 5 & 6 \end{pmatrix} \begin{pmatrix} 1 & 2 & 3 & 4 & 5 & 6 \\ 1 & 2 & 5 & 4 & 6 & 3 \end{pmatrix},$$

so each cycle can be thought of as an element of S_6.

Finally, if $\alpha = (124)(3)(56)$ and $\beta = (13)(245)(6)$ are elements of S_6, then

$$\alpha\beta = (124)(3)(56)(13)(245)(6) = (143)(256),$$

whereas

$$\beta\alpha = (13)(245)(6)(124)(3)(56) = (132)(465).$$

Returning to the 16 configurations, or colorings, in Fig. 16.5, we now examine how each element in the group G, in Fig. 16.6, acts upon these configurations. For example, $\pi_1 = \begin{pmatrix} 1 & 2 & 3 & 4 \\ 2 & 3 & 4 & 1 \end{pmatrix}$ permutes the numbers $\{1, 2, 3, 4\}$ according to a 90° clockwise rotation for the square in Fig. 16.6(a), yielding the result in Fig. 16.6(b). How does such a rotation act on $S = \{C_1, C_2, \ldots, C_{16}\}$, our set of colorings? We use π_1^* to distinguish between the 90° clockwise rotation for $\{1, 2, 3, 4\}$ and the same rotation when applied to $S = \{C_1, C_2, \ldots, C_{16}\}$. We find that

$$\pi_1^* = \begin{pmatrix} C_1 & C_2 & C_3 & C_4 & C_5 & C_6 & C_7 & C_8 & C_9 & C_{10} & C_{11} & C_{12} & C_{13} & C_{14} & C_{15} & C_{16} \\ C_1 & C_3 & C_4 & C_5 & C_2 & C_7 & C_8 & C_9 & C_6 & C_{11} & C_{10} & C_{13} & C_{14} & C_{15} & C_{12} & C_{16} \end{pmatrix}.$$

As a product of disjoint cycles,

$$\pi_1^* = (C_1)(C_2C_3C_4C_5)(C_6C_7C_8C_9)(C_{10}C_{11})(C_{12}C_{13}C_{14}C_{15})(C_{16}).$$

We note that under the action of π_1^*, no configuration is changed into one that is in another class.

As a second example, consider the reflection r_3 in Fig. 16.6(h). The action of this rigid motion on S is given by

$$r_3^* = \begin{pmatrix} C_1 & C_2 & C_3 & C_4 & C_5 & C_6 & C_7 & C_8 & C_9 & C_{10} & C_{11} & C_{12} & C_{13} & C_{14} & C_{15} & C_{16} \\ C_1 & C_2 & C_5 & C_4 & C_3 & C_7 & C_6 & C_9 & C_8 & C_{10} & C_{11} & C_{14} & C_{13} & C_{12} & C_{15} & C_{16} \end{pmatrix}$$

$$= (C_1)(C_2)(C_3C_5)(C_4)(C_6C_7)(C_8C_9)(C_{10})(C_{11})(C_{12}C_{14})(C_{13})(C_{15})(C_{16}).$$

Once again, no configuration is taken by r_3^* into one that is outside the class that it was in originally.

Using the idea of *the group G acting on the set S*, we define a relation $\mathcal{R}$ on S as follows. For colorings $C_i, C_j \in S$, where $1 \le i, j \le 16$, we write $C_i \mathcal{R} C_j$ if there is a permutation $\sigma \in G$ such that $\sigma^*(C_i) = C_j$. That is, as σ^* acts on the 16 configurations in S, C_i is transformed into C_j. This relation $\mathcal{R}$ is an equivalence relation, as we now verify.

a) (Reflexive Property) For all $C_i \in S$, where $1 \le i \le 16$, it follows that $C_i \mathcal{R} C_i$ because G contains the identity permutation. $[\pi_0^*(C_i) = C_i$ for all $1 \le i \le 16.]$

b) (Symmetric Property) If $C_i \mathcal{R} C_j$ for $C_i, C_j \in S$, then $\sigma^*(C_i) = C_j$, for some $\sigma \in G$. G is a group, so $\sigma^{-1} \in G$, and we find that $(\sigma^*)^{-1} = (\sigma^{-1})^*$. (Verify this for two choices of $\sigma \in G$.) Hence $C_i = (\sigma^{-1})^*(C_j)$, and $C_j \mathcal{R} C_i$.

c) (Transitive Property) Let $C_i, C_j, C_k \in S$ with $C_i \, \mathcal{R} \, C_j$ and $C_j \, \mathcal{R} \, C_k$. Then $C_j = \sigma^*(C_i)$ and $C_k = \tau^*(C_j)$, for some $\sigma, \tau \in G$. By closure in G, $\sigma\tau \in G$, and we find that $(\sigma\tau)^* = \sigma^*\tau^*$, where σ is applied first in $\sigma\tau$ and σ^* first in $\sigma^*\tau^*$. (Verify this for two specific permutations $\sigma, \tau \in G$.) Then $C_k = (\sigma\tau)^*(C_i)$ and $\mathcal{R}$ is transitive. [The reader may have noticed that $C_k = \tau^*(C_j) = \tau^*(\sigma^*(C_i))$ and felt that we should have written $(\sigma\tau)^* = \tau^*\sigma^*$. Once again, there has been a change in the notation for the composite function as we first defined it in Chapter 5. Here we write $\sigma^*\tau^*$ for $(\sigma\tau)^*$, and σ^* is applied first.]

Since $\mathcal{R}$ is an equivalence relation on S, $\mathcal{R}$ partitions S into equivalence classes, which are precisely the classes $c\ell(1), c\ell(2), \ldots, c\ell(6)$ of Fig. 16.5. Consequently, there are six nonequivalent configurations under the group action. So among the original 16 colorings only 6 are really distinct.

What has happened in this example generalizes as follows. With S a set of configurations, let G be a group (of permutations) that acts on S. If the relation $\mathcal{R}$ is defined on S by $x \, \mathcal{R} \, y$ if $\pi^*(x) = y$, for some $\pi \in G$, then $\mathcal{R}$ is an equivalence relation.

With only red and white disks to connect the sticks, the answer to this example could have been determined from the results in Fig. 16.5. However, we developed quite a bit of mathematical overkill to answer the question. Referring to S as the set of 2-colorings of the vertices of a square, we start to wonder about the role of 2 and seek the number of nonequivalent configurations if the disks come in three or more colors.

In addition, we might notice that the function $f(r, w) = r^4 + r^3 w + 2r^2 w^2 + r w^3 + w^4$ is the generating function (of two variables) for the number of nonequivalent configurations from S. Here the coefficient of $r^i w^{4-i}$, for $0 \le i \le 4$, yields the number of distinct 2-colorings that have i red disks and $(4 - i)$ white ones. The coefficient of $r^2 w^2$ is 2 because of the two equivalence classes $c\ell(3)$ and $c\ell(4)$. Finally, $f(1, 1) = 6$, the number of equivalence classes. This generating function $f(r, w)$ is called the *pattern inventory* for the configurations. We shall examine it in more detail in the next two sections.

For now we record an extended version of our present results in the following theorem. (A proof of this result is given on pages 136–137 of C. L. Liu [17].)

THEOREM 16.18

Burnside's Theorem. Let S be a set of configurations on which a finite group G of permutations acts. The number of equivalence classes into which S is partitioned by the action of G is then given by

$$\frac{1}{|G|} \sum_{\pi \in G} \psi(\pi^*),$$

where $\psi(\pi^*)$ is the number of configurations in S fixed under π^*.

To better accept the validity of this theorem, we first examine two examples where we already know the answers.

EXAMPLE 16.29

In Example 16.28 we find that $\psi(\pi_1^*) = 2$ because only C_1 and C_{16} are fixed, or *invariant*, under π_1^*. For $r_3 \in G$, however, $\psi(r_3^*) = 8$ because $C_1, C_2, C_4, C_{10}, C_{11}, C_{13}, C_{15}$, and C_{16} remain fixed under this group action. In like manner $\psi(\pi_2^*) = 4$, $\psi(\pi_3^*) = 2$, $\psi(\pi_0^*) = 16$,

$\psi(r_1^*) = \psi(r_2^*) = 4$, and $\psi(r_4^*) = 8$. With $|G| = 8$, Burnside's Theorem implies that the number of equivalence classes, or nonequivalent configurations, is

$$(1/8)(16 + 2 + 4 + 2 + 4 + 4 + 8 + 8) = (1/8)(48) = 6,$$

the original answer.

EXAMPLE 16.30

In how many ways can six people be arranged around a circular table if two arrangements are considered equivalent when one can be obtained from the other by means of a clockwise rotation through $i \cdot 60°$, for $0 \le i \le 5$?

Here the six distinct people are to be placed in six chairs located at a table, as shown in Fig. 16.7. Our permutation group G consists of the clockwise rotations π_i through $i \cdot 60°$, where $0 \le i \le 5$. Here reflections are not meaningful. The situation is two-dimensional, for we can rotate the circle (representing the table) only in the plane; the circle never lifts off the plane. The total number of possible configurations is 6! We find that $\psi(\pi_0^*) = 6!$ and that $\psi(\pi_i^*) = 0$, for $1 \le i \le 5$. (It's impossible to move different people and simultaneously have them stay in a fixed location.)

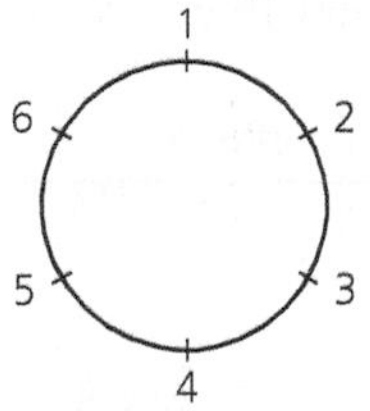

Figure 16.7

Consequently, the total number of nonequivalent seating arrangements is

$$\left(\frac{1}{|G|}\right) \sum_{\sigma \in G} \psi(\sigma^*) = \left(\frac{1}{6}\right)(6! + 0 + 0 + 0 + 0 + 0) = 5!,$$

as we found in Example 1.16 of Chapter 1.

We now examine a situation where the power of this theorem is made apparent.

EXAMPLE 16.31

In how many ways can the vertices of a square be 3-colored, if the square can be moved about in three dimensions?

Now we have the sticks of Example 16.28, along with red, white, and blue disks. Considering the group in Fig. 16.6, we find the following:

$\psi(\pi_0^*) = 3^4$, because the identity fixes all 81 configurations in the set S of possible configurations.

$\psi(\pi_1^*) = \psi(\pi_3^*) = 3$, for each of π_1^*, π_3^* leaves invariant only those configurations with all vertices the same color.

$\psi(\pi_2^*) = 9$, for π_2^* can fix only those configurations where the opposite (diagonally) vertices have the same color. Consider a square like the one shown in Fig. 16.8. There are three choices for placing a colored disk at vertex 1 and then one choice for matching it at vertex 3. Likewise, there are three choices for colors at vertex 2 and then one for vertex 4. Consequently, there are nine configurations invariant under π_2^*.

$\psi(r_1^*) = \psi(r_2^*) = 9$. In the case of r_1^*, for the square shown in Fig. 16.8 we have three choices for coloring each of the vertices 1 and 2, and then we must match the color of vertex 4 with the color of vertex 1, and the color of vertex 3 with that of vertex 2.

Finally, $\psi(r_3^*) = \psi(r_4^*) = 27$. For r_3^*, we have nine choices for coloring the two vertices at 2 and 4, and three choices for vertex 1. Then there is only one choice for vertex 3 because we must match the color of vertex 1.

By Burnside's Theorem, the number of nonequivalent configurations is

$$(1/8)(3^4 + 3 + 3^2 + 3 + 3^2 + 3^2 + 3^3 + 3^3) = 21.$$

Figure 16.8

EXERCISES 16.10

1. Consider the configurations shown in Fig. 16.5.

   a) Determine π_2^*, π_3^*, r_2^*, and r_4^*.

   b) Verify that $(\pi_1^{-1})^* = (\pi_1^*)^{-1}$ and $(r_3^{-1})^* = (r_3^*)^{-1}$.

   c) Verify that $(\pi_1 r_1)^* = \pi_1^* r_1^*$ and $(\pi_3 r_4)^* = \pi_3^* r_4^*$.

2. Express each of the following elements of S_7 as a product of disjoint cycles.

$$\alpha = \begin{pmatrix} 1 & 2 & 3 & 4 & 5 & 6 & 7 \\ 2 & 4 & 6 & 7 & 1 & 5 & 3 \end{pmatrix}$$

$$\beta = \begin{pmatrix} 1 & 2 & 3 & 4 & 5 & 6 & 7 \\ 3 & 6 & 5 & 2 & 1 & 7 & 4 \end{pmatrix}$$

$$\gamma = \begin{pmatrix} 1 & 2 & 3 & 4 & 5 & 6 & 7 \\ 2 & 3 & 1 & 7 & 5 & 4 & 6 \end{pmatrix}$$

$$\delta = \begin{pmatrix} 1 & 2 & 3 & 4 & 5 & 6 & 7 \\ 4 & 2 & 7 & 1 & 3 & 6 & 5 \end{pmatrix}$$

3. **a)** Determine the order of each of the elements in Exercise 2.

   b) State a general result about the order of an element in S_n in terms of the lengths of the cycles in its decomposition as a product of disjoint cycles.

4. **a)** Determine the number of distinct ways one can color the vertices of an equilateral triangle using the colors red and white, if the triangle is free to move in three dimensions.

   b) Answer part (a) if the color blue is also available.

5. Answer the questions in Exercise 4 for a regular pentagon.

6. **a)** How many distinct ways are there to paint the *edges* of a square with three different colors?

   b) Answer part (a) for the edges of a regular pentagon.

7. We make a child's bracelet by symmetrically placing four beads about a circular wire. The colors of the beads are red, white, blue, and green, and there are at least four beads of each color. (a) How many distinct bracelets can we make in this way, if the bracelets can be rotated but not reflected? (b) Answer part (a) if the bracelets can be rotated and reflected.

8. A baton is painted with three cylindrical bands of color (not necessarily distinct), with each band of the same length.

a) How many distinct paintings can be made if there are three colors of paint available? How many for four colors?

b) Answer part (a) for batons with four cylindrical bands.

c) Answer part (a) for batons with n cylindrical bands.

d) Answer parts (a) and (b) if adjacent cylindrical bands are to have different colors.

9. In how many ways can we 2-color the vertices of the configurations shown in Fig. 16.9 if they are free to move in (a) two dimensions? (b) three dimensions?

Figure 16.9

10. A pyramid has a square base and four faces that are equilateral triangles. If we can move the pyramid about (in three dimensions), how many nonequivalent ways are there to paint its five faces if we have paint of four different colors? How many if the color of the base must be different from the color(s) of the triangular faces?

11. **a)** In how many ways can we paint the cells of a 3×3 chessboard using red and blue paint? (The back of the chessboard is black.)

   b) In how many ways can we construct a 3×3 chessboard by joining (with paste) the edges of nine 1×1 plastic squares that are transparent and tinted red or blue? (There are nine squares of each color available.)

12. Answer Exercise 11 for a 4×4 chessboard. [Replace each "nine" in part (b) with "sixteen."]

13. In how many ways can we paint the seven (identical) horses on a carousel using black, brown, and white paint?

14. **a)** Let S be a set of configurations and G a group of permutations that acts on S. If $x \in S$, prove that $\{\pi \in G \mid \pi^*(x) = x\}$ is a subgroup of G (called the *stabilizer* of x).

   b) Determine the respective stabilizer subgroups in part (a) for each of the configurations C_7 and C_{15} in Fig. 16.5.

16.11
The Cycle Index

In applying Burnside's Theorem we have been faced with computing $\psi(\pi^*)$ for each $\pi \in G$, where G is a permutation group acting on a set S of configurations. As the number of available colors increases and the configurations get more complex, such computations can get a bit involved. In addition, it seems that if we can determine the number of 2-colorings for a set S of configurations, we should be able to use some of the work in this case to determine the number of 3-colorings, 4-colorings, and so on. We shall now find some assistance as we return to the solution of Example 16.28. This time more attention will be paid to the representation of each permutation $\pi \in G$ as a product of disjoint cycles. Our results are summarized in Table 16.10.

Table 16.10

Rigid Motions π (Elements of G)	Configurations in S that Are Invariant under π^*	Cycle Structure Representation of π	Inventory of Configurations that Are Invariant under π^*
$\pi_0 = (1)(2)(3)(4)$	2^4: All configurations in S	x_1^4	$(r+w)^4 \qquad = r^4 + 4r^3w + 6r^2w^2 + 4rw^3 + w^4$
$\pi_1 = (1234)$	$2: C_1, C_{16}$	x_4	$r^4 + w^4 \qquad = r^4 \qquad\qquad\qquad + w^4$
$\pi_2 = (13)(24)$	$2^2: C_1, C_{10}, C_{11}, C_{16}$	x_2^2	$(r^2+w^2)^2 \qquad = r^4 \qquad + 2r^2w^2 \qquad + w^4$
$\pi_3 = (1432)$	$2: C_1, C_{16}$	x_4	$r^4 + w^4 \qquad = r^4 \qquad\qquad\qquad + w^4$
$r_1 = (14)(23)$	$2^2: C_1, C_7, C_9, C_{16}$	x_2^2	$(r^2+w^2)^2 \qquad = r^4 \qquad + 2r^2w^2 \qquad + w^4$
$r_2 = (12)(34)$	$2^2: C_1, C_6, C_8, C_{16}$	x_2^2	$(r^2+w^2)^2 \qquad = r^4 \qquad + 2r^2w^2 \qquad + w^4$
$r_3 = (13)(2)(4)$	$2^3: C_1, C_2, C_4, C_{10},$ $C_{11}, C_{12}, C_{15}, C_{16}$	$x_2 x_1^2$	$(r^2+w^2)(r+w)^2 = r^4 + 2r^3w + 2r^2w^2 + 2rw^3 + w^4$
$r_4 = (1)(24)(3)$	$2^3: C_1, C_3, C_5, C_{10},$ $C_{11}, C_{12}, C_{14}, C_{16}$	$x_2 x_1^2$	$(r^2+w^2)(r+w)^2 = r^4 + 2r^3w + 2r^2w^2 + 2rw^3 + w^4$
	$P_G(x_1, x_2, x_3, x_4) =$ $\frac{1}{8}(x_1^4 + 2x_4 + 3x_2^2 + 2x_2 x_1^2)$	Complete Inventory $\Big\}$	$= 8r^4 + 8r^3w + 16r^2w^2 + 8rw^3 + 8w^4$

For π_0, the identity of G, we write $\pi_0 = (1)(2)(3)(4)$, a product of four disjoint cycles. We shall represent this cycle structure algebraically by x_1^4, where x_1 indicates a cycle of length 1. The term x_1^4 is called the *cycle structure representation* of π_0. Here we interpret "disjoint" as "independent," in the sense that whatever color is used to paint the vertices in one cycle has no bearing on the choice of color for the vertices in another cycle. As long as all the vertices in a given cycle have the same color, we shall find configurations that are invariant under π_0^*. (Admittedly, this seems like mathematical overkill again, inasmuch as π_0^* fixes all 2-colorings of the square.) In addition, since we can paint the vertices in each cycle either red or white, we have 2^4 configurations, and we find that $(r+w)^4 = r^4 + 4r^3w + 6r^2w^2 + 4rw^3 + w^4$ *generates* these 16 configurations. For example, from the term $6r^2w^2$ we find that there are six configurations with two red and two white vertices, as found in classes $c\ell(3)$ and $c\ell(4)$ of Fig. 16.5.

Turning to π_1, we find $\pi_1 = (1234)$, a cycle of length 4. This cycle structure is represented by x_4, and here there are only two invariant configurations. The fact that the cycle structure for π_1 has only one cycle tells us that for a configuration to be invariant under π_1^*, every vertex in this cycle must be painted the same color. With two colors to choose from, there are only two possible configurations, C_1 and C_{16}. In this case the term $r^4 + w^4$ generates these configurations.

Continuing with r_1, we have $r_1 = (14)(23)$, a product of two disjoint cycles of length 2; the term x_2^2 represents this cycle structure. For a configuration to be invariant under r_1^*, the vertices at 2 and 3 must be the same color; that is, we have two choices for coloring the vertices in (23). We also have two choices for coloring the vertices in (14). Consequently, we get 2^2 invariant configurations: $C_1(r^4)$, $C_7(r^2w^2)$, $C_9(r^2w^2)$, and $C_{16}(w^4)$. [$(r^2 + w^2)^2 = r^4 + 2r^2w^2 + w^4$.]

Finally, in the case of $r_3 = (13)(2)(4)$, we find that $x_2x_1^2$ indicates its decomposition into one cycle of length 2 and two of length 1. The vertices at 1 and 3 must be painted the same color if the configuration is to be invariant under r_3^*. With three cycles and two choices of color for each cycle, we find 2^3 invariant configurations. They are $C_1(r^4)$, $C_2(r^3w)$, $C_4(r^3w)$, $C_{10}(r^2w^2)$, $C_{11}(r^2w^2)$, $C_{13}(rw^3)$, $C_{15}(rw^3)$, and $C_{16}(w^4)$. These configurations are generated by $(r^2 + w^2)(r + w)^2$, for when we consider the cycle (13) we have two choices: both vertices red (r^2) *or* both vertices white (w^2). This gives us $r^2 + w^2$. For each single vertex in the two cycles of length 1, $r + w$ provides the choices for each cycle, $(r + w)^2$ the choices for the two. By the independence of choice of colors as we go from one cycle to another, $(r^2 + w^2)(r + w)^2$ generates the 2^3 configurations that are invariant under r_3^*.

Similar arguments provide the information in Table 16.10 for the permutations π_2, π_3, r_2, and r_4.

At this point we see that what determines the number of configurations that are invariant under π^*, for $\pi \in G$, depends on the cycle structure of π. Within each cycle the same color must be used, but that color can be selected from the two or more choices made available. For r_1, we had two cycles (of length 2) and 2^2 configurations. If three colors had been available, the number of invariant configurations would have been 3^2. For m colors, the number is m^2. Adding these terms for all the cycle structures that arise gives $\sum_{\pi \in G} \psi(\pi^*)$.

We now wish to place more emphasis on cycle structures, so we define the *cycle index*, P_G, for the *group* G (of permutations) as

$$P_G(x_1, x_2, x_3, x_4) = \frac{1}{|G|} \sum_{\pi \in G} (\text{cycle structure representation of } \pi).$$

In this example,

$$P_G(x_1, x_2, x_3, x_4) = (1/8)(x_1^4 + 2x_4 + 3x_2^2 + 2x_2x_1^2).$$

When each occurrence of x_1, x_2, x_3, x_4 is replaced by 2, we find that the number of non-equivalent 2-colorings is equal to

$$P_G(2, 2, 2, 2) = (1/8)(2^4 + 2(2) + 3(2^2) + 2(2)(2^2)) = 6.$$

We summarize our present findings in the following result.

THEOREM 16.19   Let S be a set of configurations that are acted upon by a permutation group G. [G is a subgroup of S_n, the group of all permutations of $\{1, 2, 3, \ldots, n\}$, and the cycle index $P_G(x_1, x_2, x_3, \ldots, x_n)$ of G is

$$(1/|G|) \sum_{\pi \in G} (\text{cycle structure representation of } \pi).]$$

The number of nonequivalent m-colorings of S is then $P_G(m, m, m, \ldots, m)$.

We close this section with an example that uses this theorem.

EXAMPLE 16.32   In how many distinct ways can we 4-color the vertices of a regular hexagon that is free to move in space?

For a regular hexagon there are twelve rigid motions: (a) the six clockwise rotations through 0°, 60°, 120°, 180°, 240°, and 300°; (b) the three reflections in diagonals through opposite vertices; and (c) the three reflections about lines passing through the midpoints of opposite edges.

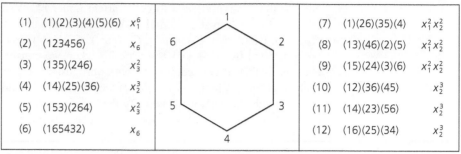

(1)	(1)(2)(3)(4)(5)(6)	x_1^6	(7)	(1)(26)(35)(4)	$x_1^2 x_2^2$
(2)	(123456)	x_6	(8)	(13)(46)(2)(5)	$x_1^2 x_2^2$
(3)	(135)(246)	x_3^2	(9)	(15)(24)(3)(6)	$x_1^2 x_2^2$
(4)	(14)(25)(36)	x_2^3	(10)	(12)(36)(45)	x_2^3
(5)	(153)(264)	x_3^2	(11)	(14)(23)(56)	x_2^3
(6)	(165432)	x_6	(12)	(16)(25)(34)	x_2^3

Figure 16.10

In Fig. 16.10 we have listed each group element as a product of disjoint cycles, together with its cycle structure representation. Here

$$P_G(x_1, x_2, x_3, x_4, x_5, x_6) = (1/12)(x_1^6 + 2x_6 + 2x_3^2 + 4x_2^3 + 3x_1^2 x_2^2),$$

and there are

$$P_G(4, 4, 4, 4, 4, 4) = (1/12)(4^6 + 2(4) + 2(4^2) + 4(4^3) + 3(4^2)(4^2)) = 430$$

nonequivalent 4-colorings of a regular hexagon. (*Note:* Even though neither x_4 nor x_5 occurs in a cycle structure representation, we may list these variables among the arguments of P_G.)

EXERCISES 16.11

1. In how many ways can we 5-color the vertices of a square that is free to move in (a) two dimensions? (b) three dimensions?

2. Answer Exercise 1 for a regular pentagon.

3. Find the number of nonequivalent 4-colorings of the vertices in the configurations shown in Fig. 16.11 when they are free to move in (a) two dimensions; (b) three dimensions.

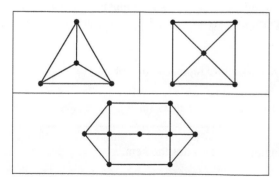

Figure 16.11

4. a) In how many ways can we 3-color the vertices of a regular hexagon that is free to move in space?

b) Give a combinatorial argument to show that for all $m \in \mathbf{Z}^+$, $(m^6 + 2m + 2m^2 + 4m^3 + 3m^4)$ is divisible by 12.

5. a) In how many ways can we 5-color the vertices of a regular hexagon that is free to move in two dimensions?

b) Answer part (a) if the hexagon is free to move in three dimensions.

c) Find two 5-colorings that are equivalent for case (b) but distinct for case (a).

6. In how many distinct ways can we 3-color the *edges* in the configurations shown in Fig. 16.11 if they are free to move in (a) two dimensions; (b) three dimensions?

7. a) In how many distinct ways can we 3-color the *edges* of a square that is free to move in three dimensions?

b) In how many distinct ways can we 3-color both the vertices and the edges of such a square?

c) For a square that can move in three dimensions, let k, m, and n denote the number of distinct ways in which we can 3-color its vertices (alone), its edges (alone), and both its vertices and edges, respectively. Does $n = km$? (Give a geometric explanation.)

16.12
The Pattern Inventory:
Polya's Method of Enumeration

In this final section we return to Example 16.28 and its continued analysis in Section 16.11. At this time we introduce the pattern inventory and how it is derived from the cycle index.

For $\pi_0 \in G$, every configuration in S is invariant. The cycle structure (representation) for π_0 is given by x_1^4, where for each cycle of length 1 we have a choice of coloring the vertex in that cycle red (r) or white (w). Using $+$ to represent *exclusive or*, we write $r + w$ to denote the two choices for that vertex (cycle of length 1). With four such cycles, $(r + w)^4$ generates the patterns of the 16 configurations.

In the case of $\pi_1 = (1234)$, x_4 denotes the cycle structure, and here all four vertices must be the same color for the configuration to remain fixed under π_1^*. Consequently, we have all four vertices red or all four vertices white, and we express this algebraically by $r^4 + w^4$.

At this point we notice that for each of the permutations we have considered, the number of factors in the expression used to generate the patterns fixed under a certain permutation equals the number of factors in the cycle structure (representation) of that permutation. Is this just a coincidence?

Continue now with $r_1 = (14)(23)$, whose cycle structure is x_2^2. For the cycle (14) we must color both of the vertices 1 and 4 either red or white. These choices are represented by $r^2 + w^2$. Since there are two such cycles of length 2, we find that $(r^2 + w^2)^2$ will generate the patterns of the configurations in S fixed under r_1^*. Once again the number of factors in the cycle structure equals the number of factors in the corresponding term used to generate the patterns.

Last, for $r_3 = (13)(2)(4)$, the cycle structure is $x_2 x_1^2 = x_1^2 x_2$. For each of the cycles (2) and (4), $r + w$ represents the choices for each of these vertices, so that $(r + w)^2$ accounts for all four colorings of the pair. The cycle (13) indicates that vertices 1 and 3 must have the same color; $r^2 + w^2$ accounts for the two possibilities. Therefore, $(r + w)^2(r^2 + w^2)$ generates the patterns of the configurations in S fixed under r_3^*, and we find three factors in both the cycle structure and the product $(r + w)^2(r^2 + w^2)$. But even more comes to light here.

Looking at the terms in the cycle structures, we see that, for $1 \leq i \leq n$, the factor x_i in the cycle structure corresponds with the term $r^i + w^i$ in the expression used to generate the patterns.

Continuing with the cycle structures for π_2, π_3, r_2, and r_4, we find that the *pattern inventory* can be obtained by replacing each x_i in $P_G(x_1, x_2, x_3, x_4)$ with $r^i + w^i$, for $1 \leq i \leq 4$. Consequently,

$$P_G(r + w, r^2 + w^2, r^3 + w^3, r^4 + w^4) = r^4 + r^3 w + 2r^2 w^2 + r w^3 + w^4.$$

(This result is (1/8)-th of the complete inventory listed in Table 16.10.)

If we had three colors (red, white, and blue), the replacement for x_i would be $r^i + w^i + b^i$, where $1 \leq i \leq 4$.

We generalize these observations in the following theorem.

THEOREM 16.20   *Polya's Method of Enumeration.* Let S be a set of configurations that are acted upon by a permutation group G, where G is a subgroup of S_n and G has cycle index $P_G(x_1, x_2, \ldots, x_n)$.

Then the pattern inventory of nonequivalent m-colorings of S is given by

$$P_G\left(\sum_{i=1}^{m} c_i, \sum_{i=1}^{m} c_i^2, \ldots, \sum_{i=1}^{m} c_i^n\right),$$

where $c_1, c_2, \ldots, c_m$ denote the m colors that are available.

One important point should be reiterated here before applying Theorem 16.20 — namely, the pattern inventory is another example of a generating function. Having made that point, we now apply this theorem in the following examples.

EXAMPLE 16.33

A child's bracelet is formed by placing three beads — red, white, and blue — on a circular piece of wire. Bracelets are considered equivalent if one can be obtained from the other by a (planar) rotation. Find the pattern inventory for these bracelets.

Here G is the group of rotations of an equilateral triangle, so $G = \{(1)(2)(3), (123), (132)\}$, where 1, 2, 3 denote the vertices of the triangle. Then $P_G(x_1, x_2, x_3) = (1/3) \cdot (x_1^3 + 2x_3)$, and the pattern inventory is given by $(1/3)[(r + w + b)^3 + 2(r^3 + w^3 + b^3)] = (1/3)[3r^3 + 3r^2w + 3r^2b + 3rw^2 + 6rwb + 3rb^2 + 3w^3 + 3w^2b + 3wb^2 + 3b^3] = r^3 + r^2w + r^2b + rw^2 + 2rwb + rb^2 + w^3 + w^2b + wb^2 + b^3$. We interpret this result as follows:

1) For each summand, other than $2rwb$, the coefficient is 1 because there is only one (distinct) bracelet of that type. That is, there is one bracelet with three red beads (for r^3), one with two red beads and one white bead (for r^2w), and so on for the other seven summands with coefficient 1.

2) The summand $2rwb$ has coefficient 2 because there are two nonequivalent bracelets with one red, one white, and one blue bead — as shown in Fig. 16.12.

If the bracelets can also be reflected, then G becomes $\{(1)(2)(3), (123), (132), (1)(23), (2)(13), (3)(12)\}$, and the pattern inventory here is the same as the one above, with one exception. Here we have rwb, instead of $2rwb$, because the nonequivalent (for rotations) patterns in Fig. 16.12 become equivalent when reflections are allowed.

Figure 16.12

EXAMPLE 16.34

Consider the 3-colorings of the configurations in Example 16.28. If the three colors are red, white, and blue, how many nonequivalent configurations have exactly two red vertices?

Given that $P_G(x_1, x_2, x_3, x_4) = (1/8)(x_1^4 + 2x_4 + 3x_2^2 + 2x_2x_1^2)$, the answer is the sum of the coefficients of r^2w^2, r^2b^2, and r^2wb in $(1/8)[(r + w + b)^4 + 2(r^4 + w^4 + b^4) + 3(r^2 + w^2 + b^2)^2 + 2(r^2 + w^2 + b^2)(r + w + b)^2]$.

In $(r + w + b)^4$, we find the term $6r^2w^2 + 6r^2b^2 + 12r^2wb$. For $3(r^2 + w^2 + b^2)^2$, we are interested in the term $6r^2w^2 + 6r^2b^2$, whereas $4r^2w^2 + 4r^2b^2 + 4r^2bw$ arises in $2(r^2 + w^2 + b^2)(r + w + b)^2$.

Then $(1/8)[6r^2w^2 + 6r^2b^2 + 12r^2wb + 6r^2w^2 + 6r^2b^2 + 4r^2w^2 + 4r^2b^2 + 4r^2bw] = 2r^2w^2 + 2r^2b^2 + 2r^2bw$, the inventory of the six nonequivalent configurations that contain exactly two red vertices.

Our next example deals with the pattern inventory for the 2-colorings of the vertices of a cube. (The colors are red and white.)

| **EXAMPLE 16.35** |

For the cube in Fig. 16.13, we find that its group G of rigid motions consists of the following.

1) The identity transformation with cycle structure x_1^8.

2) Rotations through 90°, 180°, and 270° about an axis through the centers of two opposite faces: From Fig. 16.13(a) we have

$$\begin{array}{lll} 90° \text{ rotation:} & (1234)(5678) & \text{Cycle structure:} \quad x_4^2 \\ 180° \text{ rotation:} & (13)(24)(57)(68) & \text{Cycle structure:} \quad x_2^4 \\ 270° \text{ rotation:} & (1432)(5876) & \text{Cycle structure:} \quad x_4^2 \end{array}$$

Since there are two other pairs of opposite faces, these nine rotations account for the term $3x_2^4 + 6x_4^2$ in the cycle index.

3) Rotations through 180° about an axis through the midpoints of two opposite edges: As in Fig. 16.13(b), we have the permutation $(17)(28)(34)(56)$, whose cycle structure is given by x_2^4. With six pairs of opposite edges, these rotations contribute the term $6x_2^4$ to the cycle index.

4) Rotations through 120° and 240° about an axis through two diagonally opposite vertices: From part (c) of the figure we have

$$\begin{array}{lll} 120° \text{ rotation:} & (168)(274)(3)(5) & \text{Cycle structure:} \quad x_1^2 x_3^2 \\ 240° \text{ rotation:} & (186)(247)(3)(5) & \text{Cycle structure:} \quad x_1^2 x_3^2 \end{array}$$

Here there are four such pairs of vertices, and these give rise to the term $8x_1^2 x_3^2$ in the cycle index.

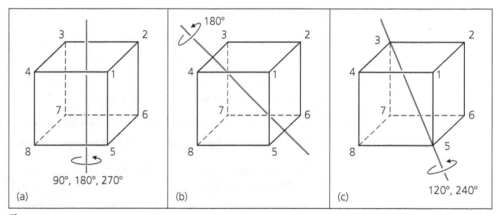

Figure 16.13

Therefore, $P_G(x_1, x_2, \ldots, x_8) = (1/24)(x_1^8 + 9x_2^4 + 6x_4^2 + 8x_1^2 x_3^2)$, and the pattern inventory for these configurations is given by the generating function

$$f(r, w) = (1/24)[(r + w)^8 + 9(r^2 + w^2)^4 + 6(r^4 + w^4)^2 + 8(r + w)^2(r^3 + w^3)^2]$$
$$= r^8 + r^7 w + 3r^6 w^2 + 3r^5 w^3 + 7r^4 w^4 + 3r^3 w^5 + 3r^2 w^6 + rw^7 + w^8.$$

Replacing r and w by 1, we find 23 nonequivalent configurations here.

Since Polya's Method of Enumeration was first developed in order to count isomers of organic compounds, we close this section with an application that deals with a certain class of organic compounds. This is based on an example by C. L. Liu. (See pp. 152–154 of reference [17].)

EXAMPLE 16.36

Here we are concerned with organic molecules of the form shown in Fig. 16.14, where C is a carbon atom and X denotes any of the following components: Br (bromine), H (hydrogen), CH_3 (methyl), or C_2H_5 (ethyl). For example, if each X is replaced by H, the compound CH_4 (methane) results. Figure 16.14 should not be allowed to mislead us. The structure of these organic compounds is three-dimensional. Consequently, we turn to the regular tetrahedron in order to model this structure. We would place the carbon atom at the center of the tetrahedron and then place our selections for X at vertices 1, 2, 3, and 4 as shown in Fig. 16.15.

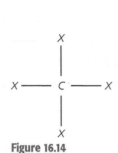

Figure 16.14

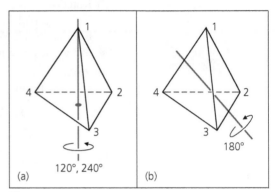

Figure 16.15

The group G acting on these configurations is given as follows:

1) The identity transformation (1)(2)(3)(4) with cycle structure x_1^4.

2) Rotations through 120° or 240° about an axis through a vertex and the center of the opposite face: As Fig. 16.15(a) shows, we have

$$120° \text{ rotation:} \quad (1)(243) \text{ with cycle structure } x_1x_3$$
$$240° \text{ rotation:} \quad (1)(234) \text{ with cycle structure } x_1x_3$$

By symmetry there are three other pairs of vertices and opposite faces, so these rigid motions account for the term $8x_1x_3$ in $P_G(x_1, x_2, x_3, x_4)$.

3) Rotations of 180° about an axis through the midpoints of two opposite edges: The case shown in part (b) of the figure is given by the permutation (14)(23) whose cycle structure is x_2^2. With three pairs of opposite edges, we get the term $3x_2^2$ in $P_G(x_1, x_2, x_3, x_4)$.

Hence $P_G(x_1, x_2, x_3, x_4) = (1/12)[x_1^4 + 8x_1x_3 + 3x_2^2]$ and $P_G(4, 4, 4, 4) = (1/12) \cdot [4^4 + 8(4^2) + 3(4^2)] = 36$, so there are 36 distinct organic compounds that can be formed in this way.

Last, if we wish to know how many of these compounds have exactly two bromine atoms, we let w, x, y, and z represent the "colors" Br, H, CH_3, and C_2H_5, respectively, and find the sum of the coefficients of w^2x^2, w^2y^2, w^2z^2, w^2xy, w^2xz, and w^2yz in the pattern inventory

$$(1/12)[(w + x + y + z)^4 + 8(w + x + y + z)(w^3 + x^3 + y^3 + z^3) + 3(w^2 + x^2 + y^2 + z^2)^2].$$

For $(w + x + y + z)^4$ the relevant term is $6w^2x^2 + 6w^2y^2 + 6w^2z^2 + 12w^2xy + 12w^2xz + 12w^2yz$. The middle summand of the pattern inventory does not give rise to any of the desired configurations, whereas in $3(w^2 + x^2 + y^2 + z^2)^2$ we find $6w^2x^2 + 6w^2y^2 + 6w^2z^2$.

Consequently that part of the pattern inventory for the compounds containing exactly two bromine atoms is

$$(1/12)[12w^2x^2 + 12w^2y^2 + 12w^2z^2 + 12w^2xy + 12w^2xz + 12w^2yz]$$

and there are six such organic compounds.

EXERCISES 16.12

1. a) Find the pattern inventory for the 2-colorings of the edges of a square that is free to move in (i) two dimensions; (ii) three dimensions. (Let the colors be red and white.)

b) Answer part (a) for 3-colorings, where the colors are red, white, and blue.

2. If a regular pentagon is free to move in space and we can color its vertices with red, white, and blue paint, how many nonequivalent configurations have exactly three red vertices? How many have two red, one white, and two blue vertices?

3. Suppose that in Example 16.35 we 2-color the faces of the cube, which is free to move in space.

a) How many distinct 2-colorings are there for this situation?

b) If the available colors are red and white, determine the pattern inventory.

c) How many nonequivalent colorings have three red and three white faces?

4. For the organic compounds in Example 16.36, how many have at least one bromine atom? How many have exactly three hydrogen atoms?

5. Find the pattern inventories for the 2-colorings of the vertices in the configurations in Fig. 16.11, when they are free to move in space. (Let the colors be green and gold.)

6. a) In how many ways can the seven (identical) horses on a carousel be painted with black, brown, and white paint in such a way that there are three black, two brown, and two white horses?

b) In how many ways would there be equal numbers of black and brown horses?

c) Give a combinatorial argument to verify that for all $n \in \mathbf{Z}^+$, $n^7 + 6n$ is divisible by 7.

7. a) In how many ways can we paint the eight squares of a 2×4 chessboard, using the colors red and white? (The back of the chessboard is black cardboard.)

b) Find the pattern inventory for the colorings in part (a).

c) How many of the colorings in part (a) have four red and four white squares? How many have six red and two white squares?

8. a) In how many ways can we 2-color the eight regions of the pinwheel shown in Fig. 16.16, using the colors black and gold, if the back of each region remains grey?

b) Answer part (a) for the possible 3-colorings, using black, gold, and blue paints to color the regions.

c) For the colorings in part (b), how many have four black, two gold, and two blue regions?

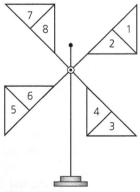

Figure 16.16

9. Let $m, n \in \mathbf{Z}^+$ with $n \geq 3$. How many distinct summands appear in the pattern inventory for the m-colorings of the vertices of a regular polygon of n sides?

16.13
Summary and Historical Review

Although the notion of a group of transformations evolved gradually in the study of geometry, the major thrust in the development of the group concept came from the study of polynomial equations.

Methods for solving quadratic equations were known to the ancient Greeks. Then in the sixteenth century, advances were made toward solving cubic and quartic polynomial equations where the coefficients were rational numbers. Continuing with polynomials of fifth and higher degree, both Leonhard Euler (1707–1783) and Joseph-Louis Lagrange (1736–1813) attempted to solve the general quintic. Lagrange realized there had to be a connection between the degree n of a polynomial equation and the permutation group S_n. However, it was Niels Henrik Abel (1802–1829) who finally proved that it was not possible to find a formula for solving the general quintic using only addition, subtraction, multiplication, division, and root extraction. During this same period, the existence of a necessary and sufficient condition for when a polynomial of degree $n \geq 5$ with rational coefficients can be solved by radicals was investigated and solved by the illustrious French mathematician Evariste Galois (1811–1832). Since the work of Galois utilizes the structures of both groups and fields, we shall say more about him in the summary of Chapter 17.

Niels Henrik Abel (1802–1829)

Examining pages 278–280 of J. Stillwell [28], one finds that the group concept, and in fact the actual word "group," first appears in Galois' work *Mémoire sur les conditions de résolubilité des équations par radicaux*, published in 1831. Associativity, the group identity, and inverses were consequences of Galois' assumptions, for he only dealt with a group of permutations of a finite set and his definition of a group required only the closure property. It was Arthur Cayley (1821–1895) (in 1854, in his paper *On the Theory of Groups, as Depending on the Symbolic Equation* $\theta^n = 1$) who first found the need to state the associative property for group elements. The first actual mention of inverses in the definition of a group occurs in the 1883 article *Gruppentheoretischen Studien* II by Walther Franz Anton von Dyck (1856–1934).

The concept of the coset, which we introduced in Section 16.3, was also developed by Evariste Galois (in 1832). The actual term was coined (in 1910) by George Abram Miller (1863–1951).

Following the accomplishments of Galois, group theory affected many areas of mathematics. During the late nineteenth century, for example, the German mathematician Felix Klein (1849–1929), in what has come to be known as the *Erlanger Programm*, attempted to codify all existing geometries according to the group of transformations under which the properties of the geometry were invariant.

Many other mathematicians, such as Augustin-Louis Cauchy (1789–1857), Arthur Cayley (1821–1895), Ludwig Sylow (1832–1918), Richard Dedekind (1831–1916), and Leopold Kronecker (1823–1891), contributed to the further development of certain types of groups. However, it was not until 1900 that lists of defining conditions were given for the general abstract group.

During the twentieth century a great deal of research took place in the attempt to analyze the structure of finite groups. For finite abelian groups, it is known that any such group is isomorphic to a direct product of cyclic groups of prime power order. However, the case of the finite nonabelian groups has turned out to be considerably more complex. Starting with the work of Galois, one finds particular attention paid to a special type of subgroup called a normal subgroup. For any group G, a subgroup H (of G) is called *normal* if, for all $g \in G$ and all $h \in H$, we have $ghg^{-1} \in H$. In an abelian group every subgroup is normal, but this is not the case for nonabelian groups. In every group G, both $\{e\}$ and G are normal subgroups, but if G has no other normal subgroups it is called *simple*. During the past six decades mathematicians have sought and determined all the finite simple groups and examined their role in the structure of all finite groups. Among the prime movers in the classification of the finite simple groups are Professors Walter Feit, John Thompson, Daniel Gorenstein, Michael Aschbacher, and Robert Griess, Jr. For more on the history and impact of this monumental work we refer the reader to the articles by J. A. Gallian [5], A. Gardiner [7], M. Gardner [9], R. Silvestri [27], and, especially, the one by D. Gorenstein [13].

There are many texts one can turn to for further study in the theory of groups. At the introductory level, the texts by J. A. Gallian [6] and V. H. Larney [16] provide further coverage beyond the introduction given in this chapter. The text by I. N. Herstein [15] is an excellent source and includes material on Galois theory.

More on the RSA public-key cryptosystem of Section 16.4 can be found in the references by T. H. Barr [2], P. Garrett [10], and W. Trappe and L. C. Washington [31]. An early description of the system is given in the article by M. Gardner [8], where a message is encrypted using, as the modulus n, the product of a 64-digit prime and a 65-digit prime. The article by G. Taubes [30] relates the effort set forth by Arjen Lenstra, Paul Leyland, Michael Graff, and Derek Atkins, along with 600 volunteers, in factoring n.

The beginnings of algebraic coding theory can be traced to 1941, when Claude Elwood Shannon began his investigations of problems in communications. These problems were prompted by the needs of the war effort. His research resulted in many new ideas and principles that were later published in 1948 in the journal article [26]. As a result of this work, Shannon is acknowledged as the founder of information theory. After this publication, results by M. J. E. Golay [11] and R. W. Hamming [14] soon followed, giving further impetus to research in this area. The 1478 references listed in the bibliography at the end of Volume II of the texts by F. J. MacWilliams and N. J. A. Sloane [18] should convey some idea of the activity in this area between 1950 and 1975.

Our coverage of coding theory followed the development in Chapter 5 of the text by L. L. Dornhoff and F. E. Hohn [4]. The texts by E. F. Assmus, Jr., and J. D. Key [1], S. W. Golomb, R. A. Scholtz, and R. E. Peile [12], V. Pless [20], and S. Roman [24] provide a nice coverage of topics at a fairly intermediate level. More advanced work in coding can be found in the books by F. J. MacWilliams and N. J. A. Sloane [18], S. Roman [25], and A. P. Street and W. D. Wallis [29]. An interesting application on the use of the pigeonhole principle in coding theory is given in Chapter XI of [29].

In Sections 10, 11, and 12 of the chapter, we came upon an enumeration technique whose development is attributed to the Hungarian mathematician George Polya (1887–1985). His article [21] provided the fundamental techniques for counting equivalence classes of chemical isomers, graphs, and trees. (To some extent, the ideas in this work were anticipated by J. H. Redfield [23].) Since then these techniques have been found invaluable for counting problems in such areas as the electronic realizations of Boolean functions. Polya's fundamental theorem was first generalized in the article by N. G. DeBruijn [3], and other extensions of these ideas can be found in the literature. The article by R. C. Read [22] relates the profound influence that Polya's Theorem has had on developments in combinatorial analysis. (The issue of the journal that contains this article also includes several other articles dealing with the life and work of George Polya.)

Our coverage of this topic follows the presentation given in the article by A. Tucker [32]. A more rigorous presentation of this method can be found in Chapter 5 of the text by C. L. Liu [17].

In dealing with Burnside's Theorem we have another instance of an inaccurate attribution. As we learn in the article by P. M. Neumann [19], the result appears in a paper by Georg Frobenius (1848–1917) that was published in 1887, as well as in some of Cauchy's work from 1845.

REFERENCES

1. Assmus, E. F., Jr., and Key, J. D. *Designs and Their Codes*. New York: Cambridge University Press, 1992.
2. Barr, Thomas H. *Invitation to Cryptology*. Upper Saddle River, N. J.: Prentice-Hall, 2002.
3. DeBruijn, Nicolaas Govert. "Polya's Theory of Counting." Chapter 5 in *Applied Combinatorial Mathematics*, ed. by Edwin F. Beckenbach. New York: Wiley, 1964.
4. Dornhoff, Larry L., and Hohn, Franz E. *Applied Modern Algebra*. New York: Macmillan, 1978.
5. Gallian, Joseph A. "The Search for Finite Simple Groups." *Mathematics Magazine* 49, 1976, pp. 163–179.
6. Gallian, Joseph A. *Contemporary Abstract Algebra*, 5th ed. Boston, Mass.: Houghton Mifflin, 2002.
7. Gardiner, Anthony. "Groups of Monsters." *New Scientist*, April 5, 1979, p. 34.
8. Gardner, Martin. "A New Kind of Cipher That Would Take Millions of Years to Break." *Scientific American* (August 1977): pp. 120–124.
9. Gardner, Martin. "The Capture of the Monster: A Mathematical Group with a Ridiculous Number of Elements." *Scientific American* 242 (6), 1980, pp. 20–32.
10. Garrett, Paul. *Making, Breaking Codes: An Introduction to Cryptology*. Upper Saddle River, N. J.: Prentice-Hall, 2001.
11. Golay, Marcel J. E. "Notes on Digital Coding." *Proceedings of the IRE* 37, 1949, p. 657.
12. Golomb, Solomon W., Scholtz, Robert A., and Peile, Robert E. *Basic Concepts in Information Theory and Coding*. New York: Plenum, 1994.
13. Gorenstein, Daniel. "The Enormous Theorem." *Scientific American* 253 (6), 1985, pp. 104–115.
14. Hamming, Richard Wesley. "Error Detecting and Error Correcting Codes." *Bell System Technical Journal* 29, 1950, pp. 147–160.

15. Herstein, Israel Nathan. *Topics in Algebra*, 2nd ed. Lexington, Mass.: Xerox College Publishing, 1975.
16. Larney, Violet H. *Abstract Algebra: A First Course*. Boston: Prindle, Weber & Schmidt, 1975.
17. Liu, C. L. *Introduction to Combinatorial Mathematics*. New York: McGraw-Hill, 1968.
18. MacWilliams, F. Jessie, and Sloane, Neil J. A. *The Theory of Error-Correcting Codes*, Volumes I and II. Amsterdam: North-Holland, 1977.
19. Neumann, Peter M. "A Lemma That Is Not Burnside's." *The Mathematical Scientist*, Vol. 4, 1979, pp. 133–141.
20. Pless, Vera. *Introduction to the Theory of Error-Correcting Codes*, 2nd ed. New York: Wiley, 1989.
21. Polya, George. "Kombinatorische Anzahlbestimmungen für Gruppen, Graphen und Chemishe Verbindungen." *Acta Mathematica* 68, 1937, pp. 145–254.
22. Read, R. C. "Polya's Theorem and Its Progeny." *Mathematics Magazine* 60, 1987, pp. 275–282.
23. Redfield, J. Howard. "The Theory of Group Reduced Distributions." *American Journal of Mathematics* 49, 1927, pp. 433–455.
24. Roman, Steven. *Introduction to Coding and Information Theory*. New York: Springer-Verlag, 1997.
25. Roman, Steven. *Coding and Information Theory*. New York: Springer-Verlag, 1992.
26. Shannon, Claude E. "The Mathematical Theory of Communication." *Bell System Technical Journal* 27, 1948, pp. 379–423, 623–656. Reprinted in C. E. Shannon and W. Weaver, *The Mathematical Theory of Communication* (Urbana: University of Illinois Press, 1949).
27. Silvestri, Richard. "Simple Groups of Finite Order." *Archive for the History of Exact Sciences* 20, 1979, pp. 313–356.
28. Stillwell, John. *Mathematics and Its History*. New York: Springer-Verlag, 1989.
29. Street, Anne Penfold, and Wallis, W. D. *Combinatorial Theory: An Introduction*. Winnipeg, Canada: The Charles Babbage Research Center, 1977.
30. Taubes, G. "Small Army of Code-breakers Conquers a 129-digit Giant." *Science* 264, 1994, pp. 776–777.
31. Trappe, Wade, and Washington, Lawrence C. *Introduction to Cryptography with Coding Theory*. Upper Saddle River, N. J.: Prentice-Hall, 2002.
32. Tucker, Alan. "Polya's Enumeration Formula by Example." *Mathematics Magazine* 47, 1974, pp. 248–256.

SUPPLEMENTARY EXERCISES

1. Let $f: G \to H$ be a group homomorphism with e_H the identity in H. Prove that

a) $K = \{x \in G | f(x) = e_H\}$ is a subgroup of G. (K is called the *kernel* of the homomorphism.)

b) if $g \in G$ and $x \in K$, then $gxg^{-1} \in K$.

2. If G, H, and K are groups and $G = H \times K$, prove that G contains subgroups that are isomorphic to H and K.

3. Let G be a group where $a^2 = e$ for all $a \in G$. Prove that G is abelian.

4. If G is a group of even order, prove that there is an element $a \in G$ with $a \neq e$ and $a = a^{-1}$.

5. Let $f: G \to H$ be a group homomorphism *onto H*. If G is a cyclic group, prove that H is also cyclic.

6. a) Consider the group $(\mathbf{Z}_2 \times \mathbf{Z}_2, \oplus)$ where, for a, b, c, $d \in \mathbf{Z}_2$, $(a, b) \oplus (c, d) = (a + c, b + d)$—the sums $a + c$ and $b + d$ are computed using addition modulo 2. What is the value of $(1, 0) \oplus (0, 1) \oplus (1, 1)$ in this group?

b) Now consider the group $(\mathbf{Z}_2 \times \mathbf{Z}_2 \times \mathbf{Z}_2, \oplus)$ where $(a, b, c) \oplus (d, e, f) = (a + d, b + e, c + f)$. (Here the sums $a + d, b + e, c + f$ are computed using addition modulo 2.) What do we obtain when we add the seven nonzero (or nonidentity) elements of this group?

c) State and prove a generalization that includes the results in parts (a) and (b).

7. Let $(G, \circ)$ be a group where

$$x \circ a \circ y = b \circ a \circ c \Rightarrow x \circ y = b \circ c,$$

for all $a, b, c, x, y \in G$. Prove that $(G, \circ)$ is an abelian group.

8. For $k, n \in \mathbf{Z}^+$ with $n \geq k \geq 1$, let $Q(n, k)$ count the number of permutations $\pi \in S_n$ where any representation of π, as a product of disjoint cycles, contains no cycle of length greater than k. Verify that

$$Q(n + 1, k) = \sum_{i=0}^{k-1} \binom{n}{i} (i!) Q(n - i, k).$$

9. For $k, n \in \mathbf{Z}^+$ where $n \geq 2$ and $1 \leq k \leq n$, let $P(n, k)$ denote the number of permutations $\pi \in S_n$ that have k cycles. [For example, $(1)(23)$ is counted in $P(3, 2)$, $(12)(34)$ is counted in $P(4, 2)$, and $(1)(23)(4)$ is counted in $P(4, 3)$.]

a) Verify that $P(n + 1, k) = P(n, k - 1) + nP(n, k)$.

b) Determine $\sum_{k=1}^n P(n, k)$.

10. For $n \geq 1$, if $\sigma, \tau \in S_n$, define the distance $d(\sigma, \tau)$ between σ and τ by

$$d(\sigma, \tau) = \max\{|\sigma(i) - \tau(i)| \mid 1 \leq i \leq n\}.$$

a) Prove that the following properties hold for d.

i) $d(\sigma, \tau) \geq 0$ for all $\sigma, \tau \in S_n$

ii) $d(\sigma, \tau) = 0$ if and only if $\sigma = \tau$

iii) $d(\sigma, \tau) = d(\tau, \sigma)$ for all $\sigma, \tau \in S_n$

iv) $d(\rho, \tau) \leq d(\rho, \sigma) + d(\sigma, \tau)$, for all $\rho, \sigma, \tau \in S_n$

b) Let ϵ denote the identity element of S_n (that is, $\epsilon(i) = i$ for all $1 \leq i \leq n$). If $\pi \in S_n$ and $d(\pi, \epsilon) \leq 1$, what can we say about $\pi(n)$?

c) For $n \geq 1$ let a_n count the number of permutations π in S_n, where $d(\pi, \epsilon) \leq 1$. Find and solve a recurrence relation for a_n.

11. Wilson's Theorem [in part (d) of Exercise 19 of Section 16.1] tells us that $(p - 1)! \equiv -1 \pmod{p}$, for p a prime.

a) Is the converse of this theorem true or false — that is, if $n \in \mathbf{Z}^+$ and $n \geq 2$, does $(n - 1)! \equiv -1 \pmod{n} \Rightarrow n$ is prime?

b) For p an odd prime, prove that

$$2(p - 3)! \equiv -1 \pmod{p}.$$

12. In how many ways can Nicole paint the eight regions of the square shown in Fig. 16.17 if

a) five colors are available?

b) she actually uses exactly four of the five available colors?

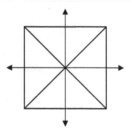

Figure 16.17

17

Finite Fields and Combinatorial Designs

It is time now to recall the ring structure of Chapter 14 as we examine rings of polynomials and their role in the construction of finite fields. We know that for every prime p, $(\mathbf{Z}_p, +, \cdot)$ is a finite field, but here we shall find other finite fields. Just as the order of a finite Boolean algebra is restricted to powers of 2, for finite fields the possible orders are p^n, where p is a prime and $n \in \mathbf{Z}^+$. Applications of these finite fields will include a discussion of such combinatorial designs as Latin squares. Finally, we shall investigate the structure of a finite geometry and discover how these geometries and combinatorial designs are interrelated.

17.1
Polynomial Rings

We recall that a ring $(R, +, \cdot)$ consists of a nonempty set R, where $(R, +)$ is an abelian group, $(R, \cdot)$ is closed under the associative operation $\cdot$, and the two operations are related by the distributive laws: $a(b + c) = ab + ac$ and $(b + c)a = ba + ca$, for all $a, b, c \in R$. (We write ab for $a \cdot b$.)

In order to introduce the formal concept of a polynomial with coefficients in R we let x denote an indeterminate — that is, a formal symbol that is not an element of the ring R. We then use this symbol x to define the following.

Definition 17.1   Given a ring $(R, +, \cdot)$, an expression of the form $f(x) = a_n x^n + a_{n-1} x^{n-1} + \cdots + a_1 x^1 + a_0 x^0$, where $a_i \in R$ for all $0 \le i \le n$, is called a *polynomial in the indeterminate x with coefficients from R.*

If a_n is not the zero element of R, then a_n is called *the leading coefficient of $f(x)$* and we say that $f(x)$ has *degree n*. Hence the degree of a polynomial is the highest power of x that occurs in a summand of the polynomial. The term $a_0 x^0$ is called the *constant*, or *constant term*, of $f(x)$.

If $g(x) = b_m x^m + b_{m-1} x^{m-1} + \cdots + b_1 x^1 + b_0 x^0$ is also a polynomial in x over R, then $f(x) = g(x)$ if $m = n$ and $a_i = b_i$ for all $0 \le i \le n$.

Finally, we use the notation $R[x]$ to represent the set of all polynomials in the indeterminate x with coefficients from R.

EXAMPLE 17.1

a) Over the ring $R = (\mathbf{Z}_6, +, \cdot)$, the expression $5x^2 + 3x^1 - 2x^0$ is a polynomial of degree 2, with leading coefficient 5 and constant term $-2x^0$. As before, here we are using a to denote $[a]$ in $\mathbf{Z}_6$. This polynomial may also be written as $5x^2 + 3x^1 + 4x^0$ since $[4] = [-2]$ in $\mathbf{Z}_6$.

b) If z is the zero element of ring R, then the zero polynomial $zx^0 = z$ is also the zero element of $R[x]$ and is said to have *no degree* and no leading coefficient. A polynomial over R that is the zero element or is of degree 0 is called a *constant polynomial*. For example, the polynomial $5x^0$ over $\mathbf{Z}_7$ has degree 0 and leading coefficient 5 and is a constant polynomial.

For a ring of coefficients $(R, +, \cdot)$, let

$$f(x) = a_n x^n + a_{n-1} x^{n-1} + \cdots + a_1 x^1 + a_0 x^0$$
$$g(x) = b_m x^m + b_{m-1} x^{m-1} + \cdots + b_1 x^1 + b_0 x^0,$$

where $a_i, b_j \in R$ for all $0 \le i \le n, 0 \le j \le m$. We introduce (closed binary) operations of addition and multiplication for these polynomials in order to obtain a new ring.

Assume that $n \ge m$. We define

$$f(x) + g(x) = \sum_{i=0}^{n} (a_i + b_i) x^i, \tag{1}$$

where $b_i = z$ for $i > m$, and

$$f(x)g(x) = (a_n b_m) x^{n+m} + (a_n b_{m-1} + a_{n-1} b_m) x^{n+m-1}$$
$$+ \cdots + (a_1 b_0 + a_0 b_1) x^1 + (a_0 b_0) x^0. \tag{2}$$

In the definition of $f(x) + g(x)$, the coefficient $(a_i + b_i)$, for each $0 \le i \le n$, is obtained from the addition of elements in R. For $f(x)g(x)$, the coefficient of x^t is $\sum_{k=0}^{t} a_{t-k} b_k$, where all additions and multiplications occur within R, and $0 \le t \le n + m$. Here is one such example to demonstrate the types of calculations that are involved.

Let $f(x) = 4x^3 + 2x^2 + 3x^1 + 1x^0$ and $g(x) = 3x^2 + x^1 + 2x^0$ be polynomials from $\mathbf{Z}_5[x]$. Here

$$a_3 = 4, \qquad a_2 = 2, \qquad a_1 = 3, \qquad a_0 = 1,$$

and

$$b_2 = 3, \qquad b_1 = 1, \qquad b_0 = 2.$$

For all $n \ge 4$ we find that $a_n = 0$. When $m \ge 3$ we have $b_m = 0$. Using the definitions in Eqs. (1) and (2), where the addition and multiplication of the coefficients are now performed modulo 5, we obtain

$$f(x) + g(x) = (4 + 0)x^3 + (2 + 3)x^2 + (3 + 1)x^1 + (1 + 2)x^0$$
$$= 4x^3 + 0x^2 + 4x^1 + 3x^0 = 4x^3 + 4x^1 + 3x^0$$

and

$$f(x)g(x) = \left(\sum_{k=0}^{5} a_{5-k} b_k \right) x^5 + \left(\sum_{k=0}^{4} a_{4-k} b_k \right) x^4 + \left(\sum_{k=0}^{3} a_{3-k} b_k \right) x^3$$
$$+ \left(\sum_{k=0}^{2} a_{2-k} b_k \right) x^2 + \left(\sum_{k=0}^{1} a_{1-k} b_k \right) x^1 + \left(\sum_{k=0}^{0} a_{0-k} b_k \right) x^0$$

$$= (0 \cdot 2 + 0 \cdot 1 + 4 \cdot 3 + 2 \cdot 0 + 3 \cdot 0 + 1 \cdot 0)x^5$$
$$+ (0 \cdot 2 + 4 \cdot 1 + 2 \cdot 3 + 3 \cdot 0 + 1 \cdot 0)x^4$$
$$+ (4 \cdot 2 + 2 \cdot 1 + 3 \cdot 3 + 1 \cdot 0)x^3$$
$$+ (2 \cdot 2 + 3 \cdot 1 + 1 \cdot 3)x^2 + (3 \cdot 2 + 1 \cdot 1)x^1 + (1 \cdot 2)x^0$$
$$= 2x^5 + 0x^4 + 4x^3 + 0x^2 + 2x^1 + 2x^0 = 2x^5 + 4x^3 + 2x^1 + 2x^0.$$

The closed binary operations defined in Eqs. (1) and (2) were designed to give us the following result.

THEOREM 17.1

If R is a ring, then under the operations of addition and multiplication given in Eqs. (1) and (2), $(R[x], +, \cdot)$ is a ring, called the *polynomial ring,* or *ring of polynomials, over R.*

Proof: The ring properties for $R[x]$ hinge upon those of R. Consequently, we shall prove the associative law of multiplication here, as an example, and shall then leave the proofs of the other properties to the reader. Let $h(x) = \sum_{k=0}^{p} c_k x^k$, with $f(x), g(x)$ as defined earlier. A typical summand in $(f(x)g(x))h(x)$ has the form Ax^t, where $0 \le t \le (m+n) + p$ and A is the sum of all products of the form $(a_i b_j)c_k$, with $0 \le i \le n, 0 \le j \le m, 0 \le k \le p$, and $i + j + k = t$. In $f(x)(g(x)h(x))$ the coefficient of x^t is the sum of all products of the form $a_i(b_j c_k)$, again with $0 \le i \le n, 0 \le j \le m, 0 \le k \le p$, and $i + j + k = t$. Since R is associative under multiplication, $(a_i b_j)c_k = a_i(b_j c_k)$ for each of these terms, and so the coefficient of x^t in $(f(x)g(x))h(x)$ is the same as it is in $f(x)(g(x)h(x))$. Hence $(f(x)g(x))h(x) = f(x)(g(x)h(x))$.

COROLLARY 17.1

Let $R[x]$ be a polynomial ring.

   a) If R is commutative, then $R[x]$ is commutative.

   b) If R is a ring with unity, then $R[x]$ is a ring with unity.

   c) $R[x]$ is an integral domain if and only if R is an integral domain.

Proof: The proof of this corollary is left for the reader.

From this point on, we shall write x instead of x^1. If R has unity u, we define $x^0 = u$, and for all $r \in R$ we write rx^0 as r.

EXAMPLE 17.2

Let $f(x), g(x) \in \mathbf{Z}_8[x]$ with $f(x) = 4x^2 + 1$ and $g(x) = 2x + 3$. Then $f(x)$ has degree 2 and $g(x)$ has degree 1. From our past experiences with polynomials, we expect the degree of $f(x)g(x)$ to be 3, the sum of the degrees of $f(x)$ and $g(x)$. Here, however, $f(x)g(x) = (4x^2 + 1)(2x + 3) = 8x^3 + 12x^2 + 2x + 3 = 4x^2 + 2x + 3$ because $[8] = [0]$ in $\mathbf{Z}_8$. So degree $f(x)g(x) = 2 < 3 = $ degree $f(x) + $ degree $g(x)$.

The cause of the phenomenon in Example 17.2 is the existence of proper divisors of zero in the ring $\mathbf{Z}_8$. This observation leads us to the following theorem.

THEOREM 17.2

Let $(R, +, \cdot)$ be a commutative ring with unity u. Then R is an integral domain if and only if for all $f(x), g(x) \in R[x]$, if neither $f(x)$ nor $g(x)$ is the zero polynomial, then

$$\text{degree } f(x)g(x) = \text{degree } f(x) + \text{degree } g(x).$$

Proof: Let $f(x) = \sum_{i=0}^{n} a_i x^i$, $g(x) = \sum_{j=0}^{m} b_j x^j$, with $a_n \neq z$, $b_m \neq z$. If R is an integral domain, then $a_n b_m \neq z$, so degree $f(x)g(x) = n + m = $ degree $f(x) + $ degree $g(x)$. Conversely, if R is not an integral domain, let $a, b \in R$ with $a \neq z, b \neq z$, but $ab = z$. The polynomials $f(x) = ax + u, g(x) = bx + u$ each have degree 1, but $f(x)g(x) = (a + b)x + u$ and degree $f(x)g(x) \leq 1 < 2 = $ degree $f(x) + $ degree $g(x)$.

Before we can proceed we need to recall an idea that was introduced in Section 14.2 — in Exercise 21. If R is a ring with unity u and $r \in R$, we define $r^0 = u, r^1 = r$, and $r^{n+1} = r^n r$ for all $n \in \mathbf{Z}^+$. [From these definitions one can show, for example, that for all $m, n \in \mathbf{Z}^+$, $(r^m)(r^n) = r^{m+n}$ and $(r^m)^n = r^{mn}$.] So now we continue as follows.

Let R be a ring with unity u and let $f(x) = a_n x^n + \cdots + a_1 x + a_0 \in R[x]$. If $r \in R$, then $f(r) = a_n r^n + \cdots + a_1 r + a_0 \in R$. We are especially interested in those values of r for which $f(r) = z$, and this interest leads us to the following concept.

Definition 17.2

Let R be a ring with unity u and let $f(x) \in R[x]$, with degree $f(x) \geq 1$. If $r \in R$ and $f(r) = z$, then r is called a *root* of the polynomial $f(x)$.

EXAMPLE 17.3

a) If $f(x) = x^2 - 2 \in \mathbf{R}[x]$, then $f(x)$ has $\sqrt{2}$ and $-\sqrt{2}$ as roots because $(\sqrt{2})^2 - 2 = 0 = (-\sqrt{2})^2 - 2$. In addition, we can write $f(x) = (x - \sqrt{2})(x + \sqrt{2})$, with $x - \sqrt{2}, x + \sqrt{2} \in \mathbf{R}[x]$. However, if we regard $f(x)$ as an element of $\mathbf{Q}[x]$, then $f(x)$ has no roots because $\sqrt{2}$ and $-\sqrt{2}$ are irrational numbers. Consequently, the existence of roots for a polynomial is dependent on the underlying ring of coefficients.

b) For $f(x) = x^2 + 3x + 2 \in \mathbf{Z}_6[x]$, we find that

$$f(0) = (0)^2 + 3(0) + 2 = 2 \qquad f(3) = (3)^2 + 3(3) + 2 = 20 = 2$$
$$f(1) = (1)^2 + 3(1) + 2 = 6 = 0 \qquad f(4) = (4)^2 + 3(4) + 2 = 30 = 0$$
$$f(2) = (2)^2 + 3(2) + 2 = 12 = 0 \qquad f(5) = (5)^2 + 3(5) + 2 = 42 = 0$$

Consequently, $f(x)$ has four roots: 1, 2, 4, and 5. This is more than we expected. In our prior experiences, a polynomial of degree 2 had at most two roots.

In this chapter we shall be primarily concerned with polynomial rings $F[x]$, where F is a field (and $F[x]$ is an integral domain). Consequently, we shall not dwell any further on situations where degree $f(x)g(x) < $ degree $f(x) + $ degree $g(x)$. In addition, unless it is stated otherwise, we shall denote the zero element of a field by 0 and use 1 to denote its unity.

As a result of Example 17.3(b), we shall now develop the concepts needed to find out when a polynomial of degree n has at most n roots.

Definition 17.3

Let F be a field. For $f(x), g(x) \in F[x]$, where $f(x)$ is not the zero polynomial, we call $f(x)$ a *divisor* (or *factor*) of $g(x)$ if there exists $h(x) \in F[x]$ with $f(x)h(x) = g(x)$. In this situation we also say that $f(x)$ *divides* $g(x)$ and that $g(x)$ is a *multiple* of $f(x)$.

This leads to the *division algorithm* for polynomials. Before proving the general result, however, we shall examine two particular examples.

EXAMPLE 17.4

Early in algebra we were taught how to perform the long division of polynomials with real coefficients. Given two polynomials $f(x)$, $g(x)$ with degree $f(x) \leq$ degree $g(x)$, we organized our work in the form

$$
\begin{array}{r}
q_1(x) + q_2(x) + \cdots + q_t(x) \; (= q(x)) \\ \hline
f(x) \overline{)\, g(x)} \\
\underline{f(x)q_1(x)} \\
g(x) - f(x)q_1(x) \\
\cdots \cdots \cdots \cdots \\ \hline
r(x)
\end{array}
$$

where we continued to divide until we found either

$$ r(x) = 0 \quad \text{or} \quad \text{degree } r(x) < \text{degree } f(x). $$

It then followed that $g(x) = q(x)f(x) + r(x)$.

For example, if $f(x) = x - 3$ and $g(x) = 7x^3 - 2x^2 + 5x - 2$, then $f(x)$, $g(x) \in \mathbf{Q}[x]$ (or $\mathbf{R}[x]$, or $\mathbf{C}[x]$), and we find

$$
\begin{array}{r}
7x^2 + 19x \;\; + 62 \quad (= q(x)) \\ \hline
x - 3 \overline{)\, 7x^3 - \;\; 2x^2 + \;\; 5x - \;\; 2} \\
\underline{7x^3 - 21x^2} \\
19x^2 + \;\; 5x - \;\; 2 \\
\underline{19x^2 - 57x} \\
62x - \;\; 2 \\
\underline{62x - 186} \\
184 \; (= r(x))
\end{array}
$$

Checking these results, we have

$$ q(x)f(x) + r(x) = (7x^2 + 19x + 62)(x - 3) + 184 = 7x^3 - 2x^2 + 5x - 2 = g(x). $$

EXAMPLE 17.5

The technique illustrated in Example 17.4 also applies when the coefficients of our polynomials are taken from a *finite field*.

If $f(x) = 3x^2 + 4x + 2$ and $g(x) = 6x^4 + 4x^3 + 5x^2 + 3x + 1$ are polynomials in $\mathbf{Z}_7[x]$, then the process of long division provides the following calculations:

$$
\begin{array}{r}
2x^2 + \;\; x \;\; + 6 \quad (= q(x)) \\ \hline
3x^2 + 4x + 2 \overline{)\, 6x^4 + 4x^3 + 5x^2 + 3x + 1} \\
\underline{6x^4 + \;\; x^3 + 4x^2} \\
3x^3 + \;\; x^2 + 3x + 1 \\
\underline{3x^3 + 4x^2 + 2x} \\
4x^2 + \;\; x + 1 \\
\underline{4x^2 + 3x + 5} \\
5x + 3 \;\; (= r(x))
\end{array}
$$

Performing all arithmetic in $\mathbf{Z}_7$, we find (as in Example 17.4) that

$$q(x)f(x) + r(x) = (2x^2 + x + 6)(3x^2 + 4x + 2) + (5x + 3)$$
$$= 6x^4 + 4x^3 + 5x^2 + 3x + 1 = g(x)$$

We turn now to the general situation.

THEOREM 17.3

Division Algorithm. Let $f(x), g(x) \in F[x]$ with $f(x)$ not the zero polynomial. There exist unique polynomials $q(x), r(x) \in F[x]$ such that $g(x) = q(x)f(x) + r(x)$, where $r(x) = 0$ or degree $r(x) <$ degree $f(x)$.

Proof: Let $S = \{g(x) - t(x)f(x) | t(x) \in F[x]\}$.

If $0 \in S$, then $0 = g(x) - t(x)f(x)$ for some $t(x) \in F[x]$. Then with $q(x) = t(x)$ and $r(x) = 0$, we have $g(x) = q(x)f(x) + r(x)$.

If $0 \notin S$, consider the degrees of the elements of S, and let $r(x) = g(x) - q(x)f(x)$ be an element in S of minimum degree. Since $r(x) \neq 0$, the result follows if degree $r(x) <$ degree $f(x)$. If not, let

$$r(x) = a_n x^n + a_{n-1}x^{n-1} + \cdots + a_2 x^2 + a_1 x + a_0, \qquad a_n \neq 0,$$
$$f(x) = b_m x^m + b_{m-1}x^{m-1} + \cdots + b_2 x^2 + b_1 x + b_0, \qquad b_m \neq 0,$$

with $n \geq m$. Define

$$h(x) = r(x) - [a_n b_m^{-1} x^{n-m}]f(x) = (a_n - a_n b_m^{-1} b_m)x^n + (a_{n-1} - a_n b_m^{-1}b_{m-1})x^{n-1}$$
$$+ \cdots + (a_{n-m} - a_n b_m^{-1}b_0)x^{n-m} + a_{n-m-1}x^{n-m-1} + \cdots + a_1 x + a_0.$$

Then $h(x)$ has degree less than n, the degree of $r(x)$. More important, $h(x) = [g(x) - q(x)f(x)] - [a_n b_m^{-1} x^{n-m}]f(x) = g(x) - [q(x) + a_n b_m^{-1} x^{n-m}]f(x)$, so $h(x) \in S$ and this contradicts the choice of $r(x)$ as having minimum degree. Consequently, degree $r(x) <$ degree $f(x)$ and we have the existence part of the theorem.

For uniqueness, let $g(x) = q_1(x)f(x) + r_1(x) = q_2(x)f(x) + r_2(x)$ where $r_1(x) = 0$ or degree $r_1(x) <$ degree $f(x)$, and $r_2(x) = 0$ or degree $r_2(x) <$ degree $f(x)$. Then $[q_2(x) - q_1(x)]f(x) = r_1(x) - r_2(x)$, and if $q_2(x) - q_1(x) \neq 0$, then degree $([q_2(x) - q_1(x)]f(x)) \geq$ degree $f(x)$, whereas $r_1(x) - r_2(x) = 0$ or degree $[r_1(x) - r_2(x)] \leq$ max{degree $r_1(x)$, degree $r_2(x)$} $<$ degree $f(x)$. Consequently, $q_1(x) = q_2(x)$, and $r_1(x) = r_2(x)$.

The division algorithm provides the following results on roots and factors.

THEOREM 17.4

The Remainder Theorem. For $f(x) \in F[x]$ and $a \in F$, the remainder in the division of $f(x)$ by $x - a$ is $f(a)$.

Proof: From the division algorithm, $f(x) = q(x)(x - a) + r(x)$, with $r(x) = 0$ or degree $r(x) <$ degree $(x - a) = 1$. Hence $r(x) = r$ is an element of F. Substituting a for x, we find $f(a) = q(a)(a - a) + r = 0 + r = r$.

THEOREM 17.5

The Factor Theorem. If $f(x) \in F[x]$ and $a \in F$, then $x - a$ is a factor of $f(x)$ if and only if a is a root of $f(x)$.

Proof: If $x - a$ is a factor of $f(x)$, then $f(x) = q(x)(x - a)$. With $f(a) = q(a)(a - a) = 0$, it follows that a is a root of $f(x)$. Conversely, suppose that a is a root of $f(x)$. By the division algorithm, $f(x) = q(x)(x - a) + r$, where $r \in F$. Since $f(a) = 0$ we have $r = 0$, so $f(x) = q(x)(x - a)$, and $x - a$ is a factor of $f(x)$.

EXAMPLE 17.6

a) Let $f(x) = x^7 - 6x^5 + 4x^4 - x^2 + 3x - 7 \in \mathbf{Q}[x]$. From the remainder theorem it follows that when $f(x)$ is divided by $x - 2$, the remainder is

$$f(2) = 2^7 - 6(2^5) + 4(2^4) - 2^2 + 3(2) - 7 = -5.$$

If we were to divide $f(x)$ by $x + 1$, then the remainder would be $f(-1) = -2$.

b) If $g(x) = x^5 + 3x^4 + x^3 + x^2 + 2x + 2 \in \mathbf{Z}_5[x]$ is divided by $x - 1$, then the remainder here is $g(1) = 1 + 3 + 1 + 1 + 2 + 2 = 0$ (in $\mathbf{Z}_5$). Consequently, $x - 1$ divides $g(x)$, and by the factor theorem,

$$g(x) = q(x)(x - 1) \qquad \text{(where degree } q(x) = 4).$$

Using the results of Theorems 17.4 and 17.5, we now establish the last major idea for this section.

THEOREM 17.6

If $f(x) \in F[x]$ has degree $n \geq 1$, then $f(x)$ has at most n roots in F.

Proof: The proof is by mathematical induction on the degree of $f(x)$. If $f(x)$ has degree 1, then $f(x) = ax + b$, for $a, b \in F$, $a \neq 0$. With $f(-a^{-1}b) = 0$, $f(x)$ has at least one root in F. If c_1 and c_2 are both roots, then $f(c_1) = ac_1 + b = 0 = ac_2 + b = f(c_2)$. By cancellation in a ring, $ac_1 + b = ac_2 + b \Rightarrow ac_1 = ac_2$. Since F is a field and $a \neq 0$, we have $ac_1 = ac_2 \Rightarrow c_1 = c_2$, so $f(x)$ has only one root in F.

Now assume the result of the theorem is true for all polynomials of degree k (≥ 1) in $F[x]$. Consider a polynomial $f(x)$ of degree $k + 1$. If $f(x)$ has no roots in F, the theorem follows. Otherwise, let $r \in F$ with $f(r) = 0$. By the factor theorem, $f(x) = (x - r)g(x)$ where $g(x)$ has degree k. Consequently, by the induction hypothesis, $g(x)$ has at most k roots in F, and $f(x)$, in turn, has at most $k + 1$ roots in F.

EXAMPLE 17.7

a) Let $f(x) = x^2 - 6x + 9 \in \mathbf{R}[x]$. Then $f(x)$ has at most two roots in $\mathbf{R}$ — namely, the roots 3, 3. So here we say that 3 is a root of *multiplicity* 2. In addition $f(x) = (x - 3)(x - 3)$, a factorization into two first-degree, or *linear*, factors.

b) For $g(x) = x^2 + 4 \in \mathbf{R}[x]$, $g(x)$ has no real roots, but Theorem 17.6 is not contradicted. (Why?) In $\mathbf{C}[x]$, $g(x)$ has the roots $2i$, $-2i$ and can be factored as $g(x) = (x - 2i)(x + 2i)$.

c) If $h(x) = x^2 + 2x + 6 \in \mathbf{Z}_7[x]$, then $h(2) = 0$, $h(3) = 0$ and these are the only roots. Also, $h(x) = (x - 2)(x - 3) = x^2 - 5x + 6 = x^2 + 2x + 6$, because $[-5] = [2]$ in $\mathbf{Z}_7$.

d) As we saw in Example 17.3(b), the polynomial $x^2 + 3x + 2$ has four roots. This is not a contradiction to Theorem 17.6 because $\mathbf{Z}_6$ is not a field. Also, $x^2 + 3x + 2 = (x + 1)(x + 2) = (x + 4)(x + 5)$, two distinct factorizations.

We close with one final remark, without proof, on the idea of factorization in $F[x]$. If $f(x) \in F[x]$ has degree n, and $r_1, r_2, \ldots, r_n$ are the roots of $f(x)$ in F (where it is possible for a root to be repeated — that is, $r_i = r_j$ for some $1 \le i < j \le n$), then $f(x) = a_n(x - r_1)(x - r_2) \cdots (x - r_n)$, where a_n is the leading coefficient of $f(x)$. This representation of $f(x)$ is unique up to the order of the first-degree factors.

EXERCISES 17.1

1. Let $f(x), g(x) \in \mathbf{Z}_7[x]$ where $f(x) = 2x^4 + 2x^3 + 3x^2 + x + 4$ and $g(x) = 3x^3 + 5x^2 + 6x + 1$. Determine $f(x) + g(x)$, $f(x) - g(x)$, and $f(x)g(x)$.

2. Determine all of the polynomials of degree 2 in $\mathbf{Z}_2[x]$.

3. How many polynomials are there of degree 2 in $\mathbf{Z}_{11}[x]$? How many have degree 3? degree 4? degree n, for $n \in \mathbf{N}$?

4. a) Find two nonzero polynomials $f(x), g(x)$ in $\mathbf{Z}_{12}[x]$ where $f(x)g(x) = 0$.

b) Find polynomials $h(x), k(x) \in \mathbf{Z}_{12}[x]$ such that degree $h(x) = 5$, degree $k(x) = 2$, and degree $h(x)k(x) = 3$.

5. Complete the proofs of Theorem 17.1 and Corollary 17.1.

6. For each of the following pairs $f(x), g(x)$, find $q(x)$, $r(x)$ so that $g(x) = q(x)f(x) + r(x)$, where $r(x) = 0$ or degree $r(x) <$ degree $f(x)$.

a) $f(x), g(x) \in \mathbf{Q}[x]$, $\quad f(x) = x^4 - 5x^3 + 7x$, $\quad g(x) = x^5 - 2x^2 + 5x - 3$

b) $f(x), g(x) \in \mathbf{Z}_2[x]$, $f(x) = x^2 + 1$, $g(x) = x^4 + x^3 + x^2 + x + 1$

c) $f(x), g(x) \in \mathbf{Z}_5[x]$, $f(x) = x^2 + 3x + 1$, $g(x) = x^4 + 2x^3 + x + 4$

7. a) If $f(x) = x^4 - 16$, find its roots and factorization in $\mathbf{Q}[x]$.

b) Answer part (a) for $f(x) \in \mathbf{R}[x]$.

c) Answer part (a) for $f(x) \in \mathbf{C}[x]$.

d) Answer parts (a), (b), and (c) for $f(x) = x^4 - 25$.

8. a) Find all roots of $f(x) = x^2 + 4x$ if $f(x) \in \mathbf{Z}_{12}[x]$.

b) Find four distinct linear polynomials $g(x), h(x), s(x)$, $t(x) \in \mathbf{Z}_{12}[x]$ so that $f(x) = g(x)h(x) = s(x)t(x)$.

c) Do the results in part (b) contradict the statements made in the paragraph following Example 17.7?

9. In each of the following, find the remainder when $f(x)$ is divided by $g(x)$.

a) $f(x), g(x) \in \mathbf{Q}[x]$, $f(x) = x^8 + 7x^5 - 4x^4 + 3x^3 + 5x^2 - 4$, $g(x) = x - 3$

b) $f(x), g(x) \in \mathbf{Z}_2[x]$, $f(x) = x^{100} + x^{90} + x^{80} + x^{50} + 1$, $g(x) = x - 1$

c) $f(x), g(x) \in \mathbf{Z}_{11}[x]$, $f(x) = 3x^5 - 8x^4 + x^3 - x^2 + 4x - 7$, $g(x) = x + 9$

10. For each of the following polynomials $f(x) \in \mathbf{Z}_7[x]$, determine all of the roots in $\mathbf{Z}_7$ and then write $f(x)$ as a product of first-degree polynomials.

a) $f(x) = x^3 + 5x^2 + 2x + 6$

b) $f(x) = x^7 - x$

11. How many units are there in the ring $\mathbf{Z}_5[x]$? How many in $\mathbf{Z}_7[x]$? How many in $\mathbf{Z}_p[x]$, p a prime?

12. Given a field F, let $f(x) \in F[x]$ where $f(x) = a_n x^n + a_{n-1}x^{n-1} + \cdots + a_2 x^2 + a_1 x + a_0$. Prove that $x - 1$ is a factor of $f(x)$ if and only if

$$a_n + a_{n-1} + \cdots + a_2 + a_1 + a_0 = 0.$$

13. Let R, S be rings, and let $g: R \to S$ be a ring homomorphism. Prove that the function $G: R[x] \to S[x]$ defined by

$$G\left(\sum_{i=0}^{n} r_i x^i\right) = \sum_{i=0}^{n} g(r_i)x^i$$

is a ring homomorphism.

14. If R is an integral domain, prove that if $f(x)$ is a unit in $R[x]$, then $f(x)$ is a constant and is a unit in R.

15. Verify that $f(x) = 2x + 1$ is a unit in $\mathbf{Z}_4[x]$. Does this contradict the result of Exercise 14?

16. For $n \in \mathbf{Z}^+, n \ge 2$, let $f(x) \in \mathbf{Z}_n[x]$. Prove that if $a, b \in \mathbf{Z}$ and $a \equiv b \pmod{n}$, then $f(a) \equiv f(b) \pmod{n}$.

17. If F is a field, let $S \subseteq F[x]$ where $f(x) = a_n x^n + a_{n-1}x^{n-1} + \cdots + a_2 x^2 + a_1 x + a_0 \in S$ if and only if $a_n + a_{n-1} + \cdots + a_2 + a_1 + a_0 = 0$. Prove that S is an ideal of $F[x]$.

18. Let $(R, +, \cdot)$ be a ring. If I is an ideal of R, prove that $I[x]$, the set of all polynomials in the indeterminate x with coefficients in I, is an ideal in $R[x]$.

17.2
Irreducible Polynomials: Finite Fields

We now wish to construct finite fields other than those of the type $(\mathbf{Z}_p, +, \cdot)$, where p is a prime. The construction will use the following special polynomials.

Definition 17.4 Let $f(x) \in F[x]$, with F a field and degree $f(x) \geq 2$. We call $f(x)$ *reducible* (over F) if there exist $g(x), h(x) \in F[x]$, where $f(x) = g(x)h(x)$ and each of $g(x), h(x)$ has degree ≥ 1. If $f(x)$ is not reducible it is called *irreducible*, or *prime*.

Theorem 17.7 contains some useful observations about irreducible polynomials.

THEOREM 17.7 For polynomials in $F[x]$,

  a) every nonzero polynomial of degree ≤ 1 is irreducible.

  b) if $f(x) \in F[x]$ with degree $f(x) = 2$ or 3, then $f(x)$ is reducible if and only if $f(x)$ has a root in the field F.

Proof: The proof is left for the reader.

EXAMPLE 17.8
  a) The polynomial $x^2 + 1$ is irreducible in $\mathbf{Q}[x]$ and $\mathbf{R}[x]$, but in $\mathbf{C}[x]$ we find $x^2 + 1 = (x + i)(x - i)$.

  b) Let $f(x) = x^4 + 2x^2 + 1 \in \mathbf{R}[x]$. Although $f(x)$ has no real roots, it is reducible because $(x^2 + 1)^2 = x^4 + 2x^2 + 1$. Hence part (b) of Theorem 17.7 is not applicable for polynomials of degree > 3.

  c) In $\mathbf{Z}_2[x]$, $f(x) = x^3 + x^2 + x + 1$ is reducible because $f(1) = 0$. But $g(x) = x^3 + x + 1$ is irreducible because $g(0) = g(1) = 1$.

  d) Let $h(x) = x^4 + x^3 + x^2 + x + 1 \in \mathbf{Z}_2[x]$. Is $h(x)$ reducible in $\mathbf{Z}_2[x]$? Since $h(0) = h(1) = 1$, $h(x)$ has no first-degree factors, but perhaps we can find $a, b, c, d \in \mathbf{Z}_2$ such that $(x^2 + ax + b)(x^2 + cx + d) = x^4 + x^3 + x^2 + x + 1$.

  By expanding $(x^2 + ax + b)(x^2 + cx + d)$ and comparing coefficients of like powers of x, we find $a + c = 1$, $ac + b + d = 1$, $ad + bc = 1$, and $bd = 1$. With $bd = 1$, it follows that $b = 1$ and $d = 1$, so $ac + b + d = 1 \Rightarrow ac = 1 \Rightarrow a = c = 1 \Rightarrow a + c = 0$. This contradicts $a + c = 1$. Consequently, $h(x)$ is irreducible in $\mathbf{Z}_2[x]$.

All of the polynomials in Example 17.8 share a common property, which we shall now define.

Definition 17.5 A polynomial $f(x) \in F[x]$ is called *monic* if its leading coefficient is 1, the unity of F.

Some of our next results (up to and including the discussion in Example 17.11) awaken memories of Chapters 4 and 14.

Definition 17.6 If $f(x), g(x) \in F[x]$, then $h(x) \in F[x]$ is *a greatest common divisor* of $f(x)$ and $g(x)$

  a) if $h(x)$ divides each of $f(x)$ and $g(x)$, and

  b) if $k(x) \in F[x]$ and $k(x)$ divides both $f(x), g(x)$, then $k(x)$ divides $h(x)$.

We now state the following results on the existence and uniqueness of what we shall call *the* greatest common divisor, which we shall abbreviate as gcd. Furthermore, there is a method for finding this gcd that is called the Euclidean algorithm for polynomials. A proof for the first result is outlined in the Section Exercises.

THEOREM 17.8

Let $f(x)$, $g(x) \in F[x]$, with at least one of $f(x)$, $g(x)$ not the zero polynomial. Then each polynomial of minimum degree that can be written as a linear combination of $f(x)$ and $g(x)$ — that is, in the form $s(x)f(x) + t(x)g(x)$, for $s(x)$, $t(x) \in F[x]$ — will be a greatest common divisor of $f(x)$, $g(x)$. If we require a gcd to be monic, then it will be unique.

THEOREM 17.9

Euclidean Algorithm for Polynomials. Let $f(x)$, $g(x) \in F[x]$ with degree $f(x) \leq$ degree $g(x)$ and $f(x) \neq 0$. Applying the division algorithm, we write

$$g(x) = q(x)f(x) + r(x), \qquad \text{degree } r(x) < \text{degree } f(x)$$
$$f(x) = q_1(x)r(x) + r_1(x), \qquad \text{degree } r_1(x) < \text{degree } r(x)$$
$$r(x) = q_2(x)r_1(x) + r_2(x), \qquad \text{degree } r_2(x) < \text{degree } r_1(x)$$
$$\vdots \qquad\qquad\qquad \vdots$$
$$r_{k-2}(x) = q_k(x)r_{k-1}(x) + r_k(x), \qquad \text{degree } r_k(x) < \text{degree } r_{k-1}(x)$$
$$r_{k-1}(x) = q_{k+1}(x)r_k(x) + r_{k+1}(x), \qquad r_{k+1}(x) = 0.$$

Then $r_k(x)$, the last nonzero remainder, is a greatest common divisor of $f(x)$, $g(x)$, and is a constant multiple of the monic greatest common divisor of $f(x)$, $g(x)$. [Multiplying $r_k(x)$ by the inverse of its leading coefficient allows us to obtain the unique monic polynomial we call *the* greatest common divisor.]

Definition 17.7

If $f(x)$, $g(x) \in F[x]$ and their gcd is 1, then $f(x)$ and $g(x)$ are called *relatively prime*.

The last results we need to construct our new finite fields provide the analog of a construction we developed in Section 14.3.

THEOREM 17.10

Let $s(x) \in F(x)$, $s(x) \neq 0$. Define relation $\mathcal{R}$ on $F[x]$ by $f(x) \mathcal{R} g(x)$ if $f(x) - g(x) = t(x)s(x)$, for some $t(x) \in F[x]$ — that is, $s(x)$ divides $f(x) - g(x)$. Then $\mathcal{R}$ is an equivalence relation on $F[x]$.

Proof: The verification of the reflexive, symmetric, and transitive properties of $\mathcal{R}$ is left for the reader.

When the situation in Theorem 17.10 occurs, we say that $f(x)$ is *congruent* to $g(x)$ *modulo* $s(x)$ and write $f(x) \equiv g(x) \pmod{s(x)}$. The relation $\mathcal{R}$ is referred to as *congruence modulo* $s(x)$.

Let us examine the equivalence classes for one such relation.

EXAMPLE 17.9

Let $s(x) = x^2 + x + 1 \in \mathbf{Z}_2[x]$. Then

a) $[0] = [x^2 + x + 1] = \{0, x^2 + x + 1, x^3 + x^2 + x, (x+1)(x^2 + x + 1), \ldots\}$
$= \{t(x)(x^2 + x + 1) | t(x) \in \mathbf{Z}_2[x]\}$

b) $[1] = \{1, x^2 + x, x(x^2 + x + 1) + 1, (x + 1)(x^2 + x + 1) + 1, \ldots\}$
$\qquad = \{t(x)(x^2 + x + 1) + 1 | t(x) \in \mathbf{Z}_2[x]\}$

c) $[x] = \{x, x^2 + 1, x(x^2 + x + 1) + x, (x + 1)(x^2 + x + 1) + x, \ldots\}$
$\qquad = \{t(x)(x^2 + x + 1) + x | t(x) \in \mathbf{Z}_2[x]\}$

d) $[x + 1] = \{x + 1, x^2, x(x^2 + x + 1) + (x + 1), (x + 1)(x^2 + x + 1)$
$\qquad\qquad + (x + 1), \ldots\} = \{t(x)(x^2 + x + 1) + (x + 1) | t(x) \in \mathbf{Z}_2[x]\}$

Are these all of the equivalence classes? If $f(x) \in \mathbf{Z}_2[x]$, then by the division algorithm $f(x) = q(x)s(x) + r(x)$, where $r(x) = 0$ or degree $r(x) <$ degree $s(x)$. Since $f(x) - r(x) = q(x)s(x)$, it follows that $f(x) \equiv r(x) \pmod{s(x)}$, so $f(x) \in [r(x)]$. Consequently, to determine all the equivalence classes, we consider the possibilities for $r(x)$. Here $r(x) = 0$ or degree $r(x) < 2$, so $r(x) = ax + b$, where $a, b \in \mathbf{Z}_2$. With only two choices for each of a, b, there are four possible choices for $r(x)$: $0, 1, x, x + 1$.

We now place a ring structure on the equivalence classes of Example 17.9. Recalling how this was accomplished in Chapter 14 for $\mathbf{Z}_n$, we define addition by $[f(x)] + [g(x)] = [f(x) + g(x)]$. Since degree $(f(x) + g(x)) \leq \max\{\text{degree } f(x), \text{degree } g(x)\}$, we can find the equivalence class for $[f(x) + g(x)]$ without too much trouble. Here, for example, $[x] + [x + 1] = [x + (x + 1)] = [2x + 1] = [1]$ because $2 = 0$ in $\mathbf{Z}_2$.

In defining the multiplication of these equivalence classes, we run into a little more difficulty. For instance, what is $[x][x]$ in Example 17.9? If, in general, we define $[f(x)][g(x)] = [f(x)g(x)]$, it is possible that degree $f(x)g(x) \geq$ degree $s(x)$, so we may not readily find $[f(x)g(x)]$ in the list of equivalence classes. However, if degree $f(x)g(x) \geq$ degree $s(x)$, then using the division algorithm, we can write $f(x)g(x) = q(x)s(x) + r(x)$, where $r(x) = 0$ or degree $r(x) <$ degree $s(x)$. With $f(x)g(x) = q(x)s(x) + r(x)$, it follows that $f(x)g(x) \equiv r(x) \pmod{s(x)}$, and we define $[f(x)g(x)] = [r(x)]$, where $[r(x)]$ *does* occur in the list of equivalence classes.

From these observations we construct Tables 17.1 and 17.2 for the addition and multiplication, respectively, of $\{[0], [1], [x], [x + 1]\}$. (In these tables we write a for $[a]$.)

Table 17.1

+	0	1	x	$x + 1$
0	0	1	x	$x + 1$
1	1	0	$x + 1$	x
x	x	$x + 1$	0	1
$x + 1$	$x + 1$	x	1	0

Table 17.2

$\cdot$	0	1	x	$x + 1$
0	0	0	0	0
1	0	1	x	$x + 1$
x	0	x	$x + 1$	1
$x + 1$	0	$x + 1$	1	x

From the multiplication table (Table 17.2), we find that these equivalence classes form not only a ring but also a field, where $[1]^{-1} = [1]$, $[x]^{-1} = [x + 1]$, and $[x + 1]^{-1} = [x]$. This field of order 4 is denoted by $\mathbf{Z}_2[x]/(x^2 + x + 1)$, and we observe that it contains (an isomorphic copy of) the *subfield* $\mathbf{Z}_2$. [In general, a subring $(R, +, \cdot)$ of a field $(F, +, \cdot)$ is called a subfield when $(R, +, \cdot)$ is a field.] In addition, for the nonzero elements of this field we find that $[x]^1 = [x]$, $[x]^2 = [x + 1]$, $[x]^3 = [1]$, so we have a cyclic group of order 3. But the nonzero elements of any field form a group under multiplication, and any group of order 3 is cyclic, so why bother with this observation? In general, the nonzero elements of *any* finite field form a cyclic group under multiplication. (A proof for this can be found in Chapter 12 of reference [10].)

The preceding construction is summarized in the following theorem. An outline of the proof is given in the Section Exercises.

THEOREM 17.11   Let $s(x)$ be a nonzero polynomial in $F[x]$.

a) The equivalence classes of $F[x]$ for the relation of congruence modulo $s(x)$ form a commutative ring with unity under the closed binary operations

$$[f(x)] + [g(x)] = [f(x) + g(x)], \qquad [f(x)][g(x)] = [f(x)g(x)] = [r(x)],$$

where $r(x)$ is the remainder obtained upon dividing $f(x)g(x)$ by $s(x)$. This ring is denoted by $F[x]/(s(x))$.

b) If $s(x)$ is irreducible in $F[x]$, then $F[x]/(s(x))$ is a field.

c) If $|F| = q$ and degree $s(x) = n$, then $F[x]/(s(x))$ contains q^n elements.

Before we continue we wish to emphasize that for $s(x)$ irreducible in $F[x]$ the elements in the field $F[x]/(s(x))$ are *not* simply polynomials (in x). But how can this be, considering the presence of the symbol x in each of the elements $[x]$ and $[x + 1]$ in the field $\mathbf{Z}_2[x]/(x^2 + x + 1)$ of Example 17.9? In order to make our point more apparent we consider an infinite example that is somewhat familiar to us.

EXAMPLE 17.10   Here we let $F = (\mathbf{R}, +, \cdot)$, the field of real numbers, and we consider the irreducible polynomial $s(x) = x^2 + 1$ in $\mathbf{R}[x]$. From part (b) of Theorem 17.11 we learn that $\mathbf{R}[x]/(s(x)) = \mathbf{R}[x]/(x^2 + 1)$ is a field.

For all $f(x) \in \mathbf{R}[x]$ it follows by the division algorithm that

$$f(x) = q(x)(x^2 + 1) + r(x), \quad \text{where } r(x) = 0 \text{ or } 0 \leq \deg r(x) \leq 1.$$

Therefore,

$$\mathbf{R}[x]/(x^2 + 1) = \{[a + bx] \mid a, b \in \mathbf{R}\},$$

where it can be shown that $[a + bx] = [a] + [bx] = [a] + [b][x]$.

Among the (infinitely many) elements of $\mathbf{R}[x]/(x^2 + 1)$ are the following:

1) $[1] = \{1 + t(x)(x^2 + 1) \mid t(x) \in \mathbf{R}[x]\}$, where we find the elements $x^2 + 2$ and $3x^3 + 3x + 1$ (from $\mathbf{R}[x]$);

2) $[r] = \{r + t(x)(x^2 + 1) \mid t(x) \in \mathbf{R}[x]\}$, where r is any (fixed) real number;

3) $[-1] = \{-1 + t(x)(x^2 + 1) \mid t(x) \in \mathbf{R}[x]\}$, where we find the polynomial $-1 + (1)(x^2 + 1) = x^2$ — so, $[x][x] = [x^2] = [-1]$; and

4) $[\sqrt{2}x - 3] = \{(\sqrt{2}x - 3) + t(x)(x^2 + 1) \mid t(x) \in \mathbf{R}[x]\}$.

Now let us consider the field $(\mathbf{C}, +, \cdot)$ of complex numbers and the correspondence

$$h: \mathbf{R}[x]/(x^2 + 1) \to \mathbf{C},$$

where $h([a + bx]) = a + bi$.

For all $[a + bx], [c + dx] \in \mathbf{R}[x]/(x^2 + 1)$, we have $[a + bx] = [c + dx] \Longleftrightarrow (a + bx) - (c + dx) = t(x)(x^2 + 1)$, for some $t(x) \in \mathbf{R}[x] \Longleftrightarrow (a - c) + (b - d)x = t(x)(x^2 + 1)$. If $t(x)$ is not the zero polynomial, then we have $(a - c) + (b - d)x$, a polynomial of degree less than 2, equal to $t(x)(x^2 + 1)$, a polynomial of degree at least 2. Consequently, $t(x) = 0$, so $a + bx = c + dx$ and $a = c$, $b = d$. This guarantees that the correspondence given by h is actually a function. In fact, h is an isomorphism of fields. (See Exercise 24 in the exercises at the end of this section.) To establish that h preserves the operation of multiplication, for example, we observe that

$$\begin{aligned}
h([a + bx][c + dx]) &= h([ac + adx + bcx + bdx^2]) \\
&= h([ac + (ad + bc)x] + [bd][x^2]) \\
&= h([ac + (ad + bc)x] + [bd][-1]) \\
&= h([ac - bd) + (ad + bc)x]) \\
&= (ac - bd) + (ad + bc)i = (a + bi)(c + di) \\
&= h([a + bx])h([c + dx]).
\end{aligned}$$

Since $\mathbf{R}[x]/(x^2 + 1)$ is isomorphic to $\mathbf{C}$, the correspondence $h([x]) = i$ makes us think of $[x]$ as a *number* in $\mathbf{R}[x]/(x^2 + 1)$ and not as a polynomial in x (in $\mathbf{R}[x]$). The number $[x]$ represents an equivalence class of polynomials in $\mathbf{R}[x]$, and this number $[x]$ behaves like the complex number i in the field $(\mathbf{C}, +, \cdot)$. We should also note that for each real number r, $h([r]) = r$, and $\{[r] | r \in \mathbf{R}\}$ is a subfield of $\mathbf{R}[x]/(x^2 + 1)$, which is isomorphic to the subfield $\mathbf{R}$ of $\mathbf{C}$.

Finally, if we identify the field $\mathbf{R}[x]/(x^2 + 1)$ with the field $(\mathbf{C}, +, \cdot)$, we can summarize what has happened above as follows: We started with the irreducible polynomial $s(x) = x^2 + 1$ in $\mathbf{R}[x]$, which had no root in the field $(\mathbf{R}, +, \cdot)$. We then enlarged $(\mathbf{R}, +, \cdot)$ to $(\mathbf{C}, +, \cdot)$ and in $\mathbf{C}$ we found the root i (and the root $-i$) for $s(x)$, which can now be factored as $(x + i)(x - i)$ in $\mathbf{C}[x]$.

Since our major concern in the chapter is with finite fields, we now examine another example of a finite field that arises by virtue of Theorem 17.11.

EXAMPLE 17.11

In $\mathbf{Z}_3[x]$ the polynomial $s(x) = x^2 + x + 2$ is irreducible because $s(0) = 2$, $s(1) = 1$, and $s(2) = 2$. Consequently, $\mathbf{Z}_3[x]/(s(x))$ is a field containing all equivalence classes of the form $[ax + b]$, where $a, b \in \mathbf{Z}_3$. These arise from the possible remainders when a polynomial $f(x) \in \mathbf{Z}_3[x]$ is divided by $s(x)$. The nine equivalence classes are $[0]$, $[1]$, $[2]$, $[x]$, $[x + 1]$, $[x + 2]$, $[2x]$, $[2x + 1]$, and $[2x + 2]$.

Instead of constructing a complete multiplication table, we examine four sample multiplications and then make two observations.

a) $[2x][x] = [2x^2] = [2x^2 + 0] = [2x^2 + (x^2 + x + 2)] = [3x^2 + x + 2] = [x + 2]$ because $3 = 0$ in $\mathbf{Z}_3$.

b) $[x + 1][x + 2] = [x^2 + 3x + 2] = [x^2 + 2] = [x^2 + 2 + 2(x^2 + x + 2)] = [2x]$.

c) $[2x + 2]^2 = [4x^2 + 8x + 4] = [x^2 + 2x + 1] = [(-x - 2) + (2x + 1)]$ since $x^2 \equiv (-x - 2) \pmod{s(x)}$. Consequently, $[2x + 2]^2 = [x - 1] = [x + 2]$.

d) Often we write the equivalence classes without brackets and concentrate on the coefficients of the powers of x. For example, 11 is written for $[x + 1]$ and 21 represents $[2x + 1]$. Consequently, $(21) \cdot (12) = [2x + 1][x + 2] = [2x^2 + 5x + 2] = [2x^2 + 2x + 2] = [2(-x - 2) + 2x + 2] = [-4 + 2] = [-2] = [1]$, so $(21)^{-1} = (12)$.

e) We also observe that

$[x]^1 = [x]$	$[x]^3 = [2x + 2]$	$[x]^5 = [2x]$	$[x]^7 = [x + 1]$
$[x]^2 = [2x + 1]$	$[x]^4 = [2]$	$[x]^6 = [x + 2]$	$[x]^8 = [1]$

Therefore the nonzero elements of $\mathbf{Z}_3[x]/(s(x))$ form a cyclic group under multiplication.

f) Finally, when we consider the equivalence classes $[0], [1], [2]$, we realize that they provide us with a subfield of $\mathbf{Z}_3[x]/(s(x))$ — a subfield we identify with the field $(\mathbf{Z}_3, +, \cdot)$.

In Example 17.9 (and in the discussion that follows it) and in Example 17.11, we constructed finite fields of orders $4 \,(= 2^2)$ and $9 \,(= 3^2)$, respectively. Now we shall close this section as we investigate other possibilities for the order of a finite field. To accomplish this we need the following idea.

Definition 17.8   Let $(R, +, \cdot)$ be a ring. If there is a least positive integer n such that $nr = z$ (the zero of R) for all $r \in R$, then we say that R has *characteristic n* and write $\operatorname{char}(R) = n$. When no such integer exists, R is said to have *characteristic 0*.

EXAMPLE 17.12

a) The ring $(\mathbf{Z}_3, +, \cdot)$ has characteristic 3; $(\mathbf{Z}_4, +, \cdot)$ has characteristic 4; in general, $(\mathbf{Z}_n, +, \cdot)$ has characteristic n.

b) The rings $(\mathbf{Z}, +, \cdot)$ and $(\mathbf{Q}, +, \cdot)$ both have characteristic 0.

c) A ring can be infinite and still have positive characteristic. For example, $\mathbf{Z}_3[x]$ is an infinite ring but it has characteristic 3.

d) The ring in Example 17.9 has characteristic 2. In Example 17.11 the characteristic of the ring is 3. Unlike the examples in part (a), the order of a finite ring can be different from its characteristic.

Examples 17.9 and 17.11, however, are more than just rings. They are fields with prime characteristic. Could this property be true for all finite fields?

THEOREM 17.12   Let $(F, +, \cdot)$ be a field. If $\operatorname{char}(F) > 0$, then $\operatorname{char}(F)$ must be prime.

Proof: In this proof we write the unity of F as u so that it is distinct from the positive integer 1. Let $\operatorname{char}(F) = n > 0$. If n is not prime, we write $n = mk$, where $m, k \in \mathbf{Z}^+$ and $1 < m < n$, $1 < k < n$. By the definition of characteristic, $nu = z$, the zero of F. Hence $(mk)u = z$. But

$$(mk)(u) = \underbrace{(u + u + \cdots + u)}_{mk \text{ summands}} = \underbrace{(u + u + \cdots + u)}_{m \text{ summands}} \underbrace{(u + u + \cdots + u)}_{k \text{ summands}} = (mu)(ku).$$

With F a field, $(mu)(ku) = z \Rightarrow (mu) = z$ or $(ku) = z$. Assume without loss of generality that $ku = z$. Then for each $r \in F$, $kr = k(ur) = (ku)r = zr = z$, contradicting the choice of n as the characteristic of F. Consequently, $\operatorname{char}(F)$ is prime.

(The proof of Theorem 17.12 actually requires that F only be an integral domain.)

If F is a finite field and $m = |F|$, then $ma = z$ for all $a \in F$ because $(F, +)$ is an additive group of order m. (See Exercise 8 of Section 16.3.) Consequently, F has positive characteristic and by Theorem 17.12 this characteristic is prime. This leads us to the following theorem.

THEOREM 17.13   A finite field F has order p^t, where p is a prime and $t \in \mathbf{Z}^+$.

Proof: Since F is a finite field, let char$(F) = p$, a prime, and let u denote the unity and z the zero element. Then $S_0 = \{u, 2u, 3u, \ldots, pu = z\}$ is a set of p distinct elements in F. If not, $mu = nu$ for $1 \le m < n \le p$ and $(n - m)u = z$, with $0 < n - m < p$. So for all $x \in F$ we now find that $(n - m)x = (n - m)(ux) = [(n - m)u]x = zx = z$, and this contradicts char$(F) = p$. If $F = S_0$, then $|F| = p^1$ and the result follows. If not, let $a \in F - S_0$. Then $S_1 = \{ma + nu \mid 0 < m, n \le p\}$ is a subset of F with $|S_1| \le p^2$. If $|S_1| < p^2$, then $m_1 a + n_1 u = m_2 a + n_2 u$, with $0 < m_1, m_2, n_1, n_2 \le p$ and at least one of $m_1 - m_2, n_2 - n_1 \ne 0$. Should $m_1 - m_2 = 0$, then $(m_1 - m_2)a = z = (n_2 - n_1)u$, with $0 < |n_2 - n_1| < p$. Consequently, for all $x \in F$, $|n_2 - n_1|x = |n_2 - n_1|(ux) = (|n_2 - n_1|u)x = zx = z$ with $0 < |n_2 - n_1| < p = $ char(F), another contradiction. If $n_1 - n_2 = 0$, then $(m_1 - m_2)a = z$ with $0 < |m_1 - m_2| < p$. Since F is a field and $a \ne z$ we know that $a^{-1} \in F$, so $|m_1 - m_2|u = |m_1 - m_2|aa^{-1} = za^{-1} = z$ with $0 < |m_1 - m_2| < p$—yet another contradiction. Hence neither $m_1 - m_2$ nor $n_1 - n_2$ is 0. Therefore, $(m_1 - m_2)a = (n_2 - n_1)u \ne z$. Choose $k \in \mathbf{Z}^+$ such that $0 < k < p$ and $k(m_1 - m_2) \equiv 1 \pmod{p}$. Then $a = k(m_1 - m_2)a = k(n_2 - n_1)u$, and $a \in S_0$, one more contradiction. Hence $|S_1| = p^2$, and if $F = S_1$ the theorem is proved. If not, continue this process with an element $b \in F - S_1$. Then $S_2 = \{\ell b + ma + nu \mid 0 < \ell, m, n \le p\}$ will have order p^3. (Prove this.) Since F is finite, we reach a point where $F = S_{t-1}$ for some $t \in \mathbf{Z}^+$, and $|F| = |S_{t-1}| = p^t$.

As a result of this theorem there can be no finite fields with orders such as 6, 10, 12, 14, 15, ... In addition, for each prime p and each $t \in \mathbf{Z}^+$, there is really only one field of order p^t. Any two finite fields of the same order are isomorphic. These fields were discovered by the French mathematician Evariste Galois (1811–1832) in his work on the nonexistence of formulas for solving general polynomial equations of degree ≥ 5 over $\mathbf{Q}$. As a result, a finite field of order p^t is denoted by $GF(p^t)$, where the letters GF stand for *Galois field*.

EXERCISES 17.2

1. Determine whether or not each of the following polynomials is irreducible over the given fields. If it is reducible, provide a factorization into irreducible factors.

  a) $x^2 + 3x - 1$ over $\mathbf{Q}, \mathbf{R}, \mathbf{C}$

  b) $x^4 - 2$ over $\mathbf{Q}, \mathbf{R}, \mathbf{C}$

  c) $x^2 + x + 1$ over $\mathbf{Z}_3, \mathbf{Z}_5, \mathbf{Z}_7$

  d) $x^4 + x^3 + 1$ over $\mathbf{Z}_2$

  e) $x^3 + 3x^2 - x + 1$ over $\mathbf{Z}_5$

2. Give an example of a polynomial $f(x) \in \mathbf{R}[x]$ where $f(x)$ has degree 6, is reducible, but has no real roots.

3. Determine all polynomials $f(x) \in \mathbf{Z}_2[x]$ such that $1 \le$ degree $f(x) \le 3$ and $f(x)$ is irreducible (over $\mathbf{Z}_2$).

4. Let $f(x) = (2x^2 + 1)(5x^3 - 5x + 3)(4x - 3) \in \mathbf{Z}_7[x]$. Write $f(x)$ as the product of a unit and three monic polynomials.

5. How many monic polynomials in $\mathbf{Z}_7[x]$ have degree 5?

6. Prove Theorem 17.7.

7. An outline for a proof of Theorem 17.8 follows.

  a) Let $S = \{s(x)f(x) + t(x)g(x) \mid s(x), t(x) \in F[x]\}$. Select an element $m(x)$ of minimum degree in S. (Recall that the zero polynomial has no degree, so it is not selected.) Can we guarantee that $m(x)$ is monic?

  b) Show that if $h(x) \in F[x]$ and $h(x)$ divides both $f(x)$ and $g(x)$, then $h(x)$ divides $m(x)$.

  c) Show that $m(x)$ divides $f(x)$. If not, use the division algorithm and write $f(x) = q(x)m(x) + r(x)$, where $r(x) \ne 0$ and degree $r(x) <$ degree $m(x)$. Then show that $r(x) \in S$ and obtain a contradiction.

  d) Repeat the argument in part (c) to show that $m(x)$ divides $g(x)$.

8. Prove Theorems 17.9 and 17.10.

9. Use the Euclidean algorithm for polynomials to find the gcd of each pair of polynomials, over the designated field F. Then write the gcd as $s(x)f(x) + t(x)g(x)$, where $s(x), t(x) \in F[x]$.

  a) $f(x) = x^2 + x - 2$, $g(x) = x^5 - x^4 + x^3 + x^2 - x - 1$ in $\mathbf{Q}[x]$

  b) $f(x) = x^4 + x^3 + 1$, $g(x) = x^2 + x + 1$ in $\mathbf{Z}_2[x]$

c) $f(x) = x^4 + 2x^2 + 2x + 2$, $g(x) = 2x^3 + 2x^2 + x + 1$ in $\mathbf{Z}_3[x]$

10. If F is any field, let $f(x), g(x) \in F[x]$. If $f(x), g(x)$ are relatively prime, prove that there is no element $a \in F$ with $f(a) = 0$ and $g(a) = 0$.

11. Let $f(x), g(x) \in \mathbf{R}[x]$ with $f(x) = x^3 + 2x^2 + ax - b$, $g(x) = x^3 + x^2 - bx + a$. Determine values for a, b so that the gcd of $f(x), g(x)$ is a polynomial of degree 2.

12. For Example 17.9, determine which equivalence class contains each of the following:

a) $x^4 + x^3 + x + 1$

b) $x^3 + x^2 + 1$

c) $x^4 + x^3 + x^2 + 1$

13. An outline for the proof of Theorem 17.11 follows.

a) Prove that the operations defined in part (a) of Theorem 17.11 are well-defined by showing that if $f(x) \equiv f_1(x) \pmod{s(x)}$ and $g(x) \equiv g_1(x) \pmod{s(x)}$, then $f(x) + g(x) \equiv f_1(x) + g_1(x) \pmod{s(x)}$ and $f(x)g(x) \equiv f_1(x)g_1(x) \pmod{s(x)}$.

b) Verify the ring properties for the equivalence classes in $F[x]/(s(x))$.

c) Let $f(x) \in F[x]$, with $f(x) \neq 0$ and degree $f(x) < $ degree $s(x)$. If $s(x)$ is irreducible in $F[x]$, why does it follow that 1 is the gcd of $f(x)$ and $s(x)$?

d) Use part (c) to prove that if $s(x)$ is irreducible in $F[x]$, then $F[x]/(s(x))$ is a field.

e) If $|F| = q$ and degree $s(x) = n$, determine the order of $F[x]/(s(x))$.

14. a) Show that $s(x) = x^2 + 1$ is reducible in $\mathbf{Z}_2[x]$.

b) Find the equivalence classes for the ring $\mathbf{Z}_2[x]/(s(x))$.

c) Is $\mathbf{Z}_2[x]/(s(x))$ an integral domain?

15. For the field in Example 17.11, find each of the following:

a) $[x + 2][2x + 2] + [x + 1]$

b) $[2x + 1]^2[x + 2]$

c) $(22)^{-1} = [2x + 2]^{-1}$

16. Let $s(x) = x^4 + x^3 + 1 \in \mathbf{Z}_2[x]$.

a) Prove that $s(x)$ is irreducible.

b) What is the order of the field $\mathbf{Z}_2[x]/(s(x))$?

c) Find $[x^2 + x + 1]^{-1}$ in $\mathbf{Z}_2[x]/(s(x))$. (*Hint:* Find a, b, c, $d \in \mathbf{Z}_2$ so that $[x^2 + x + 1][ax^3 + bx^2 + cx + d] = [1]$.)

d) Determine $[x^3 + x + 1][x^2 + 1]$ in $\mathbf{Z}_2[x]/(s(x))$.

17. For p a prime, let $s(x)$ be irreducible of degree n in $\mathbf{Z}_p[x]$.

a) How many elements are there in the field $\mathbf{Z}_p[x]/(s(x))$?

b) How many elements in $\mathbf{Z}_p[x]/(s(x))$ generate the multiplicative group of nonzero elements of this field?

18. Give the characteristic for each of the following rings:

a) $\mathbf{Z}_{11}$ **b)** $\mathbf{Z}_{11}[x]$ **c)** $\mathbf{Q}[x]$

d) $\mathbf{Z}[\sqrt{5}] = \{a + b\sqrt{5} | a, b \in \mathbf{Z}\}$, under the binary operations of ordinary addition and multiplication of real numbers.

19. In each of the following rings, the operations are componentwise addition and multiplication, as in Exercise 18 of Section 14.2. Determine the characteristic in each case.

a) $\mathbf{Z}_2 \times \mathbf{Z}_3$ **b)** $\mathbf{Z}_3 \times \mathbf{Z}_4$ **c)** $\mathbf{Z}_4 \times \mathbf{Z}_6$

d) $\mathbf{Z}_m \times \mathbf{Z}_n$, for $m, n \in \mathbf{Z}^+$, $m, n \geq 2$

e) $\mathbf{Z}_3 \times \mathbf{Z}$

20. For Theorem 17.13, prove that $|S_2| = p^3$.

21. Find the orders n for all fields $GF(n)$, where $100 \leq n \leq 150$.

22. Construct a finite field of 25 elements.

23. Construct a finite field of 27 elements.

24. a) Prove that the function h in Example 17.10 is one-to-one and onto and preserves the operation of addition.

b) Let $(F, +, \cdot)$ and $(K, \oplus, \odot)$ be two fields. If $g \colon F \to K$ is a *ring* isomorphism and a is a nonzero element of F (that is, a is a unit in F), prove that $g(a^{-1}) = [g(a)]^{-1}$. (Consequently, this function g establishes an isomorphism of fields. In particular, the function h of Example 17.10 is such a function.)

25. a) Let $\mathbf{Q}[\sqrt{2}] = \{a + b\sqrt{2} | a, b \in \mathbf{Q}\}$. Prove that $(\mathbf{Q}[\sqrt{2}], +, \cdot)$ is a subring of the field $(\mathbf{R}, +, \cdot)$. (Here the binary operations in $\mathbf{R}$ and $\mathbf{Q}[\sqrt{2}]$ are those of ordinary addition and multiplication of real numbers.)

b) Prove that $\mathbf{Q}[\sqrt{2}]$ is a field and that $\mathbf{Q}[x]/(x^2 - 2)$ is isomorphic to $\mathbf{Q}[\sqrt{2}]$.

26. Let p be a prime. (a) How many monic quadratic (degree 2) polynomials $x^2 + bx + c$ in $\mathbf{Z}_p[x]$ can we factor into linear factors in $\mathbf{Z}_p[x]$? (For example, if $p = 5$, then the polynomial $x^2 + 2x + 2$ in $\mathbf{Z}_5[x]$ would be one of the quadratic polynomials for which we should account, under these conditions.) (b) How many quadratic polynomials $ax^2 + bx + c$ in $\mathbf{Z}_p[x]$ can we factor into linear factors in $\mathbf{Z}_p[x]$? (c) How many monic quadratic polynomials $x^2 + bx + c$ in $\mathbf{Z}_p[x]$ are irreducible over $\mathbf{Z}_p$? (d) How many quadratic polynomials $ax^2 + bx + c$ in $\mathbf{Z}_p[x]$ are irreducible over $\mathbf{Z}_p$?

17.3
Latin Squares

Our first application for this chapter deals with the structure called a Latin square. Such configurations arise in the study of combinatorial designs and play a role in statistics — in the design of experiments. We introduce the structure in the following example.

EXAMPLE 17.13

A petroleum corporation is interested in testing four types of gasoline additives to determine their effects on mileage. To do so, a research team designs an experiment wherein four different automobiles, denoted A, B, C, and D, are run on a fixed track in a laboratory. Each run uses the same prescribed amount of fuel with one of the additives present. To see how each additive affects each type of auto, the team follows the schedule in Table 17.3, where the additives are numbered 1, 2, 3, and 4. This schedule provides a way to test each additive thoroughly in each type of auto. If one additive produces the best results in all four types, the experiment will reveal its superior capability.

The same corporation is also interested in testing four other additives developed for cleaning engines. A similar schedule for these tests is shown in Table 17.4, where these engine-cleaning additives are also denoted as 1, 2, 3, and 4.

Table 17.3

Auto	Day			
	Mon	**Tues**	**Wed**	**Thurs**
A	1	2	3	4
B	2	1	4	3
C	3	4	1	2
D	4	3	2	1

Table 17.4

Auto	Day			
	Mon	**Tues**	**Wed**	**Thurs**
A	1	2	3	4
B	3	4	1	2
C	4	3	2	1
D	2	1	4	3

Furthermore, the research team is interested in the combined effect of both types of additives. It requires 16 days to test the 16 possible pairs of additives (one for improved mileage, the other for cleaning engines) in every automobile. If the results are needed in four days, the research team must design the schedules so that every pair is tested once by some auto. There are 16 ordered pairs in {1, 2, 3, 4} × {1, 2, 3, 4}, so this can be done in the allotted time if the schedules in Tables 17.3 and 17.4 are superimposed to obtain the schedule in Table 17.5. Here, for example, the entry (4, 3) indicates that on Tuesday, auto C is used to test the combined effect of the fourth additive for improved mileage and the third additive for maintaining a clean engine.

Table 17.5

Auto	Day			
	Mon	**Tues**	**Wed**	**Thurs**
A	(1, 1)	(2, 2)	(3, 3)	(4, 4)
B	(2, 3)	(1, 4)	(4, 1)	(3, 2)
C	(3, 4)	(4, 3)	(1, 2)	(2, 1)
D	(4, 2)	(3, 1)	(2, 4)	(1, 3)

What has happened here leads us to the following concepts.

Definition 17.9

An $n \times n$ *Latin square* is a square array of symbols, usually $1, 2, 3, \ldots, n$, where each symbol appears exactly once in each row and each column of the array.

EXAMPLE 17.14

a) Tables 17.3 and 17.4 are examples of 4×4 Latin squares.

b) For all $n \geq 2$, we can obtain an $n \times n$ Latin square from the table of the group $(\mathbf{Z}_n, +)$ if we replace the occurrences of 0 by the value of n.

From the two Latin squares in Example 17.13 we were able to produce all of the ordered pairs in $S \times S$, for $S = \{1, 2, 3, 4\}$. We now question whether or not we can do this for $n \times n$ Latin squares in general.

Definition 17.10

Let $L_1 = (a_{ij})$, $L_2 = (b_{ij})$ be two $n \times n$ Latin squares, where $1 \leq i, j \leq n$ and each a_{ij}, $b_{ij} \in \{1, 2, 3, \ldots, n\}$. If the n^2 ordered pairs (a_{ij}, b_{ij}), $1 \leq i, j \leq n$, are distinct, then L_1, L_2 are called a *pair of orthogonal Latin squares*.

EXAMPLE 17.15

a) There is no pair of 2×2 orthogonal Latin squares because the only possibilities are

$$L_1: \begin{matrix} 1 & 2 \\ 2 & 1 \end{matrix} \quad \text{and} \quad L_2: \begin{matrix} 2 & 1 \\ 1 & 2 \end{matrix}$$

b) In the 3×3 case we find the orthogonal pair

$$L_1: \begin{matrix} 1 & 2 & 3 \\ 2 & 3 & 1 \\ 3 & 1 & 2 \end{matrix} \quad \text{and} \quad L_2: \begin{matrix} 1 & 2 & 3 \\ 3 & 1 & 2 \\ 2 & 3 & 1 \end{matrix}$$

Table 17.6

1	2	3	4
4	3	2	1
2	1	4	3
3	4	1	2

c) The two 4×4 Latin squares in Example 17.13 form an orthogonal pair. The 4×4 Latin square shown in Table 17.6 is orthogonal to each of the Latin squares in that example.

We could continue listing some larger Latin squares, but we've seen enough of them at this point to ask the following questions:

1) Is there any $n > 2$ for which there is no pair of orthogonal $n \times n$ Latin squares? If so, what is the smallest such n?

2) For $n > 1$, what can we say about the number of $n \times n$ Latin squares that can be constructed so that each pair of them is orthogonal?

3) Is there a method to assist us in constructing a pair of orthogonal $n \times n$ Latin squares for certain values of $n > 2$?

Before we can examine these questions, we need to standardize some of our results.

Definition 17.11

If L is an $n \times n$ Latin square, then L is said to be in *standard form* if its first row is $1 \quad 2 \quad 3 \quad \cdots \quad n$.

Except for the Latin square L_2 in Example 17.15(a), all the Latin squares we've seen in this section are in standard form. If a Latin square is not in standard form, it can be put in that form by interchanging some of the symbols.

EXAMPLE 17.16

The 5×5 Latin square shown in (a) is not in standard form. If, however, we replace each occurrence of 4 with 1, each occurrence of 5 with 4, and each occurrence of 1 with 5, then the result is the (standard) 5×5 Latin square shown in (b).

$$
\begin{array}{ccccc}
4 & 2 & 3 & 5 & 1 \\
1 & 3 & 5 & 4 & 2 \\
3 & 4 & 2 & 1 & 5 \\
2 & 5 & 1 & 3 & 4 \\
5 & 1 & 4 & 2 & 3 \\
\end{array}
\qquad
\begin{array}{ccccc}
1 & 2 & 3 & 4 & 5 \\
5 & 3 & 4 & 1 & 2 \\
3 & 1 & 2 & 5 & 4 \\
2 & 4 & 5 & 3 & 1 \\
4 & 5 & 1 & 2 & 3 \\
\end{array}
$$

(a)   (b)

It is often convenient to deal with Latin squares in standard form. But will this affect our results on orthogonal pairs in any way?

THEOREM 17.14

Let L_1, L_2 be an orthogonal pair of $n \times n$ Latin squares. If L_1, L_2 are standardized as L_1^*, L_2^*, then L_1^*, L_2^* are orthogonal.

Proof: The proof of this result is left for the reader.

These ideas are needed for the main results of this section.

THEOREM 17.15

In $n \in \mathbf{Z}^+$, $n > 2$, then the largest possible number of $n \times n$ Latin squares that are orthogonal in pairs is $n - 1$.

Proof: Let $L_1, L_2, \ldots, L_k$ be k distinct $n \times n$ Latin squares that are in standard form and orthogonal in pairs. We write $a_{ij}^{(m)}$ to denote the entry in the ith row and jth column of L_m, where $1 \leq i, j \leq n$, $1 \leq m \leq k$. Since these Latin squares are in standard form, we have $a_{11}^{(m)} = 1$, $a_{12}^{(m)} = 2, \ldots$, and $a_{1n}^{(m)} = n$ for all $1 \leq m \leq k$. Now consider $a_{21}^{(m)}$, for all $1 \leq m \leq k$. These entries in the second row and first column are below $a_{11}^{(m)} = 1$. Thus $a_{21}^{(m)} \neq 1$, for all $1 \leq m \leq k$, or the configuration is not a Latin square. Further, if there exists $1 \leq \ell < m \leq k$ with $a_{21}^{(\ell)} = a_{21}^{(m)}$, then the pair L_ℓ, L_m cannot be an orthogonal pair. (Why not?) Consequently, there are at best $n - 1$ choices for the a_{21} entries in any of our $n \times n$ Latin squares, and the result follows from this observation.

This theorem places an upper bound on the number of $n \times n$ Latin squares that are orthogonal in pairs. We shall find that for certain values of n, this upper bound can be attained. In addition, our next theorem provides a method for constructing these Latin squares, though initially not in standard form. The construction uses the structure of a finite field. Before proving this theorem for the general situation, however, we shall examine one special case.

EXAMPLE 17.17

Let $F = \{f_i \mid 1 \leq i \leq 5\} = \mathbf{Z}_5$ with $f_1 = 1$, $f_2 = 2$, $f_3 = 3$, $f_4 = 4$, and $f_5 = 5$, the zero of $\mathbf{Z}_5$.

For $1 \leq k \leq 4$, let L_k be the 5×5 array $(a_{ij}^{(k)})$, where $1 \leq i, j \leq 5$ and

$$a_{ij}^{(k)} = f_k f_i + f_j.$$

When $k = 1$, we construct $L_1 = (a_{ij}^{(1)})$ as follows. Here $a_{ij}^{(1)} = f_1 f_i + f_j = f_i + f_j$, for $1 \leq i, j \leq 5$. With $i = 1$, the first row of L_1 is calculated as follows:

$$a_{11}^{(1)} = f_1 + f_1 = 2 \qquad a_{12}^{(1)} = f_1 + f_2 = 3 \qquad a_{13}^{(1)} = f_1 + f_3 = 4$$
$$a_{14}^{(1)} = f_1 + f_4 = 5 \qquad a_{15}^{(1)} = f_1 + f_5 = 1$$

The entries in the second row of L_1 are computed when $i = 2$. Here we find

$$a_{21}^{(1)} = f_2 + f_1 = 3 \qquad a_{22}^{(1)} = f_2 + f_2 = 4 \qquad a_{23}^{(1)} = f_2 + f_3 = 5$$
$$a_{24}^{(1)} = f_2 + f_4 = 1 \qquad a_{25}^{(1)} = f_2 + f_5 = 2$$

Continuing these calculations, we obtain the Latin square L_1 as

$$
\begin{array}{ccccc}
2 & 3 & 4 & 5 & 1 \\
3 & 4 & 5 & 1 & 2 \\
4 & 5 & 1 & 2 & 3 \\
5 & 1 & 2 & 3 & 4 \\
1 & 2 & 3 & 4 & 5
\end{array}
$$

For $k = 2$, the entries of L_2 are given by the formula $a_{ij}^{(2)} = f_2 f_i + f_j = 2 f_i + f_j$. To obtain the first row of L_2, we set i equal to 1 and compute

$$a_{11}^{(2)} = 2 f_1 + f_1 = 3 \qquad a_{12}^{(2)} = 2 f_1 + f_2 = 4 \qquad a_{13}^{(2)} = 2 f_1 + f_3 = 5$$
$$a_{14}^{(2)} = 2 f_1 + f_4 = 1 \qquad a_{15}^{(2)} = 2 f_1 + f_5 = 2$$

When i is set equal to 2, the entries in the second row of L_2 are calculated as follows:

$$a_{21}^{(2)} = 2 f_2 + f_1 = 5 \qquad a_{22}^{(2)} = 2 f_2 + f_2 = 1 \qquad a_{23}^{(2)} = 2 f_2 + f_3 = 2$$
$$a_{24}^{(2)} = 2 f_2 + f_4 = 3 \qquad a_{25}^{(2)} = 2 f_2 + f_5 = 4$$

Similar calculations for $i = 3$, 4, and 5 result in the Latin square L_2 given by

$$
\begin{array}{ccccc}
3 & 4 & 5 & 1 & 2 \\
5 & 1 & 2 & 3 & 4 \\
2 & 3 & 4 & 5 & 1 \\
4 & 5 & 1 & 2 & 3 \\
1 & 2 & 3 & 4 & 5
\end{array}
$$

It is straightforward to check that the two Latin squares L_1 and L_2 are orthogonal. In Exercise 5 (at the end of this section) the reader will be asked to calculate L_3 and L_4. Our next result will verify that the four arrays L_1, L_2, L_3, and L_4 are Latin squares and that they are orthogonal in pairs.

THEOREM 17.16   Let $n \in \mathbf{Z}^+$, $n > 2$. If p is a prime and $n = p^t$, for $t \in \mathbf{Z}^+$, then there are $n - 1$ Latin squares that are $n \times n$ and orthogonal in pairs.

Proof: Let $F = GF(p^t)$, the Galois field of order $p^t = n$. Consider $F = \{f_1, f_2, \ldots, f_n\}$, where f_1 is the unity and f_n is the zero element.

We construct $n - 1$ Latin squares as follows.

For each $1 \leq k \leq n - 1$, let L_k be the $n \times n$ array $(a_{ij}^{(k)})$, $1 \leq i, j \leq n$, where $a_{ij}^{(k)} = f_k f_i + f_j$.

First we show that each L_k is a Latin square. If not, there are two identical elements of F in the same row or column of L_k. Suppose that a repetition occurs in a column — that is, $a_{rj}^{(k)} = a_{sj}^{(k)}$ for $1 \leq r, s \leq n$. Then $a_{rj}^{(k)} = f_k f_r + f_j = f_k f_s + f_j = a_{sj}^{(k)}$. This implies that $f_k f_r = f_k f_s$, by the cancellation for addition in F. Since $k \neq n$, it follows that $f_k \neq f_n$, the zero of F. Consequently, f_k is invertible, so $f_r = f_s$ and $r = s$. A similar argument shows that there are no repetitions in any row of L_k.

At this point we have $n - 1$ Latin squares, $L_1, L_2, \ldots, L_{n-1}$. Now we shall prove that they are orthogonal in pairs. If not, let $1 \leq k < m \leq n - 1$ with

$$a_{ij}^{(k)} = a_{rs}^{(k)}, \qquad a_{ij}^{(m)} = a_{rs}^{(m)}, \qquad 1 \leq i, j, r, s \leq n, \qquad \text{and} \qquad (i, j) \neq (r, s).$$

(Then the same ordered pair occurs twice when we superimpose L_k and L_m.) But

$$a_{ij}^{(k)} = a_{rs}^{(k)} \iff f_k f_i + f_j = f_k f_r + f_s, \qquad \text{and}$$
$$a_{ij}^{(m)} = a_{rs}^{(m)} \iff f_m f_i + f_j = f_m f_r + f_s.$$

Subtracting these equations, we find that $(f_k - f_m) f_i = (f_k - f_m) f_r$. With $k \neq m$, $(f_k - f_m)$ is not the zero of F, so it is invertible and we have $f_i = f_r$. Putting this back into either of the prior equations, we find that $f_j = f_s$. Consequently, $i = r$ and $j = s$. Therefore for $k \neq m$, the Latin squares L_k and L_m form an orthogonal pair.

The first value of n that is not a power of a prime is 6. The existence of a pair of 6×6 orthogonal Latin squares was first investigated by Leonhard Euler (1707–1783) when he sought a solution to the "problem of the 36 officers." This problem deals with six different regiments wherein six officers, each with a different rank, are selected from each regiment. (There are only six possible ranks.) The objective is to arrange the 36 officers in a 6×6 array so that in each row or column of the array, every rank and every regiment is represented exactly once. Hence each officer in the square array corresponds to an ordered pair (i, j) where $1 \leq i, j \leq 6$, with i for his regiment and j for his rank. In 1782 Euler conjectured that the problem could not be solved — that there is no pair of 6×6 orthogonal Latin squares. He went further and conjectured that for all $n \in \mathbf{Z}^+$, if $n \equiv 2 \pmod 4$, then there is no pair of $n \times n$ orthogonal Latin squares. In 1900 G. Tarry verified Euler's conjecture for $n = 6$ by a systematic enumeration of all possible 6×6 Latin squares. However, it was not until 1960, through the combined efforts of R. C. Bose, S. S. Shrikhande, and E. T. Parker, that the remainder of Euler's conjecture was proved false. They showed that if $n \in \mathbf{Z}^+$ with $n \equiv 2 \pmod 4$ and $n > 6$, then there exists a pair of $n \times n$ orthogonal Latin squares.

For more on this result and Latin squares in general, the reader should consult the chapter references.

EXERCISES 17.3

1. a) Rewrite the following 4×4 Latin square in standard form.

```
1  3  4  2
3  1  2  4
2  4  3  1
4  2  1  3
```

b) Find a 4×4 Latin square in standard form that is orthogonal to the result in part (a).

c) Apply the reverse of the process in part (a) to the result in part (b). Show that your answer is orthogonal to the given 4×4 Latin square.

2. Prove Theorem 17.14.

3. Complete the proof of the first part of Theorem 17.16.

4. The three 4×4 Latin squares in Tables 17.3, 17.4, and 17.6 are orthogonal in pairs. Can you find another 4×4 Latin square that is orthogonal to each of these three?

5. Complete the calculations in Example 17.17 in order to obtain the two 5×5 Latin squares L_3 and L_4. Rewrite each Latin square L_i, for $1 \le i \le 4$, in standard form.

6. Find three 7×7 Latin squares that are orthogonal in pairs. Rewrite these results in standard form.

7. Extend the experiment in Example 17.13 so that the research team needs three 4×4 Latin squares that are orthogonal in pairs.

8. A Latin square L is called *self-orthogonal* if L and its transpose L^{tr} form an orthogonal pair.

   a) Show that there is no 3×3 self-orthogonal Latin square.

   b) Give an example of a 4×4 Latin square that is self-orthogonal.

   c) If $L = (a_{ij})$ is an $n \times n$ self-orthogonal Latin square, prove that the elements a_{ii}, for $1 \le i \le n$, must all be distinct.

17.4
Finite Geometries and Affine Planes

In the Euclidean geometry of the real plane, we find that (a) two distinct points determine a unique line and (b) if ℓ is a line in the plane, and P a point not on ℓ, then there is a unique line ℓ' that contains P and is parallel to ℓ. During the eighteenth and nineteenth centuries, non-Euclidean geometries were developed when alternatives to condition (b) were investigated. Yet all of these geometries contained infinitely many points and lines. The notion of a finite geometry did not appear until the end of the nineteenth century in the work of Gino Fano (*Giornale di Matematiche*, 1892).

How can we construct such a geometry? To do so, we return to the more familiar Euclidean geometry. In order to describe points and lines in this plane algebraically, we introduced a set of coordinate axes and identified each point P by an ordered pair (c, d) of real numbers. This description set up a one-to-one correspondence between the points in the plane and the set $\mathbf{R} \times \mathbf{R}$. By using the idea of slope, we could uniquely represent each line in this plane by either (1) $x = a$, where the slope is infinite, or (2) $y = mx + b$, where m is the slope; a, m, and b are real numbers. We also found that two distinct lines are parallel if and only if they have the same slope. When their slopes are distinct, the lines intersect in a unique point.

Instead of using real numbers a, b, c, d, m for the point (c, d) and the lines $x = a$, $y = mx + b$, we now turn to a comparable *finite* structure, the finite field. Our objective is to construct what is called a (finite) affine plane.

Definition 17.12   Let $\mathcal{P}$ be a finite set of points, and let $\mathcal{L}$ be a set of subsets of $\mathcal{P}$, called lines. A (*finite*) *affine plane* on the sets $\mathcal{P}$ and $\mathcal{L}$ is a finite structure satisfying the following conditions.

   A1) Two distinct points of $\mathcal{P}$ are (simultaneously) in only one element of $\mathcal{L}$; that is, they are on only one line.

   A2) For each $\ell \in \mathcal{L}$, and each $P \in \mathcal{P}$ with $P \notin \ell$, there exists a unique element $\ell' \in \mathcal{L}$ where $P \in \ell'$ and ℓ, ℓ' have no point in common.

   A3) There are four points in $\mathcal{P}$, no three of which are collinear (that is, no three of these four points are in any one of the subsets $\ell \in \mathcal{L}$).

Figure 17.1

The reason for condition (A3) is to avoid uninteresting situations like the one shown in Fig. 17.1. If only conditions (A1) and (A2) were considered, then this system would be an affine plane.

We return now to our construction. Let $F = GF(n)$, where $n = p^t$ for some prime p and $t \in \mathbf{Z}^+$. In constructing our affine plane, denoted by $AP(F)$, we let $\mathcal{P} = \{(c, d)|c, d \in F\}$. Thus we have n^2 points.

How many lines should we have for the set $\mathcal{L}$?

The lines fall into two categories. For a line of infinite slope the equation is $x = a$, where $a \in F$. Thus we have n such "vertical lines." The other lines are given algebraically by $y = mx + b$, where $m, b \in F$. With n choices for each of m and b, it follows that there are n^2 lines that are not "vertical." Hence $|\mathcal{L}| = n^2 + n$.

Before we verify that $AP(F)$, with $\mathcal{P}$ and $\mathcal{L}$ as constructed, is an affine plane, we make two other observations.

First, for each line $\ell \in \mathcal{L}$, if ℓ is given by $x = a$, then there are n choices for y on $\ell = \{(a, y)|y \in F\}$. Thus ℓ contains exactly n points. If ℓ is given by $y = mx + b$, for $m, b \in F$, then for each choice of x we have y uniquely determined, and again ℓ consists of n points.

Now consider any point $(c, d) \in \mathcal{P}$. This point is on the line $x = c$. Furthermore, on each line $y = mx + b$ of finite slope m, $d - mc$ uniquely determines b. With n choices for m, we see that the point (c, d) is on the n lines of the form $y = mx + (d - mc)$. Overall, (c, d) is on $n + 1$ lines.

Thus far in our construction of $AP(F)$ we have a set $\mathcal{P}$ of points and a set $\mathcal{L}$ of lines where (a) $|\mathcal{P}| = n^2$; (b) $|\mathcal{L}| = n^2 + n$; (c) each $\ell \in \mathcal{L}$ contains n points; and (d) each point in $\mathcal{P}$ is on exactly $n + 1$ lines. We shall now prove that $AP(F)$ satisfies the three conditions to be an affine plane.

A1) Let $(c, d), (e, f) \in \mathcal{P}$. Using the two-point formula for the equation of a line, we have

$$(e - c)(y - d) = (f - d)(x - c) \tag{1}$$

as a line on which we find both (c, d) and (e, f). Each of these points is on $n + 1$ lines. Could there be a second line containing both of them?

The point (c, d) is on the line $x = c$. If (e, f) is also on this line, then $e = c$, but $f \neq d$ because the points are distinct. With $e = c$, Eq. (1) reduces to $0 = (f - d)(x - c)$, or $x = c$ because $f - d \neq 0$, and so we do not have a second line.

With $c \neq e$, if $(c, d), (e, f)$ are on a second line of the form $y = mx + b$, then $d = mc + b$, $f = me + b$, and $(f - d) = m(e - c)$. Our coefficients are taken from a field and $e \neq c$, so $m = (f - d)(e - c)^{-1}$ and $b = d - mc = d - (f - d) \cdot (e - c)^{-1}c$. Consequently, this second line containing (c, d) and (e, f) is

$$y = (f - d)(e - c)^{-1}x + [d - (f - d)(e - c)^{-1}c]$$

or, because multiplication in F is commutative, $(e - c)(y - d) = (f - d)(x - c)$, which is Eq. (1). Thus two points from $\mathcal{P}$ are on only one line, and condition (A1) is satisfied.

A2) To verify this condition, consider the point P and the line ℓ as shown in Fig. 17.2. Since there are n points on any line, let $P_1, P_2, \ldots, P_n$ be the points of ℓ. (These are the only points on ℓ, although the figure might suggest others.) The point P is not on ℓ, so P and P_i determine a unique line ℓ_i, for each $1 \leq i \leq n$. We showed earlier that each point is on $n + 1$ lines, so now there is one additional line ℓ' with P on ℓ' and with ℓ' not intersecting ℓ.

A3) The last condition uses the field F. Since $|F| \geq 2$, there is the unity 1 and the zero element 0 in F. Considering the points $(0, 0)$, $(1, 0)$, $(0, 1)$, $(1, 1)$, if line ℓ

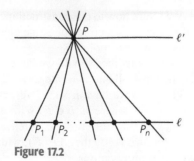

Figure 17.2

contains any three of these points, then two of the points have the form $(c, c), (c, d)$. Consequently the equation for ℓ is given by $x = c$, which is not satisfied by either (d, c) or (d, d). Hence no three of these points are collinear.

We have now shown the following.

THEOREM 17.17   If F is a finite field, then the system based on the set $\mathcal{P}$ of points and the set $\mathcal{L}$ of lines, as described above, is an affine plane denoted by $AP(F)$.

Some particular examples will indicate a connection between these finite geometries, or affine planes, and the Latin squares of the previous section.

EXAMPLE 17.18   For $F = (\mathbf{Z}_2, +, \cdot)$, we have $n = |F| = 2$. The affine plane in Fig. 17.3 has $n^2 = 4$ points and $n^2 + n = 6$ lines. For example, the line $\ell_4 = \{(1, 0), (1, 1)\}$, and ℓ_4 contains no other points that the figure might suggest. Furthermore, ℓ_5 and ℓ_6 are parallel lines in this finite geometry because they do not intersect.

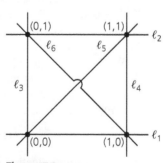

Figure 17.3

EXAMPLE 17.19   Let $F = GF(2^2)$ — the field of Example 17.9. Recall the notation of Example 17.11(d) and write $F = \{00, 01, 10, 11\}$, with addition and multiplication given by Table 17.7. We use this field to construct a finite geometry with $n^2 = 16$ points and $n^2 + n = 20$ lines. The 20 lines can be partitioned into five *parallel classes* of four lines each.

Class 1: Here we have the lines of infinite slope. These four "vertical" lines are given by the equations $x = 00$, $x = 01$, $x = 10$, and $x = 11$.

Class 2: For the "horizontal" class, or class of slope 0, we have the four lines $y = 00$, $y = 01$, $y = 10$, and $y = 11$.

Table 17.7

+	00	01	10	11
00	00	01	10	11
01	01	00	11	10
10	10	11	00	01
11	11	10	01	00

·	00	01	10	11
00	00	00	00	00
01	00	01	10	11
10	00	10	11	01
11	00	11	01	10

Class 3: The lines with slope 01 are those whose equations are $y = 01x + 00$, $y = 01x + 01$, $y = 01x + 10$, and $y = 01x + 11$.

Class 4: This class consists of the lines with equations $y = 10x + 00$, $y = 10x + 01$, $y = 10x + 10$, and $y = 10x + 11$.

Class 5: The last class contains the four lines given by $y = 11x + 00$, $y = 11x + 01$, $y = 11x + 10$, and $y = 11x + 11$.

Since each line in $AP(F)$ contains four points and each parallel class contains four lines, we shall see now how three of these parallel classes partition the 16 points of $AP(F)$.

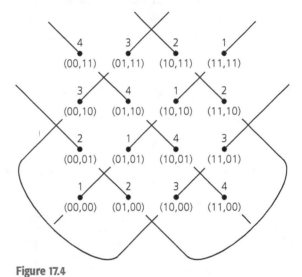

Figure 17.4

For the class with $m = 01$, there are four lines: (1) $y = 01x + 00$; (2) $y = 01x + 01$; (3) $y = 01x + 10$; and (4) $y = 01x + 11$. Above each point in $AP(F)$ we write the number corresponding to the line it is on. (See Fig. 17.4.) This configuration can be given by the following Latin square:

$$\begin{array}{cccc} 4 & 3 & 2 & 1 \\ 3 & 4 & 1 & 2 \\ 2 & 1 & 4 & 3 \\ 1 & 2 & 3 & 4 \end{array}$$

If we repeat this process for classes 4 and 5, we get the partitions shown in Figs. 17.5 and 17.6, respectively. In each class the lines are listed, for the given slope, in the same order as for Fig. 17.4. Within each figure is the corresponding Latin square.

These figures give us three 4×4 Latin squares that are orthogonal in pairs.

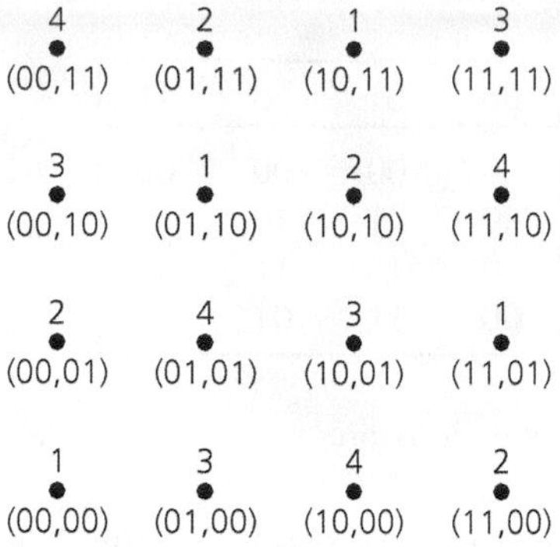

4	2	1	3
(00,11)	(01,11)	(10,11)	(11,11)
3	1	2	4
(00,10)	(01,10)	(10,10)	(11,10)
2	4	3	1
(00,01)	(01,01)	(10,01)	(11,01)
1	3	4	2
(00,00)	(01,00)	(10,00)	(11,00)

```
4 2 1 3
3 1 2 4
2 4 3 1
1 3 4 2
```

Figure 17.5

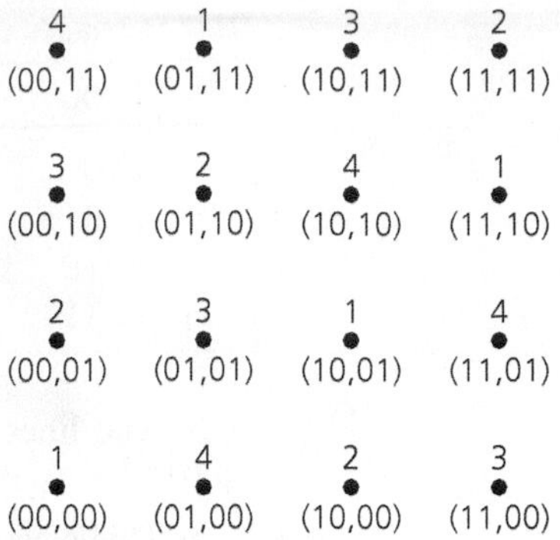

4	1	3	2
(00,11)	(01,11)	(10,11)	(11,11)
3	2	4	1
(00,10)	(01,10)	(10,10)	(11,10)
2	3	1	4
(00,01)	(01,01)	(10,01)	(11,01)
1	4	2	3
(00,00)	(01,00)	(10,00)	(11,00)

```
4 1 3 2
3 2 4 1
2 3 1 4
1 4 2 3
```

Figure 17.6

The results of this example are no accident, as demonstrated by the following theorem.

THEOREM 17.18

Let $F = GF(n)$, where $n \geq 3$ and $n = p^t$, p a prime, $t \in \mathbf{Z}^+$. The Latin squares that arise from $AP(F)$ for the $n - 1$ parallel classes, where the slope is neither 0 nor infinite, are orthogonal in pairs.

Proof: A proof of this result is outlined in the Section Exercises.

EXERCISES 17.4

1. Complete the following table dealing with affine planes.

Field	Number of Points	Number of Lines	Number of Points on a Line	Number of Lines on a Point
	25			
GF(3^2)				
		56		
				17
			31	

2. How many parallel classes do each of the affine planes in Exercise 1 determine? How many lines are in each class?

3. Construct the affine plane $AP(\mathbf{Z}_3)$. Determine its parallel classes and the corresponding Latin squares for the classes of finite nonzero slope.

4. Repeat Exercise 3 with $\mathbf{Z}_5$ taking the place of $\mathbf{Z}_3$.

5. Determine each of the following lines.

a) The line in $AP(\mathbf{Z}_7)$ that is parallel to $y = 4x + 2$ and contains $(3, 6)$.

b) The line in $AP(\mathbf{Z}_{11})$ that is parallel to $2x + 3y + 4 = 0$ and contains $(10, 7)$.

c) The line in $AP(F)$, where $F = GF(2^2)$, that is parallel to $10y = 11x + 01$ and contains $(11, 01)$. (See Table 17.7.)

6. Suppose we try to construct an affine plane $AP(\mathbf{Z}_6)$ as we did in this section.

a) Determine which of the conditions (A1), (A2), and (A3) fail in this situation.

b) Find how many lines contain a given point P and how many points are on a given line ℓ, for this "geometry."

7. The following provides an outline for a proof of Theorem 17.18.

a) Consider a parallel class of lines given by $y = mx + b$, where $m \in F$, $m \neq 0$. Show that each line in this class intersects each "vertical" line and each "horizontal" line in exactly one point of $AP(F)$. Thus the configuration obtained by labeling the points of $AP(F)$, as in Figs. 17.4, 17.5, and 17.6, is a Latin square.

b) To show that the Latin squares corresponding to two different classes, other than the classes of slope 0 or infinite

slope, are orthogonal, assume that an ordered pair (i, j) appears more than once when one square is superimposed upon the other. How does this lead to a contradiction?

17.5
Block Designs and Projective Planes

In this final section, we examine a type of combinatorial design and see how it is related to the structure of a finite geometry. The following example will illustrate this design.

EXAMPLE 17.20

Dick (d) and his wife Mary (m) go to New York City with their five children — Richard (r), Peter (p), Christopher (c), Brian (b), and Julie (j). While staying in the city they receive three passes each day, for a week, to visit the Empire State Building. Can we make up a schedule for this family so that everyone gets to visit this attraction the same number of times?

The following schedule is one possibility.

1) b, c, d **2)** b, j, r **3)** b, m, p **4)** c, j, m

5) c, p, r **6)** d, j, p **7)** d, m, r

Here the result was obtained by trial and error. For a problem of this size such a technique is feasible. However, in general, a more effective strategy is needed. Furthermore, in asking for a certain schedule, we may be asking for something that doesn't exist. In this problem, for example, each pair of family members is together on only one visit. If the family had received four passes each day, we would not be able to construct a schedule that maintained this property.

The situation in this example generalizes as follows.

Definition 17.13

Let V be a set with v elements. A collection $\{B_1, B_2, \ldots, B_b\}$ of subsets of V is called a *balanced incomplete block design*, or (v, b, r, k, λ)-*design*, if the following conditions are satisfied:

a) For each $1 \le i \le b$, the subset B_i contains k elements, where k is a fixed constant and $k < v$.

b) Each element $x \in V$ is in r $(\le b)$ of the subsets B_i, $1 \le i \le b$.

c) Every pair x, y of elements of V appears together in λ $(\le b)$ of the subsets B_i, $1 \le i \le b$.

The elements of V are often called *varieties* because of the early applications in the design of experiments that dealt with tests on fertilizers and plants. The b subsets $B_1, B_2, \ldots, B_b$ of V are called *blocks*, where each block contains k varieties. The number r is referred to as the *replication number* of the design. Finally, λ is termed the *covalency* for the design. This parameter makes the design balanced in the following sense. For general block designs we have a number λ_{xy} for each pair $x, y \in V$; if λ_{xy} is the same for all pairs of elements from

V, then λ represents this common measure and the design is called *balanced*. In this text we only deal with balanced designs.

EXAMPLE 17.21

a) The schedule in Example 17.20 is an example of a $(7, 7, 3, 3, 1)$-design.

b) For $V = \{1, 2, 3, 4, 5, 6\}$, the ten blocks

$$
\begin{array}{ccccc}
1\ \ 2\ \ 4 & 1\ \ 3\ \ 4 & 1\ \ 5\ \ 6 & 2\ \ 3\ \ 6 & 3\ \ 4\ \ 6 \\
1\ \ 2\ \ 6 & 1\ \ 3\ \ 5 & 2\ \ 3\ \ 5 & 2\ \ 4\ \ 5 & 4\ \ 5\ \ 6
\end{array}
$$

constitute a $(6, 10, 5, 3, 2)$-design.

c) If F is a finite field, with $|F| = n$, then the affine plane $AP(F)$ yields an $(n^2, n^2 + n, n + 1, n, 1)$-design. Here the varieties are the n^2 points in $AP(F)$; the $n^2 + n$ lines are the blocks of the design.

At this point there are five parameters determining our design. We now examine how these parameters are related.

THEOREM 17.19

For a (v, b, r, k, λ)-design, (1) $vr = bk$ and (2) $\lambda(v - 1) = r(k - 1)$.

Proof:

1) With b blocks in the design and k elements per block, listing all the elements of the blocks, we get bk symbols. This collection of symbols consists of the elements of V with each element appearing r times, for a total of vr symbols. Hence $vr = bk$.

2) For this property we introduce the *pairwise incidence matrix* A for the design. With $|V| = v$, let $t = \binom{v}{2}$, the number of pairs of elements in V. We construct the $t \times b$ matrix $A = (a_{ij})$ by defining $a_{ij} = 1$ if the ith pair of elements from V is in the jth block of the design; if not, $a_{ij} = 0$.

$$
\begin{array}{c}
\\
x_1 x_2 \\
x_1 x_3 \\
\vdots \\
x_1 x_v \\
x_2 x_3 \\
\vdots \\
x_{v-1} x_v
\end{array}
\begin{array}{c}
B_1 \quad\ B_2 \ \cdots\ \ B_b \\
\left[
\begin{array}{cccc}
a_{11} & a_{12} & \cdots & a_{1b} \\
a_{21} & a_{22} & \cdots & a_{2b} \\
\vdots & \vdots & \vdots & \vdots \\
a_{v-1\,1} & a_{v-1\,2} & \cdots & a_{v-1\,b} \\
a_{v\,1} & a_{v\,2} & \cdots & a_{v\,b} \\
\vdots & \vdots & \vdots & \vdots \\
a_{t\,1} & a_{t\,2} & \cdots & a_{t\,b}
\end{array}
\right]
\end{array}
$$

We now count the number of 1's in matrix A in two ways.

a) Consider the rows. Since each pair x_i, x_j, for $1 \le i < j \le v$, appears in λ blocks, it follows that each row contains λ 1's. With t rows in the matrix, the number of 1's is then $\lambda t = \lambda v(v - 1)/2$.

b) Now consider the columns. As each block contains k elements, this determines $\binom{k}{2} = k(k - 1)/2$ pairs, and this is the number of 1's in each column of matrix A. With b columns, the total number of 1's is $bk(k - 1)/2$.

Then, $\lambda v(v - 1)/2 = bk(k - 1)/2 = vr(k - 1)/2$, so $\lambda(v - 1) = r(k - 1)$.

As we mentioned earlier, when n is a power of a prime, an $(n^2, n^2 + n, n + 1, n, 1)$-design can be obtained from the affine plane $AP(F)$, where $F = GF(n)$. Here the points are the varieties and the lines are the blocks. We shall now introduce a construction that enlarges $AP(F)$ to what is called a finite projective plane. From this projective plane we can construct an $(n^2 + n + 1, n^2 + n + 1, n + 1, n + 1, 1)$-design. First let us see how these two kinds of planes compare.

Definition 17.14   If $\mathscr{P}'$ is a finite set of points and $\mathscr{L}'$ a set of lines, each of which is a nonempty subset of $\mathscr{P}'$, then the (finite) plane based on $\mathscr{P}'$ and $\mathscr{L}'$ is called a *projective plane* if the following conditions are satisfied.

P1) Two distinct points of $\mathscr{P}'$ are on only one line.

P2) Any two lines from $\mathscr{L}'$ intersect in a unique point.

P3) There are four points in $\mathscr{P}'$, no three of which are collinear.

The difference between the affine and projective planes lies in the condition dealing with the existence of parallel lines. Here the parallel lines of the affine plane based on $\mathscr{P}$ and $\mathscr{L}$ will intersect when the given system is enlarged to the projective plane based on $\mathscr{P}'$ and $\mathscr{L}'$.

The construction proceeds as follows.

EXAMPLE 17.22   Start with an affine plane $AP(F)$ where $F = GF(n)$. For each point $(x, y) \in \mathscr{P}$, rewrite the point as $(x, y, 1)$. We then think of the points as ordered triples (x, y, z) where $z = 1$. Rewrite the equations of the lines $x = c$ and $y = mx + b$ in $AP(F)$ as $x = cz$ and $y = mx + bz$, where $z = 1$. We still have our original affine plane $AP(F)$, but with a change of notation.

Add the set of points $\{(1, 0, 0)\} \cup \{(x, 1, 0) | x \in F\}$ to $\mathscr{P}$ to get the set $\mathscr{P}'$. Then $|\mathscr{P}'| = n^2 + n + 1$. Let ℓ_∞ be the subset of $\mathscr{P}'$ consisting of these new points. This new line can be given by the equation $z = 0$, with the stipulation that we never have $x = y = z = 0$. Hence $(0, 0, 0) \notin \mathscr{P}'$.

Now let us examine these ideas for the affine plane $AP(\mathbf{Z}_2)$. Here $\mathscr{P} = \{(0, 0), (1, 0), (0, 1), (1, 1)\}$, so

$$\mathscr{P}' = \{(0, 0, 1), (1, 0, 1), (0, 1, 1), (1, 1, 1)\} \cup \{(1, 0, 0), (0, 1, 0), (1, 1, 0)\}.$$

The six lines in $\mathscr{L}$ were originally

$x = 0$: $\{(0, 0), (0, 1)\}$ $y = 0$: $\{(0, 0), (1, 0)\}$ $y = x$: $\{(0, 0), (1, 1)\}$
$x = 1$: $\{(1, 0), (1, 1)\}$ $y = 1$: $\{(0, 1), (1, 1)\}$ $y = x + 1$: $\{(0, 1), (1, 0)\}$

We rewrite these as

$$x = 0 \qquad y = 0 \qquad y = x \qquad x = z \qquad y = z \qquad y = x + z$$

and add a new line ℓ_∞ defined by $z = 0$. These constitute the set $\mathscr{L}'$ of lines for our projective plane. And now at this point we consider z as a *variable*. Consequently, the line $x = z$ consists of the points $(0, 1, 0)$, $(1, 0, 1)$, and $(1, 1, 1)$. In fact, each line of $\mathscr{L}$ that contained two points will now contain three points when considered in $\mathcal{L}'$. The set $\mathcal{L}'$ consists of the following seven lines.

$x = 0$: $\{(0, 0, 1), (0, 1, 0), (0, 1, 1)\}$ $y = z$: $\{(1, 0, 0), (0, 1, 1), (1, 1, 1)\}$

$y = 0$: $\{(0, 0, 1), (1, 0, 0), (1, 0, 1)\}$ $y = x$: $\{(0, 0, 1), (1, 1, 0), (1, 1, 1)\}$

$x = z$: $\{(0, 1, 0), (1, 0, 1), (1, 1, 1)\}$ $y = x + z$: $\{(0, 1, 1), (1, 1, 0), (1, 0, 1)\}$

$z = 0$ (ℓ_∞): $\{(1, 0, 0), (0, 1, 0), (1, 1, 0)\}$

In the original affine plane the lines $x = 0$ and $x = 1$ were parallel because no point in this plane satisfied both equations simultaneously. Here in this new system $x = 0$ and $x = z$ intersect in the point $(0, 1, 0)$, so they are no longer parallel in the sense of $AP(\mathbf{Z}_2)$. Likewise, $y = x$ and $y = x + 1$ were parallel in $AP(\mathbf{Z}_2)$, whereas here the lines $y = x$ and $y = x + z$ intersect at $(1, 1, 0)$. We depict this projective plane based on $\mathcal{P}'$ and $\mathcal{L}'$ as shown in Fig. 17.7. Here the "circle" through $(1, 0, 1)$, $(1, 1, 0)$, and $(0, 1, 1)$ is the line $y = x + z$. Note that every line intersects ℓ_∞, which is often called the *line at infinity*. This line consists of the three *points at infinity*. We define two lines to be parallel in the projective plane when they intersect in a point at infinity (or on ℓ_∞).

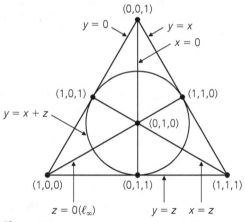

Figure 17.7

This projective plane provides us with a $(7, 7, 3, 3, 1)$-design like the one we developed by trial and error in Example 17.20.

We generalize the results of Example 17.22 as follows: Let n be a power of a prime. The affine plane $AP(F)$, for $F = GF(n)$, provides an example of an $(n^2, n^2 + n, n + 1, n, 1)$-design. In $AP(F)$ the $n^2 + n$ lines fall into $n + 1$ parallel classes. For each parallel class we add a point at infinity to $AP(F)$. The point $(0, 1, 0)$ is added for the class of lines $x = cz$, $c \in F$; the point $(1, 0, 0)$ for the class of lines $y = bz$, $b \in F$. When $m \in F$ and $m \neq 0$, then we add the point $(m^{-1}, 1, 0)$ for the class of lines $y = mx + bz$, $b \in F$. The line at infinity, ℓ_∞, is then defined as the set of $n + 1$ points at infinity. In this way we obtain the projective plane over $GF(n)$, which has $n^2 + n + 1$ points and $n^2 + n + 1$ lines. Here each point is on $n + 1$ lines, and each line contains $n + 1$ points. Furthermore, any two points in this plane are on only one line. Consequently, we have an example of an $(n^2 + n + 1, n^2 + n + 1, n + 1, n + 1, 1)$-design.

EXERCISES 17.5

1. Let $V = \{1, 2, \ldots, 9\}$. Determine the values of $v, b, r, k,$ and λ for the design given by the following blocks.

126   147   234   279   378   468
135   189   258   369   459   567

2. Find an example of a $(4, 4, 3, 3, \lambda)$-design.

3. Find an example of a $(7, 7, 4, 4, \lambda)$-design.

4. Complete the following table so that the parameters v, b, r, k, λ in any row may be possible for a balanced incomplete block design.

v	b	r	k	λ
4			3	2
9	12		3	
10		9		2
13		4	4	
	30	10		3

5. Is it possible to have a (v, b, r, k, λ)-design where (a) $b = 28, r = 4, k = 3$? (b) $v = 17, r = 8, k = 5$?

6. Given a (v, b, r, k, λ)-design with $b = v$, prove that if v is even, then λ is even.

7. A (v, b, r, k, λ)-design is called a *triple system* if $k = 3$. When $k = 3$ and $\lambda = 1$, we call the design a *Steiner triple system*.

   a) Prove that in every triple system, $\lambda(v - 1)$ is even and $\lambda v(v - 1)$ is divisible by 6.

   b) Prove that in every Steiner triple system, v is congruent to 1 or 3 modulo 6.

8. Verify that the following blocks constitute a Steiner triple system on nine varieties.

128   147   234   279   389   468
135   169   256   367   459   578

9. In a Steiner triple system with $b = 12$, find the values of v and r.

10. In each of the following, $\mathscr{P}'$ is a set of points and $\mathscr{L}'$ a set of lines, each of which is a nonempty subset of $\mathscr{P}'$. Which of the conditions (P1), (P2), and (P3) of Definition 17.14 hold for the given $\mathscr{P}'$ and $\mathscr{L}'$?

   a) $\mathscr{P}' = \{a, b, c\}$
   $\mathscr{L}' = \{\{a, b\}, \{a, c\}, \{b, c\}\}$

   b) $\mathscr{P}' = \{(x, y, z) | x, y, z \in \mathbf{R}\} = \mathbf{R}^3$
   $\mathscr{L}'$ is the set of all lines in $\mathbf{R}^3$.

   c) $\mathscr{P}'$ is the set of all lines in $\mathbf{R}^3$ that pass through $(0, 0, 0)$. $\mathscr{L}'$ is the set of all planes in $\mathbf{R}^3$ that pass through $(0, 0, 0)$.

11. Bowling teams of five students each are formed from a class of 15 college freshmen. Each of the students bowls on the same number of teams; each pair of students bowls together on two teams. (a) How many teams are there in all? (b) On how many different teams does each student bowl?

12. Mrs. Mackey gave her computer science class a list of 28 problems and directed each student to write algorithms for the solutions of exactly seven of these problems. If each student did as instructed and if for each pair of problems there was exactly one pair of students who wrote algorithms to solve them, how many students did Mrs. Mackey have in her class?

13. Consider a (v, b, r, k, λ)-design on the set V of varieties, where $|V| = v \geq 2$. If $x, y \in V$, how many blocks in the design contain either x or y?

14. In a programming class Professor Madge has a total of n students, and she wants to assign teams of m students to each of p computer projects. If each student must be assigned to the same number of projects, (a) in how many projects will each individual student be involved? (b) in how many projects will each pair of students be involved?

15. a) If a projective plane has six lines through every point, how many points does this projective plane have in all?

   b) If there are 57 points in a projective plane, how many points lie on each line of the plane?

16. In constructing the projective plane from $AP(\mathbf{Z}_2)$ in Example 17.22, why didn't we want to include the point $(0, 0, 0)$ in the set $\mathscr{P}'$?

17. Determine the values of $v, b, r, k,$ and λ for the balanced incomplete block design associated with the projective plane that arises from $AP(F)$ for the following choices of F: (a) $\mathbf{Z}_5$ (b) $\mathbf{Z}_7$ (c) $GF(8)$.

18. a) List the points and lines in $AP(\mathbf{Z}_3)$. How many parallel classes are there for this finite geometry? What are the parameters for the associated balanced incomplete block design?

   b) List the points and lines for the projective plane that arises from $AP(\mathbf{Z}_3)$. Determine the points on ℓ_∞, and use them to determine the "parallel" classes for this geometry. What are the parameters for the associated balanced incomplete block design?

17.6
Summary and Historical Review

The structure of a field was first developed in Chapter 14. In this chapter we examined polynomial rings and their role in the structure of finite fields, directing our attention to applications in finite geometries and combinatorial designs.

In Chapter 15 we saw that the order of a finite Boolean algebra could only be a power of 2. Now we find that for a finite field the order can only be a power of a prime and that for each prime p and each $n \in \mathbf{Z}^+$, there is only one field, up to isomorphism, of order p^n. This field is denoted by $GF(p^n)$, in honor of the French mathematician Evariste Galois (1811–1832).

Evariste Galois (1811–1832)

The finite fields $(\mathbf{Z}_p, +, \cdot)$, for p a prime, were obtained in Chapter 14 by means of the equivalence relation, congruence modulo p, defined on $\mathbf{Z}$. Using these finite fields, we developed here the integral domains $\mathbf{Z}_p[x]$. Then, with $s(x)$ an irreducible polynomial of degree n in $\mathbf{Z}_p[x]$, a similar equivalence relation — namely, congruence modulo $s(x)$ — gave us a set of p^n equivalence classes, denoted $\mathbf{Z}_p[x]/(s(x))$. These p^n equivalence classes became the elements of the field $GF(p^n)$. (Although we did not prove every possible result in general, it can be shown that over the finite field $\mathbf{Z}_p$, there is an irreducible polynomial of degree n for each $n \in \mathbf{Z}^+$.)

The theory of finite fields was developed by Galois in his work addressing the problem of the solutions of polynomial equations. As we mentioned in the summary of Chapter 16, the study of polynomial equations was an area of research that challenged many mathematicians from the sixteenth to the nineteenth centuries. In the nineteenth century, Niels Henrik Abel (1802–1829) first showed that the solution of the general quintic could not be given by radicals. Galois showed that for any polynomial of degree n over a field F, there is a corresponding group G that is isomorphic to a subgroup of S_n, the group of permutations of $\{1, 2, 3, \ldots, n\}$. The essence of Galois's work is that such a polynomial equation can be solved by (addition, subtraction, multiplication, division, and) radicals if its corresponding group is *solvable*. Now what makes a finite group solvable? We say that a finite group G is solvable if it has a chain of subgroups $G = K_1 \supset K_2 \supset K_3 \supset \cdots \supset K_t = \{e\}$, where for all

$2 \leq i \leq t$, K_i is a normal subgroup of K_{i-1} (that is, $xyx^{-1} \in K_i$ for all $y \in K_i$ and for all $x \in K_{i-1}$), and the quotient group K_{i-1}/K_i is abelian. One finds that all subgroups of S_i, for $1 \leq i \leq 4$, are solvable, but for $n \geq 5$ there are subgroups of S_n that are not solvable.

Though it seems that Galois theory is concerned predominantly with groups, there is a great deal more on the theory of fields that we have not mentioned. As a consequence of Galois's work, the areas of field theory and finite group theory became topics of great mathematical interest.

For more on *Galois theory*, the reader will find Chapter 6 of the text by V. H. Larney [8] and Chapter 12 in the book by N. H. McCoy and T. R. Berger [10] good places to start. Chapter 5 of I. N. Herstein [6] has more on the topic, while a detailed presentation can be found in the text by S. Roman [11] and the classic work by O. Zariski and P. Samuel [17]. Appendix E in the text by V. H. Larney [8] includes an interesting short account of the life of Galois; more on his life can be found in the somewhat fictional account by L. Infeld [7]. The article by T. Rothman [12] provides a more contemporary discussion of the inaccuracies and myths surrounding the life, and especially the death, of Galois. The biographical notes on pages 287–291 of the text by J. Stillwell [14] relate more on the life and work of this great genius.

The Latin squares, combinatorial designs, and finite geometries of the later sections of the chapter showed us how the finite field structure entered into problems of design. Dating back to the time of Leonhard Euler (1707–1783) and the problem of the "36 officers," the study of orthogonal Latin squares has been developed considerably since 1900, and especially since 1960 with the work of R. C. Bose, S. S. Shrikhande, and E. T. Parker. Chapter 7 of the monograph by H. J. Ryser [13] provides the details of their accomplishments. The text by C. L. Liu [9] includes ideas from coding theory in its discussion of Latin squares.

The study of finite geometries can be traced back to the work of Gino Fano, who, in 1892, considered a finite three-dimensional geometry consisting of 15 points, 35 lines, and 15 planes. However, it was not until 1906 that these geometries gained any notice, when O. Veblen and W. Bussey began their study of finite projective geometries. For more on this topic, the reader should find the texts by A. A. Albert and R. Sandler [1] and H. L. Dorwart [4] very interesting. The text by P. Dombowski [3] provides an extensive coverage for those seeking something more advanced.

Finally, the notion of designs was first studied by statisticians in the area called the design of experiments. Through the research of R. A. Fisher and his followers, this area has come to play an important role in the modern theory of statistical analysis. In our development, we examined conditions under which a (v, b, r, k, λ)-design could exist and how such designs were related to affine planes and finite projective planes. The text by M. Hall, Jr. [5] provides more on this topic, as does the work by A. P. Street and W. D. Wallis [15]. Chapter XIII of reference [15] includes material relating to designs and coding theory. A rather thorough coverage of the topic of designs is given in the work by W. D. Wallis [16], and the text edited by J. H. Dinitz and D. R. Stinson [2] provides the reader with a collection of more work in this area.

REFERENCES

1. Albert, A. Adrian, and Sandler, R. *An Introduction to Finite Projective Planes.* New York: Holt, 1968.
2. Dinitz, Jeffrey H., and Stinson, Douglas R., eds. *Contemporary Design Theory.* New York: Wiley, 1992.
3. Dombowski, Peter. *Finite Geometries.* New York: Springer-Verlag, 1968.

4. Dorwart, Harold L. *The Geometry of Incidence*. Englewood Cliffs, N.J.: Prentice-Hall, 1966.

5. Hall, Marshall, Jr. *Combinatorial Theory*. Waltham, Mass.: Blaisdell, 1967.

6. Herstein, Israel Nathan. *Topics in Algebra*, 2nd ed. Lexington, Mass.: Xerox College Publishing, 1975.

7. Infeld, Leopold. *Whom the Gods Love*. New York: McGraw-Hill, 1948.

8. Larney, Violet H. *Abstract Algebra: A First Course*. Boston: Prindle, Weber & Schmidt, 1975.

9. Liu, C. L. *Topics in Combinatorial Mathematics*. Mathematical Association of America, 1972.

10. McCoy, Neal H., and Berger, Thomas R. *Algebra: Groups, Rings, and Other Topics*. Boston: Allyn and Bacon, 1977.

11. Roman, Steven. *Field Theory*. New York: Springer-Verlag, 1995.

12. Rothman, Tony. "Genius and Biographers: The Fictionalization of Evariste Galois." *The American Mathematical Monthly* 89, no. 2 (1982): pp. 84–106.

13. Ryser, Herbert J. *Combinatorial Mathematics*. Carus Mathematical Monographs, Number 14, Mathematical Association of America, 1963.

14. Stillwell, John. *Mathematics and Its History*. New York: Springer-Verlag, 1989.

15. Street, Anne Penfold, and Wallis, W. D. *Combinatorial Theory: An Introduction*. Winnipeg, Canada: The Charles Babbage Research Center, 1977.

16. Wallis, W. D. *Combinatorial Designs*. New York: Marcel Dekker, Inc., 1988.

17. Zariski, Oscar, and Samuel, Pierre. *Commutative Algebra*, Vol. I. New York: Van Nostrand, 1958.

SUPPLEMENTARY EXERCISES

1. Determine n if over $GF(n)$ there are 6561 monic polynomials of degree 5 with no constant term.

2. a) Let $f(x) = a_n x^n + \cdots + a_1 x + a_0 \in \mathbf{Z}[x]$. If $r/s \in \mathbf{Q}$, with $\gcd(r, s) = 1$ and $f(r/s) = 0$, prove that $s \mid a_n$ and $r \mid a_0$.

b) Find the rational roots, if any exist, of the following polynomials over $\mathbf{Q}$. Factor $f(x)$ in $\mathbf{Q}[x]$.

   i) $f(x) = 2x^3 + 3x^2 - 2x - 3$

   ii) $f(x) = x^4 + x^3 - x^2 - 2x - 2$

c) Show that the polynomial $f(x) = x^{100} - x^{50} + x^{20} + x^3 + 1$ has no rational root.

3. a) For how many integers n, where $1 \le n \le 1000$, can we factor $f(x) = x^2 + x - n$ into the product of two first degree factors in $\mathbf{Z}[x]$?

b) Answer part (a) for $f(x) = x^2 + 2x - n$.

c) Answer part (a) for $f(x) = x^2 + 5x - n$.

d) Let $g(x) = x^2 + kx - n \in \mathbf{Z}[x]$, for $1 \le n \le 1000$. Find the smallest positive integer k so that $g(x)$ cannot be factored into two first degree factors in $\mathbf{Z}[x]$ for all $1 \le n \le 1000$.

4. Verify that the polynomial $f(x) = x^4 + x^3 + x + 1$ is reducible over every field F (finite or infinite).

5. If p is a prime, prove that in $\mathbf{Z}_p[x]$,

$$x^p - x = \prod_{a \in \mathbf{Z}_p} (x - a).$$

6. For any field F, let $f(x) = x^n + a_{n-1} x^{n-1} + \cdots + a_1 x + a_0 \in F[x]$. If $r_1, r_2, \ldots, r_n$ are the roots of $f(x)$, and $r_i \in F$ for all $1 \le i \le n$, prove that

   a) $-a_{n-1} = r_1 + r_2 + \cdots + r_n$.

   b) $(-1)^n a_0 = r_1 r_2 \cdots r_n$.

7. Four of the seven blocks in a $(7, 7, 3, 3, 1)$-design are $\{1, 3, 7\}$, $\{1, 5, 6\}$, $\{2, 6, 7\}$, and $\{3, 4, 6\}$. Determine the other three blocks.

8. Find the values of b and r for a Steiner triple system where $v = 63$.

9. a) If a projective plane has 73 points, how many points lie on each line?

b) If each line in a projective plane passes through 10 points, how many lines are there in the projective plane?

10. A projective plane is coordinatized with the elements of a field F. If this plane contains 91 lines, what are $|F|$ and $\text{char}(F)$?

11. Let $V = \{x_1, x_2, \ldots, x_v\}$ be the set of varieties and $\{B_1, B_2, \ldots, B_b\}$ the collection of blocks for a (v, b, r, k, λ)-design. We define the *incidence matrix* A for the design by

$$A = (a_{ij})_{v \times b}, \qquad \text{where } a_{ij} = \begin{cases} 1, & \text{if } x_i \in B_j \\ 0, & \text{otherwise.} \end{cases}$$

a) How many 1's are there in each row and column of A?

b) Let $J_{m \times n}$ be the $m \times n$ matrix where every entry is 1. For $J_{n \times n}$ we write J_n. Prove that for the incidence matrix A, $A \cdot J_b = r \cdot J_{v \times b}$ and $J_v \cdot A = k \cdot J_{v \times b}$.

c) Show that

$$A \cdot A^{tr} = \begin{bmatrix} r & \lambda & \lambda & \cdots & \lambda \\ \lambda & r & \lambda & \cdots & \lambda \\ \lambda & \lambda & r & \cdots & \lambda \\ \cdots & \cdots & \cdots & \cdots & \cdots \\ \lambda & \lambda & \lambda & \cdots & r \end{bmatrix}$$

$$= (r - \lambda)I_v + \lambda J_v,$$

where I_v is the $v \times v$ (multiplicative) identity.

d) Prove that

$$\det(A \cdot A^{tr}) = (r - \lambda)^{v-1}[r + (v - 1)\lambda] = (r - \lambda)^{v-1}rk.$$

12. Given a (v, b, r, k, λ)-design based on the v varieties of V, replace each of the blocks B_i, for $1 \le i \le b$, by its complement $\overline{B}_i = V - B_i$. Then the collection $\{\overline{B}_1, \overline{B}_2, \ldots, \overline{B}_b\}$ provides the blocks for a (v, b, r', k', λ')-design, also based on the set V.

a) Find this corresponding complementary (v, b, r', k', λ')-design for the design given in Exercise 1 of Section 17.5.

b) In general, how are the parameters r', k', λ' of the complementary design related to the parameters v, b, r, k, λ of the original design?

Appendix 1

Exponential and Logarithmic Functions

Throughout the study of mathematics and computer science, one confronts exponential and logarithmic functions. The function concept is introduced in Section 5.2, and in part (d) of Exercise 15 for that section we find the function $f: \mathbf{R} \to \mathbf{R}$, where $f(x) = e^x$ for $x \in \mathbf{R}$. This is an example of an exponential function. Then in Example 5.61 we come across the function $f: \mathbf{R} \to \mathbf{R}^+$, where $f(x) = e^x$ — this time in conjunction with a logarithmic function, denoted $\ln x$, where $x \in \mathbf{R}^+$. Later, in Example 5.73 of this same chapter, another logarithmic function — namely, $\log_2 n$, for $n \in \mathbf{Z}^+$ — appears in the analysis of an algorithm. And since these types of functions occur in later chapters as well, we now provide this appendix as a review of some of the fundamental properties of these two kinds of functions.

Let us start with the idea of positive integer exponents. For instance, we know that the expression 3^7 indicates the multiplication of seven 3's — that is,

$$3^7 = 3 \cdot 3 \cdot 3 \cdot 3 \cdot 3 \cdot 3 \cdot 3 = 2187.$$

In this example, the number 3 is called the *base* of 3^7; the number 7 is the *exponent*, or *power*. Generally, when the exponent is a positive integer, the base — call it b — can be any real number (including 0). In dealing with an exponent that is a negative integer, we use the following definition.

Definition A1.1	For every nonzero real number b and every $n \in \mathbf{Z}^+$, we have $b^{-n} = 1/b^n$.

EXAMPLE A1.1

From Definition A1.1 we see that

a) $3^{-7} = 1/3^7 = 1/2187$ b) $(1/2)^{-6} = 1/(1/2)^6 = 1/(1/64) = 64$

c) $(-3/5)^{-5} = 1/(-3/5)^5 = 1/(-243/3125) = -3125/243$

Finally, when our exponent is the integer 0 we define $b^0 = 1$, for any *nonzero*[†] real number b.

The preceding ideas can be summarized in the following, where we use the idea of a recursive definition (introduced in Section 2 of Chapter 4) in the first part:

For all $b \in \mathbf{R}$,

[†]The expression 0^0 is called an *indeterminate form* since its value may be different in different situations. This idea is studied in calculus and is covered in conjunction with L'Hospital's Rule.

1) $b^1 = b$, and $b^n = b \cdot b^{n-1}$, for $n \in \mathbf{Z}^+$ where $n > 1$;

2) if $b \neq 0$ and $n \in \mathbf{Z}^+$, then $b^{-n} = 1/b^n$; and

3) if $b \neq 0$, then $b^0 = 1$.

In order to proceed from integer exponents to those that are rational numbers, we recall from earlier work in algebra that if $q \in \mathbf{Z}^+$, where $q > 1$, and b is any nonnegative real number, then the expression $b^{1/q}$ denotes *the* qth root of b. Hence $b^{1/q}$ is *the* real number a where $a^q = b$. For example,

$$32^{1/5} = 2 \text{ because } 2^5 = 32, \text{ and} \qquad (1/8)^{1/3} = 1/2 \text{ because } (1/2)^3 = 1/8.$$

But when we are confronted with the equations $2^2 = 4$ and $(-2)^2 = 4$, we must ask ourselves what we shall mean here by $4^{1/2}$. The convention that is followed names the positive root as the one represented by $4^{1/2}$, so $4^{1/2} = 2$, not -2 or ± 2. Likewise, $9^{1/2} = 3$, $16^{1/2} = 4$, and for all $r \in \mathbf{R}$, $(r^2)^{1/2} = |r|$, the absolute value of r, not just plain r. Also, though $2^4 = (-2)^4 = (2i)^4 = (-2i)^4 = 16$, when the expression $16^{1/4}$ is encountered it denotes *the* positive fourth root, namely, 2.

When b is a negative real number and q is an odd positive integer, our earlier definition of $b^{1/q}$ continues to make sense. We find, for example, that $(-8)^{1/3} = -2$ since $(-2)^3 = -8$ and no other cube of a *real* number results in -8. However, for the case where $q = 2$, the expression $(-4)^{1/2}$ denotes a complex number that is not real — and so we shall avoid such situations here.

Finally, without getting into a detailed discussion on the development of irrational numbers, we shall agree that real, but irrational, numbers such as $2^{1/2} = \sqrt{2}$ and $(-5)^{1/3} = \sqrt[3]{-5}$ do exist and, in general, for $q \in \mathbf{Z}^+$ and $r \in \mathbf{R}$, the following real numbers also exist:

$$r^{1/q} = \sqrt[q]{r}, \text{ for } r \geq 0 \qquad r^{1/q} = \sqrt[q]{r}, \text{ for } r < 0 \text{ and } q \text{ odd.}$$

And now that we have settled this issue of exponents (or powers) of the form $1/q$, where q is a positive integer greater than 1, we pass to the following definition.

Definition A1.2

Let $b \in \mathbf{R}$ and let $p, q \in \mathbf{Z}^+$. Then

1) $b^{p/q} = (b^{1/q})^p$, for $b \geq 0$;

2) $b^{-p/q} = (b^{1/q})^{-p} = 1/[(b^{1/q})^p]$, for $b > 0$;

3) $b^{p/q} = (b^{1/q})^p$, for $b < 0$ and q odd; and

4) $b^{-p/q} = (b^{1/q})^{-p} = 1/[(b^{1/q})^p]$, for $b < 0$ and q odd.

This definition is illustrated in the following example.

EXAMPLE A1.2

a) $(8)^{2/3} = 8^{2/3} = (8^{1/3})^2 = 2^2 = 4 \ (= 64^{1/3} = (8^2)^{1/3})$

b) $(81)^{-3/4} = (81^{1/4})^{-3} = 3^{-3} = 1/3^3 = 1/27 \ (= (3^{-1})^3 = [(81^{1/4})^{-1}]^3 = [(81)^{-1/4}]^3)$

c) $(-1/32)^{3/5} = [(-1/32)^{1/5}]^3 = (-1/2)^3 = -1/8$

d) $(-1024)^{-2/5} = [(-1024)^{1/5}]^{-2} = (-4)^{-2} = 1/(-4)^2 = 1/16 \ (= 1/(-1024)^{2/5})$.

The last result observed in part (a) of the preceding example suggests the following, which is true in general:

$$b^{p/q} = (b^p)^{1/q}, \qquad b \geq 0, \qquad p, q \in \mathbf{Z}^+.$$

The other parts of Definition A1.2 can also be extended as

$$b^{-p/q} = (b^{-p})^{1/q} = (1/b^p)^{1/q} = (1/b^{p/q}), b > 0, p, q \in \mathbf{Z}^+.$$

$$b^{p/q} = (b^p)^{1/q}, b < 0, p, q \in \mathbf{Z}^+, q \text{ odd.}$$

$$b^{-p/q} = (b^{-p})^{1/q} = (1/b^p)^{1/q} = (1/b^{p/q}), b < 0, p, q \in \mathbf{Z}^+, q \text{ odd.}$$

EXAMPLE A1.3

Using 2 as our base, we know from Definitions A1.1 and A1.2 that

$$2^{-3} = 1/8, \qquad 2^{-2} = 1/4, \qquad 2^{-1} = 1/2, \qquad 2^0 = 1, \qquad 2^1 = 2, \qquad 2^2 = 4, \qquad 2^3 = 8$$

and that

$$2^{-3/2} = (2^{1/2})^{-3} = (\sqrt{2})^{-3} = (1/\sqrt{2})^3 = 1/(2\sqrt{2}) \doteq 0.3535534$$
$$2^{3/2} = (\sqrt{2})^3 = 2\sqrt{2} \doteq 2.8284271 (\doteq (2^3)^{1/2} = \sqrt{8}).$$

However, how do we deal with something like $2^{\sqrt{3}}$, where now an irrational power confronts us? Using the fact that $\sqrt{3} = 1.7320508\ldots$, we can evaluate the successive rational powers:

$$2^1 = 2$$
$$2^{1.7}(= 2^{17/10} = (2^{17})^{1/10} = 131072^{1/10}) \doteq 3.2490096$$
$$2^{1.73} \doteq 3.3172782$$
$$2^{1.732} \doteq 3.3218801$$
$$2^{1.7320} \doteq 3.3218801$$
$$2^{1.73205} \doteq 3.3219952$$
$$\vdots$$

With the assistance of a hand-held calculator or a computer one finds that to seven decimal places $2^{\sqrt{3}}$ is given as 3.3219971. If we want to be more precise, we can say that the real number $2^{\sqrt{3}}$ is the *limit* of the sequence $2^1, 2^{1.7}, 2^{1.73}, 2^{1.732}, 2^{1.7320}, 2^{1.73205}, \ldots$. (One studies such ideas in calculus and introductory analysis.)

In a similar way one deals with the expression b^r, where $b \in \mathbf{R}^+$ and $r \in \mathbf{R}$.

Using the results we have now learned about exponents, we state the following properties — but we do not prove any of them.

THEOREM A1.1

The Properties of Exponents. For all $a, b \in \mathbf{R}^+$ and all $x, y \in \mathbf{R}$,

    1) $(b^x)(b^y) = b^x \cdot b^y = b^{x+y}$,

    2) $(b^x)/(b^y) = b^x/b^y = b^{x-y}$,

    3) $(b^x)^y = b^{xy} = b^{yx} = (b^y)^x$, and

    4) $(ab)^x = (a^x)(b^x) = a^x \cdot b^x$.

The properties in Theorem A1.1 are illustrated in the following.

EXAMPLE A1.4

    1) $3^{5/2} \cdot 3^{3/2} = 3^{[(5/2)+(3/2)]} = 3^{8/2} = 3^4 = 81$

    2) $(7^{1/5})/(7^{11/5}) = 7^{[(1/5)-(11/5)]} = 7^{-10/5} = 7^{-2} = 1/7^2 = 1/49$

    3) $[(\sqrt{2})^3]^2 = (\sqrt{2})^6 = (2^{1/2})^6 = 2^{(1/2)6} = 2^3 = 8$

    4) $(3\sqrt{5})^4 = 3^4(\sqrt{5})^4 = (81)(25) = 2025$

We have now finished with the preliminaries needed to define an exponential function.

Definition A1.3

For a fixed positive real number b, the function $f: \mathbf{R} \rightarrow \mathbf{R}^+$ defined by $f(x) = b^x$ is called the *exponential function for base b*. [Sometimes we denote b^x by $\exp_b(x)$.]

EXAMPLE A1.5

a) In Fig. A1.1 we find the graphs of four functions:

$$f_1: \mathbf{R} \to \mathbf{R}^+, \quad f_1(x) = x^2 \qquad f_2: \mathbf{R} \to \mathbf{R}^+, \quad f_2(x) = 2^x$$
$$f_3: \mathbf{R} \to \mathbf{R}, \quad f_3(x) = x^3 \qquad f_4: \mathbf{R} \to \mathbf{R}^+, \quad f_4(x) = 3^x$$

The functions f_1 and f_3 are polynomial functions — *not* exponential functions. Hence, when we examine the exponential functions f_2 and f_4 we realize that there is a distinct difference between the expressions x^2 (for f_1) and 2^x (for f_2), and between the expressions x^3 (for f_3) and 3^x (for f_4). The exponential functions f_2 and f_4 are such that

1) $f_2(x) > 0$ and $f_4(x) > 0$, for all $x \in \mathbf{R}$ — in particular, $f_2(x) > 1$ and $f_4(x) > 1$, for all $x > 0$, while $0 < f_2(x) < 1$ and $0 < f_4(x) < 1$, for all $x < 0$.

2) for all $x, y \in \mathbf{R}$, $x < y \Rightarrow f_2(x) < f_2(y)$ [and $f_4(x) < f_4(y)$]. (This is true for every exponential function where the base $b > 1$. That is, when $b > 1$ and $x < y$, then $b^x < b^y$.)

3) if $v, w \in \mathbf{R}$ and $f_2(v) = f_2(w)$, then $v = w$. [This property is also true whenever we are dealing with an exponential function $f(x) = b^x$, for $b > 1$. So for $v, w \in \mathbf{R}$ and $b > 1$, $b^v = b^w \Rightarrow v = w$.]

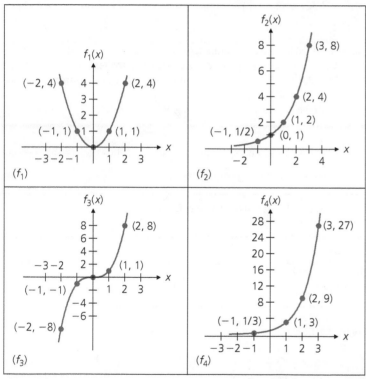

Figure A1.1

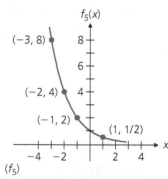

Figure A1.2

b) The graph of the function $f_5: \mathbf{R} \to \mathbf{R}^+$, defined by $f_5(x) = (1/2)^x = 2^{-x}$, is given in Fig. A1.2. This graph demonstrates the following properties, which are true for all exponential functions $f: \mathbf{R} \to \mathbf{R}^+$, where $f(x) = b^x$ for $0 < b < 1$.

1) Here $f_5(x) > 0$ for all $x \in \mathbf{R}$ — but now we find $f_5(x) > 1$ for $x < 0$ and $f_5(x) < 1$ when $x > 0$.

2) If $x, y \in \mathbf{R}$ with $x < y$, then $f_5(x) > f_5(y)$.

3) For $x, y \in \mathbf{R}$, if $f_5(x) = f_5(y)$, then $x = y$.

c) When one speaks of *the* exponential function the reference is to the function $f : \mathbf{R} \to \mathbf{R}^+$, where $f(x) = e^x$ for the irrational number $e \doteq 2.71828$. This function is shown as f_6 in Fig. A1.3, where we have used the approximations $e^2 \doteq 7.38906$ and $e^3 \doteq 20.08554$. The function f_7 (also in Fig. A1.3) is the exponential function where $f_7(x) = e^{-x}$.

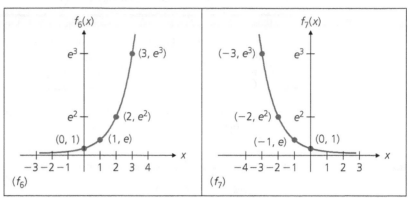

Figure A1.3

From property (3) in parts (a) and (b) of Example A1.5 we learned that for all $b \in \mathbf{R}^+$ and all $x, y \in \mathbf{R}$, if $b \neq 1$ and $b^x = b^y$ then $x = y$. This observation helps us to solve the following exponential equation.

EXAMPLE A1.6

For which real number(s) n is it true that $(1/2)^{-6n^2} = (1/8)^{-(10n+4)/3}$?

This equation can be written as $2^{6n^2} = 8^{(10n+4)/3}$ because $(1/2)^{-6n^2} = [(1/2)^{-1}]^{6n^2} = 2^{6n^2}$ and $(1/8)^{-(10n+4)/3} = [(1/8)^{-1}]^{(10n+4)/3} = 8^{(10n+4)/3}$. Then

$$2^{6n^2} = 8^{(10n+4)/3} \Rightarrow 2^{6n^2} = (2^3)^{(10n+4)/3} \Rightarrow 2^{6n^2} = 2^{(10n+4)} \Rightarrow$$

$$6n^2 = 10n + 4 \Rightarrow 3n^2 = 5n + 2 \Rightarrow$$

$$3n^2 - 5n - 2 = (3n+1)(n-2) = 0 \Rightarrow n = -1/3 \text{ or } n = 2.$$

Now that we have examined the exponential function, we shall turn our attention to a second type of function that goes hand-in-hand with the exponential function. This is the logarithm or logarithmic function. However, before we introduce this function, we shall review some of the fundamental properties of logarithms. First we consider the precise relationship between exponents and logarithms, as described in the following definition.

Definition A1.4

Let b denote a fixed positive real number other than 1. If $x \in \mathbf{R}^+$, we write $\log_b x$ to designate the *logarithm of x to the base b* (or the logarithm to the base b of x), which is the (unique) real number y that satisfies $b^y = x$.

This idea can be restated as follows: $\log_b x$ is the exponent (or power) to which we raise the base b in order to obtain x. Hence,

$$y = \log_b x \text{ if and only if } x = b^y.$$

EXAMPLE A1.7

The following results are obtained from the preceding definition:

a) Since $2^3 = 8$, we have $\log_2 8 = 3$.

b) One finds that $\log_3(1/81) = -4$ because $3^{-4} = 1/(3^4) = 1/81$.

c) For all $b \in \mathbf{R}^+$, where $b \neq 1$, it follows that

  i) $\log_b b = 1$ because $b^1 = b$,

  ii) $\log_b b^2 = 2$ because $b^2 = b^2$, and

  iii) $\log_b(1/b) = -1$ because $b^{-1} = 1/b$.

d) Since $\sqrt{7} = 7^{1/2}$, it follows that $\log_7 \sqrt{7} = 1/2$.

EXAMPLE A1.8

Suppose that $b, x \in \mathbf{R}^+$ where b is fixed and different from 1. If $\log_b x = 6$, what is $\log_{b^2} x$?

We know that $\log_b x = 6 \iff b^6 = x$, so $x = (b^2)^3$. And $x = (b^2)^3 \iff \log_{b^2} x = 3$. (In a similar manner one also finds that $\log_{b^3} x = 2$ and $\log_{b^6} x = 1$.)

In conjunction with properties (1), (2), and (3) for exponents, as found in Theorem A1.1, the following properties correspond for logarithms.

THEOREM A1.2

Let $b, r, s \in \mathbf{R}^+$ where b is fixed and other than 1. Then

1) $\log_b(rs) = \log_b r + \log_b s$,

2) $\log_b(r/s) = \log_b r - \log_b s$, and

3) $\log_b(r^s) = s \log_b r$.

Proof: We shall prove part (1) and request a proof for part (2) in the exercises at the end of this appendix. For part (3) we shall only request (in the exercises) the proof for the case where s is a nonzero integer — but we shall accept (without proof) and use the general statement given here.

Suppose that $x = \log_b r$ and $y = \log_b s$. Then, because $x = \log_b r \iff b^x = r$ and $y = \log_b s \iff b^y = s$, it follows from part (1) of Theorem A1.1 that $rs = (b^x)(b^y) = b^{x+y}$. Since $rs = b^{x+y} \iff \log_b(rs) = x + y$, we have shown that

$$\log_b(rs) = x + y = \log_b r + \log_b s.$$

In our next example we find how the three results in Theorem A1.2 can be used to calculate logarithms.

EXAMPLE A1.9

Before the advent of computers and hand-held calculators, logarithms were used to assist in calculating products, quotients, and powers and in extracting roots. Very often the base for these logarithms was 10 and tables of these numbers were available for working with logarithms. [Logarithms were invented by the Scottish mathematician John Napier (1550–1617). Navigators and astronomers used them in the seventeenth century to reduce the time it took to perform multiplication and division.]

For example, since $\log_{10} 10 = 1$ and $\log_{10} 100 = 2$, one finds that $1 < \log_{10} 31 < 2$. In fact, $\log_{10} 31 = 1.4914$. Likewise, we have $2 < \log_{10} 137 = 2.1367 < 3$. From Theorem A1.2 it then follows that

1) $\log_{10} 4247 = \log_{10}(31 \cdot 137) = \log_{10} 31 + \log_{10} 137 = 1.4914 + 2.1367 = 3.6281$,

2) $\log_{10}(137/31) = \log_{10} 137 - \log_{10} 31 = 2.1367 - 1.4914 = 0.6453$, and

3) $\log_{10} \sqrt[3]{137} = \log_{10} 137^{1/3} = (1/3) \log_{10} 137 = (1/3)(2.1367) = 0.7122$.

In calculus we find use for logarithms to the base $e \doteq 2.71828$, and these so-called *natural* logarithms are usually denoted by $\ln x$, for $x \in \mathbf{R}^+$. When dealing with the analysis of algorithms in computer science, logarithms to the base 2 often prove to be useful. But this does not mean we need to be overly concerned about dealing with logarithms in several different bases. Many hand-held calculators provide logarithms to the base 10 and the base e. And we'll find in our next result that if we can obtain logarithms in one base, we can use these to obtain logarithms in any other base.

THEOREM A1.3

The Base-Changing Formula. Let $a, b \in \mathbf{R}^+$ where neither a nor b is 1. For all $x \in \mathbf{R}^+$,

$$\log_a x = \frac{\log_b x}{\log_b a}.$$

Proof: Let $c = \log_b x$ and $d = \log_a x$. Then $b^c = x = a^d$ and $\log_b x = \log_b a^d = d \log_b a = (\log_a x)(\log_b a)$. Consequently, $\log_a x = \log_b x / \log_b a$.

EXAMPLE A1.10

From a table or hand-held calculator one finds that $\log_e 2 = \ln 2 = 0.6931$ and $\log_e 10 = \ln 10 = 2.3026$. Therefore, by virtue of Theorem A1.3, $\log_2 10 = \ln 10/\ln 2 = 2.3026/0.6931 \doteq 3.3222$.

EXAMPLE A1.11

A special formula results from Theorem A1.3 when $x = b$. In this case we find that

$$\log_a b = \frac{\log_b b}{\log_b a} = \frac{1}{\log_b a}.$$

Having reviewed the necessary preliminaries, it is time to define the logarithmic function.

Definition A1.5

Let $b \neq 1$ be a fixed positive real number. The function $g: \mathbf{R}^+ \to \mathbf{R}$ defined by $g(x) = \log_b x$ is called the *logarithmic function to the base b*.

EXAMPLE A1.12

a) Consider the logarithmic functions

$$g_1: \mathbf{R}^+ \to \mathbf{R}, \qquad g_1(x) = \log_2 x \qquad g_2: \mathbf{R}^+ \to \mathbf{R}, \qquad g_2(x) = \log_3 x.$$

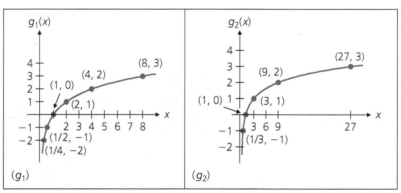

Figure A1.4

The graphs of these functions are shown in Fig. A1.4. These functions are such that

1) $g_1(x) \geq 0$ and $g_2(x) \geq 0$ for all $x \geq 1$, and $g_1(x) < 0$ and $g_2(x) < 0$ for all $x < 1$. (This is true for every logarithmic function $\log_b x$ where $b > 1$.)
2) for all $x, y \in \mathbf{R}^+$, $x < y \Rightarrow g_1(x) < g_1(y)$ [and $g_2(x) < g_2(y)$]. (Again this is true for all logarithmic functions $\log_b x$ where $b > 1$.)
3) if $u, v \in \mathbf{R}^+$ and $g_1(u) = g_1(v)$, then $u = v$. (In fact, for $b > 1$, we have $\log_b u = \log_b v \Rightarrow u = v$ because $w = \log_b u \Leftrightarrow u = b^w$, and $w = \log_b v \Leftrightarrow v = b^w$.)

b) The graph in Fig. A1.5 is for the function $g_3: \mathbf{R}^+ \to \mathbf{R}$ defined by $g_3(x) = \log_{(1/2)} x$. This graph illustrates the following properties, which are true for those logarithmic functions $\log_b x$ where $0 < b < 1$.

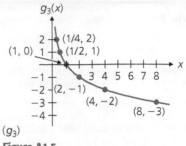

(g₃)

Figure A1.5

1) Here $g_3(x) \geq 0$ for all $x \leq 1$, while $g_3(x) < 0$ for all $x > 1$.

2) For all $x, y \in \mathbf{R}^+$, if $x < y$ then $g_3(x) > g_3(y)$.

3) If $u, v \in \mathbf{R}^+$ and $g_3(u) = g_3(v)$, then $u = v$. [The proof here is the same as that given in section (3) of part (a).]

c) In part (a) of Fig. A1.6 we have the graphs of the functions $f: \mathbf{R} \to \mathbf{R}^+$, where $f(x) = 2^x$, and $g: \mathbf{R}^+ \to \mathbf{R}$, where $g(x) = \log_2 x$. These graphs are symmetric (to each other) in the line $y = x$ — that is, if one were to fold the figure along the line $y = x$, then the graphs of f and g would coincide. Here we also observe how the points on one graph correspond with the points on the other. For instance, the point (2, 4) on the graph of f corresponds with the point (4, 2) on the graph of g. In general, each point $(x, 2^x)$ on the graph of f corresponds with the point $(2^x, x \, (= \log_2 2^x))$ on the graph of g, and $(x, \log_2 x)$ on the graph of g corresponds with $(\log_2 x, x \, (= 2^{\log_2 x}))$ on the graph of f.

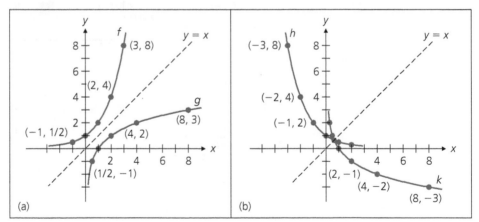

Figure A1.6

d) The graphs of the functions

$$h: \mathbf{R} \to \mathbf{R}^+, \quad h(x) = (1/2)^x \qquad k: \mathbf{R}^+ \to \mathbf{R}, \quad k(x) = \log_{(1/2)} x$$

are shown in part (b) of Fig. A1.6. As in part (c) of this example these functions are also symmetric in the line $y = x$. Here each point $(x, (1/2)^x)$ on the graph of h corresponds with the point $((1/2)^x, x \, (= \log_{(1/2)}(1/2)^x))$ on the graph of k, and $(x, \log_{(1/2)} x)$ on the graph of k corresponds with $(\log_{(1/2)} x, x \, (= (1/2)^{\log_{(1/2)} x}))$ on the graph of h. (These two graphs intersect on the line $y = x$ where $x \doteq 0.6412$.)

e) The reader may now want to examine, or reexamine, the graphs of the functions $y = e^x$ and $y = \ln x$ shown in Fig. 5.10 of Section 5.6. In that section, the relationship of symmetry of functions in the line $y = x$ [mentioned above in parts (c) and (d)] is studied in conjunction with the ideas of function composition and the inverse of a function.

EXERCISES A.1

1. Write each of the following in exponential form, for $x, y \in \mathbf{R}^+$.

  a) $\sqrt{xy^3}$ **b)** $\sqrt[4]{81x^{-5}y^3}$ **c)** $5\sqrt[3]{8x^9y^{-5}}$

2. Evaluate each of the following.

  a) $125^{-4/3}$ **b)** $0.027^{2/3}$ **c)** $(4/3)(1/8)^{-2/3}$

3. Determine each of the following.

  a) $(5^{3/4})(5^{13/4})$ **b)** $\dfrac{7^{3/5}}{7^{18/5}}$ **c)** $(5^{1/2})(20^{1/2})$

4. In each of the following find the real number(s) x for which the equation is valid.

  a) $5^{3x^2} = 5^{5x+2}$ **b)** $4^{x-1} = (1/2)^{4x-1}$

  c) $(1/25)^{1-x} = (1/125)^x$

5. Write each of the following exponential equations as a logarithmic equation.

  a) $2^7 = 128$ **b)** $125^{1/3} = 5$

  c) $10^{-4} = 1/10{,}000$ **d)** $2^a = b$

6. Find each of the following logarithms.

  a) $\log_{10} 100$ **b)** $\log_{10}(1/1000)$

  c) $\log_2 2048$ **d)** $\log_2(1/64)$

  e) $\log_4 8$ **f)** $\log_8 2$

  g) $\log_{16} 1$ **h)** $\log_{27} 9$

7. Solve for x in each of the following.

  a) $\log_x 243 = 5$ **b)** $\log_3 x = -3$

  c) $\log_{10} 1000 = x$ **d)** $\log_x 32 = 5/2$

8. Prove part (2) of Theorem A1.2.

9. Let $b, r \in \mathbf{R}^+$ where b is fixed and different from 1.

  a) For all $n \in \mathbf{Z}^+$, prove that $\log_b r^n = n \log_b r$.

  b) Prove that $\log_b r^{-n} = (-n) \log_b r$ for all $n \in \mathbf{Z}^+$.

10. Approximate each of the following on the basis that (to four decimal places) $\log_2 5 = 2.3219$ and $\log_2 7 = 2.8074$.

  a) $\log_2 10$ **b)** $\log_2 100$

  c) $\log_2(7/5)$ **d)** $\log_2 175$

11. Given that (to four decimal places) $\ln 2 = 0.6931$, $\ln 3 = 1.0986$, and $\ln 5 = 1.6094$, approximate each of the following.

  a) $\log_2 3$ **b)** $\log_5 2$ **c)** $\log_3 5$

12. Determine the value of x in each of the following.

  a) $\log_{10} 2 + \log_{10} 5 = \log_{10} x$

  b) $\log_4 3 + \log_4 x = \log_4 7 - \log_4 5$

13. Solve for x in each of the following.

  a) $\log_{10} x + \log_{10} 6 = 1$

  b) $\ln x - \ln(x - 1) = \ln 3$

  c) $\log_3(x^2 + 4x + 4) - \log_3(2x - 5) = 2$

14. Determine the value of x if

  $\log_2 x = (1/3)[\log_2 3 - \log_2 5] + (2/3) \log_2 6 + \log_2 17$.

15. Let b be a fixed positive real number other than 1. If $a, c \in \mathbf{R}^+$, prove that $a^{\log_b c} = c^{\log_b a}$.

Appendix 2

Matrices, Matrix Operations, and Determinants

Starting in Chapter 7, and then in several subsequent chapters, certain kinds of matrices have been introduced. Historically, these mathematical structures were developed and studied in the nineteenth century by the English mathematician Arthur Cayley (1821–1895) and his (English-born) American coworker James Joseph Sylvester (1814–1897). Introduced in 1858, Cayley's work in matrix algebra provides another instance where research in abstract mathematics later proved to be of importance in many applied areas — for example, in quantum theory in physics and data analysis in psychology and sociology.

For those readers who may not have studied anything about matrices in earlier coursework or who simply wish to review the matrix algebra we use in this text, the material in this appendix should prove to be helpful. (We shall not prove all of the results in general here but state many of them in conjunction with a given example. For a more rigorous development the reader should consult one of the references at the end of this appendix.)

First and foremost, we start with the following.

Definition A2.1

For $m, n \in \mathbf{Z}^+$ an $m \times n$ *matrix* is a rectangular array of mn numbers arranged in m (horizontal) rows and n (vertical) columns.

An $m \times n$ matrix A is denoted by $A = (a_{ij})_{m \times n}$, where $1 \leq i \leq m$ and $1 \leq j \leq n$, and the number a_{ij} is called the (i, j)-*entry* (that is, the entry that appears in the ith row and jth column of A). An $m \times 1$ matrix is often called a *column matrix* (or *column vector*); a $1 \times n$ matrix is referred to as a *row matrix* (or *row vector*). When $m = n$ the matrix is called *square*.

EXAMPLE A2.1

Let $A = (a_{ij})_{3 \times 2} = \begin{bmatrix} 1 & 2 \\ 0 & 3 \\ -5 & 4 \end{bmatrix}$, $B = (b_{ij})_{2 \times 4} = \begin{bmatrix} 1 & 2 & 0 & 3 \\ 1 & 2 & -1 & 7 \end{bmatrix}$, and $C = \begin{bmatrix} \pi & 0 \\ 1/2 & \sqrt{2} \end{bmatrix}$.

Here A is a 3×2 matrix where $a_{11} = 1$, $a_{12} = 2$, $a_{21} = 0$, $a_{22} = 3$, $a_{31} = -5$, and $a_{32} = 4$. The matrix B has two rows and four columns, where, for instance, one finds the entries $b_{13} = 0$ and $b_{24} = 7$. In the 2×2 square matrix C we see that the entries in a matrix may be rational numbers and even irrational numbers.

(*Note*: Although the entries in a matrix may even be complex numbers, in this appendix we shall deal only with matrices where each entry is a real number.)

As with other mathematical structures, once the structure is defined one needs to decide when two such structures are the same. The method for that decision is now addressed.

Definition A2.2

Let $A = (a_{ij})_{m \times n}$ and $B = (b_{ij})_{m \times n}$ be two $m \times n$ matrices. We say that A and B are *equal*, and we write $A = B$, when $a_{ij} = b_{ij}$ for all $1 \leq i \leq m$ and all $1 \leq j \leq n$.

EXAMPLE A2.2

In Definition A2.2 we learned that two matrices are equal when they have the same number of rows and the same number of columns and when their corresponding entries are equal. As a result, if

$$A = \begin{bmatrix} w & 2 & 0 \\ 0 & 3 & x \end{bmatrix} \quad \text{and} \quad B = \begin{bmatrix} -7 & y & 0 \\ 0 & z & 4 \end{bmatrix},$$

then for A and B to be equal we must have $w = -7$, $x = 4$, $y = 2$, $z = 3$.

Thinking back to our first encounters with arithmetic, after we learned how to count, we then started to combine integers by using addition, and then multiplication. Along the same lines we now consider how we may combine matrices.

Definition A2.3

If $A = (a_{ij})_{m \times n}$ and $B = (b_{ij})_{m \times n}$ are two $m \times n$ matrices, their *sum*, denoted $A + B$, is the $m \times n$ matrix $C = (c_{ij})_{m \times n}$, where $c_{ij} = a_{ij} + b_{ij}$, for all $1 \leq i \leq m$, $1 \leq j \leq n$.

From Definition A2.3 we see that we can only add two matrices of the same size (where they have the same number of rows and the same number of columns). Furthermore, the addition of two matrices is carried out by adding their corresponding entries.

EXAMPLE A2.3

Consider the matrices

$$A = \begin{bmatrix} 1 & 3 & 4 \\ 2 & 0 & 6 \\ 1 & 1 & 3 \end{bmatrix}, \qquad B = \begin{bmatrix} 2 & -1 & 6 \\ 3 & 1 & 7 \\ 4 & 2 & 2 \end{bmatrix}, \qquad \text{and } C = \begin{bmatrix} 1 & -1 \\ 3 & -4 \\ -7 & 6 \end{bmatrix}.$$

Here we find that $A + B = \begin{bmatrix} 1+2 & 3+(-1) & 4+6 \\ 2+3 & 0+1 & 6+7 \\ 1+4 & 1+2 & 3+2 \end{bmatrix} = \begin{bmatrix} 3 & 2 & 10 \\ 5 & 1 & 13 \\ 5 & 3 & 5 \end{bmatrix}$. In fact, we also

have $B + A = \begin{bmatrix} 3 & 2 & 10 \\ 5 & 1 & 13 \\ 5 & 3 & 5 \end{bmatrix}$, which illustrates the following general result.

For any two $m \times n$ matrices E and F, $E + F = F + E$. Hence the addition of matrices is an example of a commutative (binary) operation.

We cannot determine either of the sums $A + C$ or $B + C$ because each of A, B has three columns while C has only two. However, we can find the sum

$$C + C = \begin{bmatrix} 1 & -1 \\ 3 & -4 \\ -7 & 6 \end{bmatrix} + \begin{bmatrix} 1 & -1 \\ 3 & -4 \\ -7 & 6 \end{bmatrix} = \begin{bmatrix} 2 & -2 \\ 6 & -8 \\ -14 & 12 \end{bmatrix}.$$

In the last part of Example A2.3, we see that we could have obtained the result $C + C$ by simply multiplying each entry of C by the number 2. This leads us to the general idea we now state as follows.

Definition A2.4

If $A = (a_{ij})_{m \times n}$ and $r \in \mathbf{R}$, the *scalar product* rA is the $m \times n$ matrix where the (i, j)-entry is ra_{ij}, for all $1 \leq i \leq m$, $1 \leq j \leq n$.

EXAMPLE A2.4

a) If $A = \begin{bmatrix} 1 & 6 & 4 \\ 0 & -1 & -3 \end{bmatrix}$, then

$$3A = 3 \begin{bmatrix} 1 & 6 & 4 \\ 0 & -1 & -3 \end{bmatrix} = \begin{bmatrix} 3 \cdot 1 & 3 \cdot 6 & 3 \cdot 4 \\ 3 \cdot 0 & 3 \cdot (-1) & 3 \cdot (-3) \end{bmatrix} = \begin{bmatrix} 3 & 18 & 12 \\ 0 & -3 & -9 \end{bmatrix}.$$

b) For $A = \begin{bmatrix} 1 & 6 & 4 \\ 0 & -1 & -3 \end{bmatrix}$ and $B = \begin{bmatrix} -2 & 0 & 2 \\ -5 & 1 & 7 \end{bmatrix}$, we find $3B = \begin{bmatrix} -6 & 0 & 6 \\ -15 & 3 & 21 \end{bmatrix}$, and

$$3(A + B) = 3 \left(\begin{bmatrix} 1 & 6 & 4 \\ 0 & -1 & -3 \end{bmatrix} + \begin{bmatrix} -2 & 0 & 2 \\ -5 & 1 & 7 \end{bmatrix} \right)$$

$$= 3 \begin{bmatrix} -1 & 6 & 6 \\ -5 & 0 & 4 \end{bmatrix} = \begin{bmatrix} -3 & 18 & 18 \\ -15 & 0 & 12 \end{bmatrix} = \begin{bmatrix} 3 & 18 & 12 \\ 0 & -3 & -9 \end{bmatrix} + \begin{bmatrix} -6 & 0 & 6 \\ -15 & 3 & 21 \end{bmatrix}$$

$$= 3A + 3B.$$

c) The result in part (b) may be generalized as follows: For any two $m \times n$ matrices E, F and any $r \in \mathbf{R}$, $r(E + F) = rE + rF$. This principle is called the *Distributive Law of Scalar Multiplication over Matrix Addition*.

EXAMPLE A2.5

a) Let $A = (a_{ij})_{3 \times 2}$ represent an arbitrary 3×2 matrix, and let $Z = \begin{bmatrix} 0 & 0 \\ 0 & 0 \\ 0 & 0 \end{bmatrix}$. Then

$$A + Z = \begin{bmatrix} a_{11} & a_{12} \\ a_{21} & a_{22} \\ a_{31} & a_{32} \end{bmatrix} + \begin{bmatrix} 0 & 0 \\ 0 & 0 \\ 0 & 0 \end{bmatrix} = \begin{bmatrix} a_{11} + 0 & a_{12} + 0 \\ a_{21} + 0 & a_{22} + 0 \\ a_{31} + 0 & a_{32} + 0 \end{bmatrix} = \begin{bmatrix} a_{11} & a_{12} \\ a_{21} & a_{22} \\ a_{31} & a_{32} \end{bmatrix} = A.$$

We say that Z is the *additive identity* (or *zero element*) for all 3×2 matrices.

b) When $A = \begin{bmatrix} 1 & 1 \\ 2 & -3 \\ -4 & 5 \end{bmatrix}$ and $B = \begin{bmatrix} -1 & -1 \\ -2 & 3 \\ 4 & -5 \end{bmatrix}$, it follows that

$$A + B = \begin{bmatrix} 1 + (-1) & 1 + (-1) \\ 2 + (-2) & (-3) + 3 \\ (-4) + 4 & 5 + (-5) \end{bmatrix} = \begin{bmatrix} 0 & 0 \\ 0 & 0 \\ 0 & 0 \end{bmatrix}.$$

Consequently, we call $B = (-1)A$ the *additive inverse* of A, and also write $B = -A$.

Hopefully, what we have done so far has proved to be somewhat interesting. But what makes the study of matrices truly interesting and useful is the operation of matrix multiplication. If one tries to define this operation like the componentwise operation of matrix addition, the result is of little interest. Instead, matrix multiplication rests upon a row-and-column multiplication and summation where, for example,

$$[a_1 \quad a_2 \quad a_3] \begin{bmatrix} b_1 \\ b_2 \\ b_3 \end{bmatrix} = a_1 b_1 + a_2 b_2 + a_3 b_3 = \sum_{i=1}^{3} a_i b_i.$$

Hence, in one particular case, we have

$$[-1 \quad 4 \quad 3] \begin{bmatrix} 2 \\ 1 \\ 7 \end{bmatrix} = (-1) \cdot 2 + 4 \cdot 1 + 3 \cdot 7 = -2 + 4 + 21 = 23.$$

In general, if $a = (a_i)_{1 \le i \le n}$ is a $1 \times n$ row vector and $b = (b_i)_{1 \le i \le n}$ is an $n \times 1$ column vector, then $ab = \sum_{i=1}^{n} a_i b_i$. This result, which is a real number, is called the *scalar product* of the vectors (or matrices) a and b. This idea is the key we need for the following definition.

Definition A2.5

Given the matrices $A = (a_{ij})_{m \times n}$ and $B = (b_{jk})_{n \times p}$, the (matrix) *product* AB is the matrix $C = (c_{ik})_{m \times p}$, where

$$c_{ik} = a_{i1}b_{1k} + a_{i2}b_{2k} + \cdots + a_{in}b_{nk} = \sum_{t=1}^{n} a_{it}b_{tk}, \quad \text{for all } 1 \le i \le m, 1 \le k \le p.$$

Hence the entry c_{ik} in the ith row and kth column of the $m \times p$ matrix C is obtained from the scalar product of the ith row (vector) of A and the jth column (vector) of B.

The following demonstrates the result given by Definition A2.5.

$$AB = \begin{bmatrix} a_{11} & a_{12} & \cdots & a_{1n} \\ a_{21} & a_{22} & \cdots & a_{2n} \\ \cdots & \cdots & \cdots & \cdots \\ a_{i1} & a_{i2} & \cdots & a_{in} \\ \cdots & \cdots & \cdots & \cdots \\ a_{m1} & a_{m2} & \cdots & a_{mn} \end{bmatrix} \begin{bmatrix} b_{11} & b_{12} & \cdots & b_{1k} & \cdots & b_{1p} \\ b_{21} & b_{22} & \cdots & b_{2k} & \cdots & b_{2p} \\ \cdots & \cdots & \cdots & \cdots & \cdots & \cdots \\ \cdots & \cdots & \cdots & \cdots & \cdots & \cdots \\ \cdots & \cdots & \cdots & \cdots & \cdots & \cdots \\ b_{n1} & b_{n2} & \cdots & b_{nk} & \cdots & b_{np} \end{bmatrix}$$

$$= C = \begin{bmatrix} c_{11} & c_{12} & \cdots & c_{1k} & \cdots & c_{1p} \\ c_{21} & c_{22} & \cdots & c_{2k} & \cdots & c_{2p} \\ \cdots & \cdots & \cdots & \cdots & \cdots & \cdots \\ c_{i1} & c_{i2} & \cdots & c_{ik} & \cdots & c_{ip} \\ \cdots & \cdots & \cdots & \cdots & \cdots & \cdots \\ c_{m1} & c_{m2} & \cdots & c_{mk} & \cdots & c_{mp} \end{bmatrix}$$

EXAMPLE A2.6

a) Consider the matrices $A = (a_{ij})_{2 \times 3} = \begin{bmatrix} 1 & 2 & 1 \\ 3 & 0 & 4 \end{bmatrix}$ and $B = (b_{jk})_{3 \times 3} = \begin{bmatrix} 1 & 2 & 7 \\ 1 & 3 & 3 \\ 0 & 1 & 1 \end{bmatrix}$.

Then $AB = C = (c_{ik})_{2 \times 3} = \begin{bmatrix} c_{11} & c_{12} & c_{13} \\ c_{21} & c_{22} & c_{23} \end{bmatrix}$, where

$$c_{11} = 1 \cdot 1 + 2 \cdot 1 + 1 \cdot 0 = 3 \qquad \begin{bmatrix} 1 & 2 & 1 \\ 3 & 0 & 4 \end{bmatrix} \begin{bmatrix} 1 & 2 & 7 \\ 1 & 3 & 3 \\ 0 & 1 & 1 \end{bmatrix},$$

$$c_{12} = 1 \cdot 2 + 2 \cdot 3 + 1 \cdot 1 = 9 \qquad \begin{bmatrix} 1 & 2 & 1 \\ 3 & 0 & 4 \end{bmatrix} \begin{bmatrix} 1 & 2 & 7 \\ 1 & 3 & 3 \\ 0 & 1 & 1 \end{bmatrix},$$

$$c_{13} = 1 \cdot 7 + 2 \cdot 3 + 1 \cdot 1 = 14 \qquad \begin{bmatrix} 1 & 2 & 1 \\ 3 & 0 & 4 \end{bmatrix} \begin{bmatrix} 1 & 2 & 7 \\ 1 & 3 & 3 \\ 0 & 1 & 1 \end{bmatrix},$$

$$c_{21} = 3 \cdot 1 + 0 \cdot 1 + 4 \cdot 0 = 3 \qquad \begin{bmatrix} 1 & 2 & 1 \\ 3 & 0 & 4 \end{bmatrix} \begin{bmatrix} 1 & 2 & 7 \\ 1 & 3 & 3 \\ 0 & 1 & 1 \end{bmatrix},$$

$$c_{22} = 3 \cdot 2 + 0 \cdot 3 + 4 \cdot 1 = 10 \qquad \begin{bmatrix} 1 & 2 & 1 \\ 3 & 0 & 4 \end{bmatrix} \begin{bmatrix} 1 & 2 & 7 \\ 1 & 3 & 3 \\ 0 & 1 & 1 \end{bmatrix},$$

$$c_{23} = 3 \cdot 7 + 0 \cdot 3 + 4 \cdot 1 = 25 \qquad \begin{bmatrix} 1 & 2 & 1 \\ 3 & 0 & 4 \end{bmatrix} \begin{bmatrix} 1 & 2 & 7 \\ 1 & 3 & 3 \\ 0 & 1 & 1 \end{bmatrix}.$$

Consequently,

$$AB = C = \begin{bmatrix} 3 & 9 & 14 \\ 3 & 10 & 25 \end{bmatrix}.$$

b) With A and B as in part (a) let us try to form the matrix product $BA = \begin{bmatrix} 1 & 2 & 7 \\ 1 & 3 & 3 \\ 0 & 1 & 1 \end{bmatrix} \begin{bmatrix} 1 & 2 & 1 \\ 3 & 0 & 4 \end{bmatrix}.$

To find the entry in the first row and first column of BA we want to form the scalar product

$$[1 \quad 2 \quad 7]\begin{bmatrix} 1 \\ 3 \end{bmatrix} = 1 \cdot 1 + 2 \cdot 3 + 7 \cdot \text{``?''}.$$

Unfortunately, we do not have enough entries in the first column of A, and so we cannot form either this scalar product or the matrix product BA.

Now we may find ourselves wondering why we could form the product AB but couldn't form the product BA. Considering the product BA once again, we see that the difficulty hinges on the fact that the first column of A did not have the same number of entries as the first row of B. The number of entries in the first row of B is 3, which is the number of columns in B. The number of entries in the first column of A is 2, which is the number of rows in A. These observations lead us to the following general result.

If C is an $m \times n$ matrix and D is a $p \times q$ matrix, then the product CD can be formed when $n = p$ — that is, when the number of columns in C (the first matrix) equals the number of rows in D (the second matrix). And when $n = p$ the resulting product CD has m rows and q columns.

Let us examine matrix multiplication a little further.

EXAMPLE A2.7

a) If $A = \begin{bmatrix} 1 & 2 \\ 1 & 3 \\ 1 & 1 \end{bmatrix}$ and $B = \begin{bmatrix} 1 & 1 & 2 \\ 2 & 0 & 2 \end{bmatrix}$, then $AB = \begin{bmatrix} 5 & 1 & 6 \\ 7 & 1 & 8 \\ 3 & 1 & 4 \end{bmatrix}$ while $BA = \begin{bmatrix} 4 & 7 \\ 4 & 6 \end{bmatrix}$.

Consequently, even though it is possible to form both matrix products AB and BA, we do *not* have $AB = BA$. In fact, these products are not even of the same size.

b) For $A = \begin{bmatrix} -1 & 1 \\ 1 & -1 \end{bmatrix}$ and $B = \begin{bmatrix} 1 & 2 \\ 1 & 2 \end{bmatrix}$, one finds that $AB = \begin{bmatrix} 0 & 0 \\ 0 & 0 \end{bmatrix}$ and $BA = \begin{bmatrix} 1 & -1 \\ 1 & -1 \end{bmatrix}$.

So here AB and BA are of the same size, but $AB \neq BA$.

c) Finally, consider the matrices

$$A = \begin{bmatrix} 1 & 1 & 3 \\ 4 & -1 & 5 \end{bmatrix}, \qquad B = \begin{bmatrix} 2 & 1 \\ 0 & 1 \\ -3 & -1 \end{bmatrix}, \qquad \text{and} \qquad C = \begin{bmatrix} 1 & 2 \\ 3 & -4 \end{bmatrix}.$$

Here we find that

$$AB = \begin{bmatrix} 1 & 1 & 3 \\ 4 & -1 & 5 \end{bmatrix}\begin{bmatrix} 2 & 1 \\ 0 & 1 \\ -3 & -1 \end{bmatrix} = \begin{bmatrix} -7 & -1 \\ -7 & -2 \end{bmatrix} \qquad \text{and}$$

$$(AB)C = \begin{bmatrix} -7 & -1 \\ -7 & -2 \end{bmatrix} \begin{bmatrix} 1 & 2 \\ 3 & -4 \end{bmatrix} = \begin{bmatrix} -10 & -10 \\ -13 & -6 \end{bmatrix},$$

while

$$BC = \begin{bmatrix} 2 & 1 \\ 0 & 1 \\ -3 & -1 \end{bmatrix} \begin{bmatrix} 1 & 2 \\ 3 & -4 \end{bmatrix} = \begin{bmatrix} 5 & 0 \\ 3 & -4 \\ -6 & -2 \end{bmatrix} \quad \text{and}$$

$$A(BC) = \begin{bmatrix} 1 & 1 & 3 \\ 4 & -1 & 5 \end{bmatrix} \begin{bmatrix} 5 & 0 \\ 3 & -4 \\ -6 & -2 \end{bmatrix} = \begin{bmatrix} -10 & -10 \\ -13 & -6 \end{bmatrix}.$$

Hence, $(AB)C = A(BC)$.

In general, if $m, n, p, q \in \mathbf{Z}^+$ and $A = (a_{ij})_{m \times n}$, $B = (b_{jk})_{n \times p}$, and $C = (c_{kl})_{p \times q}$, then

$$(AB)C = A(BC),$$

so matrix multiplication is associative (when it can be performed).

From the results in parts (a) and (b) of Example A2.7 we learn two important facts:

1) The operation of matrix multiplication is not commutative in general.

2) It is possible to find two nonzero matrices $C = (c_{ij})_{m \times n} (c_{ij} \neq 0$ for some $1 \leq i \leq m, 1 \leq j \leq n)$ and $D = (d_{jk})_{n \times p}$, $(d_{jk} \neq 0$ for some $1 \leq j \leq n, 1 \leq k \leq p)$, where $CD = Z = (0)_{m \times p}$.

In short, matrix multiplication does not necessarily behave like the multiplication of real numbers.

Now that we've made some comparisons between matrix multiplication and the multiplication of real numbers, let us pursue a few more.

EXAMPLE A2.8

a) When we consider square matrices — in particular, 2×2 matrices — we learn that

$$\begin{bmatrix} a & b \\ c & d \end{bmatrix} \begin{bmatrix} 1 & 0 \\ 0 & 1 \end{bmatrix} = \begin{bmatrix} 1 & 0 \\ 0 & 1 \end{bmatrix} \begin{bmatrix} a & b \\ c & d \end{bmatrix} = \begin{bmatrix} a & b \\ c & d \end{bmatrix}.$$

Consequently, the matrix $I_2 = \begin{bmatrix} 1 & 0 \\ 0 & 1 \end{bmatrix}$ is called the *multiplicative identity* for all 2×2 matrices. In general, for a fixed positive integer $n > 1$, the matrix

$$I_n = (\delta_{ij})_{n \times n}, \qquad \text{where } \delta_{ij} = \begin{cases} 1, & \text{if } i = j \\ 0, & \text{if } i \neq j \end{cases}$$

is the *multiplicative identity* for all $n \times n$ matrices.

b) Returning to the real numbers let us recall that for each $x \in \mathbf{R}$, if $x \neq 0$, then there exists $y \in \mathbf{R}$ where $xy = yx = 1$. This real number y is termed the *multiplicative inverse* of x and is often designated by x^{-1}.

We would like to know if there is a similar situation for square matrices — and we shall concentrate on 2×2 matrices.

If $A = \begin{bmatrix} a & b \\ c & d \end{bmatrix}$, where a, b, c, d are fixed real numbers, can we find a matrix $B = \begin{bmatrix} w & x \\ y & z \end{bmatrix}$ so that $AB = BA = I_2$? (Here w, x, y, z are unknown real numbers and our objective is to determine the values of these four numbers in terms of the given real numbers a, b, c, d.)

Forming the product AB we find that

$$AB = \begin{bmatrix} a & b \\ c & d \end{bmatrix} \begin{bmatrix} w & x \\ y & z \end{bmatrix} = \begin{bmatrix} aw + by & ax + bz \\ cw + dy & cx + dz \end{bmatrix}.$$

For AB to equal I_2 — that is, for

$$\begin{bmatrix} aw + by & ax + bz \\ cw + dy & cx + dz \end{bmatrix} = \begin{bmatrix} 1 & 0 \\ 0 & 1 \end{bmatrix}$$

— it follows from the definition of equality of matrices that

(1) $aw + by = 1$ (3) $ax + bz = 0$

(2) $cw + dy = 0$ (4) $cx + dz = 1.$

Focusing on Eqs. (1) and (2), if we multiply Eq. (1) by d and Eq. (2) by b, we find that

(1)' $adw + bdy = d$ (2)' $bcw + bdy = 0.$

Subtracting Eq. (2)' from Eq. (1)', we learn that $adw - bcw = (ad - bc)w = d$, so $w = d/(ad - bc)$, if $ad - bc \neq 0$. Similar calculations yield $x = -b/(ad - bc)$, $y = -c/(ad - bc)$, $z = a/(ad - bc)$, and these formulas are also valid as long as $ad - bc \neq 0$.

[*Note*: (1) The real number $ad - bc$ is called the *determinant* of the matrix A. (2) Although we determined the values for w, x, y, and z from the equation $AB = I_2$, it can be shown that the same solutions result when we deal with the equation $BA = I_2$.]

c) Using the results in part (b), let

$$A = \begin{bmatrix} a & b \\ c & d \end{bmatrix} = \begin{bmatrix} 1 & 2 \\ 0 & 1 \end{bmatrix}.$$

Then with $ad - bc = 1 \cdot 1 - 2 \cdot 0 = 1 \ (\neq 0)$, it follows that $w = 1/1 = 1$, $x = -2/1 = -2$, $y = -0/1 = 0$, $z = 1/1 = 1$, and

$$\begin{bmatrix} 1 & 2 \\ 0 & 1 \end{bmatrix}\begin{bmatrix} 1 & -2 \\ 0 & 1 \end{bmatrix} = \begin{bmatrix} 1 & 0 \\ 0 & 1 \end{bmatrix} = \begin{bmatrix} 1 & -2 \\ 0 & 1 \end{bmatrix}\begin{bmatrix} 1 & 2 \\ 0 & 1 \end{bmatrix}.$$

Under these circumstances we write $A^{-1} = \begin{bmatrix} 1 & -2 \\ 0 & 1 \end{bmatrix}$.

d) Consider the matrix $A_1 = \begin{bmatrix} 3 & 1 \\ 1 & 2 \end{bmatrix}$, where the determinant of $A_1 = 3 \cdot 2 - 1 \cdot 1 = 5 \ (\neq 0)$. Here we find that $A_1^{-1} = \begin{bmatrix} 3 & 1 \\ 1 & 2 \end{bmatrix}^{-1} = \begin{bmatrix} 2/5 & -1/5 \\ -1/5 & 3/5 \end{bmatrix} = 1/5 \begin{bmatrix} 2 & -1 \\ -1 & 3 \end{bmatrix}$.

e) From parts (b), (c), and (d), we can say that if $A = \begin{bmatrix} a & b \\ c & d \end{bmatrix}$, then $A^{-1} = \dfrac{1}{\det(A)} \begin{bmatrix} d & -b \\ -c & a \end{bmatrix}$, when $\det(A) = $ determinant of $A = ad - bc \neq 0$.

f) For the matrix $A_2 = \begin{bmatrix} 1 & 2 \\ 3 & 6 \end{bmatrix}$, one finds that the determinant of $A_2 = 1 \cdot 6 - 2 \cdot 3 = 0$, so in this case there is no multiplicative inverse — that is, A_2^{-1} does not exist.

At this point we have developed some fundamental ideas about matrices, and the reader may be wondering how one might use these mathematical structures. Therefore we return one more time to the real numbers and some of the ideas we encountered in elementary algebra.

When the equation $2x = 3$ is solved the following list of equations may be written:

$$2x = 3 \tag{1}$$

$$\left(\tfrac{1}{2}\right)(2x) = \left(\tfrac{1}{2}\right)(3) \tag{2}$$

$$\left[\left(\tfrac{1}{2}\right)2\right]x = 3/2 \tag{3}$$

$$1 \cdot x = 3/2 \tag{4}$$

$$x = 3/2 \tag{5}$$

And in solving this equation the real number $1/2$ $(= 2^{-1})$, introduced in Eq. (2), is what we need to "get the unknown, x, by itself" as we progress through steps (3) and (4) and get to step (5). So, in general, if we start with the fixed real numbers a, b, where $a \neq 0$, then the equation $ax = b$ has the solution $x = a^{-1}b$.

Now let us consider the system of linear equations:

$$3x + y = 3 \qquad\qquad (*)$$
$$x + 2y = 7,$$

which can be represented in matrix form as

$$\begin{bmatrix} 3 & 1 \\ 1 & 2 \end{bmatrix} \begin{bmatrix} x \\ y \end{bmatrix} = \begin{bmatrix} 3 \\ 7 \end{bmatrix}.$$

[This way of representing a system of linear equations is helpful in understanding the reason behind the definition of matrix multiplication. For the left-hand side of each equation at (*) is the scalar product of a row from the matrix $\begin{bmatrix} 3 & 1 \\ 1 & 2 \end{bmatrix}$ with the column matrix $\begin{bmatrix} x \\ y \end{bmatrix}$. If we let

$$A = \begin{bmatrix} 3 & 1 \\ 1 & 2 \end{bmatrix}, \qquad X = \begin{bmatrix} x \\ y \end{bmatrix}, \qquad \text{and} \qquad B = \begin{bmatrix} 3 \\ 7 \end{bmatrix},$$

then we are seeking a solution for the (matrix) equation $AX = B$. Could the solution here be $X = A^{-1}B$, considering that it was $x = a^{-1}b$ for the earlier equation $ax = b$?

Since the determinant of $A = 3 \cdot 2 - 1 \cdot 1 = 5 \neq 0$, from part (e) of Example A2.8 we know that

$$A^{-1} = (1/5) \begin{bmatrix} 2 & -1 \\ -1 & 3 \end{bmatrix} = \begin{bmatrix} 2/5 & -1/5 \\ -1/5 & 3/5 \end{bmatrix}.$$

Then we find that

$$\begin{bmatrix} 3 & 1 \\ 1 & 2 \end{bmatrix} \begin{bmatrix} x \\ y \end{bmatrix} = \begin{bmatrix} 3 \\ 7 \end{bmatrix} \qquad\qquad (1)'$$

$$\begin{bmatrix} 2/5 & -1/5 \\ -1/5 & 3/5 \end{bmatrix} \left(\begin{bmatrix} 3 & 1 \\ 1 & 2 \end{bmatrix} \begin{bmatrix} x \\ y \end{bmatrix} \right) = \begin{bmatrix} 2/5 & -1/5 \\ -1/5 & 3/5 \end{bmatrix} \begin{bmatrix} 3 \\ 7 \end{bmatrix} \qquad\qquad (2)'$$

$$\left(\begin{bmatrix} 2/5 & -1/5 \\ -1/5 & 3/5 \end{bmatrix} \begin{bmatrix} 3 & 1 \\ 1 & 2 \end{bmatrix} \right) \begin{bmatrix} x \\ y \end{bmatrix} = \begin{bmatrix} -1/5 \\ 18/5 \end{bmatrix} \qquad\qquad (3)'$$

$$\begin{bmatrix} 1 & 0 \\ 0 & 1 \end{bmatrix} \begin{bmatrix} x \\ y \end{bmatrix} = \begin{bmatrix} -1/5 \\ 18/5 \end{bmatrix} \qquad\qquad (4)'$$

$$\begin{bmatrix} x \\ y \end{bmatrix} = \begin{bmatrix} -1/5 \\ 18/5 \end{bmatrix}. \qquad\qquad (5)'$$

From Definition A2.2 it then follows from the solution $\begin{bmatrix} x \\ y \end{bmatrix} = X = A^{-1}B = \begin{bmatrix} -1/5 \\ 18/5 \end{bmatrix}$ that $x = -1/5$ and $y = 18/5$.

In general, if $A = \begin{bmatrix} a_{11} & a_{12} \\ a_{21} & a_{22} \end{bmatrix}$ and $B = \begin{bmatrix} b_1 \\ b_2 \end{bmatrix}$ with $a_{11}, a_{12}, a_{21}, a_{22}, b_1, b_2 \in \mathbf{R}$ and $\det(A) = a_{11}a_{22} - a_{12}a_{21} \neq 0$, then the solution of the system of linear equations,

$$a_{11}x + a_{12}y = b_1$$
$$a_{21}x + a_{22}y = b_2,$$

is given by

$$X = \begin{bmatrix} x \\ y \end{bmatrix} = A^{-1}B = \frac{1}{\det(A)} \begin{bmatrix} a_{22} & -a_{12} \\ -a_{21} & a_{11} \end{bmatrix} \begin{bmatrix} b_1 \\ b_2 \end{bmatrix} = \begin{bmatrix} (1/\det(A))(a_{22}b_1 - a_{12}b_2) \\ (1/\det(A))(-a_{21}b_1 + a_{11}b_2) \end{bmatrix}.$$

Furthermore, although we cannot prove our next result, the following is true for $n \in \mathbf{Z}^+$, $n \geq 2$.

If $A = (a_{ij})_{n \times n}$ is a real matrix (which has a multliplicative inverse), and $B = (b_i)_{1 \leq i \leq n}$, $X = (x_i)_{1 \leq i \leq n}$ are $n \times 1$ column matrices (like those defined earlier for $n = 2$), then the resulting system of linear equations

$$AX = B$$

has the solution

$$X = A^{-1}B.$$

Now although we shall not deal with the inverses of any matrices larger than 2×2, we close this appendix with some further results on larger determinants.

We already know that for $A = \begin{bmatrix} a & b \\ c & d \end{bmatrix}$, the determinant of $A = \det(A) = ad - bc$. The $\det(A)$ is generally denoted by $\begin{vmatrix} a & b \\ c & d \end{vmatrix}$. In order to deal with the determinants of larger matrices we need the following idea.

Definition A2.6

Let $A = (a_{ij})_{n \times n}$ with $n \geq 3$. For all $1 \leq i \leq n$ and $1 \leq j \leq n$, the minor associated with a_{ij} is the $(n-1) \times (n-1)$ determinant obtained from matrix A after we delete the ith row and jth column of A.

EXAMPLE A2.9

a) For $A = \begin{bmatrix} 1 & 0 & 2 \\ 3 & 4 & 6 \\ -1 & 3 & 7 \end{bmatrix}$, we find that

1) the minor associated with 0 is obtained from A by deleting its first row and second column:

$$\begin{bmatrix} 1 & 0 & 2 \\ 3 & 4 & 6 \\ -1 & 3 & 7 \end{bmatrix} \quad \text{leads us to} \quad \begin{vmatrix} 3 & 6 \\ -1 & 7 \end{vmatrix}; \quad \text{and}$$

2) for $a_{23} = 6$ the minor is $\begin{vmatrix} a_{11} & a_{12} \\ a_{31} & a_{32} \end{vmatrix} = \begin{vmatrix} 1 & 0 \\ -1 & 3 \end{vmatrix}$.

b) Given the 4×4 matrix

$$B = \begin{bmatrix} 1 & 2 & 0 & 6 \\ 3 & 7 & 8 & -1 \\ 4 & -3 & -2 & 5 \\ -6 & 9 & 4 & 0 \end{bmatrix},$$

the minor associated with 3 is the 3×3 determinant

$$\begin{vmatrix} 2 & 0 & 6 \\ -3 & -2 & 5 \\ 9 & 4 & 0 \end{vmatrix},$$

obtained from the matrix B by deleting the second row and first column of B (and replacing the matrix brackets by the vertical bars for determinants).

Given a matrix $A = (a_{ij})_{3 \times 3}$, for all $1 \leq i \leq 3$, $1 \leq j \leq 3$, we shall let M_{ij} denote the minor associated with a_{ij}. Then

$$\det(A) = \begin{vmatrix} a_{11} & a_{12} & a_{13} \\ a_{21} & a_{22} & a_{23} \\ a_{31} & a_{32} & a_{33} \end{vmatrix} = a_{11}(-1)^{1+1}M_{11} + a_{12}(-1)^{1+2}M_{12} + a_{13}(-1)^{1+3}M_{13}$$

$$= a_{11}\begin{vmatrix} a_{22} & a_{23} \\ a_{32} & a_{33} \end{vmatrix} - a_{12}\begin{vmatrix} a_{21} & a_{23} \\ a_{31} & a_{33} \end{vmatrix} + a_{13}\begin{vmatrix} a_{21} & a_{22} \\ a_{31} & a_{32} \end{vmatrix},$$

and we say that we are evaluating $\det(A)$ by using an *expansion by minors*.

In this way we reduce the problem to 2×2 determinants that we know how to evaluate. Let us examine a particular example.

EXAMPLE A2.10

a) $\begin{vmatrix} 2 & 4 & -7 \\ 3 & 8 & 2 \\ 5 & 6 & 0 \end{vmatrix} = 2(-1)^{1+1}\begin{vmatrix} 8 & 2 \\ 6 & 0 \end{vmatrix} + 4(-1)^{1+2}\begin{vmatrix} 3 & 2 \\ 5 & 0 \end{vmatrix} + (-7)(-1)^{1+3}\begin{vmatrix} 3 & 8 \\ 5 & 6 \end{vmatrix}$

$$= 2(8 \cdot 0 - 2 \cdot 6) - 4(3 \cdot 0 - 2 \cdot 5) - 7(3 \cdot 6 - 8 \cdot 5)$$

$$= 2(-12) - 4(-10) - 7(-22) = 170.$$

[*Note*: In this expansion by minors we find a sum that uses each entry a_{1j}, for $1 \leq j \leq 3$, in the first row of the determinant, and each such entry is multiplied by two terms:

1) $(-1)^{1+j}$, where the exponent $1 + j$ is the sum of the row number and column number for a_{1j}; and
2) its associated minor M_{1j}.]

b) The reader may be wondering what is so special about the first row of a determinant. For suppose we expand the determinant in part (a) by the third column. The resulting expansion is

$$\sum_{i=1}^{3} a_{i3}(-1)^{i+3}M_{i3} = (-7)(-1)^{1+3}\begin{vmatrix} 3 & 8 \\ 5 & 6 \end{vmatrix} + 2(-1)^{2+3}\begin{vmatrix} 2 & 4 \\ 5 & 6 \end{vmatrix} + 0(-1)^{3+3}\begin{vmatrix} 2 & 4 \\ 3 & 8 \end{vmatrix}$$

$$= (-7)(3 \cdot 6 - 8 \cdot 5) - 2(2 \cdot 6 - 4 \cdot 5) + 0(2 \cdot 8 - 4 \cdot 3)$$

$$= (-7)(-22) - 2(-8) = 170.$$

c) What has happened in parts (a) and (b) is not just a mere conicidence. In general, for any 3×3 matrix A, the determinant of A can be evaluated by expanding along any one (fixed) row or down any one (fixed) column. And this method extends to larger square matrices — that is, for $n \in \mathbf{Z}^{+}$ where $n \geq 4$, an $n \times n$ deteminant can be expanded, along any one of its n rows or down any one of its n columns, into n summands each of which involves an $(n - 1) \times (n - 1)$ determinant.

If $A = (a_{ij})_{n \times n}$, where $n \geq 3$, then

$$\det(A) = \sum_{j=1}^{n} a_{ij}(-1)^{i+j}M_{ij} \qquad \text{[expansion across the (fixed) ith row]}$$

$$= \sum_{i=1}^{n} a_{ij}(-1)^{i+j}M_{ij} \qquad \text{[expansion down the (fixed) jth column].}$$

d) From part (c) we now realize that if $A = (a_{ij})_{n \times n}$, for any $n \geq 3$, then if A has a row or column where every entry is 0, it follows that the determinant of A is 0.

REFERENCES

The ideas presented in this appendix (and its corresponding exercises) should provide a sufficient background for what is needed in the way of matrices and determinants in this text. For the reader who would like to learn more about this area of mathematics, any one of the following should serve as a good starting point.

1. Anton, Howard, and Rorres, Chris. *Elementary Linear Algebra with Applications*. New York: Wiley, 1987.
2. Lay, David C. *Linear Algebra and Its Applications,* 3rd ed. Boston Mass.: Addison-Wesley, 2003.
3. Strang, Gilbert. *Linear Algebra and Its Applications*, 3rd ed. San Diego, Calif.: Harcourt Brace Jovanovich, 1988.

EXERCISES A.2

1. Let $A = \begin{bmatrix} 2 & 1 & 4 \\ -1 & 0 & 3 \end{bmatrix}$, $B = \begin{bmatrix} 1 & 1 & 1 \\ 1 & 2 & 4 \end{bmatrix}$, and $C = \begin{bmatrix} 0 & 1 & 2 \\ 5 & 4 & -3 \end{bmatrix}$. Find each of the following.

a) $A + B$

b) $(A + B) + C$

c) $B + C$

d) $A + (B + C)$

e) $2A$

f) $2A + 3B$

g) $2C + 3C$

h) $5C \; (= (2 + 3)C)$

i) $2B - 4C \; (= 2B + (-4)C)$

j) $A + 2B - 3C$

k) $2(3B)$

l) $(2 \cdot 3)B$

2. Solve for a, b, c, d if

$$3 \begin{bmatrix} a & b \\ c & d \end{bmatrix} + 4 \begin{bmatrix} 1 & -2 \\ -3 & -2 \end{bmatrix} = 2 \begin{bmatrix} 1 & 0 \\ 5 & 3 \end{bmatrix}.$$

3. Perform the following matrix multiplications.

a) $[\,1 \quad 3 \quad 7\,] \begin{bmatrix} -2 \\ 0 \\ 2 \end{bmatrix}$

b) $\begin{bmatrix} 1 & 1 & 2 \\ 1 & 1 & 3 \end{bmatrix} \begin{bmatrix} 1 & 4 \\ 2 & 5 \\ 3 & 6 \end{bmatrix}$

c) $\begin{bmatrix} 1 & -2 \\ 0 & 3 \end{bmatrix} \begin{bmatrix} 2 & 6 \\ 6 & 8 \end{bmatrix}$

d) $\begin{bmatrix} 1 & 1 & -1 \\ 2 & -2 & 3 \\ 4 & 0 & -5 \end{bmatrix} \begin{bmatrix} 3 & 0 & 4 \\ -1 & 0 & 6 \\ 7 & 7 & 2 \end{bmatrix}$

e) $\begin{bmatrix} 1 & 0 & 0 \\ 0 & 1 & 0 \\ 0 & 0 & 3 \end{bmatrix} \begin{bmatrix} a & b & c \\ d & e & f \\ g & h & i \end{bmatrix}$

f) $\begin{bmatrix} 1 & 0 & 0 \\ 0 & 0 & 3 \\ 0 & 1 & 0 \end{bmatrix} \begin{bmatrix} a & b & c \\ d & e & f \\ g & h & i \end{bmatrix}$

4. Let $A = \begin{bmatrix} -1 & 4 \\ 1 & 2 \\ 0 & 3 \end{bmatrix}$, $B = \begin{bmatrix} 1 & 2 & -4 \\ 1 & 3 & 5 \end{bmatrix}$, and $C = \begin{bmatrix} 0 & 3 & -2 \\ -2 & -7 & 6 \end{bmatrix}$. Show that (a) $AB + AC = A(B + C)$; and (b) $BA + CA = (B + C)A$.

[In general, if A is an $m \times n$ matrix and B, C are $n \times p$ matrices, then $AB + AC = A(B + C)$. For $n \times p$ matrices B, C and a $p \times q$ matrix A, it follows that $BA + CA = (B + C)A$. These two results are called the *Distributive Laws for Matrix Multiplication over Matrix Addition*.]

5. Find the multiplicative inverse of each of the following matrices if the multiplicative inverse exists.

a) $\begin{bmatrix} 1 & 2 \\ 3 & 1 \end{bmatrix}$

b) $\begin{bmatrix} 0 & 1 \\ 1 & 0 \end{bmatrix}$

c) $\begin{bmatrix} -3 & 1 \\ -6 & 2 \end{bmatrix}$

d) $\begin{bmatrix} 7 & -3 \\ -2 & 1 \end{bmatrix}$

6. Solve each of the following matrix equations for the 2×2 matrix A.

a) $\begin{bmatrix} 2 & 2 \\ 2 & 3 \end{bmatrix} A = \begin{bmatrix} 1 & 0 \\ 1 & -5 \end{bmatrix}$

b) $\begin{bmatrix} 2 & 3 \\ 3 & 5 \end{bmatrix} A - \begin{bmatrix} 1 & 3 \\ 2 & 1 \end{bmatrix} = \begin{bmatrix} 3 & 2 \\ -1 & 1 \end{bmatrix}$

7. If $A = \begin{bmatrix} 1 & 1 \\ 0 & 2 \end{bmatrix}$ and $B = \begin{bmatrix} -1 & 2 \\ -3 & 1 \end{bmatrix}$, determine the following.

a) A^{-1}

b) B^{-1}

c) AB

d) $(AB)^{-1}$

e) $B^{-1}A^{-1}$

8. Evaluate the following 2×2 determinants:

a) $\begin{vmatrix} 1 & 2 \\ 3 & 4 \end{vmatrix}$

b) $\begin{vmatrix} 5 & 10 \\ 3 & 4 \end{vmatrix}$

c) $\begin{vmatrix} 5 & 2 \\ 15 & 4 \end{vmatrix}$

d) $\begin{vmatrix} 5 & 10 \\ 15 & 20 \end{vmatrix}$

9. Solve the following systems of linear equations by using matrices:

a) $3x - 2y = 5$
$\quad 4x - 3y = 6$

b) $5x + 3y = 35$
$\quad 3x - 2y = 2$

10. Let $a, b, c, d \in \mathbf{R}$ with $\begin{vmatrix} a & b \\ c & d \end{vmatrix} = 7$. Determine the value of each of the following.

a) $\begin{vmatrix} 3a & 3b \\ c & d \end{vmatrix}$

b) $\begin{vmatrix} 3a & b \\ 3c & d \end{vmatrix}$

c) $\begin{vmatrix} a & b \\ 3c & 3d \end{vmatrix}$

d) $\begin{vmatrix} 3a & 3b \\ 3c & 3d \end{vmatrix}$

11. Let A be a 2×2 matrix with $\det(A) = 31$. What is $\det(2A)$? What is $\det(5A)$?

12. Expand each of the following determinants across the specified row as well as down the specified column.

a) $\begin{vmatrix} 1 & 0 & -2 \\ 3 & 1 & -1 \\ 4 & 1 & 2 \end{vmatrix}$; row 2 and column 3

b) $\begin{vmatrix} 1 & 1 & 2 \\ 2 & 3 & -4 \\ 0 & 5 & 7 \end{vmatrix}$; row 1 and column 2

13. Expand each of the following determinants across any row or down any column.

a) $\begin{vmatrix} 1 & 0 & 2 \\ 6 & -2 & 1 \\ 4 & 3 & 2 \end{vmatrix}$ **b)** $\begin{vmatrix} 4 & 7 & 0 \\ 4 & 2 & 0 \\ 3 & 6 & 2 \end{vmatrix}$ **c)** $\begin{vmatrix} 1 & 2 & -4 \\ 0 & 1 & 0 \\ 3 & 3 & 2 \end{vmatrix}$

14. a) Evaluate each of the following 3×3 determinants.

i) $\begin{vmatrix} 1 & 1 & 2 \\ 1 & 1 & 3 \\ 1 & 1 & 4 \end{vmatrix}$ **ii)** $\begin{vmatrix} a & d & a \\ b & e & b \\ c & f & c \end{vmatrix}$

iii) $\begin{vmatrix} 1 & 2 & 4 \\ 0 & 3 & -1 \\ 1 & 2 & 4 \end{vmatrix}$ **iv)** $\begin{vmatrix} d & e & f \\ a & b & c \\ a & b & c \end{vmatrix}$

b) State a general result suggested by the answers in part (a).

15. a) Evaluate each of the following 3×3 determinants.

i) $\begin{vmatrix} 1 & 2 & 1 \\ 0 & -1 & -1 \\ 2 & 3 & 0 \end{vmatrix}$

ii) $\begin{vmatrix} 5 & 2 & 1 \\ 0 & -1 & -1 \\ 10 & 3 & 0 \end{vmatrix}$

iii) $\begin{vmatrix} 5 & 2 & 5 \\ 0 & -1 & -5 \\ 10 & 3 & 0 \end{vmatrix}$

b) Let $a, b, c, d, e, f, g, h, i \in \mathbf{R}$. If $\begin{vmatrix} a & b & c \\ d & e & f \\ g & h & i \end{vmatrix} = 17$, evaluate

i) $\begin{vmatrix} 3a & b & c \\ 3d & e & f \\ 3g & h & i \end{vmatrix}$

ii) $\begin{vmatrix} 3a & b & 2c \\ 9d & 3e & 6f \\ 3g & h & 2i \end{vmatrix}$

iii) $\begin{vmatrix} 2a & 2b & 2c \\ 3d & 3e & 3f \\ 5g & 5h & 5i \end{vmatrix}$

16. Let $A = (a_{ij})_{n \times n}$ and $B = (b_{ij})_{n \times n}$ be two matrices. When the matrix product AB is formed, as defined in Definition A2.5, how many multiplications (of entries) are performed? How many additions (of entry-products) are performed?

Appendix 3
Countable and Uncountable Sets

In Example 3.2 of Section 3.1 we informally mention the ideas of what we feel are a finite set and an infinite set. This final appendix will deal with these issues in a more rigorous manner and will help us attach some meaning to $|A|$ (the size, or cardinality, of a set A) when A is an infinite set. To develop these notions more precisely let us recall the following concept that was first introduced in Section 5.6.

Definition A3.1

For any nonempty sets A, B the function $f: A \rightarrow B$ is called a *one-to-one correspondence* if f is both one-to-one and onto.

EXAMPLE A3.1

Let $A = \mathbf{Z}^+$ and $B = 2\mathbf{Z}^+ = \{2k \mid k \in \mathbf{Z}^+\} = \{2, 4, 6, \ldots\}$. The function $f: A \rightarrow B$, defined by $f(x) = 2x$, is a one-to-one correspondence:

1) For $a_1, a_2 \in A$, we have $f(a_1) = f(a_2) \Rightarrow 2a_1 = 2a_2 \Rightarrow a_1 = a_2$, so f is one-to-one.

2) If $b \in B$, then $b = 2a$ for some (unique) $a \in A$, and $f(a) = 2a = b$, making f onto.

The result in Example A3.1 now leads us to consider the following.

Definition A3.2

If A, B are two nonempty sets, we say that *A has the same size, or cardinality, as B* and we write $A \sim B$, if there exists a one-to-one correspondence $f: A \rightarrow B$.

From Example A3.1 we see that $\mathbf{Z}^+$ has the same size as $2\mathbf{Z}^+$, even though it seems that $2\mathbf{Z}^+$ has fewer elements than $\mathbf{Z}^+$ — after all, we do know that $2\mathbf{Z}^+ \subset \mathbf{Z}^+$.

If we define $g: B \rightarrow A$ (for $B = 2\mathbf{Z}^+$ and $A = \mathbf{Z}^+$) by $g(2k) = k$, then

1) $g(2k_1) = g(2k_2) \Rightarrow k_1 = k_2 \Rightarrow 2k_1 = 2k_2$, establishing that g is one-to-one; and

2) for each $k \in A$, we have $2k \in B$ with $g(2k) = k$, so g is also an onto function.

Consequently, g is a one-to-one correspondence and $B \sim A$.

So at least in the case of $A = \mathbf{Z}^+$ and $B = 2\mathbf{Z}^+$ we find that $A \sim B$ and $B \sim A$ (even though $B \subset A$). But, in reality, what has happened in this one situation holds true in general. For the function g just defined is actually the function f^{-1} for f in Example A3.1. And we learned in Theorem 5.8 that a function is invertible if and only if it is both one-to-one and onto. Consequently, whenever there are two nonempty sets A, B with $A \sim B$ then it follows from Theorem 5.8 that $B \sim A$, so we can say that *A and B have the same cardinality* and denote this by $|A| = |B|$. (*Note*: It does not necessarily follow that $A = B$.)

Let us consider another example.

EXAMPLE A3.2	For $B = 2\mathbf{Z}^+ = \{2k | k \in \mathbf{Z}^+\}$ and $C = 3\mathbf{Z}^+ = \{3k | k \in \mathbf{Z}^+\}$, the function $h: B \to C$ defined by $h(2k) = 3k$ establishes a one-to-one correspondence between B and C. Therefore we have $B \sim C$ (and $C \sim B$, and $|B| = |C|$). Furthermore, using the function $f: A \to B$ that was defined in Example A3.1, where $A = \mathbf{Z}^+$, by virtue of Theorem 5.5 we know that $h \circ f: A \to C$ is also a one-to-one correspondence. So $A \sim C$ (and $C \sim A$, and $|A| = |C|$).

What we have learned up to this point can be summarized as part of the following result.

THEOREM A3.1

For all nonempty sets A, B, C,

    a) $A \sim A$;

    b) if $A \sim B$, then $B \sim A$; and

    c) if $A \sim B$ and $B \sim C$, then $A \sim C$.

Proof:

    a) Given any nonempty set A, it follows that $A \sim A$ because the identity function $1_A: A \to A$ is a one-to-one correspondence.

    b) If $A \sim B$, then there exists a one-to-one correspondence $f: A \to B$. But then $f^{-1}: B \to A$ is also a one-to-one correspondence and we have $B \sim A$.

    c) When $A \sim B$ and $B \sim C$ there exist one-to-one correspondences $f: A \to B$ and $g: B \to C$. Since $g \circ f: A \to C$ is also a one-to-one correspondence, it follows that $A \sim C$.

We shall now use the ideas developed so far in order to define what we shall mean by a finite set and by an infinite set.

Definition A3.3

Any set A is called a *finite* set if $A = \emptyset$ or if $A \sim \{1, 2, 3, \ldots, n\}$ for some $n \in \mathbf{Z}^+$. When $A = \emptyset$ we say that A has no elements and write $|A| = 0$. In the latter case A is said to have n elements and we write $|A| = n$. When a set A is *not* finite then it is called *infinite*.

From this definition we see that if A is a nonempty finite set then there is a one-to-one correspondence $g: \{1, 2, 3, \ldots, n\} \to A$ for some $n \in \mathbf{Z}^+$. This function g provides a listing of the elements of A as $g(1), g(2), \ldots, g(n)$ — a listing where we can count (or account for) a first element, a second element, $\ldots$, and so on, up to an nth (last) element.

Also when A is an infinite set we see that there is no $n \in \mathbf{Z}^+$ for which we can find a one-to-one correspondence $f: A \to \{1, 2, 3, \ldots, n\}$. But if A, B are both infinite sets, can we conclude automatically that $|A| = |B|$ — that is, that there is a one-to-one correspondence between A and B? This is the question we shall answer, in the negative, as we continue our discussion. For now we introduce the following concept.

Definition A3.4

A set A is called *countable* (or *denumberable*) if (1) A is finite or (2) $A \sim \mathbf{Z}^+$.

We have seen that $2\mathbf{Z}^+ \sim \mathbf{Z}^+$ and $3\mathbf{Z}^+ \sim \mathbf{Z}^+$, and since $\mathbf{Z}^+ \sim \mathbf{Z}^+$, it follows that the sets $\mathbf{Z}^+, 2\mathbf{Z}^+$, and $3\mathbf{Z}^+$ are all countable sets. In fact, for all $k \in \mathbf{Z}$, $k \neq 0$, the function $f: \mathbf{Z}^+ \to k\mathbf{Z}^+$, defined by $f(x) = kx$, is a one-to-one correspondence so $k\mathbf{Z}^+$ is countable (and $|k\mathbf{Z}^+| = |\mathbf{Z}^+|$). Consequently, the set of all negative integers — that is, $(-1)\mathbf{Z}^+$ — is a countable set.

Furthermore, whenever A is infinite and $A \sim \mathbf{Z}^+$, we also have $\mathbf{Z}^+ \sim A$, so there is a one-to-one correspondence $f: \mathbf{Z}^+ \to A$ which provides a listing of the elements of A — namely, $f(1)$, $f(2)$, $f(3)$, $\ldots$ — and in this way we can *count* (but never finish counting) the elements in A.

Finally, as noted above, whenever $A \sim \mathbf{Z}^+$ we have $\mathbf{Z}^+ \sim A$. Consequently, a given set A can be shown to be countably infinite (that is, both infinite and countable) by finding either a one-to-one correspondence $f: A \to \mathbf{Z}^+$ or a one-to-one correspondence $g: \mathbf{Z}^+ \to A$.

EXAMPLE A3.3

Since $\mathbf{Z}^+$, $(-1)\mathbf{Z}^+$, and $\{0\}$ are all countable, is $\mathbf{Z} = \mathbf{Z}^+ \cup (-1)\mathbf{Z}^+ \cup \{0\}$ countable?

Consider the function $f: \mathbf{Z}^+ \to \mathbf{Z}$ defined by

$$f(x) = \begin{cases} x/2, & \text{for } x \text{ even} \\ -(x-1)/2, & \text{for } x \text{ odd.} \end{cases}$$

Here we find, for example, that

$$f(4) = 4/2 = 2 \quad \text{and} \quad f(3) = -(3-1)/2 = -2/2 = -1.$$

We claim that f is a one-to-one correspondence where $f(2\mathbf{Z}^+) = \mathbf{Z}^+$ and $f(\mathbf{Z}^+ - 2\mathbf{Z}^+) = (-1)\mathbf{Z}^+ \cup \{0\}$. For suppose that $a, b \in \mathbf{Z}^+$ with $f(a) = f(b)$.

1) If a, b are both even, then $f(a) = f(b) \Rightarrow a/2 = b/2 \Rightarrow a = b$.

2) If a, b are both odd, then $f(a) = f(b) \Rightarrow -(a-1)/2 = -(b-1)/2 \Rightarrow a - 1 = b - 1 \Rightarrow a = b$.

3) If a is even and b odd, then $f(a) = f(b) \Rightarrow a/2 = -(b-1)/2 \Rightarrow a = -b + 1 \Rightarrow a - 1 = -b$, with $a - 1 \geq 1$ and $-b < 0$. Hence this case cannot occur — nor can the case where a is odd and b even.

Consequently, the function f is at least one-to-one.

Furthermore, for all $y \in \mathbf{Z}$,

1) if $y = 0$, then $f(1) = 0$;

2) if $y > 0$, then $2y \in \mathbf{Z}^+$ and $f(2y) = 2y/2 = y$; and

3) if $y < 0$, then $-2y + 1 \in \mathbf{Z}^+$ and $f(-2y+1) = -[(-2y+1) - 1]/2 = -(-2y)/2 = y$.

So f is also an onto function and $f: \mathbf{Z}^+ \to \mathbf{Z}$ is a one-to-one correspondence. Hence $\mathbf{Z}$ is countable.

Although all of our examples of countably infinite sets have been subsets of $\mathbf{Z}$, other countably infinite sets are possible.

EXAMPLE A3.4

Let $A = \{1, 1/2, 1/3, 1/4, \ldots\} = \{1/n | n \in \mathbf{Z}^+\}$. The function $f: \mathbf{Z}^+ \to A$ defined by $f(n) = 1/n$ establishes a one-to-one correspondence between $\mathbf{Z}^+$ and A. Hence $|\mathbf{Z}^+| = |A|$ and A is countable.

In order to take our development on countable sets one step further we now introduce the following definition.

Definition A3.5

For $n \in \mathbf{Z}^+$, a *finite sequence of n terms* is a function f whose domain is $\{1, 2, 3, \ldots, n\}$. Such a sequence is usually written as an *ordered* set $\{x_1, x_2, x_3, \ldots, x_n\}$, where $x_i = f(i)$ for all $1 \leq i \leq n$.

An *infinite sequence* is a function g having $\mathbf{Z}^+$ as its domain. This type of sequence is generally denoted by the *ordered* set $\{x_i\}_{i \in \mathbf{Z}^+}$ or $\{x_1, x_2, x_3, \ldots\}$, where $x_i = g(i)$ for all $i \in \mathbf{Z}^+$.

EXAMPLE A3.5

a) The set $\{1, 1/2, 1/4, 1/8, 1/16\}$ can be thought of as a finite sequence — given by the function $f: A \to \mathbf{Q}^+$ where $A = \{1, 2, 3, 4, 5\}$ and $f(n) = 2^{-n+1}$.

b) The set A in Example A3.4 can also be expressed as $\{1/n\}_{n \in \mathbf{Z}^+}$ — an infinite sequence given by the function $g: \mathbf{Z}^+ \to \mathbf{Q}^+$, where $g(n) = 1/n$ for each $n \in \mathbf{Z}^+$.

c) The terms in a sequence need not be distinct. For instance, let $f: \mathbf{Z}^+ \to \mathbf{Z}$, where $x_n = f(n) = (-1)^{n+1}$, for each positive integer n. Then $\{x_n\}_{n \in \mathbf{Z}^+} = \{x_1, x_2, x_3, x_4, x_5, \ldots\} = \{1, -1, 1, -1, 1, \ldots\}$, but the range of f is only the two-element set $\{1, -1\}$.

Our next result ties together the concepts introduced in Definitions A3.4 and A3.5.

THEOREM A3.2

If A is a nonempty countable set, then A can be written as a sequence of distinct elements.

Proof: There are two cases to consider.

1) If A is finite, then $A \sim \{1, 2, 3, \ldots n\}$ (and $\{1, 2, 3, \ldots, n\} \sim A$) for some $n \in \mathbf{Z}^+$. Hence there is a one-to-one correspondence $f: \{1, 2, 3, \ldots, n\} \to A$.

   Define $a_i = f(i)$ for each $1 \le i \le n$. Then, since f is one-to-one and onto, $\{a_1, a_2, a_3, \ldots, a_n\}$ is a sequence of the n distinct elements of A.

2) For A infinite there is a one-to-one correspondence $g: \mathbf{Z}^+ \to A$.

   Define $a_i = g(i)$ for all $i \in \mathbf{Z}^+$. Since g is one-to-one, the elements of the infinite sequence $\{a_1, a_2, a_3, \ldots\}$ are distinct; $\{a_1, a_2, a_3, \ldots\} = A$ because g is onto.

Before moving forward let us retrace some of our steps and recall that $\mathbf{Z}^+$ is countable as are the subsets $2\mathbf{Z}^+$ and $3\mathbf{Z}^+$ (of $\mathbf{Z}^+$). This suggests that perhaps every subset of a countable set is itself countable. To deal with this possibility we introduce the next two ideas.

Definition A3.6

1) The infinite sequence $\{a_1, a_2, a_3, \ldots\} = \{a_i\}_{i \in \mathbf{Z}^+}$ is a *subsequence* of $\mathbf{Z}^+ = \{1, 2, 3, \ldots\}$ if for all $i \in \mathbf{Z}^+$, $a_i \in \mathbf{Z}^+$ and $a_i < a_{i+1}$.

2) Let $\{x_n\}_{n \in \mathbf{Z}^+}$ and $\{y_n\}_{n \in \mathbf{Z}^+}$ be two infinite sequences. We say that $\{y_n\}_{n \in \mathbf{Z}^+}$ is a *subsequence of* $\{x_n\}_{n \in \mathbf{Z}^+}$ if there exists a subsequence $\{a_k\}_{k \in \mathbf{Z}^+}$ of $\mathbf{Z}^+$ where for each $k \in \mathbf{Z}^+$ we have $y_k = x_{a_k}$.

EXAMPLE A3.6

a) $\{1, 3, 5, 7, \ldots\}$ is a subsequence of $\mathbf{Z}^+$, as is $\{1, 2, 4, 7, 11, 16, \ldots\}$. The first subsequence can be given by the function $f: \mathbf{Z}^+ \to \mathbf{Z}^+$ where $a_n = f(n) = 2n - 1$. The second subsequence can be generated recursively by

1) $c_1 = h(1) = 1$; and
2) $c_{n+1} = h(n + 1) = h(n) + n = c_n + n$, for $n \ge 1$.

b) Let $\{x_n\}_{n \in \mathbf{Z}^+}$ and $\{y_n\}_{n \in \mathbf{Z}^+}$ be two sequences where for each $n \in \mathbf{Z}^+$, $x_n = f(n) = (-1)^n + (1/n)$ and $y_n = g(n) = 1 + (1/(2n))$. So $\{x_n\}_{n \in \mathbf{Z}^+} = \{0, 3/2, -2/3, 5/4, -4/5, 7/6, -6/7, 9/8, \ldots\}$ — and $\{y_n\}_{n \in \mathbf{Z}^+} = \{3/2, 5/4, 7/6, 9/8, \ldots\}$ — and $y_n = x_{2n}$ for all $n \in \mathbf{Z}^+$. For the subsequence $\{a_k\}_{k \in \mathbf{Z}^+}$ (of $\mathbf{Z}^+$) where $a_k = 2k$ for each $k \in \mathbf{Z}^+$, we find that $y_n = x_{2n} = x_{a_n}$ for each $n \in \mathbf{Z}^+$ — and this shows us that $\{y_n\}_{n \in \mathbf{Z}^+}$ is a subsequence of $\{x_n\}_{n \in \mathbf{Z}^+}$.

c) For $n \in \mathbf{Z}^+$ let $x_n = 1/n$ and let $y_n = 1/(3n)$. Then $\{x_n\}_{n \in \mathbf{Z}^+} = \{1, 1/2, 1/3, 1/4, 1/5, 1/6, 1/7, \ldots\}$ and $\{y_n\}_{n \in \mathbf{Z}^+} = \{1/3, 1/6, 1/9, \ldots\}$. Now consider the subsequence $\{a_k\}_{k \in \mathbf{Z}^+}$ (of $\mathbf{Z}^+$) where $a_k = 3k$ for each $k \in \mathbf{Z}^+$. Then for all $n \in \mathbf{Z}^+$, $y_n = 1/(3n) = x_{3n} = x_{a_n}$, so $\{y_n\}_{n \in \mathbf{Z}^+}$ is a subsequence of $\{x_n\}_{n \in \mathbf{Z}^+}$.

And now we turn to the following result for countable sets and their subsets.

THEOREM A3.3

If S is an infinite countable set and $A \subseteq S$, then A is countable.

Proof: If A is finite, then from Definition A3.4 we know that A is countable. So assume from this point on that A is infinite. Since S is countable, we can invoke Theorem A3.2 in order to list the elements of S as an infinite sequence of distinct terms — so we write $S = \{s_1, s_2, s_3, \ldots\}$. Now define a subsequence $\{a_n\}_{n \in \mathbf{Z}^+}$ of $\mathbf{Z}^+$ as follows:

$a_1 = \min\{n | n \in \mathbf{Z}^+, \text{ and } s_n \in A\}$

$a_2 = \min\{n | n \in \mathbf{Z}^+, n > a_1 \text{ and } s_n \in A\}$

$a_3 = \min\{n | n \in \mathbf{Z}^+, n > a_2 \text{ and } s_n \in A\}$

In general, once $a_1, a_2, a_3, \ldots, a_t$ have been selected, we define $a_{t+1} = \min\{n | n \in \mathbf{Z}^+, n > a_t$ and $s_n \in A\}$. Consider the "function" $F: \mathbf{Z}^+ \to A$ given by $F(n) = s_{a_n}$. If $m, n \in \mathbf{Z}^+$, we find that $m = n \Rightarrow a_m = a_n \Rightarrow s_{a_m} = s_{a_n} \Rightarrow F(m) = F(n)$, so there is no doubt that F is a function. To complete the proof that A is countable we need to show that F is a one-to-one correspondence.

Suppose that $m, n \in \mathbf{Z}^+$ with $F(m) = F(n)$. Then $F(m) = F(n) \Rightarrow s_{a_m} = s_{a_n} \Rightarrow a_m = a_n$ because the elements of the sequence $S = \{s_1, s_2, s_3, \ldots\}$ are distinct. Furthermore, $a_m = a_n \Rightarrow m = n$ because the elements in the subsequence $\{a_n\}_{n \in \mathbf{Z}^+}$ of $\mathbf{Z}^+$ are also distinct. Consequently, this function F is one-to-one.

Now let $b \in A$. Since $A \subseteq S = \{s_1, s_2, s_3 \ldots\}$ we can write $b = s_m$ for some $m \in \mathbf{Z}^+$. If $m = a_1$ then $F(1) = s_{a_1} = s_m = b$. If $m \neq a_1$, then since $a_1 < a_2 < a_3 < \cdots$, there is a smallest $r \in \mathbf{Z}^+$ such that $a_{r-1} < m \leq a_r$. From the definition of the subsequence $\{a_n\}_{n \in \mathbf{Z}^+}$ we know that $a_r = \min\{t | t \in \mathbf{Z}^+, t > a_{r-1} \text{ and } s_t \in A\}$ — and since $m > a_{r-1}$ and $s_m \in A$, we have $a_r \leq m$. Now $a_r \leq m$ and $m \leq a_r \Rightarrow a_r = m$, and so $F(r) = s_{a_r} = s_m = b$. Consequently, the function F is also onto.

From Theorem A3.3 we deduce that a given infinite set T is countable if and only if T has the same cardinality as a subset of $\mathbf{Z}^+$. So if there is a one-to-one function $f: T \to \mathbf{Z}^+$ (not necessarily a one-to-one correspondence), then this is enough to tell us that T is countable — for $T \sim f(T)$ (or $|T| = |f(T)|$) and $f(T)$ is countable.

Up to this point every infinite set we have examined has turned out to be countable. Could it be that all infinite sets are countable — and that for all infinite sets A, B we have $|A| = |B|$? The next result settles this issue.

THEOREM A3.4

The set $(0, 1] = \{x | x \in \mathbf{R} \text{ and } 0 < x \leq 1\}$ is not a countable set.

Proof: If $(0, 1]$ were countable, then (by Theorem A3.2) we could write this set as a sequence of distinct terms: $(0, 1] = \{r_1, r_2, r_3, \ldots\}$. To avoid two representations we agree to write real numbers in $(0, 1]$ such as 0.5 as $0.499 \ldots$ — so no element in $(0, 1]$ is represented by a decimal expansion that terminates. Writing such decimal expansions for $r_1, r_2, r_3, \ldots$, we get

$$r_1 = 0.a_{11}a_{12}a_{13}a_{14} \cdots$$
$$r_2 = 0.a_{21}a_{22}a_{23}a_{24} \cdots$$
$$r_3 = 0.a_{31}a_{32}a_{33}a_{34} \cdots$$
$$\vdots$$
$$r_n = 0.a_{n1}a_{n2}a_{n3}a_{n4} \cdots$$
$$\vdots$$

where $a_{ij} \in \{0, 1, 2, 3, \ldots, 8, 9\}$ for all $i, j \in \mathbf{Z}^+$.

Now consider the real number $r = 0.b_1b_2b_3 \cdots$, where for each $k \in \mathbf{Z}^+$,

$$b_k = \begin{cases} 3, & \text{if } a_{kk} \neq 3 \\ 7, & \text{if } a_{kk} = 3. \end{cases}$$

Then $r \in (0, 1)$, but for *every* $k \in \mathbf{Z}^+$ we have $r \neq r_k$ — so $r \notin \{r_1, r_2, r_3, r_4, \ldots\}$. This contradicts our assumption that $(0, 1] = \{r_1, r_2, r_3, r_4, \ldots\}$.

The technique employed in this proof (of Theorem A3.4) is generally known as *Cantor's Diagonal Construction* in honor of the (Russian-born) German mathematician Georg Cantor (1845–1918), who introduced the idea in December of 1873.

When a set is not countable it is termed *uncountable*. So (0, 1] is uncountable. When a set A is uncountable then (1) $\mathbf{Z}^+$ and A do *not* have the same size, or cardinality, so $\mathbf{Z}^+ \not\sim A$ and the cardinality of A is greater than that of $\mathbf{Z}^+$ — that is, $|A| > |\mathbf{Z}^+|$, even though both A and $\mathbf{Z}^+$ are infinite sets. The following corollary provides another example of an uncountable set.

COROLLARY A3.1

The set $\mathbf{R}$ (of all real numbers) is an uncountable set.

Proof: If $\mathbf{R}$ were countable, then by Theorem A3.3 the subset (0, 1] of $\mathbf{R}$ would be countable.

Before continuing with anything new let us say a few more words about this notion of an uncountable set.

1) First and foremost we realize that Corollary A3.1 is a special case of the general result: For all sets A, B, if A is uncountable and $A \subseteq B$, then B is uncountable.

2) Unlike the result in Theorem A3.3 we do not find in general that nonempty subsets of uncountable sets are uncountable. We may even have an infinite subset A of an uncountable set B where A is countable — for instance, let $A = \mathbf{Z}$ and $B = \mathbf{R}$.

3) Following Theorem A3.3 we remarked that whenever we had a set A and could find a one-to-one function $f: A \to \mathbf{Z}^+$, then the set A had to be countable. We cannot reverse the roles of A and $\mathbf{Z}^+$ for the function f. If there is a one-to-one function $g: \mathbf{Z}^+ \to A$, the set A could be uncountable. Just consider $g: \mathbf{Z}^+ \to \mathbf{R}$ where $g(x) = x$ for each $x \in \mathbf{Z}^+$.

4) Consider the points in the Cartesian plane on the unit circle $x^2 + (y - 1)^2 = 1$. How large is this set $S = \{(x, y) | x, y \in \mathbf{R} \text{ and } x^2 + (y - 1)^2 = 1\}$ — that is, is S countable or uncountable?

   In Fig. A3.1 we have a unit circle (in the plane) centered at $C(0, 1)$. This circle is tangent to the real number line (or x-axis) at the point where $x = 0$. The point P, on the circumference, has coordinates (0, 2).

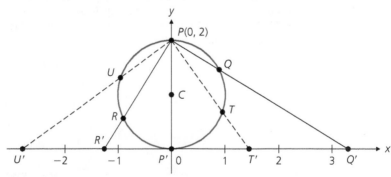

Figure A.3.1

Let (x, y) be any point on the circumference of the unit circle, other than the point $P(0, 2)$. For example, point Q is one such point, and R is another. Draw the line determined by P and Q. This line intersects the x-axis at Q'. Likewise the line determined by P and R intersects the x-axis at R'. Conversely, consider the points on the x-axis — except for the point where $x = 0$. Two such points are T' and U'. The line through P and T' intersects the unit circle at T. Point U is the point of intersection (on S) determined by the line through P and U'. Finally, correspond

P with P' (on the x-axis where $x = 0$). In this way we obtain a one-to-one correspondence between the elements of S and the set $\mathbf{R}$. Hence $|S| = |\mathbf{R}|$, so S is another uncountable set.

Summarizing what we now know about $|\mathbf{Z}|$ and $|\mathbf{R}|$ — namely, that $|\mathbf{Z}| < |\mathbf{R}|$ — we now want to determine whether $|\mathbf{Q}| = |\mathbf{Z}|$ or $|\mathbf{Q}| = |\mathbf{R}|$ or, perhaps, $|\mathbf{Z}| < |\mathbf{Q}| < |\mathbf{R}|$. In accomplishing this we shall prove something more general; to do so we start with the following.

THEOREM A3.5

The set $\mathbf{Z}^+ \times \mathbf{Z}^+$ is countable.

Proof: Define the function $f: \mathbf{Z}^+ \times \mathbf{Z}^+ \to \mathbf{Z}^+$ by $f(a, b) = 2^a 3^b$. The result will follow if we can show that f is one-to-one. For $(m, n), (u, v) \in \mathbf{Z}^+ \times \mathbf{Z}^+$, $f(m, n) = f(u, v) \Rightarrow 2^m 3^n = 2^u 3^v \Rightarrow m = u, n = v$, by the Fundamental Theorem of Arithmetic. Consequently, f is one-to-one and $\mathbf{Z}^+ \times \mathbf{Z}^+$ is countable.

Before any statements can be made about the size, or cardinality, of $\mathbf{Q}$, we first need to consider the subset $\mathbf{Q} \cap (0, 1] = \{s | s \in \mathbf{Q} \text{ and } 0 < s \leq 1\}$ of $\mathbf{Q}$.

THEOREM A3.6

The set $\mathbf{Q} \cap (0, 1]$ is countable.

Proof: First we must agree that each s in $\mathbf{Q} \cap (0, 1]$ will be written in the (unique) form p/q, where $p, q \in \mathbf{Z}^+$ and have no common divisor other than 1. Now define $f: \mathbf{Q} \cap (0, 1] \to \mathbf{Z}^+ \times \mathbf{Z}^+$ by $f(p/q) = (p, q)$, and let $K = $ range f. For $p/q, u/v \in \mathbf{Q} \cap (0, 1]$, we find that $f(p/q) = f(u/v) \Rightarrow (p, q) = (u, v) \Rightarrow p = u$ and $q = v \Rightarrow p/q = u/v$, so f is a one-to-one function. Consequently, $\mathbf{Q} \cap (0, 1] \sim K$, a subset of the countable set $\mathbf{Z}^+ \times \mathbf{Z}^+$. From Theorem A3.3 it now follows that the set $\mathbf{Q} \cap (0, 1]$ is countable.

As we continue in our efforts to determine $|\mathbf{Q}|$ we shall need the next two definitions and theorem.

Definition A3.7

Let $\mathscr{F}$ be any collection of sets from a universe $\mathscr{U}$. The *union* of all the sets in $\mathscr{F}$, written $\bigcup_{A \in \mathscr{F}} A$, is defined as $\{x | x \in \mathscr{U} \text{ and } x \in A, \text{ for some } A \in \mathscr{F}\}$.

When $\mathscr{F}$ is a countable collection — that is, $\mathscr{F} = \{A_1, A_2, A_3, \ldots\}$ — we may write $\bigcup_{A \in \mathscr{F}} A = \bigcup_{n=1}^{\infty} A_n = \bigcup_{n \in \mathbf{Z}^+} A_n$.

EXAMPLE A3.7

In each of the following the universe $\mathscr{U}$ is $\mathbf{R}$.

a) For each $n \in \mathbf{Z}^+$ let $A_n = [n - 1, n)$. Then, for example, $A_1 = [0, 1)$, $A_2 = [1, 2)$, and $A_3 = [2, 3)$. For $\mathscr{F} = \{A_1, A_2, A_3, \ldots\} = \{A_i | i \in \mathbf{Z}^+\}$ we find that $\bigcup_{A \in \mathscr{F}} A = \bigcup_{n=1}^{\infty} A_n = \bigcup_{n \in \mathbf{Z}^+} A_n = [0, +\infty)$.

b) Given any $q \in \mathbf{Q}^+$ let $A_q = (q - 1/2, q + 1/2)$. Here, for instance, $A_{1/2} = (0, 1)$, $A_4 = (7/2, 9/2)$, and $A_{11/3} = (19/6, 25/6)$. If $\mathscr{F} = \{A_q | q \in \mathbf{Q}^+\}$, then $\bigcup_{A \in \mathscr{F}} A = \bigcup_{q \in \mathbf{Q}^+} A_q = (-1/2, +\infty)$.

Definition A3.8

Let $\mathscr{F}$ be a collection of sets each taken from a universe $\mathscr{U}$. The collection $\mathscr{F}$ is called a *disjoint collection* if for all A, B in $\mathscr{F}$, when $A \neq B$ then $A \cap B = \emptyset$.

EXAMPLE A3.8

When we reexamine the two collections in Example A3.7 we find that the collection in part (a) is the only disjoint collection.

The concepts of a countable set and a disjoint collection of sets now come together in our next result.

THEOREM A3.7

Let $\mathscr{F}$ be a countable disjoint collection of sets, each of which is countable. Then $\bigcup_{A \in \mathscr{F}} A$ is also a countable set.

Proof: Since $\mathscr{F}$ is a countable disjoint collection, we may write $\mathscr{F} = \{A_1, A_2, A_3, \ldots\}$, where $A_i \cap A_j = \emptyset$ for all $i, j \in \mathbf{Z}^+$, when $i \neq j$. Furthermore, for each $n \in \mathbf{Z}^+$, A_n is countable and can be expressed as $\{a_{n1}, a_{n2}, a_{n3}, \ldots\}$, a sequence of distinct terms. In order to show that $\bigcup_{A \in \mathscr{F}} A$ is countable, consider each $x \in \bigcup_{A \in \mathscr{F}} A$.

Since $\bigcup_{A \in \mathscr{F}} A = \bigcup_{n=1}^{\infty} A_n$, we have $x \in A_n$ for some (fixed) $n \in \mathbf{Z}^+$, and this n is unique because $\mathscr{F}$ is a disjoint collection. In addition, $x \in A_n \Rightarrow x = a_{nk}$ for some $k \in \mathbf{Z}^+$ (where k is fixed and unique). Now define $f: \bigcup_{A \in \mathscr{F}} A \to \mathbf{Z}^+ \times \mathbf{Z}^+$ by $f(x) = f(a_{nk}) = (n, k)$. From Theorem A3.5 we know that $\mathbf{Z}^+ \times \mathbf{Z}^+$ is countable, so the range of f is countable. Consequently, the result will be established once we show that f is one-to-one. This readily follows, for if $x = a_{nk}$, $y = a_{pq} \in \bigcup_{A \in \mathscr{F}} A$ with $f(x) = f(y)$, then $f(a_{nk}) = f(a_{pq}) \Rightarrow (n, k) = (p, q) \Rightarrow n = p, k = q \Rightarrow a_{nk} = a_{pq} \Rightarrow x = y$.

Note that the proof of Theorem A3.7 is valid if $\mathscr{F}$ is finite (and ∞ is replaced by $|\mathscr{F}|$) or if one or more of the sets A_i, $i \in \mathbf{Z}^+$, is finite.

As a result of Theorem A3.7 we can now deal with the cardinality of $\mathbf{Q}$.

THEOREM A3.8

The set $\mathbf{Q}$ (of all rational numbers) is countable.

Proof: We start by recalling that $A_0 = \mathbf{Q} \cap (0, 1]$ is countable — from Theorem A3.6. Now for each nonzero integer n, let $A_n = \mathbf{Q} \cap (n, n+1]$ and define $f_n: A_n \to A_0$ by $f_n(q) = q - n$. Then $f_n(q_1) = f_n(q_2) \Rightarrow q_1 - n = q_2 - n \Rightarrow q_1 = q_2$, so f_n is one-to-one. Consequently, $A_n \sim f_n(A_n) \subseteq A_0$, and by Theorem A3.3 we have A_n countable. In addition, for all $m, n \in \mathbf{Z}$, $m \neq n \Rightarrow A_m \cap A_n = \emptyset$. From Example A3.3 we know that $\mathbf{Z}$ is countable, so $\mathscr{F} = \{A_0, A_1, A_{-1}, A_2, A_{-2}, \ldots\}$ is a countable disjoint collection of countable sets. Therefore, by virtue of Theorem A3.7, it follows that $\bigcup_{A \in \mathscr{F}} A = \bigcup_{n \in \mathbf{Z}} A_n = \mathbf{Q}$ is countable.

So now we know that $\mathbf{Z}^+$, $\mathbf{Z}$, and $\mathbf{Q}$ are all infinite and $\mathbf{Z}^+ \sim \mathbf{Z} \sim \mathbf{Q}$ while $\mathbf{R}$ is infinite and $\mathbf{R} \not\sim \mathbf{Z}^+$. Recall that any infinite set A, where $A \sim \mathbf{Z}^+$, is said to be countably infinite — and we shall now denote the cardinality of such a set A by writing $|A| = \aleph_0$, using the Hebrew letter aleph, with the subscripted 0, to designate the first level of infinity. The cardinality of $\mathbf{R}$ is greater than $\aleph_0$ and is usually denoted by c, for the *continuum*.

In our next theorem we shall improve upon the result in Theorem A3.7. The following lemma helps with the improvement.

LEMMA A3.1

Let $\mathscr{F} = \{A_1, A_2, A_3, \ldots\}$ be any countable collection of sets (from a universe $\mathscr{U}$). Let $\mathscr{G} = \{B_1, B_2, B_3, \ldots\}$ be the countable collection of sets where $B_1 = A_1$ and $B_n = A_n - \bigcup_{k=1}^{n-1} A_k$ for $n \geq 2$. Then $\mathscr{G}$ is a countable disjoint collection and $\bigcup_{k=1}^{\infty} A_k = \bigcup_{k=1}^{\infty} B_k$.

Proof: First we establish that the countable collection $\mathscr{G}$ is disjoint. To do so we must show that for all $i, j \in \mathbf{Z}^+$, where $i \neq j$, we have $B_i \cap B_j = \emptyset$. If not, let $i < j$ with $B_i \cap B_j \neq \emptyset$. For $x \in B_i \cap B_j$ we find that $x \in B_j = A_j - \bigcup_{k=1}^{j-1} A_k \Rightarrow x \notin A_i$, because $1 \leq i \leq j - 1$. But it also happens that $x \in B_i = A_i \bigcup_{k=1}^{i-1} A_k \Rightarrow x \in A_i$ because $A_i - \bigcup_{k=1}^{i-1} A_k \subseteq A_i$. (*Note:* $\bigcup_{k=1}^{i-1} A_k = \emptyset$ when $i = 1$.)

The contradiction — $x \notin A_i$ and $x \in A_i$ — tells us that $B_i \cap B_j = \emptyset$ for all $i, j \in \mathbf{Z}^+$, where $i \neq j$. So $\mathcal{G}$ is a disjoint countable collection of sets.

For the second part — namely, that $\bigcup_{k=1}^{\infty} A_k = \bigcup_{k=1}^{\infty} B_k$ — start with $x \in \bigcup_{k=1}^{\infty} A_k$. Then $x \in A_n$ for some $n \in \mathbf{Z}^+$, and let m denote the smallest such n. If $m = 1$, then $x \in A_1 = B_1 \subseteq \bigcup_{k=1}^{\infty} B_k$. If $m > 1$, then $x \notin A_j$ for all $1 \leq j \leq m-1$, and so $x \in A_m - \bigcup_{k=1}^{m-1} A_k = B_m \subseteq \bigcup_{k=1}^{\infty} B_k$. In either case $x \in \bigcup_{k=1}^{\infty} B_k$ and $\bigcup_{k=1}^{\infty} A_k \subseteq \bigcup_{k=1}^{\infty} B_k$. For the opposite inclusion we find that $y \in \bigcup_{k=1}^{\infty} B_k \Rightarrow y \in B_n$, for some (unique) $n \in \mathbf{Z}^+ \Rightarrow y \in A_n$, for this same $n \in \mathbf{Z}^+$, because $B_1 = A_1$ and $B_i = A_i - \bigcup_{k=1}^{i-1} A_k \subseteq A_i$, for all $i \geq 2$. Then $y \in A_n \Rightarrow y \in \bigcup_{k=1}^{\infty} A_k$, so $\bigcup_{k=1}^{\infty} B_k \subseteq \bigcup_{k=1}^{\infty} A_k$. Consequently, $\bigcup_{k=1}^{\infty} A_k = \bigcup_{k=1}^{\infty} B_k$.

As in the case of Theorem A3.7, the proof of Lemma A3.1 is valid if $\mathcal{F}$ is finite (and ∞ is then replaced by $|\mathcal{F}|$).

From Lemma A3.1 we learn that the hypothesis of Theorem A3.7 can be weakened — the countable collection $\mathcal{F}$ need not be disjoint. This is formally established as follows.

THEOREM A3.9

The union of any countable collection of countable sets is countable.

Proof: If $\mathcal{F} = \{A_1, A_2, A_3, \ldots\}$ is a countable collection of countable sets, construct the countable collection $\mathcal{G} = \{B_1, B_2, B_3, \ldots\}$ as in Lemma A3.1. For each $k \in \mathbf{Z}^+$, $B_k \subseteq A_k$, so by Theorem A3.3 each B_k is countable. Lemma A3.1 tells us that $\bigcup_{k=1}^{\infty} A_k = \bigcup_{k=1}^{\infty} B_k$, and from Theorem A3.7 we know that $\bigcup_{k=1}^{\infty} B_k$ is countable. Hence $\bigcup_{A \in \mathcal{F}} A = \bigcup_{k=1}^{\infty} A_k$ is countable.

Once again, should $\mathcal{F}$ be finite, the proof of Theorem A3.9 remains valid (upon replacing each occurrence of ∞ by $|\mathcal{F}|$).

Following Theorem A3.8 we mentioned that $|\mathbf{Z}^+| = \aleph_0$ and $|\mathbf{R}| = c$, where $\aleph_0 < c$. Although there is still a great deal more that can be said about infinite sets, we shall close this appendix by showing that these are not the only infinite cardinal numbers. In fact, there are infinitely many infinite cardinal numbers.

THEOREM A3.10

If A is any set, then $|A| < |\mathcal{P}(A)|$.

Proof: If $A = \emptyset$, then $|A| = 0$ and $|\mathcal{P}(A)| = |\mathcal{P}(\emptyset)| = |\{\emptyset\}| = 1$, so the result is true in this case. If $A \neq \emptyset$, let $f: A \to \mathcal{P}(A)$ be defined by $f(a) = \{a\}$ for each $a \in A$. The function f is a one-to-one function and it follows that $|A| = |f(A)| \leq |\mathcal{P}(A)|$. To show that $|A| \neq |\mathcal{P}(A)|$ we must prove that no function $g: A \to \mathcal{P}(A)$ can be onto. So let $g: A \to \mathcal{P}(A)$ and consider $B = \{a \mid a \in A \text{ and } a \notin g(a)\}$. Remember that $g(a) \subseteq A$ and that $B \subseteq A$. With $B \in \mathcal{P}(A)$, if g is to be an onto function there must exist $a' \in A$ such that $g(a') = B$. Now do we have $a' \in g(a')$ or $a' \notin g(a')$? Exactly one of these two results must be true.

If $a' \in g(a') = B$, then from the definition of B we have $a' \notin g(a')$ — and the contradiction: $a' \in g(a')$ and $a' \notin g(a')$. On the other hand, when $a' \notin g(a')$ then $a' \in B$ — but $B = g(a')$. Once again we get the same contradiction.

Therefore, there is no $a' \in A$ with $g(a') = B$, so g cannot be onto, and hence $|A| < |\mathcal{P}(A)|$.

As a consequence of Theorem A3.10 we find that there is no largest infinite cardinal number. For if A is any infinite set, then $|A| < |\mathcal{P}(A)| < |\mathcal{P}(\mathcal{P}(A))| < \cdots$. However, there is a smallest infinite cardinal number. As we mentioned earlier, this is $\aleph_0$.

REFERENCES

Since there is still more that can be said about countable and uncountable sets, the interested reader may want to examine one of the following for further information.

1. Enderton, Herbert B. *Elements of Set Theory*. New York: Academic Press, 1977.
2. Halmos, Paul R. *Naive Set Theory*. New York: Van Nostrand, 1960.
3. Henle, James M. *An Outline of Set Theory*. New York: Springer-Verlag, 1986.

EXERCISES A.3

1. Determine whether each of the following statements is true or false. For parts (d)–(g) provide a counterexample if the statement is false.

a) The set $\mathbf{Q}^+$ is countable.

b) The set $\mathbf{R}^+$ is countable.

c) There is a one-to-one correspondence between the sets $\mathbf{N}$ and $2\mathbf{Z} = \{2k \mid k \in \mathbf{Z}\}$.

d) If A, B are countable sets, then $A \cup B$ is countable.

e) If A, B are uncountable sets, then $A \cap B$ is uncountable.

f) If A, B are countable sets, then $A - B$ is countable.

g) If A, B are uncountable sets, then $A - B$ is uncountable.

2. a) Let $A = \{n^2 \mid n \in \mathbf{Z}^+\}$. Find a one-to-one correspondence between $\mathbf{Z}^+$ and A.

b) Find a one-to-one correspondence between $\mathbf{Z}^+$ and $\{2, 6, 10, 14, \ldots\}$.

3. Let A, B be sets with A uncountable. If $A \subseteq B$, prove that B is uncountable.

4. Let $I = \{r \in \mathbf{R} \mid r$ is irrational$\} = \mathbf{R} - \mathbf{Q}$. Is I countable or uncountable? Prove your assertion.

5. If S, T are infinite and countable, prove that $S \times T$ is countable.

6. Prove that $\mathbf{Z}^+ \times \mathbf{Z}^+ \times \mathbf{Z}^+ = \{(a, b, c) \mid a, b, c \in \mathbf{Z}^+\}$ is countable.

7. Prove that the set of all real solutions of the quadratic equations $ax^2 + bx + c = 0$, where $a, b, c \in \mathbf{Z}$, $a \neq 0$, is a countable set.

8. Determine a one-to-one correspondence between the open interval $(0, 1)$ and the open intervals (a) $(0, 3)$; (b) $(2, 7)$; and (c) (a, b), where $a, b \in \mathbf{R}$ and $a < b$.

Solutions

Chapter 1
Fundamental Principles of Counting

Sections 1.1
and 1.2−p. 11

1. a) 13   b) 40   c) The rule of sum in part (a); the rule of product in part (b)
3. a) 288   b) 24
5. $2 \times 2 \times 1 \times 10 \times 10 \times 2 = 800$ different license plates
7. 2^9 9. a) $(14)(12) = 168$ b) $(14)(12)(6)(18) = 18,144$ c) 73,156,608
11. a) $12 + 2 = 14$ b) $14 \times 14 = 196$ c) 182
13. a) $P(8, 8) = 8!$ b) 7!   6!   15. $4! = 24$
17. Class A: $(2^7 - 2)(2^{24} - 2) = 2,113,928,964$
    Class B: $2^{14}(2^{16} - 2) = 1,073,709,056$
    Class C: $2^{12}(2^8 - 2) = 1,040,384$
19. a) $7! = 5040$ b) $(4!)(3!) = 144$ c) $(5!)(3!) = 720$ d) 288
21. a) $12!/(3!\,2!\,2!\,2!)$ b) $2[11!/(3!\,2!\,2!\,2!)]$ c) $[7!/(2!\,2!)][6!/(3!\,2!)]$
23. $12!/(4!\,3!\,2!\,3!) = 277,200$ 25. a) $n = 10$ b) $n = 5$ c) $n = 5$
27. a) $(10!)/(2!\,7!) = 360$ b) 360
    c) Let x, y, and z be any real numbers and let m, n, and p be any nonnegative integers.
    The number of paths from (x, y, z) to $(x + m, y + n, z + p)$, as described in part (a), is
    $(m + n + p)!/(m!\,n!\,p!)$.
29. a) 576   b) The rule of product
31. a) $9 \times 9 \times 8 \times 7 \times 6 \times 5 = 136,080$ b) 9×10^5
        (i) (a) 68,880 (b) 450,000
        (ii) (a) 28,560 (b) 180,000
        (iii) (a) 33,600 (b) 225,000
33. a) 2^{10} b) 3^{10} 35. a) 6!   b) $2(5!) = 240$
37. $\binom{16}{10}9!\,5! = 348,713,164,800$

Section 1.3−p. 24

1. $\binom{6}{2} = 6!/(2!\,4!) = 15$. The selections of size 2 are $ab, ac, ad, ae, af, bc, bd, be, bf, cd, ce, cf,$ $de, df,$ and ef.
3. a) $C(10, 4) = 10!/(4!\,6!) = 210$ b) $\binom{12}{7} = 12!/(7!\,5!) = 792$
   c) $C(14, 12) = 91$ d) $\binom{15}{10} = 3003$
5. a) $P(5, 3) = 60$
   b) a, f, m   a, f, r   a, f, t   a, m, r   a, m, t
      a, r, t   f, m, r   f, m, t   f, r, t   m, r, t
7. a) $\binom{20}{12} = 125,970$ b) $\binom{10}{6}\binom{10}{6} = 44,100$ c) $\sum_{i=1}^{5} \binom{10}{12-2i}\binom{10}{2i}$
   d) $\sum_{i=7}^{10} \binom{10}{i}\binom{10}{12-i}$ e) $\sum_{i=8}^{10} \binom{10}{i}\binom{10}{12-i}$
9. a) $\binom{8}{2} = 28$ b) 70   c) $\binom{8}{6} = 28$ d) 37
11. a) 120   b) 56   c) 100
13. $\binom{8}{4}\left(\dfrac{7!}{4!\,2!}\right) = 7350$
15. a) $\binom{15}{2} = 105$ b) $\binom{25}{3} = 2300$; $\binom{25}{3}$; $\binom{25}{4} = 12,650$
17. a) $\sum_{k=2}^{n} \dfrac{1}{k!}$ c) $\sum_{j=1}^{7}(-1)^{j-1}j^3 = \sum_{k=1}^{7}(-1)^{k+1}k^3$ d) $\sum_{i=0}^{n} \dfrac{i+1}{n+i}$
19. $\binom{10}{3} + \binom{10}{1}\binom{9}{1} + \binom{10}{1} = 220$ $\binom{10}{4} + \binom{10}{2} + \binom{10}{1}\binom{9}{2} + \binom{10}{1}\binom{9}{1} = 705$
    $2^{10}\left(\sum_{i=0}^{5}\binom{10}{2i}\right)$

21. $\binom{n}{3}$ $\binom{n}{3} - n - n(n-4), n \geq 4$

23. **a)** $\binom{12}{9}$ **b)** $\binom{12}{9}(2^3)$ **c)** $\binom{12}{9}(2^9)(-3)^3$

25. **a)** $\binom{4}{1,1,2} = 12$ **b)** 12 **c)** $\binom{4}{1,1,2}(2)(-1)(-1)^2 = -24$

    d) -216 **e)** $\binom{8}{3,2,1,2}(2^3)(-1)^2(3)(-2)^2 = 161,280$

27. **a)** 2^3 **b)** 2^{10} **c)** 3^{10} **d)** 4^5 **e)** 4^{10}

29. $n\binom{m+n}{m} = n\frac{(m+n)!}{m!\,n!} = \frac{(m+n)!}{m!(n-1)!} = (m+1)\frac{(m+n)!}{(m+1)(m!)(n-1)!}$

$$= (m+1)\frac{(m+n)!}{(m+1)!(n-1)!} = (m+1)\binom{m+n}{m+1}$$

31. Consider the expansions of (a) $[(1+x) - x]^n$; (b) $[(2+x) - (x+1)]^n$; and (c) $[(2+x) - x]^n$.

33. **a)** $a_3 - a_0$ **b)** $a_n - a_0$ **c)** $\frac{1}{102} - \frac{1}{2} = \frac{-25}{51}$

Section 1.4–p. 34

1. **a)** $\binom{14}{10}$ **b)** $\binom{9}{5}$ **c)** $\binom{12}{8}$ **3.** $\binom{23}{20}$ **5.** **a)** 2^5 **b)** 2^n

7. **a)** $\binom{35}{32}$ **b)** $\binom{31}{28}$ **c)** $\binom{11}{8}$ **d)** 1 **e)** $\binom{43}{40}$ **f)** $\binom{31}{28} - \binom{6}{3}$

9. $n = 7$ **11.** **a)** $\binom{14}{5}$ **b)** $\binom{11}{5} + 3\binom{10}{4} + 3\binom{9}{3} + \binom{8}{2}$

13. **a)** $\binom{7}{4}$ **b)** $\sum_{i=0}^{3}\binom{9-2i}{7-2i}$ **15.** $\binom{23}{20}(24!)$ **17.** **a)** $\binom{16}{12}$ **b)** 5^{12}

19. $\binom{23}{4}$ **21.** $24,310 = \sum_{i=1}^{n} i$ [for $n = \binom{12}{3}$]

23. **a)** Place one of the m identical objects into each of the n distinct containers. This leaves $m - n$ identical objects to be placed into the n distinct containers, resulting in $\binom{n+(m-n)-1}{m-n} = \binom{m-1}{m-n} = \binom{m-1}{n-1}$ distributions.

25. **a)** 2^9 **b)** 2^4

27. **a)** $\binom{2+3-1}{3} = 4$ **b)** 10 **c)** 48 **d)** $\binom{3+4-1}{4}\binom{2+3-1}{3} + \binom{3+2-1}{2}\binom{2+5-1}{5} = 96$

    e) 180 **f)** 420

Section 1.5–p. 40

1. $\binom{2n}{n} - \binom{2n}{n-1} = \frac{(2n)!}{n!\,n!} - \frac{(2n)!}{(n-1)!(n+1)!} = \frac{(2n)!(n+1)}{(n+1)!\,n!} - \frac{(2n)!\,n}{n!(n+1)!} =$

$\frac{(2n)![(n+1)-n]}{(n+1)!\,n!} = \frac{1}{(n+1)}\frac{(2n)!}{n!\,n!} = \left(\frac{1}{n+1}\right)\binom{2n}{n}$

3. **a)** $5 (= b_3)$; $14 (= b_4)$

    b) For $n \geq 0$ there are $b_n\left(= \frac{1}{(n+1)}\binom{2n}{n}\right)$ such paths from $(0, 0)$ to (n, n).

    c) For $n \geq 0$ the first move is U and the last is R.

5. Using the results in the third column of Table 1.10 we have:

111000	110010	101010
1 2 3	1 2 5	1 3 5
4 5 6	3 4 6	2 4 6

7. There are $b_5(= 42)$ ways.

9. (i) When $n = 4$ there are $14 (= b_4)$ such diagrams.

(ii) For each $n \geq 0$, there are b_n different drawings of n semicircles on and above a horizontal line, with no two semicircles intersecting. Consider, for instance, the diagram in part (f) of Fig. 1.10. Going from left to right, write 1 the first time you encounter a semicircle and write 0 the second time that semicircle is encountered. Here we get the list 110100. The list 110010 corresponds with the drawing in part (g). This correspondence shows that the number of such drawings for n semicircles is the same as the number of lists of n 1's and n 0's where, as the list is read from left to right, the number of 0's never exceeds the number of 1's.

11. $\left(\frac{1}{7}\right)\binom{12}{6}(6!)(6!) = \left(\frac{1}{7}\right)(12!) = 68,428,800$

Supplementary Exercises–p. 43

1. $\binom{4}{1}\binom{7}{2} + \binom{4}{2}\binom{7}{4} + \binom{4}{3}\binom{7}{6}$

3. Select any four of these twelve points (on the circumference). As seen in the figure, these points determine a pair of chords that intersect. Consequently, the largest number of points of intersection for all possible chords is $\binom{12}{4} = 495$.

5. a) 10^{25} b) $(10)(11)(12) \cdots (34) = 34!/9!$ c) $(25!)\binom{24}{9}$
7. a) $C(12, 8)$ b) $P(12, 8)$ 9. a) 12 b) 49
11. $(1/11)[11!/(5! \, 3! \, 3!)]$
13. a) (i) $\binom{5}{4} + \binom{5}{2}\binom{4}{2} + \binom{4}{4}$ (ii) $\binom{8}{4} + \binom{6}{2}\binom{5}{2} + \binom{7}{4}$ (iii) $\binom{8}{4} + \binom{6}{2}\binom{5}{2} + \binom{7}{4} - 9$
    b) (i) $\binom{5}{1}\binom{4}{3} + \binom{5}{3}\binom{4}{1}$ (ii) and (iii) $\binom{5}{1}\binom{6}{3} + \binom{7}{3}\binom{4}{1}$
15. a) $2\binom{9}{4} + \binom{9}{3} = 343$ b) $[2\binom{12}{4} - 9] + [\binom{12}{3} - 1] = 1200$
17. a) $(5)(9!)$ b) $(3)(8!)$
19. a) $\binom{4}{2}7^5$ b) $2[\binom{3}{2}7^4 + \binom{4}{2}7^5]$
21. $0 = (1 + (-1))^n = \binom{n}{0} - \binom{n}{1} + \binom{n}{2} - \binom{n}{3} + \cdots + (-1)^n\binom{n}{n}$, so
    $\binom{n}{0} + \binom{n}{2} + \binom{n}{4} + \cdots = \binom{n}{1} + \binom{n}{3} + \binom{n}{5} + \cdots$
23. a) $P(20, 12) = 20!/8!$ b) $\binom{17}{9}(12!)$
25. a) $\binom{9}{1} + \binom{10}{3} + \cdots + \binom{16}{15} + \binom{17}{17} = \sum_{k=0}^{8}\binom{9+k}{1+2k}$ b) $\sum_{k=0}^{9}\binom{9+k}{2k}$
    c) $n = 2k + 1, k \geq 0: \sum_{i=0}^{k}\binom{k+1+i}{1+2i}$
       $n = 2k, k \geq 1: \sum_{i=0}^{k}\binom{k+i}{2i}$
27. a) $\binom{r+(n-r)-1}{n-r} = \binom{n-1}{n-r} = \binom{n-1}{r-1}$
    b) $\sum_{r=1}^{n}\binom{n-1}{r-1} = \binom{n-1}{0} + \binom{n-1}{1} + \cdots + \binom{n-1}{n-1} = 2^{n-1}$
29. a) $11!/(7! \, 4!)$ b) $[11!/(7! \, 4!)] - [4!/(2! \, 2!)][4!/(3! \, 1!)]$
    c) $[11!/(7! \, 4!)] + [10!/(6! \, 3! \, 1!)] + [9!/(5! \, 2! \, 2!)] + [8!/(4! \, 1! \, 3!)] + [7!/(3! \, 4!)]$ [in part (a)]
       $\{[11!/(7! \, 4!)] + [10!/(6! \, 3! \, 1!)] + [9!/(5! \, 2! \, 2!)] + [8!/(4! \, 1! \, 3!)] + [7!/(3! \, 4!)]\}$
          $- \{\{[4!/(2! \, 2!)] + [3!/(1! \, 1! \, 1!)] + [2!/2!]\} \times \{[4!/(3! \, 1!)] + [3!/(2! \, 1!)]\}\}$
       [in part (b)]
31. $\binom{9}{2}\binom{6}{2} = 540$ 33. $\binom{6}{4}(12)(11)(10)(9) = 178,200$

Chapter 2
Fundamentals of Logic

Section 2.1 – p. 54

1. The sentences in parts (a), (c), (d), and (f) are statements. The other two sentences are not.
3. a) 0 b) 0 c) 1 d) 0
5. a) If triangle ABC is equilateral, then it is isosceles.
   b) If triangle ABC is not isosceles, then it is not equilateral.
   d) Triangle ABC is isosceles, but it is not equilateral.
7. a) If Darci practices her serve daily then she will have a good chance of winning the tennis tournament.
   b) If you do not fix my air conditioner, then I shall not pay the rent.
   c) If Mary is to be allowed on Larry's motorcycle, then she must wear her helmet.
9. Statements (a), (e), (f), and (h) are tautologies.
11. a) $2^5 = 32$ b) 2^n 13. $p: 0; r: 0; s: 0$
15. a) $m = 3, n = 6$ b) $m = 3, n = 9$ c) $m = 18, n = 9$ d) $m = 4, n = 9$
    e) $m = 4, n = 9$
17. Dawn

1. a) (i)

p	q	r	$q \wedge r$	$p \to (q \wedge r)$	$p \to q$	$p \to r$	$(p \to q) \wedge (p \to r)$
0	0	0	0	1	1	1	1
0	0	1	0	1	1	1	1
0	1	0	0	1	1	1	1
0	1	1	1	1	1	1	1
1	0	0	0	0	0	0	0
1	0	1	0	0	0	1	0
1	1	0	0	0	1	0	0
1	1	1	1	1	1	1	1

(iii)

p	q	r	$q \vee r$	$p \to (q \vee r)$	$p \to q$	$\neg r \to (p \to q)$
0	0	0	0	1	1	1
0	0	1	1	1	1	1
0	1	0	1	1	1	1
0	1	1	1	1	1	1
1	0	0	0	0	0	0
1	0	1	1	1	0	1
1	1	0	1	1	1	1
1	1	1	1	1	1	1

b)

$[p \to (q \vee r)] \Longleftrightarrow [\neg r \to (p \to q)]$	From part (iii) of part (a)
$\Longleftrightarrow [\neg r \to (\neg p \vee q)]$	By the 2nd Substitution Rule, and $(p \to q) \Longleftrightarrow (\neg p \vee q)$
$\Longleftrightarrow [\neg(\neg p \vee q) \to \neg\neg r]$	By the 1st Substitution Rule, and $(s \to t) \Longleftrightarrow (\neg t \to \neg s)$ for any primitive statements s, t
$\Longleftrightarrow [(\neg\neg p \wedge \neg q) \to r]$	By DeMorgan's Law, Double Negation, and the 2nd Substitution Rule
$\Longleftrightarrow [(p \wedge \neg q) \to r]$	By Double Negation and the 2nd Substitution Rule

3. a) For any primitive statement s, $s \vee \neg s \Longleftrightarrow T_0$. Replace each occurrence of s by $p \vee (q \wedge r)$, and the result follows by the 1st Substitution Rule.

b) For any primitive statements s, t, we have $(s \to t) \Longleftrightarrow (\neg t \to \neg s)$. Replace each occurrence of s by $p \vee q$, and each occurrence of t by r, and the result is a consequence of the 1st Substitution Rule.

5. a) Kelsey placed her studies before her interest in cheerleading, but she (still) did not get a good education.

b) Norma is not doing her mathematics homework or Karen is not practicing her piano lesson.

c) Harold did pass his C++ course and he did finish his data structures project, but he did not graduate at the end of the semester.

7. a)

p	q	$(\neg p \vee q) \wedge (p \wedge (p \wedge q))$	$p \wedge q$
0	0	0	0
0	1	0	0
1	0	0	0
1	1	1	1

b) $(\neg p \wedge q) \vee (p \vee (p \vee q)) \Longleftrightarrow p \vee q$

9. a) If $0 + 0 = 0$, then $1 + 1 = 1$. (FALSE)
Contrapositive: If $1 + 1 \neq 1$, then $0 + 0 \neq 0$. (FALSE)
Converse: If $1 + 1 = 1$, then $0 + 0 = 0$. (TRUE)
Inverse: If $0 + 0 \neq 0$, then $1 + 1 \neq 1$. (TRUE)

b) If $-1 < 3$ and $3 + 7 = 10$, then $\sin\left(\frac{3\pi}{2}\right) = -1$. (TRUE)
Converse: If $\sin\left(\frac{3\pi}{2}\right) = -1$, then $-1 < 3$ and $3 + 7 = 10$. (TRUE)
Inverse: If $-1 \geq 3$ or $3 + 7 \neq 10$, then $\sin\left(\frac{3\pi}{2}\right) \neq -1$. (TRUE)
Contrapositive: If $\sin\left(\frac{3\pi}{2}\right) \neq -1$, then $-1 \geq 3$ or $3 + 7 \neq 10$. (TRUE)

11. a) $(q \to r) \vee \neg p$ **b)** $(\neg q \vee r) \vee \neg p$

13.

p	q	r	$[(p \leftrightarrow q) \wedge (q \leftrightarrow r) \wedge (r \leftrightarrow p)]$	$[(p \to q) \wedge (q \to r) \wedge (r \to p)]$
0	0	0	1	1
0	0	1	0	0
0	1	0	0	0
0	1	1	0	0
1	0	0	0	0
1	0	1	0	0
1	1	0	0	0
1	1	1	1	1

15. a) $(p \uparrow p)$ **b)** $(p \uparrow p) \uparrow (q \uparrow q)$ **c)** $(p \uparrow q) \uparrow (p \uparrow q)$ **d)** $p \uparrow (q \uparrow q)$
e) $(r \uparrow s) \uparrow (r \uparrow s)$, where r stands for $p \uparrow (q \uparrow q)$ and s for $q \uparrow (p \uparrow p)$

17.

p	q	$\neg(p \downarrow q)$	$(\neg p \uparrow \neg q)$	$\neg(p \uparrow q)$	$(\neg p \downarrow \neg q)$
0	0	0	0	0	0
0	1	1	1	0	0
1	0	1	1	0	0
1	1	1	1	1	1

19. a) $p \vee [p \wedge (p \vee q)]$ — **Reasons**
  $\Longleftrightarrow p \vee p$ — Absorption Law
  $\Longleftrightarrow p$ — Idempotent Law of $\vee$

c) $[(\neg p \vee \neg q) \to (p \wedge q \wedge r)]$ — **Reasons**
  $\Longleftrightarrow \neg(\neg p \vee \neg q) \vee (p \wedge q \wedge r)$ — $s \to t \Longleftrightarrow \neg s \vee t$
  $\Longleftrightarrow (\neg\neg p \wedge \neg\neg q) \vee (p \wedge q \wedge r)$ — DeMorgan's Laws
  $\Longleftrightarrow (p \wedge q) \vee (p \wedge q \wedge r)$ — Law of Double Negation
  $\Longleftrightarrow p \wedge q$ — Absorption Law

Section 2.3 – p. 84

1. a)

p	q	r	$p \to q$	$(p \vee q)$	$(p \vee q) \to r$
0	0	0	1	0	1
0	0	1	1	0	1
0	1	0	1	1	0
0	1	1	1	1	1
1	0	0	0	1	0
1	0	1	0	1	1
1	1	0	1	1	0
1	1	1	1	1	1

The validity of the argument follows from the results in the last row. (The first seven rows may be ignored.)

c)

p	q	r	$q \vee r$	$p \vee (q \vee r)$	$\neg q$	$p \vee r$
0	0	0	0	0	1	0
0	0	1	1	1	1	1
0	1	0	1	1	0	0
0	1	1	1	1	0	1
1	0	0	0	1	1	1
1	0	1	1	1	1	1
1	1	0	1	1	0	1
1	1	1	1	1	0	1

The results in rows 2, 5, and 6 establish the validity of the given argument. (The results in the other five rows of the table may be disregarded.)

3. a) If p has the truth value 0, then so does $p \wedge q$.

b) When $p \vee q$ has the truth value 0, then the truth value of p (and that of q) is 0.

c) If q has truth value 0, then the truth value of $[(p \vee q) \wedge \neg p]$ is 0, regardless of the truth value of p.

d) The statement $q \vee s$ has truth value 0 only when each of q, s has truth value 0. Then $(p \to q)$ has truth value 1 when p has truth value 0; $(r \to s)$ has truth value 1 when r has truth value 0. But then $(p \vee r)$ must have truth value 0, not 1.

5. a) Rule of Conjunctive Simplification

b) Invalid — attempt to argue by the converse

c) Modus Tollens

d) Rule of Disjunctive Syllogism

e) Invalid — attempt to argue by the inverse

7. 1) and 2)   Premise

3)   Steps (1) and (2) and the Rule of Detachment

4)   Premise

5)   Step (4) and $(r \to \neg q) \Longleftrightarrow (\neg\neg q \to \neg r) \Longleftrightarrow (q \to \neg r)$

6)   Steps (3) and (5) and the Rule of Detachment

7)   Premise

8)   Steps (6) and (7) and the Rule of Disjunctive Syllogism

9)   Step (8) and the Rule of Disjunctive Amplification

9. a)

1)   Premise (The Negation of the Conclusion)

2)   Step (1) and $\neg(\neg q \to s) \Longleftrightarrow \neg(\neg\neg q \vee s) \Longleftrightarrow \neg(q \vee s) \Longleftrightarrow \neg q \wedge \neg s$

3)   Step (2) and the Rule of Conjunctive Simplification

4)   Premise

5)   Steps (3) and (4) and the Rule of Disjunctive Syllogism

6)   Premise

7)   Step (2) and the Rule of Conjunctive Simplification

8)   Steps (6) and (7) and Modus Tollens

9)   Premise

10)   Steps (8) and (9) and the Rule of Disjunctive Syllogism

11)   Steps (5) and (10) and the Rule of Conjunction

12)   Step (11) and the Method of Proof by Contradiction

b) 1) $p \to q$ Premise

2) $\neg q \to \neg p$ Step (1) and $(p \to q) \Longleftrightarrow (\neg q \to \neg p)$

3) $p \vee r$ Premise

4) $\neg p \to r$ Step (3) and $(p \vee r) \Longleftrightarrow (\neg p \to r)$

5) $\neg q \to r$ Steps (2) and (4) and the Law of the Syllogism

6) $\neg r \vee s$ Premise

7) $r \to s$ Step (6) and $(\neg r \vee s) \Longleftrightarrow (r \to s)$

8) $\therefore \neg q \to s$ Steps (5) and (7) and the Law of the Syllogism

11. a) $p:1$ $q:0$ $r:1$ **c)** $p, q, r:1$ $s:0$
 b) $p:0$ $q:0$ $r:0$ or 1 **d)** $p, q, r:1$ $s:0$
 $p:0$ $q:1$ $r:1$

13. a)

p	q	r	$p \vee q$	$\neg p \vee r$	$(p \vee q) \wedge (\neg p \vee r)$	$q \vee r$	$[(p \vee q) \wedge (\neg p \vee r)] \to (q \vee r)$
0	0	0	0	1	0	0	1
0	0	1	0	1	0	1	1
0	1	0	1	1	1	1	1
0	1	1	1	1	1	1	1
1	0	0	1	0	0	0	1
1	0	1	1	1	1	1	1
1	1	0	1	0	0	1	1
1	1	1	1	1	1	1	1

From the last column of the truth table it follows that $[(p \vee q) \wedge (\neg p \vee r)] \to (q \vee r)$ is a tautology.

b) (i) Steps

	Steps	**Reasons**
1)	$p \vee (q \wedge r)$	Premise
2)	$(p \vee q) \wedge (p \vee r)$	Step (1) and the Distributive Law of $\vee$ over $\wedge$
3)	$p \vee r$	Step (2) and the Rule of Conjunctive Simplification
4)	$p \to s$	Premise
5)	$\neg p \vee s$	Step (4), $p \to s \iff \neg p \vee s$
6)	$\therefore r \vee s$	Steps (3), (5), the Rule of Conjunction, and Resolution

(iii) Steps

	Steps	**Reasons**
1)	$p \vee q$	Premise
2)	$p \to r$	Premise
3)	$\neg p \vee r$	Step (2), $p \to r \iff \neg p \vee r$
4)	$[(p \vee q) \wedge (\neg p \vee r)]$	Steps (1), (3), and the Rule of Conjunction
5)	$q \vee r$	Step (4) and Resolution
6)	$r \to s$	Premise
7)	$\neg r \vee s$	Step (6), $r \to s \iff \neg r \vee s$
8)	$[(r \vee q) \wedge (\neg r \vee s)]$	Steps (5), (7), the Commutative Law of $\vee$, and the Rule of Conjunction
9)	$\therefore q \vee s$	Step (8) and Resolution

(iv) Steps

	Steps	**Reasons**
1)	$\neg p \vee q \vee r$	Premise
2)	$q \vee (\neg p \vee r)$	Step (1) and the Commutative and Associative Laws of $\vee$
3)	$\neg q$	Premise
4)	$\neg q \vee (\neg p \vee r)$	Step (3) and the Rule of Disjunctive Amplification
5)	$[[q \vee (\neg p \vee r)] \wedge [\neg q \vee (\neg p \vee r)]]$	Steps (2), (4), and the Rule of Conjunction
6)	$(\neg p \vee r)$	Step (5), Resolution, and the Idempotent Law of $\wedge$
7)	$\neg r$	Premise
8)	$\neg r \vee \neg p$	Step (7) and the Rule of Disjunctive Amplification
9)	$[(r \vee \neg p) \wedge (\neg r \vee \neg p)]$	Steps (6), (8), the Commutative Law of $\vee$, and the Rule of Conjunction
10)	$\therefore \neg p$	Step (9), Resolution, and the Idempotent Law of $\vee$

c) Consider the following assignments.

p: Jonathan has his driver's license.
q: Jonathan's new car is out of gas.
r: Jonathan likes to drive his new car.

Then the given argument can be written in symbolic form as

$$\begin{array}{c} \neg p \vee q \\ p \vee \neg r \\ \underline{\neg q \vee \neg r} \\ \therefore \neg r \end{array}$$

Steps	Reasons
1) $\neg p \vee q$	Premise
2) $p \vee \neg r$	Premise
3) $(p \vee \neg r) \wedge (\neg p \vee q)$	Steps (2), (1), and the Rule of Conjunction
4) $\neg r \vee q$	Step (3) and Resolution
5) $q \vee \neg r$	Step (4) and the Commutative Law of $\vee$
6) $\neg q \vee \neg r$	Premise
7) $(q \vee \neg r) \wedge (\neg q \vee \neg r)$	Steps (5), (6), and the Rule of Conjunction
8) $\neg r \vee \neg r$	Step (7) and Resolution
9) $\therefore \neg r$	Step (8) and the Idempotent Law of $\vee$

Section 2.4 – p. 100

1. a) False   **b)** False   **c)** False   **d)** True   **e)** False   **f)** False

3. Statements (a), (c), and (e) are true, and statements (b), (d), and (f) are false.

5. a) $\exists x \, [m(x) \wedge c(x) \wedge j(x)]$ True
   b) $\exists x \, [s(x) \wedge c(x) \wedge \neg m(x)]$ True
   c) $\forall x \, [c(x) \rightarrow (m(x) \veebar p(x))]$ False
   d) $\forall x \, [(g(x) \wedge c(x)) \rightarrow \neg p(x)]$, or   True
       $\forall x \, [(p(x) \wedge c(x)) \rightarrow \neg g(x)]$, or
       $\forall x \, [(g(x) \wedge p(x)) \rightarrow \neg c(x)]$
   e) $\forall x \, [(c(x) \wedge s(x)) \rightarrow (p(x) \veebar e(x))]$ True

7. a)   (i) $\exists x \, q(x)$
      (ii) $\exists x \, [p(x) \wedge q(x)]$
     (iii) $\forall x \, [q(x) \rightarrow \neg t(x)]$
     (iv) $\forall x \, [q(x) \rightarrow \neg t(x)]$
      (v) $\exists x \, [q(x) \wedge t(x)]$
     (vi) $\forall x \, [(q(x) \wedge r(x)) \rightarrow s(x)]$

   b) Statements (i), (ii), (v), and (vi) are true. Statements (iii) and (iv) are false; $x = 10$ provides a counterexample for either statement.

   c)   (i) If x is a perfect square, then $x > 0$.
      (ii) If x is divisible by 4, then x is even.
     (iii) If x is divisible by 4, then x is not divisible by 5.
     (iv) There exists an integer that is divisible by 4, but it is not a perfect square.

   d) (i) Let $x = 0$.   (iii) Let $x = 20$.

9. a)   (i) True   (ii) False   Consider $x = 3$.
     (iii) True   (iv) True

   c)   (i) True   (ii) True
     (iii) True   (iv) False   For $x = 2$ or 5, the truth value of $p(x)$ is 1
                              while that of $r(x)$ is 0.

11. a) In this case the variable x is free, while the variables y, z are bound.
   b) Here the variables x, y are bound; the variable z is free.

13. a) $p(2, 3) \wedge p(3, 3) \wedge p(5, 3)$
   b) $[p(2, 2) \vee p(2, 3) \vee p(2, 5)] \vee [p(3, 2) \vee p(3, 3) \vee p(3, 5)] \vee [p(5, 2) \vee p(5, 3) \vee p(5, 5)]$

15. a) The proposed negation is correct and is a true statement.

   b) The proposed negation is wrong. A correct version of the negation is: For all rational numbers x, y, the sum $x + y$ is rational. This correct version of the negation is a true statement.

   d) The proposed negation is wrong. A correct version of the negation is: For all integers x, y, if x, y are both odd, then xy is even. The (original) statement is true.

17. a) There exists an integer n such that n is not divisible by 2 but n is even (that is, not odd).

   b) There exist integers k, m, n such that $k - m$ and $m - n$ are odd, and $k - n$ is odd.

   d) There exists a real number x such that $|x - 3| < 7$ and either $x \leq -4$ or $x \geq 10$.

19. a) *Statement*: For all positive integers m, n, if $m > n$, then $m^2 > n^2$. (TRUE)
   Converse: For all positive integers m, n, if $m^2 > n^2$, then $m > n$. (TRUE)
   Inverse: For all positive integers m, n, if $m \leq n$, then $m^2 \leq n^2$. (TRUE)
   Contrapositive: For all positive integers m, n, if $m^2 \leq n^2$, then $m \leq n$. (TRUE)

   b) *Statement*: For all integers a, b, if $a > b$, then $a^2 > b^2$. (FALSE — let $a = 1$ and $b = -2$.)
   Converse: For all integers a, b, if $a^2 > b^2$, then $a > b$. (FALSE — let $a = -5$ and $b = 3$.)
   Inverse: For all integers a, b, if $a \leq b$, then $a^2 \leq b^2$. (FALSE — let $a = -5$ and $b = 3$.)
   Contrapositive: For all integers a, b, if $a^2 \leq b^2$, then $a \leq b$. (FALSE — let $a = 1$ and $b = -2$.)

   c) *Statement*: For all integers m, n, and p, if m divides n and n divides p, then m divides p. (TRUE)
   Converse: For all integers m and p, if m divides p, then for each integer n it follows that m divides n and n divides p. (FALSE — let $m = 1$, $n = 2$, and $p = 3$.)
   Inverse: For all integers m, n, and p, if m does not divide n or n does not divide p, then m does not divide p. (FALSE — let $m = 1$, $n = 2$, and $p = 3$.)
   Contrapositive: For all integers m and p, if m does not divide p, then for each integer n it follows that m does not divide n or n does not divide p. (TRUE)

   e) *Statement*: $\forall x \, [(x^2 + 4x - 21 > 0) \rightarrow [(x > 3) \vee (x < -7)]]$ (TRUE)
   Converse: $\forall x \, [[(x > 3) \vee (x < -7)] \rightarrow (x^2 + 4x - 21 > 0)]$ (TRUE)
   Inverse: $\forall x \, [(x^2 + 4x - 21 \leq 0) \rightarrow [(x \leq 3) \wedge (x \geq -7)]]$, or $\forall x \, [(x^2 + 4x - 21 \leq 0) \rightarrow (-7 \leq x \leq 3)]$ (TRUE)
   Contrapositive: $\forall x \, [[(x \leq 3) \wedge (x \geq -7)] \rightarrow (x^2 + 4x - 21 \leq 0)]$, or $\forall x \, [(-7 \leq x \leq 3) \rightarrow (x^2 + 4x - 21 \leq 0)]$ (TRUE)

21. a) True   **b)** False   **c)** False   **d)** True   **e)** False

23. a) $\forall a \, \exists b \, [a + b = b + a = 0]$ **b)** $\exists u \, \forall a \, [au = ua = a]$ **c)** $\forall a \neq 0 \, \exists b \, [ab = ba = 1]$
   d) The statement in part (b) remains true, but the statement in part (c) is no longer true for this new universe.

25. a) $\exists x \, \exists y \, [(x > y) \wedge (x - y \leq 0)]$ **b)** $\exists x \, \exists y \, [(x < y) \wedge \forall z \, [x \geq z \vee z \geq y]]$

Section 2.5 – p. 116

1. Although we may write $28 = 25 + 1 + 1 + 1 = 16 + 4 + 4 + 4$, there is no way to express 28 as the sum of at most three perfect squares.

3.

$30 = 25 + 4 + 1$	$40 = 36 + 4$	$50 = 25 + 25$
$32 = 16 + 16$	$42 = 25 + 16 + 1$	$52 = 36 + 16$
$34 = 25 + 9$	$44 = 36 + 4 + 4$	$54 = 25 + 25 + 4$
$36 = 36$	$46 = 36 + 9 + 1$	$56 = 36 + 16 + 4$
$38 = 36 + 1 + 1$	$48 = 16 + 16 + 16$	$58 = 49 + 9$

5. a) The real number π is not an integer.

   c) All administrative directors know how to delegate authority.

   d) Quadrilateral $MNPQ$ is not equiangular.

7. a) When the statement $\exists x \, [p(x) \vee q(x)]$ is true, there is at least one element c in the prescribed universe where $p(c) \vee q(c)$ is true. Hence at least one of the statements $p(c)$, $q(c)$ has the truth value 1, so at least one of the statements $\exists x \, p(x)$ and $\exists x \, q(x)$ is true. Therefore, it follows that $\exists x \, p(x) \vee \exists x \, q(x)$ is true, and $\exists x \, [p(x) \vee q(x)] \Rightarrow \exists x \, p(x) \vee \exists x \, q(x)$.
Conversely, if $\exists x \, p(x) \vee \exists x \, q(x)$ is true, then at least one of $p(a)$, $q(b)$ has the truth value 1,

for some a, b in the prescribed universe. Assume without loss of generality that it is $p(a)$. Then $p(a) \lor q(a)$ has truth value 1 so $\exists x \, [p(x) \lor q(x)]$ is a true statement, and $\exists x \, p(x) \lor \exists x \, q(x) \Rightarrow \exists x \, [p(x) \lor q(x)]$.

b) First consider when the statement $\forall x \, [p(x) \land q(x)]$ is true. This occurs when $p(a) \land q(a)$ is true for each a in the prescribed universe. Then $p(a)$ is true [as is $q(a)$] for all a in the universe, so the statements $\forall x \, p(x)$ and $\forall x \, q(x)$ are true. Therefore, the statement $\forall x \, p(x) \land \forall x \, q(x)$ is true and $\forall x \, [p(x) \land q(x)] \Rightarrow \forall x \, p(x) \land \forall x \, q(x)$. Conversely, suppose that $\forall x \, p(x) \land \forall x \, q(x)$ is a true statement. Then $\forall x \, p(x)$, $\forall x \, q(x)$ are both true. So now let c be any element in the prescribed universe. Then $p(c)$, $q(c)$, and $p(c) \land q(c)$ are all true. And since c was chosen arbitrarily, it follows that the statement $\forall x \, [p(x) \land q(x)]$ is true, and $\forall x \, p(x) \land \forall x \, q(x) \Rightarrow \forall x \, [p(x) \land q(x)]$.

9. **1)** Premise
    2) Premise
    3) Step (1) and the Rule of Universal Specification
    4) Step (2) and the Rule of Universal Specification
    5) Step (4) and the Rule of Conjunctive Simplification
    6) Steps (5) and (3) and Modus Ponens
    7) Step (6) and the Rule of Conjunctive Simplification
    8) Step (4) and the Rule of Conjunctive Simplification
    9) Steps (7) and (8) and the Rule of Conjunction
    10) Step (9) and the Rule of Universal Generalization

11. Consider the open statements

    $w(x)$: x works for the credit union
    $\ell(x)$: x writes loan applications
    $c(x)$: x knows COBOL
    $q(x)$: x knows Excel

    and let r represent Roxe and i represent Imogene.

    In symbolic form the given argument is as follows:

    $$\forall x \, [w(x) \to c(x)]$$
    $$\forall x \, [(w(x) \land \ell(x)) \to q(x)]$$
    $$w(r) \land \neg q(r)$$
    $$\underline{q(i) \land \neg c(i)}$$
    $$\therefore \neg \ell(r) \land \neg w(i)$$

    The steps (and reasons) needed to verify this argument can now be presented.

Steps	**Reasons**
1) $\forall x \, [w(x) \to c(x)]$	Premise
2) $q(i) \land \neg c(i)$	Premise
3) $\neg c(i)$	Step (2) and the Rule of Conjunctive Simplification
4) $w(i) \to c(i)$	Step (1) and the Rule of Universal Specification
5) $\neg w(i)$	Steps (3) and (4) and Modus Tollens
6) $\forall x \, [(w(x) \land \ell(x)) \to q(x)]$	Premise
7) $w(r) \land \neg q(r)$	Premise
8) $\neg q(r)$	Step (7) and the Rule of Conjunctive Simplification
9) $(w(r) \land \ell(r)) \to q(r)$	Step (6) and the Rule of Universal Specification
10) $\neg(w(r) \land \ell(r))$	Steps (8) and (9) and Modus Tollens
11) $w(r)$	Step (7) and the Rule of Conjunctive Simplification
12) $\neg w(r) \lor \neg \ell(r)$	Step (10) and DeMorgan's Law
13) $\neg \ell(r)$	Steps (11) and (12) and the Rule of Disjunctive Syllogism
14) $\therefore \neg \ell(r) \land \neg w(i)$	Steps (13) and (5) and the Rule of Conjunction

13. a) *Contrapositive*: For all integers k and ℓ, if k, ℓ are not both odd, then $k\ell$ is not odd — OR, For all integers k and ℓ, if at least one of k, ℓ is even, then $k\ell$ is even.

Proof: Let us assume (without loss of generality) that k is even. Then $k = 2c$ for some integer c — because of Definition 2.8. Then $k\ell = (2c)\ell = 2(c\ell)$, by the associative law of multiplication for integers — and $c\ell$ is an integer. Consequently, $k\ell$ is even — once again, by Definition 2.8. (Note that this result does not require anything about the integer ℓ.)

15. *Proof*: Assume that for some integer n, n^2 is odd while n is not odd. Then n is even and we may write $n = 2a$, for some integer a — by Definition 2.8. Consequently, $n^2 = (2a)^2 = (2a)(2a) = (2 \cdot 2)(a \cdot a)$, by the commutative and associative laws of multiplication for integers. Hence, we may write $n^2 = 2(2a^2)$, with $2a^2$ an integer — and this means that n^2 is even. Thus we have arrived at a contradiction, since we now have n^2 both odd (at the start) and even. This contradiction came about from the false assumption that n is not odd. Therefore, for every integer n, it follows that n^2 odd $\Rightarrow n$ odd.

17. *Proof*:
  (1) Since n is odd, we have $n = 2a + 1$ for some integer a. Then $n + 11 = (2a + 1) + 11 = 2a + 12 = 2(a + 6)$, where $a + 6$ is an integer. So by Definition 2.8 it follows that $n + 11$ is even.
  (2) If $n + 11$ is not even, then it is odd and we have $n + 11 = 2b + 1$, for some integer b. So $n = (2b + 1) - 11 = 2b - 10 = 2(b - 5)$, where $b - 5$ is an integer, and it follows from Definition 2.8 that n is even — that is, not odd.
  (3) In this case we stay with the hypothesis — that n is odd — and also assume that $n + 11$ is not even — hence, odd. So we may write $n + 11 = 2b + 1$, for some integer b. This then implies that $n = 2(b - 5)$, for the integer $b - 5$. So by Definition 2.8 it follows that n is even. But with n both even (as shown) and odd (as in the hypothesis), we have arrived at a contradiction. So our assumption was wrong, and it now follows that $n + 11$ is even for every odd integer n.

19. This result is not true, in general. For example, $m = 4 = 2^2$ and $n = 1 = 1^2$ are two positive integers that are perfect squares, but $m + n = 2^2 + 1^2 = 5$ is not a perfect square.

21. *Proof*:
We shall prove the given result by establishing the truth of its (logically equivalent) contrapositive.

  Let us consider the negation of the conclusion — that is, $x < 50$ and $y < 50$. Then with $x < 50$ and $y < 50$ it follows that $x + y < 50 + 50 = 100$, and we have the negation of the hypothesis. The given result now follows by this indirect method of proof (by the contrapositive).

23. *Proof*: If n is odd, then $n = 2k + 1$ for some (particular) integer k. Then $7n + 8 = 7(2k + 1) + 8 = 14k + 7 + 8 = 14k + 15 = 14k + 14 + 1 = 2(7k + 7) + 1$. It then follows from Definition 2.8 that $7n + 8$ is odd.

  To establish the converse, suppose that n is not odd. Then n is even, so we can write $n = 2t$, for some (particular) integer t. But then $7n + 8 = 7(2t) + 8 = 14t + 8 = 2(7t + 4)$, so it follows from Definition 2.8 that $7n + 8$ is even — that is, $7n + 8$ is not odd. Consequently, the converse follows by contraposition.

Supplementary Exercises — p. 120

1.

p	q	r	s	$q \wedge r$	$\neg(s \vee r)$	$[(q \wedge r) \rightarrow \neg(s \vee r)]$ t	$p \leftrightarrow t$
0	0	0	0	0	1	1	0
0	0	0	1	0	0	1	0
0	0	1	0	0	0	1	0
0	0	1	1	0	0	1	0
0	1	0	0	0	1	1	0
0	1	0	1	0	0	1	0
0	1	1	0	1	0	0	1
0	1	1	1	1	0	0	1

p	q	r	s	$q \wedge r$	$\neg(s \vee r)$	$\overbrace{[(q \wedge r) \rightarrow \neg(s \vee r)]}^{t}$	$p \leftrightarrow t$
1	0	0	0	0	1	1	1
1	0	0	1	0	0	1	1
1	0	1	0	0	0	1	1
1	0	1	1	0	0	1	1
1	1	0	0	0	1	1	1
1	1	0	1	0	0	1	1
1	1	1	0	1	0	0	0
1	1	1	1	1	0	0	0

3. a)

p	q	r	$q \leftrightarrow r$	$p \leftrightarrow (q \leftrightarrow r)$	$(p \leftrightarrow q)$	$(p \leftrightarrow q) \leftrightarrow r$
0	0	0	1	0	1	0
0	0	1	0	1	1	1
0	1	0	0	1	0	1
0	1	1	1	0	0	0
1	0	0	1	1	0	1
1	0	1	0	0	0	0
1	1	0	0	0	1	0
1	1	1	1	1	1	1

It follows from the results in columns 5 and 7 that $[p \leftrightarrow (q \leftrightarrow r)] \Longleftrightarrow [(p \leftrightarrow q) \leftrightarrow r]$.

b) The truth value assignments p: 0; q: 0; r: 0 result in the truth value 1 for $[p \rightarrow (q \rightarrow r)]$ and the truth value 0 for $[(p \rightarrow q) \rightarrow r]$. Consequently, these statements are not logically equivalent.

5. (1) If Kaylyn does not practice her piano lessons, then she cannot go to the movies.

(2) If Kaylyn is to go to the movies, then she will have to practice her piano lessons.

7. a) $(\neg p \vee \neg q) \wedge (F_0 \vee p) \wedge p$

b) $(\neg p \vee \neg q) \wedge (F_0 \vee p) \wedge p$

$\Longleftrightarrow (\neg p \vee \neg q) \wedge (p \wedge p)$	$F_0 \vee p \Longleftrightarrow p$
$\Longleftrightarrow (\neg p \vee \neg q) \wedge p$	Idempotent Law of $\wedge$
$\Longleftrightarrow p \wedge (\neg p \vee \neg q)$	Commutative Law of $\wedge$
$\Longleftrightarrow (p \wedge \neg p) \vee (p \wedge \neg q)$	Distributive Law of $\wedge$ over $\vee$
$\Longleftrightarrow F_0 \vee (p \wedge \neg q)$	$p \wedge \neg p \Longleftrightarrow F_0$
$\Longleftrightarrow p \wedge \neg q$	F_0 is the identity for $\vee$.

9. a) contrapositive   **b)** inverse   **c)** contrapositive   **d)** inverse   **e)** converse

11. a)

p	q	r	$p \veebar q$	$(p \veebar q) \veebar r$	$q \veebar r$	$p \veebar (q \veebar r)$
0	0	0	0	0	0	0
0	0	1	0	1	1	1
0	1	0	1	1	1	1
0	1	1	1	0	0	0
1	0	0	1	1	0	1
1	0	1	1	0	1	0
1	1	0	0	0	1	0
1	1	1	0	1	0	1

It follows from the results in columns 5 and 7 that $[(p \veebar q) \veebar r] \Longleftrightarrow [p \veebar (q \veebar r)]$.

b) The given statements are not logically equivalent. The truth value assignments p: 1; q: 0; r: 0 provide a counterexample.

13. a) True   **b)** False   **c)** True   **d)** True   **e)** False   **f)** False   **g)** False   **h)** True

15. Suppose that the 62 squares in this 8×8 chessboard (with two opposite missing corners) can be covered with 31 dominos. The chessboard contains 30 blue squares and 32 white ones. Each domino covers one blue and one white square — for a total of 31 blue squares and 31 white ones. This contradiction tells us that we cannot cover this 62-square chessboard with the 31 dominos.

Chapter 3
Set Theory

Section 3.1 – p. 134

1. They are all the same set.
3. Parts (b) and (d) are false; the remaining parts are true.
5. **a)** $\{0, 2\}$ **b)** $\{2, 2\frac{1}{2}, 3\frac{1}{3}, 5\frac{1}{5}, 7\frac{1}{7}\}$ **c)** $\{0, 2, 12, 36, 80\}$
7. **a)** $\forall x \, [x \in A \rightarrow x \in B] \wedge \exists x \, [x \in B \wedge x \notin A]$
   b) $\exists x \, [x \in A \wedge x \notin B] \vee \forall x \, [x \notin B \vee x \in A]$
   OR, $\exists x \, [x \in A \wedge x \notin B] \vee \forall x \, [x \in B \rightarrow x \in A]$
9. **a)** $|A| = 6$ **b)** $|B| = 7$ **c)** If B has 2^n subsets of odd cardinality, then $|B| = n + 1$.
11. **a)** 31   **b)** 30   **c)** 28   **13. a)** $\binom{30}{5}$ **b)** $\binom{25}{4}$ **c)** $\binom{29}{4} + \binom{28}{4} + \binom{27}{4} + \binom{26}{4}$
15. Let $W = \{1\}$, $X = \{\{1\}, 2\}$, and $Y = \{X, 3\}$.
17. **c)** If $x \in A$, then $A \subseteq B \Rightarrow x \in B$, and $B \subset C \Rightarrow x \in C$. Hence $A \subseteq C$. Since $B \subset C$, there exists $y \in C$ with $y \notin B$. Also, $A \subseteq B$ and $y \notin B \Rightarrow y \notin A$. Consequently, $A \subseteq C$ and $y \in C$ with $y \notin A \Rightarrow A \subset C$.
   d) Since $A \subset B$, it follows that $A \subseteq B$. The result then follows from part (c).
19. **a)** For $n, k \in \mathbf{Z}^+$ with $n \geq k + 1$, consider the hexagon centered at $\binom{n}{k}$. This has the form

$$
\begin{array}{ccccc}
 & \binom{n-1}{k-1} & & \binom{n-1}{k} & \\
\binom{n}{k-1} & & \binom{n}{k} & & \binom{n}{k+1} \\
 & \binom{n+1}{k} & & \binom{n+1}{k+1} & \\
\end{array}
$$

where the two alternating triples — namely, $\binom{n-1}{k-1}, \binom{n}{k+1}, \binom{n+1}{k}$ and $\binom{n-1}{k}, \binom{n+1}{k+1}, \binom{n}{k-1}$ — satisfy $\binom{n-1}{k-1}\binom{n}{k+1}\binom{n+1}{k} = \binom{n-1}{k}\binom{n+1}{k+1}\binom{n}{k-1}$.
   b) For $n, k \in \mathbf{Z}^+$ with $n \geq k + 1$,

$$
\binom{n-1}{k-1}\binom{n}{k+1}\binom{n+1}{k} = \left[\frac{(n-1)!}{(k-1)!(n-k)!}\right]\left[\frac{n!}{(k+1)!(n-k-1)!}\right]\left[\frac{(n+1)!}{k!(n+1-k)!}\right]
$$

$$
= \left[\frac{(n-1)!}{k!(n-1-k)!}\right]\left[\frac{(n+1)!}{(k+1)!(n-k)!}\right]\left[\frac{n!}{(k-1)!(n-k+1)!}\right] = \binom{n-1}{k}\binom{n+1}{k+1}\binom{n}{k-1}.
$$

21. $n = 20$
23. The fifth, sixth, and seventh entries in the row for $n = 14$ provide the unique solution.
25. As an ordered set, $A = \{x, v, w, z, y\}$.
27. **a)** If $S \in S$, then since $S = \{A \mid A \notin A\}$ we have $S \notin S$.
   b) If $S \notin S$, then by the definition of S it follows that $S \in S$.

Section 3.2 – p. 146

1. **a)** $\{1, 2, 3, 5\}$ **b)** A **c) and d)** $\mathcal{U} - \{2\}$ **e)** $\{4, 8\}$
   f) $\{1, 2, 3, 4, 5, 8\}$ **g)** $\varnothing$ **h)** $\{2, 4, 8\}$ **i)** $\{1, 3, 4, 5, 8\}$
3. **a)** $A = \{1, 3, 4, 7, 9, 11\}$ $B = \{2, 4, 6, 8, 9\}$
   b) $C = \{1, 2, 4, 5, 9\}$ $D = \{5, 7, 8, 9\}$
5. **a)** True   **b)** True   **c)** True   **d)** False   **e)** True
   f) True   **g)** True   **h)** False   **i)** False
7. **a)** Let $\mathcal{U} = \{1, 2, 3\}$, $A = \{1\}$, $B = \{2\}$, and $C = \{3\}$. Then $A \cap C = B \cap C = \varnothing$ but $A \neq B$.
   b) For $\mathcal{U} = \{1, 2\}$, $A = \{1\}$, $B = \{2\}$, and $C = \mathcal{U}$, we have $A \cup C = B \cup C$ but $A \neq B$. [From parts (a) and (b) we see that we do not have cancellation laws for $\cap$ or $\cup$. This differs from what we know about $\mathbf{R}$, where for $a, b, c \in \mathbf{R}$ (i) $ab = ac$ and $a \neq 0 \Rightarrow b = c$; and (ii) $a + b = a + c \Rightarrow b = c$.]
   c) $x \in A \Rightarrow x \in A \cup C \Rightarrow x \in B \cup C$. So $x \in B$ or $x \in C$. If $x \in B$, then we are finished. If $x \in C$, then $x \in A \cap C = B \cap C$ and $x \in B$. In either case, $x \in B$ so $A \subseteq B$. Likewise,

$y \in B \Rightarrow y \in B \cup C = A \cup C$, so $y \in A$ or $y \in C$. If $y \in C$, then $y \in B \cap C = A \cap C$. In either case, $y \in A$ and $B \subseteq A$. Hence $A = B$.

d) Let $x \in A$. Consider two cases: (1) $x \in C \Rightarrow x \notin A \triangle C \Rightarrow x \notin B \triangle C \Rightarrow x \in B$.
(2) $x \notin C \Rightarrow x \in A \triangle C \Rightarrow x \in B \triangle C \Rightarrow x \in B$ (because $x \notin C$). In either case, $x \in B$, so $A \subseteq B$. In a similar way it follows that $B \subseteq A$ and $A = B$.

9. 7; 1

11. a) $\emptyset = (A \cup B) \cap (A \cup \overline{B}) \cap (\overline{A} \cup B) \cap (\overline{A} \cup \overline{B})$ **b)** $A = A \cup (A \cap B)$
c) $A \cap B = (A \cup B) \cap (A \cup \overline{B}) \cap (\overline{A} \cup B)$ **d)** $A = (A \cap B) \cup (A \cap \mathcal{U})$

13. a) Let $\mathcal{U} = \{1, 2, 3\}$, $A = \{1\}$, and $B = \{2\}$. Then $\{1, 2\} \in \mathcal{P}(A \cup B)$ but $\{1, 2\} \notin \mathcal{P}(A) \cup \mathcal{P}(B)$.
b) $X \in \mathcal{P}(A \cap B) \Longleftrightarrow X \subseteq A \cap B \Longleftrightarrow X \subseteq A$ and $X \subseteq B \Longleftrightarrow X \in \mathcal{P}(A)$ and $X \in \mathcal{P}(B) \Longleftrightarrow X \in \mathcal{P}(A) \cap \mathcal{P}(B)$, so $\mathcal{P}(A \cap B) = \mathcal{P}(A) \cap \mathcal{P}(B)$.

15. a) 2^6 **b)** 2^n
c) In the membership table, $A \subseteq B$ if the columns for A, B are such that whenever a 1 occurs in the column for A, there is a corresponding 1 in the column for B.
d)

A	B	C	$A \cup \overline{B}$	$(A \cap B) \cup (\overline{B \cap C})$
0	0	0	1	1
0	0	1	1	1
0	1	0	0	1
0	1	1	0	0
1	0	0	1	1
1	0	1	1	1
1	1	0	1	1
1	1	1	1	1

17. a) $A \cap (B - A) = A \cap (B \cap \overline{A}) = B \cap (A \cap \overline{A}) = B \cap \emptyset = \emptyset$
b) $[(A \cap B) \cup (A \cap B \cap \overline{C} \cap D)] \cup (\overline{A} \cap B) = (A \cap B) \cup (\overline{A} \cap B)$ by the Absorption Law $= (A \cup \overline{A}) \cap B = \mathcal{U} \cap B = B$
d) $\overline{A} \cup \overline{B} \cup (A \cap B \cap \overline{C}) = (\overline{A \cap B}) \cup [(A \cap B) \cap \overline{C}] = [(\overline{A \cap B}) \cup (A \cap B)] \cap [(\overline{A \cap B}) \cup \overline{C}] = [(\overline{A \cap B}) \cup \overline{C}] = \overline{A} \cup \overline{B} \cup \overline{C}$

19. a) $[-6, 9]$ **c)** $\emptyset$ **e)** A_7 **g)** $\mathbf{R}$

Section 3.3 – p. 150

1. 55   **3.** $2^9 + 2^8 - 2^5 = 736$ **5.** $9! + 9! - 8! = 685,440$
7. a) $24! + 24! - 22!$ **b)** $26! - [24! + 24! - 23!]$
9. $[13!/(2!)^3] - 3[12!/(2!)^2] + 3(11!/2!) - 10!$

Section 3.4 – p. 156

1. a) $3/8$ **b)** $1/2$ **c)** $1/4$ **d)** $5/8$ **e)** $5/8$ **f)** $7/8$ **g)** $1/8$
3. 6   **5. a)** $\binom{6}{2}/\binom{12}{2} = 5/22$ **b)** $7/22$ **7.** $49/99$
9. a) $1/64$ **b)** $3/32$ **c)** $15/64$ **d)** $1/2$ **e)** $11/32$ **11. a)** $55/216$ **b)** $5/54$
13. a) $\frac{14!}{15!} = \frac{1}{15}$ **b)** $2/15$ **c)** $3/35$
15. $Pr(A) = 1/3$, $Pr(B) = 7/15$, $Pr(A \cap B) = 2/15$, $Pr(A \cup B) = 2/3$; $Pr(A \cup B) = 2/3 = 1/3 + 7/15 - 2/15 = Pr(A) + Pr(B) - Pr(A \cap B)$

Section 3.5 – p. 164

1. $Pr(\overline{A}) = 0.6$; $Pr(\overline{B}) = 0.7$; $Pr(A \cup B) = 0.5$; $Pr(\overline{A \cup B}) = 0.5$; $Pr(A \cap \overline{B}) = 0.2$; $Pr(\overline{A} \cap B) = 0.1$; $Pr(A \cup \overline{B}) = 0.9$; $Pr(\overline{A} \cup B) = 0.8$
3. a) $S = \{(x, y) | x, y \in \{1, 2, 3, \ldots, 10\}, x \neq y\}$ **b)** $1/2$ **c)** $5/9$
5. 0.4   **7. a)** $11/21$ **b)** $12/21$ **c)** $9/21$ **9.** $3/16$
11. a) (i) $27/38$ (ii) $27/38$ **b)** (i) $81/361$ (ii) $18/361$
13. $11/14$ **15.** $\binom{11}{7}/\binom{80}{7} = 330/3{,}176{,}716{,}400$

17. Since $A \cup B \subseteq \mathscr{S}$, it follows from the result of the preceding exercise that $Pr(A \cup B) \leq Pr(\mathscr{S}) = 1$. So $1 \geq Pr(A \cup B) = Pr(A) + Pr(B) - Pr(A \cap B)$, and $Pr(A \cap B) \geq Pr(A) + Pr(B) - 1 = 0.7 + 0.5 - 1 = 0.2$.

Section 3.6 – p. 173

1. $1/4$ 3. $(0.80)(0.75) = 0.60$

5. In general, $Pr(A \cup B) = Pr(A) + Pr(B) - Pr(A \cap B)$. Since A, B are independent, $Pr(A \cap B) = Pr(A)Pr(B)$. So

$$Pr(A \cup B) = Pr(A) + Pr(B) - Pr(A)Pr(B) = Pr(A) + [1 - Pr(A)]Pr(B)$$
$$= Pr(A) + Pr(\overline{A})Pr(B).$$

The proof for $Pr(B) + Pr(\overline{B})Pr(A)$ is similar.

7. a) $52/85$ b) $11/26$ 9. $3/7$

11. $Pr(A \cap B) = 1/4 = (1/2)(1/2) = Pr(A)Pr(B)$, so the events A, B are independent.

13. $1/5$ 15. $(0.05)(0.02) = 0.001$ 17. $5/21$

19. Any two of the events are independent. However, $Pr(A \cap B \cap C) = 1/4 \neq 1/8 = (1/2)(1/2)(1/2) = Pr(A)Pr(B)Pr(C)$, so the events A, B, C are not independent.

21. a) $5/16$ b) $11/32$ c) $11/32$ 23. 0.6

25. a) $2^5 - \binom{5}{0} - \binom{5}{1} = 26$ b) $2^n - \binom{n}{0} - \binom{n}{1} = 2^n - (n+1)$ 27. $30/77$ 29. 0.15

Section 3.7 – p. 185

1. a) $1/4$ b) 1 c) $7/8$ d) $3/4$ e) $2/7$ f) $1/2$

3. a) $Pr(X = x) = \dfrac{\binom{10}{x}\binom{110}{5-x}}{\binom{120}{5}}$, $x = 0, 1, 2, 3, 4, 5$.

   b) $Pr(X = 4) = \dfrac{\binom{10}{4}\binom{110}{1}}{\binom{120}{5}} = 275/2,268,786$

   c) $139/1,134,393$ d) $2675/8796$

5. a) $2/3$ b) $2/3$ c) $1/4$ d) $7/2$ e) $35/12$

7. a) $c = 1/15$ b) $3/5$ c) $7/3$ d) $14/9$ 9. $n = 200$, $p = 0.35$

11. a) $(0.75)^8 \doteq 0.100113$ b) $\binom{8}{3}(0.25)^3(0.75)^5 \doteq 0.207642$

   c) $\sum_{x=6}^{8} \binom{8}{x}(0.25)^x(0.75)^{8-x} \doteq 0.004227$

   d) 0.037139 (approximately)   e) 2 f) 1.5

13. $c = 10$ 15. a) $Pr(X = 1) = 1/5$; $Pr(X = 2) = 16/95$; $Pr(X = 3) = 12/19$

   b) $7/19$ c) $19/35$ d) $231/95 \doteq 2.431579$ e) $5824/9025 \doteq 0.645319$

17. a) $E(X(X-1)) = \displaystyle\sum_{x=0}^{n} x(x-1)Pr(X=x) = \sum_{x=2}^{n} x(x-1)Pr(X=x)$

$$= \sum_{x=2}^{n} x(x-1)\binom{n}{x}p^x q^{n-x} = \sum_{x=2}^{n} \frac{n!}{x!(n-x)!}x(x-1)p^x q^{n-x}$$

$$= \sum_{x=2}^{n} \frac{n!}{(x-2)!(n-x)!}p^x q^{n-x} = p^2 n(n-1)\sum_{x=2}^{n} \frac{(n-2)!}{(x-2)!(n-x)!}p^{x-2}q^{n-x}$$

$$= p^2 n(n-1)\sum_{y=0}^{n-2} \frac{(n-2)!}{y![n-(y+2)]!}p^y q^{n-(y+2)}, \quad \text{substituting } x-2 = y,$$

$$= p^2 n(n-1)\sum_{y=0}^{n-2} \frac{(n-2)!}{y![(n-2)-y]!}p^y q^{(n-2)-y}$$

$$= p^2 n(n-1)(p+q)^{n-2}, \quad \text{by the Binomial Theorem}$$

$$= p^2 n(n-1)(1)^{n-2} = p^2 n(n-1) = n^2 p^2 - np^2$$

   b) $\text{Var}(X) = E(X)^2 - [E(X)]^2 = [E(X(X-1)) + E(X)] - [E(X)]^2 = [(n^2 p^2 - np^2) + np] - (np)^2 = n^2 p^2 - np^2 + np - n^2 p^2 = np - np^2 = np(1-p) = npq$.

19. a) $Pr(X = 2) = 1/4$; $Pr(X = 3) = 1/8$; $Pr(X = 4) = 1/4$; $Pr(X = 5) = 1/4$;
$Pr(X = 6) = 1/8$ **b)** $31/8$ **c)** $119/64$
21. $E(X) = 4$; $\sigma_X = 1$

**Supplementary
Exercises—p. 189**

1. Suppose that $(A - B) \subseteq C$ and $x \in A - C$. Then $x \in A$ but $x \notin C$. If $x \notin B$, then
$[x \in A \land x \notin B] \Rightarrow x \in (A - B) \subseteq C$. So now we have $x \notin C$ and $x \in C$. This contradiction
gives us $x \in B$, so $(A - C) \subseteq B$.
    Conversely, if $(A - C) \subseteq B$, let $y \in A - B$. Then $y \in A$ but $y \notin B$. If $y \notin C$, then
$[y \in A \land y \notin C] \Rightarrow y \in (A - C) \subseteq B$. This contradiction—that is, $y \notin B$ and $y \in B$—yields
$y \in C$, so $(A - B) \subseteq C$.

3. a) The sets $\mathcal{U} = \{1, 2, 3\}$, $A = \{1, 2\}$, $B = \{1\}$, and $C = \{2\}$ provide a counterexample.
  b) $A = A \cap \mathcal{U} = A \cap (C \cup \overline{C}) = (A \cap C) \cup (A \cap \overline{C}) = (A \cap C) \cup (A - C)$
    $= (B \cap C) \cup (B - C) = (B \cap C) \cup (B \cap \overline{C}) = B \cap (C \cup \overline{C}) = B \cap \mathcal{U} = B$

5. a) 126 (if teams wear different uniforms); 63 (if teams are not distinguishable)
    112 (if teams wear different uniforms); 56 (if teams are not distinguishable)
  b) $2^n - 2$; $(1/2)(2^n - 2)$. $2^n - 2 - 2n$; $(1/2)(2^n - 2 - 2n)$.

7. a) 128 **b)** $|A| = 8$

9. Suppose that $(A \cap B) \cup C = A \cap (B \cup C)$ and that $x \in C$. Then
$x \in C \Rightarrow x \in (A \cap B) \cup C \Rightarrow x \in A \cap (B \cup C) \subseteq A$, so $x \in A$, and $C \subseteq A$.
    Conversely, suppose that $C \subseteq A$.
    (1) If $y \in (A \cap B) \cup C$, then $y \in A \cap B$ or $y \in C$.
      (i) $y \in A \cap B \Rightarrow y \in (A \cap B) \cup (A \cap C) \Rightarrow y \in A \cap (B \cup C)$.
      (ii) $y \in C \Rightarrow y \in A$, because $C \subseteq A$. Also, $y \in C \Rightarrow y \in B \cup C$. So $y \in A \cap (B \cup C)$.
    In either case (i) or case (ii), we have $y \in A \cap (B \cup C)$, so $(A \cap B) \cup C \subseteq A \cap (B \cup C)$.
    (2) Now let $z \in A \cap (B \cup C)$. Then $z \in A \cap (B \cup C) = (A \cap B) \cup (A \cap C) \subseteq (A \cap B) \cup C$,
since $A \cap C \subseteq C$.
    From parts (1) and (2) it follows that $(A \cap B) \cup C = A \cap (B \cup C)$.

11. a) $[0, 14/3]$ **b)** $\{0\} \cup (6, 12]$ **c)** $[0, +\infty)$ **d)** $\emptyset$

13. a)

A	B	$A \cap B$
0	0	0
0	1	0
1	0	0
1	1	1

Since $A \subseteq B$, consider only rows 1, 2, and 4. For these rows, $A \cap B = A$.

c)

A	B	C	$(A \cap \overline{B}) \cup (B \cap \overline{C})$	$A \cap \overline{C}$
0	0	0	0	0
0	0	1	0	0
0	1	0	1	0
0	1	1	0	0
1	0	0	1	1
1	0	1	1	0
1	1	0	1	1
1	1	1	0	0

For $C \subseteq B \subseteq A$, consider only rows 1, 5, 7, and 8. Here $(A \cap \overline{B}) \cup (B \cap \overline{C}) = A \cap \overline{C}$.

d)

A	B	C	$A \triangle B$	$A \triangle C$	$B \triangle C$
0	0	0	0	0	0
0	0	1	0	1	1
0	1	0	1	0	1
0	1	1	1	1	0
1	0	0	1	1	0
1	0	1	1	0	1
1	1	0	0	1	1
1	1	1	0	0	0

When $A \triangle B = C$, we consider rows 1, 4, 6, and 7. In these cases, $A \triangle C = B$ and $B \triangle C = A$.

15. a) $\binom{r+1}{m}$ $(m \leq r+1)$ **b)** $\binom{n-k+1}{k}$ $(2k \leq n+1)$

17. a) 23 **b)** 8 **19.** $7^{15} - 3(3^{15}) + 3$ **21.** $\binom{12}{4}\binom{10}{3}/\binom{22}{7} \doteq 0.3483$

23. a) $\sum_{i=0}^{8} \binom{2i}{i}\binom{i+8}{8-i} = \sum_{i=0}^{8} \frac{(i+8)!}{i!\,i!(8-i)!}$

b) (i) $\binom{12}{6}\binom{14}{2}/\left[\sum_{i=0}^{8}\binom{2i}{i}\binom{i+8}{8-i}\right]$ (ii) $\binom{12}{6}\binom{13}{1}/\left[\sum_{i=0}^{8}\binom{2i}{i}\binom{i+8}{8-i}\right]$

(iii) $\left[\binom{16}{8} + \binom{12}{6}\binom{14}{2} + \binom{8}{4}\binom{12}{4} + \binom{4}{2}\binom{10}{6} + \binom{0}{0}\binom{8}{8}\right] / \left[\sum_{i=0}^{8}\binom{2i}{i}\binom{i+8}{8-i}\right]$

25. $A \cup B = [-2, 4]$, $A \cap B = \{3\}$ **27.** $135/512 \doteq 0.263672$

29. $Pr(A \cap (B \cup C)) = Pr((A \cap B) \cup (A \cap C)) = Pr(A \cap B) + Pr(A \cap C) - Pr((A \cap B) \cap (A \cap C))$. Since A, B, C are independent and $(A \cap B) \cap (A \cap C) = (A \cap A) \cap (B \cap C) = A \cap B \cap C$, $Pr(A \cap (B \cup C)) = Pr(A)Pr(B) + Pr(A)Pr(C) - Pr(A)Pr(B)Pr(C) = Pr(A)[Pr(B) + Pr(C) - Pr(B)Pr(C)] = Pr(A)[Pr(B) + Pr(C) - Pr(B \cap C)] = Pr(A)Pr(B \cup C)$, so A and $B \cup C$ are independent.

31. a) 0.99 **b)** $(0.99)^3 = 0.970299$ **33.** 3/5

35. $\binom{5}{3}(0.8)^3(0.2)^2 + \binom{5}{4}(0.8)^4(0.2) + \binom{5}{5}(0.8)^5 \doteq 0.94208$

37. 675/2048 **39. a)** $c = 1/50$ **b)** 0.82 **c)** 13/41 **d)** 2.8 **e)** 1.64

41. a) $3/\binom{47}{2}$ **b)** $\left[\binom{10}{2} - 3\right]/\binom{47}{2}$ **c)** $\left[3\binom{4}{1}\binom{4}{1} - 3\right]/\binom{47}{2}$

43. $2/[m(m+1)]$

45. a) $Pr(X = 1) = 7/16$; $Pr(X = 2) = 3/8$; $Pr(X = 3) = 3/16$

b) 7/4 **c)** $\sigma_X = 3/4$

Chapter 4
Properties of the Integers: Mathematical Induction

Section 4.1 – p. 208

1. b) Since $1 \cdot 3 = (1)(2)(9)/6$, the result is true for $n = 1$. Assume the result is true for $n = k$ (≥ 1): $1 \cdot 3 + 2 \cdot 4 + 3 \cdot 5 + \cdots + k(k+2) = k(k+1)(2k+7)/6$. Then consider the case for $n = k + 1$: $[1 \cdot 3 + 2 \cdot 4 + \cdots + k(k+2)] + (k+1)(k+3) = [k(k+1)(2k+7)/6] + (k+1)(k+3) = [(k+1)/6][k(2k+7) + 6(k+3)] = (k+1)(2k^2 + 13k + 18)/6 = (k+1)(k+2)(2k+9)/6$. Hence the result follows for all $n \in \mathbf{Z}^+$ by the Principle of Mathematical Induction.

c) $S(n)$: $\displaystyle\sum_{i=1}^{n} \frac{1}{i(i+1)} = \frac{n}{n+1}$

$S(1)$: $\displaystyle\sum_{i=1}^{1} \frac{1}{i(i+1)} = \frac{1}{1(2)} = \frac{1}{1+1}$, so $S(1)$ is true.

Assume $S(k)$: $\displaystyle\sum_{i=1}^{k} \frac{1}{i(i+1)} = \frac{k}{k+1}$. Consider $S(k+1)$.

$\displaystyle\sum_{i=1}^{k+1} \frac{1}{i(i+1)} = \sum_{i=1}^{k} \frac{1}{i(i+1)} + \frac{1}{(k+1)(k+2)} = \frac{k}{(k+1)} + \frac{1}{(k+1)(k+2)}$
$= [k(k+2) + 1]/[(k+1)(k+2)] = (k+1)/(k+2),$

so $S(k) \Rightarrow S(k+1)$ and the result follows for all $n \in \mathbf{Z}^+$ by the Principle of Mathematical Induction.

3. a) From $\sum_{i=1}^{n} i^3 + (n+1)^3 = \sum_{i=0}^{n}(i^3 + 3i^2 + 3i + 1) = \sum_{i=1}^{n} i^3 + 3\sum_{i=1}^{n} i^2 + 3\sum_{i=1}^{n} i + \sum_{i=0}^{n} 1$, we have $(n+1)^3 = 3\sum_{i=1}^{n} i^2 + 3\sum_{i=1}^{n} i + (n+1)$. Consequently,

$$3\sum_{i=1}^{n} i^2 = (n^3 + 3n^2 + 3n + 1) - 3[(n)(n+1)/2] - n - 1$$
$$= n^3 + (3/2)n^2 + (1/2)n$$
$$= (1/2)[2n^3 + 3n^2 + n] = (1/2)n(2n^2 + 3n + 1)$$
$$= (1/2)n(n+1)(2n+1), \text{ so}$$

$\sum_{i=1}^{n} i^2 = (1/6)n(n+1)(2n+1)$ (as shown in Example 4.4).

b) From $\sum_{i=1}^{n} i^4 + (n+1)^4 = \sum_{i=0}^{n}(i+1)^4 = \sum_{i=0}^{n}(i^4 + 4i^3 + 6i^2 + 4i + 1) = \sum_{i=1}^{n} i^4 + 4\sum_{i=1}^{n} i^3 + 6\sum_{i=1}^{n} i^2 + 4\sum_{i=1}^{n} i + \sum_{i=0}^{n} 1$, it follows that $(n+1)^4 = 4\sum_{i=1}^{n} i^3 + 6\sum_{i=1}^{n} i^2 + 4\sum_{i=1}^{n} i + \sum_{i=0}^{n} 1$. Consequently,

$$4\sum_{i=1}^{n} i^3 = (n+1)^4 - 6[n(n+1)(2n+1)/6] - 4[n(n+1)/2] - (n+1)$$
$$= n^4 + 4n^3 + 6n^2 + 4n + 1 - (2n^3 + 3n^2 + n) - (2n^2 + 2n) - (n+1)$$
$$= n^4 + 2n^3 + n^2 = n^2(n^2 + 2n + 1) = n^2(n+1)^2.$$

So $\sum_{i=1}^{n} i^3 = (1/4)n^2(n+1)^2$ [as shown in part (d) of Exercise 1 for this section].

From $\sum_{i=1}^{n} i^5 + (n+1)^5 = \sum_{i=0}^{n}(i+1)^5 = \sum_{i=0}^{n}(i^5 + 5i^4 + 10i^3 + 10i^2 + 5i + 1) = \sum_{i=1}^{n} i^5 + 5\sum_{i=1}^{n} i^4 + 10\sum_{i=1}^{n} i^3 + 10\sum_{i=1}^{n} i^2 + 5\sum_{i=1}^{n} i + \sum_{i=0}^{n} 1$, we have $5\sum_{i=1}^{n} i^4 = (n+1)^5 - (10/4)n^2(n+1)^2 - (10/6)n(n+1)(2n+1) - (5/2)n(n+1) - (n+1)$. So

$$5\sum_{i=1}^{n} i^4 = n^5 + 5n^4 + 10n^3 + 10n^2 + 5n + 1 - (5/2)n^4$$
$$- 5n^3 - (5/2)n^2 - (10/3)n^3 - 5n^2 - (5/3)n - (5/2)n^2 - (5/2)n - n - 1$$
$$= n^5 + (5/2)n^4 + (5/3)n^3 - (1/6)n.$$

Consequently, $\sum_{i=1}^{n} i^4 = (1/30)n(n+1)(6n^3 + 9n^2 + n - 1)$.

5. a) 7626   **b)** 627,874   **7.** $n = 10$ **9. a)** 506   **b)** 12,144

11. a) $\sum_{i=1}^{n} t_{2i} = \sum_{i=1}^{n} \frac{(2i)(2i+1)}{2} = \sum_{i=1}^{n}(2i^2 + i) = 2\sum_{i=1}^{n} i^2 + \sum_{i=1}^{n} i = 2[(n)(n+1)(2n+1)/6] + [n(n+1)/2] = [n(n+1)(2n+1)/3] + [n(n+1)/2] = n(n+1)[\frac{2n+1}{3} + \frac{1}{2}] = n(n+1)[\frac{4n+5}{6}] = n(n+1)(4n+5)/6$.

b) $\sum_{i=1}^{100} t_{2i} = 100(101)(405)/6 = 681,750$.

c)
```
begin
    sum := 0
    for i := 1 to 100 do
        sum := sum + (2 * i) * (2 * i + 1)/2
    print sum
end
```

13. a) There are 49 ($= 7^2$) 2×2 squares and 36 ($= 6^2$) 3×3 squares. In total there are $1^2 + 2^2 + 3^2 + \cdots + 8^2 = (8)(8+1)(2 \cdot 8 + 1)/6 = (8)(9)(17)/6 = 204$ squares.

b) For each $1 \le k \le n$ the $n \times n$ chessboard contains $(n - k + 1)^2$ $k \times k$ squares. In total there are $1^2 + 2^2 + 3^2 + \cdots + n^2 = n(n+1)(2n+1)/6$ squares.

15. For $n = 5$, $2^5 = 32 > 25 = 5^2$. Assume the result for $n = k$ (≥ 5): $2^k > k^2$. For $k > 3$, $k(k-2) > 1$, or $k^2 > 2k + 1$. $2^k > k^2 \Rightarrow 2^k + 2^k > k^2 + k^2 \Rightarrow 2^{k+1} > k^2 + k^2 > k^2 + (2k+1) = (k+1)^2$. Hence the result is true for $n \ge 5$ by the Principle of Mathematical Induction.

17. b) Starting with $n = 1$ we find that

$$\sum_{j=1}^{1} j H_j = H_1 = 1 = [(2)(1)/2](3/2) - [(2)(1)/4] = [(2)(1)/2]H_2 - [(2)(1)/4].$$

Assuming the truth of the given (open) statement for $n = k$, we have

$$\sum_{j=1}^{k} j H_j = [(k+1)(k)/2]H_{k+1} - [(k+1)(k)/4].$$

For $n = k + 1$ we now find that

$$\begin{aligned}
\sum_{j=1}^{k+1} j H_j &= \sum_{j=1}^{k} j H_j + (k+1)H_{k+1} \\
&= [(k+1)(k)/2]H_{k+1} - [(k+1)(k)/4] + (k+1)H_{k+1} \\
&= (k+1)[1 + (k/2)]H_{k+1} - [(k+1)(k)/4] \\
&= (k+1)[1 + (k/2)][H_{k+2} - (1/(k+2))] - [(k+1)(k)/4] \\
&= [(k+2)(k+1)/2]H_{k+2} - [(k+1)(k+2)]/[2(k+2)] - [(k+1)(k)/4] \\
&= [(k+2)(k+1)/2]H_{k+2} - [(1/4)[2(k+1) + k(k+1)]] \\
&= [(k+2)(k+1)/2]H_{k+2} - [(k+2)(k+1)/4].
\end{aligned}$$

Consequently, by the Principle of Mathematical Induction, it follows that the given (open) statement is true for all $n \in \mathbf{Z}^+$.

19. Assume $S(k)$. For $S(k+1)$, we find that $\sum_{i=1}^{k+1} i = ([k + (1/2)]^2/2) + (k+1) = (k^2 + k + (1/4) + 2k + 2)/2 = [(k+1)^2 + (k+1) + (1/4)]/2 = [(k+1) + (1/2)]^2/2$. So $S(k) \Rightarrow S(k+1)$. However, we have no first value of k where $S(k)$ is true: for all $k \geq 1$, $\sum_{i=1}^{k} i = (k)(k+1)/2$ and $(k)(k+1)/2 = [k + (1/2)]^2/2 \Rightarrow 0 = 1/4$.

21. Let $S(n)$ denote the following (open) statement: For $x, n \in \mathbf{Z}^+$, if the program reaches the top of the **while** loop, after the two loop instructions are executed $n \, (> 0)$ times, then the value of the integer variable *answer* is $x(n!)$.

First consider $S(1)$, the statement for the case where $n = 1$. Here the program (if it reaches the top of the **while** loop) will result in one execution of the **while** loop: x will be assigned the value $x \cdot 1 = x(1!)$, and the value of n will be decreased to 0. With the value of n equal to 0, the loop is not processed again and the value of the variable *answer* is $x(1!)$. Hence $S(1)$ is true.

Now assume the truth for $n = k \, (\geq 1)$: For $x, k \in \mathbf{Z}^+$, if the program reaches the top of the **while** loop, then upon exiting the loop, the value of the variable *answer* is $x(k!)$. To establish the truth of $S(k+1)$, if the program reaches the top of the **while** loop, then the following occur during the first execution:

The value assigned to the variable x is $x(k+1)$.

The value of n is decreased to $(k+1) - 1 = k$.

But then we can apply the induction hypothesis to the integers $x(k+1)$ and k, and upon exiting the **while** loop for these values, the value of the variable *answer* is $(x(k+1))(k!) = x(k+1)!$

Consequently, $S(n)$ is true for all $n \geq 1$, and we have verified the correctness of this program segment by using the Principle of Mathematical Induction.

23. b) $24 = 5+5+7+7$ $25 = 5+5+5+5+5$ $26 = 5+7+7+7$
$27 = 5+5+5+5+7$ $28 = 7+7+7+7$
Hence the result is true for all $24 \leq n \leq 28$. Assume the result true for $24, 25, 26, 27, 28, \ldots, k$, and consider $n = k + 1$. Since $k + 1 \geq 29$, we may write $k + 1 = [(k+1) - 5] + 5 = (k-4) + 5$, where $k - 4$ can be expressed as a sum of 5's and 7's. Hence $k + 1$ can be expressed as such a sum and the result follows for all $n \geq 24$ by the alternative form of the Principle of Mathematical Induction.

25.
$$E(X) = \sum_x x Pr(X = x) = \sum_{x=1}^{n} x \left(\frac{1}{n}\right) = \left(\frac{1}{n}\right) \sum_{x=1}^{n} x = \left(\frac{1}{n}\right) \left[\frac{n(n+1)}{2}\right] = \frac{n+1}{2}$$

$$E(X^2) = \sum_x x^2 Pr(X = x) = \sum_{x=1}^{n} x^2 \left(\frac{1}{n}\right) = \left(\frac{1}{n}\right) \sum_{x=1}^{n} x^2 = \left(\frac{1}{n}\right) \left[\frac{n(n+1)(2n+1)}{6}\right]$$

$$= \frac{(n+1)(2n+1)}{6}$$

$$\text{Var}(X) = E(X^2) - [E(X)]^2 = \frac{(n+1)(2n+1)}{6} - \frac{(n+1)^2}{4} = (n+1)\left[\frac{2n+1}{6} - \frac{n+1}{4}\right]$$

$$= (n+1)\left[\frac{4n+2-(3n+3)}{12}\right] = \frac{(n+1)(n-1)}{12} = \frac{n^2-1}{12}.$$

27. Let $T = \{n \in \mathbf{Z}^+ | n \geq n_0$ and $S(n)$ is false $\}$. Since $S(n_0)$, $S(n_0 + 1)$, $S(n_0 + 2)$, ..., $S(n_1)$ are true, we know that n_0, $n_0 + 1$, $n_0 + 2$, ..., $n_1 \notin T$. If $T \neq \emptyset$, then T has a least element r, because $T \subseteq \mathbf{Z}^+$. However, since $S(n_0)$, $S(n_0 + 1)$, ..., $S(r - 1)$ are true, it follows that $S(r)$ is true. Hence $T = \emptyset$ and the result follows.

Section 4.2 – p. 219

1. a) $c_1 = 7$; $c_{n+1} = c_n + 7$, for $n \geq 1$.   **b)** $c_1 = 7$; $c_{n+1} = 7c_n$, for $n \geq 1$.
c) $c_1 = 10$; $c_{n+1} = c_n + 3$, for $n \geq 1$.   **d)** $c_1 = 7$; $c_{n+1} = c_n$, for $n \geq 1$.

3. Let $T(n)$ denote the following statement: For $n \in \mathbf{Z}^+$, $n \geq 2$, and the statements $p, q_1, q_2, \ldots, q_n$,

$$p \vee (q_1 \wedge q_2 \wedge \cdots \wedge q_n) \Longleftrightarrow (p \vee q_1) \wedge (p \vee q_2) \wedge \cdots \wedge (p \vee q_n).$$

The statement $T(2)$ is true by virtue of the Distributive Law of $\vee$ over $\wedge$. Assuming $T(k)$, for some $k \geq 2$, we now examine the situation for the statements $p, q_1, q_2, \ldots, q_k, q_{k+1}$. We find that $p \vee (q_1 \wedge q_2 \wedge \cdots \wedge q_k \wedge q_{k+1})$

$$\Longleftrightarrow p \vee [(q_1 \wedge q_2 \wedge \cdots \wedge q_k) \wedge q_{k+1}]$$

$$\Longleftrightarrow [p \vee (q_1 \wedge q_2 \wedge \cdots \wedge q_k)] \wedge (p \vee q_{k+1})$$

$$\Longleftrightarrow [(p \vee q_1) \wedge (p \vee q_2) \wedge \cdots \wedge (p \vee q_k)] \wedge (p \vee q_{k+1})$$

$$\Longleftrightarrow (p \vee q_1) \wedge (p \vee q_2) \wedge \cdots \wedge (p \vee q_k) \wedge (p \vee q_{k+1}).$$

It then follows by the Principle of Mathematical Induction that the statement $T(n)$ is true for all $n \geq 2$.

5. a) (i) The intersection of A_1, A_2 is $A_1 \cap A_2$.
(ii) The intersection of A_1, A_2, ..., A_n, A_{n+1} is given by $A_1 \cap A_2 \cap \cdots \cap A_n \cap A_{n+1} = (A_1 \cap A_2 \cap \cdots \cap A_n) \cap A_{n+1}$, the intersection of the *two* sets $A_1 \cap A_2 \cap \cdots \cap A_n$ and A_{n+1}.
b) Let $S(n)$ denote the given (open) statement. Then the truth of $S(3)$ follows from the Associative Law of $\cap$. Assuming $S(k)$ true for some $k \geq 3$, consider the case for $k + 1$ sets.
(1) If $r = k$, then

$$(A_1 \cap A_2 \cap \cdots \cap A_k) \cap A_{k+1} = A_1 \cap A_2 \cap \cdots \cap A_k \cap A_{k+1},$$

from the recursive definition given in part (a).
(2) For $1 \leq r < k$, we have

$$(A_1 \cap A_2 \cap \cdots \cap A_r) \cap (A_{r+1} \cap \cdots \cap A_k \cap A_{k+1})$$

$$= (A_1 \cap A_2 \cap \cdots \cap A_r) \cap [(A_{r+1} \cap \cdots \cap A_k) \cap A_{k+1}]$$

$$= [(A_1 \cap A_2 \cap \cdots \cap A_r) \cap (A_{r+1} \cap \cdots \cap A_k)] \cap A_{k+1}$$

$$= (A_1 \cap A_2 \cap \cdots \cap A_r \cap A_{r+1} \cap \cdots \cap A_k) \cap A_{k+1}$$

$$= A_1 \cap A_2 \cap \cdots \cap A_r \cap A_{r+1} \cap \cdots \cap A_k \cap A_{k+1},$$

and by the Principle of Mathematical Induction, $S(n)$ is true for all $n \geq 3$ and all $1 \leq r < n$.

7. For $n = 2$, the truth of the result $A \cap (B_1 \cup B_2) = (A \cap B_1) \cup (A \cap B_2)$ follows by virtue of the Distributive Law of $\cap$ over $\cup$. Assuming the result for $n = k$, let us examine the case for the sets A, B_1, B_2, ..., B_k, B_{k+1}. We have $A \cap (B_1 \cup B_2 \cup \cdots \cup B_k \cup B_{k+1}) = A \cap [(B_1 \cup B_2 \cup \cdots \cup B_k) \cup B_{k+1}] = [(A \cap (B_1 \cup B_2 \cup \cdots \cup B_k)] \cup (A \cap B_{k+1}) = [(A \cap B_1) \cup (A \cap B_2) \cup \cdots \cup (A \cap B_k)] \cup (A \cap B_{k+1}) = (A \cap B_1) \cup (A \cap B_2) \cup \cdots \cup (A \cap B_k) \cup (A \cap B_{k+1})$. So the result is true for all $n \geq 2$, by the Principle of Mathematical Induction.

9. a) (i) For $n = 2$, the expression $x_1 x_2$ denotes the ordinary product of the real numbers x_1 and x_2.
(ii) Let $n \in \mathbf{Z}^+$ with $n \geq 2$. For the real numbers $x_1, x_2, \ldots, x_n, x_{n+1}$, we define

$$x_1 x_2 \cdots x_n x_{n+1} = (x_1 x_2 \cdots x_n) x_{n+1},$$

the product of the *two* real numbers $x_1 x_2 \cdots x_n$ and x_{n+1}.
b) The result holds for $n = 3$ by the Associative Law of Multiplication (for real numbers). So $x_1 (x_2 x_3) = (x_1 x_2) x_3$, and there is no ambiguity in writing $x_1 x_2 x_3$. Assuming the result true for some $k \geq 3$ and all $1 \leq r < k$, let us examine the case for $k + 1 \, (\geq 4)$ real numbers. We find that (1) if $r = k$, then $(x_1 x_2 \cdots x_k) x_{k+1} = x_1 x_2 \cdots x_k x_{k+1}$ by the recursive definition given in part (a); and (2) if $1 \leq r < k$, then $(x_1 x_2 \cdots x_r)(x_{r+1} \cdots x_k x_{k+1}) = (x_1 x_2 \cdots x_r)((x_{r+1} \cdots x_k) x_{k+1})$ $= ((x_1 x_2 \cdots x_r)(x_{r+1} \cdots x_k)) x_{k+1} = (x_1 x_2 \cdots x_r x_{r+1} \cdots x_k) x_{k+1} = x_1 x_2 \cdots x_r x_{r+1} \cdots x_k x_{k+1}$, so the result is true for all $n \geq 3$ and all $1 \leq r < n$, by the Principle of Mathematical Induction.

11. *Proof* (By the Alternative Form of the Principle of Mathematical Induction): For $n = 0, 1, 2$ we have

$$(n = 0) \quad a_{0+2} = a_2 = 1 \geq (\sqrt{2})^0;$$

$$(n = 1) \quad a_{1+2} = a_3 = a_2 + a_0 = 2 \geq \sqrt{2} = (\sqrt{2})^1; \text{ and}$$

$$(n = 2) \quad a_{2+2} = a_4 = a_3 + a_1 = 2 + 1 = 3 \geq 2 = (\sqrt{2})^2.$$

Therefore, the result is true for these first three cases, and this gives us the basis step for the proof.

Next, for some $k \geq 2$, we assume the result true for all $n = 0, 1, 2, \ldots, k$. When $n = k + 1$ we find that

$$a_{(k+1)+2} = a_{k+3} = a_{k+2} + a_k \geq (\sqrt{2})^k + (\sqrt{2})^{k-2} = [(\sqrt{2})^2 + 1](\sqrt{2})^{k-2}$$
$$= 3(\sqrt{2})^{k-2} = (3/2)(2)(\sqrt{2})^{k-2} = (3/2)(\sqrt{2})^k \geq (\sqrt{2})^{k+1},$$

because $(3/2) = 1.5 > \sqrt{2} \, (\doteq 1.414)$. This provides the inductive step for the proof.

From the basis and inductive steps it now follows by the alternative form of the Principle of Mathematical Induction that $a_{n+2} \geq (\sqrt{2})^n$ for all $n \in \mathbf{N}$.

13. *Proof* (By Mathematical Induction):
Basis Step: When $n = 1$ we find that

$$\sum_{i=1}^{1} \frac{F_{i-1}}{2^i} = F_0/2 = 0 = 1 - (2/2) = 1 - \frac{F_3}{2} = 1 - \frac{F_{1+2}}{2^1},$$

so the result holds in the first case.
Inductive Step: Assuming the given (open) statement true for $n = k$, we have $\sum_{i=1}^{k} \frac{F_{i-1}}{2^i} = 1 - \frac{F_{k+2}}{2^k}$. When $n = k + 1$, we find that

$$\sum_{i=1}^{k+1} \frac{F_{i-1}}{2^i} = \sum_{i=1}^{k} \frac{F_{i-1}}{2^i} + \frac{F_k}{2^{k+1}} = 1 - \frac{F_{k+2}}{2^k} + \frac{F_k}{2^{k+1}}$$

$$= 1 + (1/2^{k+1})[F_k - 2F_{k+2}] = 1 + (1/2^{k+1})[(F_k - F_{k+2}) - F_{k+2}]$$

$$= 1 + (1/2^{k+1})[-F_{k+1} - F_{k+2}] = 1 - (1/2^{k+1})(F_{k+1} + F_{k+2}) = 1 - (F_{k+3}/2^{k+1}).$$

From the basis and inductive steps it follows from the Principle of Mathematical Induction that

$$\forall n \in \mathbf{Z}^+ \sum_{i=1}^{n} (F_{i-1}/2^i) = 1 - (F_{n+2}/2^n).$$

15. *Proof* (By the Alternative Form of the Principle of Mathematical Induction): The result holds for $n = 0$ and $n = 1$ because

$$(n = 0) \quad 5F_{0+2} = 5F_2 = 5(1) = 5 = 7 - 2 = L_4 - L_0 = L_{0+4} - L_0; \text{ and}$$

$$(n = 1) \quad 5F_{1+2} = 5F_3 = 5(2) = 10 = 11 - 1 = L_5 - L_1 = L_{1+4} - L_1.$$

This establishes the basis step for the proof.

Next we assume the induction hypothesis — that is, for some $k \ (\geq 1)$, $5F_{n+2} = L_{n+4} - L_n$ for all $n = 0, 1, 2, \ldots, k - 1, k$. It then follows that for $n = k + 1$,

$$5F_{(k+1)+2} = 5F_{k+3} = 5(F_{k+2} + F_{k+1}) = 5(F_{k+2} + F_{(k-1)+2}) = 5F_{k+2} + 5F_{(k-1)+2}$$

$$= (L_{k+4} - L_k) + (L_{(k-1)+4} - L_{k-1}) = (L_{k+4} - L_k) + (L_{k+3} - L_{k-1})$$

$$= (L_{k+4} + L_{k+3}) - (L_k + L_{k-1}) = L_{k+5} - L_{k+1} = L_{(k+1)+4} - L_{k+1},$$

where we have used the recursive definitions of the Fibonacci numbers and the Lucas numbers to establish the second and eighth equalities.

In then follows by the alternative form of the Principle of Mathematical Induction that

$$\forall n \in \mathbf{N} \quad 5F_{n+2} = L_{n+4} - L_n.$$

17. a) Steps   **Reasons**

1) p, q, r, T_0 Part (1) of the definition
2) $(p \vee q)$ Step (1) and part (2-ii) of the definition
3) $(\neg r)$ Step (1) part (2-i) of the definition
4) $(T_0 \wedge (\neg r))$ Steps (1) and (3) and part (2-iii) of the definition
5) $((p \vee q) \rightarrow (T_0 \wedge (\neg r)))$ Steps (2) and (4) and part (2-iv) of the definition

19. a) $\binom{k}{2} + \binom{k+1}{2} = [k(k-1)/2] + [(k+1)k/2] = (k^2 - k + k^2 + k)/2 = k^2$.

c) $\binom{k}{3} + 4\binom{k+1}{3} + \binom{k+2}{3} = [k(k-1)(k-2)/6] + 4[(k+1)(k)(k-1)/6] + [(k+2) \cdot (k+1)(k)/6] = (k/6)[(k-1)(k-2) + 4(k+1)(k-1) + (k+2)(k+1)] = (k/6)[6k^2] = k^3$.

e) $k^4 = \binom{k}{4} + 11\binom{k+1}{4} + 11\binom{k+2}{4} + \binom{k+3}{4}$

In general, $k^t = \sum_{r=0}^{t-1} a_{t,r}\binom{k+r}{t}$, where the $a_{t,r}$'s are the Eulerian numbers of Example 4.21 (The given summation formula is known as Worpitzky's identity.)

Section 4.3 – p. 230

1. e) If $a|x$ and $a|y$, then $x = ac$ and $y = ad$ for some $c, d \in \mathbf{Z}$. So $z = x - y = a(c - d)$, and $a|z$. The proofs for the other cases are similar.

g) Follows from part (f) by the Principle of Mathematical Induction.

3. Since q is prime, its only positive divisors are 1 and q. With p a prime, it follows that $p > 1$. Hence $p|q \Rightarrow p = q$.

5. *Proof* (By the Contrapositive): Suppose that $a|b$ or $a|c$. If $a|b$, then $ak = b$ for some $k \in \mathbf{Z}$. But $ak = b \Rightarrow (ak)c = a(kc) = bc \Rightarrow a|bc$. A similar result is obtained if $a|c$.

7. a) Let $a = 1$, $b = 5$, $c = 2$. Another example is $a = b = 5$, $c = 3$.

b) *Proof*: $31|(5a + 7b + 11c) \Rightarrow 31|(10a + 14b + 22c)$. Also, $31|(31a + 31b + 31c)$, so $31|[(31a + 31b + 31c) - (10a + 14b + 22c)]$. Hence $31|(21a + 17b + 9c)$.

9. $[b|a \text{ and } b|(a + 2)] \Rightarrow b|[ax + (a + 2)y]$ for all $x, y \in \mathbf{Z}$. Let $x = -1$, $y = 1$. Then $b > 0$ and $b|2$, so $b = 1$ or 2.

11. Let $a = 2m + 1$ and $b = 2n + 1$, for some $m, n \in \mathbf{N}$. Then $a^2 + b^2 = 4(m^2 + m + n^2 + n) + 2$, so $2|(a^2 + b^2)$ but $4 \nmid (a^2 + b^2)$.

13. For $n = 0$ we have $7^n - 4^n = 7^0 - 4^0 = 1 - 1 = 0$, and $3|0$. So the result is true for this first case. Assuming the truth for $n = k \ (\geq 0)$, we have $3|(7^k - 4^k)$. Turning to the case for $n = k + 1$, we find that $7^{k+1} - 4^{k+1} = 7(7^k) - 4(4^k) = (3 + 4)(7^k) - 4(4^k) = 3(7^k) + 4(7^k - 4^k)$. Since $3|3$ and $3|(7^k - 4^k)$ (by the induction hypothesis), it follows from part (f) of Theorem 4.3 that $3|[3(7^k) + 4(7^k - 4^k)]$, that is, $3|(7^{k+1} - 4^{k+1})$. It now follows by the Principle of Mathematical Induction that $3|(7^n - 4^n)$ for all $n \in \mathbf{N}$.

15.

	Base 10	Base 2	Base 16
a)	22	10110	16
b)	527	1000001111	20F
c)	1234	10011010010	4D2
d)	6923	1101100001011	1B0B

17.

	Base 2	Base 10	Base 16
a)	11001110	206	CE
b)	00110001	49	31
c)	11110000	240	F0
d)	01010111	87	57

19. $n = 1, 2, 3, 6, 9, 18$

21.

	Largest Integer	Smallest Integer
a)	$7 = 2^3 - 1$	$-8 = -(2^3)$
b)	$127 = 2^7 - 1$	$-128 = -(2^7)$
c)	$2^{15} - 1$	$-(2^{15})$
d)	$2^{31} - 1$	$-(2^{31})$
e)	$2^{n-1} - 1$	$-(2^{n-1})$

23. $ax = ay \Rightarrow ax - ay = 0 \Rightarrow a(x - y) = 0$. In the system of integers, if b, $c \in \mathbf{Z}$ and $bc = 0$, then $b = 0$ or $c = 0$. Since $a(x - y) = 0$ and $a \neq 0$, it follows that $(x - y) = 0$ and $x = y$.

29. a) Since $2 | 10^t$ for all $t \in \mathbf{Z}^+$, $2|n$ if and only if $2|r_0$.

b) Follows from the fact that $4|10^t$ for $t \geq 2$.

c) Follows from the fact that $8|10^t$ for $t \geq 3$. In general,

$$2^{t+1}|n \text{ if and only if } 2^{t+1}|(r_t \cdot 10^t + \cdots + r_1 \cdot 10 + r_0).$$

Section 4.4 – p. 236

1. a) $\gcd(1820, 231) = 7 = 1820(8) + 231(-63)$

b) $\gcd(2597, 1369) = 1 = 2597(534) + 1369(-1013)$

c) $\gcd(4001, 2689) = 1 = 4001(-1117) + 2689(1662)$

3. $\gcd(a, b) = d \Rightarrow d = ax + by$, for some x, $y \in \mathbf{Z}$
$\gcd(a, b) = d \Rightarrow a/d, b/d \in \mathbf{Z}$
$1 = (a/d)x + (b/d)y \Rightarrow \gcd(a/d, b/d) = 1$.

5. *Proof*: Since $c = \gcd(a, b)$ we have $a = cx$, $b = cy$ for some x, $y \in \mathbf{Z}^+$. So $ab = (cx)(cy) = c^2(xy)$, and c^2 divides ab.

7. Let $\gcd(a, b) = h$ and $\gcd(b, d) = g$.
$\gcd(a, b) = h \Rightarrow [h|a \text{ and } h|b] \Rightarrow h|(a \cdot 1 + bc) \Rightarrow h|d$.
$[h|b \text{ and } h|d] \Rightarrow h|g$.
$\gcd(b, d) = g \Rightarrow [g|b \text{ and } g|d] \Rightarrow g|(d \cdot 1 + b(-c)) \Rightarrow g|a$.
$[g|b, g|a, \text{ and } h = \gcd(a, b)] \Rightarrow g|h.$ $h|g$, $g|h$, with g, $h \in \mathbf{Z}^+ \Rightarrow g = h$.

9. a) If $c \in \mathbf{Z}^+$, then $c = \gcd(a, b)$ if (and only if)
(1) $c|a$ and $c|b$; and
(2) $\forall d \in \mathbf{Z} \ [(d|a) \wedge (d|b)] \Rightarrow d|c$.

b) If $c \in \mathbf{Z}^+$, then $c \neq \gcd(a, b)$ if (and only if)
(1) $c \nmid a$ or $c \nmid b$; or
(2) $\exists d \in \mathbf{Z} \ [(d|a) \wedge (d|b) \wedge (d \nmid c)]$.

11. $\gcd(a, b) = 1 \Rightarrow ax + by = 1$, for some x, $y \in \mathbf{Z}$. Then $acx + bcy = c$. $a|acx$, $a|bcy$ (because $a|bc) \Rightarrow a|c$.

13. We find that for any $n \in \mathbf{Z}^+$, $(5n + 3)(7) + (7n + 4)(-5) = (35n + 21) - (35n + 20) = 1$. Consequently, it follows that $\gcd(5n + 3, 7n + 4) = 1$, or $5n + 3$ and $7n + 4$ are relatively prime.

15. One \$20 and 20 \$50 chips; six \$20 and 18 \$50 chips; eleven \$20 and 16 \$50 chips.

17. There is no solution for $c \neq 12, 18$. For $c = 12$, the solutions are $x = 118 - 165k$, $y = -10 + 14k$, $k \in \mathbf{Z}$. For $c = 18$, the solutions are $x = 177 - 165k$, $y = -15 + 14k$, $k \in \mathbf{Z}$.

19. $b = 40{,}425$ 21. $\gcd(n, n + 1) = 1$; $\text{lcm}(n, n + 1) = n(n + 1)$

1. a) $2^2 \cdot 3^3 \cdot 5^3 \cdot 11$ **b)** $2^4 \cdot 3 \cdot 5^2 \cdot 7^2 \cdot 11^2$ **c)** $3^2 \cdot 5^3 \cdot 7^2 \cdot 11 \cdot 13$

3. a) $m^2 = p_1^{2e_1} p_2^{2e_2} p_3^{2e_3} \cdots p_t^{2e_t}$ **b)** $m^2 = p_1^{3e_1} p_2^{3e_2} p_3^{3e_3} \cdots p_t^{3e_t}$

5. (The proof is similar to that given in Example 4.41.) If not, we have $\sqrt{p} = a/b$, where $a, b \in \mathbf{Z}^+$ and $\gcd(a, b) = 1$. Then $\sqrt{p} = a/b \Rightarrow p = a^2/b^2 \Rightarrow pb^2 = a^2 \Rightarrow p|a^2 \Rightarrow p|a$ (by Lemma 4.2). Since $p|a$ we know that $a = pk$ for some $k \in \mathbf{Z}^+$, and $pb^2 = a^2 = (pk)^2 = p^2k^2$, or $b^2 = pk^2$. Hence $p|b^2$ and so $p|b$. But if $p|a$ and $p|b$, then $\gcd(a, b) \geq p > 1$ — contradicting our earlier claim that $\gcd(a, b) = 1$.

7. a) 96   **b)** 270   **c)** 144   **9.** 660   **11.** There are 252 possible values for n.

13. a) *Proof*: (i) Since $10|a^2$ we have $5|a^2$ and $2|a^2$. Then by Lemma 4.2 it follows that $5|a$ and $2|a$. So $a = 5b$ for some $b \in \mathbf{Z}^+$. Further, since $2|5b$ we have $2|5$ or $2|b$ (by Lemma 4.2). Consequently, $a = 5b = 5(2c) = 10c$, and 10 divides a.

(ii) This result is false — let $a = 2$.

b) We can generalize section (i) of part (a) by replacing 10 by an integer n of the form $p_1 p_2 \cdots p_t$, a product of t distinct primes. (So n is a square-free integer — that is, no square greater than 1 divides n.)

15. 176,400   **17.** $n = 2 \cdot 3 \cdot 5^2 \cdot 7^2 = 7350$

19. a) 5   **b)** 7   **c)** 32   **d)** $7 + 7 + 5 + 25 + 20 + 20 = 84$ **e)** 84

21. 1061 ($= 512 + 256 + 293$)

23. a) From the Fundamental Theorem of Arithmetic $88,200 = 2^3 \cdot 3^2 \cdot 5^2 \cdot 7^2$. Consider the set $F = \{2^3, 3^2, 5^2, 7^2\}$. Each subset of F determines a factorization ab where $\gcd(a, b) = 1$. There are 2^4 subsets — hence, 2^4 factorizations. Since order is not relevant, this number (of factorizations) reduces to $(1/2)2^4 = 2^3$. And since $1 < a < n$, $1 < b < n$, we remove the case for the empty subset of F (or the subset F itself). This yields $2^3 - 1$ such factorizations.

b) Here $n = 2^3 \cdot 3^2 \cdot 5^2 \cdot 7^2 \cdot 11$ and there are $2^4 - 1$ such factorizations.

c) Suppose that $n = p_1^{n_1} p_2^{n_2} \cdots p_k^{n_k}$, where $p_1, p_2, \ldots, p_k$ are k distinct primes and $n_1, n_2, \ldots, n_k \geq 1$. The number of unordered factorizations of n as ab, where $1 < a < n$, $1 < b < n$, and $\gcd(a, b) = 1$, is $2^{k-1} - 1$.

25. *Proof*: (By Mathematical Induction): For $n = 2$ we find that $\prod_{i=2}^2 \left(1 - \frac{1}{i^2}\right) = \left(1 - \frac{1}{2^2}\right) = \left(1 - \frac{1}{4}\right) = 3/4 = (2+1)/(2 \cdot 2)$, so the result is true in this first case, and this establishes the basis step for our inductive proof. Next we assume the result true for some $k \in \mathbf{Z}^+$ where $k \geq 2$. This gives us $\prod_{i=2}^k \left(1 - \frac{1}{i^2}\right) = (k+1)/(2k)$. When we consider the case for $n = k + 1$, we obtain the inductive step for we find that

$$\prod_{i=2}^{k+1} \left(1 - \frac{1}{i^2}\right) = \left(\prod_{i=2}^k \left(1 - \frac{1}{i^2}\right)\right) \left(1 - \frac{1}{(k+1)^2}\right)$$

$$= [(k+1)/(2k)]\left[1 - \frac{1}{(k+1)^2}\right] = \left[\frac{k+1}{2k}\right]\left[\frac{(k+1)^2 - 1}{(k+1)^2}\right]$$

$$= \frac{k^2 + 2k}{(2k)(k+1)} = (k+2)/(2(k+1)) = ((k+1) + 1)/(2(k+1)).$$

The result now follows for all positive integers $n \geq 2$ by the Principle of Mathematical Induction.

27. a) The positive divisors of 28 are 1, 2, 4, 7, 14, and 28, and $1 + 2 + 4 + 7 + 14 + 28 = 56 = 2(28)$, so 28 is a perfect integer. The positive divisors of 496 are 1, 2, 4, 8, 16, 31, 62, 124, 248, and 496, and $1 + 2 + 4 + 8 + 16 + 31 + 62 + 124 + 248 + 496 = 992 = 2(496)$, so 496 is a perfect integer.

b) It follows from the Fundamental Theorem of Arithmetic that the divisors of $2^{m-1}(2^m - 1)$, for $2^m - 1$ prime, are $1, 2, 2^2, 2^3, \ldots, 2^{m-1}$, and $(2^m - 1), 2(2^m - 1), 2^2(2^m - 1), 2^3(2^m - 1), \ldots,$ and $2^{m-1}(2^m - 1)$. These divisors sum to $[1 + 2 + 2^2 + 2^3 + \cdots + 2^{m-1}] + (2^m - 1)[1 + 2 + 2^2 + 2^3 + \cdots + 2^{m-1}] = (2^m - 1) + (2^m - 1)(2^m - 1) = (2^m - 1)[1 + (2^m - 1)] = 2^m(2^m - 1) = 2[2^{m-1}(2^m - 1)]$, so $2^{m-1}(2^m - 1)$ is a perfect integer.

1. $a + (a + d) + (a + 2d) + \cdots + (a + (n-1)d) = na + [(n-1)nd]/2$. For $n = 1$, $a = a + 0$, and the result is true in this case. Assuming that

$$\sum_{i=1}^{k} [a + (i-1)d] = ka + [(k-1)kd]/2,$$

we have

$$\sum_{i=1}^{k+1} [a + (i-1)d] = (ka + [(k-1)kd]/2) + (a + kd) = (k+1)a + [k(k+1)d]/2,$$

so the result follows for all $n \in \mathbf{Z}^+$ by the Principle of Mathematical Induction.

3. *Conjecture*: $\sum_{i=1}^{n}(-1)^{i+1}i^2 = (-1)^{n+1}\sum_{i=1}^{n} i$, for all $n \in \mathbf{Z}^+$.

Proof (By the Principle of Mathematical Induction): If $n = 1$ the conjecture provides $\sum_{i=1}^{1}(-1)^{i+1}i^2 = (-1)^{1+1}(1)^2 = 1 = (-1)^{1+1}(1) = (-1)^{1+1}\sum_{i=1}^{1} i$, which is a true statement. And this establishes the basis step of the proof. To confirm the inductive step, we shall assume the truth of the result

$$\sum_{i=1}^{k}(-1)^{i+1}i^2 = (-1)^{k+1}\sum_{i=1}^{k} i$$

for some $k \geq 1$. When $n = k + 1$ we find that

$$\sum_{i=1}^{k+1}(-1)^{i+1}i^2 = \left(\sum_{i=1}^{k}(-1)^{i+1}i^2\right) + (-1)^{(k+1)+1}(k+1)^2 = (-1)^{k+1}\sum_{i=1}^{k} i + (-1)^{k+2}(k+1)^2$$

$$= (-1)^{k+1}(k)(k+1)/2 + (-1)^{k+2}(k+1)^2 = (-1)^{k+2}[(k+1)^2 - (k)(k+1)/2]$$

$$= (-1)^{k+2}(1/2)[2(k+1)^2 - k(k+1)] = (-1)^{k+2}(1/2)[2k^2 + 4k + 2 - k^2 - k]$$

$$= (-1)^{k+2}(1/2)[k^2 + 3k + 2] = (-1)^{k+2}(1/2)(k+1)(k+2)$$

$$= (-1)^{k+2}\sum_{i=1}^{k+1} i,$$

so the truth of the result at $n = k$ implies the truth at $n = k + 1$ — and we have the inductive step. It then follows by the Principle of Mathematical Induction that

$$\sum_{i=1}^{n}(-1)^{i+1}i^2 = (-1)^{n+1}\sum_{i=1}^{n} i,$$

for all $n \in \mathbf{Z}^+$.

5. a)

n	$n^2 + n + 41$	n	$n^2 + n + 41$	n	$n^2 + n + 41$
1	43	4	61	7	97
2	47	5	71	8	113
3	53	6	83	9	131

b) For $n = 39$, $n^2 + n + 41 = 1601$, a prime. But for $n = 40$, $n^2 + n + 41 = (41)^2$, so $S(39) \not\Rightarrow S(40)$.

7. a) For $n = 0$, $2^{2n+1} + 1 = 2 + 1 = 3$, so the result is true in this first case. Assuming that 3 divides $2^{2k+1} + 1$ for $n = k \ (\geq 0) \in \mathbf{N}$, consider the case of $n = k + 1$. Since $2^{2(k+1)+1} + 1 = 2^{2k+3} + 1 = 4(2^{2k+1}) + 1 = 4(2^{2k+1} + 1) - 3$, and 3 divides both $2^{2k+1} + 1$ and 3, it follows that 3 divides $2^{2(k+1)+1} + 1$. Consequently, the result is true for $n = k + 1$ whenever it is true for $n = k$. So by the Principle of Mathematical Induction, the result follows for all $n \in \mathbf{N}$.

9. $x = y = z = 0$ and $x = 2$, $y = 5$, $z = 5$

11. For $n = 2$ we find that $2^2 = 4 < 6 = \binom{4}{2} < 16 = 4^2$, so the (open) statement is true in this first case. Assuming the result true for $n = k \geq 2$ —that is, $2^k < \binom{2k}{k} < 4^k$, we now consider what happens for $n = k + 1$. Here we find that

$$\binom{2(k+1)}{k+1} = \binom{2k+2}{k+1} = \left[\frac{(2k+2)(2k+1)}{(k+1)(k+1)}\right]\binom{2k}{k} = 2[(2k+1)/(k+1)]\binom{2k}{k}$$
$$> 2[(2k+1)/(k+1)]2^k > 2^{k+1},$$

since $(2k+1)/(k+1) = [(k+1)+k]/(k+1) > 1$. In addition, $[(k+1)+k]/(k+1) < 2$, so $\binom{2k+2}{k+1} = 2[(2k+1)/(k+1)]\binom{2k}{k} < (2)(2)\binom{2k}{k} < 4^{k+1}$. Consequently, the result is true for all $n \geq 2$, by the Principle of Mathematical Induction.

13. First we observe that the result is true for all $n \in \mathbf{Z}^+$ where $64 \leq n \leq 68$. This follows from the calculations

$$64 = 2(17) + 6(5) \qquad 65 = 13(5) \qquad 66 = 3(17) + 3(5)$$
$$67 = 1(17) + 10(5) \qquad 68 = 4(17)$$

Now assume the result is true for all n where $68 \leq n \leq k$, and consider the integer $k + 1$. Then $k + 1 = (k - 4) + 5$, and since $64 \leq k - 4 < k$, we can write $k - 4 = a(17) + b(5)$ for some $a, b \in \mathbf{N}$. Consequently, $k + 1 = a(17) + (b + 1)(5)$, and the result follows for all $n \geq 64$, by the alternative form of the Principle of Mathematical Induction.

15. a) $r = r_0 + r_1 \cdot 10 + r_2 \cdot 10^2 + \cdots + r_n \cdot 10^n$
$\qquad = r_0 + r_1(9) + r_1 + r_2(99) + r_2 + \cdots + r_n \underbrace{(99 \ldots 9)}_{n\ 9\text{'s}} + r_n$

$\qquad = [9r_1 + 99r_2 + \cdots + (99 \cdots 9)r_n] + (r_0 + r_1 + r_2 + \cdots + r_n).$
Hence $9|r$ if and only if $9|(r_0 + r_1 + r_2 + \cdots + r_n)$.

c) $3|t$ for $x = 1$ or 4 or 7; $9|t$ for $x = 7$.

17. a) $\binom{13}{9}$ **b)** $\binom{8}{4}$

19. a) $1, 4, 9$ **b)** $1, 4, 9, 16, \ldots, k$, where k is the largest square less than or equal to n.

21. a) For all $n \in \mathbf{Z}^+$, $n \geq 3$, $1 + 2 + 3 + \cdots + n = n(n+1)/2$. If $\{1, 2, 3, \ldots, n\} = A \cup B$ with $s_A = s_B$, then $2s_A = n(n+1)/2$, or $4s_A = n(n+1)$. Since $4|n(n+1)$ and $\gcd(n, n+1) = 1$, then either $4|n$ or $4|(n+1)$.

b) Here we are verifying the converse of our result in part (a).

(i) If $4|n$, we write $n = 4k$. Here we have
$\{1, 2, 3, \ldots, k, k+1, \ldots, 3k, 3k+1, \ldots, 4k\} = A \cup B$ where $A = \{1, 2, 3, \ldots, k, 3k+1, 3k+2, \ldots, 4k-1, 4k\}$ and $B = \{k+1, k+2, \ldots, 2k, 2k+1, 3k-1, 3k\}$, with $s_A = (1+2+3+\cdots+k) + [(3k+1) + (3k+2) + \cdots + (3k+k)] = [k(k+1)/2] + k(3k) + [k(k+1)/2] = k(k+1) + 3k^2 = 4k^2 + k$, and $s_B = [(k+1) + (k+2) + \cdots + (k+k)] + [(2k+1) + (2k+2) + \cdots + (2k+k)] = k(k) + [k(k+1)/2] + k(2k) + [k(k+1)/2] = 3k^2 + k(k+1) = 4k^2 + k$.

(ii) Now we consider the case where $n + 1 = 4k$. Then $n = 4k - 1$ and we have
$\{1, 2, 3, \ldots, k-1, k, \ldots, 3k-1, 3k, \ldots, 4k-2, 4k-1\} = A \cup B$, with $A = \{1, 2, 3, \ldots, k-1, 3k, 3k+1, \ldots, 4k-1\}$ and $B = \{k, k+1, \ldots, 2k-1, 2k, 2k+1, \ldots, 3k-1\}$. Here we find $s_A = [1 + 2 + 3 + \cdots + (k-1)] + [3k + (3k+1) + \cdots + (3k + (k-1))] = [(k-1)(k)/2] + k(3k) + [(k-1)(k)/2] = 3k^2 + k^2 - k = 4k^2 - k$, and $s_B = [k + (k+1) + \cdots + (k + (k-1))] + [2k + (2k+1) + \cdots + (2k + (k-1))] = k^2 + [(k-1)(k)/2] + k(2k) + [(k-1)(k)/2] = 3k^2 + (k-1)k = 4k^2 - k$.

23. a) The result is true for $a = 1$, so consider $a > 1$. From the Fundamental Theorem of Arithmetic we can write $a = p_1^{e_1} p_2^{e_2} \cdots p_t^{e_t}$, where $p_1, p_2, \ldots, p_t$ are t distinct primes and $e_i > 0$, for all $1 \leq i \leq t$. Since $a^2|b^2$ it follows that $p_i^{2e_i}|b^2$ for all $1 \leq i \leq t$. So $b^2 = p_1^{2f_1} p_2^{2f_2} \cdots p_t^{2f_t} c^2$, where $f_i \geq e_i$ for all $1 \leq i \leq t$, and $b = p_1^{f_1} p_2^{f_2} \cdots p_t^{f_t} c = a(p_1^{f_1 - e_1} p_2^{f_2 - e_2} \cdots p_t^{f_t - e_t})c$, where $f_i - e_i \geq 0$ for all $1 \leq i \leq t$. Consequently, $a|b$.

b) This result is not necessarily true! Let $a = 8$ and $b = 4$. Then $a^2 \ (= 64)$ divides $b^3 \ (= 64)$, but a does not divide b.

25. a) Recall that

$$a^3 + b^3 = (a + b)(a^2 - ab + b^2)$$
$$a^5 + b^5 = (a + b)(a^4 - a^3b + a^2b^2 - ab^3 + b^4)$$
$$\vdots$$
$$a^p + b^p = (a + b)(a^{p-1} - a^{p-2}b + \cdots + b^{p-1})$$
$$= (a + b) \sum_{i=1}^{p} a^{p-i}(-b)^{i-1},$$

for p an odd prime.

Since k is not a power of 2 we write $k = r \cdot p$, where p is an odd prime and $r \geq 1$. Then $a^k + b^k = (a^r)^p + (b^r)^p = (a^r + b^r) \sum_{i=1}^{p} a^{r(p-i)}(-b^r)^{(i-1)}$, so $a^k + b^k$ is composite.

b) Here n is not a power of 2. If, in addition, n is not prime, then $n = r \cdot p$ where p is an odd prime. Then $2^n + 1 = 2^n + 1^n = 2^{r \cdot p} + 1^{r \cdot p} = (2^r + 1^r) \sum_{i=1}^{p} 2^{r(p-i)}(-1^r)^{(i-1)} = (2^r + 1) \sum_{i=1}^{p} (-1)^{i-1} 2^{r(p-i)}$, so $2^n + 1$ is composite — not prime.

27. *Proof*: For $n = 0$ we find that $F_0 = 0 \leq 1 = (5/3)^0$, and for $n = 1$ we have $F_1 = 1 \leq (5/3) = (5/3)^1$. Consequently, the given property is true in these first two cases (and this provides the basis step of the proof).

Assuming that this property is true for $n = 0, 1, 2, \ldots, k - 1, k$, where $k \geq 1$, we now examine what happens at $n = k + 1$. Here we find that

$$F_{k+1} = F_k + F_{k-1} \leq (5/3)^k + (5/3)^{k-1} = (5/3)^{k-1}[(5/3) + 1] = (5/3)^{k-1}(8/3)$$
$$= (5/3)^{k-1}(24/9) \leq (5/3)^{k-1}(25/9) = (5/3)^{k-1}(5/3)^2 = (5/3)^{k+1}.$$

It then follows from the alternative form of the Principle of Mathematical Induction that $F_n \leq (5/3)^n$ for all $n \in \mathbf{N}$.

29. a) There are $9 \cdot 10 \cdot 10 = 900$ such palindromes and their sum is

$\sum_{a=1}^{9} \sum_{b=0}^{9} \sum_{c=0}^{9} abcba = \sum_{a=1}^{9} \sum_{b=0}^{9} \sum_{c=0}^{9} (10001a + 1010b + 100c) =$
$\sum_{a=1}^{9} \sum_{b=0}^{9} [10(10001a + 1010b) + 100(9 \cdot 10/2)] =$
$\sum_{a=1}^{9} \sum_{b=0}^{9} (100010a + 10100b + 4500) = \sum_{a=1}^{9} [10(100010a) + 10100(9 \cdot 10/2) + 10(4500)] = 1000100 \sum_{a=1}^{9} a + 9(454500) + 9(45000) = 1000100(9 \cdot 10/2) + 4090500 + 405000 = 49,500,000.$

b) begin

```
    sum := 0
    for a := 1 to 9 do
      for b := 0 to 9 do
        for c := 0 to 9 do
          sum := sum + 10001 * a + 1010 * b + 100 * c
    print sum
  end
```

31. *Proof*: Suppose that $7|n$. We see that $7|n \Rightarrow 7|(n - 21u) \Rightarrow 7|[(n - u) - 20u] \Rightarrow 7|[10(\frac{n-u}{10}) - 20u] \Rightarrow 7|[10(\frac{n-u}{10} - 2u)] \Rightarrow 7|(\frac{n-u}{10} - 2u)$, by Lemma 4.2 since $\gcd(7, 10) = 1$. [*Note*: $\frac{n-u}{10} \in \mathbf{Z}^+$ since the units digit of $n - u$ is 0.] Conversely, if $7|(\frac{n-u}{10} - 2u)$, then since $\frac{n-u}{10} - 2u = \frac{n-21u}{10}$ we find that $7|(\frac{n-21u}{10}) \Rightarrow 7 \cdot 10 \cdot x = n - 21u$, for some $x \in \mathbf{Z}^+$. Since $7|7$ and $7|21$, it then follows that $7|n$ — by part (e) of Theorem 4.3.

33. If Catrina's selection includes any of 0, 2, 4, 6, 8, then at least two of the resulting three-digit integers will have an even unit's digit, and be even — hence, *not* prime. Should her selection include 5, then two of the resulting three-digit integers will have 5 as their unit's digit; these three-digit integers are then divisible by 5 and so, they are *not* prime. Consequently, to complete the proof we need to consider the four selections of size 3 that Catrina can make from $\{1, 3, 7, 9\}$. The following provides the selections — each with a three-digit integer that is not prime.

1) $\{1, 3, 7\}$: $713 = 23 \cdot 31$ **2)** $\{1, 3, 9\}$: $913 = 11 \cdot 83$

3) $\{1, 7, 9\}$: $917 = 7 \cdot 131$ **4)** $\{3, 7, 9\}$: $793 = 13 \cdot 61$

35. Let x denote the integer Barbara erased. The sum of the integers $1, 2, 3, \ldots, x - 1, x + 1,$ $x + 2, \ldots, n$ is $[n(n + 1)/2] - x$, so $[[n(n + 1)/2] - x]/(n - 1) = 35\frac{7}{17}$. Consequently, $[n(n + 1)/2] - x = (35\frac{7}{17})(n - 1) = (602/17)(n - 1)$. Since $[n(n + 1)/2] - x \in \mathbf{Z}^+$, it follows that $(602/17)(n - 1) \in \mathbf{Z}^+$. Therefore, from Lemma 4.2, we find that $17 | (n - 1)$ because $17 \nmid 602$. For $n = 1, 18, 35, 52$ we have:

n	$x = [n(n + 1)/2] - (602/17)(n - 1)$
1	1
18	−431
35	−574
52	−428

When $n = 69$, we find that $x = 7$ $\left[\text{and } \left(\sum_{i=1}^{69} i - 7\right)/68 = 602/17 = 35\frac{7}{17}\right]$. For $n = 69 + 17k$, $k \geq 1$, we have

$$x = [(69 + 17k)(70 + 17k)/2] - (602/17)[68 + 17k]$$
$$= 7 + (k/2)[1159 + 289k]$$
$$= [7 + (1159k/2)] + (289k^2)/2 > n.$$

Hence the answer is unique: namely, $n = 69$ and $x = 7$.

37. $(1 + m_1)(1 + m_2)(1 + m_3)$, where $m_i = \min\{e_i, f_i\}$ for $1 \leq i \leq 3$.

Chapter 5
Relations and Functions

Section 5.1 – p. 252

1. $A \times B = \{(1, 2), (2, 2), (3, 2), (4, 2), (1, 5), (2, 5), (3, 5), (4, 5)\}$
$B \times A = \{(2, 1), (2, 2), (2, 3), (2, 4), (5, 1), (5, 2), (5, 3), (5, 4)\}$
$A \cup (B \times C) = \{1, 2, 3, 4, (2, 3), (2, 4), (2, 7), (5, 3), (5, 4), (5, 7)\}$
$(A \cup B) \times C = \{(1, 3), (2, 3), (3, 3), (4, 3), (5, 3), (1, 4), (2, 4), (3, 4), (4, 4), (5, 4),$
          $(1, 7), (2, 7), (3, 7), (4, 7), (5, 7)\} = (A \times C) \cup (B \times C)$

3. a) 9   **b)** 2^9 **c)** 2^9 **d)** 2^7 **e)** $\binom{9}{5}$ **f)** $\binom{9}{7} + \binom{9}{8} + \binom{9}{9}$

5. a) Assume that $A \times B \subseteq C \times D$ and let $a \in A$ and $b \in B$. Then $(a, b) \in A \times B$, and since $A \times B \subseteq C \times D$, we have $(a, b) \in C \times D$. But $(a, b) \in C \times D \Rightarrow a \in C$ and $b \in D$. Hence, $a \in A \Rightarrow a \in C$, so $A \subseteq C$, and $b \in B \Rightarrow b \in D$, so $B \subseteq D$.

Conversely, suppose that $A \subseteq C$ and $B \subseteq D$, and that $(x, y) \in A \times B$. Then $(x, y) \in A \times B \Rightarrow x \in A$ and $y \in B \Rightarrow x \in C$ (since $A \subseteq C$) and $y \in D$ (since $B \subseteq D$) $\Rightarrow (x, y) \in C \times D$. Consequently, $A \times B \subseteq C \times D$.

b) Even if any of the sets A, B, C, D is empty, we still find that

$$[(A \subseteq C) \wedge (B \subseteq D)] \Rightarrow [A \times B \subseteq C \times D].$$

However, the converse need not hold. For example, let $A = \emptyset$, $B = \{1, 2\}$, $C = \{1, 2\}$, and $D = \{1\}$. Then $A \times B = \emptyset$ — if not, there exists an ordered pair (x, y) in $A \times B$, and this means that the empty set A contains an element x. And so $A \times B = \emptyset \subseteq C \times D$ — but $B = \{1, 2\} \not\subseteq \{1\} = D$.

7. a) 2^{20} **b)** If $|A| = m$, $|B| = n$, for $m, n \in \mathbf{N}$, then there are 2^{mn} elements in $\mathscr{P}(A \times B)$.

9. c) $(x, y) \in (A \cap B) \times C \Longleftrightarrow x \in A \cap B$ and $y \in C \Longleftrightarrow (x \in A$ and $x \in B)$ and $y \in C \Longleftrightarrow$ $(x \in A$ and $y \in C)$ and $(x \in B$ and $y \in C) \Longleftrightarrow (x, y) \in A \times C$ and $(x, y) \in B \times C \Longleftrightarrow$ $(x, y) \in (A \times C) \cap (B \times C)$

11. $(x, y) \in A \times (B - C) \Longleftrightarrow x \in A$ and $y \in B - C \Longleftrightarrow x \in A$ and $(y \in B$ and $y \notin C) \Longleftrightarrow$ $(x \in A$ and $y \in B)$ and $(x \in A$ and $y \notin C) \Longleftrightarrow (x, y) \in A \times B$ and $(x, y) \notin A \times C \Longleftrightarrow$ $(x, y) \in (A \times B) - (A \times C)$

13. a) (1) $(0, 2) \in \mathscr{R}$; and
      (2) If $(a, b) \in \mathscr{R}$, then $(a + 1, b + 5) \in \mathscr{R}$

b) From part (1) of the definition we have $(0, 2) \in \mathcal{R}$. By part (2) of the definition we then find that

    (i) $(0, 2) \in \mathcal{R} \Rightarrow (0 + 1, 2 + 5) = (1, 7) \in \mathcal{R}$;

    (ii) $(1, 7) \in \mathcal{R} \Rightarrow (1 + 1, 7 + 5) = (2, 12) \in \mathcal{R}$;

    (iii) $(2, 12) \in \mathcal{R} \Rightarrow (2 + 1, 12 + 5) = (3, 17) \in \mathcal{R}$; and

    (iv) $(3, 17) \in \mathcal{R} \Rightarrow (3 + 1, 17 + 5) = (4, 22) \in \mathcal{R}$.

Section 5.2 – p. 258

1. a) Function; range $= \{7, 8, 11, 16, 23, \ldots\}$ **b)** Relation, not a function

   c) Function; range $= \mathbf{R}$ **d)** and **e)** Relation, not a function

3. a) (1) $\{(1, x), (2, x), (3, x), (4, x)\}$ (2) $\{(1, y), (2, y), (3, y), (4, y)\}$

      (3) $\{(1, z), (2, z), (3, z), (4, z)\}$ (4) $\{(1, x), (2, y), (3, x), (4, y)\}$

      (5) $\{(1, x), (2, y), (3, z), (4, x)\}$

   b) 3^4 **c)** 0   **d)** 4^3 **e)** 24   **f)** 3^3 **g)** 3^2 **h)** 3^2

5. a) $\{(1, 3)\}$ **b)** $\{(-7/2, -21/2)\}$

   c) $\{(-8, -15)\}$ **d)** $\mathbf{R}^2 - \{(-7/2, -21/2)\} = \{(x, y) | x \neq -7/2 \text{ or } y \neq -21/2\}$

7. a) $\lfloor 2.3 - 1.6 \rfloor = \lfloor 0.7 \rfloor = 0$ **b)** $\lfloor 2.3 \rfloor - \lfloor 1.6 \rfloor = 2 - 1 = 1$

   c) $\lceil 3.4 \rceil \lfloor 6.2 \rfloor = 4 \cdot 6 = 24$ **d)** $\lfloor 3.4 \rfloor \lceil 6.2 \rceil = 3 \cdot 7 = 21$

   e) $\lfloor 2\pi \rfloor = 6$ **f)** $2\lceil \pi \rceil = 8$

9. a) $\cdots \cup [-1, -6/7) \cup [0, 1/7) \cup [1, 8/7) \cup [2, 15/7) \cup \cdots$

   b) $[1, 8/7)$ **c)** $\mathbf{Z}$ **d)** $\mathbf{R}$

11. a) $\cdots \cup (-7/3, -2] \cup (-4/3, -1] \cup (-1/3, 0] \cup (2/3, 1] \cup (5/3, 2] \cup \cdots = \bigcup_{m \in \mathbf{Z}} (m - 1/3, m]$

   b) $\cdots \cup ((-2n - 1)/n, -2] \cup ((-n - 1)/n, -1] \cup (-1/n, 0] \cup ((n - 1)/n, 1] \cup ((2n - 1)/n, 2] \cup \cdots = \bigcup_{m \in \mathbf{Z}} (m - 1/n, m]$

13. a) *Proof* (i): If $a \in \mathbf{Z}^+$, then $\lceil a \rceil = a$ and $\lfloor \lceil a \rceil / a \rfloor = \lfloor 1 \rfloor = 1$. If $a \notin \mathbf{Z}^+$, write $a = n + c$, where $n \in \mathbf{Z}^+$ and $0 < c < 1$. Then $\lceil a \rceil / a = (n + 1)/(n + c) = 1 + (1 - c)/(n + c)$, where $0 < (1 - c)/(n + c) < 1$. Hence $\lfloor \lceil a \rceil / a \rfloor = \lfloor 1 + (1 - c)/(n + c) \rfloor = 1$.

   b) Consider $a = 0.1$. Then

      (i) $\lfloor \lceil a \rceil / a \rfloor = \lfloor 0/0.1 \rfloor = \lfloor 10 \rfloor = 10 \neq 1$; and

      (ii) $\lceil \lfloor a \rfloor / a \rceil = \lceil 0/0.1 \rceil = 0 \neq 1$.

      In fact (ii) is false for all $0 < a < 1$, since $\lceil \lfloor a \rfloor / a \rceil = 0$ for all such values of a. In the case of (i), when $0 < a \leq 0.5$, it follows that $\lceil a \rceil / a \geq 2$ and $\lfloor \lceil a \rceil / a \rfloor \geq 2 \neq 1$. However, for $0.5 < a < 1$, $\lceil a \rceil / a = 1/a$ where $1 < 1/a < 2$, and so $\lfloor \lceil a \rceil / a \rfloor = 1$ for $0.5 < a < 1$.

15. a) One-to-one; the range is the set of all odd integers.

   b) One-to-one; the range is $\mathbf{Q}$.

   c) Not one-to-one; the range is $\{0, \pm 6, \pm 24, \pm 60, \ldots\} = \{n^3 - n | n \in \mathbf{Z}\}$.

   d) One-to-one; the range is $(0, +\infty)$.

   e) One-to-one; the range is $[-1, 1]$.

   f) Not one-to-one; the range is $[0, 1]$.

17. 4^2

19. a) $f(A_1 \cup A_2) = \{y \in B | y = f(x), x \in A_1 \cup A_2\} = \{y \in B | y = f(x), x \in A_1 \text{ or } x \in A_2\} = \{y \in B | y = f(x), x \in A_1\} \cup \{y \in B | y = f(x), x \in A_2\} = f(A_1) \cup f(A_2)$

   c) From part (b), $f(A_1 \cap A_2) \subseteq f(A_1) \cap f(A_2)$. Conversely, $y \in f(A_1) \cap f(A_2) \Rightarrow y = f(x_1) = f(x_2)$, for $x_1 \in A_1, x_2 \in A_2 \Rightarrow y = f(x_1)$ and $x_1 = x_2$ (because f is injective) $\Rightarrow y \in f(A_1 \cap A_2)$. So f injective $\Rightarrow f(A_1 \cap A_2) = f(A_1) \cap f(A_2)$.

21. No. Let $A = \{1, 2\}$, $X = \{1\}$, $Y = \{2\}$, $B = \{3\}$. For $f = \{(1, 3), (2, 3)\}$ we have $f|_X, f|_Y$ one-to-one, but f is *not* one-to-one.

23. a) $f(a_{ij}) = 12(i - 1) + j$ **b)** $f(a_{ij}) = 10(i - 1) + j$ **c)** $f(a_{ij}) = 7(i - 1) + j$

25. a) (i) $f(a_{ij}) = n(i - 1) + (k - 1) + j$ (ii) $g(a_{ij}) = m(j - 1) + (k - 1) + i$

   b) $k + (mn - 1) \leq r$

27. a) $A(1, 3) = A(0, A(1, 2)) = A(1, 2) + 1 = A(0, A(1, 1)) + 1 = [A(1, 1) + 1] + 1 = A(1, 1) + 2 = A(0, A(1, 0)) + 2 = [A(1, 0) + 1] + 2 = A(1, 0) + 3 = A(0, 1) + 3 = (1 + 1) + 3 = 5$

$$A(2, 3) = A(1, A(2, 2))$$

$$A(2, 2) = A(1, A(2, 1))$$

$$A(2, 1) = A(1, A(2, 0)) = A(1, A(1, 1))$$

$$A(1, 1) = A(0, A(1, 0)) = A(1, 0) + 1 = A(0, 1) + 1 = (1 + 1) + 1 = 3$$

$$A(2, 1) = A(1, 3) = A(0, A(1, 2)) = A(1, 2) + 1 = A(0, A(1, 1)) + 1$$

$$= [A(1, 1) + 1] + 1 = 5$$

$$A(2, 2) = A(1, 5) = A(0, A(1, 4)) = A(1, 4) + 1 = A(0, A(1, 3)) + 1 = A(1, 3) + 2$$

$$= A(0, A(1, 2)) + 2 = A(1, 2) + 3 = A(0, A(1, 1)) + 3 = A(1, 1) + 4 = 7$$

$$A(2, 3) = A(1, 7) = A(0, A(1, 6)) = A(1, 6) + 1 = A(0, A(1, 5)) + 1$$

$$= A(0, 7) + 1 = (7 + 1) + 1 = 9$$

b) Since $A(1, 0) = A(0, 1) = 2 = 0 + 2$, the result holds for the case where $n = 0$. Assuming the truth of the (open) statement for some k (≥ 0), we have $A(1, k) = k + 2$. Then we find that $A(1, k + 1) = A(0, A(1, k)) = A(1, k) + 1 = (k + 2) + 1 = (k + 1) + 2$, so the truth at $n = k$ implies the truth at $n = k + 1$. Consequently, $A(1, n) = n + 2$ for all $n \in \mathbf{N}$ by the Principle of Mathematical Induction.

Section 5.3–p. 265

1. a) $A = \{1, 2, 3, 4\}$, $B = \{v, w, x, y, z\}$, $f = \{(1, v), (2, v), (3, w), (4, x)\}$
 b) A, B as in (a), $f = \{(1, v), (2, x), (3, z), (4, y)\}$
 c) $A = \{1, 2, 3, 4, 5\}$, $B = \{w, x, y, z\}$, $f = \{(1, w), (2, w), (3, x), (4, y), (5, z)\}$
 d) $A = \{1, 2, 3, 4\}$, $B = \{w, x, y, z\}$, $f = \{(1, w), (2, x), (3, y), (4, z)\}$

3. a), b), c), and **f)** are one-to-one and onto.
 d) Neither one-to-one nor onto; range $= [0, +\infty)$
 e) Neither one-to-one nor onto; range $= \left[-\frac{1}{4}, +\infty\right)$

5. (For the case $n = 5$, $m = 3$):

$$\sum_{k=0}^{5} (-1)^k \binom{5}{5-k} (5-k)^3 = (-1)^0 \binom{5}{5} 5^3 + (-1)^1 \binom{5}{4} 4^3 + (-1)^2 \binom{5}{3} 3^3$$

$$+ (-1)^3 \binom{5}{2} 2^3 + (-1)^4 \binom{5}{1} 1^3 + (-1)^5 \binom{5}{0} 0^3$$

$$= 125 - 5(64) + 10(27) - 10(8) + 5 = 0$$

7. a)   (i) $2!S(7, 2)$ (ii) $\binom{5}{2}[2!S(7, 2)]$ (iii) $3!S(7, 3)$
       (iv) $\binom{5}{3}[3!S(7, 3)]$ (v) $4!S(7, 4)$ (vi) $\binom{5}{4}[4!S(7, 4)]$
 b) $\binom{n}{k}[k!S(m, k)]$

9. For each $r \in \mathbf{R}$ there is at least one $a \in \mathbf{R}$ such that $a^5 - 2a^2 + a - r = 0$, because the polynomial $x^5 - 2x^2 + x - r$ has odd degree and real coefficients. Consequently, f is onto. However, $f(0) = 0 = f(1)$, so f is not one-to-one.

11.

m \ n	1	2	3	4	5	6	7	8	9	10
9	1	255	3025	7770	6951	2646	462	36	1	
10	1	511	9330	34105	42525	22827	5880	750	45	1

13. a) Since $156{,}009 = 3 \times 7 \times 17 \times 19 \times 23$, it follows that there are $S(5, 2) = 15$ two-factor unordered factorizations of $156{,}009$, where each factor is greater than 1.
 b) $\sum_{i=2}^{5} S(5, i) = 15 + 25 + 10 + 1 = 51$ **c)** $\sum_{i=2}^{n} S(n, i)$

15. a) $n = 4: \sum_{i=1}^{4} i!S(4, i); \quad n = 5: \sum_{i=1}^{5} i!S(5, i)$
In general, the answer is $\sum_{i=1}^{n} i!S(n, i)$.
b) $\binom{15}{12} \sum_{i=1}^{12} i!S(12, i)$.

17. Let $a_1, a_2, \ldots, a_m, x$ denote the $m + 1$ distinct objects. Then $S_r(m + 1, n)$ counts the number of ways these objects can be distributed among n identical containers so that each container receives at least r of the objects.

Each of these distributions falls into exactly one of two categories:

(1) The element x is in a container with r or more other objects: Here we start with $S_r(m, n)$ distributions of $a_1, a_2, \ldots, a_m$ into n identical containers — each container receiving at least r of the objects. Now we have n *distinct* containers — distinguished by their contents. Consequently, there are n choices for locating the object x. As a result, this category provides $nS_r(m, n)$ of the distributions.

(2) The element x is in a container with $r - 1$ of the other objects: These other $r - 1$ objects can be chosen in $\binom{m}{r-1}$ ways, and then these objects — along with x — can be placed in one of the n containers. The remaining $m + 1 - r$ distinct objects can then be distributed among the $n - 1$ identical containers — where each container receives at least r of the objects — in $S_r(m + 1 - r, n - 1)$ ways. Hence this category provides the remaining $\binom{m}{r-1}S_r(m + 1 - r, n - 1)$ distributions.

19. a) We know that $s(m, n)$ counts the number of ways we can place m people — call them $p_1, p_2, \ldots, p_m$ — around n circular tables, with at least one occupant at each table. These arrangements fall into two disjoint sets: (1) The arrangements where p_1 is alone: There are $s(m - 1, n - 1)$ such arrangements; and (2) The arrangements where p_1 shares a table with at least one of the other $m - 1$ people: There are $s(m - 1, n)$ ways where $p_2, p_3, \ldots, p_m$ can be seated around the n tables so that every table is occupied. Each such arrangement determines a total of $m - 1$ locations (at all the n tables) where p_1 can now be seated — this for a total of $(m - 1)s(m - 1, n)$ arrangements. Consequently, $s(m, n) = (m - 1)s(m - 1, n) + s(m - 1, n - 1)$, for $m \geq n > 1$.

Section 5.4 – p. 272

1. Here we find, for example, that $f(f(a, b), c) = f(a, c) = c$, while $f(a, f(b, c)) = f(a, b) = a$, so f is *not* associative.

3. a), b), and **d)** are commutative and associative; **c)** is neither commutative nor associative.

5. a) 25   **b)** 5^{25} **c)** 5^{25} **d)** 5^{10}

7. a) Yes   **b)** Yes   **c)** No   **9. a)** 1216   **b)** $p^{31}q^{37}$

11. By the Well-Ordering Principle, A has a least element and this same element is the identity for g. If A is finite, then A will have a largest element, and this same element will be the identity for f. If A is infinite, then f cannot have an identity.

13. a) 5   **b)**

A_3	A_4	A_5
25	25	6
25	2	4
60	40	20
25	40	10

c) A_1, A_2

Section 5.5 – p. 277

1. The pigeons are the socks; the pigeonholes are the colors.   **3.** $26^2 + 1 = 677$

5. a) For each $x \in \{1, 2, 3, \ldots, 300\}$ write $x = 2^n \cdot m$, where $n \geq 0$ and $\gcd(2, m) = 1$. There are 150 possibilities for m: 1, 3, 5, $\ldots$, 299. When we select 151 numbers from $\{1, 2, 3, \ldots, 300\}$, there must be two numbers of the form $x = 2^s \cdot m$, $y = 2^t \cdot m$. If $x < y$, then $x | y$; otherwise $y < x$ and $y | x$.
b) If $n + 1$ integers are selected from the set $\{1, 2, 3, \ldots, 2n\}$, then there must be two integers x, y in the selection where $x | y$ or $y | x$.

7. a) Here the pigeons are the integers 1, 2, 3, $\ldots$, 25 and the pigeonholes are the 13 sets $\{1, 25\}, \{2, 24\}, \ldots, \{11, 15\}, \{12, 14\}, \{13\}$. In selecting 14 integers, we get the elements in at least one two-element subset, and these sum to 26.

b) If $S = \{1, 2, 3, \ldots, 2n + 1\}$, for n a positive integer, then any subset of size $n + 2$ from S must contain two elements that sum to $2n + 2$.

9. a) For each $t \in \{1, 2, 3, \ldots, 100\}$, we find that $1 \leq \sqrt{t} \leq 10$. When we select 11 elements from $\{1, 2, 3, \ldots, 100\}$ there must be two — say, x and y — where $\lfloor \sqrt{x} \rfloor = \lfloor \sqrt{y} \rfloor$ so that $0 < |\sqrt{x} - \sqrt{y}| < 1$.

b) Let $n \in \mathbf{Z}^+$. If $n + 1$ elements are selected from $\{1, 2, 3, \ldots, n^2\}$, then there exist two — say, x and y — where $0 < |\sqrt{x} - \sqrt{y}| < 1$.

11. Divide the interior of the square into four smaller congruent squares as shown in the figure. Each smaller square has diagonal length $1/\sqrt{2}$. Let region R_1 be the interior of square $AEKH$ together with the points on segment EK, excluding point E. Region R_2 is the interior of square $EBFK$ together with the points on segment FK, excluding points F and K. Regions R_3 and R_4 are defined in a similar way. Then if five points are chosen in the interior of square $ABCD$, at least two are in R_i for some $1 \leq i \leq 4$, and these points are within $1/\sqrt{2}$ (units) of each other.

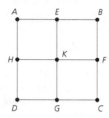

13. Consider the subsets A of S where $1 \leq |A| \leq 3$. Since $|S| = 5$, there are $\binom{5}{1} + \binom{5}{2} + \binom{5}{3} = 25$ such subsets A. Let s_A denote the sum of the elements in A. Then $1 \leq s_A \leq 7 + 8 + 9 = 24$. So by the pigeonhole principle, there are two subsets of S whose elements yield the same sum.

15. For $(\emptyset \neq) T \subseteq S$, we have $1 \leq s_T \leq m + (m - 1) + \cdots + (m - 6) = 7m - 21$. The set S has $2^7 - 1 = 128 - 1 = 127$ nonempty subsets. So by the pigeonhole principle we need to have $127 > 7m - 21$ or $148 > 7m$. Hence $7 \leq m \leq 21$.

17. a) $2, 4, 1, 3$ **b)** $3, 6, 9, 2, 5, 8, 1, 4, 7$

c) For $n \geq 2$, there exists a sequence of n^2 distinct real numbers with no decreasing or increasing subsequence of length $n + 1$. For example, consider $n, 2n, 3n, \ldots, (n - 1)n,$ $n^2, (n - 1), (2n - 1), \ldots, (n^2 - 1), (n - 2), (2n - 2), \ldots, (n^2 - 2), \ldots, 1, (n + 1),$ $(2n + 1), \ldots, (n - 1)n + 1$.

d) The result in Example 5.49 (for $n \geq 2$) is best possible — in the sense that we cannot reduce the length of the sequence from $n^2 + 1$ to n^2 and still obtain the desired subsequence of length $n + 1$.

19. *Proof*: If not, each pigeonhole contains at most k pigeons — for a total of at most kn pigeons. But we have $kn + 1$ pigeons. So we have a contradiction and the result then follows.

21. a) 1001   **b)** 2001

c) Let $n, k \in \mathbf{Z}^+$. The smallest value for $|S|$ (where $S \subset \mathbf{Z}^+$) so that there exist n elements $x_1, x_2, \ldots, x_n \in S$ where all n of these integers have the same remainder upon division by k is $k(n - 1) + 1$.

23. *Proof*: If not, then the number of pigeons roosting in the first pigeonhole is $x_1 \leq p_1 - 1$, the number of pigeons roosting in the second pigeonhole is $x_2 \leq p_2 - 1, \ldots$, and the number roosting in the nth pigeonhole is $x_n \leq p_n - 1$. Hence the total number of pigeons is $x_1 + x_2 + \cdots + x_n = (p_1 - 1) + (p_2 - 1) + \cdots + (p_n - 1) = p_1 + p_2 + \cdots + p_n - n <$ $p_1 + p_2 + \cdots + p_n - n + 1$, the number of pigeons we started with. The result now follows because of this contradiction.

Section 5.6 – p. 288

1. a) $7! - 6! = 4320$ **b)** $n! - (n - 1)! = (n - 1)(n - 1)!$
3. $a = 3, b = -1; a = -3, b = 2$

5. $g^2(A) = g(T \cap (S \cup A)) = T \cap (S \cup [T \cap (S \cup A)])$
$\quad\quad = T \cap [(S \cup T) \cap (S \cup (S \cup A))] = T \cap [(S \cup T) \cap (S \cup A)]$
$\quad\quad = [T \cap (S \cup T)] \cap (S \cup A) = T \cap (S \cup A) = g(A)$

7. a) $(f \circ g)(x) = 3x - 1$; $(g \circ f)(x) = 3(x - 1)$;

$$(g \circ h)(x) = \begin{cases} 0, & x \text{ even}; \\ 3, & x \text{ odd} \end{cases} \quad\quad (h \circ g)(x) = \begin{cases} 0, & x \text{ even}; \\ 1, & x \text{ odd} \end{cases}$$

$$(f \circ (g \circ h))(x) = f((g \circ h)(x)) = \begin{cases} -1, & x \text{ even}; \\ 2, & x \text{ odd} \end{cases}$$

$$((f \circ g) \circ h)(x) = \begin{cases} (f \circ g)(0), & x \text{ even} \\ (f \circ g)(1), & x \text{ odd} \end{cases} = \begin{cases} -1, & x \text{ even} \\ 2, & x \text{ odd} \end{cases}$$

b) $f^2(x) = f(f(x)) = x - 2$; $f^3(x) = x - 3$; $g^2(x) = 9x$; $g^3(x) = 27x$; $h^2 = h^3 = h^{500} = h$.

9. a) $f^{-1}(x) = (1/2)(\ln x - 5)$

b) For $x \in \mathbf{R}^+$,

$$(f \circ f^{-1})(x) = f((1/2)(\ln x - 5)) = e^{2((1/2)(\ln x - 5)) + 5} = e^{\ln x - 5 + 5} = e^{\ln x} = x.$$

For $x \in \mathbf{R}$,

$$(f^{-1} \circ f)(x) = f^{-1}(e^{2x+5}) = (1/2)[\ln(e^{2x+5}) - 5] = (1/2)[2x + 5 - 5] = x.$$

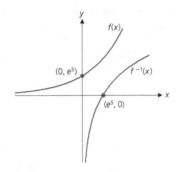

11. f, g invertible $\Rightarrow$ each of f, g is both one-to-one and onto $\Rightarrow g \circ f$ is one-to-one and onto $\Rightarrow g \circ f$ invertible. Since $(g \circ f) \circ (f^{-1} \circ g^{-1}) = 1_C$ and $(f^{-1} \circ g^{-1}) \circ (g \circ f) = 1_A$, it follows that $f^{-1} \circ g^{-1}$ is an inverse of $g \circ f$. By uniqueness of inverses, we have $f^{-1} \circ g^{-1} = (g \circ f)^{-1}$.

13. a) $f^{-1}(-10) = \{-17\}$ $\quad\quad\quad$ $f^{-1}(0) = \{-7, 5/2\}$
$\quad\quad f^{-1}(4) = \{-3, 1/2, 5\}$ $\quad\quad$ $f^{-1}(6) = \{-1, 7\}$
$\quad\quad f^{-1}(7) = \{0, 8\}$ $\quad\quad\quad\quad$ $f^{-1}(8) = \{9\}$

b) (i) $[-12, -8]$ $\quad\quad\quad\quad\quad\quad$ (ii) $[-12, -7] \cup [5/2, 3)$
$\quad\quad$ (iii) $[-9, -3] \cup [1/2, 5]$ $\quad\quad$ (iv) $(-2, 0] \cup (6, 11)$
$\quad\quad$ (v) $[12, 18)$

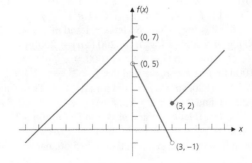

15. $3^2 \cdot 4^3 = 576$ functions

17. **a)** The range of $f = \{2, 3, 4, \ldots\} = \mathbf{Z}^+ - \{1\}$.

b) Since 1 is not in the range of f, the function is not onto.

c) For all $x, y \in \mathbf{Z}^+$, $f(x) = f(y) \Rightarrow x + 1 = y + 1 \Rightarrow x = y$, so f is one-to-one.

d) The range of g is $\mathbf{Z}^+$.   **e)** Since $g(\mathbf{Z}^+) = \mathbf{Z}^+$, the codomain of g, this function is onto.

f) Here $g(1) = 1 = g(2)$, and $1 \neq 2$, so g is not one-to-one.

g) For all $x \in \mathbf{Z}^+$, $(g \circ f)(x) = g(f(x)) = g(x + 1) = \max\{1, (x + 1) - 1\} = \max\{1, x\} = x$, since $x \in \mathbf{Z}^+$. Hence $g \circ f = 1_{\mathbf{Z}^+}$.

h) $(f \circ g)(2) = f(\max\{1, 1\}) = f(1) = 1 + 1 = 2$
$(f \circ g)(3) = f(\max\{1, 2\}) = f(2) = 2 + 1 = 3$
$(f \circ g)(4) = f(\max\{1, 3\}) = f(3) = 3 + 1 = 4$
$(f \circ g)(7) = f(\max\{1, 6\}) = f(6) = 6 + 1 = 7$
$(f \circ g)(12) = f(\max\{1, 11\}) = f(11) = 11 + 1 = 12$
$(f \circ g)(25) = f(\max\{1, 24\}) = f(24) = 24 + 1 = 25$

i) No, because the functions f, g are *not* inverses of each other. The calculations in part (h) may suggest that $f \circ g = 1_{\mathbf{Z}^+}$, since $(f \circ g)(x) = x$ for $x \geq 2$. But we also find that $(f \circ g)(1) = f(\max\{1, 0\}) = f(1) = 2$, so $(f \circ g)(1) \neq 1$, and, consequently, $f \circ g \neq 1_{\mathbf{Z}^+}$.

19. **a)** $a \in f^{-1}(B_1 \cap B_2) \Longleftrightarrow f(a) \in B_1 \cap B_2 \Longleftrightarrow f(a) \in B_1$ and $f(a) \in B_2 \Longleftrightarrow a \in f^{-1}(B_1)$ and $a \in f^{-1}(B_2) \Longleftrightarrow a \in f^{-1}(B_1) \cap f^{-1}(B_2)$

c) $a \in f^{-1}(\overline{B_1}) \Longleftrightarrow f(a) \in \overline{B_1} \Longleftrightarrow f(a) \notin B_1 \Longleftrightarrow a \notin f^{-1}(B_1) \Longleftrightarrow a \in \overline{f^{-1}(B_1)}$

21. **a)** Suppose that $x_1, x_2 \in \mathbf{Z}$ and $f(x_1) = f(x_2)$. Then either $f(x_1), f(x_2)$ are both even or they are both odd. If they are both even, then $f(x_1) = f(x_2) \Rightarrow -2x_1 = -2x_2 \Rightarrow x_1 = x_2$. Otherwise, $f(x_1), f(x_2)$ are both odd and $f(x_1) = f(x_2) \Rightarrow 2x_1 - 1 = 2x_2 - 1 \Rightarrow 2x_1 = 2x_2 \Rightarrow x_1 = x_2$. Consequently, the function f is one-to-one.

To prove that f is an onto function, let $n \in \mathbf{N}$. If n is even, then $(-n/2) \in \mathbf{Z}$ and $(-n/2) < 0$, and $f(-n/2) = -2(-n/2) = n$. For the case where n is odd we find that $(n + 1)/2 \in \mathbf{Z}$ and $(n + 1)/2 > 0$, and $f((n + 1)/2) = 2[(n + 1)/2] - 1 = (n + 1) - 1 = n$. Hence f is onto.

b) $f^{-1} : \mathbf{N} \to \mathbf{Z}$, where

$$f^{-1}(x) = \begin{cases} \left(\frac{1}{2}\right)(x + 1), & x = 1, 3, 5, 7, \ldots \\ -x/2, & x = 0, 2, 4, 6, \ldots \end{cases}$$

23. **a)** For all $n \in \mathbf{N}$, $(g \circ f)(n) = (h \circ f)(n) = (k \circ f)(n) = n$.

b) The results in part (a) do not contradict Theorem 5.7. For although $g \circ f = h \circ f = k \circ f = 1_{\mathbf{N}}$, we note that

(i) $(f \circ g)(1) = f(\lfloor 1/3 \rfloor) = f(0) = 3 \cdot 0 = 0 \neq 1$, so $f \circ g \neq 1_{\mathbf{N}}$;
(ii) $(f \circ h)(1) = f(\lfloor 2/3 \rfloor) = f(0) = 3 \cdot 0 = 0 \neq 1$, so $f \circ h \neq 1_{\mathbf{N}}$; and
(iii) $(f \circ k)(1) = f(\lfloor 3/3 \rfloor) = f(1) = 3 \cdot 1 = 3 \neq 1$, so $f \circ k \neq 1_{\mathbf{N}}$.

Consequently, none of g, h, and k is the inverse of f. (After all, since f is *not* onto, it is *not* invertible.)

Section 5.7 – p. 293

1. **a)** $f \in O(n)$ **b)** $f \in O(1)$ **c)** $f \in O(n^3)$ **d)** $f \in O(n^2)$
e) $f \in O(n^3)$ **f)** $f \in O(n^2)$ **g)** $f \in O(n^2)$

3. **a)** For all $n \in \mathbf{Z}^+$, $0 \leq \log_2 n < n$. So let $k = 1$ and $m = 200$ in Definition 5.23. Then $|f(n)| = 100 \log_2 n = 200 \left(\frac{1}{2} \log_2 n\right) < 200 \left(\frac{1}{2}n\right) = 200|g(n)|$, so $f \in O(g)$.

b) For $n = 6$, $2^n = 64 < 3096 = 4096 - 1000 = 2^{12} - 1000 = 2^{2n} - 1000$. Assuming that $2^k < 2^{2k} - 1000$ for $n = k \geq 6$, we find that $2 < 2^2 \Rightarrow 2(2^k) < 2^2(2^{2k} - 1000) < 2^2 2^{2k} - 1000$, or $2^{k+1} < 2^{2(k+1)} - 1000$, so $f(n) < g(n)$ for all $n \geq 6$. Therefore, with $k = 6$ and $m = 1$ in Definition 5.23, we find that for $n \geq k$, $|f(n)| \leq m|g(n)|$ and $f \in O(g)$.

5. To show that $f \in O(g)$, let $k = 1$ and $m = 4$ in Definition 5.23. Then for all $n \geq k$, $|f(n)| = n^2 + n \leq n^2 + n^2 = 2n^2 \leq 2n^3 = 4((1/2)n^3) = 4|g(n)|$, and f is dominated by g. To show that $g \notin O(f)$, we follow the idea given in Example 5.66, namely, that

$$\forall m \in \mathbf{R}^+ \ \forall k \in \mathbf{Z}^+ \ \exists n \in \mathbf{Z}^+ \ [(n \geq k) \wedge (|g(n)| > m|f(n)|)].$$

So no matter what the values of m and k are, choose $n > \max\{4m, k\}$. Then $|g(n)| = \left(\frac{1}{2}\right) n^3 > \left(\frac{1}{2}\right)(4m)n^2 = m(2n^2) \geq m(n^2 + n) = m|f(n)|$, so $g \notin O(f)$. Alternatively, if $g \in O(f)$, then $\exists m \in \mathbf{R}^+$ $\exists k \in \mathbf{Z}^+$ $\forall n \in \mathbf{Z}^+$ $|\left(\frac{1}{2}\right) n^3| \leq m|n^2 + n|$, or $\left(\frac{1}{2}\right) n^2 \leq m(n + 1)$. Then $\frac{n^2}{2(n+1)} \leq m \Rightarrow 0 < \frac{n^2}{4n} < \frac{n^2}{2(n+1)} \leq m \Rightarrow \frac{n}{4} \leq m$, a contradiction since n is variable and m constant.

7. For all $n \geq 1$, $\log_2 n \leq n$, so with $k = 1$ and $m = 1$ in Definition 5.23, we have $|g(n)| = \log_2 n \leq n = m \cdot n = m|f(n)|$. Hence $g \in O(f)$. To show that $f \notin O(g)$, we first observe that $\lim_{n \to \infty} \frac{n}{\log_2 n} = +\infty$. (This can be established by using L'Hospital's Rule from the calculus.) Since $\lim_{n \to \infty} \frac{n}{\log_2 n} = +\infty$, we find that for every $m \in \mathbf{R}^+$ and $k \in \mathbf{Z}^+$, there in an $n \in \mathbf{Z}^+$ such that $\frac{n}{\log_2 n} > m$, or $|f(n)| = n > m \log_2 n = m|g(n)|$. Hence $f \notin O(g)$.

9. Since $f \in O(g)$, there exists $m \in \mathbf{R}^+$, $k \in \mathbf{Z}^+$ such that $|f(n)| \leq m|g(n)|$ for all $n \geq k$. But then $|f(n)| \leq [m/|c|]|cg(n)|$ for all $n \geq k$, so $f \in O(cg)$.

11. **a)** For all $n \geq 1$, $f(n) = 5n^2 + 3n > n^2 = g(n)$. So with $M = 1$ and $k = 1$, we have $|f(n)| \geq M|g(n)|$ for all $n \geq k$ and it follows that $f \in \Omega(g)$.

   c) For all $n \geq 1$, $f(n) = 5n^2 + 3n > n = h(n)$. With $M = 1$ and $k = 1$, we have $|f(n)| \geq M|h(n)|$ for all $n \geq k$ and so $f \in \Omega(h)$.

   d) Suppose that $h \in \Omega(f)$. If so, there exist $M \in \mathbf{R}^+$ and $k \in \mathbf{Z}^+$ with $n = |h(n)| \geq M|f(n)| = M(5n^2 + 3n)$ for all $n \geq k$. Then $0 < M \leq n/(5n^2 + 3n) = 1/(5n + 3)$. But how can M be a positive *constant* while $1/(5n + 3)$ approaches 0 as n (a *variable*) gets larger? From this contradiction it follows that $h \notin \Omega(f)$.

13. **a)** For $n \geq 1$, $f(n) = \sum_{i=1}^{n} i = n(n+1)/2 = (n^2/2) + (n/2) > (n^2/2)$. With $k = 1$ and $M = 1/2$, we have $|f(n)| \geq M|n^2|$ for all $n \geq k$. Hence $f \in \Omega(n^2)$.

   b) $\sum_{i=1}^{n} i^2 = 1^2 + 2^2 + \cdots + n^2 > \lceil n/2 \rceil^2 + \cdots + n^2 > \lceil n/2 \rceil^2 + \cdots + \lceil n/2 \rceil^2 = \lceil (n+1)/2 \rceil \lceil n/2 \rceil^2 > n^3/8$. With $k = 1$ and $M = 1/8$, we have $|g(n)| \geq M|n^3|$ for all $n \geq k$. Hence $g \in \Omega(n^3)$.

   Alternatively, for $n \geq 1$, $g(n) = \sum_{i=1}^{n} i^2 = n(n+1)(2n+1)/6 = (2n^3 + 3n^2 + n)/6 > n^3/6$. With $k = 1$ and $M = 1/6$, we find that $|g(n)| \geq M|n^3|$ for all $n \geq k$ — so $g \in \Omega(n^3)$.

   c) $\sum_{i=1}^{n} i^t = 1^t + 2^t + \cdots + n^t > \lceil n/2 \rceil^t + \cdots + n^t > \lceil n/2 \rceil^t + \cdots + \lceil n/2 \rceil^t = \lceil (n+1)/2 \rceil \lceil n/2 \rceil^t > (n/2)^{t+1}$. With $k = 1$ and $M = (1/2)^{t+1}$, we have $|h(n)| \geq M|n^{t+1}|$ for all $n \geq k$. Hence $h \in \Omega(n^{t+1})$.

15. *Proof*: $f \in \Theta(g) \Rightarrow f \in \Omega(g)$ and $f \in O(g)$ (from Exercise 14 of this section) $\Rightarrow g \in O(f)$ and $g \in \Omega(f)$ (from Exercise 12 of this section) $\Rightarrow g \in \Theta(f)$.

Section 5.8 – p. 300

1. **a)** $f \in O(n^2)$ **b)** $f \in O(n^3)$ **c)** $f \in O(n^2)$ **d)** $f \in O(\log_2 n)$
   e) $f \in O(n \log_2 n)$

5. **a)** Here there are five additions and 10 multiplications.
   b) For the general case there are n additions and $2n$ multiplications.

7. For $n = 1$, we find that $a_1 = 0 = \lfloor 0 \rfloor = \lfloor \log_2 1 \rfloor$, so the result is true in this first case. Now assume the result true for all $n = 1, 2, 3, \ldots, k$, where $k \geq 1$, and consider the cases for $n = k + 1$.

   (i) $n = k + 1 = 2^m$, where $m \in \mathbf{Z}^+$: Here $a_n = 1 + a_{\lfloor n/2 \rfloor} = 1 + a_{2^{m-1}} = 1 + \lfloor \log_2 2^{m-1} \rfloor = 1 + (m - 1) = m = \lfloor \log_2 2^m \rfloor = \lfloor \log_2 n \rfloor$; and

   (ii) $n = k + 1 = 2^m + r$, where $m \in \mathbf{Z}^+$ and $0 < r < 2^m$: Here $2^m < n < 2^{m+1}$, so we have
   (1) $2^{m-1} < (n/2) < 2^m$;
   (2) $2^{m-1} = \lfloor 2^{m-1} \rfloor \leq \lfloor n/2 \rfloor < \lfloor 2^m \rfloor = 2^m$; and
   (3) $m - 1 = \log_2 2^{m-1} \leq \log_2 \lfloor n/2 \rfloor < \log_2 2^m = m$.

   Consequently, $\lfloor \log_2 \lfloor n/2 \rfloor \rfloor = m - 1$ and $a_n = 1 + a_{\lfloor n/2 \rfloor} = 1 + \lfloor \log_2 \lfloor n/2 \rfloor \rfloor = 1 + (m - 1) = m = \lfloor \log_2 n \rfloor$. Therefore it follows from the alternative form of the Principle of Mathematical Induction that $a_n = \lfloor \log_2 n \rfloor$ for all $n \in \mathbf{Z}^+$.

9. $(5/8)n + (3/8)$

11. a) procedure *LocateRepeat* (n: positive integer;
        $a_1, a_2, a_3, \ldots, a_n$: integers)
  begin
    location := 0
    i := 2
    while $j \leq n$ **and** *location* = 0 **do**
      begin
        j := 1
        while $j < i$ **and** *location* = 0 **do**
          if $a_j = a_i$ **then** *location* := i
          else j := j + 1
        i := i + 1
    end
  end {*location* is the subscript of the first array entry that
       repeats a previous array entry; *location* is 0 if the array
       contains n distinct integers.}

b) $O(n^2)$

Supplementary Exercises–p. 305

1. a) If either A or B is $\emptyset$, then $A \times B = \emptyset = A \cap B$ and the result is true. For A, B nonempty we find that:

$(x, y) \in (A \times B) \cap (B \times A) \Rightarrow (x, y) \in A \times B$ and $(x, y) \in B \times A \Rightarrow (x \in A \text{ and } y \in B)$ and $(x \in B \text{ and } y \in A) \Rightarrow x \in A \cap B$ and $y \in A \cap B \Rightarrow (x, y) \in (A \cap B) \times (A \cap B)$; and

$(x, y) \in (A \cap B) \times (A \cap B) \Rightarrow (x \in A \text{ and } x \in B)$ and $(y \in A \text{ and } y \in B) \Rightarrow (x, y) \in A \times B$ and $(x, y) \in B \times A \Rightarrow (x, y) \in (A \times B) \cap (B \times A)$.

Consequently, $(A \times B) \cap (B \times A) = (A \cap B) \times (A \cap B)$.

b) If either A or B is $\emptyset$, then $A \times B = \emptyset = B \times A$ and the result follows. If not, let $(x, y) \in (A \times B) \cup (B \times A)$. Then

$(x, y) \in (A \times B) \cup (B \times A) \Rightarrow (x, y) \in A \times B$ or $(x, y) \in (B \times A) \Rightarrow (x \in A \text{ and } y \in B)$ or $(x \in B \text{ and } y \in A) \Rightarrow (x \in A \text{ or } x \in B)$ and $(y \in A \text{ or } y \in B) \Rightarrow x, y \in A \cup B \Rightarrow (x, y) \in (A \cup B) \times (A \cup B)$.

3. a) $f(1) = f(1 \cdot 1) = 1 \cdot f(1) + 1 \cdot f(1)$, so $f(1) = 0$.   **b)** $f(0) = 0$

c) *Proof* (by Mathematical Induction): When $a = 0$ the result is true, so consider $a \neq 0$. For $n = 1$, $f(a^n) = f(a) = 1 \cdot a^0 \cdot f(a) = na^{n-1}f(a)$, so the result follows in this first case, and this establishes our basis step. Assume the result true for $n = k \ (\geq 1)$ — that is, $f(a^k) = ka^{k-1}f(a)$. For $n = k + 1$ we have $f(a^{k+1}) = f(a \cdot a^k) = af(a^k) + a^k f(a) = aka^{k-1}f(a) + a^k f(a) = ka^k f(a) + a^k f(a) = (k + 1)a^k f(a)$. Consequently, the truth of the result for $n = k + 1$ follows from the truth of the result for $n = k$. So by the Principle of Mathematical Induction the result is true for all $n \in \mathbf{Z}^+$.

5. $(x, y) \in (A \cap B) \times (C \cap D) \Longleftrightarrow x \in A \cap B, y \in C \cap D \Longleftrightarrow (x \in A, y \in C)$ and $(x \in B, y \in D) \Longleftrightarrow (x, y) \in A \times C$ and $(x, y) \in B \times D \Longleftrightarrow (x, y) \in (A \times C) \cap (B \times D)$

7. $x = 1/\sqrt{2}$ and $x = \sqrt{3/2}$

9. b) *Conjecture:* For $n \in \mathbf{Z}^+$, $f^n(x) = a^n(x + b) - b$. *Proof* (by Mathematical Induction): The formula is true for $n = 1$ — by the definition of $f(x)$. Hence we have our basis step. Assume the formula true for $n = k \ (\geq 1)$ — that is, $f^k(x) = a^k(x + b) - b$. Now consider $n = k + 1$. We find that $f^{k+1}(x) = f(f^k(x)) = f(a^k(x + b) - b) = a[(a^k(x + b) - b) + b] - b = a^{k+1}(x + b) - b$. Since the truth of the formula at $n = k$ implies the truth of the formula at $n = k + 1$, it follows that the formula is valid for all $n \in \mathbf{Z}^+$ — by the Principle of Mathematical Induction.

11. a) $(7!)/[2(7^5)]$

13. For $1 \leq i \leq 10$, let x_i be the number of letters typed on day i. Then $x_1 + x_2 + x_3 + \cdots + x_8 + x_9 + x_{10} = 84$, or $x_3 + \cdots + x_8 = 54$. Suppose that

$x_1 + x_2 + x_3 < 25$, $x_2 + x_3 + x_4 < 25$, $\ldots$, $x_8 + x_9 + x_{10} < 25$. Then
$x_1 + 2x_2 + 3(x_3 + \cdots + x_8) + 2x_9 + x_{10} < 8(25) = 200$, or $3(x_3 + \cdots + x_8) < 160$.
Consequently, we obtain the contradiction $54 = x_3 + \cdots + x_8 < \frac{160}{3} = 53\frac{1}{3}$.

15. For $\prod_{k=1}^{n}(k - i_k)$ to be odd, $(k - i_k)$ must be odd for all $1 \le k \le n$; that is, one of k, i_k must be even and the other odd. Since n is odd, $n = 2m + 1$ and in the list $1, 2, \ldots, n$ there are m even integers and $m + 1$ odd integers. Let $1, 3, 5, \ldots, n$, be the pigeons and $i_1, i_3, i_5, \ldots, i_n$ the pigeonholes. At most m of the pigeonholes can be even integers, so $(k - i_k)$ must be even for at least one $k = 1, 3, 5, \ldots, n$. Consequently, $\prod_{k=1}^{n}(k - i_k)$ is even.

17. Let the n distinct objects be $x_1, x_2, \ldots, x_n$. Place x_n in a container. Now there are two *distinct* containers. For each of $x_1, x_2, \ldots, x_{n-1}$ there are two choices, and this gives 2^{n-1} distributions. Among these there is one where $x_1, x_2, \ldots, x_{n-1}$ are in the container with x_n, so we remove this distribution and find $S(n, 2) = 2^{n-1} - 1$.

19. **a)** and **b)** $m!S(n, m)$

21. Fix $m = 1$. For $n = 1$ the result is true. Assume $f \circ f^k = f^k \circ f$ and consider $f \circ f^{k+1}$.
$f \circ f^{k+1} = f \circ (f \circ f^k) = f \circ (f^k \circ f) = (f \circ f^k) \circ f = f^{k+1} \circ f$. Hence $f \circ f^n = f^n \circ f$
for all $n \in \mathbf{Z}^+$. Now assume that for some $t \ge 1$, $f^t \circ f^n = f^n \circ f^t$. Then
$f^{t+1} \circ f^n = (f \circ f^t) \circ f^n = f \circ (f^t \circ f^n) = f \circ (f^n \circ f^t) = (f \circ f^n) \circ f^t =$
$(f^n \circ f) \circ f^t = f^n \circ (f \circ f^t) = f^n \circ f^{t+1}$, so $f^m \circ f^n = f^n \circ f^m$ for all $m, n \in \mathbf{Z}^+$.

23. *Proof*: Let $a \in A$. Then $f(a) = g(f(f(a))) = f(g(f(f(f(a))))) = f(g \circ f^3(a))$. From
$f(a) = g(f(f(a)))$ we have $f^2(a) = (f \circ f)(a) = f(g(f(f(a))))$. So $f(a) =$
$f(g \circ f^3(a)) = f(g(f(f(f(a))))) = f^2(f(a)) = f^2(g(f^2(a))) = f(f(g(f(f(a))))) =$
$f(g(f(a))) = g(a)$. Consequently, $f = g$.

25. **a)** Note that $2 = 2^1$, $16 = 2^4$, $128 = 2^7$, $1024 = 2^{10}$, $8192 = 2^{13}$, and $65536 = 2^{16}$. Consider the exponents on 2. If four numbers are selected from $\{1, 4, 7, 10, 13, 16\}$, there is at least one pair whose sum is 17. Hence if four numbers are selected from S, there are two numbers whose product is $2^{17} = 131072$.
b) Let $a, b, c, d, n \in \mathbf{Z}^+$. Let $S = \{b^a, b^{a+d}, b^{a+2d}, \ldots, b^{a+nd}\}$. If $\lceil \frac{n}{2} \rceil + 1$ numbers are selected from S then there are at least two of them whose product is b^{2a+nd}.

27. $f \circ g = \{(x, z), (y, y), (z, x)\}$; $g \circ f = \{(x, x), (y, z), (z, y)\}$; $f^{-1} = \{(x, z), (y, x), (z, y)\}$;
$g^{-1} = \{(x, y), (y, x), (z, z)\}$; $(g \circ f)^{-1} = \{(x, x), (y, z), (z, y)\} = f^{-1} \circ g^{-1}$;
$g^{-1} \circ f^{-1} = \{(x, z), (y, y), (z, x)\}$.

29. $2^3 \cdot 2^2 \cdot 3^5 = 7776$ functions

31. **a)** $(\pi \circ \sigma)(x) = (\sigma \circ \pi)(x) = x$ **b)** $\pi^n(x) = x - n$; $\sigma^n(x) = x + n$ $(n \ge 2)$
c) $\pi^{-n}(x) = x + n$; $\sigma^{-n}(x) = x - n$ $(n \ge 2)$

33. **a)** $S(8, 4)$ **b)** $S(n, m)$

35. **a)** Let $m = 1$ and $k = 1$. Then for all $n \ge k$, $|f(n)| \le 2 < 3 \le |g(n)| = m|g(n)|$, so $f \in O(g)$.

37. First note that if $\log_a n = r$, then $n = a^r$ and $\log_b n = \log_b(a^r) = r \log_b a = (\log_b a)(\log_a n)$.
Now let $m = (\log_b a)$ and $k = 1$. Then for all $n \ge k$, $|g(n)| = \log_b n = (\log_b a)(\log_a n) = m|f(n)|$, so $g \in O(f)$. Finally, with $m = (\log_b a)^{-1} = \log_a b$ and $k = 1$, we find that for all $n \ge k$, $|f(n)| = \log_a n = (\log_a b)(\log_b n) = m|g(n)|$. Hence $f \in O(g)$.

Chapter 6
Languages: Finite State Machines

Section 6.1 – p. 317

1. **a)** 25; 125   **b)** 3906   **3.** 12   **5.** 780

7. **a)** $\{00, 11, 000, 111, 0000, 1111\}$ **b)** $\{0, 1\}$
c) $\Sigma^* - \{\lambda, 00, 11, 000, 111, 0000, 1111\}$ **d)** $\{0, 1, 00, 11\}$
e) Σ^* **f)** $\Sigma^* - \{0, 1, 00, 11\} = \{\lambda, 01, 10\} \cup \{w \| \|w\| \ge 3\}$

9. **a)** $x \in AC \Rightarrow x = ac$, for some $a \in A$, $c \in C \Rightarrow x \in BD$, since $A \subseteq B$, $C \subseteq D$.
b) If $A\emptyset \ne \emptyset$, let $x \in A\emptyset$. $x \in A\emptyset \Rightarrow x = yz$, for some $y \in A$, $z \in \emptyset$. But $z \in \emptyset$ is impossible. Hence $A\emptyset = \emptyset$. [In like manner, $\emptyset A = \emptyset$.]

11. For any alphabet Σ, let $B \subseteq \Sigma$. Then, if $A = B^*$, it follows from part (f) of Theorem 6.2 that $A^* = (B^*)^* = B^* = A$.

13. a) Here A^* consists of all strings x of even length where if $x \neq \lambda$, then x starts with 0 and ends with 1, and the symbols (0 and 1) alternate.

b) In this case A^* contains precisely those strings made up of $3n$ 0's, for $n \in \mathbf{N}$.

c) Here a string $x \in A^*$ if (and only if)
  (i) x is a string of n 0's, for $n \in \mathbf{N}$; or
  (ii) x is a string that starts and ends with 0, and has at least one 1 and at least two 0's between any two 1's.

15. Let Σ be an alphabet with $\emptyset \neq A \subseteq \Sigma^*$. If $|A| = 1$ and $x \in A$, then $xx = x$ since $A^2 = A$. But $\|xx\| = 2\|x\| = \|x\| \Rightarrow \|x\| = 0 \Rightarrow x = \lambda$. If $|A| > 1$, let $x \in A$ where $\|x\| > 0$ but $\|x\|$ is minimal. Then $x \in A^2 \Rightarrow x = yz$, for $y, z \in A$. Since $\|x\| = \|y\| + \|z\|$, if $\|y\|, \|z\| > 0$, then one of y, z is in A with length smaller than $\|x\|$. Consequently, one of $\|y\|$ or $\|z\|$ is 0, so $\lambda \in A$.

17. If $A = A^2$, then it follows by the Principle of Mathematical Induction that $A = A^n$ for all $n \in \mathbf{Z}^+$. Hence $A = A^+$. By Exercise 15, $A = A^2 \Rightarrow \lambda \in A$. Hence $A = A^*$.

19. By Definition 6.11, $AB = \{ab | a \in A, b \in B\}$, and since it is possible to have $a_1 b_1 = a_2 b_2$ with $a_1, a_2 \in A, a_1 \neq a_2$, and $b_1, b_2 \in B, b_1 \neq b_2$, it follows that $|AB| \leq |A \times B| = |A| \|B\|$.

21. a) The words 001 and 011 have length 3 and are in A. The words 00011 and 00111 have length 5 and they are also in A.

b) From step (1) we know that $1 \in A$. Then by applying step (2) three times we get
  (i) $1 \in A \Rightarrow 011 \in A$;
  (ii) $011 \in A \Rightarrow 00111 \in A$; and
  (iii) $00111 \in A \Rightarrow 0001111 \in A$.

c) If 00001111 were in A, then from step (2) we see that this word would have to be generated from 000111 (in A). Likewise, 000111 in $A \Rightarrow 0011$ is in $A \Rightarrow 01$ is in A. However, there are no words in A of length 2 — in fact, there are no words of even length in A.

23. a)

Steps	Reasons
1) ( ) is in A.	Part (1) of the recursive definition
2) (( )) is in A.	Step (1) and part (2-ii) of the definition
3) (( ))( ) is in A.	Steps (1) and (2) and part (2-i) of the definition

b)

Steps	Reasons
1) ( ) is in A.	Part (1) of the recursive definition
2) (( )) is in A.	Step (1) and part (2-ii) of the definition
3) (( ))( ) is in A.	Steps (1) and (2) and part (2-i) of the definition
4) (( ))( )( ) is in A.	Steps (1) and (3) and part (2-i) of the definition

25. Length 3: $\binom{3}{0} + \binom{2}{1} = 3$ Length 4: $\binom{4}{0} + \binom{3}{1} + \binom{2}{2} = 5$

Length 5: $\binom{5}{0} + \binom{4}{1} + \binom{3}{2} = 8$ Length 6: $\binom{6}{0} + \binom{5}{1} + \binom{4}{2} + \binom{3}{3} = 13$ [Here the summand $\binom{6}{0}$ counts the strings where there are no 0's; the summand $\binom{5}{1}$ counts the strings where we arrange the symbols 1, 1, 1, 1, 00; the summand $\binom{4}{2}$ is for the arrangements of 1, 1, 00, 00; and the summand $\binom{3}{3}$ counts the arrangements of 00, 00, 00.]

27. A: (1) $\lambda \in A$
      (2) If $a \in A$, then $0a0, 0a1, 1a0, 1a1 \in A$.
  B: (1) $0, 1 \in B$.
      (2) If $a \in B$, then $0a0, 0a1, 1a0, 1a1 \in B$.

Section 6.2 – p. 324

1. a) 0010101; s_1 **b)** 0000000; s_1 **c)** 001000000; s_0

3. a) 010110   **b)**

5. a) 010000; s_2 **b)** (s_1) 100000; s_2 **c)**

(s_2) 000000; s_2

(s_3) 110010; s_2

	ν		ω	
	0	1	0	1
s_0	s_0	s_1	0	0
s_1	s_1	s_2	1	1
s_2	s_2	s_2	0	0
s_3	s_0	s_3	0	1
s_4	s_2	s_3	0	1

d) s_1 **e)** $x = 101$ (unique)

7. a) (i) 15   (ii) 3^{15} (iii) 2^{15} **b)** 6^{15}

9. a)

	ν		ω	
	0	1	0	1
s_0	s_4	s_1	0	0
s_1	s_3	s_2	0	0
s_2	s_3	s_2	0	1
s_3	s_3	s_3	0	0
s_4	s_5	s_3	0	0
s_5	s_5	s_3	1	0

b) There are only two possibilities: $x = 1111$ or $x = 0000$.

c) $A = \{111\}\{1\}^* \cup \{000\}\{0\}^*$

d) Here $A = \{11111\}\{1\}^* \cup \{00000\}\{0\}^*$.

Section 6.3 – p. 332

1. a)   **b)**

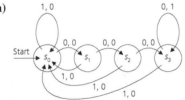 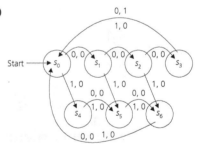

3.

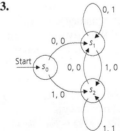

5. b) (i) 011   (ii) 0101   (iii) 00001

c) The machine outputs a 0 followed by the first $n - 1$ symbols of the n symbol input string x. Hence the machine is a unit delay.

d) This machine performs the same tasks as the one in Fig. 6.13 (but has only two states).

7. a) The transient states are s_0, s_1. State s_4 is a sink state. $\{s_1, s_2, s_3, s_4, s_5\}$, $\{s_4\}$, and $\{s_2, s_3, s_5\}$ (with the corresponding restrictions on the given function ν) constitute submachines. The strongly connected submachines are $\{s_4\}$ and $\{s_2, s_3, s_5\}$.

b) States s_2, s_3 are transient. The only sink state is s_4. The set $\{s_0, s_1, s_3, s_4\}$ provides the states for a submachine; $\{s_0, s_1\}$ and $\{s_4\}$ provide strongly connected submachines.

1. a) True   **b)** False   **c)** True   **d)** True   **e)** True   **f)** True

3. Let $x \in \Sigma$ and $A = \{x\}$. Then $A^2 = \{xx\}$ and $(A^2)^* = \{\lambda, x^2, x^4, \ldots\}$. $A^* = \{\lambda, x, x^2, x^3, \ldots\}$ and $(A^*)^2 = A^*$, so $(A^2)^* \neq (A^*)^2$.

5. $\mathbb{O}_{02} = \{1, 00\}^*\{0\}$ $\mathbb{O}_{22} = \{0\}\{1, 00\}^*\{0\}$ $\mathbb{O}_{11} = \emptyset$
$\mathbb{O}_{00} = \{1, 00\}^* - \{\lambda\}$ $\mathbb{O}_{10} = \{1\}\{1, 00\}^* \cup \{10\}\{1, 00\}^*$

7. a) By the pigeonhole principle there is a first state s that is encountered twice. Let y be the output string that resulted since s was first encountered, until we reach this state a second time. Then from that point on the output is $yyy \ldots$.

b) n **c)** n

9.
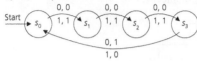

11. a)

	ν		ω	
	0	1	0	1
(s_0, s_3)	(s_0, s_4)	(s_1, s_3)	1	1
(s_0, s_4)	(s_0, s_3)	(s_1, s_4)	0	1
(s_1, s_3)	(s_1, s_3)	(s_2, s_3)	1	1
(s_1, s_4)	(s_1, s_4)	(s_2, s_4)	1	1
(s_2, s_3)	(s_2, s_3)	(s_0, s_4)	1	1
(s_2, s_4)	(s_2, s_4)	(s_0, s_3)	1	0

b) $\omega((s_0, s_3), 1101) = 1111$; M_1 is in state s_0, and M_2 is in state s_4.

Chapter 7
Relations: The Second Time Around

Section 7.1 – p. 343

1. a) $\{(1, 1), (2, 2), (3, 3), (4, 4), (1, 2), (2, 1), (2, 3), (3, 2)\}$

b) $\{(1, 1), (2, 2), (3, 3), (4, 4), (1, 2)\}$ **c)** $\{(1, 1), (2, 2), (1, 2), (2, 1)\}$

3. a) Let $f_1, f_2, f_3 \in \mathcal{F}$ with $f_1(n) = n + 1$, $f_2(n) = 5n$, and $f_3(n) = 4n + 1/n$.

b) Let $g_1, g_2, g_3 \in \mathcal{F}$ with $g_1(n) = 3$, $g_2(n) = 1/n$, and $g_3(n) = \sin n$.

5. a) Reflexive, antisymmetric, transitive   **b)** Transitive

c) Reflexive, symmetric, transitive   **d)** Symmetric   **e)** Symmetric

f) Reflexive, symmetric, transitive   **g)** Reflexive, symmetric   **h)** Reflexive, transitive

7. a) For all $x \in A$, $(x, x) \in \mathcal{R}_1, \mathcal{R}_2$, so $(x, x) \in \mathcal{R}_1 \cap \mathcal{R}_2$ and $\mathcal{R}_1 \cap \mathcal{R}_2$ is reflexive.

b) (i) $(x, y) \in \mathcal{R}_1 \cap \mathcal{R}_2 \Rightarrow (x, y) \in \mathcal{R}_1, \mathcal{R}_2 \Rightarrow (y, x) \in \mathcal{R}_1, \mathcal{R}_2 \Rightarrow (y, x) \in \mathcal{R}_1 \cap \mathcal{R}_2$ and $\mathcal{R}_1 \cap \mathcal{R}_2$ is symmetric.

(ii) $(x, y), (y, x) \in \mathcal{R}_1 \cap \mathcal{R}_2 \Rightarrow (x, y), (y, x) \in \mathcal{R}_1, \mathcal{R}_2$. By the antisymmetry of $\mathcal{R}_1$ (or $\mathcal{R}_2$), $x = y$ and $\mathcal{R}_1 \cap \mathcal{R}_2$ is antisymmetric.

(iii) $(x, y), (y, z) \in \mathcal{R}_1 \cap \mathcal{R}_2 \Rightarrow (x, y), (y, z) \in \mathcal{R}_1, \mathcal{R}_2 \Rightarrow (x, z) \in \mathcal{R}_1, \mathcal{R}_2$ (transitive property) $\Rightarrow (x, z) \in \mathcal{R}_1 \cap \mathcal{R}_2$, so $\mathcal{R}_1 \cap \mathcal{R}_2$ is transitive.

9. a) False: let $A = \{1, 2\}$ and $\mathcal{R} = \{(1, 2), (2, 1)\}$.

b) (i) Reflexive: true

(ii) Symmetric: false. Let $A = \{1, 2\}$, $\mathcal{R}_1 = \{(1, 1)\}$, and $\mathcal{R}_2 = \{(1, 1), (1, 2)\}$.

(iii) Antisymmetric and transitive: false. Let $A = \{1, 2\}$, $\mathcal{R}_1 = \{(1, 2)\}$, and $\mathcal{R}_2 = \{(1, 2), (2, 1)\}$.

d) True.

11. a) $\binom{2 + 2 - 1}{2}\binom{2 + 2 - 1}{2} = \binom{3}{2}\binom{3}{2} = 9$ **b)** 18   **c)** $\binom{4 + 2 - 1}{2}\binom{2 + 2 - 1}{2} = \binom{5}{2}\binom{3}{2} = 30$

d) 60   **e)** 81   **f)** 972

13. There may exist an element $a \in A$ such that for all $b \in A$, neither (a, b) nor (b, a) is in $\mathcal{R}$.

15. $r - n$ counts the elements in $\mathcal{R}$ of the form (a, b), $a \neq b$. Since $\mathcal{R}$ is symmetric, $r - n$ is even.

17. a) $\binom{7}{4}\binom{21}{0} + \binom{7}{2}\binom{21}{1} + \binom{7}{0}\binom{21}{2}$ **b)** $\binom{7}{5}\binom{21}{0} + \binom{7}{3}\binom{21}{1} + \binom{7}{1}\binom{21}{2}$

d) $\binom{7}{6}\binom{21}{1} + \binom{7}{4}\binom{21}{2} + \binom{7}{2}\binom{21}{3} + \binom{7}{0}\binom{21}{4}$

Section 7.2 – p. 354

1. $\mathcal{R} \circ \mathcal{S} = \{(1, 3), (1, 4)\}$; $\mathcal{S} \circ \mathcal{R} = \{(1, 2), (1, 3), (1, 4), (2, 4)\}$;
$\mathcal{R}^2 = \mathcal{R}^3 = \{(1, 4), (2, 4), (4, 4)\}$; $\mathcal{S}^2 = \mathcal{S}^3 = \{(1, 1), (1, 2), (1, 3), (1, 4)\}$

3. $(a, d) \in (\mathcal{R}_1 \circ \mathcal{R}_2) \circ \mathcal{R}_3 \Rightarrow (a, c) \in \mathcal{R}_1 \circ \mathcal{R}_2, (c, d) \in \mathcal{R}_3$ for some $c \in C \Rightarrow (a, b) \in \mathcal{R}_1$,
$(b, c) \in \mathcal{R}_2, (c, d) \in \mathcal{R}_3$ for some $b \in B, c \in C \Rightarrow (a, b) \in \mathcal{R}_1, (b, d) \in \mathcal{R}_2 \circ \mathcal{R}_3 \Rightarrow$
$(a, d) \in \mathcal{R}_1 \circ (\mathcal{R}_2 \circ \mathcal{R}_3)$, and $(\mathcal{R}_1 \circ \mathcal{R}_2) \circ \mathcal{R}_3 \subseteq \mathcal{R}_1 \circ (\mathcal{R}_2 \circ \mathcal{R}_3)$

5. $\mathcal{R}_1 \circ (\mathcal{R}_2 \cap \mathcal{R}_3) = \mathcal{R}_1 \circ \{(m, 3), (m, 4)\} = \{(1, 3), (1, 4)\}$
$(\mathcal{R}_1 \circ \mathcal{R}_2) \cap (\mathcal{R}_1 \circ \mathcal{R}_3) = \{(1, 3), (1, 4)\} \cap \{(1, 3), (1, 4)\} = \{(1, 3), (1, 4)\}$

7. This follows by the pigeonhole principle. Here the pigeons are the $2^{n^2} + 1$ integers between 0 and 2^{n^2}, inclusive, and the pigeonholes are the 2^{n^2} relations on A.

9. 2^{21}

11. Consider the entry in the ith row and jth column of $M(\mathcal{R}_1 \circ \mathcal{R}_2)$. If this entry is a 1, then there exists $b_k \in B$ where $1 \leq k \leq n$ and $(a_i, b_k) \in \mathcal{R}_1, (b_k, c_j) \in \mathcal{R}_2$. Consequently, the entry in the ith row and kth column of $M(\mathcal{R}_1)$ is 1, and the entry in the kth row and jth column of $M(\mathcal{R}_2)$ is 1. This results in a 1 in the ith row and jth column in the product $M(\mathcal{R}_1) \cdot M(\mathcal{R}_2)$.

Should the entry in row i and column j of $M(\mathcal{R}_1 \circ \mathcal{R}_2)$ be 0, then for each b_k, where $1 \leq k \leq n$, either $(a_i, b_k) \notin \mathcal{R}_1$ or $(b_k, c_j) \notin \mathcal{R}_2$. This means that in the matrices $M(\mathcal{R}_1)$, $M(\mathcal{R}_2)$, if the entry in the ith row and kth column of $M(\mathcal{R}_1)$ is 1, then the entry in the kth row and jth column of $M(\mathcal{R}_2)$ is 0. Hence the entry in the ith row and jth column of $M(\mathcal{R}_1) \cdot M(\mathcal{R}_2)$ is 0.

13. d) Let s_{xy} be the entry in row (x) and column (y) of M. Then s_{yx} appears in row (x) and column (y) of M^{tr}. $\mathcal{R}$ is antisymmetric $\iff (s_{xy} = s_{yx} = 1 \Rightarrow x = y) \iff M \cap M^{\mathrm{tr}} \leq I_n$.

15.

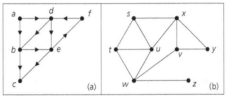

(a)   (b)

17. (i) $\mathcal{R} = \{(a, b), (b, a), (a, e), (e, a), (b, c), (c, b), (b, d), (d, b), (b, e), (e, b), (d, e),$
$(e, d), (d, f), (f, d)\}$:

$$
M(\mathcal{R}) = \begin{array}{c} \\ (a) \\ (b) \\ (c) \\ (d) \\ (e) \\ (f) \end{array}
\begin{array}{c}
\begin{array}{cccccc} (a) & (b) & (c) & (d) & (e) & (f) \end{array} \\
\left[\begin{array}{cccccc}
0 & 1 & 0 & 0 & 1 & 0 \\
1 & 0 & 1 & 1 & 1 & 0 \\
0 & 1 & 0 & 0 & 0 & 0 \\
0 & 1 & 0 & 0 & 1 & 1 \\
1 & 1 & 0 & 1 & 0 & 0 \\
0 & 0 & 0 & 1 & 0 & 0
\end{array} \right]
\end{array}
$$

For part (ii) the rows and columns of the relation matrix are indexed as in part (i).

(ii) $\mathcal{R} = \{(a, b), (b, e), (d, b), (d, c), (e, f)\}$:

$$
M(\mathcal{R}) = \left[\begin{array}{cccccc}
0 & 1 & 0 & 0 & 0 & 0 \\
0 & 0 & 0 & 0 & 1 & 0 \\
0 & 0 & 0 & 0 & 0 & 0 \\
0 & 1 & 1 & 0 & 0 & 0 \\
0 & 0 & 0 & 0 & 0 & 1 \\
0 & 0 & 0 & 0 & 0 & 0
\end{array} \right]
$$

19.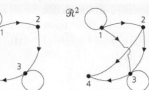

21. a) 2^{25} **b)** 2^{15}

23. a) $\mathcal{R}_1$:
$$\begin{bmatrix} 1 & 1 & 0 & 0 & 0 \\ 1 & 1 & 0 & 0 & 0 \\ 0 & 0 & 1 & 1 & 0 \\ 0 & 0 & 1 & 1 & 0 \\ 0 & 0 & 0 & 0 & 1 \end{bmatrix}$$
$\mathcal{R}_2$:
$$\begin{bmatrix} 1 & 1 & 1 & 0 & 0 \\ 1 & 1 & 1 & 0 & 0 \\ 1 & 1 & 1 & 0 & 0 \\ 0 & 0 & 0 & 1 & 1 \\ 0 & 0 & 0 & 1 & 1 \end{bmatrix}$$

b) Given an equivalence relation $\mathcal{R}$ on a finite set A, list the elements of A so that elements in the same cell of the partition (see Section 7.4) are adjacent. The resulting relation matrix will then have square blocks of 1's along the diagonal (from upper left to lower right).

25.
(s_1) a := 1
(s_2) b := 2
(s_3) a := a + 3
(s_4) c := b
(s_5) a := 2 * a − 1
(s_6) b := a * c
(s_7) c := 7
(s_8) d := c + 2

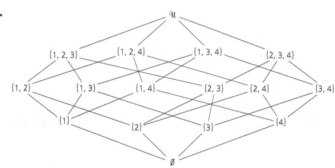

27. $n = 38$

Section 7.3 – p. 364

1.

3. For all $a \in A$, $b \in B$, we have $a \,\mathcal{R}_1\, a$ and $b \,\mathcal{R}_2\, b$, so $(a, b) \,\mathcal{R}\, (a, b)$ and $\mathcal{R}$ is reflexive.
$(a, b) \,\mathcal{R}\, (c, d), (c, d) \,\mathcal{R}\, (a, b) \Rightarrow a \,\mathcal{R}_1\, c, c \,\mathcal{R}_1\, a$ and $b \,\mathcal{R}_2\, d, d \,\mathcal{R}_2\, b \Rightarrow a = c, b = d \Rightarrow$
$(a, b) = (c, d)$, so $\mathcal{R}$ is antisymmetric. $(a, b) \,\mathcal{R}\, (c, d), (c, d) \,\mathcal{R}\, (e, f) \Rightarrow a \,\mathcal{R}_1\, c, c \,\mathcal{R}_1\, e$ and
$b \,\mathcal{R}_2\, d, d \,\mathcal{R}_2\, f \Rightarrow a \,\mathcal{R}_1\, e, b \,\mathcal{R}_2\, f \Rightarrow (a, b) \,\mathcal{R}\, (e, f)$, and this implies that $\mathcal{R}$ is transitive.

5. $\emptyset < \{1\} < \{2\} < \{3\} < \{1, 2\} < \{1, 3\} < \{2, 3\} < \{1, 2, 3\}$ (There are other possibilities.)

7. a) 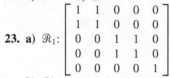 **b)** $3 < 2 < 1 < 4$ or $3 < 1 < 2 < 4$ **c)** 2

9. Let x, y both be least upper bounds. Then $x \,\mathcal{R}\, y$, since y is an upper bound and x is a least upper bound. Likewise, $y \,\mathcal{R}\, x$. $\mathcal{R}$ antisymmetric $\Rightarrow x = y$. (The proof for the glb is similar.)

11. Let $\mathcal{U} = \{1, 2\}$, $A = \mathcal{P}(\mathcal{U})$, and let $\mathcal{R}$ be the inclusion relation. Then $(A, \mathcal{R})$ is a poset but not a total order. Let $B = \{\emptyset, \{1\}\}$. Then $(B \times B) \cap \mathcal{R}$ is a total order.

13. $n + \binom{n}{2}$

15. a) The n elements of A are arranged along a vertical line. For if $A = \{a_1, a_2, \ldots, a_n\}$ where $a_1 \mathcal{R} a_2 \mathcal{R} a_3 \mathcal{R} \cdots \mathcal{R} a_n$, then the diagram can be drawn as follows:

$\bullet\ a_n$
$\bullet\ a_{n-1}$
$\bullet\ a_{n-2}$
$\vdots$
$\bullet\ a_3$
$\bullet\ a_2$
$\bullet\ a_1$

b) $n!$

17.

	lub	glb		lub	glb		lub	glb		lub	glb		lub	glb
a)	$\{1, 2\}$	$\emptyset$	**b)**	$\{1, 2, 3\}$	$\emptyset$	**c)**	$\{1, 2\}$	$\emptyset$	**d)**	$\{1, 2, 3\}$	$\{1\}$	**e)**	$\{1, 2, 3\}$	$\emptyset$

19. For each $a \in \mathbf{Z}$ it follows that $a \mathcal{R} a$ because $a - a = 0$, an even nonnegative integer. Hence $\mathcal{R}$ is *reflexive*. If $a, b, c \in \mathbf{Z}$ with $a \mathcal{R} b$ and $b \mathcal{R} c$, then

$$a - b = 2m, \qquad \text{for some } m \in \mathbf{N}$$
$$b - c = 2n, \qquad \text{for some } n \in \mathbf{N},$$

and $a - c = (a - b) + (b - c) = 2(m + n)$, where $m + n \in \mathbf{N}$. Therefore, $a \mathcal{R} c$ and $\mathcal{R}$ is *transitive*. Finally, suppose that $a \mathcal{R} b$ and $b \mathcal{R} a$ for some $a, b \in \mathbf{Z}$. Then $a - b$ and $b - a$ are both nonnegative integers. Since this can only occur for $a - b = b - a = 0$, we find that $[a \mathcal{R} b \wedge b \mathcal{R} a] \Rightarrow a = b$, so $\mathcal{R}$ is *antisymmetric*.

Consequently, the relation $\mathcal{R}$ is a partial order for $\mathbf{Z}$. But it is *not* a total order. For example, $2, 3 \in \mathbf{Z}$ and we have neither $2 \mathcal{R} 3$ nor $3 \mathcal{R} 2$, because neither -1 nor 1, respectively, is a nonnegative even integer.

21. b) & c) Here the least element (and only minimal element) is $(0, 0)$. The element $(2, 2)$ is the greatest element (and the only maximal element).

d) $(0, 0) \mathcal{R} (0, 1) \mathcal{R} (0, 2) \mathcal{R} (1, 0) \mathcal{R} (1, 1) \mathcal{R} (1, 2) \mathcal{R} (2, 0) \mathcal{R} (2, 1) \mathcal{R} (2, 2)$

23. a) False. Let $\mathcal{U} = \{1, 2\}$, $A = \mathcal{P}(\mathcal{U})$, and let $\mathcal{R}$ be the inclusion relation. Then $(A, \mathcal{R})$ is a lattice where for all $S, T \in A$, $\text{lub}\{S, T\} = S \cup T$ and $\text{glb}\{S, T\} = S \cap T$. However, $\{1\}$ and $\{2\}$ are not related, so $(A, \mathcal{R})$ is not a total order.

25. a) a **b)** a **c)** c **d)** e **e)** z **f)** e **g)** v

$(A, \mathcal{R})$ is a lattice with z the greatest (and only maximal) element and a the least (and only minimal) element.

27. a) 3 **b)** m **c)** 17 **d)** $m + n + 2mn$ **e)** 133

f) $m + n + k + 2(mn + mk + nk) + 3mnk$ **g)** 1484

h) $m + n + k + \ell + 2(mn + mk + m\ell + nk + n\ell + k\ell) + 3(mnk + mn\ell + mk\ell + nk\ell) + 4mnk\ell$

29. $429 = \left(\frac{1}{8}\right)\binom{14}{7}$ so $k = 6$, and there are $2 \cdot 7 = 14$ positive integer divisors of $p^6 q$.

Section 7.4 – p. 370

1. a) Here the collection A_1, A_2, A_3 provides a partition of A.

b) Although $A = A_1 \cup A_2 \cup A_3 \cup A_4$, we have $A_1 \cap A_2 \neq \emptyset$, so the collection A_1, A_2, A_3, A_4 does *not* provide a partition for A.

3. $\mathcal{R} = \{(1, 1), (1, 2), (2, 1), (2, 2), (3, 3), (3, 4), (4, 3), (4, 4), (5, 5)\}$

5. $\mathcal{R}$ is not transitive since $1 \mathcal{R} 2$ and $2 \mathcal{R} 3$ but $1 \not\mathcal{R} 3$.

7. a) For all $(x, y) \in A$, $x + y = x + y \Rightarrow (x, y) \mathcal{R} (x, y)$.

$(x_1, y_1) \mathcal{R} (x_2, y_2) \Rightarrow x_1 + y_1 = x_2 + y_2 \Rightarrow x_2 + y_2 = x_1 + y_1 \Rightarrow (x_2, y_2) \mathcal{R} (x_1, y_1)$.

$(x_1, y_1) \mathcal{R} (x_2, y_2), (x_2, y_2) \mathcal{R} (x_3, y_3) \Rightarrow x_1 + y_1 = x_2 + y_2, x_2 + y_2 = x_3 + y_3$, so $x_1 + y_1 = x_3 + y_3$ and $(x_1, y_1) \mathcal{R} (x_3, y_3)$. Since $\mathcal{R}$ is reflexive, symmetric, and transitive, it is an equivalence relation.

b) $[(1, 3)] = \{(1, 3), (2, 2), (3, 1)\}$; $[(2, 4)] = \{(1, 5), (2, 4), (3, 3), (4, 2), (5, 1)\}$

$[(1, 1)] = \{(1, 1)\}$

c) $A = \{(1, 1)\} \cup \{(1, 2), (2, 1)\} \cup \{(1, 3), (2, 2), (3, 1)\} \cup \{(1, 4), (2, 3), (3, 2), (4, 1)\} \cup$
$\{(1, 5), (2, 4), (3, 3), (4, 2), (5, 1)\} \cup \{(2, 5), (3, 4), (4, 3), (5, 2)\} \cup$
$\{(3, 5), (4, 4), (5, 3)\} \cup \{(4, 5), (5, 4)\} \cup \{(5, 5)\}$

9. a) For all $(a, b) \in A$ we have $ab = ab$, so $(a, b) \mathcal{R} (a, b)$ and $\mathcal{R}$ is reflexive. To see that $\mathcal{R}$ is symmetric suppose that $(a, b), (c, d) \in A$ and that $(a, b) \mathcal{R} (c, d)$. Then $(a, b) \mathcal{R} (c, d) \Rightarrow$ $ad = bc \Rightarrow cb = da \Rightarrow (c, d) \mathcal{R} (a, b)$, so $\mathcal{R}$ is symmetric. Finally, let $(a, b), (c, d),$ $(e, f) \in A$ with $(a, b) \mathcal{R} (c, d)$ and $(c, d) \mathcal{R} (e, f)$. Then $(a, b) \mathcal{R} (c, d) \Rightarrow ad = bc$ and $(c, d) \mathcal{R} (e, f) \Rightarrow cf = de$, so $adf = bcf = bde$ and since $d \neq 0$, we have $af = be$. But $af = be \Rightarrow (a, b) \mathcal{R} (e, f)$, and consequently $\mathcal{R}$ is transitive. It follows from the above that $\mathcal{R}$ is an equivalence relation on A.

b) $[(2, 14)] = \{(2, 14)\}$ $[(-3, -9)] = \{(-3, -9), (-1, -3), (4, 12)\}$
$[(4, 8)] = \{(-2, -4), (1, 2), (3, 6), (4, 8)\}$

c) There are five cells in the partition — in fact,

$$A = [(-4, -20)] \cup [(-3, -9)] \cup [(-2, -4)] \cup [(-1, -11)] \cup [(2, 14)].$$

11. a) $\left(\frac{1}{2}\right)\binom{6}{3}$ **b)** $4\binom{6}{3}$ **c)** $2\binom{6}{4}$ **d)** $\left(\frac{1}{2}\right)\binom{6}{3} + 4\binom{6}{3} + 2\binom{6}{4} + \binom{6}{5} + \binom{6}{6}$ **13.** 300

15. Let $\{A_i\}_{i \in I}$ be a partition of a set A. Define $\mathcal{R}$ on A by $x \mathcal{R} y$ if for some $i \in I$, we have $x, y \in A_i$. For each $x \in A$, $x, x \in A_i$ for some $i \in I$, so $x \mathcal{R} x$ and $\mathcal{R}$ is reflexive. $x \mathcal{R} y \Rightarrow x, y \in A_i$, for some $i \in I \Rightarrow y, x \in A_i$ for some $i \in I \Rightarrow y \mathcal{R} x$, so $\mathcal{R}$ is symmetric. If $x \mathcal{R} y$ and $y \mathcal{R} z$, then $x, y \in A_i$ and $y, z \in A_j$ for some $i, j \in I$. Since $A_i \cap A_j$ contains y and $\{A_i\}_{i \in I}$ is a partition, from $A_i \cap A_j \neq \emptyset$ it follows that $A_i = A_j$, so $i = j$. Hence $x, z \in A_i$, so $x \mathcal{R} z$ and $\mathcal{R}$ is transitive.

17. *Proof*: Since $\{B_1, B_2, B_3, \ldots, B_n\}$ is a partition of B, we have $B = B_1 \cup B_2 \cup B_3 \cup \cdots \cup B_n$. Therefore $A = f^{-1}(B) = f^{-1}(B_1 \cup \cdots \cup B_n) = f^{-1}(B_1) \cup \cdots \cup f^{-1}(B_n)$ [by generalizing part (b) of Theorem 5.10]. For $1 \leq i < j \leq n$, $f^{-1}(B_i) \cap f^{-1}(B_j) = f^{-1}(B_i \cap B_j) = f^{-1}(\emptyset) = \emptyset$. Consequently, $\{f^{-1}(B_i) | 1 \leq i \leq n, f^{-1}(B_i) \neq \emptyset\}$ is a partition of A.
Note: Part (b) of Example 7.56 is a special case of this result.

Section 7.5 – p. 376

1. a) s_2 and s_5 are equivalent.   **b)** s_2 and s_5 are equivalent.
c) s_2 and s_7 are equivalent; s_3 and s_4 are equivalent.

3. a) s_1 and s_7 are equivalent; s_4 and s_5 are equivalent.
b)   (i) 0 0 0 0
   (ii) 0
   (iii) 0 0

	ν		ω	
M:	0	1	0	1
s_1	s_4	s_1	1	0
s_2	s_1	s_2	1	0
s_3	s_6	s_1	1	0
s_4	s_3	s_4	0	0
s_6	s_2	s_1	1	0

Supplementary Exercises – p. 378

1. a) False. Let $A = \{1, 2\}$, $I = \{1, 2\}$, $\mathcal{R}_1 = \{(1, 1)\}$, and $\mathcal{R}_2 = \{(2, 2)\}$. Then $\bigcup_{i \in I} \mathcal{R}_i$ is reflexive, but neither $\mathcal{R}_1$ nor $\mathcal{R}_2$ is reflexive. Conversely, however, if $\mathcal{R}_i$ is reflexive for all (actually, at least one) $i \in I$, then $\bigcup_{i \in I} \mathcal{R}_i$ is reflexive.

3. $(a, c) \in \mathcal{R}_2 \circ \mathcal{R}_1 \Rightarrow$ for some $b \in A$, $(a, b) \in \mathcal{R}_2$, $(b, c) \in \mathcal{R}_1$. With $\mathcal{R}_1, \mathcal{R}_2$ symmetric, $(b, a) \in \mathcal{R}_2$, $(c, b) \in \mathcal{R}_1$, so $(c, a) \in \mathcal{R}_1 \circ \mathcal{R}_2 \subseteq \mathcal{R}_2 \circ \mathcal{R}_1$. $(c, a) \in \mathcal{R}_2 \circ \mathcal{R}_1 \Rightarrow (c, d) \in \mathcal{R}_2$, $(d, a) \in \mathcal{R}_1$, for some $d \in A$. Then $(d, c) \in \mathcal{R}_2$, $(a, d) \in \mathcal{R}_1$ by symmetry, and $(a, c) \in \mathcal{R}_1 \circ \mathcal{R}_2$, so $\mathcal{R}_2 \circ \mathcal{R}_1 \subseteq \mathcal{R}_1 \circ \mathcal{R}_2$ and the result follows.

5. $(c, a) \in (\mathcal{R}_1 \circ \mathcal{R}_2)^c \iff (a, c) \in \mathcal{R}_1 \circ \mathcal{R}_2 \iff (a, b) \in \mathcal{R}_1, (b, c) \in \mathcal{R}_2$, for some $b \in B \iff (b, a) \in \mathcal{R}_1^c, (c, b) \in \mathcal{R}_2^c$, for some $b \in B \iff (c, a) \in \mathcal{R}_2^c \circ \mathcal{R}_1^c$.

7. Let $\mathcal{U} = \{1, 2, 3, 4, 5\}$, $A = \mathcal{P}(\mathcal{U}) - \{\mathcal{U}, \emptyset\}$. Under the inclusion relation, A is a poset with the five minimal elements $\{x\}$, $1 \le x \le 5$, but no least element. Also, A has five maximal elements — the five subsets of $\mathcal{U}$ of size 4 — but no greatest element.

9. $n = 10$

11. a)

Adjacency List		Index List	
1	2	1	1
2	3	2	2
3	1	3	3
4	4	4	5
5	5	5	6
6	3	6	8
7	5		

b)

Adjacency List		Index List	
1	2	1	1
2	3	2	2
3	1	3	3
4	5	4	4
5	4	5	5
		6	6

c)

Adjacency List		Index List	
1	2	1	1
2	3	2	2
3	1	3	3
4	4	4	6
5	5	5	7
6	1	6	8
7	4		

13. b) The cells of the partition are the connected components of G.

15. One possible order is 10, 3, 8, 6, 7, 9, 1, 4, 5, 2, where program 10 is run first and program 2 last.

17. b) $[(0.3, 0.7)] = \{(0.3, 0.7)\}$ $[(0.5, 0)] = \{(0.5, 0)\}$ $[(0.4, 1)] = \{(0.4, 1)\}$
$[(0, 0.6)] = \{(0, 0.6), (1, 0.6)\}$ $[(1, 0.2)] = \{(0, 0.2), (1, 0.2)\}$
In general, if $0 < a < 1$, then $[(a, b)] = \{(a, b)\}$; otherwise, $[(0, b)] = \{(0, b), (1, b)\} = [(1, b)]$.

c) The lateral surface of a cylinder of height 1 and base radius $1/2\pi$.

19. $4^n - 2(3^n) + 2^n$

21. a) (i) $B \mathcal{R} A \mathcal{R} C$; (ii) $B \mathcal{R} C \mathcal{R} F$
$B \mathcal{R} A \mathcal{R} C \mathcal{R} F$ is a maximal chain. There are six such maximal chains.
b) Here $11 \mathcal{R} 385$ is a maximal chain of length 2, while $2 \mathcal{R} 6 \mathcal{R} 12$ is one of length 3. The length of a longest chain for this poset is 3.
c) (i) $\emptyset \subseteq \{1\} \subseteq \{1, 2\} \subseteq \{1, 2, 3\} \subseteq \mathcal{U}$; (ii) $\emptyset \subseteq \{2\} \subseteq \{2, 3\} \subseteq \{1, 2, 3\} \subseteq \mathcal{U}$
There are $4! = 24$ such maximal chains.
d) $n!$

23. Let $a_1 \mathcal{R} a_2 \mathcal{R} \cdots \mathcal{R} a_{n-1} \mathcal{R} a_n$ be a longest (maximal) chain in $(A, \mathcal{R})$. Then a_n is a maximal element in $(A, \mathcal{R})$ and $a_1 \mathcal{R} a_2 \mathcal{R} \cdots \mathcal{R} a_{n-1}$ is a maximal chain in $(B, \mathcal{R}')$. Hence the length of a longest chain in $(B, \mathcal{R}')$ is at least $n - 1$. If there is a chain $b_1 \mathcal{R}' b_2 \mathcal{R}' \cdots \mathcal{R}' b_n$ in $(B, \mathcal{R}')$ of length n, then this is also a chain of length n in $(A, \mathcal{R})$. But then b_n must be a maximal element of $(A, \mathcal{R})$, and this contradicts $b_n \in B$.

25. If $n = 1$, then for all $x, y \in A$, if $x \ne y$ then $x \not\mathcal{R} y$ and $y \not\mathcal{R} x$. Hence $(A, \mathcal{R})$ is an antichain, and the result follows. Now assume the result true for $n = k \ge 1$, and let $(A, \mathcal{R})$ be a poset where the length of a longest chain is $k + 1$. If M is the set of all maximal elements in $(A, \mathcal{R})$, then $M \ne \emptyset$ and M is an antichain in $(A, \mathcal{R})$. Also, by virtue of Exercise 23, $(A - M, \mathcal{R}')$, for $\mathcal{R}' = ((A - M) \times (A - M)) \cap \mathcal{R}$, is a poset with k the length of a longest chain. So by the induction hypothesis, $A - M = C_1 \cup C_2 \cup \cdots \cup C_k$, a partition into k antichains. Consequently, $A = C_1 \cup C_2 \cup \cdots \cup C_k \cup M$, a partition into $k + 1$ antichains.

27. a) n **b)** 2^{n-1} **c)** 64

Chapter 8
The Principle of Inclusion and Exclusion

Section 8.1 – p. 396

1. Let $x \in S$ and let n be the number of conditions (from among c_1, c_2, c_3, c_4) satisfied by x.

($n = 0$):   Here x is counted once in $N(\overline{c_2}\overline{c_3}\overline{c_4})$ and once in $N(\overline{c_1}\overline{c_2}\overline{c_3}\overline{c_4})$.

($n = 1$):   If x satisfies c_1 (and not c_2, c_3, c_4), then x is counted once in $N(\overline{c_2}\overline{c_3}\overline{c_4})$ and once in $N(c_1\overline{c_2}\overline{c_3}\overline{c_4})$.

If x satisfies c_i, for $i \neq 1$, then x is not counted in any of the three terms in the equation.

$(n = 2, 3, 4)$:   If x satisfies at least two of the four conditions, then x is not counted in any of the three terms in the equation.

The preceding observations show that the two sides of the given equation count the same elements from S, and this provides a combinatorial proof for the formula $N(\bar{c}_2\bar{c}_3\bar{c}_4) = N(c_1\bar{c}_2\bar{c}_3\bar{c}_4) + N(\bar{c}_1\bar{c}_2\bar{c}_3\bar{c}_4)$.

3. a) 12   **b)** 3   **5. a)** 534   **b)** 458   **c)** 76

7. 4,460,400   **9.** $\binom{37}{31} - \binom{7}{1}\binom{27}{21} + \binom{7}{2}\binom{17}{11} - \binom{7}{3}\binom{7}{1}$

11. $(15!)\left[\binom{14}{10} - \binom{5}{1}\binom{10}{6} + \binom{5}{2}\binom{6}{2}\right]$ **13.** $26! - [3(23!) + 24!] + (20! + 21!)$

15. $\left[6^8 - \binom{6}{1}5^8 + \binom{6}{2}4^8 - \binom{6}{3}3^8 + \binom{6}{2}2^8 - \binom{6}{1}\right]/6^8$

17. $9!/\left[(3!)^3\right] - 3\left[7!/[(3!)^2]\right] + 3(5!/3!) - 3!$ **19.** $651/7776 \doteq 0.08372$

21. a) 32   **b)** 96   **c)** 3200   **23. a)** 2^{n-1} **b)** $2^{n-1}(p-1)$

25. a) 1600   **b)** 4399   **27.** $\phi(17) = \phi(32) = \phi(48) = 16$

29. If 4 divides $\phi(n)$, then one of the following must hold:

(1) n is divisible by 8;

(2) n is divisible by two (or more) distinct odd primes;

(3) n is divisible by an odd prime p (such as 5, 13, and 17) where 4 divides $p - 1$; or

(4) n is divisible by 4 (and not 8) and at least one odd prime.

Section 8.2 – p. 401

1. $E_0 = 768$; $E_1 = 205$; $E_2 = 40$; $E_3 = 10$; $E_4 = 0$; $E_5 = 1$. $\sum_{i=0}^{5} E_i = 1024 = N$.

3. a) $\left[14!/(2!)^5\right] - \binom{5}{1}\left[13!/(2!)^4\right] + \binom{5}{2}\left[12!/(2!)^3\right] - \binom{5}{3}\left[11!/(2!)^2\right] + \binom{5}{4}[10!/2!] - \binom{5}{5}[9!]$

b) $E_2 = \binom{5}{2}\left[12!/(2!)^3\right] - \binom{3}{1}\binom{5}{3}\left[11!/(2!)^2\right] + \binom{4}{2}\binom{5}{4}[10!/2!] - \binom{5}{3}\binom{5}{5}[9!]$

c) $L_3 = \binom{5}{3}\left[11!/(2!)^2\right] - \binom{3}{2}\binom{5}{4}[10!/2!] + \binom{4}{2}\binom{5}{5}[9!]$

5. $E_2 = 6132$; $L_2 = 6136$

7. a) $\left[\sum_{i=0}^{3}(-1)^i\binom{4}{i}\binom{52-13i}{13}\right]/\binom{52}{13}$ **b)** $\left[\sum_{i=1}^{3}(-1)^{i+1}(i)\binom{4}{i}\binom{52-13i}{13}\right]/\binom{52}{13}$

c) $\left[\binom{4}{2}\binom{26}{13} - 3\binom{4}{3}\binom{13}{13}\right]/\binom{52}{13}$

Section 8.3 – p. 403

1. $10! - \binom{5}{1}9! + \binom{5}{2}8! - \binom{5}{3}7! + \binom{5}{4}6! - \binom{5}{5}5!$ **3.** 44

5. a) $7! - d_7$ $(d_7 \doteq (7!)e^{-1})$ **b)** $d_{26} \doteq (26!)e^{-1}$

7. $n = 11$ **9.** $(10!)d_{10} \doteq (10!)^2(e^{-1})$

11. a) $(d_{10})^2 \doteq (10!)^2 e^{-2}$ **b)** $\sum_{i=0}^{10}(-1)^i\binom{10}{i}[(10-i)!]^2$

13. For all $n \in \mathbf{Z}^+$, $n!$ counts the total number of permutations of $1, 2, 3, \ldots, n$. Each such permutation will have k elements that are deranged (that is, there are k elements $x_1, x_2, \ldots, x_k$ in $\{1, 2, 3, \ldots, n\}$ where x_1 is *not* in position x_1, x_2 is *not* in position $x_2, \ldots$, and x_k is *not* in position x_k) and $n - k$ elements that are fixed (that is, the $n - k$ elements $y_1, y_2, \ldots, y_{n-k}$ in $\{1, 2, 3, \ldots, n\} - \{x_1, x_2, \ldots, x_k\}$ are such that y_1 is in position y_1, y_2 is in position $y_2, \ldots$, and y_{n-k} is in position y_{n-k}).

The $n - k$ fixed elements can be chosen in $\binom{n}{n-k}$ ways, and the remaining k elements can then be permuted (that is, deranged) in d_k ways. Hence there are $\binom{n}{n-k}d_k = \binom{n}{k}d_k$ permutations of $1, 2, 3, \ldots, n$ with $n - k$ fixed elements (and k deranged elements). As k varies from 0 to n we count all of the $n!$ permutations of $1, 2, 3, \ldots, n$ according to the number k of deranged elements.

Consequently,

$$n! = \binom{n}{0}d_0 + \binom{n}{1}d_1 + \binom{n}{2}d_2 + \cdots + \binom{n}{n}d_n = \sum_{k=0}^{n}\binom{n}{k}d_k.$$

15. $\binom{n}{0}(n-1)! - \binom{n}{1}(n-2)! + \binom{n}{2}(n-3)! - \cdots + (-1)^{n-1}\binom{n}{n-1}(0!) + (-1)^n\binom{n}{n}$

Sections 8.4
and 8.5 – p. 410

3. a) $\binom{8}{0} + \binom{8}{1}8x + \binom{8}{2}(8 \cdot 7)x^2 + \binom{8}{3}(8 \cdot 7 \cdot 6)x^3 + \binom{8}{4}(8 \cdot 7 \cdot 6 \cdot 5)x^4 + \cdots + \binom{8}{8}(8!)x^8 = \sum_{i=0}^{8} \binom{8}{i}P(8, i)x^i$

b) $\sum_{i=0}^{n} \binom{n}{i}P(n, i)x^i$

5. a) (i) $(1 + 2x)^3$ (ii) $1 + 8x + 14x^2 + 4x^3$
(iii) $1 + 9x + 25x^2 + 21x^3$ (iv) $1 + 8x + 16x^2 + 7x^3$

b) If the board C consists of n steps, and each step has k blocks, then $r(C, x) = (1 + kx)^n$.

7. $5! - 8(4!) + 21(3!) - 20(2!) + 6(1!) = 20$ **9. a)** 20 **b)** 3/10

11. $(6!/2!) - 9(5!/2!) + 27(4!/2!) - 31(3!/2!) + 12 = 63$

Supplementary
Exercises – p. 413

1. 134 **3.** $[(24!)/(6!)^4]\left[\binom{19}{16} - \binom{4}{1}\binom{13}{10} + \binom{4}{2}\binom{7}{4}\right]$ **5.** $\sum_{i=0}^{7}(-1)^i \binom{8}{i}(8 - i)!$

7. $\sum_{k=0}^{10}(-1)^k \binom{10}{k}\binom{14 - k}{10 - k}(10 - k)! = 1{,}764{,}651{,}461$

9. Let $T = (13!)/(2!)^5$.

a) $\left(\left[\binom{5}{3}(10!)/(2!)^2\right] - \left[\binom{4}{1}\binom{5}{4}(9!)/(2!)\right] + \left[\binom{5}{2}\binom{5}{5}(8!)\right]\right)/T$

b) $[T - (E_4 + E_5)]/T$, where $E_4 = \left[\binom{4}{4}(9!)/(2!)\right] - \left[\binom{5}{1}\binom{5}{5}(8!)\right]$ and $E_5 = \binom{5}{5}(8!)$

11. a) $\binom{n - m}{r - m}$ **13.** 84

15. a) $S_1 = \{1, 5, 7, 11, 13, 17\}$ $S_2 = \{2, 4, 8, 10, 14, 16\}$
$S_3 = \{3, 15\}$ $S_6 = \{6, 12\}$
$S_9 = \{9\}$ $S_{18} = \{18\}$

b) $|S_1| = 6 = \phi(18)$ $|S_3| = 2 = \phi(6)$ $|S_9| = 1 = \phi(2)$
$|S_2| = 6 = \phi(9)$ $|S_6| = 2 = \phi(3)$ $|S_{18}| = 1 = \phi(1)$

17. a) If n is even, then by the Fundamental Theorem of Arithmetic (Theorem 4.11) we may write $n = 2^k m$, where $k \geq 1$ and m is odd. Then $2n = 2^{k+1}m$ and $\phi(2n) = (2^{k+1})(1 - \frac{1}{2})\phi(m) = 2^k \phi(m) = 2(2^k)(\frac{1}{2})\phi(m) = 2[2^k(1 - \frac{1}{2})\phi(m)] = 2[\phi(2^k m)] = 2\phi(n)$.

b) When n is odd, we find that $\phi(2n) = (2n)(1 - \frac{1}{2})\prod_{p|n}(1 - \frac{1}{p})$, where the product is taken over all (odd) primes dividing n. (If $n = 1$, then $\prod_{p|n}(1 - \frac{1}{p})$ is 1.) But $(2n)(1 - \frac{1}{2})\prod_{p|n}(1 - \frac{1}{p}) = n\prod_{p|n}(1 - \frac{1}{p}) = \phi(n)$.

19. a) $d_4(12!)^4$ **b)** $\binom{4}{1}d_3(12!)^4$ **c)** $d_4(d_{12})^4$

Chapter 9
Generating Functions

Section 9.1 – p. 417

1. a) The coefficient of x^{20} in $(1 + x + x^2 + \cdots + x^7)^4$

b) The coefficient of x^{20} in $(1 + x + x^2 + \cdots + x^{20})^2 (1 + x^2 + x^4 + \cdots + x^{20})^2$ or $(1 + x + x^2 + \cdots)^2 (1 + x^2 + x^4 + \cdots)^2$

c) The coefficient of x^{30} in $(x^2 + x^3 + x^4)(x^3 + x^4 + \cdots + x^8)^4$

d) The coefficient of x^{30} in $(1 + x + x^2 + \cdots + x^{30})^3 (1 + x^2 + x^4 + \cdots + x^{30}) \cdot (x + x^3 + x^5 + \cdots + x^{29})$ or $(1 + x + x^2 + \cdots)^3 (1 + x^2 + x^4 + \cdots)(x + x^3 + x^5 + \cdots)$

3. a) The coefficient of x^{10} in $(1 + x + x^2 + x^3 + \cdots)^6$

b) The coefficient of x^r in $(1 + x + x^2 + x^3 + \cdots)^n$

5. The answer is the coefficient of x^{31} in the generating function

$$(1 + x + x^2 + x^3 + \cdots)^3(1 + x + x^2 + \cdots + x^{10}).$$

Section 9.2 – p. 431

1. a) $(1 + x)^8$ **b)** $8(1 + x)^7$ **c)** $(1 + x)^{-1}$
d) $6x^3/(1 + x)$ **e)** $(1 - x^2)^{-1}$ **f)** $x^2/(1 - ax)$

3. a) $g(x) = f(x) - a_3 x^3 + 3x^3 = f(x) + (3 - a_3)x^3$

b) $g(x) = f(x) + (3 - a_3)x^3 + (7 - a_7)x^7$

c) $g(x) = 2f(x) + (1 - 2a_1) x + (3 - 2a_3) x^3$

d) $g(x) = 2f(x) + [5/(1 - x)] + (1 - 2a_1 - 5) x + (3 - 2a_3 - 5) x^3 + (7 - 2a_7 - 5) x^7$

5. a) $\binom{21}{7}$ **b)** $\binom{n+6}{7}$ **7.** $\binom{14}{10} - 5\binom{9}{5} + \binom{5}{2}$

9. a) 0 **b)** $\binom{14}{12} - 5\binom{16}{14}$ **c)** $\binom{18}{15} + 4\binom{17}{14} + 6\binom{16}{13} + 4\binom{15}{12} + \binom{14}{11}$

11. $\binom{99}{96} - 4\binom{64}{61} + 6\binom{29}{26}$ **13.** $\left[\binom{29}{18} - \binom{12}{1}\binom{23}{12} + \binom{12}{2}\binom{17}{6} - \binom{12}{3}\right] / \left(6^{12}\right)$

15. $(1/8) [1 + (-1)^n] + (1/4)\binom{n+1}{n} + (1/2)\binom{n+2}{n}$

17. $\left(1 - x - x^2 - x^3 - x^4 - x^5 - x^6\right)^{-1} = \left[1 - \left(x + x^2 + \cdots + x^6\right)\right]^{-1}$

$= 1 + \left(x + x^2 + \cdots + x^6\right) + \left(x + x^2 + \cdots + x^6\right)^2 + \left(x + x^2 + \cdots + x^6\right)^3 + \cdots,$

            one roll   **two rolls**   **three rolls**

where the 1 takes care of the case where the die is not rolled.

19. a) $2^4/2^7 = 1/8$ **b)** $2^{\lfloor n/2 \rfloor}/2^{n-1} = 2^{1 - \lceil n/2 \rceil}$

21. $2^{\lfloor (n-2t)/2 \rfloor}$ **23.** $2^{(n/2)-1}; 2^{(n/2)-1}$

25. a) $Pr\,(Y = y) = (5/6)^{y-1}(1/6), y = 1, 2, 3, \ldots .$

b) $E(Y) = 6$ **c)** $\sigma_Y = \sqrt{30} \doteq 5.477226$

27. $3/5$

29. a) The differences are 2, 3, 2, 7, and 0, and these sum to 14.

b) $\{3, 5, 8, 15\}$ **c)** $\{1 + a, 1 + a + b, 1 + a + b + c, 1 + a + b + c + d\}$.

31. $c_k = \sum_{i=0}^{k} i(k-i)^2 = k^2 \sum_{i=0}^{k} i - 2k \sum_{i=0}^{k} i^2 + \sum_{i=0}^{k} i^3$

$= \left(k^2\right) [k(k+1)/2] - 2k[k(k+1)(2k+1)/6] + \left[k^2(k+1)^2/4\right]$

$= (1/12) \left(k^2\right) \left(k^2 - 1\right)$

33. a) $\left(1 + x + x^2 + x^3 + x^4\right)\left(0 + x + 2x^2 + 3x^3 + \cdots\right) = \sum_{i=0}^{\infty} c_i x^i$ where $c_0 = 0, c_1 = 1$,

$c_2 = 1 + 2 = 3, c_3 = 1 + 2 + 3 = 6, c_4 = 1 + 2 + 3 + 4 = 10$, and

$c_n = n + (n-1) + (n-2) + (n-3) + (n-4) = 5n - 10$ for all $n \geq 5$.

b) $\left(1 - x + x^2 - x^3 + \cdots\right)\left(1 - x + x^2 - x^3 + \cdots\right) = 1/(1 + x)^2 = (1 + x)^{-2}$, the

generating function for the sequence $\binom{-2}{0}, \binom{-2}{1}, \binom{-2}{2}, \binom{-2}{3}, \ldots$. Hence the convolution of the

given pair of sequences is $c_0, c_1, c_2, \ldots$, where $c_n = \binom{-2}{n} = (-1)^n \binom{2+n-1}{n} = (-1)^n \binom{n+1}{n} = $

$(-1)^n (n+1), n \in \mathbf{N}$. [This is the alternating sequence 1, −2, 3, −4, 5, −6, 7, ...]

Section 9.3 – p. 435

1. $7; 6 + 1; 5 + 2; 5 + 1 + 1; 4 + 3; 4 + 2 + 1; 4 + 1 + 1 + 1; 3 + 3 + 1; 3 + 2 + 2;$
$3 + 2 + 1 + 1; 3 + 1 + 1 + 1 + 1; 2 + 2 + 2 + 1; 2 + 2 + 1 + 1 + 1; 2 + 1 + 1 + 1 + 1 + 1;$
$1 + 1 + 1 + 1 + 1 + 1 + 1$

3. The number of partitions of 6 into 1's, 2's, and 3's is 7.

5. a) and b)

$$\left(1 + x^2 + x^4 + x^6 + \cdots\right)\left(1 + x^4 + x^8 + \cdots\right)\left(1 + x^6 + x^{12} + \cdots\right) \cdots = \prod_{i=1}^{\infty} \frac{1}{1 - x^{2i}}$$

7. Let $f(x)$ be the generating function for the number of partitions of $n \in \mathbf{Z}^+$ where no summand appears more than twice. Then

$$f(x) = \prod_{i=1}^{\infty} \left(1 + x^i + x^{2i}\right).$$

Let $g(x)$ be the generating function for the number of partitions of n where no summand is divisible by 3. Here

$$g(x) = \frac{1}{1 - x} \cdot \frac{1}{1 - x^2} \cdot \frac{1}{1 - x^4} \cdot \frac{1}{1 - x^5} \cdot \frac{1}{1 - x^7} \cdots .$$

But

$$f(x) = \left(1 + x + x^2\right)\left(1 + x^2 + x^4\right)\left(1 + x^3 + x^6\right)\left(1 + x^4 + x^8\right)\cdots$$

$$= \frac{1 - x^3}{1 - x}\cdot\frac{1 - x^6}{1 - x^2}\cdot\frac{1 - x^9}{1 - x^3}\cdot\frac{1 - x^{12}}{1 - x^4}\cdots$$

$$= \frac{1}{1 - x}\cdot\frac{1}{1 - x^2}\cdot\frac{1}{1 - x^4}\cdot\frac{1}{1 - x^5}\cdot\frac{1}{1 - x^7}\cdots = g(x).$$

9. This result follows from the one-to-one correspondence between the Ferrers graphs with summands (rows) not exceeding m and the transpose graphs (also Ferrers graphs) that have m summands (rows).

Section 9.4—p. 439

1. a) e^{-x} **b)** e^{2x} **c)** e^{-ax} **d)** $e^{a^2 x}$ **e)** $ae^{a^2 x}$ **f)** xe^{2x}

3. a) $g(x) = f(x) + [(3 - a_3)/3!]\,x^3$
   b) $g(x) = f(x) + [(-1 - a_3)/3!]\,x^3 = e^{5x} - [126x^3/(3!)]$
   c) $g(x) = 2f(x) + [2 - 2a_1]\,x + [(4 - 2a_2)/2!]\,x^2$

5. $\dfrac{1}{1 - x} = 1 + x + x^2 + x^3 + \cdots = (0!)\dfrac{x^0}{0!} + (1!)\dfrac{x^1}{1!} + (2!)\dfrac{x^2}{2!} + (3!)\dfrac{x^3}{3!} + \cdots$

7. The answer is the coefficient of $\dfrac{x^{25}}{25!}$ in $\left(\dfrac{x^3}{3!} + \dfrac{x^4}{4!} + \cdots + \dfrac{x^{10}}{10!}\right)^4$.

9. a) $(1/2)\left[3^{20} + 1\right]/\left(3^{20}\right)$ **b)** $(1/4)\left[3^{20} + 3\right]/\left(3^{20}\right)$ **c)** $(1/2)\left[3^{20} - 1\right]/\left(3^{20}\right)$
   d) $(1/2)\left[3^{20} - 1\right]/\left(3^{20}\right)$ **e)** $(1/2)\left[3^{20} + 1\right]/\left(3^{20}\right)$

Section 9.5—p. 442

1. a) $\left(1 + x + x^2\right)/(1 - x)$ **b)** $\left(1 + x + x^2 + x^3\right)/(1 - x)$ **c)** $(1 + 2x)/(1 - x)^2$

5. $a_0, a_1 - a_0, a_2 - a_1, a_3 - a_2, \ldots$ **7.** $f(x) = [e^x/(1 - x)]$

Supplementary Exercises—p. 445

1. a) $6/(1 - x) + 1/(1 - x)^2$ **b)** $1/(1 - ax)$ **c)** $1/[1 - (1 + a)x]$
   d) $1/(1 - x) + 1/(1 - ax)$

3. $\left[\binom{15}{12} - \binom{4}{1}\binom{9}{6} + \binom{4}{2}\right]^2$

5. Let $f(x)$ be the generating function for the number of partitions of $n \in \mathbf{Z}^+$ in which no even summand is repeated (an odd summand may or may not be repeated). Then

$$f(x) = \left(1 + x + x^2 + x^3 + \cdots\right)\left(1 + x^2\right)\left(1 + x^3 + x^6 + x^9 + \cdots\right)\left(1 + x^4\right)\cdots$$

$$= \frac{1}{1 - x}\cdot\left(1 + x^2\right)\cdot\frac{1}{1 - x^3}\cdot\left(1 + x^4\right)\cdot\frac{1}{1 - x^5}\cdots$$

Let $g(x)$ be the generating function for the number of partitions of $n \in \mathbf{Z}^+$ where no summand occurs more than three times. Then

$$g(x) = \left(1 + x + x^2 + x^3\right)\left(1 + x^2 + x^4 + x^6\right)\left(1 + x^3 + x^6 + x^9\right)\cdots$$

$$= \left[(1 + x)\left(1 + x^2\right)\right]\left[\left(1 + x^2\right)\left(1 + x^4\right)\right]\left[\left(1 + x^3\right)\left(1 + x^6\right)\right]\cdots$$

$$= \left[\left(1 - x^2\right)/(1 - x)\right]\left(1 + x^2\right)\left[\left(1 - x^4\right)/\left(1 - x^2\right)\right]\left(1 + x^4\right)\cdot$$

$$\left[\left(1 - x^6\right)/\left(1 - x^3\right)\right]\left(1 + x^6\right)\cdots$$

$$= (1/(1 - x))\left(1 + x^2\right)\left(1/\left(1 - x^3\right)\right)\left(1 + x^4\right)\left(1/\left(1 - x^5\right)\right)\left(1 + x^6\right)\cdots = f(x).$$

7. a) $1, 5, (5)(7), (5)(7)(9), (5)(7)(9)(11), \ldots$ **b)** $a = 4, b = -\frac{7}{4}$

9. $n\left(2^{n-1}\right)$ **11. a)** $\binom{19}{8}$ **b)** $\binom{9}{4}^2/\binom{19}{8}$

13. a) $[a + (d - a)x]/(1 - x)^2$ **b)** $na + (1/2)(n)(n - 1)d$

15. a) $x^n f(x)$ **b)** $\left[f(x) - \left(a_0 + a_1 x + a_2 x^2 + \cdots + a_{n-1}x^{n-1}\right)\right]/x^n$ **17.** $(1 - p)^{m-n}$

Chapter 10
Recurrence Relations

1. a) $a_n = 5a_{n-1}, n \geq 1, a_0 = 2$ **b)** $a_n = -3a_{n-1}, n \geq 1, a_0 = 6$
c) $a_n = (2/5)a_{n-1}, n \geq 1, a_0 = 7$
3. $d = \pm(3/7)$ **5.** 141 months   **7. a)** 145   **b)** 45
9. a) 21345   **b)** 52143, 52134   **c)** 21534, 21354, 21345

1. a) $a_n = (3/7)(-1)^n + (4/7)(6)^n, n \geq 0$ **b)** $a_n = 4(1/2)^n - 2(5)^n, n \geq 0$
c) $a_n = 3\sin(n\pi/2), n \geq 0$ **d)** $a_n = (5 - n)3^n, n \geq 0$
e) $a_n = (\sqrt{2})^n[\cos(3\pi n/4) + 4\sin(3\pi n/4)], n \geq 0$
3. $a_n = (1/10)[7^n - (-3)^n], n \geq 0$
5. a) $a_n = 2a_{n-1} + a_{n-2}, n \geq 2, a_0 = 1, a_1 = 2$
$a_n = (1/2\sqrt{2})[(1 + \sqrt{2})^{n+1} - (1 - \sqrt{2})^{n+1}], n \geq 0$
b) $a_n = a_{n-1} + 3a_{n-2}, n \geq 2, a_0 = 1, a_1 = 1$
$a_n = (1/\sqrt{13})[((1 + \sqrt{13})/2)^{n+1} - ((1 - \sqrt{13})/2)^{n+1}], n \geq 0$
c) $a_n = 2a_{n-1} + 3a_{n-2}, n \geq 2, a_0 = 1, a_1 = 2$
$a_n = (3/4)(3^n) + (1/4)(-1)^n, n \geq 0$
7. a)

$$F_1 = F_2 - F_0$$
$$F_3 = F_4 - F_2$$
$$F_5 = F_6 - F_4$$
$$\vdots$$
$$F_{2n-1} = F_{2n} - F_{2n-2}$$

Conjecture: For all $n \in \mathbf{Z}^+$, $F_1 + F_3 + F_5 + \cdots + F_{2n-1} = F_{2n} - F_0 = F_{2n}$.
Proof (By Mathematical Induction): For $n = 1$ we have $F_1 = F_2$, and this is true since $F_1 = 1 = F_2$.
Consequently, the result is true in this first case (and this establishes the basis step for the proof).
Next we assume the result true for $n = k \ (\geq 1)$ — that is, we assume

$$F_1 + F_3 + F_5 + \cdots + F_{2k-1} = F_{2k}.$$

When $n = k + 1$, we then find that

$$F_1 + F_3 + F_5 + \cdots + F_{2k-1} + F_{2(k+1)-1}$$
$$= (F_1 + F_3 + F_5 + \cdots + F_{2k-1}) + F_{2k+1} = F_{2k} + F_{2k+1} = F_{2k+2} = F_{2(k+1)}.$$

Therefore, the truth for $n = k$ implies the truth at $n = k + 1$, so by the Principle of Mathematical
Induction it follows that for all $n \in \mathbf{Z}^+$

$$F_1 + F_3 + F_5 + \cdots + F_{2n-1} = F_{2n}.$$

9. $a_n = (1/\sqrt{5})[((1 + \sqrt{5})/2)^{n+1} - ((1 - \sqrt{5})/2)^{n+1}], n \geq 0$
11. a) $a_n = a_{n-1} + a_{n-2}, n \geq 3, a_1 = 2, a_2 = 3$: $a_n = F_{n+2}, n \geq 1$.
b) $b_n = b_{n-1} + b_{n-2}, n \geq 3, b_1 = 1, b_2 = 3$: $b_n = L_n, n \geq 1$.
13. $a_n = [(8 + 9\sqrt{2})/16][2 + 4\sqrt{2}]^n + [(8 - 9\sqrt{2})/16][2 - 4\sqrt{2}]^n, n \geq 0$
15. $a_n = 2^{F_n}$, where F_n is the nth Fibonacci number for $n \geq 0$
17. a) F_{n+2} **b)** (i) F_n (ii) F_{n-1} (iii) F_{n-k+2} **c)** $n + 2: 0, \ n + 3: 1$
d) These results provide a combinatorial proof that $F_{n+2} = (F_n + F_{n-1} + \cdots + F_2 + F_1) + 1$.
19. $(\alpha, \alpha), (\beta, \beta)$

21. a) *Proof* (By the alternative form of the Principle of Mathematical Induction):

$$F_3 = 2 = (1 + \sqrt{9})/2 > (1 + \sqrt{5})/2 = \alpha = \alpha^{3-2},$$
$$F_4 = 3 = (3 + \sqrt{9})/2 > (3 + \sqrt{5})/2 = \alpha^2 = \alpha^{4-2},$$

so the result is true for these first two cases (where $n = 3, 4$). This establishes the basis step. Assuming the truth of the statement for $n = 3, 4, 5, \ldots, k \, (\geq 4)$, where k is a fixed (but arbitrary) integer, we continue now with $n = k + 1$:

$$
\begin{aligned}
F_{k+1} &= F_k + F_{k-1} \\
&> \alpha^{k-2} + \alpha^{(k-1)-2} \\
&= \alpha^{k-2} + \alpha^{k-3} = \alpha^{k-3}(\alpha + 1) \\
&= \alpha^{k-3} \cdot \alpha^2 = \alpha^{k-1} = \alpha^{(k+1)-2}.
\end{aligned}
$$

Consequently, $F_n > \alpha^{n-2}$ for all $n \geq 3$ — by the alternative form of the Principle of Mathematical Induction.

23. $a_n = 2a_{n-1} + a_{n-2}, n \geq 2, a_0 = 1, a_1 = 3$:
$a_n = (1/2)[(1 + \sqrt{2})^{n+1} + (1 - \sqrt{2})^{n+1}], n \geq 0$

25. $(7/10)(7^{10}) + (3/10)(-3)^{10} = 197{,}750{,}389$

27. $a_n = a_{n-1} + a_{n-2} + 2a_{n-3}, n \geq 4, a_1 = 1, a_2 = 2, a_3 = 5$:
$a_n = (4/7)(2)^n + (3/7)\cos(2n\pi/3) + (\sqrt{3}/21)\sin(2n\pi/3), n \geq 1$

29. $x_n = 4(2^n) - 3, n \geq 0$ **31.** $a_n = \sqrt{51(4^n) - 35}, n \geq 0$

33. Since $\gcd(F_1, F_0) = 1 = \gcd(F_2, F_1)$, consider $n \geq 2$. Then

$$
\begin{aligned}
F_3 &= F_2 + F_1 \ (= 1) \\
F_4 &= F_3 + F_2 \\
F_5 &= F_4 + F_3 \\
&\vdots \\
F_{n+1} &= F_n + F_{n-1}
\end{aligned}
$$

Reversing the order of these equations, we have the steps in the Euclidean algorithm for computing the gcd of F_{n+1} and $F_n, n \geq 2$. Since the last nonzero remainder is $F_1 = 1$, it follows that $\gcd(F_{n+1}, F_n) = 1$ for all $n \geq 2$.

Section 10.3 – p. 481

1. a) $a_n = (n + 1)^2, n \geq 0$ **b)** $a_n = 3 + n(n - 1)^2, n \geq 0$
c) $a_n = 6(2^n) - 5, n \geq 0$ **d)** $a_n = 2^n + n(2^{n-1}), n \geq 0$

3. a) $a_n = a_{n-1} + n, n \geq 1, a_0 = 1$ $a_n = 1 + [n(n + 1)]/2, n \geq 0$
b) $b_n = b_{n-1} + 2, n \geq 2, b_1 = 2,$ $b_n = 2n, n \geq 1, b_0 = 1$

5. a) $a_n = (3/4)(-1)^n - (4/5)(-2)^n + (1/20)(3)^n, n \geq 0$
b) $a_n = (2/9)(-2)^n - (5/6)(n)(-2)^n + (7/9), n \geq 0$

7. $a_n = A + Bn + Cn^2 - (3/4)n^3 + (5/24)n^4$ **9.** $P = \$117.68$

11. a) $a_n = [(3/4)(3)^n - 5(2)^n + (7n/2) + (21/4)]^{1/2}, n \geq 0$ **b)** $a_n = 2, n \geq 0$

13. a) $t_n = 2t_{n-1} + 2^{n-1}, n \geq 2, t_1 = 2$:
$t_n = (n + 1)(2^{n-1}), n \geq 1$
b) $t_n = 4t_{n-1} + 3(4^{n-1}), n \geq 2, t_1 = 4$:
$t_n = (1 + 3n)4^{n-1}, n \geq 1$
c) $t_n = [1 + (r - 1)n]r^{n-1}, n \geq 1, r = |\Sigma| \geq 1$.

Section 10.4 – p. 487

1. a) $a_n = (1/2)[1 + 3^n], n \geq 0$ **b)** $a_n = 1 + [n(n - 1)(2n - 1)]/6, n \geq 0$
c) $a_n = 5(2^n) - 4, n \geq 0$ **d)** $a_n = 2^n, n \geq 0$

3. a) $a_n = 2^n(1 - 2n), b_n = n(2^{n+1}), n \geq 0$

b) $a_n = (-3/4) + (1/2)(n + 1) + (1/4)(3^n)$,
$b_n = (3/4) + (1/2)(n + 1) - (1/4)(3^n), n \geq 0$

Section 10.5—p. 493

1. $b_4 = (8!)/[(5!)(4!)] = 14$

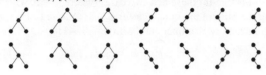

3. $\begin{pmatrix} 2n - 1 \\ n \end{pmatrix} - \begin{pmatrix} 2n - 1 \\ n - 2 \end{pmatrix} = \left[\dfrac{(2n - 1)!}{n!(n - 1)!} \right] - \left[\dfrac{(2n - 1)!}{(n - 2)!(n + 1)!} \right]$

$\qquad = \left[\dfrac{(2n - 1)!(n + 1)}{(n + 1)!(n - 1)!} \right] - \left[\dfrac{(2n - 1)!(n - 1)}{(n - 1)!(n + 1)!} \right]$

$\qquad = \left[\dfrac{(2n - 1)!}{(n + 1)!(n - 1)!} \right] [(n + 1) - (n - 1)]$

$\qquad = \dfrac{(2n - 1)!(2)}{(n + 1)!(n - 1)!} = \dfrac{(2n - 1)!(2n)}{(n + 1)! \, n!} = \dfrac{(2n)!}{(n + 1)(n!)(n!)}$

$\qquad = \dfrac{1}{(n + 1)} \begin{pmatrix} 2n \\ n \end{pmatrix}$

5. a) $(1/9)\binom{16}{8}$ **b)** $[(1/4)\binom{6}{3}]^2$ **c)** $[(1/6)\binom{10}{5}][(1/3)\binom{4}{2}]$ **d)** $(1/6)\binom{10}{5}$

7. a)

b) (iii) $(((ab)c)d)e$ (iv) $(ab)(c(de))$

9. $a_n = a_0 a_{n-1} + a_1 a_{n-2} + a_2 a_{n-3} + \cdots + a_{n-2} a_1 + a_{n-1} a_0$

Since $a_0 = 1, a_1 = 1, a_2 = 2$, and $a_3 = 5$, we find that $a_n =$ the nth Catalan number.

11. a)

x	$f_1(x)$	$f_2(x)$	$f_3(x)$	$f_4(x)$	$f_5(x)$
1	1	3	2	2	1
2	2	3	2	3	3
3	3	3	3	3	3

b) The functions in part (a) correspond with the following paths from $(0, 0)$ to $(3, 3)$.

c) The mountain ranges in Fig. 10.24 of the text.

d) For $n \in \mathbf{Z}^+$, the number of monotone increasing functions $f: \{1, 2, 3, \ldots, n\} \rightarrow \{1, 2, 3, \ldots, n\}$, where $f(i) \geq i$ for all $1 \leq i \leq n$, is $b_n = (1/(n + 1))\binom{2n}{n}$, the nth Catalan number. This follows from Exercise 3 in Section 1.5. There is a one-to-one correspondence between the paths described in that exercise and the functions being dealt with here.

13. $(1/(n + 1))\binom{2n}{n}$, the nth Catalan number

15. a) $E_3 = 2$ **b)** $E_5 = 16$

c) For each rise/fall permutation, n cannot be in the first position (unless $n = 1$); n is the second component of a rise in such a permutation. Consequently, n must be at position 2 or 4 ... or $2\lfloor n/2 \rfloor$.

d) Consider the location of n in a rise/fall permutation $x_1 x_2 x_3 \cdots x_{n-1} x_n$ of $1, 2, 3, \ldots, n$. The number n is in position $2i$ for some $1 \le i \le \lfloor n/2 \rfloor$. Here there are $2i - 1$ numbers that precede n. These can be selected in $\binom{n-1}{2i-1}$ ways and give rise to E_{2i-1} rise/fall permutations. The $(n-1) - (2i - 1) = n - 2i$ numbers that follow n give rise to E_{n-2i} rise/fall permutations. Consequently, $E_n = \sum_{i=1}^{\lfloor n/2 \rfloor} \binom{n-1}{2i-1} E_{2i-1} E_{n-2i}$, $n \ge 2$.

g) From parts (d) and (f)

$$E_n = \binom{n-1}{1} E_1 E_{n-2} + \binom{n-1}{3} E_3 E_{n-4} + \cdots + \binom{n-1}{2\lfloor n/2 \rfloor - 1} E_{2\lfloor n/2 \rfloor - 1} E_{n-2\lfloor n/2 \rfloor}$$

$$E_n = \binom{n-1}{0} E_0 E_{n-1} + \binom{n-1}{2} E_2 E_{n-3} + \cdots + \binom{n-1}{2\lfloor (n-1)/2 \rfloor} E_{2\lfloor (n-1)/2 \rfloor} E_{n-2\lfloor (n-1)/2 \rfloor - 1}$$

Adding these equations we have

$$2E_n = \sum_{i=0}^{n-1} \binom{n-1}{i} E_i E_{n-i-1} \quad \text{or} \quad E_n = (1/2) \sum_{i=0}^{n-1} \binom{n-1}{i} E_i E_{n-i-1}.$$

h) $E_6 = 61$, $E_7 = 272$

i) Consider the Maclaurin series expansions $\sec x = 1 + x^2/2! + 5x^4/4! + 61x^6/6! + \cdots$ and $\tan x = x + 2x^3/3! + 16x^5/5! + 272x^7/7! + \cdots$ One finds that $\sec x + \tan x$ is the exponential generating function of the sequence $1, 1, 1, 2, 5, 16, 61, 272, \ldots$ — namely, the sequence of Euler numbers.

Section 10.6 – p. 504

1. a) $f(n) = (5/3)(4n^{\log_3 4} - 1)$ and $f \in O(n^{\log_3 4})$ for $n \in \{3^i | i \in \mathbf{N}\}$

b) $f(n) = 7(\log_5 n + 1)$ and $f \in O(\log_5 n)$ for $n \in \{5^i | i \in \mathbf{N}\}$

3. a) $f \in O(\log_b n)$ on $\{b^k | k \in \mathbf{N}\}$ **b)** $f \in O(n^{\log_b a})$ on $\{b^k | k \in \mathbf{N}\}$

5. a) $f(1) = 0$ $f(n) = 2f(n/2) + 1$
From Exercise 2(b), $f(n) = n - 1$.

b) The equation $f(n) = f(n/2) + (n/2)$ arises as follows: There are $n/2$ matches played in the first round. Then there are $n/2$ players remaining, so we need $f(n/2)$ additional matches to determine the winner.

7. $O(1)$

9. a)

$$f(n) \le af(n/b) + cn$$
$$af(n/b) \le a^2 f(n/b^2) + ac(n/b)$$
$$a^2 f(n/b^2) \le a^3 f(n/b^3) + a^2 c(n/b^2)$$
$$\vdots \qquad \vdots$$
$$a^{k-1} f(n/b^{k-1}) \le a^k f(n/b^k) + a^{k-1} c(n/b^{k-1})$$

Hence $f(n) \le a^k f(n/b^k) + cn[1 + (a/b) + (a/b)^2 + \cdots + (a/b)^{k-1}] = a^k f(1) + cn[1 + (a/b) + (a/b)^2 + \cdots + (a/b)^{k-1}]$, because $n = b^k$. Since $f(1) \le c$ and $(n/b^k) = 1$, we have $f(n) \le cn[1 + (a/b) + (a/b)^2 + \cdots + (a/b)^{k-1} + (a/b)^k] = (cn) \sum_{i=0}^{k} (a/b)^i$.

c) For $a \neq b$,

$$cn \sum_{i=0}^{k} (a/b)^i = cn \left[\frac{1 - (a/b)^{k+1}}{1 - (a/b)} \right] = (c)(b^k) \left[\frac{1 - (a/b)^{k+1}}{1 - (a/b)} \right]$$

$$= c \left[\frac{b^k - (a^{k+1}/b)}{1 - (a/b)} \right] = c \left[\frac{b^{k+1} - a^{k+1}}{b - a} \right] = c \left[\frac{a^{k+1} - b^{k+1}}{a - b} \right].$$

d) From part (c), $f(n) \leq (c/(a - b))[a^{k+1} - b^{k+1}] = (ca/(a - b))a^k - (cb/(a - b))b^k$. But $a^k = a^{\log_b n} = n^{\log_b a}$ and $b^k = n$, so $f(n) \leq (ca/(a - b))n^{\log_b a} - (cb/(a - b))n$.

   (i) When $a < b$, then $\log_b a < 1$, and $f \in O(n)$ on $\mathbf{Z}^+$.

   (ii) When $a > b$, then $\log_b a > 1$, and $f \in O(n^{\log_b a})$ on $\mathbf{Z}^+$.

Supplementary Exercises—p. 508

1. $\binom{n}{k+1} = \dfrac{n!}{(k+1)!(n-k-1)!} = \dfrac{(n-k)}{(k+1)} \cdot \dfrac{n!}{k!(n-k)!} = \left(\dfrac{n-k}{k+1} \right) \binom{n}{k}$

3. There are two cases to consider. Case 1 (1 is a summand): Here there are $p(n-1, k-1)$ ways to partition $n - 1$ into exactly $k - 1$ summands. Case 2 (1 is not a summand): Here each summand $s_1, s_2, \ldots, s_k > 1$. For $1 \leq i \leq k$, let $t_i = s_i - 1 \geq 1$. Then $t_1, t_2, \ldots, t_k$ provide a partition of $n - k$ into exactly k summands. These cases are exhaustive and disjoint, so by the rule of sum, $p(n, k) = p(n-1, k-1) + p(n-k, k)$.

5. b) *Conjecture*: For $n \in \mathbf{Z}^+$, $A^n = \begin{bmatrix} F_{n+1} & F_n \\ F_n & F_{n-1} \end{bmatrix}$, where F_n denotes the nth Fibonacci number.

Proof: For $n = 1$, $A = A^1 = \begin{bmatrix} 1 & 1 \\ 1 & 0 \end{bmatrix} = \begin{bmatrix} F_2 & F_1 \\ F_1 & F_0 \end{bmatrix}$, so the result is true in this case.

Assume the result true for $n = k \geq 1$. That is, $A^k = \begin{bmatrix} F_{k+1} & F_k \\ F_k & F_{k-1} \end{bmatrix}$. For $n = k + 1$,

$$A^n = A^{k+1} = A^k \cdot A = \begin{bmatrix} F_{k+1} & F_k \\ F_k & F_{k-1} \end{bmatrix} \begin{bmatrix} 1 & 1 \\ 1 & 0 \end{bmatrix} = \begin{bmatrix} F_{k+1} + F_k & F_{k+1} \\ F_k + F_{k-1} & F_k \end{bmatrix} = \begin{bmatrix} F_{k+2} & F_{k+1} \\ F_{k+1} & F_k \end{bmatrix}$$

Consequently, the result is true for all $n \in \mathbf{Z}^+$, by the Principle of Mathematical Induction.

7. $(-1, 0), (\alpha, \alpha), (\beta, \beta)$

9. a) Since $\alpha^2 = \alpha + 1$, it follows that $\alpha^2 + 1 = 2 + \alpha$ and $(2 + \alpha)^2 = 4 + 4\alpha + \alpha^2 = 4(1 + \alpha) + \alpha^2 = 5\alpha^2$.

c) $\displaystyle \sum_{k=0}^{2n} \binom{2n}{k} F_{2k+m} = \sum_{k=0}^{2n} \binom{2n}{k} \left[\frac{\alpha^{2k+m} - \beta^{2k+m}}{\alpha - \beta} \right]$

$$= (1/(\alpha - \beta)) \left[\sum_{k=0}^{2n} \binom{2n}{k} (\alpha^2)^k \alpha^m - \sum_{k=0}^{2n} \binom{2n}{k} (\beta^2)^k \beta^m \right]$$

$$= (1/(\alpha - \beta))[\alpha^m (1 + \alpha^2)^{2n} - \beta^m (1 + \beta^2)^{2n}]$$

$$= (1/(\alpha - \beta))[\alpha^m (2 + \alpha)^{2n} - \beta^m (2 + \beta)^{2n}]$$

$$= (1/(\alpha - \beta))[\alpha^m ((2 + \alpha)^2)^n - \beta^m ((2 + \beta)^2)^n]$$

$$= (1/(\alpha - \beta))[\alpha^m (5\alpha^2)^n - \beta^m (5\beta^2)^n]$$

$$= 5^n (1/(\alpha - \beta))[\alpha^{2n+m} - \beta^{2n+m}] = 5^n F_{2n+m}$$

11. $c_n = F_{n+2}$, the $(n+2)$-nd Fibonacci number

13. a) F_{n+1} **b)** (i) $1 = \binom{n-0}{n-2 \cdot 0}$ (ii) $\binom{n-1}{n-2 \cdot 1}$ (iii) $\binom{n-2}{n-2 \cdot 2}$ (iv) $\binom{n-3}{n-2 \cdot 3}$ (v) $\binom{n-k}{n-2k}$

c) $F_{n+1} = \sum_{k=0}^{\lfloor n/2 \rfloor} \binom{n-k}{k} = \sum_{k=0}^{\lfloor n/2 \rfloor} \binom{n-k}{n-2k}$

15. a) For each derangement, 1 is placed in position i, where $2 \leq i \leq n$. Two things then occur. Case 1 (i is in position 1): Here the other $n - 2$ integers are deranged in d_{n-2} ways. With $n - 1$ choices for i, this results in $(n-1)d_{n-2}$ such derangements. Case 2 [i is not in position 1 (or position i)]: Here we consider 1 as the new natural position for i, so there are $n - 1$ elements to derange. With $n - 1$ choices for i, we have $(n - 1)d_{n-1}$ derangements. Since the two cases are exhaustive and disjoint, the result follows from the rule of sum.

b) $d_0 = 1$ **c)** $d_n - nd_{n-1} = d_{n-2} - (n-2)d_{n-3}$

17. a) $a_n = \binom{2n}{n}, n \geq 0$ **b)** $r = 1, s = -4, t = -1/2$

d) $b_n = (1/(2n-1))\binom{2n}{n}, n \geq 1; b_0 = 0$

19. $c = \alpha$ or $c = \beta$ **21.** $p = -\beta$

23. $a_n = a_{n-1} + a_{n-2}, n \geq 3, a_1 = 1, a_2 = 2: a_n = F_{n+1}, n \geq 1$

25. a) $(n = 0)$ $F_1^2 - F_0 F_1 - F_0^2 = 1^2 - 0 \cdot 1 - 0^2 = 1$

$(n = 1)$ $F_2^2 - F_1 F_2 - F_1^2 = 1^2 - 1 \cdot 1 - 1^2 = -1$

$(n = 2)$ $F_3^2 - F_2 F_3 - F_2^2 = 2^2 - 1 \cdot 2 - 1^2 = 1$

$(n = 3)$ $F_4^2 - F_3 F_4 - F_3^2 = 3^2 - 2 \cdot 3 - 2^2 = -1$

b) *Conjecture*: For $n \geq 0$,

$$F_{n+1}^2 - F_n F_{n+1} - F_n^2 = \begin{cases} 1, & n \text{ even} \\ -1, & n \text{ odd.} \end{cases}$$

c) *Proof*: The result is true for $n = 0, 1, 2, 3$, by the calculations in part (a). Assume the result true for $n = k \, (\geq 3)$. There are two cases to consider — namely, k even and k odd. We shall establish the result for k even, the proof for k odd being similar. Our induction hypothesis tells us that $F_{k+1}^2 - F_k F_{k+1} - F_k^2 = 1$. When $n = k + 1 \, (\geq 4)$ we find that $F_{k+2}^2 - F_{k+1} F_{k+2} - F_{k+1}^2 = (F_{k+1} + F_k)^2 - F_{k+1}(F_{k+1} + F_k) - F_{k+1}^2 = F_{k+1}^2 + 2F_{k+1}F_k + F_k^2 - F_{k+1}^2 - F_{k+1}F_k - F_{k+1}^2 = F_{k+1}F_k + F_k^2 - F_{k+1}^2 = -[F_{k+1}^2 - F_k F_{k+1} - F_k^2] = -1$. The result follows for all $n \in \mathbf{N}$, by the Principle of Mathematical Induction.

27. a) $r(C_1, x) = 1 + x$ $\qquad\qquad$ $r(C_4, x) = 1 + 4x + 3x^2$

$r(C_2, x) = 1 + 2x$ $\qquad\qquad$ $r(C_5, x) = 1 + 5x + 6x^2 + x^3$

$r(C_3, x) = 1 + 3x + x^2$ $\qquad$ $r(C_6, x) = 1 + 6x + 10x^2 + 4x^3$

In general, for $n \geq 3, r(C_n, x) = r(C_{n-1}, x) + xr(C_{n-2}, x)$.

b) $r(C_1, 1) = 2$ $\qquad$ $r(C_3, 1) = 5$ $\qquad$ $r(C_5, 1) = 13$

$r(C_2, 1) = 3$ $\qquad$ $r(C_4, 1) = 8$ $\qquad$ $r(C_6, 1) = 21$

[*Note*: For $1 \leq i \leq n$, if one "straightens out" the chessboard C_i in Fig. 10.28, the result is a $1 \times i$ chessboard — like those studied in Exercise 26.]

29. a) The partitions counted in $f(n, m)$ fall into two categories:

(1) Partitions where m is a summand. These are counted in $f(n - m, m)$, for m may occur more than once.

(2) Partitions where m is not a summand — so that $m - 1$ is the largest possible summand. These partitions are counted in $f(n, m - 1)$.

Since these two categories are exhaustive and mutually disjoint, it follows that $f(n, m) = f(n - m, m) + f(n, m - 1)$.

Chapter 11
An Introduction to Graph Theory

Section 11.1 – p. 518

1. a) To represent the air routes traveled among a certain set of cities by a particular airline.

b) To represent an electrical network. Here the vertices can represent switches, transistors, and so on, and an edge (x, y) indicates the existence of a wire connecting x to y.

c) Let the vertices represent a set of job applicants and a set of open positions in a corporation. Draw an edge (A, b) to denote that applicant A is qualified for position b. Then all open positions can be filled if the resulting graph provides a matching between a subset of the applicants and the open positions.

3. 6   **5.** 9; 3

7. a)

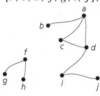

b) $\{(g, d), (d, e), (e, a)\}$; $\{(g, b), (b, c), (c, d), (d, e), (e, a)\}$

c) Two: one of $\{(b, c), (c, d)\}$ and one of $\{(b, f), (f, g), (g, d)\}$

d) No

e) Yes. Travel the path $\{(c, d), (d, e), (e, a), (a, b), (b, f), (f, g)\}$

f) Yes. Travel the trail $\{(g, b), (b, f), (f, g), (g, d), (d, b), (b, c), (c, d), (d, e), (e, a),$ $(a, b)\}$.

9. If $\{a, b\}$ is not part of a cycle, then its removal disconnects a and b (and G). If not, there is a path P from a to b, and P together with $\{a, b\}$ provides a cycle containing $\{a, b\}$. Conversely, if the removal of $\{a, b\}$ from G disconnects G, then there exist $x, y, \in V$ such that the only path P from x to y contains $e = \{a, b\}$. If e were part of a cycle C, then the edges in $(P - \{e\}) \cup (C - \{e\})$ would contain a second path connecting x to y.

11. a) Yes   **b)** No   **c)** $n - 1$

13. The partition of V induced by $\mathscr{R}$ yields the (connected) components of G.

15. The number of closed $v - v$ walks of length $n \geq 1$ is F_{n+1}, the $(n + 1)$-st Fibonacci number.

Section 11.2 – p. 528

1. a) 3   **b)** $G_1 = \langle U \rangle$, where $U = \{a, b, d, f, g, h, i, j\}$; $G_1 = G - \{c\}$

c) $G_2 = \langle W \rangle$, where $W = \{b, c, d, f, g, i, j\}$; $G_2 = G - \{a, h\}$

d)

e)

3. a) $2^9 = 512$ **b)** 3   **c)** 2^6

5. G is (or is isomorphic to) K_n, where $n = |V|$.

7. (i)

R	Y	W	B
B ⬛1 Y	R ⬛2 B	Y ⬛3 R	W ⬛4 W
W	B	Y	R

(ii) No solution

(iii)

W	B	Y	R
R ⬛1 W	W ⬛2 B	Y ⬛3 R	B ⬛4 Y
Y	R	B	W

9. a) No   **b)** Yes. Correspond a with u, b with w, c with x, d with y, e with v, and f with z.

11. a) If $G_1 = (V_1, E_1)$ and $G_2 = (V_2, E_2)$ are isomorphic, then there is a function $f: V_1 \rightarrow V_2$ that is one-to-one and onto and preserves adjacencies. If $x, y \in V_1$ and $\{x, y\} \notin E_1$, then $\{f(x), f(y)\} \notin E_2$. Hence the same function f preserves adjacencies for $\overline{G}_1, \overline{G}_2$ and can be used to define an isomorphism for $\overline{G}_1, \overline{G}_2$. The converse follows in a similar way.

b) They are not isomorphic. The complement of the graph containing vertex a is a cycle of length 8. The complement of the other graph is the disjoint union of two cycles of length 4.

13. If G is the cycle with edges $\{a, b\}$, $\{b, c\}$, $\{c, d\}$, $\{d, e\}$, and $\{e, a\}$, then $\overline{G}$ is the cycle with edges $\{a, c\}$, $\{c, e\}$, $\{e, b\}$, $\{b, d\}$, and $\{d, a\}$. Hence G and $\overline{G}$ are isomorphic. Conversely, if G is a cycle on n vertices and $G, \overline{G}$ are isomorphic, then $n = \frac{1}{2}\binom{n}{2}$, or $n = \frac{1}{4}(n)(n - 1)$, and $n = 5$.

15. a) Here f must also maintain directions. So $(a, b) \in E_1$ if and only if $(f(a), f(b)) \in E_2$.

b) They are not isomorphic. Consider vertex a in the first graph. It is incident to one vertex and incident from two other vertices. No vertex in the other graph has this property.

17. $n^2 - 3n + 3$

Section 11.3—p. 537

1. a) $|V| = 6$ **b)** $|V| = 1$ or 2 or 3 or 5 or 6 or 10 or 15 or 30

(In the first four cases, G must be a multigraph; when $|V| = 30$, G is disconnected.)

c) $|V| = 6$

3. a) 9

b)

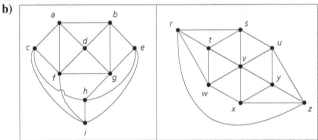

5. a) $|V_1| = 8 = |V_2|$; $|E_1| = 14 = |E_2|$

b) For V_1 we find that $\deg(a) = 3$, $\deg(b) = 4$, $\deg(c) = 4$, $\deg(d) = 3$, $\deg(e) = 3$, $\deg(f) = 4$, $\deg(g) = 4$, and $\deg(h) = 3$. For V_2 we have $\deg(s) = 3$, $\deg(t) = 4$, $\deg(u) = 4$, $\deg(v) = 3$, $\deg(w) = 4$, $\deg(x) = 3$, $\deg(y) = 3$, $\deg(z) = 4$. Hence each of the two graphs has four vertices of degree 3 and four of degree 4.

c) Despite the results in parts (a) and (b), the graphs G_1 and G_2 are *not* isomorphic.

In the graph G_2 the four vertices of degree 4 — namely, t, u, w, and z — are on a cycle of length 4. For the graph G_1 the vertices b, c, f, and g — each of degree 4 — do not lie on a cycle of length 4.

A second way to observe that G_1 and G_2 are not isomorphic is to consider once again the vertices of degree 4 in each graph. In G_1 these vertices induce a disconnected subgraph consisting of the two edges $\{b, c\}$ and $\{f, g\}$. The four vertices of degree 4 in graph G_2 induce a connected subgraph that has five edges — every possible edge except $\{u, z\}$.

7. a) 19   **b)** $\sum_{i=1}^{n} \binom{d_i}{2}$ (*Note*: No assumption about connectedness is made here.)

9. a) 16   **b)** $2^{19} = 524,288$

11. The number of edges in K_n is $\binom{n}{2} = n(n-1)/2$. If the edges of K_n can be partitioned into such cycles of length 4, then 4 divides $\binom{n}{2}$ and $\binom{n}{2} = 4t$, for some $t \in \mathbf{Z}^+$. For each vertex v that appears in a cycle, there are two edges (of K_n) incident to v. Consequently, each vertex v of K_n has even degree, so n is odd. Therefore, $n - 1$ is even and as $4t = \binom{n}{2} = n(n-1)/2$, it follows that $8t = n(n-1)$. So 8 divides $n(n-1)$, and since n is odd, it follows (from the Fundamental Theorem of Arithmetic) that 8 divides $n - 1$. Hence $n - 1 = 8k$, or $n = 8k + 1$, for some $k \in \mathbf{Z}^+$.

13. $\delta|V| \leq \sum_{v \in V} \deg(v) \leq \Delta|V|$. Since $2|E| = \sum_{v \in V} \deg(v)$, it follows that $\delta|V| \leq 2|E| \leq \Delta|V|$, so $\delta \leq 2(e/n) \leq \Delta$.

15. Start with a cycle $v_1 \rightarrow v_2 \rightarrow v_3 \rightarrow \cdots \rightarrow v_{2k-1} \rightarrow v_{2k} \rightarrow v_1$. Then draw the k edges $\{v_1, v_{k+1}\}$, $\{v_2, v_{k+2}\}, \ldots, \{v_i, v_{i+k}\}, \ldots, \{v_k, v_{2k}\}$. The resulting graph has $2k$ vertices each of degree 3.

17. (Corollary 11.1). Let $V = V_1 \cup V_2$, where $V_1 (V_2)$ contains all vertices of odd (even) degree. Then $2|E| - \sum_{v \in V_2} \deg(v) = \sum_{v \in V_1} \deg(v)$ is an even integer. For $|V_1|$ odd, $\sum_{v \in V_1} \deg(v)$ is odd.

　　(Corollary 11.2). For the converse let $G = (V, E)$ have an Euler trail with a, b as the starting and terminating vertices. Add the edge $\{a, b\}$ to G to form the larger graph $G_1 = (V, E_1)$ where G_1 has an Euler circuit. Hence G_1 is connected and each vertex in G_1 has even degree. When we remove edge $\{a, b\}$ from G_1, the vertices in G will have the same even degree except for a, b; $\deg_G(a) = \deg_{G_1}(a) - 1$, $\deg_G(b) = \deg_{G_1}(b) - 1$, so the vertices a, b have odd degree in G. Also, since the edges in G form an Euler trail, G is connected.

19. a) Let $a, b, c, x, y \in V$ with $\deg(a) = \deg(b) = \deg(c) = 1$, $\deg(x) = 5$, and $\deg(y) = 7$. Since $\deg(y) = 7$, y is adjacent to all of the other (seven) vertices in V. Therefore vertex x is not adjacent to any of the vertices a, b, and c. Since x cannot be adjacent to itself, unless we have loops, it follows that $\deg(x) \leq 4$, and we cannot draw a graph for the given conditions.

b)

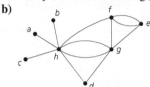

21. n odd; $n = 2$　　**23.** Yes

25. a) (i) 13　(ii) 25　(iii) 41　(iv) $2n^2 - 2n + 1$
　　b) (i) 12　(ii) 24　(iii) 40　(iv) $2n^2 - 2n$

27. In any directed graph (or multigraph), $\sum_{v \in V} \mathrm{od}(v) = |E| = \sum_{v \in V} \mathrm{id}(v)$, so $\sum_{v \in V}[\mathrm{od}(v) - \mathrm{id}(v)] = 0$. For each $v \in V$, $\mathrm{od}(v) + \mathrm{id}(v) = n - 1$, so

$$0 = (n-1) \cdot 0 = \sum_{v \in V}(n-1)[\mathrm{od}(v) - \mathrm{id}(v)]$$

$$= \sum_{v \in V}[\mathrm{od}(v) + \mathrm{id}(v)][\mathrm{od}(v) - \mathrm{id}(v)]$$

$$= \sum_{v \in V}[(\mathrm{od}(v))^2 - (\mathrm{id}(v))^2],$$

and the result follows.

29. a) and **b)**　　　　　　　　　　　　**c)**

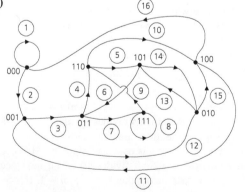

31. Let $|V| = n \geq 2$. Since G is loop-free and connected, for all $x \in V$ we have $1 \leq \deg(x) \leq n - 1$. Apply the pigeonhole principle with the n vertices as the pigeons and the $n - 1$ possible degrees as the pigeonholes.

33. a) Yes　　**b)** Yes　　**c)** No

35. No. Let each person represent a vertex for a graph. If v, w represent two of these people, draw the edge $\{v, w\}$ if the two shake hands. If the situation were possible, then we would have a graph with 15 vertices, each of degree 3. So the sum of the degrees of the vertices would be 45, an odd integer. This contradicts Theorem 11.2.

37. Assign the Gray code {00, 01, 11, 10} to the four horizontal levels: top — 00; second (from the top) — 01; second (from the bottom) — 11; bottom — 10. Likewise, assign the same code to the four vertical levels: left (or, first) — 00; second — 01; third — 11; right (or, fourth) — 10. This provides the labels for $p_1, p_2, \ldots, p_{16}$, where, for instance, p_1 has the label (00, 00), p_2 has the label (01, 00), $\ldots, p_7$ has the label (11, 01), $\ldots, p_{11}$ has the label (11, 11), $\ldots, p_{15}$ has the label (11, 10), and p_{16} has the label (10, 10).

Define the function f from the set of 16 vertices of this grid to the vertices of Q_4 by $f((ab, cd)) = abcd$. Here $f((ab, cd)) = f((a_1b_1, c_1d_1)) \Rightarrow abcd = a_1b_1c_1d_1 \Rightarrow a = a_1$, $b = b_1, c = c_1, d = d_1 \Rightarrow (ab, cd) = (a_1b_1, c_1d_1) \Rightarrow f$ is one-to-one. Since the domain and codomain of f both contain 16 vertices, it follows from Theorem 5.11 that f is also onto. Finally, let {$(ab, cd), (wx, yz)$} be an edge in the grid. Then either $ab = wx$ and cd, yz differ in one component or $cd = yz$ and ab, wx differ in one component. Suppose that $ab = wx$ and $c = y$, but $d \neq z$. Then {$abcd, wxyz$} is an edge in Q_4. The other cases follow in a similar way. Conversely, suppose that {$f((a_1b_1, c_1d_1)), f((w_1x_1, y_1z_1))$} is an edge in Q_4. Then $a_1b_1c_1d_1$, $w_1x_1y_1z_1$ differ in exactly one component — say the first. Then in the grid, there is an edge for the vertices $(0b_1, c_1d_1), (1b_1, c_1d_1)$. The arguments are similar for the other three components. Consequently, f establishes an isomorphism between the three-by-three grid and a subgraph of Q_4. (*Note:* The three-by-three grid has 24 edges while Q_4 has 32 edges.)

Section 11.4 — p. 553

1. In this situation vertex b is in the region formed by the edges {a, d}, {d, c}, {c, a}, and vertex e is outside of this region. Hence the edge {b, e} will cross one of the edges {a, d}, {d, c}, or {a, c}, (as shown).

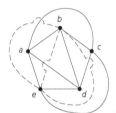

3. a)

Graph	Number of Vertices	Number of Edges
$K_{4,7}$	11	28
$K_{7,11}$	18	77
$K_{m,n}$	$m + n$	mn

b) $m = 6$

5. a) Bipartite **b)** Bipartite **c)** Not bipartite

7. a) $\binom{m}{2}\binom{n}{2}$ **b)** $m\binom{n}{2} + n\binom{m}{2} = (1/2)(mn)[m + n - 2]$
c) $(m)(n)(m - 1)(n - 1) = 4\binom{m}{2}\binom{n}{2}$

9. a) 6 **b)** $(1/2)(7)(3)(6)(2)(5)(1)(4) = 2520$ **c)** 50,295,168,000
d) $(1/2)(n)(m)(n - 1)(m - 1)(n - 2) \cdots (2)(n - (m + 1))(1)(n - m)$

11. Partition V as $V_1 \cup V_2$ with $|V_1| = m$, $|V_2| = v - m$. If G is bipartite, then the maximum number of edges that G can have is $m(v - m) = -[m - (v/2)]^2 + (v/2)^2$, a function of m. For a given value of v, when v is even, $m = v/2$ maximizes $m(v - m) = (v/2)[v - (v/2)] = (v/2)^2$. For v odd, $m = (v - 1)/2$ or $m = (v + 1)/2$ maximizes $m(v - m) = [(v - 1)/2][v - ((v - 1)/2)] = [(v - 1)/2][(v + 1)/2] = [(v + 1)/2][v - ((v + 1)/2)] = (v^2 - 1)/4 = \lfloor (v/2)^2 \rfloor < (v/2)^2$. Hence if $|E| > (v/2)^2$, then G cannot be bipartite.

13. a)

a: $\{1, 2\}$	f: $\{4, 5\}$
b: $\{3, 4\}$	g: $\{2, 5\}$
c: $\{1, 5\}$	h: $\{2, 3\}$
d: $\{2, 4\}$	i: $\{1, 3\}$
e: $\{3, 5\}$	j: $\{1, 4\}$

b) G is (isomorphic to) the Petersen graph. [See Fig. 11.52(a).]

15. mn must be even

17. a) There are 17 vertices, 34 edges, and 19 regions, and $v - e + r = 17 - 34 + 19 = 2$.

b) Here we find 10 vertices, 24 edges, and 16 regions, and $v - e + r = 10 - 24 + 16 = 2$.

19. 10

21. If not, $\deg(v) \geq 6$ for all $v \in V$. Then $2e = \sum_{v \in V} \deg(v) \geq 6|V|$ so $e \geq 3|V|$, contradicting $e \leq 3|V| - 6$ (Corollary 11.3).

23. a) $2e \geq kr = k(2 + e - v) \Rightarrow (2 - k)e \geq k(2 - v) \Rightarrow e \leq [k/(k - 2)](v - 2)$ **b)** 4

c) In $K_{3,3}$, we have $e = 9$ and $v = 6$. $[k/(k - 2)](v - 2) = (4/2)(4) = 8 < 9 = e$. Since $K_{3,3}$ is connected, it must be nonplanar.

d) Here $k = 5$, $v = 10$, $e = 15$, and $[k/(k - 2)](v - 2) = (5/3)(8) = (40/3) < 15 = e$. The Petersen graph is connected, so it must be nonplanar.

25. a) The dual for the tetrahedron [Fig. 11.59(b)] is the graph itself. For the graph (cube) in Fig. 11.59(d) the dual is the octahedron, and vice versa. Likewise, the dual of the dodecahedron is the icosahedron, and vice versa.

b) For $n \in \mathbf{Z}^+$, $n \geq 3$, the dual of the wheel graph W_n is W_n itself.

27.

29. a) As we mentioned in the remark following Example 11.18, when G_1, G_2 are homeomorphic graphs, then they may be regarded as isomorphic except, possibly, for vertices of degree 2. Consequently, two such graphs will have the same number of vertices of odd degree.

b) Now if G_1 has an Euler trail, then G_1 (is connected and) has all vertices of even degree — except two, those being the vertices at the beginning and end of the Euler trail. From part (a) G_2 is likewise connected with all vertices of even degree, except for two of odd degree. Consequently, G_2 has an Euler trail. (The converse follows in a similar way.)

c) If G_1 has an Euler circuit, then G_1 (is connected and) has all vertices of even degree. From part (a) G_2 is likewise connected with all vertices of even degree, so G_2 has an Euler circuit. (The converse follows in a similar manner.)

Section 11.5 — p. 562

1. a) **b)** **c)** **d)**

3. a) Hamilton cycle: $a \to g \to k \to i \to h \to b \to c \to d \to j \to f \to e \to a$

b) Hamilton cycle: $a \to d \to b \to e \to g \to j \to i \to f \to h \to c \to a$

c) Hamilton cycle: $a \to h \to e \to f \to g \to i \to d \to c \to b \to a$

d) Hamilton path: $a \to c \to d \to b \to e \to f \to g$

e) Hamilton path: $a \to b \to c \to d \to e \to j \to i \to h \to g \to f \to k \to l \to m \to n \to o$

f) Hamilton cycle: $a \to b \to c \to d \to e \to j \to i \to h \to g \to l \to m \to n \to o \to t \to s \to r \to q \to p \to k \to f \to a$

5. d) If we remove any one of the vertices a, b, or g, the resulting subgraph has a Hamilton cycle. For example, upon removing vertex a we find the Hamilton cycle $b \to d \to c \to f \to g \to e \to b$.

e) The following Hamilton cycle exists if we remove vertex g: $a \to b \to c \to d \to e \to j \to o \to n \to i \to h \to m \to l \to k \to f \to a$. A symmetric situation results upon removing vertex i.

7. a) $(1/2)(n-1)!$ **b)** 10   **c)** 9

9. Let $G = (V, E)$ be a loop-free undirected graph with no odd cycles. We assume that G is connected — otherwise, we work with the components of G. Select any vertex x in V, and let $V_1 = \{v \in V | d(x, v)$, the length of a shortest path between x and v, is odd$\}$ and $V_2 = \{w \in V | d(x, w)$, the length of a shortest path between x and w, is even$\}$. Note that (i) $x \in V_2$, (ii) $V = V_1 \cup V_2$, and (iii) $V_1 \cap V_2 = \emptyset$. We claim that each edge $\{a, b\}$ in E has one vertex in V_1 and the other vertex in V_2. For suppose that $e = \{a, b\} \in E$ with $a, b \in V_1$. (The proof for $a, b \in V_2$ is similar.) Let $E_a = \{\{a, v_1\}, \{v_1, v_2\}, \dots, \{v_{m-1}, x\}\}$ be the m edges in a shortest path from a to x, and let $E_b = \{\{b, v_1'\}, \{v_1', v_2'\}, \dots, \{v_{n-1}', x\}\}$ be the n edges in a shortest path from b to x. Note that m and n are both odd. If $\{v_1, v_2, \dots, v_{m-1}\} \cap \{v_1', v_2', \dots, v_{n-1}'\} = \emptyset$, then the set of edges $E' = \{\{a, b\}\} \cup E_a \cup E_b$ provides an odd cycle in G. Otherwise, let $w \, (\neq x)$ be the first vertex where the paths come together, and let $E'' =$

$$\{\{a, b\}\} \cup \{\{a, v_1\}, \{v_1, v_2\}, \dots, \{v_i, w\}\} \cup \{\{b, v_1'\}, \{v_1', v_2'\}, \dots, \{v_j', w\}\},$$

for some $1 \leq i \leq m - 1$ and $1 \leq j \leq n - 1$. Then either E'' provides an odd cycle for G or $E' - E''$ contains an odd cycle for G.

11. a)

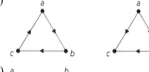

b)

od$(a) = 3$ id$(a) = 0$ od$(a) = 3$ id$(a) = 0$
od$(b) = 2$ id$(b) = 1$ od$(b) = 1$ id$(b) = 2$
od$(c) = 0$ id$(c) = 3$ od$(c) = 1$ id$(c) = 2$
od$(d) = 1$ id$(d) = 2$ od$(d) = 1$ id$(d) = 2$

od$(a) = 1$ id$(a) = 2$ od$(a) = 0$ id$(a) = 3$
od$(b) = 1$ id$(b) = 2$ od$(b) = 2$ id$(b) = 1$
od$(c) = 2$ id$(c) = 1$ od$(c) = 2$ id$(c) = 1$
od$(d) = 2$ id$(d) = 1$ od$(d) = 2$ id$(d) = 1$

13. *Proof:* If not, there exists a vertex x such that $(v, x) \notin E$ and, for all $y \in V$, $y \neq v, x$, if $(v, y) \in E$, then $(y, x) \notin E$. Since $(v, x) \notin E$, we have $(x, v) \in E$, as T is a tournament. Also, for each y mentioned earlier, we also have $(x, y) \in E$. Consequently, od$(x) \geq$ od$(v) + 1$ — contradicting od(v) being a maximum!

15. For the multigraph in the given figure, $|V| = 4$ and deg$(a) =$ deg$(c) =$ deg$(d) = 2$ and deg$(b) = 6$. Hence deg$(x) +$ deg$(y) \geq 4 > 3 = 4 - 1$ for any nonadjacent $x, y \in V$, but the multigraph has no Hamilton path.

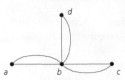

17. For $n \geq 5$, let $C_n = (V, E)$ denote the cycle on n vertices. Then C_n has (actually is) a Hamilton cycle, but for all $v \in V$, $\deg(v) = 2 < n/2$.

19. This follows from Theorem 11.9, since for all (nonadjacent) $x, y \in V$,
$\deg(x) + \deg(y) = 12 > 11 = |V|$.

21. When $n = 5$, the graphs C_5 and $\overline{C}_5$ are isomorphic, and both are Hamilton cycles on five vertices.

　　　For $n \geq 6$, let u, v denote nonadjacent vertices in $\overline{C}_n$. Since $\deg(u) = \deg(v) = n - 3$, we find that $\deg(u) + \deg(v) = 2n - 6$. Also, $2n - 6 \geq n \iff n \geq 6$, so it follows from Theorem 11.9 that the cocycle $\overline{C}_n$ contains a Hamilton cycle when $n \geq 6$.

23. a) The path $v \to v_1 \to v_2 \to v_3 \to \cdots \to v_{n-1}$ provides a Hamilton path for H_n. Since $\deg(v) = 1$, the graph cannot have a Hamilton cycle.
b) Here $|E| = \binom{n-1}{2} + 1$. (So the number of edges required in Corollary 11.6 cannot be decreased.)

25. a) (i) $\{a, c, f, h\}$, $\{a, g\}$　(ii) $\{z\}$, $\{u, w, y\}$　**b)** (i) $\beta(G) = 4$　(ii) $\beta(G) = 3$
c) (i) 3　(ii) 3　(iii) 3　(iv) 4　(v) 6　(vi) The maximum of m and n
d) The complete graph on $|I|$ vertices

Section 11.6 – p. 571

1. Draw a vertex for each species of fish. If two species x, y must be kept in separate aquaria, draw the edge $\{x, y\}$. The smallest number of aquaria needed is then the chromatic number of the resulting graph.

3. a) 3　**b)** 5

5. a) $P(G, \lambda) = \lambda(\lambda - 1)^3$
b) For $G = K_{1,n}$ we find that $P(G, \lambda) = \lambda(\lambda - 1)^n$.　$\chi(K_{1,n}) = 2$

7. a) 2　**b)** 2 (n even); 3 (n odd)
c) Figure 11.59(d): 2; Fig. 11.62(a): 3; Fig. 11.85(i): 2; Fig. 11.85(ii): 3　**d)** 2

9. a) (1) $\lambda(\lambda - 1)^2(\lambda - 2)^2$　(2) $\lambda(\lambda - 1)(\lambda - 2)(\lambda^2 - 2\lambda + 2)$
(3) $\lambda(\lambda - 1)(\lambda - 2)(\lambda^2 - 5\lambda + 7)$
b) (1) 3　(2) 3　(3) 3　**c)** (1) 720　(2) 1020　(3) 420

11. Let $e = \{v, w\}$ be the deleted edge. There are $\lambda(1)(\lambda - 1)(\lambda - 2) \cdots (\lambda - (n - 2))$ proper colorings of G_n where v, w share the same color and $\lambda(\lambda - 1)(\lambda - 2) \cdots (\lambda - (n - 1))$ proper colorings where v, w are colored with different colors. Therefore, $P(G_n, \lambda) = \lambda(\lambda - 1) \cdots (\lambda - n + 2) + \lambda(\lambda - 1) \cdots (\lambda - n + 1) = \lambda(\lambda - 1) \cdots (\lambda - n + 3)(\lambda - n + 2)^2$, so $\chi(G_n) = n - 1$.

13. a) $|V| = 2n$; $|E| = (1/2) \sum_{v \in V} \deg(v) = (1/2)[4(2) + (2n - 4)(3)] = (1/2)[8 + 6n - 12] = 3n - 2$, $n \geq 1$.
b) For $n = 1$, we find that $G = K_2$ and $P(G, \lambda) = \lambda(\lambda - 1) = \lambda(\lambda - 1)(\lambda^2 - 3\lambda + 3)^{1-1}$ so the result is true in this first case. For $n = 2$, we have $G = C_4$, the cycle of length 4, and here $P(G, \lambda) = \lambda(\lambda - 1)^3 - \lambda(\lambda - 1)(\lambda - 2) = \lambda(\lambda - 1)(\lambda^2 - 3\lambda + 3)^{2-1}$. So the result follows for $n = 2$. Assuming the result true for an arbitrary (but fixed) $n \geq 1$, consider the situation for $n + 1$. Write $G = G_1 \cup G_2$, where G_1 is C_4 and G_2 is the ladder graph for n rungs. Then $G_1 \cap G_2 = K_2$, so from Theorem 11.14 we have $P(G, \lambda) = P(G_1, \lambda) \cdot P(G_2, \lambda)/P(K_2, \lambda)$
$= [(\lambda)(\lambda - 1)(\lambda^2 - 3\lambda + 3)][(\lambda)(\lambda - 1)(\lambda^2 - 3\lambda + 3)^{n-1}]/[(\lambda)(\lambda - 1)] =$
$(\lambda)(\lambda - 1)(\lambda^2 - 3\lambda + 3)^n$. Consequently, the result is true for all $n \geq 1$, by the Principle of Mathematical Induction.

15. a) $\lambda(\lambda - 1)(\lambda - 2)$　**b)** Follows from Theorem 11.10

c) Follows by the rule of product

d)
$$P(C_n, \lambda) = P(P_{n-1}, \lambda) - P(C_{n-1}, \lambda) = \lambda(\lambda - 1)^{n-1} - P(C_{n-1}, \lambda)$$
$$= [(\lambda - 1) + 1](\lambda - 1)^{n-1} - P(C_{n-1}, \lambda)$$
$$= (\lambda - 1)^n + (\lambda - 1)^{n-1} - P(C_{n-1}, \lambda),$$

so $P(C_n, \lambda) - (\lambda - 1)^n = (\lambda - 1)^{n-1} - P(C_{n-1}, \lambda)$.

Replacing n by $n - 1$ yields

$$P(C_{n-1}, \lambda) - (\lambda - 1)^{n-1} = (\lambda - 1)^{n-2} - P(C_{n-2}, \lambda).$$

Hence

$$P(C_n, \lambda) - (\lambda - 1)^n = P(C_{n-2}, \lambda) - (\lambda - 1)^{n-2}.$$

e) Continuing from part (d),

$$P(C_n, \lambda) = (\lambda - 1)^n + (-1)^{n-3}[P(C_3, \lambda) - (\lambda - 1)^3]$$
$$= (\lambda - 1)^n + (-1)^{n-1}\left[\lambda(\lambda - 1)(\lambda - 2) - (\lambda - 1)^3\right]$$
$$= (\lambda - 1)^n + (-1)^n(\lambda - 1).$$

17. From Theorem 11.13, the expansion for $P(G, \lambda)$ will contain exactly one occurrence of the chromatic polynomial of K_n. Since no larger graph occurs, this term determines the degree as n and the leading coefficient as 1.

19. a) For $n \in \mathbf{Z}^+$, $n \geq 3$, let C_n denote the cycle on n vertices. If n is odd then $\chi(C_n) = 3$. But for each v in C_n, the subgraph $C_n - v$ is a path with $n - 1$ vertices and $\chi(C_n - v) = 2$. So for n odd C_n is color-critical.

However, when n is even we have $\chi(C_n) = 2$, and for each v in C_n, the subgraph $C_n - v$ is still a path with $n - 1$ vertices and $\chi(C_n - v) = 2$. Consequently, cycles with an even number of vertices are not color-critical.

b) For every complete graph K_n, where $n \geq 2$, we have $\chi(K_n) = n$, and for each vertex v in K_n, $K_n - v$ is (isomorphic to) K_{n-1}, so $\chi(K_n - v) = n - 1$. Consequently, every complete graph with at least one edge is color-critical.

c) Suppose that G is not connected. Let G_1 be a component of G where $\chi(G_1) = \chi(G)$, and let G_2 be any other component of G. Then $\chi(G_1) \geq \chi(G_2)$ and for all v in G_2 we find that $\chi(G - v) = \chi(G_1) = \chi(G)$, so G is not color-critical.

Supplementary Exercises – p. 576

1. $n = 17$

3. a) Label the vertices of K_6 with $a, b, \ldots, f$. Of the five edges on a, at least three have the same color, say red. Let these edges be $\{a, b\}, \{a, c\}, \{a, d\}$. If the edges $\{b, c\}, \{c, d\}, \{b, d\}$ are all blue, the result follows. If not, one of these edges, say $\{c, d\}$, is red. Then the edges $\{a, c\}, \{a, d\}, \{c, d\}$ yield a red triangle.

b) Consider the six people as vertices. If two people are friends (strangers), draw a red (blue) edge connecting their respective vertices. The result then follows from part (a).

5. a) We can redraw G_2 as

b) 72

7. a) 1260   **b)** 756

c) (Case 1: p is odd, $p = 2k + 1$ for $k \in \mathbf{N}$.) Here there are mn paths of length $p = 1$ (when $k = 0$) and $(m)(n)(m - 1)(n - 1) \cdots (m - k)(n - k)$ paths of length $p = 2k + 1 \geq 3$.

(Case 2: p is even, $p = 2k$ for $k \in \mathbf{Z}^+$.) When $p < 2m$ (i.e., $k < m$) the number of paths of length p is $(1/2)(m)(n)(m - 1)(n - 1) \cdots (n - (k - 1))(m - k) + (1/2)(n)(m)(n - 1)$ ·

$(m-1)\cdots(m-(k-1))(n-k)$. For $p=2m$ we find $(1/2)(n)(m)(n-1)(m-1)\cdots$
$(m-(m-1))(n-m)$ paths of (longest) length $2m$.

9. a) Let I be independent and $\{a,b\}\in E$. If neither a nor b is in $V-I$, then a, $b\in I$, and since they are adjacent, I is not independent. Conversely, if $I\subseteq V$ with $V-I$ a covering of G, then if I is not independent there are vertices x, $y\in I$ with $\{x,y\}\in E$. But $\{x,y\}\in E\Rightarrow$ either x or y is in $V-I$.

b) Let I be a largest maximal independent set in G and K a minimum covering. From part (a), $|K|\le|V-I|=|V|-|I|$ and $|I|\ge|V-K|=|V|-|K|$, or $|K|+|I|\ge|V|\ge|K|+|I|$.

11. $a_n=a_{n-1}+a_{n-2}, a_0=a_1=1$ $\quad a_n=F_{n+1}$, the $(n+1)$-st Fibonacci number

13. $a_n=a_{n-1}+2a_{n-2}, a_1=3, a_2=5$ $\quad a_n=(-1/3)(-1)^n+(4/3)(2^n), n\ge1$.

15. a) $\gamma(G)=2$; $\beta(G)=3$; $\chi(G)=4$.

b) G has neither an Euler trail nor an Euler circuit; G does have a Hamiltonian cycle.

c) G is not bipartite, but it is planar.

17. a) $\chi(G)\ge\omega(G)$. **b)** They are equal.

19. a) The constant term is 3, not 0. This contradicts Theorem 11.11.

b) The leading coefficient is 3, not 1. This contradicts the result in Exercise 17 of Section 11.6.

c) The sum of the coefficients is -1, not 0. This contradicts Theorem 11.12.

21. a) $a_n=F_{n+2}$, the $(n+2)$-nd Fibonacci number.

c) $H_1: 3+F_6$ $\qquad H_2: 3+F_7$ $\qquad H_3: 3+F_{n+2}$ **d)** 2^s-1+m

Chapter 12
Trees

1. a)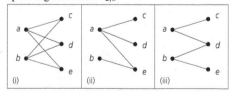

b) 5

3. a) 47 **b)** 11 **5.** Paths **7.**

9. If there is a unique path between each pair of vertices in G, then G is connected. If G contains a cycle, then there is a pair of vertices x, y with two distinct paths connecting x and y. Hence, G is a loop-free connected undirected graph with no cycles, so G is a tree.

11. $\binom{n}{2}$

13. In part (i) of the given figure we find the complete bipartite graph $K_{2,3}$. Parts (ii) and (iii) provide two nonisomorphic spanning trees for $K_{2,3}$. Up to isomorphism these are the only spanning trees for $K_{2,3}$.

15. (1) 6 (2) 36

17. a) $n\ge m+1$

b) Let k be the number of pendant vertices in T. From Theorems 11.2 and 12.3 we have $2(n-1)=2|E|=\sum_{v\in V}\deg(v)\ge k+m(n-k)$. Consequently,

$$[2(n-1)\ge k+m(n-k)]\Rightarrow[2n-2\ge k+mn-mk]$$
$$\Rightarrow[k(m-1)\ge2-2n+mn=2+(m-2)n\ge2+(m-2)(m+1)$$
$$=2+m^2-m-2=m^2-m=m(m-1)],$$

so $k\ge m$.

19. a) If the complement of T contains a cut-set, then the removal of these edges disconnects G, and there are vertices x, y with no path connecting them. Hence T is not a spanning tree for G.

b) If the complement of C contains a spanning tree, then every pair of vertices in G has a path connecting them, and this path includes no edges of C. Hence the removal of the edges in C from G does not disconnect G, so C is not a cut-set for G.

21. a) (i) 3, 4, 6, 3, 8, 4   (ii) 3, 4, 6, 6, 8, 4

b) No pendant vertex of the given tree appears in the sequence, so the result is true for these vertices. When an edge $\{x, y\}$ is removed and y is a pendant vertex (of the tree or one of the resulting subtrees), the $\deg(x)$ is decreased by 1 and x is placed in the sequence. As the process continues, either (i) this vertex x becomes a pendant vertex in a subtree and is removed but not recorded again in the sequence, or (ii) the vertex x is left as one of the last two vertices of an edge. In either case, x has been listed in the sequence $[\deg(x) - 1]$ times.

c)

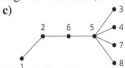

d) Input: The given Prüfer code $x_1, x_2, \ldots, x_{n-2}$

Output: The unique tree T with n vertices labeled with $1, 2, \ldots, n$. (This tree has the Prüfer code $x_1, x_2, \ldots, x_{n-2}$.)

$$C := [x_1, x_2, \ldots, x_{n-2}] \qquad \{\text{Initializes } C \text{ as a list (ordered set)}\}$$
$$L := [1, 2, \ldots, n] \qquad \{\text{Initializes } L \text{ as a list (ordered set)}\}$$
$$T := \emptyset$$

for $i := 1$ **to** $n - 2$ **do**
    $v :=$ smallest element in L not in C
    $w :=$ first entry in C
    $T := T \cup \{\{v, w\}\}$ {Add the new edge $\{v, w\}$ to the present forest.}
    delete v from L
    delete the first occurrence of w from C
$T := T \cup \{\{y, z\}\}$ {The vertices y, z are the last two remaining entries in L.}

23. a) If the tree contains $n + 1$ vertices, then it is (isomorphic to) the complete bipartite graph $K_{1,n}$ — often called the *star* graph.

b) If the tree contains n vertices, then it is (isomorphic to) a path on n vertices.

25. Let $E_1 = \{\{a, b\}, \{b, c\}, \{c, d\}, \{d, e\}, \{b, h\}, \{d, i\}, \{f, i\}, \{g, i\}\}$ and
$E_2 = \{\{a, h\}, \{b, i\}, \{h, i\}, \{g, h\}, \{f, g\}, \{c, i\}, \{d, f\}, \{e, f\}\}$.

Section 12.2 – p. 603

1. a) f, h, k, p, q, s, t **b)** a **c)** d
   d) e, f, j, q, s, t **e)** q, t **f)** 2   **g)** k, p, q, s, t

3. a) $/ + w - x y * \pi \uparrow z 3$ **b)** 0.4

5. Preorder: $r, j, h, g, e, d, b, a, c, f, i, k, m, p, s, n, q, t, v, w, u$
   Inorder: $h, e, a, b, d, c, g, f, j, i, r, m, s, p, k, n, v, t, w, q, u$
   Postorder: $a, b, c, d, e, f, g, h, i, j, s, p, m, v, w, t, u, q, n, k, r$

7. a) (i) and (iii)   (ii)

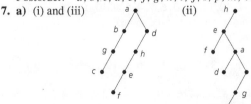

b) (i)   **(ii)**   **(iii)**

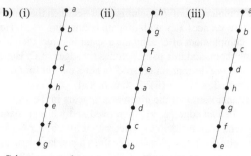

9. G is connected.

11. Theorem 12.6

a) Each internal vertex has m children, so there are mi vertices that are the children of some other vertex. This accounts for all vertices in the tree except the root. Hence $n = mi + 1$.

b) $\ell + i = n = mi + 1 \Rightarrow \ell = (m-1)i + 1$

c) $\ell = (m-1)i + 1 \Rightarrow i = (\ell - 1)/(m-1)$

$n = mi + 1 \Rightarrow i = (n-1)/m$

Corollary 12.1

Since the tree is balanced, $m^{h-1} < \ell \le m^h$ by Theorem 12.7.

$$m^{h-1} < \ell \le m^h \Rightarrow \log_m(m^{h-1}) < \log_m(\ell) \le \log_m(m^h)$$
$$\Rightarrow (h-1) < \log_m \ell \le h \Rightarrow h = \lceil \log_m \ell \rceil$$

13. a) $102; 69$

15. a)   **b)** $9; 5$ **c)** $h(m-1); (h-1) + (m-1)$

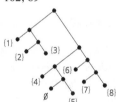

17. $21845; 1 + m + m^2 + \cdots + m^{h-1} = (m^h - 1)/(m-1)$

19.

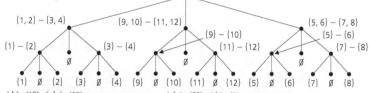

21. $\left(\frac{1}{6}\right)\binom{10}{5}\left(\frac{1}{10}\right)\binom{18}{9} = 204,204$ $\left(\frac{1}{11}\right)\binom{20}{10}\left(\frac{1}{5}\right)\binom{8}{4} = 235,144$

23. a) $1, 2, 5, 11, 12, 13, 14, 3, 6, 7, 4, 8, 9, 10, 15, 16, 17$

b) The preorder traversal of the rooted tree

Section 12.3 – p. 609

1. a) L_1: $1, 3, 5, 7, 9$ L_2: $2, 4, 6, 8, 10$

b) L_1: $1, 3, 5, 7, \ldots, 2m - 3, m + n$

L_2: $2, 4, 6, 8, \ldots, 2m - 2, 2m - 1, 2m, 2m + 1, \ldots, m + n - 1$

3. a)

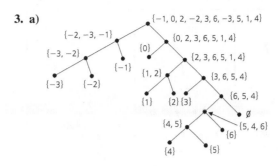

1. a) tear   **b)** tatener   **c)** rant

3. a: 111 c: 0110 e: 10 g: 11011 i: 00
b: 110101 d: 1100 f: 0111 h: 010 j: 110100

5. 55,987

7.

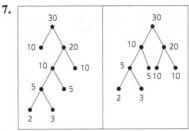

Amend part (a) of step (2) for the Huffman tree algorithm as follows. If there are n (> 2) such trees with smallest root weights w and w', then

  (i) if $w < w'$ and $n - 1$ of these trees have root weight w', select a tree (of root weight w') with smallest height; and

  (ii) if $w = w'$ (and all n trees have the same smallest root weight), select two trees (of root weight w) of smallest height.

1. The articulation points are b, e, f, h, j, k. The biconnected components are B_1: $\{\{a, b\}\}$; B_2: $\{\{d, e\}\}$; B_3: $\{\{b, c\}, \{c, f\}, \{f, e\}, \{e, b\}\}$; B_4: $\{\{f, g\}, \{g, h\}, \{h, f\}\}$; B_5: $\{\{h, i\}, \{i, j\}, \{j, h\}\}$; B_6: $\{\{j, k\}\}$; B_7: $\{\{k, p\}, \{p, n\}, \{n, m\}, \{m, k\}, \{p, m\}\}$.

3. a) T can have as few as one or as many as $n - 2$ articulation points. If T contains a vertex of degree $(n - 1)$, then this vertex is the only articulation point. If T is a path with n vertices and $n - 1$ edges, then the $n - 2$ vertices of degree 2 are all articulation points.

b) In all cases, a tree on n vertices has $n - 1$ biconnected components. Each edge is a biconnected component.

5. $\chi(G) = \max\{\chi(B_i) \mid 1 \le i \le k\}$.

7. *Proof*: Suppose that G has a pendant vertex, say x, and that $\{w, x\}$ is the (unique) edge in E incident with x. Since $|V| \ge 3$, we know that $\deg(w) \ge 2$ and that $\kappa(G - w) \ge 2 > 1 = \kappa(G)$. Consequently, w is an articulation point of G.

9. a) The first tree provides the depth-first spanning tree T for G where the order prescribed for the vertices is reverse alphabetical and the root is c.

b) The second tree provides $(\text{low}'(v), \text{low}(v))$ for each vertex v of G (and T). These results follow from step (2) of the algorithm.

For the third tree, we find $(\text{dfi}(v), \text{low}(v))$ for each vertex v. Applying step (3) of the algorithm, we find the articulation points d, f, and g, and the four biconnected components.

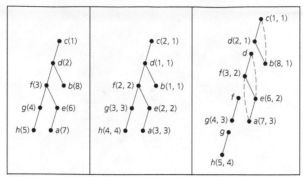

11. We always have $\text{low}(x_2) = \text{low}(x_1) = 1$. (*Note*: Vertices x_2 and x_1 are always in the same biconnected component.)

13. If not, let $v \in V$ where v is an articulation point of G. Then $\kappa(G - v) > \kappa(G) = 1$. (From Exercise 19 of Section 11.6 we know that G is connected.) Now $G - v$ is disconnected with components $H_1, H_2, \ldots, H_t$, for $t \geq 2$. For $1 \leq i \leq t$, let $v_i \in H_i$. Then $H_i + v$ is a subgraph of $G - v_{i+1}$, and $\chi(H_i + v) \leq \chi(G - v_{i+1}) < \chi(G)$. (Here $v_{t+1} = v_1$.) Now let $\chi(G) = n$ and let $\{c_1, c_2, \ldots, c_n\}$ be a set of n colors. For each subgraph $H_i + v$, $1 \leq i \leq t$, we can properly color the vertices of $H_i + v$ with at most $n - 1$ colors — and can use c_1 to color vertex v for all of these t subgraphs. Then we can join these t subgraphs together at vertex v and obtain a proper coloring for the vertices of G where we use less than $n (= \chi(G))$ colors.

Supplementary Exercises — p. 625

1. If G is a tree, consider G a rooted tree. Then there are λ choices for coloring the root of G and $(\lambda - 1)$ choices for coloring each of its descendants. The result then follows by the rule of product.

   Conversely, if $P(G, \lambda) = \lambda(\lambda - 1)^{n-1}$, then since the factor λ occurs only once, the graph G is connected. $P(G, \lambda) = \lambda(\lambda - 1)^{n-1} = \lambda^n - (n-1)\lambda^{n-1} + \cdots + (-1)^{n-1}\lambda \Rightarrow G$ has n vertices and $(n - 1)$ edges. Hence G is a tree [by part (d) of Theorem 12.5].

3. **a)** 1011001010100

   b) (i)   (ii)

   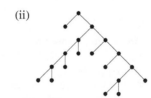

   c) Since the last two vertices visited in a preorder traversal are leaves, the last two symbols in the characteristic sequence of a complete binary tree are 00.

5. We assume that $G = (V, E)$ is connected — otherwise we work with a component of G. Since G is connected, and $\deg(v) \geq 2$ for all $v \in V$, it follows from Theorem 12.4 that G is not a tree. But every loop-free connected undirected graph that is not a tree must contain a cycle.

7. For $1 \leq i\ (< n)$, let $x_i =$ the number of vertices v where $\deg(v) = i$. Then $x_1 + x_2 + \cdots + x_{n-1} = |V| = |E| + 1$, so $2|E| = 2(-1 + x_1 + x_2 + \cdots + x_{n-1})$. But $2|E| = \sum_{v \in V} \deg(v) = (x_1 + 2x_2 + 3x_3 + \cdots + (n-1)x_{n-1})$. Solving $2(-1 + x_1 + x_2 + \cdots + x_{n-1}) = x_1 + 2x_2 + \cdots + (n-1)x_{n-1}$ for x_1, we find that $x_1 = 2 + x_3 + 2x_4 + 3x_5 + \cdots + (n-3)x_{n-1} = 2 + \sum_{\deg(v_i) \geq 3} [\deg(v_i) - 2]$.

9. **a)** G^2 is isomorphic to K_5.   **b)** G^2 is isomorphic to K_4.

   c) G^2 is isomorphic to K_{n+1}, so the number of new edges is $\binom{n+1}{2} - n = \binom{n}{2}$.

   d) If G^2 has an articulation point x, then there exists $u, v \in V$ such that every path (in G^2) from u to v passes through x. (This follows from Exercise 2 of Section 12.5.) Since G is connected, there exists a path P (in G) from u to v. If x is not on this path (which is also a path in G^2), then we contradict x being an articulation point in G^2. Hence the path P (in G) passes through x, and we can write $P: u \to u_1 \to \cdots \to u_{n-1} \to u_n \to x \to v_m \to v_{m-1} \to \cdots \to v_1 \to v$. But then in G^2 we add the edge $\{u_n, v_m\}$, and the path P' (in G^2) given by $P': u \to u_1 \to \cdots \to u_{n-1} \to u_n \to v_m \to v_{m-1} \to \cdots \to v_1 \to v$ does not pass through x. So x is not an articulation point of G^2, and G^2 has no articulation points.

11. **a)** $\ell_n = \ell_{n-1} + \ell_{n-2}$, for $n \geq 3$ and $\ell_1 = \ell_2 = 1$. Since this is precisely the Fibonacci recurrence relation, we have $l_n = F_n$, the nth Fibonacci number, for $n \geq 1$.

b) $i_n = i_{n-1} + i_{n-2} + 1$, $n \geq 3$, $i_1 = i_2 = 0$
$i_n = (1/\sqrt{5})\alpha^n - (1/\sqrt{5})\beta^n - 1 = F_n - 1$, $n \geq 1$

13. **a)** For the spanning trees of G there are two mutually exclusive and exhaustive cases: (i) The edge $\{x_1, y_1\}$ is in the spanning tree: These spanning trees are counted in b_n. (ii) The edge $\{x_1, y_1\}$ is not in the spanning tree: In this case the edges $\{x_1, x_2\}$, $\{y_1, y_2\}$ are both in the spanning tree. Upon removing the edges $\{x_1, x_2\}$, $\{y_1, y_2\}$, and $\{x_1, y_1\}$ from the original ladder graph, we now need a spanning tree for the resulting smaller ladder graph with $n - 1$ rungs. There are a_{n-1} spanning trees in this case.

b) $b_n = b_{n-1} + 2a_{n-1}$, $n \geq 2$

c) $a_n - 4a_{n-1} + a_{n-2} = 0$, $n \geq 2$
$a_n = (1/(2\sqrt{3}))[(2 + \sqrt{3})^n - (2 - \sqrt{3})^n]$, $n \geq 0$

15. **a)** (i) 3   (ii) 5

b) $a_n = a_{n-1} + a_{n-2}$, $n \geq 5$, $a_3 = 2$, $a_4 = 3$
$a_n = F_{n+1}$, the $(n + 1)$-st Fibonacci number

17. Here the input consists of

(a) the k (≥ 3) vertices of the spine — ordered from left to right as $v_1, v_2, \ldots, v_k$;

(b) $\deg(v_i)$, in the caterpillar, for all $1 \leq i \leq k$; and

(c) n, the number of vertices in the caterpillar, with $n \geq 3$.

If $k = 3$, the caterpillar is the complete bipartite graph (or star) $K_{1,n-1}$, for some $n \geq 3$. We label v_1 with 1 and the remaining vertices with 2, 3, ..., n. This provides the edge labels (the absolute value of the difference of the vertex labels) 1, 2, 3, ..., $n - 1$, a graceful labeling.

For $k > 3$ we consider the following.

$l := 2$ {l is the largest low label}
$h := n - 1$ {h is the smallest high label}
label v_1 with 1
label v_2 with n

for $i := 2$ **to** $k - 1$ **do**
    if $2\lfloor i/2 \rfloor = i$ **then**   {i is even}
       begin
          if v_i has unlabeled leaves that are not on the spine **then**
             assign the $\deg(v_i) - 2$ labels from l to $l + \deg(v_i) - 3$
             to these leaves of v_i
          assign the label $l + \deg(v_i) - 2$ to v_{i+1}
          $l := l + \deg(v_i) - 1$
       end
    else
       begin
          if v_i has unlabeled leaves that are not on the spine **then**
             assign the $\deg(v_i) - 2$ labels from $h - [\deg(v_i) - 3]$ to
             h to these leaves of v_i
          assign the label $h - \deg(v_i) + 2$ to v_{i+1}
          $h := h - \deg(v_i) + 1$
       end

19. **a)** 1, −1, 1, 1, −1, −1   1, 1, −1, 1, −1, −1   1, −1, 1, −1, 1, −1

b)

In total there are 14 ordered rooted trees on five vertices.

c) This is another example where the Catalan numbers arise. There are $\left(\frac{1}{n+1}\right)\binom{2n}{n}$ ordered rooted trees on $n+1$ vertices.

21. a) 8 **b)** 8^4 **c)** $4 \cdot 8^4$ **d)** $2(4 \cdot 8^4)$ **e)** $2(n8^n)$

Chapter 13
Optimization and Matching

Section 13.1 – p. 638

1. a) If not, let $v_i \in \overline{S}$, where $1 \le i \le m$ and i is the smallest such subscript. Then $d(v_0, v_i) < d(v_0, v_{m+1})$, and we contradict the choice of v_{m+1} as a vertex v in $\overline{S}$ for which $d(v_0, v)$ is a minimum.

b) Suppose there is a shorter directed path (in G) from v_0 to v_k. If this path passes through a vertex in $\overline{S}$, then from part (a) we have a contradiction. Otherwise, we have a shorter directed path P'' from v_0 to v_k, and P'' only passes through vertices in S. But then $P'' \cup \{(v_k, v_{k+1}), (v_{k+1}, v_{k+2}), \ldots, (v_{m-1}, v_m), (v_m, v_{m+1})\}$ is a directed path (in G) from v_0 to v_{m+1}, and it is shorter than path P.

3. a) $d(a, b) = 5$; $d(a, c) = 6$; $d(a, f) = 12$; $d(a, g) = 16$; $d(a, h) = 12$

b) $f: (a, c), (c, f)$ $g: (a, b), (b, h), (h, g)$ $h: (a, b), (b, h)$

5. False. Consider the following weighted graph.

Section 13.2 – p. 643

1. Kruskal's algorithm generates the following sequence (of forests), which terminates in a minimal spanning tree T of weight 18.

(1) $F_1 = \{\{e, h\}\}$ (2) $F_2 = F_1 \cup \{\{a, b\}\}$ (3) $F_3 = F_2 \cup \{\{b, c\}\}$

(4) $F_4 = F_3 \cup \{\{d, e\}\}$ (5) $F_5 = F_4 \cup \{\{e, f\}\}$ (6) $F_6 = F_5 \cup \{\{a, e\}\}$

(7) $F_7 = F_6 \cup \{\{d, g\}\}$ (8) $F_8 = T = F_7 \cup \{\{f, i\}\}$

(This answer is not unique.)

3. No! Consider the following counterexample:

Here $V = \{v, x, w\}$, $E = \{\{v, x\}, \{x, w\}, \{v, w\}\}$, and $E' = \{\{v, x\}, \{x, w\}\}$.

5. a) Evansville-Indianapolis (168); Bloomington-Indianapolis (51); South Bend–Gary (58); Terre Haute–Bloomington (58); South Bend–Fort Wayne (79); Indianapolis–Fort Wayne (121).

b) Fort Wayne–Gary (132); Evansville-Indianapolis (168); Bloomington-Indianapolis (51); Gary–South Bend (58); Terre Haute–Bloomington (58); Indianapolis–Fort Wayne (121).

7. a) To determine an optimal tree of maximal weight, replace the two occurrences of "small" in Kruskal's algorithm by "large."

b) Use the edges: South Bend–Evansville (303); Fort Wayne–Evansville (290); Gary-Evansville (277); Fort Wayne–Terre Haute (201); Gary-Bloomington (198); Indianapolis-Evansville (168).

9. When the weights of the edges are all distinct, in each step of Kruskal's algorithm a unique edge is selected.

1. a) $s = 2$; $t = 4$; $w = 5$; $x = 9$; $y = 4$ **b)** 18
  c) (i) $P = \{a, b, h, d, g, i\}$; $\overline{P} = \{z\}$ (ii) $P = \{a, b, h, d, g\}$; $\overline{P} = \{i, z\}$
   (iii) $P = \{a, h\}$; $\overline{P} = \{b, d, g, i, z\}$

3. (1)   (2)

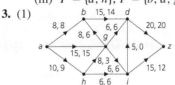 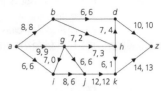

The maximum flow is 32,   The maximum flow is 23,
which is $c\{P, \overline{P}\}$ for   which is $c\{P, \overline{P}\}$ for
$P = \{a, b, d, g, h\}$ and $\overline{P} = \{i, z\}$ $P = \{a\}$ and $\overline{P} = \{b, g, i, j, d, h, k, z\}$

5. Here $c(e)$ is a positive integer for each $e \in E$, and the initial flow is defined as $f(e) = 0$ for all $e \in E$. The result follows because $\triangle_p$ is a positive integer for each application of the Edmonds-Karp algorithm and in the Ford-Fulkerson algorithm, $f(e) - \triangle_p$ will not be negative for a backward edge.

7.

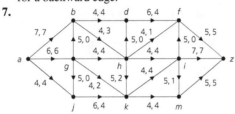

1. $5/\binom{8}{4} = 1/14$

3. Let the committees be represented as $c_1, c_2, \ldots, c_6$, according to the way they are listed in the exercise.
  a) Select the members as follows: $c_1 - A$; $c_2 - G$; $c_3 - M$; $c_4 - N$; $c_5 - K$; $c_6 - R$.
  b) Select the nonmembers as follows: $c_1 - K$; $c_2 - A$; $c_3 - G$; $c_4 - J$; $c_5 - M$; $c_6 - P$.

5. a) A one-factor for a graph $G = (V, E)$ consists of edges that have no common vertex. So the one-factor contains an even number of vertices, and since it spans G, we must have $|V|$ even.
  b) Consider the Petersen graph as shown in Fig. 11.52(a). The edges

$$\{e, a\} \quad \{b, c\} \quad \{d, i\} \quad \{g, j\} \quad \{f, h\}$$

provide a one-factor for this graph.
  c) There are $(5)(3) = 15$ one-factors for K_6.
  d) Label the vertices of K_{2n} with $1, 2, 3, \ldots, 2n - 1, 2n$. We can pair vertex 1 with any of the other $2n - 1$ vertices, and we are then confronted, in the case where $n \geq 2$, with finding a one-factor for the graph K_{2n-2}. Consequently,

$$a_n = (2n - 1)a_{n-1}, \qquad a_1 = 1.$$

We find that

$$a_n = (2n - 1)a_{n-1} = (2n - 1)(2n - 3)a_{n-2} = (2n - 1)(2n - 3)(2n - 5)a_{n-3} = \cdots$$

$$= (2n - 1)(2n - 3)(2n - 5) \cdots (5)(3)(1)$$

$$= \frac{(2n)(2n - 1)(2n - 2)(2n - 3) \cdots (4)(3)(2)(1)}{(2n)(2n - 2) \cdots (4)(2)} = \frac{(2n)!}{2^n (n!)}$$

7. Yes, such an assignment can be made by Fritz. Let X be the set of student applicants and Y the set of part-time jobs. Then for all $x \in X$, $y \in Y$, draw the edge (x, y) if applicant x is qualified for part-time job y. Then $\deg(x) \geq 4 \geq \deg(y)$ for all $x \in X$, $y \in Y$, and the result follows from Corollary 13.6.

9. a) (i) Select i from A_i for $1 \le i \le 4$.

(ii) Select $i + 1$ from A_i for $1 \le i \le 3$, and 1 from A_4.

b) 2

11. For each subset A of X, let G_A be the subgraph of G induced by the vertices in $A \cup R(A)$. If e is the number of edges in G_A, then $e \ge 4|A|$ because $\deg(a) \ge 4$ for all $a \in A$. Likewise, $e \le 5|R(A)|$ because $\deg(b) \le 5$ for all $b \in R(A)$. So $5|R(A)| \ge 4|A|$ and $\delta(A) = |A| - |R(A)| \le |A| - (4/5)|A| = (1/5)|A| \le (1/5)|X| = 2$. Then since $\delta(G) = \max\{\delta(A) | A \subseteq X\}$, we have $\delta(G) \le 2$.

13. a) $\delta(G) = 1$. A maximal matching of X into Y is given by $\{\{x_1, y_4\}, \{x_2, y_2\}, \{x_3, y_1\}, \{x_5, y_3\}\}$.

b) If $\delta(G) = 0$, there is a complete matching of X into Y, and $\beta(G) = |Y|$, or $|Y| = \beta(G) - \delta(G)$. If $\delta(G) = k > 0$, let $A \subseteq X$ where $|A| - |R(A)| = k$. Then $A \cup (Y - R(A))$ is a largest maximal independent set in G and $\beta(G) = |A| + |Y - R(A)| = |Y| + (|A| - |R(A)|) = |Y| + \delta(G)$, so $|Y| = \beta(G) - \delta(G)$.

c) Fig. 13.30(a): $\{x_1, x_2, x_3, y_2, y_4, y_5\}$; Fig. 13.32: $\{x_3, x_4, y_2, y_3, y_4\}$.

Supplementary Exercises—p. 669

1. $d(a, b) = 5$ $d(a, c) = 11$ $d(a, d) = 7$ $d(a, e) = 8$

$d(a, f) = 19$ $d(a, g) = 9$ $d(a, h) = 14$

[Note that the loop at vertex g and the edges (c, a) of weight 9 and (f, e) of weight 5 are of no significance.]

3. a) The edge e_1 will always be selected in the first step of Kruskal's algorithm.

b) Again using Kruskal's algorithm, edge e_2 will be selected in the first application of step (2) unless each of the edges e_1, e_2 is incident with the same two vertices—that is, the edges e_1, e_2 form a circuit and G is a multigraph.

5.

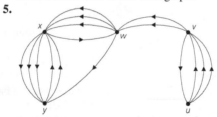

7. There are d_n, the number of derangements of $\{1, 2, 3, \ldots, n\}$.

9. The vertices [in the line graph $L(G)$] determined by E' form a maximal independent set.

Chapter 14
Rings and Modular Arithmetic

Section 14.1—p. 678

1. (Example 14.5): $-a = a, -b = e, -c = d, -d = c, -e = b$

(Example 14.6): $-s = s, -t = y, -v = x, -w = w, -x = v, -y = t$

3. a) $(a + b) + c = (b + a) + c$ Commutative Law of $+$

$\qquad\qquad = b + (a + c)$ Associative Law of $+$

$\qquad\qquad = b + (c + a)$ Commutative Law of $+$

b) $d + a(b + c) = d + (ab + ac)$ Distributive Law of $\cdot$ over $+$

$\qquad\qquad = (d + ab) + ac$ Associative Law of $+$

$\qquad\qquad = (ab + d) + ac$ Commutative Law of $+$

$\qquad\qquad = ab + (d + ac)$ Associative Law of $+$

c) $c(d + b) + ab = ab + c(d + b)$ Commutative Law of $+$

$\qquad\qquad = ab + (cd + cb)$ Distributive Law of $\cdot$ over $+$

$\qquad\qquad = ab + (cb + cd)$ Commutative Law of $+$

$\qquad\qquad = (ab + cb) + cd$ Associative Law of $+$

$\qquad\qquad = (a + c)b + cd$ Distributive Law of $\cdot$ over $+$

5. a) (i) The closed binary operation $\oplus$ is associative. For all a, b, $c \in \mathbf{Z}$ we find that

$$(a \oplus b) \oplus c = (a + b - 1) \oplus c = (a + b - 1) + c - 1 = a + b + c - 2,$$

and

$$a \oplus (b \oplus c) = a \oplus (b + c - 1) = a + (b + c - 1) - 1 = a + b + c - 2.$$

(ii) For the closed binary operation $\odot$ and all a, b, $c \in \mathbf{Z}$, we have

$$(a \odot b) \odot c = (a + b - ab) \odot c = (a + b - ab) + c - (a + b - ab)c$$
$$= a + b - ab + c - ac - bc + abc = a + b + c - ab - ac - bc + abc,$$

and

$$a \odot (b \odot c) = a \odot (b + c - bc) = a + (b + c - bc) - a(b + c - bc)$$
$$= a + b + c - bc - ab - ac + abc = a + b + c - ab - ac - bc + abc.$$

Consequently, the closed binary operation $\odot$ is also associative.

(iii) Given any integers a, b, c, we find that

$$(b \oplus c) \odot a = (b + c - 1) \odot a = (b + c - 1) + a - (b + c - 1)a$$
$$= b + c - 1 + a - ba - ca + a = a + a + b + c - 1 - ba - ca,$$

and

$$(b \odot a) \oplus (c \odot a) = (b + a - ba) \oplus (c + a - ca)$$
$$= (b + a - ba) + (c + a - ca) - 1 = a + a + b + c - 1 - ba - ca.$$

Therefore, the second distributive law holds.

c) Aside from 0 the only other unit is 2, since $2 \odot 2 = 2 + 2 - (2 \cdot 2) = 0$, the unity for $(\mathbf{Z}, \oplus, \odot)$.

d) This ring is an integral domain, but not a field. For all a, $b \in \mathbf{Z}$ we see that $a \odot b = 1$ (the zero element) $\Rightarrow a + b - ab = 1 \Rightarrow a(1 - b) = (1 - b) \Rightarrow (a - 1)(1 - b) = 0 \Rightarrow a = 1$ or $b = 1$, so there are no proper divisors of zero in $(\mathbf{Z}, \oplus, \odot)$.

7. From the previous exercise we know that we need to determine the condition(s) on k, m for which the distributive laws will hold. Since $\odot$ is commutative we can focus on just one of these laws.

If x, y, $z \in \mathbf{Z}$, then

$$x \odot (y \oplus z) = (x \odot y) \oplus (x \odot z)$$
$$\Rightarrow x \odot (y + z - k) = (x + y - mxy) \oplus (x + z - mxz)$$
$$\Rightarrow x + (y + z - k) - mx(y + z - k) = (x + y - mxy) + (x + z - mxz) - k$$
$$\Rightarrow x + y + z - k - mxy - mxz + mkx = x + y - mxy + x + z - mxz - k$$
$$\Rightarrow mkx = x \Rightarrow mk = 1 \Rightarrow m = k = 1 \text{ or } m = k = -1, \text{ since } m, k \in \mathbf{Z}.$$

9. a) We shall verify one of the distributive laws. If a, b, $c \in \mathbf{Q}$, then

$$a \odot (b \oplus c) = a \odot (b + c + 7)$$
$$= a + (b + c + 7) + [a(b + c + 7)]/7$$
$$= a + b + c + 7 + (ab/7) + (ac/7) + a,$$

while

$$(a \odot b) \oplus (a \odot c) = (a \odot b) + (a \odot c) + 7$$
$$= a + b + (ab/7) + a + c + (ac/7) + 7$$
$$= a + b + c + 7 + (ab/7) + (ac/7) + a.$$

Also, the rational number -7 is the zero element, and the additive inverse of each rational number a is $-14 - a$.

c) For each $a \in \mathbf{Q}$, $a = a \odot u = a + u + (au/7) \Rightarrow u[1 + (a/7)] = 0 \Rightarrow u = 0$, because a is arbitrary. Hence the rational number 0 is the unity for this ring. Now let $a \in \mathbf{Q}$, where $a \neq -7$, the zero element of the ring. Can we find $b \in \mathbf{Q}$ so that $a \odot b = 0$—that is, so that $a + b + (ab/7) = 0$? It follows that $a + b + (ab/7) = 0 \Rightarrow b(1 + (a/7)) = -a \Rightarrow b = (-a)/[1 + (a/7)]$. Hence every rational number, other than -7, is a unit.

11. b) $1, -1, i, -i$

13. $\begin{bmatrix} a & b \\ c & d \end{bmatrix}^{-1} = (1/(ad - bc)) \begin{bmatrix} d & -b \\ -c & a \end{bmatrix}$, $ad - bc \neq 0$

15. a) $xx = x(t + y) = xt + xy = t + y = x$
$yt = (x + t)t = xt + tt = t + t = s$
$yy = y(t + x) = yt + yx = s + s = s$
$tx = (y + x)x = yx + xx = s + x = x$
$ty = (y + x)y = yy + xy = s + y = y$

b) Since $tx = x \neq t = xt$, this ring is not commutative.

c) There is no unity and, consequently, no units.

d) The ring is neither an integral domain nor a field.

Section 14.2–p. 684

1. Theorem 14.10(a). If $(S, +, \cdot)$ is a subring of R, then $a - b, ab \in S$ for all $a, b \in S$. Conversely, since $S \neq \emptyset$, let $a \in S$. Then $a - a = z \in S$ and $z - a = -a \in S$. Also, if $b \in S$, then $-b \in S$, so $a - (-b) = a + b \in S$, and S is a subring by Theorem 14.9.

3. a) $(ab)(b^{-1}a^{-1}) = a(bb^{-1})a^{-1} = aua^{-1} = aa^{-1} = u$ and $(b^{-1}a^{-1})(ab) = b^{-1}(a^{-1}a)b = b^{-1}ub = b^{-1}b = u$, so ab is a unit. Since the multiplicative inverse of a unit is unique, $(ab)^{-1} = b^{-1}a^{-1}$.

b) $A^{-1} = \begin{bmatrix} 2 & -7 \\ -1 & 4 \end{bmatrix}$ $B^{-1} = \begin{bmatrix} 1 & -2 \\ -2 & 5 \end{bmatrix}$ $(AB)^{-1} = \begin{bmatrix} 4 & -15 \\ -9 & 34 \end{bmatrix}$

$(BA)^{-1} = \begin{bmatrix} 16 & -39 \\ -9 & 22 \end{bmatrix}$ $B^{-1}A^{-1} = \begin{bmatrix} 4 & -15 \\ -9 & 34 \end{bmatrix}$

5. $(-a)^{-1} = -(a^{-1})$

7. $z \in S, T \Rightarrow z \in S \cap T \Rightarrow S \cap T \neq \emptyset$. $a, b \in S \cap T \Rightarrow a, b \in S$ and $a, b \in T \Rightarrow a + b, ab \in S$ and $a + b, ab \in T \Rightarrow a + b, ab \in S \cap T$. $a \in S \cap T \Rightarrow a \in S$ and $a \in T \Rightarrow -a \in S$ and $-a \in T \Rightarrow -a \in S \cap T$. So $S \cap T$ is a subring of R.

9. If not, there exist $a, b \in S$ with $a \in T_1, a \notin T_2$ and $b \in T_2, b \notin T_1$. Since S is a subring of R, it follows that $a + b \in S$. Hence $a + b \in T_1$ or $a + b \in T_2$.

Assume without loss of generality that $a + b \in T_1$. Since $a \in T_1$, we have $-a \in T_1$, so by the closure under addition in T_1 we now find that $(-a) + (a + b) = (-a + a) + b = b \in T_1$, a contradiction. Therefore, $S \subseteq T_1 \cup T_2 \Rightarrow S \subseteq T_1$ or $S \subseteq T_2$.

11. b) $\begin{bmatrix} 1 & 0 \\ 0 & 1 \end{bmatrix}$ **c)** $\begin{bmatrix} 1 & 0 \\ 0 & 0 \end{bmatrix}$

d) S is an integral domain, while R is a noncommutative ring with unity.

13. Since $za = z$, it follows that $z \in N(a)$ and $N(a) \neq \emptyset$. If $r_1, r_2 \in N(a)$, then $(r_1 - r_2)a = r_1a - r_2a = z - z = z$, so $r_1 - r_2 \in N(a)$. Finally, if $r \in N(a)$ and $s \in R$, then $(rs)a = (sr)a = s(ra) = sz = z$, so $rs, sr \in N(a)$. Hence $N(a)$ is an ideal, by Definition 14.6.

15. 2

17. a) $a = au \in aR$ since $u \in R$, so $aR \neq \emptyset$. If $ar_1, ar_2 \in aR$, then $ar_1 - ar_2 = a(r_1 - r_2) \in aR$. Also, for $ar_1 \in aR$ and $r \in R$, we have $r(ar_1) = (ar_1)r = a(r_1r) \in aR$. Hence aR is an ideal of R.

b) Let $a \in R$, $a \neq z$. Then $a = au \in aR$ so $aR = R$. Since $u \in R = aR$, $u = ar$ for some $r \in R$, and $r = a^{-1}$. Hence R is a field.

19. a) $\binom{4}{2}(49)$ **b)** 7^4 **c)** Yes, the element (u, u, u, u) **d)** 4^4

21. b) If R has a unity u, define $a^0 = u$, for $a \in R$, $a \neq z$. If a is a unit of R, define a^{-n} as $(a^{-1})^n$, for $n \in \mathbf{Z}^+$.

Section 14.3 – p. 696

1. **a)** (i) Yes   (ii) No   (iii) Yes   **b)** (i) No   (ii) Yes   (iii) Yes

3. **a)** $-6, 1, 8, 15$ **b)** $-9, 2, 13, 24$ **c)** $-7, 10, 27, 44$

5. Since $a \equiv b \pmod{n}$, we may write $a = b + kn$ for some $k \in \mathbf{Z}$. And $m|n \Rightarrow n = \ell m$ for some $\ell \in \mathbf{Z}$. Consequently, $a = b + kn = b + (k\ell)m$ and $a \equiv b \pmod{m}$.

7. Let $a = 8$, $b = 2$, $m = 6$, and $n = 2$. Then $\gcd(m, n) = \gcd(6, 2) = 2 > 1$, $a \equiv b \pmod{m}$ and $a \equiv b \pmod{n}$. But $a - b = 8 - 2 = 6 \neq k(12) = k(mn)$, for any $k \in \mathbf{Z}$. Hence $a \not\equiv b \pmod{mn}$.

9. For n odd consider the $n - 1$ numbers $1, 2, 3, \ldots, n - 3, n - 2, n - 1$ as $(n-1)/2$ pairs: 1 and $(n - 1)$, 2 and $(n - 2)$, 3 and $(n - 3)$, $\ldots$, $n - \left(\frac{n-1}{2}\right) - 1$ and $n - \left(\frac{n-1}{2}\right)$. The sum of each pair is n which is congruent to 0 modulo n. Hence $\sum_{i=1}^{n-1} i \equiv 0 \pmod{n}$. When n is even we consider the $n - 1$ numbers $1, 2, 3, \ldots, (n/2) - 1, (n/2), (n/2) + 1, \ldots, n - 3, n - 2, n - 1$ as $(n/2) - 1$ pairs — namely, 1 and $n - 1$, 2 and $n - 2$, 3 and $n - 3$, $\ldots$, $(n/2) - 1$ and $(n/2) + 1$ — and the single number $(n/2)$. For each pair the sum is n, or 0 modulo n, so $\sum_{i=1}^{n-1} i \equiv (n/2) \pmod{n}$.

11. **b)** No, $2 \mathcal{R} 3$ and $3 \mathcal{R} 5$, but $5 \not{\mathcal{R}} 8$. Also, $2 \mathcal{R} 3$ and $2 \mathcal{R} 5$, but $4 \not{\mathcal{R}} 15$.

13. **a)** $[17]^{-1} = [831]$ **b)** $[100]^{-1} = [111]$ **c)** $[777]^{-1} = [735]$

15. **a)** 16 units, 0 proper zero divisors   **b)** 72 units, 44 proper zero divisors
   c) 1116 units, 0 proper zero divisors

17. $\left[\binom{334}{3} + 2\binom{333}{3} + \binom{334}{1}\binom{333}{1}^2\right] \Big/ \binom{1000}{3}$

19. **a)** For $n = 0$ we have $10^0 = 1 = 1(-1)^0$, so $10^0 \equiv (-1)^0 \pmod{11}$. [Since $10 - (-1) = 11$, $10 \equiv (-1) \pmod{11}$, or $10^1 \equiv (-1)^1 \pmod{11}$. Hence the result is also true for $n = 1$.] Assume the result true for $n = k \geq 1$ and consider the case for $k + 1$. Then, since $10^k \equiv (-1)^k \pmod{11}$ and $10 \equiv (-1) \pmod{11}$, we have $10^{k+1} = 10^k \cdot 10 \equiv (-1)^k(-1) = (-1)^{k+1} \pmod{11}$. The result now follows for all $n \in \mathbf{N}$, by the Principle of Mathematical Induction.

b) If $x_n x_{n-1} \cdots x_2 x_1 x_0 = x_n \cdot 10^n + x_{n-1} \cdot 10^{n-1} + \cdots + x_2 \cdot 10^2 + x_1 \cdot 10 + x_0$ denotes an $(n + 1)$-digit integer, then

$$x_n x_{n-1} \cdots x_2 x_1 x_0 \equiv (-1)^n x_n + (-1)^{n-1} x_{n-1} + \cdots + x_2 - x_1 + x_0 \pmod{11}.$$

Proof:

$$x_n x_{n-1} \cdots x_2 x_1 x_0 = x_n \cdot 10^n + x_{n-1} \cdot 10^{n-1} + \cdots + x_2 \cdot 10^2 + x_1 \cdot 10 + x_0$$
$$\equiv x_n(-1)^n + x_{n-1}(-1)^{n-1} + \cdots + x_2(-1)^2 + x_1(-1) + x_0$$
$$= (-1)^n x_n + (-1)^{n-1} x_{n-1} + \cdots + x_2 - x_1 + x_0 \pmod{11}.$$

21. Let $g = \gcd(a, n)$, $h = \gcd(b, n)$. $[a \equiv b \pmod{n}] \Rightarrow [a = b + kn$, for some $k \in \mathbf{Z}] \Rightarrow [g|b$ and $h|a]$. $[g|b$ and $g|n] \Rightarrow g|h$; $[h|a$ and $h|n] \Rightarrow h|g$. Since $g, h > 0$, it follows that $g = h$.

23.
(1) Plaintext	a	l	l	g	a	u	l	i	s	d	i	v	i	d	e	d
(2)	0	11	11	6	0	20	11	8	18	3	8	21	8	3	4	3
(3)	3	14	14	9	3	23	14	11	21	6	11	24	11	6	7	6
(4) Ciphertext	D	O	O	J	D	X	O	L	V	G	L	Y	L	G	H	G

	i	n	t	o	t	h	r	e	e	p	a	r	t	s
	8	13	19	14	19	7	17	4	4	15	0	17	19	18
	11	16	22	17	22	10	20	7	7	18	3	20	22	21
	L	Q	W	R	W	K	U	H	H	S	D	U	W	V

For each θ in row (2), the corresponding result below it in row (3) is $(\theta + 3)$ **mod** 26.

25. **a)** $(24)(8) = 192$ **b)** $(25)(20) = 500$ **c)** $(27)(18) = 486$ **d)** $(30)(8) = 240$

27. **a)** 9   **b)** 10, 15, 2, 13, 11, 1, 8, 5, 9

29. *Proof*: (By Mathematical Induction):
[Note that for $n \geq 1$, $(a^n - 1)/(a - 1) = a^{n-1} + a^{n-2} + \cdots + 1$, which can be computed in the ring $(\mathbf{Z}, +, \cdot)$.]

When $n = 0$, $a^0 x_0 + c[(a^0 - 1)/(a - 1)] \equiv x_0 + c[0/(a - 1)] \equiv x_0 \pmod{m}$, so the formula is true in this first basis ($n = 0$) case. Assuming the result for n (≥ 0) we have $x_n \equiv a^n x_0 + c[(a^n - 1)/(a - 1)] \pmod{m}$, $0 \leq x_n < m$. Continuing to the next case, we learn that

$$x_{n+1} \equiv ax_n + c \pmod{m}$$
$$\equiv a[a^n x_0 + c[(a^n - 1)/(a - 1)]] + c \pmod{m}$$
$$\equiv a^{n+1} x_0 + ac[(a^n - 1)/(a - 1)] + c(a - 1)/(a - 1) \pmod{m}$$
$$\equiv a^{n+1} x_0 + c[(a^{n+1} - a + a - 1)/(a - 1)] \pmod{m}$$
$$\equiv a^{n+1} x_0 + c[(a^{n+1} - 1)/(a - 1)] \pmod{m}$$

and we select x_{n+1} so that $0 \leq x_{n+1} < m$. It now follows by the Principle of Mathematical Induction that

$$x_n \equiv a^n x_0 + c[(a^n - 1)/(a - 1)] \pmod{m}, \quad 0 \leq x_n < m.$$

31. *Proof*: Let n, $n + 1$, and $n + 2$ be three consecutive integers. Then $n^3 + (n + 1)^3 + (n + 2)^3 = n^3 + (n^3 + 3n^2 + 3n + 1) + (n^3 + 6n^2 + 12n + 8) = (3n^3 + 15n) + 9(n^2 + 1)$. So we consider $3n^3 + 15n = 3n(n^2 + 5)$. If $3 | n$, then we are finished. If not, then $n \equiv 1 \pmod{3}$ or $n \equiv 2 \pmod{3}$. If $n \equiv 1 \pmod{3}$, then $n^2 + 5 \equiv 1 + 5 \equiv 0 \pmod{3}$, so $3 | (n^2 + 5)$. If $n \equiv 2 \pmod{3}$, then $n^2 + 5 \equiv 9 \equiv 0 \pmod{3}$, and $3 | (n^2 + 5)$. All cases are now covered, so we have $3 | [n(n^2 + 5)]$. Hence $9 | [3n(n^2 + 5)]$ and, consequently, 9 divides $(3n^3 + 15n) + 9(n^2 + 1) = n^3 + (n + 1)^3 + (n + 2)^3$.

33. $\displaystyle\sum_{k=0}^{n-1} p(k(n + 1), n, n) = \frac{1}{n + 1}\binom{2n}{n}$, the nth Catalan number

35. a) 112   **b)** 031-43-3464

37. a) $1, 28, 14, 34, 2, 3\ (= 2 + 1), 15\ (= 14 + 1), 4\ (= 3 + 1)$ **b)** $1, 2, 3, 4, 5$

Section 14.4 – p. 704

1. $s \to 0, t \to 1, v \to 2, w \to 3, x \to 4, y \to 5$

3. Let $(R, +, \cdot)$, $(S, \oplus, \odot)$, and $(T, +', \cdot')$ be the rings. For all $a, b \in R$, $(g \circ f)(a + b) = g(f(a + b)) = g(f(a) \oplus f(b)) = g(f(a)) +' g(f(b)) = (g \circ f)(a) +' (g \circ f)(b)$. Also, $(g \circ f)(a \cdot b) = g(f(a \cdot b)) = g(f(a) \odot f(b)) = g(f(a)) \cdot' g(f(b)) = (g \circ f)(a) \cdot' (g \circ f)(b)$. Hence $g \circ f$ is a ring homomorphism.

5. a) Since $f(z_R) = z_S$, it follows that $z_R \in K$ and $K \neq \emptyset$. If $x, y \in K$, then $f(x - y) = f(x + (-y)) = f(x) \oplus f(-y) = f(x) \ominus f(y) = z_S \ominus z_S = z_S$, so $x - y \in K$. Finally, if $x \in K$ and $r \in R$, then $f(rx) = f(r) \odot f(x) = f(r) \odot z_S = z_S$, and $f(xr) = f(x) \odot f(r) = z_S \odot f(r) = z_S$, so $rx, xr \in K$. Consequently, K is an ideal of R.

b) The kernel is $\{6n \mid n \in \mathbf{Z}\}$.

7. a)

x (in $\mathbf{Z}_{20}$)	$f(x)$ (in $\mathbf{Z}_4 \times \mathbf{Z}_5$)	x (in $\mathbf{Z}_{20}$)	$f(x)$ (in $\mathbf{Z}_4 \times \mathbf{Z}_5$)
0	(0, 0)	10	(2, 0)
1	(1, 1)	11	(3, 1)
2	(2, 2)	12	(0, 2)
3	(3, 3)	13	(1, 3)
4	(0, 4)	14	(2, 4)
5	(1, 0)	15	(3, 0)
6	(2, 1)	16	(0, 1)
7	(3, 2)	17	(1, 2)
8	(0, 3)	18	(2, 3)
9	(1, 4)	19	(3, 4)

b) (i) $f((17)(19) + (12)(14)) = (1, 2)(3, 4) + (0, 2)(2, 4) = (3, 3) + (0, 3) = (3, 1)$, and $f^{-1}(3, 1) = 11$

9. a) 4   **b)** 1   **c)** No

11. No. $\mathbf{Z}_4$ has two units, while the ring in Example 14.4 has only one unit.

13. $397 + k(648), k \in \mathbf{Z}$ **15.** $173 + k(210), k \in \mathbf{Z}$

Supplementary Exercises – p. 708

1. a) False. Let $R = \mathbf{Z}$ and $S = \mathbf{Z}^+$.   **b)** False. Let $R = \mathbf{Z}$ and $S = \{2x | x \in \mathbf{Z}\}$.

c) False. Let $R = M_2(\mathbf{Z})$ and $S = \left\{ \begin{bmatrix} a & 0 \\ 0 & 0 \end{bmatrix} \middle| a \in \mathbf{Z} \right\}$.   **d)** True.

e) False. The ring $(\mathbf{Z}, +, \cdot)$ is a subring (but not a field) in $(\mathbf{Q}, +, \cdot)$.

f) False. For any prime p, $\{a/(p^n)|a, n \in \mathbf{Z}, n \geq 0\}$ is a subring in $(\mathbf{Q}, +, \cdot)$.

g) False. Consider the field in Table 14.6.   **h)** True.

3. a) $[a + a = (a + a)^2 = a^2 + a^2 + a^2 + a^2 = (a + a) + (a + a)] \Rightarrow [a + a = 2a = z]$. Hence $-a = a$.

b) For each $a \in R$, $a + a = z \Rightarrow a = -a$. For $a, b \in R$, $(a + b) = (a + b)^2 = a^2 + ab + ba + b^2 = a + ab + ba + b \Rightarrow ab + ba = z \Rightarrow ab = -ba = ba$, so R is commutative.

5. Since $az = z = za$ for all $a \in R$, we have $z \in C$ and $C \neq \emptyset$. If $x, y \in C$, then $(x + y)a = xa + ya = ax + ay = a(x + y)$, $(xy)a = x(ya) = x(ay) = (xa)y = (ax)y = a(xy)$, and $(-x)a = -(xa) = -(ax) = a(-x)$, for all $a \in R$, so $x + y, xy, -x \in C$. Consequently, C is a subring of R.

7. b) Since m, n are relatively prime, we can write $1 = ms + nt$ where $s, t \in \mathbf{Z}$. With $m, n > 0$ it follows that one of s, t must be positive, and the other negative. Assume (without any loss of generality) that s is negative so that $1 - ms = nt > 0$.

Then $a^n = b^n \Rightarrow (a^n)^t = (b^n)^t \Rightarrow a^{nt} = b^{nt} \Rightarrow a^{1-ms} = b^{1-ms} \Rightarrow a(a^m)^{(-s)} = b(b^m)^{(-s)}$. But with $-s > 0$ and $a^m = b^m$, we have $(a^m)^{(-s)} = (b^m)^{(-s)}$. Consequently,

$$([(a^m)^{(-s)} = (b^m)^{(-s)} \neq z] \wedge [a(a^m)^{(-s)} = b(b^m)^{(-s)}]) \Rightarrow a = b,$$

since we may use the Cancellation Law of Multiplication in an integral domain.

9. Let $x = a_1 + b_1$, $y = a_2 + b_2$, for $a_1, a_2 \in A$ and $b_1, b_2 \in B$. Then $x - y = (a_1 - a_2) + (b_1 - b_2) \in A + B$. If $r \in R$ and $a + b \in A + B$, with $a \in A$ and $b \in B$, then $ra \in A$, $rb \in B$, and $r(a + b) \in A + B$. Similarly, $(a + b)r \in A + B$, and $A + B$ is an ideal of R.

11. Consider the numbers $x_1, x_1 + x_2, x_1 + x_2 + x_3, \ldots, x_1 + x_2 + x_3 + \cdots + x_n$. If one of these numbers is congruent to 0 modulo n, the result follows. If not, there exist $1 \leq i < j \leq n$ with $(x_1 + x_2 + \cdots + x_i) \equiv (x_1 + \cdots + x_i + x_{i+1} + \cdots + x_j) \pmod{n}$. Hence n divides $(x_{i+1} + \cdots + x_j)$.

13. a) 1875   **b)** 2914   **c)** 3/16

15. *Proof*: For all $n \in \mathbf{Z}$ we find that $n^2 \equiv 0 \pmod 5$ — when $5|n$ — or $n^2 \equiv 1 \pmod 5$ or $n^2 \equiv 4 \pmod 5$. Suppose that 5 does not divide any of a, b, or c. Then

(i) $a^2 + b^2 + c^2 \equiv 3 \pmod 5$ — when $a^2 \equiv b^2 \equiv c^2 \equiv 1 \pmod 5$;

(ii) $a^2 + b^2 + c^2 \equiv 1 \pmod 5$ — when each of two of a^2, b^2, c^2 is congruent to 1 modulo 5 and the other square is congruent to 4 modulo 5;

(iii) $a^2 + b^2 + c^2 \equiv 4 \pmod 5$ — when one of a^2, b^2, c^2 is congruent to 1 modulo 5 and each of the other two squares is congruent to 4 modulo 5; or

(iv) $a^2 + b^2 + c^2 \equiv 2 \pmod 5$ — when $a^2 \equiv b^2 \equiv c^2 \equiv 4 \pmod 5$.

17. $(e_1 + 1)(e_2 + 1) \cdots (e_k + 1) - 1$

Chapter 15
Boolean Algebra and Switching Functions

Section 15.1 – p. 718

1. a) 1   **b)** 1   **c)** 1   **d)** 1   **3. a)** 2^n **b)** $2^{(2^n)}$

5. a) d.n.f. $xyz + x\overline{y}z + x\overline{y}\,\overline{z} + xy\overline{z} + \overline{x}y\overline{z}$

c.n.f. $(x + y + z)(x + y + \overline{z})(x + \overline{y} + \overline{z})$

b) $f = \sum m(2, 4, 5, 6, 7) = \prod M(0, 1, 3)$

7. a) 2^{64} **b)** 2^6 **c)** 2^6

9. $m + k = 2^n$ **11. a)** $y + x\overline{z}$ **b)** $x + y$ **c)** $wx + z$

13. a) (i)

f	g	h	fg	$\overline{f}h$	gh	$fg + \overline{f}h + gh$	$fg + \overline{f}h$
0	0	0	0	0	0	0	0
0	0	1	0	1	0	1	1
0	1	0	0	0	0	0	0
0	1	1	0	1	1	1	1
1	0	0	0	0	0	0	0
1	0	1	0	0	0	0	0
1	1	0	1	0	0	1	1
1	1	1	1	0	1	1	1

Alternatively, $fg + \overline{f}h = (fg + \overline{f})(fg + h) = (f + \overline{f})(g + \overline{f})(fg + h) =$
$\mathbf{1}(g + \overline{f})(fg + h) = fgg + gh + \overline{f}fg + \overline{f}h = fg + gh + 0g + \overline{f}h = fg + gh + \overline{f}h.$
(ii) $fg + f\overline{g} + \overline{f}g + \overline{f}\,\overline{g} = f(g + \overline{g}) + \overline{f}(g + \overline{g}) = f \cdot 1 + \overline{f} \cdot 1 = f + \overline{f} = 1$

b) (i) $(f + g)(\overline{f} + h)(g + h) = (f + g)(\overline{f} + h)$
(ii) $(f + g)(f + \overline{g})(\overline{f} + g)(\overline{f} + \overline{g}) = \mathbf{0}$

15. a) $f \oplus f = \mathbf{0};\quad f \oplus \overline{f} = \mathbf{1};\quad f \oplus 1 = \overline{f};\quad f \oplus 0 = f$

b) (i) $f \oplus g = \mathbf{0} \Leftrightarrow f\overline{g} + \overline{f}g = \mathbf{0} \Rightarrow f\overline{g} = \overline{f}g = \mathbf{0}.$ [$f = 1$ and $f\overline{g} = \mathbf{0}$] $\Rightarrow g = 1.$
[$f = 0$ and $\overline{f}g = \mathbf{0}$] $\Rightarrow g = 0.$ Hence $f = g.$
(iii) $\overline{f} \oplus \overline{g} = \overline{f}\,\overline{\overline{g}} + \overline{\overline{f}}\,\overline{g} = \overline{f}g + f\overline{g} = f\overline{g} + \overline{f}g = f \oplus g$
(iv) This is the only result that is not true. When f has value 1, g has value 0 and h value 1
(or g has value 1 and h value 0), then $f \oplus gh$ has value 1 but $(f \oplus g)(f \oplus h)$ has
value 0.
(v) $fg \oplus fh = \overline{fg}fh + fg\overline{fh} = (\overline{f} + \overline{g})fh + fg(\overline{f} + \overline{h}) =$
$\overline{f}fh + f\overline{g}h + f\overline{f}g + fg\overline{h} = f\overline{g}h + fg\overline{h} = f(\overline{g}h + g\overline{h}) = f(g \oplus h)$
(vi) $\overline{f} \oplus g = \overline{f}\overline{g} + fg = fg + \overline{f}\,\overline{g} = f \oplus \overline{g}$
$\overline{f \oplus g} = \overline{f\overline{g} + \overline{f}g} = (\overline{f} + g)(f + \overline{g}) = \overline{f}\,\overline{g} + fg = \overline{f} \oplus g$

Section 15.2 – p. 727

1. a) $x \oplus y = (x + y)(\overline{xy})$

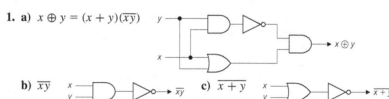

b) $\overline{xy}$  **c)** $\overline{x + y}$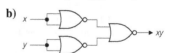

3. a) $x \multimap \overline{x}$ **b)**  $\to xy$

c) $x + y$

5. $f(w, x, y, z) = \overline{w}\,\overline{x}y\overline{z} + (w + x + \overline{y})z$

7.

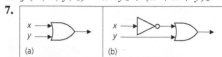

a) The output is $(x + \overline{y})(x + y) + y$. This simplifies to $x + (\overline{y}y) + y = x + 0 + y = x + y$,
and provides us with the simpler equivalent network in part (a) of the figure.

b) Here the output is $\overline{(x + \overline{y})} + (\overline{x}\,\overline{y} + y)$, which simplifies to $\overline{x}\,\overline{\overline{y}} + \overline{x}\,\overline{y} + y = \overline{x}y + \overline{x}\,\overline{y} + y = \overline{x}(y + \overline{y}) + y = \overline{x}(1) + y = \overline{x} + y$. This accounts for the simpler equivalent network in part (b) of the figure.

9. a) $f(w, x, y) = \overline{x}y + x\overline{y}$ **b)** $f(w, x, y) = x$ **c)** $f(w, x, y, z) = xz + \overline{x}\,\overline{z}$
d) $f(w, x, y, z) = w\overline{y}\,\overline{z} + x\overline{y}z + wyz + xy\overline{z}$ **e)** $f(w, x, y, z) = wy + w\overline{x}z + xyz$
f) $f(v, w, x, y, z) = \overline{v}\,\overline{w}x\overline{y}\,\overline{z} + vwx\overline{z} + \overline{v}\,\overline{x}y\overline{z} + \overline{w}\,\overline{x}z + v\overline{w}y + vyz$

11. a) 2   **b)** 3   **c)** 4   **d)** $k + 1$

13. a) $|f^{-1}(0)| = |f^{-1}(1)| = 8$ **b)** $|f^{-1}(0)| = 12, |f^{-1}(1)| = 4$
c) $|f^{-1}(0)| = 14, |f^{-1}(1)| = 2$ **d)** $|f^{-1}(0)| = 4, |f^{-1}(1)| = 12$

Section 15.3 – p. 733

1. $uv + wvy + uxz + uyz + wz$
3. a) $f(w, x, y, z) = z$ **b)** $f(w, x, y, z) = \overline{x}\,\overline{y}z + x\overline{y}z + xyz$
c) $f(v, w, x, y, z) = v\overline{y}\,\overline{z} + \overline{w}\,\overline{x}yz + \overline{v}\,\overline{w}\,\overline{z} + \overline{v}x\overline{y}$
5. $\{b, d\}, \{c, d\}, \{d, f\}, \{a, g\}, \{e, f\}, \{b, e\}, \{c, e\}, \{a, f\}, \{b, g\}, \{c, g\}$

Section 15.4 – p. 741

3. a) 30   **b)** 30   **c)** 1   **d)** 21   **e)** 30   **f)** 70
5. a) $w \leq 0 \Rightarrow w \cdot 0 = w$. But $w \cdot 0 = 0$, by part (a) of Theorem 15.3.
c) $y \leq z \Rightarrow yz = y$, and $y \leq \overline{z} \Rightarrow y\overline{z} = y$. Therefore, $y = yz = (y\overline{z})z = y(\overline{z}z) = y \cdot 0 = 0$.
7. $y \leq x$
9. From Theorem 15.5(a), with x_1, x_2 distinct atoms, if $x_1x_2 \neq 0$, then $x_1 = x_1x_2 = x_2x_1 = x_2$, a contradiction.
11. a) $f(0) = f(x\overline{x})$ for each $x \in \mathcal{B}_1$. $f(x\overline{x}) = f(x)f(\overline{x}) = f(x)\overline{f(x)} = 0$.
b) Follows from part (a) by duality.
c) $x \leq y \Longleftrightarrow xy = x \Rightarrow f(xy) = f(x) \Rightarrow f(x)f(y) = f(x) \Longleftrightarrow f(x) \leq f(y)$
13. a) $f(xy) = f(\overline{\overline{x} + \overline{y}}) = \overline{f(\overline{x} + \overline{y})} = \overline{f(\overline{x}) + f(\overline{y})} = \overline{f(\overline{x})} \cdot \overline{f(\overline{y})} = f(\overline{\overline{x}}) \cdot f(\overline{\overline{y}}) = f(x) \cdot f(y)$
b) Let $\mathcal{B}_1, \mathcal{B}_2$ be Boolean algebras with $f: \mathcal{B}_1 \to \mathcal{B}_2$ one-to-one and onto. Then f is an isomorphism if $f(\overline{x}) = \overline{f(x)}$ and $f(xy) = f(x)f(y)$ for all $x, y \in \mathcal{B}_1$. [Follows from part (a) by duality.]
15. For all $1 \leq i \leq n$, $(x_1 + x_2 + \cdots + x_n)x_i = x_1x_i + x_2x_i + \cdots + x_{i-1}x_i + x_ix_i + x_{i+1}x_i + \cdots + x_nx_i = 0 + 0 + \cdots + 0 + x_i + 0 + \cdots + 0 = x_i$, by part (b) of Theorem 15.5. Consequently, it follows from Theorem 15.7 that $(x_1 + x_2 + \cdots + x_n)x = x$ for all $x \in \mathcal{B}$. Since the one element is unique (from Exercise 10), we conclude that $1 = x_1 + x_2 + \cdots + x_n$.

Supplementary Exercises – p. 743

1. a) When $n = 2$, $x_1 + x_2$ denotes the Boolean sum of x_1 and x_2. For $n \geq 2$, we define $x_1 + x_2 + \cdots + x_n + x_{n+1}$ recursively by $(x_1 + x_2 + \cdots + x_n) + x_{n+1}$. (A similar definition can be given for the Boolean product.) For $n = 2$, $\overline{x_1 + x_2} = \overline{x_1}\,\overline{x_2}$ is true; this is one of the DeMorgan Laws. Assume the result for $n = k$ (≥ 2) and consider the case of $n = k + 1$.

$$\overline{(x_1 + x_2 + \cdots + x_k + x_{k+1})} = \overline{(x_1 + x_2 + \cdots + x_k) + x_{k+1}}$$
$$= \overline{(x_1 + x_2 + \cdots + x_k)}\,\overline{x_{k+1}}$$
$$= \overline{x_1}\,\overline{x_2} \cdots \overline{x_k}\,\overline{x_{k+1}}$$

Consequently, the result follows for all $n \geq 2$, by the Principle of Mathematical Induction.
b) Follows from part (a) by duality.
3. She can invite only Nettie and Cathy.
5. If $x \leq z$ and $y \leq z$, then from Exercise 6(b) of Section 15.4 we have $x + y \leq z + z$. And by the Idempotent Law we have $z + z = z$. Conversely, suppose that $x + y \leq z$. We find that $x \leq x + y$, because $x(x + y) = x + xy$ (by the Idempotent Law) $= x$ (by the Absorption Law). Since $x \leq x + y$ and $x + y \leq z$, we have $x \leq z$, because a partial order is transitive. (The proof that $y \leq z$ follows in a similar way.)

7. a) $x \le y \Rightarrow x + \overline{x} \le y + \overline{x} \Rightarrow 1 \le y + \overline{x} \Rightarrow y + \overline{x} = \overline{x} + y = 1$. Conversely,
$\overline{x} + y = 1 \Rightarrow x(\overline{x} + y) = x \cdot 1 \Rightarrow x\overline{x} \ (= 0) + xy = x \Rightarrow xy = x \Rightarrow x \le y$.

b) $x \le \overline{y} \Rightarrow x\overline{y} = x \Rightarrow xy = (x\overline{y})y = x(\overline{y}y) = x \cdot 0 = 0$. Conversely,
$xy = 0 \Rightarrow x = x \cdot 1 = x(y + \overline{y}) = xy + x\overline{y} = x\overline{y}$ and $x = x\overline{y} \Rightarrow x \le \overline{y}$.

9. a) $f(w, x, y, z) = \overline{w}\,\overline{x} + xy$ **b)** $g(v, w, x, y, z) = \overline{v}\,\overline{w}yz + xz + w\overline{y}\,\overline{z} + \overline{x}\,\overline{y}\,\overline{z}$

11. a) $2^{(2^{n-1})}$ **b)** $2^4;\ 2^{n+1}$

13. a) If $n = 60$, there are 12 divisors, and no Boolean algebra contains 12 elements since 12 is not a power of 2.

b) If $n = 120$, there are 16 divisors. However, if $x = 4$, then $\overline{x} = 30$ and $x \cdot \overline{x} = \gcd(x, \overline{x}) = \gcd(4, 30) = 2$, which is not the zero element. So the Inverse Laws are not satisfied.

Chapter 16
Groups, Coding Theory, and Polya's Method of Enumeration

Section 16.1 – p. 751

1. a) Yes. The identity is 1 and each element is its own inverse.

b) No. The set is not closed under addition and there is no identity.

c) No. The set is not closed under addition.

d) Yes. The identity is 0; the inverse of $10n$ is $10(-n)$ or $-10n$.

e) Yes. The identity is 1_A and the inverse of $g: A \to A$ is $g^{-1}: A \to A$.

f) Yes. The identity is 0; the inverse of $a/(2^n)$ is $(-a)/(2^n)$.

3. Subtraction is not an associative (closed) binary operation for $\mathbf{Z}$. For example, $(3 - 2) - 4 = -3 \ne 5 = 3 - (2 - 4)$.

5. Since $x, y \in \mathbf{Z} \Rightarrow x + y + 1 \in \mathbf{Z}$, the operation is a closed binary operation (or $\mathbf{Z}$ is closed under $\circ$). For all $w, x, y \in \mathbf{Z}$, $w \circ (x \circ y) = w \circ (x + y + 1) = w + (x + y + 1) + 1 = (w + x + 1) + y + 1 = (w \circ x) \circ y$, so the binary operation is associative. Furthermore, $x \circ y = x + y + 1 = y + x + 1 = y \circ x$, for all $x, y \in \mathbf{Z}$, so $\circ$ is also commutative. If $x \in \mathbf{Z}$, then $x \circ (-1) = x + (-1) + 1 = x[= (-1) \circ x]$, so -1 is the identity element for $\circ$. And finally, for each $x \in \mathbf{Z}$, we have $-x - 2 \in \mathbf{Z}$ and $x \circ (-x - 2) = x + (-x - 2) + 1 = -1 \ [= (-x - 2) \circ x]$, so $-x - 2$ is the inverse for x under $\circ$. Consequently, $(\mathbf{Z}, \circ)$ is an abelian group.

7. $U_{20} = \{1, 3, 7, 9, 11, 13, 17, 19\}$ $U_{24} = \{1, 5, 7, 11, 13, 17, 19, 23\}$

9. a) The result follows from Theorem 16.1(b) because both $(a^{-1})^{-1}$ and a are inverses of a^{-1}.

b)

$$(b^{-1}a^{-1})(ab) = b^{-1}(a^{-1}a)b = b^{-1}(e)b = b^{-1}b = e \text{ and}$$

$$(ab)(b^{-1}a^{-1}) = a(bb^{-1})a^{-1} = a(e)a^{-1} = aa^{-1} = e$$

So $b^{-1}a^{-1}$ is an inverse of ab, and by Theorem 16.1(b), $(ab)^{-1} = b^{-1}a^{-1}$.

11. a) $\{0\}; \{0, 6\}; \{0, 4, 8\}; \{0, 3, 6, 9\}; \{0, 2, 4, 6, 8, 10\}; \mathbf{Z}_{12}$

b) $\{1\}; \{1, 10\}; \{1, 3, 4, 5, 9\}; \mathbf{Z}_{11}^*$

c) $\{\pi_0\}; \{\pi_0, \pi_1, \pi_2\}; \{\pi_0, r_1\}; \{\pi_0, r_2\}; \{\pi_0, r_3\}; S_3$

13. a) There are 10: five rotations through $i(72°)$, $0 \le i \le 4$, and five reflections about lines containing a vertex and the midpoint of the opposite side.

b) For a regular n-gon ($n \ge 3$) there are $2n$ rigid motions. There are the n rotations through $i(360°/n)$, $0 \le i \le n - 1$. There are n reflections. For n odd, each reflection is about a line through a vertex and the midpoint of the opposite side. For n even, there are $n/2$ reflections about lines through opposite vertices and $n/2$ reflections about lines through the midpoints of opposite sides.

15. Since $eg = ge$ for all $g \in G$, it follows that $e \in H$ and $H \ne \emptyset$. If $x, y \in H$, then $xg = gx$ and $yg = gy$ for all $g \in G$. Consequently, $(xy)g = x(yg) = x(gy) = (xg)y = (gx)y = g(xy)$ for all $g \in G$, and we have $xy \in H$. Finally, for each $x \in H$, $g \in G$, $xg^{-1} = g^{-1}x$. So $(xg^{-1})^{-1} = (g^{-1}x)^{-1}$, or $gx^{-1} = x^{-1}g$, and $x^{-1} \in H$. Therefore, H is a subgroup of G.

17. b) (i) 216

(ii) $H_1 = \{(x, 0, 0)|x \in \mathbf{Z}_6\}$ is a subgroup of order 6

$H_2 = \{(x, y, 0)|x, y \in \mathbf{Z}_6, y = 0, 3\}$ is a subgroup of order 12

$H_3 = \{(x, y, 0)|x, y \in \mathbf{Z}_6\}$ has order 36

(iii) $-(2, 3, 4) = (4, 3, 2); -(4, 0, 2) = (2, 0, 4); -(5, 1, 2) = (1, 5, 4)$

19. a) $x = 1, x = 4$ **b)** $x = 1, x = 10$

c) $x = x^{-1} \Rightarrow x^2 \equiv 1 \pmod{p} \Rightarrow x^2 - 1 \equiv 0 \pmod{p} \Rightarrow (x - 1)(x + 1) \equiv 0 \pmod{p} \Rightarrow$
$x - 1 \equiv 0 \pmod{p}$ or $x + 1 \equiv 0 \pmod{p} \Rightarrow x \equiv 1 \pmod{p}$ or $x \equiv -1 \equiv p - 1 \pmod{p}$.

d) The result is true for $p = 2$, since $(2 - 1)! = 1! \equiv -1 \pmod 2$. For $p \geq 3$, consider the elements $1, 2, \ldots, p - 1$ in $(\mathbf{Z}_p^*, \cdot)$. The elements $2, 3, \ldots, p - 2$ yield $(p - 3)/2$ pairs of the form x, x^{-1}. (For example, when $p = 11$ we find that $2, 3, 4, \ldots, 9$ yield the four pairs 2, 6; 3, 4; 5, 9; 7, 8.) Consequently, $(p - 1)! \equiv (1)(1)^{(p-3)/2}(p - 1) \equiv p - 1 \equiv -1 \pmod{p}$.

Section 16.2 – p. 756

1. b) $f(a^{-1}) \cdot f(a) = f(a^{-1} \cdot a) = f(e_G) = e_H$ and $f(a) \cdot f(a^{-1}) = f(a \cdot a^{-1}) = f(e_G) = e_H$, so $f(a^{-1})$ is an inverse of $f(a)$. By the uniqueness of inverses (Theorem 16.1b), it follows that $f(a^{-1}) = [f(a)]^{-1}$.

3. $f(0) = (0, 0) \qquad f(1) = (1, 1) \qquad f(2) = (2, 0)$
$f(3) = (0, 1) \qquad f(4) = (1, 0) \qquad f(5) = (2, 1)$

5. $f(4, 6) = -5g_1 + 3g_2$

7. a) $o(\pi_0) = 1, o(\pi_1) = o(\pi_2) = 3, o(r_1) = o(r_2) = o(r_3) = 2$

b) (See Fig. 16.6) $o(\pi_0) = 1, o(\pi_1) = o(\pi_3) = 4, o(\pi_2) = o(r_1) = o(r_2) = o(r_3) = o(r_4) = 2$

9. a) The elements of order 10 are 4, 12, 28, and 36.

11. $\mathbf{Z}_5^* = \langle 2 \rangle = \langle 3 \rangle; \qquad \mathbf{Z}_7^* = \langle 3 \rangle = \langle 5 \rangle; \qquad \mathbf{Z}_{11}^* = \langle 2 \rangle = \langle 6 \rangle = \langle 7 \rangle = \langle 8 \rangle$

13. Let $(G, +), (H, *), (K, \cdot)$ be the given groups. For all $x, y \in G, (g \circ f)(x + y) = g(f(x + y)) = g(f(x) * f(y)) = (g(f(x))) \cdot (g(f(y))) = ((g \circ f)(x)) \cdot ((g \circ f)(y))$, since f, g are homomorphisms. Hence, $g \circ f : G \to K$ is a group homomorphism.

15. a) $(\mathbf{Z}_{12}, +) = \langle 1 \rangle = \langle 5 \rangle = \langle 7 \rangle = \langle 11 \rangle$
$(\mathbf{Z}_{16}, +) = \langle 1 \rangle = \langle 3 \rangle = \langle 5 \rangle = \langle 7 \rangle = \langle 9 \rangle = \langle 11 \rangle = \langle 13 \rangle = \langle 15 \rangle$
$(\mathbf{Z}_{24}, +) = \langle 1 \rangle = \langle 5 \rangle = \langle 7 \rangle = \langle 11 \rangle = \langle 13 \rangle = \langle 17 \rangle = \langle 19 \rangle = \langle 23 \rangle$

b) Let $G = \langle a^k \rangle$. Since $G = \langle a \rangle$, we have $a = (a^k)^s$ for some $s \in \mathbf{Z}$. Then $a^{1-ks} = e$, so $1 - ks = tn$ since $o(a) = n$. $1 - ks = tn \Rightarrow 1 = ks + tn \Rightarrow \gcd(k, n) = 1$. Conversely, let $G = \langle a \rangle$ where $a^k \in G$ and $\gcd(k, n) = 1$. Then $\langle a^k \rangle \subseteq G$. $\gcd(k, n) = 1 \Rightarrow 1 = ks + tn$, for some $s, t \in \mathbf{Z} \Rightarrow a = a^1 = a^{ks+nt} = (a^k)^s(a^n)^t = (a^k)^s(e)^t = (a^k)^s \in \langle a^k \rangle$. Hence $G \subseteq \langle a^k \rangle$. So $G = \langle a^k \rangle$, or a^k generates G.

c) $\phi(n)$.

Section 16.3 – p. 758

1. a) $\left\{ \begin{pmatrix} 1 & 2 & 3 & 4 \\ 2 & 3 & 4 & 1 \end{pmatrix}, \begin{pmatrix} 1 & 2 & 3 & 4 \\ 3 & 4 & 1 & 2 \end{pmatrix}, \begin{pmatrix} 1 & 2 & 3 & 4 \\ 4 & 1 & 2 & 3 \end{pmatrix}, \begin{pmatrix} 1 & 2 & 3 & 4 \\ 1 & 2 & 3 & 4 \end{pmatrix} \right\}$

b) $\begin{pmatrix} 1 & 2 & 3 & 4 \\ 2 & 1 & 4 & 3 \end{pmatrix} H = \left\{ \begin{pmatrix} 1 & 2 & 3 & 4 \\ 3 & 2 & 1 & 4 \end{pmatrix}, \begin{pmatrix} 1 & 2 & 3 & 4 \\ 4 & 3 & 2 & 1 \end{pmatrix}, \begin{pmatrix} 1 & 2 & 3 & 4 \\ 1 & 4 & 3 & 2 \end{pmatrix}, \begin{pmatrix} 1 & 2 & 3 & 4 \\ 2 & 1 & 4 & 3 \end{pmatrix} \right\}$

$\begin{pmatrix} 1 & 2 & 3 & 4 \\ 4 & 2 & 3 & 1 \end{pmatrix} H = \left\{ \begin{pmatrix} 1 & 2 & 3 & 4 \\ 1 & 3 & 4 & 2 \end{pmatrix}, \begin{pmatrix} 1 & 2 & 3 & 4 \\ 2 & 4 & 1 & 3 \end{pmatrix}, \begin{pmatrix} 1 & 2 & 3 & 4 \\ 3 & 1 & 2 & 4 \end{pmatrix}, \begin{pmatrix} 1 & 2 & 3 & 4 \\ 4 & 2 & 3 & 1 \end{pmatrix} \right\}$

$\begin{pmatrix} 1 & 2 & 3 & 4 \\ 1 & 2 & 4 & 3 \end{pmatrix} H = \left\{ \begin{pmatrix} 1 & 2 & 3 & 4 \\ 2 & 3 & 1 & 4 \end{pmatrix}, \begin{pmatrix} 1 & 2 & 3 & 4 \\ 3 & 4 & 2 & 1 \end{pmatrix}, \begin{pmatrix} 1 & 2 & 3 & 4 \\ 4 & 1 & 3 & 2 \end{pmatrix}, \begin{pmatrix} 1 & 2 & 3 & 4 \\ 1 & 2 & 4 & 3 \end{pmatrix} \right\}$

$\begin{pmatrix} 1 & 2 & 3 & 4 \\ 1 & 3 & 2 & 4 \end{pmatrix} H = \left\{ \begin{pmatrix} 1 & 2 & 3 & 4 \\ 2 & 4 & 3 & 1 \end{pmatrix}, \begin{pmatrix} 1 & 2 & 3 & 4 \\ 3 & 1 & 4 & 2 \end{pmatrix}, \begin{pmatrix} 1 & 2 & 3 & 4 \\ 4 & 2 & 1 & 3 \end{pmatrix}, \begin{pmatrix} 1 & 2 & 3 & 4 \\ 1 & 3 & 2 & 4 \end{pmatrix} \right\}$

$\begin{pmatrix} 1 & 2 & 3 & 4 \\ 1 & 4 & 2 & 3 \end{pmatrix} H = \left\{ \begin{pmatrix} 1 & 2 & 3 & 4 \\ 2 & 1 & 3 & 4 \end{pmatrix}, \begin{pmatrix} 1 & 2 & 3 & 4 \\ 3 & 2 & 4 & 1 \end{pmatrix}, \begin{pmatrix} 1 & 2 & 3 & 4 \\ 4 & 3 & 1 & 2 \end{pmatrix}, \begin{pmatrix} 1 & 2 & 3 & 4 \\ 1 & 4 & 2 & 3 \end{pmatrix} \right\}$

$\begin{pmatrix} 1 & 2 & 3 & 4 \\ 1 & 2 & 3 & 4 \end{pmatrix} H = H$

3. 12

5. From Lagrange's Theorem we know that $|K| = 66 (= 2 \cdot 3 \cdot 11)$ divides $|H|$ and that $|H|$ divides $|G| = 660 (= 2^2 \cdot 3 \cdot 5 \cdot 11)$. Consequently, since $K \neq H$ and $H \neq G$, it follows that $|H|$ is $2(2 \cdot 3 \cdot 11) = 132$ or $5(2 \cdot 3 \cdot 11) = 330$.

7. a) Let $\epsilon = \begin{pmatrix} 1 & 2 & 3 & 4 \\ 1 & 2 & 3 & 4 \end{pmatrix}, \alpha = \begin{pmatrix} 1 & 2 & 3 & 4 \\ 2 & 1 & 4 & 3 \end{pmatrix}, \beta = \begin{pmatrix} 1 & 2 & 3 & 4 \\ 3 & 4 & 1 & 2 \end{pmatrix}$, and $\delta = \begin{pmatrix} 1 & 2 & 3 & 4 \\ 4 & 3 & 2 & 1 \end{pmatrix}$.

·	ϵ	α	β	δ
ϵ	ϵ	α	β	δ
α	α	ϵ	δ	β
β	β	δ	ϵ	α
δ	δ	β	α	ϵ

It follows from Theorem 16.3 that H is a subgroup of G. And since the entries in the accompanying table are symmetric about the diagonal from the upper left to the lower right, we have H an abelian subgroup of G.

b) Since $|G| = 4! = 24$ and $|H| = 4$, there are $24/4 = 6$ left cosets of H in G.

c) Consider the function $f: H \to \mathbf{Z}_2 \times \mathbf{Z}_2$ defined by

$$f(\epsilon) = (0, 0), \qquad f(\alpha) = (1, 0), \qquad f(\beta) = (0, 1), \qquad f(\delta) = (1, 1).$$

This function f is one-to-one and onto, and for all $x, y \in H$ we find that

$$f(x \cdot y) = f(x) \oplus f(y).$$

Consequently, f is an isomorphism.

(*Note*: There are other possible answers that can be given here. In fact, there are six possible isomorphisms that one can define here.)

9. a) If H is a proper subgroup of G, then by Lagrange's Theorem, $|H|$ is 2 or p. If $|H| = 2$, then $H = \{e, x\}$ where $x^2 = e$, so $H = \langle x \rangle$. If $|H| = p$, let $y \in H$, $y \neq e$. Then $\circ(y) = p$, so $H = \langle y \rangle$.

b) Let $x \in G$, $x \neq e$. Then $\circ(x) = p$ or $\circ(x) = p^2$. If $\circ(x) = p$, then $|\langle x \rangle| = p$. If $\circ(x) = p^2$, then $G = \langle x \rangle$ and $\langle x^p \rangle$ is a subgroup of G of order p.

11. b) Let $x \in H \cap K$. If the order of x is r, then r must divide both m and n. Since $\gcd(m, n) = 1$, it follows that $r = 1$, so $x = e$ and $H \cap K = \{e\}$.

13. a) In $(\mathbf{Z}_p^*, \cdot)$ there are $p - 1$ elements, so by Exercise 8, for each $[x] \in (\mathbf{Z}_p^*, \cdot)$, $[x]^{p-1} = [1]$, or $x^{p-1} \equiv 1 \pmod{p}$, or $x^p \equiv x \pmod{p}$. For all $a \in \mathbf{Z}$, if $p \mid a$, then $a \equiv 0 \pmod{p}$ and $a^p \equiv 0 \equiv a \pmod{p}$. If $p \nmid a$, then $a \equiv b \pmod{p}$ where $1 \leq b \leq p - 1$, and $a^p \equiv b^p \equiv b \equiv a \pmod{p}$.

b) In the group G of units of $\mathbf{Z}_n$, there are $\phi(n)$ elements. If $a \in \mathbf{Z}$ and $\gcd(a, n) = 1$, then $[a] \in G$ and $[a]^{\phi(n)} = [1]$ or $a^{\phi(n)} \equiv 1 \pmod{n}$.

c) and **d)** These results follow from Exercises 6 and 8. They are special cases of Exercise 8.

Section 16.4 – p. 761

1. 0462   0170   1809   0462   1809   1981   0305

3. DRIVESAFELYX   **5.** $p = 157, q = 773$

Section 16.5 – p. 765

1. a) $e = 0001001$ **b)** $r = 1111011$ **c)** $c = 0101000$

3. a)   (i) $D(111101100) = 101$ (ii) $D(000100011) = 000$
   (iii) $D(010011111) = 011$

b) 000000000, 000000001, 100000000   **c)** 64

Sections 16.6 and 16.7 – p. 772

1. $S(101010, 1) = \{101010, 001010, 111010, 100010, 101110, 101000, 101011\}$
   $S(111111, 1) = \{111111, 011111, 101111, 110111, 111011, 111101, 111110\}$

3. a) $|S(x, 1)| = 11$; $|S(x, 2)| = 56$; $|S(x, 3)| = 176$

b) $|S(x, k)| = 1 + \binom{n}{1} + \binom{n}{2} + \cdots + \binom{n}{k} = \sum_{i=0}^{k} \binom{n}{i}$

5. a) The minimum distance between code words is 3. The code can detect all errors of weight $\leq$ 2 or correct all single errors.

b) The minimum distance between code words is 5. The code can detect all errors of weight $\leq$ 4 or correct all errors of weight $\leq$ 2.

c) The minimum distance is 2. The code detects all single errors but has no correction capability.

7. a) $C = \{00000, 10110, 01011, 11101\}$. The minimum distance between code words is 3, so the code can detect all errors of weight ≤ 2 or correct all single errors.

b) $H = \begin{bmatrix} 1 & 0 & 1 & 0 & 0 \\ 1 & 1 & 0 & 1 & 0 \\ 0 & 1 & 0 & 0 & 1 \end{bmatrix}$

c) (i) 01   (ii) 11   (v) 11   (vi) 10

For (iii) and (iv) the syndrome is $(111)^{tr}$, which is not a column of H. Assuming a double error, if $(111)^{tr} = (110)^{tr} + (001)^{tr}$, then the decoded received word is 01 [for (iii)] and 10 [for (iv)]. If $(111)^{tr} = (011)^{tr} + (100)^{tr}$, we get 10 [for (iii)] and 01 [for (iv)].

9. $G = [I_8 | A]$ where I_8 is the 8×8 multiplicative identity matrix and A is a column of eight 1's. $H = [A^{tr} | 1] = [11111111 | 1]$.

11. Compare the generator (parity-check) matrix in Exercise 9 with the parity-check (generator) matrix in Exercise 10.

Sections 16.8 and 16.9 — p. 779

1. $\binom{256}{2}$; 255

3. a)

Syndrome	Coset Leader			
000	00000	10110	01011	11101
110	10000	00110	11011	01101
011	01000	11110	00011	10101
100	00100	10010	01111	11001
010	00010	10100	01001	11111
001	00001	10111	01010	11100
101	11000	01110	10011	00101
111	01100	11010	00111	10001

(The last two rows are not unique.)

b)

Received Word	Code Word	Decoded Message
11110	10110	10
11101	11101	11
11011	01011	01
10100	10110	10
10011	01011	01
10101	11101	11
11111	11101	11
01100	00000	00

5. a) G is 57×63; H is 6×63 **b)** The rate is $\frac{57}{63}$.

7. a) $(0.99)^7 + \binom{7}{1}(0.99)^6(0.01)$ **b)** $[(0.99)^7 + \binom{7}{1}(0.99)^6(0.01)]^5$

Section 16.10 — p. 784

1. a) $\pi_2^* = \begin{pmatrix} C_1 & C_2 & C_3 & C_4 & C_5 & C_6 & C_7 & C_8 & C_9 & C_{10} & C_{11} & C_{12} & C_{13} & C_{14} & C_{15} & C_{16} \\ C_1 & C_4 & C_5 & C_2 & C_3 & C_8 & C_9 & C_6 & C_7 & C_{10} & C_{11} & C_{14} & C_{15} & C_{12} & C_{13} & C_{16} \end{pmatrix}$

$r_4^* = \begin{pmatrix} C_1 & C_2 & C_3 & C_4 & C_5 & C_6 & C_7 & C_8 & C_9 & C_{10} & C_{11} & C_{12} & C_{13} & C_{14} & C_{15} & C_{16} \\ C_1 & C_4 & C_3 & C_2 & C_5 & C_9 & C_8 & C_7 & C_6 & C_{10} & C_{11} & C_{12} & C_{15} & C_{14} & C_{13} & C_{16} \end{pmatrix}$

b) $(\pi_1^{-1})^* = \begin{pmatrix} C_1 & C_2 & C_3 & C_4 & C_5 & C_6 & C_7 & C_8 & C_9 & C_{10} & C_{11} & C_{12} & C_{13} & C_{14} & C_{15} & C_{16} \\ C_1 & C_5 & C_2 & C_3 & C_4 & C_9 & C_6 & C_7 & C_8 & C_{11} & C_{10} & C_{15} & C_{12} & C_{13} & C_{14} & C_{16} \end{pmatrix}$

$= (\pi_1^*)^{-1}$

c) $\pi_3^* r_4^* = \begin{pmatrix} C_1 & C_2 & C_3 & C_4 & C_5 & C_6 & C_7 & C_8 & C_9 & C_{10} & C_{11} & C_{12} & C_{13} & C_{14} & C_{15} & C_{16} \\ C_1 & C_5 & C_4 & C_3 & C_2 & C_6 & C_9 & C_8 & C_7 & C_{11} & C_{10} & C_{13} & C_{12} & C_{15} & C_{14} & C_{16} \end{pmatrix}$

$= (\pi_3 r_4)^*$

3. a) $\mathrm{o}(\alpha) = 7$; $\mathrm{o}(\beta) = 12$; $\mathrm{o}(\gamma) = 3$; $\mathrm{o}(\delta) = 6$

b) Let $\alpha \in S_n$, with $\alpha = c_1 c_2 \cdots c_k$, a product of disjoint cycles. Then $\mathrm{o}(\alpha)$ is the 1cm of $\ell(c_1), \ell(c_2), \ldots, \ell(c_k)$, where $\ell(c_i) = $ length of c_i, for $1 \le i \le k$.

5. a) 8   **b)** 39   **7. a)** 70   **b)** 55

9. Triangular figure: **a)** 8   **b)** 8   Square figure: **a)** 12   **b)** 12

11. a) 140   **b)** 102   **13.** 315

Section 16.11 – p. 788

1. a) 165   **b)** 120

3. Triangular figure: **a)** 96   **b)** 80

Square figure: **a)** 280   **b)** 220

Hexagonal figure: **a)** 131,584   **b)** 70,144

5. a) 2635   **b)** 1505

c)

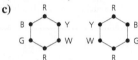

7. a) 21   **b)** 954

c) No: $k = 21$ and $m = 21$, so $km = 441 \ne 954 = n$. Here the location of a certain edge must

be considered relative to the location of the vertices. For example,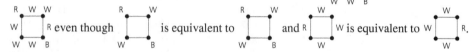

Section 16.12 – p. 793

1. a) (i) and (ii) $r^4 + w^4 + r^3 w + 2r^2 w^2 + r w^3$

b) (i) $(1/4)[(r + b + w)^4 + 2(r^4 + b^4 + w^4) + (r^2 + b^2 + w^2)^2]$

(ii) $(1/8)[(r + b + w)^4 + 2(r^4 + b^4 + w^4) + 3(r^2 + b^2 + w^2)^2$
$+ 2(r + b + w)^2 (r^2 + b^2 + w^2)]$

3. a) 10

b) $(1/24)[(r + w)^6 + 6(r + w)^2 (r^4 + w^4) + 3(r + w)^2 (r^2 + w^2)^2 + 6(r^2 + w^2)^3$
$+ 8(r^3 + w^3)^2]$

c) 2

5. Let g = green and y = gold.

Triangular figure: $(1/6)[(g + y)^4 + 2(g + y)(g^3 + y^3) + 3(g + y)^2 (g^2 + y^2)]$

Square figure: $(1/8)[(g + y)^5 + 2(g + y)(g^4 + y^4) + 3(g + y)(g^2 + y^2)^2$
$+ 2(g + y)^3 (g^2 + y^2)]$

Hexagonal figure: $(1/4)[(g + y)^9 + 2(g + y)(g^2 + y^2)^4 + (g + y)^5 (g^2 + y^2)^2]$

7. a) 136   **b)** $(1/2)[(r + w)^8 + (r^2 + w^2)^4]$ **c)** 38; 16   **9.** $\binom{m + n - 1}{n}$

Supplementary Exercises – p. 797

1. a) Since $f(e_G) = e_H$, it follows that $e_G \in K$ and $K \ne \emptyset$. If $x, y \in K$, then $f(x) = f(y) = e_H$ and $f(xy) = f(x)f(y) = e_H e_H = e_H$, so $xy \in K$. Also, for $x \in K$, $f(x^{-1}) = [f(x)]^{-1} = e_H^{-1} = e_H$, so $x^{-1} \in K$. Hence K is subgroup of G.

b) If $x \in K$, then $f(x) = e_H$. For all $g \in G$,

$$f(gxg^{-1}) = f(g)f(x)f(g^{-1}) = f(g)e_H f(g^{-1}) = f(g)f(g^{-1}) = f(gg^{-1}) = f(e_G) = e_H.$$

Hence, for all $x \in K$, $g \in G$, we find that $gxg^{-1} \in K$.

3. Let $a, b \in G$. Then $a^2 b^2 = ee = e = (ab)^2 = abab$. But $a^2 b^2 = abab \Rightarrow aabb = abab \Rightarrow ab = ba$, so G is abelian.

5. Let $G = \langle g \rangle$ and let $h = f(g)$. If $h_1 \in H$, then $h_1 = f(g^n)$ for some $n \in \mathbf{Z}$, since f is onto and G is cyclic. Therefore, $h_1 = f(g^n) = [f(g)]^n = h^n$, and $H = \langle h \rangle$.

7. For all $a, b \in G$,

$$(a \circ a^{-1}) \circ b^{-1} \circ b = b \circ b^{-1} \circ (a^{-1} \circ a) \Rightarrow$$

$$a \circ a^{-1} \circ b = b \circ a^{-1} \circ a \Rightarrow a \circ b = b \circ a,$$

and so it follows that $(G, \circ)$ is an abelian group.

9. a) Consider a permutation σ that is counted in $P(n+1, k)$. If $(n+1)$ is a cycle (of length 1) in σ, then σ (restricted to $\{1, 2, 3, \ldots, n\}$) is counted in $P(n, k-1)$. Otherwise, consider each permutation τ that is counted in $P(n, k)$. For each cycle of τ, say $(a_1 a_2 \cdots a_r)$, there are r locations in which to place $n+1$ — (1) between a_1 and a_2; (2) between a_2 and a_3; ...; $(r-1)$ between a_{r-1} and a_r; and (r) between a_r and a_1. Hence there are n locations, in total, to locate $n+1$ in τ. Consequently, $P(n+1, k) = P(n, k-1) + nP(n, k)$.

b) $\sum_{k=1}^{n} P(n, k)$ counts all of the permutations in S_n, which has $n!$ elements.

11. a) Suppose that n is composite. We consider two cases.

(1) $n = m \cdot r$, where $1 < m < r < n$: Here $(n-1)! = 1 \cdot 2 \cdots (m-1) \cdot m \cdot (m+1) \cdots (r-1) \cdot r \cdot (r+1) \cdots (n-1) \equiv 0 \pmod{n}$. Hence $(n-1)! \not\equiv -1 \pmod{n}$.

(2) $n = q^2$, where q is a prime: If $(n-1)! \equiv -1 \pmod{n}$ then $0 \equiv q(n-1)! \equiv q(-1) \equiv n - q \not\equiv 0 \pmod{n}$. So in this case we also have $(n-1)! \not\equiv -1 \pmod{n}$.

b) From Wilson's Theorem, when p is an odd prime, we find that

$$-1 \equiv (p-1)! \equiv (p-3)!(p-2)(p-1) \equiv (p-3)!(p^2 - 3p + 2) \equiv 2(p-3)! \pmod{p}.$$

Chapter 17
Finite Fields and Combinatorial Designs

Section 17.1 – p. 806

1. $f(x) + g(x) = 2x^4 + 5x^3 + x^2 + 5$

$f(x)g(x) = 6x^7 + 2x^6 + 3x^5 + 4x^4 + 2x^3 + x^2 + 4x + 4$

3. $(10)(11)^2$; $(10)(11)^3$; $(10)(11)^4$; $(10)(11)^n$

7. a) and **b)** $f(x) = (x^2 + 4)(x - 2)(x + 2)$; the roots are ± 2.

c) $f(x) = (x + 2i)(x - 2i)(x - 2)(x + 2)$; the roots are $\pm 2, \pm 2i$.

d) (a) $f(x) = (x^2 - 5)(x^2 + 5)$; there are no rational roots.

(b) $f(x) = (x - \sqrt{5})(x + \sqrt{5})(x^2 + 5)$; the roots are $\pm \sqrt{5}$.

(c) $f(x) = (x - \sqrt{5})(x + \sqrt{5})(x - \sqrt{5}i)(x + \sqrt{5}i)$; the roots are $\pm \sqrt{5}, \pm i\sqrt{5}$.

9. a) $f(3) = 8060$ **b)** $f(1) = 1$ **c)** $f(-9) = f(2) = 6$

11. 4; 6; $p - 1$

13. Let $f(x) = \sum_{i=0}^{m} a_i x^i$ and $h(x) = \sum_{i=0}^{k} b_i x^i$, where $a_i \in R$ for $0 \le i \le m$, $b_i \in R$ for $0 \le i \le k$, and $m \le k$. Then $f(x) + h(x) = \sum_{i=0}^{k}(a_i + b_i)x^i$, where $a_{m+1} = a_{m+2} = \cdots = a_k = z$, the zero of R, so $G(f(x) + h(x)) = G\left(\sum_{i=0}^{k}(a_i + b_i)x^i\right) = \sum_{i=0}^{k} g(a_i + b_i)x^i = \sum_{i=0}^{k}[g(a_i) + g(b_i)]x^i = \sum_{i=0}^{k} g(a_i)x^i + \sum_{i=0}^{k} g(b_i)x^i = G(f(x)) + G(h(x))$. Also, $f(x)h(x) = \sum_{i=0}^{m+k} c_i x^i$, where $c_i = a_i b_0 + a_{i-1}b_1 + \cdots + a_1 b_{i-1} + a_0 b_i$, and

$$G(f(x)h(x)) = G\left(\sum_{i=0}^{m+k} c_i x^i\right) = \sum_{i=0}^{m+k} g(c_i)x^i.$$

Since $g(c_i) = g(a_i)g(b_0) + g(a_{i-1})g(b_1) + \cdots + g(a_1)g(b_{i-1}) + g(a_0)g(b_i)$, it follows that

$$\sum_{i=0}^{m+k} g(c_i)x^i = \left(\sum_{i=0}^{m} g(a_i)x^i\right)\left(\sum_{i=0}^{k} g(b_i)x^i\right) = G(f(x)) \cdot G(h(x)).$$

Consequently, $G: R[x] \to S[x]$ is a ring homomorphism.

15. In $\mathbf{Z}_4[x]$, $(2x + 1)(2x + 1) = 1$, so $(2x + 1)$ is a unit. This does not contradict Exercise 14 because $(\mathbf{Z}_4, +, \cdot)$ is not an integral domain.

17. First note that for $f(x) = a_n x^n + a_{n-1}x^{n-1} + \cdots + a_2 x^2 + a_1 x + a_0$, we have $a_n + a_{n-1} + \cdots + a_2 + a_1 + a_0 = 0$ if and only if $f(1) = 0$. Since the zero polynomial is in S, the set S is *not* empty. With $f(x)$ as given here, let $g(x) = b_m x^m + b_{m-1} x^{m-1} + \cdots + b_2 x^2 + b_1 x + b_0 \in S$. (Here $m \le n$, and for $m < n$ we have $b_{m+1} = b_{m+2} = \cdots = b_n = 0$.) Then $f(1) - g(1) = 0 - 0 = 0$, so $f(x) - g(x) \in S$.

Now consider $h(x) = \sum_{i=0}^{k} r_i x^i \in F[x]$. Here $h(x) f(x) \in F[x]$ and $h(1) f(1) = h(1) \cdot 0 = 0$, so $h(x) f(x) \in S$.

Consequently, S is an ideal in $F[x]$.

Section 17.2 – p. 813

1. a) $x^2 + 3x - 1$ is irreducible over **Q**. Over **R**, **C**,

$$x^2 + 3x - 1 = [x - ((-3 + \sqrt{13})/2)][x - ((-3 - \sqrt{13})/2)].$$

b) $x^4 - 2$ is irreducible over **Q**.
Over **R**, $x^4 - 2 = (x - \sqrt[4]{2})(x + \sqrt[4]{2})(x^2 + \sqrt{2})$;
$x^4 - 2 = (x - \sqrt[4]{2})(x + \sqrt[4]{2})(x - \sqrt[4]{2}i)(x + \sqrt[4]{2}i)$ over **C**.

c) $x^2 + x + 1 = (x + 2)(x + 2)$ over $\mathbf{Z}_3$. Over $\mathbf{Z}_5$, $x^2 + x + 1$ is irreducible; $x^2 + x + 1 = (x + 5)(x + 3)$ over $\mathbf{Z}_7$.

d) $x^4 + x^3 + 1$ is irreducible over $\mathbf{Z}_2$.

e) $x^3 + 3x^2 - x + 1$ is irreducible over $\mathbf{Z}_5$.

3. Degree 1: x; $x + 1$ Degree 2: $x^2 + x + 1$ Degree 3: $x^3 + x^2 + 1$; $x^3 + x + 1$ **5.** 7^5

7. a) Yes, since the coefficients of the polynomials are from a field.

b) $h(x) | f(x), g(x) \Rightarrow f(x) = h(x)u(x), g(x) = h(x)v(x)$, for some $u(x), v(x) \in F[x]$.
$m(x) = s(x)f(x) + t(x)g(x)$ for some $s(x), t(x) \in F[x]$, so
$m(x) = h(x)[s(x)u(x) + t(x)v(x)]$ and $h(x) | m(x)$.

c) If $m(x) \nmid f(x)$, then $f(x) = q(x)m(x) + r(x)$, where $0 < \deg r(x) < \deg m(x)$.
$m(x) = s(x)f(x) + t(x)g(x)$ so $r(x) = f(x) - q(x)[s(x)f(x) + t(x)g(x)]$
$= (1 - q(x)s(x))f(x) - q(x)t(x)g(x)$, so $r(x) \in S$.
With $\deg r(x) < \deg m(x)$ we contradict the choice of $m(x)$. Hence $r(x) = 0$ and $m(x) | f(x)$.

9. a) The gcd is $(x - 1) = (1/17)(x^5 - x^4 + x^3 + x^2 - x - 1)$
$\qquad\qquad\qquad\qquad\; - (1/17)(x^2 + x - 2)(x^3 - 2x^2 + 5x - 8)$.

b) The gcd is $1 = (x + 1)(x^4 + x^3 + 1) + (x^3 + x^2 + x)(x^2 + x + 1)$.

c) The gcd is $x^2 + 2x + 1 = (x^4 + 2x^2 + 2x + 2) + (x + 2)(2x^3 + 2x^2 + x + 1)$.

11. $a = 0, b = 0; a = 0, b = 1$

13. a) $f(x) \equiv f_1(x) \pmod{s(x)} \Rightarrow f(x) = f_1(x) + h(x)s(x)$, for some $h(x) \in F[x]$, and
$g(x) \equiv g_1(x) \pmod{s(x)} \Rightarrow g(x) = g_1(x) + k(x)s(x)$, for some $k(x) \in F[x]$. Hence $f(x) + g(x) = f_1(x) + g_1(x) + (h(x) + k(x))s(x)$, so $f(x) + g(x) \equiv f_1(x) + g_1(x) \pmod{s(x)}$, and
$f(x)g(x) = f_1(x)g_1(x) + (f_1(x)k(x) + g_1(x)h(x) + h(x)k(x)s(x))s(x)$, so $f(x)g(x) \equiv f_1(x)g_1(x) \pmod{s(x)}$.

b) These properties follow from the corresponding properties for $F[x]$. For example, for the distributive law,

$$[f(x)]([g(x)] + [h(x)]) = [f(x)][g(x) + h(x)] = [f(x)(g(x) + h(x))]$$
$$= [f(x)g(x) + f(x)h(x)] = [f(x)g(x)] + [f(x)h(x)]$$
$$= [f(x)][g(x)] + [f(x)][h(x)].$$

d) A nonzero element of $F[x]/(s(x))$ has the form $[f(x)]$, where $f(x) \ne 0$ and $\deg f(x) < \deg s(x)$. With $f(x), s(x)$ relatively prime, there exist $r(x), t(x)$ with $1 = f(x)r(x) + s(x)t(x)$, so $1 \equiv f(x)r(x) \pmod{s(x)}$ or $[1] = [f(x)][r(x)]$. Hence $[r(x)] = [f(x)]^{-1}$.

e) q^n

15. a) $[2x + 1]$ **b)** $[2x + 1]$ **c)** $[2x]$ **17. a)** p^n **b)** $\phi(p^n - 1)$

19. a) 6   **b)** 12   **c)** 12   **d)** $\text{lcm}(m, n)$ **e)** 0

21. 101, 103, 107, 109, 113, 121, 125, 127, 128, 131, 137, 139, 149

23. For $s(x) = x^3 + x^2 + x + 2 \in \mathbf{Z}_3[x]$ one finds that $s(0) = 2$, $s(1) = 2$, and $s(2) = 1$. It then follows from part (b) of Theorem 17.7 and parts (b) and (c) of Theorem 17.11 that $\mathbf{Z}_3[x]/(s(x))$ is a finite field with $3^3 = 27$ elements.

25. a) Since $0 = 0 + 0\sqrt{2} \in \mathbf{Q}[\sqrt{2}]$, the set $\mathbf{Q}[2]$ is nonempty. For $a + b\sqrt{2}, c + d\sqrt{2} \in \mathbf{Q}[\sqrt{2}]$, we have

$$(a + b\sqrt{2}) - (c + d\sqrt{2}) = (a - c) + (b - d)\sqrt{2}, \text{ with } (a - c), (b - d) \in \mathbf{Q}; \text{ and}$$

$$(a + b\sqrt{2})(c + d\sqrt{2}) = (ac + 2bd) + (ad + bc)\sqrt{2}, \text{ with } ac + 2bd, ad + bc \in \mathbf{Q}.$$

Consequently, it follows from part (a) of Theorem 14.10 that $\mathbf{Q}[\sqrt{2}]$ is a subring of $\mathbf{R}$.
b) To show that $\mathbf{Q}[\sqrt{2}]$ is a subfield of $\mathbf{R}$ we need to find in $\mathbf{Q}[\sqrt{2}]$ a multiplicative inverse for each nonzero element in $\mathbf{Q}[\sqrt{2}]$. Let $a + b\sqrt{2} \in \mathbf{Q}[\sqrt{2}]$ with $a + b\sqrt{2} \neq 0$. If $b = 0$, then $a \neq 0$ and $a^{-1} \in \mathbf{Q}$—and $a^{-1} + 0 \cdot \sqrt{2} \in \mathbf{Q}[\sqrt{2}]$. For $b \neq 0$, we need to find $c + d\sqrt{2} \in \mathbf{Q}[\sqrt{2}]$ so that

$$(a + b\sqrt{2})(c + d\sqrt{2}) = 1.$$

Now $(a + b\sqrt{2})(c + d\sqrt{2}) = 1 \Rightarrow (ac + 2bd) + (ad + bc)\sqrt{2} = 1 \Rightarrow ac + 2bd = 1$ and $ad + bc = 0 \Rightarrow c = -ad/b$ and $a(-ad/b) + 2bd = 1 \Rightarrow -a^2d + 2b^2d = b \Rightarrow d = b/(2b^2 - a^2)$ and $c = -a/(2b^2 - a^2)$. (*Note*: $2b^2 - a^2 \neq 0$ because $\sqrt{2}$ is irrational.) Consequently, $(a + b\sqrt{2})^{-1} = [-a/(2b^2 - a^2)] + [b/(2b^2 - a^2)]\sqrt{2}$, with $[-a/(2b^2 - a^2)]$, $[b/(2b^2 - a^2)] \in \mathbf{Q}$. So $\mathbf{Q}[\sqrt{2}]$ is a subfield of $\mathbf{R}$.

Since $s(x) = x^2 - 2$ is irreducible over $\mathbf{Q}$, we know from part (b) of Theorem 17.11 that $\mathbf{Q}[x]/(x^2 - 2)$ is a field. Define the correspondence

$$f: \mathbf{Q}[x]/(x^2 - 2) \rightarrow \mathbf{Q}[2] \quad \text{by} \quad f([a + bx]) = a + b\sqrt{2}.$$

By an argument similar to the one given in Example 17.10 and part (a) of Exercise 24 it follows that f is an isomorphism.

Section 17.3 – p. 819

1. a)
```
1 2 3 4
2 1 4 3
4 3 2 1
3 4 1 2
```
b)
```
1 2 3 4
3 4 1 2
2 1 4 3
4 3 2 1
```
c)
```
1 3 4 2
4 2 1 3
3 1 2 4
2 4 3 1
```

3. $a_{ri}^{(k)} = a_{rj}^{(k)} \Rightarrow f_k f_r + f_i = f_k f_r + f_j \Rightarrow f_i = f_j \Rightarrow i = j$

5. L_3:
```
4 5 1 2 3
2 3 4 5 1
5 1 2 3 4
3 4 5 1 2
1 2 3 4 5
```
L_4:
```
5 1 2 3 4
4 5 1 2 3
3 4 5 1 2
2 3 4 5 1
1 2 3 4 5
```
In standard form the Latin squares L_i, $1 \leq i \leq 4$, become

L_1':
```
1 2 3 4 5
2 3 4 5 1
3 4 5 1 2
4 5 1 2 3
5 1 2 3 4
```
L_2':
```
1 2 3 4 5
3 4 5 1 2
5 1 2 3 4
2 3 4 5 1
4 5 1 2 3
```
L_3':
```
1 2 3 4 5
4 5 1 2 3
2 3 4 5 1
5 1 2 3 4
3 4 5 1 2
```
L_4':
```
1 2 3 4 5
5 1 2 3 4
4 5 1 2 3
3 4 5 1 2
2 3 4 5 1
```

7. Introduce a third factor, such as four types of transmission fluid or four types of tires.

1.

Field	Number of Points	Number of Lines	Number of Points on a Line	Number of Lines on a Point
$GF(5)$	25	30	5	6
$GF(3^2)$	81	90	9	10
$GF(7)$	49	56	7	8
$GF(2^4)$	256	272	16	17
$GF(31)$	961	992	31	32

3. There are nine points and twelve lines. These lines fall into four parallel classes.

    (i) Slope of 0: $y = 0$; $y = 1$; $y = 2$

    (ii) Infinite slope: $x = 0$; $x = 1$; $x = 2$

    (iii) Slope 1: $y = x$; $y = x + 1$; $y = x + 2$

    (iv) Slope 2 (as shown in the figure): (1) $y = 2x$ (2) $y = 2x + 1$ (3) $y = 2x + 2$

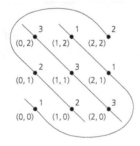

The Latin square corresponding to the fourth parallel class is

$$
\begin{array}{ccc}
3 & 1 & 2 \\
2 & 3 & 1 \\
1 & 2 & 3
\end{array}
$$

5. a) $y = 4x + 1$ **b)** $y = 3x + 10$ or $2x + 3y + 3 = 0$

    c) $y = 10x$ or $10y = 11x$

7. a) Vertical line: $x = c$. The line $y = mx + b$ intersects this vertical line at the unique point $(c, mc + b)$. As b takes on the values of F, there are no two column entries (on the line $x = c$) that are the same.

    Horizontal line: $y = c$. The line $y = mx + b$ intersects this horizontal line at the unique point $(m^{-1}(c - b), c)$. As b takes on the values of F, no two row entries (on the line $y = c$) are the same.

1. $v = 9, b = 12, r = 4, k = 3, \lambda = 1$

3. $\lambda = 2$

$$
\begin{array}{llll}
1\ \ 2\ \ 3\ \ 4 & \quad 1\ \ 3\ \ 5\ \ 7 & \quad 2\ \ 3\ \ 6\ \ 7 \\
1\ \ 2\ \ 5\ \ 6 & \quad 1\ \ 4\ \ 6\ \ 7 & \quad 2\ \ 4\ \ 5\ \ 7 & \quad 3\ \ 4\ \ 5\ \ 6
\end{array}
$$

5. a) No   **b)** No

7. a) $\lambda(v - 1) = r(k - 1) = 2r \Rightarrow \lambda(v - 1)$ is even.

    $\lambda v(v - 1) = vr(k - 1) = bk(k - 1) = b(3)(2) \Rightarrow 6 | \lambda v(v - 1)$

    b) $(\lambda = 1)$ $6 | \lambda v(v - 1) \Rightarrow 6 | v(v - 1) \Rightarrow 3 | v(v - 1) \Rightarrow 3 | v$ or $3 | (v - 1)$

    $\lambda(v - 1)$ even $\Rightarrow (v - 1)$ even $\Rightarrow v$ odd

    $3 | v \Rightarrow v = 3t$, t odd $\Rightarrow v = 3(2s + 1) = 6s + 3$ and $v \equiv 3 \pmod 6$

    $3 | (v - 1) \Rightarrow v - 1 = 3t$, t even $\Rightarrow v - 1 = 6x \Rightarrow v = 6x + 1$ and $v \equiv 1 \pmod 6$

9. $v = 9, r = 4$ **11. a)** $b = 21$ **b)** $r = 7$

13. There are λ blocks that contain both x and y. And since r is the replication number of the design, it follows that $r - \lambda$ blocks contain x, but not y. Likewise there are $r - \lambda$ blocks containing y but not x. Consequently, the number of blocks in the design that contain x or y is $(r - \lambda) + (r - \lambda) + \lambda = 2r - \lambda$.

15. a) 31   **b)** 8

17. a) $v = b = 31; r = k = 6; \lambda = 1$ **b)** $v = b = 57; r = k = 8; \lambda = 1$
  c) $v = b = 73; r = k = 9; \lambda = 1$

Supplementary
Exercises – p. 832

1. $n = 9$ **3. a)** 31   **b)** 30   **c)** 29   **d)** $k = 1000$

5. For all $a \in \mathbf{Z}_p$, $a^p = a$ [See part (a) of Exercise 13 at the end of Section 16.3.], so a is a root of $x^p - x$, and $x - a$ is a factor of $x^p - x$. Since $(\mathbf{Z}_p, +, \cdot)$ is a field, the polynomial $x^p - x$ can have at most p roots. Therefore $x^p - x = \prod_{a \in \mathbf{Z}_p} (x - a)$.

7. $\{1, 2, 4\}, \{2, 3, 5\}, \{4, 5, 7\}$ **9. a)** 9   **b)** 91

11. b) $A \cdot J_b$ is a $v \times b$ matrix whose (i, j)th entry is r, since there are r 1's in each row of A and every entry in J_b is 1. Hence $A \cdot J_b = r J_{v \times b}$. Likewise, $J_v \cdot A$ is a $v \times b$ matrix whose (i, j)th entry is k, because there are k 1's in each column of A and every entry in J_v is 1. Hence $J_v \cdot A = k \cdot J_{v \times b}$.

c) The (i, j)th entry in $A \cdot A^{\text{tr}}$ is obtained from the componentwise multiplication of rows i and j of A. If $i = j$, this results in the number of 1's in row i, which is r. For $i \ne j$, the number of 1's is the number of times x_i and x_j appear in the same block — which is given by λ. Hence $A \cdot A^{\text{tr}} = (r - \lambda)I_v + \lambda J_v$.

d)
$$\begin{vmatrix} r & \lambda & \lambda & \lambda & \cdots & \lambda \\ \lambda & r & \lambda & \lambda & \cdots & \lambda \\ \lambda & \lambda & r & \lambda & \cdots & \lambda \\ \lambda & \lambda & \lambda & r & \cdots & \lambda \\ \cdots & \cdots & \cdots & \cdots & \cdots & \cdots \\ \lambda & \lambda & \lambda & \lambda & \cdots & r \end{vmatrix}$$

$$\overset{(1)}{=} \begin{vmatrix} r & \lambda - r & \lambda - r & \lambda - r & \cdots & \lambda - r \\ \lambda & r - \lambda & 0 & 0 & \cdots & 0 \\ \lambda & 0 & r - \lambda & 0 & \cdots & 0 \\ \lambda & 0 & 0 & r - \lambda & \cdots & 0 \\ \cdots & \cdots & \cdots & \cdots & \cdots & \cdots \\ \lambda & 0 & 0 & 0 & \cdots & r - \lambda \end{vmatrix}$$

$$\overset{(2)}{=} \begin{vmatrix} r + (v - 1)\lambda & 0 & 0 & 0 & \cdots & 0 \\ \lambda & r - \lambda & 0 & 0 & \cdots & 0 \\ \lambda & 0 & r - \lambda & 0 & \cdots & 0 \\ \lambda & 0 & 0 & r - \lambda & \cdots & 0 \\ \cdots & \cdots & \cdots & \cdots & \cdots & \cdots \\ \lambda & 0 & 0 & 0 & \cdots & r - \lambda \end{vmatrix}$$

$$= [r + (v - 1)\lambda](r - \lambda)^{v-1} = (r - \lambda)^{v-1}[r + r(k - 1)] = rk(r - \lambda)^{v-1}$$

Key: (1) Multiply column 1 by -1 and add it to the other $v - 1$ columns.
   (2) Add rows 2 through v to row 1.

Appendix 1
Exponential and Logarithmic Functions

p. A-9

1. a) $\sqrt{xy^3} = x^{1/2}y^{3/2}$ **b)** $\sqrt[4]{81x^{-5}y^3} = 3x^{-5/4}y^{3/4} = \dfrac{3y^{3/4}}{x^{5/4}}$

  c) $5\sqrt[3]{8x^9y^{-5}} = 5(8^{1/3}x^{9/3}y^{-5/3}) = 5(2x^3y^{-5/3}) = \dfrac{10x^3}{y^{5/3}}$

3. a) 625   **b)** 1/343   **c)** 10

5. a) $\log_2 128 = 7$ **b)** $\log_{125} 5 = 1/3$ **c)** $\log_{10} 1/10{,}000 = -4$ **d)** $\log_2 b = a$

7. a) 3   **c)** 3

9. a) *Proof* (By Mathematical Induction):

For $n = 1$ the statement is $\log_b r^1 = 1 \cdot \log_b r$, so the result is true for this first case. Assuming the result for $n = k \ (\geq 1)$ we have $\log_b r^k = k \log_b r$. Now for the case where $n = k + 1$ we find that $\log_b r^{k+1} = \log_b (r \cdot r^k) = \log_b r + \log_b r^k$ [by part (1) of Theorem A1.2] $= \log_b r + k \log_b r$ (by the induction hypothesis) $= (1 + k)\log_b r = (k + 1)\log_b r$. Therefore, the result follows for all $n \in \mathbf{Z}^+$ by the Principle of Mathematical Induction.

b) For all $n \in \mathbf{Z}^+$, $\log_b r^{-n} = \log_b(1/r^n) = \log_b 1 - \log_b r^n$ [by part (2) of Theorem A1.2] $= 0 - n \log_b r$ [by part (a)] $= (-n)\log_b r$.

11. a) 1.5851   **b)** 0.4307   **c)** 1.4650

13. a) 5/3   **b)** 3/2   **c)** 4

15. Let $x = a^{\log_b c}$ and $y = c^{\log_a b}$. Then

$$x = a^{\log_b c} \Rightarrow \log_b x = \log_b[a^{\log_b c}] = (\log_b c)(\log_b a), \quad \text{and}$$

$$y = c^{\log_a b} \Rightarrow \log_b y = \log_b[c^{\log_b a}] = (\log_b a)(\log_b c).$$

Consequently, we find that $\log_b x = \log_b y$, from which it follows that $x = y$.

Appendix 2
Matrices, Matrix Operations, and Determinants

p. A-21

1. a) $A + B = \begin{bmatrix} 3 & 2 & 5 \\ 0 & 2 & 7 \end{bmatrix}$ **b)** $(A + B) + C = \begin{bmatrix} 3 & 3 & 7 \\ 5 & 6 & 4 \end{bmatrix}$

c) $B + C = \begin{bmatrix} 1 & 2 & 3 \\ 6 & 6 & 1 \end{bmatrix}$ **d)** $A + (B + C) = \begin{bmatrix} 3 & 3 & 7 \\ 5 & 6 & 4 \end{bmatrix}$

e) $2A = \begin{bmatrix} 4 & 2 & 8 \\ -2 & 0 & 6 \end{bmatrix}$ **f)** $2A + 3B = \begin{bmatrix} 7 & 5 & 11 \\ 1 & 6 & 18 \end{bmatrix}$

g) $2C + 3C = \begin{bmatrix} 0 & 5 & 10 \\ 25 & 20 & -15 \end{bmatrix}$ **h)** $5C = \begin{bmatrix} 0 & 5 & 10 \\ 25 & 20 & -15 \end{bmatrix}$

i) $2B - 4C = \begin{bmatrix} 2 & -2 & -6 \\ -18 & -12 & 20 \end{bmatrix}$ **j)** $A + 2B - 3C = \begin{bmatrix} 4 & 0 & 0 \\ -14 & -8 & 20 \end{bmatrix}$

k) $2(3B) = \begin{bmatrix} 6 & 6 & 6 \\ 6 & 12 & 24 \end{bmatrix}$ **l)** $(2 \cdot 3)B = \begin{bmatrix} 6 & 6 & 6 \\ 6 & 12 & 24 \end{bmatrix}$

3. a) [12], or 12   **b)** $\begin{bmatrix} 9 & 21 \\ 12 & 27 \end{bmatrix}$ **c)** $\begin{bmatrix} -10 & -10 \\ 18 & 24 \end{bmatrix}$

d) $\begin{bmatrix} -5 & -7 & 8 \\ 29 & 21 & 2 \\ -23 & -35 & 6 \end{bmatrix}$ **e)** $\begin{bmatrix} a & b & c \\ d & e & f \\ 3g & 3h & 3i \end{bmatrix}$ **f)** $\begin{bmatrix} a & b & c \\ 3g & 3h & 3i \\ d & e & f \end{bmatrix}$

5. a) $(-1/5)\begin{bmatrix} 1 & -2 \\ -3 & 1 \end{bmatrix}$ **b)** $\begin{bmatrix} 0 & 1 \\ 1 & 0 \end{bmatrix}$ **c)** The inverse does not exist.   **d)** $\begin{bmatrix} 1 & 3 \\ 2 & 7 \end{bmatrix}$

7. a) $A^{-1} = (1/2)\begin{bmatrix} 2 & -1 \\ 0 & 1 \end{bmatrix}$ **b)** $B^{-1} = (1/5)\begin{bmatrix} 1 & -2 \\ 3 & -1 \end{bmatrix}$ **c)** $AB = \begin{bmatrix} -4 & 3 \\ -6 & 2 \end{bmatrix}$

d) $(AB)^{-1} = (1/10)\begin{bmatrix} 2 & -3 \\ 6 & -4 \end{bmatrix}$ **e)** $B^{-1}A^{-1} = (1/10)\begin{bmatrix} 2 & -3 \\ 6 & -4 \end{bmatrix}$

9. b) $\begin{bmatrix} 5 & 3 \\ 3 & -2 \end{bmatrix}\begin{bmatrix} x \\ y \end{bmatrix} = \begin{bmatrix} 35 \\ 2 \end{bmatrix}$

$\begin{bmatrix} x \\ y \end{bmatrix} = \begin{bmatrix} 5 & 3 \\ 3 & -2 \end{bmatrix}^{-1}\begin{bmatrix} 35 \\ 2 \end{bmatrix} = (-1/19)\begin{bmatrix} -2 & -3 \\ -3 & 5 \end{bmatrix}\begin{bmatrix} 35 \\ 2 \end{bmatrix} = \begin{bmatrix} 4 \\ 5 \end{bmatrix}$

11. $\det(2A) = 2^2(31) = 124$, $\det(5A) = 5^2(31) = 775$

13. a) 45   **b)** -40 **c)** 14

15. a)   (i)
$$\begin{vmatrix} 1 & 2 & 1 \\ 0 & -1 & -1 \\ 2 & 3 & 0 \end{vmatrix} = 2(-1)^{3+1}\begin{vmatrix} 2 & 1 \\ -1 & -1 \end{vmatrix} + 3(-1)^{3+2}\begin{vmatrix} 1 & 1 \\ 0 & -1 \end{vmatrix}$$
$$= 2(-2 - (-1)) - 3(-1) = 2(-1) + 3 = 1.$$

(ii) 5   (iii) 25

b) (i) 51   (ii) 306   (iii) 510

Appendix 3
Countable and Uncountable Sets

p. A-32

1. a) True   **b)** False   **c)** True   **d)** True

e) False: Let $A = \mathbf{Z} \cup (0, 1]$ and $B = \mathbf{Z} \cup (1, 2]$. Then A, B are both uncountable, but $A \cap B = \mathbf{Z}$ is countable.

f) True

g) False: Let $A = \mathbf{Z}^+ \cup (0, 1]$ and $B = (0, 1]$. Then A, B are both uncountable, but $A - B = \{2, 3, 4, \ldots\}$ is countable.

3. If B were countable, then by Theorem A3.3 it would follow that A is countable. This leads us to a contradiction since we are given that A is uncountable.

5. Since S, T are countably infinite, we know from Theorem A3.2 that we can write
$S = \{s_1, s_2, s_3, \ldots\}$ and $T = \{t_1, t_2, t_3, \ldots\}$ — two (infinite) sequences of distinct terms. Define the function

$$f : S \times T \to \mathbf{Z}^+$$

by $f(s_i, t_j) = 2^i 3^j$, for all $i, j \in \mathbf{Z}^+$. If $i, j, k, \ell \in \mathbf{Z}^+$ with $f(s_i, t_j) = f(s_k, t_\ell)$, then $f(s_i, t_j) = f(s_k, t_\ell) \Rightarrow 2^i 3^j = 2^k 3^\ell \Rightarrow i = k, \, j = \ell$ (By the Fundamental Theorem of Arithmetic) $\Rightarrow s_i = s_k$ and $t_j = t_\ell \Rightarrow (s_i, t_j) = (s_k, t_\ell)$. Therefore, f is a one-to-one function and $S \times T \sim f(S \times T) \subset \mathbf{Z}^+$. So from Theorem A3.3 we know that $S \times T$ is countable.

7. The function $f : (\mathbf{Z} - \{0\}) \times \mathbf{Z} \times \mathbf{Z} \to \mathbf{Q}$ given by $f(a, b, c) = 2^a 3^b 5^c$ is one-to-one (Verify this!). So by Theorems A3.3 and A3.8 $(\mathbf{Z} - \{0\}) \times \mathbf{Z} \times \mathbf{Z}$ is countable. Now for all $(a, b, c) \in (\mathbf{Z} - \{0\}) \times \mathbf{Z} \times \mathbf{Z}$ there are at most two (distinct) real solutions for the quadratic equation $ax^2 + bx + c = 0$. From Theorem A3.9 it then follows that the set of all real solutions of the quadratic equations $ax^2 + bx + c = 0$, where $a, b, c \in \mathbf{Z}$ and $a \neq 0$, is countable.

Index

NOTATION

SPECIAL SETS OF NUMBERS		
SPECIAL SETS OF NUMBERS	$\mathbf{Z}$	the set of integers: $\{0, 1, -1, 2, -2, 3, -3, \ldots\}$
	$\mathbf{N}$	the set of nonnegative integers or natural numbers: $\{0, 1, 2, 3, \ldots\}$
	$\mathbf{Z}^+$	the set of positive integers: $\{1, 2, 3, \ldots\} = \{x \in \mathbf{Z} \mid x > 0\}$
	$\mathbf{Q}$	the set of rational numbers: $\{a/b \mid a, b \in \mathbf{Z}, b \neq 0\}$
	$\mathbf{Q}^+$	the set of positive rational numbers
	$\mathbf{Q}^*$	the set of nonzero rational numbers
	$\mathbf{R}$	the set of real numbers
	$\mathbf{R}^+$	the set of positive real numbers
	$\mathbf{R}^*$	the set of nonzero real numbers
	$\mathbf{C}$	the set of complex numbers: $\{x + yi \mid x, y \in \mathbf{R}, i^2 = -1\}$
	$\mathbf{C}^*$	the set of nonzero complex numbers
	$\mathbf{Z}_n$	$\{0, 1, 2, \ldots, n-1\}$, for $n \in \mathbf{Z}^+$
	$[a, b]$	the closed interval from a to b: $\{x \in \mathbf{R} \mid a \leq x \leq b\}$
	(a, b)	the open interval from a to b: $\{x \in \mathbf{R} \mid a < x < b\}$
	$[a, b)$	a half-open interval from a to b: $\{x \in \mathbf{R} \mid a \leq x < b\}$
	$(a, b]$	a half-open interval from a to b: $\{x \in \mathbf{R} \mid a < x \leq b\}$
ALGEBRAIC STRUCTURES	$(R, +, \cdot)$	R is a ring with binary operations $+$ and $\cdot$
	$R[x]$	the ring of polynomials over ring R
	$\deg f(x)$	the degree of the polynomial $f(x)$
	$(G, \circ)$	G is a group under the binary operation $\circ$
	S_n	the symmetric group on n symbols
	aH	a left coset of subgroup H (in group G): $\{ah \mid h \in H\}$
	$(\mathscr{B}, +, \cdot, {}^-, 0, 1)$	the Boolean algebra $\mathscr{B}$ with binary operations $+$ and $\cdot$, the unary operation $^-$, and identity elements 0 (for $+$) and 1 (for $\cdot$)
GRAPH THEORY	$G = (V, E)$	G is a graph with vertex set V and edge set E
	K_n	the complete graph on n vertices
	$\overline{G}$	the complement of graph G
	$\deg(v)$	the degree of vertex v (in an undirected graph G)
	$\operatorname{od}(v)$	the out degree of vertex v (in a directed graph G)
	$\operatorname{id}(v)$	the in degree of vertex v (in a directed graph G)
	$\kappa(G)$	the number of connected components of graph G
	Q_n	the n-dimensional hypercube: the n-cube
	$K_{m,n}$	the complete bipartite graph on $V = V_1 \cup V_2$ where $V_1 \cap V_2 = \emptyset$, $\lvert V_1 \rvert = m$, $\lvert V_2 \rvert = n$
	$\beta(G)$	the independence number of G
	$\chi(G)$	the chromatic number of G
	$P(G, \lambda)$	the chromatic polynomial of G
	$\gamma(G)$	the domination number of G
	$L(G)$	the line graph of G
	$T = (V, E)$	T is a tree with vertex set V and edge set E
	$N = (V, E)$	N is a (transport) network with vertex set V and edge set E